▤ Nonvertical line, slope m

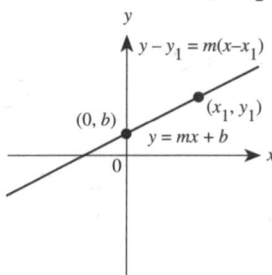

▤ Horizontal line, vertical line

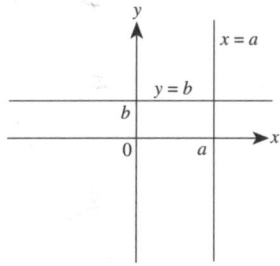

▤ Distance formula, slope m of a nonvertical line

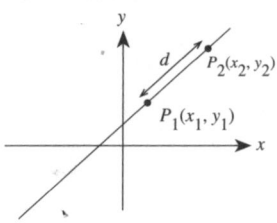

$$d = \sqrt{(x_2 - x_1)^2 + (y_2 - y_1)^2}$$

$$m = \frac{y_2 - y_1}{x_2 - x_1}$$

▤ $y = x^2$

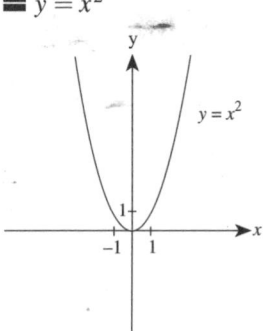

▤ $y = x^3$

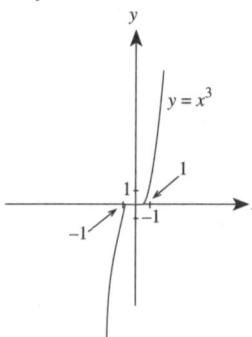

▤ $y = \sqrt{x}$

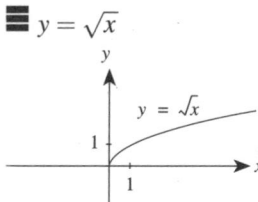

▤ $y = b^x, b > 0, b \neq 1$

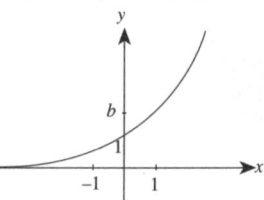

▤ $y = \log_b x, b > 1$

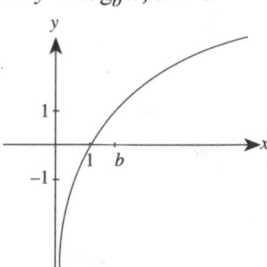

▤ $(x - h)^2 + (y - k)^2 = r^2$

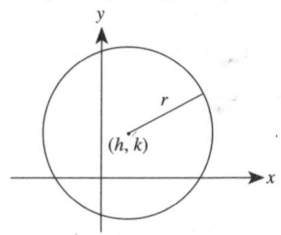

▤ $4p(y - k) = (x - h)^2, p > 0$

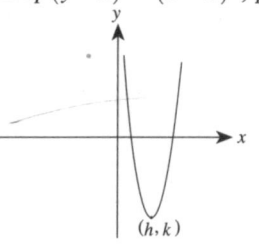

▤ $\dfrac{(x - h)^2}{a^2} + \dfrac{(y - k)^2}{b^2} = 1, a > b$

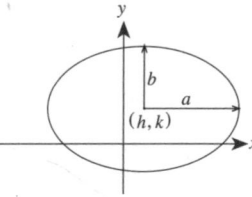

▤ $\dfrac{(x - h)^2}{a^2} - \dfrac{(y - k)^2}{b^2} = 1$

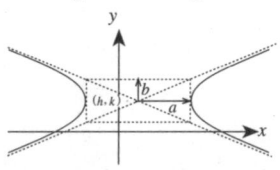

▤ $y = \sin x$

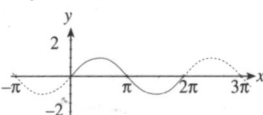

▤ $y = \cos x$

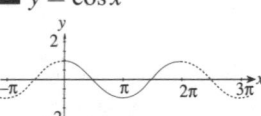

▤ $y = \tan x$

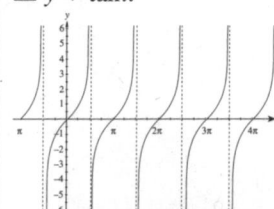

Technical Mathematics

WITH CALCULUS

Second Edition

Technical Mathematics

WITH CALCULUS
Second Edition

John C. Peterson

Chattanooga State Technical Community College

Delmar Publishers

I(T)P™   An International Thomson Publishing Company

Albany • Bonn • Boston • Cincinnati • Detroit • London • Madrid • Melbourne
Mexico City • New York • Pacific Grove • Paris • San Francisco • Singapore • Tokyo
Toronto • Washington

NOTICE TO THE READER

Publisher does not warrant or guarantee any of the products described herein or perform any independent analysis in connection with any of the product information contained herein. Publisher does not assume, and expressly disclaims, any obligation to obtain and include information other than that provided to it by the manufacturer.

The reader is expressly warned to consider and adopt all safety precautions that might be indicated by the activities described herein and to avoid all potential hazards. By following the instructions contained herein, the reader willingly assumes all risks in connection with such instructions.

The publisher makes no representations or warranties of any kind, including but not limited to, the warranties of fitness for particular purpose or merchantability, nor are any such representations implied with respect to the material set forth herein, and the publisher takes no responsibility with respect to such material. The publisher shall not be liable for any special, consequential, or exemplary damages resulting, in whole or in part, from the readers' use of, or reliance upon, this material.

Cover photo: Gary Conner

Delmar Staff
Publisher: Robert D. Lynch
Acquisitions Editor: Paul Shepherdson
Production Manager: Larry Main
Art and Design Coordinator: Nicole Reamer

COPYRIGHT © 1997
By Delmar Publishers Inc.
an International Thomson Publishing Company

The ITP logo is a trademark under license.

Printed in the United States of America

For more information, contact:

Delmar Publishers Inc.
3 Columbia Circle, Box 15015
Albany, NY 12212-5015

International Thomson Publishing Europe
Berkshire House 168-173
High Holborn
London, WC1V 7AA
England

Thomas Nelson Australia
102 Dodds Street
South Melbourne, 3205
Victoria, Australia

Nelson Canada
1120 Birchmont Road
Scarborough, Ontario
Canada, M1K 5G4

International Thomson Editores
Campos Eliseos 385, Piso 7
Col Polanco
11560 Mexico D F Mexico

International Thomson Publishing GmbH
Konigswinterer Strasse 418
53227 Bonn
Germany

International Thomson Publishing Asia
221 Henderson Road
#05-10 Henderson Building
Singapore 0315

International Thomson Publishing—Japan
Hirakawacho Kyowa Building, 3F
2-2-1 Hirakawacho
Chiyoda-ku, Tokyo 102
Japan

1 2 3 4 5 6 7 8 9 10 XXX 02 01 00 99 98 97 96

Library of Congress Cataloging-in-Publication Data

Peterson, John C. (John Charles), 1939-
    Technical mathematics with calculus / John C. Peterson. -- 2nd ed.
        p.   cm.
    Includes indexes.
    ISBN 0-8273-7243-4
    1. Mathematics.   I. Title.
QA37.2.P48 1996
510--dc20
                                          95-13516
                                             CIP

Contents

Preface

Technical Mathematics with Calculus, second edition, is a student-oriented textbook, designed to be easily read and understood by students who are preparing for technical or scientific careers. Mathematics can be a difficult subject to read or study, especially when the math terms become tangled with the technical terms used in the applications. This text offers a clear, readable development of the math concepts. But learning mathematics takes more than reading—it requires extensive practice. Therefore, *Technical Mathematics with Calculus* provides plentiful opportunities to practice problem solving. Together, these two essential elements have formed our constant focus in the development of this text: thorough, uncomplicated discussions followed by examples and problem sets that draw on real-world applications of math.

Organization and Approach

The first three chapters of this text form a thorough introduction to the real number system, algebra, and geometry. Much of the material in these chapters might be familiar to your students and can be covered quickly or skipped entirely. Chapters 4–21 focus on the precalculus areas of mathematics through trigonometry, analytic geometry, and introductory statistics. Chapters 22–34 are devoted to calculus and differential equations.

This text offers an integrated presentation of algebra, geometry, trigonometry, and analytic geometry. Practical applications of mathematics in the technical and scientific areas are provided throughout the text. Students also receive ample opportunities to solve problems using scientific calculators, graphing calculators, and microcomputers.

This textbook strives to make the reading and comprehension of its contents easier for students. The uncomplicated writing style makes difficult discussions easier to grasp. Each important concept is supported by numerous examples, applications, and practice problems.

Help for Teaching and Learning

Well-written, carefully illustrated discussions form the foundation of any textbook; but your students will form their math skills with practice. We have developed a package of learning features that will inspire students to practice hard, and this will make their efforts, and yours, effective.

Problem Solving

Students often have trouble assessing a real situation and sorting out the pertinent information. Our approach to problem solving encourages students to visualize problems, organize information, and develop their intuitive abilities. The applications, examples, and problems are intended to teach students how to interpret real-world situations. Many problems are accompanied by illustrations and photos intended to guide students in analyzing information.

Problem solving is formally introduced in Sections 2.4 and 2.5, and is developed throughout the book. Hints, Notes, and Cautions are offered to hone problem-solving techniques, and many boxed features include guidelines for solving certain types of problems.

Visualizing Mathematics

For students to develop good problem-solving skills, they must learn to visualize mathematical problems. This text offers a unique feature that will help students to see the important information in a problem. Each chapter opens with a photograph that demonstrates mathematics at work in the real world. Later in the chapter, this photograph is analyzed in an example and used to solve a problem based on the picture. A sketch placed next to the photo extracts the important mathematical elements. Where possible, this sketch is imposed over the photograph to show students how to cull the important elements from a real situation, and how to create a sketch that will help to solve a problem.

Using Technology in Problem Solving

This text emphasizes the wise use of technologies—scientific calculators, graphing calculators, and microcomputers—because they can take the drudgery out of repetitive computations and can help students learn faster. Reducing the drudgery can help students focus on the important aspects of the operations they are performing. However, students must be aware that not all solutions are aided by technology. They must also learn how to determine whether a calculator or computer has given a correct answer, a reasonable answer, or a wrong answer. Rather than indicate that only certain problems should be solved with a calculator or computer, the author has assumed that any problem can be solved with the help of one of these tools. Calculator keystroke sequences appear frequently throughout the text to help students learn to use these important tools efficiently. Some examples and exercises may be completed using a graphing calculator or a computer graphing program. These instances are indicated by the placement of a graphing calculator icon. Problems that require students to do some computer programming are indicated by a computer icon. Answers to these exercises are given in GW-BASIC. The Computer Programs section provides the debugged programs, along with additional problems to work using these programs. The use of computer programs throughout the book also allows students to explore what can and does happen when small changes are made in a problem or in generating recursive answers. For those students who do not wish to write the programs, a disk is available that contains all the programs created in the exercises.

Examples

The book contains frequent, realistic examples that show how to apply mathematical concepts. Examples cover both routine mathematical manipulations and applications. Approximately 20 percent of these examples have been selected from trade, vocational, technical, and industrial areas to demonstrate how mathematics is applied to all fields of technology.

Exercises

The second edition has increased the number of problems by 20 percent. Many of these problems show the great variety of technical and scientific applications for math.

Methods for solving most of the exercises can be found by reviewing the worked examples. Answers to the odd-numbered questions appear in the back of the book. A separate solutions manual for all problems is available to instructors, with a manual of selected solutions available to students.

Writing to Learn Math

The second edition helps instructors incorporate NCTM and AMATYC standards by offering writing problems at the end of each exercise section. These exercises appear under the heading "In Your Words," and will encourage students to work through their problem-solving processes verbally as well as numerically. The writing exercises can also be used to promote collaborative learning.

Applications

Most technical math courses are designed to prepare students to solve applied problems in their chosen technical fields. For this reason, we have taken care to present many exercises and examples that show how to apply math in a variety of technical fields. These applications are highlighted in the text with special headings that indicate the applied field. The text also offers an index of the applications by field.

Pedagogical Features

Throughout the book, you will find rules, formulas, guidelines, and hints that are boxed for easy identification. The boxes make the material easy to locate, and also make the book a valuable reference after the course is finished.

Five icons appear frequently in the margins:

The graphing calculator icon indicates that a certain example or exercise may be used with a graphing calculator. However, students may prefer to use a computer graphing program.

The computer icon indicates that a certain problem requires the student to do some computer programming.

Many years of teaching experience have shown the author where students make common errors. These places are indicated by the "CAUTION" icon. This icon not only points out the potential errors, but how to avoid them.

The "HINT" icon points out a valuable technique or learning hint that students can use to solve problems.

The "NOTE" icon refreshes students' memories about a concept, or points out interesting or unusual ideas. These notes highlight ideas that might easily be overlooked, alternative ways to solve a problem, or a possible shortcut.

New problems that require students to describe their solutions and logical processes in words have been added to the second edition.

Chapter Review

Each chapter concludes with a list of the important terms covered in the chapter, a generous set of review exercises, and a chapter test. The review exercises include both routine computations and applications. Many of these exercises draw on more than one section of the chapter, requiring students to think harder and synthesize their learning. The chapter test gives students an idea of the types of questions they might expect for a fifty-minute exam.

Supplements

The supplements package has been thoroughly revised and expanded for the second edition, offering instructors an array of products that let them tailor a course to their own student profile.

- *Instructor's Resource Guide* contains chapter-by-chapter: objectives, teaching hints, NCTM/AMATYC guidelines, hands-on math labs, and collaborative exercises.
- *Solutions Manual*, showing fully worked solutions to every text problem.
- *Computerized Test Bank*, with approximately 100 questions per chapter to choose from.
- *LabMath* is a new supplement that lets instructors customize their courses to a specific calculator. Instructors can choose among interactive workbook components that offer tutorials for the TI-82, TI-85, or Casio graphing calculators.
- The *Student Solutions Manual* has been expanded to include solutions to every odd problem.
- *Computer Software:* A disk is available containing all of the debugged computer programs that appear in the *Computer Programs* section.

Acknowledgments

Anyone who writes a book enjoys the direct and indirect assistance of many people. I would especially like to thank the late Jeff Jones, the first accuracy checker for this text. He worked every example exercise and provided valuable editorial suggestions. Rosanna Templin and Jessica Daugherty typed most of the manuscript and solutions. Melissa Savage also helped to type and scan some material. Not only did they do this with speed and efficiency, but their experience with this difficult material made my job so much easier. Special thanks go to Alan Herweyer for preparing the Solutions Manual, and to Darryl Sheppard for debugging the computer programs.

The author and publisher gratefully acknowledge the assistance of many talented reviewers and contributors. Allan Hymers, of Scarborough, Ontario, contributed several hundred problems and applications. The following reviewers also submitted applications for the text:
Calvin Holt, Franklin, Va.
Roberta Laine, West Bend, Wis.
Frank Weeks, Gresham, Oreg.
Natalie Woodrow, Sweetwater, Tex.

The following instructors provided valuable criticism during several review stages:
Haya Adner, Queensborough Community College, Bayside, N.Y.
Doris Bratcher, Northwest Iowa Technical College, Iowa
Granville Brown, Grand Rapids Community College, Grand Rapids, Mich.
Ronald Bukowski, Johnson Technical Institute, Scranton, Pa.
Ed Champy, Northern Essex Community College
Bill Ferguson, Columbus State Community College, Columbus, Ohio
Maggie Flint, Northeast State Technical Community College, Blountville, Tenn.
Marion Graziano, Montgomery County Community College, Blue Bell, Pa.
Henry Hosek, Purdue University, Calumet, Ind.
David Hutchinson, Mohawk College of Technical and Applied Arts, Hamilton, Ont.
Allan Hymers, Centennial College, Scarborough, Ont.
Stanley Koper, Triangle Tech, Belle Vernon, Pa.
Larry Lance, Columbus State Community College, Columbus, Ohio
Tom McCollow, DeVry Institute of Technology, Phoenix, Ariz.
Peter Moreno, ITT Technical Institute, San Bernardino, Calif.
Robert Opel, Waukesha County Technical College, Waukesha, Wis.
Steven B. Ottman, Southeast Community College, Nebraska
Gordon Schlafmann, Western Wisconsin Technical College
Rita Shillabeer, University at Alberta, Canada
Cathy Vollstedt, North Central Technical College, Schofield, Wis.
Anita Walker, Asheville, N.C.
Phyliss Wiegman, Ivy Tech, Fort Wayne, Ind.
Nancy Woods, Des Moines Area Community College, Des Moines, Iowa

During several rigorous accuracy checks, the following professionals examined the text and solutions in detail:
C. Lea Campbell, Lamar University, Port Arthur, Tex.
Gerry East, Tulane University, New Orleans, La.
Diane Ferris, Portland Community College, Portland, Oreg.
Alan Herweyer, Chattanooga State Technical College, Chattanooga, Tenn.

Barbara Karp, Newcastle, Ind.
Amy Relyea, Miami University, Oxford, Ohio
Gordon Schlafmann, Western Wisconsin Technical College
Randall Sowell, Central Virginia Community College, Lynchburg, Va.
John Wisthoff, Anne Arundel Community College, Md.
George Wyatt, Rensselaer Polytechnic Institute, Troy, N.Y.

Several enthusiastic instructors generously agreed to test the final pages in their classrooms:
C. Lea Campbell, Lamar University, Port Arthur, Tex.
Bill Ferguson, Columbus State Community College, Columbus, Ohio
Marion Graziano, Montgomery County Community College, Blue Bell, Pa.
Henry Hosek, Purdue University, Calumet, Ind.
Robert Opel, Waukesha County Technical College, Waukesha, Wis.
Arthur J. Varie, Trumbull County JVS, Warren, Ohio

No one could complete an effort such as this without the support, encouragement, and tolerance of an understanding family. I want to thank my wife, Dr. Marla Peterson, and our son, Matt, for giving me the time and understanding to complete this work.

John C. Peterson

CHAPTER

1

The Real Number System

In order to work mathematics, we need to be able to add, subtract, multiply, and divide real numbers. In Section 1.3, we will see how to subtract real numbers to determine the thickness of this pipe.

Courtesy of Ruby Gold

Every technician must be able to use mathematics. The science of mathematics is a universal language that helps technicians communicate and do their work. Advances in technology require that technicians know more and more mathematics. Technicians must be able to solve mathematical problems quickly and accurately.

The mathematics that are developed in this text provide the foundation for work in almost any field of technology. Perhaps as important as the actual mathematics is the ability to use calculators and computers to help work mathematics problems. As you use this book you should learn how to use calculators and computers to help you with mathematics. Basic calculator instructions are in Appendix A.

≡ 1.1
SOME SETS OF NUMBERS

Fundamental to our work in mathematics is an understanding of the different types of numbers that we will use. One of the first sets of numbers that you learned was the **natural numbers**. These numbers are 1, 2, 3, 4, and so forth. These were also the first numbers invented by people. Later, people discovered a need for the number zero (0). When 0 was added to the natural numbers, a new set of numbers, called the **whole numbers**, was formed.

Integers

People continued to use the whole numbers until they discovered that they were not able to solve some problems with the set of whole numbers. Problems such as $3 - 5$ required that a new set of numbers be invented: the **integers**. The integers included the natural numbers, zero, and a negative value for each of the natural numbers. The set of integers consists of the numbers

$$\ldots, -4, -3, -2, -1, 0, 1, 2, 3, \ldots$$

The negative numbers are called the **negative integers** and the natural numbers are called the **positive integers**. The number 0 is neither positive nor negative. Positive numbers can be written with a plus (+) sign in front as $+1, +2, +3, \ldots$ but this is not necessary.

≡ **Note**
Three points of ellipsis (...) in a series of numbers means that some numbers have been omitted. They are placed wherever the missing numbers would be.

EXAMPLE 1.1

The numbers 1, 14, 973, and 8,436 are all positive integers. They are also natural numbers.
The numbers -2, -18, and $-4,392$ are all negative integers, but not natural numbers.

Rational Numbers

The rational numbers were created to indicate when a part of something was needed. The rational numbers are used to represent division of one integer by a positive or negative integer. There are both positive and negative rational numbers. Rational numbers can be expressed in more than one way. The numbers $\frac{4}{2}$, $\frac{6}{3}$, $\frac{2}{1}$, and $\frac{142}{71}$ are all different ways of writing 2.

EXAMPLE 1.2

Some positive rational numbers are 14, $\frac{7}{3}$, $\frac{2}{5}$, $\frac{271}{18}$, $\frac{4}{2}$, and 973.
Negative rational numbers include $-\frac{2}{3}$, -5, $-\frac{9}{7}$, $-\frac{5}{23}$ and $\frac{-14}{2}$.
Zero (0) is a rational number that is neither positive nor negative.

Irrational Numbers

At one time people thought that everything could be expressed as a rational number and that every problem would have a rational number as an answer. In fact, there is a story that the first person who proved that all numbers are not rational was executed because of this proof. Perhaps because it is not rational to execute a person just because they discovered a new group of numbers, these new numbers were called the **irrational numbers**. Some of the irrational numbers are $\sqrt{2}$, $\sqrt{5}$, π, and $-\sqrt{17}$. While we will not prove it here, it is not possible to represent any of these numbers by the division of two integers.

Real Numbers

When the rational numbers are combined with the irrational numbers, a new group of numbers is formed—the **real numbers**. Most of our work in this text uses the real numbers. Later we will introduce a new group of numbers, called the imaginary numbers, and combine the real and imaginary numbers into another group, the complex numbers.

At times it is convenient to represent the real numbers on a line called the **number line**. The usual number line is a horizontal line that has been marked in equally spaced intervals. One of these marks is called the **origin** and is indicated by the number zero (0). The marks to the right are labeled using the positive integers. The negative integers are used to designate the marks to the left of the origin. A typical number line is shown in Figure 1.1.

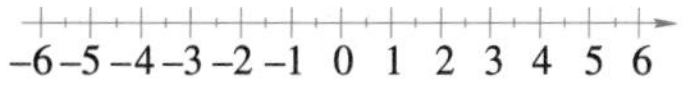

The other real numbers are located between the integers.

The fact that a number line does not have to be horizontal can be demonstrated by a thermometer. Most wall thermometers hang vertically with the positive numbers above the negative numbers and this represents a vertical number line. But the fact that some thermometers are circular shows that nothing requires a number line to be a straight line.

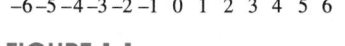

$-6\ -5\ -4\ -3\ -2\ -1\ \ 0\ \ 1\ \ 2\ \ 3\ \ 4\ \ 5\ \ 6$

FIGURE 1.1

Decimals

Each real number can be represented by a decimal number. The rational numbers are represented by repeating decimals. Irrational numbers are represented by non-repeating decimals. The decimal representations allow us to position these numbers accurately on the number line.

EXAMPLE 1.3

Repeating decimals include $\frac{1}{2} = 0.5000\ldots$ (repeats 0); $\frac{13}{4} = 3.2500\ldots$ (repeats 0), $-\frac{4}{3} = -1.333\ldots$ (repeats 3), and $\frac{-41}{11} = -3.727272\ldots$ (repeats 72). These are sometimes represented with a bar over the repeating digits. Thus $\frac{1}{2} = 0.5\overline{0}$, $\frac{13}{4} = 3.25\overline{0}$, $-\frac{4}{3} = -1.\overline{3}$, and $\frac{-41}{11} = -3.\overline{72}$. Numbers that repeat zero are called **terminating decimals** and are usually written without indicating the repeating zero. When this is done, we have $\frac{1}{2} = 0.5$ and $\frac{13}{4} = 3.25$.

EXAMPLE 1.3 (Cont.)

Irrational numbers are represented by nonrepeating and nonterminating decimals. For example, here are the decimal representations of four irrational numbers:

$$\sqrt{2} = 1.414213\ldots \qquad \sqrt{7} = 2.6457513\ldots$$

$$\pi = 3.14159265359\ldots \qquad -\frac{\sqrt{15}}{2} = -1.93649167\ldots$$

≡ Note

Here, the three points of ellipsis are placed at the end of a number to indicate that not all of the digits have been shown.

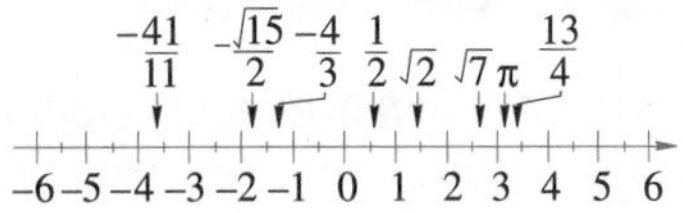

The location of each of the numbers in Example 1.3 is shown in Figure 1.2.

FIGURE 1.2

Absolute Values

Knowing the different kinds of numbers can be useful. However, what is more important is how the numbers relate to each other and how we can use them. One of the basic ideas is the distance of a number from zero or, on a number line, from the origin. The number 3 is three units from zero. So is the number -3. The number $-\frac{8}{5}$ is $\frac{8}{5}$ units from zero. The distance of a number from zero is called the **absolute value** of the number. The absolute value of a number is indicated by placing the number between vertical bars, $|\ |$.

EXAMPLE 1.4

The absolute value of 4 is 4, the absolute value of -9 is 9. These, and other absolute values, are written as

$$|4| = 4$$
$$|-9| = 9$$
$$\left|\frac{4}{3}\right| = \frac{4}{3}$$
$$\left|-\frac{18}{7}\right| = \frac{18}{7}$$
$$\left|-\sqrt{7}\right| = \sqrt{7}$$
$$|0| = 0$$

Inequalities

Some other important ideas include how numbers relate to each other. The relationship most often used is equality, or when two different numerals represent the same amount. Since $\frac{6}{3}$ and 2 both represent the same amount, we say they are equal, use the equal sign ($=$), and indicate this by writing $\frac{6}{3} = 2$.

Other relationships are concerned with numbers that are **not equal to** each other. To indicate that two numbers do not represent the same amount we use the "is not equal to" symbol, $\neq$. Thus, $3 \neq 4$ and $-5 \neq 17$ are two examples showing how to use the "is not equal to" symbol.

The $\neq$ sign is one example of a **sign of inequality**. Other signs of inequality are often used to indicate which of two numbers is the larger. A number is larger than another if it is located farther to the right on the number line. In this case we say that the number that is farther to the right on the number line **is greater than** the other number. The symbol $>$ is used to indicate that one number is greater than another. Another definition says that when the second number is subtracted from the first and the answer is positive, then the first number is greater than the second.

EXAMPLE 1.5

$7 > 5$, because $7 - 5 = 2$, a positive number
$4 > -2$, because $4 - (-2) = 6$, a positive number
$-3 > -10$, because $-3 - (-10) = 7$, a positive number

If the first number is to the left of the second number on the number line, or the answer is negative when the second number is subtracted from the first, then the first number **is less than** the second. We use the symbol $<$.

EXAMPLE 1.6

$5 < 7$, because $5 - 7 = -2$, a negative number
$-2 < 4$, because $-2 - 4 = -6$, a negative number
$-10 < -3$, because $-10 - (-3) = -7$, a negative number

There are two other inequality symbols. The symbol $\leq$ means "is less than or equal to" and the symbol $\geq$ indicates "is greater than or equal to."

EXAMPLE 1.7

$5 \geq 3$, because $5 > 3$; that is, $5 - 3 = 2$, a positive number
$3 \geq 3$, because $3 = 3$

Reciprocals

Every number, except zero, has a **reciprocal**. The reciprocal of a number is equal to 1 divided by that number. If the number is a fraction, then the reciprocal is 1 divided by the fraction or $\dfrac{1}{\text{fraction}}$. Thus, the reciprocal of $\frac{4}{3}$ is $\dfrac{1}{\frac{4}{3}} = \frac{3}{4}$.

EXAMPLE 1.8

The reciprocal of 9 is $\frac{1}{9}$.

The reciprocal of $\frac{2}{3}$ is $\dfrac{1}{\frac{2}{3}} = \frac{3}{2}$.

The reciprocal of $\sqrt{7}$ is $\dfrac{1}{\sqrt{7}}$.

The reciprocal of $\frac{-5}{2}$ is $\dfrac{1}{\frac{-5}{2}} = \frac{-2}{5}$.

≡ **Note** The product of a number and its reciprocal is 1.

In this section we have introduced some of the different sets of numbers and ways to indicate the relationship between two numbers. In the next section we will review the basic operations on, and the basic laws of, the real numbers.

Exercise Set 1.1

In Exercises 1–4, indicate all the sets of numbers to which each number belongs. Remember, a number can belong to more than one set of numbers.

1. 15

2. $\frac{-2}{3}$

3. $\frac{-\sqrt{7}}{8}$

4. 0

Evaluate each of the expressions in Exercises 5–10.

5. $|15|$

6. $\left|\frac{-2}{3}\right|$

7. $\left|\frac{-\sqrt{7}}{8}\right|$

8. $|0|$

9. $|4-7|$

10. $\left|8-\frac{7}{2}\right|$

Locate each of the numbers in Exercises 11–14 on a number line.

11. $\frac{4}{7}$

12. 2.5

13. $-\frac{8}{3}$

14. $\frac{-\pi}{3}$

In Exercises 15–26, insert the correct sign of inequality (< or >) between the given pairs of numbers.

15. 2, 3

16. 5, 3

17. −4, 7

18. 9, −7

19. −3, −8

20. −15, −7

21. $\frac{-2}{3}, -\frac{1}{2}$

22. $-\frac{7}{2}, -\frac{8}{3}$

23. 0.7, 0.5

24. −2.1, −2.0

25. $|2.3|, |-4.1|$

26. $|-5.1|, |3.7|$

Find the reciprocals of the numbers in Exercises 27–30.

27. −5

28. $\frac{1}{2}$

29. $\frac{17}{3}$

30. $\frac{-2}{\pi}$

Solve Exercises 31–40.

31. List the following numbers in numerical order from smallest to largest: $-5, -\frac{2}{3}, |-8|, \frac{16}{3}, -|4|, \frac{-1}{3}$

32. List the following numbers in numerical order from smallest to largest: $\pi, \frac{22}{7}, -\sqrt{2}, -\sqrt{7}, \sqrt{7}-\sqrt{2}, \frac{7}{5}, -\left|\frac{7}{4}\right|$

33. *Automotive technician* Front end shims come in $\frac{1}{32}$-, $\frac{1}{16}$-, and $\frac{1}{8}$-inch thicknesses.

 (a) Which is the smallest shim?
 (b) Which is the largest shim?
 (c) What is the decimal size of the $\frac{1}{16}$-inch shim?

34. *Automotive technician* The service manager records how long each technician has worked on a job in hours written as a decimal. If you spent $\frac{3}{4}$ h to change a front engine mount, $1\frac{1}{2}$ h to replace the engine oil pan, and $2\frac{1}{4}$ h to replace the clutch, how would the service manager record these times?

35. *Electricity* The voltage across an element with respect to the ground is −19.4 V initially and then it becomes −16.8 V. Determine the absolute value of the change in the voltage.

36. *Electricity*   The voltage across an element with respect to the ground is -6.5 V initially, and then it becomes to 14.7 V. Determine the absolute value of the change in the voltage.

37. *Automotive technician*   Five technicians were given the same task. By noon, Sheila had completed $\frac{5}{7}$ of her task, José had completed $\frac{3}{5}$ of his, Lamar had completed $\frac{4}{9}$ of his, Hazel had completed $\frac{2}{3}$ of hers, and Robert had completed $\frac{9}{16}$ of his task. Rank the technicians in order according to who completed most of the assigned task, with the person who had finished the most listed first.

38. *Electricity*   The voltage across an element with respect to the ground is 23.7 V initially and then it becomes -5.2 V. Determine the absolute value of the change in the voltage.

39. *Automotive technician*   Helen drives past mile marker 157 on the interstate highway. In the next hour, she drives 64 miles. At what mile marker is she now? (Hint: There are two correct answers.)

40. *Electricity*   As resistors warm, the value of their resistance increases. One resistor's value changed from 14 Ω to 19 Ω. Determine the absolute value of the change in the resistance.

 **In Your Words**

41. What is the meaning of *absolute value*?

42. Explain why you might want the absolute value of an answer rather than the actual value.

43. On a sheet of paper write an exercise that requires the use of one or more rational numbers to solve the exercise. On the back of the paper write your solution. Share your exercise with a friend. Rewrite

your exercise or solution to clarify places where your classmate had difficulty.

44. On a sheet of paper write an exercise that requires the use of absolute values to solve the exercise. Share your exercise with a friend. Rewrite your exercise to clarify places where your classmate had difficulty.

≡ 1.2
BASIC LAWS OF REAL NUMBERS

Knowing the different kinds of numbers is important. But, it is even more important that you can use these numbers. In order to use them, there are some basic laws of the real numbers that you should know. Two of these laws apply to addition and multiplication.

Commutative Laws

The first laws state that the order in which two numbers are added or multiplied does not matter. These are called the **commutative laws**. For example, the commutative law for addition would state that $5 + 2 = 2 + 5$ and also $\frac{17}{4} + \frac{-3}{7} = \frac{-3}{7} + \frac{17}{4}$. In the same way, the commutative law for multiplication guarantees that $4 \times 8 = 8 \times 4$ and also that $\dfrac{-\pi}{3} \times \dfrac{6}{\sqrt{2}} = \dfrac{6}{\sqrt{2}} \times \dfrac{-\pi}{3}$.

> **Commutative Laws for Addition and Multiplication**
>
> Symbolically, the commutative laws would be written as follows:
>
> Commutative law for addition: $a + b = b + a$.
> Commutative law for multiplication: $a \times b = b \times a$.

Caution

Subtraction and division are **not** commutative. Thus,

$$7 - 4 \neq 4 - 7 \quad \text{and} \quad 12 \div 3 \neq 3 \div 12$$

Associative Laws

The second laws state that when you have a group of three numbers it does not matter which two are added or multiplied first. These are called the **associative laws**. For example, the associative law for addition would state that $(4+2)+8 = 4+(2+8)$ or $\left(\frac{6}{5} + \frac{-2}{3}\right) + \sqrt{3} = \frac{6}{5} + \left(\frac{-2}{3} + \sqrt{3}\right)$. Similarly, the associative law for multiplication guarantees that $(6 \times 7) \times 2 = 6 \times (7 \times 2)$ and $\left(\frac{\pi}{4} \times \sqrt{5}\right) \times \frac{1}{\sqrt{7}} = \frac{\pi}{4} \times \left(\sqrt{5} \times \frac{1}{\sqrt{7}}\right)$.

> **Associative Laws for Addition and Multiplication**
>
> Symbolically, the associative laws would be written as follows:
>
> Associative law for addition: $(a+b)+c = a+(b+c)$.
> Associative law for multiplication: $(ab)c = a(bc)$.

Caution

Subtraction and division are **not** associative. Thus,

$$15 - (7-3) \neq (15-7) - 3 \quad \text{and} \quad 18 \div (6 \div 3) \neq (18 \div 6) \div 3$$

Distributive Law

A third law combines addition and multiplication and states that multiplication distributes over addition. This law, called the **distributive law**, means that if you have a problem such as $12(13+7)$ you could evaluate it two ways. One way would be to first add the numbers inside the parentheses and get $12(13+7) = 12 \times 20 = 240$. The other method would first distribute the multiplication and then add those products. For this same problem, you would get $12(13+7) = (12 \times 13) + (12 \times 7) = 156 + 84 = 240$.

> **Distributive Law**
>
> Symbolically, the distributive law is written as
>
> $$a(b+c) = ab + ac.$$

Order of Operations

Now, let's consider the order of operations. The order of operations is very important in mathematics. Mathematicians have made some basic agreements that some operations are to be performed before others. By these agreements, operations such as addition, subtraction, multiplication, division, and raising to powers are to be performed in the following order: (Remember, not all operations are in every problem.)

Order of Operations

Perform the operations in a problem in the following order.

1. Operations inside parentheses or above or below a fraction bar. Always start with the innermost parentheses and work outward.
2. Raising to a power.
3. Multiplications and divisions in the order in which they appear from left to right.
4. Additions and subtractions in the order in which they appear from left to right.

Hint

Some people use the acronym "**P**lease **E**xcuse **M**y **D**ear **A**unt **S**ally" to help remember the order of operations. Here the **P** in "Please" stands for parentheses; the **E** for exponents (raise to a power), **M** and **D** for multiplication and division, and the **A** in "Aunt" and **S** in "Sally" for addition and subtraction.

Computers and most scientific calculators are programmed to perform the operations according to the previous rules. The logic used by these calculators and computers is called **algebraic logic** or the **algebraic operating system**.

Checking: Does Your Calculator Use Algebraic Logic?

Work this problem on your calculator: $3 + 5 \times 7$. If your calculator gives the answer 38, then it uses algebraic logic.

If your calculator gives the answer 56, then it does not use algebraic logic.

If your calculator gives an answer other than 38 or 56, then you probably made a mistake. Try again.

EXAMPLE 1.9

We will consider the problem $6 + 10 \div 2$ in two different ways. Following the rules, we should do the division first:

$$6 + 10 \div 2 = 6 + 5 = 11$$

Notice that $6 + 10 \div 2$ is performed as if it were $6 + (10 \div 2)$. If you wanted to indicate that the addition should be performed first, then parentheses would have to

EXAMPLE 1.9 (Cont.)

be added:

$$(6+10) \div 2 = 16 \div 2 = 8$$

As you can see, the answers 8 and 11 are not the same. This shows the importance of performing the operations in the correct order.

≡ **Note**

$(6+10) \div 2$ is sometimes written as $\frac{6+10}{2}$. Here we perform the operation above the fraction bar (order of operation #1) before we perform the division. Thus, $\frac{6+10}{2} = \frac{16}{2} = 8$.

EXAMPLE 1.10

Solve $8 + 9 \times 2 + 16 - 12 \div 4$

Solution   Orders of operations #1 and #2 do not apply because there are no parentheses, fractions, or exponents. According to order of operation #3, we should perform the multiplication and division in order from left to right. Thus, if we use parentheses to indicate what to do first, we get

$$8 + 9 \times 2 + 16 - 12 \div 4 = 8 + (9 \times 2) + 16 - (12 \div 4)$$
$$= 8 + 18 + 16 - 3$$
$$= 39$$

EXAMPLE 1.11

Solve $6 + 12 \div 4 \times 5 - 4 \times 3$

Solution   We begin by performing the multiplication and division operations in order from left to right. First we perform $12 \div 4$ and then multiply this answer by 5 as shown.

$$6 + 12 \div 4 \times 5 - 4 \times 3 = 6 + 3 \times 5 - 4 \times 3$$
$$= 6 + 15 - 4 \times 3$$
$$= 6 + 15 - 12$$
$$= 21 - 12$$
$$= 9$$

EXAMPLE 1.12

Solve $28 - (26 - (3 - (4 - 3)))$

Solution   We begin with the innermost set of parentheses and perform the operations inside of them. So, we begin by working $4 - 3$.

$$28 - (26 - (3 - (4 - 3))) = 28 - (26 - (3 - 1))$$
$$= 28 - (26 - 2)$$

EXAMPLE 1.12 (Cont.)

$$= 28 - 24$$
$$= 4$$

Notice that at each step we performed the operations in the innermost parentheses.

Sometimes different grouping symbols are used to help a person see symbols that go together. Example 1.12 might have been written as follows

$$28 - \{26 - [3 - (4 - 3)]\}$$

You would first work the problem inside the parentheses ( ), then inside the brackets [ ], and finally, inside the braces { }.

Identity Elements

Two numbers that are very important are the **identity elements**. There is an identity element for addition and one for multiplication. The identity element for addition is zero (0) and for multiplication is one (1). When you add the identity element for addition to a number, the answer does not change. So, $4 + 0 = 4$; $\frac{-3}{8} + 0 = \frac{-3}{8}$; and $0 + \frac{\sqrt{2}}{5} = \frac{\sqrt{2}}{5}$. In the same way, when you multiply a number by the identity element for multiplication, the answer does not change. For example, $17 \times 1 = 17$; $1 \times \sqrt{5} = \sqrt{5}$; and $\frac{-9}{7} \times 1 = \frac{-9}{7}$.

> **Identity Elements**
>
> Symbolically, these are written as follows:
>
> Identity element for addition is 0: $a + 0 = 0 + a = a$.
>
> Identity element for multiplication is 1: $a \times 1 = 1 \times a = a$.

Inverses

Now that we have identity elements for addition and multiplication we can determine the inverses of each number. Every number, except zero, has two inverses—an inverse for addition and an inverse for multiplication. Zero has only an inverse for addition.

The inverse for addition, or **additive inverse**, of a number is the number that gives a sum of zero when added to the original number. Thus, the additive inverse of 5 is -5 because $5 + (-5) = 0$. (This also means that 5 is the additive inverse of -5.) The additive inverse of $\frac{-8}{7}$ is $\frac{8}{7}$ because $\frac{-8}{7} + \frac{8}{7} = 0$ and the additive inverse of $-\sqrt{5}$ is $\sqrt{5}$ because $-\sqrt{5} + \sqrt{5} = 0$.

≡ **Note**

The additive inverse of a number b is $-b$ because $b + -b = 0$. If b is a negative number, such as -8, then $-b$ is positive, $-(-8) = 8$. The additive inverse of zero is zero because $0 + 0 = 0$.

Hint

To enter a negative number on a calculator you need to press a special key. On some calculators it is the $\boxed{+/-}$ or $\boxed{\text{CHS}}$ key and it is pressed *after* the number is entered. Thus, -25 would be entered as $25 \boxed{+/-}$. Some calculators use a $\boxed{(-)}$ key and it is pressed *before* the number is entered. Here, -35 would be entered as $\boxed{(-)} 35$.

EXAMPLE 1.13

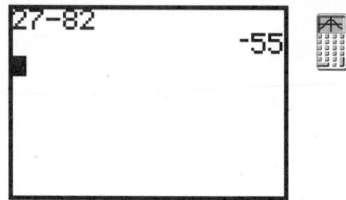

```
27-82
          -55
```

FIGURE 1.3a

```
27-82
          -55
-91-63
         -154
```

FIGURE 1.3b

```
27-82
          -55
-91-63
         -154
57--28
           85
```

FIGURE 1.3c

```
          -55
-91-63
         -154
57--28
           85
-12--38
           26
```

FIGURE 1.3d

Use a calculator to compute each of the following: (a) $27 - 82$, (b) $-91 - 63$, (c) $57 - -28$, and (d) $-12 - -38$.

Solutions The actual keystrokes given below are for a TI-82 graphics calculator. However, any graphing calculator that does not use reverse polish notation (RPN) would work in the same way.

(a) Press $27 \boxed{-} 82 \boxed{\text{ENTER}}$. The result, -55, is shown in Figure 1.3a.

(b) Press $\boxed{(-)} 91 \boxed{-} 63 \boxed{\text{ENTER}}$. The result, -154, is shown in Figure 1.3b. When you look at Figure 1.3b, you will see that the result from (a) is still displayed. This will continue to happen until you press the $\boxed{\text{CLEAR}}$ key. Notice that you had to press the $\boxed{(-)}$ key for the negative sign of -91. You pressed the $\boxed{-}$ key to subtract 63.

(c) Press $57 \boxed{-} \boxed{(-)} 28 \boxed{\text{ENTER}}$. The result, 85, is shown in Figure 1.3c. Notice that you had to press the $\boxed{-}$ key for subtraction and the $\boxed{(-)}$ key for the negative sign on -28.

(d) Press $\boxed{(-)} 12 \boxed{-} \boxed{(-)} 38 \boxed{\text{ENTER}}$. The result, 26, is shown in Figure 1.3d. Again, you should have pressed the $\boxed{(-)}$ key to indicate that -12 and -38 are negative.

Each number (except zero) also has an inverse for multiplication. The **multiplicative inverse** of a number is the number that will give a product of 1 when multiplied by the original number. So, the multiplicative inverse of 5 is $\frac{1}{5}$ because $5 \times \frac{1}{5} = 1$. (This means that the multiplicative inverse of $\frac{1}{5}$ is 5.) The multiplicative inverse of $\frac{-3}{8}$ is $\frac{8}{-3}$ because $\frac{-3}{8} \times \frac{8}{-3} = 1$ and the multiplicative inverse of $-\dfrac{\sqrt{2}}{3}$ is $-\dfrac{3}{\sqrt{2}}$ because $-\dfrac{\sqrt{2}}{3} \times (-\dfrac{3}{\sqrt{2}}) = 1$.

The multiplicative inverse of a number b is $\dfrac{1}{b}$ because $b \times \dfrac{1}{b} = 1$ provided that $b \neq 0$. Notice that a number and its multiplicative inverse have the same sign. Zero does not have a multiplicative inverse because the product of 0 and any other number is 0. The fact that $b \neq 0$ means that the denominator of a fraction, and the divisor in a division problem, cannot be 0. Since $\frac{1}{0}$ is not a number, $0 \times \frac{1}{0}$ is not defined. In particular, $0 \times \frac{1}{0} \neq 0$ and $0 \times \frac{1}{0} \neq 1$.

≡ **Note**

You might have already noticed that the multiplicative inverse and the reciprocal of a number are the same.

Inverse Elements

Symbolically, the inverse elements are written as:

Inverse element for addition of the number b is $-b$.

Inverse element for multiplication, or reciprocal, of the number b, $b \neq 0$, is $\dfrac{1}{b}$.

There are two other properties of zero that are important to remember. The product of any number multiplied by zero is zero. The quotient of zero divided by any number (except zero) is also zero. For example, $7 \times 0 = 0$ and $0 \div 9 = 0$. In symbols we would write these as follows:

Properties of 0

$b \times 0 = 0$

$0 \div b = 0$ (or $\dfrac{0}{b} = 0$) if $b \neq 0$

Exercise Set 1.2

In Exercises 1–12, determine which of the basic laws of real numbers is being used.

1. $4 + 3 = 3 + 4$
2. $8(4 \cdot 6) = (8 \cdot 4)6$
3. $9 \cdot 7 = 7 \cdot 9$
4. $4(3 + 5) = 4 \cdot 3 + 4 \cdot 5$
5. $(9 + 3) + 6 = 9 + (3 + 6)$
6. $(\pi + \sqrt{2})\sqrt{5} = \pi\sqrt{5} + \sqrt{2}\sqrt{5}$
7. $1 \times 81 = 81$
8. $\frac{3}{16} \times \frac{16}{3} = 1$
9. $-5 + 5 = 0$
10. $19 - 5 = -5 + 19$
11. $(\sqrt{3} \cdot \sqrt{4})\sqrt{5} = \sqrt{3}(\sqrt{4} \cdot \sqrt{5})$
12. $16 = 0 + 16$

Name the additive inverse of each of the numbers in Exercises 13–16.

13. 91
14. -8
15. $\sqrt{2}$
16. $\frac{1}{3}$

Name the multiplicative inverse of each of the numbers in Exercises 17–20.

17. $\frac{1}{2}$
18. -5
19. $\dfrac{\sqrt{2}}{2}$
20. $-\frac{3}{7}$

Use the correct order of operations to solve Exercises 21–30.

21. $16 - 8 \div 4$
22. $16 \div 8 + 2$
23. $24 + 3 - 10 \div 5 \times 8 + 2$
24. $13 \times 7 - 26 \div 5 + 5$
25. $(7 - 2) - (3 + 8 - 7)$
26. $7 - 2 - 3 + 8 - 7$
27. $7 \times 3 + 5 \times 2$
28. $6 \times 4 - 3 \times 5$

29. $\{-[5-(8-4)+(3-7)]-(4-2)\}$

30. $(14+3(8-6)+4(9-5))$

Use your calculator to work Exercises 31–38. Be sure to include all parentheses.

31. $-39-72$

32. $-257-63$

33. $192--78$

34. $-56--344$

35. **(a)** $243+(-691+89)$
  (b) $(243+-691)+89$

36. **(a)** $-41(63+-177)$
  (b) $-41\times63+-41\times-177$

37. **(a)** $(-342+-18)91$
  (b) $-342\times91+-18\times91$

38. **(a)** $\frac{3}{7}\times\frac{7}{3}$ **(b)** $-\frac{4}{11}+\frac{4}{11}$

In Your Words

39. Explain how you know when to use the $\boxed{(-)}$ key and when to use the $\boxed{-}$ key on a calculator.

40. **(a)** What is the inverse element for addition of -76?
  (b) What is the inverse element for multiplication of -37?
  (c) Describe how you can tell if your answers in (a) and (b) are correct.

☰ 1.3
BASIC OPERATIONS WITH REAL NUMBERS

The ability to work with numbers is very important in mathematics. In this book you will be using calculators and computers to help solve problems. In many ways, these solving technologies make it more important that you have good arithmetic skills. Both the calculator and the computer will only give the correct answer if you correctly tell the machine what to do. This section provides a brief review of arithmetic with real numbers. The first group of rules will use integers in the examples. Later in this section, we will examine the arithmetic of rational numbers.

Addition of Integers

> **Rule 1**
>
> To add two real numbers with the same sign, add the numbers and give to the sum the sign of the original numbers.

EXAMPLE 1.14

$+8+(+7)$
Solution $8+7=15$
8 and 7 both have a $+$ sign.
  So, $+8+(+7)=+15$.

EXAMPLE 1.15

$-9+(-24)$
Solution $9+24=33$
9 and 24 both have a $-$ sign.
So, $-9+(-24)=-33$.

Rule 2

To add two real numbers with different signs, take the absolute value of both numbers, subtract the smaller absolute value from the larger, and give to the answer the sign of the number with the larger absolute value.

EXAMPLE 1.16

$-8+(+5)$
Solution $|-8|=8$
$|+5|=5$
$8>5$, so the answer will be negative.
$8-5=3$ and so, $-8+(+5)=-3$.

EXAMPLE 1.17

$+14+(-9)$
Solution $|+14|=14$
$|-9|=9$
$14>9$ so the answer will be positive.
$14-9=5$ and so $+14+(-9)=+5$.

≡ Note   Remember that positive numbers are often written without a $+$ sign.

EXAMPLE 1.18

$-19+(-18)+23$
Solution $-19+(-18)+23=-37+23$, because $-19+(-18)=-37$
$=-14$

Application

EXAMPLE 1.19

When you started your car one morning, the temperature was $-16°C$. Later you hear someone say that the temperature has gone up $24°C$. What is the latest temperature?

EXAMPLE 1.19 (Cont.)

Solution To find the new temperature, we add the increase to the morning temperature:

$$-16 + 24 = 8$$

The new temperature is 8°C.

Application

EXAMPLE 1.20

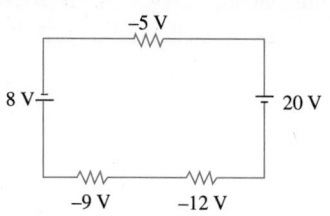

FIGURE 1.4

Figure 1.4 shows a closed circuit with two voltage sources and three resistors. The sum of voltages in such a loop must be 0. Check the voltage drops given in the figure to determine if the sum is 0.

Solution The voltage drops in Figure 1.4 are 8 V, −5 V, 20 V, −12 V, and −9 V. The sum of the voltage drops is

$$
\begin{aligned}
8 + (-5) + 20 + (-12) + (-9) &= 3 + 20 + (-12) + (-9) \\
&= 23 + (-12) + (-9) \\
&= 11 + (-9) \\
&= 2
\end{aligned}
$$

A sum of 2 V indicates an error, probably in one of the measurements.

Subtraction of Integers

> **Rule 3**
>
> To subtract one real number from another, change the sign of the number being subtracted and then add according to Rule 2.

EXAMPLE 1.21

$(-6) - (-9)$

Solution $(-6) - (-9) = (-6) + (+9)$ Change −9 to +9 and add.

$ = 3$

EXAMPLE 1.22

$(-13) - (+7)$

Solution $(-13) - (+7) = (-13) + (-7)$ Change +7 to −7 and add.

$ = -20$

Caution

Do not use the $\boxed{-}$ (subtraction) key to enter a negative number into your calculator. The $\boxed{-}$ key is used only for subtraction. A negative number is entered by using the $\boxed{+/-}$ or $\boxed{(-)}$ key.

EXAMPLE 1.23

$14 - 29$

Solution $14 - 29 = 14 + (-29)$ Change 29 to -29 and add.
$$= -15$$

Application

EXAMPLE 1.24

When performing a front-end alignment, a technician must work with the included angle, the steering axis inclination, and the camber. To find the steering axis angle, you subtract the camber angle from the included angle. If the included angle is $5°$ and the camber angle is $-2°$, what is the steering axis?

Solution   Subtracting the camber angle from the included angle, we have

$$5° - (-2°) = 5° + 2°$$
$$= 7°$$

The steering axis angle is $7°$.

Multiplication and Division of Integers

> **Rule 4**
>
> The product (or quotient) of two real numbers with the same sign is the product (or quotient) of their absolute values.

EXAMPLE 1.25

$(-8)(-9)$

Solution $(-8)(-9) = |-8| \times |-9| = 8 \times 9$
$$= 72$$

EXAMPLE 1.26

$(-27) \div (-3)$

Solution $(-27) \div (-3) = |-27| \div |-3|$
$$= 27 \div 3$$
$$= 9$$

Rule 5

The product (or quotient) of two real numbers with different signs is the additive inverse of the product (or quotient) of their absolute values.

EXAMPLE 1.27

$(-8)(9)$
Solution $\quad |-8| = 8 \qquad |9| = 9 \qquad 8 \times 9 = 72$
The additive inverse of 72 is -72.
$\qquad$ So, $(-8)(9) = -72$.

EXAMPLE 1.28

$81 \div -3$
Solution $\quad |81| = 81 \qquad |-3| = 3 \qquad 81 \div 3 = 27$
The additive inverse of 27 is -27 and so $81 \div (-3) = -27$.

Application

EXAMPLE 1.29

The total cost of a car, including finance charges, is \$9,216. This is to be repaid in 48 equal payments. How much is each payment?

Solution $\quad$ To find the size of each payment, we need to divide the total cost by the number of payments.

$$9,216 \div 48 = 192$$

Each payment is \$192.

Application

EXAMPLE 1.30

A metal plate contracts (shrinks) 0.2 mm for each degree below 60°F. If the plate is 42 mm wide at 60°F, what is its width at 35°F?

Solution $\quad$ The temperature change from 60°F to 35°F is $60° - 35° = 25°$F. A decrease, or shrinkage, of 0.2 mm can be written as -0.2. So, for a 25° change in temperature, the total shrinkage is

$$-0.2 \times 25 = -5.0$$

So, the plate will shrink 5 mm and, at 35°F, its width is

$$42 \, \text{mm} - 5 \, \text{mm} = 37 \, \text{mm}$$

Arithmetic of Rational Numbers

We will now look at the arithmetic of rational numbers. Remember, the numerator of a rational number is the top number and the denominator is the bottom number. In the rational number $\frac{-5}{8}$, -5 is the numerator and 8 is the denominator.

Sometimes a rational number is written as a mixed number that combines an integer and a rational number. Examples of mixed numbers are $2\frac{1}{2}$ and $-4\frac{2}{3}$. While it is not shown, the two parts of a mixed number are being added. Thus, $2\frac{1}{2}$ means $2 + \frac{1}{2}$ and $-4\frac{2}{3}$ means $-(4 + \frac{2}{3})$.

Every rational number can be written with either zero or one negative sign. If a rational number is written with more than one negative sign, you can rewrite it by reducing the number of negative signs by two.

EXAMPLE 1.31

The following rational number has three negative signs: $-\frac{-3}{-8}$. You can delete any two of these and get $\frac{-3}{8}$ or $-\frac{3}{8}$ or $\frac{3}{-8}$ depending on which two negative signs you delete. All three of these are equivalent to the original number.

EXAMPLE 1.32

The following rational number has two negative signs: $-\frac{9}{-5}$. If you delete both you get $\frac{9}{5}$, which is equivalent to the original number.

Rule 6

To add (or subtract) two rational numbers, change both denominators to the same positive integer (the **common denominator**). Add (or subtract) the numerators, and place the result over the common denominator. In symbols, this is written as

$$\text{Addition:} \quad \frac{a}{b} + \frac{c}{d} = \frac{ad}{bd} + \frac{bc}{bd} = \frac{ad + bc}{bd}$$

$$\text{Subtraction:} \quad \frac{a}{b} - \frac{c}{d} = \frac{ad}{bd} - \frac{bc}{bd} = \frac{ad - bc}{bd}$$

EXAMPLE 1.33

$$\frac{2}{3} + \frac{-5}{6}$$

Solution   The denominators are 3 and 6. A common denominator of 3 and 6 is 6, so $\frac{2}{3} = \frac{4}{6}$. The problem then becomes $\frac{2}{3} + \frac{-5}{6} = \frac{4}{6} + \frac{-5}{6} = \frac{4+(-5)}{6} = \frac{-1}{6}$. So, $\frac{2}{3} + \frac{-5}{6} = \frac{-1}{6}$.

EXAMPLE 1.34

$$\frac{7}{-8} + \frac{-5}{6}$$

Solution The denominators are -8 and 6. A common denominator of -8 and 6 is 24. So, $\frac{7}{-8} = \frac{-21}{24}$ and $\frac{-5}{6} = \frac{-20}{24}$. Thus,

$$\frac{7}{-8} + \frac{-5}{6} = \frac{-21}{24} + \frac{-20}{24}$$
$$= \frac{-21 + (-20)}{24}$$
$$= \frac{-41}{24}$$

≡ Note

Many common denominators of two numbers are possible. For instance, in the above example, $\frac{7}{-8} + \frac{-5}{6}$, some possible common denominators are -24, 48, -48, 72, and -72. We normally select the smallest positive common denominator because it makes computations easier.

EXAMPLE 1.35

$$\frac{3}{8} + \left(-1\frac{5}{16}\right)$$

Solution A common denominator is 16. Thus, $\frac{3}{8} = \frac{6}{16}$ and $-1\frac{5}{16} = \frac{-21}{16}$. The example then becomes $\frac{6}{16} + \frac{-21}{16} = \frac{-15}{16}$.

EXAMPLE 1.36

$$\frac{-5}{7} - \frac{-8}{3}$$

Solution This is a subtraction problem. We first change it to an addition problem using Rule 3: $\frac{-5}{7} - \frac{-8}{3} = \frac{-5}{7} + \frac{8}{3}$. A common denominator of 7 and 3 is 21: $\frac{-5}{7} = \frac{-15}{21}$ and $\frac{8}{3} = \frac{56}{21}$. Thus,

$$\frac{-5}{7} - \frac{-8}{3} = \frac{-15}{21} + \frac{56}{21}$$
$$= \frac{-15 + 56}{21}$$
$$= \frac{41}{21}$$

Application

EXAMPLE 1.37

The circular pipe shown in Figure 1.5a has an inside radius of 1.35 cm and an outside radius of 1.575 cm. What is the thickness of the pipe?

EXAMPLE 1.37 (Cont.)

Solution   An outline of the pipe is shown in Figure 1.5b. To find the thickness, we need to subtract the inner radius from the outer radius, or $1.575 - 1.35$.

$$\begin{array}{r} 1.575 \\ -1.35 \\ \hline 0.225 \end{array}$$

The pipe is 0.225 cm thick.

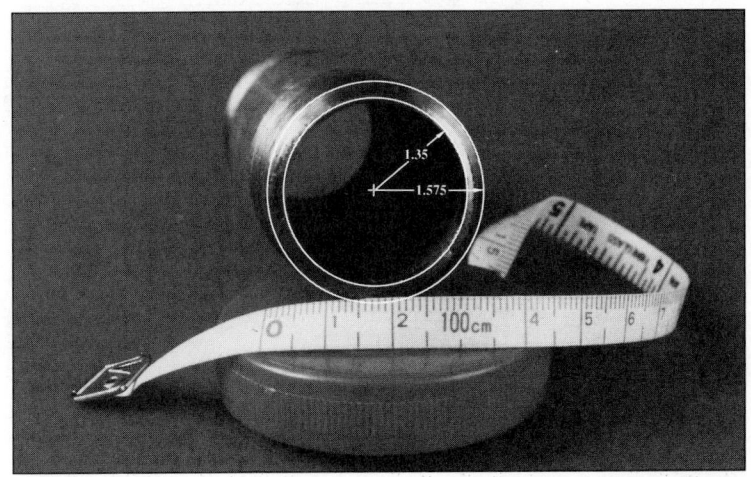

Courtesy of Ruby Gold

FIGURE 1.5a

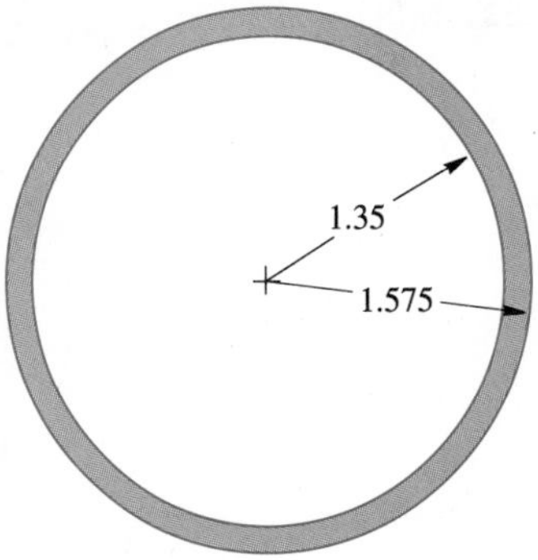

FIGURE 1.5b

Rule 7

To multiply two rational numbers, multiply the numerators and multiply the denominators. In symbols, this is written as

$$\frac{a}{b} \times \frac{c}{d} = \frac{a \times c}{b \times d} = \frac{ac}{bd}$$

EXAMPLE 1.38

$$\frac{3}{4} \times \frac{-5}{8}$$

Solution $\dfrac{3}{4} \times \dfrac{-5}{8} = \dfrac{3 \times (-5)}{4 \times 8} = \dfrac{-15}{32}$

EXAMPLE 1.39

$$\left(\frac{-7}{8}\right)(-9)$$

Solution $\left(\frac{-7}{8}\right)(-9) = \frac{-7}{8} \times \frac{-9}{1} = \frac{(-7)(-9)}{8 \times 1} = \frac{63}{8}$

 Hint

Sometimes it is possible to "cancel" factors that appear in the numerators and denominators *before* you multiply the numbers. For example, in $\frac{5}{12} \times \frac{8}{25} = \frac{5 \times 8}{12 \times 25}$, the numerator and denominator have a factor of 5×4 in common. Thus, we can write

$$\frac{5}{12} \times \frac{8}{25} = \frac{5 \times 8}{12 \times 25} = \frac{\cancel{5} \times \cancel{4} \times 2}{3 \times \cancel{4} \times \cancel{5} \times 5} = \frac{2}{3 \times 5} = \frac{2}{15}$$

Application

EXAMPLE 1.40

An auto shop charges $56 for each hour they work on your car. How much will they charge if they work on it for $3\frac{1}{4}$ hours?

Solution We need to multiply $56 \times 3\frac{1}{4}$. We begin by changing 56 and $3\frac{1}{4}$ to fractions. We write 56 as $\frac{56}{1}$ and $3\frac{1}{4}$ as $\frac{13}{4}$.

$$56 \times 3\frac{1}{4} = \frac{56}{1} \times \frac{13}{4}$$

$$= \frac{56 \times 13}{1 \times 4} \qquad \text{or} \qquad \frac{\cancel{4} \times 14 \times 13}{1 \times \cancel{4}} = \frac{14 \times 13}{1}$$

$$= \frac{728}{4} = 182 \qquad\qquad\qquad\qquad = 182$$

They will charge $182.

Caution

Be careful! $56 \times 3\frac{1}{4}$ does not mean $56 \times 3 \times \frac{1}{4} = 168 \times \frac{1}{4} = \frac{168}{4}$. Remember, $3\frac{1}{4}$ is a short way of writing $3 + \frac{1}{4}$ and $56 \times 3\frac{1}{4}$ means $56 \times \left(3 + \frac{1}{4}\right)$.

Application

EXAMPLE 1.41

To change $14°F$ to its equivalent temperature in degrees Celsius, you need to compute $\frac{5}{9}(14° - 32°)$. What is this temperature?

Solution

$$\frac{5}{9}(14° - 32°) = \frac{5}{9}(-18°)$$

$$= \frac{5}{9} \times \frac{-18}{1}$$

$$= \frac{5 \times (-18)}{9 \times 1} \qquad \text{or} \qquad \frac{5 \times (-2) \times \cancel{9}}{\cancel{9} \times 1} = \frac{-10}{1}$$

EXAMPLE 1.41 (Cont.)

$$= \frac{-90}{9} = -10 \qquad\qquad = -10$$

So, $-10°C$ is the same as $14°F$.

Rule 8

To divide one rational number by another, multiply the first number by the reciprocal of the second. In symbols, this is written as

$$\frac{a}{b} \div \frac{c}{d} = \frac{a}{b} \times \frac{d}{c} = \frac{ad}{bc}$$

EXAMPLE 1.42

$$\frac{-5}{8} \div \frac{2}{3}$$

Solution The reciprocal of $\frac{2}{3}$ is $\frac{3}{2}$ so

$$\frac{-5}{8} \div \frac{2}{3} = \frac{-5}{8} \times \frac{3}{2}$$
$$= \frac{(-5)(3)}{8 \times 2}$$
$$= \frac{-15}{16}$$

EXAMPLE 1.43

$$\frac{-7}{8} \div \frac{-9}{5}$$

Solution The reciprocal of $\frac{-9}{5}$ is $\frac{5}{-9}$ $\left(\text{or } \frac{-5}{9}\right)$ and so

$$\frac{-7}{8} \div \frac{-9}{5} = \frac{-7}{8} \times \frac{-5}{9}$$
$$= \frac{(-7)(-5)}{8 \times 9}$$
$$= \frac{35}{72}$$

Application

EXAMPLE 1.44

A tank of fuel weighs $185\frac{1}{6}$ pounds. If a gallon of gasoline weighs $7\frac{1}{3}$ pounds, how many gallons are in the tank?

| EXAMPLE 1.44 (Cont.) | **Solution** We need to divide the weight of the tank by the number of pounds for one gallon. |

$$185\frac{1}{6} \div 7\frac{1}{3} = \frac{1,111}{6} \div \frac{22}{3}$$

$$= \frac{1,111}{6} \times \frac{3}{22}$$

$$= \frac{1,111 \times 3}{6 \times 22}$$

$$= \frac{3,333}{132}$$

$$= 25\frac{33}{132} = 25\frac{1}{4}$$

The tank contains $25\frac{1}{4}$ gallons of gasoline.

Exercise Set 1.3

Perform the indicated operation in Exercises 1–40.

1. $+27 + (+23)$

2. $8 + (-19)$

3. $27 + (-13)$

4. $-9 + (-8)$

5. $7 - 16$

6. $29 - (-8)$

7. $-8 - 16$

8. $-25 - (-13)$

9. $-37 - (-49)$

10. $(-2)(6)$

11. $(-3)(-5)$

12. $(7)(-8)$

13. $-38 \div 4$

14. $-45 \div (-9)$

15. $\frac{3}{4} + \frac{-5}{8}$

16. $-1\frac{3}{4} + \frac{-2}{3}$

17. $\frac{-9}{5} + \frac{7}{3}$

18. $\frac{-2}{3} + \frac{5}{6}$

19. $\frac{2}{5} + \frac{-1}{4}$

20. $\frac{-4}{5} + \frac{-5}{6}$

21. $\frac{3}{8} - \frac{-1}{4}$

22. $1\frac{1}{3} - \frac{-5}{6}$

23. $\frac{-9}{10} - \frac{2}{3}$

24. $-\frac{5}{16} - \frac{-3}{8}$

25. $\frac{5}{32} - \frac{1}{8}$

26. $-\frac{7}{3} - \frac{-6}{7}$

27. $\frac{-2}{3} \times \frac{4}{5}$

28. $\frac{3}{4} \times \frac{-5}{8}$

29. $\frac{-1}{8} \times \frac{-3}{4}$

30. $\frac{9}{16} \times \frac{1}{2}$

31. $-\frac{4}{3} \times \frac{5}{2}$

32. $\frac{-9}{5} \times \frac{-3}{8}$

33. $-\frac{3}{4} \div \frac{-5}{8}$

34. $1\frac{3}{4} \div \frac{-2}{3}$

35. $-\frac{3}{5} \div 4$

36. $-\frac{3}{8} \div \frac{1}{4}$

37. $-\frac{7}{5} \div \left(-\frac{5}{7}\right)$

38. $\frac{2}{3} \div \left(-\frac{7}{3}\right)$

39. $\left(-2 + \frac{2}{3}\right) \times \frac{-1}{2} - \left[\frac{3}{2} \div (-3)\right] + \frac{7}{3}$

40. $\frac{4}{3} \times \frac{-7}{8} \times \frac{3}{-5} \div \frac{4}{5} + \frac{-5}{8} - \frac{7}{8}$

Solve Exercises 41–52.

41. *Recreation*   A ski resort received $17\frac{1}{2}''$ of snow in December, $15\frac{7}{8}''$ in January, $29\frac{3}{4}''$ in February, and $15\frac{3}{8}''$ in March. How much snow did the resort get during these four months?

42. *Auto technician*   The toe-in reading is $\frac{5}{32}$ on one front wheel and $\frac{3}{16}$ on the other. What is the total toe-in?

43. *Electricity*   When two or more voltages are connected in series, the total voltage is the sum of the separate voltages. When the voltages are connected in the same direction, they are all considered to be positive. When the voltages are connected in the opposite direction, one direction is considered positive and the other negative. Find the total voltage of the batteries in Figure 1.6.

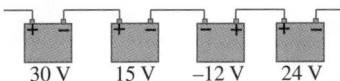

FIGURE 1.6

44. *Construction* A planer removed $\frac{1}{5}$ inch from a $1\frac{5}{8}$-inch board. Find the thickness of the finished board.

45. *Recreation* A $14\frac{1}{2}$-mi race has 3 checkpoints. The first checkpoint is $4\frac{5}{8}$ mi from the starting point. The second checkpoint is $3\frac{1}{5}$ mi from the first checkpoint. The third checkpoint is $3\frac{3}{8}$ mi from the second checkpoint.
   (a) How many miles is it from the starting point to the second checkpoint?
   (b) How many miles is it from the starting point to the third checkpoint?
   (c) How many miles is it from the third checkpoint to the finish line?

46. *Automotive technology* An automobile traveled $427\frac{1}{5}$ mi on $13\frac{3}{4}$ gal of gasoline. How many miles did it get per gallon?

47. *Electrical technician* A certain job requires 37 pieces of electrical wire that are each $2'3\frac{1}{2}''$ long. What is the total length of wire that is needed?

48. *Construction* A 4-ft by 8-ft sheet of plywood is cut into strips, each $1\frac{5}{16}$ in. by 4 ft. The saw cut is $\frac{1}{8}$-in. wide. How many strips can be cut from the sheet?

49. *Construction* Find the length indicated by the question mark in Figure 1.7.

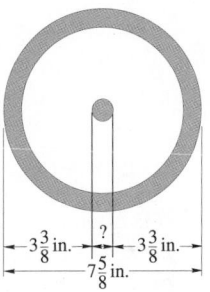

FIGURE 1.7

50. *Electricity* Determine the sum of the voltages in the closed circuit in Figure 1.8.

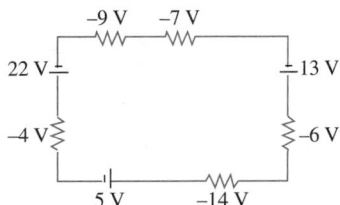

FIGURE 1.8

51. *Electricity* The voltage across an element with respect to the ground is -6.5 V initially, and then it changes to 14.7 V.
   (a) Determine the change in the voltage.
   (b) Explain the meaning of your answer in (a).

52. *Electricity* The voltage across an element with respect to the ground is 23.7 V initially, and then it changes to -5.2 V.
   (a) Determine the change in the voltage.
   (b) Explain the meaning of your answer in (a).

In Your Words

53. Explain the significance of positive and negative signs when referring to "amount of change." How does this differ from the meaning of the absolute value of the amount of change?

54. Describe how to add two rational numbers. Show your description to a classmate and see if he or she understands what you wrote. Rewrite your description to clarify areas where your classmate had difficulty.

55. Describe how to multiply two rational numbers. Show your description to a classmate and see if he or she understands what you wrote. Rewrite your description to clarify any places where your classmate had difficulty.

≡ 1.4
LAWS OF EXPONENTS

We have introduced the different types of numbers, the basic laws of arithmetic, and the basic operations with real numbers. In this section, we will learn some shorthand notation used in mathematics that will help us with the work in future sections.

Exponents

It is often necessary to multiply a number, b, by itself several times. The notation b^n is used to indicate that a total of n bs are being multiplied. The number b is called the **base**, and n is called the **exponent**. We say that b^n is the "nth power of b" or "b to the nth power." When the exponent is 1, the 1 is often not written. Thus, $19^1 = 19$.

EXAMPLE 1.45

$$3^4 = 3 \times 3 \times 3 \times 3$$
$$= 81$$

The base is 3 and the exponent is 4. This is 3 to the fourth power or the fourth power of 3.

$$7^5 = 7 \times 7 \times 7 \times 7 \times 7$$
$$= 16,807$$

The base is 7 and the exponent is 5. This is 7 to the fifth power.

Notice how much easier it is to write 7^5 than 16,807.

$$9^2 = 9 \times 9$$
$$= 81$$

The base is 9, the exponent 2. This is the second power of 9 or, as it is more frequently called, 9 squared.

$$\left(\tfrac{5}{4}\right)^3 = \tfrac{5}{4} \times \tfrac{5}{4} \times \tfrac{5}{4}$$
$$= \tfrac{125}{64}$$

The base is $\tfrac{5}{4}$, the exponent is 3. This is the third power of $\tfrac{5}{4}$ or $\tfrac{5}{4}$ cubed.

$$(-3)^4 = (-3)(-3)(-3)(-3)$$
$$= 81$$

The base is -3, the exponent 4.

At the present time the exponent must be an integer. Later we will learn how to use and understand expressions where the exponent is any real number.

Rules of Exponents

As with all of the operations we have studied so far, there are some basic rules of exponents.

Rule 1 for Exponents	$b^m b^n = b^{m+n}$

This rule states that to multiply two or more powers of the same base, you add the exponents.

EXAMPLE 1.46

$$3^5 \cdot 3^2 = 3^{5+2} = 3^7 \text{ because } 3^5 \cdot 3^2 = (3 \times 3 \times 3 \times 3 \times 3)(3 \times 3)$$
$$= 3 \times 3 \times 3 \times 3 \times 3 \times 3 \times 3$$
$$= 3^7$$

EXAMPLE 1.47

$$x^3 \cdot x^4 \cdot x = x^3 \cdot x^4 \cdot x^1 = x^{3+4+1} = x^8$$

Rule 2 for Exponents	$(b^m)^n = b^{mn}$

This means that in order to take a power of a power you multiply the exponents.

EXAMPLE 1.48

$$(3^2)^4 = 3^{2\cdot4} = 3^8 \text{ because } (3^2)^4 = (3^2)(3^2)(3^2)(3^2)$$
$$= (3 \times 3)(3 \times 3)(3 \times 3)(3 \times 3)$$
$$= 3 \times 3 \times 3 \times 3 \times 3 \times 3 \times 3 \times 3$$
$$= 3^8$$

EXAMPLE 1.49

$$(x^4)^5 = x^{4\cdot5} = x^{20}$$

Rule 3 for Exponents	$(ab)^n = a^n b^n$

EXAMPLE 1.50

$$(2 \cdot 5)^3 = (2 \cdot 5)(2 \cdot 5)(2 \cdot 5)$$
$$= (2 \cdot 2 \cdot 2)(5 \cdot 5 \cdot 5)$$
$$= 2^3 \cdot 5^3$$

EXAMPLE 1.51

$$(xy)^5 = x^5 y^5$$

Rule 4a for Exponents	$\left(\dfrac{a}{b}\right)^m = \dfrac{a^m}{b^m}, \quad$ if $b \neq 0$

This says that to raise a quotient to a power, both the numerator and denominator are raised to that power.

EXAMPLE 1.52

$$\left(\frac{2}{5}\right)^4 = \frac{2^4}{5^4} \text{ because } \left(\frac{2}{5}\right)^4 = \left(\frac{2}{5}\right)\left(\frac{2}{5}\right)\left(\frac{2}{5}\right)\left(\frac{2}{5}\right)$$
$$= \frac{2 \times 2 \times 2 \times 2}{5 \times 5 \times 5 \times 5}$$
$$= \frac{2^4}{5^4}$$

Rule 4b for Exponents	$\left(\dfrac{1}{b}\right)^m = \dfrac{1}{b^m}, \quad$ if $b \neq 0$

This is a special case of Rule 4a.

EXAMPLE 1.53

$$\left(\frac{1}{3}\right)^4 = \frac{1^4}{3^4} = \frac{1}{3^4}$$

Rule 5 for Exponents	$\dfrac{b^m}{b^n} = b^{m-n}, \quad$ if $b \neq 0$

To divide two powers with the same base, you subtract the exponent of the denominator from the exponent of the numerator.

There are really three cases to this rule depending on which is larger, m or n, or if they are the same. Let's look first at when $m > n$.

EXAMPLE 1.54

$$\frac{5^7}{5^4} = 5^{7-4} = 5^3 \text{ because } \frac{5^7}{5^4} = \frac{5 \times 5 \times 5 \times 5 \times 5 \times 5 \times 5}{5 \times 5 \times 5 \times 5}$$
$$= 5 \times 5 \times 5$$
$$= 5^3$$

EXAMPLE 1.55

$$\frac{x^9}{x^4} = \frac{x \cdot x \cdot x \cdot x \cdot x \cdot x \cdot x \cdot x \cdot x}{x \cdot x \cdot x \cdot x} = x^5$$

Next, let's look at when $m = n$. Then we have $\frac{b^n}{b^n} = b^{n-n} = b^0$. But, $\frac{b^n}{b^n} = 1$ and so $b^0 = 1$.

EXAMPLE 1.56

$$\frac{5^2}{5^2} = 5^{2-2} = 5^0 = 1$$

and $$\frac{5^2}{5^2} = \frac{5 \times 5}{5 \times 5} = \frac{25}{25} = 1$$

Finally, let's look at when $m < n$.

EXAMPLE 1.57

$$\frac{6^3}{6^7} = 6^{3-7} = 6^{-4}$$

But, $$\frac{6^3}{6^7} = \frac{6 \times 6 \times 6}{6 \times 6 \times 6 \times 6 \times 6 \times 6 \times 6} = \frac{1}{6 \times 6 \times 6 \times 6} = \frac{1}{6^4}$$

EXAMPLE 1.58

$$\frac{x^5}{x^7} = x^{5-7} = x^{-2}$$

and $$\frac{x^5}{x^7} = \frac{x \cdot x \cdot x \cdot x \cdot x}{x \cdot x \cdot x \cdot x \cdot x \cdot x \cdot x} = \frac{1}{x \cdot x} = \frac{1}{x^2}$$

Examples 1.56–1.58 have set the foundation for two more rules. From Example 1.56 we can see that

Rule 6 for Exponents	$b^0 = 1, \quad \text{if } b \neq 0$

Any nonzero number raised to the zero power has a value of one.
From Examples 1.57 and 1.58, we have

Rule 7 for Exponents	$b^{-n} = \dfrac{1}{b^n}, \quad \text{if } b \neq 0$

This rule says that any number raised to a negative exponent is the same as the reciprocal of the number raised to that positive power.

≡ **Note**   Since $n = -(-n)$, it sometimes helps to think of Rule 7 as

$$b^n = b^{-(-n)} = \frac{1}{b^{-n}}$$

EXAMPLE 1.59

$$\left(x^2 y^3 z w^0\right)^{-4} = \frac{1}{\left(x^2 y^3 z w^0\right)^4} = \frac{1}{x^8 y^{12} z^4}$$

EXAMPLE 1.60

We will use the previous note $\left(\text{that } b^n = \dfrac{1}{b^{-n}}\right)$ to solve this example.

$$\left(\frac{a^2 b}{x^3 w^2}\right)^{-5} = \frac{\left(a^2 b\right)^{-5}}{\left(x^3 w^2\right)^{-5}} = \frac{\left(x^3 w^2\right)^5}{\left(a^2 b\right)^5}$$

$$= \frac{x^{15} w^{10}}{a^{10} b^5}$$

EXAMPLE 1.61

$$\frac{\left(x^2 y^7\right)^3}{\left(x^3 y^5\right)^4} = \frac{x^6 y^{21}}{x^{12} y^{20}} = \frac{y}{x^6}$$

Application

EXAMPLE 1.62

The total resistance, R, of a series-parallel circuit is given by the formula

$$R = \left(\frac{1}{R_1} + \frac{1}{R_2}\right)^{-1} + R_3$$

Find R when $R_1 = 0.75\,\Omega$, $R_2 = 0.50\,\Omega$, and $R_3 = 0.60\,\Omega$.

Solution   We begin by adding the terms in the parentheses.

$$R = \left(\frac{1}{R_1} + \frac{1}{R_2}\right)^{-1} + R_3$$

$$= \left(\frac{1}{0.75} + \frac{1}{0.50}\right)^{-1} + 0.60$$

$$= \left(\frac{0.50}{(0.75)(0.50)} + \frac{0.75}{(0.75)(0.50)}\right)^{-1} + 0.60$$

$$= \left(\frac{0.50 + 0.75}{(0.75)(0.50)}\right)^{-1} + 0.60$$

$$= \left(\frac{1.25}{(0.75)(0.50)}\right)^{-1} + 0.60$$

EXAMPLE 1.62 (Cont.)

$$= \frac{(0.75)(0.50)}{1.25} + 0.60$$
$$= 0.3 + 0.60$$
$$= 0.9$$

The total resistance is $0.9\,\Omega$.

≡ **Note** Remember when we discussed the order of operations in Section 1.2, that order #2 was raising to a power. Thus, unless grouping symbols indicate otherwise, perform all operations of raising to a power before multiplication, division, addition, or subtraction.

Exercise Set 1.4

Evaluate Exercises 1–8

1. 5^3
2. 3.8^0
3. $\left(\frac{2}{3}\right)^{-1}$
4. $\left(\frac{3}{5}\right)^{-2}$
5. $(-4)^2$
6. $(-5)^4$
7. $\frac{7}{7^3}$
8. -3^2

In Exercises 9–42, perform the indicated operations. Leave all answers in terms of positive exponents. $\frac{1}{pr^2}$

9. $3^2 \cdot 3^4$
10. $d^8 d^5$
11. $2^4 \cdot 2^3 \cdot 2^5$
12. $f^3 f^4 f^1$
13. $\frac{2^5}{2^3}$ 2^2
14. $\frac{5^{14}}{5^3}$ $= 5^{11}$
15. $(2^3)^2$ 2^6
16. $(5^7)^3$ 5^{21}
17. $(x^4)^5$ x^{20}
18. $(xy^3)^4$
19. $(a^{-2}b)^{-3}$ $\frac{a^6}{b^3}$

20. $\left(\frac{2}{3}\right)^4$
21. $\left(\frac{x}{4}\right)^3 = \frac{x^3}{4^3}$
22. $\left(\frac{a}{b^3}\right)^5$
23. $\left(\frac{a^2b}{c^3}\right)^4$
24. 4^{-3}
25. $x^{-7} = \frac{1}{x^7}$
26. $\frac{1}{p^{-5}}$
27. $\left(\frac{1}{5}\right)^3$

28. $\frac{x^4}{x^2}$
29. $\frac{7^3}{7^8}$ $\frac{1}{7^5}$
30. $\frac{5^2}{5^{10}}$ $\frac{1}{5^8}$
31. $\frac{a^2y^3}{a^5y^7}$
32. $\frac{x^4yb^2}{xy^3b^5}$
33. $\frac{a^2p^5y^3}{a^6p^5y}$ $\frac{y^2}{a^4}$
34. $\frac{p^3q^4r^2}{p^4r^2}$ $\frac{q^4}{p}$

35. $(pr^2)^{-1}$ pr^2
36. $\left(\frac{2x^2}{y}\right)^{-1}$
37. $\left(\frac{4y^3}{5^2}\right)^{-1}$ $\frac{5^2}{4y^3}$
38. $\left(\frac{4x^3}{y^2}\right)^{-2}$
39. $\left(\frac{2b^2}{y^5}\right)^{-3}$
40. $(-8pr^2)^{-3}$
41. $(-b^4)^6$ b^{24}
42. $ap^2(-a^2p^3)^2$

Solve Exercises 43–46.

43. *Electricity* The impedance in an *RC* circuit is given by the expression

$$Z_{RC} = \sqrt{R^2 + ((2\pi fC)^{-1})^2}$$

Determine the impedance if $R = 40\,\Omega$, $f = 60\,\text{Hz}$, and $C = 8 \times 10^{-5}\,\text{F}$.

44. *Electricity*   The total resistance, R, of a certain series-parallel circuit is given by

$$R = \left(\frac{1}{R_1} + \frac{1}{R_2} \right)^{-1} + R_3$$

Find R when $R_1 = 0.8\,\Omega$, $R_2 = 0.45\,\Omega$, and $R_3 = 0.76\,\Omega$.

45. *Electricity*   The total resistance, R, of a certain series-parallel circuit is given by

$$R = R_1 + \left(\frac{1}{R_2} + \frac{1}{R_3} + \frac{1}{R_4} \right)^{-1}$$

Find R when $R_1 = 6.5\,\Omega$, $R_2 = 5\,\Omega$, $R_3 = 6\,\Omega$, and $R_4 = 7.5\,\Omega$.

46. *Business*   If P dollars are invested at an annual interest rate of r compounded n times a year, then the total amount, A, accumulated after t years is given by the formula

$$A = P \left(1 + \frac{r}{n} \right)^{nt}$$

Approximate the total amount of money accumulated after 5 years by investing $1,200 at 6.8% ($r = 0.068$) interest compounded monthly ($n = 12$).

 In Your Words

47. (a) Use your calculator to evaluate -7^2 and $(-7)^2$.
(b) Are the answers in (a) different or the same?
(c) Explain why the answers are different (or why they are the same).

48. (a) Use your calculator to evaluate $(2^3)^2$ and 2^{3^2}.
(b) Are the answers in (a) different or the same?
(c) Explain why the answers are different (or why they are the same).

≡ 1.5
SIGNIFICANT DIGITS AND ROUNDING OFF

A great deal of technical work deals with measurements. Problems in this book are in both the metric system and the customary (or English) system of measurement. If you are not familiar with the metric system of measurement, you should read Appendix B.

No measurement deals with exact numbers. For example, suppose an automotive technician says that the outside diameter of a valve stem is 7.1 mm. While the diameter may be exactly 7.1 mm, it is more than likely to be a little more or less than 7.1 mm.

The amount of precision in such a measurement depends on the measuring instrument and the person doing the measuring. In mathematical terms, we have the following definitions of precision, significant digits, and accuracy.

> **Precision, Significant Digits, and Accuracy**
>
> The **precision** of a measurement is indicated by the position of the last significant digit relative to the decimal point.
>
> The **significant digits** are those that are determined by measurement.
>
> **Accuracy** refers to the number of significant digits.

A number with five significant digits, such as 7.1043, is more accurate than a number with four significant digits, such as 7.104.

Guidelines for Determining Which Digits Are Significant

1. All nonzero digits are significant.
2. Zero digits that lie between significant digits are significant. For example, 307 has three significant digits.
3. Zero digits that lie to the right of both the decimal point and the last nonzero digit are significant. For example, .860 has three significant digits.
4. Zeros at the beginning of a decimal fraction are not significant. For example, both .045 and 0.045 have two significant digits since the zeros serve only to locate the decimal point.
5. Zeros written at the end of a whole number are significant only if there is a "tilde" (˜) or a "bar" (‾) over the last significant digit. For example, 980,0̃00 and 980,0̅00 both have four significant digits.

Perhaps a few words about notation are needed here. A number such as 15,340,000 is written in the metric system by using spaces instead of commas, as 15 340 000. The spaces are also used for numbers smaller than one, for example, 0.00002471 would be written in the metric system as 0.000 024 71. A four-digit number in the metric system does not need to be written with the space, just as we do not always use a comma in such a number. Thus, 2400 and 2 400 are both correct in the metric system and 2,400 and 2400 are both correct in the English (or standard) system. As a technician, you may encounter both metric and standard or English notations and should recognize a number in either system.

Concept of Error

Let us now return to the earlier measurement of the outside diameter of a valve stem. If the person who measured this diameter had used an instrument that measured in thousandths of millimeters and the diameter had measured 7.1 mm to the nearest thousandth, then the measurement should have been written 7.100 mm.

Since we have spent so much time describing approximate numbers and the method with which to indicate how carefully they were measured, the question might arise as to what is an exact number. **Exact numbers** result from some definition or from counting. For example, the number of spark plugs in an 8-cylinder vehicle is an exact number; the length of a board is an approximate number.

Related to the ideas of accuracy and precision are the ideas of absolute, relative, and percent error. The **absolute error** is the true value subtracted from the approximate value of a number. The absolute error can be either positive or negative depending on whether the true value is smaller or larger than the approximate value.

The **relative error** is the ratio of the absolute error to the true value. Relative error is usually expressed as a percent. When relative error is written as a percent it is often called **percent error**.

Three Types of Errors

Absolute error = approximate value − true value

$$\text{Relative error} = \frac{\text{absolute error}}{\text{true value}}$$

Percent error = Relative error × 100%

EXAMPLE 1.63

The outside diameter of a valve stem was measured as 7.127 mm when the actual diameter was 7.1346 mm. What are the absolute, relative, and percent errors?

Solution Absolute error = approximate value − true value

$$= 7.127\,\text{mm} - 7.1346\,\text{mm}$$

$$= -0.0076\,\text{mm}$$

$$\text{Relative error} = \frac{\text{absolute error}}{\text{true value}}$$

$$= \frac{-0.0076\,\text{mm}}{7.1346\,\text{mm}}$$

$$= -0.001\,065\,\text{mm}$$

$$\text{Percent error} = \text{Relative error} \times 100\%$$

$$= -0.001\,065 \times 100\%$$

$$= -0.1065\%$$

≡ Note When computing with approximate numbers, the answer will depend on the accuracy or precision of those numbers.

Working with Approximate Numbers

When you add or subtract approximate numbers, the result is only as precise as the least precise number.

EXAMPLE 1.64

Add 182.7, 43.69, 2,470.765, and 0.32, and express the answer to the correct precision.

Solution First, we will add the numbers:

$$
\begin{array}{r}
182.7 \\
43.69 \\
2{,}470.765 \\
+\quad 0.32 \\
\hline
2{,}697.475
\end{array}
$$

These four numbers have 182.7 as their least precise number, so the sum can only be significant to the tenths place. The answer would be rounded off to 2,697.5.

Application

The masses of 5 pieces of metal plate are 16.63 kg, 738.6 kg, 4.314 kg, 21.645 kg, and 0.875 2 kg. Find the total mass of the four pieces correct to 4 significant digits.

Solution These five numbers have 738.6 as their least precise number, so the sum can only be significant to the tenths place. First, we will add the numbers:

$$
\begin{array}{r}
16.63 \\
738.6 \\
4.314 \\
21.645 \\
+\ \ 0.875\,2 \\
\hline
782.064\,2
\end{array}
$$

The answer would be rounded off to 782.1 kg.

When multiplying or dividing approximate numbers, the errors are enlarged. For this reason, the result is only as accurate as the least accurate number.

If a rectangle measures 42.37 mm long and 5.81 mm wide, it has an area of

$$
\begin{array}{r}
42.37 \text{ mm} \\
\times\ \ 5.81 \text{ mm} \\
\hline
246.169\,7 \text{ mm}^2
\end{array}
$$

Since 42.37 has four significant digits and 5.81 has only three significant digits, the product is rounded off to three significant digits and the area is 246 mm^2.

Rounding off Numbers

In general, to round off a number to a certain number of significant digits, examine the digit in the next place to the right. If this digit is less than 5 (0, 1, 2, 3, or 4) then accept the digit in the last place to the left of this digit. If the digit is 5 or more (5, 6, 7, 8, or 9) then increase the digit in the last place by one.

Round off 83.427 to the nearest hundredth.

Solution The digit 2 is in the hundredths place. The next digit to the right is a 7. This is more than 5, so the 2 is increased by 1 (to 3) and the 7 is dropped. Thus, 83.427 rounded to the nearest hundredth is 83.43.

Round 5.2348 to the nearest hundredth.

Solution Here the digit in the hundredths place is a 3. The next digit to the right is a 4. Since 4 is less than 5, the 4 and all other digits to its right are dropped. So, 5.2348 rounded off to the nearest hundredth is 5.23.

EXAMPLE 1.69

Round 7,030.6 to **(a)** two significant digits, **(b)** three significant digits, and **(c)** four significant digits.

Solutions

(a) The second significant digit is a 0. The digit to the right is a 3. Since 3 is less than 5, we accept the second significant digit, 0, and a tilde is placed over it to show that it is significant. Each digit between the second significant digit and the decimal point is changed to 0. So, 7,030.6 rounded to two significant digits is 7,0̃00.

(b) The third significant digit is a 3. The digit to the right is a 0. Since 0 is less than 5, we accept the third significant digit, 3. So, 7,030.6 rounded to three significant digits is 7,030.

(c) The fourth significant digit is a 0. The digit to the right is a 6. Since 6 is greater than 5, we raise the fourth significant digit by 1 to 1. So, 7,030.6 rounded to four significant digits is 7,031.

In many scientific and engineering applications a different round-off rule is used when the test digit is 5. This rule is known as either the **Odd-Five Rule** or the **Round-to-the-Even Rule**.

The Odd-Five Rule states that if the test digit is a 5 and it is the last nonzero digit in a number, then add 1 to the round off digit if it is odd (1, 3, 5, 7, or 9) or retain the original round-off digit if it is even (0, 2, 4, 6, or 8).

EXAMPLE 1.70

Round off 43.725 to the nearest hundredth by the Odd-Five Rule.

Solution We first note that the test digit is 5 and that it is the last nonzero digit. Next, because the round-off digit 2 is even, it is retained and the digit 5 is dropped. Thus, 43.725 would round off to the nearest hundredth to 43.72, by the Odd-Five Rule.

EXAMPLE 1.71

Round off 153.835 to the nearest hundredth by using the Odd-Five Rule.

Solution Again the test digit is 5 and it is the last nonzero digit, but here the round-off digit is 3, an odd number. Using the Odd-Five Rule, we add 1 to the 3 and drop the 5, and the number is rounded off to 153.84.

 **Note**

Unless it is specified, we will **not** use the Odd-Five Rule in this book.

 **Caution**

A precaution about significant digits and electronic calculators: Most calculators display 8 or 10 digits and store 2 or 3 more for rounding off purposes. However, this is not always the case. Some calculators employ a method called **truncation** in which any digits not displayed are discarded. Thus, 489.781 truncated to tenths is

489.7. Rounded off to tenths it would be 489.8. This can result in different answers when two different calculators are used.

Estimation

As we mentioned in the introduction to Section 1.3, you will most likely be using calculators and computers in your work as a technician. You must be able to recognize when an answer provided by these tools is reasonable. The ability to estimate answers is very important in your ability to recognize when answers are reasonable. Your skill at rounding off numbers will help you estimate answers to problems.

In addition and subtraction, round off each number to one or two significant digits and add (or subtract) these rounded off figures.

EXAMPLE 1.72

Estimate the sum of $4.7 + 8.6 + 9.2 + 4.1$.

Solution   The numbers in this sum, $4.7 + 8.6 + 9.2 + 4.1$, would be rounded off to $5 + 9 + 9 + 4 = 27$. The actual sum of the original four numbers is 26.6, which rounds off to 27.

EXAMPLE 1.73

Estimate the sum of $93.74 + 182.7 + 14,325 + 83.43$.

Solution   Rounding off each number to the nearest hundred, you get the estimate $100 + 200 + 14,300 + 100 = 14,700$. The actual total is 14,684.87.

We will learn how to estimate products and quotients in Section 1.6, which involves scientific notation.

Rounding-Off with a Calculator

Most calculators display from 8 to 10 digits. It is often helpful to set your calculator so that it will only display a certain number of digits to the right of the decimal point. For this, you use the [Fix] key, or mode, of the calculator. So, if you want your answer to show 3 decimal places to the right of the decimal point, set your calculator mode to [Fix] 3. The next example shows how to get the answer of a sum to the correct amount of precision.

EXAMPLE 1.74

Use your calculator to add 182.7, 43.69, 2,470.765, and 0.32. Express the answer to the correct number of decimal places.

Solution   We worked this example earlier in Example 1.64, and found the answer to be 2,697.5. Now, let's use our calculator. The least precise number is 182.7. Set the calculator to [Fix] 1. This will cause the calculator to display our answer with only one digit to the right of the decimal point. Now we enter each number that we want to add.

EXAMPLE 1.74 (Cont.)

PRESS	DISPLAY
182.7 $\boxed{+}$	182.7
43.69 $\boxed{+}$	226.4
2470.765 $\boxed{+}$	2697.2
0.32 $\boxed{=}$	2697.5

Once again, we get the answer of 2,697.5.

Exercise Set 1.5

In Exercises 1–6, determine whether the given numbers are exact or approximate.

1. There are 27 students in class.
2. The car traveled at 88 km/h.
3. A calculator is 148 mm by 79 mm by 35 mm.
4. She bought 18 bolts for $2.79.
5. Of all the people working in the United States, 7.9% are technicians.
6. A sheet of plywood is 1 200 mm by 2 400 mm.

Give the number of significant digits for the numbers in Exercises 7–14.

7. 6.05
8. 4,030
9. 12.0
10. 0.432
11. 4,000
12. 0.00290
13. 70.06
14. 160.070

Which numbers in Exercises 15–23 are (a) more accurate and (b) more precise. (It is possible for both numbers to be accurate or for both to be precise.)

15. 6.05; 2.8
16. 0.027; 6.324
17. 0.027; 5.01
18. 19,020; 29,000
19. 27,0͠00; 37,800
20. 6,000; 0.003
21. 0.2; 86
22. 3.05; 305.00
23. 140.070; 140,070

Round off each number in Exercises 24–31 to (a) one, (b) two, and (c) three significant digits.

24. 4.362
25. 14.37
26. 4.065
27. 7.035
28. 0.006155
29. 403.2
30. 0.03725
31. 305.4

In Exercises 32–39, round off each of the numbers to (a) one, (b) two, and (c) three significant digits using the Odd-Five Rule. (These are the same numbers you rounded off in Exercises 24–31.)

32. 4.362
33. 14.37
34. 4.065
35. 7.035
36. 0.006155
37. 403.2
38. 0.03725
39. 305.4

Round off each number in Exercises 40–47 to the nearest (a) ten, (b) tenth, and (c) thousandth.

40. 25.3345
41. 89.8992
42. 125.3755
43. 237.3017
44. 96.99854
45. 437.9975
46. 12.3405
47. 78.6705

Solve Exercises 48 and 49.

48. *Construction*   The actual length of an I-beam is 12.445 m.   An engineer measures the beam as 12.45 m. What are the absolute, relative, and percent errors in this measurement?

49. *Computer science*   A microcomputer chip is supposed to measure 24 mm long, 8 mm wide, and 3 mm thick. One chip, when measured with a micrometer, is 23.72 mm long, 8.35 mm wide, and 2.98 mm thick. What are the absolute, relative, and percent errors of each of these measurements?

Estimate the answers to Exercises 50–53, then work the exercises and round off the answers using the rules for approximate numbers.

50. $4.31 + 2.015 + 18.35$

51. $97.83 - 4.378 + 5.92$

52. $243.7 + 85.37 - 62.105 + 143.8$

53. $1,342.8 + 85.32 + 173.54 + 16$

Work Exercises 54–57 and round off the answer using the rules for approximate numbers.

54. 14.3×5.7

55. 20.4×50.1

56. 0.0034×2.50

57. 3.4×5.00

Solve Exercises 58–61.

58. *Automotive technology*   The 2 front parking lamps of a car each draw a current of 0.417 A (amperes). The 2 tail lamps each draw a current of 0.457 A. The license plate lamp draws 0.736 A. Find the total current drawn by the 5 lamps. Give your answer correct to 3 significant digits.

59. *Electricity*   An inductor has 37 layers of wire. Each layer contains 132 turns. The average length of a turn is 0.072 m. Find the length of the wire in the coil. Give your answer correct to 5 significant digits.

60. *Construction*   A pile of lumber has 379 pieces with an average length of 11 ft $10\frac{1}{8}$ in. Find the total length of the lumber, in inches, correct to **(a)** three decimal places and **(b)** three significant digits.

61. *Metalworking*   A milling machine cutter with 18 teeth removes 0.086 mm of steel per tooth. The cutter makes 597.3 revolutions. What length of material is removed? Round off your answer to 4 significant digits.

In Your Words

62. On a sheet of paper, explain precision, significant digits, and accuracy. Do not look at the definitions in the book.

63. On a sheet of paper, describe how to determine absolute error and relative error. Do not look at the definitions in the book.

≡ 1.6
SCIENTIFIC AND ENGINEERING NOTATION

In scientific and technical work it is often necessary to work with very large or very small numbers. For example, the star Sirius is approximately 8 220 000 000 000 000 km from earth. On the other hand, the radius of an electron is about 0.000 000 000 000 002 82 m.

Writing Numbers in Scientific Notation

Writing numbers with all these zeros is very time consuming and increases the chance of making an error by either omitting a zero or by using too many zeros. In order to save time and reduce the chance of making a mistake, scientists adopted a method for abbreviating numerals. This method is called **scientific notation**.

Expressing a Number in Scientific Notation

To express any number in scientific notation, write the number as the product of the significant digits written with exactly one nonzero digit to the left of the decimal point and a power of 10.

EXAMPLE 1.75

$$2,400 = 2.4 \times 1,000 = 2.4 \times 10^3$$

$$38,900,000 = 3.89 \times 10,000,000 = 3.89 \times 10^7$$

$$4,07\tilde{0},000,000,000 = 4.070 \times 1,000,000,000,000 = 4.070 \times 10^{12}$$

$$100,000 = 1 \times 10^5 \quad \text{or} \quad 10^5$$

$$0.036 = 3.6 \times \frac{1}{100} = 3.6 \times \frac{1}{10^2} = 3.6 \times 10^{-2}$$

$$0.000\,000\,403 = 4.03 \times \frac{1}{10\,000\,000} = 4.03 \times \frac{1}{10^7} = 4.03 \times 10^{-7}$$

$$8.2 = 8.2 \times 10^0 \text{ or } 8.2$$

Perhaps the easiest way to remember how to express a number in scientific notation is to follow these two guidelines:

Guidelines for Expressing a Number in Scientific Notation

1. Move the decimal point to just after the first nonzero digit.
2. Next multiply the number from Step 1 by 10^n if you shifted the decimal point n places to the left or by 10^{-n} if you shifted the decimal point n places to the right.

EXAMPLE 1.76

Express $20\,500$ in scientific notation.

Solution $20\,500 = 2.05 \times 10^4$

Move decimal point
four places to the left.

EXAMPLE 1.77	Express 0.000037 in scientific notation.

Solution $0.000037 = 3.7 \times 10^{-5}$
Move decimal point
five places to the right.

Any zeros that are significant should be included when the number is written in scientific notation. For example, $0.00230 = 2.30 \times 10^{-3}$ and $847\,000 = 8.470 \times 10^{5}$.

To change a number from scientific notation to ordinary notation you just reverse the process. To change 8.37×10^{6} to ordinary notation you would move the decimal point six places to the right. So, $8.37 \times 10^{6} = 8\,370\,000$. A number like 4.61×10^{-4} is changed by moving the decimal point four places to the left. Then $4.61 \times 10^{-4} = 0.000461$.

Obviously scientific notation relies heavily on exponents. This will give us the first chance to use the rules for exponents that we learned in Section 1.4. Also, many calculators use scientific notation to express some numbers. We will now compute products and quotients using scientific notation.

Scientific Notation on a Calculator

A calculator or a computer will automatically put a number in scientific notation if the number contains too many digits to show on the screen. For example, if you multiply 250,000 and 19,700,000, the calculator displays the answer as

$$4.925\text{E}12$$

$$\text{or} \quad 4.925 \quad 12$$

which should be interpreted as 4.925×10^{12}.

Caution

Do not interpret the calculator display 4.925E12 or 4.925 12 as 4.925^{12}. It means 4.925×10^{12}.

Similarly, the answer to $0.000025 \div 800\,000$ would be displayed as

$$3.125\text{E}{-}11$$

$$\text{or} \quad 3.125 \quad -11$$

which is interpreted as 3.125×10^{-11}.

≡ Note

Some calculators place a space between the number and the exponent while other calculators use the letter "E."

To enter the number in scientific notation on a calculator, you need to use the Enter Exponent key, labeled as $\boxed{\text{EE}}$, $\boxed{\text{EEX}}$, or $\boxed{\text{EXP}}$.

EXAMPLE 1.78

Enter 4.37×10^{12} on a calculator.

Solution

PRESS	DISPLAY			
4.37 $\boxed{EE}$	4.37	00	or	4.37E
12	4.37	12	or	4.37E12

EXAMPLE 1.79

Enter 9.87×10^{-15} on a calculator.

Solution The procedure depends on the type of calculator. With some models, you use this method:

PRESS	DISPLAY	
9.87 $\boxed{EE}$	9.87	00
15 $\boxed{+/-}$	9.87	-15

With other calculators, such as the TI-8x graphics calculator, you follow these steps:

PRESS	DISPLAY
9.87 $\boxed{EE}$	9.87E
$\boxed{(-)}$ 15	9.87E-15

Hint

You might find it easier to use the $\boxed{10^x}$ key than the $\boxed{EE}$ key.

Special care has to be taken when entering a power of 10. The next example shows a correct procedure.

EXAMPLE 1.80

Enter 10^{25} into a calculator.

Solution The procedure depends on the calculator. Many graphing calculators have a $\boxed{10^x}$ key. Thus, to enter 10^{25} into such a calculator, press $\boxed{10^x}$ 25 $\boxed{ENTER}$. On a TI-8x calculator, the result is shown as 1E25.

You may also use the $\boxed{EE}$ key

PRESS	DISPLAY	
1 $\boxed{EE}$	1	00 or 1E
25	1	25 or 1E25

Caution

Remember that powers of 10, such as 10^{25}, really represent 1×10^{25}. Thus, you enter 1 $\boxed{EE}$ 25 and not 10 $\boxed{EE}$ 25. However, on a TI-8x, $\boxed{EE}$ 25 will work.

**Note**

Many calculators allow you to put the calculator in scientific notation mode. If this is done, all answers will be displayed in scientific notation.

Products and Quotients

The product of $87\,000\,000 \times 470\,000\,000\,000$ can be simplified with scientific notation. Change each number to scientific notation and then multiply.

$$
\begin{aligned}
87\,000\,000 \times 470\,000\,000\,000 &= (8.7 \times 10^7) \times (4.7 \times 10^{11}) \\
&= (8.7 \times 4.7) \times (10^7 \times 10^{11}) \\
&= 40.89 \times 10^{7+11} \\
&= 40.89 \times 10^{18} \\
&= 4.089 \times 10^{19}
\end{aligned}
$$

Notice that we had to change our first answer (40.89×10^{18}) to scientific notation because 40.89 is not between 1 and 10.

Similarly, a large and small number can be multiplied more easily by using scientific notation, as shown in the next example.

EXAMPLE 1.81

Solve $4\,100\,000\,000 \times 0.000\,002\,4$.

Solution
$$
\begin{aligned}
4\,100\,000\,000 \times 0.000\,002\,4 &= (4.1 \times 10^9) \times (2.4 \times 10^{-6}) \\
&= (4.1 \times 2.4) \times (10^9 \times 10^{-6}) \\
&= 9.84 \times 10^3 \\
&= 9\,840
\end{aligned}
$$

Division can also be simplified using scientific notation.

EXAMPLE 1.82

Solve $0.000\,000\,036 \div 0.000\,012$.

$$
\begin{aligned}
0.000\,000\,036 \div 0.000\,012 &= (3.6 \times 10^{-8}) \div (1.2 \times 10^{-5}) \\
&= \frac{3.6 \times 10^{-8}}{1.2 \times 10^{-5}} = \frac{3.6}{1.2} \times \frac{10^{-8}}{10^{-5}} \\
&= \frac{3.6}{1.2} \times 10^{-8+5} \\
&= 3 \times 10^{-8+5} \\
&= 3 \times 10^{-3} = 0.003
\end{aligned}
$$

EXAMPLE 1.83

Use a calculator to solve $4\,200\,000\,000 \div 0.000\,000\,025$.

Solution   First, we convert each of these numbers to scientific notation.

$$
\begin{aligned}
4\,200\,000\,000 &= 4.2 \times 10^9 \\
0.000\,000\,025 &= 2.5 \times 10^{-8}
\end{aligned}
$$

EXAMPLE 1.83 (Cont.)

Now, we use our calculator.

PRESS	DISPLAY	
4.2 ☐EE☐	4.2	00
9 ☐÷☐	4.2	09
2.5 ☐EE☐	2.5	00
8 ☐+/−☐	2.5	−08
☐=☐	1.68	17

So, the quotient is 1.68×10^{17}.

Application

EXAMPLE 1.84

The alternating current reactance of a circuit, X_L, is given in ohms (Ω) by the formula

$$X_L = 2\pi f L$$

where f is the frequency of the alternating current in hertz (Hz), and L is the inductance of the circuit or inductor in henrys (H). Compute the inductive reactance when $f = 10\,000\,\text{Hz}$ and $L = 0.000\,006\,5\,\text{H}$.

Solution We will use a calculator to compute this product. First, we convert $10\,000\,\text{Hz}$ and $0.000\,006\,5\,\text{H}$ to scientific notation.

$$10\,000 = 1 \times 10^4$$

$$0.000\,006\,5 = 6.5 \times 10^{-6}$$

Now, we use our calculator:

PRESS	DISPLAY	
2 ☐×☐	2	
☐π☐☐×☐	6.283185307	
1 ☐EE☐	1	00
4	1	04
☐×☐	62831.85307	
6.5 ☐EE☐	6.5	00
6 ☐+/−☐	6.5	−06
☐=☐	0.408407045	

So, the reactance is about $0.41\,\Omega$.

Using Scientific Notation for Estimates

Scientific notation can also be used to help estimate products and quotients.

EXAMPLE 1.85

Estimate the product of $362 \times 2{,}165 \times 82$.

Solution   First, round off each number so that it has only one nonzero digit. Thus, 362 is rounded off to 400, 2,165 to 2,000, and 82 to 80. Express each of these in scientific notation and multiply.

$$400 \times 2{,}000 \times 80 = 4 \times 10^2 \times 2 \times 10^3 \times 8 \times 10^1$$
$$= (4 \times 2 \times 8) \times (10^2 \times 10^3 \times 10^1)$$
$$= 64 \times 10^6$$
$$= 64{,}000{,}000$$

The actual product of $362 \times 2{,}165 \times 82$ is 64,265,860.

Much of the time the estimated product (or quotient) will not be this close to the actual product (or quotient). This method of estimation is best used to make sure that you are placing the decimal point in the correct place.

For example, $2\,965 \times 650 \times 24 = 46\,254\,000$. Using the estimation procedure previously described you would get

$$3\,000 \times 700 \times 20 = 3 \times 10^3 \times 7 \times 10^2 \times 2 \times 10$$
$$= (3 \times 7 \times 2) \times (10^3 \times 10^2 \times 10)$$
$$= 42 \times 10^6 = 42\,000\,000$$

As you can see, $46\,254\,000$ is quite a bit larger than $42\,000\,000$. Yet, the estimation procedure indicated that the correct answer would have 8 digits to the left of the decimal point.

Engineering Notation

Engineering notation is very similar to scientific notation. In engineering notation, the exponents of the 10 are always multiples of three. The main advantage of engineering notation is when SI (metric) units are used. In the metric system, the most widely used prefixes are for every third power of 10.

Expressing a Number in Engineering Notation

To express any number in engineering notation, write the number as the product of a number with one, two, or three digits to the left of the decimal point and a power of 10 where the power-of-10 exponent is a multiple of three.

EXAMPLE 1.86

Express each of the following numbers in engineering notation: **(a)** 25,300, **(b)** 120,000, **(c)** 407,000,000,000, and **(d)** 0.000 000 025.

Solutions

(a) $25{,}300 = 25.3 \times 10^3$

EXAMPLE 1.86 (Cont.)

(b) $120{,}000 = 120 \times 10^3$

(c) $407{,}\tilde{0}00{,}000{,}000 = 407.0 \times 10^9$

(d) $0.000\,000\,025 = 25 \times 10^{-9}$

≡ **Note**   Many calculators let you put the calculator in engineering notation mode. Most calculators abbreviate "engineering mode" with "Eng." If this is done, all answers will be displayed in engineering notation.

EXAMPLE 1.87

Use a TI-82 calculator to express the number 0.9988428928 in (a) "normal" mode, (b) scientific mode, and (c) engineering mode.

Solutions

(a) Key 0.9988428928 into the calculator and press ENTER. The result in "normal" mode is shown in the viewing window as .9988428928.

(b) Press MODE. The cursor is blinking on the word Normal. Press ▶ once. The word Sci should now be blinking. Press ENTER to put the calculator in scientific mode. Now press 2nd QUIT to get back to the home screen. Key in 0.9988428928, and press ENTER. The viewing window should now display 9.988428928ᴇ-1.

(c) Press MODE. The cursor is blinking on the word Normal. Press ▶ twice. The word Eng should now be blinking. Press ENTER to put the calculator in engineering mode. Now press 2nd QUIT to get back to the home screen. Key in 0.9988428928, and press ENTER. The viewing window should now display 998.8428928ᴇ-3.

Application

EXAMPLE 1.88

The rest energy, E, of an electron with rest mass, m, is given by Einstein's equation

$$E = mc^2$$

where c is the speed of light. Find E if $m \approx 9.110 \times 10^{-31}$ kg and $c \approx 2.998 \times 10^8$ m/s. Express the answer in (a) engineering notation and (b) scientific notation.

Solutions   Substituting the values of m and c into the given equation, we find

$$E \approx \left(9.110 \times 10^{-31}\right)\left(2.998 \times 10^8\right)^2$$
$$= (9.110)(2.998)^2 \left(10^{-31}\right)\left(10^8\right)^2$$
$$= (9.110)(2.998)^2 \left(10^{-31}\right)\left(10^8\right)\left(10^8\right)$$
$$= (81.8807)\left(10^{-31+8+8}\right)$$
$$= 81.8807 \times 10^{-15}$$

The rest energy of an electron is about 81.8807×10^{-15} kg·m²/s².

EXAMPLE 1.88 (Cont.)

(a) Since the exponent of 10 in the number 81.8807×10^{-15} is a multiple of three, the number 81.8807×10^{-15} is already in engineering notation.

(b) In scientific notation, we have $81.8807 \times 10^{-15} = (8.18807 \times 10) \times 10^{-15} = 8.18807 \times 10^{-14}$. Thus, in scientific notation, the rest energy of an electron is about $8.18807 \times 10^{-14}\,\text{kg·m}^2/\text{s}^2$.

Exercise Set 1.6

In Exercises 1–10, change each of the numbers to scientific notation.

1. 42 000

2. 370 000 000

3. 0.000 38

4. 0.000 007 5

5. 9 807 000 000

6. 87 000 000

7. 0.000 097 0

8. 0.400

9. 4.3

10. 2.07

In Exercises 11–20, change each of the numbers to engineering notation.

11. 74 000

12. 910 000 000

13. 0.000 47

14. 0.000 007 5

15. 9 807 000 000

16. 57 000 000

17. 0.000 053 10

18. 0.700

19. 5.6

20. 3.08

In Exercises 21–28, change each number from scientific notation to ordinary notation.

21. 4.5×10^3

22. 3.7×10^5

23. 4.05×10^7

24. 3.05×10^8

25. 6.3×10^{-5}

26. 1.87×10^{-8}

27. 7.2×10

28. 9.6×10^{-1}

In Exercises 29–36, change each number from engineering notation to ordinary notation.

29. 75×10^3

30. 19×10^6

31. 47.5×10^9

32. 317.0×10^9

33. 39.20×10^{-6}

34. 1.75×10^{-3}

35. 83.15×10^0

36. 391.25×10^{-9}

In Exercises 37–58, perform the indicated calculations by using scientific notation and/or engineering notation.

37. $760\,000 \times 20\,400\,000\,000$

38. $(43\,200\,000)(850\,000\,000)$

39. $0.000\,035 \times 0.000\,000\,76$

40. $(0.000\,42)(0.000\,075)$

41. $(840\,000\,000)(0.000\,35)$

42. $(0.000\,0042)(23\,000)$

43. $(70\,400)(0.000\,003\,2)$

44. $(0.000\,302)(4\,370\,000\,000)$

45. $(28\,800\,000\,000) \div (240\,000)$

46. $55\,500\,000\,000 \div 370\,000$

47. $375\,000 \div 150\,000\,000$

48. $79\,800 \div 840\,000\,000$

49. $0.003\,2 \div 0.000\,000\,16$

50. $0.000\,48 \div 0.000\,000\,3$

51. $0.000\,000\,36 \div 0.000\,2$

52. $0.000\,000\,009\,8 \div 0.000\,014$

53. $(8\,760\,000)(245\,000\,000)(6\,400\,000\,000)$

54. $(4\,360\,000)(625\,000\,000)(38\,700\,000\,000)$

55. $\dfrac{(250\,200)(630\,000\,000)}{0.000\,000\,006\,3}$

56. $\dfrac{(25\,200)(8\,000\,000)}{3\,970\,000\,000}$

57. $\dfrac{0.000\,005\,2 \times 480\,000\,000}{0.000\,000\,000\,006\,4}$

58. $\dfrac{96\,000\,000 \times 810\,000}{243\,000\,000\,000\,000}$

Solve Exercises 59–68.

59. Estimate the computation of $\dfrac{325 \times 86.5}{43.1}$.

60. Estimate the computation of $\dfrac{453 \times 672.4}{3.81 \times 42.3}$.

61. *Physics*   An electron moves at the rate of $300\,000\,000\,000$ mm/s (millimeters per second). How many millimeters will it move in $0.000\,003\,7$ s?

62. *Astronomy*   The speed of light is approximately $300\,000$ km/s. How many minutes will it take for a ray of light to reach the earth from the sun, if the sun is $150\,000\,000$ km from earth?

63. *Physics*   One atomic mass unit (amu) is $1.660\,6 \times 10^{-27}$ kg. If 1 carbon atom has 12 atomic mass units, what is the mass of $14\,000\,000$ carbon atoms?

64. *Physics*   One iron atom contains 55 amu. What is the mass in kilograms of $230\,000\,000$ iron atoms?

65. *Physics*   The rest mass of one electron is 9.1095×10^{-31} kg and the rest mass of one neutron is 1.6750×10^{-27} kg. Which has the larger mass—one electron or one neutron? How many times heavier is it?

66. *Electronics*   Compute the inductive reactance when $f = 10,000,000$ Hz and $L = 0.015$ H. (See Example 1.84.)

67. *Electronics*   Compute the inductive reactance when $f = 18,000,000$ Hz and $L = 0.0025$ H. Work the exercise, and express your answer in engineering notation.

68. *Physics*   The mass of the earth is about 5.975×10^{24} kg, and its volume is about 1.083×10^{21} m^3. Use engineering notation to compute the density of the earth. (Note: the density of an object is defined as its mass divided by its volume.)

✎ **In Your Words**

69. Explain why you might use scientific notation. Explain the advantages of scientific notation over ordinary notation.

70. Explain when you might use engineering notation rather than scientific notation.

≡ 1.7
ROOTS

In Section 1.4 we learned how to take powers of numbers and the different operations with exponents. In this section we will look at the reverse process of taking exponents. Remember that 3^4 meant $3 \times 3 \times 3 \times 3 = 81$. Now, suppose you were asked to find the number (or numbers) that you could raise to the 4th power and get 81. This number (or numbers) would be called the fourth root (or roots) of 81. We use the symbol $\sqrt[4]{81}$ to indicate the fourth root of 81.

In general, the symbol $\sqrt[n]{b}$ is used to indicate the *n*th root of a number *b*. When $n = 2$ it is not necessary to put the 2 in the symbol. Thus, $\sqrt{9}$ means $\sqrt[2]{9}$. The $\sqrt{}$ sign is called the **root** or **radical sign**.

EXAMPLE 1.89

$\sqrt{16}$ This is the second root of 16 or, as it is more commonly called, the square root of 16.

$\sqrt[3]{8}$ This is the third root of 8 or the cube root of 8.

$\sqrt[6]{17}$ This is the sixth root of 17.

The solution to $\sqrt{16}$, or to the question "What number squared is 16?", can readily be seen to be either 4 or -4. This presents a problem. Do you write both as the solution to $\sqrt{16}$? Again, mathematicians have decided on a way to interpret $\sqrt{16}$. They have decided that $\sqrt{16} = 4$ and not -4. The symbol $-\sqrt{16}$ is used when they want the value -4.

What we want to find as the solution to $\sqrt[n]{b}$ is the **principal nth root of b**.

Principal nth Root of b

The principal nth root of b is positive if b is positive.

The principal nth root of b is negative if b is negative and n is odd.

EXAMPLE 1.90

$$\sqrt{16} = 4 \text{ and not } -4$$
$$\sqrt[3]{8} = 2 \text{ since } 2 \times 2 \times 2 = 8$$
$$\sqrt[3]{-8} = -2 \text{ since } -2 \times -2 \times -2 = -8$$

≡ **Note**

If you want the negative number that is the square root of b, then you must write $-\sqrt{b}$. Thus, $-\sqrt{16} = -4$.

We will see later that there are n different numbers that, when raised to the nth power, will give the number b. Thus, there are n different numbers that will satisfy the definition of $\sqrt[n]{b}$. In most cases we will limit ourselves to just one of these—the principal nth root.

Rules of Roots

There are some rules of roots that are very similar to the rules of exponents.

Rule 1 of Roots $\sqrt[n]{ab} = \sqrt[n]{a} \cdot \sqrt[n]{b}$

This rule says that the root of a product is the same as the product of the root of each factor. Compare this to Rule 3 for exponents.

EXAMPLE 1.91

$$\sqrt{36} = \sqrt{4 \cdot 9} = \sqrt{4} \cdot \sqrt{9} = 2 \cdot 3 = 6$$
$$\sqrt{75} = \sqrt{25 \cdot 3} = \sqrt{25} \cdot \sqrt{3} = 5\sqrt{3}$$
$$\sqrt[3]{40} = \sqrt[3]{8 \cdot 5} = \sqrt[3]{8}\sqrt[3]{5} = 2\sqrt[3]{5}$$

Rule 2 of Roots

$$\sqrt[n]{\frac{a}{b}} = \frac{\sqrt[n]{a}}{\sqrt[n]{b}}$$

In words, this says that the root of a quotient is the root of the numerator divided by the root of the denominator.

EXAMPLE 1.92

$$\sqrt{\frac{9}{4}} = \frac{\sqrt{9}}{\sqrt{4}} = \frac{3}{2}$$

$$\sqrt[3]{\frac{5}{8}} = \frac{\sqrt[3]{5}}{\sqrt[3]{8}} = \frac{\sqrt[3]{5}}{2}$$

Rule 3 of Roots

$$\left(\sqrt[n]{b}\right)^n = b$$

This rule states that when the root of a number is raised to the same power as the root, then the result is the original number.

EXAMPLE 1.93

$$\left(\sqrt[3]{5}\right)^3 = 5$$

$$\left(\sqrt[11]{\frac{7}{3}}\right)^{11} = \frac{7}{3}$$

Fractional Exponents

Rule 4 of Roots

$$\sqrt[n]{b} = b^{1/n}$$

≡ Note   Rule 4 is very important. It means that every radical or root can be written as a fractional exponent. This means that all the rules of exponents will hold for roots. This causes Rule 3 to make even more sense because $\left(\sqrt[n]{b}\right)^n = \left(b^{1/n}\right)^n = b^{n/n} = b^1 = b$.

EXAMPLE 1.94

$$27^{1/3} = \sqrt[3]{27} = 3$$

$$16^{1/4} = \sqrt[4]{16} = 2$$

Application

EXAMPLE 1.95

According to *Blasius' formula*, the friction factor f of flow in a smooth pipe is

$$f = \frac{0.316}{\sqrt[4]{\text{Re}}}$$

where Re is *Reynold's number*, a constant that depends on the average velocity of the fluid flow, the diameter of the pipe, and the kinematic viscosity of the fluid. Calculate f when Re $= 7,340$.

Solution   We will show how to use a calculator to solve this problem. Remember, we are calculating

$$\frac{0.316}{\sqrt[4]{7,340}}$$

and also remember that $\sqrt[4]{7,340} = 7,340^{1/4}$.
   Now, we use our calculator:

PRESS	DISPLAY
.316 ÷ 7340 ∧ (1 ÷ 4) ENTER	.0341399647
or .316 ÷ 7340 ∧ 4 x^{-1} ENTER	.0341399647

So, the friction factor is about 0.0341.

Application

EXAMPLE 1.96

For a large rectangular orifice of width b and depth $h_2 - h_1$, the discharge, Q, is

$$Q = \frac{2}{3} C_Q b \left(h_2^{2/3} - h_1^{2/3} \right) \sqrt{2g}$$

where C_Q is the coefficient of discharge. Determine the discharge if $C_Q = 0.96$, $b = 0.200$ m, $h_2 = 0.512$ m, $h_1 = 0.385$ m, and $g = 9.81$ m/s^2.

Solution   We will show how to use a calculator to solve this problem. Remember, after the values above are substituted, we are calculating $\frac{2}{3}(0.96)(0.200)(0.512^{2/3} - 0.385^{2/3})\sqrt{2(9.81)}$

PRESS	DISPLAY
2 ÷ 3 × .96 × .2 × (.512	
∧ (2 ÷ 3) − .385 ∧	
(2 ÷ 3)) × 2nd	
√ (2 × 9.81) ENTER	.0628059206

So, the discharge is about 0.063 m^3/s.

Rule 4 also means that we can have numbers such as $4^{3/2} = \left(4^{1/2}\right)^3 = 2^3 = 8$. This also allows us to find some higher degree roots by simplifying them into combinations of similar roots. Thus $\sqrt[mn]{b} = \sqrt[m]{\sqrt[n]{b}}$ because, by the rules of exponents, $\sqrt[mn]{b} = b^{1/mn} = \left(b^{1/n}\right)^{1/m} = \sqrt[m]{\sqrt[n]{b}}$. It may not look as if this has been simplified. However, it has been, as Example 1.97 shows.

EXAMPLE 1.97

$$\sqrt[6]{64} = \sqrt[3]{\sqrt{64}} = \sqrt[3]{8} = 2$$

$$\sqrt[6]{729} = \sqrt[3]{\sqrt{729}} = \sqrt[3]{27} = 3$$

$$\sqrt[8]{256} = \sqrt[4]{\sqrt{256}} = \sqrt[4]{16} = 2$$

You will remember that in defining the principal nth root of a number, we said that if b was negative then n had to be an odd number. This means that, as yet, we have no definition for numbers such as $\sqrt{-1}$, $\sqrt{-9}$, or $\sqrt[4]{-16}$. In general, these are new types of numbers called imaginary numbers. They do not belong to the set of real numbers.

The special symbol j is used to represent $\sqrt{-1}$. (Some areas use i for $\sqrt{-1}$.) This allows us to represent numbers such as $\sqrt{-9}$ in terms of j, since $\sqrt{-9} = \sqrt{(9)(-1)} = \sqrt{9}\sqrt{-1} = 3j$. These are called imaginary numbers and we will study them in a later chapter. For the present time we need to remember that we cannot take an even root of a negative number.

Exercise Set 1.7

Find the principal nth root of each real number in Exercises 1–16.

1. $\sqrt{25}$ = 5
2. $\sqrt{36}$ = 6
3. $\sqrt{144}$
4. $\sqrt{121}$
5. $\sqrt[3]{8}$ ≈ 2

6. $\sqrt[3]{-64}$
7. $\sqrt[3]{-27}$
8. $\sqrt[4]{81}$
9. $\sqrt[4]{16}$ = 2
10. $\sqrt[5]{-243}$

11. $\sqrt[4]{\frac{16}{81}}$
12. $\sqrt[3]{\frac{-27}{1,000}}$
13. $\sqrt{0.04}$

14. $\sqrt{0.25}$
15. $\sqrt[3]{-0.001}$
16. $\sqrt[3]{0.125}$

Simplify each of the real numbers in Exercises 17–48. Use the rules for roots and the rules for exponents.

17. $\sqrt{3^2}$
18. $\sqrt[3]{5^3}$
19. $\sqrt[4]{8.32^4}$
20. $\sqrt[6]{7.91^6}$
21. $\sqrt[3]{5}\sqrt[3]{25}$
22. $\sqrt[5]{-3}\sqrt[5]{81}$
23. $\sqrt[4]{8}\sqrt[4]{9}$
24. $\sqrt[6]{12}\sqrt[6]{48}$

25. $\dfrac{\sqrt{75}}{\sqrt{3}}$
26. $\dfrac{\sqrt{112}}{\sqrt{7}}$
27. $\dfrac{\sqrt[3]{5}}{\sqrt[3]{40}}$
28. $\dfrac{\sqrt[3]{11}}{\sqrt[3]{297}}$
29. $\sqrt[3]{2^3} + \sqrt[4]{5^4}$

30. $\sqrt{3^2} - \sqrt[3]{2^3}$
31. $\sqrt[3]{\left(\frac{2}{3}\right)^3} - \sqrt[4]{\left(\frac{1}{3}\right)^4}$
32. $\sqrt[5]{\left(\frac{3}{4}\right)^5} + \sqrt[3]{\left(\frac{5}{4}\right)^3}$
33. $\sqrt{5^{2/3}}$
34. $\sqrt{7^{2/5}}$
35. $\sqrt[3]{16^{3/4}}$
36. $\sqrt[4]{27^{4/3}}$
37. $16^{3/4}$

38. $(-27)^{2/3}$
39. $(-8)^{2/3}$
40. $25^{3/2}$
41. $8^{-2/3}$
42. $9^{-3/2}$
43. $(-27)^{4/3}$
44. $(-32)^{3/5}$
45. $\sqrt{\dfrac{81}{(8)(0.01)}}$

46. $\sqrt[3]{\dfrac{(27)(0.008)^2}{0.027}}$ **47.** $\sqrt[3]{\dfrac{(0.125)^3\sqrt{144}}{\frac{3}{2}}}$ **48.** $\sqrt{\dfrac{64}{(0.25)(0.16)}}$

Solve Exercises 49–52.

49. *Electronics*   A wave consists of many crests and troughs. The distance between any two crests is called the wavelength. The frequency of a wave is the number of waves that pass a given point in a unit of time. The relationship between the wavelength λ of a wave, the frequency f in hertz (Hz), and the speed c at which the wave is propagated is given by the formula $\lambda = \dfrac{c}{f}$. λ and c must be in the same units of length, that is, if λ is given in meters, then c must be in meters per second (m/s) or if λ is given in feet, then c must be in feet per second (ft/s). What is the wavelength in meters of a radio wave with a frequency of 143.5 megahertz (MHz)(143.5 MHz = 143 500 000 Hz) if $c = 300\,000\,000$ m/s?

50. *Wastewater technology*   The minimum retention time (in days) of a certain waste-handling system is given by the expression below. Evaluate the given expression.

$$\dfrac{1}{0.35\left[1-\sqrt{\dfrac{8,100}{8,100+121,000}}\right]-0.045}$$

51. *Acoustics*   The speed of sound of a longitudinal sound wave in a liquid, v_L, is given by the formula

$$v_L = \sqrt{\dfrac{K}{\rho}}$$

where K is the bulk modulus and ρ is the mass density of the liquid. Determine the speed of sound in water if $K = 2.1 \times 10^9$ N/m^2 and $\rho = 1\,000$ kg/m^3.

52. *Wastewater technology*   For a large rectangular orifice of depth $h_2 - h_1$, the average discharge velocity is

$$\bar{v} = \dfrac{2C_v(h_2^{2/3} - h_1^{2/3})\sqrt{2g}}{3(h_2 - h_1)}$$

Find $\bar{v}$ (in meters per second) when C_v, the coefficient of velocity, is 0.96, $h_2 = 0.512$ m, $h_1 = 0.385$ m, and $g = 9.81$ m/s^2.

 In Your Words

53. (a) Evaluate $\sqrt{81}$ and $-\sqrt{81}$.
(b) Which of your answers in (a) is the principal square root of 81?
(c) Evaluate $\sqrt[3]{-27}$ and $-\sqrt[3]{27}$.
(d) Which of your answers in (c) is the principal cube root of -27?
(e) Explain what is meant by the principal nth root of a number.

54. Explain what is meant by $b^{m/n}$.

55. (a) Use a calculator to evaluate $64\wedge(1\div3)$.
(b) Use a calculator to evaluate $64\wedge3\boxed{x^{-1}}$.
(c) Are these two answers the same?
(d) What do they represent?

≡ CHAPTER 1 REVIEW

Important Terms and Concepts

Absolute value
Accuracy
Associative laws
Commutative laws
Decimals
Distributive law
Engineering notation
Exact numbers
Exponents
   Fractional
   Rules for
Identity elements
Inequalities
Integers
   Negative
   Positive
   Zero

Inverse elements
Irrational numbers
Natural numbers
Order of operations
Precision
Radical
Rational numbers
Real numbers
Reciprocal
Roots
Rounding off
Scientific notation
Significant digits
Zero

Review Exercises

1. To which sets of numbers do each of the following numbers belong?
   (a) -24
   (b) $\frac{15}{7}$
   (c) $\sqrt[3]{-17}$

2. Give the absolute value of each of these numbers.
   (a) $\sqrt{42}$
   (b) -16
   (c) $\frac{-5}{8}$

3. Give the reciprocal of each of these numbers.
   (a) $\frac{2}{3}$
   (b) -8
   (c) $\frac{-1}{5}$

4. Give the additive inverse of each of these numbers.
   (a) -5
   (b) 17
   (c) $4\sqrt{2}$

5. In each part, identify the basic law of real numbers that is being used.
   (a) $14+7=7+14$
   (b) $6(8+2)=6\cdot8+6\cdot2$
   (c) $14=0+14$
   (d) $\frac{2}{3}\times\frac{3}{2}=1$

In Exercises 6–20, perform the indicated operation.

6. $16+48$
7. $37+(-81)$
8. $95-42$
9. $37-(-61)$
10. $4\times(-8)$
11. $\frac{2}{3}+\frac{-5}{6}$
12. $\frac{4}{5}-\frac{5}{6}$
13. $-9\div3$
14. $12\div(-4)$
15. $\frac{2}{3}\times\frac{-3}{5}$
16. $3\frac{3}{4}\times(-4\frac{1}{3})$
17. $\frac{2}{3}\div\frac{1}{4}$
18. $\frac{1}{5}\div\frac{-2}{15}$
19. $2\frac{1}{2}\div3\frac{1}{4}$
20. $-4\frac{1}{3}\div(-3\frac{1}{6})$

Evaluate each number in Exercises 21–26.

21. 2^5

22. $(-3)^4$

23. $(-4)^3$

24. $8^{1/3}$

25. $4^{1/2}$

26. $(-64)^{1/3}$

In Exercises 27–38, perform the indicated operations. Leave all answers in terms of positive exponents.

27. $2^5 \cdot 2^3$

28. $3^5 \cdot 3^4$

29. $2^{-3} \cdot 2^5$

30. $16^{-4} \cdot 16^{-3}$

31. $\left(4^3\right)^5$

32. $\left(2^{-3}\right)^4$

33. $\left(4^{1/3}\right)^3$

34. $\left(5^{1/4}\right)^{2/3}$

35. $\dfrac{a^2 b^3}{ab^4}$

36. $\dfrac{x^2 y^3 z}{x^3 yz}$

37. $\left(ax^2\right)^{-2}$

38. $\dfrac{\left(by^{-2}\right)^2}{\left(cy^{-4}\right)^{1/4}}$

Which of the pairs of numbers in Exercises 39–42 is (a) more accurate and which is (b) more precise?

39. 2.37; 42,000

40. 0.002; 2.02

41. 7; 0.7

42. 1200; 0.0021

In Exercises 43–46, round off each number to (a) two significant digits, (b) three significant digits, (c) the nearest tenth, and (d) the nearest hundredth.

43. 7.351

44. 18.2874

45. 2.0528

46. 4.028476

Change each of the numbers in Exercises 47–54 to scientific notation.

47. 371 000 000 000

48. 2 540 000 000 000 000

49. 0.000 000 000 024

50. 0.000 000 000 000 000 000 049 1

51. 29 080 000 000 000 000

52. 753 000 000 000 000 000 000 000

53. 0.000 000 000 000 000 000 000 075

54. 0.000 000 000 000 000 193

Find the principal root of the real numbers in Exercises 55–58.

55. $\sqrt{144}$

56. $\sqrt[3]{-64}$

57. $\sqrt[4]{625}$

58. $\sqrt{\dfrac{36}{121}}$

Solve Exercises 59 and 60.

59. *Business*   A wholesale parts distributor discounts the unit cost of an item $0.14 for each time a full dozen is ordered. A manufacturer orders 418 items which normally cost $13.79 each.
(a) What was the amount of the discount?
(b) What is the total cost of the manufacturer's order?

60. *Nuclear physics*   The average kinetic energy of atomic particles in a certain location, E, can be determined from the absolute temperature, T, at that location by using the equipartition principle

$$E = \tfrac{3}{2}kT$$

where k is Boltzmann's constant $8.617\ 33 \times 10^{-5}$ eV/K.
(a) What is the average kinetic energy in electronvolts of particles at room temperature ($T = 295$ K)?
(b) What is the average kinetic energy in electronvolts of particles at the surface of the sun ($T = 5.714 \times 10^3$ K)?

≣ CHAPTER 1 TEST

1. Give the absolute value of each of the following:
   (a) $\frac{4}{3}$
   (b) $\frac{1}{2} - \frac{5}{8}$
   (c) -6

2. What is the reciprocal of $-\frac{3}{7}$?

3. What is the additive inverse of $-\frac{5}{8}$?

In Exercises 4–22, perform the indicated operation.

4. $35 + 76$

5. $-47 - 65$

6. -5×14

7. $\frac{5}{3} + 5\frac{2}{3}$

8. $\frac{7}{4} - \left(-\frac{3}{5}\right)$

9. $-4\frac{1}{2} \times 2\frac{1}{3}$

10. $\frac{-5}{7} \div \frac{15}{28}$

11. $-2\frac{1}{3} \div -4\frac{5}{6}$

12. Evaluate $(-5)^3$.

13. $4^{3/2}$

14. $2^6 \cdot 2^{-4}$

15. $\left(5^{1/4}\right)^8$

16. $3^{5/2} \div 3^{-3/2}$

17. $\dfrac{a^3 b^5}{a^4 b^2}$

18. Which is more accurate: 4.516 or 37 000?

19. Which is more precise: 0.000 51 or 51.02?

20. Write 47,500,000,000,000 in scientific notation.

21. What is the principal root of $\sqrt{\frac{49}{25}}$?

22. Write 92 500 000 000 000 in engineering notation.

2

Algebraic Concepts and Operations

In order to operate correctly, a computer relies on instructions given to it by a programmer. The computer programmer must understand algebra in order to tell the computer what to do.

Courtesy of Ruby Gold

Until now we have spent most of our time working with real numbers. We have learned the basic operations with real numbers, how to take powers, roots, and absolute values, and how to write real numbers in scientific notation. Now it is time to begin our work in algebra.

In this chapter, we will begin by looking at the operations of addition, subtraction, multiplication, and division of algebraic expressions. Once these four operations have been learned, we will use them to learn how to solve equations. Finally, we will use our equation-solving skills to solve many problems like those encountered in technical areas.

≡ 2.1
ADDITION AND SUBTRACTION

One of the first things that everyone notices about algebra is that it uses letters and other symbols to represent numbers. Algebra also uses letters and other symbols to represent unknown amounts in equations and inequalities. In the first chapter we used letters to help present some of the rules. For example, we said that $b^{-n} = \dfrac{1}{b^n}$, if $b \neq 0$.

Variables

When a letter or other symbol is used to indicate something that can be assigned any value from a given or implied set of numbers, it is called a **variable**. We have already used letters as variables in this book. For example, we used a, b, and c as variables when we said that $a(b+c) = ab + ac$. We used b and n as variables when we said that $b^{-n} = \dfrac{1}{b^n}$, if $b \neq 0$. Another place where variables are used is on a calculator. For example, the $\boxed{x^2}$, $\boxed{\sqrt{x}}$, $\boxed{1/x}$, and $\boxed{y^x}$ keys on a calculator all use the letter x as a variable and one uses the letter y.

Constants

Letters or other symbols may also be used to designate fixed, but unspecified, numbers called **constants**. One constant that you are already familiar with is π, which represents the number $3.14159265\ldots$. (The $\ldots$ at the end of this number means that there are more digits after the 5.) Usually letters near the beginning of the alphabet, like a, b, c, d, are used to represent constants. Letters near the end of the alphabet, such as w, x, y, and z, are used to indicate variables. Thus, in the expression $ax + b$, a and b represent constants and x is a variable.

Algebraic Expressions

The term **algebraic expression** is used for any combination of variables and constants that is formed using a finite number of operations. Examples of algebraic expressions include ax^2, $ax^2 + bx + c$, $\dfrac{x^3 + \sqrt{y}}{ax^2 - by}$, $\dfrac{4x^{-3} + 7x^2}{2x - 3w}$, and $1.8C + 32$.

Algebraic Terms

If an algebraic expression consists of parts connected by plus or minus signs, it is called an **algebraic sum**. Each of the parts of an algebraic sum, together with the sign preceding it, is called an **algebraic term**.

EXAMPLE 2.1

What are the algebraic terms of the algebraic sum $5x^7 + 2x^3y - \dfrac{4x}{y^2}$?

Solution   This algebraic sum has three algebraic terms: $5x^7$, $2x^3y$, and $-\dfrac{4x}{y^2}$.

An algebraic term may consist of several **factors**. The term $5xy$ has the individual factors of 5, x, and y. Other factors are products of two or more of the individual factors. Thus, $5x$, $5y$, xy, and $5xy$ are also factors of $5xy$.

EXAMPLE 2.2

What are the factors of $\dfrac{6x^2}{y}$?

Solution The individual factors are 2, 3, x, and $\dfrac{1}{y}$. Other factors are 6, $2x$, $3x$, $6x$, x^2, $2x^2$, $3x^2$, $6x^2$, $\dfrac{2}{y}$, $\dfrac{3}{y}$, $\dfrac{6}{y}$, $\dfrac{x}{y}$, $\dfrac{2x}{y}$, $\dfrac{3x}{y}$, $\dfrac{6x}{y}$, $\dfrac{x^2}{y}$, $\dfrac{2x^2}{y}$, $\dfrac{3x^2}{y}$, and $\dfrac{6x^2}{y}$.

EXAMPLE 2.3

What are the factors of $\dfrac{-10(x+y)}{z}$?

Solution The individual factors are -1, 2, 5, $x+y$, and $\dfrac{1}{z}$. The other factors are products of these factors. Notice that the negative sign was treated by using -1 as a factor.

Each term has two parts. One part is the coefficient and the other part contains the variables. The **coefficient** is the product of all of the constants. Normally the coefficient is written at the front of the term. A variable with no visible coefficient, such as x or y, is understood to have a coefficient of 1. The coefficient of the term $4x^2$ is 4; of $18y^3z$ is 18; of $\dfrac{w^2}{3}$ is $\frac{1}{3}$; and of yx^2 is 1. The coefficient of $4\pi bx$ is $4\pi b$ if we assume that b is a constant.

Terms that involve exactly the same variables raised to exactly the same power are called **like terms**. For example, $7x^2y$ and $4x^2y$ are like terms because both contain the same variables (x and y), the variable x is raised to the second power in each term and the y is raised to the first power. $4xyz^3$ and $5xyz^2$ are not like terms even though they both contain the variables x, y, and z, because one term has the variable z raised to the third power and in the other term it is raised to the second power.

Monomials and Multinomials

An algebraic expression with only one term is a **monomial**. An algebraic sum with exactly two terms is a **binomial;** an algebraic sum with three terms is a **trinomial;** and an algebraic sum with two or more terms is a **multinomial**. As you can see, binomials and trinomials are special kinds of multinomials.

EXAMPLE 2.4

$4x^2$, $\frac{7}{5}xy$, and $-8xz^3\sqrt{w}$ are all monomials.
ax^2+by, $\sqrt{3x^3z}-2w$, and $4x^2+2$ are binomials.
$3x^3-2x+1$ and $2xy+3yz-4\sqrt{2y}$ are trinomials.
$16x^3-7\sqrt{2}y^2+\frac{2}{3}xy-7^5x^3y$ is a multinomial with four terms. In this last expression the coefficient of the first term is 16, the second term has a coefficient of $-7\sqrt{2}$, the third $\frac{2}{3}$, and the fourth -7^5 (or $-16,807$).

Adding and Subtracting Multinomials

In order to add or subtract monomials, we add or subtract the coefficients of like terms. So, $3x^2+5x^2=8x^2$. We can justify this with the distributive property, since $3x^2+5x^2=(3+5)x^2=8x^2$.

≡ **Note**

Remember, you can only add or subtract like terms.

So,

$$3x^2+2xy-y^3-8xy=3x^2-6xy-y^3$$

because the only two like terms are $2xy$ and $-8xy$. When these are added, we get $-6xy$. None of the other terms can be combined because they are not like terms.

EXAMPLE 2.5

$(3x^2+2yx-5y)+(-6xy+6x^2-7y)$
$$=(3x^2+6x^2)+(2yx-6xy)+(-5y-7y)$$
$$=9x^2-4yx-12y$$
Notice that $2yx$ and $-6xy$ are like terms even though the variables are written in a different order. This is an application of the commutative law introduced in Section 1.2.

EXAMPLE 2.6

$(4y-3z+4xy)-(7y-3z-6xy)$
$$=4y-3z+4xy-7y+3z+6xy$$
$$=(4y-7y)+(-3z+3z)+(4xy+6xy)$$
$$=-3y+10xy$$
Notice that when you have a negative sign in front of a parenthesis it means to multiply *each* term inside the parentheses by -1. So,

$$-(7y-3z-6xy)=-7y+3z+6xy$$

As mentioned in Section 1.2, there are several different symbols for grouping. These symbols are parentheses ( ), brackets [ ], and braces { }. When it is necessary to simplify an expression, you should start removing grouping symbols from the inside. For example, the expression

$$5\{2x - [3x + (4x - 5) + 2] - 3x\}$$

would be simplified by first removing the parentheses and combining any like terms that are within the brackets.

$$5\{2x - [3x + (4x - 5) + 2] - 3x\} = 5\{2x - [3x + 4x - 5 + 2] - 3x\}$$
$$= 5\{2x - [7x - 3] - 3x\}$$

Next, remove the brackets. Do not forget the negative sign in front of the left bracket [. This indicates that each term inside the brackets is to be multiplied by -1 when the brackets are removed.

$$= 5\{2x - [7x - 3] - 3x\}$$
$$= 5\{2x - 7x + 3 - 3x\}$$
$$= 5\{-8x + 3\}$$

Now, remove the braces. The 5 in front of the left brace { means that each term within the braces should be multiplied by 5.

$$= 5\{-8x + 3\}$$
$$= -40x + 15$$

Application

EXAMPLE 2.7

The total resistance, R, in a parallel circuit with two resistances, R_1 and R_2, is given by the formula $\dfrac{1}{R} = \dfrac{1}{R_1} + \dfrac{1}{R_2}$. Simplify the right-hand side of this equation.

Solution In order to simplify these two fractions, we need to have a common denominator. For this fraction, the common denominator is $R_1 R_2$.

$$\frac{1}{R} = \frac{R_2}{R_1 R_2} + \frac{R_1}{R_1 R_2} = \frac{R_2 + R_1}{R_1 R_2}$$

Exercise Set 2.1

In Exercises 1–4, identify the variables and constants in each expression.

1. $4xy$

2. $\dfrac{5x^2}{3tw}$

3. $8\pi r^2$

4. $\sqrt{7}ax + by$

In Exercises 5–8, identify the algebraic terms in each expression.

5. $3x^3 + 4x$

6. $47x^4 - 9y^5$

7. $(2x^3)(5y) + \sqrt{3}ab - \dfrac{7a}{b}$

8. $-3x^5 - (4a)\left(\dfrac{b}{5}\right) + 9\sqrt{y}$

Solve Exercises 9 and 10.

9. What are the individual factors of $\dfrac{-5x^3}{y^2}$?

10. What are the individual factors of $\dfrac{15(x-y)}{x^2+3y}$?

In Exercises 11–14, identify the coefficient of each expression.

11. $47xy^5$

12. $\dfrac{2\sqrt{3}y}{x^5}$

13. $\dfrac{\pi ax}{4y}$

14. $\dfrac{\sqrt{17}byz}{\sqrt{5}at}$

In Exercises 15–20, identify the like terms in each exercise.

15. $3x^2y$, $17x^2y$, and $12xy^2$

16. $\sqrt{5}xy$, $\dfrac{2}{3}ax^5y$, and $\dfrac{2}{3}bxy$

17. $(x+y)^2$, $2(x+y)$, $4(x+y)^3$, and $5(x+y)^2$

18. $3x^2y$, $-9xy^2$, $5(xy^2)$, and $6(x^2+y)$

19. $5(a+b+c)$, $a+b-c$, and $-2(a+b-c)$

20. ax^2, a^2x, $5x^2$, and $5a^5x$

In Exercises 21–70, simplify each algebraic expression.

21. $4x+7x$

22. $5y-2y$

23. $3z-z$

24. $7w+4w-w$

25. $8x+9x^2-2x$

26. $11y-7y+6y^2$

27. $10w+w^2-8w^2$

28. y^2-6y^2+4y

29. $ax^2+a^2x+ax^2$

30. $by-by^2+by$

31. $7xy^2-5x^2y+4xy^2$

32. $12wz-8w^2z+6w^2z$

33. $(a+6b)-(a-6b)$

34. $(x-7y)-(7y-x)$

35. $(2a^2+3b)+(2b+4a)$

36. $(7c^2-8d)+(6d-8c)$

37. $(4x^2+3x)-(2x^2-3x)$

38. $(3y^2-4x)-(4y+2x)$

39. $2(6y^2+7x)$

40. $5(3a+4b)$

41. $-3(4b-2c)$

42. $-2(-6b+3a)$

43. $4(a+b)+3(b+a)$

44. $2(c+d)+8(d+c)$

45. $3(x+y)-2(x+y)$

46. $3(x^2-y)-2(y+x^2)$

47. $2(a+b+c)+3(a+b-c)$

48. $4(x-y+z)+2(x+y-z)$

49. $3[2(x+y)]$

50. $4[3(x-y)]$

51. $3(a+b)+4(a+b)-2(a+b)$

52. $2(x+y)-3(x+y)-4(x+y)$

53. $2(a+b+c)+3(a-b+c)+(a-b-c)$

54. $3(x-y+z)-2(x+y-z)+4(-x-y+z)$

55. $2(x+3y)-3(x-2y)+5(2x-y)$

56. $3(y-2a)-(a+3y)+4(a-3y)$

57. $3(x+y-z)-2(3x+2y-z)-3(x-y+4z)$

58. $5(x-y+z)-(y-x+z)+2(x-2y+z)$

59. $(x+y)-3(x-z)+4(y+4z)-2(x+y-3z)$

60. $(a+b)-2(b-c)+4(c+2d)-5(d+2c-3b-a)$

61. $x+[3x+2(x+y)]$

62. $x+[5y+3(y-x)]$

63. $y-[2z-3(y+z)+y]$

64. $2w-[4z-5(z+w)+2w]$

65. $[2x+3(x+y)-2(x-y)+y]-2x$

66. $[4a-8(a+b)+2(a-b)+3a]-3b$

67. $-\{-[2a-(3b+a)]\}$

68. $-\{-3[4x-(5x-4y)]\}$

69. $5a-2\{4[a+2(4a+b)-b]+a\}-a$

70. $7x-3\{-[x+2(x-y)-y]+2x\}-3y$

Solve the word problems in Exercises 71–74.

71. *Civil engineering*   In order to determine the cost of widening a highway, several cost comparisons were used. These led to the following expression for determining the total cost: $p + \frac{1}{2}p + \frac{2}{3}p$. Simplify by combining like terms.

72. *Construction*   A concrete mix has the following ingredients by volume: 1 part cement, 2 parts water, 3 parts aggregate, and 3 parts sand. These are used to produce the expression $x + 2x + 3x + 3x$. This expression is then used to determine the amount of each ingredient for a specific amount of concrete. Simplify by combining like terms.

73. *Electronics*   The total capacitance, C_T, is a series circuit with two capacitors, C_1 and C_2, and is given by the formula

$$\frac{1}{C_T} = \frac{1}{C_1} + \frac{1}{C_2}$$

Simplify the right-hand side of this equation.

74. *Electronics*   The total inductance, L_T, in a parallel circuit with three inductors, L_1, L_2, and L_3, is given by

$$\frac{1}{L_T} = \frac{1}{L_1} + \frac{1}{L_2} + \frac{1}{L_3}$$

Simplify the right-hand side of this equation.

 **In Your Words**

75. Explain how an algebraic term and an algebraic expression are alike and how they are different.

76. Describe what is meant by "like terms."

 **2.2**

MULTIPLICATION

In Section 1.4 we learned some rules for exponents. The very first rule stated that $b^m b^n = b^{m+n}$. We will use this rule as a foundation for learning how to multiply multinomials. We will also use the rule for multiplying real numbers from Section 1.3.

Multiplying Monomials

First, we will find the product of two or more monomials. When multiplying monomials, first multiply the numerical coefficients to get the numerical coefficient of the product. Then, multiply the remaining factors using the rules of exponents.

EXAMPLE 2.8

$$(3x^3)(-7bx^4) = -21bx^{3+4} = -21bx^7$$
$$(4ax^2)(7bx^3) = 28abx^{2+3} = 28abx^5$$
$$(-5xy^2z^3)(3x^2y^2) = -15x^3y^4z^3$$
$$(\sqrt{2}x^3y)(\sqrt{3}xy^4) = \sqrt{6}x^4y^5$$

To multiply a monomial and a multinomial, use the distributive property. Distribute the monomial over the multinomial. Thus, by the distributive property $a(b+c) = ab + ac$. Each term of the multinomial will be multiplied by the monomial.

EXAMPLE 2.9

$$4x^2y(3xy^2 + 2x^3y) = (4x^2y)(3xy^2) + (4x^2y)(2x^3y)$$
$$= 12x^3y^3 + 8x^5y^2$$

EXAMPLE 2.10

$$-7ab^2(2ax^3 - 5abx + 3b^3) = (-7ab^2)(2ax^3) + (-7ab^2)(-5abx)$$
$$+ (-7ab^2)(3b^3)$$
$$= -14a^2b^2x^3 + 35a^2b^3x - 21ab^5$$

As you gain experience, you will find that you may not always need to write all the steps. The more you practice, the better you will become, until you may be able to write the correct answer without writing the middle expression.

Multiplying Multinomials

To multiply one multinomial by another, multiply each term in one multinomial by each term in the other multinomial. Again, use the distributive property to help in the multiplication.

EXAMPLE 2.11

$$(x^2 + b)(x + c) = x^2(x + c) + b(x + c)$$
$$= x^3 + cx^2 + bx + bc$$

EXAMPLE 2.12

$$(y^2 + 2f)(y^2 - 3f) = y^2(y^2 - 3f) + 2f(y^2 - 3f)$$
$$= y^4 - 3y^2f + 2y^2f - 6f^2$$
$$= y^4 - y^2f - 6f^2$$

The preceding two examples show multiplication of two binomials. The same procedure is used for multiplying any two multinomials. The next example shows how to multiply two trinomials. This example uses an extended version of the distributive property, where $a(b + c + d) = ab + ac + ad$.

EXAMPLE 2.13

$$(2x + 3y + z)(4x - 3y - z) = 2x(4x - 3y - z) + 3y(4x - 3y - z)$$
$$+ z(4x - 3y - z)$$
$$= (8x^2 - 6xy - 2xz)$$
$$+ (12xy - 9y^2 - 3yz)$$
$$+ (4xz - 3yz - z^2)$$
$$= 8x^2 + 6xy + 2xz - 9y^2 - 6yz - z^2$$

Another method is to multiply in the same manner that you used with real numbers. This is shown in Examples 2.14 and 2.15. When using this method, like terms are put in the same column.

EXAMPLE 2.14

$(2x+5)(3x-4)$

Solution

$$
\begin{array}{r}
2x \ + \ 5 \\
\times \quad 3x \ - \ 4 \\
\hline
-8x \ - 20 \\
6x^2 + 15x \\
\hline
6x^2 + \ 7x \ - 20 \quad \text{Add the like terms } 15x \text{ and } -8x.
\end{array}
$$

EXAMPLE 2.15

$(x^3+2x^2-3x+7)(5x^2-4x+2)$

Solution

$$
\begin{array}{r}
x^3 \ + \ 2x^2 \ - \ 3x \ + \ 7 \\
\times \qquad\qquad 5x^2 \ - \ 4x \ + \ 2 \\
\hline
2x^3 \ + \ 4x^2 \ - \ 6x \ + 14 \\
-4x^4 \ - \ 8x^3 \ +12x^2 - 28x \\
5x^5 + 10x^4 - 15x^3 + 35x^2 \\
\hline
5x^5 + \ 6x^4 \ -21x^3 + 51x^2 - 34x + 14
\end{array}
$$

The FOIL Method for Multiplying Binomials

There will be many times when you will need to multiply two binomials. Because this happens so often, there are a few patterns that we should notice. The first pattern, called the **FOIL method**, is one that many people use to remember how to multiply two binomials. The letters for FOIL come from the first letters for the words First, Outside, Inside, and Last.

 Hint

Use the FOIL method to help you remember how to multiply binomials. FOIL indicates the order in which terms of two binomials could be multiplied.

F   suggests the product of the **F**irst terms.
O   suggests the product of the **O**utside terms.
I   suggests the product of the **I**nside terms.
L   suggests the product of the **L**ast terms.

EXAMPLE 2.16

Use the FOIL method to multiply $(x+4)(x+3)$.

EXAMPLE 2.16 (Cont.)

Solution

$\underbrace{(x+4)(x+3)}_{\text{F}}$ x^2 the product of the **First** terms

$\underbrace{(x+4)(x+3)}_{\text{O}}$ $3x$ the **Outside** terms

$\underbrace{(x+4)(x+3)}_{\text{I}}$ $4x$ the **Inside** terms

$\underbrace{(x+4)(x+3)}_{\text{L}}$ 12 the **Last** terms

Adding and combining the terms in the right-hand column

$$x^2 + 3x + 4x + 12 = x^2 + 7x + 12$$

gives the correct answer.

This example looks much longer than it really is. Try it! You will find that this will help "foil" errors caused by not multiplying all the terms.

The Special Product $(a+b)(a-b)$

One special product is shown by this example:

$$(x+4)(x-4) = x^2 - 4x + 4x - 16 = x^2 - 16$$

Notice that the two middle terms are additive inverses of each other and their sum is zero.

Let's look at the general case of this type of problem.

$$(a+b)(a-b) = a^2 - ab + ab - b^2 = a^2 - b^2$$

What is so special about this problem? Look at the two factors in the original problem. One factor is $a+b$ and the other is $a-b$. Notice that each factor has the same first term a, and that the second terms are additive inverses of each other, b and $-b$. Whenever you have a product of two binomials in the form $a+b$ and $a-b$, the answer is $a^2 - b^2$.

Difference of Squares	$(a+b)(a-b) = a^2 - b^2$

EXAMPLE 2.17

$(x+7)(x-7) = x^2 - 49$ Note that $49 = 7^2$.

$(3a+b)(3a-b) = 9a^2 - b^2$ Note that $(3a)^2 = 9a^2$.

$\left(\dfrac{4}{3}x - 2y\right)\left(\dfrac{4}{3}x + 2y\right) = \dfrac{16}{9}x^2 - 4y^2$

The Special Product $(a+b)^2$

The final special product of binomials is a perfect square: $(a+b)^2$. Remember that $(a+b)^2 = (a+b)(a+b)$. If we multiply these we get

$$(a+b)^2 = (a+b)(a+b) = a^2 + ab + ab + b^2 = a^2 + 2ab + b^2$$

The answer contains the squares of the first and last terms. The middle term in the answer is twice the product of the two terms, a and b. So, when you have a square of a binomial you get $(a+b)^2 = a^2 + 2ab + b^2$ or $(a-b)^2 = a^2 - 2ab + b^2$.

Square of a Binomial	$(a+b)^2 = a^2 + 2ab + b^2$
	$(a-b)^2 = a^2 - 2ab + b^2$

EXAMPLE 2.18

$$(x + 3y)^2 = x^2 + 6xy + 9y^2$$
$$(t + 7)^2 = t^2 + 14t + 49$$
$$(4a - b)^2 = 16a^2 - 8ab + b^2$$
$$\left(\frac{ax}{2} - \frac{3c}{4}\right)^2 = \frac{a^2x^2}{4} - \frac{3}{4}acx + \frac{9c^2}{16} \quad \text{Note that } 2\left(\frac{ax}{2}\right)\left(\frac{3c}{4}\right) = \tfrac{3}{4}acx.$$

Caution

Remember: $(a+b)^2 \neq a^2 + b^2$ and $(a-b)^2 \neq a^2 - b^2$.

Application

EXAMPLE 2.19

A physics equation that describes an elastic collision is $m_1(v_a - v_b)(v_a + v_b) = m_2(v_c - v_d)(v_c + v_d)$. Perform the indicated multiplications.

Solution   We will begin by multiplying the expressions in parentheses.

$$m_1(v_a - v_b)(v_a + v_b) = m_2(v_c - v_d)(v_c + v_d)$$
$$m_1(v_a{}^2 - v_b{}^2) = m_2(v_c{}^2 - v_d{}^2)$$

Next, we shall multiply through the left-hand side by m_1 and the right-hand side by m_2, producing

$$m_1 v_a{}^2 - m_1 v_b{}^2 = m_2 v_c{}^2 - m_2 v_d{}^2$$

We cannot simplify this equation any more, because none of the remaining terms are like terms.

Exercise Set 2.2

In Exercises 1–62, perform the indicated multiplication.

1. $(a^2x)(ax^2)$
2. $(by^2)(b^2y)$
3. $(3ax)(2ax^2)$
4. $(5by)(3b^2y)$
5. $(2xw^2z)(-3x^2w)$
6. $(-4ya^2b)(6y^2b)$
7. $(3x)(4ax)(-2x^2b)$
8. $(4y)(3y^2b)(-5by^2)$
9. $2(5y-6)$
10. $4(3x-5)$
11. $-5(4w-7)$
12. $-3(8+5p)$
13. $3x(7y+4)$
14. $6x(8y-7)$
15. $-5t(-3+t)$
16. $-3n(2n-5)$
17. $\frac{1}{2}a(4a-2)$
18. $\frac{1}{3}x(-21x-15)$
19. $2x(3x^2-x+4)$
20. $3y(4y^2-5y-7)$
21. $4y^2(-5y^2+2y-5+3y^{-1}-6y^{-2})$

22. $5p^2(-4p^3-3p+2+p^{-1}-7p^{-2})$
23. $(a+b)(a+c)$
24. $(s+t)(s+2t)$
25. $(x+5)(x^2+6)$
26. $(y+3)(y^2+7)$
27. $(2x+y)(3x-y)$
28. $(4a+b)(8a-b)$
29. $(2a-b)(3a-2b)$
30. $(4p+q)(3p-2q)$
31. $(b-1)(2b+5)$
32. $(4x-1)(3x-2)$
33. $(7a^2b+3c)(8a^2b-3c)$
34. $(6p^2r+2t)(5p^2r+4t)$
35. $(x+4)(x-4)$
36. $(a+8)(a-8)$
37. $(p-6)(p+6)$
38. $(b-10)(b+10)$
39. $(ax+2)(ax-2)$
40. $(xy-3)(xy+3)$
41. $(2r^2+3x)(2r^2-3x)$
42. $(4p^3-7d)(4p^3+7d)$
43. $(5a^2x^3-4d)(5a^2x^3+4d)$
44. $(3p^2st-\frac{11}{3}w^3)(3p^2st+\frac{11}{3}w^3)$

45. $\left(\frac{2}{3}pa^2f+\frac{3}{4}tb^3\right)$
    $\times\left(-\frac{2}{3}pa^2f+\frac{3}{4}tb^3\right)$
46. $\left(\frac{\sqrt{3}}{2}+\frac{7}{5}t^2u\right)$
    $\times\left(-\frac{\sqrt{3}}{2}+\frac{7}{5}t^2u\right)$
47. $(x+y)^2$
48. $(p+r)^2$
49. $(x-5)^2$
50. $(b-7)^2$
51. $(a+3)^2$
52. $(w+5)^2$
53. $(2a+b)^2$
54. $(3c+d)^2$
55. $(3x-2y)^2$
56. $(5a-6f)^2$
57. $4x(x+4)(3x-2)$
58. $5y(y-6)(2y+3)$
59. $(x+y-z)(x-y+z)$
60. $(a+b+c)(a-b-c)$
61. $(a+2)^3$
62. $(m-n)^3$

Solve Exercises 63–68.

63. *Computer science*   The amount of time required to test a computer "chip" with n cells is given by the expression $2[n(n+2)+n]$. Simplify this expression. (In this problem, asking you to simplify means that you are to perform the indicated multiplications and add the like terms.)

64. *Aeronautical engineering*   An aircraft uses its radar to measure the direct echo range R to another object. If x represents the distance to the ground echo point, then the following expression results:

$$(2R-x)^2-x^2-R^2$$

Simplify this expression.

65. *Construction*   In computing the center of mass of a plate of uniform density, the expression $\frac{1}{2}(y_2-y_1)(y_2+y_1)$ is used. Simplify this expression.

66. *Business*   After three years, an investment of $500 at an interest rate r compounded annually (once a year) is worth $500(1+r)^3$. Perform the indicated multiplications.

67. *Automotive technology*   Under certain conditions, position x of a motorcycle at time t is given by

$$x = 3(t+4)(t+1)$$

Simplify the right-hand side of this equation by multiplying the terms.

68. *Thermal science*   The quantity of heat, Q, transferred by radiation between two parallel equal areas, A, in time t is

$$Q = \bar{e}A\left\{ \left(\frac{T_1}{100} - \frac{T_2}{100}\right)\left(\frac{T_1}{100} + \frac{T_2}{100}\right) \right.$$
$$\left. \times \left[\left(\frac{T_1}{100}\right)^2 + \left(\frac{T_2}{100}\right)^2\right]\right\}10^8 t$$

where $\bar{e}$ is a radiation constant and T_1 and T_2 are absolute temperatures of the areas. Simplify this equation by multiplying the terms in the braces.

In Your Words

69. Explain how to use the FOIL method for multiplying two binomials. (Do not look at the explanation in the book.)

70. Describe how you would multiply a binomial and a trinomial.

≡ 2.3
DIVISION

To find the quotient when one multinomial is divided by another, we will again need to use the rules of exponents. Rule 5 in Section 1.4 states that $\dfrac{b^m}{b^n} = b^{m-n}$, if $b \neq 0$. This is a very helpful rule in the division of algebraic expressions.

Remember that division can be indicated by $x \div y$, $\dfrac{x}{y}$, or x/y. You must be very careful, when reading algebra problems and writing your answers, to watch how terms are grouped. For example, in $a^2 + b/c + 3$ the only division is b divided by c. Thus, $a^2 + b/c + 3$ is equivalent to $a^2 + \dfrac{b}{c} + 3$. Now, consider $(a^2 + b)/c + 3$. The parentheses indicate that the entire expression $a^2 + b$ is being divided by c. So $(a^2 + b)/c + 3 = \left(\dfrac{a^2+b}{c}\right) + 3$. This could also be written without parentheses as $\dfrac{a^2+b}{c} + 3$. Finally, $(a^2 + b)/(c + 3)$ means that the expression $a^2 + b$ is being divided by the expression $c + 3$. So, $(a^2 + b)/(c + 3) = \dfrac{a^2+b}{c+3}$.

Dividing a Monomial by a Monomial ▬▬▬

We will begin by dividing a monomial by a monomial. Consider $16x^5 \div 8x^2$. Using the laws of exponents, we have $x^5 \div x^2 = x^{5-2} = x^3$. If there are numerical coefficients, they are divided separately. So, $16x^5 \div 8x^2 = \dfrac{16}{8}\dfrac{x^5}{x^2} = 2x^{5-2} = 2x^3$.

Each variable should be treated separately. For example, $24x^3y^2 \div 8xy^2 = \dfrac{24}{8}\dfrac{x^3}{x}\dfrac{y^2}{y^2} = 3x^{3-1}y^{2-2} = 3x^2y^0 = 3x^2$.

EXAMPLE 2.20

Divide $6w^2xy^4z$ by $9wx^3y^2z$.

Solution $6w^2xy^4z \div 9wx^3y^2z = \dfrac{6w^2xy^4z}{9wx^3y^2z}$

$$= \frac{6}{9}w^{2-1}x^{1-3}y^{4-2}z^{1-1}$$

$$= \frac{2}{3}wx^{-2}y^2z^0$$

$$= \frac{2wy^2}{3x^2}$$

EXAMPLE 2.21

Divide $35a^2b^3c^{-2}$ by $-7a^{-3}bc^2$.

Solution $35a^2b^3c^{-2} \div -7a^{-3}bc^2 = \dfrac{35a^2b^3c^{-2}}{-7a^{-3}bc^2}$

$$= \frac{35}{-7}a^{2-(-3)}b^{3-1}c^{-2-2}$$

$$= -5a^5b^2c^{-4}$$

$$= \frac{-5a^5b^2}{c^4}$$

Caution

Be careful! A typical error here is cancelling noncommon factors. Do not make either of the following mistakes:

$$\frac{\cancel{x}+y}{\cancel{x}} = 1+y$$

or $\quad \dfrac{x+\cancel{4}}{\cancel{4}} = x+1$

These are *not correct,* as you can see by substituting some numerical values for x or y.

Dividing a Multinomial by a Monomial

In order to divide a multinomial by a monomial, you should divide each term of the multinomial by the monomial. You may remember that $\dfrac{a+b}{c} = \dfrac{a}{c} + \dfrac{b}{c}$. (This is part of Rule 6 in Section 1.3.)

EXAMPLE 2.22

Divide $(15x^2 + 9x)$ by $6x$.

EXAMPLE 2.22 (Cont.)

Solution $(15x^2 + 9x) \div 6x = \dfrac{15x^2 + 9x}{6x}$

$$= \dfrac{15x^2}{6x} + \dfrac{9x}{6x}$$

$$= \dfrac{5}{2}x + \dfrac{3}{2}$$

EXAMPLE 2.23

Divide $(4r^2 st^3 - 8rs^3)$ by $-2rst^2$.

Solution $(4r^2 st^3 - 8rs^3) \div -2rst^2 = \dfrac{4r^2 st^3 - 8rs^3}{-2rst^2}$

$$= \dfrac{4r^2 st^3}{-2rst^2} + \dfrac{-8rs^3}{-2rst^2}$$

$$= -2rt + 4\dfrac{s^2}{t^2}$$

Remember to divide each term in the multinomial by the monomial.

EXAMPLE 2.24

Divide $(6xy - 4xy^2 + 9x^2 y - 12x^2 y^2)$ by $3xy$.

Solution $\dfrac{6xy - 4xy^2 + 9x^2 y - 12x^2 y^2}{3xy} = \dfrac{6xy}{3xy} - \dfrac{4xy^2}{3xy} + \dfrac{9x^2 y}{3xy} - \dfrac{12x^2 y^2}{3xy}$

$$= 2 - \dfrac{4}{3}y + 3x - 4xy$$

Dividing a Multinomial by a Multinomial

The next step in the division process is to divide a multinomial by a multinomial. This is not as easy as dividing by a monomial, so we will learn it in two stages. In the first stage we will learn to divide a polynomial by a polynomial. In stage two we will apply what we learned in stage one.

Before we can learn to divide a polynomial by a polynomial we need to have a definition of a polynomial. A polynomial is a special type of multinomial. An algebraic sum is a **polynomial in** x if each term is of the form ax^n, where n is a non-negative integer. One example of a polynomial in x is $7x^3 - \sqrt{3}x^2 + \frac{1}{2}x - 2$. (Remember that -2 can be written as $-2x^0$.) Another polynomial in x is $7x^{15} + 2x^3 - 5x$.

Not every polynomial is a polynomial in x. For example, $4y^3 - y^2$ is a polynomial in y and $\dfrac{-\sqrt{3}}{2}$ is a polynomial that is a constant. The multinomial $4x^3 + 5x^2 - 3 + x^{-3}$ is *not* a polynomial because the exponent on the last term is -3, which is not a non-negative integer. The multinomial $6x + \sqrt{7x} = 6x + \sqrt{7}x^{1/2}$

is not a polynomial in x because the exponent on the variable in the last term is $\frac{1}{2}$, which is not an integer.

We will now examine how to divide one polynomial by another. Study Example 2.15. Make sure you read the comments in Example 2.15, because they explain the method we use to divide polynomials.

EXAMPLE 2.25

Divide $6x^2 - 4 - 4x + 8x^4$ by $2x + 1$.

Solution

1. Write both the dividend and the divisor in decreasing order of powers.
The dividend is $6x^2 - 4 - 4x + 8x^4$. The largest power is 4, the next largest is 2, and so the dividend should be written as $8x^4 + 6x^2 - 4x - 4$. The divisor, $2x + 1$, is already in descending order of the powers.

2. Are there any missing terms in the dividend? If so, write them with a coefficient of 0.
Here the dividend does not have an x^3 term. Rewrite the dividend as $8x^4 + 0x^3 + 6x^2 - 4x - 4$.

3. Set up the problem just as you would any long division problem.

$$2x + 1 \overline{) 8x^4 + 0x^3 + 6x^2 - 4x - 4}$$

4. Divide the first term in the dividend by the first term in the divisor.
In this example the first term in the dividend is $8x^4$ and the first term in the divisor is $2x$. The result of this division, $4x^3$, is written above the dividend directly over the term with the same power.

$$
\begin{array}{r}
4x^3 \\
2x + 1 \overline{) 8x^4 + 0x^3 + 6x^2 - 4x - 4}
\end{array}
$$

5. Multiply the divisor by this first term of the quotient. Write the product below the dividend with like terms under the like terms in the dividend. Subtract this product from the dividend. We will call this difference the "new dividend."

$$
\begin{array}{r}
4x^3 \\
2x + 1 \overline{) 8x^4 + 0x^3 + 6x^2 - 4x - 4} \\
\underline{8x^4 + 4x^3 } \\
-4x^3 + 6x^2 - 4x - 4 \quad \text{new dividend}
\end{array}
$$

Caution

Be careful doing the subtraction. A common error would be to get $+4x^3$ for the x^3 term instead of $-4x^3$.

6. Repeat the last two steps until the power of the new dividend is less than the power of the divisor. Each time the last two steps are repeated, you should divide the first term of the new dividend by the first term of the divisor as shown here.

EXAMPLE 2.25 (Cont.)

A.
$$
2x+1 \overline{\smash{\big)}\ 8x^4+0x^3+6x^2-4x-4}
$$
quotient: $4x^3-2x^2$
$$
\underline{8x^4+4x^3}
$$
$$
-4x^3+6x^2-4x-4
$$
$$
\underline{-4x^3-2x^2}
$$
$$
8x^2-4x-4 \quad \text{new dividend}
$$

B.
$$
2x+1 \overline{\smash{\big)}\ 8x^4 + 0x^3 + 6x^2 - 4x-4}
$$
quotient: $4x^3 \ -2x^2 \ +4x$
$$
\underline{8x^4 + 4x^3}
$$
$$
-4x^3 + 6x^2 - 4x-4
$$
$$
\underline{-4x^3 - 2x^2}
$$
$$
8x^2 - 4x-4
$$
$$
\underline{8x^2 + 4x}
$$
$$
-8x-4 \quad \text{new dividend}
$$

C.
$$
2x+1 \overline{\smash{\big)}\ 8x^4+0x^3+6x^2-4x-4}
$$
quotient: $4x^3-2x^2+4x-4$
$$
\underline{8x^4+4x^3}
$$
$$
-4x^3+6x^2-4x-4
$$
$$
\underline{-4x^3-2x^2}
$$
$$
8x^2-4x-4
$$
$$
\underline{8x^2+4x}
$$
$$
-8x-4
$$
$$
\underline{-8x-4}
$$
$$
0
$$

Stop! This is really $0x^0$ and its power, 0, is less than the power of the first term in the divisor, 1. So,

$$(8x^4 + 6x^2 - 4x - 4) \div (2x + 1) = 4x^3 - 2x^2 + 4x - 4$$

When you divide one polynomial by another, your work should look like the final step, C. The other steps were given to help you see what we were doing. In the next example we will show it all in one step.

EXAMPLE 2.26

Divide $4x^5 - 2x^3 + 6x^2 - 8$ by $4x^2 + 2$.

Solution The dividend and divisor are already in descending order. The dividend, $4x^5 - 2x^3 + 6x^2 - 8$, does not have an x^4 or an x^1 term. We should include those terms with coefficients of zero. The dividend now becomes $4x^5 + 0x^4 - 2x^3 + 6x^2 + 0x - 8$. We will now set the problem up in the standard long division format and divide.

EXAMPLE 2.26 (Cont.)

$$
\begin{array}{r}
x^3+0x^2-\ x+\tfrac{3}{2} \\
4x^2+2\ {\overline{\smash{\big)}\,4x^5+0x^4-2x^3+6x^2+0x-\ 8}} \\
\underline{4x^5\qquad\ +2x^3} \\
0x^4-4x^3+6x^2+0x-\ 8 \\
\underline{0x^4\qquad +0x^2} \\
-4x^3+6x^2+0x-\ 8 \\
\underline{-4x^3\qquad -2x} \\
6x^2+2x-\ 8 \\
\underline{6x^2\qquad +\ 3} \\
2x-11
\end{array}
$$

The power of the $2x$ term, 1, is less than the power of the x^2 term of the divisor, 2. This expression, $2x-11$, is the remainder.

The remainder is written as $\dfrac{2x-11}{4x^2+2}$. So,

$$(4x^5-2x^3+6x^2-8)\div(4x^2+2)=x^3+0x^2-x+\frac{3}{2}+\frac{2x-11}{4x^2+2}$$

$$=x^3-x+\frac{3}{2}+\frac{2x-11}{4x^2+2}$$

There is a faster method of division that can be used with some polynomials. This method is known as synthetic division. We will learn about synthetic division in Chapter 16.

EXAMPLE 2.27

Divide $3x^4-6x^3+x^2+12x$ by $3x^2-2$.
Solution

$$
\begin{array}{r}
x^2-\ 2x+1 \\
3x^2-2\ {\overline{\smash{\big)}\,3x^4-6x^3+\ x^2+12x+0}} \\
\underline{3x^4\qquad -2x^2} \\
-6x^3+3x^2+12x+0 \\
\underline{-6x^3\qquad +\ 4x} \\
3x^2+\ 8x+0 \\
\underline{3x^2\qquad -2} \\
8x+2
\end{array}
$$

The solution to $(3x^4-6x^3+x^2+12x)\div(3x^2-2)$ is $x^2-2x+1+\dfrac{8x+2}{3x^2-2}$.

In the next two examples, we will look at a division problem in which both the dividend and the divisor have more than one variable. The method used will be very similar to the method for dividing one polynomial by another.

EXAMPLE 2.28

Divide $27x^6 - 8y^6$ by $3x^2 - 2y^2$.

Solution

$$3x^2 - 2y^2 \overline{)\ 27x^6 \qquad\qquad\quad -8y^6}$$

quotient: $9x^4 + 6x^2y^2 + 4y^4$

$$27x^6 - 18x^4y^2$$
$$18x^4y^2 \qquad\qquad -8y^6$$
$$18x^4y^2 - 12x^2y^4$$
$$12x^2y^4 - 8y^6$$
$$12x^2y^4 - 8y^6$$
$$0$$

So, $(27x^6 - 8y^6) \div (3x^2 - 2y^2) = 9x^4 + 6x^2y^2 + 4y^4$.

EXAMPLE 2.29

Divide $2a^3 - 3a^2b + 4ab^2 - 12b^3$ by $a - 2b$.

Solution

quotient: $2a^2 + ab + 6b^2$

$$a - 2b \overline{)\ 2a^3 - 3a^2b + 4ab^2 - 12b^3}$$
$$2a^3 - 4a^2b$$
$$a^2b + 4ab^2 - 12b^3$$
$$a^2b - 2ab^2$$
$$6ab^2 - 12b^3$$
$$6ab^2 - 12b^3$$
$$0$$

So, $(2a^3 - 3a^2b + 4ab^2 - 12b^3) \div (a - 2b) = 2a^2 + ab + 6b^2$.

Exercise Set 2.3

In Exercises 1–86, perform the indicated division.

1. x^7 by x^3
2. y^8 by y^6
3. $2x^6$ by x^4
4. $3w^4$ by w^2
5. $12y^5$ by $4y^3$
6. $15a^7$ by $3a^4$
7. $-45ab^2$ by $15ab$
8. $-55xy^3$ by $-11xy$
9. $33xy^2z$ by $3xyz$
10. $65x^2yz$ by $5xyz$
11. $96a^2xy^3$ by $-16axy^2$
12. $105b^3yw^2$ by $-15b^2yw$

13. $144c^3d^2f$ by $8cf$
14. $162x^2yz^3$ by $9x^2z^2$
15. $9np^3$ by $-15n^3p^2$
16. $15rs^2t$ by $-27r^2st^3$
17. $8abcdx^2y$ by $14adxy^2$
18. $9efg^2hr$ by $24e^2fh^3r$
19. $2a^3 + a^2$ by a
20. $4x^4 - x^3$ by x^2
21. $36b^4 - 18b^2$ by $9b$
22. $49y^5 + 35y^3$ by $7y^2$
23. $42x^2 + 28x$ by 7
24. $56z^6 - 48z^3$ by 8

25. $34x^5 - 51x^2$ by $17x^2$
26. $105w^6 + 63w^4$ by $21w^2$
27. $24x^6 - 8x^4$ by $-4x^3$
28. $42y^7 - 24y^5$ by $6y^4$
29. $5x^2y + 5xy^2$ by xy
30. $7a^2b - 7ab^2$ by ab
31. $10x^2y + 15xy^2$ by $5xy$
32. $25p^2q - 15pq^2$ by $5pq$
33. $ap^2q - 2pq$ by pq
34. $bx^2w + 3xw$ by xw
35. $a^2bc + abc$ by abc
36. $x^3yz - xyz$ by xyz

37. $9x^2y^2z - 3xyz^2$ by $-3xyz$
38. $12a^2b^2c + 4abc^2$ by $-4abc$
39. $b^3x^2 + b^3$ by $-b$
40. $c^5y^3 - cy^2$ by $-c$
41. $x^2y + xy - xy^2$ by xy
42. $ab^2 - ab + a^2b$ by ab
43. $18x^3y^2z - 24x^2y^3z$ by $-12x^2yz$
44. $36a^4b^2c - 27a^2b^4c^2$ by $-27a^2bc$
45. $x^2 + 7x + 12$ by $x + 3$
46. $x^2 + x - 12$ by $x + 4$
47. $x^2 - 3x + 2$ by $x - 2$
48. $x^2 - 2x - 15$ by $x - 5$
49. $x^2 + x - 2$ by $x + 2$
50. $x^2 + x - 6$ by $x + 3$
51. $6a^2 + 17a + 7$ by $3a + 7$
52. $4b^2 + 10b - 6$ by $4b - 2$
53. $8y^2 - 8y - 6$ by $2y - 3$
54. $12t^2 + t - 6$ by $4t + 3$

55. $x^3 - 5x + 2$ by $x^2 + 2x - 1$
56. $d^3 + d^2 - 3d + 2$ by $d^2 + d + 2$
57. $6a^3 - 7a^2 + 10a - 4$ by $3a^2 - 2a + 4$
58. $9y^3 - 16y + 8$ by $3y^2 + 3y - 4$
59. $4x^3 - 3x + 4$ by $2x - 1$
60. $7p^3 + 2p - 5$ by $3p + 2$
61. $r^3 - r^2 - 6r + 5$ by $r + 2$
62. $2c^3 - 3c^2 + c - 4$ by $c - 2$
63. $x^4 - 81$ by $x - 3$
64. $y^4 - 81$ by $y + 3$
65. $23x^2 - 5x^4 + 12x^5 - 12 - 14x^3 + 8x$ by $3x^2 - 2 + x$
66. $10a + 21a^3 - 35 + 6a^5 - 25a^4 + 21a^7 - 14a^6$ by $2a + 7a^3 - 7$
67. $x^2 - y^2$ by $x - y$
68. $a^2 - b^2$ by $a + b$
69. $w^3 - z^3$ by $w - z$
70. $x^3 + y^3$ by $x + y$
71. $x^3 + xy^2 + x^2y + y^3$ by $x + y$

72. $a^3 - a^2b + ab^2 - b^3$ by $a - b$
73. $c^3d^3 - 8$ by $cd - 2$
74. $e^3f^3 + 27$ by $ef + 3$
75. $x^2 - 2xy + y^2$ by $x - y$
76. $a^2 + 6ab + 9b^2$ by $a + 3b$
77. $5p^3r - 10p^2 + 15pr^2 - p^2r^2 + 2pr - 3r^3$ by $5p - r$
78. $8x^3 - 12x^2y + 6xy^2 - y^3$ by $2x - y$
79. $a^2 - 2ad - 3d^2 + 3a - 13d - 8$ by $a - 3d - 1$
80. $2x^2 - xy - 6y^2 + x + 19y - 15$ by $2x + 3y - 5$
81. $a^2f - af^2$ by $a - f$
82. $d^3 - 2dm^2 + 2m^3 - d^2m$ by $d - m$
83. $a^2 - b^2 + 2bc - c^2$ by $a - b + c$
84. $e^2 + 2eh - f^2 + h^2$ by $e + f + h$
85. $a^4 + 2a^2 - a + 2$ by $a^2 + a + 2$
86. $x^6 - x^4 + 2x^2 - 1$ by $x^3 - x + 1$

Solve Exercises 87–90.

87. *Electronics* The expression for the total resistance of three resistances in a parallel electrical circuit is

$$\frac{R_1 R_2 R_3}{R_2 R_3 + R_1 R_3 + R_1 R_2}$$

Determine the reciprocal of this expression, and then perform the indicated division.

88. *Thermal science* The radiation constant, $\bar{e}$ (see Section 2.2, Exercise 68, page 69), is defined as

$$\bar{e} = \frac{\sigma}{\dfrac{e_1 + e_2 - e_1e_2}{e_1e_2}}$$

where e_1 and e_2 are radiation emissives of the respective areas and σ is the Stefan-Boltzmann constant. Find an equivalent formula by writing the denominator as three separate terms and simplifying those terms.

89. *Thermal science* One reference book gives the formula for the volume change of a gas under a constant pressure due to the changes in temperature from T_1 to T_2 as

$$V_2 = V_1\left(1 + \frac{T_2 - T_1}{T_1}\right)$$

where V_1 is the volume at T_1 and V_2 is the volume at T_2. Simplify the right-hand side of this equation.

90. The area of a certain rectangle is given by $x^3 + 6x^2 - 7x$, and the length of one side of the rectangle is $x + 7$.
    (a) If the area of a rectangle is the product of the length and the width, what algebraic expression describes the width of this rectangle?
    (b) If $x = 4$ ft, what are the length, width, and area of this rectangle?

In Your Words

91. Describe how to divide a multinomial by a monomial.
92. Describe how to divide a multinomial by a multinomial.

≡ 2.4
SOLVING EQUATIONS

Until now we have worked with real numbers and learned how to add, subtract, multiply, and divide algebraic expressions. In these next two sections we are going to begin learning how to use these skills to solve problems.

The ability to solve a problem often depends on the ability to write an equation for a problem and then solve that equation. We will use this section to learn how to solve an equation. In the next section we will start to learn how to write an equation for a problem.

Equations

An **equation** is an algebraic statement. It asserts that two algebraic expressions are equal. The two algebraic expressions are called the left-hand side (or left side) of the equation and the right-hand side (or right side) of the equation. As the name indicates, the left-hand side is to the left of the equal sign and the right-hand side is to the right of the equal sign.

$$\underbrace{4x^3 + 2x - \sqrt{3}}_{\text{left-hand side}} = \underbrace{7x^2 - 5x}_{\text{right-hand side}}$$

Some equations are true for all values of their variables. These equations are called **identities**. For example, $(x+2)(x-2) = x^2 - 4$ is an identity. An equation that is true for only some values of the variables and not true for the other values is a **conditional equation**. Examples of conditional equations are $4x = -24$ and $y^2 = 25$. The first is true only when $x = -6$. The second is true only when $y = 5$ or $y = -5$.

Any value that can be substituted for the variable and that makes an equation true is called a **solution** or **root** of the equation. So, 5 is a solution of $2x^2 - 5x - 5 = 20$, since $2(5^2) - 5(5) - 5 = 20$. The root of $x + 7 = 10$ is 3, and this is often indicated by the phrase, "3 satisfies the equation $x + 7 = 10$." To solve an equation means to find all of its solutions or roots.

Equivalent Equations

The purpose of this section is to help you learn how to find the solutions or roots of an equation. The techniques you learn here will be used whenever you need to solve an equation. Two equations are **equivalent** if they have exactly the same roots. There is a series of five operations that will allow you to change an equation into an equivalent equation.

To change an equation into an equivalent equation you can use any of the following five operations.

In order to solve an equation you will generally have to use a combination of these five operations. You may have to use the same operation more than once. The following examples show how these operations are used. The first four examples use one operation and the other examples use a combination of operations.

Operations for Changing Equations into Equivalent Equations

1. Add or subtract the same algebraic expression or amount to both sides of the equation.
2. Multiply or divide both sides of the equation by the same algebraic expression, provided the expression does not equal zero.
3. Combine like terms on either side of the equation.
4. Replace one side (or both sides) with an identity.
5. Interchange the two sides of the equation.

EXAMPLE 2.30

Solve $x + 7 = 15$.

Solution
$$x + 7 = 15$$
$$(x + 7) - 7 = 15 - 7 \quad \text{Subtract 7 from both sides (Operation \#1).}$$
$$x = 8$$

Substituting this value in the original equation verifies that 8 is the solution, since $8 + 7 = 15$.

EXAMPLE 2.31

Solve $3x = -15$.

Solution
$$3x = -15$$
$$\frac{1}{3}(3x) = \frac{1}{3}(-15) \quad \text{Multiply both sides by } \tfrac{1}{3}, \text{ the reciprocal of 3}$$
$$\text{(Operation \#2).}$$
$$x = -5$$

Checking the solution in the original equation, we see that -5 satisfies the equation, since $3(-5) = -15$.

EXAMPLE 2.32

Solve $5x - 3x - x = 20 - 12$.

Solution
$$5x - 3x - x = 20 - 12$$
$$x = 8 \qquad \text{Combine like terms (Operation \#3).}$$

Checking the root in the original, we see that $5 \times 8 - 3 \times 8 - 8 = 40 - 24 - 8 = 8$. This satisfies the original equation.

EXAMPLE 2.33

Solve $x^2 + 6x + 9 = 0$.

Solution $\quad x^2 + 6x + 9 = 0$

$$(x+3)^2 = 0 \quad \text{Replace } x^2 + 6x + 9 \text{ with } (x+3)^2 \text{ (Operation #4).}$$

We will stop this example here. To continue solving this problem takes some more mathematics, which we will learn later. The important thing to notice is that we used an identity to rewrite one side of the equation. ▪

The next four examples use combinations of the operations.

EXAMPLE 2.34

Solve $3x + 27 = 6x$.

Solution

$$3x + 27 = 6x$$
$$(3x + 27) - 3x = 6x - 3x \quad \text{Subtract } 3x \text{ (Operation #1).}$$
$$27 = 3x \quad \text{Combine like terms (Operation #3).}$$
$$3x = 27 \quad \text{Switch sides (Operation #5).}$$
$$\frac{1}{3}(3x) = \frac{1}{3}(27) \quad \text{Multiply by } \tfrac{1}{3} \text{ (Operation #2).}$$
$$x = 9$$

To check, replace the x in the original equation with the value we found, 9. The left-hand side is $3(9) + 27 = 27 + 27 = 54$ and the right-hand side is $6(9) = 54$. Since both sides give the same value, 9 is a solution. ▪

EXAMPLE 2.35

Solve $7y + 6 = 216 - 3y$.

Solution

$$7y + 6 = 216 - 3y$$
$$7y + 6 + 3y = 216 - 3y + 3y \quad \text{Add } 3y \text{ (Operation #1).}$$
$$10y + 6 = 216 \quad \text{Combine terms (Operation #3).}$$
$$10y + 6 - 6 = 216 - 6 \quad \text{Subtract 6 (Operation #1).}$$
$$10y = 210 \quad \text{Combine terms.}$$
$$\frac{10y}{10} = \frac{210}{10} \quad \text{Divide by 10 (Operation #2).}$$
$$y = 21$$

Check this answer in the original equation. When you replace y with 21 do you get the same value on both the left-hand and right-hand sides of the equation? You should.

Notice that in the next-to-last step we divided both sides by 10. We could have multiplied both sides by $\frac{1}{10}$ and arrived at the same result. Do you understand why? Both approaches are correct because division by 10 is the same as multiplication by $\frac{1}{10}$. ▪

EXAMPLE 2.36

Solve $3(4z - 8) + 2(3z + 5) = 4(z + 7)$.

Solution

$$3(4z - 8) + 2(3z + 5) = 4(z + 7)$$

$$12z - 24 + 6z + 10 = 4z + 28 \qquad \text{Remove parentheses by distribution.}$$

$$18z - 14 = 4z + 28 \qquad \text{Combine terms.}$$

$$14z - 14 = 28 \qquad \text{Subtract } 4z.$$

$$14z = 42 \qquad \text{Add 14.}$$

$$z = 3 \qquad \text{Divide by 14 (or multiply by } \tfrac{1}{14}).$$

Check this in the original equation. You should get 40 on each side of the equation.

EXAMPLE 2.37

Solve $\dfrac{n}{2} + 5 = \dfrac{2n}{3}$.

Solution

$$\frac{n}{2} + 5 = \frac{2n}{3}$$

$$6\left(\frac{n}{2} + 5\right) = 6\left(\frac{2n}{3}\right) \qquad \text{Multiply by 6. (6 is a common}$$

$$\text{denominator of } \frac{n}{2} \text{ and } \frac{2n}{3}.)$$

$$3n + 30 = 4n$$

$$30 = n \qquad \text{Subtract } 3n.$$

Because most people like to give the variable first and then the value of that variable, you could now use Operation #5 and rewrite this as $n = 30$.

Check by substituting 30 for n in the original equation. The left side is $\dfrac{30}{2} + 5 = 15 + 5 = 20$. The right side is $\dfrac{2(30)}{3} = \dfrac{60}{3} = 20$. So, $n = 30$ satisfies the original equation.

Hints for Solving Problems

The last four examples used a combination of the five operations used to make equivalent equations. Each of the examples showed how to use one or more of the hints for solving problems. As you continue through this book, you will get some more hints to help you become a better problem-solver.

Hints for Solving Problems

1. Eliminate fractions. Multiply both sides of the equation by a common denominator.
2. Remove grouping symbols. Perform the indicated multiplications and then combine terms to remove parentheses, brackets, and braces.
3. Combine like terms whenever possible.
4. Get all terms containing the variable on one side of the equation. All other terms should be placed on the other side of the equation.
5. Check your answer in the original equation.

Here are three more examples. See if you can work each problem before studying the solution given.

EXAMPLE 2.38

Solve $\dfrac{3p+7}{4} - \dfrac{2(p-5)}{3} = \dfrac{p-3}{2} + 6$.

Solution

$$\dfrac{3p+7}{4} - \dfrac{2(p-5)}{3} = \dfrac{p-3}{2} + 6$$

$$12\left[\dfrac{3p+7}{4} - \dfrac{2(p-5)}{3}\right] = 12\left[\dfrac{p-3}{2} + 6\right]$$ Multiply both sides by 12, a common denominator of 2, 3, and 4.

$$12\left[\dfrac{3p+7}{4}\right] - 12\left[\dfrac{2(p-5)}{3}\right] = 12\left[\dfrac{p-3}{2}\right] + 12(6)$$ Distribute the 12.

$$3[3p+7] - 4[2(p-5)] = 6[p-3] + 12(6)$$

$$9p+21 - 4[2p-10] = 6p - 18 + 72$$ Remove grouping symbols.

$$9p+21 - 8p + 40 = 6p - 18 + 72$$

$$p + 61 = 6p + 54$$ Combine terms.

$$7 = 5p$$ Use hint #4.

$$\dfrac{7}{5} = p$$ Divide by 5.

The solution is $\frac{7}{5}$, or 1.4. Check. You should get 5.2 on both sides of the equation when you evaluate each side with $p = 1.4$. This would be a good time to practice using your calculator to see if you get 5.2 on each side of the equation.

EXAMPLE 2.39

Solve $\dfrac{3w+a}{2b} - \dfrac{4(w+a)}{b} = \dfrac{2w-6}{3b} + \dfrac{2+6a}{b}$ for w.

EXAMPLE 2.39 (Cont.)

Solution   There are several letters in this equation, a, b, and w, but you are asked to solve for w. This means that w is the variable and that a and b should be treated as constants.

$$\frac{3w+a}{2b} - \frac{4(w+a)}{b} = \frac{2w-6}{3b} + \frac{2+6a}{b}$$

$$6b\left[\frac{3w+a}{2b} - \frac{4(w+a)}{b}\right] = 6b\left[\frac{2w-6}{3b} + \frac{2+6a}{b}\right]$$ Multiply both sides by a common denominator, $6b$.

$$6b\left[\frac{3w+a}{2b}\right] - 6b\left[\frac{4(w+a)}{b}\right] = 6b\left[\frac{2w-6}{3b}\right] + 6b\left[\frac{2+6a}{b}\right]$$

$$3[3w+a] - 6[4(w+a)] = 2[2w-6] + 6[2+6a]$$

$$9w + 3a - 24w - 24a = 4w - 12 + 12 + 36a$$ Remove grouping.

$$-15w - 21a = 4w + 36a$$ Combine like terms.

$$-15w - 4w = 36a + 21a$$ Hint #4.

$$-19w = 57a$$ Combine like terms.

$$w = \frac{57a}{-19}$$ Divide both sides

$$w = -3a$$ by -19.

Check. Substitute $-3a$ for w in the original equation. When you simplify each side, you get $\frac{4a}{b}$. While it is easier to substitute for w at a later step, you might have made a mistake getting to that step. Sometimes you will even make an error when you copy the problem onto your paper. Always go back to the original problem to check your work. ∎

When you work some problems you may need to modify the order of the hints, as shown in the next example.

EXAMPLE 2.40

Solve $2\left(\frac{1}{3}+x\right) = 5$.

Solution

$$2\left(\frac{1}{3}+x\right) = 5$$

$$\frac{2}{3} + 2x = 5$$ Remove grouping symbols

$$3\left(\frac{2}{3} + 2x\right) = 3(5)$$ Multiply both sides by 3

$$3\left(\frac{2}{3}\right) + 3(2x) = 3(5)$$ Distribute the 3

EXAMPLE 2.40 (Cont.)

$$2 + 6x = 15 \quad \text{Multiply}$$
$$6x = 13 \quad \text{Use hint \#4}$$
$$x = \frac{13}{6} \quad \text{Divide by 6}$$

Exercise Set 2.4

Solve the equations in Exercises 1–84.

1. $x - 7 = 32$

2. $y - 8 = 41$

3. $a + 13 = 25$

4. $b + 21 = 34$

5. $25 + c = 10$

6. $28 + d = 12$

7. $4.3 + w = 8.7$

8. $5.1 + z = 9.1$

9. $4x = 18$

10. $5y = 12$

11. $-3w = 24$

12. $6z = -42$

13. $12a = 18$

14. $15b = -25$

15. $21c = -14$

16. $24d = 16$

17. $\dfrac{p}{3} = 5$

18. $\dfrac{r}{5} = 4$

19. $\dfrac{t}{4} = -6$

20. $\dfrac{s}{-3} = -5$

21. $4a + 3 = 11$

22. $3b + 4 = 16$

23. $7 - 8d = 39$

24. $9 - 7c = 44$

25. $4x - 3 = -37$

26. $5y - 4 = -41$

27. $2.3w + 4.1 = 13.3$

28. $3.5z + 5.2 = 22.7$

29. $2x + 5x = 28$

30. $3y + 8y = 121$

31. $3x = 7 - 10$

32. $4x = 16 - 24$

33. $3a + 2(a + 5) = 45$

34. $4b + 3(7 + b) = 56$

35. $4(6 + c) - 5 = 21$

36. $5(7 + d) + 4 = 31$

37. $2(p - 4) + 3p = 16$

38. $7(n - 5) + 4n = 16$

39. $3x = 2x + 5$

40. $4y = 3y + 7$

41. $4w = 6w + 12$

42. $7z = 10z + 42$

43. $9a = 54 + 3a$

44. $8b = 55 + 3b$

45. $\dfrac{5x}{2} = \dfrac{4x}{3} - 7$

46. $\dfrac{3y}{7} = \dfrac{2y}{3} + 4$

47. $\dfrac{6p}{5} = \dfrac{3p}{2} + 4$

48. $\dfrac{5z}{3} = \dfrac{4z}{5} - 3$

49. $8n - 4 = 5n + 14$

50. $9p - 5 = 6p + 37$

51. $7r + 3 = 11r - 21$

52. $8s + 7 = 15s - 56$

53. $9t + 6 = 3t - 5$

54. $11u - 4 = 6u + 5$

55. $\dfrac{6x - 3}{2} = \dfrac{7x + 2}{3}$

56. $\dfrac{4r - 3}{3} = \dfrac{5r + 2}{2}$

57. $\dfrac{3t + 4}{4} = \dfrac{2t - 5}{2}$

58. $\dfrac{6a - 5}{3} = \dfrac{7a + 5}{6}$

59. $3(x + 5) = 2x - 3$

60. $2(y - 3) = 4 + 3y$

61. $5(w - 7) = 2w + 4$

62. $6(z + 5) = 2z - 9$

63. $\dfrac{x}{2} + \dfrac{x}{3} - \dfrac{x}{4} = 2$

64. $\dfrac{p}{2} - \dfrac{p}{3} - \dfrac{p}{4} = 3$

65. $\dfrac{4(a - 3)}{5} = \dfrac{3(a + 2)}{4}$

66. $\dfrac{5(b + 4)}{3} = \dfrac{4(b - 5)}{5}$

67. Solve $ax + b = 3ax$ for x

68. Solve $2by = 6 + 4by$ for y

69. Solve $ax - 3a + x = 5a$ for a

70. Solve $2(by - c) = 3\left(\dfrac{y}{2} - c\right)$ for y

71. $\dfrac{3}{x} + \dfrac{4}{x} = 3$

72. $\dfrac{5}{y} - \dfrac{3}{y} = 6$

73. $\dfrac{3}{4p} + \dfrac{1}{p} = \dfrac{7}{4}$

74. $\dfrac{6}{5q} - \dfrac{2}{q} = \dfrac{6}{5}$

75. $\dfrac{1}{x + 1} - \dfrac{2}{x - 1} = 0$

76. $\dfrac{3}{x + 2} - \dfrac{4}{x - 2} = 0$

77. $\dfrac{3}{2x} = \dfrac{1}{x+5}$

78. $\dfrac{4}{3x} = \dfrac{2}{x+1}$

79. $\dfrac{2x+1}{2x-1} = \dfrac{x-1}{x-3}$

80. $\dfrac{2x+3}{2x+5} = \dfrac{5x+4}{5x+2}$

81. Solve $\dfrac{6b-5a}{3} + \dfrac{5a+b}{2} = \dfrac{5(a+2b)}{4} + 5$ for a

82. Solve $\dfrac{3r+t}{5t} - \dfrac{r+5t}{3t} = \dfrac{2(r-4)}{15t} + 2$ for r

83. Solve $\dfrac{2z+a}{3x} - \dfrac{9z+a}{6x} = \dfrac{z-a}{2x} + \dfrac{4a}{3x}$ for a

84. Solve $\dfrac{3p+2x}{2y} - \dfrac{5p-3x}{3y} = \dfrac{x-2p}{y} + \dfrac{3x+p}{6y}$ for x

Solve Exercises 85 and 86.

85. *Meteorology*   The formula for converting Celsius temperatures to Fahrenheit temperatures is $F = \frac{9}{5}C + 32$, where F represents the Fahrenheit temperature and C, the Celsius temperature. Find the formula for converting Fahrenheit temperatures to Celsius by solving this equation for C.

86. *Physics*   The velocity v of a falling object after t seconds is given by the formula $v = v_0 + at$, where v_0 represents the initial velocity and a is the acceleration due to gravity.

(a) Solve this equation for t in order to determine the length of time that it takes the object to reach some velocity v.

(b) Use your answer to part a to determine how long it takes an object to strike the ground, if its initial velocity is 12 m/s, its final velocity is 97 m/s, and the acceleration is 9.8 m/s^2.

In Your Words

87. Without looking in the text, write a summary of the operations you can use for changing equations into equivalent equations.

88. Without looking in the text, write a summary of the hints for solving problems.

≡ 2.5
APPLICATIONS OF EQUATIONS

In Section 2.4 we learned how to solve some equations. But, the ability to solve equations is helpful only if you are able to take a problem and write it in the form of one or more equations. Then, once you have the equations, you can solve them to find the answer to your original problem.

Most problems you have to solve at work will be verbal problems. Someone will tell you about a problem they want you to solve or they will write part, or all, of the problem. You will have to first take this verbal problem and organize it so that it is easier to understand. Then, you will look at the problem and decide what information is important and what is not. You should then take the important information and express it in one or more equations. Once you have written the equations, the most difficult part is over. All that is left is to solve the equations and check your answers. In this section, we will focus on taking written information and writing it as an equation.

Seven Suggestions to Help Solve Word Problems ▬▬▬

Verbal, or word problems, give several numerical relationships and then ask some questions about them. You must be able to translate the word problem into equations.

The fewer equations you need, the easier it will be to solve the problem. Here are some suggestions and examples to help you.

Suggestions for Solving Word Problems

1. Read the problem carefully. Make sure you understand what the problem is asking. You may need to read the problem several times to fully understand it.
2. Clearly identify the unknown quantities. Identify each unknown quantity with a letter (or variable). Write down what each letter stands for. Use a letter for an unknown quantity that makes sense to you. For example, you might use d for distance or t for time.
3. If possible, represent all the unknowns in terms of just one variable.
4. Make a sketch (if possible) if it makes the problem clearer.
5. Analyze the problem carefully. Try to write one equation that shows how all the unknowns and knowns are related. If it is not possible to write one equation, use more. (Later we will learn how to solve several equations that show how given unknowns are related.)
6. Solve the equation.
7. Check your answer in the original problem.

EXAMPLE 2.41

In order to get an A in a word processing class, one teacher requires that you type an average of 85 words per minute for five different timings. Bill had speeds of 77, 78, 87, and 91 words per minute on his first four timings. How fast must he type on the next test in order to get an A?

Solution   The unknown quantity is the typing speed on the next timed typing. We will let s represent the unknown typing speed. The average for the five timings is the sum of these five timings divided by five or $\dfrac{77+78+87+91+s}{5}$. This average must be 85, and so we have the equation

$$\frac{77+78+87+91+s}{5} =$$

$$\frac{333+s}{5} = 85 \qquad \text{Combine terms.}$$

$$5\left(\frac{333+s}{5}\right) = 5 \times 85 \qquad \text{Multiply by 5.}$$

$$333+s = 425$$

$$333+s-333 = 425-333 \qquad \text{Subtract 333.}$$

$$s = 92$$

Bill must type 92 words per minute to have an average of 85 words per minute for the five timings, in order to get an A in the class.

EXAMPLE 2.42

At the end of a model year, a dealer advertises that the list prices of last year's models are 15% off. What was the original price of a car that is on sale for $7,990?

Solution   Suppose the original price of the car was p dollars. The discounted amount is 15% of the original price, or $0.15p$. The sale price of $7,990 is the original price p less the discounted amount, $0.15p$. This can be written as

$$p - 0.15p = 7,990$$
$$0.85p = 7,990 \quad \text{Combine terms.}$$
$$p = \frac{7,990}{0.85} \quad \text{Divide by 0.85.}$$
$$p = 9,400$$

The original price was $9,400.

You can check your answer by taking 15% of $9,400 ($0.15 \times 9,400 = 1,410$) and subtracting this from the original price ($9,400 - 1,410 = 7,990$). Is this the sale price?

We will now look at some types of problems that have applications to technology. The problems selected provide a variety of examples.

Uniform Motion Problems

The distance an object travels is governed by the rate it is traveling and the time that it travels. This is often expressed with the formula

$$d = rt$$

where d is the distance traveled, r is the uniform or average rate of travel, and t is the time spent traveling.

Application

EXAMPLE 2.43

If an airplane flies 785 km/h for 4.5 h, how far does it travel?

Solution   We use the formula $d = rt$. We are given the rate r as 785 km/h and the time t as 4.5 h. So,

$$d = rt$$
$$= 785(4.5)$$
$$= 3\,532.5 \, \text{km}$$

The plane will travel 3 532.5 km in 4.5 h.

It sometimes helps to include the units as part of your work. In Example 2.43, we could have written

$$d = rt$$
$$= 785 \, \frac{\text{km}}{\text{h}} \times 4.5 \, \text{h}$$

$$= 3\,532.5 \,\frac{\text{km} \cdot \cancel{\text{h}}}{\cancel{\text{h}}}$$

$$= 3\,532.5 \,\text{km}$$

This same formula, $d = rt$, can be used to find any one of the three values, distance, rate, or time, provided that we know the other two.

Application

EXAMPLE 2.44

If a plane travels 1,150 mi in 2.5 h, what is its average rate of speed?

Solution Again, we use the formula $d = rt$. This time we have the distance d as 1,150 mi and the time t as 2.5 h. So,

$$d = rt$$
$$1,150 = r(2.5)$$
$$\frac{1,150}{2.5} = r$$
$$460 = r$$

The plane averaged 460 mph.

Another motion problem is demonstrated in Example 2.45.

Application

EXAMPLE 2.45

A car traveling on an interstate highway leaves a rest stop at 3 p.m., traveling at 75 km/h. Another car leaves the same rest stop 15 min later, headed in the same direction. If the second car travels at 100 km/h, how long will it be before it overtakes the first car?

Solution Notice that the problem asks how long it will be before the cars meet, but does not say if this should be how much time or how many km. Let's find both.

The distance formula says that $d = rt$. If t is the amount of time (in hours) since the first car left, then for the first car the distance it has traveled, d, is $d = 75t$.

The distance traveled by the second car will be the same. The rate 100 km/h and the time will be different from those of the first car. The second car left 15 min or $\frac{1}{4}$ h later. (Notice that we had to give this time in hours.) So, the time the second car traveled is $t - \frac{1}{4}$ h. Thus, we have $d = 100\left(t - \frac{1}{4}\right)$.

Both cars traveled the same distance, which means that

$$75t = 100\left(t - \frac{1}{4}\right)$$
$$75t = 100t - 25$$
$$75t - 100t = 100t - 25 - 100t \qquad \text{Subtract } 100t.$$
$$-25t = -25 \qquad\qquad\qquad \text{Combine terms.}$$
$$t = 1 \qquad\qquad\qquad\qquad \text{Divide by } -25.$$

EXAMPLE 2.45 (Cont.)

The second car will catch the first 1 h after the first car left. This will be at 4 p.m., or 45 min after the second car leaves. The distance traveled by each car is $75(1) = 75$ km.

Work Problems

Work problems provide a different type of challenge. In these problems, we want to determine the amount of work each person or machine can do in a given amount of time.

Application

EXAMPLE 2.46

One printer can complete a certain job in 3 hours. Another printer can do the same job in 2 hours. How long would it take if both printers work on the job?

Solution We will let h stand for the hours it takes both machines to complete the job. In one hour the two machines can complete $\frac{1}{h}$ of the job. The first printer does $\frac{1}{3}$ of the job in one hour when it works alone. The second printer does $\frac{1}{2}$ of the job in one hour when it works alone, but together they complete $\frac{1}{h}$ of the job in one hour. So, $\frac{1}{h} = \frac{1}{3} + \frac{1}{2}$. A common denominator is $6h$. Multiplying by $6h$ we get

$$6 = 2h + 3h$$
$$6 = 5h$$
$$\frac{6}{5} = h$$

They can complete the job together in $\frac{6}{5}$ hours or in 1 hour 12 minutes.

FIGURE 2.1a

Mixture Problems

In a mixture problem, two quantities are combined, or mixed, to produce a third quantity. Each quantity has a different percentage of some ingredient. The problems often ask for the percent of the ingredient in the third quantity.

Application

EXAMPLE 2.47

If 100 L of gasohol containing 12% alcohol is mixed with 300 L that contains 6% alcohol, how much alcohol is in the final mixture? What is the percent of alcohol in the final mixture?

Solution Figure 2.1a shows the two given quantities on the left and the final mixture on the right. The first quantity has $0.12 \times 100 = 12$ L of alcohol. The second quantity has $0.06 \times 300 = 18$ L of alcohol.

The total amount of gasohol is $100 + 300 = 400$ L. The total amount of alcohol is $12 + 18 = 30$ L. The alcohol is $\frac{30}{400} = 7.5\%$ of the gasohol. (See Figure 2.1b.)

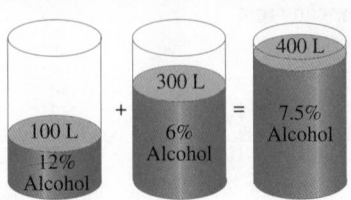

FIGURE 2.1b

Application

A 15-L cooling system contains a solution of 20% antifreeze and 80% water. In order to get the most protection, the solution should contain 60% antifreeze and 40% water. How much of the 20% solution must be replaced with pure antifreeze to get the best protection?

Solution　Let $n =$ the number of liters of the original solution that will be replaced with antifreeze. The amount of original solution that is not replaced is $(15 - n)$ L. You know that you will end up with 60% of 15 L, or 9 L of antifreeze. Thus, you can write

$$\text{antifreeze left} + \text{antifreeze added} = 9 \text{ L}$$

The antifreeze that is left is 20% of $15 - n$, so we can rewrite this equation as

$$20\%(15 - n) + n = 9$$
$$0.2(15 - n) + n = 5$$
$$3 - 0.2n + n = 9$$
$$3 + 0.8n = 9$$
$$0.8n = 6$$
$$n = \frac{6}{0.8}$$
$$n = 7.5$$

So, 7.5 L of the 20% solution must be replaced with pure antifreeze.

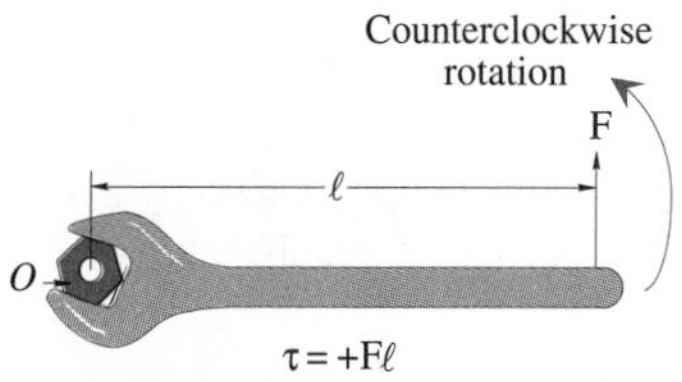

Statics Problems

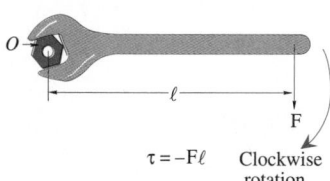

Counterclockwise rotation

$\tau = +F\ell$

FIGURE 2.2a

$\tau = -F\ell$　Clockwise rotation

FIGURE 2.2b

The **moment of force**, or **torque**, about a point O is the product of the force F and the perpendicular distance ℓ from the force to the point. The distance ℓ is also called the **moment arm** of the force about a point. The Greek letter tau, τ, is used to represent torque. Thus, we have the equation

$$\tau = F\ell$$

A torque that produces a counterclockwise rotation is considered positive, as shown in Figure 2.2a. A torque that produces a clockwise rotation is considered negative. (See Figure 2.2b.)

　　An object is in rotational equilibrium when the sum of all the torques acting on the object about any point is zero.

　　The **center of gravity** of an object is the point where the object's entire weight is regarded as being concentrated. If an object is hung from its center of gravity, it will not rotate. An object can also be balanced on its center of gravity without rotating. But, what is perhaps most important in studying the equilibrium of an object, is that its weight is considered to be a downward force from its center of gravity.

Application

EXAMPLE 2.49

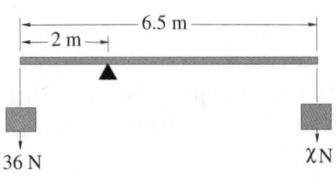

FIGURE 2.3

A 6.5-m rigid rod of negligible weight has a 36-N (newton) weight hung from one end. An unknown weight is hung from the other end. If the rod is balanced at a point 2 m from the end with the 36-N weight, how much is the unknown weight? (See Figure 2.3.)

Solution Since the rod and weights are balanced, the torques of the two weights must be equal. On the left side the moment arm is 2 m and the force is 36 N. The torque on the left-hand side, τ_1, is $\tau_1 = 36 \text{ N} \times 2\text{ m} = 72 \text{ N}\cdot\text{m}$. The moment arm on the right is $6.5\text{ m} - 2\text{ m} = 4.5\text{ m}$. We are to find the force, x. So, the torque on the right-hand side is $\tau_2 = (x \text{ N})(4.5 \text{ m}) = 4.5x \text{ N}\cdot\text{m}$. Since $\tau_1 = \tau_2$ then $72 \text{ N}\cdot\text{m} = 4.5x \text{ N}\cdot\text{m}$ or $72 = 4.5x$. Solving this, we get $x = \dfrac{72}{4.5} = 16 \text{ N}$. The unknown weight is 16 N.

Application

EXAMPLE 2.50

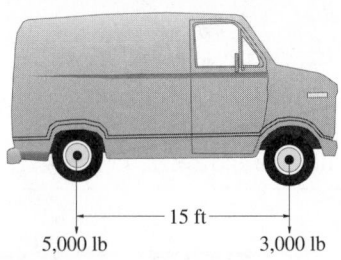

FIGURE 2.4

The front wheels of a truck together support 3,000 lb. Its rear wheels together support 5,000 lb. If the axles are 15 ft apart, where is the center of gravity? (See Figure 2.4.)

Solution Let d represent the distance from the front axle to the center of gravity. Then $15 - d$ is the distance from the center of gravity to the rear axle. The torque on the right is $\tau_1 = 3,000d \text{ ft}\cdot\text{lb}$. The torque on the left is $\tau_2 = -5,000(15 - d) \text{ ft}\cdot\text{lb}$. (Notice that τ_2 is a counterclockwise rotation around the center of gravity and so it is positive. τ_1 is a clockwise rotation around the center of gravity and so it is negative.) Since the truck is in rotational equilibrium,

$$\tau_1 + \tau_2 = 0$$
$$3,000d - 5,000(15 - d) = 0$$
$$3,000d - 75,000 + 5,000d = 0$$
$$-75,000 + 8,000d = 0$$
$$8,000d = 75,000$$
$$d = \frac{75,000}{8,000}$$
$$d = 9.375$$

The center of gravity is 9.375 ft or 9′4.5″ behind the front axle.

Exercise Set 2.5

Solve Exercises 1–32.

1. On your first three mathematics exams you received scores of 79, 85, and 74. What do you need to get on the next exam in order to have an 80 average?

2. There are three parts to a state certification exam. You must pass all three parts with an average of 75 and you cannot get below 60 on any one part. Judy's first two scores were 65 and 72. What must she get on the last part in order to pass?

3. If Raphael got scores of 85 and 82 on the first two parts of the exam described in Exercise 2, what must he get on the third part in order to pass the exam?

4. *Business*   A resort promised that the temperature would average 72°F during your four-day vacation or you would get your money back. The first three days the average temperatures were 69°, 73°, and 68°. How warm does it have to get today for the resort to be able to keep your money?

5. *Business*   A discount store sells personal computers for $920. This price is 80% of the price at a wholesale store. What is the wholesale price? How much will you save at the discount store?

6. *Business*   The personal computer discount store makes a profit of 15% based on its cost for the computer in Exercise 5. What does the computer cost the discount store?

7. An insurance company gives you $1,839 to replace a stolen car. At the time they tell you that the car was only worth 30% of its original cost. What was the original cost?

8. *Automotive*   Because of inflation, the price of automotive parts increased by 3% in January and by another 4% in May. What was last year's price of parts that cost $227.63 after the May increase?

9. *Finance*   Sally invested $4,500. Part of it was invested at 7.5% and part of it at 6%. After one year her interest was $303. How much was invested at each rate?

10. *Finance*   To help pay for his education, José worked and invested his money. Altogether he was able to invest $6,200. Some money he invested at 8.2% and the rest at 7.25%. He was able to earn $482.75 in interest during the year. How much did he invest at each rate?

11. *Business*   A factory pays time and a half for all hours over 40 hours per week. Juannita makes $8.50 an hour and one week brought home $429.25. How many hours of overtime did she work?

12. *Business*   Mladen is paid $8.20 an hour. He also gets paid time-and-a-half overtime when he works more than 40 hours a week if it is during the week. He gets double time if the extra hours are on the weekend. One week Mladen brought home $524.80 for 54 hours of work. How many hours did he work on the weekend?

13. *Transportation*   A freight train leaves Chicago traveling at an average rate of 38 mph. How far will it travel in 7 h?

14. *Transportation*   How long does it take the train in Exercise 13 to travel 475 mi?

15. *Environmental science*   An oil tanker has hit a reef and a hole has been knocked in the side of the tanker. Oil is leaking out of the tanker and forming an oil slick. The oil slick is moving toward a beach 380 km away at a rate of 12 km/d (kilometers per day). The day after the oil spill, cleanup ships leave a dock at the beach and head directly toward the oil slick at a rate of 80 km/d. How far will the ships be from the beach when they reach the oil slick?

16. A school group is traveling together on a school bus. The bus leaves a rest stop on an interstate highway at 1:00 p.m. and travels at a rate of 60 mph. One of the students did not get back on the bus before it left. A highway patrol car leaves the rest stop with the student at 1:30 p.m. and tries to catch the bus. If the patrol car averages 80 mph, at what time will it catch the bus? How far from the rest stop will the bus have traveled?

17. *Machine technology*   The worker on machine *A* can complete a certain job in 6 h. The worker on machine *B* can do the same job in 4 h. In a rush situation, how long would it take both machines to do the job?

18. *Construction* Manuel can build a house in 45 days. With Errol's help they can build the house in 30 days. How long would it take Errol to build the same house by himself?

19. A tank can be filled by Pipe *A* in 4 h. Pipe *B* fills the same tank in 2 h. How long will it take to fill the tank if both pipes are used at the same time?

20. *Energy* A solar collector can generate 70 kJ (kilojoules) in 12 min. A second solar collector can generate the same amount of kJ in 4 min. How long will it take the two of them working together to generate 70 kJ?

21. *Chemistry* To generate hydrogen in a chemistry laboratory, a 40% solution of sulfuric acid is needed. You have 50 mL of an 86% solution of sulfuric acid. How many mL of water should be added to dilute the solution to the required 40% sulfuric acid level?

22. *Petroleum engineering* A petroleum distributor has two gasohol storage tanks. One tank contains 8% alcohol and the other 14% alcohol. An order is received for 1 000 000 L of gasohol containing 9% alcohol. How many liters from each tank should be used to fill this order?

23. *Chemistry* A copper alloy that is 35% copper is to be combined with an alloy that is 75% copper. The result will be 750 kg of an alloy that is 60% copper.

How many kg of each alloy should be used?

24. *Chemistry* How many pounds of an alloy of 20% silver must be melted with 50 lb of an alloy with 30% silver in order to get an alloy of 27% silver?

25. *Civil engineering* A horizontal beam of negligible weight is 20 ft long and supported by columns at each end. A weight of 850 lb is placed at one spot on the beam. The force on one end is 500 lb. What is the force on the other end? Where is the weight located?

26. *Transportation engineering* A loaded truck weighs 140 000 N. The front wheels of the truck support 60 000 N and the rear wheels support the rest. Where is the center of gravity if the axles are 4.2 m apart?

27. *Physics* Locate the center of gravity of the beam in Figure 2.5 if the beam is 12 ft long.

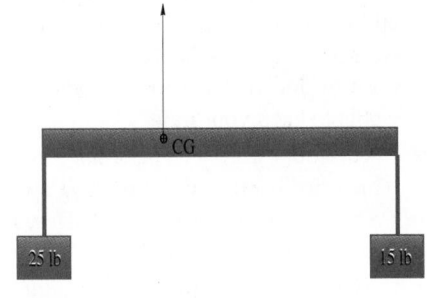

FIGURE 2.5

28. *Machine technology* Locate the center of gravity of the machine part in Figure 2.6 if it is all constructed from the same material.

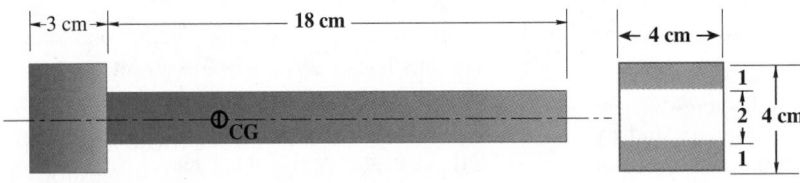

FIGURE 2.6

29. *Machine technology* Locate the center of gravity of the machine part in Figure 2.7 if it is all made of the same metal. (Hint: The volume of a cylinder is $\pi r^2 h$ or $\frac{\pi}{4} d^2 h$, where r is the radius, d is the diameter of the circular bottom, and h is the height. This machine part is made of two circular cylinders.)

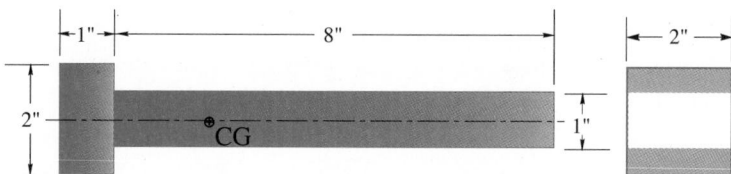

FIGURE 2.7

30. *Physics*   The center of gravity of the object in Figure 2.8 is 10 cm from the left side. What is the thickness of the right side?

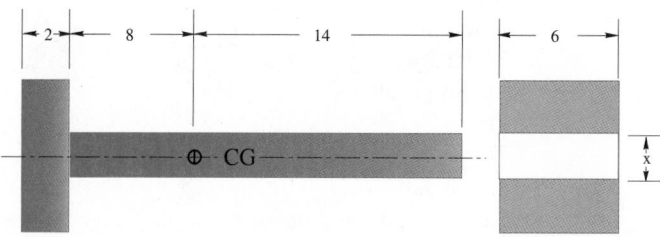

FIGURE 2.8

31. *Medical technology*   How much pure alcohol must a nurse add to 10 cc (cm³) of a 60% alcohol solution to strengthen it to a 90% solution?

32. *Chemistry*   A chemist has 300 g of a 20% hydrochloric acid solution. She wishes to drain some off and replace it with an 80% solution in order to obtain 300 g of a 25% solution. How many grams of the 20% solution must she drain and replace with the 80% solution?

![pencil] **In Your Words**

33. On a sheet of paper write a word problem. On the back of the sheet of paper, write your name and explain how to solve the problem. Give the problem you wrote to a friend and let him or her try to solve it. If your friend has difficulty understanding the problem or solving the problem, or if your friend disagrees with your solution, make any necessary changes in the problem or solution. When you have finished, give the revised problem and solution to another friend and see if he or she can solve it.

34. On a sheet of paper write a word problem that is either a uniform motion, mixture, or statics problem. (If your problem in Exercise 33 was from one of these three areas, then select a different area for this problem.) On the back of the sheet of paper, write your name and explain how to solve the problem. Give the problem you wrote to a friend and let him or her try to solve it, and follow the suggestions given in Exercise 33.

≡ CHAPTER 2 REVIEW

Important Terms and Concepts

Algebraic expression
Algebraic sum
Algebraic term
Binomial
Center of gravity
Coefficient
Conditional equation
Constant
Equation
Equivalent equations
Factors
FOIL method for multiplying binomials
Identities
Like terms
Mixture problems

Moment arm
Moment of force
Monomial
Multinomial
Polynomial
Root
Solution
Solving word problems
Statics problems
Terms
Torque
Trinomial
Uniform motion problems
Variable
Work problems

Review Exercises

Simplify the algebraic expressions in Exercises 1–48.

1. $8y - 5y$
2. $4z + 15z$
3. $7x - 4x + 2x - 8$
4. $-9a + 4a - 3a + 2$
5. $(2x^2 + 3x + 4) + (5x^2 - 3x + 7)$
6. $(3y^2 - 4y - 3) + (5y - 3y^2 + 6)$
7. $2(8x + 4)$
8. $-3(4a - 2)$
9. $-(3x - 1)$
10. $-(2z + 5)$
11. $(4x^2 + 3x) - (2x - 5x^2 + 2)$
12. $(7y^2 + 6y) - (6y - 7y^2) + 2y - 5$
13. $2(a + b) - 3(a - b) + 4(a + b)$
14. $6(c - d) - 4(d - c) + 2(c + d)$
15. $(ax^2)(a^2x)$
16. $(cy^3)(dy)$

17. $(9ax^2)(3x)$
18. $(6cy^2z)(2cz^3)$
19. $4(5x - 6)$
20. $3(12y - 5)$
21. $2x(4x - 5)$
22. $3a(6a + a^2)$
23. $(a + 4)(a - 4)$
24. $(x - 9)(x + 9)$
25. $(2a - b)(3a - b)$
26. $(4x + 1)(3x - 7)$
27. $(3x^2 + 2)(2x^2 - 3)$
28. $(4a^3 + 2)(6a - 3)$
29. $(x + 2)^2$
30. $(3 - y)^2$
31. $5x(3x - 4)(2x + 1)$
32. $6a(4a + 3)(a - 2)$
33. $a^5 \div a^2$

34. $x^7 \div x^3$
35. $8a^2 \div 2a$
36. $27b^3 \div 3b$
37. $45a^2x^3 \div (-5ax)$
38. $52b^4c^2 \div (-4b^2)$
39. $(36x^2 - 16x) \div 2x$
40. $(39b^3 + 52b^5) \div 13b^2$
41. $(x^2 - x - 12) \div (x + 3)$
42. $(x^2 - x - 30) \div (x - 6)$
43. $(x^3 - 27) \div (x - 3)$
44. $(8a^3 + 64) \div (2a + 4)$
45. $(x^2 - y^2) \div (x + y)$
46. $(y^2 - a^2) \div (y + a)$
47. $(x^3 - 2x^2y + 2y^3 - xy^2) \div (x^2 - y^2)$
48. $(a^3 - b^3 + 2ba + 3a^2b - 3ab^2) \div (a - b)$

Solve each of the equations in Exercises 49–68.

49. $x + 9 = 47$

50. $y - 19 = -32$

51. $2x = 15$

52. $-3y = 14$

53. $\dfrac{x}{4} = 9$

54. $\dfrac{y}{3} = -7$

55. $4x - 3 = 17$

56. $7 + 8y = 23$

57. $3.4a - 7.1 = 8.2$

58. $6.2b + 19.1 = 59.4$

59. $4x + 3 = 2x$

60. $7a - 2 = 2a$

61. $4b + 2 = 3b - 5$

62. $7c + 9 = 12c - 4$

63. $\dfrac{4(x-3)}{3} = \dfrac{5(x+4)}{2}$

64. $\dfrac{3(y-7)}{5} = \dfrac{5(y+4)}{2}$

65. $\dfrac{2}{a} - \dfrac{3}{a} = 5$

66. $\dfrac{4}{x} + \dfrac{5}{x} = \dfrac{1}{8}$

67. $\dfrac{3}{2a} = \dfrac{3}{a+2}$

68. $\dfrac{9}{4b} = \dfrac{12}{b+4}$

Solve Exercises 69–76.

69. A student received grades of 68, 70, and 74 on the first three exams. What does the student need to achieve on the next exam to have an average of 72?

70. *Business*   The price of a new television was $342.93. This price included 6.5% sales tax. How much was the television before the tax was added?

71. *Transportation*   An airplane leaves New York City and flies at a rate of 755 km/h. How long does it take to fly 2 718 km?

72. *Space technology*   A damaged space satellite passes over Houston at midnight and is traveling at a rate of 330 km/h. A space shuttle is attempting to overtake the satellite so it can be brought on board for repairs. The shuttle passes over Houston at 1:45 a.m., traveling in the same direction and orbit as the satellite and moving at 430 km/h. At what time does the shuttle overtake the satellite?

73. *Chemistry*   From 50 kg of solder that is half lead and half zinc, 10 kg of solder is removed and 15 kg of lead is added. How much lead is in the final mixture? What percent of the final mixture is lead?

74. *Physics*   A horizontal bar of negligible weight is supported by two columns that are 8 m apart. A load of 460 N is applied to the bar at a point x m from one end. The force on that end is 320 N. What is the force on the other end? Where was the load applied? (See Figure 2.9.)

460 N

FIGURE 2.9

75. *Physics*   A 490 N beam is suspended from two cables 10 m apart. A 2 156 N mass is placed 4 m from one end of the beam. How much tension is on each cable? (See Figure 2.10.)

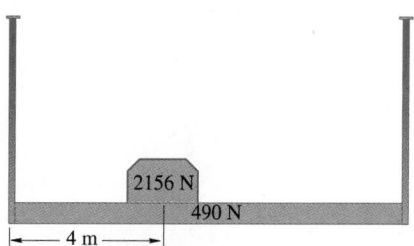

2156 N
490 N
4 m

FIGURE 2.10

76. *Medical technology*   A doctor ordered 20 g of a 52% solution of a certain medicine. The pharmacist has only bottles of 40% and 70% solutions. How much of each solution must be used to obtain the 20 g of the 52% solution?

▬ CHAPTER 2 TEST

1. Simplify $2x + 5x^2 - 5x$.

2. Simplify $(4a^3 - 2b) - (3b + a^3)$.

In Exercises 3–8, perform the indicated operation and simplify your answer.

3. $4x + [3(x + y - 2) - 5(x - y)]$

4. $(4xy^3z)(\frac{1}{2}xy^{-2}z^2)$

5. $(2b - 3)(2b + 3)$

6. $(x^3 + 3x) \div x$

7. $(6x^5 + 4x^3 - 1) \div 2x^2$

8. $(3x^3 - 2x^2 + x - 3) \div (x + 2)$

Solve Exercises 9–13.

9. $\dfrac{y}{3y + 2} - \dfrac{4}{y - 1}$

10. Solve for x: $5x - 8 = 3x$.

11. Solve for x: $\dfrac{7x + 3}{2} - \dfrac{9x - 12}{4} = 8$.

12. The price of a certain graphing calculator is $74.85. This price includes a 7% sales tax. How much was the calculator before the tax was added?

13. Your automobile's cooling system, including the heater and coolant reserve system, has a capacity of 9 qt of coolant. You know that the system currently contains 50% antifreeze and 50% water. You want to remove some of the solution and replace it with antifreeze, so that the final mixture is 60% antifreeze and 40% water. How much solution should you remove?

3

Geometry

Some oven hoods are made by combining prisms and the frustum of a pyramid. In Example 3.27, we shall see how to determine the amount of metal it takes to make an oven hood like the one shown here.

Courtesy of Best (USA) Inc.

You have been working with geometry and geometric ideas most of your life. Geometry deals with the properties and measurements of lines, angles, plane figures, and solid figures. Geometry is a very important tool in technical mathematics. In this chapter we explore the basic geometric properties and measurements needed in technology.

In Chapter 4, we will begin to combine geometry and algebra. In this chapter, we will focus on the geometric ideas and skills that you will need in the following chapters and in many technical jobs.

≡ 3.1
LINES AND ANGLES

Let's begin by discussing lines, segments, and rays. They are the building blocks for the remainder of this section. We will then look at two ways in which angles are measured. Finally, we will give a few areas in which angles and their measures are used.

Lines, Segments, and Rays

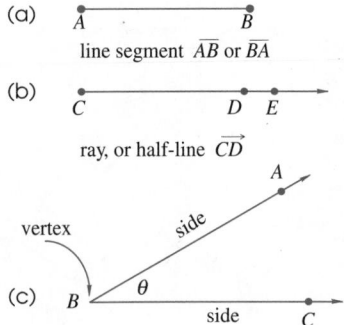

FIGURE 3.1

The basic parts of geometry are lines, angles, planes, surfaces, and the figures that they form. In this section, we will focus on the first two: lines and angles. A **line segment**, or **segment**, is a portion of a straight line between two endpoints. It is named by its endpoints, and a bar is placed over the endpoints' names. For example, the segment joining the points A and B, written $\overline{AB}$, is shown in Figure 3.1a. This segment could also be named $\overline{BA}$. A **ray** or **half-line** is the portion of a line that lies on one side of a point and includes the point. For example, in Figure 3.1b, ray $\overrightarrow{CD}$ begins at point C and passes through D. The beginning point of the ray is called the ray's **endpoint**. This second point used to name the ray can be any point on the ray other than the endpoint. Thus, in Figure 3.1b, $\overrightarrow{CD} = \overrightarrow{CE}$.

Angles

An **angle** is formed by two rays that have the same endpoint. This common endpoint is called the **vertex** of the angle and the two rays are called the **sides** of the angle. The angle in Figure 3.1c has its vertex at B and the two rays that form the sides are $\overrightarrow{BA}$ and $\overrightarrow{BC}$. This angle has several possible names, including $\angle B$, $\angle ABC$, $\angle CBA$, and $\angle \theta$, where the symbol $\angle$ means angle.

Another way to think of an angle is to think of it as generated by moving a ray from an initial position to a terminal position. One revolution is the amount a ray would move to return to its original position. If this generated angle was placed inside a circle with its vertex at the center of the circle, then you would have a figure something like the one shown in Figure 3.2.

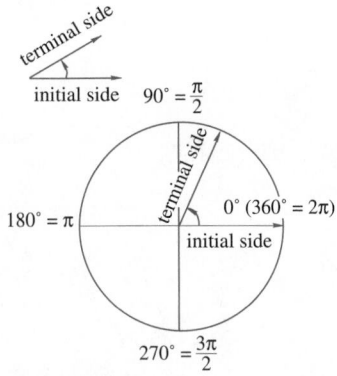

FIGURE 3.2

Degrees and Radians

Angles are measured using several different systems. The two most common units are **degrees** and **radians**. We will use both in this book. Look again at Figure 3.2. The end of the initial side is marked $0°$ ($360° = 2\pi$). This indicates that if the terminal side were to rotate completely around the circle and stop at the initial side, the size of the angle would be 360 degrees ($360°$) or 2π radians (2π rad). Later we will see a relationship between the radian and the distance around a circle.

Other important angle measures are given in relation to the distance around the circle. One-fourth of the way around the circle is $90°$ or $\frac{\pi}{2}$ rad. Halfway around the circle is $180°$ or π rad and three-fourths of the way is $270°$ or $\frac{3\pi}{2}$ rad.

≡ **Note**   If there is no degree symbol (°) after an angle measure, then assume that it is a radian measure.

Each degree is divided into 60 minutes and each minute into 60 seconds. There are 3,600 seconds in a degree. The symbol ′ is used for minutes and ″ for seconds. So, $60' = 1°$ and $60'' = 1'$. Recently the use of calculators has caused decimal values for degrees to become more popular. Thus, in decimal degrees, $15' = \left(\frac{15}{60}\right)° = 0.25°$ and $27'' = \left(\frac{27}{3600}\right)° = 0.0075°$. The increased use of calculators should cause decimal values for degrees to become even more widely used.

There are some other widely used angles besides the ones shown in Figure 3.2. These are the $30° = \frac{\pi}{6}$, $45° = \frac{\pi}{4}$, and $60° = \frac{\pi}{3}$ angles. There are times when you are going to need to convert from degrees to radians or from radians to degrees. Since $180° = \pi$ rad, then

$$1 \text{ rad} = \frac{180°}{\pi} \approx \frac{180°}{3.1416} \approx 57.296°$$

(The symbol ≈ means "is approximately equal to.")

and you also have

$$1° = \frac{\pi}{180°} \approx \frac{3.1416}{180°} \approx 0.01745 \text{ rad}$$

Therefore, we have the following conversion formulas.

Conversion Between Degrees and Radians

To change from radians to degrees, multiply the number of radians by $\dfrac{180°}{\pi}$, or 57.296°.

To change from degrees to radians, multiply by $\dfrac{\pi}{180°}$, or 0.01745 (or divide by 57.296).

EXAMPLE 3.1

Change **(a)** 1.89 and **(b)** $\frac{5\pi}{6}$ to degrees.

Solutions   Since there is no degree symbol, these must be in radians.

(a) $1.89 \text{ rad} \approx 1.89 \text{ rad} \times \dfrac{57.296°}{1 \text{ rad}} = 108.29°$

(b) $\frac{5\pi}{6} = \frac{5\pi}{6} \times \dfrac{180°}{\pi} = 150°$

EXAMPLE 3.2

Change **(a)** 120° and **(b)** 82.5° to radians.

Solutions

(a) $120° = 120° \times \dfrac{\pi}{180°} \text{ rad} = \frac{2\pi}{3} \text{ rad}$

EXAMPLE 3.2 (Cont.)

Notice here that the answer is left in terms of π and we have $120° = \frac{2\pi}{3}$, the exact answer. You may leave the answer in terms of π or compute a decimal approximation and see that $120° \approx 2.094$ rad.

(b) $82.5° \approx 82.5° \times 0.01745\,\text{rad} = 1.439625 \approx 1.44\,\text{rad}$

or $\quad 82.5° = 82.5° \div 57.296 \approx 1.440\,\text{rad}$

or $\quad 82.5° = 82.5° \times \dfrac{\pi}{180°} = \frac{11\pi}{24} \approx 1.440\,\text{rad}$

 Hint

Some people find that it is easier to convert between degrees and radians by using the proportion

$$\frac{D}{180°} = \frac{R}{\pi}$$

where D represents a known or unknown number of degrees and R represents a known or unknown number of radians.

EXAMPLE 3.3

Use the proportion $\dfrac{D}{180°} = \dfrac{R}{\pi}$ to change **(a)** $120°$ to radians and **(b)** $\frac{5\pi}{6}$ to degrees.

Solutions

(a) Since we are asked to convert $120°$ to radians, we substitute $120°$ for D in the proportion and solve it for R as follows:

$$\frac{120°}{180°} = \frac{R}{\pi}$$
$$R = \frac{120°}{180°}\pi$$
$$= \frac{2}{3}\pi$$

Thus, $120° = \frac{2}{3}\pi = \frac{2\pi}{3}$ radians.

(b) Here we want to convert $\frac{5\pi}{6}$ to degrees, so we substitute $\frac{5\pi}{6}$ for R in the proportion, and obtain

$$\frac{D}{180°} = \frac{R}{\pi}$$
$$\frac{D}{180°} = \frac{5\pi/6}{\pi}$$
$$D = \frac{5\pi}{6} \cdot \frac{180°}{\pi}$$
$$= 150°$$

As in Example 3.1(b), we obtain $\frac{5\pi}{6} = 150°$.

An easier method to work these problems would be to use a calculator that has a conversion factor built in. (Introductory material on calculators is in Appendix A.) Some calculators have a [2nd] [D·R], [2nd] [D↔R], or [2nd] [→RAD] key. This indicates that this key can be used to change from degrees to radians. In Example 3.4, we will use a calculator to work the same problems that were in Example 3.2.

EXAMPLE 3.4

Use a calculator to change **(a)** 120° and **(b)** 82.5° to radians.

Solutions

	PRESS	DISPLAY
(a)	120 [2nd] [D·R]	2.0943951
(b)	82.5 [2nd] [D·R]	1.4398966

Our answer to (b) is the same as in Example 3.2. But the answer for (a) looks different. In Example 3.2(a), we obtained $\frac{2\pi}{3}$, but the calculator gave an answer of 2.0943951. However, a little work on your calculator will show you that these are approximately equal.

Caution

Graphing calculators, such as the Texas Instruments TI-8x and the Casio fx-7700G, follow different procedures from those just shown. Consult the owner's manual for your calculator to find the exact steps you should use.

The calculator can also be used to convert from radians to degrees, but now you may need to use the [INV] key. This is the INVerse key. By using it before you use another function key, you reverse the operation of that function or key. In this example, we will need to use both the [INV] and the [2nd] keys. These can be used in either order; however, we will always use [INV] and then the [2nd]. Some calculators have a [2nd] [→DEG] key that is used to convert radians to degrees. We will now use a calculator to work the same problems you saw in Example 3.1.

EXAMPLE 3.5

Use a calculator to change **(a)** 1.89 and **(b)** $\frac{5\pi}{6}$ to degrees.

Solutions

	PRESS	DISPLAY
(a)	1.89 [INV] [2nd] [D·R]	108.28902
(b)	5 [×] [2nd] [π]	3.1415927
	[÷]	15.707963
	6 [=]	2.6179939
	[INV] [2nd] [D·R]	150

Part (b) of Example 3.4 was more complicated because we had to first convert $\frac{5\pi}{6}$ rad to its decimal approximation and then convert that approximation to degrees.

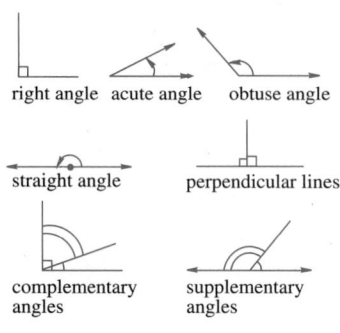

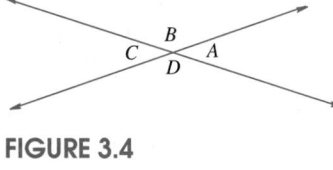

FIGURE 3.3

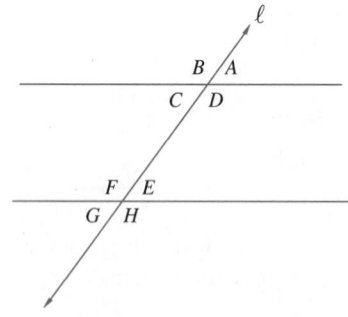

FIGURE 3.4

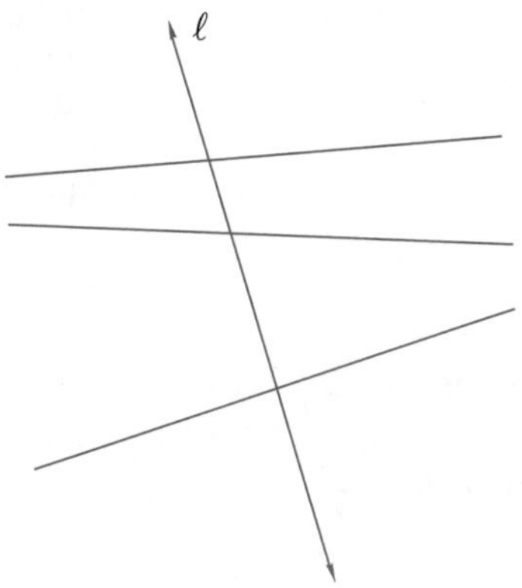

FIGURE 3.5

FIGURE 3.6

Types of Angles

Figure 3.3 shows four basic types of angles. A **right angle** has a measure of 90° or $\frac{\pi}{2}$ rad. It is usually indicated by placing a small square at the vertex of the angle. An **acute angle** measures between 0° and 90° (0 and $\frac{\pi}{2}$ rad) and an **obtuse angle** measures between 90° and 180° ($\frac{\pi}{2}$ and π rad). A **straight angle** has a measure of 180° or π rad. If two lines meet and form a right angle, then the lines are **perpendicular**. Two angles are **supplementary** if their sum is a straight angle (180° or π rad) and are **complementary** if their sum is a right angle (90° or $\frac{\pi}{2}$ rad).

Figure 3.4 shows two intersecting lines. Angles A and C are **vertical angles**. Angles B and D are also vertical angles. As you can probably see, vertical angles are the same size. In geometry, when two objects are the same size we say that they are **congruent**. Two 37° angles are said to be congruent. Two segments that are 3 cm long are congruent. So, since vertical angles are the same size, we can say that vertical angles are congruent.

Two angles that have the same vertex and a common side are called **adjacent angles**. In Figure 3.4, $\angle A$ and $\angle B$ are adjacent angles. They are also supplementary angles. There are three other pairs of adjacent angles in Figure 3.4. Can you find them?

A **transversal** is a line that intersects, or crosses, two or more lines. For example, in Figure 3.5, line ℓ is a transversal of the other three lines. In Figure 3.6, line ℓ is a transversal of the other two lines. Angles A, B, G, and H are called **exterior angles** and angles C, D, E, and F are **interior angles**. Two angles that are not adjacent but are on different sides of the transversal are called **alternate angles**. Angles A and G are both exterior angles and alternate angles, so we say that they are **alternate exterior angles**. So are angles B and H. When you refer to alternate exterior angles, you are referring to a pair of angles. Another pair of angles is called the **alternate interior angles**, such as the pair $\angle C$ and $\angle E$ or the pair $\angle D$ and $\angle F$.

If there are two nonadjacent angles on the same side of the transversal, and one is an interior angle and the other is an exterior angle, then they are **corresponding angles**. In Figure 3.6, $\angle A$ and $\angle E$ are corresponding angles. So are $\angle D$ and $\angle H$. On the other side of the transversal, $\angle B$ and $\angle F$ form one set of corresponding angles and $\angle C$ and $\angle G$ form the other.

If a transversal intersects two parallel lines, as in Figure 3.6, then some interesting things are true. There are several theorems in geometry about parallel lines and a transversal. These theorems are summarized in the following theorem.

Theorem

If two parallel lines are intersected by a transversal, then the
(a) alternate interior angles are congruent,
(b) alternate exterior angles are congruent, and
(c) corresponding angles are congruent.

One pair of alternate interior angles in Figure 3.6 is $\angle C$ and $\angle E$. Another pair is $\angle D$ and $\angle F$. Since alternate interior angles are congruent, $\angle C \cong \angle E$ and $\angle D \cong \angle F$. (The symbol $\cong$ means "is congruent to.") The alternate exterior angles in Figure 3.6 are also congruent, so $\angle A \cong \angle G$ and $\angle B \cong \angle H$.

In Figure 3.6, $\angle A \cong \angle E$ and $\angle C \cong \angle G$, because they are corresponding angles. We also know that $\angle A \cong \angle C$, because they are vertical angles. So, $\angle A \cong \angle C \cong \angle E \cong \angle G$. In the same way we can show that $\angle B \cong \angle D \cong \angle F \cong \angle H$.

Application

EXAMPLE 3.6

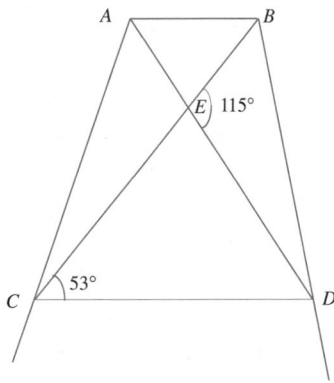

FIGURE 3.7

Some structures, such as the Eiffel Tower in Paris, France, or the Sunsphere in Knoxville, Tennessee, use a crossed girder design. A portion of a similar girder arrangement is shown in Figure 3.7, with $\overline{AB}$ parallel to $\overline{CD}$. If $\angle DCB = 53°$ and $\angle DEB = 115°$, what are the sizes of **(a)** $\angle ABC$, **(b)** $\angle CED$, and **(c)** $\angle AEC$?

Solutions

(a) $\angle ABC$ is an alternate interior angle of $\angle BCD$. Since $\angle BCD = 53°$, $\angle ABC = 53°$.

(b) $\angle CED$ is a supplementary angle of $\angle DEB$. We are given $\angle DEB = 115°$. So, $\angle CED = 180° - 115° = 65°$.

(c) $\angle AEC$ and $\angle DEB$ are vertical angles. So, $\angle AEC = \angle DEB = 115°$.

Another theorem says that if parallel lines are intersected by two transversals, then the corresponding segments are proportional. **Corresponding segments** are the parts of the transversals between the same parallel lines. In Figure 3.8, lines P_1, P_2 and P_3 are parallel lines intersected by transversals ℓ_1 and ℓ_2. The points of intersection have been marked with capital letters. Segments $\overline{AB}$ and $\overline{DE}$ are corresponding segments, as are $\overline{BC}$ and $\overline{EF}$. Likewise, $\overline{AC}$ and $\overline{DF}$ are corresponding segments. Therefore, we have the proportion

$$\frac{AB}{DE} = \frac{BC}{EF} = \frac{AC}{DF}$$

Application

EXAMPLE 3.7

Find the distance across the lake in Figure 3.9.

Solution You carefully use stakes to mark off the segments $\overline{AF}$, $\overline{BE}$, and $\overline{CD}$. The segments are parallel. A, B, C are in a line, as are D, E, and F. We know that corresponding segments are proportional, so $\dfrac{AB}{EF} = \dfrac{BC}{DE}$. Using the lengths in Figure 3.9, $\frac{100}{75} = \dfrac{BC}{168}$. Solving this, we get $BC = \dfrac{100 \times 168}{75} = 224$. It is 224 m across the lake.

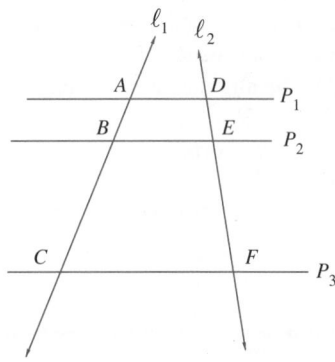

FIGURE 3.8

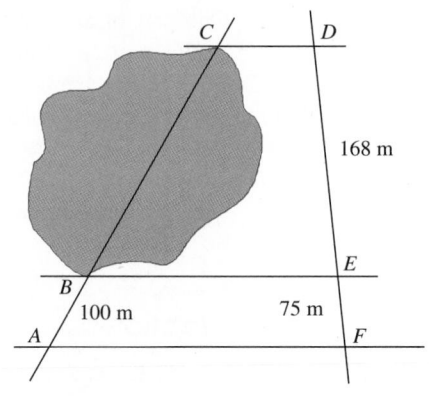

FIGURE 3.9

Exercise Set 3.1

Convert each of the angle measures in Exercises 1–14 from degrees to radians or from radians to degrees without using a calculator. (You may leave your answers in terms of π.) When you have finished, check your work with a calculator.

1. $15°$

2. $75°$

3. $210°$

4. $10°45'$

5. $85.4°$

6. $48.6°$

7. $163.5°$

8. $242°35'$

9. $\frac{4\pi}{3}$

10. $\frac{\pi}{6}$

11. 1.3π

12. 2.15

13. 0.25

14. 1.1

Solve Exercises 15 and 16.

15. Find the supplement of a $35°$ angle in **(a)** degrees and **(b)** radians.

16. Find the complement of a $65°$ angle in **(a)** degrees and **(b)** radians.

In Exercises 17–22, find the measures of angles A and B. In Exercises 19–23, lines ℓ_1 and ℓ_2 are parallel.

17.

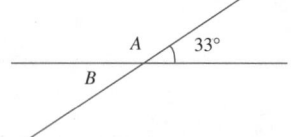

18.

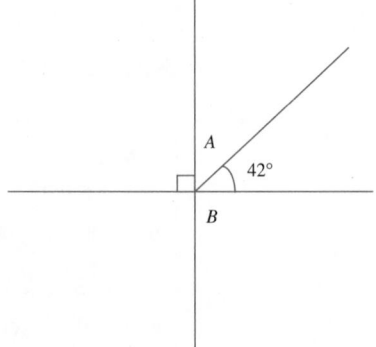

19.

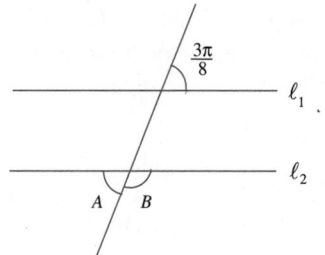

20.

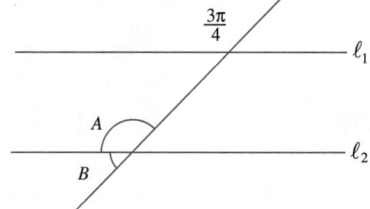

21.

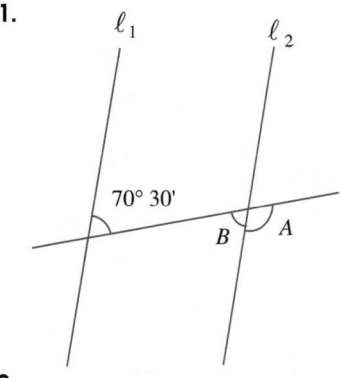

22.

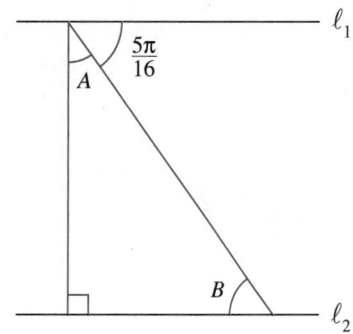

Solve Exercises 23–30.

23. What is the distance from A to B in Figure 3.10 if lines ℓ_1, ℓ_2, and ℓ_3 are parallel?

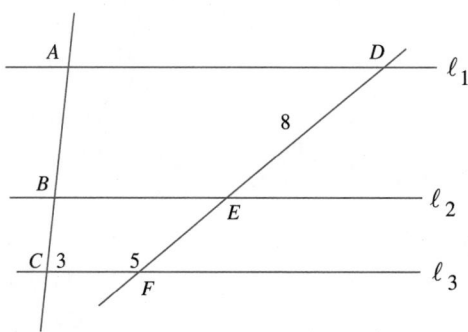

FIGURE 3.10

24. In Figure 3.11, if lines ℓ_1, ℓ_2, and ℓ_3 are parallel, what is the distance from A to B?

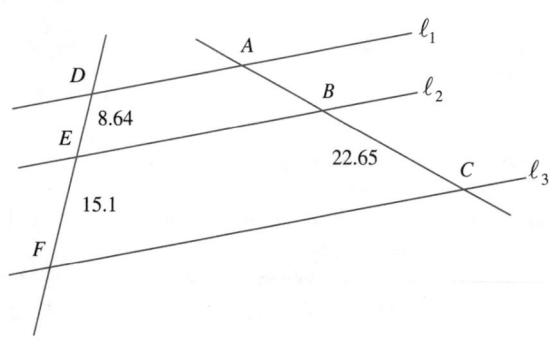

FIGURE 3.11

25. In Figure 3.12, angles A and B are the same size. Find the sizes of angles A and B.

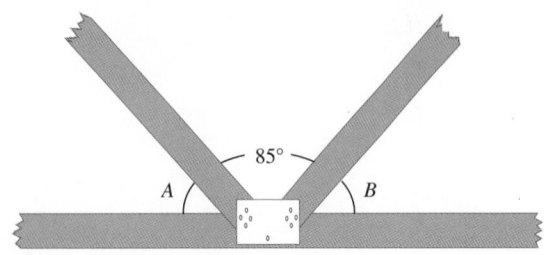

FIGURE 3.12

26. *Civil engineering*   Figure 3.13 shows part of a bridge that is made of suspended cables hung from a girder to the deck of the bridge. What is the length from *A* to *B*?

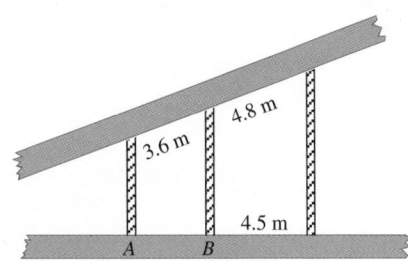

FIGURE 3.13

In Your Words

31. **(a)** Without looking in the text, describe the relationship between degrees and radians.
   (b) Explain how you can use the relationship between degrees and radians to convert between the units.

32. **(a)** Draw a diagram that shows parallel lines intersected by two transversals.

27. *Electronics*   The angular frequency ω of an ac current is $\omega = 2\pi f$ rad/s, where f is the frequency. If $f = 60$ Hz for ordinary house current, what is the angular frequency?

28. *Mechanical technology*   An angular velocity ω of a rigid object, such as a drive shaft, pulley, or wheel, is related to the angle θ through which the body rotates in a period of time t by the formula $\theta = \omega t$ or $\omega = \dfrac{\theta}{t}$.
   (a) Find the angular velocity of a gear wheel that rotates $285°$ in 0.6 s.
   (b) Find the angular velocity of a wheel that rotates $\frac{11\pi}{16}$ rad in 0.9 s.

29. *Metalworking*   A machinist needs to weld a piece of pipe parallel to two existing pipes, as shown in Figure 3.14. What is the size of angle x?

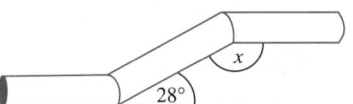

FIGURE 3.14

30. *Programming*   Use the conversion factors at the beginning of this section to write a program for your graphing calculator that will **(a)** convert degrees to radians and **(b)** convert radians to degrees.

   (b) Explain how you can use parallel lines and transversals to find some distances that are difficult to measure directly.

≡ 3.2
TRIANGLES

A **polygon** is a figure in a plane that is formed by three or more line segments, called **sides**, joined at their endpoints. The endpoints are called **vertices** and one endpoint is called a **vertex**. In this section, we will be concerned only with the one type of polygon called a triangle.

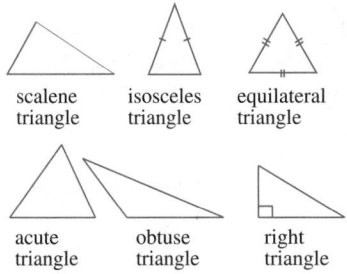

scalene triangle   isosceles triangle   equilateral triangle

acute triangle   obtuse triangle   right triangle

FIGURE 3.15

A **triangle** is a polygon that has exactly three sides. (See Figure 3.15.) Triangles are named according to a property of their sides or their angles. When classified according to its sides, a triangle is **scalene** if none of the sides are the same length, **isosceles** if two sides are the same length, and **equilateral** if all three sides are the same length. When sides are the same length, single, double, or triple marks are used to show which sides are congruent.

An **acute triangle** has three acute angles, an **obtuse triangle** has one obtuse angle, and a **right triangle** has one right angle. (See Figure 3.15.) In an equilateral triangle, all three angles are congruent. In an isosceles triangle the angles opposite the congruent sides are congruent.

The sum of the three angles of a triangle is $180°$, or π rad. Since all three angles in an equilateral triangle are congruent, the size of each angle must be $\dfrac{180°}{3} = 60°$ or $\dfrac{\pi}{3}$ rad. Since a right angle is $90°$, a right triangle must have two acute angles, because the other two angles, when added together, can have only a total of $90°$. This means that each of the angles must be less than $90°$. In a similar manner, an obtuse triangle must have two acute angles.

EXAMPLE 3.8

Find the size of the third angle of a triangle if the other two angles measure $37°$ and $96°$.

Solution   If we call the unknown angle $\angle A$, then

$$\angle A = 180° - 37° - 96° = 47°$$

The third angle of this triangle is $47°$.

The **perimeter** of a triangle is the distance around the triangle. To find the perimeter, you add the lengths of the three sides. If the lengths of the sides are a, b, and c, then the perimeter P is $P = a + b + c$.

Application

EXAMPLE 3.9

A triangular piece of land measures $42\,\text{m}$, $36.2\,\text{m}$, and $58.7\,\text{m}$ on the three sides. What is the perimeter of this plot of land?

Solution
$$
\begin{aligned}
P &= a + b + c \\
&= 42 + 36.2 + 58.7 \\
&= 136.9
\end{aligned}
$$

The perimeter is $136.9\,\text{m}$.

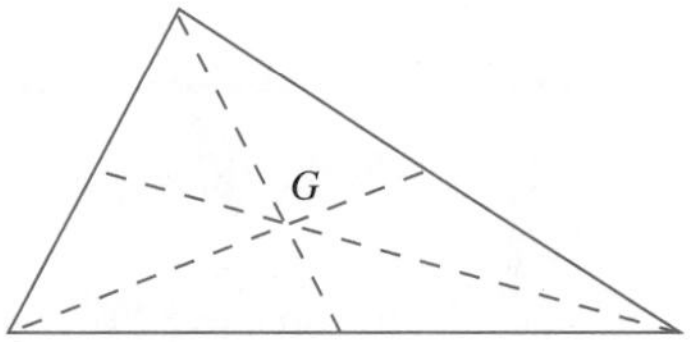

A segment drawn from a vertex to the middle (or midpoint) of the opposite side is called a **median**. If you draw all three medians of a triangle, they meet in a common point, G, called the **centroid**, as shown in Figure 3.16. The centroid is the center of gravity of a triangle.

FIGURE 3.16

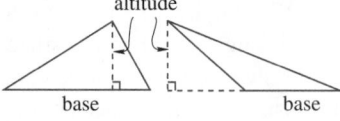

altitude

base   base

FIGURE 3.17

A segment drawn from a vertex and perpendicular to the opposite side is an **altitude** of a triangle. The opposite side is called the **base**. (See Figure 3.17.) When solving some problems, it is sometimes necessary to extend the base so that it will intersect the altitude.

The area of a triangle is found by the product of $\frac{1}{2}$ and the lengths of the base and the altitude. If b is the length of the base and h the length of the altitude, then the area A is given by the formula

$$A = \frac{1}{2}bh$$

If the base and altitude are in the same units, then the area is in square units.

EXAMPLE 3.10

Find the area of a triangle with a base of 1.2 m and an altitude of 4.5 m.

Solution $A = \frac{1}{2}bh$

$$= \frac{1}{2}(1.2)(4.5)$$

$$= 2.7$$

The area is $2.7 \, \text{m}^2$.

≡ Note

You may have noticed that m^2 was used to indicate square meters. Similar notation is used for other area units. For example, cm^2 is used for square centimeters, and ft^2 for square feet.

Sometimes the length of the altitude is not known. It is possible to find the area using **Hero's formula**, also referred to as **Heron's formula**. Instead of the altitude, you will need the lengths of the three sides and the semiperimeter. If we let a, b, and c represent the lengths of the sides, then the **semiperimeter**, s, is found using the formula $s = \dfrac{a+b+c}{2}$. We can find the area from Hero's formula:

$$A = \sqrt{s\,(s-a)\,(s-b)\,(s-c)}$$

Application ────────────────────

EXAMPLE 3.11

A triangular piece of land measures 51'9'', 47'3'', and 82'6'' on the three sides. What is the area of the lot of land?

Solution   We will let $a = 51'9'' = 51.75$ ft, $b = 47'3'' = 47.25$ ft, and $c = 82'6'' = 82.5$ ft.

EXAMPLE 3.11 (Cont.)

$$s = \frac{a+b+c}{2} = \frac{51.75+47.25+82.5}{2} = \frac{181.5}{2} = 90.75$$

$$A = \sqrt{s(s-a)(s-b)(s-c)}$$
$$= \sqrt{90.75(90.75-51.75)(90.75-47.25)(90.75-82.5)}$$
$$= \sqrt{90.75(39)(43.5)(8.25)}$$
$$= \sqrt{1,270,148.3}$$
$$\approx 1,127.0086$$

The area is about $1,127\,\text{ft}^2$.

Two triangles are congruent if the corresponding angles and sides of each triangle are congruent. For example, in Figure 3.18, sides $\overline{AB}$ and $\overline{EF}$ are both 6 units long, $\overline{BC}$ and $\overline{DF}$ are both 3 units long, and $\overline{AC}$ and $\overline{DE}$ are 7 units long. If you measure the angles you will find that $\angle A \cong \angle E$, $\angle B \cong \angle F$ and $\angle C \cong \angle D$. As you can see, **congruent triangles** have the same size and shape.

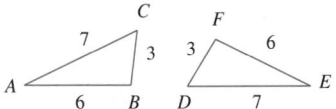

FIGURE 3.18

We indicate that the triangles in Figure 3.18 are congruent by writing $\triangle ABC \cong \triangle EFD$. When we show that two triangles are congruent, we write it so the corresponding vertices are in the same order. Thus, writing $\triangle ABC \cong \triangle EFD$ indicates that $\angle A \cong \angle E$, $\angle B \cong \angle F$, and $\angle C \cong \angle D$.

Two triangles that have the same shape are said to be **similar triangles.** In similar triangles, the corresponding angles are congruent. In Figure 3.19, $\angle A \cong \angle D$, $\angle B \cong \angle E$, and $\angle C \cong \angle F$. Since the corresponding angles are the same size, the triangles are similar. Congruent triangles are a special case of similar triangles in which the corresponding sides are the same length. In all similar triangles, the corresponding sides are proportional. In Figure 3.19, the sides of the larger triangle are three times as large as those of the smaller triangle. Thus, the ratios of the corresponding sides of the larger triangle to the smaller triangle are $\frac{12}{4} = \frac{9}{3} = \frac{6}{2} = \frac{3}{1}$.

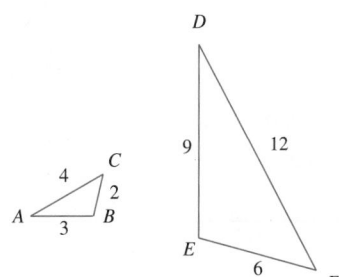

FIGURE 3.19

Application

EXAMPLE 3.12

A television tower casts a shadow that is 150 m long. At the same time, a vertical pole that is 1.6 m high casts a shadow 1.2 m long. How high is the television tower? (See Figure 3.20.)

Solution We have two similar triangles, so the corresponding sides are proportional. If we call the unknown height h, then we have

$$\frac{h}{1.6} = \frac{150}{1.2}$$
$$h = \frac{150(1.6)}{1.2}$$
$$= 200$$

The height of the tower is 200 m.

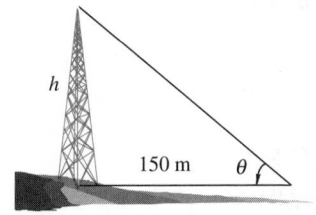

FIGURE 3.20

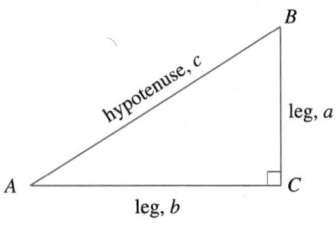

FIGURE 3.21

One of the most valuable theorems in geometry involves the right triangle. In a right triangle, the side opposite the right angle is called the **hypotenuse**, as shown in Figure 3.21. The other two sides are the **legs**. The hypotenuse is the longest side of a right triangle. If the lengths of the two legs are a and b and the length of the hypotenuse is c, then the **Pythagorean theorem** is stated as follows.

Pythagorean Theorem

$\triangle ABC$ is a right triangle with a hypotenuse of length c and legs of lengths a and b, if and only if

$$a^2 + b^2 = c^2$$

EXAMPLE 3.13

A right triangle has legs of length 6 in. and 8 in. What is the length of the hypotenuse?

Solution   If we let $a = 6$ and $b = 8$, then, by using the Pythagorean theorem, we have

$$c^2 = a^2 + b^2$$
$$= 6^2 + 8^2$$
$$= 36 + 64$$
$$= 100$$
$$c = \sqrt{100} = 10$$

The hypotenuse is 10 in. long.

EXAMPLE 3.14

The hypotenuse of a right triangle is 7.3 cm long. One leg is 4.8 cm long. What is the length of the other leg?

Solution   We are given $c = 7.3$ and $a = 4.8$, and so, by using the Pythagorean theorem, we obtain

$$c^2 = a^2 + b^2$$
$$(7.3)^2 = (4.8)^2 + b^2$$
$$(7.3)^2 - (4.8)^2 = b^2$$

or

$$b^2 = (7.3)^2 - (4.8)^2$$
$$= 53.29 - 23.04$$
$$= 30.25$$
$$b = \sqrt{30.25} = 5.5$$

The length of the remaining side is 5.5 cm.

Application

EXAMPLE 3.15

In Example 3.12 we found the height of a television tower. If a cable could be strung in a straight line from the top of the tower to a point on the ground 130 m from the base of the tower, how long would the cable have to be?

Solution   The cable, tower, and ground form a right triangle. We know that the height of the tower is 200 m. This is one leg of the triangle. The other leg is 130 m. The cable will form the hypotenuse. So,

$$c^2 = a^2 + b^2$$
$$c^2 = 130^2 + 200^2$$
$$= 16\,900 + 40\,000$$
$$= 56\,900$$
$$c = \sqrt{56\,900} \approx 238.54$$

Since the given data, 200 m and 130 m, have three significant digits, the cable would be about 239 m long.

The last example might not have been exactly realistic. It failed to account for any sag in the cable or any additional cable that would be needed to fasten it at each end. Later, we will find some methods to get more accurate answers to this problem.

Remember that the Pythagorean theorem can be used only with right triangles. The following exercises provide some additional examples that use the Pythagorean theorem.

Exercise Set 3.2

Find the indicated variables in Exercises 1–8. Identical marks on segments or angles indicate that they are congruent.

1.

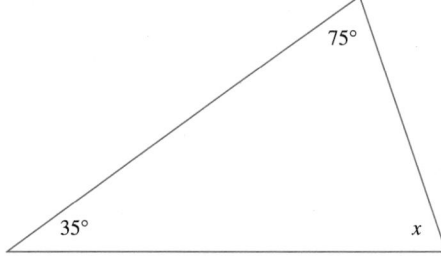

3.

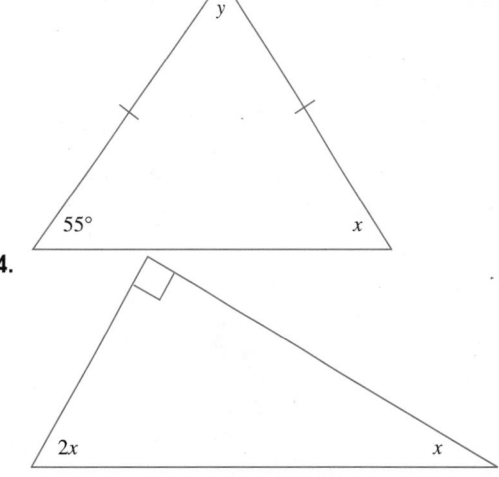

4.

2.

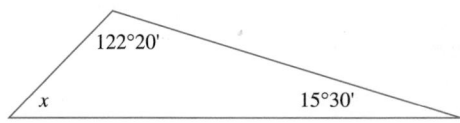

5.

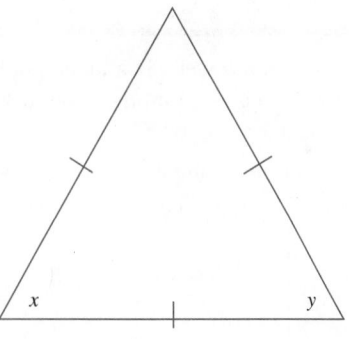

6.

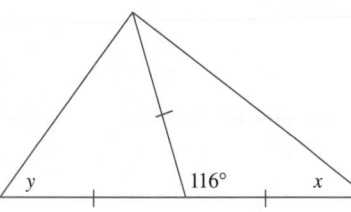

7.

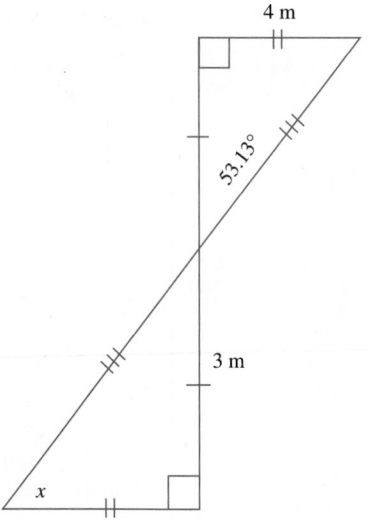

8.

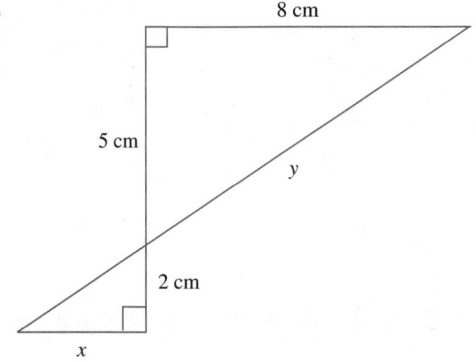

Find the area and perimeter of *ABC* in Exercises 9–12.

9.

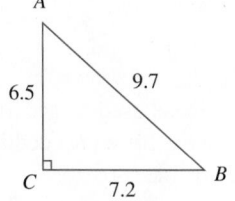

10.

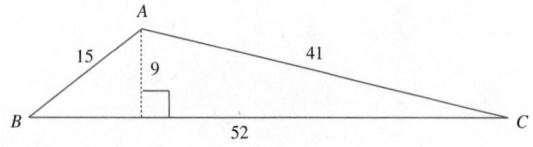

11.

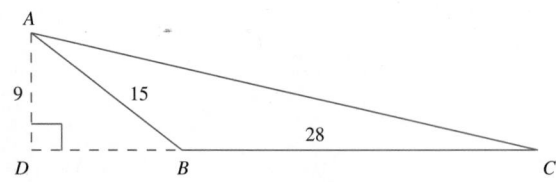

12.

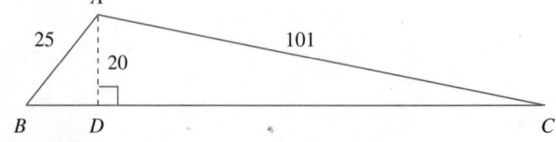

Solve Exercises 13–26.

13. *Landscape architecture*   A triangular piece of land measures 23.2 m, 47.6 m, and 62.5 m. **(a)** How much fencing is needed to enclose this land? **(b)** How many square meters of sod would be needed to sod the entire piece of land?

14. The shadow of a building is 123′6″. At the same time, the shadow of a yard stick is 2′3″. What is the height of the building?

15. A ladder 15 m long reaches the top of a building when its foot is 5 m from the building. How high is the building?

16. *Mechanical engineering*   A triangular metal plate was made by cutting a rectangular plate along a diagonal. If the rectangular piece was 23 cm long and 16 cm high, what is the area of the plate?

17. *Transportation engineering*   A traffic light support is to be suspended parallel to the ground. It reaches diagonally across the intersection of two perpendicular streets. One street is 45 ft wide and the other is 62 ft wide. Determine the length of the support.

18. *Construction*   A house is 8 m wide and the ridge is 3 m higher than the side walls. If the rafters, *r*, extend 0.5 m beyond the sides of the house, how long are the rafters? (See Figure 3.22.)

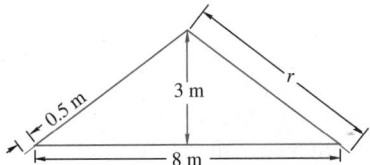

FIGURE 3.22

19. *Construction*   Several cross beams are placed across an A-frame house. The highest beam is placed 2 600 mm from the ridge and the top of this beam is 2 400 mm long. The top of the next beam is placed 2 800 mm below the bottom of the top beam and the top of the third is 2 800 mm below the bottom of the second. The bottom of the third beam is 900 mm above the ground. Each beam is 300 mm thick. **(a)** How long are the tops of the second and third

beams? **(b)** How far apart is the base of the A-frame? (See Figure 3.23.)

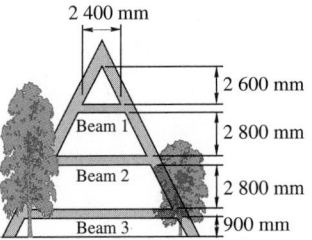

FIGURE 3.23

20. *Construction*   An antenna 175 m high is supported by cables positioned around the antenna. One cable is 50 m from the base of the antenna, one is 75 m, and the third is 100 m. How long is each cable?

21. *Electrical engineering*   A 30 m length of conduit is bent as shown in Figure 3.24. What is the length of the offset, *x*?

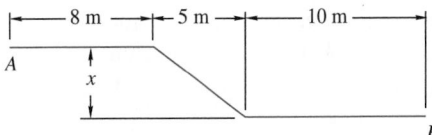

FIGURE 3.24

22. *Electrical engineering*   If it were possible to run a straight conduit from *A* to *B* in Figure 3.24, how much conduit would be saved?

23. *Construction*   A gas pipeline was constructed across a ravine by going down one side, across the bottom, and up the other side, as shown in Figure 3.25. **(a)** How much pipe was used to get from *A* to *B*? **(b)** How much would have been needed if it would have been possible to go directly from *A* to *B*?

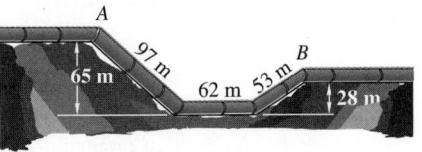

FIGURE 3.25

24. *Physics* The effect of a moving load on the stress beam is shown in the influence diagram of Figure 3.26. If the distance from A to B is 24 ft, find the lengths of $\overline{BC}$ and $\overline{CD}$. (Assume that $\overline{AF}$ is parallel to $\overline{EG}$.)

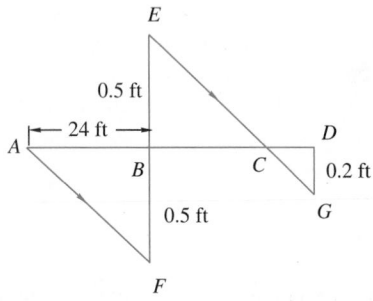

FIGURE 3.26

25. *Civil engineering* A building, located on level ground, casts a shadow of 128.45 m. At the same time, a meter stick casts a shadow of 1.75 m. What is the height of the building?

26. *Construction* Find the length of the brace in Figure 3.27.

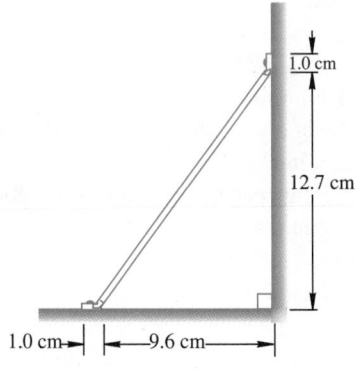

FIGURE 3.27

 **In Your Words**

27. Describe an acute, obtuse, and right triangle. How are they alike? How are they different?

28. **(a)** Explain how similar triangles differ from congruent triangles.

(b) Can similar triangles ever be congruent? Explain your answer.

≡ 3.3
OTHER POLYGONS

In Section 3.2 we looked at the simplest type of polygon, the triangle. In this section, we will look at polygons that have more than three sides.

After the triangle, the most commonly used polygons are those with four sides. A polygon with exactly four sides is a **quadrilateral**. Other polygons that we will use are the **pentagon**, which has exactly five sides, the **hexagon** with six sides, and the **octagon** with eight sides. Any polygon that has all congruent sides and congruent angles is a **regular polygon**. For example, equilateral is another name for a regular triangle.

Quadrilaterals

All quadrilaterals have four sides. Like triangles, different names are given to the quadrilaterals with special properties. The properties deal with the lengths of the sides, if the sides are parallel, and the sizes of the angles. Among the special types of quadrilaterals are the ones shown in Figure 3.28.

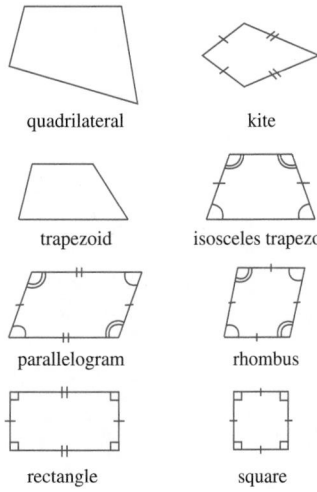

quadrilateral   kite

trapezoid   isosceles trapezoid

parallelogram   rhombus

rectangle   square

FIGURE 3.28

The **kite** has two pairs of adjacent sides congruent. A **trapezoid** has at least one pair of opposite sides parallel. If a trapezoid also has a pair of congruent sides that are not parallel, it is an **isosceles trapezoid** and has the congruent angles indicated in Figure 3.28. A **parallelogram** is a quadrilateral with both pairs of opposite sides parallel. As a result, the opposite sides are also congruent.

A **rhombus** is a parallelogram that has all four sides congruent. A **rectangle** is a parallelogram that has four right angles. A **square** is a rectangle that has four congruent sides. As you can see in Figure 3.28, a square is a special kind of rhombus.

To find the perimeter of a quadrilateral, you must add the lengths of the four sides. In the case of a rhombus or square, this is made easier by the fact that all four sides are the same length. So, if s is the length of one side of a rhombus or square, the perimeter is $4s$. A parallelogram, rectangle, and kite are not much more difficult, since each has two pairs of congruent sides. If the lengths of the sides that are not congruent are a and b, then the perimeter is $2(a+b)$.

In most cases, the area of a quadrilateral depends on the length of one or more sides and the distance between this side and its opposite side. If the height h is the distance between the sides and the length of the base is b, then the area of a rectangle, square, rhombus, or parallelogram can all be expressed as $A = bh$. In a trapezoid you must use the average length of the two parallel sides, $\frac{1}{2}(b_1 + b_2)$ and the height, h. So, for a trapezoid, $A = \frac{1}{2}h(b_1 + b_2)$. A kite is an exception, since its area can be found by multiplying $\frac{1}{2}$ by the lengths of the diagonals, d_1 and d_2. So, the area of a kite is $\frac{1}{2}d_1 d_2$. All of this is summarized in Figure 3.29, where the parts of each figure are labeled and the formulas for perimeter and area of each polygon are given.

Area and Perimeter of Other Polygons

Figure 3.29 shows the formulas for the perimeter and area of a general quadrilateral. Also included are the area and perimeter for a regular hexagon and a regular octagon. In general, with a regular polygon of n sides, you can divide it into n congruent triangles, each with height h and base with length s. The perimeter of the polygon is ns and the area is $\frac{1}{2}nsh$. In Figure 3.29, two additional formulas for the area of a regular hexagon are given. These come from the fact that it is divided into six equilateral triangles and by using the Pythagorean theorem.

| **EXAMPLE 3.16** | Find the perimeter and area of a rectangle with a length of 16 m and a width of 9 m. |

Solution $P = 2(a+b)$ $A = ab$

$\qquad\qquad\quad = 2(16+9) \qquad\quad = 16 \cdot 9$

$\qquad\qquad\quad = 2(25) = 50 \qquad\quad = 144$

The perimeter is 50 meters and the area is 144 square meters. These answers are usually written as 50 m and 144 m^2.

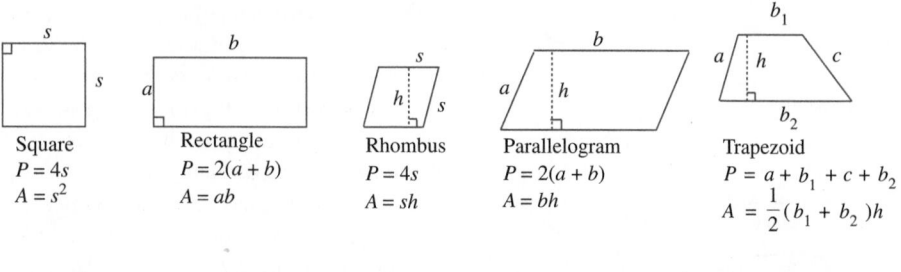

Square
$P = 4s$
$A = s^2$

Rectangle
$P = 2(a + b)$
$A = ab$

Rhombus
$P = 4s$
$A = sh$

Parallelogram
$P = 2(a + b)$
$A = bh$

Trapezoid
$P = a + b_1 + c + b_2$
$A = \dfrac{1}{2}(b_1 + b_2)h$

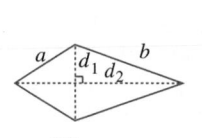

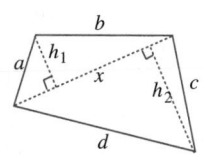

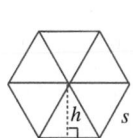

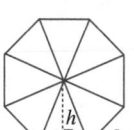

Kite
$P = 2(a + b)$
$A = \dfrac{1}{2}d_1d_2$

Quadrilateral
$P = a + b + c + d$
$A = \dfrac{1}{2}(h_1 + h_2)x$

Regular Hexagon
$P = 6s$
$A = 3sh$
$\quad = \dfrac{3\sqrt{3}}{2}s^2$
$\quad = 2\sqrt{3}\,h^2$

Regular Octagon
$P = 8s$
$A = 4sh$

FIGURE 3.29

EXAMPLE 3.17

A trapezoid has bases of 14 and 18 in. and a height of 7 in. What is its area?

Solution $A = \dfrac{1}{2}h(b_1 + b_2)$

$\qquad = \dfrac{1}{2}(7)(14 + 18)$

$\qquad = \dfrac{1}{2}(7)(32)$

$\qquad = (7)\dfrac{1}{2}(32)$

$\qquad = 7 \cdot 16 = 112$

The area is 112 square inches (often written as 112 in.^{2}).

EXAMPLE 3.18

A metal plate is made by cutting a rectangle out of an isosceles trapezoid, using the dimensions shown in Figure 3.30. **(a)** Find the area and **(b)** the perimeter of this metal plate.

EXAMPLE 3.18 (Cont.)

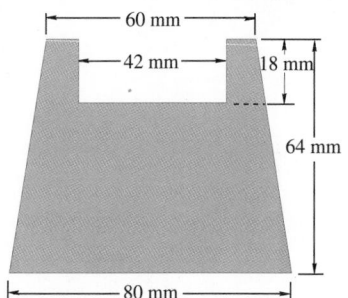

FIGURE 3.30

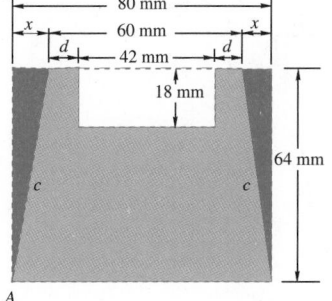

FIGURE 3.31

Solutions

(a) To find the area of this plate, we first find the area of the trapezoid and then of the rectangle. The area of the plate is the difference between the two.

$$A_{\text{trapezoid}} = \frac{1}{2}h(b_1 + b_2)$$
$$= \frac{1}{2}(64)(60+80)$$
$$= \frac{1}{2}(64)(140)$$
$$= 4{,}480$$
$$A_{\text{rectangle}} = bh$$
$$= (42)(18)$$
$$= 756$$
$$A_{\text{plate}} = A_{\text{trapezoid}} - A_{\text{rectangle}}$$
$$= 4{,}480 - 756$$
$$= 3{,}724$$

(b) We have all the measurements we need to find the perimeter except for the lengths of the two slanted sides of the trapezoid. For this we will use the Pythagorean theorem. Look at Figure 3.31. The plate has been drawn inside a rectangle. Because the plate was cut from an isosceles trapezoid, we know that the two lengths marked x are the same length. We also know that $2x + 60 = 80$, so $x = 10\,\text{mm}$.

The slanted sides of the trapezoid are the hypotenuse c of the two shaded triangles in Figure 3.31. The legs of these triangles are 10 and 64 so

$$c^2 = 10^2 + 64^2$$
$$= 100 + 4\,096$$
$$= 4\,196$$
$$c = \sqrt{4\,196} \approx 64.78$$

If we assume that the rectangular cutout is centered, then

$$d = \left(\frac{60 - 42}{2}\right) = 9\,\text{mm}$$

The perimeter can be found by starting at point A and adding the lengths of the sides in a clockwise direction:

$$P = 64.78 + 9 + 18 + 42 + 18 + 9 + 64.78 + 80$$
$$= 305.56$$

The perimeter of this plate is approximately 305.56 mm.

Exercise Set 3.3

Find the perimeter and area of each polygon in Exercises 1–10. Use the formulas in Figure 3.29.

1. 15 cm × 15 cm (square)

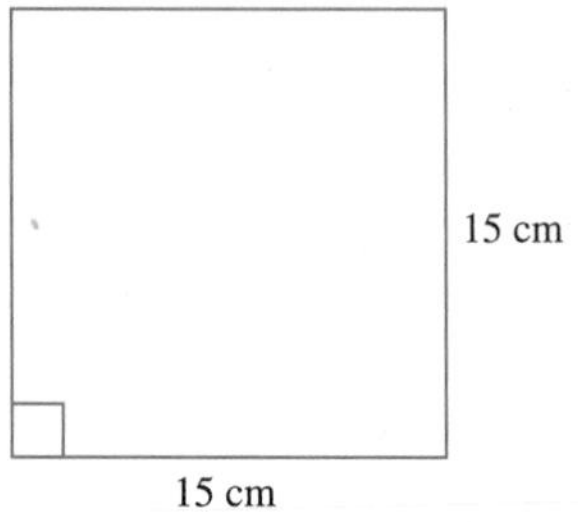

2. 9 in. × 25 in. (rectangle)

3. parallelogram: $11\frac{1}{2}$ in. (height), 33 in. (base), $15\frac{1}{4}$ in. (side)

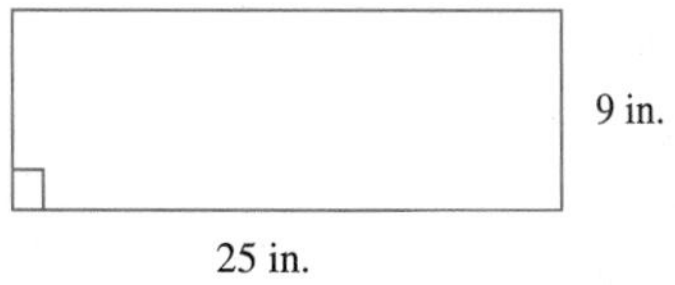

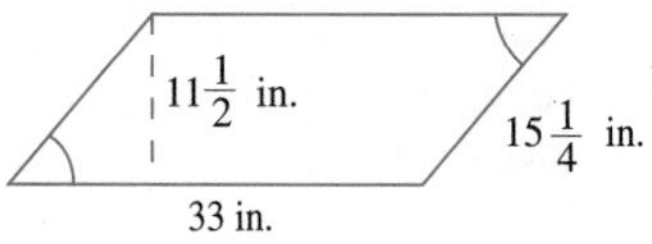

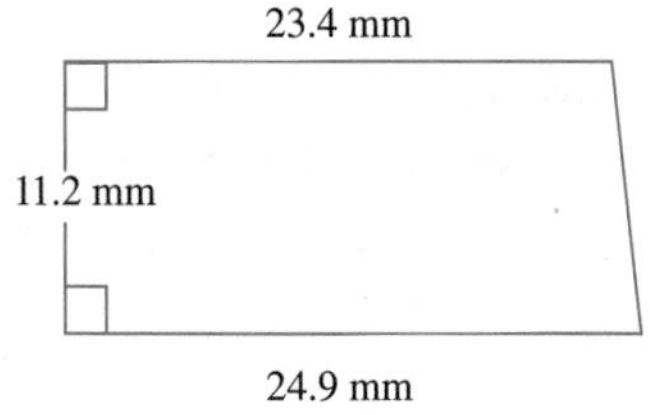

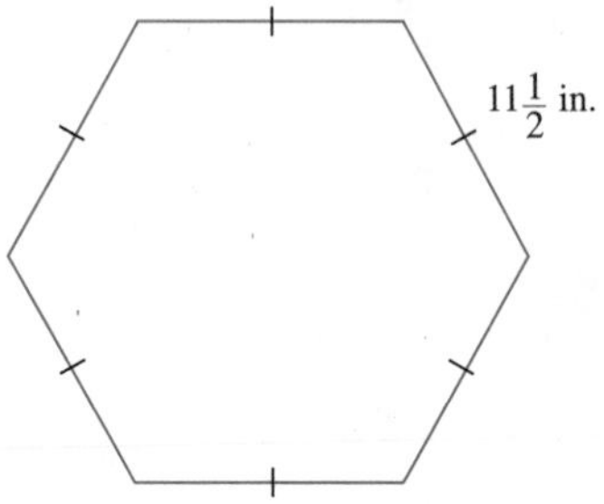

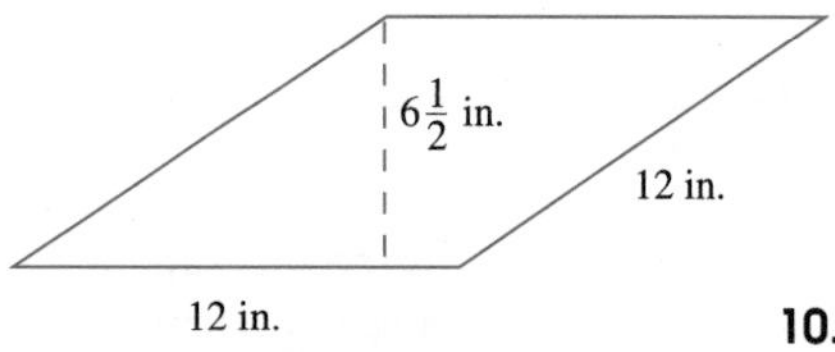

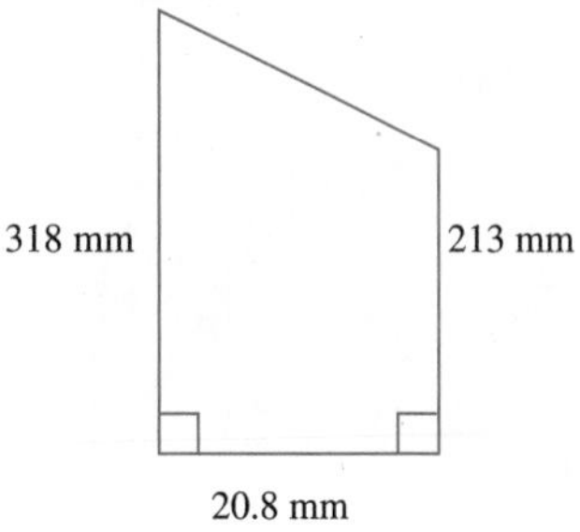

4. trapezoid: 23.4 mm (top), 11.2 mm (side), 24.9 mm (bottom)

5. hexagon: $11\frac{1}{2}$ in.

6. parallelogram: $6\frac{1}{2}$ in. (height), 12 in. (side), 12 in. (base)

7. kite: 10 cm, 16 cm, 21 cm, 12 cm

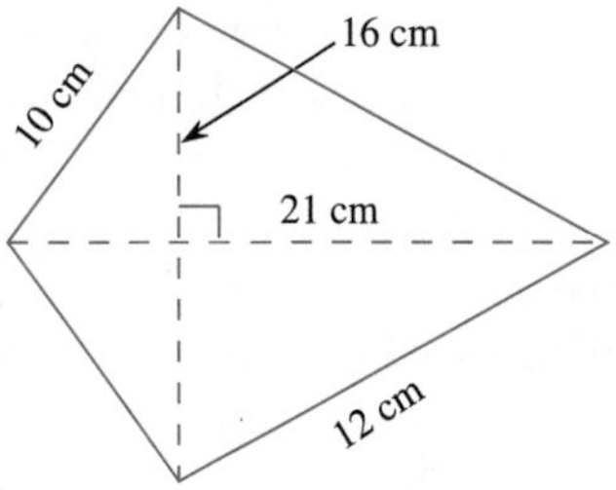

8. 318 mm, 213 mm, 20.8 mm

9. parallelogram: 15.7 mm, 21.2 mm, 13.2 mm, $14\frac{3}{4}$ in.

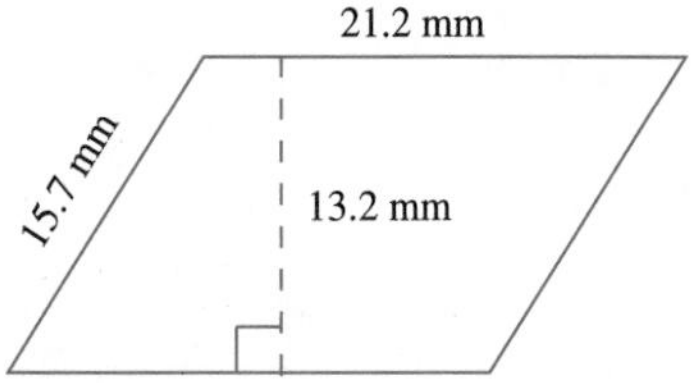

10. trapezoid: $12\frac{1}{2}$ in., 12 in., 13 in., $23\frac{1}{4}$ in.

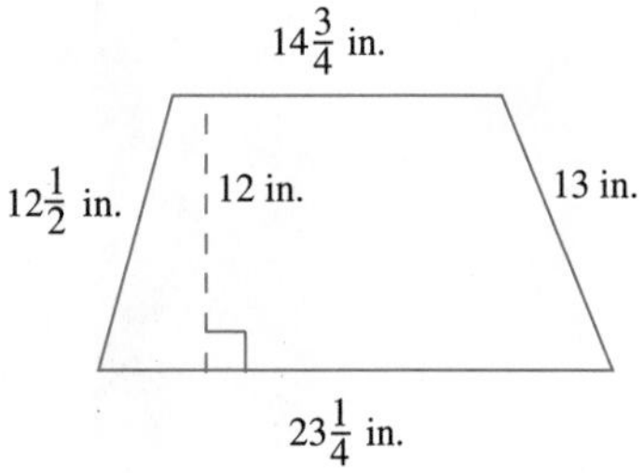

Solve Exercises 11–22.

11. *Construction*   Find the area of the L-shaped patio in Figure 3.32.

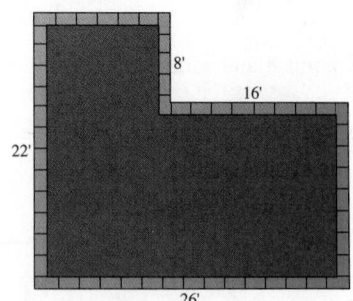

FIGURE 3.32

12. *Construction*   You plan to put a brick border around the patio in Figure 3.32. Each brick is 8 in. long and 4 in. wide. If the bricks are placed end to end, how many bricks are needed?

13. *Mechanical engineering*   What is the area of the cross-section of the I-beam in Figure 3.33a? (Hint: Think of it as a rectangle and subtract the areas of the two trapezoids. See Figure 3.33b.)

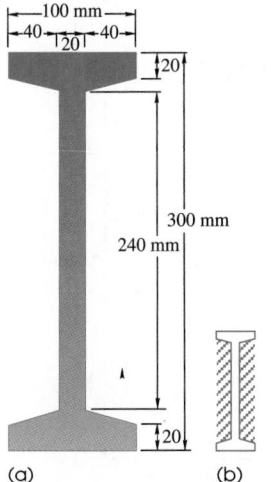

FIGURE 3.33

14. *Civil engineering*   Find the area of the concrete highway support shown in Figure 3.34.

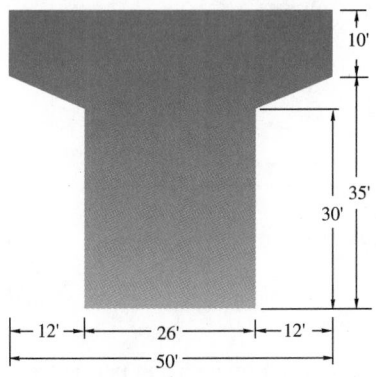

FIGURE 3.34

15. *Machine technology*   A hexagonal bolt measures $\frac{7}{8}''$ across the short distance. (See Figure 3.35.) What are the perimeter and area of this bolt?

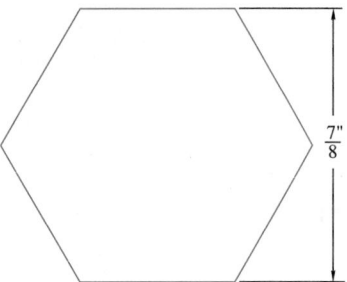

FIGURE 3.35

16. *Machine technology*   A hexagonal bolt measures 15 mm across the short distance.   What are the perimeter and area of this bolt?

17. *Civil engineering*   A river bed is going to be constructed in the shape of a trapezoid. The height of the trapezoid is designed to contain flood waters between the dykes, which form the walls. The trapezoid has the dimensions given in Figure 3.36. **(a)** What is the cross-sectional area? **(b)** How many linear feet of concrete are needed to surface the walls and the bottom?

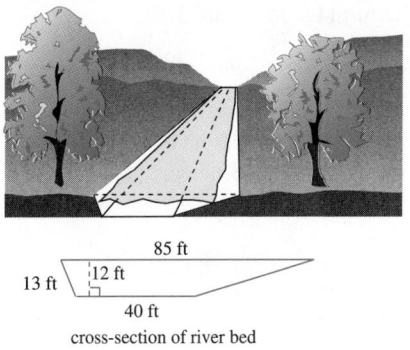

cross-section of river bed

FIGURE 3.36

18. *Interior design*   What will it cost to carpet the floor of a rectangular room that is 20′6″ by 15′3″ at $6.75 a square yard? (There are 9 ft² in 1 yd².)

19. *Civil engineering*   A structural supporting member is made in the shape of an angle as shown in Figure 3.37. What is the cross-sectional area?

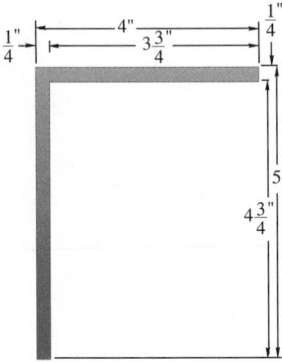

FIGURE 3.37

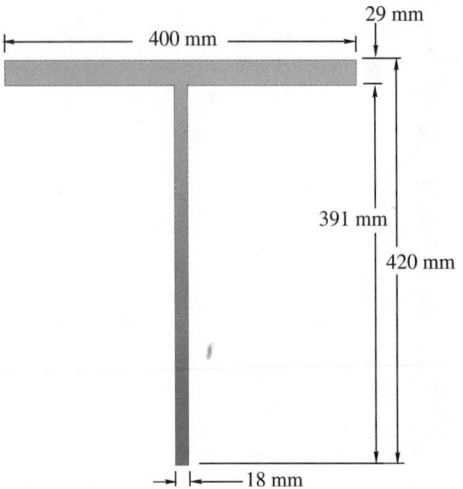

FIGURE 3.38

20. *Architecture*   How many 9″ square tiles will cover a floor 12′ by 17′3″?

21. *Civil engineering*   Find the area of the cross-section of the structural tee in Figure 3.38.

22. *Architecture*   Find the area of the side of the house in Figure 3.39.

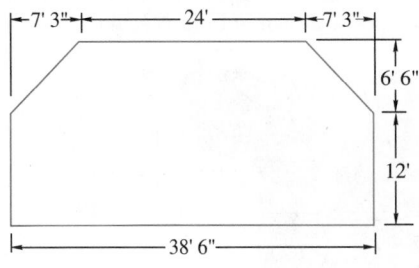

FIGURE 3.39

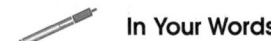   **In Your Words**

23. **(a)** Describe how a trapezoid and a parallelogram are different.
   (b) Can a trapezoid ever be a parallelogram? Explain your answer.

24. Describe what is meant by area and perimeter.

3.4
CIRCLES

Until now, the geometric figures we have studied have all been made of segments joined at their endpoints. A circle is a different type of geometric figure. Like the polygons, a circle is part of a plane. A **circle** is made up of the set of points that are all the same distance from a point called the **center**, as shown in Figure 3.40.

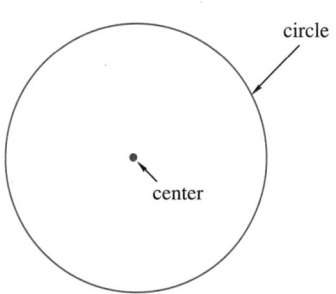

FIGURE 3.40

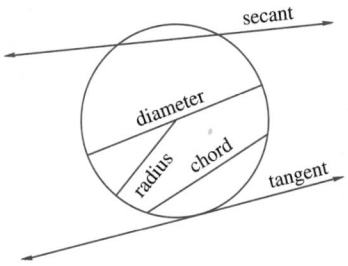

FIGURE 3.41

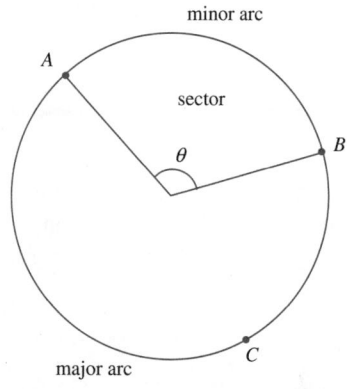

FIGURE 3.42

Parts of a Circle

There are many parts to a circle. Some of them are shown in Figure 3.41. The **radius** is a segment with one endpoint at the center and the other on the circle. A **chord** is any segment with both endpoints on the circle. A **diameter** is a chord through the center of the circle. A **secant** is a line that intersects a circle twice, and a **tangent** intersects the circle in exactly one point. A tangent is perpendicular to the radius at its point of tangency. The **circumference** of a circle is the name for the perimeter and is equal to the distance around the circle.

Figure 3.42 shows a few more parts of the circle that will be used later. A **central angle** is an angle with a vertex at the center of the circle. An **arc** is a section of a circle and is often described in terms of the size of its central angle. Thus, we might refer to a 20° arc or an arc of $\frac{\pi}{9}$ rad.

An arc with length equal to the radius is 1 rad. A central angle divides a circle into a **minor arc** and a **major arc**. We also may refer to an arc by its endpoints. The minor arc in Figure 3.42 is identified as $\overparen{AB}$. The major arc is identified as $\overparen{ACB}$, where A and B are the endpoints and C is any other point on the major arc. The length of an arc is denoted by placing an m in front of the name of the arc. Thus, $M\overparen{AB}$ is the length of $\overparen{AB}$. A **sector** is the region inside the circle and is bounded by a central angle and an arc.

Circumference and Area of Circles

The formulas for the circumference and area of a circle involve the use of the irrational number π. The circumference C of a circle is

$$C = \pi d \qquad \text{or} \qquad C = 2\pi r, \quad \text{(where } d = 2r\text{)}$$

where d is the length of a diameter and r is the length of a radius.

The area of a circle is

$$A = \pi r^2$$

Some mechanics use the formula

$$A = 0.785 d^2$$

Since $d = 2r$, you can see that $d^2 = 4r^2$ and so $r^2 = \dfrac{d^2}{4}$. The area of the circle can be written as $A = \pi \left(\dfrac{d^2}{4}\right) = \frac{\pi}{4}d^2$. Since $\frac{\pi}{4} \approx 0.785$, we have $A = 0.785d^2$.

Circumference and Area of a Circle

A circle with radius r and diameter d has a circumference C and an area A where

$$C = 2\pi r = \pi d$$
$$\text{and} \qquad A = \pi r^2$$

EXAMPLE 3.19

Find the circumference and area of a circle with **(a)** diameter 7.00 in. and **(b)** radius 8.3 mm.

Solutions   **(a)** $d = 7.00$ in.
$$C = \pi d$$
$$= \pi 7$$
$$\approx 21.99 \text{ in.}$$
$$A = \pi r^2$$
$$= \pi \left(\frac{7}{2}\right)^2$$
$$\approx 38.48 \text{ in.}^2$$

(b) $r = 8.3$ mm
$$C = 2\pi r$$
$$= 2\pi(8.3)$$
$$\approx 52.2 \text{ mm}$$
$$A = \pi r^2$$
$$= \pi(8.3)^2$$
$$\approx 216.4 \text{ mm}^2$$

This example raises a question. What value should you use for π? If you use 3.14 or $3\frac{1}{7}$, your answers may differ slightly from those shown. They were obtained using a value programmed into a calculator and recalled by pressing the $\boxed{\pi}$ key, with the resulting display of 3.1415927. As you can see, $\pi \neq 3.14$. We use 3.14 as an approximation of π.

Arc Length

The arc length s is a direct result of the size of the angle that determines the arc. An angle of 2π rad is a complete circle and has an arc length equal to the circumference, $2\pi r$. An angle of π rad is half a circle, and so its arc length is πr. Similarly, an angle of 3 rad has an arc length of $3r$. In general, a central angle of θ rad has an arc length of θr. This gives us the following formula.

Arc Length

An arc formed from a circle of radius r and central angle of θ rad has an arc length s, where

$$s = \theta r$$

Similarly, the area of a sector of a circle can be derived from the formula for the area of a circle. A complete circle has an angle of 2π rad and an area of $\pi r^2 = \frac{1}{2}(2\pi)r^2$. A semicircle, or half a circle, has an angle of π rad and an area of $\frac{1}{2}\pi r^2$. In general, a central angle of θ rad forms a sector with area $\frac{1}{2}\theta r^2 = \frac{1}{2}r^2\theta$, as given in the following box.

> **Area of a Sector**
>
> A sector formed by a circle with radius r and a central angle of θ rad has an area, A, where
>
> $$A = \frac{1}{2}r^2\theta$$

≡ Note   In both of the previous formulas, the central angle θ *must* be in radians.

Application

EXAMPLE 3.20

A landscaper is going to put some plastic edging around a pie-shaped flower bed. The flower bed is formed from a circle with radius 9.0 ft and a central angle of 105°. How much edging, in feet, will be needed?

Solution   A sketch of the flower bed is shown in Figure 3.43. The edging will go from C to A then to B and back to C around the arc of the circle. The length of edging needed is $CA + AB + m\overset{\frown}{BC} = 9.0 + 9.0 + m\overset{\frown}{BC} = 18.0 + m\overset{\frown}{BC}$. All we need to determine is the length of the arc $\overset{\frown}{BC}$.

To determine the length of the arc, we will use $s = \theta r$. Here, $r = 9.0$ and $\theta = 105°$. To use this formula, we must convert 105° to radians. Using the proportion $\dfrac{D}{180°} = \dfrac{R}{\pi}$, with $D = 105°$, we obtain $R = \frac{7\pi}{12}$. Thus,

$$
\begin{aligned}
m\overset{\frown}{BC} &= r\theta \\
&= 9.0\left(\frac{7\pi}{12}\right) \\
&= \frac{21.0\pi}{4} \approx 16.5\,\text{ft}
\end{aligned}
$$

So, the length of edging needed is $18.0 + 16.5 = 34.5$ ft.

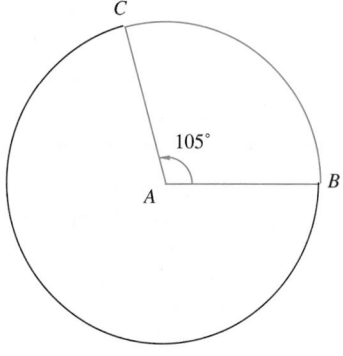

FIGURE 3.43

Application

EXAMPLE 3.21

A machine shop is installing two pulleys with radii 570 mm and 130 mm, respectively. The pulleys, as shown in Figure 3.44, are 1 250 mm apart and the length AB is 1 170 mm. If $\angle AOP = 70°$, what is the length of the driving belt?

Solution   To solve this problem we must add the lengths of the straight sections of the belt, AB and CD, and the two arc lengths, s_1 and s_2. We are given $AB = 1\,170$ mm. So, $CD = 1\,170$. To find s_1 we will use $s_1 = r\theta$. We are given $r = 570$ mm, $\theta = 360° - \angle AOD = 360° - 140° = 220° \approx 3.84$ rad. This means that

$$
\begin{aligned}
s_1 &= r\theta \\
&= 570(3.84) \\
&= 2\,188.8\,\text{mm}
\end{aligned}
$$

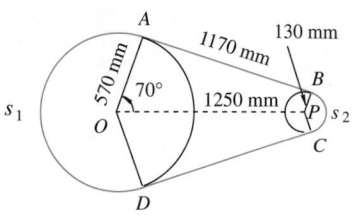

FIGURE 3.44

EXAMPLE 3.21 (Cont.)

Now, $s_2 = r\theta$, where $r_2 = 130\,\text{mm}$ and $\theta_2 = 140° \approx 2.44\,\text{rad}$. So,

$$s_2 = r_2\theta_2$$
$$= (130)(2.44)$$
$$= 317.2\,\text{mm}$$

The length of the belt is $1\,170 + 317.2 + 1\,170 + 2\,188.8 = 4\,846\,\text{mm}$.

Application

EXAMPLE 3.22

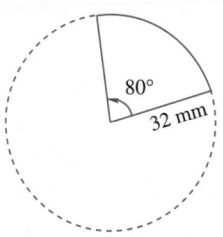

FIGURE 3.45

A pie-shaped piece is going to be cut out of a circular piece of metal. The radius of the circle is 32.0 mm and the central angle of the sector is 80°. What is the area of this sector? (See Figure 3.45.)

Solution We will use the formula $A = \frac{1}{2}r^2\theta$. We are given $r = 32.0\,\text{mm}$ and $\theta = 80° \approx 1.396\,\text{rad}$, so

$$A = \frac{1}{2}(32.0)^2(1.396) \approx 714.752$$

The area is about $714.8\,\text{mm}^2$.

Application

EXAMPLE 3.23

Two competing pizza companies sell pizza by the slice. At Checker's Pizza, a typical slice of pizza is a sector with a central angle of 60° formed from an 8″ radius pizza. A slice sells for $1.75. Pizza Plus makes each slice from a 10″ radius pizza with a central angle of 45°. A slice sells for $2.10. At which company do you get the most pizza for the price?

Solution At Checker's, a sector with $\theta = 60° = \frac{\pi}{3}$ and $r = 8″$ has an area $A = \frac{1}{2}r^2\theta = \frac{1}{2}\left(8^2\right)\frac{\pi}{3} = \frac{32\pi}{3}\,\text{in.}^2$ At $1.75 per slice, this is $1.75 \div \frac{32\pi}{3} \approx \$0.052/\text{in.}^2$

At Pizza Plus a sector with $\theta = 45° = \frac{\pi}{4}$ and $r = 10″$ has an area $A = \frac{1}{2}r^2\theta = \frac{1}{2}\left(10^2\right)\frac{\pi}{4} = \frac{25\pi}{2}\,\text{in.}^2$ At $2.10 per slice, this is $2.10 \div \frac{25\pi}{2} \approx \$0.053/\text{in.}^2$

Since the Checker's Pizza is $0.052/in.² and the Pizza Plus slice cost $0.053/in.², a slice from Checker's Pizza is the better bargain.

Exercise Set 3.4

Find the area and circumference of the circles in Exercises 1–8 with the given radius or diameter. (You may leave answers in terms of π.)

1. $r = 4\,\text{cm}$
2. $d = 16\,\text{in.}$
3. $r = 5\,\text{in.}$
4. $d = 23\,\text{mm}$
5. $r = 14.2\,\text{mm}$
6. $r = 13\frac{1}{4}\,\text{in.}$
7. $d = 24.20\,\text{mm}$
8. $d = 23\frac{1}{2}\,\text{in.}$

Solve Exercises 9–22.

9. *Interior design* **(a)** What is the area of a circular table top with a diameter of 48 in.? **(b)** How much metal edging would be needed to go around this table?

10. *Electricity* A coil of bell wire has 42 turns. The diameter of the coil is 0.5 m. How long is the wire on this coil?

11. *Industrial engineering* Two circular drums are to be riveted together, as shown in Figure 3.46, with the rivets spaced 75 mm apart. How many rivets will be needed?

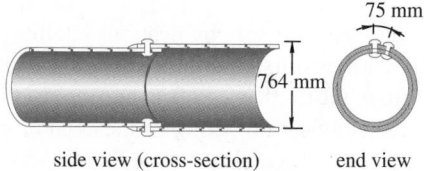

side view (cross-section)   end view

FIGURE 3.46

12. *Industrial design* Two pulleys each with a radius of $3'2''$ have their centers $11'9\frac{1}{4}''$ apart, as shown in Figure 3.47. What is the length of the belt needed for these pulleys?

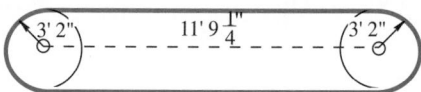

FIGURE 3.47

13. *Sheet metal technology* A sheet of copper has been cut in the shape of a sector of a circle. It is going to be rolled up to form a cone. **(a)** What is the arc length, $\overarc{AB}$, of the sector? **(b)** What is the area of this piece of copper? (See Figure 3.48.)

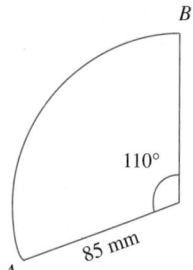

FIGURE 3.48

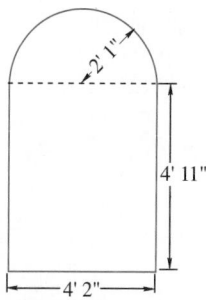

FIGURE 3.49

14. *Architecture* **(a)** Find the length of molding used around the window opening in Figure 3.49. **(b)** What is the area of the glass needed for this window?

15. *Architecture* The top of a stained glass window has the shape shown in Figure 3.50. The triangle is an equilateral triangle and the two arcs have their centers at the opposite vertices. That is, the arc from B to C is from a circle with center A. **(a)** What is the amount of molding needed for this window? **(b)** What is the area of the glass needed for this window?

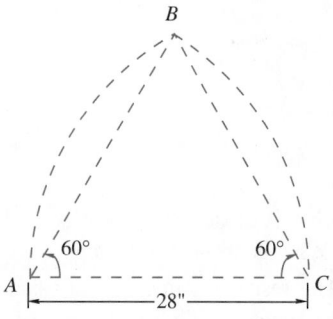

FIGURE 3.50

16. *Architecture*　(a) Find the length of molding needed to go around the window molding in Figure 3.51. (b) What is the area of the glass needed for this window? (The radii for the arcs are 2 400 mm.)

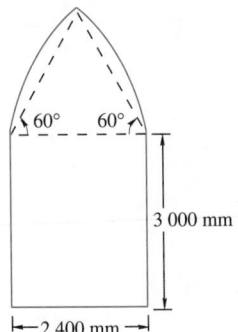

FIGURE 3.51

17. *Industrial design*　(a) What is the area of the table top in Figure 3.52? (b) How much metal edging would be required for this table?

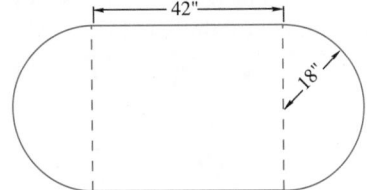

FIGURE 3.52

18. *Civil engineering*　A cross-section of pipe is shown in Figure 3.53. What is the area of the cross-section?

19. *Electronics*　The resistance in a circuit can be reduced by 50% by replacing a wire with another wire that has twice the cross-sectional area. What is the diameter of a wire that will replace one that has a diameter of 8.42 mm?

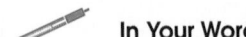　**In Your Words**

23. Explain what is meant by an arc and by a sector of a circle.
24. Write an application in your technology area of interest that requires you to use some of the circle concepts

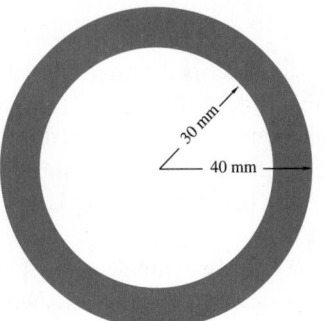

FIGURE 3.53

20. *Space technology*　A communications satellite is orbiting the Earth at a fixed altitude above the equator. If the radius of the Earth is 3,960 mi and the satellite is in direct communication with $\frac{1}{3}$ of the equator, what is the height h of the satellite? (See Figure 3.54.)

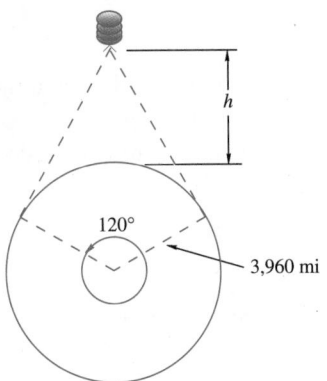

FIGURE 3.54

21. (a) What is the absolute error between $3\frac{1}{7}$ and π?
    (b) What is the percent error between $3\frac{1}{7}$ and π?
22. (a) What is the absolute error between 3.14 and π?
    (b) What is the percent error between 3.14 and π?

in this section. Give your problem to a classmate and see if he or she understands and can solve your problem. Rewrite the problem as necessary to remove any difficulties encountered by your classmate.

≡ **3.5**
GEOMETRIC SOLIDS

The geometric figures we have looked at thus far have all been plane figures; that is, figures that can be drawn in two dimensions. But, we live in a three-dimensional world. Most of the objects we work with can be characterized as three-dimensional solids. The plane geometric figures we have studied are what the objects look like on one side, when they are taken apart, or "sliced" into cross-sections.

Cylinders

A **cylinder** is a solid whose ends, or **bases**, are parallel congruent plane figures arranged in such a manner that the segments connecting corresponding points on the bases are parallel. These segments are called **elements**. In the first cylinder in Figure 3.55, $\overline{AA'}$, $\overline{BB'}$, and $\overline{CC'}$ are all elements of the cylinder. A **circular cylinder** is a cylinder in which both bases are circles. A **right circular cylinder** is the most common type of cylinder and is formed when the bases are perpendicular to the elements. The **height** or **altitude** of a cylinder is a segment that is perpendicular to both bases.

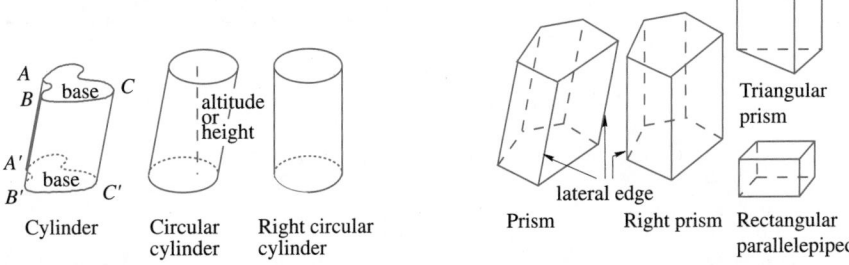

Cylinder   Circular   Right circular
               cylinder   cylinder

FIGURE 3.55

Prism   Right prism   Rectangular
                           parallelepiped

Triangular prism

lateral edge

FIGURE 3.56

Prisms

As shown in Figure 3.56, a **prism** is a solid with ends, or **bases**, that are parallel congruent polygons and with sides, called **faces** or **lateral faces**, that are parallelograms. The segments that form the intersections of the lateral faces are called the **lateral edges**. The height, or altitude, of a prism is the distance between the bases. A **right prism** has its bases perpendicular to its lateral edges; hence, its faces are rectangles.

Prisms get their names from their bases. If the bases are regular polygons, then the prism is a **regular prism**. A **triangular prism** has triangles for bases and a **rectangular prism** has rectangles for bases. The most common prisms are the right rectangular prisms, which are called **rectangular parallelepipeds** and the right square prism, more commonly known as a **cube**.

There are two kinds of areas that are usually associated with any solid figure. The **lateral area** is the sum of the areas of all the sides. The **total surface area** is the lateral area plus the area of the bases. Because the lateral surface of a right prism or right cylinder can be unfolded to form a parallelogram if it is cut along an element, the lateral area L is found by multiplying the perimeter or circumference of the base times the height. The volume of a cylinder or prism is the area of the base B times the height.

Lateral Area, Total Surface Area, and Volume of a Cylinder or Prism

The lateral area, total surface area, and volume of a cylinder or prism are given by the following formulas

Solid	Lateral Area L	Total Surface Area T	Volume V
Prism	ph	$ph + 2B$	Bh
Cylinder	$2\pi rh$	$2\pi r(r+h)$	$\pi r^2 h$

where p is the perimeter of a base of the prism, h is the height, r is the radius of a base of the cylinder, and B is the area of a base.

EXAMPLE 3.24

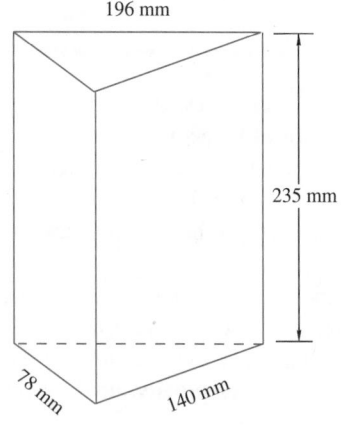

FIGURE 3.57

Find the **(a)** lateral area, **(b)** total surface area, and **(c)** volume of the triangular prism shown in Figure 3.57.

Solutions

(a) Lateral area $L = ph$, where p is the perimeter of the base and h the height of the prism. We are given $h = 235$ and add the lengths of the triangle's sides to obtain $p = 78 + 140 + 196 = 414$.

$$L = ph$$
$$= 414 \times 235$$
$$= 97\,290$$

The lateral area is $97\,290\,\text{mm}^2$.

(b) Total surface area $T = L + 2B$, where B is the area of a base. Using Hero's formula with $s = \dfrac{p}{2} = \dfrac{414}{2} = 207$, we get

$$B = \sqrt{s(s-a)(s-b)(s-c)}$$
$$= \sqrt{207(207-78)(207-140)(207-196)}$$
$$= \sqrt{207(129)(67)(11)}$$
$$= \sqrt{19\,680\,111}$$
$$\approx 4436.23$$

EXAMPLE 3.24 (Cont.)

$$T = 2B + L$$
$$= 2\,(4\,436.23) + 97\,290$$
$$= 8\,872.46 + 97\,290$$
$$= 106\,162.46$$

The total surface area is 106 162.46 mm^2.

(c) Volume $V = Bh$
$$= 4\,436.23 \times 235$$
$$= 1\,042\,514.1$$

The volume is 1 042 514.1 mm^3.

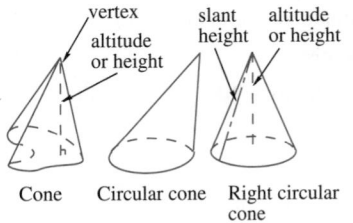

vertex, altitude or height, slant height, altitude or height

Cone   Circular cone   Right circular cone

FIGURE 3.58

Cones

A cone is formed by drawing segments from a plane figure, the **base**, to a point called the **vertex**. The vertex cannot be in the same plane as the base. The **altitude** is a segment from the vertex and perpendicular to the base. The most common cones are the **circular cone** and the **right circular cone**. Both have a base that is a circle. In a right circular cone, the altitude intersects the base at its center. The **slant height** of a right circular cone is a segment from the vertex to a point on the circumference. (See Figure 3.58.)

Pyramids

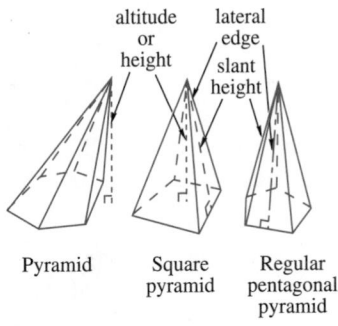

altitude or height, lateral edge, slant height

Pyramid   Square pyramid   Regular pentagonal pyramid

FIGURE 3.59

A **pyramid** is a special type of cone, which has a base that is a polygon. A typical pyramid and some of its parts are shown in Figure 3.59. Each side of a pyramid is a triangle and is called a **lateral face**. The lateral faces meet at the **lateral edges**. As with prisms, pyramids are classified according to the shape of their base. A **regular pyramid** has a regular polygon for a base and an altitude that is perpendicular to the base at its center. The slant height of a regular pyramid is the altitude of any of the lateral faces.

The volume V of a cone or pyramid is one-third the area of the base B times the height h, or $V = \frac{1}{3}Bh$. For the lateral areas we will only consider those of right circular cones and regular pyramids. The lateral area L is one-half the slant height s times the perimeter or circumference of the base. The total surface area is the lateral area plus the area of the base.

> **Lateral Area, Total Surface Area, and Volume of Cones or Pyramids**
>
> The lateral area, total surface area, and volume of a right circular cone or regular pyramid are given by the following formulas
>
> | Solid | Lateral Area L | Total Surface Area T | Volume V |
> |---|---|---|---|
> | Pyramid | $\frac{1}{2}ps$ | $\frac{1}{2}ps+B$ | $\frac{1}{3}Bh$ |
> | Cone | πrs | $\pi r(r+s)$ | $\frac{1}{3}\pi r^2 h$ |
>
> where p is the perimeter of a base of the pyramid, h is the height or altitude, s is the slant height, r is the radius of a base of the cone, and B is the area of a base.

Application

EXAMPLE 3.25

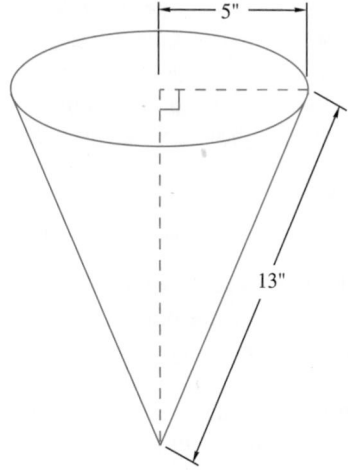

FIGURE 3.60

How many square inches of metal are needed to make a cone-shaped container like the one in Figure 3.60? What is the volume of this cone?

Solution   The amount of metal needed is the lateral area, L. We know that

$$L = \frac{1}{2}Cs$$

where C is the circumference of the base and s is the slant height of the cone.

$$C = 2\pi r \quad \text{where } r \text{ is the radius of the base}$$
$$= 2\pi 5 = 10\pi$$
$$L = \frac{1}{2}(10\pi)(13)$$
$$= 65\pi \approx 204.2$$

It will take about 204.2 in.² of metal.

$V = \frac{1}{3}\pi r^2 h$, where h is the height of the cone. The slant height, a radius, and the altitude form a right triangle with the slant height equal to the hypotenuse. So, $h^2 + 5^2 = 13^2$ or $h = 12$.

$$V = \frac{1}{3}\pi(5)^2(12)$$
$$= 100\pi \approx 314.2$$

The volume is about 314.2 in.³

Frustums

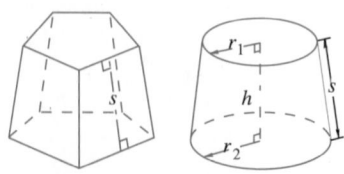

FIGURE 3.61

A **frustum** of a cone or a pyramid is formed by a plane parallel to the base that intersects the solid between the vertex and the base. We will limit our study to frustums of right circular cones or right regular pyramids, as shown in Figure 3.61.

If the height of the frustum is h and the areas of the bases are B_1 and B_2 then the volume V is given by

$$V = \frac{h}{3}\left(B_1 + B_2 + \sqrt{B_1 B_2}\right)$$

The lateral area L is given using p_1 and p_2 as the perimeter of the bases and s, the slant height:

$$L = \frac{s}{2}\left(p_1 + p_2\right)$$

For the frustum of a cone with bases of radii r_1 and r_2, p_1 and p_2 are actually the circumferences of the bases. Thus, $p_1 = 2\pi r_1$ and $p_2 = 2\pi r_2$ and the formula becomes $L = \frac{s}{2}\left(2\pi r_1 + 2\pi r_2\right)$ or

$$L = \pi s(r_1 + r_2)$$

Application

EXAMPLE 3.26

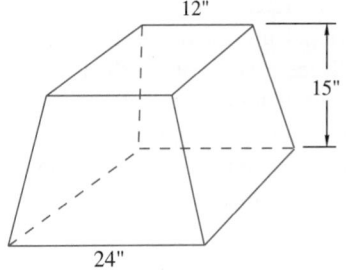

FIGURE 3.62

The concrete base of a light pole is constructed in the form of the frustum of a square pyramid, as shown in Figure 3.62. What is the volume of the base for the light pole?

Solution The volume of a frustum of a square pyramid is given by the formula $V = \frac{h}{3}\left(B_1 + B_2 + \sqrt{B_1 B_2}\right)$. We must first find the area of the bases. If B_1 is the area of the top base, then $B_1 = 12^2 = 144\,\text{in.}^2$ and $B_2 = 24^2 = 576\,\text{in.}^2$ So,

$$\begin{aligned}V &= \frac{h}{3}\left(B_1 + B_2 + \sqrt{B_1 B_2}\right)\\ &= \frac{15}{3}\left(144 + 576 + \sqrt{144 \cdot 576}\right)\\ &= \frac{15}{3}\left(144 + 576 + 288\right)\\ &= 5{,}040\end{aligned}$$

The volume is $5{,}040\,\text{in.}^3$

Application

EXAMPLE 3.27

The picture in Figure 3.63a shows the oven hood pictured at the beginning of the chapter. The hood was made by combining two prisms and the frustum of a square pyramid. In Figure 3.63b, a line drawing of the hood is shown with the measurements indicated. The bottom of the pyramid is 72 in. on each of the two longer sides and 36 in. on each of the two shorter sides. The top base measures 18 in. on each side. The slant height for the longer sides is 15 in. For the shorter sides, the slant height is 29.5 in. The top prism is 36 in. high and the bottom prism is 6 in. high. **(a)** How much metal will it take to form the outside of the part of the oven hood formed by the frustum of the pyramid? **(b)** How much metal will it take to form the outside of the entire oven hood?

EXAMPLE 3.27 (Cont.)

Courtesy of Best (USA) Inc.

FIGURE 3.63a

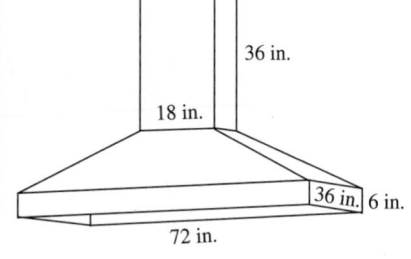

FIGURE 3.63b

Solutions

(a) The amount of metal needed can be found by determining the lateral area L of this pyramid. Notice that this is a frustum of a rectangular pyramid and not of a square pyramid. This means that we cannot use the above formula for the lateral area of the frustum of a pyramid. Since we cannot use the formula, we will think of this frustum as if it were made of four trapezoids, as shown in Figure 3.63c. These four trapezoids are two pairs of congruent trapezoids, so we need only find the areas of the left two trapezoids, marked T_1 and T_2, and double their total area to get the lateral surface area of the frustum in the oven hood.

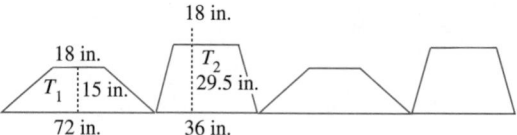

FIGURE 3.63c

The left trapezoid T_1, has an area L_1 of

$$L_1 = \frac{1}{2}(b_1 + b_2)h$$
$$= \frac{1}{2}(18 + 72)15$$
$$= 675$$

Trapezoid T_2 has an area of $L_2 = \frac{1}{2}(36 + 18)(29.5) = 796.5 \text{ in.}^2$ The total lateral surface area for the frustum and hence the amount of metal it will take to make this portion of the oven hood is $2(675 + 796.5) = 2,943 \text{ in.}^2$

EXAMPLE 3.27 (Cont.)

(b) To get the amount of metal it will take to form the outside of the entire oven hood, we add the answer to (a) to the lateral surface areas of the two prisms. The top prism has a square base that is 18 in. on each side and the prism is 36 in. high. So, it has a perimeter of $4 \times 18 = 72$ in. and its lateral area is $L_t = 36 \times 72 = 2,592$ in.^{2} The bottom prism is a rectangular prism with a perimeter of $72 + 36 + 72 + 36 = 216$ in. and a height of 6 in. Hence, its lateral surface area is $L_b = 6 \times 216 = 1,296$ in.^{2} The total amount of metal needed to form the outside of this oven hood is then $2,943 + 2,592 + 1,296 = 6,831$ in.^{2}

Spheres

The final geometric solid that we will consider is the sphere. A **sphere** consists of all the points in space that are a fixed distance from some fixed point called the center. For a sphere with radius r, the volume and surface area are given by the following formulas.

Surface Area and Volume of a Sphere

A sphere with radius r has a surface area S and volume V where

$$S = 4\pi r^2$$
$$V = \frac{4}{3}\pi r^3$$

Application

EXAMPLE 3.28

A water tower, like the one in Figure 3.64, is in the shape of a sphere on top of a cylinder. Most of the water is stored in the sphere, which has a radius of 30 ft. How much water can the tower hold?

Solution We want to determine the volume of a sphere with a radius of 30 ft. The formula for the volume of a sphere is $V = \frac{4}{3}\pi r^3$. Since $r = 30$ ft, we have

$$\begin{aligned}
V &= \frac{4}{3}\pi(30)^3 \\
&= \frac{4}{3}\pi(27,000) \\
&= 36,000\pi \\
&\approx 113,097.34
\end{aligned}$$

The water tower will hold approximately $113,097$ ft^3 of water.

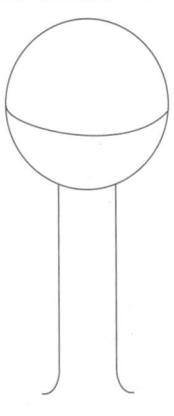

·FIGURE 3.64

A summary of all the formulas for the areas and volumes of the solid figures is given in Table 3.1.

TABLE 3.1   Areas and Volumes of Solid Figures

Solid	Lateral Area (L)	Total Surface Area (T)	Volume (V)
Rectangular prism	$2h(l+w)$	$2lw+2lh+2hw$ $=2(lw+lh+hw)$	lwh
Cube	$4s^2$	$6s^2$	s^3
Prism	ph	$ph+2B$	Bh
Cylinder	$2\pi rh$	$2\pi r(r+h)$	$\pi r^2 h$
Pyramid	$\frac{1}{2}ps$	$\frac{1}{2}ps+B$	$\frac{1}{3}Bh$
Cone	πrs	$\pi r(r+s)$	$\frac{1}{3}\pi r^2 h$
Frustum of pyramid	$\frac{s}{2}(p_1+p_2)$	$\frac{s}{2}(p_1+p_2)+B_1+B_2$	$\frac{h}{3}\left(B_1+B_2+\sqrt{B_1 B_2}\right)$
Frustum of cone	$\pi s(r_1+r_2)$	$\pi[r_1(r_1+s)+r_2(r_2+s)]$	$\frac{\pi h}{3}(r_1^2+r_2^2+r_1 r_2)$
Sphere	not applicable	$4\pi r^2$	$\frac{4}{3}\pi r^3$

Exercise Set 3.5

Find the lateral area, total surface area, and volume of each of the solids in Exercises 1–8.

1.

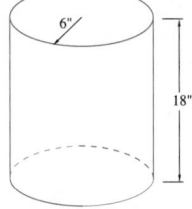

3.

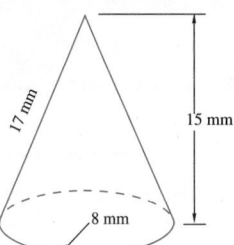

5.

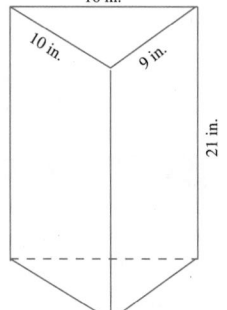

2.

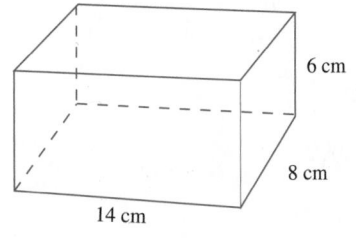

4.

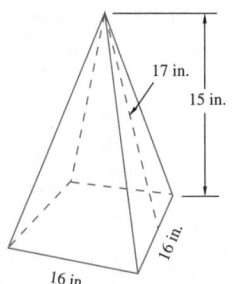

6.

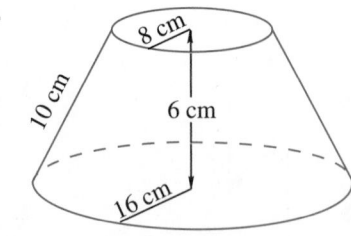

7.

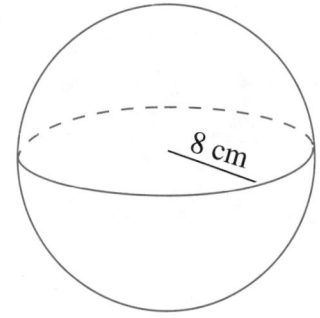

8.

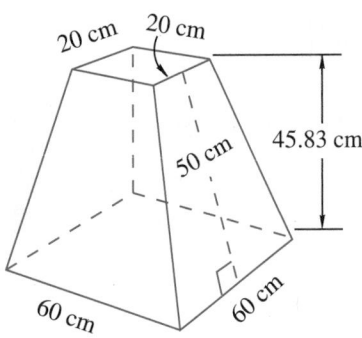

Solve Exercises 9–26.

9. *Civil engineering*   The cross-section of a road is shown in Figure 3.65. Find the number of cubic yards of concrete it will take to pave 1 mi. of this road. (There are $27\,\text{ft}^3$ in $1\,\text{yd}^3$ and 5,280 ft. in 1 mi.)

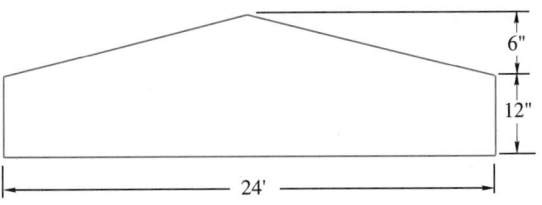

FIGURE 3.65

10. *Civil engineering*   The cross-section of an I-beam is shown in Figure 3.66. **(a)** What is the volume of this beam if it is 10 m long? **(b)** How much paint, in cm^2, will be needed for this beam? **(c)** If $1\,\text{cm}^3$ of steel has a mass of 0.008 kg what is the mass of this beam?

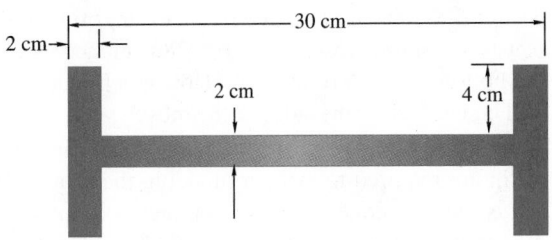

FIGURE 3.66

11. *Transportation*   A railroad container car in the shape of a rectangular parallelepiped is 30 ft. long,

10 ft. wide and 12 ft. high. **(a)** How much can the container car hold? **(b)** How many square feet of aluminum were required to make the car?

12. *Energy*   A cylindrical gas tank has a radius of 48 ft. and a height of 140 ft. What are the volume and total surface area of the tank?

13. *Product design*   A cylindrical soup can has a diameter of 66 mm and a height of 95 mm. **(a)** How much soup can the can hold (in mm^3)? **(b)** How many square millimeters of paper are needed for the label if the ends overlap 5 mm?

14. *Mechanics*   A cross-section of pipe is shown in Figure 3.67. If the pipe is 2 m long what is the volume of the material needed to make the pipe?

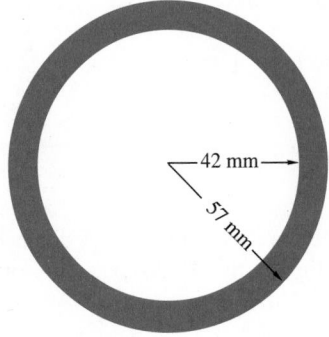

FIGURE 3.67

15. *Energy*   A spherical fuel tank has a radius of 10 m. What is its volume?

16. *Sheet metal technology*   A vent hood is made in the shape of a frustum of a square pyramid that is open at the top and bottom, as indicated in Figure 3.68. If the slant height is 730 mm, how much metal did it take to make this vent?

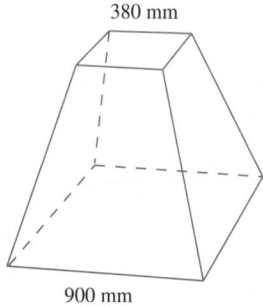

380 mm

900 mm

FIGURE 3.68

17. *Civil engineering*   The concrete highway support in Figure 3.69 is 2 ft. 6 in. thick. How many cubic yards of concrete are needed to make one support? $(27 \text{ ft}^3 = 1 \text{ yd}^3.)$

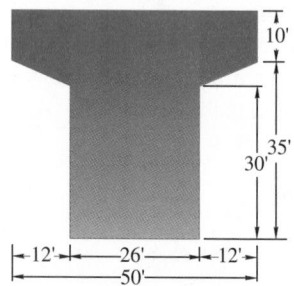

10'
35'
30'
|←12'→|←——26'——→|←12'→|
|←————50'————→|

FIGURE 3.69

18. *Construction*   How many cubic feet of dirt had to be excavated to dig a $\frac{1}{4}$-mi (1,320 ft) section of the river bed in Figure 3.70?

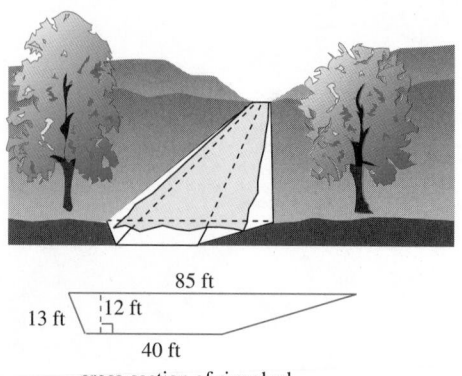

85 ft

13 ft   12 ft

40 ft

cross-section of river bed

FIGURE 3.70

19. *Sheet metal technology*   What is the volume of the cone that was made from the piece of copper in Figure 3.71?

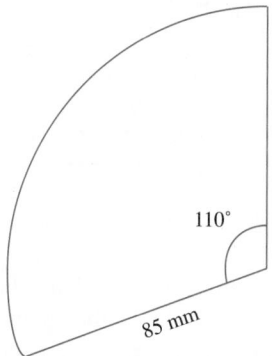

110°

85 mm

FIGURE 3.71

20. *Sheet metal technology*   A manufacturer has an order for 500 tubs in the shape of a frustum of a cone that is 490 mm across at the top, 380 mm across at the bottom, and 260 mm deep. How much material will be needed for the sides of the tubs?

21. *Sheet metal technology*   How many square inches of tin are required to make a funnel in the shape of a frustum of a cone that has a top and bottom with diameters of 3 in. and 8 in., respectively, and a slant height of 12 in.?

22. *Sheet metal technology*   The funnel in Figure 3.72 has a diameter 3 cm at *B* and 1 cm at *A*. How much metal is needed to make this funnel?

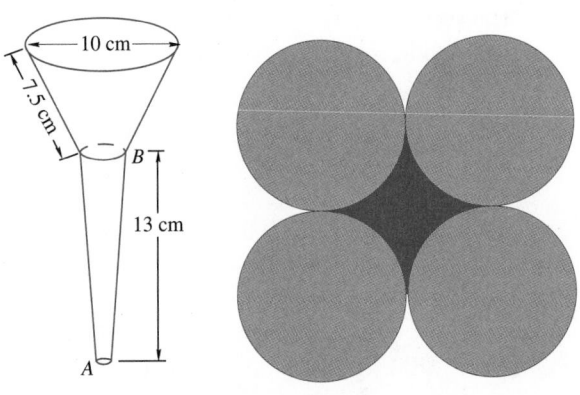

FIGURE 3.72   **FIGURE 3.73**

23. *Agricultural technology*   Figure 3.73 shows the top view of four grain elevators. When the elevators are filled, the grain will overflow to fill the space in the middle. The radius of each elevator is 3 m and the height is 10 m. What is the total volume that can be held by the four elevators and the space in the middle if the grain is leveled at the top?

24. *Automotive technology*   The piston displacement is the volume of the cylinder with a given bore (diameter) and piston stroke (height). **(a)** If the bore is 7 cm and the stroke is 8 cm, what is the displacement for this piston? **(b)** If this is a 6-cylinder engine, what is the total engine displacement?

25. *Construction*   A pile of sand falls naturally into a cone. If a pile is 4 ft. high and 12 ft. in diameter, how many cubic feet of sand are there?

26. *Construction*   A grain silo has the shape of a right circular cylinder topped by a hemisphere. The cylindrical part of the silo has a height of 40 ft. and radius of 8 ft. What is the surface area of the silo?

In Your Words

27. Describe how cylinders and prisms are alike and how they are different.

28. Describe how cylinders and cones are alike and how they are different.

≡ CHAPTER 3 REVIEW

Important Terms and Concepts

Adjacent angles
Altitude
Angle
 Acute
 Obtuse
 Right
 Straight
Arc
Arc length
Area
Central angle
Chord
Circumference
Complementary angles
Cone
Congruent angles
Congruent polygons

Corresponding angles
Cylinder
Degree
Diameter
Frustum
Hero's formula
Hexagon
Hypotenuse
Lateral angle
Line segment
Octagon
Parallel lines
Parallelogram
Pentagon
Perimeter
Perpendicular lines
Polygon

Prism
Pythagorean theorem
Quadrilateral
Radians
Radius
Rectangle
Rhombus
Secant
Similar polygons
Square
Supplementary angles
Tangent

Total surface area
Transversal
Trapezoid
Triangle
   Acute
   Equilateral
   Isosceles
   Obtuse
   Right
   Scalene
Vertex

Review Exercises

Convert each of the angle measures in Exercises 1–4 to either radians or degrees, without using a calculator. When you have finished, check your work by using a calculator.

1. $27°$ **2.** $212°$ **3.** 1.1π **4.** 0.75

Solve Exercises 5 and 6.

  5. What is the supplement of a $137°$ angle?

  6. What is the complement of a $\frac{\pi}{6}$ angle?

In Exercises 7 and 8 find the measures of angles A and B if lines ℓ_1 and ℓ_2 are parallel.

7.

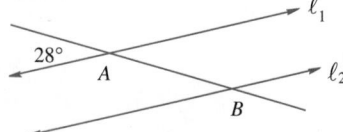

8.

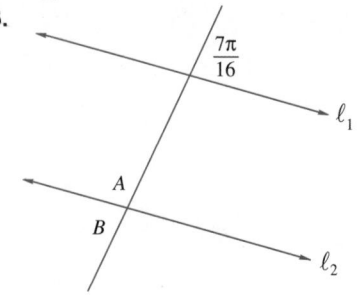

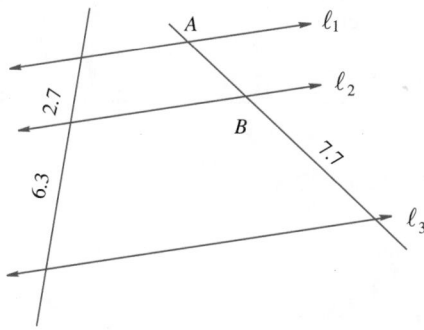

FIGURE 3.74

Find the variables indicated in Exercises 10 and 11.

10.

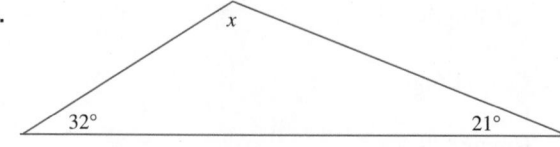

Solve Exercise 9.

  9. What is the distance from A to B in Figure 3.74, if lines ℓ_1, ℓ_2, and ℓ_3 are parallel?

11.

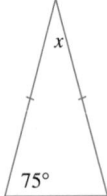

13.

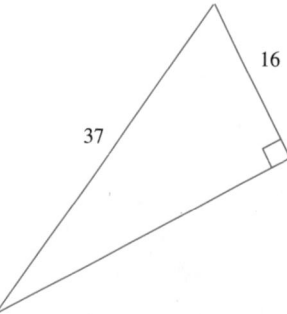

Find the length of the missing side in the right triangles in Exercises 12 and 13.

12.

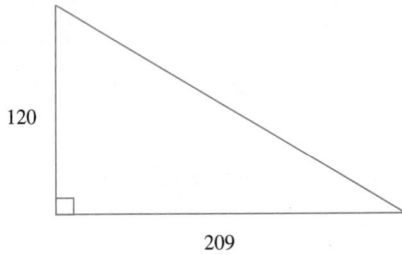

Find the area and perimeter or circumference of each of the figures in Exercises 14–21.

14.

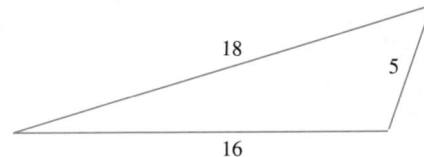

17.

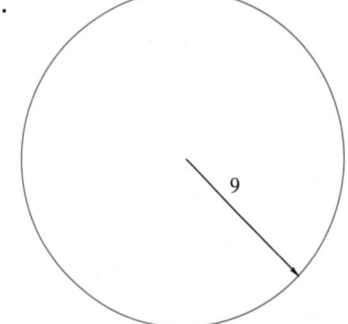

15.

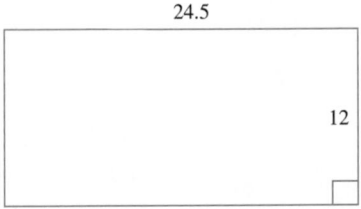

16.

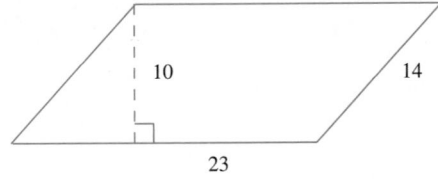

18.

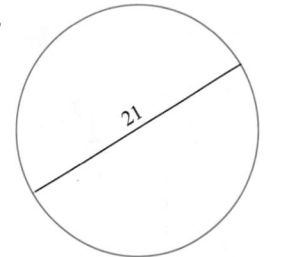

19.

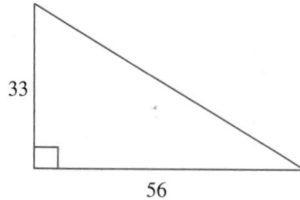

20.

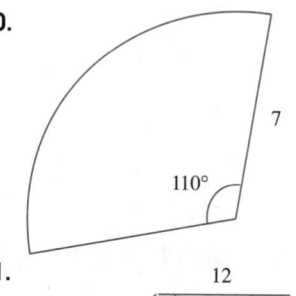

21.

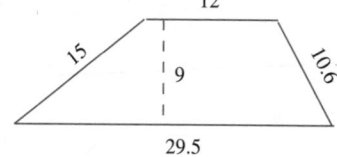

Find the lateral area, total surface area, and volume of each of the solid figures in Exercises 22–29.

22.

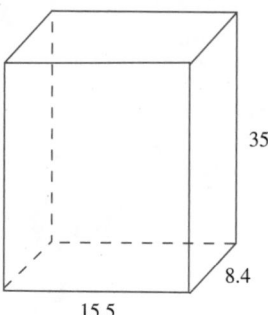

24.

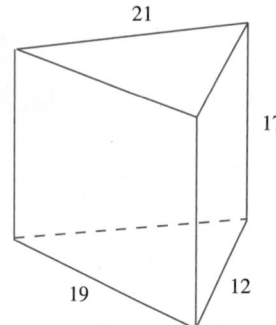

26.

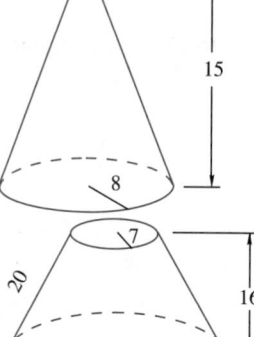

23.

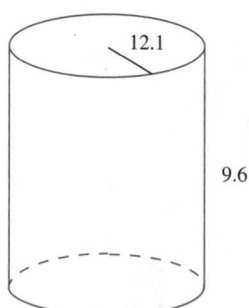

25.

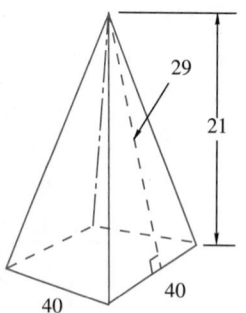

27.

28.

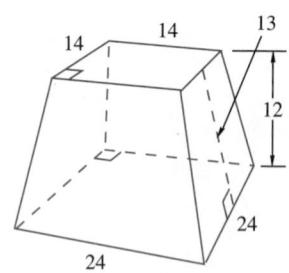

29.

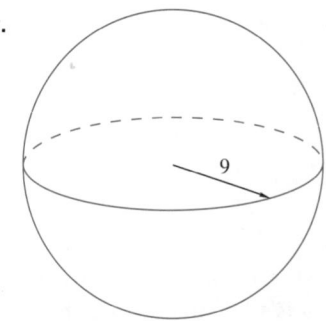

9

Solve Exercises 30–32.

30. *Construction*   An antenna 175 m high is supported by cables positioned at three positions around the antenna. At each position, four cables go to various heights of the antenna. One set of cables is attached to the antenna 50 m above the ground, one set is attached 100 m above the ground, the third set is attached 150 m above the ground, and the fourth set is attached to the top of the antenna. If each of the three positions is located 75 m from the base of the antenna, what is the total length of all the cables used to support the antenna?

31. *Civil engineering*   Find the distance across the lake in Figure 3.75.

32. *Metalworking*   A metal washer is in the shape of a circular cylinder with a (circular cylindrical) hole punched in the middle. Each washer has a diameter of 3.20 cm, the hole has a diameter of 1.05 cm, and the washer is 0.240 cm thick. The washers are made by feeding a 1 m strip of metal that is the same width and thickness as a washer into a stamping machine. For safety reasons, the last 5 cm of each strip are not fed into the machine.

(a) How many strips of metal will be needed to make (stamp) 100,000 washers?

(b) How much actual metal is required for these washers?

(c) How much scrap metal is generated in the production of these washers?

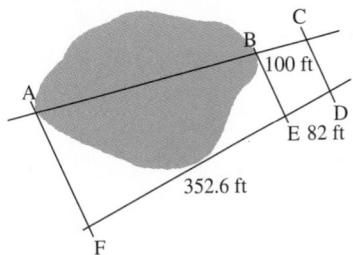

FIGURE 3.75

☰ CHAPTER 3 TEST

1. Convert 35° to radians.

2. Convert $\frac{7\pi}{15}$ to degrees.

3. What is the supplement of a 76° angle?

4. In Figure 3.76, lines ℓ_1 and ℓ_2 are parallel. What is the measure of angle A?

5. In Figure 3.77, what is the length of $\overline{AB}$, if ℓ_1, ℓ_2, and ℓ_3 are parallel?

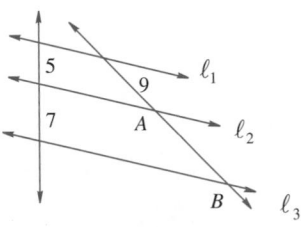

FIGURE 3.77

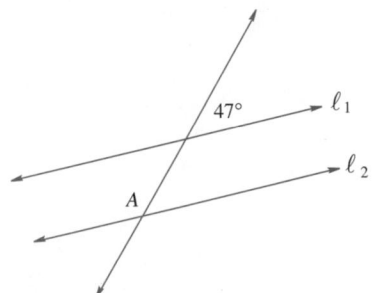

FIGURE 3.76

6. Determine the length of side a in Figure 3.78.

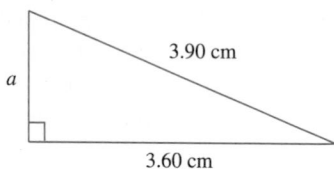

FIGURE 3.78

Find the perimeter of each of the figures in Exercises 7 and 8.

7.

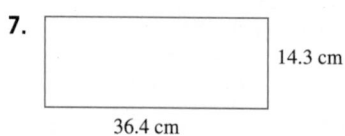

8.

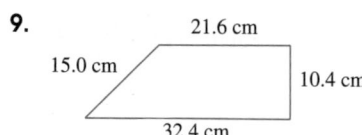

Find the area of each of the figures in Exercises 9 and 10.

9.

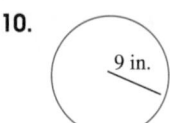

10.

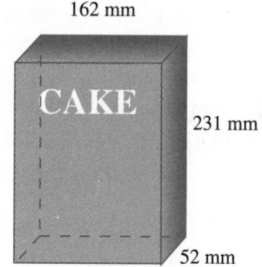

Solve Exercises 11–15.

11. A building casts a shadow of 120 m. At the same time, an antenna that is 14 m high casts a shadow of 11.6 m. How tall is the building?

12. An automobile tire has a diameter of 62 cm. How many revolutions must the tire make when the car travels 25 m in a straight line?

13. What is the volume of the box in Figure 3.79?

FIGURE 3.79

14. A spherical storage tank has a diameter of 35 ft. What is its volume?

15. The part of a cylindrical soup can that is covered by the label is 9.5 cm tall and has a diameter of 6.5 cm. What is the area of a label that covers the entire side of the can and that needs a 0.8 cm overlap to glue the ends of the label?

CHAPTER
4
Functions and Graphs

Some technicians have to take water samples at a given distance from a source of pollution. In Sections 4.4 and 4.6, we will see how to use functions and their graphs to help analyze these samples and predict the effects of pollution.

Courtesy of Janet Essman, New York State Department of Environmental Conservation

In Chapter 2, we looked at algebraic equations. We learned some ways to solve equations, and we learned how we could take some information and write an equation that showed how that information was related.

Some of the equations involved variables that stood for things related to each other. One equation used variables to show the relationship between distance, rate, and time. Torque, force, and moment arm length were other relationships in which variables were used. In each of these last two relationships, once we knew two of the variables we could determine the third.

≡ 4.1
RELATIONS AND FUNCTIONS

In this section, we will learn to use a particular kind of relationship called a function. The key to the mathematical analysis of a technical problem is your ability to recognize the relationship between the variables that describe the problem. Such a relationship often takes the form of a formula that expresses one variable as a function of another variable. We will be studying the basic ideas of a function and, later in the chapter, show how we can draw a picture or graph of a function. These basic ideas will form the foundation for much of the work we will do later in this book.

Relations

We have used the word relationship very loosely. A **relation** in mathematics is used to represent a relationship between two numbers, variables, or objects. A relation is often written as a set of ordered pairs or by a rule that describes how the items are related. In mathematics this rule may be given as an equation, an inequality, or a system of equations or inequalities. A **function** is a special kind of relation.

EXAMPLE 4.1

$a = 5b + 1$ is a relation between a and b. Pick a value for b, say $b = 3$, then $a = 5(3) + 1 = 16$.

EXAMPLE 4.2

The equation $I = 2.54C$ expresses a relation between the two variables I and C. So, if you select a value for C, say $C = 5$, you would get $I = 2.54(5) = 12.7$. You could also pick a value for I, say $I = 8.89$. Then $8.89 = 2.54C$ and $C = \frac{8.89}{2.54} = 3.5$.

EXAMPLE 4.3

The equation $v = \frac{1}{2}t^2 + 9t + 1$ describes a relation between the two variables v and t. If you let $t = 4$, then you get $v = \frac{1}{2}(4^2) + 9(4) + 1 = 45$. Let $v = 21$, then $t = 2$ or $t = -20$ will both work.

EXAMPLE 4.4

$x^2 + y^2 = 64$ is a relation between x and y. Pick a value for x between -8 and 8. If you pick $x = 0$ then $0 + y^2 = 64$ or $y^2 = 64$, which means $y = \pm 8$. If $x = 1$, then $y = \pm\sqrt{63}$. If $x = -8$, then $y = 0$. In general, $y = \pm\sqrt{64 - x^2}$. Notice that you cannot select a number smaller than -8 or greater than 8 because that would require taking the square root of a negative number.

Ordered Pairs

In addition to defining a relation by an equation, we may also define it as a set of **ordered pairs**. For example, the solutions to $I = 2.54C$ are pairs of numbers. If we agree to write the ordered pairs as (C, I), where the first number is the C-value and the second number is the I-value, then these ordered pairs also define the relation between C and I. Some of these ordered pairs are $(1, 2.54)$, $(2, 5.08)$, $(2.5, 6.35)$, $(5, 12.7)$, $(3.5, 8.89)$, and $(-3, -7.62)$.

In technical work you often work with ordered pairs of numbers. An example of a set of ordered pairs of numbers is given in Table 4.1. In this table pressures in kilopascals (kPa) and the corresponding pressure in pounds per square inch (psi) are given. For each pressure in psi there is a corresponding pressure in kPa.

Ordered pairs can be represented in table form or with each corresponding ordered pair written between parentheses. If we use the ordered pairs from Table 4.1 we would write a kPa pressure as the first number and the corresponding psi pressure as the second number. The first four ordered pairs from Table 4.1 would be written as $(140, 20)$, $(145, 21)$, $(155, 22)$, and $(160, 23)$.

TABLE 4.1 Inflation Pressure Conversion Chart (kilopascals to psi)

kPa	psi	kPa	psi
140	20	215	31
145	21	220	32
155	22	230	33
160	23	235	34
165	24	240	35
170	25	250	36
180	26	275	40
185	27	310	45
190	28	345	50
200	29	380	55
205	30	415	60

Conversion: 6.9 kPa = 1 psi

1991 Buick Skyhawk Service Manual. Flint, Michigan: Service Department, Buick Division, General Motors Corporation, 1991. Figure 3 Inflation Pressure Conversion, p. 3E-2

Whenever you have a set of ordered pairs of numbers (x, y) such that for each value of x there corresponds exactly one value of y, you have a function. You will notice that at the bottom of Table 4.1, the function is given in equation form, 6.9 kPa = 1 psi. We might rewrite this as $k = 6.9p$, where p represents a pressure in psi and k stands for a pressure in kPa. In this case, the ordered pairs are of the form (k, p).

Domain and Range

For any relation, the set of all possible first component values is called the **domain** and the set of all second component values that can result from using values in the domain is called the **range**.

EXAMPLE 4.5

The set of ordered pairs

$$\{(2,5), (3,8), (4,12), (6,-12)\}$$

expresses a relation between the set of first components, or domain, $\{2,3,4,6\}$, and the set of second components, or range, $\{5,8,12,-12\}$.

≡ **Note**

The domain and range are important ideas that will help us better understand a relation, especially when we begin graphing, and they will help us to make better use of graphing calculators and computer graphing software.

EXAMPLE 4.6

(a) In Example 4.4 where $x^2 + y^2 = 64$, both the domain and range were the real numbers between, and including, -8 and 8.

(b) In Example 4.3, the equation $v = \frac{1}{2}t^2 + 9t + 1$ expressed how the first component v was obtained from the second component t. The domain contains all the real numbers. The range would be all real numbers v greater than or equal to -39.5, or $v \geq -39.5$.

(c) If $y = \sqrt{x^2 - 81}$, then the domain would be all real numbers such that $x^2 - 81 \geq 0$. That means that the domain is the values of x larger than or equal to 9 or less than or equal to -9, or $x \leq -9$ or $x \geq 9$. The range would be the nonnegative real numbers, or $y \geq 0$.

(d) If $y = \dfrac{5}{x-3}$, the domain would be all real numbers except 3, because if $x = 3$, then $x - 3 = 0$ and $\dfrac{5}{x-3}$ is not defined. (Remember, division by zero is not defined.) The range is all real numbers except 0, because a fraction is 0 only when the numerator is 0, but this numerator is always 5 (and so, never 0).

Functions

In the relation in Example 4.4, there are two values of y for each value of x, unless $x = 8$ or $x = -8$. If a relation between x and y gives only one value of y for each value of x, then we say that y is a **function of** x. Every function is a relation. However, as Example 4.4 demonstrated, not every relation is a function.

Function

The variable y is a function of x if a relation between x and y produces exactly one value of y for each value of x.

EXAMPLE 4.7

$y = 7x - 2$ is a function described by an equation. Pick any value for x, say -4. Then $y = 7(-4) - 2 = -28 - 2 = -30$.

For each value of x there is only one value of y.

EXAMPLE 4.8

The table below is an example of a function defined by a table.

x	2	3	4	6
y	5	8	12	-12

We saw this same function earlier in Example 4.5 as a set of ordered pairs. Tables and ordered pairs are useful for describing functions that involve only a few values.

As we have seen, a function can be described by a set of ordered pairs (Example 4.5), a table of values (Example 4.8), or an equation (Example 4.7). We will see later that a function can also be described by a graph.

EXAMPLE 4.9

The equation $x^2 + y^4 = 5$ does not describe a function. Pick any value of x, say 2. Then $y = -1$ and $y = 1$ both make this equation true. Since this particular value of x, 2, produces more than one value of y that makes the equation true, we have shown that this equation is not a function of x.

 **Note**

Whenever specific values (numbers) are used to show that an equation is not true, we say that we have given a **counterexample**. The numbers we used in Example 4.9 provided a counterexample to show that $x^2 + y^4 = 5$ is *not* a function.

EXAMPLE 4.10

The set of ordered pairs $(0, 1)$, $(2, 5)$, $(2, -8)$ is not a function because there are two different second values, 5 and -8, for the first number, 2.

Independent and Dependent Variables

For the function $y = 7x - 2$, the variable x is called the **independent variable** and y is the **dependent variable**. These terms are used because the value of y depends upon the value selected for x. Since the value selected for x can be freely chosen, x is called the independent variable. These terms are arbitrary. After all, if we had written the function as $x = \frac{1}{3}(y + 5)$, then y would have been the independent variable and x the dependent variable.

Finding Domains and Ranges

Since any function can be described by a set of ordered pairs, a list, a table, an equation, or a graph, each of these can be used to determine the domain and range of the function. Here we will show how to determine the domain of a function defined by a set of ordered pairs, a list, a table, and an equation. Later in the chapter we will show how to use a graph to help determine the domain.

Finding the range is not always as easy as finding the domain. Here we will show you some techniques that you can use. You will find others as you progress through the text.

Using a Table, List, or Set of Ordered Pairs to Determine Domain and Range

A table or list can be used to determine the domain and range when it is possible to list all the ordered pairs in the function.

EXAMPLE 4.11

Determine the domain and range of the function defined by the table in Example 4.8.

Solution The table is shown below.

x	2	3	4	6
y	5	8	12	-12

Here, the first elements are in the top row of the table, the row labeled x. The numbers in this row are the domain of the function: $\{2, 3, 4, 6\}$. The second elements, in the bottom row, are the elements of the range. So the range is $\{5, 8, 12, -12\}$. Usually we like to list these in increasing order. In that case the range would be given as $\{-12, 5, 8, 12\}$.

EXAMPLE 4.12

Determine the domain and range of the function defined by the set of ordered pairs:
$$\{(-3, 30), (-2, 10), (-1, -2), (0, -6), (\tfrac{1}{2}, -5), (1, -2), (2, 10), (3, 30)\}.$$

Solution Because this function has just eight ordered pairs, it is easy to list all the first elements. This list of first elements is the domain. Thus, the domain is $\{-3, -2, -1, 0, \tfrac{1}{2}, 1, 2, 3\}$. Similarly, the range is the list of all the second elements; that is, $\{-6, -5, -2, 10, 30\}$. Notice that we did not list an element of the range the second time it was used.

To use a table for this example, you would present the data as in

x	-3	-2	-1	0	$\tfrac{1}{2}$	1	2	3
y	30	10	-2	-6	-5	-2	10	30

The domain is the set of numbers in the first row, and the range is found from the second row.

Using an Equation to Determine Domain and Range

Many times a function is defined by an equation. When this is done you will need to use algebra and the equation to help determine the domain and range.

EXAMPLE 4.13

Show how to use the equation of a function to determine the domain and range of the function $y = 2x + 7$.

Solution   It is possible to replace x with any real number and get a value of y. Thus, we conclude that the domain of this function is all real numbers.

The equation says that each value of x will be doubled and then have 7 added to it. If you want some particular value of y, say -23, you could replace y with that number and solve for x. In the case of $y = -23$, we would get $-23 = 2x + 7$ or $-30 = 2x$, and so $x = -15$. Since we can do this for any value of y, we conclude that the range is all real numbers.

Domain: All real numbers
Range: All real numbers

EXAMPLE 4.14

Show how to use the equation of a function to determine the domain and range of the function $y = x^2 - 3$.

Solution   It is possible to replace x with any real number and get a value of y. Thus, we conclude that the domain of this function is all real numbers.

The equation says that each value of x will be squared and then have -3 added to it. If you square a positive or negative number, you get a positive number. If you square 0, then you get 0. Thus, if any real number is squared it is greater than or equal to 0. If -3 is added to this number, then the sum is greater than or equal to -3. Thus, we conclude that the range is all real numbers greater than or equal to -3. We often write this as $\{y \colon y \geq -3\}$. This is read as "the set of all y's such that y is greater than or equal to -3."

Domain: All real numbers
Range: All real numbers greater than or equal to -3 or $\{y \colon y \geq -3\}$

EXAMPLE 4.15

Show how to use the equation of a function to determine the domain and range of the function $y = -\sqrt{x}$.

Solution   Because you can only take the square root of a nonnegative real number, the domain is $\{x \colon x \geq 0\}$.

The square root of a nonnegative number is nonnegative. Thus, for any x in the domain, $\sqrt{x} \geq 0$. But, we want the range of $y = -\sqrt{x}$. Since multiplying a positive number by -1 results in a negative number, we can see that the range is $\{y \colon y \leq 0\}$.

EXAMPLE 4.16

Show how to use the equation of a function to determine the domain and range of the function in Example 4.6c: $y = \sqrt{x^2 - 81}$.

Solution   To find the domain, we need to find all real numbers where $x^2 - 81 \geq 0$. These are $x \leq -9$ or $x \geq 9$. The range is all nonnegative numbers, or $y \geq 0$.

EXAMPLE 4.17

Show how to use the equation of a function to determine the domain and range of the function $y = \dfrac{5}{x-2}$.

Solution   If $x = 2$, then $\dfrac{5}{x-2}$ is not defined, so 2 is not in the domain. In fact, the domain consists of all real numbers except 2. We can write this at $\{x : x \neq 2\}$. Since the numerator is 5 it can never be zero, $y \neq 0$. Indeed, the range is all real numbers y where $y \neq 0$. We can write this at $\{y : y \neq 0\}$.

EXAMPLE 4.18

Show how to use the equation of a function to determine the domain and range of the function $y = \dfrac{x}{x+3}$.

Solution   If $x = -3$, then $\dfrac{x}{x+3}$ is not defined, so -3 is not in the domain. In fact, the domain consists of all real numbers except -3. We can write this at $\{x : x \neq -3\}$.

To help find the range, we will simplify $\dfrac{x}{x+3}$ by dividing. When we do this, we obtain $\dfrac{x}{x+3} = 1 + \dfrac{-3}{x+3}$. Since the numerator of $\dfrac{-3}{x+3}$ cannot be zero, then $\dfrac{-3}{x+3}$ cannot be zero. Hence, $\dfrac{x}{x+3} = 1 + \dfrac{-3}{x+3}$ can never be one. From this we see that the range is all real numbers y where $y \neq 1$. We can write this at $\{y : y \neq 1\}$.

 **Caution**

As indicated below, you must not simplify the equation before you determine the domain and range of a function.

EXAMPLE 4.19

The functions given by $y = \dfrac{x^2 - 4}{x - 2}$ and $y = x + 2$ agree at all values of $x \neq 2$. However, they are not equal sets of points. The reason is that they have different domains. The first function's natural domain is all real numbers $x \neq 2$, or $\{x : x \neq 2\}$. The domain of the second function is all real numbers. Thus, the functions are not equal since only the second one contains a point with x-coordinate 2, namely $(2, 4)$.

 Hint

Unless there are some other restrictions, the domain of a function defined by an equation includes all real numbers except:
- Any numbers for which a denominator is zero
- Any numbers for which the expression defined by the equation is not a real number

EXAMPLE 4.20

Create a function whose domain is $\{x: x \neq 3, x \neq -2\}$.

Solution The notation $\{x: x \neq 3, x \neq -2\}$ means that $x = 3$ cannot be in the domain and neither can $x = -2$.

The easiest way to do this is to write the function using a fraction with a denominator that is zero when $x = 3$ and $x = -2$. One solution is $y = \dfrac{1}{(x-3)(x+2)}$.

A second solution is $y = \dfrac{1}{x-3} + \dfrac{1}{x+2} = \dfrac{x^2 - x - 6}{(x-3)(x+2)}$.

Two other solutions are given by the functions $y = \dfrac{x^2}{(x-3)(x+2)}$ and

$y = \dfrac{x^2 - 4}{(x-3)(x+2)(x+2)}$.

≡ **Note**

Since no restrictions were placed on the range, the numerator can be anything that is defined for all real numbers.

EXAMPLE 4.21

Create a function whose domain is $\{x: x \geq -3\}$.

Solution Restricting the domain to $x \geq -3$ is the same as saying we want $x + 3 \geq 0$. We know that $\sqrt{x+3}$ is defined only if $x + 3 \geq 0$ and so one function with the desired domain is $y = \sqrt{x+3}$.

EXAMPLE 4.22

Create a function whose domain is $\{x: x \leq 7, x \neq -3\}$.

Solution As in Example 4.20, we can eliminate $x = -3$ by putting an $x + 3$ in the denominator.

Restricting the rest of the domain to $x \leq 7$ is the same as saying we want $x - 7 \leq 0$ (or $-(x-7) = 7 - x \geq 0$). We know that $\sqrt{7-x}$ is defined only if $7 - x \geq 0$. Let this be the numerator.

One function with domain $\{x: x \leq 7, x \neq -3\}$ is $y = \dfrac{\sqrt{7-x}}{x+3}$.

≡ **Note**

Are the conditions in Example 4.22 still satisfied if $\sqrt{7-x}$ is in the denominator, that is, if $y = \dfrac{1}{(x+3)\sqrt{7-x}}$? No? Then 7 would not be in the domain and the domain would be $\{x: x < 7, x \neq -3\}$.

Functional Notation

A function is often identified by a letter or group of letters, such as f, g, h, j, F, G, C, tan, ln, ABS, or $\cos^{-1}$. If x is the independent variable and y the dependent variable, then the number that corresponds to y is designated as $f(x)$, $g(x)$, $h(x)$, $F(x)$, or $\tan(x)$ depending on how the function is identified.

The notation $f(x)$ is read "f of x" or "f at x." We say y is a function of x and write $y = f(x)$. Thus the function in Example 4.13 could be written as $f(x) = 2x + 7$ and the function in Example 4.14 as $g(x) = x^2 - 3$.

The independent variable does not have to be x. We often use functions of other variables. Thus, $f(y) = 5y + \frac{1}{2}$ is a function of y and $h(t) = \frac{4}{3}t^2 - 2t + 1$ is a function of t.

If you want to know the value of a function at a specific point, then function notation is very useful. For example, suppose you have the function $f(x) = 2x^2 - 3$ and you want to know what value of f corresponds to $x = 4$. You would then be asking for $f(4)$ and you should substitute 4 for each x in the function. The result is

$$f(4) = 2(4^2) - 3$$
$$= 2(16) - 3$$
$$= 29$$

The value of $f(4)$ is 29. This gives the ordered pair $(4, f(4))$ or $(4, 29)$.

Caution

The notation $f(4)$ does not mean $f \cdot (4) = 4f$. The notation $f(4)$ asks for the value of the function f when $x = 4$.

EXAMPLE 4.23

If $f(x) = 3x^2 - 5x + 2$, find $f(-2)$.

Solution Substitute -2 for each value of x, with the result

$$f(-2) = 3(-2)^2 - 5(-2) + 2$$
$$= 3(4) - (-10) + 2$$
$$= 12 + 10 + 2$$
$$= 24$$

So, $f(-2) = 24$.

EXAMPLE 4.24

If $d(t) = 88t - 2.5$, find $d(4.5)$.

Solution Substitute 4.5 for the variable t, with the result

$$d(4.5) = 88(4.5) - 2.5$$
$$= 396 - 2.5$$
$$= 393.5$$

Thus, we have found that $d(4.5) = 393.5$.

EXAMPLE 4.25

If $g(x) = 3x^2 - 2x + 1$, then find $g(5+h)$.

Solution

Replace each x with $5+h$.

$$\begin{aligned} g(5+h) &= 3(5+h)^2 - 2(5+h) + 1 \\ &= 3(25 + 10h + h^2) - 2(5+h) + 1 \\ &= (75 + 30h + 3h^2) - (10 + 2h) + 1 \\ &= 75 + 30h + 3h^2 - 10 - 2h + 1 \\ &= 66 + 28h + 3h^2 \end{aligned}$$

So, $g(5+h) = 66 + 28h + 3h^2$.

Application

EXAMPLE 4.26

If 100 m of fencing is used to enclose a rectangular yard, then the resulting area of the fenced yard is given by the function

$$A(x) = x(50 - x)$$

where x is the length of the rectangle. **(a)** What is the area when the length is 10 m? **(b)** What is the area when the length is 30 m? **(c)** What restrictions must be placed on the domain x so that the problem makes physical sense?

Solutions

(a) When the length is 10 m, we have $x = 10$, and the area is

$$\begin{aligned} A(10) &= x(50 - x) \\ &= 10(50 - 10) \\ &= 10(40) = 400 \, \text{m}^2 \end{aligned}$$

(b) Here $x = 30$ m, so the area is

$$\begin{aligned} A(30) &= x(50 - x) \\ &= 30(50 - 30) \\ &= 30(20) = 600 \, \text{m}^2 \end{aligned}$$

(c) The domain should be $0 < x < 50$. If $x \leq 0$ or $x \geq 50$, we would obtain a length or width that is either 0 or a negative number. This would produce an area that is either 0 or a negative number, which would not make physical sense.

Application

EXAMPLE 4.27

The water pressure on the base of a dam is a function of the depth of the water. The weight density of water is 9 800 newtons per square meter (9 800 N/m²). The water pressure P beneath the surface of the water is $P(d) = 9800d$, where d represents the depth of the water in meters. What is the pressure at the base of a dam that is 20.5 m below the water's surface?

EXAMPLE 4.27 (Cont.)

Solution We know $P(d) = 9\,800d$ and we are given $d = 20.5$ m. Substituting, we obtain $P(d) = 9\,800(20.5) = 200\,900$. So, the pressure is $200\,900\,\text{N/m}^2$ or $200\,900$ Pa.

Very often, we do not bother to write a formula as a function. For instance, in the last example, we had the function $P(d) = 9\,800d$. Most of the time we simply write this as $P = 9\,800d$.

Explicit and Implicit Functions

When a function is expressed in the form $y = f(x)$, it is called an **explicit function**. For example, the equation $y = 5x - 7$ defines y explicitly as a function of x. If we call this function f, then we have the explicit function $f(x) = 5x - 7$.

If the relationship between x and y is not of this form, we say that x and y are related implicitly or that it is an **implicit function**. For example, if we write the above equation as $y - 5x = -7$, or $5x - y = 7$, or $5x - y - 7 = 0$, then the new equation defines y implicitly as a function of x. *Solving* for y gives the original explicit function we had before.

Some implicit functions can be used more easily if they are written as explicit functions. However, being able to write a function in terms of one (or more) explicit functions is *very* important. In most cases, in order to graph a function on a graphing calculator or with computer graphing software, you must be able to write it in explicit form.

EXAMPLE 4.28

(a) $4x - y = 2$ is an implicit equation. If we solve for y, we get an explicit function, $y = 4x - 2$.

(b) $x^2 + y^2 = 64$ is an implicit equation. It defines two explicit functions, $y = \sqrt{64 - x^2}$ and $y = -\sqrt{64 - x^2}$.

(c) $x + 3y^2 - y = 7$ is an implicit function that defines two explicit functions,
$$y = \frac{1}{6} + \sqrt{\frac{85}{36} - \frac{x}{3}} \text{ and } y = \frac{1}{6} - \sqrt{\frac{85}{36} - \frac{x}{3}}.$$

(d) $x^3 - xy + 5y^4 = 7$ is an implicit function that does not define an explicit function.

Implicit functions can be written in functional notation. For example, $4x - y = 2$ could be written as a function of the two variables x and y, by writing $f(x, y) = 4x - y - 2$ and $x^2 + y^2 = 36$ as $g(x, y) = x^2 + y^2 - 36$. There is nothing that restricts a function to two variables. Thus, we could have the implicit function $x^2 + y^2 + z = 81$ and represent it as a function of three variables, or $h(x, y, z) = x^2 + y^2 + z - 81$.

Just as we learned that if $f(x) = 7x - 2$, then $f(-1) = 7(-1) - 2 = -9$, we can substitute values for the variables in functions of more than one variable.

EXAMPLE 4.29

Let $f(x, y, z) = 2x - \frac{1}{2}y + z^2 + 4$ and find $f(1, -2, 3)$.

Solution   Here we substitute $x = 1$, $y = -2$, and $z = 3$, with the result

$$f(1, -2, 3) = 2(1) - \frac{1}{2}(-2) + 3^2 + 4$$
$$= 2 + 1 + 9 + 4$$
$$= 16$$

So, we have determined that $f(1, -2, 3) = 16$.

Exercise Set 4.1

Which of the relations in Exercises 1–8 are also functions of x?

1. $y = 10x + 2$
2. $y = 17 - 3x$
3. $y^2 = x^2 - 5$
4. $y = x^3 - 2$
5. $y = \sqrt{x} + 5$
6. $y = \sqrt[3]{x}$
7. $y^2 = 2x - 7$
8. $y = \pm\sqrt{x^2 - 5}$

Solve Exercises 9–16.

9. Is the set of ordered pairs $(0, 0)$, $(1, 1)$, $(2, 8)$, $(3, 27)$, and $(4, 64)$ a function? Why or why not?

10. Is the set of ordered pairs $(-2, 4)$, $(-1, 1)$, $(0, 0)$, $(1, 1)$, and $(2, 4)$ a function? Why or why not?

11. Is the set of ordered pairs $(4, -2)$, $(1, -1)$, $(0, 0)$, $(1, 1)$, and $(4, 2)$ a function? Why or why not?

12. Does this table describe a function? Explain.

x	-40	-30	-20	-10	0	10	20	30	40
y	-40	-22	-4	14	32	50	68	86	104

13. Give the domain and range of the function defined by these ordered pairs: $(-3, -7)$, $(-2, -5)$, $(-1, -3)$, $(0, -1)$, $(1, 1)$, and $(2, 3)$.

14. Give the domain and range of the function defined by these ordered pairs: $(-3, 3)$, $(-2, -2)$, $(-1, -5)$, $(0, -6)$, $(1, -5)$, $(2, -2)$, $(3, 3)$.

15. Give the domain and range of the function defined by the values in this table.

x	-3	-2	-1	0	1	2	3
y	-25	-6	1	2	3	10	25

16. Give the domain and range of the function defined by the values in this table.

x	-2	$-\frac{3}{2}$	-1	$-\frac{1}{2}$	0	$\frac{1}{2}$	1	$\frac{3}{2}$	2
y	20	5	-1	$-\frac{7}{3}$	-5	-2	5	25	5

In Exercises 17–28, determine the domain of each function. In Exercises 17–22, also determine the range of each function.

17. $y = x + 2$
18. $y = x - 7$
19. $y = \dfrac{x}{x - 5}$
20. $y = \dfrac{x + 1}{x + 13}$
21. $y = \sqrt{x} - 2$
22. $y = x^2$
23. $y = \dfrac{15}{(x - 2)(x + 3)}$
24. $y = \dfrac{-9}{(x + 1)(x - 5)}$
25. $y = \dfrac{x + 1}{(x + 1)^2(x - 5)}$
26. $y = \dfrac{x - 2}{(x - 2)^2(x + 3)}$
27. $y = \dfrac{x^2 - x - 2}{(x - 2)^2(x + 4)^3}$
28. $y = \dfrac{x^2 + 3x - 10}{(x - 2)^2(x + 5)^3}$

In Exercises 29–36, create a function with the given domain.

29. $\{x : x \neq 5\}$

30. $\{x : x \neq -7\}$

31. $\{x : x \neq -1, x \neq 2\}$

32. $\{x : x \neq -3, x \neq 1, x \neq 4\}$

33. $\{x : x \geq 5\}$

34. $\{x : x \leq 7\}$

35. $\{x : x \geq -1, x \neq 4\}$

36. $\{x : x > -5, x \neq 3\}$

In Exercises 37–44, given the function $f(x) = 3x - 2$, determine the following.

37. $f(0)$

38. $f(2)$

39. $f(-3)$

40. $f\left(\frac{4}{3}\right)$

41. $f(0) + f(-3)$

42. $f(b)$

43. $f(b+3)$

44. $f(x+h) - f(x)$

In Exercises 45–54, given the function $g(x) = x^2 - 5x$, determine the following.

45. $g(0)$

46. $g(-3)$

47. $g(2)$

48. $g(0.4)$

49. $g\left(\frac{2}{5}\right)$

50. $g(-x)$

51. $g(x-5)$

52. $g(3-t)$

53. $g(4m^2)$

54. $g(x+h) - g(x)$

In Exercises 55–60, given the function $F(x) = \dfrac{2-x}{x^2+2}$, determine the following.

55. $F(0)$

56. $F(2)$

57. $F(1.5)$

58. $F\left(\frac{1}{3}\right)$

59. $F(2x)$

60. $\dfrac{F(2x)}{F(x)}$

Given the function $C(f) = \frac{5}{9}(f - 32)$, determine the following in Exercises 61–64.

61. $C(32)$

62. $C(98.6)$

63. $C(-40)$

64. $C(72)$

Which of the equations in Exercises 65–68 are in explicit form and which are in implicit form? If possible, rewrite any implicit equations in explicit form.

65. $y = x^2 + 5$

66. $y + 2x = 7$

67. $y = x^2 y + x$

68. $y = x^3 - 5x + 3$

In Exercises 69–72, if $f(x,y) = 3x - 2y + 4$, then determine the indicated value.

69. $f(1, 0)$

70. $f(0, 1)$

71. $f(-1, 2)$

72. $f(2, -3)$

In Exercises 73–76, if $g(x,y) = x^2 - y^2 + 2xy$, determine the indicated value.

73. $g(-2, 0)$

74. $g(0, -2)$

75. $g(5, 4)$

76. $g(-3, 5)$

In Exercises 77–80, if $h(x,y,z) = 3x - 4y - 2z + xyz$, then determine the indicated value.

77. $h(-1, 2, 3)$

78. $h(2, -1, -3)$

79. $h(3, 2, -1)$

80. $h(a, a^2, a^3)$

Solve Exercises 81–86.

81. *Petroleum engineering*   The volume V of a cylindrical storage tank 40 ft high is a function of its radius and is described by $V = 40\pi r^2$. The total surface area S of the steel needed to construct this same tank is given by $S = 2\pi r(r + 40)$. Two tanks are designed. One has a radius of 20 ft and the other a radius of 30 ft. Compute the volume and total surface area of each tank.

82. *Dynamics*　The distance *s* that a free-falling object travels is a function of the time *t* since it started falling, and is described by the function $s(t) = 4.91t^2$, where *s* is measured in meters and *t* in seconds. **(a)** How far will an object fall in 1 s? **(b)** How far will it fall in 2 s? **(c)** How far will it fall in 5 s?

83. *Business*　The cost of renting a chain saw is $20 for the first 4 hours and $5 for each hour after 4. If $R(h)$ represents the rental charge for *h* hours, then $R(h) = 20 + 5(h-4)$. Determine the cost of renting this chain saw for 7 hours.

84. *Landscape architecture*　A rectangular yard is to be enclosed by 450 ft. of fencing.

(a) width.
(b) the width is 50 ft?
(c) the width is 150 ft?
(d) 250 ft?
(e) that the problem makes physical sense?

85. *Recreation*　The water pressure at the base of a dam is a function of the depth of the water. The weight density of water is 62.4 lb/ft³. Thus, the water pressure, *P*, beneath the surface of the water is $P(d) = 62.4d$, where *d* represents the depth of the water in feet. What is the pressure on a scuba diver that is 80 ft. below the water's surface?

86. *Medical technology*　At various times in a child's life, the average number of millions of red blood cells in each mm³ of the child's blood is given in the following table.

Age	Birth	2 days	14 days	3 mo	6 mo	1 yr	2 yr	4 yr	8–21 yr
Red cells/mm³ (in millions)	5.1	5.3	5.0	4.3	4.6	4.7	4.8	4.8	5.1

(a) What is the domain?
(b) What is the range?

 In Your Words

87. Explain how to find the domain of a function when it is described by
(a) a table
(b) a set of elements
(c) an equation

88. **(a)** Can a relation ever be a function? Explain your answer.
(b) Can a function ever be a relation? Explain your answer.

☰ 4.2
OPERATIONS ON FUNCTIONS; COMPOSITE FUNCTIONS

This section introduces inverse functions and some operations on functions. Most of the operations involve the usual operations of addition, subtraction, multiplication, and division; but a new operation, called composition, will be introduced.

Inverse Functions

We often use a pair of functions that are natural opposites of each other. By a natural opposite, we mean that one of the functions "undoes" the result of the other function, much the same as subtraction "undoes" the result of addition and division "undoes" multiplication. Two functions with this relationship are called **inverse functions**.

To determine an inverse function of $y = f(x)$, you first solve for *x*. This gives an inverse relation for *y*. If this relation is also a function, say $g(y)$, then $g(y)$ is the inverse function of $f(x)$. We use the notation $f^{-1}(x)$ to designate the inverse function of the function $f(x)$.

Figure 4.1 shows that the domain of f is the range of f^{-1} and the range of f is the domain of f^{-1}.

Do not confuse the -1 in f^{-1} with a negative exponent. The symbol f^{-1} does not represent $\dfrac{1}{f}$.

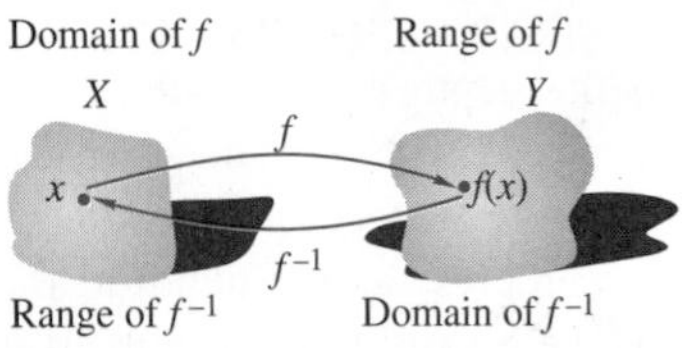

Caution

Domain of f Range of f
X Y

Range of f^{-1} Domain of f^{-1}

FIGURE 4.1

Guidelines for Finding f^{-1}

1. Solve the function $y = f(x)$ for x. Let $x = f^{-1}(y)$.
2. Exchange x and y to get $y = f^{-1}(x)$.
3. Check the domains and ranges: The domain of f and the range of f^{-1} should be the same, as should the domain of f^{-1} and the range of f.

EXAMPLE 4.30

Find the inverse function for each of the functions f: **(a)** $f(x) = 4x - 5$, **(b)** $g(x) = x^3 + 5$, and **(c)** $h(x) = x^2$.

Solutions

(a) Let $y = f(x) = 4x - 5$. Solving for x, we get $x = \dfrac{y+5}{4}$ and we write $f^{-1}(y) = \dfrac{y+5}{4}$. Since, by tradition, we usually let x represent the independent variable, we normally rewrite this as $f^{-1}(x) = \dfrac{x+5}{4}$.

(b) Let $y = g(x) = x^3 + 5$. Then $x = \sqrt[3]{y-5}$ and $g^{-1}(y) = \sqrt[3]{y-5}$ or, more traditionally, $g^{-1}(x) = \sqrt[3]{x-5}$.

(c) Let $y = h(x) = x^2$. Then $x = \pm\sqrt{y}$, but this is not the inverse function of $h(x)$. Why? Because the domain of h is all real numbers and the range of $x = \sqrt{y}$ is the nonnegative real numbers.

The range of an inverse function is the domain of the original function. Another way of saying this is that an inverse function "undoes" what the function did. In symbols, we would write $f^{-1}(f(x)) = x$. One way to see if you have the inverse of a function is to test some values in this equation.

To see what this means, let's look at the first function in Example 4.30: $f(x) = 4x - 5$. We said that $f^{-1}(x) = \dfrac{x+5}{4}$. If we try the value $x = 2$ in $f(x) = 4x - 5$, we get $f(2) = 4(2) - 5 = 3$. Now evaluate

$$f^{-1}(3) = \frac{3+5}{4}$$
$$= \frac{8}{4}$$
$$= 2$$

Good! We started and ended with 2, or $f^{-1}(f(2)) = f^{-1}(3) = 2$.

Try another number, say -10.

$$f(-10) = 4(-10) - 5$$
$$= -45$$

and

$$f^{-1}(-45) = \frac{-45 + 5}{4}$$
$$= \frac{-40}{4}$$
$$= -10$$

Again it checks! $f^{-1}(f(-10)) = f^{-1}(-45) = -10$.

Look at the third function in Example 4.30: $y = h(x) = x^2$. We solved for x and obtained $x = \sqrt{y}$. Now check some values. $h(2) = 2^2 = 4$ and $\sqrt{4} = 2$, so this checks. Try a negative number, like -10. $h(-10) = (-10)^2 = 100$ and $\sqrt{100} = 10$. It does not work. What we hoped would be the inverse returned a value of 10 instead of -10.

How can we tell if a function has an inverse? We will explore that idea later in this chapter.

Operations on Functions

Two functions, f and g, can be added, subtracted, multiplied, and divided according to the following definitions.

Basic Operations on Functions

If f and g are two functions, then their sum $(f + g)$, their difference $(f - g)$, their product $(f \cdot g)$, and their quotient $\left(f/g \text{ or } \dfrac{f}{g} \right)$ are defined by

$$(f + g)(x) = f(x) + g(x)$$
$$(f - g)(x) = f(x) - g(x)$$
$$(f \cdot g)(x) = f(x) \cdot g(x)$$
$$(f/g)(x) = \left(\frac{f}{g} \right)(x) = \frac{f(x)}{g(x)}$$

The domain of $f + g$, $f - g$, and $f \cdot g$ is the set of numbers that belong to both the domain of f and the domain of g. For the quotient function f/g, the domain also excludes all numbers where $g(x) = 0$.

EXAMPLE 4.31

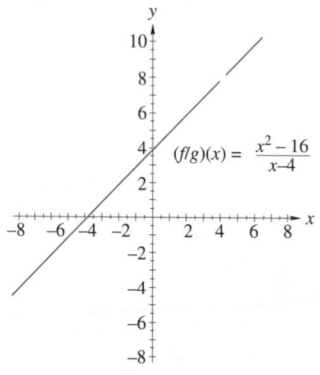

FIGURE 4.2a

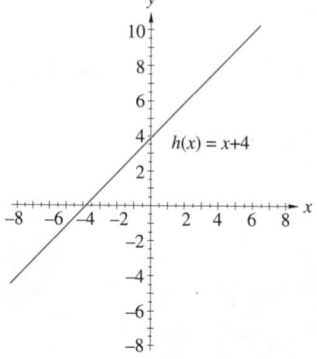

FIGURE 4.2b

Find each function, $(f+g)(x)$, $(f-g)(x)$, $(f \cdot g)(x)$, $(f/g)(x)$ and their domains, if $f(x) = x^2 - 16$ and $g(x) = x - 4$.

Solutions

$$(f+g)(x) = f(x) + g(x) = (x^2 - 16) + (x - 4)$$
$$= x^2 - 16 + x - 4$$
$$= x^2 + x - 20$$
$$(f-g)(x) = f(x) - g(x) = (x^2 - 16) - (x - 4)$$
$$= x^2 - 16 - x + 4$$
$$= x^2 - x - 12$$
$$(f \cdot g)(x) = f(x) \cdot g(x) = (x^2 - 16)(x - 4)$$
$$= x^3 - 4x^2 - 16x + 64$$
$$(f/g)(x) = \frac{f(x)}{g(x)} = \frac{x^2 - 16}{x - 4}$$
$$= \frac{(x+4)(x-4)}{x-4} = x + 4$$

The domain of all the new functions, except for f/g, is the set of real numbers. Since $g(x) = 0$ when $x = 4$, the domain of f/g is all real numbers except $x = 4$.

Notice in Example 4.31 that the function $(f/g)(x) = x + 4$, if $x \neq 4$, and the function $h(x) = x + 4$ are identical except at $x = 4$. Because $(f/g)(x)$ is not defined when $x = 4$, its graph, as shown in Figure 4.2a, looks like the line $y = x + 4$ with a "hole" when $x = 4$. The graph of $h(x) = x + 4$ is shown in Figure 4.2b.

EXAMPLE 4.32

Find each function $(f+g)(x)$, $(f-g)(x)$, $(f \cdot g)(x)$ and $(f/g)(x)$ and their domains, if $f(x) = \sqrt{x+7}$ and $g(x) = \sqrt{x-5}$.

Solutions $(f+g)(x) = \sqrt{x+7} + \sqrt{x-5}$
$$(f-g)(x) = \sqrt{x+7} - \sqrt{x-5}$$
$$(f \cdot g)(x) = \left(\sqrt{x+7}\right)\left(\sqrt{x-5}\right) = \sqrt{(x+7)(x-5)}$$
$$= \sqrt{x^2 + 2x - 35}$$
$$(f/g)(x) = \frac{\sqrt{x+7}}{\sqrt{x-5}} = \sqrt{\frac{x+7}{x-5}}$$

The domain of f is all real numbers $x \geq -7$. The domain of g is all real numbers $x \geq 5$. So, the domain of each new function is all real numbers $x \geq 5$ except for f/g; since $g(5) = 0$, the domain of f/g is all real numbers $x > 5$.

Composite Functions

There is another important way in which two functions f and g can be combined to form a new function. This new function is called the composite function. An example of a composite function is $y = (x - 7)^3$. If we let $u = g(x) = x - 7$, then $y = f(u) = u^3$. You could have also written this as $y = f(u) = f(g(x)) = (x - 7)^3$.

> **Composite Function**
>
> If f and g are two functions, the **composite function**, $f \circ g$, is defined by
>
> $$(f \circ g)(x) = f(g(x))$$
>
> where the domain of $f \circ g$ is all numbers x in the domain of g where $g(x)$ is in the domain of f.

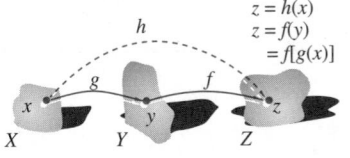

$$z = h(x)$$
$$z = f(y)$$
$$= f[g(x)]$$

FIGURE 4.3

The sketch in Figure 4.3 shows a function g, which assigns to each element x of set X, some element y of set Y. The figure also shows a function f that takes each element of set Y and assigns to it a value z of set Z. Using both f and g, an element x in X is assigned to an element z in Z. This process is a new function h, which takes an element x in X and assigns to it an element z in Z.

We do not perform composition in the same left-to-right direction as we do in reading. In the composition $f \circ g$, function g is performed first.

The definition of the domain is rather cumbersome, but it will make sense after a few examples.

EXAMPLE 4.33

Let $f(x) = x - 2$ and let $g(x)$ be defined by the following table.

x	−2	−1	0	1	2	3	4
$g(x)$	−7	3	0	−2	5	1	−7

Describe **(a)** $(f \circ g)(x)$ and **(b)** $(g \circ f)(x)$.

Solutions

(a) To find $(f \circ g)(x)$, replace each x in the expression for $f(x)$ with $g(x)$. Thus,

$$(f \circ g)(x) = f(g(x)) = g(x) - 2.$$

At this point, we can give no further description for $f \circ g$ in the form of an equation. We can, however, use a table to describe the composition.

x	−2	−1	0	1	2	3	4
$g(x)$	−7	3	0	−2	5	1	−7
$(f \circ g)(x)$	−9	1	−2	−4	3	−1	−9

This table lists all possible values for $(f \circ g)(x)$. Thus, we see that the domain of $f \circ g$ is $\{-2, -1, 0, 1, 2, 3, 4\}$, and the range of $f \circ g$ is $\{-9, -4, -2, -1, 1, 3\}$.

EXAMPLE 4.33 (Cont.)

(b) To find $(g \circ f)(x) = g(f(x)) = g(x-2)$, we use the following table.

x	-2	-1	0	1	2	3	4	5	6
$f(x) = x-2$	-4	-3	-2	-1	0	1	2	3	4
$(g \circ f)(x) = g(x-2)$			-7	3	0	-2	5	1	-7

This table lists all possible values for $(g \circ f)(x)$. Notice that two of the cells in the table are empty. Why? Consider $x = -2$. It is in the domain of f and $f(-2) = -4$. But -4 is not in the domain of g. The same argument can be made if $x = -1$. Since the domain of f is not specified, we can assume it is all real numbers. In this case, the domain of f actually is all real numbers. From the table, we see that the domain of $g \circ f$ is $\{0, 1, 2, 3, 4, 5, 6\}$, and the range of $g \circ f$ is $\{-7, -2, 0, 1, 3, 5\}$.

EXAMPLE 4.34

If $f(x) = x^2$ and $g(x) = x - 4$, find **(a)** $(f \circ g)(x)$, **(b)** $(g \circ f)(x)$, **(c)** $(f \circ g)(5)$, and **(d)** $(g \circ f)(5)$

Solutions

(a) $(f \circ g)(x) = f(g(x)) = f(x-4)$
$$= (x-4)^2$$

(b) $(g \circ f)(x) = g(f(x)) = g(x^2) = x^2 - 4$

(c) $(f \circ g)(5) = (5-4)^2 = (1)^2 = 1$

(d) $(g \circ f)(5) = 5^2 - 4 = 25 - 4 = 21$

Because the domains of both of these functions are the entire set of real numbers, the domains of both $f \circ g$ and $g \circ f$ are all real numbers.

≡ **Note**   Examples 4.34(c) and (d) show that in most cases $(f \circ g)(x) \neq (g \circ f)(x)$.

EXAMPLE 4.35

If $f(x) = x^2 - 9$ and $g(x) = \sqrt{x+4}$, then find **(a)** $(f \circ g)(x)$, **(b)** $(g \circ f)(x)$, **(c)** $(f \circ g)(2)$, **(d)** $(g \circ f)(2)$, and **(e)** the domains of $f \circ g$ and $g \circ f$.

Solutions

(a) $(f \circ g)(x) = f(g(x)) = f(\sqrt{x+4}) = (\sqrt{x+4})^2 - 9$
$$= (x+4) - 9$$
$$= x - 5$$

(b) $(g \circ f)(x) = g(f(x)) = g(x^2 - 9) = \sqrt{(x^2-9)+4}$
$$= \sqrt{x^2 - 5}$$

(c) $(f \circ g)(2) = 2 - 5 = -3$

EXAMPLE 4.35 (Cont.)

(d) $(g \circ f)(2) = \sqrt{2^2 - 5} = \sqrt{-1}$, which is not a real number. Therefore, 2 is not in the domain of $g \circ f$.

(e) The domain of f is all real numbers and the domain of g is all real numbers $x \geq -4$. The domain of $f \circ g$ will be all numbers $x \geq -4$. The domain of $g \circ f$ will be the real numbers x, where $f(x) \geq -4$ or where $x^2 - 9 \geq -4$. Simplifying this, we see that $x^2 \geq 5$, which further simplifies to $x \geq \sqrt{5}$ or $x \leq -\sqrt{5}$.

Application

EXAMPLE 4.36

An oil well in the Gulf of Mexico is leaking. There is no wind or current, so the oil is spreading on the water as a circle. At any time t, in minutes, after the leak began, the radius in meters of the circular oil slick is $r(t) = 12t$. If we know that the area of a circle of radius r is $A = \pi r^2$, **(a)** find the function $A(t)$, the area of the oil slick as a function of t, **(b)** determine the area of the slick, 15 min after the leak began, and **(c)** determine the area of the slick, 1 h after it began.

Solution

(a) We have $r(t) = 12t$ and $A(r) = \pi r^2$. Then $A(t) = (A \circ r)(t) = A(r(t)) = A(12t) = \pi(12t)^2 = 144\pi t^2 \, \text{m}^2$.

(b) When $t = 15$ min, $A(15) = 144\pi(15^2) = 32400\pi \, \text{m}^2$.

(c) When $t = 1$ h, we must use 1 h = 60 min, because $A(t)$ is for t in minutes. $A(1\,\text{h}) = A(60) = 144\pi(60^2) = 518400\pi \, \text{m}^2$.

Exercise Set 4.2

Find the inverse of each of the functions or relations in Exercises 1–8.

1. $y = x - 5$
2. $y = x + 7$
3. $y = 2x - 8$
4. $y = 5x + 12$
5. $6y = 2x - 4$
6. $9y = 3x + 5$
7. $y^2 = x - 5$
8. $y = x^3 + 7$

In Exercises 9–20, let $f(x) = x + 3$ and let $g(x)$ be defined by the following table.

x	-3	-2	-1	0	1	2	3	4
$g(x)$	-10	5	0	-1	5	1	-10	7

Determine the following.

9. $(f+g)(x)$
10. $(f-g)(x)$
11. $(g-f)(x)$
12. $(f \cdot g)(x)$
13. $(f/g)(x)$
14. $(g/f)(x)$
15. $(f \circ g)(x)$
16. $(g \circ f)(x)$
17. The domain and range of f and g
18. The domain and range of **(a)** $f+g$, **(b)** $f-g$, and **(c)** $f \cdot g$
19. The domain and range of **(a)** f/g and **(b)** g/f
20. The domain and range of **(a)** $f \circ g$ and **(b)** $g \circ f$

In Exercises 21–38, let $f(x) = x^2 - 1$ and $g(x) = 3x + 5$. Determine the following.

21. $(f + g)(x)$
22. $(f + g)(4)$
23. $(f - g)(x)$
24. $(f - g)(4)$
25. $(g - f)(x)$
26. $(g - f)(5)$

27. $(f \cdot g)(x)$
28. $(f \cdot g)(5)$
29. $(f/g)(x)$
30. $(f/g)(-2)$
31. $(g/f)(x)$
32. $(g/f)(-2)$

33. The domains of f and g
34. The domains of f/g and g/f
35. $(f \circ g)(x)$
36. $(f \circ g)(-3)$
37. $(g \circ f)(x)$
38. $(g \circ f)(-3)$

In Exercises 39–56, let $f(x) = 3x - 1$ and $g(x) = 3x^2 + x$. Determine the following.

39. $(f + g)(x)$
40. $(f + g)(4)$
41. $(f - g)(x)$
42. $(f - g)(-2)$
43. $(g - f)(x)$
44. $(g - f)(-2)$

45. $(f \cdot g)(x)$
46. $(f \cdot g)(-1)$
47. $(f/g)(x)$
48. $(f/g)(-1)$
49. $(g/f)(x)$
50. $(g/f)(-1)$

51. $(f \circ g)(x)$
52. $(f \circ g)(5)$
53. $(g \circ f)(x)$
54. $(g \circ f)(5)$
55. The domains of f and g
56. The domains of f/g and g/f

Solve Exercises 57–62.

57. *Business* The cost of manufacturing an item is the sum of the fixed costs and the variable costs. Fixed costs are the costs the business must meet just to stay in business, such as rent and insurance. Variable costs are the costs, such as material, labor, and lighting, of producing just one item. Suppose the fixed cost, in dollars, of manufacturing computer chips is $F(n) = \$7,500$/week and the variable cost is $15 per chip. **(a)** Find the cost function $C(n)$, in dollars, of manufacturing n computer chips per week. **(b)** How much will it cost to manufacture 100 chips? **(c)** How much will it cost to manufacture 1,000 chips?

58. *Business* In manufacturing a certain product, a company must pay a set-up cost of $S(n) = \dfrac{2,750,000}{n}$ dollars, where n is the number of units it produces in a production run, and with $1 \le n \le 25,000$. The fixed cost is $F(n) = \$12,500$/week and the variable cost is $8/item. **(a)** Determine the cost function, $C(n)$, of setting up and manufacturing n items in a week. **(b)** What is the cost of manufacturing 100 items? **(c)** of manufacturing 1,000 items?

59. *Business* A company's profit is equal to the revenue it generates minus the cost of producing and selling the product. The revenue function for a certain product is $R(n) = 90n$ and its cost function is $C(n) = 30n + 275$, where n is the number of units sold. **(a)** If R and C are in dollars, determine the profit function, $P(n)$, for this product. **(b)** What is the profit if you sell 50 units?

60. *Business* The demand function $n = f(p)$ relates the number of units n that can be sold to the price p charged for each unit. The revenue is equal to the price per unit multiplied by the quantity sold. A manufacturer is currently selling its calculators for p dollars each. Based on past experience, the company believes that the weekly demand for the calculator is related to its price by the equation $f(p) = 750 - 4.2p$. **(a)** What is the revenue function? **(b)** How many calculators can they expect to sell if they charge $100 for each calculator? **(c)** What is the revenue from the $100 calculators? **(d)** How many calculators can they expect to sell at $80 each? **(e)** What is the revenue for the $80 calculator?

61. *Environmental science* The population P of a certain species of large fish depends on the number n of a smaller fish on which it feeds, with

$$P(n) = 300n^2 - 50n$$

The number of smaller fish depends on the amount a of its food supply, a type of plankton, with

$$n(a) = 7a + 4$$

Find the relationship between the population P of the larger fish and the amount a of plankton available.

62. *Meteorology*   A spherical weather balloon is inflated so the radius r is changed at a constant rate of 3 cm/s.

(a) Find an algebraic expression for $V(t)$, the volume of the balloon as a function of time t in seconds.

(b) Determine the volume of the balloon after 10 s.

 **In Your Words**

63. Explain the difference between the composite functions $f \circ g$ and $g \circ f$.

64. Explain how to determine the domain of a composite function $f \circ g$.

≡ 4.3
RECTANGULAR COORDINATES

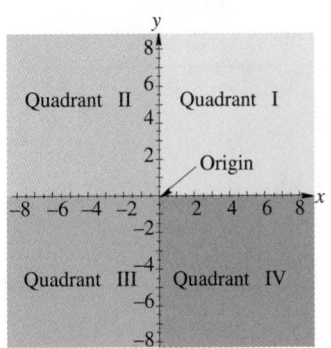

FIGURE 4.4

In Chapter 1 we graphed some points on a number line. In this section we will expand our idea of graphing to a plane.

Why is a graph so important? A graph gives us a picture of an equation. By looking at a graph, we can often get a better idea of what we can expect an equation to do. The graphs will also help us find solutions to some of our problems.

In Section 1.1 we learned that we could represent the numbers by points on a line. To get a graph in a plane, we need two number lines. The two number lines are usually drawn perpendicular to each other and intersect at the number zero as shown in Figure 4.4. The point of intersection is called the **origin**. If one of the lines is horizontal; then the other is vertical. The horizontal number line is called the **x-axis** and the vertical number line the **y-axis**. This is called the **rectangular coordinate system** or the **Cartesian coordinate system** in honor of the man who invented it, René Descartes.

On the x-axis, positive numbers are to the right of the origin and negative numbers to the left. On the y-axis, positive numbers are above the origin and negative values are below it. The two axes divide the plane into four regions called **quadrants**, with the first quadrant in the upper right section of the plane and the others numbered in a counterclockwise rotation around the origin. (See Figure 4.4.)

Suppose that P is a point in the plane. The coordinates of P can be determined by drawing a perpendicular line segment from P to the x-axis. If this perpendicular segment meets the x-axis at the value a then the x-coordinate of the point P is a. Now draw a perpendicular segment from P to the y-axis. It meets the y-axis at b, so the y-coordinate of P is b. The coordinates of P are the ordered pair (a, b).

≡ **Note**   The x-coordinate is always listed first. The order in which the coordinates are written is very important. In most cases, the point (a, b) is different than the point (b, a). (See Figure 4.5.)

EXAMPLE 4.37

The positions of $A(2, 4)$, $B(4, 2)$, $C(-4, 5)$, $D(7, 0)$, $E(-4, -3)$, $F(0, -3)$, and $G(5, -4)$ are shown in Figure 4.6. Note that $A(2, 4)$ and $B(4, 2)$ are different points as are $C(-4, 5)$ and $G(5, -4)$.

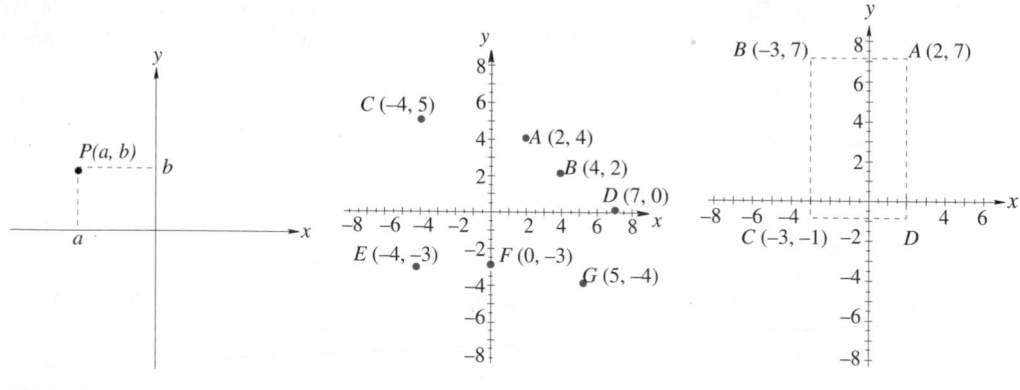

FIGURE 4.5   FIGURE 4.6   FIGURE 4.7

EXAMPLE 4.38

$A(2,7)$, $B(-3,7)$ and $C(-3,-1)$ are three vertices of a rectangle as shown in Figure 4.7. What are the coordinates of the fourth vertex, D?

Solution If we plot points A, B, and C, we can see that the missing vertex D will have the same x-coordinate as A, 2, and the same y-coordinate as C, -1. So, D has the coordinates $(2,-1)$. ▪

EXAMPLE 4.39

Some ordered pairs for the equation $y = 2x + 7$ are $(-2,3)$, $(-1,5)$, $(0,7)$, $\left(\frac{1}{2},8\right)$, $(1,9)$, and $(2,11)$. Plot these ordered pairs on a rectangular coordinate system.

Solution The ordered pairs are plotted on the graph in Figure 4.8. ▪

Application

EXAMPLE 4.40

The following table shows the results of a series of drillings used to determine the depth of the bedrock at a building site. These drillings were taken along a straight line down the middle of the lot, from the front to the back, where the building will be placed. In the table, x is the distance from the front of the parking lot and y is the corresponding depth. Both x and y are given in feet.

x	0	20	40	60	80	100	120	140	160
y	33	35	40	45	42	38	46	40	48

Plot these ordered pairs on a rectangular coordinate system. Connect the points in order to get an estimate of the profile of the bedrock.

Solution The ordered pairs are plotted on the graph in Figure 4.9. The points have been connected in order. Note that this graph is "upside down." That is, the top of the dirt is along the x-axis, while the bedrock, which is below ground, is above the x-axis. ▪

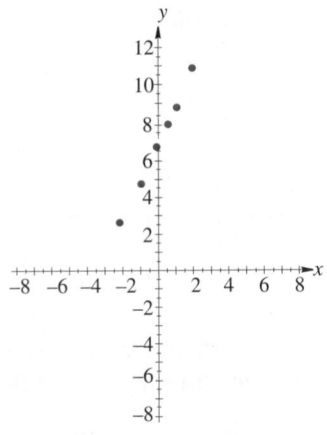

FIGURE 4.8

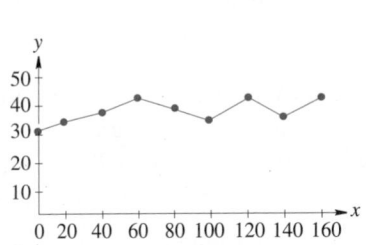

FIGURE 4.9

Exercise Set 4.3

In Exercises 1–10, plot the points on a rectangular coordinate system.

1. $(4, 5)$

2. $(1, 7)$

3. $(7, 1)$

4. $(-2, 4)$

5. $(-3, -5)$

6. $(6, -1)$

7. $(0, 0)$

8. $\left(-\frac{5}{2}, 3\right)$

9. $\left(2, -\frac{3}{2}\right)$

10. $(4.5, -1.5)$

Solve Exercises 11–28.

11. If $A(2, 5)$, $B(-1, 5)$, and $C(2, -4)$ are three vertices of a rectangle, what are the coordinates of the fourth vertex?

12. In Example 4.9, we found that $(-3, 6)$, $(-2, 1)$, $(-1, -2)$, $(0, -3)$, $(1, -2)$, $(2, 1)$, and $(3, 6)$ were ordered pairs of the equation $y = x^2 - 3$. Plot these ordered pairs on the same rectangular coordinate system.

13. Plot the ordered pairs $(-5, 3)$, $(-3, 3)$, $(0, 3)$, $(1, 3)$, $(4, 3)$, and $(6, 3)$. What do all of these points have in common?

14. Plot the ordered pairs $(-2, 6)$, $(-2, 4)$, $(-2, 1)$, $(-2, -1)$, $(-2, -3)$, and $(-2, -4)$. What do all of these points have in common?

15. *Automotive technology* Graph these ordered pairs from the conversion formula 6.9 kPa $=$

1 psi: $(138, 20)$, $(172.5, 25)$, $(207, 30)$, $(241.5, 35)$, $(276, 40)$, $(345, 50)$, $(414, 60)$. Use a ruler and draw the segment connecting $(138, 20)$ and $(414, 60)$. What seems to happen?

16. Where are all the points whose x-coordinates are 0?

17. Where are all the points whose y-coordinates are -2?

18. Where are all the points whose y-coordinates are 7?

19. Where are all the points whose x-coordinates are -5?

20. Where are all the points whose y-coordinates are 0?

21. Where are all the points (x, y) for which $x > -3$?

22. Where are all the points (x, y) for which $x > 0$ and $y < 0$?

23. Where are all the points (x, y) for which $x > 1$ and $y < -2$?

24. *Automotive technology*   If the antifreeze content of the coolant is increased, the boiling point of the coolant is also increased, as shown in the table below.

Percent antifreeze in coolant	0	10	20	30	40	50	60	70	80	90	100
Boiling temperature °F	210	212	214	218	222	228	236	246	258	271	330

(a) Plot these ordered pairs on a rectangular coordinate system.

(b) What will be the boiling point of the coolant if the system is filled with a recommended solution of 50% water and 50% antifreeze?

(c) What will be the boiling point of the coolant if the system is filled with a recommended solution of 40% water and 60% antifreeze?

25. *Automotive technology*   The temperature-pressure relationship of the refrigerant R-12 is very important to maintain proper operation and for diagnosis. The table below indicates the pressure of R-12 at various temperatures.

Temperature °F	−20	−10	0	10	20	30	40	50	60	70	80
Atmospheric pressure (psi)	2.4	4.5	9.2	14.7	21.1	28.5	37.0	46.8	57.7	70.1	84.1

(a) Plot these ordered pairs on a rectangular coordinate system. Connect the ordered pairs in order to help you answer (b) and (c).

(b) One day the early morning temperature was 45°F. What was the pressure, in psi, when the temperature was 45°F?

(c) If a technician connected a pressure gauge to an air-conditioning system filled with R-12 on a 90°F summer day, what pressure, in psi, would the gauge indicate?

26. *Automotive technology*   Recent automotive air conditioning systems use R-134a rather than R-12. This action is based on evidence which indicates that R-12 is causing depletion of the earth's protective ozone layer. As with R-12, the temperature-pressure relationship of refrigerant R-134a is very important to maintain proper operation and diagnosis. The table below indicates the pressure of R-134a at various temperatures.

Temperature (°C)	−20	−10	0	10	20	30
R-134a pressure (kPa)	31.4	99.4	191.4	312.9	469.5	666.7

(a) Plot these ordered pairs on a rectangular coordinate system. Connect the ordered pairs in order to help you answer parts (b), (c), and (d).

(b) One day the early morning temperature was 5°C. What was the pressure, in kPa, when the temperature was 5°C?

(c) If a technician connected a pressure gauge to an air conditioning system filled with R-134a on a 35°C summer day, what pressure, in kPa, would the gauge indicate?

(d) Suppose a technician connected a pressure gauge to an air conditioning system filled with R-134a and got a reading of 500 kPa. What is the temperature in °C?

27. *Machine technology*   The following table gives actual recording times and counter values as obtained on one particular cassette tape deck.

Time (minutes)	1	2	3	4	5	10	15	20	25	30
Counter reading	30	60	88	115	141	262	369	466	556	640

(a) Plot these ordered pairs on a rectangular coordinate system. Connect the ordered pairs in order to help you answer parts (b) and (c).

(b) What would you expect the counter to read after 12.5 min?

(c) How much time has passed if the counter has a value of 500?

28. *Forestry*   The following table gives the girth of a pine tree measured in feet at shoulder height and the amount of lumber in board feet that was finally obtained.

Girth (ft)	15	18	20	22	25	28	33	35	40	43
Lumber (bd ft)	9	27	42	60	90	126	198	231	324	387

(a) Plot these ordered pairs on a rectangular coordinate system. Connect the ordered pairs in order to help you answer parts (b) and (c).

(b) How many board feet of lumber would you expect to obtain from a tree with a girth of 30 ft?

(c) What was the girth of a tree that produced 275 board feet of lumber?

In Your Words

29. Look again at Example 4.40. Describe how you would have changed the graph, or the way the data was recorded, so that the graph would appear right-side up instead of upside down.

30. Gather some data from your field of interest. Write a problem that will require that data to be graphed.

Then write some questions that are answered by reading the graph. Put your answers on the back of the paper. Give your problem to a classmate and ask him or her to solve the problem. Rewrite your exercise or solution to clarify places where your classmate had difficulty.

≡ 4.4
GRAPHS

In the last section, we learned how to graph a point on a plane. Earlier, we said that a function could be represented by a graph. In this section, we will see how we can use graphs to represent functions and we will use the computer to help us draw graphs.

> **Graph of an Equation**
>
> The graph of an equation in two variables x and y is formed by all the points $P(x, y)$ whose coordinates (x, y) satisfy the given equation.

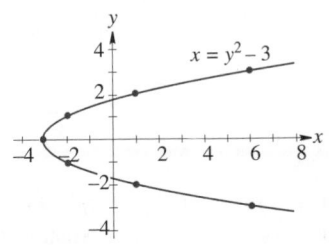

FIGURE 4.10

To graph an equation such as $x = y^2 - 3$, you can set up a table of values, plot the points in the table, and then connect the points smoothly. To fill the table, we will select values for y and use the equation to solve for the corresponding values for x. Because the ordered pairs have x listed first, we will list x as the top value in the table.

x	6	1	−2	−3	−2	1	6
y	−3	−2	−1	0	1	2	3

These points are plotted in Figure 4.10 and a smooth curve has been used to connect the points. Remember, this equation has an infinite number of possible ordered pairs. We have selected only seven of them. We must plot enough points to be confident that the points give an outline of the curve that is complete enough for us to tell what the actual curve looks like. Later, you will learn some ways to use mathematics to help determine when you have a sufficient number of points to sketch a curve.

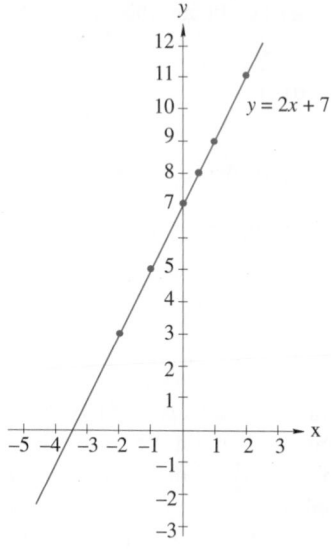

FIGURE 4.11

In Figure 4.8 we plotted six ordered pairs for the equation $y = 2x + 7$: $(-2, 3)$, $(-1, 5)$, $(0, 7)$, $\left(\frac{1}{2}, 8\right)$, $(1, 9)$, and $(2, 11)$. If we connect these points, we see that we get the straight line in Figure 4.11. In fact, any equation of the form $Ax + By + C = 0$ where A, B, and C are constants and both A and B are not 0 is a **linear equation** whose graph is a straight line. The equation $y = 2x + 7$ is also a function and can be written as $f(x) = 2x + 7$.

Intercepts

The graph of the equation $y = 2x + 7$ crosses both the x-axis and the y-axis. These points are called the **intercepts**. The graph crosses the y-axis at $(0, 7)$, so we say that the y-intercept is 7. It crosses the x-axis at $\left(-\frac{7}{2}, 0\right)$, so the x-intercept is $-\frac{7}{2}$.

 **Hint**

Any time you have a linear equation, you will only need to plot two points in order to sketch the graph of that equation. The intercepts are often the easiest two points to determine and use when graphing a linear equation.

EXAMPLE 4.41

Plot the graph of $2x + 3y = 12$.

Solution This is a linear equation, so we will use two points to determine the line. If we find the intercepts, we get the points $(6, 0)$ and $(0, 4)$. These points and the line they determine are shown in Figure 4.12.

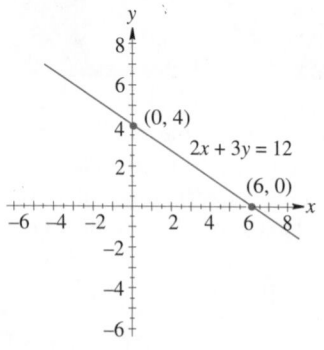

FIGURE 4.12

Slope

One important property of the graph of a straight line is its slope. The slope refers to the steepness or inclination of the line. The idea of the slope is so important that we will keep coming back to it. In fact, it is a major topic in calculus.

Slope

The **slope**, m, of a straight line is a measure of its steepness with respect to the x-axis. If (x_1, y_1) and (x_2, y_2) are two different points on a line, then the slope of the line is defined as

$$m = \frac{y_2 - y_1}{x_2 - x_1}$$

provided $x_2 - x_1 \neq 0$.

Hint

Some people like to remember the slope as the "rise over the run" or $\dfrac{\text{rise}}{\text{run}}$. Here "rise" means the vertical change and "run" the horizontal change.

EXAMPLE 4.42

What is the slope of the line $2x + 3y = 12$ in Example 4.41?

Solution We know that $(6, 0)$ and $(0, 4)$ are two points on this line. If we let $x_1 = 6$, $y_1 = 0$, $x_2 = 0$, and $y_2 = 4$, then

$$m = \frac{y_2 - y_1}{x_2 - x_1}$$
$$= \frac{4 - 0}{0 - 6} = \frac{4}{-6} = -\frac{2}{3}$$

The slope of this line is $m = -\frac{2}{3}$.

EXAMPLE 4.43

Graph $4y - 6x = 12$ and find the intercepts and the slope.

Solution This is a straight line. The intercepts are $(0, 3)$ and $(-2, 0)$. Plotting these two points and the line through them gives the graph in Figure 4.13. Letting $x_1 = 0$, $y_1 = 3$, $x_2 = -2$, and $y_2 = 0$, we determine that the slope is

$$m = \frac{0 - 3}{-2 - 0} = \frac{-3}{-2} = \frac{3}{2}$$

EXAMPLE 4.44

Graph $8x = 2 + y$, and find the intercepts and the slope.

Solution Again, this is a straight line. If we let $x = 0$, then we get $0 = 2 + y$ or $y = -2$, and so one intercept is $(0, -2)$. If we let $y = 0$, then $8x = 2$ and $x = \frac{1}{4}$. Thus, $\left(\frac{1}{4}, 0\right)$ is the other intercept. The slope is

$$m = \frac{-2 - 0}{0 - \frac{1}{4}} = \frac{-2}{-\frac{1}{4}} = 8$$

Plotting the two intercepts and drawing the line through them produces the line in Figure 4.14.

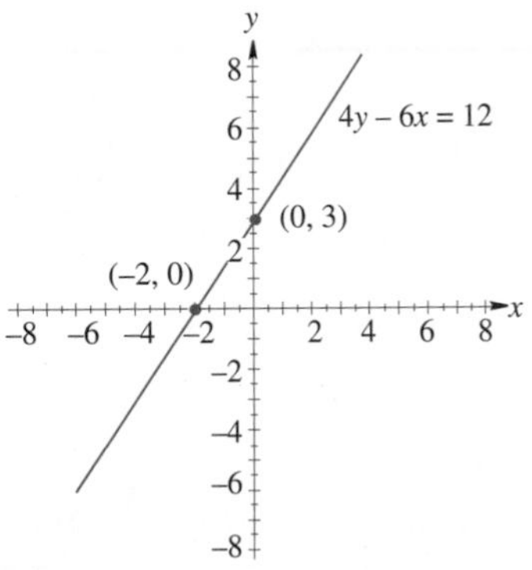

FIGURE 4.13

Look again at the last three examples. In Example 4.42, the slope of the line was $-\frac{2}{3}$. Notice that this line falls as the x-values get larger. In Example 4.43, the slope of the line was $\frac{3}{2}$. In Example 4.44, the slope of the line was 8. Both of these lines rose as the values of x got larger. Notice that the line in Example 4.44 was steeper than the line in Example 4.43.

In general, as the values of x increase, a straight line rises if it has a positive slope and falls if it has a negative slope. This is demonstrated in Figures 4.15 and 4.16.

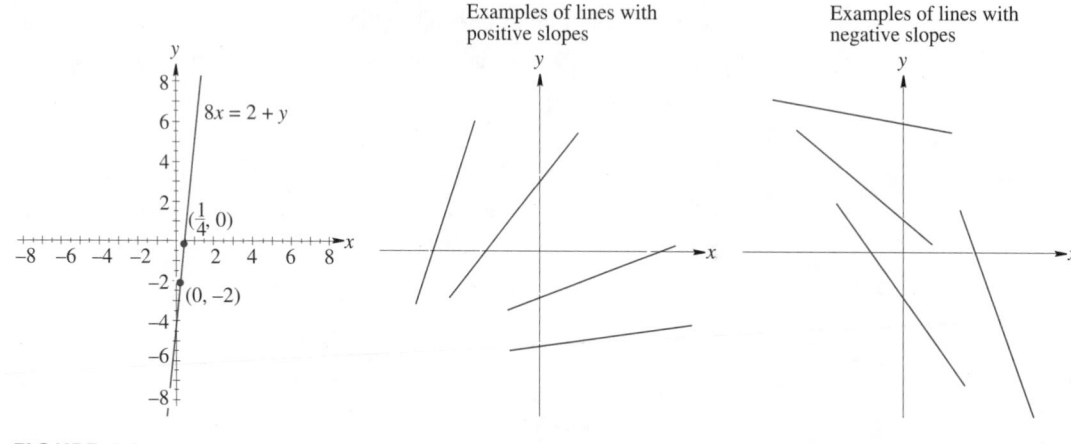

FIGURE 4.14   FIGURE 4.15   FIGURE 4.16

Graphs That Are Not Lines

Not every graph is a straight line. The next four examples show some graphs of functions that are not lines. This is followed by a test that tells how to use a graph to determine if it is a graph of a function.

EXAMPLE 4.45

Graph the function from Example 4.14, $y = x^2 - 3$.

Solution Some ordered pairs that satisfy $y = x^2 - 3$ are shown in the following table.

x	-3	-2	-1	0	1	2	3
y	6	1	-2	-3	-2	1	6

If we plot these seven points and connect them with a smooth curve, we get the graph in Figure 4.17. Notice that this graph has a y-intercept at $y = -3$ and that it has two x-intercepts, when $x = \sqrt{3}$ and when $x = -\sqrt{3}$.

EXAMPLE 4.46

Graph the function $f(x) = x^2 - 2x$.

Solution Here is a table of some of the values for this function.

x	-3	-2	-1	0	1	2	3	4	5
$f(x)$	15	8	3	0	-1	0	3	8	15

These points and the curve connecting them are shown in Figure 4.18.

EXAMPLE 4.47

Graph the function $g(x) = \dfrac{x-1}{x}$.

Solution Note that this function is not defined when $x = 0$. A partial table of values for this function follows.

x	-5	-4	-3	-2	-1	-0.4	-0.1
$g(x)$	1.2	1.25	$1.3\overline{3}$	1.5	2	3.5	11

x	0	0.1	0.4	1	2	3	4	5
$g(x)$	DNE*	-9	-1.5	0	0.5	$1.6\overline{6}$	0.75	0.80

*DNE means Does Not Exist.

The graph is shown in Figure 4.19.

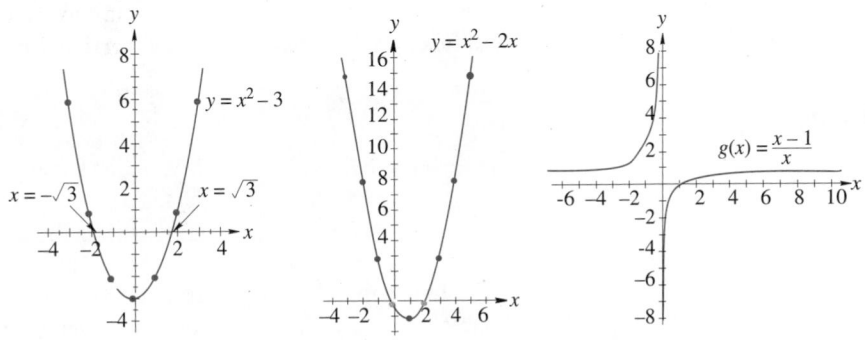

FIGURE 4.17   **FIGURE 4.18**   **FIGURE 4.19**

Another interesting graph is shown by Example 4.48.

EXAMPLE 4.48

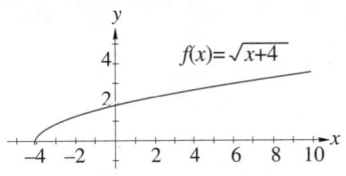

FIGURE 4.20

Graph the function $f(x) = \sqrt{x+4}$.

Solution You can only take the square root of a nonnegative number. Thus, the function is only defined for $x \geq -4$, and the domain of the function is $\{x : x \geq -4\}$. A partial table of values for this function follows. Its graph is shown in Figure 4.20.

x	-4	-3	-2	-1	0	1	2	3	4	5
y	0	1	$\sqrt{2} \approx$ 1.414	$\sqrt{3} \approx$ 1.732	2	$\sqrt{5} \approx$ 2.236	$\sqrt{6} \approx$ 2.449	$\sqrt{7} \approx$ 2.646	$\sqrt{8} \approx$ 2.828	3

Application

EXAMPLE 4.49

The total cost to manufacture x items of a certain product is given by $C(x) = \frac{8}{x} + 0.5x$, where $x > 0$. Set up a partial table of values and sketch the graph of this function.

Solution This function is only defined for $x > 0$. A partial table of values is shown. Its graph is shown in Figure 4.21.

x	1	2	3	4	5	6	7	8	9	10
y	8.5	5	$4\frac{1}{6}$	4	4.1	$4\frac{1}{3}$	$4\frac{9}{14}$	5	$5\frac{7}{18}$	5.8

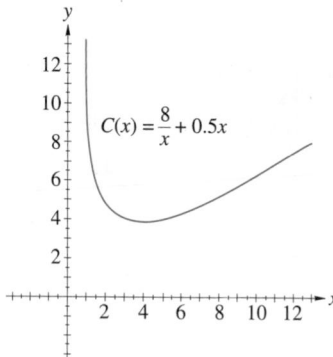

FIGURE 4.21

Vertical Line Test

A graph can provide us with an easy test to see if the equation that has been graphed is a function. This test is called the **vertical line test**.

> **Vertical Line Test**
>
> The vertical line test states that a graph is the graph of a function if no vertical line intersects the graph more than once.

To use the vertical line test, you look at a graph and determine if there are any places where a vertical line would intersect the graph more than once. If you cannot find any such places, then this is the graph of a function. If you can, then this is not the graph of a function. Figures 4.22a and 4.22b give some examples. The graph in Figure 4.22a is not the graph of a function, because the vertical line intersects the graph three times. On the other hand, the graph in Figure 4.22b is the graph of a function because no vertical line can intersect the graph more than once.

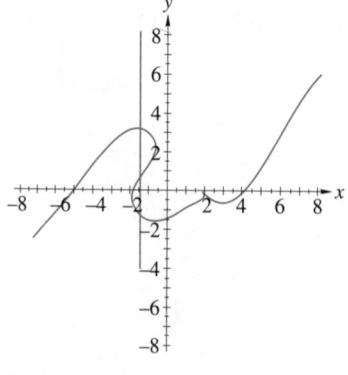

FIGURE 4.22a

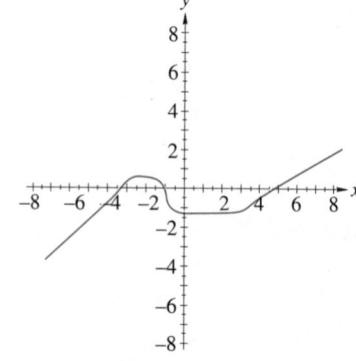

FIGURE 4.22b

Application

EXAMPLE 4.50

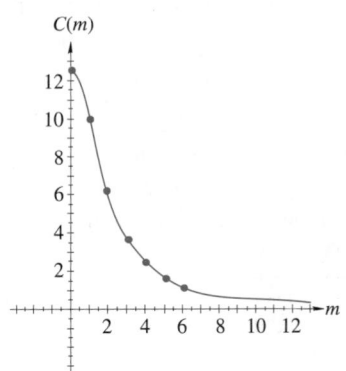

FIGURE 4.23

The concentration in parts per million (ppm) of a certain pollutant m mi from a certain factory is given by

$$C(m) = \frac{50}{m^2 + 4}, \ m \geq 0$$

Set up a partial table of values and sketch the graph of this function.

Solution   This function is defined for values of $m \geq 0$. A partial table of values follows. The graph of the function is shown in Figure 1.23. Notice that this graph passes the vertical line test.

m	0	1	2	3	4	5	6
$C(m)$	12.5	10	6.25	3.85	2.5	1.72	1.25

Using a Graph to Determine the Domain and Range of a Function

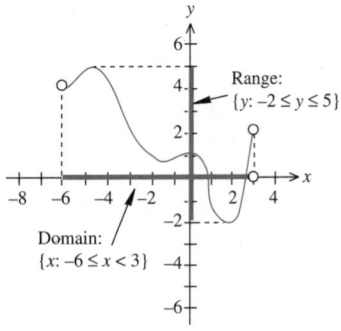

FIGURE 4.24

We just used the vertical line test to determine if a graph is the graph of a function. We can use a variation of that test to help us determine the domain of the function. To determine the domain of a function, think of vertical lines drawn from each point on the x-axis. If a vertical line intersects the graph of a function, then the x-coordinate determined by that line is in the domain. If the vertical line does not intersect the graph, then the x-coordinate is not in the domain.

We can also use a graph to help us determine the range of a function. To determine the range of a function, think of horizontal lines drawn from each point on the y-axis. If a horizontal line intersects the graph, then the y-coordinate determined by that line is in the range. If it does not intersect the graph, then the y-coordinate is not in the range.

EXAMPLE 4.51

Determine the domain and range of the function shown by the graph in Figure 1.24.

Solution   Because the graph extends from $x = -6$ to, but not including, $x = 3$, the domain is $\{x : \ -6 \leq x < 3\}$. Since the graph extends from $y = 5$ to $y = -2$, the range is $\{y : \ -2 \leq y \leq 5\}$. Notice that the range was not determined by the y-values when $x = -6$ and when $x = 3$.

EXAMPLE 4.52

Show how to use the graph of a function to determine the domain and range of the functions in Example 1.19, $y = x + 2$ and $y = \frac{x^2 - 4}{x - 2}$.

Solution   The graph of $y = x + 2$ is shown in Figure 1.25a and the graph of $y = \frac{x^2 - 4}{x - 2}$ is shown in Figure 1.25b.

EXAMPLE 4.52 (Cont.)

The graph of the equation $y = x + 2$ is the line shown in Figure 4.25a. While we cannot see the entire graph, we know that this is a line and are led to believe that any vertical line will intersect this graph exactly once. Similarly, it appears as if any horizontal line will intersect the graph. Hence, the domain and range are both the set of real numbers.

Look at the graph of $y = \dfrac{x^2 - 4}{x - 2}$ in Figure 4.25b. There appears to be a "hole" in the graph at $(2, 4)$. Because the denominator is 0 when $x = 2$, the domain of this function is $\{x : x \neq 2\}$. If you draw a vertical line through the "hole," it will hit the x-axis at $x = 2$. This graphically shows that $x = 2$ is not in the domain of this function. If you draw a horizontal line through the "hole" at $(2, 4)$, it hits the y-axis at $y = 4$. This leads us to conclude that the range is $\{y : y \neq 4\}$.

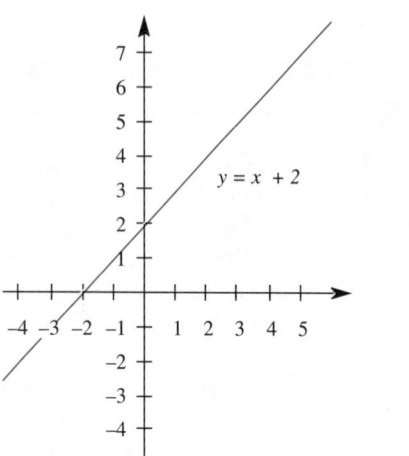

FIGURE 4.25a

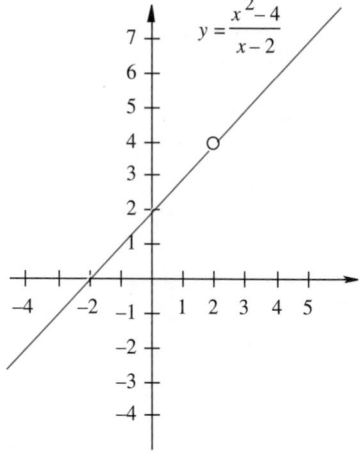

FIGURE 4.25b

Caution

You must be careful when you use the graph of a function to determine the domain and the range. Most graphs that you will draw or see show only part of the graph. You must use your algebra skills, your experience, *and* the graph to fully determine the range.

The next section will further explore how you can determine the domain and range with the aid of a graphing calculator.

Exercise Set 4.4

For each of Exercises 1–10, (a) draw the graphs of the linear equations, (b) determine the intercepts, and (c) calculate the slope.

1. $y = x$
2. $y = 2x$
3. $y = -3x$

4. $y = \frac{1}{2}x - 3$
5. $y = -4x + 2$

6. $x + y = 2$
7. $x - y = 2$
8. $2x + y = 1$

9. $3x - 6y = 9$
10. $3x - 6y = -6$

In Exercises 11–20, (a) set up a table of values, (b) graph the given functions, and (c) determine the domain and range of each function.

11. $y = x^2$
12. $y = x^2 + 3$
13. $y = x^2 - 2$

14. $y = x^2 + 2x + 1$
15. $y = x^2 - 6x + 9$
16. $y = -x^2 + 2$

17. $y = \dfrac{1}{x}$
18. $y = \dfrac{1}{x+3}$

19. $y = x^3$
20. $y = \sqrt[3]{x}$

In Exercises 21–26, set up a table of values and graph each of the equations.

21. $y^2 + x^2 = 25$
22. $y = \sqrt{25 - x^2}$

23. $y = |x + 2|$
24. $\dfrac{x^2}{4} + \dfrac{y^2}{9} = 1$

25. $\dfrac{x^2}{4} - \dfrac{y^2}{9} = 1$
26. $y = x^3 - x^2$

Which of the graphs in Exercises 27–30 are functions?

27.

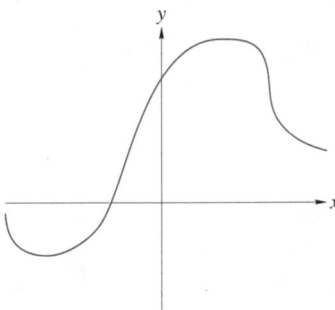

29.

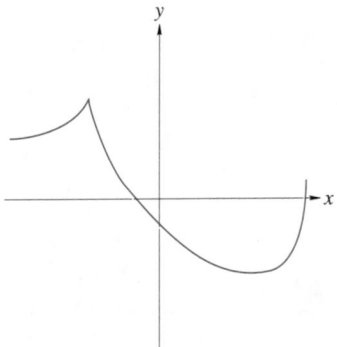

28.

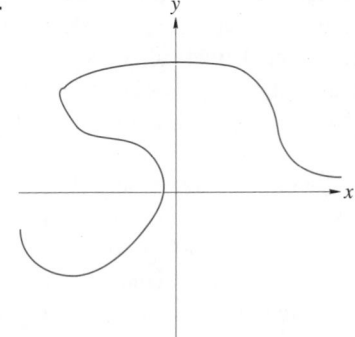

30.

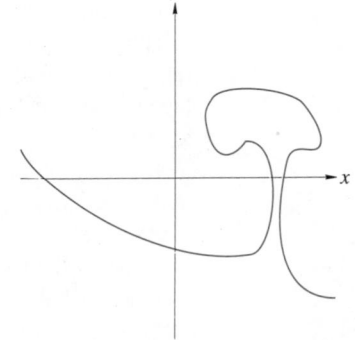

Solve Exercises 31–34.

31. *Ecology*   An ecologist is investigating the effects of air pollution from an industrial city in the plants surrounding the city. She estimates that the percentage of diseased plants is given by

$$p(k) = \frac{25}{2k+1}$$

(a) Set up a table of values and graph the equation for $0 \le k \le 9$.

(b) Does your graph pass the vertical line test?

32. *Ecology*   A second ecologist thinks that the function in the previous problem does not give the correct percentages. She thinks that the percentage of diseased plants k km from the city is given by

$$p(k) = \frac{25}{\sqrt{k+1}}$$

Make a table of values and graph the function for $0 \le k \le 9$

33. *Business*   Based on past experience, a company decides that the weekly demand (in thousands) for a new microwaveable food product is $d(p) = -p^2 + 2.5p + 10$, where p is the price (in dollars) of the product.

(a) What is the domain of this function?

(b) Make a table of values for each $0.25 change in price of $0 < p \le \$2.00$.

(c) Sketch a graph of this function.

(d) What appears to be the price that will result in the most sales?

34. *Automotive technology*   In a chrome-electroplating process, the mass m in grams of the chrome plating increases according to the formula $m = 1 + 2^{t/2}$, where t is the time in minutes.

(a) Set up a table of values and graph the equation for $1 \le t \le 15$.

(b) How long does it take to form 100 g of plating?

In Your Words

35. Describe how to use the graph of a function to help determine the range of the function.

36. Explain what the slope of a line tells you.

≡ 4.5
GRAPHING CALCULATORS AND COMPUTER-AIDED GRAPHING

The introduction of microcomputers has allowed people to use inexpensive computers in their work for writing and for performing calculations quickly and accurately; but people discovered that words and numbers were not enough, and so computer graphics were introduced.

Some of these graphing capabilities are now available on calculators. To some extent, the use of a graphing calculator or a computer removes some of the drudgery of plotting equations by hand. We will show you how you can use computer and calculator graphing capabilities to demonstrate some mathematical concepts and to relieve you of some work. However, we will also show you how to take advantage of these technological devices and show how to interpret the information they give you.

The graphics capabilities of computers and calculators vary widely. Examples in this section, and other graphics examples in this text, were run on either a Casio fx-7700G or a Texas Instruments TI-82 graphing calculator or on a Macintosh computer.

Rather than teach you how to write a graphing program for a computer, we will assume you have access to a program that can be used to graph functions. Some, such as *Master Grapher,* are specialized graphing programs; some, such as *Lotus 1–2–3,* are computer "spreadsheets" that have graphing capabilities. Other more specialized programs include *Mathematica, Derive, Microcalc,* and *Maple,* all of which might be classified as computer algebra systems.

The purpose of this book is not to explain how to use the graphing capabilities of these programs, but to help you use these computer programs to graph curves and to interpret their graphs.

Using a Graphing Calculator

We begin with a discussion on using graphing calculators. The examples and the instructions in this book are meant to supplement the user's guide for your calculator, not to replace it. Additional details on using a graphing calculator are in Appendix A and the *Student's Handbook.*

The first and third examples (Examples 4.53 and 4.55) uses the Casio fx-7700G. The second example (Example 4.54) uses a Texas Instruments TI-82 graphing calculator. Later examples will be more generic and, unless specified, will be generated using a TI-82 graphing calculator.

The first equation will be of a straight line, $y = 2x + 7$. Before we begin, we must determine the domain of this function. The range of this function is determined by the allowable values of y that can appear on the calculator's screen.

EXAMPLE 4.53

```
Range
Xmin:-4.7
  max:4.7
  scl:1.
Ymin:-3.1
  max:3.1
  scl:1.
INIT
```

FIGURE 4.26a

Use a Casio fx-7700G graphing calculator to sketch the graph of $f(x) = 2x + 7$.

Solution First, enter $\boxed{\text{SHIFT}}$ $\boxed{\text{F5}}$ $\boxed{\text{EXE}}$ to clear the graphics screen of any previous graphs that might be on the screen or in the calculator's memory.

We begin by setting the domain and range we want displayed on the screen. Press the $\boxed{\text{Range}}$ key. The screen should be similar to Figure 4.26a. Now, press the $\boxed{\text{SHIFT}}$ $\boxed{\text{F1}}$ $\boxed{\text{Range}}$ $\boxed{\text{Range}}$ keys. This clears the memory and sets the domain and range to the default values in the machine.

The part of the coordinate system that will be shown on the screen is indicated by Xmin (the smallest value of x displayed on the screen), Xmax (the corresponding largest x-value), and Xscl (the x-axis' scale or, more precisely, the distance between "tick marks" on the x-axis). The range is similarly restricted by Ymin, Ymax, and Yscl. As explained in Appendix A, these values can be changed and we will do that in later graphing calculator activities. However, for this example, let's use these preset values.

You are now ready to input a formula describing the function you want graphed, in this example $f(x) = 2x + 7$. To input $f(x) = 2x + 7$, press the following sequence of keys:

$$\boxed{\text{Graph}}\ 2\ \boxed{\text{X},\theta,\text{T}}\ \boxed{+}\ 7\ \boxed{\text{EXE}}$$

Be careful! To input the variable x, you press the $\boxed{\text{X},\theta,\text{T}}$ key or the $\boxed{\text{ALPHA}}$ $\boxed{\text{X}}$. Do not confuse the $\boxed{\text{X}}$ with the multiplication $\boxed{\times}$ key.

The result is the graph shown in Figure 4.26b. Compare this to the earlier graph of this same function in Figure 4.11.

EXAMPLE 4.53 (Cont.)

This graph is fine, but it does not show us where the graph crosses the y-axis. Let's see one way that we can change the domain and range of the display screen. Press the $\boxed{\text{SHIFT}}$ $\boxed{\text{F2}}$ $\boxed{\text{F4}}$ keys. This causes the calculator to "zoom out" from the middle of the screen. The result is the graph in Figure 4.26d. Press $\boxed{\text{Range}}$ to see how the domain and range values have changed. As you can see, they have doubled.

You still cannot see where the graph crosses the y-axis, so press the $\boxed{\text{F2}}$ $\boxed{\text{F4}}$ keys again. The calculator "zooms out" once again and produces the graph in Figure 4.26c.

You can now see where the line crosses the x- and y-axes. It is hard to determine the coordinates of these intercepts by looking at the calculator screen. In the next section, we will see how to determine the approximate coordinates of the points.

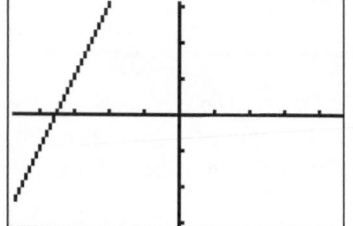

FIGURE 4.26b

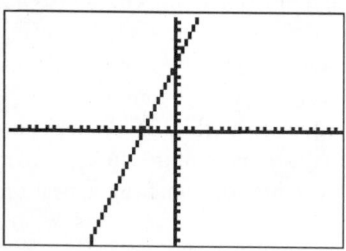

FIGURE 4.26c

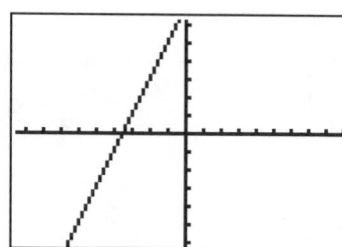

FIGURE 4.26d

EXAMPLE 4.54

```
WINDOW FORMAT
Xmin=-10
Xmax=10
Xscl=1
Ymin=-10
Ymax=10
Yscl=1
```

FIGURE 4.27a

Use a Texas Instruments TI-82 graphing calculator to sketch the graph of $f(x) = 2x + 15$.

Solution First press $\boxed{\text{ZOOM}}$ 6. You may have to wait for the screen to clear. This clears the memory and sets the domain and range to the default values in the machine. Wait until the xy-axes appear on the screen, and then press $\boxed{\text{WINDOW}}$. The screen should show an image something like that shown in Figure 4.27a.

The part of the coordinate system that will be shown on the screen is indicated by Xmin (the smallest value of x displayed on the screen), Xmax (the corresponding largest x-value), and Xscl (the x-axis scale or, more precisely, the distance between tick marks on the x-axis). The range is similarly restricted by Ymin, Ymax, and Yscl. As explained in Appendix A, these values can be changed, and we will do that in later graphing calculator activities. However, for this example, let's use these preset values.

You are now ready to input a formula describing the function you want graphed, in this example $f(x) = 2x + 15$. To input $f(x) = 2x + 15$, press the following sequence of keys:

$\boxed{\text{Y=}}$ $\boxed{\text{CLEAR}}$ 2 $\boxed{\text{X,T,}\theta}$ $\boxed{+}$ 15 $\boxed{\text{GRAPH}}$

EXAMPLE 4.54 (Cont.)

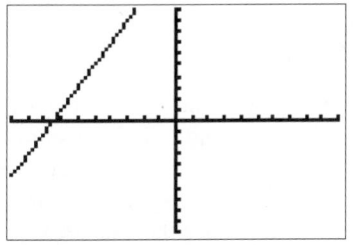

FIGURE 4.27b

≡ **Note**

EXAMPLE 4.55

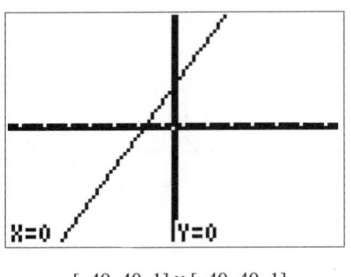

$[-40, 40, 1] \times [-40, 40, 1]$

FIGURE 4.27c

Be careful! To input the variable x, you press the $\boxed{\text{X,T},\theta}$ key or the $\boxed{\text{ALPHA}}\,\boxed{\text{X}}$ keys. Do not confuse the $\boxed{\text{X}}$ key with the multiplication, or $\boxed{\times}$, key.

The result is the graph shown in Figure 4.27b.

This graph is fine, but you will notice that it does not show us where the graph crosses the y-axis. In this case we will need to zoom out to see more of the graph. Let's see how using the calculator's "zoom-out" ability changes the domain and range of the display screen. Press the $\boxed{\text{ZOOM}}$ 3 $\boxed{\text{ENTER}}$ keys. This causes the calculator to "zoom out" from the middle of the screen. The result may look like the graph in Figure 4.27c. Press $\boxed{\text{WINDOW}}$ to see how the domain and range values have changed. In Figure 4.27c, they are Xmin= -40, Xmax= 40, Xscl= 1, Ymin= -40, Ymax= 40, and Yscl= 1.

It is hard to determine the coordinates of these intercepts by looking at the calculator screen. In the next section, we will see how to determine the approximate coordinates of the points.

Whenever we are not using the default window settings on a calculator, we will adopt the notation [Xmin, Xmax, Xscl] $\times$ [Ymin, Ymax, Yscl]. Thus, as indicated in Figure 4.27c, the viewing window is $[-40, 40, 1] \times [-40, 40, 1]$. If Xscl and Yscl are both 1 we may shorten this to $[-40, 40] \times [-40, 40]$.

Let's make the graph of a curve, which is not a straight line. In Section 4.4, we graphed $y = x^2 - 3$ with the result shown in Figure 4.17. Let's see what that looks like on a calculator.

Use a graphing calculator to sketch the graph of $y = x^2 - 3$.

Solution We will show how to do this on a Casio fx-7700G. You will have to make a few changes if you are using a different machine.

First, enter $\boxed{\text{Shift}}\,\boxed{\text{F5}}\,\boxed{\text{EXE}}$ to clear the graphics screen of the previous graphs from the screen. As before, press the $\boxed{\text{Range}}\,\boxed{\text{Shift}}\,\boxed{\text{F1}}\,\boxed{\text{Range}}\,\boxed{\text{Range}}$ keys to clear the memory and set the domain and range to the default values in the machine.

You are now ready to input a formula describing the function you want graphed. To input $y = x^2 - 3$, press the following sequence of keys:

$$\boxed{\text{GRAPH}}\,\boxed{\text{X},\theta,\text{T}}\,\boxed{\text{Shift}}\,\boxed{x^2}\,\boxed{-}\,3\,\boxed{\text{EXE}}$$

The result is the graph shown in Figure 4.28. Compare this to the earlier graph of this same function in Figure 4.17.

From this graph we can see that the curve crosses the y-axis near $y = -3$ and the x-axis around $x = \pm 1.8$. In the next section, we shall see how to obtain better approximations of these values.

By now you have probably noticed that in order to tell the calculator what you want it to graph on a Casio fx-7700G, you press the GRAPH key and on a TI-82, you press the Y= key. In each case you obtain a phrase like Casio's "Graph Y=" and TI-82's "Y1=" on the calculator's screen. This should remind you that you can only graph a function. It may be necessary to solve an equation for y before you can graph it.

EXAMPLE 4.56

Use a graphing calculator to sketch the graph of $f(x) = \dfrac{x^2+1}{x+2}$.

Solution Graphing this is very similar to the previous graphing you have done. However, you must be careful to insert enough parentheses and to insert them in the right spots.

You might want to write this equation on one line. First, since your calculator will be graphing something in the $y =$ form, you should replace the $f(x) =$ with $y =$. Then write the fraction on one line. Remember the warning at the beginning of Section 2.3. Everything over the fraction bar is to be grouped, and everything under the fraction bar is another group. Parentheses are used to group terms. So, we end up writing the equation as

$$y = (x^2+1)/(x+2)$$

You are now ready to input this equation into your graphing calculator. On a TI-82, you would press the following sequence of keys:

Y= (X,T,θ x² + 1) / (X,T,θ + 2) GRAPH

The result is shown in Figure 4.29a.

Sometimes, as in Figure 4.29a, the calculator draws in some points (usually shown as vertical lines) that are really not part of the graph. This may also happen with some computer graphing software. Sometimes it is possible to adjust the viewing window, as in Figure 4.29b, to eliminate these extraneous points. Consult the users' manual for your calculator or software program for suggestions.

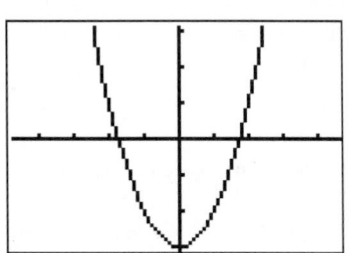

FIGURE 4.28

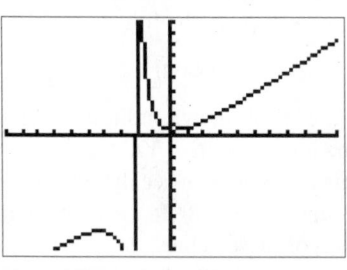

FIGURE 4.29a

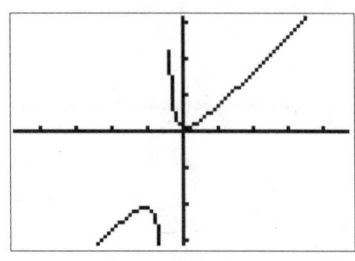

[−18.8, 18.8, 4] x [−12.4, 12.4, 4]

FIGURE 4.29b

EXAMPLE 4.57

Use a graphing calculator to sketch the graph of $y + 5 = \sqrt{x+2}$.

Solution   First, solve this equation for y, obtaining $y = \sqrt{x+2} - 5$.
You will want to put parentheses around the $x + 2$. If you do not, the same as $y = \sqrt{x} - 3$. On a TI-82 you press the following sequence of keys:

$$\boxed{Y=}\ \boxed{2nd}\ \boxed{\sqrt{\ }}\ \boxed{(}\ \boxed{X,T,\theta}\ \boxed{+}\ \boxed{2}\ \boxed{)}\ \boxed{-}\ 5\ \boxed{GRAPH}$$

The result is shown in Figure 4.30.

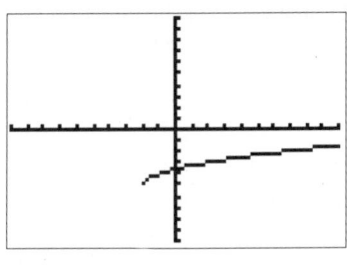

FIGURE 4.30

Exercise Set 4.5

Graph each of these functions with a calculator or on a computer. Compare the machine's graph with the graph you made of the same functions in Section 4.4.

1. $y = x$
2. $y = 2x$
3. $y = -3x$
4. $y = \frac{1}{2}x - 3$
5. $y = -4x + 2$
6. $x + y = 2$

7. $x - y = 2$
8. $2x + y = 1$
9. $3x - 6y = 9$
10. $3x - 6y = -6$
11. $y = x^2$
12. $y = x^2 + 3$

13. $y = x - 2$
14. $y = x^2 + 2x + 1$
15. $y = x^2 - 6x + 9$
16. $y = -x^2 + 2$
17. $y = \dfrac{1}{x}$

18. $y = \dfrac{1}{x+3}$
19. $y = x^3$
20. $y = \sqrt[3]{x}$

21. $y^2 + x^2 = 25$ (HINT: Solve for y, then graph two equations.)
22. $y = \sqrt{25 - x^2}$
23. $y = |x + 2|$ (HINT: Use the $\boxed{Abs}$ key. Read your manual to see how to access this on your calculator.)

24. $\dfrac{x^2}{4} + \dfrac{y^2}{9} = 1$ (HINT: Solve for y, then graph two equations.)
25. $\dfrac{x^2}{4} - \dfrac{y^2}{9} = 1$
26. $y = x^3 - x^2$

In Exercises 27–42, graph each of the given functions with a calculator or on a computer.

27. $y = x^2 - 4x$
28. $y = x^2 + 6x$
29. $y = x^2 - 10x + 37$
30. $y = 0.1x^2 - 20$
31. $y = -x^2 - 4x$

32. $y = -x^2 + 6x$
33. $y = \sqrt{x + 5}$
34. $y = \sqrt{x - 5}$
35. $y = \sqrt{9 - x}$
36. $y = -\sqrt{5 - x}$

37. $y = \dfrac{x+2}{x}$
38. $y = \dfrac{x}{x+2}$
39. $y = \dfrac{x^2+3}{x+2}$

40. $y = \dfrac{x+2}{x^2+3}$
41. $y = \dfrac{x^2-4}{x+2}$
42. $y = \dfrac{x+2}{x^2-4}$

 **In Your Words**

43. Explain the notation $[-15, 15, 2] \times [-10, 20, 4]$.
44. Write a description that tells how to zoom in and zoom out on your calculator.

≡ 4.6
USING GRAPHS TO SOLVE EQUATIONS

A graph not only gives you a picture of what an equation looks like, but it can be used to help solve the equation. In Section 4.4, we discussed the points where the graph crosses the axes. The y-intercepts were the points where the graph crossed or touched the y-axis; the x-intercepts were the points where the graph crossed or touched the x-axis. The x-intercepts are also known as the **roots** or **solutions** of the equation, because these are the values of x for which y will equal zero. If the equation is a function of x, these are called the roots (or **zeros**) of the function because they are the values of x when $f(x) = 0$.

EXAMPLE 4.58

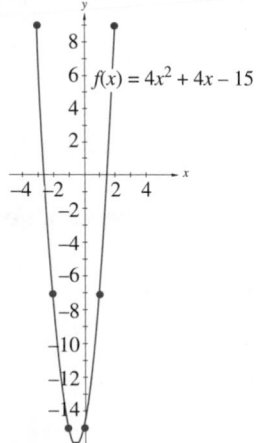

FIGURE 4.31

Find graphically the approximate roots of $4x^2 + 4x - 15 = 0$.

Solution First, set $f(x) = 4x^2 + 4x - 15$. Then set up a partial table of values for the function f.

x	-3	-2	-1	0	1	2	3
$y = f(x)$	9	-7	-15	-15	-7	9	33

These points are plotted and connected to form the curve in Figure 4.31. From the graph we can see that the x-intercepts look as if they are at $x = -2\frac{1}{2}$ and $x = 1\frac{1}{2}$. A check of the function confirms that $f\left(-2\frac{1}{2}\right) = 0$ and $f\left(1\frac{1}{2}\right) = 0$. These are the roots of this equation.

EXAMPLE 4.59

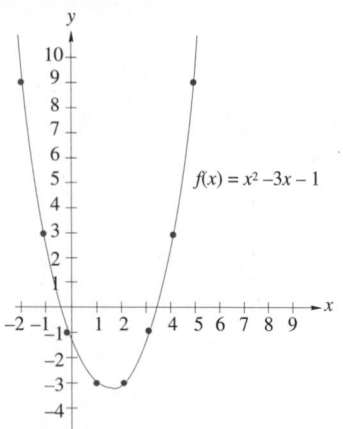

FIGURE 4.32

Find graphically the approximate roots of $f(x) = x^2 - 3x - 1$.

Solution Again, if we set up a table of values we get

x	-2	-1	0	1	2	3	4	5
$f(x)$	9	3	-1	-3	-3	-1	3	9

Graphing the curve determined by these points, we get the curve in Figure 4.32. From the graph we can see that there appear to be roots at $x = -0.25$ and $x = 3.25$. If we evaluate the function at these two values, we get $f(-0.25) = -0.1875$ and $f(3.25) = -0.1875$. This shows two things. First, it shows that -0.25 and 3.25 are *not* roots of this function, since $f(x) \neq 0$ at either of these two points. Second, it does show that the roots of this function are "close" to -0.25 and 3.25.

If we wanted more accurate approximations of the roots of the function in Example 4.59, we could substitute different values for x until we got an approximation that was as accurate as we wanted. In a later chapter, we will find out how to determine the exact roots to this function. In the next example, we will show how to use a graphing calculator to approximate this function's roots.

EXAMPLE 4.60

Use a graphing calculator or computer graphing software with "zoom" capability to find the approximate roots of $f(x) = x^2 - 3x - 1$.

Solution This is the same function we graphed in Example 4.59. We already know that the roots are near $x = -0.25$ and $x = 3.25$. Details on how to use the "zoom" features of a Casio fx-7700G and Texas Instrument's TI-81 graphing calculator are in Appendix A.

When you graph the function using the calculator's default settings for the range, you obtain the graph in Figure 4.33a on an fx-7700G or the graph in Figure 4.33b on a TI-82.

As you can see from these figures, the graph crosses the x-axis at two points; these are the x-intercepts or roots. Use the "trace" function of your calculator until the cursor is positioned near the x-intercept on the left. Figure 4.33c shows the fx-7700G screen and indicates that the cursor is at $(-0.3, -0.01)$, and Figure 4.33d shows that the TI-82 places the cursor at $(-0.212766, -0.3164328)$.

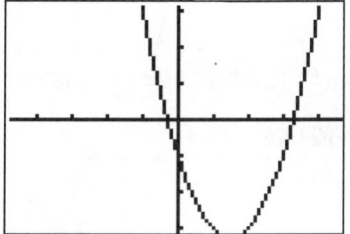

FIGURE 4.33a

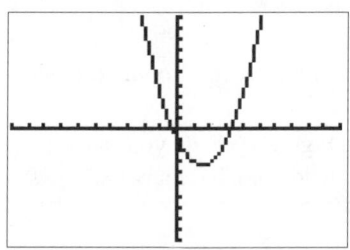

FIGURE 4.33b

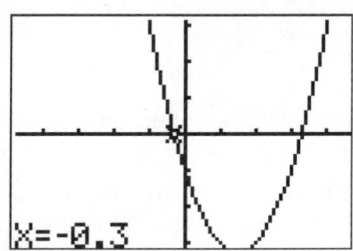

FIGURE 4.33c

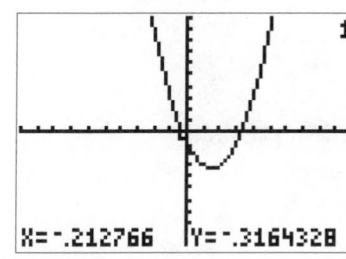

FIGURE 4.33d

For more accurate approximations of this root, you can continue to zoom in around this point on the graph. If you do, you will find that this root is $x \approx -0.302775638$. Later we will show how to find the exact value of this root.

Now, approximate the other root of this function. Return to the original graph in either Figure 4.33a or 4.33b, and zoom in around the right-hand root. If you zoom in enough, you will obtain $x \approx 3.302775638$.

EXAMPLE 4.61

Use a graphing calculator or computer graphing software with "zoom" capability to find the approximate roots of $y = x^4 - 2x^3 - 1$.

Solution The graph of this function on a TI-82 is shown in Figure 4.34a. Here Xmin $= -5$, Xmax $= 5$, Ymin $= -3$, and Ymax $= 3$. Zoom in, as in Figure 4.34b, press $\boxed{\text{Trace}}$ to move the cursor, and see that $x \approx -0.7180851$.

Notice that the second root is not on the screen. Returning to the original graph, we move the cursor over near the second root. Zooming in around this point and using the trace function, we see that the other root is approximately $x \approx 2.105$.

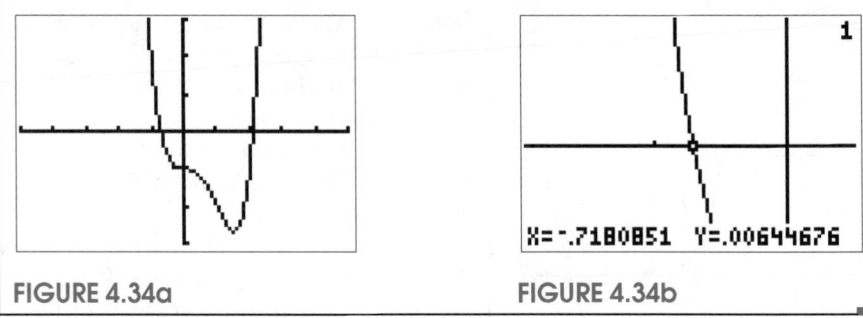

FIGURE 4.34a   FIGURE 4.34b

EXAMPLE 4.62

Use a graphing calculator or computer graphing software with "zoom" capability to determine any x-intercepts of $y = x^2 + 1.5$

Solution The graph of this function is shown in Figure 4.35. As you can see, the graph does not cross the x-axis. Thus, this function does not have any real roots.

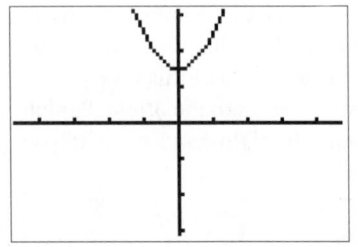

FIGURE 4.35

The graph of an equation can tell the approximate roots or, as in Example 4.62, if the equation has no real roots. There are many times when we will need more accurate values of roots and a quicker procedure than that of graphing the equation. In later chapters, we will learn some alternative methods that can be used to find the roots of an equation.

In Section 4.2, we discussed the inverse of a function. We said that two functions, f and g, are inverses of each other if $f[g(x)] = x$ for every value of x in the domain of g and $g[f(x)] = x$ for every value of x in the domain of f. We also said that if g was the inverse of f, we would write $g(x) = f^{-1}(x)$.

The graph of a function can be used to determine if the function has an inverse and to sketch the graph of the inverse function. You may remember that the vertical line test indicates that if no vertical line intersected a graph more than once, then the graph represents a function. A similar test can be used to determine if a function has an inverse.

Horizontal Line Test

The horizontal line test states that a function has an inverse function if no horizontal line intersects its graph at more than one point.

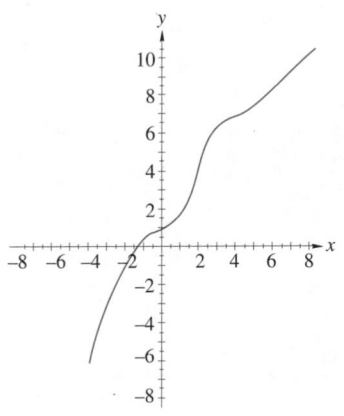

FIGURE 4.36a

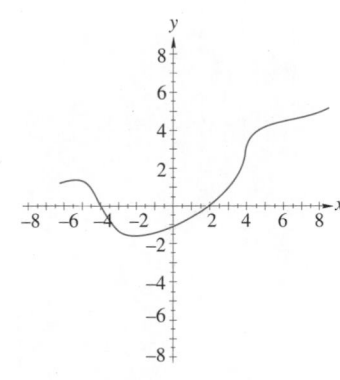

FIGURE 4.36b

For example, Figures 4.36a and 4.36b show the graphs of two functions. The function in Figure 4.36a has an inverse function, because it is not possible to draw a horizontal line that will intersect the graph at more than one point. The function in Figure 4.36b does not have an inverse because it is possible to draw a horizontal line that will intersect the graph twice. In fact, the x-axis intersects the graph in Figure 4.36b in at least two places.

Now let's use our graphing ability to graph the inverse of a function. The graph of $f(x) = \sqrt{x-1}$ is shown in Figure 4.37. Using the horizontal line test, we can see that this function has an inverse. Now, draw the line $y = x$. (In Figure 4.37 this is the dashed line.) Suppose you placed a mirror on the line $y = x$. The image you would see is indicated by the colored curve in Figure 4.37.

What this essentially does is take any point (a, b) on the function and convert it to its mirror image (b, a). Remember that if (a, b) is a point on a function f, then $f(a) = b$; and if (b, a) is a point on some function g, then $g(b) = a$. But, since $f[f^{-1}(a)] = f(b) = a$, we can see that g must be f^{-1}.

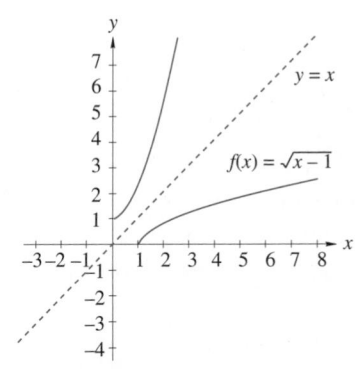

FIGURE 4.37

EXAMPLE 4.63

Graph the inverse function of $y = x^3$.

Solution   The graph of $y = x^3$ is given in Figure 4.38a and the line $y = x$ is shown by a dashed line. The reflection of $y = x^3$ in the line $y = x$ is shown in Figure 4.38b and is the graph of $y = \sqrt[3]{x}$. If $f(x) = x^3$, then $f^{-1}(x) = \sqrt[3]{x}$.

EXAMPLE 4.63 (Cont.)

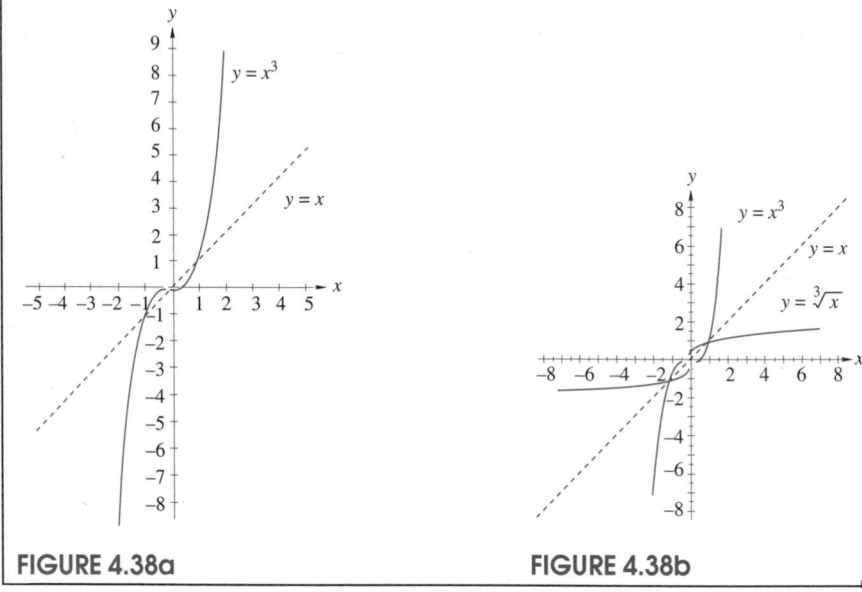

FIGURE 4.38a   FIGURE 4.38b

Application

EXAMPLE 4.64

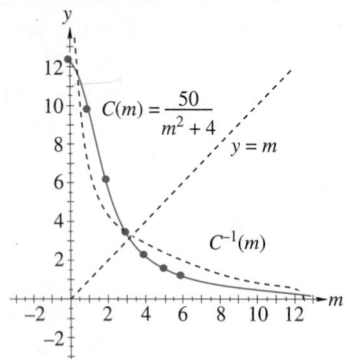

FIGURE 4.39a

In Example 4.50, we made the following partial table for the concentration in parts per million (ppm) of a certain pollutant m mi from a certain factory as given by the function

$$C(m) = \frac{50}{m^2 + 4}, \qquad m \geq 0$$

m	0	1	2	3	4	5	6
$C(m)$	12.5	10	6.25	3.85	2.5	1.72	1.25

Next, we used our table to sketch the graph of this function.

(a) Use the table to sketch the graph of the inverse of this function.

(b) Use the table or the graph to estimate how far you are from the factory, if the concentration of the pollutant is 3.08 ppm.

(c) Determine the equation that describes the inverse function of C.

(d) Use your equation for C^{-1} to determine $C^{-1}(3.08)$.

Solutions

(a) The graph of the function $y = C(m)$ is shown in Figure 4.39b. The reflection of the graph in the line $y = m$ produces the inverse function shown by the dotted curve in Figure 4.39a.

(b) We will use the graph of the original function to determine $C^{-1}(3.08)$. Move up the vertical axis for the graph of $C(m)$ until you reach the point that is approximately at 3.08 on the vertical axis. Draw a horizontal line from this

EXAMPLE 4.64 (Cont.)

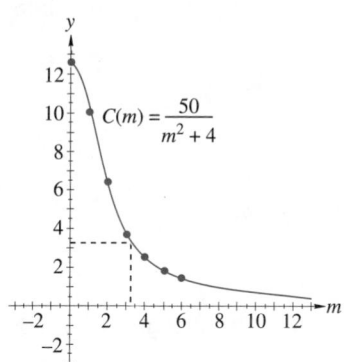

FIGURE 4.39b

EXAMPLE 4.65

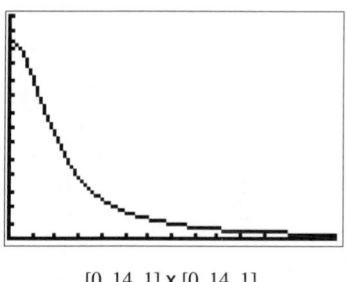

$[0, 14, 1] \times [0, 14, 1]$

FIGURE 4.40

point until it reaches the graph of C. Draw a vertical line until it crosses the horizontal axis. This should be near 3.5 as shown in Figure 4.39b. Thus, $C^{-1}(3.08) \approx 3.5$ and we conclude that the pollutant was collected about 3.5 mi from the factory.

(c) As described in Section 4.4, we will let $y = \dfrac{50}{m^2+4}$. Solving for m, we obtain

$$m = \sqrt{\dfrac{50}{y} - 4}, \text{ for } 0 < y \le 12.5. \text{ Thus, } C^{-1}(m) = \sqrt{\dfrac{50}{m} - 4}, \text{ for } 0 < m \le 12.5.$$

(d) Substituting 3.08 in the formula for C^{-1}, we obtain

$$C^{-1}(3.08) = \sqrt{\dfrac{50}{3.08} - 4} \approx 3.50$$

Thus, the formula and the graph give us approximately the same answers. ▪

Application

Use a graphing calculator to rework parts of Example 4.64.

(a) Sketch the graph of $C(m) = \dfrac{50}{m^2+4}, m \ge 0$.

(b) Use the graph and your calculator to estimate how far you are from the factory if the concentration of the pollutant is 3.08 ppm.

Solutions

(a) You must make two changes before we begin graphing. First, you must replace $C(m)$ with y. Second, your calculator will only allow you to use certain variables when you graph a function. At present the only variable you can use is x. So, rewrite the given function as $y = \dfrac{50}{x^2+4}, x \ge 0$. Don't forget the correct use of parentheses. You will want to graph this as $y = \dfrac{50}{(x^2+4)}$.

Also, to get the graph to look like the one in Figure 4.39b, change the viewing window to $[0, 14, 1] \times [0, 14, 1]$. The result is the graph shown in Figure 4.40.

(b) We want to know when the concentration of the pollutant is 3.08 ppm. This means we want to know the value of x when $y = 3.08$. Press TRACE and move the cursor until you have a y-value near 3.08. On this particular calculator, the y-values change from around 2.98 to 3.17. Zooming in around one of these points and using the trace function, we see that when $y \approx 3.08$, then $x \approx 3.4976$. Once again, we conclude that the pollutant was collected about 3.5 mi from the factory. ▪

Exercise Set 4.6

The equations in Exercises 1–18 all have at least one root between -10 and 10. Write each equation in the form $y = f(x)$. Graph each equation to find the approximate value of the roots. If possible, use a graphing calculator or a computer graphing program to help you.

1. $2x + 5 = 0$

2. $5x - 9 = 0$

3. $x^2 - 9 = 0$

4. $4x^2 - 10 = 0$

5. $x^2 = 5x$

6. $x^2 = 4x - 3$

7. $x^2 + 5x - 3 = 0$

8. $3x^2 + 5x - 3 = 0$

9. $20x^2 + 21x = 54$

10. $x^4 - 4x^3 - 25x^2 + x + 6 = 0$

11. $\sqrt{x + 1} = 0$

12. $\sqrt{x - 2.5} = 0$

13. $\sqrt{x + 1} = 3$

14. $\sqrt{x - 3} = 2$

15. $\dfrac{x}{x + 1} = -3$

16. $\dfrac{x}{x - 2} = 5$

17. $\dfrac{x}{x - 2} = 2 - x^2$

18. $\dfrac{x^2 + 5}{x - 3} = x^2 - 2$

In Exercises 19–34, graph each of the functions with a calculator or on a computer. Use the graph to help determine the domain and range of each function.

19. $y = x^2 - 4x$

20. $y = x^2 + 6x$

21. $y = x^2 - 10x + 37$

22. $y = 0.1x^2 - 20$

23. $y = -x^2 - 4x$

24. $y = -x^2 + 6x$

25. $y = \sqrt{x + 5}$

26. $y = \sqrt{x - 5}$

27. $y = \sqrt{9 - x}$

28. $y = -\sqrt{5 - x}$

29. $y = \dfrac{x + 2}{x}$

30. $y = \dfrac{x}{x + 2}$

31. $y = \dfrac{x^2 - 1}{x + 2}$

32. $y = \dfrac{x + 2}{x^2 + 3}$

33. $y = \dfrac{x^2 - 4}{x + 2}$

34. $y = \dfrac{x + 2}{x^2 - 4}$

Use the horizontal line test to determine whether or not the functions graphed in Exercises 35–38 have inverse functions.

35.

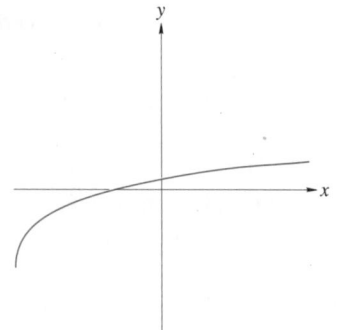

36.

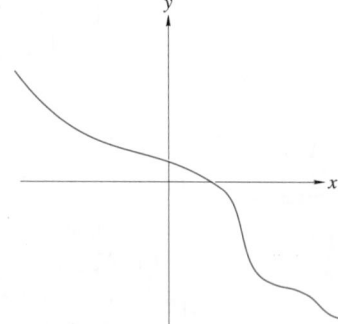

37.

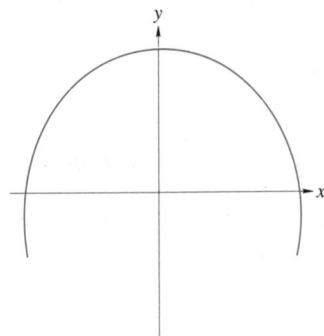

38.

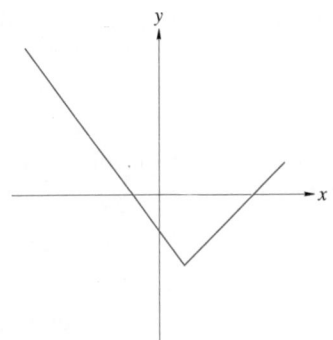

Each of the functions in Exercises 39–42 have inverses. Sketch the graph of the inverse of each function.

39.

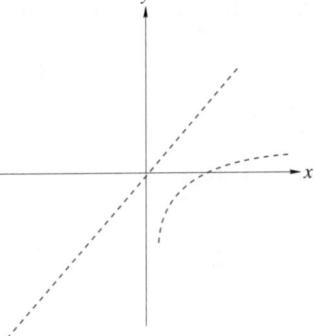

41.

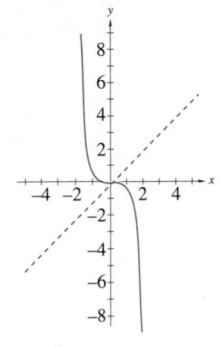

40.

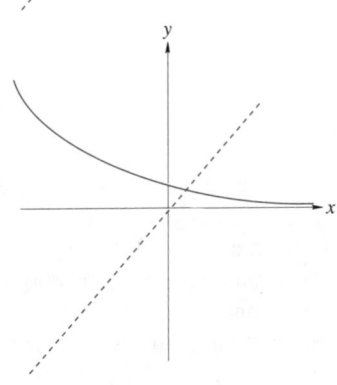

42.

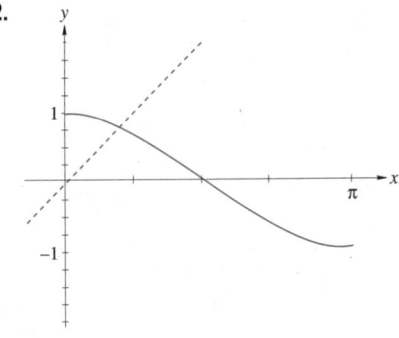

Solve Exercises 43–46.

43. *Automotive technology*   Study the following table. It shows that if the antifreeze content of a coolant is increased, the boiling point of the coolant is also increased.

Percent antifreeze in coolant	0	10	20	30	40	50	60	70	80	90	100
Boiling temperature °F	210	212	214	218	222	228	236	246	258	271	330

(a) What percent of antifreeze will be in the coolant if the boiling point of the system is 218°F?
(b) What percent of antifreeze will be in the coolant if the boiling point of the system is 236°F?
(c) What percent of antifreeze will be in the coolant if the boiling point of the system is 250°F?

44. *Environmental science* The population P of a certain species of animal depends on the number n of a smaller animal on which it feeds, with

$$P(n) = 5\sqrt{n} - 10$$

(a) Determine the inverse function for P.
(b) If the population of P is 5, how many of the small animals are there?
(c) Graph the inverse function of P.

45. *Construction* This table contains the results of a series of drillings used to determine the depth of the bedrock at a building site. Drillings were taken along a straight line down the middle of the lot, where the building will be placed. In the table, x is the distance from the front of the parking lot and y is the corresponding depth. Both x and y are given in feet.

x	0	20	40	60	80	100	120	140	160
y	33	35	40	45	42	38	46	40	48

These ordered pairs have been plotted on the following graph. Does this graph have an inverse function?

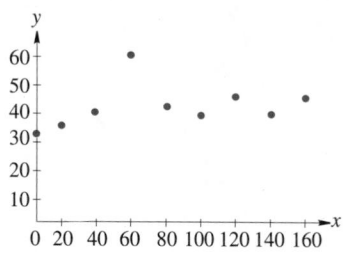

46. *Environmental science* Suppose a cost-benefit model is given by

$$C(x) = \frac{6.4x}{100 - x}$$

where $C(x)$ is the cost in millions of dollars for removing x percent of a given pollutant.
(a) What is the cost of removing 30% of the pollutant?

(b) What is the cost of removing 60% of the pollutant?
(c) What is the inverse function of C?
(d) If a community can only spend $12,000,000, what percent of the pollutant can be removed?
(e) Graph the given function and its inverse function.

In Your Words

47. Explain how to use your calculator to solve an equation.

48. On a sheet of paper, explain the horizontal line test. Do not look at the definition in the book.

CHAPTER 4 REVIEW

Important Terms and Concepts

Cartesian coordinate system
Composite function
Dependent variable
Domain
Function
Horizontal line test
Independent variable
Intercepts
Inverse function
Linear equation

Quadrants
Range
Rectangular coordinate system
Relation
Slope
Vertical line test
x-axis
x-intercept
y-axis
y-intercept

Review Exercises

For Exercises 1–6 (a) graph each of the relations, (b) determine the domain, range, x-intercept, and y-intercept of each relation, (c) use the vertical line test to determine if each relation is a function, (d) use the horizontal line test to determine if each function has an inverse function, and (e) graph each inverse function that exists.

1. $y = 8x - 7$

2. $y = 2x^2 - 4$

3. $y = \sqrt{x} - 3$

4. $y = \frac{1}{2}x^3 + 2$

5. $y = x^2 - 2x$

6. $y = x^2 + 4$

In Exercises 7–12, given the function $f(x) = 4x - 12$, determine the following.

7. $f(0)$

8. $f(-2)$

9. $f(3)$

10. $f(a)$

11. $f(a-2)$

12. $f(x+h)$

In Exercises 13–19, given the function $g(x) = \dfrac{x^2 - 9}{x^2 + 9}$, determine the following.

13. $g(0)$

14. $g(3)$

15. $g(-3)$

16. $g(-2)$

17. $g(4)$

18. $g(-5)$

19. Use your values from Exercises 13–18 to graph $g(x)$. From your graph, determine the zeros of g.

In Exercises 20–33, let $f(x) = 4x - 12$ and $g(x) = \dfrac{x^2 - 9}{x^2 + 9}$. Determine the following.

20. $(f+g)(x)$

21. $(f+g)(3)$

22. $(f-g)(x)$

23. $(f-g)(-2)$

24. $(f \cdot g)(x)$

25. $(f \cdot g)(0)$

26. $(f/g)(x)$

27. $(f/g)(5)$

28. $(g/f)(x)$

29. $(g/f)(2)$

30. $(f \circ g)(x)$

31. $(f \circ g)(4)$

32. $(g \circ f)(x)$

33. $(g \circ f)(3)$

In Exercises 34–40, graph each of these functions or relations with a calculator or computer.

34. $y = \dfrac{5}{x-4}$

35. $y = -\dfrac{7}{x+3}$

36. $y = \sqrt{2x-3}$

37. $y = \sqrt{5-4x}$

38. $y = \dfrac{x}{\sqrt{x+2}}$

39. $x^2 + y^2 = 9$

40. $\frac{1}{2}x^2 + y^2 = 16$

Each of the equations in Exercises 41–44 has a root between -10 and 10. Write each equation in the form $y = f(x)$. Graph each equation to find the approximate value of the roots.

41. $4x + 7y = 0$

42. $x^2 - 20 = 0$

43. $2x^2 + 10x + 4 = 0$

44. $8x^3 - 20x^2 - 34x + 21 = 0$

Solve Exercises 45–48.

45. *Business*   The manager of a videotape store has found that n videotapes can be sold if the price is $P(n) = 35 - \dfrac{n}{20}$ dollars.
  (a) What price should be charged in order to sell 101 videotapes? 350 videotapes? 400 videotapes?
  (b) Find an expression for the revenue from the sale of n videotapes, where revenue = demand × price.
  (c) How much revenue can be expected if 101 videotapes are sold? if 350 are sold? if 400 are sold?

46. *Business*   A videotape store has learned that the function $R(n) = 30n - \dfrac{n^2}{20}$ is a good predictor of its revenue, in dollars, from the sale of n tapes. The cost of operating the store is given by $C(n) = 550 + 10n$.
  (a) If the profit P is given by $P(n) = R(n) - C(n)$, what is the profit function?
  (b) How much profit will the store make if it sells 30 videotapes? 100 videotapes? 150 videotapes? 300 videotapes? 400 videotapes?

47. *Medical technology*   A measure of cardiac output can be determined by injecting a dye into a vein near the heart and measuring the concentration of the dye. In a normal heart, the concentration of the dye is given by the function

$$h(t) = -0.02t^4 + 0.2t^3 - 0.3t^2 + 3.2t$$

where t is the number of seconds since the dye was injected. Set up a partial table of values for $0 \leq t \leq 10$, and sketch the graph of this function.

48. *Automotive technology*   The distance s, in feet, needed to stop a car traveling v mph is given by

$$s(v) = 0.04v^2 + v$$

  (a) Set up a partial table of values for $s(v)$ with $0 \leq v \leq 70$.
  (b) What was the velocity of a car that took 265 ft. to stop?
  (c) Sketch the graph of s and s^{-1} on the same set of axes.

▤ CHAPTER 4 TEST

Use the function $f(x) = 7x - 5$ in Exercises 1 and 2, and determine the indicated value.

1. $f(-2)$

2. $f(3-a)$

In Exercises 3 and 4, let $g(x) = \dfrac{x^2 - 2x - 15}{x+3}$ and determine the indicated value.

3. $g(0)$

4. $g(5)$

Solve Exercise 5.

5. (a) Graph $h(x) = \frac{1}{2}\sqrt{x+4} - 3$.
   (b) What is the domain of h?
   (c) What is the range of h?
   (d) What is the x-intercept of h?
   (e) What is the y-intercept of h?

In Exercises 6–11, let $f(x) = 3x - 15$ and $g(x) = \dfrac{x-5}{x+5}$, and determine the indicated value.

6. $(f+g)(x)$
7. $(f-g)(x)$

8. $(f \cdot g)(x)$
9. $(f/g)(x)$

10. $(f \circ g)(x)$
11. $(g \circ f)(x)$

In Exercises 12–16, sketch the graph of each of the following functions or relations.

12. $y = 3x - 4$
13. $f(x) = x^2 - 2x + 1$

14. $g(x) = \dfrac{x+5}{x-1}$

15. $\dfrac{x^2}{x+1} = 2x - 1$
16. $\sqrt{x^2 - 1} = 3 - x$

Solve Exercises 17 and 18.

17. A researcher in physiology has decided that the function $r(s) = -s^2 + 12s - 20$ is a good mathematical model for the number of impulses fired after a nerve has been stimulated. Here, r is the number of responses per millisecond (ms) and s is the number of milliseconds since the nerve was stimulated.
    (a) Graph this function.
    (b) How many responses can be expected after 3 ms?
    (c) If there are 16 responses, how many ms have elapsed since the nerve was stimulated?
    (d) If there are 12 responses, how many ms have elapsed since the nerve was stimulated?

18. The cost of removing a certain pollutant is given by

$$C(x) = \frac{5x}{100 - x}$$

where $C(x)$ is the cost in thousands of dollars of removing x percent of the pollutant.
    (a) Graph this function.
    (b) How much will it cost to remove 50% of the pollutant?
    (c) How much will it cost to remove 90% of the pollutant?
    (d) What is the inverse function of C?
    (e) If you had only \$15,000, how much of the pollutant could you remove?

5

Systems of Linear Equations and Determinants

The ability to schedule the use of trucks is important to gaining the best use of each vehicle and for the company to make a profit. In Section 5.3, you will see how linear equations can be used to determine these factors.

Courtesy of Ruby Gold

Many technical problems require us to consider the effects of several conditions and variables simultaneously. We often need to use more than one equation to show how these variables are related. When this happens, we need to find the solutions that satisfy all of these equations.

For example, in order to determine how many computers can be made using several parts, we have to consider how many of each part are available and the number needed for each computer.

In Chapter 2, we introduced the idea of solving equations. In Chapter 4 we showed how we could use graphs to help find an equation's roots. In this chapter, we will use our algebraic skills to solve linear equations and to solve systems of two or more linear equations.

≡ 5.1
LINEAR EQUATIONS

You may remember that a linear equation is the equation of a straight line. In general, an equation is a **linear equation** if each term contains only one variable, to the first power, or the term is a constant.

EXAMPLE 5.1

The equation $4x + 5 = 25$ is a linear equation in one variable, x. We learned how to solve this equation in Chapter 2. The solution is $x = 5$. ▪

EXAMPLE 5.2

The equation $4x - 5y = 3$ is a linear equation in two variables, x and y. We learned how to solve this equation in Chapter 4. The solution to this equation is all the points that are on the line $y = \frac{4}{5}x - \frac{3}{5}$. ▪

EXAMPLE 5.3

The equation $9x + 2y - 7z = 4$ is a linear equation in three variables x, y, and z. We have not learned how to solve an equation of this type. ▪

A linear equation can have any number of variables. The previous examples show linear equations in 1, 2, and 3 variables. But, a linear equation could just as easily have 4 or 14 variables.

In the last chapter, we found that the slope of a line that went through two points (x_1, y_1) and (x_2, y_2) was given by the formula $m = \dfrac{y_2 - y_1}{x_2 - x_1}$. The slope tells us how steep the line is and whether the line is rising or falling.

Point-Slope Equation of a Line

If we know the slope of a line and we know one of the points on that line, we can then determine an equation for the line. Any point on the line is of the form (x, y). Suppose we know that a specific point (x_1, y_1) is on the line and that the slope of the line is m. (See Figure 5.1.) From the equation for the slope we know that

$$m = \frac{y - y_1}{x - x_1}$$

or $\quad m(x - x_1) = y - y_1$

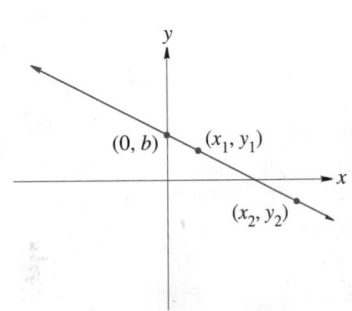

FIGURE 5.1

Rewriting this equation gives us the point-slope form of a linear equation.

The point-slope form of a linear equation

If (x_1, y_1) is a point on a line and the slope of the line is m, then

$$y - y_1 = m(x - x_1)$$

is known as the **point-slope equation** of a straight line.

EXAMPLE 5.4

Find the equation of the line through the point $(4, 5)$ with a slope of $\frac{2}{3}$.

Solution   We are told that the slope is $\frac{2}{3}$ and that a point on the line is $(4, 5)$. So we have $m = \frac{2}{3}$, $x_1 = 4$, and $y_1 = 5$. Putting these values in the point-slope form of a linear equation, we get $y - 5 = \frac{2}{3}(x - 4)$.

EXAMPLE 5.5

Find the equation of the line through the points $(2, 3)$ and $(5, 9)$.

Solution   We are not given the slope, but since we have two points we can find it. The slope is

$$m = \frac{9 - 3}{5 - 2} = \frac{6}{3} = 2$$

The point $(2, 3)$ is on the line, so we can let $x_1 = 2$, $y_1 = 3$, and using the point-slope equation of a line, we get

$$y - 3 = 2(x - 2).$$

We do not have to use the point $(2, 3)$. We can use any known point on the line. If we had used the point $(5, 9)$, then $x_1 = 5$, $y_1 = 9$, and we would get the equation

$$y - 9 = 2(x - 5)$$

We can show that this is equivalent to the previous equation.

Slope-Intercept Equation of a Line

We learned in Chapter 4 that the y-intercept is the point where the graph crosses the y-axis. The x-coordinate of this point is 0, and if the y-intercept is at b, then the point $(0, b)$ is on the line, as shown in Figure 5.1. If the slope of the line is m, then $y - b = m(x - 0)$. This simplifies to the following equation.

> **The slope-intercept form of a linear equation**
>
> The **slope-intercept form** of the equation for a line is
>
> $$y = mx + b$$
>
> where m is the slope and b is the y-intercept.

Notice that you can easily tell the slope and the y-intercept by looking at a linear equation written in the slope-intercept form.

EXAMPLE 5.6

Find the equation of the straight line with a slope of -7 and a y-intercept of 4.

Solution   Here $m = -7$ and $b = 4$.   Using the slope-intercept form, we get $y = -7x + 4$.

EXAMPLE 5.7

What are the slope and y-intercept of the line $2x - 5y - 10 = 0$?

Solution   If we solve this equation for y, it will then be in the slope-intercept form of the line. We can then determine the answers by looking at the equation for the line.

$$2x - 5y - 10 = 0$$
$$-5y = -2x + 10$$
$$y = \frac{2}{5}x - 2$$

This is now in the slope-intercept form for the line, $y = mx + b$. The coefficient of x is the slope and the constant is the y-intercept. So the slope, m, is $\frac{2}{5}$ and the y-intercept, b, is -2.

If we wanted to graph the line in Example 5.7, we would need a second point. Select a value for x and solve for y. For example, if $x = 10$, then $y = \frac{2}{5}(10) - 2 = 4 - 2 = 2$. So, the point $(10, 2)$ is on this line.

Sometimes the easiest point to use is the y-intercept. Use the idea of slope as $\frac{\text{rise}}{\text{run}}$, and, from the y-intercept $(0, -2)$, "run" 5 units to the right to $x = 5$ and then "rise" 2 units upward to $y = -2 + 2 = 0$. This gives the point $(5, 0)$, which is also on the line.

 Hint

If you use the idea of slope as $\frac{\text{rise}}{\text{run}}$ you should always "run" to the right. Then, if the slope is positive, "rise" upward, or, if the slope is negative, "rise" downward.

We could also let $y = 0$ and find the x-intercept. From the original equation we have $2x - 5(0) - 10 = 0$ or $2x - 10 = 0$ or $x = 5$. This means that the point $(5, 0)$ is also on this line. Plotting these points and drawing the line through them produces the graph in Figure 5.2.

A **horizontal line** is parallel to the x-axis and has a slope of 0. This means that any horizontal line can be written as $y = 0 \cdot x + b = b$ or simply $y = b$. Notice that all points on a horizontal line have the same y-coordinate, b.

A **vertical line** has an undefined slope. This means that a vertical line cannot be written in slope-intercept form. But, all points on a vertical line have the same x-coordinate. If we call this x-coordinate a, we can then write the equation of this vertical line as $x = a$.

Horizontal and Vertical Lines

A horizontal line is parallel to the x-axis, has a slope of 0, and can be written as $y = b$.

A vertical line is perpendicular to the x-axis, has an undefined slope, and can be written as $x = a$.

EXAMPLE 5.8

Graph the lines $x = 5$ and $y = -3$.

Solution   The line $x = 5$ is a vertical line with x-intercept 5. The line $y = -3$ is a horizontal line with y-intercept -3. These two lines are graphed in Figure 5.3.

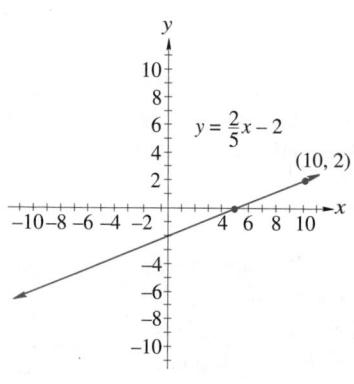

FIGURE 5.2

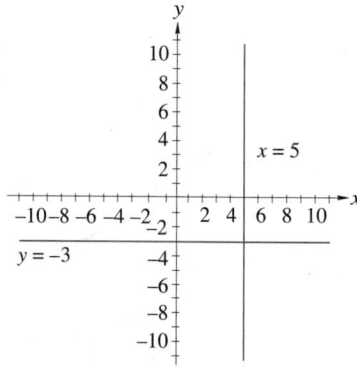

FIGURE 5.3

Exercise Set 5.1

In Exercises 1–10, determine the slope of the line through the given pair of points.

1. $(2,5), (3,8)$

2. $(4,7), (-2,1)$

3. $(1,8), (5,3)$

4. $(0,4), (5,0)$

5. $(9,3), (2,-7)$

6. $(-6,1), (2,-5)$

7. $(0,5), (4,5)$

8. $(-3,5), (-3,7)$

9. $(2,17), (2,\frac{3}{4})$

10. $(2,-7), (-7,-7)$

In Exercises 11–24, determine an equation in point-slope form for the line that satisfies the given data.

11. $m = 4$, point: $(5, 3)$

12. $m = -3$, point: $(-6, 1)$

13. $m = \frac{2}{3}$, point: $(1, -5)$

14. $m = \frac{3}{2}$, point: $(0, 5)$

15. $m = 0$, point: $(2, -5)$

16. m undefined, point: $(-3, 6)$

17. $m = -\frac{5}{3}$, point: $(2, 0)$

18. $m = -\frac{3}{4}$, point: $(-3, -1)$

19. m undefined, point: $(7, -4)$

20. $m = 0$, point: $(-16, 5)$

21. points: $(1, 5)$ and $(-3, 2)$

22. points: $(-5, 6)$ and $(1, -6)$

23. $(-5, 3), (7, 3)$

24. $(\frac{2}{3}, \frac{3}{4}), (0, \frac{25}{8})$

In Exercises 25–28, determine an equation in slope-intercept form for the line that satisfies the given information.

25. $m = 2, b = 4$

26. $m = -3, b = 5$

27. $m = 5, b = -3$

28. $m = -4, b = -2$

In Exercises 29–36, rewrite each equation in slope-intercept form, find the slope m and y-intercept b, and sketch the graph.

29. $y - 3x = 6$

30. $2x - y = 5$

31. $2y - 5x = 8$

32. $2x - 3y = 9$

33. $5x - 2y - 10 = 0$

34. $4y - 3x - 4 = 0$

35. $x = -3y + 7$

36. $3x = 5y - 6$

For each equation in Exercises 37–40, (a) give the slope and (b) graph the line determined by the equation.

37. $y = 4$

38. $x = -2$

39. $x = \frac{9}{2}$

40. $y = -7$

Solve Exercises 41–44.

41. *Machine technology*   A grinding machine operates at 1 780 rev/min. The surface speed s in cm/s is given by the formula $s = \dfrac{1780\pi d}{60}$, where d is the diameter in cm. What is the slope of this equation? (Assume that d is on the horizontal axis.)

42. *Meteorology*   The relationship between the Fahrenheit and Celsius temperatures is linear. If the Fahrenheit temperatures are put on the horizontal or x-axis and the Celsius temperatures are put on the vertical or y-axis, then two points are $(-40, -40)$ and $(32, 0)$.

    (a) What is the slope of this line?

    (b) What is the y-intercept?

    (c) Write the equation in point-slope form and in slope-intercept form.

    (d) Graph the line.

43. *Physics*   A spring coil has an unstretched or natural length of L_0, and requires a force F of kx to stretch it x units beyond its natural length. The letter k represents a constant known as the **spring constant**. The distance the spring is stretched, x, is equal to $L - L_0$, where L is the length of the spring when it is stretched.

    (a) Write an equation in slope-intercept form for the force F in terms of k, L, and L_0.

    (b) If $k = 4.5$ and $L_0 = 6$ cm, write an equation in slope-intercept form for the force F in terms of L.

    (c) Graph the equation in (b).

44. *Physics*   The pressure P at a depth h in a liquid depends on the density of the liquid, ρ. In a certain liquid at 4 ft, the pressure is 250 lb/ft². At 9 ft the pressure is 562.5 lb/ft². Write an equation in point-slope form for the pressure in terms of the depth.

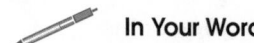 **In Your Words**

45. (a) Explain what it means for a line to have a slope of 0.
  (b) Explain what it means for a line to have an undefined slope.

46. Explain how you can determine the equation for a line if you know the coordinates of two points on that line.

≡ 5.2
GRAPHICAL AND ALGEBRAIC METHODS FOR SOLVING TWO LINEAR EQUATIONS IN TWO VARIABLES

In this section, we will begin to look at methods for solving a system of simultaneous linear equations. Simultaneous linear equations are equations containing the same variables such as

$$2x + y = 5$$
$$4x - y = 1$$

Our task in this section is to determine all the points, or ordered pairs, that these two equations have in common. We will begin by looking at a way to use graphing to help find these common points. Since we are looking for a common point of these two lines, this point will be where the two lines intersect if we graph each line. A graph of these two lines is shown in Figure 5.4.

As you can see, the lines appear to meet at point $(1, 3)$. A quick check of both equations will show that point $(1, 3)$ is on both lines. To check substitute $x = 1$ and $y = 3$ in the first equation. We obtain

$$2(1) + 3 = 2 + 3 = 5$$

Substituting $x = 1$ and $y = 3$ in the second equation, produces

$$4(1) - 3 = 4 - 3 = 1$$

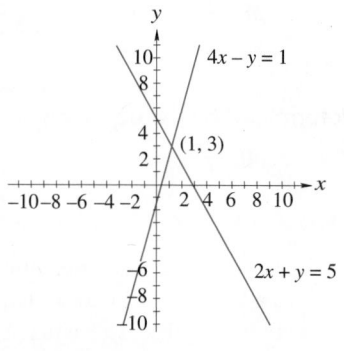

FIGURE 5.4

But, as we saw in Chapter 4, graphical methods are not always accurate ways to determine the roots to an equation. Graphical methods are also not very accurate ways to determine the common solutions of simultaneous equations. What we need are some algebraic methods for solving a system of equations. We will learn two methods in this section. Both methods involve solving for one of the variables by eliminating the other variable. These are called **elimination methods**.

Substitution Method

The first elimination method involves elimination by substitution. It is generally called the **substitution method**.

> **Substitution method for solving a system of linear equations**
>
> To use the substitution method to solve a system of linear equations:
>
> 1. Solve one equation for one of the variables.
> 2. Substitute the solution from Step 1 into the other equation and solve for the remaining variable.
> 3. Substitute the value from Step 2 into the equation from Step 1 and solve for the other variable.

In the substitution method, we change two equations in two variables into one equation in one variable. The next two examples show how to use the substitution method to solve systems of two equations in two variables.

EXAMPLE 5.9

Use the substitution method to solve the system of equations

$$\begin{cases} 2x + y = 5 & (1) \\ 4x - y = 1 & (2) \end{cases}$$

Solution   We will solve the first equation for y.

$$y = 5 - 2x \qquad (3)$$

Substitute this solution for y in equation (2). Equation (2) becomes

$$4x - (5 - 2x) = 1$$

Solve this equation for x.

$$4x - 5 + 2x = 1$$
$$6x = 6$$
$$x = 1$$

We then substitute this solution for x in equation (3) and find

$$y = 5 - 2(1)$$
$$y = 3$$

The solution is $(1, 3)$, which was the same answer we got by graphing.

EXAMPLE 5.10

Use the substitution method to solve the system of equations.

$$\begin{cases} -2x + 2y = 5 & (1) \\ x + 6y = 1 & (2) \end{cases}$$

Solution   This time we will solve equation (2) for x and get

$$x = 1 - 6y \qquad (3)$$

EXAMPLE 5.10 (Cont.)

Substituting this value, $1 - 6y$, for x in equation (1) we get

$$-2(1 - 6y) + 2y = 5$$
$$-2 + 12y + 2y = 5$$
$$-2 + 14y = 5$$
$$14y = 7$$
$$y = \tfrac{7}{14} = \tfrac{1}{2}$$

Replacing the y in equation (3) with $\frac{1}{2}$ we get

$$x = 1 - 6\left(\tfrac{1}{2}\right)$$
$$x = 1 - 3$$
$$x = -2$$

The solution appears to be $\left(-2, \tfrac{1}{2}\right)$.

◾

To be certain that we have the correct solution, we should substitute $\left(-2, \tfrac{1}{2}\right)$ into the original equations—equations (1) and (2)—and see if this solution satisfies both of these equations. It is very important that you always check your work by using the *original* equations. If you made any errors, you might not detect them unless you check your work in the original problem.

Addition Method

The second algebraic method for solving a system of linear equations is normally called the **addition method**. Technically, its name is the **elimination method by addition and subtraction**.

> **Addition method for solving a system of linear equations**
>
> To use the addition method to solve a system of linear equations:
>
> Add or subtract the two equations in order to eliminate one of the variables.
>
> It is sometimes necessary to multiply the original equations by a constant before it is possible to eliminate one of the variables by adding or subtracting the equations.

The next two examples will show how the addition method is used. These are the same examples that we worked with when using the substitution method.

EXAMPLE 5.11

Use the addition method to solve the system

$$\begin{cases} 2x + y = 5 & \qquad (1) \\ 4x - y = 1 & \qquad (2) \end{cases}$$

EXAMPLE 5.11 (Cont.)

Solution Equation (1) has a $(+y)$ term and equation (2) has a $(-y)$ term. If we add equations (1) and (2), the new equation will not have a y term.

$$2x + y = 5$$
$$4x - y = 1$$

adding $\quad 6x \quad\ \ = 6 \qquad\qquad\qquad (1)+(2)$

or $\qquad\quad x = 1 \qquad\qquad\qquad\qquad (3)$

Substituting this value of x into equation (1) we get

$$2(1) + y = 5$$
$$y = 3$$

So, the solution is $x = 1$ and $y = 3$ or the ordered pair $(1, 3)$.

EXAMPLE 5.12

Use the addition method to solve the system

$$\begin{cases} -2x + 2y = 5 & (1) \\ x + 6y = 1 & (2) \end{cases}$$

Solution Equations (1) and (2) do not have any terms that are equal so we will have to multiply at least one equation by a constant. If we multiply equation (2) by 2, the x-term will become $2x$, which is the additive inverse of the x term in equation (1). After multiplication, equation (2) becomes

$$2x + 12y = 2 \qquad\qquad\qquad (3)$$

and adding equations (1) and (3) we get

$$-2x + 2y = 5 \qquad\qquad\qquad (1)$$
$$2x + 12y = 2 \qquad\qquad\qquad (3)$$

adding $\qquad\qquad 14y = 7 \qquad\qquad (1)+(3)$

or $\qquad\qquad\qquad y = \dfrac{1}{2} \qquad\qquad\qquad (4)$

Substituting this value for y into equation (2) we get

$$x + 6\left(\tfrac{1}{2}\right) = 1$$
$$x + 3 = 1$$
$$x = -2$$

Thus, we have found $x = -2$, $y = \frac{1}{2}$, and the solution is $\left(-2, \frac{1}{2}\right)$.

You may get very strange looking results when you attempt to solve a system of equations. For example, sometimes all the variables vanish. Consider the next example.

EXAMPLE 5.13

Use the addition method to solve the system

$$\begin{cases} 2x + 3y = 6 & (1) \\ 4x + 6y = 30 & (2) \end{cases}$$

Solution If we multiply equation (1) by -2 the terms containing the variable x will be additive inverses of each other. This multiplication makes equation (1) into

$$-4x - 6y = -12 \qquad (3)$$

and adding $\underline{4x + 6y = 30} \qquad (2)$

$$0 = 18 \qquad (3)+(2)$$

Now we know that $0 \neq 18$, so something must be wrong. If we check our work, there do not appear to be any errors. Let's graph these equations. The graph in Figure 5.5 indicates the problem. The lines are parallel. They will never intersect, so there is no solution to this system of equations. ▪

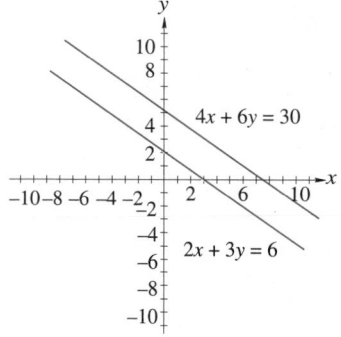

FIGURE 5.5

When two lines are parallel they will not intersect. Since there is no solution, the equations in the system are said to be **inconsistent**. When you try to solve a system of linear equations that is inconsistent, you will get an untrue equation. In Example 5.13, this equation was $0 = 18$.

In Example 5.14, we examine another kind of system of linear equations in which all the variables vanish.

EXAMPLE 5.14

Solve the system of linear equations

$$\begin{cases} 3x + 5y = 15 & (1) \\ 9x + 15y = 45 & (2) \end{cases}$$

Solution Multiplying equation (1) by -3, we get

$$-9x - 15y = -45 \qquad (3)$$

and adding this to equation (2) we have the following:

$$-9x - 15y = -45 \qquad (3)$$

$$\underline{9x + 15y = 45} \qquad (2)$$

$$0 = 0 \qquad (3)+(2)$$

It is certainly true that $0 = 0$. But, how will that help us solve this system of equations? Once again, we will turn to graphing to help us solve this system. (See Figure 5.6.) The graphs of these two equations are exactly the same. Thus, there are an unlimited number of solutions to both equations. In fact, any ordered pairs that satisfy one of the equations will satisfy the other. ▪

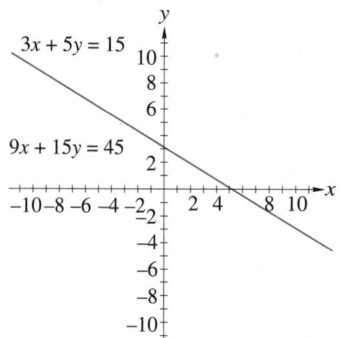

FIGURE 5.6

If the graphs of two equations coincide, we say that the equations are **dependent**. In the case of dependent equations, every solution to one equation will be a solution to the other equation. When you try to solve a system of dependent equations, you will get an equality of two constants. In Example 5.14, this was $0 = 0$.

If we solve Equation (1) in Example 5.14 for y, we get $y = -\frac{3}{5}x + 3$. (You get the same result if you solve Equation (2) for y.) Hence, every ordered pair (a, b) of the form $(a, -\frac{3}{5}a + 3)$ is a solution of the given system.

Most of the systems of equations we will work with are consistent. A system of linear equations is **consistent** if it has exactly one point as the solution.

In this section, we looked at three methods for solving systems of linear equations—the graphical method, the substitution method, and the elimination method. In the next section, we will learn about a new way of working with numbers that will help us find an easier method for solving systems of linear equations.

Exercise Set 5.2

In Exercises 1–8, graphically solve each system of equations. Estimate each answer to the nearest tenth if necessary. Check your answers, but remember that your estimate may not check exactly.

1. $\begin{cases} x + y = 6 \\ x - y = 2 \end{cases}$

3. $\begin{cases} 2x + 3y = 15 \\ 4x - 4y = 10 \end{cases}$

5. $\begin{cases} 3x + 5y = 32 \\ 10x - 5y = -32 \end{cases}$

7. $\begin{cases} 5x + 10y = 0 \\ 3x - 4y = 11 \end{cases}$

2. $\begin{cases} x + y = 8 \\ 2x - y = 1 \end{cases}$

4. $\begin{cases} 3x + 2y = 2 \\ 2x - 6y = -39 \end{cases}$

6. $\begin{cases} -5x + 11y = 11 \\ 5x + 2y = -11 \end{cases}$

8. $\begin{cases} 4x + y = -8 \\ 3x + 2y = 0 \end{cases}$

In Exercises 9–20, use the substitution method to solve each system of equations.

9. $\begin{cases} y = 3x - 4 \\ x + y = 8 \end{cases}$

13. $\begin{cases} 2x + 5y = 6 \\ x - y = 10 \end{cases}$

17. $\begin{cases} 4.8x - 1.3y = 16.9 \\ -7.2x - 2.8y = -9.2 \end{cases}$

10. $\begin{cases} x = -2y + 12 \\ x + y = 5 \end{cases}$

14. $\begin{cases} 3x - 2y = 5 \\ -7x + 4y = -7 \end{cases}$

18. $\begin{cases} 2.3x + 1.7y = 8.5 \\ -6.7x + 3.7y = 38.4 \end{cases}$

11. $\begin{cases} y = -2x - 2 \\ 3x + 2y = 0 \end{cases}$

15. $\begin{cases} 2x + 3y = 3 \\ 6x + 4y = 15 \end{cases}$

19. $\begin{cases} 4.2x + 3.7y = 10.79 \\ 6.5x - 0.3y = -15.24 \end{cases}$

12. $\begin{cases} x = 7 + 2y \\ 3x + 4y = 1 \end{cases}$

16. $\begin{cases} 2x + 2y = -3 \\ 4x + 9y = 5 \end{cases}$

20. $\begin{cases} 0.75x + 1.5y = -0.225 \\ 3.13x + 1.74y = 13.073 \end{cases}$

In Exercises 21–30, use the addition method to solve each system of equations.

21. $\begin{cases} x + y = 9 \\ x - y = 5 \end{cases}$

24. $\begin{cases} 3x - 2y = 8 \\ 5x + y = 9 \end{cases}$

27. $\begin{cases} 3x - 5y = 37 \\ 5x - 3y = 27 \end{cases}$

22. $\begin{cases} 2x + 3y = 5 \\ -2x + 5y = 3 \end{cases}$

25. $\begin{cases} 3x - 2y = -15 \\ 5x + 6y = 3 \end{cases}$

28. $\begin{cases} x + \frac{1}{2}y = 7 \\ 4x - 2y = 5 \end{cases}$

23. $\begin{cases} -x + 3y = 5 \\ 2x + 7y = 3 \end{cases}$

26. $\begin{cases} 2x - 3y = 11 \\ 6x - 5y = 13 \end{cases}$

29. $\begin{cases} 6.4x - 1.7y = 66.7 \\ -4.2x + 5.1y = -62.1 \end{cases}$

30. $\begin{cases} 1.6x - 2.9y = -2.645 \\ 2.4x + 1.4y = 11.27 \end{cases}$

In Exercises 31–40, solve each system of equations by either the substitution method or the addition method. Graph each system of equations.

31. $\begin{cases} 2x + 3y = 5 \\ x - 2y = 6 \end{cases}$

34. $\begin{cases} 6x + 12y = 7 \\ 8x - 15y = -1 \end{cases}$

37. $\begin{cases} x - 9y = 0 \\ \dfrac{x}{3} = 2y + \frac{1}{3} \end{cases}$

39. $\begin{cases} 4.9x + 1.7y = 10.6 \\ 3.6x - 1.2y = 14.4 \end{cases}$

32. $\begin{cases} 2x - 3y = -14 \\ 3x + 2y = 44 \end{cases}$

35. $\begin{cases} 10x - 9y = 18 \\ 6x + 2y = 1 \end{cases}$

38. $\begin{cases} 5x + 3y = 7 \\ \frac{3}{2}x - \frac{3}{4}y = 9\frac{1}{4} \end{cases}$

40. $\begin{cases} 3.14x + 4.57y = -7.9 \\ 2.48x + 11.84y = 15.16 \end{cases}$

33. $\begin{cases} 8x + 3y = 13 \\ 3x + 2y = 11 \end{cases}$

36. $\begin{cases} 4x - 5y = 7 \\ -8x + 10y = -30 \end{cases}$

Solve Exercises 41–52.

41. *Land management*   The perimeter of a rectangular field is 36 km. The length of the field is 8 km longer than the width. What are the length and width of the field? (Hint: To find the length and width of this rectangular field, you need to solve this system of linear equations where L represents the length of the field and w the width.)

$$\begin{cases} 2L + 2w = 36 \\ L = w + 8 \end{cases}$$

42. *Land management*   The perimeter of a rectangular field is 72 mi. The length of the field is 9 mi longer that the width. What are the length and width of the field?

43. *Land management*   The perimeter of a rectangular field is 45 km. The length is 3 times the width. What are the length and width of the field?

44. *Land management*   The perimeter of a field in the shape of an isosceles triangle is 96 yd. The length of each of the two equal sides is $1\frac{1}{2}$ times the length of the third side. What are the lengths of the three sides of this field?

45. *Petroleum technology*   Two different gasohol mixtures are available. One mixture contains 5% alcohol and the other, 13% alcohol. In order to determine how much of each mixture should be used to get 10 000 L of gasohol containing 8% alcohol, you would solve the following equations:

$$\begin{cases} x + y = 10\,000 \\ 0.05x + 0.13y = (0.08)(10\,000) \end{cases}$$

where x is the number of liters of 5% gasohol mixture and y is the number of liters of 13% gasohol mixture. Determine the number of liters of each mixture that are needed.

46. *Petroleum technology*   Two different gasohol mixtures are available. One mixture contains 4% alcohol and the other, 12% alcohol. How much of each mixture should be used to get 20 000 L of gasohol containing 9% alcohol?

47. *Physics*   Two forces are applied to the ends of a beam. The force on one end is 8 kg and the force at the other end is not known. The unknown force is 5 m from the centroid; we do not know the distance of the 8-kg force from the centroid. If an additional force of 4 kg is applied to the 8-kg force, the unknown force must be increased by 3.2 kg for equilibrium to be maintained. Find the unknown mass and distance. (See Figure 5.7.)

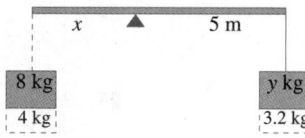

FIGURE 5.7

48. *Automotive technology* A 12-L cooling system is filled with 25% antifreeze. How many liters must be replaced with 100% antifreeze to raise the strength to 45% antifreeze?

In Exercises 49–52, use the following information.

The current that flows in each branch of a complex circuit can be found by applying **Kirchhoff's rules** to the circuit. The first rule applies to the junction of three or more wires as in Figure 5.8. The second applies to loops (circuits) or closed paths in the circuit.

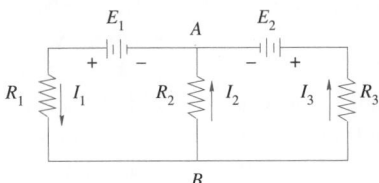

FIGURE 5.8

Rule 1: (Junction rule) The sum of the currents that flow into a junction is equal to the sum of the currents that flow out of the junction. In Figure 5.8, this means that $I_1 = I_2 + I_3$.

Rule 2: (Circuit rule) The sum of the voltages around any closed loop equals zero. In Figure 5.8, for the left loop this means that $E_1 = I_1 R_1 + I_2 R_2$ and for the right loop that $E_2 = R_2 I_2 - R_3 I_3$.

49. *Electronics* Find the currents in the three resistors of the circuit shown in Figure 5.8, given that $E_1 = 8$ V, $E_2 = 5$ V, $R_1 = 3\,\Omega$, $R_2 = 5\,\Omega$, and $R_3 = 6\,\Omega$. [Use $E_1 = I_1 R_1 + I_2 R_2$ and $E_2 = R_2 I_2 - R_3(I_1 - I_2)$.]

50. *Electronics* In Figure 5.8, if $E_1 = 10$ V, $E_2 = 15$ V, $R_1 = 2\,\Omega$, $R_2 = 4\,\Omega$, and $R_3 = 8\,\Omega$, find I_1, I_2, and I_3.

51. *Electronics* In Figure 5.9, we have $I_2 = I_1 + I_3$, $E_1 = R_1 I_1 + R_2 I_2$, and $E_2 = R_3 I_3 + R_2 I_2$. If $E_1 = 6$ V, $E_2 = 10$ V, $R_1 = 8\,\Omega$, $R_2 = 4\,\Omega$, and $R_3 = 7\,\Omega$, find I_1, I_2, and I_3.

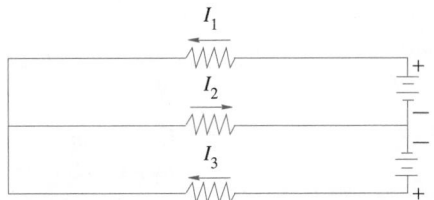

FIGURE 5.9

 52. One computer technique for solving systems of equations is called the **Gauss-Seidel method**. To apply this method, consider a system of two simultaneous linear equations.

$$\begin{cases} ax + by + c = 0 & (1) \\ dx + ey + f = 0 & (2) \end{cases}$$

Solve equation (1) for x and solve equation (2) for y to get

$$\begin{cases} x = -\dfrac{by + c}{a} & (3) \\ y = -\dfrac{dx + f}{e} & (4) \end{cases}$$

Now, guess a value for y. Since this is the first guess for y, we will call it y_1. Substitute the value y_1 in equation (3) to get a first value for x. (We will call this first x-value, x_1.) Substitute x_1 in equation (4) to get y_2. Substitute y_2 in equation (3) to get x_2, and so on. How long do you keep this up? You might continue until two consecutive values of x, say x_n and x_{n+1} are the same and until $y_n = y_{n+1}$.

Write a program to solve two linear equations in two variables using the Gauss-Seidel method. Test your program on the equations in Exercise 15. If, when you run the program, the values of x and y get large, then solve the first equation for y and the second for x.

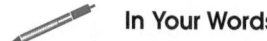
In Your Words

53. Describe the advantages and disadvantages of the substitution method compared to the addition method.

54. Describe what it means for a system of equations to be consistent or inconsistent.

55. Describe how to use the substitution method for solving a system of linear equations.

≡ 5.3
ALGEBRAIC METHODS FOR SOLVING THREE LINEAR EQUATIONS IN THREE VARIABLES

In Section 5.2 we learned to solve a system of two linear equations in two variables by a graphical method and by two algebraic methods of elimination. In this section, we will expand these algebraic techniques to allow us to solve a system of three linear variables. These methods will be used later to allow us to solve n equations with n variables.

The graph of a linear equation in three variables is a plane. Graphing a system of three equations in three variables requires the ability to graph three planes and their intersections. We cannot graph three planes so that we can tell where they intersect. As a result, we will not consider graphical solutions to a system of three equations in three variables.

Substitution Method

We will solve the same system of equations using both elimination methods. The first example uses the substitution method.

EXAMPLE 5.15

Solve this system of equations using the substitution method.

$$\begin{cases} x + 2y - 2z = 3 & (1) \\ 2x - y + 3z = -5 & (2) \\ 4x - 3y + z = 7 & (3) \end{cases}$$

Solution If we solve equation (3) for z we get

$$z = 3y - 4x + 7 \qquad (4)$$

Substituting this value of z in equation (2) changes it to

$$2x - y + 3(3y - 4x + 7) = 2x - y + 9y - 12x + 21 = -5$$

Combining terms results in

$$-10x + 8y = -26 \qquad (5)$$

Solving equation (5) for y produces

$$8y = 10x - 26$$

or $\qquad y = \tfrac{5}{4}x - \tfrac{13}{4} \qquad (6)$

Substituting the value for z from equation (4) and the value for y from equation (6)

EXAMPLE 5.15 (Cont.)

in equation (1) we get

$$x + 2\left(\tfrac{5}{4}x - \tfrac{13}{4}\right) - 2\left(3y - 4x + 7\right) = 3$$
$$x + \tfrac{5}{2}x - \tfrac{13}{2} - 6y + 8x - 14 = 3$$
$$\tfrac{23}{2}x - 6y = \tfrac{47}{2} \tag{7}$$

Substituting the value of y from equation (6) into equation (7) results in

$$\tfrac{23}{2}x - 6\left(\tfrac{5}{4}x - \tfrac{13}{4}\right) = \tfrac{47}{2}$$
$$\tfrac{23}{2}x - \tfrac{15}{2}x + \tfrac{39}{2} = \tfrac{47}{2}$$
$$\tfrac{8}{2}x = \tfrac{8}{2}$$

or

$$1x = 1$$

Using this value of $x = 1$ in equation (6) produces

$$y = \tfrac{5}{4} - \tfrac{13}{4} = -\tfrac{8}{4} = -2$$

Finally, using $x = 1$ and $y = -2$ in equation (4) we get

$$z = 3(-2) - 4(1) + 7 = -3$$

So, $x = 1$, $y = -2$, and $z = -3$.

Addition Method

There was no particular reason to begin by solving equation (3) for z in the previous example. We could just as easily have started by solving equation (1) for x or equation (2) for y or even equation (1) for y or z. Let's solve the same system of equations using the addition method.

EXAMPLE 5.16

Solve this system of equations using the addition method.

$$\begin{cases} x + 2y - 2z = 3 & (1) \\ 2x - y + 3z = -5 & (2) \\ 4x - 3y + z = 7 & (3) \end{cases}$$

Solution We will begin by eliminating one of the variables. When this is done, we will have two equations with two variables. Let's start by eliminating the variable z. To do this, we multiply equation (3) by 2 and add this to equation (1).

$$\begin{array}{ll} 8x - 6y + 2z = 14 & \text{(3) multiplied by 2} \\ \underline{x + 2y - 2z = 3} & \text{(1)} \\ 9x - 4y \phantom{{}+2z} = 17 & \text{Adding to get equation (4)} \end{array}$$

EXAMPLE 5.16 (Cont.)

Next, we multiply equation (3) by 3 and subtract equation (2) from this new equation.

$$
\begin{array}{ll}
12x - 9y + 3z = 21 & \text{(3) multiplied by 3} \\
\underline{2x - y + 3z = -5} & \text{(2)} \\
10x - 8y = 26 & \text{Subtracting to get equation (5)}
\end{array}
$$

We now have two equations, (4) and (5), with two variables, x and y.

$$
\begin{array}{ll}
9x - 4y = 17 & \text{(4)} \\
10x - 8y = 26 & \text{(5)}
\end{array}
$$

If we multiply equation (4) by 2 and subtract equation (5) from that equation, we will eliminate the variable y.

$$
\begin{array}{ll}
18x - 8y = 34 & \text{(4) multiplied by 2} \\
\underline{10x - 8y = 26} & \text{(5)} \\
8x = 8 & \text{Subtracting to get (6)}
\end{array}
$$

Solving equation (6) for x, we get $x = 1$. Substituting this value in (4), we get $9 - 4y = 17$ or $-4y = 8$, which simplifies to $y = -2$. Then, if we substitute $x = 1$ and $y = -2$ in equation (3), we get $4 + 6 + z = 7$, or $z = -3$. Again, we get the solution $x = 1$, $y = -2$, and $z = -3$. ∎

As you can see, the elimination method by addition and subtraction is often an easier method to use. We will use this method again in the next example.

Application

EXAMPLE 5.17

By volume, one alloy is 70% copper, 20% zinc, and 10% nickel. A second alloy is 60% copper and 40% nickel. A third alloy is 30% copper, 30% nickel, and 40% zinc. How much of each must be mixed in order to get $1\,000\,\text{mm}^3$ of a final alloy that is 50% copper, 18% zinc, and 32% nickel?

Solution We must first determine the equations that are needed to solve this problem. If we let a represent the volume of the first alloy in the final alloy, b the volume of the second, and c the volume of the third, then we know that the total volume of the final alloy, $1\,000\,\text{mm}^3$, is $a + b + c$.

We know that the final alloy contains 50% or $500\,\text{mm}^3$ of copper and that this is $0.7a + 0.6b + 0.3c$. Also, 18% or $180\,\text{mm}^3$ of the final solution is zinc, so $0.2a + 0.4c = 180$. This gives you a system of three linear equations in three variables.

$$
\begin{cases}
a + b + c = 1\,000 & \text{(1)} \\
0.7a + 0.6b + 0.3c = 500 & \text{(2)} \\
0.2a \phantom{{}+ 0.6b} + 0.4c = 180 & \text{(3)}
\end{cases}
$$

EXAMPLE 5.17 (Cont.)

We can also establish a fourth equation for the amount of nickel in the final solution, 32% or 320 mm^3. This is

$$0.1a + 0.4b + 0.3c = 320 \tag{4}$$

We do not need equation (4) to solve the problem, but we can use it to check our answers.

Since equation (3) does not contain variable b, we will combine equations (1) and (2) to eliminate this variable.

$$
\begin{array}{lll}
0.7a + 0.6b + 0.3c = & 500 & \text{(2)} \\
0.6a + 0.6b + 0.6c = & 600 & \text{(1) multiplied by 0.6} \\
\hline
0.1a \quad\quad\quad -0.3c = & -100 & \text{Subtract to get (5).}
\end{array}
$$

If we now multiply equation (5) by 2 and subtract this from equation (3), we will eliminate variable a.

$$
\begin{array}{lll}
0.2a + 0.4c = & 180 & \text{(3)} \\
0.2a - 0.6c = & -200 & \text{(5) multiplied by 0.2} \\
\hline
c = & 380 & \text{Subtract to get } c.
\end{array}
$$

So, alloy c is 380 mm^3. Substituting this in (5) we get

$$
\begin{aligned}
0.1a - 0.3(380) &= -100 \\
0.1a - 114 &= -100 \\
0.1a &= 14 \\
a &= 140
\end{aligned}
$$

Then, substituting these values for a and c in equation (1) we get $140 + b + 380 = 1\,000$ or $b = 480$. The answer: we need 140 mm^3 of alloy a, 480 mm^3 of alloy b, and 380 mm^3 of alloy c. If you put these values in equation (4) you will see that they check. ▪

Application

EXAMPLE 5.18

A trucking company has three sizes of trucks, large (L), medium (M), and small (S). The trucks are needed to move some packages, which come in three different shapes. We will call these three different shaped packages, A, B, and C. From experience, the company knows that these trucks can hold the combination of packages as shown in this chart.

	Size of Truck		
	Medium	Medium	Small
Package A	12	8	0
Package B	10	5	4
Package C	8	7	6

EXAMPLE 5.18 (Cont.)

The company has to deliver a total of 64 A packages, 77 B packages, and 99 C packages. How many trucks of each size are needed, if each truck is fully loaded?

Solution   From the table we can see that the 64 A packages must be arranged with 12 on each large truck, 8 on each medium truck, and 0 on each small truck. We can write this as

$$12L + 8M = 64$$

Similarly, the B packages satisfy $10L + 5M + 4S = 77$ and the C packages satisfy $8L + 7M + 6S = 99$. Thus, we have the system

$$\begin{cases} 12L & + & 8M & & & = & 64 & \qquad (1) \\ 10L & + & 5M & + & 4S & = & 77 & \qquad (2) \\ 8L & + & 7M & + & 6S & = & 99 & \qquad (3) \end{cases}$$

Since equation (1) does not contain variable S, we will combine equations (2) and (3) to eliminate it.

$$\begin{aligned} 30L + 15M + 12S &= 231 &\qquad \text{(2) multiplied by 3} \\ \underline{16L + 14M + 12S} &= \underline{198} &\qquad \text{(3) multiplied by 2} \\ 14L + M &= 33 &\qquad \text{Subtracting to get (4)} \end{aligned}$$

Now multiply equation (4) by 8 and subtract equation (1), and variable M is eliminated.

$$\begin{aligned} 112L + 8M &= 264 &\qquad \text{(4) multiplied by 8} \\ \underline{12L + 8M} &= \underline{64} &\qquad (1) \\ 100L &= 200 &\qquad \text{Subtract.} \end{aligned}$$

So, $L = \frac{200}{100} = 2$. Substituting this in (1), we get

$$24 + 8M = 64$$
$$8M = 64 - 24$$
$$8M = 40$$
$$M = 5$$

And finally, substituting these values for L and M in (2), we obtain

$$10(2) + 5(5) + 4S = 77$$
$$20 + 25 + 4S = 77$$
$$4S = 77 - 45$$
$$= 32$$
$$S = 8$$

So, a total of 2 large, 5 medium, and 8 small trucks are needed.

As with a system of two equations in two variables, it is possible to have a system of three equations with three variables that is either inconsistent or dependent. If an elimination method results in an equation of the type $0x + 0y + 0z = c$, or $0 = c$, where $c \neq 0$, then the system is **inconsistent** and has no solutions. Graphically, this would mean that the plane of one equation was parallel to the plane of another equation or the planes intersect in three pairs of parallel lines.

If the elimination method results in an equation of the type $0x + 0y + 0z = 0$, or $0 = 0$, then two or more of the equations graph the same plane and the system is **dependent**.

EXAMPLE 5.19

Solve the system

$$\begin{cases} 3x - y - z = 5 & (1) \\ x - 5y + z = 3 & (2) \\ x + 2y - z = 1 & (3) \end{cases}$$

Solution Adding equations (1) and (2) produces

$$4x - 6y = 8 \qquad (4)$$

and adding equations (2) and (3) gives

$$2x - 3y = 4 \qquad (5)$$

If we multiply equation (5) by 2 and subtract that result from equation (4), we obtain

$$0 = 0$$

Thus, we see that the given system of equations is dependent; and every ordered triple (a, b, c) of the form $(a, \frac{2}{3}a - \frac{4}{3}, \frac{7}{3}a - \frac{11}{3})$ is a solution of the given equation. ▪

Exercise Set 5.3

In Exercises 1–4, use the substitution method to solve each system of equations.

1. $\begin{cases} 2x + y + z = 7 \\ x - y + 2z = 11 \\ 5x + y - 2z = 1 \end{cases}$
2. $\begin{cases} x + y + 2z = 0 \\ 2x - y + z = 6 \\ 4x + 2y + 2z = 0 \end{cases}$
3. $\begin{cases} 2x - y - z = -8 \\ x + y - z = -9 \\ x - y + 2z = 7 \end{cases}$
4. $\begin{cases} x + y + 5z = -10 \\ x - y - 5z = 11 \\ -x + y - 5z = 13 \end{cases}$

In Exercises 5–14, use the addition method to solve each system of equations. (Exercises 5–14 are the same as Exercises 1–4.)

5. $\begin{cases} 2x + y + z = 7 \\ x - y + 2z = 11 \\ 5x + y - 2z = 1 \end{cases}$
7. $\begin{cases} 2x - y - z = -8 \\ x + y - z = -9 \\ x - y + 2z = 7 \end{cases}$
9. $\begin{cases} x + y + z = 2 \\ 8x - 2y + 4z = -3 \\ 6x - 4y - 3z = 3 \end{cases}$
11. $\begin{cases} 3x - y - 2z = 11 \\ -x + 3y + 2z = -1 \\ 2x - 2y - 4z = 17 \end{cases}$

6. $\begin{cases} x + y + 2z = 0 \\ 2x - y + z = 6 \\ 4x + 2y + 2z = 0 \end{cases}$
8. $\begin{cases} x + y + 5z = -10 \\ x - y - 5z = 11 \\ -x + y - 5z = 13 \end{cases}$
10. $\begin{cases} x + y - z = 7 \\ 8x + 4y + 2z = 21 \\ 4x + 3y + 6z = 2 \end{cases}$
12. $\begin{cases} x - 2y + z = -4 \\ 2x + y + 3z = 5 \\ 6x + 3y + 12z = 6 \end{cases}$

13.
$$\begin{cases} 2x + 3y + 3z = 9 \\ 5x - 2y + 8z = 6 \\ 4x - y + 5z = -1 \end{cases}$$

14.
$$\begin{cases} x + 2y + 3z = 4 \\ 2x - 3y - 4z = -1 \\ 3x - 4y + 5z = 6 \end{cases}$$

Solve Exercises 15–22.

15. *Electronics*   Kirchhoff's law for current states that the sum of the currents into any point equals zero. Applying this to junction A in Figure 5.10 produces the equation $I_1 - I_2 + I_3 = 0$. Kirchhoff's voltage law states that the sum of the voltages around any closed loop equals zero. Applying this first to the left loop and then the right loop in Figure 5.10 results in the equations $6I_1 + 6I_2 = 18$ and $6I_2 + I_3 = 14$. What are the values of the currents I_1, I_2, and I_3? (Note that electromotive force E equals current I times resistance or $E = IR$.)

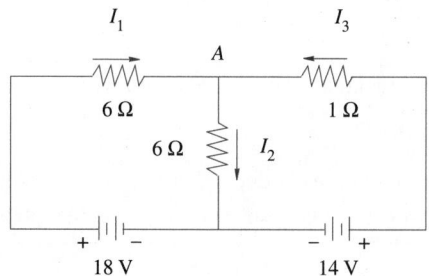

FIGURE 5.10

16. *Electronics*   Applying Kirchhoff's laws to Figure 5.11 produces the following equations.

$$\begin{cases} I_1 + I_2 - I_3 = 0 \\ 3I_1 - 5I_2 - 10 = 0 \\ 5I_2 + 6I_3 - 5 = 0 \end{cases}$$

Find the currents associated with I_1, I_2, and I_3.

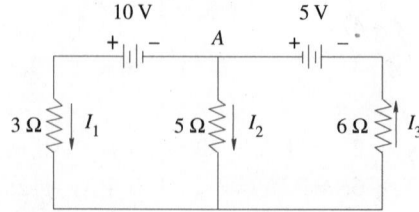

FIGURE 5.11

17. The standard equation for a circle is $x^2 + y^2 + ax + by + c = 0$. A circle passes through the points $P(5, 1)$, $Q(-2, -6)$, and $R(-1, -7)$. When the x- and y-coordinates for point P are put in the standard equation, it becomes $5^2 + 1^2 + a \cdot 5 + b \cdot 1 + c = 0$, or $5a + b + c + 26 = 0$. Use the coordinates of Q and R to obtain two more versions of the standard equation for this circle and then solve your system of equations for a, b, and c.

18. Another circle passes through the points $S(4, 16)$, $T(-6, -8)$, and $U(11, -1)$. Find the values of a, b, and c.

19. *Transportation*   A trucking company has three sizes of trucks, large (L), medium (M), and small (S). Experience has shown that the large truck can carry 7 of container A, 6 of container B, and 4 of container C. The medium truck can carry 6 of A, 3 of B, and 2 of C and the small truck can carry 8 of A, 1 of B, and 2 of C. How many trucks of the three sizes are needed to deliver 64 of container A, 33 of B, and 26 of C?

20. *Petroleum technology*   Three crude oils are to be mixed and loaded aboard a supertanker that can carry 450 000 tonnes (metric tons, t). The crudes contain the following percentages of light-, medium-, and heavy-weight oils:

	Light	Medium	Heavy
Crude oil A	10%	20%	70%
Crude oil B	30%	40%	30%
Crude oil C	43%	44%	13%

How many tonnes of each crude should be mixed so that the new mixtures contain 24% light-, 32% medium-, and 44% heavy-weight oils?

21. *Landscape architecture*   By weight, a 10-10-10 fertilizer contains 10% nitrogen, 10% phosphorous, and 10% potash. A 12-0-6 fertilizer contains 12% nitrogen, no phosphorous, and 6% potash. A landscaper has three types of fertilizer in stock: one is 10-12-15, a second is 10-0-5, and a third is 30-6-15. How much of each must be mixed in order to get 400 lb of fertilizer that is 16-3-9?

22. *Business* A company makes three types of patio furniture: chairs, tables, and recliners. Each requires the number of units of wood, plastic, and aluminum shown below:

	Wood	Plastic	Aluminum
Chair	1 unit	1 unit	2 units
Table	2 units	4 units	5 units
Recliner	1 unit	2 units	3 units

The company has in stock 500 units of wood, 900 units of plastic, and 1,300 units of aluminum. For its end-of-season production, the company wants to use all its stock. To do this, how many chairs, tables, and recliners should it make?

 **In Your Words**

23. **(a)** Which do you find easier to use: the substitution method or the addition method?

   (b) Write an explanation defending your position.

24. Under what conditions is the substitution method easier than the addition method?

25. Explain what it means for a system of equations to be dependent.

 # 5.4
DETERMINANTS

We've learned to solve two equations in two variables and to solve three equations in three variables. Many problems result in having to solve four equations in four variables or five equations in five variables.

Whenever you have to solve more than two equations in two variables, the substitution and the addition methods become very long and difficult. There are other methods that are easier, especially when used with a calculator or a computer. This section gives the background material for one of the easier methods. We begin by giving the definition of a determinant.

If a, b, c, and d are any four real numbers, then the symbol

$$\begin{vmatrix} a & b \\ c & d \end{vmatrix}$$

is called a 2×2 **determinant** or a **determinant of the second order**. The numbers a, b, c, and d are called the **elements** or **entries** of the determinant. The value of a determinant is the number $ad - cb$, so we have the following statement.

Evaluating a 2×2 Determinant

If a, b, c, and d are any four real numbers, then

$$\begin{vmatrix} a & b \\ c & d \end{vmatrix} = ad - cb$$

As a memory aid, you might want to draw the diagonals of the determinant.

$$\begin{vmatrix} a & b \\ c & d \end{vmatrix} = ad - cd$$

EXAMPLE 5.20

Evaluate the determinant $\begin{vmatrix} 7 & 5 \\ 2 & 6 \end{vmatrix}$.

Solution $\begin{vmatrix} 7 & 5 \\ 2 & 6 \end{vmatrix} = 7(6) - 2(5) = 42 - 10 = 32$

EXAMPLE 5.21

Evaluate the determinant $\begin{vmatrix} -6 & -2 \\ 3 & 4 \end{vmatrix}$.

Solution $\begin{vmatrix} -6 & -2 \\ 3 & 4 \end{vmatrix} = (-6)(4) - 3(-2) = -24 - (-6) = -18$

Determinants have many useful properties and can be used to solve simultaneous equations. In the next section, we will learn how to use determinants to solve equations. In this section, we will become familiar with their properties. We will use them later.

The determinants that we have used have all been 2×2 (read 2-by-2), or second order determinants. A 3×3 (3-by-3) determinant or a **determinant of the third order** is represented by the symbol

$$\begin{vmatrix} a_1 & a_2 & a_3 \\ b_1 & b_2 & b_3 \\ c_1 & c_2 & c_3 \end{vmatrix}$$

where $a_1, a_2, a_3, b_1, b_2, b_3, c_1, c_2$, and c_3 are any real numbers. We will learn how to evaluate a 3×3 determinant later in this section.

As we mentioned, the numbers $a_1, a_2, \ldots, c_3$, which form the determinant, are called the elements or entries. The **rows** are numbered from top to bottom.

a	b	Row 1	a_1	b_1	c_1
c	d	Row 2	a_2	b_2	c_2
		Row 3	a_3	b_3	c_3

The **columns** are numbered from left to right.

$$\begin{vmatrix} a & b \\ c & d \end{vmatrix}$$

Column 1 Column 2 Column 3

$$\begin{vmatrix} a_1 & b_1 & c_1 \\ a_2 & b_2 & c_2 \\ a_3 & b_3 & c_3 \end{vmatrix}$$

A determinant has the same number of rows and columns.

The **main** or **principal diagonal** of a determinant is the diagonal from upper left to lower right. For example, in the determinant

$$\begin{vmatrix} a & b & c \\ d & e & f \\ g & h & i \end{vmatrix}$$

the entries on the main diagonal are a, e, and i.

A determinant is in **triangular form** if all entries below, or above, the main diagonal are zero. For example, the following determinants are both in triangular form.

$$\begin{vmatrix} 2 & -3 & 5 \\ 0 & 9 & -7 \\ 0 & 0 & \frac{1}{2} \end{vmatrix} \text{ and } \begin{vmatrix} 2 & 0 & 0 \\ -5 & -\frac{3}{5} & 0 \\ 3 & \pi & 17 \end{vmatrix}$$

Property 1 for Determinants

If you interchange (or swap) any two rows or any two columns of a determinant, you change its sign.

EXAMPLE 5.22

If $\begin{vmatrix} 2 & 3 \\ 1 & 9 \end{vmatrix} = 15$ then $\begin{vmatrix} 1 & 9 \\ 2 & 3 \end{vmatrix} = -15$, because the first and second rows were swapped. Thus, $\begin{vmatrix} 2 & 3 \\ 1 & 9 \end{vmatrix} = -\begin{vmatrix} 1 & 9 \\ 2 & 3 \end{vmatrix}$.

EXAMPLE 5.23

If $\begin{vmatrix} 3 & 1 & 5 \\ 2 & 0 & -2 \\ -5 & -1 & 4 \end{vmatrix} = -14$, then $\begin{vmatrix} 5 & 1 & 3 \\ -2 & 0 & 2 \\ 4 & -1 & -5 \end{vmatrix} = 14$, because the first and third columns were swapped.

Property 2 for Determinants

If you multiply every entry in one row or column of a determinant by a constant k, the result is the same as multiplying the value of the determinant by k.

EXAMPLE 5.24

If you multiply each entry in the second column of the determinant $\begin{vmatrix} 2 & 3 \\ 1 & 9 \end{vmatrix} = 15$

by 5 you get $\begin{vmatrix} 2 & 15 \\ 1 & 45 \end{vmatrix} = 75$, and $75 = 5 \times 15$.

EXAMPLE 5.25

If each element in the first row of the determinant $\begin{vmatrix} 3 & 1 & 5 \\ 2 & 0 & -2 \\ -5 & -1 & 4 \end{vmatrix} = -14$ is

multiplied by -3, the result is the determinant $\begin{vmatrix} -9 & -3 & -15 \\ 2 & 0 & -2 \\ -5 & -1 & 4 \end{vmatrix}$. It has a value

of $(-3)(-14) = 42$.

Property 2 also means that if all the elements of a single row or column have a common factor, then it can be factored out of the entire determinant.

EXAMPLE 5.26

In the determinant $\begin{vmatrix} 2 & 9 & 8 \\ 5 & 7 & 12 \\ 9 & -2 & -16 \end{vmatrix}$, all the entries in column 3 have a common

factor of 4. So this determinant can be rewritten as $4\begin{vmatrix} 2 & 9 & 2 \\ 5 & 7 & 3 \\ 9 & -2 & -4 \end{vmatrix}$.

Property 3 for Determinants

If you add a constant multiple of the entries in any one row (or column) of a determinant to the corresponding entries in any other row (or column), the value of the determinant will not be changed.

EXAMPLE 5.27

In the determinant $\begin{vmatrix} 7 & 3 \\ -5 & 1 \end{vmatrix}$, if you multiply each number in the first row by 4 and add these new numbers to the second row, the value of the determinant will not change.

$$\begin{vmatrix} 7 & 3 \\ -5 & 1 \end{vmatrix} = 7 - (-15) = 22 \tag{5.34}$$

$$\begin{vmatrix} 7 & 3 \\ 28-5 & 12+1 \end{vmatrix} = \begin{vmatrix} 7 & 3 \\ 23 & 13 \end{vmatrix} = 91 - 69 = 22 \tag{5.36}$$

EXAMPLE 5.28

Consider $\begin{vmatrix} 3 & 1 & 5 \\ 2 & 0 & -2 \\ -5 & -1 & 4 \end{vmatrix} = -14$. Multiply the second row by 2 and add this product to the third row. The result is

$$\begin{vmatrix} 3 & 1 & 5 \\ 2 & 0 & -2 \\ 4-5 & 0-1 & -4+4 \end{vmatrix} = \begin{vmatrix} 3 & 1 & 5 \\ 2 & 0 & -2 \\ -1 & -1 & 0 \end{vmatrix} = -14$$

Property 4 for Determinants

If any two rows (or columns) of a determinant are the same, its value is zero.

If you combine Properties 2 and 4, you can see that if any row (or column) is a multiple of any other row (or column) then the value of the determinant is zero.

EXAMPLE 5.29

(a) $\begin{vmatrix} 2 & 3 & -1 \\ 9 & 8 & 7 \\ 2 & 3 & -1 \end{vmatrix}$, because rows 1 and 3 are the same.

(b) $\begin{vmatrix} 6 & 9 & 9 \\ 1 & 3 & 3 \\ -7 & 4 & 4 \end{vmatrix} = 0$, because columns 2 and 3 are the same.

(c) $\begin{vmatrix} 8 & 4 & -2 \\ -2 & -1 & \frac{1}{2} \\ 7 & 16 & -3 \end{vmatrix} = 0$, because each element in row 1 is -4 times each element in row 2.

Property 5 for Determinants

If a determinant is in triangular form, its value is the product of its main diagonal.

EXAMPLE 5.30

$$\begin{vmatrix} 9 & 3 & -1 & 2 \\ 0 & 4 & -2 & 7 \\ 0 & 0 & -\frac{1}{3} & 17 \\ 0 & 0 & 0 & -1 \end{vmatrix} = 9(4)\left(-\frac{1}{3}\right)(-1) = 12.$$

Minors

Each element in a determinant has a minor associated with it. The **minor** of a given element is the determinant that is formed by deleting the row and column in which the element lies.

EXAMPLE 5.31

Consider the determinant $\begin{vmatrix} 2 & 4 & -4 \\ 5 & 8 & 1 \\ -6 & -3 & 7 \end{vmatrix}$.

(a) The minor of the first element, 2, is found by crossing out the first row and first column.

$$\begin{vmatrix} 2 & 4 & -4 \\ 5 & 8 & 1 \\ -6 & -3 & 7 \end{vmatrix}$$

The minor of the element 2 is $\begin{vmatrix} 8 & 1 \\ -3 & 7 \end{vmatrix} = 56 + 3 = 59.$

(b) The minor of the element 8 is found by crossing out the second row and the second column.

$$\begin{vmatrix} 2 & 4 & -4 \\ 5 & 8 & 1 \\ -6 & -3 & 7 \end{vmatrix}$$

The minor of 8 is $\begin{vmatrix} 2 & -4 \\ -6 & 7 \end{vmatrix} = 14 - 24 = -10.$

(c) The minor of 1 is found by crossing out the second row and third column.

$$\begin{vmatrix} 2 & 4 & -4 \\ 5 & 8 & 1 \\ -6 & -3 & 7 \end{vmatrix}$$

The minor of 1 is $\begin{vmatrix} 2 & 4 \\ -6 & -3 \end{vmatrix} = -6 + 24 = 18.$

Cofactors

Each element also has a **cofactor**. The value of the cofactor is determined by first adding the number of the row and the number of the column where the element is located. If this sum is even, the value of the cofactor is equal to the value of the minor for that element. If the sum is odd, the value of the cofactor is then -1 times the value of the minor for that element.

EXAMPLE 5.32

We will consider the same determinant we used in Example 5.31.

(a) The cofactor of 2 is 59. The element 2 is in row 1 and column 1. Since $1 + 1$ is an even number and the minor for 2 had a value of 59. (See Example 5.31(a), its cofactor has a value of 59.)

(b) The element 8 is in row 2, column 2. Since $2 + 2 = 4$, an even number, the cofactor of 8 is -10, the same as its minor.

(c) The element 1 is in row 2, column 3 and $2 + 3 = 5$, an odd number. The value of the minor of 1 is 18, so the cofactor of 1 is $(-1)18 = -18$.

Evaluating a Determinant

To evaluate any determinant

(1) select any row or column of the determinant,

(2) multiply each element of that row or column by its cofactor, and

(3) add the results.

Notice that this procedure allows you to choose which row or column you want to use to evaluate the determinant.

EXAMPLE 5.33

Evaluate $\begin{vmatrix} 2 & 4 & -4 \\ 5 & 8 & 1 \\ -6 & -3 & 7 \end{vmatrix}$.

Solution   We will evaluate this determinant by using, or expanding, on the second row. (We picked this row because we found the cofactor of the last two elements in Example 5.32.) The cofactor of 5 is $(-1)\begin{vmatrix} 4 & -4 \\ -3 & 7 \end{vmatrix} = (-1)(28 - 12) = -16$. The cofactor of 8 is -10 and the cofactor of 1 is -18. So, the value of this determinant is

$$\begin{vmatrix} 2 & 4 & -4 \\ 5 & 8 & 1 \\ -6 & -3 & 7 \end{vmatrix} = 5(-16) + 8(-10) + 1(-18)$$

$$= -80 + -80 + -18$$

$$= -178$$

EXAMPLE 5.34

Evaluate $\begin{vmatrix} 2 & 1 & -5 & -2 \\ 1 & 0 & 4 & 5 \\ 7 & 2 & 1 & 0 \\ 5 & 0 & -3 & 2 \end{vmatrix}$.

Solution Since we can select any row or column, let's select the one that has the most zeros. The second column of this determinant has two zeros, so we shall use column 2 to evaluate it.

$$\begin{vmatrix} 2 & 1 & -5 & -2 \\ 1 & 0 & 4 & 5 \\ 7 & 2 & 1 & 0 \\ 5 & 0 & -3 & 2 \end{vmatrix} = -1\begin{vmatrix} 1 & 4 & 5 \\ 7 & 1 & 0 \\ 5 & -3 & 2 \end{vmatrix} + 0\begin{vmatrix} 2 & -5 & -2 \\ 7 & 1 & 0 \\ 5 & -3 & 2 \end{vmatrix}$$

$$-2\begin{vmatrix} 2 & -5 & -2 \\ 1 & 4 & 5 \\ 5 & -3 & 2 \end{vmatrix} + 0\begin{vmatrix} 2 & -5 & -2 \\ 1 & 4 & 5 \\ 7 & 1 & 0 \end{vmatrix}$$

$$= -1\begin{vmatrix} 1 & 4 & 5 \\ 7 & 1 & 0 \\ 5 & -3 & 2 \end{vmatrix} - 2\begin{vmatrix} 2 & -5 & -2 \\ 1 & 4 & 5 \\ 5 & -3 & 2 \end{vmatrix}$$

Now we need to evaluate each of these 3×3 determinants. We will evaluate the first determinant along column three.

$$\begin{vmatrix} 1 & 4 & 5 \\ 7 & 1 & 0 \\ 5 & -3 & 2 \end{vmatrix} = 5\begin{vmatrix} 7 & 1 \\ 5 & -3 \end{vmatrix} - 0\begin{vmatrix} 1 & 4 \\ 5 & -3 \end{vmatrix} + 2\begin{vmatrix} 1 & 4 \\ 7 & 1 \end{vmatrix}$$

$$= 5[7(-3) - 5(1)] - 0 + 2[1(1) - 7(4)]$$
$$= 5[-21 - 5] - 0 + 2[1 - 28]$$
$$= 5(-26) + 2(-27)$$
$$= -130 - 54 = -184$$

The other 3×3 determinant is evaluated along row one:

$$\begin{vmatrix} 2 & -5 & -2 \\ 1 & 4 & 5 \\ 5 & -3 & 2 \end{vmatrix} = 2\begin{vmatrix} 4 & 5 \\ -3 & 2 \end{vmatrix} - (-5)\begin{vmatrix} 1 & 5 \\ 5 & 2 \end{vmatrix} + (-2)\begin{vmatrix} 1 & 4 \\ 5 & -3 \end{vmatrix}$$

$$= 2[4(2) - (-3)5] + 5[1(2) - 5(5)] - 2[1(-3) - 5(4)]$$
$$= 2(8 + 15) + 5(2 - 25) - 2(-3 - 20)$$
$$= 2(23) + 5(-23) - 2(-23)$$
$$= 46 - 115 + 46$$
$$= -23$$

So, the value of the original determinant is

$$-1\begin{vmatrix} 1 & 4 & 5 \\ 7 & 1 & 0 \\ 5 & -3 & 2 \end{vmatrix} - 2\begin{vmatrix} 2 & -5 & -2 \\ 1 & 4 & 5 \\ 5 & -3 & 2 \end{vmatrix} = -1(-184) - 2(-23)$$

$$= 184 + 46$$
$$= 230$$

Needless to say, the last example was quite long. We will next show how to evaluate this same determinant using a calculator.

Evaluating Determinants on a Calculator

Many of today's scientific calculators allow you to evaluate a determinant quickly (and accurately). The following procedure describes how this is done on a Texas Instruments TI-82 graphics calculator.

To evaluate a determinant you will need to use the matrix features of this calculator. We will learn more about a matrix in Chapter 17.

EXAMPLE 5.35

Use a graphing calculator to evaluate the determinant in Example 5.34:

$$\begin{vmatrix} 2 & 1 & -5 & -2 \\ 1 & 0 & 4 & 5 \\ 7 & 2 & 1 & 0 \\ 5 & 0 & -3 & 2 \end{vmatrix}$$

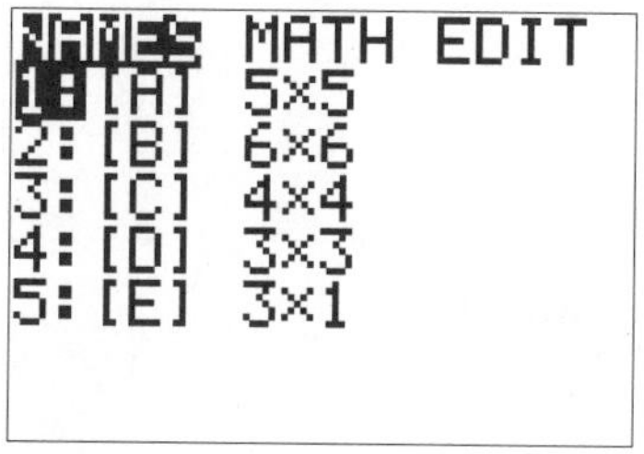

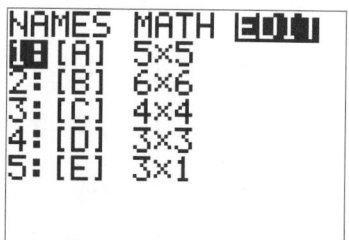

FIGURE 5.12

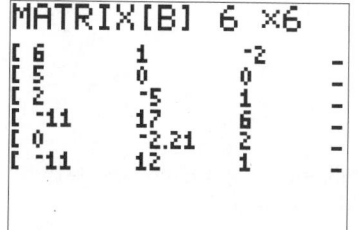

FIGURE 5.13

Solution To begin, turn on your calculator and press the [MATRX] key to access the matrix operations. You should see a display like that shown in Figure 5.12 on your screen. Across the top of the screen are the titles of three menus. The NAMES menu is highlighted. This menu shows the names and the sizes of the five matrices that can be stored in a TI-82 calculator. For example, the first matrix is named "matrix [A]." Its size is 5×5, which means that it has 5 rows and 5 columns. (What you actually see depends on whether your calculator has previously been used to enter a matrix.)

To the right of the NAMES menu is the MATH menu. It deals with several operations on matrices. We will use this menu later. The other menu, EDIT, allows us to define or modify a matrix. To evaluate a determinant, we enter it in the calculator as a matrix. To do this, you first select the EDIT menu by pressing the [▶] key twice or by pressing the [◀] key once. The screen should now look like the one in Figure 5.13.

We are now ready to define the determinant we want to evaluate. The TI-82 calculator can store five matrices, which it names [A], [B], [C], [D], and [E]. Indicate which matrix you want to define. Press the number shown at the left of each of these matrix names. We will name our matrix [B]. Pressing a [2] results in the screen display shown in Figure 5.14.

The blinking cursor is on the row dimension of the determinant. We must either accept or change the dimension on the top line. To accept the number, press [ENTER]. To change the number, enter the numbers of rows in your determinant, and then press [ENTER]. We want to evaluate a 4×4 determinant, so press 4 [ENTER] 4 [ENTER]. You should now see the display in Figure 5.15.

We are now ready to enter the elements of our determinant. Start with the element in the upper left-hand corner, 2, and press 2 [ENTER]. Next enter the second element in the top row, 1, by pressing 1 [ENTER]; the third element in the top row, −5, by pressing [(−)] 5 [ENTER]; and so on until all 16 elements have been entered.

FIGURE 5.14

EXAMPLE 5.35 (Cont.)

```
MATRIX[B]  4 ×4
[ 6     1    -2   -
[ 5     0    0    -
[ 2    -5    1    -
[ -11  17    6    -

1,1=6
```

FIGURE 5.15

When you complete a row, the calculator will go to the left-most element in the next row.

When you have finished entering the elements, return to the HOME screen by pressing 2nd QUIT.

Now, you are ready to evaluate this determinant. To evaluate a determinant, you want the MATH menu for matrices. To get this, press MATRX ►. The result is shown in Figure 5.16. There are many matrix operations listed. The one we want, determinant (or "det"), is listed first, so press 1.

If you named your matrix [B], now press MATRX 2 ENTER. The result, shown in Figure 5.17, shows that the value of the determinant of [B] is 230.

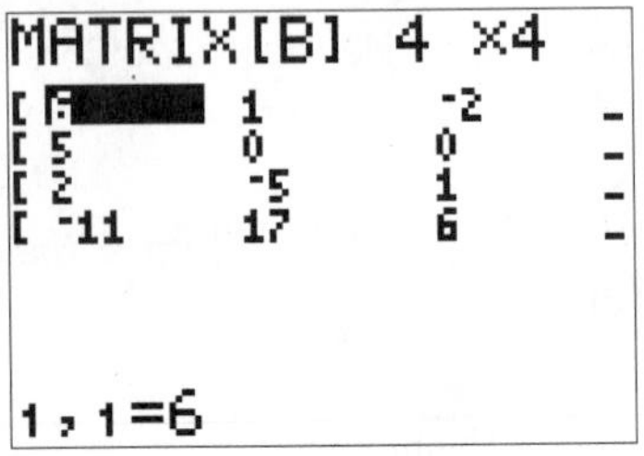

```
NAMES MATH EDIT
1▪det
2:T
3:dim
4:Fill(
5:identity
6:randM(
7↓augment(
```

FIGURE 5.16

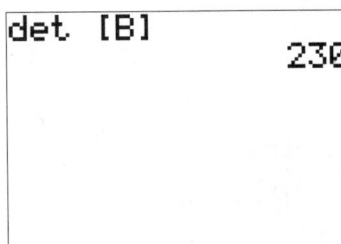

```
det [B]
            230
```

FIGURE 5.17

Exercise Set 5.4

In Exercises 1–8, evaluate each determinant.

1. $\begin{vmatrix} 2 & 3 \\ 4 & -1 \end{vmatrix}$

2. $\begin{vmatrix} 4 & -5 \\ 6 & 2 \end{vmatrix}$

3. $\begin{vmatrix} 5 & -1 \\ 8 & 1 \end{vmatrix}$

4. $\begin{vmatrix} 9 & -1 \\ -2 & 3 \end{vmatrix}$

5. $\begin{vmatrix} 4 & 7 \\ -3 & 1 \end{vmatrix}$

6. $\begin{vmatrix} -1 & -4 \\ 0 & 1 \end{vmatrix}$

7. $\begin{vmatrix} -9 & \frac{1}{2} \\ 2 & 1 \end{vmatrix}$

8. $\begin{vmatrix} 6 & -3 \\ \frac{1}{3} & -\frac{2}{3} \end{vmatrix}$

In each of Exercises 9–14, use the determinant $\begin{vmatrix} 4 & 2 & -1 \\ 3 & 7 & -4 \\ -2 & 1 & 1 \end{vmatrix}$ **and for the indicated position find (a) the element**

in that position, (b) the minor of that element, and (c) the cofactor of that element.

9. Row 1, Column 2

10. Row 2, Column 1

11. Row 3, Column 2

12. Row 2, Column 3

13. Row 3, Column 1

14. Row 1, Column 3

Solve Exercises 15 and 16.

15. Evaluate the determinant used in Exercises 9–14 by expanding on the second row.

16. Evaluate the determinant used in Exercises 9–14 by expanding on the third column.

In Exercises 17–24, use Properties 1–5 to evaluate each determinant.

17. $\begin{vmatrix} 2 & -5 & 8 \\ 16 & 4 & 3 \\ 2 & -5 & 8 \end{vmatrix}$

18. $\begin{vmatrix} 1 & -2 & 3 \\ 2 & -4 & 6 \\ 0 & 3 & 5 \end{vmatrix}$

19. $\begin{vmatrix} 1 & 2 & 5 \\ -1 & -2 & 3 \\ 3 & 6 & 15 \end{vmatrix}$

20. $\begin{vmatrix} 2 & 4 & 3 \\ 0 & 1 & 19 \\ 0 & 0 & -3 \end{vmatrix}$

21. $\begin{vmatrix} 5 & 2 & 3 \\ 4 & -5 & -6 \\ -2 & 5 & -9 \end{vmatrix}$

22. $\begin{vmatrix} 9 & 3 & 3 \\ -2 & 0 & 6 \\ 6 & 3 & 3 \end{vmatrix}$

23. $\begin{vmatrix} 4 & 3 & 9 \\ -4 & -6 & 16 \\ 2 & 3 & 2 \end{vmatrix}$

24. $\begin{vmatrix} 9 & 18 & -1 \\ 0 & -7 & 5 \\ 0 & 0 & 2 \end{vmatrix}$

Evaluate each of the determinants in Exercises 25–28. You will probably need to use a calculator. (Note: The TI–81 does not allow computations inside a matrix; the TI-82, TI-85, and Casio fx–7700G do.)

25. $\begin{vmatrix} 2.41 & -3.5 & -5.3 \\ 6.02 & -7.01 & -4.26 \\ 9.1 & -3.2 & -4.5 \end{vmatrix}$

26. $\begin{vmatrix} \sqrt{3} & \frac{1}{2} & -5 \\ \frac{1}{4} & \sqrt{2} & 4 \\ \frac{2}{5} & 7 & -1 \end{vmatrix}$

27. $\begin{vmatrix} 0.071 & -0.069 & -1.095 \\ 0.202 & -0.420 & -0.100 \\ 0.066 & -0.303 & -2.093 \end{vmatrix}$

28. $\begin{vmatrix} \sqrt{7} & -3 & \sqrt{5} \\ -1.4 & \sqrt{6} & 2 \\ \sqrt{8} & 3 & -6 \end{vmatrix}$

In Exercises 29–32, assume that $\begin{vmatrix} a & b & c \\ r & s & t \\ x & y & z \end{vmatrix} = -5.$ **Find the value of each determinant.**

29. $\begin{vmatrix} r & s & t \\ a & b & c \\ x & y & z \end{vmatrix}$

30. $\begin{vmatrix} s & t & r \\ b & c & a \\ y & z & x \end{vmatrix}$

31. $\begin{vmatrix} a & b & c \\ 3r & 3s & 3t \\ -2x & -2y & -2z \end{vmatrix}$

32. $\begin{vmatrix} a & b & c \\ r+a & s+b & t+c \\ x-r & y-s & z-t \end{vmatrix}$

Evaluate each of the determinants in Exercises 33–38.

33. $\begin{vmatrix} 1 & 2 & 3 & 4 \\ 5 & 0 & 7 & 0 \\ 9 & 10 & 11 & 12 \\ 13 & 14 & 15 & 16 \end{vmatrix}$

34. $\begin{vmatrix} 2 & 0 & 3 & -1 \\ 4 & 0 & -2 & 5 \\ 9 & -5 & 0 & -2 \\ 0 & 3 & 1 & -6 \end{vmatrix}$

35. $\begin{vmatrix} 3 & -2 & 4 & \sqrt{6} \\ 9 & 8 & \frac{1}{3} & -1 \\ 0 & 2.4 & 0 & -3.1 \\ 6 & -7 & \pi & 9 \end{vmatrix}$

36. $\begin{vmatrix} \sqrt{3} & \sqrt{2} & -\frac{1}{4} & 1 \\ 2 & 0 & -7 & -\sqrt{3} \\ \pi & 12 & -\sqrt{2} & 0 \\ -1 & 11 & -3 & 9 \end{vmatrix}$

37. $\begin{vmatrix} 2.1 & 5.7 & 3.4 & 6 \\ 0.1 & -0.3 & 7.1 & 3.5 \\ 1.05 & 2.85 & 1.7 & 3 \\ 18 & \frac{1}{3} & -6.73 & 9 \end{vmatrix}$

38. $\begin{vmatrix} -0.7 & 0.1 & 0.1 & -0.3 \\ 1.2 & 5.6 & -\frac{1}{3} & \frac{3}{17} \\ 0.01 & 0.17 & 7.4 & 3.1 \\ 9.4 & 35 & 0 & 0.9 \end{vmatrix}$

Solve Exercises 39–42.

39. Write a computer program to input and evaluate a second-order determinant. Test your program on the determinants in Exercises 1–8.

40. A third-order determinant can be evaluated using minors. Another method is to rewrite the determinant so that it has three rows and five columns. The fourth column is the same as column 1 and the fifth column is the same as column 2. You then multiply the three numbers in each diagonal. You add the product if the diagonal slopes down and subtract if it slopes up. Try this method on Exercises 21–24.

41. Write a computer program to input the elements of a third-order determinant and evaluate it using the method described in Exercise 40. Test your program on Exercises 17–28.

42. Consider the determinant $\begin{vmatrix} 1 & 2 & 3 \\ 4 & 5 & 6 \\ 7 & 8 & 9 \end{vmatrix}$.

(a) Evaluate this determinant by hand.

(b) Evaluate this determinant using your calculator.

(c) Evaluate this determinant using a different brand or model of calculator or with computer software.

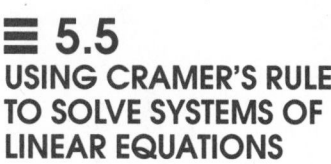

In Your Words

43. The directions for evaluating determinants on a calculator that are in the text could be adapted to fit any specific brand or model of calculator. Write directions that explain how to evaluate a determinant using your brand or model calculator or for some computer software program. Give your directions (and calculator, if necessary) to a friend and see if he or she can understand your directions. Rewrite your directions to eliminate any places where your friend had difficulties.

44. **(a)** Did you get the same answers in Exercise 42 (a), (b), and (c)?

(b) If the answers in Exercise 42 (a), (b), and (c) were different, write an explanation of why you think the two answers are different.

≡ 5.5
USING CRAMER'S RULE TO SOLVE SYSTEMS OF LINEAR EQUATIONS

In Section 5.4 we learned how to evaluate determinants. In this section we will learn how to use determinants to solve systems of linear equations. We will first work with second-order determinants.

Consider the system of linear equations

$$\begin{cases} ax + by = h \\ cx + dy = k \end{cases}$$

If we multiply the first equation by d and the second equation by b, we get

$$\begin{cases} adx + bdy = hd \\ bcx + bdy = bk \end{cases}$$

Subtracting the second equation from the first eliminates the y terms and produces the equation

$$adx - bcx = hd - bk$$

or $\quad (ad - bc)\, x = hd - bk$

Now, this last equation could have been written using determinants as $\begin{vmatrix} a & b \\ c & d \end{vmatrix} x = \begin{vmatrix} h & b \\ k & d \end{vmatrix}$, and so

$$x = \dfrac{\begin{vmatrix} h & b \\ k & d \end{vmatrix}}{\begin{vmatrix} a & b \\ c & d \end{vmatrix}} \quad \text{provided that} \quad \begin{vmatrix} a & b \\ c & d \end{vmatrix} \neq 0$$

Similarly, we can show that $y = \dfrac{\begin{vmatrix} a & h \\ c & k \end{vmatrix}}{\begin{vmatrix} a & b \\ c & d \end{vmatrix}}$.

These last two equations are cumbersome, so we'll abbreviate the determinants by letting $D = \begin{vmatrix} a & b \\ c & d \end{vmatrix}$, $D_x = \begin{vmatrix} h & b \\ k & d \end{vmatrix}$, and $D_y = \begin{vmatrix} a & h \\ c & k \end{vmatrix}$. Then we have $x = \dfrac{D_x}{D}$ and $y = \dfrac{D_y}{D}$. This is **Cramer's rule** for solving two linear equations in two variables.

Cramer's Rule for Solving a 2×2 Linear System

The unique solution to the linear system

$$\begin{cases} a_1x + b_1y = k_1 \\ a_2x + b_2y = k_2 \end{cases}$$

is

$$x = \frac{D_x}{D} = \frac{\begin{vmatrix} k_1 & b_1 \\ k_2 & b_2 \end{vmatrix}}{\begin{vmatrix} a_1 & b_1 \\ a_2 & b_2 \end{vmatrix}} \quad \text{and} \quad y = \frac{D_y}{D} = \frac{\begin{vmatrix} a_1 & k_1 \\ a_2 & k_2 \end{vmatrix}}{\begin{vmatrix} a_1 & b_1 \\ a_2 & b_2 \end{vmatrix}}$$

provided that

$$\begin{vmatrix} a_1 & b_1 \\ a_2 & b_2 \end{vmatrix} \neq 0$$

The determinant D is called the **coefficient determinant**, because its entries are the coefficients of the variables in the system of linear equations:

$$\begin{cases} a_1x + b_1y = k_1 \\ a_2x + b_2y = k_2 \end{cases} \qquad D = \begin{vmatrix} a_1 & b_1 \\ a_2 & b_2 \end{vmatrix}$$

The determinant D_x is obtained by replacing the first column of D (the coefficients of x) with the constants on the right in the system of equations. Similarly, the determinant D_y is obtained by replacing the second column of D (the coefficients of y) with the constants on the right in the system.

Shading the constants in the system, we see that D_x and D_y are obtained as follows:

$$\begin{cases} a_1x + b_1y & = & k_1 \\ a_2x + b_2y & = & k_2 \end{cases} \qquad D_x = \begin{vmatrix} k_1 & b_1 \\ k_2 & b_2 \end{vmatrix} \qquad D_y = \begin{vmatrix} a_1 & k_1 \\ a_2 & k_2 \end{vmatrix}$$

≡ Note   You cannot use Cramer's rule when $D = 0$. When $D = 0$, the equations in the system are either inconsistent or dependent.

EXAMPLE 5.36   Use Cramer's rule to solve $\begin{cases} 2x - y = 9 \\ x + 5y = 21 \end{cases}$.

EXAMPLE 5.36 (Cont.)

Solution Here we have $D = \begin{vmatrix} 2 & -1 \\ 1 & 5 \end{vmatrix} = 11$. This leads to the following two determinants. (The constants have been shaded to help you see the substitutions.)

$$D_x = \begin{vmatrix} 9 & -1 \\ 21 & 5 \end{vmatrix} = 66 \quad \text{and} \quad D_y = \begin{vmatrix} 2 & 9 \\ 1 & 21 \end{vmatrix} = 33.$$

Using Cramer's rule, we determine that $x = \dfrac{D_x}{D} = \dfrac{66}{11} = 6$ and $y = \dfrac{D_y}{D} = \dfrac{33}{11} = 3$.

EXAMPLE 5.37

Use Cramer's rule to solve $\begin{cases} 3x + y = 10 \\ 6x + 2y = 15 \end{cases}$.

Solution In this system, $D = \begin{vmatrix} 3 & 1 \\ 6 & 2 \end{vmatrix} = 0$, so Cramer's rule does not apply. Graphing these two lines would show that they are parallel, and so the system is inconsistent.

Application

EXAMPLE 5.38

A contractor mixes some aggregate with cement to make concrete. Two mixtures are on hand. One of the mixtures, which we will call Mixture A, is 40% sand and 60% aggregate. The other mixture, Mixture B, is 70% sand and 30% aggregate. How much of each should be used to get a 500-lb mixture that is 46.2% sand and 53.8% aggregate?

Solution Let a represent the amount of Mixture A and b the amount of Mixture B used in the final mixture. The final 500-lb mixture will contain 46.2%, or 231 lb, of sand and 53.8%, or 269 lb, of aggregate. Thus, we get two equations. The first

$$0.4a + 0.7b = 231 \tag{1}$$

represents the amounts of sand from each of Mixtures A and B that are needed to make the 231 lb in the final mixture.

The second equation

$$0.6a + 0.3b = 269 \tag{2}$$

represents the amount of aggregate from Mixtures A and B needed to make the aggregate in the final mixture.

Thus, we have the system

$$\begin{cases} 0.4a + 0.7b = 231 \\ 0.6a + 0.3b = 269 \end{cases}$$

Here, $D = \begin{vmatrix} 0.4 & 0.7 \\ 0.6 & 0.3 \end{vmatrix} = -0.3$, $D_a = \begin{vmatrix} 231 & 0.7 \\ 269 & 0.3 \end{vmatrix} = -119$, and

EXAMPLE 5.38 (Cont.)

$$D_b = \begin{vmatrix} 0.4 & 231 \\ 0.6 & 269 \end{vmatrix} = -31. \text{ As a result, we have } a = \frac{D_a}{D} = \frac{-119}{-0.3} \approx 396.7 \text{ and}$$

$b = \dfrac{D_b}{D} = \dfrac{-31}{-0.3} \approx 103.3.$ The final mixture should contain about 396.7 lb of Mixture A and 103.3 lb of Mixture B.

Using Cramer's Rule on a System of Three Linear Equations

To extend Cramer's rule to a system of three linear equations in three variables, you use a similar procedure, as shown in the following box.

Cramer's Rule for Solving a 3×3 Linear System

The unique solution to the linear system

$$\begin{cases} a_1 x + b_1 y + c_1 z = k_1 \\ a_2 x + b_2 y + c_2 z = k_2 \\ a_3 x + b_3 y + c_3 z = k_3 \end{cases}$$

is

$$x = \frac{D_x}{D} \qquad y = \frac{D_y}{D} \qquad \text{and} \qquad z = \frac{D_z}{D}$$

where

$$D_x = \begin{vmatrix} k_1 & b_1 & c_1 \\ k_2 & b_2 & c_2 \\ k_3 & b_3 & c_3 \end{vmatrix}, \quad D_y = \begin{vmatrix} a_1 & k_1 & c_1 \\ a_2 & k_2 & c_2 \\ a_3 & k_3 & c_3 \end{vmatrix}, \quad D_z = \begin{vmatrix} a_1 & b_1 & k_1 \\ a_2 & b_2 & k_2 \\ a_3 & b_3 & k_3 \end{vmatrix}$$

and provided that

$$D = \begin{vmatrix} a_1 & b_1 & c_1 \\ a_2 & b_2 & c_2 \\ a_3 & b_3 & c_3 \end{vmatrix} \neq 0$$

If $D = 0$, the equations are either inconsistent or dependent.

≡ Note

- If the determinant of the denominator, D, is not zero, there is a unique solution to the system of equations.
- If all determinants are zero, the system is dependent, that is, there are an infinite number of solutions.
- If the determinant of the denominator, D, is zero, and any of the determinants in the numerator is not zero, the system is inconsistent, that is, there is no solution.

EXAMPLE 5.39

Use Cramer's rule to solve the system of linear equations:

$$\begin{cases} 3x - 2y - z = 2 \\ -x - 3y + z = 10 \\ 2x + 3y - 2z = -14 \end{cases}$$

Solution For this example, we have

$$D = \begin{vmatrix} 3 & -2 & -1 \\ -1 & -3 & 1 \\ 2 & 3 & -2 \end{vmatrix}$$

$$= 3 \begin{vmatrix} -3 & 1 \\ 3 & -2 \end{vmatrix} + 2 \begin{vmatrix} -1 & 1 \\ 2 & -2 \end{vmatrix} - 1 \begin{vmatrix} -1 & -3 \\ 2 & 3 \end{vmatrix}$$

$$= 3(3) + 2(0) - 1(3)$$

$$= 6$$

Since $D \neq 0$, the system is consistent, and we can solve it by Cramer's rule. Solving for D_x, D_y, and D_z, we obtain

$$D_x = \begin{vmatrix} 2 & -2 & -1 \\ 10 & -3 & 1 \\ -14 & 3 & -2 \end{vmatrix}$$

$$= 2 \begin{vmatrix} -3 & 1 \\ 3 & -2 \end{vmatrix} + 2 \begin{vmatrix} 10 & 1 \\ -14 & -2 \end{vmatrix} - 1 \begin{vmatrix} 10 & -3 \\ -14 & 3 \end{vmatrix}$$

$$= 2(3) + 2(-6) - 1(-12)$$

$$= 6 - 12 + 12$$

$$= 6,$$

$$D_y = \begin{vmatrix} 3 & 2 & -1 \\ -1 & 10 & 1 \\ 2 & -14 & -2 \end{vmatrix}$$

$$= 3 \begin{vmatrix} 10 & 1 \\ -14 & -2 \end{vmatrix} - 2 \begin{vmatrix} -1 & 1 \\ 2 & -2 \end{vmatrix} - 1 \begin{vmatrix} -1 & 10 \\ 2 & -14 \end{vmatrix}$$

$$= 3(-6) - 2(0) - 1(-6)$$

$$= -18 - 0 + 6$$

$$= -12,$$

and $\quad D_z = \begin{vmatrix} 3 & -2 & 2 \\ -1 & -3 & 10 \\ 2 & 3 & -14 \end{vmatrix}$

$$= 3 \begin{vmatrix} -3 & 10 \\ 3 & -14 \end{vmatrix} + 2 \begin{vmatrix} -1 & 10 \\ 2 & -14 \end{vmatrix} + 2 \begin{vmatrix} -1 & -3 \\ 2 & 3 \end{vmatrix}$$

EXAMPLE 5.39 (Cont.)

$$= 3(12) + 2(-6) + 2(3)$$
$$= 36 - 12 + 6$$
$$= 30$$

So, $x = \dfrac{D_x}{D} = \dfrac{6}{6} = 1$, $y = \dfrac{D_y}{D} = \dfrac{-12}{6} = -2$, and $z = \dfrac{D_z}{D} = \dfrac{30}{6} = 5$.

Application

EXAMPLE 5.40

An applied electromotive force (emf) of 200 V produces a current of 40 mA in the simple series circuit shown in Figure 5.18. The voltage drop across resistor R_1 is 60 V more than the combined voltage drop across R_2 and R_3. The voltage drop across R_3 is one-fourth the drop across R_2. What are the values of the three resistors in ohms?

Solution   Since the voltage drop across resistor R_1 is 60 V more than the combined voltage drop across R_2 and R_3, we have

$$V_1 = 60 + V_2 + V_3 \qquad (1)$$

We also are given

$$V_3 = \frac{1}{4}V_2 \qquad (2)$$

The current of 40 mA $= 40 \times 10^{-3}$ A $= 0.040$ A. We can use Ohm's law to find the total external resistance, R_T in the circuit.

$$E = IR_T$$
$$200 = 0.040R_T$$
$$R_T = \frac{200}{0.040}$$
$$= 5\,000$$

Thus, from Kirchhoff's laws,

$$R_1 + R_2 + R_3 = 5\,000 \qquad (3)$$

If we use Ohm's law, we can use the following equations to express the voltage drop across each resistor:

$$V_1 = 0.040R_1$$
$$V_2 = 0.040R_2$$
$$V_3 = 0.040R_3$$

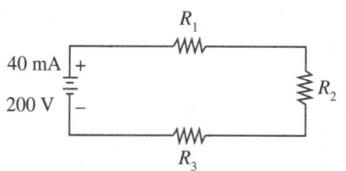

FIGURE 5.18

EXAMPLE 5.40 (Cont.)

Substituting these values of V_1, V_2, and V_3 in equations (1) and (2), we obtain

$$0.04R_1 = 60 + 0.04R_2 + 0.04R_3 \tag{4}$$

$$0.04R_3 = \frac{1}{4}0.04R_2$$

$$\text{or} \quad 4R_3 = R_2 \tag{5}$$

We are now ready to use Cramer's rule to solve the system of equations

$$\begin{cases} R_1 + R_2 + R_3 = 5,000 \\ 0.04R_1 = 60 + 0.04R_2 + 0.04R_3 \\ 4R_3 = R_2 \end{cases}$$

$$\text{or} \quad \begin{cases} R_1 + R_2 + R_3 = 5,000 \\ 0.04R_1 - 0.04R_2 - 0.04R_3 = 60 \\ R_2 - 4R_3 = 0 \end{cases}$$

Thus, we have

$$D = \begin{vmatrix} 1 & 1 & 1 \\ 0.04 & -0.04 & -0.04 \\ 0 & 1 & -4 \end{vmatrix} = 0.4$$

$$D_{R_1} = \begin{vmatrix} 5\,000 & 1 & 1 \\ 60 & -0.04 & -0.04 \\ 0 & 1 & -4 \end{vmatrix} = 1\,300$$

$$D_{R_2} = \begin{vmatrix} 1 & 5\,000 & 1 \\ 0.04 & 60 & -0.04 \\ 0 & 0 & -4 \end{vmatrix} = 560$$

$$\text{and} \quad D_{R_3} = \begin{vmatrix} 1 & 1 & 5\,000 \\ 0.04 & -0.04 & 60 \\ 0 & 1 & 0 \end{vmatrix} = 140$$

As a result, we obtain

$$R_1 = \frac{D_{R_1}}{D} = \frac{1\,300}{0.4} = 3\,250$$

$$R_2 = \frac{D_{R_2}}{D} = \frac{560}{0.4} = 1\,400$$

$$\text{and} \quad R_3 = \frac{D_{R_3}}{D} = \frac{140}{0.4} = 350$$

Thus, the three resistors have values of $3\,250\,\Omega$, $1\,400\,\Omega$, and $350\,\Omega$.

Exercise Set 5.5

In Exercises 1–18 use Cramer's rule, when applicable, to solve each system of linear equations.

1. $\begin{cases} 2x + y = 7 \\ 3x - 2y = -7 \end{cases}$

2. $\begin{cases} 3x + y = 3 \\ 2x - 3y = 13 \end{cases}$

3. $\begin{cases} 4x + 3y = 4 \\ 3x - 2y = -14 \end{cases}$

4. $\begin{cases} \frac{1}{2}x - \frac{2}{3}y = \frac{3}{4} \\ \frac{1}{3}x + 2y = \frac{5}{6} \end{cases}$

5. $\begin{cases} 1.2x + 3.7y = 9.1 \\ 4.3x - 5.2y = 8.3 \end{cases}$

6. $\begin{cases} 0.02x - 1.22y = 3.74 \\ -0.14x + 0.32y = -1.32 \end{cases}$

7. $\begin{cases} 2.5x + 3.8y = 9.3 \\ 0.5x + 0.76y = -2.44 \end{cases}$

8. $\begin{cases} 4.3x - 2.7y = -4 \\ 3.5x + 4.2y = -1 \end{cases}$

9. $\begin{cases} -5.3x + 2.1y = 4.6 \\ 6.2x - 3.1y = 6 \end{cases}$

10. $\begin{cases} 1.3x - 0.8y = 2.9 \\ 1.7x - 0.7y = 0.4 \end{cases}$

11. $\begin{cases} 6x + 2.5y = 8.2 \\ 13.8x + 5.75y = 18.86 \end{cases}$

12. $\begin{cases} 3.5x - 6.5y = 22.45 \\ 5.5x + 3.3y = -1.21 \end{cases}$

13. $\begin{cases} 4x + 3y = 2 \\ 3x - 2y = -24 \end{cases}$

14. $\begin{cases} x + 3y - z = 7 \\ 5x - 7y + z = 3 \\ 2x - y - 2z = 0 \end{cases}$

15. $\begin{cases} x - 0.5y + 1.5z = 2 \\ x + 0.5y - 3.75z = 0.25 \\ -3x - 2.5y + 4z = -0.75 \end{cases}$

16. $\begin{cases} x + y - z = -4.7 \\ 0.5x - 3.5y + 2.4z = 17.4 \\ 1.5x + 4.5y - 3.6z = -21.6 \end{cases}$

17. $\begin{cases} 2x + y + z + w = 1.2 \\ 7x + 3y + 5z + 3w = -7.2 \\ -31x - 7y + 9z - 7w = -42.3 \\ 0.4x + 2y + 4z + 5w = 30 \end{cases}$

18. $\begin{cases} 2a + 3b - c + d = -5 \\ 4a + 5b + 2c - d = 4 \\ -2a - b - c - d = 1 \\ 6a + 7b + c - 4d = 2 \end{cases}$

Solve Exercises 19–26.

19. Write a computer program that will allow you to input the numerical values for a system of two linear equations in two variables or a system of three linear equations in three variables, and then use Cramer's rule to solve this system.

20. *Electronics*   Applying Kirchhoff's laws to a certain circuit results in the following system of equations.

$$\begin{cases} I_1 + I_2 + I_3 = 0 \\ 8I_1 - 10I_3 = 8 \\ 6I_1 - 3I_2 = 12 \end{cases}$$

21. *Physics*   The displacement s of an object moving in a straight line under constant acceleration from an

initial position s_0 is given by the formula

$$s = s_0 + v_0 t + \frac{1}{2}at^2$$

where t is elapsed time in seconds, v_0 is the initial velocity, and a the acceleration. The results of three measurements are shown in the table. Find the constants s_0, v_0, and a.

Time, t (s)	2	5	7
Displacement, s (m)	212	156.5	70.5

22. *Metallurgy*   An alloy is composed of three metals A, B, and C. The percentage of each metal are given by the following system of equations

$$\begin{cases} A+B+C=100 \\ A-2B=0 \\ -4A+C=0 \end{cases}$$

Determine the percentage of each metal in the alloy.

23. *Automotive technology*   A petroleum engineer was testing three different gasoline mixtures A, B, and C in the same car and under the same driving conditions. She noticed that the car traveled 90 km further when it used mixture B than when it used mixture A. Using the fuel C, the car traveled 130 km more than when it used fuel B. The total distance traveled was 1 900 km. Find the distance traveled on the three fuels.

24. *Construction technology*   If three cables are joined at a point and three forces are applied so the system is in equilibrium, the following system of equations results.

$$\begin{cases} \frac{6}{7}F_B - \frac{2}{3}F_C = 2,000 \\ -F_A - \frac{3}{7}F_B + \frac{1}{3}F_C = 0 \\ \frac{2}{7}F_B - \frac{2}{3}F_C = 1,200 \end{cases}$$

Determine the three forces, F_A, F_B, and F_C, measured in newtons (N).

25. *Environmental science*   To control ice and protect the environment, a certain city determines that the best mixture to be spread on roads consists of 5 units of salt, 6 units of sand, and 4 units of a chemical inhibiting agent. Three companies, A, B, and C, sell mixtures of these elements according to the following table:

	Salt	Sand	Inhibiting Agent
Company A	2	1	1
Company B	2	2	2
Company C	1	5	1

(a) In what proportion should the city purchase from each company in order to spread the best mixture? (Assume that the city must buy complete truckloads.)

(b) If the city expects to need 3,297,283 units for the winter, how many units should be bought from each company?

26. *Automotive technology*   The relationship between the velocity, v, of a car (in mph) and the distance, d, (in ft) required to bring it to a complete stop is known to be of the form $d = av^2 + bv + c$, where a, b, and c are constants. Use the following data to determine the values of a, b, and c. When $v = 20$, then $d = 40$; when $v = 55$, then $d = 206.25$; and when $v = 65$, then $d = 276.25$.

In Your Words

27. Explain how to use Cramer's rule.

28. (a) Explain what it means if the coefficient determinant $D = 0$.

(b) Explain what it means if the all determinants are 0.

(c) Explain what it means if none of the determinants is 0.

CHAPTER 5 REVIEW

Important Terms and Concepts

Addition method for solving systems linear equations
Coefficient determinant
Cofactor of determinant
Consistent equations
Cramer's rule for solving linear equations
Dependent equations

Determinant
Elements of a determinant
Elimination methods
    Addition
    Substitution
Entries of a determinant

Horizontal line
   Slope of
Kirchhoff's law
Kirchhoff's rules
Linear equation
   Point-slope form
   Slope-intercept form
Linear system of equations
   Addition method for solving
   Consistent
   Cramer's rule for solving
   Dependent
   Inconsistent

Substitution method for solving
Minor of a determinant
Point-slope form of a linear equation
Rows of a determinant
Slope
   Of a horizontal line
   Of a vertical line
Slope-intercept form of a linear equation
Substitution method for solving systems of linear
   equations
Vertical line
   Slope of

Review Exercises

Evaluate each of the determinants in Exercises 1–8.

1. $\begin{vmatrix} 2 & 5 \\ -3 & 6 \end{vmatrix}$

2. $\begin{vmatrix} 4 & -7 \\ 3 & -6 \end{vmatrix}$

3. $\begin{vmatrix} 5 & 9 \\ 8 & 1 \end{vmatrix}$

4. $\begin{vmatrix} 3 & -12 \\ 0 & 5 \end{vmatrix}$

5. $\begin{vmatrix} 6 & -2 \\ -4 & 3 \end{vmatrix}$

6. $\begin{vmatrix} 8 & 9 \\ -1 & -2 \end{vmatrix}$

7. $\begin{vmatrix} 9 & 2 & 1 \\ 3 & -4 & 6 \\ 7 & 2 & 1 \end{vmatrix}$

8. $\begin{vmatrix} 9 & -2 & 3 \\ 4 & -5 & 2 \\ 3 & 2 & 1 \end{vmatrix}$

In Exercises 9–12, (a) determine the slope of the line through the given pair of points and then (b) find the equations of the line in point-slope form.

9. $(-2, 3), (5, 1)$

10. $(8, -4), (3, -2)$

11. $(9, -1), (-2, 4)$

12. $(-4, -3), (2, -1)$

In Exercises 13–16, rewrite each of the equations in slope-intercept form.

13. $4x - 2y = 6$

14. $9x + 3y = 5$

15. $4x = 5y - 8$

16. $7x + 3y - 8 = 0$

In Exercises 17–18, solve each system of equations graphically.

17. $\begin{cases} 3x + y = 4 \\ 4x + y = 10 \end{cases}$

18. $\begin{cases} 6x + 2y = 5 \\ 3x - 4y = 7 \end{cases}$

In Exercises 19–22, use the substitution method to solve each system of equations.

19. $\begin{cases} 3x + y = 5 \\ 4x - y = 16 \end{cases}$

20. $\begin{cases} 3x - 2y = 4 \\ x + 8y = 3 \end{cases}$

21. $\begin{cases} 4x - y = 7 \\ 3x + 2y = 5 \end{cases}$

22. $\begin{cases} 6x + y - 5 = 0 \\ 4y - 3x = -7 \end{cases}$

In Exercises 23–28, use the addition method to solve each system of equations.

23. $\begin{cases} x - y = -2 \\ x + y = 8 \end{cases}$

24. $\begin{cases} 6x + 5y = 7 \\ 3x - 7y = 13 \end{cases}$

25. $\begin{cases} x + \frac{1}{2}y = 2 \\ 3x - y = 1 \end{cases}$

26. $\begin{cases} \frac{1}{6}x + \frac{1}{4}y = \frac{1}{3} \\ \frac{1}{4}x - \frac{1}{2}y = 1 \end{cases}$

27. $\begin{cases} 3x + 2y + 3z = -7 \\ 5x - 3y + 2z = -4 \\ 7x + 4y + 5z = 2 \end{cases}$

28. $\begin{cases} x + y + z = 2.7 \\ 6x + 7y - 5z = -8.8 \\ 10x + 16y - 3z = -6.4 \end{cases}$

In Exercises 29–34, use Cramer's rule to solve each system of equations.

29. $\begin{cases} 3x - 2y = 4 \\ 5x + 2y = 12 \end{cases}$

31. $\begin{cases} 4x - 3y = 9 \\ 11x + 4y = 7 \end{cases}$

33. $\begin{cases} 5x + 3y - 2z = 5 \\ 3x - 4y + 3z = 13 \\ x + 6y - 4z = -8 \end{cases}$

34. $\begin{cases} 5x - 2y + 3z = 6 \\ 6x - 3y + 4z = 10 \\ -4x + 4y - 9z = 4 \end{cases}$

30. $\begin{cases} 4x + 3y = 27 \\ 2x - 5y = -19 \end{cases}$

32. $\begin{cases} 2x - 5y = -8 \\ 4x + 6y = 17 \end{cases}$

Solve Exercises 35–37.

35. *Business*   A store owner makes a special blend of coffee from Colombian Supreme costing $4.99/lb and Mocha Java costing $5.99/lb. The mixture sells for $5.39/lb. If this mixture is made in 50-lb batches, how many pounds of each type should be used?

36. *Business*   A computer company make two kinds of computers. One, a personal computer (PC), uses 4 Type A chips and 11 Type B chips. The other, a business computer (BC), uses 9 Type A chips and 6 Type B chips. The company has 670 Type A chips and 1,055 Type B chips in stock. How many PC and BC computers can the company make so that all the chips are used?

37. *Business*   An office building has 146 rooms made into 66 offices. The smallest offices each have 1 room and rent of $300 per month. The middle-sized offices have 2 rooms each and rent for $520 per month. The largest offices have 3 rooms each and rent for $730 each. If the rental income is $37,160 per month, how many of each type of offices are there?

▤ CHAPTER 5 TEST

1. **(a)** Determine the slope of the line through the points $(-4, 2)$ and $(5, 6)$.
   (b) Write the equation of the line through the points in **(a)**.

2. **(a)** Rewrite the equation $6x - 5y = 12$ in slope-intercept form.
   (b) What is the slope of this line?
   (c) What is the y-intercept?
   (d) Sketch the graph of this line.

3. Use the addition method to solve the following system.

$$\begin{cases} 2x + 3y = 1 \\ -3x + 6y = 16 \end{cases}$$

4. Evaluate the determinant.

$$\begin{vmatrix} 4 & -2 \\ 5 & 6 \end{vmatrix}$$

5. Evaluate the determinant.

$$\begin{vmatrix} 4 & 2 & -1 \\ 3 & 0 & 4 \\ -3 & 1 & 1 \end{vmatrix}$$

6. Solve the following system graphically.

$$\begin{cases} 2x + 3y = 5 \\ x - 4y = -14 \end{cases}$$

7. Solve the following system by the substitution method.

$$\begin{cases} 4x + 3y = 9 \\ 2x + y = 2 \end{cases}$$

8. Use Cramer's rule to solve the following system.

$$\begin{cases} 2x - 4y = 2 \\ -6x + 8y = -9 \end{cases}$$

9. Solve the following system:

$$\begin{cases} -x + 3y - 2z = 7 \\ 3x + 4y - 7z = -8 \\ x + 2y + z = 2 \end{cases}$$

10. Solve the following system:

$$\begin{cases} 3x_1 + 3x_2 - x_3 + x_4 = -25 \\ 4x_1 + 5x_2 - x_3 - 2x_4 = -8 \\ x_1 + x_2 - x_3 + x_4 = 1 \\ 6x_1 + 7x_2 - x_3 - 3x_4 = 17 \end{cases}$$

11. Three machine parts cost a total of $60. The first part costs as much as the other two together. The cost of the second part is $3 more than twice the cost of the third part. How much does each part cost?

12. Find the current of the system shown in Figure 5.19.

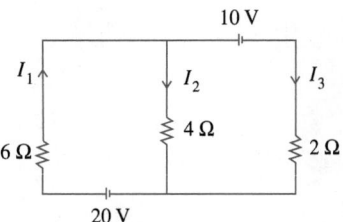

FIGURE 5.19

6

Ratio, Proportion, and Variation

An architect needs to make some scale drawings for a rectangular building. In Example 6.14, we will see how to use proportions to determine the size of the drawing on a blueprint.

Courtesy of Ruby Gold

Ratio and proportion are two ideas that we see very often in mathematics. These ideas are basic to many natural, technical, and scientific relationships. In this chapter, we will examine the concepts of ratio and proportion and look at some of their applications.

☰ 6.1
RATIO AND PROPORTION

The ideas of ratio and proportion include ratio, rate, proportion, and continued proportion. In this section, we will look at all four of these ideas.

Ratio ━━━━━━━━━━━━━━━━━━━━━━━━━━━━━━━━━━━━━━━

A **ratio** is a comparison of two quantities. If these quantities are represented by the values a and b, then the ratio comparing a to b is written $\dfrac{a}{b}$ or $a : b$. A **pure ratio** compares two quantities that have the same units. A pure ratio is written without units.

EXAMPLE 6.1

What is the ratio of **(a)** 4 m to 3 m, **(b)** 5 ft to 4 yd, and **(c)** 2 m to 35 cm?

Solutions   We will write each of these as a pure ratio.

(a) $\dfrac{4\text{ m}}{3\text{ m}} = \dfrac{4}{3}$ or $4 : 3$

(b) $\dfrac{5\text{ ft}}{4\text{ yd}} = \dfrac{5\text{ ft}}{12\text{ ft}} = \dfrac{5}{12}$ or $5 : 12$

(c) $\dfrac{2\text{ m}}{35\text{ cm}} = \dfrac{200\text{ cm}}{35\text{ cm}} = \dfrac{200}{35} = \dfrac{40}{7}$ or $40 : 7$

☰ Note

The rational numbers got their name because they indicate a ratio of two numbers.

Some ratios are always expressed as a quantity compared to 1, as shown in the next example.

Application ━━━━━━━━━━━━━━━━━━━━━━━━━━━━━━━

EXAMPLE 6.2

A driven gear has 26 teeth and the driving gear has 8 teeth. What is the gear ratio of the driven gear to the driving gear?

Solution

$$\frac{\text{driven gear}}{\text{driving gear}} = \frac{26\text{ teeth}}{8\text{ teeth}} = \frac{26}{8} = \frac{13}{4} = \frac{3.25}{1} \quad \text{or} \quad 3.25 : 1$$

Application ━━━━━━━━━━━━━━━━━━━━━━━━━━━━━━━

EXAMPLE 6.3

The **specific gravity** of a substance is the ratio of its density to the density of water. The density of aluminum is 2.7 g/cm^3 and the density of water is 1 g/cm^3. What is the specific gravity of aluminum?

Solution   specific gravity of aluminum $= \dfrac{\text{density of aluminum}}{\text{density of water}}$

$$= \frac{2.7\text{ g/cm}^3}{1\text{ g/cm}^3}$$

$$= 2.7$$

Actually, this is 2.7 : 1, but it is usually written as 2.7.

Unit Consistency and Conversion

When working an equation, the units must be dimensionally consistent. There is an old saying that "you cannot add apples and oranges." In mathematics, that means that you can only add or equate two terms if they have the same units. In calculations, units can be treated just like algebraic symbols with respect to multiplication and division.

We have used this concept already. In Example 6.1(b), in order to determine the ratio of 5 ft : 4 yd, we changed 4 yd to 12 ft. Then the calculations proceeded as $\frac{5\,\text{ft}}{12\,\text{ft}} = \frac{5\,\cancel{\text{ft}}}{12\,\cancel{\text{ft}}} = \frac{5}{12}$. Similarly, in Example 6.3, we were able to "cancel" the g/cm³ when we converted from $\frac{2.7\,\text{g/cm}^3}{1\,\text{g/cm}^3} = \frac{2.7}{1} = 2.7$.

Another example is the formula $d = rt$, or distance = rate · time. Suppose we wanted to determine how far a car moving 65 mph would travel in 2 hr. The unit mph can be thought of as mi/h or $\frac{\text{mi}}{\text{h}}$. Substituting the given data in the formula $d = rt$, we obtain

$$d = \left(\frac{65\,\text{mi}}{\cancel{\text{h}}}\right)(2\,\cancel{\text{h}})$$
$$= (65\,\text{mi})(2) = 130\,\text{mi}$$

Rate

A **rate** is a ratio that compares different kinds of units. Common rates include miles and gallons, revolutions and minutes, and miles and hours.

Application

EXAMPLE 6.4

An automobile travels 243 km in 3 h. What is the rate of speed?

Solution $\dfrac{243\,\text{km}}{3\,\text{h}} = \dfrac{81\,\text{km}}{1\,\text{h}} = 81\,\text{km/h}$

The car is traveling at the rate of 81 km/h.

Application

EXAMPLE 6.5

A crankshaft turns 9,500 revolutions (rev) in $2\frac{1}{2}$ min. How many revolutions per minute (rpm) is the crankshaft turning?

Solution $\dfrac{\text{number of revolutions}}{\text{number of minutes}} = \dfrac{9,500\,\text{rev}}{2\frac{1}{2}\,\text{min}}$

$$= \dfrac{3,800\,\text{rev}}{1\,\text{min}}$$
$$= 3,800\,\text{rpm}$$

The crankshaft is turning at the rate of 3,800 rpm.

Proportion

A statement that says two ratios are equal is called a **proportion**. If the ratio $a : b$ is equal to the ratio $c : d$, then we have the proportion $a : b = c : d$ or

$$\frac{a}{b} = \frac{c}{d}$$

which is read, "a is to b as c is to d."

The inside terms of a proportion are called the means and the outside terms are the extremes.

means
↓   ↓
$a : b = c : d$
↑   ↑
extremes

Two ratios are equal, and hence form a proportion, when the product of the means equals the product of the extremes. This is also referred to as the **cross-product**. This is stated algebraically as shown in the box.

Equal Ratios

Two ratios $a : b$ and $c : d$ are equal, if $ad = bc$.

Stated using the "fractional notation" for ratios, this says that

$$\frac{a}{b} = \frac{c}{d}, \quad \text{if and only if } ad = bc$$

The following diagram may help you to remember which terms to multiply and also to explain why this is called a cross-product.

$$\frac{a}{b} \times \frac{c}{d}$$

EXAMPLE 6.6

What is the missing number in the proportion $5 : 7 = x : 42$?

Solution   We are given $\frac{5}{7} = \frac{x}{42}$.
The product of the means is $7x$.
The product of the extremes is $5 \cdot 42 = 210$.
Since this is a proportion, the product of the means equals the product of the extremes. So, $7x = 210$ or $x = 30$, and thus we have determined that $5 : 7 = 30 : 42$.

Application

EXAMPLE 6.7

If 74.5 L of fuel are used to drive 760 km, how many liters are needed to drive 3 754 km?

Solution   We have the rates of 74.5 L to 760 km and x L to 3 754 km. We will form a proportion where the left-hand side equals the ratio $\dfrac{L}{km}$. We must then have $\dfrac{L}{km}$ on the right-hand side. If we set these two rates equal, we have

$$\frac{74.5\,\text{L}}{760\,\text{km}} = \frac{x\,\text{L}}{3\,754\,\text{km}}$$

The product of the extremes is $(74.5)(3\,754) = 279\,673$ and the product of the means is $760x$. Setting these equal to each other we get

$$760x = 279\,673$$
$$x = 367.99079$$

So, it would require about 368 L of fuel to travel 3 754 km.

Continued Proportion

A **continued proportion** is a proportion that involves six or more quantities. If there are six quantities, a continued proportion is of the form $a : b : c = x : y : z$. This is a shorthand notation for writing $\dfrac{a}{x} = \dfrac{b}{y} = \dfrac{c}{z}$. We will use continued proportions when we study similar figures in Section 6.2, and later when we study trigonometry.

EXAMPLE 6.8

Solve the proportion $2 : 5 : 7 = x : 32 : z$.

Solution   We will solve this continued proportion in stages. Remember that $2 : 5 : 7 = x : 32 : z$ is a short way of writing $\dfrac{2}{x} = \dfrac{5}{32} = \dfrac{7}{z}$. We will first find x.

Working with the proportion from the two ratios on the left, we obtain the proportion $\dfrac{2}{x} = \dfrac{5}{32}$. The product of the extremes is $2 \times 32 = 64$ and is equal to the product of the means, $5x$. If $5x = 64$, then $x = 12.8$.

Next, we work with the proportion from the two ratios on the right of the continued proportion. This proportion is $\dfrac{5}{32} = \dfrac{7}{z}$ or $5z = 7(32) = 224$, and so $z = 44.8$.

Thus, the solution to the proportion $2 : 5 : 7 = x : 32 : z$ is $2 : 5 : 7 = 12.8 : 32 : 44.8$.

EXAMPLE 6.9

A 520 mm wire is to be cut into three pieces so that the ratio of the lengths is to be $6 : 4 : 3$. How long should each piece be?

EXAMPLE 6.9 (Cont.)

Solution The length of each piece is a multiple of some unknown length, x. If we represent the length of each piece by $6x$, $4x$, and $3x$, then we have the total length of $6x + 4x + 3x = 13x$. But the length of the wire is 520 mm. Thus,

$$13x = 520$$
$$x = 40$$

The lengths must then be $6(40) = 240$, $4(40) = 160$, and $3(40) = 120$. The proportion would be $6 : 4 : 3 = 240 : 160 : 120$. ▪

Exercise Set 6.1

Express each of the rates in Exercises 1–4 as a ratio.

1. $1.38 for 16 bolts
2. 725 rpm
3. 86 L/km
4. 236 mi in 4 h

Write each of the ratios in Exercises 5–8 as a ratio compared to 1.

5. $9 : 2$
6. $\frac{7}{5}$
7. $\frac{23}{7}$
8. $37 : 4$

Use cross-products to determine whether the pairs of ratios in Exercises 9–12 are equal.

9. $\frac{2}{3}, \frac{10}{15}$
10. $\frac{1}{3}, \frac{4}{12}$
11. $145 : 25, 29 : 5$
12. $3 : 8, 12 : 33$

Solve each of the proportions in Exercises 13–24.

13. $\dfrac{7}{8} = \dfrac{c}{32}$
14. $\dfrac{3}{b} = \dfrac{9}{24}$
15. $\dfrac{124}{62} = \dfrac{158}{d}$
16. $\dfrac{a}{3.5} = \dfrac{8}{20}$
17. $\dfrac{a}{4} = \dfrac{0.16}{0.15}$
18. $\dfrac{7.5}{10.5} = \dfrac{x}{6.3}$
19. $\dfrac{20}{b} = \dfrac{8}{5.6}$
20. $\dfrac{2.4}{10.8} = \dfrac{1.6}{d}$
21. $4 : 9 = x : 4$
22. $3 : x = x : 12$
23. $2 : 5 : 9 = x : 14 : z$
24. $a : 6 : 9 = 20 : y : 45$

Solve Exercises 25–38.

25. *Mechanics* A driven gear has 54 teeth and a driving gear has 13 teeth. What is the gear ratio of the driven gear to the driving gear?

26. A room is 12 ft wide and 15 ft long. What is the ratio of length to width?

27. *Automotive technology* The **steering ratio** of an automobile can be determined by dividing the total number of degrees that the steering wheel turns by the number of degrees that the wheels turn. What is the steering ratio if the steering wheel makes $4\frac{2}{3}$ complete turns while the front wheels turn 60°?

28. If the density of gold is 19 g/cm³, what is its specific gravity?

29. *Electricity* The **capacitance** of a capacitor is the ratio of the charge on either plate of the capacitor to the potential difference between the plates. The unit of capacitance is the farad (F) where 1 F = 1 coulomb/volt, or 1 C/V. Because the farad is such a large unit the microfarad (μF) is used where 1 μF = 10^{-6} F. If a capacitor has a charge of 5×10^{-4} C when the potential difference across the plates is 300 V, what is the capacitance of this capacitor?

30. *Optics* The **linear magnification**, m, of an optical system is the ratio between the size of the image and the size of the object. What is the linear magnifica-

tion of a system if an object that is 80 mm long has an image that is 20 mm long?

31. *Automotive technology*   The **compression ratio** compares the volume of a cylinder at Bottom Dead Center (BDC) to the volume at Top Dead Center (TDC). If the compression ratio is 8.7 : 1 and the volume at BDC is 96 cm^3, what is the volume at TDC?

32. *Automotive technology*   A driver added 90 mL of fuel conditioner to 20 L of fuel. How much is needed for 54 L of fuel?

33. *Physics*   It requires 80 calories of heat to change 1 g of ice to water without raising the temperature. How many calories are needed to change 785 kg of ice to water without changing the temperature?

34. *Electricity*   In a transformer, an electric current in the primary wire induces a current in the secondary wire. The ratio of turns in the windings determines the ratio of voltages across them, and is expressed as the proportion $\dfrac{V_1}{V_2} = \dfrac{N_1}{N_2}$, where V represents voltage and N represents the number of turns in each coil. A transformer has 100 turns in its primary winding and 750 in its secondary winding. If the primary voltage is 120 V, what is the secondary voltage?

35. The ratio of mm to inches is 25.4 to 1. How many inches are there in 88.9 mm?

36. *Physics*   The acceleration due to gravity is 9.780 39 m/s^2 or 32.087 8 ft/s^2 at the equator. At the north pole, the acceleration due to gravity is 9.832 17 m/s^2. What is the value in ft/s^2 at the north pole?

37. *Automotive technology*   The ratio of gallons to liters is 0.2642 : 1. If a car's gasoline tank holds 14.2 gal, what is its capacity in liters?

38. *Nutrition*   For a certain breakfast cereal, a serving of 1.25 oz contains 140 cal and 3 g of dietary fiber. A box of this same cereal contains 14.1 oz. How many calories and how many grams of dietary fiber are in a box of this cereal?

In Your Words

39. Explain the difference between a rate and a ratio.

40. Describe how a proportion and a continued proportion are alike and how they are different.

6.2
SIMILAR GEOMETRIC SHAPES

In Chapter 3, we studied geometry. As part of that study we examined similar triangles. In this section, we will reexamine similar triangles and expand our work to include other geometric figures.

Similar Triangles

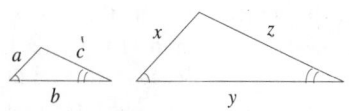

FIGURE 6.1

Two triangles that have the same shape are **similar triangles**. In similar triangles the corresponding angles are congruent and the corresponding sides are proportional. In Figure 6.1, since the triangles are similar, $\dfrac{a}{x} = \dfrac{b}{y} = \dfrac{c}{z}$ or $a : b : c = x : y : z$.

EXAMPLE 6.10

Find the lengths of the unknown sides of the similar triangles in Figure 6.2.

Solution   The corresponding sides of these triangles are proportional, thus $9.3 : 7.75 : y = 7.2 : x : 5.4$. Solving for x, produces

$$\frac{9.3}{7.2} = \frac{7.75}{x}$$
$$9.3x = (7.75)(7.2)$$
$$= 55.8$$
$$x = 6$$

Now, solving for y, we have the proportion

$$\frac{9.3}{7.2} = \frac{y}{5.4}$$
$$7.2y = (9.3)(5.4)$$
$$= 50.22$$
$$y = 6.975$$

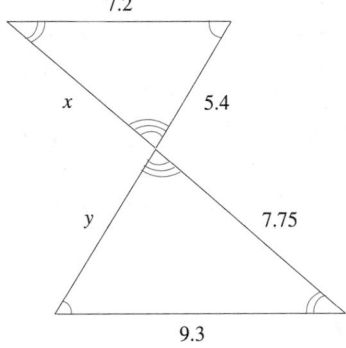

FIGURE 6.2

Other Similar Figures

Similar figures other than triangles can be more difficult to work with. If the corresponding angles of a triangle are congruent, we know that the triangles are similar. This is not true for other figures. But it is true that, for two similar figures, the distance between any two points on one figure is proportional to the distance between any two corresponding points on the other figure. This is true for any two similar figures whether they are plane figures or solid ones.

EXAMPLE 6.11

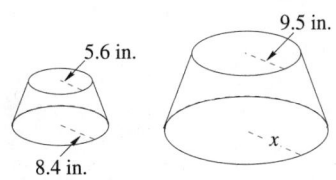

FIGURE 6.3

Figure 6.3 shows two similar frustums of cones. The radii of the bases of the smaller frustum are given as is the radius of the smaller base for the larger frustum. What is the radius of the bottom base of the larger frustum?

Solution   Since these are similar figures, we know that the ratios of the radius of the small base to the radius of the large base must be the same for each figure. Thus,

$$\frac{5.6}{8.4} = \frac{9.5}{x}$$
$$5.6x = (8.4)(9.5)$$
$$= 79.8$$
$$x = 14.25 \text{ in.}$$

The radius of the bottom base of the larger frustum is 14.25 in.

If we know that two figures are similar, then it is easy to find the area or volume of one if we know the area or volume of the other.

Areas of Similar Figures

Areas of similar figures are related to each other as the *squares* of any two corresponding dimensions. This is true for both plane and solid figures. It is also true for plane areas, lateral surface areas, total surface areas, or cross-sectional areas. For example, suppose two circles are similar. If one circle has a radius of r_1 and the other a radius of r_2, then the ratio of their areas is

$$\frac{A_1}{A_2} = \frac{\pi r_1^2}{\pi r_2^2} = \frac{r_1^2}{r_2^2}$$

Thus, the ratio of the areas of two circles is the same as the ratio of the squares of their radii.

EXAMPLE 6.12

The lateral surface area of the smaller frustum in Figure 6.3 is 220 in.^{2} What is the lateral surface area of the larger figure?

Solution   We will let L represent the lateral surface area that we are to find and use the proportion

$$\frac{220}{L} = \frac{5.6^2}{9.5^2}$$
$$5.6^2 L = (220)(9.5)^2$$
$$31.36L = 220(90.25)$$
$$= 19,855$$
$$L = 633.13138$$

The lateral surface area of the larger figure is about 633.13 in.^{2}

Volumes of Similar Figures

If two solid figures are similar, then their volumes are related to each other as the *cubes* of any two corresponding dimensions. As with the circles, two spheres are similar. If one sphere has a radius of r_1 and the other a radius of r_2, then the ratio of their volumes is

$$\frac{V_1}{V_2} = \frac{\frac{4}{3}\pi r_1^3}{\frac{4}{3}\pi r_2^3} = \frac{r_1^3}{r_2^3}$$

EXAMPLE 6.13

The volume of the smaller frustum in Figure 6.3 is 646.6 in.^{3} What is the volume of the larger figure?

EXAMPLE 6.13 (Cont.)

Solution We will let V represent the volume that we are to find and use the proportion

$$\frac{646.6}{V} = \frac{5.6^3}{9.5^3}$$
$$5.6^3 V = (646.6)(9.5)^3$$
$$175.616V = 554{,}378.67$$
$$V = 3{,}156.7663$$

The volume of the larger frustum is about 3,156.77 in.^{3}

Scale Drawings

Perhaps the most common use of similar figures applies when using scale drawings. Scale drawings are used in maps, blueprints, engineering drawings, and other figures. The ratio of distances on the drawing to corresponding distances on the actual object is called the scale of the drawing.

Application

EXAMPLE 6.14

A rectangular building $200' \times 145'$ is drawn to a scale of $\frac{1}{4}'' = 1'0''$. What is the size of the rectangle that will represent this building on the blueprint?

Solution This is the continuing proportion $\dfrac{\frac{1}{4}''}{1'0''} = \dfrac{x}{200'} = \dfrac{y}{145'}$. Solving each of these we get $x = 50''$ and $y = 36\frac{1}{4}''$.

The building will be represented by a rectangle that is $50'' \times 36\frac{1}{4}''$.

Application

EXAMPLE 6.15

A map has a scale of $1 : 24\,000$. What is the actual distance of a map distance of 32 mm?

Solution We will use the proportion $\dfrac{1}{24\,000} = \dfrac{32\ \text{mm}}{x}$.

$$x = (32\ \text{mm})(24\,000) = 768\,000\ \text{mm}$$
$$= 768\ \text{m}$$

So, 32 mm on the map represents an actual distance of 768 m.

Exercise Set 6.2

The pairs of figures in each of Exercises 1–6 are similar. Find the lengths of the unknown sides.

1.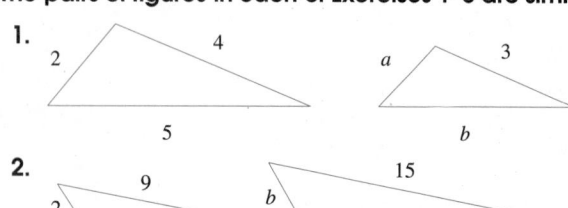

2.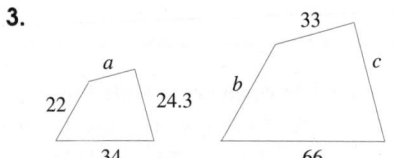

3.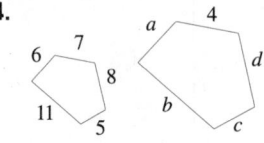

4.

5.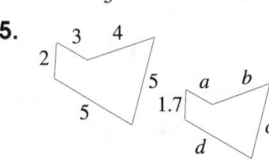

6.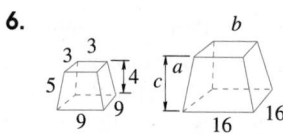

Solve Exercises 7–16.

7. *Architectural technology*   The floor plan of a building has a scale of $\frac{1}{8}$ in. $= 1$ ft. One room of the floor plan has an area of 20 in.^{2} What is the actual room area in square inches? What is the area in square feet? (Hint: 144 in.$^2 = 1$ ft^2)

8. *Wastewater technology*   A pipe at a sewage treatment plant is 75 mm in diameter and discharges 2 000 L of water in a given period of time. If a pipe is to discharge 3 000 L in the same period of time what is its diameter?

9. A square bar of steel, 38 mm on a side, has a mass of 22 kg. What is the mass of another bar of the same length that measures 19 mm on a side?

10. *Agricultural technology*   It cost $982 to fence in a circular field that has an area of 652 ft^2. What will it cost to enclose another circular field with three times as much area?

11. *Construction*   It requires 700 L to paint a spherical tank that has a radius of 20 m. How much paint will be needed to paint a tank with a radius of 35 m?

12. *Environmental technology*   A water tank that is 12 m high has a volume of 20 kiloliters (kL) or 20 000 L. What is the volume of a similar tank that is 30 m high?

13. *Sheet metal technology*   A cylinder has a capacity of 3 930 mm^3. Its diameter is 15 mm. What is the volume of a similar container with a diameter of 5 mm?

14. *Sheet metal technology*   A cylindrical container has a diameter of 4″ and a height of 5″. A similar container has a diameter of 2.5″. What is the height of the second container and the volume of each.

15. *Sheet metal technology*   A sphere with a 10-cm radius has a volume of 4 188.79 cm^3 and a surface area of 1 256.64 cm^2. Find **(a)** the surface area of a sphere with a radius of 2 cm and **(b)** the volume of a sphere with a radius of 2 cm.

16. *Product design*   A cylindrical soup can has a diameter of 66.0 mm and a height of 95.0 mm, and holds about 325.013 5 cm^3. The soup company wants to market a can that holds twice as much soup in a cylindrical can that is similar in size to the present can. What should be the dimensions of the new can?

In Your Words

17. Describe how you would use similar figures to make a scale drawing.

18. **(a)** If the lengths of each side of a triangle are twice as long as the corresponding sides of a similar triangle, how are their areas related?

(b) If the lengths of each side of a triangle are three times as long as the corresponding sides of a similar triangle, how are their areas related?

(c) Describe how the areas between any two similar triangles are related to the lengths of their sides.

≡ **6.3**
DIRECT AND INVERSE VARIATION

As we have seen, many scientific laws are given in terms of ratios or proportions. In this section, we will look at some relations between two or more variables.

Direct Variation

When two variables, x and y, are related so that their ratio $\dfrac{y}{x} = k$, where k is a constant, then y is **directly proportional** to x. Other ways of stating that y is directly proportional to x is to say that y is proportional to x, that y varies directly as x, or that y varies as x. The constant k is called the **constant of proportionality** or the **constant of variation**. The symbol $\propto$ is sometimes used to indicate that y is directly proportional to x and is written $y \propto x$. Note that $y \propto x$ is not an equation, but that $y \propto x$ means $y = kx$, where k is a constant.

Direct Variation

If x and y are variables and k is a constant, then the formula for direct variation is

$$y = kx \qquad \text{or} \qquad \frac{y}{x} = k$$

≡ **Note**

Saying two variables are in direct variation means that if one variable increases, then the other variable increases in the same proportion. Similarly, when two variables are in direct variation, if one of the variables decreases, then the other decreases by the same amount proportionally.

EXAMPLE 6.16

The circumference of a circle is directly proportional to its diameter. Here $C = kd$, where $k = \pi$.

Application

EXAMPLE 6.17

The weight (w) of an object at the earth's surface is directly proportional to the mass (m) of the object. Here $w = mg$, where g is the constant acceleration of gravity. Notice that here g is used instead of k. We saw in Exercise 36 of Exercise Set 6.1 that $g \approx 9.8$ m/s^2 or $g \approx 32$ ft/s^2.

Application

EXAMPLE 6.18

When applied to a spring, **Hooke's law** states that the force F required to stretch a spring is proportional to the distance s the spring is stretched (provided the elastic limit of the spring is not exceeded). Thus $F = ks$, where k is called the **spring constant**.

Application

EXAMPLE 6.19

A force of 10 lb is required to stretch a spring from its natural length of 5 in. to a length of 8 in. Find the spring constant.

Solution   We know from Hooke's law that $F = ks$. Here, $F = 10$ lb and $s = 8 - 5 = 3$ in. So, $10 = k \cdot 3$ and $k = \frac{10}{3}$.

It is also possible to state that y varies as the square of x, $y = kx^2$, or y varies as the cube of x, $y = kx^3$, and so on.

EXAMPLE 6.20

The surface area of a sphere varies as the square of its radius. Since $A = 4\pi r^2$, $k = 4\pi$.

EXAMPLE 6.21

The volume of a sphere varies directly as the cube of its radius. Since $V = \frac{4}{3}\pi r^3$, the constant of proportionality is $\frac{4}{3}\pi$.

Application

EXAMPLE 6.22

The distance h that an object falls when dropped from rest is directly proportional to the square of the time t that the object has fallen. If a stone is dropped 30.625 m from a bridge and takes 2.5 s to strike the water, what is the constant of proportionality? How is the constant related to the acceleration of gravity?

Solution   The statement "the distance h that an object falls when dropped from rest is directly proportional to the square of the time t the object has fallen," can be written as $h = kt^2$. We are given $h = 30.625$ m and $t = 2.5$ s.

EXAMPLE 6.22 (Cont.)

Thus, $\quad 30.625 = k(2.5)^2$
$$= k(6.25)$$
$$4.9 = k$$

The constant is $4.9\,\text{m/s}^2$. Since the acceleration of gravity g is about $9.8\,\text{m/s}^2$, then $k = \frac{1}{2}g$.

If we have two pairs of values, x_1, y_1, and x_2, y_2, from a direct variation then we know that $\dfrac{y_1}{x_1} = k$ and $\dfrac{y_2}{x_2} = k$, so $\dfrac{y_1}{x_1} = \dfrac{y_2}{x_2}$, or $y_1 x_2 = x_1 y_2$.

Application

EXAMPLE 6.23

Charles' law states that at a constant pressure the temperature of a gas varies directly as its volume. This is often expressed as $\dfrac{V_1}{T_1} = \dfrac{V_2}{T_2}$, where V_1 represents the volume of the gas at an absolute temperature T_1, and similarly for V_2 and T_2. If a gas sample of $5\,\text{m}^3$ at $0°\text{C}$ is cooled until its volume is $2\,\text{m}^3$, what is the final temperature?

Solution Charles' law is true only when we are working in degrees expressed in Kelvin (K). The official temperature unit in the SI Metric system is Kelvin but people more often use the Celsius scale. If T_C is the temperature in degrees Celsius and T_K is the temperature in Kelvin, then $T_K = T_C + 273.18$.
In this problem, $0°\text{C} = 273.18\,\text{K}$, and we have $V_1 = 5\,\text{m}^3$, $V_2 = 2\,\text{m}^3$, and $T_1 = 273.18\,\text{K}$. According to Charles' law

$$\frac{V_1}{T_1} = \frac{V_2}{T_2}$$
$$\frac{5}{273.18} = \frac{2}{T_2}$$
$$5T_2 = 546.36$$
$$T_2 = 109.272\,\text{K}$$

The temperature is $109.272\,\text{K}$ or $-163.908°\text{C}$.

EXAMPLE 6.24

The area of a circle is directly proportional to the square of its radius. If the area of a circle with radius $2\,\text{cm}$ is $4\pi\,\text{cm}^2$, what is the area of a circle with a radius of $6\,\text{cm}$?

Solution We could use the formula for the area of a circle. However, we will apply our knowledge of direct variation. If A_1, r_1, A_2, and r_2 are the areas and radii of two circles, then we know

$$\frac{A_1}{r_1^2} = \frac{A_2}{r_2^2}$$

EXAMPLE 6.24 (Cont.)

Since $A_1 = 4\pi$, $r_1 = 2$, and $r_2 = 6$, then

$$\frac{4\pi}{2^2} = \frac{A_2}{6^2}$$

or $\quad 4A_2 = (4\pi)(36)$

$$A_2 = 36\pi \text{ cm}^2$$

Inverse Variation

Another type of variation is **inverse variation**, where the product of two variables, x and y, is a constant. In inverse variation $xy = k$, or $y = \dfrac{k}{x}$, and we say that y is **inversely proportional** to x, or y **varies inversely** as x. Again k is called the constant of variation or the constant of proportionality.

> **Inverse Variation**
>
> If x and y are variables and k is a constant, then the formula for inverse variation is
>
> $$y = \frac{k}{x} \quad \text{or} \quad yx = k$$

≡ **Note**

If two variables are inversely proportional, when one variable increases, the other variable decreases in the same proportion.

Application

EXAMPLE 6.25

The electrical resistance of a wire varies inversely as its cross-sectional area. If R is the resistance of a wire and A is the area of its cross-section, then $RA = k$ or $R = \dfrac{k}{A}$.

Application

EXAMPLE 6.26

Boyle's law states that at a constant temperature the volume of a sample of gas is inversely proportional to the absolute pressure applied to the gas. Thus, if p is the gas pressure and V its volume, then $pV = k$, or $V = \dfrac{k}{p}$.

Application

EXAMPLE 6.27

If 250 L of air in a cylinder exert a gauge pressure of 1 380 kilopascals (kPa), what is the constant of proportionality?

Solution In Example 6.26, we learned that $pV = k$. In this example $p = 1\,380$ kPa and $V = 250$ L, so $k = (1\,380)(250) = 345\,000$.

As was true with direct variation, it is also possible to state that y varies inversely as the square of x, thus $yx^2 = k$ $\left(\text{or } y = \dfrac{k}{x^2}\right)$, or that y varies inversely as the cube of x, thus $yx^3 = k$ $\left(\text{or } y = \dfrac{k}{x^3}\right)$, and so on.

Application

EXAMPLE 6.28

The resistance R of a wire varies inversely as its cross-sectional area A, and so $RA = k$. The area can be determined from the diameter, d. In fact, $A = \pi r^2 = \pi \left(\dfrac{d}{2}\right)^2 = \frac{\pi}{4}d^2$, thus, because $\frac{\pi}{4}$ is a constant, we could say that the resistance of a wire varies inversely as the square of its diameter. Thus, $Rd^2 = k$, or $R = \dfrac{k}{d^2}$. Note that this constant of variation is not the same constant as in Example 6.25.

If we have two pairs of values, x_1, y_1 and x_2, y_2, from the same inverse variation we then know that, since $x_1 y_1 = k$ and $x_2 y_2 = k$, then $x_1 y_1 = x_2 y_2$ or $\dfrac{x_1}{x_2} = \dfrac{y_2}{y_1}$.

Application

EXAMPLE 6.29

Boyle's law (see Example 6.26) is often given as $p_1 V_1 = p_2 V_2$, where p_1 is the gas pressure when the volume is V_1 and p_2 is its pressure when the volume is V_2. In Example 6.27, we had 250 L of air in a cylinder with a pressure of 1 380 kPa. If the volume of the cylinder is increased to 1 m³, what is the pressure?

Solution We must first change the units for one of the volumes so that both volumes are in the same unit. Since 1 m³ $= 1\,000$ L, we can let $V_2 = 1\,000$ L, and $V_1 = 250$ L, while $p_1 = 1\,380$ kPa. So,

$$p_1 V_1 = p_2 V_2$$
$$(1\,380)(250) = p_2(1\,000)$$
$$345 = p_2$$

The pressure is 345 kPa when the volume is 1 m³.

Exercise Set 6.3

In Exercises 1–6, express each of the given quantities as variations.

1. The resistance R of a wire is directly proportional to its length l.

2. The pressure P of a column of liquid varies directly as the height h of the liquid.

3. The area A of a geometric figure varies directly as the square of any dimension, d.

4. The energy E of a photon of light is directly proportional to its frequency, f.

5. The current I in a transformer is inversely proportional to the number of turns N in the winding.

6. The intensity of illumination E produced by a source of light varies inversely as the square of the distance d from the light source.

In Exercises 7–10, give the equation relating the variables and find the constant of proportionality for the given set of values.

7. r varies directly as t and $r = 4$ when $t = 6$.

8. a varies inversely as b and $a = 8$ when $b = 3$.

9. d varies directly as the square of r and $d = 8$ when $r = 4$.

10. p varies inversely as the cube of s and $p = 8$ when $s = 2$.

In Exercises 11–16, find the required value by setting up the equation and solving.

11. Find a when $d = 18$, if a varies directly as d and $a = 4$ when $d = 8$.

12. If r varies inversely with t, and $r = 6$ when $t = 3$, find the value of r when $t = 9$.

13. If v varies directly as t and $v = 60$ m/s when $t = 20$ s, what is v when $t = 1$ min?

14. If p varies inversely with q and $p = 30$ mm^3 when $q = 10$ kg/mm^2, what is p when $q = 40$ kg/mm^2?

15. If n varies directly as d^2 and $n = 10890$ neutrons when $d = 10\,\mu$m, what is n when $d = 2\,\mu$m?

16. If m varies inversely as r^3 and $m = 25$ kPa when $r = 30$ mm, what is m when $r = 5$ mm?

Solve Exercises 17–26.

17. *Chemistry*   A balloon filled with $25\,\text{m}^3$ of air at $20°$C is heated under pressure to $85°$C. According to Charles' law, $V = kT$. Find **(a)** the value of the constant of proportionality and **(b)** the increase in volume. (Remember to change the temperatures from $°$C to K.)

18. *Navigation*   If an airplane is traveling at a uniform rate, the distance traveled will vary directly as the time it has traveled. If a plane covers 475 mi in 1 h, how long will it take to travel 1,235 mi?

19. *Electricity*   The resistance of an electrical wire varies directly with its length. If a wire 1,500 ft long has a resistance of 32 ohms what is the resistance in 5,000 ft of the same wire?

20. *Physics*   The distance d that an object falling from rest will travel varies directly as the square of the time t that the object has been falling. **(a)** If $d = 29.4$ m when $t = 3$ s, find the formula for d in terms of t. **(b)** What is d when $t = 8$ s?   **(c)** What is t when $s = 150$ m?

21. *Physics*   Using Boyle's law (see Example 6.26) and the fact that you know that the volume is 200 in.^{3} when the pressure is 40 psi, find the pressure when the volume is 50 in.^{3}

22. *Physics*   Use Exercise 6 and the fact that a given light source produces an illumination of 600 lux (1 lux = 1 lumen per m^2) on a surface 2 m away to determine how much illumination the same light source produces on a surface 6 m away?

23. *Electricity*   The current in the winding of a transformer is inversely proportional to the number of turns. A transformer has 200 turns in its primary

winding and 50 turns in its secondary winding. If the current in the secondary circuit is 0.3 A what is the current in the primary circuit?

24. *Electricity*   The voltage in a winding of a transformer is directly proportional to the number of turns in the winding.   A 120-V ac power line has 300 turns in its primary winding. If there are 40 turns in its secondary winding, what is the voltage across its secondary winding?

25. *Space technology*   The weight of a body in space varies inversely as the square of its distance from the center of earth. If a person weighs 180 lb on the surface of the earth, how much does he or she weigh 500 mi above the surface if the radius of the earth is 4,000 mi?

26. *Recreation*   The length of time required to fill a swimming pool varies inversely as the square of the diameter of the pipe used to fill the pool. If a 2-in. pipe can fill the pool in 16 h, how long will it take for a 3-in. pipe to fill the pool?

In Your Words

27. (a) Without looking at the definition in the book, explain direct variation.
(b) Find common examples, other than those in the book, that use direct variation.

28. (a) Without looking at the definition in the book, explain inverse variation.
(b) Find common examples, other than those in the book, that use inverse variation.

≡ 6.4
JOINT AND COMBINED VARIATION

In Section 6.3 we studied about direct and inverse variation. In this section, we will study two more types of variation—joint and combined.

Both direct and inverse variation are functions of one variable. In each case we had $y = f(x)$. For direct variation, $y = kx$, and for inverse variation, $y = \dfrac{k}{x}$. The types of variation we will study in this section, joint and combined variation, are functions of two or more variables. Thus joint and combined variation are of the forms $y = f(x, z)$, $y = f(x, w, z)$, or $y = f(x, w, z, p)$.

Joint Variation

Joint variation is when one quantity, y, varies directly as the product of two or more quantities. Thus, if y is jointly proportional to x and z, then $y = kxz$, or if y is jointly proportional to x, w, and z, then $y = kxwz$. The term "jointly" is sometimes omitted and we simply say that y varies as w and z or that y is proportional to w and z when $y = kwz$. Again, k is the constant of variation or the constant of proportionality.

> **Joint Variation**
>
> If x, y, and z are variables and k is a constant, then the formula for joint variation of y with x and z is
>
> $$y = kxz$$

EXAMPLE 6.30

The absolute temperature T of a perfect gas varies jointly as its pressure P and volume V. So, $T = kPV$.

EXAMPLE 6.31

The volume of a prism V varies jointly as the area of its base B and its height h. So, $V = kBh$. In this case, we know from our study of geometry in Chapter 3 that $k = 1$.

Combined Variation

If a quantity y varies with two or more variables in ways more complicated than that described in joint variation, it is then referred to as **combined variation**. Unlike direct, inverse, and joint variation, the term combined variation is seldom used in the description of the problem. In fact, a combined variation relationship will often use both "directly" and "inversely" in the description of the relationship.

> **Guidelines for Solving Variations**
>
> 1. Write the appropriate variation formula.
> 2. Use the given information to find k.
> 3. Rewrite the variation formula using the value for k found in Step 2.
>   If asked to find a particular value from other given information, then use Step 4.
> 4. Substitute this other information into the equation. Solve for the required value.

Application

EXAMPLE 6.32

The resistance R of a conductor is directly proportional to its length l, and inversely proportional to its cross-sectional area A. Thus, $R = \dfrac{kl}{A}$.

Application

EXAMPLE 6.33

The potential energy W of a capacitor is directly related to the square of its charge Q, and inversely related to its capacitance C. This relation is given by the formula $W = k\dfrac{Q^2}{C}$.

Application

EXAMPLE 6.34

The volume of a cone V is directly related to the square of the radius of its base r and its height h. Here $V = kr^2h$ and we know that the constant of proportionality $k = \frac{1}{3}\pi$.

Just as was possible with direct and inverse variation, it is possible to use a given set of values from a joint or combined variation to find an unknown value from another set.

Application

EXAMPLE 6.35

The resistance R of an electrical wire varies directly as its length l, and inversely as the square of its diameter d. If a wire 600 m long with a diameter of 5 mm has a resistance of 32 ohms (Ω), determine the resistance in a 1 500-m wire made of the same material with a diameter of 10 mm.

Solution We know that $R = \dfrac{kl}{d^2}$, where k is the constant of proportionality. From the given values of $R = 32\,\Omega$, $d = 5$ mm, and $l = 600$ m, we have $32 = \dfrac{k(600)}{5^2}$, or $k = \frac{4}{3}$. So

$$R = \frac{\frac{4}{3}(1\,500)}{10^2} = \frac{2\,000}{100} = 20$$

The resistance is $20\,\Omega$.

Application

EXAMPLE 6.36

Express the resistance, R, of an electrical wire in terms of its length, l, and cross-sectional area, A.

Solution Assuming that the cross-section is circular, then $A = \pi r^2 = \pi\left(\dfrac{d}{2}\right)^2 = \dfrac{\pi d^2}{4}$. Thus, $d^2 = \dfrac{4A}{\pi}$. Substituting this for d^2 in the formula from the previous example produces $R = \dfrac{kl}{d^2} = \dfrac{kl}{\frac{4A}{\pi}} = \dfrac{k\pi l}{4A} = \dfrac{\rho l}{A}$, where $\rho = \dfrac{k\pi}{4}$.

Application

EXAMPLE 6.37

Newton's law of universal gravitation states that the gravitational force of attraction F (in newtons, N) between two bodies is directly proportional to each of their masses m_1 and m_2 and inversely proportional to the square of the distance between them, d. Thus $F = \dfrac{km_1 m_2}{d^2}$, where k is the **gravitational constant**. Two objects that each have a mass of 1 kg and are 1 m apart exert a mutual gravitational attraction of $F = \dfrac{k(1)(1)}{1^2} = 6.67 \times 10^{-11}$ N. So, $k \approx 6.67 \times 10^{-11}$ N·m^2/kg^2. The earth's mass is approximately 5.98×10^{24} kg. What is the earth's gravitational force on a space shuttle that has a mass of 29 500 kg and is in orbit 10^8 m from the center of the earth?

EXAMPLE 6.37 (Cont.)

Solution   Since $k = 6.67 \times 10^{-11}$ N, we have $F = 6.67 \times 10^{-11} \dfrac{m_1 m_2}{d^2}$, where $m_1 = 5.98 \times 10^{24}$ kg, $m_2 = 29\,500$ kg $= 2.95 \times 10^4$ kg, and $d = 10^8$ m.

$$F = 6.67 \times 10^{-11} \frac{\left(5.98 \times 10^{24}\right)\left(2.95 \times 10^4\right)}{\left(10^8\right)^2}$$
$$= (6.67)(5.98)(2.95) \times 10^{-11+24+4-16}$$
$$= 117.67 \times 10^1 = 1\,176.7\,\text{N}$$

The earth's gravitational force on this space shuttle is about $1\,176.7$ N.

The previous examples have all used the guidelines for solving variation problems.

Exercise Set 6.4

In Exercises 1–6, express each of the given quantities as variations.

1. The kinetic energy K of a moving body is jointly proportional to its mass m and the square of its speed v.

2. The volume V of a parallelepiped varies jointly as its length l, width w, and height h.

3. The frequency f in hertz of a musical tone is directly proportional to the speed v of sound in air, and inversely proportional to the wavelength l of the sound wave.

4. The revolutions per minute r of a shaft being turned in a lathe varies directly as the cutting speed s of the shaft, and inversely as the diameter d of the shaft.

5. The voltage drop E in a wire varies directly as its length l and the current I, and inversely as the square of its diameter d.

6. The inductance of a solenoid L varies directly as the square of the number of turns of the coil N and the cross-sectional area of the solenoid, A, and inversely as its length l.

In Exercises 7–10, give the equation relating the variables and find the constant of proportionality for the given set of values.

7. a varies jointly as b and c, and $a = 20$, when $b = 4$ and $c = 2$.

8. p varies jointly as r, s, and t, and $p = 15$, when $r = 3$, $s = 10$, and $t = 25$.

9. u varies directly as v and inversely as w, and $u = 20$, when $v = 5$ and $w = 2$.

10. x varies directly as y and the cube of w, and inversely as the square of z, and $x = 175$, when $y = 3$, $w = 5$, and $z = 4$.

In Exercises 11–16, find the required value by setting up the equation and solving.

11. Find r when $s = 3$ and $t = 4$, if r varies jointly as s and t, and $r = 10$, when $s = 6$ and $t = 5$.

12. If f varies directly as r and inversely as l, and $f = 125$ when $r = 10$ and $l = 5$, then find the value of f when $r = 20$ and $l = 10$.

13. If a varies jointly as x and the square of y, and $a = 9$ when $x = 3$ and $y = 5$, then find a when $x = 6$ and $y = 15$.

14. Using the relationship from Exercise 13, find y when $x = 9$ and $a = 10$.

15. If p varies directly as r and the square root of t and inversely as w, and $p = 9$ when $r = 18$, $t = 9$, and $w = 6$, then find p when $r = 36$, $t = 4$, and $w = 2$.

16. Using the relationships from Exercise 15, find t when $p = 18$, $r = 9$, and $w = 4$.

Solve Exercises 17–26.

17. The area of a triangle varies jointly as the length of the base b and the height h of the triangle. If the area of a triangle is 12 when the base is 3 and the height is 8, what is **(a)** the constant of proportionality and **(b)** the area, when the base is 10 and the height is 5?

18. *Electricity* The energy W used by a device is directly related to the current I, the voltage V, and the length of time t, for which the energy is used. If a 240-V clothes drier draws a current of 15 A and in 45 min used 2.4 kWh (kilowatt hours) of energy, how much energy will a 120-V light bulb use, which draws a current of 1.25 A in 8 h of use?

19. *Electricity* The electric power consumed by a device is directly related to the square of the voltage and inversely related to the resistance. The 18 Ω filament of a tube is rated at 2 W and has an operating voltage of 6 V. What is the maximum voltage in a 50 Ω, 10 W resister?

20. *Electricity* The capacitance of a parallel-plate capacitor varies directly as the area of either of its plates and inversely as the distance between them. If the plates are 0.5 mm apart and each is a square that measures 4 m on a side, the capacitance is 0.3 μF. What is the capacitance if the plates are moved so they are 0.1 mm apart?

21. *Electricity* The resistance of a copper wire varies directly as its length and inversely as its cross-sectional area. If a copper wire 80 m long with a cross-sectional area of 2.5 mm² has a resistance of 0.56 Ω, what length of 0.1-mm² copper wire is needed for a resistance of 3 Ω?

22. *Electricity* The specific resistances of most metals vary with temperature. The change in the resistance of a metal is jointly proportional to the resistance of the metal and the change in temperature. If a copper wire has a resistance of 2.25 Ω at 25°C and 2.09 Ω at 10°C, what is its resistance at 50°C?

23. *Energy* The rate of heat conduction through a flat plate varies jointly as the area of one face of the plate and the difference between the temperature of the opposite faces, and inversely as the thickness of the plate. A 3-in. thick brick with a 12-in.² area on one face loses heat at the rate of 50 Btu/h when the temperature is 72°F on one side and 32°F on the other side of the brick. What is the rate of conduction if the outside temperature of 32°F drops to 10°F?

24. *Sound* The frequency in hertz (Hz) of a guitar wire varies directly as the square root of the tension and inversely as the length. If a wire 500 mm long under 25 kg tension vibrates at 256 Hz, what is the frequency of a wire 450 mm long under a tension of 30 kg?

25. *Physics* The **ideal gas law** states that the pressure varies directly as the absolute temperature and inversely as the volume. A gas sample has a volume of 5 ft³ at 70°F at a pressure of 15 lb/in.² (psi). Find the volume at 200°F at a pressure of 75 lb/in.² (The **Rankine scale** is the absolute temperature based on the Fahrenheit scale. If T_R represents a temperature on the Rankine scale and T_F represents the equivalent temperature on the Fahrenheit scale, then $T_R \approx T_F + 460°$. The freezing point of water is 32°F or about 492°R and the boiling point of water is 212°F or about 672°R.)

26. *Construction* The strength, S, of a horizontal beam supported at both ends varies jointly as the width, w, and square of the depth, d, and inversely as the length, l.
(a) Express this relationship as a variation.
(b) What is the effect on the strength of a beam if the width is doubled, the depth is halved, and the length remains the same?
(c) What change in the depth will increase the strength by 75% if the width and length remain unchanged?

In Your Words

27. (a) Describe joint variation.
    (b) Find common examples, other than those in the book, that use direct variation.
    (c) Explain how joint variation is alike and how it is different from direct variation.

28. (a) Without looking in the book, describe how you would solve a variation problem.
    (b) Compare your description in (a) with the guideline for solving variations given in the text.

CHAPTER 6 REVIEW

Important Terms and Concepts

Combined variation
Constant of proportionality
Continued proportion
Direct variation
Directly proportional
Inverse variation
Joint variation
Proportion

Rate
Ratio
Scale drawings
Similar figures
  Area
  Volume
Similar triangles

Review Exercises

Solve each of the proportions in Exercises 1–8.

1. $3 : x = 4 : 6$

2. $x : 5 = 3 : 15$

3. $\dfrac{7}{9} = \dfrac{21}{d}$

4. $\dfrac{14}{6} = \dfrac{c}{27}$

5. $4 : 8 = 19 : x$

6. $x : 12 = 15 : 32$

7. $7 : 24.5 : x = 8 : y : 42$

8. $\dfrac{12.5}{x} = \dfrac{y}{47} = \dfrac{8}{5}$

Solve Exercises 9–19.

9. *Electricity*   The **turn ratio** in a transformer is the number of turns in the primary winding to the number of turns in the secondary winding. If the turn ratio for a transformer is 25, and there are 4,000 turns in the primary winding, how many turns are in the secondary winding?

10. *Economics*   A worker's salary increase is proportional to the cost-of-living index. If the worker received a raise of $12.50/wk when the index was 6.5, how much should the raise be when the index is 9.6?

11. The longest side of triangle A is 180 mm. Triangle B has sides of 4, 5, and 8 mm. Triangles A and B are

similar. What are the lengths of the other two sides of triangle A?

12. *Physics*   The **theoretical mechanical advantage (TMA)** of an inclined plane or ramp is equal to the ratio between its length and height. A ramp 90 ft long slopes down 5 ft to the edge of a lake. What is the TMA of the ramp?

13. *Machine technology*   The efficiency of a machine is the ratio between its actual mechanical advantage (AMA) and its TMA. The AMA of the boat ramp in Exercise 12 is 16 because of the friction in a boat trailer's wheels. What is the efficiency of the ramp?

14. *Physics*   The rate of flow of a liquid through a pipe is jointly proportional to its velocity and the cross-sectional area of the pipe.   A hose with a 1-cm diameter has a liquid flow through it at 3 m/s. **(a)** If a hose with a diameter of 5 mm is attached to one end of this hose, what is the velocity of the liquid through this smaller hose? **(b)** What size hose should have been attached if the velocity was to be 30 m/s?

15. *Petroleum technology*   A horizontal pipe 0.5 m in diameter is joined to an oil pipeline 2.5 m in diameter. If the velocity of the oil in the smaller pipe is 5 m/s, what is the velocity in the large pipe? (See Exercise 14.)

16. *Energy*   The rate at which an object radiates energy, as given by the **Stefan-Boltzmann law**, varies directly as the fourth power of its absolute temperature. The sun radiates energy at the rate of 6.5×10^7 W/m^2 from its surface. If the surface temperature of the sun is 5 800 K, what is the constant of proportionality? (This is the **Stefan-Boltzmann constant**.)

17. *Electricity*   The force that one charge exerts on another is given by **Coulomb's law** and is jointly proportional to the magnitudes of the charges and inversely proportional to the square of the distance between them. If a charge of 4×10^{-9} C is 5 cm from a charge of 5×10^{-8} C, and has a force of 7.192×10^{-4} N, what is the constant of proportionality (**Coulomb's constant**)? (Hint: Change 5 cm to m.)

18. *Electricity*   The magnetic field of a straight current varies directly as the magnitude of the current and inversely as the distance from the current. A cable 5 m above the ground carrying a 100-A current has a magnetic field of 4×10^{-6} T. What is the magnetic field of a cable that carries a 150-A current and is 15 m above the ground?

19. *Light*   The **index of refraction** of a transparent medium is the ratio between the velocity of light in free space and its velocity in the medium. If the velocity of light is 3×10^8 m/s and the index of refraction in a diamond is 2.42, what is the speed of light in a diamond?

▤ CHAPTER 6 TEST

In Exercises 1–6, solve the proportions.

1. $4 : 9 = x : 51$

2. $8 : x = 68 : 93.5$

3. $\dfrac{a}{28} = \dfrac{12}{20}$

4. $\dfrac{24}{42} = \dfrac{78}{d}$

5. $6 : 8 : x = y : 14 : 24.5$

6. $\dfrac{a}{9} = \dfrac{b}{30} = \dfrac{75}{45}$

In Exercises 7–8, express in writing each of the given quantities as a variation.

7. E varies directly as the square of d.

8. h varies directly as r and inversely as the cube of t.

Solve Exercises 9–12.

9. A right triangle with sides of 5, 12, and 13 is proportional to a triangle with hypotenuse 71.5. What are the lengths of the other two sides?

10. The perimeter of a rectangular solar panel is 540 cm. The ratio of the length to the width is 3 : 2. What are the length and width?

11. The area of a geometric figure varies directly as the square of any dimension. Two similar triangles have corresponding sides of length 12 m and 18 m. The smaller triangle has an area of 72 m^2. What is the area of the other triangle?

12. The number of days d needed to erect a certain building varies inversely with the number of people working on the building p and the number of hours h they work each day. If a building requires 45 days to complete when 80 people work 8 h/day, how much time is needed if 60 people work 10 h/day?

7

Factoring and Algebraic Fractions

We often want to make a container that will have the largest possible volume while using the least possible amount of material. In Sections 7.1 and 7.2, we will learn some of the basic algebra needed to solve this type of problem.

Courtesy of Ruby Gold

This chapter uses the algebraic foundations established in Chapter 2. In Chapter 2, we learned how to multiply algebraic expressions. In this chapter we will learn how to do the reverse—to find the factors that, when multiplied together, give the original expression. Once we have the ability to factor algebraic expressions, we will use it to help solve second degree or quadratic equations.

We will also use factoring to help add, subtract, multiply, divide, and simplify algebraic fractions. In spite of all the scientific and technical examples we have used that involve linear equations, many scientific and technical situations must be described with quadratic equations or with algebraic fractions.

☰ 7.1
SPECIAL PRODUCTS

In mathematics we use certain products so often that we need to take the time to learn them. Success in factoring depends on your ability to recognize patterns in any form that they appear. All of these special products were developed in Section 2.2, but most were not stated in the general form. We will present each special product, give some examples of how it is used, and then summarize all of the special products at the end of this section.

The first special product is the distributive law of multiplication over addition, which we saw in Chapter 1.

Distributive Law	$a(x+y) = ax + ay$	#1

EXAMPLE 7.1

$$8(4m + 3n) = 8(4m) + 8(3n)$$
$$= 32m + 24n$$

Here $a = 8$, $x = 4m$, and $y = 3n$. As you can see, the variable x can represent a product of a constant and a variable.

EXAMPLE 7.2

$$3x(2y + 5ax) = (3x)(2y) + (3x)(5ax)$$
$$= 6xy + 15ax^2$$

The term a in the distributive law can represent a constant, a variable, or the product of a constant and a variable.

The next special product is formed by multiplying the sum and difference of two terms. It is equal to the square of the first term minus the square of the second term.

Difference of Two Squares	$(x+y)(x-y) = x^2 - y^2$	#2

EXAMPLE 7.3

$$(x + 6)(x - 6) = x^2 - 6^2$$
$$= x^2 - 36$$

Notice that each term to the right of the equal sign is a perfect square.

EXAMPLE 7.4

$$(2a+3\sqrt{y})(2a-3\sqrt{y}) = (2a)^2 - (3\sqrt{y})^2$$
$$= 4a^2 - 9y$$

Again, once you realize that $y = (\sqrt{y})^2$, you can see that each term to the right of the equal sign is a perfect square. As you gain practice, you will be able to omit the middle step.

The third and fourth special products will be considered together.

| **Perfect Square Trinomials** | $(x+y)^2 = x^2 + 2xy + y^2$ | #3 |
| | $(x-y)^2 = x^2 - 2xy + y^2$ | #4 |

Note that both of these demonstrate that the square of a binomial is a trinomial.

Caution

Be careful, many students make the mistake of thinking that $(x+y)^2 = x^2 + y^2$. This is not true, as you can easily verify by using the FOIL method from Chapter 2.

EXAMPLE 7.5

$$(p+7)^2 = p^2 + 2(7)p + 7^2$$
$$= p^2 + 14p + 49$$

Here $x = p$ and $y = 7$.

EXAMPLE 7.6

$$(2a - 5t^2)^2 = (2a)^2 - 2(2a)(5t^2) + (5t^2)^2$$
$$= 4a^2 - 20at^2 + 25t^4$$

Here $x = 2a$ and $y = 5t^2$.

The next special product is the product of two binomials that have the same first term. It can be verified by using the FOIL method.

| **Two Binomials, Same First Term** | $(x+a)(x+b) = x^2 + (a+b)x + ab$ | #5 |

EXAMPLE 7.7

$$(p+3)(p+7) = p^2 + (3+7)p + (3)(7)$$
$$= p^2 + 10p + 21$$

EXAMPLE 7.8

$$(r+2)(r-8) = r^2 + (2-8)r + (2)(-8)$$
$$= r^2 - 6r - 16$$

In this example, $a = 2$ and $b = -8$. Remember that $r - 8$ can be written as $r + (-8)$.

The next special product is a general version of the last one. Again, this special product can be checked by using the FOIL method.

General Quadratic Trinomial $(ax+b)(cx+d) = acx^2 + (ad+bc)x + bd$ #6

EXAMPLE 7.9

Use Special Product 6 to multiply $(4x+2)(3x+5)$.
Solution   Here $a = 4, b = 2, c = 3$, and $d = 5$, so

$$(4x+2)(3x+5) = 4 \cdot 3x^2 + (4 \cdot 5 + 2 \cdot 3)x + 2 \cdot 5$$
$$= 12x^2 + (20+6)x + 10$$
$$= 12x^2 + 26x + 10$$

EXAMPLE 7.10

Multiply $(7x+3y)(5x-2y)$.
Solution   Here $a = 7, b = 3y, c = 5$, and $d = -2y$. Using Special Product 6 produces

$$(7x+3y)(5x-2y) = (7 \cdot 5x^2) + [7(-2y) + (3y)5]x + (3y)(-2y)$$
$$= 35x^2 + (-14y + 15y)x + (-6y^2)$$
$$= 35x^2 + xy - 6y^2$$

Of course, there is nothing to prevent the combination of two or more of these special products.

≡ **Note**   Remember the order of operations from Chapter 1. Powers are executed before multiplication or division. So, in the following example, we first square $(3x-5)$ and then multiply that result by $4x$.

EXAMPLE 7.11

$$4x(3x-5)^2 = 4x(9x^2 - 30x + 25)$$
$$= 36x^3 - 120x^2 + 100x$$

EXAMPLE 7.12

$$(x^2+9)(x+3)(x-3) = (x^2+9)(x^2-9)$$
$$= x^4 - 81$$

In this example, notice that the left-hand side has the special product $(x+3)(x-3) = x^2 - 9$. When these are multiplied, we get another special product: $(x^2 + 9)(x^2 - 9)$. The time spent looking at a problem for special products can save some computation and time. It can also reduce errors.

There are four other special products that you will use less often than the ones already listed. They will be presented in pairs. The first two are the cubes of a sum or difference.

Perfect Cubes	$(x+y)^3 = x^3 + 3x^2y + 3xy^2 + y^3$	#7
	$(x-y)^3 = x^3 - 3x^2y + 3xy^2 - y^3$	#8

EXAMPLE 7.13

$$(x+5)^3 = x^3 + 3x^2(5) + 3x(5^2) + 5^3$$
$$= x^3 + 15x^2 + 75x + 125$$

EXAMPLE 7.14

$$(2a-4)^3 = (2a)^3 - 3(2a)^2(4) + 3(2a)(4^2) - (4^3)$$
$$= 8a^3 - 48a^2 + 96a - 64$$

The last two special products also deal with cubes.

Sum and Difference of Two Cubes	$(x+y)(x^2 - xy + y^2) = x^3 + y^3$	#9
	$(x-y)(x^2 + xy + y^2) = x^3 - y^3$	#10

EXAMPLE 7.15

$$(x-3)(x^2+3x+9) = x^3 - (3)^3$$
$$= x^3 - 27$$

EXAMPLE 7.16

$$(2d+5)(4d^2-10d+25) = (2d)^3 + (5)^3$$
$$= 8d^3 + 125$$

Application

An open box is going to be formed by cutting a square out of each corner of a rectangular piece of cardboard and folding up the sides as shown in Figure 7.1a. Suppose the piece of cardboard measures 10 cm × 12 cm. **(a)** Determine a function for the volume of the box. **(b)** What is the domain of this function? **(c)** Set up a partial table of values and sketch the graph of this function. **(d)** What size square seems to produce a box with the largest volume?

EXAMPLE 7.17

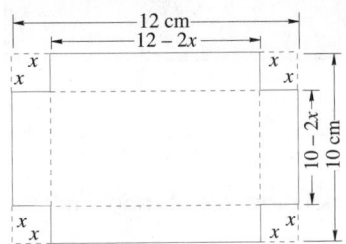

FIGURE 7.1a

Solutions

(a) If the square measures x cm × x cm, then the lengths of the sides that will be folded up are $10 - 2x$ cm and $12 - 2x$ cm. So, the volume of the box is given by

$$V(x) = x(10-2x)(12-2x)$$
$$= 120x - 44x^2 + 4x^3$$

(b) The domain of the function is all real numbers. But, we cannot cut out a square that measures 0 cm or less and we cannot cut out a square that measures 5 cm or more. So, the "realistic" domain is represented by $\{x : 0 < x < 5\}$. Notice how the factored form helped us determine the domain. The expanded form will be more useful later in this book when we get to calculus.

(c) We have the following *partial* table of values.

x	0.5	1.0	1.5	2.0	2.5	3.0	3.5	4.0	4.5
$V(x)$	49.5	80.0	94.5	96.0	87.5	72.0	52.5	32.0	13.5

A graph formed from this table is shown in Figure 7.1b.

(d) From looking at the table, or the graph, it *seems* as if a 2 cm × 2 cm square will produce a box with the largest volume. Calculus can be used to show that $x \approx 1.8$ cm will produce the largest volume: 96.768 cm³.

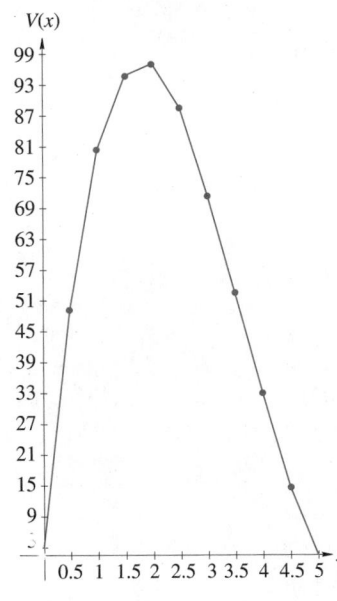

FIGURE 7.1b

Thus, we have a total of 10 special products. Study them. Become familiar with them. Learn to recognize them and to use them to help simplify your work. The ten special products are listed here together.

The Special Products	$a(x+y) = ax + ay$	#1
	$(x+y)(x-y) = x^2 - y^2$	#2
	$(x+y)^2 = x^2 + 2xy + y^2$	#3
	$(x-y)^2 = x^2 - 2xy + y^2$	#4
	$(x+a)(x+b) = x^2 + (a+b)x + ab$	#5
	$(ax+b)(cx+d) = acx^2 + (ad+bc)x + bd$	#6
	$(x+y)^3 = x^3 + 3x^2y + 3xy^2 + y^3$	#7
	$(x-y)^3 = x^3 - 3x^2y + 3xy^2 - y^3$	#8
	$(x+y)(x^2 - xy + y^2) = x^3 + y^3$	#9
	$(x-y)(x^2 + xy + y^2) = x^3 - y^3$	#10

Exercise Set 7.1

In Exercises 1–44, find the indicated products by direct use of one of the 10 special products. It should not be necessary to write intermediate steps.

1. $3(p+q)$
2. $7(2+x)$
3. $3x(5-y)$
4. $2a(4+a^2)$
5. $(p+q)(p-q)$
6. $(3a+b)(3a-b)$
7. $(2x-6p)(2x+6p)$
8. $\left(\dfrac{y}{2} + \dfrac{2p}{3}\right)\left(\dfrac{y}{2} - \dfrac{2p}{3}\right)$
9. $(r+w)^2$
10. $(q-f)^2$
11. $(2x+y)^2$
12. $(\frac{1}{2}a+b)^2$
13. $(\frac{2}{3}x+4b)^2$
14. $(a-\frac{1}{2}y)^2$
15. $(2p-\frac{3}{4}r)^2$
16. $(\frac{2}{3}r-5t)^2$
17. $(a+2)(a+3)$
18. $(x+5)(x+7)$
19. $(x-5)(x+2)$
20. $(a+9)(x-12)$

21. $(2a+b)(3a+b)$
22. $(4x+2)(3x+5)$
23. $(3x+4)(2x-5)$
24. $(3m-n)(4m+2n)$
25. $(a+b)^3$
26. $(r-s)^3$
27. $(x+4)^3$
28. $(2a+b)^3$
29. $(2a-b)^3$
30. $(3x+2y)^3$
31. $(3x-2y)^3$
32. $(4r-s)^3$
33. $(m+n)(m^2 - mn + n^2)$
34. $(a+2)(a^2 - 2a + 4)$
35. $(r-t)(r^2 + rt + t^2)$
36. $(h-3)(h^2 + 3h + 9)$
37. $(2x+b)(4x^2 - 2xb + b^2)$
38. $(3a+\frac{2}{3}c)(9a^2 - 2ac + \frac{4}{9}c^2)$
39. $(3a-d)(9a^2 + 3da + d^2)$
40. $\left(\dfrac{2e}{5} - \dfrac{5r}{4}\right)\left(\dfrac{4e^2}{25} + \dfrac{er}{2} + \dfrac{25r^2}{16}\right)$

41. $3(a+2)^2$

42. $5(2a-4)^2$

43. $\dfrac{5r}{t}\left(t+\dfrac{r}{5}\right)^2$

44. $(x^2+4)(x+2)(x-2)$

In Exercises 45–60, perform the indicated operations.

45. $(x^2-6)(x^2+6)$

46. $(\dfrac{2a}{b}+\dfrac{b}{2})^2$

47. $(\dfrac{3x}{y}-\dfrac{y}{x})(\dfrac{3x}{y}+\dfrac{y}{x})x^2y^2$

48. $(x+1)(x+2)(x+3)$

49. $(x-y)^2-(y-x)^2$

50. $(x+y)^2-(y-x)^2$

51. $(x+3)(x-3)^2$

52. $(y-4)(y+4)^2$

53. $r(r-t)^2-t(t-r)^2$

54. $5(x+4)(x-4)(x^2+16)$

55. $(5+3x)(25-15x+9x^2)$

56. $(2-\sqrt{x})(2+\sqrt{x})(4+x)$

57. $[(x+y)-(w+z)]^2$

58. $[(r+s)+(t+u)]^2$

59. $(x+y-z)(x+y+z)$

60. $(a+b+2)(a-b+2)$

Solve Exercises 61–66.

61. *Electricity*   The impedance, z, of an ac circuit at frequency f is given by the formula $z^2 = R^2 + (x_L - x_C)^2$, where R is the resistance, x_L is the inductive resistance and x_C is the capacitive resistance at the frequency f. Use a special product to expand the equation.

62. *Physics*   The kinetic energy, KE, of an object is given by the formula $KE = \frac{1}{2}mv^2$, where m is the mass of the object and v its velocity. If the velocity of an object at any time, t, is given by the equation $v = 3t + 1$, find an equation for the kinetic energy in terms of m and t and expand your result using a special formula.

63. *Physics*   The magnitude of the centripetal acceleration of a body in uniform circular motion is given by the formula $a_c = \dfrac{v^2}{r}$, where v is the velocity of the body and r is the radius of the circular path. If the velocity at any given time t is expressed as $v = 2t^2 - t$, find an equation for the centripetal acceleration in terms of r and t, and expand your result.

64. *Energy*   The work, W, done by a steam turbine in a certain period of time is given by the formula

$W = \frac{1}{2}m(v_1 - v_2)(v_1 + v_2)$, where m is the mass of the steam that passes through the turbine during that period, v_1 is the velocity of the steam when it enters, and v_2 the velocity when it leaves. Simplify this formula by multiplying the factors together.

65. *Architectural technology*   When the maximum deflections of two similar cantilever beams bearing the load at the ends are compared, the difference in deflection is given by

$$d = \left(\frac{P}{3EI}\right)(l_1 - l_2)\left(l_1^2 + l_1 l_2 + l_2^2\right)$$

Simplify this formula by multiplying the two factors containing l_1.

66. *Architectural technology*   When the maximum deflections of two similar cantilever beams bearing distributed loads are compared, the difference in deflection is given by

$$d = \left(\frac{w}{8EI}\right)(l_1 - l_2)(l_1 + l_2)\left(l_1^2 + l_2^2\right)$$

Simplify this formula by multiplying the three factors containing l_1.

In Your Words

67. Describe how you can remember the expansions of the perfect square trinomials.

68. Describe how you can remember the expansions of the perfect cubes.

≡ 7.2
FACTORING

Sometimes it is important to determine what expressions were multiplied together to form a product, instead of multiplying quantities together as we did in the special products. Recall from Section 2.1 that an algebraic term consists of several factors. For example, the term $5xy$ has factors of $1, 5, x, y, 5x, 5y, xy$, and $5xy$. In Section 7.1, we learned to multiply several factors together to form the special products. In this section, we will begin to learn how to factor an algebraic expression. Once we know how to factor we will be able to solve more complicated problems. Determining the factors of an algebraic expression is called **factoring**.

When factoring a polynomial, we will use only those factors that are also polynomials. Further, we will continue factoring a polynomial until the only remaining factors are 1 and -1. When this has been completed, we will be able to say that the polynomial has been factored completely and all of the factors will be **prime factors**.

Common Factors

The simplest type of factoring is the reverse of the distributive law of multiplication. This type of factoring is called **removing a common factor**. If each term in an expression contains the same factor, then this is a common factor and it can be factored out using the distributive law.

Common factor	$ax + ay = a(x+y)$

EXAMPLE 7.18

In the expression $6a + 3a^2$, each term contains a factor of $3a$.

$$6a + 3a^2 = 2(3a) + a(3a)$$
$$= (2+a)3a$$
$$= 3a(2+a)$$

EXAMPLE 7.19

$8y^2 + 2y = 2y(4y+1)$

Remember that $2y = (2y)(1)$. When you factor $2y$, you are left with a factor of 1.

EXAMPLE 7.20

$12x^5y + 8x^3y^2 = 4x^3y(3x^2 + 2y)$

You could have factored this in several other ways, such as $2xy(6x^4 + 4x^2y)$. This would not have been considered completely factored, since $6x^4 + 4x^2y$ can be factored further.

You may not see how to completely factor out all of the common factors in one step. For instance, in the last example you might have first written

$$12x^5y + 8x^3y^2 = 2xy(6x^4 + 4x^2y)$$

and then noticed that you could factor $6x^4 + 4x^2y$ as $2x^2(3x^2 + 2y)$. Then you would have had

$$12x^5y + 8x^3y^2 = 2xy(6x^4 + 4x^2y)$$
$$= 2xy(2x^2)(3x^2 + 2y)$$
$$= 4x^3y(3x^2 + 2y)$$

Notice that in the last step the two monomials were combined.

EXAMPLE 7.21

Factor $6x^2y + 9xy^2z - 3xyz$.

Solution   We can see that each term contains a multiple of 3, as well as an x and y factor. If we factor $3xy$ out of each term, we have $(3xy)(2x) + (3xy)(3yz) + (3xy)(-z) = 3xy(2x + 3yz - z)$. So, $6x^2y + 9xy^2z - 3xyz = 3xy(2x + 3yz - z)$.

The easiest factors to locate (and sometimes the easiest to overlook) are the monomials or common factors.

 Hint

You should always begin factoring an algebraic expression by looking for common factors. Once you have factored out all common factors, the remaining expression is easier to factor.

Using the Special Products

Special product #2 gives us our second important form of factoring. Since $(x+y)(x-y) = x^2 - y^2$, we can reverse this to get

Difference of two squares	$x^2 - y^2 = (x+y)(x-y)$

As you can see, to factor the difference between two squares you get factors that are the sum and difference of the quantities.

EXAMPLE 7.22

Factor $x^2 - 25$.
Solution $x^2 - 25 = x^2 - 5^2$
$$= (x+5)(x-5)$$
You would have had to notice that 25 was 5^2 before you could recognize that this was the difference of two squares.

EXAMPLE 7.23

Factor $9a^2 - 49b^4$.

Solution
$$9a^2 - 49b^4 = (3a)^2 - (7b^2)^2$$
$$= (3a + 7b^2)(3a - 7b^2)$$

Again, you have to recognize the perfect squares: $9a^2 = (3a)^2$ and $49b^4 = (7b^2)^2$.

EXAMPLE 7.24

Factor $27x^3 - 75xy^2$.

Solution
$$27x^3 - 75xy^2 = 3x(9x^2 - 25y^2)$$
$$= 3x(3x + 5y)(3x - 5y)$$

This example demonstrates the value of first looking for common factors. There is a common factor of $3x$. When the common factor is factored out, it is easier to see that the remaining factor is the difference of two squares.

EXAMPLE 7.25

Factor $5x^4 - 80y^4$.

Solution
$$5x^4 - 80y^4 = 5(x^4 - 16y^4)$$
$$= 5[(x^2)^2 - (4y^2)^2]$$
$$= 5(x^2 + 4y^2)(x^2 - 4y^2)$$
$$= 5(x^2 + 4y^2)(x + 2y)(x - 2y)$$

Again, you should have noticed that there was a common factor of 5 and that each of the remaining terms was a perfect square. After factoring out $x^4 - 16y^4$ you were not finished, because $x^2 - 4y^2$ is also the difference of two squares. The other factor $x^2 + 4y^2$ cannot be factored further.

≡ **Note**

Remember: $x^2 + y^2 \neq (x + y)^2$. In fact, $x^2 + y^2$ cannot be factored using real numbers.

Application

EXAMPLE 7.26

In order to shield it from stray electromagnetic radiation, an electronic device is housed in a metal canister that is shaped like a right circular cylinder. The surface area of the cylinder is given by $A = 2\pi r^2 + 2\pi rh$, where r is the radius of the base and h is the height of the cylinder. Factor the right-hand side of this formula.

Solution This formula has common factors of 2π and r. Factoring, we obtain

$$A = 2\pi r^2 + 2\pi rh$$
$$= 2\pi r(r + h)$$

Exercise Set 7.2

Completely factor each of the expressions in Exercises 1–40.

1. $6x + 6$

2. $12x + 12$

3. $12a - 6$

4. $15d - 5$

5. $4x - 2y + 8$

6. $6a + 9b - 3c$

7. $5x^2 + 10x + 15$

8. $16a^2 + 8b - 24$

9. $10x^2 - 15$

10. $14x^4 + 21$

11. $4x^2 + 6x$

12. $8a - 4a^2$

13. $7b^2y + 28b$

14. $9ax^2 + 27bx$

15. $3ax + 6ax^2 - 2ax$

16. $6by - 12b^2y + 7by^2$

17. $4ap^2 + 6a^2pq + 8apq^2$

18. $12p^2r^2 - 8p^3r + 24pr^2$

19. $a^2 - b^2$

20. $p^2 - r^2$

21. $x^2 - 4$

22. $a^2 - 16$

23. $y^2 - 81$

24. $m^2 - 49$

25. $4x^2 - 9$

26. $49y^2 - 64$

27. $9a^4 - b^2$

28. $16t^6 - a^2$

29. $25a^2 - 49b^2$

30. $121r^2 - 81t^2$

31. $144 - 25b^4$

32. $81 - 49r^4$

33. $5a^2 - 125$

34. $7x^2 - 63$

35. $28a^2 - 63b^4$

36. $81x^2 - 36t^6$

37. $a^4 - 81$

38. $b^4 - 256$

39. $16x^4 - 256y^4$

40. $25a^5 - 400ab^8$

Solve Exercises 41–48.

41. The total surface area of a cone is given by the formula $\pi r^2 + \pi r \sqrt{h^2 + r^2}$, where r is the radius of the base and h is the height of the cone. Factor this expression.

42. *Thermodynamics* The amount of heat that must be added to a metal object in order for it to melt is given by the formula $Q = mc\Delta t + mL_f$. Factor the right side of this equation.

43. *Wastewater technology* According to Bernoulli's equation, if a fluid of density d is flowing horizontally in a pipe and its pressure and velocity at one location are p_1 and v_1, respectively, and at a second location, they are p_2 and v_2, then the difference in their pressures is given by $p_1 - p_2 = \frac{1}{2}dv_2^2 - \frac{1}{2}dv_1^2$. Factor the right-hand side of this equation.

44. The cross-sectional area A of a tube can be determined from the formula $A = \pi R^2 - \pi r^2$, where R is the outside radius and r is the inside radius of the tube. Factor the right-hand side of this equation.

45. *Acoustical engineering* The angular acceleration α of a stereo turntable during a time period can be determined using the formula

$$\alpha = \frac{\omega_f^2 - \omega_0^2}{2\theta}$$

where ω_0 is the angular velocity at the beginning of the time interval, ω_f is the angular velocity at the end of the time interval, and θ is the number of radians the turntable rotated during the interval. Factor the right-hand side of this equation.

46. *Thermodynamics* A black body is a hypothetical body that absorbs, without reflection, all of the electromagnetic radiation that strikes its surface. The energy E radiated by a black body is given by $E = e\sigma T^4 - e\sigma T_0^4$, where T and T_0 are the absolute temperatures of the body and the surroundings, respectively, σ is the Stefan-Boltzmann constant, and e is the emissivity of the body. Factor the right-hand side of this equation.

47. *Energy* The work-energy equation for rotational motion is

$$W = \frac{1}{2}I\omega_2^2 - \frac{1}{2}I\omega_1^2$$

Factor the right-hand side of this equation.

48. *Optics* The *lensmaker's equation*

$$f^{-1} = nr_1^{-1} - nr_2^{-1} - r_1^{-1} + r_2^{-1}$$

gives the focal length f of a very thin lens.
(a) Rewrite the lensmaker's equation by factoring the right-hand side.
(b) Write your answer in (a) using only positive exponents.

 **In Your Words**

49. Explain how you recognize when an expression is a difference of two squares and then how you would factor it.

50. Describe what you look for when removing a common factor.

 **7.3**
FACTORING TRINOMIALS

In Section 7.2, we introduced the idea of factoring and learned how to factor two types of problems. These problems were based on the first two special products that we learned in Section 7.1.

An algebraic expression that has three terms is called a **trinomial**. Special products #3 through #6 all resulted in quadratic trinomials. In this section we will focus on factoring quadratic trinomials with the purpose of reversing one of these four special products.

Not all quadratic trinomials can be factored using real numbers. The general quadratic trinomial is of the form $ax^2 + bx + c$, where a, b, and c represent constants.

 **Note**

You can determine if a quadratic trinomial can be factored by examining the discriminant $b^2 - 4ac$. If the discriminant is a perfect square, then it is possible to factor the quadratic using rational numbers.

Discriminant

The **discriminant** of the trinomial $ax^2 + bx + c$ is $b^2 - 4ac$.

If $b^2 - 4ac > 0$, then $ax^2 + bx + c$ can be factored using real numbers.

If $b^2 - 4ac$ is a perfect square, then $ax^2 + bx + c$ can be factored using rational numbers.

If $b^2 - 4ac < 0$, then $ax^2 + bx + c$ cannot be factored using real numbers.

EXAMPLE 7.27

Can the trinomial $4x^2 + 3x - 7$ be factored?

Solution In this quadratic trinomial $a = 4$, $b = 3$, and $c = -7$, so

$$b^2 - 4ac = (3)^2 - 4(4)(-7)$$
$$= 9 - (-112)$$
$$= 121$$

Using our knowledge (or a calculator) we see that $\sqrt{121} = 11$, so this trinomial can be factored using rational numbers. Later in this section, our job will be to find its factors.

EXAMPLE 7.28

Can the quadratic trinomial $5x^2 - 3x - 7$ be factored?

Solution   Here $a = 5, b = -3$, and $c = -7$, so

$$b^2 - 4ac = (-3)^2 - 4(5)(-7)$$
$$= 9 - (-140)$$
$$= 149$$

Using a calculator, we find that $\sqrt{149} \approx 12.207$ and is not a perfect square. Thus, it is possible to factor this quadratic equation with real numbers.

We will skip special products #3 and #4 and consider special product #5: $(x + a)(x + b) = x^2 + (a + b)x + ab$. If you examine special products #3 and #4 you can see that they are just special cases of #5. In the special product where $(x + a)(x + b) = x^2 + (a + b)x + ab$, the leading coefficient (the coefficient of the x^2 term) is 1. This makes the job of factoring somewhat easier, because all we need to do is determine a and b. Notice that $a + b$ is the coefficient of the x-term and ab is the constant. Thus, we have the reverse of special product #5.

$$x^2 + (a + b)x + ab = (x + a)(x + b) \qquad \text{#5}$$

Of course, the difficulty is determining the values of a and b.

EXAMPLE 7.29

Factor $x^2 - 3x - 10$.

Solution   Since the value of the discriminant is $49 = 7^2$, we know that this can be factored using rational numbers. From the formula in special product #5, we know that it factors into $(x + a)(x + b)$ where $a + b = -3$ and $ab = -10$.

What possible choices are there for a and b? From the factors of -10 we have the following four possible pairs.

Possible pairs of factors which satisfy $ab = -10$		Sum $(a + b)$
-10 and	1	-9
10 and	-1	9
-5 and	2	-3
5 and	-2	3

As you can see, only one of these pairs adds to -3, the pair of -5 and 2. So, if we let $a = -5$ and $b = 2$, then we have

$$x^2 - 3x - 10 = (x - 5)(x + 2)$$
$$= (x + 2)(x - 5)$$

EXAMPLE 7.30

Factor $x^2 + 10x + 16$.

Solution   The discriminant is 36, a perfect square, so this can be factored using rational numbers into the form $(x + a)(x + b)$. We want to find a and b so that $a + b = 10$ and $ab = 16$. Again, we will begin with the factors of the product $ab = 16$.

Possible pairs of factors which satisfy $ab = 16$			Sum $(a+b)$
16	and	1	17
−16	and	−1	−17
8	and	2	10
−8	and	−2	
4	and	4	
−4	and	−4	

We stopped once we saw that the pairs of 8 and 2 added to 10, because we had found the pair that worked. If we let $a = 8$ and $b = 2$, we have factored the trinomial as

$$x^2 + 10x + 16 = (x+8)(x+2)$$
$$= (x+2)(x+8)$$

A check to see if the constant is positive or negative will give you some help in factoring. If the constant is positive, then the factors will have the same sign—both will be positive or both negative. If the constant term is negative, then the two factors will have different or unlike signs—one will be positive and the other negative.

EXAMPLE 7.31

Factor each trinomial completely.
Here the constant term is positive.

a. $x^2 + 7x + 12 = (x+4)(x+3)$ Both factors have positive signs.
b. $x^2 - 7x + 12 = (x-4)(x-3)$ Both factors have negative signs.

Here the constant term is negative.

c. $x^2 + x - 12 = (x+4)(x-3)$ ⎫ Factors have unlike signs—one
d. $x^2 - x - 12 = (x-4)(x+3)$ ⎬ is positive and one is negative.

Factoring a quadratic trinomial with a leading coefficient that is not 1 is not as easy. Here we are looking at the reverse of special product #6 or

$$acx^2 + (ad+bc)x + bd = (ax+b)(cx+d) \qquad \#6$$

There are three techniques used to factor these quadratic trinomials. We will look at two of them in this section and examine the third in Section 7.4.

The first technique is called trial and error. In using the trial and error method, we make use of the fact that ac is the leading coefficient, bd is the constant term, and $ad+bc$ is the middle coefficient. We can also use our knowledge of signs to help find the signs of the factors.

EXAMPLE 7.32

Factor $3x^2 - 8x + 4$.

Solution The discriminant is $16 = 4^2$, so this equation will factor using rational numbers. Since 3 is a prime number, we know that its only factors are 3 and 1. Also, since the constant is positive, both factors have the same sign. Thus, we know that the factors are $(3x+b)(x+d)$. Now, all we need to find are b and d.

Since $bd = 4$, the possible pairs of factors are $-4, -1; 4, 1; 2, 2;$ and $-2, -2$. Next, we know that $3d + b = -8$ and the only choices from the pairs of factors that satisfy this are $b = -2$ and $d = -2$. Thus, the factors of $3x^2 - 8x + 4$ are $(3x - 2)(x - 2)$. You should multiply these factors together to check that their product is the original trinomial.

EXAMPLE 7.33

Factor $6x^2 + 7x - 20$.

Solution The discriminant is $529 = 23^2$, so this equation will factor using rational numbers. The leading coefficient, 6, has factors of 6 and 1, and 2 and 3. We will try 6 and 1.

Since the constant term is negative, the factors will have unlike signs. Thus, we have $(6x+b)(x+d)$, where either b or d is negative, $bd = -20$, and $6d+b = 7$. The possible choices for the pairs b and d are $1, 20; 2, 10;$ and $4, 5$, where one is positive and the other is negative.

If we try $b = 4$ and $d = -5$, we get $6(-5) + 4 = -26$. Since this is not 7, this is not the correct solution. All other possible combinations for b and d also fail. Thus, we must make another choice for a and c.

If $a = 2$ and $c = 3$, we have $(2x+b)(3x+d)$. If $b = 5$ and $d = -4$, we get $ad + bc = 2(-4) + (5)(3) = -8 + 15 = 7$. This is the correct coefficient for the x-term, so the factors of $6x^2 + 7x - 20$ are $(2x + 5)$ and $(3x - 4)$.

As you can see, the trial and error method can be very long and frustrating, but some people learn to factor a quadratic trinomial quickly by using this method.

The other method we will look at in this section is called either the "grouping" or "split the middle" method. It is longer than the trial and error method, but it is a "sure-fire" technique. We will use the grouping method on the same problem that we just worked.

EXAMPLE 7.34

Factor $6x^2 + 7x - 20$ using the grouping method.

Solution

Step 1. Multiply the leading coefficient and the constant term: $(6)(-20) = -120$.

Step 2. Find two factors of this product whose sum is the coefficient of the middle term of the trinomial. For this problem we want to find factors p and q, where $pq = -120$ and $p + q = 7$. We get $p = 15$ and $q = -8$.

Step 3. Rewrite the trinomial by splitting the middle term into $px + qx$ and grouping the first two terms and the last two terms. In this problem, $7x = 15x - 8x$ and the trinomial becomes $(6x^2 + 15x) + (-8x - 20)$.

Step 4. Distribute the common factors from each grouping.

$$(6x^2 + 15x) + (-8x - 20) = 3x(2x + 5) - 4(2x + 5)$$

Step 5. Distribute the common factor $2x + 5$ from the entire expression.

$$3x(2x + 5) - 4(2x + 5) = (3x - 4)(2x + 5)$$

These are the required factors. Thus, we have factored $6x^2 + 7x - 20$ as $(3x - 4)(2x + 5)$.

EXAMPLE 7.35

Factor $21x^2 - 41x + 10$ using the grouping method.

Solution

Step 1: $(21)(10) = 210$

Step 2: $210 = (-35)(-6)$ and $-35 - 6 = -41$

Step 3: $(21x^2 - 35x) + (-6x + 10)$

Step 4: $7x(3x - 5) - 2(3x - 5)$

Step 5: $(7x - 2)(3x - 5)$

≡ **Note**

Remember to look for any common factors before you start to use either the trial and error or the grouping method.

EXAMPLE 7.36

Factor $20x^3 + 22x^2 - 12x$.

Solution   There is a common factor of $2x$, so

$$20x^3 + 22x^2 - 12x = 2x(10x^2 + 11x - 6)$$
$$= 2x(5x - 2)(2x + 3)$$

Not all trinomials have just one variable. The grouping method is the best method to use when factoring trinomials that have more than one variable. For example, as shown in the next example, $2x^2 - 17xy + 36y^2$ can be factored using the grouping method.

EXAMPLE 7.37

Factor $2x^2 - 17xy + 36y^2$ completely.

Solution

Step 1: $2x^2(36y^2) = 72x^2y^2$

Step 2: $72x^2y^2 = (-9xy)(-8xy)$ and $-9xy - 8xy = -17xy$

Step 3: $(2x^2 - 9xy) + (-8xy + 36y^2)$

Step 4: $x(2x - 9y) - 4y(2x - 9y)$

Step 5: $(x - 4y)(2x - 9y)$

As a check, multiplying the two factors in Step 5 gives the original expression, and so we can say that $2x^2 - 17xy + 36y^2 = (x - 4y)(2x - 9y)$.

It is also possible to factor trinomials with powers greater than 2 if one exponent is twice the other.

EXAMPLE 7.38

Factor $x^8 + 5x^4 + 6$.

Solution $x^8 + 5x^4 + 6 = (x^4)^2 + 5x^4 + 6$
$$= (x^4 + 3)(x^4 + 2)$$

Two other useful methods are based on special products #9 and #10. These are the sum and difference of two cubes.

Sum of Two Cubes	$(x + y)(x^2 - xy + y^2) = x^3 + y^3$	#9

Difference of Two Cubes	$(x - y)(x^2 + xy + y^2) = x^3 - y^3$	#10

EXAMPLE 7.39

Factor $8x^3 + 125$.

Solution $8x^3 = (2x)^3$ and $125 = 5^3$, so this is the sum of two cubes, and $8x^3 + 125 = (2x)^3 + 5^3 = (2x + 5)(4x^2 - 10x + 25)$.

Application

EXAMPLE 7.40

The volume V of a box of height x can be given by $V(x) = 4x^3 - 64x^2 + 252x$. Factor the right-hand side of this equation to determine the "realistic" domain for the height x.

Solution First, we notice that each term on the right-hand side of $V(x) = 4x^3 - 64x^2 + 252x$ has a common factor of $4x$. So, $V(x) = 4x(x^2 - 16x + 63)$. The quadratic expression factors as $x^2 - 16x + 63 = (x - 9)(x - 7)$. So, the completely factored form is

$$V(x) = 4x(x - 9)(x - 7)$$

From this, we see that the domain is $0 < x < 7$ or $x > 9$. We will see later that $x > 9$ will not satisfy many methods used to make such a box.

The six special factors are listed here. Study them, and learn to recognize them either as factors or as one of the special factors.

The Special Factors		
$ax + ay = a(x + y)$	#1
$x^2 - y^2 = (x + y)(x - y)$	#2
$x^2 + (a + b)x + ab = (x + a)(x + b)$	#5
$acx^2 + (ad + bc)x + bd = (ax + b)(cx + d)$	#6
$x^3 + y^3 = (x + y)(x^2 - xy + y^2)$	#9
$x^3 - y^3 = (x - y)(x^2 + xy + y^2)$	#10

Exercise Set 7.3

Determine if each of the trinomials in Exercises 1–6 can be factored.

1. $x^2 + 9x - 8$
2. $x^2 + 7x - 8$
3. $3x^2 - 10x - 8$
4. $2x^2 + 16x + 14$
5. $5x^2 + 23x + 18$
6. $7x^2 - 5x + 16$

Factor each of the following trinomials completely.

7. $x^2 + 7x + 10$
8. $x^2 + 8x + 15$
9. $x^2 - 12x + 27$
10. $x^2 - 14x + 33$
11. $x^2 - 27x + 50$
12. $x^2 + 19x + 48$
13. $x^2 - x - 2$
14. $x^2 - 4x - 5$

15. $x^2 - 3x - 10$
16. $p^2 - 16p + 64$
17. $r^2 + 10r + 25$
18. $v^2 - 14v + 49$
19. $a^2 + 22a + 121$
20. $e^2 + 26e + 169$
21. $f^2 - 30f + 225$
22. $3x^2 + 4x + 1$

23. $6y^2 - 7y + 1$
24. $3p^2 + 5p + 2$
25. $7t^2 + 9t + 2$
26. $5a^2 + 14a - 3$
27. $7b^2 - 34b - 5$
28. $2y^2 + y - 6$
29. $4e^2 + 19e - 5$
30. $6m^2 - 19m + 3$

31. $3u^2 + 10u + 8$
32. $7r^2 + 13r - 2$
33. $9t^2 - 25t - 6$
34. $4x^2 + 8x + 3$
35. $6x^2 + 13x - 5$
36. $8y^2 - 8y - 6$
37. $15a^2 - 16a - 15$
38. $15d^2 + 16d - 15$

39. $15e^2 + 34e + 15$

40. $14a^2 - 39a + 10$

41. $10x^2 - 19x + 6$

42. $3x^2 + 18x + 27$

43. $3r^2 - 18r - 21$

44. $15x^2 + 50x + 35$

45. $49t^4 - 105t^3 + 14t^2$

46. $2y^4 - 9y^2 + 7$

47. $6x^2 - 11xy - 10y^2$

48. $4p^2 + 20pq + 25q^2$

49. $8a^2 - 14ab - 9b^2$

50. $6d^9 + 15d^5e^2 + 6de^4$

51. $a^3 - b^3$

52. $y^3 - 8$

53. $8x^3 - 27$

54. $64p^3 + 125t^6$

Solve Exercises 55–60.

55. *Electronics*   The current i, in amperes, in a certain circuit varies with time, in seconds, according to the equation

$$i = 0.7t^2 - 2.1t - 2.8$$

Factor the right-hand side of this equation.

56. *Sheet metal technology*   A box with an open top is made from a rectangular sheet of metal by cutting equal-sized squares from the corners and folding up the sides. If the length of the side of a square that is removed is x, then the volume of this box is given by $V = 180x - 58x^2 + 4x^3$. Factor this expression.

57. *Business*   The cost C for a certain company to produce n items is given by the equation $C(n) = 0.0001n^3 - 0.2n^2 - 3n + 6,000$. Factor the right-hand side of this equation.

58. *Dynamics*   A ball is thrown upward with a speed of 48 ft/s from the edge of a cliff 448 ft above the ground. The distance s of this ball above the ground at any time t is given by $s(t) = -16t^2 + 48t + 448$. Factor the right-hand side of this equation.

59. *Ecology*   An ecology center wants to make an experimental garden. A gravel border of uniform width will be placed around the rectangular garden. The garden is 10 m long and 6 m wide. The builder has only enough gravel to cover 36 m^2 to the desired depth. In order to determine the width of the border, the equation

$$(6 + 2x)(10 + 2x) - 60 = 36$$

must be solved. **(a)** Simplify this equation and **(b)** factor your answer.

60. *Fire science*   The flow rate, Q, in a certain hose can be found by solving the equation $2Q^2 + Q - 21 = 0$. Factor the left-hand side of this equation.

In Your Words

61. Explain how to use the discriminate to find if a trinomial can be factored by **(a)** rational numbers and **(b)** real numbers.

62. Explain how you recognize when an expression is a difference of two cubes and then how you would factor it.

≡ 7.4
FRACTIONS

Working with algebraic expressions in technical situations often requires work with algebraic fractions. Working with algebraic fractions is very similar to working with fractions based on real numbers. In this section, we will work with some fundamental properties of fractions, and in Sections 7.5 and 7.6 we will use these properties on the basic operations of addition, subtraction, multiplication, and division with fractions. We begin by stating the Fundamental Principle of Fractions.

Fundamental Principle of Fractions

Multiplying or dividing both the numerator and denominator of a fraction by the same number, except zero, results in a fraction that is equivalent to the original fraction. Two fractions are equivalent if their cross-products are equal.

EXAMPLE 7.41

Since $\frac{3}{4} = \frac{3 \cdot 5}{4 \cdot 5} = \frac{15}{20}$ then the fractions $\frac{15}{20}$ and $\frac{3}{4}$ are equivalent by the Fundamental Principle of Fractions.

You can check their cross-products to verify that $\frac{3}{4}$ and $\frac{15}{20}$ are equivalent. Since $(3)(20) = 60$ and $(4)(15) = 60$, their cross-products are equal and the fractions are equivalent.

EXAMPLE 7.42

$$\frac{a}{x} = \frac{a(xy)}{x(xy)} = \frac{axy}{x^2 y}$$

Solution   We can verify that $\dfrac{a}{x} = \dfrac{axy}{x^2 y}$ by comparing their cross-products:

$$a(x^2 y) = ax^2 y \qquad \text{and} \qquad x(axy) = ax^2 y$$

Since the cross-products are equal, the fractions are equivalent as well.

EXAMPLE 7.43

$$\frac{x+3}{x-2} = \frac{(x+3)(x-3)}{(x-2)(x-3)} = \frac{x^2-9}{x^2-5x+6}$$

As long as $x \neq 3$, we can say that $\dfrac{x+3}{x-2} = \dfrac{x^2-9}{x^2-5x+6}$.

Saying that two fractions are equivalent is another way of saying that they represent the same number. Remembering the rules for signed numbers in Section 1.3, we stated that the number of negative signs can be reduced (or increased) by twos without changing the value of the expression. The Fundamental Principle of Fractions says essentially the same thing.

EXAMPLE 7.44

$$\frac{-a}{b} = \frac{-a(-1)}{b(-1)} = \frac{a}{-b}$$

One of the most important applications of the Fundamental Principle of Fractions is in reducing a fraction to **lowest terms** or **simplest form**. A fraction is in lowest terms when the numerator and denominator have no factors in common other than $+1$.

EXAMPLE 7.45

Reduce $\dfrac{14ab^2x}{7ax^3}$ to simplest form.

Solution The numerator and denominator have a common factor of $7ax$. We can then write $\dfrac{14ab^2x}{7ax^3} = \dfrac{7ax(2b^2)}{7ax(x^2)} = \dfrac{2b^2}{x^2}$. Notice that we used a part of the Fundamental Principle of Fractions that we had not used before. We *divided* both the numerator and the denominator by $7ax$, the common factor.

EXAMPLE 7.46

Reduce $\dfrac{x^2(x-3)}{x^2-9}$ to lowest terms.

Solution We need to find a common factor of both the numerator and the denominator; x^2 is *not* that common factor. While x^2 is a factor of the numerator, it is not a factor of the denominator. However, the denominator will factor into $(x-3)(x+3)$, so

$$\frac{x^2(x-3)}{x^2-9} = \frac{x^2(x-3)}{(x-3)(x+3)} = \frac{x^2}{x+3}$$

Again, we used the Fundamental Principle of Fractions and divided both the numerator and denominator by $x-3$.

≡ **Note**

Some factors differ only in sign. While this difference may not seem like much, it is often overlooked. In particular, you should note that

$$x - y = (-1)(-x+y) = -(-x+y) = -(y-x)$$

Of these four, the first and last are the ones you will use the most. Remember, $x-y$ and $y-x$ differ only in sign.

EXAMPLE 7.47

Reduce $\dfrac{5x-xy}{3y-15}$ to lowest terms.

Solution First, factor the numerator and the denominator.

$$\frac{5x-xy}{3y-15} = \frac{x(5-y)}{3(y-5)}$$

Since $5 - y = -(y-5)$, replace $5-y$ with $-(y-5)$.

$$\frac{x(5-y)}{3(y-5)} = \frac{-x(y-5)}{3(y-5)} = \frac{-x}{3}$$

| EXAMPLE 7.47 (Cont.) | Is $\dfrac{-x}{3}$ equal to $\dfrac{5x - xy}{3y - 15}$? You can always verify your answer by checking the cross-products. In this case, the cross-products are equal. |

| EXAMPLE 7.48 | Reduce $\dfrac{2x^3 + 4x^2 - 30x}{15x^2 + x^3 - 2x^4}$ to simplest form. |

Solution First remove the common factors.

$$\frac{2x^3 + 4x^2 - 30x}{15x^2 + x^3 - 2x^4} = \frac{2x(x^2 + 2x - 15)}{x^2(15 + x - 2x^2)}$$

Factor each of the expressions in parentheses.

$$\frac{2x(x^2 + 2x - 15)}{x^2(15 + x - 2x^2)} = \frac{2x(x+5)(x-3)}{x^2(5+2x)(3-x)}$$

Notice that $x - 3 = -(3 - x)$ and rewrite the numerator.

$$\frac{-2x(x+5)(3-x)}{x^2(5+2x)(3-x)}$$

Divide both numerator and denominator by $x(3 - x)$, obtaining

$$\frac{-2x(x+5)(3-x)}{x^2(5+2x)(3-x)} = \frac{-2(x+5)}{x(5+2x)}$$

and we have the final result.

$$\frac{2x^3 + 4x^2 - 30x}{15x^2 + x^3 - 2x^4} = \frac{-2(x+5)}{x(5+2x)}$$

Exercise Set 7.4

In Exercises 1–10, multiply the numerator and denominator of each fraction by the given factor. Check the cross-products to verify that your answer is equivalent to the given fraction.

1. $\frac{7}{8}$ (by 5)

2. $\frac{-5}{9}$ (by -4)

3. $\frac{x}{y}$ (by a)

4. $\frac{r}{t}$ (by z)

5. $\frac{x^2 y}{a}$ (by $3ax$)

6. $\frac{a^3 b}{ca}$ (by $3ab$)

7. $\frac{4}{x - y}$ (by $x + y$)

8. $\frac{a+b}{4}$ (by $a - 4$)

9. $\frac{a+b}{a-b}$ (by $a + b$)

10. $\frac{x-2}{x+3}$ (by $x - 3$)

In Exercises 11–18, divide the numerator and denominator of each of the fractions by the given factor. Check the cross-products to verify that your answer is equivalent to the given fraction.

11. $\frac{38}{24}$ (by 2)

12. $\frac{51}{119}$ (by 17)

13. $\frac{3x^2}{12x}$ (by $3x$)

14. $\frac{15a^3x^2}{3a^4x}$ (by $3a^3x$)

15. $\frac{4(x+2)}{(x+2)(x-3)}$ (by $x+2$)

16. $\frac{7(x-3)(x+5)}{14(x+5)(x-1)}$ [by $7(x+5)$]

17. $\frac{x^2-16}{x^2+8x+16}$ (by $x+4$)

18. $\frac{(x-a)(x-b)(x-c)}{(x-c)(x-b)(x-d)}$ [by $(x-c)(x-b)$]

In Exercises 19–40, reduce each fraction to lowest terms.

19. $\frac{4x^2}{12x}$

20. $\frac{9y}{3y^2}$

21. $\frac{x^2+3x}{x^3+5x}$

22. $\frac{y^2-4y}{2y+y^3}$

23. $\frac{6m^2-3m^3}{9m+18m^3}$

24. $\frac{4r^2+12r^3}{8r+12r^2}$

25. $\frac{x^2+3x}{x^2-9}$

26. $\frac{a^2-9a}{a^2-81}$

27. $\frac{2b^2-10b}{3b^2-75}$

28. $\frac{4e^2-196}{14e-2e^2}$

29. $\frac{z^2-9}{z^2-6z+9}$

30. $\frac{x^2-16}{x^2+8x+16}$

31. $\frac{x^2+4x+3}{x^2+7x+12}$

32. $\frac{a^2-5a+6}{a^2+5a-14}$

33. $\frac{2x^2+9x+4}{x^2+9x+20}$

34. $\frac{15m^2-22m-5}{3m^2+4m-15}$

35. $\frac{12y^3+12y^2+3y}{6y^2-3y-3}$

36. $\frac{45x^2-60x+20}{6x^2+5x-6}$

37. $\frac{x^3y^6-y^3x^6}{2x^3y^4-2x^4y^3}$

38. $\frac{x^3-y^3}{y^2-x^2}$

39. $\frac{x^2-y^2}{x+y}$

40. $\frac{y-x}{x^2-y^2}$

Solve Exercises 41 and 42.

41. *Construction* The safe load, p (in pounds), when using a drop hammer pile driver, can be determined by the formula
$$p=\frac{6whs+6wh}{3s^2+6s+3}$$
Simplify this expression.

42. *Construction* The safe load, p (in pounds), when using a steam pile driver, can be determined by the formula
$$p=\frac{2whs+2whk+2amhs+2amhk-2bhs-2bhk}{s^2+2sk+k^2}$$
Simplify this expression.

 **In Your Words**

43. Explain the Fundamental Principal of Fractions.

44. Describe how to use the Fundamental Principal of Fractions to simplify a fraction.

≣ 7.5
MULTIPLICATION AND DIVISION OF FRACTIONS

The ability to simplify fractions is a skill that will be helpful in this section and the next, as we learn to operate with fractions. After that, it is a skill that will be required throughout this book.

In Section 1.3, we learned the basic operations with real numbers. Among those were Rules 7 and 8, which dealt with multiplying and dividing rational numbers. To refresh your memory, they are repeated here.

Rule 7

To multiply two rational numbers, multiply the numerators and multiply the denominators.

EXAMPLE 7.49

$$\frac{3}{4} \times \frac{-5}{8} = \frac{3 \times (-5)}{4 \times 8} = \frac{-15}{32}$$

Rule 8

To divide one rational number by another, multiply the first by the reciprocal of the second.

EXAMPLE 7.50

Compute $\frac{-5}{8} \div \frac{2}{3}$.

Solution The reciprocal of $\frac{2}{3}$ is $\frac{3}{2}$, so

$$\frac{-5}{8} \div \frac{2}{3} = \frac{-5}{8} \times \frac{3}{2} = \frac{(-5)(3)}{(8)(2)} = \frac{-15}{16}$$

≣ **Note**

You do not need common denominators when multiplying or dividing two rational numbers.

If we express these two rules symbolically, we will have the rules for multiplying or dividing any two fractions, whether they are rational numbers or algebraic fractions. The rule for multiplying two rational numbers can be restated as the following.

Multiplying Fractions

If $\dfrac{a}{b}$ and $\dfrac{c}{d}$ are fractions, then their product is

$$\frac{a}{b} \cdot \frac{c}{d} = \frac{ac}{bd}$$

Similarly, we can state the following rule for dividing one rational number by another.

Dividing Fractions

If $\dfrac{a}{b}$ and $\dfrac{c}{d}$ are fractions, then their quotient is

$$\frac{a}{b} \div \frac{c}{d} = \frac{a}{b} \times \frac{d}{c} = \frac{ad}{bc}$$

EXAMPLE 7.51

Find the product of $\dfrac{3a^2}{x}$ and $\dfrac{7y}{5p^3}$.

Solution $\quad \dfrac{3a^2}{x} \cdot \dfrac{7y}{5p^3} = \dfrac{(3a^2)(7y)}{(x)(5p^3)} = \dfrac{21a^2y}{5xp^3}$

EXAMPLE 7.52

Find the product of $\dfrac{x-2}{x+3}$ and $\dfrac{x+2}{x-5}$.

Solution $\quad \dfrac{x-2}{x+3} \cdot \dfrac{x+2}{x-5} = \dfrac{(x-2)(x+2)}{(x+3)(x-5)}$

$$= \frac{x^2-4}{x^2-2x-15}$$

EXAMPLE 7.53

Find the quotient when $\dfrac{7x^2}{3a}$ is divided by $\dfrac{5y}{4x}$.

Solution $\quad \dfrac{7x^2}{3a} \div \dfrac{5y}{4x} = \dfrac{7x^2}{3a} \cdot \dfrac{4x}{5y}$

$$= \frac{(7x^2)(4x)}{(3a)(5y)}$$

$$= \frac{28x^3}{15ay}$$

EXAMPLE 7.54

Find $\dfrac{x-2}{x+3} \div \dfrac{x-4}{x-3}$.

Solution
$$\dfrac{x-2}{x+3} \div \dfrac{x-4}{x-3} = \dfrac{x-2}{x+3} \cdot \dfrac{x-3}{x-4} = \dfrac{(x-2)(x-3)}{(x+3)(x-4)}$$
$$= \dfrac{x^2 - 5x + 6}{x^2 - x - 12}$$

≡ Note

It is often beneficial to leave the answer in factored form. For example, in Example 7.54, you might have wanted to leave the answer as $\dfrac{(x-2)(x-3)}{(x+3)(x-4)}$. This version has 2 multiplication operations, 1 division operation, 3 subtraction operations, and 1 addition operation, for a total of 7 operations. The answer $\dfrac{x^2 - 5x + 6}{x^2 - x - 12}$ has 3 multiplication operations, 1 division operation, 3 subtractions, and 1 addition, for a total of 8 operations. In computer programming, more operations require more computer time, which costs more money.

Caution

When you divide, make sure that you multiply by the reciprocal of the *divisor*. Do not invert the dividend.

The following examples use all of the skills we have learned for simplifying fractions.

EXAMPLE 7.55

Multiply $\dfrac{x^2 - 9}{4x - 8}$ and $\dfrac{2x + 8}{x + 3}$.

Solution   If we proceed as before, we get
$$\dfrac{x^2 - 9}{4x - 8} \cdot \dfrac{2x + 8}{x + 3} = \dfrac{(x^2 - 9)(2x + 8)}{(4x - 8)(x + 3)} = \dfrac{2x^3 + 8x^2 - 18x - 72}{4x^2 + 4x - 24}$$

While this is correct, it is not the easiest approach. It is a good idea to study a problem for a few seconds before you start to work it. If we had stopped to factor these fractions we could have saved some work.
$$\dfrac{x^2 - 9}{4x - 8} \cdot \dfrac{2x + 8}{x + 3} = \dfrac{(x+3)(x-3)}{4(x-2)} \cdot \dfrac{2(x+4)}{(x+3)}$$
$$= \dfrac{2(x+3)(x-3)(x+4)}{4(x-2)(x+3)}$$
$$= \dfrac{(x-3)(x+4)}{2(x-2)}$$

or
$$= \dfrac{x^2 + x - 12}{2x - 4}$$

EXAMPLE 7.56

Compute $\dfrac{2x^2+9x-5}{3x^2-3x-60}\cdot\dfrac{3x+12}{2x+10}$.

Solution $\dfrac{2x^2+9x-5}{3x^2-3x-60}\cdot\dfrac{3x+12}{2x+10}=\dfrac{(2x-1)(x+5)(3)(x+4)}{3(x-5)(x+4)(2)(x+5)}$

$$=\dfrac{2x-1}{2(x-5)}$$

The common factor $3(x+5)(x+4)$ is easily seen using this procedure.

Caution

Be sure to factor the numerator and denominator first. Only common factors can be "canceled."

 Hint

When a polynomial is factored, all the $+$ and $-$ signs are inside parentheses.

EXAMPLE 7.57

Compute $\dfrac{x^2-9}{6x^2-21x}\div\dfrac{(x+3)^2}{2x-7}$.

Solution $\dfrac{x^2-9}{6x^2-21x}\div\dfrac{(x+3)^2}{2x-7}=\dfrac{x^2-9}{6x^2-21x}\cdot\dfrac{2x-7}{(x+3)^2}$

$$=\dfrac{(x-3)(x+3)(2x-7)}{3x(2x-7)(x+3)(x+3)}$$

$$=\dfrac{x-3}{3x(x+3)}=\dfrac{x-3}{3x^2+9x}$$

There are times when it is just as useful to leave the final answer in the factored form rather than multiplying the factors together. Thus, we could have left the last answer in the form $\dfrac{x-3}{3x(x+3)}$.

EXAMPLE 7.58

Simplify $\dfrac{\dfrac{6x}{x^2-4}}{\dfrac{2x^2+10x}{x+2}}$.

EXAMPLE 7.58 (Cont.)

Solution We have to remember that $\dfrac{a}{b}$ means $a \div b$. So, this is the division problem:

$$\frac{6x}{x^2-4} \div \frac{2x^2+10x}{x+2} = \frac{6x}{x^2-4} \cdot \frac{x+2}{2x^2+10x}$$

$$= \frac{6x}{(x-2)(x+2)} \cdot \frac{x+2}{2x(x+5)}$$

$$= \frac{6x(x+2)}{(x-2)(x+2)2x(x+5)}$$

$$= \frac{3}{(x-2)(x+5)} \cdot$$

Exercise Set 7.5

In Exercises 1–48, perform the indicated operation and simplify.

1. $\dfrac{2}{x} \cdot \dfrac{5}{y}$

2. $\dfrac{4}{y} \cdot \dfrac{x}{3}$

3. $\dfrac{4x^2}{5} \cdot \dfrac{3}{y^3}$

4. $\dfrac{7x}{6} \cdot \dfrac{5y}{2t}$

5. $\dfrac{3}{x} \div \dfrac{7}{y}$

6. $\dfrac{a}{3} \div \dfrac{b}{4}$

7. $\dfrac{2x^2}{3} \div \dfrac{7y}{4x}$

8. $\dfrac{9x}{2y} \div \dfrac{4y}{3a}$

9. $\dfrac{2x}{3y} \cdot \dfrac{5}{4x^2}$

10. $\dfrac{3xy}{7} \cdot \dfrac{14x}{5y^2}$

11. $\dfrac{3a^2b}{5d} \cdot \dfrac{25ad^2}{6b^2}$

12. $\dfrac{x^2y^2t}{abc} \cdot \dfrac{b^2c}{y^3t}$

13. $\dfrac{3y}{5x} \div \dfrac{15x^2}{8xy}$

14. $\dfrac{4y^2}{7x} \div \dfrac{8y^3}{21x}$

15. $\dfrac{3x^2y}{7p} \cdot \dfrac{15x^2p}{7y^2}$

16. $\dfrac{9xyz}{7a} \div \dfrac{3ayz}{14z}$

17. $\dfrac{4y+16}{5} \cdot \dfrac{15y}{3y+12}$

18. $\dfrac{x^2+3x}{6a} \cdot \dfrac{a^2}{x^2-9}$

19. $\dfrac{a^2-b^2}{a+3b} \cdot \dfrac{5a+15b}{a+b}$

20. $(x+y)\dfrac{x^2+2x}{x^2-y^2}$

21. $\dfrac{x^2-100}{10} \div \dfrac{2x+10}{15}$

22. $\dfrac{5a^2}{x^2-49} \div \dfrac{25ax-25a}{x^2+7x}$

23. $\dfrac{4x^2-1}{9x-3x^2} \div \dfrac{2x+1}{x^2-9}$

24. $\dfrac{x+y}{3x-3y} \div \dfrac{(x+y)^2}{x^2-y^2}$

25. $\dfrac{a^2-8a}{a-8} \cdot \dfrac{a+2}{a}$

26. $\dfrac{49-x^2}{x+y} \cdot \dfrac{x}{7-x}$

27. $\dfrac{2a-b}{4a} \cdot \dfrac{2a-b}{4a^2-4ab+b^2}$

28. $\dfrac{x^4-81}{(x-3)^2} \cdot \dfrac{x-3}{4-x^2}$

29. $\dfrac{y^2}{x^2-1} \div \dfrac{y^2}{x-1}$

30. $\dfrac{m^2-49}{m^2-5m-14} \div \dfrac{m+7}{2m^2-13m-7}$

31. $\dfrac{2y^2-y}{4y^2-4y+1} \div \dfrac{y^2}{8y-4}$

32. $\dfrac{a-1}{a^2-1} \div \dfrac{(a-1)^2}{a^2-1}$

33. $\dfrac{x^2-3x+2}{x^2+5x+6} \cdot \dfrac{x+3}{3x-6}$

34. $\dfrac{2x+2}{x^2+2x-8} \cdot \dfrac{x^2-4}{x^2+4x+4}$

35. $\dfrac{x^2+xy-6y^2}{x^2+6xy+8y^2} \cdot \dfrac{x^2-9xy+20y^2}{x^2-4xy-21y^2}$

36. $\dfrac{y^2+14xy+49x^2}{y^2-7xy-30x^2} \cdot \dfrac{y^2-100x^2}{y^3+7xy^2}$

37. $\dfrac{9x^2-25}{x^2+6x+9} \div \dfrac{3x+5}{x+3}$

38. $\dfrac{x^2-16}{x^2-6x+8} \div \dfrac{x^3+4x^2}{x^2-9x+14}$

39. $\dfrac{x^2+4xy+4y^2}{x^2-4y^2} \div \dfrac{x^2+xy-2y^2}{x^2-xy-2y^2}$

40. $\dfrac{p^3-27q^3}{3p^2+9pq+27q^2} \div \dfrac{9q^2-p^2}{6p+18q}$

41. $\dfrac{x+y}{4x-4y} \div \left[\dfrac{(x+y)^2}{x^2-y^2} \cdot \dfrac{x^3-y^3}{x^3+y^3}\right]$

42. $\dfrac{2.4m^2n}{0.8mn^2} \div \left[\dfrac{0.6m}{0.3n} \cdot \dfrac{3.6m}{2.4n}\right]$

43. $\dfrac{x^2-25}{5x^2-24x-5} \cdot \dfrac{2x^2+12x+2}{6x^2-12x} \div \dfrac{x^2+6x+1}{5x^2-9x-2}$

44. $\dfrac{a-b}{a^3-b^3} \cdot \dfrac{a^3+ab^2}{4a+4b} \div \dfrac{a^4-b^4}{8a^2-8b^2}$

45. $\left(\dfrac{2x^2-5x-3}{x^2-x-12} \div \dfrac{2x^2+5x+2}{3x+9}\right) \div \dfrac{x^2-9}{x^2-2x-8}$

46. $\dfrac{a^2+4a-5}{a^2-3a-4} \cdot \dfrac{a^2+3a-28}{(a-3)^2} \div \dfrac{a^2+12a+35}{a^2-2a-3}$

47. *Construction*   The volume strain of a beam is given by the expression

$$\frac{(a^3-a'^3)/a^3}{(a^2-a'^2)/a^2}$$

Simplify this expression.

48. *Electronics*   The charge on a capacitor is given by $Q = CV$. The energy stored in the capacitor is given by $E = \dfrac{1}{2}\dfrac{Q^2}{V}$. Find the ratio of charge to energy.

In Your Words

49. **(a)** Explain how to divide two fractions: $\dfrac{a}{b} \div \dfrac{c}{d}$.

(b) What do you think is the most common mistake people make when they divide two fractions?

50. **(a)** Explain how to multiply two fractions: $\dfrac{a}{b} \times \dfrac{c}{d}$.

(b) What do you think is the most common mistake people make when they multiply two fractions?

≡ 7.6
ADDITION AND SUBTRACTION OF FRACTIONS

In Section 7.5, we learned how to multiply and divide two fractions. In this section, we will look at two other operations with fractions—addition and subtraction.

As we mentioned earlier, much of this work with algebraic fractions is patterned after our work with rational numbers from Section 1.3 . Rule 6 dealt with the addition and subtraction of rational numbers. This rule is repeated here.

> **Rule 6**
>
> To add (or subtract) two rational numbers, change both denominators to the same positive integer (the common denominator), add (or subtract) the numerators, and place the result over the common denominator.

EXAMPLE 7.59

Perform the indicated operations and simplify **(a)** $\frac{2}{3} + \dfrac{-5}{6}$ and **(b)** $\dfrac{-5}{7} - \dfrac{-8}{3}$.

Solutions

(a) A common denominator of 3 and 6 is 6, so

$$\frac{2}{3} + \frac{-5}{6} = \frac{4}{6} + \frac{-5}{6} = \frac{4+(-5)}{6} = \frac{-1}{6}$$

EXAMPLE 7.59 (Cont.)

(b) A common denominator of 7 and 3 is 21, so

$$\frac{-5}{7} = \frac{-15}{21} \quad \text{and} \quad \frac{-8}{3} = \frac{-56}{21}$$

As a result, we obtain

$$\frac{-5}{7} - \frac{-8}{3} = \frac{-15}{21} - \frac{-56}{21} = \frac{-15 - (-56)}{21} = \frac{41}{21}$$

Before we restate Rule 6 in symbols, we will consider a special case of adding or subtracting fractions. If two fractions have the same denominator, then you need only add or subtract the numerators. Symbolically, this is represented as:

$$\frac{a}{c} + \frac{b}{c} = \frac{a+b}{c} \quad \text{and} \quad \frac{a}{c} - \frac{b}{c} = \frac{a-b}{c}$$

EXAMPLE 7.60

Simplify **(a)** $\dfrac{7}{2x} + \dfrac{9y}{2x}$ and **(b)** $\dfrac{x}{x+5} - \dfrac{2x-y}{x+5}$.

Solutions

(a) $\dfrac{7}{2x} + \dfrac{9y}{2x} = \dfrac{7+9y}{2x}$

(b) $\dfrac{x}{x+5} - \dfrac{2x-y}{x+5} = \dfrac{x-(2x-y)}{x+5} = \dfrac{x-2x+y}{x+5}$

$$= \frac{-x+y}{x+5}$$

If the denominators are not the same, then addition and subtraction become somewhat more complicated. As Rule 6 indicates, you need to find a common denominator and rewrite each fraction as an equivalent fraction with this common denominator. The quickest way to find a common denominator is to multiply the denominators together. This method is demonstrated here.

Caution

When adding or subtracting fractions, remember to multiply both the numerator and denominator of a fraction by the same quantity.

Adding and Subtracting Fractions

The sum of two fractions $\dfrac{a}{b}$ and $\dfrac{c}{d}$ is

$$\frac{a}{b}+\frac{c}{d}=\frac{ad}{bd}+\frac{bc}{bd}=\frac{ad+bc}{bd}$$

The difference of two fractions $\dfrac{a}{b}$ and $\dfrac{c}{d}$ is

$$\frac{a}{b}-\frac{c}{d}=\frac{ad}{bd}-\frac{bc}{bd}=\frac{ad-bc}{bd}$$

As you will see in Example 7.61(b), this may not be the lowest common denominator. You must also remember that both the numerator and denominator must by multiplied by the same quantity.

EXAMPLE 7.61

Simplify **(a)** $\dfrac{3}{x+5}+\dfrac{x}{x-5}$ and **(b)** $\dfrac{2x}{x+3}-\dfrac{x-4}{x^2-9}$.

Solutions

(a)
$$\frac{3}{x+5}+\frac{x}{x-5}=\frac{3}{x+5}\cdot\frac{x-5}{x-5}+\frac{x}{x-5}\cdot\frac{x+5}{x+5}$$
$$=\frac{3(x-5)}{(x+5)(x-5)}+\frac{x(x+5)}{(x+5)(x-5)}$$
$$=\frac{3(x-5)+(x^2+5x)}{(x+5)(x-5)}$$
$$=\frac{3x-15+x^2+5x}{(x+5)(x-5)}$$
$$=\frac{x^2+8x-15}{x^2-25}$$

(b)
$$\frac{2x}{x+3}-\frac{x-4}{x^2-9}=\frac{2x}{x+3}\cdot\frac{x^2-9}{x^2-9}-\frac{x-4}{x^2-9}\cdot\frac{x+3}{x+3}$$
$$=\frac{2x(x^2-9)}{(x+3)(x^2-9)}-\frac{(x-4)(x+3)}{(x+3)(x^2-9)}$$
$$=\frac{(2x^3-18x)-(x^2-x-12)}{(x+3)(x^2-9)}$$
$$=\frac{2x^3-x^2-17x+12}{x^3+3x^2-9x-27}$$

Least Common Denominator

Perhaps you wondered if the last answer was in simplest form. It is not, since

$$\frac{2x^3-x^2-17x+12}{x^3+3x^2-9x-27}=\frac{(x+3)(2x^2-7x+4)}{(x+3)(x^2-9)}=\frac{2x^2-7x+4}{x^2-9}$$

If we want our answers in the simplest form, then simply multiplying the denominators together is not the best method to use. What we need to do is determine the least common denominator, or LCD, of the fractions to be added or subtracted.

There are three steps to determining the LCD:

> **How to Find the Least Common Denominator**
>
> (1) Factor the denominator of each of the fractions in the problem,
> (2) Determine the different factors and the highest power of each factor that occurs in any denominator, and
> (3) Multiply the distinct factors from Step 2 after each has been raised to its highest power.

EXAMPLE 7.62

Find the LCD of the fractions $\dfrac{7x+1}{x^4+x^3}$, $\dfrac{14}{x^3-4x^2+4x}$, and $\dfrac{9}{2x^2-2x-4}$.

Solution

Step 1: Factor the denominator of each of the fractions: $\dfrac{7x+1}{x^3(x+1)}$, $\dfrac{14}{x(x-2)^2}$, and $\dfrac{9}{2(x+1)(x-2)}$.

Step 2: List each factor and the highest exponent of each.

factor	highest exponent	final factors
2	1	2^1
x	3	x^3
$x+1$	1	$(x+1)^1$
$x-2$	2	$(x-2)^2$

Step 3: The LCD is $2x^3(x+1)(x-2)^2$.

EXAMPLE 7.63

Find the least common denominator of $\dfrac{2x}{x^2+5x+6}$, $\dfrac{x-3}{x^3+2x^2}$, and $\dfrac{x^2+x}{x^3+6x^2+9x}$.

Solution

Step 1: Factor each denominator: $\dfrac{2x}{(x+2)(x+3)}$, $\dfrac{x-3}{x^2(x+2)}$, and $\dfrac{x^2+x}{x(x+3)^2}$.

Step 2: List each factor and the highest exponent of each.

factor	highest exponent	final factors
$x+2$	1	$x+2$
$x+3$	2	$(x+3)^2$
x	2	x^2

Step 3: The LCD is $x^2(x+2)(x+3)^2$.

Now we have the foundation for a much better way to add or subtract algebraic fractions, or any fractions. For each fraction, multiply both the numerator and denominator by a quantity that makes the denominator equal to the LCD of the fractions being added or subtracted. Then, add or subtract the numerators; place the result over the common denominator; and, if possible, simplify.

EXAMPLE 7.64

Calculate $\dfrac{2x}{x+3} - \dfrac{x-4}{x^2-9}$.

Solution This is the same difference we were asked to compute in Example 7.61(b). First we find the LCD, which is $(x+3)(x-3)$.

We rewrite the first fraction as $\dfrac{2x}{x+3} = \dfrac{2x(x-3)}{(x+3)(x-3)}$. The second fraction, $\dfrac{x-4}{x^2-9}$, is already written with the common denominator. So,

$$\frac{2x}{x+3} - \frac{x-4}{x^2-9} = \frac{2x(x-3)}{(x+3)(x-3)} - \frac{x-4}{x^2-9}$$

$$= \frac{2x(x-3)-(x-4)}{x^2-9}$$

$$= \frac{(2x^2-6x)-(x-4)}{x^2-9}$$

$$= \frac{2x^2-7x+4}{x^2-9}$$

This was the same problem we worked in Example 7.61(b). Notice how much simpler this answer looks compared to the answer we found before.

EXAMPLE 7.65

Calculate $\dfrac{2x}{x+2} + \dfrac{x}{x-2} - \dfrac{1}{x^2-4}$.

Solution The LCD of these fractions is $(x+2)(x-2)$. So,

$$\frac{2x}{x+2} = \frac{2x(x-2)}{(x+2)(x-2)},$$

$$\frac{x}{x-2} = \frac{x(x+2)}{(x-2)(x+2)},$$

and $$\frac{1}{x^2-4} = \frac{1}{(x-2)(x+2)}.$$

EXAMPLE 7.65 (Cont.)

Thus, we get

$$\frac{2x}{x+2}+\frac{x}{x-2}-\frac{1}{x^2-4} = \frac{2x(x-2)}{(x+2)(x-2)}+\frac{x(x+2)}{(x-2)(x+2)}-\frac{1}{x^2-4}$$

$$= \frac{(2x^2-4x)+(x^2+2x)-1}{x^2-4}$$

$$= \frac{3x^2-2x-1}{x^2-4}$$

Complex Fractions

A **complex fraction** is a fraction in which the numerator, the denominator, or both, contain a fraction. There are two methods that are commonly used to simplify complex fractions.

Method 1: Find the LCD of all the fractions that appear in the numerator and denominator. Multiply both the numerator and denominator by the LCD.

Method 2: Combine the terms in the numerator into a single fraction. Combine the terms in the denominator into a single fraction. Divide the numerator by the denominator.

We will work each of the next two examples using both methods. Then you will be better able to select the method you prefer.

EXAMPLE 7.66

Simplify $\dfrac{2+\dfrac{1}{x}}{x-\dfrac{2}{x^2}}$.

Solution

Method 1: The LCD of 2, $\dfrac{1}{x}$, x, and $\dfrac{2}{x^2}$ is x^2, so

$$\frac{2+\dfrac{1}{x}}{x-\dfrac{2}{x^2}} = \frac{2+\dfrac{1}{x}}{x-\dfrac{2}{x^2}}\cdot\frac{x^2}{x^2}$$

$$= \frac{2x^2+x}{x^3-2}$$

$$= \frac{x(2x+1)}{x^3-2}$$

Method 2:

$$2+\frac{1}{x} = \frac{2x}{x}+\frac{1}{x} = \frac{2x+1}{x}$$

$$x-\frac{2}{x^2} = \frac{x^3}{x^2}-\frac{2}{x^2} = \frac{x^3-2}{x^2}$$

EXAMPLE 7.66 (Cont.)

$$\frac{2+\dfrac{1}{x}}{x-\dfrac{2}{x^2}}=\frac{\dfrac{2x+1}{x}}{\dfrac{x^3-2}{x^2}}=\frac{2x+1}{x}\div\frac{x^3-2}{x^2}$$

$$=\frac{2x+1}{x}\cdot\frac{x^2}{x^3-2}=\frac{x(2x+1)}{x^3-2}$$

EXAMPLE 7.67

Simplify $\dfrac{\dfrac{1}{2x}-\dfrac{6}{y}}{\dfrac{1}{x}+\dfrac{2}{3y}}$.

Solution

Method 1: The LCD of $\dfrac{1}{2x},\dfrac{6}{y},\dfrac{1}{x}$, and $\dfrac{2}{3y}$ is $6xy$, so

$$\frac{\dfrac{1}{2x}-\dfrac{6}{y}}{\dfrac{1}{x}+\dfrac{2}{3y}}=\frac{\left(\dfrac{1}{2x}-\dfrac{6}{y}\right)}{\left(\dfrac{1}{x}+\dfrac{2}{3y}\right)}\cdot\frac{6xy}{6xy}$$

$$=\frac{3y-36x}{6y+4x}=\frac{3(y-12x)}{2(3y+2x)}$$

Method 2:

$$\frac{1}{2x}-\frac{6}{y}=\frac{y}{2xy}-\frac{12x}{2xy}=\frac{y-12x}{2xy}$$

$$\frac{1}{x}+\frac{2}{3y}=\frac{3y}{3xy}+\frac{2x}{3xy}=\frac{3y+2x}{3xy}$$

$$\frac{\dfrac{1}{2x}-\dfrac{6}{y}}{\dfrac{1}{x}+\dfrac{2}{3y}}=\frac{\dfrac{y-12x}{2xy}}{\dfrac{3y+2x}{3xy}}$$

$$=\frac{y-12x}{2xy}\div\frac{3y+2x}{3xy}$$

$$=\frac{y-12x}{2xy}\cdot\frac{3xy}{3y+2x}$$

$$=\frac{3(y-12x)}{2(3y+2x)}$$

Application

EXAMPLE 7.68

If four resistances are connected in a series-parallel circuit, the total resistance R is given by the equation

$$R = \frac{1}{R_1 + R_2} + \frac{1}{R_3 + R_4}$$

where R_1, R_2, R_3, and R_4 represent four resistances. Simplify this equation by adding the right-hand side.

Solution We find that the LCD of the right-hand side of the given equation is $(R_1 + R_2)(R_3 + R_4)$. So,

$$R = \frac{1}{R_1 + R_2} \cdot \frac{R_3 + R_4}{R_3 + R_4} + \frac{1}{R_3 + R_4} \cdot \frac{R_1 + R_2}{R_1 + R_2}$$

$$= \frac{R_3 + R_4}{(R_1 + R_2)(R_3 + R_4)} + \frac{R_1 + R_2}{(R_1 + R_2)(R_3 + R_4)}$$

$$= \frac{R_1 + R_2 + R_3 + R_4}{(R_1 + R_2)(R_3 + R_4)}$$

Exercise Set 7.6

In Exercises 1–44, perform the indicated operations and simplify.

1. $\frac{2}{7} + \frac{5}{7}$

2. $\frac{4}{5} + \frac{-11}{5}$

3. $\frac{7}{3} - \frac{5}{3}$

4. $\frac{-2}{9} - \frac{8}{9}$

5. $\frac{1}{2} + \frac{1}{3}$

6. $\frac{3}{4} + \frac{-2}{3}$

7. $\frac{4}{5} - \frac{2}{3}$

8. $-\frac{5}{7} - \frac{3}{5}$

9. $\frac{1}{x} + \frac{5}{x}$

10. $\frac{2}{y} + \frac{-5}{y}$

11. $\frac{4}{a} - \frac{3}{a}$

12. $\frac{-5}{p} - \frac{-7}{p}$

13. $\frac{2x}{y} + \frac{3x}{y}$

14. $\frac{4p}{q} - \frac{6p}{q}$

15. $\frac{3r}{2t} + \frac{-r}{2t} - \frac{5r}{2t}$

16. $\frac{3x}{2y} - \frac{5x}{2y} + \frac{x}{2y}$

17. $\frac{3}{x+2} + \frac{x}{x+2}$

18. $\frac{5}{y-3} + \frac{y}{y-3}$

19. $\frac{t}{t+1} - \frac{2}{t+1}$

20. $\frac{a}{b-3} - \frac{4}{3-b}$

21. $\frac{y-3}{x+2} + \frac{3+y}{x+2}$

22. $\frac{x+4}{x-2} + \frac{x-5}{x-2}$

23. $\frac{x+2}{a+b} - \frac{x-5}{a+b}$

24. $\frac{x+4}{y-5} - \frac{2-x}{y-5}$

25. $\frac{2}{x} + \frac{3}{y}$

26. $\frac{x}{y} + \frac{5}{x}$

27. $\frac{a}{b} - \frac{4}{d}$

28. $\frac{2x}{y} - \frac{3y}{x}$

29. $\frac{3}{x(x+1)} + \frac{4}{x^2-1}$

30. $\frac{5}{y(x+1)} + \frac{x}{y(x+2)}$

31. $\frac{2}{x^2-1} - \frac{4}{(x+1)^2}$

32. $\frac{6}{y-2} - \frac{3}{y+2}$

33. $\frac{x}{x^2-11x+30} + \frac{2}{x^2-36}$

34. $\frac{a}{a^2-9a+18} + \frac{a}{a^2-9}$

35. $\frac{2}{x^2-x-6} - \frac{5}{x^2-4}$

36. $\frac{b}{b^2-10b+21} - \frac{b}{b^2-9}$

37. $\frac{x-1}{3x^2-13x+4} + \frac{3x+1}{4x-x^2}$

38. $\frac{x-3}{x^2+3x+2} + \frac{2x-5}{x^2+x-2}$

39. $\frac{x-3}{x^2-1} + \frac{2x-7}{x^2+5x+4}$

40. $\frac{x+4}{x^2-9} - \frac{x-3}{x^2+6x+9}$

41. $\dfrac{y+3}{y^2-y-2}-\dfrac{2y-1}{y^2+2y-8}$

42. $\dfrac{1}{(a-b)(a-c)}+\dfrac{1}{(b-a)(b-c)}-\dfrac{1}{(b-c)(a-c)}$

43. $\dfrac{x}{(x^2+3)(x-1)}+\dfrac{3x^2}{(x-1)^2(x+2)}-\dfrac{x+2}{x^2+3}$

44. $\dfrac{2x-1}{x^2+5x+6}-\dfrac{x-2}{x^2+4x+3}+\dfrac{x-4}{x^2+3x+2}$

Use Method 1 to simplify each of the complex fractions in Exercises 45–50.

45. $\dfrac{1+\frac{2}{x}}{1-\frac{3}{x}}$

46. $\dfrac{x+\frac{1}{x}}{2-\frac{1}{x}}$

47. $\dfrac{x-1}{1+\frac{1}{x}}$

48. $\dfrac{x^2-25}{\frac{1}{x}-\frac{1}{5}}$

49. $\dfrac{\frac{x+y}{x}-\frac{x-y}{y}}{\frac{x}{x+y}+\frac{y}{x-y}}$

50. $\dfrac{x+3-\frac{16}{x+3}}{x-6+\frac{20}{x+6}}$

Use Method 2 to simplify each of the complex fractions in Exercises 51–56.

51. $\dfrac{1+\frac{3}{x}}{1+\frac{2}{x}}$

52. $\dfrac{y+\frac{1}{y}}{3+\frac{2}{y}}$

53. $\dfrac{t-1}{t+\frac{1}{t}}$

54. $\dfrac{x^2-36}{\frac{1}{6}-\frac{1}{x}}$

55. $\dfrac{\frac{x-y}{x-y}-\frac{y}{x+y}}{\frac{1}{x-y}+\frac{1}{x+y}}$

56. $\dfrac{t-5+\frac{25}{t-5}}{t+3+\frac{10}{t-3}}$

Solve Exercises 57–64.

57. *Electricity*   If two resistors, R_1 and R_2, are connected in parallel, the equivalent resistance of the combination can be found using $\dfrac{1}{R}=\dfrac{1}{R_1}+\dfrac{1}{R_2}$. Add the right-hand side of the equation.

58. *Optics*   The lensmakers equation states that if p is the object distance, q the image distance, and f the focal length of a lens, then $\dfrac{1}{p}+\dfrac{1}{q}=\dfrac{1}{f}$. Simplify the left-hand side of this equation.

59. *Electricity*   If three capacitors with capacitance C_1, C_2, and C_3 are connected together in series, then they can be replaced by a single capacitor of capacitance C. The value of C can be determined from the equation $\dfrac{1}{C}=\dfrac{1}{C_1}+\dfrac{1}{C_2}+\dfrac{1}{C_3}$. Simplify the right-hand side of the equation.

60. *Transportation*   A car travels the first part of a trip for a distance d_1 at velocity v_1, and the second part of the trip it travels d_2 at the velocity v_2. The average speed for these two parts of the trip is given by $\dfrac{d_1+d_2}{\frac{d_1}{v_1}+\frac{d_2}{v_2}}$. Simplify this fraction.

61. *Physics*   When an elastic body of mass m_1 collides with an elastic body of mass m_2, both moving with a preimpact velocity v toward each other, the rebound velocity of m_1 is given by

$$\left(\dfrac{m_1}{m_1+m_2}-\dfrac{m_2}{m_1+m_2}\right)v+2\left(\dfrac{m_2}{m_1+m_2}\right)v$$

Simplify this expression.

62. *Electrical engineering*   In calculating the electric intensity of a field set up by a dipole, the following expression is used.

$$\dfrac{1}{\left(r-\frac{d}{2}\right)^2}-\dfrac{1}{\left(r+\frac{d}{2}\right)^2}$$

Simplify this expression.

63. *Electrical engineering*   Millman's theorem provides a shortcut for finding the common voltage, V, across any number of parallel branches with different voltage sources. If there are three branches, then the common voltage is

$$V = \frac{\dfrac{V_1}{R_1} + \dfrac{V_2}{R_2} + \dfrac{V_3}{R_3}}{\dfrac{1}{R_1} + \dfrac{1}{R_2} + \dfrac{1}{R_3}}$$

Simplify the right-hand side of this equation.

64. *Automotive engineering*   If a car travels a distance d_1 at a rate v_1 and then another distance d_2 at a rate v_2, the average speed, v, for the entire trip is

$$v = \frac{d_1 + d_2}{\dfrac{d_1}{v_1} + \dfrac{d_2}{v_2}}$$

Simplify the right-hand side of this equation.

 **In Your Words**

65. (a) What is a common denominator?

  (b) Explain how to find the least common denominator.

66. Describe a complex algebraic fraction. Give examples of fractions that are complex algebraic fractions, and give examples of some that are not. How are they different?

▤ CHAPTER 7 REVIEW

Important Terms and Concepts

Binomial
Common factor
Denominator
Discriminant
Factor
Fractions
  Addition
  Complex

Division
Equivalent
Multiplication
Reducing
Subtraction
Least common denominator (LCD)
Numerator
Trinomial

Review Exercises

In Exercises 1–10, find the indicated products by direct use of one of the special products.

1. $5x(x - y)$
2. $(3 + x)^2$
3. $(x - 2y)^3$

4. $(x + y)(x - 6)$
5. $(2x + 3)(x - 6)$
6. $(x + 7)(x - 7)$

7. $(x^2 - 5)(x^2 + 5)$
8. $(7x - 1)(x + 5)$
9. $(2 + x)^3$

10. $(x - 7)^2$

Completely factor each of the expressions in Exercises 11–20.

11. $9 + 9y$
12. $x^2 - 4$
13. $7x^2 - 63$

14. $x^2 - 12x + 36$
15. $x^2 - 11x + 30$
16. $x^2 + 15x + 36$

17. $x^2 + 6x - 16$
18. $x^2 - 4x - 45$
19. $2x^2 - 3x - 9$

20. $8x^3 + 6x^2 - 20x$

In Exercises 21–26, reduce each fraction to lowest terms.

21. $\dfrac{2x}{6y}$

22. $\dfrac{7x^2y}{9xy^2}$

23. $\dfrac{x^2-9}{(x+3)^2}$

24. $\dfrac{x^2-4x-45}{x^2-81}$

25. $\dfrac{x^3+y^3}{x^2+2xy+y^2}$

26. $\dfrac{x^3-16x}{x^2+2x-8}$

In Exercises 27–40, perform the indicated operations and simplify.

27. $\dfrac{x^2}{y}\cdot\dfrac{3y^2}{7x}$

28. $\dfrac{x^2-9}{x+4}\cdot\dfrac{x^3-16x}{x-3}$

29. $\dfrac{4x}{3y}\div\dfrac{2x^2}{6y}$

30. $\dfrac{x^2-25}{x^2-4x}\div\dfrac{2x^2+2x-40}{x^3-x}$

31. $\dfrac{4x}{y}+\dfrac{3x}{y}$

32. $\dfrac{4}{x-y}+\dfrac{6}{x+y}$

33. $\dfrac{3(x-3)}{(x+2)(x-5)^2}+\dfrac{4(x-1)}{(x+2)^2(x-5)}$

34. $\dfrac{8a}{b}-\dfrac{3}{b}$

35. $\dfrac{x}{y+x}-\dfrac{x}{y-x}$

36. $\dfrac{2(x+3)}{(x+1)^2(x+2)}-\dfrac{3(x-1)}{(x+1)(x+2)^2}$

37. $\dfrac{x^2-5x-6}{x^2+8x+12}+\dfrac{x^2+7x+6}{x^2-4x-12}$

38. $\dfrac{2x-1}{4x^2-12x+5}-\dfrac{x+1}{4x^2-4x-15}$

39. $\dfrac{x^2-5x-6}{x^2+8x+12}\div\dfrac{x^2+7x+6}{x^2-4x-12}$

40. $\dfrac{x^2+x-2}{7a^2x^2-14a^2x+7a^2}\cdot\dfrac{14ax-28a}{1-2x+x^2}$

Simplify each of the complex fractions in Exercises 41–46.

41. $\dfrac{\dfrac{1}{x}-\dfrac{1}{y}}{\dfrac{1}{x}+\dfrac{1}{y}}$

42. $\dfrac{\dfrac{1}{x}+\dfrac{1}{y}}{x+y}$

43. $\dfrac{\dfrac{1}{x}-\dfrac{1}{y}}{\dfrac{x-y}{xy}}$

44. $\dfrac{1-\dfrac{1}{x}}{x-2+\dfrac{1}{x}}$

45. $\dfrac{\dfrac{x}{1+x}-\dfrac{1-x}{x}}{\dfrac{x}{1+x}+\dfrac{1-x}{x}}$

46. $\dfrac{x-\dfrac{xy}{x-y}}{\dfrac{x^2}{x^2-y^2}-1}$

▤ CHAPTER 7 TEST

1. Multiply $(x+5)(x-3)$.

2. Multiply $(2x-3)(2x+3)$.

3. Multiply $(3x^2-4)(2-5x)$

4. Multiply $(x-4)^3$

5. Completely factor $2x^2-128$.

6. Completely factor $x^2-12x+32$.

7. Completely factor $10x^2+x-21$.

8. Completely factor x^3-125.

9. Reduce $\dfrac{x^2-25}{x^2+6x+5}$ to lowest terms.

10. Simplify $\dfrac{3(a+b)^3-x(a+b)}{a^2-b^2}$

11. Calculate $\dfrac{3x}{x+2}\cdot\dfrac{x-1}{x+2}$.

12. Calculate $\dfrac{2x+6}{x-2}\div\dfrac{3x+9}{x^2-4}$.

13. Calculate $\dfrac{6}{x-5}+\dfrac{x^2-2x}{x-5}$.

14. Calculate $\dfrac{2x}{x+3} - \dfrac{x+4}{x-2}$.

15. Simplify $\dfrac{x - \dfrac{1}{x}}{x - \dfrac{2}{x+1}}$.

16. Reduce $\dfrac{1}{2x+1} - \dfrac{2}{4x^2+4x+1}$

17. The average rate, r, for a round trip with a one-way distance d is

$$r = \frac{2d}{\dfrac{d}{r_1} + \dfrac{d}{r_2}}$$

Simplify this complex fraction.

18. The total resistance, R, in a parallel electrical circuit with three resistances is given by

$$\frac{1}{R} = \frac{1}{R_1} + \frac{1}{R_2} + \frac{1}{R_3}$$

Express the sum on the right-hand side as a single fraction.

CHAPTER

8

Fractional and Quadratic Equations

Quadratic equations are needed to determine lengths or other dimensions. In Sections 8.3 and 8.4, you will learn two ways to determine the lengths of the rafters in this solar collector.

Courtesy of New York State Energy Research and Development Authority

In Chapter 7, we learned the special products, how to factor some algebraic expressions, and how to simplify and operate with fractions. In this chapter, we will use that knowledge to help solve problems.

You already know something about solving equations; but your knowledge has been limited to working with linear equations. In this chapter, we will venture into learning techniques for solving two new types of equations—fractional equations and quadratic equations.

≡ 8.1
FRACTIONAL EQUATIONS

An equation in which one or more terms is a fraction is called a **fractional equation**. Solving a fractional equation requires a technique that we used in solving systems of linear equations. In order to add or subtract two linear equations in Section 5.2, you often had to multiply one or both equations by a nonzero number. To solve fractional equations, we will use that same technique—we will multiply the equation by a nonzero quantity. In particular, we will multiply the equation by the LCD. This is often referred to as **clearing the equation**.

The easiest type of fractional equations to solve are those in which the variables occur only in the numerator.

EXAMPLE 8.1

Solve $\dfrac{2x}{3} - \dfrac{3x}{5} = \dfrac{1}{10}$ for x.

Solution The LCD of $\dfrac{2x}{3}$, $\dfrac{3x}{5}$, and $\dfrac{1}{10}$ is 30, so we will multiply both sides of the equation by 30.

$$30\left(\frac{2x}{3} - \frac{3x}{5}\right) = 30\left(\frac{1}{10}\right)$$
$$30\left(\frac{2x}{3}\right) - 30\left(\frac{3x}{5}\right) = 30\left(\frac{1}{10}\right)$$
$$20x - 18x = 3$$
$$2x = 3$$
$$x = \frac{3}{2}$$

If we check our answer in the original problem, we see that $\dfrac{2\left(\frac{3}{2}\right)}{3} - \dfrac{3\left(\frac{3}{2}\right)}{5} = 1 - \frac{9}{10} = \frac{1}{10}$. The answer checks.

If the variables are in the denominator, we then need to use more caution.

Caution

The original equation will not be defined for any values of the variable that give any of the denominators a value of 0. If you forget this, you may get an answer that does not satisfy the original problem. This type of answer is called an **extraneous solution** because it seems to be a solution but is not a valid one. For this reason, it is a good idea to study the equation first and to note any values that make the denominator 0.

In these next examples, we multiply the equation by the least common denominator for the fractions. Notice that we begin each solution by finding out which values make a denominator 0.

EXAMPLE 8.2

Solve $\dfrac{2}{x-5} = \dfrac{1}{4x-12}$ for x.

Solution   The LCD of $\dfrac{2}{x-5}$ and $\dfrac{1}{4x-12}$ is $4(x-5)(x-3)$. Since the LCD has a value of 0 when $x = 5$ or $x = 3$, neither of these values is a possible solution for this equation.

If we multiply both sides of the equation by the LCD, we obtain

$$4(x-5)(x-3)\left(\frac{2}{x-5}\right) = 4(x-5)(x-3)\left(\frac{1}{4x-12}\right)$$
$$4(x-3)(2) = x-5$$
$$8(x-3) = x-5$$
$$8x-24 = x-5$$
$$7x = 19$$
$$x = \frac{19}{7}$$

Thus, $x = \frac{19}{7}$ appears to be the solution. But, we should check our work to ensure that we have made no errors.

Check: The left-hand side of the equation becomes

$$\frac{2}{\dfrac{19}{7}-5} = \frac{2}{\dfrac{19}{7}-\dfrac{35}{7}} = \frac{2}{\dfrac{-16}{7}} = -\frac{7}{8}$$

The value of the right-hand side is

$$\frac{1}{4\left(\dfrac{19}{7}\right)-12} = \frac{1}{\dfrac{76}{7}-12} = \frac{1}{\dfrac{76}{7}-\dfrac{84}{7}} = \frac{1}{\dfrac{-8}{7}} = -\frac{7}{8}$$

Both sides of the equation have a value of $-\frac{7}{8}$ when $x = \frac{19}{7}$, so $x = \frac{19}{7}$ must be the correct solution. ▪

≡ **Note**

We could have used cross-multiplication to solve the equation in the last example. However, as in the next three examples, cross-multiplication would not be easy.

EXAMPLE 8.3

Solve $\dfrac{4}{x^2-1} = \dfrac{2}{x-1} - \dfrac{3}{x+1}$.

Solution   The LCD of $\dfrac{4}{x^2-1}$, $\dfrac{2}{x-1}$, and $\dfrac{3}{x+1}$ is $(x+1)(x-1) = x^2-1$. Notice that $x \neq 1$ and $x \neq -1$, because each of these values makes two of the denominators 0. Multiplying both sides of the given equation by x^2-1, we obtain

$$(x^2-1)\left(\frac{4}{x^2-1}\right) = (x^2-1)\left(\frac{2}{x-1}\right) - (x^2-1)\left(\frac{3}{x+1}\right)$$
$$4 = (x+1)2 - (x-1)3$$

EXAMPLE 8.3 (Cont.)

$$4 = 2x + 2 - (3x - 3)$$
$$4 = 2x + 2 - 3x + 3$$
$$4 = -x + 5$$
$$-1 = -x$$

or $\qquad x = 1$

Since $x = 1$ is not an allowable solution, the "solution" $x = 1$ is extraneous. *This equation has no solution.*

EXAMPLE 8.4

Solve $\dfrac{3x}{x-2} + 5 = \dfrac{7x}{x-2}$.

Solution   The LCD is $x - 2$, so $x \neq 2$. Multiplying both sides by $x - 2$, we get

$$(x-2)\left(\frac{3x}{x-2} + 5\right) = (x-2)\left(\frac{7x}{x-2}\right)$$

$$(x-2)\left(\frac{3x}{x-2}\right) + (x-2)5 = (x-2)\left(\frac{7x}{x-2}\right)$$

$$3x + 5x - 10 = 7x$$
$$8x - 10 = 7x$$
$$x = 10$$

Substituting $x = 10$ into the original equation shows that it satisfies the equation.

Application

EXAMPLE 8.5

In a lens, if the object distance is p, the image distance is q, and the focal length is f, then the relation exists where $\dfrac{1}{f} = \dfrac{1}{p} + \dfrac{1}{q}$. Solve this equation for q.

Solution   The LCD of $\dfrac{1}{f}$, $\dfrac{1}{p}$, and $\dfrac{1}{q}$ is fpq. Multiplying both sides of the equation by fpq, we obtain

$$fpq\left(\frac{1}{f}\right) = fpq\left(\frac{1}{p} + \frac{1}{q}\right)$$

$$pq = fpq\left(\frac{1}{p}\right) + fpq\left(\frac{1}{q}\right)$$

Multiplying further, we obtain

$$pq = fq + fp$$

EXAMPLE 8.5 (Cont.)

To solve for q, we put the terms containing q on the left-hand side with all other terms on the right-hand side of the equation.

$$pq - fq = fp$$

At this point, we need to determine the coefficient of q. We do this by factoring.

$$q(p - f) = fp$$

We see that $p - f$ acts as the coefficient of q. Dividing by this coefficient, we obtain the desired solution.

$$q = \frac{fp}{p - f}$$

Application

EXAMPLE 8.6

A technician can assemble an instrument in 12.5 hours. After working for 3 hours on a job, the technician is joined by another technician, who is able to assemble the instrument alone in 9.5 hours. How long does it take to assemble this instrument?

Solution The problem is similar to the work problem we solved in Chapter 2. To solve this example, let h represent the number of hours that the technicians worked together on the instrument. The time to assemble the instrument will be $h + 3$ hours because the first technician worked alone for 3 hours.

The first technician, working alone, can complete the job in 12.5 hours. So, each hour this technician works, $\frac{1}{12.5}$ of the instrument is assembled. Similarly, the second technician will assemble $\frac{1}{9.5}$ of the instrument for each hour worked. The first technician works $h + 3$ hours and is able to complete $\frac{1}{12.5}(h + 3)$ of the work. The second technician works h hours and completes $\frac{1}{9.5}(h)$ of the work. Together, they assemble the entire instrument, so we get the equation

$$\frac{1}{12.5}(h + 3) + \frac{1}{9.5}(h) = 1$$

Multiplying by the common denominator $(12.5)(9.5)$, we obtain

$$9.5(h + 3) + 12.5(h) = (9.5)(12.5)$$
$$9.5h + 28.5 + 12.5h = 118.75$$
$$22h = 90.25$$
$$h \approx 4.1$$

Thus, the two technicians will be able to completely assemble the instrument in about 7.1 hours or 7 hours 6 minutes. (Remember, the total time of 7.1 is $h + 3$ hours.)

Exercise Set 8.1

In Exercises 1–30, solve the given equations and check the results.

1. $\dfrac{x}{2} + \dfrac{x}{3} = \dfrac{1}{4}$

2. $\dfrac{x}{3} - \dfrac{x}{4} = \dfrac{1}{2}$

3. $\dfrac{y}{2} + 3 = \dfrac{4y}{5}$

4. $\dfrac{y}{5} - 5\dfrac{1}{2} = \dfrac{3y}{4}$

5. $\dfrac{x-1}{2} + \dfrac{x+1}{3} = \dfrac{x-1}{4}$

6. $\dfrac{x+2}{3} - \dfrac{x+4}{2} = \dfrac{x-1}{6}$

7. $\dfrac{1}{x} + \dfrac{2}{x} = \dfrac{1}{3}$

8. $\dfrac{3}{x} - \dfrac{4}{x} = \dfrac{2}{5}$

9. $\dfrac{7}{w-4} = \dfrac{1}{2w+5}$

10. $\dfrac{5}{y+1} = \dfrac{3}{y-3}$

11. $\dfrac{2}{2x-1} = \dfrac{5}{x+5}$

12. $\dfrac{3}{4x+2} = \dfrac{1}{x+2}$

13. $\dfrac{4x}{x-3} - 1 = \dfrac{3x}{x-3}$

14. $7 - \dfrac{3x}{x+2} = \dfrac{4x}{x+2}$

15. $\dfrac{4}{x+2} - \dfrac{3}{x-1} = \dfrac{5}{(x-1)(x+2)}$

16. $\dfrac{3}{x-3} + \dfrac{2}{2-x} = \dfrac{5}{(x-3)(x-2)}$

17. $\dfrac{x+1}{x+2} + \dfrac{x+3}{x-2} = \dfrac{2x^2+3x-5}{x^2-4}$

18. $\dfrac{x+2}{x+3} - \dfrac{x+5}{x-3} = \dfrac{2x-1}{x^2-9}$

19. $\dfrac{3}{a+1} + \dfrac{a+1}{a-1} = \dfrac{a^2}{a^2-1}$

20. $\dfrac{5}{x-4} - \dfrac{x+2}{x+4} = \dfrac{x^2}{16-x^2}$

21. $\dfrac{2}{x-1} + \dfrac{5}{x+1} = \dfrac{4}{x^2-1}$

22. $\dfrac{3x+4}{x+2} - \dfrac{3x-5}{x-4} = \dfrac{12}{x^2-2x-8}$

23. $\dfrac{5x-2}{x-3} + \dfrac{4-5x}{x+4} = \dfrac{10}{x^2+x-12}$

24. $\dfrac{2x}{x-1} - \dfrac{3}{x+2} = \dfrac{4x}{x^2+x-2} + 2$

25. $\dfrac{5}{x} + \dfrac{3}{x+1} = \dfrac{x}{x+1} - \dfrac{x+1}{x}$

26. $\dfrac{y}{y+2} + \dfrac{5}{y-1} = \dfrac{3}{y+2} + \dfrac{y}{y-1}$

27. $\dfrac{2t-4}{2t+4} = \dfrac{t+2}{t+4}$

28. $\dfrac{3x+5}{x-5} = \dfrac{3x-1}{x+3}$

29. $\dfrac{3x+1}{x-1} - \dfrac{x-2}{x+3} = \dfrac{2x-3}{x+3} + \dfrac{4}{x-1}$

30. $\dfrac{7x+2}{x+2} + \dfrac{3x-1}{x+3} = \dfrac{6x+1}{x+3} + \dfrac{4x-3}{x+2}$

Solve each of Exercises 31–38 for the indicated variable.

31. $\dfrac{1}{r} + \dfrac{1}{s} = \dfrac{1}{t}$ for s

32. $\dfrac{P_1 V_1}{T_1} = \dfrac{P_2 V_2}{T_2}$ for T_1

33. $\dfrac{1}{R} = \dfrac{1}{R_1} + \dfrac{1}{R_2} + \dfrac{1}{R_3}$ for R

34. $P = \dfrac{E^2}{R+r} - \dfrac{E^2}{(R+r)^2}$ for E^2

35. $V = 2\pi rh + 2\pi r^2$ for h

36. $\dfrac{5}{9}(F - 32) = C$ for F

37. $\dfrac{1}{f} = (n-1)\left(\dfrac{1}{R_1} + \dfrac{1}{R_2}\right)$ for R_2

38. $\dfrac{P_1}{g} + \dfrac{V_1^2}{2g} + h_1 = \dfrac{P_2}{dg} + \dfrac{V_2^2}{2g} + h_2$ for g

Solve Exercises 39–52.

39. *Electrical engineering* The capacitance, C, of a spherical capacitor is given by the formula

$$\dfrac{d}{(9 \times 10^9)C} = \dfrac{1}{R_2} - \dfrac{1}{R_1}$$

where R_2 is the outside radius of the sphere, R_1 is the inside radius, and d is the dielectric constant. Solve this equation for C.

40. *Electrical engineering*   The capacitance, C, of a circuit containing three capacitances C_1, C_2, and C_3 in series is given by

$$\frac{1}{C} = \frac{1}{C_1} + \frac{1}{C_2} + \frac{1}{C_3}$$

Solve this equation for C.

41. *Optics*   The *lensmaker's equation*

$$\frac{1}{f} = (n-1)\left(\frac{1}{r_1} - \frac{1}{r_2}\right)$$

gives the focal length, f, of a very thin lens. Solve this equation for f.

42. *Optics*   An important equation in optics is

$$\frac{1}{f} = \frac{1}{p} + \frac{1}{q}$$

where f is the focal length of the lens, p is the distance to the object from the lens, and q is the distance to the image from the lens. Solve this equation for p.

43. *Mechanical engineering*   A formula relating the depth, h, of a gear tooth to the major diameter, D, of the gear and the minor diameter, d, of the gear may be expressed as

$$\frac{h}{D-d} = 2$$

Solve this equation for D.

44. *Automotive engineering*   The formula for the efficiency of a diesel engine is given by

$$\text{Eff} = 1 - \frac{T_4 - T_1}{\alpha(T_3 - T_2)}$$

Solve this equation for T_1.

45. *Business*   One computer can process a company's payroll in 10 h, while a newer computer can do the same job in 6 h. Working together at these rates, how long would it take to complete the payroll?

46. *Wastewater technology*   Working alone, one pipe can fill a tank in 12 h, while a second pipe can fill the tank in 15 h. If both pipes are opened at the same time, how long does it take to fill the tank?

47. *Wastewater technology*   Pipe A can fill a tank in 6 h and Pipe B can fill it in 4 h. If Pipe A is opened 1 h before Pipe B is opened, how long does it take to fill the tank?

48. *Computer technology*   One microprocessor can process a set of data in 5 μs (microseconds) and a second microprocessor can process the same amount of data in 8 μs. If they process the data together, how many microseconds should it take?

49. *Electricity*   A generator can charge a group of batteries in 18 h. It begins charging the batteries, and 4 h later a second generator starts charging the same set of batteries. If the second generator alone could charge the batteries in 12 h, how long will it take both to charge the batteries?

50. *Transportation*   An airplane traveling against the wind travels 500 km in the same time it takes it to travel 650 km with the wind. If the wind speed is 20 km/h, find the speed of the airplane in still air.

51. *Wastewater technology*   Working alone, it takes one large pipe 3 h to fill a tank. Two smaller pipes are used to drain the tank. Each of the smaller pipes requires 4 h to drain the tank. By mistake, one of the small pipes is left open when the tank is being filled. Assuming that no one notices the mistake, how long does it take to fill the tank?

52. *Solar energy*   Solar collector A can absorb 12,000 Btu in 8 h. A second collector, B, is added. Together collectors A and B collect 45,000 Btu in 6 h. How long would it take collector B alone to collect 45,000 Btu?

 **In Your Words**

53. **(a)** What is an extraneous solution? **(b)** Are they good or bad? **(c)** How can you tell if you have an extraneous solution? **(d)** What should you do if you get an extraneous solution?

54. Describe what is meant by clearing an equation.

≡ 8.2
QUADRATIC EQUATIONS AND FACTORING

Until now, all the equations we have solved have been first degree, or linear, equations and systems of linear equations. Many technical problems require the ability to solve more complicated equations. In the remainder of this chapter, we will focus on second-degree, or quadratic, equations. As we continue through the book, we will learn how to solve more types of equations.

Quadratic Equations

We worked with quadratic trinomials and binomials in Chapter 7. A polynomial equation of the second degree is a **quadratic equation**.

> **Quadratic Equation**
>
> If a, b, and c are constants and $a \neq 0$, then
> $$ax^2 + bx + c = 0$$
> is the **standard quadratic equation**.

EXAMPLE 8.7

The following are all quadratic equations written in the standard form:

(a) $2x^2 - 3x + 5 = 0$ $a = 2$, $b = -3$, $c = 5$
(b) $4x^2 + 7x = 0$ $a = 4$, $b = 7$, $c = 0$
(c) $5x^2 - 125 = 0$ $a = 5$, $b = 0$, $c = -125$
(d) $(p+3)x^2 + px - p + 2 = 0$ $a = p+3$, $b = p$, $c = 2 - p$

EXAMPLE 8.8

The following are also quadratic equations, but are not in the standard form.

(a) $x^2 = 49$ $a = 1$, $b = 0$, $c = -49$
(b) $8 + 2x = \dfrac{7x^2}{2}$ $a = \dfrac{7}{2}$, $b = -2$, $c = -8$

≡ **Note**

Quadratic equations that contain fractions are often simplified by writing them without fractions and with $a > 0$. This simplification is achieved by multiplying the equation by the LCD of the coefficients. (We could write the equation in Example 8.8(b) as $-\frac{7}{2}x^2 + 2x + 8 = 0$, but it would be better to multiply the equation by -2 and obtain the equivalent equation $7x^2 - 4x - 16 = 0$, since integer coefficients are usually easier to use.)

EXAMPLE 8.9

The following are not quadratic equations.

(a) $2x^3 - x^2 + 5 = 0$ This equation has a term of degree 3. The highest degree of any term in a quadratic equation is 2.

(b) $4x + 5 = 0$ This does not have a term of degree 2. This means that $a = 0$, which contradicts part of the definition of a quadratic equation.

In order to solve a quadratic equation, we need another property for the real numbers. We have not needed the **zero-product rule** until now.

> **Zero-Product Rule for Real Numbers**
>
> If a and b are numbers and $ab = 0$, then $a = 0$, $b = 0$, or both a and b are 0.

This is a very simple but powerful statement, as shown by the next example.

EXAMPLE 8.10

Use the zero-product rule to solve $(x - 1)(x + 5) = 0$.

Solution Here a has the value $x - 1$ and b has the value $x + 5$. According to the zero-product rule, either $x - 1 = 0$ or $x + 5 = 0$ (or both). If $x - 1 = 0$, then $x = 1$. If $x + 5 = 0$, then $x = -5$. These are the roots (also called solutions or zeros) of the equation.

Substitute 1 into the original equation. Did you get 0? Now substitute -5 and you will get 0 again. So, both of these answers check. The solutions are $x = -5$ and $x = 1$.

Roots of Quadratic Equations

The zero-product rule indicates that if we can factor a quadratic equation, then we can find its roots or solutions. A quadratic equation will never have more than two roots. Normally, all quadratic equations are considered to have two roots. But, there are times when both of these roots are the same number. In this case, the roots are referred to as **double roots**. There are also times when there will be no real numbers that are roots of a quadratic equation. In this case, the roots will be imaginary numbers. (We will talk more about this later in this chapter and again in Chapter 14.)

Finding Roots by Factoring

Let's begin by looking at the general idea behind the method of factoring to find roots of a quadratic equation. Suppose we have a general quadratic equation $ax^2 + bx + c = 0$ and that this equation can be factored as follows.

$$ax^2 + bx + c = (rx + t)(sx + v) = 0$$

From the zero-product rule, we know that $rx + t = 0$ or $sx + v = 0$. Solving each of these linear equations, we get $x = \dfrac{-t}{r}$ and $x = \dfrac{-v}{s}$, which are the roots of the quadratic equation. Now, let's look at some examples showing how to use this method.

EXAMPLE 8.11

Find the roots of $x^2 - x - 6 = 0$.

Solution   This quadratic equation factors to $(x - 3)(x + 2)$. So, from the zero-product rule, we have

$$x^2 - x - 6 = 0$$
$$(x - 3)(x + 2) = 0$$
$$x - 3 = 0 \text{ and } x = 3$$
or $\qquad\qquad x + 2 = 0 \text{ and } x = -2$

EXAMPLE 8.12

Find the roots of $x^2 + 6x + 9 = 0$.

Solution   Factoring $x^2 + 6x + 9$, we get $(x + 3)^2$, so

$$x^2 + 6x + 9 = 0$$
$$(x + 3)(x + 3) = 0$$
$$x + 3 = 0 \text{ and } x = -3$$

This is a double root. Both roots are the same: -3.

EXAMPLE 8.13

Find the roots of $6x^2 - 11x - 35 = 0$.

Solution $\qquad 6x^2 - 11x - 35 = 0$
$$(3x + 5)(2x - 7) = 0$$
$$3x + 5 = 0 \text{ and } x = -\frac{5}{3}$$
or $\qquad\qquad 2x - 7 = 0 \text{ and } x = \frac{7}{2}$

It is very important that one side of the equation is equal to 0.

Caution

The property that we are using, the zero-product rule, says that if $ab = 0$, then $a = 0$ or $b = 0$ (or both). It is guaranteed to work *only* if the right-hand side of the equation is 0. For example, $ab = 1$ does *not* imply that $a = 1$ or $b = 1$.

Suppose you had a problem such as $(x - 2)(x + 4) = 6$, where the right-hand side of the equation was not 0, and you tried to use the zero-product rule to solve the equation. If $(x - 2)(x + 4) = 6$, you would say $x - 2 = 6$ or $x + 4 = 6$. In the first case, $x - 2 = 6$, we get a possible solution of $x = 8$. In the second case, $x + 4 = 6$, we obtain an answer of $x = 2$.

Now check these answers.

If $x = 8$, then $(x-2)(x+4) = (8-2)(8+4) = (6)(12) = 72$, which is certainly not 6. This answer does not check. Let's try the other solution.

If $x = 2$, then $(x-2)(x+4) = (2-2)(2+4) = (0)(6) = 0$. Again, we do not get an answer of 6.

This example was intended to show you that it is important to make sure that the right-hand side of the equation is 0 before applying the zero-product rule. The next example will show you what to do when the right-hand side of the equation is not 0.

EXAMPLE 8.14

Find the roots of $4x^2 - 10 = 3x$.

Solution Before we can factor this problem, we have to get all the terms on the left-hand side of the equation; then the right-hand side will be 0.

$$4x^2 - 10 = 3x$$
$$4x^2 - 3x - 10 = 0$$
$$(4x+5)(x-2) = 0$$
$$4x+5 = 0 \text{ and } x = -\frac{5}{4}$$
$$\text{or} \qquad x - 2 = 0 \text{ and } x = 2$$

So, the two roots of this equation are $x = -\frac{5}{4}$ and $x = 2$.

It is also possible to solve some fractional equations by factoring. You will need to first multiply the equation by the LCD, and then put all the nonzero terms on the left-hand side of the equation before you begin to factor the equation. The next example shows how to do this.

EXAMPLE 8.15

Find the roots of $\dfrac{6}{x(3-2x)} - \dfrac{4}{3-2x} = 1$.

Solution The LCD is $x(3-2x)$, so $x \neq 0$ and $x \neq \frac{3}{2}$. Multiplying both sides of the equation by $x(3-2x)$ provides

$$6 - 4x = x(3-2x)$$
$$6 - 4x - x(3-2x) = 0$$
$$6 - 4x - 3x + 2x^2 = 0$$
$$2x^2 - 7x + 6 = 0$$
$$(2x-3)(x-2) = 0$$
$$2x - 3 = 0 \text{ and } x = \frac{3}{2}$$
$$\text{or} \qquad x - 2 = 0 \text{ and } x = 2$$

Since $x \neq \frac{3}{2}$, the only root is $x = 2$.

This section will not place much emphasis on graphing. But, you should realize that a quadratic function and its roots can be graphically represented. Figures 8.1 and 8.2 show the graphs for the quadratic functions $f(x) = x^2 - x - 6$ and $f(x) = x^2 + 6x + 9$, formed by the quadratic equations in Examples 8.11 and 8.12. As we will learn in Chapter 15, these are both examples of graphs called "parabolas."

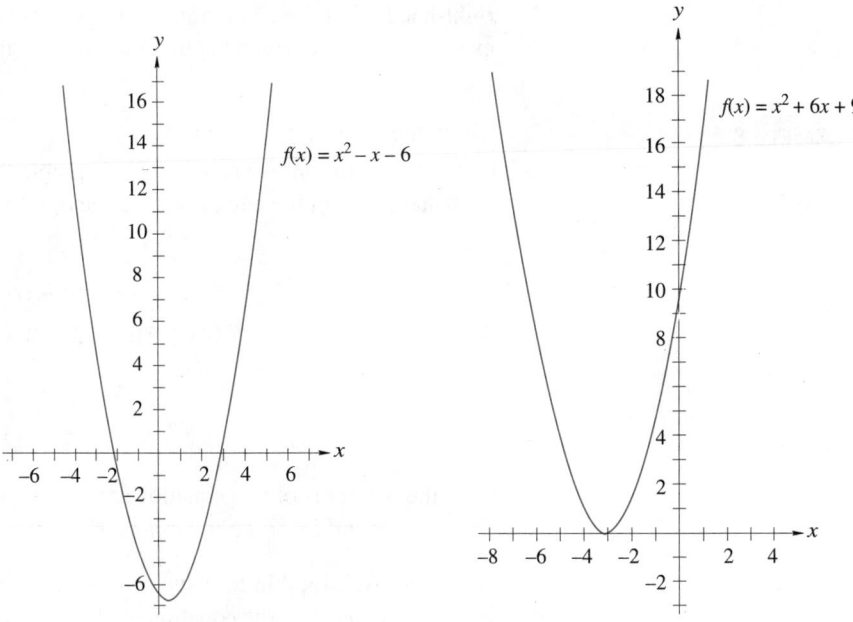

FIGURE 8.1   FIGURE 8.2

In Example 8.11, we found that the roots were -2 and 3 and that these are the points where the graph crosses the x-axis. In Example 8.12, the quadratic equation $x^2 + 6x + 9 = 0$ had a double root, $x = -3$. The graph of the quadratic function for Example 8.12 intersects the x-axis at exactly one point, and that one point is when $x = -3$.

EXAMPLE 8.16

Find the roots of $x^2 = 16$.

Solution This equation is equivalent to $x^2 - 16 = 0$. Factoring, we obtain $(x - 4)(x + 4) = 0$ and $x = 4$ or $x = -4$. We could have solved this problem with less work if we had taken the square root of both sides.

$$x^2 = 16$$
$$x = \pm\sqrt{16}$$
$$x = \pm 4$$

In general, if $c \geq 0$ and $x^2 = c$, then $x = \pm\sqrt{c}$.

EXAMPLE 8.17

Find the roots of $9x^2 = 25$.

Solution $9x^2 = 25$

$$x^2 = \frac{25}{9}$$

$$x = \pm\sqrt{\frac{25}{9}} = \pm\frac{5}{3}$$

EXAMPLE 8.18

Find the roots of $(x-4)^2 = 25$.

Solution   There are two ways to work this problem.

Method 1: First, we will square the left-hand side of the equation and collect like terms.

$$(x-4)^2 = 25$$
$$x^2 - 8x + 16 = 25$$
$$x^2 - 8x + 16 - 25 = 0$$
$$x^2 - 8x - 9 = 0$$
$$(x-9)(x+1) = 0$$
$$x = 9 \text{ or } x = -1$$

Method 2: This is a short cut. It will save time, but you have to be careful that you do not make errors. We will first take the square root of both sides. Once this is done, we will solve the linear equation for both values of x.

$$(x-4)^2 = 25$$
$$\sqrt{(x-4)^2} = \pm\sqrt{25}$$
$$x - 4 = \pm 5$$
$$x = 4 + 5 = 9$$
or
$$x = 4 - 5 = -1$$

This is the same answer we got using the first method.

Application

EXAMPLE 8.19

A rectangular piece of aluminum is to be used to form a box. (See Figure 8.3a.) A 2-in. square is to be cut from each corner and the ends are to be folded up to form an open box. (See Figure 8.3b.) If the original piece of aluminum was twice as long as it was wide, and the volume of the box is 672 in.^{3}, what were the dimensions of the original rectangle?

EXAMPLE 8.19 (Cont.)

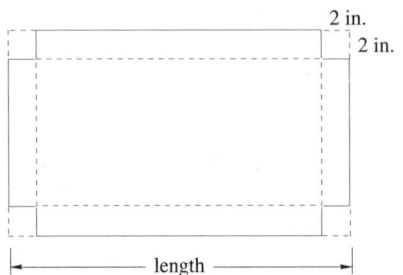

FIGURE 8.3a

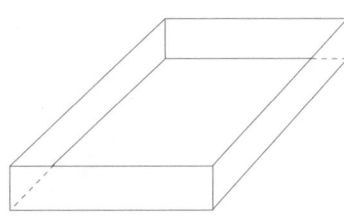

FIGURE 8.3b

Solution   If the width of the original rectangle is w, then the length is $2w$. The width of the box is $w - 4$ and the length of the box is $2w - 4$. The volume of the box is the product of the width, length, and height or

$$wlh = V$$
$$(w - 4)(2w - 4)2 = 672$$
$$4w^2 - 24w + 32 = 672 \qquad \text{Multiply.}$$
$$4w^2 - 24w - 640 = 0 \qquad \text{Put equation in standard form.}$$
$$w^2 - 6w - 160 = 0 \qquad \text{Divide both sides by 4.}$$
$$(w - 16)(w + 10) = 0 \qquad \text{Factor.}$$

If $w - 16 = 0$, then $w = 16$, and if $w + 10 = 0$, then $w = -10$. The last answer does not make sense. We cannot have a rectangle with a width of -10 in. So, the width of the original rectangle is 16 in. and the length is 32 in.

Exercise Set 8.2

In Exercises 1–44, solve the quadratic equations by factoring.

1. $x^2 - 9 = 0$

2. $x^2 - 100 = 0$

3. $x^2 + x - 6 = 0$

4. $x^2 - 6x - 7 = 0$

5. $x^2 - 11x - 12 = 0$

6. $x^2 - 5x + 4 = 0$

7. $x^2 + 2x - 8 = 0$

8. $x^2 + 2x - 15 = 0$

9. $x^2 - 5x = 0$

10. $x^2 + 10x = 0$

11. $x^2 + 12 = 7x$

12. $x^2 = 7x - 10$

13. $2x^2 - 3x - 14 = 0$

14. $2x^2 + x - 15 = 0$

15. $2x^2 + 12 = 11x$

16. $2x^2 + 18 = 15x$

17. $3x^2 - 8x - 3 = 0$

18. $3x^2 - 4x - 4 = 0$

19. $4x^2 - 24x + 35 = 0$

20. $6x^2 - 13x + 6 = 0$

21. $6x^2 + 11x - 35 = 0$

22. $10x^2 + 9x - 9 = 0$

23. $10x^2 - 17x + 3 = 0$

24. $14x^2 - 29x - 15 = 0$

25. $6x^2 = 31x + 60$

26. $15x^2 = 23x - 4$

27. $(x - 1)^2 = 4$

28. $(x + 2)^2 = 9$

29. $(5x - 2)^2 = 16$

30. $(3x + 2)^2 = 64$

31. $\dfrac{x}{x + 1} = \dfrac{x + 2}{3x}$

32. $\dfrac{4x}{x - 1} = \dfrac{7x + 2}{x}$

33. $(x + 5)^3 = x^3 + 1385$

34. $(x - 3)^3 = x^3 - 63$

35. $(x - 4)^3 - x^3 = -316$

36. $(x + 2)^3 - x^3 = 56$

37. $\dfrac{1}{x - 3} + \dfrac{1}{x + 4} = \dfrac{1}{12}$

38. $\dfrac{1}{x - 5} + \dfrac{1}{x + 3} = \dfrac{1}{3}$

39. $\dfrac{1}{x - 1} + \dfrac{1}{x - 2} = \dfrac{7}{12}$

40. $\dfrac{1}{x + 1} - \dfrac{1}{x + 2} = \dfrac{1}{20}$

41. $\dfrac{2}{x - 4} + \dfrac{1}{x - 9} = \dfrac{1}{6}$

42. $\dfrac{x}{x + 1} + \dfrac{1}{x} = \dfrac{13}{12}$

43. $\dfrac{x}{x+1} - \dfrac{2x}{x+3} = -\dfrac{1}{15}$

44. $\dfrac{3x}{x-1} - \dfrac{9x}{x+2} = 8.4$

Solve Exercises 45–58.

45. *Dynamics*   A ball is thrown vertically upward into the air from the roof of a building 192 ft high. The height of the ball above the ground is a function of the time in seconds and the initial velocity of the ball. If the initial velocity is 64 ft/s, then the height is given by $h(t) = -16t^2 + 64t + 192$. How many seconds will it take for the ball to return to the roof? (That is, when will $h(t) = 192$?)

46. *Dynamics*   If the ball in Exercise 45 misses the building when it comes down, how long will it take for the ball to hit the ground? (When will $h(t) = 0$?)

47. The length of a rectangle is 5 cm more than its width. Find the length and width, if the area is 104 cm².

48. *Dynamics*   The World Trade Center is 411 m high. A ball dropped from the top will fall according to $4.9t^2$ m. How long will it take for this ball to strike the ground? (Hint: Solve $4.9t^2 = 411$.)

49. *Solar Energy*   In order to support a solar collector at the correct angle, the roof trusses for a building are designed as right triangles, as shown in Figure 8.4. The rafter on the same side of the solar collector is 7 m shorter than the other rafter and the base of each truss is 13 m long. What are the lengths of the rafters?

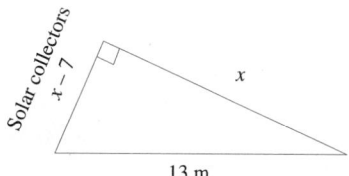

FIGURE 8.4

50. *Computer technology*   Working alone, computer A can complete a data-processing job in 6 h less than computer B working alone. Together the two computers can complete the job in 4 h. How long would it take each computer by itself?

51. *Construction*   A rectangular concrete pipe is constructed with a 10.00 in. by 16.00 in. interior channel, as shown in Figure 8.5. What uniform width w of

concrete must be formed on all sides if the total cross-sectional area of the concrete must be 192 in.²?

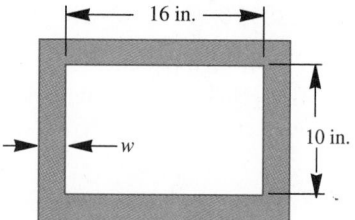

FIGURE 8.5

52. *Construction*   A rectangular parking lot at a shopping mall is 50.0 m wide and 80.0 m long. The developers of the mall want to expand the parking area to 18 000 m². They plan to do this by adding an equal length, in meters, to the length and to the width. What are the dimensions of the new parking lot?

53. *Recreation*   A swim club has a circular swimming pool with a diameter of 24 m. The club wants to build a deck of uniform width around the pool. Because of the financial condition of the club and the cost of materials, they can only afford to build a deck of area 432π m² around the pool. How wide should they build the deck?

54. *Recreation*   A swim club has a rectangular swimming pool 30.0 ft long and 22 ft wide. The club wants to build a deck of uniform width around the pool. Because of the financial condition of the club and the cost of materials, they can only afford to build a deck of 480 ft² around the pool. How wide should they build the deck?

55. *Wastewater technology*   Working together, two pipes, A and B, can fill a tank in 4 h. It takes pipe A 6 h longer than pipe B to fill the tank alone. How long would it take each pipe alone to fill the tank?

56. *Electricity*   In a certain circuit, the power (in watts) is measured using current, I (in amperes, A), voltage E (in volts, V), and resistance, R (in ohms, Ω), by $RI^2 + EI = 18\,000$. If the resistance is 10.0 Ω and the voltage is 510 V, find the current in amperes.

57. *Forestry* Volume estimates, V, in ft^3, for shortleaf pine trees are based on D, the d.b.h. (diameter at breast height) in inches; top d.i.b., the diameter inside the bark at the top of the tree, in inches; and H, the height of the tree in ft. One formula for trees with a 3-in. top d.i.b. is

$$V = 0.002837D^2H - 0.127248$$

Determine D for a 75-ft-high tree that has a volume estimate of 47.75 ft^3. (Note: This formula does not require any unit conversions.)

58. *Forestry* Volume estimates, V, in ft^3, for shortleaf pine trees are based on D, the d.b.h. (diameter at breast height) in inches; top d.i.b., the diameter inside the bark at the top of the tree, in inches; and H, the height of the tree in ft. One formula for trees with a 4-in. top d.i.b. is

$$V = 0.002835D^2H - 0.337655$$

Determine D for an 85-ft-high tree that has a volume estimate of 61.35 ft^3.

 **In Your Words**

59. **(a)** Give an example of a quadratic equation.
   (b) How do you know that your equation is quadratic?

60. **(a)** What is a root of a quadratic equation?
   (b) What is the zero-product rule for real numbers?

(c) How does the zero-product rule help to find roots of a quadratic equation?

☰ 8.3
COMPLETING THE SQUARE

Factoring is one method that can be used to solve quadratic equations. However, it is very difficult to factor some equations. In fact, most equations cannot be factored using integers or with rational numbers. We are going to introduce a technique called **completing the square**, which we can use to solve these quadratic equations.

In Section 8.2, we worked Example 8.18 using two different methods. We will begin our work in this section by reviewing the second method that we used in that example.

$$(x-4)^2 = 25$$
$$\sqrt{(x-4)^2} = \pm\sqrt{25}$$
$$x - 4 = \pm 5$$
$$x = 4 + 5 = 9$$

or

$$x = 4 - 5 = -1$$

Suppose that this was a slightly different problem: $(x-4)^2 = 26$. Let's solve it the same way.

$$(x-4)^2 = 26$$
$$\sqrt{(x-4)^2} = \pm\sqrt{26}$$
$$x-4 = \pm\sqrt{26}$$
$$x = 4 \pm \sqrt{26}$$
$$x = 4 + \sqrt{26}$$
or $$x = 4 - \sqrt{26}$$

Now, let's look at some special products—those that are perfect squares. The general form is $(x+k)^2 = x^2 + 2kx + k^2$. Notice that the constant, k^2, is the square of one-half of $2k$, the coefficient of x. If we combine the method for solving quadratic equations and the special product for perfect squares, we can develop the completing-the-square method.

Suppose you had the equation $x^2 + 6x - 10 = 0$. A quick check of the discriminant $(b^2 - 4ac)$ shows that it is $6^2 - 4(1)(-10) = 36 + 40 = 76$. Since 76 is not a perfect square, we cannot factor this equation. Rewrite the equation so that the variables are on the left-hand side and the constant term is on the right-hand side. (We have placed an empty box on each side of the equation to show that we will add something to both sides when we complete the square.)

$$x^2 + 6x + \square = 10 + \square$$

Complete the square on the left-hand side by taking one-half of the coefficient of the x-term, squaring it, and adding this number to both sides of the equation. The coefficient of x is 6, half of that is 3, and $3^2 = 9$. This is added to both sides of the equation and placed inside the empty boxes.

$$x^2 + 6x + \boxed{9} = 10 + \boxed{9}$$
or $$x^2 + 6x + 9 = 19$$

The left-hand side is now a perfect square.

$$(x+3)^2 = 19$$

We can solve this equation by using the second method from Example 8.18.

$$\sqrt{(x+3)^2} = \pm\sqrt{19}$$
$$x+3 = \pm\sqrt{19}$$
$$x = -3 \pm \sqrt{19}$$
$$x = -3 + \sqrt{19} \text{ or } x = -3 - \sqrt{19}$$

Check these answers. (The easiest way is to use your calculator.) You should see that they check when they are substituted in the original equation $x^2 + 6x - 10 = 0$.

EXAMPLE 8.20

Find the roots of $x^2 - 9x - 5 = 0$.

Solution

$$x^2 - 9x - 5 = 0$$

$$x^2 - 9x + \boxed{} = 5 + \boxed{}$$

$$x^2 - 9x + \boxed{\left(\tfrac{9}{2}\right)^2} = 5 + \boxed{\left(\tfrac{9}{2}\right)^2}$$

$$= 5 + \frac{81}{4}$$

$$= \frac{101}{4}$$

$$\left(x - \frac{9}{2}\right)^2 = \frac{101}{4}$$

$$\sqrt{\left(x - \frac{9}{2}\right)^2} = \pm\sqrt{\frac{101}{4}} = \pm\frac{\sqrt{101}}{2}$$

$$x - \frac{9}{2} = \pm\frac{\sqrt{101}}{2}$$

$$x = \frac{9}{2} \pm \frac{\sqrt{101}}{2}$$

$$x = \frac{9 + \sqrt{101}}{2} \text{ or } x = \frac{9 - \sqrt{101}}{2}$$

Before you complete the square, the coefficient on the x^2-term must be 1. One way to do this is shown in the next example.

EXAMPLE 8.21

Find the roots of $2x^2 - 8x + 3 = 0$.

Solution This is slightly complicated by the fact the coefficient of $2x^2$ is not 1. Our first step will be to divide the equation by 2 and then proceed as we have before.

$$2x^2 - 8x + 3 = 0$$

$$x^2 - 4x + \frac{3}{2} = 0$$

$$x^2 - 4x + \boxed{} = -\frac{3}{2} + \boxed{}$$

$$x^2 - 4x + \boxed{2^2} = -\frac{3}{2} + \boxed{2^2}$$

$$(x - 2)^2 = -\frac{3}{2} + 4 = \frac{5}{2}$$

$$x - 2 = \pm\sqrt{\frac{5}{2}}$$

$$x = 2 \pm \sqrt{\frac{5}{2}}$$

Application

EXAMPLE 8.22

The photograph in Figure 8.6a shows a solar collector. The roof trusses for a solar collector are often designed as right triangles, so the solar collector will be supported at the correct angle. Rafters form the legs of the right triangle and the base of the truss forms the hypotenuse. Suppose the rafter along the back of the solar collector is 3.5 m shorter than the other rafter and that the base of each truss is 6.5 m long. What is the length of each rafter?

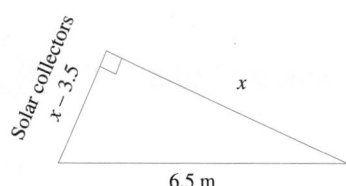

Courtesy of New York State Energy Research and Development Authority.
FIGURE 8.6a   **FIGURE 8.6b**

Solution   A triangle has been drawn over the photograph of the solar collector in Figure 8.6a. The triangle is labeled with the given information and shown by itself in Figure 8.6b. Since this is a right triangle, we can use the Pythagorean theorem. Thus, we have

$$x^2 + (x - 3.5)^2 = 6.5^2$$

Squaring both sides produces

$$x^2 + (x^2 - 7x + 12.25) = 42.25$$

or
$$2x^2 - 7x = 30$$

Dividing by 2, we get

$$x^2 - 3.5x = 15$$

EXAMPLE 8.22 (Cont.)

Next, we complete the square.

$$x^2 - 3.5x + \boxed{} = 15 + \boxed{}$$

$$x^2 - 3.5x + \boxed{\left(\frac{3.5}{2}\right)^2} = 15 + \boxed{\left(\frac{3.5}{2}\right)^2}$$

$$= 15 + \frac{12.25}{4}$$

$$= \frac{72.25}{4}$$

$$\left(x - \frac{3.5}{2}\right)^2 = \frac{72.25}{4}$$

$$\sqrt{\left(x - \frac{3.5}{2}\right)^2} = \pm\sqrt{\frac{72.25}{4}} = \pm\frac{8.5}{2}$$

$$x - \frac{3.5}{2} = \pm\frac{8.5}{2}$$

$$x = \frac{3.5}{2} \pm \frac{8.5}{2}$$

$$= 6 \text{ or } x = -2.5$$

It would not make sense for a rafter to be -2.5 m long. The lengths of the rafters must be 6 m and $6 - 3.5 = 2.5$ m long.

Exercise Set 8.3

Find the roots of each of the quadratic equations in Exercises 1–14 by completing the square.

1. $x^2 + 6x + 8 = 0$

2. $x^2 - 7x - 8 = 0$

3. $x^2 - 10x = 11$

4. $2x^2 - 3x = 14$

5. $x^2 + 6x + 3 = 0$

6. $x^2 + 8x + 10 = 0$

7. $x^2 - 5x + 5 = 0$

8. $x^2 - 7x + 11 = 0$

9. $2x^2 - 6x - 10 = 0$

10. $3x^2 + 12x - 18 = 0$

11. $4x^2 - 12x - 18 = 0$

12. $3x^2 - 9x = 33$

13. $x^2 + 2kx + c = 0$

14. $px^2 + 2qx + r = 0$

Solve Exercises 15–20.

15. *Physics* When an object is dropped from a building that is 196 ft tall, its height h at any time t after it was dropped is given by the function $h(t) = 196 - 16t^2$. Find the time it takes for the object to strike the ground.

16. *Business* Total revenue from selling a certain object is the product of the price and the quantity of the object sold. If q represents the quantity sold and the price is $\$2,520 - 3q$, find the quantity that produces a total revenue of $\$140,400$.

17. *Business* Total revenue from selling a certain object is the product of the price and the quantity of the object sold. If q represents the quantity sold and the price of the object is $\$1,560 - 4q$, find the quantity that produces a total revenue of $\$29,600$.

18. *Agriculture* A farmer has 1,500 ft of fencing to enclose a rectangular field. Find the dimensions of the field so that the enclosed area will be 137,600 ft^2.

19. *Agriculture*   A farmer has 2,700 ft of fencing to enclose two adjacent rectangular fields. Find the dimensions of the fields, so that the enclosed area will be 270,000 ft^2.

20. *Electricity*   When two resistors, R_1 and R_2, are connected in parallel, their combined resistance, R, is given by $\dfrac{1}{R} = \dfrac{1}{R_1} + \dfrac{1}{R_2}$. Two certain resistors that are connected in parallel have a combined resistance of 6 Ω, and the resistance of one of them is 5 Ω more than that of the other. What is the resistance of each resistor?

In Your Words

21. Without looking at the definition in the book, describe the technique called "completing the square."

22. Write a word problem in your technology area of interest that requires you to use a quadratic equation. On the back of the sheet of paper, write your name and explain how to solve the problem by using factoring or completing the square. Give the problem you wrote to a friend and let him or her try to solve it. If your friend has difficulty understanding the problem or solving the problem, or disagrees with your solution, make any necessary changes in the problem or solution. When you have finished, give the revised problem and solution to another friend and see if he or she can solve it.

≡ 8.4
THE QUADRATIC FORMULA

In Section 8.3, we learned how to solve a quadratic equation by completing the square. In this section, we will develop a general formula that can be used to find the roots of any quadratic equation.

   If there is a formula that will allow us to solve a quadratic equation with very little difficulty, why did we take the time to learn how to complete the square? There are two reasons. The first reason is that we will use completing the square to develop the quadratic formula. The second reason is that it is a tool that we will need when working with conic sections.

   Suppose we have a standard quadratic equation

$$ax^2 + bx + c = 0, \quad \text{with } a \neq 0$$

What are the roots of this quadratic equation? If we complete the square, we can find out.

$$ax^2 + bx + c = 0$$

$$x^2 + \frac{b}{a}x + \frac{c}{a} = 0 \qquad\qquad \text{Divide both sides by } a.$$

$$x^2 + \frac{b}{a}x = -\frac{c}{a} \qquad\qquad \text{Add } -\frac{c}{a} \text{ to both sides.}$$

$$x^2 + \frac{b}{a}x + \left(\frac{b}{2a}\right)^2 = -\frac{c}{a} + \left(\frac{b}{2a}\right)^2 \qquad \text{Complete the square by adding } \left(\frac{b}{2a}\right)^2$$
$$\text{to both sides.}$$

$$\left(x + \frac{b}{2a}\right)^2 = \frac{b^2}{4a^2} - \frac{4ac}{4a^2} \qquad\qquad \text{Factor the left-hand side; reverse terms on the right-hand side.}$$

$$= \frac{b^2 - 4ac}{4a^2} \qquad \text{Collect terms on the right-hand side.}$$

$$x + \frac{b}{2a} = \pm\sqrt{\frac{b^2 - 4ac}{4a^2}} \qquad \text{Take the square root of both sides.}$$

$$= \pm\frac{\sqrt{b^2 - 4ac}}{2a}$$

$$x = \frac{-b}{2a} \pm \frac{\sqrt{b^2 - 4ac}}{2a} \qquad \text{Solve for } x.$$

$$= \frac{-b \pm \sqrt{b^2 - 4ac}}{2a}$$

This is the **quadratic formula**.

Quadratic Formula

The solutions of the equation $ax^2 + bx + c = 0$, $(a \neq 0)$, are given by

$$x = \frac{-b \pm \sqrt{b^2 - 4ac}}{2a}$$

The expression $b^2 - 4ac$ is called the **discriminant**.

To solve a quadratic equation by using the quadratic formula, write the equation in the standard form, identify a, b, and c, and substitute these numbers into the equation. The quadratic formula is a very useful tool, but many equations are easier to solve by factoring.

You may recognize something we have used before in part of the quadratic formula. The quantity $b^2 - 4ac$ is the discriminant. We used it to help tell us when a quadratic equation can be factored. Now we can use it to tell us something else. Since the quadratic formula takes the square root of the discriminant, a quadratic equation will have only real numbers as roots when $b^2 - 4ac \geq 0$. In Chapter 14, when we study complex numbers, we will consider quadratic equations in which the discriminant is negative.

EXAMPLE 8.23

Solve $x^2 + 7x - 8 = 0$.

Solution In this equation, $a = 1$, $b = 7$, and $c = -8$. Putting these values in the quadratic formula we get

$$x = \frac{-7 \pm \sqrt{7^2 - 4(1)(-8)}}{2(1)}$$

$$= \frac{-7 \pm \sqrt{49 + 32}}{2}$$

$$= \frac{-7 \pm \sqrt{81}}{2}$$

EXAMPLE 8.23 (Cont.)

$$= \frac{-7 \pm 9}{2}$$

$$x = \frac{-7 + 9}{2} = \frac{2}{2} = 1$$

$$\text{or} \quad x = \frac{-7 - 9}{2} = \frac{-16}{2} = -8$$

The roots are 1 and -8. This is an equation that we could have solved by factoring. ∎

EXAMPLE 8.24

Solve $2x^2 + 5x - 3 = 0$.

Solution In this equation $a = 2$, $b = 5$, and $c = -3$. Putting these values in the quadratic formula we get

$$x = \frac{-5 \pm \sqrt{5^2 - 4(2)(-3)}}{2(2)}$$

$$= \frac{-5 \pm \sqrt{25 + 24}}{4}$$

$$= \frac{-5 \pm \sqrt{49}}{4}$$

$$= \frac{-5 \pm 7}{4}$$

So,

$$x = \frac{-5 + 7}{4} = \frac{2}{4} = \frac{1}{2} \text{ or } x = \frac{-5 - 7}{4} = \frac{-12}{4} = -3$$

The roots are $\frac{1}{2}$ and -3. ∎

EXAMPLE 8.25

Solve $9x^2 + 49 = 42x$.

Solution This equation is not in the standard form. If we subtract $42x$ from both sides, we get $9x^2 - 42x + 49 = 0$, with $a = 9$, $b = -42$, and $c = 49$. Substituting these values for a, b, and c in the quadratic formula, we obtain

$$x = \frac{42 \pm \sqrt{(-42)^2 - 4(9)(49)}}{2(9)}$$

$$= \frac{42 \pm \sqrt{1764 - 1764}}{18}$$

$$= \frac{42 \pm 0}{18}$$

$$= \frac{7}{3}$$

This is a double root; in this case both roots are $\frac{7}{3}$. ∎

EXAMPLE 8.26

Find the roots of $3x^2 + 7x + 3 = 0$.

Solution In this equation, $a = 3$, $b = 7$, and $c = 3$, so

$$x = \frac{-7 \pm \sqrt{7^2 - 4(3)(3)}}{2(3)}$$
$$= \frac{-7 \pm \sqrt{49 - 36}}{6}$$
$$= \frac{-7 \pm \sqrt{13}}{6}$$

So, $x = \dfrac{-7 + \sqrt{13}}{6}$ or $x = \dfrac{-7 + \sqrt{13}}{6}$

Notice that we really needed the quadratic formula to find the factors.

 **Hint**

Unless you quickly see that an equation can be factored, it may be best to use the quadratic formula.

Application

EXAMPLE 8.27

The length of a rectangular piece of cardboard is 4 in. more than its width. A 3-in. square is removed from each corner as shown in Figure 8.7. The remaining cardboard is bent to form an open box. If the volume of the box is 420 in.³, what are the dimensions of the original piece of cardboard?

Solution We will begin by letting x be the width, in inches, of the original piece of cardboard. The length of this cardboard is $x + 4$ in. After the squares are removed and the sides have been folded up, the dimensions of the box are

$$\text{length} = (x + 4) - 2 \cdot 3 = x - 2$$
$$\text{width} = x - 6$$
$$\text{height} = 3$$

The volume, 420 in.³, is: length × width × height, or $V = lwh$, and so we obtain

$$(x - 2)(x - 6)3 = 420$$
$$(x - 2)(x - 6) = 140$$
$$x^2 - 8x + 12 = 140$$
$$x^2 - 8x - 128 = 0$$
$$(x - 16)(x + 8) = 0$$
$$x = 16 \text{ or } x = -8$$

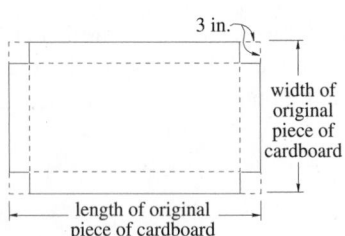

FIGURE 8.7

It would make no sense for the width of a piece of cardboard to be −8 in., so we reject $x = -8$ as an answer.

Thus, the piece of cardboard must have a width of 16 in. and a length of $16 + 4 = 20$ in.

Application

EXAMPLE 8.28

A ball is thrown upward from the top of a building that is 555 ft high with an initial velocity of 64 ft/s. The height of the ball in feet above the ground at any time t is given the formula $s(t) = 555 + 64t - 16t^2$. When will the ball hit the ground?

Solution When the ball hits the ground, its height will be 0. So, we want to solve the equation

$$0 = 555 + 64t - 16t^2$$

for t. First write the equation in standard form and then use the quadratic formula to solve it. Multiplying the equation by -1 produces

$$16t^2 - 64t - 555 = 0$$

Here $a = 16$, $b = -64$, and $c = -555$. Substituting these into the quadratic formula we get

$$
\begin{aligned}
t &= \frac{-b \pm \sqrt{b^2 - 4ac}}{2a} \\
&= \frac{-(-64) \pm \sqrt{(-64)^2 - 4(16)(-555)}}{2(16)} \\
&= \frac{64 \pm \sqrt{4096 + 35{,}520}}{32} \\
&= \frac{64 \pm \sqrt{39{,}616}}{32} \\
&\approx \frac{64 \pm 199.03768}{32} \\
&\approx 2 \pm 6.22
\end{aligned}
$$

Thus, $t \approx 8.22$ s or $t \approx -4.22$ s. The second answer makes no sense, because this would mean that the ball struck the ground before it was thrown. The first answer, about 8.22 s, checks when it is substituted into the original equation, and so it is the correct answer.

Exercise Set 8.4

In Exercises 1–38, use the quadratic formula to find the roots of each equation.

1. $x^2 + 3x - 4 = 0$
2. $x^2 - 8x - 33 = 0$
3. $3x^2 - 5x - 2 = 0$
4. $7x^2 + 5x - 2 = 0$
5. $7x^2 + 6x - 1 = 0$
6. $2x^2 - 3x - 20 = 0$
7. $2x^2 - 5x - 7 = 0$
8. $3x^2 + 4x - 7 = 0$

9. $3x^2 + 2x - 8 = 0$
10. $9x^2 - 6x + 1 = 0$
11. $9x^2 + 12x + 4 = 0$
12. $3x^2 + 3x - 7 = 0$
13. $2x^2 - 3x - 1 = 0$
14. $2x^2 - 5x + 1 = 0$
15. $x^2 + 5x + 2 = 0$
16. $3x^2 - 6x - 2 = 0$

17. $2x^2 + 6x - 3 = 0$
18. $5x^2 + 2x - 1 = 0$
19. $x^2 - 2x - 7 = 0$
20. $x^2 + 3 = 0$
21. $2x^2 - 3 = 0$
22. $2x^2 = 5$
23. $3x^2 + 4 = 0$
24. $3x^2 + 1 = 5x$

25. $\frac{2}{3}x^2 - \frac{1}{9}x + 3 = 0$
26. $\frac{1}{2}x^2 - 2x + \frac{1}{3} = 0$
27. $0.01x^2 + 0.2x = 0.6$
28. $0.16x^2 = 0.8x - 1$
29. $\frac{1}{4}x^2 + 3 = \frac{5}{2}x$
30. $\frac{3}{2}x^2 + 2x = \frac{7}{2}$
31. $1.2x^2 = 2x - 0.5$
32. $1.4x^2 + 0.2x = 2.3$

33. $3x^2 + \sqrt{3}x - 7 = 0$

34. $2x^2 - \sqrt{89}x + 5 = 0$

35. $\dfrac{x-3}{7} = 2x^2$

36. $\dfrac{x-5}{3} = 5x^2$

37. $\dfrac{2}{x-1} + 3 = \dfrac{-2}{x+1}$

38. $\dfrac{3x}{x+2} + 2x = \dfrac{2x^2-1}{x+1}$

Solve Exercises 39–56.

39. *Dynamics*   The World Trade Center is 411 m high. If a ball is dropped from the top of the World Trade Center, its height at time t is given by the formula $h(t) = -4.9t^2 + 411$. The ball will hit the ground when $h(t) = 0$. How long does it take for the ball to fall to the ground?

40. *Dynamics*   If the ball in the previous problem is thrown downward with an initial velocity of 20 m/s, its height at time t is given by $h(t) = -4.9t^2 - 20t + 411$. When does it hit the ground?

41. *Dynamics*   If the ball in the previous problem is thrown upward with an initial velocity of 20 m/s, its height at time t is given by $h(t) = -4.9t^2 + 20t + 411$. How long will it take to hit the ground?

42. *Sheet metal technology*   An open box is to be made from a square piece of aluminum by cutting out a 4-cm square from each corner and folding up the sides. If the box is to have a volume of $100\,\text{cm}^3$, find the dimensions of the piece of aluminum that is needed.

43. *Sheet metal technology*   An open box is to be made from a rectangular piece of aluminum. A 3-cm square is to be cut from each corner and the sides will be folded up. If the original piece of aluminum was 1.5 times as long as its width and the volume of the box is $578\,\text{cm}^3$, what were the dimensions of the original rectangle?

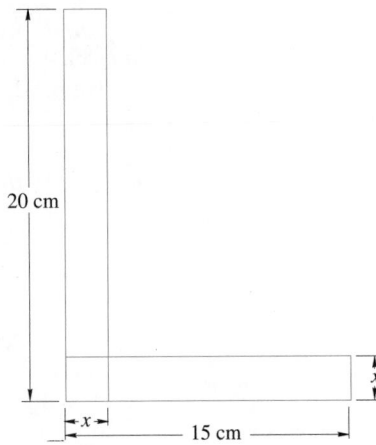

20 cm

x

←x→   15 cm

FIGURE 8.8

44. *Construction*   An angle beam is to be constructed as shown in Figure 8.8. If the cross-sectional area of the angle beam is $81.25\,\text{cm}^2$, what is the thickness of the beam?

45. *Sheet metal technology*   A gutter is to be made by folding up the edges of a strip of metal as shown in Figure 8.9. If the metal is 12 in. wide and the cross-sectional area of the gutter is to be $16\frac{7}{8}\,\text{in.}^2$, what are the width and depth of the gutter?

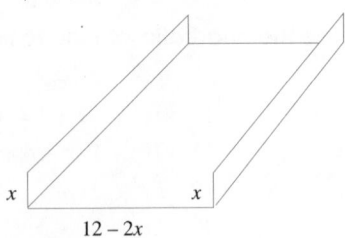

x x

$12 - 2x$

FIGURE 8.9

46. *Package design*   A cylindrical container is to be made with a height of 10 cm. If the total surface area of the container is to be 245 cm², what is the radius of the base?

47. *Business*   An oil distributor has 1,000 commercial customers who each pay a base rate of $30 per month for oil. The distributor figures that for each $1-per-month increase in the base rate, 5 customers will convert to coal. The distributor needs to increase its monthly income to $45,000 and lose as few customers as possible. How much should the base rate be increased?

48. *Electricity*   In an ac circuit that contains resistance, inductance, and capacitance in series, the applied voltage V can be found by solving $V^2 = V_R^2 + (V_L - V_C)^2$. If $V = 5.8\,\text{V}$, $V_R = 5\,\text{V}$, and $V_C = 10\,\text{V}$, find V_L.

49. *Electricity*   In an ac circuit that contains resistance, inductance, and capacitance in series, the impedance of the circuit Z is related to the resistance R, the inductive reactance X_L, and the capacitive reactance X_C, by the formula, $Z^2 = R^2 + (X_L - X_C)^2$. If a circuit has $Z = 610\,\Omega$, $R = 300\,\Omega$, and $X_C = 531\,\Omega$, find X_L.

50. *Construction*   For a simply supported beam of length l having a distributed load of w kg/m, the binding moment M at any distance x from one end is given by

$$M = \frac{1}{2}wlx - \frac{1}{2}wx^2$$

At which locations is the binding moment zero?

51. *Construction*   A rectangular concrete pipe is constructed with a 10.00 in. by 16.00 in. interior channel, as shown in Figure 8.10. What uniform width w of concrete must be formed on all sides if the total cross-sectional area of the concrete must be 245 in.²?

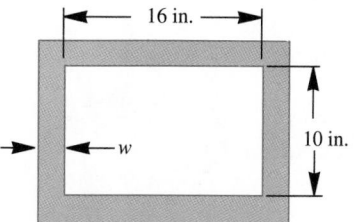

FIGURE 8.10

52. *Construction*   A rectangular parking lot at a shopping mall is 50.0 m wide and 80.0 m long. The developers of the mall want to double the parking area. They plan to do this by adding an equal distance, in meters, to the length and to the width of the lot. What are the dimensions of the new parking lot?

53. *Forestry*   The Scribner log-rule equation for 16-ft logs is $V = 0.79D^2 - 2D - 4$, where V is the volume, in board ft, of the log and D is the diameter, in in., of the small end of a log inside the bark. If a certain 16-ft log has a volume of 926.0 board ft, what is the diameter of the small end of the log inside the bark?

54. *Forestry*   Use the Scribner log-rule equation (see Exercise 53), to determine the diameter of the small end of the log inside the bark of a 16.0-ft log that has a volume of 1078.0 board ft.

55. Write a computer program to solve a quadratic equation by using the quadratic formula.

56. Write a program for your calculator to use the quadratic formula to solve a quadratic equation.

 **In Your Words**

57. **(a)** Use factoring to solve $3x^2 + 5x - 2 = 0$.

   (b) Use the quadratic formula to solve $3x^2 + 5x - 2 = 0$. (You should get the same answers you got in (a).)

   (c) Explain to a classmate how to use the quadratic formula. Check to see how well your classmate understands your explanation by asking him or her to solve $3x^2 + 5x - 2 = 0$.

   (d) Under what conditions is the quadratic formula easier to use than factoring?

58. **(a)** What is the discriminant?

   (b) What does the discriminant tell you if it is non-negative?

☰ CHAPTER 8 REVIEW

Important Terms and Concepts

Completing the square   Quadratic equation
Discriminant   Quadratic formula
Fractional equation   Zero-product rule
General quadratic equation

Review Exercises

Solve each of Exercises 1–6 for x.

1. $\dfrac{x}{3} + \dfrac{x}{2} = 5$

2. $\dfrac{2}{x} - \dfrac{3}{x} = \dfrac{1}{5}$

3. $\dfrac{x-2}{4} - \dfrac{x+2}{5} = \dfrac{x}{2}$

4. $\dfrac{2}{x-1} + \dfrac{3}{x-2} = \dfrac{4}{x^2-3x+2}$

5. $\dfrac{3x}{x-1} - \dfrac{3x+2}{x} = 3$

6. $\dfrac{4x}{2x-3} + \dfrac{1}{x-1} = \dfrac{2x+1}{x-1}$

Solve each of Exercises 7–10 for the indicated variable.

7. $A = 2lw + 2(l+w)h$, for h

8. $A = 2lw + 2(l+w)h$, for l

9. $\dfrac{1}{f} = \dfrac{1}{p} + \dfrac{1}{q}$, for f

10. $X = wL - \dfrac{1}{wc}$, for c

Use factoring to find the roots of each of the quadratic equations in Exercises 11–22.

11. $x^2 - 8x + 7 = 0$

12. $x^2 + 4x + 3 = 0$

13. $x^2 - 11x + 10 = 0$

14. $x^2 + 15x + 14 = 0$

15. $x^2 + 15x + 56 = 0$

16. $x^2 - 8x + 12 = 0$

17. $2x^2 - 5x + 3 = 0$

18. $3x^2 + 10x + 7 = 0$

19. $6x^2 + 7x - 10 = 0$

20. $8x^2 + 22x + 9 = 0$

21. $4x^2 - 9 = 0$

22. $9x^2 - 16 = 0$

Find the roots of each of the quadratic equations in Exercises 23–28 by completing the square.

23. $x^2 + 4x - 5 = 0$

24. $x^2 - 7x + 6 = 0$

25. $x^2 + 19x = 11$

26. $2x^2 + 7x = 15$

27. $3x^2 + 5x - 14 = 0$

28. $4x^2 + 2x - 5 = 0$

Use the quadratic formula to find the roots of each of the equations in Exercises 29–40.

29. $x^2 - 8x + 5 = 0$

30. $x^2 + 7x - 6 = 0$

31. $2x^2 + 3x - 5 = 0$

32. $2x^2 - 7x + 4 = 0$

33. $3x^2 + 2x - 4 = 0$

34. $4x^2 - 3x = 1$

35. $5x^2 + 2 = 8x$

36. $6x^2 + 2x = 3$

37. $3x^2 - 8x + 10 = 0$

38. $8x^2 = 4x + 3$

39. $\dfrac{x}{x-1} + \dfrac{2}{x+1} = 3$

40. $\dfrac{2}{x} - \dfrac{3}{x+2} = 4$

Solve Exercises 41–44.

41. *Electricity* Figure 8.11 shows two currents flowing in a single resistor R. The total current in the resistor is $i_1 + i_2$ and the power dissipated P is

$$P = (i_1 + i_2)^2 R$$

If $R = 50\,\Omega$ and $i_2 = 0.4\,\text{A}$, find the current i_1 needed to produce a power of $12\,\text{W}$.

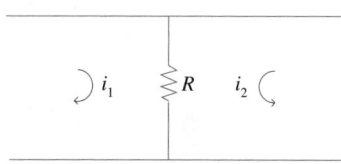

FIGURE 8.11

42. John can mow a field in 7 h and Matt can mow the same field in 8 h. If they work together, how long will it take to mow the field?

43. *Dynamics* A stone is dropped from the edge of a cliff that is 144 ft high. If the stone's height at time t is given by $L(t) = 144 - 16t^2$, how long does it take for the stone to reach the bottom of the cliff?

44. *Industrial design* A wedge is being designed in the shape of a right triangle. The hypotenuse will be 27 mm long and one leg will be 6 mm shorter than the other leg. What are the lengths of the sides of the triangle?

≡ CHAPTER 8 TEST

In Exercises 1–10, solve each equation for x.

1. $\dfrac{x}{8} + \dfrac{3x}{4} = \dfrac{7}{2}$

2. $\dfrac{12}{x^2 - 9} = \dfrac{x}{x-3} - \dfrac{x}{x+3}$

3. $x^2 - 4x - 5 = 0$

4. $3x^2 + x - 2 = 0$

5. $16x^2 + 9 = 24x$

6. $2x^2 + 3x + 3 = 0$

7. $x^2 + 3x + 1 = 0$

8. Solve $x^2 - 8x + 5 = 0$ by completing the square.

9. Solve $x^2 + 7x - 30 = 0$ by factoring.

10. Solve $3x^2 + 5x - 28 = 0$ by factoring.

Solve Exercises 11 and 12.

11. Solve $V = 3D^2H - 5$ for H.

12. Solve $V = 3D^2H - 5$ for D.

Solve Exercises 13 and 14.

13. A rectangular work area is to be 3 m longer than it is wide and will have an area of $46.75\,\text{m}^2$. What are the dimensions of the work area?

14. The position s at time t of an object moving rectilinearly with an initial velocity of v and an initial acceleration of a is

$$s = vt + \frac{at^2}{2}$$

Solve this equation for v in terms of a, t, and s.

9

Trigonometric Functions

Trigonometry allows us to measure distances when we cannot measure them directly. In Section 9.3, we will determine the height that this balloon is above the ground.

Courtesy of Michael A. Gallitelli, Metroland Photo Inc.

Until now, most of the functions we have used involved polynomials. Even our work with systems of linear equations and with quadratic equations dealt with polynomials. But much of the mathematics used by people working in technical areas cannot be solved by using polynomial functions.

In this chapter, we will learn about a new type of function—the trigonometric function. Trigonometric functions were originally developed to describe the relationship between the sides and angles of triangles. But, as so often happens in mathematics, other uses were discovered for these functions. Technical and scientific areas rely a great deal on trigonometric functions.

We will begin our study of trigonometry with the study of angles and angular measurements. We will then define the six trigonometric functions and their inverse functions, and study how we can use trigonometry to solve problems involving right triangles. (Chapter 10 will expand the applications to all types of triangles.) This chapter also provides some good opportunities to use our calculators and computers in new ways.

☰ 9.1
ANGLES, ANGLE MEASURE, AND TRIGONOMETRIC FUNCTIONS

At the beginning of this section, we will review our knowledge of angles and how they are measured. From this introduction we will quickly move into a study of the trigonometric functions.

Positive and Negative Angles

In Chapter 3, we gave two definitions of an angle. One of these was to think of generating a ray from an initial position to a terminal position. One revolution is the amount a ray would turn to return to its original position. As can be seen in Figure 9.1, if the rotation of the terminal side from the initial side is counterclockwise, the angle is a positive angle, but if the rotation is clockwise, the angle is a negative angle.

Degrees and Radians

Angles are measured using several different systems. The two most common are degrees and radians. In Chapter 3, we discussed how to change from degrees to radians and from radians to degrees. This is an important skill. We will briefly review this technique, but for more details you should refer to Section 3.1.

A degree is $\frac{1}{360}$ of a circle and the symbol ° is used to indicate degree(s). A radian is $\frac{1}{2\pi}$ of a circle. An entire circle contains 360° or 2π rad.

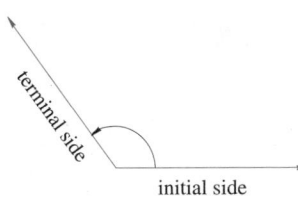

terminal side

initial side

Positive Angle
(counterclockwise rotation)

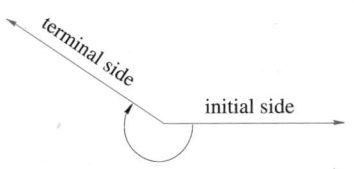

terminal side

initial side

Negative Angle
(clockwise rotation)

FIGURE 9.1

Degree-Radian Conversions

To convert from degrees to radians, multiply the number of degrees by $\frac{\pi}{180°} \approx$ 0.01745, or divide by its reciprocal, 57.296.

To convert from radians to degrees, multiply the number of radians by $\frac{180°}{\pi} \approx$ 57.296 or divide by its reciprocal, 0.01745.

Some scientific calculators have a key that will allow you to convert between degrees and radians. (Consult your owner's manual to see if your calculator can do this. Many graphing calculators cannot.)

EXAMPLE 9.1

Convert 72° to radians.

Solution $72° \approx 72 \times 0.01745 \, \text{rad}$

$72° = 1.2564 \, \text{rad}$

or $72° = 72° \left(\dfrac{\pi}{180°} \right)$

$= \dfrac{72\pi}{180}$

$= \dfrac{2\pi}{5} \, \text{rad}$

EXAMPLE 9.2

Convert 0.62 rad to degrees

Solution $0.62 \, \text{rad} = 0.62 \times 57.296°$

$= 35.52352°$

EXAMPLE 9.3

Convert $\frac{3\pi}{4}$ rad to degrees.

Solution   Because this angle is given in terms of π, we will multiply by $\dfrac{180°}{\pi}$ rather than 57.296°.

$$\frac{3\pi}{4} = \frac{3\pi}{4} \times \frac{180°}{\pi}$$
$$= \frac{3}{4} \times 180°$$
$$= 135°$$

EXAMPLE 9.4

Use a calculator to convert 110° to radians.

Solution

	PRESS	DISPLAY
Algebraic calculator:	110 [2nd] [D·R]	1.9198622
RPN calculator:	110 [2nd] [→RAD]	1.919862177
Graphing calculator:	110 [×] [π] [÷] 180	1.919862177

EXAMPLE 9.5

Use a calculator to convert 1.6 rad to degrees.

Solution

	PRESS	DISPLAY
Algebraic calculator:	1.6 [INV] [2nd] [D·R]	91.673247
RPN calculator:	1.6 [2nd] [→DEG]	91.67324722
Graphing calculator:	1.6 [×] 180 [÷] [π]	91.67324722

If you cannot remember these conversion values and if your calculator is not handy, use your knowledge of proportions. Since $180° = \pi$ rad, we can use the proportion given in the box.

Converting Between Degrees and Radians

To convert d degrees to r radians (or vice versa), use the proportion

$$\frac{d}{180°} = \frac{r}{\pi}$$

When you use this ratio, you will know either d or r and want to find the other.

EXAMPLE 9.6

Use this ratio to convert $72°$ to radians.

Solution   Here $d = 72°$, so

$$\frac{72°}{180°} = \frac{r}{\pi}$$

and $\dfrac{72°}{180°}\pi = \frac{2}{5}\pi \approx 1.25664$ gives you the same answer we got in Example 9.1. Notice that $72°$ is exactly $\frac{2}{5}\pi$, while 1.25664 is an approximation.

■

Notice that the degree-radian method used in Example 9.1 gave an answer of $72° \approx 1.2564$, while the proportion method used in Example 9.6 produced $72° = \frac{2}{5}\pi \approx 1.25664$. The difference in these two results is caused by the multiplication of an approximate number in the first method. As you can see, the proportion method used in Example 9.6, and used next in Example 9.7, produces more accurate results.

EXAMPLE 9.7

Use the ratio to convert $\dfrac{3\pi}{4}$ to degrees.

Solution   Method I:
This is the same problem we worked in Example 9.3. Here $r = \dfrac{3\pi}{4}$, so

$$\frac{d}{180°} = \frac{3\pi/4}{\pi}$$

and

$$d = \frac{3\pi/4}{\pi} \cdot 180°$$

$$= \frac{3\pi}{4} \cdot \frac{1}{\pi} \cdot 180°$$

$$= \frac{3\cancel{\pi}}{4} \cdot \frac{1}{\cancel{\pi}} \cdot 180°$$

EXAMPLE 9.7 (Cont.)

$$= \frac{3}{4} \cdot 180°$$
$$= 135°.$$

Thus, we see that $\frac{3\pi}{4} = 135°$.

Method II:

This can be called the UFO method (UFO = Useful Form of One). A useful form of one is $\frac{180°}{\pi}$. Multiplying $\frac{3\pi}{4}$ by $\frac{180°}{\pi}$ we get $3 \times 45° = 135°$.

≡ **Note**   If you are converting from degrees to radians, the UFO method uses $\frac{\pi}{180°}$.

Coterminal Angles

If two angles have the same initial side and the same terminal side, they are **coterminal angles**. An example of coterminal angles is given in Figure 9.2. One way to find a coterminal angle of a given angle is to add 360° to the original angle. In Figure 9.2, the original angle is 50° and one coterminal angle is $50° + 360° = 410°$. In fact, you could add any integer multiple of 360° to the original angle to find a coterminal angle. So, $50° + 4(360°) = 50° + 1,440° = 1,490°$ is another coterminal angle of a 50° angle.

In the same way, you could subtract an integer multiple of 360° from the original angle to get a coterminal angle. Thus, $50° - 360° = -310°$ is a coterminal angle of a 50° angle, and so is $50° - 2(360°) = 50° - 720° = -670°$.

An angle is in **standard position** if its vertex is at the origin of a rectangular coordinate system and its initial side coincides with the positive x-axis. The angle is determined by the position of the terminal side. The angle is said to be in a certain quadrant if its terminal side lies in that quadrant. If the terminal side coincides with one of the coordinate axes, the angle is a **quadrantal angle**.

Consider an angle θ in a standard position and let $P(x, y)$ be a fixed point on the terminal side of θ, as in Figure 9.3. We will call r the distance from O to P. From the Pythagorean theorem we know $r = \sqrt{x^2 + y^2}$. The length r is also called the **radius vector**. Suppose $Q(x_1, y_1)$ is any other point on the terminal side θ. If PR and QS are both perpendicular to the x-axis, then $\triangle POR$ and $\triangle QOS$ are similar. As you remember from our chapter on proportions (Chapter 6), the corresponding sides of similar triangles are proportional. So, the ratios of the corresponding sides are equal. For example, $\frac{y}{r} = \frac{y_1}{r_1}$ and $\frac{x}{y} = \frac{x_1}{y_1}$. As long as θ does not change, these ratios will not change.

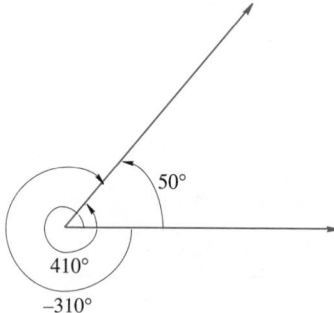

FIGURE 9.2

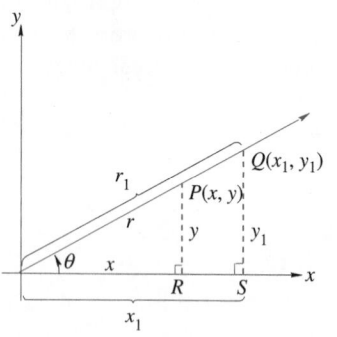

FIGURE 9.3

The Trigonometric Functions

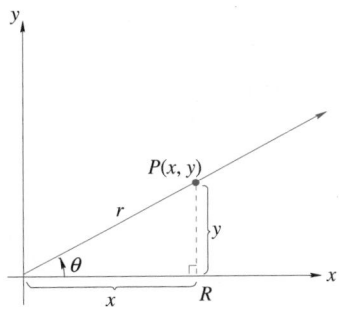

FIGURE 9.4

There are six possible ratios of two sides of a triangle. Each of these ratios has been given a name. These names (and their abbreviations) are sine (sin), cosine (cos), tangent (tan), cosecant (csc), secant (sec), and cotangent (cot). Because these ratios depend on the size of angle θ, they are written $\sin\theta$, $\cos\theta$, and so on. So, what you have are ratios that are functions of θ—the trigonometric functions. The six trigonometric, or trig, functions are defined using the triangle in Figure 9.4.

Trigonometric Functions		
	$\sin\theta = \dfrac{y}{r}$	$\csc\theta = \dfrac{r}{y}$
	$\cos\theta = \dfrac{x}{r}$	$\sec\theta = \dfrac{r}{x}$
	$\tan\theta = \dfrac{y}{x}$	$\cot\theta = \dfrac{x}{y}$

EXAMPLE 9.8

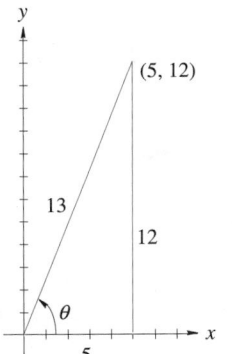

FIGURE 9.5

Given the point $(5, 12)$ on the terminal side of an angle θ, find the six trigonometric functions of θ.

Solution   A sketch of the angle is in Figure 9.5. Since $x = 5$, and $y = 12$, we can find the radius vector r by using the Pythagorean theorem.

$$r = \sqrt{x^2 + y^2} = \sqrt{5^2 + 12^2} = \sqrt{25 + 144} = \sqrt{169} = 13$$

$$\sin\theta = \frac{y}{r} = \frac{12}{13} \qquad \csc\theta = \frac{r}{y} = \frac{13}{12}$$

$$\cos\theta = \frac{x}{r} = \frac{5}{13} \qquad \sec\theta = \frac{r}{x} = \frac{13}{5}$$

$$\tan\theta = \frac{y}{x} = \frac{12}{5} \qquad \cot\theta = \frac{x}{y} = \frac{5}{12}$$

≡ **Note**

The values of x and y may be positive, negative, or zero; but r is always positive or zero. Remember, r is never negative because $r = \sqrt{x^2 + y^2}$ and the expression $\sqrt{x^2 + y^2}$ is never negative.

   Of course, there is nothing to restrict the terminal side to the first quadrant. In the next two examples, the terminal side is in Quadrants II and IV, respectively.

EXAMPLE 9.9

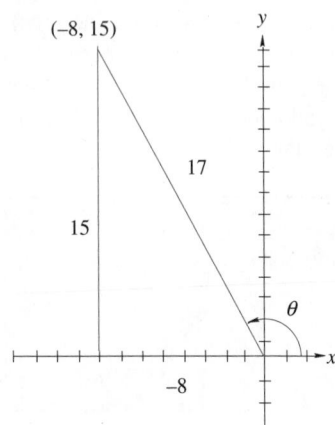

FIGURE 9.6

Given the point $(-8, 15)$ on the terminal side of an angle θ, find the six trigonometric functions of θ.

Solution Here we have $x = -8$ and $y = 15$, so the radius vector r is given by $\sqrt{(-8)^2 + (15)^2} = \sqrt{64 + 225} = \sqrt{289} = 17$ as shown in Figure 9.6. The six trigonometric functions are

$$\sin\theta = \frac{15}{17} \qquad \csc\theta = \frac{17}{15}$$

$$\cos\theta = \frac{-8}{17} \qquad \sec\theta = \frac{-17}{8}$$

$$\tan\theta = -\frac{15}{8} \qquad \cot\theta = -\frac{8}{15}$$

As a final example, we will consider a case where the radius vector is not an integer. Note that none of the sides of the triangle have to be integers.

EXAMPLE 9.10

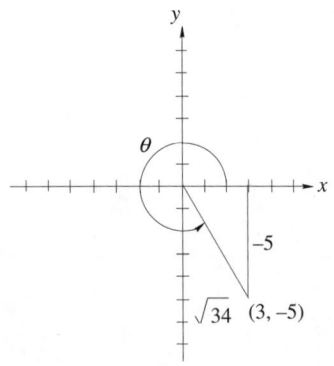

FIGURE 9.7

Given the point $(3, -5)$ on the terminal side of an angle θ, find the six trigonometric functions of θ.

Solution Since $x = 3$ and $y = -5$ the radius vector, r, is $\sqrt{9 + 25} = \sqrt{34}$, as shown in Figure 9.7. The trigonometric functions of θ are

$$\sin\theta = \frac{-5}{\sqrt{34}} \approx -0.8575 \qquad \csc\theta = -\frac{\sqrt{34}}{5} \approx -1.1662$$

$$\cos\theta = \frac{3}{\sqrt{34}} \approx 0.5145 \qquad \sec\theta = \frac{\sqrt{34}}{3} \approx 1.9437$$

$$\tan\theta = -\frac{5}{3} \approx -1.6667 \qquad \cot\theta = -\frac{3}{5} = -0.6$$

Exercise Set 9.1

Convert each of the angle measures in Exercises 1–8 from degrees to radians.

1. $90°$

2. $45°$

3. $80°$

4. $-15°$

5. $155°$

6. $-235°$

7. $215°$

8. $180°$

Convert each of the angle measures in Exercises 9–16 from radians to degrees.

9. 2   **11.** 1.5   **13.** $\frac{\pi}{3}$ **15.** $-\frac{\pi}{4}$

10. 3   **12.** π **14.** $\frac{5\pi}{6}$ **16.** -1.3

In Exercises 17–20, (a) draw each of the angles in standard position; (b) draw an arrow to indicate the rotation; (c) for each angle, find two other angles, one positive and one negative, which are coterminal with the given angle. (Note: There are many possible correct answers.)

17. $150°$ **18.** $315°$ **19.** $-135°$ **20.** $-30°$

Each point in Exercises 21–30 is on the terminal side of angle θ in standard position. Find the six trigonometric functions of the angle θ associated with each of these points.

21. $(4, 3)$ **24.** $(-20, 21)$ **27.** $(10, -8)$ **30.** $(-6, \sqrt{13})$

22. $(-6, -8)$ **25.** $(1, 2)$ **28.** $(-5, -2)$

23. $(8, -15)$ **26.** $(-2, 4)$ **29.** $(\sqrt{11}, 5)$

Find the trigonometric functions that exist for each of the quadrantral angles θ when drawn in standard position for each of the points in Exercises 31–34.

31. $(3, 0)$ **32.** $(0, -4)$ **33.** $(-5, 0)$ **34.** $(0, 6)$

For each of Exercises 35–40, there is a point P on the terminal side of an angle θ in standard position. From the information given, determine the six trigonometric functions of θ.

35. $x = 6, r = 10, y > 0$ **37.** $x = -20, r = 29, y > 0$ **39.** $x = -7, r = 8, y < 0$

36. $y = -9, r = 15, x > 0$ **38.** $y = 5, r = 13, x < 0$ **40.** $y = 5, r = 30, x < 0$

  In Your Words

41. **(a)** Write an explanation of how to convert from degrees to radians.
   (b) Ask a classmate to use your written explanation to convert $50°$ to radians.
   (c) Did your classmate get $50° \approx 0.87266$? If not, either your classmate did not follow your directions or your explanation needs to be rewritten. Decide where the error was made and make the necessary corrections.

42. **(a)** Draw a right triangle on the xy-coordinate system with an acute angle at the origin, one leg on the positive x-axis, and the other leg vertical. Label the angle at the origin, θ, the hypotenuse, r, the horizontal leg, x, and the vertical leg, y.
   (b) Define each of the trigonometric functions of θ in terms of x, y, and r.
   (c) Compare your drawing in (a) with Figure 9.4. Compare your definitions in (b) with those in the "Trigonometric Functions" box near Figure 9.4.

≡ 9.2
VALUES OF THE TRIGONOMETRIC FUNCTIONS

We have learned how to calculate the values of the trigonometric functions for an angle θ in standard position when we are given a point on the terminal side of θ. This is not always the most convenient way to find the values of the trigonometric functions for θ.

If we have an angle θ in the standard position and draw the triangle as we did in Figure 9.4, we get a picture that helps us to determine the values of the trigonometric functions for θ. If we look at just $\triangle POR$, we get a figure much like the one in Figure 9.8. The length of the hypotenuse is r, the length y is for the side opposite angle θ, and the other side x is the side adjacent to angle θ. We can use these descriptions of the sides to rephrase our definitions for the trigonometric function.

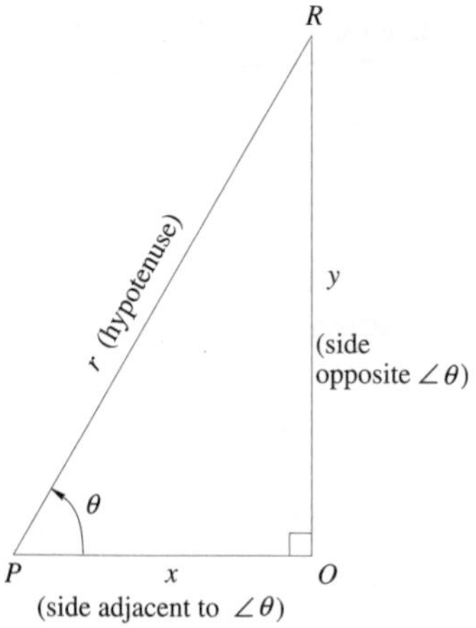

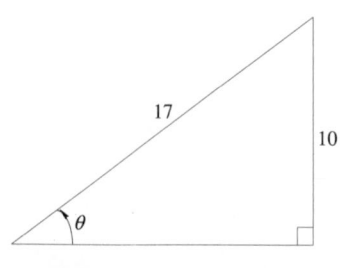

FIGURE 9.8

Trigonometric Functions

$$\sin\theta = \frac{y}{r} = \frac{\text{side opposite } \theta}{\text{hypotenuse}} \qquad \csc\theta = \frac{r}{y} = \frac{\text{hypotenuse}}{\text{side opposite } \theta}$$

$$\cos\theta = \frac{x}{r} = \frac{\text{side adjacent to } \theta}{\text{hypotenuse}} \qquad \sec\theta = \frac{r}{x} = \frac{\text{hypotenuse}}{\text{side adjacent to } \theta}$$

$$\tan\theta = \frac{y}{x} = \frac{\text{side opposite } \theta}{\text{side adjacent to } \theta} \qquad \cot\theta = \frac{x}{y} = \frac{\text{side adjacent to } \theta}{\text{side opposite } \theta}$$

With these relationships, we can find the trigonometric functions of any angle of a right triangle.

EXAMPLE 9.11

Determine the values of the trigonometric functions for an angle of a triangle with a hypotenuse of 17 and the opposite side of length 10, as shown in Figure 9.9.

Solution The length of the adjacent side x is missing. Use the Pythagorean theorem, $x = \sqrt{r^2 - y^2}$. Since $r = 17$ and $y = 10$, $x = \sqrt{17^2 - 10^2} = \sqrt{189} = 3\sqrt{21}$. The trigonometric functions are

$$\sin\theta = \frac{\text{side opposite}}{\text{hypotenuse}} = \frac{10}{17} \qquad \csc\theta = \frac{\text{hypotenuse}}{\text{side opposite}} = \frac{17}{10}$$

$$\cos\theta = \frac{\text{side adjacent}}{\text{hypotenuse}} = \frac{3\sqrt{21}}{17} \qquad \sec\theta = \frac{\text{hypotenuse}}{\text{side adjacent}} = \frac{17}{3\sqrt{21}}$$

$$\tan\theta = \frac{\text{side opposite}}{\text{side adjacent}} = \frac{10}{3\sqrt{21}} \qquad \cot\theta = \frac{\text{side adjacent}}{\text{side opposite}} = \frac{3\sqrt{21}}{10}$$

FIGURE 9.9

We could have worked the problem had we been given the value of one of the trigonometric functions rather than the lengths of two of the sides.

EXAMPLE 9.12

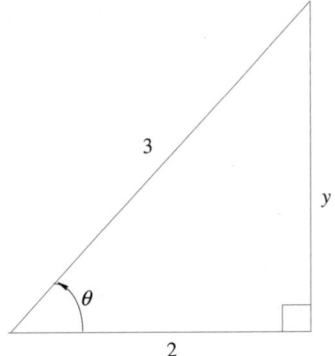

FIGURE 9.10

In Figure 9.10, if θ is an angle of a triangle and $\cos\theta = \frac{2}{3}$, what are the values of the other trigonometric functions?

Solution The length y of the side opposite θ is $\sqrt{3^2 - 2^2} = \sqrt{5}$. Thus, we have the values for the other functions

$$\sin\theta = \frac{\sqrt{5}}{3} \qquad \csc\theta = \frac{3}{\sqrt{5}} \qquad \sec\theta = \frac{3}{2}$$

$$\tan\theta = \frac{\sqrt{5}}{2} \qquad \cot\theta = \frac{2}{\sqrt{5}}$$

Look at Figure 9.11. This is the same triangle that was in Figure 9.8. There is another acute angle in that triangle, labeled ϕ. Angles θ and ϕ are complementary. Remember, complementary angles are two angles that measure 90° when added. So, $\theta + \phi = 90°$. What are the trigonometric functions for ϕ? The side opposite ϕ is x and the side adjacent to ϕ is y, so

$$\sin\phi = \frac{x}{r} \qquad \csc\phi = \frac{r}{x}$$

$$\cos\phi = \frac{y}{r} \qquad \sec\phi = \frac{r}{y}$$

$$\tan\phi = \frac{x}{y} \qquad \cot\phi = \frac{y}{x}$$

From this, we get the principle of cofunctions of complementary angles.

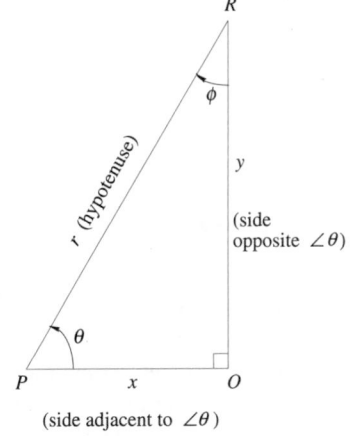

FIGURE 9.11

Trigonometric Functions of Complementary Angles

If θ and ϕ are complementary angles, then

$$\sin\theta = \cos\phi \qquad \tan\theta = \cot\phi \qquad \csc\theta = \sec\phi$$

$$\cos\theta = \sin\phi \qquad \cot\theta = \tan\phi \qquad \sec\theta = \csc\phi$$

You may have begun to notice a pattern with the trigonometric functions. The values of some of the trig functions are the reciprocals of other functions. These are known as the **reciprocal identities**. There are three reciprocal identities. (An identity is an equation that is true for every value in the domain of its variables.)

Reciprocal Identities

$$\csc\theta = \frac{1}{\sin\theta} \qquad \sec\theta = \frac{1}{\cos\theta} \qquad \cot\theta = \frac{1}{\tan\theta}$$

There is also a relationship between the $\sin\theta$, $\cos\theta$, and the $\tan\theta$. This relationship, and its reciprocal, are known as the **quotient identities**.

Quotient Identities	$\tan\theta = \dfrac{\sin\theta}{\cos\theta}$ $\cot\theta = \dfrac{\cos\theta}{\sin\theta}$

None of these identities is true when the denominator is zero.

EXAMPLE 9.13

If $\sin\theta = \frac{1}{2}$ and $\cos\theta = \dfrac{\sqrt{3}}{2}$, find the values of the other four trigonometric functions.

Solution   Using the reciprocal and quotient identities we have

$$\tan\theta = \frac{\sin\theta}{\cos\theta} = \frac{\dfrac{1}{2}}{\dfrac{\sqrt{3}}{2}} = \frac{1}{\sqrt{3}}$$

$$\cot\theta = \frac{1}{\tan\theta} = \frac{1}{\dfrac{1}{\sqrt{3}}} = \frac{\sqrt{3}}{1} = \sqrt{3}$$

$$\csc\theta = \frac{1}{\sin\theta} = \frac{1}{\dfrac{1}{2}} = \frac{2}{1} = 2$$

$$\sec\theta = \frac{1}{\cos\theta} = \frac{1}{\dfrac{\sqrt{3}}{2}} = \frac{2}{\sqrt{3}}$$

Trigonometric Values

Until now, we have been finding the values of trigonometric functions without knowing the size of the angle θ. But, many times we are given the value of θ and are asked to determine the values of the trigonometric functions for that angle.

There are two basic ways that are used to find values of trigonometric functions. By far, the easiest way is with a calculator or a computer. The other basic method is with a table of trigonometric values. There are some trigonometric angles that seem to be used quite frequently. Some people learn the values of these basic angles so they can readily use these values when they are needed. We will concentrate on calculators and computers for determining the value of a trigonometric function.

Before you use a calculator or a computer, you need to decide if the angle is measured in degrees or radians. Most calculators can compute the trigonometric functions of an angle hether it is in degrees or radians, as long as the calculator is set to work in that mode.

FIGURE 9.12

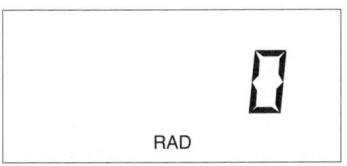

FIGURE 9.13

FIGURE 9.14

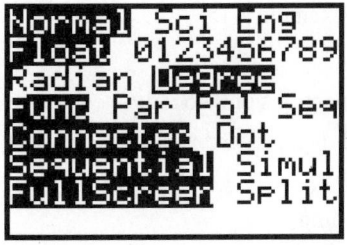

FIGURE 9.15

Nongraphing Calculators

Some calculators have a $\boxed{\text{DRG}}$ button. When you turn on the calculator, a 0 is the only figure displayed, as shown in Figure 9.12. The calculator is in degree mode when it is turned on.

Press the $\boxed{\text{DRG}}$ button. Now the screen displays a 0 and the abbreviation RAD, as in Figure 9.13. The calculator is now ready to work with angles in radians.

Press the $\boxed{\text{DRG}}$ again. The RAD disappears and the word GRAD appears, as shown in Figure 9.14. GRAD stands for another type of angle measure in which the circle is divided into 400 equal parts. One grad is $\frac{1}{400}$ of a circle. We will not use grads in this book.

Press the $\boxed{\text{DRG}}$ again. The calculator returns to the degree mode shown in Figure 9.12. You can reverse the order by using the $\boxed{\text{INV}}$ key. Thus, to change from the radian mode to the degree mode press $\boxed{\text{INV}}$ $\boxed{\text{DRG}}$.

Some calculators have separate keys for each function. Again, the calculator will be in degree mode when it is turned on. To get the calculator to the radian mode, press $\boxed{\text{2nd}}$ $\boxed{\text{RAD}}$. Unfortunately, you may not be able to see any difference in the display.

Graphing Calculators

Other calculators have different symbols or different locations for the symbols that let you know which mode the calculator is in. For example, on a graphing calculator, such as the Casio fx-7700G or Texas Instruments TI-82, press $\boxed{\text{MODE}}$. The TI-82 will display a screen like that in Figure 9.15. (The calculator displayed in Figure 9.15 is in degree mode.)

Use the cursor keys $\boxed{\blacktriangle}$, $\boxed{\blacktriangledown}$, $\boxed{\blacktriangleleft}$, and $\boxed{\blacktriangleright}$ to darken the appropriate degree or radian mode. Press $\boxed{\text{ENTER}}$ or $\boxed{\text{EXE}}$ to set the calculator in that mode. Make sure that you consult the owner's manual for your calculator.

Trigonometric Values on a Calculator

Some calculators have three trigonometric function keys— $\boxed{\text{sin}}$, $\boxed{\text{cos}}$, and $\boxed{\text{tan}}$. If you want one of the other trigonometric functions, you will have to use one of these keys and the $\boxed{1/x}$ key. Here you need to know the reciprocal identities discussed earlier in this section. You must press the $\boxed{1/x}$ or $\boxed{x^{-1}}$ key *after* you have pressed the trigonometric function key.

If you are using a nongraphing calculator to find the trigonometric function for a particular angle in degrees, enter the measure of the angle and press the appropriate function key to obtain sin, cos, or tan. To obtain csc, sec, or cot, press the $\boxed{1/x}$ or $\boxed{x^{-1}}$ key *after* you have pressed the appropriate key for its reciprocal function. On graphing calculators, you press the function key first.

Make sure that your calculator is set in the correct degree or radian mode before you begin to work the problem.

EXAMPLE 9.14

Determine the trigonometric functions of 48.9°.

Solution Put the calculator in degree mode, and then proceed as follows.

Function	ENTER (Algebraic)	ENTER (Graphing)	DISPLAY
$\sin 48.9°$	48.9 $\boxed{\sin}$	$\boxed{\sin}$ 48.9 $\boxed{\text{ENTER}}$	0.753563
$\cos 48.9°$	48.9 $\boxed{\cos}$	$\boxed{\cos}$ 48.9 $\boxed{\text{ENTER}}$	0.6573752
$\tan 48.9°$	48.9 $\boxed{\tan}$	$\boxed{\tan}$ 48.9 $\boxed{\text{ENTER}}$	1.1463215
$\csc 48.9°$	48.9 $\boxed{\sin}$ $\boxed{1/x}$	$\boxed{(}$ $\boxed{\sin}$ 48.9 $\boxed{)}$ $\boxed{x^{-1}}$ $\boxed{\text{ENTER}}$	1.3270284
$\sec 48.9°$	48.9 $\boxed{\cos}$ $\boxed{1/x}$	$\boxed{(}$ $\boxed{\cos}$ 48.9 $\boxed{)}$ $\boxed{x^{-1}}$ $\boxed{\text{ENTER}}$	1.5212012
$\cot 48.9°$	48.9 $\boxed{\tan}$ $\boxed{1/x}$	$\boxed{(}$ $\boxed{\tan}$ 48.9 $\boxed{)}$ $\boxed{x^{-1}}$ $\boxed{\text{ENTER}}$	0.8723556

Hint

You do not need to change modes to find the trigonometric function of an angle measured in degrees. There is a degree symbol on the $\boxed{\text{MATH}}$ menu of a TI-81, on the $\boxed{\text{ANGLE}}$ menu of a TI-82, and on the $\boxed{\text{MATH}}$ $\boxed{\text{F3}}$ menu of a TI-85 graphics calculator. You can use the calculator's degree symbol to get the correct answer from either degree or radian mode. To get $\sin 48.9°$ on a TI-82, press $\boxed{\sin}$ 48.9 $\boxed{\text{2nd}}$ $\boxed{\text{ANGLE}}$ 1 $\boxed{\text{ENTER}}$. The calculator should display the result .7535633923.

To find the values of the trigonometric functions for an angle in radians, put the calculator in the radian mode. Then proceed as in Example 9.15.

EXAMPLE 9.15

Use a calculator to determine the trigonometric functions of 0.65 rad.

Solution Put the calculator in radian mode and then proceed as follows.

Function	ENTER (Algebraic)	ENTER (Graphing)	DISPLAY
$\sin 0.65$	0.65 $\boxed{\sin}$	$\boxed{\sin}$ 0.65 $\boxed{\text{ENTER}}$	0.6051864
$\cos 0.65$	0.65 $\boxed{\cos}$	$\boxed{\cos}$ 0.65 $\boxed{\text{ENTER}}$	0.7960838
$\tan 0.65$	0.65 $\boxed{\tan}$	$\boxed{\tan}$ 0.65 $\boxed{\text{ENTER}}$	0.7602044
$\csc 0.65$	0.65 $\boxed{\sin}$ $\boxed{1/x}$	$\boxed{(}$ $\boxed{\sin}$ 0.65 $\boxed{)}$ $\boxed{x^{-1}}$ $\boxed{\text{ENTER}}$	1.6523834
$\sec 0.65$	0.65 $\boxed{\cos}$ $\boxed{1/x}$	$\boxed{(}$ $\boxed{\cos}$ 0.65 $\boxed{)}$ $\boxed{x^{-1}}$ $\boxed{\text{ENTER}}$	1.2561492
$\cot 0.65$	0.65 $\boxed{\tan}$ $\boxed{1/x}$	$\boxed{(}$ $\boxed{\tan}$ 0.65 $\boxed{)}$ $\boxed{x^{-1}}$ $\boxed{\text{ENTER}}$	1.3154357

Hint

Just as you do not need to change modes to find the trigonometric function of an angle measured in degrees, you also do not need to change modes to find the trigonometric function of an angle measured in radians. The radian symbol on the TI-8x calculators is an r. You can use the calculator's degree symbol to get the correct answer from either degree or radian mode. To get $\sin 0.65$ on a TI-82, press $\boxed{\sin}$ .65 $\boxed{\text{2nd}}$ $\boxed{\text{ANGLE}}$ 3 $\boxed{\text{ENTER}}$. The calculator should display $\sin$ .65^r on one line, with the result, .6051864057, on the next line.

Trigonometric Values on a Computer

Computers operate in much the same way as do calculators. There are two main differences. Computers normally operate in radians. If you want to work in degrees, you will need to write a program that will convert radians to degrees. The second difference is that you must put parentheses around the size of the angle. Like calculators, computers do not have the secant, cosecant, and cotangent built into them. In Example 9.16, we will use the same angle that we used in Example 9.15 and we will only find the values of the sin, tan, and sec.

EXAMPLE 9.16

Use a computer to determine the sin, tan, and sec of 0.65 rad.

Solution

Function	ENTER	DISPLAY
sin .65	PRINT SIN(.65)	.605186406
tan .65	PRINT TAN(.65)	.760204399
sec .65	PRINT 1/COS(.65)	1.25614917

Trigonometric Tables

The other principal way to determine the values of trigonometric functions is with a table of trigonometric functions. Until the advent of electronic calculators, this was the main way people determined the value of a trigonometric function. Trigonometric tables can be found in many mathematics books or in special books of mathematical tables. We have not included the trigonometric tables in the back of this book, because we assume that you will use a calculator or a computer to obtain values of trigonometric functions.

Exercise Set 9.2

In Exercises 1–6, find the six trigonometric functions for angle θ for the indicated sides of $\triangle ABC$. (See Figure 9.16.)

1. $a = 5, c = 13$
2. $a = 7, b = 8$
3. $a = 1.2, c = 2$
4. $a = 2.1, b = 2.8$
5. $a = 1.4, b = 2.3$
6. $c = 3.5, b = 2.1$

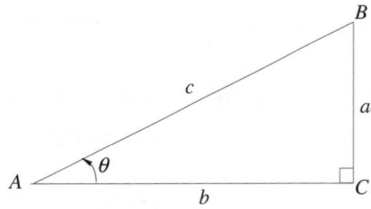

FIGURE 9.16

In Exercises 7–12, let θ be an angle of a right triangle with the given trigonometric function. Find the values of the other trigonometric functions.

7. $\cos\theta = \frac{8}{17}$

8. $\tan\theta = \frac{21}{20}$

9. $\sin\theta = \frac{3}{5}$

10. $\sec\theta = \frac{10}{6}$

11. $\csc\theta = \frac{29}{20}$

12. $\csc\theta = \frac{15}{12}$

In Exercises 13–16, the values of two of the trigonometric functions of angle θ are given. Use this information to determine the values of the other four trigonometric functions for θ.

13. $\sin\theta = 0.866$, $\cos\theta = 0.5$

14. $\sin\theta = 0.975$, $\cos\theta = 0.222$

15. $\sin\theta = 0.085$, $\sec\theta = 1.004$

16. $\csc\theta = 4.872$, $\cos\theta = 0.979$

Use a calculator to determine the values of each of the indicated trigonometric functions in Exercises 17–32.

17. $\sin 18.6°$

18. $\cos 38.4°$

19. $\tan 18.3°$

20. $\sin 20°15'$

21. $\tan 76°32'$

22. $\sec 24°14'$

23. $\cot 82.6°$

24. $\csc 19°50'$

25. $\sin 0.25\,\text{rad}$

26. $\cos 0.4\,\text{rad}$

27. $\tan 0.63\,\text{rad}$

28. $\sec 1.35\,\text{rad}$

29. $\cot 1.43\,\text{rad}$

30. $\sin 1.21\,\text{rad}$

31. $\csc 0.21\,\text{rad}$

32. $\tan 1.555\,\text{rad}$

Solve Exercises 33–38.

33. *Dynamics* If there is no air resistance, a projectile fired at an angle θ above the horizontal with an initial velocity of v_0 has a range R of

$$R = \frac{v_0{}^2}{g}\sin 2\theta$$

A football is thrown with initial velocity of 15 m/s at an angle of 40° above the horizontal. If $g = 9.8\,\text{m/s}^2$, how far will the football travel?

34. *Dynamics* If the football in Exercise 33 had been thrown at an angle of 45° above the horizontal, how much further, or shorter, would the ball have traveled?

If the angle had been 48°, how would the results have differed?

35. *Electricity* In an ac circuit that contains resistance, inductance, and capacitance in series, the angle of the applied voltage v and the voltage drop across the resistance V_R is the **phase angle** ϕ, and $V_R = V\cos\phi$. If the phase angle is 32° and the applied voltage is 5.8 V, what is the effective voltage across the resistor V_R?

36. *Electricity* Consider an ac circuit that contains resistance, inductance, and capacitance in series, as in Exercise 35. If the phase angle is 71° and the applied voltage is 11.25 V, what is the effective voltage across the resistor, V_R?

You will need the following information in order to work Exercises 37 and 38.

Tables of trigonometric functions are usually computed using the Maclaurin series. For the sin x and cos x the series are

$$\sin x = x - \frac{x^3}{3!} + \frac{x^5}{5!} - \frac{x^7}{7!} + \cdots \quad \text{and} \quad \cos x = 1 - \frac{x^2}{2!} + \frac{x^4}{4!} - \frac{x^6}{6!} + \frac{x^8}{8!} - \cdots$$

where x is the angle in radians, and $3! = 3\cdot 2\cdot 1 = 6, 4! = 4\cdot 3\cdot 2\cdot 1 = 24, 5! = 5\cdot 4\cdot 3\cdot 2\cdot 1 = 120$, and so on. (Some calculators have an $\boxed{x!}$ key.) The symbols $x!$, $5!$, $2!$, and so on, are read x factorial, 5 factorial, 2 factorial, and so on.

37. Write a program to compute and print the sines and cosines of an angle from 0 rad to 1.50 rad every 0.1 rad, using the first four terms of the series.

 38. Convert your program from Exercise 37 so that it will compute the sines and cosines from 0° to 90° every 1°.

In Your Words

39. (a) Draw a right triangle with acute angle θ. Label the hypotenuse, adjacent side, and opposite side.
(b) Define each of the trigonometric functions of θ in terms of the hypotenuse, the adjacent side, and the opposite side.

(c) Compare your drawing in (a) with Figure 9.8. Check your definitions in (b) with those in the box near Figure 9.8.

40. (a) What are the reciprocal identities?
(b) What are the quotient identities?

≡ 9.3
THE RIGHT TRIANGLE

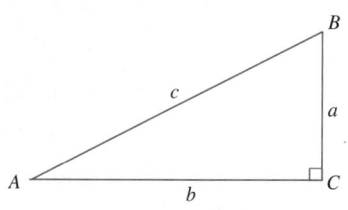

FIGURE 9.17

In Section 9.2, we learned how to determine trigonometric functions given the lengths of two sides of a right triangle. We are now ready to begin finding the unknown parts of a triangle if we know the length of one side and the size of one of the acute angles. Once we can do that, we will begin to apply our knowledge to solving some problems that involve triangles.

Look at the right triangle in Figure 9.17. The right angle is labeled C and the other two vertices A and B. The side opposite each of these vertices is labeled with the lowercase version of the same letter. Thus, side a is opposite $\angle A$, side b is opposite $\angle B$, and side c is opposite $\angle C$. Since this is a right triangle, angles A and B are complementary.

We now have the tools to solve right triangles. To solve a triangle means to find the sizes of all unknown sides and angles.

EXAMPLE 9.17

Given $A = 55°$ and $b = 7.92$, solve the right triangle ABC.

Solution This data chart contains the information that we know and shows what we have left to find. Notice that, since this is a right triangle, we wrote that $C = 90°$.

sides	angles
$a = \underline{\quad}$	$A = 55°$
$b = 7.92$	$B = \underline{\quad}$
$c = \underline{\quad}$	$C = 90°$

Since $A = 55°$, then $B = 90° - 55° = 35°$. If you look at Figure 9.18 you see that $\tan A = \dfrac{a}{b}$ or $a = b \tan A$. We know $A = 55°$ and $b = 7.92$, so $a = 7.92 \tan 55° =$

EXAMPLE 9.17 (Cont.)

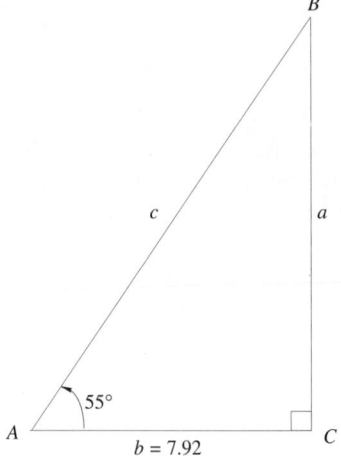

FIGURE 9.18

$(7.92)(1.428148) = 11.310932$ or about 11.31. Now $\sec A = \dfrac{c}{b}$, so $c = b\sec A = (7.92)(1.7434468) = 13.808099$ or 13.81. Since your calculator does not have a sec key, it may be better to find c using cos. Now,

$$\cos A = \frac{b}{c}$$

so
$$c = \frac{b}{\cos A}$$
$$= \frac{7.92}{\cos 55°} \approx \frac{7.92}{0.5736}$$
$$\approx 13.81$$

Look at the data chart again. This time all of the parts of the chart have been filled in, so we know that we have solved for all the missing parts of this triangle. From the completed chart, we can easily see the measures of any part of the triangle.

sides	angles
$a \approx 11.31$	$A = 55°$
$b = 7.92$	$B = 35°$
$c \approx 13.81$	$C = 90°$

We could have used our derived value of a and the Pythagorean theorem to find c, or we could have used a and one of the other trigonometric functions.

Hint

It is usually best not to use the derived values because of errors that can be introduced by rounding.

By the way, if you use the Pythagorean theorem to check the results in Example 9.17, you will see that $(7.92)^2 + (11.31)^2 \neq (13.81)^2$. The Pythagorean theorem says that these should be equal, but because we are using rounded-off values, they are not. However, they are equal to three significant digits.

Using the [2nd] or [INV] Key

Until now, we have calculated the values of a trigonometric function from the size of the angle. But, what if we know the value of a trigonometric function and want to know the size of the angle? Here you press the [INV] or [2nd] key on the calculator before you enter the function key.

EXAMPLE 9.18

Use a graphing calculator to determine the angles that have each of the following values of trigonometric functions: $\sin A = 0.785$, $\cos B = 0.437$, $\tan C = 4.213$, and $\csc D = 8.915$.

EXAMPLE 9.18 (Cont.)

Solution

Function	ENTER	DISPLAY
$\sin A = 0.785$	2nd sin 0.785	51.720678
$\cos B = 0.437$	2nd cos 0.437	64.087375
$\tan C = 4.213$	2nd tan 4.213	75.547345
$\csc D = 8.915$	2nd sin 8.915 x^{-1}	6.4404506
or	2nd sin (1 ÷ 8.915)	6.4404506

So, $A \approx 51.72°$, $B \approx 64.09°$, $C \approx 76.65°$, and $D \approx 6.44°$. Notice that to find angle D, we had to use the 2nd sin keys with the reciprocal of 8.915. This is because $\csc D = 8.915$ is the same as $\dfrac{1}{\sin D} = 8.915$ or $\sin D = \dfrac{1}{8.915}$. Hence, $D = \sin^{-1}\dfrac{1}{8.915}$ which may be entered in your calculator as either

$$2nd \ \ sin \ \ 8.915 \ \ x^{-1}$$

or $\quad$ 2nd sin (1 ÷ 8.915)

In BASIC computer language, you are able to find an angle, in radian measure, if you know its tangent. Thus if $\tan B = 4.213$, to find the size of B, you would enter PRINT ATN (4.213) and get the display 1.33774853. This is in radians and converts to 76.64°. ATN is the BASIC function for arctan or INV tan. If you want to find an angle in BASIC where you know $\sin x$ or $\cos x$, you need to define special functions to do this. We discuss this definition of special functions later in this chapter.

EXAMPLE 9.19

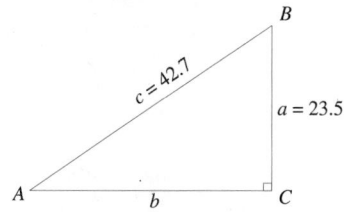

FIGURE 9.19

Solve right triangle ABC ($\triangle ABC$), if $a = 23.5$ and $c = 42.7$, as shown in Figure 9.19.

Solution The following data chart shows the information we are starting with and what we have to find.

sides	angles
$a = 23.5$	$A = $ ___
$b = $ ___	$B = $ ___
$c = 42.7$	$C = 90°$

From the Pythagorean theorem we know that $b^2 = c^2 - a^2 = (42.7)^2 - (23.5)^2 = 1271.04$, so $b = 35.7$. Now, $\sin A = \frac{23.5}{42.7} \approx 0.5503513$. Pressing 2nd sin 0.5503513, we get 33.391116. Thus, $A \approx 33.4°$ and $B \approx 90° - 33.4° = 56.6°$. The completed data chart follows.

sides	angles
$a = 23.5$	$A \approx 33.4°$
$b \approx 35.7$	$B \approx 56.6°$
$c = 42.7$	$C = 90°$

Hint

Two possible shortcuts could have been taken in Example 9.19 when you determined the size of A.

1. After you calculated $\sin A = \dfrac{23.5}{42.7} \approx 0.5503513$, you could have used the ability of a graphing calculator to keep the last answer in memory. Thus, rather than press [2nd] [sin] 0.5503513, you could have pressed

[2nd] [sin] [2nd] [ANS] [ENTER]

2. An even faster method would have performed the calculation in one step:

[2nd] [sin] [(] 23.5 [÷] 42.7 [)] [ENTER]

EXAMPLE 9.20

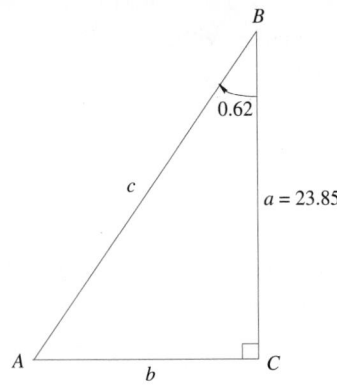

FIGURE 9.20

Solve for the right triangle ABC if $B = 0.62$ rad and $a = 23.85$, as shown in Figure 9.20.

Solution   The data chart for this triangle is

sides	angles
$a = 23.85$	$A = \underline{}$
$b = \underline{}$	$B = 0.62$
$c = \underline{}$	$C = \dfrac{\pi}{2} \approx 1.57$

In this example, we can use angle B as our reference angle. So, $\tan B = \dfrac{b}{a}$ and $b = a \tan B = (23.85)\tan 0.62 = (23.85)(0.713909) = 17.02673$, or $b \approx 17.03$. (Did you remember to put your calculator in radian mode?)

Also, $\cos B = \dfrac{a}{c}$, so

$$
\begin{aligned}
c &= \frac{a}{\cos B} \\
&= \frac{23.85}{\cos 0.62} \\
&= \frac{23.85}{0.81387846} \\
&= 29.30413
\end{aligned}
$$

or ≈ 29.30

Finally, we have $\tan A = \frac{23.85}{17.03} = 1.4004698$, and pressing [2nd] [tan] 1.4004698, we get 0.9507055, or $A \approx 0.95$ rad.

Remember from Exercise 1 in Exercise Set 9.1 that $90° \approx 1.57$ rad. If you used your calculator to convert $90°$ to radians, you could then have found A as $1.57 - 0.62 = 0.95$.

The completed data chart for this example is

sides	angles
$a = 23.85$	$A \approx 0.95$
$b \approx 17.03$	$B = 0.62$
$c \approx 29.30$	$C = \dfrac{\pi}{2} \approx 1.57$

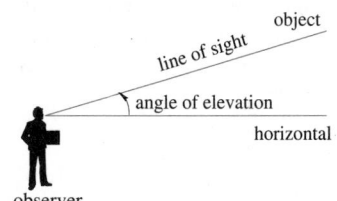

FIGURE 9.21

Angles of Elevation or Depression

Frequently when solving problems involving trigonometry, we have to use the **angle of elevation** (see Figure 9.21), which is the angle, measured from the horizontal, through which an observer would have to elevate his or her line of sight in order to see an object. Similarly, the **angle of depression** is the angle, measured from the horizontal, through which an observer has to lower his or her line of sight in order to see an object. (See Figure 9.22.)

EXAMPLE 9.21

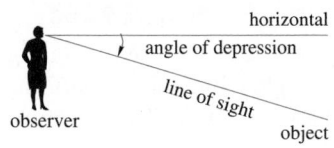

FIGURE 9.22

Application

A person is standing 50 m from the base of a tower. (See Figure 9.23.) The angle of elevation to the top of the tower is 76°. How high is the tower?

Solution We have a right triangle as sketched in Figure 9.23. The height of the tower is labeled x, so we have $\tan 76° = \dfrac{x}{50}$, or

$$x = 50 \tan 76°$$
$$= 50(4.0107809)$$
$$= 200.53905$$

The tower is about 200.5 m high.

Application

EXAMPLE 9.22

Two people are in a hot air balloon. One of them is able to get a sighting from the gondola of the balloon as it passes over one end of a football field, as shown in Figure 9.24. The angle of depression to the other end of the football field is 53.8°. This person knows that the length of the football field, including the end zones, is 120 yd. How high was the balloon when it went over the football field?

Solution We want to determine the height of the right triangle. The height has been labeled h in Figure 9.24. So, we have $\tan 53.8° = \dfrac{h}{120}$, or

$$h = 120 \tan 53.8°$$
$$= 120(1.366326733)$$
$$= 163.9592079$$

The balloon was about 164 yd high.

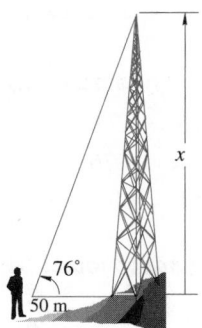

FIGURE 9.23

Application

EXAMPLE 9.23

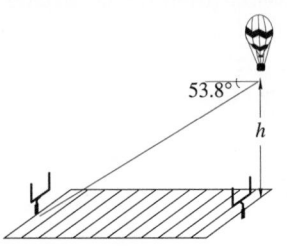

FIGURE 9.24

A surveyor marks off a right-angle corner of a rectangular house foundation. In sighting on the diagonally opposite corner of the foundation, the line of sight has to move through an angle of 35° as shown in Figure 9.25a. If the length of the short side of the foundation is 46.3 ft, what is the length of the long side?

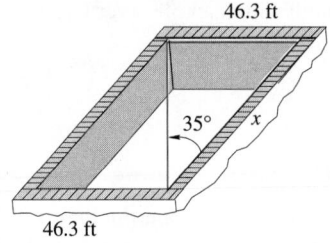

FIGURE 9.25a

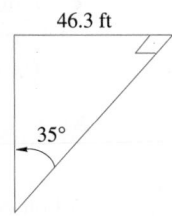

FIGURE 9.25b

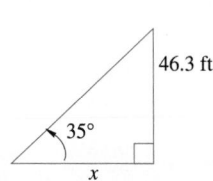

FIGURE 9.25c

Solution   We have the situation sketched in Figure 9.25b. (Notice that in Figure 9.25b, the right angle does not look like a right angle.) But, this is confusing; so in Figure 9.25c the triangle has been rotated to the more "typical" position with the right angle along the bottom. Since $\tan 35° = \dfrac{46.3}{x}$, we have

$$x = \frac{46.3}{\tan 35°}$$
$$= \frac{46.3}{0.70020754}$$
$$= 66.123253$$

So the long side is about 66.1 ft.

Exercise Set 9.3

Find the acute angles for the trigonometric functions given in Exercises 1–12. (Give answers to the nearest 0.01.)

1. $\sin A = 0.732$ **4.** $\sin D = 0.049$ **7.** $\sec G = 3.421$ **10.** $\sec J = 1.734$

2. $\cos B = 0.285$ **5.** $\cos E = 0.839$ **8.** $\csc H = 1.924$ **11.** $\csc K = 4.761$

3. $\tan C = 4.671$ **6.** $\tan F = 0.539$ **9.** $\cot I = 0.539$ **12.** $\cot L = 4.021$

In Exercises 13–30, sketch each right triangle and solve it. Use your knowledge of significant figures to round off appropriately.

13. $A = 16.5°, a = 7.3$ **17.** $A = 43°, b = 34.6$ **21.** $A = 0.15\,\text{rad}, c = 18$

14. $B = 53°, b = 9.1$ **18.** $B = 67°, c = 32.4$ **22.** $B = 0.23\,\text{rad}, a = 9.7$

15. $A = 72.6°, c = 20$ **19.** $A = 0.92\,\text{rad}, a = 6.5$ **23.** $A = 1.41\,\text{rad}, b = 40$

16. $B = 12.7°, a = 19.4$ **20.** $B = 1.13\,\text{rad}, b = 24$ **24.** $A = 1.15\,\text{rad}, c = 18$

25. $a = 9, b = 15$

26. $a = 19.3, c = 24.4$

27. $b = 9.3, c = 18$

28. $a = 14, b = 9.3$

29. $a = 20, c = 30$

30. $b = 15, c = 25$

Solve Exercises 31–42.

31. *Electricity*   In an ac circuit that has inductance x_L and resistance R, the phase angle can be determined from the equation $\tan \phi = \dfrac{x_L}{R}$. If $x_L = 12.3\,\Omega$ and $R = 19.7\,\Omega$, what is the phase angle?

32. From the top of a lighthouse, the angle of depression to the waterline of a boat is $23.2°$. If the lighthouse is 222 ft high, how far away is the ship from the bottom of the lighthouse?

33. *Navigation*   An airplane is flying at an altitude of 700 m when the copilot spots a ship in distress at an angle of depression of $37.6°$. How far is it from the plane to the ship?

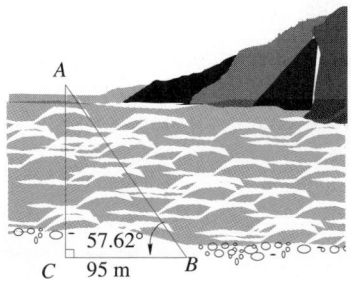

FIGURE 9.26

34. *Construction*   A bridge is to be constructed across a river. As shown in Figure 9.26, a piling is to be placed at point A and another at C. To find the distance between A and C a surveyor locates point B exactly 95 m from C so that $\angle C$ is a right angle. If $\angle B$ is $57.62°$, how far apart are the pilings?

35. A vector is usually resolved into component vectors that are perpendicular to each other. Thus, you get a rectangle with one side F_x, the horizontal component and the other F_y, the vertical component, as shown in Figure 9.27. The original vector F is the diagonal of the rectangle. If $F = 10\,\text{N}$ and $\theta = 30°$, what are the horizontal and vertical components of this force?

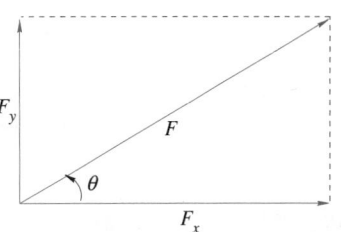

FIGURE 9.27

36. Two forces, one of 20 lb and the other of 12 lb, act on a body in directions perpendicular to each other. What is the magnitude of the resultant of these forces? What is the size of the angle the resultant of these forces makes with the smaller force?

37. *Transportation engineering*   Highway curves are usually banked at an angle ϕ. The proper banking angle for a car making a turn of radius r at velocity v is $\tan \phi = \dfrac{v^2}{gr}$, where $g = 32\,\text{ft/s}^2$ or $9.8\,\text{m/s}^2$. Find the proper banking angle for a car moving at 55 mph to go around a curve 1,200 ft in radius. (First, convert 55 mph to 80.67 ft/s.)

38. *Transportation engineering*   Find the proper banking angle for a car moving 88 km/h to go around a curve 500 m in radius. (First, convert 88 km/h to 24.44 m/s.)

39. *Transportation engineering*   Two straight highways A and B intersect at an angle $67°$. A service station is located on highway A 300 m from the intersection. What is the location of the point on highway B that is closest to the service station? How near is it?

40. *Air traffic control*   One way of measuring the ceiling, or height of the bottom of the clouds, at an airport is to focus a light beam straight up at the clouds at a known distance from an observation point. The angle of elevation from the observation point to the light beam on the clouds is θ. Find the ceiling height if the observation point is 950 ft from the source of the light beam and $\theta = 58.6°$.

41. *Forestry* Using a clinometer and tape measure, a forest ranger locates the top of a tree at a 67.3° angle of elevation from a point 25.75 ft from the base of the tree and on the same level as the base. How tall is the tree?

42. *Fire science* A 32.0 m ladder on a fire truck can be extended at an angle of 65.0°. The base of the ladder is located at a point that is 2.7 m above the ground.

(a) How close to the base of a building must the base of the ladder be located for the end of the ladder to reach the building?

(b) A person needing to be rescued is located 35.85 m above the ground. Can a 1.92 m tall firefighter reach the person if the firefighter (foolishly) stands on the very top rung of the ladder?

In Your Words

43. Describe what it means to solve a triangle.

44. What is the difference between an angle of elevation and an angle of depression? How can you tell one from another?

≡ 9.4
TRIGONOMETRIC FUNCTIONS OF ANY ANGLE

In Sections 9.2 and 9.3, we have concentrated on the trigonometric functions for angles between 0° and 90° or 0 and $\frac{\pi}{2}$ rad. We will now examine other angles.

If you remember, we originally defined the trigonometric functions in terms of a point on the terminal side of an angle in standard position. We want to return to that definition as we look at angles that are not acute.

Reference Angles

Consider angle θ in Figure 9.28. Point P is on the terminal side and is in the second quadrant. If a perpendicular line is dropped from P to the x-axis, a triangle is formed with an acute angle θ_{Ref}. If we use the definition of the trigonometric functions from Section 9.1, we seem to get the same values we got for angle θ of the triangle.

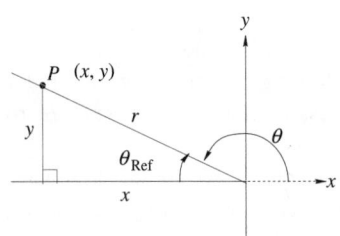

FIGURE 9.28

> **Reference Angle**
>
> The acute angle, θ_{Ref}, between the terminal side of θ and the x-axis is called the **reference angle** for θ. Thus, we can see that the trigonometric functions of θ are numerically the same as those of its reference angle, θ_{Ref}. The reference angle θ_{Ref} for any angle θ in each of four quadrants is shown in Figures 9.29a–d.

EXAMPLE 9.24

Find the reference angle θ_{Ref} for each angle θ: **(a)** $\theta = 75°$, **(b)** $\theta = 218°$, **(c)** $\theta = 320°$, **(d)** $\theta = \frac{3\pi}{4}$, **(e)** $\theta = \frac{7\pi}{6}$.

Solutions The solutions will use the guidelines demonstrated in Figure 9.29.

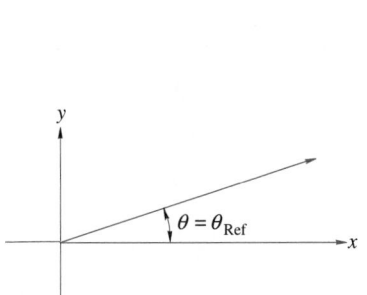

FIGURE 9.29a

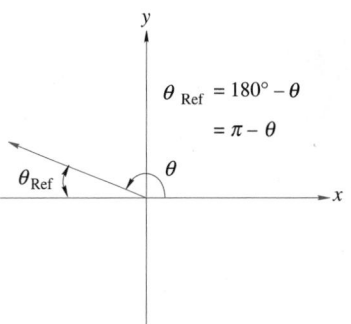

FIGURE 9.29b

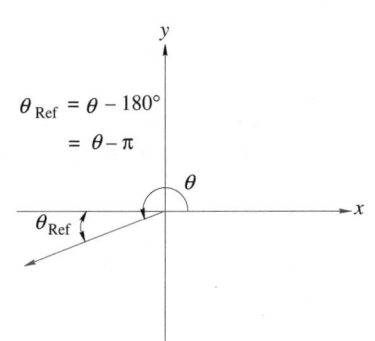

FIGURE 9.29c

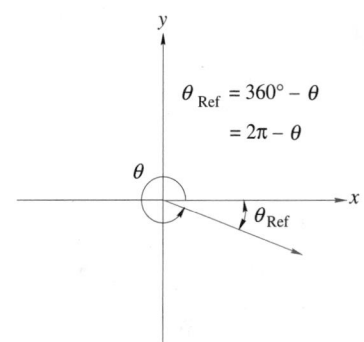

FIGURE 9.29d

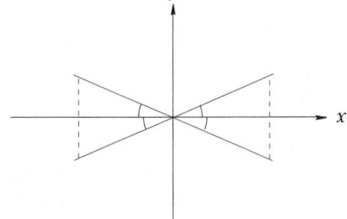

FIGURE 9.30

(a) $75°$ is in Quadrant I, so $\theta_{\text{Ref}} = 75°$.

(b) $218°$ is Quadrant III, so $\theta_{\text{Ref}} = 218° - 180° = 38°$.

(c) $320°$ is in Quadrant IV, so $\theta_{\text{Ref}} = 360° - 320° = 40°$.

(d) $\frac{3\pi}{4}$ is in Quadrant II, so $\theta_{\text{Ref}} = \pi - \frac{3\pi}{4} = \frac{4\pi}{4} - \frac{3\pi}{4} = \frac{\pi}{4}$.

(e) $\frac{7\pi}{6}$ is in Quadrant III, so $\theta_{\text{Ref}} = \frac{7\pi}{6} - \pi = \frac{7\pi}{6} - \frac{6\pi}{6} = \frac{\pi}{6}$.

 Hint

The reference angle is always measured from the x-axis and never from the y-axis. The "bowtie" in Figure 9.30 may help you remember how to determine the reference angle.

EXAMPLE 9.25

A point on the terminal side of an angle θ has the coordinates $(-5, -12)$. Write the six trigonometric functions of θ to three decimal places.

Solution   A sketch of the angle is in Figure 9.31. This angle is in the third quadrant. From the Pythagorean theorem we get

$$r = \sqrt{(-5)^2 + (-12)^2} = \sqrt{25 + 144} = \sqrt{169} = 13$$

So, the trigonometric functions of θ are

$$\sin\theta = \frac{y}{r} = \frac{-12}{13} \approx -0.923 \qquad \csc\theta = \frac{r}{y} = \frac{13}{-12} \approx -1.083$$

$$\cos\theta = \frac{x}{r} = \frac{-5}{13} \approx -0.385 \qquad \sec\theta = \frac{r}{x} = \frac{13}{-5} = -2.6$$

$$\tan\theta = \frac{y}{x} = \frac{-12}{-5} = 2.4 \qquad \cot\theta = \frac{x}{y} = \frac{-5}{-12} \approx 0.417$$

As you can see, the sine and cosine and their reciprocals are negative. The tangent, and its reciprocal are positive.

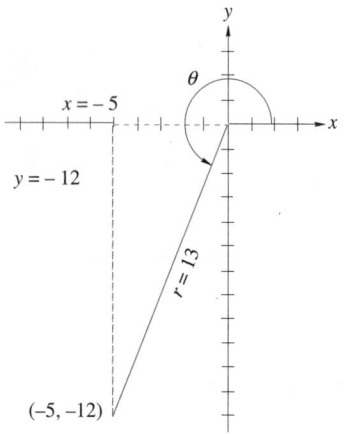

FIGURE 9.31

In our previous work with right triangles, we found that the values of the trigonometric functions were always positive. You can see here and from our work in Section 9.1 that they can sometimes be negative. Whenever the angle is in the second, third, or fourth quadrant, some of the trigonometric functions will be negative.

EXAMPLE 9.26

Use a calculator to find the values of the following trigonometric functions: $\sin 215°$, $\cos 110°$, $\tan 332°$, $\csc 163°$, $\sec 493°$, and $\cot(-87°)$.

Solution   We find these the same way we used the calculator to find the values of angles between $0°$ and $90°$:

Function	ENTER (Algebraic)	ENTER (Graphing)	DISPLAY
$\sin 215°$	215 sin	sin 215 ENTER	−0.5735764
$\cos 110°$	110 cos	cos 110 ENTER	−0.3420201
$\tan 332°$	332 tan	tan 332 ENTER	−0.5317094
$\csc 163°$	163 sin 1/x	(sin 163) x^{-1} ENTER	3.4203036
$\sec 493°$	493 cos 1/x	(cos 493) x^{-1} ENTER	−1.4662792
$\cot(-87°)$	87 +/− tan 1/x	(tan (−) 87) x^{-1} ENTER	−0.0524078

There are times when we need to know the sign of an angle and we may not have our calculator handy. One case might be if you are using the table of trigonometric functions. But you will need it more often when you have to find the size of an angle and you know the value of a trigonometric function. You will see in Section 9.5 that the calculator will give you an answer, but it may not be the correct answer to the problem. You will have to determine the correct answer from the calculator's answer and from your knowledge of the signs of the trigonometric functions.

To find the sign of a trigonometric function, make a sketch of the angle. You do not need a very accurate sketch, but make sure you get the angle in the correct quadrant. Draw the triangle for the reference angle and note the signs of x and y. Remember that the radius vector r is always positive. Then set up the ratio using the two values from x, y, and r that are appropriate for this angle.

EXAMPLE 9.27

What is the sign of $\sec 215°$?

Solution   Examine the sketch in Figure 9.32. We have drawn a right triangle for the reference triangle and labeled the sides with the sign in parentheses after the letter x, y, or r. Since $\sec 215° = \dfrac{r}{x}$ where r is positive and x is negative, the sign of the $\sec 215°$ is negative.

Figure 9.33 summarizes the signs of the trigonometric functions in each quadrant. If a function is not mentioned in a quadrant, then the function is negative. Thus in Quadrant II, since they are not mentioned, the cos, tan, cot, and sec are all negative.

We will finish this section with a discussion of how to find the values of trigonometric functions from a table. If an angle is not an acute angle, you need to use the reference angle and the coterminal angle.

If an angle is negative, or if it is a positive angle with measure more than 360° or 2π rad, replace it with a coterminal angle that is between 0° and 360° or 0 and 2π rad. To do this, you will add or subtract multiples of 360° if the angle is in degrees, or 2π if it is in radians, until you have the desired coterminal angle.

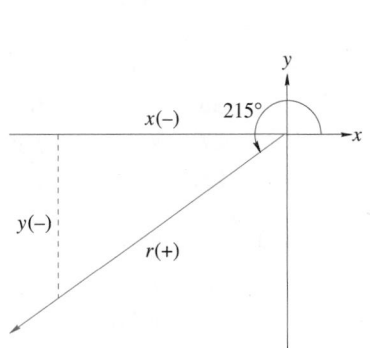

FIGURE 9.32

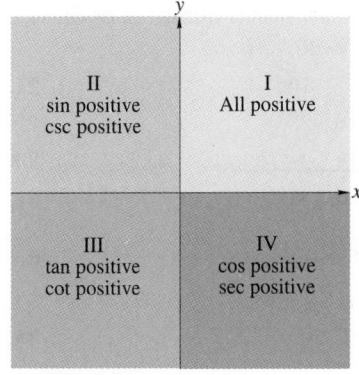

FIGURE 9.33

EXAMPLE 9.28

For each angle find a coterminal angle between 0° and 360° or 0 and 2π rad: **(a)** $1,292°$, **(b)** $-683°$, **(c)** $\frac{17}{4}\pi$, **(d)** $-\frac{5\pi}{3}$, **(e)** -8.7 rad.

EXAMPLE 9.28 (Cont.)	**Solutions**

(a) $1,292° - 3(360°) = 1,292° - 1,080° = 212°$

(b) $-683° + 2(360°) = -683° + 720° = 37°$

(c) $\frac{17}{4}\pi - 2(2\pi) = \frac{17}{4}\pi - 2\left(\frac{8\pi}{4}\right) = \frac{17\pi}{4} - \frac{16\pi}{4} = \frac{\pi}{4}$

(d) $-\frac{5\pi}{3} + 1(2\pi) = \frac{-5\pi}{3} + \frac{6\pi}{3} = \frac{\pi}{3}$

(e) $-8.7 + 2(2\pi) = -8.7 + 12.566371 = 3.8663706 \text{ rad}$

The value of any trigonometric function of an angle θ is the same as the value of the function for the reference angle θ_{Ref}, except for an occasional change in the algebraic sign. If the angle is not between $0°$ and $360°$ or 0 and 2π rad, first find the coterminal angle in that interval and then find the reference angle of the coterminal angle.

Exercise Set 9.4

Give the reference angle for each of the angles in Exercises 1–8.

1. $87°$ **3.** $137°$ **5.** $\frac{9\pi}{8}$ rad   **7.** 4.5 rad

2. $200°$ **4.** $298°$ **6.** 2.1 rad   **8.** 5.85 rad

Give the coterminal angle for each of the angles in Exercises 9–16.

9. $518°$ **11.** $-871°$ **13.** 7.3 rad   **15.** -2.17 rad

10. $1,673°$ **12.** $-137°$ **14.** $\frac{16\pi}{3}$ rad   **16.** -8.43 rad

State which quadrant or quadrants the terminal side of θ is in for each of the angles or expressions in Exercises 17–30.

17. $165°$ **21.** $250°$ **25.** $\tan\theta$ is negative.   **29.** $\csc\theta$ is negative and $\cos\theta$ is positive.

18. $285°$ **22.** $197°$ **26.** $\cos\theta$ is positive.

19. $-47°$ **23.** $98°$ **27.** $\sec\theta$ is negative.   **30.** $\tan\theta$ and $\sec\theta$ are negative.

20. $312°$ **24.** $\sin\theta$ is positive.   **28.** $\cot\theta$ is negative.

State whether each of the expressions in Exercises 31–38 is positive or negative.

31. $\sin 105°$ **33.** $\tan 372°$ **35.** $\cos 1.93$ rad   **37.** $\sin 215°$

32. $\sec 237°$ **34.** $\cos(-53°)$ **36.** $\cot 4.63$ rad   **38.** $\csc 5.42$ rad

Find the values of each of the trigonometric functions in Exercises 39–62.

39. $\sin 137°$ **43.** $\tan 164.2°$ **47.** $\tan 6.1$ rad   **51.** $\sin 415.5°$

40. $\cos 263°$ **44.** $\csc 197.3°$ **48.** $\sec 4.32$ rad   **52.** $\tan 512.1°$

41. $\tan 293°$ **45.** $\sin 2.4$ rad   **49.** $\tan 1.37$ rad   **53.** $\cot -87.4°$

42. $\sin 312°$ **46.** $\cos 1.93$ rad   **50.** $\sin 3.2$ rad

54. $\cos 372.1°$

55. $\csc -432.4°$

56. $\cos 357.3°$

57. $\sin 6.5 \,\text{rad}$

58. $\sec(-4.3 \,\text{rad})$

59. $\tan 8.35 \,\text{rad}$

60. $\sin 9.42 \,\text{rad}$

61. $\cos -0.43 \,\text{rad}$

62. $\sec 9.34 \,\text{rad}$

Solve Exercises 63–68.

63. *Electronics*   The intensity, I, of a sinusoidal current in an ac circuit is given by $I = I_{max} \sin\theta$, where I_{max} is the maximum intensity of the current. Find I when $I_{max} = 32.65$ mA and $\theta = 132.0°$.

64. *Electronics*   The potential voltage, V, of a sinusoidal current in an ac circuit is given by $V = V_{max} \sin\theta$, where V_{max} is the maximum potential. Find V when $V_{max} = 115.2$ V and $\theta = 113.4°$.

65. *Metalworking*   Find the length of piece of metal ABC in Figure 9.34.

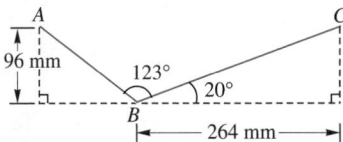

FIGURE 9.34

66. *Metalworking*   Figure 9.35 shows the outline of an isosceles trapezoid that is to be cut from a metal sheet. Find the size of $\angle\theta$.

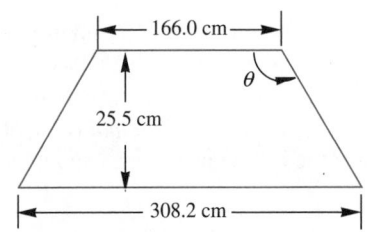

FIGURE 9.35

In Your Words

69. Explain what is meant by a reference angle.

67. *Landscape architecture*   Figure 9.36 shows the outline of some ground that needs to be sodded. How many square feet of sod are needed to completely cover the ground.

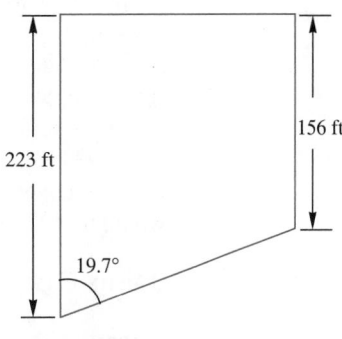

FIGURE 9.36

68. *Metalworking*   Find the length of side AB on the metal sheet shown in Figure 9.37.

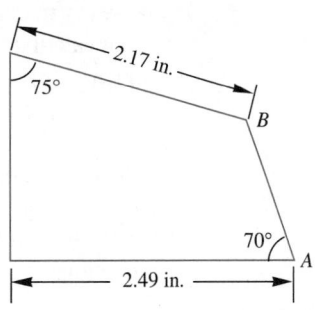

FIGURE 9.37

70. Describe how you can tell the signs of the six trigonometric functions of a particular angle if you know the quadrant in which the angle lies.

≡ 9.5
INVERSE TRIGONOMETRIC FUNCTIONS

In Section 9.3, we were introduced to the inverse trigonometric functions. For example, when we knew that $\sin\theta = 0.2437$ we could enter 0.2437 $\boxed{\text{INV}}$ $\boxed{\text{sin}}$ or $\boxed{\text{2nd}}$ $\boxed{\text{sin}}$ 0.2437 into a calculator and find that $\theta = 14.105021°$. In this section, we look more closely at these functions.

Inverse Trigonometric Functions

In our earlier work with inverse trigonometric functions, we were concerned with right triangles. Thus, all the angles we had to find were acute angles. In Section 9.4, we learned that, simply knowing the value of a trigonometric function for an angle did not allow us to determine the size of the angle. For example, if $\sin\theta = 0.2437$, we know that θ must be in Quadrant I or II. These are the only quadrants where sine is positive. This means that $\theta = 14.105021°$ or $\theta = 180° - 14.105021° = 165.894979°$. It is possible that θ could be any angle coterminal with either of these two angles.

In a similar manner, if $\tan\theta = -1.5$, then using a calculator you obtain, $\theta = -56.309932°$. This is coterminal with $\theta = 303.690068°$. This angle is in Quadrant IV. The tangent is also negative for values in Quadrant II. In Quadrant II, the angle would be $123.69007°$.

Using Calculators for Inverse Trigonometric Functions

Using a calculator as we did in the two previous examples, is certainly fast and accurate. But, it may not always give us the angle that we want. For example, it will never give an angle in Quadrant III. Why is this? It goes back to the basic idea of a function—the idea that for each different value of x there is exactly one value of y.

Because people wanted inverse trigonometric *functions* they had to restrict the answers to exactly one value of y for each value of x. That is why, when you use your calculator to determine inverse sin 0.5, you obtain 30° by pressing 0.5 $\boxed{\text{INV}}$ $\boxed{\text{sin}}$ or $\boxed{\text{2nd}}$ $\boxed{\text{sin}}$ 0.5.

This is nothing new. At one time we studied a function $f(x) = x^2$ and its inverse $f^{-1}(x) = \sqrt{x}$. We knew that $f(-5) = f(5) = 5^2 = 25$, but that $f^{-1}(25) = \sqrt{25} = 5$. The inverse function of x^2, $\sqrt{x}$, only gave the principal value of the square root. In the same way, the inverse trigonometric functions have only the *principal value* of the angle defined for each value of x. Thus, we have the following inverse trigonometric functions.

Inverse Trigonometric Functions

If $\sin\theta = x$, then $\arcsin x = \theta$, where $-90° \le \theta \le 90°$
or $\quad -\frac{\pi}{2} \le \theta \le \frac{\pi}{2}$

If $\cos\theta = x$, then $\arccos x = \theta$, where $0° \le \theta \le 180°$
or $\quad 0 \le \theta \le \pi$

If $\tan\theta = x$, then $\arctan x = \theta$, where $-90° \le \theta \le 90°$
or $\quad \frac{-\pi}{2} < \theta < \frac{\pi}{2}$

If $\csc\theta = x$, then $\text{arccsc } x = \theta$, where $-90° \le \theta \le 90°, \theta \ne 0°$
or $\quad -\frac{\pi}{2} < \theta < \frac{\pi}{2}, \theta \ne 0$

If $\sec\theta = x$, then $\text{arcsec } x = \theta$, where $0° \le \theta \le 180°, \theta \ne 90°$
or $\quad 0 \le \theta \le \pi, \theta \ne \frac{\pi}{2}$

If $\cot\theta = x$, then $\text{arccot } x = \theta$, where $0° < \theta < 180°$
or $\quad 0 < \theta < \pi$

As you can see, for positive values you always get an angle in the first quadrant. If you want the inverse value of a negative number, then you will get an angle in the fourth quadrant for INV sin and INV tan and an angle in the second quadrant for INV cos. Working with a calculator will also show you that the angles in the fourth quadrant are given as negative angles. Thus on a calculator, $\arcsin(-0.5) = -30°$ and not $330°$.

In the previous list, we used the symbol $\arcsin x$ to represent the inverse of the $\sin\theta$. Arcsin is an accepted symbol for the inverse of the sine function. Another symbol that is often used is $\sin^{-1} x$. Just as we used $f^{-1}(x) = \sqrt{x}$; here we use $f^{-1}(x) = \sin^{-1} x$. Similarly, $\arccos x = \cos^{-1} x$, $\arctan x = \tan^{-1} x$, and so on.

Finding All Angles for Inverse Trigonometric Functions ▬▬

Since inverse trigonometric functions only give one answer to problems like $\sin\theta = \frac{1}{2}$ where $\theta = \sin^{-1}\frac{1}{2}$, we must develop a method for finding other angles that satisfy this equation. We will use our knowledge of reference angles and the sign of the functions in the four quadrants to do this.

To find the reference angle of an equation of the form $\sin\theta = n$, where n is a number in the range of sine, we will use the absolute value of n, $|n|$. Then the desired answers can be found by adding or subtracting the reference angle from $180°$ (or π) or subtracting from $360°$ (or 2π).

EXAMPLE 9.29

Find all angles θ in the interval $[0°, 360°)$ where $\cos\theta = -0.42$.

Solution   Here, we have $n = -0.42$, so $|n| = |-0.42| = 0.42$ and $\theta_{\text{Ref}} = \cos^{-1} 0.42 = 65.2°$. Since n was negative, we have answers in Quadrants II and III, the quadrants where cosine is negative.

$$\theta_{\text{II}} = 180° - \theta_{\text{Ref}} = 180° - 65.2° = 114.8°$$
$$\theta_{\text{III}} = 180° + \theta_{\text{Ref}} = 180° + 65.2° = 245.2°$$

The answers are $114.8°$ and $245.2°$.

EXAMPLE 9.30

Find all angles θ in the interval $[0, 2\pi)$ where $\cot\theta = -1.73$.

Solution   Here, we have $n = -1.73$, so $|n| = |-1.73| = 1.73$ and $\theta_{\text{Ref}} = \cot^{-1} 1.73 = \tan^{-1}\frac{1}{1.73} = 0.5241$. Since n was negative, we have answers in Quadrants II and IV, the quadrants where cotangent is negative.

$$\theta_{\text{II}} = \pi - \theta_{\text{Ref}} = \pi - 0.5241 \approx 2.6175$$
$$\theta_{\text{IV}} = 2\pi - \theta_{\text{Ref}} = 2\pi - 0.5241 \approx 5.7591$$

The answers are 2.6175 rad and 5.7591 rad.

In the previous two examples, we only found two answers. There are, however, an infinite number of angles coterminal with these answers. If we want to represent all these answers, we proceed as follows.

For Example 9.29:

$$114.8° + 360°k \text{ and } 245.2° + 360°k, \text{ where } k \text{ is an integer}$$

For Example 9.30:

$$2.6175 + 2\pi k \text{ rad and } 5.7591 + 2\pi k \text{ rad, where } k \text{ is an integer}$$

 Using Computers for Inverse Trigonometric Functions

A computer operates much like a calculator. If you want to find the arctangent of a value, you use the BASIC function ATN. The answer will be in radians. For example, if you would enter PRINT ATN (4.213) and press ⟨RETURN⟩, the computer would display 1.33774853.

Some BASIC languages do not have BASIC functions equivalent to the inverses of the other trigonometric functions. (Some languages, such as True BASIC™, have libraries that do contain these functions.) You can define the inverses of the other trigonometric functions in terms of ATN(x), which follows. In a program, you must check to see that the value of x, is in the domains of the functions previously listed. For example, for the $\sin^{-1}(x)$ you have to make sure that x is not larger than $\frac{\pi}{2} = 1.57079633$ or that x is not less than $-\frac{\pi}{2}$. For $\csc^{-1} x$ you must also check to make sure that $x \neq 0$.

```
ARCSIN(X)=ATN(X/SQR(-X*X+1))
ARCCOS(X)=-ATN(X/SQR(-X*X+1))+1.57079633
ARCCSC(X)=ATN(1/SQR(X*X-1))+(SGN(X)-1)
*1.57079633
```

Note $|x| > 1$

```
ARCSEC(X)=ATN(SQR(X*X-1))+(SGN(X)-1)
*1.57079633
ARCCOT(X)=-ATN(X)+1.57079633
```

A new function was used. The SGN function determines if a value is positive, negative, or zero. $\text{SGN}(X) = +1$ if x is positive, $\text{SGN}(X) = -1$ if x is negative, and $\text{SGN}(X) = 0$ if x is zero.

Exercise Set 9.5

State in which quadrants the angles θ lie for the given trigonometric functions in Exercises 1–8.

1. $\sin\theta = \frac{1}{2}$

2. $\tan\theta = \frac{3}{4}$

3. $\cot\theta = -2$

4. $\cos\theta = -0.25$

5. $\sec\theta = 4.3$

6. $\cos\theta = 0.8$

7. $\csc\theta = -6.1$

8. $\sin\theta = -\dfrac{\sqrt{3}}{2}$

In Exercises 9–16 find, to the nearest tenth of a degree, all angles θ, where $0° \leq \theta < 360°$, with the given trigonometric function.

9. $\sin\theta = \frac{1}{2}$

10. $\tan\theta = \frac{3}{4}$

11. $\cot\theta = -2$

12. $\cos\theta = -0.25$

13. $\sec\theta = 4.3$

14. $\cos\theta = 0.8$

15. $\csc\theta = -6.1$

16. $\sin\theta = -\dfrac{\sqrt{3}}{2}$

In Exercises 17–24 find, to the nearest hundredth of a radian, all angles θ, where $0 \leq \theta < 2\pi$, with the given trigonometric function.

17. $\sin\theta = 0.75$

18. $\tan\theta = 1.6$

19. $\cot\theta = -0.4$

20. $\cos\theta = 0.25$

21. $\csc\theta = 4.3$

22. $\sin\theta = -0.08$

23. $\sec\theta = 2.7$

24. $\cos\theta = -0.95$

Evaluate each of the functions in Exercises 25–32. Write each answer to the nearest tenth of a degree.

25. $\arcsin 0.84$

26. $\arccos(-0.21)$

27. $\arctan 4.21$

28. $\text{arccot}(-0.25)$

29. $\sin^{-1} 0.32$

30. $\cos^{-1} 0.47$

31. $\tan^{-1}(-0.64)$

32. $\csc^{-1}(-3.61)$

Evaluate each of the functions in Exercises 33–40. Write each answer to the nearest hundredth of a radian.

33. $\arccos(-0.33)$

34. $\arctan 1.55$

35. $\arccos 0.29$

36. $\text{arcsec}(-3.15)$

37. $\sin^{-1} 0.95$

38. $\cos^{-1}(-0.67)$

39. $\tan^{-1} 0.25$

40. $\cot^{-1}(-0.75)$

Solve Exercises 41–48.

41. *Electronics* The phase angle, θ, in an RC circuit between the capacitive reactance, X_C, and the resistance, R, when resistance is in series with capacitive reactance, can be found by using $\theta = \tan^{-1}\left(-\dfrac{X_C}{R}\right)$. Find θ when $X_C = 33\,\Omega$ and $R = 40\,\Omega$.

42. *Electronics* The phase angle, θ, in an inductive circuit between the impedance, Z, and the resistance, R, can be found by using $\theta = \cos^{-1}\left(\dfrac{Z}{R}\right)$. Find θ when $Z = 30\,\Omega$ and $R = 50\,\Omega$.

43. *Automotive technology* Find the angle, θ, of the diesel engine valve face in Figure 9.38.

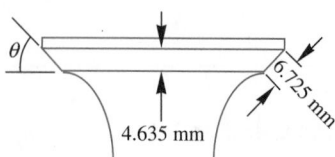

FIGURE 9.38

44. *Mechanical engineering* The maximum height, h, of a projectile with initial velocity, v_0, at angle θ is given by

$$h = \frac{v_0^2 \sin^2 \theta}{2g}$$

where g is the acceleration due to gravity.
(a) Solve for θ.
(b) If $v_0 = 172.5$ m/s and $g = 9.8$ m/s^2, at what angle was the projectile shot if its maximum height was 431.1 m?

 45. Write a computer program that will allow you to input a value of x and that will print x and $\arctan x$.

 46. Write a computer program that will allow you to input a value of x. Check to see if it is an acceptable value. Then print x and $\arccos x$.

 47. Write a computer program that will allow you to input a value of x and that will print x and all possible inverse trigonometric functions for that value of x.

 48. Adapt each of the programs that you just wrote to print the answers in degrees instead of radians.

 **In Your Words**

49. Describe the difference between $\sin^{-1}\theta$ and $(\sin\theta)^{-1}$.
50. Describe the difference between $\cos^{-1}\theta$ and $(\sec\theta)^{-1}$.

≡ 9.6
APPLICATIONS OF TRIGONOMETRY

In earlier sections, we looked at some applications of trigonometry, particularly applications dealing with right triangles. In this section, we will look at some additional applications of trigonometry.

We defined a radian as $\frac{1}{2\pi}$ of a circle. Another way to look at a radian is to examine the circle in Figure 9.39. If r is the radius of the circle and s is the length of the arc opposite θ, then $\theta = \dfrac{s}{r}$ or $\theta r = s$, where θ is in radians.

We can use this relationship to find the area of the sector of a circle. For the shaded sector in Figure 9.39, the area $A = \dfrac{rs}{2} = \dfrac{r^2\theta}{2} = \dfrac{1}{2}r^2\theta$. Both of these formulas work if the angles are in radians.

EXAMPLE 9.31

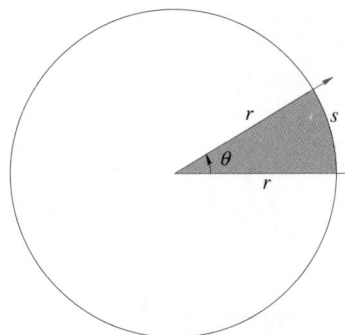

FIGURE 9.39

Find the arc length and area of a sector of a circle, with radius 2.7 cm and central angle 1.3 rad.

Solution   To find the arc length we use $s = \theta r$ with $\theta = 1.3$ rad and $r = 2.7$ cm.

$$s = \theta r$$
$$= (1.3)(2.7)$$
$$= 3.51 \text{ cm}$$

The area is found using $A = \frac{1}{2}r^2\theta$.

$$A = \frac{1}{2}r^2\theta$$
$$= \frac{1}{2}(2.7)^2(1.3)$$
$$= 4.7385 \text{ or } 4.74 \text{ cm}^2$$

Radian measure was developed with the aid of a circle and is used to measure angles. One application of radian measure involves rotational motion. When we work with motion in a straight line, we use the formula $d = vt$ or distance = velocity (or rate) × time. We assume that the rate is constant.

Suppose instead that we had an object moving around a circle at a constant rate of speed. If we let s represent the distance around the circle and v the velocity (or rate), then $s = vt$ or $v = \frac{s}{t}$. Since $s = \theta r$, where r is the radius of the circle, we have $v = \frac{\theta r}{t}$. The centripetal acceleration is $a_c = \frac{v^2}{r} = \frac{\theta^2 r}{t^2}$.

Application

EXAMPLE 9.32

A ball is spun in a horizontal circle 60 cm in radius at the rate of one revolution every 3 s. What is the ball's velocity and centripetal acceleration?

Solution   The velocity $v = \frac{s}{t}$. We know that $t = 3$ s, but we have to find s. In one revolution, the ball covers one complete circle, so $\theta = 2\pi$, and since $r = 60$ cm, we have $s = \theta r = 120\pi$. Thus,

$$v = \frac{s}{t}$$
$$= \frac{120\pi \text{ cm}}{3 \text{ s}}$$
$$= 40\pi \text{ cm/s} = 125.66 \text{ cm/s}$$

The centripetal acceleration is

$$a_c = \frac{v^2}{r} = \frac{(40\pi \text{ cm/s})^2}{60 \text{ cm}}$$
$$= 263.19 \text{ cm/s}^2$$

The angular distance D (measured as the earth's center) between two points P_1 and P_2 on the earth's surface is determined by

$$\cos D = \sin L_1 \sin L_2 + \cos L_1 \cos L_2 \cos(M_1 - M_2)$$

where L_1, L_2 and M_1, M_2 are the respective latitudes and longitudes of two points. The latitude is the angle measured at the earth's surface between a point on the earth and the equator on the same meridian. Meridians are imaginary lines on the surface of the earth, which pass through both the north and south poles. (See Figure 9.40.) The longitude is the angle between the meridian passing through a point on the earth and the $0°$ meridian passing through Greenwich, England.

Application

EXAMPLE 9.33

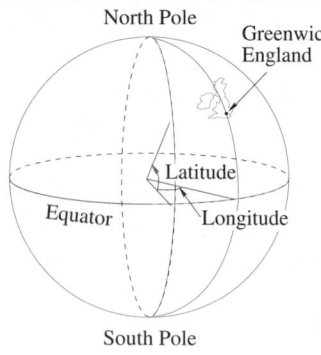

FIGURE 9.40

If Knoxville, Tennessee, is at latitude $36°$ N and longitude $83°55'$ W, and Denver, Colorado, is at latitude $39°44'$ N and longitude $104°59'$ W, and the radius of the earth is 3,960 mi, how far is it from Knoxville to Denver?

Solution We have $L_1 = 36°$, $L_2 = 39°44'$, $M_1 = 83°55'$, and $M_2 = 104°59'$, so $M_1 - M_2 = -21°04'$.

$$
\begin{aligned}
-\cos D &= (\sin 36°)[\sin(39°44')] + (\cos 36°)(\cos 39°44')[\cos(-21°04')] \\
&= (0.5878)(0.6392) + (0.8090)(0.7690)(0.9332) \\
&= 0.3757 + 0.5806 \\
&= 0.9563 \\
D &= \cos^{-1}(0.9563) \\
&= 17.00° \\
&= 0.2967 \text{ rad}
\end{aligned}
$$

Notice that we had to convert our answer for D from degrees to radians.

Since $r = 3,960$ mi, the distance from Knoxville to Denver is

$$s = \theta r = (0.2967)(3,960) = 1,174.93 \text{ mi}$$

This would, of course, be air miles and not the distance you would travel by automobile.

One use of radian measure involves rotational motion. Remember that for motion in a straight line, distance = velocity (or rate) × time or $d = vt$, where we assumed that the rate remains constant. Now, if the motion is in a circular direction instead of a straight line, we would have the formula $s = vt$. Since $s = \theta r$ and $v = \dfrac{s}{t}$, then $v = \dfrac{\theta r}{t} = r\left(\dfrac{\theta}{t}\right)$. This ratio, $\dfrac{\theta}{t}$, is called the **angular velocity** and is usually denoted by the Greek letter ω (omega). Thus it is known that angular velocity ω of an object that turns through the angle θ in time t is given by $\omega = \dfrac{\theta}{t}$. The linear velocity v of a point that moves in a circle of radius r with uniform angular velocity

ω is $v = \omega r$. A rotating body with an angular velocity that changes from ω_0 to ω_f in the time interval t has an angular acceleration of $\alpha = \dfrac{\omega_f - \omega_0}{t}$.

Application

EXAMPLE 9.34

A steel cylinder 8 cm in diameter is to be machined in a lathe. If the desired linear velocity of the cylinder's surface is to be 80.5 cm/s, at how many rpm should it rotate?

Solution We have $v = \omega r$. We want to find ω and $\omega = \dfrac{v}{r}$, where $v = 80.5$ cm/s and $r = 4$ cm.

$$\omega = \frac{80.5\,\text{cm/s}}{4\,\text{cm}}$$
$$= 20.125\,\text{rad/s}$$

Now, 1 rpm is 2π rad/60 s, so using the proportion

$$\frac{2\pi\,\text{rad}}{60\,\text{s}} = \frac{x\,\text{rad}}{1\,\text{s}}$$

we get 1 rpm $= 0.1047$ rad/s. Thus

$$\omega = \frac{20.125\,\text{rad/s}}{(0.1047\,\text{rad/s})/\text{rpm}}$$

$$= 20.125\,\frac{\text{rad}}{\text{s}}\left(\frac{1\,\text{rpm}}{0.1047\,\frac{\text{rad}}{\text{s}}}\right)$$

$$= \frac{20.125}{0.1047}\,\text{rpm}$$

$$\approx 192.22\,\text{rpm}$$

Application

EXAMPLE 9.35

An engine requires 6 s to go from its idling speed of 600 rpm to 1 500 rpm. What is its angular acceleration?

Solution The initial velocity of the engine is

$$\omega_0 = 600\,\text{rpm}$$
$$= 600(0.1047)$$
$$= 62.82\,\text{rad/s}$$

and the final velocity is

$$\omega_f = 1\,500\,\text{rpm}$$
$$= 1\,500(0.1047)$$
$$= 157.05\,\text{rad/s}$$

EXAMPLE 9.35 (Cont.)

The angular acceleration is

$$\alpha = \frac{\omega_f - \omega_0}{t}$$

$$= \frac{157.05 - 62.82}{6}$$

$$= \frac{94.23 \text{ rad/s}}{6 \text{ s}}$$

$$= 15.705 \text{ rad/s}$$

Exercise Set 9.6

Solve Exercises 1–26.

1. *Transportation engineering*   A circular highway curve has a radius of 1 050.250 m and a central angle of 47° measured to the center line of the road. What is the length of the curve?

2. *Physics*   An object is moving around a circle of radius 8 in. with an angular velocity of 5 rad/s. What is the linear velocity?

3. *Physics*   A flywheel rotates with an angular velocity of 30 rpm. If its radius is 12 in., what is the linear velocity of the rim?

4. *Industrial technology*   A pulley belt 4 m long takes 2 s to complete one revolution. If the radius of the pulley is 180 mm, what is the angular velocity of a point on the rim of the pulley?

5. In a circle of radius 15 cm, what is the length of the arc intercepted by a central angle of $\frac{5\pi}{8}$.

6. In a circle of radius 235 mm, what is the length of the arc intercepted by a central angle of 85°?

7. What is the area of the sector in Exercise 5?

8. What is the area of the sector in Exercise 6?

9. *Physics*   A clock pendulum 1.2 m long oscillates through an angle of 0.07 rad on each side of the vertical. What is the distance the end of the pendulum travels from one end to the other?

10. *Space technology*   A communication satellite is in orbit at 22,300 mi above the surface of the earth. The satellite is in permanent orbit above a certain point on the equator. Thus, the satellite makes one revolution every 24 h. What is the angular velocity of the satellite? What is its linear velocity? (The radius of the earth at the equator is 3,963 mi.)

11. *Acoustical engineering*   A phonograph record is 175.26 mm in diameter and rotates at 45 rpm. **(a)** What is the linear velocity of a point on the rim? **(b)** How far does this point travel in 1 min?

12. *Acoustical engineering*   A phonograph turntable rotating at 4.2 rad/s makes 4 complete turns before it stops after 11.92 s. What is its angular acceleration?

13. *Navigation*   City A is at 35°30′ N, 78°40′ W and City B is at 40°40′ N, 88°50′ W. What is the angular distance between the two cities? How far is it between the cities?

14. *Mechanical engineering*   A 1,200-rpm motor is directly connected to a 12-in.-diameter circular saw blade. What is the linear velocity of the saw's teeth in ft/min?

15. *Computer science*   The outer track on a $5\frac{1}{4}''$ diameter diskette for a microcomputer is $2\frac{1}{2}''$ from the center of the diskette. If 8 bytes of a data can be stored on $\frac{1}{4}''$ of track, how many bytes can be stored in the length of the track subtended by $\frac{\pi}{3}$ rad?

16. *Computer science*   Some diskettes are divided into 16 sectors per track. What is the length of one sector of the track in Exercise 15?

17. *Mechanical engineering*   A flywheel makes 850 rev in 1 min. How many degrees does it rotate in 1 s?

18. *Mechanical engineering* If the flywheel in Exercise 17 has a 15 in. diameter, what is the linear velocity of a point on its outer edge?

19. *Energy technology* A wind generator has blades 4.2 m long. What is the speed of a blade tip when the blades are rotating in 20 rpm?

20. *Physics* A 2,500 newton weight is resting on an inclined plane that makes an angle of 23° with the horizontal. Find the component F_x and F_y of the weight parallel to and perpendicular to the surface of the plane, as shown in Figure 9.41.

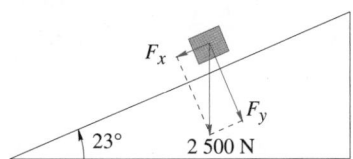

FIGURE 9.41

21. *Navigation* The arctic circle is approximately at latitude 66.5° N. If the radius of the earth is 6,370 km, how far is the arctic circle from the equator?

22. *Optics* When unpolarized light strikes an interface between two materials, **Brewster's law** states that the reflected light is completely polarized perpendicular to the plane of incidence if the angle of incidence, θ, is given by

$$\tan\theta = \frac{n_b}{n_a}$$

If sunlight is reflected off commercial plate glass, at what angle of reflection is the light completely polarized? Here, for air $n_a = 1.00$ and for commercial plate glass, $n_b = 1.516$.

 **In Your Words**

27. Write a word problem in your technology area of interest that requires you to use trigonometry. On the back of the sheet of paper, write your name and explain how to solve the problem by using factoring or completing the square. Give the problem you wrote to a friend and let him or her try to solve it. If your friend has difficulty understanding the problem or solving the problem, or disagrees with your solu-

23. *Optics* The intensity, I, of a light beam transmitted by a pair of polarizing filters is given by the equation

$$I = I_m \cos^2\theta$$

where I_m is the maximum intensity and θ is the angle between the filters. At what angle does the intensity drop to three-fourths of the maximum intensity?

24. *Construction* A roof that slopes at 23° to the horizontal is 14.5 m long and has a slant height of 9.2 m. How large an area does the roof actually cover? (Hint: Find the area of rectangle $ABCD$ in Figure 9.42)

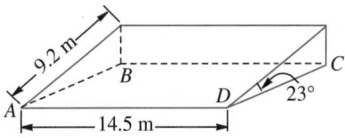

FIGURE 9.42

25. *Space technology* The global positioning system (GPS) consists of 24 satelites, each orbiting the earth in a circular orbit at an altitude of 11,000 mi. Each satellite orbits the earth once every 12 hr. If the earth's radius is 3,960 mi, determine:
(a) The angular speed of each GPS satellite in rad/s
(b) The linear speed of each GPS satellite in mi/s

26. *Navigation* A nautical mile may be defined as the arc length intercepted on the surface of the earth by a central angle with measure 1′. Suppose that the radius of the earth is 3,960 (statute) miles and that one statute mile is 5,280 ft.
(a) Determine the number of feet in a nautical mile.
(b) How much longer (in ft) is a nautical mile than a statute mile?

tion, make any necessary changes in the problem or solution. When you have finished, give the revised problem and solution to another friend and see if he or she can solve it.

28. As in Exercise 27, write a word problem in your technology area that requires you to use trigonometry. This problem should use a different trigonometric function than the one in Exercise 27.

☰ CHAPTER 9 REVIEW

Important Terms and Concepts

Angle(s)
    Complementary
    Coterminal
    Negative
    of Depression
    of Elevation
    Positive
    Quadrantal
    Reference
    Standard position
Degrees
Inverse trigonometric functions
    Arccos or $\cos^{-1}$
    Arccot or $\cot^{-1}$

Arccsc or $\csc^{-1}$
Arcsec or $\sec^{-1}$
Arcsin or $\sin^{-1}$
Arctan or $\tan^{-1}$
Quotient identities
Radians
Reciprocal identities
Trigonometric functions
    Cosecant (csc)
    Cosine (cos)
    Cotangent (cot)
    Secant (sec)
    Sine (sin)
    Tangent (tan)

Review Exercises

Convert each of the angle measures in Exercises 1–6 from degrees to radians.

1. $60°$
2. $198°$
3. $325°$
4. $180°$
5. $-115°$
6. $435°$

Convert each of the angle measures to Exercises 7–12 from radians to degrees.

7. $\frac{3\pi}{4}$ rad
8. 1.10 rad
9. 2.15 rad
10. $\frac{7\pi}{3}$ rad
11. -4.31 rad
12. 5.92 rad

For Exercises 13–24, (a) tell which quadrant the angle is in, (b) give the reference angle for the given angle, and (c) give two coterminal angles, one positive and one negative.

13. $60°$
14. $198°$
15. $325°$
16. $180°$
17. $-115°$
18. $435°$
19. $\frac{3\pi}{4}$ rad
20. 1.10 rad
21. 2.15 rad
22. $\frac{7\pi}{3}$ rad
23. -4.31 rad
24. 5.92 rad

Each point in Exercises 25–30 is on the terminal side of an angle θ in standard position. Find the six trigonometric functions of the angle θ associated with each of these points.

25. $(3, -4)$ **27.** $(-20, 21)$ **29.** $(7, 1)$

26. $(5, 12)$ **28.** $(-4, -7)$ **30.** $(-12, 8)$

In Exercises 31–36, find the trigonometric functions of angle θ at vertex A of triangle ABC for the indicated sides.

31. $a = 8, c = 17$ **33.** $b = 8, c = \frac{40}{3}$ **35.** $b = 7, c = 18.2$

32. $a = 5, b = 12$ **34.** $a = 6, c = 7.5$ **36.** $a = 42, b = 44.1$

In Exercises 37–39, let θ be an angle of a triangle with the given trigonometric function. Find the values of the other trigonometric functions.

37. $\sin\theta = \dfrac{12.8}{27.2}$ **38.** $\tan\theta = \dfrac{16}{16.8}$ **39.** $\sec\theta = \dfrac{4}{2.5}$

In Exercises 40–42, the values of two trigonometric functions of an angle θ are given. Use this information to determine the values of the other four trigonometric functions for θ.

40. $\sin\theta = 0.532, \tan\theta = 0.628$ **41.** $\sin\theta = 0.5, \cos\theta = 0.866$ **42.** $\cos\theta = 0.680, \csc\theta = 1.364$

Use a calculator to determine the values of each of the indicated trigonometric functions in Exercises 43–50.

43. $\sin 45°$ **46.** $\sec(-81°)$ **49.** $\tan(-3.2)\,\text{rad}$

44. $\cos 82.5°$ **47.** $\cos 2.3\,\text{rad}$ **50.** $\csc 0.21\,\text{rad}$

45. $\tan 213.5°$ **48.** $\sin 4.75\,\text{rad}$

In Exercises 51–54, find to the nearest tenth of a degree, all angles θ, where $0° \le \theta < 360°$, with the given trigonometric function.

51. $\sin\theta = 0.5$ **52.** $\tan\theta = 2.5$ **53.** $\cos\theta = -0.75$ **54.** $\csc\theta = 3.0$

In Exercises 55–58, find to the nearest hundredth of a radian, all angles θ, where $0 \le \theta < 2\pi$, with the given trigonometric function.

55. $\cos\theta = -0.5$ **56.** $\sin\theta = 0.717$ **57.** $\tan\theta = -0.95$ **58.** $\sec\theta = 2.25$

Evaluate each of the functions in Exercises 59–64. Write each answer to the nearest tenth of a degree.

59. $\arcsin 0.866$ **61.** $\arctan(-1)$ **63.** $\sin^{-1} 0.385$

60. $\arccos 0.5$ **62.** $\cos^{-1}(-0.707)$ **64.** $\cot^{-1}(3.5)$

Solve Exercises 65–80.

65. *Optics*   The index of refraction n of a medium is the ratio of the speed of light c in air to the speed of light in the medium c_m. According to Snell's law, the angles of incidence i and refraction r (see Figure 9.43) of a light ray are related by the formula

$$n = \frac{\sin i}{\sin r}$$

The index of refraction for water is 1.33. If a light beam enters a lake at an angle of incidence of 30°, what is the angle of refraction?

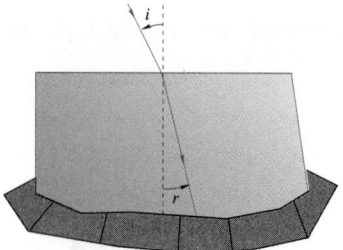

FIGURE 9.43

66. *Optics*   A beam of light enters a pane of glass at an angle of incidence of 0.90 rad. If the angle of refraction is 0.55 rad, what is the index of refraction of the glass?

67. *Electricity*   The current in an ac circuit varies with time t and the maximum value of the current I_m according to the formula

$$I = I_m \sin \omega t$$

where ω is the angular frequency of the alternating current. Find I, if $I_m = 12.6$ A and ωt is $\frac{\pi}{5}$ rad.

68. *Machine technology*   An engine valve is shown in Figure 9.44. What is the angle θ of the valve face, in degrees?

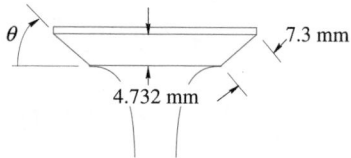

FIGURE 9.44

69. *Electricity*   If resistor and capacitor are connected in series to an ac power source, the phase angle θ is given by the formula

$$\tan \theta = \frac{V_L}{V_R}$$

where θ is the phase angle between the total voltage V and the resistive voltage V_R. If V_L, the voltage across the inductor, is 44 V and $V_R = 50$ V, what is the phase angle θ? What is the total voltage V? (See Figure 9.45.)

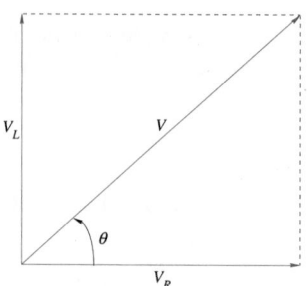

FIGURE 9.45

70. *Machine technology*   A tapered shaft in the shape of an equilateral trapezoid is shown in Figure 9.46. What is the height of the shaft?

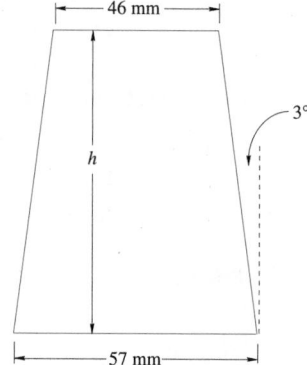

FIGURE 9.46

71. *Transportation engineering*   A guard rail is going to be constructed around the outer edge of a highway curve. If the curve has a radius of 900 ft and an angle of 37°, how many feet of railing are needed?

72. *Computer technology*   The drive speed of a disk drive for a microcomputer is 300 rpm. How many radians does it rotate in 1 s? What is the linear velocity of a point on the rim of a $5\frac{1}{4}''$ diskette?

73. *Space technology*   The planet Earth is approximately 93,000,000 mi from the sun and revolves around the sun in an almost circular orbit every 365 days. What is the approximate linear speed in miles per hour of Earth in its orbit?

74. *Construction*   A guy wire for an electric pole is anchored to the ground at a point 15 ft from the base of the pole. The wire makes an angle of 68° with the level ground. How high up the pole is the wire attached?

75. Find the size of angle x in Figure 9.47 to the nearest tenth of a degree.

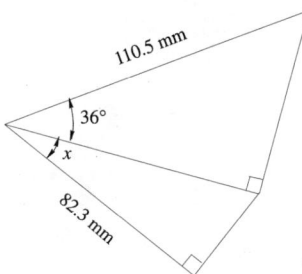

FIGURE 9.47

76. *Forestry*   A tree 56 ft tall casts a shadow 43 ft long. What is the angle of the elevation of the sun at that moment?

77. *Construction*   The light on the top of a tower is at an angle of elevation of 53.7°, when a person is 150 ft from the base of the tower. How high is the light?

78. Determine the size of angle θ in Figure 9.48 to the nearest tenth of a degree?

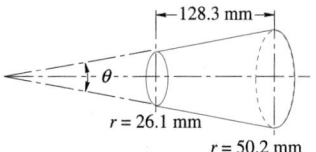

FIGURE 9.48

79. *Automotive technology*   A 3,500 lb automobile rests on a ramp that makes an angle of 34.3° with the horizontal. Determine the components F_x and F_y, as shown in Figure 9.49.

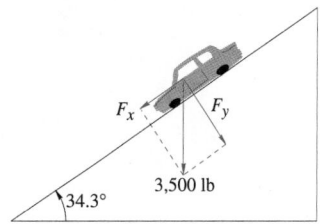

FIGURE 9.49

80. *Construction*   Two guy wires from the top of a tower are anchored in the ground 35 m apart and in direct line with the tower. If the wires make angles of 36.4° and 23.7° with the top of the tower as shown in Figure 9.50, what is the height of the tower?

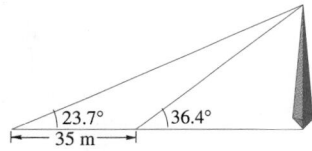

FIGURE 9.50

▦ CHAPTER 9 TEST

1. Convert 50° to radians.

2. Convert $\frac{4\pi}{3}$ to degrees.

3. (a) In what quadrant is a 237° angle?
   (b) Give the reference angle for a 237° angle.

4. The point $(-5, 12)$ is on the terminal side of an angle θ in standard position. Give the six trigonometric functions of this angle.

5. If angle θ is at vertex A of right $\triangle ABC$ with sides $a = 9$ and $c = 23.4$, then determine
   (a) $\sin\theta$
   (b) $\tan\theta$

6. If $\cot\theta = \frac{6.75}{30}$, then determine
   (a) $\tan\theta$
   (b) $\sin\theta$

7. Determine each of the indicated values:
   (a) $\sin 53°$
   (b) $\tan(-112°)$
   (c) $\sec 127°$
   (d) $\cos 4.2$

8. Find, to the nearest tenth of a degree, all angles θ, where $0° \le \theta < 360°$, with the given function.
   (a) $\tan \theta = 1.2$
   (b) $\sin \theta = -0.72$

9. A guy wire for a radio tower is anchored to the ground 135 ft from the base of the tower. The wire makes an angle of 53° with the level ground. How high up the tower is the wire attached?

10. The drive speed of a computer hard disk drive is 3,600 rpm. How many radians does it rotate in 1 s?

CHAPTER

10

Vectors and Trigonometric Functions

A police officer uses a wheel tape to measure the length of skid marks at an accident. In Section 10.1, we will learn how to determine the horizontal and vertical forces the officer exerts.

Courtesy of Michael A. Gallitelli, Metroland Photo Inc.

Many applications of mathematics involve quantities that have both a magnitude (or size) and a direction. The quantities include force, displacement, velocity, acceleration, torque, and magnetic flux density. In this chapter, we will focus on

377

ways to represent these quantities and how they can be used in conjunction with trigonometry to solve problems. These ways will require that you learn about vectors and how to use them with trigonometry.

≡ 10.1
INTRODUCTION TO VECTORS

Our study will begin with the introduction of two quantities: scalars and vectors. After we finish this introductory material, you will be ready to learn some of the basic operations with scalars and vectors. We will use these operations to help solve some applied problems.

Scalars

Quantities that have size, or magnitude, but no direction are called **scalars**. Time, volume, mass, speed, distance, and temperature are some examples of scalars.

Vectors

A **vector** is a quantity that has both magnitude and direction. The magnitude of the vector is indicated by its length. The direction is often given by an angle. A vector is usually pictured as an arrow with the arrowhead pointing in the direction of the vector. Force, velocity, acceleration, torque, and electric and magnetic fields are all examples of vectors.

Vectors are usually represented with boldface letters such as **a** or **A** or by letters with arrows over them, $\vec{a}$ or $\vec{A}$. If a vector extends from a point A, called the **initial point**, to a point B, called the **terminal point**, then the vector can be represented by $\overrightarrow{AB}$ with the arrowhead over the terminal point B. In written work, it is hard to write in boldface, so you should use $\vec{a}$ or $\vec{b}$ or $\overrightarrow{AB}$.

The magnitude of a vector **A** is usually denoted by $|\mathbf{A}|$ or A. The magnitude of a vector is never negative. Two vectors, $\overrightarrow{AB}$ and $\overrightarrow{CD}$, are **equal**, or **equivalent**, if they have the same magnitude and direction, and we write $\overrightarrow{AB} = \overrightarrow{CD}$. In Figure 10.1, $\overrightarrow{AB} = \overrightarrow{CD}$ but $\overrightarrow{AB} \neq \overrightarrow{EF}$ because, while they have the same magnitude, they are not in the same direction. Also, $\overrightarrow{AB} \neq \overrightarrow{GH}$, because they do not have the same magnitude, even though they are in the same direction.

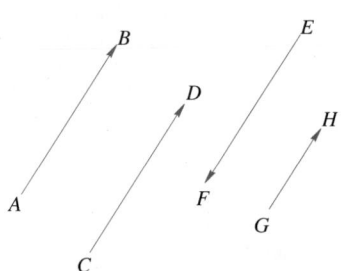

FIGURE 10.1

Resultant Vectors

If $\overrightarrow{AB}$ is the vector from A to B and $\overrightarrow{BC}$ is the vector from B to C, then the vector $\overrightarrow{AC}$ is the **resultant vector** and represents the sum of $\overrightarrow{AB}$ and $\overrightarrow{BC}$.

$$\overrightarrow{AC} = \overrightarrow{AB} + \overrightarrow{BC}$$

The sum of two vectors is shown geometrically in Figure 10.2a. It is also possible to add vectors such as $\overrightarrow{AB} + \overrightarrow{AD}$, as shown in Figure 10.2b. Here $\overrightarrow{AC} = \overrightarrow{AB} + \overrightarrow{AD}$. As you can probably guess from the shape of Figure 10.2b, this method of adding vectors is called the **parallelogram method**.

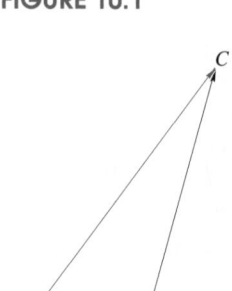

FIGURE 10.2a

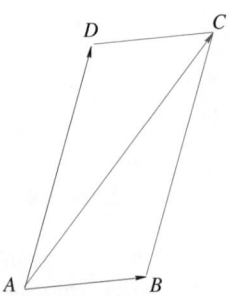

FIGURE 10.2b

The vectors that are being added do not need an endpoint in common. For example, in Figure 10.3 to find $\overrightarrow{AB} + \overrightarrow{EF}$ we would find the vector $\overrightarrow{BC}$, which is equivalent to $\overrightarrow{EF}$. That is, $\overrightarrow{BC} = \overrightarrow{EF}$. We are then back to the first method and $\overrightarrow{AB} + \overrightarrow{EF} = \overrightarrow{AB} + \overrightarrow{BC} = \overrightarrow{AC}$.

Vectors can also be subtracted. Here you need to realize that the vector $-\mathbf{V}$ has the opposite direction of the vector $\mathbf{V}$. An example of vector subtraction, $\overrightarrow{AB} - \overrightarrow{CD} = \overrightarrow{R}$ is shown in Figure 10.4.

Another way to represent subtraction of vectors is shown in Figure 10.5. If we want to subtract $\overrightarrow{AB} - \overrightarrow{CD}$ as in Figure 10.4, we could draw $\overrightarrow{AE}$, where $\overrightarrow{AE} = \overrightarrow{CD}$. Then $\overrightarrow{EB}$ would be the vector that represents $\overrightarrow{AB} - \overrightarrow{CD} = \overrightarrow{AB} - \overrightarrow{AE}$. Notice that in Figure 10.5, $\overrightarrow{EB}$ is drawn from the terminal point of the second vector in the difference $(\overrightarrow{AE})$ to the terminal point of the first vector $(\overrightarrow{AB})$.

A vector can be multiplied by a scalar. Thus, $2\mathbf{a}$ has twice the magnitude but the same direction as $\mathbf{a}$ and $5\mathbf{a}$ has five times the magnitude and the same direction as $\mathbf{a}$. (See Figure 10.6.)

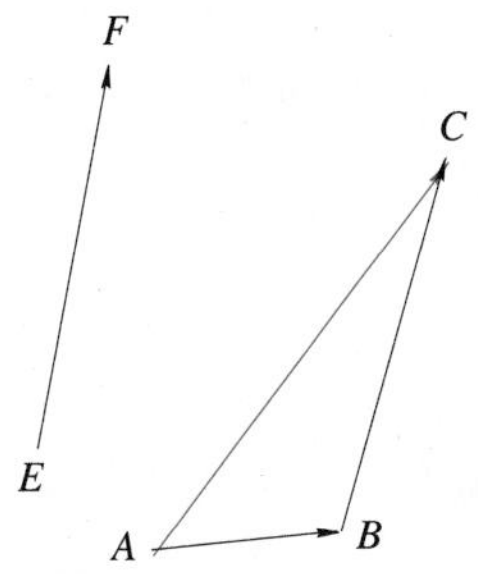

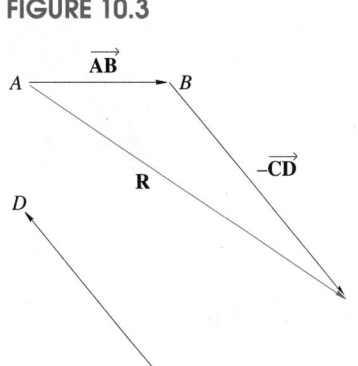

FIGURE 10.3

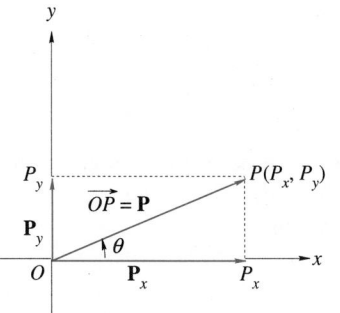

FIGURE 10.4

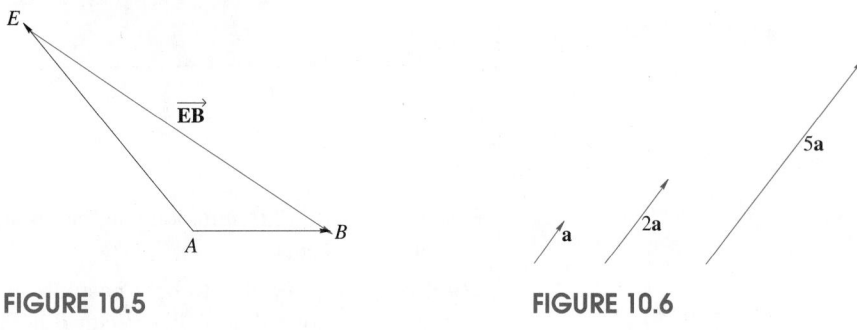

FIGURE 10.5　　　　　　　　　**FIGURE 10.6**

Component Vectors

In addition to adding two vectors together to find a resultant vector, we often need to reverse the process and think of a vector as the sum of two other vectors. Any two vectors that can be added together to give the original vector are called **component vectors**. To resolve a vector means to replace it by its component vectors. Usually a vector is resolved into component vectors that are perpendicular to each other.

Position Vector

If P is any point in a coordinate plane and O is the origin, then $\overrightarrow{OP}$ is called the **position vector** of P. It is relatively easy to resolve a position vector into two component vectors by using the x- and y-axes. These are called the **horizontal** (or x-)**component** and the **vertical** (or y-)**component**. In Figure 10.7, $\mathbf{P}_x$ is the horizontal component of $\overrightarrow{OP}$ and $\mathbf{P}_y$ is the vertical component.

FIGURE 10.7

Finding Component Vectors

If you study Figure 10.7, you can see that the coordinates of P are (P_x, P_y). (Remember that P_x is the magnitude of vector $\mathbf{P}_x$.) Since every position vector has O as its initial point, it is easier to refer to a position vector by its terminal point. From now on we will refer to $\overrightarrow{OP}$ as $\mathbf{P}$.

If the angle that a position vector $\mathbf{P}$ makes with the positive x-axis is θ, then the components of $\mathbf{P}$ are found as follows.

Components of a Vector

A position vector $\mathbf{P}$ that makes an angle θ with the positive x-axis can be resolved into component vectors $\mathbf{P}_x$ and $\mathbf{P}_y$ along the x- and y-axis respectively, with magnitudes P_x and P_y, where

$$\mathbf{P}_x = P \cos\theta$$
$$\text{and} \quad \mathbf{P}_y = P \sin\theta$$

EXAMPLE 10.1

Resolve a vector 12.0 units long and at an angle of 150° into its horizontal and vertical components.

Solution Consider this to be a position vector and put the initial point at the origin. The vector will look like vector $\mathbf{P}$ in Figure 10.8. We are told that $P = 12$ and $\theta = 150°$, so

$$\mathbf{P}_x = 12\cos 150°$$
$$\approx 12(-0.866)$$
$$= -10.392$$
$$\text{and} \quad \mathbf{P}_y = 12\sin 150°$$
$$= 12(0.5)$$
$$= 6$$

We have resolved $\mathbf{P}$ into two component vectors. One component is along the negative x-axis and has approximate magnitude 10.4 units. The other component is along the positive y-axis and has magnitude 6.00 units.

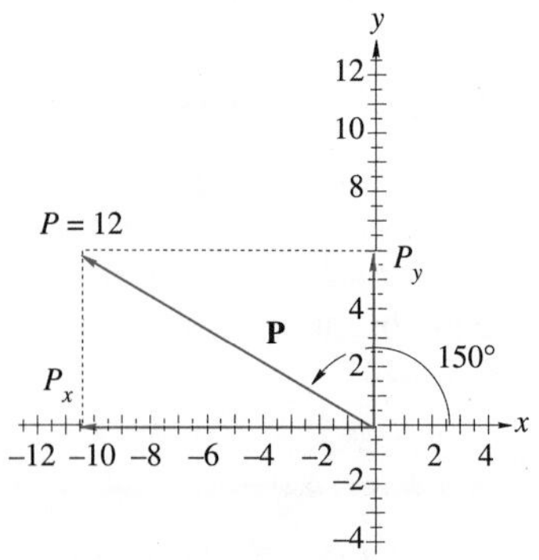

FIGURE 10.8

EXAMPLE 10.2

Vector $\mathbf{P}$ is shown in Figure 10.9a. Resolve $\mathbf{P}$ into its horizontal and vertical components.

EXAMPLE 10.2 (Cont.)

Solution This vector is in Quadrant III, so both components will be negative. The reference angle is 67°. We want to know the angle that this vector makes with the positive x-axis. Since the vector is in Quadrant III, $\theta = 180° + 67° = 247°$. We see that $P = 130$, so we determine

$$\mathbf{P}_x = 130\cos 247°$$
$$\approx 130(-0.3907)$$
$$\approx -50.7950$$
$$\approx -50.8$$
$$\mathbf{P}_y = 130\sin 247°$$
$$\approx 130(-0.9205)$$
$$\approx -119.6656$$
$$\approx -120$$

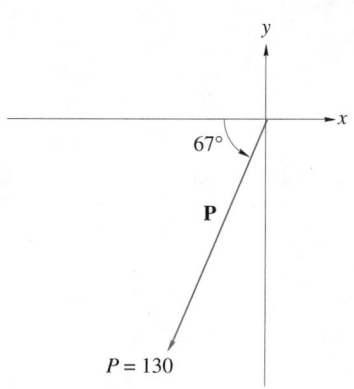

FIGURE 10.9a

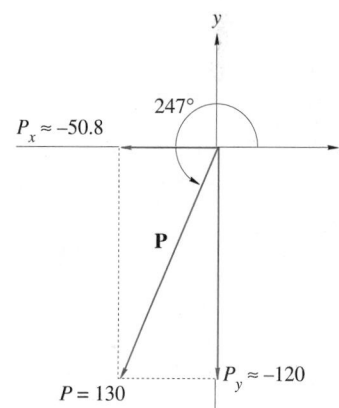

FIGURE 10.9b

Thus, $\mathbf{P}$ has been resolved into component vectors $\mathbf{P}_x$ of magnitude 50.8 units along the negative x-axis and $\mathbf{P}_y$ of magnitude 120 units along the negative y-axis, as shown in Figure 10.9b.

Application

EXAMPLE 10.3

The police officer in Figure 10.10a is measuring skid marks at an accident scene by pushing a wheel tape with a force of 10 lb and holding the handle at an angle of 46° with the ground. Resolve this into its horizontal and vertical component vectors.

EXAMPLE 10.3 (Cont.)

Solution A vector diagram has been drawn over the photograph (Figure 10.10a) of the police officer operating the wheel tape. The vector diagram is shown alone in Figure 10.10b. The initial point of the vector is at the officer's hand and the terminal point is at the hub, or axle, of the wheel tape. We have placed the origin of our

Courtesy of Michael A. Gallitelli, Metroland Photo Inc.

FIGURE 10.10a

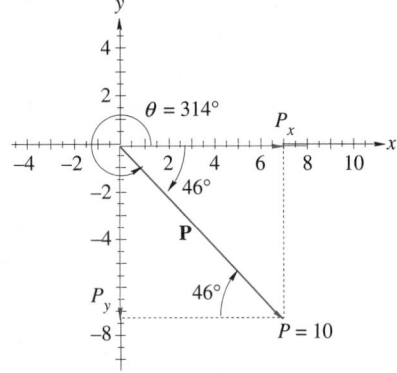

FIGURE 10.10b

coordinate system at the initial point of the vector, and the horizontal, or x-axis, is parallel to the ground.

The vector makes an angle of $360° - 46° = 314°$ with the positive x-axis. Thus, we have $P = 10$ and $\theta = 314°$, so

$$\mathbf{P}_x = 10\cos 314°$$
$$\approx -6.9466$$

and

$$\mathbf{P}_y = 10\sin 314°$$
$$\approx -7.1934$$

The police officer exerts a horizontal force of about 6.9 lb and a vertical force of approximately 7.2 lb.

Application

EXAMPLE 10.4

A cable supporting a television tower exerts a force of 723 N at an angle of 52.7° with the horizontal, as shown in Figure 10.11. Resolve this force into its vertical and horizontal components.

EXAMPLE 10.4 (Cont.)

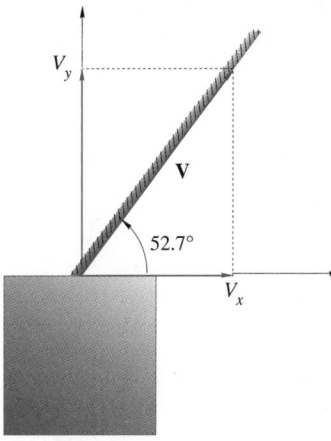

FIGURE 10.11

Solution A vector diagram has been drawn in Figure 10.11. If the cable is represented by vector **V**, then the horizontal component is $\mathbf{V}_x$ and the vertical component is $\mathbf{V}_y$. In this example, $\theta = 52.7°$, and so,

$$V_x = 723 \cos 52.7°$$
$$\approx 438.1296$$

and $\quad V_y = 723 \sin 52.7°$
$$\approx 575.1273$$

We see that this cable exerts a horizontal force of approximately 438 N and a vertical force of about 575 N.

Finding the Magnitude and Direction of Vectors

If we have the horizontal and vertical components of a vector **P**, then we can use the components to determine the magnitude and direction of the resultant vector.

Magnitude and Direction of a Vector

If $\mathbf{P}_x$ is the horizontal component of vector **P** and $\mathbf{P}_y$ is its vertical component, then

$$|\mathbf{P}| = P = \sqrt{P_x^2 + P_y^2}$$

and $\quad \theta_{\text{Ref}} = \tan^{-1} \left| \dfrac{P_y}{P_x} \right|$

EXAMPLE 10.5

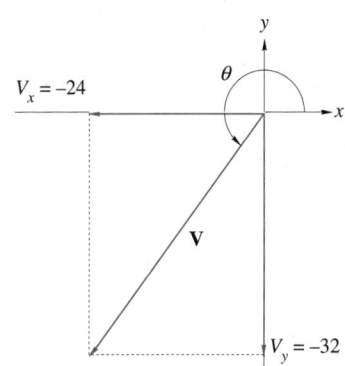

FIGURE 10.12

A position vector **V** has its horizontal component $\mathbf{V}_x = -24$ and its vertical component $\mathbf{V}_y = -32$. What are the direction and magnitude of **V**?

Solution A sketch of this problem in Figure 10.12 shows that **V** is in the third quadrant. Since $\theta_{\text{Ref}} = \tan^{-1} \left| \frac{-32}{-24} \right| \approx 53.13°$, we see that $\theta \approx 180° + 53.13° = 233.13°$.

We know that

$$V = \sqrt{V_x^2 + V_y^2}$$
$$= \sqrt{(-24)^2 + (-32)^2} = \sqrt{1,600}$$
$$= 40$$

So, the magnitude of **V** is 40 and **V** is at an angle of $233.13°$.

Application

EXAMPLE 10.6

A pilot heads a jet plane due east at a ground speed of 425.0 mph. If the wind is blowing due north at 47 mph, find the true speed and direction of the jet.

Solution A sketch of this situation is shown in Figure 10.13. If **V** represents the vector with components $\mathbf{V}_x$ and $\mathbf{V}_y$, then we are given $V_x = 425.0$ and $V_y = 47$. Thus,

$$V = \sqrt{V_x{}^2 + V_y{}^2}$$
$$= \sqrt{425.0^2 + 47^2}$$
$$\approx 427.59$$

and $\quad \theta = \tan^{-1}\left(\dfrac{47}{425}\right)$

$$= \tan^{-1} 0.110588$$

and so, $\quad \theta = 6.31°$

The jet is flying at a speed of approximately 427.6 mph in a direction 6.31° north of due east.

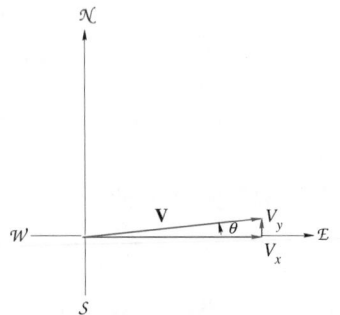

FIGURE 10.13

Exercise Set 10.1

In Exercises 1–4, add the given vectors by drawing the resultant vector.

1.

2.

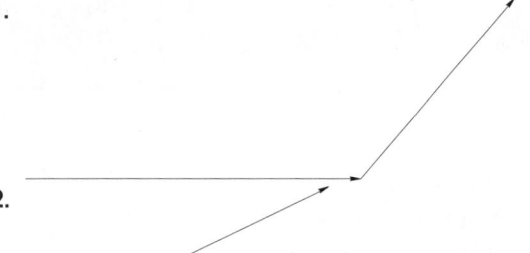

3.

4.

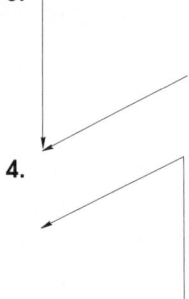

For Exercises 5–20, trace each of the vectors A–D in Figure 10.14. Use these vectors to find each of the indicated sums or differences.

 A B 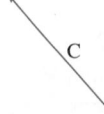 C D

FIGURE 10.14

5. $A + B$	9. $A - B$	13. $A + 3B$	17. $A + 2B - C$
6. $B + D$	10. $C - D$	14. $C + 2D$	18. $A + 3C - D$
7. $A + B + C$	11. $B - A$	15. $2C - D$	19. $2B - C + 2.5D$
8. $B + C + D$	12. $D - C$	16. $3B - A$	20. $4C - 3B - 2D$

In Exercises 21–26, use trigonometric functions to find the horizontal and vertical components of the given vectors.

21. Magnitude 20, $\theta = 75°$

22. Magnitude 16, $\theta = 212°$

23. Magnitude 18.4, $\theta = 4.97$ rad

24. Magnitude 23.7, $\theta = 2.22$ rad

25. $V = 9.75, \theta = 16°$

26. $P = 24.6, \theta = 317°$

In Exercises 27–30, the horizontal and vertical components are given for a vector. Find the magnitude and direction of each resultant vector.

27. $A_x = -9; A_y = 12$

28. $B_x = 10; B_y = -24$

29. $C_x = 8; C_y = 15$

30. $D_x = -14; D_y = -20$

Solve Exercises 31–38.

31. *Navigation*   A ship heads into port at 12.0 km/h. The current is perpendicular to the ship at 5 km/h. What is the resultant velocity of the ship?

32. *Navigation*   A pilot heads a jet plane due east at a ground speed of 756.0 km/h. If the wind is blowing due north at 73 km/h, find the true speed and direction of the jet.

33. *Construction*   A cable supporting a tower exerts a force of 976 N at an angle of 72.4° with the horizontal. Resolve this force into its vertical and horizontal components.

34. *Construction*   A sign of mass 125.0 lb hangs from a cable. A worker is pulling the sign horizontally by a force of 26.50 lb. Find the force and the angle of the resultant force on the sign.

35. *Medical technology*   A ramp for the physically challenged makes an angle of 12° with the horizontal. A woman and her wheelchair weigh 153 lb. What are the components of this weight parallel and perpendicular to the ramp?

36. *Electricity*   If a resistor, a capacitor, and an inductor are connected in series to an ac power source, then the effective voltage of the source is given by the vector V, where V is the sum of the vector quantities V_R and $V_L - V_C$, as shown in Figure 10.15. The phase angle, ϕ, is the angle between V and V_R, where

$\tan\phi = \dfrac{V_L - V_C}{V_R}$. If the effective voltages across the circuit components are $V_R = 12$ V, $V_C = 10$ V, and $V_L = 5$ V, determine the effective voltage and the phase angle.

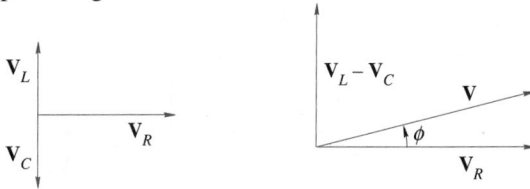

FIGURE 10.15

37. *Electricity*   A resistor, capacitor, and inductor are connected in series to an ac power source. If the effective voltages across the circuit components are $V_R = 15.0$ V, $V_C = 17.0$ V, and $V_L = 8.0$ V, determine the effective voltage and the phase angle.

38. *Electricity*   A resistor, capacitor, and inductor are connected in series to an ac power source. If the effective voltages across the circuit components are $V_R = 22.6$ V, $V_C = 15.2$ V, and $V_L = 28.3$ V, determine the effective voltage and phase angle.

In Your Words

39. **(a)** Distinguish between scalars and vectors. How are they alike and how are they different?

(b) What does it mean for two vectors to be equal?

40. **(a)** Describe how to use the components of a vector to determine the vector's magnitude and direction.

(b) Explain how to use the magnitude and direction of a vector to determine its component vectors.

☰ 10.2
ADDING AND SUBTRACTING VECTORS

In Section 10.1, we learned how to use diagrams to add and subtract vectors. We also learned how to use trigonometry to determine the horizontal and vertical components of a vector. In this section, we will learn how to use trigonometry and the Pythagorean theorem to add and subtract vectors.

We will begin by looking at two special cases. In the first case, both vectors are on the same axis. In the second case, the vectors are on different axes.

EXAMPLE 10.7

FIGURE 10.16

Add vectors **A** and **B**, where $A = 9.6$, $\theta_A = 0°$ and $B = 4.3$, $\theta_B = 180°$.

Solution If we find the horizontal and vertical components of these two vectors we see that

$$\mathbf{A}_x = 9.6\cos 0° = 9.6$$
$$\mathbf{A}_y = 9.6\sin 0° = 0 - 9999$$
$$\mathbf{B}_x = 4.3\cos 180° = -4.3$$
$$\mathbf{B}_y = 4.3\sin 180° = 0$$

The resultant vector **R** has horizontal component $\mathbf{R}_x = \mathbf{A}_x + \mathbf{B}_x = 9.6 + -4.3 = 5.3$ and a vertical component $\mathbf{R}_y = 0 + 0 = 0$. The angle of the resultant vector θ_R is found using $\tan\theta_R = \dfrac{R_y}{R_x} = \dfrac{0}{5.3} = 0$. Since $R_x > 0$, we know that θ_R is $0°$ rather than $180°$. (See Figure 10.16.)

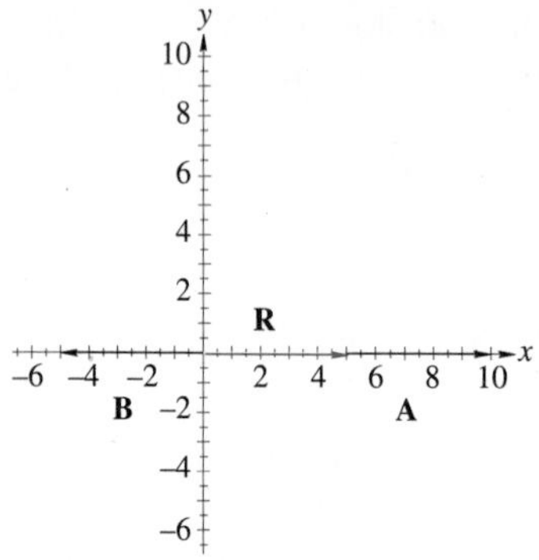

In this first case, the component method required a little extra work. But, the example gave us a foundation for the next example.

EXAMPLE 10.8

Find the resultant of **C** and **D**, when $C = 6.2$, $\theta_C = 270°$, $D = 12.4$, and $\theta_D = 180°$, as shown in Figure 10.17.

Solution Since **C** and **D** are perpendicular, the length of the resultant vector can be found by using the Pythagorean theorem.

$$R = \sqrt{C^2 + D^2} = \sqrt{(6.2)^2 + (12.4)^2} \approx 13.9$$

We also know that $\tan\theta = \dfrac{C}{D} = \dfrac{6.2}{12.4} = \dfrac{1}{2}$ and that $\tan^{-1}\tfrac{1}{2} = 26.57°$. From Figure 10.17, we can see that **R** is in the third quadrant. So, $\theta_R = 206.57° \approx 207°$.

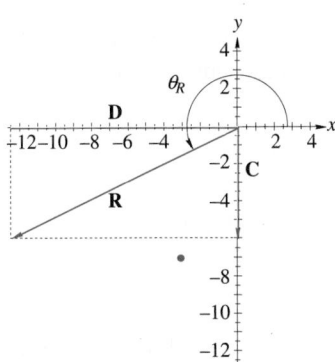

FIGURE 10.17

If we want to find the resultant of two vectors that are not at right angles, the method takes a bit longer. We first resolve each vector into its horizontal and vertical components and then add the horizontal and vertical components, as outlined in the following box.

Adding Vectors Using Components

1. Resolve each vector into its horizontal and vertical components.
2. Add the horizontal components. This sum is R_x, the horizontal component of the resultant vector.
3. Add the vertical components. This sum is R_y, the vertical component of the resultant vector.
4. Use R_x and R_y to determine the magnitude R and direction θ of the resultant vector, where

$$R = \sqrt{R_x{}^2 + R_y{}^2} \text{ and } \tan\theta = \frac{R_y}{R_x}$$

Calculators yield the same answer for $\tan^{-1}\left(\dfrac{R_y}{R_x}\right)$ and $\tan^{-1}\left(\dfrac{-R_y}{-R_x}\right)$. Similarly, they also give the same answer to $\tan^{-1}\left(\dfrac{-R_y}{R_x}\right)$ and $\tan^{-1}\left(\dfrac{R_y}{-R_x}\right)$. You must be careful to check the quadrant in which **R** is located and then, if necessary, add π (or 180°) to the resultant angle.

EXAMPLE 10.9

Find the resultant of two vectors **E** and **F**, where $E = 109, \theta_E = 33.4°, F = 125$ and $\theta_F = 69.4°$. (See Figure 10.18a.)

Solution Resolving **E** into its horizontal and vertical components we get

$$\mathbf{E}_x = 109\cos 33.4° = 91$$

and $\mathbf{E}_y = 109\sin 33.4° = 60$

The components for vector **F** are

$$\mathbf{F}_x = 125\cos 69.4° = 44$$
$$\mathbf{F}_y = 125\sin 69.4° = 117$$

EXAMPLE 10.9 (Cont.)

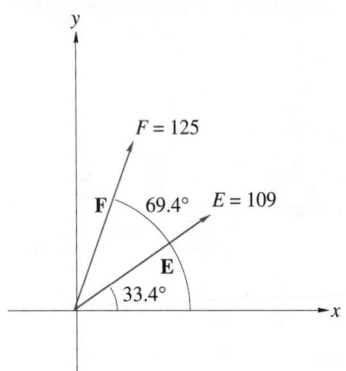

FIGURE 10.18a

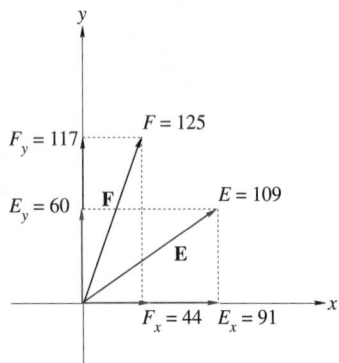

FIGURE 10.18b

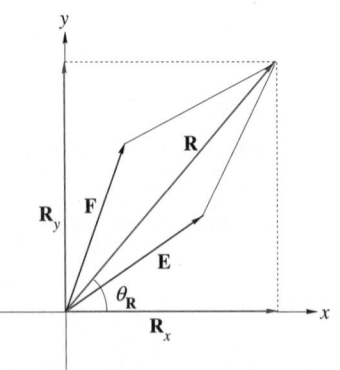

FIGURE 10.18c

These components are shown in Figure 10.18b. The components of the resultants vector are found by adding the horizontal and vertical components of $\mathbf{E}$ and $\mathbf{F}$.

$$\mathbf{R}_x = \mathbf{E}_x + \mathbf{F}_x$$
$$= 91 + 44$$
$$= 135$$
$$\mathbf{R}_y = \mathbf{E}_y + \mathbf{F}_y$$
$$= 60 + 117$$
$$= 177$$

These two component vectors and their resultant vector are shown in Figure 10.18c.

$$R = \sqrt{R_x{}^2 + R_y{}^2}$$
$$= \sqrt{135^2 + 177^2}$$
$$\approx 222.61$$

and

$$\tan\theta_R = \frac{R_y}{R_x}$$
$$\tan\theta_R = \frac{177}{135}$$
$$\theta_R = \tan^{-1}\frac{177}{135}$$

so

$$\theta_R = 52.7°$$

It is often helpful to use a table to help keep track of vectors and their components. It is an effective way to organize this information. A table for this example follows. It lists the horizontal and vertical components of each vector and the resultant vector.

Vector	Horizontal component	Vertical component
$\mathbf{E}$	$\mathbf{E}_x = 109\cos33.4° = \underline{91}$	$\mathbf{E}_y = 109\sin33.4° = \underline{60}$
$\mathbf{F}$	$\mathbf{F}_x = 125\cos69.4° = \underline{44}$	$\mathbf{F}_y = 125\sin69.4° = \underline{117}$
$\mathbf{R}$	$\mathbf{R}_x \qquad\qquad\quad 135$	$\mathbf{R}_y \qquad\qquad\quad 177$

From these values for R_x and R_y, we can determine $R \approx 223$ and $\theta_R = 52.7°$. ▪

EXAMPLE 10.10

Find the resultant of two vectors **G** and **H**, where $G = 449$, $\theta_G = 128.6°$, $H = 521$, and $\theta_H = 327.6°$. (See Figure 10.19a.)

Solution Again, we will resolve each vector into its horizontal and vertical components.

$$\mathbf{G}_x = 449 \cos 128.6°$$
$$= -280.1$$
$$\mathbf{G}_y = 449 \sin 128.6°$$
$$= 350.9$$
$$\mathbf{H}_x = 521 \cos 327.6°$$
$$= 439.9$$
$$\mathbf{H}_y = 521 \sin 327.6°$$
$$= -279.2$$

Figure 10.19b shows each vector and its horizontal and vertical components.

Adding the horizontal components gives the horizontal component of the resultant vector.

$$\mathbf{R}_x = \mathbf{G}_x + \mathbf{H}_x$$
$$= -280.1 + 439.9$$
$$= 159.8$$

Similarly, we can determine the vertical component of the resultant vector.

$$\mathbf{R}_y = \mathbf{G}_y + \mathbf{H}_y$$
$$= 350.9 - 279.2$$
$$= 71.7$$

The resultant vector is shown in Figure 10.19c.

The magnitude of the resultant vector is

$$R = \sqrt{R_x{}^2 + R_y{}^2}$$
$$= \sqrt{159.8^2 + 71.7^2}$$
$$\approx 175.1$$

and the direction of the resultant vector is found from

$$\tan \theta_R = \frac{R_y}{R_x}$$
$$= \frac{71.7}{159.8}$$
$$= 0.4487$$

so

$$\theta_R = \tan^{-1} 0.4487$$
$$= 24.2°$$

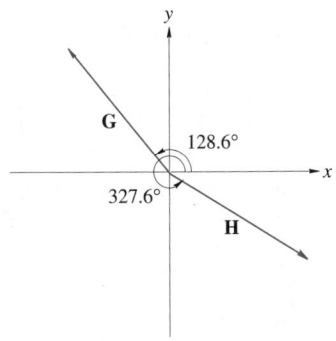

FIGURE 10.19a

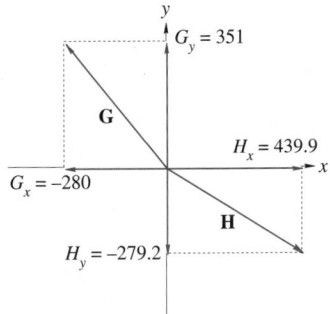

FIGURE 10.19b

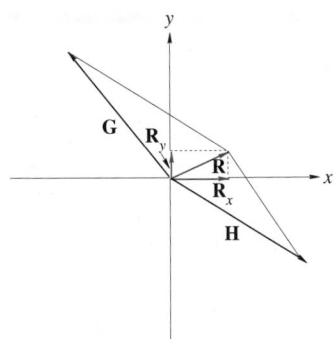

FIGURE 10.19c

The table method for this example would appear as follows.

Vector	Horizontal component	Vertical component
G	$G_x = 449\cos 128.6° = -280.1$	$G_y = 449\sin 128.6° = 350.9$
H	$H_x = 521\cos 327.6° = 439.9$	$H_y = 521\sin 327.6° = -279.2$
R	R_x 159.8	R_y 71.7

These values for R_x and R_y can be used as before to find $R = 175.1$ and $\theta_R = 24.2°$.

EXAMPLE 10.11

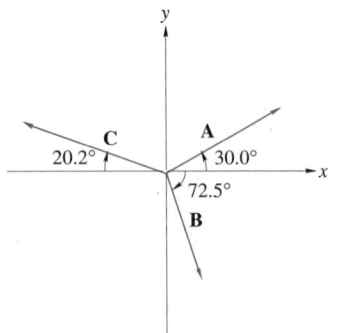

FIGURE 10.20

Find the resultant of the three vectors shown in Figure 10.20, if $A = 137$, $B = 89.4$, and $C = 164.6$.

Solution The table that follows lists the horizontal and vertical components of each of the vectors and the resultant vector **R**.

Vector	Horizontal component	Vertical component
A	$A_x = 137\cos 30° = 118.6$	$A_y = 137\sin 30° = 68.5$
B	$B_x = 89.4\cos 287.5° = 26.9$	$B_y = 89.4\sin 287.5° = -85.3$
C	$C_x = 164.6\cos 159.8° = -154.5$	$C_y = 164.6\sin 159.8° = 56.8$
R	R_x -9.0	R_y 40.0

Once again, the values for R_x and R_y can be used to find

$$R = \sqrt{R_x^2 + R_y^2}$$
$$= \sqrt{(-9)^2 + 40^2}$$
$$= 41$$

and

$$\theta_R = \tan^{-1}\frac{40}{-9}$$
$$\theta_R = 102.7°$$

Application

EXAMPLE 10.12

A sign has been lifted into position by two cranes, as shown in Figure 10.21a. If the sign weighs 420 lb, what is the tension in each of the three cables?

Solution We will draw the coordinate axes and label the angles as shown in Figure 10.21b. The origin is placed at the ring where the three cables meet, because the tension on all three cables acts on this ring.

As usual, we make a table listing the horizontal and vertical components of each vector.

Vector	Horizontal component	Vertical component
A	$A_x = A\cos 145°$	$A_y = A\sin 145°$
B	$B_x = B\cos 40°$	$B_y = B\sin 40°$
C	$C_x = C\cos 270° = 0$	$C_y = C\sin 270° = -C = -420$
R	R_x	R_y

EXAMPLE 10.12 (Cont.)

FIGURE 10.21a

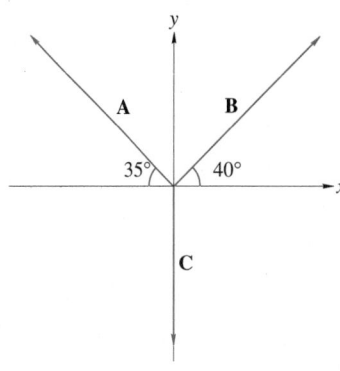

FIGURE 10.21b

The ring is at rest as a result of these three forces, so $R_x = 0$ and $R_y = 0$. Thus, we have

$$R_x = A\cos 145° + B\cos 40° = 0$$
$$R_y = A\sin 145° + B\sin 40° - 420 = 0$$

Evaluating the trigonometric functions in these two equations, we are led to the following system of two equations in two variables:

$$\begin{cases} -0.81915A + 0.76604B &= 0 \\ 0.57358A + 0.64279B &= 420 \end{cases}$$

Solving this system of two equations in two variables (by Cramer's rule), we get

$$A \approx 333.0862\,\text{lb}$$

and $\quad B \approx 356.1792\,\text{lb}$

The tension in the cable on the left is about 333 lb and the tension in the cable on the right side is about 356 lb.

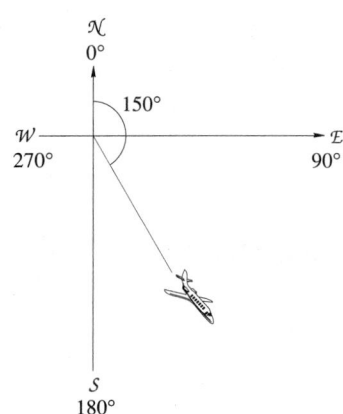

FIGURE 10.22

≡ **Note**   In flight terminology, the **heading** of an aircraft is the direction in which the aircraft is pointed. Usually the wind is pushing the aircraft so that it is actually moving in a different direction, called the **track** or **course**. The angle between the heading and the course is the **drift angle**. The **air speed** is the speed of the plane relative to the air. The **ground speed** is the speed of the aircraft relative to the ground.

≡ **Note**   In navigation, directions are usually given in terms of the size of the angle measured clockwise from true north. For example, the airplane in Figure 10.22 has a heading of 150°.

Application

EXAMPLE 10.13

An airplane is flying at 340.0 mph with a heading of 210°. If a 50 mph wind is blowing from 165°, find the ground speed, drift angle, and course of the airplane.

Solution In Figure 10.23a, $\overrightarrow{OA}$ represents the airspeed of 340 mph with a heading of 210°, and $\overrightarrow{OW}$ represents a wind of 50 mph from 165°. In Figure 10.23b, we have completed the parallelogram to obtain the vector $\overrightarrow{OR} = \overrightarrow{OA} + \overrightarrow{OW}$. Notice that, while the wind is from 165°, it has a heading of $180° + 165° = 345°$.

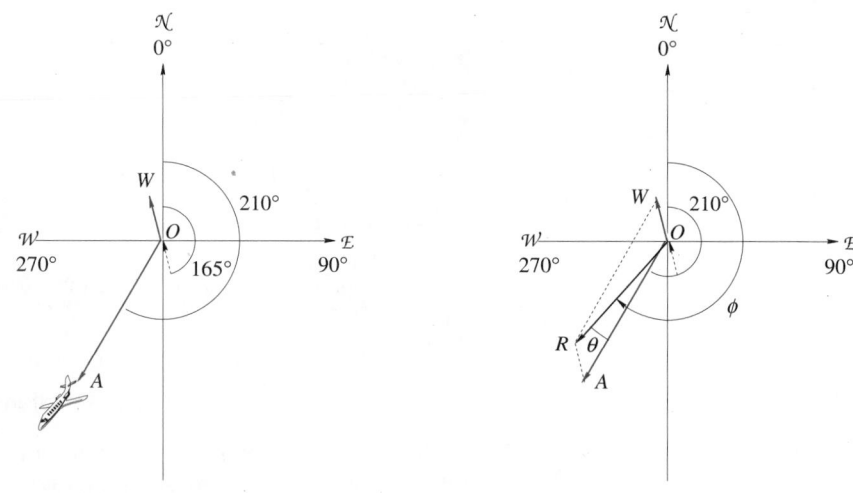

FIGURE 10.23a   **FIGURE 10.23b**

The length $\overrightarrow{OR}$ represents the ground speed, θ is the drift angle, and ϕ is the track or course. The following table lists the horizontal and vertical components of each vector $\overrightarrow{OA}$ and $\overrightarrow{OW}$ and the resultant vector $\overrightarrow{OR}$.

Vector	Horizontal component	Vertical component
$\overrightarrow{OA} = \mathbf{A}$	$\mathbf{A}_x = 340\cos 210° \approx -294.45$	$\mathbf{A}_y = 340\sin 210° = -170.00$
$\overrightarrow{OW} = \mathbf{W}$	$\mathbf{W}_x = 50\cos 345° \approx \underline{48.30}$	$\mathbf{W}_y = 50\sin 345° \approx \underline{-12.94}$
$\overrightarrow{OR} = \mathbf{R}$	$\mathbf{R}_x -246.15$	$\mathbf{R}_y -182.94$

Using these values for R_x and R_y, we see that the length of the resultant vector is $\overrightarrow{OR} = \sqrt{(-246.15)^2 + (-182.94)^2} \approx 306.69$ and $\phi \approx \tan^{-1}\left(\frac{-182.94}{-246.15}\right) = 36.62°$. Since ϕ is in the third quadrant, $\phi = 180° + 36.62° = 216.62°$.

So, the plane has a ground speed of 306.69 mph, a drift angle of 6.62°, and a course of 216.62°.

Exercise Set 10.2

In Exercises 1–4, vectors *A* and *B* are both on the same axis. Find the magnitude and direction of the resultant vector.

1. $A = 20.0, \theta_A = 0°, B = 32.5, \theta_B = 180°$

2. $A = 14.3, \theta_A = 90°, B = 7.2, \theta_B = 90°$

3. $A = 121.7, \theta_A = 270°, B = 86.9, \theta_B = 90°$

4. $A = 63.1, \theta_A = 180°, B = 43.5, \theta_B = 180°$

In Exercises 5–8, vectors *C* and *D* are perpendicular. Find the magnitude and direction of the resultant vectors. It may help if you draw vectors *C*, *D*, and the resultant vector.

5. $C = 55, \theta_C = 90°, D = 48, \theta_D = 180°$

6. $C = 65, \theta_C = 270°, D = 72, \theta_C = 180°$

7. $C = 81.4, \theta_C = 0°, D = 37.6, \theta_D = 90°$

8. $C = 63.4, \theta_C = 270°, D = 9.4, \theta_D = 0°$

In Exercises 9–16, find the magnitude and direction of the vector with the given components.

9. $A_x = 33, A_y = 56$

10. $B_x = 231, B_y = 520$

11. $C_x = 11.7, C_y = 4.4$

12. $D_x = 31.9, D_y = 36.0$

13. $E_x = 6.3, E_y = 1.6$

14. $F_x = 5.1, F_y = 14.0$

15. $G_x = 8.4, G_y = 12.6$

16. $H_x = 15.3, H_y = 9.2$

In Exercises 17–30, add the given vectors by using the trigonometric functions and the Pythagorean theorem.

17. $A = 4, \theta_A = 60°, B = 9, \theta_B = 20°$

18. $C = 12, \theta_C = 75°, D = 15, \theta_D = 37°$

19. $C = 28, \theta_C = 120°, D = 45, \theta_D = 210°$

20. $E = 72, \theta_E = 287°, F = 65, \theta_F = 17°$

21. $A = 31.2, \theta_A = 197.5°, B = 62.1, \theta_B = 236.7°$

22. $C = 53.1, \theta_C = 324.3°, D = 68.9, \theta_D = 198.6°$

23. $E = 12.52, \theta_E = 46.4°, F = 18.93, \theta_F = 315°$

24. $G = 76.2, \theta_G = 15.7°, H = 89.4, \theta_H = 106.3°$

25. $A = 9.84, \theta_A = 215°30', B = 12.62, \theta_B = 105°15'$

26. $C = 79.63, \theta_C = 262°45', D = 43.72, \theta_D = 196°12'$

27. $E = 42.0, \theta_E = 3.4\,\text{rad}, F = 63.2, \theta_F = 5.3\,\text{rad}$

28. $G = 37.5, \theta_G = 0.25\,\text{rad}, H = 49.3, \theta_H = 1.92\,\text{rad}$

29.

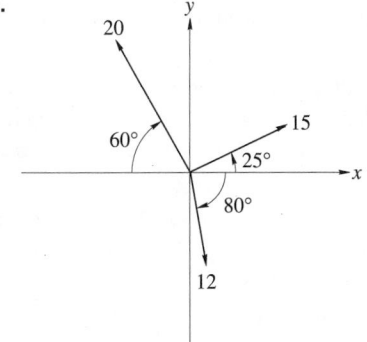

30.

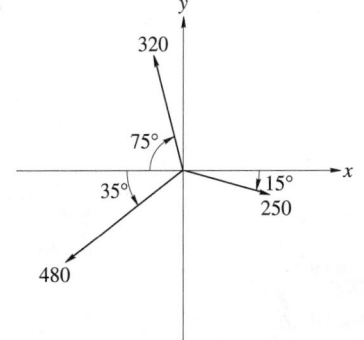

Solve Exercises 31–40.

31. *Construction technology*   A sign is held in position by three cables, as shown in Figure 10.24. If the sign has a weight of 215 N, what is the tension in each of the three cables?

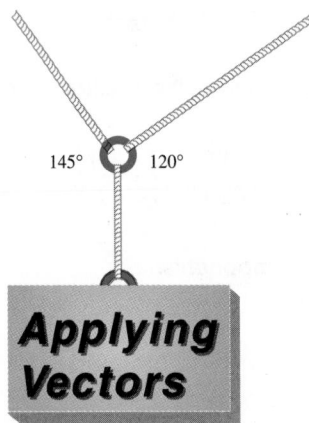

FIGURE 10.24

32. *Construction technology*   A 175 lb sign is supported from a wall by a cable inclined 53° with the horizontal, and a brace perpendicular to the wall, as shown in Figure 10.25. Find the magnitudes of the forces in the cable and the brace that will keep the sign in equilibrium.

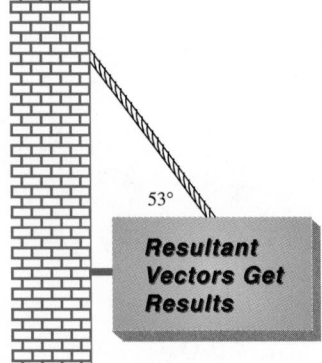

FIGURE 10.25

33. *Construction technology*   A 235 lb sign is supported from a wall by a cable inclined 37° with the horizontal, and a brace perpendicular to the wall, similar to that shown in Figure 10.25. Find the tension in the cable and the compression in the boom.

34. *Construction technology*   An 85 kg sign is held in position by three ropes, as shown in Figure 10.26. What is the tension in each of the three ropes?

FIGURE 10.26

35. *Navigation*   An airplane is flying at 480 mph with a heading of 63°. A 45.0 mph wind is blowing from 325°. Find the ground speed, course, and drift angle of the airplane.

36. *Navigation*   An airplane is flying at 320.0 mph with a heading of 172°. A 72.0 mph wind is blowing from 137°. Find the ground speed, course, and drift angle of the airplane.

37. *Electricity*   A circuit has two ac voltages that are 68° out of phase. Each voltage is 196 V. What is the total voltage?

38. *Electricity*   A 2-phase generator produces two voltages that are 120° out of phase. The first voltage is 86 V. The second voltage is 110 V. What is the total voltage?

39. Write a computer program to determine the component vectors for a position vector. Remember, compute the sines and cosines of the given angles in radians.

 40. Write a computer program to determine the resultant vector when two or more vectors are added.

 In Your Words

41. Without looking in the text, describe how you can use a vector's components to add vectors.

42. Flight terminology uses several technical terms. Write a brief definition of each of the following terms. Draw a figure to illustrate your explanations.

(a) Heading
(b) Drift angle
(c) Course
(d) Ground speed
(e) Wind speed

≡ 10.3
APPLICATIONS OF VECTORS

Vectors are used in science and technology, as well as in mathematics. In this section, we will look at some of those applications.

Application

EXAMPLE 10.14

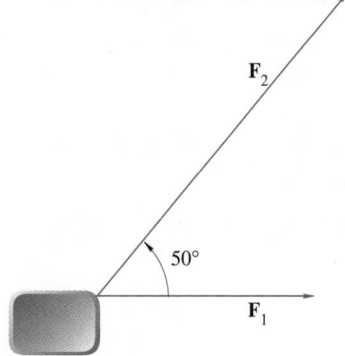

FIGURE 10.27

Two forces, F_1 and F_2, act on an object. If F_1 is 40 lb, F_2 is 75 lb, and the angle θ between them is $50°$, find the magnitude and direction of the resultant force.

Solution   We sketch the two forces as vectors and place the object at the origin with $\mathbf{F_1}$ along the positive x-axis as in Figure 10.27. We will use a table similar to the one in the last example to find the components of $\mathbf{F_1}, \mathbf{F_2}$, and the resultant vector.

Vector	Horizontal component		Vertical component	
$\mathbf{F_1}$		40.0		0.0
$\mathbf{F_2}$	$75\cos 50° =$	48.2	$75\sin 50° =$	57.5
$\mathbf{R}$	R_x	88.2	$\mathbf{R_y}$	57.5

So, $R_x = 88.2$ and $R_y = 57.5$. The magnitude of $\mathbf{R}$ is $R = \sqrt{88.2^2 + 57.5^2} = 105.3$ lb and $\theta_R = \tan^{-1} \frac{57.5}{88.2} = 33.1°$.

Application

EXAMPLE 10.15

A truck weighing 22,500 lb is on a $25°$ hill. Find the components of the truck's weight parallel and perpendicular to the road.

Solution   The weight of an object (truck, car, building, etc.) is the gravitational force with which earth attracts it. This force always acts vertically downward and

EXAMPLE 10.15 (Cont.)

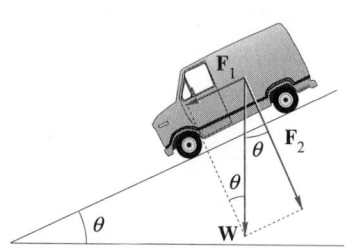

FIGURE 10.28

is indicated by the vector **W** in Figure 10.28. The components of **W** have been labeled F_1 and F_2. Because **W** is vertical and F_2 is perpendicular to the road, the angle θ between **W** and F_2 is equal to the angle the road makes with the horizon. Using the trigonometric functions, we have

$$F_1 = W \sin\theta = 22{,}500 \sin 25° = 9{,}509 \, \text{lb}$$

$$F_2 = W \cos\theta = 22{,}500 \cos 25° = 20{,}392 \, \text{lb}$$

Thus, the components of the truck's weight are 9,509 lb parallel to the road and 20,392 lb perpendicular to the road.

Application

EXAMPLE 10.16

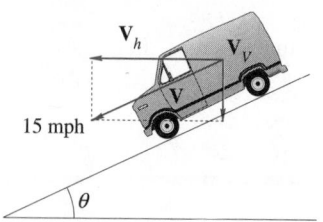

FIGURE 10.29

If the truck in Example 10.15 rolls down the hill at 15 mph, find the magnitudes of the horizontal and vertical components of the truck's velocity.

Solution   The velocity vector **V** is shown in Figure 10.29. The horizontal and vertical components of **V** are marked V_h and V_v.

$$V_v = V \sin 25° = 15 \sin 25° = 6.3 \, \text{mph}$$

$$V_h = V \cos 25° = 15 \cos 25° = 13.6 \, \text{mph}$$

Application

EXAMPLE 10.17

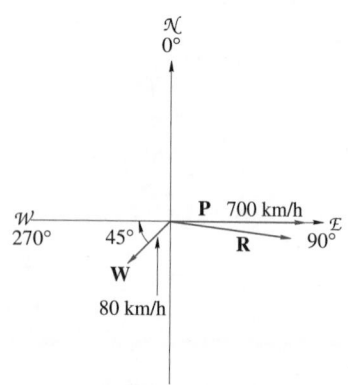

FIGURE 10.30

A jet plane is traveling due east at an airspeed of 700 km/h. If a wind of 80 km/h is blowing due southwest, find the magnitude and direction of the plane's resultant velocity. (See Figure 10.30.)

Solution   A table of the components for the wind's vector **W** and the plane's vector **P** allows us to quickly find the resultant vector **R**.

Vector	Horizontal component		Vertical component	
W	$80 \cos 225° =$	-56.6	$80 \sin 225° =$	-56.6
P	$700 \cos 0° =$	700.0		0.0
R		643.4		-56.6

The magnitude of **R** is $R = \sqrt{643.4^2 + (-56.6)^2} = 645.9$ km/h. This is the ground speed of the plane. The plane's direction is $\theta_R = \tan^{-1}\left(\frac{-56.6}{643.4}\right) = -5.03°$ or $5.03°$ south of east.

Application

EXAMPLE 10.18

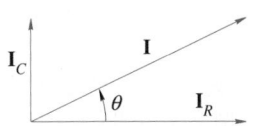

FIGURE 10.31

In a parallel RC (resistance-capacitance) circuit, the current $\mathbf{I}_C$ through the capacitance leads the current $\mathbf{I}_R$ through the resistance by 90°, as shown in Figure 10.31. If $I_C = 0.5$ A and $I_R = 1.2$ A, find the total current $\mathbf{I}$ in the circuit and the phase angle θ of the circuit.

Solution $I = \sqrt{I_R{}^2 + I_C{}^2} = \sqrt{(1.2)^2 + (0.5)^2} = 1.3$ A

$$\theta = \tan^{-1}\left(\frac{I_C}{I_R}\right)$$

$$\theta = \tan^{-1}\left(\frac{0.5}{1.2}\right) = 22.6°$$

Application

EXAMPLE 10.19

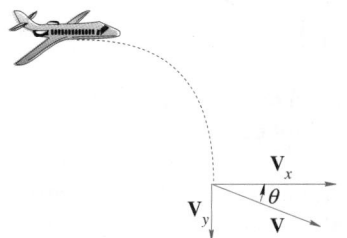

FIGURE 10.32

An airplane in level flight drops an object. The plane was traveling at 180 m/s at a height of 7500 m. The vertical component of the dropped object is given by $\mathbf{V}_y = -9.8t$ m/s. What is the magnitude of the velocity of the object after 8 s? At what angle with the ground is the object moving at this time?

Solution   The horizontal velocity of the object will have no effect on the vertical motion. Hence, V_x will remain at 180 m/s. We are told that $\mathbf{V}_y = -9.8t$. So, when $t = 8$, $\mathbf{V}_y = -78.4$. The magnitude of the velocity of the object when $t = 8$ is

$$V = \sqrt{(-78.4)^2 + 180^2} = 196.3 \text{ m/s}$$

If θ is the angle that the object makes with the ground, then

$$\theta = \tan^{-1}\frac{V_y}{V_x} = \tan^{-1}\left(\frac{-78.4}{180}\right) \approx -23.5°$$

or 23.5° below the horizontal. (See Figure 10.32.)

Exercise Set 10.3

Solve Exercises 1–24.

1. *Physics*   Two forces act on an object. One force is 70 lb and the other is 50 lb. If the angle between the two forces is 35°, find the magnitude and direction of the resultant force.

2. *Physics*   A person pulling a cart exerts a force of 35 lb on the cart at an angle of 25° above the horizontal. Find the horizontal and vertical components of this force.

3. *Physics*   A person pushes a lawn mower with a force of 25 lb. The handle of the lawn mower is 55° above the horizontal. **(a)** How much downward force is being exerted on the ground? **(b)** How much horizontal forward force is being exerted? **(c)** How do these forces change if the handle is lowered to 40° with the horizontal?

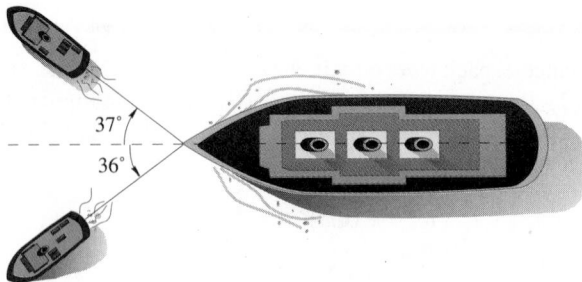

FIGURE 10.33

4. *Navigation* Two tugboats are pulling a ship. As shown in Figure 10.33, the first tug exerts a force of 1 500 N on a cable, making an angle of 37° with the axis of the ship. The second tug pulls on a cable, making an angle of 36° with the axis of the ship. The resultant force vector is in line with the axis of the ship. What is the force being exerted by the second tug?

5. *Navigation* A tugboat is turning a barge by pushing with a force of 1 700 N at an angle of 18° with the axis of the barge. What are the forward and sideward forces in newtons exerted by the tug on the barge? (See Figure 10.34.)

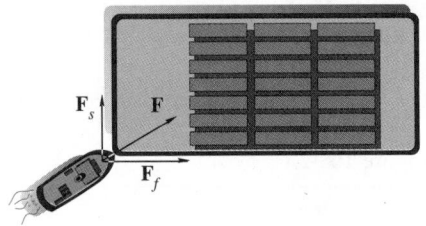

FIGURE 10.34

6. *Navigation* A ship heads due northwest at 15 km/h in a river that flows east at 7 km/h. What is the magnitude and direction of the ship's velocity relative to Earth's surface?

7. *Navigation* A plane is heading due south at 370 mph with a wind from the west at 40 mph. What are the ground speed and the true direction of the plane?

8. *Navigation* On a compass, due north is 0°, east is 90°, south 180°, and so on. A plane has a compass heading of 115° and is traveling at 420 mph. The wind is at 62 mph and blowing at 32°. What are the ground speed and true direction of the plane?

9. *Physics* A car with a mass of 1 200 kg is on a hill that makes an angle of 22° with the horizon. Which components of the car's mass are parallel and perpendicular to the road?

10. *Physics* Find the force necessary to push a 30-lb ball up a ramp that is inclined 15° with the horizon. (You want to find the component parallel to the ramp.)

11. *Construction* A guy wire runs from the top of a utility pole 35 ft high to a point on the ground 27 ft from the base of the pole. The tension in the wire is 195 lb. What are the horizontal and vertical components?

12. *Electricity* In a parallel RC circuit, the current I_C through the capacitance leads the current through the resistance I_R by 90°. If $I_C = 2.4$ A and $I_R = 1.6$ A, find the magnitude of the total current in the circuit and the phase angle.

13. *Electricity* If the total current I in a parallel RC circuit is 9.6 A and I_R is 7.5 A, what are I_C and the phase angle?

14. *Electricity* In a parallel RL (resistance-inductance) circuit, the current I_L through the inductance lags the current I_R through the resistance by 90°, as shown in Figure 10.35. Find the magnitude of the total current I and the negative phase angle θ, when $I_L = 6.2$ A and $I_R = 8.4$ A.

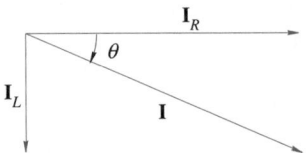

FIGURE 10.35

15. *Electricity* If the total current I in a parallel RL circuit is 12.4 A and I_R is 6.3 A, find I_L and the negative phase angle.

16. *Electricity* The total impedance Z of a series ac circuit is the resultant of the resistance R, the inductive reactance X_L, and the capacitive reactance X_C, as shown in Figure 10.36. Find Z and the phase angle, when $X_C = 50\,\Omega$, $X_L = 90\,\Omega$, and $R = 12\,\Omega$.

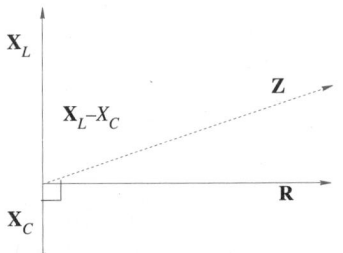

FIGURE 10.36

17. *Electricity* Find the total impedance Z and the phase angle of a series ac circuit, when $X_C = 60\,\Omega$, $X_L = 40\,\Omega$, and $R = 24\,\Omega$.

18. *Electricity* Find the total impedance and phase angle, when $X_L = 38\,\Omega$, $X_C = 265\,\Omega$, and $R = 75\,\Omega$.

19. *Electricity* In a synchronous ac motor, the current leads the applied voltage, and in an induction ac electric motor, the current lags the applied voltage. A circuit with a synchronous motor A connected in parallel with an induction motor B and a purely resistive load C has a current diagram as shown in Figure 10.37. If $I_A = 20$ A, $\theta_A = 35°$, $I_B = 15$ A, $\theta_B = -20°$, and $I_C = 25$ A, find the total current I and the phase angle between the total current and the applied voltage.

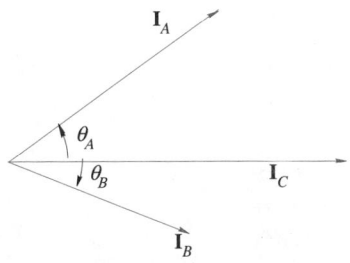

FIGURE 10.37

20. *Navigation* A ship is sailing at a speed of 12 km/h in the direction of 10°. A strong wind is exerting enough pressure on the ship's superstructure to move it in the direction of 270° at 2 km/h. A tidal current is flowing in the direction of 140° at the rate of 6 km/h. What is the ship's velocity and direction relative to Earth's surface?

21. *Navigation* An airplane in level flight drops an object. The plane was traveling at 120 m/s at a height of 5 000 m. The vertical component of the dropped object is $V_y = -9.8t$ m/s. What is the magnitude of the velocity of the object after 4 s and at what angle with the ground is the object moving at this time?

22. *Navigation* At any time t the object in Exercise 21 will have fallen $4.9t^2$ m. How long will it take for it to strike the ground? What is its velocity at this time? At what angle will it strike the ground?

23. *Construction* A parallelogram with adjacent sides of lengths $1'9''$ and $2'3''$ is to be cut from a rectangular piece of plywood. The parallelogram contains a 40° angle. What are the dimensions of the smallest piece of plywood from which this parallelogram can be cut?

24. *Construction* A parallelogram with adjacent sides of lengths 15 cm and 32 cm is to be cut from a rectangular piece of plywood. The parallelogram contains a 35° angle. What are the dimensions of the smallest piece of plywood from which this parallelogram can be cut?

In Your Words

25. Write a word problem in your technology area of interest that requires you to use vectors. On the back of the sheet of paper, write your name and explain how to solve the problem by using vectors. Give the problem you wrote to a friend and let him or her try to solve it. If your friend has difficulty understanding the problem or solving the problem, or if he or she disagrees with your solution, make any necessary changes in the problem or solution. When you have finished, give the revised problem and solution to another friend and see if he or she can solve it.

26. Write a word problem in your technology area of interest that requires you to use at least three vectors and their components in its solution. Follow the same procedures described in Exercise 25 for writing, sharing, and revising your problem.

10.4
OBLIQUE TRIANGLES: LAW OF SINES

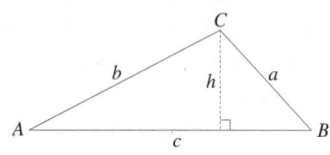

FIGURE 10.38

The triangles we have examined until now have been right triangles. At first, trigonometry and the trigonometric functions dealt only with right triangles. Mathematicians quickly discovered that they needed to work with triangles that did not have a right angle. These triangles, the ones with no right angle, were named **oblique triangles**.

The trigonometric methods for solving right triangles do not work with oblique triangles. There are two methods that are usually used with oblique triangles. One of these, the Law of Sines, will be studied in this section. The other, the Law of Cosines, will be studied in Section 10.5.

The Law of Sines can be developed without too much difficulty. Consider the triangle in Figure 10.38. Select one of the vertices, in this case vertex C. Drop a perpendicular to the opposite side. From the resulting right triangles we have

$$\sin A = \frac{h}{b} \text{ and } \sin B = \frac{h}{a}$$

or $$h = b \sin A \text{ and } h = a \sin B$$

Since both of these are equal to h, then $b \sin A = a \sin B$ and $\frac{a}{\sin A} = \frac{b}{\sin B}$. If we dropped the perpendicular from vertex B we would get $\frac{a}{\sin A} = \frac{c}{\sin C}$.

Putting these together, we get the Law of Sines

The Law of Sines

The **Law of Sines** or **Sine Law** is a continued proportion and states that if $\triangle ABC$ is a triangle with sides of lengths a, b, and c and opposite angles A, B, and C, then

$$\frac{a}{\sin A} = \frac{b}{\sin B} = \frac{c}{\sin C}.$$

You may recognize that this can also be written as the continued proportion $a : b : c = \sin A : \sin B : \sin C$.

The Law of Sines can be used to solve a triangle when the parts of a triangle are known in either of the following two cases.

Case 1 (SSA) The measure of two sides and the angle opposite one of them is known.

Case 2 (AAS) The measure of two angles and one side is known.

As is true with any continued proportion, you work with just two of the ratios at any one time. You should also remember that the angles of a triangle add up to $180°$ or π rad.

Let's look at an example that uses the Law of Sines. This example will fit the description of a Case 1 triangle or SSA.

EXAMPLE 10.20

In a triangle, $A = 33°$, $a = 9.4$, and $c = 14.3$. Solve the triangle.

Solution We are to solve this triangle. That means that we are to find the length of the third side and the sizes of the other two angles. The following data chart shows the parts that we know and those we are to determine.

sides	angles
$a = 9.4$	$A = 33°$
$b = \_\_$	$B = \_\_$
$c = 14.3$	$C = \_\_$

Since two of the known parts have the same letter, a, one of the ratios we should use is $\dfrac{a}{\sin A}$. The other known part has the letter c so the other ratio should be $\dfrac{c}{\sin C}$. By the Law of Sines

$$\frac{a}{\sin A} = \frac{c}{\sin C}$$

$$\frac{9.4}{\sin 33°} = \frac{14.3}{\sin C}$$

$$\sin C = \frac{(14.3)(\sin 33°)}{9.4}$$

$$\approx 0.8285466$$

$$C \approx 55.95°$$

Since $C \approx 55.95°$ and $A = 33°$, then

$$B \approx 180 - 33° - 55.95°$$
$$= 91.05°$$

We can use this information to find the length of the third side of the triangle, b.

$$\frac{a}{\sin A} = \frac{b}{\sin B}$$

$$\frac{9.4}{\sin 33°} = \frac{b}{\sin 91.05°}$$

$$b = \frac{(9.4)(\sin 91.05°)}{\sin 33°}$$

$$= 17.26$$

EXAMPLE 10.20 (Cont.)

We can now complete the data chart and show all the parts of this triangle:

sides	angles
$a = 9.4$	$A = 33°$
$b = 17.26$	$B = 91.05°$
$c = 14.3$	$C = 55.95°$

The last part of Example 10.20 was somewhat like the situation in Case 2. You knew two of the angles. Once you found the size of the third angle, you were ready to continue solving the problems.

EXAMPLE 10.21

Solve the triangle ABC, if $A = 82.17°$, $B = 64.43°$, and $c = 9.12$.

Solution The beginning data chart is

sides	angles
$a = \_\_$	$A = 82.17°$
$b = \_\_$	$B = 64.43°$
$c = 9.12$	$C = \_\_$

This is a triangle in that it satisfies Case 2. (It is an AAS triangle.) Since we know that $A + B + C = 180°$ and are given that $A + B = 82.17° + 64.43° = 146.60°$, we know that $C = 180° - 146.60° = 33.40°$.

We can now use the Sine Law to find either a or b. We will first find a.

$$\frac{a}{\sin A} = \frac{c}{\sin C}$$

$$\frac{a}{\sin 82.17°} = \frac{9.12}{\sin 33.40°}$$

$$a = \frac{(9.12)(\sin 82.17°)}{\sin 33.40°}$$

$$\approx 16.41$$

Now we use the Sine Law to find b.

$$\frac{b}{\sin B} = \frac{c}{\sin C}$$

$$\frac{b}{\sin 64.43°} = \frac{9.12}{\sin 33.40°}$$

$$b = \frac{(9.12)(\sin 64.43°)}{\sin 33.40°}$$

$$\approx 14.94$$

EXAMPLE 10.21 (Cont.)

The completed data chart is

sides	angles
$a = 16.41$	$A = 82.17°$
$b = 14.94$	$B = 64.43°$
$c = 9.12$	$C = 33.40°$

■

It is possible to find 0, 1, or 2 correct solutions to the triangle if the given information includes two sides and one angle. When this ambiguous case occurs, carefully consider the practical application of the solution. We will now examine two situations that produce ambiguous results.

EXAMPLE 10.22

Solve $\triangle ABC$, if $a = 20$, $b = 24$, and $A = 55.4°$.

Solution   As usual, we begin with the data chart showing the given and unknown measurements.

sides	angles
$a = 20$	$A = 55.4°$
$b = 24$	$B = \underline{\ \ }$
$c = \underline{\ \ }$	$C = \underline{\ \ }$

We will now find B.

$$\frac{a}{\sin A} = \frac{b}{\sin B}$$

$$\frac{20}{\sin 55.4°} = \frac{24}{\sin B}$$

$$\sin B = \frac{24(\sin 55.4°)}{20}$$

$$\approx 0.9877636$$

$$B = \sin^{-1}(0.9877636)$$

So, either $B = 81.03°$ or $B = 98.97°$. Remember, the sine is positive in both the first and second quadrants. Whenever you use the Sine Law and get an equation of the form $B = \sin^{-1} n$ or $\sin B = n$, where $0 < n < 1$, then there are two possible values for B.

Will both of these answers satisfy the given parts of the triangle? Let's call the two answers for angle B, B_1, and B_2. If $B_1 = 81.03°$, and since $A = 55.4°$, then $C_1 = 180° - 81.03° - 55.4° = 43.57°$. If $B_2 = 98.97°$, then $C_2 = 180° - 98.97° - 55.4° = 25.63°$.

EXAMPLE 10.22 (Cont.)

We will now use these angles to find the length of side c. If $C_1 = 43.57°$, then

$$\frac{a}{\sin A} = \frac{c_1}{\sin C_1}$$

$$\frac{20}{\sin 55.4°} = \frac{c_1}{\sin 43.57°}$$

$$c_1 = \frac{20(\sin 43.57°)}{\sin 55.4°}$$

$$= 16.75$$

If we use $C_2 = 25.63°$, then we get

$$\frac{a}{\sin A} = \frac{c_2}{\sin C_2}$$

$$\frac{20}{\sin 55.4°} = \frac{c_2}{\sin 25.63°}$$

$$c_2 = \frac{20(\sin 25.63°)}{\sin 55.4°}$$

$$= 10.51$$

The data chart is now complete and written as two data charts.

sides	angles	sides	angles
$a = 20$	$A = 55.4°$	$a = 20$	$A = 55.4°$
$b = 24$	$B_1 = 81.03°$	$b = 24$	$B_2 = 98.97°$
$c_1 = 16.75$	$C_1 = 43.57°$	$c_2 = 10.51$	$C_2 = 25.63°$

The triangles formed with these two solutions are shown in Figure 10.39. Both are correct.

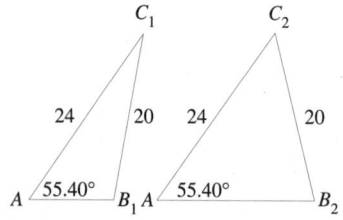

FIGURE 10.39

The other problem that can develop with the ambiguous case is that there may be no solution. Consider the situation in Example 10.23.

EXAMPLE 10.23

Solve for triangle ABC, when $a = 20, b = 27$, and $A = 70°$.

Solution The beginning data chart is:

sides	angles
$a = 20$	$A = 70°$
$b = 27$	$B = \underline{\ \ }$
$c = \underline{\ \ }$	$C = \underline{\ \ }$

EXAMPLE 10.23 (Cont.)

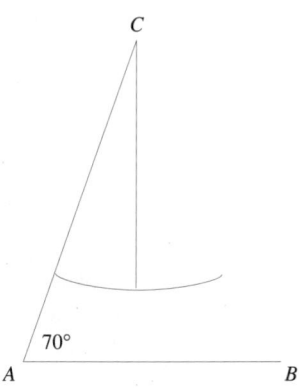

FIGURE 10.40

By the Sine Law $\dfrac{a}{\sin A} = \dfrac{b}{\sin B}$, so

$$\frac{20}{\sin 70°} = \frac{27}{\sin B}$$

$$\sin B = \frac{27(\sin 70°)}{20}$$

$$= 1.27$$

The sine of an angle is never larger then 1. This triangle is not possible. Perhaps if you look at Figure 10.40 you can get a better idea why this is not a legitimate triangle.

As with all trigonometry, there are many applications of the Law of Sines in technical areas. The problem of the technician is to recognize when the application requires trigonometry and then to select the appropriate method to solve it. The next example and the problems in the exercise set will help you make the correct selection.

Application

EXAMPLE 10.24

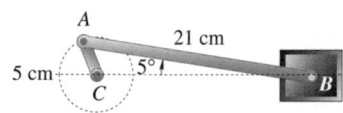

FIGURE 10.41

The crankshaft $\overline{CA}$ of an engine is 5 cm long and the connecting rod $\overline{AB}$ is 21 cm long. Find the size of $\angle ACB$ when the size of $\angle ABC$ is 5°.

Solution A sketch of this crankshaft is in Figure 10.41. This problem falls into the SSA case. We know the lengths of two sides, AC and AB, and of the angle opposite one of them, $\angle B$. Since $AC = b$ and $AB = c$, we can use the Sine Law with

$$\frac{b}{\sin B} = \frac{c}{\sin C}$$

$$\frac{5}{\sin 5} = \frac{21}{\sin C}$$

$$\sin C = \frac{21 \sin 5°}{5} = 0.3660541$$

$$C = 21.47° \text{ or } 158.53°$$

There are two possible solutions. By looking at Figure 10.41, you can see that both are acceptable.

Exercise Set 10.4

In Exercises 1–20, solve each triangle with the given parts. Check for the ambiguous cases.

1. $A = 19.4°, B = 85.3°, c = 22.1$

2. $a = 12.4, B = 62.4°, C = 43.9°$

3. $a = 14.2, b = 15.3, B = 97°$

4. $A = 27.42°, a = 27.3, b = 35.49$

5. $A = 86.32°, a = 19.19, c = 18.42$
6. $B = 75.46°, b = 19.4, C = 44.95°$
7. $B = 39.4°, b = 19.4, c = 35.2$
8. $A = 84.3°, b = 9.7, C = 12.7°$
9. $A = 45°, a = 16.3, b = 19.4$
10. $a = 10.4, c = 5.2, C = 30°$
11. $a = 42.3, B = 14.3°, C = 16.9°$
12. $A = 105.4°, B = 68.2°, c = 4.91$

13. $b = 19.4, c = 12.5, C = 35.6°$
14. $a = 121.4, A = 19.7°, c = 63.4$
15. $a = 19.7, b = 8.5, B = 78.4°$
16. $b = 9.12, B = 1.3 \, \text{rad}, C = 0.67 \, \text{rad}$
17. $b = 8.5, c = 19.7, C = 1.37 \, \text{rad}$
18. $b = 19.7, c = 36.4, C = 0.45 \, \text{rad}$
19. $A = 0.47 \, \text{rad}, b = 195.4, C = 1.32 \, \text{rad}$
20. $a = 29.34, A = 1.23 \, \text{rad}, C = 1.67 \, \text{rad}$

Solve Exercises 21–30.

21. *Civil engineering*   Two high-tension wires are to be strung across a river. There are two towers, A and B, on one side of the river. These two towers are 360 m apart. A third tower, C, is on the other side of the river. If $\angle ABC$ is 67.4° and $\angle BAC$ is 49.3°, what are the distances between towers A and C and towers B and C?

22. *Civil engineering*   A tunnel is to be dug between points A and B on opposite sides of a hill. A point C is chosen 250 m from A and 275 m from B. If $\angle BAC$ measures 43.62°, find the length of the tunnel.

23. *Navigation*   A plane leaves airport A with a heading of 313°. Several minutes later the plane is spotted from airport B at a heading of 27°. Airport B is due west of airport A and the two airports are 37 mi apart. How far had the airplane flown?

24. *Civil engineering*   From a point on the top of one end of a football stadium, the angle of depression to the 40-yard marker is 10.45°. The angle of depression to the 50-yard marker is 11.36°. How high is that end of the stadium above the playing field?

25. *Civil engineering*   Two technicians release a balloon containing a radio-controlled camera. In order for the camera's photographs to cover enough territory, they plan to start the camera when the balloon reaches a certain height. The technicians are 400 m apart. One study triggers the balloon when the angle of elevation at that spot is 47°. At that instant, the angle of elevation for the other technician is 67°. If the balloon is directly above the line connecting the two technicians, what is its height?

26. *Automotive technology*   The angles between the three holes in a seat bracket are shown in Figure 10.42. If $AB = 15.6$ cm, determine AC and BC.

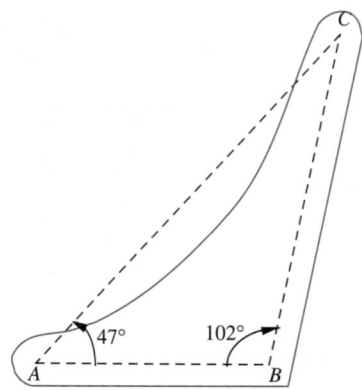

FIGURE 10.42

27. *Civil engineering*   A 225.0-ft antenna mast stands on the edge of the roof of a building. A 6′0″-tall observer on the ground at some point away from the building determines that the angles of elevation to the top and bottom of the mast are 67.4° and 57.8°, respectively. How high is the building? (The observer's eyes are 5′7″ above the ground.)

28. *Machine technology*   Holes are to be drilled in a metal plate at five equally spaced locations around a circle with a radius of 6.25 in. Find the distance between two adjacent holes.

29. *Machine technology*   Holes are to be drilled in a metal plate at 12 equally spaced locations around a circle with a radius of 16.40 cm. Find the distance between two adjacent holes.

30. *Civil engineering*   A guy wire to the top of a pole makes a 63.75° angle with level ground. At a point 8.2 m farther from the pole than the guy wire, the angle of elevation of the top of the pole is 42.5°. How long is the guy wire?

 In Your Words

31. Without looking in the text, write the Sine Law and describe how to use it.

32. Describe how you can tell when to use the Sine Law.

33. When using the Sine Law you have to be careful on the ambiguous case.

(a) Explain what is meant by the ambiguous case.
(b) How do you know that you might be working a problem that involves the ambiguous case?
(c) What should you do if you are working a problem that may include the ambiguous case?

≡ 10.5
OBLIQUE TRIANGLES: LAW OF COSINES

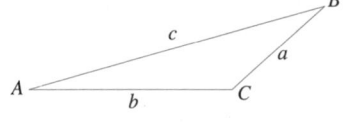

FIGURE 10.43

In Section 10.4, we learned the Law of Sines and when to use it. We were able to use the Sine Law in two cases. In Case 1, we knew the measures for two sides and the angle opposite one of them. This we named the SSA case. Case 2 existed when we knew the measures of two angles and one side of a triangle. This we called the AAS case.

There are two other cases that lead to solvable triangles. We will learn to solve the following two cases in this section.

Case 3 (SAS) The measure of two sides and the included angle are known.

Case 4 (SSS) The measure of three sides are known.

The Law of Cosines is used to help solve both Cases 3 and 4. Using the general oblique triangle in Figure 10.43, the Law of Cosines can be stated as follows.

The Law of Cosines

If $\triangle ABC$ is a triangle with sides of lengths a, b, and c, and opposite angles A, B, C, then by the **Law of Cosines** or **Cosine Law**,

$$a^2 = b^2 + c^2 - 2bc \cos A$$
$$b^2 = a^2 + c^2 - 2ac \cos B$$
or $$c^2 = a^2 + b^2 - 2ab \cos C$$

Notice that there are three versions of the Cosine Law. Each version simply restates the law so that different parts of the triangle are used.

 Hint

In the Cosine Law, the side on the left-hand side of the equation has the same letter as the angle on the right-hand side of the equation.

EXAMPLE 10.25

If $b = 14.7, c = 9.3$, and $A = 46.3°$, solve the triangle.

Solution The data chart is

sides	angles
$a = \underline{}$	$A = 46.3°$
$b = 14.7$	$B = \underline{}$
$c = 9.3$	$C = \underline{}$

This is Case 3 or the SAS type of problem. Since we know the size of angle A, we first use the Law of Cosines to find the length of side a.

$$
\begin{aligned}
a^2 &= b^2 + c^2 - 2bc\cos A \\
&= (14.7)^2 + (9.3)^2 - 2(14.7)(9.3)\cos 46.3° \\
&= 216.09 + 86.49 - 273.42(0.6908824) \\
&= 216.09 + 86.49 - 188.90107 \\
&= 113.67893
\end{aligned}
$$

So, $a = 10.662032$ or $a \approx 10.7$.

At this point, you have a choice as to which method to use to solve the remainder of the problem. You know the size of one angle so you could use the Sine Law. We will use an alternate version of the Cosine Law, which takes advantage of the fact that you know the lengths of all three sides. This alternate version of the Cosine Law is used for the SSS type of problem. It will be the next example, after we state the alternate version of the Cosine Law.

The Law of Cosines (Alternate Version)

If $\triangle ABC$ is a triangle with sides of lengths a, b, and c, and opposite angles A, B, C, then by the **Law of Cosines**,

$$\cos A = \frac{b^2 + c^2 - a^2}{2bc}$$

$$\cos B = \frac{a^2 + c^2 - b^2}{2ac}$$

or $\quad \cos C = \dfrac{a^2 + b^2 - c^2}{2ab}$

You may notice that the alternate version of the Law of Cosines is just the original versions solved for cos A, cos B, and cos C, respectively.

EXAMPLE 10.26

If $a = 10.7$, $b = 14.7$, and $c = 9.3$, find the sizes of the three angles.

Solution We will use the Cosine Law to find the size of angle B.

$$b^2 = a^2 + c^2 - 2ac\cos B$$

$$\cos B = \frac{(10.7)^2 + (9.3)^2 - (14.7)^2}{2(10.7)(9.3)}$$

$$\cos B = \frac{114.49 + 86.49 - 216.09}{199.02} \approx -0.07592201793$$

$$B = 94.354201 \approx 94.4°$$

Since this is a continuation of Example 10.25, we know that $A = 46.3°$, so $C = 180° - 46.3° - 94.4° = 39.3°$. The completed data chart is:

sides	angles
$a = 10.7$	$A = 46.3°$
$b = 14.7$	$B = 94.4°$
$c = 9.3$	$C = 39.3°$

Application

EXAMPLE 10.27

An electric transmission line is planned to go directly over a swamp. The power line will be supported by towers at points A and B in Figure 10.44. A surveyor measures the distance from B to C as 573 m, the distance from A to C as 347 m, and $\angle BCA$ as 106.63°. What is the distance from tower A to tower B?

Solution $BC = a = 573$ m; $AC = b = 347$ m; $\angle BCA = \angle C = 106.63°$. This is an SAS type of problem. We want to find $AB = c$. Using the Cosine Law, we have

$$c^2 = a^2 + b^2 - 2ab\cos C$$
$$= (573)^2 + (347)^2 - 2(573)(347)\cos 106.63$$
$$= 562\,544.93$$
$$c \approx 750$$

The distance between the towers will be about 750 m.

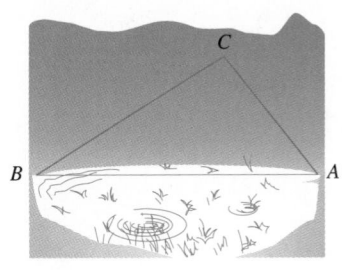

FIGURE 10.44

Exercise Set 10.5

In Exercises 1–20, solve each triangle with the given parts. The angles in Exercises 1–10 are in degrees and those in Exercises 11–20 are in radians.

1. $a = 9.3$, $b = 16.3$, $C = 42.3°$
2. $A = 16.25°$, $b = 29.43$, $c = 36.52$
3. $a = 47.85$, $B = 113.7°$, $c = 32.79$
4. $a = 19.52$, $b = 63.42$, $c = 56.53$
5. $a = 29.43$, $b = 16.37$, $c = 38.62$
6. $A = 121.37°$, $b = 112.37$, $c = 93.42$
7. $a = 63.92$, $B = 92.44°$, $c = 78.41$
8. $a = 19.53$, $b = 7.66$, $C = 32.56°$
9. $a = 4.527$, $b = 6.239$, $c = 8.635$
10. $A = 7.53°$, $b = 37.645$, $c = 42.635$

11. $a = 8.5$, $b = 15.8$, $C = 0.82\,\text{rad}$
12. $A = 0.31\,\text{rad}$, $b = 15.8$, $c = 38.47$
13. $a = 52.65$, $B = 1.98\,\text{rad}$, $c = 35.8$
14. $a = 43.5$, $b = 63.4$, $c = 37.3$
15. $a = 36.27$, $b = 24.55$, $c = 44.26$
16. $A = 2.41\,\text{rad}$, $b = 153.21$, $c = 87.49$
17. $a = 54.8$, $B = 1.625\,\text{rad}$, $c = 38.33$
18. $a = 7.621$, $b = 3.429$, $C = 0.183\,\text{rad}$
19. $a = 2.317$, $b = 1.713$, $c = 1.525$
20. $A = 0.09\,\text{rad}$, $b = 40.75$, $c = 50.25$

Solve Exercises 21–28.

21. *Space technology*   A tracking antenna is aimed 34.7° above the horizon. The distance from the antenna to a spacecraft is 12325 km. If the radius of the earth is 6335 km, how high is the spacecraft above the surface of the earth? (See Figure 10.45.) (Hint: Find the length of BD.)

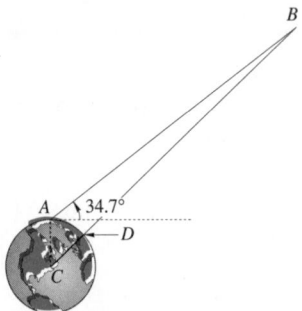

FIGURE 10.45

22. *Navigation*   A ship leaves port at noon and travels due north at 21 km/h. At 3 p.m., the ship changes direction to a heading of 37°. How far from the port is the ship at 7 p.m.? What is the bearing of the ship from the port?

23. *Construction*   A hill makes an angle of 12.37° with the horizontal. A 75-ft antenna is erected on the top of the hill. A guy wire is to be strung from the top of

the antenna to a point on the hill that is 40 ft from the base of the antenna. How long is the guy wire?

24. *Physics*   Two forces are acting on an object. The magnitude of one force is 35 lb and the magnitude of the second force is 50 lb. If the angle between the two forces is 32.15°, what is the magnitude of the resultant force?

25. *Physics*   Figure 10.46 shows two forces represented by vector $\overrightarrow{AB}$ and $\overrightarrow{BC}$. If $AB = 12\,\text{N}$, $BC = 23\,\text{N}$, and $\angle ABC = 121.27°$, find the magnitude of **R** and the size of θ.

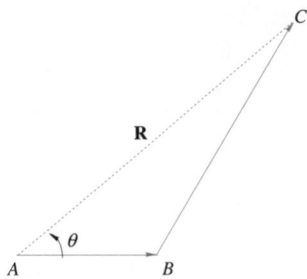

FIGURE 10.46

26. *Physics*   Two forces of 15.5 lb and 36.4 lb are acting on an object. The resultant force is 30.1 lb. What is the size of the angle between the original two forces?

27. *Electronics*   An inductive reactance, X_L, of 56 kΩ occurs in a circuit that has a resistance, R, of 38 kΩ. Power losses cause X_L to be 74° out of phase with R as shown in Figure 10.47. What is the impedance, Z, of the circuit?

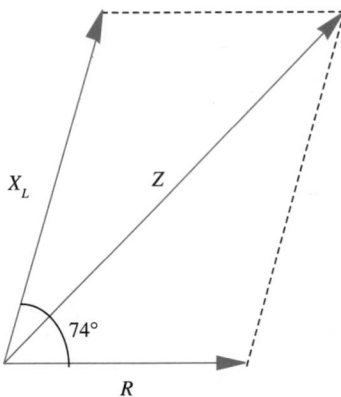

FIGURE 10.47

28. *Civil engineering*   Find the length represented by x the metal truss shown in Figure 10.48.

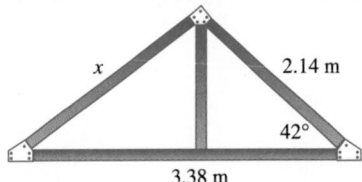

FIGURE 10.48

    In Your Words

29. Without looking in the text, write the Cosine Law and describe how to use it.

30. Describe the types of information about a triangle that you need in order to use the Cosine Law.

31. Explain how you decide whether to use the Sine Law or the Cosine Law.

≡ **CHAPTER 10 REVIEW**

Important Terms and Concepts

Adding vectors
   By adding components
   By the parallelogram method
Component vectors
Cosine Law
Direction vector
Initial point
Magnitude of a vector
Oblique triangle

Parallelogram method
Position vector
Resultant vector
Scalar
Sine Law
   Ambiguous case
Terminal point
Vectors

Review Exercises

In Exercises 1–4, add the given vectors by drawing the resultant vector.

1.

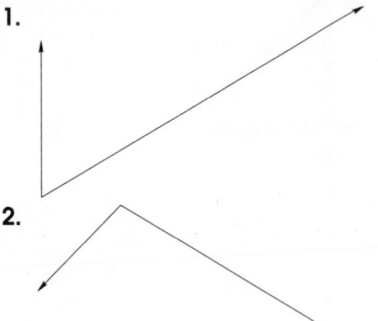

2.

3.

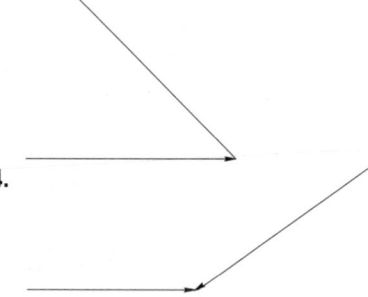

4.

In Exercises 5–8, find the horizontal and vertical components of the given vectors.

5. Magnitude 35, $\theta = 67°$

6. Magnitude 19.7, $\theta = 237°$

7. Magnitude 23.4, $\theta = 172.4°$

8. Magnitude 14.5, $\theta = 338°$

In Exercises 9 and 10, the horizontal and vertical components are given for a vector. Find each of the resultant vectors.

9. $A_x = 16, A_y = -8$

10. $B_x = -27, B_y = 32$

In Exercises 11–14, find the components of the indicated vectors.

11. $A = 38, \theta_A = 15°$

12. $B = 43.5, \theta_B = 127°$

13. $C = 19.4, \theta_C = 1.25$

14. $D = 62.7, \theta_D = 5.37$

In Exercises 15–18, add the given vectors by using the trigonometric functions and the Pythagorean theorem.

15. $A = 19, \theta_A = 32°, B = 32, \theta_B = 14°$

16. $C = 24, \theta_C = 57°, D = 35, \theta_D = 312°$

17. $E = 52.6, \theta_E = 2.53, F = 41.7, \theta_F = 3.92$

18. $G = 43.7, \theta_G = 4.73, H = 14.5, \theta_H = 4.42$

Solve each of the triangles in Exercises 19–26.

19. $a = 14, b = 32, c = 27$

20. $a = 43, b = 52, B = 86.4°$

21. $b = 87.4, B = 19.57°, c = 65.3$

22. $A = 121.3°, b = 42.5, c = 63.7$

23. $a = 127.35, A = 0.12, b = 132.6$

24. $b = 84.3, c = 95.4, C = 0.85$

25. $a = 67.9, b = 54.2, C = 2.21$

26. $a = 53.1, b = 63.2, c = 74.3$

Solve Exercises 27–32.

27. *Physics*   Two tow trucks are attempting to right an overturned vehicle. One truck is exerting a force of 1 650 kg. Its tow chain makes an angle of 68° with the axis of the vehicle. The other truck is exerting a force of 1 325 kg. Its chain makes an angle of 76° with the axis of the vehicle. What is the magnitude and direction of the resultant force vector? (See Figure 10.49.)

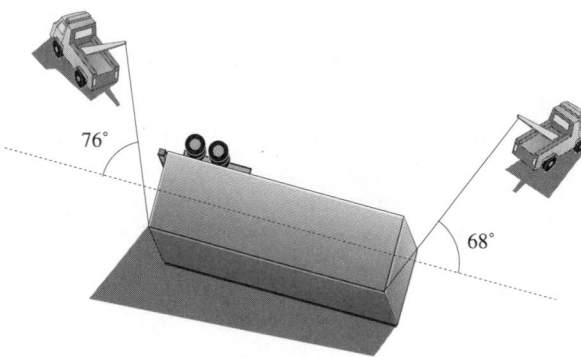

FIGURE 10.49

28. *Civil engineering*   A highway engineer has to decide whether to go over or to cut through a hill. The top of the hill makes an angle of 72.4° with the sides. One side of the hill is 2,342 ft and other side is 3,621 ft. It will cost 2.3 times as much per foot to cut through the hill and take the alternate route in Figure 10.50. How long is the alternate route? Which route is less expensive?

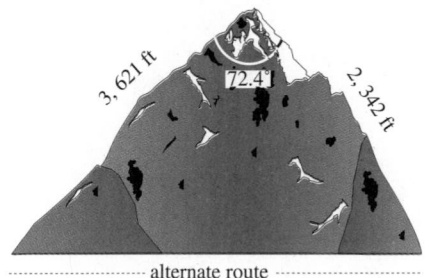

FIGURE 10.50

29. *Physics*   A block is resting on a ramp that makes an angle of 31.7° with the horizontal. The block weighs 126.5 lb. Find the components of the block's weight that are parallel and perpendicular to the road. (See Figure 10.51.)

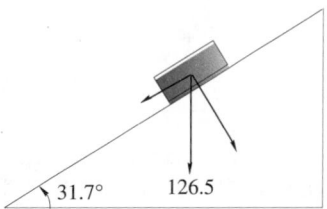

FIGURE 10.51

30. *Electricity*   Find the total impedance Z and the phase angle of a series ac circuit when $X_C = 72\,\Omega$, $X_L = 52\,\Omega$, and $R = 35\ \Omega$.

31. *Civil engineering*   A wire is to be strung across a valley. The wire will run from Tower A to Tower B. A surveyor is able to set up a position at a point C on the same side of the valley as Tower A, as shown in Figure 10.52. The distance from A to C is 73 m and $\angle BAC = 123.4°$ and $\angle ACB = 42.1°$. What is the distance from Tower A to Tower B?

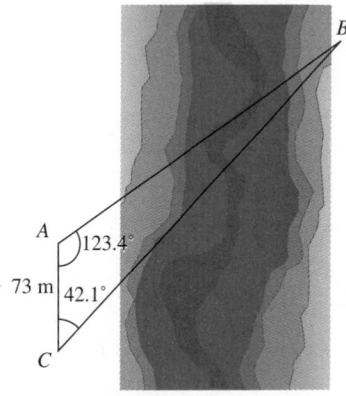

FIGURE 10.52

32. *Surveying*   A surveyor needs to determine the distance across a swamp. From a point C in Figure 10.53, she locates a point B on one side of the swamp. The distance from B to C is $1\,235$ m. Point A is directly across the swamp from B. The distance from A to C is 962 m and $\angle BCA$ is 52.57°. How far is it across the swamp from A to B?

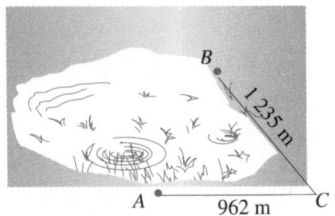

FIGURE 10.53

☰ CHAPTER 10 TEST

1. Determine the horizontal and vertical components of a vector **V** of magnitude 47 and direction $\theta = 117°$.

2. If $A_x = 12.91$ and $A_y = -14.36$, determine the magnitude and direction of vector **A**.

3. Add the given vectors by using the trigonometric functions and the Pythagorean theorem: $A = 25$, $\theta_A = 64°$, $B = 40$, $\theta_B = 112°$

4. Use the Sine Law to determine the length of side b in $\triangle ABC$, if $a = 9.42$, $\angle A = 35.6°$, and $\angle B = 67.5°$.

5. Use the Cosine Law to determine the length of side a of $\triangle ABC$, if $b = 4.95$, $c = 6.24$, and $\angle A = 113.4°$.

6. Solve $\triangle ABC$, if $A = 24°$, $b = 36.5$, and $C = 97°$.

7. Two walls meet at an angle of 97° to form the sides of a triangular corner cupboard. If the sides of the cupboard along each wall measure 30 in. and 36 in., what is the length of the front of the cupboard?

11

Graphs of Trigonometric Functions

The screen of an oscilloscope displays the graph of an electrical signal. This oscilloscope displays a Lissajous figure. In Section 11.6, we will see how to produce such a figure from two trigonometric functions.

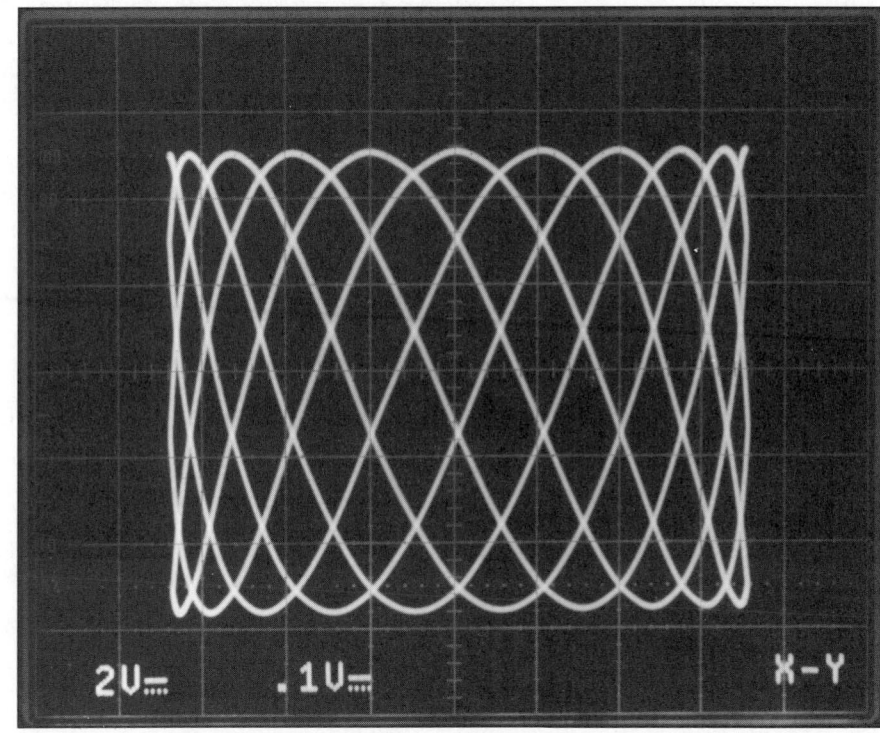

Trigonometry originally helped people solve triangles. For many years this was the main use of trigonometry. When people began graphing equations, it became possible to obtain pictures or graphs of the trigonometric functions.

Later, when people discovered light and sound waves, they found that these waves repeated at regular intervals—just like the graphs of trigonometric functions.

In fact, light, sound, electricity, ocean waves, and spring action are all examples of things that can be displayed as a graph of one or more trigonometric function.

☰ 11.1
SINE AND COSINE CURVES: AMPLITUDE AND PERIOD

As you know, the trigonometric functions repeat their values on a regular basis. Thus, $\sin 15° = \sin(360° + 15°) = \sin(720° + 15°)$, and so on. This means that the graphs of the trigonometric functions repeat. This repeating nature helps us apply these graphs to many physical and technical situations. In this chapter, we will use a graphics calculator to help draw the graphs, and show how to apply these graphs to your technical area.

Sine Curves

Sine and cosine curves are used in applications of mathematics that involve circular motion. In this chapter, we will first look at ways to graph the sine and cosine functions. Later in the chapter, we will look at some applications.

When we began graphing algebraic equations, we started by making a table of values. We shall adopt the same approach in graphing trigonometric functions. The following table gives the values of x in both degrees and radians of the function $y = \sin x$. The values of x only go from 0° to 360° (0 to 2π rad), since the values of y begin to repeat after 360° (or 2π rad). The graph of these values is shown in Figure 11.1.

x (degrees)	0	30	45	60	90	120	135	150	180
x (radians)	0	$\frac{\pi}{6}$	$\frac{\pi}{4}$	$\frac{\pi}{3}$	$\frac{\pi}{2}$	$\frac{2\pi}{3}$	$\frac{3\pi}{4}$	$\frac{5\pi}{6}$	π
$y = \sin x$	0	0.5	0.707	0.866	1	0.866	0.707	0.5	0

x (degrees)	180	210	225	240	270	300	315	330	360
x (radians)	π	$\frac{7\pi}{6}$	$\frac{5\pi}{4}$	$\frac{4\pi}{3}$	$\frac{3\pi}{2}$	$\frac{5\pi}{3}$	$\frac{7\pi}{4}$	$\frac{11\pi}{6}$	2π
$y = \sin x$	0	−0.5	−0.707	−0.866	−1	−0.866	−0.707	−0.5	0

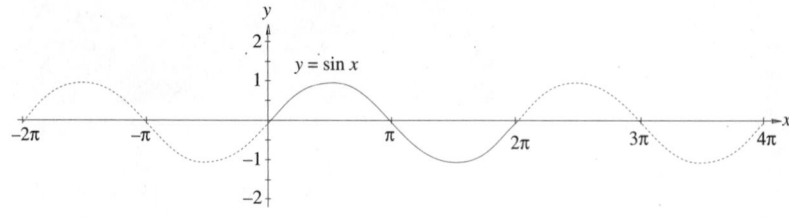

FIGURE 11.1

If we were to continue plotting the values for x larger than 2π rad or less than 0, we would get the points indicated by the dotted line in Figure 11.1. As you can see, the graph, like the values for $\sin x$, begins to repeat. The graph of the sine function has a very distinctive wave shape. A graph that has the general shape of the sine function is called a **sine wave**, a **sine curve**, or a **sinusoidal curve**.

Amplitude

One term used to help describe some functions is amplitude.

> **Amplitude**
>
> The **amplitude** A is one-half the difference between the maximum and minimum values for a periodic function.

For $y = \sin x$, the maximum value is 1 and the minimum value is -1. The difference between these two values is $1 - (-1) = 2$ and half of that is 1. So, the amplitude of $y = \sin x$ is 1.

Let's examine another trigonometric function. Suppose we want to graph $y = 3 \sin x$. This is the same as taking each of the values for $y = \sin x$ in the previous table and multiplying it by 3. The resulting table would be as follows and the graph would be like the one in Figure 11.2.

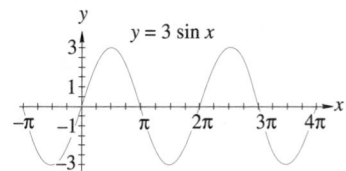

FIGURE 11.2

x	0	$\frac{\pi}{6}$	$\frac{\pi}{3}$	$\frac{\pi}{2}$	$\frac{2\pi}{3}$	$\frac{5\pi}{6}$	π	$\frac{7\pi}{6}$	$\frac{4\pi}{3}$	$\frac{3\pi}{2}$	$\frac{5\pi}{3}$	$\frac{11\pi}{6}$	2π
$y = 3\sin x$	0	1.5	2.6	3	2.6	1.5	0	-1.5	-2.6	-3	-2.6	-1.5	0

Notice that the maximum value of $y = 3 \sin x$ is 3 and the minimum value is -3. The amplitude of $y = 3 \sin x$ is $\frac{3-(-3)}{2} = 3$. In fact, you should be able to see that the amplitude of $y = A \sin x$ is $|A|$.

Period and Frequency

There are some things we should notice about these first two curves. They are examples of periodic functions. Each of them repeats its shape or completes one wave every 2π units, so we say these curves each have a period of 2π rad.

> **Periodic Function**
>
> For a function f, if p is the smallest positive constant where $f(x) = f(x+p)$ for all values of x, we say that f is a **periodic function** and that it has **period** p. The portion of a graph that falls within one period is called a **cycle**.

The sine function is a periodic function with period 2π (or 360°), because $\sin x = \sin(x + 2\pi)$.

Frequency

The **frequency** of a periodic function is defined as $\dfrac{1}{\text{period}}$.

Thus, the sine function with period 2π has a frequency of $\frac{1}{2\pi}$.

Now, let's graph a slightly different version of the sine curve, $y = \sin 4x$. Again, we will begin by using a table. This table will have values for x and $4x$, as well as for $y = \sin 4x$. The table follows, and the graph of $y = \sin 4x$ is in Figure 11.3.

x	0	$\frac{\pi}{24}$	$\frac{\pi}{12}$	$\frac{\pi}{8}$	$\frac{\pi}{6}$	$\frac{5\pi}{24}$	$\frac{\pi}{4}$	$\frac{7\pi}{24}$	$\frac{\pi}{3}$	$\frac{3\pi}{8}$	$\frac{5\pi}{12}$	$\frac{11\pi}{24}$	$\frac{\pi}{2}$
$4x$	0	$\frac{\pi}{6}$	$\frac{\pi}{3}$	$\frac{\pi}{2}$	$\frac{2\pi}{3}$	$\frac{5\pi}{6}$	π	$\frac{7\pi}{6}$	$\frac{4\pi}{3}$	$\frac{3\pi}{2}$	$\frac{5\pi}{3}$	$\frac{11\pi}{6}$	2π
$y = \sin 4x$	0	0.5	0.866	1	0.866	0.5	0	-0.5	-0.866	-1	-0.866	-0.5	0

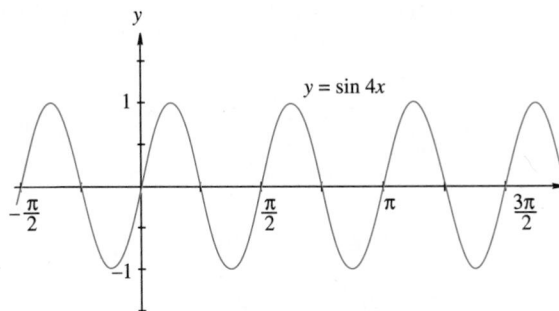

FIGURE 11.3

This graph is interesting in that it looks like a sine curve, with an amplitude of 1, but it does not have a period of 2π. In fact, its period is $\frac{\pi}{2}$. Notice that $\frac{\pi}{2} = 2\pi \div 4$. In general, we have the following statement for both the sine and the cosine functions.

Period and Frequency

The **period** and **frequency** of $y = \sin Bx$ and $y = \cos Bx$ are

Period	Frequency	
$\dfrac{2\pi}{\lvert B\rvert}$	$\dfrac{\lvert B\rvert}{2\pi}$	if measured in radians
$\dfrac{360°}{\lvert B\rvert}$	$\dfrac{\lvert B\rvert}{360°}$	if measured in degrees

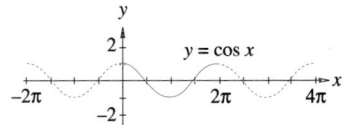

FIGURE 11.4

Cosine Curves

As yet, we have graphed only the sine curve. It is time to graph the cosine curve. A table of values follows and the graph of $y = \cos x$ is shown in Figure 11.4.

x	0	$\frac{\pi}{6}$	$\frac{\pi}{3}$	$\frac{\pi}{2}$	$\frac{2\pi}{3}$	$\frac{5\pi}{6}$	π	$\frac{7\pi}{6}$	$\frac{4\pi}{3}$	$\frac{3\pi}{2}$	$\frac{5\pi}{3}$	$\frac{11\pi}{6}$	2π
$y = \cos x$	1	0.87	0.5	0	-0.5	-0.87	-1	-0.87	-0.5	0	0.5	0.87	1

Here, as with the sine curve in Figure 11.1, we have shown the values graphed from the table with a solid line. The dashed line indicates what the graph would look like if we continued to plot values for $x > 2\pi$ or $x < 0$. The solid line represents one cycle. As you can see, the cosine function is periodic; $y = \cos x$ has period of 2π and amplitude of 1. A careful examination of the curves in Figures 11.1 and 11.4 shows that the shapes of the curves are exactly the same but are displaced along the horizontal axis relative to one another.

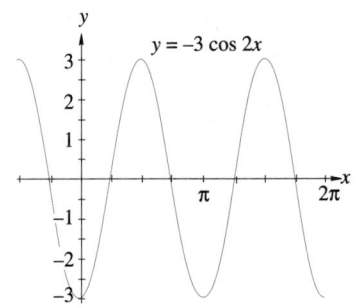

FIGURE 11.5

As our final example, we will graph $y = -3\cos 2x$. The period of $y = \cos x$ is 2π, so the period of $y = -3\cos 2x$ is $2\pi \div 2 = \pi$. The amplitude of $y = A\cos x$ is $|A|$, so the amplitude of $y = -3\cos 2x$ is $|-3| = 3$. To sketch this curve, we need to find only a few values of x. These x values will be at the maximum and minimum values and at the x-intercepts (when $y = 0$). This table follows and the graph of $y = -3\cos 2x$ is in Figure 11.5.

x	0	$\frac{\pi}{4}$	$\frac{\pi}{2}$	$\frac{3\pi}{4}$	π
$y = -3\cos 2x$	-3	0	3	0	-3

Study the graph of $y = -3\cos 2x$. This is the graph of $y = 3\cos 2x$ only "flipped" over the x-axis.

≡ **Note**

The technical expression for indicating that a graph is "flipped over the x-axis" is that it is "reflected across the x-axis."

EXAMPLE 11.1

Sketch the graph of $y = 4\sin 3x$.

Solution We can see that the amplitude is 4 and the period is $\frac{2\pi}{3}$. The frequency is $\frac{1}{\frac{2\pi}{3}} = \frac{3}{2\pi}$. From this we know that $y = 0$ when $x = \frac{2\pi}{3}$, $x = \frac{\pi}{3}$, and $x = 0$. The maximum and minimum values will occur halfway between these values of x, or when $x = \frac{\frac{\pi}{3}+0}{2} = \frac{\pi}{6}$ and $x = \frac{\frac{2\pi}{3}+\frac{\pi}{3}}{2} = \frac{\pi}{2}$. The table for these values follows.

x	0	$\frac{\pi}{6}$	$\frac{\pi}{3}$	$\frac{\pi}{2}$	$\frac{2\pi}{3}$
$y = 4\sin 3x$	0	4	0	-4	0

Based on these values and our knowledge of the shape of the sine curves, we get the graph shown in Figure 11.6. Notice how this graph is "compressed" or "shrunk" over a period of $\frac{2\pi}{3}$.

EXAMPLE 11.2

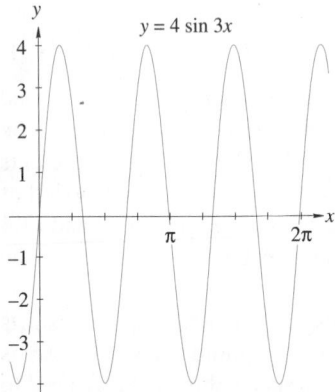

FIGURE 11.6

Sketch the graph of $y = 1.5 \cos \dfrac{x}{2}$.

Solution   We can see the amplitude is 1.5 and the period is $2\pi \div \frac{1}{2} = 4\pi$ or $720°$. Since the frequency is $\dfrac{1}{\text{period}}$, it is $\dfrac{1}{4\pi}$. The maximum values will occur when $x = 0$ and $x = 720°$. The minimum value will be halfway between these two points or when $x = 360°$. The zeros are at the midpoints between these three points or when $x = 180°$ and $x = 540°$. These findings are shown in this table.

x	0°	180°	360°	540°	720°
$y = 1.5 \cos \dfrac{x}{2}$	1.5	0	−1.5	0	1.5

Based on these values and our knowledge of the shape of the cosine curve, we get the graph in Figure 11.7. Notice how this curve is "stretched out" over a period of 4π or $720°$.

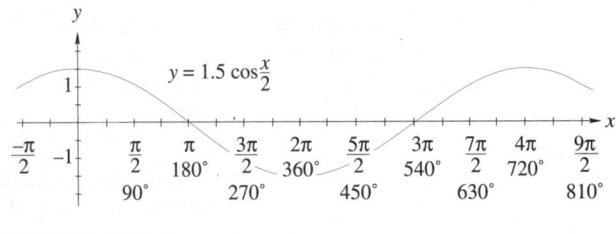

FIGURE 11.7

The following box summarizes this section's discussion on amplitude and period.

Summary

A function of the form $y = A \sin Bx$ and $y = A \cos Bx$ has

Amplitude	Period	Frequency	
$\lvert A\rvert$	$\dfrac{2\pi}{\lvert B\rvert}$	$\dfrac{\lvert B\rvert}{2\pi}$	if measured in radians
$\lvert A\rvert$	$\dfrac{360°}{\lvert B\rvert}$	$\dfrac{\lvert B\rvert}{360°}$	if measured in degrees

If $A < 0$, the graph is "flipped over" (or "reflected across") the x-axis.

Exercise Set 11.1

In Exercises 1–14, find the period, amplitude, and frequency of the given functions.

1. $y = 3\sin 2x$

2. $y = 5\sin 6x$

3. $y = 2\cos x$

4. $y = 7\cos 3x$

5. $y = 8\cos 2\pi x$

6. $y = 7\cos \pi x$

7. $y = \frac{1}{2}\sin 4x$

8. $y = 5\cos \frac{1}{3}x$

9. $y = \dfrac{\cos \frac{1}{2}x}{3}$

10. $y = \dfrac{\cos 3x}{5}$

11. $y = -3\sin \dfrac{x}{4}$

12. $y = -5\cos \dfrac{x}{3}$

13. $y = \dfrac{-\cos x}{2}$

14. $y = \dfrac{\sin 3x}{-2}$

For Exercises 15–28, sketch one cycle of each of the functions in Exercises 1–14.

Solve Exercise 29–32.

29. *Electronics*   The current, I (in amperes), in an ac circuit is given by $I = 6.5\sin 120\pi t$, where t is the time in seconds. **(a)** Find the period, amplitude, and frequency for this function. **(b)** Plot I as a function of t for $0 \le t \le 0.1$ s.

30. *Electronics*   In a simple generator, the electomotive force, E (measured in volts), can be expressed as a function of time, t, as $E = 12\sin 120\pi t$.
    (a) Find the amplitude, period, and frequency for this function.
    (b) Sketch the graph of E for two periods. Make sure that both axes are labeled correctly.

31. *Solar energy*   When the receiving surface of a solar radiation detector is not perpendicular to the direction of the sun, the incident radiation per unit area, G_i, will be reduced by the $\cos i$, which is the angle between the rays of the sun and a line perpendicular

to the receiving surface. The equation $G_i = G_n \cos i$ gives the incident radiation per unit area where G_n is the radiant energy incident upon a surface on the earth placed normal (perpendicular) to the rays of the sun. If $G_n = 1094$ W/h/m^2.
    (a) Find the amplitude, period, and frequency for this function.
    (b) Sketch the graph of G_i for two periods. Make sure that both axes are labeled correctly.

32. *Oceanography*   Under certain conditions, the height of the tide above its mean level is given by $y = 2.1\cos 0.45t$, where y is in meters and t is in hours.
    (a) Find the amplitude, period, and frequency for this function.
    (b) Sketch the graph of y for two periods. Make sure that both axes are labeled correctly.

In Your Words

33. If you were shown the graphs of $y = \sin x$, $y = 2\sin x$, and $y = \sin 2x$ on the same axes, describe how you could distinguish among the graphs.

34. **(a)** Describe how you can determine the amplitude of a sine or cosine function.
    (b) How can you use the amplitude to help you draw the graph of $y = 5\sin x$?
    (c) Describe how you can determine the period and frequency of a sine or cosine function.

☰ 11.2
SINE AND COSINE CURVES: HORIZONTAL DISPLACEMENT OR PHASE SHIFT AND VERTICAL DISPLACEMENT

The graphs of the sine and cosine curves that we have examined have all had one thing in common. That is, the sine curves all contained the point $(0, 0)$ and the cosine curves all contained $(0, A)$, where $|A|$ was the amplitude. In this section, we will shift the graphs away from these points.

Phase Shift or Horizontal Displacement

EXAMPLE 11.3

Sketch the cycle of the graph of $y = \sin\left(x + \frac{\pi}{3}\right)$.

Solution   As usual, we make a table of values. In this table, we show three rows. The first row contains values of x, the next row contains values of $x + \frac{\pi}{3}$, and the third contains the values for $y = \sin\left(x + \frac{\pi}{3}\right)$.

x	0	$\frac{\pi}{6}$	$\frac{\pi}{3}$	$\frac{\pi}{2}$	$\frac{2\pi}{3}$	$\frac{5\pi}{6}$	π
$x + \frac{\pi}{3}$	$\frac{\pi}{3}$	$\frac{\pi}{2}$	$\frac{2\pi}{3}$	$\frac{5\pi}{6}$	π	$\frac{7\pi}{6}$	$\frac{4\pi}{3}$
$y = \sin\left(x + \frac{\pi}{3}\right)$	0.866	1	0.866	0.5	0	-0.5	-0.866

x	π	$\frac{7\pi}{6}$	$\frac{4\pi}{3}$	$\frac{3\pi}{2}$	$\frac{5\pi}{3}$	$\frac{11\pi}{6}$	2π
$x + \frac{\pi}{3}$	$\frac{4\pi}{3}$	$\frac{3\pi}{2}$	$\frac{5\pi}{3}$	$\frac{11\pi}{6}$	2π	$\frac{13\pi}{6}$	$\frac{8\pi}{3}$
$y = \sin\left(x + \frac{\pi}{3}\right)$	-0.866	-1	-0.866	-0.5	0	0.5	0.866

The graph of one cycle of this curve is given in Figure 11.8. Notice that this looks just like a sine curve except that it has to be shifted $\frac{\pi}{3}$ units to the left.

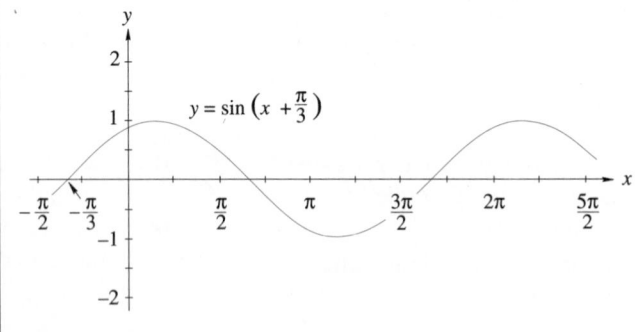

FIGURE 11.8

Let's look at a more general case.

Application

EXAMPLE 11.4

Sketch one cycle of the graph of $y = \sin\left(2x - \frac{\pi}{6}\right)$.

Solution   The following table contains three rows, x, $2x - \frac{\pi}{6}$, and $\sin\left(2x - \frac{\pi}{6}\right)$.

x	0	$\frac{\pi}{12}$	$\frac{\pi}{6}$	$\frac{\pi}{4}$	$\frac{\pi}{3}$	$\frac{5\pi}{12}$	$\frac{\pi}{2}$
$2x - \frac{\pi}{6}$	$-\frac{\pi}{6}$	0	$\frac{\pi}{6}$	$\frac{\pi}{3}$	$\frac{\pi}{2}$	$\frac{2\pi}{3}$	$\frac{5\pi}{6}$
$y = \sin\left(2x - \frac{\pi}{6}\right)$	-0.5	0	0.5	0.866	1	0.866	0.5

x	$\frac{\pi}{2}$	$\frac{7\pi}{12}$	$\frac{2\pi}{3}$	$\frac{3\pi}{4}$	$\frac{5\pi}{6}$	$\frac{11\pi}{12}$	π
$2x - \frac{\pi}{6}$	$\frac{5\pi}{6}$	π	$\frac{7\pi}{6}$	$\frac{4\pi}{3}$	$\frac{3\pi}{2}$	$\frac{5\pi}{3}$	$\frac{11\pi}{6}$
$y = \sin\left(2x - \frac{\pi}{6}\right)$	0.5	0	-0.5	-0.866	-1	-0.866	-0.5

One cycle of this graph is shown in Figure 11.9. Notice that it has a period of π and has been shifted $\frac{\pi}{12}$ units to the right.

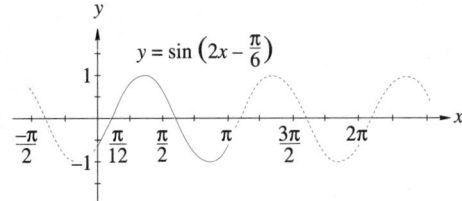

FIGURE 11.9

Graphs of $y = A\sin\left(Bx + C\right)$ and $y = A\cos\left(Bx + C\right)$

A curve of the form $y = A\sin(Bx + C)$ or $y = A\cos(Bx + C)$ has

Amplitude	Period	Frequency	Phase shift or horizontal displacement	If measured in
$|A|$	$\dfrac{2\pi}{|B|}$	$\dfrac{|B|}{2\pi}$	$-\dfrac{C}{B}$	radians
$|A|$	$\dfrac{360°}{|B|}$	$\dfrac{|B|}{360°}$	$-\dfrac{C}{B}$	degrees

If the phase shift is negative, the entire curve will be shifted to the left. When the phase shift is positive, the entire curve will be shifted to the right. The phase shift gives you the x-coordinate at which to begin drawing one period of $y = A\sin(Bx + C)$ or $y = A\cos(Bx + C)$.

If $A < 0$, the graph is "flipped over" (or "reflected across") the x-axis.

EXAMPLE 11.5

Sketch one cycle of the curve $y = \frac{1}{2}\cos(3x + \pi)$.

Solution Using the information contained in the equation, $A = \frac{1}{2}$, $B = 3$, and $C = \pi$, so the amplitude is $\frac{1}{2}$, the period is $\frac{2\pi}{3}$, and the phase shift is $-\frac{\pi}{3}$.

Since the phase shift is $-\frac{\pi}{3}$, we start drawing the graph at $x = -\frac{\pi}{3}$. The period of $\frac{2\pi}{3}$ indicates that one complete cycle will end at $x = -\frac{\pi}{3} + \frac{2\pi}{3} = \frac{\pi}{3}$. For cosine, the maximum values occur at these two points. The minimum value will occur midway between the maximum values, or at $x = 0$. The zeros are midway between the maximum and minimum values, or at $x = -\frac{\pi}{6}$ and $x = \frac{\pi}{6}$. A sketch of $y = \frac{1}{2}\cos(3x + \pi)$ is in Figure 11.10 with the one complete cycle we just described in solid and additional cycles dashed.

EXAMPLE 11.6

Sketch one cycle of the graph of $y = -4\sin\left(\frac{1}{2}x - \frac{\pi}{8}\right)$.

Solution Here $A = -4$, $B = \frac{1}{2}$, and $C = -\frac{\pi}{8}$. The function has an amplitude of $|-4| = 4$ and a period of $2\pi \div \frac{1}{2} = 4\pi$. The phase shift is $-\left(-\frac{\pi}{8} \div \frac{1}{2}\right) = \frac{\pi}{4}$. We begin sketching the curve at $x = \frac{\pi}{4}$. The period of 4π indicates that one complete cycle will end at $x = \frac{\pi}{4} + 4\pi = \frac{17\pi}{4}$. The zeros of this sine function occur when $x = \frac{\pi}{4}$, $x = \frac{17\pi}{4}$, and midway between these two points, at $x = \frac{9\pi}{4}$. The minimum value is at $x = \frac{5\pi}{4}$ and the maximum when $x = \frac{13\pi}{4}$. The graph of $y = -4\sin\left(\frac{1}{2}x - \frac{\pi}{8}\right)$ is shown in Figure 11.11.

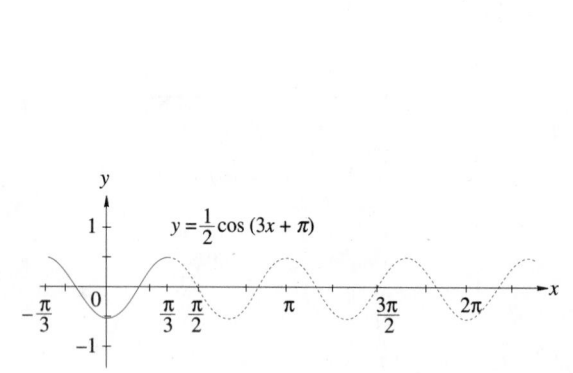

FIGURE 11.10

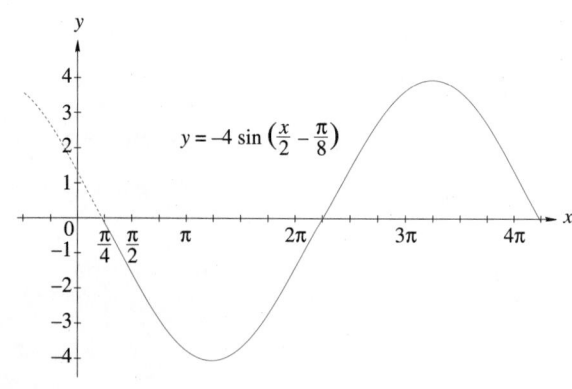

FIGURE 11.11

Vertical Displacement

We have seen how graphs can be shifted horizontally. They can also be shifted vertically.

EXAMPLE 11.7

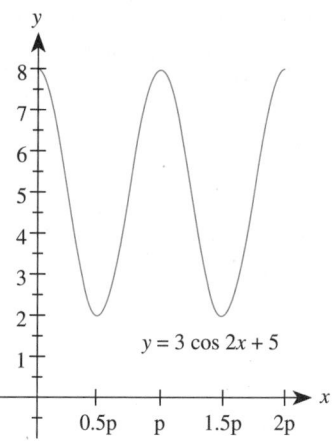

$y = 3 \cos 2x + 5$

FIGURE 11.12

Sketch two cycles of the graph of $y = 3\cos 2x + 5$.

Solution This equation is in the form $y = A\sin(Bx + C) + D$, with $A = 3$, $B = 2$, $C = 0$, and $D = 5$. The function has an amplitude of $|3| = 3$ and a period of $2\pi \div 2 = \pi$. Since $C = 0$, the horizontal phase shift is 0. What effect does D have on this graph? Let's look at a table of values.

x	0	$\frac{\pi}{4}$	$\frac{\pi}{2}$	$\frac{3\pi}{4}$	π	$\frac{5\pi}{4}$	$\frac{3\pi}{2}$	$\frac{7\pi}{4}$	2π
$y = 3\cos 2x$	3	0	−3	0	3	0	−3	0	3
$y = 3\cos 2x + 5$	8	5	2	5	8	5	2	5	8

The value of D in this example, 5, moves or displaces the graph 5 units vertically. This can be seen by the graphs in Figure 11.12. If you moved the graph of $y = 3\cos 2x$ up 5 units it would overlap the graph of $y = 3\cos 2x + 5$.

From Example 11.7 it seems as if the effect of D is to shift the graph vertically. When we add this information to the previous information about graphing, we get the following summary box. It includes everything we have discussed in the last two sections, so this one box can replace all of the other boxes in these two sections.

Graphs of $y = A\sin(Bx + C) + D$ **and** $y = A\cos(Bx + C) + D$

A curve of the form $y = A\sin(Bx + C) + D$ or $y = A\cos(Bx + C) + D$ has

Amplitude	Period	Frequency	Phase shift or horizontal displacement	Vertical displacement	If measured in
$|A|$	$\dfrac{2\pi}{|B|}$	$\dfrac{|B|}{2\pi}$	$-\dfrac{C}{B}$	D	radians
$|A|$	$\dfrac{360°}{|B|}$	$\dfrac{|B|}{360°}$	$-\dfrac{C}{B}$	D	degrees

If the phase shift is negative, the entire curve will be shifted to the left. When the phase shift is positive, the entire curve will be shifted to the right. The phase shift gives you the x-coordinate at which to begin drawing one period of $y = A\sin(Bx + C) + D$ or $y = A\cos(Bx + C) + D$.

If $A < 0$, the graph is "flipped" over the line $y = D$.

EXAMPLE 11.8

Sketch one cycle of the graph of $y = -\frac{1}{2}\sin(-4x + \pi) - 3$.

Solution Here $A = -\frac{1}{2}$, $B = -4$, $C = \pi$, and $D = -3$. The function has an amplitude of $\left|-\frac{1}{2}\right| = \frac{1}{2}$ and a period of $2\pi \div |(-4)| = \frac{\pi}{2}$. Since $C = \pi$, the horizontal phase shift is $-\left(\pi \div (-4)\right) = \frac{\pi}{4}$. We begin sketching the curve at $x = \frac{\pi}{4}$. The period of $\frac{\pi}{2}$ indicates that one complete cycle will end at $x = \frac{\pi}{4} + \frac{\pi}{2} = \frac{3\pi}{4}$. The zeros of $y = -\frac{1}{2}\sin(-4x + \pi)$ will occur when $x = \frac{\pi}{4}$ and $x = \frac{3\pi}{4}$, and midway between these

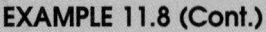

EXAMPLE 11.8 (Cont.)

two points at $x = \frac{\pi}{2}$. The minimum value is at $x = \frac{3\pi}{8}$ and the maximum value is at $x = \frac{5\pi}{8}$. The vertical displacement is -3, which means that the graph will be shifted 3 units downward. The graph of $y = -\frac{1}{2}\sin(-4x+\pi)-3$ is shown in Figure 11.13. ▪

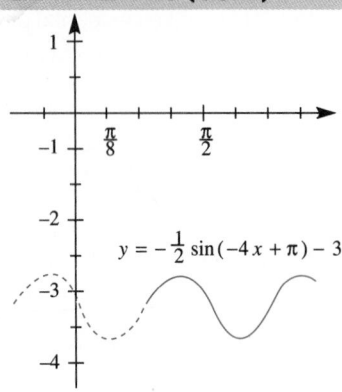

$y = -\frac{1}{2}\sin(-4x+\pi)-3$

FIGURE 11.13

Exercise Set 11.2

In Exercises 1–20, give the amplitude, period, phase shift, and vertical displacement, and then graph the given functions.

1. $y = 2\sin\left(x + \frac{\pi}{4}\right)$

2. $y = 1.5\cos\left(x - \frac{\pi}{3}\right)$

3. $y = 2.5\sin(3x - \pi)$

4. $y = 3\cos\left(2x + \frac{\pi}{2}\right)$

5. $y = 6\cos(1.5x + 180°)$

6. $y = 8\sin(3.5x - 90°)$

7. $y = -2\cos(3x - 90°)$

8. $y = -4\sin(5x + 600°)$

9. $y = 4.5\sin(6x - 8)$

10. $y = 9\cos(5x + 3)$

11. $y = 0.5\cos(\pi x + \frac{\pi}{8})$

12. $y = -0.25\sin\left(\frac{x}{\pi} - \frac{1}{4}\right)$

13. $y = -0.75\sin\left(\pi x + \frac{\pi^2}{3}\right)$

14. $y = \pi\cos\left(\frac{1}{\pi}x + \frac{2}{3}\right)$

15. $y = 2\cos\left(\pi x + \frac{1}{3}\right) + 4$

16. $y = 3.25\sin(3x - 4) - \frac{7}{2}$

17. $y = -\frac{1}{2}\sin(\pi x + 1) + \frac{3}{2}$

18. $y = -1.75\cos(2x - \pi) + 2.25$

19. $y = -2\cos(2x - 45°) + 3.5$

20. $y = 3\sin(6x + 510°) - 2.5$

Solve Exercises 21–26.

21. *Electronics*   The current, I (in amperes), in an ac circuit is given by $I = 65\cos\left(120\pi t + \frac{\pi}{2}\right)$, where t is the time in seconds. **(a)** Determine the amplitude, period, and phase shift for this function. **(b)** Sketch one complete cycle of this function.

22. *Oceanography*   At a certain point in the ocean, the vertical displacement of the water due to wave action is given by $y = 3\sin\left[\frac{\pi}{6}(t - 5)\right]$, where y is measured in meters and t is in seconds. **(a)** Determine the amplitude, period, and phase shift for the graph of this wave action. **(b)** Sketch one complete cycle of this function.

23. *Oceanography*   Tides rise and fall every 12.4 h. As a result of the tides, the depth in a certain harbor ranges from 6.2 m to 11.8 m. The depth of the water at a given time can be approximated by a sine function. **(a)** Write an equation that gives the approximate depth of the harbor at any given time, t, if midnight is at $t = 0$ and high tide is at 4:00 a.m. **(b)** Sketch the graph of this function for 24 hours, starting at midnight.

24. *Electronics*   In an ac circuit with only a constant inductance, the voltage is given by

$$V = 12\sin(120\pi t + 0.5\pi)$$

(a) Determine the amplitude, period, phase shift, and vertical displacement of this function.

(b) Sketch the graph of this function for one period. Make sure that you label the times on the x-axis where the graph crosses the axis and where it is at its highest and lowest points.

25. *Meteorology*   The average weekly temperature at a particular location can usually be determined from the appropriate sine function. In most cases, the yearly average occurs around March 21. In Minneapolis, the maximum and minimum temperatures are 72°F and 11°F, respectively.

In Your Words

27. If you were shown the graphs of $y = \sin x$, $y = \sin x + 2$, and $y = \sin(x + 2)$ on the same axes, describe how you could distinguish among the graphs.

28. **(a)** Describe how you can determine the phase shift of a sine or cosine function.

(b) Describe how you can determine the vertical displacement of a sine or cosine function.

(a) Write an equation that gives the approximate weekly temperature of any week in the year, starting with January 1. (March 21 will occur in the 12th week of the year.)

(b) Sketch the graph of this function for one year, starting with the first week of January.

26. *Meteorology*   The average weekly temperature at a particular location can usually be determined from the appropriate sine function. In most cases, the yearly average occurs around March 21. In Chattanooga, Tennessee, the maximum and minimum temperatures are 89.3°F and 29.2°F, respectively.

(a) Write an equation that gives the approximate weekly temperature of any week in the year, starting with January 1. (March 21 will occur in the 12th week of the year.)

(b) Sketch the graph of this function for one year, starting with the first week of January.

(c) How can you use the amplitude and the vertical displacement to help you draw the graph of $y = 5\sin x$?

(d) How can you use the amplitude and the vertical displacement to help you draw the graph of $y = 5\sin x + 7$?

≡ 11.3
COMPOSITE SINE AND COSINE CURVES

In earlier sections, we looked at adding and subtracting functions. At that time, the functions were all algebraic. For example, when $f(x) = x^2$ and $g(x) = x$, we had little difficulty graphing $(f + g)(x)$. In a case such as this, we simply graphed $x^2 + x$ as if it were one function. Graphing composite trigonometric curves is not as easy.

In this section, we will learn how to graph trigonometric curves that combine two or more functions, using the ability to graph that we have learned earlier in this book. After we are competent with this technique, we will begin graphing trigonometric functions with the help of a calculator. Later in this chapter, we will see how these curves are used in technical areas.

The easiest way to learn how to graph a combination or composite curve is through an example.

EXAMPLE 11.9

Sketch the graph of $y = \sin x + \cos x$.

Solution   We will begin with a table of values. Both of the functions $y_1 = \sin x$ and $y_2 = \cos x$ have amplitude of 1 and period of 2π. The period of a composite curve is an integral multiple of the least common multiple of the period of each of

EXAMPLE 11.9 (Cont.)

the individual curves. In this case, since y_1 and y_2 each have periods of 2π, the least common multiple is also 2π. Since the amplitudes of $\sin x$ and $\cos x$ are both 1, the largest possible amplitude for $y = \sin x + \cos x$ is 2. There is no vertical displacement. We can use this estimate for the amplitude and the amount of vertical displacement to help determine reasonable values for the y-axis. (It turns out that the amplitude is actually $\sqrt{2}$.)

The following table contains values for y_1, y_2, and $y = y_1 + y_2$. [Notice that $\sin\frac{\pi}{4} = \frac{\sqrt{2}}{2} \approx 0.71$ and $\cos\frac{\pi}{4} = \frac{\sqrt{2}}{2} \approx 0.71$, so $\sin\frac{\pi}{4} + \cos\frac{\pi}{4} = \frac{\sqrt{2}}{2} + \frac{\sqrt{2}}{2} = \sqrt{2} \approx$ 1.41. However, adding the values in the table for $\sin\frac{\pi}{4}$ and $\cos\frac{\pi}{4}$, we get $0.71 + 0.71 = 1.42$. The difference in these two answers is the result of adding exact values $\left(\dfrac{\sqrt{2}}{2} + \dfrac{\sqrt{2}}{2}\right)$ or approximate values $(0.71 + 0.71)$. It helps to show why you should wait until you have finished all calculations before you round off any numbers.]

x	0	$\frac{\pi}{6}$	$\frac{\pi}{4}$	$\frac{\pi}{3}$	$\frac{\pi}{2}$	$\frac{2\pi}{3}$	$\frac{3\pi}{4}$	$\frac{5\pi}{6}$	π
$y_1 = \sin x$	0.0	0.5	0.71	0.87	1.0	0.87	0.71	0.5	0.0
$y_2 = \cos x$	1.0	0.87	0.71	0.5	0.0	−0.5	−0.71	−0.87	−1.0
$y = y_1 + y_2$	1.0	1.37	1.42	1.37	1.0	0.37	0.0	−0.37	−1.0

x	π	$\frac{7\pi}{6}$	$\frac{5\pi}{4}$	$\frac{4\pi}{3}$	$\frac{3\pi}{2}$	$\frac{5\pi}{3}$	$\frac{7\pi}{4}$	$\frac{11\pi}{6}$	2π
$y_1 = \sin x$	0.0	−0.5	−0.71	−0.87	−1.0	−0.87	−0.71	−0.5	0.0
$y_2 = \cos x$	−1.0	−0.87	−0.71	−0.5	0.0	0.5	0.71	0.87	1.0
$y = y_1 + y_2$	−1.0	−1.37	−1.42	−1.37	−1.0	−0.37	0.0	0.37	1.0

A sketch of all three curves is shown in Figure 11.14. At any point x, the value for y was obtained by adding the values for y_1 and y_2. The dotted line indicates how this was done graphically.

EXAMPLE 11.10

Sketch the graph of $y = 2\sin x + \frac{1}{2}\sin 5x$.

Solution Once again we begin with a table of values. The function $y_1 = 2\sin x$ has amplitude 2 and period 2π. The function $y_2 = \frac{1}{2}\sin 5x$ has amplitude $\frac{1}{2}$ and period $\frac{2\pi}{5}$. Because the least common multiple of 2π and $\frac{2\pi}{5}$ is 2π, the period of y will be 2π. Since the amplitude of $2\sin x$ is 2 and the amplitude of $\frac{1}{2}\sin 5x$ is $\frac{1}{2}$, the largest possible amplitude for $y = 2\sin x + \frac{1}{2}\sin 5x$ is $2 + \frac{1}{2} = \frac{5}{2}$. There is no vertical displacement. We use this estimate for the amplitude and the amount of vertical displacement to help determine reasonable values for the y-axis. We plan for y-values of this graph to range from $-\frac{5}{2}$ to $\frac{5}{2}$.

The table contains values for y_1 and y_2 as well as for $y = y_1 + y_2$.

The sketch of the graphs for all three functions is in Figure 11.15. Notice at $\frac{\pi}{2}$, or 90°, the value y was obtained by adding the graphical heights of y_1 and y_2. This same method of adding the graphical distances or heights could be performed at any value of x to get the value of y at that point.

EXAMPLE 11.10 (Cont.)

x	0	$\frac{\pi}{6}$	$\frac{\pi}{3}$	$\frac{\pi}{2}$	$\frac{2\pi}{3}$	$\frac{5\pi}{6}$	π
$y_1 = 2\sin x$	0	1.00	1.73	2.00	1.73	1.00	0
$y_2 = \frac{1}{2}\sin 5x$	0	0.25	−0.43	0.50	−0.43	0.25	0
$y = y_1 + y_2$	0	1.25	1.30	2.50	1.30	1.25	0

x	π	$\frac{7\pi}{6}$	$\frac{4\pi}{3}$	$\frac{3\pi}{2}$	$\frac{5\pi}{3}$	$\frac{11\pi}{6}$	2π
$y_1 = 2\sin x$	0	−1.00	−1.74	−2.00	−1.74	−1.00	0
$y_2 = \frac{1}{2}\sin 5x$	0	−0.25	0.43	−0.50	0.43	−0.25	0
$y = y_1 + y_2$	0	−1.25	−1.31	−2.50	−1.31	−1.25	0

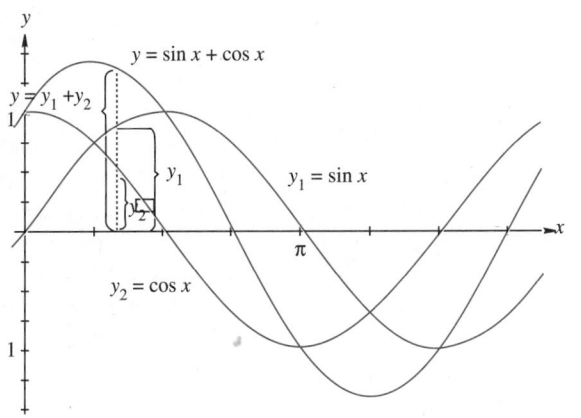

FIGURE 11.14

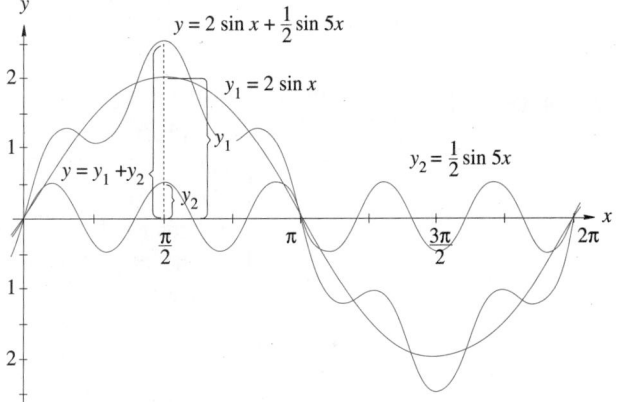

FIGURE 11.15

A different graph can be obtained by changing the two functions y_1 and y_2. If we shift y_1 to the left $\frac{\pi}{6}$ rad, we would get $y_1 = 2\sin\left(x + \frac{\pi}{6}\right)$. If y_2 remains the same, we have the graph of $y = 2\sin\left(x + \frac{\pi}{6}\right) + \frac{1}{2}\sin 5x$. The following table shows some of the values for this curve. Not only can you add y_1 and y_2 in the table to get the value of y, but this can be done by adding values on the graph shown in Figure 11.16.

x	0	$\frac{\pi}{6}$	$\frac{\pi}{3}$	$\frac{\pi}{2}$	$\frac{2\pi}{3}$	$\frac{5\pi}{6}$	π
$y_1 = 2\sin\left(x + \frac{\pi}{6}\right)$	1.00	1.73	2.00	1.73	1.0	0	−1.00
$y_2 = \frac{1}{2}\sin 5x$	0	0.25	−0.43	0.50	−0.43	0.25	0
$y = y_1 + y_2$	1.00	1.98	1.57	2.23	0.57	0.25	−1.00

x	π	$\frac{7\pi}{6}$	$\frac{4\pi}{3}$	$\frac{3\pi}{2}$	$\frac{5\pi}{3}$	$\frac{11\pi}{6}$	2π
$y_1 = 2\sin\left(x + \frac{\pi}{6}\right)$	−1.00	−1.74	−2.00	−1.74	−1.00	0	1.00
$y_2 = \frac{1}{2}\sin 5x$	0	−0.26	0.43	−0.50	0.43	−0.25	0
$y = y_1 + y_2$	−1.00	−2.00	−1.57	−2.23	−0.57	−0.25	1.00

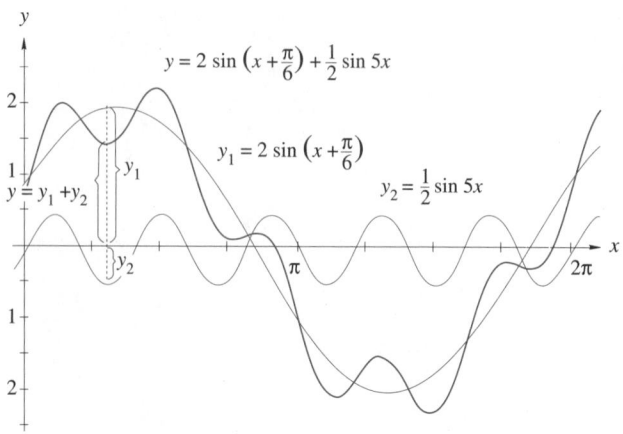

FIGURE 11.16

EXAMPLE 11.11

Sketch the graph of $y = 3\cos 2x - \sin 4x + 5$.

Solution   Before we begin with a table of values, we will analyze this function. The function $y_1 = 3\cos 2x$ has amplitude 3, period π, and vertical displacement 0. The function $y_2 = -\sin 4x + 5$ has amplitude 1, period $\frac{\pi}{2}$, and vertical displacement 5. Because the least common multiple of π and $\frac{\pi}{2}$ is π, the period for this function is π. The amplitudes are 3 and 1, so the largest possible amplitude is 4. With a vertical displacement of 5, the range of this function is in the interval $[5-4, 5+4] = [1, 9]$. Note that this is probably *not* the actual range of the function. This range is just an estimate to help us determine what values we will need on the y-axis. This estimate will be very important when we do calculator graphing. (The actual range is about $(1.515, 8.485)$.)

Let's look at a table of values:

x	0	$\frac{\pi}{8}$	$\frac{\pi}{4}$	$\frac{3\pi}{8}$	$\frac{\pi}{2}$	$\frac{5\pi}{8}$	$\frac{3\pi}{4}$	$\frac{7\pi}{8}$	π
$y_1 = 3\cos 2x$	3	2.12	0	-2.12	-3	-2.12	0	2.12	3
$y_2 = -\sin 4x + 5$	5	4	5	6	5	4	5	6	5
$y = y_1 + y_2$	8	6.12	5	3.88	2	1.88	5	8.12	8

The graphs of y_1, y_2, and $y = y_1 + y_2$ are in Figure 11.17.

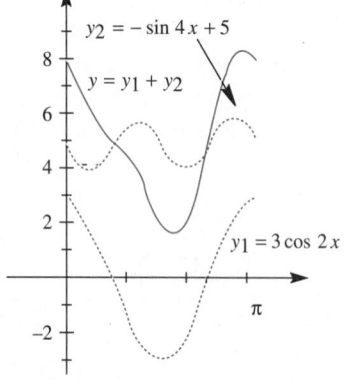

FIGURE 11.17

As the final sample graph, let's involve something more complicated than adding and subtracting two trigonometric functions. In this example, we will use a familiar curve, $y_1 = \sin 2x$, and a new curve, $y_2 = \dfrac{\sin x}{x}$.

EXAMPLE 11.12

Sketch the graph of $y = y_1 y_2$, where $y_1 = \sin 2x$, and $y_2 = \dfrac{\sin x}{x}$.

Solution The graph of y_2 is something like a sine curve. But you will notice that y_2 is not defined when $x = 0$. Also, since the values of $\sin x$ are always between $+1$ and -1, as the value of x gets larger and larger, the amplitude of y_2 will get smaller and smaller. The graph of $y_2 = \dfrac{\sin x}{x}$ is shown in Figure 11.18.

If we multiply these two functions, we get a new function, y, which is $y = (y_1)(y_2) = (\sin 2x)\left(\dfrac{\sin x}{x}\right) = \dfrac{(\sin 2x)(\sin x)}{x}$. A table of values for y_1, y_2 and $y = y_1 \cdot y_2$ follows. The curves for all three functions are shown in Figure 11.19.

x	0	$\frac{\pi}{3}$	$\frac{2\pi}{3}$	π	$\frac{4\pi}{3}$	$\frac{5\pi}{3}$	2π	$\frac{7\pi}{3}$	$\frac{8\pi}{3}$	3π
$y_1 = \sin 2x$	0	0.87	-0.87	0.02	0.87	-0.87	0.02	0.87	-0.87	0.02
$y_2 = \frac{\sin x}{x}$	*	0.83	0.41	0	-0.21	-0.17	0	0.12	0.10	0
$y = y_1 \cdot y_2$	*	0.71	-0.36	0	-0.18	0.14	0	0.10	-0.09	0

Calculator Graphics

Graphing these curves by hand takes a great deal of time. Using a calculator or a computer to draw the graphs of trigonometric curves would save time. We will briefly describe how to use a graphing calculator to graph trigonometric functions. All of the directions and illustrations are specific to a Texas Instruments TI-82 graphics calculator. If you have another brand or model then you may have to vary the directions slightly.

While you can graph trigonometric functions with your calculator in degree mode, it is easier if you first set it in radian mode.

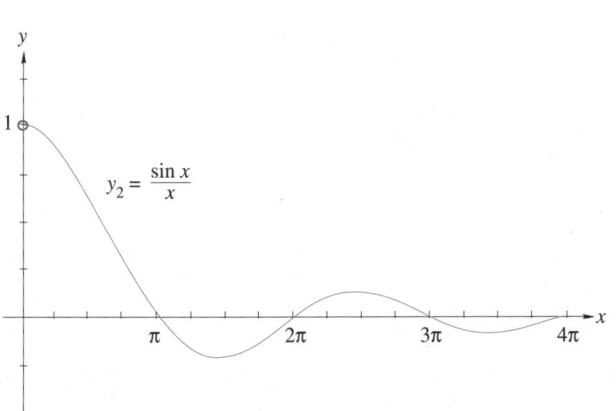

FIGURE 11.18

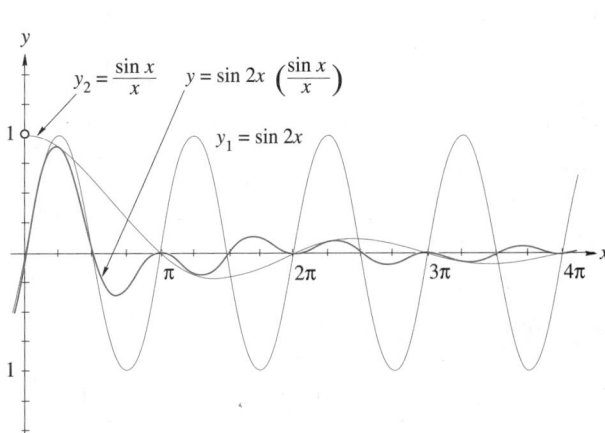

FIGURE 11.19

EXAMPLE 11.13

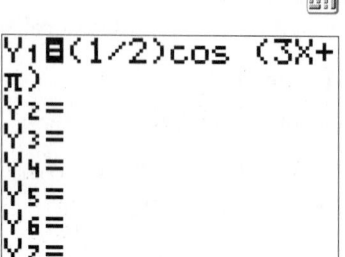

FIGURE 11.20a

Use a graphing calculator to graph $y = \frac{1}{2}\cos(3x + \pi)$.

Solution Make sure that your calculator is in radian mode. Set your calculator to the "standard" setting by pressing [ZOOM] 6 [ENTER]. Now, press [Y=] [CLEAR]. You are now ready to enter the function you want to graph. Press the following sequence of keys:

[(] 1 [÷] 2 [)] [COS] [(] 3 [X|T] [+] [2nd] [π] [)]

The calculator screen should look like the one shown in Figure 11.20a. (Note that we could have typed 0.5 instead of [(] 1 [÷] 2 [)].) Press [GRAPH] and you should obtain the graph shown in Figure 11.20b. Compare this graph to the one in Figure 11.10.

This graph is very difficult to see. We need to change the scale. We know that the amplitude of this function is $\frac{1}{2}$ and that it has a period of $\frac{2\pi}{3} \approx 2.09$. Let's change the scale as follows: Xmin $= -1$, Xmax $= 3$, Xscl $= \frac{\pi}{4} \approx 0.7854$, Ymin $= -1$, Ymax $= 1$, Yscl $= 0.5$. Why did we select these values for Xscl and Yscl? We selected Yscl $= 0.5$ to help us see where the highest and lowest values of the graph are. Now press [GRAPH]. The result is shown in Figure 11.20c.

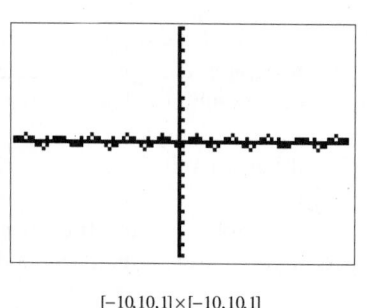

$[-10,10,1] \times [-10,10,1]$

FIGURE 11.20b

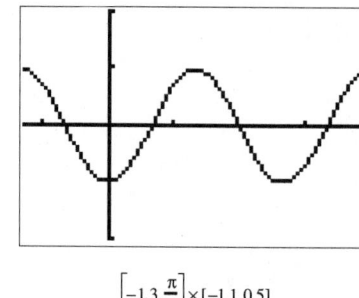

$\left[-1,3,\frac{\pi}{4}\right] \times [-1,1,0.5]$

FIGURE 11.20c

EXAMPLE 11.14

Use a graphing calculator to graph the product $y = (\sin 2x)\left(\frac{\sin x}{x}\right)$.

Solution This is the same function we graphed in Example 11.12. We will work this so we can separately graph $y_1 = \sin 2x$ and $y_2 = \frac{\sin x}{x}$ and then graph their product $y = y_1 y_2$.

Based on our work in Example 11.12, we choose Xmin $= 0$, Xmax $= 10$, Xscl $= 0.7854$, Ymin $= -1.1$, Ymax $= 1.1$, and Yscl $= 0.5$. Now press [Y=] [CLEAR] [SIN] 2 [X|T] This should appear on line Y1.

EXAMPLE 11.14 (Cont.)

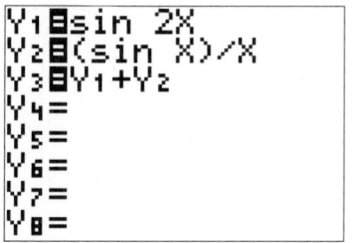

FIGURE 11.21a

To enter $y_2 = \dfrac{\sin x}{x}$, press $\boxed{\text{ENTER}}$. This moves the cursor to the next line, Y2. Now press

$$\boxed{(}\;\boxed{\text{SIN}}\;\boxed{\text{X}|\text{T}}\;\boxed{)}\;\boxed{\div}\;\boxed{\text{X}|\text{T}}$$

We will put $y = y_1 y_2$ on line Y3. Rather than entering both y_1 and y_2 again, we will let the calculator do this for us. Press $\boxed{\text{ENTER}}$ to move the cursor to line Y3 and press the following key sequence:

$$\boxed{\text{2nd}}\;\boxed{\text{Y-VARS}}\;1\;1\;\boxed{\text{2nd}}\;\boxed{\text{Y-VARS}}\;1\;2$$

Notice that line Y3 now reads Y3=Y1Y2. The calculator screen should look like that in Figure 11.21a. Press $\boxed{\text{GRAPH}}$. You should see each of the functions y_1, y_2, and y graphed in sequence, with the final result like that shown in Figure 11.21b.

This is too cluttered. Let's change it so we see only the graph of $y = y_1 y_2$. Press $\boxed{\text{Y=}}\;\boxed{\blacktriangleleft}$ to move the cursor on top of the equals sign. The equals sign is white on a black background, indicating that the calculator will graph this function. Press $\boxed{\text{ENTER}}$. Now the equals sign is black on a white background. This shows that the calculator will not graph Y1. Press $\boxed{\blacktriangledown}$ to move the cursor to the equals sign on line Y2 and press $\boxed{\text{ENTER}}$ to tell the calculator not to graph Y2. Now press $\boxed{\text{GRAPH}}$.

The result, shown in Figure 11.21c, is the graph of $y = y_1 y_2 = (\sin 2x)\left(\dfrac{\sin x}{x}\right)$.

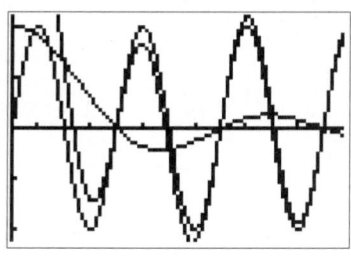

$$\left[0, 10, \frac{\pi}{4}\right] \times [-1.1, 1.1, 0.5]$$

FIGURE 11.21b

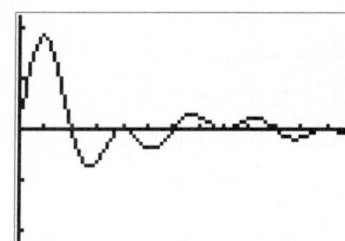

$$\left[0, 10, \frac{\pi}{4}\right] \times [-1.1, 1.1, 0.5]$$

FIGURE 11.21c

EXAMPLE 11.15

Use a graphing calculator to graph two cycles of $y = 3\cos 2x - \sin 4x + 5$.

Solution This is the same function we graphed in Example 11.11, so we can use much of the information from our work on that example to help us here. The two important pieces are the period and the range. These help us decide on the viewing window for the graph. The period for this function is π. The range of this function is in the interval $[5 - 4, 5 + 4] = [1, 9]$. Since the period is π, an interval from $x = 0$ to $x = 2\pi$ will include two cycles. We will set the viewing

EXAMPLE 11.15 (Cont.)

window to $\left[0, 2\pi, \frac{\pi}{8}\right] \times [0, 9, 1]$. Even though the range is no more than $[1, 9]$, we set the y-values in the viewing window to $[0, 9]$ so the x-axis would be included in the graph. The graph of $y = 3\cos 2x - \sin 4x + 5$ is in Figure 11.22. Compare it to the graph we got in Figure 11.17.

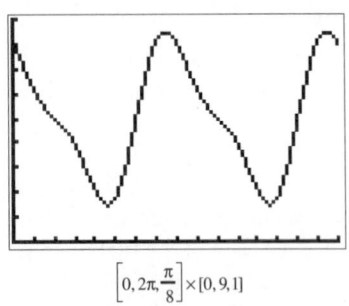

$\left[0, 2\pi, \dfrac{\pi}{8}\right] \times [0, 9, 1]$

FIGURE 11.22

Exercise Set 11.3

For each of the given functions in Exercises 1–24, (a) determine its period and maximum amplitude (if they exist), (b) determine a reasonable viewing window, and (c) sketch the graph by hand, with a graphing calculator or with computer graphing software.

1. $y = \sin x - \cos x$
2. $y = \sin 2x + 2\cos x$
3. $y = \frac{1}{3}\sin 4x + \sin 2x$
4. $y = 3\sin\frac{x}{3} + \cos 2x$
5. $y = x + \sin x$
6. $y = \sin x - \sin 2x$
7. $y = \cos 2x - \cos x$
8. $y = x^2 + \cos x$
9. $y = \sin \pi x - \cos 2x$
10. $y = \cos\frac{\pi}{2}x + 2\sin 3x$
11. $y = \cos 2x + \sin\left(x - \frac{\pi}{3}\right)$
12. $y = \frac{1}{4}\cos\left(x - \frac{\pi}{6}\right) + 4\sin\left(x + \frac{\pi}{6}\right)$

13. $y = 4\cos 2\pi x + \sin\left(\frac{\pi x}{2} + \frac{\pi}{4}\right)$
14. $y = \frac{\cos x}{x^2 + 1} - \sin x$
15. $y = \frac{\sin x}{x^2 + 1} + 2\cos\left(\frac{x}{3} - \frac{\pi}{2}\right)$
16. $y = 2\sin\left(3x + \frac{\pi}{4}\right) + 5\cos\left(2x - \frac{\pi}{3}\right)$
17. $y = x\sin x$
18. $y = 3(x + \pi)\cos\left(x - \frac{\pi}{4}\right)$
19. $y = 3\sin(2x + 1) + 2\cos(4x - 1) + 2$
20. $y = 2\sin(3x - \pi) + 5\cos(3x + 1) - 4$
21. $y = -2\sin(3x - 2) + 4\cos(5x + 1) - 3$
22. $y = -3\sin(4x - \pi) + \frac{1}{2}\cos\left(5x + \frac{\pi}{2}\right) + 6$
23. $y = 2.5\sin\left(2\pi x + \frac{\pi}{4}\right) + 3\sin\left(x - \frac{\pi}{4}\right) - 3.5$
24. $y = 0.5\sin(x - \pi) + \frac{1}{3}\cos\left(x + \frac{\pi}{2}\right) + 2.1$

Solve Exercises 25 and 26.

25. *Music* Plucking a guitar string produces a triangular wave. It is possible to represent such a triangular wave by adding several sinusoidal functions, such as $y_1 = A\sin kx$, $y_2 = -\frac{A}{3^2}\sin 3kx$, and $y_3 = \frac{A}{5^2}\sin 5kx$. Graph $y = y_1 + y_2 + y_3$ with $A = 1$ and $k = 1$.

26. *Electronics* The equation for a half-wave rectified pulsating waveform can be approximated by

$$V = 0.318V_m + 0.500V_m\cos\alpha$$
$$+ 0.212V_m\cos 2\alpha - 0.0424V_m\cos 4\alpha$$

Graph V from $x = 0$ to $x = 4\pi$ with $V_m = 1$.

In Your Words

27. Explain how to use the amplitude and period to help determine a useful viewing window.

28. Write a word problem in your technology area of interest that requires you to add two or more trigonometric functions. Give the problem you wrote to a friend and let him or her try to solve it. If your friend has difficulty understanding or solving the problem, or disagrees with your solution, make any necessary changes in the problem or solution. When you have finished, give the revised problem and solution to another friend and see if he or she can solve it.

≡ 11.4
GRAPHS OF THE OTHER TRIGONOMETRIC FUNCTIONS

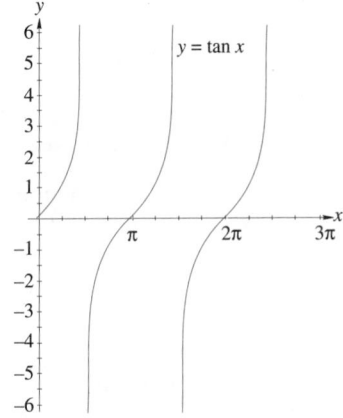

FIGURE 11.23

We have learned how to graph two of the trigonometric functions, $y = \sin x$ and $y = \cos x$. We can change the frequency, amplitude, and displacement of each of them. In this section, we will look at the graphs of the other four trigonometric functions.

The sin and cos functions that we have graphed were each bounded. That means that there was a limit, or bound, on how large or small the value of each function could get. For the sine and cosine functions, this bound was called the amplitude.

The four remaining trigonometric functions (tangent, cotangent, secant, cosecant) are not bounded. Thus, there is no limit to how large or small each function can become.

Graph of the Tangent Function

We will begin with the last of the basic trigonometric functions, the tangent function. A table of values for the tangent function, $y = \tan x$, follows. A graph based on these values is in Figure 11.23.

x	0	$\frac{\pi}{6}$	$\frac{\pi}{3}$	$\frac{\pi}{2}$	$\frac{2\pi}{3}$	$\frac{5\pi}{6}$	π			
$y = \tan x$	0	0.58	1.73	$*$	-1.73	-0.58	0			

x	π	$\frac{7\pi}{6}$	$\frac{4\pi}{3}$	$\frac{3\pi}{2}$	$\frac{5\pi}{3}$	$\frac{11\pi}{6}$	2π	$\frac{13\pi}{6}$	$\frac{7\pi}{3}$	$\frac{5\pi}{2}$
$y = \tan x$	0	0.58	1.73	$*$	-1.73	-0.58	0	0.58	1.73	$*$

*The tangent at this point is not defined.

Notice that the function is not defined when $x = \frac{\pi}{2}, \frac{3\pi}{2}, \frac{5\pi}{2}$, and so on. The vertical lines at each of these values of x that the graph of tangent approaches but never touches are called the **vertical asymptotes**. You might also note that the period of the tangent curve is π. (Remember, the sine and cosine have periods of 2π.)

Graphs of Cotangent, Secant, and Cosecant Functions

In a similar procedure, we could graph the three remaining trigonometric functions: cotangent, secant, and cosecant. The graphs for each of these functions are given in Figures 11.24, 11.25, 11.26, respectively. Notice that the period of the cotangent

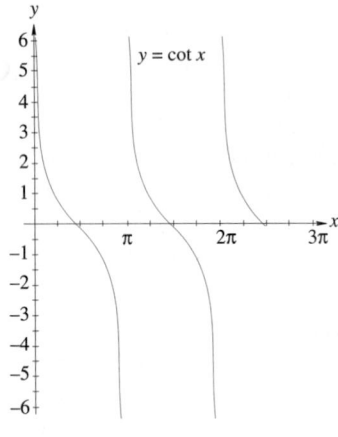

FIGURE 11.24

is also π. The period of the secant and cosecant functions is 2π. (Remember that $\csc x = \dfrac{1}{\sin x}$ and $\sec x = \dfrac{1}{\cos x}$, and both sine and cosine have period 2π.)

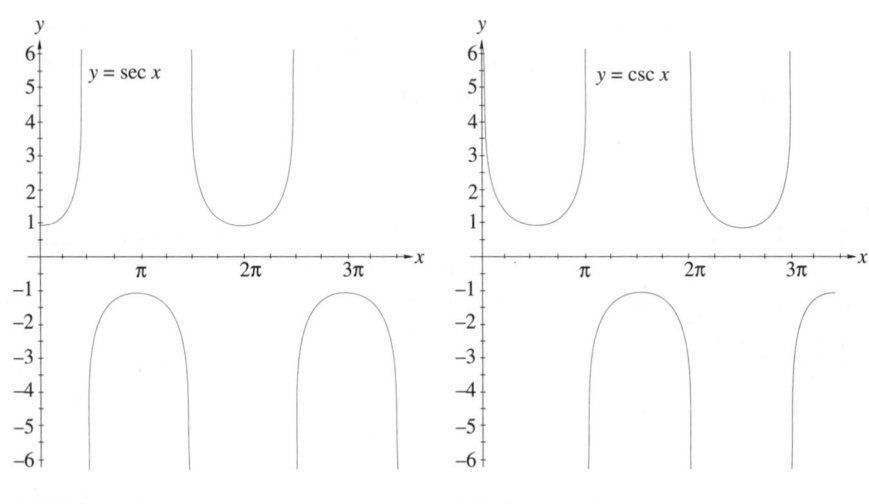

FIGURE 11.25   **FIGURE 11.26**

Variations in the Graphs of Tan, Cot, Sec, and Csc ▬▬▬

The terms period and frequency have the same meaning for these four functions as they did for the sine and cosine functions. The tangent and cotangent each have period π. The secant and cosecant have period 2π. Amplitude has no meaning for these functions. Each of these graphs has vertical asymptotes in which the function is not defined.

Sketching a function such as $y = 3\tan x$ or $y = -2\sec x$ is not difficult if you know the shapes of the six basic trigonometric functions. Knowing the shape means knowing where each graph crosses the x-axis (if it does), where its high and low points are, and where the function is not defined.

We will sketch both $y = 3\tan x$ and $y = -2\sec x$ to give you an idea of how to proceed. The graph for $y = 3\tan x$ is shown in Figure 11.27. The graph for $y = -2\sec x$ is shown in Figure 11.28. First, sketch the basic curve for each function. For $y = 3\tan x$, sketch $y = \tan x$. For $y = -2\sec x$, sketch $y = \sec x$. Now, multiply the y-values of each curve by the appropriate value and plot these points. For $y = \tan x$, multiply each y-value by 3 to get $y = 3\tan x$. For $y = \sec x$, multiply each y-value by -2 to get $y = -2\sec x$.

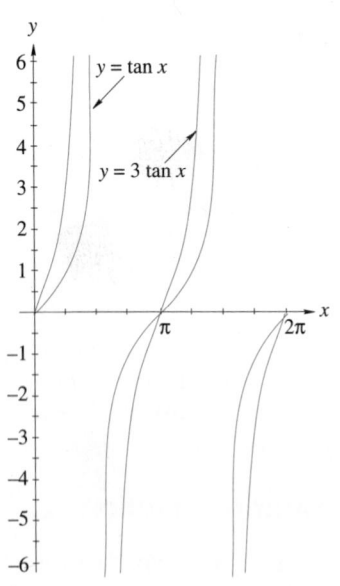

FIGURE 11.27

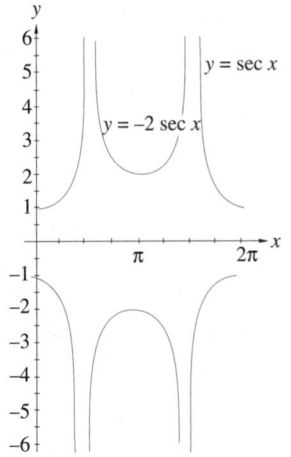

FIGURE 11.28

A summary of the period, amplitude, phase shift, and vertical asymptotes for all of the six trigonometric functions is given in Table 11.1.

TABLE 11.1   Summary for All Six Trigonometric Functions

Function	Period	Amplitude	Phase Shift	Vertical asymptotes
$y = A\sin(Bx + C)$	$\dfrac{2\pi}{|B|}$	$|A|$	$-\dfrac{C}{B}$	none
$y = A\cos(Bx + C)$	$\dfrac{2\pi}{|B|}$	$|A|$	$-\dfrac{C}{B}$	none
$y = A\tan(Bx + C)$	$\dfrac{\pi}{|B|}$	*	$-\dfrac{C}{B}$	$Bx + C = n\pi + \dfrac{\pi}{2}$
$y = A\cot(Bx + C)$	$\dfrac{\pi}{|B|}$	*	$-\dfrac{C}{B}$	$Bx + C = n\pi$
$y = A\sec(Bx + C)$	$\dfrac{2\pi}{|B|}$	*	$-\dfrac{C}{B}$	$Bx + C = n\pi + \dfrac{\pi}{2}$
$y = A\csc(Bx + C)$	$\dfrac{2\pi}{|B|}$	*	$-\dfrac{C}{B}$	$Bx + C = n\pi$

*Not defined

Calculator Graphics

EXAMPLE 11.16

Use a graphing calculator to sketch the graph of $y = -\frac{1}{3}\tan\left(2x + \frac{\pi}{4}\right)$.

Solution   Make sure your calculator is in radian mode. From Table 11.1, we see that the period is $\frac{\pi}{2} \approx 1.57$. We would like to see two periods of this graph, so we set Xmin $= -2$, Xmax $= 2$, Xscl $= \frac{\pi}{4} \approx 0.7854$, Ymin $= -4$, Ymax $= 4$, and Yscl $= 1$. Now press $\boxed{\text{Y=}}$ $\boxed{\text{CLEAR}}$ and then the following sequence of keys:

$\boxed{(-)}$ $\boxed{(}$ $\boxed{1}$ $\boxed{\div}$ $\boxed{3}$ $\boxed{)}$ $\boxed{\text{TAN}}$ $\boxed{(}$ $\boxed{2}$ $\boxed{\text{X|T}}$ $\boxed{+}$ $\boxed{(}$ $\boxed{\text{2nd}}$ $\boxed{\pi}$ $\boxed{\div}$ $\boxed{4}$ $\boxed{)}$ $\boxed{)}$

The calculator screen should now look like the one shown in Figure 11.29a. Press $\boxed{\text{GRAPH}}$ and you should obtain the graph in Figure 11.29b.

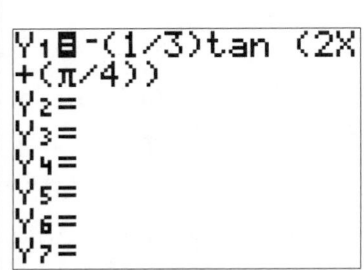

FIGURE 11.29a

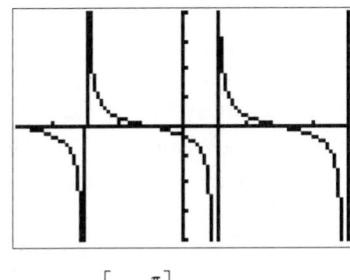

$\left[-1, 2, \frac{\pi}{4}\right] \times [-4, 4, 1]$

FIGURE 11.29b

Notice that the calculator seems to have drawn in the vertical asymptotes. Look closely. These are not vertical lines. The portion above the x-axis is not directly above the portion below the x-axis. These apparent vertical asymptotes are put in by some graphing calculators and computer graphing programs.

EXAMPLE 11.17

Use a graphing calculator to sketch $y = 1.5 \sec (x + 2)$.

Solution We set Xmin $= -1$, Xmax $= 7$, Xscl $= 1$, Ymin $= -5$, Ymax $= 5$, and Yscl $= .5$. Press $\boxed{\text{Y=}}$ $\boxed{\text{CLEAR}}$ to erase the previously graphed function. A calculator does not have a secant key, so before you graph this function you will need to use the reciprocal identity $\sec x = \dfrac{1}{\cos x}$ to rewrite this function as $y = \dfrac{1.5}{\cos (x + 2)}$. Now, press the following sequence of keys:

$$1.5 \boxed{\div} \boxed{(} \boxed{\text{COS}} \boxed{(} \boxed{\text{X|T}} \boxed{+} 2 \boxed{)} \boxed{)}$$

Press $\boxed{\text{GRAPH}}$ and you should obtain the graph in Figure 11.30. Once again, you can see that the calculator seems to have drawn in the vertical asymptotes. ▪

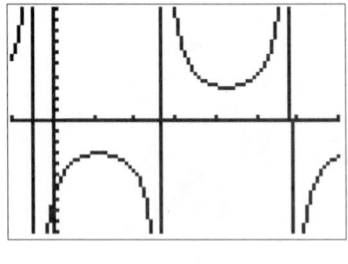

$[-1,7,1] \times [-5,5,0.5]$

FIGURE 11.30

Exercise Set 11.4

For each of the given functions in Exercises 1–20, (a) determine a reasonable viewing window, and (b) sketch the graph by hand, with a graphing calculator or with computer graphing software.

1. $y = 3 \tan x$
2. $y = -4 \tan x$
3. $y = \frac{1}{2} \sec x$
4. $y = 3 \csc x$
5. $y = -\frac{1}{2} \cot x$
6. $y = -2 \csc x$
7. $y = \tan \left(x + \frac{\pi}{4}\right)$

8. $y = \sec \left(x - \frac{\pi}{3}\right)$
9. $y = \cot \left(x + \frac{\pi}{6}\right)$
10. $y = \csc \left(x - \frac{2\pi}{5}\right)$
11. $y = \tan 2x$
12. $y = \sec 3x$
13. $y = \csc 4x$
14. $y = \cot 3x$

15. $y = 4 \tan \left(2x + \frac{\pi}{3}\right)$
16. $y = -3 \cot \left(5x - \frac{\pi}{6}\right)$
17. $y = \frac{1}{2} \sec \left(3x + \frac{\pi}{4}\right)$
18. $y = -\frac{1}{3} \csc \left(2x + \frac{\pi}{6}\right)$
19. $y = 2 \tan \left(0.5x + \frac{\pi}{4}\right) - 2$
20. $y = -2 \sec \left(\pi x - \frac{\pi}{6}\right) + 3$

Solve Exercises 21 and 22.

21. *Machine technology* The small-end diameter, d, of a tapered hole is a function of the angle, θ, of taper, and is given by

$$d = -2h \tan \theta + D$$

where h is the depth of the tapered hole and D is the large-end diameter. A certain tapered hole, measured in cm, has $h = 8$ and $D = 4.5$.

(a) What is the small-end diameter when $\theta = 10°$?

(b) What is the smallest angle that will produce a small-end diameter of $0°$?

(c) Sketch the graph of the equation for d when $0° \leq \theta < 30°$.

22. *Broadcasting* A television camera will be used to broadcast a parade. The camera, C, is located on a platform 75.0 ft from a point on the street, P, where $\overline{CP}$ is perpendicular to the street. Because of several obstructions, the camera can only show the parade between the angles of 40° to the left and 55° to the right.

(a) Determine the maximum distance to the left of P that this camera can televise.
(b) Determine the maximum distance to the right of P that this camera can televise.
(c) Sketch the graph of the distance from the camera to the parade over these viewing angles.

In Your Words

23. Explain how to graph a function such as $y = \csc(3x - 1)$ on your graphing calculator or with computer graphing software. Make sure that you show what keys need to be pressed.

24. What is a vertical asymptote? How can you tell when a trigonometric function will have a vertical asymptote?

≡ 11.5
APPLICATIONS OF TRIGONOMETRIC GRAPHS

There are many applications of trigonometric functions and their graphs. In this section we will examine some of them.

Periodic and Simple Harmonic Motion

One area in which period, frequency, amplitude, and displacement are applied is physics. **Periodic motion** is when an object repeats a certain motion indefinitely, always returning to its starting point after a constant time interval and then beginning a new cycle.

Simple harmonic motion is one type of periodic motion. One form of simple harmonic motion is a straight line. The acceleration is proportional to the displacement and in the opposite direction. Examples of simple harmonic motion include a weight bobbing on a spring, a simple pendulum, and a boat bobbing on water.

Simple Harmonic Motion

If a point moves so that its displacement varies with time t, measured in seconds, the point is in equilibrium when $t = 0$. Its displacement at any time t is given by

$$y = A \sin 2\pi f t$$

or $y = A \sin \omega t$

where $\omega = 2\pi f$. Then the point is in **simple harmonic motion**.

As you can see from these equations, the amplitude is $|A|$. The Greek letter omega, ω, represents the angular velocity in radians per second. The period is $\frac{2\pi}{\omega}$ and the frequency f is $\frac{\omega}{2\pi}$. The frequency is measured in cycles per second and the units are hertz (Hz), kilohertz (kHz), and megahertz (MHz). One **hertz** is equivalent to 1 cycle per second. Period is measured in seconds.

Application

EXAMPLE 11.18

An object is suspended from a spring. The object is pulled down 10 cm and released. It oscillates in simple harmonic motion. If it takes 0.2 s for it to complete one oscillation, and if we let $t = 0$ when it is at rest, find the amplitude, period, frequency, and angular velocity. Write an equation to describe the motion. (See Figure 11.31.)

Solution   We are given $A = 10$ cm and period $= 0.2$ s. Since the frequency is $\frac{1}{\text{period}}$, we can determine that the frequency is $\frac{1}{0.2} = 5$ oscillations/second. Now, the angular velocity ω is $2\pi f = 2\pi(5) = 10\pi$ rad/s. Since $\omega = 10\pi$, and $A = 10$, the equation that describes this motion is $y = 10\sin 10\pi t$.

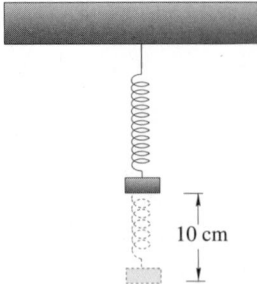

10 cm

FIGURE 11.31

Simple harmonic motion can also be described using the cosine function. If a point is at its maximum displacement when $t = 0$, and the rest of the previously mentioned conditions are met, then the displacement of the point at any time t can be expressed by

$$y = A\cos 2\pi f t$$

or $$y = A\cos \omega t$$

If a point is in simple harmonic motion, but is not in equilibrium or at its maximum value when $t = 0$, the sine or cosine curves can be used with the appropriate phase shift. Thus, $y = A\sin(\omega t + \phi)$ could be the equation for an object in simple harmonic motion with a phase shift of ϕ.

Alternating Current

Another common use of trigonometric curves is in the study of alternating current ac in electricity. The voltage of an ac generator varies with time t. The curve of the output voltage for a generator is a sine curve. Voltage varies with time according to these formulas.

$$V = V_{\text{max}} \sin 2\pi f t$$

or $$V = V_{\text{max}} \sin \omega t$$

In these formulas, V_{max} is the maximum value of the voltage and corresponds to the amplitude of the sine wave. The **angular frequency** of the alternating current is represented by $\omega = 2\pi f$ and is in rad/s. As before, f represents the frequency of the voltage in hertz.

In a similar manner, the current in an ac circuit varies with time in the same manner as the voltage. If the maximum value of the current is I_{max}, then the current I is represented by

$$I = I_{\text{max}} \sin 2\pi f t$$

or $$I = I_{\text{max}} \sin \omega t$$

Application

EXAMPLE 11.19

Maximum voltage across a power line is 170 V at a frequency of 60 Hz. What is the amplitude, period, and angular frequency of this voltage?

Solution The amplitude is $|V_{max}| = 170$ V. Since the frequency is 60 Hz or 60 cycles/s and the period is $1/f$, the period is $\frac{1}{60}$ s/cycle. The angular frequency is $\omega = 2\pi f = 2\pi(60) = 120\pi$ rad/s.

Application

EXAMPLE 11.20

If household current has a frequency of 60 Hz and a maximum of 2.5 A, find the amplitude of the current.

Solution The amplitude is $|I_{max}| = 2.5$ A.

Application

EXAMPLE 11.21

In the household current of Example 11.20, what is the current when $t = 1.32$ s?

Solution From Example 11.20 we know that $I_{max} = 2.5$ A and $f = 60$ Hz. So, $I = I_{max}\sin 2\pi f t = 2.5\sin 120\pi t$ describes the current. When $t = 1.32$ s, $I = 2.378$ A.

There are three different relationships that exist between a voltage V and a current I in an ac circuit. These three relationships are shown in Figure 11.32. In an ac circuit that contains only resistance, the current and voltage are **in phase** with each other. This means that both are 0 at the same time and both reach their maximum value at the same time. This is shown in Figure 11.32a.

In an ac circuit containing only inductance, the voltage leads the current by $\frac{1}{4}$ cycle, as shown in Figure 11.32b. In this case, the current and voltage are said to be $90°$ ($\frac{\pi}{2}$ rad) **out of phase**, since $90°$ is $\frac{1}{4}$ cycle. This is also referred to by stating that the current **lags** the voltage by $90°$.

Figure 11.32c demonstrates the relationship between the current and the voltage in a circuit that contains only capacitance. The current **leads** the voltage by $90°$ ($\frac{\pi}{2}$ rad) or $\frac{1}{4}$ cycle.

In an ac circuit that contains resistance, inductance, and capacitance in series, the instantaneous voltages across the circuit elements are V_R, V_L, and V_C, respectively. At any moment, the applied voltage $V = V_R + V_L + V_C$. Because V_R, V_L, and V_C are out of phase with one another, this formula holds only for instantaneous voltages. The phase angle ϕ can be determined from the relationship $\tan\phi = \dfrac{V_L - V_C}{V_R}$ or

$$\phi = \tan^{-1}\left(\frac{V_L - V_C}{V_R}\right).$$

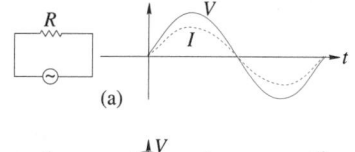

(a)

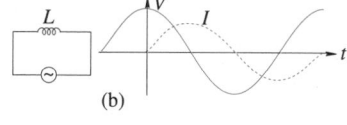

(b)

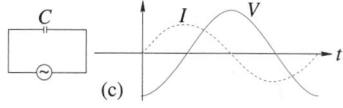

(c)

FIGURE 11.32

Application

A capacitor, inductor, and resistor are connected in series with a 120-V, 60-Hz power source. If $V_{R_{max}} = 135$ V, $V_{L_{max}} = 190$ V, and $V_{C_{max}} = 90$ V, graph the instantaneous voltage curve for the capacitance, inductance, resistance, and applied voltage and determine the phase angle.

Solution   The equation for the resistance is determined from the equation $V_R = V_{R_{max}} \sin \omega t$. Since $V_{R_{max}} = 135$ V and $\omega = 2\pi f = 120\pi$, we have $V_R = 135 \sin 120\pi t$. The inductance leads the current by 90° or $\frac{\pi}{2}$ rad, so $V_L = 190 \sin\left(120\pi + \frac{\pi}{2}\right)$.

Similarly, since the capacitance lags the current by 90°, or $\frac{\pi}{2}$, we have $V_C = 90 \sin\left(120\pi - \frac{\pi}{2}\right)$. Each of these curves and $V = V_R + V_L + V_C$ are shown in Figure 11.33. The phase angle ϕ is determined from $\tan\phi = \dfrac{V_L - V_C}{V_R} = \dfrac{190 - 90}{135} = 0.74$, so $\phi = \tan^{-1} 0.74 = 36.5°$ or 0.638 rad.

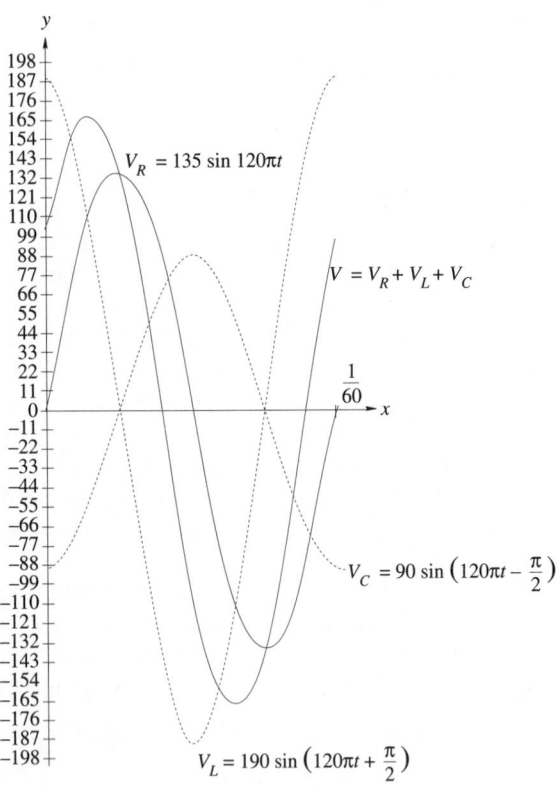

FIGURE 11.33

Exercise Set 11.5

Solve Exercises 1–22.

1. *Mechanics* A weight vibrates on a spring in simple harmonic motion. The amplitude is 10 cm and the frequency is 4 Hz. Write an equation for the weight's position y, at any time t, if $y = 0$ when $t = 0$ s.

2. *Mechanics* Write an equation for the position of the weight in Exercise 1 at any time t if $y = 8$ when $t = 0.1$ s.

3. *Oceanography* A raft bobs on the water in simple harmonic motion. Write a formula for the raft's position if the amplitude is 0.8 m, $\omega = \frac{\pi}{3}$ rad/s, and $y = 0$ when $t = 0$ s.

4. *Oceanography* What is the frequency of the raft in Exercise 3?

5. *Physics* The acceleration a of a pendulum is given by $a = -g \sin \theta$, where g is gravitational acceleration and θ is the angular displacement from the vertical. What is the acceleration of a pendulum, when $\theta = 5°$ and $g = 32$ ft/s^2?

6. *Physics* What is the acceleration of a pendulum when $\theta = 0.05$ rad and $g = 9.8$ m/s^2?

7. *Mechanics* A weight hanging from a spring vibrates in simple harmonic motion according to the equation $y = 8.5 \cos 2.8t$, where y is the position of the weight in centimeters and t is time in seconds. (a) What is the amplitude of the vibrating weight? (b) What is its period? (c) What is its frequency? (d) Sketch one complete cycle of this curve.

8. *Automotive technology* Each piston in a certain engine has stroke or total travel distance of 10.50 cm. (a) If the engine is operating at 3 000 rpm, what is the frequency of oscillation for each piston? (b) Write an equation to represent the position of the piston at any second t, if we assume the piston was at its maximum height (TDC), when $t = 0$. (c) If the engine speed is increased to 4 500 rpm, what equation would represent the position of the piston in (b) at any time t?

9. *Electricity* In an ac circuit, the current I is given by the relation $I = 10 \sin 120\pi t$. What are the amplitude, period, frequency, and angular velocity?

10. *Electricity* In an ac circuit, write the equation of the sine curve for the current when $I_{max} = 6.8$ A and $f = 80$ Hz.

11. *Electricity* In the circuit in Exercise 10, what is the equation of the sine curve when the phase angle is $\frac{\pi}{3}$ rad?

12. *Electricity* What is the equation of the sine curve for the voltage in an ac circuit, when $V_{max} = 220$ V and $f = 40$ Hz?

13. *Electricity* What is the equation of the sine curve for the voltage in Exercise 12, when the phase angle is $-\frac{\pi}{3}$ rad?

14. *Electricity* What is the equation of the cosine curve for the voltage in Exercise 12?

15. *Electricity* What is the equation of the cosine curve for the voltage in Exercise 13?

16. *Medical technology* The voltage for an x-ray can have a maximum value of 250 000 V. If the frequency of an x-ray is 10^{19} Hz, write the equation of the sine curve for the voltage.

17. *Physics* A gamma ray can be described by the equation

$$y = 10^{-12} \sin 2\pi 10^{23} t$$

What are the frequency and amplitude of this gamma ray?

18. *Electricity* In an ac circuit containing only a constant capacitance, the current leads the voltage by $\frac{1}{4}$ cycle. If $V = V_{max} \sin 2\pi f t$, $I_{max} = 3.6$ A and $f = 60$ Hz, write the equation for the current. Sketch one cycle of the curve.

19. *Electricity* In a resistance-inductance circuit, $I = I_{max} \sin(2\pi f t + \phi)$. If $I_{max} = 1.2$ A, $f = 400$ Hz, and $\phi = 37°$, sketch I vs t for one cycle.

20. *Electricity* A capacitor, inductor, and resistor are connected in series with a 120-V, 60-Hz power source. If $V_R = 120$ V, $V_L = 80$ V, and $V_C = 130$ V, graph the instantaneous voltage curves for the capacitance, inductance, resistance, and applied voltage, and determine the phase angle.

21. *Medical technology*   An x-ray is described by the equation

$$y = 10^{-10} \sin\left(2\pi \times 10^{18} t + \pi\right)$$

where y is in meters and t is in seconds.
(a) Determine the period and amplitude of this x-ray.
(b) Sketch one cycle of this function. Make sure that you label the axes correctly.

22. *Ecology*   The population of a certain species of animal in a region can be determined at any given time by the equation

$$P = 25{,}000 + 7{,}250 \cos\left(\frac{\pi}{12} t\right)$$

where P is the population after t months.
(a) What are the maximum and minimum sizes in the population?
(b) What is the population after one year?
(c) What is a reasonable viewing window for a graph of this function on your calculator?
(d) Graph this function for a three-year period.

 **In Your Words**

23. You may have had difficulty using a graphing calculator or computer graphing software to obtain a graph of the x-ray in Exercise 21. Explain why the calculator or software may have caused this problem.

24. Write a word problem in your technology area of interest that requires you to use a trigonometric function. On the back of the sheet of paper, write your name and explain how to solve the problem. Give the problem you wrote to a friend and let him or her try to solve it. If your friend has difficulty understanding or solving the problem, or disagrees with your solution, make any necessary changes in the problem or solution. When you have finished, give the revised problem and solution to another friend and see if he or she can solve it.

25. Write a word problem in your technology area of interest that requires you to use a trigonometric function different from the one you used in Exercise 24. Follow the same procedures described in Exercise 24 for writing, sharing, and revising your problem.

☰ 11.6
PARAMETRIC EQUATIONS

Graphing an equation in terms of two variables x and y is often done by setting up a table of values. We usually select a value for x or y and solve for the other variables. In the case of a function such as $y = x^2$ or $y = x^3 + 2x$, this has not been a problem. We have had difficulty with equations that were not functions, such as $x^2 + y^2 = 9$.

Parametric Equations

One solution is to rewrite this equation by expressing x and y as functions of a third variable, called a **parameter**. These equations are called **parametric equations**.

EXAMPLE 11.23

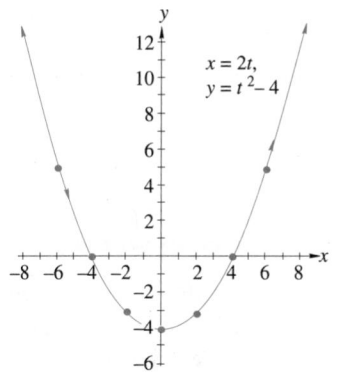

FIGURE 11.34

Describe and sketch the curve represented by the parametric equations

$$x = 2t, \ y = t^2 - 4$$

Solution The parameter for these equations is t. We will set up a table of values for t, x, and y. The graph is shown in Figure 11.34. We connect the points in order of the increasing values of t, as indicated by the arrows in Figure 11.34.

t	-4	-3	-2	-1	0	1	2	3	4
x	-8	-6	-4	-2	0	2	4	6	8
y	12	5	0	-3	-4	-3	0	5	12

Does the curve look familiar? It looks very much like some of the equations that we graphed earlier. It is possible to eliminate the parameter and write the equation in rectangular form. In these two equations, since $x = 2t$, then $t = \dfrac{x}{2}$. Substituting this value for t in the equation $y = t^2 - 4$, we get $y = \dfrac{x^2}{4} - 4$ or $y = \frac{1}{4}x^2 - 4$.

When eliminating the parameter, you must be very careful about the domain. Consider the following example.

EXAMPLE 11.24

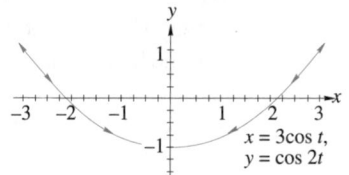

FIGURE 11.35

Describe and sketch the curve represented by the parametric equations $x = 3\cos t$ and $y = \cos 2t$.

Solution We will set up a table of values for t, x, and y.

t	0	$\frac{\pi}{6}$	$\frac{\pi}{4}$	$\frac{\pi}{3}$	$\frac{\pi}{2}$	$\frac{2\pi}{3}$	$\frac{3\pi}{4}$	$\frac{5\pi}{6}$	π	$\frac{7\pi}{6}$	$\frac{5\pi}{4}$	$\frac{4\pi}{3}$	$\frac{3\pi}{2}$
x	3	2.6	2.1	1.5	0	-1.5	-2.1	-2.6	-3	-2.6	-2.1	-1.5	0
y	1	0.5	0	-0.5	-1	-0.5	0	0.5	1	0.5	0	-0.5	-1

Notice that these values begin to repeat once we get to $t = \pi$. The sketch of this curve is shown in Figure 11.35. Again, we get a shape similar to the one in Example 11.23, except that the curve does not continue in each direction. It oscillates back and forth. Using techniques we will develop later, we can eliminate the parameter to form the rectangular equation $y = \dfrac{2x^2}{9} - 1$, where $-3 \le x \le 3$. Notice the restriction on the domain of x.

A simpler example demonstrating how the domain is often restricted when parametric equations are written in the rectangular form is shown by the next example.

EXAMPLE 11.25

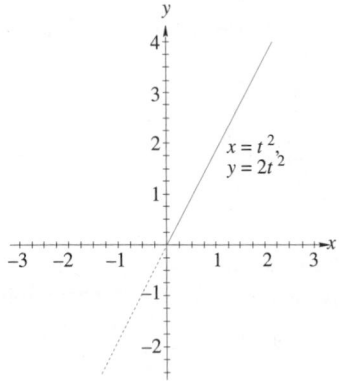

FIGURE 11.36

Describe and sketch the curve represented by the parametric equations $x = t^2$ and $y = 2t^2$.

Solution Since $x = t^2$ and $y = 2t^2$, we see that the rectangular form is $y = 2x$. This is the equation of a straight line. Notice, however, that for all values of t, $x \geq 0$ and $y \geq 0$. The graph formed by these parametric equations is shown in Figure 11.36 as a solid line. The remainder of the curve $y = 2x$ is indicated by the dashed line.

Lissajous Figures

When the parametric equations of a point describe simple harmonic motion, the resulting curve is called a **Lissajous figure**. When voltages of different frequencies are applied to the vertical and horizontal plates of an oscilloscope, a Lissajous figure results.

EXAMPLE 11.26

Sketch the graph of the parametric equations $x = \cos t$ and $y = \sin 3t$.

Solution The Lissajous curve for these parametric equations is shown in Figure 11.37.

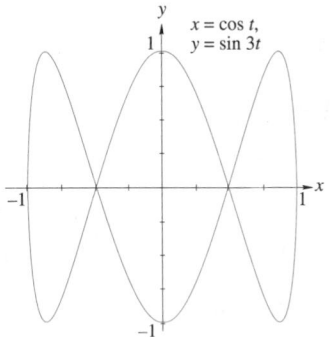

FIGURE 11.37

Examine the Lissajous figure in Figure 11.37. There are three loops along the top edge of the figure and one loop along the side. This is a ratio of 3:1. Now look at the frequencies of the parametric equations that generated this curve. The frequency of $x = \cos t$ is $\frac{1}{2\pi}$ and the frequency of $y = \sin 3t$ is $\frac{3}{2\pi}$. The ratio of the frequencies is 3:1. In Exercise Set 11.6 we will predict the number of loops on the top and side from the parametric equations and then graph the curve. While this is only an exercise in this text, it is a technique that can be used to calibrate signal generators.

Using a Calculator to Graph Parametric Equations

Before you can use a graphing calculator to draw the graphs of parametric equations, you need to put the calculator in parametric mode. On a TI-82 this is done by pressing MODE ▼ ▼ ▼ ▶ ENTER . The screen on a TI-82 should look like the one shown in Figure 11.38. On a Casio fx-7700G, press the key sequence MODE SHIFT × , to set the calculator in parametric mode.

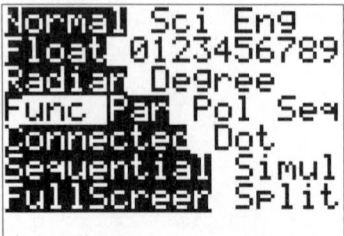

FIGURE 11.38

EXAMPLE 11.27

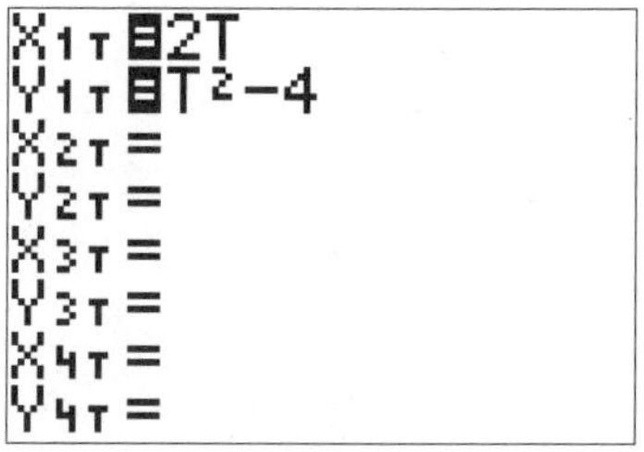

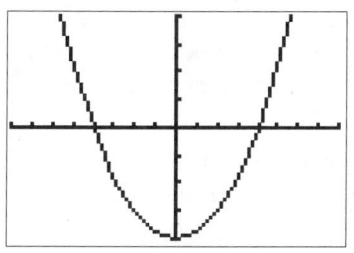

FIGURE 11.39a

$[-8, 8, 1] \times [-4, 4, 1]$

FIGURE 11.39b

Use a graphing calculator to sketch the curve represented by the parametric equations $x = 2t$ and $y = t^2 - 4$.

Solution   These are the same parametric equations we graphed in Example 11.23. We will use the following table from that example to help graph these equations.

t	-4	-3	-2	-1	0	1	2	3	4
x	-8	-6	-4	-2	0	2	4	6	8
y	12	5	0	-3	-4	-3	0	5	12

Press WINDOW. On a TI-82, you are first asked for Tmin, Tmax, and Tstep. Based on the table, we will let Tmin $= -4$ and Tmax $= 4$. The value of Tstep determines how much the value of t should be increased before calculating new values for x and y. We will pick Tstep $= 0.5$. You may want to try different values. Based on this table, let Xmin $= -8$, Xmax $= 8$, Xscl $= 1$, Ymin $= -4$, Ymax $= 4$, and Yscl $= 1$. A Casio fx-7700G calculator asks for these same values, but Tmin, Tmax, and Tptch are requested at the end, rather than at the beginning.

Now press Y=. On the first line, enter the right-hand side of the parametric equation for x. Press

$$2 \;\boxed{\text{X,T,}\theta}$$

Notice that when you pressed the $\boxed{\text{X,T,}\theta}$ key a T is displayed on the screen. This will always happen when the calculator is in parametric mode.

Next, enter the parametric equation for y. Press ENTER to move the cursor to the line labeled Y₁ₜ and press the key sequence

$$\boxed{\text{X,T,}\theta}\;\boxed{x^2}\;\boxed{-}\;4$$

The result in displayed in Figure 11.39a. Now press GRAPH and you should see the result in Figure 11.39b.

EXAMPLE 11.28

Use a graphing calculator to sketch the curve represented by the parametric equations $x = 2\cos t$ and $y = \sin t$.

Solution   Because the periods of sin and cos are 2π, we will let Tmin $= 0$, Tmax $= 6.3$, and Tstep $= 0.1$. Since the range of $x = 2\cos t$ is $[-2, 2]$, we choose Xmin $= -2$, Xmax $= 2$, and Xscl $= 1$. The range of $y = \sin t$ is $[-1, 1]$. We let Ymin $= -1.5$, Ymax $= 1.5$, and Yscl $= 1$. Now press Y= and enter the parametric equations by pressing

$$2 \;\boxed{\text{COS}}\;\boxed{\text{X,T,}\theta}\;\boxed{\text{ENTER}}\;\boxed{\text{SIN}}\;\boxed{\text{X,T,}\theta}\;\boxed{\text{GRAPH}}$$

The result is displayed in Figure 11.40.

EXAMPLE 11.28 (Cont.)

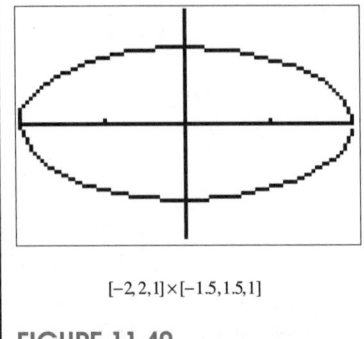

$[-2,2,1] \times [-1.5,1.5,1]$

FIGURE 11.40

Exercise Set 11.6

Graph each curve given by the parametric equations in Exercises 1–14. Make a table of values for at least six values of t. Show how the figure was drawn by drawing arrows on it. Eliminate the parameters in Exercises 1–6 and write the equation as a function of y.

1. $x = t,\ y = 3t$
2. $x = 2t,\ y = 4t + 1$
3. $x = t,\ y = \dfrac{1}{t}$
4. $x = t + 5,\ y = 3t - 2$

5. $x = 3 - t,\ y = t^2 - 9$
6. $x = t + 2,\ y = t^2 - t$
7. $x = 2\sin t,\ y = 2\cos t$
8. $x = 5\sin t,\ y = 2\cos t$
9. $x = 5\sin t,\ y = 3\sin t$

10. $x = 2\cos t,\ y = 6\cos t$
11. $x = t - \sin t,\ y = 1 - \cos t$
12. $x = \tan t,\ y = 6\cot t$
13. $x = \sec t,\ y = 2\csc t$
14. $x = 3\sec t,\ y = \tan t$

Examine the pairs of parametric equations in Exercises 15–24. For each pair, (a) predict the ratio of the loops along the top to the number of loops along the side, then (b) graph each of these curves. If possible, use a graphing calculator or computer graphing program to graph the curves.

15. $x = \sin t,\ y = \cos t$
16. $x = \sin 2t,\ y = \cos t$
17. $x = \sin t,\ y = \cos 2t$
18. $x = \sin 3t,\ y = \cos t$

19. $x = \sin 4t,\ y = \cos t$
20. $x = \sin 4t,\ y = \cos 2t$
21. $x = \sin 3t,\ y = \cos 2t$
22. $x = \sin 4t,\ y = \cos 3t$

23. $x = \sin 5t,\ y = \cos 2t$
24. $x = \sin 5t,\ y = \cos 3t$

Exercises 25–30 are Lissajous curves. For these curves, we have changed the amplitude. Graph each curve.

25. $x = 2\sin t,\ y = \cos t$
26. $x = 4\sin t,\ y = \cos t$

27. $x = 3\sin 2t,\ y = 2\cos t$
28. $x = 3\sin 3t,\ y = \cos t$

29. $x = \sin 3t,\ y = 4\cos t$
30. $x = 2\sin 2t,\ y = 5\cos 3t$

Solve Exercises 31 and 32.

31. *Automotive technology* The motion of a piece of gravel leaving a (rear) wheel at an angle α with speed V, in ft/s, can be described by

$$x = (V\cos\alpha)t$$
$$y = (V\sin\alpha)t - \frac{1}{2}gt^2$$

where $g \approx 32\,\text{ft/s}^2$.
(a) Assume that a car is traveling 30 mph (44 ft/s) and that three pieces of gravel leave its rear tire, one at $\alpha = 30°$, one at $\alpha = 45°$, and one at $\alpha = 50°$. Graph the path of each of these three pieces of gravel on the same set of axes.
(b) Rewrite these parametric equations as one equation in rectangular form.
(c) Use your answer to (b) to determine how far the gravel travels before hitting the road.

32. *Business* After a new consumer electronics product is introduced, sales rise quickly and the price gradually decreases. Let t be the number of years since a product was introduced (so, $t = 0$ is when the product was introduced). Suppose the unit price at time t, in hundreds of dollars, is $p(t) = \dfrac{t^2+20}{t^2+5}$, and the monthly sales, in 100,000 units, are $s(t) = \dfrac{t^2+3t}{t^2+1}$.
(a) Graph this pair of parametric equations for 5 years with $s(t)$ on the horizontal axis and $p(t)$ on the vertical axis.
(b) What was the price when the product was introduced? After 1 year? After 5 years?
(c) What were the monthly sales during the 12th month? After 5 years?

 **In Your Words**

33. Distinguish between rectangular equations and parametric equations.

34. Describe situations in which it is more helpful to use parametric equations than rectangular equations for graphing equations.

 **11.7**
POLAR COORDINATES

Every time we have represented a point in a plane, we used the rectangular coordinate system. Each point has an x- and a y-coordinate. There is another type of coordinate system that is used to represent points. This system is called the **polar coordinate system**.

To introduce a system of polar coordinates in a plane, we begin with a fixed point O called the **pole** or **origin**. From the pole, we shall draw a half-line that has O as its endpoint. This half-line will be called the **polar axis**.

Polar Coordinates

Consider any point in the plane different from point O. Call this point P. Let the polar axis form the initial side of an angle and $\overrightarrow{OP}$ the terminal side. This angle has measure θ. If the distance from O to P is r, we can say that the **polar coordinates** of P are (r, θ), as shown in Figure 11.41.

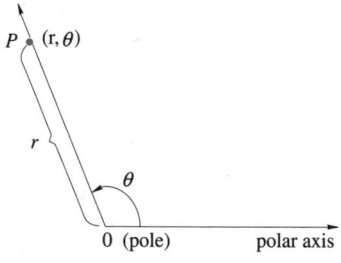

FIGURE 11.41

We have some of the same understanding about polar coordinates as we do about the angles from trigonometry. If the angle is generated by a counterclockwise rotation of the polar axis, the angle is positive. If it is generated by a clockwise rotation, θ is negative. If r is negative, the terminal side of the angle is extended in the opposite direction through the pole and is measured off r units on this extended side.

A special type of graph paper, called **polar coordinate paper**, is used to graph polar coordinates. This paper has concentric circles with their centers at the pole. The distance between any two consecutive circles is the same. Lines are drawn through the pole and correspond to some of the common angles. An example of polar coordinate paper is shown in Figure 11.42.

EXAMPLE 11.29

Plot the points with the following polar coordinates: $\left(5, \frac{\pi}{4}\right)$, $\left(-5, \frac{\pi}{4}\right)$, $(3, 150°)$, $(3, -70°)$, $(6, \pi)$, $(4, 0°)$, and $(6, 450°)$.

Solution   The points are plotted in Figure 11.43.

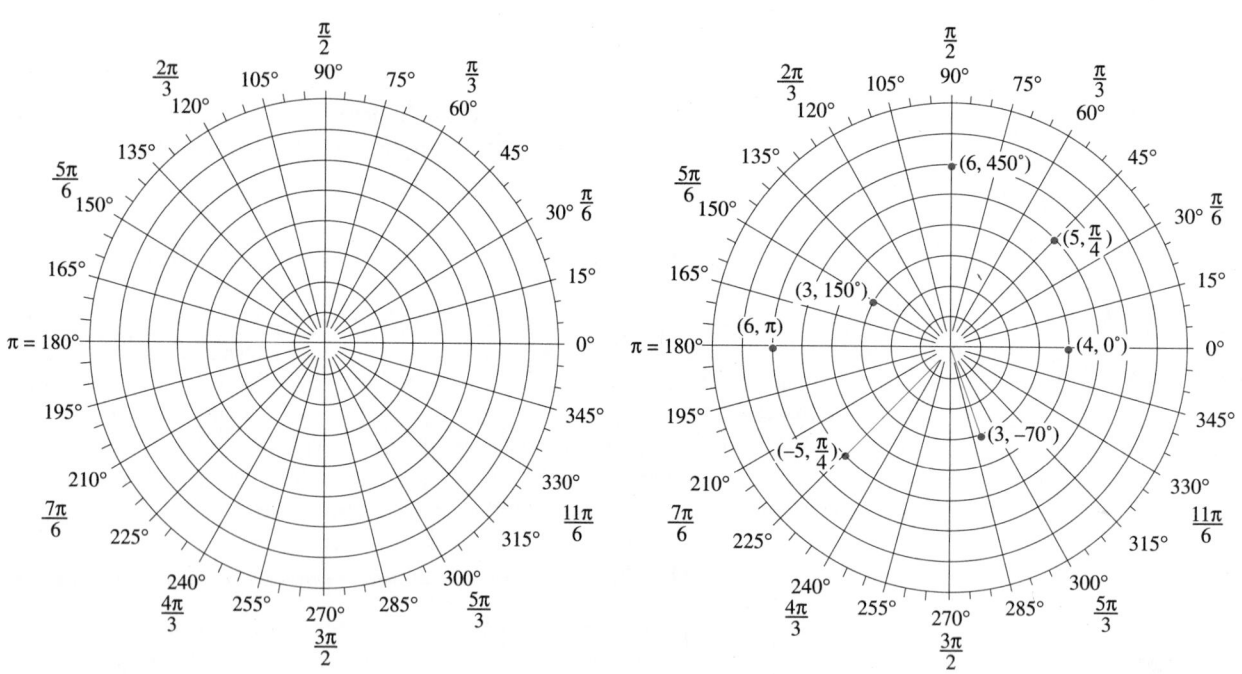

FIGURE 11.42   FIGURE 11.43

Notice that there is nothing unique about these points. For example, the point $(6, 450°)$ would be the same as the points $(6, 90°)$, $(6, -270°)$, or $(-6, 270°)$. The pole has the polar coordinate $(0, \theta)$, where θ can be any angle.

Converting Between Polar and Rectangular Coordinates ▪

Converting between the polar coordinate system and the rectangular coordinate system requires the use of trigonometry.

Converting Polar Coordinates to Rectangular Coordinates

If the point P has polar coordinates (r, θ) and rectangular coordinates (x, y), then

$$x = r\cos\theta$$
$$y = r\sin\theta$$

This converts the polar coordinates to rectangular coordinates.

EXAMPLE 11.30

Find the rectangular coordinates of the point with the polar coordinates $\left(8, \frac{5\pi}{6}\right)$.

Solution   From the equations previously given, we have $x = r\cos\theta$ and $y = r\sin\theta$. In this example, $r = 8$ and $\theta = \frac{5\pi}{6}$. So, $x = 8\left(\dfrac{-\sqrt{3}}{2}\right) = -4\sqrt{3} \approx -6.93$ and $y = 8\left(\frac{1}{2}\right) = 4$. The rectangular coordinates are $(-4\sqrt{3}, 4)$.

To convert from rectangular to polar coordinates, we need to use the Pythagorean theorem.

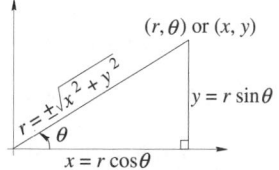

FIGURE 11.44

Converting Rectangular Coordinates to Polar Coordinates

The equations

$$r^2 = x^2 + y^2 \qquad \text{or} \qquad r = \pm\sqrt{x^2 + y^2}$$
$$\tan\theta = \frac{y}{x}, x \neq 0$$

will convert rectangular coordinates to polar coordinates. (See Figure 11.44.)

EXAMPLE 11.31

Find polar coordinates of $\left(-4\sqrt{3}, 4\right)$.

Solution   This is the reverse of Example 11.30, so we know that the answer should be $\left(8, \frac{5\pi}{6}\right)$. Let's practice using the two conversion formulas $r^2 = x^2 + y^2$ and $\tan\theta = \dfrac{y}{x}$.

EXAMPLE 11.31 (Cont.)

Here, $x = -4\sqrt{3}$ and $y = 4$.

$$r^2 = x^2 + y^2$$
$$= \left(-4\sqrt{3}\right)^2 + 4^2$$
$$= 48 + 16 = 64$$

and $\qquad r = \pm 8$

$$\tan\theta = \frac{4}{-4\sqrt{3}} = \frac{-1}{\sqrt{3}}$$

$$\theta = \tan^{-1}\left(\frac{-1}{\sqrt{3}}\right)$$

$$\approx 0.5236$$

and so $\qquad \approx -\dfrac{\pi}{6}$

It looks as if the answer is $\left(8, -\frac{\pi}{6}\right)$ or $\left(-8, -\frac{\pi}{6}\right)$. Now $\left(-8, -\frac{\pi}{6}\right)$ is correct, but $\left(8, -\frac{\pi}{6}\right)$ is not correct. What happened? Plot $\left(-4\sqrt{3}, 4\right)$. It is in Quadrant II. Now plot $= \left(8, \frac{-\pi}{6}\right)$. It is in Quadrant IV. Remember that $\tan^{-1}$ will only give angles in Quadrants I and IV. You should first determine which quadrant a point is in. If it is in Quadrants II or III, you will need to add π (or $180°$) to the answer you get using the . conversion formula. $\left(\text{Because } \theta \text{ may be in Quadrant II or III, we write } \tan\theta = \frac{y}{x}, \text{ rather than } \tan^{-1}\frac{y}{x} = \theta.\right)$ If we add π to $\frac{-\pi}{6}$, we get $\frac{5\pi}{6}$. Thus, two possible answers are $\left(-8, -\frac{\pi}{6}\right)$ and $\left(8, \frac{5\pi}{6}\right)$. Both of these are polar coordinates of the point with Cartesian coordinates $\left(-4\sqrt{3}, 4\right)$.

Polar Equations

Equations can also be written using the variables r and θ. These are called **polar equations**. A polar equation states a relationship between all the points (r, θ) that satisfy the equation. In the remaining part of this section, we will graph some polar equations.

As we usually do when we graph an equation, we will use a table of values, plot the points in the table, then connect these points in order as the values of θ increase.

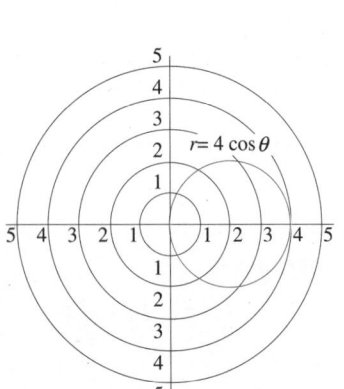

FIGURE 11.45

EXAMPLE 11.32

Graph the function $r = 4\cos\theta$.

Solution A table of values follows. The graph of the points is shown in Figure 11.45

θ	0°	30°	60°	90°	120°	150°	180°	210°	240°	270°	300°	330°	360°
r	4	3.46	2	0	−2	−3.46	−4	−3.46	−2	0	2	3.46	4

Notice that this is a circle with radius 2 centered at $(2, 0°)$.

> **Circles**
>
> The graph of any equation of the type $r = a\cos\theta$ is a circle of radius $\left|\frac{a}{2}\right|$ and center $\left(\frac{a}{2}, 0°\right)$. An equation of the type $r = a\sin\theta$ is a circle with radius $\left|\frac{a}{2}\right|$ and center $\left(\frac{a}{2}, 90°\right)$.

EXAMPLE 11.33

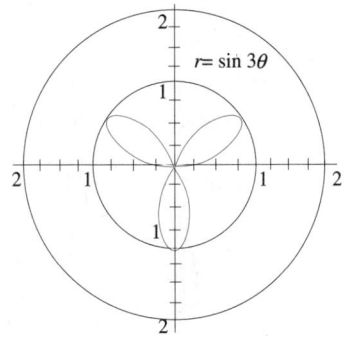

FIGURE 11.46

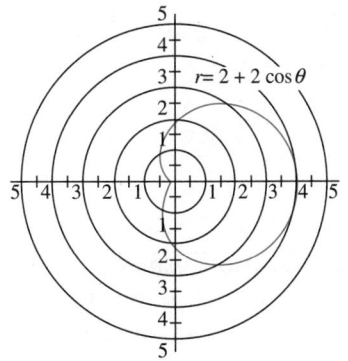

FIGURE 11.47

Graph the function $r = \sin 3\theta$.

Solution The table of values follows. The graph of these points is in Figure 11.46.

θ	0°	10°	20°	30°	40°	50°	60°	70°	80°	90°
$r = \sin 3\theta$	0	0.5	0.87	1	0.87	0.5	0	−0.5	−0.87	−1

θ	90°	100°	110°	120°	130°	140°	150°	160°	170°	180°
$r = \sin 3\theta$	−1	−0.87	−0.5	0	0.5	0.87	1	0.87	0.5	0

A curve of this type is called a **rose**. This rose has three petals.

> **Roses**
>
> Any polar equation of the form $r = \sin n\theta$ or $r = \cos n\theta$, where n is a positive integer, is a rose. If n is an odd number, the rose will have n petals. If n is an even number, the rose will have $2n$ petals.

For many functions, polar equations are much simpler to work with than are rectangular equations (and vice versa). A good example of this will be seen in a later chapter.

Another type of polar graph is demonstrated by the graph of the polar equation $r = 2 + 2\cos\theta$. The graph of this equation is shown in Figure 11.47.

The curve in Figure 11.47 is called a **cardioid** because of its heart shape. In Exercise Set 11.7, we will graph other polar equations. If the curve has a special name, we will give that name.

Using a Calculator to Graph Polar Equations

You can use a graphing calculator or a computer to help you graph many of these curves. We will describe how it is done using a graphics calculator. The procedures for a TI-82 differ quite a bit from those for a Casio fx-7700G. We will use a TI-82 for the first example and a Casio fx-7700G for the second example.

EXAMPLE 11.34

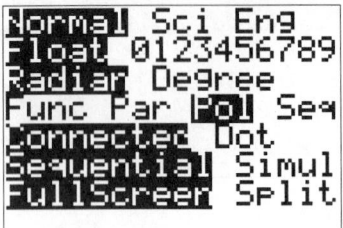

FIGURE 11.48a

Use a TI-82 graphing calculator to graph $r = \sin^2 3\theta$.

Solution First, put your calculator in polar mode. On a TI-82 this is done by pressing MODE ▼ ▼ ▼ ► ► ENTER. The screen on a TI-82 should look like the one shown in Figure 11.48a.

Press WINDOW to set the size of the viewing window. The function $r = \sin^2 3\theta$ has a period of 2π, so it will take no more than from $\theta = 0$ to $\theta = 2\pi$ to completely sketch this graph, so set $\theta\text{min} = 0$, $\theta\text{max} = 2\pi$, and $\theta\text{step} = \pi/24$. Since the function has an amplitude of 1, a window of $[-1.5, 1.5] \times [1, 1]$ should be right. You are now ready to enter the function into the calculator.

Now press Y=. On the first line, enter the right-hand side of the parametric equation for r. Press

$$(\boxed{\sin} \; 3 \; \boxed{\text{X,T,}\theta} \;) \; \boxed{x^2}$$

Notice that when you pressed the X,T,θ key a θ was displayed on the screen. This will always happen when the calculator is in polar mode. Remember, that $\sin^2 \theta$ is entered into a calculator or computer as $(\sin \theta)^2$. The result in displayed in Figure 11.48b. Now press GRAPH and you should see the result in Figure 11.48c.

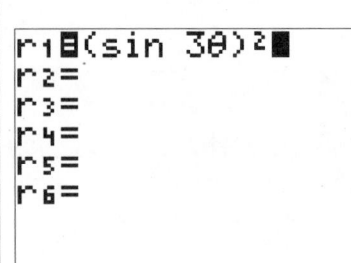

FIGURE 11.48b

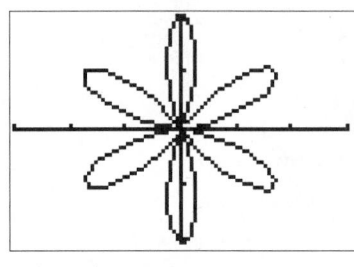

$[-1.5, 1.5, 0.5] \times [-1, 1, 0.5]$

FIGURE 11.48c

EXAMPLE 11.35

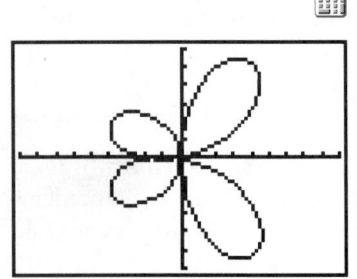

FIGURE 11.49

Use a Casio fx-7700G graphing calculator to graph $r = 5\sin 2\theta - 2\cos 3\theta$.

Solution First, put your calculator in polar mode. Press MODE SHIFT −. Next, set the range. Since the ranges of sine and cosine are both $[-1, 1]$, we know that $-7 \le r \le 7$. Because the screen is longer horizontally than it is vertically, we set Xmin -9, Xmax $= 9$, Xscl $= 1$, Ymin $= -6$, Ymax $= 6$, and Yscl $= 1$. After you press EXE, you get a new screen, where you enter the values of θ. The period of $\sin 2\theta$ is π; the period of $\cos 3\theta$ is $\frac{2}{3}\pi$. Since the least common multiple of π and $\frac{2}{3}\pi$ is 2π, the period of r is a multiple of 2π. We will set $\theta\text{min} = 0$, $\theta\text{max} = 2\pi$ (press 2 SHIFT π), and $\theta\text{ptch} = 0.1$. Press EXE Graph and then the following:

$$5 \; \boxed{\text{SIN}} \; 2 \; \boxed{\text{X,}\theta\text{,T}} \; \boxed{-} \; 3 \; \boxed{\text{X,}\theta\text{,T}} \; \boxed{\text{EXE}}$$

The result is a butterfly-shaped curve as shown in Figure 11.49.

Exercise Set 11.7

In Exercises 1–12, plot the points with the given polar coordinates.

1. $\left(2, \frac{\pi}{4}\right)$
2. $(4, 60°)$
3. $(3, 90°)$

4. $(5, \pi)$
5. $\left(-4, \frac{2\pi}{3}\right)$
6. $(-6, 30°)$

7. $(-4, 0°)$
8. $\left(-2, \frac{5\pi}{6}\right)$
9. $(5, -30°)$

10. $\left(3, -\frac{3\pi}{4}\right)$
11. $\left(-7, \frac{11\pi}{6}\right)$
12. $(-2, -270°)$

In Exercises 13–24, convert each polar coordinate to its equivalent rectangular coordinates.

13. $\left(4, \frac{\pi}{3}\right)$
14. $(5, 75°)$
15. $(2, 135°)$

16. $\left(3, \frac{3\pi}{2}\right)$
17. $(-6, 20°)$
18. $\left(-3, \frac{5\pi}{3}\right)$

19. $(-2, 4.3)$
20. $(-5, 255°)$
21. $(3, -170°)$

22. $\left(4, -\frac{\pi}{8}\right)$
23. $(-6, -2.5)$
24. $(-3, -195°)$

In Exercises 25–36, convert each rectangular coordinate to an equivalent polar coordinate with $r > 0$.

25. $(4, 4)$
26. $(3, 6)$
27. $(4, 3)$

28. $(5, 12)$
29. $(-20, 21)$
30. $(-12, 5)$

31. $(-3, 4)$
32. $(9, -5)$
33. $(-7, -10)$

34. $(-8, -3)$
35. $(2, 9)$
36. $(-6, 1)$

In Exercises 37–58, graph the polar equations.

37. $r = 4$
38. $r = -6$
39. $r = 3\sin\theta$
40. $r = -5\cos\theta$
41. $r = 4 - 4\sin\theta$ (cardioid)
42. $r = 1 + 3\cos\theta$ (limacon)
43. $r = 3\cos 5\theta$ (five-petaled rose)
44. $r = 3\sin 2\theta$ (four-petaled rose)
45. $r = 5\sec\theta$
46. $r = -7\csc\theta$
47. $r = \theta$ (Let θ get larger than 4π.)
48. $r = 3^\theta$ (spiral)
49. $r = 4 + 4\sec\theta$

50. $r = \dfrac{1}{\theta}, \theta > 0$
51. $r = 3 + \cos\theta$
52. $r^2 = 16\sin 2\theta$
53. $r = 2 + 5\sin\theta$
54. $r = 1 + 4\sec\theta$
55. $r = \dfrac{6}{3 + 2\sin\theta}$
56. $r = \dfrac{6}{1 + 3\cos\theta}$
57. $r = \dfrac{3}{2 + 2\cos\theta}$
58. $r = \dfrac{4\sec\theta}{2\sec\theta - 1}$

Solve Exercises 59–62.

59. *Space technology*   The polar equation for a certain satellite's orbit is given by

$$r = \frac{6000}{1.4} - 0.25\cos\theta$$

Sketch the graph of this satellite.

60. *Architecture*   In the design of a geodesic dome, an architect uses the equation

$$r^2 = \frac{E^2}{E^2\cos^2\theta + \sin^2\theta}$$

where E is a constant.

(a) What is the rectangular form of this equation?

(b) Let $E = 0.5$ and graph this function on your calculator.

61. *Electronics*   The field strength, r, in μV/m, of a broadcast station 1 mi from the antenna is given by

$$r = 2 + 5\cos 2\theta + \sin\theta$$

Use your calculator to sketch the graph of this antenna pattern.

62. *Robotics*   The path that an industrial robot must follow in a certain production process is described by

$$r = 1 - 3\sin\theta$$

(a) Convert this equation into rectangular form.

(b) Use your calculator to sketch the graph of this robot's path.

In Your Words

63. Distinguish among polar equations, rectangular equations, and parametric equations.

64. Describe situations in which polar equations are more helpful than rectangular equations for graphing equations.

≡ CHAPTER 11 REVIEW

Important Terms and Concepts

Amplitude
Cosine curve
Displacement
Frequency
Harmonic motion
Horizontal displacement
Lissajous figures
Parametric equations

Period
  Periodic motion
Phase shift
Polar coordinates
Polar equations
Sine curves
Sinusoidal curve
Vertical displacement

Review Exercises

In Exercises 1–10, find the period, amplitude, frequency, and displacement of the given functions and graph one cycle of each.

1. $y = 8\cos 4x$

2. $y = 3\sin 2x$

3. $y = 2\tan 3x$

4. $y = 3\sin\left(2x + \frac{\pi}{2}\right)$

5. $y = \frac{1}{2}\cos\left(3x - \frac{\pi}{3}\right)$

6. $y = \frac{1}{3}\sec\left(2x - \frac{\pi}{6}\right)$

7. $y = -4\cot\left(x + \frac{\pi}{4}\right)$

8. $y = -\frac{1}{2}\csc\left(\frac{x}{3} - \frac{\pi}{5}\right)$

9. $y = 2\sin\left(\frac{2}{3}x + \frac{\pi}{6}\right)$

10. $y = \frac{-1}{4}\cos\left(-\frac{x}{2} + \frac{2\pi}{3}\right)$

In Exercises 11–20, sketch the curves of the given equations.

11. $y = \sin 2x + \cos 3x$

12. $y = 4\sin \dfrac{x}{4} - \cos 2x$

13. $y = \cos 2x + \sin 3x$

14. $y = x + \sin\left(3x + \frac{\pi}{4}\right)$

15. $x = 3t, y = 5t - 2$

16. $x = t + 2, y = t^2 + t$

17. $x = 4\sin t, y = 3\cos t$

18. $x = \sin t, y = \cos 4t$

19. $r = \cos 7\theta$

20. $r = 3 + 2\sin\theta$

Solve Exercises 21–24.

21. *Mechanics*   A weight vibrates on a spring in simple harmonic motion. The amplitude is 85 mm and the frequency is 5 Hz. Write an equation for the weight's position y at any time t, if $y = 0$ when $t = 0$.

22. *Oceanography*   A raft is bobbing on the water in simple harmonic motion according to the equation $y = 1.7\sin 3.4t$, where y is the position of the raft in meters and t is the time in seconds. **(a)** What is the amplitude of the raft? **(b)** What is its period?

(c) What is its frequency? **(d)** Sketch one complete cycle for this curve.

23. *Electronics*   In an ac circuit containing only a constant capacitance, the current leads the voltage by $\frac{1}{4}$ cycle. If $V = V_{max}\sin 2\pi f t$ V, $I_{max} = 4.8$ A, and $f = 60$ Hz, write the equation for the current. Sketch one cycle of the curve.

24. *Electronics*   In a resistance-inductance circuit, $I = I_{max}\sin(2\pi f t + \phi)$. If $I_{max} = 1.5$ A, $f = 360$ Hz, and $\phi = -31°$, sketch I vs t for one cycle.

≡ CHAPTER 11 TEST

1. For the following trigonometric functions, find the period, amplitude, frequency, and displacement of the given functions.
   (a) $y = -3\sin 5x$
   (b) $y = 2.4\cos\left(3x - \frac{\pi}{4}\right)$
   (c) $y = 1.5\tan\left(2x + \frac{\pi}{5}\right)$

2. Sketch each of the given curves.
   (a) $y = -3\sin 5x$
   (b) $y = \frac{1}{2}\sec\left(x + \frac{\pi}{2}\right)$
   (c) $y = \sin 2x + 2\cos 3x$

3. Sketch $x = 2t, y = t^2 + 1$.

4. Sketch $r = 1 + 2\cos\theta$.

5. Convert $(2, 55°)$ to rectangular coordinates.

6. Convert $(5, -12)$ to polar coordinates.

7. In an ac electric circuit, write the equation of the sine curve for the current, when $I_{max} = 5.7$ A and $f = 60$ Hz.

12

Exponents and Radicals

When an object is placed between two light sources so the illuminance is the same from each source, an equation involving radicals is used to determine the intensity of each light. In Section 12.4, we will learn how to determine this intensity.

Courtesy of Michael A. Gallitelli, Metroland Photo Inc.

Exponents and radicals were introduced in Chapter 1. In this chapter, we will explore the topic further and examine the close relationship between exponents and radicals.

≡ 12.1
FRACTIONAL EXPONENTS

The basic rules for exponents were given in Section 1.4. They are repeated here to refresh your memory.

Basic Rules for Exponents	
	Rule 1: $b^m b^n = b^{m+n}$
	Rule 2: $(b^m)^n = b^{mn}$
	Rule 3: $(ab)^n = a^n b^n$
	Rule 4: $\left(\dfrac{a}{b}\right)^m = \dfrac{a^m}{b^m}, b \neq 0$
	Rule 5: $\dfrac{b^m}{b^n} = b^{m-n}, b \neq 0$
	Rule 6: $b^0 = 1, b \neq 0$
	Rule 7: $b^{-n} = \dfrac{1}{b^n}, b \neq 0$

When we studied these rules in Section 1.4, m and n had to be integers. In Section 1.7, we examined the meaning of $b^{1/n}$ and found that it meant $\sqrt[n]{b}$. This makes sense, since $(b^{1/n})^n = b^{n/n} = b$. This gives us a new rule:

Rule 8 for Exponents	$b^{1/n} = \sqrt[n]{b}$

EXAMPLE 12.1

Evaluate (a) $27^{1/3}$, (b) $16^{1/2}$, and (c) $625^{1/4}$.

Solutions   (a) $27^{1/3} = \sqrt[3]{27} = 3$

            (b) $16^{1/2} = \sqrt{16} = 4$

            (c) $625^{1/4} = \sqrt[4]{625} = 5$

If we combine Rules 2 and 8, we obtain the following new rule.

Rule 9 for Exponents	$b^{m/n} = (b^m)^{1/n} = \sqrt[n]{b^m}$

EXAMPLE 12.2

Evaluate (a) $8^{2/3}$, (b) $64^{5/2}$, (c) $81^{5/4}$, and (d) $9^{-3/2}$.

Solutions   (a) $8^{2/3} = \sqrt[3]{8^2} = \sqrt[3]{64} = 4$

            (b) $64^{5/2} = \sqrt{64^5} = \sqrt{1{,}073{,}741{,}824} = 32{,}768$

            (c) $81^{5/4} = \sqrt[4]{81^5} = \sqrt[4]{3{,}486{,}784{,}401} = 243$

            (d) $9^{-3/2} = \dfrac{1}{\sqrt{9^3}} = \dfrac{1}{\sqrt{729}} = \dfrac{1}{27}$

It is often easier to find the root before raising the number to a power. Thus, we could rewrite Rule 9.

Rule 9 for Exponents	$b^{m/n} = \left(b^{1/n}\right)^m = \left(\sqrt[n]{b}\right)^m$

We will rework the problems in Example 12.2 using this variation of Rule 9.

EXAMPLE 12.3

Evaluate **(a)** $8^{2/3}$, **(b)** $64^{5/2}$, **(c)** $81^{5/4}$, **(d)** $9^{-3/2}$, **(e)** $(-8)^{2/3}$, **(f)** $\left(\frac{8}{27}\right)^{-1/3}$, **(g)** $\left(\frac{1}{16}\right)^{-5/4}$, and **(h)** $\left(-\frac{64}{125}\right)^{-2/3}$.

Solutions

(a) $8^{2/3} = \left(\sqrt[3]{8}\right)^2 = 2^2 = 4$

(b) $64^{5/2} = \left(\sqrt{64}\right)^5 = 8^5 = 32,768$

(c) $81^{5/4} = \left(\sqrt[4]{81}\right)^5 = 3^5 = 243$

(d) $9^{-3/2} = \dfrac{1}{\left(\sqrt{9}\right)^3} = \dfrac{1}{3^3} = \dfrac{1}{27}$

(e) $(-8)^{2/3} = \left(\sqrt[3]{-8}\right)^2 = (-2)^2 = 4$

(f) $\left(\dfrac{8}{27}\right)^{-1/3} = \left(\dfrac{27}{8}\right)^{1/3} = \sqrt[3]{\dfrac{27}{8}} = \dfrac{3}{2}$

(g) $\left(\dfrac{1}{16}\right)^{-5/4} = 16^{5/4} = \left(\sqrt[4]{16}\right)^5 = 2^5 = 32$

(h) $\left(-\dfrac{64}{125}\right)^{-2/3} = \left(-\dfrac{125}{64}\right)^{2/3} = \left(\sqrt[3]{-\dfrac{125}{64}}\right)^2 = \left(-\dfrac{5}{4}\right)^2 = \dfrac{25}{16}$

All of the rules for integer exponents apply to fractional exponents. This is demonstrated in the following example.

EXAMPLE 12.4

(a) $x^{1/2}x^{2/3} = x^{1/2+2/3} = x^{7/6}$

(b) $(y^{2/3})^{4/5} = y^{2/3 \cdot 4/5} = y^{8/15}$

(c) $\left(\dfrac{x}{y}\right)^{4/5} = \dfrac{x^{4/5}}{y^{4/5}}$

(d) $\dfrac{x^{3/4}}{x^{5/3}} = x^{3/4-5/3} = x^{-11/12} = \dfrac{1}{x^{11/12}}$

(e) $\dfrac{4x^2}{x^{2/3}} = 4x^{2-2/3} = 4x^{4/3}$

(f) $(8y^3)^{5/3} = 8^{5/3}(y^3)^{5/3} = 32y^5$

EXAMPLE 12.4 (Cont.)

(g) $\left(\dfrac{x^{15}}{y^9}\right)^{-1/3} = \left(\dfrac{y^9}{x^{15}}\right)^{1/3} = \dfrac{y^3}{x^5}$

(h) $\left(\dfrac{x^{1/3}y^{2/5}}{z^{3/5}}\right)^{15} = \dfrac{(x^{1/3})^{15}(y^{2/5})^{15}}{(z^{3/5})^{15}} = \dfrac{x^5 y^6}{z^9}$

(i) $x^{2/5}x^{-1/3} = x^{2/5-1/3} = x^{1/15}$

Evaluating Exponents Using a Calculator

You might want to review Appendix A and the work with calculators, where it describes how to use the $\boxed{x^2}$ key to calculate the square, or second power, of a number and how to use the $\boxed{x^y}$, $\boxed{y^x}$, or $\boxed{\wedge}$ key for the value of any power. To use the $\boxed{x^y}$ key on an algebraic calculator, you must also use the $\boxed{=}$ key before the answer is displayed.

EXAMPLE 12.5

Evaluate $4.7^{3.42}$

Solution

	PRESS	DISPLAY
Algebraic calculator:	4.7 $\boxed{x^y}$ 3.42 $\boxed{=}$	198.8724729
RPN calculator:	4.7 $\boxed{\text{ENTER}}$ 3.42 $\boxed{y^x}$	198.8724729
TI-82 calculator:	4.7 $\boxed{\wedge}$ 3.42 $\boxed{\text{ENTER}}$	198.8724729

Some calculators have a $\boxed{\sqrt[x]{\ }}$ key which you access through a menu. To use this key to evaluate $\sqrt[3]{8}$ you press 3 $\boxed{\sqrt[x]{\ }}$ 8 $\boxed{\text{ENTER}}$. Check the manual for your calculator to see if it has such an option and which menu you use to access it.

EXAMPLE 12.6

Evaluate $\sqrt[7]{943.2}$

Solution

	PRESS	DISPLAY
Algebraic calculator:	943.2 $\boxed{x^y}$ $\boxed{(}$ 1 $\boxed{\div}$ 7 $\boxed{)}$ $\boxed{=}$	2.660378313
RPN calculator:	943.2 $\boxed{\text{ENTER}}$ 1 $\boxed{\text{ENTER}}$ 7 $\boxed{\div}$ $\boxed{y^x}$	2.660378314
TI-82 calculator:	943.2 $\boxed{\wedge}$ $\boxed{(}$ 1 $\boxed{\div}$ 7 $\boxed{)}$ $\boxed{\text{ENTER}}$	2.660378313
or	7 $\boxed{\sqrt[x]{\ }}$ 943.2 $\boxed{\text{ENTER}}$	2.660378313
or	943.2 $\boxed{\wedge}$ 7 $\boxed{x^{-1}}$ $\boxed{\text{ENTER}}$	2.660378313

Hint

To evaluate $b^{m/n}$ where n is odd and b is a negative number, your calculator must "think" of this stated as $\left(b^m\right)^{1/n}$ or, equivalently as $\left(b^{1/n}\right)^m$.

EXAMPLE 12.7

Evaluate $12^{5/3}$

Solution

	PRESS	DISPLAY
Algebraic calculator:	12 $\boxed{x^y}$ $\boxed{(}$ 5 $\boxed{\div}$ 3 $\boxed{)}$ $\boxed{=}$	62.89779346
RPN calculator:	12 $\boxed{\text{ENTER}}$ 5 $\boxed{\text{ENTER}}$ 3 $\boxed{\div}$ $\boxed{y^x}$	62.89779351
TI-82 calculator:	12 $\boxed{\wedge}$ $\boxed{(}$ 5 $\boxed{\div}$ 3 $\boxed{)}$ $\boxed{\text{ENTER}}$	62.89779351
or	3 $\boxed{\sqrt[x]{\ }}$ 12 $\boxed{\wedge}$ 5 $\boxed{\text{ENTER}}$	62.89779351

EXAMPLE 12.8

Evaluate $(-8)^{5/3}$

Solution   You have to treat $(-8)^{5/3}$ as $\left[(-8)^5\right]^{1/3}$ or as $\left[(-8)^{1/3}\right]^5$.

	PRESS	DISPLAY
Algebraic calculator:	8 $\boxed{+/-}$ $\boxed{x^y}$ 5 $\boxed{\text{2nd}}$ $\boxed{x^y}$ $\boxed{(}$ 1 $\boxed{\div}$ 3 $\boxed{=}$	-32
RPN calculator:	8 $\boxed{\text{ENTER}}$ 5 $\boxed{\text{ENTER}}$ 3 $\boxed{\div}$ $\boxed{y^x}$ $\boxed{\text{CHS}}$	-32.00000002
TI-82 calculator:	$\boxed{(}$ $\boxed{(-)}$ 8 $\boxed{)}$ $\boxed{\wedge}$ $\boxed{(}$ 1 $\boxed{\div}$ 3 $\boxed{)}$ $\boxed{\wedge}$ 5 $\boxed{\text{ENTER}}$	-32
or	$\boxed{(}$ $\boxed{(-)}$ 8 $\boxed{)}$ $\boxed{\wedge}$ 5 $\boxed{\wedge}$ $\boxed{(}$ 1 $\boxed{\div}$ 3 $\boxed{)}$ $\boxed{\text{ENTER}}$	-32
or	$\boxed{(}$ 3 $\boxed{\sqrt[x]{\ }}$ $\boxed{(-)}$ 8 $\boxed{)}$ $\boxed{\wedge}$ 5 $\boxed{\text{ENTER}}$	-32

≡ **Note**   Remember, you cannot take an even root of a negative number. That is, you cannot take the square root of a negative number, you cannot take the 4th root of a negative number, you cannot take the 6th root of a negative number, and so on.

Caution   If you are using a TI-85 calculator, use the above directions for the TI-82 calculator. If you *do* calculate $(-8)^{5/3}$ on a TI-85 calculator by pressing $(-8) \wedge (5/3)$, you get an answer of $(16, -27.7128129211)$. This is the TI-85's way to represent the complex number $16 - 27.7128129211 j$. We discuss complex numbers in Chapter 14, and in Section 14.5 we will show why this answer makes sense.

Application

EXAMPLE 12.9

One study of a lake found that the light intensity was reduced 15% through a depth of 25 cm. The formula $I = (0.85)^{d/25}$ gives the approximate fraction of surface light intensity at a depth d, in centimeters. Find I at a depth of 60 cm.

Solution   Since $d = 60$ cm, we want to evaluate $I = (0.85)^{60/25}$ Using an algebraic calculator, we enter

$$0.85 \boxed{x^y} \boxed{(} 60 \boxed{\div} 25 \boxed{)} \boxed{=}$$

and obtain 0.677026116. So, the light intensity at 60 cm is about 67.7% of the surface light intensity.

Exercise Set 12.1

In Exercises 1–20, evaluate the given expression without the use of a calculator.

1. $25^{1/2}$
2. $27^{1/3}$
3. $64^{1/3}$
4. $125^{1/3}$
5. $25^{-1/2}$
6. $81^{-1/4}$

7. $32^{-1/5}$
8. $64^{-1/3}$
9. $27^{2/3}$
10. $81^{3/4}$
11. $125^{2/3}$
12. $32^{3/5}$

13. $16^{-3/4}$
14. $(-8)^{2/3}$
15. $(-8)^{-1/3}$
16. $(-27)^{-2/3}$
17. $\left(\frac{1}{8}\right)^{1/3}$

18. $\left(\frac{1}{25}\right)^{3/2}$
19. $\left(\frac{1}{16}\right)^{-5/4}$
20. $\left(\frac{-1}{27}\right)^{4/3}$

In Exercises 21–70, express each of the given expressions in the simplest form containing only positive exponents.

21. $3^2 \cdot 3^5$
22. $5^9 5^8$
23. $7^6 7^{-2}$
24. $11^9 11^{-6}$
25. $x^4 x^6$
26. $y^7 y^9$
27. $y^6 y^{-4}$
28. $x^8 x^{-2}$
29. $(9^5)^2$
30. $(11^8)^{-5}$
31. $(x^7)^3$
32. $(p^9)^{-5}$
33. $(xy)^5$
34. $(yt)^3$
35. $(ab)^{-5}$
36. $(xyz)^{-9}$
37. $\dfrac{x^{10}}{x^2}$
38. $\dfrac{p^9}{p^3}$
39. $\dfrac{x^2}{x^8}$
40. $\dfrac{a^3}{a^{12}}$
41. $x^{1/2} x^{3/2}$
42. $a^{1/3} a^{4/3}$

43. $r^{3/4} r$
44. $a^2 a^{2/3}$
45. $a^{1/2} a^{1/3}$
46. $b^{2/3} b^{1/4}$
47. $d^{2/3} d^{-1/4}$
48. $x^{3/5} y^{-2/3}$
49. $\dfrac{a^2 b^5}{a^5 b^2}$
50. $\dfrac{x^3 y^2}{x^7 y}$
51. $\dfrac{r^5 s^2 t}{t r^3 s^5}$
52. $\dfrac{a^2 b c^3}{(abc)^3}$
53. $\dfrac{(xy^2 z)^4}{x^4 (yz^2)^2}$
54. $\dfrac{m^6 n^7}{(m^2 n)^3}$
55. $\left(\dfrac{a}{b^2}\right)^3 \left(\dfrac{a}{b^3}\right)^2$
56. $\left(\dfrac{x^2}{y}\right)^4 \left(\dfrac{x}{y^2}\right)^2$
57. $\dfrac{(xy^2 b^3)^{1/2}}{(x^{1/4} b^4 y)^2}$
58. $\dfrac{(x^{1/3} y^3)^3}{(y^{10} x^5)^{1/5}}$

59. $\left(\dfrac{2x}{p^2}\right)^{-2} \left(\dfrac{p}{4}\right)^{-1}$
60. $\left(\dfrac{5a^2}{6b}\right)^{-2} \left(\dfrac{6}{a}\right)^{-4}$
61. $\dfrac{(x^2 y^{-1} z)^{-2}}{(xy)^{-4}}$
62. $\left\{\left[(2x^2)^3\right]^{-4}\right\}^{-1}$
63. $\dfrac{(27x^6 y^3)^{-1/3}}{(16x^4 y^{12})^{-1/4}}$
64. $\dfrac{(64x^6 y^{12})^{-5/6}}{(9x^4 y^2)^{-3/2}}$
65. $\left(\dfrac{x^2}{y^3}\right)^{-1/5} \left(\dfrac{y^2}{x^5}\right)^{1/3}$
66. $\left(\dfrac{a^2 x^3}{b^5 y}\right)^{-4/3} \left(\dfrac{axy^{-2}}{b}\right)^{5/3}$
67. $\left(\dfrac{9x^3}{t^5}\right)^{-1/2} \left(\dfrac{8t^2}{x^5}\right)^{-1/3}$
68. $\left(\dfrac{125a^6}{8b^3}\right)^{-2/3} \left(\dfrac{8b}{5a^2}\right)^{-1/2}$
69. $\dfrac{(9x^4 y^{-6})^{-3/2}}{(8x^{-6} y^3)^{-2/3}}$
70. $\dfrac{(81x^5 y^{-8} z)^{-3/4}}{(64xy^{-3} z^2)^{-5/3}}$

Use a calculator or computer to determine the value of each of the numbers in Exercises 71–80.

71. $8.3^{2/3}$

72. $7.3^{2/5}$

73. $92.47^{5/7}$

74. $81.94^{3/5}$

75. $(-81.52)^{2/7}$

76. $(-78.64)^{1/3}$

77. $432.61^{1/4}$

78. $(-537.15)^{2/3}$

79. $(-32.35)^{-3/7}$

80. $(-0.1439074804)^{-3/5}$

Solve Exercises 81–86.

81. *Physics*   The distance in meters traveled by a falling body starting from rest is $9.8t^2$, where t is the time, in seconds, the object has been falling. In 4.75 s, the distance fallen will be $9.8(4.75)^2$ m. Evaluate this quantity.

82. The volume of a sphere is $\frac{4}{3}\pi r^3$. If the radius r of a sphere is 19.25 in., what is the volume?

83. *Physics*   When 5 m³ of helium at a temperature of 315 K and a pressure of 15 N/m² is adiabatically compressed to 0.5 m³, its new pressure p and temperature T are given by

$$p = (5)^5 \left(\frac{15}{0.5}\right)^{5/3}$$

and $$T = 315 \left(\frac{15}{0.5}\right)^{5/3}$$

Evaluate p and T.

84. *Nuclear technology*   Radium has a half-life of approximately 1,600 years. It decays according to the formula $q(t) = q_0 2^{-t/1,600}$, where q_0 is the original quantity of radium and $q(t)$ is the amount left at time t. In this problem, t is given in years. If you begin with 75 mg of pure radium, how much will remain after 2,000 years?

85. *Aeronautical engineering*   The efficiency of a turbojet engine is given by the expression

$$E = 1 - \left(\frac{p_1}{p_2}\right)^{\alpha/(1-\alpha)}$$

where $\dfrac{p_1}{p_2}$ is the compression ratio. Simplify the expression for $\alpha = 1.4$.

86. *Finance*   The periodic payment, R, for a debt, A, at the annual interest rate, r, with n payments per year for t years is given by

$$R = A \left[\frac{\dfrac{r}{n}}{1 - \left(1 + \dfrac{r}{n}\right)^{-nt}} \right]$$

A person borrowed $15,000 for an automobile at an annual rate of 9%. How much is each monthly payment if she borrowed the money for 5 years?

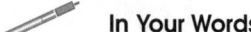
In Your Words

87. Describe how to use your calculator to evaluate $(1.44)^{3/2}$.

88. (a) Use your calculator to evaluate $\left(-\frac{1}{8}\right)^{-2/3}$, $-\dfrac{1}{8^{-2/3}}$, $-8^{2/3}$, $(-8)^{2/3}$, and $(-8)^{-2/3}$.

(b) Explain why some answers in (a) are the same and some are different.

☰ 12.2
LAWS OF RADICALS

In Section 1.3, we introduced the concept of roots. We used them in Section 12.1 for the discussion of fractional exponents. The more general name for roots is **radicals**. Fractional exponents can be used for any operation that requires radicals.

Laws of Radicals

A **radical** is any number of the form $\sqrt[n]{b}$. The number under the radical b, is called the **radicand**. The number indicating the root n is called the **order** or **index**.

In Section 1.7, we introduced four basic rules or laws of radicals. These rules are listed in the following box.

Basic Rules for Radicals

Rule 1: $\sqrt[n]{ab} = \sqrt[n]{a}\sqrt[n]{b}$

Rule 2: $\sqrt[n]{\dfrac{a}{b}} = \dfrac{\sqrt[n]{a}}{\sqrt[n]{b}}$

Rule 3: $(\sqrt[n]{b})^n = b^{n/n} = b$

Rule 4: $\sqrt[n]{b} = b^{1/n}$

We assume in each of these rules that if n is even, neither a nor b are negative real numbers.

EXAMPLE 12.10

Simplify **(a)** $\sqrt[6]{27}$, **(b)** $\sqrt[5]{96}$, **(c)** $\sqrt{x^6y^8}$, **(d)** $\sqrt[3]{\dfrac{125y^6}{x^9}}$, and **(e)** $\sqrt[3]{54x^8}$

Solutions

(a) $\sqrt[6]{27} = \sqrt[6]{3^3} = 3^{3/6} = 3^{1/2} = \sqrt{3}$

(b) $\sqrt[5]{96} = \sqrt[5]{32 \cdot 3} = \sqrt[5]{32}\sqrt[5]{3} = \sqrt[5]{2^5}\sqrt[5]{3} = 2\sqrt[5]{3}$

(c) $\sqrt{x^6y^8} = (x^6y^8)^{1/2} = x^{6/2}y^{8/2} = x^3y^4$

(d) $\sqrt[3]{\dfrac{125y^6}{x^9}} = \sqrt[3]{\dfrac{5^3y^6}{x^9}} = \dfrac{5^{3/3}y^{6/3}}{x^{9/3}} = \dfrac{5y^2}{x^3}$

(e) $\sqrt[3]{54x^8} = \sqrt[3]{2 \cdot 3^3 x^6 x^2}$
$= 2^{1/3}3^{3/3}x^{6/3}x^{2/3}$
$= 2^{1/3} \cdot 3 \cdot x^2 x^{2/3}$
$= 3x^2 2^{1/3} x^{2/3}$
$= 3x^2 \left(2x^2\right)^{1/3}$
$= 3x^2 \sqrt[3]{2x^2}$

Notice that several examples started with as many factors as possible with nth roots that could be easily found. Thus, in Example 12.10(b), we used $\sqrt[n]{ab} = \sqrt[n]{a}\sqrt[n]{b}$, when we wrote $\sqrt[5]{96} = \sqrt[5]{32} \cdot \sqrt[5]{3}$.

 Hint

In general, we try to express a radical so that the exponent of any factor in the radicand is less than the index of the radical. Thus, in Example 12.10(e), we wrote $\sqrt[3]{x^8}$ as $x^2\sqrt[3]{x^2}$.

Rationalizing Denominators

When a radicand is a fraction, a variation of Rule 2, $\sqrt[n]{\dfrac{a}{b}} = \dfrac{\sqrt[n]{a}}{\sqrt[n]{b}}$, is used to eliminate the radical in the denominator. This technique is called **rationalizing the denominator**.

> **Rationalizing the Denominator**
>
> To rationalize a denominator of the form $\sqrt[n]{x^r}$, multiply the denominator by another radical with the same radicand and the same index. In $\sqrt[n]{x^s}$, $r + s$ is a multiple of n.

The process of rationalizing the denominator makes the denominator a perfect power of x and eliminates the radical in the denominator.

≡ **Note**   Remember that whenever you multiply the denominator by something other than 1, you must also multiply the numerator by the same quantity.

EXAMPLE 12.11

Rationalize the denominators of **(a)** $\sqrt{\dfrac{3}{5}}$, **(b)** $\dfrac{1}{\sqrt[3]{2}}$, and **(c)** $\sqrt[5]{\dfrac{3}{8x^2}}$.

Solutions

(a) Rewrite $\sqrt{\dfrac{3}{5}}$ as $\dfrac{\sqrt{3}}{\sqrt{5}}$.

Multiply both the numerator and denominator by $\sqrt{5}$ because $5 \cdot 5 = 5^2$.

$$\sqrt{\frac{3}{5}} = \frac{\sqrt{3}}{\sqrt{5}} = \frac{\sqrt{3}}{\sqrt{5}} \cdot \frac{\sqrt{5}}{\sqrt{5}}$$

$$= \frac{\sqrt{3 \cdot 5}}{\sqrt{5 \cdot 5}} = \frac{\sqrt{15}}{\sqrt{5^2}} = \frac{\sqrt{15}}{5}$$

(b) Multiply both the numerator and denominator by $\sqrt[3]{2^2}$ because $2 \cdot 2^2 = 2^3$, a perfect cube.

$$\frac{1}{\sqrt[3]{2}} = \frac{1}{\sqrt[3]{2}} \cdot \frac{\sqrt[3]{2^2}}{\sqrt[3]{2^2}}$$

$$= \frac{\sqrt[3]{2^2}}{\sqrt[3]{2^3}} = \frac{\sqrt[3]{4}}{2}$$

(c) Rewrite $\sqrt[5]{\dfrac{3}{8x^2}}$ as $\dfrac{\sqrt[5]{3}}{\sqrt[5]{8x^2}}$.

EXAMPLE 12.11 (Cont.)

$\sqrt[5]{8x^2} = \sqrt[5]{2^3 x^2}$, so we will multiply both the numerator and denominator by $\sqrt[5]{2^2 x^3}$ because $2^3 x^2 \cdot 2^2 x^3 = 2^5 x^5$.

$$\sqrt[5]{\frac{3}{8x^2}} = \frac{\sqrt[5]{3}}{\sqrt[5]{8x^2}} = \frac{\sqrt[5]{3}}{\sqrt[5]{8x^2}} \cdot \frac{\sqrt[5]{2^2 x^3}}{\sqrt[5]{2^2 x^3}}$$

$$= \frac{\sqrt[5]{3 \cdot 2^2 x^3}}{\sqrt[5]{8x^2 2^2 x^3}} = \frac{\sqrt[5]{3 \cdot 2^2 x^3}}{\sqrt[5]{2^5 x^5}}$$

$$= \frac{\sqrt[5]{12x^3}}{2x}$$

Rationalizing the denominator was originally developed as a way to help computations, but the increased use of calculators and computers reduces its importance in calculations. Rationalizing the denominator, however, is often a useful way to write numbers.

Another helpful rule is used with radicals.

Rule 5 for Radicals
$$\sqrt[m]{\sqrt[n]{b}} = \sqrt[mn]{b}$$

EXAMPLE 12.12

(a) $\sqrt[3]{\sqrt{27}} = \sqrt[6]{27}$, **(b)** $\sqrt[4]{\sqrt[5]{914}} = \sqrt[20]{914}$, **(c)** $\sqrt[5]{\sqrt[3]{-82}} = \sqrt[15]{-82}$, and

(d) $\sqrt[4]{\sqrt[3]{x}} = \sqrt[12]{x}$

Sometimes it is possible to reduce the index of a radical. For example,

$$\sqrt[6]{y^2} = y^{2/6} = y^{1/3} = \sqrt[3]{y}$$

Here the index was reduced from 6 to 3. Another example can be seen from 12.10(a).

$$\sqrt[3]{\sqrt{27}} = \sqrt[3]{\sqrt{3^3}} = \sqrt[6]{3^3} = 3^{3/6} = 3^{1/2} = \sqrt{3}$$

The last version, $\sqrt{3}$, is a simpler version with which to work.

EXAMPLE 12.13

Reduce the index of **(a)** $\sqrt[8]{16}$, **(b)** $\sqrt[6]{16x^2}$, and **(c)** $\sqrt[12]{27x^6 y^3}$.
Solutions

(a) $\sqrt[8]{16} = \sqrt[8]{2^4} = 2^{4/8} = 2^{1/2} = \sqrt{2}$

(b) $\sqrt[6]{16x^2} = \sqrt[6]{2^4 x^2} = 2^{4/6} x^{2/6} = 2^{2/3} x^{1/3} = \sqrt[3]{4x}$

(c) $\sqrt[12]{27x^6 y^3} = \sqrt[12]{3^3 x^6 y^3} = 3^{3/12} x^{6/12} y^{3/12} = 3^{1/4} x^{2/4} y^{1/4} = \sqrt[4]{3x^2 y}$

Notice that, in the last example, we could have factored the x^2 out of the radical using Rule 1 and written $\sqrt[4]{3x^2y} = \sqrt[4]{3y}\sqrt[4]{x^2} = \sqrt[4]{3y}\sqrt{x}$.

Simplifying Radicals

Simplifying a radical makes it easier to work with. The following three steps are used to simplify a radical.

Steps for Simplifying Radicals

A radical is simplified when all of the following steps are finished.

Step 1: All possible factors have been removed from the radicand.

Step 2: All denominators are rationalized.

Step 3: The index has been reduced as much as possible.

EXAMPLE 12.14

Simplify the following: **(a)** $\sqrt{\dfrac{x^3}{y}}$, **(b)** $\sqrt[3]{x^5y^{10}}$, **(c)** $\sqrt[5]{x^3y^{-7}}$, and **(d)** $\sqrt[3]{\dfrac{x}{y^2} + \dfrac{5y^2}{x}}$.

Solutions

(a) $\sqrt{\dfrac{x^3}{y}} = \sqrt{\dfrac{x^3}{y}}\sqrt{\dfrac{y}{y}} = \dfrac{\sqrt{x^3y}}{\sqrt{y^2}} = \dfrac{\sqrt{x^3y}}{y} = \dfrac{x\sqrt{xy}}{y}$

(b) $\sqrt[3]{x^5y^{10}} = \sqrt[3]{x^3x^2y^9y} = \sqrt[3]{x^3y^9} \cdot \sqrt[3]{x^2y} = xy^3\sqrt[3]{x^2y}$

(c) $\sqrt[5]{x^3y^{-7}} = \sqrt[5]{\dfrac{x^3}{y^7}} = \sqrt[5]{\dfrac{x^3}{y^7} \cdot \dfrac{y^3}{y^3}} = \dfrac{\sqrt[5]{x^3y^3}}{\sqrt[5]{y^{10}}} = \dfrac{\sqrt[5]{x^3y^3}}{y^2}$

(d) $\sqrt[3]{\dfrac{x}{y^2} + \dfrac{5y^2}{x}} = \sqrt[3]{\dfrac{x \cdot x}{y^2x} + \dfrac{5y^2y^2}{xy^2}} = \sqrt[3]{\dfrac{x^2 + 5y^4}{xy^2}}$

$= \sqrt[3]{\dfrac{(x^2 + 5y^4)x^2y}{(xy^2)x^2y}} = \dfrac{\sqrt[3]{x^4y + 5x^2y^5}}{\sqrt[3]{x^3y^3}}$

$= \dfrac{\sqrt[3]{x^4y + 5x^2y^5}}{xy}$

Notice that in Example 12.14(d) we had to find a common denominator before we could add the fractions. After the fractions were added, we were able to rationalize the denominator.

Application

EXAMPLE 12.15

The frequency of oscillation f of a simple pendulum is given by

$$f = \frac{1}{2\pi} \sqrt{\frac{g}{L}}$$

where $g \approx 9.8 \text{ m/s}^2$ is the acceleration due to gravity in the metric system and L is the length of the pendulum in meters. **(a)** Express f in simplest form, when $L = 0.35 \text{ m}$. **(b)** Evaluate f to the nearest hundredth.

Solution **(a)** Given $g \approx 9.8 \text{ m/s}^2$ and $L = 0.35 \text{ m}$,

$$f = \frac{1}{2\pi} \sqrt{\frac{g}{L}}$$

$$= \frac{1}{2\pi} \sqrt{\frac{9.8}{0.35}}$$

$$= \frac{1}{2\pi} \sqrt{28}$$

$$= \frac{\sqrt{7}}{\pi}$$

(b) Evaluating $\dfrac{\sqrt{7}}{\pi}$ with an algebraic calculator, we press $7\ \boxed{\sqrt{\ }}\ \boxed{\div}\ \boxed{\pi}\ \boxed{=}$ and obtain 0.8421687987. Thus, the pendulum oscillates about once every 0.84 s.

Exercise Set 12.2

Use the rules for radicals to express each of Exercises 1–60 in simplest radical form.

1. $\sqrt[3]{16}$
2. $\sqrt[3]{81}$
3. $\sqrt{45}$
4. $\sqrt[3]{40}$
5. $\sqrt[3]{y^{12}}$
6. $\sqrt[4]{p^8}$
7. $\sqrt[5]{a^7}$
8. $\sqrt[7]{b^{10}}$
9. $\sqrt{x^2 y^7}$
10. $\sqrt[3]{x^5 y^3}$
11. $\sqrt[4]{a^5 b^3}$
12. $\sqrt[5]{p^{12} y^8}$
13. $\sqrt[3]{8x^4}$
14. $\sqrt[4]{81 y^9}$

15. $\sqrt{27 x^3 y}$
16. $\sqrt[3]{32 a^5 b^2}$
17. $\sqrt[3]{-8}$
18. $\sqrt[5]{-243}$
19. $\sqrt[3]{a^2 b^4} \sqrt[3]{ab^5}$
20. $\sqrt[5]{x^3 y^2 z^4} \sqrt[5]{x^2 y^8 z}$
21. $\sqrt[4]{p^3 q^2 r^6} \sqrt[4]{pq^6 r}$
22. $\sqrt[6]{m^3 n^2 e^7} \sqrt[6]{m^2 n^4 e^5}$
23. $\sqrt[3]{\dfrac{8x^3}{27}}$
24. $\sqrt[4]{\dfrac{81 y^8}{16}}$
25. $\sqrt[5]{\dfrac{x^5 y^{10}}{z^5}}$

26. $\sqrt[3]{\dfrac{a^3 b^9}{c^6}}$
27. $\sqrt[3]{\dfrac{16 x^3 y^2}{z^6}}$
28. $\sqrt{\dfrac{125 a^5 b^2}{c^4}}$
29. $\sqrt{\dfrac{64 x^3 y^4}{9 z^4 p^2}}$
30. $\sqrt[3]{\dfrac{8 a^5 b^3}{27 r^6 s^9}}$
31. $\sqrt{\dfrac{16}{3}}$
32. $\sqrt{\dfrac{4}{5}}$
33. $\sqrt[3]{\dfrac{27}{4}}$

34. $\sqrt[3]{\dfrac{16}{25}}$
35. $\sqrt{\dfrac{25}{2x}}$
36. $\sqrt[3]{\dfrac{8}{5 y^2}}$
37. $\sqrt[4]{\dfrac{81}{32 z^2}}$
38. $\sqrt[3]{\dfrac{-2}{25 r^2}}$
39. $\sqrt[3]{\dfrac{16 x^2 y}{x^5}}$
40. $\sqrt[4]{\dfrac{25 a^3 b^5}{a^7 b}}$
41. $\sqrt[3]{\dfrac{-8 x^3 yz}{27 b^2 z^4}}$

42. $\sqrt[4]{\dfrac{25a^2b^3}{16c^3b^6}}$

43. $\sqrt{4 \times 10^4}$

44. $\sqrt{9 \times 10^6}$

45. $\sqrt{25 \times 10^3}$

46. $\sqrt{16 \times 10^7}$

47. $\sqrt{4 \times 10^7}$

48. $\sqrt{9 \times 10^9}$

49. $\sqrt[3]{1.25 \times 10^{10}}$

50. $\sqrt[5]{3.2 \times 10^{14}}$

51. $\sqrt{\dfrac{x}{y} + \dfrac{y}{x}}$

52. $\sqrt{\dfrac{a}{b} - \dfrac{b}{a}}$

53. $\sqrt{a^2 + 2ab + b^2}$

54. $\sqrt{x^2 - 2xy + y^2}$

55. $\sqrt{\dfrac{1}{a^2} + \dfrac{1}{b}}$

56. $\sqrt{\dfrac{x}{y^2} + \dfrac{y}{x^2}}$

57. $\sqrt[6]{\sqrt[3]{27x^2}}$

58. $\sqrt[4]{\sqrt[6]{125x^3y^5}}$

59. $\sqrt[3]{\sqrt[7]{-624.2x^{15}y^{10}z}}$

60. $\sqrt{\sqrt[4]{9x^8y^3}}$

Solve Exercises 61–66.

61. *Music*   Many musical instruments contain strings, which vibrate to produce music. The frequency of vibration f of a string of length L fixed at both ends and vibrating in its fundamental mode, is given by

$$f = \frac{1}{2L}\sqrt{\frac{T}{\mu}}$$

where μ is the mass per unit length and T is the tension in the string. Rationalize the right-hand side of this equation.

62. *Music*   In the equation in Exercise 61, what happens to the frequency when the tension is quadrupled?

63. *Electricity*   The impedance Z of a certain circuit is given by the equation

$$Z = \frac{1}{\sqrt{\dfrac{1}{x^2} + \dfrac{1}{R^2}}}$$

Rationalize the denominator in order to simplify this equation.

64. *Chemistry*   The distance between ion layers of a sodium chloride crystal is given by the expression

$$\sqrt[3]{\frac{M}{2\rho N}}$$

where M is the molecular weight, N is Avogadro's number, and ρ is the density.   Express this in simplest form.

65. *Mechanical engineering*   The formula

$$k = 1 + \sqrt{1 + \frac{2h}{m}}$$

is used to calculate the impact factor for dynamic loading. Rewrite the right-hand side in simplest radical form. Make sure you rationalize the denominator.

66. *Civil engineering*   Frequency f, in Hz, of an object attached to the end of a cantilever beam of length l is given by

$$f = \frac{1}{2\pi}\sqrt{\frac{3EI}{ml^3}}$$

Express the formula in simplest radical form.

In Your Words

67. Without looking in the text, explain the meaning of each of the following terms: **(a)** radical, **(b)** radicand, and **(c)** index.

68. Explain how to rationalize the denominator of the form $\sqrt[n]{x^r}$.

≡ 12.3
BASIC OPERATIONS WITH RADICALS

In this section, we will study the basic operations of addition, subtraction, multiplication, and division of radicals. Adding and subtracting radicals is similar to adding and subtracting algebraic expressions. However, there are two cases to consider when multiplying or dividing radicals. The first case is when the radicals have the same index; the second case is when they have different indices. Each of these cases will be considered in this section.

Addition and Subtraction of Radicals

When we learned how to add and subtract algebraic expressions, we found that only like terms can be added or subtracted. Addition and subtraction of radicals is very similar.

≡ **Note**   Radicals can only be added or subtracted if the radicands are identical and the indices are the same.

Trying to add or subtract two radicals such as $\sqrt{2}$ and $\sqrt{5}$ is similar to adding and subtracting x and y. Example 12.16 shows the similarity between combining radicals and combining algebraic expressions.

EXAMPLE 12.16

	Combining Radicals	Combining Algebraic Expressions
(a)	$\sqrt{2}+\sqrt{5}+\sqrt{2}$ $=2\sqrt{2}+\sqrt{5}$	$x+y+x=2x+y$, where $x=\sqrt{2}$ and $y=\sqrt{5}$
(b)	$2\sqrt{3}+4\sqrt{7}+6\sqrt{3}$ $=8\sqrt{3}+4\sqrt{7}$	$2a+4b+6a=8a+4b$, where $a=\sqrt{3}$ and $b=\sqrt{7}$
(c)	$3\sqrt[3]{6}-7\sqrt[3]{6}=-4\sqrt[3]{6}$	$3p-7p=-4p$, where $p=\sqrt[3]{6}$
(d)	$\dfrac{5\sqrt{3}}{2}-\dfrac{\sqrt{3}}{2}=\dfrac{4\sqrt{3}}{2}=2\sqrt{3}$	$\dfrac{5x}{2}-\dfrac{x}{2}=\dfrac{4x}{2}=2x$, where $x=\sqrt{3}$

In each case, only the similar radicals are combined. Radicals that are not similar remain as separate terms and the addition or subtraction is only indicated. Sometimes it is possible to combine radicals that do not appear to be similar by first simplifying the radicals. For example $\sqrt{8}$ and $\sqrt{2}$ can be combined, since $\sqrt{8}=\sqrt{4}\sqrt{2}=2\sqrt{2}$.

EXAMPLE 12.17

Simplify and combine similar radicals.
(a) $\sqrt{2}+\sqrt{8}=\sqrt{2}+2\sqrt{2}=3\sqrt{2}$
(b) $3\sqrt{8}+5\sqrt{18}=6\sqrt{2}+15\sqrt{2}=21\sqrt{2}$
  Note that $\sqrt{18}=\sqrt{9}\sqrt{2}=3\sqrt{2}$ and $5\sqrt{18}=5\sqrt{9}\sqrt{2}=5\cdot 3\sqrt{2}=15\sqrt{2}$.
(c) $\sqrt{98x}+\sqrt{32x}=7\sqrt{2x}+4\sqrt{2x}=11\sqrt{2x}$
(d) $\sqrt{20x^3}+\sqrt{8x^2}-\sqrt{45x}=2x\sqrt{5x}+2x\sqrt{2}-3\sqrt{5x}=(2x-3)\sqrt{5x}+2x\sqrt{2}$
  Note that we can only indicate the difference of $2x\sqrt{5x}-3\sqrt{5x}$ as $(2x-3)\sqrt{5x}$. We factored $\sqrt{5x}$ out of each term.

EXAMPLE 12.17 (Cont.)

(e) $\sqrt{\dfrac{5}{3}} + \dfrac{\sqrt{3}}{\sqrt{125}} = \dfrac{\sqrt{5}}{\sqrt{3}} + \dfrac{\sqrt{3}}{5\sqrt{5}} = \dfrac{\sqrt{5}\sqrt{3}}{\sqrt{3}\sqrt{3}} + \dfrac{\sqrt{3}\sqrt{5}}{5\sqrt{5}\sqrt{5}}$

$\qquad\qquad = \dfrac{\sqrt{15}}{3} + \dfrac{\sqrt{15}}{25}$

$\qquad\qquad = \dfrac{25\sqrt{15}}{75} + \dfrac{3\sqrt{15}}{75}$

$\qquad\qquad = \dfrac{28\sqrt{15}}{75}$

Here, we rationalized the denominators and then added the fractions by finding the common denominator. We could have rationalized the second fraction by multiplying by $\sqrt{125}$, but it was easier to first simplify the denominator to $5\sqrt{5}$ then multiply by $\sqrt{5}$.

Multiplying Radicals with the Same Index

Multiplying and dividing radicals that have the same index use two of the rules we have discussed.

$$\sqrt[n]{a}\sqrt[n]{b} = \sqrt[n]{ab}$$

$$\text{and} \qquad \dfrac{\sqrt[n]{a}}{\sqrt[n]{b}} = \sqrt[n]{\dfrac{a}{b}}, \qquad b \neq 0$$

EXAMPLE 12.18

Multiply the following radical expressions: **(a)** $\sqrt{2}\sqrt{5}$, **(b)** $\sqrt[3]{5}\sqrt[3]{10}$, **(c)** $\sqrt{xy}\sqrt{2x}$, **(d)** $\sqrt[3]{\dfrac{3}{2}}\sqrt[3]{\dfrac{5x}{4}}$, **(e)** $\sqrt{5x}(\sqrt{5x}+\sqrt{10x^3})$, and **(f)** $(\sqrt{x}+\sqrt{y})(\sqrt{x}-\sqrt{y})$.

Solutions

(a) $\qquad\qquad \sqrt{2}\sqrt{5} = \sqrt{2\cdot 5} = \sqrt{10}$

(b) $\qquad\qquad \sqrt[3]{5}\sqrt[3]{10} = \sqrt[3]{5\cdot 10} = \sqrt[3]{50}$

(c) $\qquad \sqrt{xy}\sqrt{2x} = \sqrt{(xy)(2x)} = \sqrt{2x^2 y} = x\sqrt{2y}$

(d) $\quad \sqrt[3]{\dfrac{3}{2}}\sqrt[3]{\dfrac{5x}{4}} = \sqrt[3]{\left(\dfrac{3}{2}\right)\left(\dfrac{5x}{4}\right)} = \sqrt[3]{\dfrac{15x}{8}}$

$\qquad\qquad\qquad = \dfrac{\sqrt[3]{15x}}{2}$

(e) $\quad \sqrt{5x}\left(\sqrt{5x}+\sqrt{10x^3}\right) = \sqrt{5x}\sqrt{5x} + \sqrt{5x}\sqrt{10x^3}$

$\qquad\qquad\qquad\qquad = \sqrt{25x^2} + \sqrt{50x^4}$

$\qquad\qquad\qquad\qquad = 5x + 5x^2\sqrt{2}$

(f) $\quad (\sqrt{x}+\sqrt{y})(\sqrt{x}-\sqrt{y}) = (\sqrt{x})^2 - (\sqrt{y})^2 = x - y$ using the special product for the difference of two squares.

Multiplying Radicals with Different Indices

Radicals with different indices can be multiplied if they are rewritten so that they have the same index. The easiest way to do this is with fractional exponents.

EXAMPLE 12.19

Multiply the following: **(a)** $\sqrt[3]{2}\sqrt{7}$, **(b)** $\sqrt[4]{5x^2}\sqrt[3]{2x}$, and **(c)** $\sqrt[3]{4ab^2}\sqrt[5]{16a^4b^2}$.

Solutions

(a)
$$\sqrt[3]{2}\sqrt{7} = 2^{1/3}7^{1/2} = 2^{2/6}7^{3/6} = (2^2 7^3)^{1/6}$$
$$= \sqrt[6]{2^2 7^3}$$
$$= \sqrt[6]{1,372}$$

(b)
$$\sqrt[4]{5x^2}\sqrt[3]{2x} = (5x^2)^{1/4}(2x)^{1/3} = (5x^2)^{3/12}(2x)^{4/12}$$
$$= \sqrt[12]{(5x^2)^3}\,\sqrt[12]{(2x)^4}$$
$$= \sqrt[12]{(5x^2)^3(2x)^4}$$
$$= \sqrt[12]{5^3 x^6 2^4 x^4}$$
$$= \sqrt[12]{2,000x^{10}}$$

(c)
$$\sqrt[3]{4ab^2}\sqrt[5]{16a^4b^2} = (4ab^2)^{1/3}(16a^4b^2)^{1/5}$$
$$= (4ab^2)^{5/15}(16a^4b^2)^{3/15}$$
$$= \sqrt[15]{(4ab^2)^5}\,\sqrt[15]{(16a^4b^2)^3}$$
$$= \sqrt[15]{(2^2 ab^2)^5}\,\sqrt[15]{(2^4 a^4 b^2)^3}$$
$$= \sqrt[15]{2^{10}a^5 b^{10}}\,\sqrt[15]{2^{12}a^{12}b^6}$$
$$= \sqrt[15]{2^{22}a^{17}b^{16}}$$
$$= 2ab\,\sqrt[15]{2^7 a^2 b}$$
$$= 2ab\,\sqrt[15]{128a^2 b}$$

Division of Radicals

Division of radicals with the same index is done using Rule 2 for radicals. The result is usually simplified by rationalizing the denominator.

EXAMPLE 12.20

Divide each of the following: **(a)** $\dfrac{\sqrt{10}}{\sqrt{3}}$, **(b)** $\dfrac{\sqrt[3]{4x}}{\sqrt[3]{2x}}$, **(c)** $\dfrac{\sqrt[3]{x^2y}}{\sqrt[3]{z}}$, and **(d)** $\dfrac{\sqrt{3xy}}{\sqrt{7x^5y}}$.

Solutions

(a) $\dfrac{\sqrt{10}}{\sqrt{3}} = \dfrac{\sqrt{10}}{\sqrt{3}} \cdot \dfrac{\sqrt{3}}{\sqrt{3}} = \dfrac{\sqrt{30}}{3}$

(b) $\dfrac{\sqrt[3]{4x}}{\sqrt[3]{2x}} = \sqrt[3]{\dfrac{4x}{2x}} = \sqrt[3]{2}$

(c) $\dfrac{\sqrt[3]{x^2y}}{\sqrt[3]{z}} = \dfrac{\sqrt[3]{x^2y}}{\sqrt[3]{z}} \cdot \dfrac{\sqrt[3]{z^2}}{\sqrt[3]{z^2}} = \dfrac{\sqrt[3]{x^2yz^2}}{z}$

(d) $\dfrac{\sqrt{3xy}}{\sqrt{7x^5y}} = \dfrac{\sqrt{3xy}}{x^2\sqrt{7xy}} = \dfrac{1}{x^2} \cdot \sqrt{\dfrac{3xy}{7xy}} = \dfrac{1}{x^2}\sqrt{\dfrac{3}{7}}$

We now rationalize the denominator.

$$= \dfrac{1}{x^2}\sqrt{\dfrac{3}{7}}\sqrt{\dfrac{7}{7}} = \dfrac{\sqrt{21}}{7x^2}$$

Conjugate of a Radical Expression

If a, b, $\sqrt{m}$, and $\sqrt{n}$ are real numbers, then radical expressions of the form $a\sqrt{m}+b\sqrt{n}$ and $a\sqrt{m}-b\sqrt{n}$ are **conjugates** of each other.

Sometimes the denominator of a fraction is the sum of the difference of two square roots. Examples of this are $\sqrt{2}+\sqrt{5}$ and $\sqrt{x}-\sqrt{y}$. In this case, the numerator and denominator can be multiplied by its conjugate in order to make the denominator the difference of two squares. For example, if the denominator is $\sqrt{2}+\sqrt{5}$, then multiply the numerator and denominator by $\sqrt{2}-\sqrt{5}$. If the denominator is $\sqrt{x}-\sqrt{y}$, then multiply both numerator and denominator by $\sqrt{x}+\sqrt{y}$. Notice that in each case the denominator is then in the form a^2-b^2, where a or b is a radical.

EXAMPLE 12.21

Rationalize each of these denominators: **(a)** $\dfrac{1}{\sqrt{2}+\sqrt{5}}$, **(b)** $\dfrac{a}{\sqrt{x}-\sqrt{y}}$, and

(c) $\dfrac{\sqrt{x+y}}{1+\sqrt{x+y}}$.

EXAMPLE 12.21 (Cont.)

Solutions   **(a)** The conjugate of $\sqrt{2}+\sqrt{5}$ is $\sqrt{2}-\sqrt{5}$.

$$\frac{1}{\sqrt{2}+\sqrt{5}} = \frac{1}{\sqrt{2}+\sqrt{5}} \cdot \frac{\sqrt{2}-\sqrt{5}}{\sqrt{2}-\sqrt{5}}$$

$$= \frac{\sqrt{2}-\sqrt{5}}{(\sqrt{2})^2 - (\sqrt{5})^2}$$

$$= \frac{\sqrt{2}-\sqrt{5}}{2-5}$$

$$= -\frac{\sqrt{2}-\sqrt{5}}{3} = \frac{\sqrt{5}-\sqrt{2}}{3}$$

(b) The conjugate of $\sqrt{x}-\sqrt{y}$ is $\sqrt{x}+\sqrt{y}$.

$$\frac{a}{\sqrt{x}-\sqrt{y}} = \frac{a}{\sqrt{x}-\sqrt{y}} \cdot \frac{\sqrt{x}+\sqrt{y}}{\sqrt{x}+\sqrt{y}}$$

$$= \frac{a(\sqrt{x}+\sqrt{y})}{(\sqrt{x})^2-(\sqrt{y})^2}$$

$$= \frac{a\sqrt{x}+a\sqrt{y}}{x-y}, \quad x \neq y$$

(c) The conjugate of $1+\sqrt{x+y}$ is $1-\sqrt{x+y}$.

$$\frac{\sqrt{x+y}}{1+\sqrt{x+y}} = \frac{\sqrt{x+y}}{1+\sqrt{x+y}} \cdot \frac{1-\sqrt{x+y}}{1-\sqrt{x+y}}$$

$$= \frac{\sqrt{x+y}(1-\sqrt{x+y})}{(1)^2-(\sqrt{x+y})^2}$$

$$= \frac{\sqrt{x+y}-(\sqrt{x+y})^2}{1-(x+y)}$$

$$= \frac{\sqrt{x+y}-x-y}{1-x-y}, \quad x+y \neq 1$$

Finally, to find the quotient of two radicals with different indices, we use fractional exponents in the same manner as when we multiplied radicals with different indices.

EXAMPLE 12.22

Find the following quotients: **(a)** $\dfrac{\sqrt[3]{15}}{\sqrt{15}}$ and **(b)** $\dfrac{\sqrt{2x}}{\sqrt[4]{8x^3}}$.

Solutions **(a)**

$$\frac{\sqrt[3]{15}}{\sqrt{15}} = \frac{(15)^{1/3}}{(15)^{1/2}} = \frac{(15)^{2/6}}{(15)^{3/6}} = \frac{\sqrt[6]{15^2}}{\sqrt[6]{15^3}} = \frac{\sqrt[6]{15^2}}{\sqrt[6]{15^3}} \cdot \frac{\sqrt[6]{15^3}}{\sqrt[6]{15^3}}$$

$$= \frac{\sqrt[6]{15^5}}{15}$$

(b)

$$\frac{\sqrt{2x}}{\sqrt[4]{8x^3}} = \frac{(2x)^{1/2}}{(8x^3)^{1/4}} = \frac{(2x)^{2/4}}{(8x^3)^{1/4}} = \frac{\sqrt[4]{(2x)^2}}{\sqrt[4]{8x^3}} = \frac{\sqrt[4]{4x^2}\,\sqrt[4]{2x}}{\sqrt[4]{8x^3}\,\sqrt[4]{2x}}$$

$$= \frac{\sqrt[4]{8x^3}}{2x}$$

Exercise Set 12.3

In Exercises 1–60, perform the indicated operations and express the answers in simplest form.

1. $2\sqrt{3}+5\sqrt{3}$

2. $5\sqrt{6}-3\sqrt{6}$

3. $\sqrt[3]{9}+4\sqrt[3]{9}$

4. $\sqrt[4]{8}+3\sqrt[4]{8}$

5. $2\sqrt{3}+4\sqrt{2}+6\sqrt{3}$

6. $5\sqrt{3}-6\sqrt{5}-9\sqrt{3}$

7. $\sqrt{5}+\sqrt{20}$

8. $\sqrt{8}+\sqrt{2}$

9. $\sqrt{7}-\sqrt{28}$

10. $\sqrt{8}-\sqrt{32}$

11. $\sqrt{60}-\sqrt{\dfrac{5}{3}}$

12. $\sqrt{84}+\sqrt{\dfrac{3}{7}}$

13. $\sqrt{\dfrac{1}{2}}-\sqrt{\dfrac{9}{2}}$

14. $\sqrt{\dfrac{4}{3}}-\sqrt{\dfrac{25}{3}}$

15. $\sqrt{x^3y}+2x\sqrt{xy}$

16. $\sqrt{a^5b^3}-3ab\sqrt{a^3b}$

17. $\sqrt[3]{24p^2q^4}+\sqrt[3]{3p^8q}$

18. $\sqrt[4]{16a^2b}-\sqrt[4]{81a^6b}$

19. $\sqrt{\dfrac{x}{y^3}}-\sqrt{\dfrac{y}{x^3}}$

20. $\sqrt{\dfrac{a^3}{b^3}}+\sqrt{\dfrac{b}{a^5}}$

21. $a\sqrt{\dfrac{b}{3a}}+b\sqrt{\dfrac{a}{3b}}$

22. $x\sqrt{\dfrac{y}{5x}}-y\sqrt{\dfrac{x}{5y}}$

23. $\sqrt{5}\sqrt{8}$

24. $\sqrt[5]{-7}\sqrt[5]{11}$

25. $\sqrt{3x}\sqrt{5x}$

26. $\sqrt[3]{7x^2}\sqrt[3]{3x}$

27. $(\sqrt{4x})^3$

28. $(\sqrt[3]{2x^2y})^4$

29. $\sqrt{\dfrac{5}{8}}\sqrt{\dfrac{9}{10}}$

30. $\sqrt{\dfrac{7}{6}}\sqrt{\dfrac{12}{3}}$

31. $\sqrt{2}(\sqrt{x}+\sqrt{2})$

32. $\sqrt{3}(\sqrt{12}-\sqrt{y})$

33. $(\sqrt{x}+\sqrt{y})^2$

34. $(\sqrt{a}+3\sqrt{b})^2$

35. $\sqrt[3]{\dfrac{5}{2}}\sqrt[3]{\dfrac{2}{7}}$

36. $\sqrt{\dfrac{ab}{5c}}\sqrt{\dfrac{abc}{5}}$

37. $(\sqrt{a}+\sqrt{b})(\sqrt{a}-\sqrt{b})$

38. $(\sqrt{x}-2\sqrt{y})(\sqrt{x}+5\sqrt{y})$

39. $\sqrt[3]{x}\sqrt{x}$

40. $\sqrt{5x}\sqrt[3]{2x^2}$

41. $\sqrt[3]{49x^2}\sqrt[4]{3x}$

42. $\sqrt[3]{9y}\sqrt[5]{-8x^2}$

43. $\dfrac{\sqrt{32}}{\sqrt{2}}$

44. $\dfrac{\sqrt[3]{a^2}}{\sqrt[3]{4a}}$

45. $\dfrac{\sqrt[3]{4b^2}}{\sqrt[3]{16b}}$

46. $\dfrac{\sqrt{5a^3b}}{\sqrt{15ab^3}}$

47. $\dfrac{\sqrt{3x^2}}{\sqrt[3]{9x}}$

48. $\dfrac{\sqrt[3]{15x}}{\sqrt{5x}}$

49. $\dfrac{\sqrt[3]{16a^2b}}{\sqrt[4]{2ab^3}}$

50. $\dfrac{\sqrt[5]{75a^3x^2}}{\sqrt[3]{-15ax^2}}$

51. $\dfrac{\sqrt[3]{4}}{\sqrt{2}}$

52. $\dfrac{\sqrt[5]{-8}}{\sqrt[3]{4}}$

53. $\dfrac{1}{x+\sqrt{5}}$

54. $\dfrac{1}{x+\sqrt{3}}$

55. $\dfrac{\sqrt{5}-\sqrt{3}}{\sqrt{5}+\sqrt{3}}$

56. $\dfrac{\sqrt{5}+\sqrt{7}}{\sqrt{7}-\sqrt{5}}$

57. $\dfrac{\sqrt{x+1}}{\sqrt{x-1}}+\dfrac{\sqrt{x-1}}{\sqrt{x+1}}$

58. $\dfrac{\sqrt{x+3}}{\sqrt{x-3}}-\dfrac{\sqrt{x-3}}{\sqrt{x+3}}$

59. $\dfrac{\sqrt{x+y}}{\sqrt{x-y}-\sqrt{x}}$

60. $\dfrac{\sqrt{1+y}}{\sqrt{1-y}+\sqrt{y}}$

Solve Exercises 61–66.

61. Use the quadratic equation to find the roots of $ax^2 + bx + c = 0$. What is the sum of the two roots?

62. *Oceanography* The velocity v of a small water wave is given by

$$v = \sqrt{\frac{\pi}{4\lambda d}} + \sqrt{\frac{4\pi}{\lambda d}}$$

Simplify and combine this equation.

63. *Electricity* The equivalent resistance R of two resistors, R_1 and R_2, connected in parallel is expressed

$$R = \frac{R_1 R_2}{R_1 + R_2}$$

In a given circuit, $R_1 = x^{3/2}$ and $R_2 = \sqrt{x}$. **(a)** Express R in terms of x. Make sure you simplify the answer. **(b)** If $x = 20\,\Omega$, what is R?

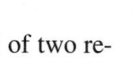

In Your Words

67. Explain how to add two radicals. What precautions do you need to take regarding the radicands and the indices?

64. *Sound* The theory of waves in wires uses the equation

$$\frac{\sqrt{d_1} - \sqrt{d_2}}{\sqrt{d_1} + \sqrt{d_2}}$$

Simplify this expression.

65. What is the product of the two roots of the quadratic equation $ax^2 + bx + c = 0$?

66. *Civil Engineering* The potential energy in a truss is given by

$$C = \frac{3}{8\sqrt[3]{2}} \cdot \frac{KA}{\sqrt[3]{L}}$$

Simplify this expression.

68. Describe how to multiply two radicals
   (a) With the same index
   (b) With different indices.

12.4
EQUATIONS WITH RADICALS

You should now know the laws of radicals. However, the laws of radicals are rather dry without some technical and algebraic application. Because technical situations can involve solving equations that contain radicals, we now learn how to use radicals in equations and to solve those equations.

Radical Equations; Extraneous Roots

Working with radicals often means that we must work with radical equations. An equation in which the variable occurs under a radical sign or has a fractional exponent is called a **radical equation**. In order to solve radical equations, we need to eliminate the radicals or the fractional exponents. This is done by raising both sides of the equation to some power that will eliminate the radical. The equation that results may not be equivalent to the original equation. In fact, the new equation may have more roots than the original one. The roots of the new equation that are not roots of the original equation are called **extraneous roots**. Check all solutions to make sure they are actual roots. Reject any extraneous roots.

EXAMPLE 12.23

Solve the radical equation $\sqrt{x+4}-9=0$.

Solution   This radical equation has only one radical term. So, isolate this radical term on one side of the equation.

$$\sqrt{x+4}=9$$
$$\left(\sqrt{x+4}\right)^2=9^2 \qquad\qquad \text{Square both sides.}$$
$$x+4=81$$
$$x=77$$

Substituting 77 in the original equation, we get $\sqrt{77+4}-9=\sqrt{81}-9=9-9=0$. The answer checks, so 77 is the solution.

EXAMPLE 12.24

Solve $\sqrt{x+5}-\sqrt{x-3}=10$.

Solution   This equation contains two radicals. Rewrite the equation with one radical on each side of the equals sign.

$$\sqrt{x+5}=\sqrt{x-3}+10$$
$$\left(\sqrt{x+5}\right)^2=\left(\sqrt{x-3}+10\right)^2 \qquad\qquad \text{Square both sides.}$$
$$x+5=\left(\sqrt{x-3}\right)^2+20\sqrt{x-3}+10^2 \qquad\qquad \text{``FOIL''}$$
$$x+5=x-3+20\sqrt{x-3}+100$$
$$x+5=x+97+20\sqrt{x-3}$$

This is now an equation with one radical. Isolate the term with the radical so it is on one side of the equation, and use the method of squaring both sides.

$$-92=20\sqrt{x-3}$$
$$(-92)^2=\left(20\sqrt{x-3}\right)^2$$
$$8{,}464=400\,(x-3)$$
$$8{,}464=400x-1{,}200$$
$$9{,}664=400x$$
$$x=\frac{9{,}664}{400}=\frac{604}{25}$$
$$=24.16$$

Replacing $x=24.16$ in the left-hand side of the original equation gives $\sqrt{29.16}-\sqrt{21.16}=0.8$. This does not equal the right-hand side of the original equation, 10. Since we have not made any mistakes, we can only conclude that $x=24.16$ is an extraneous root and that there is no solution to this problem.

If we had not wanted to show how you can get extraneous roots when you square both sides, we could have solved the previous example with less work. Notice the equation $-92 = 20\sqrt{x-3}$. Since $\sqrt{x-3}$ is never negative, the product of $20\sqrt{x-3}$ can never be negative. So, there is no solution to the equation $-92 = 20\sqrt{x-3}$. The next example also involves an extraneous root.

EXAMPLE 12.25

Solve $\sqrt{2x-3} - \sqrt{x+7} = 4$.

Solution   As in the last example, this equation contains two radicals, so we will first rewrite the problem with one radical on each side of the equal sign.

$$\sqrt{2x-3} = \sqrt{x+7} + 4$$
$$\left(\sqrt{2x-3}\right)^2 = \left(\sqrt{x+7}+4\right)^2 \qquad \text{Square both sides.}$$
$$\left(\sqrt{2x-3}\right)^2 = \left(\sqrt{x+7}\right)^2 + 8\sqrt{x+7} + 4^2$$
$$2x-3 = x+7 + 8\sqrt{x+7} + 16$$
$$x - 26 = 8\sqrt{x+7} \qquad \text{Simplify.}$$
$$(x-26)^2 = \left(8\sqrt{x+7}\right)^2 \qquad \text{Square again.}$$
$$x^2 - 52x + 676 = 64(x+7)$$
$$x^2 - 52x + 676 = 64x + 448$$
$$x^2 - 116x + 228 = 0$$
$$(x-2)(x-114) = 0$$

So, $x = 2$ or $x = 114$.

Replacing 2 for x in the original equation, we get $\sqrt{2\cdot 2 - 3} - \sqrt{2+7} = \sqrt{1} - \sqrt{9} = 1 - 3 = -2$. Since $-2 \neq 4$, the number 2 must be an extraneous root.

Next we replace 114 for x in the original equation, and we get $\sqrt{2\cdot 114 - 3} - \sqrt{114+7} = \sqrt{225} - \sqrt{121} = 15 - 11 = 4$. This checks. The only solution to this problem is 114.

EXAMPLE 12.26

Solve $\sqrt{x+3} = \sqrt[3]{x+21}$.

Solution   This problem also involves two radicals, but the indices of the radicals are not the same. If these were written with fractional exponents, you would have $(x+3)^{1/2} = (x+21)^{1/3}$. Raise both sides of the equation to the least common denominator of these powers. In this problem, the LCD is 6, so we would have

$$\left(\sqrt{x+3}\right)^6 = \left(\sqrt[3]{x+21}\right)^6$$
$$(x+3)^3 = (x+21)^2$$
$$x^3 + 9x^2 + 27x + 27 = x^2 + 42x + 441$$
$$x^3 + 8x^2 - 15x - 414 = 0$$

As you will learn in Chapter 18, the only real number solution to this problem is $x = 6$. Check this in the original problem to see if it is a root.

EXAMPLE 12.27

Solve $\sqrt{(x+1)^3} = 64$.

Solution $\sqrt{(x+1)^3}$ can be written as $(x+1)^{3/2}$. Raising both sides to the $\frac{2}{3}$ power we get

$$\left((x+1)^{3/2}\right)^{2/3} = 64^{2/3}$$
$$x+1 = 16$$
$$x = 15$$

Checking this in the original problem confirms that it is a root.

Application

EXAMPLE 12.28

At some point P between two light sources, the illuminance from each source is the same. If d_1 represents the distance from the light source with intensity of I_1 and d_2 is the distance from the source with intensity I_2, then

$$\frac{d_1}{d_2} = \frac{\sqrt{I_1}}{\sqrt{I_2}}$$

Find I_1 and I_2 in candelas (cd), when $d_1 = 15$ m, $d_2 = 6$ m, and $I_2 = 5I_1 - 11$.

Solution Substituting the given values in the previous equation produces

$$\frac{d_1}{d_2} = \frac{\sqrt{I_1}}{\sqrt{I_2}}$$
$$\frac{15}{6} = \frac{\sqrt{I_1}}{\sqrt{5I_1 - 11}}$$

Squaring both sides, we obtain

$$\frac{225}{36} = \frac{I_1}{5I_1 - 11}$$

or $\quad 225(5I_1 - 11) = 36I_1$
$$1\,125I_1 - 2\,475 = 36I_1$$
$$1\,125I_1 - 36I_1 = 2\,475$$
$$1\,089I_1 = 2\,475$$
$$I_1 = \frac{2\,475}{1\,089} \approx 2.27$$

So, I_1 is approximately 2.27 cd.

Substituting the exact value $I_1 = \frac{2\,475}{1\,089}$ in the equation $I_2 = 5I_1 - 11$, we determine

$$I_2 = 5\left(\frac{2\,475}{1\,089}\right) - 11 = \left(\frac{12\,375}{1\,089}\right) - 11 \approx 0.36$$

The illuminance at I_2 is approximately 0.36 cd.

Exercise Set 12.4

In Exercises 1–20, solve the radical equations. Check all roots.

1. $\sqrt{x+3} = 5$

2. $\sqrt{x-16} = 5$

3. $\sqrt{2x+4} - 7 = 0$

4. $\sqrt{3y-9} - 18 = 0$

5. $\sqrt{x^2+24} = x - 4$

6. $\sqrt{x^2-75} = x + 5$

7. $(y+12)^{1/3} = 3$

8. $(r+9)^{1/4} = 7$

9. $\left(\frac{x}{2}+1\right)^{1/2} = 3$

10. $\left(\frac{y}{3}+4\right)^{1/3} = \frac{5}{2}$

11. $(x-1)^{3/2} = 27$

12. $(x+1)^{3/4} = 8$

13. $\sqrt{x+1} = \sqrt{2x-1}$

14. $\sqrt{x^2-4} = \sqrt{2x-1}$

15. $\sqrt{x+1} + \sqrt{x-5} = 7$

16. $\sqrt{x-3} - \sqrt{x+2} = 19$

17. $\sqrt{x-1} + \sqrt{x+5} = 2$

18. $\sqrt{x+3} - \sqrt{2x-4} = 3$

19. $\sqrt{\dfrac{1}{x}} = \sqrt{\dfrac{4}{3x-1}}$

20. $\sqrt{\dfrac{2}{x-1}} = \sqrt{\dfrac{5}{x+1}}$

Solve Exercises 21–26.

21. *Civil engineering*   The maximum speed at which it is possible to negotiate a curve without slipping is given by

$$v = \sqrt{\mu_s g R}$$

Solve this for R.

22. *Physics*   The velocity v of an object falling under the influence of gravity g is given by $v = \sqrt{v_0^2 - 2gh}$, where v_0 is the initial velocity and h is the height fallen. Solve this equation for h.

23. *Lighting technology*   The illuminance from two light sources is the same at some point P between them. Suppose d_A is the distance to P from light A and d_B is the distance from P to light B. If the luminous intensity at A is I_A and at B is I_B, then we have the equation

$$\frac{d_A}{d_B} = \frac{\sqrt{I_A}}{\sqrt{I_B}}$$

If $d_A = 11\,\text{m}$, $d_B = 20\,\text{m}$, and $I_A = 3.5I_B - 15$, then find I_A and I_B.

24. *Lighting technology*   The intensity of a light source can be determined with a photometer. The relationship between the luminous intensity I_x of an unknown source is found by comparing it with a standard source of known intensity I_y. The distances from each source are adjusted until a grease spot is equally illuminated by each source. If r_x is the distance to the unknown source and r_y is the distance to the standard, then the equation

$$\frac{\sqrt{I_x}}{r_x} = \frac{\sqrt{I_y}}{r_y}$$

can be used. (a) Solve this equation for I_x. (b) If $I_y = 30\,\text{cd}$ and $r_x = 2r_y + 3$, then solve for I_x.

25. *Physics*   If an object is dropped, the time t it takes for it to fall s ft is given by $t = \frac{1}{4}\sqrt{s}$. An object is dropped from the top of a building. It takes 8.6 s for the object to hit the ground. How tall is the building?

26. *Electronics*   The impedance, Z, in an ac circuit can be determined by the formula $Z = \sqrt{R^2 + X^2}$, where R is the resistance in Ω and X is the reactance in Ω. Find the reactance of a circuit with a resistance of $25\,\Omega$ and an impedance of $43\,\Omega$.

In Your Words

27. **(a)** What is an extraneous root? **(b)** How can you tell if a root is extraneous?

28. Describe the general approach you should use to solve a radical equation.

▤ CHAPTER 12 REVIEW

Important Terms and Concepts

Adding radicals	Multiplying radicals
Conjugate	Radical
Dividing radicals	Radical equations
Extraneous roots	Radicand
Fractional exponent	Rationalizing denominators
Index	Subtracting radicals

Review Exercises

In Exercises 1–6, evaluate the given expression without the use of a calculator. Check your answers with a calculator.

1. $49^{1/2}$
2. $81^{1/4}$
3. $25^{3/2}$
4. $64^{2/3}$
5. $9^{-3/2}$
6. $125^{-4/3}$

In Exercises 7–18, express each of the given expressions in the simplest form containing only positive exponents.

7. $5^4 5^9$
8. $13^6 13^{-9}$
9. $\left(2^4\right)^8$
10. $\left(5^3\right)^6$
11. $\left(x^2 y^3\right)^4$
12. $\left(y^6 x\right)^4$
13. $\left(\dfrac{a^3}{ya^4}\right)^{-5}$
14. $\left(\dfrac{a^5}{ab^4}\right)^3$
15. $x^{1/2} x^{1/3}$
16. $y^{2/3} y^{3/4}$
17. $\dfrac{\left(xy^2 a\right)^5}{x^4 \left(ya^3\right)^2}$
18. $\dfrac{\left(ab^3 c^2\right)^4}{\left(a^2 bc\right)^3}$

Use the laws of radicals to express each of Exercises 19–42 in simplest radical form.

19. $\sqrt[3]{-27}$
20. $\sqrt[4]{81^3}$
21. $\sqrt{a^2 b^5}$
22. $\sqrt[4]{a^3 b^7}$
23. $\sqrt[3]{\dfrac{-8x^3 y^6}{z}}$
24. $\sqrt[4]{\dfrac{32a^6 b^8}{cd^2}}$
25. $\sqrt{\dfrac{a}{b^2} - \dfrac{b}{a^2}}$

26. $\sqrt{a^2 + 8a + 16}$
27. $2\sqrt{5} + 6\sqrt{5}$
28. $3\sqrt{7} - 6\sqrt{7}$
29. $3\sqrt{6} + 5\sqrt{6} - 2\sqrt{3}$
30. $\sqrt{6} + \sqrt{24}$
31. $\sqrt[4]{16a^3 b} + \sqrt[4]{81a^7 b}$
32. $\sqrt[4]{32x^5} - x^2 \sqrt[4]{2x}$
33. $a\sqrt{\dfrac{a}{7b}} - b\sqrt{\dfrac{b}{7a}}$
34. $\sqrt{\dfrac{x-2}{x+2}} + \sqrt{\dfrac{x+2}{x-2}}$

35. $\sqrt{3}\sqrt{5}$
36. $\sqrt{7}\sqrt{6}$
37. $\sqrt{2x}\sqrt{7x}$
38. $\sqrt[3]{5y}\sqrt[3]{25y}$
39. $\left(\sqrt{a} + \sqrt{2b}\right)\left(\sqrt{a} - \sqrt{2b}\right)$
40. $\left(\sqrt{x} - 3\sqrt{y}\right)\left(\sqrt{x} + 3\sqrt{y}\right)$
41. $\dfrac{\sqrt[3]{2a^2}}{\sqrt[3]{16a}}$
42. $\dfrac{\sqrt[3]{5a^2 b}}{\sqrt[3]{25ab^2}}$

In Exercises 43–50, solve the radical equations. Check your answers for extraneous roots.

43. $\sqrt{x-1} = 9$

44. $\sqrt{2x-7} - 5 = 0$

45. $\sqrt{x^2 - 7} = 2x - 4$

46. $\sqrt{\dfrac{3x}{4} - 1} = 5x - 3$

47. $\sqrt{x+1} + \sqrt{x-5} = 8$

48. $\sqrt{x+3} + \sqrt{x-2} = 20$

49. $\sqrt{\dfrac{1}{x}} = \sqrt{\dfrac{3x}{4} - 1}$

50. $\sqrt{\dfrac{3}{x-1}} = \sqrt{\dfrac{4}{x+1}}$

Solve Exercises 51–54.

51. *Nuclear technology*   The half-life of tritium is 12.5 yr. It decays according to the formula $q(t) = q_0 2^{-t/12.5}$, where q_0 is the original amount of tritium and $q(t)$ is the amount left at time t, where t is given in years. If you begin with 50 mg of tritium, **(a)** how much will remain after 10 years? **(b)** after 20 years?

52. *Nuclear technology*   The velocity of a proton can be described in terms of its kinetic energy, KE, and mass, m, with the formula

$$v = \sqrt{\dfrac{2KE}{m}}$$

(a) Express this in simplest radical form. **(b)** Solve the equation for m.

53. *Navigation*   Figure 12.1 shows the vector diagram for the resultant velocity v of a missile with horizontal component $\mathbf{v}_x$ and vertical component $\mathbf{v}_y$. If $\mathbf{v}_x = 3t + 2$ and $\mathbf{v}_y = 4t - 1$, express v in terms of t in simplest radical form.

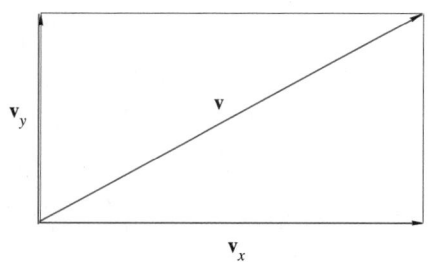

FIGURE 12.1

54. *Energy technology*   Two sources, A and B, are 12 m apart as shown in Figure 12.2. Each emit waves of wavelength $\lambda = 5\,m$. **Constructive interference** occurs when the joint effect of the two waves results in a wave of larger amplitude. Here, constructive interference of the waves will occur at point C, located x units from A when

$$BC - AC = \lambda$$

How far is point C from A? How far from B?

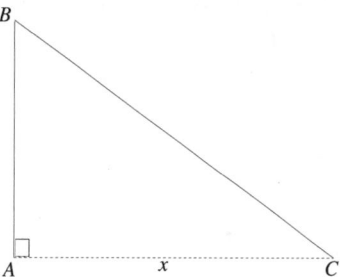

FIGURE 12.2

≡ CHAPTER 12 TEST

In Exercises 1–4, express each of the given expressions in the simplest form containing only positive exponents.

1. $8^5 8^{-7}$

2. $\left(3^5\right)^4$

3. $\left(\dfrac{x^2}{xy^3}\right)^5$

4. $x^{1/4} x^{4/3}$

In Exercises 5–13, use the laws of radicals to express each of the following in simplest form.

5. $\sqrt[3]{-64}$

6. $\sqrt[4]{81}$

7. $\sqrt{x^4 y^3}$

8. $\sqrt[3]{\dfrac{-27x^6 y}{z^2}}$

9. $\sqrt{x^2 + 6x + 9}$

10. $3\sqrt{6} + 5\sqrt{24}$

11. $\sqrt{2x}\sqrt{6x}$

12. $\left(3\sqrt{x} + \sqrt{5y}\right)\left(3\sqrt{x} - \sqrt{5y}\right)$

13. $\dfrac{\sqrt[3]{4x^2 y}}{\sqrt[3]{2x^2 y^2}}$

In Exercises 14–17, solve the radical equations.

14. $\sqrt{x+3} = 6$

15. $\sqrt{2x+1} = 5$

16. $\sqrt{2x^2 - 7} = x + 3$

17. $\sqrt{\dfrac{5}{x+3}} = \sqrt{\dfrac{4}{x-3}}$

Solve the following exercises.

18. The diameter d of a cylindrical column necessary for it to resist a crushing force F is

$$d = \left(1.12 \times 10^{-2}\right)\sqrt[3]{Fl}$$

where l is the length of the column. **(a)** Solve for F. **(b)** Determine F, when $d = 1.2\,\text{m}$ and $l = 5\,\text{m}$.

13

Exponential and Logarithmic Functions

Medical technologists and people working in microbiology often need to determine how many of a certain type of bacteria they can expect after a certain period of time. This type of growth tends to be exponential. In Section 13.2, you will learn how to determine exponential growth.

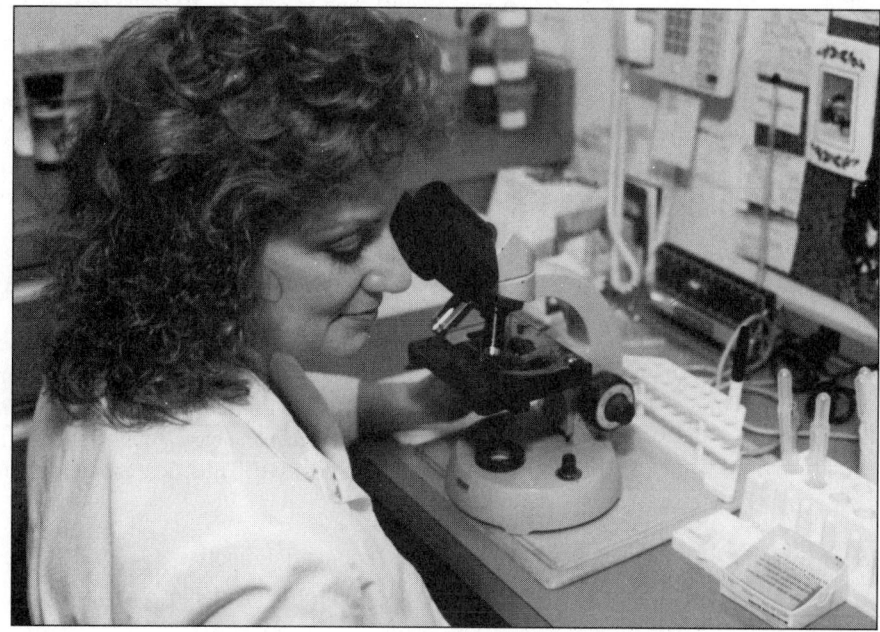

Courtesy of Ruby Gold

The first functions we studied in this text were algebraic functions. Then we introduced the trigonometric functions, examples of functions that were not algebraic. In this chapter, we will introduce two new functions. These functions are not algebraic functions, but they are very important in many technical areas, such as business, finance, nuclear technology, acoustics, electronics, and astronomy. Many of the applications in this chapter will involve growth or decay.

≡ 13.1
EXPONENTIAL
FUNCTIONS

We will begin this section with the definition of an exponential function and with some examples of exponential functions.

Exponential Function

An **exponential function** is any function of the form

$$f(x) = b^x$$

where $b > 0$, $b \neq 1$, and x is any real number. The number b is called the **base**.

EXAMPLE 13.1

Some examples of exponential functions are: $f(x) = 2^x$, $g(x) = 3^x$, $h(x) = \pi^x$, $j(x) = 4.2^x$, and $k(x) = (\sqrt{3})^x$. The following exponential functions contain constants, represented by a and b: $y(x) = (5b)^{-x}$, $h(x) = (2a)^{x+1}$, and $k(x) = 3a^{x/3}$.

The functions $f(x) = (-2)^x$, $g(x) = 0^x$, and $h(x) = 1^x$ are not exponential functions. In $f(x) = (-2)^x$ the base, -2, is less than zero. Therefore, it is not an exponential function. We can see that if $x = \frac{1}{2}$, then $(-2)^{1/2} = \sqrt{-2}$ is not a real number. The function $g(x) = 0^x$ is not an exponential function, because the base is 0. Also, $h(x) = 1^x$ is not an exponential function, since the base is 1.

Graphing Exponential Functions; Asymptotes

Graphing an exponential function is done in the same manner in which we have graphed other functions. We will choose values for x and determine the corresponding values for $f(x)$. We will then plot these points and connect them in order to get the graph. Notice that, since the base b of an exponential function is positive, all powers of that base are also positive. Thus, an exponential function is positive for all values of x.

EXAMPLE 13.2

Graph the exponential function $f(x) = 2^x$.

Solution The following table gives integer values of x from -3 to 5.

x	-3	-2	-1	0	1	2	3	4	5
$f(x) = 2^x$	0.125	0.25	0.5	1	2	4	8	16	32

As x gets larger, $f(x)$ increases at a faster and faster rate. As x gets smaller, $f(x)$ keeps getting closer to 0, but never reaches 0.

The graph of $f(x) = 2^x$ is shown in Figure 13.1. Notice how the curve keeps getting closer to the negative x-axis. As a curve approaches a line when a variable approaches a certain value, the curve is said to be asymptotic to the line. The line is an **asymptote** of the curve. In this example, $y = 2^x$ is asymptotic to the negative x-axis.

Either a calculator or a computer is helpful for finding values of an exponential function. On a calculator, you will use the $\boxed{y^x}$ key. (On some calculators, this key is

x^y and on some graphics calculators it is $\boxed{\wedge}$.) On the $\boxed{y^x}$ key, the base is indicated by the letter y.

EXAMPLE 13.3

Determine $f(4.5)$, when $f(x) = 2^x$.

Solution

	PRESS	DISPLAY
Algebraic calculator:	2 $\boxed{x^y}$ 4.5 $\boxed{=}$	22.62741700
RPN calculator:	2 $\boxed{ENTER}$ 4.5 $\boxed{y^x}$	22.62741700
TI-81 calculator:	2 $\boxed{\wedge}$ 4.5 $\boxed{ENTER}$	22.627417

The most common bases are those that are larger than 1. In fact, when the base is less than 1 (and greater than 0), you get an interesting effect as shown in the next example.

EXAMPLE 13.4

Graph the exponential function $g(x) = \left(\frac{1}{2}\right)^x$.

Solution   Again, we will make a table of values.

x	−5	−4	−3	−2	−1	0	1	2	3
$g(x) = \left(\frac{1}{2}\right)^x$	32	16	8	4	2	1	0.5	0.25	0.125

The graph of these points is shown in Figure 13.2. This graph would be identical to the graph of $f(x) = 2^x$, if it were reflected in the y-axis.

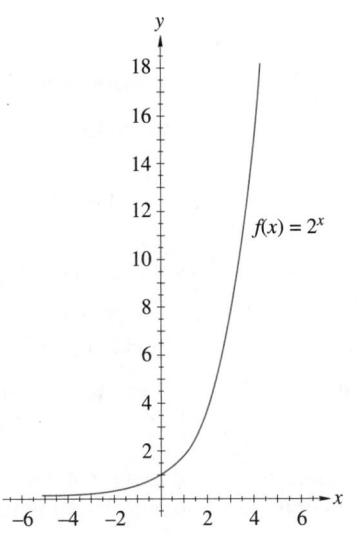

FIGURE 13.1

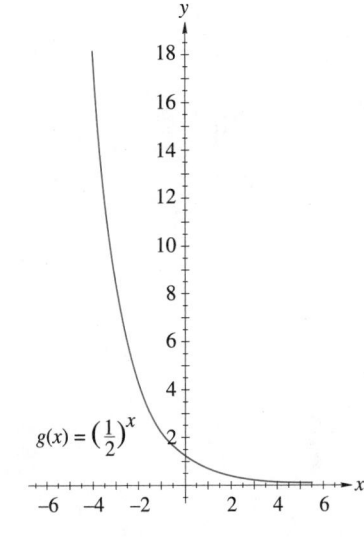

FIGURE 13.2

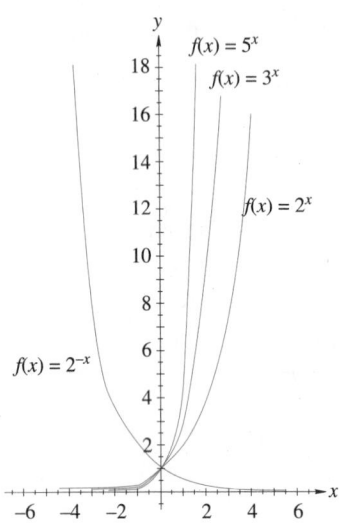

FIGURE 13.3

Study the function again. Remember that $\frac{1}{2} = 2^{-1}$, so $\left(\frac{1}{2}\right)^x = 2^{-x}$. These two examples show the differences between a function of the form b^x and one of the form b^{-x}. If $b > 1$, the graph of b^x rises and the graph of b^{-x} falls, as we move from left to right.

Figure 13.3 shows the graph of several exponential functions on the same set of coordinates. These functions, and all exponential functions, have the following three features in common.

> **Common Features of Exponential Functions**
>
> If $f(x) = b^x$, then
>
> 1. The y-intercept is 1.
> 2. If $b > 1$, the negative x-axis is an asymptote; if $b < 1$, the positive x-axis is an asymptote.
> 3. If $b > 1$, the curves all rise as the values of x increase; if $b < 1$, the curves all fall as the values of x increase.

Business Applications

One application of exponential functions has to do with money. When an amount of money P, called the principal, is invested and interest is compounded annually (once a year) at the interest rate of r per year, then the amount of money at the end of t yr is given by the formula $S = P(1 + r)^t$. In this formula, r is expressed as a decimal. So, at 5% interest, $r = 0.05$ and at $6\frac{1}{4}$% interest, $r = 0.0625$.

Application

EXAMPLE 13.5

If $800 is invested at 6% compounded annually, how much is the total value after 10 yr?

Solution   Here $P = \$800, r = 0.06$, and $t = 10$. As a result we have

$$S = P(1 + r)^t$$
$$= 800(1 + 0.06)^{10}$$
$$= 800(1.790847697)$$
$$\approx 1{,}432.68$$

After 10 yr, the total value of this investment is $1,432.68.

If interest is compounded more than once a year, then the formula is changed. If interest is compounded semiannually, or twice a year, the formula becomes

$$S = P\left(1 + \frac{r}{2}\right)^{2t}$$

If interest is compounded quarterly, or four times a year, the formula is changed to

$$S = P\left(1 + \frac{r}{4}\right)^{4t}$$

In general, we have the following compound interest formula.

Compound Interest Formula

If an amount of money P, is invested and interest is compounded k times a year at an interest rate of r per year, then the amount of money S at the end of t yr is

$$S = P\left(1 + \frac{r}{k}\right)^{kt}$$

Application

EXAMPLE 13.6

If \$800 is invested in a savings account paying 6% interest compounded monthly, how much is this money worth after 10 yr?

Solution   In this example, $P = \$800, r = 0.06, k = 12,$ and $t = 10$, so

$$\begin{aligned}
S &= P\left(1 + \frac{r}{k}\right)^{kt} \\
&= 800\left(1 + \frac{0.06}{12}\right)^{12\cdot10} \\
&= 800(1.005)^{120} \\
&= 800(1.819396734) \\
&\approx 1,455.52
\end{aligned}$$

Compare this to the result we obtained in Example 13.5. Compounding monthly increased the total by \$22.84.

In Section 13.2, we will examine what happens if you compound interest daily or continuously. We will see that there is a limit to the amount of money you will receive.

Exercise Set 13.1

In Exercises 1–6, use a calculator to approximate the given numbers. Round off each answer to four decimal places.

1. $3^{\sqrt{2}}$

2. 4^{π}

3. π^3

4. $(\sqrt{3})^{\pi}$

5. $(\sqrt{3})^{\sqrt{5}}$

6. $(\sqrt{4})^{4/3}$

In Exercises 7–16, make a table of values and draw the graph of each function.

7. $f(x) = 4^x$

8. $g(x) = 3^x$

9. $h(x) = 1.5^x$

10. $k(x) = 2.5^x$

11. $f(x) = 3^{-x}$

12. $g(x) = 5^{-x}$

13. $h(x) = 2.4^{-x}$

14. $k(x) = (\sqrt{3})^x$

15. $f(x) = 3^{(x+1/2)}$

16. $g(x) = 2^{(x-1/4)}$

Solve Exercises 17–26.

17. *Finance*   The sum of $1,000 is placed in a savings account at 6% interest. If interest is compounded **(a)** annually, **(b)** semiannually, **(c)** quarterly, and **(d)** monthly, what is the total after 5 yr?

18. *Finance*   The sum of $2,000 is placed in a savings account at 8% interest. If interest is compounded **(a)** annually, **(b)** semiannually, **(c)** quarterly, and **(d)** monthly, what is the total after 10 yr?

19. *Finance*   One bank offers 6% interest compounded semiannually. A second bank offers the same interest but compounded monthly. How much more income will result by depositing $1,000 in the second account for 5 yr than by depositing $1,000 in the first account for 5 yr?

20. *Finance*   A bank offers 8% interest compounded annually. A second bank offers 8% interest compounded quarterly. How much more income will result by depositing $2,000 in the second bank for 10 yr than in the first?

21. *Biology*   The number of a certain type of bacteria is given by the equation $Q = Q_0 2^t$, where Q_0 is the initial number of bacteria (that is, the number of bacteria when $t = 0$) and t is the time in hours since the initial count was taken.
(a) If $Q = 200{,}000$ when $t = 2$, find Q_0.
(b) Find the number of bacteria present at the end of 4 h.
(c) How long does it take for Q to become twice as large as Q_0?
(d) How long does it take for Q to become eight times as large as Q_0?

22. *Medical technology*   A pharmaceutical company is growing an organism to be used in a vaccine. The or-

ganism's growth is given by the equation $Q = Q_0 3^t$, where Q_0 is the initial number of bacteria (that is, the number of bacteria when $t = 0$) and t is the time in hours since the initial count was taken. When $t = 1$, we know that $Q = 2{,}760$.
(a) Find Q_0.
(b) What are the number of organisms at the end of 5 h?
(c) How long will it take for Q to become 3 times as large as Q_0?

23. *Medical technology*   A pharmaceutical company is growing an organism to be used in a vaccine. The organism's growth is given by the equation $Q = Q_0 3^{kt}$, where Q_0 is the initial number of bacteria, t is the time in hours since the initial count was taken, and k is a constant. When $t = 0$, we know that $Q = 400$, and when $t = 2\frac{1}{2}$ h, $Q = 1{,}200$.
(a) Find Q_0.
(b) Find k.
(c) What are the number of organisms present at the end of 5 h?

24. *Ecology*   In 1995, the world population was growing at the rate of approximately 1.8% per year. This can be expressed mathematically as $P = P_0 (1.018)^t$, where $P_0 = 5.7 \times 10^9$ and P is the population t years after 1995.
(a) What will the world's approximate population be in the year 2000?
(b) What will the world's approximate population be in the year 2050?
(c) Graph $P = P_0 (1.018)^t = 5.7 \times 10^9 (1.018)^t$.
(d) Use the graph from (c) to approximate the year that the world's population will reach $10{,}000{,}000{,}000 = 10^{10}$?

25. *Lighting technology*   Each 1 mm thickness of a certain translucent material reduces the intensity of a light beam passing through it by 12%. This means that the intensity, I, is a function of the thickness, T, of the material as given by $I = 0.88^T$.
  (a) What is the intensity when the thickness is 1.75 mm?
  (b) Graph I as a function of T.
  (c) Use the graph from (b) to approximate the thickness that produces an intensity of 0.50?

26. *Electronics*   An important triode formula is

$$I_p + I_g = K \left(V_g + \frac{V_p}{\mu} \right)^{3/2}$$

where I_p is the plate current in amperes, I_g is the grid current in amperes, V_p is the plate voltage, V_g is the grid voltage, and μ is an amplification factor. Calculate $I_p + I_g$ if $K = 0.0005$, $V_p = 350$ V, $V_g = 8$ V, and $\mu = 14$.

In Your Words

27. The text listed three common features of exponential functions. Without looking in the text, list each of these features. You may want to draw some graphs to help your explanation.

 **28.**   **(a)** Graph the two functions $f(x) = x^2$ and $g(x) = 2^x$ on your calculator.

  (b) Explain how the two graphs are alike and how they are different.
  (c) Describe how you would help someone learn which was the graph of $f(x) = x^2$ and which was the graph of $g(x) = 2^x$.

≡ 13.2
THE EXPONENTIAL
FUNCTION e^x

At the end of Section 13.1, we were working with an equation that determined the amount of interest gathered in an account. A variation of that formula is $\left(1 + \dfrac{1}{n}\right)^n$.

Let's examine the values of $\left(1 + \dfrac{1}{n}\right)^n$ as n gets larger. Use your calculator to check these numerical values.

n	$\left(1 + \dfrac{1}{n}\right)^n$
1	2
10	2.59374246
100	2.704813829
1,000	2.716923932
10,000	2.718145927
100,000	2.718268237
1,000,000	2.718280469
10,000,000	2.718281693
100,000,000	2.718281815
1,000,000,000	2.718281827

Mathematicians have been able to prove that as the value of n gets larger, the value of $\left(1 + \dfrac{1}{n}\right)^n$ also continues to get larger, but has a limit on how large it can get. This limit is such a special number that it has been given its own symbol, e. The first nine digits of e are the same ones that we got in the previous table, 2.71828182.

As we will see later in this section, the number e is a very important number. Some calculators have two special numbers marked on them—π and e. Later in this section, we will need to determine values for the exponential function $f(x) = e^x$. Both the calculator and the computer can be used to find these values.

Finding e^x with a Calculator

Some calculators have an $\boxed{e^x}$ key. If you want to determine e^5, use this method.

	PRESS	DISPLAY
Algebraic calculator:	5 $\boxed{e^x}$	148.4131591
RPN calculator:	5 $\boxed{\text{ENTER}}$ $\boxed{e^x}$	148.4131591
TI-82 calculator:	$\boxed{\text{2nd}}$ $\boxed{e^x}$ 5 $\boxed{\text{ENTER}}$	148.4131591

Some calculators do not have a key marked e^x. On these calculators, you must use two keys. You will need to use the $\boxed{\text{INV}}$ key followed by the key $\boxed{\ln x}$. This may seem strange, but you will understand more clearly when we cover the next section. So, to determine e^5, you would use this method.

PRESS	DISPLAY
5 $\boxed{\text{INV}}$ $\boxed{\ln x}$	148.41316

Business and Finance

Let's return to a problem involving compound interest. In Section 13.1, we learned that if a certain principal amount P is invested and interest is compounded k times a year at an interest rate of r, the amount after t yr would be

$$S = P\left(1 + \frac{r}{k}\right)^{kt}$$

If we let $n = \dfrac{k}{r}$, then $k = nr$, and this formula becomes

$$S = P\left(1 + \frac{r}{nr}\right)^{nrt}$$
$$= P\left[\left(1 + \frac{1}{n}\right)^n\right]^{rt}$$

Now, if interest is compounded continuously, then the expression inside the brackets, $\left(1 + \dfrac{1}{n}\right)^n$, is equal to the number represented by e. So, the amount accumulated after t yr at the interest rate of r would be

$$S = Pe^{rt}$$

Application

EXAMPLE 13.7

If $800 is invested in a savings account paying interest compounded continuously at 6%, how much has accumulated after 10 yr?

Solution   Here $P = 800$, $r = 0.06$, and $t = 10$, so

$$S = 800e^{(0.06)10}$$
$$= 800e^{0.6}$$
$$\approx 800(1.8221188)$$
$$= 1,457.70$$

After 10 yr, this $800 investment has increased to $1,457.70.

When this is compared to the result in Examples 13.5 and 13.6, we see that, after 10 yr, continuous compounding of interest has provided an additional $2.18.

Exponential Growth and Decay

The number represented by e is used in the two areas known as exponential growth and exponential decay.

Exponential growth can be explained by letting y represent the size of a quantity at time t. If this quantity grows or decays exponentially, it obeys the exponential growth formula given here.

Exponential Growth and Decay

The basic formula for the exponential growth or decay of a quantity is

$$y = ce^{kt}$$

where y represents the size of the quantity at time t, and c and k are positive real number constants.
If $k > 0$, this is an **exponential growth** function.
If $k < 0$, this is an **exponential decay** function.

Application

EXAMPLE 13.8

A culture of bacteria originally numbers 500. After 4 h, there are 8,000 bacteria in the culture. If we assume that these bacteria grow exponentially, how many will there be after 10 h?

Solution   Since the number of bacteria grows exponentially, their growth obeys the formula $y = ce^{kt}$. When $t = 0$, we are told that $y = 500$, and so $500 = ce^{k \cdot 0} = c$. (Remember, $k \cdot 0 = 0$, $e^{k \cdot 0} = e^0 = 1$.) The formula is now $y = 500e^{kt}$. When $t = 4$ h, we have $y = 8,000$ and from the formula,

$$8,000 = 500e^{k4}$$
$$16 = e^{k4}$$

EXAMPLE 13.8 (Cont.)

However, the example asks us to determine y, when $t = 10$. It is possible to do this without determining k. (In Section 13.3, we will learn how to find the value of k.)

$$y = 500e^{k \cdot 10}$$
$$= 500(e^{k4})^{2.5}, \text{ since } 10 = (4)(2.5)$$
$$= 500(16)^{2.5}, \text{ since } e^{k4} = 16$$
$$= 500(1{,}024)$$
$$= 512{,}000$$

After 10 h, the number of bacteria increased from 500 to 512,000.

Exponential decay is seen most often in radioactive substances. A common measure of the rate of decay is the **half-life** of a substance. The half-life is the amount of time needed for a substance to diminish to one-half its original size.

Application

EXAMPLE 13.9

The half-life of copper-67 is 62 h. How much of 100 g will remain after 15 days?

Solution　As in Example 13.8, we have some basic information. When $t = 0$, we know that $y = 100$ g, so

$$100 = ce^{k \cdot 0}$$
$$\text{or} \quad 100 = c$$

and the exponential decay formula for copper-67 becomes

$$y = 100e^{kt}$$

Now, since the half-life is 62 h, we know that, when $t = 62$, there is only half as much copper-67, or $y = 50$ g, so

$$50 = 100e^{k(62)}$$
$$\frac{1}{2} = e^{k(62)}$$

We want to determine the amount when $t = 15$ days. Because the half-life is given in hours, we first convert 15 days to $15 \times 24 = 360$ h. Since $360 \div 62 \approx 5.8$, we can write the formula as

$$y = 100e^{k \cdot 360}$$
$$\approx 100(e^{k(62)})^{5.8}$$

and since $e^{k(62)} = \frac{1}{2}$, we have

$$y \approx 100 \left(\frac{1}{2}\right)^{5.8}$$
$$= 1.79 \text{ g}$$

So, of the original 100 g of copper-67, about 1.79 g remain after 15 days.

Application

EXAMPLE 13.10

When a charged capacitor is discharged through a resistance, as in Figure 13.4, the decrease in charge Q is given by the formula

$$Q = Q_0 e^{-t/T}$$

where $Q_0 = CV$, the initial charge, $T = RC$, and C is the capacitance, V the battery voltage, and R the resistance. The product RC is called the time constant of the circuit. If a 15-μF capacitor is charged by being connected to a 60-V battery through a circuit with a resistance of $12\,000\,\Omega$, what is the charge on the capacitor 9 s after the battery is disconnected?

Solution   We are given $C = 15\,\mu\text{F} = 15 \times 10^{-6}$ F, $V = 60$ V, and $R = 12\,000\,\Omega$. Thus

$$
\begin{aligned}
Q_0 &= CV \\
&= 15 \times 10^{-6}\,\text{F} \times 60\,\text{V} \\
&= 900 \times 10^{-6} \\
&= 9 \times 10^{-4}\,\text{C}
\end{aligned}
$$

and

$$
\begin{aligned}
T &= RC \\
&= 12\,000\,\Omega \times 15 \times 10^{-6}\,\text{F} \\
&= 180\,000 \times 10^{-6}\,\text{F} \\
&= 0.18\,\text{s}
\end{aligned}
$$

Substituting these values for Q_0 and T into the formula $Q = Q_0 e^{-t/T}$, we obtain

$$Q = 9 \times 10^{-4} e^{-t/0.18} \tag{*}$$

We want to find the value of Q when $t = 9$ s, or

$$\frac{t}{T} = \frac{9\,\text{s}}{0.18\,\text{s}} = 50$$

Thus, returning to equation (*), with $\dfrac{t}{T} = 50$, we get

$$
\begin{aligned}
Q &= 9 \times 10^{-4} e^{-50} \\
&= 1.74 \times 10^{-25}\,\text{C}
\end{aligned}
$$

So, the charge is about 1.74×10^{-25} C, 9 s after the battery is disconnected.

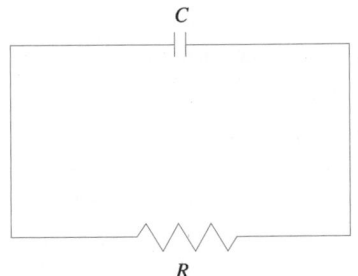

C

R

FIGURE 13.4

Exercise Set 13.2

Use a calculator or a computer to evaluate each of the numbers in Exercises 1–8.

1. e^3

2. e^7

3. $e^{4.65}$

4. $e^{5.375}$

5. e^{-4}

6. e^{-9}

7. $e^{-2.75}$

8. $e^{-0.25}$

Make a table of values and graph each of the functions in Exercises 9–12 over the given domains of x.

9. $f(x) = 4e^x$, $\{x : -2 \le x \le 4\}$

10. $g(x) = 3.5e^{5x}$, $\{x : -1 \le x \le 4\}$

11. $h(x) = 4e^{-x}$, $\{x : -4 \le x \le 2\}$

12. $k(x) = 8.5e^{-6x}$, $\{x : -4 \le x \le 2\}$

Solve Exercises 13–28.

13. *Nuclear technology*   The number of milligrams of a radioactive substance present after t yr is given by $Q = 125e^{-0.375t}$. **(a)** How many milligrams were present at the beginning? **(b)** How many milligrams were present after 1 yr? **(c)** How many milligrams are present after 16 yr?

14. *Nuclear technology*   Radium decays exponentially and has a half-life of 1,600 yr. How much of 100 mg will be left after 2,000 yr?

15. *Biology*   The number of bacteria in a certain culture increases from 5,000 to 15,000 in 20 h. If we assume these bacteria grow exponentially, **(a)** how many will be there after 10 h? **(b)** How many can we expect after 30 h? **(c)** How many can we expect after 3 days?

16. *Finance*   If $5,000 is invested in an account that pays 5% interest compounded continuously, how much can we expect to have after 10 yr?

17. *Biology*   The population of a certain city is increasing at the rate of 7% per year. The present population is 200,000. **(a)** What will be the population in 5 yr? **(b)** What can the population be expected to reach in 10 yr?

18. *Medical technology*   A pharmaceutical company is growing an organism to be used in a vaccine. The organism grows at a rate of 4.5% per hour. How many units of this organism must they begin with in order to have 1,000 units at the end of 7 days?

19. *Thermodynamics*   According to **Newton's law of cooling,** the rate at which a hot object cools is proportional to the difference between its temperature

and the temperature of its surroundings. The temperature T of the object after a period of time t is

$$T = T_m + (T_0 - T_m)e^{-kt}$$

where T_0 is the initial temperature and T_m is the temperature of the surrounding medium. An object cools from 180°F to 150°F in 20 min when surrounded by air at 60°F. What is the temperature at the end of 1 h of cooling?

20. *Thermodynamics*   A piece of metal is heated to 150°C and is then placed in the outside air, which is 30°C. After 15 min the temperature of the metal is 90°C. What will its temperature be in another 15 min? (See Exercise 19.)

21. *Thermodynamics*   You like your drinks at 45°F. When you arrive home from the store, the cans of drink you bought are 87°F. You place the cans in a refrigerator. The thermostat is set at 37°F. When you open the refrigerator 25 min later, the drinks are at 70°F. How long will it take for the drinks to get to 45°F? (See Exercise 19.)

22. *Electricity*   The circuit in Figure 13.5 contains a resistance R, a voltage V, and an inductance L. The current I at t s after the switch is closed is given by

$$I = I_0 \left(1 - e^{-t/T}\right)$$

where I_0 is the steady state current $\dfrac{V}{R}$ and $T = \dfrac{L}{R}$. If a circuit has $V = 120$ V, $R = 40\,\Omega$, and $L = 3.0$ H (henrys), determine the current in the circuit **(a)** 0.01 s, **(b)** 0.1 s, and **(c)** 1.0 s after the connection is made.

23. *Electricity*   A 130-μF capacitor is charged by being connected to a 120-V circuit. The resistance is 4 500 Ω. What is the charge on the capacitor 0.5 s after the circuit is disconnected?

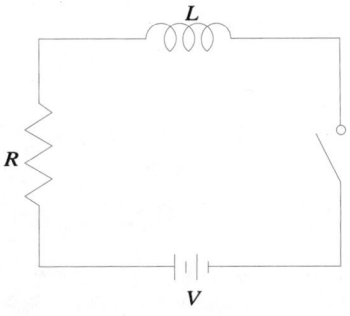

FIGURE 13.5

24. *Nuclear technology*   The half-life of tritium is 12.5 yr. How much of 100 g will remain after 40 yr?

25. *Medical technology*   A radioactive material used in radiation therapy has a half-life of 5.4 days. This means that the radioactivity decreases by one-half each 5.4 days. A hospital gets a new supply that measures 1 200 microcuries (μCi). How much of this material will still be radioactive after 30 days?

26. *Medical technology*   The average doubling time for a breast cancer cell is 100 days. It takes about 9 yr before a lump can be seen on a mammogram and 10 yr before a breast cancer is large enough to be felt as a lump. (Use 365.25 days = 1 year.)
(a) How many cancer cells are in a lump that is just large enough to be seen on a mammogram?
(b) How many cancer cells are in a lump that is just large enough to be felt?

In Your Words

29. Suppose that you are shown the graphs of $y = e^x$, $y = 2^x$, $y = x^2$, and $y = x^3$, but the graphs are not labeled. Explain how you would distinguish among the graphs.

(c) If a lump must be approximately 1 cm in diameter before it can be felt, how large is each cancer cell? (Assume the lump is a perfect sphere.)

27. *Electronics*   In a capacitive circuit the equation for the current in amperes is given by

$$i = \frac{V}{R} e^{-t/RC}$$

where t is any elapsed time in seconds after the switch is closed, V is the impressed voltage, C is the capacitance of the circuit in F, and R is the circuit resistance in Ω. A capacitance of 500 μF in series with 1 kΩ is connected across a 50-V generator.
(a) What is the value of the current at the instant the switch is closed?
(b) What is the value of the current 0.02 s after the switch is closed?
(c) What is the value of the current 0.04 s after the switch is closed?

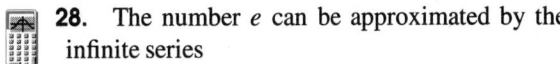
28.   The number e can be approximated by the infinite series

$$e = 2 + \frac{1}{2} + \frac{1}{2 \cdot 3} + \frac{1}{2 \cdot 3 \cdot 4} + \frac{1}{2 \cdot 3 \cdot 4 \cdot 5} \cdots$$
$$= 2 + \frac{1}{2!} + \frac{1}{3!} + \frac{1}{4!} + \frac{1}{5!}$$

(Remember, from the exercises in Section 9.2, that $3! = 3 \cdot 2 \cdot 1 = 6$ and $4! = 4 \cdot 3 \cdot 2 \cdot 1 = 24$. In general, $n! = n \cdot (n-1) \cdot (n-2) \cdots 4 \cdot 3 \cdot 2 \cdot 1$.) Calculate e to four figures by adding the first six terms of this series.

30. Explain what is meant by the term "half-life."

☰ 13.3
LOGARITHMIC FUNCTIONS

If you look at the graphs of exponential functions, such as the ones in Figures 13.1, 13.2, and 13.3, you can see that they would pass the horizontal line test for the inverse of a function. The inverse of the exponential function $f(x) = b^x$, $b > 0$, $b \neq 1$, is called the **logarithmic function**. The symbol for the logarithmic function is $\log_b x$. So, if $f(x) = b^x$, we have

$$f^{-1}(x) = \log_b x, \quad b > 0, \ b \neq 1$$

which is read "log to the base b of x." This provides for the following definition.

Logarithmic Function

A **logarithmic function** is any function of the form

$$f(x) = \log_b x$$

where $x > 0$, $b > 0$, and $b \neq 1$. If $y = \log_b x$, then $x = b^y$. The number represented by b is called the **base**.

Remember that for any function f, with an inverse f^{-1}, we have the following relationship.

$$f\left(f^{-1}(x)\right) = x \qquad \text{and} \qquad f^{-1}(f(x)) = x$$

If $f(x) = b^x$ and $f^{-1}(x) = \log_b x$, then we see that

$$b^{\log_b x} = x \qquad \text{and} \qquad \log_b b^x = x$$

Since $y = \log_b x$ is equivalent to $b^y = x$, we can express each logarithm in exponential form.

EXAMPLE 13.11

Use the fact that $y = \log_b x$ is equivalent to $b^y = x$ to rewrite each logarithm in exponential form.

	Logarithmic form	Exponential form
(a)	$\log_5 125 = 3$	$3 = 125$
(b)	$\log_7 49 = 2$	$7^2 = 49$
(c)	$\log_2 128 = 7$	$2^7 = 128$
(d)	$\log_5\left(\frac{1}{25}\right) = -2$	$5^{-2} = \frac{1}{25}$
(e)	$\log_2 0.125 = -3$	$2^{-3} = 0.125 = \frac{1}{8}$
(f)	$\log_8 16 = \frac{4}{3}$	$8^{4/3} = 16$
(g)	$\log_b 1 = 0$	$b^0 = 1$

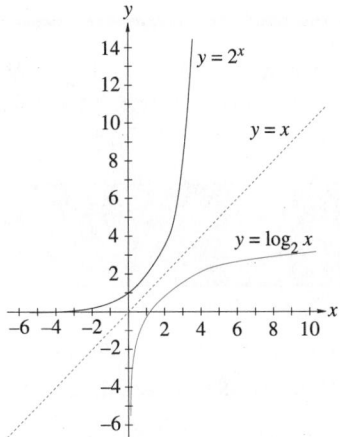

FIGURE 13.6

Figure 13.6 shows the graph of $y = 2^x$. The line $y = x$ is shown as a dashed line. The reflection of $y = 2^x$ in the line $y = x$ is shown by the colored curve and is the graph of $y = \log_2 x = f^{-1}(2^x)$.

This is a rather awkward method to graph the curve of $y = \log_b x$. Let's find another way to evaluate $\log_b x$ so that we can set up a table of values and plot points, and to connect the points to sketch the graph.

Common Logs; Natural Logs

There is another way to evaluate $\log_b x$, but the ease of doing it depends on the base of the logarithm. There are two bases that are used most often. These are 10 and e. Logarithms that have a base of 10 are called **common logs** and those with a base of e are called **natural logs**.

Because the bases 10 and e are used so often, they have special symbols. The symbol **log**, written with no indicated base, shows that common logs, or base 10 logs, are being used. If natural logs are involved, the symbol **ln** is used.

There are three major ways to find the values of a logarithm. These ways include a calculator, a computer, and a table. We will explain how to use a calculator to find the logarithm of a number. Tables are seldom used today. If, or when, you need to use a table of logarithms, you should carefully read the instructions for its use.

Finding Logs with a Calculator

Look at your calculator. It probably has two keys on it that we have seldom used. One key is $\boxed{\log}$ or $\boxed{\text{LOG}}$, and the other is $\boxed{\ln x}$ or $\boxed{\text{LN}}$. Now, since $\log_{10} x$ means $\log x$, we can use the $\boxed{\log}$ key to determine $\log_{10} x$. Similarly, we can use the $\boxed{\ln x}$ or $\boxed{\text{LN}}$ key to determine values of $\log_e x = \ln x$.

EXAMPLE 13.12

Use a calculator to evaluate **(a)** $\log 2$, **(b)** $\log 100$, **(c)** $\log 9.53$, **(d)** $\ln 2$, **(e)** $\ln 12.4$, **(f)** $\ln 1$, and **(g)** $\ln 32.4$.

Solution

	PRESS	DISPLAY
(a)	$\boxed{\text{LOG}}$ 2 $\boxed{\text{ENTER}}$	0.3010299957
(b)	$\boxed{\text{LOG}}$ 100 $\boxed{\text{ENTER}}$	2
(c)	$\boxed{\text{LOG}}$ 9.53 $\boxed{\text{ENTER}}$	0.9790929006
(d)	$\boxed{\text{LN}}$ 2 $\boxed{\text{ENTER}}$	0.6931471806
(e)	$\boxed{\text{LN}}$ 12.4 $\boxed{\text{ENTER}}$	2.517696473
(f)	$\boxed{\text{LN}}$ 1 $\boxed{\text{ENTER}}$	0
(g)	$\boxed{\text{LN}}$ 32.4 $\boxed{\text{ENTER}}$	3.478158423

 **Note**

If you want use a nongraphing calculator to determine a logarithm such as $\ln 32.4$, first press 32.4 and then press the $\boxed{\ln x}$ key.

Logs of Different Bases

Now we see that we can use a calculator to help us get values of $\ln x$ and $\log x$. We can also use calculators and computers to help us find the value of $\log_b x$. To do this, we use the following relationship.

Change of Base Formula	$\log_b x = \dfrac{\ln x}{\ln b}$

The relationship $\log_b x = \dfrac{\ln x}{\ln b}$ seems to be an unusual one. It uses some properties of logarithms that we will learn in Section 13.4. But, it works! In fact, it works for any base a of the logarithms on the right-hand side. So, it is true that $\log_b x = \dfrac{\log_a x}{\log_a b}$.

Since $5^3 = 125$, we know $\log_5 125 = 3$. By this formula, $\log_5 125 = \dfrac{\ln 125}{\ln 5}$ should also be 3. Try it on your calculator.

PRESS	DISPLAY
125 $\boxed{\ln x}$	4.8283137
$\boxed{\div}$ 5 $\boxed{\ln x}$	1.6094379
$\boxed{=}$	3

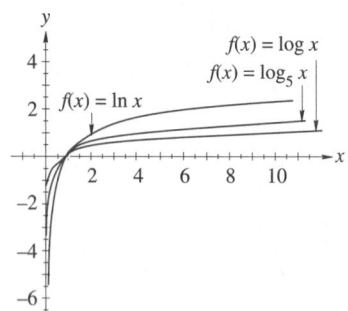

FIGURE 13.7

Plot the graphs of $\log x$, $\ln x$, and $\log_5 x$.

Solution   A table of values follows and the graphs of the functions are shown in Figure 13.7.

x	0.50	1	1.50	2.00	2.50	3.00	3.50	4.00	4.50	5.00	5.50	6.00
$\log x$	-0.30	0	0.18	0.30	0.40	0.48	0.54	0.60	0.65	0.70	0.74	0.78
$\ln x$	-0.69	0	0.41	0.69	0.92	1.10	1.25	1.39	1.50	1.61	1.70	1.79
$\log_5 x$	-0.43	0	0.25	0.43	0.57	0.68	0.78	0.86	0.93	1.00	1.06	1.11

Notice that all three curves cross the x-axis at the point $(1, 0)$. If you look back at Example 13.12(f), you will see that $\log_b 1 = 0$. All logarithmic curves cross the x-axis at the point $(1, 0)$.

Exercise Set 13.3

In Exercises 1–10, rewrite each logarithm in exponential form.

1. $\log_6 216 = 3$

2. $\log_9 6,561 = 4$

3. $\log_4 16 = 2$

4. $\log_7 16,807 = 5$

5. $\log_{1/7} \frac{1}{49} = 2$

6. $\log_{1/2} \frac{1}{64} = 6$

7. $\log_2 \frac{1}{32} = -5$

8. $\log_3 \frac{1}{243} = -5$

9. $\log_9 2,187 = \frac{7}{2}$

10. $\log_8 2,048 = \frac{11}{3}$

In Exercises 11–20, rewrite each exponential in logarithmic form.

11. $5^4 = 625$

12. $3^5 = 243$

13. $2^7 = 128$

14. $4^3 = 64$

15. $7^3 = 343$

16. $11^2 = 121$

17. $5^{-3} = \frac{1}{125}$

18. $3^{-5} = \frac{1}{243}$

19. $4^{7/2} = 128$

20. $125^{5/3} = 3,125$

In Exercises 21–32, use a calculator or a computer to evaluate each of these logarithms.

21. $\ln 5$

22. $\ln 19$

23. $\ln 4.751$

24. $\ln 35.62$

25. $\log 4$

26. $\log 23$

27. $\log 12.67$

28. $\log 78.143$

29. $\log_5 8$

30. $\log_3 20$

31. $\log_{12} 16.4$

32. $\log_8 691.45$

In Exercises 33–39, make a table of values and sketch the graph of each function.

33. $f(x) = \ln x$

34. $g(x) = \log x$

35. $h(x) = \log_2 x$

36. $k(x) = \log_3 x$

37. $f(x) = \log_{12} x$

38. $g(x) = \log_{1/2} x$

39. $h(x) = \log_{1/4} x$

Solve Exercises 40–44.

40. *Seismology* The **Richter scale** is used to measure the magnitude of earthquakes. The formula for the Richter scale is $R = \log I$, where R is the Richter number, and I is the intensity of the earthquake. Express the Richter scale in exponential form.

41. *Acoustical engineering* The **decibel (dB) scale** is used for sound intensity. This scale is used because the response of the human ear to sound intensity is not proportional to the intensity. The intensity, $I_0 = 10^{-12}$ W/m^2 (watts/m^2), is just audible and so is given a value of 0 dB. A sound 10 times more intense is given the value 10 dB; a sound $100 = 10^2$ times more intense than 0 dB is given the value 20 dB. Continuing in this manner gives the formula

$$\beta = 10 \log \frac{I}{I_0}$$

where β is the intensity in decibels and I is the intensity in W/m^2. **(a)** Express the decibel scale in exponential notation. **(b)** What is the intensity in decibels of a sound that measures 10^{-7} W/m^2? **(c)** What is the intensity of a heavy truck passing a pedestrian at the side of a road if the sound wave intensity of the truck I is 10^{-3} W/m^2?

42. *Land management* To plan for the future needs of a city, the city engineer uses the function

$$P(t) = 47,000 + 9,000 \ln(0.7t + 1)$$

to estimate the population t years from now. **(a)** What is the city's current population? **(b)** What is the expected population in 10 years? **(c)** How long will it be until the population reaches 75,000?

43. *Forestry* The yield, Y, in total ft^3/acre, of thinned stands of yellow poplar can be predicted by the equation

$$\ln Y = 5.36437 - 101.16296S^{-1} - 22.00048A^{-1}$$
$$+ 0.97116\ln BA$$

where S is the site index, A is the current age of the trees, and BA is the basal area. What is the predicted yield of a 40-year-old stand growing on a site with an index of 110 and a basal area of 90 ft^2 per acre?

44. *Computer technology* In designing a PC board, the impedance of a trace depends on the trace width, conductor thickness, and PC board material in a strip-line configuration, as shown by the equation

$$Z_0 = \frac{87}{\sqrt{\epsilon' + 1.4}} \ln\left(\frac{6H}{0.8W + t}\right)$$

Solve this equation for t.

In Your Words

45. Explain how the graph of a logarithmic function differs from an exponential function.

46. **(a)** What are common logarithms?
(b) What are natural logarithms?
(c) How are common logarithms and natural logarithms alike and how are they different?

≡ 13.4
PROPERTIES OF LOGARITHMS

In Section 13.3, we saw that logarithms were the inverses of exponentials. We also saw that we could write each logarithm as an exponential. It is not unexpected that the properties of logarithms must be related to the rules of exponents. We will examine three properties of logarithms and mention three others. These properties used to be very important in helping to calculate logarithms. Today, we use calculators and computers for these calculations, but the properties will be important later when we need to solve equations that involve logarithms or exponents.

Property 1

Let's consider two numbers, x and y, and consider $\log_b xy$. Suppose we know that $\log_b x = m$ and $\log_b y = n$, or, in exponential form, $b^m = x$ and $b^n = y$. Then,

$$xy = b^m b^n = b^{m+n}$$

Rewriting $xy = b^{m+n}$ in logarithmic form, we get $\log_b xy = m + n = \log_b x + \log_b y$. We have established the first property of logarithms.

> **Logarithms: Property 1**
>
> If x and y are positive real numbers, $b > 0$, and $b \neq 1$, then
>
> $$\log_b xy = \log_b x + \log_b y$$

EXAMPLE 13.14

Use the first property of logarithms to rewrite each of the following: **(a)** $\log_4 35$, **(b)** $\log 21$, **(c)** $\ln 18$, **(d)** $\log_3 5x$, **(e)** $\log_8 2 + \log_8 5$, **(f)** $\log 5 + \log 2 + \log 6$, **(g)** $\ln 3 + \ln 13$, and **(h)** $\log 5 + \log a$.

Solutions

(a) $\log_4 35 = \log_4 (5 \cdot 7) = \log_4 5 + \log_4 7$

(b) $\log 21 = \log(3 \cdot 7) = \log 3 + \log 7$

(c) $\ln 18 = \ln(2 \cdot 3 \cdot 3) = \ln 2 + \ln 3 + \ln 3$

(d) $\log_3 5x = \log_3 5 + \log_3 x$

(e) $\log_8 2 + \log_8 5 = \log_8 (2 \cdot 5) = \log_8 10$

(f) $\log 5 + \log 2 + \log 6 = \log(5 \cdot 2 \cdot 6) = \log 60$

(g) $\ln 3 + \ln 13 = \ln(3 \cdot 13) = \ln 39$

(h) $\log 5 + \log a = \log 5a$

Property 2

For the second property, let's consider $\log_b \dfrac{x}{y}$. Again, if we let $x = b^m$ and $y = b^n$, we have $\dfrac{x}{y} = \dfrac{b^m}{b^n} = b^{m-n}$. Now, $\log_b b^{m-n} = m - n = \log_b x - \log_b y$. This gives us the following property of logarithms.

> **Logarithms: Property 2**
>
> If x and y are positive real numbers, $b > 0$, and $b \neq 1$, then
> $$\log_b \frac{x}{y} = \log_b x - \log_b y$$

Caution

The symbol $\dfrac{\log_b x}{\log_b y}$ is not the same as $\log_b \dfrac{x}{y}$. In particular, $\dfrac{\log_b x}{\log_b y} \neq \log_b x - \log_b y$.

EXAMPLE 13.15

Use the second property of logarithms to rewrite each of the following: **(a)** $\log \frac{3}{5}$, **(b)** $\ln \frac{5}{4}$, **(c)** $\log_5 \dfrac{7}{x}$, **(d)** $\ln 8 - \ln 2$ **(e)** $\log_3 5 - \log_3 2$, and **(f)** $\log 8x^2 - \log 2x$.

Solutions

(a) $\log \frac{3}{5} = \log 3 - \log 5$

(b) $\ln \frac{5}{4} = \ln 5 - \ln 4$

(c) $\log_5 \dfrac{7}{x} = \log_5 7 - \log_5 x$

(d) $\ln 8 - \ln 2 = \ln \frac{8}{2} = \ln 4$

EXAMPLE 13.15 (Cont.)

(e) $\log_3 5 - \log_3 2 = \log_3 \frac{5}{2}$

(f) $\log 8x^2 - \log 2x = \log \dfrac{8x^2}{2x} = \log 4x$

Property 3

Now let's consider x^p. If $m = \log_b x$, then $x = b^m$ and $x^p = (b^m)^p = b^{mp}$. So, if $x^p = b^{mp}$, then $\log_b x^p = mp$, and since $m = \log_b x$, we have the following third property of logarithms.

> **Logarithms: Property 3**
>
> If x and y are positive real numbers, $b > 0$, and $b \neq 1$, then
>
> $$\log_b x^p = p \log_b x.$$

EXAMPLE 13.16

Use the third property of logarithms to rewrite each of the following: **(a)** $\log 5^3$, **(b)** $\log_2 16^5$, **(c)** $\log 100^{3.4}$, **(d)** $\log \sqrt[3]{25}$, and **(e)** $2\log 5 + 3\log 4 - 4\log 2$.

Solutions

(a) $\log 5^3 = 3\log 5$

(b) $\log_2 16^5 = 5\log_2 16$

(c) $\log 100^{3.4} = \log(10^2)^{3.4} = \log 10^{6.8} = 6.8\log 10 = 6.8$

(d) $\log \sqrt[3]{25} = \log 25^{1/3} = \frac{1}{3}\log 25$

(e) $2\log 5 + 3\log 4 - 4\log 2 = \log 5^2 + \log 4^3 - \log 2^4$

$$= \log 25 + \log 64 - \log 16$$

$$= \log \frac{25 \cdot 64}{16} = \log 100$$

$$= 2$$

Properties 4, 5, and 6

In addition to these properties, there are three other properties of logarithms. These three properties are listed in the following box.

> **Logarithms: Properties 4, 5, and 6**
>
> If x and y are positive real numbers, $b > 0$, and $b \neq 1$, then
>
> $$\log_b 1 = 0 \qquad \qquad \text{Property 4}$$
> $$\log_b b = 1 \qquad \qquad \text{Property 5}$$
> $$\log_b b^n = n \qquad \qquad \text{Property 6}$$

The properties of logarithms allow us to simplify the logarithms of products, quotients, powers, and roots.

Caution

There is no way to simplify logarithms of sums or differences. You cannot change $\log_b(x + y)$ to $\log_b x + \log_b y$, except when $x = y$. This is very tempting, but do not make this error. Remember, $\log_b(x + y) \neq \log_b x + \log_b y$.

Logarithms were originally developed to help people compute. Slide rules were developed as a computational tool based on logarithms. At one time, every technician had, and knew how to use, a slide rule. Electronic calculators and the microcomputers have replaced slide rules and reduced the importance of logarithms as a help in computing.

However, logarithms are still an important tool in working many problems. In the remainder of this section, we will demonstrate the properties of logarithms. Values are sometimes obtained from a table of logarithms. We will not expect you to get values from tables. These exercises will prepare you for the next section where you will solve equations that involve logarithms.

EXAMPLE 13.17

If $\log 2 = 0.3010$, $\log 3 = 0.4771$, and $\log 5 = 0.6990$, determine each of the following: **(a)** $\log 6$, **(b)** $\log 81$, **(c)** $\log 1.5$, **(d)** $\log \sqrt{5}$, and **(e)** $\log 50$.

Solutions

(a) $\log 6 = \log(2 \cdot 3) = \log 2 + \log 3 = 0.3010 + 0.4771 = 0.7781$

(b) $\log 81 = \log 3^4 = 4 \log 3 = 4(0.4771) = 1.9084$

(c) $\log 1.5 = \log \frac{3}{2} = \log 3 - \log 2 = 0.4771 - 0.3010 = 0.1761$

(d) $\log \sqrt{5} = \log 5^{1/2} = \frac{1}{2} \log 5 = \frac{1}{2}(0.6990) = 0.3495$

(e) $\log 50 = \log(5 \cdot 10) = \log 5 + \log 10 = 0.6990 + 1 = 1.6990$

Use a calculator to check each of these answers.

Exercise Set 13.4

In Exercises 1–12, write each logarithm as the sum or difference of two or more logarithms.

1. $\log \frac{2}{3}$

2. $\log \frac{5}{4}$

3. $\log 14$

4. $\log 55$

5. $\log 12$

6. $\log 28$

7. $\log \frac{150}{7}$

8. $\log \frac{588}{5x}$

9. $\log 2x$

10. $\log \frac{1}{5x}$

11. $\log \frac{2ax}{3y}$

12. $\log \frac{4bc}{3xy}$

Express each of Exercises 13–24 as a single logarithm.

13. $\log 2 + \log 11$

14. $\log 3 + \log 13$

15. $\log 11 - \log 3$

16. $\log 17 - \log 23$

17. $\log 2 + \log 2 + \log 3$

18. $\log 3 + \log 3 + \log 5$

19. $\log 4 + \log x - \log y$

20. $\log 5 + \log 7 - \log 11$

21. $5 \log 2 + 3 \log 5$

22. $7 \log 3 - 2 \log 8$

23. $\log \frac{2}{3} + \log \frac{6}{7}$

24. $\log \frac{3}{24} + \log \frac{4}{7} - \log \frac{1}{3}$

If log 2 = 0.3010, log 3 = 0.4771, and log 5 = 0.6990, determine the value of each logarithm in Exercises 25–36.

25. $\log 8$

26. $\log 9$

27. $\log 12$

28. $\log 36$

29. $\log 15$

30. $\log 30$

31. $\log \frac{75}{2}$

32. $\log \frac{45}{8}$

33. $\log 200$

34. $\log 150$

35. $\log 5{,}000$

36. $\log 4{,}500$

If $\log_b 2 = 0.3869$, $\log_b 3 = 0.6131$, and $\log_b 5 = 0.8982$, determine the value of each logarithm in Exercises 37–40.

37. $\log_b 16$

38. $\log_b 15$

39. $\log_b \frac{5}{3}$

40. $\log_b \frac{24}{5}$

Solve Exercises 41–44.

41. Use the change of base formula to determine the base for the logarithms in Exercises 33–40. (Hint: The answer is an integer. You may not get an integer, because the values have been rounded off to four decimal places.)

42. *Electronics*   The gain or loss of power can be determined by the formula

$$N = 20(\log I_1 - \log I_2) + 10(\log R_1 - \log R_2)$$

Simplify this formula.

43. *Acoustical engineering*   Decibel gain or loss, ΔdB, can be computed by the formula

$$\Delta\text{dB} = 20 \left(\log I_2 + \frac{1}{2} \log R_2 - \log I_1 - \frac{1}{2} \log R_1 \right)$$

Simplify the right-hand side of this equation.

44. *Ecology*   As a result of pollution, the population of fish in a certain river decreases according to the formula $\ln \left(\dfrac{P}{P_0} \right) = -0.0435t$, where P is the population after t years and P_0 is the original population.

(a) After how many years will there be only 50% of the original fish population remaining?

(b) After how many years will the the original population be reduced by 90%?

(c) What percentage will die in the first year of pollution?

In Your Words

45. Explain Properties 4, 5, and 6 in your own words. Why do you think these three properties are especially useful?

46. (a) Explain the differences between $\log_b x - \log_b y$, $\log_b(x - y)$, and $\log_b x - y$.

(b) Graph $y_1 = \ln x - \ln 4x$, $y_2 = \ln(x - 4x)$, and $y_3 = \ln x - 4x$.

(c) Do your graphs in (b) support your explanations in (a)? If not, examine your work in both parts (a) and (b) and make any necessary changes.

≡ 13.5
EXPONENTIAL AND LOGARITHMIC EQUATIONS

We mentioned in Section 13.4 that logarithms were originally developed as an aid for computation. The wide use of electronic calculators decreased the importance of logarithms as a help for calculation. In this section, we will focus on ways we can use logarithms to help solve equations.

Exponential Equations

We will start with an equation that is relatively easy. For one thing, we already know the answer. This will help us see that the method works.

Exponential Equations

An **exponential equation** is any equation in which the variable is an exponent.

Steps for Solving Exponential Equations

In solving an exponential equation, you should:
1. Use the properties of exponents to rewrite each side of the equation in terms of exponents with the same base.
2. Use the fact that $x = b^y$ is equivalent to $\log_b x = y$ to rewrite the equation.
3. Solve the resulting equation.

EXAMPLE 13.18

Solve $3^x = 81$.

Solution   To solve this equation, we will take the logarithm of both sides. We can use any base for the logarithm, but it is easiest to use one of the bases that is on a calculator. We will first use base 10 and then solve the same equation using base e to show that it will work either way.

$$3^x = 81$$
$$\log 3^x = \log 81$$

EXAMPLE 13.18 (Cont.)

Since $\log 3^x = x \log 3$, we have

$$x \log 3 = \log 81$$

$$x = \frac{\log 81}{\log 3}$$

$$\approx \frac{1.908485}{0.4771213} = 4$$

≡ Note

Notice that $\log 81 \div \log 3 = \dfrac{\log 81}{\log 3} = 4$ and that $\log \frac{81}{3} = \log 81 - \log 3 \approx 1.431$. Since $4 \neq 1.431$, we see that $\log 81 \div \log 3 = \dfrac{\log 81}{\log 3} \neq \log \frac{81}{3}$.

Let's solve this same equation using natural logarithms.

$$3^x = 81$$
$$\ln 3^x = \ln 81 \qquad \text{Take ln of both sides.}$$
$$x \ln 3 = \ln 81 \qquad \text{Use Property 3 of logs.}$$
$$x = \frac{\ln 81}{\ln 3} \qquad \text{Divide.}$$
$$\approx \frac{4.3944492}{1.0986123} = 4.$$

EXAMPLE 13.19

Solve $5^{2x+1} = 25^{4x-1}$.

Solution   Notice that $25 = 5^2$, so we can write this equation as

$$5^{2x+1} = 25^{4x-1}$$
$$= \left(5^2\right)^{4x-1}$$
$$= 5^{8x-2}$$

Taking $\log_5$ of both sides produces

$$\log_5(5^{2x+1}) = \log_5(5^{8x-2})$$
$$\text{or} \qquad (2x+1)\log_5 5 = (8x-2)\log_5 5$$

which simplifies to

$$2x + 1 = 8x - 2$$
$$3 = 6x$$
$$\tfrac{1}{2} = x$$

Logarithmic Equations

Just as we used logarithms to solve equations that involved exponentials, we can use exponentials to solve problems that involve logarithms.

Logarithmic Equations

A **logarithmic equation** is an equation that contains a logarithm of the variable.

Steps for Solving Logarithmic Equations

In solving a logarithmic equation, you should:
1. Use the properties of logarithms to combine all the logarithms into one.
2. Use the fact that $\log_b x = y$ is equivalent to $x = b^y$ to rewrite the equation.
3. Solve the resulting equation.

EXAMPLE 13.20

Solve $\ln x = 7$.

Solution   The base in this equation is e, so $\ln x$ is the same as $\log_e x$ and this is equivalent to $\log_e x = 7$. Using the fact that $\log_e x = y$ is the same as $x = e^y$, we get

$$x = e^7$$
$$\approx 1{,}096.6$$

It is important that you remember to use the correct base when you solve a logarithmic equation. The next example will show you how to proceed with a more complicated situation. It also uses a different base; in this case, the base is 10.

EXAMPLE 13.21

Solve $\log(3x - 5) + 2 = \log 4x$.

Solution   We will first combine this into one logarithm

$$\log(3x - 5) + 2 = \log 4x$$
$$\log(3x - 5) - \log 4x = -2$$
$$\log\left(\frac{3x - 5}{4x}\right) = -2$$

EXAMPLE 13.21 (Cont.)

We will now use the fact that $\log_b x = y$ is equivalent to $x = b^y$. Since we are using common logarithms, the base is 10.

$$\frac{3x - 5}{4x} = 10^{-2}$$

$$\frac{3x - 5}{4x} = 0.01$$

We solve this as we would any fractional equation.

$$\frac{3x - 5}{4x} = 0.01$$

$$3x - 5 = 0.04x$$

$$2.96x = 5$$

$$x = \frac{5}{2.96} \approx 1.69$$

≡ Note

The approximate answer 1.69 does not check, but the exact answer $\frac{5}{2.96}$ does check.

Business and Finance

An interesting equation results from the work with exponentials and we can use logarithms to solve it. If a certain amount of money P was invested at 5% compounded continuously, the investment would be worth

$$S = Pe^{rt}$$

after t yr. How long does it take for the money to double?

If it doubles, then it is worth $2P$ and we have

$$2P = Pe^{rt}$$

$$\text{or} \quad 2 = e^{rt}$$

Taking the natural logarithm of both sides, we obtain

$$\ln 2 = \ln e^{rt}$$

$$\text{or} \quad \ln 2 = rt$$

Solving for t results in

$$t = \frac{\ln 2}{r}$$

It will take $\dfrac{\ln 2}{r}$ yr for the money to double. This same formula will apply to anything that is growing or decaying at an exponential rate.

Application

EXAMPLE 13.22

How long will it take for $1,000 to double at 5% compounded continuously?

Solution We use the formula

$$t = \frac{\ln 2}{r}$$

and since

$$r = 5\% = 0.05,$$

$$t = \frac{\ln 2}{0.05} \approx 13.86$$

Application

EXAMPLE 13.23

What is the half-life of a radioactive material that decays at the rate of 1% per year?

Solution The same formula will apply; only here $r = 1\% = 0.01$.

$$t = \frac{\ln 2}{0.01} \approx 69.31 \text{ yr}$$

Exercise Set 13.5

In Exercises 1–24, solve each equation for the indicated variable. You may want to use a calculator.

1. $5^x = 29$
2. $6^y = 32$
3. $4^{6x} = 119$
4. $3^{-7x} = 4$
5. $3^{y+5} = 16$
6. $2^{y-7} = 67$
7. $e^{2x-3} = 10$
8. $4^{3y+5} = 30$

9. $3^{4x+1} = 9^{3x-5}$
10. $2^{3-2x} = 8^{1+2x}$
11. $e^{3x-1} = 5e^{2x}$
12. $3^{2-5x} = 8(9^{x-1})$
13. $\log x = 2.3$
14. $\ln x = 5.4$
15. $\log(x - 5) = 17$
16. $\ln(y + 3) = 19$

17. $\ln 2x + \ln x = 9$
18. $2\ln x = \ln 4$
19. $2\ln 3x + \ln 2 = \ln x$
20. $2\ln(x + 3) = \ln x + 4$
21. $\log(x - 1) = 2$
22. $\log(x - 4) = \log x - 4$
23. $2\log(x - 4) = 2\log x - 4$
24. $\ln x = 1 + 3\ln x$

Solve Exercises 25–42.

25. *Finance* The amount of $5,000 is placed in a savings account, where interest is compounded continuously at the rate of 6% per year. How long will it take for this amount to be doubled?

26. *Finance* How long will it take the money in Exercise 25 to double if the rate is 8% per year?

27. *Nuclear energy* A radioactive substance decays at the rate of 0.5% per year. What is the half-life of this substance?

28. *Nuclear energy* Another radioactive substance decays at the rate of 3% per year. What is its half-life?

29. *Nuclear energy* The half-life of tritium is 12.5 yr. What is its annual rate of decay?

30. *Nuclear energy* The half-life of the sodium isotope $^{24}_{11}\text{Na}$ against beta decay is 15 h. What is the rate of decay?

31. *Finance* A person has some money to invest and would like to double the investment in 8.5 yr. What annual rate of interest, compounded continuously, will be needed in order for this to be accomplished?

32. *Automotive technology* In a chrome-electroplating process, the mass m in grams of the chrome plating increases according to the formula $m = 200 - 2^{t/2}$, where t is the time in minutes. How long does it take to form 100 g of plating?

33. *Environmental science* In chemistry, the pH of a substance is defined by $pH = -\log[H^+]$, where $[H^+]$ is the concentration of hydrogen ions in the substance measured in moles per liter. The pH of distilled water is 7. A substance with a pH less than 7 is known as an **acid**. A substance with a pH greater than 7 is a **base**. Rain and snow have a natural concentration of $[H^+] = 2.5 \times 10^{-6}$ moles per liter. What is the natural pH of rain and snow?

34. *Environmental science* The pH of some acid rain is 5.3. What is the concentration of hydrogen ions in acid rain?

35. *Meteorology* The barometric equation,

$$H = (30T + 8{,}000) \ln \frac{P_0}{P}$$

relates the height H in meters above sea level, the air temperature T in degrees Celsius, the atmospheric pressure P_0 in centimeters of mercury at sea level, and the atmospheric pressure P in centimeters of mercury at height H. Atmospheric pressure at the summit of Mt. Whitney in California on a certain day is 43.29 cm of mercury. The average air temperature is $-5°C$ and the atmospheric pressure at sea level is 76 cm of mercury. What is the height of Mt. Whitney?

36. *Environmental science* In Exercise Set 13.3, Exercise 41, we saw that the loudness in decibels β of a noise is given by the formula $\beta = 10 \log \dfrac{I}{I_0}$, where $I_0 = 10^{-12}$ W/m^2, and I is the intensity of the noise in W/m^2. At takeoff, a certain jet plane has a noise level of 105 dB. What is the intensity I of the sound wave produced by this airplane?

37. *Electronics* The formula for the exponential decay of electric current is given by the formula $Q = Q_0 e^{-t/T}$, where $T = RC$. (See Example 13.10.) If $Q_0 = 0.40$ A, $R = 500\,\Omega$, and $C = 100\,\mu$F, what is t when $Q = 0.05$ A?

38. *Thermodynamics* The temperature T of an object after a period of time t is given by $T = T_0 + ce^{-kt}$, where T_0 is the temperature of the surrounding medium, and $c = T_I - T_0$, where T_I is the initial temperature of the object. A steel bar with a temperature of $1\,200°C$ is placed in water with a temperature of $20°C$. If the rate of cooling k is 8% per hour, how long will it take for the steel to reach a temperature of $40°C$?

39. *Electronics* In an ac circuit, the current I at any time t, in seconds, is given by $I = I_0 \left(1 - e^{-Rt/L}\right)$, where I_0 is the maximum current, L is the inductance, and R the resistance. If a circuit has a 0.2 H inductor, a resistance of $4\,\Omega$, and a maximum current of 1.5 A, at what instant does the current reach 1.4 A?

40. *Forestry* The yield, Y, in total ft^3/acre, of thinned stands of yellow poplar can be predicted by the equation

$$\ln Y = 5.36437 - 101.16296 S^{-1} - 22.00048 A^{-1}$$
$$+ 0.97116 \ln BA$$

where S is the site index, A is the current age of the trees, and BA is the basal area. What is the basal area if the predicted yield of a 60-year-old stand growing on a site with an index of 110 is 4,720 ft^3/acre?

41. *Medical technology* An implantable pacemaker normally has a capacitive output. The amplitude of emitted pulses declines with time according to the equation

$$E = \frac{U^2 C}{2} \left(1 - e^{-2t/RC}\right)$$

where C is the capacitance of the pacemaker's output capacitor in farads (F), delivering the impulse to the heart, U is the pulse voltage in volts (V), E is the energy in joules (J), I is the pulse current in amperes (A), and t is the pulse duration in seconds (s). Solve this equation for t.

42. *Medical technology* In Exercise 41, the following formula was given for the amplitude of emitted pulses of an implantable pacemaker

$$E = \frac{U^2 C}{2} \left(1 - e^{-2t/RC}\right)$$

Solve this equation for R.

43. Without looking in the text, describe the steps for solving logarithmic equations.

44. Explain the difference between a logarithmic equation and an exponential equation.

≡ 13.6

GRAPHS USING SEMILOGARITHMIC AND LOGARITHMIC PAPER

Exponential functions often produce some very large numbers. This makes graphing exponential functions difficult to show on normal graph paper. Look at the graph of $f(x) = 2^x$ in Figure 13.1. By the time x is 5, y is 32; when $x = 6$, $y = 64$. This presents a problem when trying to represent a large portion of an exponential graph. Semilogarithmic and logarithmic graph papers have been developed to allow for the plotting of a large range of values.

Semilogarithmic Graph Paper

Semilogarithmic or **semilog graph paper** has one of the axes (usually the y-axis) marked off in distances proportional to the logarithms of numbers. Thus, the distances between lines on this axis are not equally spaced. The other axis has the distance between lines equally spaced. The result of all this is that the graph of an exponential function $y = b^x$ is a straight line when it is drawn on semilog paper.

EXAMPLE 13.24

Sketch the graph of $y = 2^x$ on semilog paper.

Solution   Since this is the same curve we graphed in Figure 13.1, we will use the same table of values from that example. The table is reproduced here.

x	-3	-2	-1	0	1	2	3	4	5
$f(x) = 2^x$	0.125	0.25	0.5	1	2	4	8	16	32

Along the x-axis, we label the points -3 through 5. Along the y-axis, the points are labeled as indicated in Figure 13.8. Notice that each unit in the colored interval has been labeled 1, 2, 3, . . . , 10. The interval directly above it has each unit representing a multiple of 10 (10, 20, 30, etc.). In the next interval, the units would be multiples of 100 (100, 200, 300, etc.). The interval directly below the colored one has each unit represented in tenths (0.1, 0.2, 0.3, etc.), and the interval below that would be in hundredths (0.01, 0.02, 0.03, etc.).

EXAMPLE 13.25

Sketch the graph of $y = 5(4^x)$ on semilog paper.

Solution   A table of values from $x = -2$ to $x = 5$ is given here.

x	-2	-1	0	1	2	3	4	5
$f(x) = 5(4^x)$	0.31	1.25	5	20	80	320	1,280	5,120

The graph of this curve on semilog paper is shown in Figure 13.9. You should study the scale of the y-axis to be sure that you understand how the units are marked.

Study Figures 13.8 and 13.9. Both of these graphs are straight lines. Yet, in both cases we started with exponential equations of the form $y = b^x$.

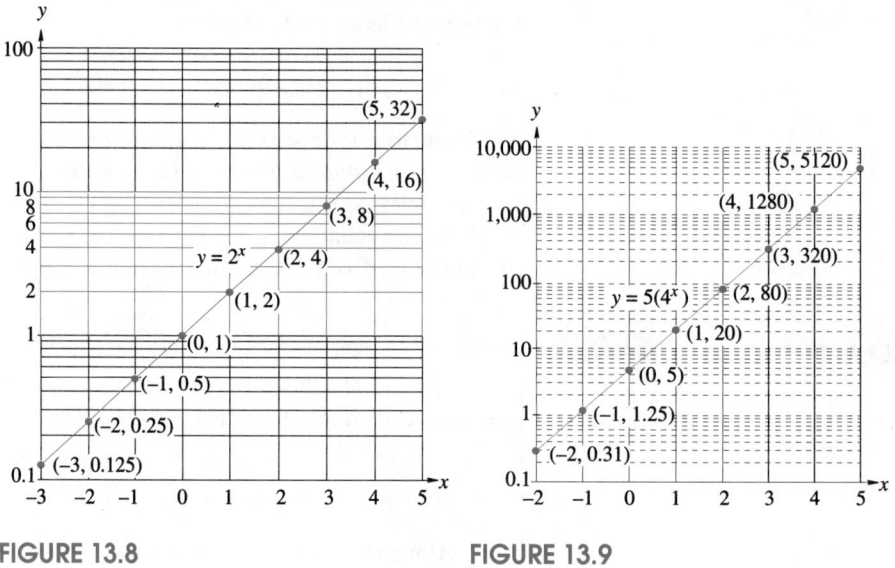

FIGURE 13.8   FIGURE 13.9

≡ **Note**

One of the reasons for using semilog paper is that graphs of exponential functions of the form $y = b^x$ will be straight lines when they are drawn on semilogarithmic paper.

Logarithmic Graph Paper

Use semilog paper when you want to indicate a large range of values for one of the variables. When this is needed for both variables, use **logarithmic** or **log-log graph paper**. Both axes on log-log paper are marked off in logarithmic scales. A function of the type $y^b = x^a$ or $y = x^{a/b}$ graphs as a straight line on log-log paper.

EXAMPLE 13.26

Sketch the graph of $y^3 = x^2$ on log-log paper.

Solution   The equation $y^3 = x^2$ is equivalent to $y = x^{2/3}$. If we select values of x that are perfect cubes, we then get integer values for y. The following table gives some values.

x	-1	0	$\frac{1}{8} = 0.125$	1	8	27	64	125	216	$1{,}000$
$y = x^{2/3}$	1	0	$\frac{1}{4} = 0.25$	1	4	9	16	25	36	100

In Figure 13.10, we show the sketch of this curve. Notice the way in which both axes are labeled.

≡ **Note**

Negative values of x and the point $(0, 0)$ do not appear on the graph. Since a logarithmic scale contains only positive values, the coordinates that are not positive cannot be plotted.

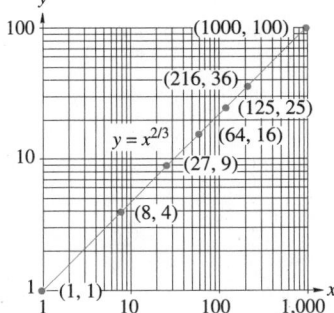

FIGURE 13.10

Semilog and log-log paper can be used to graph functions that may not graph as a straight line. This happens when a large range of values needs to be graphed, particularly when the data are exponential in nature or when the data are gathered from an experiment.

Both Semilogarithmic and Logarithmic Graph Paper

Sometimes it is necessary to graph the data on both semilog and log-log graph paper.

Using Graphs to Help Determine an Equation

If the graph of data is a straight line on semilog paper, then the data are related by a function of the form $y = ab^x$.

If the graph of data is a straight line with slope m on log-log paper, then the data are related by a function of the form $y = ax^m$. A function of this form, $y = ax^m$, is called a **power function**.

Application

EXAMPLE 13.27

Data from a certain experiment result in the following table.

x	1	2	3	4	5	6
y	5	40	135	320	625	1,080

Determine an equation that relates y as a function of x.

Solution   We will plot this data on both semilog graph paper and log-log graph paper. When the points are connected, we get the curves shown in Figures 13.11a and 13.11b.

The curve in Figure 13.11b is a straight line. This curve was drawn on log-log graph paper, so we conclude that these data satisfy a power function of the form $y = ax^m$.

EXAMPLE 13.27 (Cont.)

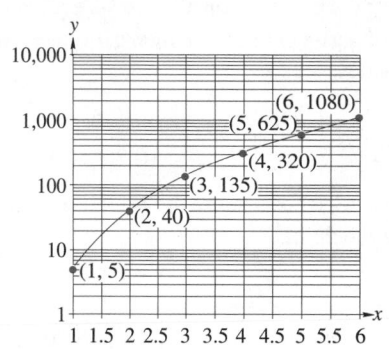

FIGURE 13.11a

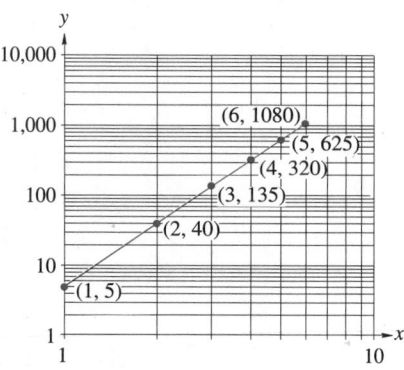

FIGURE 13.11b

The slope of this straight line is m, and if (x_1, y_1) and (x_2, y_2) are two points on the curve, we have

$$m = \frac{\log y_2 - \log y_1}{\log x_2 - \log x_1}$$

If we use the points $(1, 5)$ and $(2, 40)$, we will get

$$
\begin{aligned}
m &= \frac{\log 40 - \log 5}{\log 2 - \log 1} \\
&\approx \frac{1.602059991 - 0.698970004}{0.301029996 - 0} \\
&= \frac{0.903089987}{0.301029996} \\
&= 3
\end{aligned}
$$

Thus, we know that these data are of the form $y = ax^3$.

To determine a, we select one pair of values from the data $(3, 135)$ and get

$$
\begin{aligned}
y &= ax^3 \\
135 &= a(3)^3 \\
&= 27a \\
a &= \frac{135}{27} \\
&= 5
\end{aligned}
$$

The data in the table satisfy the equation $y = 5x^3$.

Unlike Example 13.27, data from an actual experiment seldom fit a curve exactly. In that case, we select points on a curve that seem to "best fit" the actual data points. We will explore this in more detail in Chapter 21.

Exercise Set 13.6

In Exercises 1–12, sketch the graphs of the given functions on semilog graph paper.

1. $y = 3^x$
2. $y = 5^x$
3. $y = 3(6^x)$

4. $y = 5(7^x)$
5. $y = 2^{-x}$
6. $y = 8^{-x}$

7. $y = x^4$
8. $y = x^5$
9. $y = 8x^3$

10. $y = 5x^4$
11. $y = 3x^2 + 4x$
12. $y = 8x^3 + 2x^2 + x$

In Exercises 13–24, sketch the graphs of the given functions on log-log graph paper.

13. $y = x^{1/2}$
14. $y = x^{1/3}$
15. $y = 2x^3$

16. $y = 3x^5$
17. $y = x^2 + \sqrt{x}$
18. $y = x^3 + x$

19. $y = 5x^{-1}$
20. $y = 8x^{-3}$
21. $y^2 = 4x^3$

22. $y^4 = 6x^5$
23. $x^2 y^3 = 8$
24. $x^3 y = 100$

In Exercises 25–32, sketch the indicated graphs.

25. *Meteorology*   The atmospheric pressure P at a given altitude h, in feet, is given by $P = P_0 e^{-kh}$, where P_0 and k are constants. On semilog paper, plot P in atmospheres and h, the altitude, in feet, for $0 \le h \le 10^4$, $P_0 = 1$ atmosphere, and at 0°C, $k = 1.25 \times 10^{-4}$.

26. *Meteorology*   The atmospheric pressure in kilopascals (kPa) is given approximately by $P = 100e^{-0.3h}$, where h is the altitude in kilometers. Graph P vs h on semilog paper.

27. *Electricity*   The resistance R of a copper wire varies inversely as the square of its cross-sectional diameter D. Thus $RD^2 = k$. If $k = 0.9\,\Omega\cdot\text{mm}$, graph R vs D on log-log paper, for $D = 0.1$ mm to 10 mm.

28. *Physics*   Boyle's law states that at a constant temperature, the volume V of a sample of gas is inversely proportional to the absolute pressure applied to the gas P. Thus, $PV = C$, a constant. If $C = 5$ atm·ft^3, use log-log paper to plot the graph of P in atmo-

spheres vs V in cubic feet. Let values of P range from 0.1 to 10.

29. An experimenter gathered the data in the following table. Plot the data on semilog and log-log papers and determine the equation that relates y as a function of x.

x	1	2	5	10
y	1,000	250	40	10

30. Repeat Exercise 29 for the data in this table.

x	1	2	3	4	5
y	16	80	400	2 000	10 000

31. *Electronics*   For a certain electric circuit, the voltage is given by $V = e^{-0.25t}$. Graph V vs t on semilog graph paper for $0 \le t \le 5$.

32. *Electronics*   For a certain electric circuit, the voltage is given by $V = e^{-0.35t}$. Graph V vs t on semilog graph paper for $0 \le t \le 10$.

In Your Words

33. Explain the difference between semilog and log-log graph paper.

34. Explain how a graph of a function drawn on semilog or log-log graph paper can help you determine the equation that describes that function.

☰ CHAPTER 13 REVIEW

Important Terms and Concepts

Exponential equation
Exponential function
Logarithmic equation

Logarithmic function
Logarithmic graph paper
Semilogarithmic graph paper

Review Exercises

In Exercises 1–8, use a calculator to evaluate each number.

1. e^5

2. e^{-7}

3. $e^{4.67}$

4. $e^{-3.91}$

5. $\log 8$

6. $\log 196.5$

7. $\ln 81.3$

8. $\ln 325.6$

In Exercises 9–16, make a table of values and graph each of these functions on ordinary graph paper over the indicated domains.

9. $f(x) = 5e^x$ $(x = -2 \text{ to } 4)$

10. $g(x) = 9e^{-0.5x}$ $(x = -4 \text{ to } 2)$

11. $h(x) = 5^x$ $(x = -3 \text{ to } 2)$

12. $k(x) = 2.1^{-x}$ $(x = -4 \text{ to } 2)$

13. $f(x) = \log x$ $(0 \le x \le 10)$

14. $g(x) = 3\log 2x$ $(0 \le x \le 10)$

15. $h(x) = \ln(2x+1)$ $\left(-\frac{1}{2} \le x \le 10\right)$

16. $k(x) = 2 + 3\ln x$ $(0 \le x \le 10)$

In Exercises 17–24, express each as a sum, difference, or multiple of logarithms.

17. $\log \frac{3}{4}$

18. $\log 77$

19. $\log 5x$

20. $\ln \dfrac{7x}{3}$

21. $\ln(4x)^3$

22. $\ln 4x^3$

23. $\log \sqrt{48}$

24. $\log \dfrac{\sqrt[3]{2x^5}}{4x}$

In Exercises 25–30, express each as a single logarithm.

25. $\log 5 + \log 9$

26. $\log 19 - \log 11$

27. $4\log x + \log x$

28. $\log 3a + \log b - \log a$

29. $7\log x + 2\log x - \log x$

30. $\ln \frac{2}{3} + \ln \frac{15}{8}$

In Exercises 31–38, solve each equation for the indicated variable. Use a calculator if you feel it is necessary.

31. $4^x = 28$

32. $5^{3x} = 250$

33. $e^{5x} = 11$

34. $e^{5x-1} = 2e^x$

35. $\ln x = 9.1$

36. $4\log x = 49$

37. $\log(2x+1) = 50$

38. $2\ln 4x = 100 + \ln 2x$

In Exercises 39–42, sketch the graphs of the given functions on semilog paper.

39. $y = 4(2^x)$

40. $y = 8x^4$

41. $y = 8(4^x)$

42. $y = 3(4^{x-1})$

In Exercises 43–46, sketch the graphs of the given functions on log-log paper.

43. $y = x^{1/4}$

44. $y = 5x^3$

45. $x^3 y^4 = 2$

46. $y^3 = 10x^2$

Solve Exercises 47–54.

47. *Finance* One savings institution offers 7.5% interest compounded semiannually. A second offers the same interest rate, but it is compounded monthly. A third compounds that same interest continuously. How much income will result if $1,000 is deposited in each institution for 10 yr? How long will it take for the $1,000 to double in the third bank?

48. *Nuclear engineering* If radium has a half-life of 1,600 yr, how much of 100 mg will remain after 100 yr? after 1,000 yr?

49. *Biology* The amount of bacteria in a culture increases from 4,000 to 25,000 in 24 h. If we assume these bacteria grow exponentially, **(a)** how many will be there after 12 h? **(b)** How many can we expect there to be after 48 h?

50. *Seismology* The San Francisco earthquake of 1906 registered 8.3 on the Richter scale. If $R = \log I$, where I is the relative intensity of the shock, determine the number of times the 1906 earthquake was greater than **(a)** the San Francisco quake of 1979, which registered a 6.0, and **(b)** the San Francisco quake of 1989, which registered a 7.1.

51. Data from an experiment produced a straight line on semilog paper. Two of the data points were (2, 15.76) and (4, 620.84). Find an equation that approximates this scale.

52. Data from an experiment produced a straight line on log-log paper. Two of the data points were (2, 8.75) and (4, 2.1875). Find an equation that approximates this data.

53. *Astronomy* The brightness of a star perceived by the naked eye is measured in terms of a quantity called **magnitude**. The brightest stars are of magnitude 1 and the dimmest of magnitude 6. The magnitude M is given by $M = 6 - 2.5 \log \dfrac{I}{I_0}$, where I_0 is the intensity of light from a just-visible star, and I is the actual intensity from a star. Even though the sun is a star, it has a magnitude of -27, and the moon has a magnitude of -12.5. **(a)** Express this equation in exponential form by solving for I. **(b)** How many times more intense is the moon than Polaris, which has a magnitude of 2.0?

54. *Biology* When the size of a colony of bacteria is limited because of a lack of room or nutrients, the growth is described by the law of logistic growth:

$$Q = \frac{m Q_0}{Q_0 + (m - Q_0) e^{-kmt}}$$

where Q_0 is the initial quantity present, m the maximum size, and k a positive constant. If $Q_0 = 400$, $m = 2,000$, and $Q = 800$ when $t = \frac{1}{2}$, find k.

☰ CHAPTER 13 TEST

In Exercises 1–3, use a calculator to evaluate each number.

1. $e^{5.3}$

2. $\log 715.3$

3. $\ln 72.35$

In Exercises 4–7, express each as a sum, difference, or multiple of logarithms.

4. $\log \frac{9}{15}$

5. $\log 65x$

6. $\ln(7x)^{-2}$

7. $\log \dfrac{\sqrt[5]{5x^3}}{9x}$

In Exercises 8–10, express each as a single logarithm.

8. $\log 11 + \log 3$

9. $\log 17 - 2 \log x$

10. $\ln 4 - \ln 5 + \ln 15 - \ln 8$

In Exercises 11–16, solve each equation for the indicated variable. Use a calculator if you feel it is necessary.

11. $\ln x = 17.2$

12. $3 \log x = 55$

13. $3^x = 35$

14. $e^{4x+1} = 2e^x$

15. $\log (2x^2 + 4x) = 5 + \log 2x$

16. Make a table of values and graph $f(x) = 3^x$ on ordinary graph paper over the domain $x = -3$ to 3.

Solve Exercises 17–21.

17. Make a table of values and graph $y = 0.4(3^x)$ on semilog paper from $x = 1$ to 7.

18. Graph $y = 2x^{1/5}$ on log-log paper.

19. The yield, Y, in total ft^3/acre, of thinned stands of yellow poplar can be predicted by the equation

$$\ln Y = 5.36437 - 101.16296 S^{-1} - 22.00048 A^{-1}$$
$$+ 0.97116 \ln BA$$

where S is the site index, A is the current age of the

trees, and BA is the basal area. What is the predicted yield of a 60-year-old stand growing on a site with an index of 120 and a basal area of 130 ft^2 per acre?

20. A bank offers 5.5% interest compounded monthly. If $5,000 is deposited in this bank, how much will it be worth in 8 years?

21. A radioactive substance decays at the rate of 0.75% per year. What is the half-life of the substance?

14
Complex Numbers

Complex numbers have many important applications in electronics. In Section 14.6, we will see how to use complex numbers to solve problems that involve alternating current.

Courtesy of DeVry Institutes

All the numbers we have used until now have been real numbers. Several times in the history of mathematics, problems developed that people could not solve with the number system they had. These problems led to the invention of new numbers. The first time this happened was when the number 0 had to be invented. Later, negative numbers were needed. Each time, new numbers were invented to allow people to solve new kinds of problems. At last, the set of real numbers was developed, and with it we can solve many problems.

Not all problems, however, can be solved with the real numbers. People learned that they could take the cube root -1 or of -8. We know that $\sqrt[3]{-1} = -1$ because $(-1)^3 = -1$ and that $\sqrt[3]{-8} = -2$ because $(-2)^3 = -8$. But, there is no real number for the square root of -1, or -4, or for the square root of any negative number. People had problems that could be worked only if it were possible to take the square root of a negative number. As a result, they invented the numbers we will begin using with this chapter—the complex numbers.

14.1
IMAGINARY AND COMPLEX NUMBERS

Complex numbers have many important uses in technology. They make it much easier to work with vectors and problems that involve alternating current (ac). We will see many uses for complex numbers as we work through this chapter.

Imaginary Unit

As we stated in the chapter introduction, the need for complex numbers arose because people had to solve problems that involved the square roots of negative numbers. In order to solve this dilemma, a new number was invented to correspond to the square root of -1. The name for this number is the **imaginary unit**, and it is represented by the symbol j. Thus, we have

$$j = \sqrt{-1}$$

Another popular name for the imaginary unit is the **j operator**.

≡ Note

Many mathematics books use the symbol i instead of j. But, because i is used to represent electrical current, people in science and technology use j for the imaginary unit.

One of the basic steps we learn in working with imaginary numbers allows us to represent the square root of a negative number as the product of a real number and the imaginary unit, j. The square root of a negative number is called a **pure imaginary number** and is defined in the following box. Remember, if b is a real number, the symbol $\sqrt{b}$ represents the **principal square root** of b and is never negative. Thus, $\sqrt{9} = 3$, $\sqrt{25} = 5$, and $\sqrt{\frac{16}{9}} = \frac{4}{3}$.

Pure Imaginary Number

If b is a real number, $b \geq 0$, then $\sqrt{-b}$ is a **pure imaginary number** and we have
$$\sqrt{-b} = \sqrt{(-1)b} = \sqrt{-1}\sqrt{b} = j\sqrt{b}$$
where $j = \sqrt{-1}$.

We call $j\sqrt{b}$ or $\sqrt{b}\,j$ the **standard form for a pure imaginary number**.

EXAMPLE 14.1

Simplify and express each of the following radicals in the standard form for a pure imaginary number: (a) $\sqrt{-9}$, (b) $\sqrt{-0.25}$, (c) $\sqrt{-3}$, (d) $-\sqrt{-18}$, and (e) $\sqrt{\frac{-4}{9}}$.

Solutions

(a) $\sqrt{-9} = \sqrt{9}\sqrt{-1} = 3j$

(b) $\sqrt{-0.25} = \sqrt{0.25}\sqrt{-1} = 0.5j$

(c) $\sqrt{-3} = \sqrt{3}\sqrt{-1} = \sqrt{3}j$

(d) $-\sqrt{-18} = -\sqrt{18}\sqrt{-1} = -\sqrt{9}\sqrt{2}\sqrt{-1} = -3\sqrt{2}j = -3j\sqrt{2}$

(e) $\sqrt{\frac{-4}{9}} = \sqrt{\frac{4}{9}}\sqrt{-1} = \frac{\sqrt{4}}{\sqrt{9}}\sqrt{-1} = \frac{2}{3}j$

≡ **Note**

Many people write the symbol j in front of a radical sign in order to reduce the danger of thinking that it is under the radical. Thus, you might prefer to write the answers to (c) and (d) as $j\sqrt{3}$ and $-3j\sqrt{2}$.

Since $j = \sqrt{-1}$, we have some interesting relationships.

$$j^2 = -1$$
$$j^3 = j^2 j = (-1)j = -j$$
$$j^4 = j^2 j^2 = (-1)(-1) = 1$$

Any larger power of j can be reduced to one of these basic four. Thus,

$$j^5 = j^{4+1} = j^4 j^1 = 1 \cdot j = j$$
$$j^{15} = j^{4+4+4+3} = j^4 j^4 j^4 j^3 = 1 \cdot 1 \cdot 1 \cdot (-j) = -j$$

≡ **Note**

The powers of j are cyclic, as can be seen above and in the table below.

$$
\begin{array}{ccccccccccc}
1 & = & j^0 & = & j^4 & = & j^8 & = & j^{12} & = & \cdots \\
j & = & j^1 & = & j^5 & = & j^9 & = & j^{13} & = & \cdots \\
-1 & = & j^2 & = & j^6 & = & j^{10} & = & j^{14} & = & \cdots \\
-j & = & j^3 & = & j^7 & = & j^{11} & = & j^{15} & = & \cdots
\end{array}
$$

This cyclic nature of imaginary numbers and of the trigonometric functions allow us to connect imaginary and complex numbers to cyclic applications, such as alternating electrical current.

We need to be careful when we work with imaginary numbers. Consider the problem $\sqrt{-9}\sqrt{-4}$. We know that $\sqrt{-9} = 3j$ and $\sqrt{-4} = 2j$, so $\sqrt{-9}\sqrt{-4} = (3j)(2j) = 6j^2 = -6$. But, we have gotten used to using the property $\sqrt{a}\sqrt{b} = \sqrt{ab}$. What if we tried to use it on this problem: $\sqrt{-9}\sqrt{-4} = \sqrt{(-9)(-4)} = \sqrt{36} = 6$? If we are going to have a successful set of numbers, we cannot get two different answers when we multiply imaginary numbers.

Caution

Remember that whenever you work with square roots of negative numbers, express each number in terms of j before you proceed.

Thus, the correct answer to $\sqrt{-9}\sqrt{-4}$ is -6.

Multiplication of Radicals

If a and b are real numbers, then

$$\sqrt{a}\sqrt{b} = \sqrt{ab} \text{ if } a \geq 0 \text{ and } b \geq 0$$

If either $a < 0$ or $b < 0$ (or both a and b are negative), then convert the radical to "j form" before multiplying.

EXAMPLE 14.2

Simplify the following: **(a)** $(\sqrt{-4})^2$, **(b)** $\sqrt{-3}\sqrt{-12}$, **(c)** $\sqrt{2}\sqrt{-8}$, **(d)** $\sqrt{-0.5}\sqrt{-7}$, and **(e)** $(2\sqrt{-5})(\sqrt{-7})(3\sqrt{-14})$.

Solutions

(a) $(\sqrt{-4})^2 = (2j)^2$

$\qquad = 4j^2$

$\qquad = -4$

(b) $\sqrt{-3}\sqrt{-12} = (j\sqrt{3})(2j\sqrt{3})$

$\qquad\qquad = 2j^2\sqrt{3^2}$

$\qquad\qquad = 2(-1)(3) = -6$

(c) $\sqrt{2}\sqrt{-8} = (\sqrt{2})(j\sqrt{8})$

$\qquad\qquad = j\sqrt{16}$

$\qquad\qquad = 4j$

(d) $\sqrt{-0.5}\sqrt{-7} = (j\sqrt{0.5})(j\sqrt{7})$

$\qquad\qquad = j^2\sqrt{(0.5)(7)}$

$\qquad\qquad = -\sqrt{3.5}$

(e) $(2\sqrt{-5})(\sqrt{-7})(3\sqrt{-14}) = (2j\sqrt{5})(j\sqrt{7})(3j\sqrt{14})$

$\qquad\qquad\qquad = 6j^3\sqrt{5 \cdot 7 \cdot 14}$

$\qquad\qquad\qquad = 42j^3\sqrt{10}$

$\qquad\qquad\qquad = -42j\sqrt{10}$

Note that $\sqrt{5 \cdot 7 \cdot 14} = 7\sqrt{10}$ and that $j^3 = -j$.

Complex Numbers

When an imaginary number and a real number are added, we get a complex number. A **complex number** is of the form $a + bj$, where a and b are real numbers. When $a = 0$ and $b \neq 0$, we have a number of the form bj, which is a **pure imaginary number**. When $b = 0$, we get a number of the form a, which is a real number.

Rectangular Form of a Complex Number

The form $a + bj$ is known as the **rectangular form** of a complex number, where a is the **real part** and b is the **imaginary part**.

Two complex numbers are equal if both the real parts are equal and the imaginary parts are equal. Symbolically, we express this as follows.

Equality of Complex Numbers

If $a + bj$ and $c + dj$ are two complex numbers, then $a + bj = c + dj$, if and only if $a = c$ and $b = d$.

Hint

Two complex numbers, $a + bj$ and $c + dj$, are equal if the real parts are equal, that is, if $a = c$, **and** if the imaginary parts are equal, that is, if $b = d$.

EXAMPLE 14.3

Solve $4 + 3j = 7j + x + 2 + yj$ for x and y.

Solution Here we need to determine both x and y. The best way is to rearrange the terms so that the known values are on one side of the equation and the variables are on the other.

$$4 + 3j = 7j + x + 2 + yj$$
$$4 + 3j - (2 + 7j) = x + yj$$
$$\text{or} \qquad x + yj = 4 + 3j - (2 + 7j)$$
$$x + yj = 2 - 4j$$

So, $x = 2$ and $y = -4$, since the real parts must be equal and the imaginary parts must also be equal.

EXAMPLE 14.4

Simplify and express in the form $a + bj$: (a) $7(3 + 2j)$, (b) $j(5 - 3j)$, and (c) $\dfrac{4 - \sqrt{-12}}{2}$.

Solutions

(a) $7(3 + 2j) = 21 + 14j$

(b) $j(5 - 3j) = 5j - 3j^2 = 5j - 3(-1) = 3 + 5j$

(c) $\dfrac{4 - \sqrt{-12}}{2} = \dfrac{4 - 2j\sqrt{3}}{2} = \dfrac{4}{2} - \dfrac{2j\sqrt{3}}{2} = 2 - j\sqrt{3}$

Notice that, in this last example, we had to divide *each* term of the numerator by 2 in order to get the final answer in the form $a + bj$.

Conjugates of Complex Numbers

Every complex number has a conjugate. As you will see in Section 14.2, conjugates are particularly useful when you are dividing by a complex number.

> **Conjugate of a Complex Number**
>
> The conjugate of a complex number $a + bj$ is the complex number $a - bj$.

To form the conjugate of a complex number, you need to change only the sign of the imaginary part of the complex number.

EXAMPLE 14.5

(a) The conjugate of $3 + 4j$ is $3 - 4j$.

(b) The conjugate of $5 - 2j$ is $5 + 2j$.

(c) The conjugate of $-7j$ is $7j$, since $-7j = 0 - 7j$ and its conjugate is $0 + 7j = 7j$.

(d) The conjugate of 15 is 15, since $15 = 15 + 0j$ and its conjugate is $15 - 0j = 15$.

≡ **Note**

The conjugate of $a + bj$ is $a - bj$ and the conjugate of $a - bj$ is $a + bj$. Thus, each number is the conjugate of the other.

EXAMPLE 14.6

Use the quadratic formula to solve $x^2 + 2x + 10 = 0$.

Solution The discriminant is $b^2 - 4ac$. Here $a = 1$, $b = 2$, and $c = 10$, so the discriminant is $2^2 - 4(1)(10) = 4 - 40 = -36$. Since the discriminant is negative, this trinomial has no real number roots.

EXAMPLE 14.6 (Cont.)

The quadratic formula states that the solutions of $ax^2 + bx + c = 0$ are $x = \dfrac{-b \pm \sqrt{b^2 - 4ac}}{2a}$. Use the values of $a = 1$, $b = 2$, and $c = 10$ with the quadratic formula. From above we know that $b^2 - 4ac = -36$, so

$$x = \frac{-2 \pm \sqrt{-36}}{2}$$

$$= \frac{-2}{2} \pm \frac{\sqrt{36}}{2} j$$

$$= -1 \pm \frac{6}{2} j = -1 \pm 3j$$

Thus, the roots are $x = -1 + 3j$ and $x = -1 - 3j$. Notice that these roots are complex conjugates of each other.

EXAMPLE 14.7

Use the quadratic formula to solve $2x^2 + 3x + 5 = 0$.

Solution Here $a = 2$, $b = 3$, and $c = 5$, and the discriminant is -31. Thus, this quadratic equation has no real number solutions. Using the quadratic formula, we obtain

$$x = \frac{-3 \pm \sqrt{-31}}{4} = \frac{-3 \pm j\sqrt{31}}{4}$$

Writing these in the form $a + bj$, we get $x = \dfrac{-3}{4} + \dfrac{\sqrt{31}}{4} j$ and $x = \dfrac{-3}{4} - \dfrac{\sqrt{31}}{4} j$. As you can see, these roots are complex conjugates of each other.

Application

EXAMPLE 14.8

The susceptance, B, in siemans (S) of an ac circuit that contains R resistance and X reactance, both in Ω, is given by $B = \dfrac{X}{R^2 + X^2}$. Find the value of X when $B = 0.1$ S and $R = \sqrt{34} \approx 5.831\,\Omega$.

Solution Substituting the values for B and R in the given equation, we get

$$0.1 = \frac{X}{\left(\sqrt{34}\right)^2 + X^2} = \frac{X}{34 + X^2}$$

Multiplying both sides by the denominator produces the equation

$$0.1\left(34 + X^2\right) = X$$

$$0.1\left(34 + X^2\right) - X = 0$$

EXAMPLE 14.8 (Cont.)

$$3.4 + 0.1X^2 - X = 0$$
$$0.1X^2 - X + 3.4 = 0$$

This is a quadratic equation with $a = 0.1$, $b = -1$, and $c = 3.4$. Substituting these values in the quadratic formula we obtain

$$X = \frac{-b \pm \sqrt{b^2 - 4ac}}{2a}$$

$$= \frac{-(-1) \pm \sqrt{(-1)^2 - 4(0.1)(3.4)}}{2(0.1)}$$

$$= \frac{1 \pm \sqrt{-0.36}}{0.2}$$

$$= \frac{1 \pm \sqrt{0.36}\, j}{0.2}$$

$$= \frac{1}{0.2} \pm \frac{\sqrt{0.36}\, j}{0.2}$$

$$= \frac{1}{0.2} \pm \frac{0.6\, j}{0.2}$$

$$= 5 \pm 3j$$

The reactance is either $5 + 3j\,\Omega$ or $5 - 3j\,\Omega$.

In Chapter 18, we will show that if a polynomial has only real number coefficients and has one complex root, then the conjugate of that complex number is also a root.

Exercise Set 14.1

In Exercises 1–12, simplify and express each radical in terms of j.

1. $\sqrt{-25}$
2. $\sqrt{-81}$
3. $\sqrt{-0.04}$
4. $\sqrt{-1.44}$

5. $\sqrt{-75}$
6. $-\sqrt{-72}$
7. $-3\sqrt{-20}$
8. $5\sqrt{-30}$

9. $\sqrt{-\frac{9}{16}}$
10. $\sqrt{-\frac{25}{36}}$

11. $-4\sqrt{-\frac{9}{16}}$
12. $-3\sqrt{-\frac{10}{81}}$

In Exercises 13–38, simplify each problem.

13. $(\sqrt{-11})^2$
14. $(\sqrt{-7})^2$
15. $(3\sqrt{-2})^2$
16. $(-2\sqrt{-3})^2$
17. $\sqrt{-4}\sqrt{-25}$
18. $\sqrt{-9}\sqrt{-16}$
19. $(\sqrt{-49})(2\sqrt{-9})$

20. $(3\sqrt{-16})(\sqrt{-36})$
21. $(\sqrt{-5})(-\sqrt{5})$
22. $(-\sqrt{-7})(\sqrt{7})$
23. j^7
24. j^{26}
25. $\sqrt{-\frac{1}{4}}\sqrt{\frac{16}{9}}$

26. $\sqrt{\frac{4}{25}}\sqrt{-\frac{49}{9}}$
27. $\sqrt{-\frac{3}{25}}\sqrt{\frac{3}{16}}$
28. $\sqrt{\frac{75}{36}}\sqrt{-\frac{49}{5}}$
29. $\sqrt{-\frac{25}{36}}\sqrt{-\frac{9}{16}}$
30. $\sqrt{-\frac{16}{81}}\sqrt{-\frac{9}{64}}$

31. $(\sqrt{-0.5})(\sqrt{0.5})$

32. $(\sqrt{0.8})(\sqrt{-0.2})$

33. $(-2\sqrt{2.7})(\sqrt{-3})$

34. $(-4\sqrt{1.6})(2\sqrt{-0.4})$

35. $(\sqrt{-5})(\sqrt{-6})(\sqrt{-2})$

36. $(\sqrt{-3})(\sqrt{-9})(\sqrt{-15})$

37. $(-\sqrt{-7})^2(\sqrt{-2})^2 j^3$

38. $(\sqrt{-3})^2(\sqrt{-5})^2 j^2(\sqrt{-2})$

In Exercises 39–48, solve each problem for the variables x and y.

39. $x + yj = 7 - 2j$

40. $x + yj = -9 + 2j$

41. $x + 5 + yj = 15 - 3j$

42. $6j - x + yj = 4 + 2j$

43. $x - 5j + 2 = 4 - 3j + yj$

44. $2x - 4j = 6j + 4 - yj$

45. $\frac{1}{2}x + \frac{3}{4}j = 2j - \frac{1}{4} + yj$

46. $\frac{2}{5}x - \frac{2}{3}j + 1 = 5 - \frac{4}{3}j - yj$

47. $1.2x + 3 + yj = 7.2 - 4.3j$

48. $3.7j - 1.5x = -4.8 + 2.4j + 0.5yj$

In Exercises 49–70, simplify each problem and express it in the form a + bj.

49. $2(4 + 5j)$

50. $3(2 - 4j)$

51. $-5(2 + j)$

52. $-3(4 + 7j)$

53. $j(3 - 2j)$

54. $j(5 + 4j)$

55. $2j(4 + 3j)$

56. $-3j(2 - 5j)$

57. $\frac{1}{2}(6 - 8j)$

58. $\frac{2}{3}(6 + 9j)$

59. $\dfrac{5 - 10j}{5}$

60. $\dfrac{6 + 12j}{3}$

61. $\dfrac{6 + \sqrt{-18}}{3}$

62. $\dfrac{7 - \sqrt{-98}}{7}$

63. $\dfrac{8 - \sqrt{-24}}{4}$

64. $\dfrac{9 + \sqrt{-27}}{6}$

65. $\frac{2}{3}\left(\frac{3}{4} - \frac{1}{3}j\right)$

66. $\frac{1}{2}\left(\frac{8}{5} + \frac{9}{4}j\right)$

67. $-\frac{5}{3}\left(-\frac{3}{8} + \frac{6}{15}j\right)$

68. $-\frac{7}{4}\left(\frac{8}{21} - \frac{16}{35}j\right)$

69. $1.5(2.4 - 3j)$

70. $-7.2(-0.25 + 1.75j)$

In Exercises 71–80, write the conjugate of the given numbers.

71. $7 + 2j$

72. $9 + \frac{1}{2}j$

73. $6 - 5j$

74. $\frac{1}{2} - 9j$

75. 19

76. $7j$

77. $-8j$

78. -11

79. $\sqrt{2} + 7.3j$

80. $-4.5 - \sqrt{19}j$

In Exercises 81–88, use the quadratic formula to solve each of the problems. Express your answers in the form a + bj.

81. $x^2 + x + 2.5 = 0$

82. $x^2 + 2x + 5 = 0$

83. $x^2 + 9 = 0$

84. $x^2 + 25 = 0$

85. $2x^2 + 3x + 7 = 0$

86. $2x^2 + 7x + 9 = 0$

87. $5x^2 + 2x + 5 = 0$

88. $3x^2 + 2x + 10 = 0$

Solve Exercises 89–92.

89. *Electronics* The susceptance, B, in siemans (S) of an ac circuit that contains R resistance and X reactance, both in Ω, is given by $B = \dfrac{X}{R^2 + X^2}$. Find the value of X when $B = 0.08$ S and $R = 50\,\Omega$.

90. *Electronics* The susceptance, B, in siemans (S) of an ac circuit that contains R resistance and X reactance, both in Ω, is given by $B = \dfrac{X}{R^2 + X^2}$. Find the value of X when $B = 0.05$ S and $R = 12\,\Omega$.

91. *Electronics* The susceptance, B, in siemans (S) of an ac circuit that contains R resistance and X reactance, both in Ω, is given by $B = \dfrac{X}{R^2 + X^2}$. Solve the equation for R.

92. *Electronics* The susceptance, B, in siemans (S) of an ac circuit that contains R resistance and X reactance, both in Ω, is given by $B = \dfrac{X}{R^2 + X^2}$. Find the general formula in terms of X.

✎ **In Your Words**

93. Explain what it means for two complex numbers to be equal.

94. Describe how to determine the conjugate of a complex number.

☰ 14.2
OPERATIONS WITH COMPLEX NUMBERS

As with all number systems, we want to be able to perform the four basic operations of addition, subtraction, multiplication, and division. These operations are performed after all complex numbers have been expressed in terms of j.

Addition and Subtraction ▬▬▬

We will begin by giving the definitions for addition and subtraction of complex numbers. After each definition, we will provide several examples showing how to use the definition.

> **Addition of Complex Numbers**
>
> If $a+bj$ and $c+dj$ are any two complex numbers, then their sum is defined as
>
> $$(a+bj)+(c+dj)=(a+c)+(b+d)j$$

In words, this says to add the real parts of the complex numbers and add their imaginary parts.

EXAMPLE 14.9

Find each of these sums: **(a)** $(9+2j)+(8+6j)$, **(b)** $(6+3j)+(5-7j)$, **(c)** $(-2\sqrt{3}+4j)+(5-6j)$, and **(d)** $(-4+3j)+(-1-\sqrt{-4})$.

Solutions

(a) $(9+2j)+(8+6j)=(9+8)+(2+6)j$
$$= 17+8j$$

(b) $(6+3j)+(5-7j)=(6+5)+(3-7)j$
$$= 11-4j$$

(c) $(-2\sqrt{3}+4j)+(5-6j)=(-2\sqrt{3}+5)+(4-6)j$
$$= 5-2\sqrt{3}-2j$$

Notice that the real part of this complex number is $5-2\sqrt{3}$ and the imaginary part is $-2j$.

(d) $(-4+3j)+(-1-\sqrt{-4})=(-4+3j)+(-1-2j)$
$$=(-4-1)+(3-2)j$$
$$=-5+j$$

Application

EXAMPLE 14.10

In an ac circuit, if two sections are connected in series and have the same current in each section, the voltage V is given by $V = V_1 + V_2$. Find the total voltage in a given circuit if the voltages in the individual sections are $8.9 - 2.4j$ and $11.2 + 6.3j$.

Solution To find the total voltage in this circuit, we need to add the voltages in the individual sections.

$$\begin{aligned} V &= V_1 + V_2 \\ &= (8.9 - 2.4j) + (11.2 + 6.3j) \\ &= (8.9 + 11.2) + (-2.4 + 6.3)j \\ &= 20.1 + 3.9j \end{aligned}$$

The voltage in this circuit is $20.1 + 3.9j$ V.

Subtraction of Complex Numbers

If $a + bj$ and $c + dj$ are complex numbers, then their difference is defined as

$$(a + bj) - (c + dj) = (a - c) + (b - d)j$$

EXAMPLE 14.11

Find each of the following differences: **(a)** $(3 + 4j) - (2 + j)$, **(b)** $(5 + 7j) - (3 - 10j)$, **(c)** $(-8 + 4j) - (3 + 10j)$, and **(d)** $(9 + \sqrt{-18}) - (6 + \sqrt{-2})$.

Solutions

(a) $\begin{aligned}[t] (3 + 4j) - (2 + j) &= (3 - 2) + (4 - 1)j \\ &= 1 + 3j \end{aligned}$

(b) $\begin{aligned}[t] (5 + 7j) - (3 - 10j) &= (5 - 3) + [7 - (-10)]j \\ &= 2 + 17j \end{aligned}$

(c) $\begin{aligned}[t] (-8 + 4j) - (3 + 10j) &= (-8 - 3) + (4 - 10)j \\ &= -11 - 6j \end{aligned}$

(d) $\begin{aligned}[t] (9 + \sqrt{-18}) - (6 + \sqrt{-2}) &= (9 + 3j\sqrt{2}) - (6 + j\sqrt{2}) \\ &= (9 - 6) + (3\sqrt{2} - \sqrt{2})j \\ &= 3 + 2j\sqrt{2} \end{aligned}$

Notice that in Examples 14.9(d) and 14.11(d), we had to first write some of the numbers in the $a + bj$ form. In Example 14.9(d), we changed $\sqrt{-4}$ to $2j$, and in Example 14.11(d) we changed $\sqrt{-18}$ to $3j\sqrt{2}$ and $\sqrt{-2}$ to $j\sqrt{2}$. Do not overlook this step.

Multiplication

Multiplication of complex numbers uses the FOIL method, which was introduced in Chapter 2. As you can see, you will have to replace j^2 with -1 to obtain a simplified answer.

$$(a+bj)(c+dj) = ac+adj+bcj+bdj^2$$
$$= ac+adj+bcj-bd$$
$$= (ac-bd)+(ad+bc)j$$

Multiplication of Complex Numbers

If $a+bj$ and $c+dj$ are any two complex numbers, then their product $(a+bj)(c+dj)$ is defined as

$$(a+bj)(c+dj) = (ac-bd)+(ad+bc)j$$

EXAMPLE 14.12

Multiply and write each answer in the form $a+bj$: **(a)** $(2+5j)(3-4j)$, **(b)** $(5+3j)^2$, **(c)** $(7+3j)(7-3j)$, and **(d)** $(a+bj)(a-bj)$.

Solutions We have used the FOIL method rather than the definition to determine these products. By using the FOIL method, you do not have to remember the rule. However, remember that $j^2 = -1$.

(a) $(2+5j)(3-4j) = 2 \cdot 3 + 2(-4)j + 5j(3) + 5(-4)j^2$
$$= 6-8j+15j-20(-1)$$
$$= 6-8j+15j+20$$
$$= 26+7j$$

(b) $(5+3j)^2 = (5+3j)(5+3j)$
$$= 5^2 + 2(5)(3)j + 9j^2$$
$$= 25+30j+9(-1)$$
$$= 25+30j-9$$
$$= 16+30j$$

(c) Here we can use the difference of squares.
$$(7+3j)(7-3j) = 7^2 - 3^2j^2$$
$$= 49-9(-1)$$
$$= 49+9$$
$$= 58$$

(d) $(a+bj)(a-bj) = a^2 + abj - abj - b^2j^2$
$$= a^2 - b^2j^2$$
$$= a^2 + b^2$$

Application

EXAMPLE 14.13

In an ac circuit, the formula $V = ZI$ relates the voltage V, to impedance Z and current I. Find the voltage in a given circuit if the impedance is $8 - 2j\,\Omega$ and the current is $11 + 6j$ A.

Solution   To find the voltage, we need to multiply the given values of Z and I. As before, we will use the FOIL method.

$$\begin{aligned}
V &= ZI \\
&= (8 - 2j)(11 + 6j) \\
&= 8 \cdot 11 + 8(6j) + (-2j)(11) + (-2j)6j \\
&= 88 + 48j - 22j + 12 \\
&= 100 + 26j
\end{aligned}$$

The voltage in this circuit is $100 + 26j$ V.

Division

In Examples 14.12(c) and (d) we multiplied a complex number and its conjugate. The following note will be helpful when we divide complex numbers.

≡ Note

The product of a complex number and its conjugate is a real number.

Division of Complex Numbers

If $a + bj$ and $c + dj$ are complex numbers, then the quotient $(a + bj) \div (c + dj) = \dfrac{a + bj}{c + dj}$, is defined as

$$\begin{aligned}
\frac{a + bj}{c + dj} &= \frac{a + bj}{c + dj} \cdot \frac{c - dj}{c - dj} = \frac{(ac + bd) + (bc - ad)j}{c^2 + d^2} \\
&= \frac{ac + bd}{c^2 + d^2} + \frac{bc - ad}{c^2 + d^2}\, j
\end{aligned}$$

This looks very complicated, but the following hint gives an important idea to remember whenever you divide by a complex number.

💡 Hint

In division of complex numbers, you multiply both the numerator and the denominator by the conjugate of the denominator.

The four problems in Example 14.14 show how to use the conjugate to divide.

EXAMPLE 14.14

Divide and express each answer in the form $a + bj$: **(a)** $\dfrac{10 - 4j}{1 + j}$, **(b)** $\dfrac{8 + 6j}{2 - j}$, **(c)** $\dfrac{3 - 2j}{4 + 2j}$, and **(d)** $\dfrac{0.5 + j\sqrt{3}}{2.4j}$.

Solutions

(a)
$$\frac{10 - 4j}{1 + j} = \frac{10 - 4j}{1 + j} \cdot \frac{1 - j}{1 - j}$$
$$= \frac{10 - 10j - 4j + 4j^2}{1 + 1}$$
$$= \frac{10 - 10j - 4j - 4}{2}$$
$$= \frac{6 - 14j}{2} = 3 - 7j$$

(b)
$$\frac{8 + 6j}{2 - j} = \frac{8 + 6j}{2 - j} \cdot \frac{2 + j}{2 + j}$$
$$= \frac{16 + 8j + 12j + 6j^2}{4 + 1}$$
$$= \frac{16 + 8j + 12j - 6}{5}$$
$$= \frac{10 + 20j}{5} = 2 + 4j$$

(c)
$$\frac{3 - 2j}{4 + 2j} = \frac{3 - 2j}{4 + 2j} \cdot \frac{4 - 2j}{4 - 2j}$$
$$= \frac{12 - 6j - 8j - 4}{16 + 4}$$
$$= \frac{8 - 14j}{20} = \frac{8}{20} - \frac{14}{20}j$$
$$= \frac{2}{5} - \frac{7}{10}j = 0.4 - 0.7j$$

(d)
$$\frac{0.5 + j\sqrt{3}}{2.4j} = \frac{0.5 + j\sqrt{3}}{2.4j} \cdot \frac{-2.4j}{-2.4j}$$
$$= \frac{(0.5)(-2.4j) + (j\sqrt{3})(-2.4j)}{(2.4j)(-2.4j)}$$
$$= \frac{-1.2j + 2.4\sqrt{3}}{5.76}$$
$$= \frac{2.4\sqrt{3} - 1.2j}{5.76}$$
$$= \frac{2.4\sqrt{3}}{5.76} - \frac{1.2}{5.76}j = \frac{\sqrt{3}}{2.4} - \frac{1}{4.8}j$$

Application

EXAMPLE 14.15

In an ac circuit, use the formula $V = ZI$ to find the impedance in a given circuit if the voltage is $95 + 9j$ V and the current is $7 - 3j$ A.

Solution We want to find the impedance Z, given V and I. Using the formula $V = ZI$, we see that $Z = \dfrac{V}{I}$.

$$
\begin{aligned}
Z &= \frac{V}{I} \\
&= \frac{95 + 9j}{7 - 3j} \\
&= \frac{95 + 9j}{7 - 3j} \cdot \frac{7 + 3j}{7 + 3j} \\
&= \frac{95(7) + 95(3j) + (9j)7 + (9j)(3j)}{7^2 - (3j)^2} \\
&= \frac{665 + 285j + 63j + 27j^2}{49 - 9j^2} \\
&= \frac{638 + 348j}{58} \\
&= 11 + 6j
\end{aligned}
$$

The impedance in this circuit is $11 + 6j\,\Omega$.

Using Calculators with Complex Numbers

Some calculators can be used for working with complex numbers. One way to tell if your calculator can work with complex numbers is to look in the index of the owner's manual. Another way is to see if your calculator has an $\boxed{\text{Img}}$, an $\boxed{\text{I}}$, or a $\boxed{\theta}$ key. If it does, then it can probably be used to calculate with complex numbers.

Many graphing calculators have a θ on one of the keys. For example, a TI-82 has an $\boxed{\text{X,T,}\theta}$ key and a Casio fx-7700 has an $\boxed{\text{X,}\theta\text{,T}}$ key. Neither of these calculators can be used to calculate with complex numbers. On the other hand, the Sharp EL-9300C (which has an $\boxed{\text{X/}\theta\text{/T}}$ key), the TI-85, and the Hewlett-Packard HP-48 can all be used to calculate with complex numbers.

If you are going to use complex numbers on your algebraic calculator, you must first put the calculator in the complex mode. Turn on your calculator and press the $\boxed{\text{Img}}$ key. The display should show 0 on the right and CMPLX in the lower left corner. (See Figure 14.1.) The calculator is now in the complex mode. To get the calculator out of the complex mode press $\boxed{\text{INV}}$ $\boxed{\text{Img}}$.

To enter a complex number of the form $a + bj$ into an algebraic calculator, first enter the imaginary part b and press $\boxed{\text{Img}}$. Next, enter the real part, a. With an RPN calculator, first enter the real part a, press $\boxed{\text{ENTER}}$, and then enter the imaginary part b and press $\boxed{\text{2nd}}$ $\boxed{\text{I}}$.

To enter a complex number of the form $a + bj$ into a TI-85 graphing calculator, you enter the number as an ordered pair (a, b). With a Sharp EL-9200 graphing calculator, the calculator must first be put in complex mode. You do this by pressing

CMPLX

FIGURE 14.1

MENU A 4. The complex number $a+bj$ can be put into the calculator as $a+bi$. However, for most computations it is better to enter the number as $(a+bi)$.

EXAMPLE 14.16

Enter **(a)** $9+2j$ and **(b)** $7-4j$ into a calculator.

Solutions

		PRESS	DISPLAY
(a)	Algebraic calculator:	2 ⬚Img	0
		9	9
	RPN calculator:	9 ⬚ENTER	0
		2 ⬚2nd ⬚I	9
	TI-85	⬚(9 ⬚, 2 ⬚)	(9,2)
	Sharp EL-9300	9 ⬚+ 2 ⬚2nd ⬚i	9+2i
	or	⬚(9 ⬚+ 2 ⬚2nd ⬚i ⬚)	(9+2i)
(b)	Algebraic calculator:	4 ⬚Img ⬚+/−	0
		7	7
	RPN calculator:	7 ⬚ENTER	7
		4 ⬚CHS ⬚2nd ⬚I	7
	TI-85	⬚(7 ⬚, ⬚(−) 4 ⬚)	(7,-4)
	Sharp EL-9300	7 ⬚− 4 ⬚2nd ⬚i	7-4i
	or	⬚(7 ⬚− 4 ⬚2nd ⬚i ⬚)	(7-4i)

If you want to determine the value stored in the imaginary part of the algebraic calculator, press ⬚EXC ⬚Img and, in the case of Example 14.16(b), the display will show −4 and the symbol CMPLX will blink. The blinking indicates that this is the imaginary part of the number. Enter ⬚EXC ⬚Img again and you will see the real part, 7, of the number. The CMPLX symbol stops blinking and is a steady display to show that this is the real part of the display. You must return to the real part of the display before you can perform any calculations.

To view the value stored in the imaginary part of an RPN calculator, press ⬚2nd ⬚(i). In the case of Example 14.16(b), the display will show −4 for about two seconds and then the real part of the complex number, 7, will again be displayed. To see the imaginary value for longer than two seconds, press the ⬚2nd ⬚(i) keys and hold down the ⬚(i) key until you are finished viewing the imaginary part of the display.

Calculations with a calculator are done in the same manner in which they are done with real numbers. The following example shows how each of these is done. Examples of complex number computation with graphing calculators will use TI-85 keystrokes. However, other graphing calculators that allow computations with non-real complex numbers use similar procedures.

EXAMPLE 14.17		

Use a calculator to perform each of the following: **(a)** $(3 + 5j) + (7 - 8j)$, **(b)** $(2 + j)(-6 - 3j)$, and **(c)** $\dfrac{10 - 4j}{1 + j}$.

Solutions

	PRESS	DISPLAY	
(a)	Algebraic calculator:		
	5 Img 3 +	3	
	8 +/− Img 7 =	10	steady CMPLX
	EXC Img	−3	blinking CMPLX
	RPN calculator:		
	3 ENTER 5 2nd I	3	
	7 ENTER 8 CHS 2nd I	7	
	+	10	
	2nd (i)	−3	
	Graphing calculator:		
	(3 , 5)	(3,5)	
	+ (7 , (−) 5) ENTER	(10,-3)	

So, $(3 + 5j) + (7 - 8j) = 10 - 3j$

	PRESS	DISPLAY	
(b)	Algebraic calculator:		
	1 Img 2 ×	2	
	3 +/− Img 6 +/− =	−9	steady CMPLX
	EXC Img	−12	blinking CMPLX
	RPN calculator:		
	2 ENTER 1 2nd I	2	
	6 CHS ENTER 3 2nd I	−6	
	×	−9	
	2nd (i)	−12	
	Graphing calculator:		
	(2 , 1)	(2,1)	
	((−) 6 , (−) 3) ENTER	(-9,-12)	

From this we see that $(2 + j)(-6 - 3j) = -9 - 12j$.

	PRESS	DISPLAY	
(c)	Algebraic calculator:		
	4 +/− Img 10 ÷	10	
	1 Img 1 =	3	steady CMPLX
	EXC Img	−7	blinking CMPLX
	RPN calculator:		
	10 ENTER 4 CHS 2nd I	10	
	1 ENTER 1 2nd I	1	
	÷	3	
	2nd (i)	−7	

EXAMPLE 14.17 (Cont.)

Graphing calculator:

(10 , (−) 4) $(10, -4)$

÷ (1 , 1) ENTER $(3, -7)$

And, as we saw in Example 14.14(a), $(10 - 4j) \div (1 + j) = 3 - 7j$.

Exercise Set 14.2

In Exercises 1–48, perform the indicated operations. Express all answers in the form $a + bj$. If possible, use a calculator to check your answers.

1. $(5 + 2j) + (-6 + 5j)$
2. $(9 - 7j) + (6 - 8j)$
3. $(11 - 4j) + (-6 + 2j)$
4. $(21 + 3j) + (-7 - 6j)$
5. $(4 + \sqrt{-9}) + (3 - \sqrt{-16})$
6. $(-11 + \sqrt{-4}) + (9 + \sqrt{-36})$
7. $(2 + \sqrt{-9}) + (8j - \sqrt{5})$
8. $(3 + \sqrt{-8}) + (3 - \sqrt{8})$
9. $(14 + 3j) - (6 + j)$
10. $(-8 + 3j) - (4 - 3j)$
11. $(9 - \sqrt{-4}) - (\sqrt{-16} + 6)$
12. $(\sqrt{-25} - 3) - (3 - \sqrt{-25})$
13. $(4 + 2j) + j + (3 - 5j)$
14. $(2 - 3j) - j - (6 + \sqrt{-81})$
15. $(2 + j)3j$
16. $(5 - 3j)2j$
17. $(9 + 2j)(-5j)$
18. $(11 - 4j)(-3j)$
19. $(2 + j)(5 + 3j)$
20. $(3 - 2j)(4 + 5j)$

21. $(6 - 2j)(5 + 3j)$
22. $(4 - 2j)(7 - 3j)$
23. $(2\sqrt{-9} + 3)(5\sqrt{-16} - 2)$
24. $(6\sqrt{-25} - 4)(-3 - 2\sqrt{-49})$
25. $(\sqrt{-3})^4$
26. $(\sqrt{-9})^3$
27. $(1 + 2j)^2$
28. $(3 + 4j)^2$
29. $(7 - j)^2$
30. $(4 - 3j)^2$
31. $(5 + 2j)(5 - 2j)$
32. $(7 + 3j)(7 - 3j)$
33. $\dfrac{6 - 4j}{1 + j}$
34. $\dfrac{4 - 8j}{2 - 2j}$
35. $\dfrac{6 - 3j}{1 + 2j}$
36. $\dfrac{5 - 10j}{1 - 2j}$
37. $\dfrac{4 + 2j}{1 - 2j}$

38. $\dfrac{9 + 5j}{3 + j}$
39. $\dfrac{2j}{5 + j}$
40. $\dfrac{5j}{6 - j}$
41. $\dfrac{\sqrt{3} - \sqrt{-6}}{\sqrt{-3}}$
42. $\dfrac{\sqrt{5} + \sqrt{-10}}{\sqrt{-5}}$
43. $\dfrac{(5 + 2j)(3 - j)}{4 + j}$
44. $\dfrac{(6 - j)(2 + 3j)}{-1 + 3j}$
45. $(1 + j)^4$
46. $(1 - j)^4$
47. $\left(\dfrac{1}{2} + \dfrac{\sqrt{3}}{2}j\right)^2$
48. $\dfrac{4 + j}{(3 - 2j) + (4 - 3j)}$

Solve Exercises 49–66.

49. Show that the sum of a complex number and its conjugate is a real number.

50. Show that the difference of a complex number $a + bj$, with $b \neq 0$ and its conjugate is a pure imaginary number.

51. Show that the product of a complex number and its conjugate is a real number.

52. *Electricity* In an ac circuit, if two sections are connected in series and have the same current in each section, the voltage V is given by $V = V_1 + V_2$. Find

the total voltage in a given circuit if the voltages in the individual sections are $9.32 - 6.12j$ and $7.24 + 4.31j$.

53. *Electricity*   Find the total voltage in an ac series circuit that has the same current in each section, if the voltages in the individual sections are $6.21 - 1.37j$ and $4.32 - 2.84j$.

54. *Electricity*   If two sections of an ac series circuit have the same current in each section, what is the voltage in one section if the total voltage is $19.2 - 3.5j$ and the voltage in the other section is $12.4 + 1.3j$?

55. *Electricity*   If two sections of an ac series circuit have the same current in each section, what is the voltage in one section if the total voltage is $7.42 + 1.15j$ and the voltage in the other section is $2.34 - 1.73j$?

56. *Electricity*   The total impedance Z of an ac circuit containing two impedances Z_1 and Z_2 in series is $Z = Z_1 + Z_2$. If $Z_1 = 0.25 + 0.20j\,\Omega$ and $Z_2 = 0.15 - 0.25j\,\Omega$, what is Z?

57. *Electricity*   If the total impedance of an ac series circuit containing two impedances is $Z = 9.13 - 4.27j\,\Omega$ and one of the impedances is $3.29 - 5.43j\,\Omega$, what is the impedance of the other circuit?

58. *Electricity*   If an ac circuit contains two impedances Z_1 and Z_2 in parallel, then the total impedance Z is given by

$$Z = \frac{Z_1 Z_2}{Z_1 + Z_2}$$

What is Z when $Z_1 = 6\,\Omega$ and $Z_2 = 3j\,\Omega$?

59. *Electricity*   What is the total impedance in an ac circuit that contains two impedances Z_1 and Z_2 in parallel, if $Z_1 = 20 + 10j\,\Omega$ and $Z_2 = 10 - 20j\,\Omega$?

60. *Electricity*   In an ac circuit the voltage V, current I, and impedance Z are related by $V = IZ$. If $I = 12.3 + 4.6j$ A and $Z = 16.4 - 9.0j\,\Omega$, what is the voltage?

61. *Electricity*   What is the current when $V = 5.2 + 3j$ V and $Z = 4 - 2j\,\Omega$? (See Exercise 57.)

62. *Electricity*   What is the impedance when $V = 10.6 - 6.0j$ V and $I = 4 + j$ A?

63.   Write a computer program that will add or subtract two complex numbers.

64.   Write a computer program that will multiply two complex numbers.

65.   Write a computer program that will divide two complex numbers.

66. *Programming*   If your calculator cannot be used to calculate with complex numbers, write a program for your calculator that will **(a)** allow you to enter a complex number and **(b)** add, subtract, multiply, and divide two complex numbers.

In Your Words

67. Describe how to multiply two complex numbers.

68. Describe how to divide two complex numbers.

69. If your calculator can be used to calculate with complex numbers, explain how to enter a complex number into your calculator and how to divide a complex number by another complex number. Give your explanation to a classmate and ask him or her to follow your directions. Rewrite your directions to clarify places where your classmate had difficulty.

≡ 14.3
GRAPHING COMPLEX NUMBERS; POLAR FORM OF A COMPLEX NUMBER

We have been able to graph real numbers since Chapter 4. It would be helpful if we could also represent complex numbers as points in a plane. The fact that each complex number has a real part and an imaginary part makes it possible to graph complex numbers.

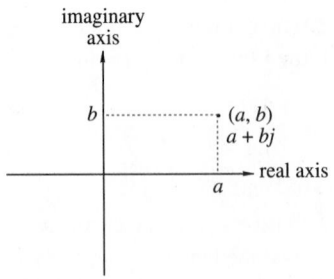

FIGURE 14.2

Complex Plane

Each complex number can be written in the form $a + bj$. We can represent this as the point (a, b) in the plane, as shown in Figure 14.2. Notice that the origin corresponds to the point $(0, 0)$, or $0 + 0j$. Since the points in the plane are representing complex numbers, this is called the **complex plane**. It is also referred to as the **Argand plane**, after the French mathematician Argand (1768–1822). In the complex plane, the horizontal axis acts as the real axis and the vertical axis as the **imaginary axis**.

EXAMPLE 14.18

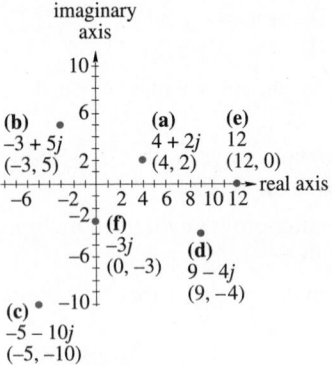

FIGURE 14.3

Graph the following complex numbers: **(a)** $4 + 2j$, **(b)** $-3 + 5j$, **(c)** $-5 - 10j$, **(d)** $9 - 4j$, **(e)** 12, and **(f)** $-3j$.

Solutions   The solutions are shown in Figure 14.3.

The complex number $a + bj$ can also be represented in the plane by the position vector **OP** from the origin to the point $P(a, b)$. We now have three correct and interchangeable ways to refer to the complex number: $a + bj$, the point (a, b) on the complex (Argand) plane, and the vector $\mathbf{a} + \mathbf{bj}$. (See Figure 14.4.)

If we have two complex numbers on a graph, their sum or difference can be represented in the same way in which we add or subtract vectors. To add two complex numbers graphically, locate the point corresponding to one of them and draw the position vector for that point. Repeat the process for the second point. Finally, use the parallelogram method to add these two vectors. The sum will be the diagonal of the parallelogram that has the origin as an endpoint.

EXAMPLE 14.19

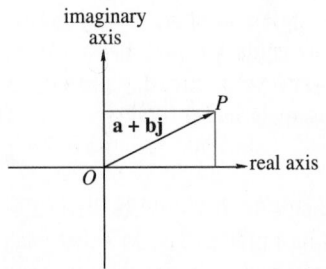

FIGURE 14.4

Graphically add the complex numbers $-9 + 5j$ and $4 - 7j$.

Solution   The solution is shown in Figure 14.5.

As you can see, the graphical solution agrees with the method given in Section 14.2.

$$(-9 + 5j) + (4 - 7j) = (-9 + 4) + (5 - 7)j$$
$$= -5 - 2j$$

Polar Form of a Complex Number

The rectangular form is not the only way to represent complex numbers. In Figure 14.6, the angle θ that the vector **OP** makes with the positive real axis is called the **argument** or **amplitude** of the complex number $a + bj$. The length r of **OP** is called the **absolute value** or **modulus** of $a + bj$. The absolute value is a real number and is never negative. The absolute value is always positive or 0. Using the Pythagorean theorem, we see that the length of r is $\sqrt{a^2 + b^2}$.

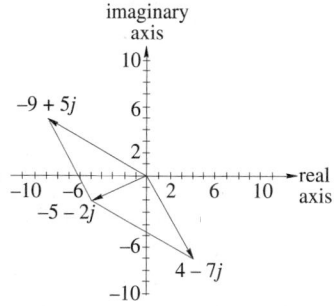

imaginary axis

$-9 + 5j$

$-5 - 2j$

real axis

$4 - 7j$

FIGURE 14.5

Since $a + bj$ can be considered a vector in the complex plane, $|a + bj|$ is the magnitude of the vector.

Absolute Value of a Complex Number

The **absolute value of a complex number** $a + bj$ is denoted $|a + bj|$ and has the value

$$|a + bj| = \sqrt{a^2 + b^2}$$

A careful examination of Figure 14.6 reveals four useful relationships. First, from our definitions of the trigonometric functions, we see that $\cos\theta = \dfrac{a}{r}$ and $\sin\theta = \dfrac{b}{r}$. From these, we see that

$$a = r\cos\theta \tag{1}$$

$$\text{and} \qquad b = r\sin\theta \tag{2}$$

The other two relationships require not only our knowledge of trigonometry but of the Pythagorean theorem. These two relationships state that

$$\tan\theta = \frac{b}{a} \tag{3}$$

$$\text{and} \qquad r = \sqrt{a^2 + b^2} \tag{4}$$

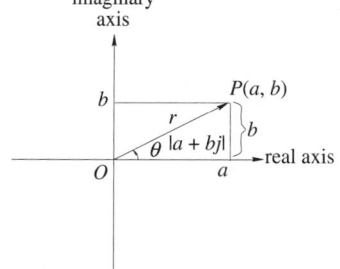

imaginary axis

$P(a, b)$

r

θ | $|a + bj|$

b

b

O a real axis

FIGURE 14.6

These four equations will be very valuable to us. From the first two, we see that

$$a + bj = r\cos\theta + jr\sin\theta = r(\cos\theta + j\sin\theta)$$

The expression $r(\cos\theta + j\sin\theta)$ is often abbreviated as $r\operatorname{cis}\theta$ or $r\,\underline{/\theta}$. In the abbreviation $r\operatorname{cis}\theta$, the c represents cosine, the s represents sine, and the i represents the mathematician's symbol for j. The symbol $r\,\underline{/\theta}$ is read "r at angle θ." The right-hand side of the previous equation, $r(\cos\theta + j\sin\theta)$, is called the **polar** or **trigonometric form** of a complex number.

Changing Complex Numbers from Polar to Rectangular Form

A complex number written in polar form as

$$r\,\underline{/\theta} \qquad \text{or} \qquad r\operatorname{cis}\theta \qquad \text{or} \qquad r(\cos\theta + j\sin\theta)$$

has the rectangular coordinates $a + bj$, where

$$a = r\cos\theta \qquad \text{and} \qquad b = r\sin\theta$$

EXAMPLE 14.20

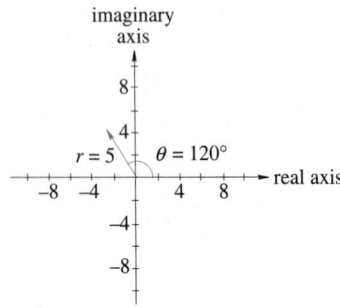

FIGURE 14.7

Locate the point $5(\cos 120° + j \sin 120°)$ in the complex plane and convert the number to rectangular form.

Solution   The graphical representation is given in Figure 14.7.
In this example, $r = 5$ and $\theta = 120°$, and so

$$a = r \cos\theta = 5 \cos 120° = 5(-0.5) = -2.5$$
$$b = r \sin\theta = 5 \sin 120° \approx 5(0.8660) = 4.3301$$

Thus, $5 \operatorname{cis} 120° \approx -2.5 + 4.3301 j$.

The previous example was worked using a calculator. The exact value for b would have been represented by $\sin\theta = \frac{\sqrt{3}}{2}$, and so $b = \frac{5\sqrt{3}}{2}$. Even though a calculator gives values to more than four decimal places, we will give degrees to the nearest tenth and trigonometric functions to four decimal places.

Do not round off numbers until all calculations have been finished. Rounding off numbers before you complete the problem can make your final results differ from the degree of accuracy you are seeking.

EXAMPLE 14.21

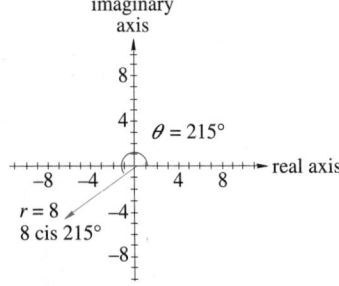

FIGURE 14.8

Locate the point $8 \,\underline{/215°}$ in the complex plane and convert the number to rectangular form.

Solution   The graphical representation is given in Figure 14.8.
In this example, $r = 8$ and $\theta = 215°$, so

$$a = r \cos\theta = 8 \cos 215° \approx 8(-0.8192) = -6.5532$$
$$b = r \sin\theta = 8 \sin 215° \approx 8(-0.5736) = -4.5886$$

Thus, $8 \,\underline{/215°} \approx -6.5532 - 4.5886 j$.

Changing Complex Numbers from Rectangular to Polar Form

A complex number written in rectangular form as

$$a + bj$$

has the polar coordinate forms

$$r \,\underline{/\theta}, \qquad r \operatorname{cis}\theta, \qquad \text{or} \qquad r(\cos\theta + j \sin\theta),$$

where

$$r = \sqrt{a^2 + b^2} \qquad \text{and} \qquad \tan\theta = \frac{b}{a}$$

≡ **Note**   While there may be many values for θ that satisfy the given conditions, we will normally select the smallest positive value.

EXAMPLE 14.22

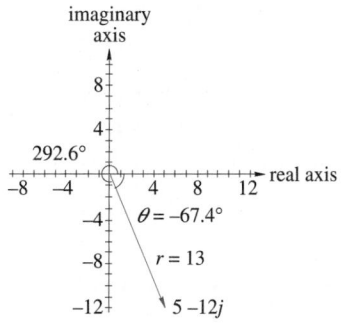

FIGURE 14.9

Represent $5 - 12j$ graphically and convert it to polar form.

Solution   Graphically, this point is shown as a vector in Figure 14.9.
Here $a = 5$ and $b = -12$, so

$$r = \sqrt{a^2 + b^2}$$
$$= \sqrt{5^2 + (-12)^2} = \sqrt{25 + 144}$$
$$= \sqrt{169} = 13$$

From the graph we can see that this point is in Quadrant IV.

$$\tan\theta = \frac{b}{a}$$
$$= \frac{-12}{5}$$

so
$$\theta = \tan^{-1}\left(\frac{-12}{5}\right)$$
$$\approx -67.4°$$

Notice that we used the arctan function. We can let θ be either $-67.4°$ or $-67.4° + 360° = 292.6°$. So, $5 - 12j \approx 13\,\underline{/292.6°} = 13\,\underline{/-67.4°}$. But, as we just stated, we will usually use the smallest positive value for θ that satisfies the given conditions. In this case, $\theta = 292.6°$.

EXAMPLE 14.23

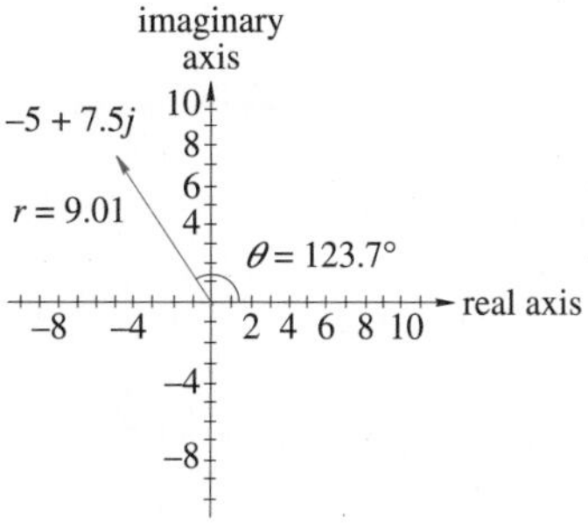

FIGURE 14.10

Locate the point $-5 + 7.5j$ in the complex plane and convert it to polar form.

Solution   The point $-5 + 7.5j$ is shown in the graph in Figure 14.10 and we see that it is in Quadrant II.
In this example $a = -5$ and $b = 7.5$, so

$$r = \sqrt{(-5)^2 + 7.5^2}$$
$$= \sqrt{25 + 56.25}$$
$$= \sqrt{81.25} \approx 9.01$$

$$\tan\theta = \frac{7.5}{-5} = -1.5$$

so
$$\theta \approx 123.7°$$

Thus, $-5 + 7.5j \approx 9.01\,\underline{/123.7°}$.

Caution

Remember that using a calculator to work with inverse trigonometric functions may not give the desired angle. You must determine which quadrant the point is in.

In Example 14.23, we had $\tan\theta = -1.5$. If you use a calculator (in degree mode) to determine $\theta = \tan^{-1}(-1.5)$, you get

	PRESS	DISPLAY
Algebraic calculator:	1.5 [+/−] [INV] [tan]	- 56.309932
Graphics calculator:	[2nd] [tan] [(−)] 1.5	- 56.30993247

This value for θ, $-56.3°$, is an angle in the fourth quadrant. We can see, from Figure 14.10, that the point is in Quadrant II. So, $\theta = -56.3° + 180° = 123.7°$

Finding the polar form of a real number or a pure imaginary number is relatively easy, as shown in Example 14.24.

EXAMPLE 14.24

Express each of the following complex numbers in polar form: **(a)** 8, **(b)** −9, **(c)** $3j$, and **(d)** $-4j$.

Solutions

(a) $8 = 8 + 0j$ lies on the positive real axis, so $\theta = 0°$ and $r = 8$; thus, $8 = 8(\cos 0° + j\sin 0°) = 8\operatorname{cis} 0°$.

(b) $-9 = -9 + 0j$ lies on the negative real axis so, $\theta = 180°$ and $r = 9$; thus, $-9 = 9\operatorname{cis} 180°$.

(c) $3j = 0 + 3j$ lies on the positive imaginary axis, so $\theta = 90°$ and $r = 3$; $3j = 3\underline{/90°}$.

(d) $-4j = 0 - 4j$ lies on the negative imaginary axis, so $\theta = 270°$ and $r = 4$; $-4j = 4(\cos 270° + j\sin 270°) = 4\operatorname{cis} 270°$.

≡ Note

Many graphing calculators can convert a complex number from polar form to rectangular form or from rectangular form to polar form. Since each make and model does this in a different way, consult the user's manual for your calculator to see how this is done.

Application

EXAMPLE 14.25

The voltage in an ac circuit is represented by the complex number $V = 28.4 - 65.7j$ V. Express this complex number in polar form.

Solution First, we find r. Here, $a = 28.4$ and $b = -65.7$, so

$$r = \sqrt{28.4^2 + (-65.7)^2} = \sqrt{806.56 + 4{,}316.49} = \sqrt{5{,}123.05} \approx 71.58$$

Next, we find the angle θ.

$$\tan\theta = \frac{-65.7}{28.4}$$

EXAMPLE 14.25 (Cont.)	

$$\theta = \tan^{-1}\left(\frac{-65.7}{28.4}\right)$$

$$\theta \approx -66.6°$$

Since V is in the fourth quadrant, $\theta \approx -66.6°$. If we want to express θ as an angle in the interval $[0, 360°)$, then $\theta = -66.6° + 360° = 293.4°$.

So, in polar form, voltage $V = 28.4 - 65.7j \approx 71.6\underline{/-66.6°} = 71.6\underline{/293.4°}$ V.

Application

EXAMPLE 14.26

Express current $i = 4.5\underline{/40°}$ A in rectangular form.

Solution Here, $r = 4.5$ and $\theta = 40°$, so $a = r\cos\theta = 4.5\cos40° \approx 3.45$ and $b = r\sin\theta = 4.5\sin40° \approx 2.89$. In rectangular form the current in amps, A, is $i = 4.5\underline{/40°} \approx 3.45 + 2.89j$.

Exercise Set 14.3

For each number in Exercises 1–16, locate the point in the complex plane and express the number in rectangular form.

1. $4(\cos30° + j\sin30°)$
2. $10(\cos45° + j\sin45°)$
3. $5(\cos135° + j\sin135°)$
4. $8(\cos305° + j\sin305°)$
5. $7\operatorname{cis}260°$
6. $3\operatorname{cis}340°$
7. $2\operatorname{cis}115°$
8. $5\operatorname{cis}285°$
9. $3\underline{/25°}$
10. $4\underline{/240°}$
11. $5\underline{/340°}$
12. $6\underline{/90°}$
13. $4.5\underline{/245°}$
14. $6.8\underline{/10°}$
15. $2.5\underline{/180°}$
16. $5.9\underline{/270°}$

For each number in Exercises 17–32, locate the point in the complex plane and express the number in polar form.

17. $5+5j$
18. $6+3j$
19. $4-8j$
20. $8-2j$
21. $-4+7j$
22. $-9+3j$
23. $-6-2j$
24. $-10-2j$
25. 6
26. $1.2+7.3j$
27. $4.2-6.3j$
28. $9j$
29. $-5.8+0.2j$
30. $-7j$
31. -2.7
32. $-4.7-1.1j$

In Exercises 33–40, perform the indicated operations graphically.

33. $(5+j)+(3+2j)$
34. $(3-4j)+(4-2j)$
35. $(-4+2j)+(2-8j)$
36. $(8-j)+(9+6j)$
37. $(-8+7j)+(-5-3j)$
38. $(4+2j)+(-3-9j)$
39. $(-5+3j)+(4-8j)$
40. $(-3+2j)+(-8-5j)$

Solve Exercises 41–47.

41. *Electronics* The current in amps, A, of an ac circuit is given by the complex number $4.7 - 6.5j$. Write this number in polar form.

42. *Electronics* The current in amps, A, of an ac circuit is given by the complex number $8.2 + 5.4j$. Write this number in polar form.

43. *Electronics* Convert the voltage given by $V = -110.4 + 46.1j$ V to polar form.

44. *Electronics* Convert the voltage given by $V = 108.5 - 57.6j$ V to polar form.

45. *Electronics* Express the current $i = 2.5\,\underline{/-50°}$ A in rectangular form.

46. *Electronics* Express the impedance $Z = 8.5\,\underline{/2\pi/3}\ \Omega$ in rectangular form.

 47. Write a computer program that will change a complex number from the rectangular form to polar form, or vice versa.

 In Your Words

48. **(a)** What is the absolute value or modulus of a complex number?
   (b) What is the amplitude or argument of a complex number?

49. Describe how to change a complex number in rectangular form to its equivalent complex number in polar form.

50. Describe how to change a complex number in polar form to its equivalent complex number in rectangular form.

 51. Explain how to use your calculator to change a complex number from rectangular form to polar form, or vice versa.

≡ 14.4
EXPONENTIAL FORM OF A COMPLEX NUMBER

There is yet another way in which complex numbers are often represented. It is called the **exponential form of a complex number**, because it involves exponents of the number e. (Remember from Section 13.2, that $e \approx 2.718281828\ldots$.) If $z = r\,\underline{/\theta}$ is a complex number, then we know that $z = r(\cos\theta + j\sin\theta)$.

Exponential Form of a Complex Number

The **exponential form** of a complex number uses **Euler's formula**, $e^{j\theta} = \cos\theta + j\sin\theta$, and states that

$$z = re^{j\theta}$$

where θ is in radians.

While θ can have any value, we will express answers with $0 \le \theta < 2\pi$.

When using the exponential form, you must remember that θ is in radians.

Caution

EXAMPLE 14.27

Write the complex number $6(\cos 180° + j\sin 180°)$ in exponential form.

Solution In this example, $r = 6$ and $\theta = 180°$. The exponential form requires that θ be expressed in radians: $180° = \pi$ rad and so $\theta = \pi$. Thus $6(\cos 180° + j\sin 180°) = 6e^{j\pi}$.

EXAMPLE 14.28

Write the complex number $8\,\underline{/225°}$ in exponential form.

Solution   In this example, $r = 8$ and $\theta = 225° = \frac{5\pi}{4} \approx 3.927\,\text{rad}$.

$$8\,\underline{/225°} = 8e^{\frac{5\pi}{4}j} = 8e^{5j\pi/4} \approx 8e^{3.927j}$$

All of the last three versions are correct. Because you will probably be using a calculator to convert from degrees to radians, the last version is the one that you will most likely use. But remember, $8e^{3.927j}$ is rounded off and therefore is the least accurate.

EXAMPLE 14.29

Express $-4+3j$ in exponential form.

Solution   This example is in the form $a+bj$. We must first determine r and θ. Now $r = \sqrt{a^2+b^2} = \sqrt{(-4)^2+3^2} = 5$ and $\tan\theta = \frac{3}{-4} = -0.75$. With a graphing calculator in radian mode, we can determine θ.

PRESS	DISPLAY
2nd | tan | (−) 0.75 + π ENTER	2.498091545

The last two steps were needed because the arctan of a negative number produces an angle in Quadrant IV. The point $-4+3j$ is in Quadrant II, so we added π to the original answer. Thus, we can see that $\theta \approx 2.4981$ and $-4+3j \approx 5e^{2.4981j}$.

EXAMPLE 14.30

Express $-5-8j$ in exponential form.

Solution $r = \sqrt{a^2+b^2} = \sqrt{(-5)^2+(-8)^2} = \sqrt{25+64} = \sqrt{89} \approx 9.4340$. $\tan\theta = \frac{-8}{-5} = 1.6$. Since $-5-8j$ is in Quadrant III, $\theta = \tan^{-1}1.6+\pi \approx 4.1538$. Thus, we see that $-5-8j = 9.4340e^{4.1538j}$.

EXAMPLE 14.31

Express $4.6e^{5.7j}$ in polar and rectangular form.

Solution   Here, $r = 4.6$ and $\theta = 5.7$, so we write the given expression in polar form as $4.6e^{5.7j} = 4.6(\cos 5.7 + j\sin 5.7)$. With a calculator in radian mode, we get

$$a = 4.6\cos 5.7 \approx 3.8397$$
$$b = 4.6\sin 5.7 \approx -2.5332$$

We have determined that $4.6e^{5.7j} = a+bj \approx 3.8397 - 2.5332j$.

Application

EXAMPLE 14.32

The voltage in an ac circuit is represented by the complex number $V = -85.6 + 72.3j$ V. Express this complex number in exponential form.

EXAMPLE 14.32 (Cont.)

Solution First, we find r.

$$r = \sqrt{(-85.6)^2 + 72.3^2} = \sqrt{7327.36 + 5227.29} \approx 112.0$$

Next, we find the angle θ.

$$\tan\theta = \frac{72.3}{-85.6}$$
$$\theta \approx -0.70137$$

Since V is in the second quadrant, $\theta = -0.70137 + \pi \approx 2.4402$. So, in exponential form, we find that voltage $V = -85.6 + 72.3j \approx 112e^{2.4402j}$ V.

Application

EXAMPLE 14.33

Express the current $I = 2.5\,\underline{/-50°}$ A in exponential and rectangular forms.

Solution Here, $r = 2.5$ and $\theta = -50° \approx -0.8727$ rad, so in exponential form we have $I = re^{j\theta} = 2.5e^{-0.8727j}$.

To convert the given number to rectangular form, we have $x = 2.5\cos(-50°) \approx 1.6070$ and $y = 2.5\sin(-50°) \approx -1.9151$. Thus, in rectangular form, the current is $I = 1.6070 - 1.9151j$ A.

Multiplying and Dividing in Exponential Form

One advantage of using the exponential form for complex numbers is that complex numbers written in exponential form obey the laws of exponents. There are three properties of exponents that are of interest. These three properties concern multiplication, division, and powers, and all of them use four basic rules introduced in Section 1.4.

$$b^m b^n = b^{m+n}$$
$$\frac{b^m}{b^n} = b^{m-n}$$
$$(b^m)^n = b^{mn}$$
$$(ab)^m = a^m b^m$$

If we have two complex numbers $z_1 = r_1 e^{j\theta_1}$ and $z_2 = r_2 e^{j\theta_2}$, we can then multiply, divide, or take powers of them using the preceding rules. For example,

$$z_1 z_2 = (r_1 e^{j\theta_1})(r_2 e^{j\theta_2}) = r_1 r_2 e^{j(\theta_1 + \theta_2)}$$

and $\qquad \dfrac{z_1}{z_2} = \dfrac{r_1 e^{j\theta_1}}{r_2 e^{j\theta_2}} = \dfrac{r_1}{r_2} e^{j(\theta_1 - \theta_2)}$

EXAMPLE 14.34

Multiply $7e^{4.2j}$ and $2e^{1.5j}$.

Solution $(7e^{4.2j})(2e^{1.5j}) = 7 \cdot 2e^{(4.2+1.5)j} = 14e^{5.7j}$

If we want to divide, then

$$\frac{z_1}{z_2} = \frac{r_1 e^{j\theta_1}}{r_2 e^{j\theta_2}} = \frac{r_1}{r_2} e^{j(\theta_1 - \theta_2)}$$

EXAMPLE 14.35

Divide $9e^{3.2j}$ by $2e^{4.3j}$.

Solution $\dfrac{9e^{3.2j}}{2e^{4.3j}} = \frac{9}{2} e^{(3.2-4.3)j} = 4.5e^{-1.1j}$

If you want to express the angle in the answer between 0 and 2π $(0 \le \theta < 2\pi)$, then let $\theta = 2\pi - 1.1$, or, using a calculator,

PRESS	DISPLAY
2 $\boxed{\times}$ $\boxed{\pi}$ $\boxed{-}$	6.2831853
1.1 $\boxed{=}$	5.1831853

Thus, $4.5e^{-1.1j}$ could be expressed as $4.5e^{5.2j}$.

The last property involves raising a complex number to a power and the properties $(b^m)^n = b^{mn}$ and $(ab)^m = a^m b^m$.

$$z^n = (re^{j\theta})^n = r^n e^{jn\theta}$$

EXAMPLE 14.36

Calculate $(4e^{2.3j})^5$.

Solution $(4e^{2.3j})^5 = 4^5 e^{5(2.3)j}$

$$= 1{,}024e^{11.5j}$$

$$\approx 1{,}024e^{5.2j}$$

Since 11.5 is greater than 2π, we subtracted 2π to get an angle of 5.2, which is between 0 and 2π.

Application

EXAMPLE 14.37

Given that $I = 5.8e^{-0.4363j}$ A and $Z = 8.5e^{1.047j}$ Ω, calculate $V = IZ$.

Solution Since $V = IZ$, we want to determine the product of the given numbers.

$$V = (5.8e^{-0.4363j})(8.5e^{1.047j}) = (5.8)(8.5)e^{-0.4363j + 1.047j}$$
$$= 49.3e^{0.6107j}$$

The voltage is $49.3e^{0.6107j}$ V.

Application

EXAMPLE 14.38

If $V_C = 78.3e^{-.3725j}$ V and $X_C = 87.0e^{-1.6500j}$ Ω, and $V_C = IX_C$, determine I.

EXAMPLE 14.38 (Cont.)

Solution   Since $V_C = IX_C$, then $I = \dfrac{V_C}{X_C}$, and so we need to divide.

$$I = \frac{V_C}{X_C} = \frac{78.3e^{-.3725j}}{87.0e^{-1.6500j}}$$
$$= \frac{78.3}{87.0}e^{-.3725j-(-1.6500j)}$$
$$= 0.9e^{1.2775j}$$

So, we have determined that $I = 0.9e^{1.2775j}$ A. ▪

Exercise Set 14.4

In Exercises 1–12, express each complex number in exponential form.

1. $3\left(\cos\frac{3\pi}{2} + j\sin\frac{3\pi}{2}\right)$
2. $7(\cos 1.4 + j\sin 1.4)$
3. $2(\cos 60° + j\sin 60°)$
4. $11(\cos 320° + j\sin 320°)$

5. $1.3(\cos 5.7 + j\sin 5.7)$
6. $9.5(\cos 2.1 + j\sin 2.1)$
7. $3.1(\cos 25° + j\sin 25°)$
8. $10.5(\cos 195° + j\sin 195°)$

9. $8 + 6j$
10. $12 - 5j$
11. $-9 + 12j$
12. $-8 - 12j$

In Exercises 13–16, express each number in rectangular and polar form.

13. $5e^{0.5j}$
14. $8e^{1.9j}$
15. $2.3e^{4.2j}$
16. $4.5e^{\frac{7\pi}{6}j}$

In Exercises 17–34, perform each of the indicated operations.

17. $2e^{3j} \cdot 6e^{2j}$
18. $3e^{j} \cdot 4e^{2j}$
19. $e^{1.3j} \cdot 2.4e^{4.6j}$
20. $1.5e^{4.1j} \cdot 0.2e^{1.7j}$
21. $7e^{4.3j} \cdot 4e^{5.7j}$

22. $3.6e^{5.4j} \cdot 2.5e^{6.1j}$
23. $8e^{3j} \div 2e^{j}$
24. $28e^{5j} \div 7e^{2j}$
25. $17e^{4.3j} \div 4e^{2.8j}$
26. $8.5e^{3.4j} \div 2e^{5.3j}$

27. $(3e^{2j})^4$
28. $(4e^{3j})^5$
29. $(2.5e^{1.5j})^4$
30. $(7.2e^{2.3j})^5$
31. $(4e^{6j})^{1/2}$

32. $(16e^{6j})^{1/4}$
33. $(6.25e^{4.2j})^{1/2}$
34. $(1.728e^{2.1j})^{1/3}$

Solve Exercises 35–44.

35. *Electricity*   The voltage in an ac circuit is represented by the complex number $V = 56.5 + 24.1j$ V. Express this complex number in exponential form.

36. *Electricity*   Express the current $I = 4.90 - 4.11j$ A in exponential form.

37. *Electricity*   Express the impedance $Z = 135 \times \underline{/-52.5°}$ Ω in exponential and rectangular forms.

38. *Electricity*   Express the capacitive reactance $X_C = 40.5\underline{/-\pi/2}$ Ω in exponential and rectangular forms.

39. *Electricity*   Given that $I = 12.5e^{-0.7256j}$ A and $Z = 6.4e^{1.4285j}$ Ω, find $V = IZ$.

40. *Electricity*   Given that $I = 4.24e^{0.5627j}$ A and $X_L = 28.5e^{-1.5708j}$ Ω, find $V_L = IX_L$.

41. *Electricity*   If $V = 115e^{-0.2145j}$ V and $Z = 2.5e^{0.5792j}$ Ω, find I, given that $V = IZ$.

42. *Electricity*   If $V_R = 35.1e^{1.3826j}$ V and $I = 0.78e^{1.3826j}$ A, find R, given that $V_R = IR$.

43. *Electricity*   If $V = 122.4e^{0.2551j}$ V and $I = 36e^{-0.8189j}$ A, find Z, given that $V = IZ$.

44. *Electricity*   If $V_L = IX_L$ with $V_L = 119.7e^{0.7254j}$ V and $I = 4.2e^{-0.1246j}$ A, find X_L.

In Your Words

45. Describe how to change a complex number in rectangular form to its equivalent complex number in exponential form.

46. Describe how to change a complex number in exponential form to its equivalent complex number in rectangular form.

≡ 14.5
OPERATIONS IN POLAR FORM; DeMOIVRE'S FORMULA

Multiplication, division, and powers of complex numbers are easily performed when the numbers are written in exponential form. They are just as easily performed when the numbers are in polar form. Remember the relationship between the exponential and polar forms.

$$re^{j\theta} = r(\cos\theta + j\sin\theta)$$

Multiplication

Again, we will let $z_1 = r_1 e^{j\theta_1}$ and $z_2 = r_2 e^{j\theta_2}$ be two complex numbers. We know that

$$z_1 z_2 = r_1 r_2 e^{j(\theta_1 + \theta_2)}$$

so in polar form, this would be

$$r_1(\cos\theta_1 + j\sin\theta_1) \cdot r_2(\cos\theta_2 + j\sin\theta_2) = r_1 r_2[\cos(\theta_1 + \theta_2) + j\sin(\theta_1 + \theta_2)]$$

(In Chapter 19, you will develop skills with trigonometry that will allow you to verify this trigonometric identity.)

Using the alternative way of writing the polar form of a complex number, we have

$$r_1 \underline{/\theta_1} \cdot r_2 \underline{/\theta_2} = r_1 r_2 \underline{/\theta_1 + \theta_2}$$

Notice that the angles do not have to be written in radians. Angles can be in either degrees or radians.

Product of Two Complex Numbers

The product of two complex numbers $z_1 = r_1 \operatorname{cis}\theta_1 = r_1 \underline{/\theta_1} = r_1 e^{j\theta_1}$ and $z_2 = r_2 \operatorname{cis}\theta_2 = r_2 \underline{/\theta_2} = r_2 e^{j\theta_2}$ is

$$z_1 z_2 = (r_1 \operatorname{cis}\theta_1)(r_2 \operatorname{cis}\theta_2) = r_1 r_2 \operatorname{cis}(\theta_1 + \theta_2)$$
$$= r_1 \underline{/\theta_1} \, r_2 \underline{/\theta_2} = r_1 r_2 \underline{/\theta_1 + \theta_2}$$
$$= r_1 r_2 e^{j(\theta_1 + \theta_2)}$$

EXAMPLE 14.39

Find each of the following products.
(a) $2(\cos 15° + j \sin 15°) \cdot 5(\cos 80° + j \sin 80°)$, (b) $6\underline{/30°} \cdot 3\underline{/60°}$,
(c) $4\underline{/2.3} \cdot 1.5\underline{/0.5}$, and (d) $(3e^{1.2j})(5e^{0.3j})$.

Solutions

(a) $2(\cos 15° + j \sin 15°) \cdot 5(\cos 80° + j \sin 80°)$
$$= 2 \cdot 5[\cos(15° + 80°) + j \sin(15° + 80°)]$$
$$= 10(\cos 95° + j \sin 95°)$$

(b) $6\underline{/30°} \cdot 3\underline{/60°} = 6 \cdot 3\underline{/30° + 60°} = 18\underline{/90°}$
$$= 18(\cos 90° + j \sin 90°)$$
$$= 18(0 + j \cdot 1) = 18j$$

(c) $4\underline{/2.3} \cdot 1.5\underline{/0.5} = 4(1.5)\underline{/2.3 + 0.5} = 6\underline{/2.8}$

(d) $(3e^{1.2j})(5e^{0.3j}) = 3(5)e^{(1.2+0.3)j} = 15e^{1.5j}$

≣ **Note**

When complex numbers are written in exponential form, $z = re^{j\theta}$, you *must* have θ in radians. Complex numbers written in polar form, $z = r \operatorname{cis} \theta$, may have θ in either degrees or radians.

Division

Division uses a similar process. Again, we will use the exponential form from Section 14.4 to show that

$$\frac{z_1}{z_2} = \frac{r_1 e^{j\theta_1}}{r_2 e^{j\theta_2}} = \frac{r_1}{r_2} e^{j(\theta_1 - \theta_2)}$$

The results for both the exponential and polar forms are summarized as follows.

Quotient of Two Complex Numbers

The quotient of two complex numbers $z_1 = r_1 \operatorname{cis} \theta_1 = r_1\underline{/\theta_1} = r_1 e^{j\theta_1}$ and $z_2 = r_2 \operatorname{cis} \theta_2 = r_2\underline{/\theta_2} = r_2 e^{j\theta_2}$ is

$$\frac{z_1}{z_2} = \frac{r_1(\cos\theta_1 + j\sin\theta_1)}{r_2(\cos\theta_2 + j\sin\theta_2)}$$
$$= \frac{r_1}{r_2}[\cos(\theta_1 - \theta_2) + j\sin(\theta_1 - \theta_2)]$$

or

$$= \frac{r_1\underline{/\theta_1}}{r_2\underline{/\theta_2}}$$
$$= \frac{r_1}{r_2}\underline{/\theta_1 - \theta_2}$$

or

$$= \frac{r_1 e^{j\theta_1}}{r_2 e^{j\theta_2}} = \frac{r_1}{r_2} e^{j(\theta_1 - \theta_2)}$$

EXAMPLE 14.40

Find each of the following quotients.

(a) $\dfrac{12(\cos 45° + j\sin 45°)}{2(\cos 15° + j\sin 15°)}$, (b) $\dfrac{15\underline{/135°}}{3\underline{/75°}}$, and (c) $4e^{2.1j} \div 8e^{1.7j}$.

Solutions

(a) $\dfrac{12(\cos 45° + j\sin 45°)}{2(\cos 15° + j\sin 15°)} = \dfrac{12}{2}[\cos(45° - 15°) + j\sin(45° - 15°)]$

$$= 6(\cos 30° + j\sin 30°)$$

(b) $\dfrac{15\underline{/135°}}{3\underline{/75°}} = \dfrac{15}{3}\underline{/135° - 75°} = 5\underline{/60°}$

(c) $4e^{2.1j} \div 8e^{1.7j} = \frac{4}{8}e^{(2.1-1.7)j} = \frac{1}{2}e^{0.4j}$

Application

EXAMPLE 14.41

Ohm's law for ac circuits is $V = IZ$. If $V = 15\underline{/39°}$ V and $Z = 8\underline{/26°}$ Ω, what is the current?

Solution Substituting the given values into the formula $V = IZ$, we obtain $15\underline{/39°} = I \cdot 8\underline{/26°}$. Solving for I produces

$$I = \frac{V}{Z}$$

$$= \frac{15\underline{/39°}}{8\underline{/26°}}$$

$$= \frac{15}{8}\underline{/39° - 26°}$$

$$= \frac{15}{8}\underline{/13°}$$

The current is $\frac{15}{8}\underline{/13°} = 1.875\underline{/13°}$ A.

Powers, Roots, and DeMoivre's Formula

Finding powers of complex numbers in polar form also uses the same process we developed for powers in the exponential form.

DeMoivre's Formula

For any complex number $z = r\operatorname{cis}\theta = r\underline{/\theta} = re^{j\theta}$

$$z^n = (re^{j\theta})^n = r^n e^{j\theta n} \quad \text{or} \quad [r(\cos\theta + j\sin\theta)]^n = r^n(\cos n\theta + j\sin n\theta)$$

This formula is known as **DeMoivre's formula**.

EXAMPLE 14.42

Use DeMoivre's formula to find each of the following powers. Convert your answers to rectangular form.
(a) $[3(\cos 30° + j \sin 30°)]^6$, **(b)** $(2\,\underline{/135°})^5$, **(c)** $(16\,\underline{/225°})^{1/4}$, and **(d)** $(2e^{0.4j})^7$.

Solutions

(a)
$$[3(\cos 30° + j \sin 30°)]^6 = 3^6(\cos 6 \cdot 30° + j \sin 6 \cdot 30°)$$
$$= 729(\cos 180° + j \sin 180°)$$
$$= -729$$

(b)
$$(2\,\underline{/135°})^5 = 2^5\,\underline{/5 \cdot 135°} = 32\,\underline{/675°}$$
$$a = 32\cos 675° \approx 22.6274$$
$$b = 32\sin 675° \approx -22.6274$$
so,
$$(2\,\underline{/135°})^5 = 22.6274 - 22.6274\,j$$

(c)
$$(16\,\underline{/225°})^{1/4} = 16^{1/4}\,\underline{/\tfrac{225°}{4}} = 2\,\underline{/56.25°}$$
$$a = 2\cos 56.25° \approx 1.1111$$
$$b = 2\sin 56.25° = 1.6629$$
so,
$$(16\,\underline{/225°})^{1/4} \approx 1.1111 + 1.6629\,j$$

There are three other 4th roots of $16\,\underline{/255°}$. We will soon see how to use DeMoivre's formula to find these other three roots.

(d)
$$(2e^{0.4j})^7 = (2)^7 e^{(0.4)7j} = 128e^{2.8j}$$
$$\approx -120.6045 + 42.8785\,j$$

DeMoivre's formula can be used to help find all of the roots of a complex number. For example, the equation $z^3 = -1$ has three roots. One of the roots is -1. What are the other two? DeMoivre's formula can be used to find those roots.

First, we will write -1 in polar form.

$$-1 = 1(\cos 180° + j \sin 180°)$$

Using DeMoivre's formula with $n = \frac{1}{3}$, we get

$$(-1)^{1/3} = 1^{1/3}\left(\cos\frac{180°}{3} + j\sin\frac{180°}{3}\right)$$
$$= 1(\cos 60° + j \sin 60°)$$
$$= 0.5000 + \frac{\sqrt{3}}{2}j$$
$$\approx 0.5000 + 0.8660\,j$$

We can see that this is not -1, the answer that we had before. A check would verify that $(0.5 + 0.8660\,j)^3 = -1$. So, this is a correct answer.

Why did we get a different answer? If you divide any number between $0°$ and $1{,}080°$ (or 0 and 6π) by 3, you find an angle between $0°$ and $360°$ (or between 0 and 2π). Now, $180°$, $540°$, and $900°$ all have the same terminal side. So, we could have written -1 as $1(\cos 540° + j \sin 540°)$ or as $1(\cos 900° + j \sin 900°)$. Let's find the cube root of each of these numbers.

$$^{1/3} = 1^{1/3}\left(\cos\frac{540°}{3} + j\sin\frac{540°}{3}\right)$$
$$= 1(\cos 180° + j\sin 180°)$$
$$= -1$$

$$[1(\cos 900° + j\sin 900°)]^{1/3} = 1^{1/3}\left(\cos\frac{900°}{3} + j\sin\frac{900°}{3}\right)$$
$$= 1(\cos 300° + j\sin 300°)$$
$$= 0.5000 - 0.8660j$$

We have found three different cube roots of -1. The first and last are conjugates of each other. We have also given an example of a process we can use to find all of the nth roots of a number, where n is a positive integer.

Roots of a Complex Number

If $z = r(\cos\theta + j\sin\theta)$, then the nth roots of z are given by the formula

$$w_k = \sqrt[n]{r}\left[\cos\left(\frac{\theta}{n} + \frac{360° \cdot k}{n}\right) + j\sin\left(\frac{\theta}{n} + \frac{360° \cdot k}{n}\right)\right] \qquad (*)$$

where $k = 0, 1, 2, \ldots, n-1$.

If θ is in radians, then substitute 2π for $360°$.

EXAMPLE 14.43

Find the five fifth roots of $32j$.

Solution The five fifth roots will be called w_0, w_1, w_2, w_3, and w_4. Using $32j = 32(\cos 90° + j\sin 90°)$ and applying formula $(*)$, we obtain the following.

$$w_0 = \sqrt[5]{32}\left[\cos\left(\frac{90°}{5} + \frac{360° \cdot 0}{5}\right) + j\sin\left(\frac{90°}{5} + \frac{360° \cdot 0}{5}\right)\right]$$
$$= 2(\cos 18° + j\sin 18°)$$
$$\approx 1.9021 + 0.6180j$$

$$w_1 = \sqrt[5]{32}\left[\cos\left(\frac{90°}{5} + \frac{360° \cdot 1}{5}\right) + j\sin\left(\frac{90°}{5} + \frac{360° \cdot 1}{5}\right)\right]$$
$$= 2(\cos 90° + j\sin 90°)$$
$$= 2j$$

EXAMPLE 14.43 (Cont.)

$$w_2 = \sqrt[5]{32}\left[\cos\left(\frac{90°}{5} + \frac{360° \cdot 2}{5}\right) + j\sin\left(\frac{90°}{5} + \frac{360° \cdot 2}{5}\right)\right]$$
$$= 2(\cos 162° + j\sin 162°)$$
$$\approx -1.9021 + 0.6180 j$$

$$w_3 = \sqrt[5]{32}\left[\cos\left(\frac{90°}{5} + \frac{360° \cdot 3}{5}\right) + j\sin\left(\frac{90°}{5} + \frac{360° \cdot 3}{5}\right)\right]$$
$$= 2(\cos 234° + j\sin 234°)$$
$$\approx -1.1756 - 1.6180 j$$

$$w_4 = \sqrt[5]{32}\left[\cos\left(\frac{90°}{5} + \frac{360° \cdot 4}{5}\right) + j\sin\left(\frac{90°}{5} + \frac{360° \cdot 4}{5}\right)\right]$$
$$= 2(\cos 306° + j\sin 306°)$$
$$\approx 1.1756 - 1.6180 j$$

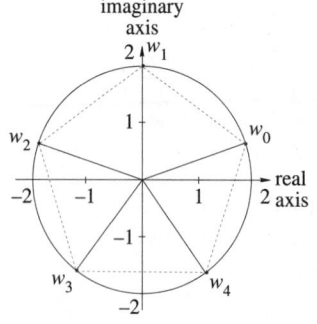

FIGURE 14.11

The five roots are shown in Figure 14.11. Notice the symmetry of these five roots around the circle. Any time you graph the n roots of a number, they should be equally spaced around a circle with the center at the origin in the complex plane. ▪

EXAMPLE 14.44

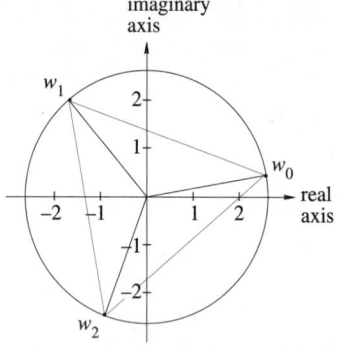

FIGURE 14.12

Find the three cube roots of $2\sqrt{11} + 10j$.

Solution We first write this number in polar form. $r = \sqrt{(2\sqrt{11})^2 + 10^2} = \sqrt{144} = 12$ and $\tan\theta = \dfrac{10}{2\sqrt{11}} \approx 1.5075567$, so $\theta \approx 56.4°$. We will find the three cube roots w_0, w_1, and w_2.

$$w_0 = \sqrt[3]{12}\left[\cos\left(\frac{56.4°}{3} + \frac{360° \cdot 0}{3}\right) + j\sin\left(\frac{56.4°}{3} + \frac{360° \cdot 0}{3}\right)\right]$$
$$= \sqrt[3]{12}(\cos 18.8° + j\sin 18.8°)$$
$$\approx 2.1673 + 0.7378 j$$

$$w_1 = \sqrt[3]{12}\left[\cos\left(\frac{56.4°}{3} + \frac{360° \cdot 1}{3}\right) + j\sin\left(\frac{56.4°}{3} + \frac{360° \cdot 1}{3}\right)\right]$$
$$= \sqrt[3]{12}(\cos 138.8° + j\sin 138.8°)$$
$$\approx -1.7230 + 1.5076 j$$

$$w_2 = \sqrt[3]{12}\left[\cos\left(\frac{56.4°}{3} + \frac{360° \cdot 2}{3}\right) + j\sin\left(\frac{56.4°}{3} + \frac{360° \cdot 2}{3}\right)\right]$$
$$= \sqrt[3]{12}(\cos 258.8° + j\sin 258.8°)$$
$$\approx -0.4447 - 2.2459 j$$

Geometrically, these three points are shown on the graph in Figure 14.12.

EXAMPLE 14.45

Find the three cube roots of $(-8)^5$.

Solution We first write -8 in polar form. Since $r = 8$ and $\theta = 180°$, we see that $-8 = 8$ cis $180°$. Using DeMoivre's formula to raise this to the 5th power, we obtain

$$(-8)^5 = 8^5 \operatorname{cis}(5 \cdot 180°)$$
$$= 32,768 \operatorname{cis} 900°$$
$$= 32,768 \operatorname{cis} 180°$$

We will find the three cube roots, w_0, w_1, and w_2, of 32,768 cis 180°.

$$w_0 = \sqrt[3]{32,768} \left[\cos \left(\frac{180°}{3} + \frac{360° \cdot 0}{3} \right) + j \sin \left(\frac{180°}{3} + \frac{360° \cdot 0}{3} \right) \right]$$
$$= 32 \left(\cos 60° + j \sin 60° \right)$$
$$\approx 16 + 27.71 j$$

$$w_1 = \sqrt[3]{32,768} \left[\cos \left(\frac{180°}{3} + \frac{360° \cdot 1}{3} \right) + j \sin \left(\frac{180°}{3} + \frac{360° \cdot 1}{3} \right) \right]$$
$$= 32 \left(\cos 180° + j \sin 180° \right)$$
$$= -32$$

$$w_2 = \sqrt[3]{32,768} \left[\cos \left(\frac{180°}{3} + \frac{360° \cdot 2}{3} \right) + j \sin \left(\frac{180°}{3} + \frac{360° \cdot 2}{3} \right) \right]$$
$$= 32 \left(\cos 300° + j \sin 300° \right)$$
$$\approx 16 - 27.71 j$$

Thus, the three cube roots of $(-8)^5$ are $16 + 27.71 j$, -32, and $16 - 27.71 j$.

In the caution immediately following Example 12.8, we commented that pressing $(-8) \wedge (5/3)$ on a TI-85 produces an answer of $(16, -27.7128129211)$. This is the value of w_2 that we found above. On the other hand, if you pressed $(-8) \wedge 5 \wedge (1/3)$ a TI-85 produces -32, which is our value for w_1.

Exercise Set 14.5

In Exercises 1–20, perform the indicated operations and give the answers in polar form.

1. $3(\cos 46° + j \sin 46°) \cdot 5(\cos 23° + j \sin 23°)$

2. $4(\cos 135° + j \sin 135°) \cdot 5(\cos 63° + j \sin 63°)$

3. $2.5(\cos 1.43 + j \sin 1.43) \cdot 4(\cos 2.67 + j \sin 2.67)$

4. $6.4(\cos 0.25 + j \sin 0.25) \cdot 3.5(\cos 1.1 + j \sin 1.1)$

5. $\dfrac{8(\cos 85° + j \sin 85°)}{2(\cos 25° + j \sin 25°)}$

6. $\dfrac{6(\cos 273° + j \sin 273°)}{3(\cos 114° + j \sin 114°)}$

7. $\dfrac{9 \,/\, 137°}{2 \,/\, 26°}$

8. $\dfrac{18 \,/\, 3.52}{5 \,/\, 2.14}$

9. $(3 \,/\, 2.7)(4 \,/\, 5.3)$

10. $[3(\cos 20° + j \sin 20°)]^4$

11. $[5(\cos 84° + j \sin 84°)]^6$

12. $[2.5(\cos 118° + j \sin 118°)]^3$

13. $[10.4(\cos 3.42 + j \sin 3.42)]^3$

14. $(2 \underline{/1.38})^5$

15. $(3.4 \underline{/5.3})^4$

16. $(4.41 \underline{/124°})^{1/2}$

17. $(4e^{2.1j})(3e^{1.7j})$

18. $6e^{1.5j} \div 4e^{0.9j}$

19. $(0.5e^{0.3j})^3$

20. $(0.0625e^{4.2j})^{1/2}$

In Exercises 21–28, use DeMoivre's formula to find the indicated roots. Give the answers in rectangular form.

21. cube roots of 1

22. cube roots of j

23. cube roots of $-8j$

24. fourth roots of -16

25. fourth roots of $-16j$

26. fourth roots of $1 - j$

27. fifth roots of $1 + j$

28. sixth roots of $-1 - j$

In Exercises 29–32, solve the given equations. Express your answers in rectangular form.

29. $x^3 = -j$

30. $x^3 = 125j$

31. $x^6 - 64j = 0$

32. $x^6 - 1 = j$

Solve Exercises 33–43.

33. *Electricity*   Ohm's law for alternating current states that for a current with voltage V, current I, and impedance Z, $V = IZ$. If the current is $12 \underline{/-23°}$ and the impedance is $9 \underline{/42°}$, what is the voltage?

34. *Electricity*   If an ac circuit has a voltage of $20 \underline{/30°}$ and a current of $5 \underline{/40°}$, what is the impedance?

35. *Electronics*   The voltage divider rule in ac circuits is

$$V_x = \frac{Z_x E}{Z_T}$$

where V_x is the voltage across one or more elements in series that have total impedance Z_x, E is the total voltage appearing across the series circuit, and Z_T is the total impedance of the series circuit. Find the voltage across the element V_x given that $Z_x = 4 \underline{/-90°}$ Ω, $E = 100 \underline{/0°}$ V, and $Z_T = 4 \underline{/-90°} + 3 \underline{/0°}$ Ω

36. *Electronics*   Use the voltage divider rule in ac circuits to find the voltage across the element V_x given that $Z_x = 6 \underline{/0°}$ Ω, $E = 50 \underline{/30°}$ V, and $Z_T = 6 \underline{/0°} + 9 \underline{/90°} + 17 \underline{/-90°}$ Ω

37. *Electronics*   The impedance in a series RLC circuit is given by

$$Z = \frac{(1+j)^2(1-j)^2}{(3-4j)^2}$$

Evaluate Z by changing the expression to polar form.

38. *Electronics*   The admittance Y in an ac circuit is measured in siemens (S) and is given by $Y = Z^{-1}$,

where Z is the impedance. If $Z = 12 - 5j$, find Y by using

(a) polar form and DeMoivre's formula. (Give the answer in polar form.)

(b) rectangular form.

39. *Electronics*   The admittance, Y, in siemens (S) of an ac circuit is given by $Y = Z^{-1}$, where Z is the impedance. If $Z = 4.68 \underline{/20.56°}$ Ω,

(a) find Y by using DeMoivre's formula. (Give the answer in polar form.)

(b) convert your answer in (a) to rectangular form.

40. *Electronics*   The admittance Y in an ac circuit is given by $Y = Z^{-1}$, where Z is the impedance. If $Y = 0.087 - 0.034j$ S,

(a) find Z by using DeMoivre's formula. (Give the answer in polar form.)

(b) convert your answer in (a) to rectangular form.

 41. Modify your program for multiplying complex numbers to also work when the numbers are entered in polar form. (See Exercise Set 14.2, Exercise 64.)

 42. Modify your program for dividing two complex numbers to also work when the numbers are entered in polar form. (See Exercise Set 14.2, Exercise 65.)

 43. Write a computer program to take the power, or find all the roots, of a complex number by using DeMoivre's formula.

In Your Words

44. Without looking in the text, describe how to multiply or divide two complex numbers in **(a)** exponential form and **(b)** polar form.

45. Explain how to use DeMoivre's formula to find all the n nth roots of the complex number z.

≡ 14.6
COMPLEX NUMBERS IN AC CIRCUITS

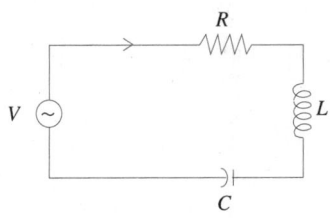

FIGURE 14.13

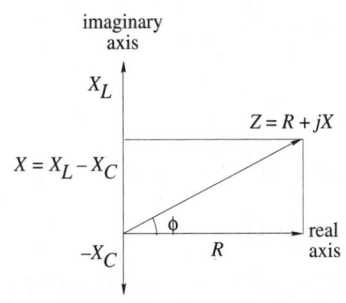

FIGURE 14.14

In direct current (dc) circuits, the basic relation between voltage and current is given by Ohm's law. If V is the voltage across a resistance R, and I is the current flowing through the resistor, then Ohm's law states that

$$V = IR$$

In alternating current (ac) circuits, there is a very similar equation. If V is the voltage across an impedance Z, and I is the current flowing through the impedance, then we have the relationship

$$V = IZ$$

The main difference in these equations is that the dc circuits are expressed as real numbers. Using complex numbers with the ac equations allows them to take the same simple form as the dc equations, except that all quantities are complex numbers. We will examine some of those relationships in this section.

Impedance is the opposition to alternating current produced by a resistance R, an inductance L, a capacitance C, or any combination of these. When a sinusoidal voltage V of a given frequency f is applied to a circuit of constant resistance R, constant capacitance C, and constant inductance L, the circuit, like the one in Figure 14.13, is an RLC circuit.

In Figure 14.13, the opposition to the current produced by the inductance is called the **inductive reactance** X_L and the opposition to the current produced by the capacitance is the **capacitive reactance** X_C. The **reactance** X is a measure of how much the capacitance and inductance retard the flow of current in an ac circuit and is the difference between the capacitive reactance and the inductive reactance. Thus,

$$X = X_L - X_C$$

The impedance, resistance, and reactance can be represented by a **vector impedance triangle**. The angle ϕ between Z and R is the **phase angle**. In the complex plane, we can represent the resistance along the real axis and the reactance along the imaginary axis as shown in Figure 14.14 Thus,

$$Z = R + jX = R + j(X_L - X_C)$$
$$= Z \underline{/\phi}$$

From our study of complex numbers, you can see that the **magnitude** of the . impedance is $|Z| = \sqrt{R^2 + X^2}$ and $\tan \phi = \dfrac{X}{R}$.

Application

A circuit has a resistance of $8\,\Omega$ in series with a reactance of $5\,\Omega$. What are the magnitude of the impedance and its phase angle?

Solution Using a vector impedance triangle, we can see that the impedance can be represented by

$$Z = 8 + 5j$$

The magnitude of the impedance is

$$|Z| = \sqrt{R^2 + X^2} = \sqrt{8^2 + 5^2} = \sqrt{89} \approx 9.43\,\Omega$$

The phase angle ϕ is given by

$$\phi = \arctan \tfrac{5}{8} = 32° \text{ or } 0.56\,\text{rad}$$

Application

In the RLC circuit in Figure 14.14, $R = 60\,\Omega$, $X_L = 75\,\Omega$, $X_C = 30\,\Omega$, and $I = 2.25\,\text{A}$. Find **(a)** the magnitude and phase angle of Z, and **(b)** the voltage across the circuit.

Solutions

(a) The reactance $X = X_L - X_C = 75 - 30 = 45\,\Omega$. The impedance $Z = 60 + 45j$, so the magnitude of the impedance is

$$|Z| = \sqrt{R^2 + X^2} = \sqrt{60^2 + 45^2} = 75\,\Omega$$

The phase angle ϕ is given by

$$\phi = \arctan \tfrac{45}{60} \approx 36.87° \text{ or } 0.64\,\text{rad}$$

(b) Since the current is 2.25 A and the impedance is $75\,\Omega$, the voltage $V = IZ = (2.25)(75) = 168.75\,\text{V}$, or approximately 169 V.

An alternating current is produced by a coil of wire rotating through a magnetic field. If the angular velocity of the wire is ω, the capacitive reactance X_C and the inductive reactance X_L are given by the formulas

$$X_C = \frac{1}{\omega C} \qquad \text{and} \qquad X_L = \omega L$$

Since $\omega = 2\pi f$, where f is the frequency of the current, these are also expressed as

$$X_C = \frac{1}{2\pi f C} \qquad \text{and} \qquad X_L = 2\pi f L$$

From these formulas you can see that if C, L, and either ω or f are known, the reactance of the circuit may be determined.

Application

EXAMPLE 14.48

If $R = 40\,\Omega$, $L = 0.1\,\text{H}$, $C = 50\,\mu\text{F}$, and $f = 60\,\text{Hz}$, determine the impedance and phase difference between the current and voltage.

Solution   Converting $C = 50\,\mu\text{F}$ to farads produces $C = 50 \times 10^{-6}\,\text{F}$.

$$X_C = \frac{1}{2\pi f C} = \frac{1}{2\pi(60)(50 \times 10^{-6})} \approx 53\,\Omega$$
$$X_L = 2\pi f L = 2\pi(60)(0.1) \approx 38\,\Omega$$
$$Z = R + jX = R + j(X_L - X_C) = 40 + (38 - 53)j = 40 - 15j$$
$$|Z| = \sqrt{R^2 + X^2} = \sqrt{40^2 + (-15)^2} \approx 42.7\,\Omega$$
$$\phi = \arctan \tfrac{-15}{40} \approx -20.6°$$

So, the impedance is $42.7\,\Omega$ and the phase difference is $-20.6°$.

In the study of dc circuits, you learn that if several resistors are connected in series, their total resistance is the sum of the individual resistances. Thus, if two resistors R_1 and R_2 are connected in series, their total resistance, R equals $R_1 + R_2$. If R_1, R_2, and R_3 are connected in series, then $R = R_1 + R_2 + R_3$.

If the resistors are connected in parallel, the relationship is more complicated. The total resistance is the reciprocal of the sum of the reciprocals of the resistances. What this means is that if two resistors R_1 and R_2 are connected in parallel, then the total resistance $R = \dfrac{1}{\dfrac{1}{R_1} + \dfrac{1}{R_2}}$. This can be simplified to $R = \dfrac{R_1 R_2}{R_1 + R_2}$. If three resistors, R_1, R_2, and R_3 are connected in parallel, then $R = \dfrac{1}{\dfrac{1}{R_1} + \dfrac{1}{R_2} + \dfrac{1}{R_3}}$. This can be rewritten as

$$R = \frac{R_1 R_2 R_3}{R_1 R_2 + R_1 R_3 + R_2 R_3}$$

Corresponding formulas hold for complex impedances. Thus, if two impedances Z_1 and Z_2 are connected in series, the total impedance is $Z = Z_1 + Z_2$. If there are three impedances in series, Z_1, Z_2, and Z_3, then $Z = Z_1 + Z_2 + Z_3$.

If complex impedances are connected in parallel, we then have the more complicated formulas. If two impedances are connected in parallel, then the total impedance $Z = \dfrac{1}{\dfrac{1}{Z_1} + \dfrac{1}{Z_2}} = \dfrac{Z_1 Z_2}{Z_1 + Z_2}$. If three impedances are connected in parallel, then the total impedance is $Z = \dfrac{1}{\dfrac{1}{Z_1} + \dfrac{1}{Z_2} + \dfrac{1}{Z_3}} = \dfrac{Z_1 Z_2 Z_3}{Z_1 Z_2 + Z_1 Z_3 + Z_2 Z_3}$.

Application

EXAMPLE 14.49

If $Z_1 = 2 + 3j\,\Omega$ and $Z_2 = 1 - 6j\,\Omega$, what is the total impedance if these are connected **(a)** in series and **(b)** in parallel?

Solutions

(a) If they are connected in series,

$$\begin{aligned} Z &= Z_1 + Z_2 \\ &= (2 + 3j) + (1 - 6j) \\ &= 3 - 3j\,\Omega \end{aligned}$$

(b) If they are connected in parallel,

$$\begin{aligned} Z &= \frac{1}{\dfrac{1}{Z_1} + \dfrac{1}{Z_2}} = \frac{Z_1 Z_2}{Z_1 + Z_2} \\[2mm] &= \frac{(2 + 3j)(1 - 6j)}{(2 + 3j) + (1 - 6j)} \\[2mm] &= \frac{20 - 9j}{3 - 3j} \qquad \text{multiply by } \frac{3 + 3j}{3 + 3j} \\[2mm] &= \frac{87 + 33j}{18} \\[2mm] &= \frac{87}{18} + \frac{33}{18}j\,\Omega \\[2mm] &= \frac{29}{6} + \frac{11}{6}j\,\Omega \end{aligned}$$

Application

EXAMPLE 14.50

If $Z_1 = 3.16\,\underline{/18.4^\circ}\ \Omega$ and $Z_2 = 4.47\,\underline{/63.4^\circ}\ \Omega$, what is the total impedance if these are connected **(a)** in series and **(b)** in parallel?

Solutions

(a) If they are connected in series,

$$Z = Z_1 + Z_2$$

Since we cannot add complex numbers in polar form, we need to change these to rectangular form.

$$\begin{aligned} Z_1 &= 3.16(\cos 18.4^\circ + j\sin 18.4^\circ) \approx 2.998 + 0.997j \\ Z_2 &= 4.47(\cos 63.4^\circ + j\sin 63.4^\circ) \approx 2.001 + 3.997j \\ Z &= (2.998 + 0.997j) + (2.001 + 3.997j) \\ &= 4.999 + 4.994j\,\Omega \\ &\approx 7.07\,\underline{/45.0^\circ}\ \Omega \end{aligned}$$

EXAMPLE 14.50 (Cont.)

(b) If they are in parallel,

$$Z = \frac{Z_1 Z_2}{Z_1 + Z_2}$$
$$= \frac{(3.16 \underline{/18.4°})(4.47 \underline{/63.4°})}{7.07 \underline{/45.0°}}$$

Using our knowledge of multiplying and dividing complex numbers in polar form, we get

$$Z = \frac{(3.16)(4.47)}{7.07} \underline{/18.4° + 63.4° - 45.0°}$$
$$= 2.0 \underline{/36.8°}\ \Omega$$

As you can see, some problems are easier to work if you use the rectangular form and some are easier if you use the polar form. They would all be easier to work if you had a computer program to solve these problems, such as the one in Exercise Set 14.6, Exercise 35.

Exercise Set 14.6

In Exercises 1–12, find the total impedance if the given impedances are connected (a) in series and (b) in parallel.

1. $Z_1 = 2 + 3j, Z_2 = 1 - 5j$
2. $Z_1 = 4 - 7j, Z_2 = -3 + 4j$
3. $Z_1 = 1 - j, Z_2 = 3j$
4. $Z_1 = 2 + j, Z_2 = 4 - 3j$
5. $Z_1 = 2.19 \underline{/18.4°}, Z_2 = 5.16 \underline{/67.3°}$
6. $Z_1 = 2\sqrt{3} \underline{/30°}, Z_2 = 2 \underline{/120°}$
7. $Z_1 = 3\sqrt{5} \underline{/\frac{\pi}{7}}, Z_2 = 1.5 \underline{/0.45}$
8. $Z_1 = 2.57 \underline{/0.25}, Z_2 = 1.63 \underline{/1.38}$
9. $Z_1 = 4 + 3j, Z_2 = 3 - 2j, Z_3 = 5 + 4j$
10. $Z_1 = 3 - 4j, Z_2 = 1 + 5j, Z_3 = -2j$
11. $Z_1 = 1.64 \underline{/38.2°}, Z_2 = 2.35 \underline{/43.7°}, Z_3 = 4.67 \underline{/-39.6°}$
12. $Z_1 = 0.15 \underline{/0.95}, Z_2 = 2.17 \underline{/1.39}, Z_3 = 1.10 \underline{/0.40}$

In Exercises 13–18, use the formula $V = IZ$ to determine the missing unit.

13. $I = 4 - 3j$ A, $Z = 8 - 15j\ \Omega$
14. $V = 5 + 5j$ V, $I = 4 + 3j$ A
15. $Z = 1 - j\ \Omega, I = 1 + j$ A
16. $V = 3 + 4j$ V, $Z = 5 - 12j\ \Omega$
17. $V = 7 \underline{/36.3°}$ V, $I = 2.5 \underline{/12.6°}$ A
18. $V = 3 \underline{/1.37}$ V, $Z = 4 \underline{/0.16}\ \Omega$

In Exercises 19–24, determine the inductive reactance, capacitive reactance, impedance, and phase difference between the current and voltage.

19. $R = 38\ \Omega, L = 0.2$ H, $C = 40\ \mu$F, and $f = 60$ Hz
20. $R = 35\ \Omega, L = 0.15$ H, $C = 80\ \mu$F, and $f = 60$ Hz
21. $R = 20\ \Omega, L = 0.4$ H, $C = 60\ \mu$F, and $f = 60$ Hz
22. $R = 12\ \Omega, L = 0.3$ H, $C = 250\ \mu$F, and $\omega = 80$ rad/s
23. $R = 28\ \Omega, L = 0.25$ H, $C = 200\ \mu$F, and $\omega = 50$ rad/s
24. $R = 2\ 000\ \Omega, L = 3.0$ H, $C = 0.5\ \mu$F, and $\omega = 1\ 000$ rad/s

In the *RLC* circuits in Exercises 25–28, find (a) the magnitude and phase angle of *Z* and (b) the voltage across the circuit.

25. $R = 75\,\Omega$, $X_L = 60\,\Omega$, $X_C = 40\,\Omega$, and $I = 3.50$ A

26. $R = 40\,\Omega$, $X_L = 30\,\Omega$, $X_C = 60\,\Omega$, and $I = 7.50$ A

27. $R = 3.0\,\Omega$, $X_L = 6.0\,\Omega$, $X_C = 5.0\,\Omega$, and $I = 2.85$ A

28. $R = 12.0\,\Omega$, $X_L = 11.4\,\Omega$, $X_C = 2.4\,\Omega$, and $I = 0.60$ A

Solve Exercises 29–38.

29. *Electronics*   Figure 14.15 indicates part of an electrical circuit. Kirchhoff's law implies that $I_2 = I_1 + I_3$. If $I_1 = 7 + 2j$ A and $I_2 = 9 - 7j$ A, find I_3.

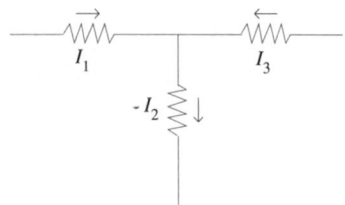

FIGURE 14.15

30. *Electronics*   **Resonance** in an *RLC* circuit occurs when $X_L = X_C$. Under these conditions, the current and voltage are in phase. Resonance is required for the tuning of radio and television receivers. If $\omega = 100\,\text{rad/s}$ and $L = 0.500$ H, what is the value of C if the system is in resonance?

31. *Electronics*   What is the frequency in Hertz for a circuit in resonance if $L = 2.5$ H and $C = 20.0\,\mu\text{F}$?

32. *Electronics*   The **admittance** Y is the reciprocal of the impedance. If $Z = 3 - 2j\,\Omega$, what is the admittance?

33. *Electronics*   If the admittance is $4 + 3j\,\Omega$, what is the impedance?

34. *Electronics*   The **susceptance** B of an ac circuit of reactance X and impedance Z is defined as $B = \dfrac{X}{Z^2}j$. If the reactance is $4\,\Omega$ and $Z = 8 + 7j$, what is the susceptance?

    35. Write a computer program that will compute the total impedance of a group of impedances in an ac circuit that are connected either in series or in parallel. The program should allow the impedance to be entered in either rectangular or polar form.

    36. Using the computer program from Exercise 35, rework Exercises 1–12.

In Your Words

37. Write an electrical application that requires you to use complex numbers and some of the calculation techniques of this chapter. Give your problem to a classmate and see if he or she understands and can solve your problem. Rewrite the problem as necessary to remove any difficulties encountered by your classmate.

38. Suppose that you are applying for a job with an electronics company. One question on the application form asks you to explain how to perform a specific type of calculation involving complex numbers. It then asks you to give a specific example that uses this type of calculation. Write your answer using complete sentences.

☰ CHAPTER 14 REVIEW

Important Terms and Concepts

Changing complex numbers from
    Polar form to rectangular form
    Rectangular form to polar form
Complex numbers
    Absolute value
    Addition
    Conjugate
    Division
    Exponential form
    Imaginary part

Multiplication
Polar form
Power
Real part
Rectangular part
Roots
Subtraction
Complex plane
DeMoivre's formula
Imaginary unit

Review Exercises

In Exercises 1–6, simplify each number in terms of *j*.

1. $\sqrt{-49}$

2. $\sqrt{-36}$

3. $\sqrt{-54}$

4. $(2j^3)^3$

5. $\sqrt{-2}\sqrt{-18}$

6. $\sqrt{-9}\sqrt{-27}$

In Exercises 7–14, perform the indicated operation. Express each answer in the form *a + bj*.

7. $(2-j)+(7-2j)$

8. $(9+j)(4+7j)$

9. $(5+j)-(6-3j)$

10. $\dfrac{1}{11-j}$

11. $(6+2j)(-5+3j)$

12. $\dfrac{2-5j}{6+3j}$

13. $(4-3j)(4+3j)$

14. $\dfrac{-4}{\sqrt{3}+2j}$

In Exercises 15–20, graph each complex number and change each number from rectangular form to polar form, or vice versa.

15. $9-6j$

16. $-8+2j$

17. $4-4j$

18. $4\operatorname{cis}60°$

19. $6.5\underline{/2.3}$

20. $10\underline{/20°}$

In Exercises 21–28, perform the indicated operation and express each answer in rectangular form.

21. $(2\operatorname{cis}30°)(5\operatorname{cis}150°)$

22. $\dfrac{3\operatorname{cis}20°}{6\operatorname{cis}80°}$

23. $\left(3\operatorname{cis}\frac{5\pi}{4}\right)^{14}$

24. $(324\operatorname{cis}225°)^{1/5}$

25. $\left(3\underline{/\frac{\pi}{4}}\right)\left(9\underline{/\frac{2\pi}{3}}\right)$

26. $44\underline{/125°}\div4\underline{/97°}$

27. $\left(2\underline{/\frac{\pi}{6}}\right)^{12}$

28. $(2048\underline{/330°})^{1/11}$

For Exercises 29–32, find all roots and express in rectangular form.

29. $\sqrt[3]{-j}$

30. $\sqrt[4]{16}$

31. $\sqrt{16\operatorname{cis}120°}$

32. $\sqrt[3]{27j}$

In Exercises 33–36, change each number to the exponential form and perform the indicated operation.

33. $(3+2j)(5-j)$

34. $\dfrac{4-7j}{3+j}$

35. $(5+3j)^5$

36. $(-7-2j)^{1/3}$

Solve Exercises 37–40.

37. *Physics*   A force vector in the complex plane is given by $7.3 - 1.4j$. What is the magnitude and direction (argument) of this vector?

38. *Electronics*   Given an RLC circuit with $R = 3.0\,\Omega$, $X_L = 7.0\,\Omega$, $X_C = 4.5\,\Omega$, and $I = 1.5\,A$, **(a)** what is the magnitude and phase angle of Z, and **(b)** what is the voltage across the circuit?

39. *Electronics*   What are the inductive reactance, capacitive reactance, impedance, and phase difference between the current and voltage, if $R = 55\,\Omega$, $L = 0.3$ H, $C = 50\,\mu$F, and $f = 60$ Hz?

40. *Electronics*   If $Z_1 = 3+5j$ and $Z_2 = 6-3j$, then what is Z, if Z_1 and Z_2 are connected **(a)** in series and **(b)** in parallel?

☰ CHAPTER 14 TEST

1. Write $\sqrt{-80}$ in terms of j.

2. Change $7 - 2j$ from rectangular form to polar form.

3. Change $8\operatorname{cis}150°$ from polar form to rectangular form.

In Exercises 4–11, perform the indicated operation. Express each answer in the form *a + bj*.

4. $(5+2j)+(8-6j)$

5. $(-5+2j)-(8-6j)$

6. $(2+3j)(4-5j)$

7. $\dfrac{6+5j}{-3-4j}$

8. $(7\operatorname{cis}75°)(2\operatorname{cis}105°)$

9. $\dfrac{4\operatorname{cis}115°}{3\operatorname{cis}25°}$

10. $\left(9\operatorname{cis}\frac{2\pi}{3}\right)^{5/2}$

11. $(27\underline{/129°})^{1/3}$

Solve Exercises 12–16.

12. Find all four roots of $\sqrt[4]{j}$.

13. If $Z_1 = 4+2j$ and $Z_2 = 5-3j$, what is Z, if Z_1 and Z_2 are connected in **(a)** series and **(b)** parallel?

14. The admittance, Y, in an ac circuit is given by $Y = Z^{-1}$, where Z is the impedance. If $Z = 9 - 3j\,\Omega$, what is the admittance?

15. The voltage in a certain ac circuit is $5+2j$ V and the current is $3 - 4j$ A. Use Ohm's law, $V = IZ$, to determine the impedance Z.

16. Two ac circuits with impedances Z_1 and Z_2 have total impedance Z, where

$Z = Z_1 + Z_2$ if they are connected in series, and

$Z = \dfrac{1}{\dfrac{1}{Z_1} + \dfrac{1}{Z_2}}$ if they are connected in parallel

If $Z_1 = 9 + 3j\,\Omega$ and $Z_2 = 7 - 2j\,\Omega$, determine the total impedance if they are connected **(a)** in series and **(b)** in parallel.

CHAPTER
15

An Introduction to Plane Analytic Geometry

A lithotripter is a medical instrument. Its design is based on an ellipse. In Section 15.4, we will see how the ellipse makes this instrument effective.

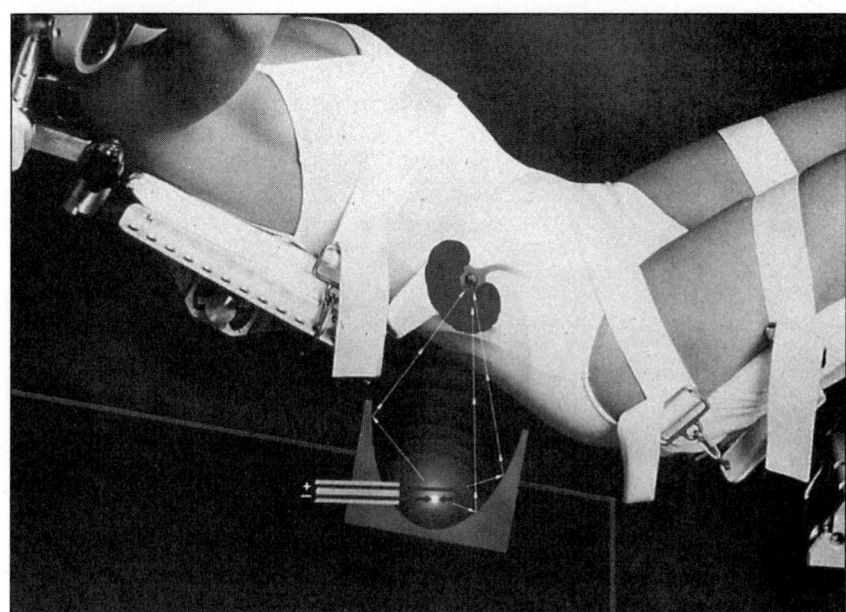

Courtesy of Dornier Medical Systems Inc.

Many scientific and technical applications make use of four special curves: the circle, parabola, ellipse, and hyperbola. These are four of the seven conic sections that can be formed by a plane intersecting a cone. (The others are a point, a line, and two intersecting lines.)

One medical instrument that uses an ellipse is a lithotripter, an instrument that uses sound waves to break up kidney stones. An ellipse has two foci. (Foci is the plural of focus.) A sound wave transmitter is placed at one focus and the patient's kidney is placed at the other. The elliptical shape of the lithotripter causes sound waves to be reflected off the ellipse and focused on the kidney stones.

In Chapter 4, the rectangular coordinate system was introduced. We began by graphing equations on the Cartesian coordinate system. The process of using algebra to describe geometry is called analytic geometry. In this chapter, we first review the line and then look at circles, parabolas, ellipses, and hyperbolas.

☰ 15.1
BASIC DEFINITIONS AND STRAIGHT LINES

We introduced graphing in Chapter 4 and have worked with graphs in almost every chapter since then. In this section, we will introduce some new ideas and review some of those that were introduced earlier.

If we have two points in a plane, we often want to know some things about them. Besides the location of the points, some of the things we might want to know include the distance between them, the point halfway between them, and the equation of the line through those two points, including its slope and intercepts. Let's look at these and a few other ideas, one at a time.

The Distance Formula

For most of this discussion we will be using two points, P_1 and P_2. Point P_1 has coordinates (x_1, y_1) and P_2 has coordinates (x_2, y_2). Using the Pythagorean theorem, we can find the distance d between these points.

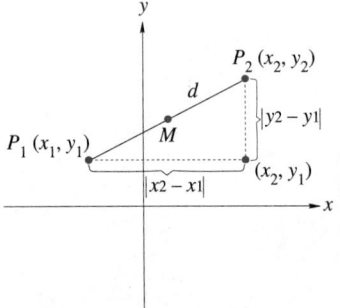

FIGURE 15.1

The Distance Formula

The distance d between any two points $P_1(x_1, y_1)$ and $P_2(x_2, y_2)$ in a plane is given by the **distance formula**

$$d = \sqrt{(x_1 - x_2)^2 + (y_1 - y_2)^2}$$

If there are several points, the symbol $d(P_1, P_2)$ would then represent the two points that are being used. (See Figure 15.1.)

EXAMPLE 15.1

Find the lengths of the sides of the triangle with vertices $A(-1, -3)$, $B(6, 1)$, and $C(2, -7)$.

Solutions

$$d(C, A) = \sqrt{(-1-2)^2 + (-3+7)^2} = \sqrt{(-3)^2 + 4^2}$$
$$= \sqrt{9 + 16} = \sqrt{25} = 5$$

$$d(B, A) = \sqrt{(-1-6)^2 + (-3-1)^2} = \sqrt{(-7)^2 + (-4)^2}$$
$$= \sqrt{49 + 16} = \sqrt{65}$$

$$d(C, B) = \sqrt{(6-2)^2 + (1+7)^2} = \sqrt{4^2 + 8^2}$$
$$= \sqrt{16 + 64} = \sqrt{80} = 4\sqrt{5}$$

The Midpoint Formula

If you want to find the coordinates of the point M halfway between two points P_1 and P_2, then you use the **midpoint formula**.

The Midpoint Formula

The coordinates of the midpoint M between any two points $P_1(x_1, y_1)$ and $P_2(x_2, y_2)$ are given by the **midpoint formula**

$$M = \left(\frac{x_1 + x_2}{2}, \frac{y_1 + y_2}{2} \right)$$

The midpoint formula gives you the coordinates of the point on the segment $\overline{P_1 P_2}$ halfway between P_1 and P_2.

EXAMPLE 15.2

Find the midpoints of the sides of the triangle with vertices $A(-1, -3)$, $B(6, 1)$, and $C(2, -7)$. Note that this is the same triangle we used in Example 15.1.

Solutions
$$M(A, B) = \left(\frac{-1 + 6}{2}, \frac{-3 + 1}{2} \right) = \left(\frac{5}{2}, \frac{-2}{2} \right) = \left(\frac{5}{2}, -1 \right)$$

$$M(A, C) = \left(\frac{-1 + 2}{2}, \frac{-3 - 7}{2} \right) = \left(\frac{1}{2}, \frac{-10}{2} \right) = \left(\frac{1}{2}, -5 \right)$$

$$M(B, C) = \left(\frac{6 + 2}{2}, \frac{1 - 7}{2} \right) = \left(\frac{8}{2}, \frac{-6}{2} \right) = (4, -3)$$

Slope

As we have discussed before, the slope of a line measures the steepness of the line. The slope also indicates whether, as values of x increase, a line is rising (positive slope), falling (negative slope), or constant (slope of 0). A formula for determining the slope of the line through two points follows.

Slope of a Line

If a line through two points $P_1(x_1, y_1)$ and $P_2(x_2, y_2)$ is not vertical, then the **slope**, m, of this line is

$$m = \frac{y_2 - y_1}{x_2 - x_1}$$

All lines except vertical lines have a slope. The slope of a vertical line is undefined. The slope of a horizontal line is 0.

EXAMPLE 15.3

What is the slope of the line through the points $A(2, -5)$ and $B(-4, 7)$?

Solution

$$m = \frac{7 - (-5)}{-4 - 2} = \frac{7 + 5}{-4 - 2} = \frac{12}{-6} = -2$$

Any line that is not parallel to the x-axis must eventually cross the x-axis. The angle measure in a positive direction from the x-axis to a line is called the **inclination** of the line. The inclination of a line parallel to the x-axis is defined as 0. The inclination provides us with an alternative definition for the slope. If α is the inclination that a line makes with the x-axis, as shown in Figures 15.2a and 15.2b, then the slope is

$$m = \tan \alpha, \qquad 0° \leq \alpha < 180° \qquad \text{or} \qquad 0 \leq \alpha < \pi$$

Of course, since $\tan 90°$ and $\tan \frac{\pi}{2}$ are undefined, the inclination of a vertical line is undefined.

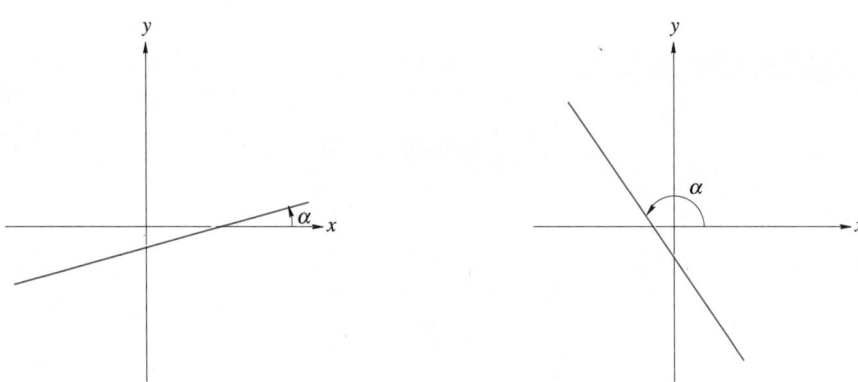

FIGURE 15.2a   FIGURE 15.2b

 **Note**

Many technical areas have their own terms that refer to the slope of a line or surface. For example, construction workers talk about the **pitch** of a roof, truck drivers refer to the **grade** of a hill, and, in some fields, **declination** refers to a line with negative slope.

EXAMPLE 15.4

What is the slope of a line with an inclination of **(a)** 30°, **(b)** 0.85 rad, and **(c)** 115°?

Solutions

(a) $m = \tan 30° = 0.5773503$

(b) $m = \tan 0.85 \text{ rad} = 1.1383327$

(c) $m = \tan 115° = -2.1445069$

EXAMPLE 15.5

The line in Example 15.3 has a slope of −2. What is the inclination of this line?

Solution

$$m = \tan \alpha = -2 \text{ so } \alpha = \tan^{-1}(-2) \approx 116.56505°$$

[If you used a calculator, $\tan^{-1}(-2) \approx -63.43498°$. To obtain the inclination, you need to add 180°, since the inclination must be between 0° and 180°, with the result $180° + (-63.43498°) = 116.56505°$.]

If two lines are parallel, they have the same inclination, and so parallel lines have the same slope. If two lines are perpendicular, they intersect at a 90° angle. This means that their inclinations must differ by 90°. It also means that the slope of one of these lines is the negative reciprocal of the other.

Slopes of Perpendicular Lines

If m_1 is the slope of a line and m_2 is the slope of a line perpendicular to the first line, then

$$m_1 = -\frac{1}{m_2}$$

This can also be written as $m_1 m_2 = -1$.

Caution

The above rule does not apply if one of the lines is horizontal because the other line would then be vertical and not have a slope.

EXAMPLE 15.6

What is the slope of a line that is perpendicular to a line with a slope of −2?

Solution If $m_1 = -2$, then the slope of the perpendicular line is $m_2 = -\frac{1}{-2} = \frac{1}{2}$

In Chapter 5, we introduced several formulas for the equation of a line. If b represents the y-intercept of a line, then the line crosses the y-axis at the point $(0, b)$. If the slope of this line is m, then the **slope-intercept form for the equation of a line** is

$$y = mx + b$$

EXAMPLE 15.7

Write an equation for a line that has a slope of $\frac{1}{4}$ and a y-intercept of 5.

Solution We have $m = \frac{1}{4}$ and $b = 5$, so $y = \frac{1}{4}x + 5$. Writing this without fractions, we get $4y = x + 20$.

EXAMPLE 15.8

What are the slope and y-intercept of the line $4x + 6y - 3 = 0$?

Solution We need to rewrite this equation in the form $y = mx + b$.

$$4x + 6y - 3 = 0$$
$$6y = -4x + 3$$
$$y = -\frac{4}{6}x + \frac{3}{6}$$
$$= -\frac{2}{3}x + \frac{1}{2}$$

The slope is $-\frac{2}{3}$ and the y-intercept is $\frac{1}{2}$.

If you know the slope of a line is m and $P(x_1, y_1)$ is a point on the line, then the equation for the line can be written in the **point-slope form for the equation of a line**.

$$y - y_1 = m(x - x_1)$$

EXAMPLE 15.9

What is the equation of the line through the point $(2, -3)$ and with a slope of 5?

Solution In this example, $x_1 = 2$, $y_1 = -3$, and $m = 5$. Substitute these values into the equation $y - y_1 = m(x - x_1)$.

$$y - (-3) = 5(x - 2)$$
$$y + 3 = 5x - 10$$
$$y = 5x - 13$$

Every straight line can be written in the **general form for the equation of a line**. This form is represented by the equation

$$Ax + By + C = 0$$

where A, B, and C represent constants and A and B are not both 0.

EXAMPLE 15.10

What are the slope and intercepts of the line $4x + 3y - 24 = 0$?

Solution We know that the intercepts are the points where the line crosses the axes. The line crosses the x-axis at $(a, 0)$ and the y-axis at $(0, b)$.

If we let $y = 0$, we then get $x = 6$, and if $x = 0$, then $y = 8$; so the x-intercept is $(6, 0)$ and the y-intercept is $(0, 8)$. We will use these two points to determine the following slope.

$$m = \frac{8 - 0}{0 - 6} = \frac{8}{-6} = -\frac{4}{3}$$

The three forms for the equation of a line are summarized in the following box.

Different Forms for the Equation of a Line

If a line has a slope of m, a y-intercept of $(0, b)$, and $P(x_1, y_1)$ is a point on the line, then the equation for the line can be written in any of these three forms:

Type of Equation	Equation for Line
Point-slope form	$y - y_1 = m(x - x_1)$
Slope-intercept form	$y = mx + b$
General form	$Ax + By + C = 0$

Application

EXAMPLE 15.11

A mechanic charges \$96 for a job that takes 2 h to finish and \$135 if the job is completed in 5 h. **(a)** Find a linear equation that describes how much the mechanic should charge for a job of x hours. **(b)** Use your equation to determine how much should be charged for a job that takes 7 h 15 min.

Solutions   **(a)** We begin by thinking of the given information as points on a line. We give each point as an ordered pair of the form (x, C), where x represents the number of hours a job takes to complete and C stands for the amount the mechanic charges for this job.

A 2-h job with a charge of \$96 can be thought of as the point $(2, 96)$, and the 5-h job as the point $(5, 135)$. From this information we determine that the slope of this line is

$$m = \frac{135 - 96}{5 - 2} = \frac{39}{3} = 13$$

Using this value for the slope and selecting the point $(2, 96)$ we can use the point-slope form for the equation of a line to determine the needed equation.

$$C - 96 = 13(x - 2)$$
$$C - 96 = 13x - 26$$

or $\qquad C = 13x + 70$

Thus the desired equation is $C = 13x + 70$. **(b)** A 7 h 15 min job can be thought of as a 7.25-h job. Substituting 7.25 for x in the equation $C = 13x + 70$, we obtain $C = 13(7.25) + 70 = 94.25 + 70 = 164.25$. The mechanic should charge \$164.25 for a job that will take 7 h and 15 min.

Application

The resistance of a circuit element varies directly with its temperature. The resistance at $0°C$ is found to be $5.00\,\Omega$. Further tests show that the resistance increases by $1.5\,\Omega$ for every $1°C$ increase in temperature. **(a)** Determine the equation for the resistance R in terms of the temperature T. **(b)** Use the equation from part (a) to find the resistance at $27°C$.

Solutions **(a)** In order to find R in terms of T, we should think of R as the dependent variable. Since resistance varies directly with temperature, the two quantities have a linear relationship. From the given information, we notice that since $R = 5.00$ when $T = 0$, we have the R-intercept. As a result, we shall use the slope-intercept form for the equation of a line.

$$R = mT + b$$
$$R = mT + 5.00$$

To find the slope m, we will use

$$m = \frac{\text{Change in } R}{\text{Change in } T} = \frac{1.5\,\Omega}{1°C} = 1.5\,\Omega/°C$$

Substituting 1.5 for m in the slope-intercept equation, we are able to complete the equation.

$$R = 1.5T + 5.00$$

(b) To find the resistance at $27°C$, we will substitute $27°C$ for T in the equation from part (a).

$$R = 1.5(27) + 5.00 = 45.5\,\Omega$$

Exercise Set 15.1

In Exercises 1–8, find the distance between the given pairs of points.

1. $(2, 4)$ and $(7, -9)$

2. $(-3, 5)$ and $(0, 1)$

3. $(5, -6)$ and $(-3, -5)$

4. $(-8, -4)$ and $(6, 10)$

5. $(12, 1)$ and $(3, -13)$

6. $(11, -2)$ and $(4, -8)$

7. $(-2, 5)$ and $(5, 5)$

8. $(-4, -4)$ and $(-4, 7)$

In Exercises 9–16, find the midpoints of the given pairs of points in Exercises 1–8.

In Exercises 17–24, find the slopes of the lines through the points in Exercises 1–8.

In Exercises 25–32, use the point-slope form of the equation for the line to write the equation of the lines through the points in Exercises 1–8.

In Exercises 33–36, find the slopes of the lines with the given inclinations.

33. $45°$ **34.** $175°$ **35.** $0.15\,\text{rad}$ **36.** $1.65\,\text{rad}$

In Exercises 37–40, find the inclinations of the lines with the given slopes.

37. 2.5 **38.** 0.30 **39.** -0.50 **40.** -1.475

In Exercises 41–44, determine the slope of a line that is perpendicular to a line of the given slope.

41. 3 **42.** $\frac{2}{5}$ **43.** $-\frac{1}{2}$ **44.** -5

In Exercises 45–56, find the equation of each line with the given properties.

45. Passes through $(2, -5)$ with a slope of 6
46. Passes through $(6, -2)$ with a slope of $-\frac{1}{2}$
47. Passes through $(2, 3)$ and $(4, -2)$
48. Passes through $(-5, 2)$ and $(-3, -4)$
49. Passes through $(-2, -4)$ with an inclination of $60°$
50. Passes through $(5, 1)$ with an inclination of $0.75\,\text{rad}$
51. Has an inclination of $20°$ and a y-intercept of $(0, 3)$

52. Has an inclination of $2.5\,\text{rad}$ and a y-intercept of $(0, -2)$
53. Passes through $(2, 5)$ and is parallel to $2x - 3y + 4 = 0$
54. Passes through $(-3, 2)$ and is parallel to $3x + 4y = 12$
55. Passes through $(4, -1)$ and is perpendicular to $2x + 5y = 20$
56. Passes through $(-1, -6)$ and is perpendicular to $8x - 3y = 24$

In Exercises 57–60, determine the slope and intercepts of each line.

57. $3x + 2y = 12$ **58.** $5x - 3y = 15$ **59.** $x - 3y = 9$ **60.** $6x + y = 9$

Solve Exercises 61–68.

61. *Physics* The instantaneous velocity v of an object under constant acceleration a during an elapsed time t is given by $v = v_0 + at$, where v_0 is its initial velocity. If an object has an initial velocity of 2.6 m/s and a velocity of 5.8 m/s after 8 s of constant acceleration, write the equation relating velocity to time.

62. *Physics* The **coefficient of linear expansion** α is the change in length of a solid due to the change in temperature.
  (a) Determine the coefficient of linear expansion of a copper rod that is 1.000 000 cm at 10°C and expands to 1.000 084 cm at 15°C.
  (b) Determine the coefficient of linear expansion of a copper rod that is 4.000 000 cm at 10°C and is 4.000 336 cm at 15°C.
  (c) The coefficient of linear expansion for any specific solid is a constant. This means that the answers for parts (a) and (b) should have been the same. Assume that the answer in part (a) is cor-

rect. What changes need to be made in the way you determine the coefficient of linear expansion?
  (d) Determine the coefficient of linear expansion of a glass rod that expands from 72.000 024 cm at 10°C to 72.005 208 cm at 18°C.

63. *Electronics* In a dc circuit, when the internal resistance of the voltage source is taken into account, the voltage E is a linear function of the current I and is given by $E = IR + Ir$, where R is the circuit resistance and r is the internal resistance. If the resistance of the circuit $R = 4.0\,\Omega$ and $I = 2.5\,A$ when $E = 12.0\,V$, find r.

64. *Electronics* The resistance of a circuit element is found to increase by $0.006\,\Omega$ for every 1°C increase in temperature over a wide range of temperatures. If the resistance is $7.000\,\Omega$ at 0°C, **(a)** write the equation relating the resistance R to the temperature T, and **(b)** find R, when $T = 17°C$.

65. *Industrial management*   The production of a certain computer component has fixed costs of $1,225 and additional costs of $1.25 per component manufactured.
   (a) Write an equation relating the total cost, C, to the number of components produced, n.
   (b) What is the cost of producing 20,000 components?

66. *Accounting*   The linear method for determining the depreciated value, V, of a deductible item is given by

$$V = C\left(1 - \frac{n}{N}\right)$$

where C is the original cost, n is the number of years since the depreciation began, and N is the usable life of the item. Graph V as a function of C when $n = 5$ and $N = 8$.

67. *Environmental science*   One estimate of the concentration, C, of carbon dioxide (CO_2) in parts per million (ppm) in year t is given by

$$C = 1.44t + 282.88$$

where t is the number of years after 1940.

(a) What was the CO_2 concentration in 1995?
(b) Predict the CO_2 concentration for the year 2050.
(c) Predict the CO_2 concentration for the current year.

68. *Energy technology*   The width, W, and height, H, of a roof overhang and the latitude, L, of a north-south facing room determine whether the room will get any summer sunshine. The overhang dimensions that give no summer sunshine are $W = H \tan B$, where $B = L - 23.5°$, is the angle of inclination of the sun at summer solstice. (For houses at either latitude 37° N or 37° S, let $L = 37$.)
   (a) Chattanooga, Tennessee is at latitude 31°2′ N. What overhang width on an 8′-high room would ensure that it gets no summer sunshine?
   (b) The latitude of Vancouver, British Columbia, is 49°16′ N. What overhang width on a 2.7 m-high room would ensure that it gets no summer sunshine?
   (c) The latitude of Anchorage, Alaska, is 61°13′ N. Will a room with an overhang that is 8′6″ high and 6′3″ wide get any summer sunshine?

In Your Words

69. Describe how to determine if two lines are perpendicular.

70. Compare and contrast the point-slope, slope-intercept, and general forms for the equation of a line. How are they alike and how are they different?

≣ 15.2
THE CIRCLE

In Section 15.1, we developed a general equation for a straight line. In this section, we will work with the circle.

   A **circle** is the set of all points in a plane that are at some fixed distance from a fixed point in the plane. The fixed point is called the **center** of the circle and the fixed distance is called the **radius**. In general, suppose we have a circle with radius r and center (h, k). An seen in Figure 15.3, if (x, y) is any point on the circle, then the distance from (x, y) to (h, k) must be r. Using the distance formula, we have the equation

$$\sqrt{(x - h)^2 + (y - k)^2} = r$$

Squaring both sides, we have

$$(x-h)^2 + (y-k)^2 = r^2$$

Thus, we have established the following standard equation of a circle.

> **Standard Equation of a Circle**
>
> The standard equation of a circle with center at (h, k) and radius r is
>
> $$(x-h)^2 + (y-k)^2 = r^2$$

EXAMPLE 15.13

The equation $(x-5)^2 + (y+3)^2 = 36$ is a circle with center at $(5, -3)$ and a radius of 6. Remember, since the equation states that $y + 3 = y - k$, you obtain $k = -3$. Note very carefully the signs of the numbers in the equation and the signs of the coordinates of the center. This circle is shown in Figure 15.4.

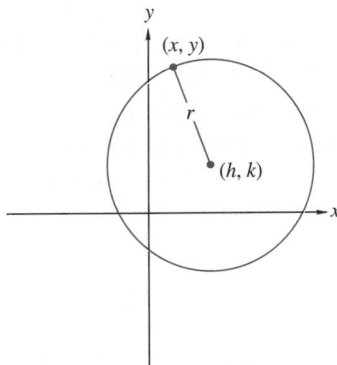

FIGURE 15.3

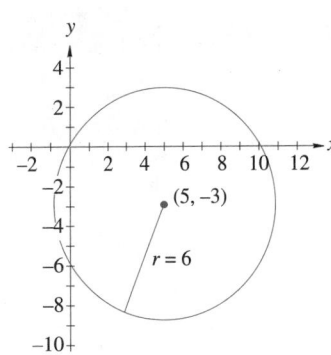

FIGURE 15.4

EXAMPLE 15.14

Write the equation for the circle with center at $(-2, 4)$ and radius 7.

Solution   The circle is shown in Figure 15.5. Since the center is $(-2, 4)$, we have $h = -2$ and $k = 4$, and since the radius is 7, then $r = 7$. So,

$$(x-h)^2 + (y-k)^2 = r^2$$
$$[x-(-2)]^2 + (y-4)^2 = 7^2$$
$$(x+2)^2 + (y-4)^2 = 49$$

Notice that if the center is at the origin, then $h = 0$ and $k = 0$. The equation becomes

$$x^2 + y^2 = r^2$$

If we expand the general equation for a circle, we get

$$(x - h)^2 + (y - k)^2 = r^2$$
$$x^2 - 2hx + h^2 + y^2 - 2ky + k^2 = r^2$$

Since h^2, k^2, and r^2 are all constants, we will let $h^2 + k^2 - r^2 = F$. If we let the constants $-2h = D$ and $-2k = E$, then the equation for a circle can be written as the following general equation.

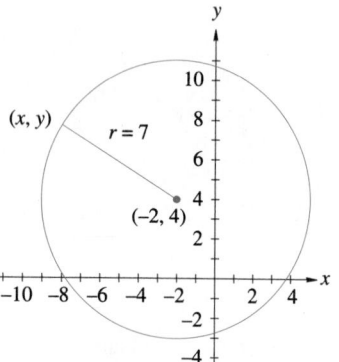

y

10
8
6
$r = 7$
(x, y)
4
$(-2, 4)$
2

−10 −8 −6 −4 −2   2   4   x
−2
−4

FIGURE 15.5

General Equation of a Circle

If A, D, E, and F are any real number constants, $A \neq 0$, then the general equation for a circle is

$$Ax^2 + Ay^2 + Dx + Ey + F = 0$$

EXAMPLE 15.15

Write $x^2 + y^2 + 4x - 8y + 4 = 0$ in the standard form for the equation of a circle and determine the center and radius.

Solution   To convert the equation to the standard form, we need to complete the square. We first write the constant on the right-hand side and group the terms containing x together and then group those containing y. The ◯ and ☐ show the missing constants that we must determine in order to complete the square. (See Section 8.3.)

$$x^2 + y^2 + 4x - 8y + 4 = 0$$
$$\left(x^2 + 4x + \bigcirc\right) + \left(y^2 - 8y + \square\right) = -4 + \bigcirc + \square$$

The coefficient of the x-term is 4. If we take half of that and square it, we can add the result, 4, to both sides of the equation. Similarly, half of -8 is -4, and we also add $(-4)^2 = 16$ to both sides. The equation then becomes

$$\left(x^2 + 4x + \boxed{4}\right) + \left(y^2 - 8y + \boxed{16}\right) = -4 + 4 + 16$$
$$(x + 2)^2 + (y - 4)^2 = 16$$

Since $16 = 4^2$, the radius is 4, and the center is $(-2, 4)$.

EXAMPLE 15.16

Write the equation $x^2 + y^2 + x - 5y + 2 = 0$ in the standard form for the equation of a circle and determine the center and radius.

EXAMPLE 15.16 (Cont.)

Solution   Again we will complete the square and use a ◯ and a ☐ to show where the constants need to be placed.

$$x^2 + y^2 + x - 5y + 2 = 0$$
$$(x^2 + x + \bigcirc) + (y^2 - 5y + \square) = -2 + \bigcirc + \square$$
$$\left(x^2 + x + \frac{1}{4}\right) + \left(y^2 - 5y + \frac{25}{4}\right) = -2 + \frac{1}{4} + \frac{25}{4}$$
$$\left(x + \frac{1}{2}\right)^2 + \left(y - \frac{5}{2}\right)^2 = \frac{18}{4}$$

The center is $\left(-\frac{1}{2}, \frac{5}{2}\right)$ and the radius is $\sqrt{\frac{18}{4}} = \frac{3\sqrt{2}}{2}$.

EXAMPLE 15.17

Write the equation $4x^2 + 4y^2 - 8x - 24y - 9 = 0$ in the standard form for the equation of a circle and determine the center and radius.

Solution   Again, we will complete the square. First, move the constant to the right-hand side and group the x-terms and the y-terms.

$$4x^2 - 8x + 4y^2 - 24y = 9$$

Next, factor the 4 out of the x-terms and the y-terms.

$$4(x^2 - 2x) + 4(y^2 - 6y) = 9$$

We now use a ◯ and a ☐ to indicate the missing constants. Notice that a $4\bigcirc$ and a $4\square$ are added on the right-hand side of the equation to show where the constants need to be placed.

$$4(x^2 - 2x + \bigcirc) + 4(y^2 - 6y + \square) = 9 + 4\bigcirc + 4\square$$
$$4(x^2 - 2x + 1) + 4(y^2 - 6y + 9) = 9 + 4(1) + 4(9)$$
$$4(x - 1)^2 + 4(y - 3)^2 = 49$$
$$(x - 1)^2 + (y - 3)^2 = \frac{49}{4}$$

The center is $(1, 3)$ and the radius $r = \sqrt{\frac{49}{4}} = \frac{7}{2}$.

While we could have started by dividing this entire equation by 4, we waited until the end in order to delay working with fractions. This example uses a process that we will need in the sections on ellipses and hyperbolas.

A circle with an equation of the form $(x - h)^2 + y^2 = r^2$ has its center on the x-axis [at $(h, 0)$] and thus is symmetrical with the x-axis. In the same way, a circle with an equation of the form $x^2 + (y - k)^2 = r^2$ has its center on the y-axis [at $(0, k)$] and is symmetrical with the y-axis.

If $(x-h)^2 + (y-k)^2 = 0$, the circle has a radius of 0. This is sometimes called a **point circle**. If the radius is 1, the circle is often referred to as a **unit circle**.

Finally, if $r^2 < 0$, then the equation does not define a circle and this equation would have no graph on the Cartesian plane.

Application

EXAMPLE 15.18

A square concrete post is reinforced axially with eight rods arranged symmetrically around a circle, much like those shown in the photo in Figure 15.6a. The cross section in Figure 15.6b provides a schematic drawing of the post and rods. **(a)** Determine the equation of the circle using the upper left-hand corner of the post as the origin. **(b)** Find the coordinates of each rod's location.

Solution **(a)** While it is not specifically stated, we will assume that the eight rods are placed around a circle that shares the center of the square post. This would place the center at the point $(15, -15)$. The radius of the circle is $2.925 + 7.075\,\text{cm} = 10\,\text{cm}$. Thus, the equation for this circle is $(x-15)^2 + (y+15)^2 = 100$. **(b)** The coordinates of rods 1, 3, 5, and 7 should be fairly easy to determine from the given information. Determining the coordinates of the four remaining rods is a little more difficult. We will show how to determine the coordinates for Rod 2 and then use the symmetry of the rod's placement to determine the coordinates of the other three rods.

The radius of the circle on which the rods are placed is 10 cm. Thus, each rod is located 10 cm from the center of the post. Rod 2 is located directly above a point that is 7.075 cm from the center of the post. Using the Pythagorean theorem, we find that this rod is placed $\sqrt{10^2 - 7.075^2} = \sqrt{100 - 50.055625} = \sqrt{49.944375} \approx 7.067\,\text{cm}$ above the line through Rods 7 and 3. Then the y-coordinate of Rod 2 is $-15 + 7.067 = -7.933$. This places Rod 2 at the coordinates $(22.075, -7.933)$. The eight rods have the coordinates $(15, -5)$, $(22.075, -7.933)$, $(25, -15)$, $(22.075, -22.067)$, $(15, -25)$, $(7.925, -22.067)$, $(5, -15)$, and $(7.925, -7.933)$.

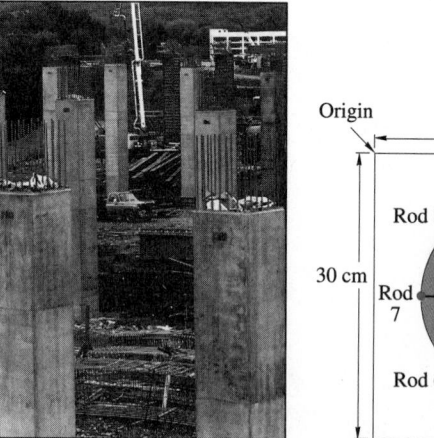

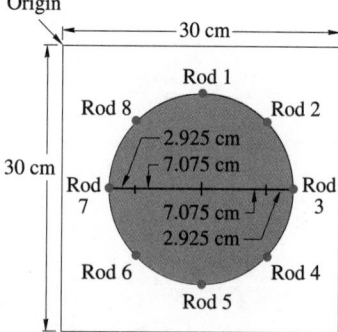

Courtesy of Barry, Bette, and Led Duke Inc.

FIGURE 15.6a　　　　　　　　**FIGURE 15.6b**

Exercise Set 15.2

In Exercises 1–8, find the standard and general form of an equation for the circle with the given center C and radius r.

1. $C = (2, 5), r = 3$
2. $C = (4, 1), r = 8$
3. $C = (-2, 0), r = 4$
4. $C = (0, -7), r = 2$
5. $C = (-5, -1), r = \frac{5}{2}$
6. $C = (-4, -5), r = \frac{5}{4}$
7. $C = (2, -4), r = 1$
8. $C = (5, -2), r = \sqrt{3}$

In Exercises 9–14, give the center and radius of the circle described by each equation. Sketch the circle.

9. $(x - 3)^2 + (y - 4)^2 = 9$
10. $(x - 7)^2 + (y + 5)^2 = 25$
11. $\left(x + \frac{1}{2}\right)^2 + \left(y + \frac{13}{4}\right)^2 = 7$
12. $\left(x + \frac{7}{2}\right)^2 + \left(y - \frac{7}{3}\right)^2 = \frac{11}{6}$
13. $x^2 + \left(y - \frac{7}{3}\right)^2 = 6$
14. $(x + 3)^2 + y^2 = 1.21$

In Exercises 15–26, describe the graph of each equation. If it is a circle, give the center and radius.

15. $x^2 + y^2 + 4x - 6y + 4 = 0$
16. $x^2 + y^2 - 10x + 2y + 22 = 0$
17. $x^2 + y^2 + 10x - 6y - 47 = 0$
18. $x^2 + y^2 + 2x - 12y - 27 = 0$
19. $x^2 + y^2 - 2x + 2y + 3 = 0$
20. $x^2 + y^2 + 2x + 2y - 3 = 0$
21. $x^2 + y^2 + 6x - 16 = 0$
22. $x^2 + y^2 - 8y - 9 = 0$
23. $x^2 + y^2 + 5x - 9y = 9.5$
24. $x^2 + y^2 - 7x + 3y + 2.5 = 0$
25. $9x^2 + 9y^2 + 18x - 15y + 27 = 0$
26. $25x^2 + 25y^2 - 10x + 30y + 1 = 0$

Solve Exercises 27–34.

27. *Physics*   When a particle with mass m and charge q enters a magnetic field of induction β with a velocity v at right angle to β, the path of the particle is a circle. The radius of the path is $\sigma = \dfrac{mv}{\beta q}$. If a proton of mass 1.673×10^{-27} kg and charge 1.5×10^{-19} C enters a magnetic field with induction 4×10^{-4} T (tesla $= \text{V} \cdot \text{s} \cdot \text{m}^2$) with a velocity of 1.186×10^7 m/s, find the equation of the path of the electron.

28. *Astronomy*   The earth's orbit around the sun is approximately a circle of radius 1.495×10^8 km. The moon's orbit around earth is approximately a circle of radius 3.844×10^5 km. If the sun is placed at the center of a coordinate system, what is the equation of the earth's orbit?

29. *Astronomy*   If the sun is placed at the center of a coordinate system and the earth on the positive x-axis, what is the equation of the moon's orbit around the earth?

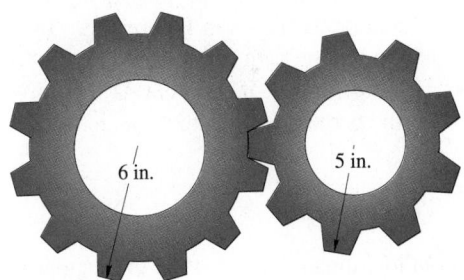

FIGURE 15.7

30. *Industrial design*   A drafter is drawing two gears that intermesh. They are represented by two intersecting circles, as shown in Figure 15.7. The first circle has a radius of 6 in. and the second has a radius of 5 in. The section of intersection has a maximum depth of 1 in. What is the equation of each circle, if the center of the first circle is at the origin and the positive x-axis passes through the center of the second circle?

31. *Industrial design*   In designing a tool, an engineer calls for a hole to be drilled with its center 0.9 cm directly above a specified origin. If the hole must have a diameter of 1.1 cm, find the equation of the hole.

32. *Industrial engineering*   In designing dies for blanking a sheet-metal piece, the configuration shown in Figure 15.8 is used. Write the equation of the small circle with its center at B, using the point marked A as the origin. Use the following values: $r = 3$ in., $a = 3.25$ in., $b = 1.0$ in., and $c = 0.25$ in.

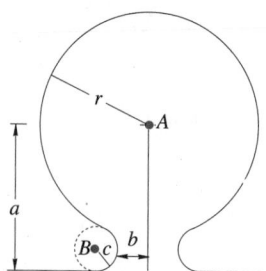

FIGURE 15.8

33. *Industrial engineering*   A flywheel that is 26 cm in diameter is to be mounted so that its shaft is 5 cm above the floor, as shown in Figure 15.9. **(a)** Write an equation for the path followed by a point on the rim. Use the surface of the floor as the horizontal axis and the perpendicular line through the center as the vertical axis. **(b)** Find the width of the opening in the floor, allowing a 2-cm clearance on both sides of the wheel.

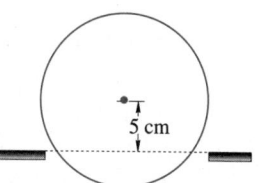

FIGURE 15.9

34. *Electricity*   The impedance, Z, in an ac circuit is given by

$$Z = \sqrt{R^2 + (X_L - X_C)^2}$$

where R is the resistance, X_L the inductive reactance, and X_C the inductive capacitance.
(a) If $R = 6\,\Omega$, $X_L = 10\,\Omega$, and $X_C = 12\,\Omega$, determine Z.
(b) If $R = 8\,\Omega$ and $X_C = 12.5\,\Omega$, what are the possible values for Z?
(c) If $Z = 40\,\Omega$ and $X_C = 50\,\Omega$, what are the possible values for X_L?
(d) Use your result from (c) to graph the relationship between resistance and inductive reactance when $Z = 40\,\Omega$ and $X_C = 50\,\Omega$. (Put R on the horizontal axis.)

In Your Words

35. Without looking in the text, write the standard equation of a circle. Explain what each constant represents.

36. Describe the procedures you would use to change the general equation of a circle to the standard equation, and vice versa.

☰ 15.3
THE PARABOLA

The second curve we will study in this chapter is the parabola. A **parabola** is the set of points in a plane for which the distance from a fixed line is equal to its distance from a fixed point not on the line. The given line is called the **directrix** and the given point is the **focus**.

In Figure 15.10, we have indicated the line ℓ is the directrix, and F is the focus. The line through F perpendicular to the directrix is called the **axis** of the parabola. Examination of the parabola in Figure 15.10 indicates that the axis of the parabola is the only axis of symmetry for the parabola. The point V on the axis that is halfway between the focus and the directrix is the **vertex**.

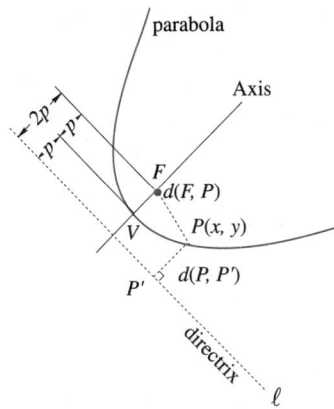

FIGURE 15.10

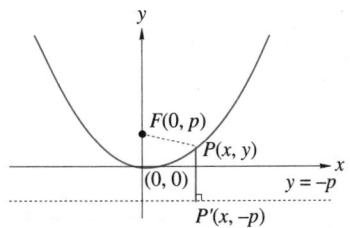

FIGURE 15.11

Let P be any point on the parabola. Draw a line through P that is perpendicular to the directrix at a point P'. According to the definition, the distance from P to P' is the same as the distance from P to F.

Let's set up a similar drawing on a coordinate system so that we can develop a formula for a parabola. We will let the vertex be at the origin and the focus F at $(0, p)$. Since the vertex is midway between the focus and the directrix, the directrix is the line $y = -p$, as shown in Figure 15.11. If $P(x, y)$ is any point on the parabola, then the distance from F to P is, by the distance formula,

$$\sqrt{(x-0)^2 + (y-p)^2}$$

Since P' is on the directrix, its coordinates are $(x, -p)$. So the distance from P to P' is

$$\sqrt{(x-x)^2 + (y+p)^2}$$

Since these two distances are equal, we have

$$\sqrt{(x-0)^2 + (y-p)^2} = \sqrt{(x-x)^2 + (y+p)^2}$$

Squaring both sides and simplifying, we obtain

$$x^2 + (y-p)^2 = (y+p)^2$$
$$x^2 + y^2 - 2py + p^2 = y^2 + 2py + p^2$$
$$x^2 = 4py$$

We have established the following standard equation of a vertical parabola. The term vertical parabola is used because the axis is a vertical line. Notice that $|p|$ is the distance from the focus to the vertex.

Standard Equation of a Vertical Parabola

The standard equation of a vertical parabola with focus F at $(0, p)$ and directrix $y = -p$ is

$$x^2 = 4py$$

If $p > 0$, the parabola opens upward, as shown in Figure 15.11.

If $p < 0$, the parabola opens downward.

EXAMPLE 15.19

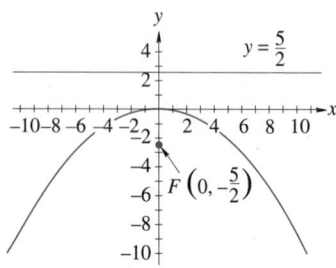

FIGURE 15.12

Find the focus and directrix of the parabola with the equation $x^2 = -10y$. Sketch its graph.

Solution   Since $x^2 = 4py$, then $4p = -10$ and $p = -\frac{5}{2}$. The focus is at $\left(0, -\frac{5}{2}\right)$. The equation of the directrix is $y = -p$ or $y = \frac{5}{2}$. The graph is sketched in Figure 15.12. Notice that $p < 0$ and that this parabola opens downward.

EXAMPLE 15.20

Find the equation of the parabola that has its vertex at the origin, opens upward, and passes through $(-5, 9)$.

Solution   The general form of the equation is $x^2 = 4py$. Since it opens upward, $p > 0$, and since $(-5, 9)$ is on the parabola, then $(-5)^2 = 4p(9)$. Solving for p, we get $p = \frac{25}{36}$. Substituting this value for p in the equation, we get

$$x^2 = 4\left(\frac{25}{36}\right)y$$

$$x^2 = \frac{25}{9}y$$

$$\text{or} \qquad 9x^2 = 25y$$

If we had used the x-axis as the axis of the parabola, and the vertex was still at the origin, the focus would then have been at $(p, 0)$, and the directrix would have had the equation $x = -p$. Using the same method that we used earlier, we would get the following standard equation for a horizontal parabola. It is called a horizontal parabola because the axis is horizontal.

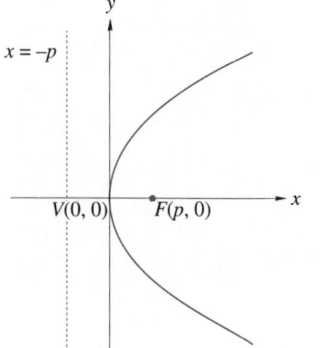

FIGURE 15.13

Standard Equation of a Horizontal Parabola

The standard equation of a horizontal parabola with focus F at $(p, 0)$ and directrix $x = -p$ is

$$y^2 = 4px$$

If $p > 0$, the parabola opens to the right, as shown in Figure 15.13.

If $p < 0$, the parabola opens to the left.

EXAMPLE 15.21

Find the equation of the parabola with its vertex at the origin and focus at $(5, 0)$.

Solution Since the focus is at $(5, 0)$, $p = 5$. A parabola with focus on the x-axis and vertex at the origin is of the form $y^2 = 4px$. Since $p = 5$, the equation is $y^2 = 20x$.

EXAMPLE 15.22

Sketch the graphs of one horizontal and one vertical parabola, each of which has its vertex at the origin and passes through the point $(3, -6)$. Determine the equation for each parabola.

Solution The graphs are in Figure 15.14. One graph is of the form $x^2 = 4py$ and the other is of the form $y^2 = 4px$. Substituting 3 for x and -6 for y in each equation, we get

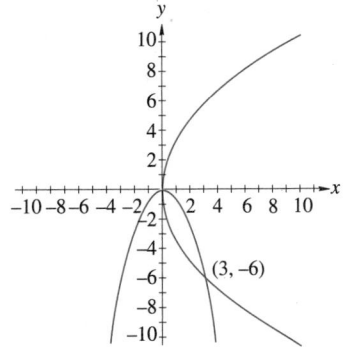

$$x^2 = 4py \qquad\qquad y^2 = 4px$$
$$3^2 = 4p(-6) \qquad\qquad (-6)^2 = 4p(3)$$
$$9 = -24p \qquad\qquad 36 = 12p$$
$$p = -\frac{9}{24} = -\frac{3}{8} \qquad\qquad p = 3$$

$$x^2 = 4\left(-\frac{3}{8}\right)y \qquad\qquad y^2 = 4(3)x$$
$$x^2 = -\frac{3}{2}y \qquad\qquad y^2 = 12x$$
$$\text{or} \quad 2x^2 = -3y$$

FIGURE 15.14

The vertical parabola has the equation $x^2 = -\frac{3}{2}y$ and the horizontal parabola has the equation $y^2 = 12x$.

Reflective Properties of Parabolas

A three-dimensional object called a **paraboloid of revolution** is formed when a parabola is revolved around its axis of symmetry. (See Figure 15.15.) Paraboloids of revolution have many uses based on the following two facts.

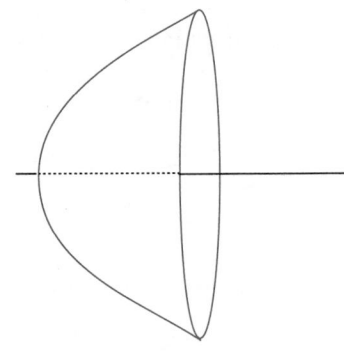

1. Rays entering a parabola along lines parallel to its axis are all reflected through its focus. Many examples exist for the different types of energy rays. Radio telescopes, radar antennae, and satellite television dishes used as downlinks are all examples of paraboloids of revolution. Parabolic reflectors are sometimes used at sports events to pick out specific sounds from background noises. In each of these cases, a ray or wave that is directed toward the dish is reflected off the sides of the paraboloid through its focus.

2. Rays drawn through a parabola's focus are all reflected along lines parallel to the parabola's axis. This uses rays in the reverse of the first property. Applications include flashlights, automobile headlights, search lights, and satellite dishes used as uplinks.

FIGURE 15.15

Paraboloids are sometimes used to both send and receive information. For example, a radar transmitter alternately sends and receives waves.

Application

EXAMPLE 15.23

The television satellite dish shown in Figure 15.16a is in the shape of a paraboloid of revolution measuring 0.4 m deep and 2.5 m across at its opening. Where should the receiver be placed in order to pick up the incoming signals?

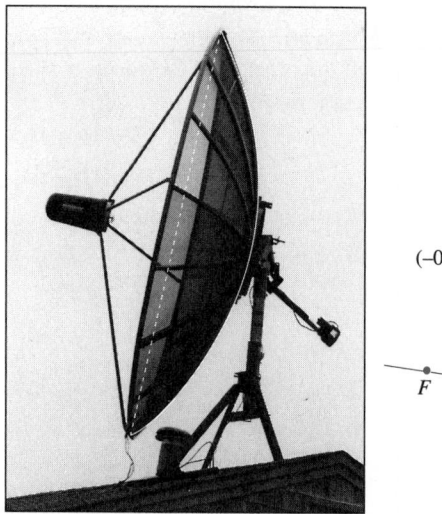

Courtesy of Michael A. Gallitelli, Metroland Photo Inc.
FIGURE 15.16a

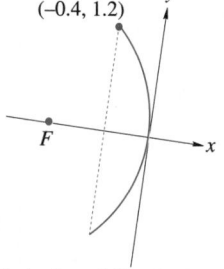

FIGURE 15.16b

Solution A cross-section through the axis of this satellite dish has been drawn over the photograph of the dish in Figure 15.16a. Because the cross-section is a parabola, the receiver should be placed at the parabola's focus. We will next determine the location of that focus.

If we place the drawing of this cross-section on a coordinate system so that the vertex is at the origin and the focus is placed on the x-axis, we get Figure 15.16b. We know that the parabola in Figure 15.16b has an equation of the form $y^2 = 4px$. From the given data, the point $(-0.4, 1.2)$ is on the parabola. This means that $p = \dfrac{y^2}{4x} = \dfrac{1.2^2}{-1.6} = -0.9$. Since the focus is at the point $(0, -p)$, this means that we can place the focus 0.9 m from the vertex.

Using a Graphing Calculator

Graphing calculators can be used to sketch the graph of any of the curves in this chapter. You must remember, however, that these calculators will only sketch the graphs of functions. This means that you will first need to solve the equation algebraically for y and then graph the one or two equations that result. The next two examples will demonstrate how this is done.

EXAMPLE 15.24

Use a graphing calculator to sketch the graphs of the parabolas **(a)** $y = \frac{3}{4}x^2$ and **(b)** $y = \frac{3}{4}x^2 - 3$.

Solution As usual, after you turn on the calculator, clear the screen of any previous graphs.

Now you are ready to graph $y = \frac{3}{4}x^2$. Since this equation has already been solved for y, we need only to press the following key sequence on a Casio fx-7700G.

$$\boxed{\text{Graph}}\,\boxed{(}\,\boxed{3}\,\boxed{\div}\,\boxed{4}\,\boxed{)}\,\boxed{\text{X},\theta,\text{T}}\,\boxed{\text{SHIFT}}\,\boxed{x^2}\,\boxed{\text{EXE}}$$

The result is the graph in Figure 15.17a.

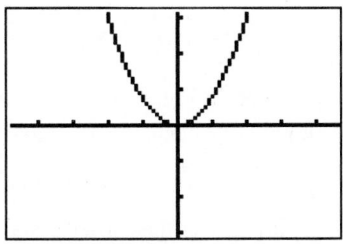
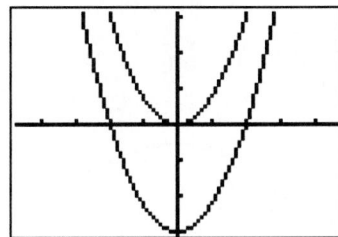

FIGURE 15.17a   **FIGURE 15.17b**

If you want to graph $y = \frac{3}{4}x^2 - 3$, enter the key sequence

$$\boxed{\text{Graph}}\,\boxed{(}\,\boxed{3}\,\boxed{\div}\,\boxed{4}\,\boxed{)}\,\boxed{\text{X},\theta,\text{T}}\,\boxed{\text{SHIFT}}\,\boxed{x^2}\,\boxed{-}\,\boxed{3}\,\boxed{\text{EXE}}$$

The result is the graph in Figure 15.17b. (If you did not clear the screen after you graphed $y = \frac{3}{4}x^2$, you will see both graphs on your screen.) Notice how the graph of $y = \frac{3}{4}x^2 - 3$ is simply the graph of $y = \frac{3}{4}x^2$ shifted down 3 units.

EXAMPLE 15.25

Use a TI-82 graphing calculator to sketch the graph of the circle $x^2 + y^2 + 4x - 8y + 4 = 0$.

Solution This is the same equation we used in Example 15.15. This equation can be simplified to

$$(x+2)^2 + (y-4)^2 = 16$$

and represents the equation of a circle with center $(-2, 4)$ and radius $r = 4$.

In order to use the graphing calculator, we need to solve the equation for y.

$$(x+2)^2 + (y-4)^2 = 16$$
$$(y-4)^2 = 16 - (x+2)^2$$

EXAMPLE 15.25 (Cont.)

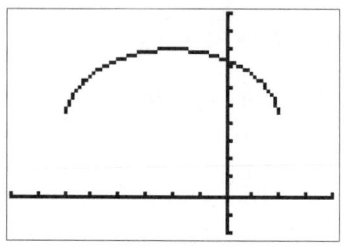

[–8, 4] x [–2, 10]

FIGURE 15.18a

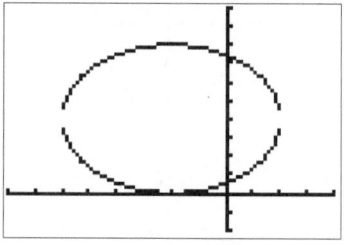

[–8, 4] x [–2, 10]

FIGURE 15.18b

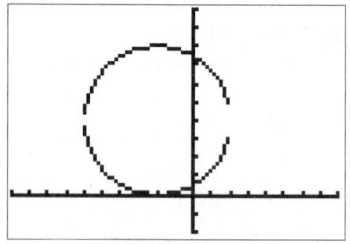

[–10, 8] x [–2, 10]

FIGURE 15.18c

$$y - 4 = \pm\sqrt{16 - (x+2)^2}$$
$$y = 4 \pm \sqrt{16 - (x+2)^2}$$

Since this result has two solutions, we will have to graph the two equations $y = 4 + \sqrt{16 - (x+2)^2}$ and $y = 4 - \sqrt{16 - (x+2)^2}$.

Now, we know that since this circle has its center at $(-2, 4)$ and a radius of 4, the x-values will be from $x = -2 - 4 = -6$ to $x = -2 + 4 = 2$ and the y-values will be from $y = 4 - 4 = 0$ to $y = 4 + 4 = 8$. We will want to set the calculator screen's "range" values so that both of these appear on the screen. We will let x be over the interval $[-8, 4]$ and y be over the interval $[-2, 10]$.

The graph of $y = 4 + \sqrt{16 - (x+2)^2}$ is obtained by first pressing $\boxed{Y=}$ and then pressing the key sequence

4 $\boxed{+}$ $\boxed{\sqrt{}}$ $\boxed{(}$ 16 $\boxed{-}$ $\boxed{(}$ $\boxed{X,T,\theta}$ $\boxed{+}$ 2 $\boxed{)}$ $\boxed{\text{SHIFT}}$ $\boxed{x^2}$ $\boxed{)}$ $\boxed{\text{GRAPH}}$

The result is the graph in Figure 15.18a. Now graph $y = 4 - \sqrt{16 - (x+2)^2}$ by pressing $\boxed{Y=}$, moving the cursor to Y2, and pressing the sequence

4 $\boxed{-}$ $\boxed{\sqrt{}}$ $\boxed{(}$ 16 $\boxed{-}$ $\boxed{(}$ $\boxed{X,T,\theta}$ $\boxed{+}$ 2 $\boxed{)}$ $\boxed{\text{SHIFT}}$ $\boxed{x^2}$ $\boxed{)}$ $\boxed{\text{GRAPH}}$

This graph is displayed over the earlier graph as shown in Figure 15.18b.

This does not look like a circle. First, the top and bottom parts do not touch, and it does not have a circular shape. The fact that parts do not touch is a common error of graphing calculators and computer graphing programs. The noncircular shape is caused by the fact that the calculator screen's ranges for both the x- and y-values are 12 units long. But, the calculator does not have a square screen. In fact, the ratio of the screen's horizontal to vertical size on calculators such as the Casio fx-7700, TI-81, and TI-82 is $3 : 2$; on the TI-85 the ratio is approximately $7 : 4$. So, if we want this graph of a circle to *look* like a circle, we need to have the ratio of the TI-82's calculator screen's range of x-values to the range of y-values be $3 : 2$.

Let's leave the y-values at $[-2, 10]$. This is 12 units, so the x-values should cover 18 units. We will change them to $[-10, 8]$. Now press $\boxed{\text{GRAPH}}$ and the graphs should be redrawn as shown in Figure 15.18c.

If you use a Texas Instruments TI-82 graphics calculator, the following procedure makes the units on the axes the same length. After you have graphed the two functions and get a graph like the one in Figure 15.18b, press the $\boxed{\text{ZOOM}}$ button on the top row of the calculator. You should see a screen similar to the one in shown in Figure 15.18d. This zoom menu presents you with nine options. Pressing a $\boxed{5}$ selects option 5: ZSquare. This option adjusts the current range values so the width of the dots on the x- and y-axes are equalized.

Some calculators will draw a circle if you enter the coordinates of the center and the radius. The next example illustrates this on a TI-82.

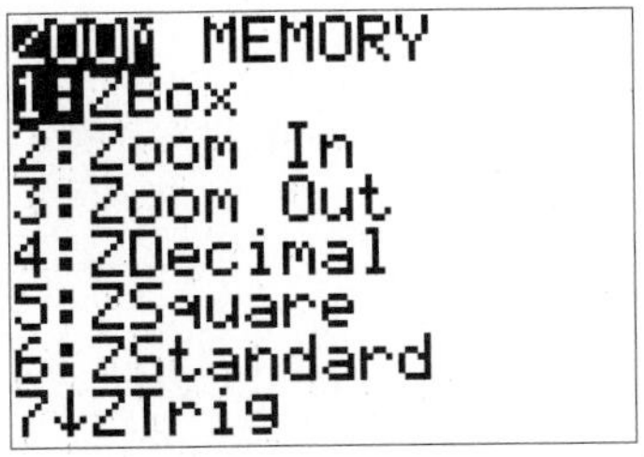

EXAMPLE 15.26

Use a TI-82 to draw the circle

$$(x+2)^2 + (y-4)^2 = 25$$

Solution   Here $h = -2$, $k = 4$, and $r = 5$. The "width" of this circle will stretch from $h - r = -7$ to $h + r = 3$ and the "height" from $k - r = -1$ to $k + r = 9$. A reasonable viewing window that will show this entire circle and will show a circular-looking circle is $[-9.4, 9.4] \times [-2.2, 10.2]$. (Notice that the ratio of width to height is $18.8 : 12.4 \approx 3 : 2$. Press $\boxed{\text{2nd}}$ $\boxed{\text{DRAW}}$ 9. Your screen should look like that shown in Figure 15.19a. Next, enter the values of h, k, and r separated by commas, by pressing $\boxed{(-)}$ 2 $\boxed{,}$ 4 $\boxed{,}$ $5 \boxed{)}$. Press $\boxed{\text{ENTER}}$ and you should get a result similar to the one in Figure 15.19b.

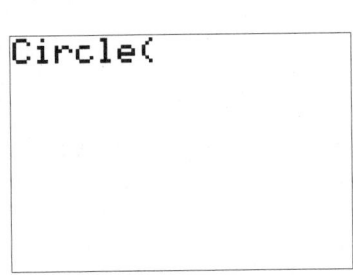

FIGURE 15.18d

FIGURE 15.19a

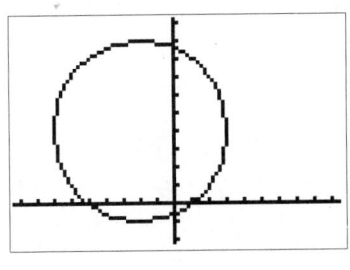

[-9.4, 9.4] x [-2.2, 10.2]

FIGURE 15.19b

EXAMPLE 15.27

Use a graphing calculator to draw the parabola

$$y^2 = -12x$$

Solution   In order to put this in our calculator, we will have to solve this for y, getting $y = \pm\sqrt{-12x}$. This will have to be graphed as two separate functions, $y_1 = \sqrt{-12x}$ and $y_2 = -\sqrt{-12x}$. Make sure you graph both y_1 and y_2. The graph of y_1 is the "top half" of the parabola, and the graph of y_2 is the "bottom half." The result is shown in Figure 15.20.

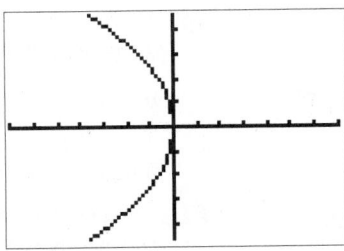

[-14.1, 14.1, 2] x [-9.3, 9.3, 2]

FIGURE 15.20

Exercise Set 15.3

In Exercises 1–12, determine the coordinates of the focus and the equation of the directrix of each parabola. State the direction in which each parabola opens and sketch each curve.

1. $x^2 = 4y$ **4.** $y^2 = -20x$ **7.** $y^2 = 10x$ **10.** $x^2 = -17y$

2. $x^2 = 12y$ **5.** $x^2 = -8y$ **8.** $y^2 = -14x$ **11.** $y^2 = -21x$

3. $y^2 = -4x$ **6.** $x^2 = -16y$ **9.** $x^2 = 2y$ **12.** $y^2 = 5x$

In Exercises 13–20, determine the equation of the parabola satisfying the given conditions.

13. Focus $(0, 4)$, directrix $y = -4$

14. Focus $(0, -5)$, directrix $y = 5$

15. Focus $(-6, 0)$, directrix $x = 6$

16. Focus $(3, 0)$, directrix $x = -3$

17. Focus $(2, 0)$, vertex $(0, 0)$

18. Focus $(0, -7)$, vertex $(0, 0)$

19. Vertex $(0, 0)$, directrix $x = -\frac{3}{2}$

20. Vertex $(0, 0)$, directrix $y = \frac{3}{4}$

Solve Exercises 21–28.

21. *Civil engineering*   In a suspension bridge, the main cables are in a parabolic shape. This is because a parabola is the only shape that will bear the total weight load evenly. The twin towers of a certain bridge extend 90 m above the road surface and are 360 m apart, as shown in Figure 15.21. The cables are suspended from the tops of the towers and are tangent to the road surface at a point midway between the towers. What is the height of the cable above the road surface at a point 100 m from the center of the bridge?

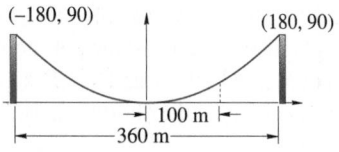

$(-180, 90)$ $(180, 90)$

⟶ 100 m ⟵

⟵――360 m――⟶

FIGURE 15.21

22. *Optics*   A parabolic reflector is a mirror formed by rotating a parabola around its axis. If a light is placed at the focus of the mirror, all the light rays starting from the focus will be reflected off the mirror in lines parallel to the axis. A radar antenna is constructed in the shape of a parabolic reflector. The receiver is placed at the focus. If the reflector has a diameter of 2 m and a depth of 0.4 m, what is the location of the receiver?

23. *Electronics*   When rays from a distant source strike a parabolic reflector, they will be reflected to a single point—the focus. A parabolic antenna is used to catch the television signals from a satellite. The antenna is 5 m across and 1.5 m deep. If the receiver is located at the focus, what is its location?

24. *Energy*   The rate at which heat is dissipated in an electric current is referred to as the power loss. This loss, P, is given by the relationship $P = I^2 R$, where I is the current and R is the resistance. If the resistance is $10\,\Omega$, sketch a graph of the power loss as a function of the current.

25. *Electronics*   For a simple linear resistance R in Ω, the power, P, in W dissipated in the circuit depends on the current, I, in A according to the equation $P = RI^2$. Graph P as a function of I for $0 \leq I \leq 6.0$ when $R = 0.25\,\Omega$.

26. *Thermodynamics*   The heat, H, in joules (J) produced by a voltage, V, across a heating coil is given by $H = \dfrac{V^2}{R}t$, where t is the time and R the resistance. If $t = 2\,\text{h}$ and $R = 15\,\Omega$, sketch the graph of H as a function of V from $V = 0$ to $V = 120\,\text{V}$.

27. *Broadcasting*   A cable television "dish" is 2 ft deep and measures 10 ft across at its opening. A technician must place the receiver at the focus. How far is this from the vertex?

28. *Astronomy* Astronomers and optical experts have been exploring the possibility of making mirrors from liquids. When a liquid is spun in a container, the surface of the liquid takes on a parabolic shape. If a liquid metal, such as mercury, is used, then the optical quality is about as good as that of more expensive ground-glass mirrors. The parabolic shape is kept as long as the mirror is kept spinning. The focal length of the mirror is proportional to the square of the rotation period, p. (The rotation period is the reciprocal of the rotational frequency.) If a mirror spinning at 45 rpm has a focal length of 22.1 cm, then

(a) determine the focal length of a mirror spinning at $33\frac{1}{3}$ rpm

(b) determine the focal length of a mirror spinning at 15 rpm

In Your Words

29. Describe how you can tell by looking at a standard equation for a parabola whether it is a horizontal or a vertical parabola.

30. Explain how to tell from the equation if a parabola opens upward, downward, left, or right.

≡ 15.4
THE ELLIPSE

The third curve we will study in this chapter is the ellipse. An **ellipse** is the set of points in a plane that have the sum of their distances from two fixed points a constant. Each of the fixed points is called a **focus**. The **major axis** of an ellipse is the line segment through the two foci with endpoints on the ellipse. The endpoints of the major axis are called the **vertices**. In Figure 15.22, the foci are labeled F and F' and the vertices are V and V'. The **center** C of the ellipse is the midpoint of the segment joining the foci. The segment through the center and perpendicular to the major axis is called the **minor axis**. The endpoints of the minor axis are on the ellipse. In Figure 15.22, the endpoints of the minor axis are M and M'. The major axis is always longer than the minor axis.

If we were to draw an ellipse with its center at the origin and its major axis along the x-axis, we would get a figure like the one in Figure 15.23. We will let the vertices have the coordinates $V(a,0)$ and $V'(-a,0)$. The foci will have the coordinates $F(c,0)$ and $F'(-c,0)$. If $P(x,y)$ is any point on the ellipse, then the distance from P to F, d_1, plus the distance from P to F', d_2, is a constant, which we will call $2a$. The sum of these distances, $d_1 + d_2 = (a-c) + (a+c) = 2a$.

The points F, F', and M form an isosceles triangle with the lengths of two equal sides, $\overline{FM}$ and $\overline{F'M}$, being a. Thus, F, M, and C form a right triangle with legs of lengths c and b and hypotenuse of length a. As a result, we see that $a^2 = b^2 + c^2$.

Applying the distance formula, we see that $d_1 = \sqrt{(x-c)^2 + y^2}$ and $d_2 = \sqrt{(x+c)^2 + y^2}$. Since $d_1 + d_2 = 2a$, we have $\sqrt{(x-c)^2 + y^2} + \sqrt{(x+c)^2 + y^2} = 2a$. Using the techniques from Chapter 12 for solving radical equations, we get

$$\sqrt{(x-c)^2 + y^2} = 2a - \sqrt{(x+c)^2 + y^2}$$

$$\left[\sqrt{(x-c)^2 + y^2}\right]^2 = \left[2a - \sqrt{(x+c)^2 + y^2}\right]^2 \qquad \text{Square both sides.}$$

$$(x-c)^2 + y^2 = (2a)^2 - 4a\sqrt{(x+c)^2 + y^2} + (x+c)^2 + y^2$$

$$4a\sqrt{(x+c)^2 + y^2} = 4a^2 + 4cx$$

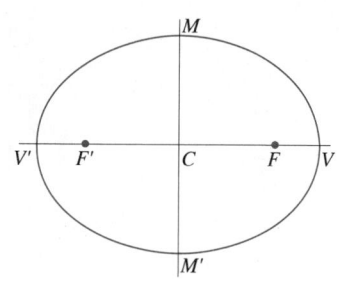

FIGURE 15.22

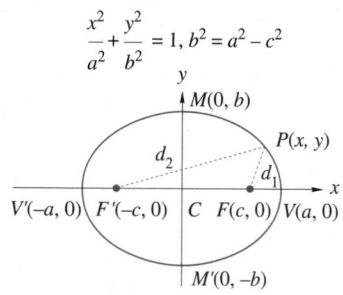

$$\frac{x^2}{a^2} + \frac{y^2}{b^2} = 1, \ b^2 = a^2 - c^2$$

FIGURE 15.23

$$a\sqrt{(x+c)^2 + y^2} = a^2 + cx$$

$$\left[a\sqrt{(x+c)^2 + y^2}\right]^2 = \left[a^2 + cx\right]^2 \qquad \text{Square again.}$$

$$a^2\left[(x+c)^2 + y^2\right] = a^4 + 2a^2cx + c^2x^2$$

$$a^2x^2 + 2a^2cx + a^2c^2 + a^2y^2 = a^4 + 2a^2cx + c^2x^2$$

$$(a^2 - c^2)x^2 + a^2y^2 = a^2(a^2 - c^2)$$

But, $a^2 = b^2 + c^2$, so $a^2 - c^2 = b^2$. The equation becomes

$$b^2x^2 + a^2y^2 = a^2b^2$$

Dividing both sides by a^2b^2 and simplifying results in one of the standard equations for an ellipse:

$$\frac{x^2}{a^2} + \frac{y^2}{b^2} = 1$$

Standard Equation: Ellipse with a Horizontal Major Axis

The standard equation of an ellipse with center at $(0, 0)$ and the major axis on the x-axis is

$$\frac{x^2}{a^2} + \frac{y^2}{b^2} = 1$$

where $a > b$.

The vertices are $(-a, 0)$ and $(a, 0)$.

The endpoints of the minor axis are $(0, -b)$ and $(0, b)$.

The foci are at $(-c, 0)$ and $(c, 0)$, where $c^2 = a^2 - b^2$.

EXAMPLE 15.28

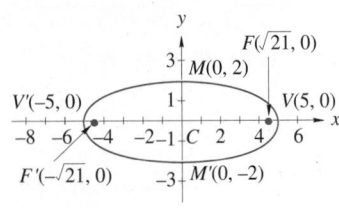

FIGURE 15.24

Sketch the ellipse $4x^2 + 25y^2 = 100$. Give the coordinates of the vertices, foci, and endpoints of the minor axis.

Solution   This equation is not in the standard form, since the right-hand side is not 1. To get the equation into the standard form, we must divide both sides of the equation by 100. This produces the equation

$$\frac{x^2}{25} + \frac{y^2}{4} = 1 \qquad .$$

From this, we see that $a^2 = 25$, so $a = 5$ and $b^2 = 4$ or $b = 2$. Since $c^2 = a^2 - b^2 = 25 - 4 = 21$, we see that $c = \sqrt{21}$. So, the vertices are $V(5, 0)$ and $V'(-5, 0)$, and the foci are $F(\sqrt{21}, 0)$ and $F'(-\sqrt{21}, 0)$. The endpoints of the minor axis are $M(0, 2)$ and $M'(0, -2)$. A sketch of this ellipse is shown in Figure 15.24.

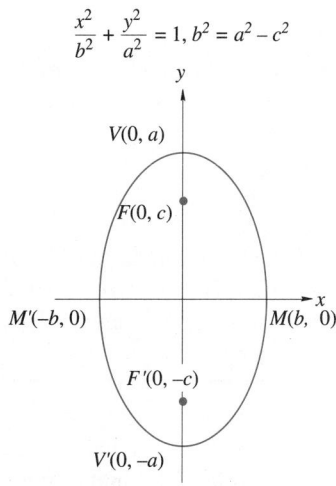

$$\frac{x^2}{b^2} + \frac{y^2}{a^2} = 1, b^2 = a^2 - c^2$$

FIGURE 15.25

Not all ellipses have their major axis along the x-axis. For some situations, it is more convenient if the center is at the origin and the major axis is along the y-axis. When that is the case, the vertices are $V(0, a)$ and $V'(0, -a)$, the foci are $F(0, c)$ and $F'(0, -c)$, and the endpoints of the minor axis are $M(b, 0)$ and $M'(-b, 0)$. In this case, we have the following standard form of the equation for an ellipse. A typical graph of an ellipse with center at the origin and a vertical major axis is shown in Figure 15.25.

Standard Equation: Ellipse with a Vertical Major Axis

The standard equation of an ellipse with center at $(0, 0)$ and the major axis on the y-axis is

$$\frac{x^2}{a^2} + \frac{y^2}{b^2} = 1$$

where $a < b$.

The vertices are $(0, -b)$ and $(0, b)$.

The endpoints of the minor axis are $(-a, 0)$ and $(a, 0)$.

The foci are at $(0, -c)$ and $(0, c)$, where $c^2 = b^2 - a^2$.

If we put these two types of ellipses together, we see that the equation of an ellipse with center at the origin and foci on a coordinate axis can always be written in the form

$$\frac{x^2}{p^2} + \frac{y^2}{q^2} = 1 \qquad \text{or} \qquad q^2 x^2 + p^2 y^2 = p^2 q^2$$

where p and q are both positive. If $p^2 > q^2$, then the major axis is along the x-axis. If $p^2 < q^2$, then the major axis is along the y-axis.

≡ Note

The x-axis is the major, or longer, axis if the denominator of x^2 is the larger denominator, and the y-axis is the major axis if the denominator of y^2 is the larger denominator.

Hint

Remember, c^2 is found by subtracting the smaller denominator from the larger. You may want to remember this as $c^2 = |a^2 - b^2|$.

EXAMPLE 15.29

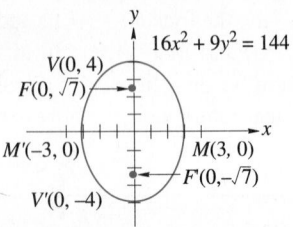

FIGURE 15.26

Discuss and graph the equation $16x^2 + 9y^2 = 144$.

Solution Dividing both sides of the equation by 144, we get the standard equation

$$\frac{16x^2}{144} + \frac{9y^2}{144} = \frac{144}{144}$$

$$\frac{x^2}{9} + \frac{y^2}{16} = 1$$

So, $a^2 = 9$, $b^2 = 16$, and $c^2 = b^2 - a^2 = 7$. Since $a = 3$ and $b = 4$, we see that $a < b$, so the major axis is along the y-axis. The vertices are $V(0, 4)$ and $V'(0, -4)$, the foci are $F(0, \sqrt{7})$ and $F'(0, -\sqrt{7})$, and the endpoints of the minor axis are $M(3, 0)$ and $M'(-3, 0)$. The sketch of this ellipse is shown in Figure 15.26.

EXAMPLE 15.30

Find the equation of the ellipse with center $(0, 0)$, one focus at $(5, 0)$, and a vertex at $(-8, 0)$.

Solution The second focus is at $(-5, 0)$ and the other vertex is at $(8, 0)$. Since the major axis is horizontal (along the x-axis), the equation is of the form

$$\frac{x^2}{a^2} + \frac{y^2}{b^2} = 1$$

where $c^2 = a^2 - b^2$, $a > b$.

Since one focus is at $(5, 0)$, $c = 5$ and the fact that one vertex is at $(8, 0)$ tells us that $a = 8$. So, $5^2 = 8^2 - b^2$ or $b^2 = 64 - 25 = 39$, and $b = \sqrt{39}$. The equation of this ellipse is $\dfrac{x^2}{64} + \dfrac{y^2}{39} = 1$.

Reflective Properties of Ellipses

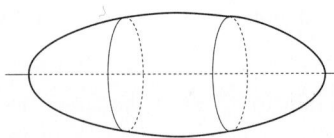

FIGURE 15.27

A three-dimensional object called a **ellipsoid of revolution** is formed when an ellipse is revolved around one of its axes of symmetry. In Figure 15.27, the ellipse has been revolved around its longer axis.

These ellipsoids of revolution have uses based on the fact that rays leaving or passing through one focus of an ellipse are all reflected by the ellipse so that they pass through the other focus of the ellipse. One architectural application of these ellipsoids of revolution is in the construction of whispering galleries, such as the one in Statuary Hall of the Capitol in Washington, D.C. or St. Paul's Cathedral in London, England. If you position yourself at one focus in a whispering gallery, you will be able to hear anything said at the other focus, regardless of the direction in which the speakers address their remarks.

Application

EXAMPLE 15.31

A lithotripter is a medical device that is used to break up kidney stones with shock waves through water. An ellipsoid is cut in half and careful measurements are used to place the patient's kidney stones at one focus of the ellipse. A source

EXAMPLE 15.31 (Cont.)

that produces ultra–high-frequency shock waves is placed at the other focus. (See Figures 15.28a and b.) The shock waves are reflected by the ellipsoid to the other focus, where they break up the kidney stones.

If the endpoints of the major axis are located 6 in. from the center of the ellipse that is rotated to make a lithotripter and one end of the minor axis is 2.5 in. from the center, what are the locations of the lithotripter's foci?

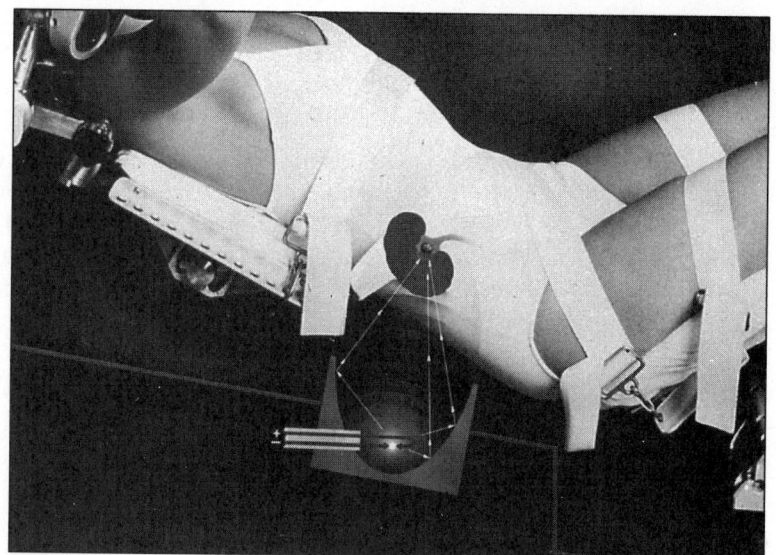

Courtesy of Dornier Medical Systems Inc.

FIGURE 15.28a   **FIGURE 15.28b**

Solution   This appears to be a vertically oriented ellipse. If we place the drawing of the cross-section of the ellipse in Figure 15.28a on a coordinate system so that the center is at the origin and the major axis is on the y-axis, we get Figure 15.28b. The solid curve in Figure 15.28b represents the lithotripter. The dotted curve represents the portion of the ellipse that has been removed. From the given information, we know that the ellipse in Figure 15.28b has endpoints on the major axis at $(0, -6)$ and $(0, 6)$, and that the endpoints of the minor axis are $(-2.5, 0)$ and $(2.5, 0)$. From the given data, we know that the desired ellipse has the equation $\frac{x^2}{a^2} + \frac{y^2}{b^2} = \frac{x^2}{2.5^2} + \frac{y^2}{6^2} = 1$. Thus, $a = 2.5$ and $b = 6$ and, since $c^2 = b^2 - a^2$, we have $c^2 = 6^2 - 2.5^2 = 29.75$ and so $c = \sqrt{29.75} \approx 5.45$. Thus, the foci are at $(0, -5.45)$ and $(0, 5.45)$.

Using a Graphing Calculator

When using a graphing calculator, choose a viewing window that places the endpoints of the major and minor axes are in the window. Again, since the calculator will graph only a function, you will need to solve the equation algebraically for y and then graph the two equations that result.

EXAMPLE 15.32

Use a graphing calculator to graph $\dfrac{x^2}{9} + \dfrac{y^2}{64} = 1$.

Solution   Since this is in standard form, we see that this is a vertically oriented ellipse. Here $a^2 = 9$, so $a = 3$ and $b^2 = 64$, so $b = 8$. A viewing window that will include the entire ellipse would set Xmin $= -3$, Xmax $= 3$, Ymin $= -8$, and Ymax $= 8$. As we will see, this does not graph the ellipse so that it appears to be vertically oriented.

Now, solve the equation for y.

$$\frac{x^2}{9} + \frac{y^2}{64} = 1$$

$$\frac{y^2}{64} = 1 - \frac{x^2}{9}$$

$$y^2 = 64\left(1 - \frac{x^2}{9}\right)$$

$$y = \pm\sqrt{64\left(1 - \frac{x^2}{9}\right)}$$

$$= \pm 8\sqrt{1 - \frac{x^2}{9}}$$

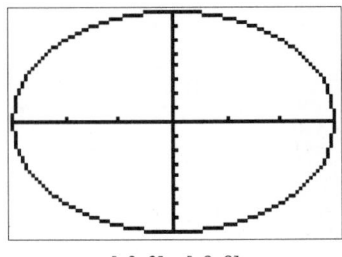

[–3, 3] × [–8, 8]

FIGURE 15.29a

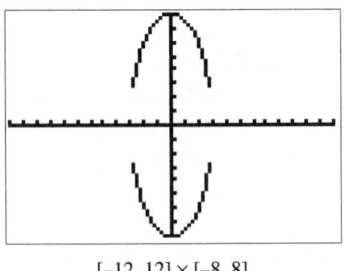

[–12, 12] × [–8, 8]

FIGURE 15.29b

We will graph $y_1 = 8\sqrt{1 - \dfrac{x^2}{9}}$ and $y_2 = -8\sqrt{1 - \dfrac{x^2}{9}}$. This result is shown in Figure 15.29a.

This ellipse does not look as if it is a vertical ellipse. If we set the ratio $(\text{Xmax} - \text{Xmin}) : (\text{Ymax} - \text{Ymin}) = 3 : 2$ and leave the Ymin and Tmax settings alone, we get

$$\frac{3}{2} = \frac{\text{Xmax} - \text{Xmin}}{16}$$

$$24 = \text{Xmax} - \text{Xmin}$$

This ellipse is centered at the origin, so Xmin and Xmax should be the same distance from the center. Let Xmin $= -12$ and Xmax $= 12$. The resulting graph is shown in Figure 15.29b.

Exercise Set 15.4

In Exercises 1–8, find the equation of the ellipse with the stated properties. Each ellipse has its center at (0,0).

1. Focus at $(4, 0)$, vertex at $(6, 0)$
2. Focus at $(9, 0)$, vertex at $(12, 0)$
3. Focus at $(0, 2)$, vertex at $(0, -4)$
4. Focus at $(0, -3)$, vertex at $(0, 5)$

5. Focus at $(3, 0)$, vertex at $(5, 0)$
6. Focus at $(0, -2)$, vertex at $(0, -3)$
7. Length of the minor axis 6, vertex at $(4, 0)$
8. Length of the minor axis 10, vertex at $(0, -6)$

In Exercises 9–16, give the coordinates of the vertices, foci, and endpoints of the minor axis, and sketch each curve.

9. $\dfrac{x^2}{4} + \dfrac{y^2}{9} = 1$

10. $\dfrac{x^2}{9} + \dfrac{y^2}{4} = 1$

11. $\dfrac{x^2}{4} + \dfrac{y^2}{1} = 1$

12. $\dfrac{x^2}{16} + \dfrac{y^2}{36} = 1$

13. $4x^2 + y^2 = 4$
14. $9x^2 + 4y^2 = 36$
15. $25x^2 + 36y^2 = 900$

16. $9x^2 + y^2 = 18$

Solve Exercises 17–24.

17. *Astronomy*   The orbit of the earth is an ellipse with the sun at one focus. The major axis has a length of 2.992×10^8 km and the ratio of $\dfrac{c}{a} = \dfrac{1}{60}$. What is the length of the minor axis? $\left(\text{The ratio } \dfrac{c}{a} \text{ is called the } \textbf{eccentricity.}\right)$

18. *Civil engineering*   In order to support a bridge, an arch in the shape of the upper half of an ellipse is built. As shown in Figure 15.30, the bridge is to span a river 80 ft wide and the center of the arch is to be 24 ft above the water. If the water level is used as the x-axis and the y-axis passes through the center of the bridge, write an equation for the ellipse that forms the arch of the bridge.

FIGURE 15.30

19. *Mechanical engineering*   A circular pipe has an inside diameter of 10 in. One end of the pipe is cut at an angle of $45°$, as shown in Figure 15.31. What are the lengths of the major and minor axes of the elliptical opening?

FIGURE 15.31

20. *Mechanical engineering*   Elliptical gears are used in some machinery to provide a quick-return mechanism or a slow power stroke (for heavy cutting) in each revolution. Figure 15.32 shows two congruent gears that remain in contact as they rotate around the indicated foci. If the distance between the foci used at the centers of rotation is 6 cm and the shortest distance from a focus to the edge of a gear is 1 cm, what is the length of the minor axis?

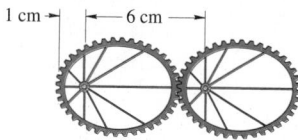

FIGURE 15.32

21. *Astronomy*   The comet Kahoutek has major and minor axes of lengths 3 600 and 44 astronomical units. One astronomical unit is about 1.495×10^8 km. What is the eccentricity of the comet's orbit?

22. *Space technology*   A satellite orbits the earth in an elliptical path with focus at the center of the earth. The altitude of the satellite ranges from 764.4 km to 3 017.5 km. If the radius of the earth is approximately 6 373 km, what is the equation of the path of the orbit?

23. *Mechanical engineering*   Two circular pipes that are each 8.0 in. in diameter are joined at a 45° angle as shown in Figure 15.33. Find the length of the major axis for the elliptical intersection.

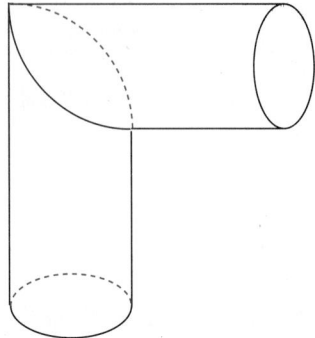

FIGURE 15.33

24. *Transportation engineering*   A road passes through a tunnel whose cross-section is a semiellipse 52 ft wide and 12.5 ft high at the center. How tall is the tallest vehicle that can pass under the tunnel at a point 14 ft from the center?

In Your Words

25. Describe how you can tell by inspecting the standard equation of an ellipse whether it is a horizontal or a vertical ellipse.

26. Explain how to use the endpoints of the axes to locate the foci of an ellipse.

≡ 15.5
THE HYPERBOLA

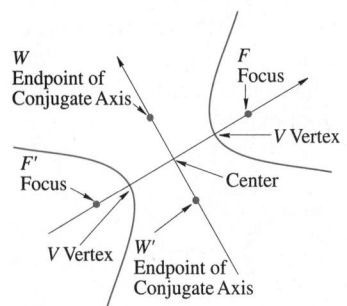

FIGURE 15.34

The fourth and last conic section that we will study is the hyperbola. A **hyperbola** is the set of points in a plane for which the difference of the distances of the points from two fixed points is a constant. As with the ellipse, each of the fixed points is called a **focus**. The **transverse axis** is the line segment through the two foci with its endpoints on the hyperbola. The endpoints of the transverse axis are called the **vertices**.

A hyberbola is shown by the blue curve in Figure 15.34. In Figure 15.34, the foci are labeled F and F' and the vertices V and V'. The **center** C of the hyperbola is the midpoint of the foci. The line segment through the center that is perpendicular to the transverse axis is called the **conjugate axis**. The endpoints of the conjugate axis, labeled W and W' in Figure 15.34, are not on the hyperbola. We will learn how to determine the length of the conjugate axis later in this section.

If we draw a hyperbola with its center at the origin and its transverse axis along the x-axis, we get a figure like the one in Figure 15.35. If $P(x, y)$ is any point on the hyperbola, the distance from P to F, d_1, minus the distance from P to F', d_2, is a constant. The difference of the distances is $|d_1 - d_2| = |(c - a) - (c + a)| = 2a$, $a > 0$.

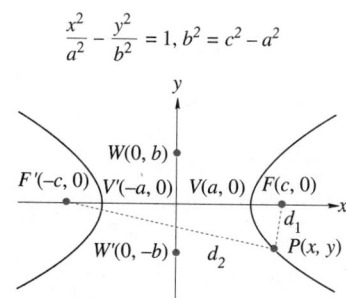

$$\frac{x^2}{a^2} - \frac{y^2}{b^2} = 1, b^2 = c^2 - a^2$$

FIGURE 15.35

Here we will use the distance formulas $d_1 = \sqrt{(x - c)^2 + y^2}$ and $d_2 = \sqrt{(x + c)^2 + y^2}$. As a result, we compute the difference of the distances as

$$\left| \sqrt{(x - c)^2 + y^2} - \sqrt{(x + c)^2 + y^2} \right| = 2a$$

or $$\sqrt{(x - c)^2 + y^2} - \sqrt{(x + c)^2 + y^2} = \pm 2a$$

Using the same technique we used on the ellipse, we get

$$c^2 x^2 - a^2 x^2 - a^2 y^2 = a^2 c^2 - a^4$$

or $$x^2(c^2 - a^2) - a^2 y^2 = a^2(c^2 - a^2)$$

If we let $b^2 = c^2 - a^2$, this can be written as $b^2 x^2 - a^2 y^2 = a^2 b^2$. Dividing both sides by $a^2 b^2$, the equation becomes

$$\frac{x^2}{a^2} - \frac{y^2}{b^2} = 1$$

which is one of the standard equations for a hyperbola. It can be shown that the lines $y = \frac{b}{a}x$ and $y = -\frac{b}{a}x$ are the hyperbola's asymptotes.

Standard Equation: Hyperbola with a Horizontal Major Axis

The standard equation of a hyperbola with center at $(0, 0)$ and the major axis on the x-axis is

$$\frac{x^2}{a^2} - \frac{y^2}{b^2} = 1$$

The vertices are $(-a, 0)$ and $(a, 0)$.

The endpoints of the conjugate axis are $(0, -b)$ and $(0, b)$.

The foci are at $(-c, 0)$ and $(c, 0)$, where $c^2 = a^2 + b^2$.

The lines $y = \frac{b}{a}x$ and $y = -\frac{b}{a}x$ are **asymptotes** of this hyperbola.

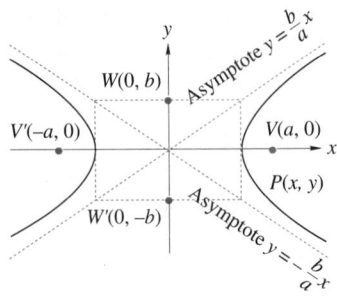

FIGURE 15.36

The asymptotes provide convenient guidelines for drawing hyperbolas. Asymptotes are easily drawn as the diagonals of a rectangle. The midpoints of the sides of the rectangles are the vertices and the endpoints of the conjugate axis, as shown in Figure 15.36.

If the transverse axis is along the y-axis and the conjugate axis is along the x-axis, the hyperbola has a second standard form.

Standard Equation: Hyperbola with a Vertical Major Axis

The standard equation of a hyperbola with center at $(0,0)$ and the major axis on the y-axis is

$$\frac{y^2}{a^2} - \frac{x^2}{b^2} = 1$$

The vertices are $(0, -a)$ and $(0, a)$.

The endpoints of the conjugate axis are $(-b, 0)$ and $(b, 0)$.

The foci are at $(0, -c)$ and $(0, c)$, where $c^2 = a^2 + b^2$.

The lines $y = \dfrac{a}{b}x$ and $y = -\dfrac{a}{b}x$ are **asymptotes** of this hyperbola.

Caution

The formulas relating a, b, and c for the hyperbola are different from the corresponding formulas for the ellipse.

For the ellipse
$$c^2 = |a^2 - b^2| \quad \text{(See Figure 15.37a.)} \qquad a > c \quad \text{(See Figure 15.37b.)}$$

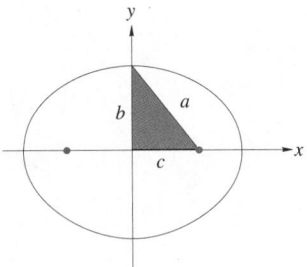

FIGURE 15.37a

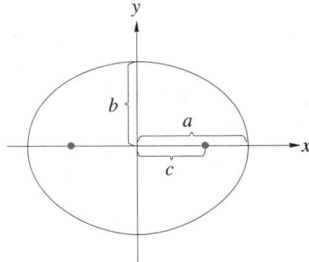

FIGURE 15.37b

For the hyperbola

$$c^2 = a^2 + b^2 \quad \text{(See Figure 15.38a.)} \qquad c > a \quad \text{(Figure 15.38b.)}$$

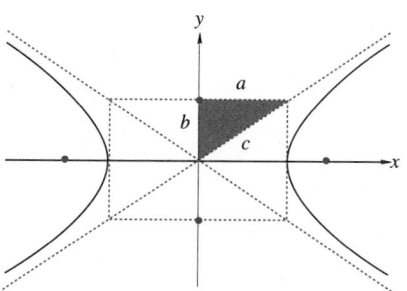

FIGURE 15.38a

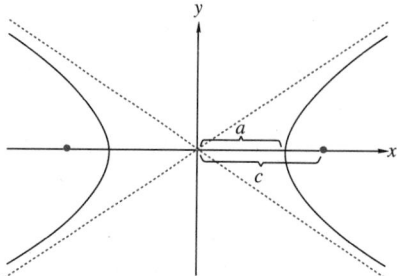

FIGURE 15.38b

EXAMPLE 15.33

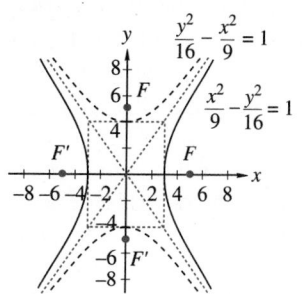

FIGURE 15.39

Discuss and graph the equation $16x^2 - 9y^2 = 144$.

Solution   To get the equation in standard form, we divide both sides by 144 and obtain the equation

$$\frac{x^2}{9} - \frac{y^2}{16} = 1$$

This is the equation of a hyperbola with center $(0,0)$. The transverse axis is the x-axis. Also, $a^2 = 9$ and $b^2 = 16$, so $c^2 = a^2 + b^2 = 9 + 16 = 25$. The foci are at $F(5,0)$ and $F'(-5,0)$, the vertices at $V(3,0)$ and $V'(-3,0)$, and the endpoints of the conjugate axis at $W(0,4)$ and $W'(0,-4)$. If we form the rectangle with midpoints of its sides V, V', W, and W' and draw its diagonals, we get the asymptotes $y = \frac{4}{3}x$ and $y = -\frac{4}{3}x$ shown by the dotted lines in Figure 15.39. The sketch of the graph is shown by the solid curve in Figure 15.39.

EXAMPLE 15.34

Discuss and graph the equation $9y^2 - 16x^2 = 144$.

Solution   Again, we divide both sides by 144 to get

$$\frac{y^2}{16} - \frac{x^2}{9} = 1$$

This is the equation of a hyperbola with center $(0,0)$. The transverse axis is the y-axis. Also, $a^2 = 16$ and $b^2 = 9$, so $c^2 = 25$. The foci are at $F(0,5)$ and $F'(0,-5)$, the vertices at $V(0,4)$ and $V'(0,-4)$, and the endpoints of the conjugate axis at $W(3,0)$ and $W'(-3,0)$. This hyperbola has the same asymptotes as the hyperbola in Example 15.33. The sketch of this hyperbola is shown by the dotted curves in Figure 15.39. The hyperbolas in Examples 15.33 and 15.34 are called **conjugate hyperbolas**, because they have the same set of asymptotes.

EXAMPLE 15.35

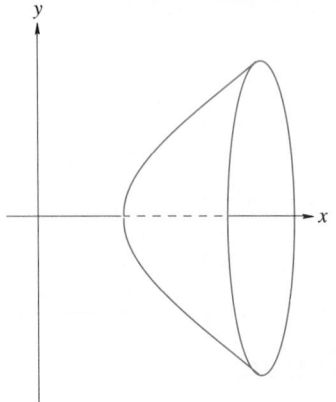

FIGURE 15.40

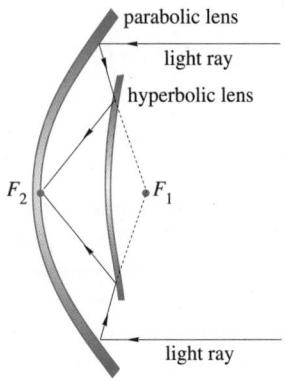

FIGURE 15.41

Find the equation of a hyperbola with center $(0,0)$, a focus at $(6,0)$, and a vertex at $(-3,0)$.

Solution   The second focus is at $(-6,0)$, and the second vertex at $(3,0)$. Since the foci and vertices are on the x-axis, it is the transverse axis. The equation is

$$\frac{x^2}{a^2} - \frac{y^2}{b^2} = 1$$

Since $a = 3$ and $c = 6$, then $b^2 = c^2 - a^2 = 36 - 9 = 27$. The equation of this hyperbola is

$$\frac{x^2}{9} - \frac{y^2}{27} = 1$$

or $27x^2 - 9y^2 = 243$.

Reflective Properties of Hyperbolas

Like the other conic sections, hyperbolas are used in many of today's technological applications. A three-dimensional object called a **hyperboloid of revolution** is formed when a hyperbola is revolved around its transverse axis of symmetry. (See Figure 15.40.) Notice that a hyperbola of revolution normally uses only one branch of the hyperbola. Uses of these hyperboloids of revolution are based on the fact that rays aimed at one focus of a hyperbolic reflector are reflected so that they pass through the other focus.

Hyperboloids are frequently used to locate the source of a sound or radio signal. In Exercise Set 15.5, Exercise 18 describes how a system called LORAN (for LOng RAnge Navigation) uses hyperbolas as aids for navigation.

A telescope can use both a hyperbolic lens and a parabolic lens. The main lens is parabolic with its focus at F_1 and vertex at F_2. A second lens is hyperbolic with its foci at F_1 and F_2, as shown in Figure 15.41. The eye is positioned at F_2. The key to this arrangement is that the same point, F_1, is both the focus of the parabola and a focus of the hyperbola.

Application

EXAMPLE 15.36

As we will see in Section 15.6, many telescopes use both a parabolic reflector and a hyperbolic reflector. In this example, we shall examine a simple hyperbolic reflector. In Section 15.6, we shall consider an application that involves both parabolic and hyperbolic reflectors.

A hyperbolic reflector is designed to fit in a confined area. It is possible to get the foci so they are 8 ft apart and one vertex is 3 ft from a focus. What is the equation for this hyperbolic reflector?

Solution   We shall arrange this reflector on a coordinate system, so that the transverse axis is on the x-axis and the center is at the origin. A cross-section through the axis of this hyperbolic reflector has been drawn on the desired coordinate system in Figure 15.42.

EXAMPLE 15.36 (Cont.)

Since the foci are 8 ft apart, this means that the foci are at $(-4, 0)$ and $(4, 0)$. The given vertex is 3 ft from one of the foci (and hence, 5 ft from the other), so it is at $(-1, 0)$ or $(1, 0)$. We have $a = 1$ and $c = 4$, so $b^2 = c^2 - a^2 = 4^2 - 1^2 = 15$. Thus, the desired equation is $\dfrac{x^2}{1} - \dfrac{y^2}{15} = 1$.

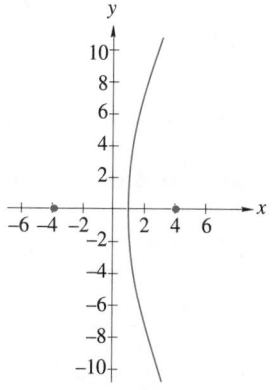

FIGURE 15.42

Using a Graphing Calculator

When using a graphing calculator, choose a viewing window that places the endpoints of the transverse and conjugate axes in substantially from the edge of the window. The next example will give a way to estimate where to put the viewing window. Again, since the calculator will graph only a function, you will need to solve the equation algebraically for y and then graph the two equations that result.

EXAMPLE 15.37

Use a graphing calculator to graph $\dfrac{x^2}{25} - \dfrac{y^2}{81} = 1$.

Solution Since this is in standard form, we see that this is a horizontally oriented hyperbola. Here $a^2 = 25$, so $a = 5$ and $b^2 = 81$, so $b = 9$. A viewing window that sets Xmin $= -5$, Xmax $= 5$, Ymin $= -9$, and Ymax $= 9$ would only show the vertices of the hyperbola. Thus, we need a wider window. As rough rule, we will use $2a$ and $2b$ to set the window settings. For this hyperbola, we will use a viewing window that sets Xmin $= -10$, Xmax $= 10$, Ymin $= -18$, and Ymax $= 18$.

Next, solve the equation for y.

$$\frac{x^2}{25} - \frac{y^2}{81} = 1$$

$$\frac{y^2}{81} = \frac{x^2}{25} - 1$$

$$y^2 = 81\left(\frac{x^2}{25} - 1\right)$$

$$y = \pm\sqrt{81\left(\frac{x^2}{25} - 1\right)}$$

$$y = \pm 9\sqrt{\frac{x^2}{25} - 1}$$

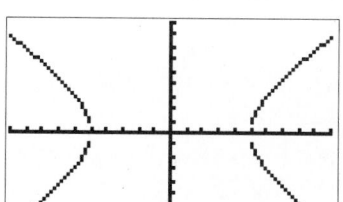

$[-10, 10, 1] \times [-18, 18, 2]$

FIGURE 15.43

We will graph $y_1 = 9\sqrt{\dfrac{x^2}{25} - 1}$ and $y_2 = -9\sqrt{\dfrac{x^2}{25} - 1}$. This result is shown in Figure 15.43.

Exercise Set 15.5

In Exercises 1–8, find the equation of the hyperbola with the stated properties. Each hyperbola has its center at (0,0).

1. Focus at $(6,0)$, vertex at $(4,0)$

2. Focus at $(12,0)$, vertex at $(9,0)$

3. Focus at $(0,5)$, vertex at $(0,-3)$

4. Focus at $(0,-4)$, vertex at $(0,2)$

5. Focus at $(5,0)$, vertex at $(3,0)$

6. Focus at $(0,-3)$, vertex at $(0,-2)$

7. Focus at $(4,0)$, length of conjugate axis, 6

8. Focus at $(0,-6)$, length of conjugate axis, 10

In Exercises 9–16, give the coordinates of the vertices, foci, and endpoints of the conjugate axis and sketch each curve.

9. $\dfrac{x^2}{4} - \dfrac{y^2}{9} = 1$

10. $\dfrac{x^2}{9} - \dfrac{y^2}{4} = 1$

11. $\dfrac{y^2}{4} - \dfrac{x^2}{1} = 1$

12. $\dfrac{y^2}{36} - \dfrac{x^2}{16} = 1$

13. $4x^2 - y^2 = 4$

14. $9x^2 - 4y^2 = 36$

15. $36y^2 - 25x^2 = 900$

16. $y^2 - 9x^2 = 18$

Solve Exercises 17–26.

17. The **eccentricity** e of the hyperbola is defined as $\dfrac{c}{a}$. Since $c > a$, then $e > 1$. (Remember, for the ellipse, $0 < e < 1$.) What is the equation of a hyperbola with center at the origin, vertex at $(7,0)$, and $e = 1.5$?

18. *Navigation*   Hyperbolas are used in long-range navigation as part of the LORAN system of navigation. A transmitter is located at each focus and radio signals are sent to the navigator simultaneously from each station. The difference in time at which the signals are received allows the navigator to determine the position. Suppose the transmitting towers are 1 000 km apart on an east-west line. A ship on the line between two towers determines its position to be 250 km from the west tower.

(a) Sketch the signal hyperbola on which the ship is located. Place the center at $(0,0)$ and let the x-axis be the transverse axis.

(b) Find the equation of the hyperbola.

19. A hyperbola for which $a = b$ is called an **equilateral hyperbola**. Find the eccentricity of an equilateral hyperbola.

20. *Civil engineering*   A proposed design for a cooling tower at a nuclear power plant is a branch of a hyperbola rotated about a conjugate axis. Find the equation of the hyperbola passing through the point $(238, -616)$ with center $(0,0)$ and one vertex at $(153, 0)$.

21. *Electronics*   For any given voltage the product of current I and resistance R in a simple circuit is constant. If the current in a circuit is 3.2 A when the resistance is 18 Ω, sketch the graph of current as a function of resistance for this value of the voltage.

22. *Civil engineering*   The silhouette of the cooling tower of a nuclear reactor forms a hyperbola, like the one in Figure 15.44. If the asymptotes are the lines given by $y = 1.25x$ and $y = -1.25x$, find the equation of the hyperbola.

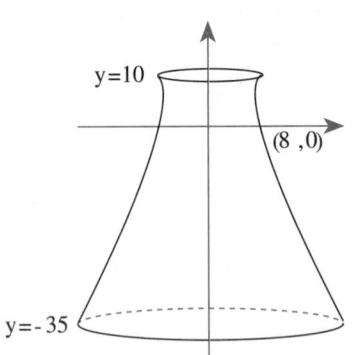

y=10

(8 ,0)

y=-35

FIGURE 15.44

23. *Aeronautical engineering*   A supersonic jet plane flying at a constant speed produces a shock wave in the shape of a cone. If the plane's path is parallel to the ground, the intersection of the cone with the ground is a hyperbola. Find the equation of the hyperbola if the center is at the origin of a coordinate system, one vertex is at $(-42, 0)$, and the hyperbola passes through the point $(-126, 30\sqrt{3})$.

24. *Electronics*   Ohm's law states that voltage V, current I, and resistance R are related by the formula $V = IR$. If $V = 14.0\,\text{V}$, sketch the graph of I as a function of positive values of R.

25. *Wastewater technology*   The formula $A = 8.34FC$ is used to determine the amount, A, of chlorine (in lb) to add to a basin with a flow rate of F million gallons per day to achieve a chlorine concentration of C parts per million (ppm). If 2250 lb of chlorine is added to a certain reservoir, sketch the graph of flow rate as a function of chlorine concentration for this amount of added chlorine.

26. *Wastewater technology*   The formula $A = 8.34FC$ is used to determine the amount, A, of chlorine in kg to add to a basin with a flow rate of F million liters per day to achieve a chlorine concentration of C parts per million (ppm). If 750 kg of chlorine is added to a certain reservoir, sketch the graph of flow rate as a function of chlorine concentration for this amount of added chlorine.

In Your Words

27. Describe how you can tell by inspecting the standard equation of a hyperbola whether it is a horizontal or a vertical hyperbola.

28. Explain how to use the endpoints of the axes to locate the foci of an hyperbola.

29. In Exercise 19, it was stated that an equilateral hyperbola is one in which $a = b$. Why is this called an equilateral hyperbola?

☰ 15.6
TRANSLATION OF AXES

So far, the equations that we have used for the ellipse and the hyperbola have all been centered at the origin. You may remember from our study of circles that some circles had their center at the origin and some did not. In this section, we will look at cases where one of the axes of an ellipse is parallel to one of the coordinate axes. To do this, we will use translation axes.

The method we will use to translate axes will be very similar to the one we used when we were solving the equation for a circle. We took each equation that was not centered at the origin and completed the square. This told us the location of the center and the radius.

The equations that we will be working with are all in the form

$$Ax^2 + Cy^2 + Dx + Ey + F = 0$$

where A and C cannot both be zero at the same time. The process can best be explained by following the next example.

EXAMPLE 15.38

Discuss and sketch the graph of $9x^2 + 4y^2 - 36x + 40y + 100 = 0$.

Solution   As we just mentioned, we will complete the square. First we will group the terms, and use ◯ and ▢ to indicate the missing constants that we must determine

EXAMPLE 15.38 (Cont.)

in order to complete the square.

$$(9x^2 - 36x) + (4y^2 + 40y) + 100 = 0$$
$$9(x^2 - 4x + \bigcirc) + 4(y^2 + 10y + \square) = -100 + 9\bigcirc + 4\square$$

Next, we complete the square of each expression within parentheses by adding 4 to the x terms and 25 to the y terms. Remember that on the right-hand side of the equation, each of these numbers (4 and 25) must be multiplied by the number outside the parentheses (9 and 4):

$$9(x^2 - 4x + 4) + 4(y^2 + 10y + 25) = -100 + 9(4) + 4(25)$$

or $$9(x - 2)^2 + 4(y + 5)^2 = 36$$

Dividing both sides by 36, we get

$$\frac{(x-2)^2}{4} + \frac{(y+5)^2}{9} = 1$$

This looks like the equation for an ellipse. If we let $x' = x - 2$ and $y' = y + 5$, then the equation can be written as

$$\frac{(x')^2}{4} + \frac{(y')^2}{9} = 1$$

This is an ellipse with center at $(x', y') = (0', 0')$. If we draw a new coordinate system, the $x'y'$-system, with its origin at $(2, -5)$, we can draw our ellipse on this new coordinate system. Thus, this ellipse is centered at $(0', 0')$, the vertices are at $V(0', 3')$ and $V'(0', -3')$, the endpoints of the minor axis at $M(2', 0')$ and $M'(-2', 0')$, and the foci are at $F(0', \sqrt{5}')$ and $F'(0', -\sqrt{5}')$. Notice that we have used $3', 0', 2'$, and so on, to show that these are points on the $x'y'$-axes. A sketch of this graph is shown in Figure 15.45. The $x'y'$-axes have been traced over the xy-system.

We can use a table to show the coordinates in both systems. Since we know the coordinates in the $x'y'$-system and we know that $x' = x - 2$ and $y' = y + 5$, to find the coordinates in the xy-system, we solve these equations for x and y: $x = x' + 2$ and $y = y' - 5$. The table follows.

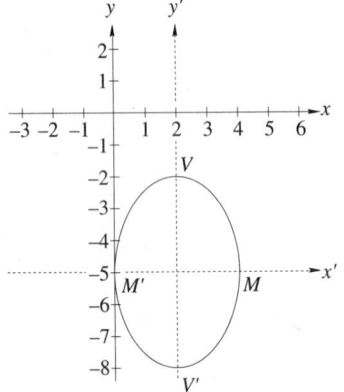

FIGURE 15.45

	$x'y'$-system	xy-system
center	$(0', 0')$	$(2, -5)$
vertices	$(0', \pm 3')$	$(2, -2), (2, -8)$
foci	$(0', \pm\sqrt{5}')$	$(2, \sqrt{5} - 5), (2, -\sqrt{5} - 5)$
endpoints of minor axis	$(\pm 2', 0')$	$(4, -5), (0, -5)$

In general, if we want the point (h, k) in the xy-coordinate system to become identified as the origin of the $x'y'$-coordinate system, we use the substitution

$$x' = x - h \qquad \text{and} \qquad y' = y - k$$

This procedure is called a **translation of axes**. As a result of a translation of axes, new axes are formed that are parallel to the old ones.

EXAMPLE 15.39

Discuss and sketch the graph of $x^2 - 18y^2 + 6x - 36y - 45 = 0$.

Solution   Again, we will complete the square.

$$(x^2 + 6x) + (-18y^2 - 36y) = 45$$
$$(x^2 + 6x + \bigcirc) - 18(y^2 + 2y + \square) = 45 + \bigcirc - 18\square$$

We add 9 to complete the square of $x^2 + 6x$ and 1 to complete $y^2 + 2y$.

$$(x^2 + 6x + 9) - 18(y^2 + 2y + 1) = 45 + 9 - 18(1)$$
$$(x + 3)^2 - 18(y + 1)^2 = 36$$

Dividing both sides by 36, we get

$$\frac{(x+3)^2}{36} - \frac{(y+1)^2}{2} = 1$$

If $x' = x + 3$ and $y' = y + 1$, then the equation becomes

$$\frac{x'^2}{36} - \frac{y'^2}{2} = 1$$

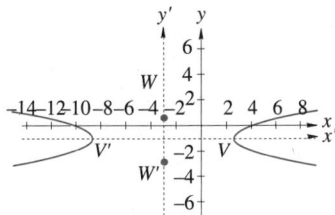

FIGURE 15.46

which is a hyperbola with center at $(0', 0')$ and vertices at $(\pm 6', 0')$ of the $x'y'$-system. The origin of the $x'y'$-system is found at $(-3, -1)$ of the xy-system. The sketch of this hyperbola on both coordinate systems is shown in Figure 15.46. Corresponding coordinates in the two systems are shown in the following table.

	$x'y'$-system	xy-system
center	$(0', 0')$	$(-3, -1)$
vertices	$(\pm 6', 0')$	$(-9, -1), (3, -1)$
foci	$(\pm\sqrt{38}, 0')$	$(-3 + \sqrt{38}, -1),$
		$(-3 - \sqrt{38}, -1)$
endpoints of conjugate axis	$(0', \pm\sqrt{2})$	$(-3, -1 + \sqrt{2}), (-3, -1 - \sqrt{2})$

EXAMPLE 15.40

Discuss and sketch the graph of $y^2 + 12x + 2y - 23 = 0$.

Solution   There is no x^2-term, so this is the equation for a parabola. We will complete the square on just the y-terms. We will also put the y-terms on one side of the equation and all the other terms on the other side.

$$y^2 + 2y = -12x + 23$$
$$y^2 + 2y + 1 = -12x + 23 + 1$$
$$(y + 1)^2 = -12(x - 2)$$

EXAMPLE 15.40 (Cont.)

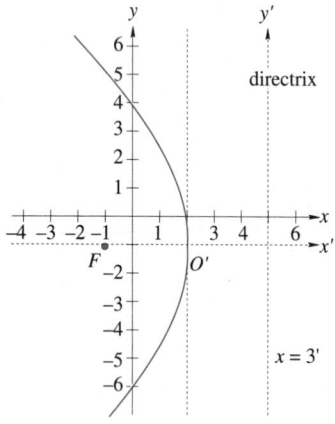

FIGURE 15.47

If $y' = y + 1$ and $x' = x - 2$, the equation becomes

$$y'^2 = -12x'$$

which is a parabola with vertex at $(0', 0')$, a horizontal axis, and that opens to the left. Since $4p = -12$, $p = -3$ and the directrix is the line $x = 3'$. The focus is at $(-3', 0')$. The sketch of this curve is shown in Figure 15.47.

EXAMPLE 15.41

Find the equation of the hyperbola with vertices $(3, 1)$ and $(-5, 1)$, and focus $(5, 1)$.

Solution The center of a hyperbola is at the midpoint between the vertices. Using the midpoint formula, we can determine that the center is at $\left(\dfrac{3 + (-5)}{2}, \dfrac{1 + 1}{2} \right) = (-1, 1)$. The distance from the vertex to the center is 4, so $a = 4$. The distance from the focus to the center is 6, so $c = 6$. This is a hyperbola, so $b^2 = c^2 - a^2 = 6^2 - 4^2 = 20$. Finally, this is a hyperbola with foci on the x'-axis and the equation is

$$\frac{x'^2}{16} - \frac{y'^2}{20} = 1$$

Since $x' = x - h$ and $y' = y - k$, we have $x' = x + 1$ and $y' = y - 1$, so the equation becomes

$$\frac{(x+1)^2}{16} - \frac{(y-1)^2}{20} = 1$$

Application

EXAMPLE 15.42

A telescope is shown in Figure 15.48a. The focus of one branch of the hyperbolic lens is 20.4 cm from the vertex of the other branch of this same lens. The parabolic lens is 63 cm deep and measures 168 cm across. Determine **(a)** the location of the focus for the parabolic lens **(b)** an equation for the parabola, and **(c)** an equation for the hyperbola.

Solution A cross-section through the axis of this telescope has been drawn on a coordinate system in Figure 15.48b. Because the focus of one branch of the hyper-

EXAMPLE 15.42 (Cont.)

bolic lens is also the vertex of the parabolic lens, we have placed the given focus at the origin. The given vertex of the other branch has been located at $(20.4, 0)$. The parabolic lens is 63 cm deep and 168 cm across and so the point $(63, 84)$ is on this lens.

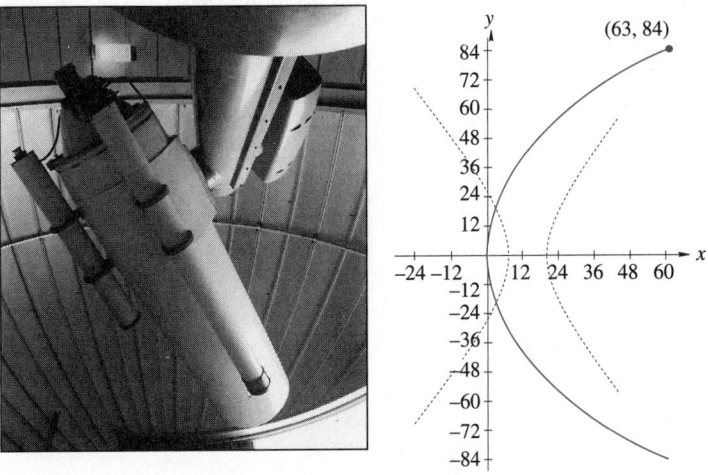

FIGURE 15.48a   FIGURE 15.48b

This is a parabola. Its equation is of the form $y^2 = 4px$. We know that $(63, 84)$ is a point on the parabola, hence we have $p = \dfrac{y^2}{4x} = \dfrac{84^2}{4(63)} = 28$. Thus, the parabola's focus is at $(28, 0)$ or 28 cm from the vertex and the equation of the parabola is $y^2 = 4(28)x = 112x$.

The foci of the hyperbola are at $(0, 0)$ and $(28, 0)$. Thus, the center of the hyperbola is at $(14, 0)$, and so $c = 28 - 14 = 14$. Since one of the hyperbola's vertices is at $(20.4, 0)$, $a = 20.4 - 14 = 6.4$. Using these values of a and c, we can determine that $b^2 = c^2 - a^2 = 14^2 - 6.4^2 = 155.04$. Thus, the standard equation of the hyperbola for this lens is

$$\frac{(x - 14)^2}{6.4^2} - \frac{y^2}{155.04} = 1$$

Using a Graphing Calculator

When using a graphing calculator, choose a viewing window that puts the center of the conic section at the center of the window. Then use the guidelines presented earlier for determining the viewing window. The next example shows how to do this with a translated ellipse. Again, since the calculator will graph only a function, you will need to solve the equation algebraically for y and then graph the two equations that result.

EXAMPLE 15.43

Use a graphing calculator to graph $9x^2 + 4y^2 - 36x + 40y + 100 = 0$.

Solution This is the same curve we graphed in Example 15.42. From that example, we know that the conic is an ellipse with standard equation $\dfrac{(x-2)^2}{4} + \dfrac{(y+5)^2}{9} = 1$, center at $(2, -5)$, $a = 2$, and $b = 3$. For the viewing window we will use the following settings, where c_x and c_y represent the x- and y-coordinates of the center.

$$
\begin{aligned}
\text{Xmin} &= c_x - a &&= 2 - 2 = 0 \\
\text{Xmax} &= c_x + a &&= 2 + 2 = 4 \\
\text{Ymin} &= c_y - b &&= -5 - 3 = -8 \\
\text{Ymax} &= c_y + b &&= -5 + 3 = -2
\end{aligned}
$$

Next, solve the equation for y.

$$\frac{(x-2)^2}{4} + \frac{(y+5)^2}{9} = 1$$

$$\frac{(y+5)^2}{9} = 1 - \frac{(x-2)^2}{4}$$

$$(y+5)^2 = 9\left(1 - \frac{(x-2)^2}{4}\right)$$

$$y+5 = \pm\sqrt{9\left(1 - \frac{(x-2)^2}{4}\right)}$$

$$y = -5 \pm \sqrt{9\left(1 - \frac{(x-2)^2}{4}\right)}$$

$$y = -5 \pm 3\sqrt{1 - \frac{(x-2)^2}{4}}$$

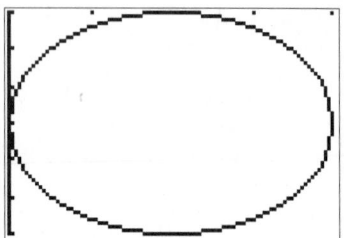

FIGURE 15.49a

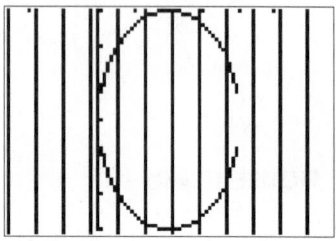

FIGURE 15.49b

We will graph $y_1 = -5 + 3\sqrt{1 - \dfrac{(x-2)^2}{4}}$ and $y_2 = -5 - 3\sqrt{1 - \dfrac{(x-2)^2}{4}}$. This result is shown in Figure 15.49a. Pressing $\boxed{\text{ZOOM}}$ 5, we obtain the proportionally correct graph shown in Figure 15.49b.

Exercise Set 15.6

Sketch the graphs in Exercises 1–12, after making suitable translations of axes.

1. $\dfrac{(x-4)^2}{9} + \dfrac{(y+3)^2}{4} = 1$

2. $\dfrac{(x+5)^2}{16} - \dfrac{(y+3)^2}{25} = 1$

3. $\dfrac{(y-3)^2}{36} - (x+4)^2 = 1$

4. $(y+5)^2 = 12(x+1)$

5. $100(x+5)^2 - 4y^2 = 400$

6. $(y+3)^2 = 8(x-2)$

7. $16x^2 + 4y^2 + 64x - 12y + 57 = 0$

8. $x^2 + y^2 - 2x + 2y - 2 = 0$

9. $25x^2 + 4y^2 - 250x - 16y + 541 = 0$

10. $100x^2 - 180x - 100y + 81 = 0$

11. $2x^2 - y^2 - 16x + 4y + 24 = 0$

12. $9x^2 - 36x + 16y^2 - 32y - 524 = 0$

In Exercises 13–22, determine the equation of each of the curves described by the given information.

13. Parabola, vertex at $(2, -3)$, $p = 8$, axis parallel to y-axis

14. Hyperbola, vertex at $(2, 1)$, foci at $(-6, 1)$ and $(8, 1)$

15. Ellipse, center at $(4, -3)$, focus at $(8, -3)$, vertex at $(10, -3)$

16. Ellipse, center at $(-2, 0)$, focus at $(6, 0)$, vertex at $(9, 0)$

17. Hyperbola, center at $(-3, 2)$, focus at $(-3, 7)$, transverse axis 6 units

18. Ellipse, center at $(3, 5)$, focus at $(3, 8)$, minor axis 2 units

19. Parabola, vertex at $(-5, 1)$, $p = -4$, axis parallel to x-axis

20. Parabola, vertex at $(-3, -6)$, $p = -12$, axis parallel to y-axis

21. Ellipse, center at $(-4, 1)$, focus at $(-4, 9)$, minor axis 12 units

22. Hyperbola, center at $(2, 8)$, vertex at $(6, 8)$, conjugate axis 10 units

Solve Exercises 23–34.

23. *Physics* The height s of a ball thrown vertically upward is given by the equation $s = 29.4t - 4.9t^2$, where s is in meters and t is the elapsed time in seconds. Graph this curve. Discuss the curve. Determine the maximum height of the ball.

24. *Astronomy* Satellites often orbit earth in an elliptical path with the center of the earth at one of the foci. For a certain satellite, the maximum altitude is 140 mi above the surface of the earth and the minimum is 90 mi. If the radius of earth is approximately 4,000 mi, find the equation of the orbit.

25. *Navigation* Two navigational transmitting towers A and B are 1 000 km apart along an east-west line. Radio signals are sent (traveling at 300 m/μs) simultaneously from both towers. An airplane is located somewhere north of the line joining the two towers. The signal from A arrives at the plane 600 μs after the signal from B. The signal sent from B and reflected by the plane takes a total of 8 000 μs to reach A. What is the location of the plane?

26. *Astronomy* The orbit of Halley's comet is an ellipse with the sun at one focus. The major and minor semi-axes of this orbit are 18.09 A.U. and 4.56 A.U., respectively. (A **semi-axis** is half an axis. One A.U. is one astronomical unit or about 1.496×10^8 km.) What is the equation of this orbit, if the sun is at the origin and the major axis is along the x-axis? What

are the maximum and minimum distances from the sun to Halley's comet?

27. *Mechanical engineering* A cantilever beam is a beam fixed at one end. Under a uniform load, the beam assumes a parabolic curve with the fixed end as the vertex. For a cantilever beam 2.00 m long, the equation of the load parabola is approximately $x^2 + 200y - 4.00 = 0$, where x is the distance in meters from the fixed end and y is the displacement. See Figure 15.50.

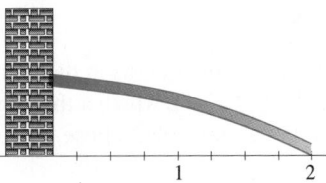

FIGURE 15.50

(a) Change this equation to the standard equation for a parabola.

(b) How far is the free end of the beam displaced from its no-load position?

28. *Mechanical engineering* A cantilever beam is a beam fixed at one end. Under a uniform load the beam assumes a parabolic curve, with the fixed end as the vertex. For a cantilever beam 8.0 ft long, the equation of the load parabola is approximately $x^2 + 100y - 4.00 = 0$, where x is the distance in feet from the fixed end and y is the displacement.

(a) Change this equation to the standard equation.

(b) How far is the free end of the beam displaced from its no-load position?

29. *Air traffic control*   An airplane starts at the origin and flies in a straight line northeast, that is, up and to the right along the line $y = x$. Transmitters at coordinates (in miles) $(0, 0)$ and $(0, 250)$ send out synchronized radio signals, and instruments in the plane measure the difference in their arrival times. If the speed of the radio signal is known, then it is possible to locate the set of points where a plane is 50 mi farther from $(0, 0)$ than from $(0, 250)$ is one branch of a hyperbola.

(a) Find the equation of the hyperbola and identify the correct branch.

(b) Find the location of the airplane when it is 50 mi farther from $(0, 0)$ than from $(0, 250)$.

30. *Air traffic control*   In the same setting as Exercise 29, a second plane flies along the line $y = 2x$. Comparison of radio signals shows that it is 50 mi farther from $(0, 0)$ than from $(0, 250)$. Find the location of the airplane.

31. *Automotive technology*   The design of an automobile cam consists of the graph of the circle $0.25x^2 +$

$0.25y^2 = 0.0625$ and the ellipse $0.4x^2 + 0.16y^2 = 0.1$.

(a) Solve each of these equations so their graphs, when drawn on the same pair of axes, will produce this cam.

(b) Use your equations from (a) to sketch this cam.

32. *Police science*   The stopping distance, d, in ft of a car moving at v mph is given approximately by the equation $d = v + \frac{1}{20}v^2$. Plot d as a function of v for $0 \le v \le 80$.

33. *Industrial management*   A factory normally packages items in lots of 24. It costs $0.90/item to package a lot of exactly 24 items. However, the cost per item is reduced by $0.01 for each item over 24.

(a) Determine a function for the cost in terms of the number of items produced.

(b) Graph your function from (a).

(c) Using your graph in (b), determine the lot size that will produce the highest packaging cost.

(d) What is the highest packaging cost?

34. *Electronics*   For a certain temperature-sensitive electronic device, the voltage, V, is given by $V = 5.0T - 0.025T^2$, where T is the ambient temperature in °C.

(a) Plot V as a function of t for values of $V \ge 0$.

(b) At what temperature is $V = 0$?

In Your Words

35. Explain how you can tell by looking at the equation $Ax^2 + Cy^2 + Dx + Ey + F = 0$ whether the graph of the equation is a circle, parabola, ellipse, or hyperbola.

36. Describe how to change an equation of the form $Ax^2 + Cy^2 + Dx + Ey + F = 0$ into the standard form for its conic section.

≡ 15.7
ROTATION OF AXES; THE GENERAL SECOND-DEGREE EQUATION

The conic sections that we have considered so far have all had their axes on a coordinate axis or parallel to a coordinate axis. All of these could be written as second-degree equations of the form

$$Ax^2 + Cy^2 + Dx + Ey + F = 0$$

Notice that this equation does not contain an xy-term.

In this section, we will work with curves in which an axis is not parallel to a coordinate axis. However, the axes are still perpendicular to each other. These equations will all have an xy-term. In order to recognize its graph, we will change to a new coordinate system. In Section 15.6, we changed to a new coordinate system through a translation of axes. In this section, we will use a rotation of axes and we will work with the general second-degree equation.

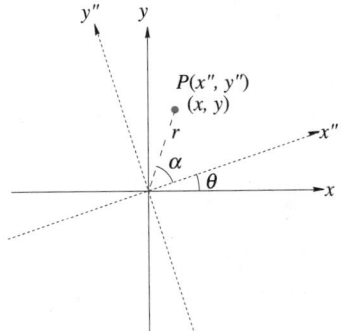

FIGURE 15.51

General Form of a Second-Degree Equation

The **general form of a second-degree equation** is

$$Ax^2 + Bxy + Cy^2 + Dx + Ey + F = 0$$

where A, B, and C are not all 0.

In a rotation, the origin remains fixed, while the x-axis and y-axis are rotated through a positive acute angle θ. These rotated axes will be labeled the x''-axis and the y''-axis. (See Figure 15.51.) If $P(x, y)$ is any general point in the xy-coordinate system, then that same point would have the coordinates $P(x'', y'')$ in the $x''y''$-coordinate system.

We can convert from one coordinate system to the other with the help of a pair of equations that express x and y in terms of x'' and y''.

$$x = x'' \cos\theta - y'' \sin\theta$$
$$y = y'' \cos\theta + x'' \sin\theta$$

These appear to be terrible equations to work with, but the following examples should show that they are not as difficult to use as they seem.

EXAMPLE 15.44

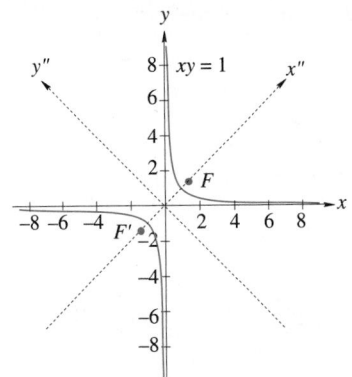

FIGURE 15.52

Transform the equation $xy = 1$ by rotating the axes through an angle of $45°$.

Solution Since $\theta = 45°$, the conversion equations become

$$x = x'' \cos 45° - y'' \sin 45° = 0.7071x'' - 0.7071y''$$
$$y = y'' \cos 45° + x'' \sin 45° = 0.7071y'' + 0.7071x''$$

Substituting these into the original equation, $xy = 1$, we get

$$\cdot [0.7071(y'' + x'')] = 1$$
$$0.5(x''^2 - y''^2) = 1$$
$$\frac{x''^2}{2} - \frac{y''^2}{2} = 1$$

This is an equation of a hyperbola with center at the origin of the $x''y''$-coordinate system and transverse axis along the x''-axis. Here $a = b = \sqrt{2}$ and $c = 2$. The asymptotes of this hyperbola happen to be the original xy-axes. A graph of this curve is shown in Figure 15.52.

Any equation of the type in Example 15.44 is called an **equilateral hyperbola** or a **rectangular hyperbola** because the asymptotes are perpendicular. Equilateral hyperbolas are of the form $xy = k$, where k is a constant.

Now, suppose we have an equation that is a general second-degree equation and $B \neq 0$. What we want to do is rotate the axes so that we will get an equation of the form

$$A''(x^2) + B''xy + C''(y^2) + D''x + E''y + F'' = 0$$

where $B'' \neq 0$. Then we will be able to put the equation in the $x''y''$-coordinate system in a recognizable form of one of the conic sections.

We can cause B'' to be 0 if we let θ be the unique acute angle where

$$\cot 2\theta = \frac{A - C}{B} \qquad 0 < \theta < 90° \qquad 0 < \theta < \frac{\pi}{2}$$

If you use a calculator, you have to be careful. First there is no $\boxed{\text{COT}}$ key on your calculator, so you will have to take the reciprocal of the value. Since $\dfrac{1}{\cot \theta} = \tan \theta$, you can then use the $\boxed{\text{INV}}$ $\boxed{\text{tan}}$ keys. But, the arctan function will give answers between $-\frac{\pi}{2}$ and $\frac{\pi}{2}$. If you get a negative angle, you will have to add π rad or 180°. Let's see how it works in the next example.

EXAMPLE 15.45

Determine the graph of the equation $29x^2 + 24xy + 36y^2 - 54x - 72y - 135 = 0$.

Solution We will first determine the angle of rotation using the formula $\cot 2\theta = \dfrac{A - C}{B}$. Here $A = 29$, $B = 24$, and $C = 36$. The following description shows how to use a graphing calculator to determine θ.

PRESS	DISPLAY	
$\boxed{(}\ 29\ \boxed{-}\ 36\ \boxed{)}\ \boxed{\div}\ 24\ \boxed{\text{ENTER}}$	-0.2916666667	
$\boxed{x^{-1}}\ \boxed{\text{ENTER}}$	-3.428571429	Changes from $\cot 2\theta$ to $\tan 2\theta$
$\boxed{\text{2nd}}\ \boxed{\text{TAN}}\ \boxed{\text{2nd}}\ \boxed{\text{ANS}}\ \boxed{\text{ENTER}}$	-73.73979529	2θ in degrees
$\boxed{+}\ 180$	106.2602047	Add 180° because 2θ is negative
$\boxed{\div}\ 2$	53.13010235	This is θ

So, using our conversion formulas and the fact that $\sin \theta = 0.8$ and $\cos \theta = 0.6$, we get

$$x = x'' \cos \theta - y'' \sin \theta = 0.6x'' - 0.8y''$$
$$y = y'' \cos \theta + x'' \sin \theta = 0.6y'' + 0.8x''$$

Substituting these values for x and y in the given equation, $29x^2 + 24xy + 36y^2 - 54x - 72y - 135 = 0$, we obtain

$$29(0.6x'' - 0.8y'')^2 + 24(0.6x'' - 0.8y'')(0.6y'' + 0.8x'')$$
$$+ 36(0.6y'' + 0.8x'')^2 - 54(0.6x'' - 0.8y'')$$

EXAMPLE 15.45 (Cont.)

$$-72(0.6y'' + 0.8x'') - 135 = 0$$
$$29(0.36x''^2 - 0.96x''y'' + 0.64y''^2)$$
$$+ 24(0.48x''^2 + 0.36x''y'' - 0.48y''^2 - 0.64x''y'')$$
$$+ 36(0.36y''^2 + 0.96x''y'' + 0.64x''^2)$$
$$- 54(0.6x'' - 0.8y'') - 72(0.6y'' + 0.8x'') - 135 = 0$$
$$10.44x''^2 - 27.84x''y'' + 18.56y''^2 + 11.52x''^2 + 8.64x''y'' - 11.52y''^2$$
$$- 15.36x''y'' + 12.96y''^2 + 34.56x''y'' + 23.04x''^2 - 32.4x''$$
$$+ 43.2y'' - 43.2y'' - 57.6x'' - 135 = 0$$
$$45x''^2 + 0x''y'' + 20y''^2 - 90x'' + 0y'' - 135 = 0$$
$$45x''^2 + 20y''^2 - 90x'' - 135 = 0$$

Completing the square on the x''-terms, we get

$$45(x''^2 - 2x'' + 1) + 20y''^2 = 135 + 45$$
$$45(x'' - 1)^2 + 20y''^2 = 180$$

Dividing both sides by 180 we obtain

$$\frac{(x'' - 1)^2}{4} + \frac{y''^2}{9} = 1$$

This is the equation of an ellipse with a major axis of 6 ($a = 3$) and a minor axis of 4 ($b = 2$) with the major axis along the vertical axis. The center of this ellipse is at $(1'', 0'')$. The graph of this ellipse is shown in Figure 15.53.

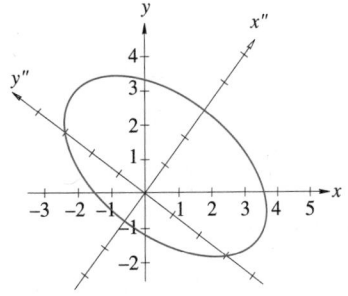

FIGURE 15.53

The Discriminant

There is an easy method for determining the nature of the curve described by a general second-degree equation. It turns out that the **discriminant**, $B^2 - 4AC$, will tell the type of curve described by the equation.

Classifying the Graph of a General Quadratic Equation

The graph of a general quadratic equation

$$Ax^2 + Bxy + Cy^2 + Dx + Ey + F = 0$$

is either a conic or a degenerate conic. If it is a conic, then the discriminant, $B^2 - 4AC$, can be used to classify the graph of the equation by using the following:

If $B^2 - 4AC > 0$, the curve is a hyperbola.
If $B^2 - 4AC = 0$, the curve is a parabola.
If $B^2 - 4AC < 0$, the curve is an ellipse.

The majority of the problems we will work are not as long as the last example. We are more likely to get problems similar to the one in the next example.

Application

EXAMPLE 15.46

When the power P in an electric circuit is constant, the voltage V is inversely proportional to the current I, as indicated by the equation $P = IV$. If the power is 120 W, sketch the relationship of I vs V.

Solution The equation is $IV = 120$. A table of values is

I	1	2	4	6	8	10	12	15	20	30	60	120
V	120	60	30	20	15	12	10	8	6	4	2	1

Negative values do not apply. The graph is one branch of an equilateral hyperbola, as shown in Figure 15.54.

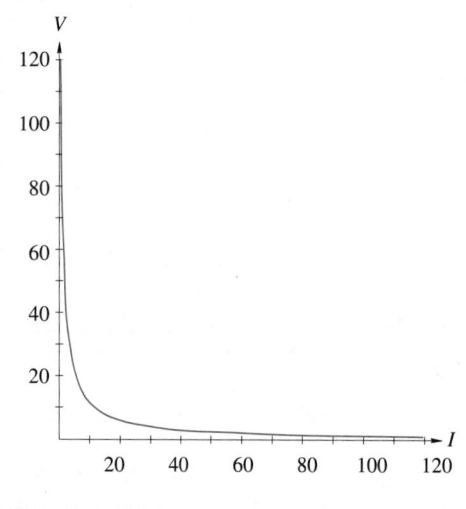

FIGURE 15.54

Using a Graphing Calculator

The most difficult part about using a graphing calculator to graph a rotated conic is getting the equation for the conic in a form that the calculator can use. This requires completing the square in a way you have probably not seen.

EXAMPLE 15.47

Use a graphing calculator to graph $29x^2 + 24xy + 36y^2 - 54x - 72y - 135 = 0$.

Solution This is the same curve we graphed in Example 15.45. From that example, we know that the conic is an ellipse rotated about $53.13°$ and with standard equation on the rotated axes $\dfrac{(x'' - 1)^2}{4} + \dfrac{y''^2}{9} = 1$ on the rotated axes.

EXAMPLE 15.47 (Cont.)

We will use the original equation and complete the square in order to solve the equation for y.

$$29x^2 + 24xy + 36y^2 - 54x - 72y - 135 = 0$$

$$24xy + 36y^2 - 72y = 135 - 29x^2 + 54x$$

$$36y^2 + 24xy - 72y = 135 + 54x - 29x^2$$

$$36y^2 + 24y(x-2) = 135 + 54x - 29x^2$$

$$36\left[y^2 + \frac{2}{3}y(x-2)\right] = 135 + 54x - 29x^2$$

$$y^2 + \frac{2}{3}y(x-2) = \frac{1}{36}(135 + 54x - 29x^2)$$

$$y^2 + \frac{2}{3}y(x-2) + \left[\frac{1}{3}(x-2)\right]^2 = \frac{1}{36}(135 + 54x - 29x^2) + \left[\frac{1}{3}(x-2)\right]^2$$

$$\left[y + \frac{1}{3}(x-2)\right]^2 = \frac{1}{36}(151 + 38x - 25x^2)$$

$$y + \frac{1}{3}(x-2) = \pm\frac{1}{6}\sqrt{(151 + 38x - 25x^2)}$$

$$y = -\frac{1}{3}(x-2) \pm \frac{1}{6}\sqrt{(151 + 38x - 25x^2)}$$

We will graph the equations $y_1 = -\frac{1}{3}(x-2) + \frac{1}{6}\sqrt{(151 + 38x - 25x^2)}$ and $y_2 = -\frac{1}{3}(x-2) - \frac{1}{6}\sqrt{(151 + 38x - 25x^2)}$. This result is shown in Figure 15.55. Compare this to the earlier result in Figure 15.53.

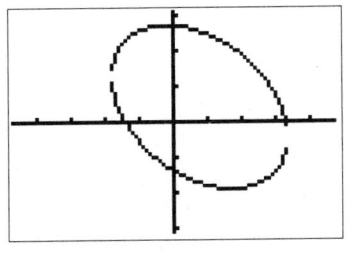

FIGURE 15.55

Exercise Set 15.7

In Exercises 1–12, (a) use the discriminant to identify the graph of the given equation, (b) determine the angle θ needed to rotate the coordinate axes to remove the xy-term, (c) rotate the axes through the angle θ, (d) determine the equation of the conic in the $x''y''$-coordinate system, and (e) sketch the graph.

1. $xy = -9$
2. $xy = 9$
3. $x^2 - 6xy + y^2 - 8 = 0$
4. $x^2 + 4xy - 2y^2 - 6 = 0$
5. $52x^2 - 72xy + 73y^2 - 100 = 0$
6. $11x^2 + 10\sqrt{3}xy + y^2 - 4 = 0$

7. $x^2 - 2xy + y^2 + x + y = 0$
8. $3x^2 + 2\sqrt{3}xy + y^2 - 2x + 2\sqrt{3}y = 0$
9. $2x^2 + 12xy - 3y^2 - 42 = 0$
10. $7x^2 - 20xy - 8y^2 + 52 = 0$
11. $6x^2 - 5xy + 6y^2 - 26 = 0$
12. $9x^2 - 6xy + y^2 - 12\sqrt{10}x - 36\sqrt{10}y = 0$

Solve Exercises 13–16.

13. *Physics* Boyle's law states that the volume of a gas is inversely proportional to its pressure, provided that the mass and temperature are constant. Thus, if P is the pressure and V the volume, $PV = k$, where k is a constant. Draw the graph of P vs V, if 15 mm^3 of gas is under a pressure of 400 kPa at a constant temperature.

14. *Electricity*　For a given alternating current circuit, the capacitance C and the capacitive reactance X_C are related by the equation $X_C = \dfrac{1}{\omega C}$, where ω is the angular frequency. What kind of curve is this? Sketch a graph of the equation if $\omega = 280\,\text{rad/s}$.

15. *Thermodynamics*　When a cross-section of a pipe is suddenly enlarged, as shown in Figure 15.56, the loss of heat of the fluid through the pipe is related by the formula $19.62 h_L = (\overline{v_1} - \overline{v_2})^2$, where h_L is the heat loss and $\overline{v_1}$ and $\overline{v_2}$ are the average velocities in the two pipes. If h_L is to be held to less that $5°C$, describe this curve. Sketch a graph of the curve with

$(\overline{v_1} - \overline{v_2})$ on one axis and h_L on the other.

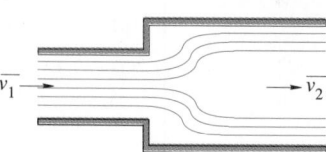

FIGURE 15.56

16. The sides of a rectangle are $3x$ and y and the diagonal is $x + 10$. What kind of curve is represented by the equation relating x and y? Sketch the curve.

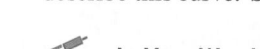

In Your Words

17. **(a)** What is the discriminant of the general quadratic equation

$$Ax^2 + Bxy + Cy^2 + Dx + Ey + F = 0$$

(b) Explain how to use the discriminant to classify the graph of a general quadratic equation.

18. Describe how you can graph an equation of the form $Ax^2 + Bxy + Cy^2 + Dx + Ey + F = 0$ on your graphing calculator.

☰ 15.8
CONIC SECTIONS IN POLAR COORDINATES

In Section 11.7, we studied graphs in the polar coordinate system. Some of the special curves we were able to draw were the rose, the cardioid, and the limaçon. In this section, we will return to the polar coordinate system and see how we can graph the conic sections using polar coordinates.

As we progressed through this chapter, you saw that the equations of the conic sections have very simple forms if the center or vertex is at the origin. But, we had to learn a different form of each conic section. Polar coordinates will allow us to use one equation to represent each conic except the circle. Each of these conics will have a focus at the origin and one axis will be a coordinate axis.

When we defined the parabola, we said that it was the set of points that was an equal distance from the focus and the directrix. The ratio of these two distances, since the distances are the same, is 1. This ratio is called the **eccentricity**. In the problems for the ellipse and the hyperbola, we also used the eccentricity. Each conic can be defined in terms of a point on the curve and the ratio of its distance from a focus and a line called the directrix. This ratio is the eccentricity, e.

Caution

Do not confuse the e used to denote the eccentricity with the number $e \approx 2.71828182846$.

Each type of conic is determined by the eccentricity and leads to the general definition of a conic section that follows.

General Definition of a Conic Section

Let ℓ be a fixed line (the directrix) and F a fixed point (**focus**) not on ℓ. A **conic section** is the set of all points P in the plane such that

$$\frac{d(P, F)}{d(P, \ell)} = e$$

where $d(P, F)$ is the distance from P to F, and $d(P, \ell)$ is the perpendicular distance from P to ℓ. The constant e is called the **eccentricity** and if

$$0 < e < 1, \text{ the conic is an ellipse}$$
$$e = 1, \text{ the conic is a parabola}$$
$$e > 1, \text{ the conic is a hyperbola}$$

≡ **Note**

Since $d(P, F)$ and $d(P, \ell)$ are distances, they are not negative, and so e cannot be negative.

Now, if $P(r, \theta)$ is a point on a conic with focus O, directrix $x = p$ ($p > 0$), and eccentricity e (see Figure 15.57), then

$$\frac{d(P, O)}{d(P, Q)} = e$$

where Q is a point on the directrix.

But, $d(P, O) = r$ and $d(P, Q) = p + r \cos \theta$, and so

$$\frac{r}{p + r \cos \theta} = e$$

Solving this for r, we obtain

$$r = \frac{pe}{1 - e \cos \theta}$$

If we had chosen the directrix as $x = -p$ ($p > 0$), we would have obtained the equation

$$r = \frac{pe}{1 + e \cos \theta}$$

If the directrix is $y = \pm p (p > 0)$, the equations would be

$$r = \frac{pe}{1 \pm e \sin \theta}$$

Thus, we have two sets of equations, and by determining e, we can determine the type of curve. The previous results are summarized in the following box.

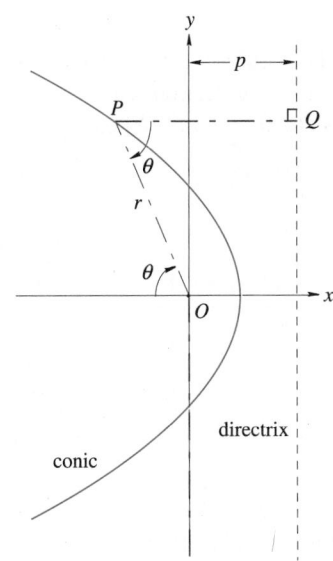

FIGURE 15.57

> **Polar Equations of Conic Sections**
>
> A polar equation that has one of the four forms
>
> $$r = \frac{pe}{1 \mp e \cos\theta} \qquad r = \frac{pe}{1 \pm e \sin\theta}$$
>
> is a conic section. The conic is a parabola if $e = 1$, an ellipse if $0 < e < 1$, or a hyperbola if $e > 1$.

EXAMPLE 15.48

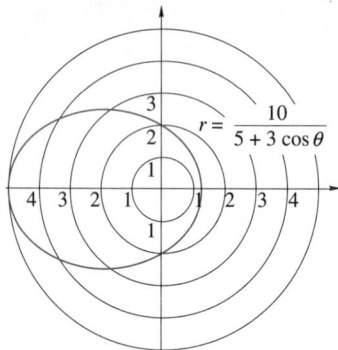

FIGURE 15.58

Describe and sketch the graph of the equation

$$r = \frac{10}{5 + 3\cos\theta}$$

Solution Since the constant term in the denominator must be 1, we divide both numerator and denominator by 5. The equation then becomes

$$r = \frac{2}{1 + \frac{3}{5}\cos\theta}$$

From this, we see that $e = \frac{3}{5}$. Since $\frac{3}{5} < 1$, the conic is an ellipse. The denominator contains the cosine function and so the major axis of this conic section is horizontal. The vertices can be determined by setting θ equal to 0 and π. When $\theta = 0$, $r = \frac{2}{8/5} = \frac{10}{8} = 1.25$ and when $\theta = \pi$, $r = \frac{2}{2/5} = 5$. So $2a = 5 + 1.25 = 6.25$, or $a = 3.125$. The eccentricity $e = \frac{c}{a}$, so $\frac{3}{5} = \frac{c}{3.125}$ and we get $c = 1.875$. Finally, $b^2 = a^2 - c^2 = 6.25$, thus $b = 2.5$. The sketch of this ellipse is given in Figure 15.58.

EXAMPLE 15.49

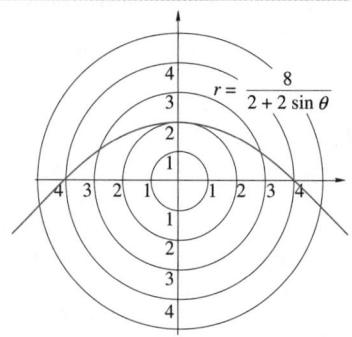

FIGURE 15.59

Describe and sketch the graph of the equation

$$r = \frac{8}{2 + 2\sin\theta}$$

Solution We divide the numerator and denominator by 2, obtaining

$$r = \frac{4}{1 + 1\sin\theta}$$

From this we see that $e = 1$, and the curve is a parabola. Also, since $pe = 4$, $p = 4$. This curve has a vertical axis, so the directrix is $y = 4$. If we plot the points that correspond to the x- and y-intercepts, we get the following table and the curve in Figure 15.59.

θ	0	$\frac{\pi}{2}$	π	$\frac{3\pi}{2}$
r	4	2	4	Not defined

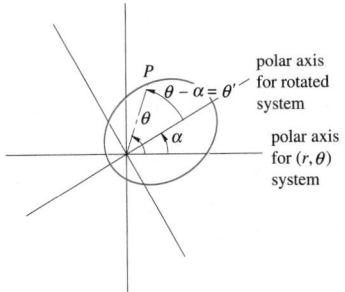

FIGURE 15.60

EXAMPLE 15.50

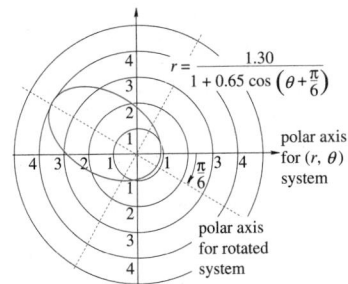

FIGURE 15.61

In Section 15.7, we learned a very complicated process for rotating conic sections on a rectangular coordinate system. The polar coordinate equation of a rotated conic is quite simple. Consider the ellipse in Figure 15.60. It has been rotated through a positive angle α about its focus at the origin O. In the rotated system with polar coordinates (r', θ'), the ellipse has the equation

$$r' = \frac{pe}{1 - e\cos\theta'}$$

But, $r' = r$ and $\theta' = \theta - \alpha$, so the equation of this ellipse in the original (unrotated) polar coordinate system is

$$r = \frac{pe}{1 - e\cos(\theta - \alpha)}$$

Since, in this example, the conic is an ellipse, $0 < e < 1$. But if $e = 1$, the conic would be a parabola, and if $e > 1$, the conic would be a hyperbola.

Discuss and sketch the graph of the equation

$$r = \frac{1.30}{1 + 0.65\cos\left(\theta + \frac{\pi}{6}\right)}$$

Solution This equation has a denominator that begins with the number 1, so we can immediately see that $e = 0.65$. Since $e < 1$, the conic is an ellipse. The angle is $\theta + \frac{\pi}{6}$, so the major axis is rotated $\alpha = -\frac{\pi}{6}$. The denominator contains the cosine function, thus the major axis is the horizontal axis, which has been rotated $-\frac{\pi}{6}$ rad. Again, the vertices can be determined by setting $\theta = 0$ and $\theta = \pi$. This gives $V = 0.83$ and $V' = 2.97$, so $2a = 3.80$ and $a = 1.90$. The eccentricity $e = \frac{c}{a}$, so $c = ea = (0.65)(1.90) = 1.235$. Thus, $b^2 = a^2 - c^2 = 2.085$ or $b \approx 1.444$. The sketch of this ellipse is shown in Figure 15.61. ▪

We will see some applications of this polar equation for a rotated conic section in the following exercise set and again in Chapter 20.

Using a Graphing Calculator

In order to graph a conic section in polar coordinates, you must make sure that your calculator is in polar mode. (Not all graphing calculators have a polar mode, so check your user's manual.) Once your calculator is in polar mode, you graph a conic section by entering the function $r(\theta)$ into the calculator, setting an appropriate viewing window, and graphing the function.

EXAMPLE 15.51

Use a graphing calculator to graph $r = \dfrac{1.30}{1 + 0.65 \cos\left(\theta + \dfrac{\pi}{6}\right)}$.

Solution   This is the same function we graphed in Example 15.50, so the result should look much like the graph in Figure 15.61.

Make sure that the calculator is in both radian and polar modes. On a TI-82 press $\boxed{\text{Y=}}$ and enter the right-hand side of the equation, as shown in Figure 15.62a. Using the window settings θmin $= 0$, θmax $= 2\pi$, θstep $= 0.1$, Xmin $= -4.7$, Xmax $= 4.7$, Xscl $= 1$, Ymin $= -3.1$, Ymax $= 3.1$, Yscl $= 1$, you should get the result in Figure 15.62b.

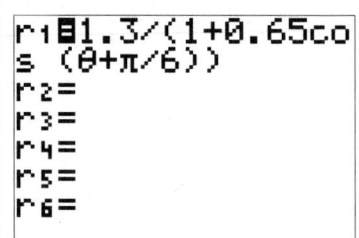

FIGURE 15.62a   **FIGURE 15.62b**

Exercise Set 15.8

In Exercises 1–10, identify and sketch the conic section with the given equation.

1. $r = \dfrac{6}{1 + 3\cos\theta}$

2. $r = \dfrac{8}{1 - 2\sin\theta}$

3. $r = \dfrac{12}{3 - \cos\theta}$

4. $r = \dfrac{15}{3 + 9\sin\theta}$

5. $r = \dfrac{12}{4 - 4\cos\theta}$

6. $r = \dfrac{8}{6 + 6\sin\theta}$

7. $r = \dfrac{12}{3 + 2\cos\theta}$

8. $r = \dfrac{12}{2 - 3\cos\theta}$

9. $r = \dfrac{12}{2 - 4\cos\left(\theta + \frac{\pi}{3}\right)}$

10. $r = \dfrac{6}{2 + \sin\left(\theta + \frac{\pi}{6}\right)}$

In Exercises 11–16, find a polar equation of the conic with focus at the origin and the given eccentricity and directrix.

11. Directrix $x = 4$; $e = \frac{3}{2}$

12. Directrix $x = -2$; $e = \frac{3}{4}$

13. Directrix $y = -5$; $e = 1$

14. Directrix $y = 3$; $e = 2$

15. Directrix $x = 1$; $e = \frac{2}{3}$

16. Directrix $x = 5$; $e = 1$

Solve Exercises 17–20.

17. *Astronomy*   The planet Mercury travels around the sun in an elliptical orbit given approximately by

$$r = \dfrac{3.442 \times 10^7}{1 - 0.206\cos\theta}$$

where r is measured in miles and the sun is at the pole. Determine Mercury's greatest and shortest distance from the sun.

18. *Mechanical engineering*   An engine gear is tested for dynamic balance by rotating it and tracing a polar graph of a point on its rim. Draw the graph given by the equation $r = \dfrac{4}{1 - 0.05 \sin\theta}$ and determine the state of balance of the gear.

19. *Mechanical engineering*   A cam is shaped such that the equation of the upper half is given by $r = 2 + \cos\theta$ and the equation of the lower half is given by $r = \dfrac{3}{2 - \cos\theta}$. Sketch the shape of this cam.

20. *Astronomy*   A certain comet is following a parabolic path, with the sun as the focus. When the comet is 50 000 000 km from the center of the sun, the line from the comet to the sun makes an angle of 45° with the axis of the parabola.
    (a) Write an equation that describes the path of this comet.
    (b) Sketch the path of this comet.
    (c) How close will the comet come to the center of the sun, that is, how far is the vertex of this parabola from the sun?

  In Your Words

21. Describe how to tell if a polar equation of a conic section is a parabola, ellipse, or hyperbola.

22. Explain the differences between the e used to denote the eccentricity and the number $e \approx 2.71828182846$.

 CHAPTER 15 REVIEW

Important Terms and Concepts

Angle of inclination
Circle
   Center
   Radius
Degenerate conics
Distance
Eccentricity
Ellipse
   Center
   Foci
   Major axis
   Minor axis
   Vertices
Hyperbola
   Asymptotes
   Center

   Conjugate axis
   Foci
   Transverse axis
   Vertices
Midpoint
Parabola
   Axis
   Directrix
   Foci
   Vertices
Polar equations for conic sections
Rotation of axes
Slope
Translation of axes
x-intercept
y-intercept

Review Exercises

For each pair of points in Exercises 1–8, find (a) the distance between them, (b) their midpoint, (c) the slope of the line through the points, and (d) the equation of the line through the points.

1. $(2, 5)$ and $(-1, 9)$

2. $(-2, -5)$ and $(10, -10)$

3. $(1, -4)$ and $(3, 6)$

4. $(2, -5)$ and $(-6, 3)$

5. For each line in Exercises 1–8, find the slope of one of its perpendiculars.

6. For each pair of points in Exercises 1–8, write the equation for the line passing through the midpoint of each pair and perpendicular to the line through them.

7. What is the equation of the line that passes through the point $(-3, 5)$ and is parallel to $2y + 4x = 9$?

8. What is the equation of the line through $(2, -7)$ with a slope of 4?

Sketch each of the conic sections in Exercises 9–28.

9. $x^2 + y^2 = 16$

10. $x^2 - y^2 = 16$

11. $x^2 + 4y^2 = 16$

12. $y^2 = 16x$

13. $(x - 2)^2 + (y + 4)^2 = 16$

14. $(x - 2)^2 - (y + 4)^2 = 16$

15. $(x - 2)^2 + 4(y + 4)^2 = 16$

16. $(y + 4)^2 = 16(x - 2)$

17. $x^2 + y^2 + 6x - 10y + 18 = 0$

18. $x^2 - 4y^2 + 6x + 40y - 107 = 0$

19. $x^2 + 4y^2 + 6x - 40y + 93 = 0$

20. $x^2 + 6x - 4y + 29 = 0$

21. $2x^2 + 12xy - 3y^2 - 42 = 0$

22. $5x^2 - 4xy + 8y^2 - 36 = 0$

23. $3x^2 + 2\sqrt{3}xy + y^2 + 8x - 8\sqrt{3}y = 32$

24. $2x^2 - 4xy - y^2 = 6$

25. $r = \dfrac{16}{5 - 3\cos\theta}$

26. $r = \dfrac{9}{3 - 5\cos\theta}$

27. $r = \dfrac{9}{3 + 3\sin\theta}$

28. $r = \dfrac{2}{1 + \cos\theta}$

Solve Exercises 29 and 30.

29. (a) Graph $y^2 + 6y = 16x + 13$.
    (b) Identify the graph.
    (c) Specify all the "important" points, such as foci, vertices, etc.

30. (a) Graph $16x^2 - 9y^2 = 144$.
    (b) Identify the graph.
    (c) Specify all the "important" points, such as foci, vertices, etc.

Solve Exercises 31–34.

31. *Civil engineering*   A concrete bridge for a highway overpass is constructed in the shape of half an elliptic arch, as shown in Figure 15.63. The arch is 100 m long. In order to have enough clearance for tall vehicles, the arch must be 6.1 m high at a point 5 m from the end of the arch. What is the equation of the ellipse for this arch, if the origin is at the midpoint of the major axis?

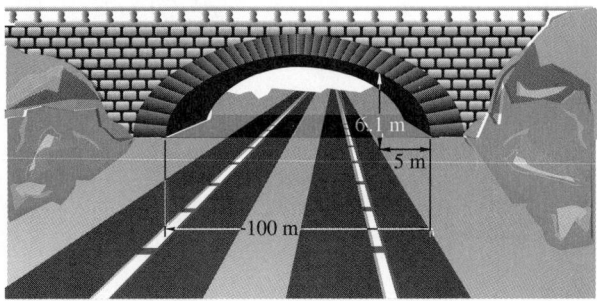

FIGURE 15.63

the road surface and are 400 m apart. If the cable's lowest point is 10 m above the road surface, what is the equation of the parabola for the main cable?

33. *Navigation* An airplane sends out an impulse that travels at the speed of sound (320 m/μs). The plane is 50 km south of the line connecting two receiving stations. The stations are on an east-west line with station A 400 km west of station B. Station A receives the signal from the plane 500 μs after station B. What is the location of the plane?

34. *Astronomy* A map of the solar system is drawn so that the surface of earth is represented by the equation $x^2 + y^2 - 2x + 4y - 6{,}361 = 0$. A satellite orbits earth in a circular orbit 0.8 units above earth. What is the equation of the satellite's orbit on this map?

32. *Civil engineering* The main cables on a suspension bridge approximate a parabolic shape. The twin towers of a suspension bridge are to be 120 m above

CHAPTER 15 TEST

1. Find the focus and the directrix of the parabola $y^2 = -18x$, and sketch the graph.

2. Graph the hyperbola $9x^2 - 4y^2 = 36$. Specify the foci, vertices, and endpoints of the conjugate axis.

3. (a) Graph $16x^2 + 9y^2 = 144$.
   (b) Identify the graph.
   (c) Specify all the "important" points, such as foci, vertices, etc.

4. Determine the distance between the points $(-2, 4)$ and $(5, -6)$.

5. What is the equation of the line that is perpendicular to the line through the points $(4, -5)$ and $(2, -8)$ and passes through their midpoint?

6. The x-intercept of a line is 3, and its angle of inclination is $30°$. Write the equation of the line in slope-intercept form.

7. (a) Graph $14(x + 1)^2 - 9(y - 2)^2 = 36$.
   (b) Identify the graph.

   (c) Specify all the "important" points, such as foci, vertices, etc.

8. (a) Write the equation $3x^2 + 4y^2 - 6x + 16y + 7 = 0$ in the appropriate standard form. **(b)** Determine the type of conic section described by this equation. **(c)** Determine the coordinates of all "significant" points, such as vertices and foci. **(d)** Graph the equation.

9. (a) Determine the angle needed to rotate the coordinate axes so that the transformed equation of $x^2 + xy + y^2 = 4$ has no xy-term. **(b)** Identify the graph.

10. Identify and sketch the graph of the polar equation $r = \dfrac{2}{1 + 4\cos\theta}$.

11. A cable hangs in a parabolic curve between two vertical supports that are 120 ft apart. At a distance 48 ft in from each support, the cable is 3.0 ft above its lowest point. How high up is the cable attached on each support?

CHAPTER

16
Systems of Equations and Inequalities

Every week, an electronics company manufactures CD players and television sets. In Section 16.5, we will learn how to analyze variables, such as cost of materials and the profit on each manufactured item, to determine at which point the company will make the most profit.

In Chapter 5, we solved systems of linear equations using four methods: graphing, elimination by substitution, elimination by addition and subtraction, and Cramer's rule. In this chapter, we will use some of these same methods to solve systems, where one or both equations are of second degree.

We will also begin to explore inequalities. Many problems in technology, industry, business, science, engineering, and mathematics require the use of inequalities. Some of the same techniques we used on equations will be used to solve inequalities.

Finally, we will study linear programming and use this powerful technique in applied situations.

16.1
SOLUTIONS OF NONLINEAR SYSTEMS OF EQUATIONS

We will look at two types of nonlinear systems of equations in this section. A **nonlinear system** is a system of equations in which one or more of the equations is not linear. The first nonlinear system we will look at contains one linear and one quadratic equation. Next, we will look at a nonlinear system of two quadratic equations. With each type, we will begin with a graphical method of solution and then consider algebraic techniques.

For our first system of nonlinear equations, we will use the following:

$$\begin{cases} x^2 + y^2 = 25 \\ x - y = 1 \end{cases}$$

Graphically, we can see in Figure 16.1 that these two curves seem to intersect at the points $(4, 3)$ and $(-3, -4)$. In the next example, we will see that most graphical solutions are not solved as accurately as this one.

We will now examine how these equations can be solved algebraically using elimination by substitution. When you have a nonlinear system of equations that contains one linear equation, first solve the linear equation for one of the variables and substitute this solution into the other equation.

EXAMPLE 16.1

Use the substitution method to solve the system

$$\begin{cases} x^2 + y^2 = 25 & (1) \\ x - y = 1 & (2) \end{cases}$$

Solution Solve equation (2) for one of the variables. If you solve equation (2) for x, you get

$$x = y + 1 \qquad (3)$$

Substitute the value of x from equation (3) into equation (1) and then solve for y.

$$(y+1)^2 + y^2 = 25$$
$$y^2 + 2y + 1 + y^2 = 25$$
$$2y^2 + 2y = 24$$
$$y^2 + y - 12 = 0$$
$$(y+4)(y-3) = 0$$

So, $y = -4$ or $y = 3$

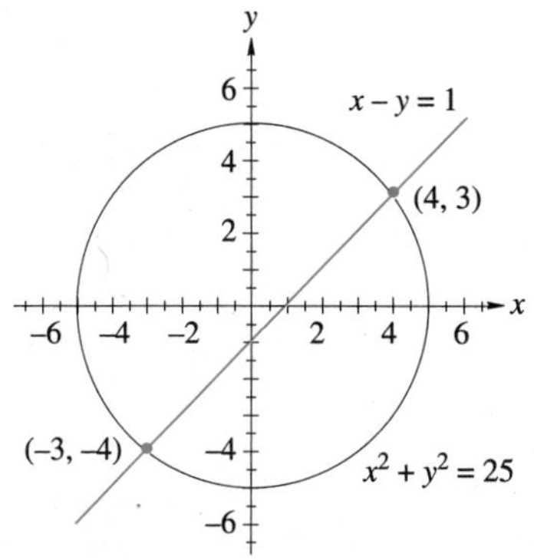

FIGURE 16.1

EXAMPLE 16.1 (Cont.)

Substituting these values for y in equation (3), we get the corresponding values for x.

$$\begin{aligned} \text{If } y = -4, \text{ then } x &= -4+1 \\ x &= -3 \\ \text{and if } y = 3, \text{ then } x &= 3+1 \\ x &= 4 \end{aligned}$$

The solutions are the points $(-3, -4)$ and $(4, 3)$. These are the same solutions we got when we graphed this system of equations.

Now, let's try this technique on another system of nonlinear equations that contains one linear equation.

EXAMPLE 16.2

Solve.

$$\begin{cases} x^2 - 4x - 4y = 16 & (1) \\ x - 2y + 5 = 0 & (2) \end{cases}$$

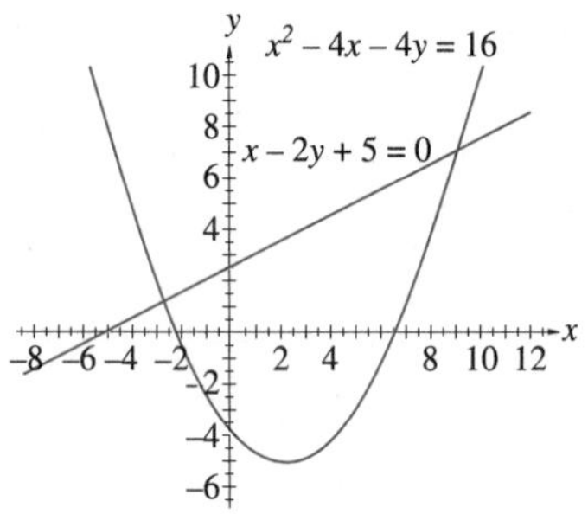

FIGURE 16.2

Solution The graphs of these two curves are shown in Figure 16.2. We need to determine where they intersect.
We solve equation (2) for x, and obtain

$$x = 2y - 5$$

Substituting this value of x into equation (1), we get

$$\begin{aligned} (2y-5)^2 - 4(2y-5) - 4y &= 16 \\ 4y^2 - 20y + 25 - 8y + 20 - 4y &= 16 \\ 4y^2 - 32y + 29 &= 0 \end{aligned}$$

Using the quadratic formula, $y = \dfrac{-b \pm \sqrt{b^2 - 4ac}}{2a}$, we find that

$$\begin{aligned} y &= \frac{32 \pm \sqrt{32^2 - 4(4)(29)}}{2(4)} \\ &= \frac{32 \pm \sqrt{1024 - 464}}{8} \\ &= \frac{32 \pm \sqrt{560}}{8} \\ &\approx \frac{32 \pm 23.66}{8} \end{aligned}$$

$$\text{So,} \quad y \approx \frac{32 + 23.66}{8} \approx 6.96$$

$$\text{and} \quad y \approx \frac{32 - 23.66}{8} \approx 1.04$$

EXAMPLE 16.2 (Cont.)

Substituting these values of y into equation (2) we get

$$x - 2(6.96) + 5 = 0 \text{ or } x = 8.92$$

$$\text{and} \quad x - 2(1.04) + 5 = 0 \text{ or } x = -2.92$$

The approximate solutions are $(-2.92, 1.04)$ and $(8.92, 6.96)$.

When neither of the equations is a linear equation, you may then use either of the elimination methods: substitution, or addition and subtraction.

EXAMPLE 16.3

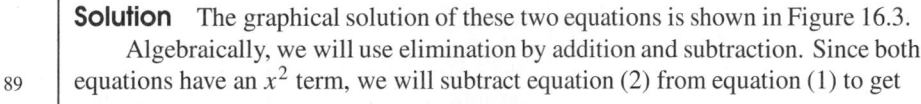

FIGURE 16.3

Solve.

$$\begin{cases} x^2 + 5y^2 = 89 & (1) \\ x^2 + y^2 = 25 & (2) \end{cases}$$

Solution   The graphical solution of these two equations is shown in Figure 16.3. Algebraically, we will use elimination by addition and subtraction. Since both equations have an x^2 term, we will subtract equation (2) from equation (1) to get

$$4y^2 = 64 \qquad (3)$$

$$y^2 = 16$$

$$y = \pm 4$$

Since $y = \pm 4$, we will substitute these values in equation (2) to determine x. For both substitutions we get

$$x^2 + 16 = 25$$

$$x^2 = 9$$

$$x = \pm 3$$

This gives a total of four solutions: $(3, 4)$, $(3, -4)$, $(-3, 4)$, and $(-3, -4)$, and these appear to be the four points in Figure 16.3 where the two curves intersect.

In Example 16.3, a circle and an ellipse intersected in four points. It is also possible that a circle and an ellipse would not intersect at any points, at one point, at two points, or at three points.

EXAMPLE 16.4

Solve.

$$\begin{cases} x^2 - 3y^2 = 22 & (1) \\ xy = -5 & (2) \end{cases}$$

EXAMPLE 16.4 (Cont.)

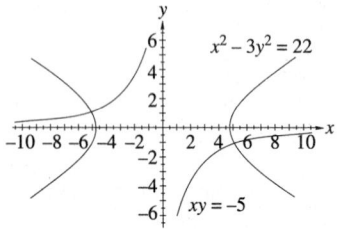

FIGURE 16.4

Solution The graphical solution is shown in Figure 16.4.

Algebraically, we will use elimination by substitution. We will solve equation (2) for y and substitute this value into equation (1).

$$y = \frac{-5}{x} \tag{3}$$

$$x^2 - 3\left(\frac{-5}{x}\right)^2 = 22$$

$$x^2 - \frac{75}{x^2} = 22$$

Multiplying both sides of the equation by x^2, we get

$$x^4 - 75 = 22x^2$$

or $\quad x^4 - 22x^2 - 75 = 0$

which factors into

$$(x^2 - 25)(x^2 + 3) = 0$$

Now, $x^2 - 25 = 0$ means that $x = \pm 5$ and $x^2 + 3 = 0$ means that $x = \pm j\sqrt{3}$. Substituting these values of x into equation (3), we get $y = \pm 1$ (when $x = \pm 5$) and $y = \pm\dfrac{5}{j\sqrt{3}} = \pm\frac{5}{3}j\sqrt{3}$ (when $x = \pm j\sqrt{3}$).

The two real solutions are $(5, -1)$ and $(-5, 1)$. These are shown on the graph in Figure 16.4. There are two imaginary roots $\left(j\sqrt{3}, \frac{5}{3}j\sqrt{3}\right)$ and $\left(-j\sqrt{3}, -\frac{5}{3}j\sqrt{3}\right)$. These two solutions are not on the graph, because this graph is in the real number plane and not the complex plane.

In our examples, all the real roots were at integer values. However, it would not be unexpected if the solutions were any real (or complex) number, as shown in the next example.

Application

EXAMPLE 16.5

Two satellites are in orbit in the same plane. One orbit is described by $5(x + 10)^2 + 3y^2 = 3\,000$. The other is described by $5x^2 + 4y^2 = 4\,000$. What are the points where the two orbits intersect?

Solution We want to solve the system that contains the equations of two ellipses.

$$\begin{cases} 5(x + 10)^2 + 3y^2 = 3\,000 & (1) \\ 5x^2 + 4y^2 = 4\,000 & (2) \end{cases}$$

EXAMPLE 16.5 (Cont.)

We will multiply equation (1) by 4 and equation (2) by 3. This will give the y^2-terms the coefficient of 12.

$$\begin{cases} 20(x+10)^2 + 12y^2 = 12\,000 \\ 15x^2 + 12y^2 = 12\,000 \end{cases}$$

Subtracting, we get

$$20(x+10)^2 - 15x^2 = 0$$

$$\text{or} \qquad 5x^2 + 400x + 2\,000 = 0$$

$$x^2 + 80x + 400 = 0$$

Using the quadratic formula we get

$$x = \frac{-80 \pm \sqrt{80^2 - 4(400)}}{2} = \frac{-80 \pm \sqrt{4800}}{2}$$
$$= -40 \pm 20\sqrt{3}$$

Substituting $x = -40 + 20\sqrt{3} \approx -5.36$ into equation (2), we get

$$143.65 + 4y^2 = 4\,000$$
$$y^2 = 964.09$$
$$y \approx \pm 31.05$$

Substituting $x = -40 - 20\sqrt{3} \approx -74.64$ into equation (2), we get

$$27\,855.65 + 4y^2 = 4\,000$$
$$y^2 \approx -5\,963.91$$

Since $y^2 < 0$, these roots are imaginary.

The satellites will intersect at two points: $(-5.36, 31.05)$ and $(-5.36, -31.05)$.

Exercise Set 16.1

In Exercises 1–10, graph the systems of equations and then solve each system algebraically for x and y. Be sure to include real and complex roots.

1. $\begin{cases} x - 2y = 5 \\ x^2 - 4y^2 = 45 \end{cases}$

2. $\begin{cases} x^2 + y^2 = 13 \\ 2x - y = 4 \end{cases}$

3. $\begin{cases} x^2 + 4y^2 = 32 \\ x + 2y = 0 \end{cases}$

4. $\begin{cases} x - y = 4 \\ x^2 - y^2 = 32 \end{cases}$

5. $\begin{cases} x^2 + y^2 = 4 \\ x^2 - 2y = 1 \end{cases}$

6. $\begin{cases} x^2 - y^2 = 4 \\ 2x - 3y = 10 \end{cases}$

7. $\begin{cases} x^2 + y^2 = 7 \\ y^2 = 6x \end{cases}$

8. $\begin{cases} x^2 - y^2 + 6 = 0 \\ y^2 = 5x \end{cases}$

9. $\begin{cases} xy = 3 \\ 2x^2 - 3y^2 = 15 \end{cases}$

10. $\begin{cases} y = x^2 - 4 \\ x^2 + 3y^2 + 4y - 6 = 0 \end{cases}$

Solve Exercises 11–24.

11. *Transportation*   A truck driver travels the first 100 mi of a 120-mi trip in light traffic. For the last 20 mi, traffic is heavy enough that the average speed is reduced by 10 mph. The total time for the trip is 3 h. What was the average speed for each part of the trip?

12. *Agricultural technology*   The perimeter of a rectangular field is 160 m and the area is $1\,200\,\text{m}^2$. What are the dimensions of the field?

13. *Electricity*   In a direct current (dc) circuit, the equivalent resistance R of two resistors R_1 and R_2 connected in series is $R = R_1 + R_2$. If the resistors are connected in parallel, then $R = \dfrac{R_1 R_2}{R_1 + R_2}$. When two resistors are connected in series, the equivalent resistance is $100\,\Omega$. When they are connected in parallel, $R = 24\,\Omega$. Find R_1 and R_2.

14. *Electricity*   When two resistors are connected in series, the equivalent resistance is $10\,\Omega$. When they are connected in parallel, the equivalent resistance is $2.1\,\Omega$. What are each of the resistances? (See Exercise 13.)

15. *Physics*   When a 40-kg object traveling at a velocity of v_1 m/s collides with a 60-kg object on a frictionless surface traveling in the same direction at v_2 m/s, the final total momentum is $450\,\text{kg} \cdot \text{m/s}$ and is given by the formula

$$40v_1 + 60v_2 = 450$$

The relationship between the initial and final kinetic energy $(KE = \tfrac{1}{2}mv^2)$ is given by the equation

$$20v_1{}^2 + 30v_2{}^2 = 1\,087.5\,\text{J}$$

where J stands for joule. Determine the velocity of each object.

16. *Space technology*   Two satellites are in orbits in the same plane. One orbit is described by $20x^2 + 40y^2 = 8\,000$. The other is described by $30x^2 + 20y^2 = 6\,000$. What are the points where these two orbits intersect?

17. *Physics*   A 200-kg object 0.8 m from the fulcrum of a lever can be lifted by a certain minimum force at the other end of the lever. If that same force is applied 1 m closer to the fulcrum, an additional 50 kg are needed to lift the object. If the minimum force

is F and its original distance from the fulcrum is ℓ, then we know that

$$F\ell = (200)(0.8)$$
$$\text{and} \qquad (F+50)(\ell - 1) = (200)(0.8)$$

Find F and ℓ.

18. *Physics*   The vertical distance in meters that an object falls is given by $y = 15t - 4.9t^2$ m. When the horizontal distance equals twice the vertical distance in meters then $y = 10t$. What values of y and t satisfy these conditions?

19. *Industrial design*   One of the surfaces of a surgical tool is a circle with a radius of x cm. A square hole y cm on each side is cut from the center of the tool, leaving $303.22\,\text{mm}^2$ of metal. If $x - y = 5.59\,\text{cm}$, find x and y.

20. *Electrical engineering*   In determining the Thomson electromotive force of a certain thermocouple, it is necessary to solve the system of equations

$$0.02\left(T_2{}^2 - T_1{}^2\right) = 4280$$
$$T_2 - T_1 = 250$$

Solve this system of equations for T_1 and T_2 in degrees Kelvin.

21. *Transportation engineering*   In the track for a rapid transit system, the total length of the merge and demerge region is related to the radius, r, of the curve and the offset of the distance, x, by the equation $L = 4rx$. The radius is also related to the velocity of the vehicle by the equation $r = 0.334v^2$. Combine these two equations in order to express L in terms of r and v only.

22. *Space technology*   A space probe is designed to separate into two parts when a spring-loaded connecting bolt is fired. The motion of the two parts, with mass $m_1 = 52\,\text{kg}$ and $m_2 = 78\,\text{kg}$, satisfies the energy equation $\left(\tfrac{1}{2}\right)(52)\left(v_1{}^2\right) + \left(\tfrac{1}{2}\right)(78)\left(v_2{}^2\right) = 23\,400$ and the momentum equation $52v_1 + 78v_2 = 0$, where v_1 and v_2 are the velocities of the parts in m/s. Find v_1 and v_2.

23. *Business*   In business, the break-even point for a product is where the revenue from selling the product equals the cost of producing the item. A company

sells two cold remedies, one a generic remedy and the other a "name" brand. The company has determined that the cost of producing the two brands is given by $4g^2 + 15n = 4291$. Meanwhile, the revenue is given by $8g^2 + 15g - 5n = 8607$. In both equations, g is the number of items (in millions) of the generic brand and n is the number of items (in millions) of the name brand. What is the break-even point?

24. *Business*   In business, the break-even point for a product is where the revenue from selling the product equals the cost of producing the item. A company produces and sells x units of product A and y units of product B. The company has determined that the cost of producing the two products is given by $x^2 + 1.5y^2 - xy = 2,891,189$. Meanwhile, the revenue is give by $80x + 100y = 252,640$. What is the break-even point?

 **In Your Words**

25. In Example 16.3, a circle and an ellipse intersected in four points. It is also possible that a circle and an ellipse can intersect in 0, 1, 2, or 3 points. Describe how each of these can happen.

26. Explain how you decide whether to use the substitution method or the addition and subtraction method to solve a system of nonlinear equations.

≡ 16.2
PROPERTIES OF INEQUALITIES; LINEAR INEQUALITIES

In Section 1.1, we introduced the inequality symbols. Since that time, we have not used them except to indicate when conditions were being placed on a group of numbers. The fact that we have not used the inequality symbols should not indicate that they are seldom used. In fact, in some parts of mathematics, inequalities are used as often as equations. In the remainder of this chapter, we will focus on inequalities.

An **inequality** is formed whenever two expressions are separated by one of the inequality symbols: $>$, $<$, $\geq$, and $\leq$. The two expressions are called the **sides**, or **members** of the inequality. A **compound inequality** has more than two expressions separated by inequality symbols.

EXAMPLE 16.6

$$2x < 5,$$
$$3x + 7 \geq 8,$$
$$16x - 8 \leq 5x - 2,$$
$$\text{and} \quad 5x^2 - 3x - 2 > x - 7$$

are all inequalities.

$$3x < 4x - 5 < 9,$$
$$x < y \leq 2z,$$
$$-8 \leq 3x + 1 \leq 4,$$
$$\text{and} \quad 1 > 2x - 3 \geq 17$$

are all compound inequalities.

A compound inequality can always be written as a combination of simple inequalities.

EXAMPLE 16.7

The compound inequality $3x < 4x - 5 < 9$ is the same as the two simple inequalities $3x < 4x - 5$ and $4x - 5 < 9$.

$-8 \leq 3x + 1 \leq 4$ is the same as the two simple inequalities $-8 \leq 3x + 1$ and $3x + 1 \leq 4$.

There are three types of inequalities. Most of the inequalities we will use are conditional inequalities. A **conditional inequality** is true for some real numbers and false for others. For example, $x \geq 5$ is a conditional inequality. It is true if x is 5, $6\frac{1}{2}$, 19, or any other number larger than 5. On the other hand, it is false for values of x such as 4, -3, $2\frac{1}{2}$, or any number smaller than 5.

An **absolute inequality** is true for all real numbers. An example of an absolute inequality is $x + 1 > x$. No matter which real number you select, this is a true statement. The opposite of an absolute inequality is a contradictory inequality. A **contradictory inequality** is false for all real numbers. For example $x + 1 < x$ is a contradictory inequality, because it is never true.

The **solution** of an inequality consists of those values of the variable that make the inequality true. To **solve an inequality** is to determine the solutions of the inequality.

Properties of Inequalities

There are six basic properties or rules of inequalities that are used to solve an inequality. In all of the explanations and examples of these properties, we will use the $<$ symbol. We could just as easily have used any of the other three inequality symbols.

Property 1 of Inequalities

If a, b, and c are real numbers, with $a < b$, then $a + c < b + c$.

In words, this property says that the same algebraic expression can be added or subtracted to both sides of an inequality without changing the direction of the inequality symbol.

EXAMPLE 16.8

(a) $x + 3 < 7$ is equivalent to $x < 4$.   Subtract 3 from both sides.
(b) $3x + 4 < 6 - 2x$ is equivalent to $5x < 2$.   Add $2x$ to both sides, and subtract 4 from both sides.

Property 2 of Inequalities

If a, b, and c are real numbers, with $a < b$ and $c > 0$, then $ac < bc$ or $\dfrac{a}{c} < \dfrac{b}{c}$

In words, this states that you can multiply or divide both sides of an inequality by the same *positive* expression without changing the direction of the inequality symbol.

EXAMPLE 16.9

(a) $\dfrac{x}{3} - \dfrac{4}{3} < 2x + 1$ is equivalent to $x - 4 < 6x + 3$. Multiply both sides by 3.

(b) $3x < 21$ is equivalent to $x < 7$. Divide both sides by 3.

Property 3 of Inequalities

If a, b, and c are real numbers, with $a < b$ and $c < 0$, then $ac > bc$ or $\dfrac{a}{c} > \dfrac{b}{c}$

≡ **Note**

If you multiply or divide both sides of an inequality by the same *negative* expression, then the direction of the inequality symbol is reversed.

EXAMPLE 16.10

(a) $-3 < 7$ is equivalent to $6 > -14$. Multiply both sides by -2.

(b) $\dfrac{-x}{3} < 5$ is equivalent to $x > -15$. Multiply both sides by -3.

(c) $-4x < 20$ is equivalent to $x > -5$. Divide both sides by -4.

Property 4 of Inequalities

If a, b, and n are positive real numbers and $a < b$, then $a^n < b^n$ and $\sqrt[n]{a} < \sqrt[n]{b}$

If both sides of an inequality are positive and n is a positive number, then the nth power or root of both sides retains the direction of the inequality.

EXAMPLE 16.11

(a) If $3 < 5$ and $n = 6$, then $3^6 < 5^6$.

(b) If $2 < 10$ and $n = 8$, then $\sqrt[8]{2} < \sqrt[8]{10}$.

Property 5 of Inequalities

If x and a are real numbers, $a > 0$, and $|x| < a$, then $-a < x < a$

EXAMPLE 16.12

(a) $|x+3| < 5$ is equivalent to $-5 < x+3 < 5$.
(b) $|2x-7| < 10$ is equivalent to $-10 < 2x-7 < 10$.

Property 6 of Inequalities

If x and a are real numbers, $a > 0$, and $|x| > a$, then $x > a$ or $x < -a$

EXAMPLE 16.13

(a) $|x-5| > 2$ is equivalent to $x-5 > 2$ or $x-5 < -2$.
(b) $|7x+4| > 9$ is equivalent to $7x+4 > 9$ or $7x+4 < -9$.

Solving Linear Inequalities

We can use these six properties to help us solve problems that use inequalities. As is true when you solve an equation, you often use several of the properties in each problem. Each of the problems is an example of a linear inequality.

Linear Inequality

A **linear inequality** in one variable, x, is an inequality of the form $ax+b < 0$, where a and b are constants and $a \neq 0$.

First, we will solve each of the linear inequalities in the following examples algebraically and then graph the solution. Since these are inequalities in one variable, each solution will be graphed on a number line.

EXAMPLE 16.14

Solve $6x+3 \geq 21$.

Solution $\quad 6x+3 \geq 21$
$\qquad\qquad 6x \geq 18 \quad$ Subtract 3 from both sides.
$\qquad\qquad\; x \geq 3 \quad$ Divide both sides by 6.

The solution to $6x+3 \geq 21$ is all numbers greater than or equal to 3. This is shown by the shaded section of the number line in Figure 16.5. The solid circle at 3 indicates that 3 is included in the solution.

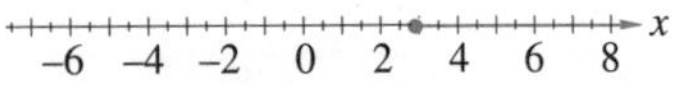

FIGURE 16.5

EXAMPLE 16.15

Solve $\frac{3}{4} - \frac{5}{6}x > \frac{-1}{2}x - \frac{11}{12}$.

Solution

$$\frac{3}{4} - \frac{5}{6}x > \frac{-1}{2}x - \frac{11}{12}$$

$9 - 10x > -6x - 11$ Multiply by 12, the LCD of the denominators.

$9 - 4x > -11$ Add $6x$ to both sides.

$-4x > -20$ Subtract 9 from both sides.

$x < 5$ Divide both sides by -4 and reverse the inequality sign.

The solution is $x < 5$. This is shown graphically in Figure 16.6. The open circle at 5 indicates that the solution does not include 5.

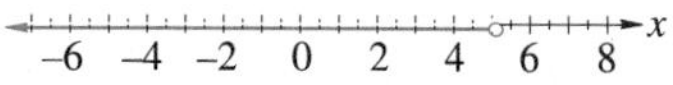

FIGURE 16.6

EXAMPLE 16.16

Solve $|3x - 5| < 13$.

Solution $|3x - 5| < 13$

$-13 < 3x - 5 < 13$ Apply Property 5.

$-8 < 3x < 18$ Add 5 to all three expressions.

$-\frac{8}{3} < x < 6$ Divide by 3.

This solution is shown graphically in Figure 16.7. Notice that $-\frac{8}{3} < x < 6$ is equivalent to $-\frac{8}{3} < x$ and $x < 6$.

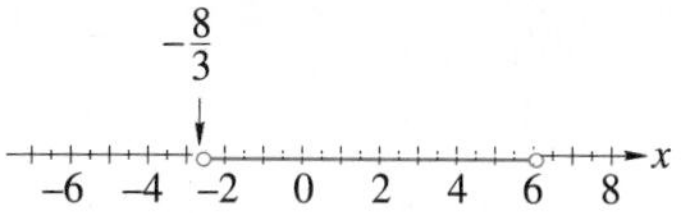

FIGURE 16.7

EXAMPLE 16.17

Solve $|3x + 2| \geq 8$.

Solution $|3x + 2| \geq 8$

$3x + 2 \geq 8$ or $3x + 2 \leq -8$ Apply Property 6.

$3x \geq 6$ or $3x \leq -10$ Subtract 2.

$x \geq 2$ or $x \leq -\frac{10}{3}$ Divide by 3.

The solution is the numbers less than or equal to $-\frac{10}{3}$ or those greater than or equal to 2. This is shown graphically in Figure 16.8.

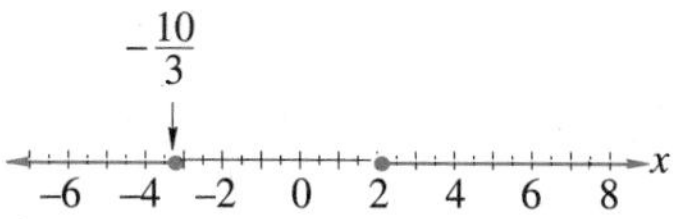

FIGURE 16.8

Application

EXAMPLE 16.18

A technician determines that an electronic circuit fails to operate because the resistance between A and B, $600\,\Omega$, exceeds the specifications. See Figure 16.9. The specifications state that the resistance must be between $200\,\Omega$ and $500\,\Omega$. If a shunt

EXAMPLE 16.18 (Cont.)

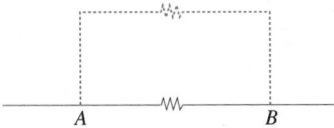

FIGURE 16.9

resistor of $R\,\Omega$, $R > 0$, is added, the circuit will satisfy the specifications. The equivalent resistance for the parallel connection is $\dfrac{1}{\dfrac{1}{600} + \dfrac{1}{R}}$ or $\dfrac{600R}{600 + R}\,\Omega$. What are the possible values for R?

Solution To satisfy the specifications, the following inequality must be true.

$$200 < \frac{600R}{600 + R} < 500$$

Since $R > 0$, $600 + R > 0$, and we can multiply the inequality by $600 + R$ without changing the direction of the inequality sign.

$$200(600 + R) < 600R < 500(600 + R)$$
$$120\,000 + 200R < 600R < 300\,000 + 500R$$

We will now write this compound inequality as two simple inequalities:

$$
\begin{array}{lcl}
120\,000 + 200R < 600R600R & \text{and} & < 300\,000 + 500R \\
120\,000 < 400R100R & \text{and} & < 300\,000 \\
300\ \ < R\ \ R & \text{and} & < 3\,000
\end{array}
$$

Thus, $300 < R < 3\,000$ and the shunt resistance must be between $300\,\Omega$ and $3\,000\,\Omega$.

Exercise Set 16.2

In Exercises 1–8, using the inequality $x < 10$, state the inequality that results when the operations given are performed on each side. Assume $x > 0$.

1. Add 5.

2. Subtract 7.

3. Multiply by 3.

4. Multiply by -4.

5. Divide by -2.

6. Divide by 5.

7. Square both.

8. Take the square root of both.

In Exercises 9–32, solve each of the given inequalities algebraically and graphically.

9. $3x > 6$

10. $4x < -8$

11. $2x + 5 > -7$

12. $3x - 4 \leq 8$

13. $2x - 5 < x$

14. $3x + 4 \geq x$

15. $4x - 7 \geq 2x + 5$

16. $7 - 3x \leq 3 + 5x$

17. $\frac{2}{3}x - 4 < \frac{1}{3}x + 2$

18. $\frac{1}{5}x - \frac{2}{5} > \frac{3}{5}x + \frac{4}{5}$

19. $\dfrac{x-2}{4} \leq \dfrac{3}{8}$

20. $\dfrac{x+2}{3} \geq \dfrac{5}{6}$

21. $\dfrac{x-3}{4} \leq \dfrac{2x}{3}$

22. $\dfrac{2x+5}{3} > \dfrac{3x-1}{2}$

23. $|x + 1| < 5$

24. $|2x - 1| < 7$

25. $|x + 4| > 6$

26. $|3x + 9| \geq 6$

27. $|x + 5| < -3$

28. $|3x + 2| < 12$

29. $-7 \leq 3x + 5 < 26$

30. $-6 < 5 - 3x \leq 17$

31. $3x + 1 < 5 < 2x - 3$

32. $4x - 6 \leq 11 < 9x + 1$

For Exercises 33–44, express the answer in terms of an inequality.

33. *Astronomy*   The radius of the Earth at the equator is 6 378 km and the polar radius is 6 357 km. What is the range of the radius r?

34. *Astronomy*   The moon has a maximum distance of 407 000 km from the Earth and a minimum distance of 357 000 km. What is the range of the moon's distance, d, from the Earth?

35. *Automotive technology*   The company specifications state that the camber angle c of the wheels should be $+0.60° \pm 0.50°$. Express the range for c as an inequality.

36. *Automotive technology*   The amount of R-12 in an automotive air-conditioning system may vary from 0.9 kg to 1.8 kg (2 to 4 lb). Knowing that there are 16 oz/lb, express the range of R-12 in ounces.

37. *Electricity*   If the resistance between A and B in Example 16.18 has been 800 Ω, what are the possible values for the shunt resistance?

38. *Machine technology*   A certain welding operation must be performed at temperatures between 1 800° and 2 200°C. What is the temperature range in degrees Fahrenheit? [Use $C = \frac{5}{9}(F - 32)$.]

39. *Business*   The weekly cost of manufacturing x microcomputers of a certain type is given by $C = 1,500 + 25x$. The revenue from selling these is given by $R = 60x$. At least how many microcomputers must be made and sold each week to produce a profit?

40. *Energy technology*   A rectangular solar collector is to have a height of 1.5 m. The collector will supply 600 W per m^2 and is to supply a total between 2 400 and 4 000 W. What range of values can the length of the collector have in order to provide this voltage?

41. *Electronics*   The instrument error, ϵ, of a wattmeter is less than 0.6 W. When the pointer is exactly on the 9-W mark, a technician records the power, P, as 9.0 W.
    (a) Express the error as an inequality, using absolute value.
    (b) Write an absolute value inequality that indicates the possible true power the technician should have recorded.
    (c) What is the possible range of values for P?

42. *Medical technology*   The medical dosage for a very young child is sometimes calculated by the formula $c = \dfrac{Ad}{150}$, where d is the adult dose, c is the child's dose, and A is the child's age in months. For what ages is the child's dosage between 25% and 50% of an adult's dose?

43. *Business*   A publishing company has found that the cost of publishing each copy of a certain magazine is $0.38. The company gets $0.35/copy from magazine dealers. Advertising revenue is 10% of the revenue received from dealers after the first 10,000 copies have been sold. What is the least number of copies which must be sold in order for the company to make a profit on this magazine?

44. *Computer science*   A certain computer diskette must be kept with a temperature range given by $|x - 29.4°C| \le 3.7°C$. What are the minimum and maximum temperatures of the range?

✎ **In Your Words**

45. Properties 2 and 3 of inequalities concern multiplying or dividing both sides of an inequality by a nonzero expression. Explain each property emphasizing how they are alike and how they are different.

46. Properties 5 and 6 of inequalities concern when an absolute value is less than or greater than a positive expression. Explain each property emphasizing how they are alike and how they are different.

≡ **16.3**
NONLINEAR
INEQUALITIES

In this section, we will learn how to solve nonlinear inequalities in one variable. There are two generally accepted methods for solving these kinds of inequalities. We will briefly show you one of those methods and then concentrate on the other method. You can use the one that seems easier.

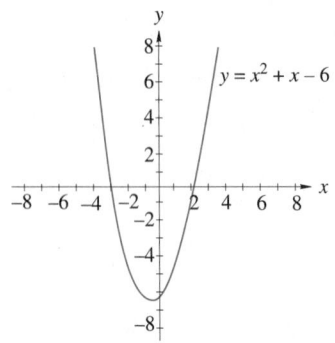

FIGURE 16.10

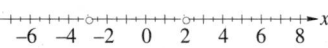

FIGURE 16.11a

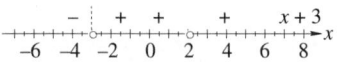

FIGURE 16.11b

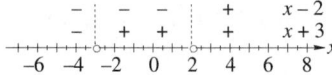

FIGURE 16.11c

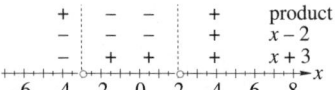

FIGURE 16.11d

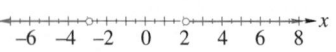

FIGURE 16.11e

Let's consider the inequality $x^2 + x - 6 > 0$. This is not linear, since it has an x^2-term. It is an inequality in one variable, x.

The first method for solving this inequality is to graph the equation $y = f(x) = x^2 + x - 6$, as shown in Figure 16.10. Look at the graph. When is $y > 0$? That is, when is the graph of $y = x^2 + x - 6$ above the x-axis? You can see that $y > 0$ when $x < -3$ or $x > 2$. Thus, the solution of $x^2 + x - 6 > 0$ is $x < -3$ or $x > 2$. This is a fairly easy approach. The fact that you have to graph the function requires time. While a graphing calculator can speed up the graphing, it is not always easy to determine *exactly* where the graph crosses the x-axis. We need to find a faster and more accurate method.

We will use the same inequality to demonstrate the second method. We begin by finding the roots of the corresponding equation $x^2 + x - 6 = 0$. Since $x^2 + x - 6 = (x + 3)(x - 2) = 0$, the roots are -3 and 2. Draw a number line and mark the roots on the number line, as in Figure 16.11a.

At one of these roots, put an open circle on top of the mark for that root. For example, in Figure 16.11a we have put an open circle on top of the mark for -3. Draw a dashed line above the circle at -3. The line is dashed because -3 is not a solution of this inequality. On the right side of this circle $x + 3 > 0$, and on the other side $x + 3 < 0$. Put a row of $+$ signs on the side where $x + 3$ is positive and a row of $-$ signs on the other side. Write $x + 3$ at one end to remind you which factor is indicated by that row. Your number line should now look like the one in Figure 16.11b.

At the second root, put an open circle at the mark for the root at 2 and draw a dashed line upward from this circle. Indicate the side where $x - 2$ is positive and the side where $x - 2$ is negative, as in Figure 16.11c.

We need one more row of signs to complete the process. The last row is the product row. The original inequality is $x^2 + x - 6 > 0$ or $(x + 3)(x - 2) > 0$. The product row will indicate the sign of $(x + 3)(x - 2)$. Remember, whenever you multiply an even number of negative numbers together, you get a positive answer. If you multiply an odd number of negative numbers, you get a negative number.

In Figure 16.11d, to the right of 2 both the $x - 2$ and the $x + 3$ rows are positive, so their product is positive. We have put $+$ signs in the product row to the right of the dashed line above the 2. The next interval is the section between -3 and 2. Here $x - 2$ is negative, and $x + 3$ is positive, so the product is negative. Thus, we have put $-$ signs in the product row between the dashed lines above -3 and 2. The last interval is to the left of -3 or the numbers less than -3. In this interval both factors are negative, so the product is positive. This is indicated by the $+$ signs in the product row to the left of the dashed line above the -3, as shown in Figure 16.11d.

By inspecting the product row in Figure 16.11d, we can see that the solution is $x < -3$ or $x > 2$. This is shown graphically in Figure 16.11e. This is the same answer we got using the other method. This seems like a longer and more complicated method, but it is actually easier. If you are solving an inequality that has the sign $\leq$ or $\geq$, you would fill in the circles that mark the roots and draw solid lines instead of dashed lines above the circles.

EXAMPLE 16.19

Solve $(x-2)(x+3)(x-5)(x+1) \le 0$.

Solution   This is already factored. The roots are $2, -3, 5$, and -1. The sign rows for each of these factors are given in Figure 16.12a. The product row is given in Figure 16.12b. Because this inequality has the sign $\le$, we have filled in the circles to indicate that the roots are included. From inspection, we see that the product row is negative when $-3 \le x \le -1$ or $2 \le x \le 5$. This is the solution, and it is shown graphically in Figure 16.12c.

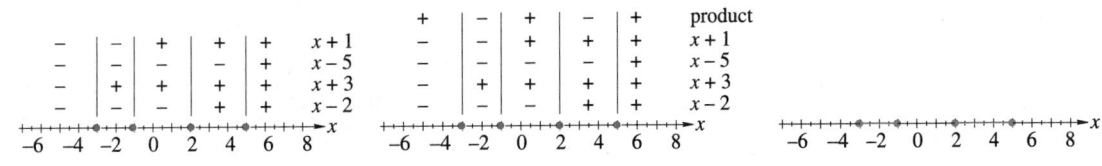

FIGURE 16.12a   **FIGURE 16.12b**   **FIGURE 16.12c**

Do not be misled into thinking that the signs in each row are always positive to the right of the root. You need to solve each linear inequality to determine the signs, as the next example demonstrates.

EXAMPLE 16.20

Solve: $x(x-2)(5-x) > 0$.

Solution   The roots are $0, 2$, and 5. The sign row for each root is given in Figure 16.13a. The circles are not filled in because the inequality sign does not include an equal sign. Also, notice the sign row for $5-x$. The product in Figure 16.13b indicates that the solution is $x < 0$ or $2 < x < 5$. Figure 16.13c shows the answer graphically.

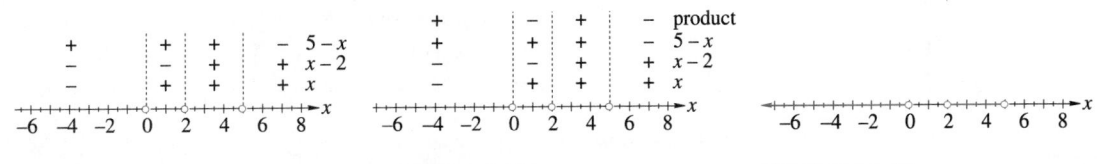

FIGURE 16.13a   **FIGURE 16.13b**   **FIGURE 16.13c**

The procedure for finding the solutions to a division problem can be determined the same way. You must check to ensure that any numbers that make the quotient zero are not in the solutions.

EXAMPLE 16.21

Solve: $\dfrac{x^2 + 6x + 5}{x+2} \le 0$.

Solution   The numerator factors into $(x+5)(x+1)$ and so the numerator is zero at -5 and -1; the denominator is zero at $x = -2$. These three points are indicated in Figure 16.14a with the sign rows for $x+5$, $x+1$, and $x+2$. Notice that the circles are filled at -5 and -1 because the inequality includes an equal sign. The circle at -2 is not filled in because the denominator is zero at -2. Thus, -2 is not a

EXAMPLE 16.21 (Cont.)

solution. If we remember that $\dfrac{x^2+6x+5}{x+2} = (x+5)(x+1)\left(\dfrac{1}{x+2}\right)$ and realize

that $x+2$ and $\dfrac{1}{x+2}$ have the same signs, then the signs of the answer are given by a product row. The product row is shown in Figure 16.14b and the graphical solution is given in Figure 16.14c. From this, we see that the solution is $x \leq -5$ or $-2 < x \leq -1$.

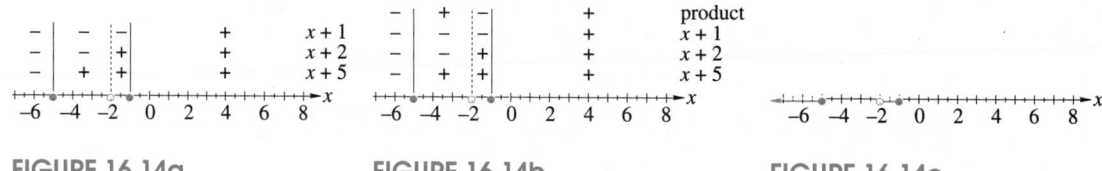

FIGURE 16.14a   FIGURE 16.14b   FIGURE 16.14c

Using a Graphing Calculator

A graphing calculator can be used to help solve inequalities. You must first determine the real roots of the associated function. Next, graph the function.

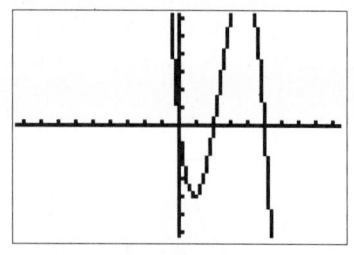

[–9.4, 9.4, 1] × [–6.2, 6.2, 1]

FIGURE 16.15

> **Solving Inequalities with a Graphing Calculator**
>
> 1. Determine the real roots and the domain of the equation associated with the inequality.
> 2. Graph the function associated with the equation or inequality.
> 3. **(a)** If the given inequality is of the form $f(x) > 0$, the solutions are the intervals on the x-axis where the graph is above the x-axis.
>   **(b)** If the given inequality is of the form $f(x) < 0$, the solutions are the intervals on the x-axis where the graph is below the x-axis.

The next two examples rework Examples 16.20 and 16.21 using a graphing calculator.

EXAMPLE 16.22

Use a graphing calculator to solve $x(x-2)(5-x) > 0$.

Solution   The equation associated with this inequality is $x(x-2)(5-x) = 0$. It has real roots $x = 0$, $x = 2$, and $x = 5$. The domain is all real numbers.

The function associated with this inequality is $f(x) = x(x-2)(5-x)$. Its graph, as it appears on a TI-82, is shown in Figure 16.15. Notice that the graph intersects the x-axis at each of the roots.

Since the inequality is of the form $f(x) > 0$, we want the intervals where the graph is above the x-axis. The graph is above the x-axis for the intervals $(-\infty, 0)$ and $(2, 5)$. Thus, the solution is $x < 0$ or $2 < x < 5$.

EXAMPLE 16.23

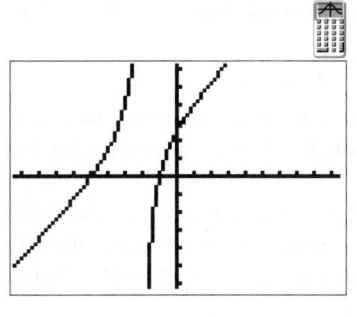

[−9.4, 9.4, 1] x [−6.2, 6.2, 1]

FIGURE 16.16

Use a graphing calculator to solve $\dfrac{x^2+6x+5}{x+2} \leq 0$.

Solution   The function associated with this inequality is $f(x) = \dfrac{x^2+6x+5}{x+2}$.

From Example 16.21, we know that $f(x) = 0$ has real roots at $x = -5$ and $x = -1$ and is undefined at $x = -2$.

The graph of f is shown in Figure 16.16. Notice that $y = f(x)$ intersects the x-axis at each of the roots and is not graphed when $x = -2$.

Because this inequality is $\leq$, we want the intervals where the graph is below or touching the x-axis. By looking at the graph we see that these occur when $x \leq -5$ or $-2 < x \leq -1$. Remember, since $f(x)$ is undefined at $x = -2$, the solution does not contain -2.

Application

EXAMPLE 16.24

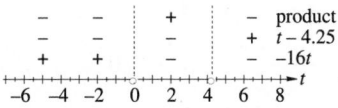

FIGURE 16.17

From the top of a 120-ft building an object is thrown upward with an initial velocity of 68 ft/s. Its distance d above the ground at any time t is given by the equation $d = 120+68t-16t^2$. For what period of time is the object higher than the building?

Solution   Since the building is 120 ft high, this is really asking when

$$120+68t-16t^2 > 120$$

$$\text{or} \qquad -16t^2+68t > 0$$

This factors into $-16t(t-4.25) > 0$. This inequality is zero at $t = 0$ and $t = 4.25$. As shown in Figure 16.17, the sign row for $-16t$ is negative for $t > 0$ and positive for $t < 0$. According to the product row, the object is above the building for the first 4.25 s after it is thrown.

In each of these examples, we were concerned with solutions in the real numbers. If the nonlinear equation does not have any real roots, the problem will be an absolute inequality and every real number will satisfy the inequality or it will be a contradictory inequality and no real number will satisfy the inequality. The next example illustrates one of these situations.

EXAMPLE 16.25

Solve: $x^2+4x+5 > 0$.

Solution   Using the quadratic formula, we see that the roots are $\dfrac{-4\pm\sqrt{16-20}}{2} = -2\pm j$. Both of these are complex numbers. Since there are only two roots to $x^2+4x+5 = 0$, and neither of them is a real number, this inequality is either absolute or contradictory. If it is absolute, any real number will satisfy it. Select any real number, say 0 and we get $0^2+4(0)+5 = 5 > 0$. So, 0 is a solution. This is an absolute inequality and all real numbers satisfy it.

The graph of $y = x^2+4x+5$ is shown in Figure 16.18. From this graph, you can see that y is always larger than 0. This confirms our solution.

EXAMPLE 16.26

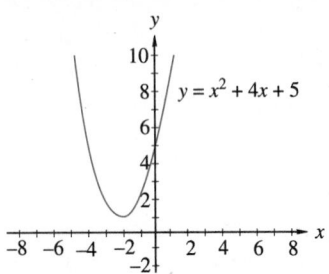

FIGURE 16.18

Solve: $x^2 + 4x + 5 < 0$.

Solution Check the graph of $y = x^2 + 4x + 5$ in Figure 16.18. As we saw in the previous example, y is always positive. This is a contradictory inequality and no real numbers will make it true.

We could have also solved this in the same manner as described in Example 16.25. Select a real number, say 2. Since $2^2 + 4(2) + 5 = 17$, this inequality has no real number solutions.

The techniques described in Examples 16.25 and 16.26 depend on first determining that there are no real roots. At present, we can only do this for quadratics. In Chapter 18, we will learn how to determine the roots of higher degree polynomials.

Exercise Set 16.3

Solve the inequalities in Exercises 1–40.

1. $(x+1)(x-3) > 0$

2. $(x-4)(x+2) \geq 0$

3. $(x-1)(x+4) \leq 0$

4. $(x-2)(x+5) < 0$

5. $x^2 - 1 < 0$

6. $x^2 \leq 16$

7. $x^2 - 2x - 15 \leq 0$

8. $x^2 + x - 2 > 0$

9. $x^2 - x - 2 \geq 0$

10. $x^2 - 3x - 10 < 0$

11. $x^2 - 5x > -6$

12. $x^2 + 7x \leq -12$

13. $2x^2 + 7x + 3 < 0$

14. $2x^2 - 5x + 2 \geq 0$

15. $2x^2 - x < 1$

16. $2x^2 - x \leq 3$

17. $4x^2 + 2x > x^2 - 1$

18. $2x^2 + 5x < 2 - x^2$

19. $(x+1)(x-2)(x+3) < 0$

20. $(x-1)(x+2)(x-3) \leq 0$

21. $(x+3)(2x-5)(x+4) > 0$

22. $(x+4)(3x-1)(x-5) \geq 0$

23. $(x-2)^2(x+4) < 0$

24. $(x+3)^2(x-5) \geq 0$

25. $x^3 - 4x > 0$

26. $x^4 - 9x^2 \leq 0$

27. $x^2 + 2x + 3 \leq 0$

28. $x^2 + x + 1 > 0$

29. $\dfrac{(x-2)(x-5)}{x+1} < 0$

30. $\dfrac{(x+2)(x-4)}{x-2} \geq 0$

31. $\dfrac{x}{(x+1)(x-2)} > 0$

32. $\dfrac{x(x-3)}{(x+1)(x-4)} \leq 0$

33. $\dfrac{(3x+1)(x-3)}{2x-1} \leq 0$

34. $\dfrac{(x-1)(x+6)}{(x+1)(x-3)} > 0$

35. $\dfrac{4}{x-1} < \dfrac{5}{x+1}$

36. $\dfrac{-3}{x+2} \geq \dfrac{6}{x-3}$

37. $\dfrac{x^2 + 2x + 3}{x-1} \geq 0$

38. $\dfrac{x^2 + x + 1}{x+2} < 0$

39. $|x^2 + 3x + 2| \leq 4$

40. $|x^2 - 3x + 2| > 5$

Solve Exercises 41–48.

41. *Lighting technology* The intensity I in candelas (cd) of a certain light is $I = 75d^2$, where d is the distance in meters from the source. For what range of distances will the intensity be between 75 and 450 cd?

42. *Physics* An object is thrown straight upward from the ground with an initial velocity of 34.3 m/s. Its distance d above the ground at any time t is given by the equation $d = 34.3t - 4.9t^2$. For what time period is the object more than 49 m above the ground?

43. *Architecture* The deflection of a beam d is given by $x^2 - 1.1x + 0.2$, where x is the distance from one end. For what values of x is $d > 0.08$?

44. *Architecture*   The load L that can be safely supported by a wooden beam of length ℓ with rectangular cross-section of width w and depth d is given by the formula

$$L = \frac{kwd^2}{\ell}$$

where k is a constant, depending on the type of wood. If $k = 90$ and the beam is to be 18 ft long and 6 in. wide, what are the acceptable values for d, if the beam must support at least 3,000 lb and $d < 12$ in.? (Do not change the length measurements to the same units.)

45. *Dynamics*   A ball is thrown vertically upward into the air from the roof of a 452-ft building. If the initial velocity of the ball is 64 ft/s, the height of the ball above the ground is given by the function $h(t) = -16t^2 + 64t + 452$.
(a) For what times t will the height be greater than 500 ft?
(b) For what times t will the height be less than 395 ft?

46. *Forestry*   Volume estimates V, in cubic feet, for shortleaf pine trees are based on D, the d.b.h. (diameter at breast height) in inches; top d.i.b., the diameter, in inches, inside the bark at the top of the tree; and H, the height of the tree in feet. One formula for trees with a 3-in. top d.i.b. is

$$V = 0.002837D^2H - 0.127248.$$

Determine D for a 75 ft tree that has a volume estimate greater than 47.75 ft^3.

47. *Forestry*   The Scribner log-rule equation for 16 ft logs is $V = 0.79D^2 - 2D - 4$, where V is the volume, in board feet, of the log and D is the diameter, in inches, of the small end of a log inside the bark. What diameter of the small end of the log inside the bark is needed for a 16 ft log to have a volume of at least 926.0 board ft?

48. *Material science* · The range of temperatures, T (in degrees Kelvin), for 100.00 g of silver when the silver receives 2.00×10^4 cal of heat is given by $|-2.8T^2 + 820T - 44,000| < 10,000$. Solve for T to the nearest 0.1 K.

    In Your Words

49.   Describe how you can use your calculator to solve a nonlinear inequality.

50.   Explain how to use a sign chart to solve an inequality.

≡ 16.4
INEQUALITIES IN TWO VARIABLES

One variable is not enough to describe some real problems. As a result, we often work with equations of two or more variables. In this section, we will consider inequalities with two variables.

Solutions to equations and inequalities in one variable are represented on a number line. Solutions to equations are represented by points and solutions to inequalities by intervals.

Equations in two variables are represented by a graph in a plane. Inequalities in two variables are also represented graphically in a plane. The graph, however, is not a line or a curve.

We will begin by looking at linear inequalities in two variables.

> **Linear Inequalities in Two Variables**
>
> A **linear inequality in two variables**, x and y, is an inequality of the form $ax + by + c < 0$, where $a, b,$ and c are constants and a and b are not both zero.

≡ **Note**

The use of the $<$ symbol in this definition is not meant to restrict this to only inequalities that use $<$. This same definition applies for $>$, $\leq$, and $\geq$.

Every nonvertical line can be written in the slope-intercept form, $y = mx + b$. This line divides the plane into three separate regions, as shown in Figure 16.19.

1. The points that satisfy $y = mx + b$ (the line)
2. The region that satisfies $y < mx + b$ (the points below the line)
3. The region that satisfies $y > mx + b$ (the points above the line)

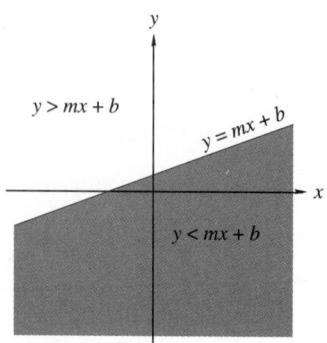

FIGURE 16.19

A vertical line is of the form $x = a$, where a is a constant. Vertical lines also divide the plane into three separate regions, as shown in Figure 16.20: the line $(x = a)$, the points to the right of the line $(x > a)$, and the points to the left of the line $(x < a)$.

To illustrate how this works, we will solve $2x + y < 4$. We begin by solving the inequality for y, getting $y < -2x + 4$. Next, we graph the line $y = -2x + 4$. This inequality does not include the points where $y = -2x + 4$, so we will make this a dotted line rather than a solid line. The plane is now divided into three regions. We want the region below the line. This area has been shaded in Figure 16.21.

It is always a good idea to check your results. Select any point in the shaded region. For example, select the point $(-1, 4)$. Substitute these values in the original inequality and, on the left-hand side, you get $2(-1) + 4 = -2 + 4 = 2$. This is certainly less than 4, so this point checks. Obviously checking one value does not mean that you have not made a mistake. This answer has an infinite number of solutions and we checked only one. But, checking one value at least *helps* us see if we may be correct.

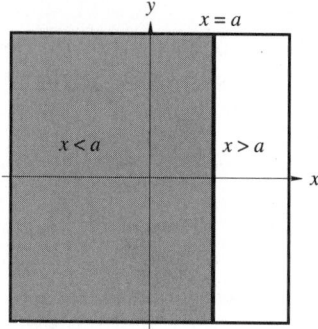

FIGURE 16.20

≡ **Note**

Notice that if the inequality does not include equality, the curve dividing the region will be a broken line. A solid line will indicate that the points on the line are included in the solution.

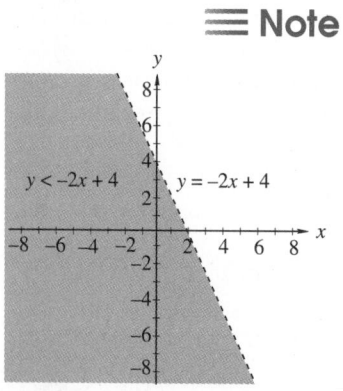

FIGURE 16.21

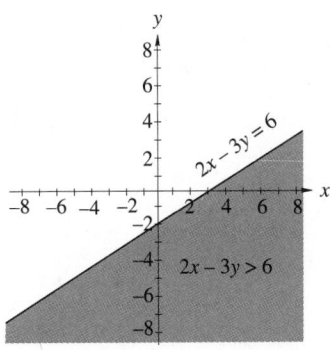

FIGURE 16.22

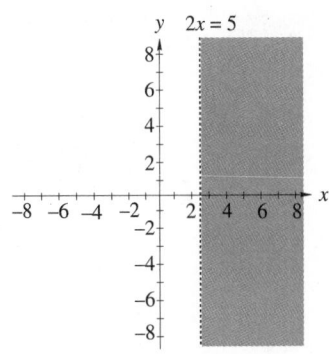

FIGURE 16.23

EXAMPLE 16.27

Find the region described by $2x - 3y \geq 6$.

Solution We first solve the inequality for y.

$$2x - 3y \geq 6$$
$$-3y \geq -2x + 6$$
$$y \leq \frac{2}{3}x - 2$$

Notice that, when we divided by -3, the direction of the inequality symbol reversed.

Next, graph the line $y = \frac{2}{3}x - 2$, as shown in Figure 16.22. Because this problem involves equality ($\leq$), we make it a solid line. When we solved the inequality for y, we got $y \leq \frac{2}{3}x - 2$ and this indicates that we want the region below the line, as is shaded in Figure 16.22. This shaded region and the line are the solutions for this problem.

EXAMPLE 16.28

Find the solution to $2x > 5$.

Solution If you solve this for x, you get $x > \frac{5}{2}$. Graph the line $x = \frac{5}{2}$ with a dotted line. The desired region is to the right of this line. The solution is the shaded region in Figure 16.23.

Not all equations in two variables are equations of a line. In the same way, not all inequalities in two variables are of regions separated by a line.

EXAMPLE 16.29

Find the solution to $y \leq x^2 - 4$.

Solution   The graph of $y = x^2 - 4$ is the parabola in Figure 16.24. All points on the parabola or below it satisfy the given inequality, as indicated in Figure 16.24.

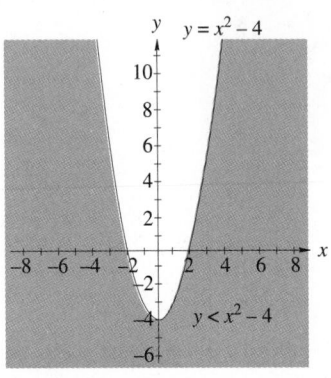

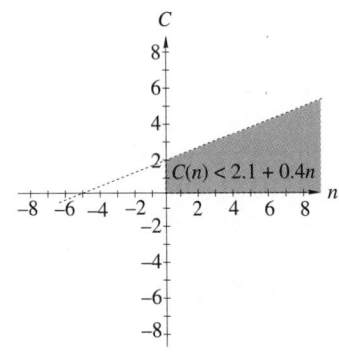

FIGURE 16.24   **FIGURE 16.25**

Application

EXAMPLE 16.30

A company has determined that it can make a profit on its microcomputers if the production costs C satisfies the inequality $C(n) < 2.1 + 0.4n$, where n is the number of computers produced. Graph this inequality.

Solution   We will let n be on the horizontal axis and C the vertical axis. We graph the line $C(n) = 2.1 + 0.4n$, and make the line dashed to show that the solution does not include the points on the line. Since we want $C(n) < 2.1 + 0.4n$, we shade the points below the line. We only shade points in the first quadrant, because $n > 0$. (After all, the company cannot make less than 0 microcomputers.) We have also $C(n) > 0$, because it will cost something to produce the computers. The solution is shown by the graph in Figure 16.25.

Exercise Set 16.4

Graph each inequality in Exercises 1–30.

1. $y > 2$
2. $y \leq -3$
3. $x \geq 5$
4. $y < 2$
5. $x + y > 3$
6. $y - x \leq 4$
7. $2x + y < 4$
8. $3x + y > -4$

9. $x + 2y \leq 4$
10. $2y - x > -3$
11. $2x + 3y < 6$
12. $3x + 2y \geq 6$
13. $4x - 3y \leq -6$
14. $3y + 4x \geq 9$
15. $0.5x + 1.5y > 2.5$
16. $0.3x - 0.9y \leq 1.2$

17. $1.4x - 0.7y < 1.0$
18. $0.2x + 0.6y \leq 2.0$
19. $y \leq x^2 + 2$
20. $y \geq x^2 - 3$
21. $y \geq 3x^2 - 1$
22. $y \leq 4x^2 - 5$
23. $y < x^2 + 2x + 1$
24. $y \geq x^2 - 4x + 4$

25. $x^2 + y^2 \leq 16$
26. $x^2 + y^2 > 9$
27. $4x^2 + 9y^2 \leq 36$
28. $9x^2 + y^2 > 9$
29. $|x + 1| < 4$
30. $|x - 5| \leq 3$

Solve Exercises 31–34.

31. *Chemistry*   The temperature (°C) and pressure (kPa) of a controlled chemical reaction must satisfy $1.8P + T < 270$. Graph this inequality for $T > 0$ and $P \geq 0$.

32. *Business*   In order to make a profit, the cost c of producing a computer chip in a certain factory must satisfy the inequality $C(n) < n^2 + 10n + 35$, where n is the number of chips produced. Graph this inequality. Remember that $n \geq 0$.

33. *Business*   A certain bank's return on a home loan is 8.5% and 6.75% on a commercial loan. The manager of the loan department wants to earn a return of at least $3.6 million on these loans. Let x represent the amount (in million dollars) for home loans and y the amount for commercial loans. Graph this inequality.

34. *Nutrition*   A brand-X multivitamin tablet contains 15 mg of iron. One brand-Y multivitamin tablet contains 9 mg of iron. A nutritionist advises a patient to take at least 90 mg of iron each day. Graph this inequality.

 **In Your Words**

35. Describe how you can decide which region of the plane should be shaded.

36. What determines whether the curve dividing the region will be a broken line or a solid line?

≡ 16.5
SYSTEMS OF INEQUALITIES; LINEAR PROGRAMMING

Several times in this book we have worked with systems of equations. In fact, the first section of this chapter was devoted to solving systems of quadratic equations. You can just as easily have systems of inequalities. In this section, you will learn how to solve linear inequalities and then learn how to apply this to a technique called linear programming.

The **graph of a system of inequalities** consists of all the points in the plane that satisfy every inequality in the system. To get this graph you will graph on the same coordinate system the region determined by each of the inequalities. The region where all these regions overlap is the graph of the system of inequalities.

EXAMPLE 16.31

Graph the solution of the following system.

$$\begin{cases} x + y \leq 5 \\ x - 3y < 3 \end{cases}$$

Solution   This system is equivalent to

$$\begin{cases} y \leq -x + 5 \\ y > \dfrac{x}{3} - 1 \end{cases}$$

As we did in Section 16.4, we first sketch the lines $y = -x + 5$ and $y = \dfrac{x}{3} - 1$. Use broken lines when you sketch them. Next, lightly shade the regions that satisfy each inequality, as shown in Figures 16.26a and 16.26b.

EXAMPLE 16.31 (Cont.)

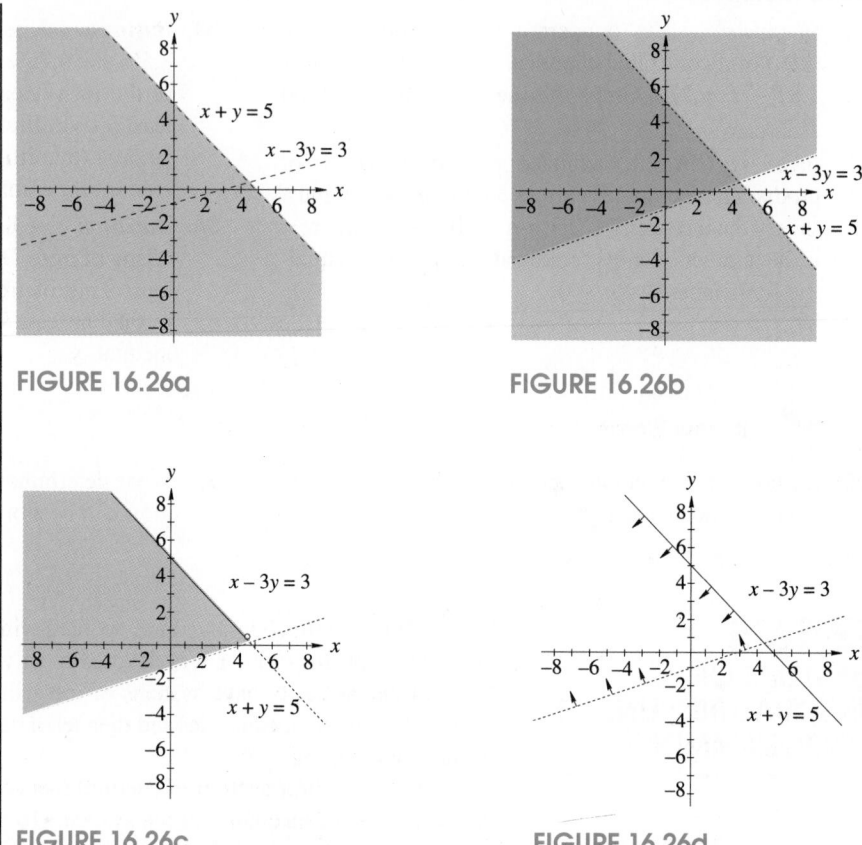

FIGURE 16.26a

FIGURE 16.26b

FIGURE 16.26c

FIGURE 16.26d

The part where both of the shaded regions overlap is part of the solution. Shade this overlapped region more heavily. The first inequality used the $\leq$ symbol, so the part of the line $y = -x + 5$ that borders the answer region should be made solid. The area you have shaded is the solution and should look like the one in Figure 16.26c. Remember to work with broken lines until you can determine which section of the line is included in the solution.

Notice the "hole" where the two boundaries intersect. This point of intersection $(4.5, 0.5)$ does not satisfy the second inequality and so it is not part of the solution for this linear system. In essence, the intersection of a solid line and a broken line results in a "hole."

Another approach is just to place arrows on the boundaries with the arrows pointing toward the region that satisfies the inequality. This is shown in Figure 16.26d.

EXAMPLE 16.32

Graph the solution of this system.

$$\begin{cases} y + x^2 < 5 \\ x \geq -1 \\ x + 2y \geq -2 \end{cases}$$

Solution   This system is equivalent to

$$\begin{cases} y < -x^2 + 5 \\ x \geq -1 \\ y \geq -\frac{1}{2}x - 1 \end{cases}$$

Each curve is sketched using broken lines. The region defined by each inequality is lightly shaded and the region common to all three is then shaded more darkly. The borders of this common region that were formed by the last two inequalities are changed from broken lines to solid lines. The final region is shown in Figure 16.27. If you want to verify that this is the correct region, then select a point in the region and verify that it satisfies all of these inequalities. A good point to test is $(0, 0)$ and you can readily see that it satisfies all three inequalities.

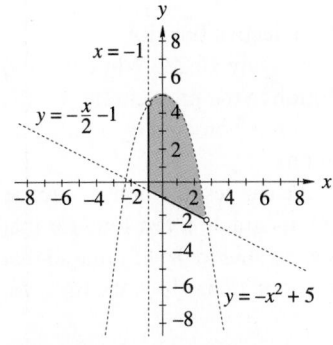

FIGURE 16.27

One very important area of discipline that uses graphs of systems of inequalities is called **linear programming**. With the use of linear programming it is possible to solve problems in which a quantity is to be maximized or minimized, subject to certain limitations called **constraints**.

Linear programming is used with functions of two or more variables. We will work only with functions or equations of two variables. In linear programming, you establish a system of linear inequalities and graph its solution. This region is the set of **feasible solutions** to the problem and is called the **feasible region**. Any point in this region is called a **feasible point**. Finally, the **objective function** is formed, which describes the quantity to be maximized or minimized. The maximum and minimum values of the objective function will occur on the boundary of the feasible region. This will always be a vertex or a segment that connects two vertices of the feasible region.

Application

EXAMPLE 16.33

A company manufactures two types of microcomputers: Model A with 256K RAM and Model B with 128K RAM. The company can make a maximum of 75 model A micros per day and 50 Model B micros per day. It takes 8 h to manufacture a Model A micro and 3 h to manufacture a Model B machine. The number of employees can provide a total of 630 h of work each day. The profit on each Model A computer is $20 and the profit on each Model B is $27. How many of each type should be made each day to give the maximum profit?

EXAMPLE 16.33 (Cont.)

Solution   If x is the number of Model A microcomputers made each day and y the number of Model B, we then have the following constraints.

$$0 \leq x \leq 75 \quad \text{They can make no more than 75 of Model } A.$$
$$0 \leq y \leq 50 \quad \text{They can make a maximum of 50 Model } B.$$
$$8x + 3y \leq 630 \quad \text{They can only work 630 h a day.}$$

The profit P is given by $P = 20x + 27y$. This is the objective function.

The set of feasible points is the shaded region in Figure 16.28. While it is too lengthy to prove here, it can be shown that the solution to the problem lies on the boundary of the feasible region. Thus, since this region is bordered by a polygon, the solution will be a vertex or one side of the polygon.

There are two ways to approach this solution. We can test the object value at each vertex or we can select a value for P and draw its graph, a test line, for that equation. The answer will be found by drawing a line parallel to the test line so that every point in the feasible region is either on the line or on one side of the line. We will begin with the first method.

The feasible region has five vertices. If we evaluate the objective function $P = 20x + 27y$, at each of these vertices we get the following table of values.

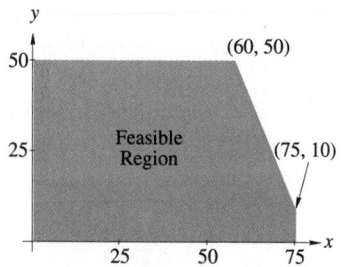

FIGURE 16.28

Vertex	P
$(0,0)$	0
$(0,50)$	1,350
$(60,50)$	2,550
$(75,10)$	1,770
$(75,0)$	1,500

From this table, we can clearly see that the most profit will occur when they make 60 Model A and 50 Model B microcomputers each day.

With the second method, you can select a value for P, say 1,080, and graph $1,080 = 20x + 27y$, as shown in Figure 16.29. Now find a line parallel to this, where the entire feasible region is on the same side of the line except for the point, or points, where the line intersects the region. (We know that this occurs when $P = 2,550$.) This happens when $y = 50$ and $8x + 3y = 630$ intersect. Solving these two simultaneous equations gives us the same solution, $x = 60$ and $y = 50$, and the maximum daily profit of $2,550.

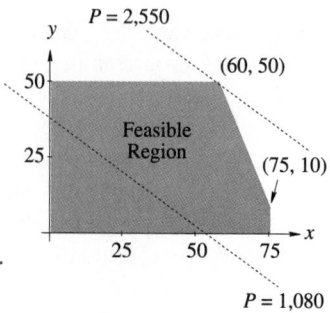

FIGURE 16.29

Application

EXAMPLE 16.34

Each week, an electronics company makes home entertainment sets. Among those that they make are compact disc players (CDs) and television sets (TVs). The number of hours it takes to manufacture each set, the cost of the materials, and the profit are shown in the following table. The last line of the table shows the maximum number of each item that are available each week. How many of each type of set should the company make in order to make the largest profit?

	Manufacturing time (h)	Material Cost ($)	Profit ($)
CD	3.5	112	67
TV	4.0	76	58
Available	488.0	13,500	

Solution   Since the company makes a profit of $67 on each CD and only $58 on each TV, you might suggest that they make only CD players. Since there are 488 h of work time available each week, and since each CD takes 3.5 h to make, they could make a total of

$$488 \div 3.5 = 139.4 \text{ CD players}$$

It requires $112 of materials to make each CD, so a total of 139 CDs would require $112 \times 139 = \$15,568$. Unfortunately, there is only $13,500 available for materials each week.

   Using a similar argument, we can see that it would not be wise to make as many CDs as possible with the available money for materials. Since each CD requires $112 worth of materials and we have $13,500 available, then why not make

$$\$13,500 \div \$112 = 120.5 \text{ CD players}$$

Thus, we could make 120 CD players each week (and make a weekly profit of $120 \times \$67 = \$8,040$). It would require $120 \times 3.5 = 420$ h to make these sets. That would leave 68 h when no manufacturing was taking place, but no more money to make any TV sets. The company is committed to making both CD players and TV sets, so this solution is not acceptable.

   Now, let's use linear programming to see how many CDs and TVs should be made in order to make the largest profit.

   If we let c represent the number of CDs and t the number of TVs we can make in one week, then our weekly profit P will be

$$P = 67c + 58t$$

This equation, $P = 67c + 58t$, is the objective function.

   The manufacturing time cannot be more than 488 h, so

$$3.5c + 4t \leq 488$$

The cost of materials cannot be more than $13,500, so

$$112c + 76t \leq 13,500$$

EXAMPLE 16.34 (Cont.)

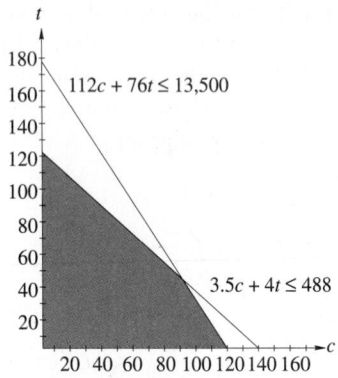

FIGURE 16.30

Finally, we assume that $c \geq 0$ and $t \geq 0$. Both of these last two inequalities contain the equal signs, because we want to allow for the possible (but unexpected) answer that we should make all CD players and no TVs or all TVs and no CDs. We now have our objective function and four constraints. The constraints are graphed in Figure 16.30.

Three vertices of the feasible region are $(0, 0)$, $(0, 120)$ and $(120, 0)$. We find the fourth vertex of the feasible region by simultaneously solving the systems of equations

$$\begin{cases} 3.5c + 4t = 488 \\ 112c + 76t = 13{,}500 \end{cases}$$

Using the techniques from Chapter 5, such as Cramer's rule, we see that $c \approx 92.92$ and $t \approx 40.69$. These are not integers, so the solution is still not clear. If we round both of these numbers off to the nearest whole number, we would conclude that we should make 93 CDs and 41 TVs. But if we do, this would require 489.5 h of manufacturing time and \$13,532 of materials a week. Both of these are more than we are allowed.

The alternative is to round one of these numbers up and the other one down. This gives the following two possibilities.

$$c = 92 \text{ and } t = 41$$

or $\quad c = 93 \text{ and } t = 40$

In order to determine the answer, we substitute each pair of values in the objective function $P = 67c + 58t$ and see which pair of numbers produces the higher profit. When

$$c = 92 \text{ and } t = 41, \ P = 67(92) \text{ and } 58(41)$$
$$= 8{,}542$$

and when $c = 93$ and $t = 40$, $P = 67(93)$ and $58(40)$
$$= 8{,}551$$

Thus, the company will make the most profit, \$8,551, when they manufacture 93 CDs and 40 TVs.

For many companies, other factors might have entered into this example. One factor is the size of the company's warehouse. Suppose that the company in Example 16.34 only ships once a week. That means that they must store all the work from one week in the warehouse. Further suppose that there was only enough room for 125 boxes in the warehouse (assuming that the TV and CD boxes are the same size). Then our answer of 93 CDs and 40 TVs, or 133 total products might not be acceptable. Other factors might be considered, such as the amount of time available to test each new product or the number of people that are trained to assemble each type of equipment.

Exercise Set 16.5

In Exercises 1–14, give the graphical solution to each system of inequalities.

1. $\begin{cases} x - y > 0 \\ y - 2x < 4 \end{cases}$

2. $\begin{cases} x + y \geq 6 \\ y - 3x < 6 \end{cases}$

3. $\begin{cases} x + y \geq 5 \\ y > 1 \end{cases}$

4. $\begin{cases} x + 2y \leq 4 \\ y < 1 \end{cases}$

5. $\begin{cases} 2x + 3y \leq 7 \\ 3x - y \leq 5 \end{cases}$

6. $\begin{cases} -2x + y > 6 \\ x + 3y \leq 9 \end{cases}$

7. $\begin{cases} x + 2y - 3 \leq 0 \\ 2x + y - 4 > 0 \end{cases}$

8. $\begin{cases} -2x - 3y < 6 \\ -4x + 3y \geq 12 \end{cases}$

9. $\begin{cases} 3x + y \geq 4 \\ y - 2x \geq -1 \end{cases}$

10. $\begin{cases} -4x + 7y \leq 28 \\ 2x + 3y < 6 \\ y > -3 \end{cases}$

11. $\begin{cases} x + 2y < 4 \\ x - 2y < 4 \\ y \leq 3 \end{cases}$

12. $\begin{cases} x > 1 \\ y < 2 \\ 8x + 3y \geq 24 \end{cases}$

13. $\begin{cases} -x + 2y \leq 6 \\ 3x + y \leq 9 \\ x > -1 \\ y \geq 1 \end{cases}$

14. $\begin{cases} x - 2y > -10 \\ 4x + 3y \geq 26 \\ 3x - 4y < 7 \\ x + y \leq 14 \end{cases}$

In Exercises 15–20, find the maximum or minimum value (as specified) of the objective function that is subject to the given constraints.

15. Maximum $P = 2x + y$ and $\begin{cases} 3x + 4y \leq 24 \\ x \geq 2 \\ y \geq 3 \end{cases}$

16. Minimum $C = 9x + 5y$ and $\begin{cases} 3x + 4y \geq 25 \\ x + 3y \geq 15 \\ x \geq 0 \\ y \geq 0 \end{cases}$

17. Maximum $F = 15x + 10y$ and $\begin{cases} 3x + 2y \leq 80 \\ 2x + 3y \leq 70 \\ x \geq 0 \\ y \geq 0 \end{cases}$

18. Minimum $F = 6x + 9y$ and $\begin{cases} 2x + 5y \leq 50 \\ x + y \leq 12 \\ 2x + y \leq 20 \\ y - x \geq 2 \\ x \geq 0 \end{cases}$

19. Maximum $P = 8x + 10y$ and $\begin{cases} 5x + 10y \leq 180 \\ 10x + 5y \leq 180 \\ x \geq 0 \\ y \geq 3 \end{cases}$

20. Minimum $M = 10x - 8y + 15$ and $\begin{cases} y \geq -3 \\ x - y \geq -5 \\ x + y \geq -5 \\ x \leq 2 \end{cases}$

Solve Exercises 21–28.

21. *Business*   A company manufactures two products. Each product must pass two inspection points, A and B. Each unit of Product X requires 30 min at A and 45 min at B. Product Y requires 15 min at A and 10 min at B. There are enough trained people to provide 100 h at A and 80 h at B. The company makes a profit of \$10 and \$8 on each of X and Y, respectively. What numbers of each should be manufactured to make the most profit? How much is the most profit? If more people could be added at A or B (but not both), where should they be added to increase profits?

22. *Business*   An electronics company manufactures two models of computer chips. Model A requires 1 unit of labor and 4 units of parts. Model B requires 1 unit of labor and 3 units of parts. If 120 units of labor and 390 units of parts are available, and if the company makes a profit of \$7.00 on each model A chip and \$5.50 on each model B, how many should it manufacture to maximize its profits?

23. *Business*   The company in Exercise 22 raises its prices so that it makes a profit of $8.00 on each model A chip and $9.50 on each model B. Now, how many should it manufacture to maximize profits?

24. *Business*   A computer company manufactures a personal computer (PC) and a business computer (BC). Each computer uses two types of chips, an AB chip and an EP chip. The number of chips needed for each computer and the profit for each are given in the following table.

Computer	AB	EP	Profit
PC	2	3	145
BC	3	8	230

Because of a shortage in chips, the company has 200 AB chips and 450 EP chips in stock. **(a)** How many of each computer should be made in order to maximize profits? **(b)** What is that maximum profit?

25. *Business*   The loan department of a bank has set aside $50 million for commercial and home loans. The bank's policy is to allocate at least five times as much money to home loans as to commercial loans. The policy also states that it must make some commercial loans. A certain bank's return on a home loan is 8.5% and 6.75% on a commercial loan. The manager of the loan department wants to earn a return of at least $3.6 million on these loans. What is the fewest of each kind of loan the bank should make in order to reach its goal of at least $3.6 million dollars?

26. *Nutrition*   A brand-X multivitamin tablet contains 15 mg of iron and 10 mg of zinc. One brand-Y multivitamin tablet contains 9 mg of iron and 12 mg of zinc. A nutritionist advises a patient to take at least 90 mg of iron and 90 mg of zinc each day. If brand-X pills cost 5 cents each and brand-Y pills cost 3.5 cents each, how many of each should the patient take per day to fulfill the prescription at minimum cost?

27. *Industrial management*   The chemistry department of a company decides to stock at least 8,000 small test tubes and 5,000 large test tubes. It can take advantage of a special price if it buys at least 15,000 test tubes. The small tubes are broken twice as often as the large tubes, so the company plans to order at least twice as many small tubes as large ones. If the small test tubes cost 15 cents each and the large ones, made of cheaper glass, cost 12 cents each, how many of each size should they order to minimize cost?

28. *Industrial management*   Two types of industrial machines are produced by a certain manufacturer. Machine A requires 3 h of labor for the body and 1 h for wiring. Machine B requires 2 h of labor for the body and 2 h for wiring. The profit on machine A is $32, and the profit on machine B is $45. The body shop can provide 120 h of time per week, and the wiring area can furnish 80 h. How many of each type of machine should be manufactured each week in order to maximize profit?

In Your Words

29. Explain what is meant by a feasible region and a feasible point? How are they alike? How are they different?

30. Describe how to use the objective function to determine the solution to a linear system.

▬ CHAPTER 16 REVIEW

Important Terms and Concepts

Feasible region
Feasible solution
Inequalities
   Linear
   Linear system
   Nonlinear

Properties
Linear programming
Nonlinear system of equations
Objective function
Substitution method

Review Exercises

In Exercises 1–8, solve each inequality both algebraically and graphically.

1. $3x < -12$

2. $4x + 5 < 6$

3. $2x - 7 \geq 15$

4. $7 - 5x < 2 + 3x$

5. $\dfrac{2x+5}{4} < \dfrac{4x-1}{3}$

6. $|2x - 1| \leq 5$

7. $|3x + 2| > 7$

8. $2x + 3 < 13 \leq 3x - 9$

In Exercises 9–13, solve each inequality graphically.

9. $-2x + y < 5$

10. $2x + 3y \geq 6$

11. $4x - 3y < 3$

12. $y > 2x^2 - 5$

13. $y \leq x^2 - 6x + 9$

In Exercises 14–21, graph each system of equations or inequalities.

14. $\begin{cases} x + y = 8 \\ x^2 + y^2 = 49 \end{cases}$

15. $\begin{cases} x^2 + y^2 = 9 \\ x^2 = 5y \end{cases}$

16. $\begin{cases} 2x^2 - y^2 = 8 \\ x^2 + 2y^2 = 4 \end{cases}$

17. $\begin{cases} 2x + 4y^2 = 16 \\ xy = 10 \end{cases}$

18. $\begin{cases} x + 3y < 5 \\ x - 4y \leq 8 \end{cases}$

19. $\begin{cases} 4x - y \leq 4 \\ y + x > -2 \\ y - x < 1 \end{cases}$

20. $\begin{cases} 2x + y - 3 < 0 \\ x - 2y - 4 \geq 0 \end{cases}$

21. $\begin{cases} 3x - y \leq -2 \\ 4 - x - y \geq 0 \\ x > -3 \\ 2x + y > -4 \end{cases}$

In Exercises 22–23, find the maximum or minimum value (as specified) of the objective function that is subject to the given constraints.

22. Maximum $P = 3x + 5y$ and $\begin{cases} x + y \leq 5 \\ y - x \geq -2 \\ x \geq -1 \\ x \leq 3 \end{cases}$

23. Minimum $F = 4x - 3y$ and $\begin{cases} 2x + y \leq 6 \\ y \geq \dfrac{x}{2} - 3 \\ x \geq -1 \\ y \leq 4x - 1 \end{cases}$

Solve Exercises 24–27.

24. The length of a computer chip is given as $7.0\,\text{mm} < \ell < 7.5\,\text{mm}$ and the width as $1.4\,\text{mm} < w < 1.6\,\text{mm}$. What is the range in the area?

25. *Electricity*   Two resistances R_1 and R_2 connected in parallel must have a total resistance R of at least $4\,\Omega$. If $R_1 R_2 = 20\,\Omega$, what are acceptable values for $R_1 + R_2$? Remember $R = \dfrac{R_1 R_2}{R_1 + R_2}$.

26. *Transportation*   A trucker was delayed 30 min because of trouble loading the shipment. To make up for the delay, the trucker increased the speed and took advantage of interstate highways. The speed was increased by an average of 4 mph, and as a result, the trucker was on time, 7 h and 30 min after leaving. Find the usual average speed of the trucker and the distance the truck traveled.

27. *Business* A robotics manufacturing company makes two types of robots: a cylindrical coordinate robot (Model *C*) and a spherical (polar) coordinate robot (Model *S*). Each robot has two assembly stations. The number of hours needed at each station and the profit of each robot are given in the following table.

Robot	Station 1 (hours)	Station 2 (hours)	Profit ($)
C	12	15	1,250
S	18	10	1,620

During the next month, the workers will take their vacations. As a result, the company will only have enough workers for 220 h at Station 1 and 180 h at Station 2. How many of each robot should be manufactured in order to make the most profit?

☰ CHAPTER 16 TEST

1. Solve the inequality $5x > 30$, both algebraically and graphically.

2. Solve $|4x + 5| < 17$, both algebraically and graphically.

3. Solve $\dfrac{4x - 5}{2} \le \dfrac{7x - 2}{3}$, both algebraically and graphically.

4. Solve $(x - 2)(x - 5) > 0$, both algebraically and graphically.

5. Solve $5x^2 + 3x \le 2x^2 - x - 1$, both algebraically and graphically.

6. Graph this system of inequalities.
$$\begin{cases} x + 2y > 4 \\ 3x - y \le 3 \end{cases}$$

7. A brand *X* multivitamin tablet contains 12 mg of iron and 10 mg of zinc. Each brand *Y* tablet contains 5 mg of iron and 8 mg of zinc. A nutritionist suggests that a patient take at least 80 mg of iron and 90 mg of zinc each day. The brand *X* tablets cost 4¢ each and brand *Y* pills are 6¢ each.

(a) How many of each pill should the patient take each day to satisfy the suggestion at the minimum cost?

(b) What is the minimum cost?

CHAPTER
17
Matrices

Kirchhoff's laws can be used to write a system of linear equations. In Section ??, we will learn how to use matrices to solve this linear system and determine the currents in each electric circuit's path.

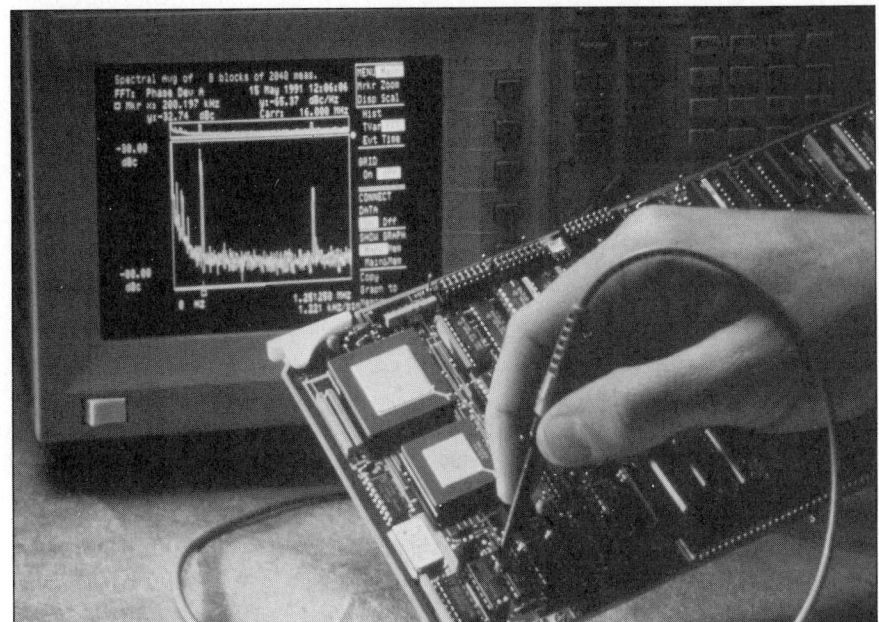

Courtesy of Hewlett-Packard Company

In Chapter 5, we studied systems of equations and we learned that determinants can be used to help solve systems of linear equations. In this chapter, another technique for solving systems of equations will be introduced. This technique uses matrices. The introduction of computers has led to increased applications of matrices in engineering, physics, biology, information science, transportation, and other technical areas. We will use calculators and computer programs to help us work with matrices.

☰ 17.1
MATRICES

Dimension

A **matrix** is a rectangular array of numbers arranged in rows and columns. A matrix with m rows and n columns is an $m \times n$ matrix. ($m \times n$ is read m by n.) The **dimensions** of a matrix are given in the form $m \times n$, where m represents the number of rows and n the number of columns.

A matrix is often used for presenting numerical data in a condensed form.

EXAMPLE 17.1

$$\begin{bmatrix} 4 & 3 & 7 \\ 8 & 19 & -11 \end{bmatrix}$$ is a matrix with dimension 2×3.

$$\begin{bmatrix} 5 & 8 & -9 & 12 \\ 4 & 0 & -3 & 5.7 \\ 9 & 3 & 21 & -6 \end{bmatrix}$$ is a 3×4 matrix.

$$\begin{bmatrix} 2 & 4 & 6 & 8 & 9 \end{bmatrix}$$ is a 1×5 matrix.

$$\begin{bmatrix} 10 \\ 8 \\ 7 \\ 6 \\ 3 \end{bmatrix}$$ is a 5×1 matrix.

Suppose a company makes two types of robots, C and S. Model C requires 78 h to assemble and contains 14 Type I computer chips, 7 Type II chips, and 16 m of wiring. Model S requires 65 h to assemble and contains 12 Type I chips, 9 Type II chips, and 11 m of wiring. The company makes a profit of \$1,250 on each Model C it sells and \$1,500 on each Model S. All this information can be presented in a rectangular array or matrix like the following.

	Hours	Type I Chip	Type II Chip	Wiring (meters)	Profit ($)
Model C	78	14	7	16	1,250
Model S	65	12	9	11	1,500

The numbers in the first row indicate the data for the Model C robot and those in the second row indicate the Model S data. This is a simple example of a matrix that has two rows and five columns.

Row and Column Vectors

A matrix of dimension $1 \times n$ is a **row vector** and a matrix of dimension $m \times 1$ is a **column vector**.

EXAMPLE 17.2

Some row vectors are:

$$\begin{bmatrix} 2 & 4 & 6 \end{bmatrix} \qquad \begin{bmatrix} 1 & -3 & 2 & 8 \end{bmatrix}$$

$$\begin{bmatrix} 4 & 6 & 0 & -7 & 0 \end{bmatrix} \qquad [9]$$

Some column vectors are:

$$\begin{bmatrix} 2 \\ 4 \\ 6 \end{bmatrix} \quad \begin{bmatrix} 1 \\ -3 \\ 2 \\ 8 \end{bmatrix} \quad \begin{bmatrix} 4 \\ 6 \\ 0 \\ -7 \\ 0 \end{bmatrix} \quad [9]$$

Notice that [9] is both a row vector and a column vector.

Zero Matrix

A matrix is a **zero matrix** if all the elements are zero. The matrix

$$\begin{bmatrix} 0 & 0 \\ 0 & 0 \\ 0 & 0 \end{bmatrix}$$

is a zero matrix of dimension 3×2.

Square Matrix

A matrix with the same number, n, of rows and columns is a **square matrix of order n**.

EXAMPLE 17.3

The matrix

$$\begin{bmatrix} 2 & 1 \\ 5 & 7 \end{bmatrix}$$

is a square matrix of order 2 and the matrix

$$\begin{bmatrix} 5 & -7 & 8 \\ 2 & -1 & 0 \\ 0 & 3 & 5 \end{bmatrix}$$

is a square matrix of order 3.

A double-subscript notation has been developed to allow us to refer to specific elements. The first numeral in the subscripts refers to the row in which the element lies and the second numeral refers to the column. An example of this notation is matrix A.

$$A = \begin{bmatrix} a_{11} & a_{12} & a_{13} & a_{14} \\ a_{21} & a_{22} & a_{23} & a_{24} \\ a_{31} & a_{32} & a_{33} & a_{34} \\ a_{41} & a_{42} & a_{43} & a_{44} \end{bmatrix}$$

The elements $a_{11}, a_{22}, a_{33}, a_{44}$ are the **main diagonal** of this square matrix of order 4.

Coefficient Matrix

When working with a system of linear equations, a **coefficient matrix** can be formed from the coefficients of the equations. For example, the system

$$\begin{cases} 2x + 3y - 5z = 8 \\ 4x - 7y + 22z = 15 \end{cases}$$

has the coefficient matrix

$$\begin{bmatrix} 2 & 3 & -5 \\ 4 & -7 & 22 \end{bmatrix}$$

Augmented Matrix

If the column vector formed by the constants to the right of the equal symbols, $\begin{bmatrix} 8 \\ 15 \end{bmatrix}$, is adjoined to the right of the coefficient matrix, we get a new matrix called the **augmented matrix** of the system of linear equations. For example, for the system used previously,

$$\begin{cases} 2x + 3y - 5z = 8 \\ 4x - 7y + 22z = 15 \end{cases}$$

the augmented matrix would be

$$\left[\begin{array}{ccc:c} 2 & 3 & -5 & 8 \\ 4 & -7 & 22 & 15 \end{array} \right]$$

The dashed line between columns 3 and 4 is not needed but is often used to indicate an augmented matrix.

Two matrices are **equal**, if they have the same dimension and their corresponding elements are equal.

EXAMPLE 17.4

If $\begin{bmatrix} 6 & 4 & 3 & x \\ 2 & 9 & y & -3 \end{bmatrix} = \begin{bmatrix} z & 4 & 3 & -11 \\ 2 & w & 7 & -3 \end{bmatrix}$, then $x = -11$, $y = 7$, $z = 6$, and $w = 9$.

EXAMPLE 17.5

For the matrices $A = \begin{bmatrix} 1 & 2 & 6 \\ 7 & 8 & 9 \end{bmatrix}$, $B = \begin{bmatrix} 1 & 2 & 6 \\ 7 & 8 & 5 \end{bmatrix}$, and $C = \begin{bmatrix} 1 & 2 \\ 7 & 8 \end{bmatrix}$, $A \neq B$ because the corresponding elements $a_{23} = 9$ and $b_{23} = 5$ are not equal. $A \neq C$ because they have different dimensions.

Addition and Subtraction of Matrices

If A and B are two matrices, each of dimension $m \times n$, then their sum (or difference) is defined to be another matrix, C, also of dimension $m \times n$, where every element of C is the sum (or difference) of the corresponding elements A and B.

EXAMPLE 17.6

(a) $\begin{bmatrix} 2 & 3 & 5 \\ 6 & 7 & 8 \end{bmatrix} + \begin{bmatrix} 10 & 11 & 12 \\ 14 & 16 & 22 \end{bmatrix} = \begin{bmatrix} 2+10 & 3+11 & 5+12 \\ 6+14 & 7+16 & 8+22 \end{bmatrix}$

$= \begin{bmatrix} 12 & 14 & 17 \\ 20 & 23 & 30 \end{bmatrix}$

(b) $\begin{bmatrix} 4 & 9 \\ -2 & 8 \\ 6 & 1 \end{bmatrix} + \begin{bmatrix} 16 & 21 \\ 2 & -8 \\ 9 & 14 \end{bmatrix} = \begin{bmatrix} 4+16 & 9+21 \\ -2+2 & 8+(-8) \\ 6+9 & 1+14 \end{bmatrix}$

$= \begin{bmatrix} 20 & 30 \\ 0 & 0 \\ 15 & 15 \end{bmatrix}$

(c) $\begin{bmatrix} 1 & -5 \\ 4 & 3 \\ 2 & -1 \end{bmatrix} - \begin{bmatrix} 4 & -6 \\ 2 & 1 \\ -3 & 9 \end{bmatrix} = \begin{bmatrix} 1-4 & -5-(-6) \\ 4-2 & 3-1 \\ 2-(-3) & -1-9 \end{bmatrix}$

$= \begin{bmatrix} -3 & 1 \\ 2 & 2 \\ 5 & -10 \end{bmatrix}$

Caution

You can only add or subtract two matrices if they have the same dimension.

EXAMPLE 17.7

If $A = \begin{bmatrix} 4 & 2 & 7 & 9 \\ 5 & 4 & 6 & 1 \end{bmatrix}$ and $B = \begin{bmatrix} 5 & 3 & 8 & 1 \\ 4 & 6 & 2 & 4 \end{bmatrix}$, then determine $A + B$, $A - B$, and $B - A$.

EXAMPLE 17.7 (Cont.)

Solutions $A + B = \begin{bmatrix} 4+5 & 2+3 & 7+8 & 9+1 \\ 5+4 & 4+6 & 6+2 & 1+4 \end{bmatrix}$

$= \begin{bmatrix} 9 & 5 & 15 & 10 \\ 9 & 10 & 8 & 5 \end{bmatrix}$

$A - B = \begin{bmatrix} 4-5 & 2-3 & 7-8 & 9-1 \\ 5-4 & 4-6 & 6-2 & 1-4 \end{bmatrix}$

$= \begin{bmatrix} -1 & -1 & -1 & 8 \\ 1 & -2 & 4 & -3 \end{bmatrix}$

$B - A = \begin{bmatrix} 5-4 & 3-2 & 8-7 & 1-9 \\ 4-5 & 6-4 & 2-6 & 4-1 \end{bmatrix}$

$= \begin{bmatrix} 1 & 1 & 1 & -8 \\ -1 & 2 & -4 & 3 \end{bmatrix}$

Application

EXAMPLE 17.8

At the beginning of a laboratory experiment, two groups of four mice were timed to see how long it took them to go through a maze. The mice in the group that went through the maze at night took 8, 6, 5, and 7 s. The mice in the group that went through the maze during the day took 22, 14, 18, and 12 s. **(a)** Write a 2 × 4 matrix using this information.

A week later, the mice were sent through the maze again. They went in the same order as they did the week before. This time the night group completed the maze in 4, 3, 5, and 3 s and the day group took 9, 8, 10, and 8 s. **(b)** Write this information as a 2 × 4 matrix and use matrix subtraction to write a matrix that shows the amount of change each mouse made in its time to complete the maze.

Solutions

(a) We will put the night group in the top row and the day group of mice in the bottom row. The result is the following matrix.

$\begin{matrix} \text{Night group} \\ \text{Day group} \end{matrix} \begin{bmatrix} 8 & 6 & 5 & 7 \\ 22 & 14 & 18 & 12 \end{bmatrix}$

(b) The matrix for the data at the end of the week is

$\begin{matrix} \text{Night group} \\ \text{Day group} \end{matrix} \begin{bmatrix} 4 & 3 & 5 & 3 \\ 9 & 8 & 10 & 8 \end{bmatrix}$

To find the change in the amount of time each mouse took going through the maze, we subtract the last matrix from the matrix in **(a)**.

$\begin{matrix} \text{Night group} \\ \text{Day group} \end{matrix} \begin{bmatrix} 8 & 6 & 5 & 7 \\ 22 & 14 & 18 & 12 \end{bmatrix} - \begin{bmatrix} 4 & 3 & 5 & 3 \\ 9 & 8 & 10 & 8 \end{bmatrix}$

$= \begin{bmatrix} 4 & 3 & 0 & 4 \\ 13 & 6 & 8 & 4 \end{bmatrix}$

Scalar Multiplication

If A is an $m \times n$ matrix and k a real number, then kA is an $m \times n$ matrix B, where $b_{ij} = ka_{ij}$ for each element of A. This is referred to as **scalar multiplication** and the real number k is called a **scalar**.

EXAMPLE 17.9

If $A = \begin{bmatrix} 3 & 5 \\ 9 & 2 \end{bmatrix}$ and $k = 4$, then

$$kA = 4 \begin{bmatrix} 3 & 5 \\ 9 & 2 \end{bmatrix} = \begin{bmatrix} 4 \cdot 3 & 4 \cdot 5 \\ 4 \cdot 9 & 4 \cdot 2 \end{bmatrix}$$
$$= \begin{bmatrix} 12 & 20 \\ 36 & 8 \end{bmatrix}$$

EXAMPLE 17.10

If $B = \begin{bmatrix} 2 & 4 \\ -3 & 7 \end{bmatrix}$ and $k = -\frac{1}{2}$, then

$$kB = -\frac{1}{2} \begin{bmatrix} 2 & 4 \\ -3 & 7 \end{bmatrix} = \begin{bmatrix} -\frac{1}{2} \cdot 2 & -\frac{1}{2} \cdot 4 \\ -\frac{1}{2}(-3) & -\frac{1}{2} \cdot 7 \end{bmatrix}$$
$$= \begin{bmatrix} -1 & -2 \\ \frac{3}{2} & -\frac{7}{2} \end{bmatrix}$$

These two examples help show that $A + A = 2A$, $B + B + B = 3B$, and so on.

Application

EXAMPLE 17.11

A computer retailer sells three types of computers: the personal computer (PC), the business computer (BC), and the industrial computer (IC). The retailer has two stores. The matrix below shows the number of each computer model in stock at each store.

$$\begin{array}{c} \\ \text{Store A} \\ \text{Store B} \end{array} \begin{array}{ccc} \text{PC} & \text{BC} & \text{IC} \end{array} \\ \begin{bmatrix} 6 & 18 & 12 \\ 10 & 22 & 28 \end{bmatrix}$$

The store plans to have a sale in the near future. In order to have enough computers in stock at each store when the sale begins, it plans to order 1.5 times the current number. How many of each computer should be ordered for each store?

EXAMPLE 17.11 (Cont.)

Solution Here the scalar is 1.5, so we want to multiply each element of this matrix by 1.5. The result is

$$1.5 \begin{pmatrix} & \text{PC} & \text{BC} & \text{IC} \\ \text{Store A} & \begin{bmatrix} 6 & 18 & 12 \\ \text{Store B} & 10 & 22 & 28 \end{bmatrix} \end{pmatrix} = \begin{matrix} & \text{PC} & \text{BC} & \text{IC} \\ \text{Store A} & \begin{bmatrix} 9 & 27 & 18 \\ \text{Store B} & 15 & 33 & 42 \end{bmatrix} \end{matrix}$$

Thus, we see that the retailer needs to order 9 PCs for Store A and 15 for Store B, 27 BCs for Store A and 33 for Store B, and 18 ICs for Store A and 42 for Store B.

We can summarize the work in this section by listing the following properties of matrices.

Matrix Properties

If A, B, and C are three matrices, all of the same dimension, 0 is the zero matrix, and k is a real number, then

$A + B = B + A$ (commutative law)

$A + 0 = 0 + A$ (identity for addition)

$A + (B + C) = (A + B) + C$ (associative law)

$k(A + B) = kA + kB$

Exercise Set 17.1

Give the dimension of the matrices in Exercises 1–6. Solve Exercises 7–10.

1. $\begin{bmatrix} 2 & 4 & 5 \\ 3 & 2 & 1 \end{bmatrix}$

2. $\begin{bmatrix} 3 & 4 & 6 \\ 2 & 5 & 9 \\ 8 & 7 & 2 \end{bmatrix}$

3. $\begin{bmatrix} 2 & 1 & 0 & 7 & 9 & 6 \\ 3 & 2 & 4 & 8 & 7 & 2 \\ 9 & 6 & 4 & 5 & 0 & 2 \end{bmatrix}$

4. $\begin{bmatrix} 3 \\ 2 \\ 1 \\ 4 \end{bmatrix}$

5. $\begin{bmatrix} 3 & 2 \\ 4 & 1 \\ 5 & 3 \\ 7 & 9 \end{bmatrix}$

6. $\begin{bmatrix} 1 & 2 & 4 & 6 & 9 & 11 & 12 \end{bmatrix}$

7. Given matrix A, determine the values of elements $a_{11}, a_{24}, a_{21},$ and a_{32}.

$$A = \begin{bmatrix} 1 & 2 & 5 & 7 \\ 8 & 9 & 11 & 13 \\ 4 & 6 & 12 & 15 \end{bmatrix}$$

8. Given matrix B, determine the values of elements $b_{12}, b_{21}, b_{23},$ and b_{32}.

$$B = \begin{bmatrix} -5 & 0 & 2 \\ 4 & 7 & 9 \\ 5 & -8 & 11 \end{bmatrix}$$

9. Given that $\begin{bmatrix} 4 & 3 & 2 & x \\ 5 & 9 & 7 & 11 \\ y & 8 & 3 & 4 \end{bmatrix} = \begin{bmatrix} 4 & 3 & z & 12 \\ w & 9 & 7 & 11 \\ -6 & 8 & 3 & 4 \end{bmatrix}$,

determine the values of $x, y, z,$ and w.

10. Given that

$$\begin{bmatrix} 3 & 2 & x-2 \\ 4 & y+5 & 9 \\ w+6 & 11 & 10 \end{bmatrix} = \begin{bmatrix} 3z & 2 & 7 \\ 4 & 3 & 9 \\ 7 & 11 & 2p \end{bmatrix},$$

determine the values of $x, y, z, w,$ and p.

In Exercises 11–14, find the indicated sum or difference.

11. $\begin{bmatrix} 3 & 2 & 1 & -4 \\ 9 & 7 & 6 & -1 \\ 4 & 3 & 5 & 2 \end{bmatrix} + \begin{bmatrix} -6 & 4 & 5 & -4 \\ 3 & 2 & 11 & 1 \\ 5 & -3 & 0 & 8 \end{bmatrix}$

12. $\begin{bmatrix} 2 & 4 \\ -5 & 1 \\ 3 & 4 \end{bmatrix} + \begin{bmatrix} 5 & -6 \\ 2 & 9 \\ 8 & 2 \end{bmatrix}$

13. $\begin{bmatrix} 3 & 2 & -1 \\ 9 & 12 & 4 \end{bmatrix} - \begin{bmatrix} 1 & 3 & -5 \\ 4 & 3 & 6 \end{bmatrix}$

14. $\begin{bmatrix} 3 & 2 & -1 & 5 \\ 4 & 3 & 11 & 9 \\ 16 & 4 & 3 & -8 \\ 12 & 5 & 4 & 6 \\ 2 & 9 & 18 & 7 \end{bmatrix} - \begin{bmatrix} 1 & 2 & -4 & 2 \\ 1 & 3 & 2 & 8 \\ 2 & -4 & 3 & 6 \\ 2 & -5 & 2 & -4 \\ 2 & 3 & -5 & 3 \end{bmatrix}$

In Exercises 15–22, use matrices A, B, and C to determine the indicated matrix, where

$$A = \begin{bmatrix} 4 & 3 & 2 \\ 5 & 0 & 7 \end{bmatrix}, B = \begin{bmatrix} -5 & 4 & 2 \\ 3 & 1 & 8 \end{bmatrix}, \text{ and } C = \begin{bmatrix} 5 & 8 & 4 \\ 6 & -2 & 9 \end{bmatrix}.$$

15. $A+B$ **17.** $3A$ **19.** $2B+C$ **21.** $2B-3A$

16. $B-C$ **18.** $4B$ **20.** $4A-3C$ **22.** $4C+2B$

In Exercises 23–30, use matrices D, E, and F to determine the indicated matrix, where

$$D = \begin{bmatrix} 4 \\ 2 \\ -1 \\ 5 \end{bmatrix}, E = \begin{bmatrix} 7 \\ 9 \\ -4 \\ 10 \end{bmatrix}, \text{ and } F = \begin{bmatrix} 14 \\ 18 \\ -8 \\ 5 \end{bmatrix}.$$

23. $D-E$ **25.** $3E$ **27.** $3D+2E$ **29.** $7D-2F$

24. $E+F$ **26.** $F-2E$ **28.** $2F-7D$ **30.** $5E-2F$

Solve Exercises 31–36.

31. *Business*   In keeping inventory records, a company uses a matrix. At one storage point, they have 18 of computer chip A, 7 of computer chip B, 9 EPROMS, 7 keyboards, 11 motherboards, and 4 disk drives. This is represented by the matrix $\begin{bmatrix} 18 & 7 & 9 \\ 7 & 11 & 4 \end{bmatrix}$. At a second storage point, the inventory is $\begin{bmatrix} 9 & 5 & 6 \\ 11 & 4 & 12 \end{bmatrix}$. How many of each item do they have on hand?

32. *Business* Wrecker's Auto Supply Company has three stores, each of which carries five sizes of a certain tire model. The following matrix represents the present inventories.

Store	195/70SR-14	205/70SR-14	185/70SR-15	205/70SR-15	215/70SR-15
A	12	15	28	7	16
B	25	19	11	40	11
C	15	29	21	17	17

At the beginning of the next month, they would like to have the following inventory.

Store	195/70SR-14	205/70SR-14	185/70SR-15	205/70SR-15	215/70SR-15
A	20	48	60	24	44
B	36	64	24	40	32
C	44	72	48	60	40

Write a matrix that represents the number of each size of tire that must be ordered for each store, so as to achieve their goal. (Assume that no more tires are sold.)

33. *Medical technology* A drug company tested 400 patients to see if a new medicine is effective. Half of the patients received the new drug and half received a placebo. The results for the first 200 patients are shown in the following matrix.

	New drug	Placebo
Effective	70	40
Not effective	30	60

Using the same matrix format, the results for a second 200 patients were $\begin{bmatrix} 65 & 42 \\ 35 & 58 \end{bmatrix}$. What were the results for the entire test group?

34. *Business* A computer supply company has its inventory of four types of computer chips in three warehouses. At the beginning of the month they had the following inventory.

Chip type:	286	386	486	586
Warehouse A	1,200	3,200	4,800	900
Warehouse B	1,650	4,580	7,200	700
Warehouse C	1,120	5,100	6,200	1,200

The sales for the month were

Chip type:	286	386	486	586
Warehouse A	270	2,130	3,210	265
Warehouse B	1,120	4,230	3,124	75
Warehouse C	320	3,126	2,743	1,012

Write a matrix that shows the inventory at the end of the month.

35. *Business* The computer supply company in Exercise 34 expects sales to increase by 10% during the next month. What are next month's projected sales? (Round off any fractional answers.) Remember, last month the company sold

Chip type:	286	386	486	586
Warehouse A	270	2,130	3,210	265
Warehouse B	1,120	4,230	3,124	75
Warehouse C	320	3,126	2,743	1,012

36. *Business* The following matrix represents the normal monthly order of a retail store for three models of exercise suits in four different sizes.

	S	M	L	XL
Jogging	5	4	3	4
Sweating	7	12	15	7
Walking	4	8	12	14

During its spring sale, the store expects to do four times the usual volume of business. Determine the matrix that represents their order for the month of the sale.

37. Write a computer program to add or subtract two $m \times n$ matrices.

 38. Write a computer program to multiply a matrix by a scalar.

 39. Write a computer program that will allow you to multiply two matrices by a scalar and then add or subtract the results.

 In Your Words

40. Describe how to enter a matrix into your calculator. (Hint: Review the procedures in Section 5.4.)

41. Describe how to use your calculator to **(a)** add or subtract two matrices and **(b)** multiply a matrix by a scalar.

42. Explain what is needed for two matrices to be equal.

43. What is an augmented matrix?

≡ 17.2
MULTIPLICATION OF MATRICES

Learning to add and subtract matrices was straightforward. We had to be careful that each matrix had the same number of rows and the same number of columns. Multiplying matrices is more involved.

Multiplying Matrices

If A and B are two matrices, then in order to define the product AB, the number of columns in matrix A must be the same as the number of rows in matrix B. Thus, matrix A must be of dimension $m \times n$ and matrix B dimension $n \times p$. The product will have dimension $m \times p$.

We will begin by multiplying a row vector and a column vector.

> **Product of a Row Vector and a Column Vector**
>
> If $A = \begin{bmatrix} a_{11} & a_{12} & a_{13} & \cdots & a_{1n} \end{bmatrix}$ is a row vector, and $B = \begin{bmatrix} b_{11} \\ b_{21} \\ b_{31} \\ \vdots \\ b_{n1} \end{bmatrix}$ is a column vector, then
>
> $$AB = [a_{11}b_{11} + a_{12}b_{21} + a_{13}b_{31} + \cdots + a_{1n}b_{n1}]$$

EXAMPLE 17.12

If $A = \begin{bmatrix} 3 & 4 \end{bmatrix}$ and $B = \begin{bmatrix} 5 \\ 1 \end{bmatrix}$, then

$$AB = [3 \cdot 5 + 4 \cdot 1] = [15 + 4] = [19]$$

Since A is 1×2 and B is 2×1, the product AB is 1×1.

EXAMPLE 17.13

If $C = \begin{bmatrix} 1 & 2 & 3 & x \end{bmatrix}$, $D = \begin{bmatrix} 3 \\ -1 \\ 1 \\ 2 \end{bmatrix}$, and $CD = [22]$, find x.

Solution
$$\begin{aligned} CD &= [1(3) + 2(-1) + 3(1) + 2x] \\ &= [3 - 2 + 3 + 2x] \\ &= [4 + 2x] = [22] \end{aligned}$$
These two matrices are equal only if $4 + 2x = 22$ or $2x = 18$, so $x = 9$.

We will use the following method of multiplying a row vector and a column vector to multiply larger matrices.

Matrix Multiplication

If A is an $m \times n$ matrix and B is an $n \times p$ matrix, then the product matrix $C = AB$ is an $m \times p$ matrix, where element c_{ij} in the ith row and jth column is formed by multiplying the elements in the ith row of A by the corresponding elements in the jth row of B and adding the results. Thus,

$$c_{ij} = a_{i1}b_{1j} + a_{i2}b_{2j} + \cdots + a_{in}b_{nj}$$

We will try to illustrate this definition by showing how you get an element in the following product. Suppose. $A = \begin{bmatrix} 1 & 2 & -1 & 0 \\ 3 & 5 & 2 & -3 \end{bmatrix}$ and $B = \begin{bmatrix} 2 & 1 & 3 \\ 0 & 4 & -2 \\ 1 & 2 & 0 \\ 1 & 3 & -2 \end{bmatrix}$.

You can see that A is a 2×4 matrix and B is a 4×3 matrix, so AB will have dimension 2×3. If we let $C = AB$, we will show how to find c_{12} or the element in row 1, column 2 of matrix C. To get this element, we multiply row 1 of matrix A and column 2 of matrix B. This is the same as multiplying a row vector and a column vector. Thus,

$$c_{12} = \begin{bmatrix} 1 & 2 & -1 & 0 \end{bmatrix} \begin{bmatrix} 1 \\ 4 \\ 2 \\ 3 \end{bmatrix}$$
$$= 1(1) + 2(4) - 1(2) + 0(3)$$
$$= 7$$

EXAMPLE 17.14

Multiply $\begin{bmatrix} 1 & 2 & -1 & 0 \\ 3 & 5 & 2 & -3 \end{bmatrix} \begin{bmatrix} 2 & 1 & 3 \\ 0 & 4 & -2 \\ 1 & 2 & 0 \\ 1 & 3 & -2 \end{bmatrix}$.

Solution $\begin{bmatrix} 1 & 2 & -1 & 0 \\ 3 & 5 & 2 & -3 \end{bmatrix} \begin{bmatrix} 2 & 1 & 3 \\ 0 & 4 & -2 \\ 1 & 2 & 0 \\ 1 & 3 & -2 \end{bmatrix}$

$= \begin{bmatrix} 1(2)+2(0)-1(1)+0(1) & 1(1)+2(4)-1(2)+0(3) \\ 3(2)+5(0)+2(1)-3(1) & 3(1)+5(4)+2(2)-3(3) \end{bmatrix}$
$\begin{bmatrix} 1(3)+2(-2)-1(0)+0(-2) \\ 3(3)+5(-2)+2(0)-3(-2) \end{bmatrix}$

$= \begin{bmatrix} 2+0-1+0 & 1+8-2+0 & 3-4-0+0 \\ 6+0+2-3 & 3+20+4-9 & 9-10+0+6 \end{bmatrix}$

$= \begin{bmatrix} 1 & 7 & -1 \\ 5 & 18 & 5 \end{bmatrix}$

≡ **Note**

In general, $AB \neq BA$. In fact, if AB is defined, then BA may not be defined. The only time that both AB and BA are defined is when both A and B are square matrices with the same dimensions. Thus, the commutative law does not hold for multiplication of matrices.

The following properties will hold, assuming that the dimensions of the matrices allow the products or sums to be defined.

$$A(BC) = (AB)C \qquad \text{associative law}$$
$$A(B+C) = AB + AC \qquad \text{left distributive law}$$
$$(B+C)A = BA + CA \qquad \text{right distributive law}$$

Because the commutative law does not hold, the left and right distributive laws usually give different results.

Application

EXAMPLE 17.15

A computer supply company has its inventory of four types of computer chips in three warehouses. The sales of each chip from each warehouse are shown in matrix S.

Chip type:	286	386	486	586
Warehouse A	270	2,130	3,210	265
$S=$ Warehouse B	1,120	4,230	3,124	75
Warehouse C	320	3,126	2,743	1,012

EXAMPLE 17.15 (Cont.)

Matrix P below shows the selling price and the profit for each type of chip.

$$P = \begin{array}{c} \text{Chip} \\ 286 \\ 386 \\ 486 \\ 586 \end{array} \begin{bmatrix} \text{Selling Price} & \text{Profit} \\ 25 & 10 \\ 35 & 13 \\ 52 & 17 \\ 97 & 33 \end{bmatrix}$$

(a) How much money in sales did each warehouse generate during this month?
(b) How much profit did each warehouse make?

Solution   Since matrix S is a 3×4 matrix and P is a 4×2, we can determine the solution from the 3×2 matrix that results when we multiply SP.

$$SP = \begin{bmatrix} 270 & 2{,}130 & 3{,}210 & 265 \\ 1{,}120 & 4{,}230 & 3{,}124 & 75 \\ 320 & 3{,}126 & 2{,}743 & 1{,}012 \end{bmatrix} \cdot \begin{bmatrix} 25 & 10 \\ 35 & 13 \\ 52 & 17 \\ 97 & 33 \end{bmatrix}$$

$$SP = \begin{array}{c} \\ \text{Warehouse A} \\ \text{Warehouse B} \\ \text{Warehouse C} \end{array} \begin{bmatrix} \text{Sales} & \text{Profit} \\ 273{,}925 & 93{,}705 \\ 345{,}773 & 121{,}773 \\ 358{,}210 & 123{,}865 \end{bmatrix}$$

From this matrix we can see, for example, that Warehouse A sold \$273,925 in chips for a profit of \$93,705. ▪

Exercise Set 17.2

In Exercises 1–10, find the indicated products.

1. $\begin{bmatrix} 2 & 3 \\ 1 & -4 \end{bmatrix} \begin{bmatrix} 4 & -3 \\ 6 & 5 \end{bmatrix}$

2. $\begin{bmatrix} 1 & -2 & 4 \end{bmatrix} \begin{bmatrix} 3 \\ 9 \\ -7 \end{bmatrix}$

3. $\begin{bmatrix} 5 & 1 & 6 & 2 \end{bmatrix} \begin{bmatrix} 2 \\ 9 \\ -8 \\ 3 \end{bmatrix}$

4. $\begin{bmatrix} 2 & 3 & 1 \\ -4 & 2 & 0 \end{bmatrix} \begin{bmatrix} 4 & 1 \\ -5 & 2 \\ 3 & 0 \end{bmatrix}$

5. $\begin{bmatrix} 1 & 2 & 4 \\ 2 & 3 & 6 \end{bmatrix} \begin{bmatrix} 5 & -7 & 4 \\ 4 & 6 & 0 \\ -4 & 2 & 1 \end{bmatrix}$

6. $\begin{bmatrix} 4 & 7 \\ 2 & 3 \\ 6 & 2 \\ 9 & 1 \end{bmatrix} \begin{bmatrix} 3 & 7 & 6 & 1 & -5 \\ 4 & -3 & 2 & 0 & 10 \end{bmatrix}$

7. $\begin{bmatrix} 1 & 2 & 3 \\ 6 & 5 & 4 \\ -1 & 0 & 2 \end{bmatrix} \begin{bmatrix} 9 & 8 & 7 \\ -2 & 3 & -1 \\ 0 & 1 & 0 \end{bmatrix}$

8. $\begin{bmatrix} 9 & 8 & 7 \\ -2 & 3 & -1 \\ 0 & 1 & 0 \end{bmatrix} \begin{bmatrix} 1 & 2 & 3 \\ 6 & 5 & 4 \\ -1 & 0 & 2 \end{bmatrix}$

9. $\begin{bmatrix} 3 & 4 & -2 \\ 7 & 8 & 10 \end{bmatrix} \begin{bmatrix} 0 & 4 & 1 \\ 1 & 0 & 2 \\ 8 & -1 & 0 \end{bmatrix}$

10. $\begin{bmatrix} 1 & 2 & -1 \\ 2 & -1 & 3 \end{bmatrix} \begin{bmatrix} 0 & 1 & 0 & 2 \\ 1 & 0 & 2 & -1 \\ 3 & 4 & -1 & 0 \end{bmatrix}$

In Exercises 11–20, use matrices A, B, and C to determine the indicated answer, where

$$A = \begin{bmatrix} 1 & 2 \\ 3 & 4 \end{bmatrix}, B = \begin{bmatrix} 3 & 6 \\ 2 & 5 \end{bmatrix}, \text{ and } C = \begin{bmatrix} 4 & -2 \\ -6 & 2 \end{bmatrix}.$$

11. AB

12. BA

13. AC

14. BC

15. $A(BC)$

16. $(AB)C$

17. $A(B+C)$

18. $(B+C)A$

19. $(2A)(3B)$

20. $\left(2A+\frac{1}{2}C\right)B$

Solve Exercises 21–30.

21. Suppose $A = \begin{bmatrix} 2 & 10 \\ 3 & 15 \end{bmatrix}$ and $B = \begin{bmatrix} -10 & 25 \\ 2 & -5 \end{bmatrix}$. What is AB? Can you conclude that if $AB = 0$, then either $A = 0$ or $B = 0$?

22. If $\begin{bmatrix} 2 & x \\ y & 5 \end{bmatrix} \begin{bmatrix} 4 & 6 \\ 9 & -1 \end{bmatrix} = \begin{bmatrix} 35 & 9 \\ 49 & 1 \end{bmatrix}$ determine x and y.

23. *Business* A computer supply company has its inventory of four types of computer chips in three warehouses. One month they had the following sales.

Chip type:	286	386	486	586
Warehouse A	270	2,130	3,210	265
Warehouse B	1,120	4,230	3,124	75
Warehouse C	320	3,126	2,743	1,012

The 286 chips sell for $25, the 386 chips sell for $35, the 486 chips sell for $55, and the 586 chips sell for $95. How much money in sales did each warehouse generate during this month?

24. *Construction* A contractor builds three sizes of houses (ranch, bi-level, and two story) in two different models, A and B. The contractor plans to build 100 new homes in a subdivision. Matrix P shows the number of each type of house planned for the subdivision.

	Ranch	Bi-level	Two story
Model A	30	20	15
Model B	10	10	15

$$P = $$

The amounts of each type of exterior material needed for each house are shown in matrix A. Here, concrete is in cubic yards, lumber in 1,000 board feet, bricks in 1,000s, and shingles in units of $100\,\text{ft}^2$.

	Concrete	Lumber	Bricks	Shingles
Ranch	10	2	2	3
Bi-level	15	3	4	4
Two story	25	5	6	3

$$A = $$

The cost of each of the units for each kind of material is given by matrix C.

$$
C = \begin{array}{c} \text{Concrete} \\ \text{Lumber} \\ \text{Brick} \\ \text{Shingles} \end{array}
\overset{\text{Cost per unit}}{\begin{bmatrix} 25 \\ 210 \\ 75 \\ 40 \end{bmatrix}}
$$

(a) How much of each type of material will the contractor need for each house model?

(b) What will it cost to build each size of house?

(c) What will it cost to build each house model?

(d) What will it cost to build the entire subdivision?

25. *Physics*   The **Pauli spin matrices** in quantum mechanics are $A = \begin{bmatrix} 0 & 1 \\ 1 & 0 \end{bmatrix}$, $B = \begin{bmatrix} 0 & -i \\ i & 0 \end{bmatrix}$, and $C = \begin{bmatrix} 1 & 0 \\ 0 & -1 \end{bmatrix}$, where $i = \sqrt{-1}$. Show that $A^2 = B^2 = C^2 = I$, where $I = \begin{bmatrix} 1 & 0 \\ 0 & 1 \end{bmatrix}$.

26. *Physics*   Show that the Pauli spin matrices **anticommute**. That is, show that $AB = -BA$, $AC = -CA$, and $BC = -CB$.

27. *Physics*   The **commutator** of two matrices P and Q is $PQ - QP$. For the Pauli spin matrices, show that (a) the commutator of A and B is $2iC$, (b) the commutator of A and C is $-2iB$, and (c) the commutator of B and C is $2iA$, where $i = \sqrt{-1}$.

28. Show, by multiplying the matrices, that the following equation represents an ellipse.

$$
\begin{bmatrix} x & y \end{bmatrix} \begin{bmatrix} 5 & -7 \\ 7 & 3 \end{bmatrix} \begin{bmatrix} x \\ y \end{bmatrix} = [30]
$$

29.   Write a computer program to multiply two matrices.

30. *Cartography*   A figure defined by the four points $P_1(x_1, y_1)$, $P_2(x_2, y_2)$, $P_3(x_3, y_3)$, and $P_4(x_4, y_4)$ is rotated through an angle θ by using the product

$$
\begin{bmatrix} x_1 & y_1 \\ x_2 & y_2 \\ x_3 & y_3 \\ x_4 & y_4 \end{bmatrix} \begin{bmatrix} \cos\theta & \sin\theta \\ -\sin\theta & \cos\theta \end{bmatrix}
$$

(a) Find the resulting matrix if the points $P_1(1, 2)$, $P_2(5, 4)$, $P_3(-2, 7)$, and $P_4(-1, -5)$ are rotated through an angle $\theta = \pi$.

(b) Find the resulting matrix if the points $P_1(1, 2)$, $P_2(5, 4)$, $P_3(-2, 7)$, and $P_4(-1, -5)$ are rotated through an angle $\theta = \frac{\pi}{2}$.

(c) What product would be used if five points $P_1(x_1, y_1)$, $P_2(x_2, y_2)$, $P_3(x_3, y_3)$, $P_4(x_4, y_4)$, and $P_5(x_5, y_5)$ were rotated through an angle θ?

31. The text states that the commutative law does not hold for the multiplication of matrices.

(a) If $A = \begin{bmatrix} 1 & 2 & 5 \\ 6 & 8 & 9 \end{bmatrix}$ and $B = \begin{bmatrix} 5 & 4 & 2 \\ 0 & 1 & 2 \\ 7 & 2 & 0 \end{bmatrix}$,

does $AB = BA$? Explain why or why not.

(b) If $C = \begin{bmatrix} 2 & 5 \\ 3 & 4 \end{bmatrix}$ and $D = \begin{bmatrix} 5 & 4 \\ -2 & 0 \end{bmatrix}$, does $CD = DC$? Explain why or why not.

32. Describe how to multiply two matrices on your calculator.

≡ 17.3
INVERSES OF MATRICES

One use of matrices is to help solve a system of linear equations. For example, if you had the system of linear equations

$$\begin{cases} a_1 x_1 + a_2 x_2 + a_3 x_3 = k_1 \\ b_1 x_1 + b_2 x_2 + b_3 x_3 = k_2 \\ c_1 x_1 + c_2 x_2 + c_3 x_3 = k_3 \end{cases}$$

you could write these as the matrices:

$$\begin{bmatrix} a_1 & a_2 & a_3 \\ b_1 & b_2 & b_3 \\ c_1 & c_2 & c_3 \end{bmatrix} \begin{bmatrix} x_1 \\ x_2 \\ x_3 \end{bmatrix} = \begin{bmatrix} k_1 \\ k_2 \\ k_3 \end{bmatrix}$$

Another way to look at this would be to let the coefficients of the linear equations be represented by the matrix

$$A = \begin{bmatrix} a_1 & a_2 & a_3 \\ b_1 & b_2 & b_3 \\ c_1 & c_2 & c_3 \end{bmatrix}$$

If X represents the column vector of variables, and K represents the column vector of constants, you have

$$X = \begin{bmatrix} x_1 \\ x_2 \\ x_3 \end{bmatrix} \text{ and } K = \begin{bmatrix} k_1 \\ k_2 \\ k_3 \end{bmatrix}$$

and the system of equations could be represented by $AX = K$.

With normal equations you would divide both sides by A to get the values of X, but these are matrices and there is no commutative law for multiplication. We will return to this dilemma after we introduce a new matrix.

In Section 17.2, we introduced matrix multiplication. A square $n \times n$ matrix with a 1 in each position of the main diagonal and zeros elsewhere is called the $n \times n$ **identity matrix** and is denoted as I_n. Thus,

$$I_2 = \begin{bmatrix} 1 & 0 \\ 0 & 1 \end{bmatrix} \text{ and } I_3 = \begin{bmatrix} 1 & 0 & 0 \\ 0 & 1 & 0 \\ 0 & 0 & 1 \end{bmatrix}$$

are both identity matrices.

If the size of the identity matrix is understood, we simply write I.

Now, back to our problem of $AX = K$. What we need is a matrix A^{-1}, where $A^{-1}A = I$. We would then find that

$$X = A^{-1}K$$

When A is an $n \times n$ matrix, then A is an **invertible** or **nonsingular matrix**, if there exists another $n \times n$ matrix, A^{-1}, where

$$A^{-1}A = AA^{-1} = I$$

If A^{-1} exists, it is called the **inverse** of matrix A. A matrix is called a **singular** matrix if it does not have an inverse.

There are two procedures for finding the inverse of a matrix. The first method may be easier on 2×2 matrices.

Steps for Finding the Inverse of a 2 × 2 Matrix

If $A = \begin{bmatrix} a & b \\ c & d \end{bmatrix}$ is a 2×2 matrix, then use the following four steps to determine A^{-1}:

1. Interchange the elements on the main diagonal.
2. Change the signs of the elements that are not on the main diagonal.
3. Find the determinant of the original matrix.
4. Divide each element at the end of Step 2 by the determinant of the original matrix.

≡ **Note**

Symbolically, we can summarize the four steps for finding the inverse of a 2×2 matrix as

$$A^{-1} = \frac{1}{ad - bc} \begin{bmatrix} d & -b \\ -c & a \end{bmatrix}$$

EXAMPLE 17.16

Find the inverse of the matrix $A = \begin{bmatrix} 1 & 2 \\ 4 & 10 \end{bmatrix}$.

Solution   **Step 1:** $\begin{bmatrix} 10 & 2 \\ 4 & 1 \end{bmatrix}$ Interchange elements on main diagonal.

Step 2: $\begin{bmatrix} 10 & -2 \\ -4 & 1 \end{bmatrix}$ Change signs of elements not on main diagonal.

Step 3: $\begin{vmatrix} 1 & 2 \\ 4 & 10 \end{vmatrix} = 10 - 8 = 2$ Find the determinant of the original matrix.

Step 4: $\begin{bmatrix} \frac{10}{2} & \frac{-2}{2} \\ \frac{-4}{2} & \frac{1}{2} \end{bmatrix}$ Divide each element from Step 2 by the determinant (Step 3).

EXAMPLE 17.16 (Cont.)

The inverse is $A^{-1} = \begin{bmatrix} 5 & -1 \\ -2 & \frac{1}{2} \end{bmatrix}$. Check your answer. Is $A^{-1}A = I$?

$$\begin{bmatrix} 5 & -1 \\ -2 & \frac{1}{2} \end{bmatrix} \begin{bmatrix} 1 & 2 \\ 4 & 10 \end{bmatrix} = \begin{bmatrix} 5-4 & 10-10 \\ -2+2 & -4+5 \end{bmatrix} = \begin{bmatrix} 1 & 0 \\ 0 & 1 \end{bmatrix} = I$$

The second method for determining the inverse of a matrix is slightly more complicated for a 2×2 matrix, but it will work for a matrix of any size.

With the second method, we are going to transform the given matrix into the identity matrix. At the same time, we will change an identity matrix into the inverse of the given matrix.

Steps for Finding the Inverse for Any $n \times n$ Matrix

The idea in this method for finding A^{-1} is to begin with the matrix $[A \;\vdots\; I]$ and use the following two steps to change it to $[I \;\vdots\; A^{-1}]$.

1. Multiply or divide all elements in a row by a nonzero constant.
2. Add a constant multiple of the elements of one row to the corresponding elements of another row.

We demonstrate this method in Example 17.17 with the same matrix used in Example 17.16.

EXAMPLE 17.17

Find the inverse of $A = \begin{bmatrix} 1 & 2 \\ 4 & 10 \end{bmatrix}$.

Solution   First, we form the augmented matrix $[A \;\vdots\; I]$.

$$\begin{bmatrix} 1 & 2 & \vdots & 1 & 0 \\ 4 & 10 & \vdots & 0 & 1 \end{bmatrix}$$

In the following demonstration, we let R_1 mean Row 1 and R_2 mean Row 2 for this 2×4 matrix. The arrows point to the row that we changed and that row is always listed last. Everything we do is designed to change the left half of the matrix to the identity matrix.

$$\begin{bmatrix} 1 & 2 & \vdots & 1 & 0 \\ 4 & 10 & \vdots & 0 & 1 \end{bmatrix}$$

EXAMPLE 17.17 (Cont.)

There is already a 1 in the a_{11} position. We need to get a 0 below it. First, multiply R_1 by -4 and add that to R_2 to get a new Row 2.

$$\left[\begin{array}{cc:cc} 1 & 2 & 1 & 0 \\ 0 & 2 & -4 & 1 \end{array}\right] \quad \leftarrow -4R_1 + R_2$$

Now that the first column is $\left[\begin{array}{c} 1 \\ 0 \end{array}\right]$, we move to the second column to get it in the form $\left[\begin{array}{c} 0 \\ 1 \end{array}\right]$.

Next, multiply R_2 by -1 and add that to R_1. This is the new row 1.

$$\left[\begin{array}{cc:cc} 1 & 0 & 5 & -1 \\ 0 & 2 & -4 & 1 \end{array}\right] \quad \leftarrow -R_2 + R_1$$

Now, all that is left is to divide Row 2 by 2 (or multiply it by $\frac{1}{2}$).

$$\left[\begin{array}{cc:cc} 1 & 0 & 5 & -1 \\ 0 & 1 & -2 & \frac{1}{2} \end{array}\right] \quad \leftarrow \frac{1}{2}R_2$$

The left half of the matrix is the identity matrix and the right half is the inverse that we wanted.

$$A^{-1} = \left[\begin{array}{cc} 5 & -1 \\ -2 & \frac{1}{2} \end{array}\right]$$

This is the same answer we got for Example 17.16.

We will work another 2×2 example and then work a 3×3 example. Before beginning, you should always check to see if the matrix is invertible. It will not be invertible if every element in any row or column is zero. Thus

$$\left[\begin{array}{ccc} 1 & 2 & 3 \\ 0 & 0 & 0 \\ 4 & 5 & 6 \end{array}\right] \text{ and } \left[\begin{array}{ccc} 1 & 2 & 0 \\ 4 & 5 & 0 \\ 7 & 8 & 0 \end{array}\right]$$

cannot be inverted. A matrix cannot be inverted if every element in one row (or column) is a constant multiple of every corresponding element in another row (or column). Thus $\left[\begin{array}{cc} 4 & 2 \\ 12 & 6 \end{array}\right]$ cannot be inverted, because each element in row 2 is 3 times its corresponding element in row 1. Notice that a matrix that cannot be inverted has a determinant of 0.

EXAMPLE 17.18

Find the inverse of $B = \begin{bmatrix} 8 & 4 \\ 3 & 2 \end{bmatrix}$.

Solution

$$\left[\begin{array}{cc:cc} 8 & 4 & 1 & 0 \\ 3 & 2 & 0 & 1 \end{array}\right]$$

$$\left[\begin{array}{cc:cc} 8 & 4 & 1 & 0 \\ 0 & 4 & -3 & 8 \end{array}\right] \quad \leftarrow -3R_1 + 8R_2$$

$$\left[\begin{array}{cc:cc} 8 & 0 & 4 & -8 \\ 0 & 4 & -3 & 8 \end{array}\right] \quad \leftarrow -R_2 + R_1$$

$$\left[\begin{array}{cc:cc} 1 & 0 & \frac{1}{2} & -1 \\ 0 & 1 & -\frac{3}{4} & 2 \end{array}\right] \quad \begin{array}{l} \leftarrow R_1 \div 8 \\ \leftarrow R_2 \div 4 \end{array}$$

The inverted matrix is $B^{-1} = \begin{bmatrix} \frac{1}{2} & -1 \\ -\frac{3}{4} & 2 \end{bmatrix}$.

EXAMPLE 17.19

Find the inverse of $C = \begin{bmatrix} 7 & -8 & 5 \\ -4 & 5 & -3 \\ 1 & -1 & 1 \end{bmatrix}$.

Solution

$$\left[\begin{array}{ccc:ccc} 7 & -8 & 5 & 1 & 0 & 0 \\ -4 & 5 & -3 & 0 & 1 & 0 \\ 1 & -1 & 1 & 0 & 0 & 1 \end{array}\right]$$

$$\left[\begin{array}{ccc:ccc} 7 & -8 & 5 & 1 & 0 & 0 \\ -4 & 5 & -3 & 0 & 1 & 0 \\ 0 & -1 & -2 & 1 & 0 & -7 \end{array}\right] \quad \leftarrow R_1 - 7R_3$$

$$\left[\begin{array}{ccc:ccc} 7 & -8 & 5 & 1 & 0 & 0 \\ 0 & 3 & -1 & 4 & 7 & 0 \\ 0 & -1 & -2 & 1 & 0 & -7 \end{array}\right] \quad \leftarrow 4R_1 + 7R_2$$

$$\left[\begin{array}{ccc:ccc} 7 & -8 & 5 & 1 & 0 & 0 \\ 0 & 3 & -1 & 4 & 7 & 0 \\ 0 & 0 & -7 & 7 & 7 & -21 \end{array}\right] \quad \leftarrow R_2 + 3R_3$$

EXAMPLE 17.19 (Cont.)

$$\begin{bmatrix} 7 & -8 & 5 & \vdots & 1 & 0 & 0 \\ 0 & -21 & 0 & \vdots & -21 & -42 & -21 \\ 0 & 0 & -7 & \vdots & 7 & 7 & -21 \end{bmatrix} \quad \leftarrow R_3 - 7R_2$$

$$\begin{bmatrix} 7 & -8 & 5 & \vdots & 1 & 0 & 0 \\ 0 & 1 & 0 & \vdots & 1 & 2 & 1 \\ 0 & 0 & 1 & \vdots & -1 & -1 & 3 \end{bmatrix} \quad \begin{matrix} \leftarrow R_2 \div -21 \\ \leftarrow R_3 \div -7 \end{matrix}$$

$$\begin{bmatrix} 7 & -8 & 0 & \vdots & 6 & 5 & -15 \\ 0 & 1 & 0 & \vdots & 1 & 2 & 1 \\ 0 & 0 & 1 & \vdots & -1 & -1 & 3 \end{bmatrix} \quad \leftarrow -5R_3 + R_1$$

$$\begin{bmatrix} 7 & 0 & 0 & \vdots & 14 & 21 & -7 \\ 0 & 1 & 0 & \vdots & 1 & 2 & 1 \\ 0 & 0 & 1 & \vdots & -1 & -1 & 3 \end{bmatrix} \quad \leftarrow 8R_2 + R_1$$

$$\begin{bmatrix} 1 & 0 & 0 & \vdots & 2 & 3 & -1 \\ 0 & 1 & 0 & \vdots & 1 & 2 & 1 \\ 0 & 0 & 1 & \vdots & -1 & -1 & 3 \end{bmatrix} \quad \leftarrow R_1 \div 7$$

The inverse of C is $C^{-1} = \begin{bmatrix} 2 & 3 & -1 \\ 1 & 2 & 1 \\ -1 & -1 & 3 \end{bmatrix}$.

Inverting a Matrix with a Graphing Calculator

Using a graphing calculator to invert a matrix is much simpler than the process used in Examples 17.16 and 17.17.

Using a Graphing Calculator to Invert a Matrix

1. Define the matrix as described in Section 5.4
2. Enter the name of the matrix
3. Press $\boxed{x^{-1}}$ $\boxed{\text{ENTER}}$

The next example shows how to find the inverse of a matrix by using a graphing calculator.

EXAMPLE 17.20

Use a graphing calculator to find the inverse of $B = \begin{bmatrix} 8 & 4 \\ 3 & 2 \end{bmatrix}$.

Solution   Enter matrix B as described in the user's guide for your calculator, or as described in Section 5.4. Let this matrix be named B.

Quit the edit matrix mode and return to the home screen. Now press $\boxed{B}\ \boxed{x^{-1}}$ $\boxed{\text{ENTER}}$. The result should be similar to the one shown in Figure 17.1a.

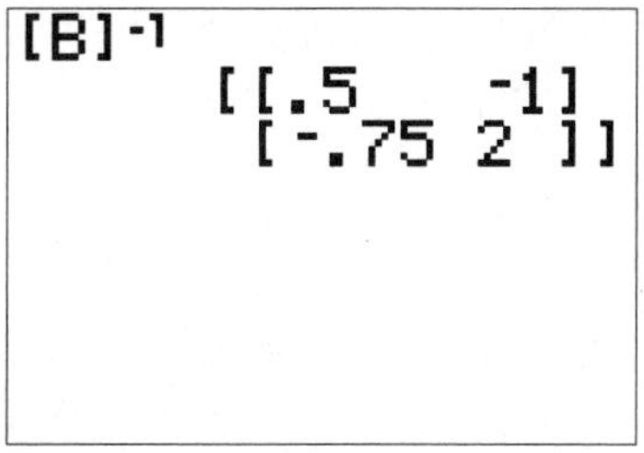

```
[B]⁻¹
        [[.5    -1]
         [-.75 2 ]]
```

```
[B]⁻¹
        [[.5    -1]
         [-.75 2 ]]
Ans▶Frac
        [[1/2   -1]
         [-3/4 2 ]]
```

FIGURE 17.1a 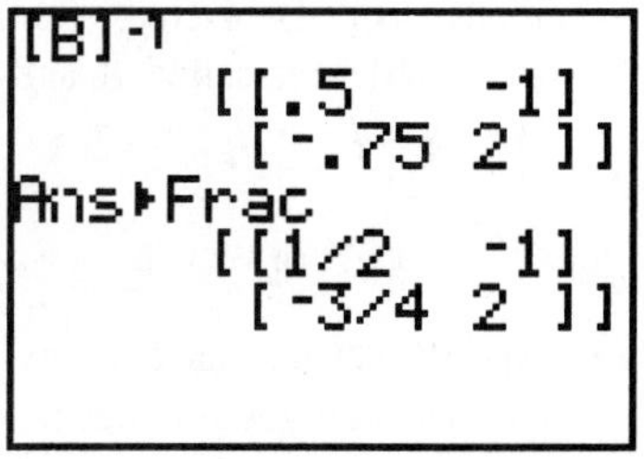   **FIGURE 17.1b**

If your calculator can convert decimals to fractions, you might want to convert B^{-1} to fractional form. For example, on a TI-82 or TI-85, press the $\boxed{\text{▶Frac}}$ key. The result, shown in Figure 17.1b, matches the result we got in Example 17.18.

Exercise Set 17.3

In Exercises 1–20, if the given matrix is invertible find its inverse.

1. $\begin{bmatrix} 6 & 1 \\ 5 & 1 \end{bmatrix}$

2. $\begin{bmatrix} 5 & 1 \\ 9 & 2 \end{bmatrix}$

3. $\begin{bmatrix} 10 & 4 \\ 8 & 3 \end{bmatrix}$

4. $\begin{bmatrix} 9 & 6 \\ 4 & 3 \end{bmatrix}$

5. $\begin{bmatrix} 8 & -6 \\ -6 & 4 \end{bmatrix}$

6. $\begin{bmatrix} 12 & 9 \\ -9 & -7 \end{bmatrix}$

7. $\begin{bmatrix} 15 & 10 \\ 4 & 3 \end{bmatrix}$

8. $\begin{bmatrix} 15 & 10 \\ 3 & 4 \end{bmatrix}$

9. $\begin{bmatrix} 0 & -3 & 0 \\ 1 & 0 & 0 \\ 0 & 0 & 4 \end{bmatrix}$

12. $\begin{bmatrix} 4 & 5 & 1 \\ 1 & 0 & 1 \\ 4 & 5 & 1 \end{bmatrix}$

15. $\begin{bmatrix} 3 & -1 & 0 \\ -6 & 2 & 0 \\ 1 & 0 & 5 \end{bmatrix}$

18. $\begin{bmatrix} 3 & 1 & -1 \\ 1 & -2 & 0 \\ 0 & 3 & 1 \end{bmatrix}$

10. $\begin{bmatrix} 1 & 0 & 0 \\ 0 & 4 & 0 \\ 0 & 0 & 2 \end{bmatrix}$

13. $\begin{bmatrix} 8 & 7 & -1 \\ -5 & -5 & 1 \\ -4 & -4 & 1 \end{bmatrix}$

16. $\begin{bmatrix} 1 & -1 & 1 \\ 7 & -8 & 5 \\ -4 & 5 & -3 \end{bmatrix}$

19. $\begin{bmatrix} 1 & -1 & 1 \\ 0 & 2 & -1 \\ 2 & 3 & 0 \end{bmatrix}$

11. $\begin{bmatrix} 1 & 2 & 6 \\ 0 & 0 & 2 \\ -3 & -6 & -9 \end{bmatrix}$

14. $\begin{bmatrix} 1 & 2 & 3 \\ 2 & 5 & 7 \\ 1 & 1 & 1 \end{bmatrix}$

17. $\begin{bmatrix} 2 & 0 & 0 \\ 2 & 2 & 0 \\ 2 & 2 & 2 \end{bmatrix}$

20. $\begin{bmatrix} 1 & 2 & -1 \\ 2 & -2 & 1 \\ 6 & 4 & 3 \end{bmatrix}$

Solve Exercises 21 and 22.

21. Check your answers to 1–20 by multiplying the given matrix and its inverse. Remember, if the matrix in the problem is called A and your answer A^{-1}, then $AA^{-1} = I$.

22. Write a computer program to determine the inverse of an $n \times n$ matrix.

In Your Words

23. Not all matrices have an inverse. List the conditions that will tell you if a matrix is not invertible.

24. If your calculator can work with complex numbers, (a) describe how to find the inverse of the matrix $A = \begin{bmatrix} i & 4+3i \\ 3 & 2-2i \end{bmatrix}$ and (b) compute A^{-1}.

☰ 17.4
MATRICES AND LINEAR EQUATIONS

Earlier in this chapter, we said that a system of equations could be represented by matrices. For example, the system

$$\begin{cases} 2x + 3y = 5 \\ 3x + 5y = 9 \end{cases}$$

would be represented by three matrices A, X, and K, where $AX = K$. To do this, let A be the matrix formed by the coefficients of the variables, let X be a column vector of the variables, and let K be a column vector of the constants. Thus, for this system, we have

$$A = \begin{bmatrix} 2 & 3 \\ 3 & 5 \end{bmatrix} \qquad X = \begin{bmatrix} x \\ y \end{bmatrix} \qquad K = \begin{bmatrix} 5 \\ 9 \end{bmatrix}$$

Now that we can invert a matrix, we can find A^{-1}. Since $A^{-1}A = I$ then $A^{-1}(AX) = (A^{-1}A)X = IX = X$. With this knowledge, we can find the solution to our system of equations.

$$AX = K$$
$$A^{-1}(AX) = A^{-1}K$$
$$X = A^{-1}K$$

EXAMPLE 17.21

Use the inverse of the coefficient matrix to solve the linear system

$$\begin{cases} 2x + 3y = 5 \\ 3x + 5y = 9 \end{cases}$$

Solution   Since $A = \begin{bmatrix} 2 & 3 \\ 3 & 5 \end{bmatrix}$, and $|A| = \begin{vmatrix} 2 & 3 \\ 3 & 5 \end{vmatrix} = 1$, we have, by the first

method of inverting a matrix, $A^{-1} = \begin{bmatrix} 5 & -3 \\ -3 & 2 \end{bmatrix}$. So

$$\begin{aligned} X = A^{-1}K &= \begin{bmatrix} 5 & -3 \\ -3 & 2 \end{bmatrix} \cdot \begin{bmatrix} 5 \\ 9 \end{bmatrix} \\ &= \begin{bmatrix} 5(5) - 3(9) \\ -3(5) + 2(9) \end{bmatrix} \\ &= \begin{bmatrix} 25 - 27 \\ -15 + 18 \end{bmatrix} \\ &= \begin{bmatrix} -2 \\ 3 \end{bmatrix} \end{aligned}$$

and $x = -2$ and $y = 3$.

Let's try this method for solving a system of linear equations on a system of three equations with three variables.

EXAMPLE 17.22

Use the inverse of the coefficient matrix to solve the linear system

$$\begin{cases} 7x - 8y + 5z = 18 \\ -4x + 5y - 3z = -11 \\ x - y + z = 1 \end{cases}$$

Solution   This linear system can be represented by the three matrices A, X, and K, where $AX = K$. Again, A is the matrix formed by the coefficients, X is the column vector of the variables, and K is the column vector of the constants. From this system we have

$$A = \begin{bmatrix} 7 & -8 & 5 \\ -4 & 5 & -3 \\ 1 & -1 & 1 \end{bmatrix} \qquad X = \begin{bmatrix} x \\ y \\ z \end{bmatrix} \qquad K = \begin{bmatrix} 18 \\ -11 \\ 1 \end{bmatrix}$$

EXAMPLE 17.22 (Cont.)

We will first need A^{-1}. In Example 17.19, we found that

$$A^{-1} = \begin{bmatrix} 2 & 3 & -1 \\ 1 & 2 & 1 \\ -1 & -1 & 3 \end{bmatrix}$$

Thus,

$$
\begin{aligned}
X = A^{-1}K &= \begin{bmatrix} 2 & 3 & -1 \\ 1 & 2 & 1 \\ -1 & -1 & 3 \end{bmatrix} \begin{bmatrix} 18 \\ -11 \\ 1 \end{bmatrix} \\[2mm]
&= \begin{bmatrix} 2(18) & + & 3(-11) & - & 1(1) \\ 1(18) & + & 2(-11) & + & 1(1) \\ -1(18) & - & 1(-11) & + & 3(1) \end{bmatrix} \\[2mm]
&= \begin{bmatrix} 36 & - & 33 & - & 1 \\ 18 & - & 22 & + & 1 \\ -18 & + & 11 & + & 3 \end{bmatrix} \\[2mm]
&= \begin{bmatrix} 2 \\ -3 \\ -4 \end{bmatrix}
\end{aligned}
$$

From this, we see that $x = 2$, $y = -3$, and $z = -4$.

This is a lot of work. It would be nice if we had a computer program to help solve a system of equations. If fact, the design of such a program is included as an exercise in the following exercise set.

Application

EXAMPLE 17.23

Apply Kirchhoff's laws to the circuit in Figure 17.2 and determine the currents in I_1, I_2, and I_3.

Solution   If we apply Kirchhoff's laws to the circuit in Figure 17.2, we obtain the following system of linear equations.

$$
\begin{cases}
I_1 - I_2 + I_3 = 0 \\
5I_1 + 8I_2 = 4 \\
8I_2 + 2I_3 = 38
\end{cases}
$$

The coefficient matrix for this system is $A = \begin{bmatrix} 1 & -1 & 1 \\ 5 & 8 & 0 \\ 0 & 8 & 2 \end{bmatrix}$. (Notice that

EXAMPLE 17.23 (Cont.)

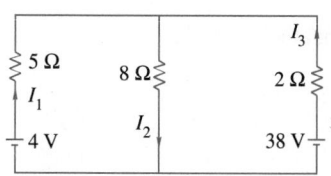

FIGURE 17.2

the coefficients of the missing variables are 0.) The inverse of A is

$$A^{-1} = \begin{bmatrix} \frac{8}{33} & \frac{5}{33} & -\frac{4}{33} \\ -\frac{5}{33} & \frac{1}{33} & \frac{5}{66} \\ \frac{20}{33} & -\frac{4}{33} & \frac{13}{66} \end{bmatrix} = \frac{1}{66} \begin{bmatrix} 16 & 10 & -8 \\ -10 & 2 & 5 \\ 40 & -8 & 13 \end{bmatrix}$$

If the constant matrix is $B = \begin{bmatrix} 0 \\ 4 \\ 38 \end{bmatrix}$ and $X = \begin{bmatrix} I_1 \\ I_2 \\ I_3 \end{bmatrix}$, then

$$X = A^{-1}B$$

$$= \frac{1}{66} \begin{bmatrix} 16 & 10 & -8 \\ -10 & 2 & 5 \\ 40 & -8 & 13 \end{bmatrix} \begin{bmatrix} 0 \\ 4 \\ 38 \end{bmatrix}$$

$$= \frac{1}{66} \begin{bmatrix} -264 \\ 198 \\ 462 \end{bmatrix}$$

$$= \begin{bmatrix} -4 \\ 3 \\ 7 \end{bmatrix}$$

Thus, the three currents are $I_1 = -4$ A, $I_2 = 3$ A, and $I_3 = 7$ A.

Exercise Set 17.4

In Exercises 1–10, solve the given systems of equations by using the inverse of the coefficient matrix. The numbers in parentheses refer to exercises in Exercise Set 17.3, where you determined the inverse of the coefficient matrix.

1. $\begin{cases} 6x + y = -4 \\ 5x + y = -3 \end{cases}$ (Exercise Set 17.3, Exercise 1)

2. $\begin{cases} 10x + 4y = 8 \\ 8x + 3y = 7 \end{cases}$ (Exercise Set 17.3, Exercise 3)

3. $\begin{cases} 8x - 6y = -27 \\ -6x + 4y = 19 \end{cases}$ (Exercise Set 17.3, Exercise 5)

4. $\begin{cases} 15x + 10y = -5 \\ 4x + 3y = 0 \end{cases}$ (Exercise Set 17.3, Exercise 7)

5. $\begin{cases} x + 2y + 6z = 12 \\ 2z = 2 \\ -3x - 6y - 9z = -27 \end{cases}$ (Exercise Set 17.3, Exercise 11)

6. $\begin{cases} 8x + 7y - z = 9 \\ -5x - 5y + z = -1 \\ -4x - 4y + z = 0 \end{cases}$ (Exercise Set 17.3, Exercise 13)

7. $\begin{cases} x + 2y + 3x = 4 \\ 2x + 5y + 7z = 7.5 \\ x + y + z = 1 \end{cases}$ (Exercise Set 17.3, Exercise 14)

8. $\begin{cases} x - y + z = -6.6 \\ 7x - 8y + 5z = -43 \\ -4x + 5y - 3z = 26.9 \end{cases}$ (Exercise Set 17.3, Exercise 16)

9. $\begin{cases} 2x = 11 \\ 2x + 2y = 4 \\ 2x + 2y + 2z = -9 \end{cases}$ (Exercise Set 17.3, Exercise 17)

10. $\begin{cases} x - y + z = 22 \\ 2y - z = -23 \\ 2x + 3y = -11.2 \end{cases}$ (Exercise Set 17.3, Exercise 19)

In Exercises 11–20, solve each system of equations by using the inverse of the coefficient matrix.

11. $\begin{cases} 2x + y = 1 \\ -3x + 2y = 16 \end{cases}$

12. $\begin{cases} 4x + 5y = 2 \\ 3x - 2y = 13 \end{cases}$

13. $\begin{cases} 2x + 2y = 4 \\ 4x + 3y = 1 \end{cases}$

14. $\begin{cases} 3x + y = -5 \\ 4x + 3y = 2 \end{cases}$

15. $\begin{cases} 1.5x + 2.5y = 0.3 \\ 3.2x + 2.6y = 7.2 \end{cases}$

16. $\begin{cases} 7x + 2y + z = 2 \\ 3x - 2y + 4z = 13 \\ 4x + 5y - z = 1 \end{cases}$

17. $\begin{cases} 5x + 2y + 3z = 1 \\ -3x + 2y - 8z = 6 \\ 4x - 2y + 9z = -7 \end{cases}$

18. $\begin{cases} 2x + 4y + z = 10 \\ 4x + 2y + z = 8 \\ 6x + 4y + 7z = -2 \end{cases}$

19. $\begin{cases} x + 2y + 4z = 7 \\ 3x + y + 4z = -2 \\ 2x + 9y - 2z = 10 \end{cases}$

20. $\begin{cases} 2x - y + z = 7 \\ 4x - 2y + z = 5 \\ 6x - 3y + 5z = -3 \end{cases}$

Solve Exercises 21–31.

21. *Electricity* If Kirchhoff's laws are applied to the circuit in Figure 17.3, the following equations are obtained. Determine the indicated currents.

$$I_A - I_B - I_C = 0$$
$$16I_A + 4I_C = 100$$
$$12I_B - 4I_C = 60$$

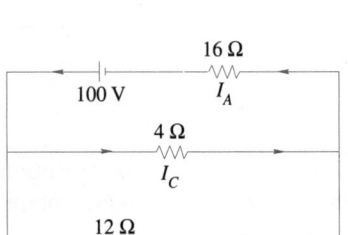

FIGURE 17.3

22. *Electricity* Applying Kirchhoff's laws to the circuit in Figure 17.4 results in the following equations. Determine the indicated currents.

$$I_3 - I_1 - I_2 = 0$$
$$20I_1 + 0.5I_1 - 15I_2 - 0.4I_2 = 120 - 80$$
$$15I_2 + 0.4I_2 + 10I_3 + 0.6I_3 = 140$$

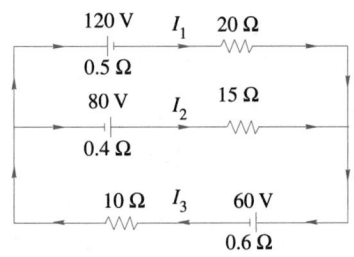

FIGURE 17.4

23. *Electricity* The currents through the resistors in Figure 17.5 produce the following equations. Determine the currents.

$$7.18I_1 - I_2 + 2.2I_3 = 10$$
$$-I_1 + 5.8I_2 + 1.5I_3 = 15$$
$$2.2I_1 + 1.5I_2 + 8.4I_3 = 20$$

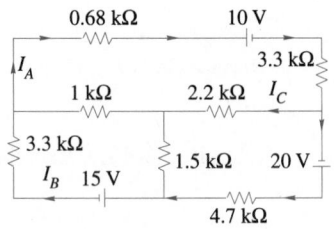

FIGURE 17.5

24. *Business*   A company manufactures robotic controls. Their current models are the *RC*-1 and *RC*-2. Each *RC*-1 unit requires 8 transistors and 4 integrated circuits. Each *RC*-2 unit uses 9 transistors and 5 integrated circuits. Each day the company receives 1,596 transistors and 860 integrated circuits. How many units of each model can be made if all parts are used?

 25. Write a computer program that uses matrices to solve a system of n linear equations with n variables.

26. *Metallurgy*   An alloy is composed of three metals, *A*, *B*, and *C*. The percentages of each metal are indicated by the following system of equations:

$$\begin{cases} A + B + C = 100 \\ A - 2B = 0 \\ -4A + C = 0 \end{cases}$$

Determine the percentage of each metal in the alloy.

27. *Automotive technology*   A petroleum engineer was testing three different gasoline mixtures, *A*, *B*, and *C*, in the same car and under the same driving conditions. She noticed that the car traveled 90 km farther when it used mixture *B* than when it used mixture *A*. Using fuel *C*, the car traveled 130 km more than when it used fuel *B*. The total distance traveled was 1 900 km. Find the distance traveled on the three fuels.

28. *Construction technology*   If three cables are joined at a point and three forces are applied so the system is in equilibrium, the following system of equations results.

$$\begin{cases} \tfrac{6}{7}F_B - \tfrac{2}{3}F_C = 2,000 \\ -F_A - \tfrac{3}{7}F_B + \tfrac{1}{3}F_C = 0 \\ \tfrac{2}{7}F_B + \tfrac{2}{3}F_C = 1,200 \end{cases}$$

Determine the three forces, F_A, F_B, and F_C, measured in newtons (N).

29. *Environmental science*   To control ice and protect the environment, a certain city determines that the best mixture to be spread on roads consists of 5 units of salt, 6 units of sand, and 4 units of a chemical inhibiting agent. Three companies, *A*, *B*, and *C*, sell mixtures of these elements according to the following table

	Salt	Sand	Inhibiting Agent
Company *A*	2	1	1
Company *B*	2	2	2
Company *C*	1	5	1

(a) In what proportion should the city purchase from each company in order to spread the best mixture? (Assume that the city must buy complete truckloads.)

(b) If the city expects to need 3,630,000 units for the winter, how many units should be bought from each company.

30. *Automotive technology*   The relationship between the velocity, v, of a car (in mph) and the distance, d (in ft), required to bring it to a complete stop is known to be of the form $d = av^2 + bv + c$, where a, b, and c are constants. Use the following data to determine the values of a, b, and c. When $v = 20$, then $d = 40$; when $v = 55$, then $d = 206.25$; and when $v = 65$, then $d = 276.25$.

31. *Metallurgy*   An alloy is composed of four metals, *A*, *B*, *C*, and *D*. The percentages of each metal are indicated by the following system of equations:

$$\begin{cases} A + B + C + D = 100 \\ A + B - C = 0 \\ -1.64A + D = 0 \\ 3A - 2C + 2D = -1 \end{cases}$$

Determine the percentage of each metal in the alloy.

In Your Words

32. What conditions are necessary in order to use matrices to solve a system of equations.

33. Describe how to use your calculator to solve the system of linear equations $AX = K$, where A, X, and K are matrices.

 CHAPTER 17 REVIEW
_____

Important Terms and Concepts

Augmented matrix
Coefficient matrix
Cofactor
Column vector
Dimension
Identity matrix
Inverse of a matrix
Invertible matrix
Matrix
   Addition

Multiplication
Scalar multiplication
Subtraction
Nonsingular matrix
Row vector
Scalar
Singular matrix
Square matrix
Zero matrix

Review Exercises

In Exercises 1–6, use the following matrices to determine the indicated matrix.

$$A = \begin{bmatrix} 4 & 3 & 2 & 5 \\ 6 & 7 & -1 & 4 \\ 9 & 10 & -8 & 3 \end{bmatrix} \quad B = \begin{bmatrix} 3 & -2 & 1 & 0 \\ 5 & -1 & 2 & 0 \\ 4 & 3 & -2 & 0 \end{bmatrix} \quad C = \begin{bmatrix} 0 & 1 & 0 & 2 \\ 3 & 0 & 4 & 0 \\ 0 & -5 & 0 & -6 \end{bmatrix}$$

1. $A + C$

2. $B + C$

3. $A - B$

4. $C - B$

5. $2A - 3C$

6. $4A + B - 2C$

In Exercises 7–10, find the indicated products.

7. $\begin{bmatrix} 3 & 2 \\ 4 & 5 \\ 1 & 0 \end{bmatrix} \begin{bmatrix} 1 & 2 \\ 0 & 1 \end{bmatrix}$

8. $\begin{bmatrix} 1 & -4 \\ 5 & 1 \end{bmatrix} \begin{bmatrix} 3 & 4 & 1 \\ 2 & 5 & 0 \end{bmatrix}$

9. $\begin{bmatrix} 3 \\ 2 \\ -1 \end{bmatrix} \begin{bmatrix} 4 & 5 & 1 \end{bmatrix}$

10. $\begin{bmatrix} 4 & 5 & 1 \end{bmatrix} \begin{bmatrix} -2 \\ 7 \\ -3 \end{bmatrix}$

In Exercises 11–14, if the given matrix is invertible, find its inverse.

11. $\begin{bmatrix} 2 & 3 \\ -4 & -5 \end{bmatrix}$

12. $\begin{bmatrix} 2 & -1 \\ 0 & 4 \end{bmatrix}$

13. $\begin{bmatrix} -2 & 1 & 0 \\ 0 & 4 & 0 \\ 1 & 0 & 1 \end{bmatrix}$

14. $\begin{bmatrix} 2 & 3 & -1 \\ 1 & 2 & 1 \\ -1 & -1 & 3 \end{bmatrix}$

In Exercises 15–18, solve the system of equations by using the inverse of the coefficient matrix.

15. $\begin{cases} 12x + 5y = -2 \\ 3x + y = 1.1 \end{cases}$

16. $\begin{cases} 4x + y = -4 \\ -3x + 2y = 14 \end{cases}$

17. $\begin{cases} 2x + 3y + 5z = 20 \\ -2x + 3y + 5z = 12 \\ 5x - 3y - 2z = 9 \end{cases}$

18. $\begin{cases} x + y + 6z = -3 \\ -2x + 2y + 4z = -2 \\ 3x + 2y + 4z = 14 \end{cases}$

Solve Exercises 19–24.

19. *Physics*   Masses of 9 kg and 11 kg are attached to a cord that passes over a frictionless pulley. When the masses are released, the acceleration a of each mass, in meters per second squared (m/s^2), and the tension T in the cord, in newtons (N), are related by the system

$$\begin{cases} T - 75.4 = 9a \\ 100.0 - T = 11a \end{cases}$$

Find a and T.

20. *Business*   A computer company makes three types of computers—a personal computer (PC), a business computer (BC), and a technical computer (TC). There are three parts in each computer that they have difficulty getting: RAM chips, EPROMS, and transistors. The number of each part needed by each computer is shown in this table.

	RAM	EPROM	Transistor
PC	4	2	7
BC	8	3	6
TC	12	5	11

If the company is guaranteed 1,872 RAM chips, 771 EPROMS, and 1,770 transistors each week, how many of each computer can be made?

21. Given the equations

$$\begin{cases} x' = \tfrac{1}{2}\left(x + y\sqrt{3}\right) \\ y' = \tfrac{1}{2}\left(-x\sqrt{3} + y\right) \end{cases}$$

and

$$\begin{cases} x'' = \tfrac{1}{2}\left(-x' + y'\sqrt{3}\right) \\ y'' = -\tfrac{1}{2}\left(x'\sqrt{3} + y'\right) \end{cases}$$

write each set as a matrix equation and solve for x'' and y'' in terms of x and y by multiplying matrices.

22. The equations in Exercise 21 represent rotations of axes in two directions. In Section 15.7, we found that, if the angle of rotation is θ, then

$$\begin{cases} x'' = x\cos\theta + y\sin\theta \\ y'' = -x\sin\theta + y\cos\theta \end{cases}$$

What was the rotation angle for the equations in Exercise 21?

23. *Optics*   The following matrix product is used in discussing two thin lenses in air:

$$M = \begin{bmatrix} 1 & -\dfrac{1}{f_2} \\ 0 & 1 \end{bmatrix} \cdot \begin{bmatrix} 1 & 0 \\ d & 1 \end{bmatrix} \cdot \begin{bmatrix} 1 & -\dfrac{1}{f_1} \\ 0 & 1 \end{bmatrix}$$

where f_1 and f_2 are the focal lengths of the lenses and d is the distance between them. Evaluate M.

24. *Optics*   In Exercise 23, element M_{12} of M is $-\dfrac{1}{f}$, where f is the focal length of the combination. Determine $\dfrac{1}{f}$.

≣ CHAPTER 17 TEST

1. Given $A = \begin{bmatrix} 8 & 0 & -4 \\ 16 & -6 & 2 \end{bmatrix}$ and $B = \begin{bmatrix} -1 & 5 & -3 \\ 3 & 0 & 4 \end{bmatrix}$, find
   (a) $A + B$
   (b) $3A - 2B$

2. Calculate the product $\begin{bmatrix} 1 & -2 & 3 \end{bmatrix} \begin{bmatrix} -4 \\ -6 \\ 8 \end{bmatrix}$.

3. If $C = \begin{bmatrix} 4 & 6 \\ -10 & 4 \end{bmatrix}$ and $D = \begin{bmatrix} \frac{1}{2} & 0 \\ -\frac{3}{2} & 4 \end{bmatrix}$, calculate
   (a) CD and (b) DC.

4. If $E = \begin{bmatrix} 2 & -3 \\ 7 & 9 \end{bmatrix}$, (a) find E^{-1}. (b) Use E^{-1} to solve $EX = F$, where $F = \begin{bmatrix} -31 \\ 28 \end{bmatrix}$.

5. Solve the following system of equations by using the inverse of the coefficient matrix.
$$\begin{cases} x + 3y + z = -2 \\ 2x + 5y + z = -5 \\ x + 2y + 3z = 6 \end{cases}$$

6. Three machine parts cost a total of $60. The first part cost as much as the other two together. The cost of the second part is $3 more than twice the cost of the third part. How much does each part cost?

7. Find the current of the system in Figure 17.6.

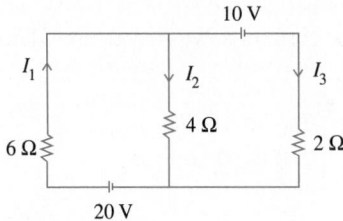

FIGURE 17.6

18

Higher Degree Equations

Package designers often need to design a box that will hold a specific volume. In Section 18.3, we will see how to design such a box made from a given sheet of material.

The majority of equations that we have solved have been linear or quadratic equations, or systems of linear equations. We were able to solve higher degree equations when they could be factored. It seems, however, that most equations do not factor. When you get to equations of degree higher than two, there is no easy formula, such as the quadratic formula, to help solve them. In this chapter, we will explore some methods for solving polynomial equations of degree higher than two.

☰ 18.1
THE REMAINDER AND FACTOR THEOREM

We will begin this section with the definition of a polynomial.

Polynomial

A **polynomial** is a function of the form

$$P(x) = a_n x^n + a_{n-1} x^{n-1} + \cdots + a_2 x^2 + a_1 x + a_0$$

where $a_0, a_1, a_2, \ldots, a_n$ are numbers, n is a non-negative integer, and $a_n \neq 0$. This polynomial has **degree** n.

a_n is called the **leading coefficient** of the polynomial.

EXAMPLE 18.1

The following are some examples of polynomials.

$$P(x) = 3x^2 + 2x - 1 \text{ has degree 2.}$$
$$P(x) = 5x^6 + \frac{7}{3}x^3 - 4x + 1 \text{ has degree 6.}$$
$$P(x) = -\sqrt{7} \text{ has degree 0.}$$
$$P(x) = \frac{2}{3}x^4 - \frac{1}{2}x^3 + x^2 \text{ has degree 4.}$$

EXAMPLE 18.2

The following are not polynomials.
$$P(x) = 4x^{5/2} + 2x^2 - 3,$$ because one of the exponents, $\frac{5}{2}$, is not an integer.
$$P(x) = 5x^{-3},$$ because the exponent is not positive.

The third polynomial in Example 18.1, $P(x) = -\sqrt{7}$, is an example of a constant polynomial.

Constant and Zero Polynomials

A **constant polynomial function** is of the form $P(x) = c$, where c is a real number and $c \neq 0$. The degree of a constant polynomial function is zero.

$P(x) = 0$ is called the **zero polynomial**. It has no degree.

The roots, solutions, or zeros of a polynomial function $P(x)$ are the values of x for which $P(x) = 0$. In earlier chapters, we learned that, if $P(x) = ax + b$, then $P(x) = 0$, when $x = \dfrac{-b}{a}$. We also learned the quadratic formula, which states that if $P(x)$ is of degree 2 and $P(x) = ax^2 + bx + c$, then $P(x) = 0$, when $x = \dfrac{-b \pm \sqrt{b^2 - 4ac}}{2a}$.

EXAMPLE 18.3

Find the roots of (a) $P(x) = 7x - \sqrt{2}$ and (b) $P(x) = 4x^2 - 3x - 5$.

Solutions

(a) The roots occur when $P(x) = 7x - \sqrt{2} = 0$. That is, when $7x = \sqrt{2}$ or $x = \frac{1}{7}\sqrt{2}$.

(b) Here, we want the solution to be $4x^2 - 3x - 5 = 0$. Using the quadratic formula, we get

$$x = \frac{-(-3) \pm \sqrt{(-3)^2 - 4(4)(-5)}}{2(4)}$$
$$= \frac{3 \pm \sqrt{9 + 80}}{8} = \frac{3 \pm \sqrt{89}}{8}$$

The roots are $x = \frac{3}{8} + \frac{1}{8}\sqrt{89}$ and $x = \frac{3}{8} - \frac{1}{8}\sqrt{89}$.

We will now look at some ways we can solve polynomial functions of degree larger than two.

Remainder Theorem

Whenever we divide a polynomial $P(x)$ by $x - a$ we get a quotient $Q(x)$ that is a polynomial and a remainder R that is a constant. In fact, when $P(x)$ is divided by $x - a$, the remainder $R = P(a)$. This is the **remainder theorem**.

Remainder Theorem

If a polynomial $P(x)$ is divided by $x - a$ until a remainder that does not contain x is obtained, then the remainder $R = P(a)$.

EXAMPLE 18.4

Determine the remainder, when $P(x) = x^3 - 4x^2 + 2x + 5$ is divided by $x - 3$ and $x + 2$.

Solution When $P(x) = x^3 - 4x^2 + 2x + 5$ is divided by $x - 3$, then $a = 3$. So, $P(3) = 3^3 - 4(3)^2 + 2(3) + 5 = 27 - 36 + 6 + 5 = 2$. When $P(x)$ is divided by $x - 3$, the remainder is 2.

When we divide $P(x)$ by $x + 2$, we have $a = -2$, so $P(-2) = (-2)^3 - 4(-2)^2 + 2(-2) + 5 = -8 - 16 - 4 + 5 = -23$. So, when $P(x)$ is divided by $x + 2$, the remainder is -23.

We can check to see if these are true by dividing $P(x)$ by $x - 3$ and by $x + 2$.

$$
\begin{array}{r}
x^2 - x - 1 \\
x - 3 \,\overline{)\, x^3 - 4x^2 + 2x + 5} \\
\underline{x^3 - 3x^2} \\
-x^2 + 2x + 5 \\
\underline{-x^2 + 3x} \\
-x + 5 \\
\underline{-x + 3} \\
2
\end{array}
\qquad
\begin{array}{r}
x^2 - 6x + 14 \\
x + 2 \,\overline{)\, x^3 - 4x^2 + 2x + 5} \\
\underline{x^3 + 2x^2} \\
-6x^2 + 2x + 5 \\
\underline{-6x^2 - 12x} \\
14x + 5 \\
\underline{14x + 28} \\
-23
\end{array}
$$

From this we see that, when $P(x) = x^3 - 4x^2 + 2x + 5$ is divided by $x - 3$, we get a quotient $Q(x) = x^2 - x - 1$ and a remainder of $R = 2$. When $P(x)$ is divided by $x + 2$, the quotient is $Q(x) = x^2 - 6x + 14$ with a remainder of $R = -23$. These remainders are the same ones we found in Example 18.4.

Factor Theorem

In some cases, $R = 0$, and we see that $P(x) = Q(x)(x - a) + R$ can be rewritten as

$$P(x) = Q(x)(x - a)$$

That is, $x - a$ is a factor of $P(x)$. This leads to our second theorem for this section.

Factor Theorem

The **factor theorem** states that a polynomial $P(x)$ contains $x - a$ as a factor if and only if $P(a) = 0$.

The factor theorem means that a is a root of $P(x)$ if $x - a$ is a factor of $P(x)$.

EXAMPLE 18.5

Use the factor theorem to determine if $x + 1$ and $x - 2$ are factors of $P(x) = x^4 + x^3 - 3x^2 - 4x - 1$.

Solutions   To see if $x + 1$ is a factor, we evaluate $P(x)$ at $x = -1$. Since $P(-1) = 0$, then $x + 1$ is a factor of P. [In fact, $P(x) = (x + 1)(x^3 + 3x - 1)$.]

To see if $x - 2$ is a factor of P, we evaluate $P(x)$ at $x = 2$. Since $P(2) = 3$, we know that $x + 2$ is not a factor of $x^4 + x^3 - 3x^2 - 4x - 1$.

Synthetic Division

What we need is a shorter method to determine the quotient and remainder of a polynomial. Both of the divisions we worked earlier took a lot of time and a lot of space. A process has been developed that allows us to shorten the division. This process is called **synthetic division**.

Consider the problem we worked earlier when $x^3 - 4x^2 + 2x + 5$ was divided by $x + 2$. We had the following work.

$$
\begin{array}{r}
x^2 - 6x + 14 \\
x + 2 \overline{)\; x^3 - 4x^2 + 2x + 5} \\
\underline{x^3 + 2x^2} \\
-6x^2 + 2x + 5 \\
\underline{-6x^2 - 12x} \\
14x + 5 \\
\underline{14x + 28} \\
-23
\end{array}
$$

In synthetic division, we do not write the x's and we do not repeat terms. So, if we erase all the x's and the terms that repeat, we get a group of coefficients that look like this.

$$
\begin{array}{r}
1 \;\; -6 \;\; 14 \\
2 \overline{)\; 1 \;\; -4 \;\; 2 \;\; 5} \\
2 \\
\hline
-6 \\
-12 \\
\hline
14 \\
28 \\
\hline
-23
\end{array}
$$

We can write all the numbers below the dividend in two lines and get

$$
\begin{array}{r}
1 \;\;\;\; -6 \;\;\;\; 14 \\
2 \overline{)\; 1 \;\; -4 \;\;\;\; 2 \;\;\;\; 5} \\
2 \;\; -12 \;\;\;\; 28 \\
\hline
-6 \;\;\;\; 14 \;\; -23
\end{array}
$$

If we repeat the leading coefficient on the bottom line, we see that the numbers on the third line represent the quotient and the remainder. Finally, we will change the sign of the divisor. Changing the sign of the divisor forces us to change the signs in the second row. This allows us to add the first two rows.

$$
\begin{array}{r|rrrr}
-2 & 1 & -4 & 2 & 5 \\
   &   & -2 & 12 & -28 \\
\hline
   & 1 & -6 & 14 & -23
\end{array}
$$

$$\underbrace{}_{\text{quotient}} \quad \underbrace{}_{\text{remainder}}$$

The next example shows you how to use synthetic division.

EXAMPLE 18.6

Use synthetic division to determine the quotient and remainder when $x^4 - 3x^2 + 10x - 5$ is divided by $x + 3$.

Solution

$$
\begin{array}{r|rrrrr}
-3 & 1 & 0 & -3 & 10 & -5 \\
\end{array}
$$

Step 1: If the divisor is $x - a$, write a in the box. Arrange the coefficients of the dividend by descending powers of x. Use a zero coefficient when a power is missing. In this example, there is no x^3 term and $a = -3$.

$$
\begin{array}{r|rrrrr}
-3 & 1 & 0 & -3 & 10 & -5 \\
& & & & & \\
\hline
& 1 & & & & \\
\end{array}
$$

Step 2: Copy the leading coefficient in the third row.

$$
\begin{array}{r|rrrrr}
-3 & 1 & 0 & -3 & 10 & -5 \\
& & -3 & & & \\
\hline
& 1 & -3 & & & \\
\end{array}
$$

Step 3: Multiply the last entry in the third row by the number in the box and write the result in the second row under the second coefficient. Add the numbers in that column.

$$
\begin{array}{r|rrrrr}
-3 & 1 & 0 & -3 & 10 & -5 \\
& & -3 & 9 & & \\
\hline
& 1 & -3 & 6 & & \\
\end{array}
$$

Step 4a: Repeat the process from Step 3, but write the results under the third coefficient.

$$
\begin{array}{r|rrrrr}
-3 & 1 & 0 & -3 & 10 & -5 \\
& & -3 & 9 & -18 & 24 \\
\hline
& 1 & -3 & 6 & -8 & 19 \\
\end{array}
$$

quotient   remainder

Step 4b: Repeat the process from Step 3 until there are as many entries in row 3 as there are in row 1. The last number in row 3 is the remainder. The other numbers are the coefficients of the quotient.

From this we see that the quotient is $Q(x) = x^3 - 3x^2 + 6x - 8$ and the remainder is $R = 19$.

EXAMPLE 18.7

Use synthetic division to determine the quotient and remainder when $2x^6 - 11x^4 + 17x^2 - 20$ is divided by $x - 2$.

Solution Since there are no x^5, x^3, or x terms, we will replace them with 0 coefficients. The synthetic division looks like this.

$$
\begin{array}{r|rrrrrrr}
2 & 2 & 0 & -11 & 0 & 17 & 0 & -20 \\
& & 4 & 8 & -6 & -12 & 10 & 20 \\
\hline
& 2 & 4 & -3 & -6 & 5 & 10 & 0 \\
\end{array}
$$

quotient   remainder

The quotient is $2x^5 + 4x^4 - 3x^3 - 6x^2 + 5x + 10$ and the remainder is 0. So, $x - 2$ is a factor of the polynomial $2x^6 - 11x^4 + 17x^2 - 20$.

EXAMPLE 18.8

Use synthetic division to determine if $3x - 2$ and $x - 1$ are factors of $18x^4 - 12x^3 - 45x^2 + 57x - 18$.

Solution   The remainder and factor theorems both refer to division by $x - a$. We are to check $3x - 2$. But $3x - 2 = 3\left(x - \frac{2}{3}\right)$, and if $x - \frac{2}{3}$ divides the polynomial, then $3x - 2$ will also divide it. Our synthetic division follows.

$$
\begin{array}{r|rrrrr}
\frac{2}{3} & 18 & -12 & -45 & 57 & -18 \\
 & & 12 & 0 & -30 & 18 \\
\hline
 & 18 & 0 & -45 & 27 & 0
\end{array}
$$

Since the remainder is zero, $3x - 2$ is a factor of $18x^4 - 12x^3 - 45x^2 - 57x - 18$. We will now see if $x - 1$ is a factor. Instead of checking the original equation, we will use the result from the synthetic division.

When we divided $18x^4 - 12x^3 - 45x^2 + 57x - 18$ by $x - \frac{2}{3}$, we got $18 \quad 0 \quad -45 \quad 27$ in the third row. This told us that $18x^4 - 12x^3 - 45x^2 + 57x - 18 = \left(x - \frac{2}{3}\right)$ $(18x^3 - 45x + 27)$. The factor $18x^3 - 45x + 27$ is called a **depressed equation** of the original equation. If $x - 1$ is a factor of the depressed equation $18x^3 - 45x + 27$, it is a factor of the original equation. Thus we can use the third row from our earlier synthetic division as the first row when we check $x - 1$.

$$
\begin{array}{r|rrrrr}
\frac{2}{3} & 18 & -12 & -45 & 57 & -18 \\
 & & 12 & 0 & -30 & 18 \\
1 & 18 & 0 & -45 & 27 & 0 \\
 & & 18 & 18 & -27 & \\
\hline
 & 18 & 18 & -27 & 0 &
\end{array}
$$

Thus, $x - 1$ is a factor of the depressed equation $18x^3 - 45x + 27$, so $x - 1$ is also a factor of $18x^4 - 12x^3 - 45x^2 + 57x - 18$.

So,

$$18x^4 - 12x^3 - 45x^2 + 57x - 18$$
$$= \left(x - \frac{2}{3}\right)(x - 1)(18x^2 + 18x - 27)$$
$$= 9\left(x - \frac{2}{3}\right)(x - 1)(2x^2 + 2x - 3)$$

We can use the quadratic formula on the last depressed equation to find that the remaining roots are $x = \dfrac{-1 + \sqrt{7}}{2}$ and $x = \dfrac{-1 - \sqrt{7}}{2}$.

Exercise Set 18.1

In Exercises 1–8, find the value of $P(x)$ for the given value of x.

1. $P(x) = 3x^2 - 2x + 1$; $x = 2$
2. $P(x) = 2x^3 - 4x + 5$; $x = -1$
3. $P(x) = x^4 + x^3 + x^2 - x + 1$; $x = -1$
4. $P(x) = x^4 - 2x^2 + x$; $x = -2$

5. $P(x) = 5x^3 - 4x + 7$; $x = 3$
6. $P(x) = 7x^4 - 5x^2 + x - 7$; $x = -2$
7. $P(x) = x^5 - x^4 + x^3 + x^2 - x + 1$; $x = -1$
8. $P(x) = 3x^5 + 4x^2 - 3$; $x = -3$

In Exercises 9–16, use the remainder theorem to find the remainder R, when $P(x)$ is divided by $x - a$.

9. $P(x) = x^3 + 2x^2 - x - 2$; $x - 1$
10. $P(x) = x^3 + 2x^2 - 12x - 9$; $x - 3$
11. $P(x) = x^3 - 3x^2 + 2x + 5$; $x - 3$
12. $P(x) = x^3 - 9x^2 + 23x - 15$; $x - 1$

13. $P(x) = 4x^4 + 13x^3 - 13x^2 - 40x + 12$; $x + 2$
14. $P(x) = 2x^4 - 2x^3 - 6x^2 - 14x - 7$; $x - 7$
15. $P(x) = 3x^4 - 12x^3 - 60x + 4$; $x - 5$
16. $P(x) = 4x^3 - 4x^2 - 10x + 8$; $x - \frac{1}{2}$

In Exercises 17–24, use the factor theorem to determine whether or not the second factor is a factor of the first.

17. $x^3 + 2x^2 - 12x - 9$; $x - 3$
18. $x^4 - 9x^3 + 18x^2 - 3$; $x + 1$
19. $2x^5 - 6x^3 + x^2 + 4x - 1$; $x + 1$
20. $2x^5 - 6x^3 + x^2 + 4x - 1$; $x - 1$

21. $3x^5 + 3x^4 - 14x^3 + 4x^2 - 24x$; $x + 3$
22. $3x^5 + 3x^4 - 14x^3 + 4x^2 - 24x$; $x - 2$
23. $6x^4 - 15x^3 - 8x^2 + 20x$; $2x - 5$
24. $20x^4 + 12x^3 + 10x + 9$; $5x + 3$

In Exercises 25–32, use synthetic division to determine the quotient and remainder when each polynomial is divided by the given $x - a$.

25. $x^5 - 17x^3 + 75x + 9$; $x - 3$
26. $2x^5 - x^2 + 8x + 44$; $x + 2$
27. $5x^3 + 7x^2 + 9$; $x + 3$
28. $x^3 + 3x^2 - 2x - 4$; $x - 2$

29. $8x^5 - 4x^3 + 7x^2 - 2x$; $x - \frac{1}{2}$
30. $9x^5 + 3x^4 - 6x^3 - 2x^2 + 6x + 1$; $x + \frac{1}{3}$
31. $4x^4 - 12x^3 + 9x^2 - 8x + 12$; $2x - 3$
32. $4x^3 + 7x^2 - 3x - 15$; $4x - 5$

✏️ **In Your Words**

33. Suppose that $P(x)$ is a polynomial and you determine that $P(3) = -5$. What does this mean?

34. Explain what it means for $P(r) = 0$ for some polynomial $P(x)$.

35. What precautions must you take when using synthetic division?

≡ **18.2**
ROOTS OF AN EQUATION

In Section 18.1, we learned the factor theorem to help us determine if a number is a root of a polynomial. We also learned how to use synthetic division to quickly find the quotient and remainder when a polynomial is divided by a first degree polynomial, $x - a$. In this section, we shall learn some theorems that determine the number of roots of the equation $P(x) = 0$.

In working with first degree polynomials, we were always able to find one root. With second degree polynomials, we could find two roots. At times, as in $P(x) = x^2 + 6x + 9$, both of these roots were the same. (Both roots of $x^2 + 6x + 9 = 0$ were -3.) As you might expect, every polynomial of degree n has exactly n roots. The **fundamental theorem of algebra** states that every polynomial equation of degree $n > 0$ has at least one (real or complex) root.

Combining the fundamental theorem of algebra with the factor theorem leads to the **linear factorization theorem**, which states the following:

Linear Factorization Theorem

If $P(x)$ is a polynomial function of degree $n > 1$, then there is a non-zero number a and there are numbers, $r_1, r_2, \ldots, r_n$, such that

$$P(x) = a(x - r_1)(x - r_2)\cdots(x - r_n)$$

The proof of this theorem is fairly easy. If $P(x)$ is a polynomial and $P(x) = 0$, then by the fundamental theorem there is a number r_1 such that $P(r_1) = 0$. From the factor theorem, we know $P(x) = (x - r_1)P_1(x)$ where $P_1(x)$ is a polynomial.

Again, the fundamental theorem states that there is a number r_2 such that $P_1(r_2) = 0$, so $P_1(x) = (x - r_2)P_2(x)$ and $P(x) = (x - r_1)(x - r_2)P_2(x)$.

We continue until one of the quotients is a constant a. At that time, we have

$$P(x) = a(x - r_1)(x - r_2)\cdots(x - r_n)$$

A linear factor appears each time a root is found. Since the degree of $P(x)$ is n, there are n linear factors and n roots. These roots are $r_1, r_2, r_3, \ldots, r_n$. Now, as we have seen, these roots may not all be different. Yet, even if they are not distinct, each root is counted.

EXAMPLE 18.9

Use the linear factorization theorem to determine the roots of the polynomial $P(x) = 3x^4 - 8x^3 - 11x^2 + 28x - 12$.

Solution $\quad 3x^4 - 8x^3 - 11x^2 + 28x - 12 = (x - 1)(3x^3 - 5x^2 - 16x + 12)$

$$= (x - 1)(x - 3)(3x^2 + 4x - 4)$$
$$= (x - 1)(x - 3)(x + 2)(3x - 2)$$
$$= 3(x - 1)(x - 3)(x + 2)\left(x - \frac{2}{3}\right)$$

The roots are 1, 3, -2, and $\frac{2}{3}$.

Since $P(x)$ was of degree 4, it should have four roots. It does. Notice that the constant factor is the leading coefficient.

Just as it is not necessary that all roots be distinct, it is not necessary that all roots are real numbers. Remember that complex numbers can be roots to quadratic equations.

EXAMPLE 18.10

Determine the roots of $P(x) = x^2 + 6x + 25$.

Solution Using the quadratic formula, we determine that $P(x) = x^2 + 6x + 25$ has the following roots.

$$
\begin{aligned}
x &= \frac{-6 \pm \sqrt{6^2 - 4(25)}}{2} \\
&= \frac{-6 \pm \sqrt{36 - 100}}{2} \\
&= \frac{-6 \pm \sqrt{-64}}{2} \\
&= \frac{-6 \pm 8j}{2} \\
&= -3 \pm 4j
\end{aligned}
$$

The roots are $r_1 = -3 + 4j$ and $r_2 = -3 - 4j$.

EXAMPLE 18.11

Determine the roots of $P(x) = (x-2)^3(x^2 + 6x + 25)$.

Solution From the previous example we know that the roots of $x^2 + 6x + 25$ are $r_1 = -3 + 4j$ and $r_2 = -3 - 4j$. Thus, the roots of $P(x) = (x-2)^3(x^2 + 6x + 25)$ would be 2, 2, 2, $-3 + 4j$, and $-3 - 4j$. Notice that there are five roots, but three of them are the same.

One basic property of complex roots follows.

> **Complex Roots Theorem**
>
> If $P(x)$ is a polynomial with real coefficients and $a + bj$ is a root of $P(x)$, then its conjugate, $a - bj$, is also a root.

This would not be true if $P(x)$ was a polynomial with complex coefficients, but it is true when all the coefficients of P are real numbers.

 Hint

Remember, when solving polynomials, if you find enough roots so the remaining factor is quadratic, you can always find the last two roots by using the quadratic formula.

EXAMPLE 18.12

Solve the equation $4x^3 - 9x^2 - 25x - 12$, given the fact that $-\frac{3}{4}$ is a root.

Solution   Using synthetic division, we get

$$
\begin{array}{r|rrrr}
-\frac{3}{4} & 4 & -9 & -25 & -12 \\
 & & -3 & 9 & 12 \\
\hline
 & 4 & -12 & -16 & 0
\end{array}
$$

So,

$$
\begin{aligned}
4x^3 - 9x^2 - 25x - 12 &= \left(x + \frac{3}{4}\right)(4x^2 - 12x - 16) \\
&= 4\left(x + \frac{3}{4}\right)(x^2 - 3x - 4)
\end{aligned}
$$

We can factor $x^2 - 3x - 4$ as $(x-4)(x+1)$, and so

$$
4x^3 - 9x^2 - 25x - 12 = 4\left(x + \frac{3}{4}\right)(x-4)(x+1)
$$

The roots of $P(x) = 4x^3 - 9x^2 - 25x - 12$ are $-\frac{3}{4}$, 4, and -1.

EXAMPLE 18.13

Solve $P(x) = x^4 + 5x^3 + 10x^2 + 20x + 24$ if you know that $2j$ is a root.

Solution   Since $2j$ is a root and the coefficients of P are all real, its conjugate $-2j$ is also a root. So, two of the linear factors are $x - 2j$ and $x + 2j$. We can use synthetic division twice or divide the original polynomial by $(x - 2j)(x + 2j) = x^2 + 4$. We will use synthetic division twice.

$$
\begin{array}{r|rrrrr}
2j & 1 & 5 & 10 & 20 & 24 \\
 & & +2j & -4+10j & -20+12j & -24 \\
\hline
-2j & 1 & (5+2j) & (6+10j) & (0+12j) & 0 \\
 & & -2j & -10j & -12j & \\
\hline
 & 1 & 5 & 6 & 0 &
\end{array}
$$

Thus,

$$
\begin{aligned}
x^4 + 5x^3 + 10x^2 + 20x + 24 &= (x-2j)(x+2j)(x^2+5x+6) \\
&= (x-2j)(x+2j)(x+3)(x+2)
\end{aligned}
$$

The four roots are $2j$, $-2j$, -3, and -2.

Notice that the second time we performed synthetic division it was done on the depressed equation that resulted from the first synthetic division.

EXAMPLE 18.14

Solve $2x^4 - 17x^3 + 49x^2 - 51x + 9$ if 3 is a double root.

Solution   Since 3 is a double root, we know that two of the linear factors are $x - 3$ and $x - 3$. Using synthetic division twice, we get

$$
\begin{array}{r|rrrrr}
3 & 2 & -17 & 49 & -51 & 9 \\
  &   & 6   & -33 & 48 & -9 \\
\hline
3 & 2 & -11 & 16 & -3 & 0 \\
  &   & 6   & -15 & 3   &   \\
\hline
  & 2 & -5 & 1   & 0   &   \\
\end{array}
$$

The last factor is $2x^2 - 5x + 1$. Using the quadratic formula, we see that its roots are

$$
x = \frac{5 \pm \sqrt{(-5)^2 - 4(2)(1)}}{2(2)}
$$

$$
= \frac{5 \pm \sqrt{17}}{4}
$$

Thus, the roots are 3, 3, $\dfrac{5 + \sqrt{17}}{4}$, and $\dfrac{5 - \sqrt{17}}{4}$.

Exercise Set 18.2

In Exercises 1–22, solve the equations using synthetic division and the given roots.

1. $5x^3 - 8x + 3$, $r_1 = 1$
2. $2x^3 + 5x^2 - 11x + 4$, $r_1 = -4$
3. $9x^3 - 3x^2 - 81x + 27$, $r_1 = \frac{1}{3}$
4. $10x^3 - 4x^2 - 40x + 16$, $r_1 = \frac{2}{5}$
5. $x^4 - 3x^2 - 4$, $r_1 = j$
6. $3x^4 + 6x^2 - 189$, $r_1 = -3j$
7. $x^4 + 2x^3 - 4x^2 - 18x - 45$, $r_1 = -1 + 2j$
8. $2x^4 + 4x^3 + 2x^2 - 16x - 40$, $r_1 = -1 - 2j$
9. $x^4 - 3x^3 - 3x^2 + 7x + 6$, $r_1 = 2, r_2 = 3$
10. $x^4 - 3x^3 - 12x^2 + 52x - 48$, $r_1 = 2, r_2 = -4$
11. $3x^4 + 12x^3 + 6x^2 - 12x - 9$, $r_1 = r_2 = -1$ (a double root)
12. $2x^4 + 6x^3 - 12x^2 - 24x + 16$, $r_1 = 2, r_2 = -2$
13. $6x^4 + 25x^3 + 33x^2 + x - 5$, $r_1 = -\frac{1}{2}, r_2 = \frac{1}{3}$

14. $12x^4 - 47x^3 + 55x^2 + 9x - 5$, $r_1 = \frac{1}{4}, r_2 = -\frac{1}{3}$
15. $3x^4 - 2x^3 - 3x + 2$, $r_1 = \frac{2}{3}, r_2 = 1$
16. $4x^4 + 3x^3 - 32x - 24$, $r_1 = 2, r_2 = -\frac{3}{4}$
17. $x^5 - x^4 + x^3 - 7x^2 + 10x - 4$, $r_1 = r_2 = r_3 = 1$ (a triple root)
18. $x^5 - 5x^4 + 7x^3 - 2x^2 + 4x - 8$, $r_1 = r_2 = r_3 = 2$
19. $3x^5 - 2x^4 - 24x^3 + x^2 + 28x - 12$, $r_1 = 3, r_2 = -2$, $r_3 = \frac{2}{3}$
20. $3x^5 - 4x^4 + 5x^3 - 18x^2 - 28x - 8$, $r_1 = -\frac{2}{3}, r_2 = 2j$
21. $x^6 - 6x^5 + 3x^4 - 60x^3 - 61x^2 - 54x - 63$, $r_1 = j$, $r_2 = -3j$
22. $6x^6 - 19x^5 + 63x^4 - 152x^3 + 216x^2 - 304x + 240$, $r_1 = r_2 = 2j$

In Your Words

23. The polynomial $P(x) = x^3 + ix^2 - 4x - 4i$ has two real roots, 2 and -2, and one nonreal complex root, i. Explain why this does not violate the complex roots theorem.

24. Explain how to create a polynomial if you know its leading coefficient and its roots.

≡ 18.3
RATIONAL ROOTS

In Section 18.2, we determined the number of roots of a polynomial. We also learned that complex roots come in pairs and that once we are able to reduce the nonlinear factor to degree 2, we can use the quadratic formula.

What we need is some help in locating roots when the degree is larger than two. Those roots can be complex numbers, irrational numbers, or rational numbers. In this section, we will focus on ways to determine the rational roots of a polynomial.

Rational Root Test

Let's consider the general polynomial:

$$P(x) = a_n x^n + a_{n-1} x^{n-1} + \cdots + a_2 x^2 + a_1 x + a_0$$

We begin with a test to help determine if a rational number is a root of P.

Rational Root Test

If the rational number $\dfrac{c}{d}$ is in lowest terms and is a root of $P(x) = a_n x^n + a_{n-1} x^{n-1} + \cdots + a_2 x^2 + a_1 x + a_0$, then c is a factor of a_0 and d is a factor of a_n.

One consequence of this test is that if $a_n = 1$, then all rational roots are integers. The next three examples show how to use the rational root test.

EXAMPLE 18.15

Find all rational roots of $P(x) = 4x^3 - 5x^2 - 2x + 3$.

Solution In this polynomial, $4x^3 - 5x^2 - 2x + 3$, $n = 3$, so $a_n = a_3 = 4$ and $a_0 = 3$. The factors of $a_n = 4$ are $d : \pm 1, \pm 2, \pm 4$, and the factors of $a_0 = 3$ are $c : \pm 1, \pm 3$. Any rational roots must be among these combinations of $\dfrac{c}{d}$: $\pm 1, \pm 3$, $\pm\frac{1}{2}, \pm\frac{3}{2}, \pm\frac{1}{4}, \pm\frac{3}{4}$. A check will verify that the roots are $-\frac{3}{4}$, 1, and 1.

EXAMPLE 18.16

Find all rational roots of $P(x) = 5x^6 - 4x^5 - 41x^4 + 32x^3 + 43x^2 - 28x - 7$.

Solution   In this polynomial, $n = 6$, so $a_n = a_6 = 5$ and $a_0 = -7$. The factors of a_6 are $d : \pm 1, \pm 5$. The factors of a_0 are $c : \pm 1, \pm 7$. Any rational roots must be among these combinations of $\frac{c}{d}$: ± 1, ± 7, $\pm \frac{1}{5}$, and $\pm \frac{7}{5}$. A check will verify that the rational roots are -1, $-\frac{1}{5}$, 1, and 1. The other two roots, $\sqrt{7}$ and $-\sqrt{7}$, are not rational numbers.

EXAMPLE 18.17

Find all rational roots of $P(x) = x^5 - 4x^4 - 5x^3 + 8x^2 - 32x - 40$.

Solution   In this polynomial, $n = 5$, so $a_n = a_5 = 1$ and $a_0 = -40$. Since $a_5 = 1$, any rational roots will be integers. The factors of a_0 are $c : \pm 1, \pm 2, \pm 4, \pm 5, \pm 8, \pm 10, \pm 20, \pm 40$. And these are the only 16 possible rational roots. A check will verify that the rational roots are -2, -1, and 5 and that the complex roots are $1 + j\sqrt{3}$ and $1 - j\sqrt{3}$.

In Example 18.15, we were able to narrow the number of possible rational roots to 12, in Example 18.16 we had only eight possible rational roots, and in Example 18.17 we had 16 possible rational roots. It would be beneficial if there were some additional methods we could use to help us in our search for roots.

Descartes' Rule of Signs

Descartes' rule of signs gives us a little more help. This rule gives an idea of the number of positive and negative roots there are in a polynomial. The rule applies to all real number roots.

> **Descartes' Rule of Signs**
>
> **Descartes' rule of signs** states that the number of positive roots of a polynomial $P(x) = 0$ cannot be more than the number of changes of sign in the terms of $P(x)$ and, if there are less, it will be reduced by a multiple of two. The equation $P(x) = 0$ has at most as many negative roots as the number of changes in sign of $P(-x)$ and, if there are less, it will be reduced by a multiple of two.

EXAMPLE 18.18

What information does Descartes' rule of signs tell us about the roots of $P(x) = 3x^5 - 2x^4 + x^3 - x - 7$?

Solution   The signs of the terms of this polynomial equation are marked below each term.

$$P(x) = \underbrace{3x^5 - 2x^4}_{1} \underbrace{+ x^3 - x}_{2} \underbrace{- x}_{3} - 7$$

EXAMPLE 18.18 (Cont.)

There are three places where the signs change. By Descartes' rule of signs, there can be no more than three positive real roots. If there are less than three, then there is one. So, this equation has either one or three positive real roots.

To find the number of negative real roots, we perform the same procedure on $P(-x)$.

$$P(-x) = 3(-x)^5 - 2(-x)^4 + (-x)^3 - (-x) - 7$$
$$= -3x^5 - 2x^4 - x^3 + x - 7$$

There are two sign changes and so the equation has either two or zero negative real roots.

Thus, this polynomial equation $3x^5 - 2x^4 + x^3 - x - 7 = 0$ has no more than three positive and two negative roots.

Caution

Remember, a polynomial may have complex number roots. This means that you cannot find the number of negative roots by subtracting the number of positive roots from the total number of roots.

Upper and Lower Bounds

We spent a lot of time in Section 18.2 learning synthetic division because it is a quick and easy way to determine if a given number is a root. What we will develop next is a way to tell when to stop looking for roots larger or smaller than the root we just checked.

Before we give this next technique, we need to learn two new terms. A real number U is an **upper bound** of the real roots of a polynomial if no root is larger than the number U. A real number L is a **lower bound** of the real roots of a polynomial if no root is smaller than L.

Let $P(x) = a_n x^n + a_{n-1} x^{n-1} + \cdots + a_2 x^2 + a_1 x + a_0$ be a polynomial with real coefficients and $a_n > 0$. If $k_1 > 0$ and if the terms in the third row of synthetic division of $P(x)$ by $x - k_1$ are all positive or zero, then k_1 is an upper bound of the real roots of $P(x)$. If $k_2 \leq 0$ and if the terms in the third row of synthetic division of $P(x)$ by $x - k_2$ alternate in sign, then k_2 is a lower bound of the real roots of $P(x)$.

EXAMPLE 18.19

Find an upper and lower bound for the roots of $P(x) = 2x^4 - 8x^3 + 5x^2 + 12x - 12$.

Solution If $\dfrac{c}{d}$ is a rational root, then c must be a factor of -12 and d must be a factor of 2. The factors of -12 are ±1, ±2, ±3, ±4, ±6, and ±12. The factors of 2 are ±1 and ±2. The possible rational roots are ±1, ±2, ±3, ±4, ±6, ±12, $\pm\frac{1}{2}$, and $\pm\frac{3}{2}$.

If we use synthetic division to determine if 4 is a root, we get

$$
\begin{array}{r|rrrrr}
4 & 2 & -8 & 5 & 12 & -12 \\
  &   & 8 & 0 & 20 & 128 \\
\hline
  & 2 & 0 & 5 & 32 & 116
\end{array}
$$

All are positive or zero.

EXAMPLE 18.19 (Cont.)

From this we see that 4 is not a root, but it is an upper bound. Thus, none of the numbers larger than 4 (6 and 12) can be roots of this polynomial.

Now, we will check to see if -2 is a root.

$$\begin{array}{r|rrrrr} -2 & 2 & -8 & 5 & 12 & -12 \\ & & -2 & 20 & -25 & 26 \\ \hline & 2 & -10 & 25 & -13 & 14 \end{array}$$ Signs alternate.

Again, -2 is not a root, but the numbers in the third row alternate sign. So, -2 is a lower bound and none of the numbers less than -2 (that is, -3, -6, and -12) can be roots.

You may want to check and verify that the actual roots are 2 (multiplicity 2), $\frac{1}{2}\sqrt{6}$, and $-\frac{1}{2}\sqrt{6}$. Thus,

$$P(x) = 2x^4 - 8x^3 + 5x^2 + 12x - 12$$
$$= 2(x-2)^2 \left(x - \frac{1}{2}\sqrt{6}\right)\left(x + \frac{1}{2}\sqrt{6}\right)$$
$$= (x-2)^2(2x^2 - 3).$$

Summary

Now let's summarize what we have learned in the last two sections and then work an example that uses everything. We have developed a list of six hints at the top of the next page to help you determine the roots of a polynomial. See the box that follows.

Hints for Finding Roots of Polynomials

If $P(x) = 0$ is a polynomial equation of degree n:

1. There are n roots.
2. Complex roots appear in conjugate pairs.
3. Any rational root of the form $\frac{c}{d}$, when in lowest terms, has c as a factor of the constant term, a_0, and d as a factor of the leading coefficient, a_n.
4. The maximum number of positive roots is the number of sign changes in $f(x)$; the maximum number of negative roots is the number of sign changes in $f(-x)$.
5. If we use synthetic division to determine whether or not a non-negative number c is a root and the third row has all non-negative numbers, then c is an upper bound; if c is a non-positive number and the third row alternates signs, then c is a lower bound.
6. Once we determine $n-2$ roots, the remaining roots can be found by using the quadratic formula or by factoring.

EXAMPLE 18.20

Find the roots of $P(x) = 6x^4 + 23x^3 + 3x^2 - 32x + 12$.

Solution $P(x)$ has degree 4, so there are four roots. There are two sign changes in $P(x)$, so there are two or zero positive real roots. Now, $P(-x) = 6x^4 - 23x^3 + 3x^2 + 32x + 12$ also has two sign changes, so $P(x)$ has two or zero negative real roots.

The leading coefficient $a_4 = 6$ and the constant $a_0 = 12$. The factors of a_4 are ± 1, ± 2, ± 3, and ± 6. The factors of a_0 are ± 1, ± 2, ± 3, ± 4, ± 6, and ± 12. The possible rational roots of $P(x)$ are ± 1, ± 2, ± 3, ± 4, ± 6, ± 12, $\pm\frac{1}{2}$, $\pm\frac{1}{3}$, $\pm\frac{1}{6}$, $\pm\frac{2}{3}$, $\pm\frac{3}{2}$, and $\pm\frac{4}{3}$.

We will now check these to see if any are roots of $P(x)$. Using synthetic division we first check 1.

$$
\begin{array}{r|rrrrr}
1 & 6 & 23 & 3 & -32 & 12 \\
  &   & 6 & 29 & 32 & 0 \\
\hline
  & 6 & 29 & 32 & 0   & 12 \\
\end{array}
$$

We found two pieces of information from this synthetic division. First, 1 is not a root. But, since the numbers in the third row are all nonnegative, 1 is an upper bound. Thus, none of the numbers larger than 1 (2, 3, 4, 6, 12, $\frac{3}{2}$, and $\frac{4}{3}$) are roots. The only possible positive rational roots must be selected from $\frac{1}{6}$, $\frac{1}{3}$, $\frac{1}{2}$, and $\frac{2}{3}$.

Instead of continuing to look for positive roots, we will check for negative roots. We will first check -1.

$$
\begin{array}{r|rrrrr}
-1 & 6 & 23 & 3 & -32 & 12 \\
   &   & -6 & -17 & 14 & 18 \\
\hline
   & 6 & 17 & -14 & -18 & 30 \\
\end{array}
$$

Since -1 is neither a root nor a lower bound, we will check -2.

$$
\begin{array}{r|rrrrr}
-2 & 6 & 23 & 3 & -32 & 12 \\
   &   & -12 & -22 & 38 & -12 \\
\hline
   & 6 & 11 & -19 & 6 & 0 \\
\end{array}
$$

We have found our first root, -2. If you check -2 again, you will see it is not a double root. It is also not a lower bound. We know that there is another negative root. We next check -3, but this time on the depressed equation in the third row.

$$
\begin{array}{r|rrrr}
-3 & 6 & 11 & -19 & 6 \\
   &   & -18 & 21 & -6 \\
\hline
   & 6 & -7 & 2 & 0 \\
\end{array}
$$

We see that -3 is also a root. The depressed equation from the third row is $6x^2 - 7x + 2$, which factors into $(3x - 2)(2x - 1)$. The roots of $P(x) = 6x^4 + 23x^3 + 3x^2 - 32x + 12$ are -3, -2, $\frac{1}{2}$, and $\frac{2}{3}$.

Application

A company needs a box like the one shown in Figure 18.1a. The box is to be made from a sheet of metal that measures 45 cm by 30 cm by cutting a square from each corner and bending up the sides. If the box is to hold $3\,500\,\text{cm}^3$, what are the lengths of the sides of the squares that are cut out of each corner?

Solution A sketch of this situation is shown in Figure 18.1b. The lengths of the sides of the squares have been labeled x. Once these squares have been removed, the part of the remaining sheet that will form the length of the box is $45 - 2x$ cm. Similarly, the box's width is $30 - 2x$ cm. The width of the box must be at least 0 and less than 30 cm. So, $0 < 30 - 2x < 30$ or $0 < x < 15$.

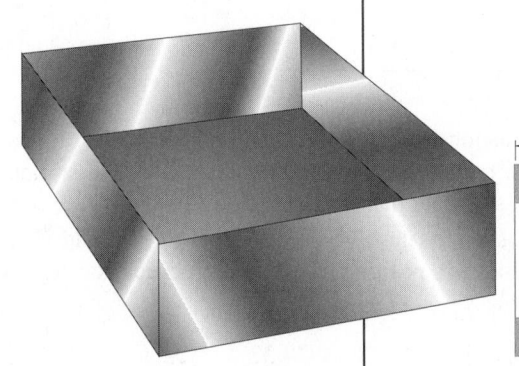

FIGURE 18.1a

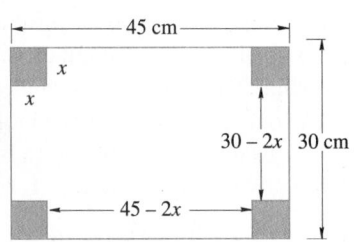

FIGURE 18.1b

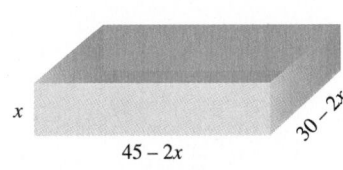

FIGURE 18.1c

When the metal is bent to form the sides of the box, a box like the one in Figure 18.1c is obtained. This box is a rectangular prism, so its volume is $V = (45 - 2x)(30 - 2x)x$. We are told that this box is to have a volume of $3\,500\,\text{cm}^3$, so

$$(45 - 2x)(30 - 2x)x = 3\,500$$

Multiplying the factors on the left-hand side produces

$$1\,350x - 150x^2 + 4x^3 = 3\,500$$

or $\quad 4x^3 - 150x^2 + 1\,350x - 3\,500 = 0$

There are three sign changes, so there is either 1 or 3 positive real roots. $P(-x) = -4x^3 - 150x^2 - 1\,350x - 3\,500$ has no sign changes, so there are zero negative real roots.

The leading coefficient is $a_3 = 4$ and the constant is $a_0 = 3\,500$. Since we know there are no negative real roots (and since x must be positive), we can eliminate the negative factors. The factors of a_3 are $d : 1, 2, 4$. The positive factors of a_0 are $c : 1, 2, 4, 5, 7, 10, 14, 20, 25, 28 \ldots$. (Since $x < 15$, we can disregard any potential roots of a_0 that are 15 or larger.) Thus, possible roots less than 15 include $\dfrac{c}{d}$: 1, 2, 4, 5, 7, 10, 14, $\frac{1}{2}$, $\frac{1}{4}$, $\frac{5}{2}$, $\frac{5}{4}$, $\frac{7}{2}$, $\frac{7}{4}$, $\frac{25}{2}$, and $\frac{25}{4}$.

Since $P(1) = -2\,296$, 1 is not a root. Similarly, $P(2) = -1\,368$ and $P(4) = -244$ indicate that neither 2 nor 4 are roots. But, $P(5) = 0$, so 5 is a root.

EXAMPLE 18.21 (Cont.)

Using synthetic division, we see that

$$
\begin{array}{r|rrrr}
5 & 4 & -150 & 1\,350 & -3\,500 \\
  &   & 20   & -650   & 3\,500 \\
\hline
  & 4 & -130 & 700   & 0
\end{array}
$$

and so

$$4x^3 - 150x^2 + 1\,350x - 3\,500 = (x-5)(4x^2 - 130x + 700)$$

Using the quadratic formula on the depressed equation $4x^2 - 130x + 700$, we obtain the other two roots.

$$
\begin{aligned}
x &= \frac{130 \pm \sqrt{130^2 - 4(4)(700)}}{8} \\
  &= \frac{130 \pm \sqrt{5\,700}}{8}
\end{aligned}
$$

So, $x = \dfrac{130 + \sqrt{5\,700}}{8} \approx 25.7$

or $x = \dfrac{130 - \sqrt{5\,700}}{8} \approx 6.8$

One of these possible solutions, 25.7 is too large since $x < 15$.

Thus, there are two possible ways to cut this metal to obtain a box with the desired volume. If $x = 5\,$cm, the box has a length of $45 - 2(5) = 35\,$cm and a width of $30 - 2(5) = 20\,$cm. Checking, we see that $(35)(20)5 = 3\,500$.

If $x \approx 6.8127\,$cm, the length is approximately $31.3746\,$cm and the width is about $16.3746\,$cm. Multiplying, we get a volume of $(31.3746)(16.3746)(6.8127) = 3\,500.000952 \approx 3\,500\,$cm^3.

This problem has two correct solutions. The solution the company uses will depend on other factors, such as which is easier (or less expensive) to make, which one does a better job of holding the product, and which shape is more appealing to the customer.

Exercise Set 18.3

In Exercises 1–10, use Descartes' rule of signs to determine the number of possible positive and negative real roots of the polynomial equation. List the possible rational roots of each equation.

1. $x^3 - 3x - 2 = 0$
2. $x^3 - 2x^2 - x + 2 = 0$
3. $x^3 - 8x^2 - 17x + 6 = 0$
4. $x^3 + 2x^2 - x - 2 = 0$
5. $x^3 - 2x^2 - 5x + 6 = 0$

6. $3x^3 - x^2 - 2x + 2 = 0$
7. $2x^4 - x^3 - 5x^2 + x + 3 = 0$
8. $2x^4 - 7x^3 - 10x^2 + 33x + 18 = 0$
9. $6x^4 - 7x^3 - 13x^2 + 4x + 4 = 0$
10. $5x^6 - x^5 - 5x^4 + 6x^3 - x^2 - 5x + 1 = 0$

In Exercises 11–26, find all rational roots of the polynomial equation. If possible, find all roots.

11. $x^3 - 9x^2 + 23x - 15 = 0$

12. $x^3 - 3x^2 - 4x + 12 = 0$

13. $x^3 - 3x - 2 = 0$

14. $x^3 - x^2 - 8x + 12 = 0$

15. $4x^3 - 5x^2 + 10x + 12 = 0$

16. $3x^3 - 4x^2 - 17x + 6 = 0$

17. $x^4 - 4x^3 + 7x^2 - 12x + 12 = 0$

18. $4x^4 + 4x^3 + 9x^2 + 8x + 2 = 0$

19. $4x^4 - 20x^3 + x^2 + 18x + 6 = 0$

20. $x^5 - x^3 + 27x^2 - 27 = 0$

21. $x^4 - 8x^3 + 39x^2 + 2x - 10 = 0$

22. $8x^3 - 36x^2 + 54x - 27 = 0$

23. $2x^5 - 13x^4 + 26x^3 - 22x^2 + 24x - 9 = 0$

24. $8x^5 - 4x^4 + 6x^3 - 3x^2 - 2x + 1 = 0$

25. $x^5 - 5x^4 + 6x^3 + 11x^2 - 43x + 30 = 0$

26. $2x^5 - x^4 - 6x^3 - 18x^2 + 4x + 40 = 0$

Solve Exercises 27–34.

27. *Sheet metal technology* A rectangular sheet of metal was made from a box by cutting identical squares from the four corners and bending up the sides. If the piece of sheet metal originally measured 8.0 in. by 10.0 in. and the volume of the box is 48 in.³, what was the length of each side of the squares that were removed?

28. *Drafting* A rectangular box is constructed so that its width is 2.5 cm longer than its height and the length is 4 cm longer than the width. If the box has a volume of 210 cm³, what are its dimensions?

29. *Agriculture* A grain silo has the shape of a right circular cylinder with a hemisphere on top, as shown in Figure 18.2. The total height of the silo is 34 ft. Determine the radius of the cylinder if the total volume is 2,511π ft³.

30. *Petroleum engineering* A propane gas storage tank is in the shape of a right circular cylinder of height 6 m with a hemisphere attached at each end. Determine the radius r so that the volume of the tank is 18π m³.

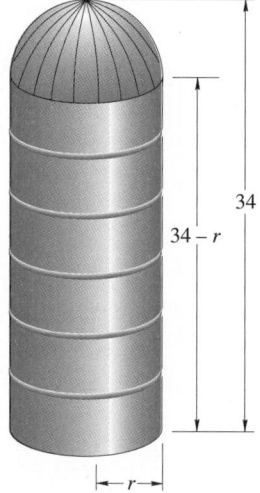

FIGURE 18.2

31. *Electricity* Three electric resistors are connected in parallel. The second resistor is 4 Ω more than the first and the third resistor is 1 Ω larger than the first. The total resistance is 1 Ω. In order to find the first resistance R, we must solve the equation

$$\frac{1}{R} + \frac{1}{R+4} + \frac{1}{R+1} = 1$$

What are the values of the resistances?

32. *Electricity*   Suppose that the resistors in Exercise 31 had been related such that the second was $4\,\Omega$ more than the first and the third $9\,\Omega$ more than the first. If the total resistance was $3\,\Omega$, then

$$\frac{1}{R} + \frac{1}{R+4} + \frac{1}{R+9} = \frac{1}{3}$$

What are the values of these resistances?

33. *Industrial technology*   A rectangular box is made from a piece of metal $12\,\text{cm}$ by $19\,\text{cm}$ by cutting a square from each corner and bending up the sides and welding the seams. If the volume of the box is $210\,\text{cm}^3$, what is the size of the square that is cut from each corner?

34. *Architecture*   The bending moment of a beam is given by $M(d) = 0.1d^4 - 2.2d^3 + 15.2d^2 - 32d$, where d is the distance in meters from one end. Find the values of d, where the bending moment is zero. (Hint: First multiply by 10 to eliminate the decimals.)

In Your Words

35. Some people think that the wording of Exercise 28 is not clear. Rewrite the exercise so that you think it is easier to understand. Hand it to a classmate for his or her comments. Did your classmate really think your wording was easier to understand than the wording in the textbook?

36. Without looking in the textbook or at your notes, write the hints for finding roots of polynomials.

≡ 18.4
IRRATIONAL ROOTS

In Section 18.3, we concentrated on rational roots of polynomials. In this section, we expand our search for roots to include the remaining real numbers—the irrational numbers.

We have no difficulty in locating irrational roots of quadratic equations, because we can use the quadratic formula. Difficulties arise with polynomials of degrees larger than two.

In Section 18.3, we learned some rules that apply to all real numbers, not just rational numbers. Descartes' rule of signs applies to all real numbers. The upper- and lower-bound tests also apply to all real numbers. Suppose that we have a polynomial of degree 4 and Descartes' rule of signs tells us that there are some real roots. We have checked all possible rational roots and none of them satisfy the equation. The real roots must be irrational numbers. How can we find out what they are?

Locator Theorem

Because we are working with polynomials, the function is continuous. There is a theorem, called the **locator theorem**, that we can use.

Locator Theorem

If $P(x)$ is a real polynomial and a and b are two real numbers such that $P(a)$ and $P(b)$ have different signs, then there is at least one real root between a and b. Thus, there is a real number c between a and b, where $P(c) = 0$.

EXAMPLE 18.22

Locate the roots of $P(x) = x^3 + x^2 - 7x + 3$.

Solution   Descartes' rule of signs tells us there are 0 or 2 positive real roots and one negative real root. The only possible rational roots are ± 1 and ± 3. By using synthetic division, we see that none of these are roots, but 3 is an upper bound.

A table of values for $P(x)$ will let us use the locator theorem. Since 3 is an upper bound we do not have to check values of x larger than 3.

x	−5	−4	−3	−2	−1	0	1	2	3
$P(x)$	−62	−17	6	13	10	3	−2	1	18

According to the locator theorem, there is a root between −4 and −3, because $P(-4) = -17 < 0$ and $P(-3) = 6 > 0$. There is also a root between 0 and 1 and one between 1 and 2. Notice that while −3 was the smallest possible rational root, there is an irrational root that is smaller.

Linear Interpolation

Once the locator theorem has given us an idea where to look for roots, we need a method to help get approximations of their values. We will use a process called **linear interpolation**. There are other methods that are more efficient, but they all use calculus. This method is easy to remember, but, as you will see, using a calculator or a computer is a definite advantage.

We have three roots. We will demonstrate linear interpolation for the root between −4 and −3. We will approximate the graph of $P(x)$ by drawing the line joining $(-4, -17)$ and $(-3, 6)$, as shown in Figure 18.3. The x-intercept of this line approximates the root of $P(x)$. The slope of this line is 23, $\left(\frac{6+17}{-3+4}\right)$, and its equation is

$$y + 17 = 23(x + 4)$$

$$\text{or} \qquad x = -4 + \frac{y + 17}{23}$$

The x-intercept occurs where $y = 0$ or $x = -4 + \frac{17}{23} \approx -3.261$. $P(-3.261) \approx 1.7806$. Thus, by the locator theorem, the actual root is between −4 and −3.261. We can repeat this process on these two values to find a closer approximation.

The process can be summarized as follows. Suppose we found that a root of $P(x)$ lies between a and b. The line joining $(a, P(a))$ and $(b, P(b))$, has the equation

$$y - P(a) = m(x - a)$$

$$\text{where} \qquad m = \frac{P(b) - P(a)}{b - a}.$$

The x-intercept of this line provides an approximation of the root. This intercept occurs where $y = 0$ in the previous equation or where

$$x = a - \frac{P(a)}{m}$$

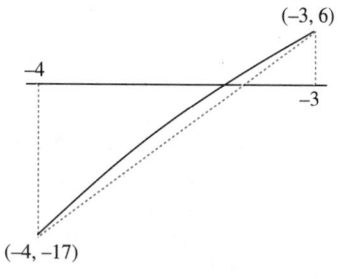

(−3, 6)

−4

−3

(−4, −17)

FIGURE 18.3

We will call this value c. Repeating the process will give closer approximations to the roots. You can continue the process until one of two things happens: (1) two consecutive values of c have the same value to the nearest tenth, hundredth, or whatever you select; or (2) the absolute value of $f(c)$ is below a certain value. Remember, since each value of c is closer to the actual root, the values of $f(c)$ are getting closer to zero.

The following table shows the approximate values that we found for the root between $x = -4$ and $x = -3$ for $P(x) = x^3 + x^2 - 7x + 3$, when we continued until $|P(x)| < 0.01$. Notice that in each step b assumes the value of c from the previous step.

a	$P(a)$	b	$P(b)$	$c = a - \dfrac{P(a)}{m}$	$P(c)$
-4	-17	-3.000	6.000	-3.2609	1.7806
-4	-17	-3.261	1.786	-3.3311	0.4508
-4	-17	-3.331	0.451	-3.3484	0.1090
-4	-17	-3.348	0.109	-3.3526	0.0261
-4	-17	-3.353	0.026	-3.3535	0.0062

Application

EXAMPLE 18.23

The stability of a molecule can be determined by solving its characteristic polynomial, $P(x)$. The roots of the equation $P(x) = 0$ determine the molecule's pi electrons. The characteristic polynomial of naphthalene $C_{10}H_8$ is

$$P(x) = x^{10} - 11x^8 + 41x^6 - 65x^4 + 43x^2 - 9$$

Determine the energies of naphthalene's 10 pi electrons.

Solution Descartes' rule of signs tell us that there are 1, 3, or 5 positive real roots and 1, 3, or 5 negative real roots. In fact, P is an even function and so it is symmetric about the y-axis. Thus, for each positive root, its additive inverse is also a root.

The only possible rational roots are ± 1, ± 3, and ± 9. Using synthetic division, we would determine that $x = 1$, and so $x = -1$, are roots. The resulting depressed polynomial is

$$P_1(x) = x^8 - 10x^6 + 31x^4 - 34x^2 + 9$$

A table of values of $P_1(x)$ will let us use the locator theorem. Because P_1 is even, we only need to check positive values of x.

x	0	0.5	1	1.5	2.0	2.5
$P_1(x)$	9	2.3	-3	1.2	-1.5	91.9

According to the locator theorem, there is a root of P_1 (and so also a root of P) between 0.5 and 1, a root between 1 and 1.5, a root between 1.5 and 2, and a root between 2 and 2.5.

				EXAMPLE 18.23 (Cont.)	

EXAMPLE 18.23 (Cont.)

Using linear interpolation to approximate the root between $x = 0.5$ and $x = 1$, we obtain the following table.

a	$P_1(a)$	b	$P_1(b)$	$c = a - \dfrac{P_1(a)}{m}$	$P_1(c)$
0.5	2.2852	1.000000	−3.0000	0.716186	−1.5638
0.5	2.2852	0.716186	−1.5638	0.628351	−0.1827
0.5	2.2852	0.628351	−0.1827	0.618847	−0.0145
0.5	2.2852	0.618847	−0.0145	0.618096	−0.0011
0.5	2.2852	0.618096	−0.0011	0.618039	−0.0001

One of the roots is $x \approx 0.618039$, so $x \approx -0.618039$ is another root. Similarly, we can determine that the other six roots are approximately ± 1.303776, ± 1.618034, and ± 2.302776. ■

Notice that we could have obtained similar results to Example 18.23 if we had used a graphing calculator or a computer graphing program with its zoom and trace features.

Exercise Set 18.4

In Exercises 1–6, approximate the specified irrational root to the nearest 0.01.

1. The positive root of $x^3 + 5x - 3 = 0$

2. The largest root of $x^3 - 3x + 1 = 0$

3. The root of $x^4 + 2x^3 - 5x^2 + 1 = 0$, which is between 0 and 1

4. The root of $x^4 + 2x^3 - 5x^2 + 1 = 0$, which is between 1 and 2

5. The root of $x^3 + x^2 - 7x + 3 = 0$, which is between 0 and 1

6. The root of $x^3 + x^2 - 7x + 3 = 0$, which is between 1 and 2

In Exercises 7–16, find to two decimal places all rational and irrational roots of the given polynomial equation.

7. $x^4 - x - 2 = 0$
8. $x^5 - 2x^2 + 4 = 0$
9. $x^4 + x^3 - 2x^2 - 7x - 5 = 0$
10. $x^4 - 2x^3 - 3x + 4 = 0$
11. $2x^4 + 3x^3 - x^2 - 2x - 2 = 0$
12. $3x^4 + 3x^3 + x^2 + 4x + 3 = 0$
13. $2x^5 - 5x^3 + 2x^2 + 4x - 1 = 0$
14. $x^5 + 3x^4 - 5x^3 - 2x^2 + x + 2 = 0$
15. $8x^4 + 6x^3 - 15x^2 - 12x - 2 = 0$
16. $9x^4 + 15x^3 - 20x^2 - 20x + 16 = 0$

Solve Exercises 17–23.

17. *Petroleum engineering* The pressure drop P in pounds per square inch (psi) in a particular oil reservoir is a function of the number of years t that the reservoir has been in operation. The pressure drop is approximated by the equation

$$P = 150t - 20t^2 + t^3$$

How many years will it take for the pressure to drop 400 psi?

18. *Industrial design*   A cylindrical storage tank 12 ft high contains 674 ft³. Determine the thickness of the tank, if the outside radius is 4.5 ft and the sides, top, and bottom have the same thickness.

19. *Nuclear technology*   A cylindrical container for storing radioactive waste is to be constructed from lead. The sides, top, and bottom of the cylinder of the container must be at least 15.5 cm thick. **(a)** If the volume of the outside cylinder is $1\,000\,000\pi\,cm^3$, and the height of the inside cylinder is twice the radius of the inside cylinder (as shown in Figure 18.4), determine the radius of the inside cylinder. **(b)** What is the volume of the inside cylinder?

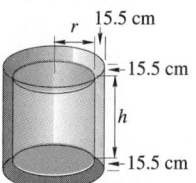

FIGURE 18.4

 **In Your Words**

24. Explain how the locator theorem helps you determine where to look for roots.

20. *Mechanical engineering*   The characteristic polynomial for a certain material is

$$S^3 - 6S^2 - 78S + 108 = 0$$

Find the approximate value(s) of the stress that lies between 1 and 2 psi.

21. *Chemistry*   The reference polynomial for naphthalene $C_{10}H_8$ is given by

$$R(x) = x^{10} - 11x^8 + 41x^6 - 61x^4 + 31x^2 - 3$$

Solve this reference polynomial.

 22.   Write a program that uses linear interpolation to approximate a root of a polynomial.

23. *Medical technology*   A pharmaceutical company is growing an organism to be used in a vaccine. The number of bacteria in millions at any given time t, in hours, is given by $Q(t) = -0.009t^5 + t^3 + 5.5$. Determine to the nearest tenth of a minute the value of t for which $Q(t) = 32.6$.

25. Describe how to use linear approximation to determine a root.

☰ 18.5
RATIONAL FUNCTIONS

When you multiply two polynomials, you get another polynomial. In this section, we will study what happens when we divide two polynomials.

Rational Function

If $P(x)$ and $Q(x)$ are two polynomials and if $Q(x) \neq 0$, then a function of the form

$$R(x) = \frac{P(x)}{Q(x)}$$

is called a **rational function**.

EXAMPLE 18.24

Each of the following are rational functions. Notice the restrictions placed on x so that the denominator is not zero.

(a) $R(x) = \dfrac{2x^2+3x-5}{x+2}, \quad x \neq -2$

(b) $f(x) = \dfrac{4x^3-7x+3}{3x-4}, \quad x \neq \dfrac{4}{3}$

(c) $g(x) = \dfrac{4}{x+3}, \quad x \neq -3$

(d) $h(x) = \dfrac{9x^2+4x-5}{x^2+6x+5}, \quad x \neq -1,\ x \neq -5$

Let's first look at solving these functions and then at graphing them. In order to solve a rational function, we need to remember that any fraction $\dfrac{a}{b} = 0$, if and only if $a = 0$ and $b \neq 0$. Using this rule, we will determine when each of the functions in Example 18.24 is zero.

EXAMPLE 18.25

Solve: $\dfrac{2x^2+3x-5}{x+2} = 0$.

Solution This equation will be true only when $2x^2+3x-5 = 0$ and $x+2 \neq 0$. The left-hand equation factors to $2x^2+3x-5 = (2x+5)(x-1) = 0$, and so $x = 1$, or $x = -\frac{5}{2}$. Neither of these make $x+2 = 0$, so 1 and $-\frac{5}{2}$ are both roots of the original equation.

EXAMPLE 18.26

Solve: $\dfrac{4x^3-7x+3}{3x-4} = 0$.

Solution This equation will be true only when $4x^3-7x+3 = 0$ and $3x-4 \neq 0$. The left-hand equation has a root at 1, so $4x^3-7x+3 = (x-1)(4x^2+4x-3) = (x-1)(2x-1)(2x+3)$; thus $x = 1, \frac{1}{2}, -\frac{3}{2}$. None of these make $3x-4 = 0$, so all three are roots of the original equation.

EXAMPLE 18.27

Solve: $\dfrac{4}{x+3} = 0$.

Solution Since the numerator, 4, is never zero, this equation has no solution. That is, there are no numbers that make it true.

EXAMPLE 18.28

Solve: $\dfrac{9x^2+4x-5}{x^2+6x+5}=0$.

Solution   This equation will be true when $9x^2+4x-5=0$, if $x^2+6x+5 \neq 0$ for the same values of x. The numerator factors to $9x^2+4x-5=(9x-5)(x+1)=0$, so the numerator is zero when $x=-1$ or $x=\frac{5}{9}$. The denominator is zero when $x^2+6x+5=(x+5)(x+1)=0$, or $x=-5$ or $x=-1$. Notice that -1 makes both the numerator and the denominator zero. Since the denominator cannot be zero, -1 is not a solution. The only solution of the original equation is $\frac{5}{9}$.

Graphing Rational Functions

Some interesting things occur when we graph rational functions. Let's graph some of the examples we just worked.

EXAMPLE 18.29

Sketch the graph of $g(x)=\dfrac{4}{x+3}$.

Solution   We already know that this is defined everywhere except when $x=-3$. We also know that it has no real roots, so it does not cross the x-axis. Some of the values for $g(x)$ are shown in this table.

x	-11	-10	-9	-8	-7	-6	-5	-4
$g(x)$	-0.5	-0.57	$-\frac{2}{3}$	-0.8	-1	$-\frac{4}{3}$	-2	-4

x	-2	-1	0	1	2	3	4	5
$g(x)$	4	2	$\frac{4}{3}$	1	0.8	$\frac{2}{3}$	0.57	0.5

Notice that $x=-3$ is not in the table. Let's see what happens when we get values of x closer to -3, as shown in this table.

x	-4	-3.8	-3.6	-3.4	-3.2	-3.1
$g(x)$	-4	-5	$-6\frac{2}{3}$	-10	-20	-40

x	-2.9	-2.8	-2.6	-2.4	-2.2	-2.0
$g(x)$	40	20	10	$6\frac{2}{3}$	5	4

From this second table, you can see that as x gets closer to -3 from the left, $x+3$ gets closer to zero and $g(x)$ gets smaller. As x gets closer to -3 from the right, $x+3$ again gets closer to zero, but this time $g(x)$ gets larger. Because $g(x)$ is not defined at $x=-3$, the graph never crosses the line $x=-3$. Thus $x=-3$ is a **vertical asymptote** for this graph.

In the same way, the x-axis is a horizontal asymptote for the graph. The graph of $g(x)$ is shown in Figure 18.5.

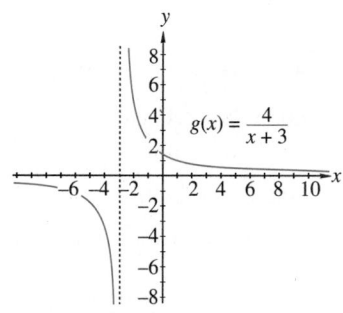

$g(x)=\dfrac{4}{x+3}$

FIGURE 18.5

> **Vertical Asymptotes**
>
> The graph of any rational function $R(x) = \dfrac{P(x)}{Q(x)}$, $Q(x) \neq 0$, where P and Q have no common factors, will have vertical asymptotes at the values of x where $Q(x) = 0$.

In Example 18.29, the denominator was 0 when $x = -3$ and the line $x = -3$ was a vertical asymptote.

EXAMPLE 18.30

Determine the vertical asymptotes, if any, for the graph of $R(x) = \dfrac{2x^2 + 3x - 5}{x + 2}$.

Solution   The denominator is zero when $x = -2$ and this is the vertical asymptote, as shown in Figure 18.6.

EXAMPLE 18.31

Sketch the graph of $h(x) = \dfrac{9x^2 + 4x - 5}{x^2 + 6x + 5}$.

Solution   The denominator is zero when $x = -5$ or $x = -1$. These are not both vertical asymptotes. Both the numerator and denominator have a factor of $x + 1$, so, $\dfrac{9x^2 + 4x - 5}{x^2 + 6x + 5} = \dfrac{9x - 5}{x + 5}$, except when $x = -1$. The graph of these functions will be the same, except one is defined at $x = -1$ and the other is not. Thus, the vertical asymptote is $x = -5$. The following is a table of values and the graph is given in Figure 18.7.

x	-10	-9	-8	-7	-6	-4	-3	-2	-1
$h(x)$	19	21.5	$25\frac{2}{3}$	34	59	-41	-16	$-7\frac{2}{3}$	$\boxed{-3.5}$
x	0	1	2	3	4	5	10		
$h(x)$	-1	$\frac{2}{3}$	1.86	2.75	3.44	4	$5\frac{2}{3}$		

The value of $h(x) = -3.5$ is boxed because the function is not defined at $x = -1$. The value of -3.5 is the value of the simplified version of the function, $h(x) = \dfrac{9x - 5}{x + 5}$, when $x \neq -1$. In Figure 18.7, an open circle at the point $(-1, -3.5)$ emphasizes the fact that $h(x)$ is not defined at $x = -1$. This is called a point of discontinuity.

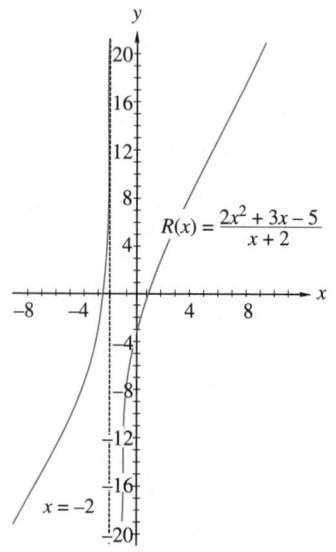

$R(x) = \dfrac{2x^2 + 3x - 5}{x + 2}$

$x = -2$

FIGURE 18.6

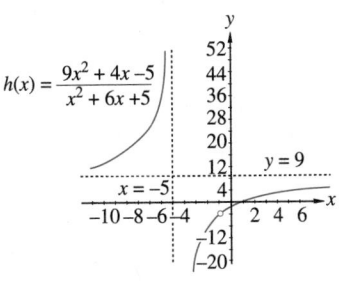

$h(x) = \dfrac{9x^2 + 4x - 5}{x^2 + 6x + 5}$

FIGURE 18.7

It appears as if the function h shown in Figure 18.7 has another asymptote, a horizontal one; and, in fact, it does. This asymptote is $y = 9$.

Horizontal Asymptote

- If two polynomials, $P(x)$ and $Q(x)$, $Q(x) \neq 0$, are of the same degree, then the rational function $R(x) = \dfrac{P(x)}{Q(x)}$ has a **horizontal asymptote** at $y = \dfrac{a}{b}$, where a is the leading coefficient of $P(x)$ and b is the leading coefficient of $Q(x)$.
- If the degree of P is less than the degree of Q, then $y = 0$ is a horizontal asymptote.

In Example 18.31, $P(x) = 9x^2 + 4x - 5$ with a leading coefficient of 9 and $Q(x) = x^2 + 6x + 5$ with a leading coefficient of 1. So, $y = \frac{9}{1} = 9$ is the horizontal asymptote.

EXAMPLE 18.32

Determine the asymptotes of $f(x) = \dfrac{8x^2 - 22x - 6}{3x^2 - 2x - 1}$.

Solution The numerator and denominator have the same degree, 2, so there is a horizontal asymptote. The leading coefficient of the numerator, $8x^2 - 22x - 6$, is 8; the leading coefficient of the denominator, $3x^2 - 2x - 1$, is 3. The horizontal asymptote is $y = \frac{8}{3}$.

The denominator factors into $(3x + 1)(x - 1)$, so it is zero when $x = -\frac{1}{3}$ or $x = 1$. Since neither of these makes the numerator zero, they are both vertical asymptotes. The graph of f is shown in Figure 18.8 with the asymptotes indicated by dashed lines.

EXAMPLE 18.33

Determine the asymptotes and sketch the graph of $g(x) = \dfrac{5x^2 + 10}{x^2 + 7x + 10}$.

Solution The numerator and denominator are both degree 2, so there is a horizontal asymptote at $y = 5$.

The denominator factors as $x^2 + 7x + 10 = (x + 5)(x + 2)$, so it is zero when $x = -5$ and $x = -2$. Neither of these makes the numerator zero, so they are vertical asymptotes. The graph of g is shown in Figure 18.9 with the asymptotes indicated by dashed lines. Notice that the graph actually crosses the asymptote $y = 5$ at $x \approx -1.143$.

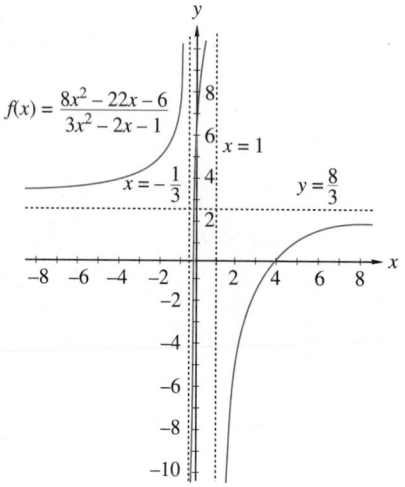

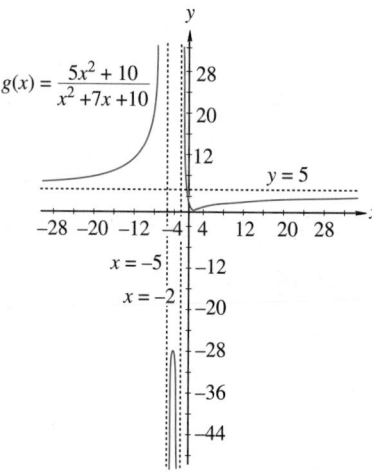

FIGURE 18.8   **FIGURE 18.9**

Solving Rational Functions

Now let's return to solving rational functions. We want to solve an equation of the form $R(x) = T(x)$, where R and T are both rational functions. We solve this in the same way that we work with fractions. First we write $R(x) - T(x)$ as a fraction and then solve $R(x) - T(x) = 0$.

EXAMPLE 18.34

Solve $\dfrac{6}{x+1} = \dfrac{4}{x+2}$.

Solution   First, we get both rational functions on the same side of the equal sign.

$$\frac{6}{x+1} = \frac{4}{x+2}$$

or $\qquad \dfrac{6}{x+1} - \dfrac{4}{x+2} = 0$

The lowest common denominator (LCD) of these two rational functions is $(x+1)(x+2)$. Writing each of these rational functions with this common denominator produces

$$\frac{6}{x+1} = \frac{6(x+2)}{(x+1)(x+2)}$$

and $\qquad \dfrac{4}{x+2} = \dfrac{4(x+1)}{(x+1)(x+2)}$

So,

$$\frac{6}{x+1} - \frac{4}{x+2} = \frac{6(x+2)}{(x+1)(x+2)} - \frac{4(x+1)}{(x+1)(x+2)}$$

EXAMPLE 18.34 (Cont.)

$$= \frac{(6x+12)-(4x+4)}{(x+1)(x+2)}$$

$$= \frac{2x+8}{(x+1)(x+2)} = 0$$

This is zero when $2x+8 = 0$ or $x = -4$. Since $x = -4$ does not make the denominator zero, the solution is $x = -4$.

Exercise Set 18.5

In Exercises 1–10, determine the horizontal and vertical asymptotes, if any, of the function. Sketch the graph.

1. $f(x) = \dfrac{3}{x+2}$

2. $g(x) = \dfrac{4}{x^2-9}$

3. $g(x) = \dfrac{2x}{x+4}$

4. $k(x) = \dfrac{5x}{x^2+x-6}$

5. $j(x) = \dfrac{3x^2}{x^2+6x+8}$

6. $f(x) = \dfrac{x^2-4}{x^2-16}$

7. $g(x) = \dfrac{x(x+1)}{(x+1)(x+2)}$

8. $h(x) = \dfrac{x^2+1}{2x^2}$

9. $j(x) = \dfrac{4x^2-12x-27}{x^2-6x+8}$

10. $k(x) = \dfrac{4x^2-9}{2x^2+7x+3}$

In Exercises 11–20, solve the given equation.

11. $\dfrac{(x+2)(x-3)}{(x+5)(x+4)} = 0$

12. $\dfrac{x^2-2x-1}{x(x+1)} = 0$

13. $\dfrac{x^2-1}{x^2+2x+1} = 0$

14. $\dfrac{3x-4}{6x^2-5x-4} = 0$

15. $\dfrac{4}{x-3} = \dfrac{2}{x+1}$

16. $\dfrac{5}{x+7} = \dfrac{2}{x-8}$

17. $\dfrac{5}{2x} - \dfrac{3}{4x} = \dfrac{2}{3}$

18. $\dfrac{2x-3}{x+5} = 2$

19. $\dfrac{3}{x-2} + \dfrac{5}{x+2} = \dfrac{20}{x^2-4}$

20. $\dfrac{4}{x+3} - \dfrac{6}{x-3} = \dfrac{10}{x^2-9}$

Solve Exercises 21–24.

21. *Medical technology*   When medicine is injected into the bloodstream, its concentration t h after injection is given by

$$C(t) = \frac{3t^2+t}{t^3+50}$$

The concentration is greatest when $3t^4+2t^3-300t-50 = 0$. Approximate this time to the nearest hundredth of an hour.

22. *Medical technology*   One of Poiseuille's laws states that the resistance, R, encountered by blood flowing through a blood vessel is given by $R(r) = C\dfrac{L}{r^4}$, where C is a positive constant determined by the

viscosity of the blood, L is the length of the blood vessel, and r is the radius of the blood vessel. If $C = 1$, $L = 100\,\text{mm}$, and $R = 6.25/\text{mm}^3$, determine r.

23. *Business*   The average cost in dollars of producing x sets of golf clubs is given by $A(x) = 73 + \dfrac{7{,}250}{x}$. What is the lowest average cost that the manufacturer can expect?

24. *Business*   It has been determined that after t weeks of training, a certain student in a word processing class can keyboard at the rate of $W(t) = 65 + \dfrac{70t^2}{t^2 + 15}$ words per minute. What is the most words per minute that we can expect this student to keyboard?

In Your Words

25. **(a)** Explain how to determine if a rational function has a horizontal asymptote.
　　(b) Explain how to determine the horizontal asymptote of a rational function.

26. **(a)** Explain how to determine if a rational function has a vertical asymptote.
　　(b) Explain how to determine the vertical asymptotes of a rational function.

CHAPTER 18 REVIEW

Important Terms and Concepts

Complex roots theorem
Constant polynomial
Descartes' rule of signs
Factor of a polynomial
Factor theorem
Fundamental theorem of algebra
Horizontal asymptote
Imaginary root
Irrational root
Linear factorization theorem
Linear interpolation
Lower bound

Multiple roots
Point of discontinuity
Polynomial function
Rational function
Rational root test
Remainder theorem
Root of an equation
Synthetic division
Upper bound
Vertical asymptote
Zero of a function
Zero polynomial

Review Exercises

In Exercises 1–4, find the value of $P(x)$ for the given value of x and give the degree of $P(x)$.

1. $P(x) = 9x^2 - 4x + 2$, $x = 2$

2. $P(x) = 2x^3 - 4x^2 + 7$, $x = 3$

3. $P(x) = 4x^3 - 5x^2 + 7x + 3$, $x = -1$

4. $P(x) = 5x^2 + 6x - 3$, $x = -3$

In Exercises 5–8, find the quotient and remainder when $P(x)$ is divided by $x - a$.

5. $P(x) = 4x^2 - 6x - 3$; $x + 3$

6. $P(x) = 6x^3 - 2x + 1$; $x - 1$

7. $P(x) = 9x^4 - 5x^3 + 2x^2 - 7x + 1$; $x - 1$

8. $P(x) = 7x^3 - 3x^2 - 12$; $x + 2$

In Exercises 9–16, find all the roots of the given equation. Approximate irrational roots to 3 decimal places.

9. $x^3 - 3x^2 + 3x - 1 = 0$

10. $x^4 - 5x^3 + 9x^2 - 7x + 2 = 0$

11. $x^4 - 6x^2 - 8x - 3 = 0$

12. $x^4 + 10x^3 + 25x^2 - 16 = 0$

13. $x^4 + x^3 - 6x^2 + x + 1 = 0$

14. $2x^5 - 3x^4 - 18x^3 + 75x^2 - 104x + 48 = 0$

15. $x^6 + 2x^5 + 3x^4 + 4x^3 + 3x^2 + 2x + 1 = 0$

16. $x^6 - 2x^5 - 4x^4 + 6x^3 - x^2 + 8x + 4 = 0$

In Exercises 17–20, determine the horizontal and vertical asymptotes, if any, and solve the equation.

17. $\dfrac{5x}{x^2 + 3x - 4} = 0$

18. $\dfrac{7x^3}{x^3 + 1} = 0$

19. $\dfrac{4}{x - 2} = \dfrac{5}{x + 2}$

20. $\dfrac{3}{x - 1} + \dfrac{2}{x + 1} = \dfrac{4}{x^2 - 1}$

Solve Exercises 21 and 22.

21. *Medical technology* Psychologists have developed mathematical models, called **learning curves**, that are used to predict the performance as a function of the number of trials n for a certain task. One learning curve is

$$P(n) = \frac{0.5 + 0.8(n - 1)}{1.5 + 0.8(n - 1)}, \quad n > 0$$

where P is the percentage, expressed as a decimal, of the correct responses after n trials. According to this model, what is the limiting percentage of correct responses as n increases; that is, what is the horizontal asymptote?

22. *Medical technology* An experiment on learning retention is conducted by the training people at a certain company. Each trainee is given 1 day to memorize a list of 30 rules. At the end of the day, the list is turned in. Each day after that, each trainee is asked to turn in a written list of as many rules as he or she can remember. It is found that

$$N(t) = \frac{5t^2 + 6t + 15}{t^2}, \quad t > 0$$

provides a close approximation of the average number of rules, $N(t)$, remembered after t days. Determine the limiting number of rules that the trainees remembered; that is, find the horizontal asymptote of N.

≡ CHAPTER 18 TEST

1. What is the degree of $P(x) = 7x^5 - 2x^4 - 5$?

2. If $P(x) = 4x^3 - 5x^2 + 2x - 8$, what is $P(-2)$?

3. Find the quotient and remainder when $P(x) = x^4 - 8x^3 + 3x^2 + 44x + 75$ is divided by $x - 6$.

4. List the possible rational roots of $P(x) = 3x^4 - 7x^2 + 2x - 2$.

5. Find all the roots of $3x^3 + 5x^2 - 4x - 4 = 0$.

6. Find all the roots of $x^6 + 2x^5 - 4x^4 - 10x^3 - 41x^2 - 72x - 36 = 0$.

7. Determine the horizontal and vertical asymptotes, if any, and solve the equation

$$\frac{3x^2 - 2x - 1}{x^2 + 5x + 6} = 0$$

8. A rectangular piece of cardboard 12.0 in. by 22 in. is formed into a box by cutting a square from each corner and folding up the sides. If the volume of the box is 175 in.^{3}, what is the length of each side of the squares that were cut out?

9. The pressure drop in lb/in.^{2} in a certain oil reservoir is given by $P(t) = 250t - 22t^2 + t^3$ where t is the time in years. After how much time does the pressure drop by exactly 500 lb/in.^{2}?

CHAPTER
19

Sequences, Series, and the Binomial Formula

How much insulation is needed to reduce fuel consumption by 20%? The solution to this question requires knowledge of geometric sequences, a topic we will study in Section 19.2.

Sequences have played an important role in advanced mathematics. Many natural and physical patterns can be described by a sequence of numbers. In this chapter, we will study two specific kinds of sequences—arithmetic and geometric.

724

We will also study the sum of a sequence, which is called a series. Both sequences and series are studied in more detail in calculus.

≡ 19.1
SEQUENCES

A **sequence** is a set of numbers arranged in some order. Each number in the sequence is labeled with a variable, such as a. The variable is indexed with a natural number that tells its position in the sequence. The numbers a_1, a_2, a_3, and so on are the **terms** of the sequence. So, the first term in the sequence is a_1, the second term a_2, the third term a_3, and so on.

EXAMPLE 19.1

The sequence 3, 7, 11, 15, 19, 23 has $a_1 = 3$, $a_2 = 7$, $a_3 = 11$, $a_4 = 15$, $a_5 = 19$, and $a_6 = 23$. This sequence has six terms. ∎

Many sequences follow some sort of pattern. The pattern is usually described by the nth term of the sequence. This term, a_n, is called the **general term** of the sequence. A **finite sequence** has a specific number of terms and so it has a last term. An **infinite sequence** does not have a last term. The notation $\{a_n\}$ is often used to present the nth term of a sequence. The { } indicate that it is a sequence.

EXAMPLE 19.2

Find the first six terms of the sequence $a_n = 4n + 1$.
Solution
$$a_1 = 4(1) + 1 = 5$$
$$a_2 = 4(2) + 1 = 9$$
$$a_3 = 4(3) + 1 = 13$$
$$a_4 = 4(4) + 1 = 17$$
$$a_5 = 4(5) + 1 = 21$$
$$a_6 = 4(6) + 1 = 25$$
The first six terms of this sequence are 5, 9, 13, 17, 21, and 25. ∎

EXAMPLE 19.3

Find the first five terms of the sequence $\{n^2 - 3\}$.
Solution
$$a_1 = 1^2 - 3 = 1 - 3 = -2$$
$$a_2 = 2^2 - 3 = 4 - 3 = 1$$
$$a_3 = 3^2 - 3 = 9 - 3 = 6$$
$$a_4 = 4^2 - 3 = 16 - 3 = 13$$
$$a_5 = 5^2 - 3 = 25 - 3 = 22$$
The first five terms of this sequence are -2, 1, 6, 13, and 22. ∎

EXAMPLE 19.4

Find the first seven terms of the sequence $\left\{\dfrac{n}{n+2}\right\}$.

Solution $a_1 = \dfrac{1}{1+2} = \dfrac{1}{3}$

$a_2 = \dfrac{2}{2+2} = \dfrac{2}{4} = \dfrac{1}{2}$

$a_3 = \dfrac{3}{3+2} = \dfrac{3}{5}$

$a_4 = \dfrac{4}{4+2} = \dfrac{4}{6} = \dfrac{2}{3}$

$a_5 = \dfrac{5}{5+2} = \dfrac{5}{7}$

$a_6 = \dfrac{6}{6+2} = \dfrac{6}{8} = \dfrac{3}{4}$

$a_7 = \dfrac{7}{7+2} = \dfrac{7}{9}$

The first seven terms of this sequence are $\frac{1}{3}, \frac{1}{2}, \frac{3}{5}, \frac{2}{3}, \frac{5}{7}, \frac{3}{4}$, and $\frac{7}{9}$.

Recursion Formula

A **recursion formula** defines a sequence in terms of one or more previous terms. A sequence that is specified by giving the first term, or the first few terms, and a recursion formula is said to be **defined recursively**.

EXAMPLE 19.5

Find the first five terms of the sequence defined recursively by $a_1 = 2$ and $a_n = na_{n-1}$.

Solution   We are told that $a_1 = 2$. By the recursion formula

$$a_2 = 2(a_1) = 2(2) = 4$$
$$a_3 = 3(a_2) = 3(4) = 12$$
$$a_4 = 4(a_3) = 4(12) = 48$$
$$a_5 = 5(a_4) = 5(48) = 240$$

The first five terms are 2, 4, 12, 48, and 240.

EXAMPLE 19.6

Give the first six terms of the sequence defined by $a_1 = 1$, $a_n = -2a_{n-1}$.

Solution $a_1 = 1$

$$a_2 = -2(a_1) = -2(1) = -2$$
$$a_3 = -2(a_2) = -2(-2) = 4$$
$$a_4 = -2(a_3) = -2(4) = -8$$
$$a_5 = -2(a_4) = -2(-8) = 16$$
$$a_6 = -2(a_5) = -2(16) = -32$$

The first six terms are 1, −2, 4, −8, 16, and −32.

Application

EXAMPLE 19.7

A contractor is preparing a bid for constructing an office building. The first floor will cost $275,000. Each floor above the first will cost $15,000 more than the floor below it. **(a)** How much will the fifth floor cost? **(b)** What is the total cost for the first five floors?

Solutions

(a) Notice that this is a recursive sequence with $a_1 = 275,000, a_n = a_{n-1} + 15,000$. The first five terms of the sequence are

$$a_1 = 275,000$$
$$a_2 = a_1 + 15,000 = 275,000 + 15,000 = 290,000$$
$$a_3 = a_2 + 15,000 = 290,000 + 15,000 = 305,000$$
$$a_4 = a_3 + 15,000 = 305,000 + 15,000 = 320,000$$
$$a_5 = a_4 + 15,000 = 320,000 + 15,000 = 335,000$$

The cost of the fifth floor is $335,000.

(b) The total cost for the first 5 floors is

$$a_1 + a_2 + a_3 + a_4 + a_5 = 275,000 + 290,000 + 305,000$$
$$+ 320,000 + 335,000$$
$$= 1,525,000$$

The cost for the first five floors in $1,525,000.

Exercise Set 19.1

In Exercises 1–12, find the first six terms with the given specified general term.

1. $a_n = \dfrac{1}{n+1}$

2. $b_n = \dfrac{(-1)^n}{n}$

3. $a_n = (n+1)^2$

4. $a_n = \dfrac{1}{n(n+1)}$

5. $a_n = (-1)^n n$

6. $a_n = \dfrac{1+(-1)^n}{1+4n}$

7. $\left\{ \dfrac{2n-1}{2n+1} \right\}$

8. $\left\{ \left(\dfrac{-1}{2} \right)^n \right\}$

9. $\left\{ \left(\dfrac{2}{3} \right)^{n-1} \right\}$

10. $\{ n \cos n\pi \}$

11. $\left\{ \left(\dfrac{n-1}{n+1} \right)^2 \right\}$

12. $\left\{ \dfrac{n^2-1}{n^2+1} \right\}$

In Exercises 13–24, find the first six terms of the recursively defined sequence.

13. $a_1 = 1, a_n = n a_{n-1}$

14. $a_1 = 3, a_n = a_{n-1} + n$

15. $a_1 = 5, a_n = a_{n-1} + 3$

16. $a_1 = 2, a_n = (a_{n-1})^n$

17. $a_1 = 2, a_n = (a_{n-1})^{n-1}$

18. $a_1 = 1, a_n = \left(\dfrac{-1}{n} \right) a_{n-1}$

19. $a_1 = 1, a_n = n^{a_{n-1}}$

20. $a_1 = \frac{1}{2}, a_n = (a_{n-1})^{-n}$

21. $a_1 = 1, a_2 = \frac{1}{2}, a_n = (a_{n-1})(a_{n-2})$

22. $a_1 = 1, a_2 = 1, a_n = a_{n-2} + a_{n-1}$

23. $a_1 = 0, a_2 = 2, a_n = a_{n-1} - a_{n-2}$

24. $a_1 = 1, a_2 = 2, a_n = (a_{n-1})(a_{n-2})$

Solve Exercises 25–28.

25. *Environmental Science*   An ocean beach is eroding at the rate of 3 in. per year. If the beach is currently 25 ft wide, how wide will the beach be in 25 y? (Hint: $a_1 = 25$ ft -3 in.; find a_{25}.)

26. *Construction*   A contractor is preparing a bid for constructing an office building. The foundation and basement will cost $750,000. The first floor will cost $320,000. The second floor will cost $240,000. Each floor above the second will cost $12,000 more than the floor below it. **(a)** How much will the 10th floor cost? **(b)** How much will the 20th floor cost?

27. *Business*   You are offered a job with a starting salary of $26,000 per year with a guaranteed raise of $1,100 per year. What can you expect as an annual salary in your **(a)** 5th year? **(b)** 10th year?

28. *Architecture*   The first row of an auditorium has 60 seats. Each row after the first has 4 more seats than the row in front of it. How many seats are in the 10th row?

In Your Words

29. What is a recursion formula? Support your explanation by giving an example different from those in the text.

30. What is a sequence? How does a finite sequence differ from an infinite sequence?

≡ 19.2
ARITHMETIC AND GEOMETRIC SEQUENCES

In Section 19.1, we began our study of sequences. In this section, we will learn about two special sequences, the arithmetic and geometric sequences.

Arithmetic Sequence ▬▬▬▬▬

The first special sequence we will study is the arithmetic sequence. Its definition depends on a recursion formula.

Arithmetic Sequence

An **arithmetic sequence**, or **arithmetic progression**, is a sequence where each term is obtained from the preceding term by adding a fixed number called the **common difference**. If the common difference is d, then an arithmetic sequence follows the recursion formula

$$a_n = a_{n-1} + d$$

The terms of an arithmetic sequence follow the pattern a_1, $a_2 = a_1 + d$, $a_3 = a_2 + d = a_1 + 2d$, $a_4 = a_3 + d = a_1 + 3d$, ..., $a_n = a_{n-1} + d = a_1 + (n-1)d$. We have developed a formula for finding the nth term of an arithmetic sequence.

> **Terms of an Arithmetic Sequence**
>
> If a_1 is the first term of an arithmetic sequence, a_n the nth term and d the common difference, then
>
> $$a_n = a_1 + (n-1)d$$
>
> If a_{n-1} and a_n are consecutive terms of an arithmetic sequence, then
>
> $$d = a_n - a_{n-1}$$

EXAMPLE 19.8

Find the 15th term of $3, 7, 11, 15 \ldots$.

Solution   Since $a_1 = 3$ and $a_2 = 7$, then $d = 7 - 3 = 4$. We want the 15th term, so $n = 15$, and we have

$$\begin{aligned} a_{15} &= a_1 + (15-1)d \\ &= 3 + 14 \cdot 4 \\ &= 59 \end{aligned}$$

The 15th term of this sequence is 59.

EXAMPLE 19.9

If the 20th term is 122 and the first term is 8, what is the common difference?

Solution   Since we know the 20th term and the first term, we let $n = 20$. Then

$$\begin{aligned} a_n &= a_1 + (n-1)d \\ a_{20} &= a_1 + (20-1)d \\ 122 &= 8 + (19)d \\ 114 &= 19d \\ 6 &= d \end{aligned}$$

The common difference is 6.

EXAMPLE 19.10

If the first term is 5, the last (nth) term is -139, and $d = -6$, how many terms are there?

Solution
$$\begin{aligned} a_n &= a_1 + (n-1)d \\ -139 &= 5 + (n-1)(-6) \\ -144 &= -6n + 6 \\ -150 &= -6n \\ 25 &= n \end{aligned}$$
There are 25 terms.

Geometric Sequence

The second special sequence that we will study is the geometric sequence. Like the arithmetic sequence, it follows a recursion formula.

Geometric Sequence

A **geometric sequence**, or **geometric progression**, is a sequence where each term is obtained by multiplying the preceding term by a fixed number called the **common ratio**. If the common ratio is r, then a geometric sequence follows the recursion formula:

$$a_n = ra_{n-1}$$

The terms of a geometric sequence follow the pattern a_1, $a_2 = ra_1$, $a_3 = ra_2 = r^2 a_1$, $a_4 = ra_3 = r^3 a_1$, ..., $a_n = ra_{n-1} = r^{n-1} a_1$.

Terms of a Geometric Sequence

If a_1 is the first term of a geometric sequence, a_n the nth term, and r the common ratio, then

$$a_n = r^{n-1} a_1$$

If a_{n-1} and a_n are consecutive terms of a geometric sequence, then

$$r = \frac{a_n}{a_{n-1}}$$

EXAMPLE 19.11

Find the eighth term of the geometric sequence 4, 20, 100, ...

Solution   Since we know that this is a geometric sequence, the common ratio is

$$r = \frac{a_2}{a_1} = \frac{20}{4} = 5$$

and since $a_1 = 4$, then

$$a_8 = r^{8-1} a_1$$
$$= 5^7 \cdot 4$$
$$= 312{,}500$$

The eighth term is 312,500.

EXAMPLE 19.12

If the 12th term is $\frac{-3}{2,048}$ and the first term is -3, what is the common ratio?

Solution Here we let $n = 12$, and so

$$a_{12} = r^{11}a_1$$

$$\frac{-3}{2,048} = r^{11}(-3)$$

$$\frac{1}{2,048} = r^{11}$$

$$r = \sqrt[11]{\frac{1}{2,048}}$$

$$= \frac{1}{2}$$

The common ratio is $\frac{1}{2}$.

EXAMPLE 19.13

In a geometric sequence, if the first term is -2, the last (nth) term is $-1,062,882$, and $r = 3$, how many terms are there?

Solution

$$a_n = r^{n-1}a_1$$

$$-1,062,882 = 3^{n-1}(-2)$$

$$531,441 = 3^{n-1}$$

$$\ln 531,441 = (n-1)\ln 3$$

$$n - 1 = \frac{\ln 531,441}{\ln 3}$$

$$= 12$$

$$n = 13$$

There are 13 terms in this sequence.

Application

EXAMPLE 19.14

Each 2-in. layer of insulation like that shown in the photograph in Figure 19.1, reduces energy consumption by 4%. How many layers would be needed to reduce consumption by 20%, from 850 kW to 680 kW?

Solution This is a geometric sequence with $a_1 = 850$ and $a_n = 680$. Each 2-in. layer reduces energy consumption by 4%, so $r = 1 - 4\% = 1 - 0.04 = 0.96$. We need to determine n, the number of 2-in. layers of insulation.

Substituting the given values into the formula

$$a_n = a_1 r^{n-1}$$

we get $\quad 680 = 850(0.96)^{n-1}$

or $\quad 0.80 = (0.96)^{n-1}$

EXAMPLE 19.14 (Cont.)

Courtesy of Owens-Corning Fiberglass
FIGURE 19.1

Using logarithms, we obtain

$$\ln 0.80 = \ln \left[(0.96)^{n-1} \right]$$
$$= (n-1)\ln 0.96$$
$$\frac{\ln 0.80}{\ln 0.96} = n-1$$
$$5.47 \approx n-1$$
so, $\qquad n \approx 6.47$

This means that 14 in. of insulation (7 layers of 2-in. insulation) are needed to reduce energy consumption by 20%.

You should notice in Example 19.14, that even though the answer (6.47) would round off to 6, we would have to use 7 layers of insulation. The actual energy used after adding 7 layers of insulation is

$$a_7 = a_1 r^6$$
$$= 850(0.96)^6$$
$$\approx 665.34 \, \text{kW}$$

This represents a savings of $850 - 665.34 = 184.66 \, \text{kW}$, which is $\frac{184.66}{850} \approx 0.217$ or about 21.7%.

You may have noticed in the last two solutions that we used a technique that we had not used for several chapters—the technique of taking the logarithm of both sides of an equation.

Exercise Set 19.2

In Exercises 1–20, determine whether the given sequence is an arithmetic sequence, a geometric sequence, or neither. For the arithmetic and geometric sequences, find the common difference or ratio and the indicated term.

1. $1, 9, 17, \ldots$ (9th term)

2. $3, -5, -13, \ldots$ (7th term)

3. $8, -4, 2, \ldots$ (10th term)

4. $4, 1, \frac{1}{4}, \ldots$ (8th term)

5. $-1, 4, -16, \ldots$ (6th term)

6. $-1, 4, 9, \ldots$ (12th term)

7. $3, -2, -7, \ldots$ (8th term)

8. $1, -3, 9, \ldots$ (7th term)

9. $3, -2, 1, \ldots$ (10th term)

10. $1, \sqrt{2}, 2, \ldots$ (12th term)

11. $0.4, 0.8, 1.2, \ldots$ (9th term)

12. $4, 2, \sqrt{2}, \ldots$ (11th term)

13. $\frac{2}{3}, \frac{1}{2}, \frac{1}{3}, \ldots$ (8th term)

14. $\frac{2}{3}, \frac{1}{3}, \frac{1}{6}, \ldots$ (8th term)

15. $5, 0.5, 0.05, \ldots$ (6th term)

16. $6, -2, \frac{2}{3}, \ldots$ (7th term)

17. $1 + \sqrt{3}, 3 + \sqrt{3}, 5 + \sqrt{3}, \ldots$ (9th term)

18. $1.02, 10.2, 102, \ldots$ (10th term)

19. $4.3, -3.2, 2.1, \ldots$ (6th term)

20. $3.4, -6.8, 13.6, \ldots$ (9th term)

Solve Exercises 21–36.

21. For an arithmetic sequence $a_5 = 9$ and $a_6 = 24$, what is a_1?

22. For an arithmetic sequence $a_7 = 11$ and $a_9 = 15$, what is a_1?

23. For a geometric sequence $a_4 = 9$ and $a_5 = 3$, what is a_1?

24. For a geometric sequence $a_8 = \frac{1}{3}$ and $a_9 = 5$, what is a_1?

25. In an arithmetic sequence $a_1 = 5$, $a_n = 81$, and $d = 4$, what is n?

26. In an arithmetic sequence $a_1 = 20$, $a_n = -31$, and $d = -3$, what is n?

27. In a geometric sequence $a_1 = 3$, $a_n = \frac{1}{559,872}$ and $r = \frac{1}{6}$, find n.

28. In a geometric sequence $a_1 = -4$, $a_n = \frac{-1}{4,096}$, and $r = \frac{1}{2}$, what is n?

29. If the first term of an arithmetic sequence is -5 and the 7th term is 4, what is the second term?

30. If the first term of a geometric sequence is $\frac{1}{3}$ and the 8th term is 729, what is the third term?

31. *Medical technology* A group of people make out an exercise program. The first day they will jog $\frac{1}{2}$ mi. After that they will jog a certain amount more each day until on the 61st day, 8 mi are jogged. How much was the distance increased each day?

32. *Physics* A ball is dropped to the ground from a height of 80 m. Each time it bounces it goes $\frac{7}{8}$ as high as the previous bounce. How high does it go on the sixth bounce?

33. *Physics* A pendulum swings 15 cm on its first swing. Each subsequent swing is reduced by 0.3 cm. How far is the 10th swing? How many times will the pendulum swing before it comes to rest?

34. *Aeronautical engineering* The atmospheric pressure at sea level is approximately 100 kPa and decreases 12.5% for each km increase in altitude. What is the atmospheric pressure at the top of Mt. Everest, which is about 8.8 km high?

35. *Environmental science* A chemical spill pollutes a river. A monitor located 1 mi downstream from the spill measures 940 parts of the chemical for every million parts of water. (This is written as 940 ppm.) The readings decrease 16.2% for each mile farther downstream. **(a)** How far downstream from the spill will it be before the concentration is reduced to 100 ppm? **(b)** The water is considered safe for human consumption at 1.5 ppm. How far downstream from the spill will you have to go before the water is considered safe for humans to drink?

36. *Environmental science*   The level of a particular metal pollutant in a certain lake was found to be increasing geometrically. In January, there were 3.50 parts per billion (ppb) and in April there were 18.76 ppb.

(a) By what ratio is the level increasing each month?

(b) What will be the pollution level in December?

 **In Your Words**

37. Explain how an arithmetic sequence and a geometric sequence are alike and how they are different.

38. In Exercises 1–20, you were asked to determine if each given sequence is an arithmetic sequence, a geometric sequence, or neither. Describe how you made your decision.

≡ 19.3
SERIES

There are many times when we need to add the terms of a sequence. The sum of the terms of a sequence is called a **series**. The series

$$a_1 + a_2 + a_3 + a_4 + a_5$$

is a **finite series** with five terms. A finite series can have $1, 2, 5, 19, 437$, or any number of terms as long as the number is a natural number. A series of the form

$$a_1 + a_2 + a_3 + a_4 + \cdots$$

is an **infinite series**. An infinite series has an infinite or endless number of terms.

Summation Notation

A more compact notation is often used to indicate a series. This notation is referred to as **summation** or **sigma notation** because it uses the capital Greek letter sigma ($\sum$). In general, the sigma notation means

$$\sum_{k=1}^{n} a_k = a_1 + a_2 + a_3 + \cdots + a_n$$

Here $\sum$ indicates a sum. The letter k is called an **index of summation**. The summation begins with $k = 1$ as is indicated below the $\sum$ and ends with $k = n$ as is indicated above the $\sum$.

Sometimes it is useful to indicate that a series begins with the zeroth, second, or other term. If it starts with the 0th term, then

$$\sum_{k=0}^{n} a_k = a_0 + a_1 + a_2 + \cdots + a_n$$

and if it starts with the second term,

$$\sum_{k=2}^{n} a_k = a_2 + a_3 + a_4 + \cdots + a_n$$

The numbers below and above the $\sum$ are the **limits of summation**. Here they are 2 and n.

EXAMPLE 19.15

Evaluate $\displaystyle\sum_{k=1}^{5}(-3)k$.

Solution $\displaystyle\sum_{k=1}^{5}(-3)k = (-3)1+(-3)2+(-3)3+(-3)4+(-3)5$

$$= -3+(-6)+(-9)+(-12)+(-15)$$
$$= -45$$

```
sum seq(-3X,X,1,
5,1)
              -45
```

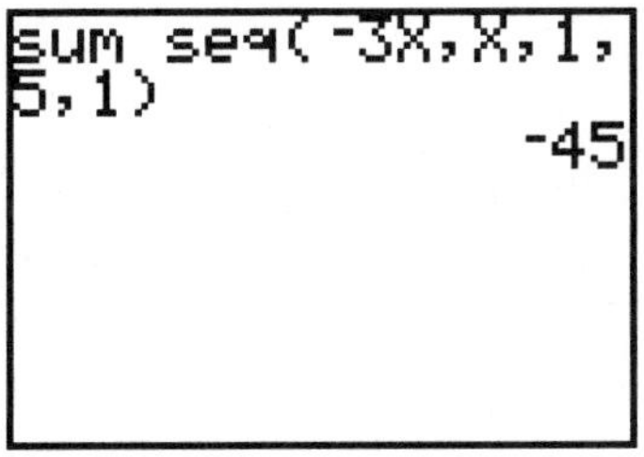

FIGURE 19.2

You can use some calculators to add sequences. For instance, you can use a TI-82 to evaluate $\displaystyle\sum_{k=1}^{5}(-3)k$ by using the keystrokes [2nd] [LIST] [MATH] 5(sum) [2nd] [LIST] 5(seq() [(−)] 3 [X,T,θ] [,] [X,T,θ] [,] 1 [,] 5 [,] 1 [)] [ENTER]. The result, −45, as shown in Figure 19.2, is the same result we obtained in Example 19.15.

EXAMPLE 19.16

Evaluate $\displaystyle\sum_{k=1}^{4}(2k-3)$.

Solution $\displaystyle\sum_{k=1}^{4}(2k-3) = (2\cdot1-3)+(2\cdot2-3)+(2\cdot3-3)+(2\cdot4-3)$

$$= (2-3)+(4-3)+(6-3)+(8-3)$$
$$= -1+1+3+5$$
$$= 8$$

EXAMPLE 19.17

Evaluate $\displaystyle\sum_{k=0}^{3}\frac{2^k}{3k-5}$.

Solution $\displaystyle\sum_{k=0}^{3}\frac{2^k}{3k-5} = \frac{2^0}{3\cdot0-5}+\frac{2^1}{3\cdot1-5}+\frac{2^2}{3\cdot2-5}+\frac{2^3}{3\cdot3-5}$

$$= \frac{1}{0-5}+\frac{2}{3-5}+\frac{4}{6-5}+\frac{8}{9-5}$$
$$= \frac{1}{-5}+\frac{2}{-2}+\frac{4}{1}+\frac{8}{4}$$
$$= 4\tfrac{4}{5}$$

Partial Sums

Informally speaking, the expression $\sum_{k=0}^{n} a_k$ directs us to find the "sum" of the terms $a_1, a_2, a_3, \ldots, a_n$. To carry out this process, we proceed as follows. We let S_n denote the sum of the first n terms of the series. Thus,

$$S_1 = a_1$$
$$S_2 = a_1 + a_2$$
$$S_3 = a_1 + a_2 + a_3$$
$$\vdots$$
$$S_n = a_1 + a_2 + a_3 + \cdots + a_n = \sum_{k=1}^{n} a_k$$

The number S_n is called the *n*th **partial sum** of the series. The sequence $S_1, S_2, S_3, \ldots, S_n$ is called the **sequence of partial sums**.

EXAMPLE 19.18

Find the first five partial sums of the series $\sum_{k=1}^{n} k^2$.

Solution
$$S_1 = 1^2 = 1$$
$$S_2 = 1^2 + 2^2 = 5$$
$$S_3 = 1^2 + 2^2 + 3^2 = 14$$
$$S_4 = 1^2 + 2^2 + 3^2 + 4^2 = 30$$
$$S_5 = 1^2 + 2^2 + 3^2 + 4^2 + 5^2 = 55$$

Arithmetic Series

Now let's look at the series formed from the two special sequences that we studied in Section 19.2. An arithmetic sequence has a first term of a_1 and a common difference d. The sum of the first n terms of an arithmetic sequence is the *n*th partial sum of the sequence:

$$S_n = a_1 + (a_1 + d) + (a_1 + 2d) + (a_1 + 3d) + \cdots + [a_1 + (n-1)d] \tag{1}$$

If we write the *n*th term a_n first, then S_n could be written as

$$S_n = a_n + (a_n - d) + (a_n - 2d) + (a_n - 3d) + \cdots + [a_n - (n-1)d] \tag{2}$$

If we add the corresponding terms of equations (1) and (2), we get

$$2S_n = (a_1 + a_n) + (a_1 + a_n) + (a_1 + a_n) + (a_1 + a_n) + \cdots + (a_1 + a_n)$$

The right-hand side has n terms that are all the same, $a_1 + a_n$. Thus, we have the following sum S_n of the first n terms of an arithmetic sequence.

> **Sum of First n Terms of an Arithmetic Sequence**
>
> The sum S_n of the first n terms of an arithmetic sequence is
>
> $$S_n = \frac{n(a_1 + a_n)}{2}$$
>
> where a_1 is the first term and a_n is the nth term.

EXAMPLE 19.19

Find the sum of the first eight terms of the arithmetic sequence 3, 8, 13, 18, 23, 28, 33, 38, ...

Solution We will use the formula $S_n = \dfrac{n(a_1 + a_n)}{2}$ with $n = 8$, $a_1 = 3$, and $a_8 = 38$.

$$S_8 = \frac{8(3 + 38)}{2} = 164$$

EXAMPLE 19.20

Find the sum of the first 14 terms of the arithmetic sequence 9, 3, −3, ...

Solution In this sequence, $d = -6$, $a_1 = 9$, and $n = 14$. From Section 19.2, we know that

$$\begin{aligned} a_{14} &= a_1 + (14 - 1)d \\ &= 9 + 13(-6) \\ &= -69 \end{aligned}$$

We can now find the sum of the first 14 terms.

$$\begin{aligned} S_{14} &= \frac{14(a_1 + a_n)}{2} \\ &= \frac{14(9 - 69)}{2} \\ &= -420 \end{aligned}$$

The sum of the first 14 terms of this sequence is −420.

Geometric Series

Now let's see if we can develop a similar formula for the first n terms of a geometric series. The sum of the first n terms of a geometric sequence is

$$S_n = a_1 + a_1 r + a_1 r^2 + a_1 r^3 + \cdots + a_1 r^{n-1} \tag{3}$$

Multiplying both sides of equation (3) by r, we get

$$r S_n = a_1 r + a_1 r^2 + a_1 r^3 + a_1 r^4 + \cdots + a_1 r^{n-1} + a_1 r^n \tag{4}$$

If we subtract equation (4) from equation (3), we get

$$S_n - rS_n = a_1 - a_1 r^n$$

Factoring each side $S_n(1-r) = a_1(1-r^n)$ and solving for S_n, we have the following formula for the first n terms of a geometric sequence.

Sum of First n Terms of a Geometric Sequence

The sum S_n of the first n terms of a geometric sequence is

$$S_n = a_1 \frac{1-r^n}{1-r}$$

where a_1 is the first term and r is the common ratio.

EXAMPLE 19.21

Find the sum of the first 10 terms of the geometric sequence $4, -8, 16, -32, \ldots$.

Solution   In this sequence, $a_1 = 4$, $r = -2$, and $n = 10$.

$$\begin{aligned}
S_{10} &= a_1 \frac{1-r^{10}}{1-r} \\
&= 4 \left(\frac{1-(-2)^{10}}{1-(-2)} \right) \\
&= 4 \left(\frac{1-1{,}024}{3} \right) \\
&= 4 \left(\frac{-1{,}023}{3} \right) \\
&= -1{,}364
\end{aligned}$$

The sum of the first 10 terms of $4 - 8 + 16 - 32 + \cdots$ is $-1{,}364$.

EXAMPLE 19.22

Find the sum of the geometric series $\frac{1}{2} + \frac{1}{3} + \frac{2}{9} + \cdots + \frac{1}{2}(\frac{2}{3})^8$.

Solution   In this series, $a_1 = \frac{1}{2}$, $r = \frac{2}{3}$, and $n = 9$.

$$S_9 = \frac{1}{2} \left(\frac{1-(\frac{2}{3})^9}{1-\frac{2}{3}} \right) = \frac{1}{2} \left(\frac{1-\frac{512}{19{,}683}}{1-\frac{2}{3}} \right) = \frac{1}{2} \left(\frac{\frac{19{,}171}{19{,}683}}{\frac{1}{3}} \right) = \frac{57{,}513}{39{,}366}$$

$$\approx 1.46$$

An Application of the Geometric Series

One application of the geometric series deals with compound interest. Suppose you deposit $100 in a savings account that pays 8% compounded annually. The amount of money after 1 year would be

$$100 + 100(0.08) = 100 + 8$$
$$= \$108$$

After 2 years the amount would be

$$108 + 108(0.08) = 108 + 8.64$$
$$= \$116.64$$

As you can see, the amount after each year forms a geometric sequence

$$100 + 108 + 116.64 + \cdots$$

where $a_1 = 100$ and $r = 1.08$. Thus $r = 1 + i$, where i is the interest rate. From this, we get the formula for compound interest.

Compound Interest

$$A = P(1+i)^n$$

where A is the amount after n interest periods, P is the principal or initial amount invested, and i is the interest rate per interest period expressed as a decimal.

Thus A is similar to S_n. Since the initial investment is P, this formula begins with $n = 0$.

The interest rate is given for the interest period. This means that in order to determine i, you must divide the annual interest rate by the number of interest periods. An 8% interest rate compounded quarterly (four times a year) would have $i = \frac{0.08}{4} = 0.02$. (Remember 8% = 0.08.) If the 8% interest rate was compounded monthly, or 12 times a year, then $i = \frac{0.08}{12} = 0.0067$.

Application

EXAMPLE 19.23

If $100 is deposited in a savings account paying 6% annually, what is the amount after 1 year if interest is compounded **(a)** quarterly, or **(b)** monthly?

Solutions   We will use $A = P(1+i)^n$ in both parts with $P = 100$.

(a) Since interest is compounded quarterly

$$i = \frac{0.06}{4} = 0.015 \text{ and } n = 4$$
$$A = 100(1 + 0.015)^4$$
$$= 106.14$$

The total is $106.14 at the end of 1 year.

EXAMPLE 19.23 (Cont.)

(b) In this example, the interest is compounded monthly, so

$$i = \frac{0.06}{12} = 0.005 \text{ and } n = 12$$
$$A = 100(1.005)^{12}$$
$$= 106.17$$

The total is $106.17 at the end of 1 year.

Application

EXAMPLE 19.24

Suppose the money in Example 19.23 were left in the savings account for 4 more years (a total of 5). What would it then be worth?

Solutions

(a) At quarterly compounding, i is still 0.015, but n is now 4×5 (4 times a year for 5 years).

$$A = 100(1.015)^{20}$$
$$= 134.69$$

(b) At monthly compounding, $i = 0.005$ and $n = 12 \times 5 = 60$.

$$A = 100(1.005)^{60}$$
$$= 134.89$$

The totals are $134.69 if the money is compounded quarterly for 5 years and $134.89 if it is compounded monthly.

Some older calculators have a $\boxed{\Sigma}$ or $\boxed{\Sigma+}$ key. This key is used to sum a group of numbers. The total is placed in memory and the number of terms is displayed. On some calculators, it is possible to recall this total from memory and on others it is not. The $\boxed{\Sigma+}$ key on a calculator is primarily used for statistical problems.

 **Caution**

Check the owner's manual for your calculator to see if you can use your calculator to sum a sequence.

Exercise Set 19.3

In Exercises 1–4, write the first four terms of the indicated series.

1. $\displaystyle\sum_{k=1}^{20} 3\left(\tfrac{1}{2}\right)^{k}$

2. $\displaystyle\sum_{n=1}^{50} [4+(n-1)3]$

3. $\displaystyle\sum_{n=0}^{20} (-1)^{n}\frac{3^{n}}{n+1}$

4. $\displaystyle\sum_{k=1}^{60} (-1)^{k}$

In Exercises 5–12, evaluate the given sum.

5. $\displaystyle\sum_{k=1}^{5} (k-3)$

7. $\displaystyle\sum_{k=1}^{6} k^2$

9. $\displaystyle\sum_{n=1}^{5} n^3$

11. $\displaystyle\sum_{i=3}^{8} \frac{2^i}{3i+1}$

6. $\displaystyle\sum_{k=0}^{4} (2k+1)$

8. $\displaystyle\sum_{k=2}^{8} \frac{k+2}{(k-1)k}$

10. $\displaystyle\sum_{n=1}^{8} \frac{(-1)^n(n^2+1)}{n}$

12. $\displaystyle\sum_{i=0}^{7} \frac{(-1)^{i+1}(i+1)^2}{2i+1}$

In Exercises 13–28, determine whether the terms of the given series form an arithmetic or geometric series and find the indicated sum.

13. $3+6+9+12+\cdots; S_{10}$

14. $5+1-3+\cdots; S_{12}$

15. $1+2+4+\cdots; S_8$

16. $\frac{1}{2}+2+8+\cdots; S_9$

17. $-6-2+2+\cdots; S_{12}$

18. $1-\frac{1}{2}+\frac{1}{4}+\cdots; S_{10}$

19. $0.5+0.75+1.0+\cdots; S_{20}$

20. $0.5+0.2+0.08+\cdots; S_{15}$

21. $\frac{3}{4}-\frac{1}{4}+\frac{1}{12}+\cdots; S_{16}$

22. $0.4+0.04+0.004+\cdots; S_6$

23. $0.4+1.6+2.8+\cdots; S_{14}$

24. $16+8+4+\cdots; S_{10}$

25. $2\sqrt{3}+6+6\sqrt{3}+\cdots; S_{15}$

26. $\sqrt{5}+10+20\sqrt{5}+\cdots; S_{12}$

27. $3+9+27+\cdots; S_8$

28. $5-25+125+\cdots; S_{10}$

Solve Exercises 29–36.

29. *Physics*   A ball is dropped to the ground from a height of 80 m, and each time it bounces 80% as high as it did on the previous bounce. How far has it traveled when it hits the ground the fourth time? The tenth time?

30. *Physics*   A pendulum swings a distance of 50 m initially from one side to the other. After the first swing, each swing is only 0.8 of the distance of the previous swing. What is the total distance covered by the pendulum in 10 swings?

31. *Construction*   The end of a gable roof is in the shape of a triangle. If the span is 32 ft (as shown in Figure 19.3), the height of the gable is 8 ft, and the studs are placed every 16 in., what is the total length of the studs?

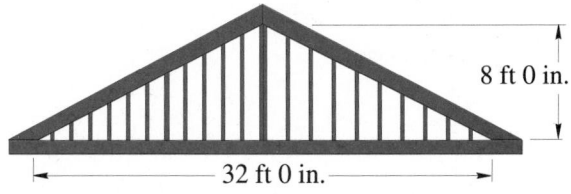

8 ft 0 in.

32 ft 0 in.

FIGURE 19.3

32. *Finance*   How much will $2,000 earn in 3 years if it is invested in a savings account that pays 8% compounded quarterly?

33. *Finance*   You have the opportunity to invest the $2,000 in Exercise 32 in a savings account that pays 8% compounded monthly. How much would this total after 3 years? How much more was this than the amount in Exercise 32?

34. *Finance*   Your final chance is to invest the money in Exercise 32 at 8% compounded continuously. What would this total after 3 years? (See Section 13.2.)

35. *Construction*   In Exercise 26, Exercise Set 19.1, a contractor was preparing a bid for the construction of an office building. The foundation and basement will cost $750,000 and the first floor will cost $320,000. The second floor will cost $240,000 and each floor above the second will cost $12,000 more than the floor below it. **(a)** If the 10th floor will cost $336,000, how much will a 10-story building cost? **(b)** If the 20th floor will cost $456,000, how much will it cost to build a 20-story building?

36. *Finance*   At the beginning of each year, the owner of a automotive repair shop deposits $3,000 into a retirement account paying 5.25% compounded annually. How much money will be in the account when the owner retires 25 years later? (Note: The owner retires at the end of the 25th year.)

In Your Words

37. Explain how a series and a sequence are different.

38. What is a partial sum? How you you determine the sequence of partial sums?

39. Describe how to determine the sum of the first n terms of an arithmetic sequence.

40. Describe how to determine the sum of the first n terms of a geometric sequence.

≡ 19.4
INFINITE GEOMETRIC SERIES

In this section, we will continue discussing geometric series. All of the series that we discussed in Section 19.3 were finite series. A finite series has a first term and a last term. In this section, we will discuss series that are not finite. In particular, those that do not have a last term. A series that does not have a last term is called an **infinite series**.

If we have a sequence of the form

$$a_1, a_2, a_3, \ldots, a_n, \ldots$$

then this is an **infinite sequence** because it does not have a last term. The corresponding infinite series is designed by

$$\sum_{k=1}^{\infty} a_k = a_1 + a_2 + a_3 + \cdots + a_n + \cdots$$

The symbol ∞ is read, "infinity."

We have seen examples of infinite numbers before and have always found ways of working with these numbers. But now we want to find a way to determine the sum of an infinite series.

One way that we can determine the sum of an infinite series is with our earlier definition of partial sum. Remember, each partial sum S_n is the sum of the first n term in the series. That means that each partial sum is the sum of a finite number of terms. So,

$$S_1 = a_1$$
$$S_2 = a_1 + a_2$$
$$S_3 = a_1 + a_2 + a_3$$
$$\vdots$$
$$S_n = a_1 + a_2 + a_3 + \cdots + a_n$$

S_n is called the n**th partial sum** of the series $\sum_{k=1}^{\infty} a_k$. The **sequence of partial sums** is $S_1, S_2, S_3, \ldots, S_n$. If it happens that, as n gets larger and larger, S_n seems to be approaching some number that we will call S, we then say that S is the sum of the infinite series.

The idea of the number S is much the same as when we studied asymptotes. If you remember a curve kept getting closer and closer to an asymptote, but never reached it. In the same way, as n gets larger, S_n will get closer and closer to S, but never reach it. We say that S is the limit of the sequence S_n and write it symbolically as

$$S = \lim_{n \to \infty} S_n = \sum_{n=1}^{\infty} a_n$$

The symbol $\lim_{n \to \infty} S_n$ is read "the limit of S_n as n goes to infinity." Limits are studied in detail in calculus.

Now, let's consider a couple of infinite geometric sequences. The first one we will consider is

$$1 + \frac{1}{2} + \frac{1}{4} + \frac{1}{8} + \cdots + \frac{1}{2^{n-1}} + \cdots$$

The partial sums are

$$S_1 = 1$$
$$S_2 = 1 + \frac{1}{2} = \frac{3}{2}$$
$$S_3 = 1 + \frac{1}{2} + \frac{1}{4} = \frac{7}{4}$$
$$S_4 = 1 + \frac{1}{2} + \frac{1}{4} + \frac{1}{8} = \frac{15}{8}$$
$$S_5 = 1 + \frac{1}{2} + \frac{1}{4} + \frac{1}{8} + \frac{1}{16} = \frac{31}{16}$$

From the Section 19.3 we know that

$$S_n = a_1 \left(\frac{1 - r^n}{1 - r} \right) = 1 \left[\frac{1 - (\frac{1}{2})^n}{1 - \frac{1}{2}} \right] = 2 \left[1 - \left(\frac{1}{2} \right)^n \right]$$

The sequence of partial sums

$$1, \frac{3}{2}, \frac{7}{4}, \frac{15}{8}, \frac{31}{16}, \dots, 2 \left[1 - \left(\frac{1}{2} \right)^n \right], \dots$$

seems to be approaching 2 as a limit. Since $\left(\frac{1}{2} \right)^n$ keeps getting closer and closer to 0 as n gets larger, it seems that

$$\lim_{n \to \infty} S_n = \lim_{n \to \infty} 2 \left[1 - \left(\frac{1}{2} \right)^n \right] = 2(1 - 0) = 2$$

Not all partial sums of infinite series approach a finite limit. If the partial sums of an infinite series approach a finite limit, we say that the series **converges** or is a **convergent series**. A series that does not converge is said to **diverge** or to be a **divergent series**. All arithmetic series diverge.

In general, an infinite geometric series of the form

$$\sum_{n=1}^{\infty} a_1 r^{n-1} = a_1 + a_1 r + a_1 r^2 + a_1 r^3 + \cdots$$

converges under certain circumstances. We know from Section 19.3 that the nth partial sum of this series is

$$S_n = \frac{a_1(1 - r^n)}{1 - r}$$

When does this have a limit? The key is in the term r^n. Pick a number, such as 2, and take larger powers of 2. What happens? $2^2 = 4$, $2^4 = 16$, $2^8 = 256$, ... Notice that as n gets larger, 2^n keeps getting larger. Now select a smaller number such as 0.5. $(0.5)^2 = 0.25$, $(0.5)^4 = 0.0625$, $(0.5)^8 = 0.00390625$. As n gets larger, $(0.5)^n$ gets smaller and smaller. In fact, for any value r that is between -1 and 1, as n gets larger r^n gets closer to zero. In symbols, using our new limit notation, this is

$$\text{If } |r| < 1, \text{then } \lim_{n \to \infty} r^n = 0$$

This means that for an infinite geometric series, if $|r| < 1$, then

$$\lim_{n \to \infty} S_n = \lim_{n \to \infty} a_1 \left(\frac{1 + r^n}{1 - r}\right) = \frac{a_1}{1 - r}$$

This provides us with a formula for the sum of an infinite geometric series.

Sum of an Infinite Geometric Series

If $|r| < 1$, then the infinite geometric series

$$a_1 + a_1 r + a_1 r^2 + a_1 r^3 + \cdots + a_1 r^n + \cdots$$

has the sum $S = \dfrac{a_1}{1 - r}$.

If $|r| \geq 1$, then the series diverges.

EXAMPLE 19.25

Show that the series $\displaystyle\sum_{n=1}^{\infty} \left(\frac{-2}{5}\right)^n$ converges and find its sum.

Solution   This is a geometric series with $r = \frac{-2}{5}$. Since $|r| = \left|\frac{-2}{5}\right| = \frac{2}{5}$, and $|r|$ is less than one, the sum of this geometric series is

$$\frac{a_1}{1 - r} = \frac{-\frac{2}{5}}{1 - \frac{-2}{5}} = \frac{-\frac{2}{5}}{\frac{7}{5}} = \frac{-2}{7}$$

Always remember that we can never reach $-\frac{2}{7}$ exactly, no matter how many terms in the series we add; but, by definition, $-\frac{2}{7}$ is the sum of this series.

EXAMPLE 19.26

Show that each of the following series is divergent:

(a) $1 + \frac{4}{3} + \frac{16}{9} + \cdots$; **(b)** $\displaystyle\sum_{k=1}^{\infty} (-1)^k$

EXAMPLE 19.26 (Cont.)

Solutions

(a) Since $r = \frac{4}{3}$, we can see that $r > 1$ and so this series is divergent.

(b) The expanded series is $-1 + 1 - 1 + 1 - 1 + 1 - 1 + \cdots$ which has partial sums

$$S_1 = -1, \ S_2 = 0, \ S_3 = -1, \ S_4 = 0, \ S_5 = -1, \ldots$$

These partial sums do not get arbitrarily large, but they also do not approach some specific limit. Since $r = -1$, we see that $|r| = |-1| = 1$, and so this series diverges.

EXAMPLE 19.27

Find the fraction that has the repeating decimal form $0.232323\ldots$.

Solution This decimal can be thought of as the series $0.23 + 0.0023 + 0.000023 + \cdots$, which is the geometric series $\sum_{n=0}^{\infty} 0.23(0.01)^n$. In this series $a_1 = 0.23$ and $r = 0.01$, so

$$S = \frac{a_1}{1-r} = \frac{0.23}{1-0.01} = \frac{0.23}{0.99} = \frac{23}{99}$$

Thus, the decimal $0.232323\ldots$ is equivalent to the fraction $\frac{23}{99}$.

EXAMPLE 19.28

Express $4.2315315315\ldots$ as a rational number.

Solution The number $4.2315315315\ldots$ repeats the digits 315. We can write the number as $4.2 + (0.0315 + 0.0000315 + 0.0000000315 + \cdots)$. The portion inside the parentheses is an infinite geometric series with $a_1 = 0.0315$ and $r = 0.001$. So, $4.2315315\ldots = 4.2 + S$, where

$$S = \frac{0.0315}{1-0.001} = \frac{0.0315}{0.999} = \frac{315}{9,990} = \frac{35}{1,110}$$

Since $4.2 = \frac{42}{10}$, we have

$$4.2315315\ldots = \frac{42}{10} + \frac{35}{1,110} = \frac{4,662 + 35}{1,110} = \frac{4,697}{1,110}.$$

Exercise Set 19.4

Determine which of the infinite geometric series in Exercises 1–22 converge and which diverge. For those that converge, find the sum.

1. $\frac{1}{2} - \frac{1}{4} + \frac{1}{8} - \frac{1}{16} + \cdots$

2. $1 + \frac{3}{4} + \frac{9}{16} + \cdots$

3. $1 - \frac{2}{3} + \frac{4}{9} - \frac{8}{27} + \cdots$

4. $1.5 + 2.25 + 3.375 + \cdots$

5. $3 + \frac{3}{4} + \frac{3}{16} + \cdots$

6. $4 - 1 + \frac{1}{4} + \cdots$

7. $\frac{5}{4} + \frac{1}{8} + \frac{1}{80} + \cdots$

8. $\frac{3}{4} + \frac{3}{4^2} + \frac{3}{4^3} + \cdots$

9. $0.03 + 0.003 + 0.0003 + \cdots$

10. $1.4 + 0.014 + 0.00014 + \cdots$

11. $\sum\limits_{n=1}^{\infty} (-0.8)^{n-1}$

12. $\sum\limits_{n=1}^{\infty} \left(\frac{3}{4}\right)^n$

13. $\sum\limits_{n=1}^{\infty} \left(\frac{5}{4}\right)^{n+1}$

14. $\sum\limits_{n=1}^{\infty} (-1.1)^n$

15. $\sum\limits_{n=1}^{\infty} 0.3(10)^n$

16. $\sum\limits_{n=1}^{\infty} 2^{-n}$
[Hint: Let $2^{-n} = (2^{-1})^n$.]

17. $\sum\limits_{n=1}^{\infty} \left(\frac{3}{2}\right)^{-n}$

18. $\sum\limits_{n=1}^{\infty} 6\left(-\frac{2}{3}\right)^n$

19. $\sum\limits_{n=1}^{\infty} \left(\frac{1}{\sqrt{2}}\right)^n$

20. $\sum\limits_{n=1}^{\infty} \left(\frac{\sqrt{3}}{3}\right)^n$

21. $\sum\limits_{n=1}^{\infty} \left(\sqrt{5}\right)^n$

22. $\sum\limits_{n=1}^{\infty} \left(\sqrt{7}\right)^{1-n}$

In Exercises 23–30, use infinite series to find the rational number corresponding to the given decimal number.

23. $0.444\ldots$

24. $0.777\ldots$

25. $0.575757\ldots$

26. $0.848484\ldots$

27. $1.352135213521\ldots$

28. $4.12341234123\ldots$

29. $6.3021021021\ldots$

30. $2.1906906906\ldots$

Solve Exercises 31–34.

31. *Physics* A ball is dropped from a height of 5 m. After the first bounce, the ball reaches a height of 4 m, after the second, 3.2 m, and so on. What is the total distance traveled by the ball before it comes to rest?

32. *Physics* An object suspended on a spring is oscillated up and down. The first oscillation was 100 mm and each oscillation after that was $\frac{9}{10}$ that of the preceding one. What is the total distance that the object traveled?

33. *Machine technology* When a motor is turned off, a flywheel attached to the motor coasts to a stop. In the first second, the flywheel makes 250 revolutions. Each of the following seconds, it revolves $\frac{8}{10}$ of the number in the preceding second. What are the total number of revolutions made by the flywheel when it stops?

34. *Physics* A pendulum swings a distance of 50 cm initially from one side to the other. After the first swing, each swing is 0.85 the distance of the previous swing. What is the total distance covered by the pendulum when it comes to a rest?

 **In Your Words**

35. What is the difference between a convergent series and a divergent series?

36. Describe how to determine if an infinite geometric series converges.

≣ 19.5
THE BINOMIAL THEOREM

In Section 7.1, we learned several special products. Among them were

$$(x+y)^2 = x^2 + 2xy + y^2$$
$$(x+y)^3 = x^3 + 3x^2y + 3xy^2 + y^3$$
$$(x-y)^2 = x^2 - 2xy + y^2$$

and $\quad (x-y)^3 = x^3 - 3x^2y + 3xy^2 - y^3$

To find the larger powers of $x + y$ or $x - y$ would require repeated multiplication by $x + y$ or $x - y$. In this section, we will develop the binomial theorem. This theorem allows us to expand $x + y$ or $x - y$ to any power without direct multiplication.

Expansions of $(x + y)^n$ and Pascal's Triangle

By direct multiplication, we can obtain the following expansions of $x + y$:

$$(x+y)^0 = 1$$
$$(x+y)^1 = x+y$$
$$(x+y)^2 = x^2 + 2xy + y^2$$
$$(x+y)^3 = x^3 + 3x^2y + 3xy^2 + y^3$$
$$(x+y)^4 = x^4 + 4x^3y + 6x^2y^2 + 4xy^3 + y^4$$
$$(x+y)^5 = x^5 + 5x^4y + 10x^3y^2 + 10x^2y^3 + 5xy^4 + y^5$$
$$(x+y)^6 = x^6 + 6x^5y + 15x^4y^2 + 20x^3y^3 + 15x^2y^4 + 6xy^5 + y^6$$

If you look at these expansions you may notice some patterns.

1. There are $n + 1$ terms in a binomial raised to the nth power.
2. The powers of x decrease by 1, each term beginning with x^n and ending with x^0.
3. The powers of y increase by 1, each term beginning with y^0 and ending with y^n.
4. The first term is x^n and the last term is y^n.
5. In each term, the sum of the exponents of x and y is n. For example, the third term of $(x + y)^6$ contains x^4y^2 and $4 + 2 = 6$, and the fifth term contains $xy^5 = x^1y^5$ and $1 + 5 = 6$.
6. The coefficients of terms equidistant from the ends are equal.
7. If the coefficient of any term is multiplied by the exponent of x in that term, and this product is divided by the exponent of y in the next term, we get the coefficient of the next term.
8. The coefficients form a pattern known as **Pascal's triangle** in which each coefficient is the sum of the two nearest coefficients in the row above.

$n = 0$							1							
$n = 1$						1		1						
$n = 2$					1		2		1					
$n = 3$				1		3		3		1				
$n = 4$			1		4		6		4		1			
$n = 5$		1		5	+	10		10		5		1		
$n = 6$	1		6		15		20		15	+	6		1	
$n = 7$	1	7		21		35		35		21		7		1

You can see that the first and last number in each row is 1 and the second and next to last numbers are n. The pattern in Pascal's triangle can be continued forever. It would take a long time to develop row 26 if you needed the coefficients for $(x + y)^{26}$. That is why pattern number 7 is easier to use.

EXAMPLE 19.29

Use Pascal's triangle to expand $(3a - 2b)^6$.

Solution Here $x = 3a$ and $y = -2b$. The coefficients from row 6 of Pascal's triangle are 1, 6, 15, 20, 15, 6, and 1, and so we get

$$1(3a)^6 + 6(3a)^5(-2b) + 15(3a)^4(-2b)^2 + 20(3a)^3(-2b)^3$$
$$+ 15(3a)^2(-2b)^4 + 6(3a)(-2b)^5 + 1(-2b)^6$$
$$= 729a^6 - 2{,}916a^5b + 4{,}860a^4b^2 - 4{,}320a^3b^3 + 2{,}160a^2b^4$$
$$- 576ab^5 + 64b^6$$

Notice that the signs alternate when the second term in the binomial is negative.

Binomial Formula

If we were to apply pattern number 7 to a general binomial expansion $(x + y)^n$, we would get the following expansion known as the **binomial formula**.

$$(x + y)^n = x^n + nx^{n-1}y + \frac{n(n-1)}{2}x^{n-2}y^2 + \frac{n(n-1)(n-2)}{2 \cdot 3}x^{n-3}y^3 + \cdots + y^n$$

Mathematicians have developed a way to abbreviate these coefficients. They use $\binom{n}{r}$ to represent $\frac{n!}{r!(n-r)!}$, where $n!$, read "n factorial," is the product

$$n! = n(n-1)(n-2)(n-3)\cdots(3)(2)(1)$$
$$\text{and} \quad 0! = 1$$

EXAMPLE 19.30

Determine 5! and 8!

Solutions $5! = 5 \cdot 4 \cdot 3 \cdot 2 \cdot 1 = 120$

$8! = 8 \cdot 7 \cdot 6 \cdot (5!) = 40{,}320$

Some calculators have a key marked $\boxed{n!}$ or $\boxed{x!}$. On some of these calculators you have to use the $\boxed{\text{2nd}}$ key in order to use the $\boxed{x!}$. As you can see from comparing 5! and 8! in the last example, factorials get very large, very quickly. To determine 10! with a calculator you would

PRESS DISPLAY

$10\,\boxed{n!}$ 3628800

Now,

$$\binom{n}{2} = \frac{n!}{2!(n-2)!}$$
$$= \frac{n(n-1)(n-2)\cdots(3)(2)(1)}{(2)(1)(n-2)(n-3)\cdots(2)(1)}$$
$$= \frac{n(n-1)}{2 \cdot 1}$$

This is the coefficient we got for the $x^{n-2}y^2$ term. A similar pattern holds for each term. This allows us to rewrite the binomial formula as follows.

Binomial Formula

$$(x+y)^n = x^n + \binom{n}{1}x^{n-1}y + \binom{n}{2}x^{n-2}y^2 + \binom{n3}{x}^{n-3}y^3 + \cdots$$

$$+ \binom{n}{n-1}xy^{n-1} + y^n$$

$$= \sum_{r=0}^{n} \binom{n}{r}x^{n-r}y^r$$

where the general term is $\binom{n}{r}x^{n-r}y^r$ and

$$\binom{n}{r} = \frac{n!}{(n-r)!r!} = \frac{n(n-1)(n-2)\cdots(n-r+1)}{r!}$$

EXAMPLE 19.31

Find the first four terms $(x+y)^{15}$.

Solution The first two coefficients are 1 and 15. The next two terms are

$$\binom{15}{2} = \frac{15!}{2!13!}$$

$$= \frac{15 \cdot 14 \cdot 13 \cdots 3 \cdot 2 \cdot 1}{(2 \cdot 1)(13 \cdot 12 \cdots 3 \cdot 2 \cdot 1)}$$

$$= \frac{15 \cdot 14}{2} = 105$$

and $$\binom{15}{3} = \frac{15!}{3!12!}$$

$$= \frac{15 \cdot 14 \cdot 13 \cdots 3 \cdot 2 \cdot 1}{(3 \cdot 2 \cdot 1)(12 \cdot 11 \cdots 2 \cdot 1)}$$

$$= \frac{15 \cdot 14 \cdot 13}{3 \cdot 2 \cdot 1} = 455$$

So, the first four terms of $(x+y)^{15}$ are

$$x^{15} + 15x^{14}y + 105x^{13}y^2 + 455x^{12}y^3$$

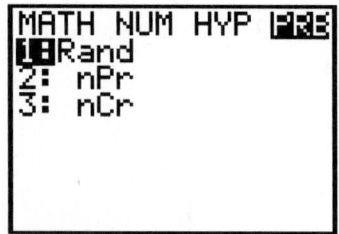

FIGURE 19.4a

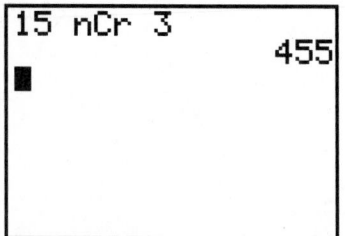

FIGURE 19.4b

You could have used the $\boxed{n!}$ key on a calculator to determine these coefficients. For example, to determine $\binom{15}{3}$

PRESS	DISPLAY
15 $\boxed{n!}$	1.3076744 12
$\boxed{\div}$ 12 $\boxed{n!}$	4.790016 08
$\boxed{\div}$	2730
3 $\boxed{n!}$	6
$\boxed{=}$	455

Some calculators have a $\boxed{Cy,x}$ or $\boxed{nC_r}$ key that can be used to determine this value. For example, on an RPN calculator, $\binom{15}{3}$ could be evaluated by using the keystroke sequence $\boxed{\text{ENTER}}$ 3 $\boxed{\text{2nd}}$ $\boxed{Cy,x}$ and the answer 455 will be displayed.

On a TI-82, you first press the top number in this expression, 15, then press $\boxed{\text{MATH}}$ to access that Math menu. Next press $\boxed{\blacktriangleright}$ three times so that the PRB at the top of the Math menu is highlighted. (PRB stands for probability.) There are three operations listed on the PRB screen. The third operation is listed as "3: nCr," as shown in Figure 19.4a. Press "3" (or $\boxed{\blacktriangledown}$ three times) and the second number, 3. Press $\boxed{\text{ENTER}}$. The screen now displays the answer, 455, as shown in Figure 19.4b.

EXAMPLE 19.32

Find the eighth term of $(x-2)^{17}$.

Solution The eighth term will be

$$\binom{17}{7}x^{10}(-2)^7 = \frac{17!}{7!10!}x^{10}(-2)^7$$
$$= 19{,}448x^{10}(-128)$$
$$= -2{,}489{,}344x^{10}$$

Binomial Series

If we rewrite the binomial in the form $1+x$, then we obtain the **binomial series**

$$(1+x)^n = 1+nx+\binom{n}{2}x^2+\binom{n}{3}x^3+\cdots+\binom{n}{n-1}x^{n-1}+x^n$$

It can be shown that the binomial series is valid for any real number n if $|x| < 1$. If n is negative or a rational number, we then get an infinite series.

$$(1+x)^n = 1+nx+\frac{n(n-1)}{2!}x^2+\frac{n(n-1)(n-2)}{3!}x^3\cdots$$

Normally, with an infinite binomial series, we calculate as many terms as are needed. One application of the binomial series is to find numerical approximations, particularly if you want more accuracy than you can get from a calculator or computer.

EXAMPLE 19.33

Find the value of $\sqrt[7]{0.875}$ to three significant digits.

Solution We can rewrite $\sqrt[7]{0.875} = (1 - 0.125)^{1/7}$. We will use the binomial series with $n = \frac{1}{7}$ and $x = -0.125$.

$$\sqrt[7]{0.875} = 1 + \frac{1}{7}(-0.125) + \frac{\frac{1}{7}\left(\frac{-6}{7}\right)}{2}(-0.125)^2 + \frac{\frac{1}{7}\left(\frac{-6}{7}\right)\left(\frac{-13}{7}\right)}{6}(-0.125)^3 + \cdots$$

We can omit the terms after the fourth one, because they are not significant.

$$\sqrt[7]{0.875} \approx 1 - 0.0179 - 0.0010 - 0.0001 = 0.981.$$

Checking this on a calculator, you get $0.9811 \ldots$, which is the same to three significant digits.

EXAMPLE 19.34

Show that $\displaystyle\lim_{n\to\infty}\left(1 + \frac{1}{n}\right)^n = e$.

Solution This is a binomial series and

$$\left(1 + \frac{1}{n}\right)^n = 1 + n\left(\frac{1}{n}\right) + \frac{n(n-1)}{2!}\left(\frac{1}{n}\right)^2 + \frac{n(n-1)(n-2)}{3!}\left(\frac{1}{n}\right)^3 + \cdots$$

$$= 1 + 1 + \frac{1}{2!}\left(\frac{n-1}{n}\right) + \frac{1}{3!}\left(\frac{n-1}{n}\right)\left(\frac{n-2}{n}\right) + \cdots$$

$$= 1 + 1 + \frac{1}{2!}\left(1 - \frac{1}{n}\right) + \frac{1}{3!}\left(1 - \frac{1}{n}\right)\left(1 - \frac{2}{n}\right) + \cdots$$

Since $\displaystyle\lim_{n\to\infty}\frac{1}{n} = 0$, this is equivalent to

$$1 + 1 + \frac{1}{2!} + \frac{1}{3!} + \frac{1}{4!} + \cdots$$

If you calculate these values, you will get 2.7083333. Including more values will show that

$$\lim_{n\to\infty} 1 + 1 + \frac{1}{2!} + \frac{1}{3!} + \cdots + \frac{1}{n!} + \cdots = 2.71828\cdots = e.$$

Exercise Set 19.5

In Exercises 1–8, expand and simplify the given expression by use of Pascal's triangle.

1. $(a+1)^4$

2. $(2x+b)^5$

3. $(3x-1)^4$

4. $(x-2y)^5$

5. $\left(\dfrac{x}{2}+d\right)^6$

6. $\left(xy-\dfrac{a}{3}\right)^6$

7. $\left(\dfrac{a}{2}-\dfrac{4}{b}\right)^5$

8. $\left(\dfrac{x}{3}+\dfrac{2}{y}\right)^4$

In Exercises 9–16, expand and simplify the given expression by use of the binomial formula.

9. $(a+b)^7$

10. $(a+b)^9$

11. $(t-a)^8$

12. $(x-2a)^7$

13. $(2a-1)^6$

14. $(a^2-3)^9$

15. $\left(x^2y+\dfrac{a}{2}\right)^7$

16. $\left(\dfrac{a}{3}-\dfrac{xy}{2}\right)^6$

In Exercises 17–22, find the first four terms of each binomial expansion.

17. $(x+y)^{12}$

18. $(x-5)^{10}$

19. $(1-a)^{-2}$

20. $(1+x)^{1/2}$

21. $(1+b)^{-1/3}$

22. $(1-y)^{-1/4}$

In Exercises 23–30, approximate the values of the given expression to three decimal places using the binomial series. Check your result with a calculator.

23. $(1.1)^4=(1+0.1)^4$

24. $(0.85)^5=(1-0.15)^5$

25. $\sqrt{1.1}$

26. $\sqrt[3]{1.01}$

27. $\sqrt[5]{1.04}$

28. $\sqrt[6]{0.98}$

29. $\sqrt[3]{0.95}$

30. $\sqrt[4]{0.925}$

In Exercises 31–36, find the indicated term of the given binomial expression.

31. The sixth term of $(x+y)^{15}$

32. The eighth term of $(a+b)^{20}$

33. The fifth term of $(2x-y)^{12}$

34. The fourth term of $(3x-2y)^{10}$

35. The term involving b^4 in $(a+b)^{14}$

36. The term involving x^5 in $(x+y)^{15}$

Solve Exercises 37–40.

37. *Nuclear physics*   The energy of an electron traveling at speed v in special relativity theory, is
$$mc^2\left(1-\frac{v^2}{c^2}\right)^{-1/2}$$, where m is the electron mass and c is the speed of light. The factor mc^2 is called the **rest mass energy** (or the energy when $v=0$). **(a)** Find the first three terms of the expansion of $\left(1-\dfrac{v^2}{c^2}\right)^{-1/2}$.
(b) Multiply the result of **(a)** by mc^2 to get the energy at speed v.

38. *Nuclear physics*   The velocity v of electrons from a high-energy accelerator is very near the velocity of light, c. Given the voltage V of the accelerator, we often need to calculate the ratio v/c. This ratio can be calculated from the formula
$$\frac{v}{c}=\sqrt{1-\frac{1}{4V^2}}, \quad V=\text{number of million volts}$$
(a) Find the first two terms of the expansion of
$$\sqrt{1-\frac{1}{4V^2}}$$

(b) Use your answer from **(a)** to determine $1-\dfrac{v}{c}$, given the following values of V. (Remember, V is the number of million volts.)
(i) 100 million volts
(ii) 500 MV (1 MV = 1 megavolt = 1 million volts)
(iii) 125 billion volts
(iv) 250 GV (1 GV = 1 gigavolt = 1 billion volts)

39. *Physics*   At a point in a magnetic field, the field strength is given by
$$H=\frac{2m\ell}{(r^2+\ell^2)^{3/2}}$$
Use the binomial series to find the first three terms of the expression $(r^2+\ell^2)^{3/2}$ by expressing it as
$$(r^2)^{3/2}\left(1+\frac{\ell^2}{r^2}\right)^{3/2}$$

40. In the formula for a derivative in calculus, the following expression may occur:
$$\frac{(x+h)^4-x^4}{h}$$
Expand and simplify this expression.

 **In Your Words**

41. Describe the binomial formula. What is the binomial formula used for?

42. What is Pascal's triangle and what does it have to do with the binomial formula?

▤ CHAPTER 19 REVIEW

Important Terms and Concepts

Arithmetic sequence
   Common difference
   Sum of first n terms
Binomial theorem
Geometric sequence
   Common ratio
   Sum of first n terms
Limit
Partial sum
Pascal's triangle
Recursion formula
Sequence
   Arithmetic

Finite
Geometric
Infinite
Series
   Arithmetic
   Binomial
   Convergent
   Divergent
   Finite
   Geometric
   Infinite
Summation notation

Review Exercises

In Exercises 1–4, find the first six terms with the specified general term.

1. $a_n = \dfrac{1}{n+2}$

2. $b_n = \dfrac{3}{2n-1}$

3. $\left\{ \dfrac{(-1)^n}{n(2n+1)} \right\}$

4. $\left\{ \dfrac{n^2+1}{3n-1} \right\}$

In Exercises 5–8, find the first six terms of the recursively defined sequence.

5. $a_1 = 1,\, a_n = \dfrac{a_{n-1}}{n}$

6. $a_1 = 5,\, a_n = a_{n-1} - n$

7. $a_1 = 1,\, a_2 = 3,\, a_n = a_{n-1} + a_{n-2}$

8. $a_1 = 1,\, a_2 = 2,\, a_n = (a_{n-1})(a_{n-2})$

In Exercises 9–14, determine whether the given sequence is an arithmetic sequence, a geometric sequence, or neither. For the arithmetic and geometric sequences, find the common difference or ratio and the indicated term.

9. $10, 7, 4, \ldots$ (10th term)

10. $8, 2, \frac{1}{2}, \ldots$ (8th term)

11. $-1, 3, 7, \ldots$ (7th term)

12. $7, 4, 0, -5, \ldots$ (9th term)

13. $3, -\frac{1}{2}, \frac{1}{12}, \ldots$ (10th term)

14. $2.05, 20.5, 205, \ldots$ (7th term)

Solve Exercises 15–17.

15. For an arithmetic sequence, $a_7 = 12$ and $a_8 = 18$. What is a_1?

16. For a geometric sequence, $a_6 = 9$ and $a_7 = \frac{9}{7}$. What is a_8?

17. Write the first four terms of

(a) $\displaystyle\sum_{k=1}^{30} 4\left(\frac{1}{3}\right)^k$ and (b) $\displaystyle\sum_{k=0}^{25} \frac{k+2}{k+1}$.

In Exercises 18–20, evaluate the given sum.

18. $\displaystyle\sum_{k=1}^{4} (k+1)$

19. $\displaystyle\sum_{k=0}^{5} (3k-1)$

20. $\displaystyle\sum_{n=1}^{4} \frac{(-1)^n n}{2^n+1}$

In Exercises 21–24, determine whether the terms of the given series form an arithmetic or geometric series and find the indicated sum.

21. $4+9+14+\cdots$; S_{10}

22. $2-5-12, \cdots$; S_{12}

23. $1+\frac{1}{3}+\frac{1}{9}+\cdots$; S_{14}

24. $\sqrt{5}+1+2\sqrt{5}+2+3\sqrt{5}+3\cdots$; S_8

In Exercises 25–30, determine which of the infinite geometric series converge and which diverge. For those that converge, find the sum.

25. $\frac{1}{3}+\frac{1}{6}+\frac{1}{12}+\cdots$

26. $8+6+4.5+\cdots$

27. $0.03+0.3+3+\cdots$

28. $1.5-0.15+0.015+\cdots$

29. $\displaystyle\sum_{n=1}^{\infty} \left(\frac{1}{\sqrt{3}}\right)^n$

30. $\displaystyle\sum_{n=0}^{\infty} \left(\sqrt{2}\right)^{-n}$

In Exercises 31 and 32, use infinite series to find the rational numbers corresponding to the given decimal number.

31. $0.185185\ldots$

32. $0.611111\ldots$

In Exercises 33–36, expand and simplify the given binomial expression.

33. $(a+2)^5$

34. $(3x-y)^6$

35. $\left(\frac{x}{2}-3y^2\right)^6$

36. $\left(\frac{2a^3}{5}+\frac{5b}{2}\right)^5$

In Exercises 37–40, find the first four terms of the given binomial expansion.

37. $(2x+y)^{15}$

38. $(1+ax^2)^{10}$

39. $(1-x)^{-5}$

40. $(1+b)^{-1/4}$

Solve Exercises 41–50.

41. What is the seventh term of $(x+2y)^{20}$?

42. What is the fifth term of $(a-3b)^{12}$?

43. Approximate $\sqrt[5]{1.02}$ to three decimal places using the binomial expansion. Check your work with a calculator.

44. Approximate $\sqrt[7]{0.98}$.

45. *Finance* If you invest $400 at 6% compounded monthly for 10 years, how much will you have at the end of that time?

46. Some copying machines make reduced copies. Suppose that you copy a page 20 cm wide and it comes out $\frac{3}{4}$ as wide. Then you copy this, and so on. How wide is the fifth copy?

47. If you lay the original and five copies from Exercise 46 side by side on the floor, how far will they extend?

48. Suppose that you continued the copying process that you began in Exercise 46 indefinitely. If you lay the original and all the copies side by side on the floor, how far would they extend?

49. *Physics*   A ball starting from rest rolls down a uniform incline so that it covers 10 in. during the first second, 25 in. during the second second, 40 in. during the third second, and so on. How long will it take until it covers 250 ft in 1 s?

50. *Business*   A company had sales of $250,000 during its first year of operation. Sales increased by $40,000 per year during each successive year. What were the sales of the company in the 10th year? What were the total sales of the company for the first 10 years?

▤ CHAPTER 19 TEST

1. Determine the first six terms of $a_n = \dfrac{5}{2n-3}$.

2. Determine the first six terms of $a_1 = -2$, $a_n = 3 + na_{n-1}$.

3. Determine whether the given sequence is an arithmetic sequence, geometric sequence, or neither. If appropriate, find the common difference or ratio and the indicated term.

$$15, 12.5, 10, \dots \text{ (10th term)}$$

4. For an arithmetic sequence with $a_5 = 10$ and $a_6 = 14$, what is a_1?

5. Evaluate $\displaystyle\sum_{k=1}^{5} (2k-1)$.

6. Determine whether this infinite geometric series converges or diverges. If it converges, find its sum.

$$\frac{1}{5} - \frac{1}{10} + \frac{1}{20} + \cdots$$

7. Determine the rational number that corresponds to $0.435435\dots$.

8. Find the first four terms of $(3x - 2y)^{12}$.

9. A computer software company predicts that it will sell 2,000 copies of a new type of software during the first month and that each month sales will be 1,500 copies more than the sales of the previous month. How many copies can they expect to sell during the first year?

10. A transistor's leakage current increases 1% for each increase of one degree Celsius. If a certain transistor's leakage is 1.40 nA at 20°C, how much will it be at 30°C? (Hint: An increase of 1% means that $r = 1.10$.)

CHAPTER

20

Trigonometric Formulas, Identities, and Equations

A highway engineer is designing the curve at an intersection where two highways intersect at angle θ. In Section 20.3, we will learn how to use trigonometry to help design this curve.

Courtesy of Ruby Gold

In Chapters 9 through 11, we discussed the fundamental reciprocal and quotient identities of trigonometry. We also learned how to use trigonometry to solve both right and oblique triangles and how to graph trigonometric functions. In this chapter, we will return to the study of trigonometry, establishing the remaining standard trigonometric identities. Among these will be identities for the sums, differences, and multiples of angles. The identities you will learn in this chapter are used in advanced mathematics, particularly calculus, and in engineering, physics, and technical areas to simplify complicated expressions and to help solve equations that involve trigonometry.

≡ 20.1
BASIC IDENTITIES

An **identity** is an equation that is true for all values of the variable. In Section 9.2, we introduced two groups of trigonometric identities. One group was called **reciprocal identities** and consisted of

$$\csc\theta = \frac{1}{\sin\theta} \qquad \sec\theta = \frac{1}{\cos\theta} \qquad \cot\theta = \frac{1}{\tan\theta}$$

The second group was the **quotient identities**.

$$\tan\theta = \frac{\sin\theta}{\cos\theta} \qquad \cot\theta = \frac{\cos\theta}{\sin\theta}$$

Pythagorean Identity

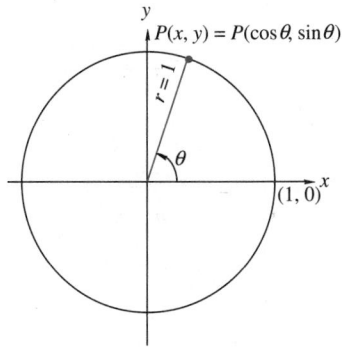

FIGURE 20.1

Now, suppose that we have a unit circle as shown in Figure 20.1. (Remember, a unit circle is a circle with radius 1.) If this circle is centered at the origin and $P(x, y)$ is any point on the circle on the terminal side of an angle θ in standard position, then $x = \cos\theta$ and $y = \sin\theta$. By the Pythagorean theorem, $x^2 + y^2 = r^2$, and since $r = 1$, we get the first of the **Pythagorean identities**.

$$\sin^2\theta + \cos^2\theta = 1$$

≡ Note

The term $\sin^2\theta$ is an abbreviation for $(\sin\theta)^2$. Similarly, $\cos^2\theta$ is an abbreviation for $(\cos\theta)^2$ and $\tan^2\theta$ is an abbreviation for $(\tan\theta)^2$. While we write $\sin^2\theta$, $\cos^2\theta$, or $\tan^2\theta$, you must enter them in a calculator as $(\sin\theta)^2$, $(\cos\theta)^2$, or $(\tan\theta)^2$.

If we divide both sides of this identity by $\cos^2\theta$, we get

$$\frac{\sin^2\theta}{\cos^2\theta} + \frac{\cos^2\theta}{\cos^2\theta} = \frac{1}{\cos^2\theta}$$

or $\qquad \tan^2\theta + 1 = \sec^2\theta$

This is the second Pythagorean identity. The third, and last, Pythagorean identity is produced by dividing $\sin^2\theta + \cos^2\theta = 1$ by $\sin^2\theta$, with the result

$$1 + \cot^2\theta = \csc^2\theta$$

Thus, there are three Pythagorean identities.

Pythagorean Identities	$\sin^2\theta + \cos^2\theta = 1$
	$\tan^2\theta + 1 = \sec^2\theta$
	$1 + \cot^2\theta = \csc^2\theta$

Proving Identities

The eight basic identities, the reciprocal, quotient, and Pythagorean identities, can be used to develop and prove other identities. Your ability to prove an identity depends greatly on your familiarity with the eight basic identities.

To prove that an identity is true, you change either side, or both sides, until the sides are the same. Each side must be worked separately. Since you do not know that the two sides are equal (which is what you are trying to prove), you cannot transpose terms from one side to the other. Some people draw a vertical line between the two sides until they can show that the sides are equal. The vertical line acts as a reminder that you should work on each side separately.

EXAMPLE 20.1

Prove the identity $\csc\theta = \dfrac{\cot\theta}{\cos\theta}$.

Solution We will change the right-hand side of the identity until it looks like the left-hand side.

$$
\begin{array}{c|ll}
\csc\theta & \dfrac{\cot\theta}{\cos\theta} & \\[2ex]
 & \dfrac{\frac{\cos\theta}{\sin\theta}}{\cos\theta} & \text{Change } \cot\theta \text{ to } \dfrac{\cos\theta}{\sin\theta}. \\[3ex]
 & \dfrac{\cos\theta}{\sin\theta}\cdot\dfrac{1}{\cos\theta} & \text{Change the division problem to a multiplication problem.} \\[3ex]
 & \dfrac{1}{\sin\theta} & \text{Multiply.} \\[3ex]
 & \csc\theta & \text{Reciprocal identity.}
\end{array}
$$

So, $\csc\theta = \dfrac{\cot\theta}{\cos\theta}$.

EXAMPLE 20.2

Prove the identity: $|\sin x| = \dfrac{|\tan x|}{\sqrt{1+\tan^2 x}}$.

Solution The right-hand side is more complicated than the left-hand side, so we will simplify the right-hand side until it matches the left-hand side.

EXAMPLE 20.2 (Cont.)

$$|\sin x| \quad \left| \quad \frac{|\tan x|}{\sqrt{1+\tan^2 x}} \right.$$

$$\frac{|\tan x|}{\sqrt{\sec^2 x}}$$ Use the Pythagorean identity to replace $1+\tan^2 x$ with $\sec^2 x$.

$$\left| \frac{\tan x}{\sec x} \right|$$ Take the square root. Notice that $\sqrt{\sec^2 x} = |\sec x|$.

$$\left| \frac{\frac{\sin x}{\cos x}}{\frac{1}{\cos x}} \right|$$ Express $\tan x$ and $\sec x$ in terms of $\sin x$ and $\cos x$.

$$\left| \frac{\sin x}{\cos x} \cdot \frac{\cos x}{1} \right|$$ Change the division problem to a multiplication problem.

$$|\sin x|$$ Multiply.

Thus, we have shown that $|\sin x| = \dfrac{|\tan x|}{\sqrt{1+\tan^2 x}}$.

Notice that we had to use the properties of absolute value to prove this identity.

EXAMPLE 20.3

Prove the identity $\sec\theta - \sec\theta \sin^2\theta = \cos\theta$.

Solution In this example, we start with the more complicated left-hand side and simplify it until it matches the right-hand side.

$$\sec\theta - \sec\theta \sin^2\theta \quad \left| \quad \cos\theta \right.$$
$$\sec\theta(1 - \sin^2\theta) \qquad\qquad \text{Factor.}$$
$$\sec\theta(\cos^2\theta) \qquad\qquad \text{Pythagorean identity.}$$
$$\frac{1}{\cos\theta}(\cos^2\theta) \qquad\qquad \text{Reciprocal identity.}$$
$$\cos\theta \qquad\qquad \text{Multiply.}$$

And so, $\sec\theta - \sec\theta \sin^2\theta = \cos\theta$.

In this example, we used a different version of a Pythagorean identity. We gave you the identity $\sin^2\theta + \cos^2\theta = 1$. You should also recognize the two variations of this identity: $\sin^2\theta = 1 - \cos^2\theta$ and $\cos^2\theta = 1 - \sin^2\theta$. There are two variations of each of the other Pythagorean identities.

EXAMPLE 20.4

Prove the identity: $\sec^2\theta\csc^2\theta = \sec^2\theta + \csc^2\theta$.

Solution Here we simplify the right-hand side until it matches the left-hand side.

$\sec^2\theta\csc^2\theta$	$\sec^2\theta + \csc^2\theta$	
	$\dfrac{1}{\cos^2\theta} + \dfrac{1}{\sin^2\theta}$	Reciprocal identities.
	$\dfrac{\sin^2\theta}{\cos^2\theta\sin^2\theta} + \dfrac{\cos^2\theta}{\cos^2\theta\sin^2\theta}$	Rewrite with a common denominator.
	$\dfrac{\sin^2\theta + \cos^2\theta}{\cos^2\theta\sin^2\theta}$	Add.
	$\dfrac{1}{\cos^2\theta\sin^2\theta}$	Pythagorean identity.
	$\dfrac{1}{\cos^2\theta} \cdot \dfrac{1}{\sin^2\theta}$	Factor.
	$\sec^2\theta\csc^2\theta$	Reciprocal identities.

So, $\sec^2\theta\csc^2\theta = \sec^2\theta + \csc^2\theta$.

Application

EXAMPLE 20.5

Malus's law concerns light incident on a polarizing plate and describes the amount of light transmitted I in terms of the angle of incidence θ and the maximum intensity of light transmitted, M. Malus's law can be written as

$$I = M - M\tan^2\theta\cos^2\theta$$

Express the right-hand side of this equation in terms of the $\cos\theta$.

Solution We have

$$I = M - M\tan^2\theta\cos^2\theta$$
$$= M - M\left(\frac{\sin^2\theta}{\cos^2\theta}\right)\cos^2\theta$$
$$= M - M\sin^2\theta$$
$$= M(1 - \sin^2\theta)$$
$$= M\cos^2\theta$$

So, Malus's law can be more simply expressed as

$$I = M\cos^2\theta$$

Using Graphs to Help Verify Identities

A graphing calculator or graphing software can also be used to prove or disprove an indentity. With this technique you separately graph each side of the proposed identity over an interval that contains at least one complete period. If both graphs are identical, then you can conclude that the identity is true whenever the functions are defined. A major difficulty arises because neither calculators nor computers define the cot, sec, and csc functions.

EXAMPLE 20.6

Use a graphing calculator to "prove" the identity $\sec^2\theta\csc^2\theta = \sec^2\theta + \csc^2\theta$.

Solution We know that this identity is true because it is the same identity that we proved in Example 20.4.

The graph of the left-hand side of this identity is shown in Figure 20.2a and the graph of the right-hand side is shown in Figure 20.2b. You can see that these two graphs are identical, hence this is a valid identity.

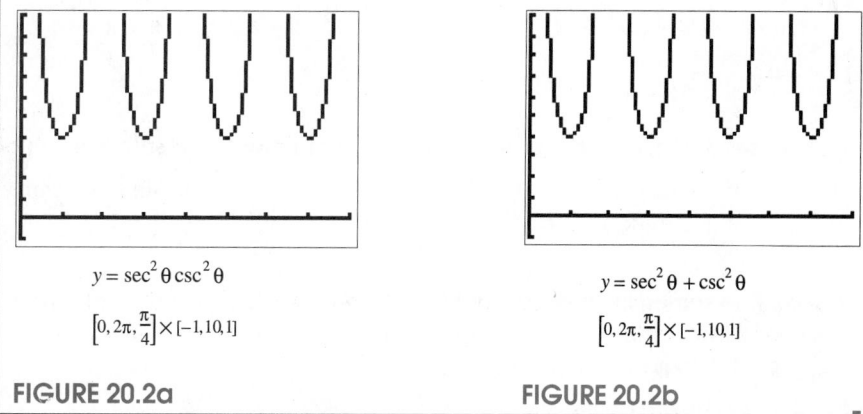

$$y = \sec^2\theta\csc^2\theta \qquad\qquad y = \sec^2\theta + \csc^2\theta$$

$$\left[0, 2\pi, \tfrac{\pi}{4}\right] \times [-1, 10, 1] \qquad\qquad \left[0, 2\pi, \tfrac{\pi}{4}\right] \times [-1, 10, 1]$$

FIGURE 20.2a   **FIGURE 20.2b**

Caution

In practice, when you graph the two sides of an identity, they both appear on the same screen. You have to watch carefully as the second graph is drawn to see if any new points appear.

Exercise Set 20.1

1. Prove the Pythagorean identity $1 + \cot^2\theta = \csc^2\theta$ from the identity $\sin^2\theta + \cos^2\theta = 1$.

Prove each of the identities in Exercises 2–30.

2. $\tan x \cot x = 1$

3. $\sin\theta\sec\theta = \tan\theta$

4. $\cos\theta(\tan\theta + \sec\theta) = \sin\theta + 1$

5. $\dfrac{\sin\theta}{\cot\theta} = \sec\theta - \cos\theta$

6. $\tan x = \dfrac{\sec x}{\csc x}$

7. $(1 - \sin^2\theta)(1 + \tan^2\theta) = 1$

8. $\dfrac{\sin A}{\csc A} + \dfrac{\cos A}{\sec A} = 1$

9. $1 - \dfrac{\sin A}{\csc A} = \cos^2 A$

10. $(1 + \tan\theta)(1 - \tan\theta) = 2 - \sec^2\theta$

11. $(1 + \cos x)(1 - \cos x) = \sin^2 x$

12. $\sec^4 x - \sec^2 x = \tan^4 x + \tan^2 x$

13. $2\csc\theta = \dfrac{\sin\theta}{1 + \cos\theta} + \dfrac{1 + \cos\theta}{\sin\theta}$

14. $\cos x = \sin x \cot x$

15. $(\sin\theta + \cos\theta)^2 = 1 + 2\sin\theta\cos\theta$

16. $\csc^2 x(1 - \cos^2 x) = 1$

17. $\dfrac{\tan\theta + \cot\theta}{\tan\theta - \cot\theta} = \dfrac{\tan^2\theta + 1}{\tan^2\theta - 1}$

18. $\dfrac{1 - \sin x}{\cos x} = \dfrac{\cos x}{1 + \sin x}$

19. $\dfrac{\sec\theta - \csc\theta}{\sec\theta + \csc\theta} = \dfrac{\tan\theta - 1}{\tan\theta + 1}$

20. $(\sin A + \cos A)^2 + (\sin A - \cos A)^2 = 2$

21. $\tan^2 x \cos^2 x + \cot^2 x \sin^2 x = 1$

22. $\tan\theta + \dfrac{\cos\theta}{1 + \sin\theta} = \sec\theta$

23. $\sec^4 x - \sec^2 x = \tan^2 x \sec^2 x$

24. $\cos^2 A - \sin^2 A = 2\cos^2 A - 1$

25. $\dfrac{\tan\theta - \sin\theta}{\sin^3\theta} = \dfrac{\sec\theta}{1 + \cos\theta}$

26. $\dfrac{\sin x - \cos x + 1}{\sin x + \cos x - 1} = \dfrac{\sin x + 1}{\cos x}$

27. $\tan^2\theta \csc^2\theta \cot^2\theta \sin^2\theta = 1$

28. $\tan x \sin x + \cos x = \sec x$

29. $\dfrac{\sec A + \csc A}{\tan A + \cot A} = \sin A + \cos A$

30. $\dfrac{\sin^3 x + \cos^3 x}{\sin x + \cos x} = 1 - \sin x \cos x$

In Exercises 31–34, use your graphing calculator to prove or disprove each of the following identities.

31. $2\csc 2x = \sec x \csc x$

32. $\cos 2x + 1 = 2\cos^2 x$

33. $\sin \tfrac{1}{2} x = \tfrac{1}{2} \sin x$

34. $1 - \cos 2x = \sin^2 x$

To show that something is not an identity, all you need is one counterexample. A counterexample is an example that shows that something is not true. In Exercises 35–39, use the indicated angles as a counterexample to show that the relation is not an identity.

35. $2\sin\theta \neq \sin 2\theta;\ \theta = 90°$

36. $\dfrac{\tan A}{2} \neq \tan\left(\dfrac{A}{2}\right);\ A = 60°$

37. $\cos(\theta^2) \neq (\cos\theta)^2;\ \theta = \pi$

38. $\sin(x - y) \neq \sin x - \sin y;\ x = 60°,\ y = 30°$

39. $\sin x \neq \dfrac{\tan x}{\sqrt{1 + \tan^2 x}};\ x = 120°$

Solve Exercises 40–44.

40. In finding the rate of change of $\cot x$, you get the expression $\dfrac{(\sin x)(-\sin x) - (\cos x)(\cos x)}{\sin^2 x}$. Show that this is equal to $-\csc^2 x$.

41. In finding the rate of change of $\cot^2 x$, you get the expression $-2\cot x \csc^2 x$. Show that this is equal to $-2\cos x \csc^3 x$.

42. In calculus, in order to determine the integral of $\sin^5 x$, we need to show that it is identical to $(1 - 2\cos^2 x + \cos^4 x)\sin x$. Prove this identity.

43. *Electricity*   In electric circuit theory, we use the expression

$$\frac{(1.2\sin\omega t - 1.6\cos\omega t)^2 + (1.6\sin\omega t + 1.2\cos\omega t)^2}{2L}$$

Show that this is identical to $\dfrac{2.0}{L}$.

44. *Physics*   The range, R, of a projectile fired at an acute angle θ with the horizontal at an initial velocity v is given by

$$R = \frac{2v^2\cos\theta\sin\theta}{g}$$

where g is the constant of gravitational acceleration. Rewrite the right-hand side using the sine function but not the cosine function.

In Your Words

45. Describe how to develop the other two Pythagorean identities from the identity $\sin^2 x + \cos^2 x = 1$.

46. Explain the process you use to prove an identity is true.

≡ 20.2
THE SUM AND DIFFERENCE IDENTITIES

As we saw in Exercises 35 and 38 in Exercise Set 20.1, $2\sin\theta \neq \sin 2\theta$ and $\sin(x - y) \neq \sin x - \sin y$. In this section, we will develop some identities for the sum and differences of the trigonometric functions. These identities are important for further studies in mathematics, such as in calculus. They are also important in wave mechanics, electric circuit theory, and in theory for other technical areas.

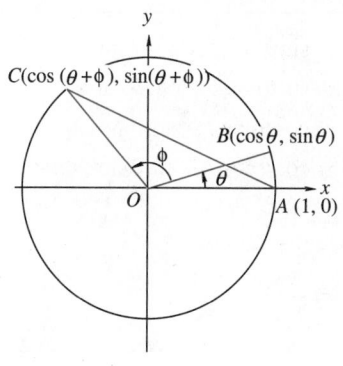

FIGURE 20.3a

Cos($\theta + \phi$)

We will begin with a rather lengthy development of the identity for the $\cos(\theta + \phi)$. Even this lengthy proof does not include all cases, but it will serve our purposes.

In Figures 20.3a and 20.3b, we have drawn two unit circles. In Figure 20.3a, $\angle AOB$ is θ and $\angle BOC$ is ϕ, so $\angle AOC$ is $\theta + \phi$. The coordinates of A, B, and C are also given. Since A is on the x-axis, its coordinates are $(1, 0)$. The coordinates of B are given in terms of θ and those of C are given in terms of $\theta + \phi$.

In Figure 20.3b, we have rotated $\triangle AOC$ through the angle $-\theta$ to get $\triangle DOF$. The coordinates of D and F are given in terms of θ and ϕ. Now, since $\triangle AOC$ is congruent to $\triangle DOF$, the distance from A to C must be the same as the distance from D to F. According to the distance formula from Section 15.1, the distance from A to C is

$$d(A, C) = \sqrt{(\cos(\theta + \phi) - 1)^2 + (\sin(\theta + \phi) - 0)^2}$$

Squaring both sides we get

$$\begin{aligned}
[d(A, C)]^2 &= (\cos(\theta + \phi) - 1)^2 + (\sin(\theta + \phi) - 0)^2 \\
&= \cos^2(\theta + \phi) - 2\cos(\theta + \phi) + 1 + \sin^2(\theta + \phi) \\
&= 2 - 2\cos(\theta + \phi)
\end{aligned}$$

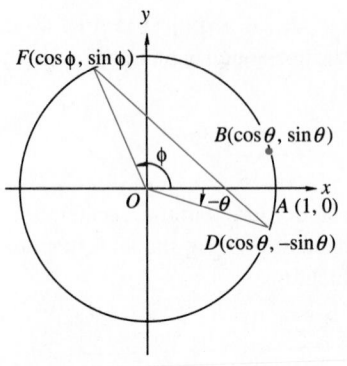

FIGURE 20.3b

In a similar manner, the distance from D to F is

$$d(D, F) = \sqrt{(\cos\phi - \cos\theta)^2 + (\sin\phi + \sin\theta)^2}$$

Again, squaring both sides, we get

$$
\begin{aligned}
[d(D, F)^2] &= (\cos\phi - \cos\theta)^2 + (\sin\phi + \sin\theta)^2 \\
&= \cos^2\phi - 2\cos\theta\cos\phi \\
&\quad + \cos^2\theta + \sin^2\phi + 2\sin\theta\sin\phi + \sin^2\theta \\
&= (\cos^2\phi + \sin^2\phi) + (\cos^2\theta + \sin^2\theta) \\
&\quad - 2\cos\theta\cos\phi + 2\sin\theta\sin\phi \\
&= 2 - 2\cos\theta\cos\phi + 2\sin\theta\sin\phi.
\end{aligned}
$$

Since $d(A, C) = d(D, F)$, we have

$$2 - 2\cos(\theta + \phi) = 2 - 2\cos\theta\cos\phi + 2\sin\theta\sin\phi$$

or

$$\cos(\theta + \phi) = \cos\theta\cos\phi - \sin\theta\sin\phi$$

Cos($\theta - \phi$)

If we substitute $-\phi$ for ϕ in the previous formula and remember the two identities $\cos(-\phi) = \cos\phi$ and $\sin(-\phi) = -\sin\phi$, we could show that

$$\cos(\theta - \phi) = \cos\theta\cos\phi + \sin\theta\sin\phi$$

Identities for the sum and difference of the sine and tangent functions can also be developed in a similar manner. The result is a total of six sum and difference identities.

Sum and Difference Identities	
	$\sin(\theta + \phi) = \sin\theta\cos\phi + \cos\theta\sin\phi$
	$\sin(\theta - \phi) = \sin\theta\cos\phi - \cos\theta\sin\phi$
	$\cos(\theta + \phi) = \cos\theta\cos\phi - \sin\theta\sin\phi$
	$\cos(\theta - \phi) = \cos\theta\cos\phi + \sin\theta\sin\phi$
	$\tan(\theta + \phi) = \dfrac{\tan\theta + \tan\phi}{1 - \tan\theta\tan\phi}$
	$\tan(\theta - \phi) = \dfrac{\tan\theta - \tan\phi}{1 + \tan\theta\tan\phi}$

EXAMPLE 20.7

If $\sin\theta = \frac{4}{5}$, $\cos\phi = \frac{-12}{13}$, θ is in Quadrant I, and ϕ is in Quadrant II, find
(a) $\sin(\theta + \phi)$, **(b)** $\cos(\theta - \phi)$, and **(c)** $\tan(\theta + \phi)$.

Solutions   If we draw reference triangles for θ and ϕ, and use the Pythagorean theorem to determine the missing side, we can determine the values of $\cos\theta$, $\tan\theta$, $\sin\phi$, and $\tan\phi$. These triangles are shown in Figure 20.4. From them, we can

EXAMPLE 20.7 (Cont.)

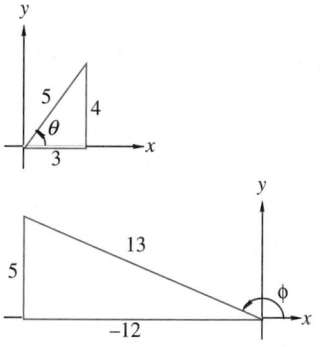

FIGURE 20.4

determine that $\cos\theta = \frac{3}{5}$, $\tan\theta = \frac{4}{3}$, $\sin\phi = \frac{5}{13}$, and $\tan\phi = \frac{-5}{12}$. We are now ready to apply the formulas.

(a) $\sin(\theta + \phi) = \sin\theta\cos\phi + \cos\theta\sin\phi$

$$= \left(\frac{4}{5}\right)\left(\frac{-12}{13}\right) + \left(\frac{3}{5}\right)\left(\frac{5}{13}\right)$$

$$= \frac{-48}{65} + \frac{15}{65} = -\frac{33}{65}$$

(b) $\cos(\theta - \phi) = \cos\theta\cos\phi + \sin\theta\sin\phi$

$$= \left(\frac{3}{5}\right)\left(\frac{-12}{13}\right) + \left(\frac{4}{5}\right)\left(\frac{5}{13}\right)$$

$$= \frac{-36}{65} + \frac{20}{65} = -\frac{16}{65}$$

(c) $\tan(\theta + \phi) = \dfrac{\tan\theta + \tan\phi}{1 - \tan\theta\tan\phi}$

$$= \frac{\frac{4}{3} + \frac{-5}{12}}{1 - \left(\frac{4}{3}\right)\left(\frac{-5}{12}\right)}$$

$$= \frac{\frac{16}{12} - \frac{5}{12}}{\frac{56}{36}} = \frac{\frac{11}{12}}{\frac{56}{36}}$$

$$= \frac{11}{12} \cdot \frac{36}{56} = \frac{33}{56}$$

So, $\sin(\theta+\phi) = -\frac{33}{65}$, $\cos(\theta-\phi) = -\frac{16}{65}$, and $\tan(\theta+\phi) = \frac{33}{56}$. Since $\sin(\theta+\phi)$ is negative and $\tan(\theta+\phi)$ is positive, $\theta+\phi$ is in the third quadrant.

EXAMPLE 20.8

If $\sin\alpha = 0.25$ and $\cos\beta = 0.65$, α in Quadrant II and β in Quadrant I, find **(a)** $\sin(\alpha - \beta)$ and **(b)** $\tan(\alpha + \beta)$.

Solutions Using the Pythagorean theorem and the fact that α is in Quadrant II, we determine that $\cos\alpha = -\sqrt{1 - 0.25^2} \approx -0.97$. Similarly, we find that $\sin\beta \approx 0.76$. We can now apply the formulas.

(a) $\sin(\alpha - \beta) = \sin\alpha\cos\beta - \cos\alpha\sin\beta$

$$\approx (0.25)(0.65) - (-0.97)(0.76)$$

$$= 0.1625 + 0.7372$$

$$= 0.8997$$

$$\approx 0.90$$

EXAMPLE 20.8 (Cont.)

(b) $\tan(\alpha + \beta) = \dfrac{\tan \alpha + \tan \beta}{1 - \tan \alpha \tan \beta}$

$\approx \dfrac{\dfrac{0.25}{-0.97} + \dfrac{0.76}{0.65}}{1 - \left(\dfrac{0.25}{-0.97}\right)\left(\dfrac{0.76}{0.65}\right)}$

$\approx \dfrac{-0.2577 + 1.1692}{1 - (-0.2577)(1.1692)}$

$= 0.7005$

EXAMPLE 20.9

Find the exact value of $\sin 75°$ by using the trigonometric values for $30°$ and $45°$.

Solution Since $75° = 30° + 45°$, we will use $\sin 75° = \sin(30° + 45°)$. Now $\sin 30° = \frac{1}{2}$, $\cos 30° = \dfrac{\sqrt{3}}{2}$, and $\sin 45° = \cos 45° = \dfrac{\sqrt{2}}{2}$. This gives

$$\sin 75° = \sin 30° \cos 45° + \cos 30° \sin 45°$$
$$= \frac{1}{2} \cdot \frac{\sqrt{2}}{2} + \frac{\sqrt{3}}{2} \cdot \frac{\sqrt{2}}{2}$$
$$= \frac{\sqrt{2}}{4} + \frac{\sqrt{6}}{4}$$
$$= \frac{\sqrt{2} + \sqrt{6}}{4}$$

≡ Note

We realize that the use of calculators makes it very unlikely that you will use such procedures to evaluate a given trigonometric value. We did these examples and have included exercises to give you practice using the sum and difference identities with numbers that you can verify on your calculator. This practice will also help you remember the identities later when you need them.

EXAMPLE 20.10

Simplify $\sin\left(x + \frac{\pi}{2}\right)$.

Solution $\sin\left(x + \dfrac{\pi}{2}\right) = \sin x \cos \dfrac{\pi}{2} + \cos x \sin \dfrac{\pi}{2}$

$= \sin x \cdot 0 + \cos x \cdot 1$

$= \cos x$

EXAMPLE 20.11

Verify that $\sin(\alpha+\beta)+\sin(\alpha-\beta)=2\sin\alpha\cos\beta$.

Solution

$\sin(\alpha+\beta)+\sin(\alpha-\beta)$	$2\sin\alpha\cos\beta$
$\sin\alpha\cos\beta+\cos\alpha\sin\beta$ $+\sin\alpha\cos\beta-\cos\alpha\sin\beta$	Expand $\sin(\alpha+\beta)$ and $\sin(\alpha-\beta)$.
$2\sin\alpha\cos\beta$	Collect terms.

So, $\sin(\alpha+\beta)+\sin(\alpha-\beta)=2\sin\alpha\cos\beta$.

Caution

Remember, $\sin(\alpha+\beta)\neq\sin\alpha+\sin\beta$. Make sure that you rewrite $\sin(\alpha+\beta)$ as $\sin\alpha\cos\beta+\cos\alpha\sin\beta$. In a similar way, you can show that $\cos(\alpha+\beta)\neq\cos\alpha+\cos\beta$ and $\tan(\alpha+\beta)\neq\tan\alpha+\tan\beta$.

Application

EXAMPLE 20.12

The displacement d of an object oscillating in simple harmonic motion can be determined by the expression

$$d = a\sin 2\pi ft\cos\beta + a\cos 2\pi ft\sin\beta$$

Express the right-hand side as a single term.

Solution If we factor an a out of both terms, then

$$d = a(\sin 2\pi ft\cos\beta + \cos 2\pi ft\sin\beta)$$

If we let $\alpha = 2\pi ft$, then the expression in parentheses is in the form $\sin\alpha\cos\beta + \cos\alpha\sin\beta = \sin(\alpha+\beta)$. So, the desired expression is

$$d = a\sin(2\pi ft + \beta)$$

Exercise Set 20.2

In Exercises 1–8, use the fact that $\sin 30° = \cos 60° = \frac{1}{2}$, $\sin 60° = \cos 30° = \frac{\sqrt{3}}{2}$, and $\sin 45° = \cos 45° = \frac{\sqrt{2}}{2}$, along with the other facts you know from trigonometry to determine the following.

1. $\sin 15°$
2. $\cos 75°$
3. $\sin 120°$
4. $\cos(-15°)$
5. $\tan 15°$
6. $\tan 135°$
7. $\sin 150°$
8. $\cos 105°$

In Exercises 9–16, simplify the given expression.

9. $\sin(x+90°)$
10. $\cos(x+\pi)$
11. $\cos(x+\frac{\pi}{2})$
12. $\sin(\frac{\pi}{2}-x)$
13. $\cos(\pi-x)$
14. $\tan(x-\frac{\pi}{4})$
15. $\sin(180°-x)$
16. $\tan(180°+x)$

If α and β are first quadrant angles, sin $\alpha = \frac{3}{4}$, and cos $\beta = \frac{7}{8}$, evaluate the given expressions in Exercises 17–24.

17. $\sin(\alpha + \beta)$

18. $\cos(\alpha + \beta)$

19. $\tan(\alpha + \beta)$

20. $\sin(\alpha - \beta)$

21. $\cos(\alpha - \beta)$

22. $\tan(\alpha - \beta)$

23. In what quadrant is $\alpha + \beta$?

24. In what quadrant is $\alpha - \beta$?

If α is a second quadrant angle, and β is a third quadrant angle with sin $\alpha = \frac{3}{4}$ and cos $\beta = \frac{-7}{8}$, determine each of the given expressions in Exercises 25–32.

25. $\sin(\alpha + \beta)$

26. $\cos(\alpha + \beta)$

27. $\tan(\alpha + \beta)$

28. $\sin(\alpha - \beta)$

29. $\cos(\alpha - \beta)$

30. $\tan(\alpha - \beta)$

31. In what quadrant is $\alpha + \beta$?

32. In what quadrant is $\alpha - \beta$?

Simplify the given expression in Exercises 33–40.

33. $\sin 47° \cos 13° + \cos 47° \sin 13°$

34. $\sin 47° \sin 13° + \cos 47° \cos 13°$

35. $\cos 32° \cos 12° - \sin 32° \sin 12°$

36. $\dfrac{\tan 40° + \tan 15°}{1 - \tan 40° \tan 15°}$

37. $\cos(\alpha + \beta)\cos \beta + \sin(\alpha + \beta)\sin \beta$

38. $\sin(x - y)\cos y + \cos(x - y)\sin y$

39. $\cos(x + y)\cos(x - y) - \sin(x + y)\sin(x - y)$

40. $\sin A \cos(-B) + \cos A \sin(-B)$

Prove each of the identities in Exercises 41–46.

41. $\sin(x + y)\sin(x - y) = \sin^2 x - \sin^2 y$

42. $(\sin A \cos B - \cos A \sin B)^2 + (\cos A \cos B + \sin A \sin B)^2 = 1$

43. $\cos \theta = \sin(\theta + 30°) + \cos(\theta + 60°)$

44. $\dfrac{\sin(x + y)}{\cos(x - y)} = \dfrac{\tan x + \tan y}{1 + \tan x \tan y}$

45. $\tan x - \tan y = \dfrac{\sin(x - y)}{\cos x \cos y}$

46. $\cos(A + B)\cos(A - B) = 1 - \sin^2 A - \sin^2 B$

In Exercises 47–50, use a calculator to show that the statements are true.

47. $\sin(20° + 37°) = \sin 20° \cos 37° + \cos 20° \sin 37°$

48. $\cos(15° + 63°) = \cos 15° \cos 63° - \sin 15° \sin 63°$

49. $\tan(0.2 + 1.3) = \dfrac{\tan 0.2 + \tan 1.3}{1 - (\tan 0.2)(\tan 1.3)}$

50. $\sin(2.3 - 1.1) = (\sin 2.3)(\cos 1.1) - (\cos 2.3)(\sin 1.1)$

In Exercises 51–54, with the help of a calculator use the given angles as a counterexample to show that each relation is not an identity.

51. $\sin(x + y) \neq \sin x + \sin y; x = 55°, y = 37°$

52. $\cos(x - y) \neq \cos x - \cos y; x = 68°, y = 24°$

53. $\cos(x + y) \neq \cos x + \cos y; x = 40°, y = 35°$

54. $\tan(x - y) \neq \tan x - \tan y; x = 76°, y = 37°$

Solve Exercises 55–58.

55. In Chapter 14, we learned that when two complex numbers are written in polar form their product is

$$[r_1(\cos\theta_1 + j\sin\theta_1)][r_2(\cos\theta_2 + j\sin\theta_2)]$$
$$= r_1 r_2[\cos(\theta_1 + \theta_2) + j\sin(\theta_1 + \theta_2)]$$

   Prove this formula.

56. *Physics*   A spring vibrating in harmonic motion described by the equation $y_1 = A_1\cos(\omega t + \pi)$ is subjected to another harmonic motion described by $y_2 = A_2\cos(\omega t - \pi)$. Show that

$$y_1 + y_2 = -(A_1 + A_2)\cos\omega t$$

57. *Optics*   When a light beam passes from one medium through another medium and exists in a third medium of the same density as the first, the displacement d of the light beam is $d = \dfrac{h}{\cos\theta_r}\sin(\theta_i - \theta_r)$, where θ_r is the angle of refraction, θ_i is the angle of incidence, and h is the thickness of the medium. Show that $d = h(\sin\theta_i - \cos\theta_i\tan\theta_r)$.

58. *Electricity*   The angle between voltage and current in an RC circuit is 45°. Develop an expression in terms of ω for $i(t)$ in milliamperes (mA) using $i(t) = I_p\sin(\theta - \omega t)$, if $I_p = 14.8$ mA.

In Your Words

59. Describe how to develop $\sin(\theta - \phi)$ if you know that $\sin(\theta + \phi) = \sin\theta\cos\phi + \cos\theta\sin\phi$.

60. Describe how to develop the $\tan(\theta + \phi)$ identity from the identities for $\sin(\theta + \phi)$ and $\cos(\theta + \phi)$.

≡ 20.3
THE DOUBLE- AND HALF-ANGLE IDENTITIES

In the Section 20.2, we studied the sum and difference identities. We can use these identities to develop double-angle identities. The double-angle identities can then be used to develop some half-angle identities.

Double-Angle Identities

The identity for $\sin(\theta + \phi)$ can be used to develop an identity for $\sin 2\theta$. To do this, calculate $\sin(\theta + \theta)$.

$$\begin{aligned}
\sin 2\theta &= \sin(\theta + \theta)\\
&= \sin\theta\cos\theta + \cos\theta\sin\theta\\
&= 2\sin\theta\cos\theta
\end{aligned}$$

In the same manner, we can develop $\cos 2\theta$.

$$\begin{aligned}
\cos 2\theta &= \cos(\theta + \theta)\\
&= \cos\theta\cos\theta - \sin\theta\sin\theta\\
&= \cos^2\theta - \sin^2\theta
\end{aligned}$$

This last identity has two other forms. If we use the Pythagorean identity, $\sin^2\theta + \cos^2\theta = 1$, we get

$$\begin{aligned}
\cos 2\theta &= \cos^2\theta - \sin^2\theta\\
&= (1 - \sin^2\theta) - \sin^2\theta\\
&= 1 - 2\sin^2\theta
\end{aligned}$$

We can replace $\sin^2\theta$ with $1 - \cos^2\theta$ and get a third version of this formula.

$$\cos 2\theta = \cos^2\theta - \sin^2\theta$$
$$= \cos^2\theta - (1 - \cos^2\theta)$$
$$= 2\cos^2\theta - 1$$

Once again, if we evaluate $\tan(\theta + \theta)$, we get the third double-angle identity:

$$\tan 2\theta = \tan(\theta + \theta)$$
$$= \frac{\tan\theta + \tan\theta}{1 - \tan\theta\tan\theta}$$
$$= \frac{2\tan\theta}{1 - \tan^2\theta}$$

This completes the list of double-angle identities.

Double-Angle Identities

$$\sin 2\theta = 2\sin\theta\cos\theta$$
$$\cos 2\theta = \cos^2\theta - \sin^2\theta$$
$$= 2\cos^2\theta - 1$$
$$= 1 - 2\sin^2\theta$$
$$\tan 2\theta = \frac{2\tan\theta}{1 - \tan^2\theta}$$

EXAMPLE 20.13

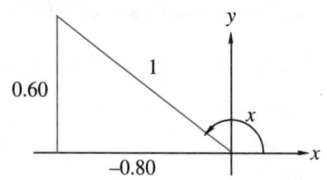

FIGURE 20.5

If $\sin x = 0.60$ and x is in the second quadrant, then determine **(a)** $\sin 2x$, **(b)** $\cos 2x$, and **(c)** $\tan 2x$.

Solutions We will first determine the value of $\cos x$. Since $\sin x = 0.60$, $\cos^2 x = 1 - \sin^2 x = 1 - (0.60)^2 = 0.64$ and $\cos x = \pm\sqrt{0.64} = \pm 0.80$. Since x is in Quadrant II (see Figure 20.5), $\cos x = -0.80$.

(a) $\sin 2x = 2\sin x\cos x$
$$= 2(0.60)(-0.80) = -0.96$$

(b) $\cos 2x = \cos^2 x - \sin^2 x$
$$= (-0.8)^2 - (0.6)^2$$
$$= 0.64 - 0.36 = 0.28$$

(c) Since $\sin x = 0.60$, $\cos x = -0.8$, and $\tan x = \dfrac{\sin x}{\cos x}$, we have $\tan x = \dfrac{0.60}{-0.80} = -\dfrac{3}{4}$. Thus,

$$\tan 2x = \frac{2\tan x}{1 - \tan^2 x}$$
$$= \frac{2\left(-\frac{3}{4}\right)}{1 - \left(-\frac{3}{4}\right)^2}$$

EXAMPLE 20.13 (Cont.)

$$= \frac{-\frac{3}{2}}{1 - \frac{9}{16}}$$

$$= \frac{-\frac{3}{2}}{\frac{7}{16}}$$

$$= \frac{-24}{7}$$

$$\approx -3.43$$

Caution

Don't forget that $\sin 2\alpha \neq 2\sin\alpha$ and $\cos 2\alpha \neq 2\cos\alpha$. We now know that

$$\sin 2\alpha = 2\sin\alpha\cos\alpha$$

and that

$$\cos 2\alpha = \cos^2\alpha - \sin^2\alpha$$

EXAMPLE 20.14

Rewrite $\cos 4x$ in terms of $\cos x$.

Solution Using the double-angle identity for $\cos 2\theta$, if we let $\theta = 2x$, then we get $\cos 4x = 2\cos^2 2x - 1$. Now using the double-angle identity $\cos 2x = 2\cos^2 x - 1$, we get

$$\cos 4x = 2(2\cos^2 x - 1)^2 - 1$$
$$= 2(4\cos^4 x - 4\cos^2 x + 1) - 1$$
$$= 8\cos^4 x - 8\cos^2 x + 1$$

Half-Angle Identities

If we solve $\cos 2x = 2\cos^2 x - 1$ for the $\cos x$, we get another identity—a half-angle identity.

$$\cos 2x = 2\cos^2 x - 1$$

$$\text{or} \quad \cos^2 x = \frac{1 + \cos 2x}{2}$$

and, taking the square root of both sides,

$$\cos x = \pm\sqrt{\frac{1 + \cos 2x}{2}}$$

If we let $x = \frac{\theta}{2}$, then

$$\cos\frac{\theta}{2} = \pm\sqrt{\frac{1 + \cos\theta}{2}}$$

If we solve $\cos^2 x = 1 - 2\sin^2 x$ for $\sin x$, we obtain

$$\sin^2 x = \frac{1 - \cos 2x}{2}$$

or $\qquad \sin x = \pm\sqrt{\frac{1 - \cos 2x}{2}}$

Again, if $x = \dfrac{\theta}{2}$, then

$$\sin \frac{\theta}{2} = \pm\sqrt{\frac{1 - \cos\theta}{2}}$$

Since $\tan x = \dfrac{\sin x}{\cos x}$, we can show that

$$\tan \frac{\theta}{2} = \pm\sqrt{\frac{1 - \cos\theta}{1 + \cos\theta}}$$
$$= \frac{\sin\theta}{1 + \cos\theta} = \frac{1 - \cos\theta}{\sin\theta}$$

We have developed the following three half-angle identities.

Half-Angle Identities

$$\sin \frac{\theta}{2} = \pm\sqrt{\frac{1 - \cos\theta}{2}}$$

$$\cos \frac{\theta}{2} = \pm\sqrt{\frac{1 + \cos\theta}{2}}$$

$$\tan \frac{\theta}{2} = \pm\sqrt{\frac{1 - \cos\theta}{1 + \cos\theta}} = \frac{\sin\theta}{1 + \cos\theta}$$
$$= \frac{1 - \cos\theta}{\sin\theta}$$

EXAMPLE 20.15

If $\cos\theta = \frac{-5}{13}$ and θ is in the third quadrant, find the values of **(a)** $\sin\dfrac{\theta}{2}$, **(b)** $\cos\dfrac{\theta}{2}$, and **(c)** $\tan\dfrac{\theta}{2}$.

Solution We need to determine which quadrant $\dfrac{\theta}{2}$ is in. We know θ is in Quadrant III, or $\pi < \theta < \frac{3\pi}{2}$, and so $\frac{\pi}{2} < \frac{\theta}{2} < \frac{1}{2}\left(\frac{3\pi}{2}\right)$ or $\frac{\pi}{2} < \frac{\theta}{2} < \frac{3\pi}{4}$. This means that $\dfrac{\theta}{2}$ is in Quadrant II. In Quadrant II, the sine is positive, cosine is negative, and tangent is negative.

(a) $\sin\dfrac{\theta}{2} = +\sqrt{\dfrac{1 - \cos\theta}{2}} = +\sqrt{\dfrac{1 - \left(-\frac{5}{13}\right)}{2}} = \sqrt{\dfrac{9}{13}} \approx 0.832$

EXAMPLE 20.15 (Cont.)

(b) $\cos\dfrac{\theta}{2} = -\sqrt{\dfrac{1+\cos\theta}{2}} = -\sqrt{\dfrac{1+\left(-\frac{5}{13}\right)}{2}} = -\sqrt{\dfrac{4}{13}} \approx -0.555$

(c) $\tan\dfrac{\theta}{2} = -\sqrt{\dfrac{1-\cos\theta}{1+\cos\theta}} = -\sqrt{\dfrac{1-\left(-\frac{5}{13}\right)}{1+\left(-\frac{5}{13}\right)}} = -\sqrt{\dfrac{\frac{18}{13}}{\frac{8}{13}}} = -\sqrt{\dfrac{9}{4}}$

$\qquad\qquad = -\dfrac{3}{2} = -1.500$

We could have determined $\tan\dfrac{\theta}{2}$ by using $\dfrac{\sin\frac{\theta}{2}}{\cos\frac{\theta}{2}}$. With the values above, we would have gotten -1.499. The error of 0.001 was caused by the use of approximations in parts **(a)** and **(b)**.

EXAMPLE 20.16

Prove the identity $2\sin\dfrac{x}{2}\cos\dfrac{x}{2} = \sin x$.

Solution

$2\sin\dfrac{x}{2}\cos\dfrac{x}{2}$	$\sin x$
$2\left(\pm\sqrt{\dfrac{1-\cos x}{2}}\right)\left(\pm\sqrt{\dfrac{1+\cos x}{2}}\right)$	Replace with half-angle identities.
$2\sqrt{\dfrac{1-\cos^2 x}{4}}$	Multiply.
$2\sqrt{\dfrac{\sin^2 x}{4}}$	Pythagorean identity.
$2\left(\dfrac{\sin x}{2}\right)$	Take the square root.
$\sin x$	Simplify.

Caution

Be careful when you use the double- and half-angle formulas. Begin by calculating the necessary values of θ, 2θ, and $\dfrac{\theta}{2}$ before they are substituted into the formula.

Application

EXAMPLE 20.17

A highway engineer is designing the curve at an intersection like the one shown by the photograph in Figure 20.6a. These two highways intersect at an angle θ. The edge of the highway is to join the two points A and B with an arc or a circle that is tangent to the highways at these two points. Determine the relationship between the radius of the arc r, the distance d of A and B from the intersection, and angle θ.

Courtesy of Michael A. Gallitelli, Metroland Photo Inc.

FIGURE 20.6a

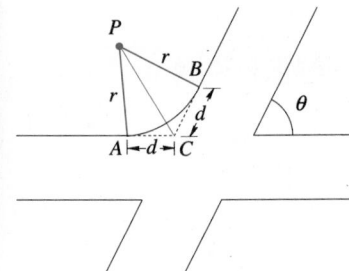

FIGURE 20.6b

Solution We begin by noticing in Figure 20.6b, that $\angle BCA$ and θ are supplementary angles. So, $m\angle BCA = 180° - \theta$. If the center of the circle is at P, then $\overline{PA} \perp \overline{AC}$, because a tangent to a circle, in this case $\overline{AC}$, is perpendicular to a radius at the point of tangency. Now, $\overline{PC}$ bisects $\angle BCA$, so $m\angle PCA = \frac{1}{2}m\angle BCA = 90° - \frac{\theta}{2}$. Since $\triangle PAC$ is a right triangle with right angle at A, $\tan \angle PCA = \frac{r}{d}$; so

$$d = \frac{r}{\tan \angle PCA} = r \cot \angle PCA = r \cot \left(90° - \frac{\theta}{2} \right) = r \tan \frac{\theta}{2}. \text{ Thus, we have shown}$$

that $d = r \tan \frac{\theta}{2}$.

Application

EXAMPLE 20.18

Two highways meet at an angle of 34°. The curb is to join points A and B located 45 ft from the beginning of the intersection. **(a)** Approximate the radius of the arc joining A and B. **(b)** Determine the length of the arc.

EXAMPLE 20.18 (Cont.)	

Solutions (a) From Example 20.17, we have the formula

$$d = r \tan \frac{\theta}{2}$$

In this example, $d = 45$ ft and $\theta = 34°$. We are to determine r.

$$r = \frac{d}{\tan \dfrac{\theta}{2}}$$

$$= \frac{45}{\tan\left(\frac{34}{2}\right)°}$$

$$= \frac{45}{\tan 17°}$$

$$\approx \frac{45}{0.3057}$$

$$\approx 147.19\,\text{ft}$$

(b) We want the length of the arc that forms the curb. As we saw in Section 9.6, the arc length s of a circle with radius r, formed by an angle θ, is

$$s = r\theta$$

provided that θ is in radians.

In this example, $\theta = 34° = \frac{34\pi}{180} = \frac{17\pi}{90}$ rad. So,

$$s = r\theta$$

$$= (147.19)\left(\frac{17\pi}{90}\right)$$

$$\approx 87.34\,\text{ft}$$

Exercise Set 20.3

In Exercises 1–14, if $\sin 30° = \cos 60° = \frac{1}{2}$, $\sin 60° = \cos 30° = \dfrac{\sqrt{3}}{2}$, and $\sin 45° = \cos 45° = \dfrac{\sqrt{2}}{2}$, determine the exact values of the given trigonometric function.

1. $\cos 15°$	**5.** $\sin 105°$	**9.** $\tan 22\frac{1}{2}°$	**12.** $\sin 37\frac{1}{2}°$
2. $\sin 75°$	**6.** $\cos 210°$	**10.** $\cos 67\frac{1}{2}°$	**13.** $\sin 127\frac{1}{2}°$
3. $\sin 15°$	**7.** $\cos 7\frac{1}{2}°$	**11.** $\cos 75°$	**14.** $\tan(-15°)$
4. $\cos 105°$	**8.** $\tan 15°$		

Use the information given in each of Exercises 15–20 to determine $\sin 2x$, $\cos 2x$, $\tan 2x$, $\sin\dfrac{x}{2}$, $\cos\dfrac{x}{2}$, and $\tan\dfrac{x}{2}$.

15. $\sin x = \frac{7}{25}$, x in Quadrant II

16. $\cos x = \frac{8}{17}$, x in Quadrant IV

17. $\sec x = \frac{29}{20}$, x in Quadrant I

18. $\csc x = \frac{-41}{9}$, x in Quadrant III

19. $\tan x = \frac{35}{12}$, x in Quadrant III

20. $\cot x = \frac{-45}{28}$, x in Quadrant II

Prove the identities in Exercises 21–32 for all angles in the domains of the functions.

21. $\cos^2 x = \sin^2 x + \cos 2x$

22. $\cos^2 4x - \sin^2 4x = \cos 8x$

23. $\cos 4x = 1 - 8\sin^2 x \cos^2 x$

24. $\cos 3x = 4\cos^3 x - 3\cos x$

25. $\dfrac{1 + \tan^2 \alpha}{1 - \tan^2 \alpha} = \sec 2\alpha$

26. $\sin 2x \cos 2x = \frac{1}{2}\sin 4x$

27. $1 - 2\sin^2 3x = \cos 6x$

28. $\dfrac{2\tan 3x}{1 - \tan^2 3x} = \tan 6x$

29. $\sin^2 x \cos^2 x = \frac{1}{4}\sin^2 2x$

30. $1 + \tan \beta \tan \dfrac{\beta}{2} = \dfrac{1}{\cos \beta}$

31. $\tan\left(\dfrac{\alpha + \beta}{2}\right)\cot\left(\dfrac{\alpha - \beta}{2}\right) = \dfrac{(\sin \alpha + \sin \beta)^2}{\sin^2 \alpha - \sin^2 \beta}$

32. $\cot \theta = \frac{1}{2}\left(\cot\dfrac{\theta}{2} - \tan\dfrac{\theta}{2}\right)$

In Exercises 33–36, use a calculator and the given angles as counterexamples to show that the following are not identities.

33. $\cos 2\theta \neq 2\cos \theta$; $\theta = 45°$

34. $\tan\dfrac{\theta}{2} \neq \dfrac{\tan \theta}{2}$; $\theta = 80°$

35. $\cot 2\alpha \neq 2\cos \alpha$; $\alpha = 150°$

36. $\sin\dfrac{x}{2} \neq \dfrac{\sin x}{2}$; $x = 210°$

Solve Exercises 37–40.

37. *Optics*   The index of refraction n of a prism whose apex angle is α and whose angle of minimum deviation is ϕ is given by

$$n = \dfrac{\sin\left(\dfrac{\alpha + \phi}{2}\right)}{\sin\dfrac{\alpha}{2}} \quad \text{with } n > 0$$

Show that

$$n = \sqrt{\dfrac{1 - \cos \alpha \cos \phi + \sin \alpha \sin \phi}{1 - \cos \alpha}}$$

38. *Optics*   Show that an equivalent expression for the index of refraction described in Exercise 37 is

$$n = \sqrt{\dfrac{1 + \cos \phi}{2}} + \left(\cot\dfrac{\alpha}{2}\right)\sqrt{\dfrac{1 - \cos \phi}{2}}$$

39. *Electricity*   In an ac circuit containing reactance, the instantaneous power is given by

$$P = V_{\max} I_{\max} \cos \omega t \sin \omega t$$

Show that $P = \dfrac{V_{\max} I_{\max}}{2}\sin 2\omega t$.

40. *Physics*   A cable vibrates with a decreased amplitude that is given by $A = \sqrt{e^{-2x}(1 + \sin 2x)}$. Show that $A = e^{-x}(\sin x + \cos x)$.

41. One of the identities for $\cos 2\theta$ is $\cos 2\theta = \cos^2 \theta - \sin^2 \theta$. There are two other identities for $\cos 2\theta$. Describe how you would develop each of them from the one above.

42. How would you establish the identity for $\tan 2\theta$.

43. The only difference between the formula for $\sin \dfrac{\theta}{2}$ and $\cos \dfrac{\theta}{2}$ is a $+$ or $-$ sign. Explain how you can tell which identity is for $\sin \dfrac{\theta}{2}$ and which is for $\cos \dfrac{\theta}{2}$.

44. The identities for $\sin \dfrac{\theta}{2}$ and $\cos \dfrac{\theta}{2}$ can both be developed from the identities for $\cos 2\theta$. Describe how you would do this.

≡ 20.4
TRIGONOMETRIC EQUATIONS

For Sections 20.1 through 20.3, we studied different types of trigonometric identities. Many people find proving and developing trigonometric identities to be very interesting. Our main interest in them, however, was to give you some skills for solving equations that involve trigonometric functions.

A **trigonometric equation** is an equation involving trigonometric functions of unknown angles. If these equations have been true for all angles, then we have called them identities. A trigonometric equation that is not an identity is a **conditional equation**. A conditional equation is true for some values for the angle and not true for others. To **solve** a conditional trigonometric equation means to find all values of the angle for which the equation is true. To solve a trigonometric equation, you must use both algebraic and trigonometric identities.

Solving a trigonometric equation of the type $2\tan x = 1$ would produce an infinite number of answers. As you remember from our earlier study, the trigonometric functions are periodic. Thus, the solution to this equation would not only be true when $x = 26.565°$ but also for $x = 26.565° + 180°n$, where n is any integer. Usually, it is sufficient to give only the **primary solutions** or **principal values**, which are the solutions for x, where $0° \leq x < 360°$ or $0 \leq x < 2\pi$.

EXAMPLE 20.19

Solve $2\tan x = 1$.
Solution $2\tan x = 1$

$$\tan x = \frac{1}{2}$$

We know that $x = \arctan \frac{1}{2} = \tan^{-1}(\frac{1}{2})$. Using a calculator, we see that $x \approx 26.565°$. But, we know that the tangent function is also positive in Quadrant III, so $x = 26.565° + 180° = 206.565°$. The primary solutions are $26.565°$ and $206.565°$.

EXAMPLE 20.20

Solve $\cos 4x = \dfrac{\sqrt{2}}{2}$, where $0 \le x < 2\pi$.

Solution A natural way to proceed would be to use the double-angle identities to rewrite this as an equation in x.

A little foresight will save a lot of work. We will let $\theta = 4x$ and solve for θ. Once we have the value for θ we can then solve for x. But, since $x = \dfrac{\theta}{4}$ and $0 \le x < 2\pi$, we must solve for $0 \le \theta < 8\pi$.

$$\cos\theta = \frac{\sqrt{2}}{2}$$

$$\theta = \cos^{-1}\left(\frac{\sqrt{2}}{2}\right) = \frac{\pi}{4}$$

Since the cosine is also positive in Quadrant IV, we see that $\theta = \frac{7\pi}{4}$. If we keep adding 2π to each of these answers until we exceed 8π, we will get the other solutions for θ:

$$\frac{\pi}{4} + 2\pi = \frac{9\pi}{4},$$

$$\frac{\pi}{4} + 4\pi = \frac{17\pi}{4},$$

$$\frac{\pi}{4} + 6\pi = \frac{25\pi}{4};$$

and we also get

$$\frac{7\pi}{4} + 2\pi = \frac{15\pi}{4},$$

$$\frac{7\pi}{4} + 4\pi = \frac{23\pi}{4},$$

and $$\frac{7\pi}{4} + 6\pi = \frac{31\pi}{4}.$$

The solutions then for $0 \le \theta < 8\pi$ are $4x = \theta = \frac{\pi}{4}, \frac{7\pi}{4}, \frac{9\pi}{4}, \frac{15\pi}{4}, \frac{17\pi}{4}, \frac{23\pi}{4}, \frac{25\pi}{4},$ and $\frac{31\pi}{4}$, and so the values of x are $x = \frac{\pi}{16}, \frac{7\pi}{16}, \frac{9\pi}{16}, \frac{15\pi}{16}, \frac{17\pi}{16}, \frac{23\pi}{16}, \frac{25\pi}{16},$ and $\frac{31\pi}{16}$.

EXAMPLE 20.21

Solve $\sin\theta \tan\theta = \sin\theta$ for $0 \le \theta < 360°$.

Solution We begin by collecting terms and factoring:

$$\sin\theta \tan\theta = \sin\theta$$
$$\sin\theta \tan\theta - \sin\theta = 0$$
$$\sin\theta(\tan\theta - 1) = 0$$

We now determine when each of these factors can be 0.

EXAMPLE 20.21 (Cont.)

So $\sin\theta = 0$ or $\tan\theta - 1 = 0$

$\qquad \sin\theta = 0 \qquad\qquad\qquad \tan\theta - 1 = 0$

$\qquad\quad \theta = 0°, 180° \qquad\qquad\quad \tan\theta = 1$

$\qquad\qquad\qquad\qquad\qquad\qquad\qquad \theta = 45°, 225°$

The solutions are $0°, 45°, 180°$, and $225°$.

EXAMPLE 20.22

Solve $2\sin^2 x - \cos x - 1 = 0$ for $0 \le x < 2\pi$.

Solution   We will use one of the Pythagorean identities to replace $\sin^2 x$ in the given expression.

$$2(1 - \cos^2 x) - \cos x - 1 = 0$$
$$2 - 2\cos^2 x - \cos x - 1 = 0$$
$$-2\cos^2 x - \cos x + 1 = 0$$
$$\text{or} \qquad 2\cos^2 x + \cos x - 1 = 0$$

Factoring, we get

$$(2\cos x - 1)(\cos x + 1) = 0$$

Solving each factor, we have

$$2\cos x - 1 = 0$$
$$\cos x = \frac{1}{2}$$
$$x = \frac{\pi}{3}, \frac{5\pi}{3}$$
$$\text{and,} \qquad \cos x + 1 = 0$$
$$\cos x = -1$$
$$x = \pi$$

The solutions are $\frac{\pi}{3}$, π, and $\frac{5\pi}{3}$.

EXAMPLE 20.23

Solve $\tan\theta + \sec\theta = 1$ for $0 \le \theta < 2\pi$.

Solution   We will follow a procedure similar to the one we followed when solving radical equations with two radicals.

This equation has two different trigonometric functions and there does not seem to be any identity that relates them. We will rewrite the equation with the functions on different sides of the equation and then square both sides. When we

EXAMPLE 20.23 (Cont.)

solved identities, we could not square both sides. However, in solving a conditional equation, it is possible to square both sides, if you check for extraneous roots later.

$$\tan\theta + \sec\theta = 1$$

$$\begin{array}{ll} \sec\theta = 1 - \tan\theta & \text{Rewrite.} \\ \sec^2\theta = (1 - \tan\theta)^2 & \text{Square both sides.} \\ \sec^2\theta = 1 - 2\tan\theta + \tan^2\theta & \text{Simplify.} \\ 1 + \tan^2\theta = 1 - 2\tan\theta + \tan^2\theta & \text{Pythagorean identity.} \\ 0 = -2\tan\theta & \text{Subtract } 1 + \tan^2\theta. \\ 0 = \tan\theta & \text{Divide by } -2. \end{array}$$

The solutions to $\tan\theta = 0$ are 0 and π. Now we need to check each of these in the *original* equation to see if they are actual roots or extraneous roots.

We first check $\theta = 0$:

$$\tan 0 + \sec 0 = 0 + 1 = 1$$

This checks so $\theta = 0$ is a root.

Next we check $\theta = \pi$:

$$\tan\pi + \sec\pi = 0 + -1 = -1$$

This is not equal to 1, so π is not a solution. (π is an extraneous root.)

Thus, $\theta = 0$ is the only solution.

Application

EXAMPLE 20.24

The range r of a projectile thrown at an angle of elevation θ at a velocity v is given by

$$r = \frac{2v^2 \cos\theta \sin\theta}{g}$$

If v is in ft/s, then g is $32\,\text{ft/s}^2$ and if v is in m/s, then g is $9.8\,\text{m/s}^2$. A projectile is fired with a velocity of $750\,\text{m/s}$ with the purpose of hitting an object $20\,000\,\text{m}$ away. Determine the angle θ at which the projectile should be fired.

Solution Here $v = 750\,\text{m/s}$, $g = 9.8\,\text{m/s}^2$, and $r = 20\,000\,\text{m}$. Substituting these values in the given equation, we obtain

$$20\,000 = \frac{2(750)^2 \cos\theta \sin\theta}{9.8}$$

Now, $2\cos\theta\sin\theta = 2\sin\theta\cos\theta = \sin 2\theta$, and the equation becomes

$$20\,000 = \frac{(750)^2 \sin 2\theta}{9.8}$$

or $$\sin 2\theta = \frac{20\,000(9.8)}{(750)^2}$$

$$\approx 0.3484$$

EXAMPLE 20.24 (Cont.)

so

$$2\theta \approx 20.39° \text{ or } 159.61°$$

and $\quad \theta \approx 10.195° \text{ or } 79.805°$

So, the projectile should be fired at an angle of $10.195°$ or $79.805°$. ∎

Exercise Set 20.4

Solve each equation in Exercises 1–30 for nonnegative angles less than 360° or 2π. You may want to use a calculator.

1. $2\cos\theta = 0$

2. $2\sin\theta = -1$

3. $\sqrt{3}\tan x = 1$

4. $\sqrt{3}\sec x = -2$

5. $4\sin\theta = -3$

6. $2\cos x = 3$

7. $4\tan\alpha = 5$

8. $3\csc x = 1$

9. $\cos 2x = -1$

10. $\sin 2x = \frac{1}{2}$

11. $\tan\dfrac{\theta}{4} = 1$

12. $\cos\dfrac{\theta}{3} = -1$

13. $\sin^2\alpha = \sin\alpha$

14. $\cos^2\beta = \frac{1}{2}\cos\beta$

15. $\sin x \cos x = 0$

16. $\dfrac{\sec\theta}{\csc\theta} = -1$

17. $3\tan^2 x = 1$

18. $\sec^2\theta = 2$

19. $4\sin\alpha\cos\alpha = 1$

20. $\sin^2\beta = \frac{1}{2}\sin\beta$

21. $\sin\theta - \cos\theta = 0$

22. $\tan\theta = \csc\theta$

23. $\sin 6\theta + \sin 3\theta = 0$

24. $4\tan^2 x = 3\sec^2 x$

25. $2\cos^2 x - 3\cos 2x = 1$

26. $\sin^2 4\alpha = \sin 4\alpha + 2$

27. $\sec^2\theta + \tan\theta = 1$

28. $\tan 2x + \sec 2x = 1$

29. $\sin 2x = \cos x$

30. $\csc^2\theta - \cot\theta = 1$

Solve Exercises 31–36.

31. *Optics* The second law of refraction (Snell's law) states that as a light ray passes from one medium to a second, the ratio of the sine of the angle of incidence θ_i to the sine of the angle of refraction θ_r is a constant, μ, called the **index of refraction**, with respect to the two mediums. (See Figure 20.7.) Thus, we have

$$\frac{\sin\theta_i}{\sin\theta_r} = \mu$$

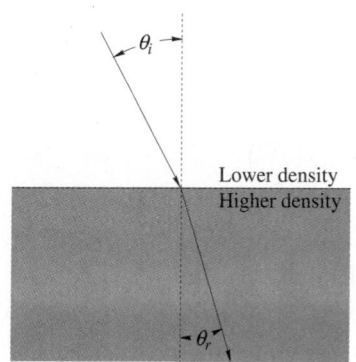

Lower density
Higher density

FIGURE 20.7

The index of refraction of general epoxy relative to air is $\mu = 1.61$. Determine the angle of refraction θ_r of a ray of light that strikes some general epoxy with an angle of incidence $\theta_i = 35°$.

32. *Optics* The index of refraction of glass silicone relative to air is 1.43. If the angle of incidence is 27°, what is the angle of refraction?

33. *Electronics* An oscillating signal voltage is given by $E = 125\cos(\omega t - \phi)$ millivolts, where the angular frequency is $\omega = 120\pi$, phase angle is $\phi = \frac{\pi}{2}$, and t is time in seconds. The triggering mechanism of an oscilloscope starts the sweep when $E = 60\,\text{mV}$. What is the smallest positive value of t for which the triggering occurs?

34. *Agriculture* As shown in Figure 20.8, an irrigation ditch has a cross-section in the shape of an isosceles trapezoid with the smaller base on the bottom. The area A of the trapezoid is given by

$$A = a\sin\theta(b + a\cos\theta)$$

If $a = 3\,\text{m}$, $b = 3.4\,\text{m}$, and $A = 10\,\text{m}^2$, find θ to the nearest tenth of a degree if $\sin\theta\cos\theta \approx 0.4675$.

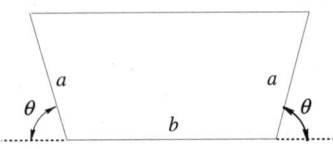

FIGURE 20.8

35. *Automotive technology* The displacement d of a piston is given by

$$d = \sin\omega t + \frac{1}{2}\sin 2\omega t$$

For what primary solutions of ωt less than 2π is $d = 0$?

36. *Optics* Refraction causes a submerged object in a liquid to appear closer to the surface than it actually is. The relation between the true depth a and the apparent depth b is

$$\frac{a}{b} = \sqrt{\frac{\mu^2 - \sin^2\theta_i}{\cos^2\theta_i}}$$

where μ is the index of refraction for the two media and θ_i is the angle of incidence. An object that is $14\,\text{ft}$ under water appears to be only $10\,\text{ft}$ under water. If the index of refraction of water is 1.333, what is the angle of incidence?

 In Your Words

37. Equations with multiple angles have more than two solutions. Explain how you can determine that you have found all solutions.

38. Describe how you would use the methods for solving a trigonometric equation to solve a trigonometric inequality.

 CHAPTER 20 REVIEW

Important Terms and Concepts

Double-angle identities
Half-angle identities
Pythagorean identities
Quotient identities

Reciprocal identities
Sum and difference identities
Trigonometric equations

Review Exercises

Prove the identities in Exercises 1–10.

1. $\dfrac{\sin(x+y)}{\cos x \cos y} = \tan x + \tan y$

2. $(\sin x + \cos x)^2 = 1 + \sin 2x$

3. $\dfrac{\sin 3x}{\sin x} - \dfrac{\cos 3x}{\cos x} = 2$

4. $\cos(\theta + \phi)\cos(\theta - \phi) = \cos^2\phi - \sin^2\theta$

5. $\sin(\alpha - \beta)\cos\beta - \cos(\alpha + \beta)\sin\beta = \sin\alpha$

6. $\tan 2x = \dfrac{2\cos x}{\csc x - 2\sin x}$

7. $\cos^4 x - \sin^4 x = \cos 2x$

8. $\dfrac{\sin 2x - \sin x}{\cos 2x + \cos x} = \tan\dfrac{x}{2}$

9. $\sin 3\theta = 2\sin\theta\cos 2\theta + \sin\theta$

10. $\dfrac{\cos 3x - \cos 5x}{\sin 3x + \sin 5x} = \tan x$
(Hint: Let $3x = 4x - x$ and $5x = 4x + x$.)

Solve the equations in Exercises 11–20 for nonnegative values less than 360° or 2π.

11. $2\tan x = -\sqrt{3}$

12. $3\sin x = -2$

13. $\cos 2x + \cos x = -1$

14. $\cos x - \sin 2x - \cos 3x = 0$

15. $\sin 4x - 2\sin 2x = 0$

16. $\sin(30° + x) - \cos(60° + x) = -\dfrac{\sqrt{3}}{2}$

17. $2\sin\theta = \sin 2\theta$

18. $\sin^2\alpha + 5\cos^2\alpha = 3$

19. $\sin^2 x = 1 + \sin x$

20. $\sin\theta - 2\csc\theta = -1$

In Exercises 21–24, use the facts that $\sin x = \frac{5}{13}$, $\cos x = -\frac{12}{13}$, and x is in Quadrant II to determine the exact value of the given function.

21. $\sin 2x$

22. $\cos\dfrac{x}{2}$

23. $\sin\dfrac{x}{2}$

24. $\tan 2x$

In Exercises 25–30, use the facts that x is in Quadrant III, $\sin x = \frac{-5}{13}$, and $\cos x = \frac{-12}{13}$, and y is in Quadrant II, $\sin y = \frac{8}{17}$, and $\cos y = \frac{-15}{17}$ to determine the exact value of the given function.

25. $\sin(x + y)$

26. $\cos(x - y)$

27. $\cos(x + y)$

28. $\tan(x + y)$

29. $\cos(y - x)$

30. $\sin(x - y)$

Solve Exercises 31–33.

31. *Automotive technology*   The acceleration of a piston is given by

$$a = 5.0(\sin\omega t + \cos 2\omega t)$$

For what primary solutions of ωt does $a = 0$?

32. *Acoustical engineering*   When two sinusoidal sound waves that are close together in frequency are superimposed, the resultant disturbance exhibits beats.   The two waves are represented by $y_1 = A_1\cos\omega_1 t$ and $y_2 = A_2\cos(\omega_2 t + \phi)$, where ω_2 is slightly larger than ω_1 and ϕ is a phase constant. If

$\phi = 0$, $\alpha = \dfrac{\omega_1 + \omega_2}{2}$, and $\beta = \dfrac{\omega_2 - \omega_1}{2}$, show that

$$y = y_1 + y_2 = A_1\cos(\alpha - \beta)t + A_2\cos(\alpha + \beta)t$$

33. *Mechanics*   Consider a machine that is mounted on four springs with a known stiffness and on four dampers with a known damping constant . If this system is initially at rest and a certain force is applied, then under certain conditions the system has a time-displacement equation of $x = 0.01e^{-6t}(\cos 8t + \sin 8t)$. Another version gives the time-displacement equation at $x = \dfrac{\sqrt{2}}{100}e^{-6t}\cos(8t - \frac{\pi}{4})$. Show that these two equations are identical.

☰ CHAPTER 20 TEST

In Exercises 1–8, use the fact that $\sin\alpha = \frac{4}{5}$ and α is in Quadrant II and that $\cos\beta = -\frac{12}{13}$, with β in Quadrant III to determine the exact value of the given function.

1. $\sin(\alpha + \beta)$

2. $\cos(\alpha + \beta)$

3. $\sin(\alpha - \beta)$

4. $\cos(\alpha - \beta)$

5. $\sin 2\alpha$

6. $\cos 2\beta$

7. $\sin\dfrac{\alpha}{2}$

8. $\cos\dfrac{\beta}{2}$

784 ■ CHAPTER 20. TRIGONOMETRIC FORMULAS, IDENTITIES, AND EQUATIONS

Solve Exercises 9–13.

9. Write $8\cos 6x \sin 6x$ using a single trigonometric function.

10. Prove the identity $\tan x = \dfrac{\sec x}{\csc x}$

11. Prove the identity $\dfrac{\sin x}{1 - \cos x} + \dfrac{\sin x}{1 + \cos x} = 2\csc x.$

12. Solve $6\cos^2 x + \cos x = 2$ for x with either $0° \le x < 360°$ or $0 \le x < 2\pi$.

13. The equation $x = 4r\sin^2\left(\dfrac{\theta}{2}\right)$ arises when engineers design the merging region for a rapid transit system. Solve this equation for x in terms of $\cos\theta$.

CHAPTER
21
Statistics and Empirical Methods

A police officer is measuring car speed in a residential area. In this chapter, we will see some ways in which statistics can be used to help interpret the information gathered by this officer.

In this chapter, we will begin exploring a new area of mathematics—the world of probability, statistics, and empirical methods. Probability theory has many applications in the physical sciences and technology. It is of basic importance in statistical mechanics. Probability is needed in any problem dealing with large numbers of variables where it is impossible or impractical to have complete information.

In many technological settings, information is needed about an operation. When it is not possible to gather information about the entire operation, information is gathered on a part, or sample, of the operation. This information is then analyzed and decisions are made as a result of this analysis. Statistics are the basis of this analysis, and in this chapter we will learn how to conduct some statistical analyses and how to use that information to help us make decisions.

21.1 PROBABILITY

The word "probably" is used often in everyday life. For example, you might tell someone that "the test will probably be hard" or that "it will probably rain today." The word "probably" indicates that we do not know what will happen and are predicting what we think *might* happen. The theory of probability tries to express more precisely just how much we do know.

In the theory of probability, a numerical value between 0 and 1 is given to the likelihood that some particular event will happen. When we do this, we assume that all events are equally likely to occur, unless we have some special information that indicates otherwise.

Outcome and Sample Space

In working with probability, we will be concerned with the possible outcomes of experiments. An experiment can be something as simple as tossing a coin or something more complicated such as determining the number of bad computer chips at a production station. A result of an experiment is called an **outcome**. The group of all possible outcomes for an experiment is the **sample space**. Each performance of an experiment is called a **trial**.

EXAMPLE 21.1

(a) If a coin is tossed, the sample space contains the two possible outcomes of heads (H) or tails (T) and can be written as $\{H, T\}$.

(b) If a die is rolled, the sample space of the number of dots on the upper face is $\{1, 2, 3, 4, 5, 6\}$. Each trial will produce one of these numbers.

(c) If two coins are tossed, the sample space has the following four possible outcomes $\{HH, HT, TH, TT\}$.

Event

An event is something that may or may not occur from the possible sample space. In particular, an **event** is some part (or all) of the sample space. For example, in rolling a die we might consider the event that an odd number is obtained. Another possible event would be that a 6 is obtained. A third possible event would be that a number larger than 2 is obtained. The outcomes for which an event occurs are said to be **favorable** to the event. For example, the outcomes 1, 3, and 5 are favorable to the event of "rolling an odd number."

Three Types of Probability

If E is an event from a sample space with n equally likely elements and k is the number of ways in which this event can happen, then the **probability of** E, $P(E)$, is

$$P(E) = \frac{k}{n}$$

This definition is known as the **classical** or **a priori** approach. In this approach, no experimenting is conducted and the probability is based on our knowledge of the nature of the event.

A second approach is the **empirical** or **frequency** approach, and it is based on the results of an actual experiment or of previous experience. In this definition, a series of N trials are made and the event is observed to occur K times. The empirical definition says that the probability of this event occurring would be

$$P(E) = \frac{K}{N}$$

There is a third approach to probability called the subjective approach, and it is based on a person's belief that an event will occur. The subjective approach is normally employed when there is no past history to use for an empirical approach and no basis for using the classical approach.

Application

EXAMPLE 21.2

Find the probability of drawing a 9 or 10 from a well-shuffled deck of cards.

Solution   A deck has 52 cards. Since this is well shuffled, the likelihood of getting any one card is the same as the likelihood of getting any other card. There are four 9s and four 10s in the deck. There are 8 cards in this event and so $P(9 \text{ or } 10) = \frac{8}{52} = \frac{2}{13}$. This is an example of the classical approach to probability.

Application

EXAMPLE 21.3

Samples have shown that, out of every 1,000 discs it copies, a certain machine will produce 75 discs with errors. The probability of a disc being bad if copied on this machine is $\frac{75}{1,000} = \frac{3}{40}$. This is an example of the empirical approach to probability.

As you can see, since $0 \leq K \leq N$, we can make the following five points about the probability of any event:

- The probability that a given event will happen ranges from 0 to 1 or $0 \leq P(E) \leq 1$.
- The higher the probability (the closer to 1), the more likely it is that the event will happen.
- If the probability is $\frac{1}{2}$, then it is equally likely that an event will happen (or will not happen).
- A probability of 1 indicates that the event is certain to occur.
- A probability of 0 means that the event will not happen.

If E represents a certain event, then E' represents all the elements in the sample space that are not in E. So, if the probability that E will happen is $P(E)$, then the probability that E will not happen (or that E' will happen) is $P(E') = 1 - P(E)$.

Application

EXAMPLE 21.4	What is the probability that the disc-copying machine in Example 21.3 will make a good copy?

Solution   The probability that a copy is bad is $\frac{3}{40}$, so the probability that a copy is good is $1 - \frac{3}{40} = \frac{37}{40}$.

Sometimes we are interested in the probability of an event when we know that another event has already occurred. If the probability that the second event will occur is not affected by what happens to the first event, we then say that the events are **independent**.

Probability of Two Independent Events

If events A and B are independent, then the probability that both A and B will occur is $P(A \text{ and } B) = P(A) \cdot P(B)$.

EXAMPLE 21.5	Two cards are to be drawn from a well-shuffled deck of cards. After the first card is drawn, it is replaced and the deck is shuffled again. What is the probability that we will draw two diamonds?

Solution   There are 13 diamonds in the deck, so the probability of getting a diamond is $\frac{13}{52} = \frac{1}{4}$. Since the first card is replaced and the deck is shuffled, the probability of getting a diamond on the second draw is independent of what was drawn on the first card. The probability of getting a diamond on the second card is $\frac{1}{4}$ and the probability of getting two diamonds is $\frac{1}{4} \cdot \frac{1}{4} = \frac{1}{16}$.

How would the results of Example 21.5 have been different if we had not replaced the first card? In this instance, we are working with **conditional probability**. If A is an event that has already occurred, and we want to know the probability that a second event B will occur, we call this the **probability of B given A**. This is often written $P(B|A)$. To determine $P(B|A)$, we will use the following reasoning: We know that event A has already occurred. This reduces the sample space to the elements in A. We now count the number of times that event B is in this reduced sample space. Then $P(B|A)$ are the number of elements in B that are also in A divided by the number of elements in A. Thus, we have the following.

Probability of Two Events That Are Not Independent

$$P(A \text{ and } B) = P(A) \cdot P(B|A) \quad \text{or} \quad P(B|A) = \frac{P(A \text{ and } B)}{P(A)}$$

EXAMPLE 21.6

Two cards are going to be drawn from a well-shuffled deck of cards. The first card will not be replaced after it is drawn. What is the probability that both cards will be diamonds?

Solution Let A be the event that the first card is a diamond. Let B be the event that the second card is a diamond. We want the probability of A and B. We know from Example 21.5 that $P(A) = \frac{1}{4}$. Since the first card is not replaced, the deck now has 51 cards. Of these 51 cards, 12 of them are diamonds. So, $P(B|A) = \frac{12}{51} = \frac{4}{17}$ and

$$P(A \text{ and } B) = P(A) \cdot P(B|A)$$
$$= \frac{1}{4} \cdot \frac{4}{17} = \frac{1}{17}$$

In this section, we have given you a very brief introduction to probability. As you might imagine, there are many more complicated situations that could happen, and determining their probabilities is much more involved. We have, however, given you a foundation for the work with statistics that is in the remainder of this chapter.

Exercise Set 21.1

In Exercises 1–10, let the sample space be an ordinary deck of well-shuffled playing cards. What is the probability of drawing each of the following?

1. a heart

2. a six

3. a queen

4. a two or a seven

5. a face card (jack, queen, or king)

6. a black card

7. two black cards on successive draws, if the first card is replaced before the second card is drawn

8. a black card and then a face card, if the first card is replaced before the second card is drawn

9. a red card and then a black card, if the first card is not replaced before the second draw

10. a face card and then an ace, if the first card is not replaced before the second draw

In Exercises 11–18, assume that you are rolling a single die. What is the probability of rolling the following?

11. 6

12. 1 or 6

13. even number

14. 4 or more

15. two successive 6s

16. not a 5

17. at least one 5 on two successive rolls (First, determine the probability of not getting a 5.)

18. three successive 6s

In Exercises 19–20, assume that you are rolling two dice. There are 36 possible ways for the dice to fall.

19. What is the sample space?

20. What is the probability of getting two 6s?

When rolling a pair of dice, what is the probability of obtaining each of the totals in Exercises 21–30?

21. 12

22. 2

23. 1

24. 3

25. 7

26. 7 or 11

27. less than 5

28. 7 on two successive rolls

29. 12 on two successive rolls

30. at least 7

In Exercises 31–34, use the fact that a loaded (or unfair) die has probabilities of $\frac{1}{21}$, $\frac{2}{21}$, $\frac{3}{21}$, $\frac{4}{21}$, $\frac{5}{21}$, and $\frac{6}{21}$ on showing a 1, 2, 3, 4, 5, and 6, respectively.

31. What is the probability of rolling two 6s in succession?

32. What is the probability of rolling a 3 and then a 4?

33. What is the probability that the number on the die is even?

34. What is the probability of a total of 12 when two dice are rolled?

Solve Exercises 35–42.

35. *Medical technology* A certain medication is known to cure a specific illness for 75% of the people who have the illness. If two people with the illness are selected at random and take the medicine, what is the probability that
(a) both will be cured?
(b) neither will be cured?
(c) only one will be cured?

36. *Aeronautics* On a two-engine jet plane, the probability of either engine failing is 0.001. If one engine fails, then the probability that the second will fail is 0.005. What is the probability that both engines will fail?

37. *Insurance* Insurance company tables show that for a married couple of a certain age group the probability that the husband will be alive in 25 years is 0.7.

The probability that the wife will be alive in 25 years is 0.8.
(a) What is the probability that both are alive in 25 years?
(b) What is the probability that both are dead in 25 years?
(c) What is the probability that in 25 years only one is alive?

38. *Industrial technology* A machine produces 25 defective parts out of every 1,000. What is the probability of a defective part being produced?

39. *Industrial technology* What is the probability of testing two parts from the machine in Exercise 38 and finding that one of them is defective?

40. *Industrial technology* What is the probability of testing four parts from the machine in Exercise 38 and finding that all four are not defective?

41. *Medical technology* Blood groups for a certain sample of people are shown in the following table.

Blood Group	Frequency
O	110
A	64
B	20
AB	6

If one person from this sample of people is randomly selected, what is the probability that he or she has type B blood?

42. *Insurance* Among 800 randomly selected drivers in the 20–24 age bracket, 316 were in a car accident during the last year. If a driver of that age bracket is randomly selected, what is the probability that he or she will be in a car accident during the next year?

In Your Words

43. What is an event?

44. What is meant by the probability of an event?

45. Explain how the classical or *a priori* approach to determine the probability of an event is different from the empirical or frequency approach.

46. What are independent events?

≡ 21.2
MEASURES OF CENTRAL TENDENCY

In Section 21.1, we stated that there were three types of probability. One of these was the empirical or frequency approach. This approach is based on past experience or on the results of an actual experiment. The experiment might consist of a series of trials. For each of these trials, the person directing the experiment will collect data, analyze the data, and then interpret it. In the remainder of this chapter, we will look at some of the ways in which the experimenter might gather and analyze this data.

In statistics, we deal with numbers. Sometimes it is possible to deal with entire sets of numbers. For example, if your teacher wants to determine how much your class has learned about probability, a test could be given to the class. Businesses also are sometimes able to test an entire group of items. For example, a company might be able to test each television set to see that it works. These situations describe **populations**.

There are times when it is not possible to check each item and so a **sample** of items must be selected. For example, it is not possible to check every electric fuse to see that it provides the proper protection. It also might not be practical to check every part made on a machine that produces 10,000 parts a day. In these last two cases, a sample is selected and tested. They might check every 100th fuse or every 1,000th part made by a machine. No matter how the data are gathered, it needs to be organized in a way that makes it easier to understand.

≡ **Note**

A sample is a subset of the population and is used to provide *estimates* about the population.

Frequency Distribution

One way to organize the information is in a **frequency distribution**. In a frequency distribution, one line contains a list of possible values and a second line contains the number of times each value was observed in a particular time.

Application

EXAMPLE 21.7

People who live near an interstate highway have complained about traffic noise. A measuring instrument is placed along the highway and every 15 min it measures the noise level in decibels. A frequency distribution is made of the readings for the first day.

Decibels	50–54	55–59	60–64	65–69	70–74	75–79	80–84
Frequency	2	2	4	4	4	6	8

Decibels	85–89	90–94	95–99	100–104	105–109	110–114	115–119
Frequency	8	10	12	14	10	8	4

Graphs

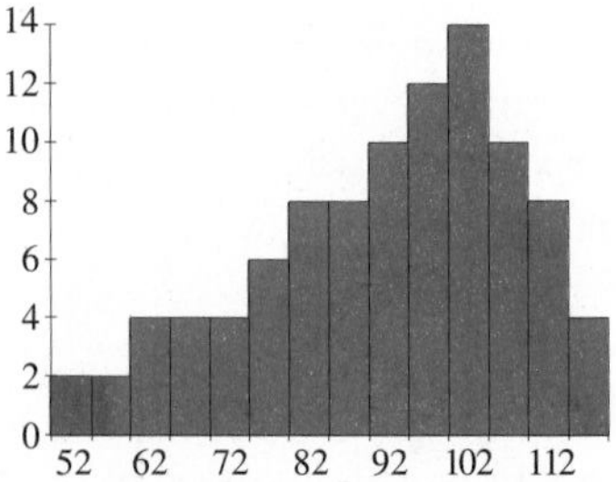

Many times it is difficult for us to understand the numbers in a table such as a frequency distribution. We find that a chart or graph helps us to get a better idea of these numbers. One type of graph is the **histogram**. Another type of graph is the **frequency polygon** or **broken line graph**. A histogram for the data in the above frequency table showing the noise level along an interstate highway from the previous example is given in Figure 21.1. The numbers along the bottom mark the middle of each interval. A broken line graph for the same information is shown in Figure 21.2.

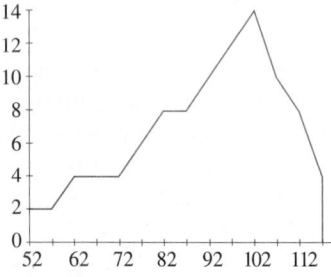

FIGURE 21.1

Sometimes a histogram can provide very important information. For example, the histogram in Figure 21.3 shows the results when inspectors measured the diameters in millimeters of 500 steel rods. The dotted line marked LAL is the lowest acceptable limit. Rods smaller than 10.00 mm are too loose in the bearing and would be thrown out or rejected in a later operation. When a rod is rejected, the company loses all the labor and material cost that went into making that rod.

The histogram in Figure 21.3 contains some very interesting information. We can see that 40 rods were rejected because they were below the LAL; but, there is a gap in the histogram at the interval just below the LAL. The gap was unexpected and is a result of inspectors passing rods that were barely below the LAL. This gap is one that the inspectors could have corrected.

FIGURE 21.2

Median, Quartiles, and Box Plots

Tables and graphs can give us some general ideas about the data, but we often need more information. Among this information are some indications for the location of the center of the distribution. We would also like some measure of how the data are spread. This gives us some numerical descriptions of the data and helps us compare groups of data.

The first types of information that we will examine are the ones that provide some indication of the center. These are known as the **measures of central tendency** and there are three of them: the median, the mean, and the mode.

The first measure of central tendency we will examine is the median.

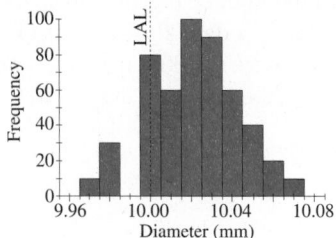

FIGURE 21.3

> **Median**
>
> The **median** is the middle number of the numbers in the distribution. One-half of the values are larger than the median, and one-half are smaller. To find the median,
>
> 1. Arrange the numbers in increasing (or decreasing) order.
> 2. If there is an odd number of items, the median is the value of the middle item.
> 3. If there is an even number of items, the median is the number half-way between the two middle items.

EXAMPLE 21.8

Given the numbers 9, 8, 3, 2, 4, determine the median.

Solution   First arrange the numbers in increasing order: 2, 3, 4, 8, 9. There are five numbers, so the middle number is the third number. The third number is 4, so the median is 4.

EXAMPLE 21.9

Find the median of 11, 12, 15, 18, 20, 20.

Solution   These number are already in numerical order. There are six numbers. The median will be half-way between the third and fourth numbers. The third number is 15 and the fourth is 18. Midway between these is 16.5, thus the median is 16.5.

The median divides the items into two equally sized parts. In the same way, the **quartiles** Q_1, Q_2, and Q_3, divide the numbers into four equally sized parts when the numbers are arranged in increasing (or decreasing) order. As we will see, quartiles provide a quick way to graphically see how the numbers are distributed.

> **Finding Quartiles**
>
> 1. Arrange the numbers in increasing order.
> 2. Q_2 is the median. It divides the numbers into a lower half and an upper half.
> 3. Q_1 is the median of the lower half of the numbers.
> 4. Q_3 is the median of the upper half of the numbers.

EXAMPLE 21.10

Determine the quartiles of 12, 15, 42, 37, 61, 14, 14, 9, 25, 32, 27, and 30.

Solution   We begin by arranging the numbers in increasing order:

$$9, 12, 14, 14, 15, 25, 27, 30, 32, 37, 42, 61$$

EXAMPLE 21.10 (Cont.)

There are 12 numbers, so the second quartile, Q_2 (or median), is midway between the sixth and seventh numbers. The sixth number is 25. The seventh number is 27.

$$Q_2 = \frac{25+27}{2} = 26$$

Q_1 is the median of the lower half, and so it is the median of the smallest six numbers. Q_1 is midway between the third and fourth numbers. These are both 14, so $Q_1 = 14$.

Q_3 is the median of the upper half. The upper half of the items is 27, 30, 32, 37, 42, and 61. The median of these is midway between 32 and 37, so

$$Q_3 = \frac{32+37}{2} = 34.5$$

The three quartiles, the lowest number, and the highest number are used to make a diagram called a **box plot** or a **box and whisker diagram**. The following steps describe how to make a box and whisker diagram.

Making a Box and Whisker Diagram

1. Arrange the numbers in increasing order.
2. Find the lowest and highest scores, Q_1, Q_2, and Q_3.
3. Draw a scale that will include the lowest and highest numbers.
4. Draw a box with the ends of the box at Q_1 and Q_3.
5. Draw a line through the box at Q_2.
6. Draw a whisker from the lowest number to the box and draw another whisker from the highest number to the box.

≡ **Note**

The box of a box and whisker diagram contains the middle 50% of the data, one whisker shows the bottom 25% of the data, and the other whisker shows the top 25% of the data.

Application

EXAMPLE 21.11

At an automobile engine plant, a quality control technician pulls crankshafts from the assembly line at regular intervals. The technician measures a critical dimension on each of these crankshafts. Even though the dimension is supposed to be 182.000 mm, some variation will occur during production. Here are the measurements for one morning's sample:

182.120	182.005	182.025	181.987	181.898	182.034
181.960	181.940	182.055	181.897	181.935	182.063
182.015	182.026	181.965	181.985	182.362	181.998
182.107	181.934	181.991	182.005	182.012	181.984

EXAMPLE 21.11 (Cont.)

Make a box plot for these measurements.

Solution   We begin by listing the numbers in numerical order.

$$
\begin{array}{cccccc}
181.897 & 181.898 & 181.934 & 181.935 & 181.940 & 181.960 \\
181.965 & 181.984 & 181.985 & 181.987 & 181.991 & 181.998 \\
182.005 & 182.005 & 182.012 & 182.015 & 182.025 & 182.026 \\
182.034 & 182.055 & 182.063 & 182.107 & 182.120 & 182.362
\end{array}
$$

From this arrangement of the 24 numbers, we can see that the lowest is 181.897 mm and the highest is 182.362 mm. Notice that the numbers are arranged in four rows with six numbers in each row. Thus, Q_1 will be between the last number in the first row and the first number in the second row, or $Q_1 = \frac{181.960 + 181.965}{2} = 191.9625$. In a similar way, we find $Q_2 = \frac{181.998 + 182.005}{2} = 182.0015$; and $Q_3 = \frac{182.026 + 182.034}{2} = 182.03$. The box plot for these data is shown in Figure 21.4.

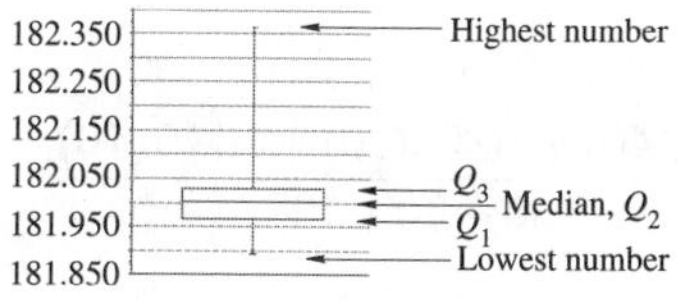

FIGURE 21.4

Mean and Mode

The second measure of central tendency is the **mean**, sometimes called the **arithmetic mean**.

Mean

To determine the mean, you add all the values and divide by the number of values. The symbol $\overline{x}$ is used to represent the mean. In symbols,

$$
\overline{x} = \frac{\sum\limits_{i=1}^{n} x_i}{n}
$$

EXAMPLE 21.12

Find the mean of the values in **(a)** Example 21.8 and **(b)** Example 21.9.

Solutions

(a) In Example 21.8, the five numbers were 9, 8, 3, 2, and 4, so

$$
\overline{x} = \frac{9 + 8 + 3 + 2 + 4}{5} = \frac{26}{5} = 5.2
$$

(b) In Example 21.9, the six numbers were 11, 12, 15, 18, 20, and 20. The mean of these six numbers is

$$
\overline{x} = \frac{11 + 12 + 15 + 18 + 20 + 20}{6} = \frac{96}{6} = 16
$$

If you want to find the mean of a large number of values, and if some appear more than once, then there is a quicker method to get the total. Consider the following frequency distribution for the sample of 500 steel rods in Figure 21.3.

Diameter	9.97	9.98	9.99	10.00	10.01	10.02	10.03	10.04	10.05	10.06	10.07
Frequency	10	30	0	80	60	100	90	60	40	20	10

It would take a long time to add these 500 values. Instead, if we multiply each diameter by the frequency for that diameter, we reduce the 500 additions to ten multiplications and nine additions. Thus,

$$\overline{x} = \frac{\begin{array}{l}9.97(10) + 9.98(30) + 9.99(0) + 10.00(80) + 10.01(60) + 10.02(100) \\ + 10.03(90) + 10.04(60) + 10.05(40) + 10.06(20) + 10.07(10)\end{array}}{10 + 30 + 80 + 60 + 100 + 90 + 60 + 40 + 20 + 10}$$
$$= \frac{5,010.7}{500}$$
$$= 10.0214$$

Using the sigma notation from Chapter 19, the mean is given by the following formula.

Mean

$$\overline{x} = \frac{\sum_{i=1}^{n} x_i f_i}{\sum_{i=1}^{n} f_i}, \text{ where } x_i \text{ represents the } i\text{th value and } f_i \text{ the frequency for that value.}$$

In Example 21.12, there were 11 values: $x_1 = 9.97$ and $f_1 = 10$, $x_2 = 9.98$ and $f_2 = 30$, and so on until $x_{11} = 10.07$ and $f_{11} = 10$.

The third, and final, measure of central tendency is the mode.

Mode

The **mode** is the value that has the greatest frequency. A set of numbers can have more than one mode. If there are two modes, the data are said to be **bimodal**.

EXAMPLE 21.13

What is the mode for the data in Example 21.9: 11, 12, 15, 18, 20, and 20?

Solution The mode is 20, because that value occurs twice and all the other values occur once.

Application

EXAMPLE 21.14

What is the mode for the sample of 500 steel rods in Figure 21.3?

Solution   The 500 steel rods had a mode of 10.20, because the 100 measures for that diameter were more than for any other.

Using a Graphing Calculator

A graphing calculator can be a great help in finding many statistical measures. In most graphing calculators, there are two modes for working with statistics. One mode is for data entry. In this mode you enter the raw data, usually giving each score its frequency. The other mode is used to perform the calculations using the data stored during the data entry mode.

There are several types of statistical calculations that will be performed. All calculators will compute the mean and sample size from this section as well as the variance and standard deviation from the next section. Some calculators will also give the median, mode, and quartiles, and draw a histogram. All of these calculations are performed under the category of single-variable (or 1-var) statistics. Because so much diversity exists among calculator brands and models, no attempt will be made here to give directions. Be sure to study the user's guide for your calculator.

Exercise Set 21.2

Use the following four sets of numbers for Exercises 1–20.
  A: 4, 2, 6, 3, 7, 4, 6, 2, 4, 4
  B: 50, 52, 52, 54, 54, 54, 54, 56, 58, 58
  C: 80, 77, 82, 73, 92, 89, 100, 96, 96, 94, 74, 94, 94, 96, 83, 84, 96, 87, 84, 96
  D: 100, 98, 96, 94, 93, 90, 89, 85, 82, 78, 76, 66, 64, 64, 78, 89, 93, 96, 98, 96, 93, 64, 96

In Exercises 1–4, set up a frequency distribution table for the numbers in the indicated set.

1. Set A   **2.** Set B   **3.** Set C   **4.** Set D

In Exercises 5–8, set up a frequency distribution table for each of the given intervals in the indicated sets.

5. 1–2, 3–4, 5–6, 7–8 in Set A
6. 50–52, 53–55, 56–58 in Set B
7. 71–75, 76–80, 81–85, 86–90, 91–95, 96–100 in Set C

8. 61–65, 66–70, 71–75, 76–80, 81–85, 86–90, 91–95, 96–100 in Set D

In Exercises 9–12, draw histograms for the data in the given exercises.

9. Exercise 1   **10.** Exercise 2   **11.** Exercise 7   **12.** Exercise 8

In Exercises 13–16, draw broken line graphs for the data in the indicated exercises.

13. Exercise 1   **14.** Exercise 2   **15.** Exercise 7   **16.** Exercise 8

In Exercises 17–20, determine the mean, median, mode, and quartiles of the given set. Then draw a box plot of the data.

17. Set A   **18.** Set B   **19.** Set C   **20.** Set D

Solve Exercises 21–38.

 21. Write a computer program that will allow you to enter individual values and their frequencies and to determine the mean of the data.

22. *Energy technology*   The following data are based on the energy consumption for one household's electric bills for 36 two-month periods.

Energy (kWh)	700–719	720–739	740–759	760–779	780–799
Frequency	2	2	4	5	3

Energy (kWh)	800–819	820–839	840–859	860–879	880–899
Frequency	4	7	5	2	2

Construct a histogram that corresponds to this frequency table.

23. *Energy technology*   A sample of 100 batteries was selected from the day's production for a machine. The batteries were tested to see how long they would operate a flashlight with these results:

Hours	211–215	216–220	221–225	226–230
Frequency	4	9	19	23

Hours	231–235	236–240	241–245	246–250
Frequency	16	14	10	5

Form a histogram for this data.

24. *Energy technology*   Form a histogram for the data in Exercise 23, using the intervals 211–220, 221–230, etc.

25. *Industrial technology*   A technician was measuring the thickness of a plastic coating on some pipe and obtained the following data:

Thickness (mm)	0.01	0.02	0.03	0.04	0.05
Frequency	1	5	40	50	36

Thickness (mm)	0.06	0.07	0.08	0.09	
Frequency	30	25	10	3	

Form a histogram of this data.

26. *Industrial technology*   Draw a broken-line graph for the data in Exercise 25.

27. *Industrial technology*   Form a histogram for the data in Exercise 25 over the intervals 0.01–0.02, 0.03–0.04, etc.

28. *Industrial technology*   Determine the mean, median, mode, and quartiles for the data in Exercise 25.

29. *Industrial technology*   Draw a box plot for the data in Exercise 25.

30. *Electrical technology*   A technician tested an electric circuit and found the following values in milliamperes on successive trials:

5.24, 5.31, 5.42, 5.26, 5.31, 5.47, 5.27, 5.29, 5.35, 5.44, 5.35, 5.31, 5.45, 5.46, 5.39, 5.34, 5.35, 5.46, 5.26, 5.27, 5.47, 5.34, 5.28, 5.39, 5.34, 5.42, 5.43, 5.46, 5.34, 5.29

Form a frequency distribution for this data.

31. *Electrical technology*   For the data in Exercise 30, draw a histogram for the intervals 5.21–5.25, 5.26–5.30, etc.

32. *Electrical technology*   Determine the mean, median, and mode for the data in Exercise 30.

33. *Electrical technology*   Determine the quartiles and draw a box plot for the data in Exercise 30.

34. *Police science*   A patrol officer using a laser gun recorded the following speeds for motorists driving through a 55 mph speed zone:

52 57 62 59 67 54
55 64 65 59 63 72

(a) Determine the mean, median, mode, and quartiles for the given data.
(b) Draw a box plot for the given data.

35. *Environmental science*   An environmental officer measured the carbon monoxide emissions (in g/m) for several vehicles. The results are shown in the following table:

| 5.02 | 12.36 | 13.46 | 6.92 | 7.44 | 8.52 | 12.82 |
| 11.92 | 14.32 | 12.06 | 8.02 | 11.34 | 6.66 | 9.28 |

(a) Determine the mean, median, mode, and quartiles for the given data.

(b) Draw a box plot for the given data.

36. *Energy science*   The carbon monoxide emissions (in g/m) were measured for several vehicles. The results are shown in the following table:

892	673	534	437	449	524
627	735	892	923	1024	905
865	704	624	535	432	495
572	625	655	684	532	484

Determine the mean, median, mode, and quartiles for the given data.

37. *Insurance*   The blood alcohol content of 15 drivers involved in fatal accidents and then convicted with jail sentences are given below:

0.14	0.16	0.21	0.10	0.13
0.19	0.26	0.22	0.13	0.09
0.11	0.18	0.12	0.24	0.27

Determine the mean, median, mode, and quartiles for the given data.

38. *Business*   The daily sales in dollars for one 31-day month at a store are shown below:

24,562	38,646	43,988	15,122	14,321	17,479	19,478
25,625	39,476	45,353	15,972	13,793	17,457	18,681
20,562	38,606	53,788	15,122	10,321	13,037	17,038
25,625	37,036	45,353	15,732	13,373	13,053	18,681
21,903	41,775	52,117				

Determine the mean, median, mode, and quartiles for the given data.

In Your Words

39. What are the mean, median, and mode? What do they measure? How are they determined?

40. What are the quartiles? Describe how you find them.

41. Describe how to make a box plot.

42. What is a histogram? Describe how a histogram is constructed.

≡ 21.3
MEASURES OF DISPERSION

In Section 21.2, we looked at the three measures of central tendency. In this section, we will learn a technique that will tell us how close together the information is distributed. **Measures of dispersion** tell how close together or how spread out the data are.

Variance and Standard Deviation

There are several measures of dispersion. We will look at two of them: variance and standard deviation.

Variance and Standard Deviation

The **variance** $v = s^2$ of a set of numbers is given by the formula

$$v = s^2 = \frac{\sum\limits_{i=1}^{n} (x_i - \overline{x})^2}{n - 1}$$

and the **standard deviation** s is the square root of the variance. Thus,

$$s = \sqrt{s^2}$$

EXAMPLE 21.15

Find the variance and standard deviation of the numbers $9, 8, 3, 2, 4$.

Solution   This is the same set of numbers we used in Example 21.8. In Example 21.12, we determined that the mean was 5.2. We can set up a table to help with the calculations.

x	$x - \overline{x}$	$(x - \overline{x})^2$
9	3.8	14.44
8	2.8	7.84
3	−2.2	4.84
2	−3.2	10.24
4	−1.2	1.44
		38.80

$$v = \frac{38.8}{4} = 9.70$$

$$s = \sqrt{9.70} \approx 3.11$$

In some sets of data, some values occur more than once. The formula for the variance becomes

$$s^2 = \frac{\sum f_i (x_i - \overline{x})^2}{n - 1}$$

What you would do is add two more columns to your table, one for f_i and the other for $f_i (x_i - \overline{x})^2$, as is shown by the next example. Note that $n = \sum f_i$.

EXAMPLE 21.16

Determine the mean and standard deviation for the following values: $39, 41, 39, 44, 39, 40, 39, 40, 37, 42, 37, 43, 44, 38, 38, 38, 43, 38, 41, 39$.

Solution   We first make a frequency distribution table and then add columns for $x_i f_i$, $x_i - \overline{x}$, $(x_i - \overline{x})^2$, and $(x_i - \overline{x})^2 f_i$.

EXAMPLE 21.16 (Cont.)

x_i	f_i	$x_i f_i$	$x_i - \bar{x}$	$(x_i - \bar{x})^2$	$(x_i - \bar{x})^2 f_i$
37	2	74	−2.95	8.7025	17.4050
38	4	152	−1.95	3.8025	15.2100
39	5	195	−0.95	0.9025	4.5125
40	2	80	0.05	0.0025	0.0050
41	2	82	1.05	1.1025	2.2050
42	1	42	2.05	4.2025	4.2025
43	2	86	3.05	9.3025	18.6050
44	2	88	4.05	16.4025	32.8050
Totals	20	799			94.95

The mean is $\bar{x} = \dfrac{\sum x_i f_i}{(\sum f_i) - 1} = \dfrac{799}{19} = 42.05$ and the variance is $v = \dfrac{\sum (x_i - \bar{x})^2 f_i}{(\sum f_i) - 1} = \dfrac{94.95}{19} = 4.9974$. The standard deviation is $s = \sqrt{4.9974} \approx 2.24$.

Standard Deviation on a Calculator

Again, a calculator will do much of the work for you. Follow the procedures for finding the mean as discussed in the user's guide for your calculator. In most cases, a graphing calculator will calculate both the variance and standard deviation at the same time as it computes the mean.

Caution

Some calculators report two standard deviation scores. For example, the TI-8x graphing calculators report an Sx and a σx. The symbol Sx, or S_x, represents the **sample standard deviation** and the symbol σx, or σ_x, represents the **population standard deviation**. In most cases you will want the higher of the two scores, Sx, or S_x, the sample standard deviation.

The importance of different values of the standard deviation are interpreted mostly from experience. But a small standard deviation indicates that the values are closely clustered about the mean. On the other hand, a large standard deviation indicates that the values are spread out widely from the mean.

Normal Curve

When we make a large number of measurements, we usually expect that a majority of them are near the mean. One of the most important frequency curves is called the **normal curve** or the **normal distribution curve**. The curve is bell shaped and symmetric about the mean. The normal curve was first discovered by DeMoivre, whose theorem for complex numbers you used in Chapter 14.

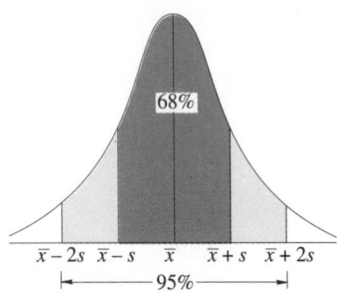

FIGURE 21.5

The standard form of the normal curve considers the mean to be the origin $(\overline{x} = 0)$ and the standard deviation as 1, and has the equation

$$y = \frac{1}{\sqrt{2\pi}}e^{-x^2/2}$$

Some results of this curve are that 68% of the values should be within one standard deviation of the mean. (While 68.26% is closer, we will use 68%.) Thus, 68% of the values are in the interval from $\overline{x} - s$ to $\overline{x} + s$. A total of about 95% of the values (it is actually closer to 95.44%) are supposed to be within two standard deviations of the mean or from $\overline{x} - 2s$ to $\overline{x} + 2s$. (See Figure 21.5.) You may want to round off the values $\overline{x} - s$, $\overline{x} + s$, $\overline{x} - 2s$, and $\overline{x} + 2s$ before you count the values in each interval.

EXAMPLE 21.17

In Example 21.16, there were 20 values. The mean was 39.95 and the standard deviation was 2.18. Thus, the interval from $\overline{x} - s$ to $\overline{x} + s$ is from 37.77 to 42.13. This rounds off to the interval from 38 to 42 and there are 14 values in this interval. This is 70% of the values and is the closest you can get to 68% with a sample of 20 values.

A z-score is used to make it easier to tell whether the score is above or below the mean and how far the score is from the mean with respect to the standard deviation.

Changing an x-value to a z-score

Population		Sample
$z = \dfrac{x - \mu}{\sigma}$	or	$z = \dfrac{x - \overline{x}}{s}$

EXAMPLE 21.18

For a set of data with $\overline{x} = 45$ and $s = 8$, find the z-score corresponding to **(a)** $x = 35$ and **(b)** $x = 51$.

Solutions

(a) $z = \dfrac{x - \overline{x}}{s} = \dfrac{35 - 45}{8} = \dfrac{-10}{8} = -1.25$

(b) $z = \dfrac{x - \overline{x}}{s} = \dfrac{51 - 45}{8} = \dfrac{6}{8} = 0.75$

Application

For men between the ages of 18 and 24 years, serum cholesterol levels, in mg/100 ml, have a mean of 178.1 and a standard deviation of 40.7. Find the z-score for a 22-year-old man who has a serum cholesterol level of 237.5.

Solution $\quad z = \dfrac{237.5 - 178.1}{40.7} = \dfrac{59.4}{40.7} \approx 1.48$

When a distribution is described in intervals, and you have no other information, then you must make some decisions before counting or not counting the points in an interval. In computing the mean and standard deviation, the midpoint of each interval is used as the "value" for that interval. The same criteria are used when counting the number of points within one or more standard deviations of the mean. For example, if $\bar{x} + s$ is at or past the midpoint of an interval, then include all the points of that interval. On the other hand, if $\bar{x} + s$ does not reach the midpoint of the interval, then do not count any of the points in the interval.

Exercise Set 21.3

In Exercises 1–12, use the following sets of numbers.

    A: 4, 2, 6, 3, 7, 4, 6, 2, 4, 4
    B: 50, 52, 52, 54, 54, 54, 54, 56, 58, 58
    C: 80, 77, 82, 73, 92, 89, 100, 96, 96, 94, 74, 94, 94, 96, 83, 84, 96, 87, 84, 96
    D: 100, 98, 96, 94, 93, 90, 89, 85, 82, 78, 76, 66, 64, 64, 78, 89, 93, 96, 98, 96, 93, 64, 96

In Exercises 1–4 find the variance v for the indicated set of numbers.

1. Set A   **2.** Set B   **3.** Set C   **4.** Set D

In Exercises 5–8, find the standard deviation for the indicated set of numbers.

5. Set A   **6.** Set B   **7.** Set C   **8.** Set D

In Exercises 9–12, determine the number of values within (a) one standard deviation and (b) two standard deviations of the mean.

9. Set A   **10.** Set B   **11.** Set C   **12.** Set D

Solve Exercises 13 and 14.

13. Write a computer program to calculate the mean, variance, and standard deviation. For the variance, you might want to use the formula

$$s^2 = \frac{n \sum f_i x^2{}_i - \left(\sum x^2{}_i\right)}{n(n-1)}$$

14. Modify your program in Exercise 13 to have the program determine the total number of values and the number and percentage within one and two standard deviations of the mean.

In Exercises 15–18, assume the given data are normally distributed, (a) find the standard deviation for the indicated sets of numbers, (b) determine the number of data points within one standard deviation of the mean, and (c) those within two standard deviations of the mean.

15. *Machine technology*   The 500 steel rods in Figure 21.3. (See frequency distribution following Example 21.12.)

16. *Energy technology*   The data from the frequency distribution for the batteries in Exercise 23 of Exercise Set 21.2. (Use the midpoint of each interval as the value for that interval.)

17. *Machine technology*   The data from the frequency distribution for the plastic pipe coating in Exercise 25 of Exercise Set 21.2.

18. *Electrical technology*   The data from the electric circuit tests in Exercise 30, Exercise Set 21.2.

In Exercises 19–24, (a) find the mean and standard deviation for each of the indicated sets of numbers, (b) determine the percent of the given data within one standard deviation of the mean, and (c) determine the percent of the given data within two standard deviations of the mean.

19. *Police science*   A patrol officer using a laser gun recorded the following speeds for motorists driving through a 55 mph speed zone:

    52   57   62   59   67   54
    55   64   65   59   63   72

20. *Environmental science*   An environmental officer measured the carbon monoxide emissions (in g/m) for several vehicles. The results are shown in the following table:

    5.02   12.36   13.46   6.92   7.44   8.52   12.82
    11.92   14.32   12.06   8.02   11.34   6.66   9.28

21. *Energy science*   The electrical energy officer measured the carbon monoxide emissions (in g/m) for several vehicles. The results are shown in the following table:

    892   673   534   437   449   524
    627   735   892   923   1024   905
    865   704   624   535   432   495
    572   625   655   684   532   484

22. *Insurance*   The blood alcohol content of 15 drivers involved in fatal accidents and then convicted with jail sentences are given below:

    0.14   0.16   0.21   0.10   0.13
    0.19   0.26   0.22   0.13   0.09
    0.11   0.18   0.22   0.24   0.16

23. *Business*   The daily sales in dollars for one 31-day month at a store are shown below:

    24,562 38,646 43,988 15,122 14,321 17,479 19,478
    25,625 39,476 45,353 15,972 13,793 17,457 18,681
    20,562 38,606 53,788 15,122 10,321 13,037 17,038
    25,625 37,036 45,353 15,732 13,373 13,053 18,681
    21,903 41,775 52,117

24. *Energy technology*   The following data are based on the energy consumption for one household's electric bills for 36 two-month periods. (Hint: Use the midpoint of each class as the value for that class.)

Energy (kWh)	700–719	720–739	740–759	760–779	780–799
Frequency	2	2	4	5	3

Energy (kWh)	800–819	820–839	840–859	860–879	880–899
Frequency	4	7	5	2	2

In Your Words

25. Explain the standard deviation and the variance. How are they alike? How are they different?

26. What is a normal curve?

≡ 21.4
FITTING A LINE TO DATA

Until now, when we have dealt with the relationship between two variables we have been given the functional relationship between them. But, many times we do not know what the relationship is or even if it exists. Sometimes we have some pairs of values that were obtained from an experiment or from observation. In Sections 21.4 and 21.5, we will take these pairs of values and plot them. Then we will try to find a function $y = f(x)$ that is suggested by the data.

In this section, we will show how to "fit" a line to a given set of points. As so often happens, when we are dealing with information about people or with the results of an experiment, the results do not exactly match what theory says they should. When you are conducting an experiment, you are often trying to determine the type of curve, and its equation, that fits the data. In this section, we will be interested in fitting a straight line to the data. In order to fit a straight line, you are going to find the line that best fits these data.

We will assume that a relationship does exist between the variables in each of the examples and in all of the exercises. Our job is to find the straight line that comes the closest to fitting these data.

Application

EXAMPLE 21.20

Below are the heights, in inches, and the weights, in pounds, for 12 students.

Student	Height	Weight
A	66	130
B	67	140
C	68	180
D	69	160
E	70	185
F	71	190
G	72	200
H	70	195
I	72	195
J	71	210
K	72	210
L	73	180

EXAMPLE 21.20 (Cont.)

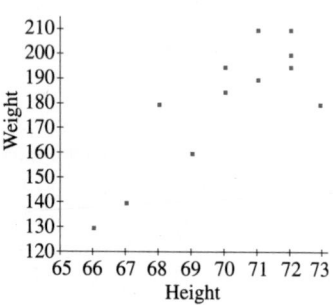

FIGURE 21.6

A graph of these 12 points is shown in Figure 21.6. A quick glance at this graph tells you that these points do not all lie on the same straight line. But, in general, the taller a person is, the more he or she weighs. Thus, we think that it is possible that there might be some straight line that would come close to these points and from which none of the points differ by very much.

▪

A graph like the one in Figure 21.6, which plots individual data points on a coordinate system, is called a **scatter diagram**.

Method of Least Squares

There are several ways to determine the line that best fits these points. We will use a method known as the **method of least squares**. The idea behind this method is that the sum of the squares of the deviation of all points from the best line is a minimum. The deviation we refer to is the difference between the y-value of the line and the y-value for a given value of x.

Suppose that the line $y = mx + b$ "passes through" these data points and that the point (x_1, y_1) is a point from the data. As the line passes through these points, it may or may not intersect any of them. But, some points will lie on one side of the line and other points on the other side of the line. The difference between the observed y-value and the one on the line is the deviation. In Figure 21.7, this is indicated by the letter d.

Now by finding these deviations for all the points of observation and squaring them, we will be finding the minimum of the sum of the squares. If

$$y = mx + b$$

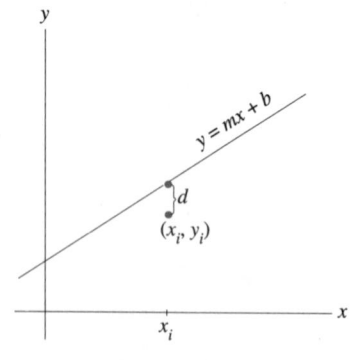

FIGURE 21.7

is the equation of the least squares line, then it can be shown that

$$m = \frac{\dfrac{\Sigma x_i \Sigma y_i}{N} - \Sigma x_i y_i}{\dfrac{(\Sigma x_i)^2}{N} - \Sigma x_i^2}$$

and $\quad b = \bar{y} - m\bar{x}$

The formula for the slope looks very complicated. Some people prefer to write it in the somewhat simpler form

$$m = \frac{n(\Sigma xy) - (\Sigma x)(\Sigma y)}{n(\Sigma x^2) - (\Sigma x)^2}$$

We will now determine the least square line for Example 21.20.

EXAMPLE 21.21

Determine the least square line for the data in Example 21.20.

Solution   The heights will be the x-scores and the weights the y-scores.

Student	Height x	Weight y	xy	x^2
A	66	130	8,580	4,356
B	67	140	9,380	4,489
C	68	180	12,240	4,624
D	69	160	11,040	4,761
E	70	185	12,950	4,900
F	71	190	13,490	5,041
G	72	200	14,400	5,184
H	70	195	13,650	4,900
I	72	195	14,040	5,184
J	71	210	14,910	5,041
K	72	210	15,120	5,184
L	73	180	13,140	5,329
Totals	841	2175	152,940	58,993

$$\bar{x} = \frac{841}{12} = 70.0833$$

$$\bar{y} = \frac{2,175}{12} = 181.25$$

We now use the simplified form of the equation for the slope of the least squares line.

$$m = \frac{n\left(\Sigma xy\right) - \left(\Sigma x\right)\left(\Sigma y\right)}{n\left(\Sigma x^2\right) - \left(\Sigma x\right)^2}$$

$$= \frac{12\left(152,940\right) - \left(841\right)\left(2,175\right)}{12\left(58,993\right) - \left(841\right)^2}$$

$$= \frac{6,105}{635} \approx 9.61$$

$$b = \bar{y} - m\bar{x}$$

$$= 181.25 - \left(9.61\right)\left(70.08\right)$$

$$\approx -492.22$$

Thus, the equation of the line is $y = 9.61x - 492.22$.

Using a Calculator

Again, the calculator can give you a great deal of help. In fact, it can be used to calculate the slope and intercept of the least squares line.

Here you will use the statistical function called **linear regression**. Be sure to read the user's manual for your calculator because calculators use the same variables to represent different quantities. In most cases, the calculator will give you the variables a, b, and r. Variable r represents the correlation coefficient which is discussed in the next subsection. Variables a and b represent the slope and intercept of the least squares line. You will use these values of a and b to get the least squares line $y = ax + b$ or $y = a + bx$. One calculator, the TI-82, gives you the option of which form you want.

EXAMPLE 21.22

Use a calculator to determine the least square line for the data from Example 21.20.

Solution The actual keystrokes given below are for a TI-82 graphics calculator. Enter the data for the height in the column headed L1 and the data for the weight in the column headed L2. When you have finished entering both columns of data, the calculator screen should look like the one in Figure 21.8a.

To determine the equation for the least squares line, press STAT ▶ 5 ENTER. The result, as shown in Figure 21.8b, shows that the equation is of the form $y = ax + b$ with $a \approx 9.614173228$ and $b \approx -492.5433071$. Thus, the equation of the least square line for the data from Example 21.20 is $y = 9.6142x - 492.5433$.

L₁	L₂	L₃
71	190	
72	200	
70	195	
72	195	
71	210	
72	210	
73	180	

L₂(12)=180

FIGURE 21.8a

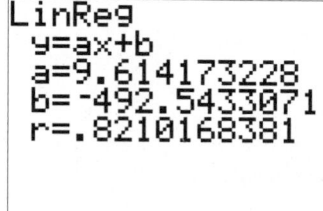

FIGURE 21.8b

Correlation Coefficient

Another statistical value that is often used is the **correlation coefficient** r, where $-1 \le r \le 1$. A value of r close to $+1$ indicates that the relationship between x and y is close to being linear and both variables are increasing. A value close to -1 also indicates a nearly linear relationship, but one where one variable increases as the other decreases. If $-0.75 < r < 0.75$, the correlation is considered to be poor. Graphs of points with different values of r are shown in Figures 21.9a–d.

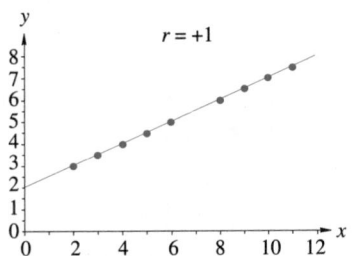

FIGURE 21.9a

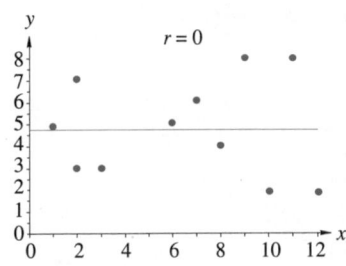

FIGURE 21.9b

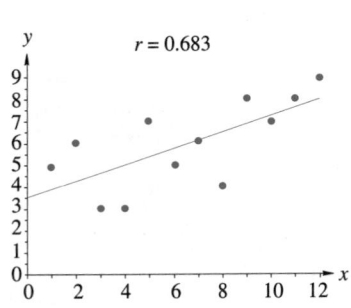

FIGURE 21.9c

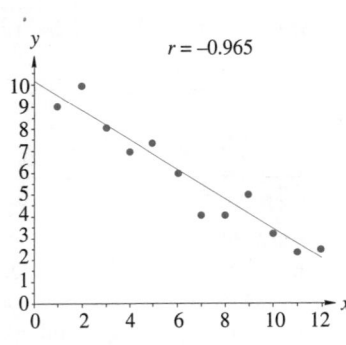

FIGURE 21.9d

One formula for the correlation coefficient is given by $r = m\dfrac{s_x}{s_y}$, where m is the slope of the least square line, s_x is the standard deviation of the x-values, and s_y is the standard deviation of the y-values.

Application

EXAMPLE 21.23

Determine the correlation coefficient for the data on the students in Example 21.20.

Solution   In Example 21.21, we determined that $m = 9.61$. We can calculate $s_x = 2.19$ and $s_y = 25.68$, so

$$r = m\frac{s_x}{s_y}$$

$$= 9.61\left(\frac{2.19}{25.68}\right)$$

$$\approx 0.82$$

It is also possible to determine the correlation coefficient with the help of some calculators. A calculator that can determine the linear regression will also report the correlation coefficient. For example, if you look at Figure 21.8a, you will see that the last item on the calculator screen is $r = .8210168381$. This rounds off to the same value we got in Example 21.23.

Exercise Set 21.4

In Exercises 1–8, find the equation of the least square line and the correlation coefficient for the given data. For each problem, graph the given data and the least square line.

1.

x	5	6	8	9	12	14
y	2	3	7	7	8	10

2.

x	1	3	5	7	9
y	16	45	86	104	132

3.

x	2	4	6	8	10
y	30	55	93	122	132

4.

x	68.5	67.2	67.7	63.8	69.8
y	33.6	35.0	30.2	30.0	33.2
x	64.7	66.4	69.1	65.3	64.8
y	30.8	30.2	33.3	32.9	37.3

5. *Machine technology*   A tensile ring is to be calibrated by measuring the deflection in thousandths of an inch at various loads in thousands of pounds. The following results were obtained by applying increasingly larger load forces from 1,000 to 12,000 lb.

load force (lb $\times$ 1,000)	1	2	3	4	5	6
deflection (in $\times$ 0.001)	16	34	44	76	84	98

load force	7	8	9	10	11	12
deflection	108	126	137	158	165	185

6. *Business*   A vending machine company studied the relationship between maintenance cost and dollar

sales for their machines. Here are the results from eight machines:

maintenance cost	95	105	85	130	145
sales	1,100	1,250	600	890	1,450

maintenance cost	125	90	110	90
sales	1,500	800	1,300	870

7. *Energy technology*   An engineering company studied the relationship between the air velocity and the evaporation of burning fuel droplets in an impulse engine. The following results were obtained:

air velocity (mm/sec)	200	600	1,000
evaporation coefficient (mm^2/s)	0.18	0.37	0.39

air vel.	1,400	1,800	2,200	2,600	3,000
evap. coeff.	0.75	0.82	0.93	1.15	1.42

8. *Automotive engineering*   An automobile company studied the relationship between highway mileage and automobile weight.

highway mpg	42	37	45	40	36	35
weight (lb $\times$ 100)	21	23	24	22	25	24

highway mpg	45	43	30	25
weight (lb $\times$ 100)	19	20	25	26

Solve Exercise 9.

9. Write a computer program that will allow you to input pairs of values and then compute the slope and y-intercept of the least square line and the correlation coefficient.

In Exercises 10–14, find the equation of the least square line and the correlation coefficient for the given data.

10. *Police science*   The following information was collected on the age, in years, and the BAC (blood alcohol count) when convicted DUI jail inmates were first arrested for driving under the influence of alcohol:

Age	16.8	53.4	35.6	45.3	18.9	20.0	37.4	56.6	22.6	26.8
BAC	0.19	0.21	0.22	0.16	0.22	0.26	0.14	0.16	0.19	0.25

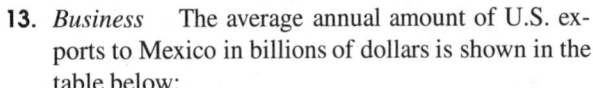

11. *Insurance*   An insurance company noted the following age and vehicle speed for motorists stopped for speeding in a 55 mph speed zone:

Age	42	63	37	74	32	27	18	21	32	16
Speed	62	57	62	59	67	64	65	72	61	82

Age	16	20	42	63	37	28	24	32	27	19
Speed	75	64	65	59	63	72	62	62	67	74

12. *Environmental science*   The average annual electricity use, in kWh/yr, of a refrigerator is shown in the following table.

Year	1970	1974	1978	1982	1986	1990	1994
Annual electricity used (kWh/yr)	1740	1600	1450	1175	1000	950	625

13. *Business*   The average annual amount of U.S. exports to Mexico in billions of dollars is shown in the table below:

Year	1990	1991	1992	1993	1994
U.S. exports to Mexico	28	32	40	41	49

14. *Automotive technology*   Several different makes and models of automobiles were selected. The displacement in liters of each engine and the average miles per gallon (mpg) of the that automobile are given in the table below:

Engine displacement	1.5	1.7	1.9	2.2	2.3	2.8
Average mpg	43.1	41.9	38.7	36.1	34.7	33.0

Engine displacement	3.5	3.8	4.3	4.5	4.9
Average mpg	32.1	28.6	26.5	27.1	26.1

 In Your Words

15. What is the equation of the least square line? What does it tell you?

16. Describe how you can use your calculator to determine the equation of the least square line.

17. Explain the correlation coefficient. How do you find it? What does it tell you? What are the possible values?

18. Describe how you can use your calculator to determine the correlation coefficient.

☰ 21.5
FITTING NONLINEAR CURVES TO DATA

In Section 21.4, we studied the case where the regression curve of y on x is linear. Thus, the curve was of the form $y = mx + b$ and, if we knew the value of x, we could predict a value of y. In the same manner, we could predict the value of x for any value of y.

In this section, we will investigate cases where the regression curve is not linear, but where we can still use the methods we learned in Section 21.4. We will then look at some examples of polynomial regression.

Engineers and technicians often plot data on different kinds of graph paper. In Section 13.6, we plotted data on semilogarithmic and logarithmic graph paper. We learned that if the graph of data is a straight line with slope m on log-log (logarithmic) paper, then the data are related by a power function, which is of the form $y = ax^m$. If the graph of the data is a straight line on semilog (semilogarithmic) paper, then the data are related by a function of the form $y = ab^x$.

Suppose that you plot several pairs of points on regular, semilog, and log-log graph paper. If the points "straighten out" on one of them, say the semilog paper, you then have an idea of the nature of the curve. If it is "straight" on the semilog paper, you know it is of the form $y = ab^x$. If you take the logarithm of both sides, you get

$$\log y = \log a + x \log b$$

We can get estimates of $\log a$ and $\log b$ using the methods from Section 21.4. Of course, once we know $\log a$ and $\log b$ we can find a and b.

Application

EXAMPLE 21.24

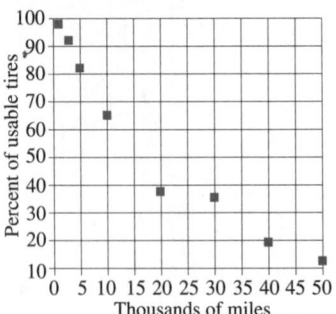

FIGURE 21.10a

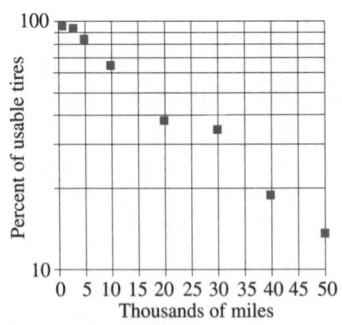

FIGURE 21.10b

A tire company collected the following data on the percentage of radial tires that are still usable after having been driven any given number of miles. The data are shown in the following table.

Miles driven ($\times 1,000$)	1	2	5	10	20	30	40	50
Percent usuable	98.3	92.1	82.1	64.9	38.0	35.2	19.3	13.5

Solution The plot of these data are shown in Figure 21.10a. As you can see, when semilog graph paper was used in Figure 21.10b, the points appear to be in a straight line.

Since we think that we have a straight line of the form $\log y = \log a + x \log b$, we expand the data table to include the values for $\log y$.

Miles (x)	Percent (y)	$\log y$
1	98.3	1.99255
2	92.1	1.96426
5	82.1	1.91434
10	64.9	1.81224
20	38.0	1.57978
30	35.2	1.54654
40	19.3	1.28556
50	13.5	1.13033

Using a calculator, or the technique from Section 21.4, we find that $\log a = 1.994676$ and $\log b = -0.0172898$, so $a \approx 98.78$ and $b \approx 0.96$, and we see that $y = (98.78)(0.96)^x$. The correlation of these data are $r = -0.99$ (which indicates a strong validity for the equation).

If the graph appeared to be a straight line on log-log paper, the function would then be of the form $y = ax^m$. Again, you would take the logarithm of both sides. This time you would find the least square line that fits

$$\log y = \log a + m \log x$$

If we have reason to believe that the data would best fit a polynomial of degree p, then the least square curve is of the form

$$y = A_0 + A_1 x + A_2 x^2 + \cdots + A_{p-1} x^{p-1} + A_p x^p$$

Advanced mathematics can be used to derive $p+1$ **normal equations** that can be used to determine approximations for the A_i values. These normal equations are of the following form, where a_i represents the approximation for A_i that we will obtain.

$$\sum y = na_0 + a_1 \sum x + a_2 \sum x^2 + \cdots + a_p \sum x^p$$

$$\sum xy = a_0 \sum x + a_1 \sum x^2 + a_2 \sum x^3 + \cdots + a_p \sum x^{p+1}$$

$$\sum x^2 y = a_0 \sum x^2 + a_1 \sum x^3 + a_2 \sum x^4 + \cdots + a_p \sum x^{p+2}$$

$$\vdots$$

$$\sum x^p y = a_0 \sum x^p + a_1 \sum x^{p+1} + a_2 \sum x^{p+2} + \cdots + a_p \sum x^{2p}$$

Notice that this is a system of $p+1$ linear equations in the $p+1$ variables a_0, a_1, a_2, ..., a_p. Your choice is to select the degree of the polynomial that you think best approximates the plotting of the data.

EXAMPLE 21.25

A plastics company tested a new additive to determine the relation of the amount of additive to the time it takes the plastic to set. Find the polynomial that best fits the following data.

Amount of additive (g)	0	1	2	3	4	5	6	7	8
Setting time (hr)	7.5	5	3.5	2.5	2	2.5	2.5	4	5

Solution The plot of these points is shown in Figure 21.11. The points give the appearance of a parabola, so we will see which quadratic equation best fits these data. Calculating the sums needed for the normal equations, we get

$$\sum x = 36 \qquad \sum x^2 = 204 \qquad \sum x^3 = 1{,}296 \qquad \sum x^4 = 8{,}772$$
$$\sum y = 34.5 \qquad \sum xy = 123 \qquad \sum x^2 y = 742$$

With this information, we can solve the following system of three linear equations in three variables.

$$34.5 = 9a_0 + 36a_1 + 204a_2$$
$$123 = 36a_0 + 204a_1 + 1{,}296a_2$$
$$742 = 204a_0 + 1{,}296a_1 + 8{,}772a_2$$

Solving these, we get $a_0 = 7.258$, $a_1 = -2.328$, and $a_2 = 0.260$, and we obtain the least squares quadratic :

$$y = 7.258 - 2.328x + 0.260x^2$$

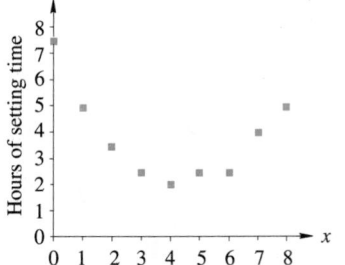

FIGURE 21.11

If you think you know which curve the data will fit, you can take a slightly different approach. Suppose the data seems to fit a function of the type $f(x)$. Then you can solve for the least squares line

$$y = mf(x) + b$$

Here, you first calculate $f(x)$ and then treat the problem as the least square line was treated in Section 21.4.

EXAMPLE 21.26

The resonant frequency of an electric circuit containing a $4\,\mu F$ capacitor was measured as a function of an inductance in the circuit, and the following data were found.

f (Hz)	490	370	260	200	175	150
L (H)	1.0	2.0	4.0	6.0	8.0	10.0

Find the least square curve of $f = m\left(\dfrac{1}{\sqrt{L}}\right) + b$.

Solution   We are given $f(x) = \dfrac{1}{\sqrt{L}}$. We will solve the system of normal equations from Section 21.4:

$$\sum y = nb + m\sum f(x)$$
$$\sum f(x)y = b\sum f(x) + m\sum [f(x)]^2$$

Here we use $f(x)$ instead of x. Since $f(x) = \dfrac{1}{\sqrt{L}}$, we let y represent the frequency and get the following sums for our system of equations.

$$\sum y = 1{,}645 \qquad\qquad \sum f(x) = 3.285$$
$$\sum f(x)y = 1{,}072.585 \qquad\qquad \sum [f(x)]^2 = 2.142$$

Thus, we have the system of equations

$$1{,}645 = 6b + 3.285m$$
$$1{,}072.585 = 3.285b + 2.142m$$

Solving these, we find that

$$m = 500.630 \text{ and } b = 0.0720$$

and so the most accurate curve is

$$f = \frac{500.630}{\sqrt{L}} + 0.0720$$

Using a Calculator

Once again, a graphing calculator can do a lot of the work for you. As we saw in the last section, many graphing calculators will give you the slope and the y-intercept of the least squares line. Many will also find the best logarithmic, exponential, power, quadratic, cubic, or quartic equation to fit a set of given data.

As when you determined the least squares line, all of these regression equations are accessed through the statistical menu on your calculator. Be sure to read the user's manual for your calculator to determine which regression equations you can access.

EXAMPLE 21.27

Use a calculator to determine the cubic polynomial that best fits the data in Example 21.25.

Solution The actual keystrokes given below are for a TI-82 graphics calculator. Enter the data for the amount of additive in grams in the column headed L1 and the data for the setting time in hours in the column headed L2.

To determine the equation for the cubic polynomial that best fits these data, press [STAT] [▶] [7] [ENTER]. The result, as shown in Figure 21.12a, shows that the equation is of the form $y = ax^3 + bx^2 + cx + d$ with $a \approx -0.0126262626$, $b \approx 0.4112554113$, $c \approx -2.784992785$, and $d \approx 7.46969697$. Thus, the equation of the cubic that best fits the given data is

$$y = -0.0126262626x^3 + 0.4112554113x^2 - 2.784992785x + 7.46969697.$$

A calculator-generated graph of these nine points and this cubic is shown in Figure 21.12b.

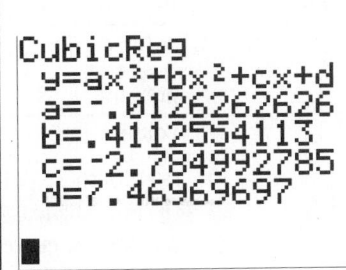

FIGURE 21.12a

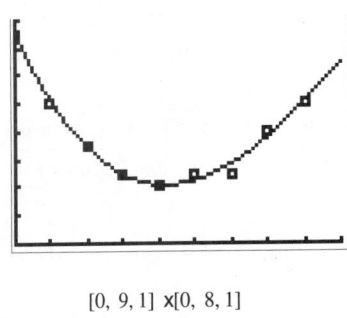

[0, 9, 1] x[0, 8, 1]

FIGURE 21.12b

Exercise Set 21.5

1. *Chemistry* The following data gives the amount of chlorine residual in a swimming pool at different times after it has been treated with chemicals.

Number of hours (x)	12	24	36	48	60
Chlorine residual (y) [parts per million (ppm)]	1.80	1.47	1.25	1.10	1.01

(a) Plot the data on semilog graph paper.

(b) Use the method of least squares to fit a curve of the form $y = ab^x$.

2. *Biology* The following data represents the growth of a colony of bacteria in a culture medium:

Days since inoculation (x)	3	6	9	12
Bacteria count ($\times 1,000$) (y)	115	147	240	354

Days since inoculation (x)	15	18
Bacteria count ($\times 1,000$) (y)	580	866

(a) Plot the data on semilog graph paper.

(b) Use the method of least squares to fit a curve of the form $y = ab^x$.

3. *Business*   A company compiled the following data on the numbers of pipes in a shipment and the number of defective pipes in the shipment.

Pipes shipped (× 100) (x)	5	10	4	10	7
Defective pipe (y)		30	51	26	52	40

(x)	8	8	5	10	5	12	6
(y)	43	45	31	52	30	59	36

(a) Plot these points on regular graph paper.
(b) Find the best fit quadratic equation for these points.
(c) Graph the curve you found in (b) on the same graph you used for (a).

4. *Machine technology*   The following data pertains to the amount of hydrogen present in core drillings made at 100-mm intervals along the length of a vacuum-cast ingot.

Core location (mm from base) (x)	1	2	3
Hydrogen present (ppm) (y)		1.28	1.47	1.00

(x)	4	5	6	7	8	9	10
(y)	0.85	0.72	0.63	0.89	0.91	1.05	1.08

(a) Plot these points.
(b) Find a parabola by the method of least squares.
(c) Graph the parabola on your graph from (a).

5. *Physics*   The following data were gathered on the atmospheric pressure, in atmospheres, at various altitudes, in meters.

altitude (×1,000 m), x	1	2	3	4
pressure (atm), y	0.879	0.782	0.687	0.601

altitude (×1,000 m), x	5	6	7	8
pressure (atm), y	0.540	0.475	0.420	0.370

(a) Plot these points on semilog paper.
(b) Find the least square curve of $y = ae^{-kx}$.
(c) Plot the curve from (b) on your graph from (a).

In Your Words

9. Describe how you might be able to determine whether a given set of data can best be approximated by a logarithmic or power function.

6. *Electrical technology*   The following data for the resistance of a copper wire and a cross-sectional diameter were gathered.

diameter (mm), x	0.1	0.2	0.3	0.4	0.5
resistance (Ω), y	90	22	10	5.6	3.8

diameter (mm), x	0.6	0.7	0.8	0.9	1.0
resistance (Ω), y	2.5	1.8	1.4	1.1	0.8

(a) Plot these points on log-log paper.
(b) Fit the least square curve of $yx^2 = k$.
(c) Plot the curve from (b) on your graph from (a).

7. *Physics*   The following are data for the pressure and volume of a gas at a constant temperature.

pressure	1	2	3	4	5	6	7
volume	5.3	2.7	1.8	1.4	0.9	0.8	0.7

pressure	8	9	10
volume	0.6	0.5	0.5

(a) Plot these points on log-log paper.
(b) Find the least square curve of $pv = k$.
(c) Plot the curve from (b) on your graph for (a).

8. *Energy technology*   An experimenter is testing a new windmill design. She gathers the data in the following table showing the power P generated by the windmill for various wind velocities v.

v	10	20	30	40	50
P	18	78	390	2,050	9,900

(a) Sketch the data on semilog graph paper.
(b) Sketch the data on log-log graph paper.
(c) Find the least square curve that relates P as a function of v.
(d) Graph your curve from (c) on the graph from (a).

 10.   Describe how you can use your calculator to determine the equation for fitting a nonlinear curve to a given set of data.

☰ CHAPTER 21 REVIEW

Important Terms and Concepts

Box and whisker diagram	Mode
Box plot	Normal curve
Conditional probability	Normal equations
Correlation	Outcome
Dependent events	Population
Event	Probability
Frequency distribution	Sample
Histogram	Sample space
Independent events	Scatter diagram
Linear regression	Standard deviation
Mean	Trial
Median	Variance
Method of least squares	Z-score

Review Exercises

In Exercises 1–8, find the probability of the event.

1. Getting a "tail" when a coin is tossed.
2. Drawing a joker from a deck of 54 cards.
3. Drawing a green ball, blindfolded, from a bag containing 6 red and 10 green balls.
4. Not drawing a queen from a deck of 52 cards.
5. Rolling a 5 or higher with a single die.
6. Rolling a sum of 4 or less with a pair of die.
7. Rolling a 5 and then a 2 with a single die.
8. Drawing two aces from a well-shuffled deck of cards, if the first card is not replaced.

Solve Exercises 9–22.

9. *Quality control*   A quality control engineer inspects a random sample of three disc drives from each lot of 50 that is ready to be shipped. If such a lot contains four drives with slight defects, find the probability that the inspector's sample will include only drives with no defects.

10. *Quality control*   An inspector at a computer disc-drive manufacturing company has collected the following information on the revolutions per minute of the disc drives: 300.1, 300.3, 298.9, 299.6, 300.1, 300.4, 300.5, 297.8, 299.8, 300.7, and 301.2. Determine the median of these numbers.

11. *Quality control*   Determine the quartiles of the numbers in Exercise 10.

12. *Quality control*   Draw a box plot of the data in Exercise 10.

13. *Quality control*   Determine the mode of the numbers in Exercise 10.

14. *Quality control*   Determine the mean of the data in Exercise 10.

15. *Quality control*   Determine the standard deviation of the data in Exercise 10.

16. *Quality control*   What percent of the data is in the interval $\bar{x} - s.d., \bar{x} + s.d.$ for the data in Exercise 10?

17. *Quality control*   What percent of the data in Exercise 10 is in the interval $\bar{x} - 2s.d., \bar{x} + 2s.d.$?

18. *Machine technology*   Some forged alloy bars were twisted until they broke. The number of twists for the bars tested were 33, 24, 37, 48, 26, 52, 37, 45, 23, 43, 39, 42, 32, 50, and 38. Find (a) the mean, (b) the median, (c) the quartiles, and (d) the standard deviation for the data.

19. *Business*   A microcomputer leasing company conducted periodic maintenance visits to the places where they rented their microcomputers. They collected the following data on the number of microcomputers at each location and the total time, in minutes, needed to complete the maintenance.

No. of microcomputers	4	6	2	5
Time for maintenance (min)	205	282	93	237

No. of microcomputers	7	7	6	3
Time for maintenance (min)	324	336	277	153

No. of microcomputers	8	5	3	1
Time for maintenance (min)	368	242	140	75

(a) Graph these data.
(b) Find the equation of the least square line for the given data.
(c) Graph the least square line.
(d) What is the correlation coefficient?

20. *Physics*   An experiment was conducted to determine the specific heat ratio α for a certain gas. In the experiment, the gas was compressed adiabatically to several predetermined volumes V and the corre-

sponding pressure P was measured. The results of the experiment are shown in the following table.

P (lb/in.^{2})	16.3	22.4	33.5	59.3	157.2	414.3
V (in.^{3})	50	40	30	20	10	5

(a) Plot these points on log-log graph paper.
(b) Find the least square curve of $pV^\alpha = K$.
(c) Plot the curve from (b) on your graph for (a).

21. *Business*   A company finds that when the price of its product is p dollars per unit, the number of units sold, n, is indicated in the following table.

price, p	10	30	40	50	60	70
number sold, n	70	68	63	50	43	30

(a) Plot these data.
(b) Find the equation of the least square line and the correlation coefficient for the given data.
(c) Graph the line from (b) on the graph for (a).

22. For the points in the following table, find the least square curve $y = m\sqrt{x} + b$.

x	0	2	4	6	8	10
y	6	8.3	9.1	9.5	10.0	11.0

x	12	14	16	18	20
y	10.6	11.3	11.5	11.9	12.3

▃ CHAPTER 21 TEST

1. What is the probability of drawing a red queen from a deck of 52 cards?

2. A quality control engineer inspects a random sample of 5 carburetors from each lot of 100 that is ready to

be shipped. If such a lot of 100 carburetors contains 2 with a defect, find the probability that the engineer's sample will include only carburetors with no defects.

Use the following information to solve Exercises 3–11.

An inspector at a disc-drive manufacturing company has collected the following information on the revolutions per minute of 20 hard disc drives: 3,600.1, 3,600.4, 3,599.9, 3,599.2, 3,598.9, 3,601.2, 3,600.8, 3,598.7, 3,599.6, 3,600.4, 3,598.2, 3,600.1, 3,599.6, 3,600.4, 3,599.2, 3,601.6, 3,598.6, 3,600.4, 3,600.2, and 3,601.4.

3. Determine the median of these numbers.

4. Determine the quartiles for these numbers.

5. Draw a box plot for these data.

6. Determine the mode of these numbers.

7. Determine the mean of these data.

8. Determine the standard deviation.

9. What percent of the data is in the interval $\bar{x} - s.d.$, $\bar{x} + s.d.$?

10. What percent of the data is in the interval $\bar{x} - 2s.d., \bar{x} + 2s.d.$?

11. An automobile rental agency wanted to see if there was a relationship between the minimum cost to rent a car and the number of cars rented. The follow-ing data were collected on the minimum cost and the number of cars rented each week.

Minimum cost ($)	19.95	20.95	18.95	19.95
Number of cars rented	265	232	293	237

Minimum cost ($)	19.95	21.95	22.95	18.95
Number of cars rented	224	186	167	346

Minimum cost ($)	19.95	20.95	22.95	19.95
Number of cars rented	318	242	140	275

(a) Graph these data.

(b) Find the equation of the least square line for the given data.

(c) Graph the least square line.

(d) What is the correlation coefficient?

22

An Introduction to Calculus

In order to pave this parking lot, we must figure out how much asphalt is needed. In this chapter, we will see how calculus can be used to find this answer.

Courtesy of Michael A. Gallitelli, Metroland Photo Inc.

In the middle of the seventeenth century, scientists began to study speed, motion, and rates of change. From this study evolved a new branch of mathematics called calculus. For the next 250 years, most of the important developments in mathematics and science were connected with calculus. Two men, Isaac Newton (1642–1727) and Gottfried Wilhelm Leibniz (1646–1716) are given credit for its discovery.

The development of calculus was a result of attempts to answer several questions in geometry. One question dealt with the slope of the tangent line to a curve, which led to differential calculus. A second question, concerning the area enclosed by the graph of a function and the x-axis, led to integral calculus. The solutions to these questions, the tangent question, and the area question will be explored in Sections 22.1 and 22.2. In Chapters 23 and 25 we will begin a more detailed look at ways to solve these questions.

This chapter forms a foundation for your study of calculus. As you proceed through these next few chapters, you will be given opportunities to use calculus in such areas as heat, light, sound, electricity, and magnetism. Computer programs and calculators will help your study of the ideas behind this powerful field of mathematics.

22.1
THE TANGENT QUESTION

In Chapter 15, we studied the slope of a straight line. In particular, we found that if $P(x_1, y_1)$ and $Q(x_2, y_2)$ are two points on a line, then the slope of the line is given by the formula

$$m = \frac{y_2 - y_1}{x_2 - x_1}$$

But, what about curves that are not straight lines?

Consider the curve in Figure 22.1. As you can see, this is not a straight line, but it is the curve of a function. Slope is an indication of the "steepness" of a line. The slope of a straight line is the same at every point on the line. The curve in Figure 22.1 gives the impression that it gets steeper as the values of x increase. We might expect that the slope of a nonlinear curve would be different at different points on the curve. We would like a way to measure the steepness, or slope, of a nonlinear curve at any particular point on that curve.

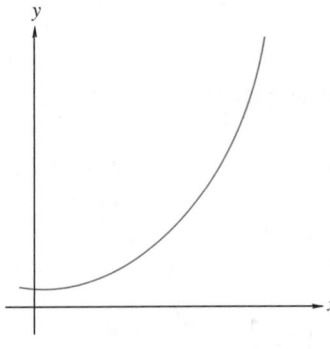

FIGURE 22.1

One way to think of the slope of a curve at some point is to draw the tangent line to the curve at that point. Look at the curve in Figure 22.2. We have drawn the tangent to this curve at point P. We have also drawn the tangent to the curve at point Q. We can tell at a glance that the slope of the tangent at Q is greater than the slope of the tangent at P. Thus, we can let the slope of the tangent to a curve at some point be used for the slope of the curve at that point.

This is a helpful idea. Now, all we have to do is figure out some way to determine the slope of the tangent to a curve at any point. At present we do not have the background to determine the slope of a tangent line to a curve. We will substitute another idea until we have the necessary background.

Look at the graph of the function in Figure 22.3. This is the same curve we used in the previous example. As in Figure 22.2, P and Q are different points on the curve. P has the coordinates (x_1, y_1) and Q has the coordinates (x_2, y_2). The line that passes through points P and Q is called a **secant line**. The slope of the secant line through points P and Q is given by

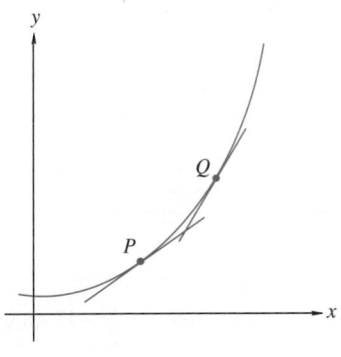

FIGURE 22.2

$$m_{PQ} = \frac{y_2 - y_1}{x_2 - x_1}$$

Average Slope

This is an important idea. We have found that the slope of the secant line is a way to approximate the slope of the tangent line. What the secant line really gives is the **average rate of change** or **average slope** of the curve over the interval from (x_1, y_1) to (x_2, y_2). As we did in Chapter 21, we will use a bar to indicate an average. Thus, the symbol for the average slope will be $\bar{m}$, the symbol for the average velocity $\bar{v}$, and so on. The average rate of change through any two points such as $P(x_1, y_1)$ and

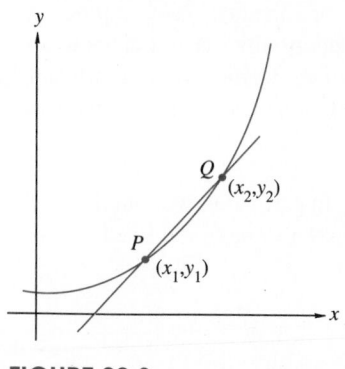

FIGURE 22.3

$Q(x_2, y_2)$ is the same as the slope of the line through those two points, and is given by

$$\bar{m} = \frac{y_2 - y_1}{x_2 - x_1}$$

or, since $y_1 = f(x_1)$ and $y_2 = f(x_2)$,

$$\bar{m} = \frac{f(x_2) - f(x_1)}{x_2 - x_1}$$

EXAMPLE 22.1

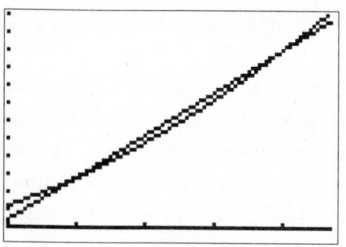

[2, 6.7, 1] x [−2, 60, 5]

FIGURE 22.4

Consider the function $f(x) = x^2 + 3x - 4$. Find the average slope $\bar{m}$ of the curve over the interval from $x = 3$ to $x = 6$.

Solution   If we let $x_1 = 3$ and $x_2 = 6$, then, since $f(x) = x^2 + 3x - 4$, we have $f(3) = 14$ and $f(6) = 50$. Figure 22.4 shows a portion of the graph of $f(x) = x^2 + 3x - 4$ and the secant line through the points $(3, 14)$ and $(6, 50)$. Thus, the average slope of $\bar{m}$ is derived as follows:

$$\begin{aligned}
\bar{m} &= \frac{f(x_2) - f(x_1)}{x_2 - x_1} \\
&= \frac{f(6) - f(3)}{6 - 3} \\
&= \frac{50 - 14}{6 - 3} \\
&= \frac{36}{3} \\
&= 12
\end{aligned}$$

The average slope of $f(x) = x^2 + 3x - 4$ from $x = 3$ to $x = 6$ is 12.

EXAMPLE 22.2

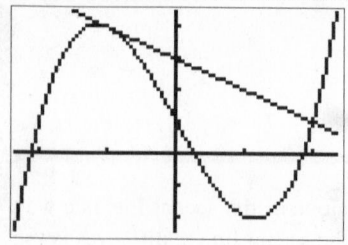

[−2.35, 2.35, 1] x [−2.5, 4.5, 1]

FIGURE 22.5

Find the average slope of the function $g(x) = x^3 - 4x + 1$ from $x = -1$ to $x = 2$.

Solution   We begin by letting $x_1 = -1$ and $x_2 = 2$, and evaluating $g(-1)$ and $g(2)$.

$$g(-1) = (-1)^3 - 4(-1) + 1$$
$$= 4$$

and $g(2) = 2^3 - 4(2) + 1$
$$= 1$$

Figure 22.5 shows the graph of $g(x) = x^3 - 4x + 1$ over the interval $[-2.35, 2.35]$ and the secant line through the points $(-1, 4)$ and $(2, 1)$. Thus,

$$\bar{m} = \frac{g(2) - g(-1)}{2 - (-1)}$$

EXAMPLE 22.2 (Cont.)

$$= \frac{1-4}{2+1}$$

$$= \frac{-3}{3}$$

$$= -1$$

The average slope of $g(x) = x^3 - 4x + 1$ from $x = -1$ to $x = 2$ is -1.

EXAMPLE 22.3

Find the average slope of the curve $h(x) = 2x^2 + 5x - 6$ from $x = -4$ to $x = x_1$, where $x_1 \neq -4$.

Solution This is a more general version of the average slope problem. When $x = -4$, $h(x) = h(-4) = 6$ and when $x = x_1$, $h(x_1) = 2x_1^2 + 5x_1 - 6$. Figure 22.6 shows the graph of h and a line through the points $(-4, 6)$ and $(x_1, h(x_1))$. The average slope is

$$\bar{m} = \frac{h(x_1) - h(-4)}{x_1 - (-4)}$$

$$= \frac{(2x_1^2 + 5x_1 - 6) - 6}{x_1 + 4}$$

$$= \frac{2x_1^2 + 5x_1 - 12}{x_1 + 4}$$

$$= \frac{(2x_1 - 3)(x_1 + 4)}{x_1 + 4}$$

$$= 2x_1 - 3$$

since $x_1 \neq -4$.

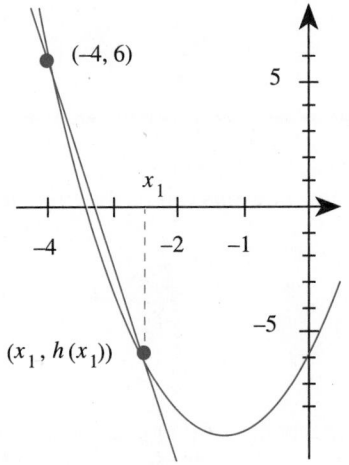

FIGURE 22.6

This last example provides a general formula for finding the average slope from $x = -4$ to any other point on the curve. For example, if we want the average slope from -4 to -5, we can use this formula with $x_1 = -5$ and get $\bar{m} = 2x_1 - 3 = 2(-5) - 3 = -13$.

The idea for finding the average slope can be applied to finding any type of average change. This can best be shown by the following example.

Application

EXAMPLE 22.4

A tank is filled with water by opening a valve on an inlet pipe. The volume V in liters of water in the tank t min after the valve is opened is given by the formula $V(t) = 5t^2 + 4t$. **(a)** What is the average rate of increase in the volume from the first minute to the second minute? **(b)** What is the average rate of increase in the volume during the next 30 seconds?

EXAMPLE 22.4 (Cont.)

Solution   We want to find the average rate of change in the volume.
(a) When $t = 2$, $V(t) = V(2) = 28$ and when $t = 1$, $V(t) = V(1) = 9$. If R_{Vol} is the rate of change in the volume, then

$$\bar{R}_{Vol} = \frac{V(2) - V(1)}{2 - 1} = \frac{28 - 9}{2 - 1} = 19$$

The volume changed at the rate of 19 L/min from the first to the second minute.
(b) The next 30 seconds will be from $t = 2.0$ min to $t = 2.5$ min. When $t = 2.5$, we determine that $V(t) = V(2.5) = 41.25$. Thus,

$$\bar{R}_{Vol} = \frac{V(2.5) - V(2)}{2.5 - 2.0} = \frac{41.25 - 28}{2.5 - 2.0} = \frac{13.25}{0.5} = 26.5$$

Thus, the volume changed at the rate of 26.5 L/min from minutes 2 to 2.5. ∎

Using a Graphing Calculator

Many graphing calculators can be used to approximate the slope of a tangent line at a particular point on a curve. Example 22.5 will show how this can be done on the function used in Example 22.3.

EXAMPLE 22.5

Use a graphing calculator to find the slope of the tangent line to $h(x) = 2x^2 + 5x - 6$ at $x = -4$.

Solution   Begin by graphing h. The result of graphing h on a TI-82 as $y_1 = 2x^2 + 5x - 6$ is shown in Figure 22.7a.

Next, press $\boxed{\text{TRACE}}$ and move the cursor to $x = -4$ and then press $\boxed{\text{2nd}}$ $\boxed{\text{DRAW}}$ 5 $\boxed{\text{ENTER}}$[1]. (Note that pressing $\boxed{\text{2nd}}$ $\boxed{\text{DRAW}}$ 5 accesses the "Draw-Tangent" option on the calculator.) The result is shown in Figure 22.7b. In the lower left-hand corner of the calculator screen shown in Figure 22.7b, you should see $dy/dx = -11$. The notation $dy/dx = -11$ indicates that the slope of this tangent line is -11. In Example 22.3, we found that the average slope of the tangent line to $h(x) = 2x^2 + 5x - 6$ near $x = -4$ was $\overline{m} = 2x_1 - 3$. If you let $x_1 = -4$, then $\overline{m} = 2(-4) - 3 = -11$, the value of dy/dx.

[1]On a TI-85, press $\boxed{\text{GRAPH}}$ $\boxed{\text{MORE}}$ $\boxed{\text{MATH}}$ $\boxed{\text{MORE}}$ $\boxed{\text{MORE}}$ $\boxed{\text{TANLN}}$ and then move the cursor to (or near) $x = -4$ and press $\boxed{\text{ENTER}}$. Setting xMin=-6.3 and xMax=6.3 will allow the cursor to stop at $x = -4$.

EXAMPLE 22.5 (Cont.)

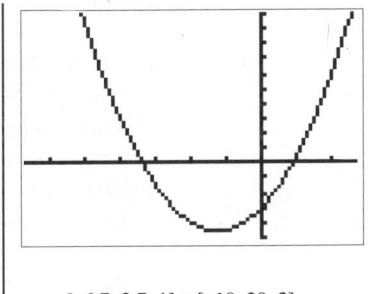

$[-6.7, 2.7, 1] \times [-10, 20, 2]$

FIGURE 22.7a

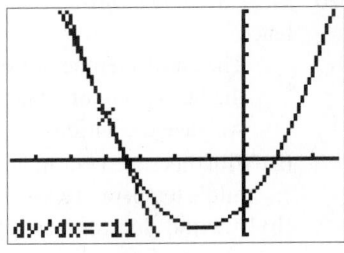

$[-6.7, 2.7, 1] \times [-10, 20, 2]$

FIGURE 22.7b

Exercise Set 22.1

Find the average slope over the indicated intervals for the functions in Exercises 1–14.

1. $f(x) = x^2 + 5x$ from $x = 1$ to $x = 3$

2. $g(x) = 5x - 3$ from $x = 4$ to $x = 7$

3. $h(x) = 7x + 1$ from $x = -3$ to $x = -1$

4. $j(x) = 4x^2 - 5$ from $x = -2$ to $x = 0$

5. $k(x) = 3x^3 + 7x - 1$ from $x = -2$ to $x = -1$

6. $m(x) = 2x^4 - 5x^2 + x - 1$ from $x = 0$ to $x = 1$

7. $f(x) = \dfrac{5}{x}$ from $x = 2$ to $x = 5$

8. $g(x) = \dfrac{8}{x+2}$ from $x = -6$ to $x = -3$

9. $h(x) = x^2$ from $x = 1$ to $x = 0$

10. $j(x) = x^2 + 3x + 5$ from $x = -2$ to $x = b$, $b \neq -2$

11. $k(x) = x^2 + x - 7$ from $x = 1$ to $x = x_1$, $x_1 \neq 1$

12. $m(x) = x^3 + 4$ from $x = 1$ to $x = x_1$, $x_1 \neq 1$

13. $f(x) = x^2 + 1$ from $x = x_1$ to $x = x_1 + h$, $h \neq 0$

14. $g(x) = x^2 + 4x - 1$ from $x = x_1$ to $x = x_1 + h$, $h \neq 0$

Solve Exercises 15–22.

15. *Machine technology*   What was the average rate of change in the volume for the tank in Example 22.4 **(a)** from the second to the third minute, **(b)** from the third to the fourth minute, and **(c)** from the second to the fourth minute?

16. *Physics*   A stone is dropped into a pool of water and causes a ripple that travels in the shape of a circle out from the point of impact at a rate of 2 m/s. What is the average change in the area within this circle from the third to the fourth second?

17. Let $f(x) = 3x^2 - 5$. Find the average slope of the curve from $x_1 = 4$ to **(a)** $x_2 = 6$, **(b)** $x_3 = 5$, **(c)** $x_4 = 4.5$, **(d)** $x_5 = 4.25$, **(e)** $x_6 = 4.1$, and **(f)** $x_7 = 4.05$.

18. For the same function in Exercise 17, $f(x) = 3x^2 - 5$, find the average slope of the curve from $x_1 = 4$ to **(a)** $x_2 = 2$, **(b)** $x_3 = 3$, **(c)** $x_4 = 3.5$, **(d)** $x_5 = 3.75$, **(e)** $x_6 = 3.9$, and **(f)** $x_7 = 3.95$.

 19. Write a computer program to determine the average slope between any two points on a curve. Use your program to check your results from Exercises 17 and 18.

20. *Medical technology*   Consider the following sentence:

    > The child's temperature has been rising for the last two hours, but not as rapidly since we gave the antibiotic an hour ago.

    (a) With this statement in mind, sketch a graph of the child's temperature as a function of time.
    (b) How did you (or how can you) use tangents to your graph to show that your graph is consistent with the statement?

21. *Transportation*   On a 50-minute trip, a car travels for 15 min with an average velocity of 20 mph and then 35 min with an average velocity of 54 mph. Determine (a) the total distance traveled and (b) the average velocity for the entire trip.

22. *Agriculture*   The graph in Figure 22.8 gives the number of $f(t)$ farms in a certain Iowa county t years after 1980.

(a) Calculate the average rate of change in the number of farms from 1980 to 1985.
(b) Calculate the average rate of change in the number of farms from 1990 to 1995.
(c) During what two-year period was the number of farms decreasing most rapidly?
(d) What was the rate of change in the number of farms for the two years you answered in (c)?

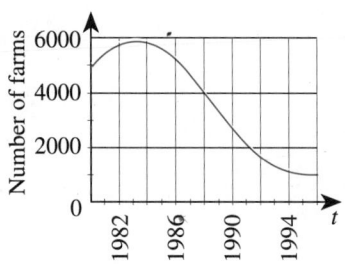

FIGURE 22.8

In Your Words

23. You are in an automobile that is traveling from your home to school. Explain how you can determine the average rate that the car is traveling.

24. What is the difference between a secant line and a tangent line?

☰ 22.2

THE AREA QUESTION

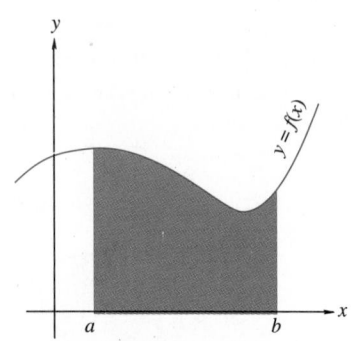

FIGURE 22.9

In Section 22.1, we examined one of the basic questions that led to the development of calculus. In this section, we will look at the other basic question—the area question.

The area question concerns the area between the graph of a function and the x-axis. There are some restrictions on this problem that can best be described by examining Figure 22.9. The function $f(x)$ should be nonnegative over a closed interval; that is, it should not fall below the x-axis in some closed interval. If the closed interval is $[a, b]$, then the desired area lies between the curve of $f(x)$, the x-axis, and the vertical lines $x = a$ and $x = b$. This is indicated by the shading in Figure 22.9.

We have used several different formulas for the area. Most of these were introduced in Chapter 4. We will use two of them to help determine the shaded area in Figure 22.9.

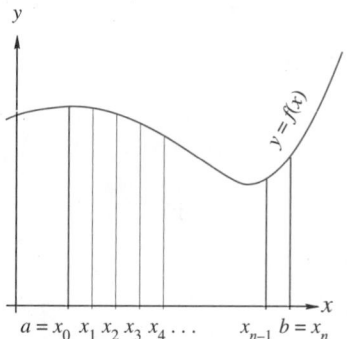

FIGURE 22.10

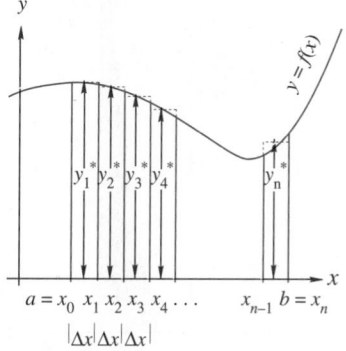

FIGURE 22.11

Rectangular Approximation

The first method we will use will approximate the area by adding the areas of several rectangles. To begin, we will divide the interval $[a, b]$ into n segments. While each segment can be a different length, we will make them all the same length. If we let $a = x_0$ and $b = x_n$, the first segment is $[x_0, x_1]$; the second segment $[x_1, x_2]$; the third segment $[x_2, x_3]$; and so on. The last segment is $[x_{n-1}, x_n]$. In general, the ith segment would be $[x_{i-1}, x_i]$. At each of these points, $x_0, x_1, x_2, \ldots, x_n$, we will erect a line perpendicular to the x-axis. This creates a group of n strips or bands, as shown in Figure 22.10.

We want to draw a line segment parallel to the x-axis across the top of each of these strips. Where we draw these segments is up to us. We would like to draw them so that we can get a close approximation to the area under the curve. What we will do is use an approach that should give a reasonable approximation.

Look at Figure 22.11. We have selected a point in the middle of each strip and drawn the line segment so that it went through the y-coordinate at that point. For example, in the first strip, the midpoint between x_0 and x_1 is the point we will call x_1^*. The height of the curve at x_1^* is $f(x_1^*)$, which we will call y_1^*. Now, since each segment is the same width, the width of each rectangle is the same. This width, which we will call Δx, is equal to $\dfrac{x_n - x_0}{n}$. This means that the area of the first rectangle is $(\Delta x)(y_1^*)$. In the same way, the area of the second rectangle would be $(\Delta x)(y_2^*)$ and the area of the last rectangle would be $(\Delta x)(y_n^*)$.

We can now get an idea of the area under this curve by adding all the areas of these rectangles. Thus, the area under the curve, A, can be written as

$$A \approx (y_1^*)(\Delta x) + (y_2^*)(\Delta x) + (y_3^*)(\Delta x) + \cdots + (y_n^*)(\Delta)$$
$$= (y_1^* + y_2^* + y_3^* + \cdots + y_n^*)(\Delta x)$$
$$= \left(\sum_{i=1}^{n} y_i^* \right) \Delta x$$

EXAMPLE 22.6

Suppose $f(x) = x^2 + 3$. Find an approximation of the area between this curve and the x-axis from $x = 1$ to $x = 6$.

Solution We will divide the interval $[1, 6]$ into 5 equal segments. (The number of segments is purely arbitrary and we selected 5 because it means that each segment has a length of 1, or $\Delta x = 1$.) The midpoints of the segments are at 1.5, 2.5, 3.5, 4.5, and 5.5 as can be seen in Figure 22.12. The value of the function at each of these points can be seen in the following table.

x^*	1.50	2.50	3.50	4.50	5.50
$f(x^*) = y^*$	5.25	9.25	15.25	23.25	33.25

EXAMPLE 22.6 (Cont.)

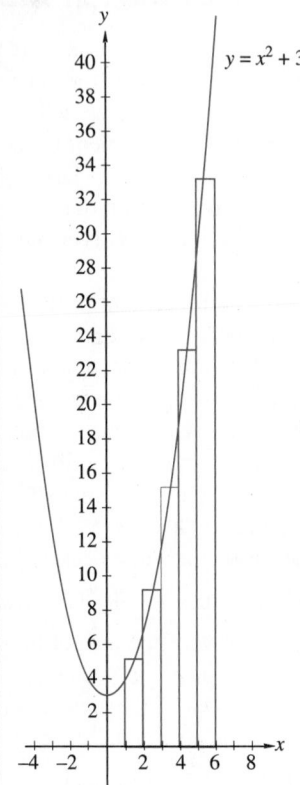

FIGURE 22.12

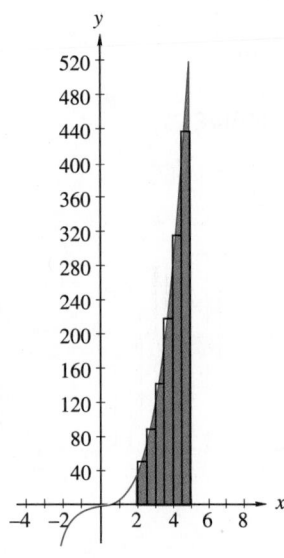

FIGURE 22.13

The area then is found using our new formula for area:

$$
\begin{aligned}
A &\approx (y_1^*)(\Delta x) + (y_2^*)(\Delta x) + (y_3^*)(\Delta x) + \cdots + (y_n^*)(\Delta x) \\
&\approx (y_1^* + y_2^* + y_3^* + \cdots + y_n^*)(\Delta x) \\
&= (5.25 + 9.25 + 15.25 + 23.25 + 33.25)(1) \\
&= 86.25
\end{aligned}
$$

The area is approximately 86.25 square units.

We will work another example, only this time we will make the lengths of the segments something other than 1.

EXAMPLE 22.7

Approximate the area under the curve $g(x) = 4x^3 + 2x - 1$ from $x = 2$ to $x = 5$.

Solution   The curve is shown in Figure 22.13 with the shaded region indicating the area we want to approximate. The total length of the interval is 3 units. We will divide this into 6 segments, so each segment will be 0.5 units long or $\Delta x = 0.5$. The following table gives the values of the function at the midpoints of each of these intervals.

EXAMPLE 22.7 (Cont.)

x^*	2.25	2.7500	3.2500	3.7500	4.2500	4.7500
y^*	49.0625	87.6875	142.8125	217.4375	314.5625	437.1875

Again, using our formula for approximating the area we get

$$A \approx (y_1^* + y_2^* + y_3^* + \cdots + y_n^*)(\Delta x)$$
$$= (49.0625 + 87.6875 + 142.8125 + 217.4375$$
$$+ 314.5625 + 437.1875)(0.5)$$
$$= (1,248.75)(0.5)$$
$$= 624.375$$

The approximate area under this curve from $x = 2$ to $x = 5$ is 624.375 square units.

Application

EXAMPLE 22.8

The following table gives the results of a series of drillings to determine the depth of the bedrock at a building site. These drillings were taken along a straight line down the middle of the lot where the building will be placed. In the table, x is the distance from the front of the parking lot and y is the corresponding depth. Both x and y are given in feet.

x	0	20	40	60	80	100	120	140	160
y	33	35	40	45	42	38	46	40	48

Approximate the area of this cross-section.

Solution The total length of the interval is 160 ft. This has been divided into 8 segments, with each segment 20 ft long, so $\Delta x = 20$. Since we do not know the function, we will have to approximate the depth at the midpoints of the intervals. We will do this by taking the average of the values at each end of an interval. Thus, the length at the midpoint of the first interval is $y_1^* = \dfrac{y_1 + y_2}{2} = \dfrac{33+35}{2} = 34$. The next table gives the approximate values at the midpoints of these intervals.

x^*	10	30	50	70	90	110	130	150
y^*	34	37.5	42.5	43.5	40	42	43	44

Once again, using our formula for approximating the area, we get

$$A \approx (y_1^* + y_2^* + y_3^* + \cdots + y_8^*)\Delta x$$
$$= (34 + 37.5 + 42.5 + 43.5 + 40 + 42 + 43 + 44)(20)$$
$$= (326.5)(20)$$
$$= 6,530 \, \text{ft}^2$$

In Section 25.5, we will find another method to approximate the area of this cross-section, at which point we will get a more accurate approximation of $6,460 \, \text{ft}^2$.

Application

EXAMPLE 22.9

Before the parking lot for the building in Example 22.8 is completed it will have to be paved. A photograph of the completed lot is shown in Figure 22.14a. In order to pave this parking lot, its area had to be determined. This area was then multiplied by the thickness of the asphalt to obtain the total volume of the asphalt that was needed.

In order to find the area, the contractor drew a base line through the "middle" of the parking lot and then drew lines perpendicular to the base line every 50 feet. Each of these lines were drawn until they reached the other side of the parking lot, as shown in Figure 22.14b. The following table shows the lengths labeled y (in feet) of each of these lines.

x	0	50	100	150	200	250	300	350	400	450	500	550	600
y	72	104	160	200	200	200	190	190	190	180	180	180	180

Use the data in the table to approximate the area of the parking lot.

Courtesy of Michael A. Gallitelli, Metroland Photo Inc.

FIGURE 22.14a

base line

◄─┤├─ 50 ft

FIGURE 22.14b

Solution We are given that the length of each interval is 50 ft, and so $\Delta x = 50$. As in Example 22.8, we will approximate the lengths at the midpoints by taking the average of the values at each end of the intervals. The next table gives the approximate values at the midpoints of these intervals.

x^*	25	75	125	175	225	275	325	375	425	475	525	575
y^*	88	132	180	200	200	195	190	190	185	180	180	180

EXAMPLE 22.9 (Cont.)

As before, we use our formula for approximating the area and obtain

$$A \approx (y_1^* + y_2^* + y_3^* + \cdots + y_{12}^*)\Delta x$$
$$= (88 + 132 + 180 + 200 + 200 + 195 + 190$$
$$+ 190 + 185 + 180 + 180 + 180)(50)$$
$$= (2,100)(50)$$
$$= 105,000$$

Thus, we see that the area of this parking lot is approximately $105,000\,\text{ft}^2$.

Exercise Set 22.2

In Exercises 1–10, find the area under the graph of the function from *a* to *b* using the method described in this section. The value of *n* in each problem indicates the number of segments into which you should divide the interval.

1. $f(x) = 3x + 2; a = 1, b = 7, n = 6$
2. $g(x) = 7 - 4x; a = -4, b = 1, n = 5$
3. $h(x) = x^2 + 1; a = 0, b = 3, n = 6$
4. $k(x) = 3x^2 - 2; a = 1, b = 3, n = 4$
5. $j(x) = 4x^2 + 3x - 5; a = 1, b = 5, n = 8$
6. $m(x) = x^3 + 2; a = -1, b = 2, n = 6$
7. $f(x) = 16 - x^2; a = -2, b = 4, n = 8$
8. $g(x) = 4x - x^3; a = 0, b = 2, n = 10$
9. $h(x) = 3x^2 - 2x + 7; a = -5, b = -1, n = 8$
10. $k(x) = x^4 - 3; a = 2, b = 4, n = 8$

In Exercises 11–12, find the approximate area under the curve defined by graphing the sets of experimental data.

11.

x	5	6	7	8	9	10	11
y	4.2	3.9	3.8	4.0	3.5	3.4	3.9

x	12	13
y	4.1	4.3

12.

x	1.00	1.25	1.50	1.75	2.00
y	16.32	16.48	16.73	16.42	16.38

x	2.25	2.50
y	16.29	16.25

Solve Exercises 13–18.

13. Write a computer program to use the rectangular method described in this section to approximate the area under a curve. Your program should be general enough that you can enter the function, and input the endpoints of the interval and the number of segments. Check your program by using it to solve Exercises 1–10.

14. Use the program you wrote in Exercise 13 to get more accurate approximations of the area under the curve in Exercises 9–10. Make the value of *n* (**a**) 15, (**b**) 24, (**c**) 48, and (**d**) 96.

15. *Business*   The marginal profit for a certain type of coat is given by $P'(x) = 78 - 0.8x$ dollars per coat, where *x* is the number of coats produced and sold weekly. The profit for the first *n* coats that are produced and sold is determined by finding the area under the graph of P' from $a = 0$ to $b = n$. What is the profit for the first 40 coats produced and sold?

16. *Environmental science*   The level of pollution in San Juan Pedro Bay, due to a sewage spill, is estimated to be $f(t) = \dfrac{1200t}{\sqrt{t^2 + 10}}$ parts per million, where t is the time in days since the spill occurred. Find the total amount of pollution during the first 5 days of the oil spill by using the method described in this section and 10 divisions.

17. *Electronics*   The charge on a capacitor in millicoulombs can be estimated by finding the area under the graph of $f(t) = 0.6t^2 - 0.2t^3$ from $t = 1$ to $t = 3$. Estimate the charge on this capacitor by using the method described in this section and 8 divisions.

18. *Ecology*   A town wants to drain and fill the swamp shown in Figure 22.15.
   (a) What is the surface area of the swamp?

 **In Your Words**

19. Describe how to use the procedures of this chapter to find the area of an irregular-shaped region.

20. What changes would you have to make in the procedures of this chapter if the rectangles were not all the same width?

(b) If the swamp has an average depth of 6 ft, how many cubic yards of dirt will it take to fill the "hole" that is left after the swamp is drained?

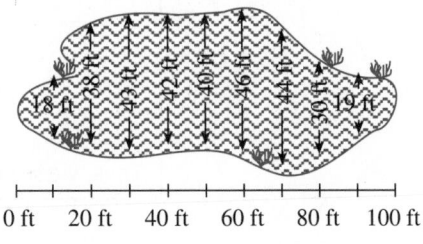

| 0 ft | 20 ft | 40 ft | 60 ft | 80 ft | 100 ft |

FIGURE 22.15

21. Use the procedures described in Example 22.19 on an irregular-shaped region such as a parking lot or a lake to determine its area.

≡ 22.3
LIMITS: AN INTUITIVE APPROACH

In Sections 22.1 and 22.2, we have looked at ways in which we could attempt to solve two of the questions that led to the development of calculus. These two questions are known as the tangent question and the area question. So far we have laid a foundation for answering them. In this section, we will learn about a third part of the foundation—the limit.

To begin our introduction to the limit, we will return to our look at the tangent question. Consider the curve of the function $f(x) = x^2 - 4$. Suppose that we want to find the slope of the tangent to this curve at the point $(3, 5)$. In Section 22.1, we found out how to determine the slope of a secant line to this curve through the point $(3, 5)$. Let $(x, f(x))$ be any point, except the point $(3, 5)$, on the graph of $f(x) = x^2 - 4$. The secant line through these two points is shown in Figure 22.16. The slope of the secant line is given by

$$\bar{m} = \frac{f(x) - 5}{x - 3} = \frac{(x^2 - 4) - 5}{x - 3} = \frac{x^2 - 9}{x - 3}$$

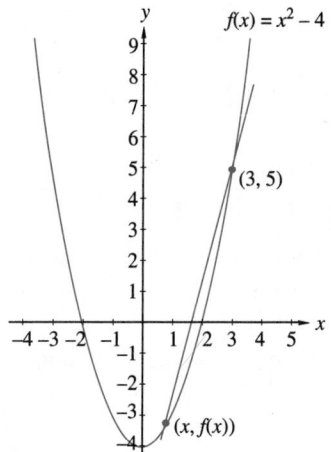

$f(x) = x^2 - 4$

(3, 5)

$(x, f(x))$

FIGURE 22.16

Since the point $(x, f(x))$ is different from $(3, 5)$, we know that $x \neq 3$ and so we can simplify our equation to

$$\bar{m} = \frac{(x+3)(x-3)}{x-3} = x+3$$

So far so good! Everything that we have done is just like what we did in Section 22.1. But now, we want to look at the problem a little closer. What happens as x gets closer to 3? The values of $\bar{m} = x + 3$ get closer to 6. We put this idea into symbols by writing

$$\bar{m}_{\tan} = \lim_{x \to 3} \frac{x^2 - 9}{x - 3} = \lim_{x \to 3} (x+3) = 6$$

Since we know that the slope of the tangent line at the point $(3, 5)$ is 6, we are able to determine that the equation of the tangent is

$$y - 5 = 6(x - 3)$$
$$y = 6x - 13$$

This is the basic idea behind a limit. We will now begin to look at a more detailed explanation.

Limit of a Function

If a function $y = f(x)$ approaches the real number L as x approaches a particular value a, then we say that L is the **limit** of f as x approaches a. The notation for this concept is

$$\lim_{x \to a} f(x) = L.$$

The $\lim_{x \to a} f(x)$ may or may not exist for any function at any particular a. The following examples use numerical calculations to locate limits.

The Numerical Approach to a Limit

To get a better idea of a limit, we will return to Exercises 17–18 from Exercise Set 22.1.

In these two exercises $f(x) = 3x^2 - 5$. We will set up two tables of values. Each table will contain values for x and $\bar{m} = \dfrac{f(x) - f(4)}{x - 4} = \dfrac{f(x) - 43}{x - 4}$. In the first, we will put the values for Exercise 17, and in the second, the values for Exercise 18.

In Exercise 18, all of the values of x were larger than 4 with the first one being the largest, 6, and each one getting smaller. The last value was the smallest, 4.05. We will continue this process by selecting additional numbers larger than 4 (and smaller than 4.05), each one getting closer to 4.

x	6	5	4.5	4.25	4.10	4.05	4.01	4.001	4.0001
m	30	27	25.5	24.75	24.30	24.15	24.03	24.003	24.0003

The table of values for Exercise 18 contains values of x that are all smaller than 4 with the largest being 3.95. Again, we will continue the process. This time the numbers will all be smaller than 4, each getting increasingly closer to 4.

x	2	3	3.5	3.75	3.9	3.95	3.99	3.999	3.9999
m	18	21	22.5	23.75	23.70	23.85	23.97	23.997	23.9997

These tables give us a numerical method for approximating the $\lim_{x \to 4} (3x^2 - 5)$. As you can see from the two tables, when x is 4.0001, $m \approx 24.0004$ and when x is 3.9999, $m \approx 23.9995$. It appears that if we were to pick values of x that were closer to 4, we could then get a value of $f(x) = 3x^2 - 5$ that was closer to 24.

EXAMPLE 22.10

Use a numerical approach to find $\lim_{x \to 1} \dfrac{x^3 - 1}{x - 1}$.

Solution   As in the previous explanation, we will construct a table (using the program from Exercise 19, Exercise Set 22.1). This time, since we want to find the limits as x gets closer to 1, we will select values of x that keep getting closer to 1. Some of the values selected are less than 1, and some are larger than 1. The results are shown in the following table.

x	0.500	0.900	0.990	0.999	1.001	1.010	1.100	1.500
$\dfrac{x^3 - 1}{x - 1}$	1.75	2.71	2.9701	2.997	3.003	3.0301	3.31	4.75

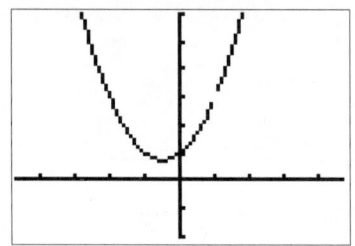

[−4.7, 4.7, 1] x [−2, 6, 1]

FIGURE 22.17

It appears from the table that $\lim_{x \to 1} \dfrac{x^3 - 1}{x - 1} = 3$.

This result is confirmed by looking at the graph of $y = \dfrac{x^3 - 1}{x - 1}$, as shown in Figure ??. The "hole" in the graph appears to be at $(1, 3)$.

You may remember that one of the factors that we had in Chapter 7 was that $x^3 - 1 = (x - 1)(x^2 + x + 1)$. We can use this knowledge to see that as long as $x \neq 1$, $g(x) = \dfrac{x^3 - 1}{x - 1} = x^2 + x + 1$. This makes it a lot easier to see that as x gets closer to 1, the value of the function gets closer to 3.

This next example will show that this type of simplification does not work for all limits.

EXAMPLE 22.11

Use a numerical approach to find $\lim\limits_{x \to 0} \dfrac{\sin x}{x}$.

Solution Since we cannot factor an x out of $\sin x$, the simplification method we have been using will not work. But, if we carefully construct a table of values for x and $g(x) = \dfrac{\sin x}{x}$ for values of x near 0, we can get an idea of what this limit will be. The following table shows this. Notice that the values of x are in radians.

x (radians)	0.200	0.100	0.010	−0.010	−0.100	−0.200
$\sin x$	0.1987	0.0998	0.0099	−0.0099	−0.0998	−0.1987
$\dfrac{\sin x}{x}$	0.9933	0.9983	0.9999	0.9999	0.9983	0.9933

From this table of values it appears that

$$\lim_{x \to 0} \frac{\sin x}{x} = 1$$

The graph in Figure 22.18 seems to reinforce this numerical method that shows $\lim\limits_{x \to 0} \dfrac{\sin x}{x} = 1$. Notice that if you $\boxed{\text{TRACE}}$ this curve, there is no y-value when $x = 0$. That is because 0 is not in the domain of the function.

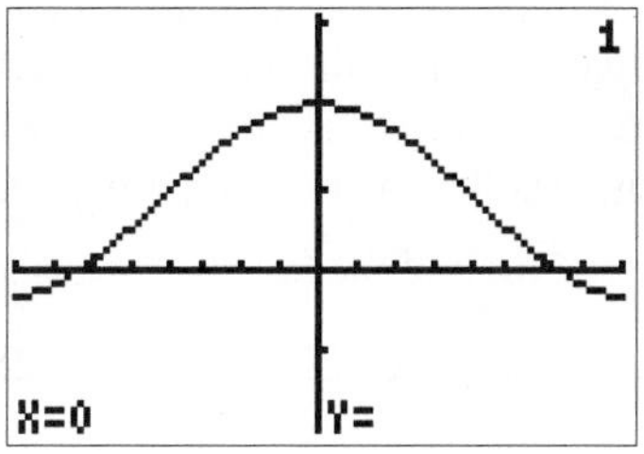

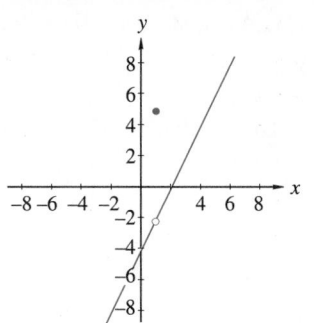

$[-4, 4, 0.5] \times [-1, 1.5, 1]$

FIGURE 22.18

The Graphical Approach to a Limit

The numerical approach is often a good way to approximate a limit. There are some times, however, when a graphical approach will be easier. We will use the graphical approach in the next two examples.

EXAMPLE 22.12

Use a graphical approach to find $\lim\limits_{x \to 1} h(x)$, where

$$h(x) = \begin{cases} 2x - 4 & \text{if } x \neq 1 \\ 5 & \text{if } x = 1 \end{cases}$$

Solution The graph of this function is shown in Figure 22.19. As you can see, as the values of x get close to 1 the values of $h(x)$ get very close to -2. In fact, if we select x close enough to 1, we can get $h(x)$ as close to -2 as we want. We decide, based on our observation, that $\lim\limits_{x \to 1} h(x) = -2$. Notice, however, that when $x = 1$, $h(x) = 5$. Thus, $\lim\limits_{x \to 1} h(x) \neq h(1)$.

FIGURE 22.19

≡ **Note** While Example 22.12 is somewhat artificial, you can see that the limit of a function at a given point is not always the same as the value of the function at that point.

The limit of a function at a certain point may not always exist. Until now, all of the functions we have examined have approached a certain number as the values of x got close to a particular value. The next example will show that this does not always happen.

EXAMPLE 22.13

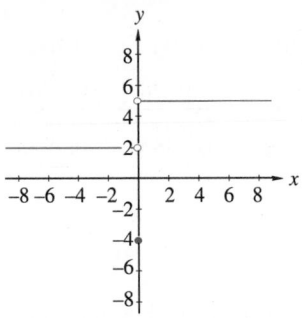

FIGURE 22.20

Use a graphical approach to find $\lim_{x \to 0} k(x)$, where

$$k(x) = \begin{cases} 2 & \text{if } x < 0 \\ -4 & \text{if } x = 0 \\ 5 & \text{if } x > 0 \end{cases}$$

Solution The graph of this function is shown in Figure 22.20. As you can see, if the values of x are close to 0 and negative, the value of $k(x)$ is 2. On the other hand, for values of x that are close to 0 and positive, the value of $k(x)$ is 5. There is no specific number that the values of $k(x)$ are near when x is close to 0. We must conclude that the $\lim_{x \to 0} k(x)$ does not exist. The fact that the function is defined when $x = 0$, $k(0) = -4$, has no effect on whether the limit of the function exists at that point.

We will use the graphical approach to show two basic rules for limits:

Rules for Limits

Rule 1: $\lim_{x \to c} A = A$, where A and c are real numbers

Rule 2: $\lim_{x \to c} x = c$, where c is a real number

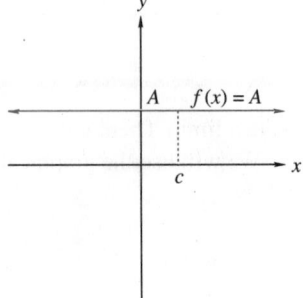

FIGURE 22.21

The graph for the first rule is shown in Figure 22.21. As you can see, the value of $f(x)$ is always A and so, as the values of x approach c, $f(x) = A$, thus we have $\lim_{x \to c} f(x) = A$.

The second rule is demonstrated in Figure 22.22. The graph of $f(x) = x$ is a straight line. For any value of c, as x gets close to c, the values of $f(x)$ are just as close to c, since $f(x) = x$. This shows that $\lim_{x \to c} f(x) = c$.

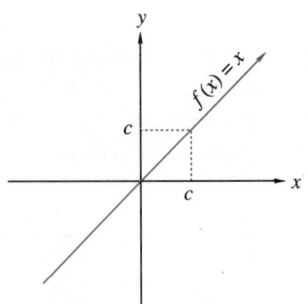

FIGURE 22.22

EXAMPLE 22.14

Evaluate each of the following limits: **(a)** $\lim_{x \to 3} 6$, **(b)** $\lim_{x \to -2} 7$, **(c)** $\lim_{x \to -5} x$, and **(d)** $\lim_{x \to 7} x$.

Solution The first two, (a) and (b), follow rule 1, $\lim_{x \to c} A = A$. In (a), $A = 6$ so, $\lim_{x \to 3} 6 = 6$ and in (b), $A = 7$ and $\lim_{x \to -2} 7 = 7$.

The last two parts, (c) and (d), use rule 2, $\lim_{x \to c} x = c$. In (c) we have $\lim_{x \to -5} x = -5$ and in (d) $\lim_{x \to 7} x = 7$.

In this section, we have tried to give you a foundation for the idea of a limit. The following exercises are designed to further develop this foundation. In Section 22.4 we will take a more detailed look at the limit concept.

Exercise Set 22.3

In Exercises 1–8, complete each table and use it to determine the indicated limit.

1.

x	0.9	0.99	0.999	0.9999	1.0001	1.001	1.01	1.1
$f(x) = 3x$								

$\lim_{x \to 1} 3x = ?$

2.

x	2.9	2.99	2.999	2.9999	3.0001	3.001	3.01	3.1
$g(x) = x - 4$								

$\lim_{x \to 3} (x - 4) = ?$

3.

x	−1.1	−1.01	−1.001	−1.0001	−0.9999	−0.999	−0.99	−0.9
$h(x) = x^2 + 2$								

$\lim_{x \to -1} (x^2 + 2) = ?$

4.

x	−2.1	−2.01	−2.001	−2.0001	−1.9999	−1.999	−1.99	−1.9
$k(x) = \dfrac{x^2 - 4}{x + 2}$								

$\lim_{x \to -2} \dfrac{x^2 - 4}{x + 2} = ?$

5.

x	−0.1	−0.01	−0.001	−0.0001	0.0001	0.001	0.01	0.1
$f(x) = \dfrac{\tan x}{x}$								

$\lim_{x \to 0} \dfrac{\tan x}{x} = ?$

6.

x	−0.1	−0.01	−0.001	−0.0001	0.0001	0.001	0.01	0.1
$g(x) = \dfrac{1 - \cos x}{x^2}$								

$\lim_{x \to 0} \dfrac{1 - \cos x}{x^2} = ?$

7.

x	0.9	0.99	0.999	0.9999	1.0001	1.001	1.01	1.1
$h(x) = \dfrac{x}{x-1}$								

$$\lim_{x \to 1} \frac{x}{x-1} = ?$$

8.

x	−0.1	−0.01	−0.001	−0.0001	0.0001	0.001	0.01	0.1
$k(x) = \dfrac{1}{x^2}$								

$$\lim_{x \to 0} \frac{1}{x^2} = ?$$

Solve Exercises 9 and 10.

9. For the function f in the given figure, find the following limits or state that the limit does not exist. Give a reason for your answers. **(a)** $\lim\limits_{x \to 0} f(x)$ **(b)** $\lim\limits_{x \to 2} f(x)$ **(c)** $\lim\limits_{x \to 3} f(x)$

10. For the function g in the given figure, find the following limits or state that the limit does not exist. Give reasons for your answers. **(a)** $\lim\limits_{x \to -2} g(x)$ **(b)** $\lim\limits_{x \to 0} g(x)$ **(c)** $\lim\limits_{x \to 3} g(x)$

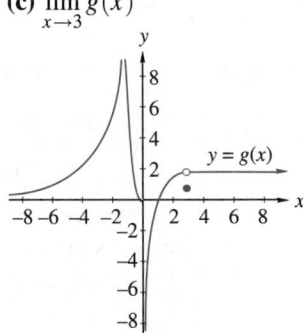

In Exercises 11–18, sketch the graphs of the functions and find out if $\lim\limits_{x \to c}$ exists at the given value of c or state that the limit does not exist.

11. $f(x) = \begin{cases} 3x - 2 & \text{if } x \neq 1 \\ 0 & \text{if } x = 1 \end{cases}$ $\quad c = 1$

12. $g(x) = \begin{cases} x + 1 & \text{if } x < 0 \\ x^2 + 1 & \text{if } x \geq 0 \end{cases}$ $\quad c = 0$

13. $h(x) = \begin{cases} 5 - 2x & \text{if } x < 0 \\ 0 & \text{if } x = 0 \\ 4x + 5 & \text{if } x > 0 \end{cases}$ $\quad c = 0$

14. $k(x) = \begin{cases} x + 1 & \text{if } x < 1 \\ \text{not defined} & \text{if } x = 1 \\ 2x & \text{if } x > 1 \end{cases}$ $\quad c = 1$

15. $f(x) = \begin{cases} 1 - 3x & \text{if } x < 0 \\ 3x + 1 & \text{if } x \geq 0 \end{cases}$ $\quad c = 0$

16. $g(x) = \begin{cases} 3x + 1 & \text{if } x \leq -1 \\ 1 - 3x & \text{if } x > -1 \end{cases}$ $\quad c = -1$

17. $h(x) = \begin{cases} 3x + 1 & \text{if } x < 0 \\ 4 & \text{if } x = 0 \\ 1 - 3x & \text{if } x > 0 \end{cases}$ $\quad c = 0$

18. $k(x) = \begin{cases} 3x + 1 & \text{if } x < 0 \\ 2 & \text{if } x = 0 \\ x + 1 & \text{if } x > 0 \end{cases}$ $\quad c = 0$

Evaluate the limits in Exercises 19–24.

19. $\lim\limits_{x\to 2}(-15)$

20. $\lim\limits_{x\to 0} x$

21. $\lim\limits_{x\to -4} x$

22. $\lim\limits_{x\to -7} 19$

23. $\lim\limits_{x\to 8} x$

24. $\lim\limits_{x\to -5}(-9)$

In Exercises 25–28, use a graphing calculator or graphing software to guess whether each of the following limits exist. If the limit does exist, estimate its value.

25. $\lim\limits_{x\to 0} \dfrac{2x^2}{x^2 + \sin x}$

26. $\lim\limits_{x\to 0} \dfrac{5x^2}{2x^2 + \sin^2 x}$

27. $\lim\limits_{x\to 0} \dfrac{2x^2}{x^2 + \cos x - 1}$

28. $\lim\limits_{x\to 0} \dfrac{3x^2}{x^2 + \tan x}$

 **In Your Words**

29. Explain the notation $\lim\limits_{x\to a} f(x) = L$.

30. Suppose that $y = g(x)$ is a function and $\lim\limits_{x\to 2} g(x) = 4$ and $g(2) = -3$.
(a) Sketch a possible graph for g.

(b) Use your graph to explain what is happening to g for values of x close to $x = 2$.

31. Suppose that the function $y = h(x)$ is not defined at $x = 4$. Is it possible for $\lim\limits_{x\to 4} h(x)$ to exist? Explain your answer.

☰ 22.4
ONE-SIDED LIMITS

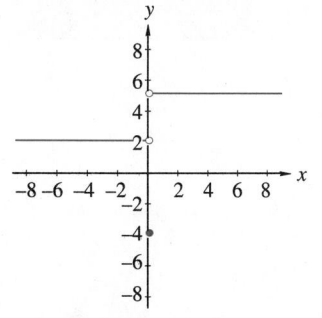

FIGURE 22.23

In using both the numerical and the graphical approaches to finding $\lim\limits_{x\to c} f(x)$, we studied the behavior of the function f on both sides of c. There are times when it is necessary to investigate the limit on just one side of c. When this is done, we are investigating one-sided limits.

There are two types of **one-sided limits**. One, the **left-hand limit**, is written as $\lim\limits_{x\to c^-} f(x)$, and the other, the **right-hand limit**, is written as $\lim\limits_{x\to c^+} f(x)$.

To get a better idea of one-sided limits, we will return to an example we worked earlier. In Example 22.13, we wanted to find $\lim\limits_{x\to 0} k(x)$, where

$$k(x) = \begin{cases} 2 & \text{if } x < 0 \\ -4 & \text{if } x = 0 \\ 5 & \text{if } x > 0 \end{cases}$$

The graph of the function is shown in Figure 22.23.

We know that $\lim\limits_{x\to 0} k(x)$ does not exist, because when x is positive, the value of $k(x)$ is 5, and when x is negative, the value of $k(x)$ is 2. What we have really said is that on the right-hand side of 0 the limit of $k(x)$ is 5 and on the left-hand side, the limit is 2. These are examples of one-sided limits.

In general, the **one-sided limits** of a function $f(x)$ at a point c are the **left-hand limit**, $\lim\limits_{x\to c^-} f(x)$, and the **right-hand limit**, $\lim\limits_{x\to c^+} f(x)$. In the function in Example 22.13, $\lim\limits_{x\to 0^-} k(x) = 2$ and $\lim\limits_{x\to 0^+} k(x) = 5$.

EXAMPLE 22.15

Determine the one-sided limits of $f(x)$ at $x = 3$, if

$$f(x) = \begin{cases} 2x+1 & \text{if } x < 1 \\ 4-x & \text{if } 1 \le x \le 3 \\ x & \text{if } x > 3 \end{cases}$$

Solution The way in which $f(x)$ behaves depends on whether $x < 3$ or $x > 3$, as shown in Figure 22.24. We have different definitions for $f(x)$ for each one-sided limit. When $x < 3$ (but close to 3), f is defined by $f(x) = 4 - x$, and so the left-hand limit is

$$\lim_{x \to 3^-} f(x) = \lim_{x \to 3^-} (4-x) = 1$$

But, when $x > 3$, f is defined by $f(x) = x$, and so the right-hand limit is

$$\lim_{x \to 3^+} f(x) = \lim_{x \to 3^+} x = 3$$

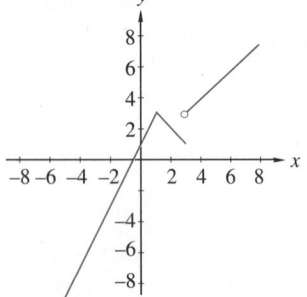

FIGURE 22.24

One-sided limits provide a way to determine whether or not the limit of a function exists at a point. The limit of a function $f(x)$ will exist at a point c, if and only if the right-hand limit of $f(x)$ at c and the left-hand limit of $f(x)$ at c both equal the same real number. Symbolically, this relationship is written as follows.

> **Relationship Between One-Sided and Two-Sided Limits**
>
> $\lim_{x \to c} f(x) = L$, if and only if $\lim_{x \to c^-} = L$ and $\lim_{x \to c^+} f(x) = L$, where L is a real number.

EXAMPLE 22.16

Determine the one-sided limits of $f(x)$ at $x = 1$, where

$$f(x) = \begin{cases} 2x+1 & \text{if } x < 1 \\ 4-x & \text{if } 1 \le x \le 3 \\ x & \text{if } x > 3 \end{cases}$$

Solution This is the same function we studied in Example 22.15. We will take the one-sided limits at $x = 1$. Here, when $x < 1$, f behaves as if $f(x) = 2x+1$, and we get

$$\lim_{x \to 1^-} f(x) = \lim_{x \to 1^-} (2x+1) = 3$$

When $x > 1$ (but close to 1), we use $f(x) = 4 - x$, with the result that

$$\lim_{x \to 1^+} f(x) = \lim_{x \to 1^+} (4-x) = 3$$

Since $\lim_{x \to 1^-} f(x) = \lim_{x \to 1^+} f(x) = 3$, we can say that $\lim_{x \to 1} f(x) = 3$.

We just specified that in order for the limit to exist, both of the one-sided limits must equal the same real number. The next example will show why we must make sure that it is a real number.

EXAMPLE 22.17

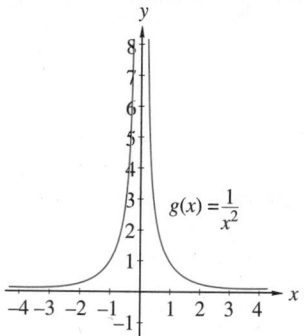

$g(x) = \dfrac{1}{x^2}$

FIGURE 22.25

Find the one-sided limits of $g(x) = \dfrac{1}{x^2}$, where $x = 0$.

Solution   We will use a table of values and a graph to give us an indication of these limits. The following table and the graph in Figure 22.25 seem to indicate that as $x \to 0$, from both the left and the right, $g(x)$ increases without bound or without limit.

x	± 1	± 0.5	± 0.1	± 0.01	± 0.001	± 0.0001
$g(x)$	1	4	100	10,000	1,000,000	100,000,000

So, there is no limit for $g(x)$ at $x = 0$. But, because both one-sided limits increase without bound, we say that $g(x)$ becomes positively infinite as x approaches 0. Symbolically, we write

$$\lim_{x \to 0} \frac{1}{x^2} = \infty$$

Caution

Be careful! The fact that we used an equal sign in this last limit does not mean that the limit exists. In fact, just the opposite is true. The infinity symbol ∞ here means that the limit does not exist and shows this because the one-sided limits increase without bound.

A somewhat different result is shown in the next example, where we look at $\dfrac{1}{x}$ as it gets close to 0.

EXAMPLE 22.18

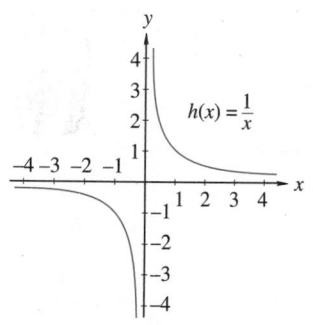

$h(x) = \dfrac{1}{x}$

FIGURE 22.26

Find the one-sided limits of $h(x) = \dfrac{1}{x}$ as $x \to 0$.

Solution   Again, we will use a table of values and a graph to demonstrate our answer. The table follows and the graph of $h(x)$ is in Figure 22.26.

x	-1	-0.1	-0.01	-0.001	0.001	0.01	0.1	1
$h(x)$	-1	-10	-100	$-1,000$	1,000	100	10	1

As we can see, $\lim\limits_{x \to 0^+} \dfrac{1}{x} = \infty$. But, what happens when x is negative? In this case, the values of $h(x)$ become negatively infinite, so

$$\lim_{x \to 0^-} \frac{1}{x} = -\infty$$

Thus we see two reasons that $\lim\limits_{x \to 0} \frac{1}{x}$ does not exist. One reason is that the left-hand and right-hand limits are not equal; the other is that infinity, either positive or negative, is not a real number.

Limits at Infinity

What happens to $h(x)$ as x gets increasingly larger? The following table gives some indication of what happens when the values of x get larger.

x	100	1,000	10,000	100,000	1,000,000
$h(x) = \dfrac{1}{x}$	0.01	0.001	0.0001	0.00001	0.000001

This table and the graph in Figure 22.26 indicate that

$$\lim_{x \to \infty} \frac{1}{x} = 0$$

That is, as x gets larger without bound, the value of $\dfrac{1}{x}$ approaches 0. Similarly, as x gets smaller without bound, the value of $\dfrac{1}{x}$ approaches 0. This is written as

$$\lim_{x \to -\infty} \frac{1}{x} = 0$$

These two results are summarized in the following box.

Limits at Infinity

$$\lim_{x \to \infty} \frac{1}{x} = 0$$

and

$$\lim_{x \to -\infty} \frac{1}{x} = 0$$

Exercise Set 22.4

Solve Exercises 1 and 2.

1. For the function f given in the figure below, find the following limits or state that the limit does not exist.

   (a) $\lim_{x \to 2^-} f(x)$ (c) $\lim_{x \to 3^-} f(x)$

   (b) $\lim_{x \to 2^+} f(x)$ (d) $\lim_{x \to 3^+} f(x)$

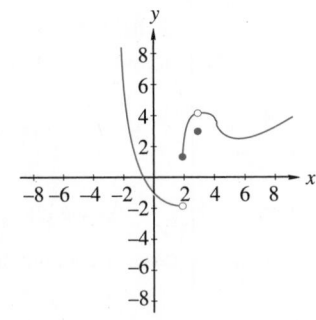

2. For the function g given in the figure below, find the following limits or state that the limit does not exist.

(a) $\lim\limits_{x \to -1^-} g(x)$ **(c)** $\lim\limits_{x \to 0^-} g(x)$ **(e)** $\lim\limits_{x \to \infty} g(x)$

(b) $\lim\limits_{x \to -1^+} g(x)$ **(d)** $\lim\limits_{x \to 0^+} g(x)$ **(f)** $\lim\limits_{x \to -\infty} g(x)$

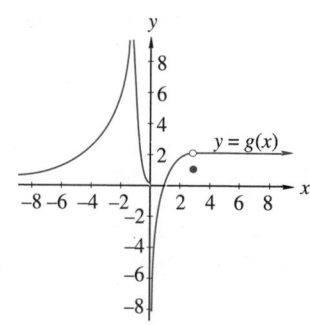

In Exercises 3–18, use a table or a graph to determine the one-sided limits.

3. $\lim\limits_{x \to 5^+} (x - 2)$

4. $\lim\limits_{x \to 2^-} (x^2 - 1)$

5. $\lim\limits_{x \to 4^+} \dfrac{x}{x - 4}$

6. $\lim\limits_{x \to 3^-} \dfrac{2x}{x - 3}$

7. $\lim\limits_{x \to -2^+} \dfrac{x^2 - 4}{x + 2}$

8. $\lim\limits_{x \to -5^-} \dfrac{x^2 + 25}{x + 5}$

9. $\lim\limits_{x \to 1^-} \dfrac{|x - 1|}{x - 1}$

10. $\lim\limits_{x \to 1^+} \dfrac{|x - 1|}{x - 1}$

11. $\lim\limits_{x \to 2^-} \dfrac{|x - 2|}{x - 2}$

12. $\lim\limits_{x \to 2^+} \dfrac{|x - 2|}{x - 2}$

13. $\lim\limits_{x \to 2^+} \sqrt{x - 2}$

14. $\lim\limits_{x \to -4^+} \sqrt{x + 4}$

15. $\lim\limits_{x \to \infty} \dfrac{9}{3x}$

16. $\lim\limits_{x \to -\infty} \dfrac{-8}{4x}$

17. $\lim\limits_{x \to -\infty} \dfrac{1}{2x + 1}$

18. $\lim\limits_{x \to \infty} \dfrac{-5}{9x - 4}$

In Exercises 19–24, use a graphing calculator or graphing software to guess whether each of the following limits exists. If the limit does exist, estimate its value.

19. $\lim\limits_{x \to \infty} \dfrac{2x^2}{x^2 + \sin x}$

20. $\lim\limits_{x \to \infty} \dfrac{5x^2}{2x^2 + \sin^2 x}$

21. $\lim\limits_{x \to 0^-} \dfrac{2x^2}{x^2 - \cos x}$

22. $\lim\limits_{x \to 0^+} \dfrac{2x^2}{x^2 - \tan x}$

23. $\lim\limits_{x \to \infty} \dfrac{5^x - 5}{2^x - 2}$

24. $\lim\limits_{x \to -\infty} \dfrac{5^x - 5}{2^x - 2}$

Solve Exercises 25–28.

25. *Medical technology*   The concentration of a drug in a patient's bloodstream t hours after it was injected is given by

$$C(t) = \frac{0.25t}{t^2 + 5}$$

Find each of the following. **(a)** $C(0.5)$, **(b)** $C(2)$, **(c)** $\lim\limits_{t \to \infty} C(t)$

26. *Business*   It has been determined that after t weeks of training, a certain student in a word processing class can keyboard at the rate of $W(t) = 65 + \dfrac{70t^2}{t^2 + 15}$ words per minute.   What are the most words per minute that we can expect this student to keyboard, that is, what is $\lim\limits_{t \to \infty} \left(65 + \dfrac{70t^2}{t^2 + 15}\right)$?

27. *Business*   The cost, in dollars, of an overseas telephone call is given by the function

$$C(t) = \begin{cases} 4.75 & \text{if } 0 < t \le 3 \\ 0.65t + 2.80 & \text{if } t > 3 \end{cases}$$

where t is the length of the call in minutes. Determine each of the following:
(a) $\lim\limits_{t \to 3^-} C(t)$
(b) $\lim\limits_{x \to 3^+} C(t)$
(c) $\lim\limits_{x \to 3} C(t)$

28. *Finance*   A certain state income tax schedule can be given by the function

$$T(x) = \begin{cases} 0.04x & \text{if } 0 < x \le 12{,}500 \\ 0.03x + 125 & \text{if } 12{,}500 < x \le 35{,}000 \\ 0.02x + 465 & \text{if } x > 35{,}000 \end{cases}$$

where x is the taxable income in dollars, $0 < x$, and $T(x)$ is in dollars. Determine each of the following:

(a) $\lim\limits_{x \to 12{,}500^-} T(x)$

(b) $\lim\limits_{x \to 12{,}500^+} T(x)$

(c) $\lim\limits_{x \to 12{,}500} T(x)$

(d) $\lim\limits_{x \to 35{,}000^-} T(x)$

(e) $\lim\limits_{x \to 35{,}000^+} T(x)$

(f) $\lim\limits_{x \to 35{,}000} T(x)$

In Your Words

29. (a) Describe the differences among $\lim\limits_{x \to a^-} f(x)$, $\lim\limits_{x \to a^+} f(x)$, and $\lim\limits_{x \to a} f(x)$.

(b) Are all three of the limits in (a) ever the same? If so, what does that mean?

(c) Can two of the limits in (a) ever be the same (and the third one different)? If so, what does that mean?

(d) Can all three of the limits in (a) be different? If so, what does that mean?

30. Explain the difference between $\lim\limits_{x \to c} f(x) = \infty$ and $\lim\limits_{x \to \infty} f(x) = c$.

≡ 22.5
ALGEBRAIC TECHNIQUES FOR FINDING LIMITS

Until now, we have found limits by using graphs or using tables of values. In many cases it would be a lot easier to use our knowledge of algebra to determine a limit. The following rules of limits will help us find a limit with much less difficulty.

In Section 22.3, we had the following two rules for limits:

Rules 1 and 2 for Limits

Rule 1: $\lim\limits_{x \to c} A = A$, where A and c are real numbers

Rule 2: $\lim\limits_{x \to c} x = c$, where c is a real number

For all of the following rules, we will assume that f and g are two functions and that $\lim\limits_{x \to c} f(x) = L$ and $\lim\limits_{x \to c} g(x) = M$, where L and M are both real numbers.

Rule 3: Limit of a Sum or Difference $\lim\limits_{x \to c}[f(x) \pm g(x)] = \lim\limits_{x \to c} f(x) \pm \lim\limits_{x \to c} g(x)$

$$= L \pm M$$

Thus, in order to find the limit of the sum (or difference) of two functions, you can find the sum (or difference) of their limits.

EXAMPLE 22.19

Evaluate $\lim\limits_{x \to 9}(x+6)$.

Solution

$$\lim_{x \to 9}(x+6) = \lim_{x \to 9} x + \lim_{x \to 9} 6 \qquad \text{(rule 3)}$$
$$= 9+6 \qquad \text{(rules 1 and 2)}$$
$$= 15$$

EXAMPLE 22.20

Evaluate $\lim\limits_{x \to 2}(3-x)$.

Solution

$$\lim_{x \to 2}(3-x) = \lim_{x \to 2} 3 - \lim_{x \to 2} x \qquad \text{(rule 3)}$$
$$= 3-2 \qquad \text{(rules 1 and 2)}$$
$$= 1$$

Rule 4: Limit of a Product $\qquad \lim\limits_{x \to c}[f(x)g(x)] = \left[\lim\limits_{x \to c} f(x)\right]\left[\lim\limits_{x \to c} g(x)\right]$
$$= LM$$

This rule states that the limit of the product of two functions is the product of their limits.

EXAMPLE 22.21

Evaluate $\lim\limits_{x \to -3} 7x$.

Solution

$$\lim_{x \to -3} 7x = \left(\lim_{x \to -3} 7\right)\left(\lim_{x \to -3} x\right) \qquad \text{(rule 4)}$$
$$= (7)(-3) \qquad \text{(rules 1 and 2)}$$
$$= -21$$

EXAMPLE 22.22

Evaluate $\lim\limits_{x \to 4} 6x^2$.

Solution

$$\lim_{x \to 4} 6x^2 = \left(\lim_{x \to 4} 6x\right)\left(\lim_{x \to 4} x\right) \qquad \text{(rule 4)}$$

EXAMPLE 22.22 (Cont.)

$$= 6\left(\lim_{x \to 4} x\right)\left(\lim_{x \to 4} x\right) \qquad \text{(rules 1 and 4)}$$

$$= 6 \cdot 4 \cdot 4 \qquad \text{(rule 2)}$$

$$= 96$$

Rule 5: Limit of a Quotient

$$\lim_{x \to c} \frac{f(x)}{g(x)} = \frac{\lim\limits_{x \to c} f(x)}{\lim\limits_{x \to c} g(x)} = \frac{L}{M}, \quad \text{if } M \neq 0$$

If the limit of the denominator is not 0, then the limit of the quotient of two functions is the quotient of their limits.

EXAMPLE 22.23

Evaluate $\lim\limits_{x \to 5} \dfrac{2x+4}{x-1}$.

Solution

$$\lim_{x \to 5} \frac{2x+4}{x-1} = \frac{\lim\limits_{x \to 5}(2x+4)}{\lim\limits_{x \to 5}(x-1)} \qquad \text{(rule 5)}$$

$$= \frac{14}{4} = \frac{7}{2}$$

Rule 6: Limit of $[f(x)]^n$ or $\sqrt[n]{f(x)}$.

If n is a positive integer,

$$\lim_{x \to c} [f(x)]^n = \left[\lim_{x \to c} f(x)\right]^n = L^n$$

and $\qquad \lim\limits_{x \to c} \sqrt[n]{f(x)} = \sqrt[n]{\lim\limits_{x \to c} f(x)} = \sqrt[n]{L}$

EXAMPLE 22.24

Evaluate $\lim\limits_{x \to -2}(2x-1)^3$.

Solution

$$\lim_{x \to -2}(2x-1)^3 = \left[\lim_{x \to -2}(2x-1)\right]^3 \qquad \text{(rule 6)}$$

$$= (-5)^3 = -125$$

EXAMPLE 22.25

Evaluate $\lim\limits_{x \to 3} \sqrt[6]{(3x-1)^2}$.

Solution

$$
\begin{aligned}
\lim_{x \to 3} \sqrt[6]{(3x-1)^2} &= \sqrt[6]{\lim_{x \to 3}(3x-1)^2} \qquad &\text{(rule 6)}\\
&= \sqrt[6]{[\lim_{x \to 3}(3x-1)]^2} \qquad &\text{(rule 6)}\\
&= \sqrt[6]{(8)^2} = \sqrt[6]{64} = 2
\end{aligned}
$$

EXAMPLE 22.26

Evaluate $\lim\limits_{x \to 2} \dfrac{x^2 - 5x + 6}{x^2 - 4}$.

Solution Checking the denominator, we see that $\lim\limits_{x \to 2}(x^2 - 4) = 0$. This means that we cannot use rule 5. But, this does not mean the limit does not exist. If we factor both the numerator and denominator, we see that

$$
\frac{x^2 - 5x + 6}{x^2 - 4} = \frac{(x-2)(x-3)}{(x-2)(x+2)}
$$

We are interested in the limits as x approaches 2 and not when x has the value of 2. Thus, the factor $x - 2$ is not 0 for these values and we can cancel this common factor. We then get

$$
\begin{aligned}
\lim_{x \to 2} \frac{x^2 - 5x + 6}{x^2 - 4} &= \lim_{x \to 2} \frac{(x-2)(x-3)}{(x-2)(x+2)}\\
&= \lim_{x \to 2} \frac{x-3}{x+2}\\
&= \frac{2-3}{2+2} = \frac{-1}{4}
\end{aligned}
$$

Caution

In Example 22.26, substituting 2 for x in the expression $\dfrac{x^2 - 5x + 6}{x^2 - 4}$ resulted in $\frac{0}{0}$. Whenever substitution results in $\frac{0}{0}$, we must do more work to determine whether a limit exists.

Until now, we have applied the six rules for limits only for the limit of a function as x approaches a specific point. As the next examples show, we often have to be more careful when we find limits at infinity.

EXAMPLE 22.27

Evaluate $\lim\limits_{x\to\infty} \dfrac{1}{x^n}$, $n > 0$.

Solution

$$\lim_{x\to\infty} \frac{1}{x^n} = \left(\lim_{x\to\infty} \frac{1}{x}\right)^n = 0^n = 0$$

Notice from this example that as $x \to \infty$, $\dfrac{1}{x^2}$, $\dfrac{1}{x^3}$, $\dfrac{1}{x^4}$, etc., all have limits of 0. Using this with rule 3, we see that

$$\lim_{x\to\infty} \frac{c}{x^n} = \left(\lim_{x\to\infty} c\right)\left(\lim_{x\to\infty} \frac{1}{x^n}\right) = c \cdot 0 = 0$$

Thus, we have the following.

Limits at Infinity

If c is a constant, then

$$\lim_{x\to\infty} \frac{c}{x^n} = 0$$

We will now show some techniques for finding limits at infinity.

EXAMPLE 22.28

Evaluate $\lim\limits_{x\to\infty} \dfrac{4x^3 + 2x^2 + 5}{5x^3 + x}$.

Solution We will first divide both the numerator and denominator by the largest power of x in the denominator, in this case x^3. This will make each term into a constant or a term with a variable in the denominator and allow us to use the properties for limits at infinity.

$$\lim_{x\to\infty} \frac{4x^3 + 2x^2 + 5}{5x^3 + x} = \lim_{x\to\infty} \frac{\dfrac{4x^3 + 2x^2 + 5}{x^3}}{\dfrac{5x^3 + x}{x^3}}$$

$$= \lim_{x\to\infty} \frac{\dfrac{4x^3}{x^3} + \dfrac{2x^2}{x^3} + \dfrac{5}{x^3}}{\dfrac{5x^3}{x^3} + \dfrac{x}{x^3}}$$

$$= \lim_{c\to\infty} \frac{4 + \dfrac{2}{x} + \dfrac{5}{x^3}}{5 + \dfrac{1}{x^2}}$$

EXAMPLE 22.28 (Cont.)

Now, according to the properties for limits at infinity, all the terms with an x in the denominator have a limit of 0. So, we now have

$$\lim_{x\to\infty} \frac{4 + \dfrac{2}{x} + \dfrac{5}{x^3}}{5 + \dfrac{1}{x^2}} = \frac{4+0+0}{5+0} = \frac{4}{5}$$

 Hint

When you want to find a limit at infinity, divide both the numerator and denominator by the largest power of the variable in the denominator.

EXAMPLE 22.29

Evaluate $\displaystyle\lim_{x\to\infty} \frac{7x^5 - 2x^2 + 1}{4x^6 + 5x^3 + x - 5}$.

Solution As in Example 22.28, we will divide both the numerator and denominator by the largest power of x in the denominator, in this case x^6.

$$\lim_{x\to\infty} \frac{7x^5 - 2x^2 + 1}{4x^6 + 5x^3 + x - 5} = \lim_{x\to\infty} \frac{\dfrac{7x^5 - 2x^2 + 1}{x^6}}{\dfrac{4x^6 + 5x^3 + x - 5}{x^6}}$$

$$= \lim_{x\to\infty} \frac{\dfrac{7x^5}{x^6} - \dfrac{2x^2}{x^6} + \dfrac{1}{x^6}}{\dfrac{4x^6}{x^6} + \dfrac{5x^3}{x^6} + \dfrac{x}{x^6} - \dfrac{5}{x^6}}$$

$$= \lim_{x\to\infty} \frac{\dfrac{7}{x} - \dfrac{2}{x^4} + \dfrac{1}{x^6}}{4 + \dfrac{5}{x^3} + \dfrac{1}{x^5} - \dfrac{5}{x^6}}$$

$$= \frac{0 - 0 + 0}{4 + 0 + 0 - 0} = \frac{0}{4} = 0$$

Thus, $\displaystyle\lim_{x\to\infty} \frac{7x^5 - 2x^2 + 1}{4x^6 + 5x^3 + x - 5} = 0$.

Exercise Set 22.5

In Exercises 1–34, evaluate the given limit using algebraic techniques.

1. $\lim_{x \to 5} 8$

2. $\lim_{x \to -4} x$

3. $\lim_{x \to 8} (x - 3)$

4. $\lim_{x \to -7} (x + 4)$

5. $\lim_{x \to -6} (x + 7)$

6. $\lim_{x \to 6} (9 - x)$

7. $\lim_{x \to 2} 3x$

8. $\lim_{t \to -3} 5t$

9. $\lim_{x \to 6} \frac{2}{3}x + 5$

10. $\lim_{x \to 4} \left(\frac{3}{2}x - 2\right)$

11. $\lim_{p \to 3} \frac{p^2 + 6}{p}$

12. $\lim_{x \to -2} \frac{x^2 - 6}{x}$

13. $\lim_{s \to 5} \frac{3s^2 + 5}{2s}$

14. $\lim_{x \to 6} \frac{2x^2 - 3}{3 - x}$

15. $\lim_{x \to 2} (x + 1)^3$

16. $\lim_{y \to 5} (y - 1)^3$

17. $\lim_{x \to -1} (2x^2 - 1)^2$

18. $\lim_{t \to 4} (3t^2 + 2)^2$

19. $\lim_{x \to 3} \sqrt{x + 3}$

20. $\lim_{x \to -6} \sqrt{6 - 7x}$

21. $\lim_{x \to 2} \frac{x^2 - 2x}{x - 2}$

22. $\lim_{w \to 3} \frac{w^2 - 1}{w + 1}$

23. $\lim_{x \to -1} \frac{x^2 - 1}{x + 1}$

24. $\lim_{y \to 3} \frac{y^2 - 7y + 12}{y - 3}$

25. $\lim_{x \to -4} \frac{x^2 + 2x - 8}{x^2 + 5x + 4}$

26. $\lim_{x \to \frac{1}{2}} \frac{2x^2 + 5x - 3}{4x^2 - 2x}$

27. $\lim_{t \to -7} \frac{2t^2 + 15t + 7}{t^2 + 5t - 14}$

28. $\lim_{x \to 0} \frac{(x + 3)^2 - 9}{x}$

29. $\lim_{h \to 0} \frac{(4 + h)^2 - 4^2}{h}$

30. $\lim_{x \to c} \frac{x^3 - c^3}{x - c}$

31. $\lim_{x \to \infty} \frac{2x^2 + 4}{5x^2 + 3x + 2}$

32. $\lim_{x \to \infty} \frac{6x^3 + 9x + 1}{2x^3 + 4x + 8}$

33. $\lim_{x \to \infty} \frac{7x^4 + 5x^2}{x^5 + 2}$

34. $\lim_{x \to \infty} \frac{4x^5 + 3x}{2x^4 + 1}$

In Your Words

35. Suppose that $\lim_{x \to 1} f(x) = 0$ and $\lim_{x \to 1} g(x) = \infty$.
   (a) Give an example of two functions f and g so that $\lim_{x \to 1} [f(x) + g(x)] = \infty$.
   (b) Give an example of two functions f and g so that $\lim_{x \to 1} [f(x) \cdot g(x)] = 0$.
   (c) Give an example of two functions f and g so that $\lim_{x \to 1} [f(x) \cdot g(x)] = 5$.

36. Suppose that neither $\lim_{x \to 2} f(x)$ and $\lim_{x \to 2} g(x)$ exists. Give an example of two functions f and g so that $\lim_{x \to 2} [f(x) + g(x)]$ does exist or explain why it is not possible.

≡ 22.6
CONTINUITY

In Sections 22.4 and 22.5, we have talked about two of the important ideas leading up to calculus. In this section, we will look at the last new idea we need before we begin calculus—**continuity at a point**.

Let's consider the two functions

$$f(x) = x$$

$$\text{and} \quad g(x) = \begin{cases} x & \text{if } x \neq 2 \\ 3 & \text{if } x = 2 \end{cases}$$

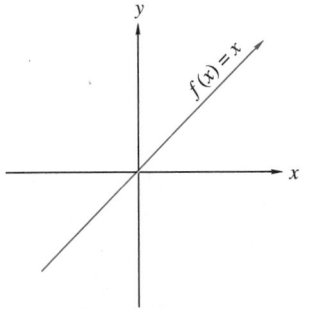

FIGURE 22.27

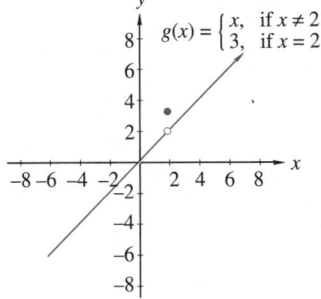

$$g(x) = \begin{cases} x, & \text{if } x \neq 2 \\ 3, & \text{if } x = 2 \end{cases}$$

FIGURE 22.28

The graphs of these functions appear in Figures 22.27 and 22.28, respectively. Looking at either the graphs or the algebraic definitions, you can see that the two functions are exactly the same, except at one point. At that point, when $x = 2$, there is a "break" in the graph of g, but no "break" in the graph of f. The function g is an example of a function that is not continuous, while f demonstrates a continuous function.

The break in g is used to determine that the function is not continuous. One way of thinking of a continuous function is that it is a function you can draw without lifting your pencil from the paper. Look again at the graphs of f and g. You could draw f without lifting your pencil. You could not draw g without lifting your pencil.

Continuity at a Point

As we saw above, we can describe a function as either continuous or noncontinuous by looking at how it is drawn. We can use the idea of limits to provide an algebraic definition of continuity. Look again at the definitions of f and g. What is $\lim_{x \to 2} f(x)$? What are $f(2)$ and $g(2)$?

As you can see, $\lim_{x \to 2} f(x) = f(2) = 2$. We know that f is a continuous function at $x = 2$. But $\lim_{x \to 2} g(x) = 2 \neq g(2) = 3$, thus g is not a continuous function, or is **discontinuous** at $x = 2$. This provides us with a more formal definition:

Continuity at a Point

A function f is **continuous** at $x = c$, if and only if the following three conditions are satisfied:

(a) $f(x)$ is defined at $x = c$,

(b) $\lim_{x \to c} f(x)$ exists, and

(c) $\lim_{x \to c} f(x) = f(c)$.

A function that is not continuous at $x = c$ is **discontinuous** at that point.

≡ **Note** A function is discontinuous at $x = c$ if any one of these three conditions are not satisfied.

EXAMPLE 22.30

Discuss the continuity of the function f at $x = 4$, where

$$f(x) = \begin{cases} \dfrac{x^2 - 16}{x - 4} & \text{if } x \neq 4 \\ 8 & \text{if } x = 4 \end{cases}$$

Solution   In order to determine if f is continuous, we need to check the three conditions in the definition: **(a)** The function is defined at $x = 4$, since $f(4) = 8$,
(b) $\displaystyle\lim_{x \to 4} \frac{x^2 - 16}{x - 4} = \lim_{x \to 4} \frac{(x + 4)(x - 4)}{x - 4} = 8$, and **(c)** $\displaystyle\lim_{x \to 4} \frac{x^2 - 16}{x - 4} = 8 = f(4)$.
The three conditions are satisfied, so this function is continuous at $x = 4$.

EXAMPLE 22.31

Discuss the continuity of each of the following functions at the indicated points:

(a) $\quad g(x) = \dfrac{1}{x}$ $\qquad\qquad\qquad$ at $x = 0$

(b) $\quad h(x) = \begin{cases} x - 1 & \text{if } x < 2 \\ x + 1 & \text{if } x \geq 2 \end{cases}$ $\qquad$ at $x = 2$

(c) $\quad j(x) = \begin{cases} x^2 + 2 & \text{if } x \neq 0 \\ 0 & \text{if } x = 0 \end{cases}$ $\qquad$ at $x = 0$

Solutions   None of these are continuous at the designated points.
(a) $g(x)$ is not defined when $x = 0$, so it does not satisfy the first condition.
(b) $h(x)$ is defined at $x = 2$, but we can see that $\displaystyle\lim_{x \to 2^-} h(x) = 1$ and $\displaystyle\lim_{x \to 2^+} h(x) = 3$,
so the $\displaystyle\lim_{x \to 2} h(x)$ does not exist and the second condition is not satisfied.
(c) $j(x)$ is defined at 0, $j(0) = 0$. $\displaystyle\lim_{x \to 0} j(x) = 2$. The third condition is not satisfied,
since $\displaystyle\lim_{x \to 0} j(x) = 2 \neq 0 = j(0)$.

The graphs in these three functions are shown in Figures 22.29a, 22.29b, and 22.29c, respectively.

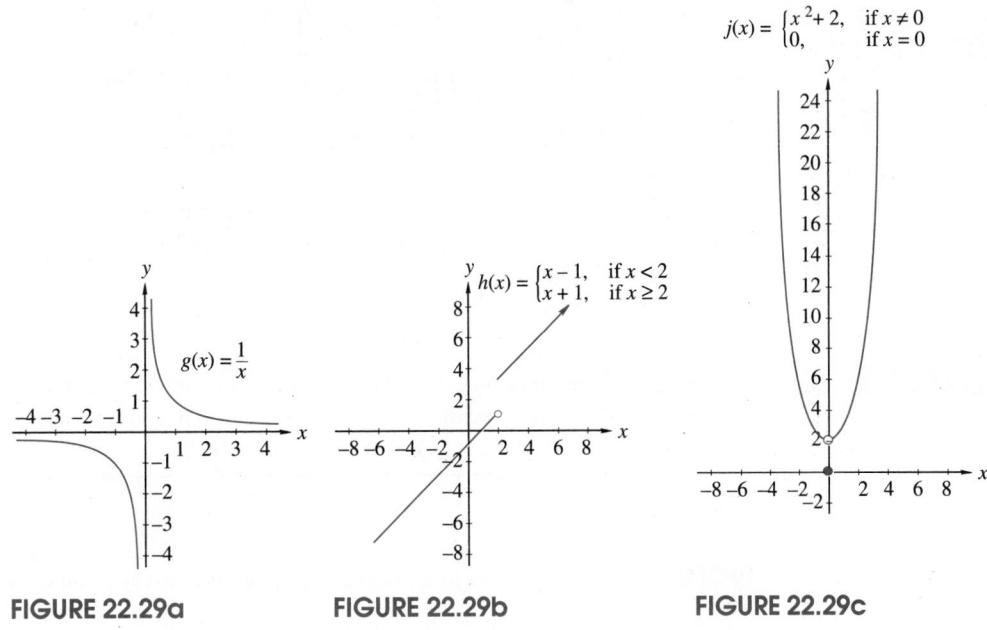

FIGURE 22.29a   FIGURE 22.29b   FIGURE 22.29c

Continuity on Intervals

When we talk about functions, we usually consider them over an interval. An **interval** is a set of all points on the number line between a and b. Because the endpoints are not always included in the intervals, notation has been developed to make these distinctions. We will use the following notations and terms:

Interval Notation	
	$(a,b) : a < x < b$
	$[a,b) : a \leq x < b$
	$(a,b] : a < x \leq b$
	$[a,b] : a \leq x \leq b$

The interval (a,b) is called an **open interval** and does not include either endpoint. The interval $[a,b]$ is a **closed interval** and includes both endpoints. The other two intervals, $[a,b)$ and $(a,b]$, are **half-open intervals**. Because $+\infty$ and $-\infty$ are not numbers, any interval that includes either of these is open on that end. For example $a < x < \infty$, or simply $a < x$, is the interval, (a,∞) and $-\infty < x \leq b$, or simply $x \leq b$, is the interval $(-\infty, b]$.

The definition of continuity we just gave is for a function continuous at a point. We are usually more interested in the continuity of a function over an interval.

Continuity Over an Interval

We say that a function is **continuous on an interval** if it is continuous at each point in the interval.

If you are concerned with the continuity of a function on an interval that includes an endpoint of the interval, then you need to examine the one-sided limits at that endpoint.

EXAMPLE 22.32

The function $k(x) = \sqrt{x-2}$ is continuous on the interval $[2, \infty)$.
The function $m(x) = \sqrt{3-x}$ is continuous on the interval $(-\infty, 3]$.
The function $f(x) = \sqrt[3]{x}$ is continuous on $(-\infty, \infty)$.

≡ Note All polynomial functions are continuous at every point or on the interval $(-\infty, \infty)$.

Functions that are continuous at all points, such as a polynomial, are said to be **continuous everywhere**, or continuous, and are referred to as **continuous functions**. Rational functions are continuous over their domains.

EXAMPLE 22.33

Find any points of discontinuity for each of the following functions:

(a) $f(x) = \dfrac{x^3 - 3x^2 + 17x - 5}{x^2 - 1}$

(b) $g(x) = \begin{cases} x^2 & \text{if } x < 2 \\ x^2 + 5 & \text{if } x \geq 2 \end{cases}$

(c) $h(x) = \begin{cases} 6 & \text{if } x < 3 \\ x + 3 & \text{if } x > 3 \end{cases}$

Solutions

(a) Possible points of discontinuity occur when the denominator is 0. The denominator is 0 when x is ± 1, so f is not defined at these two points. By condition (a), if the function is not defined at ± 1, it is not continuous at those two points. These are the only points of discontinuity for f.

(b) Possible points of discontinuity occur when $x = 2$, since this is the only place where a break may occur. When $x < 2$, $\lim\limits_{x \to 2^-} g(x) = \lim\limits_{x \to 2^-} x^2 = 4$ and when $x > 2$, $\lim\limits_{x \to 2^+} g(x) = \lim\limits_{x \to 2^+} (x^2 + 5) = 9$. The limit does not exist at $x = 2$, so this is a point of discontinuity.

(c) The function h is not defined at $x = 3$, so the function is not continuous at this point. It is continuous for all other values of x.

Properties of Continuous Functions

The formal definition that we gave for a continuous function was based on a limit. The rules that apply to limits help us develop three properties that apply to continuous functions.

Properties of Continuous Functions

Property 1: If f and g are both continuous at a point c, then so are $f + g$, $f - g$, and $f \cdot g$.

Property 2: If f and g are both continuous at c and $g(c) \neq 0$, then $f/g = \dfrac{f}{g}$ is also continuous at c.

Property 3: If f is continuous at $g(c)$ and g is continuous at c, then $f \circ g = f(g(x))$ is continuous at c.

We used property 2 in the solution of Example 22.33a. There, we let f be the polynomial $x^3 - 3x^2 + 17x - 5$ and g the polynomial $x^2 - 1$. Since f and g are both polynomials they are continuous everywhere, so f/g is continuous where $g \neq 0$, namely when $x \neq \pm 1$.

This concludes this section on continuity and completes your foundation for calculus. Limits and continuity will both provide necessary tools as we continue to develop our knowledge of calculus.

Exercise Set 22.6

In Exercises 1–10, determine where the function is continuous at c. If the function is not continuous, give a reason.

1. $f(x) = \begin{cases} 3x + 1 & \text{if } x \leq 4 \\ x^2 - 3 & \text{if } x > 4 \end{cases} \quad c = 4$

2. $g(x) = \begin{cases} 2x - 1 & \text{if } x \leq 1 \\ 2x & \text{if } x > 1 \end{cases} \quad c = 1$

3. $h(x) = \begin{cases} 3x + 1 & \text{if } x < 1 \\ 3 & \text{if } x = 1 \\ 4x & \text{if } x > 1 \end{cases} \quad c = 1$

4. $j(x) = \begin{cases} 5x - 1 & \text{if } x < 1 \\ 4 & \text{if } x = 1 \\ 4x & \text{if } x > 1 \end{cases} \quad c = 1$

5. $k(x) = \begin{cases} x^2 + 2 & \text{if } x > 2 \\ 3x & \text{if } x < 2 \end{cases} \quad c = 2$

6. $f(x) = \begin{cases} x^3 + 1 & \text{if } x > 2 \\ 0 & \text{if } x = 2 \\ x^2 + 5 & \text{if } x < 2 \end{cases} \quad c = 2$

7. $g(x) = \begin{cases} x^2 & \text{if } x < -2 \\ 3 & \text{if } x = -2 \\ -4x + 2 & \text{if } x > -2 \end{cases} \quad c = -2$

8. $h(x) = \begin{cases} x^2 & \text{if } x \leq -2 \\ 2x & \text{if } x > -2 \end{cases} \quad c = -2$

9. $j(x) = \begin{cases} x^3 - 4 & \text{if } x < 2 \\ 5 & \text{if } x = 2 \\ x^2 & \text{if } x > 2 \end{cases} \quad c = 2$

10. $k(x) = \dfrac{x^3 - 1}{x - 1}$ $c = 1$

In Exercises 11–20, find all points of discontinuity.

11. $f(x) = 2x^2 - 5$

12. $g(x) = \dfrac{4}{x - 5}$

13. $h(x) = \dfrac{x^2 + 3x + 2}{x + 1}$

14. $j(x) = 7t - 4$

15. $k(x) = \dfrac{x + 1}{x^2 + x - 6}$

16. $f(x) = \dfrac{x^2 + x - 6}{x^2 + 5x + 6}$

17. $g(x) = \dfrac{x}{x^2 + 1}$

18. $h(x) = \dfrac{x}{x}$

19. $j(x) = \begin{cases} x^2 + 2 & \text{if } x < 2 \\ \sqrt{x - 2} & \text{if } x \ge 2 \end{cases}$

20. $k(x) = \begin{cases} \dfrac{1}{x} & \text{if } x \ne 5 \\ 3 & \text{if } x = 5 \end{cases}$

In Exercises 21–30, determine the intervals over which the given function is continuous.

21. $f(x) = \dfrac{x}{x + 3}$

22. $g(x) = \dfrac{x}{x^2 - 1}$

23. $h(x) = \dfrac{x + 2}{x^2 - 4}$

24. $j(x) = \dfrac{x^2 + 3x - 4}{x + 4}$

25. $k(x) = \sqrt{x + 3}$

26. $f(x) = \sqrt{5 - x}$

27. $g(x) = \sqrt{x^2 - 9}$

28. $h(x) = \sqrt{x^2 + 4x + 4}$

29. $j(x) = \dfrac{1}{x}$

30. $k(x) = \dfrac{1}{\sqrt{x}}$

Solve Exercises 31–34.

31. *Physics*   The acceleration due to gravity, g, varies with the height above the surface of the Earth. The acceleration varies in a different way below the Earth's surface. It has been determined that, as a function of r, the distance from the center of the Earth, g is described by

$$g(r) = \begin{cases} \dfrac{GMr}{R^3} & \text{for } r < R \\ \dfrac{GM}{r^2} & \text{for } r \ge R \end{cases}$$

where $R \approx 6.371 \times 10^3$ km is the radius of the Earth, $M \approx 5.975 \times 10^{24}$] kg is the mass of the Earth, and $G \approx 6.672 \times 10^{-11}$ N·m²/kg² is the gravitational constant.

(a) Is g a continuous function of r? Explain your answer.

(b) Sketch a graph of g near $r = R$.

32. *Business*   The cost, in dollars, of an overseas telephone call is given by the function

$$C(t) = \begin{cases} 8.50 & \text{if } 0 < t \le 3 \\ 0.85t + 5.95 & \text{if } t > 3 \end{cases}$$

where t is the length of the call in minutes. Is C a continuous function at $t = 3$?

33. *Finance*   A recent federal income tax schedule can be given by the function

$$T(x) = \begin{cases} 0.15x & \text{if } 0 < x \le 23{,}900 \\ 0.28x - 3{,}107 & \text{if } 23{,}900 < x \le 61{,}650 \\ 0.33x - 6{,}189.50 & \text{if } 61{,}650 < x \le 123{,}790 \end{cases}$$

where x is the taxable income in dollars, $0 < x \le 123{,}790$, and $T(x)$ is in dollars. Determine each of the following:

(a) $\lim\limits_{x \to 23{,}900^-} T(x)$

(b) $\lim\limits_{x \to 23{,}900^+} T(x)$

(c) $\lim\limits_{x \to 23{,}900} T(x)$

(d) Is T continuous at $x = 23{,}900$?

(e) $\lim\limits_{x \to 61{,}650^-} T(x)$

(f) $\lim\limits_{x \to 61{,}650^+} T(x)$

(g) $\lim\limits_{x \to 61{,}650} T(x)$

(h) Is T continuous at $x = 61{,}650$?

34. *Business*   The local gas company uses the following function for computing their customers' monthly gas bills:

$$C(t) = \begin{cases} 0.47x + 2.95 & \text{if } 0 < x \le 24 \\ 0.89x - 7.13 & \text{if } x > 24 \end{cases}$$

where x is the number of thermal units (therms) used by the customer and $C(x)$ is the cost in dollars. Determine each of the following:

(a) $\lim\limits_{x \to 24^-} C(x)$

(b) $\lim\limits_{x \to 24^+} C(x)$

(c) $\lim\limits_{x \to 24} C(x)$

(d) Is C continuous at $x = 24$?

 In Your Words

35. The definition of continuity at a point has three conditions that must be satisfied if the function is to be continuous at a specific point.
   (a) List each of these conditions.
   (b) For each condition, give an example of a function which is not continuous because it does not satisfy that condition but does satisfy any previous conditions.

36. Suppose that f and g are functions and that neither is continuous as $x = 5$. Explain how $f + g$ can be continuous at $x = 5$.

CHAPTER 22 REVIEW

Important Terms and Concepts

Area under a graph
Average slope
Continuity
   At a point
   Over an interval
Continuous function
Interval notation

Limits
   At infinity
   Left-hand
   One-sided
   Right-hand
Properties of continuous functions
Rules for limits

Review Exercises

Find the average slope over the indicated intervals for the given functions in Exercises 1–6.

1. $f(x) = x^2 - 7$ from $x = 0$ to $x = 2$

2. $g(x) = 4x + 5$ from $x = -3$ to $x = -1$

3. $h(x) = 2x^2 + 1$ from $x = -3$ to $x = -2$

4. $j(x) = \dfrac{8}{x+4}$ from $x = 0$ to $x = \frac{1}{2}$

5. $k(x) = x^2 + 2x$ from $x = -1$ to $x = b$

6. $m(x) = x^2 - 5x$ from $x = 6$ to $x = x_1$

In Exercises 7–10, find the areas under the graph of the given function over the indicated interval. The value of n indicates the number of segments into which each interval should be divided.

7. $f(x) = -(4x+7)$ over $[-5, -2]$, $n = 6$

8. $g(x) = 2x^2 + 5$ over $[0, 4]$, $n = 8$

9. $h(x) = 3x^2 + 2x - 1$ over $[2, 4]$, $n = 8$

10. $j(x) = 5 - 2x^2$ over $[-1, 1]$, $n = 8$

In Exercise 11–20, use either the graphical or algebraic approach to determine the limit of the given function at the indicated point, or state that the limit does not exist.

11. $\lim\limits_{x \to 1} (3x^2 + 2x - 5)$

12. $\lim\limits_{x \to -2} \dfrac{3x^2 - 12}{x + 2}$

13. $\lim\limits_{x \to 0} \dfrac{3x^2 - 12}{x - 2}$

14. $\lim\limits_{x \to -3} \dfrac{x^2 - 9}{3x + 9}$

15. $\lim\limits_{x \to \infty} \dfrac{6x^2 + 3}{2x^2}$

16. $\lim\limits_{x \to \infty} \dfrac{9x^3 + 2x - 1}{3x^2 + x + 1}$

17. $\lim\limits_{x \to 5^+} \sqrt{x - 5}$

18. $\lim\limits_{x \to 4^-} \dfrac{x + 4}{x^2 - 16}$

19. $\lim\limits_{x \to \infty} \dfrac{x + 4}{2x^2 + 1}$

20. $\lim\limits_{x \to 1} f(x)$

if $f(x) = \begin{cases} x^2 - 1 & \text{if } x < 1 \\ 1 - x & \text{if } x > 1 \end{cases}$

In Exercises 21–24, determine the points of discontinuity for the given function.

21. $f(x) = \dfrac{x}{x - 5}$

22. $g(x) = \dfrac{3}{x^3}$

23. $h(x) = \begin{cases} x + 5 & \text{if } x < -3 \\ 5 - x & \text{if } x \geq -3 \end{cases}$

24. $g(x) = \begin{cases} \dfrac{x}{x + 1} & \text{if } x < 1 \\ \dfrac{1}{3 - x} & \text{if } x \geq 1 \end{cases}$

In Exercises 25–28, determine the intervals where the given function is continuous.

25. $f(x) = \dfrac{5}{x + 7}$

26. $g(x) = \dfrac{x^2 - 4}{x^2 + 5x + 6}$

27. $h(x) = \dfrac{x + 5}{\sqrt{x^2 - 25}}$

28. $j(x) = \sqrt{\dfrac{3 - x}{3 + x}}$

▤ CHAPTER 22 TEST

Solve Exercises 1 and 2.

1. Find the average slope of $f(x) = 2x^3 - 3$ from $x = 0$ to $x = 1$.

2. Find the area under the graph of $g(x) = 3x^2 - 2x$ over the interval $[1, 5]$, when it is divided into $n = 4$ subintervals.

In Exercises 3–6, determine the limit of the given function or state that the limit does not exist.

3. $\lim_{x \to 2} (5x^2 - 7)$

4. $\lim_{x \to 2} \left(\dfrac{x^2 - 4}{3x - 6} \right)$

5. $\lim_{x \to 1^+} f(x)$ where $f(x) = \begin{cases} x^3 - 1 & \text{if } x < 1 \\ 4 & \text{if } x = 1 \\ 1 + x & \text{if } x > 1 \end{cases}$

6. $\lim_{x \to \infty} \dfrac{5x^3 - 4x + 1}{7x^3 - 2}$

Solve Exercises 7 and 8.

7. Determine the points of discontinuity for $h(x) = \begin{cases} x + 7 & \text{if } x < 2 \\ \dfrac{1}{x^2 - 9} & \text{if } x \geq 2. \end{cases}$

8. Determine the intervals where $j(x) = \dfrac{x^2 - 9}{x^2 - x - 12}$ is continuous.

CHAPTER
23
The Derivative

When working with lenses, optical technicians need to be able to determine the line perpendicular to the surface of the lens at the point where the light enters the lens. Section 23.5 explains how to determine the equations of these normal lines.

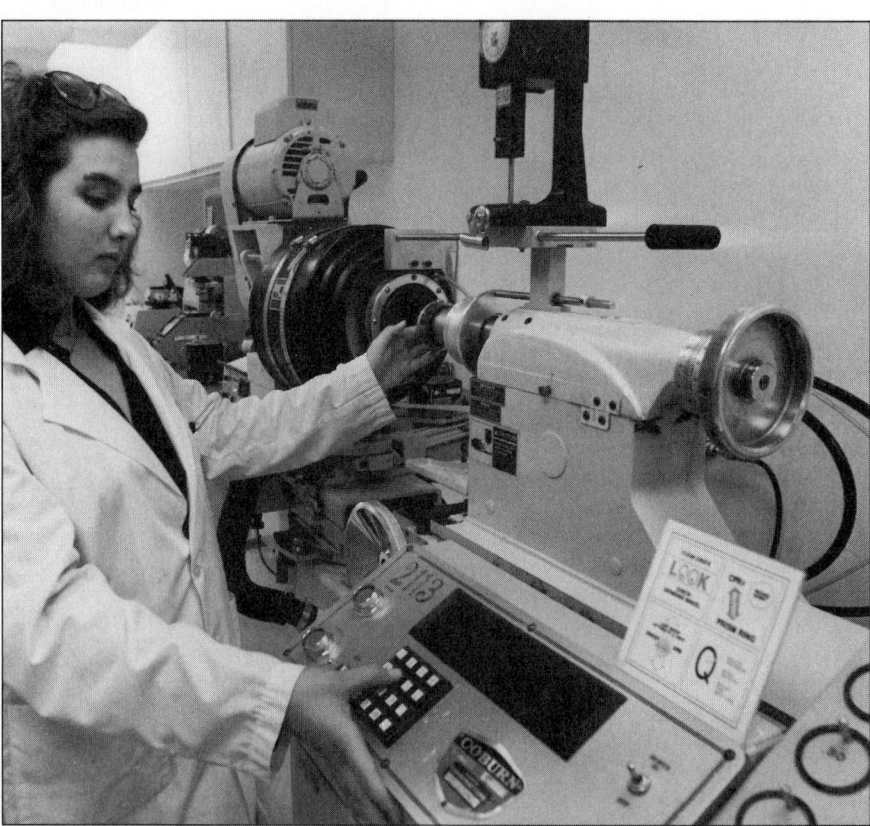

In Chapter 22, we looked at one of the questions that led to the development of calculus—the tangent question. The answer to that question depends on the use of limits. As you pursue the answer to the tangent question, you will develop an understanding of a part of calculus called the derivative. In this chapter, you will not only develop an understanding of a derivative but you will develop techniques for finding derivatives of functions.

≡ 23.1

THE TANGENT QUESTION AND THE DERIVATIVE

We will begin with the graph of the function we examined in Section 22.1. The graph is shown again in Figure 23.1. The curve has two points, P and Q. P has the coordinates (x_1, y_1) and Q the coordinates (x_2, y_2). The slope of the secant line through P and Q is

$$m_{PQ} = \frac{y_2 - y_1}{x_2 - x_1}$$

But, we are not really interested in the slope of a secant line. What we want to determine is the slope of the line that is tangent to this curve at the point P. To do this we are going to look at the various secant lines that we get by leaving P fixed. Figure 23.2 contains the same curve as in Figure 23.1, again with the points P and Q. A series of points Q_1, Q_2, Q_3, and Q_4 have been selected with each of these points lying between P and Q, each closer to P than the previous point. As each point gets closer to P, the slope of the secant line through that point and P approaches the slope of the tangent to the curve at P.

A similar thing happens if points Q^{I}, Q^{II}, Q^{III}, and Q^{IV} are on the curve but on the opposite side of P, as shown in Figure 23.3. As each point gets closer to P, the slope of the secant line through that point and P approaches the slope of the tangent line through P. Both of these examples lead us to the idea that the slope of a curve at a point P is the slope of the tangent to the curve at P.

It looks as if both of these examples involve the idea of a limit. Let's look at a more general case for the slope of these secant lines. Again, we will examine a function $y = f(x)$ as shown by the curve in Figure 23.4. We want to find an expression for the slope of the curve, or the slope of the tangent to the curve, at the point P.

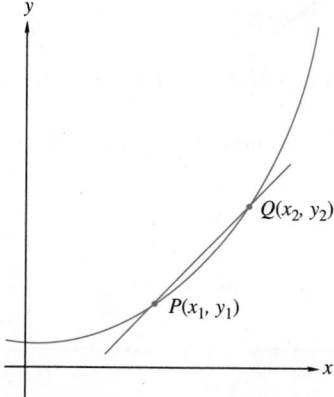

FIGURE 23.1

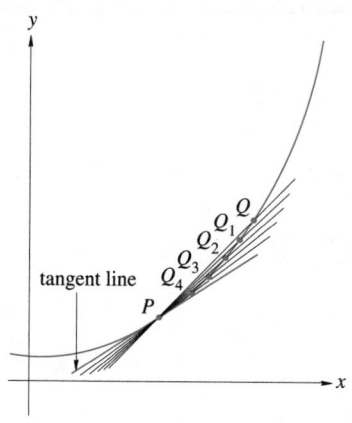

FIGURE 23.2

FIGURE 23.3

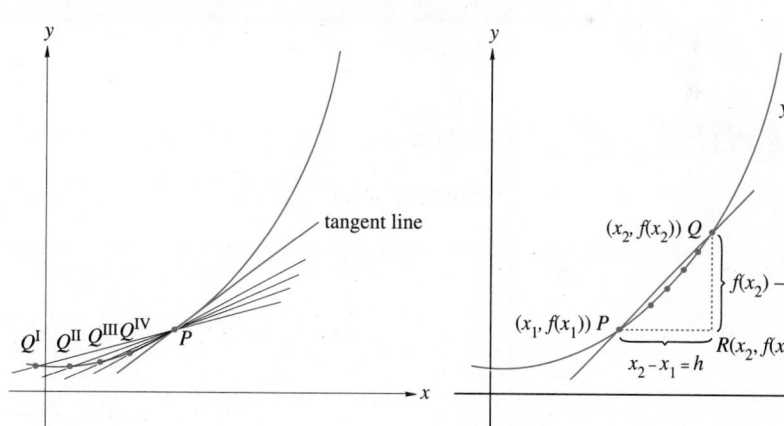

FIGURE 23.4

The point P has the coordinates $(x_1, y_1) = (x_1, f(x_1))$. We will let Q be any other point on the curve and let its coordinates be $(x_2, f(x_2))$. The slope of the secant line through P and Q is

$$m_{PQ} = \frac{f(x_2) - f(x_1)}{x_2 - x_1}$$

Now draw a line through Q parallel to the y-axis and a line through P parallel to the x-axis. These two lines intersect at the point R, which has the coordinates $(x_2, f(x_1))$. The distance from P to R is $x_2 - x_1$. If we let $h = x_2 - x_1$, then $x_2 = x_1 + h$. This allows us to write the slope of the secant line through P and Q as

$$m_{PQ} = \frac{f(x_1 + h) - f(x_1)}{(x_1 + h) - x_1}$$
$$= \frac{f(x_1 + h) - f(x_1)}{h}$$

We know that h is not 0 because if $x_1 = x_2$, either we would not have two distinct points or we would not have a function.

As Q is picked closer to P, x_2 approaches x_1. This means that h is getting closer to 0. The slope of the tangent line will be the limit

$$\lim_{h \to 0} \frac{f(x_1 + h) - f(x_1)}{h}$$

This formula gives us the slope of the curve at the point $(x_1, f(x_1))$.

Slope of a Curve

The slope of a curve $y = f(x)$ at the point $(x_1, f(x_1))$ is

$$\lim_{h \to 0} \frac{f(x_1 + h) - f(x_1)}{h}$$

provided that this limit exists.

EXAMPLE 23.1

Find the slope of $f(x) = x^2$ at the point $(3, 9)$.

Solution
$$\lim_{h \to 0} \frac{f(3 + h) - f(3)}{h} = \lim_{h \to 0} \frac{(3 + h)^2 - 3^2}{h}$$
$$= \lim_{h \to 0} \frac{9 + 6h + h^2 - 9}{h}$$
$$= \lim_{h \to 0} \frac{6h + h^2}{h}$$
$$= \lim_{h \to 0} \frac{h(6 + h)}{h}$$
$$= \lim_{h \to 0} (6 + h) = 6$$

So, the slope of the tangent line to $y = x^2$ at $(3, 9)$ is 6.

The process that we have been describing will determine the slope of a tangent to a curve or the slope of the curve at a point. This process can be used to describe how any function is changing at a particular point. In other words, we have been using the method for finding the average change in a function and then the limit of that average change as the denominator approaches 0. This gives the instantaneous change of the function at that point. We will see this by taking a different look at Example 22.4.

Application

EXAMPLE 23.2

A tank is filled with water by opening an inlet pipe. The volume V in liters of water in the tank t minutes after the valve is opened is given by $V(t) = 5t^2 + 4t$. What is the instantaneous rate of increase in the volume at the first minute?

Solution We want to find the instantaneous rate of change R in the volume when $t = 1$. We can do this by using the formula for the rate of change $R_{\text{Vol}} = \dfrac{V(t_2) - V(t_1)}{t_2 - t_1}$ and letting $t_1 = 1$ and $t_2 = t_1 + h = 1 + h$, then finding the limit when h approaches 0.

$$
\begin{aligned}
R_{\text{Vol}} &= \lim_{h \to 0} \frac{V(1+h)^2 - V(1)}{h} \\
&= \lim_{h \to 0} \frac{[5(1+h)^2 + 4(1+h)] - [5(1)^2 + 4(1)]}{(1+h) - 1} \\
&= \lim_{h \to 0} \frac{[5(1+2h+h^2) + 4(1+h)] - [5+4]}{h} \\
&= \lim_{h \to 0} \frac{(5 + 10h + 5h^2 + 4 + 4h) - (9)}{h} \\
&= \lim_{h \to 0} \frac{9 + 14h + 5h^2 - 9}{h} \\
&= \lim_{h \to 0} \frac{14h + 5h^2}{h} \\
&= \lim_{h \to 0} (14 + 5h) = 14
\end{aligned}
$$

When $t = 1$, the volume of the tank is increasing at the rate of 14 L/min.

The Derivative

All of this discussion about instantaneous change of a function leads us to the following definition of the derivative.

Derivative of a Function f at a Number c

If $y = f(x)$ defines a function f and if the point c is in the domain of f, then the **derivative of f at c**, written as $f'(c)$, is defined as

$$f'(c) = \lim_{h \to 0} \frac{f(c+h) - f(c)}{h}$$

provided that this limit exists.

We used the notation $f'(c)$ to indicate the derivative of the function at the point c. In general, the derivative of the function is the following.

Derivative of a Function	$f'(x) = \lim\limits_{h \to 0} \dfrac{f(x+h) - f(x)}{h}$

≡ **Note**

The notation $f'(x)$ is not a unique notation for the derivative. Other notations include $D_x y$, $\dfrac{dy}{dx}$, $\dfrac{d}{dx}[f(x)]$, y', $Df(x)$, $\dfrac{df}{dx}$, and $D_x[f(x)]$. There are times when a meaning is clearer if one of these notations is used instead of $f'(x)$.

Caution

$\dfrac{dy}{dx}$, $\dfrac{df}{dx}$, and $\dfrac{d}{dx}$ are not fractions but are symbols for a derivative.

Four-Step Method

We can calculate the derivative of a function f at the point c by using the following four steps.

Four Steps for Calculating $f'(c)$

Step 1: Find $f(c+h)$.

Step 2: Subtract $f(c)$ from $f(c+h)$.

Step 3: Divide the result in Step 2 by h to get the difference quotient.

Step 4: Find the limit, if it exists, of the difference quotient in Step 3, as h approaches 0.

EXAMPLE 23.3

Find the derivative of $f(x) = x^3$ at $x = 3$ by using the four-step method.

Solution

Step 1: $f(3+h) = (3+h)^3 = 27 + 27h + 9h^2 + h^3$

Step 2: Since $f(3) = 3^3 = 27$, $f(3+h) - f(3) = [27 + 27h + 9h^2 + h^3] - 27$
$$= 27h + 9h^2 + h^3$$

Step 3: $\dfrac{f(3+h) - f(3)}{h} = \dfrac{27h + 9h^2 + h^3}{h} = 27 + 9h + h^2$

Step 4: $\lim\limits_{h \to 0} (27 + 9h + h^2) = 27$.

So, the derivative of $f(x) = x^3$ at $x = 3$ is $f'(3) = 27$.

We can also use the four-step method to find the derivative of a function at any point x, rather than at a particular point.

EXAMPLE 23.4

Find the derivative of $g(x) = 3x^2 + x$ using the four-step method.

Solution

Step 1: $g(x+h) = 3(x+h)^2 + (x+h)$

$= 3(x^2 + 2xh + h^2) + (x+h)$

$= 3x^2 + 6xh + 3h^2 + x + h$

Step 2: $g(x+h) - g(x) = (3x^2 + 6xh + 3h^2 + x + h) - (3x^2 + x)$

$= 6xh + 3h^2 + h$

Step 3: $\dfrac{g(x+h) - g(x)}{h} = \dfrac{6xh + 3h^2 + h}{h} = 6x + 3h + 1u$

Step 4: $\displaystyle\lim_{h\to 0} \dfrac{g(x+h) - g(x)}{h} = \lim_{h\to 0}(6x + 3h + 1) = 6x + 1$

So, the derivative of $g(x) = 3x^2 + x$ is $g'(x) = 6x + 1$.

Once you have the derivative of a function, you can evaluate it at any particular point.

EXAMPLE 23.5

If $g(x) = 3x^2 + x$, what are **(a)** $g'(2)$, **(b)** $g'(0)$, and **(c)** $g'(-3)$?

Solutions In Example 23.4, we learned that the derivative of g is $g'(x) = 6x + 1$. As a result, we have

(a) $g'(2) = 6(2) + 1 = 13$

(b) $g'(0) = 6(0) + 1 = 1$

(c) $g'(-3) = 6(-3) + 1 = -17$

EXAMPLE 23.6

Use the four-step method to find the derivative of $y = \dfrac{1}{x+2}$.

Solution

Step 1: $y = f(x)$, so $f(x+h) = \dfrac{1}{x+h+2}$

Step 2: $f(x+h) - f(x) = \dfrac{1}{x+h+2} - \dfrac{1}{x+2}$.

In order to conduct the subtraction, we need to first find a common denominator and then subtract. The common denominator is $(x+2)(x+h+2)$.

$$\dfrac{1}{x+h+2} - \dfrac{1}{x+2} = \dfrac{1}{x+h+2}\left(\dfrac{x+2}{x+2}\right) - \dfrac{1}{x+2}\left(\dfrac{x+h+2}{x+h+2}\right)$$

EXAMPLE 23.6 (Cont.)

$$= \frac{x+2}{(x+h+2)(x+2)} - \frac{x+h+2}{(x+h+2)(x+2)}$$

$$= \frac{(x+2)-(x+h+2)}{(x+h+2)(x+2)}$$

$$= \frac{-h}{(x+h+2)(x+2)}$$

Step 3: $\dfrac{f(x+h)-f(x)}{h} = \dfrac{\frac{-h}{(x+h+2)(x+2)}}{h}$

$$= \frac{-h}{h(x+h+2)(x+2)}$$

$$= \frac{-1}{(x+h+2)(x+2)}$$

Step 4: $\displaystyle\lim_{h\to 0} \frac{-1}{(x+h+2)(x+2)} = \frac{-1}{(x+2)^2}$

So, $y' = \dfrac{-1}{(x+2)^2}$.

EXAMPLE 23.7

Use the four-step method to find the derivative of $f(x)=\sqrt{x-3}$ at $x=4$.

Solution   We will first find the general expression for $f'(x)$ and then determine $f'(4)$.

Step 1: $f(x+h)=\sqrt{x+h-3}$

Step 2: $f(x+h)-f(x)=\sqrt{x+h-3}-\sqrt{x-3}$

Step 3: $\dfrac{f(x+h)-f(x)}{h} = \dfrac{\sqrt{x+h-3}-\sqrt{x-3}}{h}$

In order to simplify this expression, we will rationalize the numerator by multiplying both numerator and denominator by $\sqrt{x+h-3}+\sqrt{x-3}$.

$$\frac{\sqrt{x+h-3}-\sqrt{x-3}}{h} = \frac{(\sqrt{x+h-3}-\sqrt{x-3})(\sqrt{x+h-3}+\sqrt{x-3})}{h(\sqrt{x+h-3}+\sqrt{x-3})}$$

$$= \frac{(x+h-3)-(x-3)}{h(\sqrt{x+h-3}+\sqrt{x-3})}$$

$$= \frac{h}{h(\sqrt{x+h-3}+\sqrt{x-3})}$$

$$= \frac{1}{\sqrt{x+h-3}+\sqrt{x-3}}$$

EXAMPLE 23.7 (Cont.)

Step 4: $\displaystyle\lim_{h\to 0}\frac{1}{\sqrt{x+h-3}+\sqrt{x-3}}=\frac{1}{\sqrt{x-3}+\sqrt{x-3}}$

$$=\frac{1}{2\sqrt{x-3}}$$

Now that we know $f'(x)=\dfrac{1}{2\sqrt{x-3}}$, we can find $f'(4)=\dfrac{1}{2\sqrt{4-3}}=\dfrac{1}{2}$.

Thus, the derivative of $f(x)=\sqrt{x-3}$ at $x=4$ is $\frac{1}{2}$. ▪

Exercise Set 23.1

In Exercises 1–24, use the definition of the derivative (that is, use the four-step method) to differentiate the given function.

1. $f(x)=3x+2$
2. $g(x)=5x-7$
3. $f(x)=4-3x$
4. $g(t)=6-4t$
5. $y=3x^2$
6. $y=-5t^2$
7. $y=2x^2+5$

8. $j(x)=3x^2-x+2$
9. $k(x)=3-4x^2$
10. $g(x)=2x-7x^2$
11. $y=-4x^3$
12. $y=x^2-2x^3$
13. $y=t-3t^3$
14. $f(x)=\dfrac{2x+1}{3}$

15. $f(x)=\dfrac{1}{x+1}$
16. $g(x)=\dfrac{5}{2x-3}$
17. $j(x)=\dfrac{4}{1-x^2}$
18. $f(x)=\dfrac{9}{x^2+2x}$

19. $s=16t^2-6t+3$
20. $q=t^3-4t^2+5t-2$
21. $y=\sqrt{x^2+4}$
22. $y=\sqrt{5-x}$
23. $y=\sqrt{2x-x^2}$
24. $y=\sqrt{5x-4x^3}$

In Exercises 25–36, evaluate the derivative at each of the indicated points.

25. $f(x)=3x+7$; $f'(2)$, $f'(9)$
26. $g(x)=x^2-2$; $g'(2)$, $g'(9)$
27. $s(t)=16t^2+2t$; $s'(3)$, $s'(2)$, $s'(0)$
28. $s(t)=2t^2-1$; $s'(1)$, $s'(5)$
29. $q(t)=t^3-4t^2$; $q'(0)$, $q'(3)$
30. $j(x)=\dfrac{1}{\sqrt{x}}$; $j'(4)$, $j'(9)$
31. $k(x)=\dfrac{1}{x^2-4}$; $k'(5)$, $k'(-5)$

32. $f(x)=\dfrac{4}{x^2+1}$; $f'(0)$, $f'(3)$, $f'(-5)$
33. $j(t)=\sqrt{5t-t^2}$; $j'(1)$, $j'(4)$
34. $g(x)=\dfrac{1}{\sqrt{x+4}}$; $g'(0)$, $g'(5)$, $g'(-2)$
35. $f(x)=\dfrac{5}{\sqrt{25-x^2}}$; $f'(0)$, $f'(3)$, $f'(-4)$
36. $h(x)=\sqrt{x^2+4x+9}$; $h'(-4)$, $h'(0)$, $h'(4)$

 **In Your Words**

37. Without looking in the book, write the four steps used to calculate $f'(c)$.

38. Explain the relationship between the slope of a curve at some point x_0 and the derivative of the function that graphs that curve at x_0.

≡ 23.2
DERIVATIVES OF POLYNOMIALS

At this point, the definition by the four-step method is our only way to take derivatives. The four-step method is a very important process and a method that we will use again and again. But because of its length, it is time consuming and there are many opportunities to make mistakes. In this section, we are going to use the definition to develop some formulas for taking the derivative of any polynomial. In the remainder of the chapter, we will develop some rules for taking more involved derivatives.

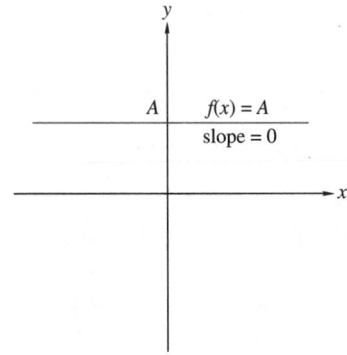

FIGURE 23.5

Constant Function

This first polynomial we will study is the constant function, $f(x) = k$. A look at the graph of this function in Figure 23.5 will show us what to expect. The slope of a constant function is 0, therefore, since the derivative is a measure of the slope of the tangent to the curve, we should expect the derivative to be 0. Let's use the four-step method to confirm our theory.

Step 1: $f(x+h) = k$

Step 2: $f(x+h) - f(x) = k - k = 0$

Step 3: $\dfrac{f(x+h) - f(x)}{h} = \dfrac{0}{h} = 0$

Step 4: $\lim\limits_{h \to 0} \dfrac{f(x+h) - f(x)}{h} = \lim\limits_{h \to 0} 0 = 0$

Thus, we have determined the following derivative of a constant function.

Rule 1: Derivative of $f(x) = k$

For the function $f(x) = k$, where k is a constant, the derivative is $f'(x) = 0$.

Linear Function

The next polynomial is the linear function $f(x) = mx + b$. Notice that we have written this function in the slope-intercept form of the equation. We know that the slope of this line is m. The derivative is supposed to give the slope of a tangent to the curve. Since this curve is a line, and its tangent is the same line, we expect that its derivative is also m. We use the four-step method:

Step 1: $f(x+h) = m(x+h) + b = mx + mh + b$

Step 2: $f(x+h) - f(x) = (mx + mh + b) - (mx + b) = mh$

Step 3: $\dfrac{f(x+h) - f(x)}{h} = \dfrac{mh}{h} = m$

Step 4: $\lim\limits_{h \to 0} \dfrac{f(x+h) - f(x)}{h} = \lim\limits_{h \to 0} m = m$

Again, we confirm our expectation and have shown the following for the derivative of a straight line.

> **Rule 2: Derivative of** $f(x) = mx + b$
>
> For the linear function $f(x) = mx + b$, the derivative is $f'(x) = m$.

EXAMPLE 23.8

Find the derivative of **(a)** $f(x) = -8$ and **(b)** $g(x) = 9x - 4$.

Solution

(a) $f(x) = -8$ is a constant function, so $f'(x) = 0$.

(b) $g(x) = 9x - 4$ is a linear function, so $g'(x) = 9$.

Polynomials of the Form $f(x) = x^n$

We will now consider higher-degree polynomials. This will be done in three steps. The first step will consider two specific forms of x^n, where n is 2 and 3. The second step will generalize this for any real number n. The third step will consider the sums and differences of these terms.

In order to develop the rule for the derivative of x^n, we will determine derivatives for x^2 and x^3. We begin with $f(x) = x^2$.

Step 1: $f(x+h) = (x+h)^2 = x^2 + 2xh + h^2$

Step 2: $f(x+h) - f(x) = \left(x^2 + 2xh + h^2\right) - x^2$
$$= 2xh + h^2$$

Step 3: $\dfrac{f(x+h) - f(x)}{h} = \dfrac{2xh + h^2}{h} = 2x + h$

Step 4: $\displaystyle\lim_{h \to 0} \dfrac{f(x+h) - f(x)}{h} = \lim_{h \to 0}(2x + h) = 2x$

So, the derivative of $f(x) = x^2$ **is** $f'(x) = 2x$.
Next, we find the derivative of $g(x) = x^3$.

Step 1: $g(x+h) = (x+h)^3 = x^3 + 3x^2h + 3xh^2 + h^3$

Step 2: $g(x+h) - g(x) = \left(x^3 + 3x^2h + 3xh^2 + h^3\right) - x^3$
$$= 3x^2h + 3xh^2 + h^3$$

Step 3: $\dfrac{g(x+h) - g(x)}{h} = \dfrac{3x^2h + 3xh^2 + h^3}{h} = 3x^2 + 3xh + h^2$

Step 4: $\displaystyle\lim_{h \to 0} \dfrac{g(x+h) - g(x)}{h} = \lim_{h \to 0}(3x^2 + 3xh + h^2) = 3x^2$

Thus, the derivative of $g(x) = x^3$ **is** $g'(x) = 3x^2$.

If we continued this process, we could show that the derivative of x^4 is $4x^3$, the derivative of x^5 is $5x^4$, the derivative of x^6 is $6x^5$, and so on. We could also show that the derivative of x^{-7} is $-7x^{-8}$, the derivative of $x^{5/2}$ is $\frac{5}{2}x^{3/2}$, and the derivative of $x^{-7/3}$ is $-\frac{7}{3}x^{-10/3}$.

What we have demonstrated is how to find the derivative of x^n, when n is a rational number. Thus, we now have

---

Rule 3: Derivative of $f(x) = x^n$

The derivative of any function $f(x) = x^n$, n a rational number, is $f'(x) = nx^{n-1}$.

EXAMPLE 23.9

Find the derivative of **(a)** $f(x) = x^7$, **(b)** $g(x) = x^{-3}$, **(c)** $h(x) = x^{2/3}$, and **(d)** $j(x) = x^{-5/2}$.

Solution

(a) $f(x) = x^7$ and so $f'(x) = 7x^{7-1} = 7x^6$

(b) $g(x) = x^{-3}$ and $g'(x) = -3x^{-3-1} = -3x^{-4}$

(c) $h(x) = x^{2/3}$ and $h'(x) = \frac{2}{3}x^{\frac{2}{3}-1} = \frac{2}{3}x^{-\frac{1}{3}}$

(d) $j(x) = x^{-5/2}$ and $j'(x) = -\frac{5}{2}x^{-\frac{5}{2}-1} = -\frac{5}{2}x^{-7/2}$

The next example demonstrates the usefulness of fractional exponents—both positive and negative.

EXAMPLE 23.10

Find the derivatives of **(a)** $y = x^2\sqrt{x}$ and **(b)** $f(x) = \dfrac{1}{x^{5/2}}$.

Solution

(a) $y = x^2\sqrt{x}$. To find the derivative of y, or $D_x y$, we must first rewrite $x^2\sqrt{x}$ as $x^2 x^{1/2} = x^{5/2}$. So, $y = x^{5/2}$ and

$$D_x y = \frac{5}{2}x^{3/2} = \frac{5}{2}x\sqrt{x}$$

(b) $f(x) = \dfrac{1}{x^{5/2}}$. To apply rule 3, first write $f(x)$ as $f(x) = x^{-5/2}$. In Example 23.9(d), we showed that $f'(x) = -\frac{5}{2}x^{-7/2}$.

Caution

Be careful. Do not just differentiate the denominator. If you do, you get

$$D_x \left(\frac{1}{x^{5/2}} \right) \neq \frac{1}{\frac{5}{2}x^{3/2}} = \frac{2}{5}x^{-3/2}$$

which is not the correct answer.

Three General Formulas

Suppose we have a function $F(x) = kf(x)$, where k is a constant and $f'(x)$ exists. [This might be a function such as $G(x) = 9x^3$, where $k = 9$ and $g(x) = x^3$.] We will use the four-step method to find the derivative of $F(x)$.

Step 1: $F(x+h) = kf(x+h)$

Step 2: $F(x+h) - F(x) = kf(x+h) - kf(x) = k[f(x+h) - f(x)]$

Step 3: $\dfrac{F(x+h) - F(x)}{h} = \dfrac{k[f(x+h) - f(x)]}{h}$

$$= k\left[\frac{f(x+h) - f(x)}{h}\right]$$

Step 4: $\displaystyle\lim_{h \to 0} \frac{F(x+h) - F(h)}{h} = \lim_{h \to 0} k\left[\frac{f(x+h) - f(x)}{h}\right]$

$$= k \lim_{h \to 0} \frac{f(x+h) - f(x)}{h}$$

$$= kf'(x)$$

Rule 4: Derivative of $F(x) = kf(x)$

If $F(x) = kf(x)$, k is a constant and $f'(x)$ exists, then

$$F'(x) = kf'(x)$$

EXAMPLE 23.11

Find the derivative of **(a)** $f(x) = 9x^3$, **(b)** $g(x) = 7x^{-5}$, and **(c)** $h(x) = \frac{2}{3}x^{5/2}$.

Solutions

(a) $f(x) = 9x^3$ so $f'(x) = 9(3x^2) = 27x^2$

(b) $g(x) = 7x^{-5}$ and $g'(x) = 7(-5x^{-6}) = -35x^{-6}$

(c) $h(x) = \frac{2}{3}x^{5/2}$ and $h'(x) = \frac{2}{3}(\frac{5}{2}x^{3/2}) = \frac{5}{3}x^{3/2}$

The next two formulas have to do with the sum and difference of two functions. These last two rules will allow us to find derivatives of higher-degree polynomials.

If f and g are two functions that can be differentiated, then what is the derivative of their sum or difference? It turns out that the derivative of a sum (or difference) is the sum (or difference) of the derivative of each term. These are stated formally in the following box, which contains the last two rules for this section.

> **Rules 5 and 6: Derivative of** $F(x) = f(x) \pm g(x)$
>
> Rule 5: If $F(x) = f(x) + g(x)$ and $f'(x)$ and $g'(x)$ exist, then
> $$F'(x) = f'(x) + g'(x)$$
>
> Rule 6: If $F(x) = f(x) - g(x)$ and $f'(x)$ and $g'(x)$ exist, then
> $$F'(x) = f'(x) - g'(x)$$

EXAMPLE 23.12

Find the derivative of $f(x) = 7x^5 - 4x^3 + 2x - 8$.

Solution   According to the last two rules, the derivative of a sum (or difference) is the sum (or difference) of the individual derivatives. We will differentiate this term by term:

$$f'(x) = 7(5x^4) - 4(3x^2) + 2(1) - 0$$
$$= 35x^4 - 12x^2 + 2$$

Exercise Set 23.2

In Exercises 1–37, find the derivative of the function by using the six rules given in this section.

1. $f(x) = 19$

2. $g(x) = -4$

3. $h(x) = 7x - 5$

4. $j(x) = -3x + 7$

5. $k(x) = 9x^2$

6. $m(x) = 15x^3$

7. $f(x) = \frac{1}{3}x^{15}$

8. $g(x) = \frac{2}{5}x^{10}$

9. $h(x) = 5x^{-2}$

10. $j(x) = -3x^{-4}$

11. $k(x) = -\frac{2}{3}x^{3/2}$

12. $j(x) = \frac{4\sqrt{x}}{3}$

13. $h(x) = 4\sqrt{3x}$

14. $m(x) = \frac{3}{4}x^{-5/3}$

15. $g(x) = \frac{3}{\sqrt[3]{x}}$

16. $f(x) = 4x^3 - 5x^2 + 2$

17. $g(x) = 9x^2 + 3x - 4$

18. $n(x) = 7x^7 - 5x^5$

19. $h(x) = \frac{1}{3}x^3 + \frac{1}{2}x^2 - 5x + \frac{2}{3}x^{-2} + 5$

20. $j(x) = \frac{3}{4}x^4 - \frac{2}{3}x^3 + \frac{5}{4}x^2 - \frac{1}{2}x + 2x^{-3} + x^{-4} + 7$

21. $f(x) = \frac{4}{5}x^5 - \frac{3}{2}x^3 - 7x^0 + \frac{1}{2}x^{-2} - \frac{2}{3}x^{-3}$

22. $k(x) = \sqrt{2}x^5 - \sqrt{3}x^3 + 3x + \pi$

23. $\ell(x) = 3\sqrt{3}x^4 + 3\sqrt{5}x + \sqrt{4x^4}$

24. $m(x) = \sqrt{7}x^6 - \sqrt{4}x^3 + 5\sqrt{3}x^2 - \sqrt{7}$

25. $s(t) = 16t^2 - 32t + 5$

26. $q(t) = 4.3t^3 - 2.7t + 3.0$

27. $\alpha(t) = 30 - 4.0t^2 + 2t^{1/2}$

28. $s(t) = \frac{5}{3}t^4 + 7\sqrt{2}t^3 - 8t^{2/3} + 7t^{-1/2}$

29. $q(t) = \sqrt{6t} - 4t\sqrt{3t} + \sqrt[3]{t^2}$

30. $f(x) = 3x\sqrt{x}$

31. $g(x) = 5x^2\sqrt[3]{x}$

32. $j(x) = \frac{3x^3}{\sqrt[3]{x}}$

33. $h(x) = \frac{4}{x^2}$

34. $j(x) = 7x^2 + 2x\sqrt{x} + \dfrac{5}{x^3}$

35. $K(x) = \dfrac{4\sqrt[3]{x}}{\sqrt{x}}$

36. $M(x) = \dfrac{3\sqrt{x^3}}{\sqrt{3x}}$

37. $f(x) = \dfrac{4\sqrt{3x} + 3x^2 - 7\sqrt{x^3} + 2x^{-1}}{\sqrt{x}}$

Solve Exercises 38–44.

38. Derive rule 6.

39. Find all values of x where $f'(x) = 0$, if $f(x) = x^3 + 3x^2$.

40. Find all values of t where $\theta'(t) = 0$, if $\theta(t) = 3t^3 - t$.

41. Find all values of x where $g'(x) = 0$, if $g(x) = 2x^3 + x^2 - 4x$.

42. If $\omega(t) = 4.0t^2 - 2.0\sqrt{t}$, find the values of t for which $\omega'(t) = 0$.

43. If $j(t) = 4t^3 + \dfrac{1}{t}$, for what values of t is $j'(t) = 0$?

44. If $h(x) = 4 + \dfrac{1}{x} - \dfrac{8}{x^3}$, for what values of x is $h'(x) = 0$?

Solve Exercises 45–56.

45. *Demography* The population of a certain city t months from now can be estimated by the formula $P(t) = 146{,}000 + 64t - 0.04t^2$. Find

  (a) The population of the city six months from now

  (b) The rate of change in the city's population six months from now

46. *Business* A manufacturer estimates that the revenue in dollars from the sale of c cellular telephone is $R(c) = 120c - c^2$. Find

  (a) The revenue from the sale of 30 telephones

  (b) The rate of change in the revenue when $c = 30$

47. *Physics* A ball is rolled down an incline. The distance in feet of the ball from the starting point after t seconds is given by $s(t) = 15t + 7t^2$. Find

  (a) The distance the ball has traveled when $t = 4\,\text{s}$

  (b) The velocity (rate of change) of the ball when $t = 4\,\text{s}$

48. *Petroleum engineering* When an oil tank is drained for cleaning, the volume, V, of oil in gallons left in the tank t minutes after the drain valve is opened is given by the formula $V(t) = 75{,}000 - 3{,}000t + 30t^2$.

  (a) How much oil is left in the tank 20 min after the drain is opened?

  (b) What is the average rate that the oil drains during the first 20 min?

  (c) What is the rate of change that the oil is draining when $t = 20$?

49. *Demography* A certain city's population t years from now can be estimated by using the function $P(t) = 125{,}000 + 1{,}500t - 36\sqrt{t}$. Find

  (a) The city's population six years from now

  (b) The rate at which the city's population is growing six years from now

  (c) The city's population ten years from now

  (d) The rate at which the city's population is growing ten years from now

50. *Business* The total cost, C, in dollars, of producing n units of a certain product is given by $C(n) = 5{,}000 + 25n - 0.4n^2$, where $0 \le n \le 60$.

  (a) Determine the rate of change in the cost when $n = 10$.

  (b) Determine the rate of change in the cost when $n = 30$.

  (c) Determine the rate of change in the cost when $n = 50$.

51. *Electronics* The power, P, in watts, delivered to a $25\,\Omega$ resistor is $P = 25i^2$. Find the instantaneous rate of change of P with respect of i when $i = 2.4\,\text{A}$.

52. *Material science* For certain temperatures, the tensile strength, S, in pounds, of a piece of material as a function of its temperature is described by $S(T) = 520 - 0.000085T^2$. Find the rate of change of S at the instant that $T = 135°\text{F}$.

53. *Business*   A manufacturer has determined that the revenue in dollars from the sale of n cellular telephones is given by $R(n) = 78n - 0.025n^2$. The cost of producing n telephones is $C(n) = 9,800 + 22.5n$.
(a) If the profit, P, is the difference between the revenue and the cost, what is the profit function?
(b) Find the marginal profit function, P'.
(c) Find $P'(50)$.

54. *Electronics*   The output voltage, V, in volts, of a circuit increases gradually with time t, in seconds, according to the function $V(t) = 30.0 + 0.04t^2$. Find an expression for the instantaneous rate of change of the voltage with time.

55. *Physics*   The altitude, h, of a rocket fired from an airplane flying over the ocean can be described by the function $h(t) = -t^4 + 12t^3 - 16t^2 + 605$, where t is in seconds and the altitude is in feet above sea level. The velocity of the rocket at time t is $v(t) = h'(t)$.
(a) From what altitude was the rocket launched?
(b) At what time does the rocket fall into the ocean?
(c) What is the rocket's altitude and velocity at $t = 2$ s?

(d) What is the rocket's altitude and velocity at $t = 4$ s?
(e) What is the rocket's altitude and velocity at $t = 6$ s?
(f) What is the rocket's altitude and velocity at $t = 10$ s?
(g) What is the rocket's velocity when it falls into the ocean?

56. *Navigation*   Upon reversing the engines, the distance traveled in meters by a supertanker in a straight line before coming to a stop is given by $s(t) = 84.5t - \frac{1}{6}t^3$, where t is in minutes.
(a) If the velocity of the supertanker at time t is $v(t) = s'(t)$, find an equation for the velocity.
(b) What was the velocity of the tanker when the engines were put in reverse?
(c) How long does it take for the supertanker to stop? That is, when is $v(t) = 0$?
(d) How far does the supertanker travel from the time the engines are reversed until it comes to a stop?

In Your Words

57. Describe how to find the derivative of $f(x) = kx^n$.

58. Describe how to find the derivative of the sum of two functions.

▤ 23.3
DERIVATIVES OF PRODUCT AND QUOTIENTS

In this section, we will learn general formulas for finding the derivatives of functions that are products or quotients.

The Product Rule ▬▬▬▬▬▬▬▬

As before, we will assume that we have two functions, $f(x)$ and $g(x)$, and that their derivatives $f'(x)$ and $g'(x)$ exist. If we let F be the product of f and g, then $F(x) = f(x)g(x)$. We want to find $F'(x)$. Again, we will use the four-step method. The first two steps will be quite straightforward. In the third step, we will have to employ some more complicated algebra, which will be explained at that time.

Step 1: $F(x+h) = f(x+h)g(x+h)$

Step 2: $F(x+h) - F(x) = f(x+h)g(x+h) - f(x)g(x)$

Step 3: $\dfrac{F(x+h) - F(x)}{h} = \dfrac{f(x+h)g(x+h) - f(x)g(x)}{h}$

This does not look too promising. But, what we will do is add a fancy version of 0 to the numerator. We will add $-f(x+h)g(x)+f(x+h)g(x)$ to the numerator. Since this is equal to 0, the value of the numerator is not changed and we get

$$\frac{f(x+h)g(x+h)-f(x+h)g(x)+f(x+h)g(x)-f(x)g(x)}{h}$$

$$= f(x+h)\frac{g(x+h)-g(x)}{h}+\frac{f(x+h)-f(x)}{h}g(x)$$

Now we will go to Step 4 and take the limit of the quantity. Remember that the limit of a product is the product of the limits.

Step 4: $\displaystyle\lim_{h\to0}\frac{F(x+h)-F(x)}{h} = \lim_{h\to0}f(x+h)\lim_{h\to0}\frac{g(x+h)-g(x)}{h}$

$$+ \lim_{h\to0}\frac{f(x+h)-f(x)}{h}\lim_{h\to0}g(x)$$

$$= f(x)g'(x)+f'(x)g(x)$$

This gives us the product rule for derivatives.

Product Rule for Derivatives

If $F(x) = f(x)g(x)$ and $f'(x)$ and $g'(x)$ exist, then

$$F'(x) = f(x)g'(x)+f'(x)g(x)$$

In words, this can be stated as "the derivative of the product of two functions, whose derivatives exist, is equal to the first function times the derivative of the second plus the second function times the derivative of the first."

≡ Note

Note carefully that the derivative of a product does not behave as nicely as the limit of a product. **The derivative of a product is not the product of the derivatives.**

EXAMPLE 23.13

Find the derivative of $F(x) = (x^2+x)(x^3-1)$.

Solution Here the function $F(x)$ is the product of $f(x) = x^2+x$ and $g(x) = x^3-1$. We calculate the derivatives of each of these two functions as $f'(x) = 2x+1$ and $g'(x) = 3x^2$. According to the new rule

$$F'(x) = f(x)g'(x)+f'(x)g(x)$$
$$= (x^2+x)(3x^2)+(2x+1)(x^3-1)$$
$$= 3x^4+3x^3+2x^4+x^3-2x-1$$
$$= 5x^4+4x^3-2x-1$$

Caution

Once again we stress that $F'(x) \neq f'(x)g'(x)$. In Example 23.13 we found that if $F(x) = (x^2 + x)(x^3 - 1)$, then $F'(x) = 5x^4 + 4x^3 - 2x - 1$. But, $f'(x)g'(x) = (2x + 1)(3x^2) = 6x^3 + 3x^2 \neq F'(x)$.

EXAMPLE 23.14

Find $G'(x)$, if $G(x) = (x^3 + 2x^2 + x)(x^2 - 5x)$.

Solution   We could multiply these two polynomials and then take the derivative of that product, but we will use the product rule instead. The time will come when we will have to use the product rule, so we might as well get some practice. Also, the product rule makes the task somewhat easier. We will let

$$f(x) = x^3 + 2x^2 + x \text{ and } g(x) = x^2 - 5x$$

so

$$f'(x) = 3x^2 + 4x + 1 \text{ and } g'(x) = 2x - 5$$

$$G'(x) = f(x)g'(x) + f'(x)g(x)$$

$$= (x^3 + 2x^2 + x)(2x - 5) + (3x^2 + 4x + 1)(x^2 - 5x)$$

$$= 5x^4 - 12x^3 - 27x^2 - 10x$$

The Quotient Rule

Suppose now that $F(x) = \dfrac{f(x)}{g(x)}$, where $g(x) \neq 0$, and $f'(x)$ and $g'(x)$ both exist. What is $F'(x)$? There are a couple of ways to develop this formula. One way uses the four-step method. We are going to use a slightly different approach that depends on the assumption that $F'(x)$ exists.

If $F(x) = \dfrac{f(x)}{g(x)}$, then $F(x)g(x) = f(x)$. Now, using the product rule on the second equation, we see that

$$F(x)g'(x) + F'(x)g(x) = f'(x)$$

Solving this for $F'(x)$ we get

$$F'(x) = \frac{f'(x) - F(x)g'(x)}{g(x)}$$

But, $F(x) = \dfrac{f(x)}{g(x)}$, and so

$$F'(x) = \frac{f'(x) - \dfrac{f(x)}{g(x)}g'(x)}{g(x)} = \frac{g(x)f'(x) - f(x)g'(x)}{[g(x)]^2}$$

And so we have the second, and last, rule of this section, the quotient rule.

Quotient Rule for Derivatives

If $F(x) = \dfrac{f(x)}{g(x)}$, $g(x) \neq 0$ and both $f'(x)$ and $g'(x)$ exist, then

$$F(x) = \frac{g(x)f'(x) - f(x)g'(x)}{[g(x)]^2}$$

EXAMPLE 23.15

If $F(x) = \dfrac{x^2 + 1}{3x - 2}$, find $F'(x)$.

Solution   According to the quotient rule,

$$F'(x) = \frac{g(x)f'(x) - f(x)g'(x)}{[g(x)]^2}$$

We will let $f(x) = x^2 + 1$ and $g(x) = 3x - 2$, so $f'(x) = 2x$ and $g'(x) = 3$. Thus,

$$F'(x) = \frac{(3x - 2)(2x) - (x^2 + 1)(3)}{(3x - 2)^2}$$

$$= \frac{(6x^2 - 4x) - (3x^2 + 3)}{(3x - 2)^2}$$

$$= \frac{3x^2 - 4x - 3}{(3x - 2)^2}$$

≡ Note

Note that the derivative of the quotient of two functions is **not** the quotient of the derivatives. Thus, if $F(x) = \dfrac{f(x)}{g(x)}$, then $F'(x) \neq \dfrac{f'(x)}{g'(x)}$. In Example 23.15,

$$\frac{f'(x)}{g'(x)} = \frac{2x}{3} \neq F'(x).$$

EXAMPLE 23.16

If $y = \dfrac{1}{x^3}$, find y'.

Solution   Here y can be considered a quotient with $f(x) = 1$ and $g(x) = x^3$. This means that $f'(x) = 0$ and $g'(x) = 3x^2$. Using the quotient rule, we get

$$y' = \frac{x^3(0) - (1)(3x^2)}{(x^3)^2}$$

$$= \frac{0 - 3x^2}{x^6} = \frac{-3}{x^4}$$

EXAMPLE 23.16 (Cont.)

We could have used rule 3 to solve this by rewriting $y = \dfrac{1}{x^3} = x^{-3}$. Then we would have gotten $y' = -3x^{-4} = \dfrac{-3}{x^4}$.

EXAMPLE 23.17

If $f(x) = \dfrac{4x^2 - 8\sqrt{x}}{x^3 + \sqrt{x^5}}$, find $f'(x)$.

Solution   We begin by writing the radicals with fractional exponents. Thus, we get

$$f(x) = \frac{4x^2 - 8x^{1/2}}{x^3 + x^{5/2}}$$

We now use the quotient rule:

$$
\begin{aligned}
f'(x) &= \frac{(x^3 + x^{5/2})(8x - 4x^{-1/2}) - (4x^2 - 8x^{1/2})(3x^2 + \frac{5}{2}x^{3/2})}{(x^3 + x^{5/2})^2} \\[2mm]
&= \frac{(8x^4 - 4x^{5/2} + 8x^{7/2} - 4x^2) - (12x^4 + 10x^{7/2} - 24x^{5/2} - 20x^2)}{(x^3 + x^{5/2})^2} \\[2mm]
&= \frac{-4x^4 - 2x^{7/2} + 20x^{5/2} + 16x^2}{(x^3 + x^{5/2})^2}
\end{aligned}
$$

Normal Lines

Until now we have concentrated on finding the slope of a line that is tangent to a curve at a particular point. We are also interested in the line that is perpendicular to the curve at a given point. This is called the **normal line**. If the slope of the tangent line is m_1 and the slope of the normal line is m_2, then, since the lines are perpendicular, we know that $m_1 = \dfrac{-1}{m_2}$.

EXAMPLE 23.18

Find an equation of the tangent and normal lines to $y = \dfrac{1}{x^3}$ at the point $\left(2, \frac{1}{8}\right)$.

Solution   In Example 23.16, we found that the derivative of this function is $y' = \dfrac{-3}{x^4}$. The derivative is used to find the slope of a line tangent to y. Evaluating y' at the point $\left(2, \frac{1}{8}\right)$, we find that the line tangent to y at this point will have slope $m = \dfrac{-3}{2^4} = \frac{-3}{16}$. We now know that the slope is $-\frac{3}{16}$ at $\left(2, \frac{1}{8}\right)$, so the equation of the tangent line is

$$y - y_0 = m(x - x_0)$$
$$y - \frac{1}{8} = -\frac{3}{16}(x - 2)$$

EXAMPLE 23.18 (Cont.)

or $\qquad y = -\dfrac{3}{16}x + \dfrac{1}{2}$

The slope of the normal line is $-\dfrac{1}{-3/16} = \frac{16}{3}$. The equation of the normal line to $y = \dfrac{1}{x^3}$ at the point $\left(2, \frac{1}{8}\right)$ is

$$y - \frac{1}{8} = \frac{16}{3}(x - 2)$$

or $\qquad y = \dfrac{16}{3}x - \dfrac{253}{24}$

The next example will require the use of the quotient rule to find the slope of a normal line to a curve.

EXAMPLE 23.19

Determine the equation for the normal line to $f(x) = \dfrac{x^2 + 3x - 1}{x + 4}$ at $(3, 2)$.

Solution We begin by finding the slope of the tangent line at this point. To do this, we find the derivative of f. Using the quotient rule,

$$f'(x) = \frac{(x+4)(2x+3) - (x^2 + 3x - 1)(1)}{(x+4)^2}$$

To find the slope of f at $(3, 2)$, we evaluate f' at $x = 3$. Notice that it is not necessary to simplify the derivative in order to evaluate it.

$$f'(3) = \frac{(7)(9) - (17)(1)}{7^2} = \frac{46}{49}$$

Since the slope of the tangent line is $\frac{46}{49}$, the slope of the normal line is $-\dfrac{1}{46/49} = -\dfrac{49}{46}$. We now know the slope of the normal line to f at $(3, 2)$, and so the equation of that normal line is

$$y - 2 = -\frac{49}{46}(x - 3)$$

or $\qquad y = -\dfrac{49}{46}x + \dfrac{239}{46}$

EXAMPLE 23.20

Find the equation of the tangent and normal lines to the curve

$$y = \frac{(x+1)(x^2 + 2x + 5)}{3 - x}$$

at the point $(1, 8)$.

Solution We have one point on the line, $(1, 8)$. What we need to find is the slope. To do this, we will find the derivative y' and evaluate it at the point $(1, 8)$. While y

EXAMPLE 23.20 (Cont.)

is in the form $\dfrac{f(x)}{g(x)}$, the function f is a product of two functions. We will use the quotient rule, but indicate the derivatives that need to be taken by using the notation D_x. This way we get

$$y' = \frac{(3-x)D_x[(x+1)(x^2+2x+5)] - (x+1)(x^2+2x+5)D_x(3-x)}{(3-x)^2}$$

We use the product rule to find

$$D_x[(x+1)(x^2+2x+5)] = (x+1)(2x+2) + (1)(x^2+2x+5)$$

We also have $D_x(3-x) = -1$, so

$$y' = \frac{(3-x)[(x+1)(2x+2)+(x^2+2x+5)] - (x+1)(x^2+2x+5)(-1)}{(3-x)^2}$$

To find the slope of y at $(1,8)$, we need to evaluate y' when $x = 1$. It is not necessary to simplify the derivative to do this. When $x = 1$,

$$y' = \frac{(2)[(2)(4)+8] - (2)(8)(-1)}{2^2} = \frac{2(16)+16}{4} = \frac{48}{4} = 12$$

We now know that the slope of the tangent line is 12 and a point on the line is $(1,8)$, so an equation of the line is

$$y - 8 = 12(x-1)$$

The slope of the normal line is $\dfrac{-1}{12}$, so an equation of that line is $y-8 = \dfrac{-1}{12}(x-1)$.

Exercise Set 23.3

In Exercises 1–38, differentiate the function.

1. $f(x) = (3x+1)(2x-7)$
2. $g(x) = (6x-2)(5-4x)$
3. $h(x) = (2x^2+x-1)(3x-5)$
4. $k(x) = (4x^3-1)(x^2-7x)$
5. $j(x) = (4-3x)(6x^2+6x-4)$
6. $s(t) = (6-4t)(3t^2-7t)$
7. $q(t) = 4(t^2-3t)^2$
8. $r(p) = (4p-3p^2)(2p-4)$
9. $f(w) = (3w^3-4w^2+2w-5)(w^2-w^{-1})$
10. $g(r) = (2r^4-4r^2+2r)(r^2+2r-1)$
11. $f(x) = \dfrac{4x-1}{2x+3}$
12. $h(y) = (y^3-1)(y^2-1)(y-1)$

13. $g(x) = \dfrac{9x^2+2}{4x-1}$
14. $h(x) = 4x^2 - \dfrac{4}{x^2}$
15. $j(s) = 6s^2 - \dfrac{1}{6s^2}$
16. $K(s) = \dfrac{3-4s}{5s-2s^2}$
17. $f(t) = 4t^3 - \dfrac{2t}{t-2}$
18. $j(x) = \dfrac{1+2x}{1-2x}$
19. $H(x) = \dfrac{x^3-1}{x-1}$

20. $k(x) = \dfrac{x^4+4}{3x}$

21. $f(\phi) = \dfrac{\phi^2}{3\phi^2-1}$

22. $L(t) = \dfrac{t^2+3t+2}{t^2-4t-4}$

23. $h(x) = \dfrac{3x^3-x+1}{x^3-3x-1}$

24. $m(x) = \dfrac{x^2+5x+6}{x^2-5x+6}$

25. $f(t) = \dfrac{3t^2-t-1}{\sqrt[3]{t}}$

26. $F(x) = \dfrac{2x^{3/2}-4x^{1/2}}{4x^{1/2}-2}$

27. $g(w) = 12w^2 + \dfrac{w-1}{w+1}$

28. $j(t) = \dfrac{t^2+1}{2t+1} + \dfrac{t-1}{2t+1}$

29. $h(x) = \dfrac{1}{x^2+1}$

30. $k(x) = \dfrac{5}{x^3-1}$

31. $H(x) = \dfrac{4-x^3}{(2-x^2)(3x-x^3)}$

32. $L(x) = \dfrac{(2x+1)(3x^2-4x)}{7x-1}$

33. $n(s) = \dfrac{2s^3}{(s^2-1)(s-1)}$

34. $m(z) = \dfrac{(z-5)(z^2+7)}{z^2+3z}$

35. $y = \dfrac{(2x-3)(x^2-4x+1)}{3x^3+1}$

36. $y = \dfrac{5x}{5-x}(25-x^2)$

37. $y = \dfrac{(t^2-1)}{(2t+1)} \cdot \dfrac{(t-1)}{(2t+1)}$

38. $y = \left(\dfrac{1-x}{x}\right)(1-x^2)$

Solve Exercises 39 and 40.

39. Find the slope of the curve $y = (3x^2+2x-1) \times (x^3-x+1)$ at $(1,4)$.

40. Find the slope of the curve $y = \dfrac{x^3}{x^2+1}$ at $(-1,-\frac{1}{2})$.

In Exercises 41–46, find the equation of the tangent and normal lines to the curves at the given points.

41. $y = \dfrac{8}{x+1}$ at $(3,2)$

42. $y = \dfrac{4x+1}{5x}$ at $(1,1)$

43. $y = \dfrac{5x+4}{2x+1}$ at $(1,3)$

44. $y = (x^2-3x-6)(2x-6)$ at $(-1,16)$

45. $y = \dfrac{9x+2}{x^2(x+4)}$ at $(-2,-2)$

46. $y = (x^3+5x^2-1)(x^2-2x)$ at $(1,-5)$

Solve Exercises 47–52.

47. *Demography* It is estimated that t years from now the population of a certain city, in thousands of people, will be $P(t) = (0.8t-6)(0.5t+9)+87$. How fast will the population be growing in 5 years?

48. *Business* The profit from the sale of n items of a certain product is given by $P(n) = (4-0.2n)(2.6n+7)$, where $P(n)$ is in hundreds of dollars and $1 \le n \le 20$. If the marginal profit is $P'(n)$, find the marginal profit when $n = 7$.

49. *Electronics* In a voltage divider circuit the output, V, in volts, depends on the value of a variable resistor, R, in ohms, according to the function $V = \dfrac{110R}{R+60}$. Find $V' = \dfrac{dV}{dR}$.

50. *Electronics*   The impedance, Z, in ohms, of a Wein bridge depends on the resistance, R, in ohms, of one arm of the bridge so that $Z = \dfrac{(1+8R)^2 + 8R}{8(1+8R)}$. Find $Z' = \dfrac{dZ}{dR}$.

51. *Medical technology*   It is estimated by a medical research team that the population of a bacterial culture after t hours is approximately $N(t) = \dfrac{t^2 - 2t + 1}{3\sqrt{t} + 2}$,

where $N(t)$ is in thousands and $0 \le t \le 12$. **(a)** Find $N'(t)$. **(b)** Find the rate of growth after 5 hours.

52. *Optics*   For thin lenses the object distance, s, is given by $s = \dfrac{s' f}{s' - f}$, where f is the focal length and s' is the image distance. If f is a constant, find the rate of change of s with respect to s', that is, find $\dfrac{ds}{ds'}$. (Note that in this standard optics formula, s' does not indicate the derivative of s.)

In Your Words

53. Without looking in the text, explain how to determine the derivative of the product of two functions.

54. Without looking in the text, explain how to determine the derivative of the quotient of two functions.

≡ 23.4
DERIVATIVES OF COMPOSITE FUNCTIONS

None of the functions that we have differentiated have involved composite functions. In this section, we will learn how to find the derivative of a function that is the composite of two differentiable functions. The rule for finding these derivatives is called the chain rule.

We will begin by briefly reviewing composite functions. We will then look at a special case of the chain rule, called the power rule.

Suppose that f and g are two functions. The composite function of f and g is $(f \circ g)(x) = f(g(x))$, where the domain of f is the range of g.

EXAMPLE 23.21

(a) If $f(x) = x^2$ and $g(x) = x + 1$, then $(f \circ g)(x) = (x+1)^2$

(b) If $f(x) = 2x^3 + 5x$ and $g(x) = x^4$, then

$$(f \circ g)(x) = 2(x^4)^3 + 5(x^4)$$
$$= 2x^{12} + 5x^4$$

(c) If $g(w) = \sin 2w$ and $h(w) = 3w^2 + 1$, then

$$(g \circ h)(w) = \sin[2(3w^2 + 1)]$$
$$= \sin(6w^2 + 2)$$

It is just as important to be able to recognize a composite function and to decompose it into its individual functions. In this case, you begin with $y = f \circ g$ and determine f and g.

EXAMPLE 23.22

(a) $y = (f \circ g)(x) = (x^2 - 2x + 1)^3$, where $f(x) = x^3$ and $g(x) = x^2 - 2x + 1$

(b) $y = (f \circ g)(x) = (x^5 - 4)^{-2}$, where $f(x) = x^{-2}$ and $g(x) = x^5 - 4$

(c) $y = (f \circ g)(x) = \sqrt{x^3 - 2x}$, where $f(x) = \sqrt{x} = x^{1/2}$ and $g(x) = x^3 - 2x$

(d) $y = (f \circ g)(x) = \ln x^5$, where $f(x) = \ln x$ and $g(x) = x^5$

(e) $y = (f \circ g)(x) = \tan 2x$, where $f(x) = \tan x$ and $g(x) = 2x$

The Power Rule

We will begin our exploration of the chain rule by looking at some special cases. In each of these cases, $f(x) = x^n$ and so $(f \circ g)(x) = [g(x)]^n$. We will use the product rule to look at the derivatives when $n = 2$, 3, and 4. Because it is easier, we will use the $\dfrac{d}{dx}$ notation to indicate what is being differentiated.

$$n = 2 \qquad \frac{d}{dx}[g(x)]^2 = \frac{d}{dx}[g(x)g(x)]$$
$$= g(x)g'(x) + g'(x)g(x)$$
$$= 2g(x)g'(x)$$

$$n = 3 \qquad \frac{d}{dx}[g(x)]^3 = \frac{d}{dx}\{[g(x)]^2 g(x)\}$$
$$= [g(x)]^2 g'(x) + \frac{d}{dx}[g(x)]^2 g(x)$$

From the example when $n = 2$, we know that

$$\frac{d}{dx}[g(x)]^2 = 2g(x)g'(x)$$

and so we get

$$\frac{d}{dx}[g(x)]^3 = [g(x)]^2 g'(x) + 2g(x)g'(x)g(x)$$
$$= 3[g(x)]^2 g'(x)$$

$$n = 4 \qquad \frac{d}{dx}[g(x)]^4 = \frac{d}{dx}\{[g(x)]^3 g(x)\} = [g(x)]^3 g'(x) + \left\{\frac{d}{dx}[g(x)]^3\right\} g(x)$$
$$= [g(x)]^3 g'(x) + 3[g(x)]^2 g'(x)g(x)$$
$$= 4[g(x)]^3 g'(x)$$

Look at these results:

$$\frac{d}{dx}[g(x)]^2 = 2g(x)g'(x)$$
$$\frac{d}{dx}[g(x)]^3 = 3[g(x)]^2 g'(x)$$
$$\frac{d}{dx}[g(x)]^4 = 4[g(x)]^3 g'(x)$$

These results seem to indicate a pattern. It is possible to show that the pattern results in the following rule, called the **general power rule**.

Rule 9: General Power Rule for Derivatives

If g is a function and its derivative g' exists, then

$$\frac{d}{dx}[g(x)]^n = n[g(x)]^{n-1}g'(x).$$

Notice that this is very similar to rule 3, which states that if $f(x) = x^n$, then $f'(x) = nx^{n-1}$. The main difference is the last factor, $g'(x)$. If $g(x) = x$, then $g'(x) = 1$, and rule 9 is the same as rule 3.

EXAMPLE 23.23

Find the derivatives of **(a)** $(x^2 + 1)^3$, **(b)** $(4 - x^3)^5$, and **(c)** $(x^3 - 2x^2 + 1)^{-4}$.

Solutions

(a) If $y = (x^2 + 1)^3$, then this is $[g(x)]^3$, where $g(x) = x^2 + 1$. Since $g'(x) = 2x$ we have

$$y' = 3[g(x)]^2 g'(x) = 3(x^2 + 1)^2(2x)$$

(b) Here $y = (4 - x^3)^5 = [g(x)]^5$, where $g(x) = 4 - x^3$ and $g'(x) = -3x^2$.

$$\begin{aligned} y' &= 5[g(x)]^4 g'(x) \\ &= 5(4 - x^3)^4(-3x^2) \\ &= -15x^2(4 - x^3)^4 \end{aligned}$$

(c) $y = (x^3 - 2x^2 + 1)^{-4} = [g(x)]^{-4}$, where $g(x) = x^3 - 2x^2 + 1$ and $g'(x) = 3x^2 - 4x$.

$$\begin{aligned} y' &= -4[g(x)]^{-5} g'(x) \\ &= -4(x^3 - 2x^2 + 1)^{-5}(3x^2 - 4x) \end{aligned}$$

The general power rule is not always used alone. It can be used with other rules, such as the product rule or the quotient rule.

EXAMPLE 23.24

Find the derivatives of **(a)** $f(x) = 3x(x^3 - 1)^4$ and **(b)** $f(x) = \left(\dfrac{4x + 1}{x^2 - 2}\right)^3$.

Solutions

(a) Here we have a product of $3x$ and $(x^3 - 1)^4$. We will first use the product rule.

$$f'(x) = 3x\left[\frac{d}{dx}(x^3 - 1)^4\right] + (x^3 - 1)^4(3)$$

EXAMPLE 23.24 (Cont.)

Now we use the general power rule to determine that

$$\frac{d}{dx}(x^3-1)^4 = 4(x^3-1)^3(3x^2) = 12x^2(x^3-1)^3$$

Putting this value into the formula for $f'(x)$, we get

$$\begin{aligned} f'(x) &= 3x[12x^2(x^3-1)^3] + (x^3-1)^4(3) \\ &= 36x^3(x^3-1)^3 + 3(x^3-1)^4 \\ &= [36x^3 + 3(x^3-1)](x^3-1)^3 \\ &= (39x^3 - 3)(x^3-1)^3 \end{aligned}$$

(b) Here $g(x) = \dfrac{4x+1}{x^2-2}$ and $f(x) = [g(x)]^3$

Using the general power rule, we see that

$$\begin{aligned} f'(x) &= 3[g(x)]^2 g'(x) \\ &= 3\left(\frac{4x+1}{x^2-2}\right)^2 g'(x) \end{aligned}$$

We will use the quotient rule to get

$$\begin{aligned} g'(x) &= \frac{(x^2-2)(4) - (4x+1)(2x)}{(x^2-2)^2} \\ &= \frac{(4x^2-8) - (8x^2+2x)}{(x^2-2)^2} \\ &= \frac{-4x^2 - 2x - 8}{(x^2-2)^2} \end{aligned}$$

Putting this value for $g'(x)$ into the answer for $f'(x)$, we get

$$\begin{aligned} f'(x) &= 3\left(\frac{4x+1}{x^2-2}\right)^2 \left(\frac{-4x^2-2x-8}{(x^2-2)^2}\right) \\ &= -3\frac{(4x+1)^2(4x^2+2x+8)}{(x^2-2)^4} \end{aligned}$$

≡ **Note**

In Example 23.24(a), after we found the derivative as $f'(x) = 3x[12x^2(x^3-1)^3] + (x^3-1)^4(3)$, we continued until it was factored as $f'(x) = (39x^3-3)(x^3-1)^3$ because the factored form is more useful in applications.

The Chain Rule

The power rule is a special case of the chain rule. Another way of writing the composition function $y = (f \circ g)(x) = f(g(x))$ is

$$y = f(u) \quad \text{where } u = g(x)$$

The major change here is that y is now a function of u and u is a function of x. By substituting, we can get y as a function of x.

Rule 10: Chain Rule

If $y = f(u)$ and $u = g(x)$, and both $f'(u)$ and $g'(x)$ exist, then

$$y' = f'(u)g'(x)$$

or

$$y' = f'[g(x)] \cdot g'(x)$$

≡ **Note**

The chain rule is often written as

$$\frac{dy}{dx} = \frac{dy}{du} \cdot \frac{du}{dx}$$

EXAMPLE 23.25

Find y', if $y = (x^3 - 2x^2 + 1)^{-4}$.

Solution   If $y = (x^3 - 2x^2 + 1)^{-4}$, then we will let $y = f(u) = u^{-4}$ and $u = g(x) = x^3 - 2x^2 + 1$. $f'(u) = -4u^{-5}$ and $g'(x) = 3x^2 - 4x$, so

$$y' = f'(u)g'(u)$$
$$= -4u^{-5}(3x^2 - 4x)$$

Substituting for u, we get

$$y' = -4(x^3 - 2x^2 + 1)^{-5}(3x^2 - 4x)$$

This is the same answer we got when we worked this problem in Example 23.23(c).

EXAMPLE 23.26

Find y', if $y = \sqrt[3]{x^5 - 4x}$.

Solution $y = f(u) = \sqrt[3]{u} = u^{1/3}$ and $u = g(x) = x^5 - 4x$. $f'(u) = \frac{1}{3}u^{-2/3}$ and $g'(x) = 5x^4 - 4$, and so

$$y = f'(u)g'(x)$$
$$= \frac{1}{3}u^{-2/3}(5x^4 - 4)$$
$$= \frac{1}{3}(x^5 - 4x)^{-2/3}(5x^4 - 4) \qquad \text{Substitute for } u.$$

Hint

It is possible to extend the chain rule. For example, suppose $y = f(u)$, $u = g(v)$, and $v = h(x)$. Then the composite function

$$y = (f \circ g \circ h)(x) = f(g(h(x)))$$

has a derivative

$$y' = f'(u)g'(v)h'(x)$$

or
$$= \frac{dy}{du}\frac{du}{dv}\frac{dv}{dx}$$

EXAMPLE 23.27

If $y = \left[6\left(\dfrac{5}{x^2}\right)^4 - 7\right]^{10}$, find y'.

Solution This looks more difficult than it is. First, if we let $v = h(x) = \dfrac{5}{x^2} = 5x^{-2}$, then we have $y = (6v^4 - 7)^{10}$. Now let $u = g(v) = 6v^4 - 7$ and we have $y = f(u) = u^{10}$. So, if $f(u) = u^{10}$, $g(v) = 6v^4 - 7$, and $h(x) = 5x^{-2}$, then

$$y' = f'(u)g'(v)h'(x)$$
$$= 10u^9(24v^3)(-10x^{-3})$$
$$= 10(6v^4 - 7)^9(24v^3)(-10x^{-3}) \qquad \text{Substitute for } u.$$
$$= 10[6(5x^{-2})^4 - 7]^9[24(5x^{-2})^3](-10x^{-3}) \qquad \text{Substitute for } v.$$
$$= 10\left[6\left(\frac{625}{x^8}\right) - 7\right]^9\left(24 \cdot \frac{125}{x^6}\right)\left(\frac{-10}{x^3}\right)$$
$$= -\frac{300,000}{x^9}\left[6\left(\frac{625}{x^8}\right) - 7\right]^9$$

Application

EXAMPLE 23.28

After a sewage spill, the level of pollution in Herd Weyer Bay is estimated by $P(t) = \dfrac{100t + 25}{\sqrt{5t^2 + 10}}$ parts per million (ppm), where t is the time in days since the spill occurred.

(a) What is the rate of change in the level of pollution?

(b) What is the rate of change in the level of pollution after 1.5 days?

(c) What is the rate of change in the level of pollution after 2.5 days?

EXAMPLE 23.28 (Cont.)

Solution

(a) We find the rate of change in the level of pollution by finding the derivative of the level of pollution function $P(t) = \dfrac{100t + 25}{\sqrt{5t^2 + 10}} = \dfrac{100t + 25}{(5t^2 + 10)^{1/2}}$.

$$P'(t) = \frac{(5t^2 + 10)^{1/2}(100) - (100t + 25)\dfrac{1}{2}(5t^2 + 10)^{-1/2}}{\left((5t^2 + 10)^{1/2}\right)^2}$$

$$= \frac{(5t^2 + 10)^{1/2}(100) - \left(50t + \dfrac{25}{2}\right)(5t^2 + 10)^{-1/2}}{5t^2 + 10}$$

$$= \frac{(5t^2 + 10)^{1/2}(100) - \left(50t + \dfrac{25}{2}\right)(5t^2 + 10)^{-1/2}}{5t^2 + 10} \cdot \frac{(5t^2 + 10)^{1/2}}{(5t^2 + 10)^{1/2}}$$

$$= \frac{(5t^2 + 10)(100) - \left(50t + \dfrac{25}{2}\right)}{(5t^2 + 10)^{3/2}}$$

$$= \frac{500t^2 + 1000 - 50t + \dfrac{25}{2}}{(5t^2 + 10)^{3/2}} = \frac{500t^2 - 50t + 1012.5}{(5t^2 + 10)^{3/2}}$$

Thus, the pollution is changing at the rate of $P'(t) = \dfrac{500t^2 - 50t + 1012.5}{(5t^2 + 10)^{3/2}}$ ppm/day.

(b) The rate of change in the level of pollution after 1.5 days is

$$P'(1.5) = \frac{500(1.5)^2 - 50(1.5) + 1012.5}{(5(1.5)^2 + 10)^{3/2}} \approx 21.055 \text{ ppm/day.}$$

(c) The rate of change in the level of pollution after 2.5 days is

$$P'(2.5) = \frac{500(2.5)^2 - 50(2.5) + 1012.5}{(5(2.5)^2 + 10)^{3/2}} \approx 15.145 \text{ ppm/day.}$$

So far, we have had ten rules for derivatives. These are summarized in the following box.

Ten Rules for Derivatives

1. If $f(x) = k$, k a constant, then $f'(x) = 0$.
2. If $f(x) = mx + b$, then $f'(x) = m$.
3. If $f(x) = x^n$, n a rational number, then $f'(x) = nx^{n-1}$.
4. If $F(x) = kf(x)$, k a constant, and $f'(x)$ exists, then $F'(x) = kf'(x)$.
5. If $F(x) = f(x) + g(x)$ and $f'(x)$ and $g'(x)$ exist, then $F'(x) = f'(x) + g'(x)$.
6. If $F(x) = f(x) - g(x)$ and both $f'(x)$ and $g'(x)$ exist, then $F'(x) = f'(x) - g'(x)$.
7. Product rule: If $F(x) = f(x)g(x)$, and both $f'(x)$ and $g'(x)$ exist, then

$$F'(x) = f(x)g'(x) + f'(x)g(x)$$

8. Quotient rule: If $F(x) = f(x)/g(x)$, $g(x) \neq 0$, and both $f'(x)$ and $g'(x)$ exist, then

$$F'(x) = \frac{g(x)f'(x) - f(x)g'(x)}{[g(x)]^2}$$

9. General power rule: If $g'(x)$ exists, then

$$\frac{d}{dx}[g(x)]^n = n[g(x)]^{n-1}g'(x)$$

10. Chain rule: If $y = f(u)$, $u = g(x)$, and both $f'(u)$ and $g'(x)$ exist, then $y' = f'(u)g'(x)$ or

$$\frac{dy}{dx} = \frac{dy}{du} \cdot \frac{du}{dx}$$

Exercise Set 23.4

In Exercises 1–22, find the derivative of the given function by using the general power rule.

1. $f(x) = (3x - 6)^4$
2. $g(x) = (7 - 2x)^3$
3. $h(x) = (5x - 7)^{-5}$
4. $k(x) = (9x + 5)^{-3}$
5. $f(x) = (x^2 + 3x)^4$
6. $h(x) = \sqrt{4x^2 + 7}$
7. $H(x) = \dfrac{1}{\sqrt{4x^2 + 7}}$
8. $g(x) = (x^7 - 9)^6$
9. $s(t) = (t^4 - t^3 + 2)^3$
10. $g(t) = (3t^5 + 2t^3 - t)^{10}$
11. $f(u) = (u^3 + 2u^{-4})^3$

12. $f(u) = (4u^2 - 3u^{-1})^2$
13. $g(x) = \left[(x - 2)(3x^2 - x)\right]^3$
14. $j(t) = 4t^3\sqrt{t^2 - 4}$
15. $h(v) = (v^2 + 1)^2(2v - 5)^3$
16. $g(x) = \left[(x^2 - 4x)(2x^3 - 7)\right]^3$
17. $h(x) = \left[(x^3 - 6x^2)(5x - 6x^2 + x^3)\right]^4$
18. $j(w) = (w^3 + 2w)^4(3w - 5)^{-2}$
19. $k(x) = (3x^2 + 2)^3\sqrt{x^3 - 7x}$
20. $h(x) = \left(\dfrac{3x + 4}{2x^2 - 1}\right)^4$

21. $g(v) = \dfrac{(v^2 - 4v)^3}{(v^3 - 9)^2}$

22. $f(u) = \left(\dfrac{4u^2 + 5}{6u^3 - 3u}\right)^5$

In Exercises 23–42, find the derivative of the given function by using the chain rule.

23. $y = f(u) = u^6, u = g(x) = x^5 + 4$

24. $y = f(u) = u^3, u = g(x) = 2x^3 - 5x$

25. $y = f(u) = \sqrt{u}, u = g(x) = 4x^2 - 5$

26. $y = f(u) = 4u^5, u = g(x) = \sqrt{3x^2 + 5x}$

27. $y = f(u) = \sqrt[3]{u^2}, u = g(x) = 7x^3 - 9x$

28. $y = f(u) = (u+1)^2, u = g(x) = \dfrac{2}{x}$

29. $y = f(u) = (u^2 + 1)^3, u = g(x) = \dfrac{1}{x+3}$

30. $y = f(u) = u^3 - 4u, u = x^4 + 5$

31. $y = f(u) = \sqrt[3]{u^2 - 2u}, u = x^3 + 4$

32. $y = f(u) = 2u^3 - 8u, u = 6x^2 - 5x - 1$

33. $y = g(x) = 4x^4 - 3x, x = 3t^2 - 4t$

34. $y = f(u) = 5u^3 - 7u^2 - 5u + 1, u = 6t^2 - 8t$

35. $y = (9x^2 + 4x)^6$

36. $y = 3(2x^2 - 5x + 1)^{5/3}$

37. $y = (11x^5 - 2x + 1)^{10}$

38. $y = \dfrac{8}{(x^3 - 5x + 2)^4}$

39. $y = \dfrac{7}{(9x^2 - 4)^8}$

40. $y = \left(\sqrt[3]{2x^5 - x^3}\right)^4$

41. $y = \sqrt[4]{(2x^2 - 5)^3}$

42. $y = \dfrac{1}{\sqrt[5]{(7x - 4x^3)^2}}$

Solve Exercises 43–56.

43. Find $\dfrac{dy}{dx}$, if $y = u^4, u = 2v^3 - 1$, and $v = \dfrac{4}{x^2}$.

44. Find $\dfrac{dy}{dt}$, if $y = 4u^2 - u, u = x^3 - 8$, and $x = 6t + 4$.

45. Find $\dfrac{dy}{dx}$, if $y = 3u^2, u = \dfrac{4}{v}$, and $v = x^5$.

46. Find an equation of the tangent line to y at $(4, 1)$, if $y = \sqrt[3]{(x - 5)^2}$.

47. Find an equation of the tangent line to y at $(2, -8)$, if $y = (4x^2 - 18)^3$.

48. Find an equation for the normal line to y at $(-3, 1)$, if $y = \dfrac{4}{\sqrt{25 - x^2}}$.

49. *Energy technology* The energy output of an electrical heater varies with time t according to the equation $E = 6(1 + 4t^2)^3$.
   (a) Find the power, P, in watts, generated by the heater if $P = E' = \dfrac{dE}{dt}$.
   (b) The answer in (a) is in watts. Rewrite your answer so that it is in kilowatts.
   (c) Find the power, in kilowatts, at $t = 2.5$ s

50. *Demography* It is estimated that t years from now the population of a certain city will be $P(t) = 10(45 + 3.5t)^2 - 1750t$.

(a) What is the rate of change in the population?
(b) How fast will the population be growing in 5 years?

51. *Environmental science* After a sewage spill, the level of pollution in San Juan Pedro Bay is estimated by $P(t) = \dfrac{250t^2}{\sqrt{t^2 + 15}}$ parts per million (ppm), where t is the time in days since the spill occurred.
   (a) What is the rate of change in the level of pollution?
   (b) What is the rate of change in the level of pollution after 5.0 days?

52. *Environmental science* It has been estimated that t years from now the level of pollution in the air will be $P(t) = \dfrac{0.6\sqrt{8t^2 + 11t + 60}}{(t+1)^2}$ ppm.
   (a) What is the rate of change in the level of pollution?
   (b) Find the rate of change in the level of pollution in 10.0 years?

53. *Optics* When light passes from air into a transparent medium, the ratio of reflected light to incident light is given approximately by $R = \left(\dfrac{n-1}{n+1}\right)^2$, where n is the index of reflection for the medium. Find the derivative of R with respect to n.

54. *Electronics*   The charge on a capacitor over a short interval of time, t, in seconds, is given by $q = 5t^2 + \sqrt{2 - 5t}$. The formula for the current is the derivative of the formula for the charge, that is, $I = q'$.
(a) What is the formula for the current?
(b) Find the value of I when $t = 0.25\,\text{s}$?

55. *Electronics*   In an RC circuit the current, I, is given by $I = \dfrac{V}{\sqrt{R^2 + X_C^2}}$, where V is the voltage in volts, R is the resistance in ohms, and X_C is the capacitative resistance in ohms. For this exercise, assume that $V = 120\,\text{V}$ and $R = 30\,\Omega$.

(a) What is the rate of change in the current with respect to the capacitative resistance?
(b) Find the value of I' when $X_C = 25\,\Omega$?

56. *Environmental technology*   Some studies have shown that the average level of certain pollutants in the air is given by $L = 1 + 0.25x + 0.001x^2$ ppm when the population is x thousand people. It is estimated that t years from now the population will be $x = \dfrac{200}{\sqrt{9 - 0.4t}}$, in thousands of people.
(a) Find the rate of change in the level of pollutants with respect to the number of years from now.
(b) Find the rate of change in the level of pollutants when $t = 5$ years.
(c) Find the rate of change in the level of pollutants when $t = 10$ years.

In Your Words

57. Explain how to use the chain rule to differentiate the composition of two functions.

58. The general power rule for differentiation is a special case of the chain rule. Describe a function where you could use the chain rule but not the general power rule to find its derivative.

≡ 23.5
IMPLICIT
DIFFERENTIATION

So far, we have considered only functions in the explicit form $y = f(x)$. If the functional relationship between the independent variable x and the dependent variable y is not of this form, they we say that x and y are related implicitly. Examples of x and y being implicitly related are

$$4x - 8y + 6 = 0 \qquad 4x^2 + 9y^2 - 36 = 0 \qquad y \le 0 \qquad xy + 9 = 0$$

Sometimes it is possible to form an explicit function from an expression where x and y are related implicitly. When this can be done, the relationship is solved for y, and a function $y = f(x)$ results. For example, the implicit equation $4x - 8y + 6 = 0$ results in the explicit function $y = \frac{1}{2}x + \frac{3}{4}$. In this section, we will learn how to take derivatives of implicit functions. The result of differentiating an implicit function is called an **implicit derivative** and the process is called **implicit differentiation**.

For many equations it is very difficult, or even impossible, to solve for y in terms of x. In these cases, implicit differentiation is a much easier method of finding the derivative. We will begin by taking the derivative in both the explicit and the implicit form.

We will begin with the implicit function $4x - 8y + 6 = 0$, which we write in the explicit form $y = \frac{1}{2}x + \frac{3}{4}$. The derivative of the explicit form is $y' = \frac{1}{2}$.

In taking an implicit derivative, it is not necessary to solve for y in terms of x. Instead, we will use the general power rule or the chain rule. One important realization is the different way to take the derivative of x^2 and of y^2. At present we know that $\dfrac{dx^2}{dx} = 2x$. But $\dfrac{dy^2}{dx} = 2y\dfrac{dy}{dx}$ because we are differentiating with respect to x. This is an application of the general power rule with $g(x)$ replaced by y. In general, if $y = f(x)$, then

$$\frac{d}{dx}[f(x)]^n = n[f(x)]^{n-1}\frac{d}{dx}[f(x)]$$
$$= ny^{n-1}\frac{dy}{dx}$$
$$= ny^{n-1}y'$$

Let's return to our example, $4x - 8y + 6 = 0$. We will differentiate each term with respect to x and get

$$\frac{d}{dx}(4x) - \frac{d}{dx}(8y) + \frac{d}{dx}(6) = \frac{d}{dx}(0)$$
$$4 - 8\frac{dy}{dx} + 0 = 0$$

Solving this for $\dfrac{dy}{dx}$ we get

$$\frac{dy}{dx} = y' = \frac{-4}{-8} = \frac{1}{2}$$

This is the same answer we got before. Notice that we used the chain rule to differentiate $8y$. Thus $\dfrac{d}{dx}(8y) = \dfrac{d}{dy}(8y)\dfrac{dy}{dx} = 8\dfrac{dy}{dx}$.

In general, use the following four steps in order to differentiate implicitly.

Four Steps for Implicit Differentiation

In the list below, we assume that x is the independent variable and that we are trying to determine dy/dx.

1. Differentiate both sides of the equation with respect to the independent variable, x.
2. Collect the terms with dy/dx on one side of the equation and place the remaining terms on the other side of the equals sign.
3. Factor out dy/dx.
4. Solve for dy/dx.

In the next example, we will again find the derivative using both the explicit and implicit methods.

EXAMPLE 23.29

Find $\dfrac{dy}{dx}$ of $4x^2 + 9y^2 - 36 = 0$.

Solution

Explicit form

This is the equation of an ellipse. If we solve this for y in terms of x, we get

$$y = \pm\sqrt{\frac{36 - 4x^2}{9}} = \pm\frac{1}{3}\sqrt{36 - 4x^2}$$

This can be written as two expressions:

$$y = \frac{1}{3}\sqrt{36 - 4x^2} \text{ and } y = -\frac{1}{3}\sqrt{36 - 4x^2}$$

Each separate expression is now a function. As you can see in Figure 23.6, $y = \frac{1}{3}\sqrt{36 - 4x^2}$ is the upper half of the ellipse and $y = -\frac{1}{3}\sqrt{36 - 4x^2}$ is the lower half.

If you want to find the slope of the tangent at a given point, you need to first determine which half of the ellipse the point is on and then differentiate the appropriate equation. We will differentiate the equation for the upper half:

$$y = \frac{1}{3}\sqrt{36 - 4x^2} = \frac{1}{3}(36 - 4x^2)^{1/2}$$
$$y' = \frac{1}{6}(36 - 4x^2)^{-1/2}(-8x)$$
$$= \frac{-4x}{3\sqrt{36 - 4x^2}}$$

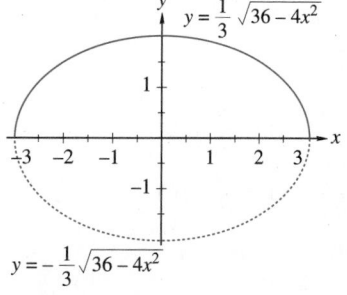

$y = \frac{1}{3}\sqrt{36 - 4x^2}$

$y = -\frac{1}{3}\sqrt{36 - 4x^2}$

FIGURE 23.6

Let's compare this process to the implicit method.

Implicit form

The original equation is $4x^2 + 9y^2 - 36 = 0$. We will differentiate both sides with respect to x.

$$\frac{d}{dx}(4x^2 + 9y^2 - 36) = \frac{d}{dx}(0)$$
$$8x + 18y\frac{dy}{dx} = 0$$
$$8x + 18yy' = 0$$

Solving for y', we get

$$y' = \frac{-8x}{18y} = \frac{-4x}{9y}$$

This is certainly a much simpler equation. But, do we get the same results?

The easiest way to find out is to substitute one of the other equations for y. You will see that when $y = \frac{1}{3}\sqrt{36 - 4x^2}$ is substituted, you get

$$y = \frac{-4x}{3\sqrt{36 - 4x^2}}$$

It works! The results are not only the same, but by using implicit differentiation, we got an easier result to use.

EXAMPLE 23.30

Use implicit differentiation to find the derivative of $9x^2 - 4y^2 = 2x$.

Solution　We differentiate both sides of the equation with respect to x.

$$\frac{d}{dx}(9x^2 - 4y^2) = \frac{d}{dx}(2x)$$

$$18x - 8y\frac{dy}{dx} = 2$$

$$\frac{dy}{dx} = \frac{2 - 18x}{-8y}$$

$$\frac{dy}{dx} = \frac{18x - 2}{8y} = \frac{9x - 1}{4y}, \text{ if } y \neq 0$$

EXAMPLE 23.31

Find $\dfrac{dy}{dx}$, if $xy^2 - x^2y + 4x - 5y = 16$.

Solution

$$\frac{d}{dx}(xy^2 - x^2y + 4x - 5y) = \frac{d}{dx}(16)$$

$$\frac{d}{dx}(xy^2) - \frac{d}{dx}(x^2y) + \frac{d}{dx}(4x) - \frac{d}{dx}(5y) = \frac{d}{dx}(16)$$

We must use the product rule on the first two terms.

$$x\frac{d}{dx}(y^2) + y^2\frac{d}{dx}(x) - x^2\frac{d}{dx}y - y\frac{d}{dx}(x^2) + \frac{d}{dx}(4x) - \frac{d}{dx}(5y) = \frac{d}{dx}(16)$$

$$x\left(2y\frac{dy}{dx}\right) + y^2(1) - x^2\frac{dy}{dx} - y(2x) + 4 - 5\frac{dy}{dx} = 0$$

If $\dfrac{dy}{dx} = y'$, then we get

$$2xyy' + y^2 - x^2y' - 2xy + 4 - 5y' = 0$$

$$(2xy - x^2 - 5)y' = 2xy - y^2 - 4$$

$$y' = \frac{2xy - y^2 - 4}{2xy - x^2 - 5}$$

if the denominator is not 0.

EXAMPLE 23.32

Find y', if $\sqrt[3]{x^2 - y} = x$.

Solution

$$\frac{d}{dx}(x^2 - y)^{1/3} = \frac{d}{dx}(x)$$

$$\frac{1}{3}(x^2 - y)^{-2/3}\left[\frac{d}{dx}(x^2 - y)\right] = \frac{d}{dx}(x)$$

EXAMPLE 23.32 (Cont.)

$$\frac{1}{3}(x^2-y)^{-2/3}\left[2x-\frac{dy}{dx}\right]=1$$

And if $(x^2-y)^{-2/3}\neq 0$,

$$2x-y'=\frac{1}{\frac{1}{3}(x^2-y)^{-2/3}}$$
$$=3(x^2-y)^{2/3}$$
$$y'=2x-3(x^2-y)^{2/3}$$

EXAMPLE 23.33

Find the slope of the tangent line to the graph of $x^3+3xy+12y^2=10$ at the point $(-2,-1)$.

Solution   We first find the derivative, y'.

$$\frac{d}{dx}(x^3+3xy+12y^2)=\frac{d}{dx}(10)$$
$$3x^2+\frac{d}{dx}(3xy)+12\frac{d}{dx}y^2=0$$
$$3x^2+(3xy'+3y)+24yy'=0$$
$$(3x+24y)y'=-3x^2-3y$$
$$y'=\frac{-3x^2-3y}{3x+24y}$$
$$=-\frac{x^2+y}{x+8y}$$

Now, at the point $(-2,-1)$, we see that

$$y'=-\frac{(-2)^2+(-1)}{-2+8(-1)}=-\frac{4-1}{-2-8}=\frac{3}{10}$$

Thus, the slope of the tangent line to the graph of $x^3+3xy+12y^2=10$ at the point $(-2,-1)$ is $\frac{3}{10}$.

In Section 23.3, we learned how to use derivatives to calculate the equation of tangent and normal lines to a curve. When light enters a lens, the light forms an angle with the line perpendicular to the surface of the lens at the light's entry point.

The perpendicular line is **normal** to the surface of the lens. The next example shows how to use implicit differentiation to determine the equation of this normal line.

Application

EXAMPLE 23.34

A profile of a lens is shown in Figure 23.7a. This particular lens is described by the quadratic equation $x^2-xy+y^2=13$. Find the equation of the normal line to this lens at the point $(-1,3)$.

EXAMPLE 23.34 (Cont.)

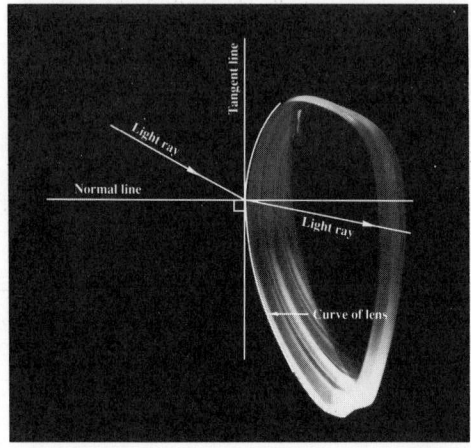

Courtesy of Ruby Gold

FIGURE 23.7a　　　　　　　　　　　**FIGURE 23.7b**

Solution　The sketch in Figure 23.7a shows a lens, the tangent and normal lines, where the normal line is the path of a "typical" light ray. In Figure 23.7b, we have sketched the lens on a coordinate system and shown the tangent and normal lines at the point $(-1, 3)$. We begin by implicitly differentiating the equation that describes the lens.

$$\frac{d}{dx}(x^2 - xy + y^2) = \frac{d}{dx}(13)$$

$$\frac{d}{dx}(x^2) - \frac{d}{dx}(xy) + \frac{d}{dx}y^2 = \frac{d}{dx}(13)$$

Using the product rule on the middle term, we obtain

$$\frac{d}{dx}(x^2) - \left[x\frac{d}{dx}(y) + y\frac{d}{dx}(x)\right] + \frac{d}{dx}y^2 = \frac{d}{dx}(13)$$

$$2x - x\frac{dy}{dx} - y + 2y\frac{dy}{dx} = 0$$

Letting $\frac{dy}{dx} = y'$, produces

$$2x - xy' - y + 2yy' = 0$$

or　　　　　　　　$$(2y - x)y' = y - 2x$$

and so,　　　　　　$$y' = \frac{y - 2x}{2y - x}$$

Now, at the point $(-1, 3)$, we see that

$$y' = \frac{3 - 2(-1)}{2(3) - (-1)} = \frac{3 + 2}{6 + 1} = \frac{5}{7}$$

EXAMPLE 23.34 (Cont.)	Thus, the slope of the tangent line to the lens at $(-1, 3)$ is $\frac{5}{7}$, so the slope of the normal line at this point is $-\frac{7}{5}$. Using the point-slope form for the line, we see that the equation of the normal line is

$$y - 3 = -\frac{7}{5}(x+1)$$

or $7x + 5y = 8$

Exercise Set 23.5

In Exercises 1–34, use implicit differentiation to find $\dfrac{dy}{dx}$.

1. $4x + 5y = 0$

2. $6x - 7y = 3$

3. $x - y^2 = 4$

4. $x^2 + y^2 = 9$

5. $x^2 - y^2 = 16$

6. $x^2 - 2y^2 = 2x$

7. $9x^2 + 16y^2 = 144$

8. $16x^2 - 9y^2 = 144$

9. $4x^2 - y^2 = 4y$

10. $x^2y + xy^2 = x$

11. $xy + 5xy^2 = x^2$

12. $x^2y - 3xy^2 = y^3$

13. $x^2 + 4xy + y^2 = y$

14. $x^3y - 3xy^2 = 2x$

15. $x + 5x^2 - 10y^2 = 3y$

16. $x^2 - 2xy + y^2 = x$

17. $4x^2 + y^3 = 9$

18. $y^4 - 9x^2 = 9$

19. $6y^3 + x^3 = xy$

20. $x^2y = 7y$

21. $x^2y = x + 1$

22. $x^2 + \dfrac{1}{y} = 2x$

23. $\dfrac{1}{x} + \dfrac{1}{y} = 16$

24. $\dfrac{1}{x^2} - \dfrac{1}{y^2} = xy$

25. $\dfrac{3}{x+1} + x^2y = 5$

26. $y^2 = \dfrac{x}{y+1}$

27. $xy + \dfrac{y}{x} = x$

28. $x^3 - 8x^2y^2 + y = 9x$

29. $x^4 - 6x^2y^2 + y^2 = 10x$

30. $(x + 2y)^2 = 4x$

31. $\sqrt{x^2 + y^2} = \dfrac{2y}{x}$

32. $\sqrt{x + xy} = x^2y^2$

33. $(x^2 + y^2)^3 = y$

34. $(x^2y + 4)^2 = 3x$

Solve Exercises 35–44.

35. Find the equation of the line tangent to the circle $x^2 + y^2 = 25$ at the point $(3, -4)$.

36. Find the equation of the line normal to the circle $x^2 + y^2 - 2x - 4y = 20$ at the point $(-2, 6)$.

37. Find the equation of the line tangent to the ellipse $9x^2 + 16y^2 = 100$ at the point $(-2, 2)$.

38. Find the slopes of the lines tangent and normal to the parabola $y^2 = -16x$ at the point $(-1, 4)$.

39. Find the slopes of the lines tangent and normal to the hyperbola $12x^2 - 16y^2 = 192$ at the point $(-8, -6)$.

40. *Physics*   The position of a particle at time t is described by the relation $s^3 - 4st + 2t^3 - 5t = 0$. Find the velocity, ds/dt, of the particle.

41. *Automotive technology*   An automobile's position, s, in miles, on a track at time t, in minutes, is described by the relation $2s^2 + \sqrt{st} - 3t = 0$.

(a) Find $\dfrac{ds}{dt}$, the velocity of the car.

(b) What is the position of the car when $t = 2$?

(c) What is the velocity of the car when $t = 2$?

42. *Economics*   The **Cobb-Douglas production formula**, $P = Cx^ay^{1-a}$, is frequently used by economists to relate the cost of a production process, P, to labor and capital, where C is a constant. Suppose that a firm's level of production is given by $P = 20x^{1/5}y^{4/5}$, where x represents the units of labor and y the units of capital. Currently, the company is using 32 units of labor and 243 units of capital. If labor is increasing by 25% per month, what change in units of capital are needed to maintain the same level of production?

43. *Economics*   The manager of an electronics store has determined that the number of digital satellites and the number of television sets sold weekly are related by the equation $0.7y^2 = 12x + xy$, where x is the number of digital satellites and y is the number of television sets.

(a) Find $\dfrac{dy}{dx}$ when $x = 10$ and $y = 25$.

(b) Interpret your answer.

44. *Business*   The number of pairs of trousers, x, and the number of shirts, y, sold at a clothing store are related by the equation $48x = 12y + 0.01x^2y$.

(a) Find $\dfrac{dy}{dx}$ when $x = 10$ and $y = 40$.

(b) Interpret your answer.

 **In Your Words**

45. Describe how to differentiate an implicit function.

46. Explain how an implicit function differs from an explicit function.

≡ 23.6
HIGHER ORDER DERIVATIVES

The derivative of a function, such as $f(x)$, is also a function, $f'(x)$, called the **derivative function** f'. At present, the only application of the derivative that we have examined is that of the slope of the curve or of a tangent to the curve. In Chapter 24, we will look at more applications of the derivative. As we work these applications there will be times when we need to take the derivative of a derivative.

When it exists, the derivative of the derivative f' is called the **second derivative** and is denoted by f''. It is possible to continue this process: The derivative of the second derivative is the **third derivative**, f''', and its derivative is the **fourth derivative**, $f^{(4)}$. The process can be continued indefinitely, as long as the resulting function has a derivative.

In general, if n is a positive integer, then $f^{(n)}$ represents the nth derivative of f and is found by starting with f and differentiating successively, n times. The integer n is called the **order** of the derivative $f^{(n)}$ and, as a group, these are known as **higher order derivatives**.

As with the derivative, there are many different notations. If $y = f(x)$, then

$$y'' = f''(x) = D_x^2 y = \frac{d^2 y}{dx^2} = \frac{d^2}{dx^2} f(x)$$

all represent second derivatives,

$$y''' = f'''(x) = D_x^3 y = \frac{d^3 y}{dx^3} = \frac{d^3}{dx^3} f(x)$$

all represent third derivatives,

$$y^{(4)} = f^{(4)}(x) = D_x^4 y = \frac{d^4 y}{dx^4} = \frac{d^4}{dx^4} f(x)$$

all represent fourth derivatives, and in general,

$$y^{(n)} = f^{(n)}(x) = D_x^n y = \frac{d^n y}{dx^n} = \frac{d^n}{dx^n} f(x)$$

all represent the nth derivative.

EXAMPLE 23.35

Find the first four derivatives of $f(x) = 2x^5 - 3x$.

Solution $f'(x) = 10x^4 - 3$
$$f''(x) = 40x^3$$
$$f'''(x) = 120x^2$$
$$f^{(4)}(x) = 240x$$

EXAMPLE 23.36

Find the first three derivatives of $f(x) = 4x^3 + \dfrac{2}{x}$.

Solution   We will first rewrite $f(x)$ as $4x^3 + 2x^{-1}$.
$$f'(x) = 12x^2 - 2x^{-2}$$
$$f''(x) = 24x + 4x^{-3}$$
$$f'''(x) = 24 - 12x^{-4}$$

EXAMPLE 23.37

Use implicit differentiation to find y' and y'' of $x^2 + 2xy - y^2 = 25$.

Solution   Using implicit differentiation, we get
$$\frac{d}{dx}(x^2 + 2xy - y^2) = \frac{d}{dx}(25)$$
$$\frac{d}{dx}(x^2) + \frac{d}{dx}(2xy) - \frac{d}{dx}(y^2) = \frac{d}{dx}(25)$$
$$2x + 2y + 2xy' - 2yy' = 0$$
$$y' = -\frac{2x+2y}{2x-2y} = \frac{x+y}{y-x}, \quad \text{if } y \neq x$$

To find y'', we will use the quotient rule:
$$y'' = \frac{d}{dx}\left(\frac{x+y}{y-x}\right)$$
$$= \frac{(y-x)\dfrac{d}{dx}(x+y) - (x+y)\dfrac{d}{dx}(y-x)}{(y-x)^2}$$
$$= \frac{(y-x)(1+y') - (x+y)(y'-1)}{(y-x)^2}$$
$$= \frac{2y - 2xy'}{(y-x)^2}$$

EXAMPLE 23.37 (Cont.)

Substituting $\dfrac{x+y}{y-x}$ for y', we get

$$y'' = \dfrac{2y - 2x\left(\dfrac{x+y}{y-x}\right)}{(y-x)^2}$$

$$= \dfrac{\dfrac{2y(y-x) - 2x(x+y)}{y-x}}{(y-x)^2}$$

$$= \dfrac{2y^2 - 2yx - 2x^2 - 2xy}{(y-x)^3}$$

$$= \dfrac{2y^2 - 4xy - 2x^2}{(y-x)^3}$$

$$= \dfrac{-2(x^2 + 2xy - y^2)}{(y-x)^3}$$

$$= \dfrac{-2(25)}{(y-x)^3} = \dfrac{-50}{(y-x)^3}$$

This concludes our introduction to derivatives. In Chapter 24, we will look at many of the applications of derivatives.

Exercise Set 23.6

In Exercises 1–19, find the indicated higher derivatives of the given function.

1. $y = 4x^3 - 6x^2 + 3x - 10$; find y''.

2. $y = 9x^4 + x^3 - 10$; find y'''.

3. $y = 7x^5 - 3x^3 + x$; find $y^{(4)}$.

4. $f(x) = 10x - 1$; find $f''(x)$.

5. $f(x) = x + \dfrac{1}{x}$; find $f''(x)$.

6. $f(t) = t^3 - \dfrac{1}{t^2}$; find $f'''(t)$.

7. $f(t) = t^2 + \dfrac{1}{t+1}$; find $f''(t)$.

8. $h(x) = (x^2 + 2)(x - 1)$; find $h''(x)$.

9. $g(u) = \dfrac{u-1}{u^2}$; find $g'''(u)$.

10. $y = 3\sqrt[3]{x^2 + 1}$; find $\dfrac{d^2y}{dx^2}$.

11. $y = \sqrt{x+1}$; find $\dfrac{d^2y}{dx^2}$.

12. $y = (4x - x^2)^2$; find $\dfrac{d^2y}{dx^2}$.

13. $y = (2x + 1)^3$; find $\dfrac{d^3y}{dx^3}$.

14. $y = x^4 - 3x^2 + 2$; find $\dfrac{d^3y}{dx^3}$.

15. $y = \dfrac{1}{\sqrt[3]{x^2}}$; find $\dfrac{d^2y}{dx^2}$.

16. $y = \dfrac{x^3}{\sqrt[3]{x}}$; find $\dfrac{d^3y}{dx^3}$.

17. $f(x) = \dfrac{x}{x+1}$; find $D_x^4 f(x)$.

18. $f(t) = \dfrac{t+1}{t+2}$; find $D_t^3 f(t)$.

19. $f(w) = (3w^2 + 1)^{-2}$; find $D_w^2 f(w)$.

In Exercises 20–27, find the indicated derivatives.

20. $\dfrac{d^5}{dx^5}(x^4 - 3x^2 + 1)$

21. $\dfrac{d^4}{dx^4}(x^7 - 4x^5 - x^3)$

22. $\dfrac{d^8}{dx^8}(x^8 - 3x^2 + 2)$

23. $\dfrac{d^3}{dx^3}\left(x - \dfrac{1}{x^2}\right)$

24. $\dfrac{d^2}{dx^2}\left(\dfrac{x^2}{2x + 3}\right)$

25. $\dfrac{d^2}{dx^2}\left(\dfrac{x^3}{3x - 1}\right)$

26. $\dfrac{d^2}{dx^2}\left[(x^2 + 2x)(x - 1)\right]$

27. $\dfrac{d^2}{dx^2}\left(\dfrac{x}{\sqrt{x + 1}}\right)$

In Exercises 28–34, find y' and y'' in terms of x and y.

28. $x^2 + y^2 = 9$

29. $x^2 - 4y^2 = x$

30. $x^2 - y^2 = 16$

31. $xy^2 + 4x = y$

32. $xy^2 - 5x = 4y$

33. $x^2 y = 4$

34. $xy + y^2 = 2x^2$

Solve Exercises 35–38.

35. *Police science* Due to the rapid increase in major crimes, the mayor of a large city plans to organize a major crime task force. It is estimated that for each 1,000 people in the city, the number of major crimes will be $N(t) = 62 + 5t^2 - 0.08t^{5/2}$, where t is the number of months after the task force has been organized and $0 \leq t \leq 12$. **(a)** Find $N'(t)$. **(b)** Find $N''(t)$.

36. *Meteorology* Meteorological results for a certain city indicate that for the month of April, the daily temperature in °F between midnight and 6:00 P.M. can be approximated by $T(t) = -0.04t^3 + 1.16t^2 - 9.3t + 56$, where t is the number of hours after midnight and $0 \leq t \leq 18$. **(a)** Find $T'(t)$. **(b)** Find $T''(t)$.

37. *Meteorology* A weather balloon is released from the ground and rises in the atmosphere. The balloon's height above the ground, in meters, from the moment it is released is given by $h(t) = 120t - 12t^2$, where the time, t, varies from 0 to 6 min. **(a)** Find $h'(t)$. **(b)** Find $h''(t)$.

38. *Automotive* A car accelerates from rest to a maximum velocity at $t = 45$ s and then decelerates to a stop at $t = 120$ s. The distance traveled in meters is given by $s(t) = 3t^2 - \frac{1}{60}t^3$. **(a)** Find $s'(t)$. **(b)** Find $s''(t)$. (Note: $s'(t)$ is the velocity and $s''(t)$ is the acceleration.)

In Your Words

39. If the derivative of a function describes the rate of change of the function, what information does the second derivative give?

40. Explain the difference in the meanings of the notations $f^{(4)}(x)$ and $f^4(x)$.

CHAPTER 23 REVIEW

Important Terms and Concepts

Chain rule
Derivative
  At a point
  Four-step method
  Of a curve
  Of polynomials
Higher order derivatives
Implicit differentiation

Normal line
Power rule
Product rule
Quotient rule
Rules for derivatives
  Chain rule
  Constant function
  Of a composite function

Linear function   Quotient rule
Of a sum or difference   Second derivative
Power rule   Third derivative
Product rule

Review Exercises

In Exercises 1–6, use the four-step method to differentiate the given function.

1. $f(x) = 2x^2 - x$

2. $g(x) = x^3 - 1$

3. $f(x) = x - 2x^3$

4. $k(x) = \dfrac{4}{x^2}$

5. $m(x) = \sqrt{x}$

6. $j(t) = \sqrt{t+4}$

In Exercises 7–32, find the first derivative of the given function.

7. $f(x) = 3x^2 + 2x - 1$

8. $g(x) = 4x^{-5} + 2x$

9. $H(x) = 3x^5 + \dfrac{4}{x^3} + 3\sqrt{x}$

10. $y = \dfrac{1}{x^2 + 1}$

11. $y = \dfrac{3x}{x+3}$

12. $y = (2x+1)(x^2+3)$

13. $y = (x-4)(x^3 - 3x)$

14. $F(x) = \dfrac{\sqrt{x+1}}{x}$

15. $g(x) = \dfrac{x+1}{\sqrt{x+1}}$

16. $G(x) = \dfrac{x^2}{\sqrt{x+1}}$

17. $h(x) = (x^2 + 4)(x - 2)(x^3 - 2)$

18. $f(x) = (3x^2 - 4x)(5x^3 + 2x - 1)$

19. $F(x) = \sqrt{(x^3 + 7)(x+5)}$

20. $g(x) = (x+5)^3$

21. $y = (x^2 + 5x)^2 (x^3 + 1)^{-2}$

22. $h(x) = (x^2 + 7x - 1)^4$

23. $y = (x+1)^2 (x-1)^3$

24. $y = (x+4)^3 \sqrt{x-1}$

25. $y = \dfrac{5}{(x^2 + 1)^3}$

26. $y = \dfrac{4x}{(3x+4)^2}$

27. $y = \dfrac{x^2}{\sqrt{(x^2 - 2)^3}}$

28. $f(t) = \dfrac{4t^3 - 3}{t^2 - 2}$

29. $g(u) = (u^2 + \dfrac{1}{u} - \dfrac{4}{u^2})^{-4}$

30. $R(t) = \sqrt{t^2 - 2t + 1}\sqrt[3]{t^2 + 1}$

31. If $y = u^3 - u$, $u = x^2 + 1$, find $\dfrac{dy}{dx}$.

32. If $y = (u+4)^3$, $u = 2t - 5$, find $\dfrac{dy}{dt}$.

In Exercises 33–38, find the implicit derivatives.

33. $x^2 + 4xy + 4y^2 = 16$

34. $x^3 - 2xy = y^2$

35. $2xy - x^2 = y^2 x$

36. $x^2 y + yx^3 = 10$

37. $x^2 y^3 + \dfrac{1}{y} = x$

38. $4x^2 + xy^2 + y^{-1} = 2x$

In Exercises 39–49, find the indicated derivative.

39. $f(x) = x^7 + 2x^4 + 3x$; find $f''(x)$.

40. $h(x) = \sqrt{x} - \dfrac{1}{\sqrt{x}}$; find $h'''(x)$.

41. $g(x) = 4x^3 - 5x + \dfrac{2}{x}$; find $g'''(x)$.

42. $y = 4x^5 - 2x^{-3}$; find $\dfrac{d^2 y}{dx^2}$.

43. $y = 7x^3 + \dfrac{7}{x^3}$; find $\dfrac{d^3 y}{dx^3}$.

44. $y = x^{-3} + 4x^{-1}$; find $\dfrac{d^3 y}{dx^3}$.

45. $f(x) = 5x^3 + \sqrt{x}$; find $D_x^3 f(x)$.

46. $g(x) = (x^2 + 1)^3$; find $\dfrac{d^3}{dx^3} g(x)$.

47. $x^2 y = y + 2x$; find y''.

48. $y^2 = xy^2 + x^3$; find y'.

49. $xy = x^2 + y^2$; find y' and y''.

Solve Exercises 50–60.

50. Find the slopes of the tangent and the normal to the curve $y = 4x^2 - 9x$ at $(3, 9)$.

51. Find the slope of the tangent to the curve $f(x) = 7x^3 - 3x + 6$ at $(-1, 2)$

52. Find the slope of the tangent to the ellipse $4x^2 + 9y^2 = 40$ at $(-1, 2)$

53. Find the slopes of the tangent and the normal to the curve $4x^2 + 3y^2 - 5xy - 4x + 10y - 11 = 0$ at $(2, -1)$.

54. If $f(x) = x^3 - 3x - 1$, find all values of x for which $f'(x) = 0$.

55. If $g(x) = \dfrac{x^2}{x+1}$, find all values of x for which $g'(x) = 0$.

56. Find the equation of the tangent to the curve $y = x^2 + 4$ at $(-1, 5)$.

57. Find the equation of the normal to the ellipse $4x^2 + 5y^2 = 36$ at $(-2, 2)$.

58. If $y = \sqrt{x+4}$, find all values of x for which $y' = 0$.

59. Find all values of x where $f''(x) = 0$ if $f(x) = 3x^4 - 6x^2 + 2$.

60. If $y = x^3 - 9x^2 - 4$, find the value(s) of x for which $y'' = 0$.

☰ CHAPTER 23 TEST

In Exercises 1–8, find the first derivative of the given functions.

1. $f(x) = 5x^3 - 4$

2. $g(x) = 3x^{-4} + 5x^2$

3. $h(x) = \sqrt[3]{x^6 - 2}$

4. $j(x) = (2x+1)(3x^2 - x)$

5. $k(x) = \dfrac{2x+1}{\sqrt{4x^2+1}}$

Solve Exercises 6–10.

6. Find the second and third derivatives of $f(x) = 5x^4 - 4x^{-3}$.

7. Find y', if $2xy + y^3 x = x^3$.

8. Find the equation of the tangent to the curve $g(x) = 4x^3 - 2x^2 + 4$ at $(1, 6)$.

9. A particle moving in a straight line is at a distance of $s(t) = 4.5t^2 + 27t$ ft from its starting point after

t s. If the velocity $v(t)$ of the particle at time, t, is $v(t) = s'(t)$, find the velocity of this particle.

10. The population of a certain bacterial culture after t hours is approximated by $P(t) = \dfrac{t^2 - 5t}{2\sqrt{t} + 7}$, where $P(t)$ is in thousands of bacteria. **(a)** Find an expression for the rate of growth of these bacteria. **(b)** Determine the rate of growth after 5 hours.

24

Applications of Derivatives

A telephone company is asked to provide service to the house shown here. In Section 24.4, you will learn where to connect this line so that it will cost the least.

Courtesy of Michael A. Gallitelli, Metroland Photo Inc.

In Chapter 23, we learned how to take derivatives of many functions. In this chapter, we will look at some of the applications of derivatives. We will begin by looking at motion applications. We will then look at ways to determine maxima and minima of functions. Finally, we will learn two tests that use derivatives.

≡ 24.1
RATES OF CHANGE

One idea that is important to the study of mathematics is **change**. As a technician, one of your responsibilities may be to observe something and measure it at certain time intervals. When these observations are completed, it will be necessary to see if any changes resulted and how those changes occurred. One of the powerful aspects of calculus is that it can be used to determine the rate at which something changes.

Velocity, Speed, and Accéleration ▬▬▬▬▬

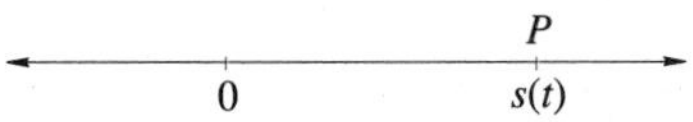

Suppose the distance x of an object p from the origin at any time t is given by $s(t)$. It is often convenient to consider **rectilinear motion** where the object or particle moves along a straight line. Usually we use a horizontal axis to represent rectilinear motion and select the origin as the initial position of the particle. Thus, $x = s(t)$ represents the distance of the object p from the origin 0 at t seconds. (See Figure 24.1.) Because $x = s(t)$ gives the position of p at time t, it is often referred to as a **position function**.

In order to measure the velocity of p at some given time c, we look at the average velocity of p from c to $c+h$. Now, the average velocity is the distance traveled divided by the time it took to go that distance. In this case, the average velocity.

FIGURE 24.1

$$\bar{v} = \frac{s(c+h) - s(c)}{h}$$

As you can see, if we take the limit as h approaches 0, we will get the **instantaneous velocity** of p at $t = c$ or $v(c)$. Thus,

$$v(c) = \lim_{h \to 0} \frac{s(c+h) - s(c)}{h}$$

But, the expression on the right-hand side is the derivative of s and so

$$v(c) = s'(c)$$

Thus, the **velocity function** for p is the derivative of the position function s. In other words,

$$v(t) = s'(t) = D_t s = \frac{ds}{dt}$$

The units of s and t determine the units for the velocity. If s is in miles and t in hours, the velocity is in miles per hour (mph). If s is in meters or feet and t is in seconds, the velocity is in meters per second (m/s) or feet per second (ft/s).

We should probably point out that the velocity of an object is not the same as the speed of an object.

≡ **Note**   **Speed** indicates the magnitude of the velocity and is defined as the absolute value of velocity, $|v(t)|$. Thus, speed is always nonnegative. Velocity can be positive, negative, or zero.

If the velocity is positive, the object is moving in a positive direction. If it is negative, the object is moving in the opposite direction. Thus, speed tells us how fast the object is moving while velocity tells us both the speed and the direction.

Now, if we differentiate the velocity, we get $v'(t) = s''(t)$. This measures the instantaneous rate of change of the velocity and is known as the **acceleration**. If we let $a(t)$ denote the acceleration, then

$$a(t) = v'(t) = s''(t) = \frac{d^2s}{dt^2}$$

The units for the acceleration are cm/s^2 (centimeters per second per second), m/s^2 (meters per second per second), mi/h^2 (miles per hour per hour), and so on.

EXAMPLE 24.1

The position function s of a particle p on a coordinate line is given by

$$s(t) = t^3 - 9t^2 + 24t - 10$$

where t is in seconds and $s(t)$ in millimeters. Where is the particle when $t = 3$ and what is its velocity and acceleration at that time?

Solution   The position is given by

$$\begin{aligned} s(3) &= 3^3 - 9(3^2) + 24(3) - 10 \\ &= 27 - 81 + 72 - 10 \\ &= 8 \text{ mm or 8 mm to the right of the origin} \end{aligned}$$

The first derivative of s, $s'(t)$, is the velocity function

$$v(t) = s'(t) = 3t^2 - 18t + 24$$

and when $t = 3$,

$$\begin{aligned} v(t) &= 3(3^2) - 18(3) + 24 \\ &= 27 - 54 + 24 = -3 \text{ mm/s} \end{aligned}$$

The object is moving with a speed of 3 mm/s in a negative direction. The acceleration function is the second derivative of the position function, and so

$$a(t) = s''(t) = 6t - 18$$

and when $t = 3$,

$$a(3) = 6(3) - 18 = 0 \text{ mm/s}^2$$

The following table shows the values for the position, velocity, and acceleration of the particle in Example 24.1, each second from $t = 0$ through $t = 6$.

t	0	1	2	3	4	5	6
s(t)	−10	6	10	8	6	10	26
v(t)	24	9	0	−3	0	9	24
a(t)	−18	−12	−6	0	6	12	18

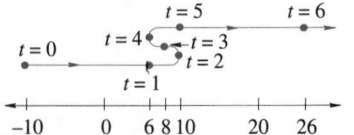

FIGURE 24.2

The motion of an object such as the particle in Example 24.1 can often be represented by a drawing like that shown in Figure 24.2. This curve does not show the path of the particle, but only shows how the particle moves. The direction of the motion is indicated by the arrows.

Not all motion is horizontal. The motion of a projectile fired straight upward is another example of rectilinear motion. In this instance, the axis is vertical rather than horizontal, the origin is at ground level, and the positive direction is upward.

Application

EXAMPLE 24.2

A projectile is fired straight upward with a velocity of 192 ft/s. Its distance above the ground at t seconds is given by $s(t) = -16t^2 + 192t$. Find the time when the projectile hits the ground. What is its velocity and acceleration at the time of impact?

Solution We are given the distance function as $s(t) = -16t^2 + 192t$. When $s(t) = 0$, the projectile will be at ground level, so we need to solve $s(t) = -16t^2 + 192t = 0$. We can factor t out of the expression to obtain

$$t(-16t + 192) = 0$$

and so $s(t) = 0$ when $t = 0$ or when $t = 12$. Since $t = 0$ when the projectile is fired, it must strike the ground when $t = 12$.

The velocity is represented by $v(t) = s'(t) = -32t + 192$ and the acceleration by $a(t) = s''(t) = v'(t) = -32$. When $t = 12$, $v(12) = -192$ and $a(t) = -32$. So, the velocity is 192 ft/s downward and the acceleration is 32 ft/s^2 downward.

Electricity

Time is not the only variable that can be used to study rates of change. In the next example, we will look at the rate at which the current changes with respect to the resistance.

Application

EXAMPLE 24.3

The current I in a certain electrical circuit is given by the formula $I = \dfrac{120}{R}$, where R represents the resistance. What is the rate of change of I with respect to R when the resistance is 60 Ω?

Solution We are given $I = \dfrac{120}{R}$ and want to find $\dfrac{dI}{dR}$ when $R = 60$.

$$I = \frac{120}{R} = 120R^{-1}$$
$$\frac{dI}{dR} = -120R^{-2}$$
$$= \frac{-120}{R^2}$$

So, when $R = 60$, $\dfrac{dI}{dR} = \dfrac{-120}{60^2} = \frac{-1}{30}$. Thus, when $R = 60\,\Omega$, the current is decreasing at a rate of $\frac{1}{30}$ A/Ω (amperes per ohm).

Exercise Set 24.1

just find the accel. & the velocity.

In Exercises 1–8, position functions of points moving rectilinearly are defined. Find the functions for the velocity and acceleration at time t. Make a table of values for each of the three functions at each second. (You might want to write a computer program to help you.) Illustrate the motion by means of a diagram like the one shown in Figure 24.2.

1. $s(t) = 3t^2 - 12t + 5$ from $t = 0$ to $t = 5$
2. $s(t) = 3t^2 - 18t + 10$ from $t = 0$ to $t = 5$
3. $s(t) = t^3 - 12t + 2$ from $t = -4$ to $t = 4$
4. $s(t) = t^3 - 27t + 20$ from $t = -4$ to $t = 4$
5. $s(t) = t + 4/t$ from $t = 1$ to $t = 4$

6. $s(t) = t + 8/t$ from $t = 1$ to $t = 4$
7. $s(t) = 2\sqrt{t} + \dfrac{1}{\sqrt{t}}$ from $t = 1$ to $t = 4$
8. $s(t) = t^3 - 8$ from $t = 0$ to $t = 4$

In Exercises 9–12, an object is fired straight upward. Its height is represented by the given formula. (a) When will it hit the ground? (b) What are its velocity and acceleration when it hits the ground? (c) What was its height when it was fired?

9. $s(t) = 144t - 16t^2$ ft
10. $s(t) = 250 + 256t - 16t^2$ ft

11. $s(t) = 29.4t - 4.9t^2$ m
12. $s(t) = 180 + 98t - 4.9t^2$ m

Solve Exercises 13–28.

13. *Electronics*   The repulsion F, in dynes, between a certain pair of electrical charges varies with the distance apart, s, in centimeters.
    (a) If $F(s) = \dfrac{50}{s^2}$, what is the rate of change of F with respect to s?
    (b) What is the rate of change when $s = 3$ cm?

14. *Electronics*   The current at a given point in a circuit i, in amperes (A), is defined as the instantaneous rate of change of the charge q, in coulombs (C), at any given time t, in seconds (s). The expression for the charge is given by
    $$q = 25 + 10t - 2.0t^2$$
    (a) What is the general expression for the current?
    (b) What is the initial charge; that is, what is the charge when $t = 0.0$ s?
    (c) What is the current at $t = 0.0$ s and $t = 4.0$ s?

15. *Physics*   If the relationship between F, the temperature in degrees Fahrenheit (°F), and C, the temperature in degrees Celsius (°C), is given by $F = \frac{9}{5}C + 32$, find (a) $\dfrac{dF}{dC}$, (b) $\dfrac{dC}{dF}$, and (c) $\dfrac{dC}{dF}$ at $C = 20$°C.

16. *Thermodynamics*   When metal is heated it expands. The area of a circle is related to its radius.

    (a) If a circular metal plate is heated, what is the rate of change of the area with respect to the radius?
    (b) What is the rate when $r = 1.5$ cm?
    (c) What is the rate when $r = 3$ cm?

17. *Hydrology*   As water leaks out of a tank, the quantity Q of water in gallons remaining in the tank after t min is given by the formula $Q(t) = 900 - 50t + 0.5t^2$.
    (a) What is the rate of change in the quantity of water?
    (b) What is the rate of change when $t = 4$ min?
    (c) What is the rate of change when $t = 8$ min?

18. *Physics*   An object rolls down a ramp. The distance in millimeters it rolls in t seconds is given by the formula $s(t) = 5t^2 + 2t$.
    (a) What is the formula for its velocity?
    (b) What is the formula for its acceleration?
    (c) What distance has it rolled after 3 s, and what are its velocity and acceleration at that time?
    (d) At what time is its velocity 46 mm/s?

19. *Physics*   The position of a ball that is projected upward on a frictionless inclined plane is given by $s(t) = 8.0t - 2.0t^2$, where $s(t)$ is in meters (m) and t is time in seconds (s).
    (a) What is the initial velocity?

(b) How far will the ball move up the inclined plane before it stops and starts to roll back down the plane?

(c) How long will it take for the ball to roll up the plane and return to the initial position?

20. Find the rate of change of the area A of a circle with respect to its circumference C.

21. *Electronics* The induced voltage in a coil v, in volts (V), is given by $v = -N\dfrac{d\phi}{dt}$, where N is the number of turns in the coil, ϕ is the magnetic flux in webers (Wb), and t is time in seconds (s). If a coil of 50 turns is connected by a magnetic flux of $\phi = 2.4t^{1/2} - 0.4t^2$, find

   (a) the general expression for the induced voltage and

   (b) the induced voltage at $t = 1.0$ s and $t = 9.0$ s.

22. The height of a cone h is twice the radius r of the base of the cone.

   (a) Find the instantaneous rate of change of the volume V with respect to the radius.

   (b) What is $\dfrac{dV}{dr}$ when $r = 24.3$ mm?

23. *Hydrology* A stone dropped in a pond makes a circular ripple that travels out from the point of impact at 1.7 m/s. At what rate is the area of the circle increasing when $t = 8$ s?

24. *Electricity* The current i, in amperes (A), in a resistance of $r\,\Omega$ when the voltage V, in volts (V), is given by $i = \dfrac{V}{r}$. The resistance changes with time t in seconds (s) according to $r = 3.0t^{1/2}$ and the voltage changes according to $V = t^2 - 1.2$.

   (a) Give an expression for i in terms of t.

   (b) Find $\dfrac{di}{dt}$.

   (c) Find $\dfrac{di}{dt}$ at $t = 2.4$ s.

In Your Words

29. Describe how the position, velocity, and acceleration of an object are related.

25. *Physics* The pressure of a gas P in kilopascals (kPa) varies inversely with the volume V in cubic meters. If the pressure of the gas is $P = 5.21$ kPa when the volume is $1.23\,\text{m}^3$,

   (a) state the equation for P, and

   (b) find the instantaneous rate of change of the pressure with respect to the volume.

26. *Physics* The angular displacement of a rotating wheel θ in radian measure (rad) is given by $\theta(t) = 3.4t - 1.2t^2$, where t is time in seconds (s).

   (a) Find the angular velocity, $\omega(t)$.

   (b) Find the angular acceleration, $\alpha(t)$.

   (c) What is the initial angular velocity and acceleration?

27. *Automotive* A car accelerates from rest to a maximum velocity at $t = 45$ s and then decelerates to a stop at $t = 120$ s. The distance traveled in meters is given by $s(t) = 3t^2 - \frac{1}{60}t^3$.

   (a) What is the velocity as a function of time t of the car during this period?

   (b) What is the acceleration as a function of time t of the car during this period?

   (c) What is the speed of the car during this period?

28. *Meteorology* A weather balloon is released from the ground and rises in the atmosphere. The balloon's height above the ground, in meters, for the first six minutes after it is released is given by $h(t) = 120t - 12t^2$.

   (a) What is the velocity of the balloon during this period?

   (b) What is the acceleration of the balloon during this period?

30. What is the relationship between the speed and the velocity of an object?

24.2
EXTREMA AND THE FIRST DERIVATIVE TEST

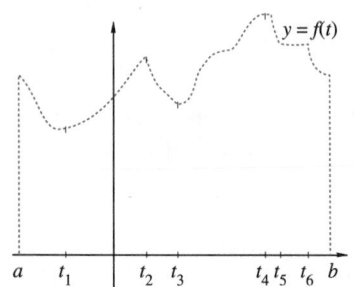

FIGURE 24.3

In Section 24.1, we looked at several applications of derivatives, particularly as they apply to the distance function. We were unable to determine when the distance, velocity, or acceleration is the greatest or least, or when the distance or speed stops increasing and begins decreasing. In this section, we will look at methods for using calculus to determine this type of information.

We will begin by examining the graph of a function f shown in Figure 24.3. This graph might have resulted when a recording instrument was used to measure something over a period of time. Several times have been marked because they are of interest. For example, the graph decreased in the time interval $[a, t_1]$ and increased in the interval $[t_1, t_2]$, decreased in $[t_2, t_3]$, and so on.

You can also see that, in the time period immediately surrounding t_1, the function is at a low point at t_1. Similarly, you can see that the function reaches a high point at t_2, another low point at t_3, and another high point at t_4. Because t_1 represents a place where the graph is at its lowest in an interval around the point, it is referred to as a **local minimum** or **relative minimum**. Another local minimum is at t_3. Similarly, t_2 and t_4 each represent a **local maximum**, because each one is a place where the graph is at its highest over some interval. Now let's look at all of these ideas more closely.

Increasing and Decreasing Functions

The first thing we noticed in Figure 24.3 was that over, some intervals the function increased, over some it decreased, and on one interval it neither increased nor decreased.

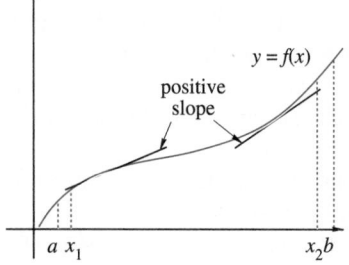

FIGURE 24.4

Look at the graph of $y = f(x)$ in Figure 24.4. As the values of x increase (going from left to right) on the interval $[a, b]$, the corresponding values of $f(x)$ also increase and the curve rises. More formally we would state the following.

> **Increasing Function**
>
> If x_1 and x_2 are any two points in the interval $[a, b]$ with $x_1 < x_2$ and if $f(x_1) < f(x_2)$, then f is an **increasing function** on the interval.

As you can see, the curve has a positive slope at every point in this interval and so $f'(x) > 0$ for all values of x in $[a, b]$.

In a similar manner, we should look at the function in Figure 24.5. The graph shows a decreasing function. In particular, as the values of x increase on the interval $[c, d]$, the corresponding values of f decrease and the curve falls.

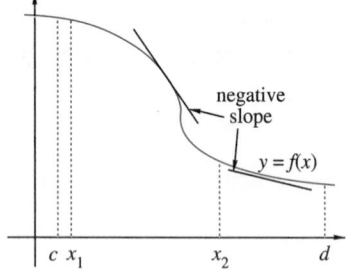

FIGURE 24.5

> **Decreasing Function**
>
> If x_1 and x_2 are any two points in the interval $[a, b]$ with $x_1 < x_2$ and if $f(x_1) > f(x_2)$, then f is a **decreasing function** on the interval.

The tangent lines of a decreasing function all have a negative slope and so $f'(x) < 0$ for all points in the interval $[c, d]$.

This leads us to an important rule dealing with increasing and decreasing functions.

Rule 1: Test to See if a Function is Increasing or Decreasing

If $f'(x) < 0$ on an interval, then f is a decreasing function on that interval.

If $f'(x) > 0$ on an interval, then f is an increasing function on that interval.

If $f'(x) = 0$ on an interval, then f is a constant function on that interval.

Determining the intervals when a derivative is positive or negative will often require your ability to solve inequalities. (See Chapter 16.) The next two examples will show how to determine when a function is increasing and when it is decreasing.

EXAMPLE 24.4

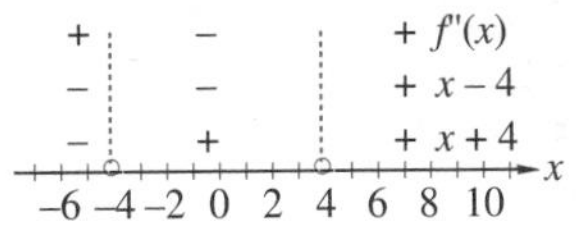

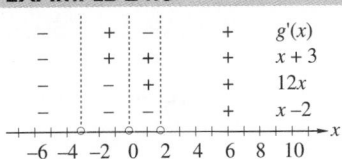

FIGURE 24.6

Find the intervals on which $f(x) = \frac{4}{3}x^3 - 64x$ is increasing or decreasing.

Solution   Rule 1 says that we can determine when f is decreasing or increasing by looking at f'. The derivative of f is

$$f'(x) = 4x^2 - 64$$
$$= 4(x^2 - 16)$$
$$= 4(x+4)(x-4)$$

The sign chart in Figure 24.6 shows where $f'(x)$ is positive and where it is negative. Since $f'(x)$ is negative on the interval $(-4, 4)$, we can say the f decreases on $(-4, 4)$. Similarly, f increases over the intervals $(-\infty, -4)$ and $(4, \infty)$.

EXAMPLE 24.5

−	+	−	+	$g'(x)$
−	+	+	+	$x+3$
−	−	+	+	$12x$
−	−	−	+	$x-2$

$$-6 \quad -4 \quad -2 \quad 0 \quad 2 \quad 4 \quad 6 \quad 8 \quad 10 \qquad x$$

FIGURE 24.7

Find the intervals on which $g(x) = 3x^4 + 4x^3 - 36x^2$ is increasing or decreasing.

Solution   The derivative of $g(x)$ is

$$g'(x) = 12x^3 + 12x^2 - 72x$$
$$= 12(x^3 + x^2 - 6x)$$
$$= 12x(x^2 + x - 6)$$
$$= 12x(x+3)(x-2)$$

From Figure 24.7 we can see that $g'(x) < 0$ on the intervals $(-\infty, -3)$ and $(0, 2)$, so $g(x)$ is decreasing on those intervals. Similarly, since $g'(x) > 0$ on $(-3, 0)$ and $(2, \infty)$, $g(x)$ is increasing on these two intervals.

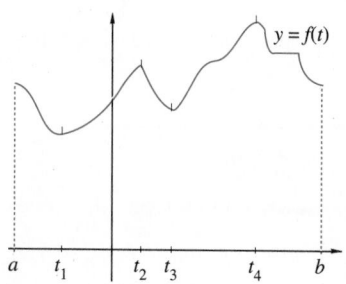

FIGURE 24.8

Extrema and Critical Points

Let's look again at the graph that was used to begin this section. The graph is repeated in Figure 24.8. At t_2 and t_4 we see that the function is higher than any other "nearby" points on the curve. The function is lower than any other nearby points at t_1 and t_3. Since these may not be the highest (or lowest) points on the entire curve, we say that f has a local maximum or a relative maximum at t_2 and t_4. If we can show that the curve does have a highest point then it is called the **absolute maximum**. Just as we described a function as increasing or decreasing over an interval, we say that a function has a local maximum at a point if the value of the function at that point is the highest value in the interval. When finding the maxima or minima over an interval you must always check the endpoints of the interval.

As you look at the local maxima and minima in Figure 24.8, there is something else you should notice. To the left of each relative minimum the function decreases; to the right, the function increases. Just the opposite is true for relative maxima.

EXAMPLE 24.6

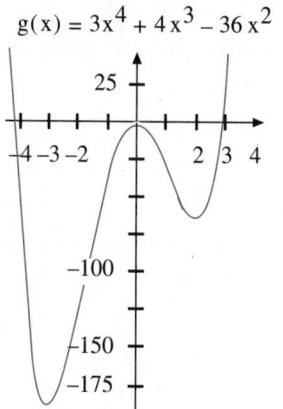

FIGURE 24.9

Determine the relative maxima and minima of

$$g(x) = 3x^4 + 4x^3 - 36x^2$$

Solution This is the same function we examined in Example 24.5. We found that on

$(-\infty, -3)$, g is decreasing.
$(-3, 0)$, g is increasing.
$(0, 2)$, g is decreasing.
$(2, \infty)$, g is increasing.

Thus, we have a minimum when $x = -3$, a maximum when $x = 0$, and a minimum when $x = 2$.

The graph of g, as shown in Figure 24.9, supports the conclusions we reached using calculus.

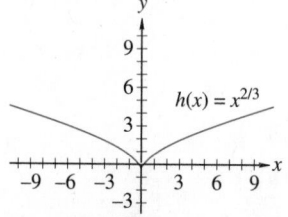

FIGURE 24.10

In the previous example, we began a practice we will follow from now on. We will let maximum refer to either an absolute or a relative maximum. We will also use minimum for either an absolute or relative minimum. Collectively, maximum and minimum values are referred to as **extreme values** or **extrema**. (A single maximum or minimum is called an **extremum**.)

Consider the function $h(x) = x^{2/3}$. Its graph is shown in Figure 24.10. Looking at the graph, we can see that $h(x)$ has a minimum when $x = 0$. But,

$$h'(x) = \frac{2}{3}x^{-1/3} = \frac{2}{3x^{1/3}}$$

When $x = 0$, $h'(x)$ is not defined.

The example above leads to a definition of critical values (and critical points) for a function.

Critical Values and Critical Points

Extrema may occur at values of x in the domain of a function $y = f(x)$ for which $f'(x) = 0$ or $f'(x)$ is not defined. These values of x are referred to as **critical values**. If x is a critical value, then the point $(x, f(x))$ is called a **critical point**.

First Derivative Test

We are now ready to talk about a method for testing critical values to see if they are extrema. This method is called the **first derivative test** for $f(x)$ and it has four steps.

First Derivative Test

1. Find $f'(x)$.
2. Determine the critical values.
3. Use the critical values to determine intervals when f is decreasing $(f'(x) < 0)$ or increasing $(f'(x) > 0)$.
4. At each critical value c, determine if $f'(x)$ changes sign as x increases through c. To do this, select a value of x less than c and one greater than c. If, as x increases,
   (a) $f'(x)$ changes from + to −, then $f(c)$ is a maximum;
   (b) $f'(x)$ changes from − to +, then $f(c)$ is a minimum;
   (c) the sign of $f'(x)$ does not change, then $f(c)$ is neither a maximum nor a minimum.

EXAMPLE 24.7

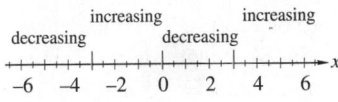

FIGURE 24.11

If $f(x) = x^4 - 18x^2$, use the first derivative test to find the intervals where f is increasing or decreasing and to locate all extrema.

Solution

1. $f'(x) = 4x^3 - 36x = 4x(x+3)(x-3)$
2. Letting $f'(x) = 0$ gives the critical values when $x = 0, -3$, and 3.
3. There are four intervals to consider. Using the same procedures we used in Examples 24.4 and 24.5, we find that on

$$(-\infty, -3), \ f'(x) < 0, \text{ so } f \text{ is decreasing}$$
$$(-3, 0), \ f'(x) > 0, \text{ so } f \text{ is increasing}$$
$$(0, 3), \ f'(x) < 0, \text{ so } f \text{ is decreasing}$$
$$(3, \infty), \ f'(x) > 0, \text{ so } f \text{ is increasing}$$

This is summarized in Figure 24.11.

EXAMPLE 24.7 (Cont.)

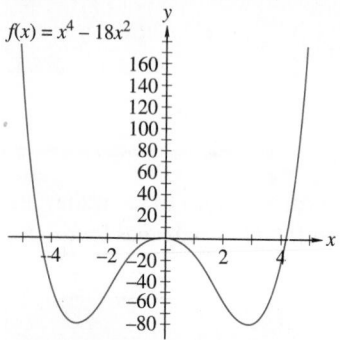

$f(x) = x^4 - 18x^2$

FIGURE 24.12

4. At $x = -3$, there is a minimum since $f'(x)$ changes from $-$ to $+$. When $x = 0$, there is a maximum since $f'(x)$ changes from $+$ to $-$. At $x = 3$, there is a minimum since $f'(x)$ changes from $-$ to $+$.

The graph of this function, $f(x) = x^4 - 18x^2$, is shown in Figure 24.12.

EXAMPLE 24.8

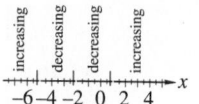

FIGURE 24.13

If $g(x) = x + \dfrac{9}{x+2}$, use the first derivative test to find all extrema.

Solution   Notice that -2 is not in the domain of g and that the graph of g has a vertical asymptote at $x = -2$.

1. $g'(x) = 1 - \dfrac{9}{(x+2)^2}$

$$= \frac{(x+2)^2}{(x+2)^2} - \frac{9}{(x+2)^2}$$

$$= \frac{(x+2)^2 - 9}{(x+2)^2}$$

$$= \frac{x^2 + 4x + 4 - 9}{(x+2)^2}$$

$$= \frac{x^2 + 4x - 5}{(x+2)^2}$$

$$= \frac{(x+5)(x-1)}{(x+2)^2}$$

2. $g'(x) = 0$ when $(x+5)(x-1) = 0$ or when $x = -5$ or $x = 1$. $g'(x)$ is undefined when $(x+2)^2 = 0$ or when $x = -2$. However, since -2 is not in the domain of g, it is not a critical value. The only critical values are $x = -5$ and $x = 1$.

3. There are four intervals to consider. Remember that $(x+2)^2$ is never negative. Solving we get the following results.

$$(-\infty, -5),\ g'(x) > 0,\ \text{so } g \text{ is increasing}$$
$$(-5, -2),\ g'(x) < 0,\ \text{so } g \text{ is decreasing}$$
$$(-2, 1),\ g'(x) < 0,\ \text{so } g \text{ is decreasing}$$
$$(1, \infty),\ g'(x) > 0,\ \text{so } g \text{ is increasing}$$

This is summarized in Figure 24.13.

EXAMPLE 24.8 (Cont.)

4. At $x = -5$, there is a maximum since $g'(x)$ changes from $+$ to $-$. At $x = 1$, there is a minimum since $g'(x)$ changes from $-$ to $+$. The value $x = -2$ is not in the domain of g. There is no extrema at that point. The graph of g is shown in Figure 24.14.

■

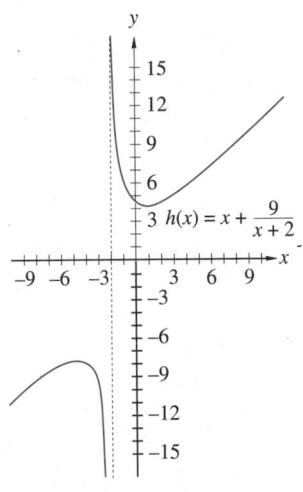

FIGURE 24.14

Exercise Set 24.2

In Exercises 1–26, determine when the function is increasing or decreasing, locate all critical values, and determine all extrema. Do not sketch the graph of the function.

1. $f(x) = x^2 - 8x$
2. $g(x) = 10x - x^2$
3. $m(x) = 16 - x^2$
4. $h(x) = 3x^2 - 12x + 10$
5. $j(x) = 15 + 16x - 4x^2$
6. $k(x) = x^2 - 4x + 11$
7. $f(x) = 4x^2 - 16x + 1$
8. $s(t) = -t^2 + 7t + 4$
9. $q(t) = t^3 - 3t$
10. $v(t) = 21t + 9t^2 - t^3$

11. $y = \frac{1}{4}x^4 - 18x^2 + 16$
12. $t = 3.2 - 12.5w^2 + 0.25w^4$
13. $f(z) = (z+4)^3$
14. $g(x) = (2x^2 - 8)^2$
15. $g(v) = 3v^4 - 4v^3 - 12v^2 + 12$
16. $y = \frac{5}{x}$
17. $s(t) = \frac{6}{t+1}$
18. $h(x) = x + \frac{1}{x}$

19. $g(x) = x^2 - \frac{2}{x}$
20. $j(x) = x + \frac{1}{x+3}$
21. $k(x) = \frac{x^2}{1-x}$
22. $m(b) = 4\sqrt{b}$
23. $f(t) = t\sqrt{1-t^2}$
24. $g(s) = (s+2)^2(s-1)^{2/3}$
25. $y = x(x-1)^{2/3}$
26. $f(x) = 9\sqrt[3]{x} - 4x$

Solve Exercises 27–36.

27. *Physics*   An object is fired straight upward with its height, given in feet, represented by $s(t) = 250 + 256t - 16t^2$. **(a)** At what time is the altitude a maximum? **(b)** What is the maximum altitude?

28. *Physics*   An object is fired straight upward with its height, given in meters, represented by $s(t) = 180 + 98t - 4.9t^2$. **(a)** At what time is its height a maximum? **(b)** What is the maximum height?

29. *Electricity*   The power P in watts (W) in a resistor is given by $P = 4.7(t - 2.0)^2$, where t is time in seconds (s).
    (a) What is the lowest power in the resistor?
    (b) When does this occur?

30. *Electricity*   The current i in amperes (A) at a given point in a circuit is given by $i = 4.8t - 1.2t^2$ where t is time in seconds (s).
    (a) At what time is the current maximum?
    (b) What is the maximum current?

31. *Electricity*   When two resistors R_1 and R_2 are connected in parallel, their total resistance R is

$$R = \frac{R_1 R_2}{R_1 + R_2}$$

If $R_1 = 9\,\Omega$ and R_2 is a variable resistor, then

$$R = \frac{9R_2}{9 + R_2}$$

    (a) What values of R_2 provide the minimum and maximum values of R?
    (b) What are the minimum and maximum values of R?

32. *Physics*   A particle moves along the x-axis according to the position function $x(t) = 8t^3 - t^4$. Find the velocity at the instant the acceleration is a maximum.

33. *Business*   The profit from the sale of n items of a certain product is given by $P(n) = (4 - 0.2n)(2.5n + 7)$, where $P(n)$ is in hundreds of dollars and $1 \leq n \leq 20$.

(a) If the marginal profit is $P'(n)$, find the marginal profit function.
(b) How many items must be sold to make the most profit?
(c) What is the maximum profit?

34. *Energy technology*   The energy output in joules of an electrical heater varies with time, t, according to the equation $E = 0.01\left(6t - t^2\right)^3$, where $0 \leq t \leq 6$.
    (a) When is the energy output a maximum?
    (b) What is the maximum energy output?
    (c) When are the energy outputs at their lowest levels?

35. *Police science*   Due to the rapid increase in major crimes, the mayor of a large city plans to organize a major crime task force. It is estimated that for each 1,000 people in the city, the number of major crimes will be $N(t) = 132 + 4\sqrt{t} + 2t^{3/2} - 0.8t^{5/2}$, where t is the number of months after the task force has been organized and $0 \leq t \leq 6$. (a) Find $N'(t)$. (b) When is the number of crimes at its peak?

36. *Meteorology*   Meteorological results for a certain city indicate that for the month of April, the daily temperature in °F between midnight and 6:00 P.M. can be approximated by $T(t) = -0.04t^3 + 1.26t^2 - 9.3t + 56$, where t is the number of hours after midnight and $0 \leq t \leq 18$.
    (a) Find $T'(t)$.
    (b) When is the temperature the highest?
    (c) When is the temperature the lowest?

✎ In Your Words

37. Explain the first derivative test and what information it gives about a function.

38. (a) What determines whether a function is increasing or decreasing?

(b) How can you use calculus to determine whether a function is increasing or decreasing?

☰ 24.3
CONCAVITY AND THE SECOND DERIVATIVE TEST

In Section 24.2, we used the first derivative to determine when a function is increasing or decreasing and to locate its extrema. In this section, we will learn some uses for the second derivative. In order to be able to use the second derivative, we need to take a closer look at a curve and its shape.

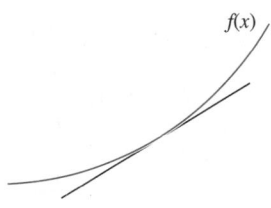

FIGURE 24.15a

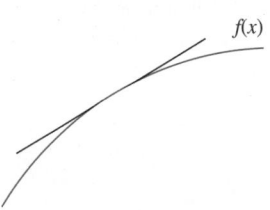

FIGURE 24.15b

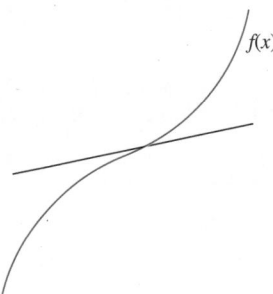

FIGURE 24.15c

Concavity

We will begin by looking at the three curves $f(x)$ in Figures 24.15a–c. In all three cases a tangent line has been drawn. Looking at the tangents, we can see that they all have positive slopes and so $f'(x) > 0$. These figures illustrate the three possible situations. In Figure 24.15a, the curve lies above the tangent line; in Figure 24.15b, the curve lies below the tangent line; and in Figure 24.15c, the curve crosses the tangent line. This example is not unique. We could have the same three situations if the first derivative were negative or if it were zero.

The three situations demonstrated in Figures 24.15a–c give us the chance to introduce some new terminology. If there is an open interval around a point where the graph of a function f is above the tangent lines for all points in that interval, then the graph of f is **concave up** at that point. This is demonstrated in Figure 24.15a and in Figure 24.16. There is an open interval (a, b) that contains the point c. The graph of $y = f(x)$ passes through $(c, f(c))$ and is above the tangent line through that point. Pick any other point in (a, b) and draw the tangent lines through that point. If the curve is above that tangent line, then the curve is concave up. In Figure 24.16, the points x_1 and x_2 represent any two points in (a, b), except c. The tangent lines to the curve at each of the points $(x_1, f(x_1))$ and $(x_2, f(x_2))$ are drawn. In both cases the curve is above the tangent line and so the curve is concave up.

We have a similar situation if the curve is below the tangent lines as in Figure 24.15b. In this case, if there is an open interval around a point where the graph of a function is below the tangent lines for all points in that interval, then the graph of the function is **concave down** at that point.

Test for Concavity

It would be rather cumbersome to have to examine the tangent lines to a graph every time we wanted to determine if the graph was concave up or down. Fortunately, a test has been developed that allows us to determine the concavity of a curve. The test of concavity uses the second derivative and states the following.

> **Test for Concavity**
>
> The graph of a function f is
>
> **(a)** concave up on any interval where $f''(x) > 0$
> **(b)** concave down on any interval where $f''(x) < 0$

💡 **Hint**

You might want to use these happy and sad faces to help remember how to use the second derivative to check the concavity of a function.

 If the second derivative is positive, the graph is concave upward.

 If the second derivative is negative, the graph is concave downward.

EXAMPLE 24.9

Determine when $f(x) = x^3 - 6x^2 + 3x + 5$ is concave up and when it is concave down.

Solution   We find the second derivative of f is $f''(x) = 6x - 12$. Solving this, we see that $f''(x) < 0$ when $x < 2$ and $f''(x) > 0$ when $x > 2$. Thus, $f(x)$ is concave down on the interval $(-\infty, 2)$ and concave up on $(2, \infty)$.

We have two items that have not been examined. What happens when a curve crosses a tangent line as in Figure 24.15c? What happens when $f''(x) = 0$?

It turns out that we can use Figure 24.15c to help answer both of these questions. To the left of the tangent line the curve is concave down and so the second derivative is negative. To the right of the tangent line the curve is concave up and so the second derivative is positive. At the tangent line in Figure 24.15c the curve changes from concave down to concave up. A point on the graph at which the concavity changes is called an **inflection point**.

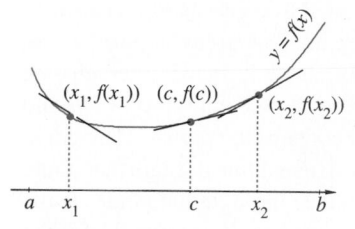

FIGURE 24.16

> **Locating Inflection Points**
>
> There are two ways to locate possible inflection points.
>
> - Determine when the second derivative is zero.
> - Determine when the second derivative is undefined.
>
> Once you have found all of these points, look at the sign of the second derivative on both sides of each point. If the signs are different, then the concavity changes and it is a point of inflection.

EXAMPLE 24.10

Locate all inflection points of $f(x) = x^3 - 6x^2 + 3x + 5$.

Solution   The second derivative is $f''(x) = 6x - 12$. Solving this, we see that $f''(x) = 0$ when $x = 2$. In Example 24.9, we showed that $f''(x) < 0$ when $x < 2$ and $f''(x) > 0$ when $x > 2$. Since the sign of f'' is negative to the left of 2 and positive to the right of 2, the point $(2, f(2)) = (2, -5)$ is an inflection point. The graph of f is shown in Figure 24.17.

The fact that the second derivative is zero at some point does not mean that it is an inflection point. Consider the graph of $f(x) = x^4$. The second derivative is $f''(x) = 12x^2$ and $f''(x) = 0$ when $x = 0$. But $f''(x) > 0$ for all $x \neq 0$. The sign of the second derivative is the same on both sides of 0. Thus, there is no inflection point when $x = 0$, as can be seen in the graph of $f(x) = x^4$ in Figure 24.18.

Second Derivative Test

The second derivative can sometimes tell us more than the concavity of a curve and the location of possible inflection points. Suppose a curve is concave up on some interval. If there is a point c in this interval with a first derivative of zero, then it must

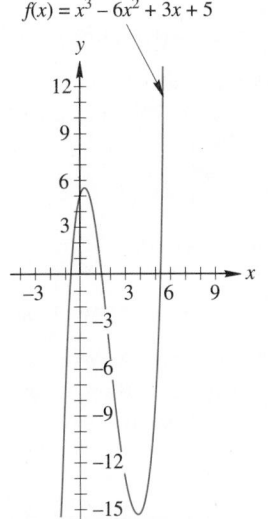

$f(x) = x^3 - 6x^2 + 3x + 5$

FIGURE 24.17

EXAMPLE 24.11

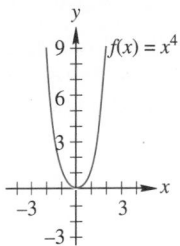

$f(x) = x^4$

FIGURE 24.18

EXAMPLE 24.12

have a horizontal tangent line. Since the curve is concave up, the tangent line is below the curve. Thus the point is minimum. This forms the basis for what is known as the **second derivative test**.

Second Derivative Test

If f is a function whose second derivative exits, and if there is a point c where $f'(c) = 0$:

(a) if $f''(c) < 0$, then $f(c)$ is a maximum,
(b) if $f''(c) > 0$, then $f(c)$ is a minimum,
(c) if $f''(c) = 0$, then the second derivative will not tell if $f(c)$ is a maximum or minimum.

Use the second derivative test to find all extrema of $g(x) = 2x^3 - 15x^2 + 36x - 25$.

Solution Using the first derivative $g'(x) = 6x^2 - 30x + 36$, we find that $x = 2$ and $x = 3$ are critical values. The second derivative is $g''(x) = 12x - 30$. Evaluating the second derivative at each critical value, we get $g''(2) = -6$ and $g''(3) = 6$. Since $g''(2) < 0$, there is a maximum at 2 and since $g''(3) > 0$, there is a minimum at 3. Remember that the minimum occurs when $x = 3$ so the location of the minimum is at $(3, \ g(3))$ or $(3, 2)$. Similarly, the maximum is at $(2, g(2))$ or $(2, 3)$.

In tabular form, we get the following results from the first derivative:

Interval	$g'(x)$	$g(x)$
$(-\infty, 2)$	$+$	increasing
$(2, 3)$	$-$	decreasing
$(3, \infty)$	$+$	increasing

Setting $g''(x) = 0$, we see that the only possible inflection point is when $x = \frac{30}{12} = \frac{5}{2} = 2.5$. Using the concavity test, we get the following results:

Interval	$g''(x)$	$g(x)$
$(-\infty, 2.5)$	$-$	concave down
$(2.5, \infty)$	$+$	concave up

The graph of g is shown in Figure 24.19. ▪

Sketch the graph of $h(x) = 3 + x + 3x^{2/3}$.

Solution $h'(x) = 1 + \dfrac{2}{x^{1/3}} = \dfrac{x^{1/3} + 2}{x^{1/3}}$. The critical values are when $h'(x) = 0$ and when $h'(x)$ is not defined. Now, since a fractional expression is 0 only when the

EXAMPLE 24.12 (Cont.)

$g(x) = 2x^3 - 15x^2 + 36x - 25$

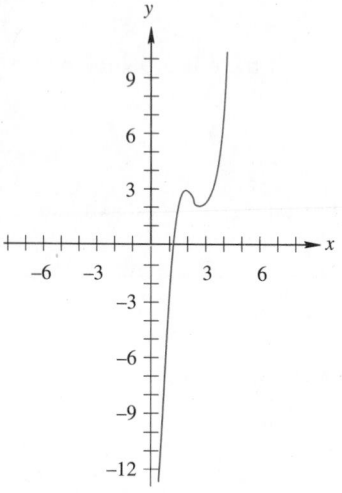

FIGURE 24.19

numerator is 0, then $h'(x) = \dfrac{x^{1/3}+2}{x^{1/3}} = 0$ is equivalent to $x^{1/3}+2=0$ or $x^{1/3}=-2$ and $x=-8$. Now, $h'(x)$ is not defined whenever the denominator is 0, or when $x^{1/3}=0$, which is when $x=0$. Since $x=0$ is in the domain of h, the critical values are $x=-8$ where $h'(x)=0$ and $x=0$ where $h'(x)$ is not defined.

The second derivative is $h''(x) = -\dfrac{2}{3x^{4/3}}$. When $x=-8$, then $h''(-8) = -\dfrac{2}{3(-8^{4/3})} = -\dfrac{2}{48} < 0$, and so $(-8, 7)$ is a maximum. When $x=0$, then $h''(0) = -\dfrac{2}{3(0^{4/3})}$, which is not defined and the second derivative test fails. Using the first derivative test, we see that when $x > 0$, $h'(x) > 0$ and so h is increasing. When x is in the interval $(-8, 0)$, $h'(x) < 0$ and so h is decreasing. Thus, $x = 0$ produces the minimum $(0, h(0))$ or $(0, 3)$. Summarizing our findings and adding those of the test for concavity, we get the following results.

Interval	$h'(x)$	h	Interval	$h''(x)$	h
$(-\infty, -8)$	+	increasing	$(-\infty, 0)$	−	concave down
$(-8, 0)$	−	decreasing	$(0, \infty)$	−	concave down
$(0, \infty)$	+	increasing			

The graph of this function is shown in Figure 24.20.

In Example 24.12, the domain of $h(x) = 3 + x + 3x^{2/3}$ was all real numbers. It is important to notice the domain. For example, graphing `h(x) = 3 + x + 3x^(2/3)` on a graphing calculator or a computer graphing program may result in something like the graph in Figure 24.21a. However, graphing h as `h(x) = 3 + x + 3(x^2)^(1/3)` produces the correct graph shown in Figure 24.21b.

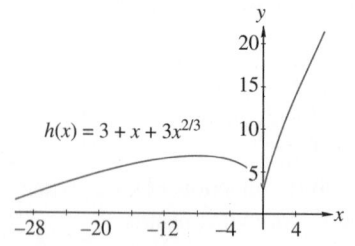

$h(x) = 3 + x + 3x^{2/3}$

FIGURE 24.20

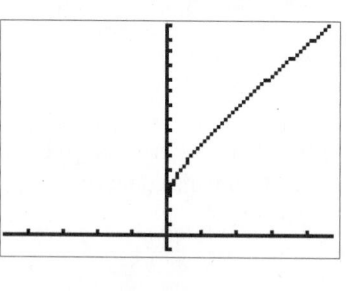

[−3.1, 3.1, 1] x [−1, 16, 1]

FIGURE 24.21a

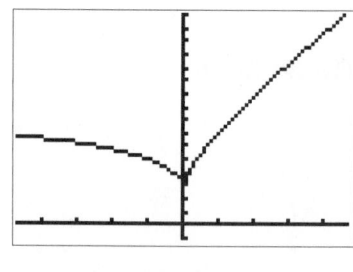

[−3.1, 3.1, 1] x [−1, 16, 1]

FIGURE 24.21b

In Section 18.5, we learned how to find some horizontal and vertical asymptotes. Basically, in order to test for horizontal asymptotes of $f(x)$, we evaluate $\lim\limits_{x \to \infty} f(x)$ and $\lim\limits_{x \to -\infty} f(x)$. Vertical asymptotes occur when the function $f(x)$ is not defined. Consider the next example.

EXAMPLE 24.13

Sketch the graph of $k(x) = \dfrac{x}{x^2+4}$.

Solution   We will first check for asymptotes.

$$\lim_{x \to \infty} \frac{x}{x^2+4} = \lim_{x \to \infty} \frac{1/x}{1+4/x^2} = \frac{0}{1} = 0$$

and $\displaystyle\lim_{x \to -\infty} \frac{x}{x^2+4} = 0$. So, the x-axis, or $y = 0$, is a horizontal asymptote.

Finding the derivatives of $k(x)$, we get

$$k'(x) = \frac{4-x^2}{(x^2+4)^2} \text{ and } k''(x) = \frac{2x^3 - 24x}{(x^2+4)^3} = \frac{2x(x^2-12)}{(x^2+4)^3}$$

From the first derivative we see that the critical values are ± 2. In tabular form, we get the following results from the first derivative.

Interval	$k'(x)$	$k(x)$
$(-\infty, -2)$	$-$	decreasing
$(-2, 2)$	$+$	increasing
$(2, \infty)$	$-$	decreasing

Thus, when $x = -2$, we have a minimum at $(-2, -0.25)$, and at $x = 2$, we have a maximum at $(2, 0.25)$.

Setting $k''(x) = 0$, we see that the possible inflection points are when $x = 0$ and $x = \pm\sqrt{12} = \pm 2\sqrt{3}$. Using the concavity test, we get the following results.

Interval	$k''(x)$	$k(x)$
$(-\infty, -2\sqrt{3})$	$-$	concave down
$(-2\sqrt{3}, 0)$	$+$	concave up
$(0, 2\sqrt{3})$	$-$	concave down
$(2\sqrt{3}, \infty)$	$+$	concave up

Thus, we have points of inflection at all three points. The sketch of the graph is shown in Figure 24.22.

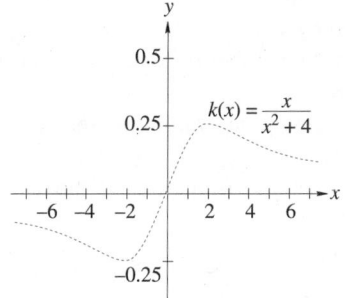

FIGURE 24.22

Exercise Set 24.3

In Exercises 1–28, find the extrema, determine the concavity, locate the inflection points, and sketch the graph.

1. $f(x) = -3x^2 + 12x$
2. $g(x) = 3x^2 - 6x + 5$
3. $g(x) = x^3 - 12x + 12$
4. $f(x) = x^3 + 3x^2 - 5$
5. $j(x) = x^3 - 6x^2 + 12x - 4$
6. $k(x) = x^3 - 6x^2 - 20$
7. $h(x) = -2x^3 + 3x^2 - 12x + 5$
8. $m(x) = x^4 - 4x^3 + 15$

9. $j(x) = 4x^2 - \frac{4}{3}x^3 - x^4$
10. $f(x) = 3x^4 - 4x^3 + 8$
11. $h(x) = (1-x)^3$
12. $g(x) = 3x^5 - 5x^3$
13. $h(x) = \sqrt[3]{x} + 2$
14. $j(x) = x - x^{2/3}$
15. $k(x) = x + \dfrac{4}{x}$

16. $m(x) = x - \dfrac{9}{x}$
17. $n(x) = x^2 - \dfrac{2}{x}$
18. $f(x) = \dfrac{1}{x-2}$
19. $g(x) = \dfrac{x^3}{3} - \dfrac{x^2}{2} - 6x$
20. $h(x) = x^4 - 32x + 48$

21. $j(x) = 2\sqrt{x} - x$

22. $k(x) = -\sqrt{x^2 - 4}$

23. $g(x) = 2 + x^{2/3}$

24. $h(x) = 2x - 3x^{2/3}$

25. $f(x) = x^{2/3}(x - 10)$

26. $g(x) = \dfrac{2}{x^2 - 4}$

27. $h(x) = \dfrac{x}{x^2 + 1}$

28. $j(x) = \dfrac{x^{2/3}}{x - 4}$

Solve Exercises 29–32.

29. *Agriculture*   The catfish population of a commercial catfish farm can be approximated by $F(t) = 30 + 0.621t^2 - 0.009t^3$, where t is the number of weeks since this pond was started, $0 \le t \le 52$. To produce the maximum yield of catfish, the best time to harvest the catfish is at the point of the fastest growth in the fish population. This point is an inflection point. When is the best time to harvest this fish population?

30. *Economics*   When a company starts a successful advertising campaign, sales and the rate at which sales grow both increase. At the **point of diminishing returns**, even though more money is spent on advertising, sales continue to grow but at a slower rate. The point of diminishing returns occurs at an inflection point where the sales curve changes from concave upward to concave downward. Find the point of diminishing returns for the sales function $S(x) = 97 + 2.1x^2 - 0.1x^3$, where x represents thou-

sands of dollars on advertising, $0 \le x \le 10$, and S is sales in thousands of dollars.

31. *Police science*   Due to the rapid increase in major crimes, the mayor of a large city plans to organize a major crime task force. It is estimated that for each 1,000 people in the city, the number of major crimes will be $N(t) = 67.4 + 3t^2 - 0.8t^{5/2}$, where t is the number of months after the task force has been organized and $0 \le t \le 12$.
(a) Find the maximum of $N(t)$.
(b) When is the maximum rate of increase in $N(t)$?

32. *Meteorology*   Meteorological results for a certain city indicate that for the month of April, the daily temperature in °F between midnight and 6:00 P.M. can be approximated by $T(t) = -0.04t^3 + 1.26t^2 - 9.3t + 56$, where t is the number of hours after midnight and $0 \le t \le 18$.
(a) What are the maximum and minimum temperatures?
(b) What is the maximum rate of increase in the temperature?

In Your Words

33. Explain the second derivative test and what information it gives about a function.

34. What is an inflection point?

≡ 24.4
APPLIED EXTREMA PROBLEMS

Up to this point, most of the applications for extrema have been made to curve sketching. We have used derivatives to help determine the high and low points on the curve, when it is increasing or decreasing, and other useful information. A curve is a picture of a function and functions are often used to represent real situations. In this section, we will look at some ways to solve problems that involve using derivatives.

As each example is worked, you should make sure that you understand what was done. The most important part will be in understanding how to set up the function; then, being aware of how derivatives are used to find possible solutions and to determine the correct answer should be understood.

Application

EXAMPLE 24.14

A company needs to erect a fence around a rectangular storage yard next to a warehouse. They have 400 m of available fencing. If they do not fence in the side next to the warehouse, what is the largest area that can be enclosed?

Solution   The clue to the desired answer is "largest area." We need to develop a function for the area and then determine when it is a maximum.

Look at Figure 24.23. The fenced-in rectangular yard has dimensions that have been labeled x and y. The area of the fenced-in region is $A = xy$. We need to write A in terms of a single variable. The total amount of fencing is $400\,\text{m} = y + x + y$. Thus, we have

$$2y + x = 400$$

or $x = 400 - 2y$

If we substitute this in the area formula, we obtain

$$\begin{aligned} A &= xy \\ &= (400 - 2y)y \\ &= 400y - 2y^2 \end{aligned}$$

The area is now expressed in terms of one variable, y, and so we have

$$A(y) = 400y - 2y^2$$

We want to know when $A(y)$ is a maximum. Taking the first derivative, we get

$$A'(y) = 400 - 4y$$

The only critical value occurs when $A'(y) = 0$, which is when $y = 100$.

Is the area a maximum when $y = 100$? Using the second derivative test, we see that $A''(y) = -4$. Since $A''(100) < 0$, the area is a maximum when $y = 100$.

We have one other item to check—is this an acceptable solution? We know that $y > 0$. We can also determine that the largest possible value for y is 200 m. Thus, the answer must be in the interval $[0, 200]$.

Now, when $y = 100$, $x = 200$, and so the maximum area is $A = (200)(100) = 20\,000\,\text{m}^2$.

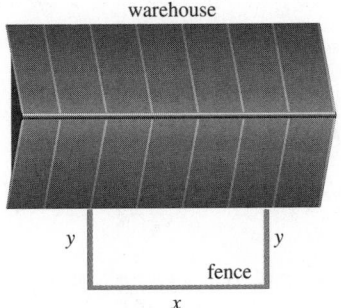

warehouse

y y

fence

x

FIGURE 24.23

Application

EXAMPLE 24.15

When a battery or generator of voltage V and an internal resistance r is connected to an external resistance R, the total resistance in the circuit is $R + r$ and the current flow is $I = \dfrac{V}{R+r}$. The power P dissipated in the load R is $P = I^2 R$. If the voltage is 100 V and $r = 5\,\Omega$, what value of R will produce the maximum load power? (See Figure 24.24.)

EXAMPLE 24.15 (Cont.)

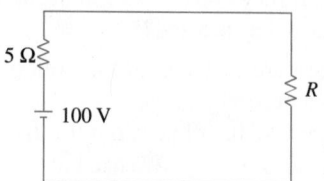

FIGURE 24.24

Solution We need to develop a formula in terms of the external resistance R. Since $I = \dfrac{V}{R+r}$, we have $I = \dfrac{100}{R+5}$. Substituting this into the power formula, we get

$$P = I^2 R = \left(\frac{100}{R+5}\right)^2 R = \frac{10\,000R}{(R+5)^2}$$

Differentiating produces

$$\begin{aligned} P' &= \frac{(R+5)^2(10\,000) - 10\,000R(2)(R+5)}{(R+5)^4} \\ &= \frac{(R+5)10\,000 - 20\,000R}{(R+5)^3} \\ &= \frac{50\,000 - 10\,000R}{(R+5)^3} \end{aligned}$$

Since $R > 0$, the only critical value occurs when $P' = 0$ or when $R = 5$. Using the first derivative test, we see that this is a maximum and so the maximum load power occurs when $R = 5\,\Omega$.

We could generalize this last example to show that the maximum load power occurs when the external load is equal to the internal resistance.

Application

EXAMPLE 24.16

A container is to be made in the form of a right circular cylinder and is to contain 1 L. What dimensions of the container will require the least amount of material?

Solution We know that the volume of a right circular cylinder is given by the formula $V = \pi r^2 h$ and the total surface area is $A = 2\pi rh + 2\pi r^2$. We will assume that the thickness of the material is negligible. We want to minimize the area. The formula for the area is given in terms of two variables, h and r. We know that $V = 1\,\text{L} = 1\,000\,\text{cm}^3$. We can solve the equation $1\,000 = \pi r^2 h$ for either r or h and substitute the result into the area formula. If we solve for h we get

$$h = \frac{1\,000}{\pi r^2}$$

Substituting, we get the area as a function of r:

$$\begin{aligned} A(r) &= 2\pi r \left(\frac{1\,000}{\pi r^2}\right) + 2\pi r^2 \\ &= \frac{2\,000}{r} + 2\pi r^2 \end{aligned}$$

Differentiating with respect to r, we have

$$A'(r) = -2\,000r^{-2} + 4\pi r$$

EXAMPLE 24.16 (Cont.)

Critical values occur when $r = 0$, because $A'(r)$ is not defined when $r = 0$, and when $A'(r) = 0$. If we set the derivative equal to zero, we obtain

$$\frac{2\,000}{r^2} = 4\pi r$$

$$\text{or} \qquad r^3 = \frac{2\,000}{4\pi} = \frac{500}{\pi}$$

$$\text{and so} \qquad r = \sqrt[3]{\frac{500}{\pi}}$$

Using the second derivative test, we see that

$$A''(r) = 4\,000 r^{-3} + 4\pi$$

When $r = 0$, $A''(r)$ is not defined, but we know that $r > 0$. (Why? What happens if $r = 0$?)

When $r = \sqrt[3]{\dfrac{500}{\pi}}$, $A''(r) > 0$, so the area is a minimum.

When $r = \sqrt[3]{\dfrac{500}{\pi}} \approx 5.4193\,\text{cm}$, $h = \dfrac{1\,000}{\pi r^2} \approx 10.8385\,\text{cm}$ and the area is $A \approx 553.5810\,\text{cm}^2$.

Application

EXAMPLE 24.17

A telephone company is asked to provide telephone service to a customer whose house is 0.5 km away from the road along which the telephone lines run. The nearest telephone box is 5 km down the road. The cost to connect the telephone line is $80 per kilometer along the road and $100 per kilometer away from the road. At which point along the road should the company connect the line in order to minimize the cost?

Solution Examine the sketch of this problem in Figure 24.25. We let $5 - x$ represent the distance from the box B to the place where the connection is made, A. The distance x is from the connection A to the point on the road closest to the house, H. The telephone line will go from B to A and then from A to H. The length AH is $\sqrt{x^2 + 0.5^2} = \sqrt{x^2 + 0.25}$. The total cost C is $C(x) = 80(5 - x) + 100\sqrt{x^2 + 0.25}$. We want $C(x)$ to be a minimum.

$$C'(x) = -80 + \frac{100x}{\sqrt{x^2 + 0.25}}$$

$$= \frac{-80\sqrt{x^2 + 0.25} + 100x}{\sqrt{x^2 + 0.25}}$$

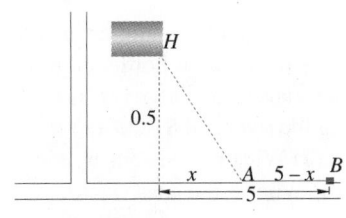

FIGURE 24.25

EXAMPLE 24.17 (Cont.)

The critical value occurs when the numerator is 0; that is, when

$$80\sqrt{x^2 + 0.25} = 100x$$
$$6\,400(x^2 + 0.25) = 10\,000x^2$$
$$1\,600 = 3\,600x^2$$
$$\frac{16}{36} = x^2$$

or when

$$x = \pm\frac{2}{3}$$

When $x = \frac{2}{3}$ km, the cost will be a minimum. The minimum cost is \$430.

When solving extrema problems, there are some helpful procedures to use. If you go back and look at the examples in this section, you will see that we used each of them.

Hint

Hints for Solving Extrema Problems

1. When possible, draw a figure to illustrate the problem. Label all the parts that are important to the problem.
2. Write an equation for the quantity that is going to be a maximum or a minimum.
3. If more than one variable is involved, eliminate all but one of them.
4. Take the derivative and locate the critical values.
5. Use the first and/or second derivative test on each critical value to see if it provides a maximum or a minimum.
6. If the function is defined for a limited range of values, examine the endpoints for possible extrema values.

Exercise Set 24.4

Solve Exercises 1–36.

1. *Agriculture* A company has 10,000 ft of available fencing. It wants to use the fencing to enclose a rectangular field. One side of the field is bordered by a river. If no fencing is placed on the side next to the river, what is the largest area that can be enclosed?

2. *Agriculture* If the company in Exercise 1 decides to fence the side bordering the river, what is the largest area that can be enclosed?

3. *Agriculture* A rectangular field contains an area of 6,000,000 ft². One side of the field borders a river. Because of difficulties caused by the river, it costs \$2.00/ft to fence along the river and \$1.00/ft to fence the other three sides. **(a)** What dimensions will give the minimum cost? **(b)** What is the cost?

4. *Agriculture*   A 200-m piece of link fencing is to be cut so that one piece encloses a square section of garden and the other piece encloses a circular section. Where should the fencing be cut if the enclosed areas are to be a maximum?

5. *Industrial design*   An open box is to be formed by cutting a square from each corner of a square piece of cardboard 24 cm on a side. After the squares are removed from each corner, the sides are folded up as shown in the following figure. How large a square must be removed from each corner in order for the box to have the largest volume?

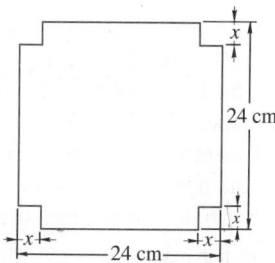

6. *Business*   A rectangular sign is to be painted on the end of a semicircular-shaped building that has a diameter of 20.00 m. What dimensions will produce a sign with the largest area?

7. What is the largest rectangle that can be inscribed in the first quadrant of the ellipse $9x^2 + 16y^2 = 144$?

8. *Industrial design*   An open rectangular box is to be made from a piece of cardboard 8 in. wide and 15 in. long by cutting a square from each corner and bending up the sides. How large a square should be removed from each corner in order for the box to have the largest volume?

9. *Industrial design*   A box with a square base and an open top is to hold 32 in.³ What dimensions will require the least amount of material?

10. *Industrial design*   If the box in Exercise 9 is to be closed on top, what dimensions will require the least amount of material?

11. *Industrial design*   The total surface area of a right-circular cylindrical tin is 480 cm². What are the dimensions of the cylinder that has the largest volume?

12. *Mechanical engineering*   The cutting speed s (in meters per minute) of a saw used to cut a certain type

of material is given by $s(t) = \sqrt{t - 9t^2}$, where t is the time in seconds. What is the maximum speed of this product?

13. *Printing*   A page of a book is to have 580 cm² with a 3-cm margin at the bottom and sides and a 2-cm margin at the tope. Find the dimensions of the page that will allow the largest printed area.

14. Find the point on the curve $y = 2x^2$ that is closest to the point $(5, 1)$.

15. *Construction*   The strength of a rectangular beam is directly proportional to the width of the beam and the square of its depth. What are the dimensions of the strongest beam that can be cut from a log whose cross-section is a circle of radius 15 in.? What are the dimensions of the strongest beam cut from a circular log whose cross-section has a radius of r?

16. *Sheet metal technology*   A metal trough is to be made by bending a rectangular sheet of metal in the middle to form a "V." If the piece of metal is 300 mm wide, what angle between the sides will give the maximum capacity?

17. *Sheet metal technology*   A gutter is to be made from a long sheet of metal 320 mm wide by bending up equal widths along the edges into vertical positions. What dimensions will give the largest capacity?

18. *Electronics*   The current I in amperes (A) through a resistor of $R\,\Omega$ with a constant voltage E in volts (V), is given by

$$I = \frac{E}{\sqrt{R^2 + X^2}}$$

(a) Assuming that R is constant, what value of X will make the current in the resistor a maximum?
(b) What is this maximum current?

19. *Electricity*   The output of a battery is given by the formula $P = VI - RI^2$, where V is the voltage, I the current, and R the resistance. Find the current for which the output is a maximum if $V = 10$ V and $R = 4.0\,\Omega$.

20. *Chemistry*   The rate of a certain autocatalytic reaction is given by $v(x) = 5x(a - x)$ for x in $[0, a]$, where a is the original amount of the substance produced by the reaction. Determine when the reaction is a maximum?

21. *Construction*   A power company wishes to install a temporary power line. The purchase price of the wire is proportional to the cross-sectional area A of the wire. The total cost C of the wire is given by the formula $C(A) = bA + \dfrac{c}{A}$, where b and c are constants. What cross-sectional area will produce the least cost?

22. *Architecture*   A window above the entrance to a room that has a cathedral ceiling is to be designed in the shape of a rectangle with a semicircle on top. If the area of the rectangular part is to be $2.00\,\text{m}^2$, what is the diameter of the semicircle that will result in a window with the smallest perimeter?

23. *Architecture*   What would be the diameter of the semicircular part of the window in Exercise 22 if the perimeter is $8.00\,\text{m}$ and the total area of the window is to be a maximum?

24. *Business*   The cost of operating a truck on an interstate highway is $0.35 + \dfrac{s}{300}$ dollars per mile, where s is the speed of the truck in miles per hour. The driver's salary of \$12 per hour is extra. (a) At what speed should the truck driver operate the truck to make an 800-mi trip for the least cost? (b) What is the cost of this 800-mi trip?

25. *Electricity*   The power P in watts (W) in a circuit is given by $P = I^2 R$, where I is current in amperes (A) and R is resistance in ohms (Ω). If $R = 25.4\,\Omega$ and $I = 4.5 - t$, where t in time is seconds (s), find (a) the time when there is minimum power, and (b) the minimum power.

26. *Electricity*   The current in a circuit is given by $\dfrac{dq}{dt}$, where q is the charge in coulombs (C) and t is the time in seconds (s). If $q = -4.36t^2 + 2.14t^3$, find the minimum current in the circuit and the time at which this occurs.

27. *Lighting technology*   The intensity of illumination at a point varies inversely as the square of the distance between the point and the light source. Two lights are 8 m apart. One of the lights has an intensity four times that of the other. At which point between the lights is the illumination the least?

28. *Aerodynamics*   The drag D on an airplane traveling at velocity v is $D = av^2 + \dfrac{b}{v^2}$, where a and b are positive constants. At what speeds does the airplane have the least drag?

29. *Lighting technology*   The deflection Y of a beam of length L at a horizontal distance x from one end is given by $Y = k(2x^4 - 5Lx^3 + 3L^2x^2)$, where k is a positive constant. For what value of x does the maximum deflection occur?

30. *Mechanical engineering*   The efficiency E of a screw is given by $E = \dfrac{T(1 - T\mu)}{T + \mu}$ where μ is the coefficient of friction and T is the tangent of the pitch angle of the screw. For what value of T is the efficiency the greatest?

31. *Civil engineering*   A proposed tunnel has a fixed cross-sectional area. The tunnel will have a horizontal floor, vertical walls of equal height, and the ceiling will be a semicircle. The ceiling cost four times as much per square meter as the floor and walls. What ratio of the diameter of the semicircular ceiling to the height of the walls will produce the least cost?

32. A person in a rowboat is 4 km offshore, where the nearest point P is on a straight shoreline of the lake. A town is 20 km down the shore road from P. This person can row 2.5 km/h and can walk 4.5 km/h. At which point should the boat be landed in order to arrive in town in the shortest time?

33. *Petroleum engineering*   An oil company wants to lay a pipeline from its offshore drilling rig to a storage tank on shore. The rig is 5 mi offshore, and the storage tank is 12 mi down the straight shoreline. The cost of laying pipe underwater is \$1,200 per mi and along the shoreline the cost is \$600 per mi. How far from the storage tank should they locate the connection of the underwater pipe to the shoreline pipe?

34. *Package design*   A manufacturer is designing a closed rectangular shipping crate with a square base. The volume of the crate is to be $36\,\text{ft}^3$. Material for the top costs \$1 per square foot, material for the sides costs \$0.75/ft^2, and material for the bottom costs \$1.50/ft^2.
    (a) Find the dimensions of the box that will minimize the total cost of material.
    (b) Determine the minimum cost.

35. *Construction*   A rectangular warehouse is being constructed with a total floor area of $4{,}200\,\text{ft}^2$. A single interior wall will partition the area into storage space and office space. It costs \$175/ft to construct an exterior wall and \$115/ft to erect the interior wall.
   (a) What are the dimensions of the warehouse that will minimize the cost?
   (b) Determine the minimum cost.

36. *Package design*   The U.S. Postal Service has a limit of 108 in. on the combined length and girth of a rectangular package that can be mailed. Find the dimensions of the package of maximum volume that has a square cross-section. (The girth is defined as the smallest perimeter of a rectangular cross-section of the box.)

In Your Words

37. Without looking in the text, describe how you solve a problem that asks you to find a maximum or a minimum.

38. Write an application in your technology area of interest that requires you to find a maximum or a mini-

mum. Give your problem to a classmate and see if he or she understands and can solve your problem using the techniques of this and earlier chapters. Rewrite the problem as necessary to remove any difficulties encountered by your classmate.

≡ 24.5
RELATED RATES

It is not unusual to have two or more quantities that vary with time and an equation that expresses some relationship between them. Normally, the values of these quantities at some time are given. We are also given all but one of the rates of change. The problem is to find the rate of change that is not given. We do this by taking the derivative of the expression (with respect to time) that relates the variables. The following examples show the basic method for solving these problems.

EXAMPLE 24.18

A ladder 30 ft long leans against a vertical wall. The foot of the ladder is slipping away from the wall at the rate of 4 ft/s. How fast is the top of the ladder sliding down the wall at the instant when the foot of the ladder is 18 ft from the wall?

Solution   Figure 24.26 shows the ladder against the wall. The constants are the length of the ladder, 30 ft, and the rate at which the foot of the ladder moves, 4 ft/s. The quantities that change are the height of the top of the ladder, h, and the distance of the foot of the ladder from the wall, x.

We want to find a rate. If h denotes the height of the top of the ladder above the ground in feet and t denotes the time in seconds, then we want to find dh/dt: the rate at which the height of the ladder is changing (dh) with respect to time (dt). In particular, we want to find dh/dt when $x = 18$ ft.

We have a right triangle with the right angle where the wall meets the ground. The ladder is the hypotenuse of the triangle. Thus, $x^2 + h^2 = 30^2$. Using implicit differentiation with respect to time t, we get

$$\frac{d(x^2)}{dt} + \frac{d(h^2)}{dt} = 0$$

or $$2x\frac{dx}{dt} + 2h\frac{dh}{dt} = 0$$

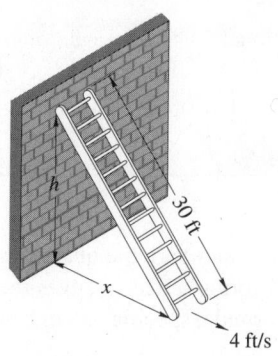

FIGURE 24.26

EXAMPLE 24.18 (Cont.)

Thus, the rate we want to find is

$$\frac{dh}{dt} = \left(\frac{-x}{h}\right)\left(\frac{dx}{dt}\right)$$

When $x = 18$, $h = 24$, and $\dfrac{dx}{dt} = 4$ ft/s, we obtain

$$\frac{dh}{dt} = -\frac{18}{24}(4) = -3 \, \text{ft/s}$$

Thus, the top of the ladder is falling at the rate of 3 ft/s. ▪

We solved Example 24.18 for the general case when the foot of the ladder was x-feet from the bottom of the wall. Once we had this answer, we solved it for the specific case when $x = 18$. Thus, by first solving for x we are then able to determine the rate for any value of x.

As with the extrema problems, there are some guidelines to help us solve related rate problems. Some of the guidelines are virtually the same as those for extrema, but are restated here to keep the guidelines as a group.

Hint

Hints for Solving Related Rates Problems

1. Make a diagram of the problem. Which quantities change; which remain constant?
2. Write down what you are to find. (In Example 24.18, it was a rate.) Express it in terms of a variable (in Example 24.18, dh/dt). Show the variable in the sketch (h in Figure 24.26).
3. Identify the other variables in the problem. Show them in the sketch. Write down any numerical information that is given about them.
4. Write equations that relate the variables.
5. Use substitution and differentiation to establish additional relationships among the variables and their derivatives.
6. Substitute numerical values for the variables and the derivatives. Solve for the unknown rate.

Caution

A common error is to introduce specific values for the rates and variable quantities too early in the solution. Remember to first get a general formula that involves the rates of change at any time t. After you have the general formula, specific values can be substituted to get the desired solution.

Application

EXAMPLE 24.19

A weather balloon in the shape of a sphere is being inflated at the rate of $12\,\text{m}^3/\text{min}$. Find the rate at which the surface area of the balloon is changing when the radius is $6\,\text{m}$.

Solution   The variables in this problem are:

$$t = \text{time in minutes}$$
$$r = \text{radius in meters of the balloon at time } t$$
$$V = \text{volume in cubic meters at time } t$$
$$S = \text{surface area in square meters at time } t$$

We have the following rates of changes:

$$\frac{dr}{dt} = \text{rate of change of radius with respect to time}$$

$$\frac{dV}{dt} = \text{rate of change of volume with respect to time}$$

$$\frac{dS}{dt} = \text{rate of change of surface area with respect to time}$$

We are told that $\dfrac{dV}{dt} = 12\,\text{m}^3/\text{min}$. We want to find $\dfrac{dS}{dt}$ when $r = 6\,\text{m}$.

The volume of a sphere is $V = \frac{4}{3}\pi r^3$ and the surface area is $S = 4\pi r^2$. Differentiating each of these with respect to t, we get

$$\frac{dV}{dt} = 4\pi r^2 \frac{dr}{dt}$$

$$\text{and} \qquad \frac{dS}{dt} = 8\pi r \frac{dr}{dt}$$

Since we want to find $\dfrac{dS}{dt}$ and we know the value of $\dfrac{dV}{dt}$, we will solve the $\dfrac{dV}{dt}$ equation for $\dfrac{dr}{dt}$ and substitute this into the $\dfrac{dS}{dt}$ equation. Solving for $\dfrac{dr}{dt}$ we get

$$\frac{dr}{dt} = \frac{1}{4\pi r^2} \cdot \frac{dV}{dt}$$

Substituting, this gives

$$\frac{dS}{dt} = 8\pi r \left(\frac{1}{4\pi r^2}\right) \frac{dV}{dt} = \frac{2}{r} \frac{dV}{dt}$$

When $r = 6\,\text{m}$ and $\dfrac{dV}{dt} = 12\,\text{m}^3/\text{min}$, we get

$$\frac{dS}{dt} = \frac{2}{6} \cdot 12 = 4\,\text{m}^2/\text{min}$$

The surface area is increasing at a rate of $4\,\text{m}^2/\text{min}$ when the radius is $6\,\text{m}$.

Application

EXAMPLE 24.20

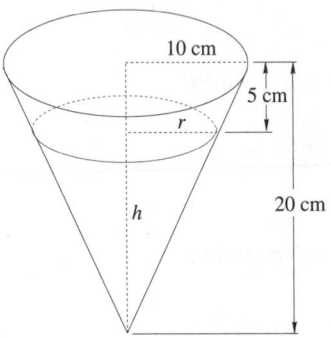

FIGURE 24.27

Water is running out of a conical funnel at a rate of $5\,\text{cm}^3/\text{s}$. If the radius of the base of the funnel is $10\,\text{cm}$ and the altitude is $20\,\text{cm}$, find the rate at which the water level is dropping when it is $5\,\text{cm}$ from the top.

Solution A sketch of this problem is given in Figure 24.27. The variables are:

$$t = \text{time in seconds}$$
$$r = \text{radius in cm of water level at time } t$$
$$h = \text{height in cm of surface of water at time } t$$
$$V = \text{volume in cm}^3 \text{ of water in cone at time } t$$

We have the following rates of change:

$$\frac{dh}{dt} = \text{rate of change of height of water with respect to time}$$

$$\frac{dV}{dt} = \text{rate of change of volume of water with respect to time}$$

By similar triangles, we have $\dfrac{r}{10} = \dfrac{h}{20}$ or $r = \frac{1}{2}h$. The volume of a cone is $V = \frac{1}{3}\pi r^3$, and substituting for r, we get

$$V = \frac{1}{3}\pi\left(\frac{1}{2}h\right)^3 = \frac{1}{24}\pi h^3$$

Differentiating with respect to t we get

$$\frac{dV}{dt} = \frac{dV}{dh}\cdot\frac{dh}{dt} = \frac{1}{8}\pi h^2\frac{dh}{dt}$$

We are given $\dfrac{dV}{dt} = -5$. This is negative because the volume is decreasing. When the water level is $5\,\text{cm}$ from the top, $h = 20 - 5 = 15\,\text{cm}$ and so we have

$$-5 = \frac{1}{8}\pi(15)^2\frac{dh}{dt}$$

or $\qquad \dfrac{dh}{dt} = -\dfrac{40}{225\pi} = \dfrac{-8}{45\pi}\,\text{cm/s}$

Application

24.21

When air changes volume adiabatically (without the addition of heat), the pressure P and volume V satisfy the formula $PV^{1.4} = k$, where k is a constant. At a certain instant, the pressure is $75\,\text{lb/in.}^2$ and the volume is $25\,\text{in.}^3$ and is decreasing at the rate of $3\,\text{in.}^3/\text{s}$. How rapidly is the pressure changing at that time?

EXAMPLE 24.21 (Cont.)

Solution Since the equation is given, a drawing is not needed for this problem. The variables are:

$$t = \text{time in seconds}$$
$$P = \text{pressure in lb/in.}^2$$
$$V = \text{volume in in.}^3$$

We have the following rates of change:

$$\frac{dV}{dt} = \text{rate of change of volume with respect to time}$$

$$\frac{dP}{dt} = \text{rate of change of pressure with respect to time}$$

From the formula $PV^{1.4} = k$ and the fact that when $P = 75$, $V = 25$, we can determine that $k \approx 6{,}794.81$.

We are given $\dfrac{dV}{dt} = -3$ when $V = 25$ and need to find $\dfrac{dP}{dt}$. We know that $P = kV^{-1.4}$ or

$$\frac{dP}{dt} = \frac{dP}{dV} \cdot \frac{dV}{dt} = -1.4kV^{-2.4}\frac{dV}{dt}$$

$$= \frac{-1.4(6{,}794.81)(-3)}{(25)^{2.4}}$$

$$= 12.6$$

The pressure is increasing at the rate of 12.6 lb/in.^{2}

Exercise Set 24.5

Solve Exercises 1–44.

1. Gas is escaping from a balloon at the rate of 20 L/min. How fast is the surface area changing when the radius is 300 cm? (1 liter = 1000 cm^3)

2. *Construction* Sand is emptied down a chute at the rate of 10 ft^3/s forms a conical pile whose height is always twice the radius. At what rate is the radius changing when the height is 4 ft?

3. Water is emptied from a spherical tank of radius 24 ft. The water in the tank is 8 ft deep and decreasing at the rate of 2 ft/min. At this time, what is the rate of change of the radius of the top of the water?

4. *Environmental technology* A circular oil slick of uniform thickness is caused by a spill of 10 m^3 of oil. The thickness of the oil slick is decreasing at the rate of 1 mm/h. At what rate is the radius of the oil slick changing when the radius is 12 m?

5. *Transportation* A helicopter flies parallel to the ground at an altitude of 0.75 km and at a speed of 3 km/min. If the helicopter flies in a straight line that will pass directly over you, at what rate is the distance between you and the helicopter changing 1 min after it passes over you?

6. *Recreation* A rectangular swimming pool 15 m long and 10 m wide is 4 m deep at one end and 1 m deep at the other, with a constant drop from the shallow end to the deep end. Water is pumped into the pool at the rate of 50 L/min. At what rate is the water rising when it is 2 m deep at the deep end? (1 000 L = 1 m^3)

7. *Landscaping*   A fish pond is in the shape of the bottom half of a sphere with a radius of 5 ft. Water is pumped into the pond at the rate of 10 ft³/min. (a) At what rate is the water rising when it is 2 ft deep? (b) What is the rate when the water is 3 ft deep? (c) What is the rate when the water is 4 ft deep? [Note: $V_{liquid} = \frac{1}{3}\pi h^2(3R-h) = \frac{1}{6}\pi h(3r^2+h^2)$.]

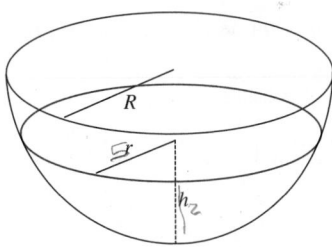

8. *Landscaping*   When the pond in Exercise 7 is drained, the water drains at the rate of 45 ft³/min. (a) At what rate does the water drop when it is 4 ft deep? (b) What is the rate when it is 2 ft deep?

9. A horizontal trough 10.00 m long has a cross-section in the shape of an equilateral triangle with a top width of 0.500 m. If the trough is filled with water at the rate of 25.0 L/min, how fast is the surface of the water rising when the depth is 0.175 m?

10. The base of a triangle is increasing at the rate of 5.00 cm/s and the height is decreasing at the rate of 4.00 cm/s. At what rate is the area of the triangle changing when the base is 2.625 m and the height is 1.375 m?

11. *Environmental science*   Air is pumped into a spherical weather balloon at the rate of 60 cm³/s. How fast is the surface area changing when the radius is 8 cm?

12. The height of a right-circular cone with radius of 2.43 m is increasing at the rate of 0.425 m/h. How fast is the lateral area of the cone changing when the height is 4.27 m?

13. A balloon is in the shape of a cylinder with a hemisphere on each end. The height of the cylinder is four times its radius. Air is pumped into the balloon at the rate of 60 cm³/s. How fast is the surface area changing when the radius is 8 cm?

14. *Thermodynamics*   When a metal plate is heated it expands. If the shape of the plate is circular and the

radius is increasing at the rate of 0.2 mm/s, at what rate is the area of the top surface changing when the radius is 20 mm?

15. *Thermodynamics*   If a metal plate in the shape of a square is heated and the length of one side is increasing at the rate of 0.3 mm/s, at what rate is the area of the top surface changing when the length of one side is 40 mm?

16. At a certain instant when air is changing adiabatically, the pressure is 60 dynes/cm² and is increasing at a rate of 4 dynes/cm²/s. If the volume at that time is 100 cm³, what is the rate at which the volume is changing?

17. *Physics*   The pressure P in kilopascals (kPa) and volume V in cubic meters of a gas is given by $PV^{1.2} = 400$. If the pressure is increasing at a rate of 10% of itself per hour, find the rate at which the volume is changing expressed as a percent of itself.

18. *Meteorology*   A weather balloon is rising vertically at the rate of 5 ft/s. An observer is 500 ft from a point on the ground directly below the balloon. At what rate is the distance between the balloon and the observer changing when the altitude of the balloon is 2,000 ft?

19. *Space technology*   A radar station is following the path of a rocket which has been launched vertically from a site 2.430 km away. How fast is the rocket rising when it is 3.750 km high, if the distance between the rocket and the radar site is increasing at the rate of 325.0 m/s?

20. *Physics*   **Boyle's law** for gases states that $PV = k$, where P is the pressure and V the volume, and k is a constant. At a certain instant the volume is 50 in.³, the pressure 20 lb/in.², and the pressure is increasing at the rate of 1.5 lb/in.² every minute. At what rate is the volume changing at this instant?

21. *Physics*   The displacement s in meters and the velocity v of an object in meters per second are related according to $v^2 = 1200 - 36.0s$. (a) Find the acceleration of the object. (b) What is the acceleration when $s = 2.45$ m?

22. A 10-m ladder is leaning against a vertical wall. If the base of the ladder slips away from the wall at the rate of 0.5 m/s, how fast is the top of the ladder going down the wall when the base of the ladder is (**a**) 2 m from the wall? (**b**) 3 m from the wall? (**c**) 4 m from the wall? (**d**) 6 m from the wall?

23. *Navigation*   Two radar stations are at points A and B. Point B is 8 mi east of A. The two stations are tracking a ship. At a certain instant, the ship is 8 mi from A and this distance is increasing at the rate of 14 mph. At the same instant, the ship is also 6 mi from B and this distance is increasing at the rate of 2 mph. (**a**) Where is the ship? (**b**) How fast is it moving? (**c**) In what direction is it moving?

24. *Transportation*   A car leaves an intersection and travels due north with a speed of 80.0 km/h. Another car leaves the same intersection 0.5 h later going due west at a speed of 100.0 km/h. How fast are the two cars separating
(**a**) 1.00 h after the second car started; and
(**b**) 2.00 h after the second car started?

25. A person 6 ft tall walks at the rate of 5 ft/s away from a street light that is 20 ft above the ground. (**a**) At what rate is the top of the person's shadow moving? (**b**) At what rate is the length of the shadow changing when the person is 10 ft from the base of the light? (**c**) What is the rate of change of the shadow when the person is 15 ft from the base of the light?

26. *Physics*   A 75-lb weight is attached to a rope 50 ft long, which passes over a pulley at P, 20 ft above the ground. The other end of the rope is attached to the bumper of a truck at a point 2 ft above the ground. If the truck moves at the rate of 8 ft/s, how fast is the weight rising when it is 10 ft above the ground?

27. *Electricity*   The electric resistance R of a certain resistor is a function of the temperature T, as shown by $R = 6 + 0.008T^2$, where R is in ohms and T in degrees Celsius. If the temperature is increasing at the rate of 0.01°C/s, how fast is the resistance changing when $T = 40$°C?

28. *Transportation*   Two cars leave the same point, traveling on straight roads that are perpendicular. One car is traveling east at the rate of 30 mph and the other car is traveling north at the rate of 40 mph. How fast is the distance between them changing (**a**) after 3 min (0.05 h) and (**b**) after 6 min (0.10 h)?

29. *Electronics*   The impedance Z in ohms (Ω) of a circuit is given by

$$Z = \frac{RX}{R+X}$$

If R is constant at $3.00\,\Omega$ and X is increasing at a rate of $1.45\,\Omega$/min, what is the rate at which Z is changing when $X = 1.05\,\Omega$?

30. *Electronics*   The impedance Z in a series circuit is given by $Z^2 = R^2 + X^2$, where X is the reactance. If $X = 12\,\Omega$ and R increases at the rate of $3.0\,\Omega$/s, what is the rate at which Z is changing when $R = 6.0\,\Omega$?

31. *Electricity*   If two variable resistances R_1 and R_2 are linked in parallel, then the effective resistance R is given by $\dfrac{1}{R} = \dfrac{1}{R_1} + \dfrac{1}{R_2}$. At a certain instant, R_1 is $4.00\,\Omega$ and increasing at the rate of $0.50\,\Omega$/s. At the same time, R_2 is $5.00\,\Omega$ and increasing at the rate of $0.40\,\Omega$/s. What is the rate of change of R?

32. *Electricity*   The power P, in watts, in a circuit varies according to $P = RI^2$. If $R = 80.00\,\Omega$ and I varies at the rate of 0.24 A/s, what is the rate of change of P when $I = 2.5$ A?

33. *Electronics*   The resistance in a resistor is given by $R = 35.0 + 0.0174T^2$, where R is in ohms and T is in degrees Celcius. How is the resistance changing when the temperature of the resistor is 47.0°C and is decreasing at the rate of 1.25°C/min?

34. *Aeronautical engineering*   In the air, the speed of sound v, in m/s, is given by the formula $v = 331\sqrt{\dfrac{T}{273}}$, where T is the temperature in degrees Kelvin. (n degrees Celsius $= n + 273$ degrees Kelvin.) As an airplane increases altitude, the outside temperature decreases at the rate of 5.8°C/km. What is the rate of change in the speed of sound when the temperature outside the airplane is -30°C?

35. *Acoustical engineering* The **Doppler effect** is the apparent change in frequency of a sound source when there is relative motion between the source and the observer. When a person stands still, the apparent change in frequency f in hertz (Hz) of an object moving toward the person is given by the formula

$$f = f_s \frac{v_L}{v_L - v_s}$$

where f_s is the frequency of sound, in Hz, from the source; v_s is the velocity of the source in m/s; and v_L is the speed of sound in m/s. At 21°C, the speed of sound is about 343 m/s. A train is moving at a speed of 29 m/s toward an observer at the instant it sounds it horn at a frequency of 200 Hz. **(a)** What is the frequency of sound to the observer? **(b)** What is the apparent rate of change of sound to the observer?

36. *Medical technology* Blood flows faster the closer it is to the center of a blood vessel. According to **Poiseuille's laws**, the velocity, V, of blood is given by $V = k(R^2 - r^2)$, where R is the radius of the blood vessel, r is the distance of a layer of blood flow from the center of the vessel, and k is a constant.
   (a) Find $\dfrac{dV}{dt}$ when $k = 475$. Treat r as a constant.
   (b) Suppose that a cross-country skier's blood vessel has radius $R = 0.015$ mm and that the cold weather is causing the vessel to contract at a rate of $\dfrac{dR}{dt} = -0.001$ mm/min. How fast is the velocity of the blood changing?

37. *Medical technology* One of the treatments for high blood pressure is to give a patient some medication to dilate the blood vessels. Use **Poiseuille's law** with $k = 637.5$ to find the rate of change of the blood velocity when $R = 0.0250$ mm and R is changing at 0.002 mm/min.

38. *Police science* Crime rates are influenced by temperature. In a certain town with a population of 100,000, the crime rate has been approximated at $C = 0.1(T - 47)^2 + 95$, where C is the number of crimes per month and T is the average monthly temperature. The average temperature was 51°F, and by the end of the month the average monthly temperature was rising at the rate of 7°F/mo. How fast is the crime rate rising at the end of April?

39. *Electronics* The impedance of a circuit containing a resistance and an inductance in parallel is given by $Z = \dfrac{RX}{R + X}$. If $R = 7.5\,\Omega$ and X is decreasing at the rate of 2.5 Ω/s, at what rate is Z changing when $X = 4.0\,\Omega$?

40. *Petroleum engineering* An oil spill on the open sea in calm weather spreads in a circular pattern. If the radius of the oil slick increases at the rate of 0.9 m/s, how fast is the area of the spill increasing when its radius is 225 m?

41. *Electronics* The impedance of a circuit containing a resistance and an inductance in parallel is given by $Z = \dfrac{RX}{R + X}$. If $R = 7.5\,\Omega$ and Z is increasing at the rate of 2.0 Ω/s, at what rate is X changing when $X = 3.5\,\Omega$?

42. *Petroleum engineering* Oil is leaking from a tanker at the rate of 5 000 L/min. The leak results in an oil spill that is circular on the ocean's surface. The depth of the spill varies linearly from a maximum of 5.0 cm at the origin of the spill to a minimum of 0.5 cm at the outside edge of the oil slick. How fast is the radius of the slick increasing 4 h after the tanker started leaking?

43. *Heating technology* A boiler pipe has a circular cross-section. The inner radius of the pipe was originally 1 in. but is decreasing at the rate of 0.1 in./yr due to mineralization. At what rate is the cross-sectional area of its opening decreasing at the instant when the radius is 0.5 in.?

44. *Construction technology* A concrete pier is being constructed to support a bridge. The pier is in the shape of the frustum of a square pyramid. The bottom base of the pier measures 3 m on each side, the top base measures 2 m on a side, and the height is 2 m. If the concrete is poured at a constant rate of 0.01 m³/s, at what rate is the height h of the poured concrete rising when $h = 1$ m?

In Your Words

45. Without looking in the book, describe the steps used to solve a related rate problem.

46. Explain how you can tell a problem involves related rates.

≣ 24.6
NEWTON'S METHOD

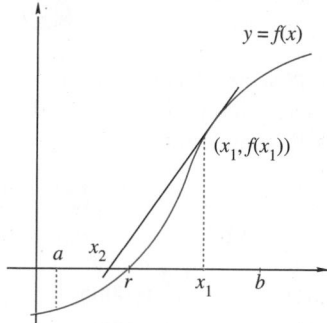

FIGURE 24.28

In Section 18.4, we learned a method for finding the roots of a polynomial by using linear interpolation. In this section, we will learn a faster method that uses derivatives. This new technique is known as **Newton's method** or the **Newton-Raphson method** for finding roots.

In using the linear interpolation method, we used the locator theorem to determine that a root existed between the real numbers a and b. We will begin Newton's method the same way. For a continuous function f over an interval, we have found two real numbers, a and b, in the interval, such that one of the values of $f(a)$ and $f(b)$ is positive and the other is negative.

Assume that $a < b$. We will select some point x_1 in the interval $[a, b]$. We know that the tangent to the graph of f at x_1 is given by $f'(x_1)$. We also know that the equation of this tangent line is $y - f(x_1) = f'(x_1)(x - x_1)$ or $y = f(x_1) + f'(x_1)(x - x_1)$. Suppose that x_2 is the x-intercept of this tangent line as shown in Figure 24.28. Then we know that $(x_2, 0)$ is a point on the tangent line. Thus, the tangent line satisfies the equation

$$0 = f(x_1) + f'(x_1)(x_2 - x_1)$$

If $f'(x_1) \neq 0$, we can rewrite this equation as

$$x_2 = x_1 - \frac{f(x_1)}{f'(x_1)}$$

The process can be continued. We can use x_2 to determine a new point x_3, and so on until we get a point x_n, where $f(x_n)$ is within some desired distance of zero. In general, if x_n is the nth approximation, then

$$x_{n+1} = x_n - \frac{f(x_n)}{f'(x_n)}$$

provided $f'(x_n) \neq 0$.

EXAMPLE 24.22

Use Newton's method to approximate the root between -4 and -3, if $P(x) = x^3 + x^2 - 7x + 3$.

Solution We will let our first guess, x_1, be -3. The results of the successive approximations are given in the following table. $P'(x) = 3x^2 + 2x - 7$

i	x_i	$P(x_i)$	$P'(x_i)$	$x_{i+1} = x_i - \dfrac{P(x_i)}{P'(x_i)}$	$P(x_{i+1})$
1	−3.0000	6.0000	14.0000	−3.4286	−1.5487
2	−3.4286	−1.5487	21.4087	−3.3563	−0.0490
3	−3.3563	−0.0490	20.0816	−3.3539	−0.0009

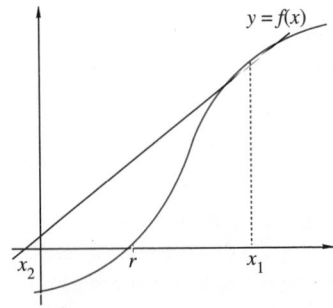

FIGURE 24.29

In Section 18.4, we worked this same problem. It took five steps to get $x_6 = -3.3535$ and $P(x_6) = 0.0062$. Newton's method took three steps to get $x_4 = -3.3539$ and $P(x_4) = -0.0009$. It appears that Newton's method is almost twice as fast as the linear interpolation method.

Newton's method does not always work. In particular, if the first choice of x_1 is not sufficiently close to the root r, then the second approximation, x_2, will be farther away from r instead of closer. This is indicated in Figure 24.29. In general, you do not want to select a point where the derivative is close to zero. As a rule of thumb, whenever $|f'(x_1)| \leq 1$, then select a different choice for x_1.

EXAMPLE 24.23

Use Newton's method to approximate $\sqrt{11}$.

Solution This is equivalent to approximating the positive real root of $f(x) = x^2 - 11$. Since $f(3) = -2$ and $f(4) = 5$, we know there is a root between 3 and 4. Using $f'(x) = 2x$ and selecting $x_1 = 3$, we get the following results.

i	x_i	$P(x_i)$	$P'(x_i)$	$x_{i+1} = x_i - \dfrac{P(x_i)}{P'(x_i)}$	$P(x_{i+1})$
1	3.0000	−2.0000	6.0000	3.3333	0.1109
2	3.3333	0.1109	6.6667	3.3167	0.0005
3	3.3167	0.0005	6.6334	3.3166	−0.0002

Thus, $\sqrt{11} \approx 3.3166$. ∎

Exercise Set 24.6

1. Modify the computer program in Section 18.4 to find roots using Newton's method.

In Exercises 2–8, use Newton's method to approximate the real roots to four decimal places.

2. $x^3 + 5x - 8 = 0$

3. $x^3 - 4x + 2 = 0$

4. $x^4 - x - 4 = 0$

5. $x^4 - 2x^3 - 3x + 2 = 0$

6. $x^4 = 125$

7. $x^3 = 91$

8. $x^5 = 111$

Solve Exercises 9–10.

9. *Sheet metal technology* An open-topped barrel in the shape of a right circular cylinder is to be fabricated from sheet metal. The manufacturer has specified that it holds a volume of $V = 10\,000\,\text{cm}^3$ and has a total surface area of $A = 2\,400\,\text{cm}^2$.
   (a) Show that the radius r of the barrel must satisfy the equation $2V + \pi r^3 - Ar = 0$, where V and A are the area and volume given above.
   (b) Use Newton's method to find the radius of the barrel accurate to 4 decimal places.

10. *Air traffic control* One method for clearing fog from an airport runway is to ignite long troughs filled with fuel that have been placed along the sides of the runway. A trough is to be formed from a rectangular sheet of metal that measures $20\,\text{m} \times 0.7\,\text{m}$. The cross-section of the trough is an arc of a circle with radius r and subtended angle θ. If each trough is to hold $0.85\,\text{m}^3$ of fuel, use Newton's method to determine the size of θ.

In Your Words

11. What is Newton's method and how is it used?

12. What are the advantages and disadvantages of Newton's method compared to using a graphing calculator?

≣ 24.7
DIFFERENTIALS

In studying the derivative of a function $y = f(x)$, we used the notations $f'(x)$ and dy/dx to represent the derivative. The symbols dy and dx are called **differentials**. They may be given their own meaning as we'll see in this section.

Since $f'(x) = \dfrac{dy}{dx}$, we can solve this equation for dy and get $dy = f'(x)dx$, which is called the **differential of** f. In this equation, $dx \neq 0$.

EXAMPLE 24.24

Find the differential of the function $f(x) = 3 + x + 14x^{2/7}$.

Solution   Here $f'(x) = 1 + 4x^{-5/7}$. Thus, the differential is

$$dy = f'(x)\,dx = \left(1 + 4x^{-5/7}\right)dx$$

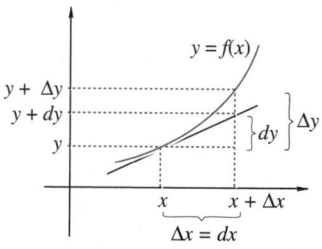

FIGURE 24.30

Figure 24.30 shows the change in height of a point that moves along the tangent line to the function $y = f(x)$ at the point $(x, f(x))$, as x changes from x to $x + \Delta x$. This change in height is called dy. The actual change in the value of y is the increment Δy. If we think of x as a fixed point, then dy is a linear function of the increment Δx. For this reason, dy is called the **linear approximation** to the actual increment Δy. We can approximate $f(x + \Delta x)$ by writing dy in place of Δy. Thus, we see that

$$f(x + \Delta x) = y + \Delta y \approx y + dy$$

Since $y = f(x)$ and $dy = f'(x)\Delta x$, we get the **linear approximation formula**:

$$f(x + \Delta x) \approx f(x) + f'(x)\Delta x$$

EXAMPLE 24.25

If $y = x^3$, determine Δy and dy, if $x = 2$ and $\Delta x = 0.01$.

Solution   Since $f(x + \Delta x) = y + \Delta y$ and $y = f(x)$, we can see that $\Delta y = f(x + \Delta x) - f(x)$. Thus, when $x = 2$ and $\Delta x = 0.01$, we have

$$\begin{aligned}
\Delta y &= (x + \Delta x)^3 - x^3 \\
&= (2.01)^3 - 2^3 \\
&= 8.1206 - 8 = 0.1206
\end{aligned}$$

We also know that $dy = f'(x)dx$. If $f(x) = x^3$, then $f'(x) = 3x^2$. Since $dx = \Delta x$, we have

$$dy = 3(2)^2(.01) = 0.12$$

As you can see, dy is a good approximation for Δy.

Application

EXAMPLE 24.26

A spherical ball-bearing has a radius of 8 mm when it is new. What is the approximate volume of metal lost after it wears down to a radius of 7.981 mm?

Solution The exact volume of the metal lost is the change in volume ΔV of the sphere, where $V = \frac{4}{3}\pi r^3$. The change in radius is $\Delta r = 7.981 - 8.000 = -0.019$ mm. The change in volume is approximated by dV, where

$$dV = 4\pi r^2 dr$$
$$= 4\pi(8^2)(-0.019) \approx -15.281$$

The approximate loss in volume is 15.281 mm^3.

Application

EXAMPLE 24.27

A hemispherical dome has a radius of 25 ft. It is going to be given a coat of paint 0.01 in. thick. What is the approximate volume of paint needed for this job?

Solution What we are looking at is the change in volume of this hemisphere when the radius changes from 25.000 ft to 25.00083 ft (0.01 in. $\approx$ 0.00083 ft) or $\Delta r = 0.00083$ ft. The volume of a hemisphere is one-half that of a sphere, or $V = \frac{2}{3}\pi r^3$. Again, using differentials we see that $dV = 2\pi r^2 dr$. When $r = 25$ and $dr = 0.00083$, we have

$$dv = 2\pi(25^2)(0.00083)$$
$$\approx 3.2594$$

About 3.26 ft^3 of paint are needed.

Differentials can be used to estimate errors in results.

Application

EXAMPLE 24.28

A cylindrical can with an open top has an inside height of exactly 100 mm and a radius of 25 mm with a maximum error of 0.1 mm. Estimate the maximum error in the calculated volume of this can.

Solution If r denotes the radius and h the height, then $V = \pi r^2 h$. If the maximum error in the radius is denoted as dr, then we can see that $dV = 2\pi rh \, dr$. If $r = 25, h = 100$, and $dr = 0.1$, then

$$dV = 2\pi(25)(100)(0.1)$$
$$= 500\pi \approx 1\,570.8 \text{ mm}^3$$

Thus, the maximum error in volume due to the error on the radius is $1\,570.8$ mm^3.

If there is an error in measurement, we define the **relative** or **average error** as the ratio of the actual error to the actual size of the quantity. This is often written in differential form as

$$\text{relative error} = \frac{dy}{y}$$

The relative error is usually expressed as a percentage.

In Example 24.28, the actual error in the radius was 0.1 mm when the radius was 25 mm. Thus the relative error in measuring the radius is $\dfrac{dr}{r} = \frac{0.1}{25} = 0.004 = 0.4\%$. The actual volume of the can is $62\,500\pi$ with a possible error of 500π. The relative error is $\dfrac{500\pi}{62\,500\pi} = 0.008 = 0.8\%$.

Exercise Set 24.7

In Exercises 1–6, find the differential of the given functions.

1. $f(x) = x^4 - x^2$

2. $g(x) = 3x^2 + 2x$

3. $h(x) = 5x^3 - x^2 + x$

4. $j(x) = (x^2 - 4)^3$

5. $k(x) = \sqrt[3]{4 - 2x}$

6. $m(x) = \dfrac{x}{x - 4}$

In Exercises 7–12, find Δy and dy for the given values of x and Δx.

7. $y = x^2 - x$, $x = 3$, $\Delta x = 0.2$

8. $y = x^3 + x$, $x = 2$, $\Delta x = 0.1$

9. $f(x) = 3x^2 - 4x + 1$, $x = 5$, $\Delta x = 0.15$

10. $g(x) = x^4$, $x = 4.1$, $\Delta x = 0.1$

11. $h(x) = \dfrac{1}{x^3}$, $x = 3$, $\Delta x = 0.2$

12. $j(x) = \dfrac{x}{x + 1}$, $x = 0.5$, $\Delta x = 0.1$

Solve Exercises 13–36.

13. *Physics* A tank filled with water is in the shape of a right-circular cone. The radius of the cone is measured to be 2 m with a maximum error of 0.01 m. The height is exactly 4 m. **(a)** What is the approximate volume of the water in the tank? **(b)** What is the maximum error in this volume? **(c)** What is the relative error in the volume?

14. *Physics* The radius of a circular disk is estimated to be 200 mm with a maximum error of 1.5 mm. **(a)** What is the maximum error in calculating the area of one side of the disc? **(b)** What is the relative error in the radius? **(c)** What is the relative error in the area?

15. The side of a square is measured to be 24 in. with a possible error of 0.02 in. **(a)** What is the maximum error in the area of the square? **(b)** What is the relative error?

16. A cube has an edge measuring 1.452 m. If the error in measurement is ±0.5 mm, what is the approximate error in the volume?

17. The radius of a hemispherical dome is measured as 100 m with a maximum error of 10 mm. What is the maximum error in calculating its surface area?

18. With what accuracy must the radius of the dome in Exercise 17 be measured in order to have a percentage error of at most 0.01% in its calculated surface area?

19. The diameter of a sphere is measured to be 12.4 m. If measurement error is plus or minus half the unit of the most precise digit, that is ± 0.05 m, (a) what is the approximate error in the volume of the sphere due to measurement error? (b) What is the approximate relative error?

20. If the diameter in Exercise 19 is given as 12.40 m, what is the approximate error in the volume and the relative error in volume due to measurement error?

21. *Energy technology* The **Stefan-Bolzmann law** for the emission of radiant energy from the surface of a body is given by $R = \sigma T^4$, where R is the rate of emission per unit area, T is the temperature in degrees Kelvin, and σ is the Stefan-Bolzmann constant. If a relative error of 0.02 is made in T, what is the maximum error in R?

22. *Electricity* A wire has a resistance of R ohms given by $R = \dfrac{250}{r^2}$, where r is the radius of the wire. There is a possible error in the radius of $\pm 0.4\%$.
    (a) What is the approximate error in the resistance? (Express your answer in terms of r.)
    (b) What is the relative error?

23. *Electricity* In a dc circuit, $V = IR$. If a circuit has a constant voltage of 30 V, what is the approximate change in the current if the resistance changes from 10.0 to 10.1 Ω?

24. *Electricity* The equivalent resistance R when a resistance of 20 Ω and a variable resistance R_v are connected in parallel is
$$R = \frac{20R_v}{20 + R_v}$$
If the variable resistance is changed from 10.0 to 10.1 Ω, what is the approximate change in R?

25. *Electricity* The resistance in a resistor is given by $R = 35.0 + 0.0174T^2$, where R is in ohms (Ω) and T is the temperature in degrees Celsius (°C). If the error in reading the temperature at $T = 125$°C is $\pm - 0.5$°C, what is the approximate error in the resistance?

26. *Construction* A coat of paint 0.5 mm thick is applied to the surface of a spherical storage tank with a radius of 20 m. How many liters of paint are needed?

27. *Mechanical engineering* A company manufactures stainless-steel, spherical ball-bearings with a radius

of 8 mm. The percentage error in the radius must be no more than 0.5%. What is the approximate percentage error for the surface area of the ball bearing?

28. *Material science* Stainless steel has a mass of 0.00794 g/mm^3.
    (a) What is the mass of one of the ball-bearings in Exercise 27?
    (b) What is the maximum error in the mass?

29. *Industrial technology* In a precision manufacturing process, ball bearings must be made with a radius of 0.6 mm, with a maximum error in the radius of ± 0.015 mm. Estimate the maximum error in the volume of the ball bearing.

30. *Pharmacology* The concentration of a certain drug in the bloodstream x hours after being administered is approximately $C(x) = \dfrac{5x}{9 + x^2}$. Use differentials to approximate the changes in concentration as x changes from (a) 1 h to 1.5 h, (b) 2 h to 2.25 h, and (c) 3 h to 3.25 h.

31. *Medical technology* A tumor is approximately spherical in shape. If the radius of the tumor changes from 15 mm to 15.5 mm, use differentials to find the approximate change in volume of the tumor.

32. *Petroleum engineering* The top of an oil slick is in the shape of a circle. Find the approximate increase in the area of the slick's top if its radius changes from 2.1 km to 2.3 km.

33. *Business* A company estimates that the revenue in dollars from the sale of n computers is given by
$$R(n) = 2345 + 0.02n^2 + 0.00015n^3$$
Use differentials to approximate the change in revenue from the sale of one more computer when 500 have been sold.

34. *Meteorology* A spherical weather balloon is being inflated. Use differentials to approximate the change in the volume of the balloon as the radius changes from 20.0 in. to 21.5 in.

35. *Machine technology* A plate in the shape of a right circular cylinder 3.2 cm thick was originally made 30.0 cm in diameter. If 0.2 cm is trimmed all around its edge, use differentials to approximate the volume of the metal that was trimmed.

36. *Acoustical engineering*   The frequency, f, in Hertz, of the fundamental note on a guitar string is related to the tension, T, in the string; the length, L, of the string; and the linear density, μ, of the string

according to the formula

$$f = \frac{1}{2L}\sqrt{\frac{T}{\mu}}$$

Estimate the amount by which the frequency will change if the tension is adjusted 1% during tuning. Let $L = 1.15$ m, $\mu = 0.0036$ kg/m, and $T = 620$ N.

In Your Words

37. What is meant by the differential terms dy and dx?

38. Explain the difference between dy and Δy.

≡ 24.8
ANTIDERIVATIVES

Until now, we have always differentiated a function to arrive at a solution. For example, if we were given a position function, we took its first derivative to get its velocity function and its second derivative to gets its acceleration function. Is it possible to reverse this process? For example, if we know the velocity function for an object, could we determine its position function? As we will see in this section, it is sometimes possible to reverse the differentiation process.

Suppose we have a function F and the derivative of F is f. Thus, $F' = f$. In this case, F is called the **antiderivative** of f.

EXAMPLE 24.29

Find an antiderivative of $f(x) = 5x^3$.

Solution   We want to find $F(x)$, where $F'(x) = f(x) = 5x^3$. When we differentiate a polynomial, we reduce the power of x by 1. We also multiply the coefficient by the power of x. Thus, the derivative of x^n is nx^{n-1}. Since the power of f is 3, then this must be one less than the power of F. Thus, F is of the form $F(x) = kx^4$. Now $F'(x) = 4kx^3 = 5x^3$ and so $k = \frac{5}{4}$. Thus $F(x) = \frac{5}{4}x^4$.

But, we have not found a unique solution. For example, the derivative of $\frac{5}{4}x^4 + 9$ is also $5x^3$, and the derivative of $\frac{5}{4}x^4 - 7$ is $5x^3$. In fact, since the derivative of a constant C is 0, the antiderivative of $5x^3$ is $\frac{5}{4}x^4 + C$.

What we have discovered is that, in general, if $f(x) = kx^n$, then

$$F(x) = \frac{k}{n+1}x^{n+1} + C, \text{ if } n \neq -1$$

Notice that

$$F'(x) = \frac{k(n+1)}{n+1}x^{n+1-1} = kx^n = f(x).$$

So F is the antiderivative of f. A special case occurs when $n = 0$. Then $f(x) = k$ and $F(x) = kx + C$.

Antiderivatives Differ by a Constant

If G and F are both antiderivatives of the same function, f, then

$$G(x) = F(x) + C \qquad \text{where } C \text{ is a constant.}$$

EXAMPLE 24.30

Find the antiderivative of $g(x) = 3x^2 - 2x + 5$.

Solution Using this new method to take the antiderivative of each term, we get

$$G(x) = x^3 - x^2 + 5x + C$$

You should check to see that $G'(x) = g(x)$.

EXAMPLE 24.31

Find the antiderivative of $h(x) = 4x^2 + 2x^{-2}$.

Solution The antiderivative of $4x^2$ is $\frac{4}{3}x^3$ and of $2x^{-2}$ is $-2x^{-1}$, so $H(x) = \frac{4}{3}x^3 - 2x^{-1} + C$.

We can apply antiderivatives to many problems. For example, we know that the acceleration due to the Earth's gravity is about $32\,\text{ft/s}^2$ or $9.8\,\text{m/s}^2$ downward. We also know that the acceleration is the derivative of the velocity or that the velocity function is an antiderivative of the acceleration. Similarly, the position function is an antiderivative of the velocity function. With enough information we can determine the value of the constant of the antiderivative. This is demonstrated in the next example.

Application

EXAMPLE 24.32

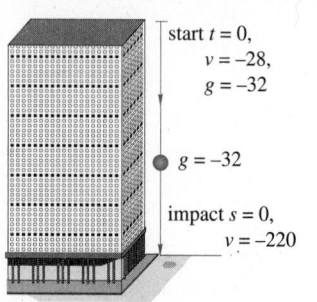

start $t = 0$,
$v = -28$,
$g = -32$

$g = -32$

impact $s = 0$,
$v = -220$

FIGURE 24.31

An object is thrown straight downward from the top of a building with an initial speed of $28\,\text{ft/s}$. The object strikes the ground with a speed of $220\,\text{ft/s}$. How tall is the building?

Solution First, let's set up a coordinate system with the ground level at $y = 0$, as shown in Figure 24.31. We know that acceleration is given by $a(t) = -32$. The velocity is the antiderivative of the acceleration and so $v(t) = -32t + C$. When the object is thrown, $t = 0$ and we are told that $v(0) = -28$. (It is negative because the velocity is downward.) Thus, we can see that $C = -28$ and $v(t) = -32t - 28$.

The object strikes the ground with a velocity of $220\,\text{ft/s}$ downward or -220. Thus, when it strikes the ground, $v(t) = -32t - 28 = -220$. Solving for t, we see that it strikes the ground when $t = 6\,\text{s}$.

The height of the object is given by the position function $s(t)$, and is the antiderivative of the velocity. Thus, $s(t) = -16t^2 - 28t + k$. When the object

EXAMPLE 24.32 (Cont.)	hits the ground, $t = 6$ and $s(t) = 0$, so $-16(6^2) - 28(6) + k = 0$. Solving, we get $k = 744$ and $s(t) = -16t^2 - 28t + 744$. At $t = 0$ when the object is released, $s(0) = 744$ and so the height of the building is 744 ft.

Exercise Set 24.8

In Exercises 1–16, find the antiderivative of the given function.

1. $f(x) = 7$

2. $g(x) = 4x$

3. $h(x) = 2 - 3x^2$

4. $j(x) = x^2 - 3x + 5$

5. $k(x) = 4x^3 - 3x^2 + 2x + 9$

6. $m(x) = \frac{4}{5}x^3 - 6x^2$

7. $f(x) = x^{-4} + 2x^{-3} - x^{-2} + 5$

8. $g(x) = \frac{1}{x^3} + 2$

9. $h(x) = \sqrt{x} + x - \frac{1}{x^2}$

10. $j(x) = 3x^2 - \frac{1}{\sqrt[3]{x}} - \frac{4}{x^2}$

11. $k(x) = 2x^{-1.4} + 3.5x^{-6} + \frac{1.2}{x^{1.3}}$

12. $m(x) = \frac{2}{\sqrt{x}} + 3x^2\sqrt{x}$

13. $s(t) = t^2 + 2t$

14. $a(t) = \frac{2}{3}t - \frac{3}{4}t^2 + \sqrt{t^3}$

15. $v(t) = 42t - 5$

16. $p(v) = \frac{1}{3}v^3 - \frac{3}{v^4}$

Solve Exercises 17–36.

17. *Physics*   A ball is dropped from a building that is 600 ft high. **(a)** How long does it take it to reach the ground? **(b)** What is its velocity when it hits the ground?

18. *Physics*   A ball is dropped from a building that is 140 m high. **(a)** How long does it take to reach the ground? **(b)** What is its velocity when it hits the ground?

19. *Physics*   A ball is thrown straight down from a building with an initial speed of 25 ft/s. It hits the ground with a speed of 185 ft/s. **(a)** How long was the ball in the air? **(b)** How high is the building?

20. *Physics*   A ball is thrown straight down from a building with an initial speed of 12.1 m/s. It hits the ground with a speed of 66 m/s. **(a)** How long was the ball in the air? **(b)** How high is the building?

21. *Physics*   A ball is thrown straight upward from the top of a building with an initial speed of 160 ft/s. It

strikes the ground with a speed of 384 ft/s. **(a)** How long is it in the air? **(b)** How high is the building?

22. *Physics*   A ball is thrown straight upward from the top of a building with an initial speed of 45 m/s. It strikes the ground with a speed of 120 m/s. **(a)** How long is it in the air? **(b)** How high is the building?

23. *Transportation*   A car starting from rest has a constant acceleration of 3.2 m/s². How far has the car traveled when the speed is 32 m/s?

24. *Automotive engineering*   A car's brakes are applied when it is going 45 mph and provides a constant deceleration of 20 ft/s². **(a)** How long does it take the car to stop? **(b)** How far does it travel before it stops? (Change 45 mph to ft/s.)

25. *Aeronautical engineering*   A runway is 500 m long. If a plane lands at an end of the runway with a speed of 50 m/s, what is the constant acceleration that would bring the plane to a stop at the other end of the runway?

26. *Automotive engineering*   The angular acceleration of a wheel is given by $\alpha(t) = 4 - t$ rad/s². Angular velocity is given by $\omega(t)$ and angular displacement by $\theta(t)$. If $\theta(0) = 0$ rad, and $\omega(0) = 10$ rad/s, find
(a) an expression for angular displacement,
(b) the angular displacement when the wheel turns in the opposite direction, and
(c) the time when the angular velocity is -14 rad/s.

27. *Machine technology*   When the motor turning a flywheel is shut off, the angular velocity of the wheel is given by $\omega(t) = 12.4 - 3.4t + 0.30t^2$ rad/s.
(a) During the first 3 s after the motor has been shut off, how many revolutions does the wheel complete?
(b) When the angular acceleration is $\alpha(t) = -1$, how many revolutions has the wheel completed?

28. *Physics*   The acceleration of a ball starting from rest and moving in a straight line is proportional to the time t in seconds (s). If the ball moves 24 m in the first 4 s, find an expression for the displacement $s(t)$.

29. *Electronics*   An expression for the current i in a circuit in amperes (A) is given by $i = 4.4t - 2.1t^2$, where t is in seconds. If $i = \dfrac{dq}{dt}$, where q is the charge in coulombs (C), find (a) an expression for q at $t = 0.0$ s, $q = 5.0$ C, and (b) the charge at $t = 3.2$ s.

30. *Electronics*   The current $i(t)$, in amperes (A), flowing to a capacitor is given by $i(t) = 6\sqrt{t}$, where t is time in seconds (s). If $i(t) = q'(t)$, where q is the charge in coulombs (C), find the charge on the capacitor at $t = 0.25$ s, if $q(0.16) = 0.347$ C.

31. *Electronics*   The voltage in volts (V) across a coil is given by $v(t) = -N\phi'(t)$, where N is the number of turns in the coil, $\phi(t)$ is the magnetic flux in webers (Wb), and t is the time in seconds (s). If $v(t) = 2t - 4t^{1/3}$ and $\phi(0) = 0.020$ Wb when N is 200 turns, find (a) a general expression for $\phi(t)$ and (b) $\phi(0.729)$.

32. A general expression for the slope of a tangent at any point on a certain curve is given by $m = 4x + 2$. If the curve goes through the point $(-1, 5)$, state the equation of the curve.

33. The slope of a normal line to a curve $y = f(x)$ is given by $m = \sqrt{x}$. Find $f(x)$, if $f(4) = -3$.

34. *Industrial engineering*   A stamping machine produces computer cases at a rate that varies over the working day according to the equation $\dfrac{dN}{dt} = 175 + 2t - 0.2t^2$, where t is the time in hours. (a) Find an expression for N and (b) calculate N for an 8-hour day if $N = 0$ at $t = 0$.

35. *Thermodynamics*   The temperature, T, in °C in an industrial furnace varies from its center to a point outside. The rate of temperature change is given by the equation

$$\frac{dT}{dx} = -\frac{5750}{(x+1)^3}$$

Find an expression for T at a distance of x meters from the center if $T = 2900$°C at $x = 0$.

36. *Ecology*   Biologists are treating a stream contaminated with bacteria. The level of contamination is changing at a rate of $\dfrac{dN}{dt} = -\dfrac{870}{t^2} - 210$ bacteria/cm³/day, where t is the number of days since the treatment began. Find a function $N(t)$ to estimate the level of contamination if the level after 1 day was about 7320 bacteria/cm³.

✏️   **In Your Words**

37. Explain how an antiderivative differs from a derivative.

38. A function has an infinite number of antiderivatives. How can you determine which is the "correct" antiderivative for a particular function?

☰ CHAPTER 24 REVIEW

Important Terms and Concepts

Acceleration
Antiderivative
Concave downward
Concave upward
Concavity
Critical point
Critical value
Decreasing function
Differentials
Extrema
First derivative test
Increasing function
Inflection point
Linear approximation formula
Maximum
    Absolute

Local
    Relative
Minimum
    Absolute
    Local
    Relative
Newton's method
Position function
Rectilinear motion
Related rates
Second derivative test
Speed
Velocity
    Average velocity
    Instantaneous velocity

Review Exercises

In Exercises 1–10, find critical values, the extrema, the intervals on which the curve is concave up and concave down, the inflection points, and the asymptotes. Sketch the curve.

1. $f(x) = x^4 - x^2$
2. $g(x) = x^4 - 32x$
3. $h(x) = 18x^2 - x^4$
4. $j(x) = 3x^4 - 4x^3 + 1$
5. $k(x) = \sqrt[5]{x}$
6. $f(x) = x\sqrt[3]{4-x}$
7. $g(x) = \dfrac{x^2+1}{x^2-4}$
8. $h(x) = \dfrac{x-1}{x+2}$
9. $j(x) = \dfrac{x}{x^2+x-2}$
10. $k(x) = \dfrac{x^3}{x^2-9}$

In Exercises 11 and 12, position functions of points moving rectilinearly are given. Find the function for the velocity and acceleration at time t. When are the position, velocity, and acceleration maximum and minimum for the given interval?

11. $s(t) = t^3 - 9t^2 + t, [-4, 4]$
12. $s(t) = t + \dfrac{4}{t}, [1, 4]$

In Exercises 13–16, find the antiderivative of each function.

13. $f(x) = 3x^2 - 4x$
14. $g(x) = 2x^{-3} + 5x^6$
15. $h(x) = \sqrt{x} + x^2 - \dfrac{3}{x^2}$
16. $s(t) = 4.9t^2 - 3.6t + 14$

Solve Exercises 17–29.

17. *Physics* An object is thrown directly upward with a velocity of 288 ft/s. Its height $s(t)$ in feet above the ground after t s is given by $s(t) = 288t - 16t^2$. **(a)** What are the velocity and acceleration after t s? **(b)** What are the height, velocity, and acceleration after 4 s? **(c)** What is the maximum height? **(d)** At what time is the height a maximum? **(e)** When does the object strike the ground? **(f)** What is its velocity when it strikes the ground?

18. *Physics* An object rolls down a ramp such that the distance in centimeters after t s is given by $s(t) = 15t^2 + 5$. **(a)** What is its velocity after t s? **(b)** What is its velocity after 2 s? **(c)** When will its velocity be 105 cm/s?

19. *Business* Suppose it costs $1 + 0.00058v^{3/2}$ dollars/mi to operate a truck at v mph. Additional costs, including the driver's salary, amount to $25 per hour. What speed will minimize the total cost of a 1,000 mi trip?

20. *Construction* A cylindrical barrel is to be constructed to hold 246π ft^3 of liquid. The cost per ft^2 of constructing the side of the barrel is three times the cost of constructing the top and the bottom. What are the dimensions of the barrel that will cost the least to construct?

21. *Agriculture* What is the maximum rectangular area that can be enclosed with 1 200 m of fencing?

22. *Petroleum engineering* A storage tank is in the shape of a sphere with a radius of 8 ft. If oil is pumped into the tank at the rate of 20 gal/min, how fast is the oil level rising when it has reached a height of 11 ft? (1 gal ≈ 0.1335805 ft^3.)

23. The radius of a sphere is 1.5 m with a possible error in measurement of 0.01m=1 cm. What is the maximum error you would expect in **(a)** the calculation of the volume? **(b)** the calculation of the surface area?

24. *Sheet metal technology* A gutter is to be formed by bending up equal widths of the opposite sides of a sheet of metal that is 380 mm wide. If the bent sides are perpendicular to the base, what dimensions will give the largest capacity?

25. *Thermodynamics* The voltage of a certain thermocouple as a function of temperature is given by $v = 4.5T + 0.0003T^3$. What is the approximate change in voltage when the temperature changes from 100°C to 101°C?

26. *Navigation* Two ships leave the same port at the same time. One travels west at the rate of 12 km/h and the other travels south at the rate of 5 km/h. At what rate is the distance between them changing 3 h after they leave the port?

27. A right-circular cylinder is to have a volume of 20 kL. What dimensions will make the total surface area a minimum?

28. *Navigation* A motorist is in a desert 8 mi from point A, the nearest point on a long straight road. The motorist wants to get to an aid station at point B on the road 10 mi from A. The car can travel 15 mph in the desert and 55 mph on the road. At what point should the motorist meet the road to get to B in the shortest possible time?

29. *Physics* A ball is dropped from the top of a building. It takes 7 s for the ball to reach the ground. **(a)** How high is the building? **(b)** What is its speed when the ball strikes the ground?

≡ CHAPTER 24 TEST

1. For the function $f(x) = x^4 - 8x^2$, determine **(a)** all critical values, **(b)** all extrema, **(c)** intervals on which f is concave upward, **(d)** intervals on which f is concave downward, and **(e)** sketch the curve.

2. For the function $g(x) = \dfrac{x^2}{9 - x^2}$, determine **(a)** all critical values, **(b)** all extrema, **(c)** intervals on which f is concave upward, **(d)** intervals on which f is concave downward, **(e)** asymptotes, and **(f)** sketch the curve.

3. Use Newton's method to approximate the positive root of $x^2 - 7 = 0$.

4. Determine the antiderivative of $f(x) = 5x^4 - 7x$.

5. The position function of points moving rectilinearly is given by the function $s(t) = t^3 - 12t^2 + 5$. **(a)** Find the function for the velocity and acceleration at time t. **(b)** When are the position, velocity, and acceleration maximum and minimum for the interval $[-2, 4]$?

6. A rectangular page is to contain 24 in.^{2} of print. The margins at the top and bottom of the page are $1\frac{1}{2}$ in., the margin toward the binding is $1\frac{1}{4}$ in., and the margin on the outside edge is $\frac{3}{4}$ in. What should the dimensions of the page be so that the least amount of paper is used?

7. A farm silo consists of a right circular cylinder of radius r and height h topped by a hemisphere. The volume of the silo is $V = \pi r^2 h + \frac{2}{3}\pi r^3$. If $h = 20$ m and $V = 800$ m^3, use Newton's method to find r to two decimal places.

25

Integration

Integration is used to determine area and volume. In Section 25.5, you will learn a numerical method for determining the cross-sectional area of the amount of dirt that must be excavated at a building site.

Courtesy of Mike Nelson

In Chapters 23 and 24, we looked at one of the two main parts of calculus—differential calculus. We have not finished looking at differential calculus, but it is time to take a look at the other main part of calculus—integral calculus.

In Chapter 22, we spent time looking at some of the foundations of calculus. As we discovered, calculus was developed in order to solve two types of problems. One of the problems, called the tangent question, formed the basis for our study of differential calculus. The other type of problem, which we called the area question, gives the foundation for our study of integral calculus.

≣ 25.1
THE AREA QUESTION AND THE INTEGRAL

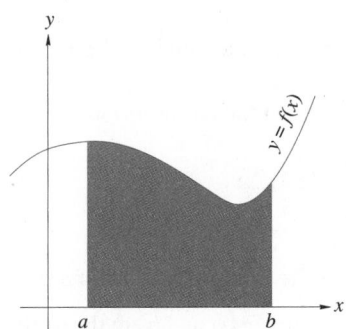

FIGURE 25.1

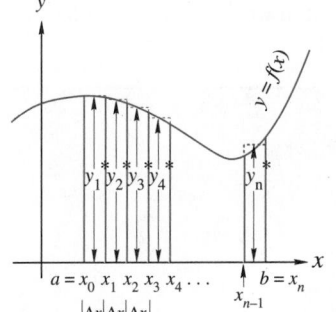

FIGURE 25.2

In Section 22.2, we looked at the area question. We began by looking at the shaded area in Figure 25.1. The area was enclosed by the curve $f(x)$, the x-axis, and the vertical lines $x = a$ and $x = b$.

We began our effort to find the area by dividing the interval $[a, b]$ into n equal segments. We let $a = x_0$ and $b = x_n$ with the first segment $[x_0, x_1]$, the second $[x_1, x_2]$, the third $[x_2, x_3]$, and so on. The midpoint of the ith interval was designated x_i^* and we located the height of the curve at that point, labeling it $y_i^* = f(x_i^*)$. Thus, in the interval $[x_0, x_1]$, the midpoint was x_1^* and the height was $y_1^* = f(x_1^*)$, as shown in Figure 25.2.

These y_i^* were used as altitudes of rectangles. The width of each rectangle was $\Delta x = \dfrac{x_n - x_0}{n}$. So, the area of the first rectangle is $(y_1^*)(\Delta x)$, the area of the second rectangle is $(y_2^*)(\Delta x)$, the area of the third rectangle is $(y_3^*)(\Delta x)$, and so on. Thus, the area A under the curve was approximated by

$$A \approx (y_1^*)(\Delta x) + (y_2^*)(\Delta x) + (y_3^*)(\Delta x) + \cdots + (y_n^*)(\Delta x)$$
$$= (y_1^* + y_2^* + y_3^* + \cdots + y_n^*)\Delta x$$
$$= [f(x_1^*) + f(x_2^*) + f(x_3^*) + \cdots + f(x_n^*)]\Delta x$$

Using the summation or sigma notation from Chapter 19, we have

$$A = \sum_{i=1}^{n} f(x_i^*)\Delta x$$

In this problem, we made things easy by letting each interval be the same size and by selecting the midpoints of the intervals. This did not have to be done. We can let each interval be a different size. The width of the ith interval would be $\Delta x_i = x_i - x_{i-1}$. We can also pick any point in the interval, say x_i', and let $f(x_i')$ be the altitude of the rectangle at that point. The area would then be approximated by the formula

$$A \approx \sum_{i=1}^{n} f(x_i')\Delta x_i.$$

In the first method, the more intervals that are selected the closer we will get to the actual area. Taking this further, we can see that

$$A = \lim_{n \to \infty} \sum_{i=1}^{n} f(x_i^*)\Delta x$$

In the second method, as the widths of the intervals get smaller, we get a closer approximation to the actual area. If we let $\|\Delta x\|$ represent the width of the largest Δx_i, then we can see that

$$A = \lim_{\|\Delta x\| \to 0} \sum_{i=1}^{n} f(x_i')\Delta x_i$$

In fact, a combination of these two is the easiest to use. We will let the number of equal-sized intervals increase and we will select any convenient point in each interval to get

$$A = \lim_{n \to \infty} \sum_{i=1}^{n} f(x_i') \Delta x$$

Notice that, since each interval is the same size, increasing the number of intervals has the effect of decreasing the width of the interval.

We will adopt a more convenient notation to represent the above formula:

$$A = \int_a^b f(x)\, dx = \lim_{n \to \infty} \sum_{i=1}^{n} f(x_i') \Delta x$$

The middle expression, $\int_a^b f(x)\, dx$, is called the **definite integral of f from a to b.** The function $f(x)$ is known as the **integrand** and the numbers a and b are the **limits of integration**. More specifically, a is the **lower limit** and b is the **upper limit**. The symbol $\int$ is an elongated "S" for "sum" and is called the **integral sign**.

The following summation formulas will be helpful in working the examples and exercises in this section.

Summation Formulas

$$\sum_{i=1}^{n} c = nc$$

$$\sum_{i=1}^{n} cf(x) = c \sum_{i=1}^{n} f(x)$$

$$\sum_{i=1}^{n} i = 1 + 2 + 3 + 4 + \cdots + n = \frac{n(n+1)}{2}$$

$$\sum_{i=1}^{n} i^2 = 1^2 + 2^2 + 3^2 + 4^2 + \cdots + n^2$$

$$= 1 + 4 + 9 + 16 + \cdots + n^2 = \frac{n(n+1)(2n+1)}{6}$$

$$\sum_{i=1}^{n} i^3 = 1^3 + 2^3 + 3^3 + 4^3 + \cdots + n^3$$

$$= 1 + 8 + 27 + 64 + \cdots + n^3 = \left[\frac{n(n+1)}{2} \right]^2$$

EXAMPLE 25.1

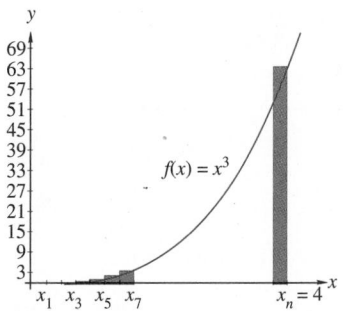

FIGURE 25.3

Evaluate $\int_0^4 x^3 \, dx$.

Solution The interval is $[0, 4]$ and we will divide it into n subintervals of equal length. Thus, each subinterval has a length $\Delta x = \dfrac{4}{n}$. While we can select any point in the subinterval, we will select the right-hand endpoint. The values of x_i, as shown in Figure 25.3, are $x_1 = 0 + \dfrac{4}{n} = \dfrac{4}{n}$, $\left(x_1 = 0 + \dfrac{4}{n}\right.$, because 0 is the left endpoint of the interval $[0, 4]\Big)$, $x_2 = 0 + 2\left(\dfrac{4}{n}\right) = \dfrac{8}{n}$, $x_3 = 0 + 3\left(\dfrac{4}{n}\right) = \dfrac{12}{n}$, ...,

$$x_i = 0 + i\left(\frac{4}{n}\right) = \frac{4i}{n}, \ldots, x_n = 0 + n\left(\frac{4}{n}\right) = 4.$$

The altitude of each rectangle is computed using $f(x) = x^3$, and so

$$f(x_1) = \left(\frac{4}{n}\right)^3 = \frac{64}{n^3}$$

$$f(x_2) = \left[2\left(\frac{4}{n}\right)\right]^3 = 2^3 \cdot \frac{64}{n^3}$$

$$f(x_3) = \left[3\left(\frac{4}{n}\right)\right]^3 = 3^3 \cdot \frac{64}{n^3}$$

$$\vdots$$

$$f(x_i) = \left[i\left(\frac{4}{n}\right)\right]^3 = i^3 \cdot \frac{64}{n^3}$$

$$\vdots$$

and $f(x_n) = 4^3 = 64$

The sum of the areas of the rectangles can be written as

$$\sum_{i=1}^{n} f(x_i)\Delta x = \sum_{i=1}^{n} i^3 \frac{64}{n^3} \cdot \frac{4}{n} = \frac{256}{n^4} \sum_{i=1}^{n} i^3$$

since $\dfrac{256}{n^4}$ is a constant. Now $\displaystyle\sum_{i=1}^{n} i^3 = \left[\dfrac{n(n+1)}{2}\right]^2$, and so

$$\frac{256}{n^4} \sum_{i=1}^{n} i^3 = \frac{256}{n^4}\left[\frac{n(n+1)}{2}\right]^2$$

$$= \frac{256}{n^4} \frac{n^2(n+1)^2}{4}$$

$$= \frac{64(n+1)^2}{n^2}$$

$$= \frac{64(n^2 + 2n + 1)}{n^2}$$

EXAMPLE 25.1 (Cont.)

We are now ready to find the integral.

$$\int_0^4 x^3\, dx = \lim_{n\to\infty} \frac{64\left(n^2+2n+1\right)}{n^2}$$

$$= \lim_{n\to\infty} \frac{64\left(1+\dfrac{2}{n}+\dfrac{1}{n^2}\right)}{1}$$

$$= 64$$

Thus, the area of the region is 64 square units.

EXAMPLE 25.2

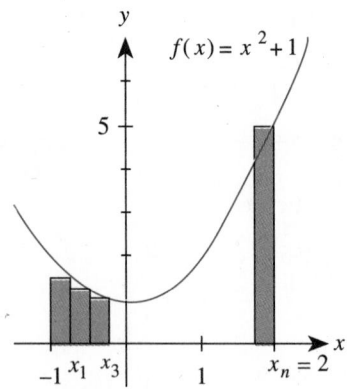

FIGURE 25.4

Evaluate $\displaystyle\int_{-1}^{2}\left(x^2+1\right)dx$.

Solution The interval $[-1,2]$ has length 3. Using equal subintervals, each subinterval will have length $\Delta x = \dfrac{3}{n}$.

Again, using the right-hand endpoint of each subinterval as shown in Figure 25.4, $x_i = -1 + \dfrac{3i}{n} = \dfrac{3i-n}{n}$. Since $f(x) = x^2+1$, we see that $f(x_i) = \left(\dfrac{3i-n}{n}\right)^2 + 1 = \dfrac{9i^2-6in+n^2}{n^2}+1 = \dfrac{9i^2}{n^2}-\dfrac{6i}{n}+2$.

Thus,

$$\sum_{i=1}^{n} f(x_i)\Delta x = \sum_{i=1}^{n}\left(\frac{9i^2}{n^2}-\frac{6i}{n}+2\right)\left(\frac{3}{n}\right)$$

$$= \sum_{i=1}^{n}\left(\frac{27i^2}{n^3}-\frac{18i}{n^2}+\frac{6}{n}\right)$$

$$= \frac{27}{n^3}\sum_{i=1}^{n}i^2 - \frac{18}{n^2}\sum_{i=1}^{n}i + \frac{6}{n}\sum_{i=1}^{n}1$$

$$= \frac{27}{n^3}\left[\frac{n(n+1)(2n+1)}{6}\right] - \frac{18}{n^2}\left[\frac{n(n+1)}{2}\right] + \frac{6}{n}\cdot n$$

$$= \frac{27}{n^3}\left(\frac{2n^3+3n^2+n}{6}\right) - \frac{18}{n^2}\left(\frac{n^2+n}{2}\right) + 6$$

$$= \left(9+\frac{27}{2n}+\frac{9}{2n^2}\right) - \left(9+\frac{9}{n}\right) + 6$$

EXAMPLE 25.2 (Cont.)

We find the integral by taking the limit of this sum as follows:

$$\int_{-1}^{2} (x^2 + 1)\, dx = \lim_{n \to \infty} \left[\left(9 + \frac{27}{2n} + \frac{9}{2n^2} \right) - \left(9 + \frac{9}{n} \right) + 6 \right]$$
$$= (9 + 0 + 0) - (9 + 0) + 6 = 6$$

Thus, the area of the region is 6 square units.

Exercise Set 25.1

In Exercises 1–18, use summation notation and limits to evaluate the definite integrals.

1. $\int_{1}^{3} x\, dx$

2. $\int_{0}^{2} 2x\, dx$

3. $\int_{0}^{4} (3x + 1)\, dx$

4. $\int_{1}^{3} (2x + 2)\, dx$

5. $\int_{1}^{3} (2x - 2)\, dx$

6. $\int_{0}^{4} (1 - 3x)\, dx$

7. $\int_{1}^{4} (3 - 2x)\, dx$

8. $\int_{0}^{4} (x^2)\, dx$

9. $\int_{0}^{2} (x^2 - 1)\, dx$

10. $\int_{0}^{4} (x^2 + 2)\, dx$

11. $\int_{0}^{2} (2x^2 + 1)\, dx$

12. $\int_{0}^{4} (2 - x^2)\, dx$

13. $\int_{1}^{3} x^3\, dx$

14. $\int_{1}^{3} (1 - x^3)\, dx$

15. $\int_{0}^{4} (x^3 + 2)\, dx$

16. $\int_{0}^{3} (2x^3 - 1)\, dx$

17. $\int_{0}^{2} (x^2 + x)\, dx$

18. $\int_{0}^{1} x^4\, dx$

Solve Exercise 19.

19. Use the computer program from Exercise 11 of Exercise Set 22.2 to solve Exercises 1--18.

Input different values of n to see how close the computer approximations get to the actual value.

In Your Words

20. In the expression $\int_{a}^{b} f(x)\, dx$, what is the integrand, the lower limit of integration, and the upper limit of integration? What does each of them tell you?

21. What are the differences in the meaning between $\sum_{1}^{10} f(x) \Delta x$ and $\int_{1}^{10} f(x)\, dx$?

☰ 25.2
THE FUNDAMENTAL THEOREM OF CALCULUS

In this section, we will examine the relationship between the derivative and the definite integral. We will then look at three properties of integrals. One of the most important tools we will attain in this section will be a general method for evaluating a definite integral.

Figure 25.5 shows the graph of $y = f(x)$. We will assume that $f(x)$ is continuous on the closed interval $[a, b]$ and that its graph does not fall below the x-axis. From Section 25.1, we know that the area of the shaded region is given by $\int_{a}^{b} f(x)\, dx$. We

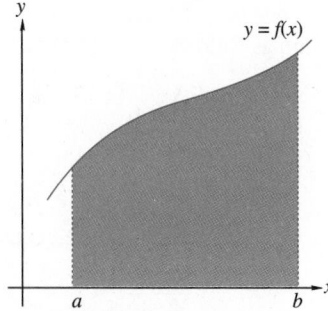

FIGURE 25.5

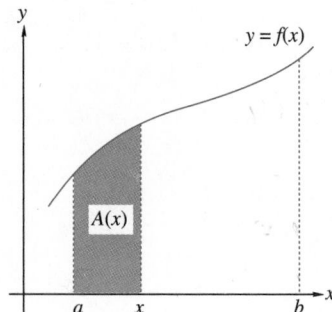

FIGURE 25.6

found that area by dividing the region into smaller and smaller rectangles, finding the sum of the areas of these rectangles, and taking the limit of this sum as the number of rectangles became infinitely large. We will now look at another way to determine the area.

Suppose there is a function $A(x)$, which we will refer to as an "area" function. The function $A(x)$ gives the area of the region below the graph of f and above the x-axis from a to x, where a is a fixed point and $a \le x \le b$. The region is shaded in Figure 25.6.

We already know two properties of $A(x)$:

1. $A(a) = 0$, since there is no area from a to a.

2. $A(b) = \displaystyle\int_a^b f(x)\,dx$.

Now, suppose that x is increased by h units as shown in Figure 25.7 and $A(x+h)$ is the area of the new shaded region. The difference in the area of the new region and the area of the original region is shown in Figure 25.8, and is written as $A(x+h) - A(x)$.

The area of the region in Figure 25.9 is the same as the area of a rectangle whose base is h and whose height is some value $\overline{y}$. Since the value of $\overline{y}$ depends on h, $\overline{y}$ is a function of h. The area of the rectangle is $A(x+h) - A(x)$ and is also represented as $\overline{y}h$. Thus, we have the equation

$$A(x+h) - A(x) = \overline{y}h$$

Dividing by h:
$$\frac{A(x+h) - A(x)}{h} = \overline{y}$$

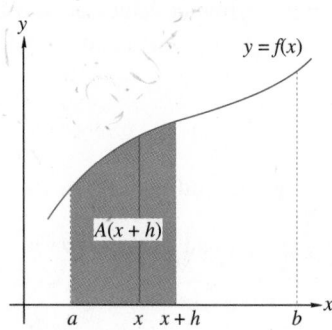

FIGURE 25.7

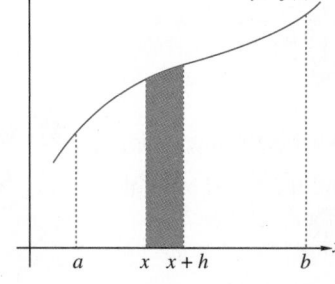

FIGURE 25.8

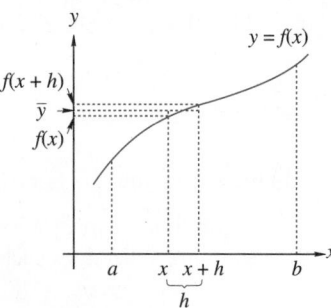

FIGURE 25.9

Now, what happens as the value of h approaches zero? One thing that will happen is that the value of $\overline{y}$ will get closer to the value of $f(x)$, and so

$$\lim_{h \to 0} \frac{A(x+h) - A(x)}{h} = \lim_{h \to 0} \overline{y} = f(x)$$

But, the left-hand side of this equation is the derivative of $A(x)$, thus we have

$$A'(x) = f(x)$$

What this means is that the area function $A(x)$ has the property that its derivative A' is the function f. That means that A is an antiderivative of f.

In Section 24.7, we found that the antiderivative of $f(x)$ was $F(x)+C$, where C is a constant. Thus, we know that $A(x) = F(x)+C$. Now, $A(a) = 0$, so when $x = a$, we have $0 = F(a)+C$ or $C = -F(a)$. Thus, we have the equation

$$A(x) = F(x) - F(a)$$

If $x = b$, then this equation becomes

$$A(b) = F(b) - F(a)$$

But, $A(b) = \displaystyle\int_a^b f(x)\,dx$. Combining these last two equations we get

$$\int_a^b f(x)\,dx = F(b) - F(a)$$

As we can see, there is a relationship between a definite integral and differentiation. All we need to do to find $\displaystyle\int_a^b f(x)\,dx$ is find an antiderivative of f, which we call F, and subtract the value of F at the lower limit from its value at the upper limit. This result is known as the **Fundamental Theorem of Calculus**. It was discovered independently by Newton and Leibnitz.

Fundamental Theorem of Calculus

If f is continuous on the interval $[a, b]$, then

$$\int_a^b f(x)\,dx = F(x)\,\Big|_a^b = F(b) - F(a)$$

where F is a function such that $F' = f$.

In stating the Fundamental Theorem of Calculus, we introduced some new notation, $F(x)\,\big|_a^b$. This notation is often used to help keep track of the limits of integration after we have integrated but before the antiderivative is evaluated.

EXAMPLE 25.3

Evaluate $\displaystyle\int_0^3 x^3\,dx$.

Solution This is the same function we evaluated in Example 25.1. We know that the antiderivative of x^3 is $\dfrac{x^4}{4}+C$, so

$$\int_0^3 x^3\,dx = \left.\frac{x^4}{4}\right|_0^4$$

$$= \frac{4^4}{4} - \frac{0^4}{4} = 64 - 0 = 64$$

≡ Note

When using antiderivatives with the Fundamental Theorem of Calculus, the constant is not used. If the antiderivative of $f(x)$ is $F(x)+C$, then

$$\int_a^b f(x)\,dx = \left. F(x)+C\ \right|_a^b$$

$$= [F(b)+C] - [F(a)+C]$$

$$= F(b) - F(a)$$

As you can see, the constant C does not have to be included since $C - C = 0$. From now on we will call the process of finding an antiderivative by its more common title of **integration**.

EXAMPLE 25.4

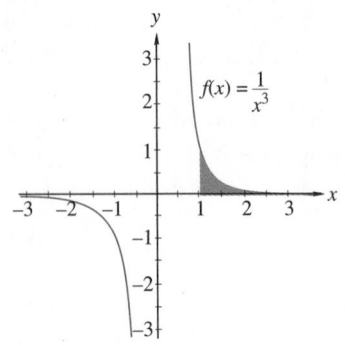

FIGURE 25.10

Find the area under the curve of $f(x) = \dfrac{1}{x^3}$ from $x = 1$ to $x = 3$.

Solution The desired area is shown in Figure 25.10. Note that f is continuous from $x = 1$ to $x = 3$. Since the curve lies above the x-axis in the interval $[-1, 3]$, the area can be found by evaluating $\displaystyle\int_1^3 f(x)\,dx$. Using the Fundamental Theorem of Calculus, we have

$$\int_1^3 \frac{1}{x^3}\,dx = \int_1^3 x^{-3}\,dx$$

$$= \left.\frac{1}{-2}x^{-2}\right|_1^3$$

$$= \left(\frac{1}{-2\,(3^2)}\right) - \left(\frac{1}{-2\cdot 1^2}\right)$$

$$= \frac{-1}{18} + \frac{1}{2}$$

$$= \frac{8}{18} = \frac{4}{9} \text{ square units}$$

At the present time we can only find the area when a curve is above the x-axis.

EXAMPLE 25.5

Evaluate $\int_1^3 (3x-2)\, dx$.

Solution $\displaystyle \int_1^3 (3x-2)\, dx = \left(\frac{3}{2}x^2 - 2x\right)\Big|_1^3$

$$= \left[\frac{3}{2}(3^2) - 2(3)\right] - \left[\frac{3}{2}(1^2) - 2(1)\right]$$

$$= \left(\frac{27}{2} - 6\right) - \left(\frac{3}{2} - 2\right) = 8$$

The following are three properties, or rules, of definite integrals.

Properties of Definite Integrals

Rule 1: $\displaystyle \int_a^b kf(x)\, dx = k \int_a^b f(x)\, dx$, where k is a constant

Rule 2: $\displaystyle \int_a^b [f(x) \pm g(x)]\, dx = \int_a^b f(x)\, dx \pm \int_a^b g(x)\, dx$

Rule 3: If f is continuous on an interval containing a, b, and c, then

$$\int_a^b f(x)\, dx = \int_a^c f(x)\, dx + \int_c^b f(x)\, dx$$

EXAMPLE 25.6

Evaluate $\int_0^2 \left(3x^2 - \sqrt{x}\right) dx$.

Solution $\displaystyle \int_0^2 \left(3x^2 - \sqrt{x}\right) dx = \int_0^2 \left(3x^2 - x^{1/2}\right) dx$

$$= \left(x^3 - \frac{2}{3}x^{3/2}\right)\Big|_0^2$$

$$= \left[2^3 - \frac{2}{3}(2)^{3/2}\right] - \left[0^3 - \frac{2}{3}(0)^{3/2}\right]$$

$$= 8 - \frac{4\sqrt{2}}{3}$$

EXAMPLE 25.7

Evaluate $\int_1^2 4x^2 \, dx$.

Solution According to rule 1,

$$\int_1^2 4x^2 \, dx = 4 \int_1^2 x^2 \, dx = 4 \left(\frac{1}{3} x^3 \Big|_1^2 \right)$$

$$= 4 \left(\frac{8}{3} - \frac{1}{3} \right)$$

$$= \frac{28}{3}$$

EXAMPLE 25.8

Evaluate $\int_1^2 (x^2 - x) \, dx$.

Solution Using rule 2, we see that

$$\int_1^2 (x^2 - x) \, dx = \int_1^2 (x^2) \, dx - \int_1^2 x \, dx$$

$$= \frac{1}{3} x^3 \Big|_1^2 - \frac{1}{2} x^2 \Big|_1^2$$

$$= \frac{7}{3} - \frac{3}{2} = \frac{5}{6}$$

EXAMPLE 25.9

Evaluate $\int_{-1}^1 3x^2 \, dx$.

Solution Using rule 3, we know that

$$\int_{-1}^1 3x^2 \, dx = \int_{-1}^0 3x^2 \, dx + \int_0^1 3x^2 \, dx$$

$$= x^3 \Big|_{-1}^0 + x^3 \Big|_0^1$$

$$= (0+1) + (1-0) = 2$$

At present these three rules are not particularly needed. But, we are setting the foundation for later, when they will play an important part in our continued development and use of calculus.

Application

EXAMPLE 25.10

From the top of a building 192 ft high, a ball is thrown upward with an initial velocity of 64 ft/s. **(a)** Find the position function giving the height s as a function of the time t. **(b)** When does the ball strike the ground?

Solutions

(a) We will let $t = 0$ be the initial time. We are given the initial conditions that

$$s(0) = 192 \quad \text{(initial height is 192 ft)}$$
$$s'(0) = 64 \quad \text{(initial velocity is 64 ft/s)}$$

Since acceleration due to gravity is -32 ft/s², we have

$$s''(t) = -32$$

and integrating, we obtain

$$s'(t) = \int s''(t)\,dt$$
$$= \int -32\,dt = -32t + C_1$$

Since $s'(0) = 64 = -32(0) + C_1$, we have $C_1 = 64$. Thus, $s'(t) = -32t + 64$. Now, by integrating $s'(t)$, we have

$$s(t) = \int s'(t)\,dt = \int (-32t + 64)\,dt = -16t^2 + 64t + C_2$$

Using the initial height $s(0) = 192 = -16(0^2) + 64(0) + C_2$, we obtain $C_2 = 192$. We have now found the position function

$$s(t) = -16t^2 + 64t + 192$$

(b) The ball will hit the ground when $s(t) = 0$.

$$s(t) = -16t^2 + 64t + 192 = 0$$
$$-16(t^2 - 4t - 12) = 0$$
$$-16(t - 6)(t + 2) = 0$$
$$t = -2, 6$$

Since t must be positive, we conclude that the ball hits the ground 6 s after it was thrown.

Exercise Set 25.2

In Exercises 1–37, use the Fundamental Theorem of Calculus to evaluate the given definite integral.

1. $\int_0^5 6\,dx$

2. $\int_1^2 (x - 6)\,dx$

3. $\int_1^5 2x\,dx$

4. $\int_{-1}^2 (6 - 2x)\,dx$

5. $\int_{-1}^3 \frac{5}{3}x^3\,dx$

6. $\int_0^5 4x^2\,dx$

7. $\int_1^2 \frac{8}{5}x^3\,dx$

8. $\int_{-2}^2 6x\,dx$

9. $\displaystyle\int_0^2 (4x+5)\,dx$

10. $\displaystyle\int_1^3 (-4)\,dt$

11. $\displaystyle\int_2^0 (4x+5)\,dx$

12. $\displaystyle\int_{-1}^2 (2t-t^2)\,dt$

13. $\displaystyle\int_1^3 (-4x)\,dx$

14. $\displaystyle\int_2^5 (5w^2+2w)\,dw$

15. $\displaystyle\int_0^3 (t^2+2t)\,dt$

16. $\displaystyle\int_{-2}^0 (5-t^3)\,dt$

17. $\displaystyle\int_{-1}^3 (3y^2-2y)\,dy$

18. $\displaystyle\int_0^4 (s^2-4s+16)\,ds$

19. $\displaystyle\int_1^4 (4-2w^2)\,dw$

20. $\displaystyle\int_{-2}^0 -dt$

21. $\displaystyle\int_2^3 (x^2+2x+1)\,dx$

22. $\displaystyle\int_{-3}^0 (3y-y^3)\,dy$

23. $\displaystyle\int_3^2 (x^2+2x+1)\,dx$

24. $\displaystyle\int_1^4 \left(\sqrt{x}-\frac{1}{\sqrt{x}}\right)dx$

25. $\displaystyle\int_6^8 dx$

26. $\displaystyle\int_{1/2}^1 \frac{4\,dx}{x^2}$

27. $\displaystyle\int_{-4}^{-2} (3y^2-y-1)\,dy$

28. $\displaystyle\int_1^4 \left(\frac{3}{x^3}-\frac{2}{x^2}\right)dx$

29. $\displaystyle\int_2^4 (x^2-2x+1)\,dx$

30. $\displaystyle\int_0^4 (\sqrt{x})^3\,dx$

31. $\displaystyle\int_1^2 -3x^{-4}\,dx$

32. $\displaystyle\int_0^1 3x^{5/6}\,dx$

33. $\displaystyle\int_{-2}^{-1} \frac{t^{-2}}{3}\,dt$

34. $\displaystyle\int_0^2 (x^4-x^3)\,dx$

35. $\displaystyle\int_0^2 \sqrt{x}\,dx$

36. $\displaystyle\int_1^3 x\sqrt[3]{x}\,dx$

37. $\displaystyle\int_{1/2}^3 \left(\frac{1}{\sqrt{x}}-2\right)dx$

Solve Exercises 38–48.

38. The Fundamental Theorem of Calculus *seems* to say that

$$\int_{-1}^1 x^{-2}\,dx = -x^{-1}\Big|_{-1}^1 = -2$$

But x^{-2} is always positive and we would expect the area under this curve to be positive. Why does the Fundamental Theorem of Calculus appear to give a wrong answer?

39. *Physics*   A ball is projected upward along a frictionless inclined plane. The velocity of the ball is given by $v(t)=4.00t-1.00t^2$ in meters per second (m/s). What is the change in displacement in the first 4.00 s?

40. *Mechanical technology*   The angular velocity in radians per second (rad/s) for a gear is given by $\omega(t)=1.00t^3-27.0t$. What is the change in angular displacement between $t=1.00$ s and $t=3.00$ s?

41. *Dynamics*   From the top of a building 180 m high, a ball is thrown down with an initial velocity of 10 m/s. **(a)** How fast is the ball moving after 3 s? **(b)** How far does the ball move in the first 3 s? **(c)** How long does it take the ball to hit the ground?

42. *Aeronautics*   A hot-air balloon is rising slowly, so the balloonist drops a sandbag from the bottom of the

balloon. If the sandbag was dropped at an elevation of 840 ft, what was the speed of the balloon at the time the sandbag was dropped if it took 8 s for the sandbag to hit the ground?

43. *Nuclear technology*   A charged particle enters a linear accelerator. Its initial velocity is 500 m/s and its velocity increases with a constant acceleration to a velocity of 10 500 m/s in 0.01 s. **(a)** What is the acceleration? **(b)** How far does the particle travel in that 0.01 s?

44. *Electricity*   The current i in amperes (A) to a capacitor is given by $i=1.00t+1.00\sqrt{t}$, where t is time in seconds (s). If $i=\dfrac{dq}{dt}$, where q is the charge in coulombs, find the change in the charge on the capacitor from $t=1.00$ s to $t=9.00$ s.

45. *Environmental science*   The manager of a wildlife preserve has started a management program to control the preserve's moose population. It is estimated that the population will continue to grow according to the function $N'(t)=25-6t^{1/2}$ per year, where t is the number of years since the plan was implemented. Find the change in the population during the first 9 years of the program.

46. *Industrial engineering*   A stamping machine produces computer cases at a rate that varies over the working day according to the equation $\dfrac{dN}{dt} = 175 + 2t - 0.2t^2$, where t is the time in h.
 (a) Find an expression for $N(t)$ if $N(0) = 0$.
 (b) Find the number of computer cases the machine produces in the first 4 hours of the day.
 (c) Find the number of computer cases the machine produces in the last 4 hours of the 8-hour day.

47. *Forestry*   A certain type of hardwood tree will grow at the rate of $G'(t) = 0.5 + 4t^{-3}$ ft/yr, where t is time in years after the tree is 20 years old.

 (a) How much will the tree grow in the second year after it is 20 years old?
 (b) How much will the tree grow in the fifth year after it is 20 years old?
 (c) How much will the tree grow in years 1 through 5 after it is 20 years old?

48. *Ecology*   Pollution from a factory is entering a lake. The rate of concentration of the pollutant at time t is given by $P'(t) = 91t^{5/2}$, where t is the number of years since the factory began introducing pollutants into the lake. Ecologists estimate that the lake can accept a total level of pollution of 5,720 units before all the fish are killed. How long can the factory pollute the lake before all the fish are killed?

 **In Your Words**

49. Explain the fundamental theorem of calculus.
50. Three properties of definite integrals were given in the text. Explain what each one means and how you can use it.

≡ 25.3
THE INDEFINITE INTEGRAL

The Fundamental Theorem of Calculus showed us a relationship between definite integrals and antiderivatives. In fact, we found that we were able to evaluate $\displaystyle\int_a^b f(x)\,dx$ by finding an antiderivative of f. The general form of the antiderivative is called the **indefinite integral**. The integral symbol, $\displaystyle\int$, without any limits is used to indicate an indefinite integral. Thus, the notation $\displaystyle\int f(x)\,dx$ is used to indicate all the antiderivatives of f.

EXAMPLE 25.11

$$\int x^2\,dx = \frac{1}{3}x^3 + C$$

$$\int 4x^3\,dx = x^4 + C$$

$$\int (2x^{-3} + 5)\,dx = -x^{-2} + 5x + C$$

In each of these examples, C represents an arbitrary constant called the **constant of integration**.

≡ Note   It is important to point out the difference between a definite integral and an indefinite integral. The definite integral is a number; the indefinite integral is a family of functions.

The value of the definite integral depends on the limits of integration. For example,

$$\int_1^2 3x^2\,dx = x^3\ \Big|_1^2 = 7 \qquad \int_1^3 3x^2\,dx = x^3\ \Big|_1^3 = 26$$

Notice that in each case the definite integral is a number.

However, the indefinite integral has no limits to affect the answer. Thus,

$$\int 3x^2\,dx = x^3 + C$$

and C, or the constant of integration, is used to show that this integral represents a family of functions and not just one particular value.

Properties of Indefinite Integrals

At the end of Section 25.2, we gave three rules for definite integrals. We are now going to give several rules for indefinite integrals. Two of these rules are the indefinite integral versions of previous rules. All of them are based on the six rules for derivatives we gave in Section 23.2.

Because an indefinite integral of f is defined as an antiderivative of f, we know that we can differentiate the indefinite integral and get the original function. Thus,

Rule 1
$$\frac{d}{dx} \int f(x)\,dx = f(x)$$

EXAMPLE 25.12
$$\frac{d}{dx} \int \sqrt[3]{x^2 - 5}\,dx = \sqrt[3]{x^2 - 5}$$

The next rule is similar to one we had for definite integrals.

Rule 2
$$\int kf(x)\,dx = k \int f(x)\,dx, \text{ where } k \text{ is a real number}$$

This rule states that we can, in effect, factor a constant out of a function before we integrate it.

The next rule is also similar to one we had for definite integrals.

Rule 3
$$\int [f(x) \pm g(x)]\, dx = \int f(x)\, dx \pm \int g(x)\, dx$$

Thus, the indefinite integral of a sum (or difference) is equal to the sum (or difference) of the indefinite integrals.

The next rule is the antiderivative of rule 3 from Section 23.2.

Rule 4
$$\int x^n\, dx = \frac{1}{n+1} x^{n+1} + C,\, n \neq -1$$

EXAMPLE 25.13

$$\int (3x^2 + 4x - 5)\, dx = \int 3x^2\, dx + \int 4x\, dx - \int 5\, dx \qquad \text{By rule 3}$$

$$= 3\int x^2\, dx + 4\int x\, dx - 5\int dx \qquad \text{By rule 2}$$

$$= 3\left(\frac{x^3}{3}\right) + 4\left(\frac{x^2}{2}\right) - 5x + C \qquad \text{By rule 4}$$

$$= x^3 + 2x^2 - 5x + C$$

We used three of the four rules in Example 25.13.

Method of Substitution

All of the rules just given are useful, but they do not show us how to integrate $\int (3x-7)^{10}\, dx$, $\int (x^4+6)^3 4x^3\, dx$, or $\int \sqrt[3]{4x^3 - 8x^2}\, dx$. The first integral could be found by expanding $(3x-7)^{10}$ and then integrating, but it would be very long and inefficient. Also, this method will not work on the last two integrals.

Some integrals that cannot be evaluated directly by using the previously stated rules can sometimes be evaluated using the method of substitution. This method involves the introduction of a function that changes the integrand, such that the given rules will work when integrating. Examples 25.14 and 25.15 show how to use the method of substitution on the last two integrals given.

EXAMPLE 25.14

Use the method of substitution to determine

$$\int (x^4+6)^3 4x^3 \, dx$$

Solution In this integral, we will first use the substitution $u = x^4 + 6$. Differentiating with respect to x, we get $\dfrac{du}{dx} = 4x^3$ or $du = 4x^3 \, dx$. We then substitute u for $x^4 + 6$ and du for $4x^3 \, dx$ in the original integral, with the following result:

$$\int (x^4+6)^3 4x^3 \, dx = \int u^3 \, du \qquad \text{Let } x^4 + 6 = u \text{ and } 4x^3 \, dx = du.$$

$$= \frac{1}{4} u^4 + C \qquad\qquad \text{Integrating with respect to } u.$$

$$= \frac{1}{4} \left(x^4+6\right)^4 + C \quad \text{Substituting for } u, \text{ we obtain this.}$$

EXAMPLE 25.15

Use the method of substitution to find $\int \sqrt[3]{4x^3 - 8}\, x^2 \, dx$.

Solution We will let $u = 4x^3 - 8$, and so $\dfrac{du}{dx} = 12x^2$. Thus, $du = 12x^2 \, dx$ and $x^2 \, dx = \frac{1}{12} du$. Substituting these values in the given integral, we get the following solution:

$$\int \sqrt[3]{4x^3 - 8}\, x^2 \, dx = \int \left(4x^3 - 8\right)^{1/3} x^2 \, dx \quad \text{Let } 4x^3 - 8 = u \text{ and } x^2 \, dx = \frac{1}{12} du$$

$$= \int \left(u^{1/3}\right) \frac{1}{12} \, du$$

$$= \frac{1}{12} \int u^{1/3} \, du$$

$$= \frac{1}{12} \frac{u^{4/3}}{4/3} + C$$

$$= \frac{1}{16} u^{4/3} + C$$

$$= \frac{1}{16} \left(4x^3 - 8\right)^{4/3} + C$$

≡ **Note**   We substituted not only for the integrand, but for the differential. The most difficult part is selecting the correct substitution. Once a substitution is chosen, its differential is determined. In all cases, we are attempting to rewrite the integral so as to get a more general form of rule 4:

$$\int u^n \, du = \frac{u^{n+1}}{n+1}, \qquad n \neq -1$$

The substitution we select must allow us to rewrite $f(x)\, dx$ as $u^n \, du$ with the possible exception of a multiplicative constant.

EXAMPLE 25.16

Evaluate $\displaystyle\int \frac{dx}{\sqrt{x^2+5}}$.

Solution   If we let $u = x^2 + 5$, then $du = 2x\, dx$. But $2x$ is not a constant, so we cannot write this integral as a constant multiple of $u^n \, du$. Thus, we cannot use the substitution method and we cannot integrate this problem using the present methods.

EXAMPLE 25.17

Determine $\displaystyle\int (x^3 + 5)^2 \, dx$.

Solution   If we let $u = x^3 + 5$, then $du = 3x^2 \, dx$. Again the substitution method cannot be used because the given integrand is not a constant multiple of $u^n \, du$. However, we can expand the given problem and solve it.

$$\int (x^3 + 5)^2 \, dx = \int (x^6 + 10x^3 + 25)\, dx$$

$$= \frac{1}{7}x^7 + \frac{5}{2}x^4 + 25x + C$$

Application

EXAMPLE 25.18

The work, W, done by a variable force, F, over a distance, x, is given by $W = \int F\, dx$. What is the formula for the work done if $F = 15(4x+1)^2$ and $W = 4$ ft·lb when $x = 0$ ft?

EXAMPLE 25.18 (Cont.)

Solution   Since $W = \int F \, dx$, we have $W = \int 15(4x+1)^2 \, dx$. Let $u = 4x + 1$

and then $\dfrac{du}{dx} = 4$ and $4 \, dx = du$. Making these substitutions, we obtain the following

solution. Notice in the second line of the solution that we rewrite 15 as $\dfrac{15}{4} \cdot 4$ in

order to get $4 \, dx$.

$$
\begin{aligned}
W &= \int 15(4x+1)^2 \, dx \\
&= \int \frac{15}{4}(4x+1)^2 \, 4 \, dx \qquad \text{Let } 4x+1 = u \text{ and } 4 \, dx = du \\
&= \frac{15}{4} \int u^2 \, du \\
&= \frac{15}{4} \frac{u^3}{3} + C \\
&= 1.25 \, (4x+1)^3 + C
\end{aligned}
$$

Thus, the formula for the work is $W = 1.25 \, (4x+1)^3 + C$. But, we are given some additional information. We are told that $W = 4 \, \text{ft·lb}$ when $x = 0 \, \text{ft}$. Substituting $x = 0$ in our formula for W produces $W = 1.25 \, (4(0)+1)^3 + C = 1.25(0+1)^3 + C = 1.25 + C$. Since $W = 4$ when $x = 0$, then $1.25 + C = 4$ and $C = 4 - 1.25 = 2.75$. With this value for C we get the desired formula for the work as:

$$W = 1.25 \, (4x+1)^3 + 2.75$$

Exercise Set 25.3

In Exercises 1–56, find the given indefinite integral.

1. $\displaystyle \int 9 \, dx$

2. $\displaystyle \int 3x \, dx$

3. $\displaystyle \int 6x^2 \, dx$

4. $\displaystyle \int \sqrt{x} \, dx$

5. $\displaystyle \int x^2 \sqrt{x} \, dx$

6. $\displaystyle \int x^{-2/3} \, dx$

7. $\displaystyle \int (t^3 + 1) \, dt$

8. $\displaystyle \int (2 - 4s^3) \, ds$

9. $\displaystyle \int (y^2 + 4y - 3) \, dy$

10. $\displaystyle \int \left(4\sqrt{y} + \frac{1}{4\sqrt{y}} \right) dy$

11. $\displaystyle \int \left(\sqrt[3]{x} - \frac{3}{\sqrt[3]{x^2}} \right) dx$

12. $\displaystyle \int \left(\frac{1}{x^3} - \frac{3}{\sqrt{x}} + \frac{\sqrt{x}}{3} \right) dx$

13. $\displaystyle \int (x^2 + 3) 2x \, dx$

14. $\displaystyle \int (3x^2 - 5) 6x \, dx$

15. $\displaystyle \int (4 - 2x^2) 4x \, dx$

16. $\displaystyle \int (2x^3 + 1)^4 6x^2 \, dx$

17. $\displaystyle \int (3 - x^2)^3 2x \, dx$

18. $\displaystyle \int (x^2 - 3)^4 x \, dx$

19. $\displaystyle \int \sqrt{x^2 + 4x} \, dx$

20. $\displaystyle \int (x^3 + 1)^5 x^2 \, dx$

21. $\displaystyle \int \frac{(\sqrt{x} - 1)^3}{\sqrt{x}} \, dx$

22. $\displaystyle \int (x^4 - 5)^7 x^3 \, dx$

23. $\int (3x^3+1)^4 x^2\,dx$

24. $\int (3-6x^2)^{3/2}\,2x\,dx$

25. $\int \dfrac{x\,dx}{\sqrt{x^2+3}}$

26. $\int \dfrac{x\,dx}{\sqrt{(4x^2-1)^3}}$

27. $\int \dfrac{x\,dx}{(x^2+3)^3}$

28. $\int (x^2+3)^2\,dx$

29. $\int (3x^2-1)^2\,dx$

30. $\int \dfrac{x^2-1}{x+1}\,dx$

31. $\int \dfrac{x^2-x-6}{x+2}\,dx$

32. $\int \dfrac{3x\,dx}{(x^2+4)^5}$

33. $\int \dfrac{5x\,dx}{\sqrt[3]{x^2-1}}$

34. $\int \dfrac{dx}{\sqrt{4-x}}$

35. $\int (1+3x^2)^2 x\,dx$

36. $\int 2x\sqrt{5x^2+3}\,dx$

37. $\int 3x^2(4x^3-5)^{2/3}\,dx$

38. $\int \dfrac{2x^2\,dx}{(5-3x^3)^{2/3}}$

39. $\int (1+3x^2)^2\,dx$

40. $\int (2x^2+4x)^3(4x+4)\,dx$

41. $\int \sqrt{4x^2+2x}\,(4x+1)\,dx$

42. $\int (x^2+x-5)^4(2x+1)\,dx$

43. $\int \dfrac{x-1}{(x-1)^3}\,dx$

44. $\int \dfrac{2x+1}{(x^2+x-5)^3}\,dx$

45. $\int (2x^4-3)^2 8x\,dx$

46. $\int \dfrac{x+3}{\sqrt[3]{x^2+6x}}\,dx$

47. $\int (x^3-3x)^{2/3}(x^2-1)\,dx$

48. $\int (x^3-2)\sqrt{x^4-8x}\,dx$

49. $\int (3x+1)^3\,dx$

50. $\int 7x(x^3-2)^2\,dx$

51. $\int (2x^3-1)\sqrt[4]{x^4-2x}\,dx$

52. $\int \dfrac{x^2}{(4-x^3)^{3/2}}\,dx$

53. $\int (t+7)^{1/2}\,dt$

54. $\int \dfrac{x^4-1}{x-1}\,dx$

55. $\int \dfrac{y^3-8}{y-2}\,dy$

56. $\int (6x-7)^8\,dx$

In Exercises 57–66, evaluate the given definite integral.

57. $\displaystyle\int_0^2 (x^2-4)2x\,dx$

58. $\displaystyle\int_0^4 x(4x^2+2)^3\,dx$

59. $\displaystyle\int_1^2 6x(3x^2-7)\,dx$

60. $\displaystyle\int_1^3 2x^2(3x^3-1)^2\,dx$

61. $\displaystyle\int_0^2 2x(20-3x^2)\,dx$

62. $\displaystyle\int_0^5 x\sqrt{x^2+144}\,dx$

63. $\displaystyle\int_0^3 (x^3+2)x^2\,dx$

64. $\displaystyle\int_1^2 \dfrac{dx}{(x+4)^2}$

65. $\displaystyle\int_1^4 (3x^2-2)^4 x\,dx$

66. $\displaystyle\int_{-1}^0 (6x^2-1)^3 x\,dx$

Solve Exercises 67–72.

67. *Environmental science* The air quality control office estimates that for a population of x thousand people, the level of pollutant in the air is increasing at a rate of $L'(x)=0.3+0.002x$ parts per million (ppm) per thousand people. Find a function, $L(x)$, that estimates the level of the pollutants if the level is 7.2 ppm when the population is 20,000 people.

68. *Ecology* Biologists are treating a stream contaminated with bacteria. The level of contamination is changing at the rate of $N'(t)=-\dfrac{750}{t^2}-120$ bacteria/cm^3/day, where t is the number of days since the treatment began. Find a function, $N(t)$, to estimate the level of contamination if the level after 1 day was about 7,320 bacteria/cm^3.

69. *Business*   The marginal profit of a fast-food restaurant is given by $P'(x) = 4x + 35$, where x is the sales volume in thousands of hamburgers. The "profit" is $\$ -150$ when no hamburgers are sold. Determine the profit function, $P(x)$.

70. *Medical technology*   A medical laboratory estimates that t hours after some certain bacteria are introduced into a culture, the population will be increasing at the rate of $P'(t) = \dfrac{1800}{(18 - 0.5t)^{1/2}}$ bacteria per hour. Find the increase in the population during the first 8 hours.

71. *Thermodynamics*   The temperature, T, in °C, in an industrial furnace varies from its center to a point outside. The rate of temperature change is given by the equation

$$\frac{dT}{dx} = -\frac{5750}{(x+1)^3}$$

Find an expression for T at a distance of x m from the center if $T(0) = 2900$°C.

72. *Electronics*   The current, I, and the charge, q, are related by $I = \dfrac{dq}{dt}$. At $t = 0$ s, the current to a discharged capacitor is $I = 0.005$ A. If the current remains constant, what is the charge, q, in coulombs on the capacitor?

 **In Your Words**

73. What are the differences in the meaning of $\displaystyle\int f(x)\,dx$ and $\displaystyle\int_a^b f(x)\,dx$?

74. Describe how to use the method of substitution.

≡ 25.4
THE AREA BETWEEN TWO CURVES

We introduced the definite integral as an answer to one of the two basic questions that lead to the development of calculus—the area question. In this section, we will continue our study of area.

Until now, we have only been concerned with finding the area of a curve that is above the x-axis. Suppose we want to find the area for a curve that goes below the x-axis, such as the one shown in Figure 25.11. If we were to compute its area, since $f(x) < 0$, we would get $\displaystyle\int_a^b f(x)\,dx < 0$ or a negative area. This does not make sense. The idea of area means that the area of something is a nonnegative number. Thus, we need a better idea for finding the area. This will become clearer as we look at the next example.

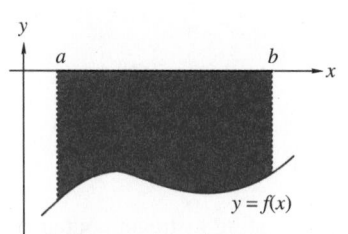

FIGURE 25.11

EXAMPLE 25.19

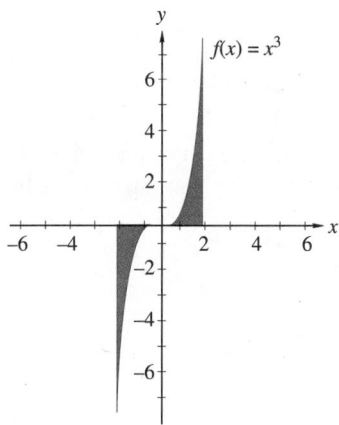

FIGURE 25.12

Find the area between the curve $f(x) = x^3$ and the x-axis from $x = -2$ to $x = 2$.

Solution The graph of the desired region is shaded in Figure 25.12. Using the definite integral we see that

$$\int_{-2}^{2} x^3 \, dx = \frac{1}{4} x^4 \Big|_{-2}^{2} = \frac{1}{4}(2^4) - \frac{1}{4}(-2)^4$$

$$= \frac{1}{4}(16) - \frac{1}{4}(16)$$

$$= 4 - 4$$

$$= 0$$

This does not make sense. As we can see, the shaded area of Figure 25.12 is not zero. Because this shaded area is symmetric about the origin, we can see that the area above the graph is the same as the area below the graph. Thus, we can see that the total area is $2 \int_{0}^{2} x^3 \, dx = 2(4) = 8$.

The integral $\int_{a}^{b} f(x) \, dx$ defines the area between the curve $y = f(x)$ and the x-axis only when $f(x)$ is a nonnegative continuous function for all x in the interval $[a, b]$. If $f(x)$ is continuous but $f(x) < 0$ for all x in the interval $[a, b]$, then $\int_{a}^{b} f(x) \, dx < 0$ and we define the area to be $\left| \int_{a}^{b} f(x) \, dx \right|$. If $f(x)$ is continuous and assumes both positive and negative values on the interval $[a, b]$, then the area is found by separating the integral into two or more parts.

One of the rules for definite integrals states that if a, b, and c are three points in an interval and if f is continuous on that interval, then

$$\int_{a}^{b} f(x) \, dx = \int_{a}^{c} f(x) \, dx + \int_{c}^{b} f(x) \, dx$$

Until now we have used this rule only once, in Example 25.9. Here, in Example 25.19, we have the opportunity to use it again. Since the curve crosses the x-axis at $x = 0$, we can compute the area as

$$\int_{-2}^{2} x^3 \, dx = \left| \int_{-2}^{0} x^3 \, dx \right| + \int_{0}^{2} x^3 \, dx$$

$$= \left| \frac{1}{4} x^4 \Big|_{-2}^{0} \right| + \frac{1}{4} x^4 \Big|_{0}^{2}$$

$$= \left| \frac{1}{4} 0^4 - \frac{1}{4}(-2)^4 \right| + \left[\frac{1}{4} 2^4 - \frac{1}{4} 0^4 \right]$$

$$= |0 - 4| + (4 - 0)$$

$$= 4 + 4 = 8$$

Caution

As you can see, it is very important that you sketch the graph when you are doing an area problem.

EXAMPLE 25.20

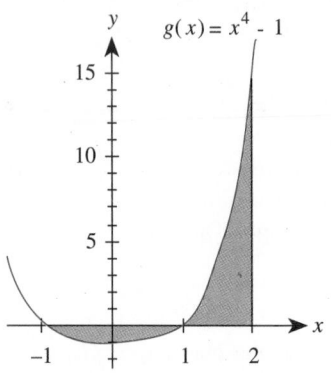

FIGURE 25.13

Find the area between the curve $g(x) = x^4 - 1$ and the x-axis from $x = -1$ to $x = 2$.

Solution The desired area is shown in Figure 25.13. The curve lies below the x-axis for the interval $[-1, 1]$ and above the x-axis for $[1, 2]$. The total area is the sum of these two separate areas. Thus,

$$\text{Area} = \left| \int_{-1}^{1} (x^4 - 1)\, dx \right| + \int_{1}^{2} (x^4 - 1)\, dx$$

Now,

$$\int_{-1}^{1} (x^4 - 1)\, dx = \left[\frac{1}{5}x^5 - x \right]_{-1}^{1}$$
$$= \left(\frac{1}{5} - 1 \right) - \left(-\frac{1}{5} + 1 \right) = -\frac{8}{5}$$

and

$$\int_{1}^{2} (x^4 - 1)\, dx = \left[\frac{1}{5}x^5 - x \right]_{1}^{2}$$
$$= \left(\frac{32}{5} - 2 \right) - \left(\frac{1}{5} - 1 \right) = \frac{26}{5}$$

Thus, the total area of the region is $\left| -\frac{8}{5} \right| + \frac{26}{5} = \frac{34}{5} = 6.8$ square units.

EXAMPLE 25.21

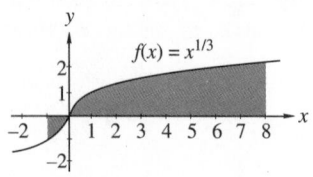

FIGURE 25.14

Find the area between the curve $f(x) = x^{1/3}$ and the x-axis from $x = -1$ to $x = 8$.

Solution The desired area is the shaded region of Figure 25.14. The graph of $f(x) = x^{1/3}$ crosses the x-axis at $x = 0$, so the area A will be

$$A = \left| \int_{-1}^{0} x^{1/3}\, dx \right| + \int_{0}^{8} x^{1/3}\, dx$$
$$= \left| \frac{3}{4}x^{4/3} \right|_{-1}^{0} + \frac{3}{4}x^{4/3} \Big|_{0}^{8}$$
$$= \left| \frac{3}{4}(0)^{4/3} - \frac{3}{4}(-1)^{4/3} \right| + \left[\frac{3}{4}(8)^{4/3} - \frac{3}{4}(0)^{4/3} \right]$$
$$= \left| 0 - \frac{3}{4} \right| + (12 - 0)$$
$$= \frac{3}{4} + 12 = 12\frac{3}{4}$$

≡ Note

The direct evaluation of $\int_{-1}^{8} x^{1/3}\,dx = \frac{3}{4}x^{4/3}\,\big|_{-1}^{8} = 12 - \frac{3}{4} = 11\frac{1}{4}$. This is not the area of the shaded region in Figure 25.14. It is, however, the value of the definite integral $\int_{-1}^{8} x^{1/3}\,dx$. You will need to be careful when evaluating a definite integral to proceed using the method that is correct for the desired meaning.

Let's expand our thinking to include a different type of area. Until now, each area has been for some region between a curve and the x-axis. Now we want to find the area between two curves.

Consider the functions f and g as shown in Figure 25.15. We want to find the area between the curves $y = f(x)$ and $y = g(x)$ and the lines $x = a$ and $x = b$. We will return to finding the area by dividing the region into rectangular regions of thickness dx. The height of each of these rectangles is $f(x) - g(x)$, regardless of the position of f and g. As a result, the area is given by

$$\int_{a}^{b} [f(x) - g(x)]\,dx$$

The result works whether the graphs lie above the x-axis, partially above and partially below the x-axis, or both lie below the x-axis.

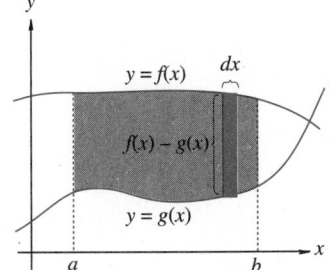

FIGURE 25.15

EXAMPLE 25.22

Find the area enclosed by the graphs of $f(x) = 12x - 3x^2$ and $g(x) = 3x - 12$.

Solution The graph of the two functions is shown in Figure 25.16. Before we can compute the area, we need to locate the points of intersection of the two graphs. Thus, we need to determine when $f(x) = g(x)$ or when

$$12x - 3x^2 = 3x - 12$$
$$3x^2 - 9x - 12 = 0$$
$$3(x - 4)(x + 1) = 0$$

and so the graphs intersect when $x = -1$ and $x = 4$.

From Figure 25.16, we can see that $f(x) \geq g(x)$ on $[-1, 4]$, so the area we want is

$$\int_{-1}^{4} [f(x) - g(x)]\,dx = \int_{-1}^{4} [(12x - 3x^2) - (3x - 12)]\,dx$$
$$= \int_{-1}^{4} (9x - 3x^2 + 12)\,dx$$
$$= \left(\frac{9}{2}x^2 - x^3 + 12x\right)\bigg|_{-1}^{4}$$
$$= (72 - 64 + 48) - \left(\frac{9}{2} + 1 - 12\right)$$
$$= 56 - (-6.5) = 62.5$$

The desired area is 62.5 square units.

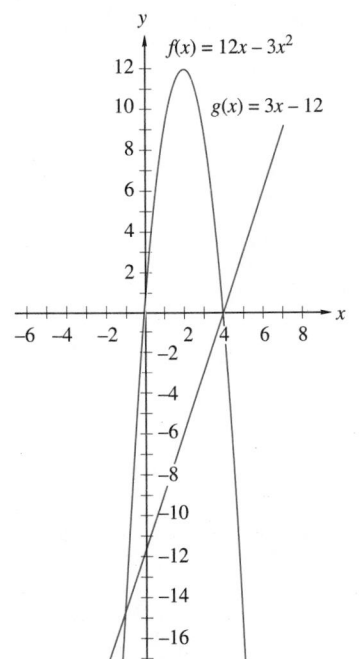

FIGURE 25.16

In the next example, we will combine the techniques that we used in the previous two examples.

EXAMPLE 25.23

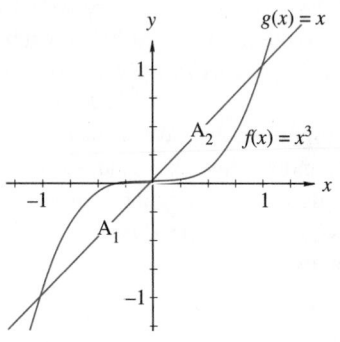

FIGURE 25.17

Find the area enclosed by the graphs of $f(x) = x^3$ and $g(x) = x$.

Solution   The graph of the two functions is shown in Figure 25.17. The points of intersection occur when

$$f(x) = g(x)$$

or when
$$x^3 = x$$

$$x^3 - x = 0$$

$$x(x^2 - 1) = x(x - 1)(x + 1) = 0$$

The two graphs intersect when $x = 0$, $x = 1$, and $x = -1$.

Study the graph of the two curves. On the interval $[-1, 0]$, $f(x) \geq g(x)$, and so the area A_1 enclosed by the curves in this interval is

$$A_1 = \int_{-1}^{0} [f(x) - g(x)]\, dx = \int_{-1}^{0} (x^3 - x)\, dx$$

$$= \left(\frac{1}{4}x^4 - \frac{1}{2}x^2 \right) \Big|_{-1}^{0}$$

$$= 0 - \left(\frac{1}{4} - \frac{1}{2} \right) = \frac{1}{4}$$

On the interval $[0, 1]$, $g(x) \geq f(x)$, and so the area A_2 enclosed by the curves in this interval is

$$A_2 = \int_{0}^{1} [g(x) - f(x)]\, dx = \int_{0}^{1} (x - x^3)\, dx$$

$$= \left(\frac{1}{2}x^2 - \frac{1}{4}x^4 \right) \Big|_{0}^{1}$$

$$= \left(\frac{1}{2} - \frac{1}{4} \right) - 0 = \frac{1}{4}$$

The total area is $A_1 + A_2 = \frac{1}{4} + \frac{1}{4} = \frac{1}{2}$.

≡ Note   Again, we need to point out the importance of graphing. If you had applied the rule for the area between two curves without graphing, you could easily have reached the incorrect result shown below:

$$\int_{-1}^{1} [f(x) - g(x)] \, dx = \int_{-1}^{1} (x^3 - x) \, dx$$

$$= \left(\frac{1}{4} x^4 - \frac{1}{2} x^2 \right) \Big|_{-1}^{1}$$

$$= \left(\frac{1}{4} - \frac{1}{2} \right) - \left(\frac{1}{4} - \frac{1}{2} \right)$$

$$= 0$$

EXAMPLE 25.24

Find the area enclosed by the graphs of $f(x) = \sqrt{9 - 9x}$, $g(x) = \sqrt{9 - x}$, and the x-axis.

Solution   The graphs of these functions are shown in Figure 25.18. The shading indicates the area we want to measure.

Again, we need to exercise caution. In the interval $[0, 1]$, the shaded region A_1 is between $f(x)$ and $g(x)$ with $g(x) \geq f(x)$. Thus

$$A_1 = \int_{0}^{1} [g(x) - f(x)] \, dx$$

$$= \int_{0}^{1} \left(\sqrt{9 - x} - \sqrt{9 - 9x} \right) dx$$

$$= \int_{0}^{1} \sqrt{9 - x} \, dx - \int_{0}^{1} \sqrt{9 - 9x} \, dx$$

$$= -\frac{2}{3} (9 - x)^{3/2} \Big|_{0}^{1} - \frac{-2}{27} (9 - 9x)^{3/2} \Big|_{0}^{1}$$

$$= \left[-\frac{2}{3} (8)^{3/2} + \frac{2}{3} (9)^{3/2} \right] - \left[\frac{-2}{27} (0) + \frac{2}{27} (9)^{3/2} \right]$$

$$= \left(-\frac{16}{3} \sqrt{8} + 18 \right) - (0 + 2)$$

$$= 16 - \frac{16}{3} \sqrt{8}$$

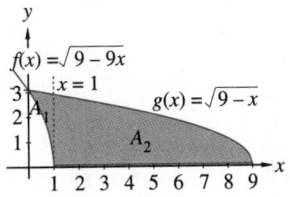

FIGURE 25.18

In the interval $[1, 9]$, the shaded region A_2 is between $g(x)$ and the x-axis, with $g(x) \geq 0$, and so

$$A_2 = \int_{1}^{9} \sqrt{9 - x} \, dx = -\frac{2}{3} (9 - x)^{3/2} \Big|_{1}^{9} = \frac{16}{3} \sqrt{8}$$

The total area is $A_1 + A_2 = 16 - \frac{16}{3} \sqrt{8} + \frac{16}{3} \sqrt{8} = 16$.

Partitions Along the *y*-axis

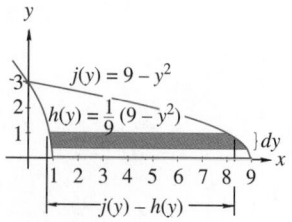

FIGURE 25.19

Until now, we have always integrated with respect to x. We have done this because the functions have always been given in terms of x and the x-axis. Let's look again at Example 25.24 and rewrite the functions in terms of y.

We had $f(x) = \sqrt{9 - 9x}$ or $y = \sqrt{9 - 9x}$. If we solve this for x, we get $y^2 = 9 - 9x$ or $x = \frac{1}{9}(9 - y^2)$. Let's call this function $h(y) = \frac{1}{9}(9 - y^2)$. In the same way, if we solve $g(x)$ for y, we get $y = \sqrt{9 - x}$ or $x = 9 - y^2$. We will call this function $j(y)$, and so $j(y) = 9 - y^2$.

Look at Figure 25.19. This time we have partitioned the y-axis and we get rectangles of width dy and length $j(y) - h(y)$. We want the area A between these curves and the x-axis (or when $y = 0$). Thus,

$$A = \int_0^3 [j(y) - h(y)]dy$$

$$= \int_0^3 \left[(9 - y^2) - \frac{1}{9}(9 - y^2)\right]dy$$

$$= \int_0^3 \left(8 - \frac{8}{9}y^2\right)dy$$

$$= \left(8y - \frac{8}{27}y^3\right)\Big|_0^3$$

$$= 8(3) - \frac{8}{27}(3)^3 = 24 - 8 = 16$$

As you can see in this case, a partition of the y-axis made the integration simpler. There is no clear method to tell which way to partition the graph. Usually, the graph will give you some indication.

Improper Integrals

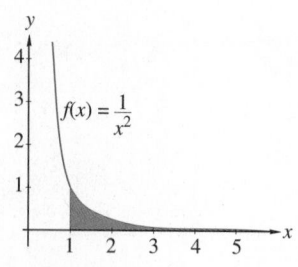

FIGURE 25.20

In the next example, we want to determine the area between a curve and the x-axis. In this case, we want the area between $f(x) = \dfrac{1}{x^2}$, the x-axis, and to the right of $x = 1$, as shown in Figure 25.20. We know that the x-axis is an asymptote of $\dfrac{1}{x^2}$, so the curve never intersects the x-axis. Thus, it looks as if the area is determined by

$$\int_1^\infty \frac{1}{x^2}dx$$

When we defined the definite integral at the beginning of this chapter, it was over a closed interval $[a, b]$. This integral is on the half-open interval $[a, \infty)$. To get around this problem, we will define

$$\int_1^\infty \frac{1}{x^2}dx = \lim_{b \to \infty} \int_1^b \frac{1}{x^2}dx, \text{ if the limit exists}$$

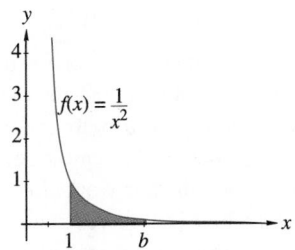

FIGURE 25.21

This is an example of an **improper integral**. Geometrically, we are finding the area under the graph on the interval $[1, b]$, as shown in Figure 25.21, and then we let $b \to \infty$. If the limit exists, the limit is defined to be the area. For this problem,

$$\int_1^\infty \frac{1}{x^2}\, dx = \lim_{b \to \infty} \int_1^b \frac{1}{x^2}\, dx$$

$$= \lim_{b \to \infty} \left(-\frac{1}{x} \right) \Big|_1^b$$

$$= \lim_{b \to \infty} \left(-\frac{1}{b} + 1 \right) = 1$$

Exercise Set 25.4

In Exercises 1–42, find the area of the region bounded by the given curves. Sketch the graphs and shade the desired area first.

1. $f(x) = x$, $x = 4$, x-axis
2. $f(x) = -2x$, $x = -2$, x-axis
3. $f(x) = x^2$, $x = 4$, x-axis
4. $h(x) = x^2 + 4$, $x = -1$, $x = 1$, x-axis
5. $g(x) = 3x + 1$, $x = 10$, x-axis
6. $h(x) = 10 - 2x$, $x = 0$, x-axis
7. $k(x) = 2x + 5$, $x = -2$, $x = 1$, x-axis
8. $j(x) = 2x^2$, x-axis, $x = 1$, $x = 2$
9. $k(x) = 2x^2 - x$, x-axis, $x = -2$
10. $j(x) = x^3 + 1$, $x = 1$, x-axis
11. $m(x) = x^2 - 4x$, x-axis
12. $n(x) = x^2 - 4$, x-axis
13. $f(x) = 9 - x^2$, x-axis
14. $h(x) = 6 - x - x^2$, x-axis
15. $g(x) = x^3 - 4x$, x-axis
16. $f(x) = x^3 - x^2$, x-axis
17. $h(x) = \sqrt{x + 9}$, x-axis, $x = 0$
18. $j(x) = \sqrt{x}$, $x = 9$, x-axis
19. $f(x) = x^2$, $g(x) = 2x$
20. $g(x) = 4 - x^2$, $h(x) = x + 2$
21. $f(x) = x^2 + 3$, $g(x) = 9$

22. $f(x) = x^2 + 4$, $x = -2$, $x = 2$, $y = -1$
23. $f(x) = x - 4$, $g(x) = \sqrt{2x}$
24. $f(x) = \sqrt{x}$, $x = 0$, $y = 4$
25. $g(x) = 4 - x^2$, $h(x) = -3x$
26. $f(x) = x^2 - 4$, $g(x) = 8 - 2x^2$
27. $h(x) = 2x - \frac{1}{2}x^2$, $j(x) = \frac{1}{2}x - 2$
28. $m(x) = x^4$, $n(x) = x^2$
29. $j(x) = 8 - x^2$, $k(x) = x^2$, $x = -2$, $x = 1$
30. $f(x) = 2\sqrt{x}$, $y = 4$, $x = -2$, x-axis
31. $f(x) = x^3$, $g(x) = 4x$
32. $h(x) = x + 3$, $k(x) = 9 - x^2$, $x = 3$
33. $g(x) = 4x - x^2 + 8$, $h(x) = x^2 - 2x$
34. $f(x) = \sqrt{x}$, $g(x) = -x$, $h(x) = x - 2$
35. $f(x) = \sqrt{4x}$, $y = 3$, $x = 0$
36. $h(x) = x^3 - 3x + 2$, $k(x) = x + 2$
37. $y = \sqrt{16 - 2x}$, $y = \sqrt{16 - 4x}$, x-axis
38. $f(x) = x^{1/3}$, $g(x) = -2$, $x = 8$
39. $f(x) = x^{-5/3}$, x-axis, to the right of $x = 8$
40. $f(x) = \frac{1}{x^2}$, $g(x) = 4$, $x = -2$, $x = 2$, x-axis
41. $x^4 y = 1$, x-axis, to the right of $x = 1$
42. $g(x) = x^4 - x^2$, x-axis

Solve Exercises 43–50.

43. *Product design*   A tile manufacturing company plans to produce enamel tiles with the design and color scheme shown in Figure 25.22. Will more blue or white enamel be needed?

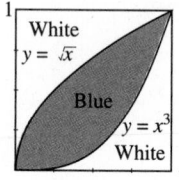

FIGURE 25.22

44. *Environmental science*   A new smog-control device will reduce the output of sulfur oxides from automobile exhausts but increase the production of nitrous oxides. It is estimated that using this device will produce a rate of savings to the community of $S(x) = -x^2 + 11x + 15$, in millions of dollars after x years of using the device. The increase in the production of nitrous oxides will result in a rate of additional costs to the community of $C(x) = \frac{5}{4}x^2$, in millions of dollars. If $S(x) - C(x)$ represents the net savings, **(a)** For how many years will it pay to use this new device? **(b)** What will be the net total savings for this period of time?

45. *Industrial design*   A double-channel trough used on an assemblyline is constructed with end plates defined by the curve $y = x^4 - 2x^2 + 1$ and the line $y = 9$, where x and y are measured in cm. What is the volume of a trough that is 10 m long and has these end plates?

46. *Machine technology*   Find the area of the cam outlined by the curves $y = 10x - x^2$ and $4y = x^2$, where x and y are measured in cm.

47. *Automotive engineering*   The work cycle, or **Otto cycle**, of an internal combustion engine may be approximated by the pressure-volume (or pV) diagram shown in Figure 25.23. The area of the region between the curves is proportional to the work done by the engine during one cycle. The curves represent adiabatic, or zero, heat input/output processes, where $pV^\gamma = k$, where k is a constant and $\gamma \approx 1.33$ for gasoline vapor. Find the work done by the cycle shown in Figure 25.23. (Note: the answer will be in British thermal units, Btu.)

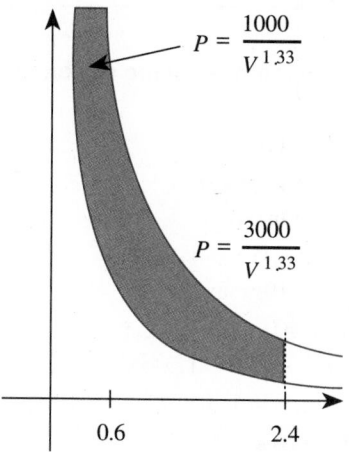

FIGURE 25.23

48. *Automotive engineering*   The work cycle, or **Diesel cycle**, of a diesel engine may be approximated by the pV-diagram shown in Figure 25.24. The area of the region between the curves is proportional to the work done by the engine during one cycle. The curves represent adiabatic, or zero, heat input/output processes, where $pV^\gamma = k$, where k is a constant, and $\gamma \approx 1.40$ for diesel vapor. Find the work done by the cycle shown in Figure 25.24. (Note: the answer will be in British thermal units, Btu.)

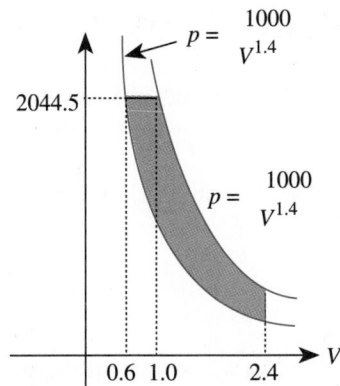

FIGURE 25.24

In Your Words

51. Describe the process used to find the area between two curves.

49. *Mechanical engineering* A heavy flywheel initially rotating on its axis is slowing because of friction in its bearings. If its angular acceleration, α, is given by $\alpha = \dfrac{d\omega}{dt} = \dfrac{100}{\sqrt{(t+1)^3}}$, find its angular velocity, ω, at time $t = 10\,\text{s}$, if $\omega = 0$ at $t = 0$.

50. *Electronics* The magnetic potential, V, at a point on the axis of a certain coil is given by an equation of the form $V = l \displaystyle\int_a^\infty \dfrac{x}{\left(R^2 + x^2\right)^{3/2}}\, dx$, where k and R are constants. Integrate in order to find an expression for V.

52. What is an indefinite integral?

☰ 25.5
NUMERICAL INTEGRATION

The Fundamental Theorem of Calculus provides a useful method for finding the area under a curve. It is not always an easy task to find the antiderivative of a function. It is also possible that we do not know how to find the antiderivative of a certain function. In this section, we will look at two methods for approximating the value of a definite integral to as many decimal places as desired. All techniques of this type come under the heading of **numerical integration**. The two methods that we will examine are called the **trapezoidal rule** and **Simpson's rule**. As with most numerical techniques, your work with these can be eased by the use of a calculator or a computer.

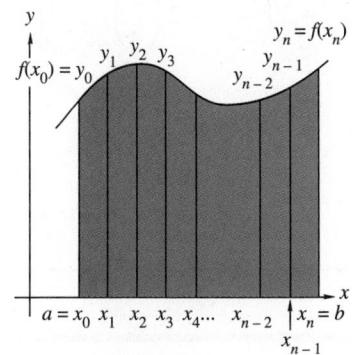

FIGURE 25.25

The Trapezoidal Rule

In both the trapezoidal rule and Simpson's rule, we will use some of the work we did in developing the definite integral back in Section 25.1. First we will assume that we want to find the area under some curve, which we will describe by $y = f(x)$. We will also assume that we want to find the area under the curve $f(x)$, above the x-axis, and between the lines $x = a$ and $x = b$, with $a < b$.

We will divide the interval $[a, b]$ into n equal parts of length $dx = \dfrac{b-a}{n}$, where $a = x_0 < x_1 < x_2 < x_3 < \cdots < x_{n-1} < x_n = b$. The corresponding points on the curve are $y_0 = f(x_0)$, $y_1 = f(x_1)$, $y_2 = f(x_2)$, $y_3 = f(x_3)$, $\ldots$, $y_{n-1} = f(x_{n-1})$, $y_n = f(x_n)$. If we connect these points, we obtain n trapezoids as shown in Figure 25.25.

The area of a trapezoid is equal to one-half the product of the altitude and the sum of its bases. A typical trapezoid is shown in Figure 25.26a. If you draw the trapezoid as shown in Figure 25.26b, you will find it easier to see that the altitude of

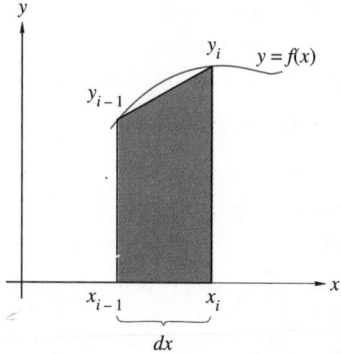

FIGURE 25.26a

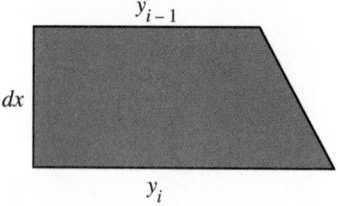

FIGURE 25.26b

this trapezoid is dx and the lengths of its bases are y_{i-1} and y_i. Thus, the area of the trapezoid in Figure 25.26a is $\frac{1}{2}(y_{i-1} + y_i)\,dx$.

The area under the curve in Figure 25.25 can be approximated by adding the areas of all the trapezoids. Thus,

$$\int_a^b f(x)\,dx \approx \frac{1}{2}(y_0 + y_1)\,dx + \frac{1}{2}(y_1 + y_2)\,dx + \frac{1}{2}(y_2 + y_3)\,dx + \cdots$$
$$+ \frac{1}{2}(y_{n-1} + y_n)\,dx$$
$$= \frac{1}{2}(y_0 + 2y_1 + 2y_2 + 2y_3 + \cdots + 2y_{x-1} + y_n)\,dx$$

Using the more conventional function notation, $y = f(x)$, we can rewrite the formula as follows:

Trapezoidal Rule $\quad\int_a^b f(x)\,dx \approx \dfrac{dx}{2}[f(x_0) + 2f(x_1) + 2f(x_2) + 2f(x_3) + \cdots$
$$+ 2f(x_{n-1}) + f(x_n)]$$

where $dx = \dfrac{b-a}{n}$.

EXAMPLE 25.25

Use the trapezoidal rule with $n = 6$ to approximate the value of $\displaystyle\int_0^2 x^3\,dx$.

Solution For $n = 6$, we have $dx = \frac{2-0}{6} = \frac{1}{3}$. Thus, $x_0 = 0$, $x_1 = \frac{1}{3}$, $x_2 = \frac{2}{3}$, $x_3 = 1$, $x_4 = \frac{4}{3}$, $x_5 = \frac{5}{3}$, and $x_6 = 2$. Using the trapezoidal rule, we have

$$\int_0^2 x^3\,dx \approx \frac{1/3}{2}\left[0^3 + 2\left(\frac{1}{3}\right)^3 + 2\left(\frac{2}{3}\right)^3 + 2(1)^3 + 2\left(\frac{4}{3}\right)^3 + 2\left(\frac{5}{3}\right)^3 + 2^3\right]$$
$$= \frac{1}{6}\left(0 + \frac{2}{27} + \frac{16}{27} + 2 + \frac{128}{27} + \frac{250}{27} + 8\right)$$
$$= \frac{37}{9}$$
$$\approx 4.111$$

The exact value obtained by integration is 4.

Simpson's Rule

For the second approximation technique, we will use the same curve. Again, the interval $[a, b]$ is divided into n equal parts, but this time n must be an even number. Instead of the straight lines we drew connecting y_0, y_1, y_2, y_3, and so on, we are

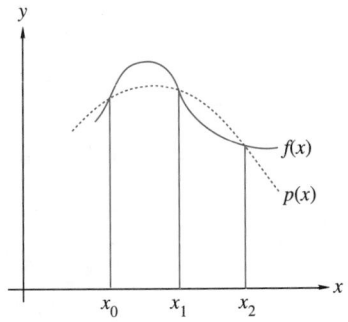

FIGURE 25.27

going to draw parabolas through each group of three consecutive points, as shown in Figure 25.27. A lengthy algebraic manipulation will show that

$$\int_{x_0}^{x_2} f(x)\,dx = \frac{dx}{3}\left[f(x_0) + 4f(x_1) + f(x_2)\right]$$

Now, if we did this for each subinterval $[x_0, x_2], [x_2, x_4], [x_4, x_6], \ldots, [x_{n-2}, x_n]$, we would get the result known as Simpson's rule.

Simpson's Rule
$$\int_{a}^{b} f(x)\,dx \approx \frac{dx}{3}\left[f(x_0) + 4f(x_1) + 2f(x_2) + 4f(x_3) + 2f(x_4)\right.$$
$$\left. + \cdots + 2f(x_{n-2}) + 4f(x_{n-1}) + f(x_n)\right]$$

where n is an even number and $dx = \dfrac{b-a}{n}$.

EXAMPLE 25.26

Use Simpson's rule with $n = 6$ to approximate the value of $\int_{0}^{2} x^3\,dx$.

Solution   Since $n = 6$, n is an even number. This is the same integral we approximated in Example 25.25, so the values of x and dx are the same.

$$\int_{0}^{2} x^3\,dx \approx \frac{1/3}{3}\left[0^3 + 4\left(\frac{1}{3}\right)^3 + 2\left(\frac{2}{3}\right)^3 + 4(1)^3 + 2\left(\frac{4}{3}\right)^3 + 4\left(\frac{5}{3}\right)^3 + 2^3\right]$$

$$= \frac{1}{9}\left(0 + \frac{4}{27} + \frac{16}{27} + 4 + \frac{128}{27} + \frac{500}{27} + 8\right)$$

$$= \frac{1}{9}(36)$$

$$= 4$$

This happens to be a better approximation than we got using the trapezoidal rule. In fact, for this function it gives the exact value of the integral.

Both the trapezoidal rule and Simpson's rule are much easier to use if one can take advantage of a calculator or computer.

Some scientific calculators use Simpson's rule to evaluate numeric integrals of functions. The next five examples demonstrate how to do this. Examples 25.27 and 25.28 use a Casio fx-7700G and Examples 25.29, 25.30, and 25.31 use a TI-82.

EXAMPLE 25.27

Use a Casio fx-7700G to evaluate $\int_0^2 x^3\,dx$.

Solution   Press the following sequence of keys:

$\boxed{\text{SHIFT}}\ \boxed{\int dx}\ \boxed{\text{X},\theta,\text{T}}\ \boxed{x^y}\ 3\ \boxed{\text{SHIFT}}\ \boxed{,}\ 0\ \boxed{\text{SHIFT}}\ \boxed{,}\ 2\ \boxed{)}\ \boxed{\text{EXE}}$

After a slight pause, the calculator will return the solution 4, the same as that which we got in Example 25.26.

EXAMPLE 25.28

Use a Casio fx-7700G to evaluate $\int_1^5 (2x^5 - 3x)\,dx$.

Solution   Press:

$\boxed{\text{SHIFT}}\ \boxed{\int dx}\ 2\ \boxed{\text{X},\theta,\text{T}}\ \boxed{x^y}\ 5\ \boxed{-}\ 3\ \boxed{\text{X},\theta,\text{T}}\ \boxed{\text{SHIFT}}\ \boxed{,}\ 1\ \boxed{\text{SHIFT}}\ \boxed{,}\ 5\ \boxed{)}\ \boxed{\text{EXE}}$

After a fairly long pause, we get the result 5,172. So, according to the calculator, $\int_1^5 (2x^5 - 3x)\,dx = 5{,}172$. How does this compare with the actual value?

$$\int_1^5 (2x^5 - 3x)\,dx = \frac{1}{3}x^6 - \frac{3}{2}x^2\ \bigg|_1^5$$

$$= \left(\frac{15{,}625}{3} - \frac{75}{2}\right) - \left(\frac{1}{3} - \frac{3}{2}\right)$$

$$= 5{,}172$$

EXAMPLE 25.29

Use a TI-82 to evaluate $\int_0^2 x^3\,dx$.

Solution   There are two ways to use a TI-82 to evaluate a definite integral. We will show the first method in this example and the other in the next example.

In the first method you press the following sequence of keys:

$\boxed{\text{MATH}}\ 9\ \boxed{\text{X},\text{T},\theta}\ \boxed{\wedge}\ 3\ \boxed{,}\ \boxed{\text{X},\text{T},\theta}\ \boxed{,}\ 0\ \boxed{,}\ 2\ \boxed{)}\ \boxed{\text{ENTER}}$

After a slight pause, the calculator will return the solution 4, the same as we got in Example 25.27. (Note that pressing $\boxed{\text{MATH}}$ 9 accesses the "fnInt(" option on the calculator.)

EXAMPLE 25.30

Use a TI-82 to evaluate $\int_1^5 (2x^5 - 3x)\,dx$.

Solution   In this method you begin as if your are going to graph the integrand. So, for this integral begin by entering the integrad as $y_1 = 2x^5 - 3x$. Then, after you access the fnInt(option on the calculator, you recall the name of the function, in this case, y_1. The remainder of the procedure is exactly the same as shown in

EXAMPLE 25.30 (Cont.)

Example 25.28. Press:

$\boxed{\text{MATH}}\ 9\ \boxed{\text{2nd}}\ \boxed{\text{VARS}}\ 1\quad 1\ \boxed{,}\ \boxed{\text{X,T,}\theta}\ \boxed{,}\ 1\ \boxed{,}\ 5\ \boxed{)}\ \boxed{\text{ENTER}}$

After a brief pause, we get the result 5172.

According to the calculator, $\displaystyle\int_1^5 (2x^5 - 3x)\,dx = 5172$. How does this compare

with the actual value? As we showed in Example 25.27, $\displaystyle\int_1^5 (2x^5 - 3x)\,dx = 5{,}172$,

so the calculator gave the exact value of this definite integral.

It is also possible to get the calculator to both shade the region being studied and evaluate the integral. Since an integral is a measure of the area under a curve, this can provide a graphical and numerical interpretation of an integral.

EXAMPLE 25.31

Use a TI-82 to graph the region, and determine the area under the curve of $f(x) = x^2 + \dfrac{1}{x^3} - 4x + 6$ from $x = 1$ to $x = 4.2$.

Solution As in the previous method, you begin as if you are going to graph the integrand. So, for this integral begin by entering the integrand as $y_1 = x^2 + \dfrac{1}{x^3} - 4x + 6$. You may want to graph the curve, in order to select a viewing window that shows the graph over the entire interval from $x = 1$ to $x = 4.2$. In this example, a viewing window of $[0, 4.7] \times [-1, 8]$ was selected.

Now access $\boxed{\text{2nd}}\ \boxed{\text{CALC}}$. (On a TI-82, the $\boxed{\text{CALC}}$ key is above the $\boxed{\text{TRACE}}$ key.) Select the $\int f(x)\,dx$ option on the calculator, by pressing $\boxed{7}$. The result in Figure 25.28a shows the graph of the function, puts a blinking cursor on the curve at the x-value in the middle of the window, asks for a Lower Limit?, and gives the coordinates of the cursor. Use the left or right arrow key, $\boxed{\blacktriangleleft}$ or $\boxed{\blacktriangleright}$, to move the cursor to the lower limit, in this case $x = 1$, and press the $\boxed{\text{ENTER}}$ key. The result is shown in Figure 25.28b. Notice that the calculator is now asking for the Upper Limit?. Use the $\boxed{\blacktriangleright}$ key to move the cursor to the x-value that indicates the upper limit as shown in Figure 25.28c. Press $\boxed{\text{ENTER}}$. You should see the calculator shade the area of the region being intergrated and then, at the bottom of the screen, give an approximation of the area as 10.754322, as shown in Figure 25.28d.

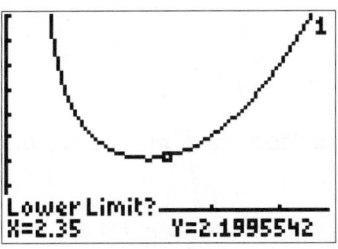

$[0, 4.7, 1] \times [-1, 8, 1]$

FIGURE 25.28a

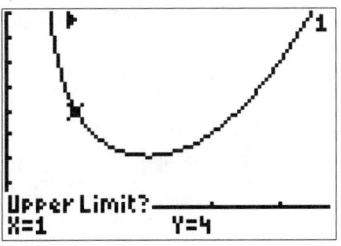

$[0, 4.7, 1] \times [-1, 8, 1]$

FIGURE 25.28b

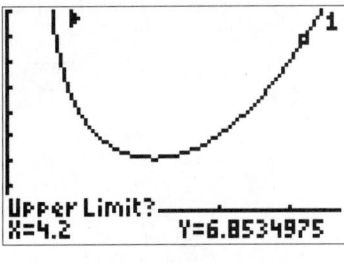

$[0, 4.7, 1] \times [-1, 8, 1]$

FIGURE 25.28c

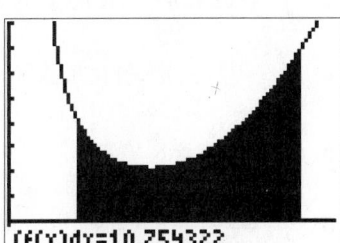

$[0, 4.7, 1] \times [-1, 8, 1]$

FIGURE 25.28d

It is also possible to use both of these formulas to obtain integrals of functions that are available only in graphical or tabular form, as shown by the next example.

Application

EXAMPLE 25.32

The following table gives the results of a series of drillings meant to determine the depth of the bedrock at a building site. These drillings were taken along a straight line down the middle of the lot where the building will be placed. In the table, x is the distance from the front of the parking lot and y is the corresponding depth. Both x and y are given in feet.

x	0	20	40	60	80	100	120	140	160
y	33	35	40	45	42	38	46	40	48

Approximate the area of this cross-section by using Simpson's rule.

Solution We first determine that $n = 8$, an even number, and $dx = 20$. By Simpson's rule,

$$\int_0^{160} f(x)\,dx \approx \frac{20}{3}[33 + 4(35) + 2(40) + 4(45) + 2(42) + 4(38)$$
$$+ 2(46) + 4(40) + 48]$$
$$= \frac{20}{3}(969)$$
$$= 6,460\,\text{ft}^2$$

Exercise Set 25.5

In Exercises 1–16, find the approximate value of the given integrals with the specified value of n by (a) the trapezoidal rule and (b) Simpson's rule.

1. $\int_1^2 x^4\,dx$, $n = 6$ (Check by integration.)

2. $\int_0^2 2x^5\,dx$, $n = 8$ (Check by integration.)

3. $\int_0^1 \sqrt{x}\,dx$, $n = 4$ (Check by integration.)

4. $\int_1^2 x^{1/3}\,dx$, $n = 4$ (Check by integration.)

5. $\int_1^5 \frac{dx}{x}$, $n = 8$

6. $\int_1^4 \frac{dx}{x^2}$, $n = 6$ (Check by integration.)

7. $\int_1^5 \frac{dx}{1+x}$, $n = 8$

8. $\int_0^2 x^\pi\,dx$, $n = 4$ (Check by integration.)

9. $\int_0^1 \frac{4}{1+x^2}$, $n = 10$

10. $\int_0^2 e^x\,dx$, $n = 4$

11. $\int_0^2 \sqrt{1+x^3}\,dx$, $n = 8$

12. $\int_0^{0.2} \sin x\,dx$, $n = 4$ (x in radian measure)

13. $\int_{-2}^2 \sqrt{x+4}\,dx$, $n = 8$

14. $\int_1^3 \ln x\,dx$, $n = 8$

15. $\int_0^4 \dfrac{\sqrt{x}}{1+\sqrt{x}}\,dx, n = 8$

16. $\int_0^1 \dfrac{e^x + e^{-x}}{2}\,dx, n = 4$

In Exercises 17–19, use the trapezoidal rule and Simpson's rule to find the approximate area under the curve defined by the given sets of experimental data.

17.

x	5.0	6.0	7.0	8.0	9.0	10.0
y	4.2	3.9	3.8	4.0	3.5	3.4

x	11.0	12.0	13.0
y	3.9	4.1	4.3

18.

x	2.00	2.10	2.20	2.30	2.40	2.50	2.60
y	4.32	4.57	5.14	5.78	6.84	6.62	6.51

19.

x	1.00	1.25	1.50	1.75	2.00
y	16.32	16.48	16.73	16.42	16.38

x	2.25	2.50
y	16.29	16.25

Solve Exercises 20–26.

20. *Environmental science* Find the area of the pond in Figure 25.29.

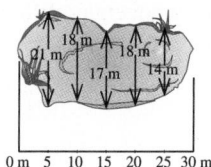

FIGURE 25.29

21. *Electronics* The current i in amperes (A) to a capacitor is given by $i = 4.0e^{-10t} - 4.0e^{-20t}$, where t is time in seconds (s). If $i = \dfrac{dq}{dt}$ where q is the charge in coulombs (C), use **(a)** the trapezoid rule and **(b)** Simpson's rule to find the approximate change in charge on the capacitor from $t = 1.0$ s to $t = 2.0$ s, using four intervals.

22. *Light* The graph in Figure 25.30 shows the light in lumens given off by the flash of a flashbulb. A typical focal plane shutter is open from 10 milliseconds to 60 milliseconds after the camera button is pressed. What is the total amount of light given by the flashbulb during this interval?

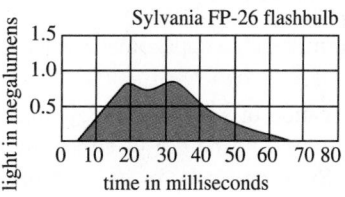

FIGURE 25.30

23. Write a computer program to determine the approximate area under a curve. The program should allow you to select whether to approximate the area by **(a)** the trapezoidal rule, **(b)** Simpson's rule, or **(c)** both rules. Your program should also allow for the input of the endpoints and the number of intervals. If Simpson's rule is used, the program should check to make sure that the number of intervals in even. The program should also allow for either a function $y = f(x)$ to be given or a set of observed data points to be used. Check your program by comparing results with Examples 25.25 and 25.26.

24. *Business* The marginal revenue for the sale of computers by a store is given by the function $R'(x) = 25\sqrt{x}e^{-0.004x}$, where $0 \le x \le 600$. Use the trapezoidal rule with $n = 10$ to determine the approximate revenue from the sale of the first 600 computers sold by the store.

25. *Agriculture*   To determine the amount of fertilizer needed for an irregularly shaped field, a farmer needed to estimate the number of acres in the field. The farmer drew a line across the field and then, at 100 ft intervals, measured across the land perpendicular to the farmer's line. The farmer's measurements are shown in the following table:

x	0	100	200	300	400	500
y	300	300	350	400	500	600

x	600	700	800	900	1000	
y	500	600	550	650	600	

Use both the trapezoidal rule and Simpson's rule to estimate the number of acres if 1 acre $= 43{,}500\,\text{ft}^2$.

26. *Pharmacology*   The reaction rate to a new drug is given by $r = e^{-t^2} + \dfrac{1}{t}$ where t is the time measured in hours after the drug is administered. Use both the trapezoidal rule and Simpson's rule with $n = 8$ to estimate the total reaction to the drug from $t = 1$ to $t = 9$.

✏️ **In Your Words**

27. Explain the purpose of the trapezoidal rule and Simpson's rule.

28. Write an application in your technology area of interest that requires you the use of either the trapezoidal rule or Simpson's rule for its solution. Give your problem to a classmate and see if he or she understands and can solve your problem. Rewrite the problem as necessary to remove any difficulties encountered by your classmate.

▬ CHAPTER 25 REVIEW

Important Terms and Concepts

Area between curves
Definite integral
  Properties of
Fundamental Theorem of Calculus
Improper integrals
Indefinite integrals
  Properties of

Numerical integration
Simpson's rule
Substitution method
Summation formulas
Trapezoidal rule

Review Exercises

In Exercises 1–30, evaluate the given definite or indefinite integral.

1. $\displaystyle\int x^5\,dx$

2. $\displaystyle\int (\sqrt[3]{x}+4)\,dx$

3. $\displaystyle\int_0^3 (x^2+4x-6)\,dx$

4. $\displaystyle\int_1^4 (x+4)^2\,dx$

5. $\displaystyle\int_1^8 \frac{dt}{\sqrt[3]{t}}$

6. $\displaystyle\int_1^4 (\sqrt{x}+4)\,dx$

7. $\displaystyle\int \frac{dt}{(t+5)^4}$

8. $\displaystyle\int \frac{x\,dx}{\sqrt[3]{x^2+4}}$

9. $\displaystyle\int_1^8 x\sqrt[3]{x}\,dx$

10. $\displaystyle\int_1^4 \frac{4\,dx}{3\sqrt{x^3}}$

11. $\displaystyle\int (u^4 - u^{-4})\,du$

12. $\displaystyle\int \frac{x^4-1}{x^3}\,dx$

13. $\displaystyle\int 2x\sqrt{x^2-5}\,dx$

14. $\displaystyle\int_0^4 x\sqrt{x^2+9}\,dx$

15. $\displaystyle\int x\sqrt[3]{6-x^2}\,dx$

16. $\displaystyle\int 3x\left(4-x^2\right)^3 dx$

17. $\displaystyle\int_{-1}^{2}(x^2+4)(x^3-3)\,dx$

18. $\displaystyle\int_1^2 x\,(x+2)^2\,dx$

19. $\displaystyle\int \frac{y^2\,dy}{(y^3-5)^{2/3}}$

20. $\displaystyle\int \frac{5-6x+x^2}{1-x}\,dx$

21. $\displaystyle\int_0^1 (1-x^2)^3\,dx$

22. $\displaystyle\int \frac{x+2}{(x^2+4x)^2}\,dx$

23. $\displaystyle\int\left(\sqrt{2x}-\frac{1}{\sqrt{2x}}\right)dx$

24. $\displaystyle\int (x^3-4x)\,(3x^2-4)\,dx$

25. $\displaystyle\int_1^4 \frac{9-t^3}{3t^2}\,dt$

26. $\displaystyle\int_0^1 \sqrt{x}\,(3\sqrt{x}-5)\,dx$

27. $\displaystyle\int_{-1}^8 \frac{4\sqrt[3]{u}\,du}{(1+u^{4/3})^3}$

28. $\displaystyle\int_1^2 \frac{3x^3-2x^{-3}}{x^2}\,dx$

29. $\displaystyle\int_1^\infty \frac{dx}{x^4}$

30. $\displaystyle\int_0^\infty \frac{4x\,dx}{(3x^2+1)^5}$

In Exercises 31–38, find the areas bounded by the given curves.

31. $f(x)=4x-6$, x-axis, $x=-3$, $x=4$

32. $g(x)=2-x-x^2$, x-axis

33. $f(x)=x^2-x-12$, $g(x)=-6$

34. $h(x)=\sqrt{x+2}$, x-axis, $x=7$

35. $f(x)=4x^3$, $g(x)=x$

36. $f(x)=x^3-3x^2$, $g(x)=x-3$

37. $y^2+x=0$, $y=2x+1$

38. $f(x)=2x^3$, $g(x)=x^4$

In Exercises 39–44, find the approximate value of the given integrals with the specified values of n by (a) the trapezoidal rule and (b) Simpson's rule.

39. $\displaystyle\int_0^1 \sqrt{1-x^2}\,dx$, $n=10$

40. $\displaystyle\int_1^4 \frac{1}{3x}\,dx$, $n=6$

41. $\displaystyle\int_2^{2.5} \sqrt{x^3-1}\,dx$, $n=10$

42. $\displaystyle\int_{-1}^4 \sqrt{1+x^5}\,dx$, $n=10$

43. $\displaystyle\int_0^1 (e^x-e^{-x})\,dx$, $n=4$

44.

x	1.00	1.25	1.50	1.75	2.00
y	4.25	5.72	5.13	3.19	2.10

x	2.25	2.50	2.75	3.00
y	0.15	1.65	3.10	3.70

Solve Exercises 45–48.

45. *Dynamics*   An object is moving rectilinearly with a velocity $v(t)=t+\dfrac{1}{\sqrt{1+t}}$ m/s. **(a)** If the object began at the origin, what is its position when $t=15$? **(b)** What is its acceleration when $t=15$?

46. *Electricity*   The power P in watts (W) in a circuit is given by $P=\dfrac{d\omega}{dt}$, where ω is dissipated energy in joules (J) and t is time in seconds (s). If the power in a circuit is given by $P=10t^3$, find the energy dissipated from $t=3.0$ s to $t=5.0$ s.

47. *Electricity*   The induced voltage v in volts (V) in an inductor is given by $v=-L\dfrac{di}{dt}$, where i is the current in amperes (A) and L is the inductance in henries (H). If v is given by $v=2.0-1.0t^2$, where t is time in seconds (s) and $L=2$ H, find the change in the current to the inductor from $t=1.0$ s to $t=6.0$ s.

48. *Highway safety* A car moving initially at 88 ft/s, decelerates at the constant rate of 4 ft/s². **(a)** How long does it take to stop the car? **(b)** How far does it travel from the time the brakes are applied until it stops?

≡ CHAPTER 25 TEST

In Exercises 1–7, evaluate the given integral.

1. $\displaystyle\int_0^1 2x\,dx$

2. $\displaystyle\int_1^4 (x+1)(x^2+1)\,dx$

3. $\displaystyle\int (x^2+3)^2\,dx$

4. $\displaystyle\int 4(x^4+7)^5 x^3\,dx$

5. $\displaystyle\int \frac{x^2+2}{x^2}\,dx$

6. $\displaystyle\int (x^2+8x)^3(x+4)\,dx$

7. $\displaystyle\int_1^2 (3x^2\sqrt{x^3+x}+\sqrt{x^3+x})\,dx$

Solve Exercises 8–11.

8. Find the total area of the region between the curve $y=4x^3-4$ and the x-axis over the interval $[-1,2]$.

9. Find the area of the region enclosed by the curve $y=x^2-2$ and the line $x+y=4$.

10. Use the trapezoidal and Simpson's rules to determine $\displaystyle\int_2^4 \frac{1}{x}\,dx$ with $n=8$.

11. For a particular circuit, the current, in amperes, after time t, in seconds, at a certain point, P, is given by $i=0.0004(t^{-1/2}+t^{1/2})$. Find the charge, in C, that passes point P during the first second.

Applications of Integration

A tank truck, like the milk truck shown here, is used to carry liquids. In Section 26.7, you will learn how to determine the force on one end of this truck when it is full of milk.

Courtesy of H.P. Hood Inc.

We introduced the integral as a method for determining the area under a curve. Later, we showed that integrals could be used to determine the velocity and position functions from the acceleration. We also used integrals to find the area between two curves. In this chapter, we will see some of the many other applications for the integral.

We begin with the average value and root mean square of a function. The average value will be used to determine the amount of heating oil that will be used during one winter day. Examples will use the root mean square to determine the effective current and voltage in an electric circuit.

Next, we look at methods for determining the volume of a three-dimensional object. Two different methods will be examined to allow you to select the better method for a particular problem. Then we will look at methods for finding the length of a curve and the area of a surface formed by revolving that curve around a line.

We will end the chapter with three sections that look at several physical applications—an object's mass, its center of gravity, its moment of inertia, the amount of work needed to move something, and the pressure exerted by a liquid.

≡ 26.1
AVERAGE VALUES AND OTHER ANTIDERIVATIVE APPLICATIONS

In this section, we will look at two ideas: One is the idea of the average value of a function; the other idea looks at some additional applications of the antiderivative.

Average Value of a Function

The concept of an average is a common one to most of us. Sometimes the term is applied to a finite number of items such as the average grade on a test by a group of students. Sometimes it is based on a continuous function. There are times when that function is known and times when it is not.

One example that most of us hear every day is the reference to the average temperature. The weather bureau assumes that there is some function f that will give the temperature T at some time t. Thus, $T = f(t)$. Normally, what is done is to take a temperature reading every hour. The average or mean temperature $\bar{T}$ is the sum of these temperatures divided by the number of hours. Thus, for one day, we have

$$\bar{T} = \frac{1}{24} \sum_{t=1}^{24} f(t)$$

If we divide the day into smaller intervals and take a temperature reading at each of those intervals, we get a more accurate reading. So, if we take n temperature readings we have

$$\bar{T} = \frac{1}{n} \sum_{i=1}^{n} f(t_i)$$

where t_i represents the time of the ith temperature reading.

Naturally, the larger that n becomes the more accurate is the average. If we wanted the true average, we would take the limit of $\frac{1}{n} \sum_{i=1}^{n} f(t_i)$ as n approaches infinity or

$$\bar{T} = \lim_{n \to \infty} \frac{1}{n} \sum_{i=1}^{n} f(t_i)$$

The right-hand side of this equation looks familiar. If dt represents the width of a time interval, $\frac{b-a}{n}$, where a is the beginning of the first interval and b the end of the last interval, and if we write $\frac{1}{n}$ as $\frac{b-a}{n} \cdot \frac{1}{b-a}$, then

$$\bar{T} = \lim_{n \to \infty} \frac{1}{b-a} \sum_{n=1}^{n} f(t_i) \frac{b-a}{n}$$

$$= \frac{1}{b-a} \lim_{n \to \infty} \sum_{i=1}^{n} f(t_i) dt$$

$$= \frac{1}{b-a} \int_{a}^{b} f(t) dt$$

If we assume that f is continuous, then we can use this idea for the following definition.

Average Value

If f is a continuous function over the interval $[a, b]$, then the **average value** $\bar{y}$ is

$$\bar{y} = \frac{1}{b-a} \int_a^b f(x)dx$$

Application

EXAMPLE 26.1

A ball is dropped from a height of 490 m. Find its average velocity between the time it is dropped and the time it hits the ground.

Solution If acceleration is -9.8 m/s^2 and, since $v(0) = 0$, then we have $v(t) = -9.8t$. We also have $s(0) = 490$, and so $s(t) = -4.9t^2 + 490$. Solving for t, we see that it takes 10 s to hit the ground. The average velocity is

$$\begin{aligned}
\bar{v} &= \frac{1}{10-0} \int_0^{10} v(t)dt \\
&= \frac{1}{10}[s(t)]\Big|_0^{10} \\
&= \frac{1}{10}(-4.9t^2 + 490)\Big|_0^{10} \\
&= \frac{1}{10}(-490) = -49
\end{aligned}$$

The average velocity is 49 m/s downward.

Application

EXAMPLE 26.2

The amount of heating oil required to heat a house for one day during the winter can be approximated by multiplying the difference between 20°C and the average daily temperature by 0.8 gal. During one particular cold spell, the temperature T, at any time of the day t h after midnight, was given by $T(t) = 2 - \frac{1}{15}(t-15)^2$. Determine the level of oil consumption during one of these days.

Solution If we let A represent the amount of oil required and $\bar{T}$ the average temperature during a day, then we have

$$A = (20°C - \bar{T})(0.8\,\text{gal/}°C)$$

We will first compute the average temperature during one of these days. We will let t range from midnight at the start of the day, $t = 0$, to midnight at the end of the

EXAMPLE 26.2 (Cont.)

day, $t = 24$. Thus, for our formula, $a = 0$ and $b = 24$

$$\bar{T} = \frac{1}{24-0} \int_0^{24} T(t)\,dt$$

$$= \frac{1}{24-0} \int_0^{24} \left[2 - \frac{1}{15}(t-15)^2 \right] dt$$

$$= \frac{1}{24} \left[2t - \frac{1}{45}(t-15)^3 \right]_0^{24}$$

$$= \frac{1}{24}(31.80 - 75) = \frac{1}{24}(-43.2) = -1.8$$

So, the average temperature during this particular 24-hour period was $-1.8°C$. We can now determine the amount of fuel oil consumed during this period.

$$A = (20°C - \bar{T})(0.8\,\text{gal/}°C)$$
$$= [20°C - (-1.8°C)](0.8\,\text{gal/}°C)$$
$$= (21.8°C)(0.8\,\text{gal/}°C)$$
$$= 17.44\,\text{gal}$$

A total of 17.44 gal of heating oil were used during this 24-h period.

Root Mean Square

The **root mean square**, rms, of a function $y = f(x)$ over the interval $[a, b]$ is defined to be the square root of the average value of y^2 on the interval.

> **Root Mean Square**
>
> If a function f is continuous on $[a, b]$, then the **root mean square**, f_{rms}, is
>
> $$f_{rms} = \sqrt{\frac{1}{b-a} \int_a^b [f(x)]^2\,dx}$$

EXAMPLE 26.3

Find the root mean square of $f(x) = x^4$ on $[-1, 3]$.
Solution

$$f_{rms} = \sqrt{\frac{1}{3-(-1)} \int_{-1}^3 (x^4)^2\,dx}$$

$$= \sqrt{\frac{1}{4} \int_{-1}^3 x^8\,dx}$$

EXAMPLE 26.3 (Cont.)

$$= \sqrt{\left. \frac{1}{4} \cdot \frac{1}{9} x^9 \right|_{-1}^{3}}$$

$$= \sqrt{\frac{1}{36}[19{,}683 - (-1)]}$$

$$= \sqrt{\frac{19{,}684}{36}} \approx 23.38$$

Because an alternating current changes continuously, its maximum value, $\pm I_{\max}$, does not indicate its ability to do work or to produce heat as does the magnitude of a direct current. For this reason, it is customary to refer to the **effective current**. The same is true for the voltage, where it is more common to refer to the **effective voltage** rather than $V_{\max}$.

≡ **Note**   The effective current and the effective voltage are the same as the root mean square of the current and voltage. In symbols, this means that $i_{\text{eff}} = i_{\text{rms}}$ and $V_{\text{eff}} = V_{\text{rms}}$.

If the resistance is constant, then the average or mean power, in watts, produced by the current is

$$P = i_{\text{rms}}^2 \cdot R$$

Application

EXAMPLE 26.4

If a current $i = t - 2t^2$ A flows through a resistor of $20\,\Omega$ from $t = 0$ to $t = 3$ s, find i_{eff} and the power generated.

Solution

$$i_{\text{eff}} = i_{\text{rms}} = \sqrt{\frac{1}{3} \int_0^3 (t - 2t^2)^2 \, dt}$$

$$= \sqrt{\frac{1}{3} \int_0^3 (t^2 - 4t^3 + 4t^4) \, dt}$$

$$= \sqrt{\frac{1}{3} \left[\frac{1}{3} t^3 - t^4 + \frac{4}{5} t^5 \right]_0^3}$$

$$= \sqrt{\frac{1}{3}(9 - 81 + 194.4)}$$

$$= \sqrt{40.8} \approx 6.39 \, \text{A}$$

and

$$P = \left(\sqrt{40.8} \right)^2 \cdot 20$$

$$\approx 816 \, \text{W}$$

Current and Power

The current i, in amperes, in an electric circuit is defined as the time rate of change of the charge q, in coulombs, which passes a given point in the circuit. This is written as

$$i = \frac{dq}{dt}$$

When this is placed in the differential form, we get $dq = i\,dt$. Integrating both side of this equation, we get

$$q = \int i\,dt$$

Thus we can find the total charge transmitted in a circuit.

The charge in a capacitor is given by

$$q = CV_c$$

where C is the capacitance in farads and V_c represents the voltage between the capacitor terminals. Substituting for q from the previous formula and solving for V_c, we get

$$V_c = \frac{1}{C}q$$

$$\text{or} \quad V_c = \frac{1}{C}\int i\,dt$$

Application

EXAMPLE 26.5

The current in a circuit is given by the equation $i = 4t^3 + 2$ A. How many coulombs are transmitted in 3 s?

Solution We have $q = \int i\,dt$ and from the information given we can see that

$$q = \int_0^3 (4t^3 + 2)dt$$
$$= (t^4 + 2t)|_0^3$$
$$= 87\,\text{C}$$

≡ **Note** The C at the end of Example 26.5 stands for coulombs. Do not confuse it with the C for capacitance, with the constant of integration, or with °C for degrees Celsius.

Application

EXAMPLE 26.6

A 1-μF capacitor receives a charge such that it has a terminal-to-terminal voltage of 33 V. We connect this capacitor at $t = 0$ s to a source that sends a current $i = t^2$ A into it. What is the voltage across the capacitor when $t = 0.2$ s?

EXAMPLE 26.6 (Cont.)

Solution We will need the formula $V_c = \dfrac{1}{C} \displaystyle\int i \, dt$. We are given $C = 1\,\mu F$ $= 10^{-6}\,F$ and $i = t^2$, so

$$V_c = \frac{1}{10^{-6}} \int t^2 \, dt = 10^6 \cdot \frac{1}{3} t^3 + k$$

(We will let k be the constant of integration.) Now when $t = 0$, $V_c = 33$ and $k = 33$ and so

$$V_c = 10^6 \cdot \frac{1}{3} t^3 + 33$$

When $t = 0.2$, we get

$$V_c = 10^6 \cdot \frac{1}{3}(0.2)^3 + 33$$
$$\approx 2,699.67\,V$$
$$\approx 2,700\,V$$

Application

EXAMPLE 26.7

A certain capacitor has a voltage of 150 V across it. A current given by $i = 0.08\sqrt{t}$ is sent through the circuit. After 0.36 s, the voltage across the capacitor is measured at 200 V. What is the capacitance of this capacitor?

Solution We will use the equation $V_c = \dfrac{1}{C} \displaystyle\int i \, dt$. Since $i = 0.08\sqrt{t}$, we have

$$V_c = \frac{1}{C} \int 0.08 t^{1/2} \, dt$$
$$= \frac{0.08}{C} \int t^{1/2} \, dt$$
$$= \frac{0.16}{3C} t^{3/2} + k$$

When $t = 0$, we know that $V_c = 150$ and so $k = 150$. Thus,

$$V_c = \frac{0.16}{3C} t^{3/2} + 150$$

Now $V_c = 200\,V$ when $t = 0.36\,s$, and so

$$200 = \frac{0.16}{3C}(0.36)^{3/2} + 150$$
$$200 = \frac{0.01152}{C} + 150$$

Solving for C, we get

$$C = 0.0002304\,F$$
$$= 2.304 \times 10^{-4}\,F$$
$$= 230.4\,\mu F$$

As a last application in this section, the amount of energy (**work**), W, expended in an electrical system is given by

$$W = \int P \, dt$$

Here W is in joules, the power P is in watts, and t is in seconds.

Application

EXAMPLE 26.8

The power in a certain electrical system is changed according to $P = 5t^{1/4}$ W. What is the formula for energy in this circuit?

Solution   Using the formula $W = \int P \, dt$ with $P = 5t^{1/4}$, we have

$$W = \int 5t^{1/4} \, dt$$
$$= 4t^{5/4} + k \, \text{J}$$

The **instantaneous power** in a circuit can be represented by $P = Vi$, $P = i^2 R$, or $P = V^2/R$, and any of these can be substituted for P in the formula $W = \int P \, dt$.

Exercise Set 26.1

In Exercises 1–12, find the average value and the root mean square of the given function over the indicated interval.

1. $f(x) = x^2$, over $[0, 1]$
2. $g(x) = 3x^2$, over $[2, 4]$
3. $h(x) = x^2 - 1$, over $[1, 3]$
4. $j(x) = 9 - x^2$, over $[-3, 3]$
5. $f(x) = \sqrt{x+4}$, over $[0, 5]$
6. $k(x) = x^3 - 1$, over $[1, 3]$

7. $f(t) = t^2 + 4$, over $[0, 4]$
8. $g(t) = 16t^2 - 25$, over $[2, 4]$
9. $f(x) = 6x - x^2$, over $[1, 2]$
10. $h(t) = 16t^2 - 8t + 1$, over $[0, 3]$
11. $j(t) = 4.9t^2 - 2.8t + 4$, over $[0, 2.5]$
12. $k(t) = 4.9t^2 - 4.2t + 9$, over $[0, 1.6]$

Solve Exercises 13–32.

13. *Meteorology*   The rainfall per day, measured in centimeters, x days after the beginning of the year is $0.0002(4991 + 366x - x^2)$. What is the average rainfall for the first 90 days? For the year (365 days)?

14. *Material science*   The mass density of a metal bar of length 4 m is given by $\rho(x) = 1 + x - \sqrt{x}$ kg/m³, where x is the distance in meters from one end of the bar. What is the average mass density over the length of the bar?

15. *Physics*   If a ball is dropped from a height of 360 ft, find its average height and average velocity between the time it is dropped and the time it hits the ground.

16. A 20 000-L water tank takes 10 min to drain. After t min, the amount of water left in the tank is given by $V(t) = 200(100 - t^2)$. What is the average amount of water in the tank while it is draining?

1.94 $V_{rms}/R = i^2 r_{ms}$

17. *Meteorology*   One day the temperature of the air t h after noon was given by $60 + 4t - t^{2/3}\,°\text{F}$.
    (a) What was the average temperature between noon and 6 p.m.?
    (b) What was the average temperature between 3 p.m. and 9 p.m.?

18. *Business*   A laboratory receives a shipment of 900 cases of test tubes every 30 days. The number of cases on hand t days after the shipment arrives is given by $I(t) = 900 - 15\sqrt{120t}$. What is the average daily inventory?

19. *Electricity*   Find the average current from $t = 0$ to $t = 3$ s, if $i = 1.0t + \sqrt{t}$ A.

20. *Electricity*   Find the effective current for the current in Exercise 19.

21. *Electricity*   Find the average current from $t = 0$ to $t = 4$ s, if $i = 4t\sqrt{t^2 + 1}$ A.

22. *Electricity*   What is the effective current for the current in Exercise 21?

23. *Electricity*   If the current in Exercise 21 flows through a 30-Ω resistor, what is its average power?

24. *Electricity*   A current $i = t^3 - 2t^2$ A flows through a resistance of 5.0 Ω from $t = 0.0$ to $t = 0.25$ s. Find the average power developed over that interval.

25. *Electricity*   The voltage across a 10-Ω resistor is given by $V = 4\sqrt{t} - 2t$. **(a)** What is the effective voltage during the time interval from $t = 0$ to $t = 6$ s?

(b) What is the average power generated during this time interval?

26. *Electricity*   A voltage of $v = 8.0t^{1/2} - 16t^{3/2}$ V is applied across a resistance of 9.75 Ω. Find the average power developed during the period when v is greater than 0?

27. *Electricity*   A capacitor contains a charge of 0.01 C. If it is supplied with a current of $i = 2t$ A, what charge resides in the capacitor after 0.1 s?

28. *Electricity*   An 80-μF capacitor is charged to 100 V. The capacitor is then supplied a current $i = 0.04t^3$ A. After what time interval does the capacitor voltage reach 225 V?

29. *Electricity*   The voltage applied to a circuit was $V = 2t + 1$ V. If the current was $i = 0.03t$ A, what was the energy delivered from $t = 0$ to $t = 50$ s?

30. *Electricity*   The voltage across a 90-μF capacitor is zero. What is the voltage after 0.001 s, if a current of 0.20 A charges the capacitor?

31. *Electricity*   The voltage across a 7.5-μF capacitor is zero. What is the voltage after 0.005 s, if a current of $i = 0.4t$ A charges the capacitor?

32. *Electricity*   The current in a certain wire changes with time according to the relation $i = \sqrt[3]{1 + 5t}$. How many coulombs of charge pass a certain point in the first 3 s?

🖋 **In Your Words**

33. Describe how to calculate the average value of a function.

34. What is the root mean square? Give three applications of the root mean square.

≡ 26.2
VOLUMES OF REVOLUTION: DISK AND WASHER METHODS

Consider the area under the graph of a continuous nonnegative function, $y = f(x)$, over the interval $[a, b]$. If this area is revolved around the x-axis, then we get a solid of revolution such as the one shown in Figure 26.1. Common examples of solids of revolution include a cone, a cylinder, a soft drink bottle, and a table leg formed by turning a piece of wood on a lathe.

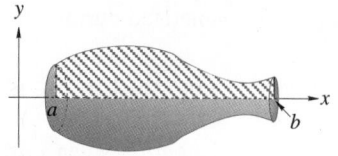

FIGURE 26.1

A cone is formed by revolving the area under a line of the form $f(x) = mx$ from 0 to h around the x-axis, as shown in Figure 26.2a. A cylinder is formed when the area under a horizontal line from a to b is revolved around the x-axis as shown in Figure 26.2b. A soft drink bottle is formed by revolving the area under a curve, such as the one shown in Figure 26.2c.

Disks

In this section and in Section 26.3, we will develop ways for finding the volume of solids of revolution such as these. As when we developed the definite integral, we will let $y = f(x)$ be a continuous function with $f(x) \geq 0$ on the interval $[a, b]$. If we rotate the area under this curve around the x-axis, we generate a solid of revolution. If we cut this solid with parallel planes all perpendicular to the x-axis, we will see that a typical cross-section is a circle and a typical slice is a disk as shown in Figure 26.3a.

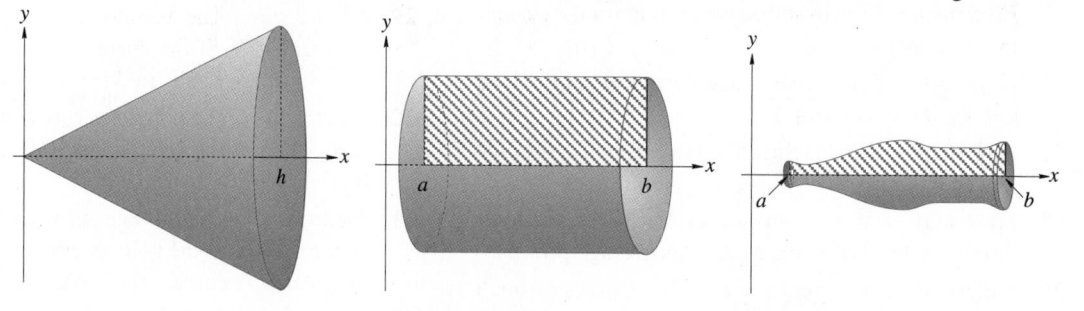

FIGURE 26.2a   **FIGURE 26.2b**   **FIGURE 26.2c**

If we divide $[a, b]$ into subintervals and let x_i be a point in one of the subintervals, then a typical disk is a cylinder with height dx and radius $f(x_i)$ as shown in Figure 26.3b. The volume of a cylinder is $\pi r^2 h$, so the volume of this disk is $\pi[f(x_i)]^2 dx$. Summing the volume of all of these disks over the interval $[a, b]$, we get the total volume as shown in the following box.

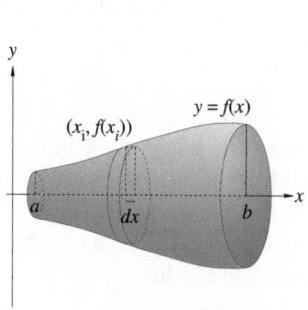

FIGURE 26.3a

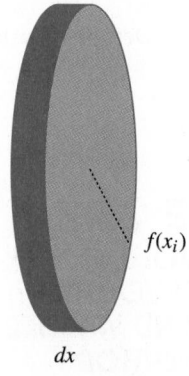

FIGURE 26.3b

Disk Method for Finding a Volume of Revolution (x-Axis)

The volume V of the solid generated by revolving the area between the graph of $y = f(x)$ and the x-axis from $x = a$ to $x = b$ around the x-axis is

$$V = \pi \int_a^b [f(x)]^2 \, dx$$

 Hint

It is often easier to remember the disk method by using the following guidelines:
1. Sketch the graph over the limits of integration.
2. Draw a typical disk by slicing the solid perpendicular to the axis of revolution.
3. Determine the radius and thickness of the disk.
4. Determine the volume of a typical disk by using $\pi \, (\text{radius})^2 \cdot (\text{thickness})$.
5. Use the following integration formula to find the total volume:

$$V = \pi \int_a^b (\text{radius})^2 \cdot (\text{thickness})$$

EXAMPLE 26.9

Find the volume of the solid of revolution generated by revolving the area under the graph of $f(x) = \sqrt{x}$ from $x = 0$ to $x = 4$ around the x-axis.

Solution   The graph of $f(x) = \sqrt{x}$ on $[0, 4]$ is given in Figure 26.4a. A rectangle has been drawn to show a typical element.

The solid that results from revolving the graph of f around the x-axis is shown in Figure 26.4b, with a disk formed by the typical element. The radius of the disk is $y = \sqrt{x}$, and its thickness is dx. A typical disk is shown in Figure 26.4c. Its volume is

$$\pi \, (\text{radius})^2 \cdot (\text{thickness}) = \pi \left(\sqrt{x} \right)^2 dx$$

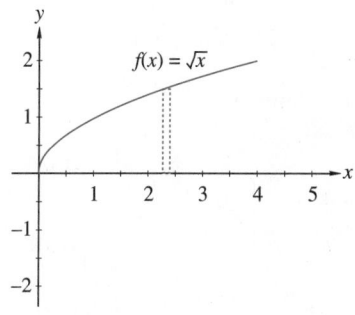

FIGURE 26.4a

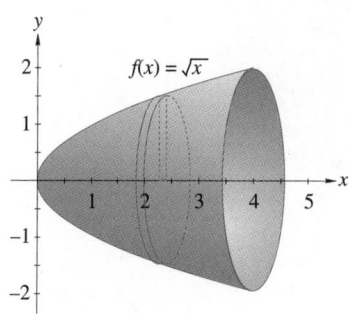

FIGURE 26.4b

EXAMPLE 26.9 (Cont.)

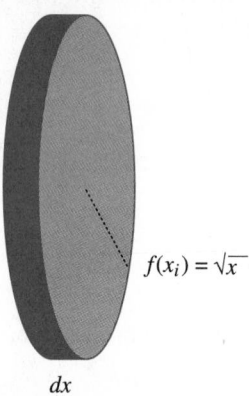

$f(x_i) = \sqrt{x}$

dx

FIGURE 26.4c

EXAMPLE 26.10

Thus, we get the following integral.

$$V = \pi \int_0^4 \left(\sqrt{x}\right)^2 dx$$

$$= \pi \int_0^4 x\, dx$$

$$= \pi \left(\frac{1}{2}x^2\right)\Big|_0^4 = 8\pi$$

The volume is 8π units^3.

Application

A wing tank for an airplane is a solid of revolution formed by rotating the region bounded by the graph of $y = \frac{1}{16}x\sqrt{4-x}$ and the x-axis from $x = 0$ to $x = 4$ around the x-axis, where x and y are measured in meters, as shown in Figure 26.5a. Find the volume of the tank.

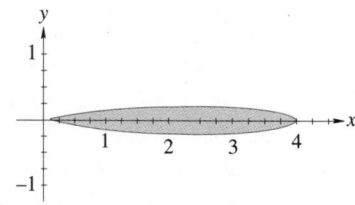

FIGURE 26.5a

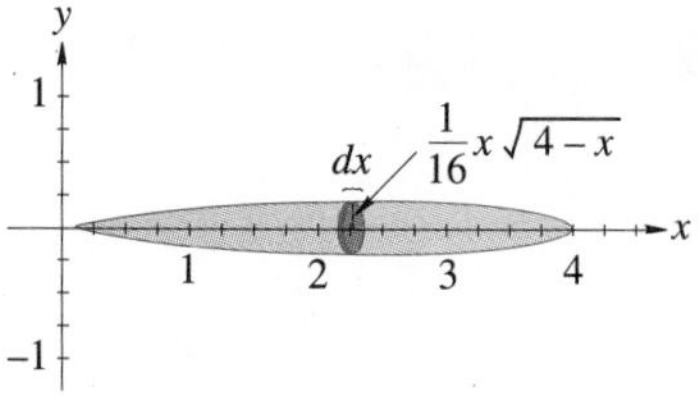

FIGURE 26.5b

Solution A disk formed from a typical element is shown in Figure 26.5b. The radius of the disk is $\frac{1}{16}x\sqrt{4-x}$ and its thickness is dx. Thus the volume of a typical disk is

$$\pi\,(\text{radius})^2 \cdot (\text{thickness}) = \pi \left(\frac{1}{16}x\sqrt{4-x^2}\right)^2 dx$$

and the total volume is

$$V = \pi \int_0^4 \left(\frac{1}{16}x\sqrt{4-x}\right)^2 dx$$

$$= \pi \int_0^4 \frac{1}{256}x^2\,(4-x)\,dx$$

$$= \frac{\pi}{256} \int_0^4 \left(4x^2 - x^3\right) dx$$

$\frac{1}{16}x = \sqrt{4-x}$

dx

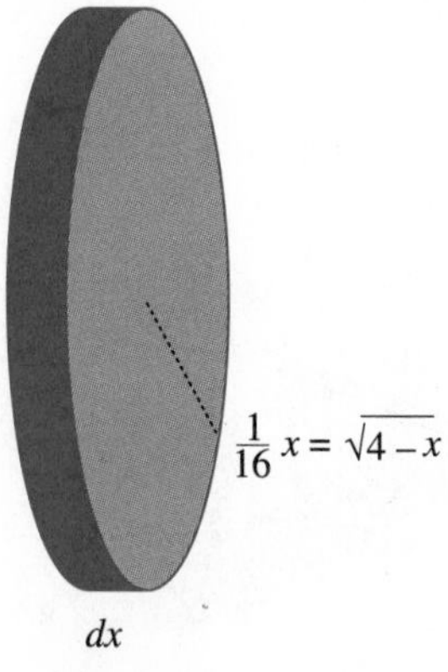

FIGURE 26.5c

EXAMPLE 26.10 (Cont.)

$$= \frac{\pi}{256} \left(\frac{4}{3}x^3 - \frac{1}{4}x^4 \right) \Big|_0^4$$

$$= \frac{\pi}{256} \left(\frac{256}{3} - 64 \right) = \frac{\pi}{12} \approx 0.262$$

The volume of this wing tank is approximately $0.262 \, \text{m}^3$.

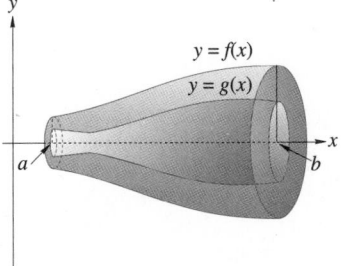

FIGURE 26.6a

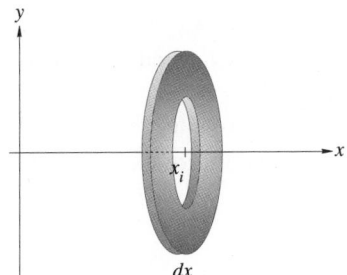

FIGURE 26.6b

Washers

In the disk method, we revolved the area between one curve and the x-axis around the x-axis. But, what happens if we revolve the area between two curves around the x-axis? In Figure 26.6a, we revolved the area enclosed by $y = f(x)$ and $y = g(x)$, where $f(x) \geq g(x)$, and between the lines $x = a$ and $x = b$.

If we sliced this as we did in the disk method, we would get a disk with a hole in the middle. Thus, the slice would look like a washer, as shown in Figure 26.6b. As before, the height of this washer is dx. Its volume could be determined by finding the volume of the solid disk formed by $y = f(x)$ and subtracting the disk that forms the hole in the washer, $y = g(x)$. The area of the entire solid disk is $\pi[f(x_i)]^2 dx$ and of the hole is $\pi[g(x_i)]^2 dx$. Thus, the area of the washer is

$$\pi[f(x)]^2 dx - \pi[g(x)]^2 dx = \pi[f(x)]^2 - [g(x)]^2 dx$$

Summing this over the entire interval, we get a second method for finding a volume of revolution.

Washer Method for Finding a Volume of Revolution

The volume V of the solid generated by revolving the area between the graph of $y = f(x)$ and the graph of $y = g(x)$ from $x = a$ to $x = b$ around the x-axis is

$$V = \pi \int_a^b \left\{ [f(x)]^2 - [g(x)]^2 \right\} dx$$

where $f(x) =$ the outer radius and $g(x) =$ the inner radius of the washer.

EXAMPLE 26.11

Find the volume of the solid of revolution generated by revolving the area enclosed by $y = x^2$ and $y = 4x$ around the x-axis.

EXAMPLE 26.11 (Cont.)

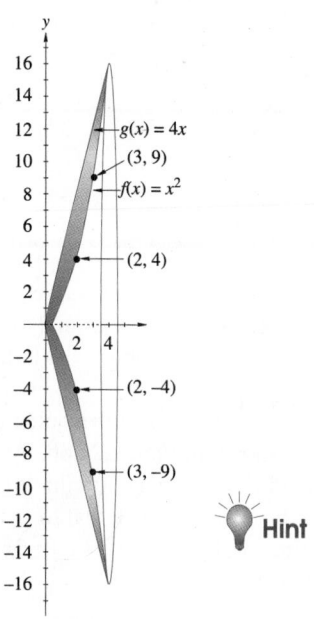

FIGURE 26.7

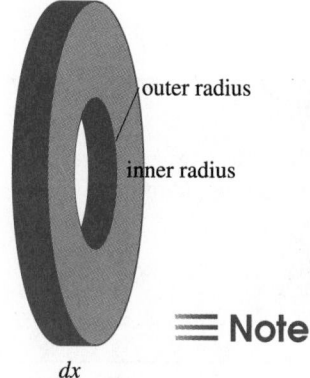

outer radius

inner radius

dx

FIGURE 26.8

EXAMPLE 26.12

Solution Let $f(x) = x^2$ and $g(x) = 4x$. The graph of these two functions is shown in Figure 26.7. The two graphs intersect when $f(x) = g(x)$ or when $x^2 = 4x$, which is at $x = 0$ and $x = 4$. On the interval $[0, 4]$, $4x \geq x^2$, the outer radius is $f(x) = x^2$, and the inner radius is $g(x) = 4x$. We determine the volume as

$$V = \pi \int_0^4 \left[(4x)^2 - (x^2)^2 \right] dx$$

$$= \pi \int_0^4 (16x^2 - x^4)dx$$

$$= \pi \left(\frac{16}{3}x^3 - \frac{1}{5}x^5 \right)\Big|_0^4$$

$$= \pi \left(\frac{2{,}048}{15} \right) \approx 136.53\pi$$

The volume is about 136.53π units^3.

💡 **Hint**

The following guidelines are a variation of those for the disk method. Following them should make it easier to remember the washer method.

1. Sketch the graphs of the two functions and find the limits of integration.
2. Draw a typical washer, as in Figure 26.8, and remember to slice the solid perpendicular to the axis of revolution.
3. Find the outer and inner radii for the washer.
4. Determine the thickness of the washer.
5. The volume of a typical washer is

$$\pi \left[(\text{outer radius})^2 - (\text{inner radius})^2 \right] \cdot (\text{thickness})$$

6. Use the following integration formula to find the total volume:

$$V = \pi \int_a^b \left[(\text{outer radius})^2 - (\text{inner radius})^2 \right] \cdot (\text{thickness})$$

☰ **Note**   Be sure to label the volume in appropriate cubic units.

Application

A drafter is designing a plastic bead that can be strung to help form a necklace. The bead is in the shape of a sphere with a hole drilled through the center, as shown in Figure 26.9a. If the sphere has a radius of 0.25 in. and the hole a radius of 0.07 in., how much plastic will be needed to make one bead?

EXAMPLE 26.12 (Cont.)

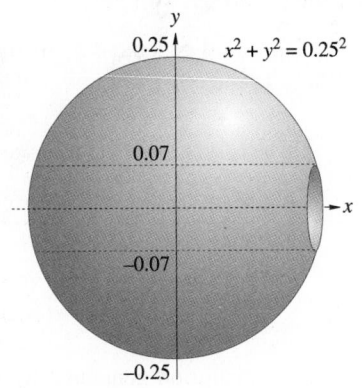

FIGURE 26.9a

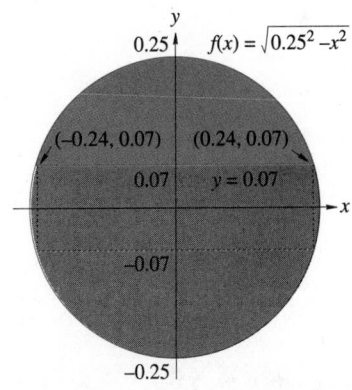

FIGURE 26.9b

Solution   We will sketch this so the origin is at the center of the sphere. A cross-section of the bead, as in Figure 26.9b, shows that the sphere is formed by rotating the top half of the circle $x^2 + y^2 = 0.25^2$, so the outer radius is $f(x) = \sqrt{0.25^2 - x^2}$. The hole is formed by rotating the region formed by the line $y = 0.07$, so the inner radius is $g(x) = 0.07$. If we let $y = 0.7$ and we solve the equation $x^2 + y^2 = 0.25^2$, we get the points of intersection at $x = \pm 0.24$ in. Since the thickness is dx, we have all the information we need, and the volume is given by

$$V = \pi \int_{-0.24}^{0.24} \left[\left(\sqrt{0.25^2 - x^2} \right)^2 - (0.07)^2 \right] dx$$

$$= \pi \int_{-0.24}^{0.24} \left(0.0625 - x^2 - 0.0049 \right) dx$$

$$= \pi \int_{-0.24}^{0.24} \left(0.0576 - x^2 \right) dx$$

$$= \pi \left[0.0576x - \frac{x^3}{3} \right]_{-0.24}^{0.24}$$

$$= \frac{\pi}{3} (0.055296) = 0.018432\pi \approx 0.0579$$

It will take about 0.0579 in.³ of plastic to make each bead.

Revolving Around the y-Axis

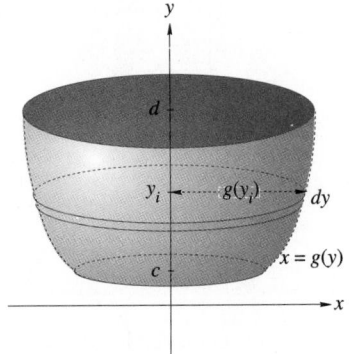

FIGURE 26.10

Until now we have found the volumes by revolving regions around the x-axis. It is possible to revolve a region around any line. We will look at revolving the area enclosed by $x = g(y)$, $y = c$, and $y = d$, where $g(y) \geq 0$, around the y-axis, as shown in Figure 26.10. The process is the same as when a region is revolved around the x-axis. Again, slices are taken, only this time they are perpendicular to the y-axis. A typical disk is shown in Figure 26.10. The height of the disk is dy and its radius is $g(y_i)$, so the volume of the disk is $\pi[g(y_i)]^2 dy$ and the total volume is given in the following box.

Disk Method for Finding a Volume of Revolution (y-Axis)

The volume V of the solid generated by revolving the area between the graph of $x = g(y)$ and the y-axis from $y = c$ to $y = d$ around the y-axis is

$$V = \pi \int_c^d (\text{radius})^2 \cdot (\text{thickness}) = \pi \int_c^d [g(y)]^2 \, dy$$

EXAMPLE 26.13

Find the volume of revolution generated by revolving the area enclosed by $y = \frac{1}{2}x^2$, $y = 0$, and $y = 8$ around the y-axis.

Solution Since the region is to be revolved around the y-axis, we will draw, horizontally, the rectangle that forms a typical disk as shown in Figure 26.11. Since this is revolved around the y-axis, we need to describe the curve as a function of y. Solving $y = \frac{1}{2}x^2$ for x, we get $x = g(y) = \sqrt{2y}$. Thus, the volume is

$$V = \pi \int_0^8 (\sqrt{2y})^2 \, dy = \pi \int_0^8 2y \, dy = \pi y^2 \big|_0^8 = 64\pi$$

The volume is 64π units^3.

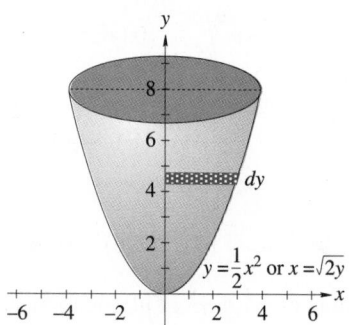

FIGURE 26.11

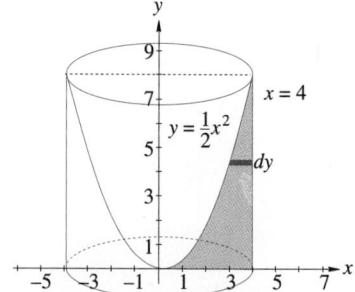

FIGURE 26.12

EXAMPLE 26.14

Find the volume of revolution generated by revolving the area enclosed by $y = \frac{1}{2}x^2$, $x = 0$, $x = 4$, and the x-axis around the y-axis.

Solution The graph of the region that is being revolved is shown in Figure 26.12. The point of intersection of $x = 4$ and $y = \frac{1}{2}x^2$ is $y = 8$. A typical element would

EXAMPLE 26.14 (Cont.)

form a washer, so its volume is $[4^2 - (\sqrt{2y})^2]dy$ and the total volume is

$$V = \pi \int_0^8 \left[(4)^2 - (\sqrt{2y})^2 \right] dy = \pi \int_0^8 (16 - 2y)dy$$
$$= \pi \left(16y - y^2 \right)\Big|_0^8 = 64\pi$$

The volume is 64 units^3.

Application

EXAMPLE 26.15

A hemispherical water tank, as shown in Figure 26.13a, has a radius of 4 m. The water is 1.5 m deep at the center. How much water is in the tank?

Solution We will sketch a cross-section of the tank with the center at the origin of the quarter-circle used to generate the hemisphere, as shown in Figure 26.13b. A typical element has been sketched in the figure. The length of this element is determined by solving the equation for the circle, $x^2 + y^2 = 4^2$ for x, with the result $x = \sqrt{4^2 - y^2}$. The thickness of the element is dy. Since the elements will move from the bottom of the tank, when $y = -4$, to the top of the water, when $y = -2.5$, we have the limits of integration. We see that the volume is

$$V = \pi \int_{-4}^{-2.5} \left(\sqrt{4^2 - y^2} \right)^2 dy$$
$$= \pi \int_{-4}^{-2.5} \left(16 - y^2 \right) dy$$
$$= \pi \left[16y - \frac{y^3}{3} \right]_{-4}^{-2.5}$$
$$= \pi \left(-\frac{835}{24} + \frac{128}{3} \right) = \left(\frac{189}{24} \right) \pi = \frac{63}{8}\pi \approx 24.74$$

There is about 24.74 m^3 of water in the tank.

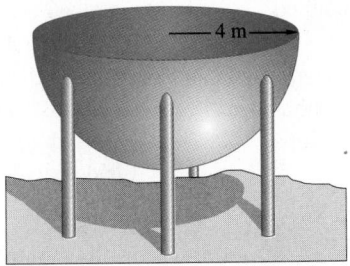

FIGURE 26.13a

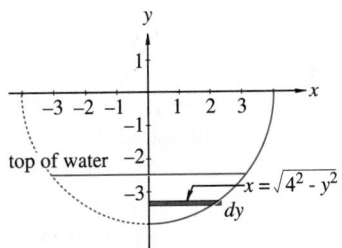

FIGURE 26.13b

Exercise Set 26.2

TRY EXES. Not Homework

In Exercises 1–19, find the volume of the solid of revolution generated by revolving the enclosed region about the indicated axis.

1. $y = 4x$, $x = 1$, $x = 5$, $y = 0$, about the x-axis
2. $y = x^2$, $x = 0$, $x = 2$, $y = 0$, about the x-axis
3. $y = x$, $x = 4$, $y = 0$, about the x-axis
4. $y = 6 - x$, $x = 0$, $y = 0$, about the x-axis
5. $y = \sqrt{x+1}$, $x = 1$, $x = 7$, $y = 0$, about the x-axis
6. $y = x^3$, $x = 2$, $y = 1$, about the x-axis
7. $y = x^2$, $y = 4$, $x = 0$, about the y-axis
8. $x + y = 5$, $y = 0$, $x = 0$, about the y-axis
9. $y = x$, $y = 4 - x$, $x = 0$, about the x-axis
10. $y = \sqrt[3]{x}$, $x = 0$, $y = 8$, about the y-axis

$dV = \pi$
$X = 8$
$y = 3\sqrt[3]{x}$
$\sqrt[3]{x} = x$

11. $y = \sqrt[3]{x}$, $x = 8$, $y = 0$, about the y-axis
12. $y = \sqrt{x}$, $y = x^2$, about the y-axis
13. $y = 4 - x^2$, $y = 8 - 2x^2$, about the x-axis
14. $y = x^2$, $y = x^3$, about the y-axis
15. $y = x^2$, $y = x^3$, about the x-axis
16. $y = x^2$, $y = 4x$, about the x-axis
17. $y = x^2$, $y = 4x$, about the y-axis
18. $y = \sqrt{8 - x^3}$, $x = 0$, $y = 0$, about the x-axis
19. $y = \sqrt{8 - x^2}$, $x = 0$, $y = 0$, about the y-axis

Solve Exercises 20–28.

20. Derive the formula for the volume of a cone of radius r and height h.

21. Derive the formula for the volume of a sphere of radius r.

22. The equation of an ellipse centered at the origin is

$$\frac{x^2}{a^2} + \frac{y^2}{b^2} = 1$$

Calculate the volume of the solid generated by revolving this ellipse around the x-axis.

23. *Electronics* A parabolic reflector is formed by rotating a parabola around its axis. Suppose a satellite television antenna is a parabolic reflector with a diameter of 3 m and a depth of 1 m. (a) Find the equation of this parabola. (b) What is the volume contained by the reflector?

24. *Electronics* If the "skin" of the parabolic reflector in Exercise 19 has a constant 0.4 cm thickness, what is the volume of the skin? (Assume the dimensions given in Exercise 18 are the interior dimensions.)

25. *Nuclear engineering* The cooling towers at a nuclear power plant are designed in the shape of the hyperbola $g(y) = \sqrt{(147)^2 + 0.16y^2}$. The base of the tower is 442 ft below the vertex and the top is 123 ft above the vertex as shown in Figure 26.14. If we assume the hyperbola is rotated around the y-axis, what is the volume of the cooling tower?

26. *Nuclear engineering* If the walls of the cooling tower in Exercise 25 are a constant 6 in. thick, then the interior walls are in the shape of a hyperbola with the equation $x = \sqrt{146.5^2 + 0.16y^2}$ that is rotated around the y-axis. What is the volume of the walls of the cooling tower?

27. *Wastewater technology* The tank on a water tower is a sphere of radius 18 m.
   (a) Determine the volume of water when the water is 4 m deep at the center.
   (b) Determine the volume of water when the water is 24 m deep at the center.

28. *Mechanical technology* A flange is formed in the shape of a solid of revolution generated by rotating about the x-axis the region bounded by a graph of $y = x^2 + 5$, where x and y are in centimeters; the y-axis; and the lines $y = 2$ and $x = 4$. Determine the mass of this flange if it is made of a material of density $0.016 \, \text{kg/cm}^3$. (Remember: mass is the product of the density and volume.)

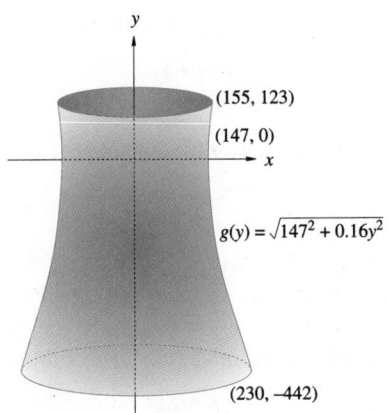

FIGURE 26.14

 **In Your Words**

29. Describe how to find the volume of a solid of revolution by using the disk method.

30. Describe how to find the volume of a solid of revolution by using the washer method.

31. Describe the changes that have to be made when a volume of revolution is formed by revolving a region around the y-axis rather than around the x-axis.

☰ 26.3
VOLUMES OF REVOLUTION: SHELL METHOD

In Section 26.2, we found the volume of a solid of revolution by the disk and washer methods. In this section, we will learn how to find these volumes by another method— the shell method. The shell method is based on finding the volume of cylindrical shells and adding those volumes to get the total volume.

An example of a cylindrical shell is a water pipe. A cylindrical shell has some thickness to it. In order to find its volume, we need to find the inner radius r_1 of the shell and the outer radius r_2, as shown in Figure 26.15. If the shell has height h, then its volume is

$$V = \pi(r_2)^2 h - \pi(r_1)^2 h = \pi h(r_2{}^2 - r_1{}^2)$$

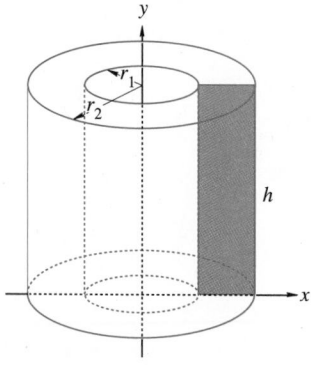

FIGURE 26.15

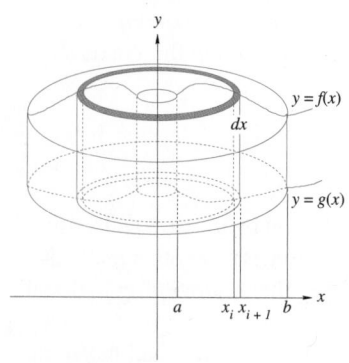

FIGURE 26.16

Some extra algebraic work will change this to a more useful formula.

$$V = \pi h(r_2{}^2 - r_1{}^2) = \pi h(r_2 + r_1)(r_2 - r_1) = 2\pi h\left(\frac{r_2 + r_1}{2}\right)(r_2 - r_1)$$

This formula is very involved, but we will show you a way to remember it after we work Example 26.16. In this last statement, $\frac{r_2 + r_1}{2}$ represents the average radius of the shell and $r_2 - r_1$ represents the thickness.

If we have two curves, $y = f(x)$ and $y = g(x)$, with $f(x) \geq g(x)$, that are continuous over the interval $[a, b]$ with $a \geq 0$, and we revolve the area between these curves around the y-axis, we get a solid of revolution similar to the one in Figure 26.16. Next we divide the interval $[a, b]$ into subintervals. Each subinterval will have the width $r_2 - r_1 = dx$, height $f(x_i) - g(x_i)$, and average radius $\frac{x_{i-1} + x_i}{2}$. Thus, when this subinterval is revolved around the y-axis, it has a volume of

$$2\pi[f(x_i) - g(x_i)]\left(\frac{x_{i-1} + x_i}{2}\right)dx$$

If dx is small enough $\left(\frac{x_{i-1} + x_i}{2} \approx x_i\right)$, and if we sum all the partitions over the interval $[a, b]$, we get the following formula.

Shell Method for Finding a Volume of Revolution (y-Axis)

The volume V of the solid generated by revolving the area between the graph of $y = f(x)$ and $y = g(x)$ from $x = a$ to $x = b$, where a and b are both on the same side of the y-axis around the y-axis, is

$$V = 2\pi \int_a^b x[f(x) - g(x)]\, dx$$

EXAMPLE 26.16

Find the volume of the solid generated by revolving around the y-axis the region between $y = 5x^3 - x^4$ and the x-axis.

Solution The graph of this region is shown in Figure 26.17. (Notice that the scale is not the same on the x- and y-axes.) These two curves, $y = 0$ and $y = 5x^3 - x^4$,

EXAMPLE 26.16 (Cont.)

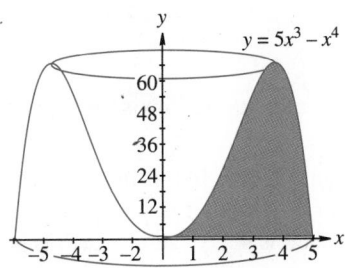

FIGURE 26.17

intersect at $x = 0$ and at $x = 5$; so, $a = 0$ and $b = 5$. Using the new formula, we see that

$$V = 2\pi \int_0^5 x[(5x^3 - x^4) - 0]\,dx$$

$$= 2\pi \int_0^5 (5x^4 - x^5)\,dx = 2\pi \left(x^5 - \frac{1}{6}x^6 \right)\Big|_0^5$$

$$= 2\pi \left(3{,}125 - \frac{15{,}625}{6} \right) = 1{,}041\frac{2}{3}\pi$$

The volume is $1{,}041\frac{2}{3}\pi$ units^3.

It would have been possible to solve Example 26.16 by using the washer method. But, that would have meant solving the equation $y = 5x^3 - x^4$ for x in order to get a function of y. This would have been a very difficult and complicated task. The shell method provides a much easier alternative.

 **Hint**

It is easier to remember the shell method by using the following guidelines:
1. Sketch the graph over the limits of integration.
2. Draw a typical shell parallel to the axis of revolution.
3. Determine the radius, height, and thickness of the shell.
4. The volume of a typical shell is $2\pi(\text{radius}) \cdot (\text{height}) \cdot (\text{thickness})$.
5. Use the following integration formula to determine the total volume:

$$V = 2\pi \int_a^b (\text{radius}) \cdot (\text{height}) \cdot (\text{thickness})$$

EXAMPLE 26.17

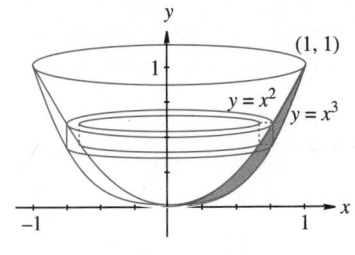

FIGURE 26.18

Find the volume of the solid generated by revolving around the y-axis the region in the first quadrant that is between $y = x^2$ and $y = x^3$.

Solution These two curves intersect at $(0, 0)$ and $(1, 1)$ as shown in Figure 26.18, so $a = 0$ and $b = 1$. On this interval, $x^2 \geq x^3$ and so, using the shell method, the volume is

$$V = 2\pi \int_0^1 x(x^2 - x^3)\,dx$$

$$= 2\pi \int_0^1 (x^3 - x^4)\,dx$$

$$= 2\pi \left(\frac{1}{4}x^4 - \frac{1}{5}x^5 \right)\Big|_0^1$$

$$= 2\pi \left(\frac{1}{4} - \frac{1}{5} \right) = \frac{\pi}{10}$$

The volume is $\dfrac{\pi}{10}$ units^3.

We can just as easily use the shell method to revolve a region around the x-axis. When this is done, the function will have to be written to ensure that it is a function of y.

EXAMPLE 26.18

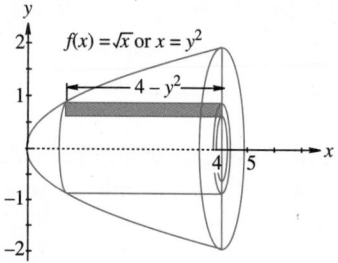

FIGURE 26.19

Find the volume of the solid generated by revolving around the x-axis the region bounded by $y = \sqrt{x}$, $y = 0$, and $x = 4$.

Solution   The graph of this region is shown in Figure 26.19. In this instance, we have two curves that must be written as functions of y. The first is $x = 4$ and the other, $y = \sqrt{x}$, is $x = y^2$. Thus, the height of each shell is $4 - y^2$ and the radius is y. Then the volume is

$$V = 2\pi \int_0^2 y(4 - y^2)dy$$

$$= 2\pi \int_0^2 (4y - y^3)dy$$

$$= 2\pi \left(2y^2 - \frac{1}{4}y^4\right)\Big|_0^2$$

$$= 2\pi(8 - 4) = 8\pi$$

The volume is 8π units^3.

Compare the result of Example 26.18 to that of Example 26.9, where we found the volume of the same solid by using the disk method.

Rotation about a Noncoordinate Axis

As we said, it is possible to rotate a region around any line in order to get a desired volume of revolution. In the next example, we will rotate a region around a line parallel to the y-axis. The formula for the shell method will not work. Instead, we need to rely on the basic concept behind the shell method. In that concept we saw that

$$V = 2\pi(\text{radius})(\text{height})(\text{thickness})$$

EXAMPLE 26.19

Find the volume of the solid formed by revolving the region bounded by $y = x^2 + 2$ and $y = x + 8$ around the line $x = 4$.

Solution   The region is shown in Figure 26.20. Since this region is revolved around the line $x = 4$, the radius of a typical shell is $4 - x$. The two curves, $y = x + 8$ and $y = x^2 + 2$, intersect at $(-2, 6)$ and $(3, 11)$. In the interval $[-2, 3]$ we can see that

EXAMPLE 26.19 (Cont.)

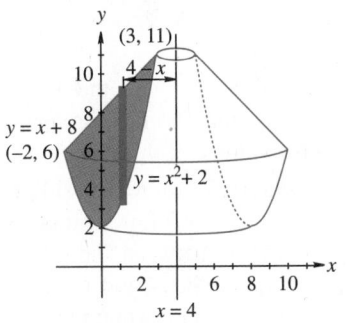

FIGURE 26.20

$x+8 \geq x^2+2$ and so the height of a typical shell is $(x+8)-(x^2+2)=6+x-x^2$. Thus, the volume of a typical shell is

$$V = 2\pi \int_{-2}^{3} (6+x-x^2)(4-x)dx$$

$$= 2\pi \int_{-2}^{3} (24-2x-5x^2+x^3)dx$$

$$= 2\pi \left(24x - x^2 - \frac{5}{3}x^3 + \frac{1}{4}x^4 \right) \Big|_{-2}^{3}$$

$$\approx 2\pi [38.25 - (-34.67)] = 145.84\pi$$

The volume is 145.84π units^3.

Summary of Disk and Shell Methods

In Sections 26.2 and 26.3, we have shown two methods for finding the volume of a solid of revolution. These two methods are summarized in Table 26.1.

TABLE 26.1   Summary of Disk and Shell Methods

	Disk or Washer Method	Shell Method
Revolution about x-axis	Partition the x-axis. Use vertical strips of width dx. Integrate with respect to x.	Partition the y-axis. Use horizontal strips of width dy. Integrate with respect to y.
Revolution about y-axis	Partition the y-axis. Use horizontal strips of width dy. Integrate with respect to y.	Partition the x-axis. Use vertical strips of width dx. Integrate with respect to x.

Exercise Set 26.3

In Exercises 1–18, use the shell method to obtain the volume of the solid of revolution generated by revolving the enclosed region about the indicated axis.

1. $y = x^2$, $y = 0$, $x = 2$, about the y-axis
2. $x = y^3$, $x = 8$, about the y-axis
3. $y = x^2 - 1$, $x = 1$, $y = 8$, about the y-axis
4. $y = x^3 + x$, $x = 0$, $x = 2$, about the y-axis
5. $y = 25 - x^2$, $y = 0$, about the y-axis
6. $y = x^2$, $y = 8 - x^2$, about the y-axis
7. $y = x^3$, $x = 0$, $y = 8$, about the x-axis

8. $y = x^4$, $y = x^3$, about the x-axis
9. $y = \sqrt[3]{x}$, $x = 0$, $y = 2$, about the x-axis
10. $x = \sqrt[3]{y^2}$, $x = 0$, $y = 8$, about the x-axis
11. $y = x^{1/2} - x^{3/2}$, $y = 0$, about the y-axis
12. $x = y$, $x + 2y = 3$, $y = 0$, about the x-axis
13. $y = 2x^3$, $y^2 = 4x$, about the x-axis
14. $x = 4y - y^3$, $x = 0$, about the x-axis

15. $y = x^2$, $y = 2x$, about the line $y = 6$

16. $y = 2 - x$, $y = x^2$, $y = 1$, about the x-axis (the area above $y = 1$)

17. $y = 2 - x$, $y = x^2$, about $x = 1$

18. $y = 4 - x^2$, $y = 0$, about the line $x = 4$

Solve Exercises 19–22.

19. A hole of radius 2 is drilled vertically through the center of a sphere of radius 5. Find the volume removed by the drilling.

20. *Civil engineering* A water tank is in the shape of a sphere with a radius of 10 m. **(a)** What is the volume of water in the tank if it is filled to a depth of 3 m? **(b)** If the density of water is 1 000 kg/m³, what is the mass of the water in the tank?

21. *Sheet metal technology* The inside of a bundt pan, used for baking, is designed by revolving the region

between $y = -(x - 3.25)^4 + \left(\frac{5}{4}\right)^4$ and the x-axis around the y-axis. Assume that the measurements are in inches, and find the volume of this pan.

22. *Industrial design* The glass reflector for a flashlight is made in the shape of a solid of revolution. It is obtained by revolving the region between the curves $x = 0.5y^2$, $x = 0.5y^2 + 0.1$, the x-axis, and $y = 2$ for $0 \le x \le 2$ around the x-axis. If all measurements are in centimeters, what is the volume of the reflector?

In Your Words

23. Describe how to find the volume of a solid of revolution by using the shell method.

24. Describe the changes that have to be made when a volume of revolution is formed using the shell method rather than either the disk or washer method.

25. Draw an example of a region for which the washer method is more appropriate than the shell method

when the region is revolved around the y-axis. Explain why the washer method is more appropriate.

26. Draw an example of a region for which the shell method is more appropriate than the washer method when the region is revolved around the y-axis. Explain why the shell method is more appropriate.

≡ 26.4
ARC LENGTH AND SURFACE AREA

In this section, we will look at two related ideas. One is the length of a curve; the second is the area of the surface that is formed when that curve is rotated around an axis.

One useful application for the length of a curve is in determining the lengths of the cables that are used to support a suspension bridge. The cables between the towers are parabolic in shape and so this results in finding the length of a parabola between two values.

Consider a curve such as $y = f(x)$, as shown in Figure 26.21. We would like to find the length of this curve from $x = a$ to $x = b$. As before, we will partition the interval $[a, b]$ into n subintervals. With each subinterval $x_0, x_1, x_2, \ldots, x_n$ there is a corresponding point on the graph $P_0, P_1, P_2, \ldots, P_n$. Each point P_i has the coordinates $(x_i, f(x_i))$. If we connect these points in succession, we get a series of line segments that approximate the curve. The sum of the lengths of these segments

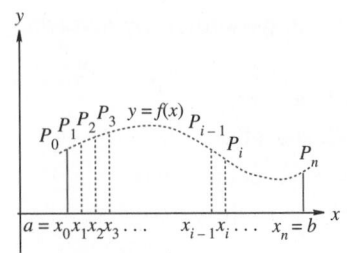

FIGURE 26.21

will approximate the length of the curve. This sum L can be written as

$$L = d(P_0, P_1) + d(P_1, P_2) + d(P_2, P_3) + \cdots + d(P_{n-1}, P_n)$$

$$= \sum_{i=1}^{n} d(P_{i-1}, P_i)$$

where $d(P_{i-1}, P_i)$ is the length of the segment from P_{i-1} to P_i.

Using the distance formula, we get

$$d(P_{i-1}, P_i) = \sqrt{(x_i - x_{i-1})^2 + [f(x_i) - f(x_{i-1})]^2}$$

$$= \sqrt{(\Delta x_i)^2 + (\Delta y_i)^2}$$

Factoring out $(\Delta x_i)^2$ produces

$$= \sqrt{1 + \left(\frac{\Delta y_i}{\Delta x_i}\right)^2} \, \Delta x_i$$

where $\Delta x_i = x_i - x_{i-1}$ and $\Delta y_i = f(x_i) - f(x_{i-1})$. Thus, we can rewrite the formula for L as

$$L = \sum_{i=1}^{n} \sqrt{1 + \left(\frac{\Delta y_i}{\Delta x_i}\right)^2} \, \Delta x_i$$

We are going to assume that f has a derivative; then $\displaystyle\lim_{x_i \to 0} \frac{\Delta y_i}{\Delta x_i} = \frac{dy_i}{dx_i} = f'(x_i)$. With this change, the length of the curve L becomes

$$L = \sum_{i=1}^{n} \sqrt{1 + [f'(x_i)]^2} \, dx_i$$

Now, as the length of each subinterval gets smaller and smaller, we get the desired formula as a limit.

Length of a Graph

If f is a function with a continuous derivative on $[a, b]$, then the length of the graph of $y = f(x)$ on $[a, b]$ is

$$L = \int_a^b \sqrt{1 + [f'(x)]^2} \, dx$$

EXAMPLE 26.20

Find the length of $y = 3 + x^{2/3}$ from $x = 1$ to $x = 8$.

Solution   If $y = f(x) = 3 + x^{2/3}$, then $f'(x) = \dfrac{2}{3x^{1/3}}$. Thus,

$$
\begin{aligned}
L &= \int_1^8 \sqrt{1 + \left[\frac{2}{3x^{1/3}}\right]^2}\, dx \\
&= \int_1^8 \sqrt{1 + \frac{4}{9x^{2/3}}}\, dx \\
&= \int_1^8 \sqrt{\frac{9x^{2/3} + 4}{9x^{2/3}}}\, dx \\
&= \int_1^8 \frac{\sqrt{9x^{2/3} + 4}}{3x^{1/3}}\, dx
\end{aligned}
$$

We will use the substitution method to solve this. If $u = 9x^{2/3} + 4$, then $du = 6x^{-1/3}dx$ and so $\frac{1}{3}x^{-1/3}\,dx = \frac{1}{18}\,du$. Thus,

$$
\begin{aligned}
L &= \int \frac{\sqrt{u}\,du}{18} \\
&= \frac{1}{18} \int \sqrt{u}\,du \\
&= \frac{1}{18} \cdot \frac{2}{3}u^{3/2} = \frac{1}{27}u^{3/2}
\end{aligned}
$$

Since $u = 9x^{2/3} + 4$, we have

$$
\begin{aligned}
L &= \frac{1}{27}\left(9x^{2/3} + 4\right)^{3/2}\Big|_1^8 \\
&= \frac{1}{27}\left(40^{3/2} - 13^{3/2}\right) \\
&\approx 7.63
\end{aligned}
$$

We should point out that if the curve is a function of y, such as $x = g(y)$, where g' is continuous on $[c, d]$, then the length of L, of the graph of $x = g(y)$ from c to d, is given by

$$
L = \int_c^d \sqrt{1 + [g'(y)]^2}\, dy
$$

EXAMPLE 26.21

Find the length of $x = y^{3/2}$ from $y = 1$ to $y = 4$.

Solution Since $x' = \frac{3}{2}y^{1/2}dy$,

$$
\begin{aligned}
L &= \int_1^4 \sqrt{1 + \left(\frac{3}{2}y^{1/2}\right)^2}\, dy \\
&= \int_1^4 \sqrt{1 + \frac{9}{4}y}\, dy \\
&= \frac{4}{9}\int_1^4 \frac{9}{4}\sqrt{1 + \frac{9}{4}y}\, dy \\
&= \frac{2}{3}\cdot\frac{4}{9}\left(1 + \frac{9}{4}y\right)^{3/2}\Bigg|_1^4 \\
&= \frac{8}{27}\left[(10)^{3/2} - \left(\frac{13}{4}\right)^{3/2}\right] \approx 7.63
\end{aligned}
$$

Surface Area

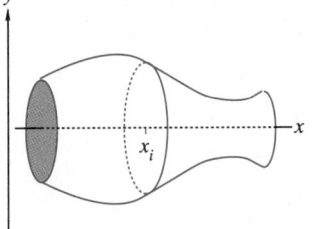

FIGURE 26.22

Now suppose that the curve in Figure 26.21 is rotated around the x-axis. Then a surface is formed. If we want to find the surface area of this figure, we need to think of it as a series of circles. In Figure 26.22, the circle with center at x_i has a circumference of $2\pi f(x_i)$, or $2\pi y_i$, where $y_i = f(x_i)$. If we add the circumferences of all the circles over the entire length of the curve from a to b, we get a total surface area, S.

Surface Area

If f is a function with a continuous derivative on $[a, b]$, then the area of the surface generated by revolving the graph of $y = f(x)$ about the x-axis is

$$
S = \int_a^b 2\pi y\sqrt{1 + [f'(x)]^2}\, dx
$$

EXAMPLE 26.22

Find the surface area of the solid generated by rotating $y = \frac{1}{3}x^3$ from $x = 0$ to $x = 2$ around the x-axis.

Solution The formula for the surface is

$$
S = \int_a^b 2\pi y\sqrt{1 + [f'(x)]^2}\, dx
$$

EXAMPLE 26.22 (Cont.)

Here, $y = \frac{1}{3}x^3$ and $f'(x) = x^2$, so we have

$$S = \int_0^2 \frac{2}{3}\pi x^3 \sqrt{1+(x^2)^2}\,dx = \frac{2}{3}\pi \int_0^2 x^3 \sqrt{1+x^4}\,dx$$

If we let $u = 1+x^4$, then $du = 4x^3\,dx$, and we have

$$\frac{2}{3}\pi \cdot \frac{1}{4}\int \sqrt{u}\,du = \frac{\pi}{6}\cdot\frac{2}{3}u^{3/2}$$
$$= \frac{\pi}{9}(1+x^4)^{3/2}\Big|_0^2 = \frac{\pi}{9}(17^{3/2}-1^{3/2})$$
$$\approx 7.68\pi$$

The surface area is about 7.68π units^2.

At present, our ability to solve many of these problems is limited by our lack of ability in integration techniques. We will be using arc length and surface areas as we improve our skills at integration.

Exercise Set 26.4

In Exercises 1–6, find the lengths of the curves over the indicated interval.

1. $y = \frac{1}{3}(x^2+2)^{3/2}$, over $[0,3]$
2. $y = x^{3/2}$, over $[0,8]$
3. $9x^2 = 4y^3$, from the point $(0,0)$ to the point $(2\sqrt{3},3)$
4. $y = \frac{x^4}{4} + \frac{1}{8x^2}$, over $[1,2]$

5. $y = \frac{x^3}{6} + \frac{1}{2x}$, over $[1,3]$
6. $x = \frac{y^3}{3} + \frac{1}{4y}$, over $[1,3]$

In Exercises 7–12, find the surface area of the solid generated by rotating the given curves around the indicated axis.

7. $y = 4x$ from $x=0$ to $x=2$ about the x-axis
8. $y = \frac{12}{5}x$ from $x=0$ to $x=8$ about the x-axis
9. $y = x$ from $(0,0)$ to $(2,2)$ about the x-axis
10. $y^2 = 4x$ from $x=0$ to $x=8$ about the x-axis
11. $x = 4y$ from $y=1$ to $y=3$ about the x-axis
12. $x = \sqrt{y}$ from $y=0$ to $y=6$ about the y-axis

Solve Exercises 13 and 14.

13. *Construction* A suspension bridge has its roadbed supported by vertical cables that are connected to a cable that is strung between two towers and from each tower to the ground. Each cable from the top of the tower to the ground fits the equation $y = \frac{2}{75}x^{3/2}$. Find the length of one of the cables, if it meets the ground 225 m from the base of the tower. (In Chapter 28, we will learn how to determine the length of the cable between the towers.)

14. *Interior design* A lamp shade is formed by rotating the graph of $y = (4-x)^3$ over $[2,4]$, around the x-axis, where x and y are in inches. What is the surface area of the lamp shade?

In Your Words

15. Explain how to find the length of a graph.
16. Write an application in your technology area of interest that requires you to find either the arc length of a (nonlinear) curve or the surface area of an irreg-

ular shape. Give your problem to a classmate and see if he or she understands and can solve your problem. Rewrite the problem as necessary to remove any difficulties encountered by your classmate.

≡ 26.5
CENTROIDS

Thus far we have used integration to study moving objects as if they were particles that had mass but no size. Objects, however, are made up of many particles. If an object is moved, then its particles may have different kinds of behavior. This type of situation can be handled by studying what happens to one particular point of the object. This point is known as the center of mass or the centroid. In this section, we will study how to find the centroid for areas and for solids of revolution.

Suppose we had a metal bar 4 m long that we place along the x-axis with the center of the bar at the origin, one end at 2, and the other end at -2. Next, suppose we put a 10-kg mass at $x = 2$ and a 4-kg mass at $x = -2$. We would like to find the center of mass for this system. The **center of mass** of a system is the point where all the mass seems to be concentrated. If the system has a constant density, then the center of mass is called the **centroid**.

Before we solve this example, we will look at the idea behind the centroid. Suppose we had a system much like the one in Figure 26.23. The point 0 indicates the origin of the system. A mass m_1 is located x_1 units to the left of the origin. A second mass m_2 is located x_2 units to the right of the origin.

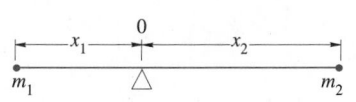

FIGURE 26.23

Associated with each mass is its moment with respect to a certain axis. In Figure 26.23, the axis is at the origin. Each **moment** is the product of the mass and its distance from the axis. For the system in Figure 26.23, the moment for mass 1 is $m_1 d_1$ and for mass 2 the moment is $m_2 d_2$. The **first moment** of this system is the sum of the individual moments. In this case, the first moment for the system is $m_1 x_1 + m_2 x_2$. We now select a new point $\bar{x}$ with the property that $(m_1 + m_2)\bar{x}$ is equal to the first moment of the system. The location of $\bar{x}$ is the center of mass. Since $(m_1 + m_2)\bar{x} = m_1 x_1 + m_2 x_2$, we have

$$\bar{x} = \frac{m_1 x_1 + m_2 x_2}{m_1 + m_2}$$

Continuing this process for more than two points gives the general formula for n points:

$$\bar{x} = \frac{\sum_{i=1}^{n} m_i x_i}{\sum_{i=1}^{n} m_i}$$

In this equation, the numerator is the first moment of the system and the denominator is the total mass of the system.

EXAMPLE 26.23

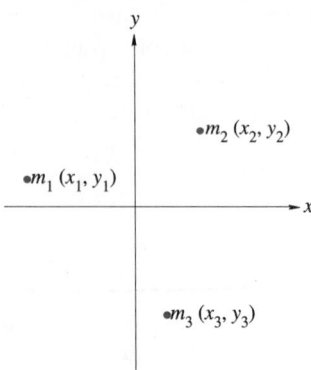

FIGURE 26.24

Four particles of masses 2, 4, 3, and 5 kg, respectively, are located on the x-axis at coordinates 3, -2, -3, and 6 m. Where is the center of mass of this system?

Solution

$$\bar{x} = \frac{m_1 x_1 + m_2 x_2 + m_3 x_3 + m_4 x_4}{m_1 + m_2 + m_3 + m_4}$$

$$= \frac{2(3) + 4(-2) + 3(-3) + 5(6)}{2 + 4 + 3 + 5} = \frac{19}{14} \text{ m}$$

The center of mass is at the coordinate $\frac{19}{14}$ m on the x-axis.

If we have several point masses located in a plane, then we locate the center of mass by separately finding the center of mass for the x-coordinates and for the y-coordinates. Thus, for the system in Figure 26.24, the center of mass would have coordinates

$$\bar{x} = \frac{m_1 x_1 + m_2 x_2 + m_3 x_3}{m_1 + m_2 + m_3} \quad \text{and} \quad \bar{y} = \frac{m_1 y_1 + m_2 y_2 + m_3 y_3}{m_1 + m_2 + m_3}$$

In general, we have the formulas in the box for the center of mass for n points.

Center of Mass for n Points

If the point masses $m_1, m_2, \ldots, m_n$ are located at (x_1, y_1), (x_2, y_2), $\ldots$, (x_n, y_n), respectively, then the center of mass has the coordinates

$$\bar{x} = \frac{\displaystyle\sum_{i=1}^{n} m_i x_i}{m} \quad \text{and} \quad \bar{y} = \frac{\displaystyle\sum_{i=1}^{n} m_i y_i}{m}$$

or

$$\bar{x} = \frac{M_y}{m} \quad \text{and} \quad \bar{y} = \frac{M_x}{m}$$

where

$$M_y = \sum_{i=1}^{n} m_i x_i \text{ is the first moment around the } y\text{-axis,}$$

$$M_x = \sum_{i=1}^{n} m_i y_i \text{ is the first moment around the } x\text{-axis, and}$$

$$m = \sum_{i=1}^{n} m_i \text{ is the total mass of the system.}$$

EXAMPLE 26.24

Calculate the center of mass of the system with four particles of masses 10 g, 5 g, 3 g, and 8 g located respectively at the points $(2, -1)$, $(3, 2)$, $(-4, 1)$, and $(-3, 4)$ (measured in cm).

Solution

$$\bar{x} = \frac{10(2) + 5(3) + 3(-4) + 8(-3)}{10 + 5 + 3 + 8}$$

$$= \frac{-1}{26} \text{ cm}$$

$$\bar{y} = \frac{10(-1) + 5(2) + 3(1) + 8(4)}{10 + 5 + 3 + 8}$$

$$= \frac{35}{26} \text{ cm}$$

The center of mass is at $\left(-\frac{1}{26}, \frac{35}{26}\right)$.

EXAMPLE 26.25

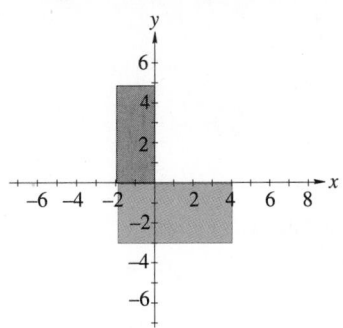

FIGURE 26.25

Find the center of mass of the area in Figure 26.25 if it has uniform density ρ.

Solution We will first divide the area into two rectangles. This division will be along the x-axis with region I above the x-axis and region II below. The geometric center of region I is $\left(-1, \frac{5}{2}\right)$ and of region II, $\left(1, -\frac{3}{2}\right)$. The mass of each region is the product of the density ρ (mass/unit area) and the area. The area of region I is 10 units and so its mass is 10ρ. The area of region II is 18 units and so its mass is 18ρ. The total mass of this figure is

$$m = 10\rho + 18\rho = 28\rho$$

The coordinates of the center of mass will be determined by adding the product of the coordinate of the centroid for each region by its mass. Thus, the x-coordinate is

$$\bar{x} = \frac{(-1)(10\rho) + (1)(18\rho)}{28\rho} = \frac{8\rho}{28\rho} = \frac{2}{7}$$

and the y-coordinate is

$$\bar{y} = \frac{\left(\frac{5}{2}\right)(10\rho) + \left(-\frac{3}{2}\right)(18\rho)}{28\rho} = \frac{-2\rho}{28\rho} = -\frac{1}{14}$$

The center of mass is at $\left(\frac{2}{7}, -\frac{1}{14}\right)$.

It is possible for the center of mass to be outside the system. You may have noticed that the density ρ was not used to locate the center of mass. This will happen as long as the object has uniform density, as will be true throughout the remainder of this text.

Now let's apply this above technique to an arbitrary region. To begin, we will let the region be found by $y = f(x)$, $x = a$, $x = b$, and the x-axis, as shown in Figure 26.26. As we have done before, we will partition the interval $[a, b]$ into n subintervals as indicated in Figure 26.26.

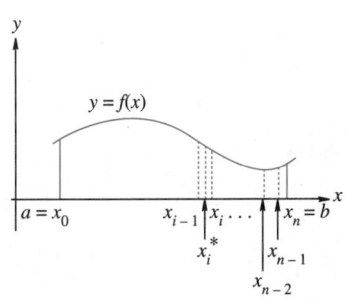

FIGURE 26.26

Look at the shaded region in Figure 26.26. It is almost rectangular in shape and so its center is the midpoint of the interval $[x_{i-1}, x_i]$, which we will call $x_i{}^*$, and half the height of the rectangle, or $\frac{1}{2}f(x_i{}^*)$. The area of this rectangle is $f(x_i{}^*)(x_i - x_{i-1}) = f(x_i{}^*)\Delta x$, and so the mass of this rectangle is $\rho f(x_i{}^*)\Delta x$, where ρ is the density. The first moment of this shaded region is given by $\rho x_i{}^* f(x_i{}^*)\Delta x$. The first moment for all of these rectangular strips with respect to the y-axis is given by

$$\sum_{i=1}^{n} \rho x_i{}^* f(x_i{}^*)\Delta x$$

So, if the number of intervals increases, we get a first moment of

$$M_y = \lim_{\Delta x \to 0} \sum_{i=1}^{n} \rho x_i{}^* f(x_i{}^*)\Delta x$$

or $\quad M_y = \int_a^b \rho x f(x)dx$

In a similar way, we can show that the first moment with respect to the x-axis is given by

$$M_x = \frac{1}{2}\int_a^b \rho[f(x)]^2 dx$$

The mass m of the region is

$$m = \int_a^b \rho f(x)dx$$

Combining these, we get the coordinate of the centroid as $(\bar{x}, \bar{y})$ where

$$\bar{x} = \frac{M_y}{m} \text{ and } \bar{y} = \frac{M_x}{m}$$

If $\rho = 1$, then the mass will be the same as the area.

EXAMPLE 26.26

Find the centroid of the plane region bounded by $y = 4 - x^2$ and $y = 0$.

Solution   These two curves intersect at $x = -2$ and $x = 2$. Thus, we have

$$M_y = \int_{-2}^{2} \rho x \left(4 - x^2\right) dx$$

$$= \int_{-2}^{2} \rho \left(4x - x^3\right) dx$$

$$= \rho \left(2x^2 - \frac{1}{4}x^4\right)\Big|_{-2}^{2}$$

$$= \rho[(8-4)-(8-4)] = 0$$

$$M_x = \frac{1}{2}\int_{-2}^{2} \rho \left(4 - x^2\right)^2 dx$$

EXAMPLE 26.26 (Cont.)

$$= \frac{1}{2} \int_{-2}^{2} \rho \left(16 - 8x^2 + x^4\right) dx$$

$$= \frac{\rho}{2} \left(16x - \frac{8}{3}x^3 + \frac{x^5}{5} \right)\Big|_{-2}^{2}$$

$$= \frac{\rho}{2} \left(\frac{256}{15} + \frac{256}{15} \right)$$

$$= \frac{256}{15}\rho$$

$$m = \int_{-2}^{2} \rho(4 - x^2) dx$$

$$= \rho \left(4x - \frac{x^3}{3} \right)\Big|_{-2}^{2}$$

$$= \rho \left(\frac{16}{3} + \frac{16}{3} \right) = \frac{32}{3}\rho$$

Thus we have the following coordinates:

$$\bar{x} = \frac{M_y}{m} = \frac{0}{32\rho/3} = 0$$

$$\bar{y} = \frac{M_x}{m} = \frac{256\rho/15}{32\rho/3} = \frac{8}{5}$$

The centroid of this region is $\left(0, \frac{8}{5}\right)$.

If the region is between two curves, $f(x)$ and $g(x)$, with $f(x) \geq g(x)$, we have the following, more general, formulas. Notice that if $g(x) = 0$, we have the previous equations.

Moments, Mass, and Centroid of a Plane Region

If f and g are continuous functions with $f(x) \geq g(x)$ on the interval $[a, b]$, then the moments, mass, and centroid of the region of uniform density ρ bounded by the graphs of $y = f(x)$ and $y = g(x)$, on $[a, b]$, are

$$M_x = \frac{1}{2} \int_a^b \rho([f(x)]^2 - [g(x)]^2) dx$$

$$M_y = \int_a^b \rho x[f(x) - g(x)] dx$$

$$m = \int_a^b \rho[f(x) - g(x)] dx$$

$$\bar{x} = \frac{M_y}{m} \text{ and } \bar{y} = \frac{M_x}{m}$$

EXAMPLE 26.27

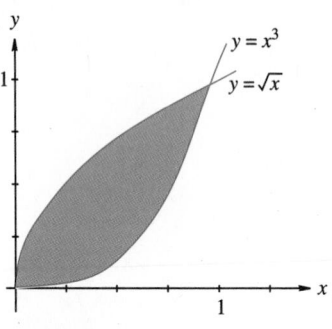

FIGURE 26.27

Find the centroid of the region bounded by $y = x^3$ and $y = \sqrt{x}$.

Solution These two curves intersect at $x = 0$ and $x = 1$. As we can see in Figure 26.27, $\sqrt{x} \geq x^3$ in this interval. Using the new formulas, we get

$$M_y = \int_0^1 \rho x \left(\sqrt{x} - x^3 \right) dx$$

$$= \int_0^1 \rho \left(x^{3/2} - x^4 \right) dx$$

$$= \rho \left[\frac{2}{5} x^{5/2} - \frac{1}{5} x^5 \right]_0^1 = \frac{1}{5} \rho$$

$$M_x = \frac{1}{2} \int_0^1 \rho \left[\left(\sqrt{x} \right)^2 - \left(x^3 \right)^2 \right] dx$$

$$= \frac{1}{2} \int_0^1 \rho \left(x - x^6 \right) dx$$

$$= \rho \frac{1}{2} \left(\frac{1}{2} x^2 - \frac{1}{7} x^7 \right) \Big|_0^1 = \frac{5}{28} \rho$$

$$m = \int_0^1 \rho \left(\sqrt{x} - x^3 \right) dx$$

$$= \rho \left(\frac{2}{3} x^{3/2} - \frac{1}{4} x^4 \right) \Big|_0^1 = \frac{5}{12} \rho$$

Thus, we have

$$\bar{x} = \frac{M_y}{m} = \frac{1/5\rho}{5/12\rho} = \frac{12}{25}$$

$$\bar{y} = \frac{M_x}{m} = \frac{5/28\rho}{5/12\rho} = \frac{12}{28} = \frac{3}{7}$$

The centroid is at $\left(\frac{12}{25}, \frac{3}{7} \right)$.

EXAMPLE 26.28

A thin plate covers a region bounded by the x-axis, $y = x^2 + 1$, and $x = 2$. The plate's density at the point (x, y) is $\rho = 5x$. Find the total mass and the center of mass of the plate.

Solution

$$M_y = \int_0^2 \rho x \left(x^2 + 1 \right) dx$$

$$= \int_0^2 (5x) x \left(x^2 + 1 \right) dx$$

$$= \int_0^2 \left(5x^4 + 5x^2 \right) dx = x^5 + \frac{5}{3} x^3 \Big]_0^2 = \frac{136}{3}$$

EXAMPLE 26.28 (Cont.)

$$M_x = \frac{1}{2}\int_0^2 \rho\left(x^2+1\right)^2 dx$$

$$= \frac{1}{2}\int_0^2 5x\left(x^4+2x^2+1\right) dx$$

$$= \frac{1}{2}\int_0^2 \left(5x^5+10x^3+5x\right) dx$$

$$= \frac{1}{2}\left(\frac{5}{6}x^6+\frac{5}{2}x^4+\frac{5}{2}x^2\right)\Big]_0^2 = \frac{153}{3}$$

$$m = \int_0^2 \rho\left(x^2+1\right) dx$$

$$= \int_0^2 \left(5x^3+5x\right) dx = \frac{5}{4}x^4+\frac{5}{2}x^2\Big]_0^2 \; 30$$

$$\overline{x} = \frac{M_y}{m} = \frac{136/3}{30} = \frac{136}{90} = \frac{68}{45}$$

$$\overline{y} = \frac{M_x}{m} = \frac{155/3}{30} = \frac{31}{18}$$

Thus, the total mass is 30 units and the center of mass is at $\left(\frac{68}{45}, \frac{31}{18}\right)$.

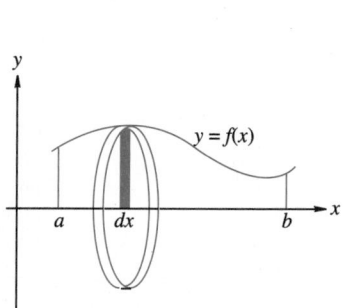

FIGURE 26.28

Centroid of a Solid of Revolution

This method can be extended to solids of revolution. As in the past, we will revolve the region bounded by the curve $y = f(x)$, $x = a$, $x = b$, and the x-axis around the x-axis, as shown in Figure 26.28. Because of the symmetry of the rotation, we know that the centroid is on the axis of rotation. In this case $\overline{y} = 0$. We need only locate $\overline{x}$.

The value of $\overline{x}$ will be $\dfrac{M_y}{m}$, where M_y is the first moment of the solid with respect to the y-axis and m is the mass. We already know that

$$m = \rho V = \pi \int_a^b \rho[f(x)]^2 \, dx$$

where V is the volume.

The moment of a typical disk, as shown in Figure 26.28, is the product of x, ρ, and the volume of the disk, $\pi [f(x)]^2 \, dx$. If we sum these for all the disks, we obtain

$$M_y = \pi \int_a^b \rho x [f(x)]^2 dx$$

Putting these together produces the following result.

Centroid of a Solid of Revolution

If f is a continuous function on the interval $[a, b]$, then the centroid of the solid generated by revolving the region bounded by $y = f(x)$, $x = a$, $x = b$, and the x-axis about the x-axis is at $(\bar{x}, 0)$, where

$$\bar{x} = \frac{M_y}{m}$$

when $M_y = \pi \int_a^b \rho x \, [f(x)]^2 \, dx$ and $m = \rho V = \pi \int_a^b \rho [f(x)]^2 \, dx$.

EXAMPLE 26.29

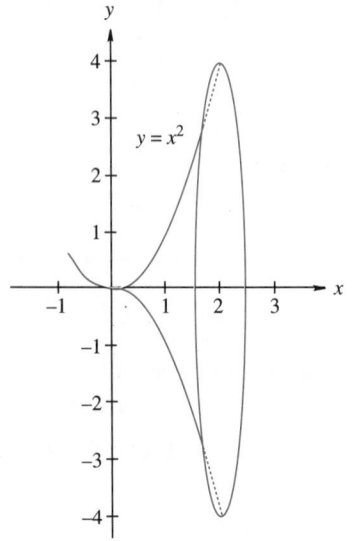

FIGURE 26.29

Find the centroid of the solid formed by revolving the region bounded by $y = x^2$, $y = 0$, and $x = 2$ about the x-axis, as shown in Figure 26.29.

Solution Because this is revolved around the x-axis, $\bar{y} = 0$. To determine $\bar{x}$, we need m and M_y. We get m from V.

$$V = \pi \int_0^2 [x^2]^2 \, dx$$

$$= \pi \int_0^2 x^4 \, dx = \pi \frac{x^5}{5} \Big|_0^2 = \frac{32}{5} \pi$$

$$m = \rho V = \frac{32}{5} \rho \pi$$

$$M_y = \pi \int_0^2 \rho x \, [x^2]^2 \, dx$$

$$= \pi \int_0^2 \rho x^5 \, dx = \rho \pi \frac{x^6}{6} \Big|_0^2 = \frac{32}{3} \rho \pi$$

$$\bar{x} = \frac{M_y}{m} = \frac{32/3 \rho \pi}{32/5 \rho \pi} = \frac{5}{3}$$

The centroid is at $\left(\frac{5}{3}, 0\right)$.

≡ **Note** If a solid is revolved around the y-axis, we would follow very similar procedures, only the centroid would be at $(0, \bar{y})$.

EXAMPLE 26.30

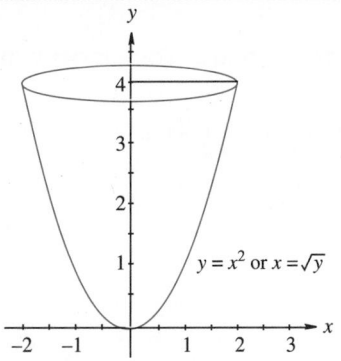

FIGURE 26.30

Find the centroid of the solid formed by revolving the region bounded by $y = x^2$, $y = 4$, and the y-axis around the y-axis as in Figure 26.30.

Solution Because it is revolved around the y-axis, $\bar{x} = 0$. To find $\bar{y}$, we need M_x and m. Again, we get m from V.

$$V = \pi \int_0^4 [\sqrt{y}]^2 \, dy$$

$$= \pi \int_0^4 y \, dy = \pi \frac{y^2}{2} \Big|_0^4 = 8\pi$$

$$m = \rho V = 8\rho\pi$$

$$M_x = \pi \int_0^4 \rho y [\sqrt{y}]^2 \, dy$$

$$= \pi \int_0^4 \rho y^2 \, dy = \pi\rho \frac{y^3}{3} \Big|_0^4 = \frac{64}{3}\rho\pi$$

$$\bar{y} = \frac{M_x}{m} = \frac{64/3\rho\pi}{8\rho\pi} = \frac{8}{3}$$

The centroid is at $\left(0, \frac{8}{3}\right)$.

≡ **Note**

If a region is bounded by two functions $f(x)$ and $g(x)$ or $h(y)$ and $k(y)$, then the washer method should be used.

EXAMPLE 26.31

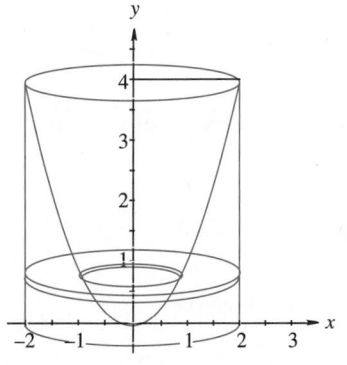

FIGURE 26.31

Find the centroid of the solid formed by revolving the region in Example 26.29 around the y-axis.

Solution Because it is revolved around the y-axis, we know that $\bar{x} = 0$.

We will use the washer method, as shown in Figure 26.31. To determine $\bar{y}$, we need M_x and m.

$$V = \pi \int_0^4 [2^2 - (\sqrt{y})^2] \, dy = \pi \int_0^4 (4 - y) dy$$

$$= \pi \left(4y - \frac{y^2}{2}\right) \Big|_0^4 = 8\pi$$

$$m = \rho V = 8\pi\rho$$

$$M_x = \pi \int_0^4 \rho y [2^2 - (\sqrt{y})^2] \, dy = \pi \int_0^4 \rho(4y - y^2) dy$$

$$= \rho\pi \left(2y^2 - \frac{y^3}{3}\right) \Big|_0^4 = \frac{32}{3}\pi\rho$$

$$\bar{y} = \frac{M_x}{m} = \frac{32/3\rho\pi}{8\pi\rho} = \frac{4}{3}$$

The centroid is at $\left(0, \frac{4}{3}\right)$.

Exercise Set 26.5

In Exercises 1–4, find the centroid of the system with masses located at the indicated points. Each mass is in grams and the units are in centimeters.

1. $m_1 = 4$ at $(5, 2)$ and $m_2 = 6$ at $(-3, 7)$

2. $m_1 = 3$ at $(-1, 4)$, $m_2 = 4$ at $(-2, -5)$, and $m_3 = 13$ at $(1, 2)$

3. $m_1 = 2$ at $(1, 5)$, $m_2 = 3$ at $(-1, 4)$, and $m_3 = 5$ at $(6, -4)$

4. $m_1 = 1$ at $(1, 1)$, $m_2 = 2$ at $(2, -2)$, $m_3 = 3$ at $(-3, 3)$, and $m_4 = 4$ at $(-4, -4)$

In Exercises 5–8, find the coordinates of the centroid for the indicated figures.

5.
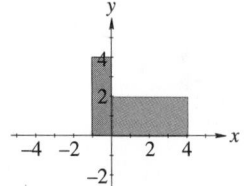

6.
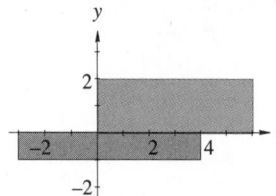

7. **8.**
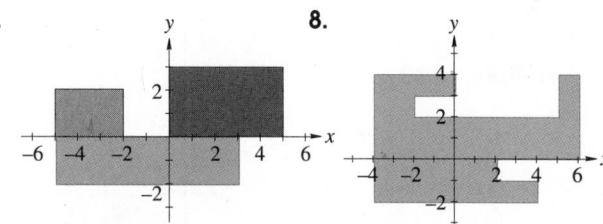

In Exercises 9–14, calculate the centroid of the plane region bounded by the given curve and the x-axis over the indicated interval.

9. $y = 2x + 3$, $[0, 3]$

10. $y = x^3$, $[0, 2]$

11. $y = x^{1/3}$, $[0, 8]$

12. $y = x^4$, $[-1, 2]$

13. $y = \sqrt{x + 4}$, $[0, 5]$

14. $y = x^2 + 16$, $[0, 4]$

In Exercises 15–22, find the centroids of the plane regions bounded by the given curves and lines.

15. $y = x^2$, $y = 4x$, $x = 0$, $x = 2$

16. $y = x^3$, $y = 2x$, $x = 0$, $x = 1$

17. $y = 2 - x$, $y = x^2$

18. $y = x^{3/2}$, $y = x$

19. $y = x^2$, $y = 18 - x^2$

20. $y = \sqrt{x}$, $y = x$

21. $y = x$, $y = 12 - x^2$

22. $y = x^2$, $y = x^3$

In Exercises 23–34, find the centroids of the solid of revolution formed by rotating the enclosed region around the indicated axis.

23. $y = x^3$, $y = 0$, $x = 2$, about the x-axis

24. $y = x^3$, $x = 0$, $y = 8$, about the y-axis

25. $y = x^4$, $y = 1$, about the y-axis

26. $y = (x - 1)^2$, $y = 1$, about the y-axis

27. $x + y = 6$, $x = 3$, $y = 0$, $x = 0$, about the x-axis

28. $x + y = 8$, $x = 4$, $y = 0$, $x = 8$, about the x-axis

29. $x + y = 8$, $x = 4$, $y = 0$, about the y-axis

30. $x^2 + y^2 = 16$, $x = 0$, $y = 0$, about the x-axis

31. $x^2 - y^2 = 1$, $x = 3$, about the x-axis

32. $x^2 - y^2 = 1$, $x = 3$, $y = 0$, about the y-axis

33. $y = x^2$, $y = \sqrt{x}$, about the x-axis

34. $y = x^4$, $y = x^2$, about the y-axis

Solve Exercises 35–38.

35. *Construction* The ends of two thin steel rods of equal lengths are welded together to make a right-angle frame, as shown in Figure 26.32.

FIGURE 26.32

(a) Locate the frame's centroid, if the rods are each 1 m long and 1 cm wide.

(b) Locate the frame's centroid, if the rods are each h units long and w units wide.

36. *Product design* A plastic pylon used at work sites is in the shape of a frustum of a cone on a square base, as shown in Figure 26.33. The radii of the bases of the cone are $2\frac{1}{2}$ in. and 7 in. and the height between the bases is 28 in.

(a) Determine the centroid of the portion of the pylon that is in the shape of a frustum of a cone.

(b) If the square base is 1 in. thick and 18 in. on each side, determine the centroid of the entire pylon.

37. Determine the mass and the center of mass of a rod of length 40 cm, where the density ρ of the rod at a point x cm from one end is $5x + 1$ g/cm.

38. Determine the mass and the center of mass of a rod of length 8 m, where the density ρ of the rod at a point x m from one end is $x + 5$ kg/m.

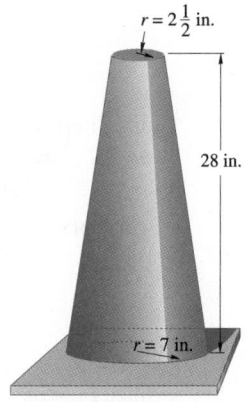

$r = 2\frac{1}{2}$ in.

28 in.

$r = 7$ in.

FIGURE 26.33

 **In Your Words**

39. Explain the meaning of the moments, mass, and centroid of a plane region.

40. Describe how to determine the centroid of an irregular-shaped plane region.

☰ 26.6
MOMENTS OF INERTIA

In Section 26.5, the first moment around the y-axis (the line $x = 0$) of n masses was defined as

$$M_y = \sum_{i=1}^{n} m_i x_i$$
$$= m_1 x_1 + m_2 x_2 + m_3 x_3 + \cdots + m_n x_n$$

For other problems, other kinds of moments are more useful. In this section, we will study the moments of inertia. The **moment of inertia** is the measure of the tendency of an object to resist a change in motion. The moment of inertia is also called the **second moment** and for a system of n masses around the y-axis it is defined as

$$I_y = \sum_{i=1}^{n} m_i x_i{}^2$$
$$= m_1 x_1{}^2 + m_2 x_2{}^2 + m_3 x_3{}^2 + \cdots + m_n x_n{}^2$$

If all of these masses were at the same distance r from the axis of rotation, we would have

$$I_y = m_1 r^2 + m_2 r^2 + m_3 r^2 + \cdots + m_n r^2$$
$$= (m_1 + m_2 + m_3 + \cdots + m_n) r^2$$
$$= mr^2$$

where m is the total mass of the system. The number r is called the **radius of gyration**.

EXAMPLE 26.32

Find the moment of inertia and the radius of gyration about the x- and y-axes for three masses of 5 kg, 2 kg, and 4 kg located, respectively, at $(2, 1)$, $(4, 0)$, and $(-3, -5)$, where the units are in meters.

Solution

$$I_y = \sum_{i=1}^{3} m_i x_i^2 = 5(2)^2 + 2(4)^2 + 4(-3)^2 = 88\,\text{kg}\cdot\text{m}^2$$

Since $I_y = m r_y{}^2$, we know that $r_y{}^2 = \dfrac{I_y}{m}$ or $r_y = \sqrt{\dfrac{I_y}{m}}$. The total mass of the system is $m = 5 + 2 + 4 = 11$ kg. Thus, $r_y = \sqrt{\frac{88}{11}} = \sqrt{8} = 2\sqrt{2} \approx 2.83$ m.

With respect to the x-axis, we get the results

$$I_x = 5(1)^2 + 2(0)^2 + 4(-5)^2 = 105\,\text{kg}\cdot\text{m}^2$$

and $\quad r_x = \sqrt{\dfrac{105}{11}} \approx 3.09\,\text{m}$

These answers mean that a mass of 11 kg placed on the line $x = 2.83$ (or $x = -2.83$) will have the same rotational inertia about the y-axis as the three objects. Similarly, placing an 11-kg mass on the line $y = 3.09$ (or $y = -3.09$) will produce the same rotational inertia about the x-axis as the three objects.

It is often of interest to determine the radius of gyration for a system that is rotated about the origin, r_0. In this case, $r_0 = \sqrt{r_x{}^2 + r_y{}^2}$. In Example 26.32, the radius of gyration with respect to the origin is $r_0 = \sqrt{8 + \frac{105}{11}} \approx 4.19$. Thus, placing an 11-kg mass at 4.19 m from the origin will produce the same rotational inertia about the origin as the three objects in Example 26.32.

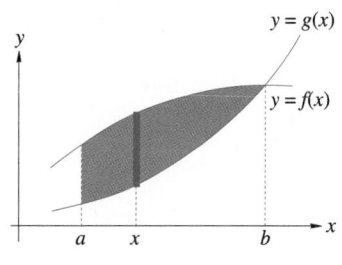

FIGURE 26.34

Moments of Inertia for a Region

Suppose an area is bounded by the curves of two functions such as $y = f(x)$ and $y = g(x)$ and the lines $x = a$ and $x = b$ with $f(x) \geq g(x)$, as shown in Figure 26.34. The moment of inertia for this region with respect to the y-axis is the sum of the moments of inertia for the individual elements of the region. The mass of each element is $\rho[f(x) - g(x)]\,dx$, where ρ is the density (mass per unit area) and $[f(x) - g(x)]\,dx$ is the area of the element. If this element is x-units from the y-axis, then the moment of inertia for the element is $\rho x^2[f(x) - g(x)]\,dx$. If we did this for all the elements, then their sum would give us the moment of inertia.

Moment of Inertia

If a region is bounded by the curves of two functions such as $y = f(x)$ and $y = g(x)$ and by the lines $x = a$ and $x = b$ with $f(x) \geq g(x)$, then the **moment of inertia** of this area with respect to the y-axis, I_y, is

$$I_y = \rho \int_a^b x^2[f(x) - g(x)]\,dx$$

where ρ is the density of the region.

If we have the moment of inertia with respect to the y-axis and the mass of the region, then we can determine the radius of gyration of the region as shown in the following box.

Radius of Gyration

The radius of gyration r_y with respect to the y-axis is

$$r_y = \sqrt{\frac{I_y}{m}}$$

where m is the mass of the region. ($m = \rho A$ where A is the area of the region.)

EXAMPLE 26.33

Find the moment of inertia and the radius of gyration for the region bounded by $y = \sqrt[3]{x}$, $x = 8$, and the x-axis with respect to the y-axis, if the region has a constant density of $3\,\text{kg/m}^2$. Assume all dimensions are meters.

EXAMPLE 26.33 (Cont.)

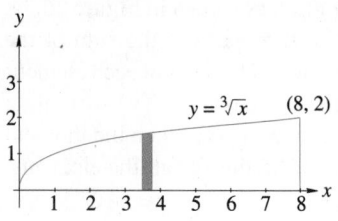

FIGURE 26.35

Solution   The region is sketched in Figure 26.35. Here

$$\rho = 3\,\text{kg/m}^2, \quad f(x) = \sqrt[3]{x}, \text{ and } g(x) = 0$$

$$I_y = 3 \int_0^8 x^2(\sqrt[3]{x} - 0)\,dx$$

$$= 3 \int_0^8 x^{7/3}\,dx$$

$$= 3 \left(\frac{3}{10}\right) x^{10/3} \Big|_0^8 = 921.6\,\text{kg} \cdot \text{m}^2$$

$$A = \int_0^8 \sqrt[3]{x}\,dx$$

$$= \left(\frac{3}{4}\right) x^{4/3} \Big|_0^8 = 12\,\text{m}^2$$

Since $m = \rho A$, we have $m = 3A = 36\,\text{kg}$.

$$r_y = \sqrt{\frac{I_y}{m}}$$

$$= \sqrt{\frac{921.6}{36}} = \sqrt{25.6} \approx 5.06\,\text{m}$$

EXAMPLE 26.34

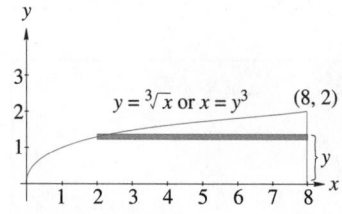

FIGURE 26.36

Find the moment of inertia and the radius of gyration of the region in Example 26.33 with respect to the x-axis.

Solution   The region is the same as before, only this time a typical element is parallel to the x-axis, as shown in Figure 26.36. Each element is y units from the x-axis and has the area $(8 - y^3)dy$. Thus, we have

$$I_x = 3 \int_0^2 y^2(8 - y^3)\,dy = 3 \int_0^2 (8y^2 - y^5)\,dy$$

$$= 3 \left(\frac{8}{3}y^3 - \frac{1}{6}y^6\right) \Big|_0^2$$

$$= 3 \left(\frac{64}{3} - \frac{64}{6}\right) = 32\,\text{kg} \cdot \text{m}^2$$

$$A = \int_0^2 (8 - y^3)\,dy$$

$$= \left(8y - \frac{1}{4}y^4\right) \Big|_0^2 = (16 - 4) = 12\,\text{kg}$$

$$m = \rho A = 3A = 36\,\text{kg}$$

$$r_x = \sqrt{\frac{I_x}{m}}$$

$$= \sqrt{\frac{32}{36}} = \frac{\sqrt{8}}{3} \approx 0.94\,\text{m}$$

Moments of Inertia of a Solid of Revolution ————

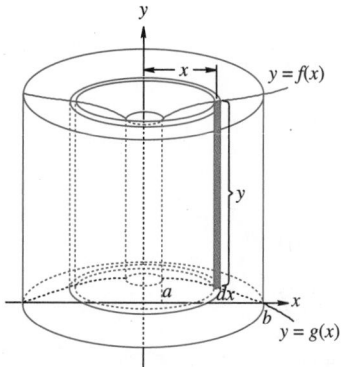

FIGURE 26.37

We next move to moments of inertia and radii of gyration for solids of revolution. For this we will use the shell method. Consider a region bounded by the curve of the function $y = f(x)$, $x = a$, $x = b$, and $y = g(x)$. The volume of a typical element, such as the one in Figure 26.37, is $2\pi xy\,dx$, where x represents the radius of the cylindrical shell, y, or $f(x) - g(x)$ [assuming $f(x) \geq g(x)$], is its height, and dx is its thickness. The second moment of this shell is $\rho(2\pi xy\,dx)x^2$. Expressing the sums of the second moments as an integral, we get the moment of inertia with respect to the y-axis, I_y, as shown in the following box.

Moment of Inertia; Radius of Gyration

The **moment of inertia** of a solid of revolution formed by generating a region around the y-axis is

$$I_y = 2\pi\rho \int_a^b x^3 y\,dx = 2\pi\rho \int_a^b x^3 [f(x) - g(x)]\,dx$$

and the **radius of gyration** is

$$r_y = \sqrt{\frac{I_y}{m}}$$

where m is the mass of the volume of the solid of revolution.

A similar formula exists for the moment of inertia with respect to the x-axis. This second moment, I_x, is

$$I_x = 2\pi\rho \int_c^d y^3 x\,dy = 2\pi\rho \int_c^d y^3 [h(y) - k(y)]\,dy$$

where the region is bounded by $x = h(y)$, $x = k(y)$, $y = c$, and $y = d$, and where $h(y) \geq k(y)$ on $[c, d]$.

EXAMPLE 26.35

Find the moment of inertia and the radius of gyration of the solid formed by rotating the region bounded by $y = \sqrt[3]{x}$, $x = 8$ cm, and $y = 0$ around the y-axis. This solid will be made from aluminum, which has a density of $2.70\,\text{g/cm}^3$.

Solution For this problem $a = 0$, $b = 8$, $f(x) = \sqrt[3]{x}$, $g(x) = 0$, and $\rho = 2.70$.

$$I_y = 2\pi\rho \int_0^8 x^3 \left(\sqrt[3]{x} - 0\right) dx$$

$$= 5.40\pi \int_0^8 x^{10/3}\,dx$$

EXAMPLE 26.35 (Cont.)

$$= 5.40\pi \left(\frac{3}{13}\right) x^{13/3} \Big|_0^8$$

$$= \frac{16.2\pi}{13}(8192) \approx 32\,070.924\,\text{g}\cdot\text{cm}^2$$

The mass is

$$m = \rho V = 2\pi\rho \int_0^8 x\left(\sqrt[3]{x} - 0\right) dx$$

$$= 5.40\pi \int_0^8 x^{4/3}\,dx$$

$$= 5.40\pi \left(\frac{3}{7}\right) x^{7/3} \Big|_0^8$$

$$= \frac{16.2\pi}{7}(128) \approx 930.630\,\text{g}$$

The radius of gyration is

$$r_y = \sqrt{\frac{I_y}{m}}$$

$$= \sqrt{\frac{32\,070.924}{930.630}} \approx 5.87\,\text{cm}$$

EXAMPLE 26.36

Find the moment of inertia and the radius of gyration of the solid formed by rotating the region bounded by $y = \sqrt[3]{x}$, $x = 8$ cm, and $y = 0$ around the x-axis. This solid will be made from nylon, which has a density of 1.14 g/cm^3.

Solution This is rotated around the x-axis, so we will use functions of y. Here $h(y) = 8$, $k(y) = y^3$, $c = 0$, $d = 2$, and $\rho = 1.14$. We will not use the value of ρ until we have finished.

$$I_x = 2\pi\rho \int_0^2 y^3\left(8 - y^3\right) dy$$

$$= 2\pi\rho \int_0^2 (8y^3 - y^6)\,dy$$

$$= 2\pi\rho \left(2y^4 - \frac{1}{7}y^7\right)\Big|_0^2$$

$$\approx 27.43\pi\rho \approx 86.17\rho\,\text{g}\cdot\text{cm}^2$$

Since $\rho = 1.14$, $I_x = 98.23$ g·cm^2. The mass is ρV and so

$$m = 2\pi\rho \int_0^2 y(8 - y^3)\,dy$$

EXAMPLE 26.36 (Cont.)

$$= 2\pi\rho \int_0^2 (8y - y^4)dy$$

$$= 2\pi\rho \left(4y^2 - \frac{1}{5}y^5\right)\Big|_0^2 = 19.2\pi\rho \approx 60.32\rho\,\text{g}$$

The radius of gyration is

$$r_x = \sqrt{\frac{I_x}{m}}$$

$$= \sqrt{\frac{27.43\pi\rho}{19.2\pi\rho}} = \sqrt{\frac{27.43}{19.2}} \approx 1.20\,\text{cm}$$

Exercise Set 26.6

In Exercises 1–4, find the moment of inertia and the radius of gyration with respect to the origin, r_0, of the given masses at the given points.

1. 3 g at $(4,0)$ and 5 g at $(-3,0)$ cm
2. 6 g at $(0,2)$, 3 g at $(0,-5)$, and 1 g at $(0,4)$

3. 4 kg at $(2,1)$ and 3 kg at $(-1,4)$ m
4. 3 g at $(2,3)$, 4 g at $(-2,-4)$, and 3 g at $(-4,5)$

In Exercises 5–14, find the moment of inertia and radius of gyration about the given axis. Unless specified, let ρ represent the density.

5. The region bounded by $y = x^2$, the y-axis, $y = 2$ about y-axis
6. The region of Exercise 5 about x-axis
7. The region bounded by $x = 0$, $y = 0$, $x = 3$, $y = 5$ about the x-axis
8. The region in Exercise 7 about the y-axis
9. The region bounded by $y = x^3$ and $y = x^2$ about the x-axis

10. The region in Exercise 9 about the y-axis
11. The region bounded by $y = 4x - x^2$ and the x-axis, $\rho = 5$ g/cm^2 about y-axis
12. The region bounded by $y^3 = x^2$, $y = 4$ and the y-axis, $\rho = 4$ g/cm^2 about the x-axis
13. The region bounded by $y = \frac{1}{x^2}$, $y = x^2$, $x = 2$, $x = 1$, $\rho = 8$ g/cm^2 about the y-axis
14. The region in Exercise 13 about the x-axis

In Exercises 15–24, find the moment of inertia and the radius of gyration for the solid formed by rotating the region about the given axis. Unless specified, let ρ represent the density.

15. The region bonded by $y = x^2$, the y-axis, and $y = 2$ about the y-axis
16. The region in Exercise 15, but rotated about the x-axis
17. The region bounded by $y = x^2$ and $y = 4x$ about the y-axis
18. The region in Exercise 17 about the x-axis
19. The region bounded by $x = 0$, $y = 0$, $x = 4$ and $y = 6$ about the x-axis

20. The region in Exercise 19, but rotated about the y-axis
21. The region bounded by $y = 4x - x^2$ and $x + y = 4$, $\rho = 3$ g/cm^3 about y-axis
22. The region bounded by $y = \sqrt{4-x}$ and $x + 2y = 4$, $\rho = 5$ g/cm^3, about the x-axis.

23. The region bounded by $y = \dfrac{1}{x}$, $y = x^3$, $x = 1$, $x = 2$, $\rho = 5$ g/cm^3, about the y-axis

24. The region in Exercise 23 about the x-axis

In Your Words

25. Without looking in the text, explain the concept of moment of inertia.

26. What is meant by a radius of gyration?

≣ 26.7
WORK AND FLUID PRESSURE

The last applications of the integral that we will consider in this chapter deal with the topic of work. **Work** is defined as the product of the force exerted on an object and the distance the object is moved by that force. For example, if a force of 10 lb is lifted a distance of 5 ft, the work performed is $(10\,\text{lb}) \times (5\,\text{ft}) = 50\,\text{ft·lb}$. In the same way, if a force of 20 dynes is lifted 9 cm the worked performed is $(20\,\text{dynes}) \times (9\,\text{cm}) = 180$ dyne·cm= 180 ergs. Just as an erg is a dyne·cm, the joule is the equivalent of the Newton-meter (N · m).

Most work problems are not as simple as these two. If they were, we certainly would not be discussing them in a chapter about applications of integral calculus.

Hooke's Law

Consider the work done to stretch a spring. When you begin to stretch the spring, it takes very little force. The further the spring is stretched, more force is required. Hooke's law states that the force F required to stretch a spring a distance x is proportional to x. Thus, if k is the constant of proportionality, $F = kx$ and so the force is a function of x or $F(x)$.

We want to determine the amount of work that is done to move the spring from point a to point b. As usual, we partition the interval $[a, b]$ into n subintervals of lengths Δx. If x_i^* is the midpoint of one of the intervals, then the work done to move the spring through that interval is approximately $\Delta W = F(x_i^*)\Delta x$. Adding the work over all of the intervals and taking the limits as $\Delta x \to 0$, we get the following formula.

> **Work**
>
> The **work** W done by a continuous force directed along the x-axis from $x = a$ to $x = b$ is
>
> $$W = \int_a^b F(x)\,dx$$

Application

EXAMPLE 26.37

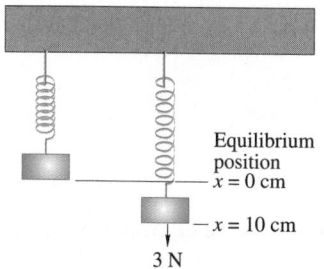

Equilibrium
position
$x = 0$ cm

$-x = 10$ cm

3 N

FIGURE 26.38

A spring is stretched 10 cm by a force of 3 N, as shown in Figure 26.38. How much work is needed to stretch the spring 40 cm?

Solution   According to Hooke's law, $F = kx$. We are given $F = 3$ N when $x = 10$ cm $= 0.1$ m. Substituting these into the formula for Hooke's law produces

$$F = kx$$
$$3 = k(0.1)$$
$$30 = k$$

Thus, for this particular spring, $F = 30x$ or the function $F(x) = 30x$. The amount of work needed to stretch this spring 40 cm $= 0.4$ m is

$$W = \int_0^{0.4} F(x)\,dx$$
$$= \int_0^{0.4} 30x\,dx = 15x^2 \Big|_0^{0.4} = 2.4\,\text{N}\cdot\text{m}$$

The work needed to stretch the spring is $2.4\,\text{N}\cdot\text{m} = 2.4\,\text{J}$. (The symbol "J" is used for joule or joules.)

Application

EXAMPLE 26.38

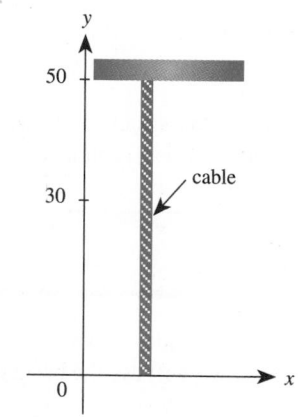

cable

FIGURE 26.39

Find the work done in winding 30 m of a 50-m cable, if the cable has a density of 4.5 kg/m.

Solution   First we let y denote the length of the cable that has been wound up at any time. The amount of cable that has yet to be wound is $50 - y$. As usual, we will subdivide the interval $[0, 30]$ into n subdivisions. A typical subdivision is shown in Figure 26.39. This subdivision has a length dy and is y units from the top. The mass of a subdivision is $\rho\,dy$ and its weight is $\rho g\,dy$, where ρ is the density and g the gravitational constant. In this example, ρ was given as 4.5 kg/m and, since this problem is in the metric system, $g = 9.8$ m/s^2. Thus, the force needed at any point is

$$(4.5)(9.8)(\text{amount of cable left to be wound}) = (4.5)(9.8)(50 - y)$$
$$= 44.1(50 - y)$$

The work needed to move this element from its initial position to the top is $44.1(50 - y)dy$. Since only 30 m are going to be wound, the work is

$$W = \int_0^{30} 44.1(50 - y)dy$$
$$= 44.1\left(50y - \frac{1}{2}y^2\right)\Big|_0^{30}$$
$$= 44.1(1\,050)\,\text{J}$$
$$= 46\,305\,\text{J}$$

Electrical Charges

According to Coulomb's law, the electric force between two charges, q_1 and q_2, in a vacuum is directly proportional to the product of these two changes and inversely proportional to the square of the distance r between them. Thus, the force is a function of r and we have

$$F(r) = k\frac{q_1 q_2}{r^2}$$

where $k = 8.988 \times 10^9\,\text{N·m}^2/\text{C}^2$. Using the formula for work

$$W = \int F(r)\,dr$$

we can determine the work done when two charges move toward or away from each other.

Application

EXAMPLE 26.39

Find the work done when two protons, $q = 1.6 \times 10^{-19}\,\text{C}$ each, move until they are 100 nm apart, if they were originally separated by 1 m.

Solution   We have $q_1 = q_2 = 1.6 \times 10^{-19}$. The particles move from 1 m apart to $100\,\text{nm} = 100 \times 10^{-9}\,\text{m} = 10^{-7}\,\text{m}$ apart.

$$\begin{aligned}
W &= \int_1^{10^{-7}} \frac{(8.988 \times 10^9)(1.6 \times 10^{-19})(1.6 \times 10^{-19})}{r^2}\,dr \\
&= 2.301 \times 10^{-28} \int_1^{10^{-7}} \frac{dr}{r^2} \\
&= 2.301 \times 10^{-28} \left(\frac{-1}{r}\right)\Big|_1^{10^{-7}} \\
&= -2.301 \times 10^{-21}\,\text{J}
\end{aligned}$$

The negative sign means that the force must be done *on* the system to move the particles together. Thus, it takes 2.301×10^{-21} J acting on these protons to move them from 1 m apart to 100 nm apart. ▪

Pumping Liquids

As a third example of a problem where the force is variable, consider the work needed to pump a liquid out of, or into, a tank. In these examples, we will use the basic idea that work = (weight of object) × (distance moved).

Application

EXAMPLE 26.40

An oil tank is in the shape of a right-circular cylinder with a height of 40 ft, a radius of 10 ft, and is filled with oil to a depth of 30 ft. How much work is required to pump the oil out of the top of the tank, if oil weighs 54.8 lb/ft^3?

Solution   We will position the tank as shown in Figure 26.40. The position of the top of the tank is at $x = 0$ and the bottom at $x = 40$. The work needed to pump a

EXAMPLE 26.40 (Cont.)

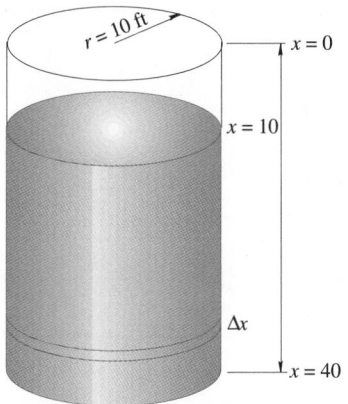

FIGURE 26.40

certain volume of oil over the top depends on the weight of this volume of oil and its distance from the top.

The oil fills the tank from $x = 10$ to $x = 40$. We will partition the interval $[10, 40]$ into subintervals and let Δx be the length of one interval. The volume at this interval is $\pi r^2 h$, and since $r = 10$ and $h = \Delta x$, we have the volume as $100\pi \Delta x$. We are told that oil weighs $54.8\,\text{lb/ft}^3$, so the weight of this layer is $(54.8)(100\pi \Delta x)$. If the interval is located at x_i^*, then the work done to lift this layer is $x_i^*(54.8)(100\pi \Delta x)$.

The layers of oil occur from 10 ft to 40 ft, thus the work required is given by

$$\int_{10}^{40} (54.8)(100\pi)x\,dx = 5480\pi \left(\frac{1}{2}x^2\right)\Big|_{10}^{40}$$
$$= 4{,}110{,}000\pi\,\text{ft}\cdot\text{lb}$$
$$\approx 12{,}911{,}946\,\text{ft}\cdot\text{lb}$$

It takes a total of $12{,}911{,}946$ ft·lb to pump the oil out of the top of the tank.

The procedure used in Example 26.40 follows the following guidelines for finding the work needed to pump a liquid out of a tank.

Guidelines for Finding Work Done During Pumping

1. Draw a picture with a coordinate system.
2. Determine the mass of a thin horizontal slab of the liquid.
3. Find an expression for the work needed to lift this slab to its destination.
4. Integrate the work expression from step 3 from the bottom of the liquid to the top.

Application

EXAMPLE 26.41

Suppose a hemispherical tank of radius 10 ft is on a platform so that the bottom is 30 ft off the ground. How much work is done filling the tank from a source of water at ground level, if the water is pumped through a hole in the bottom?

Solution We will put the origin of the vertical axis at ground level. Then the bottom of the tank is at 30 ft and the top is at 40 ft, as shown in Figure 26.41a.

Divide the interval $[30, 40]$ into n subintervals. Suppose a typical interval is x_i units from the top of the tank, as shown in Figure 26.41b. The slice of water taken at this point has a radius y_i. Because the tank is a hemisphere, we know that the radius r of the tank is the hypotenuse of the right triangle formed by x_i and y_i. This means that $x^2 + y^2 = r^2$. The volume of this slice of water is $\Delta V = \pi y^2 \Delta x = \pi(r^2 - x^2)\Delta x$. The force required to lift this slice is

$$\Delta w = 62.4\pi(r^2 - x^2)\Delta x$$

EXAMPLE 26.41 (Cont.)

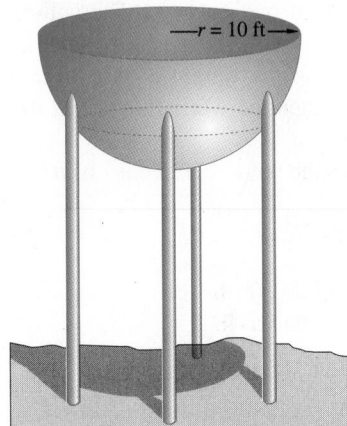

FIGURE 26.41a

where the specific weight of water is $w = 62.4\,\text{lb/ft}^3$. Now all we need is the distance this slice was lifted. Since it was lifted from $x = 0$ to a point x_i units below $x = 40$, it was lifted a total of $40 - x_i$ units. Putting this together with the fact that $r = 10$, we get

$$
\begin{aligned}
W &= \int_0^{10} 62.4\pi(10^2 - x^2)(40 - x)\,dx \\
&= 62.4\pi \int_0^{10} (4{,}000 - 100x - 40x^2 + x^3)\,dx \\
&= 62.4\pi \left(4{,}000x - 50x^2 - \frac{40}{3}x^3 + \frac{1}{4}x^4 \right) \Big|_0^{10} \\
&\approx 4{,}737{,}521.7\,\text{ft}\cdot\text{lb}
\end{aligned}
$$

Fluid Pressure

When a container is built to hold a fluid, it is important to know the force F caused by the liquid pressure on the sides of the container. The pressure P exerted by a liquid of mass density ρ at depth h units below the surface of the liquid is defined as

$$P = \rho g h$$

where g is the acceleration of gravity, $32\,\text{ft/s}^2 \approx 9.8\,\text{m/s}^2$.

≡ **Note** In the English system we are given the weight density w as lb/ft^3. Here $w = \rho g$.

If a flat plate is submerged in a fluid and placed in a horizontal position, the pressure on the plate is the same at all points. The total force on one side of the plate is the product of the pressure P and the area A of the side $F = PA = \rho g h A$.

Application

EXAMPLE 26.42

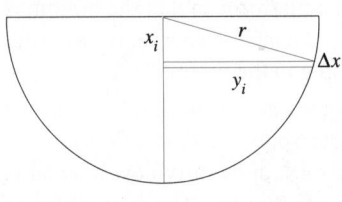

FIGURE 26.41b

A tank 3 m long and 2 m wide is filled with 1.5 m of water. What is the force exerted on the bottom of the tank?

Solution The density of water is about $\rho = 1\,000\,\text{kg/m}^3$. The area of the bottom of the tank is $A = (3\,\text{m})(2\,\text{m}) = 6\,\text{m}^2$. The pressure is $P = (1\,000)(9.8)(1.5) = 14\,700\,\text{N/m}^2$. $F = PA = (14\,700)(6) = 88\,200\,\text{N}$.

If a flat plate submerged in a fluid of density ρ is placed in a vertical position, then the pressure varies as the depth changes. If the plate is subdivided into n horizontal strips, as shown by the typical strip in Figure 26.42, then the pressure is almost constant on each strip.

As shown in Figure 26.42, each horizontal strip has a width dy and length $L(y)$. So, the area is $L(y)\,dy$. We introduce a coordinate system, as shown in the figure. If the plate extends over the interval $[c, d]$ on the y-axis, then for each y in this interval, the depth of each strip is $h(y)$. If dy is small, the pressure at any point in the horizontal

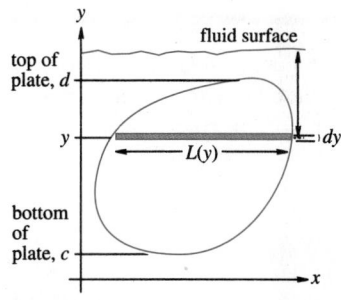

FIGURE 26.42

strip is $\rho h(y)$. Thus, the force on each rectangular strip is given by the product of the pressure on any point in the strip and the area of the strip, or $\rho h(y)L(y)dy$. Integrating this over the depth of the plate, we get the following formula.

Fluid Force

The total force F exerted by a fluid of constant density ρ on a submerged vertical plate from depth $y = c$ to $y = d$ is given by

$$F = \rho \int_c^d h(y)L(y)dy$$

where $h(y)$ is the depth of the fluid at y and $L(y)$ is the horizontal length of the plate at y.

Application

EXAMPLE 26.43

A tank has a cross-section in the shape of a trapezoid with a lower base of 4 ft, and upper base 8 ft. It is filled with 2.5 ft of water. What is the total pressure due to the water on one end of the tank?

Solution   A sketch of the tank is shown in Figure 26.43. An xy-coordinate system has been placed on the figure to help in the calculations. The y-axis is partitioned into n subintervals. A typical strip formed by this partition is shaded in the figure. The force against this ith strip is $P_i A_i = 62.4(2.5 - y_i)A_i$, where $h(y_i) = 2.5 - y_i$ is the depth of the strip and A_i is its area.

The length of the strip is approximately $2x_i$. But the edge of the tank, line BC, passes through $(2, 0)$ and $(4, 2.5)$. This line has the equation $y = \frac{2.5}{2}x - 2.5$ and so $x = \frac{2}{2.5}y + 2 = \frac{4}{5}y + 2$. Thus the length of the ith strip is $2\left(\frac{4}{5}y_i + 2\right)$ and its area $A_i = 2\left(\frac{4}{5}y_i + 2\right)\Delta y = \left(\frac{8}{5}y_i + 4\right)\Delta y_i$.

We know that the force F_i on this ith strip is

$$F_i = 62.4(2.5 - y_i)\left(\frac{8}{5}y_i + 4\right)\Delta y_i$$

Integrating, we get the total force

$$F = \int_0^{2.5} 62.4(2.5 - y)\left(\frac{8}{5}y + 4\right)dy$$

$$= 62.4 \int_0^{2.5}\left(10 - \frac{8}{5}y^2\right)dy$$

$$= 62.4\left(10y - \frac{8}{15}y^3\right)\Big|_0^{2.5}$$

$$= 62.4(16.67) \approx 1,040\,\text{lb}$$

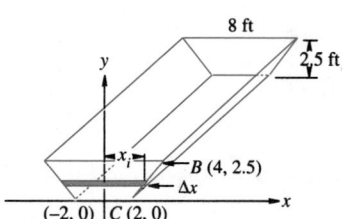

FIGURE 26.43

Application

EXAMPLE 26.44

A tank truck, like the milk truck in the photograph in Figure 26.44a, is used to carry liquids. If the tank on this truck is in the shape of a circular cylinder with a radius of 3 ft and length 35 ft, determine the force on one end of this truck when it is full of milk, if milk has a density of 64.5 lb/ft^3.

Courtesy of H.P. Hood Inc.

FIGURE 26.44a

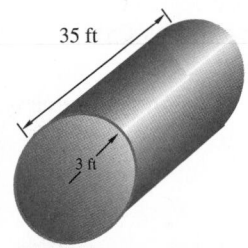

FIGURE 26.44b

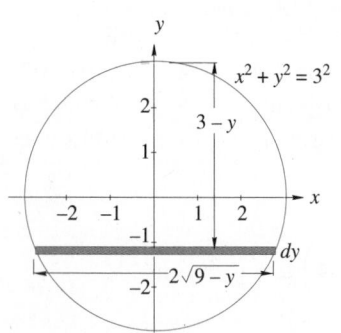

FIGURE 26.44c

Solution A sketch of the tank is shown in Figure 26.44b. A cross-section of an end of the tank is shown in Figure 26.44c. A coordinate system and a typical horizontal strip have been drawn in the figure. Since this is a circular cylinder and the origin of the coordinate system is placed at the center of the base, the circle is described by the equation $x^2 + y^2 = 3^2$, so the length of the strip is $x = 2\sqrt{9 - y^2}$. Since the width of the strip is dy, its area is $A_i = 2\sqrt{9 - y^2}\,dy$. As can be seen in Figure 26.44c, the depth of this strip is $3 - y$. Thus, the force on this strip is $F_i = \rho(3 - y)2\sqrt{9 - y^2}\,dy$ and the total force on the end of the tank is

$$F = 2\rho \int_{-3}^{3} (3 - y)\sqrt{9 - y^2}\,dy$$

$$= 6\rho \int_{-3}^{3} \sqrt{9 - y^2}\,dy + 2\rho \int_{-3}^{3} (-y)\sqrt{9 - y^2}\,dy$$

While we do not yet know how to integrate the left-hand integral, we do know that its integral will be the area of one-half a circle of radius 3. From geometry, we know that the area of a complete circle of radius 3 is $\pi r^2 = \pi 3^2 = 9\pi$ and so the desired integral will be one-half that amount, or $\frac{9}{2}\pi$. Substituting this in the previous equation, we have

$$F = 6\rho \frac{9}{2}\pi + 2\rho \int_{-3}^{3} (-y)(9 - y^2)^{1/2}\,dy$$

$$= 27\pi\rho + \frac{2}{3}\,\rho(9 - y^2)^{3/2}\Big|_{-3}^{3} = 27\pi\rho$$

Since milk has a density of $\rho = 64.5$ lb/ft^3, the total force on the end of the tank is $1{,}741.5\pi \approx 5{,}471.08$ lb/ft^3. Notice that the length of the tank had no effect on the force exerted on the ends.

Exercise Set 26.7

Solve Exercises 1–36.

1. *Physics* A force of 3 lb stretches a spring 4 in. How much work is done in stretching the spring 10 in.?

2. *Physics* If the spring in Exercise 1 is stretched from 1 ft to 2 ft, how much work is done?

3. *Physics* A force of 6 N stretches a spring 20 cm. How much work is done to stretch the spring 1 m?

4. *Physics* How much work is done to stretch the spring in Exercise 3 from 10 cm to 20 cm?

5. *Physics* If the work done in stretching a spring from rest at 0.50 m to a length of 0.60 m is 0.20 J, find the force needed to keep the spring at this length.

6. *Physics* A spring has a natural (or unstretched) length of 10 in. A force of 5 lb compresses the spring to a length of 8 in. Find the work needed to compress it from 9 in. to 6 in.

7. *Physics* If 105 ft·lb of work was needed to stretch the spring in Exercise 1 from its equilibrium position, how far was it stretched?

8. *Mechanical engineering* A cable 50 ft long and weighting 2 lb/ft is hanging from a winch. How much work is needed to wind 20 ft of it up?

9. *Mechanical engineering* A cable 30 m long weighing 2.5 N/m is hanging from a winch. How much work is needed to wind all of the cable onto the winch?

10. *Mechanical engineering* A chain with a mass of 4.7 kg/m has a length of 15 m. If there is a load of 87 kg at the end of the chain, what work is done in winding the chain and load up 10 m?

11. *Mechanical engineering* A chain 12 ft long is lying on the floor. If the chain weighs 0.75 lb/ft, how much work is needed to raise one end to a height of 20 ft?

12. *Mechanical engineering* A chain 5 m long with a density of 0.4 kg/m is lying on the floor. How much work is needed to raise one end to a height of 10 m?

13. *Mechanical engineering* A tank with a mass of 45 kg is filled with 1 000 L of water. The water is used to spray chemical cleaner off the side of a building. If the water is used at the rate of 20 L/min and

the tank is raised at 0.50 m/min, how much work is done in raising the tank 20 m up the wall?

14. *Nuclear physics* Find the work needed to move a charge of $+5 \times 10^{-7}$ C and a charge of -2×10^{-7} C until they are 1 cm apart, if they were originally 3 mm apart. (Note: Change all distances to meters.)

15. *Nuclear physics* A charge of $+4 \times 10^{-19}$ C is 5 cm from a charge of $+5 \times 10^{-8}$ C. How much work is needed to move them until they are 10 nm apart? ($1 \text{ nm} = 10^{-9}$ m.)

16. *Nuclear physics* A proton has a charge of $+1.6 \times 10^{-18}$ C and an electron has a charge of -1.6×10^{-19} C. How much work is needed to move them from 530 nm apart to 1 mm apart?

17. *Physics* According to Newton's law of gravitation, the gravitational force of attraction between two objects that are r units apart is given by $F(r) = \dfrac{k}{r^2}$. If two objects are 1 cm apart, how much work is needed to separate them by a distance of 1 m? (Leave answer in terms of k.)

18. *Physics* If two objects are 1 ft apart, how much work is needed to separate them until they are 20 ft apart? (Leave answer in terms of k.)

19. *Space technology* How much work is required to lift a 1,000-lb satellite from the surface of the Earth to an orbit of 1,000 mi above the Earth's surface? (See Exercise 17 with $k = 16 \times 10^9$ mi²·lb and the radius of the Earth as 4,000 mi.)

20. *Petroleum engineering* A cylindrical tank 4 ft in radius and 8 ft high is filled with petroleum. How much work is needed to pump the oil out over the top? (The density of oil is about 54.8 lb/ft³.)

21. *Petroleum engineering* How much work is required to pump all the oil over the top of a full cylindrical tank that is 4 m in diameter and 8 m high? (The mass density of petroleum varies. For this, and subsequent exercises, let the mass density $\rho = 880$ kg/m³.)

22. *Recreation* A swimming pool is 2 m deep, 10 m long, and 7 m wide. If it is filled with water, how much work is needed to empty the pool, if the water is pumped over the top?

23. *Recreation* How much work is needed to empty the pool in Exercise 22, if the water was only 1.5 m deep?

24. *Petroleum engineering* The oil in a cylindrical tank that is 10 ft high and has a diameter of 6 ft is to be pumped to a height 5 ft above the top of the tank. How much work is required?

25. *Petroleum engineering* An upright conical tank is filled with oil from a reservoir that is 15.0 m below the bottom of the tank. If the top of the tank has a radius of 2.40 m and the height of the tank is 3.60 m, what work is done in filling the tank when the density of the oil is 847 kg/m^3?

26. *Wastewater technology* A water tank is in the shape of a hemisphere with a radius of 4 m. If the tank is already filled with water, how much work is needed to pump all the water over the top of the tank? (The density of water is 1000 kg/m^3.)

27. *Wastewater technology* How much work is required to pump the top 2 m of water out the top of the tank in Exercise 26?

28. *Wastewater technology* If the tank in Exercise 26 is placed on a platform so that the bottom of the tank is 10 m above the ground, how much work is required to fill the tank with water to the depth of 3 m from ground level?

29. *Recreation* (a) What is the pressure of the water on the bottom of the tank in Exercise 22? (b) What is the force of the water on the bottom of this tank?

30. *Petroleum engineering* (a) What is the pressure of the petroleum on the bottom of the tank in Exercise 24? (b) What is the force of the petroleum on the bottom of this tank?

31. *Recreation* If the pool in Exercise 22 is full of water, what is the total force due to water pressure on one short side of the pool?

32. *Recreation* For the pool in Exercise 22, what is the total force due to water pressure on a long side of the pool?

33. *Civil engineering* A vertical flood gate is in the shape of an isosceles triangle which has a base of 3 m and an altitude of 2 m. If the base of the floodgate is on the surface of the water, what is the force on its face?

34. *Recreation* A rectangular swimming pool is 40.0 m long, 1.00 m deep at the shallow end, and 3.00 m deep at the deep end. Assuming that all of the sides are vertical, find the total force on the long side of the swimming pool when it is full.

35. *Transportation* A tanker truck is in the shape of a right-circular cylinder. The radius of the cylinder is 1.6 m and the density of gasoline is about $\rho = 680$ kg/m^3. If the tanker is half filled, what is the total force due to the pressure of the gasoline on one end of the cylinder?

36. *Transportation* A tanker truck is in the shape of a right-circular cylinder with a radius of 4 ft. If the tank is half full of gasoline, and the weight of gasoline is about 42.5 lb/ft^3, what is the total force due to the pressure of the gasoline on one end of the tanker?

✎ **In Your Words**

37. Explain what is meant by the concept of work as used in this text.

38. Without looking in the text, describe the four guidelines for finding the work done during pumping.

CHAPTER 26 REVIEW

Important Terms and Concepts

Arc length	Moment of inertia	Surface area
Average value	Moments	Volumes of revolution
Centroid	First	Disk method
Fluid pressure	Second	Shell method
Hooke's law	Radius of gyration	Washer method
Mass	Root mean square	Work

Review Exercises

In Exercises 1–4, find the average value and the root mean square of the function over the given interval.

1. $f(x) = 4x$, over $[1, 6]$

2. $g(x) = x^3$, over $[0, 2]$

3. $h(x) = x^2 - 4$, over $[2, 4]$

4. $k(t) = 4.9t^2 - 2.8t - 4$, over $[0, 2.5]$

In Exercises 5–12, find (a) the volume and (b) the centroid of the solid of revolution generated by revolving the enclosed region about the indicated axis.

5. $y = 6x$, $x = 1$, $x = 5$, $y = 0$, about the x-axis

6. $y = -x$, $x = 1$, $x = 4$, $y = 0$, about the y-axis

7. $y = x^4$, $x = 1$, $x = 2$, $y = 0$, about the x-axis

8. $y = \sqrt[3]{x^2}$, $x = 0$, $x = 1$, $y = 0$, about the x-axis

9. $y = x^3$, $y = \sqrt[3]{x}$, about the y-axis

10. $y = x^3$, $y = \sqrt[3]{x}$, about the x-axis

11. $y = x^2$, $y = 9 - x^2$, about the y-axis

12. $y = x^3$, $y = 4x$, about the line $x = 5$

In Exercises 13–14, find the lengths of the curves over the indicated interval.

13. $x = \sqrt{y^3}$ from $y = 0$ to $y = 4$

14. $y = \sqrt{8x^3}$ from $x = 0$ to $x = 2$

In Exercises 15–16, find (a) the lengths of the curves over the indicated interval and (b) the surface area of the solid generated by rotating the curve around the indicated axis.

15. $y = \dfrac{x^4}{4} + \dfrac{1}{8x^2}$ from $x = 1$ to $x = 3$, about the x-axis

16. $x = \dfrac{y^6}{6} + \dfrac{1}{16y^4}$ from $y = 1$ to $y = 2$, about the y-axis

Solve Exercises 17–8.

17. *Physics* An unstretched spring 0.6 m long is stretched to a length of 0.8 m by a force of 5 N. How much work is needed to stretch the spring to a length of 1.2 m?

18. *Electricity* The voltage across a 60-μF capacitor is zero. What is the voltage after 0.001 s, if a current of 0.40 A charges the capacitor?

19. *Electricity* The electric current, as a function of time, for a certain circuit is $I = 4t - t^3$. What is the average value of the current, with respect to time, for the first 2 s?

20. *Physics* If a ball is dropped from the top of the Washington Monument, which is 555 ft high, what is its average height and average velocity between the time it is dropped and the time it strikes the ground?

21. *Construction*   A steel beam weighing 1,000 lb hangs from a 50-ft cable. The cable weighs 5 lb/ft. How much work is done in winding 30 ft of cable with a winch?

22. *Civil engineering*   A dam is shaped like a trapezoid with a height of 100 ft. It is 400 ft long at the top and 200 ft long at the bottom. When the water level behind the dam is level with the top, what is the total force that the water exerts on the dam?

23. *Petroleum engineering*   A tank is in the shape of a right-circular cylinder with a radius of 6 m and a height of 16 m. If it is filled with petroleum, how much work is needed to pump it out the top? (Let $\rho = 880 \, \text{kg/m}^3$.)

24. *Waste technology*   A tank is in the shape of an inverted cone with a radius of 6 ft and a height of 16 ft. If it is filled with water, how much work is needed to pump it out the top?

≡ CHAPTER 26 TEST

1. Find the average value of $f(x) = 3x^2 + 1$ over the interval $[1, 5]$.

2. Determine the root mean square of $g(x) = x^3 - 1$ over $[1, 3]$.

3. Find the volume of the solid generated by revolving the region bounded by $y = \sqrt{9x + 1}$, $x = 0$, $x = 7$, and the x-axis about the x-axis.

4. Find the centroid of the solid of revolution generated by revolving the region bounded by $y = x^2 + 1$, $x = 0$, $x = 4$, and $y = 0$, about the x-axis.

5. Find the length of the curve $y = 1 + x^{3/2}$ from $x = 1$ to $x = 6$.

6. Find the surface area of the solid generated by rotating the curve $y = \frac{1}{6}x^3$ from $x = 1$ to $x = 3$ around the x-axis.

7. How much work does it take to pump the water from a full upright circular-cylindrical tank of radius 5 m

and height 10 m to a point 4 m above the top of the tank?

8. A vertical gate is in the shape of a right triangle with its base on top (and the right angle at the bottom), as shown in Figure 26.45. If the base is 5 ft below the surface of the water and the altitude of the gate is 3 ft, what is the force against the gate?

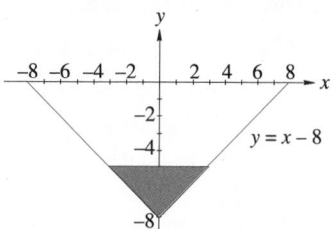

FIGURE 26.45

CHAPTER
27
Derivatives of Transcendental Functions

How far back from a movie screen like the one shown should you sit in order to have the best viewing angle? In Section 27.4, we will show how to use calculus to answer this question.

$\mathbf{W}$e have learned how to differentiate and integrate polynomials and certain other algebraic functions. A function that is not algebraic is called transcendental. The transcendental functions include the trigonometric, logarithmic, exponential, and inverse trigonometric functions. In this chapter, we will learn how to differentiate these four types of transcendental functions.

In Sections 27.1 and 27.2 we develop the derivatives of the six trigonometric functions and in Section 27.3 we develop the derivatives of the inverse trigonometric functions. Section 27.4 looks at some applications that use these derivatives.

Sections 27.5–27.6 examine derivatives of exponential and logarithmic functions, and in Section 27.7, we consider several applications that involve the derivatives of these functions.

≡ 27.1
DERIVATIVES OF THE SINE AND COSINE FUNCTIONS

In Chapter 23, we developed some rules for differentiating. These rules allowed us to find derivatives much faster than using the four-step method. In this section, we are going to develop rules for the derivatives of the sine and cosine functions. In order to do this, we return to the four-step method.

Before we develop derivatives for the sine and cosine, we must review three ideas. In Section 24.3 we showed that

$$\lim_{x \to 0} \frac{\sin x}{x} = 1$$

where x is measured in radians.

In Section 20.2 we showed that

$$\sin(A + B) = \sin A \cos B + \cos A \sin B$$

and

$$\sin(A - B) = \sin A \cos B - \cos A \sin B$$

There is one additional limit that we will need: $\lim\limits_{x \to 0} \dfrac{\cos x - 1}{x}$. We will determine its limit using algebra.

$$\lim_{x \to 0} \frac{\cos x - 1}{x} = \lim_{x \to 0} \frac{\cos x - 1}{x} \cdot \frac{\cos x + 1}{\cos x + 1}$$

$$= \lim_{x \to 0} \frac{\cos^2 x - 1}{x(\cos x + 1)}$$

$$= \lim_{x \to 0} \frac{-\sin^2 x}{x(\cos x + 1)}$$

$$= \lim_{x \to 0} \frac{\sin x}{x} \cdot \lim_{x \to 0} \frac{-\sin x}{\cos x + 1} = 1 \cdot \frac{0}{2} = 0$$

Thus, we have shown the following:

$$\lim_{x \to 0} \frac{\cos x - 1}{x} = 0$$

Derivative of Sine Function

We are now ready to find the derivative of $f(x) = \sin x$ by the four-step method.

Step 1: $f(x+h) = \sin(x+h)$

Step 2: $f(x+h) - f(x) = \sin(x+h) - \sin x$

$$= \sin x \cos h + \cos x \sin h - \sin x$$

$$= \sin x (\cos h - 1) + \cos x \sin h$$

Step 3: $\dfrac{f(x+h) - f(x)}{h} = \dfrac{\sin x (\cos h - 1)}{h} + \dfrac{\cos x \sin h}{h}$

$$= (\sin x) \left(\frac{\cos h - 1}{h} \right) + (\cos x) \left(\frac{\sin h}{h} \right)$$

Step 4: $\displaystyle\lim_{h \to 0} \dfrac{f(x+h) - f(x)}{h} = \lim_{h \to 0} (\sin x) \left(\frac{\cos h - 1}{h} \right) + \lim_{h \to 0} (\cos x) \left(\frac{\sin h}{h} \right)$

$$= (\sin x)(0) + (\cos x)(1) = \cos x$$

Since, $\displaystyle\lim_{h \to 0} \dfrac{f(x+h) - f(x)}{h} = (\sin x)(0) + (\cos x)(1) = \cos x$, we have the following derivative for the sine function.

Derivative of the Sine Function $\dfrac{d}{dx}(\sin x) = \cos x$

A more general form of the derivative of the sine function is developed by using the chain rule. In Section 23.4, we found that if $y = f(u)$ and $u = g(x)$, then

$$y' = f'(u)g'(x)$$

or, in differential notation,

$$\frac{dy}{dx} = \frac{dy}{du} \cdot \frac{du}{dx}$$

Thus, we have the following.

Derivative of the Sine Function $\dfrac{d}{dx}(\sin u) = (\cos u)\dfrac{du}{dx}$

We will use the chain rule in the following examples.

EXAMPLE 27.1

If $y = \sin x^2$, what is y'?

Solution   First, we need to remember that $\sin x^2 = \sin(x^2)$. We will use the chain rule with $y = \sin u$ and $u = x^2$. Then

$$y' = f'(u)g'(x) = (\cos u)(2x)$$

Since $u = x^2$, we have

$$y' = (\cos x^2)(2x)$$
$$\text{or} \qquad = 2x \cos x^2$$

EXAMPLE 27.2

If $y = (\sin x)^3$, find y'.

Solution   We can solve this by either using the general power rule or the chain rule. We will show how to do it both ways.

Power rule:

$$y' = 3(\sin x)^2 \left(\frac{d}{dx} \sin x \right)$$
$$= 3(\sin x)^2 (\cos x)$$
$$= 3 \sin^2 x \cos x$$

[Remember, $\sin^2 x = (\sin x)^2$.]
Chain rule: Let $u = \sin x$, so $y = u^3$

$$y' = f'(u)g'(x)$$
$$= 3u^2 \cos x$$
$$= 3(\sin x)^2 \cos x = 3 \sin^2 x \cos x$$

EXAMPLE 27.3

Differentiate $f(x) = \sin \sqrt{3x^2 + 1}$.

Solution   We let $u = \sqrt{3x^2 + 1} = (3x^2 + 1)^{1/2}$. So,

$$\frac{du}{dx} = \frac{1}{2}(3x^2 + 1)^{-1/2}(6x) = \frac{3x}{\sqrt{3x^2 + 1}}$$

Now $f(u) = \sin u$ and so

$$f'(x) = \cos u \frac{du}{dx}$$
$$= \left(\cos \sqrt{3x^2 + 1} \right) \left(\frac{3x}{\sqrt{3x^2 + 1}} \right)$$
$$= \frac{3x}{\sqrt{3x^2 + 1}} \cos \sqrt{3x^2 + 1}$$
$$= \frac{3x \cos \sqrt{3x^2 + 1}}{\sqrt{3x^2 + 1}}$$

Derivative of Cosine Function

Next we find the derivative of $y = \cos x$. We will use the identity $\cos x = \sin(\frac{\pi}{2} - x)$ and the chain rule with $y = \sin u$ and $u = \frac{\pi}{2} - x$.

$$y = \cos x = \sin\left(\frac{\pi}{2} - x\right) = \sin u$$

$$y' = \frac{dy}{du} \cdot \frac{du}{dx} = (\cos u)(-1) = -\cos u$$

$$= -\cos\left(\frac{\pi}{2} - x\right) = -\sin x$$

Again, using the chain rule, we get the more general form.

Derivative of the Cosine Function	$\dfrac{d}{dx}(\cos u) = -\sin u \dfrac{du}{dx}$

EXAMPLE 27.4

If $y = \cos(3x + 1)$, determine y'.

Solution   If we let $u = 3x + 1$, then $y = \cos u$ and

$$y' = f'(u)g'(x)$$
$$= (-\sin u)(3) = -3\sin u$$
$$= -3\sin(3x + 1)$$

EXAMPLE 27.5

Find the derivative of $y = x^2 \cos(3x + 1)$.

Solution   Using the product rule we have

$$y' = x^2 \frac{d}{dx}[\cos(3x + 1)] + [\cos(3x + 1)]\frac{d}{dx}(x^2)$$

Since $\dfrac{d}{dx}[\cos(3x + 1)] = -3\sin(3x + 1)$, we have

$$y' = -3x^2 \sin(3x + 1) + 2x \cos(3x + 1)$$

EXAMPLE 27.6

Differentiate: $f(x) = \dfrac{\sin 3x}{\cos^2 x}$.

Solution We will use the quotient rule.

$$f'(x) = \frac{\cos^2 x \dfrac{d}{dx}(\sin 3x) - \sin 3x \dfrac{d}{dx}\cos^2 x}{(\cos^2 x)^2}$$

$$= \frac{\cos^2 x(3\cos 3x) - \sin 3x(2\cos x)(-\sin x)}{\cos^4 x}$$

$$= \frac{3\cos^2 x \cos 3x + 2\sin x \sin 3x \cos x}{\cos^4 x}$$

$$= \frac{3\cos x \cos 3x + 2\sin x \sin 3x}{\cos^3 x}$$

Application

EXAMPLE 27.7

The voltage, V, in volts, available from the wall outlet of a residential house is given by $V = 120\sin 377t$, where t is the time, in seconds. Find the rate of change of the voltage with respect to time when $t = 4.50\,\text{ms}$.

Solution The rate of change of the voltage with respect to time means that we want to find $\dfrac{dV}{dt}$. Differentiating, we obtain $\dfrac{dV}{dt} = V' = 120(377)\cos 377t = 45240\cos 377t$.

Converting $t = 4.50\,\text{ms}$ to seconds, we obtain $t = 4.50\,\text{ms} = 0.00450\,\text{s}$. Substituting this value for t in our derived formula for V', we obtain $V' = 45240 \times \cos 377(0.0045) = -5671.869$. The voltage is changing at the rate of $-5671.869\,\text{V/s}$.

Exercise Set 27.1

Differentiate the functions in Exercises 1–30.

1. $y = \sin 3x$
2. $y = \cos 4x$
3. $y = 3\cos 2x$
4. $y = 5\sin 6x$
5. $y = \sin(x^2 + 1)$
6. $y = \cos(x^3 - 5)$
7. $y = 4\sin^2 3x$
8. $y = 5\cos^3 2x$

9. $y = \cos(3x^2 - 2)$
10. $y = \sin(2x^3 + 1)$
11. $y = \sin\sqrt{x}$
12. $y = \cos x^{3/2}$
13. $y = \cos\sqrt{2x^3 - 4}$
14. $y = \sin^2\sqrt{x+1}$
15. $y = x^2 + \sin^2 x$
16. $y = \cos x + \sin x$

17. $y = \sin x \cos x$
18. $y = (\sin x - \cos x)^2$
19. $y = \dfrac{2\cos x}{\sin 2x}$
20. $y = \dfrac{\sin 2x}{2x}$
21. $y = x^2 \sin x$
22. $y = x^3 \cos x$
23. $y = \sqrt{x}\sin x$

24. $y = \dfrac{\sin^3 x}{x}$
25. $y = (\sin 2x)(\cos 3x)$
26. $y = x^2\cos(3x^2 - 1)$
27. $y = \sin^3(x^4)$
28. $y = \sqrt{\cos\sqrt{x}}$
29. $y = \sin^2 x + \cos^2 x$
30. $y = \sin^2 x - 2\cos^2 x$

Solve Exercises 31–48.

31. If $f(x) = \sin x$, what is $f''(x)$?

32. If $g(x) = \cos x$, what is $g''(x)$?

33. Find y''' if $y = \sin x$.

34. Find y''' if $y = \cos x$.

35. Find $\dfrac{d^4}{dx^4}(\sin x)$.

36. Find $\dfrac{d^4}{dx^4}(\cos x)$.

37. **(a)** Determine the equation of the line tangent to $f(x) = \sqrt{2}\cos x$ at $x = \frac{\pi}{4}$.

    (b) Graph f and the tangent line to f at $x = \dfrac{\pi}{4}$ on the same coordinate system.

38. **(a)** Determine the equation of the line tangent to $g(x) = 2\cos^2 x$ at $x = \dfrac{2\pi}{3}$.

    (b) Graph g and the tangent line to g at $x = \dfrac{2\pi}{3}$ on the same coordinate system.

39. **(a)** Determine the equation of the normal line to $h(x) = \sin 4x$ at $\left(\dfrac{\pi}{2}, 0\right)$.

    (b) Graph h and the normal line to h at $\left(\dfrac{\pi}{2}, 0\right)$ on the same coordinate system.

40. **(a)** Determine the equation of the normal line to $j(x) = \sin x \cos 2x$ at $\left(\dfrac{\pi}{4}, 0\right)$.

    (b) Graph j and the normal line to j at $\left(\dfrac{\pi}{4}, 0\right)$ on the same coordinate system.

41. *Electronics* The voltage, V, in volts, available from the wall outlet of a residential house is given by $V = 120\sin 377t$, where t is the time, in seconds. Find the rate of change of the voltage with respect to time when $t = 1.571\,\text{s}$.
$-3470\,\text{V}$

42. *Automotive engineering* The motion of an automobile engine piston approximates simple harmonic

motion given by $s = 4.8\cos 75t$, where y is the displacement in centimeters and t is in seconds. Find the velocity and acceleration when $t = 0.01\,\text{s}$.

43. *Electronics* The charge across a capacitor is $q(t) = 0.25\cos(t - 1.45)$. Find the current to the capacitor at $t = 7.6\,\text{s}$.

44. *Navigation* A nautical mile varies with latitude. The equation $l = 6077 - 31\cos 2\theta$ gives the length of a nautical mile in feet at latitude θ, in radians. Find the rate at which l changes with respect to θ when $\theta = \frac{\pi}{3}$.

45. *Optics* According to Malus's law the amount of light transmitted, I, in terms of the angle of incidence, θ, and the maximum intensity of light transmitted, M, is given by $I = M\cos^2\theta$. Find the rate at which I changes with respect to θ when $\theta = \frac{\pi}{3}$.

46. *Lighting technology* The intensity of reflected light, I_{L_r}, increases with the angle of incidence, θ, as given by

$$I_{L_r} = I_{L_i} - I_{L_i}(1 - \mu_r)\cos\theta_i$$

Find the rate at which I_{L_r} changes with respect to θ_i when $\theta_i = \dfrac{\pi}{6}$, $\mu_r = 0.6$, and $I_{L_i} = 100\,\text{cd}$.

47. *Electricity* The current in an electric circuit is given by $I = 2\sin t + \cos t$. Find the maximum and minimum values of the current for one cycle.

48. *Electronics* The magnetic field, B, in tesla (T), produced by a rotating armature varies with the angle of rotation according to the equation

$$B = \frac{2\sin\theta}{1 + \cos^2(2\theta)}$$

Find the rate at which B changes with respect to θ when $\theta = \dfrac{\pi}{6}$.

In Your Words

49. The derivatives of the sine and cosine functions are very similar. Explain how you can tell which derivative has the minus sign.

50. Many objects move in circles or, like pistons, have linear motion that is the result of moving something in a circular motion. Think of a place in your tech-

nology area of interest where this is the case. Write an application that uses derivatives based on your observations. Give your problem to a classmate and see if he or she understands and can solve your problem. Rewrite the problem as necessary to remove any difficulties encountered by your classmate.

▤ 27.2
DERIVATIVES OF THE OTHER TRIGONOMETRIC FUNCTIONS

In Section 27.1, we developed the formulas for the derivatives of the sine and cosine functions. In this section, we will use those formulas to develop the derivatives for the other trigonometric functions.

Derivative of Tangent Function

In order to find the derivatives of the tangent function, we use one of the basic trigonometric identities.

$$\tan u = \frac{\sin u}{\cos u}$$

Using this, we get

$$\frac{d}{dx}\tan u = \frac{d}{dx}\left(\frac{\sin u}{\cos u}\right)$$

$$= \frac{\cos u \dfrac{d}{dx}\sin u - \sin u \dfrac{d}{dx}\cos u}{\cos^2 u}$$

$$= \frac{\cos^2 u \dfrac{du}{dx} + \sin^2 u \dfrac{du}{dx}}{\cos^2 u}$$

$$= \frac{(\cos^2 u + \sin^2 u)\dfrac{du}{dx}}{\cos^2 u}$$

Since $\cos^2 u + \sin^2 u = 1$, we have

$$\frac{d}{dx}\tan u = \frac{1}{\cos^2 u}\frac{du}{dx}$$
$$= \sec^2 u \frac{du}{dx}$$

So, we have the derivative of the tangent function.

Derivative of Tangent Function $\dfrac{d}{dx}\tan u = \sec^2 u \dfrac{du}{dx}$

Derivatives of Secant, Cosecant, and Cotangent Functions

In a similar manner, using $\sec u = \dfrac{1}{\cos u}$, we have

$$\frac{d}{dx}\sec u = \frac{d}{dx}\left(\frac{1}{\cos u}\right)$$

$$= \frac{\cos u\dfrac{d}{dx}(1) - (1)\dfrac{d}{dx}\cos u}{\cos^2 u}$$

$$= \frac{0 - (-\sin u)\dfrac{du}{dx}}{\cos^2 u}$$

$$= \frac{\sin u}{\cos^2 u}\frac{du}{dx}$$

$$= \frac{1}{\cos u}\frac{\sin u}{\cos u}\frac{du}{dx}$$

$$= \sec u \tan u \frac{du}{dx}$$

Thus, we have the derivative of the secant function.

Derivative of Secant Function	$\dfrac{d}{dx}\sec u = \sec u \tan u \dfrac{du}{dx}$

The development of the derivatives for the remaining two trigonometric functions will be done as exercises.

Derivative of Cotangent and Cosecant Functions	$\dfrac{d}{du}\cot u = -\csc^2 u \dfrac{du}{dx}$
	$\dfrac{d}{du}\csc u = -\csc u \cot u \dfrac{du}{dx}$

EXAMPLE 27.8

Differentiate $y = \sqrt{\tan 2x}$.

Solution We rewrite this as $y = (\tan 2x)^{1/2}$ and use the general power rule:

$$y' = \frac{1}{2}(\tan 2x)^{-1/2}\frac{d}{dx}(\tan 2x)$$

$$= \frac{1}{2}(\tan 2x)^{-1/2}2\sec^2 2x$$

$$= \frac{\sec^2 2x}{\sqrt{\tan 2x}}$$

EXAMPLE 27.9

Differentiate $y = x\csc^2(4x)$.

Solution We will use the product rule.

$$\begin{aligned}
y' &= x\frac{d}{dx}[\csc^2(4x)] + \csc^2(4x)\frac{d}{dx}(x) \\
&= x(2\csc 4x)\frac{d}{dx}(\csc 4x) + \csc^2 4x \\
&= 2x\csc 4x(-\csc 4x\cot 4x)4 + \csc^2 4x \\
&= -8x\csc^2 4x\cot 4x + \csc^2 4x \\
&= (\csc^2 4x)(1 - 8x\cot 4x)
\end{aligned}$$

EXAMPLE 27.10

Differentiate $f(x) = \sec^3\sqrt{x}$.

Solution
$$\begin{aligned}
f'(x) &= 3\sec^2\sqrt{x}\frac{d}{dx}\left(\sec\sqrt{x}\right) \\
&= 3\sec^2\sqrt{x}\left(\sec\sqrt{x}\tan\sqrt{x}\right)\frac{d}{dx}\sqrt{x} \\
&= \frac{3}{2\sqrt{x}}\sec^3\sqrt{x}\tan\sqrt{x}
\end{aligned}$$

EXAMPLE 27.11

Differentiate $g(x) = \sin 2x\tan x^2$.

Solution We will use the product rule.

$$\begin{aligned}
g'(x) &= \sin 2x\frac{d}{dx}\tan x^2 + \tan x^2\frac{d}{dx}\sin 2x \\
&= \sin 2x\sec^2 x^2\frac{d}{dx}x^2 + \tan x^2\cos 2x\frac{d}{dx}(2x) \\
&= 2x\sin 2x\sec^2 x^2 + 2\tan x^2\cos 2x
\end{aligned}$$

EXAMPLE 27.12

Use implicit differentiation to find y', if $y = x + \cot(xy)$.

Solution Using the implicit differentiation and the chain rule produces

$$\begin{aligned}
y' &= 1 + [-\csc^2(xy)]\frac{d}{dx}(xy) \\
&= 1 - [\csc^2(xy)](xy' + y) \\
y'[1 + x\csc^2(xy)] &= 1 - y\csc^2(xy) \\
y' &= \frac{1 - y\csc^2(xy)}{1 + x\csc^2(xy)}, \text{ if } x\csc^2 xy \neq -1
\end{aligned}$$

The derivatives of the six trigonometric functions are summarized below.

Derivatives of the Trigonometric Functions

$$\frac{d}{dx}\sin u = \cos u \frac{du}{dx}$$

$$\frac{d}{du}\csc u = -\csc u \cot u \frac{du}{dx}$$

$$\frac{d}{dx}\cos u = -\sin u \frac{du}{dx}$$

$$\frac{d}{du}\sec u = \sec u \tan u \frac{du}{dx}$$

$$\frac{d}{dx}\tan u = \sec^2 u \frac{du}{dx}$$

$$\frac{d}{du}\cot u = -\csc^2 u \frac{du}{dx}$$

Exercise Set 27.2

Differentiate each of the functions in Exercises 1–30.

1. $y = \tan^2 \sqrt{x}$
2. $y = \tan 4x$
3. $y = \sec 5x$
4. $y = \cot(1+2x)$
5. $y = \csc(2x-1)$
6. $y = \tan \dfrac{x}{2}$
7. $y = \sin x \tan x$
8. $y = \sin x \cot x$

9. $y = \tan^3 x$
10. $y = \cot^3 4x$
11. $y = \sec^4(x^2)$
12. $y = \csc^3 \sqrt{x^3}$
13. $y = \tan x \cot x$
14. $y = \sec x \csc x$
15. $y = \sin^2 x \cot x$
16. $y = \sin 5x^2$

17. $y = \tan \dfrac{1}{x}$
18. $y = \cot \sqrt{x^2+1}$
19. $y = \sqrt{1+\tan x^2}$
20. $y = \sqrt{\tan x + \cot x}$
21. $y = \dfrac{\tan x}{1+\sec x}$
22. $y = (\csc x + \cot x)^3$
23. $y = \dfrac{\cot x}{1-\csc x}$

24. $y = \sqrt{x} + \tan \sqrt{x}$
25. $y = (\csc x + 2\tan x)^3$
26. $y = \sqrt{1+\cot^2 x}$
27. $y = (\tan 2x)^{3/5}$
28. $y = x \csc x$
29. $y = \dfrac{\sec x}{1+\tan x}$
30. $y = \dfrac{\cos x}{1+\sec^2 x}$

Solve Exercises 31–32.

31. Develop the formula $\dfrac{d}{dx}\cot u = -\csc^2 u \dfrac{du}{dx}$.

32. Develop the formula $\dfrac{d}{dx}\csc u = -\csc u \cot u \dfrac{du}{dx}$.

In Exercises 33–36, find the second derivative of the given function.

33. $y = \tan 2x$
34. $y = \sec 3x$
35. $y = x \tan x$
36. $y = \dfrac{\cos x}{x}$

In Exercises 37–40, use implicit differentiation to find y'.

37. $y = x \sin y$
38. $\sin xy + xy = 0$
39. $x + y = \sin(x+y)$
40. $y^2 = \tan 4xy$

Solve Exercises 41–44.

41. *Optics* The index of refraction of the medium that light enters is given by $n = \csc \theta$, where θ is the critical angle. Find the rate of change of n with respect to θ when $\theta = \dfrac{\pi}{8}$.

42. *Aeronautics* An airplane that is flying horizontally at a constant speed of 360 mph is approaching a control tower at a height of 6,250 ft above an observer in the tower. At what rate is the angle of elevation changing when $\theta = 37°$?

43. *Electrical engineering*   The strength of a radar signal, in millivolts, varies with the orientation of the antenna according the the formula $I = 16\cot^2(2\theta)$. Find the rate of change of signal strength when $\theta = \dfrac{7\pi}{16}$.

44. *Physics*   If the displacement in meters of an object at time t in seconds is given by the equation $s = 8t^2\tan t^2$, $0 \le t \le 1.5$, find its velocity and acceleration when $t = 0.8$ s.

 In Your Words

45. Describe how you can tell the difference between the derivative of the tangent and the derivative of cotangent functions.

46. Describe how you can tell the difference between the derivative of the tangent and the derivative of secant functions.

≡ 27.3
DERIVATIVES OF INVERSE TRIGONOMETRIC FUNCTIONS

In Section 9.5, we studied the inverse trigonometric functions. As a typical example, the inverse sine function was written as either arcsin or $\sin^{-1}$. If $y = \arcsin x$, then $\sin y = x$. Because the trigonometric functions are periodic, the values of y lie in a certain interval. In review, we have the following inverse trigonometric functions.

Inverse Trigonometric Functions		
Function	Domain	Range
$y = \sin^{-1} x$	$-\dfrac{\pi}{2} \le y \le \dfrac{\pi}{2}$	$-1 \le x \le 1$
$y = \cos^{-1} x$	$0 \le y \le \pi$	$-1 \le x \le 1$
$y = \tan^{-1} x$	$-\dfrac{\pi}{2} < y < \dfrac{\pi}{2}$	x is any real number.
$y = \cot^{-1} x$	$0 < y < \pi$	x is any real number.
$y = \sec^{-1} x$	$0 \le y < \dfrac{\pi}{2}$ or $\pi \le y \le \dfrac{3\pi}{2}$	$\lvert x \rvert \ge 1$
$y = \csc^{-1} x$	$0 < y \le \dfrac{\pi}{2}$ or $\pi < y \le \dfrac{3\pi}{2}$	$\lvert x \rvert \ge 1$

To find the derivative of $y = \sin^{-1} u$, we write it as $u = \sin y$ and differentiate implicitly.

$$\frac{du}{dx} = \left(\frac{d}{dy}\sin y\right)\frac{dy}{dx}$$
$$\frac{du}{dx} = \cos y \frac{dy}{dx}$$

Solving for $\dfrac{dy}{dx}$ we get

$$\frac{dy}{dx} = \frac{1}{\cos y} \frac{du}{dx}$$

One of the Pythagorean identities states that $\sin^2 y + \cos^2 y = 1$. Solving this for $\cos y$, we get $\cos y = \pm\sqrt{1 - \sin^2 y}$. Since y is in $\left[-\frac{\pi}{2}, \frac{\pi}{2}\right]$, we know $\cos y \ge 0$, and so we select the positive square root. Thus, $\cos y = \sqrt{1 - \sin^2 y}$. But, $u = \sin y$ and so $\cos y = \sqrt{1 - u^2}$. Thus, we have the derivative of the inverse sine:

Derivative of Sin^{-1}
$$\frac{d}{dx} \sin^{-1} u = \frac{1}{\sqrt{1 - u^2}} \frac{du}{dx}, \qquad (|u| < 1)$$

EXAMPLE 27.13

Differentiate $y = \sin^{-1}(3x^2)$.

Solution We will use the chain rule with $y = \sin^{-1} u$ and $u = 3x^2$. Then

$$y' = \frac{1}{\sqrt{1 - (3x^2)^2}} \left[\frac{d}{dx}(3x^2)\right]$$

$$= \frac{6x}{\sqrt{1 - 9x^4}}$$

EXAMPLE 27.14

Find y', if $y = \sqrt{\sin^{-1}(4x - 3)}$.

Solution We will use an expanded version of the chain rule where $y' = \dfrac{dy}{dx} = \dfrac{dy}{du} \cdot \dfrac{du}{dv} \cdot \dfrac{dv}{dx}$ with $y = \sqrt{u} = u^{1/2}$, $u = \sin^{-1} v$, and $v = 4x - 3$.

$$y' = \left(\frac{d}{du} u^{1/2}\right) \left(\frac{d}{dv} \sin^{-1} v\right) \left[\frac{d}{dx}(4x - 3)\right]$$

$$= \left(\frac{1}{2} u^{-1/2}\right) \left(\frac{1}{\sqrt{1 - v^2}}\right)(4)$$

$$= \frac{2}{\sqrt{u}\sqrt{1 - v^2}}$$

$$= \frac{2}{\sqrt{\sin^{-1}(4x - 3)}\sqrt{1 - (4x - 3)^2}}$$

Through a process similar to the one we used to develop the derivative of $y = \sin^{-1} x$, we can determine the following derivative of $\cos^{-1}$.

Derivative of Cos^{-1} $\dfrac{d}{dx}\cos^{-1} u = \dfrac{-1}{\sqrt{1-u^2}}\dfrac{du}{dx}$, $(|u| < 1)$

≡ **Note**   The derivative of the arccos function is the negative of the derivative of the arcsin function.

EXAMPLE 27.15

Find y', if $y = \cos^{-1}(2x^3 + x)$.

Solution   Using the chain rule, we let $y = \cos^{-1} u$ and $u = 2x^3 + x$. Then,

$$y' = \frac{dy}{du} \cdot \frac{du}{dx}$$

$$= \frac{-1}{\sqrt{1-u^2}}(6x^2 + 1)$$

$$= \frac{-(6x^2 + 1)}{\sqrt{1-(2x^3 + x)^2}}$$

To find a formula for the derivative of $y = \tan^{-1} u$, we write $u = \tan y$ and differentiate implicitly with respect to x.

$$\frac{du}{dx} = \frac{d}{dy}\tan y \frac{dy}{dx}$$

$$= \sec^2 y \frac{dy}{dx}$$

Solving for $\dfrac{dy}{dx}$, we get

$$\frac{dy}{dx} = \frac{1}{\sec^2 y}\frac{du}{dx}$$

Since $\sec^2 y = 1 + \tan^2 y$ and $u = \tan y$, we have

$$\frac{dy}{dx} = \frac{1}{1 + \tan^2 y}\frac{du}{dx} = \frac{1}{1 + u^2}\frac{du}{dx}$$

We conclude the following.

Derivative of Tan^{-1}	$\dfrac{d}{dx}\tan^{-1}u = \dfrac{1}{1+u^2}\dfrac{du}{dx}$

In a similar manner, we can show the following.

Derivative of Cot^{-1}	$\dfrac{d}{dx}\cot^{-1}u = \dfrac{-1}{1+u^2}\dfrac{du}{dx}$

≡ **Note**   The derivative of the arccot function is the negative of the derivative of the arctan function.

EXAMPLE 27.16

Differentiate $y = (\tan^{-1}3x)^2$.

Solution   We will use the chain rule with $y = u^2$, $u = \tan^{-1}v$, and $v = 3x$. With these substitutions,

$$\frac{dy}{dx} = \frac{dy}{dy}\cdot\frac{du}{dv}\cdot\frac{dv}{dx}$$

$$= \frac{d}{du}(u^2)\cdot\frac{d}{dv}(\tan^{-1}v)\frac{d}{dx}(3x)$$

$$= 2u\left(\frac{1}{1+v^2}\right)3 = \frac{6u}{1+v^2}$$

$$= \frac{6(\tan^{-1}3x)}{1+9x^2}$$

EXAMPLE 27.17

Differentiate $y = x^4\arctan x^2$.

Solution   We need to use the product rule to solve this problem. Remember that $\arctan x^2 = \tan^{-1}x^2$.

$$y' = x^4\frac{d}{dx}\arctan x^2 + \left(\frac{d}{dx}x^4\right)\arctan x^2$$

$$= x^4\left(\frac{1}{1+x^4}\right)2x + 4x^3\arctan x^2$$

$$= \frac{2x^5}{1+x^4} + 4x^3\arctan x^2$$

We will have less opportunity to use the derivatives of the other two inverse trigonometric functions. However, we present them here in a summary of the derivatives of all six inverse trigonometric functions.

Derivatives of Inverse Trigonometric Functions

$$\frac{d}{dx}\sin^{-1}u = \frac{1}{\sqrt{1-u^2}}\frac{du}{dx}, \qquad |u| < 1$$

$$\frac{d}{dx}\cos^{-1}u = \frac{-1}{\sqrt{1-u^2}}\frac{du}{dx}, \qquad |u| < 1$$

$$\frac{d}{dx}\tan^{-1}u = \frac{1}{1+u^2}\frac{du}{dx}$$

$$\frac{d}{dx}\cot^{-1}u = \frac{-1}{1+u^2}\frac{du}{dx}$$

$$\frac{d}{dx}\sec^{-1}u = \frac{1}{u\sqrt{u^2-1}}\frac{du}{dx}, \qquad |u| > 1$$

$$\frac{d}{dx}\csc^{-1}u = \frac{-1}{u\sqrt{u^2-1}}\frac{du}{dx}, \qquad |u| > 1$$

Exercise Set 27.3

Differentiate the given function in Exercises 1–28.

1. $y = \sin^{-1}2x$
2. $y = \cos^{-1}4x$
3. $y = \tan^{-1}\dfrac{x}{2}$
4. $y = \sin^{-1}\left(\dfrac{x}{3}\right)$
5. $y = \cos^{-1}(1-x^2)$
6. $y = \sin^{-1}\sqrt{x}$
7. $y = \sin^{-1}(1-2x)$

8. $y = \tan^{-1}(4x-1)$
9. $y = \cos^{-1}(x^3-x)$
10. $y = \sin^{-1}(x-3)^2$
11. $y = \sec^{-1}(4x+2)$
12. $y = \csc^{-1}(3x+1)$
13. $y = x\sin^{-1}x$
14. $y = \sin^{-1}\sqrt{x+1}$
15. $y = x^2\cos^{-1}x$

16. $y = \cos^{-1}\sqrt{1-x^2}$
17. $y = \tan^{-1}\left(\dfrac{2x-1}{2x}\right)$
18. $y = \csc^{-1}\sqrt{x}$
19. $y = x\tan^{-1}(x+1)$
20. $y = (1+\tan^{-1}x)^2$
21. $y = \sqrt{\sin^{-1}(1-x^2)}$
22. $y = \sqrt{1-x^2}\sin^{-1}x$

23. $y = x\cot^{-1}(1+x^2)$
24. $y = \sin^{-1}\left(\dfrac{x-1}{x+1}\right)$
25. $y = \dfrac{\sin^{-1}x}{x}$
26. $y = \dfrac{\arccos x^2}{x}$
27. $y = \sqrt[3]{\arcsin x}$
28. $y = \sqrt[3]{\arctan x^2}$

Solve Exercises 29–32.

29. Show that $\dfrac{d}{du}\left(\cos^{-1}u\right) = \dfrac{-1}{\sqrt{1-u^2}}\dfrac{du}{dx}$.

30. Show that $\dfrac{d}{du}\left(\cot^{-1}u\right) = \dfrac{-1}{1+u^2}\dfrac{du}{dx}$.

31. *Photography* A race car is traveling at a speed of 320 ft/s as it passes a television camera at trackside 42 ft away. Find $\dfrac{d\theta}{dt}$, the rate at which the camera must turn to follow the car when $\theta = 15°$.

32. *Aeronautics* The angle of elevation between an observer and an airplane that is flying horizontally directly toward the observer is changing. If the airplane is flying at a constant speed of 270.0 mph at a height of 7,920 ft above an observer, at what rate is the angle of elevation changing when the plane is at a horizontal distance of 6 mi from the observer?

 **In Your Words**

33. The answer to Exercise 31 is 7.1086 rad/s. Is it practical to place a television camera at this location? If not, what are some alternative locations? Justify your answers by determining the rate at which the camera must turn to follow the car when it covers $\theta = 15°$ at each of these locations.

34. The derivatives for the inverse trigonometric functions seem naturally to group themselves into three groups of two. Which derivatives would you group? Why did you decide to put these in one group? How are you going to remember the derivatives of the functions in the other groups? Are there any other ways you could group these functions that might help you remember their derivatives?

≡ 27.4
APPLICATIONS

In this section, we have the opportunity to apply the derivatives of the trigonometric and inverse trigonometric functions. The manner in which these functions are applied is essentially the same as for algebraic functions. We will use derivatives to find slopes, tangents and normal lines, maxima and minima, related rates, and differential problems.

Application

EXAMPLE 27.18

If $f(x) = 2\sqrt{3}\sin x + \cos 2x$, find the extrema and sketch the graph of one complete cycle of f.

Solution Since the period of $\sin x$ is 2π and the period of $\cos 2x$ is π, the period of f is 2π. We will graph one period from $x = 0$ to $x = 2\pi$. Differentiating, we get $f'(x) = 2\sqrt{3}\cos x - 2\sin 2x$.

In Section 20.2, we found that $\sin 2x = 2\sin x \cos x$. Making this substitution, we get

$$f'(x) = 2\sqrt{3}\cos x - 4\sin x \cos x$$
$$= \cos x (2\sqrt{3} - 4\sin x)$$

The critical values occur when $f'(x) = 0$. This happens when $\cos x = 0$ or when $2\sqrt{3} - 4\sin x = 0$. Now, $\cos x = 0$ at $\frac{\pi}{2}$ and $\frac{3\pi}{2}$ on $[0, 2\pi]$, and $2\sqrt{3} - 4\sin x = 0$ if $\sin x = \dfrac{\sqrt{3}}{2}$ or when $x = \frac{\pi}{3}$ or $\frac{2\pi}{3}$.

The second derivative is $f''(x) = -2\sqrt{3}\sin x - 4\cos 2x$. If we evaluate the second derivative at each of the critical values, we find that $f''\left(\frac{\pi}{2}\right) > 0$ and $f''\left(\frac{3\pi}{2}\right) > 0$ while $f''\left(\frac{\pi}{3}\right) < 0$ and $f''\left(\frac{2\pi}{3}\right) < 0$. Thus, we have minima at $\frac{\pi}{2}$ and $\frac{3\pi}{2}$, and maxima at $\frac{\pi}{3}$ and $\frac{2\pi}{3}$.

EXAMPLE 27.18 (Cont.)

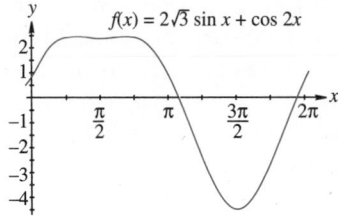

FIGURE 27.1

In order to determine inflection points, we determine when the second derivative is 0. If we substitute $\cos 2x = 1 - 2\sin^2 x$, we get

$$-2\sqrt{3}\sin x - 4\cos 2x = -2\sqrt{3}\sin x - 4 + 8\sin^2 x = 0$$

Using the quadratic formula, we obtain

$$\sin x = \frac{2\sqrt{3} \pm \sqrt{12 + 128}}{16} = \frac{2\sqrt{3} \pm \sqrt{140}}{16} = \frac{\sqrt{3} \pm \sqrt{35}}{8}$$

Solving this for x over $[0, 2\pi]$, we obtain $x \approx 1.28$, 1.87, 3.69, and 5.73 rad as possible critical values. (In degrees, these values for x are approximately $72.9°$, $107.1°$, $211.5°$, and $328.5°$.)

Summarizing we get the following results.

Minima: $\left(\frac{\pi}{2}, 2.46\right)$ and $\left(\frac{3\pi}{2}, -4.46\right)$

Maxima: $\left(\frac{\pi}{3}, 2.5\right)$ and $\left(\frac{2\pi}{3}, 2.5\right)$

Inflection points: $(1.27, 2.48)$, $(1.87, 2.48)$, $(3.69, -1.35)$, and $(5.73, -1.37)$ or, in degrees, $(73°, 2.48)$, $(107°, 2.48)$, $(211.5°, -1.35)$, and $(328.5°, -1.37)$

The graph of this function over $[0, 2\pi]$ is shown in Figure 27.1. ■

Application

EXAMPLE 27.19

Find the equation of the tangent and normal lines to $f(x) = x^2 + \sin x$ at $(1, 1.84)$.

Solution   The slopes of the tangent and normal lines are found by evaluating the derivative of $f'(x)$ at $x = 1$. (Even though it is not stated, it should be understood that all trigonometric functions will be evaluated in radians.) Taking the derivative of f, we obtain

$$f'(x) = 2x + \cos x$$

and when $x = 1$ rad, $f'(x) \approx 2.54$.

Using the point-slope form for the equation of a line, we have $y - y_1 = f'(x)(x - x_1)$. We know that $(x_1, y_1) = (1, 1.84)$ and so $y - 1.84 = 2.54(x - 1)$ or $y = 2.54x - 0.70$.

A normal line is perpendicular to a tangent to $f(x)$; its slope is $\dfrac{-1}{f'(x)}$.

Again using the point-slope form for the equation of a line, we have

$$y - 1.84 = \frac{-1}{2.54}(x - 1)$$
$$= -0.39(x - 1)$$
$$\text{or} \quad y = -0.39x + 2.23$$

We have shown that the equation of the tangent to $f(x) = x^2 + \sin x$ at $(1, 1.84)$ is $y = 2.54x - 0.70$ and the equation of the normal line at this same point is $y = -0.39x + 2.23$. ■

Application

EXAMPLE 27.20

A balloon is 500 m above the ground. It is moving horizontally at the rate of 10 m/s away from an observer. If the distance from the observer to a point directly below the balloon is 100 m, how quickly is the angle of elevation changing?

Solution   A sketch of the problem is shown in Figure 27.2. We have $\tan\theta = \dfrac{500}{x}$, where x is the distance from the observer to the point directly under the balloon and θ is the angle of elevation. Because we are asking "how quickly," we will differentiate θ with respect to time t. That means we want to determine $\dfrac{d\theta}{dt}$. Taking the derivative, we get $\sec^2\theta\,\dfrac{d\theta}{dt} = \dfrac{-500}{x^2}\cdot\dfrac{dx}{dt}$ and so, solving for $\dfrac{d\theta}{dt}$, we obtain

$$\frac{d\theta}{dt} = \frac{-500}{x^2\sec^2\theta}\cdot\frac{dx}{dt}$$
$$= \frac{-500}{x^2}\cos^2\theta\cdot\frac{dx}{dt}$$

We let $\tan\theta = \dfrac{500}{x}$, and so $\theta = \arctan\dfrac{500}{x}$. When $x = 100$ m, $\theta = \arctan\dfrac{500}{x} = \arctan 5 \approx 1.37\ (78.7°)$, so $\cos^2\theta = 0.038$. Now, since we were given $\dfrac{dx}{dt} = 10$ m/s, we have

$$\frac{d\theta}{dt} = \frac{-500}{(100)^2}(0.038)(10) = -0.019$$

The angle of elevation is decreasing at a rate of 0.019 rad/s $\approx$ 1.09 degrees/s.

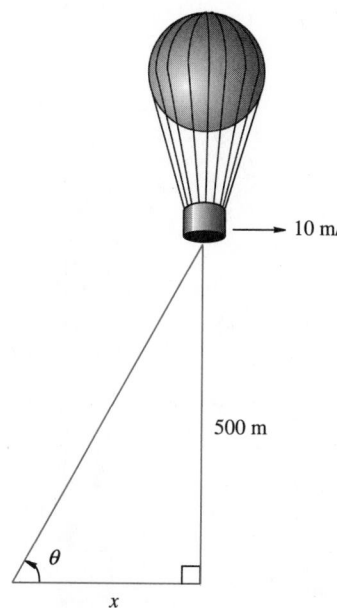

500 m

10 m/s

θ

x

FIGURE 27.2

Application

EXAMPLE 27.21

The screen in a movie theater, like the one in Figure 27.3a, is 30 ft high and its bottom edge is 10 ft above the level of a seated person's eyes. How far from a point directly below the screen should a person sit in order for the angle formed by the top of the screen, the person's eyes, and the bottom of the screen to be a maximum?

Solution   A sketch of the problem is shown in Figure 27.3b. The desired angle that we want maximized is θ. We have let x represent the distance from the viewer to the point on the floor below the screen. From the figure, we can see that $\tan(\theta + \phi) = \dfrac{40}{x}$ and $\tan\phi = \dfrac{10}{x}$. This means that

$$\tan\theta = \tan[(\theta + \phi) - \phi]$$
$$= \frac{\tan(\theta + \phi) - \tan\phi}{1 + \tan(\theta + \phi)\tan\phi}$$

EXAMPLE 27.21 (Cont.)

$$= \frac{\dfrac{40}{x} - \dfrac{10}{x}}{1 + \dfrac{40}{x} \cdot \dfrac{10}{x}} = \frac{\dfrac{30}{x}}{1 + \dfrac{400}{x^2}}$$

$$= \frac{30x}{x^2 + 400}$$

FIGURE 27.3a

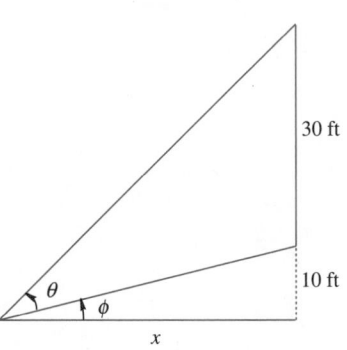

30 ft

10 ft

x

FIGURE 27.3b

Thus, $\theta = \arctan\left(\dfrac{30x}{x^2 + 400}\right)$. We want θ to be a maximum, so we take the derivative of arctan:

$$\frac{d\theta}{dx} = \frac{1}{1 + \left(\dfrac{30x}{x^2 + 400}\right)^2}\left[\frac{d}{dx}\left(\frac{30x}{x^2 + 400}\right)\right]$$

$$= \left[\frac{(x^2 + 400)^2}{x^4 + 800x^2 + 900x^2 + 160{,}000}\right]\left[\frac{(x^2 + 400)30 - 30x(2x)}{(x^2 + 400)^2}\right]$$

$$= \frac{12{,}000 - 30x^2}{x^4 + 1700x^2 + 160{,}000}$$

The derivative is 0 when $12{,}000 - 30x^2 = 30(400 - x^2) = 0$. The critical value is $x = 20$ and, by the first derivative test, it is a maximum. Thus, the viewer should sit 20 ft from the front of the screen.

Exercise Set 27.4

1. Show that $\sin x$ is increasing on the interval $\left[-\frac{\pi}{2}, \frac{\pi}{2}\right]$.

2. Show that $\tan x$ is increasing for all values where it is defined.

3. Show that $\arccos x$ is decreasing for all $|x| \leq 1$.

4. Show that $\arctan x$ is increasing for all values of x.

In Exercises 5–14 find all extrema and points of inflection and sketch the graph for $0 \leq x \leq 2\pi$.

5. $f(x) = 2\sin x + \cos 2x$

6. $g(x) = 2\cos x + \sin 2x$

7. $h(x) = \cos x - \sin x$

8. $j(x) = \cos^2 x - \sin x$

9. $k(x) = \cos^2 x + \sin x$

10. $m(x) = \sec x - \tan x$

11. Find an equation of the line tangent to $y = \sin x$ at $x = \frac{\pi}{6}$.

12. Find an equation of the line normal to $y = \sin x$ at $x = \frac{\pi}{6}$.

13. Find an equation of the line tangent to $y = \tan^3 x$ at $x = \frac{\pi}{4}$.

14. Find an equation of the line normal to $y = \tan^3 x$ at $x = \frac{\pi}{4}$.

Solve Exercises 15–26.

15. *Navigation*   A searchlight is following an airplane flying at an altitude of 6,000 ft in a straight line over the light. The plane's velocity is 400 mph. At what rate is the searchlight turning when the distance between the light and the plane is 10,000 ft?

16. *Navigation*   A ship is moving in a straight line past a lighthouse. At its closest it is 300 m from the lighthouse. The lighthouse keeper is watching the ship through a telescope. How fast does the telescope have to turn if the ship's velocity is 10 km/h and it is 400 m from the ship to the lighthouse?

17. *Navigation*   The lighthouse in Exercise 16 is located 500 m from the base of a cliff that runs along a straight shoreline. The searchlight turns at the rate of 1 rpm. How fast is the light beam moving along the cliff when it is at a point on the cliff that is 500 m from the point that is closest to the lighthouse?

18. *Electronics*   The power P in a certain ac circuit is to remain constant. As the impedance phase angle θ is varied, the **apparent power**, P_{app}, is defined as $P_{app} = P \sec \theta$. If $P = 15\,W$, find the time-rate change of P_{app}, if θ is changing at the rate of 0.6 rad/min when $\theta = \frac{\pi}{3}$.

19. *Navigation*   A radar antenna is rotating at a rate of 30 rpm. If the antenna is aboard a ship that is 8 mi from a straight shore line, how fast does the radar

beam travel along the shore when the beam makes an angle of 40° with the shore?

20. *Physics*   A weight hangs on a spring that is 2 m long when it is at rest with the weight attached. The weight is pulled down 0.75 m and then released. The weight oscillates up and down. The length x of the spring at any time after t s have elapsed is given by $x = 2 + 0.75 \cos 2\pi t$.
    (a) What is the length of the spring when $t = 0, \frac{1}{2}$, and 1?
    (b) What is the velocity function of the weight?
    (c) What is the weight's velocity when $t = 0, \frac{1}{2}$, and 1?
    (d) What is the acceleration function of the weight?
    (e) When is the velocity a maximum?

21. *Environmental science*   The amount of fluid flowing across a weir (or dam) with a V-shaped notch is approximately

$$\bar{Q} = 1.33h^{5/2} \tan \frac{\theta}{2}$$

where $\bar{Q}$ is the volume of flow given in m³/s, h is the height in meters, and θ is the angle of the notch, as shown in Figure 27.4. Assuming that h is kept constant at 2 m while θ is increasing, how fast is $\bar{Q}$ changing when (a) $\theta = 30$°? (b) $\theta = 40$°? (c) At what angle θ is $\bar{Q}$ a maximum, if h is constant?

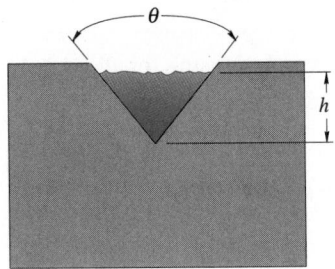

FIGURE 27.4

22. *Civil engineering* A guy wire is attached to a ring 12 m above the ground on an antenna pole. If the length s of the wire is changed, the angle between the pole and the wire is also changed. If the wire is shortened at the rate of 0.4 m/min, what is the rate of change of this angle when s is 20 m?

23. *Electricity* The current in a certain electric circuit is given by $I = \sin^2 3t$ A. At what rate is I changing when $t = 1.5$ s?

24. *Interior design* A 4-ft high picture is hung with the bottom 2.5 ft above eye level. How far away should a person stand so that the angle formed by the top of the picture, the eye, and the bottom of the picture is a maximum?

25. *Physics* A 2-kg mass is attached to a spring and is pulled out and released. At any time t, the length of the spring is $x(t) = \frac{1}{4}\cos 2t$ m. What is the maximum speed of this mass?

26. *Sheet metal technology* A gutter is to be made from a sheet of metal 30 cm wide by turning up strips of width 10 cm along each side that make equal angles θ, as shown in Figure 27.5. What angle θ will make the cross-sectional area a maximum?

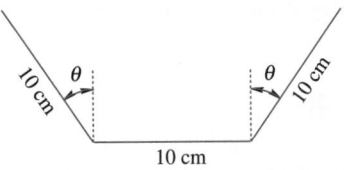

FIGURE 27.5

Use Newton's method to approximate the real roots to four decimal places in Exercises 27–30.

27. $f(x) = \cos x + x - 2$

28. $f(x) = \sin x - x^2$

29. $g(x) = \sin x + \dfrac{x^3}{4} - 2x$

30. $h(x) = \cos x + x^2 - 3$

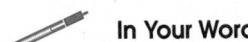

In Your Words

31. Write an application in your technology area of interest that requires you to use derivatives of a trigonometric function. Give your problem to a classmate and see if he or she understands and can solve your problem. Rewrite the problem as necessary to remove any difficulties encountered by your classmate.

32. As in Exercise 31, write a word problem in your technology area that requires you to use an inverse trigonometric function. This problem should use a different trigonometric function than the one in Exercise 31.

≡ 27.5
DERIVATIVES OF LOGARITHMIC FUNCTIONS

In Sections 27.5 and 27.6, we will study the derivatives of two other transcendental functions—the logarithmic and exponential functions.

When we studied Section 13.2, we had not yet studied limits. We did however use the idea of a limit and developed

$$\lim_{n \to \infty} \left(1 + \frac{1}{n}\right)^n = e$$

or, similarly, letting $x = \dfrac{1}{n}$

$$\lim_{x \to 0} (1 + x)^{1/x} = e$$

Any symbol can be used in place of x. For example, if we used $\dfrac{h}{x}$, then we would have

$$\lim_{\frac{h}{x} \to 0} \left(1 + \frac{h}{x}\right)^{x/h} = e$$

Now let's look at the derivative of $f(x) = \log_a x$. Using the four-step method produces the following result.

Step 1: $f(x + h) = \log_a(x + h)$

Step 2: $f(x + h) - f(x) = \log_a(x + h) - \log_a x$

$$= \log_a\left(\frac{x + h}{x}\right)$$

$$= \log_a\left(1 + \frac{h}{x}\right)$$

Step 3: $\dfrac{f(x + h) - f(x)}{h} = \dfrac{\log_a(1 + h/x)}{h}$

$$= \frac{1}{h} \log_a\left(1 + \frac{h}{x}\right)$$

This last expression does not appear to be of much help since it cannot be simplified by our usual methods. But, we can rewrite $\dfrac{1}{h}$ as $\dfrac{1}{x} \cdot \dfrac{x}{h}$, and this gives us the help we need.

$$\frac{1}{h} \log_a\left(1 + \frac{h}{x}\right) = \frac{1}{x} \cdot \frac{x}{h} \log_a\left(1 + \frac{h}{x}\right) = \frac{1}{x} \log_a\left(1 + \frac{h}{x}\right)^{x/h}$$

We are now ready for the last step.

Step 4: $\displaystyle\lim_{h \to 0} \frac{f(x + h) - f(x)}{h} = \lim_{h \to 0} \frac{1}{x} \log_a\left(1 + \frac{h}{x}\right)^{x/h}$

$$= \frac{1}{x} \lim_{h \to 0} \log_a\left(1 + \frac{h}{x}\right)^{x/h}$$

Since x is a fixed number, if $h \to 0$, then $\dfrac{h}{x} \to 0$ and we have

$$\frac{1}{x} \lim_{h \to 0} \log_a \left(1 + \frac{h}{x} \right)^{x/h} = \frac{1}{x} \lim_{\frac{h}{x} \to 0} \log_a \left(1 + \frac{h}{x} \right)^{x/h}$$

$$= \frac{1}{x} \log_a e$$

From the change of base formula in Section 13.3, we have $\log_a x = \dfrac{\ln x}{\ln a}$, so $\log_a e = \dfrac{\ln e}{\ln a} = \dfrac{1}{\ln a}$. Thus we have the following results for the derivative of the logarithm for the general base a.

Derivative of $y = \log_a x$

$$\frac{d}{dx} \log_a x = \frac{1}{x} \log_a e$$

$$= \frac{1}{x} \cdot \frac{1}{\ln a}$$

Now $\ln x = \log_e x$ and, since $\log_e e = 1$, we have

$$\frac{d}{dx} \ln x = \frac{1}{x} \log_e e = \frac{1}{x}$$

Using the chain rule, we have the following two derivatives.

Derivative of Log Function

$$\frac{d}{du} \log_a u = \frac{1}{u} \log_a e \frac{du}{dx}$$

$$= \frac{1}{u} \cdot \frac{1}{\ln a} \frac{du}{dx}$$

and

$$\frac{d}{du} \ln u = \frac{1}{u} \frac{du}{dx}$$

EXAMPLE 27.22

Differentiate $y = \log_3 x^2$.

Solution   We have $a = 3$, and if we use properties of logarithms, then we can rewrite this as $y = \log_3 x^2 = 2 \log_3 x$. Differentiating, we obtain $y' = 2 \left(\dfrac{1}{x \ln 3} \right) = \dfrac{2}{x \ln 3}$.

EXAMPLE 27.23

Differentiate $y = \log_4(x^2 + 1)$.

Solution   Here $a = 4$ and $u = x^2 + 1$, so

$$y = \frac{1}{u}\left(\frac{1}{\ln 4}\right)\frac{d}{dx}(x^2 + 1)$$

$$= \frac{1}{x^2 + 1}\left(\frac{1}{\ln 4}\right)(2x)$$

$$= \frac{2x}{(x^2 + 1)\ln 4}$$

EXAMPLE 27.24

Differentiate $y = \ln\sqrt{x}$.

Solution   We will find this derivative by two different methods. The second method uses our definition for the derivative of $\ln u$.

Method 1: In the first method we will use some of the properties of logarithms.

$$y = \ln\sqrt{x} = \ln(x^{1/2}) = \frac{1}{2}\ln x$$

Differentiating, we get

$$y' = \frac{1}{2}\cdot\frac{1}{x} = \frac{1}{2x}$$

Method 2: If we let $u = \sqrt{x}$, then the given equation is $y = \ln u$, and we have

$$y' = \frac{1}{u}\frac{du}{dx}$$

$$= \frac{1}{\sqrt{x}}\cdot\frac{1}{2\sqrt{x}}$$

$$= \frac{1}{2x}$$

The second method was not much simpler than the first, but it gave us a method that we can use on more difficult problems.

EXAMPLE 27.25

Differentiate $y = \ln \sqrt{\dfrac{\sin^2 x}{x^3}}$.

Solution   We first simplify this by using the properties of logarithms.

$$y = \ln \sqrt{\frac{\sin^2 x}{x^3}}$$

$$= \ln \left(\frac{\sin^2 x}{x^3}\right)^{1/2} = \frac{1}{2} \ln \left(\frac{\sin^2 x}{x^3}\right)$$

$$= \frac{1}{2}(\ln \sin^2 x - \ln x^3)$$

$$= \frac{1}{2}(2 \ln \sin x - 3 \ln x)$$

$$y = \ln \sin x - \frac{3}{2} \ln x$$

Now, we will differentiate the function $y = \ln \sin x - \frac{3}{2} \ln x$.

$$y' = \frac{1}{\sin x} \frac{d}{dx}(\sin x) - \frac{3}{2}\left(\frac{1}{x}\right)$$

$$= \frac{\cos x}{\sin x} - \frac{3}{2x}$$

$$= \cot x - \frac{3}{2x}$$

Exercise Set 27.5

In Exercises 1–20, differentiate the given function and simplify the resulting derivative.

1. $y = \log 5x$
2. $y = \log_5 x^3$
3. $y = \ln(x^2 + 4x)$
4. $y = \ln \sqrt{x^2 + 4x}$
5. $y = \sqrt{\ln(x^2 + 4x)}$
6. $y = \sqrt{\ln x^2 + 4x}$
7. $y = \ln \dfrac{1}{x}$

8. $y = \ln \cos x$
9. $y = \ln \tan x$
10. $y = \log_4 \sqrt{x}$
11. $y = \dfrac{\ln x}{x}$
12. $y = \ln \left(\dfrac{x}{1+x}\right)$

13. $y = \sqrt{\ln \left(\dfrac{1+x}{1-x}\right)}$
14. $y = \frac{1}{2} \ln \left(\dfrac{1+x^2}{1-x^2}\right)$
15. $y = \ln \dfrac{4x^3}{\sqrt{x^2 + 4}}$

16. $y = \dfrac{\ln x}{x^3}$
17. $y = (\ln x)^2$
18. $y = \sin(\ln x)$
19. $y = \ln(\ln x)$
20. $y = \dfrac{x}{\ln x}$

In Exercises 21–30 find the second derivative of the given function.

21. $y = \ln x$
22. $y = x \ln x$
23. $y = \dfrac{1}{x} \ln x$

24. $y = \sqrt{\ln x}$
25. $y = \ln \sqrt{\dfrac{\sin^2 x}{x^3}}$

26. $y = \ln \sqrt[3]{\dfrac{\sin^3 x}{x}}$
27. $y = \sqrt{\ln x^2}$

28. $y = \ln(x + \cos x)$
29. $y = x^2 \ln(\sin 2x)$
30. $y = (\cos x)(\ln x)$

Solve Exercises 31–34.

31. *Agriculture* Mediterranean fruit flies, better known as medflies, have been discovered in a citrus orchard. Based on their experience, the Department of Agriculture believes that the population of medflies t hours after the orchard has been sprayed with a pesticide is about $N(t) = 72 - 5t \ln(0.045t) - 0.5t$, where $0 \le t \le 36$.
(a) What will be the maximum number of medflies in the orchard?
(b) Will the medfly infestation be eradicated within the 36 h period? If so, how long did it take to eliminate all the medflies? If not, how many flies remain at the end of the 36 h period?

32. *Computer technology* The speed of the signal, s, in an ethernet line is given by $s = -kt^2 \ln t$, where k is a constant and the variable, t, is a function of the thickness of the line.
(a) Determine $\dfrac{ds}{dt}$.

(b) What is the maximum value of s in terms of k?

33. *Environmental science* The level of ozone in a major city is a function of the time of the day or night. It has been estimated that t h after midnight in a major city the ozone level P in parts per million (ppm) in the air is about $P(t) = 0.014 - 0.008t \ln(0.025t)$.
(a) At what time is the maximum ozone level?
(b) What is the maximum level?

34. *Electronics* A cylindrical capacitor consists of a pair of oppositely charged coaxial cylinders. Its capacitance is given by the equation $C = \dfrac{kL}{\ln(r_b/r_a)}$ where r_b is the radius of the outer cylinder and r_a is the radius of the inner cylinder. Find $\dfrac{dC}{dr_a}$ when $r_b = 1.5\,\text{cm}$, $r_a = 1.0\,\text{cm}$, $L = 6.0\,\text{m}$, and $k = 55.6\,\text{pF/m}$. (Note: $1\,\text{pF} = 10^{-12}\,\text{F}$.)

In Your Words

35. Describe how to differential $\log_b x$.

36. If you use the change of base formula $\log_b x = \dfrac{\ln x}{\ln b}$,

then you only need one rule for differentiating a logarithmic function. Explain how this is done.

≡ 27.6
DERIVATIVES OF EXPONENTIAL FUNCTIONS

The logarithmic and exponential functions are inverses of each other. As a result, we can use our knowledge of differentiating logarithmic functions to show how to find the derivatives of exponential functions.

Suppose we have the exponential function $y = f(x) = a^u$. If we take the logarithm of both sides we have

$$\ln y = \ln a^u = u \ln a$$

Using implicit differentiation, we have

$$\frac{d}{dx} \ln y = \frac{du}{dx} \ln a$$

$$\frac{1}{y} \frac{dy}{dx} = \frac{du}{dx} \ln a$$

Solving for $\dfrac{dy}{dx}$, we get

$$\frac{dy}{dx} = y \frac{du}{dx} \ln a$$

But $y = a^u$, and so

$$\frac{dy}{dx} = a^u \frac{du}{dx} \ln a,$$

or

Derivative of $y = a^u$	$\dfrac{d}{dx} a^u = a^u \dfrac{du}{dx} \ln a$

EXAMPLE 27.26

Differentiate $y = 3^x$.

Solution   Here $a = 3$. Notice that $u = x$ and so $\dfrac{du}{dx} = \dfrac{dx}{dx} = 1$.
Then $y' = 3^x \cdot 1 \cdot \ln 3 = 3^x \ln 3$.

EXAMPLE 27.27

Differentiate $y = 5^{x^2+1}$.

Solution   Here $a = 5$ and $u = x^2 + 1$, so $\dfrac{du}{dx} = 2x$.

$$y' = 5^{x^2+1}(2x) \ln 5$$

Now lets look at the special case when $y = e^u$. According to the rule for differentiating $y = a^u$, we obtain

$$y = e^u \frac{du}{dx} \ln e$$

But, $\ln e = 1$ and so

Derivative of $y = e^u$	$\dfrac{d}{dx} e^u = e^u \dfrac{du}{dx}$

EXAMPLE 27.28

Differentiate $y = e^x$.

Solution   Here, $u = x$, so $\dfrac{du}{dx} = \dfrac{dx}{dx} = 1$

$$y' = e^x \frac{du}{dx} = e^x$$

EXAMPLE 27.29

Differentiate $y = e^{\sqrt{x}}$.

Solution We have $u = \sqrt{x}$ and so $\dfrac{du}{dx} = \dfrac{1}{2\sqrt{x}}$. Differentiating, we get

$$y' = e^{\sqrt{x}}\left(\frac{d}{dx}\sqrt{x}\right)$$
$$= e^{\sqrt{x}}\frac{1}{2\sqrt{x}} = \frac{e^{\sqrt{x}}}{2\sqrt{x}}$$

EXAMPLE 27.30

Differentiate $y = xe^x$.

Solution We will have to use the product rule.

$$y' = x\frac{d}{dx}e^x + e^x\frac{d}{dx}x$$
$$= xe^x + e^x$$
$$= e^x(x+1)$$

EXAMPLE 27.31

Differentiate $y = x^3 3^x$.

Solution Again, we use the product rule.

$$y' = x^3\frac{d}{dx}3^x + 3^x\frac{d}{dx}x^3$$
$$= x^3 \cdot 3^x \ln 3 + 3^x 3x^2$$
$$= x^2 \cdot 3^x(x\ln 3 + 3)$$

EXAMPLE 27.32

Differentiate $y = e^{\sin x}$.

Solution $y' = e^{\sin x}\dfrac{d}{dx}(\sin x)$

$$= \cos x\, e^{\sin x}$$

EXAMPLE 27.33

Differentiate $y = \ln \tan e^{5x}$.

Solution Using the derivative of the logarithm, we get

$$
\begin{aligned}
y' &= \frac{1}{\tan e^{5x}} \frac{d}{dx}(\tan e^{5x}) \\
&= \frac{1}{\tan e^{5x}} \sec^2 e^{5x} \frac{d}{dx} e^{5x} \\
&= \frac{\sec^2 e^{5x}}{\tan e^{5x}} e^{5x} \frac{d}{dx}(5x) \\
&= \frac{5e^{5x} \sec^2 e^{5x}}{\tan e^{5x}}
\end{aligned}
$$

Logarithmic Differentiation

In developing the rule for the derivative of the exponential function, we took the logarithm of both sides. This method, known as **logarithmic differentiation**, can often simplify derivatives when both the base and the power are variables, as in the next example.

EXAMPLE 27.34

Differentiate $y = x^{\sin x}$.

Solution Taking the logarithm of both sides and simplifying gives

$$
\ln y = \sin x \ln x
$$

Using implicit differentiation, we get

$$
\begin{aligned}
\frac{1}{y} y' &= \frac{1}{x} \sin x + (\ln x)(\cos x) \\
y' &= y \left[\frac{\sin x}{x} + (\ln x)(\cos x) \right] \\
&= x^{\sin x} \left[\frac{\sin x}{x} + (\ln x)(\cos x) \right]
\end{aligned}
$$

Hyperbolic Functions

Some combinations of functions happen often enough in applications that they are given special names. The **hyperbolic functions** form such a group. Each hyperbolic function is abbreviated by placing an h at the end of the symbol representing its corresponding circular trigonometric function. Thus hyperbolic sine is abbreviated sinh, hyperbolic cosine is abbreviated cosh, and so on. The hyperbolic functions are defined as follows.

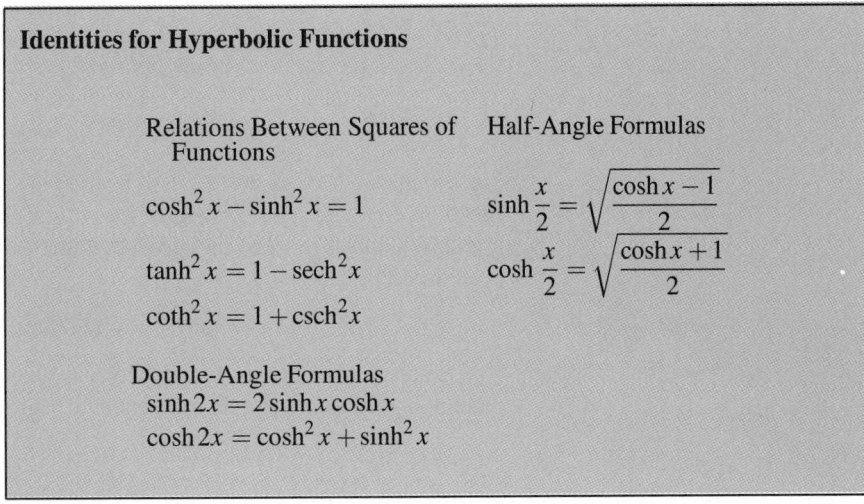

Definitions of Hyperbolic Functions

Hyperbolic sine	$\sinh x =$	$\dfrac{e^x - e^{-x}}{2}$
Hyperbolic cosine	$\cosh x =$	$\dfrac{e^x + e^{-x}}{2}$
Hyperbolic tangent	$\tanh x =$	$\dfrac{\sinh x}{\cosh x} = \dfrac{e^x - e^{-x}}{e^x + e^{-x}}$
Hyperbolic cotangent	$\coth x =$	$\dfrac{\cosh x}{\sinh x} = \dfrac{e^x + e^{-x}}{e^x - e^{-x}}$
Hyperbolic secant	$\operatorname{sech} x =$	$\dfrac{1}{\cosh x} = \dfrac{2}{e^x + e^{-x}}$
Hyperbolic cosecant	$\operatorname{csch} x =$	$\dfrac{1}{\sinh x} = \dfrac{2}{e^x - e^{-x}}$

Just as with the trigonometric functions, hyperbolic functions satisfy their own set of identities. As you can see in the following box, except for differences in sign, these are similar to the identities for the trigonometric functions. While we will not develop these, some are stated here because they are especially useful to help solve problems involving hyperbolic functions.

Identities for Hyperbolic Functions

Relations Between Squares of   Half-Angle Formulas
Functions

$$\cosh^2 x - \sinh^2 x = 1 \qquad \sinh \frac{x}{2} = \sqrt{\frac{\cosh x - 1}{2}}$$

$$\tanh^2 x = 1 - \operatorname{sech}^2 x \qquad \cosh \frac{x}{2} = \sqrt{\frac{\cosh x + 1}{2}}$$

$$\coth^2 x = 1 + \operatorname{csch}^2 x$$

Double-Angle Formulas
$$\sinh 2x = 2 \sinh x \cosh x$$
$$\cosh 2x = \cosh^2 x + \sinh^2 x$$

We are now ready to examine derivatives of hyperbolic functions. We will develop the derivative of the hyperbolic sine function, and then state the derivatives and integrals of the remaining functions.

EXAMPLE 27.35

Differentiate $y = \sinh x$.

Solution $y = \sinh x = \dfrac{1}{2}(e^x - e^{-x})$

$$y' = \frac{1}{2}\left(e^x - e^{-x}\frac{d}{dx}(-x)\right)$$

$$= \frac{1}{2}(e^x + e^{-x})$$

$$= \cosh x$$

Derivatives of Hyperbolic Functions

$$\frac{d}{dx}(\sinh u) = \cosh u \frac{du}{dx} \qquad \frac{d}{dx}(\coth u) = -\operatorname{csch}^2 u \frac{du}{dx}$$

$$\frac{d}{dx}(\cosh u) = \sinh u \frac{du}{dx} \qquad \frac{d}{dx}(\operatorname{sech} u) = -\operatorname{sech} u \tanh u \frac{du}{dx}$$

$$\frac{d}{dx}(\tanh u) = \operatorname{sech}^2 u \frac{du}{dx} \qquad \frac{d}{dx}(\operatorname{csch} u) = -\operatorname{csch} u \coth u \frac{du}{dx}$$

It can be shown that any cable that hangs freely between two supports hangs in the shape of a hyperbolic cosine curve. Examples of such curves include telephone lines and electric power cables strung from one pole or tower to another. In Section 28.3, we will examine applications involving hyperbolic functions.

Inverse Hyperbolic Functions

Since $\sinh x$ and $\cosh x$ are defined in terms of e^x, we might expect that their inverse functions are expressed in terms of the natural logarithm function. As the following box shows, this is indeed the case. As with the inverse trigonometric functions, you need to recall that

$$y = \sinh^{-1} x, \text{ if and only if } \sinh y = x$$

Since each of the inverse hyperbolic functions can be defined in terms of its natural logarithm, we should have little difficulty finding their derivatives.

Formulas for Inverse Hyperbolic Functions

$$\sinh^{-1} x = \ln\left(x + \sqrt{x^2 + 1}\right), \qquad -\infty < x < \infty$$

$$\cosh^{-1} x = \ln\left(x + \sqrt{x^2 - 1}\right), \qquad x \geq 1$$

$$\tanh^{-1} x = \frac{1}{2}\ln\frac{1+x}{1-x}, \qquad |x| < 1$$

$$\coth^{-1} x = \frac{1}{2}\ln\frac{1+x}{1-x}, \qquad |x| > 1$$

$$\operatorname{sech}^{-1} x = \ln\left(\frac{1 + \sqrt{1-x^2}}{x}\right), \qquad 0 < x \leq 1$$

$$\operatorname{csch}^{-1} x = \ln\left(\frac{1}{x} + \frac{\sqrt{1-x^2}}{|x|}\right), \qquad x \neq 0$$

EXAMPLE 27.36

If $y = \sinh^{-1} x$, determine y'.

Solution Rewriting $y = \sinh^{-1} x$ as $y = \ln(x + \sqrt{x^2 + 1})$ and using the derivative of the natural logarithm produces

$$y' = \left(\frac{1}{x + \sqrt{x^2+1}}\right)\frac{d}{dx}\left(x + \sqrt{x^2+1}\right)$$

$$= \frac{1}{x + \sqrt{x^2+1}}\left(1 + \frac{x}{\sqrt{x^2+1}}\right)$$

$$= \frac{\sqrt{x^2+1} + x}{\left(x + \sqrt{x^2+1}\right)\sqrt{x^2+1}}$$

$$= \frac{1}{\sqrt{x^2+1}}$$

Using similar procedures, we can develop all of the following derivatives.

Derivatives of Inverse Hyperbolic Functions

$$\frac{d}{dx}\left(\sinh^{-1}u\right) = \frac{1}{\sqrt{u^2+1}}\frac{du}{dx}$$

$$\frac{d}{dx}\left(\cosh^{-1}u\right) = \frac{1}{\sqrt{u^2-1}}\frac{du}{dx},\ u \geq 1$$

$$\frac{d}{dx}\left(\tanh^{-1}u\right) = \frac{1}{1-u^2}\frac{du}{dx},\ |u| < 1$$

$$\frac{d}{dx}\left(\coth^{-1}u\right) = \frac{1}{1-u^2}\frac{du}{dx},\ |u| > 1$$

$$\frac{d}{dx}\left(\text{sech}^{-1}u\right) = \frac{-1}{u\sqrt{1-u^2}}\frac{du}{dx},\ 0 < u < 1$$

$$\frac{d}{dx}\left(\text{csch}^{-1}u\right) = \frac{-1}{|u|\sqrt{1+u^2}}\frac{du}{dx},\ u \neq 1$$

EXAMPLE 27.37

If $f(x) = \sinh^{-1}\sqrt{x^4-9}$, determine $f'(x)$.

Solution Using the formula for the derivative of $\sinh^{-1}$ and the chain rule, we obtain

$$f'(x) = \frac{1}{\sqrt{\left(\sqrt{x^4-9}\right)^2 + 1}}\left[\frac{1}{2}(x^4-9)^{-1/2}(4x^3)\right]$$

$$= \frac{1}{\sqrt{x^4-9+1}}\left[\frac{2x^3}{\sqrt{x^4-9}}\right]$$

$$= \frac{2x^3}{\sqrt{x^4-8}\sqrt{x^4-9}}$$

In Section 28.5, we will examine applications involving the inverse hyperbolic functions.

Exercise Set 27.6

In Exercises 1–26, differentiate and simplify the given function.

1. $y = 4^x$

2. $y = e^{x^2}$

3. $y = 5^{\sqrt{x}}$

4. $y = 6^{x^2+x}$

5. $y = 2^{\sin x}$

6. $y = e^{5x+3}$

7. $y = e^{x^2+x}$

8. $y = e^{2x}$

9. $y = 4^{x^4}$

10. $y = x + e^x$

11. $y = \dfrac{e^x}{x^2}$

12. $y = x^3 e^x$

13. $y = \dfrac{1+e^x}{x^2}$

14. $y = \dfrac{1+e^x}{e^x}$

15. $y = e^{\tan x}$

16. $y = e^{\cos 3x}$

17. $y = \sin e^{x^2}$

18. $y = \cos 2^{x^2}$

19. $y = 3^x(x^3-1)$

20. $y = e^{x^2 - \ln x}$

21. $y = x^{\cos x}$

22. $y = x^{x^3}$

23. $y = (\sin x)^x$

24. $y = e^{\arctan x}$

25. $y = \ln \sin e^{3x}$

26. $y = e^{\sin x} \ln \sqrt{x}$

Solve Exercises 27–32.

27. Find y' implicitly, if $y = e^y + y + x$.

28. Find y' implicitly, if $ye^x - xe^y = 1$

29. Show that $\dfrac{d}{dx}(\cosh x) = \sinh x$.

30. Verify that $\cosh^2 x - \sinh^2 x = 1$.

31. Verify that $\cosh x + \sinh x = e^x$.

32. Verify that $\sinh 2x = 2 \sinh x \cosh x$.

In Exercises 33–42, find the derivative of the given function.

33. $y = e^{\sinh 3x}$

34. $f(x) = \sinh(2x + 3)$

35. $g(x) = \tanh \sqrt{x^3 + 4}$

36. $h(x) = \sinh \ln(x^2 + 3x)$

37. $j(x) = \cosh^3(x^4 + \sin x)$

38. $k(x) = \tanh^2 x + \operatorname{sech}^2 x$

39. $f(x) = \sinh^{-1} 7x$

40. $g(x) = \cosh^{-1} \sqrt{x}$

41. $j(x) = x \sinh^{-1} \dfrac{1}{x}$

42. $k(x) = \tanh^{-1}(\sin x),\ -\frac{\pi}{2} < x < \frac{\pi}{2}$

Solve Exercises 43–48.

43. *Advertising* An electronics dealer estimates that the total number people who have heard one of her radio advertisements is approximated by $N(t) = 615\left(1 - e^{-0.02t}\right)$, where t is the number of days since the advertisement began. How fast is N growing after 7 days?

44. *Medical technology* At the beginning of an experiment, a culture of 100 bacteria is growing in a medium that will allow a maximum of 10,000 bacteria to survive. The population, N, at time t, in hours, is estimated to be $N(t) = \dfrac{10{,}000}{1 + 99e^{-0.14t}}$ after the experiment has begun.
(a) What was the initial population?
(b) Find a formula for the rate of change of the bacteria population.
(c) How fast is the population changing at the end of 5 h?
(d) When is the population growing the fastest?
(e) What is the fastest rate of population growth?

45. *Electronics* The charge across a capacitor is $q(t) = 5e^{-t} \cos 2.5t$.
(a) What is the current in the capacitor $I = q'(t)$?
(b) What is the current at $t = 0.45$ s?

46. *Meteorology* If we assume that the temperature of the atmosphere is a constant, then the formula $14.7e^{-0.0003655h}$ gives the atmospheric pressure, P, in pounds per square inch, at altitude h, in feet above sea level, for $0 \le h \le 50$ mi.
(a) Find the rate of change of P with respect to h.
(b) What is $\dfrac{dP}{dh}$ for an airplane that is flying at 35,000 ft?

47. *Biology* Consider a culture of bacteria growing in a glass jar. The number of bacteria present after t h is estimated to be $N(t) = 10^5 + 10^6 \tanh(0.1t)$.
(a) What is the rate of growth of these bacteria?
(b) What is the rate of growth of these bacteria at $t = 6.93$ h?
(c) How many bacteria are there at $t = 6.93$ h?

48. *Environmental science*   Radon gas can readily diffuse through solid materials such as brick and concrete. If the direction of diffusion in a basement wall is perpendicular to the surface, then the radon concentration $C(x)$ in J/cm^3 in the air-filled pores within the wall at distance x from the outside surface, can be approximated by

$$C(x) = A\sinh(qx) + B\cosh(qx) + k$$

where constant q depends on the porosity of the wall, the half-life of radon, and a diffusion coefficient; constant k is the maximum radon concentration in the air-filled pores; and A and B are constants that depend on the initial conditions. Find the rate at which the concentration is changing with respect to distance x cm.

In Your Words

49. You can use the change of base formula $\log_b x = \dfrac{\ln x}{\ln b}$ so that you need but one rule for differentiating a logarithmic function. Will this also work to learn the derivatives of e^x and a^x? If you answered yes, explain how it is done. If you answered no, explain why it will not work.

50. Explain how the derivatives of the sine and cosine functions differ from those of the hyperbolic sine and hyperbolic cosine functions, respectively.

▤ 27.7
APPLICATIONS

In this section, we will look at applications that require the derivative of logarithmic and exponential functions.

Application

EXAMPLE 27.38

Sketch the graph of $f(x) = \dfrac{x}{e^x}$. Locate all extrema and inflection points.

Solution   One thing that we notice before we even begin with calculus is that the function $\dfrac{1}{e^x} = e^{-x}$ is always positive. This means that when $x < 0$, $f(x) = \dfrac{x}{e^x} < 0$ and when $x > 0$, $f(x) > 0$.

Taking the derivative of $f(x) = \dfrac{x}{e^x} = xe^{-x}$, we get $f'(x) = -xe^{-x} + e^{-x} = e^{-x}(1-x)$. The only critical value occurs when $x = 1$. Since $e^{-x} > 0$, the sign of $f'(x)$ depends on the sign of $1-x$. When $x < 1$, we see that $1-x > 0$ and when $x > 1$, then $1-x < 0$. Thus, by the first derivative test, there is a maximum when $x = 1$ or at the point $\left(1, \dfrac{1}{e}\right) \approx (1, 0.368)$.

Taking the second derivative, we get $f''(x) = xe^{-x} - e^{-x} - e^{-x} = (x-2)e^{-x}$. The second derivative is zero when $x = 2$. If $x > 2$, then $f''(x) > 0$, so $f(x)$ is concave up on $(2, \infty)$. If $x < 2$, then $f''(x) < 0$, and $f(x)$ is concave down on $(-\infty, 2)$. Thus, the point $\left(2, \dfrac{2}{e^2}\right) \approx (2, 0.271)$ is a point of inflection.

A sketch of this curve is shown in Figure 27.6.

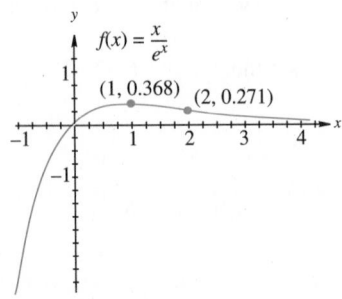

FIGURE 27.6

Application

EXAMPLE 27.39

Find an equation of the tangent and the normal lines to the graph of $y = \ln x^3$ at $(1, 0)$.

Solution In order to determine these lines, we need the slope, which means we need to find the first derivative and evaluate it when $x = 1$. Using properties of logarithms, we have $y = \ln x^3 = 3 \ln x$. Thus, $y' = \dfrac{3}{x}$.

When $x = 1$, $y' = 3$ and so the slope of the tangent line is 3 and the slope of the normal line is $\frac{-1}{3}$. Using the slope-intercept form for the equation of a line, we get the tangent line

$$y - 0 = 3(x - 1)$$
$$y = 3x - 3$$

Similarly, since the slope of the normal line is $-\frac{1}{3}$, the equation of the normal line is

$$y - 0 = -\frac{1}{3}(x - 1)$$
$$y = -\frac{1}{3}x + \frac{1}{3}$$

or $\quad 3y + x = 1$

Application

EXAMPLE 27.40

A 1-kg object suspended from a spring stretches it 15 cm. An unknown velocity is given to the object at $t = 0$. Because of a damping factor, the position s of the object at time t is given by $s(t) = -2e^{-2t} \sin 4t$. **(a)** What is the velocity given the object? **(b)** What is the maximum displacement of the object? **(c)** Sketch the graph of this function.

Solution In order to find the velocity, we need to differentiate $s(t)$.

$$v(t) = s'(t) = -2(4e^{-2t} \cos 4t - 2e^{-2t} \sin 4t)$$
$$= -4e^{-2t}(2 \cos 4t - \sin 4t)$$

We can now answer the first two parts. **(a)** The initial velocity occurs when $t = 0$ and $v(0) = -8$, so the initial velocity is 8 cm/s downward. **(b)** The maximum displacement will occur when $s'(t) = 0$. Since $e^{-2t} > 0$ for all values of t, $s'(t) = 0$ when $2 \cos 4t - \sin 4t = 0$, or when $\sin 4t = 2 \cos 4t$. Dividing both sides by $\cos 4t$, we get $\tan 4t = 2$ or $4t = \tan^{-1} 2$. Using a calculator, we see that $4t = 1.107 + \pi n$, so $t = 0.277 + \frac{\pi}{4}n$ rad. When $n = 0$, we have $t = 0.277$ and $s(t) \approx -1.028$ cm. When $n = 1$, we obtain $t = 0.277 + \frac{\pi}{4}$ and we get $s(t) = 0.214$ cm. The maximum displacement is at $(0.277, -1.028)$. **(c)** The graph of this function is shown in Figure 27.7.

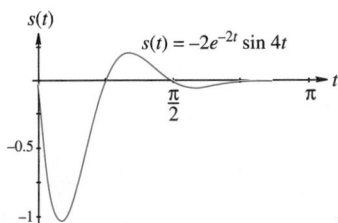

FIGURE 27.7

Application

EXAMPLE 27.41

A submarine telephone cable consists of a conducting circular core surrounded by a layer of insulation. If x is the ratio of the radius of the core to the thickness of the insulation, then the speed v of the signal is proportional to $x^2 \ln\left(\dfrac{1}{x}\right)$. Thus $v(x) = kx^2 \ln\dfrac{1}{x}$, where k is the constant of proportionality. For what value of x is the speed of the signal a maximum?

Solution   We take the first derivative of $v(x)$.

$$v'(x) = kx^2 \left(\frac{-1/x^2}{1/x}\right) + 2kx \ln\frac{1}{x}$$

$$= -kx + 2kx \ln\frac{1}{x}$$

$$= kx \left(2\ln\frac{1}{x} - 1\right)$$

We see that $v'(x)$ is undefined when $x = 0$, and that $v'(x) = 0$ when $\ln\dfrac{1}{x} = \dfrac{1}{2}$ or when $\ln x = \frac{-1}{2}$. Solving for x, we see that $v'(x) = 0$ when $x = e^{-1/2} \approx 0.607$. For x in the interval $\left(0, e^{-1/2}\right)$, we have $v'(x) > 0$, and for x in $\left(e^{-1/2}, \infty\right)$, we see that $v'(x) < 0$. Thus, $s(x)$ is a maximum when $x = e^{-1/2}$.

Exercise Set 27.7

In Exercises 1–10, find all extrema and points of inflection and sketch the graph.

1. $y = x \ln x$

2. $y = \dfrac{\ln x}{x}$

3. $y = xe^x$

4. $y = x^2 e^x$

5. $y = \ln\dfrac{1}{x^2 + 1}$

6. $y = \dfrac{4 \ln^2 x}{x}$

7. $y = \sinh x$

8. $y = \cosh x$

9. $y = e^{-x} \cos x$ for $-2\pi \le x \le 2\pi$

10. $y = \dfrac{\sin x}{\ln(x + 2)}$ for $0 \le x \le 4\pi$

Solve Exercises 11–24.

11. Find an equation of the line tangent to $y = x^3 \ln x$ at $(1, 0)$.

12. Find an equation of the line normal to $y = x^3 \ln x$ at $(1, 0)$.

13. Find an equation of the line tangent to $y = x^2 e^x$ at $(1, e)$.

14. Find an equation of the line normal to $y = x^2 e^x$ at $(1, e)$.

15. *Physics*   A particle moves on a coordinate line with its position at t s given by $s(t) = e^{-3t}$.
    (a) What is its velocity function?
    (b) When is its velocity a maximum?

16. *Physics*   A particle moves along a coordinate line with its position given by $s(t) = t^2 + 4\ln(t+1)$, where t is in seconds.
    (a) What is the velocity function?
    (b) What is the acceleration function?
    (c) When is the velocity a minimum?
    (d) What is the minimum velocity?

17. *Physics*   A particle moves on a coordinate line with its position given by $s(t) = \sin(e^t)$.
    (a) What are the velocity and acceleration functions?
    (b) When does the particle first have a velocity of 0 and what is its acceleration at that time?

18. *Biology*   The number of bacteria in a particular colony at time t hours is given by

    $$x = 5000e^{0.5t}$$

    What is the rate of growth of this colony?

19. *Material science*   The density of a 5-m bar is given by $\rho(x) = xe^{-x^{2/3}}$ kg/m, where x is in meters from the left of the bar. At what point is the bar the most dense?

20. *Meteorology*   The atmospheric pressure of a height of x m above sea level is given by $P(x) = 10^4 e^{-0.0012x}$ kg/m^2. What is the rate of change of the pressure with respect to the height?

21. *Electricity*   The current in a series RL circuit is given by $i = 1 - e^{-t^2/2L}$. When is the current a maximum?

22. *Electronics*   The instantaneous charge on a capacitor for a certain electric circuit is given by $Q(t) = e^{-6t}(4\cos 8t + 3\sin 8t) - 0.4\cos 10t$.
    (a) What is the rate of change of this charge?
    (b) What is the rate of change when $t = \frac{\pi}{16}$.

23. *Space technology*   The power supply P in watts in a satellite is given by $P = 100e^{-0.015t}$, where t is measured in days. What is the time-rate of change of power when $t = 50$ days?

24. *Refrigeration technology*   A warm object is placed in a refrigerator to cool. The temperature in degrees Celsius, at any given time t in minutes, is given by $T = 10 + 15e^{-0.875t}$. What is the rate of change of the temperature when $t = 30$ min?

In Exercises 25–28, use Newton's method to approximate the real roots of the given function.

25. $f(x) = x - \ln x - 2$

26. $g(x) = \ln x - \sin x$

27. $h(x) = e^x \cdot \cos x$

28. $j(x) = e^x + 2x - 2$

In Your Words

29. Write an application in your technology area of interest that requires you to use derivatives of a logarithmic function. Give your problem to a classmate and see if he or she understands and can solve your problem. Rewrite the problem as necessary to remove any difficulties encountered by your classmate.

30. Write an application in your technology area of interest that requires you to use derivatives of an exponential function. Give your problem to a classmate and see if he or she understands and can solve your problem. Rewrite the problem as necessary to remove any difficulties encountered by your classmate.

▤ CHAPTER 27 REVIEW

Important Terms and Concepts

Hyperbolic functions
Inverse hyperbolic functions

Logarithmic differentiation

Review Exercises

In Exercises 1–24, differentiate and simplify the given functions.

1. $y = \sin 2x + \cos 3x$
2. $y = \tan 3x^2$
3. $y = \tan^2 3x$
4. $y = \sqrt{\sin 2x}$
5. $y = x^2 \sin x$
6. $y = \dfrac{\cos 3x}{3x}$
7. $y = \sin^{-1}(3x - 2)$
8. $y = \cos^{-1} x^2$

9. $y = \arctan 3x^2$
10. $y = \arctan\left(\dfrac{1+x}{1-x}\right)$
11. $y = \arcsin\sqrt{1-x^2}$
12. $y = \arccos\sqrt{-3+4x-x^2}$
13. $y = \log_4(3x^2 - 5)$
14. $y = \ln(x+4)^3$
15. $y = \ln^2(x^3+4)$
16. $y = \ln\cos x$

17. $y = \ln\dfrac{x^3}{(4x-3)^2}$
18. $y = \ln\arctan x$
19. $y = e^{4x^2}$
20. $y = e^{\ln x^2}$
21. $y = e^{\sin x^2}$
22. $y = xe^{x^2} + \sin x$
23. $y = e^{-x}\sin 5x$
24. $y = e^{-x}\ln x$

In Exercises 25–30, locate all the extrema and the points of inflection for the given function. Sketch the graph.

25. $y = x^2 e^x$
26. $y = 4e^{-9x^2}$
27. $y = e^{-1/2x}\sin 2x$ on $[0, 2\pi]$
28. $y = e\sin 3x - e^{-1}\cos x$ on $[0, 2\pi]$
29. $y = x^3\ln x,\, x > 0$
30. $y = 2\arccos\dfrac{x}{4}$

In Exercises 31–36, find an equation of (a) the line tangent and (b) the line normal to the given curve at the indicated point.

31. $y = \arctan x$ at $\left(1, \frac{\pi}{4}\right)$
32. $y = \sin^3 x$ at $\left(\frac{\pi}{4}, \frac{\sqrt{2}}{4}\right)$
33. $y = \ln 3x$ at $(1, 0)$

34. $y = \dfrac{e^x}{x}$ at $(1, e)$
35. $y = \ln(\sin x)$ at $\left(\frac{\pi}{4}, -\ln\sqrt{2}\right)$
36. $y = \sin(\ln x)$ at $(1, 0)$

Solve Exercises 37–40.

37. *Thermodynamics* An object is heated to 95°C and allowed to cool in a room with air temperature of 20°C. The temperature T after t min is given by the equation $T = 75e^{-0.05109t} + 20$.
    (a) What is the temperature of the object after 10 min?
    (b) What is the rate of change of the temperature after 10 min?

38. *Physics* A person parachutes from an airplane flying at 10,000 ft and opens the parachute immediately. The height of the parachutist t s after jumping is given by $s(t) = 10{,}000 - \frac{400}{3}\ln(\cosh 1.6t)$.
    (a) What is the velocity of the parachutist?

(b) What is the velocity when $t = 60$ s?

39. *Advertising* A billboard parallel to a highway is to be 40 ft high and its bottom will be 8 ft above the eye of a passing motorist. How far from the highway should the billboard be placed in order to maximize the angle it subtends at the motorist's eyes?

40. *Meteorology* A weather balloon is rising from the ground at the rate of 2 m/s from a point 100 m from an observer. Determine the rate of increase of the angle of inclination of the observer's line of sight when the balloon is 30 m high.

☰ CHAPTER 27 TEST

In Exercises 1–10, differentiate the given functions.

1. $f(x) = \sin 7x$

2. $g(x) = \tan(3x^2 + 2x)$

3. $h(x) = e^{2x}$

4. $j(x) = \ln 5x^2$

5. $k(x) = \cos^{-1}\left(e^{x^2}\right)$

6. $f(x) = \dfrac{\sin 4x}{1 + e^{2x}}$

7. $g(x) = (7x^2 + 3x)\tan^2 5x$

8. $h(x) = e^{\sin x}\ln\sqrt{x}$

9. The charge across a capacitor is $q(t) = 3e^{-t}\cos 2.5t$. Find the current to the capacitor when $t = 0.65$ s.

10. A particle is moving along the x-axis so that its distance x (in meters) from the origin at time t s is $x(t) = 3.2\sqrt{\tan 5.0t}$. Find the velocity at time $t = 1.5$ s.

28

Techniques of Integration

The main cable joining two towers of a suspension bridge, like the one shown, takes the shape of a parabola. In Section 28.7, we will learn how to determine the length of this cable.

Courtesy of New York Convention and Visitors Bureau Inc.

In Chapters 23 and 27, we learned to differentiate almost any function. What is perhaps even more important is that we have developed a method that will help us find the derivative of almost any conceivable function. The same cannot be said about integration.

In this chapter, we will learn more about integration. We will begin by expanding our use of the power formula to include transcendental functions. Next we will look at the integrals for some logarithmic, exponential, and trigonometric functions. Finally, we will learn some techniques that can be used to integrate some additional functions.

When you have finished this chapter, you will have a better idea of which functions can be integrated and how to proceed in order to find the indefinite integral.

≡ 28.1
THE GENERAL POWER FORMULA

In Section 25.3, the fourth rule of integration stated that, for any variable x, $\int x^n \, dx = \frac{1}{n+1} x^{n+1} + C$. If we let u represent any function, then this can be restated as the **general power formula**.

General Power Formula for Integration

$$\int u^n \, du = \frac{1}{n+1} u^{n+1} + C, n \neq -1$$

Until now we have only applied the general power formula to algebraic functions, but it can be applied to transcendental functions as well, as is shown in the next five examples.

EXAMPLE 28.1

Find $\int \sin^2 x \cos x \, dx$.

Solution   If we let $u = \sin x$, then $du = \cos x \, dx$ and the integral is as follows.

$$\int \sin^2 x \cos x \, dx = \int u^2 \, du$$
$$= \frac{1}{3} u^3 + C$$
$$= \frac{1}{3} \sin^3 x + C$$

EXAMPLE 28.2

Find $\int \frac{\ln^4 x}{x} \, dx$.

Solution   If $u = \ln x$, then $du = \frac{1}{x} \, dx$ and we have

$$\int \frac{\ln^4 x}{x} \, dx = \int (\ln x)^4 \left(\frac{1}{x} \right) dx$$
$$= \int u^4 \, du$$
$$= \frac{1}{5} u^5 + C$$
$$= \frac{1}{5} \ln^5 x + C$$

≡ **Note**   Recall that integration can be checked by differentiation. Thus,

$$D_x \left(\frac{1}{5} \ln^5 x + C \right) = \frac{1}{5} 5 \ln^4 x \left(\frac{1}{x} \right)$$

$$= \frac{\ln^4 x}{x}$$

EXAMPLE 28.3

Determine $\int \sec^5 x \tan x \, dx$.

Solution   Let $u = \sec x$ and then $du = \sec x \tan x \, dx$. If we write $\sec^5 x \tan x \, dx$ as $\sec^4 x (\sec x \tan x \, dx)$, then we have

$$\int \sec^5 x \tan x \, dx = u^4 \, du$$

$$= \frac{1}{5} u^5 + C$$

$$= \frac{1}{5} \sec^5 x + C$$

EXAMPLE 28.4

Determine $\int \dfrac{e^{3x} \, dx}{\sqrt{1 + e^{3x}}}$.

Solution   We will let $u = 1 + e^{3x}$ and so $du = 3e^{3x} \, dx$. We do not have du in our problem, but we do have $\frac{1}{3} du$. We will multiply the problem by $\frac{3}{3}$ in order to get du into the problem.

$$\int \frac{e^{3x} \, dx}{\sqrt{1 + e^{3x}}} = \frac{3}{3} \int \frac{e^{3x} \, dx}{\sqrt{1 + e^{3x}}}$$

$$= \frac{1}{3} \int \frac{3e^{3x} \, dx}{\sqrt{1 + e^{3x}}}$$

$$= \frac{1}{3} \int \frac{du}{\sqrt{u}}$$

$$= \frac{1}{3} \int u^{-1/2} \, du = \left(\frac{1}{3} \right) \frac{2}{1} u^{1/2} + C$$

$$= \frac{2}{3} u^{1/2} = \frac{2}{3} \left(1 + e^{3x} \right)^{1/2} + C$$

EXAMPLE 28.5

Evaluate $\displaystyle\int_0^1 \frac{\arcsin x}{\sqrt{1-x^2}}\,dx$.

Solution We first determine the improper integral by letting $u = \arcsin x$ and $du = \dfrac{dx}{\sqrt{1-x^2}}$. Then

$$\int \frac{\arcsin x}{\sqrt{1-x^2}}\,dx = \int u\,du$$

$$= \frac{1}{2}u^2 + C$$

$$= \frac{1}{2}(\arcsin x)^2 + C$$

Evaluating this, we get the proper integral

$$\int_0^1 \frac{\arcsin x}{\sqrt{1-x^2}\,dx} = \left.\frac{1}{2}(\arcsin x)^2\right|_0^1$$

$$= \frac{1}{2}\left(\frac{\pi}{2}\right)^2 = \frac{\pi^2}{8}$$

Application

EXAMPLE 28.6

The charge q stored in a capacitor decreases at the rate of

$$\frac{dq}{dt} = e^{-t}\left(e^{-t}-4\right)^2$$

What is the charge dissipated from $t = 0.5$ s to $t = 2$ s?

Solution As before, we will first determine the improper integral. If $u = e^{-t}-4$, then $du = -e^{-t}\,dt$. Substituting these into the original integral, produces

$$\int e^{-t}\left(e^{-t}-4\right)^2 dt = -\int u^2\,du$$

$$= -\frac{1}{3}u^3 + C$$

$$= -\frac{1}{3}\left(e^{-t}-4\right)^3 + C$$

Now we can evaluate this over the time interval form $t = 0.5$ s to $t = 2$ s.

$$\int_{0.5}^2 e^{-t}\left(e^{-t}-4\right)^2 dt = \left.-\frac{1}{3}\left(e^{-t}-4\right)^3\right|_{0.5}^2$$

$$\approx 2.1611 - .9019 = 1.2592$$

About 1.26 C is dissipated during this time interval.

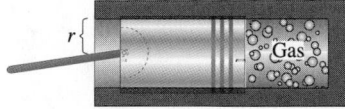

FIGURE 28.1

Consider a piston of radius r in a cylindrical casing, such as the one shown in Figure 28.1. As the gas in the cylinder expands, the piston moves and work is done. If we let P represent the pressure of the gas against the piston head and V the volume of the gas, then as the volume of the gas expands from V_0 to V_1, the temperature remains constant. The work done in moving the piston is given by the formula

$$W = \int_{V_0}^{V_1} P \, dV$$

If we assume that the pressure of the gas is inversely proportional to its volume, then $P = \dfrac{k}{V}$ and the work done in moving the piston is expressed as

$$W = \int_{V_0}^{V_1} \frac{k}{V} \, dV$$

Application

EXAMPLE 28.7

A quantity of gas with an initial volume of $0.100 \, \text{m}^3$ and pressure 300 kPa expands to a volume of $0.200 \, \text{m}^3$. Find the work done by the gas, assuming that the pressure of the gas is inversely proportional to its volume.

Solution Since the pressure of the gas is inversely proportional to its volume, we know that $P = \dfrac{k}{V}$. We are given that when the volume is $0.100 \, \text{m}^3$, the pressure is 300 kPa, which means that $300 = \dfrac{k}{0.100}$. Solving for k, we find that $k = 30$. Thus, the work is

$$W = \int_{V_0}^{V_1} \frac{k}{V} \, dV$$

$$= \int_{0.100}^{0.200} \frac{30}{V} \, dV$$

$$= 30 \ln V \Big]_{0.100}^{0.200}$$

$$= 30[\ln 0.2 - \ln 0.1] = 30 \ln \left(\frac{0.2}{0.1} \right)$$

$$= 30 \ln 2 \approx 20.7944$$

So, the work is about $20.7944 \, \text{N·m}$.

Exercise Set 28.1

In Exercises 1–28, find the given integral.

1. $\displaystyle \int \sin^3 x \cos x \, dx$

2. $\displaystyle \int \cos^4 x \sin x \, dx$

3. $\displaystyle \int \sqrt{\sin^3 x} \cos x \, dx$

4. $\displaystyle \int \sin 2x \cos 2x \, dx$

5. $\displaystyle \int x \sec^2 x^2 \, dx$

6. $\displaystyle \int \cos^3 x \sin x \, dx$

7. $\displaystyle \int \sec^2 x \tan x \, dx$

8. $\displaystyle \int \frac{e^{2x}}{(1 + e^{2x})^{1/2}} \, dx$

9. $\displaystyle \int \frac{\arccos x}{\sqrt{1 - x^2}} \, dx$

10. $\displaystyle\int \frac{\arctan x}{1+x^2}\,dx$

11. $\displaystyle\int \frac{1}{x}\sqrt{\frac{\operatorname{arccsc}x}{x^2-1}}\,dx$

12. $\displaystyle\int \frac{\arctan 6x}{1+36x^2}\,dx$

13. $\displaystyle\int \frac{(\arcsin 2x)^3}{\sqrt{1-4x^2}}\,dx$

14. $\displaystyle\int \frac{(\ln x)^3}{x}\,dx$

15. $\displaystyle\int \frac{[\ln(x+4)]^4}{x+4}\,dx$

16. $\displaystyle\int \frac{[5+2\ln x]^3}{x}\,dx$

17. $\displaystyle\int \frac{e^x\,dx}{(e^x+4)^2}$

18. $\displaystyle\int (e^x+e^{-x})^{1/3}(e^x-e^{-x})\,dx$

19. $\displaystyle\int (e^{2x}-1)^3 e^{2x}\,dx$

20. $\displaystyle\int (3e^{3x}+1)^{2/3}e^{3x}\,dx$

21. $\displaystyle\int_0^{\pi/4} \sin^3 x\cos x\,dx$

22. $\displaystyle\int_{\pi/4}^{\pi/3} \tan^3 x\sec^2 x\,dx$

23. $\displaystyle\int_1^2 \frac{x[\ln(x^2+1)]^3}{x^2+1}\,dx$

24. $\displaystyle\int_1^{\sqrt{3}} \frac{(\arctan x)^3}{1+x^2}\,dx$

25. $\displaystyle\int_1^2 \frac{e^{2x}\,dx}{(e^{2x}-1)^2}$

26. $\displaystyle\int_0^2 8(e^{2x}-1)^3 e^{2x}\,dx$

Solve Exercises 27–32.

27. Find the area under the curve $y = \sin^2 x\cos x$ from $x = 0$ to $x = \pi$.

28. Find the area under the curve $y = (e^{3x}+1)e^{3x}$ from $x = 0$ to $x = 1$.

29. *Electricity*   The charge q stored in a capacitor decreases at the rate of

$$\frac{dq}{dt} = e^{-t}\left(e^{-t}-2\right)^2$$

What is the charge dissipated from $t = 0.7$ s to $t = 3$ s?

30. *Electricity*   The charge q stored in a capacitor decreases at the rate of

$$\frac{dq}{dt} = e^{-t}\left(e^{-t}-5\right)^2$$

What is the charge dissipated from $t = 0.25$ s to $t = 2.50$ s?

31. *Ecology*   Ecologists are treating a stream contaminated with bacteria. The level of contamination is changing at the rate of $\dfrac{dN}{dt} = \dfrac{1020}{t^2} - 340$ bacteria/cm^3/day, where t is the number of days since the treatment began.
(a) Find a function $N(t)$ that will estimate the level of contamination if $N(1) = 9{,}750$ bacteria/cm^3.
(b) Determine $N(10)$.

32. *Quality control*   A quality control engineer has determined that the rate of expenditure in dollars/year for maintaining a certain machine at a specified level is given by $M'(t) = \sqrt{3t^2+18t}\,(6t+18)$, where t is time measured in years. Maintenance expenditures for the third year were \$945.
(a) Find the total maintenance function, $M(t)$.
(b) According to the engineer, the machine should be replaced when the total expenses reach \$7,200. How long should the company keep this machine?

In Your Words

33. In the general power formula for integration, what is the role of du?

34. In the general power formula for integration, how do you determine which part of the integrand is u and which part is du?

≡ 28.2
BASIC LOGARITHMIC AND EXPONENTIAL INTEGRALS

The general power formula from Section 28.1 is valid for all values of n except $n = -1$. This gap can be filled by taking advantage of the derivative of the natural logarithm function and the fact that the integral is the inverse operation of the derivative.

We know from Chapter 27 that

$$\frac{d}{dx}\ln u = \frac{1}{u}\frac{du}{dx}$$

In differential form this is written as

$$d(\ln u) = \frac{1}{u}du$$

Thus, we see that by using antiderivatives, we obtain

$$\int \frac{1}{u}du = \ln u + C,\ u > 0$$

The logarithmic function is defined for positive values of u. What do we do if u is negative? Well, if u is negative, then $u < 0$ and $-u > 0$, so

$$\frac{d}{dx}\ln(-u) = \frac{-1}{u}\left(-\frac{du}{dx}\right) = \frac{1}{u}\frac{du}{dx}$$

From this we see that

$$\int \frac{1}{u}du = \int \frac{-1}{u}(-du) = \ln(-u) + C,\ \text{if } u < 0$$

We can combine these two cases if we remember that

$$|u| = \begin{cases} -u & \text{if } u < 0 \\ u & \text{if } u > 0 \end{cases}$$

This means that

$$\int \frac{du}{u} = \ln|u| + C$$

We have now filled the gap in the general power formula.

EXAMPLE 28.8

Determine $\int \frac{x\,dx}{x^2+4}$.

Solution If we let $u = x^2 + 4$, then $du = 2x\,dx$. If we multiply the above integral by $\frac{2}{2}$, we get

$$\int \frac{x\,dx}{x^2+4} = \frac{2}{2}\int \frac{x\,dx}{x^2+4}$$
$$= \frac{1}{2}\int \frac{2x\,dx}{x^2+4}$$

EXAMPLE 28.8 (Cont.)

$$= \frac{1}{2} \int \frac{du}{u}$$

$$= \frac{1}{2} \ln|u| + C$$

$$= \frac{1}{2} \ln|x^2 + 4| + C$$

In this case, since $x^2 \geq 0$, $x^2 + 4 > 0$, so $|x^2 + 4| = x^2 + 4$ and we can write the answer as $\frac{1}{2} \ln(x^2 + 4) + C$.

We also learned in Chapter 27 that $\dfrac{d}{dx} e^u = e^u \dfrac{du}{dx}$. Reversing this, we get the integral of $e^u \, du$.

$$\int e^u \, du = e^u + C$$

EXAMPLE 28.9

Find $\displaystyle\int e^{\sin x} \cos x \, du$.

Solution If $u = \sin x$, then $du = \cos x \, du$. Thus

$$\int e^u \, du = e^u + C$$

$$= e^{\sin x} + C$$

If we are working with bases other than e, we get the following integral.

$$\int a^u \, du = a^u \log_a e + C$$

$$= \frac{a^u}{\ln a} + C$$

EXAMPLE 28.10

Find $\int 3^x\, dx$.

Solution   We let $u = x$ and so $du = dx$.

$$\int 3^x\, dx = \int 3^u\, du$$
$$= \frac{3^u}{\ln 3} + C$$
$$= \frac{3^x}{\ln 3} + C$$

EXAMPLE 28.11

Determine $\int x^2 e^{x^3}\, dx$.

Solution   If $u = x^3$, then $du = 3x^2\, dx$. We will need to multiply by $\frac{3}{3}$ to get

$$\int x^2 e^{x^3}\, dx = \frac{1}{3} \int 3x^2 e^{x^3}\, dx$$
$$= \frac{1}{3} \int e^u\, du$$
$$= \frac{1}{3} e^u + C$$
$$= \frac{1}{3} e^{x^3} + C$$

Hint

When integrating e raised to a function, the substitution $u =$ the exponent of e usually leads to a recognizable integration.

EXAMPLE 28.12

Evaluate $\int_1^2 \frac{e^{1/x}}{x^2}\, dx$.

Solution   If $u = \frac{1}{x} = x^{-1}$, then $du = -x^{-2}\, dx$ and $\frac{1}{x^2} = -du$. This produces

$$\int \frac{e^{1/x}\, dx}{x^2} = -\int e^u\, du = -e^u + C$$

Since $u = \frac{1}{x}$, we have

$$\int_1^2 \frac{e^{1/x}}{x^2}\, dx = -e^{1/x}\Big|_1^2$$
$$= -e^{1/2} - (-e^1) = e - \sqrt{e}$$
$$\approx 1.0696$$

Application

EXAMPLE 28.13

An object weighing 128 lb is thrown from a height of 200 ft with an initial velocity of 16 ft/s. If the air resistance is proportional to the velocity of the body and the limiting velocity is 128 ft/s, the velocity of the object at time t is given by $v(t) = 112e^{-t/4} - 128$. What is the position of the object at any time t?

Solution The velocity function is the derivative of the position function, $s(t)$. This means that the position function is

$$s(t) = \int v(t)\,dt$$
$$= \int (112e^{-t/4} - 128)\,dt$$
$$= -448e^{-t/4} - 128t + C$$

Now at $t = 0$, we know that the object is 200 ft above the ground. So, $s(0) = 200$. Since $s(0) = -448e^0 - 128(0) + C = -448 + C$, we find that $C = 648$. Thus, the position function of this object at time t is $s(t) = -448e^{-t/4} - 128t + 648$ ft.

Application

EXAMPLE 28.14

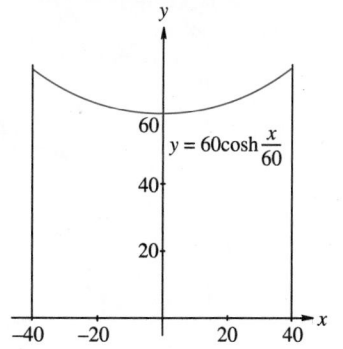

FIGURE 28.2

Any free hanging cable suspended from two towers, like the one shown in Figure 28.2, takes the shape of a **catenary** whose equation is

$$y = a\cosh\frac{x}{a}$$

for some real number a. If an electric cable is hung between two towers that are 80.00 m apart so that the lowest part of the cable is $a = 60.00$ m above the ground, the cable takes the equation

$$y = 60\cosh\frac{x}{60}$$

Find the arc length of the cable between the two towers.

Solution A sketch of the cable with a coordinate system has been drawn in Figure 28.2. Since $y = 60\cosh\dfrac{x}{60}$, we know from Section 27.6 that $y = 60\cosh\dfrac{x}{60} = 30\left(e^{x/60} + e^{-x/60}\right)$. To find the arc length, we need $(y')^2$. We can determine that $y' = \frac{1}{2}\left(e^{x/60} - e^{-x/60}\right)$ and so

$$(y')^2 = \left[\frac{1}{2}\left(e^{x/60} - e^{-x/60}\right)\right]^2$$
$$= \frac{1}{4}\left(e^{x/30} - 2 + e^{-x/30}\right)$$

EXAMPLE 28.14 (Cont.)

We are now ready to determine the arc length L of the cable.

$$L = \int_a^b \sqrt{1+(y')^2}\, dx$$

$$= \int_{-40}^{40} \sqrt{1+\frac{1}{4}\left(e^{x/30} - 2 + e^{-x/30}\right)}\, dx$$

$$= \int_{-40}^{40} \sqrt{\frac{1}{4}\left(e^{x/30} + 2 + e^{-x/30}\right)}\, dx$$

$$= \int_{-40}^{40} \sqrt{\left[\frac{1}{2}\left(e^{x/60} + e^{-x/60}\right)\right]^2}\, dx$$

$$= \int_{-40}^{40} \frac{1}{2}\left(e^{x/60} + e^{-x/60}\right) dx$$

$$= 30\left(e^{x/60} - e^{-x/60}\right)\Big|_{-40}^{40}$$

$$= 60\left(e^{2/3} - e^{-2/3}\right) \approx 86.0590$$

The cable is about 86.06 m long.

Exercise Set 28.2

Evaluate the integrals in Exercises 1–28.

1. $\displaystyle\int \frac{dx}{4x+1}$

2. $\displaystyle\int \frac{dx}{9x-5}$

3. $\displaystyle\int e^{-3x}\, dx$

4. $\displaystyle\int e^{6x}\, dx$

5. $\displaystyle\int \frac{\sin x}{\cos x}\, dx$

6. $\displaystyle\int \frac{10x+3}{5x^2+3x-7}\, dx$

7. $\displaystyle\int \frac{2\sec^2 x}{\tan x}\, dx$

8. $\displaystyle\int \frac{2e^x}{e^x+4}\, dx$

9. $\displaystyle\int 4^x\, dx$

10. $\displaystyle\int (e^x - e^{-x})\, dx$

11. $\displaystyle\int (e^x + e^{-x})\, dx$

12. $\displaystyle\int \sec^2 x\, e^{\tan x}\, dx$

13. $\displaystyle\int \frac{e^{2x}}{1+e^{2x}}\, dx$

14. $\displaystyle\int \frac{x}{1+x^2}\, dx$

15. $\displaystyle\int \frac{\ln(1/x)}{x}\, dx$

16. $\displaystyle\int e^{8\sqrt{x}} \frac{\sqrt{x}}{}\, dx$

17. $\displaystyle\int e^{\frac{\sqrt[3]{x}}{x^{2/3}}}\, dx$

18. $\displaystyle\int \frac{(\ln x)^{3/4}}{x}\, dx$

19. $\displaystyle\int_0^4 \frac{x}{x^2+1}\, dx$

20. $\displaystyle\int_0^1 \frac{dx}{e^{2x}}$

21. $\displaystyle\int_1^3 \frac{1}{\ln e^x}\, dx$

22. $\displaystyle\int_0^3 \frac{x^2}{x^3+1}\, dx$

23. $\displaystyle\int_0^1 x^3 e^{x^4}\, dx$

24. $\displaystyle\int_{\pi/4}^{\pi/3} \frac{dx}{(x^2+1)\arctan x}$

25. $\displaystyle\int_2^4 \left(\tfrac{1}{5}\right)^{x/2}\, dx$

26. $\displaystyle\int_{\ln 2}^{\ln 5} \frac{e^x}{e^x+1}\, dx$

27. Find the area between the curve $y = e^{x+1}$, the y-axis, $y = x^2$, and $x = 1$.

28. Find the area bounded by $y = 0$, $x = 0$, $y = e^{-x}$, and $x = 4$.

Solve Exercises 29–38.

29. *Physics*   An object of mass 24 kg is thrown verti-cally upward from the ground into the air with an initial velocity of 10 m/s. If the object encounters air resistance proportional to its velocity, the velocity in m/s of the object at any time t is given by

$$v(t) = 394e^{0.025t} - 384$$

   (a) What is the position of the object at any time t?
   (b) At what time does the object reach its maximum height?
   (c) What is the maximum height?

30. *Electricity*   Find the average value of the current $i = e^{(-2/3)t}$ in a circuit from $t = 0$ to $t = 2$.

31. Find the volume of the solid generated by rotating the region bounded by $y = e^{-x}$, $y = 0$, $x = 0$, and $x = 2$ around the x-axis.

32. *Electricity*   The voltage across an inductor varies according to $v = 200e^{-50t}$.
   (a) Find the average voltage over the inductor from $t = 0$ to $t = 0.02$ s.
   (b) Determine the effective voltage from $t = 0$ to $t = 0.02$ s.

33. *Electronics*   The signal voltage V across a tele-phone circuit changed with the distance s mi along the circuit at the rate $\dfrac{dV}{ds} = -0.45e^{-0.15s}$. If the applied voltage at the sending end was 3 V, at what distance along the line is the voltage 1.5 V?

34. *Physics*   A quantity of gas with an initial volume of $0.200\,\text{m}^3$ and pressure 500 kPa expands to a volume of $0.800\,\text{m}^3$. Find the work done by the gas, assuming that the pressure of the gas is inversely proportional to its volume.

35. *Physics*   A quantity of gas with an initial volume of $0.400\,\text{ft}^3$ and pressure 20 psi expands to a volume of $1.500\,\text{ft}^3$. Find the work done by the gas, assuming that the pressure of the gas is inversely proportional to its volume.

36. *Pharmacology*   If x mg of a certain drug is admin-istered to a person, the rate of change in the person's temperature, in °F, with respect to the dosage is given by $T'(x) = \frac{4}{2x+9}$. A dose of 1 mg raises the person's temperature 2.4°F. Determine the function that gives total change in body temperature.

37. *Nuclear engineering*   The integral $F(x) = \displaystyle\int_0^x 12te^{-3t^2}\,dt$ appears in the analysis of failure rates and reliability in nuclear reactor design. Integrate to find $F(x)$.

38. *Medical technology*   The rate of change of the con-centration $C(t)$ at time t of a radioactive tracer drug is $C'(t) = 2^{-t}$, where t is measured in hours. If the initial concentration is 1 µg/L, determine the concen-tration at time t.

In Your Words

39. You can use the change of base formula $\log_b x = \dfrac{\ln x}{\ln b}$ so that you need but one rule for differentiating a log-arithmic function. Will this also work to integrate $\ln u$ and $\log_b u$? If you answered yes, explain how it is done. If you answered no, explain why it will not work.

40. You can use the change of base formula $\log_b x = \dfrac{\ln x}{\ln b}$ so that you need but one rule for differentiating a log-arithmic function. Will this also work to integrate e^x and a^x? If you answered yes, explain how it is done. If you answered no, explain why it will not work.

≡ 28.3
BASIC TRIGONOMETRIC AND HYPERBOLIC INTEGRALS

In this section, we will discuss the basic trigonometric integrals. This will include the integrals of the six trigonometric functions. Integration is the inverse of differentiation. Integrating the derivatives of the six trigonometric functions gives us the following formulas.

Basic Trigonometric Integrals

$$\int \cos u \, du = \sin u + C$$

$$\int \sin u \, du = -\cos u + C$$

$$\int \sec^2 u \, du = \tan u + C$$

$$\int \csc^2 u \, du = -\cot u + C$$

$$\int \sec u \tan u \, du = \sec u + C$$

$$\int \csc u \cot u \, du = -\csc u + C$$

 **Hint**   The substitution $u =$ the angle will usually allow an easier integration.

EXAMPLE 28.15

Determine $\int 2\cos 2x \, dx$.

Solution   Here the angle is $2x$, so we will let $u = 2x$. Then $du = 2\,dx$ and

$$\int 2\cos 2x \, dx = \int \cos u \, du$$
$$= \sin u + C$$
$$= \sin 2x + C$$

EXAMPLE 28.16

Find $\int \sec(3x+4)\tan(3x+4)\,dx$.

Solution   If we let $u = 3x+4$, then $du = 3\,dx$ and $dx = \frac{1}{3}\,du$. We get the following result:

$$\int \sec(3x+4)\tan(3x+4)\,dx = \frac{1}{3}\int \sec u \tan u \, du$$
$$= \frac{1}{3}\sec u + C$$
$$= \frac{1}{3}\sec(3x+4) + C$$

EXAMPLE 28.17

Evaluate $\int \tan^2 x\, dx$.

Solution This does not seem to fit in one of the six basic equations. But, if we remember that $\tan^2 x = \sec^2 x - 1$, we can see that

$$\int \tan^2 x\, dx = \int (\sec^2 x - 1)\, dx$$
$$= \int \sec^2 x\, dx - \int dx$$
$$= \tan x - x + C$$

We have yet to determine the integrals for four of the trigonometric functions—tangent, cotangent, secant, and cosecant. We will develop them next.

For the tangent and cotangent functions, we use the fact that $\tan x = \dfrac{\sin x}{\cos x}$ and $\cot x = \dfrac{\cos x}{\sin x}$.

$\int \tan u\, du = \int \dfrac{\sin u}{\cos u}\, du$. If we let $v = \cos u$, then $dv = -\sin u\, du$. Multiplying the integral by $\frac{-1}{-1}$, we get $\int \dfrac{\sin u}{\cos u}\, du = -\int \dfrac{-\sin u}{\cos u}\, du = -\int \dfrac{dv}{v} = -\ln|v| + C.$ But $v = \cos u$, so

$$\int \tan u\, du = -\ln|\cos u| + C$$

Remembering that $-\ln|\cos u| = \ln(|\cos u|)^{-1} = \ln\left|\dfrac{1}{\cos u}\right| = \ln|\sec u|$, we get the following two alternative results for $\int \tan u\, du$.

$$\int \tan u\, du = -\ln|\cos u| + C$$
$$= \ln|\sec u| + C$$

In the exercises you will be asked to show that

$$\int \cot u\, du = \ln|\sin u| + C$$

Developing the integral of the secant and cosecant functions are a little more complicated. We will develop the integral of the secant function and let you show that our integral of the cosecant function is true. The integral of the secant function

requires us to multiply the function by an unusual form of the number 1: $\dfrac{\sec u + \tan u}{\sec u + \tan u}$.
We now proceed as follows.

$$\int \sec u \, du = \int \frac{(\sec u)(\sec u + \tan u)}{\sec u + \tan u} \, du$$

$$= \int \frac{\sec^2 u + \sec u \tan u}{\sec u + \tan u} \, du$$

If we let $v = \sec u + \tan u$, then $dv = (\sec u \tan u + \sec^2 u) \, du$. This means that our integral can be rewritten as

$$\int \frac{dv}{v} = \ln|v| + C = \ln|\sec u + \tan u| + C$$

Thus, we have the result we were seeking. A similar effort would produce the integral for the cosecant function.

> **Integrals of Secant and Cosecant**
> $$\int \sec u \, du = \ln|\sec u + \tan u| + C$$
> $$\int \csc u \, du = \ln|\csc u - \cot u| + C$$

EXAMPLE 28.18

Find $\displaystyle\int \tan(7x - 5)\, dx$.

Solution   If we let $u = 7x - 5$, then $du = 7\,dx$ and $dx = \frac{1}{7}\,du$. This produces

$$\int \tan(7x - 5)\, dx = \frac{1}{7} \int \tan u \, du$$

$$= \frac{1}{7} \ln|\sec u| + C$$

$$= \frac{1}{7} \ln|\sec(7x - 5)| + C$$

EXAMPLE 28.19

Evaluate $\displaystyle\int x \sec 3x^2 \, dx$.

Solution   Let $u = 3x^2$, then $du = 6x\,dx$ and $x\,dx = \frac{1}{6}\,du$. These substitutions produce

$$\int x \sec 3x^2 \, dx = \frac{1}{6} \int \sec u \, du$$

$$= \frac{1}{6} \ln|\sec u + \tan u| + C$$

$$= \frac{1}{6} \ln|\sec 3x^2 + \tan 3x^2| + C$$

Application

EXAMPLE 28.20

The current in a particular circuit is given by the equation $i = 100\cos(120t - \sqrt{2})$. What is the average current as t ranges from 0 to 0.5 s?

Solution The average current $\bar{i}$ is given by

$$\bar{i} = \frac{1}{b-a} \int_a^b i(t)\, dt$$

$$= \frac{1}{0.5-0} \int_0^{0.5} 100\cos(120t - \sqrt{2})\, dt$$

$$= \frac{1}{0.5} \frac{100}{120} \sin\left(120t - \sqrt{2}\right)\Big|_0^{0.5}$$

$$\approx \frac{5}{3}[0.8932 - (-0.9878)]$$

$$\approx 3.135$$

The average current is 3.135 A.

Integrals of Hyperbolic Functions

In Section 27.6, we defined and found derivatives of the hyperbolic and inverse hyperbolic functions. We will next examine the integrals of the hyperbolic functions. In Section 28.5, we will integrate the inverse hyperbolic functions.

Based on the derivatives of the hyperbolic functions, we have the following integrals.

Integrals Involving Hyperbolic Functions

$$\int \sinh u\, du = \cosh u + C \qquad\qquad \int \operatorname{csch}^2 u\, du = -\coth u + C$$

$$\int \cosh u\, du = \sinh u + C \qquad\qquad \int \operatorname{sech} u \tanh u\, du = -\operatorname{sech} u + C$$

$$\int \operatorname{sech}^2 u\, du = \tanh u + C \qquad\qquad \int \operatorname{csch} u \coth u\, du = -\operatorname{csch} u + C$$

EXAMPLE 28.21

Evaluate $\int \cosh 9x\, dx$.

Solution   If we let $u = 9x$, then $du = 9\,dx$. We multiply the integral by $\frac{9}{9}$, and obtain the result

$$\int \cosh 9x\, dx = \frac{9}{9} \int \cosh 9x\, dx$$
$$= \frac{1}{9} \int 9 \cosh 9x\, dx$$
$$= \frac{1}{9} \sinh 9x + C$$

EXAMPLE 28.22

Evaluate $\int \coth 17x\, dx$.

Solution   Here we rewrite $\coth 17x$ as $\dfrac{\cosh 17x}{\sinh 17x}$. Now, let $u = \sinh 17x$, then $du = 17 \cosh 17x\, dx$. We use these results as we evaluate the following integral.

$$\int \coth 17x\, dx = \int \frac{\cosh 17x}{\sinh 17x}\, dx$$
$$= \frac{1}{17} \int \frac{du}{u}$$
$$= \frac{1}{17} \ln |u| + C$$
$$= \frac{1}{17} \ln |\sinh 17x| + C$$

In Example 28.14, we went through a rather long procedure to solve a problem similar to this. Now that we are able to integrate $\cosh x$, we can use the much simpler process demonstrated by the next example.

Application

EXAMPLE 28.23

A cable suspended between two poles takes the shape $y = 5 \cosh \dfrac{x}{5}$ for $-10.00 \le x \le 10.00$. Find the length of the cable, if x and y are in meters.

Solution   Recalling from Section 26.4, the formula for the arc length is $L = \int_a^b \sqrt{1 + [f'(x)]^2}\, dx$. We begin by differentiating the given function, with the result

$$y' = \sinh \left(\frac{x}{5} \right)$$

EXAMPLE 28.23 (Cont.)

Substituting this into the arc length formula, we have

$$L = \int_{-10}^{10} \sqrt{1 + \left[\sinh\left(\frac{x}{5}\right)\right]^2}\, dx$$

$$= \int_{-10}^{10} \sqrt{\cosh^2\left(\frac{x}{5}\right)}\, dx$$

$$= \int_{-10}^{10} \cosh\left(\frac{x}{5}\right) dx$$

$$= 5\sinh\frac{x}{5}\Big]_{-10}^{10}$$

$$\approx 5[3.63 - (-3.63)]$$

$$= 36.30$$

This cable is about 36.30 m long.

Exercise Set 28.3

In Exercises 1–40, determine the given integral.

1. $\displaystyle\int \sin\frac{x}{2}\, dx$

2. $\displaystyle\int \cos 3x\, dx$

3. $\displaystyle\int \tan 5x\, dx$

4. $\displaystyle\int x\csc x^2\, dx$

5. $\displaystyle\int \frac{1}{\cos x}\, dx$

6. $\displaystyle\int \frac{1}{\sin x}\, dx$

7. $\displaystyle\int \frac{1}{\cos^2 x}\, dx$

8. $\displaystyle\int \frac{\csc x}{\sin x}\, dx$

9. $\displaystyle\int \frac{\tan\sqrt{x}}{\sqrt{x}}\, dx$

10. $\displaystyle\int \tan^2 6x\, dx$

11. $\displaystyle\int \sin(4x - 1)\, dx$

12. $\displaystyle\int \frac{\sin x + \cos x}{\cos x}\, dx$

13. $\displaystyle\int \frac{\sin x}{\cos^2 x}\, dx$

14. $\displaystyle\int \frac{1 + \sin x}{\cos x}\, dx$

15. $\displaystyle\int \frac{\sec 3x \tan 3x}{5 + 2\sec 3x}\, dx$

16. $\displaystyle\int \cos^2 x \sin x\, dx$

17. $\displaystyle\int \sec^4 3x \tan 3x\, dx$

18. $\displaystyle\int \tan 3x \sec^2 3x\, dx$

19. $\displaystyle\int \frac{3\, dx}{\tan 3x}$

20. $\displaystyle\int \frac{5\, dx}{\sin^2 5x}$

21. $\displaystyle\int (\sec x + 2)^2\, dx$

22. $\displaystyle\int (\tan x + 3)^2\, dx$

23. $\displaystyle\int_0^{\pi/2} (\sec 0.5x + 5)^2\, dx$

24. $\displaystyle\int_{\pi/12}^{\pi/6} (\cot 3x - 4)^2\, dx$

25. $\displaystyle\int \frac{1 + \cos 4x}{\sin^2 4x}\, dx$

26. $\displaystyle\int x\csc^2 x^2\, dx$

27. $\displaystyle\int \tan\frac{x}{4}\, dx$

28. $\displaystyle\int x\sec^2(x^2 + 1)\tan(x^2 + 1)\, dx$

29. $\displaystyle\int_{\pi/4}^{\pi/2} \frac{1 + \cot^2 x}{\csc^2 x}\, dx$

30. $\displaystyle\int_0^{\pi/2} \frac{\cos x}{1 + \sin x}\, dx$

31. $\displaystyle\int_{\pi/4}^{\pi/2} \frac{\csc\sqrt{x}\cot\sqrt{x}}{\sqrt{x}}\, dx$

32. $\displaystyle\int_0^{\pi/8} \sec^5 2x \tan 2x\, dx$

33. $\displaystyle\int_{\pi/6}^{\pi/2} \frac{\cos^2 x}{\sin x}\, dx$

34. $\displaystyle\int_0^{5\pi/12} \frac{\sec^2 x}{2\tan x + 4}\, dx$

35. $\int x \sinh x^2 \, dx$

36. $\int \dfrac{\sinh x}{1 + \cosh x} \, dx$

37. $\int 3x \cosh x^2 \sqrt{\sinh x^2} \, dx$

38. $\int \operatorname{sech}^2 5x \, dx$

39. $\int \sinh^3 x \cosh^2 x \, dx$

40. $\int \tanh 3x \operatorname{sech} 3x \, dx$

Solve Exercises 41–52.

41. Show that $\int \cot u \, du = \ln \sin u + C$

42. Show that $\int \csc u \, du = \ln \csc u - \cot u + C$

43. Find the area of the region bounded by $y = \sin 2x$, $y = 0$, $x = 0$, and $x = \dfrac{\pi}{2}$.

44. Find the area of the region bounded by $y = \sec x$, $y = x$, $x = 0$, and $x = \dfrac{\pi}{4}$.

45. Find the average value of $y = \tan x$ on the interval $\left[0, \dfrac{\pi}{4}\right]$

46. *Physics* A certain particle has its velocity described by $v(t) = 5 \sin 2t$. If we know that at $t = 0$ it was located at $s = 4$, what is its distance function $s(t)$?

47. *Electricity* The electromotive force E of a particular electrical circuit is given by

$$E = 5 \sin 4t$$

where E is measured in volts and t is in seconds. Find the average value of E during the first second.

48. *Electricity* A 60-Hz alternator generates a current given by $i = 6.5 \sin(377t)$ A and flows through a 40-Ω resistor. The rate P at which heat is produced in the resistor is given by $P = 40i^2$. Determine the average rate of heat produced over one complete cycle. (Hint: Use the half-angle identity for the sine.)

✎ **In Your Words**

53. Explain how the integrals of the sine and cosine functions differ from those of the hyperbolic sine and hyperbolic cosine functions, respectively.

49. *Construction* A power cable suspended between two poles takes the shape $y = 12 \cosh \dfrac{x}{12}$ for $-18.0 \le x \le 18.0$. Find the length of the cable, if x and y are in meters.

50. *Construction* Electric wires suspended between two towers form a catenary with the equation $y = 60 \cosh \dfrac{x}{60}$, where x is measured in feet. Find the length of the suspended wire, if the towers are 120 ft apart.

51. *Construction* Electric wires suspended between two towers have the equation $y = 80 \cosh \dfrac{x}{80}$, where x is measured in feet. Find the length of the suspended wire, if the towers are 300 ft apart.

52. *Construction* The Gateway Arch in St. Louis has the shape of an inverted catenary. The shape of the arch can by approximated by

$$y = -127.7 \cosh \dfrac{x}{127.7} + 757.7$$

for $-315.0 \le x \le 315.0$, where all units are in feet.
(a) Determine the maximum height of the arch.
(b) Determine the width of the opening of the arch at its base.
(c) Approximate the total open area under the arch.
(d) Approximate the total length of the arch.

54. Explain how to use the method for developing $\int \tan u \, du$ and your knowledge of $\int \sinh u \, du$ and $\int \cosh u \, du$ to develop $\int \tanh u \, du$.

☰ 28.4
MORE TRIGONOMETRIC INTEGRALS

In Section 28.3, we learned the basic trigonometric integrals. In Section 28.1, we learned how to integrate some trigonometric functions by using the general power formula. In this section, we will learn some additional methods for integrating trigonometric functions. These methods will use the Pythagorean identities and half-angle formulas. Those equations are given below, so that you will be able to use them more easily.

Pythagorean Identities	$\sin^2\theta + \cos^2\theta = 1$
	$\tan^2\theta + 1 = \sec^2\theta$
	$1 + \cot^2\theta = \csc^2\theta$

Half-Angle Formulas	$\sin^2\theta = \dfrac{1 - \cos 2\theta}{2}$
	$\cos^2\theta = \dfrac{1 + \cos 2\theta}{2}$

The integrals discussed in this section fall into four types. We will give examples on integrating each type.

Type I: $\displaystyle\int \sin^n x \, dx \qquad \int \cos^n x \, dx,$ n a positive integer

This first type is really a special case of the second type. However, we will use this as a good learning experience. We will consider two cases: (a) n is odd and (b) n is even.

When n is an odd positive integer such as 3, 5, or 15, then $n-1$ is an even integer. We will first rewrite the function as $\sin^n x = \sin^{n-1} x \sin x$. Since $n-1$ is even, we can use the identity $\sin^2 x = 1 - \cos^2 x$ and then use the general power formula.

EXAMPLE 28.24

Evaluate $\displaystyle\int \sin^7 x \, dx$.

Solution

$$\int \sin^7 x \, dx = \int \sin^6 x \sin x \, dx$$

$$= \int (\sin^2 x)^3 \sin x \, dx$$

$$= \int (1 - \cos^2 x)^3 \sin x \, dx$$

If we let $u = \cos x$, then $du = -\sin x \, dx$ and the integral becomes

$$\int \sin^7 x \, dx = \int (1 - \cos^2 x)^3 \sin x \, dx$$

$$= -\int (1 - u^2)^3 \, du$$

$$= -\int (1 - 3u^2 + 3u^4 - u^6) \, du$$

$$= -\left(u - u^3 + \frac{3}{5}u^5 - \frac{1}{7}u^7\right) + C$$

$$= -\cos x + \cos^3 x - \frac{3}{5}\cos^5 x + \frac{1}{7}\cos^7 x + C$$

We would use the same technique to find $\int \cos^n x \, dx$ if n was odd, except that we would use the identity $\cos^2 x = 1 - \sin^2 x$ and the substitution $u = \sin x$.

If n is an even positive integer, we will use the half-angle formulas to help simplify the integral.

EXAMPLE 28.25

Determine $\int \sin^2 3x \, dx$.

Solution   Use the half-angle formula $\sin^2 \theta = \frac{1}{2}(1 - \cos \theta)$ with $\theta = 2x$. Then

$$\int \sin^2 3x \, dx = \frac{1}{2} \int (1 - \cos 6x) \, dx$$

$$= \frac{1}{2}x - \frac{1}{2} \int \cos 6x \, dx$$

$$= \frac{1}{2}x - \frac{1}{12} \sin 6x + C$$

EXAMPLE 28.26

Find $\int \cos^4 5x \, dx$.

Solution $\displaystyle \int \cos^4 5x \, dx = \int (\cos^2 5x)^2 \, dx$

$$= \int \left[\frac{1}{2}(1 + \cos 10x) \right]^2 dx$$

$$= \frac{1}{4} \int (1 + 2\cos 10x + \cos^2 10x) \, dx$$

$$= \frac{1}{4} \int \left[1 + 2\cos 10x + \frac{1}{2}(1 + \cos 20x) \right] dx$$

$$= \frac{1}{4}x + \frac{1}{20} \sin 10x + \frac{1}{8}x + \frac{1}{160} \sin 20x + C$$

$$= \frac{3}{8}x + \frac{1}{20} \sin 10x + \frac{1}{160} \sin 20x + C$$

Application

EXAMPLE 28.27

Find the area under the graph of $y = \cos^4 5x$ and above the x-axis from $x = 0$ to $x = 2\pi$.

EXAMPLE 28.27 (Cont.)

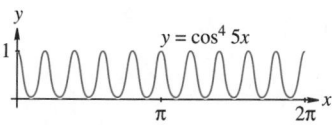

FIGURE 28.3

Solution The graph of this curve is shown in Figure 28.3. In Example 28.26, we found that $\int \cos^4 5x\, dx = \frac{3}{8}x + \frac{1}{20}\sin 10x + \frac{1}{160}\sin 20x + C$. Thus, the area under this curve over the indicated interval is

$$A = \int_0^{2\pi} \cos^4 5x\, dx$$

$$= \left[\frac{3}{8}x + \frac{1}{20}\sin 10x + \frac{1}{160}\sin 20x \right]_0^{2\pi}$$

$$\approx 2.3562 - 0 = 2.3562$$

The desired area is about 2.3562 square units.

Type II: $\int \sin^m x \cos^n x\, dx$

Again, we will consider two cases: **(a)** either m or n is an odd positive integer and **(b)** both m and n are even positive integers. Since Type I is a special case of Type II, where $m = 0$ or $n = 0$, we can use the same techniques we used on Type I integrals.

EXAMPLE 28.28

Find $\int \sin^{1/3} x \cos^5 x\, dx$.

Solution The power of the cosine function is an odd positive integer. We will rewrite $\cos^5 x$ as $(\cos^2 x)^2 \cos x$ and let $u = \sin x$.

$$\int \sin^{1/3} x \cos^5 x\, dx = \int \sin^{1/3} x (\cos^2 x)^2 \cos x\, dx$$

$$= \int \sin^{1/3} x (1 - \sin^2 x)^2 \cos x\, dx$$

$$= \int u^{1/3} (1 - u^2)^2\, du$$

$$= \int u^{1/3} (1 - 2u^2 + u^4)\, du$$

$$= \int (u^{1/3} - 2u^{7/3} + u^{13/3})\, du$$

$$= \frac{3}{4}u^{4/3} - \frac{3}{5}u^{10/3} + \frac{3}{16}u^{16/3} + C$$

$$= \frac{3}{4}\sin^{4/3} x - \frac{3}{5}\sin^{10/3} x + \frac{3}{16}\sin^{16/3} x + C$$

If both m and n are even numbers, then we can use the Pythagorean identity $\sin^2 x + \cos^2 x = 1$.

EXAMPLE 28.29

Determine $\int \sin^4 x \cos^2 x \, dx$.

Solution Substituting $1 - \sin^2 x$ for $\cos^2 x$, we obtain

$$\int \sin^4 x \cos^2 x \, dx = \int \sin^4 x (1 - \sin^2 x) \, dx$$
$$= \int (\sin^4 x - \sin^6 x) \, dx$$

The integrand is now in the form of a Type I integral and we can use those techniques to solve this integral. ■

Application

EXAMPLE 28.30

Find the volume of the solid of revolution generated by rotating the region bounded by $y = \sin^{1/6} x \cos^{5/2} x$, $y = 0$, $x = 0$, and $x = \dfrac{\pi}{2}$ around the x-axis.

Solution The graph of this region is shown in Figure 28.4a and the solid of revolution formed by rotating it around the x-axis is shown in Figure 28.4b. A typical disk has been drawn in the solid. The volume of this solid is

$$V = \pi \int_0^{\pi/2} \left(\sin^{1/6} x \cos^{5/2} x \right)^2 dx$$
$$= \pi \int_0^{\pi/2} \sin^{1/3} x \cos^5 x \, dx$$

From Example 28.28, we know that

$$\int \sin^{1/3} x \cos^5 x \, dx = \frac{3}{4} \sin^{4/3} x - \frac{3}{5} \sin^{10/3} x + \frac{3}{16} \sin^{16/3} x + C$$

Thus,

$$V = \pi \left[\frac{3}{4} \sin^{4/3} x - \frac{3}{5} \sin^{10/3} x + \frac{3}{16} \sin^{16/3} x \right]_0^{\pi/2}$$
$$= \pi \left(\frac{3}{4} - \frac{3}{5} + \frac{3}{16} \right)$$
$$= \frac{27}{80} \pi$$

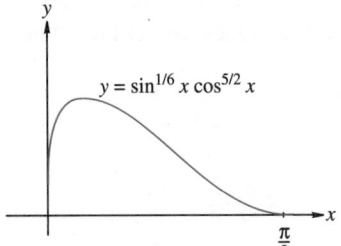

FIGURE 28.4a

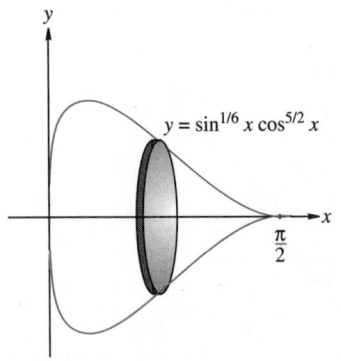

FIGURE 28.4b

Type III: $\displaystyle \int \tan^m x \sec^n x \, dx \qquad \int \cot^m x \csc^n x \, dx$

In this type, we will look first at the case where n is even. We will factor a $\sec^2 x$ out of the integrand and rewrite the remainder of the integrand in terms of the tangent by using the identity $\sec^2 x = 1 + \tan^2 x$. (If the integral involves the cotangent and cosecant functions, you would factor a $\csc^2 x$ out of the integrand and use $\csc^2 x = \cot^2 x + 1$.)

EXAMPLE 28.31

Find $\int \tan^3 x \sec^4 x \, dx$.

Solution
$$\int \tan^3 x \sec^4 x \, dx = \int \tan^3 x (\sec^2 x) \sec^2 x \, dx$$
$$= \int \tan^3 x (\tan^2 x + 1) \sec^2 x \, dx$$
$$= \int (\tan^5 x + \tan^3 x) \sec^2 x \, dx$$

If we let $u = \tan x$, then $du = \sec^2 x \, dx$, and we can rewrite this as

$$\int (u^5 + u^3) \, du = \frac{1}{6} u^6 + \frac{1}{4} u^4 + C$$
$$= \frac{1}{6} \tan^6 x + \frac{1}{4} \tan^4 x + C$$

If m is an odd number, then we will factor $\tan x \sec x$ out of the integrand and once again use the identity $\tan^2 x = \sec^2 x - 1$ (or factor $\csc x \cot x$ out of the integrand and use $\cot^2 x = \csc^2 x - 1$).

EXAMPLE 28.32

Evaluate $\int \tan^3 x \sec^3 x \, dx$.

Solution
$$\int \tan^3 x \sec^3 x \, dx = \int \tan^2 x \sec^2 x (\tan x \sec x) \, dx$$
$$= \int (\sec^2 x - 1) \sec^2 x (\tan x \sec x) \, dx$$
$$= \int (\sec^4 x - \sec^2 x)(\tan x \sec x) \, dx$$

If $u = \sec x$, then $du = \tan x \sec x \, du$, and we get

$$\int (u^4 - u^2) \, du = \frac{1}{5} u^5 - \frac{1}{3} u^3 + C$$
$$= \frac{1}{5} \sec^5 x - \frac{1}{3} \sec^3 x + C$$

Similar methods can be used for $\int \cot^m x \csc^n x \, dx$.

Type IV: $\int \tan^n x \, dx$ $\int \sec^n x \, dx$, n even

Both of these require the use of the identity $\tan^2 x = \sec^2 x - 1$. In Section 28.7, we will learn how to integrate $\sec^n x$ when n is odd.

EXAMPLE 28.33

Find $\int \tan^4 x \, dx$.

Solution $\int \tan^4 x \, dx = \int \tan^2 x \tan^2 x \, dx$

$$= \int \tan^2 x (\sec^2 x - 1) \, dx$$

$$= \int (\tan^2 x \sec^2 x - \tan^2 x) \, dx$$

$$= \int (\tan^2 x \sec^2 x - \sec^2 x + 1) \, dx$$

$$= \frac{1}{3} \tan^3 x - \tan x + x + C$$

EXAMPLE 28.34

Evaluate $\int \sec^4 3x \, dx$.

Solution $\int \sec^4 3x \, dx = \int \sec^2 3x \sec^2 3x \, dx$

$$= \int (1 + \tan^2 3x) \sec^2 3x \, dx$$

$$= \int (\sec^2 3x + \tan^2 3x \sec^2 3x) \, dx$$

$$= \frac{1}{3} \tan 3x + \frac{1}{9} \tan^3 3x + C$$

Exercise Set 28.4

In Exercises 1–30, evaluate the given integral.

1. $\int \sin^2 3x \cos 3x \, dx$

2. $\int \sin 2x \cos^2 2x \, dx$

3. $\int \cos x \sin^3 x \, dx$

4. $\int \sin^3 4x \cos^2 4x \, dx$

5. $\int \cos 5x \sin^4 5x \, dx$

6. $\int \cos^2 y \sin^5 y \, dy$

7. $\int \sin^2 x \cos^4 x \, dx$

8. $\int \sin^8 x \, dx$

9. $\int \cos^6 3x \, dx$

10. $\int \dfrac{\cos x}{\sin^3 x} \, dx$

11. $\int \sin^2 2\theta \cos^4 2\theta \, d\theta$

12. $\int \sin^4 3\theta \cos^2 3\theta \, d\theta$

13. $\int \sec^2 x \tan^2 x \, dx$

14. $\int \sec^4 y \tan^3 y \, dy$

15. $\int \csc^4 x \cot x \, dx$

16. $\int \dfrac{\sin^2 \theta}{\cos^4 \theta} \, d\theta$

17. $\int \sin^{1/2} 3\theta \cos^3 3\theta \, d\theta$

18. $\int \csc^6 2x \cot^2 2x \, dx$

19. $\int \csc x \cot^3 x \, dx$

20. $\int \sec t \tan^5 t \, dt$

21. $\int \tan^3 x \sec^2 x \, dx$

22. $\int \cot 2x \csc^4 2x \, dx$

23. $\int \tan^6 x \sec^2 x \, dx$

24. $\int \tan^5 x \, dx$

25. $\int \cot^6 x \, dx$

26. $\int \sec^6 2\theta \, d\theta$

27. $\int_0^{\pi/4} \tan^2 x \, dx$

28. $\displaystyle\int_{\pi/6}^{\pi/3} \tan^3 2t\, dt$

29. $\displaystyle\int_0^{\pi/2} \sin^4 x\, dx$

30. $\displaystyle\int_0^{\pi/6} \sec^3 2x \tan 2x\, dx$

Solve Exercises 31–36.

31. Find the volume of the solid of revolution generated by rotating the region bounded by $y = \cos x$, $y = 0$, $x = 0$, and $x = \dfrac{\pi}{3}$ around the x-axis.

32. *Electricity* Find the root mean square of the voltage for one period, if the voltage is given by $V = 80\sin 120\pi t$.

33. *Physics* The integral $I = \dfrac{mr^2}{4}\displaystyle\int_0^{2\pi} \sin^2\theta\, d\theta$ gives the moment of inertia about a diameter of a circular disk of radius r and mass m.
(a) Integrate to find I.

(b) Find the moment of inertia for a disk of radius 0.1 m and mass 12.4 kg.

34. *Physics* Under certain conditions, the rate of radiation by an accelerated charge is given by $R = \displaystyle\int \cos^3\theta\, d\theta$. Find an expression for the rate of radiation.

35. *Electronics* Find the effective current from $t = 0$ to $t = 0.5$ s, if $i = 2\sin t\sqrt{\cos t}$.

36. *Electronics* Find the rms of the current $i = 4\sin t\cos t$ from $t = 0$ to $t = \pi$ s.

In Your Words

37. Explain how to evaluate $\displaystyle\int \sin^m x \cos^n x\, dx$ when m is an odd positive integer and n is an even positive integer.

38. Explain how to evaluate $\displaystyle\int \sin^m x \cos^n x\, dx$ when m is even and n is odd.

39. Explain how to evaluate $\displaystyle\int \sin^m x \cos^n x\, dx$ when m and n are even positive integers.

≡ 28.5
INTEGRALS RELATED TO INVERSE TRIGONOMETRIC AND INVERSE HYPERBOLIC FUNCTIONS

In this section, we will discuss the integrals of certain algebraic functions that result from derivatives of inverse trigonometric functions.

Integrals Related to Inverse Trigonometric Functions ▬▬

In Section 27.3, we learned the derivatives of inverse trigonometric functions. In particular, we learned that

$$\frac{d}{dx}\sin^{-1}u = \frac{1}{\sqrt{1-u^2}}\frac{du}{dx}, \quad |u| < 1$$

$$\frac{d}{dx}\tan^{-1}u = \frac{1}{1+u^2}\cdot\frac{du}{dx}$$

$$\frac{d}{dx}\sec^{-1}u = \frac{1}{u\sqrt{u^2-1}}\frac{du}{dx}, \quad |u| > 1$$

The derivatives for the $\arccos x$, $\operatorname{arccot} x$, and $\operatorname{arccsc} x$ were similar except for the signs of the derivatives. We will look at ways of reversing these differentiation formulas.

The first integration formula we will consider is

$$\int \frac{du}{\sqrt{a^2-u^2}} = \arcsin\frac{u}{a} + C$$

In this formula, we assume that a is a constant.

EXAMPLE 28.35

Find $\int \dfrac{dx}{\sqrt{16-x^2}}$.

Solution

$$\int \frac{dx}{\sqrt{16-x^2}} = \int \frac{dx}{\sqrt{4^2-x^2}} = \arcsin\frac{x}{4} + C$$

The second integration formula we will consider is

$$\int \frac{du}{a^2+u^2} = \frac{1}{a}\arctan\frac{u}{a} + C$$

EXAMPLE 28.36

Find $\int \dfrac{3\,dx}{7+9x^2}$.

Solution $7+9x^2$ does not seem to be in the form a^2+x^2. If we let $u=3x$, then $u^2=9x^2$ and $du=3\,dx$, and if $a=\sqrt{7}$, then $a^2=7$. Thus $7+9x^2 = (\sqrt{7})^2 + (3x)^2$ and

$$\int \frac{du}{7+9x^2} = \frac{1}{\sqrt{7}}\arctan\frac{3x}{\sqrt{7}} + C$$

The last integral is

$$\int \frac{du}{u\sqrt{u^2-a^2}} = \frac{1}{a}\operatorname{arcsec}\frac{u}{a} + C$$

EXAMPLE 28.37

Evaluate $\displaystyle\int \frac{dx}{x\sqrt{16x^2-9}}$.

Solution   If $u = 4x$ and $a = 3$, then $du = 4\,dx$ and

$$\int \frac{dx}{x\sqrt{16x^2-9}} = \int \frac{4\,dx}{4x\sqrt{16x^2-9}}$$

$$= \int \frac{du}{u\sqrt{u^2-a^2}}$$

$$= \frac{1}{a}\operatorname{arcsec}\frac{u}{a} + C$$

$$= \frac{1}{3}\operatorname{arcsec}\frac{4x}{3} + C$$

EXAMPLE 28.38

Find $\displaystyle\int \frac{dx}{x^2+2x+10}$.

Solution   This does not appear to fit any of the methods we have yet seen. But if we complete the square, as in Section 8.3, we will see that

$$x^2 + 2x + 10 = x^2 + 2x + 1 + 9$$
$$= (x+1)^2 + 3^2$$

If we let $u = x + 1$ and $du = dx$, then

$$\int \frac{dx}{x^2+2x+10} = \int \frac{du}{u^2+3^2}$$

$$= \frac{1}{3}\arctan\frac{u}{3} + C$$

$$= \frac{1}{3}\arctan\frac{x+1}{3} + C$$

EXAMPLE 28.39

Determine $\displaystyle\int \frac{2x-7}{4x^2+9}\,dx$.

Solution   We will rewrite this as the sum of two integrals:

$$\int \frac{2x-7}{4x^2+9}\,dx = \int \frac{2x}{4x^2+9}\,dx - \int \frac{7}{4x^2+9}\,dx$$

EXAMPLE 28.39 (Cont.)

The first integral is in logarithmic form with $u = 4x^2 + 9$ and $du = 8x\, dx$. Multiplying by $\frac{4}{4}$, we get $\frac{1}{4} \int \frac{8x\, dx}{4x^2 + 9} = \frac{1}{4} \int \frac{du}{u}$. The second integral is of the form $\int \frac{dv}{a^2 + v^2}$ with $v = 2x$, $dv = 2\, dx$, and $a = 3$. Thus we have

$$\int \frac{2x - 7}{4x^2 + 9}\, dx = \int \frac{2x\, dx}{4x^2 + 9} - 7 \int \frac{dx}{4x^2 + 9}$$

$$= \frac{1}{4} \int \frac{du}{u} - 7 \int \frac{dv}{v^2 + 3^2}$$

$$= \frac{1}{4} \ln|u| - \frac{7}{3} \arctan \frac{v}{3} + C$$

$$= \frac{1}{4} \ln(4x^2 + 9) - \frac{7}{3} \arctan \frac{2x}{3} + C$$

Caution

Many integrals may look like they are of the inverse trigonometric form when they are not. Some integrals that use the general power formula or are of a logarithmic form look very similar to the integrals we studied in this chapter. It is important that you recognize the correct integration form to use. The exercises in this section will include integral forms from previous sections.

As an example of some of the confusion that is possible, look at the following integrals.

$$\textbf{(a)} \int \frac{dx}{1 + x^2} \qquad \textbf{(b)} \int \frac{x\, dx}{1 + x^2} \qquad \textbf{(c)} \int \frac{dx}{\sqrt{1 - x^2}} \qquad \textbf{(d)} \int \frac{x\, dx}{\sqrt{1 - x^2}}$$

Integrals **(a)** and **(b)** are *almost* identical. So are integrals **(c)** and **(d)**. Look carefully! Integral **(a)** is an inverse tangent integral, but integral **(b)** is in the logarithmic form. Integral **(c)** is an inverse sine integral, but integral **(d)** uses the general power formula.

Integrals Related to Inverse Hyperbolic Functions ▬▬▬

We present some integrals of the inverse hyperbolic functions in the box on the facing page. While it is not a complete list, it is sufficient for our purposes. You can verify each integral by differentiating the right-hand side of each equation.

EXAMPLE 28.40

Evaluate $\displaystyle\int_{-1}^{4} \frac{1}{\sqrt{9+4x^2}}\,dx$.

Solution This is in the form $\displaystyle\int \frac{1}{\sqrt{u^2+a^2}}\,du$ with $u=2x$, $a=3$, and $du=2\,dx$.
Thus, we have

$$\int_{-1}^{4} \frac{1}{\sqrt{9+4x^2}}\,dx = \frac{1}{2}\int_{-1}^{4} \frac{2}{\sqrt{9+4x^2}}\,dx$$

$$= \frac{1}{2}\sinh^{-1}\frac{2x}{3}\Bigg]_{-1}^{4}$$

$$\approx \frac{1}{2}[1.7074-(-0.6251)] \approx 1.1663$$

Integrals Involving Inverse Hyperbolic Functions

$$\int \frac{1}{\sqrt{u^2+a^2}}\,du = \sinh^{-1}\frac{u}{a}+C$$

$$\int \frac{1}{\sqrt{u^2-a^2}}\,du = \cosh^{-1}\frac{u}{a}+C$$

$$\int \frac{1}{a^2-u^2}\,du = \frac{1}{a}\tanh^{-1}\frac{u}{a}+C$$

$$\int \frac{1}{u\sqrt{a^2-u^2}}\,du = \frac{1}{a}\operatorname{sech}^{-1}\frac{u}{a}+C$$

Exercise Set 28.5

In Exercises 1–30, find the given integral.

1. $\displaystyle\int \frac{dx}{\sqrt{4-x^2}}$

2. $\displaystyle\int \frac{dx}{\sqrt{1-9x^2}}$

3. $\displaystyle\int \frac{dx}{\sqrt{4-9x^2}}$

4. $\displaystyle\int \frac{8\,dx}{1+16x^2}$

5. $\displaystyle\int \frac{-x\,dx}{\sqrt{9-x^2}}$

6. $\displaystyle\int \frac{dx}{x\sqrt{x^2-16}}$

7. $\displaystyle\int \frac{dx}{3x\sqrt{9x^2-4}}$

8. $\displaystyle\int \frac{x\,dx}{\sqrt{1-25x^2}}$

9. $\displaystyle\int \frac{dx}{1+(3-x)^2}$

10. $\displaystyle\int \frac{dx}{(2x+1)\sqrt{(2x+1)^2-4}}$

11. $\displaystyle\int \frac{4x-6}{4x^2+25}\,dx$

12. $\displaystyle\int \frac{dx}{(x+3)\sqrt{(x+3)^2-1}}$

13. $\displaystyle\int \frac{dx}{x^2+6x+10}$

14. $\displaystyle\int \frac{dx}{\sqrt{-4x^2+4x+15}}$

15. $\displaystyle\int \frac{x\,dx}{1+x^4}$

16. $\displaystyle\int \frac{\sec^2 x\,dx}{4+\tan^2 x}$

17. $\displaystyle\int \frac{\sin x\,dx}{\sqrt{1-\cos^2 x}}$

18. $\displaystyle\int \frac{x-5}{\sqrt{x^2-10x+16}}\,dx$

19. $\displaystyle\int_0^{\pi/4} \frac{\cos x\,dx}{1+\sin^2 x}$

20. $\displaystyle\int_3^4 \frac{dx}{\sqrt{-x^2+8x-15}}$

21. $\displaystyle\int \frac{dx}{\sqrt{x^2-25}}$

22. $\displaystyle\int \frac{dx}{\sqrt{(x-3)^2+16}}$

23. $\displaystyle\int \frac{dx}{\sqrt{25+9x^2}}$

24. $\displaystyle\int \frac{e^{2x}}{25-e^{4x}}\,dx$

25. $\displaystyle\int_0^{0.5} \frac{dx}{\sqrt{1-x^2}}$

26. $\displaystyle\int_0^{\sqrt{3}} \frac{x\,dx}{\sqrt{1+x^2}}$

27. $\displaystyle\int_1^5 \frac{dx}{x^2-4x+13}$

28. $\displaystyle\int_{-5}^{-1} \frac{dx}{\sqrt{-x^2-7x-6}}$

29. $\displaystyle\int_0^{\pi/6} \frac{\sec^2 x\,dx}{1+16\tan^2 x}$

30. $\displaystyle\int_0^1 \frac{x\,dx}{\sqrt{1+x^4}}$

Solve Exercises 31–38.

31. Find the area enclosed by $y=\dfrac{1}{1+x^2}$, $y=0$, $x=0$, and $x=1$.

32. Find the area enclosed by $y=\dfrac{1}{\sqrt{1-4x^2}}$, $x=-\frac14$, $x=\frac14$, and $y=0$.

33. The region bounded by the graphs of $y=e^x$, $y=\dfrac{1}{\sqrt{x^2+1}}$, and $x=1$ is rotated around the x-axis. Find the volume of the solid that is generated.

34. Find the moment M_y with respect to the y-axis of the region bounded by $y=\dfrac{1}{1+x^4}$, $y=0$, $x=1$, and $x=2$. Assume $\rho=1$.

35. Find the length of the curve $y=\sqrt{4-x^2}$ from $x=-1$ to $x=1$.

36. Find the centroid of the region bounded by $y=\dfrac{4}{\sqrt{4-x^2}}$, $x=0$, $x=1$, and $y=0$.

$\left[\text{Hint: }\displaystyle\int \frac{1}{4-x^2}\,dx = \frac14 \ln\left|\frac{x+2}{x-2}\right|+C\right]$

37. *Transportation* When a tractor trailer turns a corner at the intersection of two perpendicular streets, its rear wheels follow a curve known as a **tractrix**. If the origin is the middle of the intersection, the tractix can be shown to be the graph of $y=f(x)$, where

$$f'(x)=-\frac{1}{x\sqrt{1-x^2}}+\frac{x}{\sqrt{1-x^2}}$$

Find an equation for y, given the fact that $f(1)=0$.

38. *Robotics* During one cycle, the velocity in m/s of a robotic welding device is given by $v(t)=4t-\dfrac{12}{4-t^2}$, where t is time in seconds since the cycle began. Find an expression for the position $s(t)$, in meters, at time t if $s(0)=0$.

![pencil icon] **In Your Words**

39. In the caution in this section, you were warned that many integrals look alike. In particular, you were shown

(a) $\displaystyle\int \frac{dx}{1+x^2}$ (b) $\displaystyle\int \frac{x\,dx}{1+x^2}$

(c) $\displaystyle\int \frac{dx}{\sqrt{1-x^2}}$ (d) $\displaystyle\int \frac{x\,dx}{\sqrt{1-x^2}}$

and told that integrals a and b are *almost* identical as are integrals c and d. Explain how you determine which type of integral each one is.

40. Describe how to determine the integrals of $\displaystyle\int \frac{dx}{1+x^2}$ and $\displaystyle\int \frac{dx}{1-x^2}$.

☰ 28.6
TRIGONOMETRIC SUBSTITUTION

In Section 28.5, we looked at integrals that had square roots in the denominator. In this section, we will look at integrals that contain one of the forms of $\sqrt{a^2 - x^2}$, $\sqrt{a^2 + x^2}$, or $\sqrt{x^2 - a^2}$ in the integrand. In each case, we will use an appropriate trigonometric substitution to eliminate the radical and, hopefully, change the integrand into a trigonometric integral that we have already studied.

For each of the cases in this section, we will make a different substitution. The substitutions can be memorized, but you will probably prefer to learn the technique and derive the necessary relationships. Before we present the three substitutions, let's work an example.

EXAMPLE 28.41

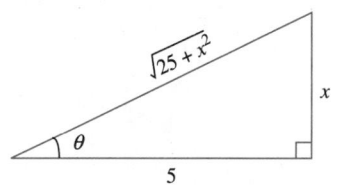

FIGURE 28.5

Evaluate $\displaystyle\int \frac{1}{(x^2 + 25)^{3/2}}\,dx$.

Solution   If we use the substitution $x = 5\tan\theta$, then $dx = 5\sec^2\theta\,d\theta$. Making these substitutions in the original integral we get

$$\int \frac{dx}{(x^2 + 25)^{3/2}} = \int \frac{5\sec^2\theta\,d\theta}{[(5\tan\theta)^2 + 25]^{3/2}}$$
$$= \int \frac{5\sec^2\theta\,d\theta}{(25\tan^2\theta + 25)^{3/2}}$$

But, $25\tan^2\theta + 25 = 25(\tan^2\theta + 1) = 25\sec^2\theta$, and so

$$\int \frac{5\sec^2\theta\,d\theta}{(25\sec^2\theta)^{3/2}} = \int \frac{5\sec^2\theta\,d\theta}{(5\sec\theta)^3}$$
$$= \int \frac{d\theta}{25\sec\theta}$$
$$= \frac{1}{25}\int \cos\theta\,d\theta$$
$$= \frac{1}{25}\sin\theta + C$$

This result is in terms of θ and not in terms of x. In order to get it in terms of x, we need to return to our substitution $x = 5\tan\theta$. In Figure 28.5, we have drawn a right triangle with an acute angle θ. Two sides are marked so that $\tan\theta = \dfrac{x}{5}$. The length of the third side was determined by using the Pythagorean theorem. Studying this triangle, we can see that $\sin\theta = \dfrac{x}{\sqrt{25 + x^2}}$. Putting this in the previous answer changes

$$\frac{1}{25}\sin\theta + C \text{ to } \frac{x}{25\sqrt{25 + x^2}} + C$$

Thus, we get the result

$$\int \frac{1}{(x^2 + 25)^{3/2}}\,dx = \frac{x}{25\sqrt{25 + x^2}} + C$$

≡ **Note**

This process had two parts. One part was the substitution to get the integrand written in terms of θ. The second part involved a right triangle with acute angle θ and the sides in terms of x. The three cases and the proper substitutions are given in Table 28.1.

TABLE 28.1

If the integrand involves	Use the substitution	Use the identity	And use the right triangle
$\sqrt{a^2 - u^2}$	$u = a\sin\theta$	$\cos^2\theta = 1 - \sin^2\theta$	
$\sqrt{a^2 + u^2}$	$u = a\tan\theta$	$\sec^2\theta = 1 + \tan^2\theta$	
$\sqrt{u^2 - a^2}$	$u = a\sec\theta$	$\tan^2\theta = \sec^2\theta - 1$	

EXAMPLE 28.42

Find $\displaystyle\int \frac{dx}{x^2\sqrt{4-9x^2}}$.

Solution This integral involves $\sqrt{a^2-u^2}$, where $u=3x$ and $a=2$. We will substitute $u=2\sin\theta$ for $3x$. Thus $x=\frac{2}{3}\sin\theta$ and $dx=\frac{2}{3}\cos\theta\,d\theta$, so we have

$$\int \frac{dx}{x^2\sqrt{4-9x^2}} = \int \frac{\frac{2}{3}\cos\theta\,d\theta}{\left(\frac{2}{3}\sin\theta\right)^2 \sqrt{4-(2\sin\theta)^2}}$$

$$= \int \frac{\frac{2}{3}\cos\theta\,d\theta}{\left(\frac{4}{9}\sin^2\theta\right)\sqrt{4(1-\sin^2\theta)}}$$

$$= \int \frac{\cos\theta\,d\theta}{\frac{2}{3}\sin^2\theta\sqrt{4\cos^2\theta}}$$

$$= \int \frac{\cos\theta\,d\theta}{\frac{2}{3}(\sin^2\theta)(2\cos\theta)}$$

EXAMPLE 28.42 (Cont.)

$$= \int \frac{3\,d\theta}{4\sin^2\theta}$$

$$= \int \frac{3}{4}\csc^2\theta = \frac{-3}{4}\cot\theta + C$$

From the triangle in Figure 28.6, we see that

$$\cot\theta = \frac{\sqrt{4-9x^2}}{3x}$$

and $\qquad -\dfrac{3}{4}\cot\theta + C = \dfrac{-3}{4}\dfrac{\sqrt{4-9x^2}}{3x} + C$

$$= -\frac{\sqrt{4-9x^2}}{4x} + C$$

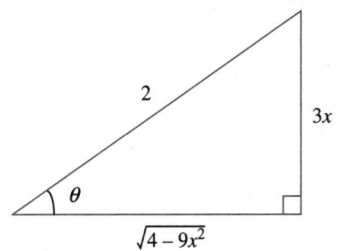

FIGURE 28.6

EXAMPLE 28.43

Determine $\displaystyle\int_0^4 \sqrt{16-x^2}\,dx$.

Solution This is in the form $\sqrt{a^2-u^2}$, so we'll let $u = x = 4\sin\theta$ and $dx = 4\cos\theta\,d\theta$. Substituting, we obtain

$$\int_0^4 \sqrt{16-x^2}\,dx = \int_{x=0}^{x=4} \sqrt{16-16\sin^2\theta}\,(4\cos\theta)\,d\theta$$

$$= \int_{x=0}^{x=4} 4\sqrt{1-\sin^2\theta}\,(4\cos\theta)\,d\theta$$

$$= 16\int_{x=0}^{x=4} \sqrt{\cos^2\theta}\,\cos\theta\,d\theta$$

$$= 16\int_{x=0}^{x=4} \cos^2\theta\,d\theta$$

From Section 28.5, we know that to integrate this we let $\cos^2\theta = \frac{1}{2}(1+\cos 2\theta)$ and get

$$16\int_{x=0}^{x=4} \frac{1}{2}(1+\cos 2\theta)\,d\theta = 8\left(\theta + \frac{1}{2}\sin 2\theta\right)\Big|_{x=0}^{x=4}$$

From the triangle in Figure 28.7, we see that if $\theta = \arcsin\dfrac{x}{4}$ and since $\sin 2\theta = 2\sin\theta\cos\theta = \dfrac{x\sqrt{16-x^2}}{8}$, we have

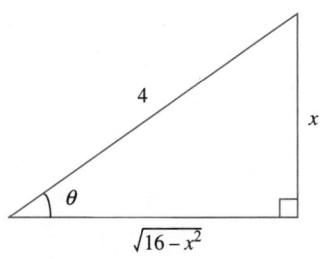

FIGURE 28.7

$$8\left(\theta + \frac{1}{2}\sin 2\theta\right)\Big|_{x=0}^{x=4} = 8\left(\arcsin\frac{x}{4} + \frac{x\sqrt{16-x^2}}{16}\right)\Big|_0^4$$

$$= 8\left(\frac{\pi}{2} + 0\right) - (0+0) = 4\pi$$

EXAMPLE 28.44

Find $\displaystyle\int \frac{dx}{x^2\sqrt{x^2+49}}$.

Solution We have the term $\sqrt{x^2+a^2}$ with $a=7$. We will use the substitution $x=7\tan\theta$ with $dx=7\sec^2\theta\,d\theta$. The problem then becomes

$$\int \frac{dx}{x^2\sqrt{x^2+49}} = \int \frac{7\sec^2\theta\,d\theta}{(7\tan\theta)^2\sqrt{49\tan^2\theta+49}}$$

$$= \int \frac{7\sec^2\theta\,d\theta}{49\tan^2\theta\sqrt{49(\tan^2\theta+1)}}$$

$$= \int \frac{7\sec^2\theta\,d\theta}{49\tan^2\theta\sqrt{49\sec^2\theta}}$$

$$= \int \frac{7\sec^2\theta\,d\theta}{49(\tan^2\theta)(7\sec\theta)}$$

$$= \int \frac{\sec\theta\,d\theta}{49\tan^2\theta}$$

$$= \frac{1}{49}\int \cot^2\theta\sec\theta\,d\theta$$

$$= \frac{1}{49}\int \frac{\cos^2\theta}{\sin^2\theta}\frac{1}{\cos\theta}\,d\theta$$

$$= \frac{1}{49}\int \frac{\cos\theta}{\sin^2\theta}\,d\theta$$

$$= \frac{1}{49}\left(\frac{-1}{\sin\theta}\right)+C = \frac{-1}{49}\csc\theta+C$$

From the triangle in Figure 28.8, we see that $\csc\theta = \dfrac{\sqrt{49+x^2}}{x}$, and so

$$\int \frac{dx}{x^2\sqrt{x^2+49}} = \frac{-\sqrt{49+x^2}}{49x}+C$$

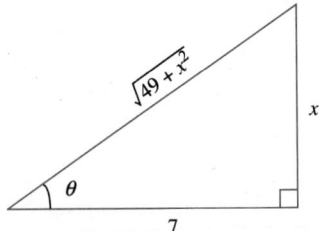

FIGURE 28.8

Exercise Set 28.6

In Exercises 1–36, determine the given integral.

1. $\displaystyle\int \frac{x}{\sqrt{9-x^2}}\,dx$

2. $\displaystyle\int \frac{dx}{\sqrt{9-x^2}}$

3. $\displaystyle\int \frac{x^2}{\sqrt{9-x^2}}\,dx$

4. $\displaystyle\int (9-x^2)^{3/2}\,dx$

5. $\displaystyle\int x^3\sqrt{9-x^2}\,dx$

6. $\displaystyle\int \frac{x}{\sqrt{x^2-9}}\,dx$

7. $\displaystyle\int \frac{dx}{\sqrt{x^2-9}}$

8. $\displaystyle\int \frac{x^3}{\sqrt{x^2-9}}\,dx$

9. $\displaystyle\int (x^2-9)^{1/2}\,dx$

10. $\displaystyle\int x^3\sqrt{x^2-9}\,dx$

11. $\displaystyle\int \frac{x}{\sqrt{x^2+9}}\,dx$

12. $\displaystyle\int \frac{dx}{\sqrt{x^2+9}}$

13. $\displaystyle\int \frac{dx}{x^2+9}$

14. $\displaystyle\int \frac{x^2}{x^2+9}\,dx$

15. $\displaystyle\int x^3\sqrt{9+x^2}\,dx$

16. $\displaystyle\int \frac{\sqrt{1-x^2}}{x^2}\,dx$

17. $\displaystyle\int \frac{\sqrt{4-3x^2}}{x^4}\,dx$

18. $\displaystyle\int \frac{\sqrt{4x^2-9}}{x^3}\,dx$

19. $\displaystyle\int \frac{x^3}{\sqrt{9x^2+4}}\,dx$

20. $\displaystyle\int \frac{dx}{(3x^2+6)^{3/2}}$

21. $\displaystyle\int \frac{\sqrt{x^2+1}}{x^2}\,dx$

22. $\displaystyle\int \frac{dx}{\sqrt{4-(x-1)^2}}$

23. $\displaystyle\int \sqrt{4-(x-1)^2}\,dx$

24. $\displaystyle\int \frac{dx}{\sqrt{4+(x-1)^2}}$

25. $\displaystyle\int \frac{dx}{\sqrt{(x-1)^2-4}}$

26. $\displaystyle\int \sqrt{x^2-2x+5}\,dx$

27. $\displaystyle\int \sqrt{x^2-2x-3}\,dx$

28. $\displaystyle\int \frac{dx}{(x^2-2x+10)^{3/2}}$

29. $\displaystyle\int \frac{dx}{(x-3)\sqrt{x^2-6x+25}}$

30. $\displaystyle\int \frac{x-3}{\sqrt{x^2-6x+25}}\,dx$

31. $\displaystyle\int_1^3 \frac{\sqrt{9-x^2}}{x^2}\,dx$

32. $\displaystyle\int_1^2 \frac{dx}{(8-x^2)^{3/2}}$

33. $\displaystyle\int_4^8 \frac{dx}{x\sqrt{x^2-4}}$

34. $\displaystyle\int_1^4 \frac{dx}{x^4\sqrt{x^2+4}}$

35. $\displaystyle\int_0^6 x^2\sqrt{36-x^2}\,dx$

36. $\displaystyle\int_0^{1.5\sqrt{3}} \frac{x^2\,dx}{\sqrt{9-x^2}}$

Solve Exercises 37–44.

37. Find the area under the graph of $y = \dfrac{x^3}{\sqrt{16-x^2}}$ from $x = 0$ to $x = 3$.

38. Find the arc length of the portion of the parabola $y = 10x - x^2$ that is above the x-axis.

39. Find the area bounded by the graph of $y = \sqrt{9+x^2}$, $y = 0$, $x = 0$ and $x = 4$.

40. *Civil engineering* The density of a 6-m bar is given by $\rho(x) = \sqrt{108 - 3x^2}$. What is the total mass of the bar?

41. An ellipsoid of revolution is obtained by revolving the ellipse $\dfrac{x^2}{25} + \dfrac{y^2}{4} = 1$ around the x-axis. What is the surface area of this ellipsoid?

42. *Electricity* The current in a circuit varies according to $i = t\sqrt{t+2}$ A. Find the charge transmitted during the interval from $t = 0$ to $t = 1$ s.

43. *Electricity* The current in a transformer is given by $i = \displaystyle\int \frac{\sqrt{t^2+1}}{9t^2}\,dt$. Find the current from $t = 0.5$ to $t = 1$ s.

44. *Petroleum engineering* A 10-ft-long cylindrical oil tank is lying on its side. The base of the tank is an ellipse described by the equation $x^2 + 4y^2 = 36$. If all measurements are in feet, find the volume of oil in the tank when the depth of the oil in the center is 2.5 ft.

In Your Words

45. What trigonometric substitution do you use for integrands of the form $\sqrt{a^2+u^2}$? Explain how this substitution is used.

46. In Exercise 45, you stated the trigonometric substitution you would use for integrands of the form $\sqrt{a^2+u^2}$. Will this same substitution work for integrands of the form $(a^2+u^2)^n$? If so, describe how you would use it. If not, tell how you would determine $\displaystyle\int (a^2+u^2)^n\,du$.

≡ 28.7
INTEGRATION BY PARTS

One of the most useful integration techniques is a direct result of the product rule of differentiation. You should remember that the product rule states that if f and g are two functions whose derivatives exist, then

$$D_x[f(x)g(x)] = f(x)g'(x) + g(x)f'(x)$$

Rewriting this, we get

$$f(x)g'(x) = D_x[f(x)g(x)] - g(x)f'(x)$$

Integrating both sides, we have

$$\int f(x)g'(x)\,dx = \int D_x[f(x)g(x)]\,dx - \int g(x)f'(x)\,dx$$

Since $\int D_x[f(x)g(x)]\,dx = f(x)g(x)$, this can be written as

$$\int f(x)g'(x)\,dx = f(x)g(x) - \int g(x)f'(x)\,dx$$

It is customary in this equation to let $u = f(x)$ and $v = g(x)$, so $du = f'(x)\,dx$ and $dv = g'(x)\,dx$. The formula then becomes what is called the integration by parts formula.

> **Integration by Parts**
>
> $$\int u\,dv = uv - \int v\,du$$

In order to apply this formula, we need to split the integrand into two parts. One part is labeled u and the other part is labeled dv. Unfortunately, there are no firm rules for selecting the part that is labeled u and the part that is dv. In some problems, there are several possible choices and only experience or trial and error will help you with the correct choices. There are, however, three general guidelines that should be followed.

> **Guidelines for Integrating by Parts**
>
> 1. dx is always part of dv.
> 2. You must be able to integrate dv.
> 3. $v\,du$ is easier to integrate than $u\,dv$.

 **Note**

As an aid, we will make a table for each problem that uses integration by parts. The table will be in the following form:

u	v
du	dv

Two of the cells in this table will be filled in when we select u and dv. We will have to determine the values for the other two cells before we can apply the integration by parts formula.

EXAMPLE 28.45

Determine $\int xe^x\,dx$.

Solution   We will select $u = x$ and $dv = e^x\,dx$. So, the table looks like

u	x	v	
du		dv	$e^x\,dx$

We will complete the table by taking the derivative of u and by integrating dv. The completed table now looks like this:

u	x	v	e^x
du	dx	dv	$e^x\,dx$

The integration by parts formula states that $\int u\,dv = uv - \int v\,du$. Using the above values produces

$$\int xe^x\,dx = xe^x - \int e^x\,dx$$
$$= xe^x - e^x + C$$
$$= (x-1)e^x + C$$

 Hint

When deciding which function to select for u, remember the *LIATE* rule: u should be the first available type of function from the following list:

Logarithmic
Inverse trigonometric
Algebraic
Trigonometric
Exponential

EXAMPLE 28.46

Evaluate $\int x\cos x\,dx$.

Solution   We set up our table with $u = x$ and $dv = \cos x\,dx$.

u	x	v	$\sin x$
du	dx	dv	$\cos x\,dx$

Then

$$\int x\cos x\,dx = uv - \int v\,du$$
$$= x\sin x - \int \sin x\,dx$$
$$= x\sin x - (-\cos x) + C$$
$$= x\sin x + \cos x + C$$

EXAMPLE 28.47

Find $\int x^3 e^{x^2}\, dx$.

Solution   The natural split is to let $u = x^3$ and $dv = e^{x^2}\, dx$. The difficulty with this is that we cannot integrate $e^{x^2}\, dx$. If we let $dv = xe^{x^2}\, dx$, then dv can be integrated. This, of course, means that $u = x^2$.

u	x^2	v	$\frac{1}{2}e^{x^2}$
du	$2x\, dx$	dv	$xe^{x^2}\, dx$

$$\int x^3 e^{x^2}\, dx = \frac{1}{2}x^2 e^{x^2} - \int xe^{x^2}\, dx$$
$$= \frac{1}{2}x^2 e^{x^2} - \frac{1}{2}e^{x^2} + C$$
$$= \frac{1}{2}e^{x^2}(x^2 - 1) + C$$

There are times when it is necessary to use integration by parts more than once on the same problem. This is demonstrated in the next example.

EXAMPLE 28.48

Evaluate $\int x^2 e^x\, dx$.

Solution   If we let $u = x^2$ and $dv = e^x\, dx$, then the table is as follows.

u	x^2	v	e^x
du	$2x\, dx$	dv	$e^x\, dx$

$$\int x^2 e^x\, dx = x^2 e^x - \int 2xe^x\, dx$$
$$= x^2 e^x - 2\int xe^x\, dx$$

The integral on the right-hand side can be obtained by integration by parts. (In fact, we did this in Example 28.45.) Using integration by parts, we obtain

$$\int x^2 e^x\, dx = x^2 e^x - 2(xe^x - e^x) + C$$
$$= e^x(x^2 - 2x + 2) + C$$

There are times that integration by parts seem to be leading us in the wrong direction when, in fact, it is not. This next example illustrates one of these times.

EXAMPLE 28.49

Find $\displaystyle\int e^x \sin x \, dx$.

Solution We will let $u = e^x$ and $dv = \sin x \, dx$. The table is

u	e^x	v	$-\cos x$
du	$e^x \, dx$	dv	$\sin x \, dx$

$$\int e^x \sin x \, dx = -e^x \cos x - \int -e^x \cos x \, dx$$
$$= -e^x \cos x + \int e^x \cos x \, dx$$

We will use integration by parts a second time.

u	e^x	v	$\sin x$
du	$e^x \, dx$	dv	$\cos x \, dx$

$$\int e^x \sin x \, dx = -e^x \cos x + e^x \sin x - \int e^x \sin x \, dx$$

It looks as if we have gone around in a circle. In some respects we have, because we have the same integral on the right-hand side that we started with. Adding this integral, $\displaystyle\int e^x \sin x \, dx$, to both sides of the previous equation produces

$$2\int e^x \sin x \, dx = -e^x \cos x + e^x \sin x$$
$$= -e^x(\cos x - \sin x)$$

Dividing both sides by 2 gives us the answer we are after.

$$\int e^x \sin x \, dx = -\frac{1}{2}e^x(\cos x - \sin x) + C$$
$$= \frac{1}{2}e^x(\sin x - \cos x) + C$$

EXAMPLE 28.50

Find $\displaystyle\int \sec^3 \theta \, d\theta$.

Solution We will let $u = \sec\theta$ and $dv = \sec^2\theta \, d\theta$. The table is

u		$\sec\theta$	v	$\tan\theta$
du		$\sec\theta\tan\theta \, d\theta$	dv	$\sec^2\theta \, d\theta$

EXAMPLE 28.50 (Cont.)

Applying integration by parts, we obtain

$$\int \sec^3 \theta \, d\theta = \sec\theta\tan\theta - \int \sec\theta\tan^2\theta \, d\theta$$

$$= \sec\theta\tan\theta - \int \sec\theta \left(\sec^2\theta - 1\right) d\theta$$

$$= \sec\theta\tan\theta - \int \sec^3\theta \, d\theta + \int \sec\theta \, d\theta$$

$$= \sec\theta\tan\theta - \int \sec^3\theta \, d\theta + \ln|\sec\theta + \tan\theta| + C$$

Adding $\int \sec^3 \theta \, d\theta$ to both sides, we get

$$2\int \sec^3\theta \, d\theta = \sec\theta\tan\theta + \ln|\sec\theta + \tan\theta| + C$$

and dividing by 2 produces

$$\int \sec^3\theta \, d\theta = \frac{1}{2}\sec\theta\tan\theta + \frac{1}{2}\ln|\sec\theta + \tan\theta| + C$$

Application

EXAMPLE 28.51

Find the volume of the region generated by rotating the area between the curves $y = \sqrt{\ln x}$, $y = x$, and the lines $x = 1$ and $x = e$ around the x-axis.

Solution A sketch of this region has been shaded in Figure 28.9. As you can see, $x > \sqrt{\ln x}$ for all values of $x > 1$. So, using the washer method, we obtain

$$V = \pi \int_1^e \left[x^2 - \left(\sqrt{\ln x}\right)^2 \right] dx$$

$$= \pi \int_1^e \left(x^2 - \ln x \right) dx$$

$$= \pi \int_1^e x^2 \, dx - \int_1^e \ln x \, dx$$

We will need to use integration by parts on $\int \ln x \, dx$. If we let $u = \ln x$ and $dv = dx$, then we can complete the table as follows.

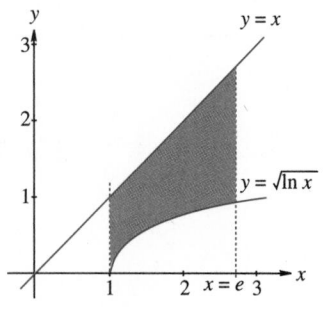

FIGURE 28.9

u	$\ln x$	v	x
du	$\dfrac{1}{x}dx$	dv	dx

EXAMPLE 28.51 (Cont.)

Thus, $\int \ln x \, dx = x \ln x - \int dx = x(\ln x - 1)$, so we are ready to complete the work for determining the volume.

$$V = \pi \int_1^e x^2 \, dx - \int_1^e \ln x \, dx$$

$$= \pi \left[\frac{1}{3}x^3 - x(\ln x - 1) \right]_1^e$$

$$= \pi \frac{e^3 - 4}{3} \approx 16.8447$$

The volume of this region when it is rotated around the x-axis is approximately 16.8447 cubic units.

 Hint

When a problem involves repeated integration by parts, a tabular method can often be used to organize the work. The technique works well for integrals of the form

$$\int P(x) \sin ax \, dx, \qquad \int P(x) \cos ax \, dx, \qquad \text{or} \qquad \int P(x) e^{ax} \, dx$$

where $P(x)$ is a polynomial. In each case, left $u = P(x)$. The next examples will demonstrate how this method can be used.

EXAMPLE 28.52

Find $\int x^3 \cos 2x \, dx$.

Solution   Here $P(x) = x^3$, so we will let $u = x^3$ and $dv = \cos 2x \, dx$. Next, we construct a table with three columns, as follows:

Alternate signs	u and its derivatives	dv and its antiderivatives
$+$	x^3	$\cos 2x$
$-$	$3x^2$	$\frac{1}{2}\sin 2x$
$+$	$6x$	$-\frac{1}{4}\cos 2x$
$-$	6	$-\frac{1}{8}\sin 2x$
$+$	0	$\frac{1}{16}\cos 2x$

$\uparrow$

Differentiate until you obtain a 0 as a derivative.

EXAMPLE 28.52 (Cont.)

The solution is given by adding the *signed* products of the diagonal entries.

$$\int x^3 \cos 2x \, dx = (x^3)\frac{1}{2}\sin 2x - (3x^2)\left(-\frac{1}{4}\cos 2x\right) + (6x)\left(-\frac{1}{8}\sin 2x\right)$$

$$- (6)\frac{1}{16}\cos 2x + C$$

$$= \frac{1}{2}x^3 \sin 2x + \frac{3}{4}x^2 \cos 2x - \frac{3}{4}x \sin 2x - \frac{3}{8}\cos 2x + C$$

EXAMPLE 28.53

Find $\int (5x^2 - 3x + 7)e^{4x} \, dx$.

Solution   Here $P(x) = 5x^2 - 3x + 7$, so $u = 5x^2 - 3x + 7$ and $dv = e^{4x} \, dx$. Again, we construct a table with three columns.

Alternate signs	u and its derivatives	dv and its antiderivatives
$+$	$5x^2 - 3x + 7$	e^{4x}
$-$	$10x - 3$	$\frac{1}{4}e^{4x}$
$+$	10	$\frac{1}{16}e^{4x}$
$-$	0	$\frac{1}{64}e^{4x}$

The solution is given by adding the *signed* products of the diagonal entries.

$$\int (5x^2 - 3x + 7)e^{4x} \, dx = (5x^2 - 3x + 7)\frac{1}{4}e^{4x} - (10x - 3)\frac{1}{16}e^{4x}$$

$$+ (10)\frac{1}{64}e^{4x} + C$$

$$= \left(\frac{5}{4}x^2 - \frac{11}{8}x + \frac{67}{32}\right)e^{4x} + C$$

Application

EXAMPLE 28.54

A suspension bridge, like the one shown in Figure 28.10a, has its roadbed supported by vertical cables that are connected to a main cable strung between two towers. The main cable does not take the shape of a catenary, but takes instead the shape of a parabola. If the towers are 1,000 ft apart and the main cable is connected to the towers 100 ft above the roadbed, determine the arc length of the main cable.

Solution   We will put the origin at the roadbed midway between the towers; that is, at the vertex of the parabola, as shown in Figure 28.10b. This means that the cables are attached to the towers at the points $(-500, 100)$ and $(500, 100)$. By placing the vertex at the origin, we know that this parabola is of the form $y = 4px^2$. Since $(500, 100)$ is a point on this parabola, we see that $100 = 4p(500)^2$, so $p = \dfrac{1}{10,000}$

EXAMPLE 28.54 (Cont.)

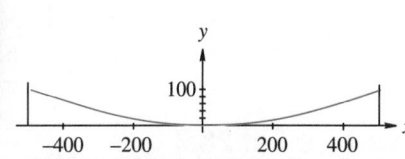

Courtesy of New York Convention and Visitors Bureau Inc.

FIGURE 28.10a   **FIGURE 28.10b**

and the parabola has the equation $y = \dfrac{4}{10,000}x^2 = \dfrac{1}{2,500}x^2$. For the arc length we need the derivative of y, which is $y' = \dfrac{2}{2,500}x = \dfrac{1}{1,250}x$. So, the length of the cable is

$$L = \int_{-500}^{500} \sqrt{1 + \left(\frac{1}{1,250}x\right)^2}\, dx = \frac{1}{1,250}\int_{-500}^{500} \sqrt{1,250^2 + x^2}\, dx$$

Using the trigonometric substitutions of $x = 1,250\tan\theta$ and $dx = 1,250\sec^2\theta\, d\theta$, the indefinite integral becomes

$$\frac{1}{1,250}\int \sqrt{1,250^2 + x^2}\, dx$$
$$= \frac{1}{1,250}\int \sqrt{1,250^2 + 1,250^2\tan^2\theta}\,\left(1,250\sec^2\theta\right) d\theta$$
$$= 1,250 \int \sqrt{1 + \tan^2\theta}\,\sec^2\theta\, d\theta$$
$$= 1,250 \int \sec^3\theta\, d\theta$$

From Example 28.50, we know that

$$\int \sec^3\theta\, d\theta = \frac{1}{2}\sec\theta\tan\theta + \frac{1}{2}\ln|\sec\theta + \tan\theta| + C$$

and so

$$L = \frac{1,250}{2}\left[\sec\theta\tan\theta + \ln|\sec\theta + \tan\theta|\right] + C$$
$$= 625\left[\sec\theta\tan\theta + \ln|\sec\theta + \tan\theta|\right] + C$$

EXAMPLE 28.54 (Cont.)

Now, since $\tan\theta = \dfrac{x}{1{,}250}$, we determine that $\sec\theta = \dfrac{\sqrt{1{,}250^2 + x^2}}{1{,}250}$ and we have

$$L = 625\left[\frac{\sqrt{1{,}250^2 + x^2}}{1{,}250}\,\frac{x}{1{,}250} + \ln\left|\frac{\sqrt{1{,}250^2 + x^2}}{1{,}250} + \frac{x}{1{,}250}\right|\right]_{-500}^{500}$$

$$= 625\left[\frac{2\sqrt{29}}{25} + \ln\left|\frac{\sqrt{29}+2}{5}\right| - \left(-\frac{2\sqrt{29}}{25} + \ln\left|\frac{\sqrt{29}-2}{5}\right|\right)\right]$$

$$= 625\left(\frac{4\sqrt{29}}{25} + \ln\left|\frac{\sqrt{29}+2}{\sqrt{29}-2}\right|\right) \approx 1{,}026.0606$$

Thus, the length of the main cable is approximately 1,026.0606 ft.

Exercise Set 28.7

In Exercises 1–24, find the given integral.

1. $\displaystyle\int x\ln x\,dx$

2. $\displaystyle\int \tan^{-1}x\,dx$

3. $\displaystyle\int x\sin 2x\,dx$

4. $\displaystyle\int x^5\ln x\,dx$

5. $\displaystyle\int x^2\ln x\,dx$

6. $\displaystyle\int x\sqrt{1+x}\,dx$

7. $\displaystyle\int x\sin^{-1}x\,dx$

8. $\displaystyle\int x^2\sin x\,dx$

9. $\displaystyle\int x^3 e^{2x}\,dx$

10. $\displaystyle\int x\tan^{-1}x\,dx$

11. $\displaystyle\int x\sqrt{4x+1}\,dx$

12. $\displaystyle\int x^2\cos 3x\,dx$

13. $\displaystyle\int x^2 e^{x/4}\,dx$

14. $\displaystyle\int e^{4x}\sin 2x\,dx$

15. $\displaystyle\int e^x\cos x\,dx$

16. $\displaystyle\int x^4 e^{2x}\,dx$

17. $\displaystyle\int x^3\cos x^2\,dx$

18. $\displaystyle\int x^2 e^{-x}\,dx$

19. $\displaystyle\int x^3 e^{-x}\,dx$

20. $\displaystyle\int \sin^2 x\,dx$

21. $\displaystyle\int e^{-x}\cos x\,dx$

22. $\displaystyle\int x4^x\,dx$

23. $\displaystyle\int x\cos 4x\,dx$

24. $\displaystyle\int e^{4x}\cos 2x\,dx$

Solve Exercises 25–38.

25. The region bounded by $y = \cos x$, $y = 0$, $x = 0$, and $x = \dfrac{\pi}{2}$ is rotated around the x-axis. What is the volume of the solid that is generated?

26. Find the volume of the solid generated by rotating the region in Exercise 25 around the y-axis.

27. *Electricity*   During a certain interval the current in a circuit varied according to $i = 4t\sin 2t$ A. What was the charge transferred during the period from $t = 0$ to $t = 4$ s?

28. The density of a 6-m bar is given by $\rho = xe^{-x}$ kg/m, where x is measured in meters. What is the total mass of the bar?

29. Find the centroid of the area bounded by $y = x^2 e^x$, $y = 0$, $x = -1$ and $x = 2$.

30. Find the root mean square of $y = \sqrt{\cos^{-1}x}$ from $x = 0$ to $x = 1$.

31. *Physics*   The force acting on a point on a coordinate axis is given by $F(x) = x^3\cos x\,dx$. Find the work required to move the point from $x = 0$ to $x = \dfrac{\pi}{2}$.

32. *Physics*   A particle moves along a coordinate axis with a velocity of $v(t) = t^5(1 - t^3)^{1/2}$. How far does it travel from the time $t = 0$ to $t = 1$, where t is in seconds?

33. *Construction* Determine the length of the main cable of a suspension bridge, if the towers are 1,000 ft apart and the main cable is connected to the towers 200 ft above the roadbed.

34. *Pharmacology* The rate of reaction to a drug is given by $r'(t) = 3t^2 e^{-t}$, where t is the time in hours since the drug was administered.
(a) Find an expression for $r(t)$, the total reaction to the drug, if $r(0) = 0$.
(b) Find the total reaction to the drug for the first 8 hours.

35. *Automotive technology* Friction between spinning circular surfaces is important in the design of clutch mechanisms. In one type of clutch mechanism, two flat circular disks of radius R are mounted so they can be brought into contact causing a net frictional force. In one case the contact pressure can be given by $P(r) = P_0 e^{-kr}$ where k is a positive constant, P_0 is the initial pressure, and r is the distance from the center of the discs.
(a) Evaluate the integral $F = 2\pi \int_0^R r \cdot P(r)\,dr$ in order to determine the total contact force between the discs.

(b) Evaluate the integral $T = 2\pi\mu \int_0^R r^2 \cdot P(r)\,dr$ in order to determine the total torque force between the discs, where μ is the coefficient of friction.

36. *Petroleum engineering* The production of a new oil field in thousands of barrels per month is estimated to be $R(t) = 10te^{-0.1t}$, where t is in months.
(a) Integrate R in order to determine the total production, P, of the well for the first t months.
(b) Estimate the total production in the first year of operation.
(c) Estimate the total production in the second year of operation.

37. *Environmental science* The concentration of particulate matter in ppm t hours after a factory ceases operation for the day is given by $C(t) = \dfrac{40\ln(t+2)}{(t+2)^2}$. Find the average concentration for the time period from $t = 0$ to $t = 6$.

38. *Electronics* When a periodic electromotive force is applied to an RL circuit, the current can be calculated using $i(t) = \dfrac{V}{L}\int e^{kt}\sin\omega t\,dt$. Determine the current.

In Your Words

39. Without looking in the text, write the integration by parts formula and describe how to use it.

40. Without looking in the text, summarize the guidelines for integrating by parts and explain how to use them.

41. A table was given in the note following the the guidelines for integrating by parts. Explain how to use this table.

≡ 28.8
USING INTEGRATION TABLES

There are many types of integrals that we encounter in applications. It would be almost impossible for anyone to remember how to find each of these integrals. For this reason, extensive tables of integrals have been prepared. One book contains a list of over 700 integrals. A much shorter list is given in Appendix C at the back of this book.

In the first seven sections of this chapter, we learned how to integrate many different types of functions. All of the integrals in Appendix C can be obtained from one of these methods, but you can often save a lot of time by using the tables for one of these integrals.

EXAMPLE 28.55

$$\int 5\cos^2 5x\, dx.$$

Solution   This integral involves a trigonometric function, so we look in the section of the table headed Trigonometric Forms. Note that this fits Formula 57 if we let $u = 5x$ and $du = 5\, dx$. Then by direct substitution, we have

$$\int 5\cos^2 5x\, dx = \frac{5x}{2} + \frac{1}{2}\sin 5x \cos 5x + C$$

Notice that Formula 57 has a second version, so we could have also written

$$\int 5\cos^2 5x\, dx = \frac{5x}{2} + \frac{1}{4}\sin 10x + C$$

Not all integrals can be solved by using direct substitution. Sometimes we have to multiply by a form of 1 in order to get the problem in the correct form.

EXAMPLE 28.56

Solve $\displaystyle\int \frac{2\, dx}{x(25x^2 - 9)^{3/2}}$.

Solution   The closest version to this form is Formula 45. Here $u = 5x$, $a = 3$, and $du = 5\, dx$. With these substitutions, Formula 45 becomes

$$\int \frac{du}{u(u^2 - a^2)^{3/2}} = \int \frac{5\, dx}{5x(25x^2 - 9)^{3/2}}$$
$$= \int \frac{dx}{x(25x^2 - 9)^{3/2}}$$

This is almost the integral we have been asked to solve. If we multiply by $\frac{2}{2}$, we get

$$\frac{1}{2}\int \frac{2\, dx}{x(25x^2 - 9)^{3/2}} = \frac{1}{2}\left(\frac{-1}{9\sqrt{25x^2 - 9}} - \frac{1}{27}\operatorname{arcsec}\frac{5x}{3} \right) + C$$

by Formula 45.

Caution

The integral you are trying to solve does not always bear a great resemblance to the integrals in the table. You will have to recognize the necessary substitutions.

EXAMPLE 28.57

Find $\displaystyle\int \frac{\tan 2x\, dx}{3 + 5\cos 2x}$.

Solution   A look through all of the integrals in the Trigonometric Forms section does not reveal any formula that even remotely resembles this one. In fact, there are no formulas that contain both the tangent and the cosine functions. That is a clue. Let's rewrite this entirely in terms of sine and cosine.

$$\int \frac{\tan 2x\, dx}{3 + 5\cos 2x} = \int \frac{\sin 2x\, dx}{\cos 2x(3 + 5\cos 2x)}.$$

EXAMPLE 28.57 (Cont.)

If we let $u = \cos 2x$, then $du = -2\sin 2x$ and we can rewrite this as

$$\frac{-1}{2}\int \frac{-2\sin 2x\, dx}{\cos 2x(3+5\cos 2x)} = \frac{-1}{2}\int \frac{du}{u(3+5u)}$$

This looks like Formula 48 with $a = 3$ and $b = 5$, and we see that

$$\int \frac{\tan 2x\, dx}{3+5\cos 2x} = \frac{-1}{2}\left(\frac{1}{3}\ln\left|\frac{\cos 2x}{3+5\cos 2x}\right|\right)+C$$

$$= \frac{-1}{6}\ln\left|\frac{\cos 2x}{3+5\cos 2x}\right|+C$$

As the last example, we will consider one of the formulas that has an integral in the answer.

EXAMPLE 28.58

Determine $\int \sin^4 x\, dx$.

Solution This is definitely of the form in Formula 64 with $n = 4$. According to that formula,

$$\int \sin^4 x\, dx = -\frac{1}{4}\sin^3 x\cos x + \frac{3}{4}\int \sin^2 x\, dx$$

The integral in the answer is also of the form of formula 64 with $n = 2$. It also fits formula 56. In this case, we have a choice. We will use formula 56 to get the final answer.

$$\int \sin^4 x\, dx = -\frac{1}{4}\sin^3 x\cos x + \frac{3}{4}\left(\frac{1}{2}x - \frac{1}{4}\sin 2x\right)+C$$

$$= -\frac{1}{4}\sin^3 x\cos x + \frac{3}{8}x - \frac{3}{16}\sin 2x + C$$

Any formula for an integral that contains an integral on the right-hand side of the formula is known as a **reduction formula**. You will notice that the function being integrated on the right-hand side in a reduction formula is the same as the original function, but is of a lower degree.

Exercise Set 28.8

In Exercises 1–24, use the table of integrals in Appendix C to integrate the given function.

1. $\int (1+\tan 3x)^2\, dx$

2. $\int \frac{\sqrt{16+25x^2}}{x}\, dx$

3. $\int \frac{dx}{x(4x-3)}$

4. $\int \frac{5x\, dx}{3+7x}$

5. $\int \frac{x^2\, dx}{x^6\sqrt{16+x^6}}$

6. $\int (25-4x^2)^{3/2}\, dx$

7. $\int \frac{x^2\, dx}{\sqrt{9-x^2}}$

8. $\int x^2\sqrt{9x^2-49}\, dx$

9. $\displaystyle\int \cos^3 x\, dx$

13. $\displaystyle\int e^{10x} \cos 6x\, dx$

17. $\displaystyle\int \arcsin 4x\, dx$

21. $\displaystyle\int x^3 e^{2x}\, dx$

10. $\displaystyle\int \sin 5x \sin 2x\, dx$

14. $\displaystyle\int x^2 \tan^{-1} x\, dx$

18. $\displaystyle\int \frac{dx}{(4-3x^2)^{3/2}}$

22. $\displaystyle\int x^4 \ln 2x\, dx$

11. $\displaystyle\int \sin^6 3x\, dx$

15. $\displaystyle\int x^7 \ln x\, dx$

19. $\displaystyle\int \frac{\sqrt{9+x^2}}{x}\, dx$

23. $\displaystyle\int x^3 \sin 2x\, dx$

12. $\displaystyle\int e^{-6x} \sin 10x\, dx$

16. $\displaystyle\int \frac{\sqrt{7-9x^2}}{x}\, dx$

20. $\displaystyle\int e^{\sin x} \sin x \cos x\, dx$

24. $\displaystyle\int \frac{\sqrt{\tan^2 2x - 9}}{\cos^2 2x}\, dx$

Solve Exercises 25–28.

25. *Electronics*   Find the average value of the voltage $V(t) = t^2 e^{5t}$ from $t = 0$ s to $t = 2.5$ s.

26. *Robotics*   The angular velocity, $\omega = \dfrac{d\theta}{dt}$, of a certain rotating system depends on the time according to the equation $\omega(t) = te^{0.25t}$, where t is in seconds.
    umber of revolutions, R, during the first n s is
    by $R = \dfrac{1}{2\pi} \displaystyle\int_0^n \omega(t)\, dt.$
    etermine the number of revolutions during the
    ırst n s.
    (b) Determine the number of revolutions during the first 10 s.

27. *Mechanical engineering*   The force in N acting on an object is given by $F = \dfrac{1}{49 - 9x^2}$, where x is the distance in meters from the initial position.   Find the work done in moving the object from $x = 0$ to $x = 2.00$.

28. *Electronics*   The current across capacitance $C = 0.02$ F in an electric circuit is given by $I = \dfrac{1}{t^2 + t}$, where $t \geq 0$ and t in s. The voltage across the capacitance at any time t is given by $V = \dfrac{1}{C} \displaystyle\int I\, dt$. If the voltage across this capacitance is 0 when $t = 1$, find the voltage across the capacitance at any time t.

✏️   **In Your Words**

29. Find a function in an earlier section that you were not able to integrate. Describe how you would use integration tables to now complete this integral and then use the tables to integrate the function.

30. Write an application in your technology area of interest that requires you to integrate a function. Give

your problem to a classmate and see if he or she understands and can solve your problem using any of the techniques in this chapter. Rewrite the problem as necessary to remove any difficulties encountered by your classmate.

▦ CHAPTER 28 REVIEW

Important Terms and Concepts

Catenary
Integration
   By parts
   By general power rule
   Of exponential functions
   Of inverse trigonometric functions

   Of logarithmic functions
   Of trigonometric functions
   By trigonometric substitution
   Using tables
Trigonometric substitution

Review Exercises

In Exercises 1–44, integrate the given function without the use of the table of integrals in Appendix C.

1. $\displaystyle\int xe^{3x}\,dx$

2. $\displaystyle\int \sin^3 x \cos^2 x\,dx$

3. $\displaystyle\int \frac{x}{\sqrt{25-x^2}}\,dx$

4. $\displaystyle\int x^3\sqrt{25-x^2}\,dx$

5. $\displaystyle\int \sin^4 2x \cos 2x\,dx$

6. $\displaystyle\int 2(e^x - e^{-x})\,dx$

7. $\displaystyle\int \frac{dx}{9x+5}$

8. $\displaystyle\int \tan^2 8x\,dx$

9. $\displaystyle\int \sin(7x+2)\,dx$

10. $\displaystyle\int \sin^3 2x \cos 2x\,dx$

11. $\displaystyle\int \sin^5 3x \cos^2 3x\,dx$

12. $\displaystyle\int \frac{dx}{x^2+4x+20}$

13. $\displaystyle\int \frac{dx}{(4x^2+49)^{3/2}}$

14. $\displaystyle\int x^3 \ln x\,dx$

15. $\displaystyle\int x^2 e^{x^3}\,dx$

16. $\displaystyle\int \frac{dx}{\sqrt{4x^2+49}}$

17. $\displaystyle\int \frac{dx}{\sqrt{4x^2+49}}$

18. $\displaystyle\int \cot 5x \csc^4 5x\,dx$

19. $\displaystyle\int \frac{\sec 4x \tan 4x}{9+2\sec 4x}\,dx$

20. $\displaystyle\int \sin^6 \frac{3x}{2}\,dx$

21. $\displaystyle\int \frac{[\ln(2x+1)]^5}{2x+1}\,dx$

22. $\displaystyle\int \frac{\arctan 7x}{1+49x^2}\,dx$

23. $\displaystyle\int \frac{x^2}{x^3+4}\,dx$

24. $\displaystyle\int 4x^3 e^{x^4}\,dx$

25. $\displaystyle\int \tan\frac{x}{5}\,dx$

26. $\displaystyle\int x\sin^3 x\,dx$

27. $\displaystyle\int e^{\cos x}\sin x\,dx$

28. $\displaystyle\int x^4 e^{-x}\,dx$

29. $\displaystyle\int \frac{\cos x\,dx}{\sin^2 x+9}$

30. $\displaystyle\int \frac{e^x\,dx}{e^x+16}$

31. $\displaystyle\int_1^e x^3 \ln x^2\,dx$

32. $\displaystyle\int e^{8x}\cos 2x\,dx$

33. $\displaystyle\int \sin^{1/3} 4x \cos^5 4x\,dx$

34. $\displaystyle\int \frac{\sin^3 x\,dx}{\cos^4 x}$

35. $\displaystyle\int \frac{\cos x\,dx}{\sqrt{16-4\sin^2 x}}$

36. $\displaystyle\int \frac{e^{5x}}{4-e^{5x}}\,dx$

37. $\displaystyle\int x^5 e^{x^2}\,dx$

38. $\displaystyle\int \frac{e^{3x}}{(e^{3x}-1)^2}\,dx$

39. $\displaystyle\int \frac{(\arctan 2x)^4}{1+4x^2}\,dx$

40. $\displaystyle\int \frac{\sec^2 5x\,dx}{2\tan 5x+9}$

41. $\displaystyle\int_0^{\ln 49} \sqrt{9+e^x}\,dx$

42. $\displaystyle\int_0^1 \arcsin\left(\frac{x}{2}\right)\,dx$

43. $\displaystyle\int_0^3 \frac{x^3\,dx}{\sqrt{9+x^2}}$

44. $\displaystyle\int_0^{\pi/2} \sin^3 x \cos^3 x\,dx$

▌ CHAPTER 28 TEST

In Exercises 1–6, integrate the given function.

1. $\displaystyle\int \frac{e^x\,dx}{\sqrt{9-e^x}}$

2. $\displaystyle\int \tan^2 4x \cos^4 4x\,dx$

3. $\displaystyle\int \frac{5\,dx}{x^2+1}$

4. $\displaystyle\int \frac{4x\,dx}{(x^2+1)^3}$

5. $\displaystyle\int xe^{4x}\,dx$

6. $\displaystyle\int \frac{e^{\tan x}}{\cos^2 x}\,dx$

Solve Exercises 7 and 8.

7. Find the volume of the solid generated by revolving about the x-axis the region bounded by the x-axis and the curve $y=\sqrt{x}\,e^x$ from $x=0$ to $x=1$.

8. Use the table of integrals in Appendix C to find the arc length of $y=x^2$ from $x=0$ to $x=2$.

Parametric Equations and Polar Coordinates

The position, velocity, and acceleration of a projectile, like this football, can be described by pairs of parametric equations. In Section 29.1, we will learn how to use these equations to describe the flight of such a projectile.

Courtesy of Eric Lars Bakke/Denver Broncos

In Chapter 11, we introduced parametric equations and polar coordinates. Later, in Chapter 14, we were able to use polar coordinates in our work with complex numbers. In this chapter, we will look at the calculus of these two types of equations.

29.1
DERIVATIVES OF PARAMETRIC EQUATIONS

The graph of a function such as $y = f(x)$ is intersected no more than once by a vertical line. To study more complicated graphs that are not graphs of functions, we need to use a different method. One of these methods is to introduce a third variable and to use a pair of equations. In this way, we let the equations

$$x = f(t) \quad \text{and} \quad y = g(t)$$

define a curve in the plane in terms of the variable t. These equations are called **parametric equations** and for each value of t they give us a value of x and a value of y.

EXAMPLE 29.1

Graph the parametric equations $x = \sin t$ and $y = 2\cos t$ for $0 \le t \le 2\pi$.

Solution A table of values follows with the graph shown in Figure 29.1.

t	0	$\frac{\pi}{6}$	$\frac{\pi}{3}$	$\frac{\pi}{2}$	$\frac{2\pi}{3}$	$\frac{5\pi}{6}$	π	$\frac{7\pi}{6}$	$\frac{4\pi}{3}$	$\frac{3\pi}{2}$	$\frac{5\pi}{3}$	$\frac{11\pi}{6}$	2π
$x = f(t)$	0	0.500	0.866	1	0.866	0.500	0	−0.500	−0.866	−1	−0.866	−0.500	0
$y = g(t)$	2	1.732	1.000	0	−1.000	−1.732	−2	−1.732	−1.000	0	1.000	1.732	2

As we see, the curve in Example 29.1 is an ellipse. In order to eliminate the parameter t and write this as an equation in terms of x and y, we use the fact that

$$\sin^2 t + \cos^2 t = 1$$

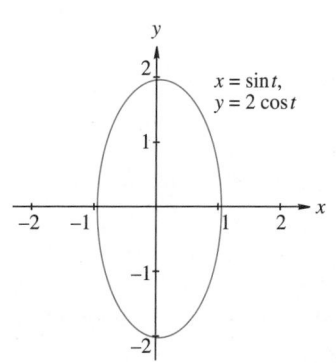

From the equation $x = \sin t$ and $y = 2\cos t$, we obtain $x = \sin t$ and $\frac{y}{2} = \cos t$. Substituting these into $\sin^2 t + \cos^2 t = 1$ produces

$$x^2 + \frac{y^2}{4} = 1$$

which is indeed the equation of an ellipse centered at the origin with a vertical major axis of 4 and minor axis of 2.

Now, suppose that we are interested in finding the slope of this graph, or the slope of the graph of any curve with equations in parametric form. This means we want the derivative $\frac{dy}{dx}$. The following statement shows how to find the derivative of the parametric equations for a curve.

FIGURE 29.1

> **Derivative of Parametric Equations**
>
> If a curve is described by the parametric equations $x = f(t)$ and $y = g(t)$, and if $\dfrac{dx}{dt} \neq 0$, then the derivative of these parametric equations is
>
> $$\frac{dy}{dx} = \frac{dy/dt}{dx/dt}$$

EXAMPLE 29.2

Find the derivative of the parametric equations in Example 29.1: $x = \sin t$, $y = 2\cos t$.

Solution To find this derivative, we need to recall that $\dfrac{d}{dt}(\sin t) = \cos t$ and $\dfrac{d}{dt}(\cos t) = -\sin t$.

$$\frac{dy}{dx} = \frac{dy/dt}{dx/dt} = \frac{-2\sin t}{\cos t} = -2\tan t$$

If you want to find a second derivative, then you take the derivative of the first derivative. Because the first derivative is a function of t and we want to take the derivative with respect to x, we must use the chain rule.

$$\frac{d}{dx}\left(\frac{dy}{dx}\right) = \frac{d}{dt}\left(\frac{dy/dt}{dx/dt}\right)$$

$$= \frac{\dfrac{d}{dx}\left(\dfrac{dy}{dt}\right)}{dx/dt}$$

$$= \frac{\dfrac{d}{dt}\left(\dfrac{dy}{dx}\right)}{dx/dt}$$

Thus, we have the following rule for $\dfrac{d^2y}{dx^2}$, the second derivative of parametric equations.

Second Derivative of Parametric Equations

If a curve is described by the parametric equations $x = f(t)$ and $y = g(t)$, and if $\dfrac{dx}{dt} \neq 0$, then the second derivative of these parametric equations is

$$\frac{d^2y}{dx^2} = \frac{\dfrac{d}{dt}\left(\dfrac{dy}{dx}\right)}{\dfrac{dx}{dt}}$$

Caution

The notation $\dfrac{d}{dt}\left(\dfrac{dy}{dx}\right)$ is used to indicate that we are finding the derivative of $\dfrac{dy}{dx}$ with respect to t, *not* $\dfrac{d}{dt}$ multiplied by $\dfrac{dy}{dx}$.

Similarly, the third derivative is $\dfrac{d^3y}{dx^3} = \dfrac{\dfrac{d}{dt}\left(\dfrac{d^2y}{dx^2}\right)}{\dfrac{dx}{dt}}$. Notice that the denominator for each derivative is $\dfrac{dx}{dt}$.

EXAMPLE 29.3

Find the second derivative of the parametric equations $x = \sin t$ and $y = 2\cos t$.

Solution   In Example 29.2, we found that $\dfrac{dx}{dt} = \cos t$ and $\dfrac{dy}{dx} = -2\tan t$. We need to first determine $\dfrac{d}{dt}\left(\dfrac{dy}{dx}\right) = \dfrac{d}{dt}(-2\tan t) = -2\sec^2 t$. Thus, the second derivative is

$$\begin{aligned}
\frac{d^2y}{dx^2} &= \frac{\dfrac{d}{dt}\left(\dfrac{dy}{dx}\right)}{\dfrac{dx}{dt}} \\
&= \frac{-2\sec^2 t}{\cos t} = -2\sec^3 t
\end{aligned}$$

EXAMPLE 29.4

Determine the first and second derivatives of $x = 4(t - \sin t)$ and $y = 4(1 - \cos t)$, then sketch the graph of the curve described by these parametric equations.

EXAMPLE 29.4 (Cont.)

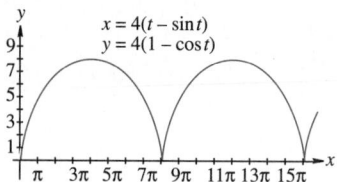

FIGURE 29.2

Solution   The first derivative is

$$\frac{dy}{dx} = \frac{dy/dt}{dx/dt} = \frac{4\sin t}{4(1-\cos t)}$$
$$= \frac{\sin t}{1-\cos t}$$

For the second derivative, we need $\dfrac{d}{dt}\left(\dfrac{dy}{dx}\right) = \dfrac{d}{dt}\left(\dfrac{\sin t}{1-\cos t}\right)$. We use the quotient rule to obtain this derivative. Thus, the second derivative is

$$\frac{d^2y}{dx^2} = \frac{\dfrac{d}{dt}\left(\dfrac{dy}{dx}\right)}{dx/dt} = \frac{\dfrac{(1-\cos t)\cos t - \sin t \sin t}{(1-\cos t)^2}}{4(1-\cos t)}$$
$$= \frac{\cos t - \cos^2 t - \sin^2 t}{4(1-\cos t)^3}$$
$$= \frac{\cos t - 1}{4(1-\cos t)^3}$$
$$= \frac{-1}{4(1-\cos t)^2}$$

You might have noticed that, except when t is an odd multiple of π, the second derivative is always negative. This indicates that, except for odd multiples of π, this curve is always concave downward. Maximum points occur when t is an odd multiple of π (π, 3π, etc.). Minima occur when t is an even multiple of π (0, 2π, 4π, etc.). When $t = \pi$, we get the maximum at $x = 4\pi$, $y = 8$. The graph is shown in Figure 29.2.

EXAMPLE 29.5

Consider the curve defined by the parametric equations $x = 4\cos t + 1$ and $y = 2\sin t$. **(a)** Find the equation for the line tangent to this curve at $t = \frac{\pi}{4}$, **(b)** find the equation for the line normal to this curve at $t = \frac{\pi}{4}$, **(c)** find all the points in the interval $[0, 2\pi]$ where the curve has a horizontal or vertical tangent line, and **(d)** sketch the graph of this curve and the tangent to the curve at $t = \frac{\pi}{4}$.

Solution

(a) To find the equation for the tangent to this curve, we evaluate the derivative at $t = \frac{\pi}{4}$.

$$\frac{dy}{dx} = \frac{dy/dt}{dx/dt}$$
$$= \frac{2\cos t}{-4\sin t} = -\frac{1}{2}\cot t$$

EXAMPLE 29.5 (Cont.)

When $t = \frac{\pi}{4}$, the slope of the tangent line is $-\frac{1}{2}\cot\left(\frac{\pi}{4}\right) = -\frac{1}{2}$. The x- and y-coordinates of this point are

$$x = 4\cos\left(\frac{\pi}{4}\right) + 1 = 4\left(\frac{\sqrt{2}}{2}\right) + 1 = 2\sqrt{2} + 1 \approx 3.828$$

$$y = 2\sin\left(\frac{\pi}{4}\right) = 2\left(\frac{\sqrt{2}}{2}\right) = \sqrt{2} \approx 1.414$$

Thus, the equation of the tangent line at this point is

$$y - y_0 = m(x - x_0)$$
$$y - \sqrt{2} = -\frac{1}{2}\left[x - \left(2\sqrt{2} + 1\right)\right]$$
$$= -\frac{1}{2}x + \sqrt{2} + \frac{1}{2}$$
$$y = -\frac{1}{2}x + 2\sqrt{2} + \frac{1}{2}$$

(b) Since the tangent line to the curve at $t = \frac{\pi}{4}$ has a slope of $-\frac{1}{2}$, the slope of the normal line to the curve at this point is $\dfrac{-1}{-1/2} = 2$. The equation of the normal line is

$$y - y_0 = m(x - x_0)$$
$$y - \sqrt{2} = 2\left[x - \left(2\sqrt{2} + 1\right)\right]$$
$$= 2x - 4\sqrt{2} - 2$$
$$y = 2x - 3\sqrt{2} - 2$$

(c) This curve has horizontal tangent lines when $\dfrac{dy}{dx} = 0$, and it has vertical tangent lines when $\dfrac{dy}{dx}$ is undefined. In (a), we found that $\dfrac{dy}{dx} = -\frac{1}{2}\cot t$.

Now, $\cot t = 0$ when $t = \frac{\pi}{2}$ and when $t = \frac{3\pi}{2}$. Thus, this curve has a horizontal tangent line when $t = \frac{\pi}{2}$, which is the point $x = 4\cos\left(\frac{\pi}{2}\right) + 1 = 1$, $y = 2\sin\left(\frac{\pi}{2}\right) = 2$ or $(1, 2)$. The other horizontal tangent line is when $t = \frac{3\pi}{2}$, or at the point $(1, -2)$.

The curve has vertical tangent lines when $t = 0$, or at the point $(5, 0)$, and when $t = \pi$ at the point $(-3, 0)$.

(d) The graph of the curve and the tangent when $t = \frac{\pi}{4}$, as drawn on a TI-82, are shown in Figure 29.3. The graph of the tangent line was drawn by using the parametric equations $x = t$, $y = -\frac{1}{2}t + 2\sqrt{2} + \frac{1}{2}$.

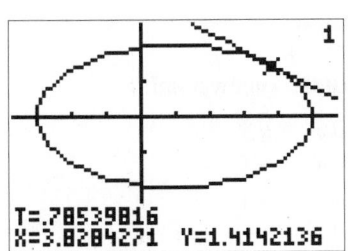

T=.78539816
X=3.8284271   Y=1.4142136

[−3.7, 5.7,1] × [−3.6, 2.6, 1]

FIGURE 29.3

Application

EXAMPLE 29.6

A certain projectile has its position at time t described by the parametric equations $s_x(t) = 6{,}250t \cos \frac{7\pi}{15}$ and $s_y(t) = 6{,}250t \sin \frac{7\pi}{15} - 16t^2$, where $s_x(t)$ represents the horizontal component of its position in feet at time t and $s_y(t)$ is the corresponding vertical component. When $t = 6$ s, determine **(a)** the horizontal and vertical components of the projectile's position, **(b)** its horizontal and vertical velocity, and **(c)** its horizontal and vertical acceleration.

Solutions **(a)** When $t = 6$, the horizontal and vertical components are $s_x(6) = 6{,}250(6) \cos \frac{7\pi}{15} \approx 3{,}919.82$ ft and $s_y(6) = 6{,}250(6) \sin \frac{7\pi}{15} - 16(6)^2 \approx 36{,}718.57$ ft.
(b) The horizontal and vertical velocities are found by taking the derivatives of the horizontal and vertical position components with respect to time. Thus, the horizontal velocity is $v_x(t) = s'_x(t) = 6{,}250 \cos \frac{7\pi}{15}$. When $t = 6$, we get $v_x(6) = 6{,}250 \cos \frac{7\pi}{15} \approx 653.30$ ft/s. The vertical velocity is $v_y(t) = s'_y(t) = 6{,}250 \sin \frac{7\pi}{15} - 32t$, and when $t = 6$, we obtain $v_y(6) = 6{,}250 \sin \frac{7\pi}{15} - 32(6) \approx 6{,}023.76$ ft/s.
(c) The horizontal and vertical accelerations are found by taking the derivatives of the horizontal and vertical velocity components with respect to time. Thus, the horizontal acceleration is $a_x(t) = v'_x(t) = 0$. The vertical acceleration is $a_y(t) = v'_y(t) = -32$ ft/s^2.

Exercise Set 29.1

In Exercises 1–10, find the first and second derivatives of the given parametric equations. Locate all extrema and any inflection points.

1. $x = t^2 + t$, $y = t + 1$
2. $x = t + 3$, $y = t^2 + t$
3. $x = t^2 - 6t + 12$, $y = t + 4$
4. $x = t^2 + 6t + 12$, $y = t + 4$
5. $x = t^2 + t$, $y = t^2 - t$

6. $x = 4t + 1$, $y = 9t^2$
7. $x = 3 + 4\cos t$, $y = -1 + \cos t$
8. $x = 3 + 4\cos t$, $y = 1 - \sin t$
9. $x = 2 + \sin t$, $y = -1 + \cos t$
10. $x = -2 + 4e^t$, $y = 3 + 2e^{-t}$

In Exercises 11–16, find an equation of the lines tangent and normal to the curve at the given point.

11. $x = 2t - 1$, $y = 4t^2 - 2t$, $t = 1$
12. $x = t - 4$, $y = t^3 + 2t^2 - 5t - 2$, $t = 1$
13. $x = t^3$, $y = t^2$, $t = -3$

14. $x = 2\cos t$, $y = 3\sin t$, $t = \frac{\pi}{4}$.
15. $x = 2 + \cos t$, $y = 2\sin t$, $t = \frac{\pi}{2}$
16. $x = e^t + 1$, $y = e^t + e^{-t}$, $t = 1$

In Exercises 17–20, find all the points where the curve has a horizontal or vertical tangent.

17. $x = t + 3$, $y = t^2 - 4t$
18. $x = t - 4$, $y = (t^2 + t)^2$

19. $x = 3\cos t$, $y = 5\sin t$
20. $x = t^2 + 1$, $y = \cos t$

Solve Exercises 21–26.

21. *Physics*　A certain projectile has its position, in feet, at time t described by the parametric equations $s_x(t) = 3{,}780t \cos \frac{4\pi}{15}$ and $s_y(t) = 3{,}780t \sin \frac{4\pi}{15} - 16t^2$. When $t = 5$ s, determine

    (a) the horizontal and vertical components of the projectile's position,

    (b) its horizontal and vertical velocity, and

    (c) its horizontal and vertical acceleration.

22. *Physics*　A certain projectile has its position, in meters, at time t described by the parametric equations $s_x(t) = 1{,}250t \cos \frac{3\pi}{11}$ and $s_y(t) = 1{,}250t \sin \frac{3\pi}{11} - 4.9t^2$. When $t = 8$ s, determine

    (a) the horizontal and vertical components of the projectile's position,

    (b) its horizontal and vertical velocity, and

    (c) its horizontal and vertical acceleration.

23. *Electronics*　An electron in an electric field moves in a path described by the parametric equations

    $$x = \frac{100}{\sqrt{t^2+1}} \quad \text{and} \quad y = \frac{100t}{\sqrt{t^2+1}}$$

    where x and y are in kilometers and t is in seconds.

    (a) Find the magnitude and direction of this electron when $t = 3.0$ s.

    (b) Find the horizontal and vertical components of this electron's velocity.

    (c) Find the magnitude and direction of the velocity of this electron when $t = 3.0$ s.

    (d) Find the horizontal and vertical components of this electron's acceleration.

    (e) Find the magnitude and direction of this electron's acceleration when $t = 3.0$ s.

In Your Words

27. Describe a situation when you might want to know the derivatives of the horizontal and vertical components of a function.

24. *Automotive technology*　The motion of a piece of gravel leaving a (rear) wheel at angle α with speed V, in feet per second, can be described by

    $$x = (V \cos \alpha)t$$
    $$y = (V \sin \alpha)t - \frac{1}{2}gt^2$$

    where $g \approx 32$ ft/s^2. Assume that a car is traveling 30 mph (44 ft/s) and that three pieces of gravel leave its rear tire, one at $\alpha = 30°$, one at $\alpha = 45°$, and one at $\alpha = 50°$.

    (a) Determine the horizontal and vertical components of each piece's velocity.

    (b) Find the magnitude and direction of the velocity of each piece of gravel at $t = 1$ s.

25. *Space technology*　During the first 100 s after launch, a spacecraft moves in a path described by the parametric equations

    $$x = 10\sqrt{t^4+1} - 1 \quad \text{and} \quad y = 40t^{3/2}$$

    where x and y are in meters and t is in seconds.

    (a) Find the horizontal and vertical components of this spacecraft's velocity.

    (b) Determine the equations for the magnitude and direction of this spacecraft's velocity at time t.

    (c) Find the magnitude and direction of the velocity of this spacecraft when $t = 5.0$ s.

26. *Space technology*　Consider the spacecraft in Exercise 25.

    (a) Find the horizontal and vertical components of this spacecraft's acceleration.

    (b) Determine the equations for the magnitude and direction of this spacecraft's acceleration at time t.

    (c) Find the magnitude and direction of the acceleration of this spacecraft when $t = 5.0$ s.

28. Without looking in the text, explain how to take the first and second derivatives of a function described by the parametric equations $x = f(t)$ and $y = g(t)$.

☰ 29.2
DIFFERENTIATION IN POLAR COORDINATES

Parametric equations use the rectangular or xy-coordinate system. Polar coordinates do not. They use the polar coordinate system, where every point in the plane can be represented as a pair of numbers (r, θ) with $r \geq 0$ and $0 \leq \theta \leq 2\pi$. The number represented by r is the distance of the point from the origin (or pole) and the number represented by θ is the direction.

The polar and rectangular coordinate systems are related by the equations

$$x = r \cos \theta \text{ and } y = r \sin \theta$$

Similarly, we have

$$r = \sqrt{x^2 + y^2} \text{ and } \tan \theta = \frac{y}{x}, \text{ if } x \neq 0$$

If we are given a polar equation, $r = f(\theta)$, it is a simple matter to find the derivative $\frac{dr}{d\theta}$. The problem is that this does not represent the same thing as the derivative $\frac{dy}{dx}$, where $y = g(x)$. As you remember, $\frac{dy}{dx}$ represents the slope of the tangent to the curve $y = g(x)$. The expression $\frac{dr}{d\theta}$ represents the change in the radius of the curve with respect to angle θ.

Let us return to our original equation, $r = f(\theta)$. We want to be able to use the derivative as a measure of the slope of the graph of this equation. This means we need to express $r = f(\theta)$ in terms of x and y or $x = r \cos \theta$ and $y = r \sin \theta$. Since $r = f(\theta)$, we get the equations

$$x = f(\theta) \cos \theta \text{ and } y = f(\theta) \sin \theta$$

These are parametric equations in terms of θ rather than t. Thus, using the parametric differentiation from Section 29.1,

$$\frac{dy}{dx} = \frac{dy/d\theta}{dx/d\theta}$$

$$= \frac{f'(\theta) \sin \theta + f(\theta) \cos \theta}{f'(\theta) \cos \theta - f(\theta) \sin \theta}$$

$$= \frac{r' \sin \theta + r \cos \theta}{r' \cos \theta - r \sin \theta}$$

This provides the following definition.

Derivative of an Equation in Polar Form

If $r = f(\theta)$ is a differentiable function, then the slope of the tangent line to the graph of r at the point (r, θ) is

$$\frac{dy}{dx} = \frac{r' \sin \theta + r \cos \theta}{r' \cos \theta - r \sin \theta}$$

provided $\frac{dx}{d\theta} \neq 0$ at (r, θ).

EXAMPLE 29.7

Find the derivative $\dfrac{dy}{dx}$ of $r = 1 + 2\cos\theta$.

Solution First we determine that $r' = \dfrac{dr}{d\theta} = -2\sin\theta$. Then

$$\begin{aligned}
\frac{dy}{dx} &= \frac{r'\sin\theta + r\cos\theta}{r'\cos\theta - r\sin\theta} \\[2mm]
&= \frac{(-2\sin\theta)(\sin\theta) + (1 + 2\cos\theta)\cos\theta}{(-2\sin\theta)(\cos\theta) - (1 + 2\cos\theta)\sin\theta} \\[2mm]
&= \frac{-2\sin^2\theta + \cos\theta + 2\cos^2\theta}{-4\sin\theta\cos\theta - \sin\theta} \\[2mm]
&= \frac{2(\cos^2\theta - \sin^2\theta) + \cos\theta}{-4\sin\theta\cos\theta - \sin\theta} \\[2mm]
&= \frac{2\cos 2\theta + \cos\theta}{-2\sin 2\theta - \sin\theta}
\end{aligned}$$

That is a rather messy formula. Fortunately, we will usually want to know the value of the derivative at a specific point, as shown in the next example.

EXAMPLE 29.8

Find the slope of the curve $r = 1 + \sin\theta$ at $\theta = \frac{2\pi}{3}$.

Solution $r' = \cos\theta$ and so

$$\begin{aligned}
\frac{dy}{dx} &= \frac{r'\sin\theta + r\cos\theta}{r'\cos\theta - r\sin\theta} \\[2mm]
&= \frac{\cos\theta\sin\theta + (1 + \sin\theta)\cos\theta}{\cos\theta\cos\theta - (1 + \sin\theta)\sin\theta} \\[2mm]
&= \frac{\cos\theta(1 + 2\sin\theta)}{\cos^2\theta - \sin\theta - \sin^2\theta}
\end{aligned}$$

When $\theta = \frac{2\pi}{3}$, $\sin\theta = \dfrac{\sqrt{3}}{2}$ and $\cos\theta = \dfrac{-1}{2}$, thus

$$\begin{aligned}
\frac{dy}{dx} &= \frac{\dfrac{-1}{2}\left[1 + 2\left(\dfrac{\sqrt{3}}{2}\right)\right]}{\left(\dfrac{-1}{2}\right)^2 - \dfrac{\sqrt{3}}{2} - \left(\dfrac{\sqrt{3}}{2}\right)^2} \\[3mm]
&= \frac{\dfrac{-1}{2}(1 + \sqrt{3})}{\dfrac{1}{4} - \dfrac{\sqrt{3}}{2} - \dfrac{3}{4}} = \frac{\dfrac{-1}{2}(1 + \sqrt{3})}{\dfrac{-\sqrt{3}}{2} - \dfrac{1}{2}} = \frac{1 + \sqrt{3}}{1 + \sqrt{3}} = 1
\end{aligned}$$

The slope of $r = 1 + \sin\theta$ at $\theta = \frac{2\pi}{3}$ is 1.

EXAMPLE 29.9

Find an equation of the normal and tangent lines to $r = 2\sin 2\theta$ at $\theta = \frac{\pi}{4}$.

Solution $r' = 4\cos 2\theta$ and so

$$
\begin{aligned}
\frac{dy}{dx} &= \frac{r'\sin\theta + r\cos\theta}{r'\cos\theta - r\sin\theta} \\
&= \frac{4\cos 2\theta \sin\theta + 2\sin 2\theta \cos\theta}{4\cos 2\theta \cos\theta - 2\sin 2\theta \sin\theta} \\
&= \frac{2\cos 2\theta \sin\theta + \sin 2\theta \cos\theta}{2\cos 2\theta \cos\theta - \sin 2\theta \sin\theta}
\end{aligned}
$$

When $\theta = \frac{\pi}{4}$, we have

$$
\begin{aligned}
\frac{dy}{dx} &= \frac{2\left(\cos\frac{\pi}{2}\right)\left(\sin\frac{\pi}{4}\right) + \left(\sin\frac{\pi}{2}\right)\left(\cos\frac{\pi}{4}\right)}{2\left(\cos\frac{\pi}{2}\right)\left(\cos\frac{\pi}{4}\right) - \left(\sin\frac{\pi}{2}\right)\left(\sin\frac{\pi}{4}\right)} \\
&= \frac{0 + \dfrac{\sqrt{2}}{2}}{0 - \dfrac{\sqrt{2}}{2}} = -1
\end{aligned}
$$

So, the slope of the tangent line is -1.

Next, we determine the x- and y-coordinates when $\theta = \frac{\pi}{4}$. When $\theta = \frac{\pi}{4}$, we have $r = 2\sin 2\theta = 2\sin\frac{\pi}{2} = 2$. So, the point $\left(2, \frac{\pi}{4}\right)$ is equivalent to the point (x, y), where $x = r\cos\theta = 2\cos\frac{\pi}{4} = \sqrt{2}$ and $y = r\sin\theta = \sqrt{2}$. Thus, the tangent line passing through the point with x- and y-coordinates $(\sqrt{2}, \sqrt{2})$ and with a slope of -1, has the equation $y - \sqrt{2} = -(x - \sqrt{2})$ or $y = -x + 2\sqrt{2}$.

The normal line has a slope of 1 and has the equation

$$y - \sqrt{2} = x - \sqrt{2}$$

or $\qquad y = x$

EXAMPLE 29.10

Find the extrema of $r = 1 + \sin\theta$.

Solution In Example 29.8 we found that

$$\frac{dy}{dx} = \frac{\cos\theta(1 + 2\sin\theta)}{\cos^2\theta - \sin\theta - \sin^2\theta}$$

The numerator is 0 if $\cos\theta = 0$ or if $\sin\theta = \frac{-1}{2}$. Thus, $\theta = \frac{\pi}{2}, \frac{3\pi}{2}, \frac{7\pi}{6}$, and $\frac{11\pi}{6}$ are critical values.

The denominator is 0 if

$$\cos^2\theta - \sin\theta - \sin^2\theta = 0$$

Since $\cos^2\theta = 1 - \sin^2\theta$, we can write this as

$$1 - \sin^2\theta - \sin\theta - \sin^2\theta = 0$$

or $\qquad 2\sin^2\theta + \sin\theta - 1 = 0$

EXAMPLE 29.10 (Cont.)

$$(2\sin\theta - 1)(\sin\theta + 1) = 0$$

This is a true equation when $\sin\theta = \frac{1}{2}$ or $\sin\theta = -1$, so $\frac{\pi}{6}$, $\frac{5\pi}{6}$, and $\frac{3\pi}{2}$ are critical values.

Checking these by the first derivative test, we see that $(r, \theta) = \left(2, \frac{\pi}{2}\right)$ and $\left(0, \frac{3\pi}{2}\right)$ are maxima and that $(r, \theta) = \left(\frac{1}{2}, \frac{7\pi}{6}\right)$ and $\left(\frac{1}{2}, \frac{11\pi}{6}\right)$ are minima.

As you can see from the graph of this curve in Figure 29.4, $\frac{5\pi}{6}$ and $\frac{\pi}{6}$ are not where r (or y) is a maximum or minimum. The first derivative test does indicate that they are extreme points and indeed the maximum value of x occurs when $\theta = \frac{\pi}{6}$ and the minimum x-value occurs when $\theta = \frac{5\pi}{6}$. The tangent lines are vertical at the points $\left(\frac{3}{2}, \frac{\pi}{6}\right)$ and $\left(\frac{3}{2}, \frac{5\pi}{6}\right)$.

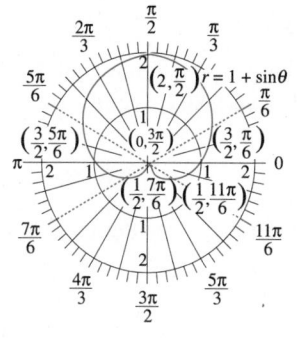

FIGURE 29.4

Exercise Set 29.2

In Exercises 1–8, find $\dfrac{dy}{dx}$.

1. $r = 3\sin\theta$
2. $r = -2\cos\theta$
3. $r = 1 + \cos\theta$
4. $r = \cos 3\theta$
5. $r = 1 + \cos 3\theta$
6. $r = \csc\theta$
7. $r = \tan\theta$
8. $r = e^{\theta}$

In Exercises 9–16, find the equations (in rectangular coordinates) of the tangent and normal lines to the given curve at the given point. The coordinates of the given points are in polar coordinates, (r, θ).

9. $r = \sin\theta$, $\left(1, \frac{\pi}{2}\right)$
10. $r = 2\cos\theta$, $\left(\sqrt{2}, \frac{\pi}{4}\right)$
11. $r = 5\sin 3\theta$, $\left(\frac{5}{\sqrt{2}}, \frac{\pi}{12}\right)$
12. $r = 2 - 3\sin\theta$, $\left(\frac{1}{2}, \frac{5\pi}{6}\right)$
13. $r = 6\sin^2\theta$, $\left(4.5, \frac{2\pi}{3}\right)$
14. $r = 2 + 3\sec\theta$, $\left(-4, \frac{2\pi}{3}\right)$
15. $r = e^{\theta}$, $\left(2.8497, \frac{\pi}{3}\right)$
16. $r = \tan\theta$, $\left(-1, \frac{3\pi}{4}\right)$

In Exercises 17–20, find all extrema.

17. $r = 3\cos 2\theta$
18. $r = 1 - \cos\theta$
19. $r = 1 + 2\cos\theta$
20. $r = \sin^2\theta$

Solve Exercises 21 and 22.

21. *Robotics*   A robot arm joint moves in the elliptical path described by the equation $r = \dfrac{10}{5 + 3\cos\theta}$, where r is in meters and $0 \le \theta < 2\pi$.

(a) Find $\dfrac{dr}{d\theta}$.

(b) Determine the value of θ where r is a maximum.

22. *Robotics*   Consider the robot arm joint described in Exercise 21.

(a) Find expressions for the x- and y-coordinates of r in terms of θ.

(b) Determine $\dfrac{dy}{dx}$.

(c) Determine where the graph has horizontal tangents and where it has vertical tangents.

In Your Words

23. What is a physical interpretation of the expression $\dfrac{dr}{d\theta}$?

24. What is a physical interpretation of the expression $\dfrac{dy}{dx}$ when $x = r\cos\theta$ and $y = r\sin\theta$ and $r = f(\theta)$ is a polar equation?

≡ 29.3
ARC LENGTH AND SURFACE AREA REVISITED

In Section 26.4, we studied the arc length of a curve and how the arc length could be used to find the surface area of a solid generated around a line. The curves we studied before were all expressed in rectangular coordinates. In this section, we will learn how to determine and use the arc length of a curve that is in polar form or given as parametric equations.

Arc Length of Parametric Equations

If we have a curve described by the function $y = f(x)$, then the length L of the curve from $x = a$ to $x = b$ is given by the equation

$$L = \int_a^b \sqrt{1 + [f'(x)]^2}\,dx = \int_a^b \sqrt{1 + \left(\frac{dy}{dx}\right)^2}\,dx$$

If we differentiate this equation with respect to x, we get

$$\frac{dL}{dx} = \sqrt{1 + \left(\frac{dy}{dx}\right)^2}$$

Squaring both sides results in

$$\left(\frac{dL}{dx}\right)^2 = 1 + \left(\frac{dy}{dx}\right)^2$$

or

$$(dL)^2 = (dx)^2 + (dy)^2$$

Now, suppose that x and y are both functions of a third variable t. Then they are parametric equations of the form $x = f(t)$ and $y = g(t)$. Differentiating the previous equation with respect to t we obtain

$$\left(\frac{dL}{dt}\right)^2 = \left(\frac{dx}{dt}\right)^2 + \left(\frac{dy}{dt}\right)^2$$

or

$$\frac{dL}{dt} = \sqrt{\left(\frac{dx}{dt}\right)^2 + \left(\frac{dy}{dt}\right)^2}$$

Integrating from $t = t_1$ to $t = t_2$, we get the formula for the arc length of a curve represented by parametric equations.

Arc Length of a Curve Represented by Parametric Equations

$$L = \int_{t_1}^{t_2} \sqrt{\left(\frac{dx}{dt}\right)^2 + \left(\frac{dy}{dt}\right)^2}\, dt$$

EXAMPLE 29.11

Find the length of the curve $x = 2t^2 + 5$, $y = 3t^3 - 1$ from $t = 0$ to $t = 2$.

Solution   To use the new formula, we need to first differentiate x and y. $\dfrac{dx}{dt} = 4t$ and $\dfrac{dy}{dt} = 9t^2$. Thus, we have

$$L = \int_0^2 \sqrt{(4t)^2 + (9t^2)^2}\, dt$$

$$= \int_0^2 \sqrt{16t^2 + 81t^4}\, dt$$

$$= \int_0^2 t\sqrt{16 + 81t^2}\, dt$$

If we let $u = 16 + 81t^2$, then $du = 162t\, dt$ and we have

$$\frac{1}{162}\int \sqrt{u}\, du = \frac{1}{243}u^{3/2} = \frac{1}{243}\left(16 + 81t^2\right)^{3/2}\Big|_0^2$$

$$= \frac{1}{243}\left(340\sqrt{340} - 64\right) \approx 25.536$$

The arc length of this curve is about 25.536 units.

EXAMPLE 29.12

Find the length of $x = e^t \sin t$, $y = e^t \cos t$ from $t = 0$ to $t = \pi$.

Solution   Again, we begin by finding

$$\frac{dx}{dt} = e^t \cos t + e^t \sin t = e^t(\cos t + \sin t)$$

and

$$\frac{dy}{dt} = -e^t \sin t + e^t \cos t = e^t(\cos t - \sin t)$$

Thus

$$
\begin{aligned}
L &= \int_0^\pi \sqrt{[e^t(\cos t + \sin t)]^2 + [e^t(\cos t - \sin t)]^2}\, dt \\
&= \int_0^\pi \sqrt{e^{2t}(\cos^2 t + 2\cos t \sin t + \sin^2 t) + e^{2t}(\cos^2 t - 2\cos t \sin t + \sin^2)}\, dt \\
&= \int_0^\pi \sqrt{e^{2t}(1 + 2\cos t \sin t) + e^{2t}(1 - 2\cos t \sin t)}\, dt \\
&= \int_0^\pi \sqrt{2e^{2t}}\, dt = \sqrt{2}\int_0^\pi e^t\, dt = \sqrt{2}e^t \Big|_0^\pi = \sqrt{2}(e^\pi - 1) \approx 31.312
\end{aligned}
$$

So, the arc length of this curve is approximately 31.312 units.

Arc Length of a Polar Equation

Suppose we have a curve given by the polar equation $r = f(\theta)$, where f is continuous on the interval $a \le \theta \le b$. We know that we can express this in rectangular coordinates using

$$x = r\cos\theta \qquad\qquad \text{and} \qquad y = r\sin\theta$$
$$= f(\theta)\cos\theta \qquad\qquad\qquad\qquad = f(\theta)\sin\theta$$

We can find the arc length of this curve using the form for parametric equations. We have $\dfrac{dx}{d\theta} = f'(\theta)\cos\theta - f(\theta)\sin\theta$ and $\dfrac{dy}{d\theta} = f'(\theta)\sin\theta + f(\theta)\cos\theta$. Squaring these and adding produces the following. (Notice that we reversed the order in which the terms were written.)

$$\left(\frac{dx}{d\theta}\right)^2 = [f(\theta)]^2 \sin^2\theta - 2f'(\theta)f(\theta)\sin\theta\cos\theta + [f'(\theta)]^2 \cos^2\theta$$

$$\left(\frac{dy}{d\theta}\right)^2 = [f(\theta)]^2 \cos^2\theta + 2f'(\theta)f(\theta)\sin\theta\cos\theta + [f'(\theta)]^2 \sin^2\theta$$

Adding, we get

$$\left(\frac{dx}{d\theta}\right)^2 + \left(\frac{dy}{d\theta}\right)^2 = [f(\theta)]^2 (\sin^2\theta + \cos^2\theta) + [f'(\theta)]^2 (\cos^2\theta + \sin^2\theta)$$

$$= [f(\theta)]^2 + [f'(\theta)]^2$$

Thus, the length L of a curve in polar form $r = f(\theta)$ from $\theta = a$ to $\theta = b$ is given by the following formula.

> **Arc Length of a Curve in Polar Form:** $r = f(\theta)$
>
> If $r = f(\theta)$ has a continuous first derivative for $a \leq \theta \leq b$ and if $r = f(\theta)$ is traced exactly once for $a \leq \theta \leq b$, then the length of the curve is
>
> $$L = \int_a^b \sqrt{[f(\theta)]^2 + [f'(\theta)]^2}\, d\theta$$
>
> $$= \int_a^b \sqrt{r^2 + \left(\frac{dr}{d\theta}\right)^2}\, d\theta$$

EXAMPLE 29.13

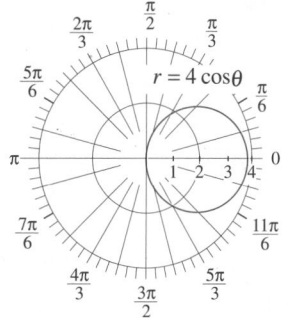

FIGURE 29.5

Find the arc length of $r = 4\cos\theta$ from $\theta = 0$ to $\theta = \pi$.

Solution A sketch of this curve is shown in Figure 29.5. We need $r' = -4\sin\theta$, then

$$L = \int_0^\pi \sqrt{(4\cos\theta)^2 + (-4\sin\theta)^2}\, d\theta$$

$$= \int_0^\pi \sqrt{16\cos^2\theta + 16\sin^2\theta}\, d\theta$$

$$= \int_0^\pi \sqrt{16(\cos^2\theta + \sin^2\theta)}\, d\theta = \int_0^\pi \sqrt{16}\, d\theta$$

$$= \int_0^\pi 4\, d\theta = 4\theta \Big|_0^\pi = 4\pi$$

The arc length of $r = 4\cos\theta$ from $\theta = 0$ to $\theta = \pi$ is 4π units. As you can see from the graph in Figure 29.5, this curve is a circle of radius 2. Thus, the arc length is its circumference, $C = 2\pi r = 2\pi(2) = 4\pi$ units.

EXAMPLE 29.14

Find the arc length of $r = 1 + \sin\theta$ from $\theta = 0$ to $\theta = 2\pi$.

Solution The sketch of this curve is in Figure 29.6.
Here $r' = \cos\theta$, and so the arc length is

$$L = \int_0^{2\pi} \sqrt{(1 + \sin\theta)^2 + (\cos\theta)^2}\, d\theta$$

$$= \int_0^{2\pi} \sqrt{1 + 2\sin\theta + \sin^2\theta + \cos^2\theta}\, d\theta$$

$$= \int_0^{2\pi} \sqrt{2 + 2\sin\theta}\, d\theta$$

EXAMPLE 29.14 (Cont.)

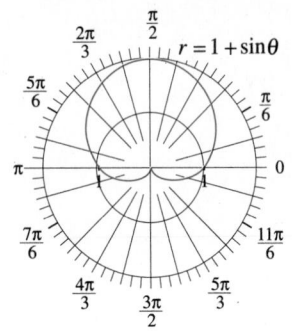

FIGURE 29.6

Using the trigonometric identity from Section 20.3, $1 + \cos\theta = 2\cos^2\left(\frac{\theta}{2}\right)$, this becomes

$$L = \int_0^{2\pi} \sqrt{4\cos^2\left(\frac{\theta}{2}\right)} \, d\theta$$

$$= \int_0^{2\pi} 2\left|\cos\left(\frac{\theta}{2}\right)\right| \, d\theta$$

Now, for $0 \leq \theta \leq \pi$, we know that $\cos\left(\frac{\theta}{2}\right) \geq 0$ and for $\pi \leq \theta \leq 2\pi$, we know $\cos\left(\frac{\theta}{2}\right) \leq 0$ which means that for $\pi \leq \theta \leq 2\pi$, we have $\left|\cos\left(\frac{\theta}{2}\right)\right| = -\cos\left(\frac{\theta}{2}\right)$.
We are now ready to evaluate the integral.

$$L = \int_0^{\pi} 2\cos\left(\frac{\theta}{2}\right) d\theta - \int_{\pi}^{2\pi} 2\cos\left(\frac{\theta}{2}\right) d\theta$$

$$= 4\sin\left(\frac{\theta}{2}\right)\Big]_0^{\pi} - 4\sin\left(\frac{\theta}{2}\right)\Big]_{\pi}^{2\pi}$$

$$= 4\,[1-0] - 4\,[0-1] = 8$$

The arc length of this cardioid is 8 units.

Surface Area

As we saw in Section 26.4, the area of a surface of revolution is given by the following formulas.

> **Surface Area of a Curve Revolved around an Axis**
>
> If a curve, written in rectangular form $y = f(x)$ from $x = a$ to $x = b$, is revolved around the x-axis, the area of its surface of revolution is
>
> $$S = 2\pi \int_a^b y\sqrt{1 + [f'(x)]^2}\, dx$$
>
> If it is revolved around the y-axis, then
>
> $$S = 2\pi \int_a^b x\sqrt{1 + [f'(x)]^2}\, dx$$

The corresponding formulas for parametric and polar equations are given as follows.

Surface Area of a Curve Revolved around an Axis (Parametric Form)

If the curve, written in parametric form, $x = f(t)$, $y = g(t)$, $a \leq t \leq b$, is revolved around the x-axis, the surface area is

$$S = 2\pi \int_a^b y \sqrt{\left(\frac{dx}{dt}\right)^2 + \left(\frac{dy}{dt}\right)^2}\, dt$$

where $y \geq 0$. If it is revolved around the y-axis, then

$$S = 2\pi \int_a^b x \sqrt{\left(\frac{dx}{dt}\right)^2 + \left(\frac{dy}{dt}\right)^2}\, dt$$

where $x \geq 0$.

EXAMPLE 29.15

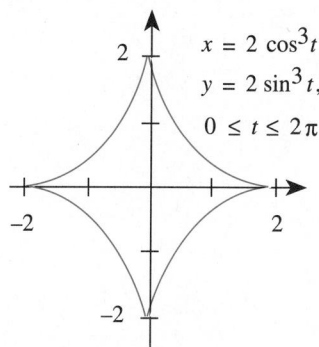

$x = 2\cos^3 t,$
$y = 2\sin^3 t,$
$0 \leq t \leq 2\pi$

FIGURE 29.7a

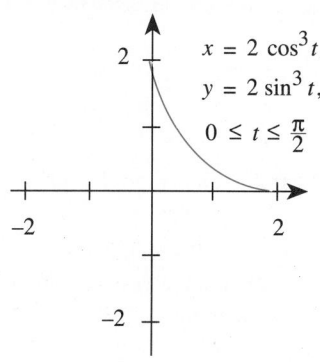

$x = 2\cos^3 t,$
$y = 2\sin^3 t,$
$0 \leq t \leq \frac{\pi}{2}$

FIGURE 29.7b

Find the surface area of the curve with the parametric equations $x = 2\cos^3 t$, $y = 2\sin^3 t$ from $0 \leq t \leq 2\pi$ when it is revolved around the x-axis.

Solution The graph of $x = 2\cos^3 t$, $y = 2\sin^3 t$ for $0 \leq t \leq 2\pi$ is shown in Figure 29.7a. Since we are required to have $y \geq 0$, we will restrict this to $0 \leq t \leq \pi$. We will also take advantage of the symmetry of the figure and find the surface area when the portion in the first quadrant is revolved around the x-axis and then double that answer. Restricting t to the interval $\left[0, \frac{\pi}{2}\right]$ produces the desired graph, as shown in Figure 29.7b.

When the curve in Figure 29.7b is revolved around the x-axis, the surface area is

$$S = 2\pi \int_0^{\pi/2} y \sqrt{\left(\frac{dx}{dt}\right)^2 + \left(\frac{dy}{dt}\right)^2}\, dt$$

$$= 2\pi \int_0^{\pi/2} \left(2\sin^3 t\right) \sqrt{\left(-6\cos^2 t \,\sin t\right)^2 + \left(6\sin^2 t \,\cos t\right)^2}\, dt$$

$$= 2\pi \int_0^{\pi/2} \left(2\sin^3 t\right) \sqrt{36\cos^4 t \,\sin^2 t + 36\sin^4 t \,\cos^2 t}\, dt$$

$$= 2\pi \int_0^{\pi/2} \left(2\sin^3 t\right) \sqrt{36\cos^2 t \,\sin^2 t \left(\cos^2 t + \sin^2 t\right)}\, dt$$

$$= 2\pi \int_0^{\pi/2} \left(2\sin^3 t\right) \sqrt{36\cos^2 t \,\sin^2 t}\, dt$$

$$= 2\pi \int_0^{\pi/2} \left(2\sin^3 t\right) \left(6\cos t \,\sin t\right)\, dt$$

EXAMPLE 29.15 (Cont.)

$$= 2\pi \int_0^{\pi/2} 12\sin^4 t \cos t \, dt$$

$$= 2\pi \frac{12}{5} \sin^5 t \Big]_0^{\pi/2} = \frac{24\pi}{5}$$

Doubling this result produces a surface area of $2\left(\dfrac{24\pi}{5}\right) = \dfrac{48\pi}{5} \approx 30.1593$ units^2.

Surface Area of a Curve Revolved around an Axis (Polar Form)

If the curve, written in polar form $r = f(\theta)$ is traced exactly once from $\theta = a$ to $\theta = b$, and is revolved around the polar (or x) axis, then

$$S = 2\pi \int_a^b y\sqrt{[f(\theta)]^2 + [f'(\theta)]^2} \, d\theta$$

$$= 2\pi \int_a^b y\sqrt{r^2 + \left(\frac{dr}{d\theta}\right)^2} \, d\theta$$

where $y = r\sin\theta$.

If the curve is revolved around the $\frac{\pi}{2}$ (or y) axis, then

$$S = 2\pi \int_a^b x\sqrt{[f(\theta)]^2 + [f'(\theta)]^2} \, d\theta$$

$$= 2\pi \int_a^b x\sqrt{r^2 + \left(\frac{dr}{d\theta}\right)^2} \, d\theta$$

where $x = r\cos\theta$.

EXAMPLE 29.16

The curve $r = e^\theta$ for $0 \le \theta \le \pi$ is revolved around the polar axis. Find the surface area of the generated figure.

Solution   The sketch of this curve is shown in Figure 29.8a. When it is revolved around the polar axis, the surface shown in Figure 29.8b results. This curve is in polar form. According to the formula, the surface area is given by

$$S = 2\pi \int_0^\pi y\sqrt{[f(\theta)]^2 + [f'(\theta)]^2} \, d\theta$$

where $y = r\sin\theta$.

EXAMPLE 29.16 (Cont.)

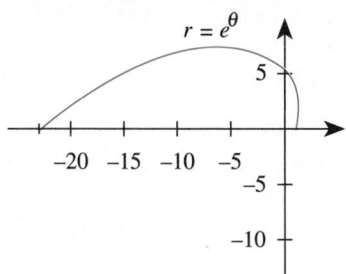

$r = e^\theta$

FIGURE 29.8a

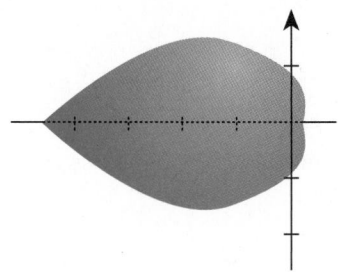

FIGURE 29.8b

We have $y = e^\theta \sin\theta$ and $f'(\theta) = e^\theta$, so

$$S = 2\pi \int_0^\pi e^\theta \sin\theta \sqrt{(e^\theta)^2 + (e^\theta)^2}\, d\theta$$

$$= 2\pi \int_0^\pi e^\theta \sin\theta \sqrt{2e^{2\theta}}\, d\theta$$

$$= 2\pi \int_0^\pi e^\theta \sin\theta (e^\theta \sqrt{2})\, d\theta$$

$$= 2\sqrt{2}\pi \int_0^\pi e^{2\theta} \sin\theta\, d\theta$$

Integration by parts (or Formula 87 in Appendix C) produces

$$= 2\sqrt{2}\pi \left[\frac{e^{2\theta}}{5}(2\sin\theta - \cos\theta)\right]\Big|_0^\pi$$

$$= \frac{2}{5}\sqrt{2}\pi[e^{2\pi} + e^0]$$

$$= \frac{2}{5}\sqrt{2}(e^{2\pi} + 1)\pi \approx 953.428$$

This surface has an area of approximately 953.428 square units.

Exercise Set 29.3

In Exercises 1–10, find the lengths of the given curves for the given intervals.

1. $x = 4t^3$, $y = 3t^2$ from $t = 0$ to $t = 1$
2. $x = \sin t$, $y = \cos t$ from $t = 0$ to $t = \pi$
3. $x = 3\cos t$, $y = 3\sin t$ from $t = 0$ to $t = 2\pi$
4. $x = \cos^3 t$, $y = \sin^3 t$ from $t = 0$ to $t = \frac{\pi}{2}$
5. $x = \cos t + t\sin t$, $y = \sin t - t\cos t$ from $t = 0$ to $t = \pi$
6. $r = \theta$ from $\theta = 0$ to $\theta = \frac{\pi}{2}$
7. $r = 1 + \cos\theta$ from $\theta = 0$ to $\theta = \pi$
8. $r = \cos^2\theta$ from $\theta = 0$ to $\theta = \frac{\pi}{2}$
9. $r = \sin^2\theta$ from $\theta = 0$ to $\theta = \frac{\pi}{2}$
10. $r = e^{\theta/2}$ from $\theta = 0$ to $\theta = 4$

In Exercises 11–18, find the area of the surface of revolution.

11. $x = t+4$, $y = t^3$ from $t = 0$ to $t = 2$ around the x-axis
12. $x = t$, $y = 4 - t^2$ from $t = 0$ to $t = 2$ around the y-axis
13. $x = \cos t$, $y = \sin t$ from $t = 0$ to $t = \frac{\pi}{2}$ around the x-axis
14. $x = 1 + \sin t$, $y = \cos t$ from $t = 0$ to $t = \frac{\pi}{2}$ around the y-axis
15. $x = 1 + \sin t$, $y = \cos t$ from $t = 0$ to $t = \frac{\pi}{2}$ around the x-axis
16. $r = \sin\theta$, from $\theta = 0$ to $\theta = \frac{\pi}{2}$ around the polar axis
17. $r = 1 + \cos\theta$, from $\theta = 0$ to $\theta = \pi$ around the polar axis
18. $r = e^{\theta/2}$ from $\theta = 0$ to $\theta = \pi$ around the polar axis

Solve Exercises 19 and 20.

19. *Environmental science* An **Archimedean spiral** has a polar equation of the form $r = a + b\theta$. Each successive loop of this spiral is the same distance $d = 2\pi b$ from the adjacent loops. One brand of mosquito coil is designed as an Archimedean spiral described by the equation $r = b\theta$, where r is in cm and $0 \le \theta < 2n\pi$. Here n determines the number of complete coils.
   (a) What is the length of this coil if there are four complete coils that are 1 cm apart?
   (b) If the coil burns at the rate of 5 cm/h, how long before the coil is all burned?

20. *Electronics* An electric heater coil is constructed from nichrome wire of resistance $4\,\Omega/\text{m}$. If L is the length of nichrome used, the total resistance, R, of the heating element is $R = 4L\,\Omega$. The shape of the coil is described by the equation $r = \dfrac{\theta}{2\pi}$, where $2\pi \le \theta \le 8\pi$. If the heater is designed to be connected across a voltage, V, of 110 V
   (a) Find the length, L, of nichrome wire needed for this coil.
   (b) Find the power needed for this length of coil.
   (Remember, $P = \dfrac{V^2}{R}$.)

![pencil icon] **In Your Words**

21. Without looking in the text, describe the method used to determine the arc length of a curve in either polar or parametric form.

22. Without looking in the text, describe the method used to determine the surface area of a curve revolved around an axis if the equation of the curve is in either polar or parametric form.

≡ 29.4
INTERSECTION OF GRAPHS OF POLAR COORDINATES

In Section 29.5, we will be finding the area of a region of the plane enclosed by one or more graphs of polar equations. In order to do, this we will need to know where these graphs intersect. Previously, when we needed to find where the graphs of two equations in rectangular coordinates intersected, we solved the equations simultaneously. This does not always work when we are working in polar coordinates.

EXAMPLE 29.17

Find the points of intersection of $r = 1$ and $r = 2\cos\theta$.

Solution Substituting $r = 1$ from the first equation into the second equation, we get

$$1 = 2\cos\theta \text{ or } \cos\theta = \frac{1}{2}$$

Solving this for θ, we get $\theta = \frac{\pi}{3}$ or $\frac{5\pi}{3}$. Thus the points of intersection are $\left(1, \frac{\pi}{3}\right)$ and $\left(1, \frac{5\pi}{3}\right)$. The graphs of these two curves, and their points of intersection, are shown in Figure 29.9.

We solved these two equations simultaneously and got the points of intersection. But, as we said, this does not always work. Consider the next example.

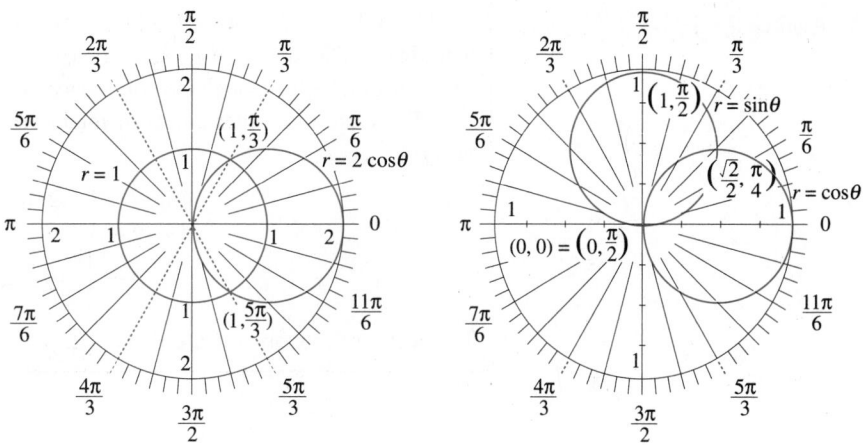

FIGURE 29.9   **FIGURE 29.10**

EXAMPLE 29.18

Find the points of intersection of $r = \sin\theta$ and $r = \cos\theta$.

Solution   If we solve these simultaneously, we get $\sin\theta = \cos\theta$ or $\dfrac{\sin\theta}{\cos\theta} = \tan\theta = 1$. This is true when $\theta = \frac{\pi}{4} + n\pi$ or, in the range $0 \le \theta < 2\pi$, when $\theta = \frac{\pi}{4}$ or $\frac{5\pi}{4}$. Thus, we get the points $\left(\dfrac{\sqrt{2}}{2}, \frac{\pi}{4}\right)$ and $\left(-\dfrac{\sqrt{2}}{2}, \frac{5\pi}{4}\right)$. We have a problem. These are not different points. They are just different ways of writing the same point.

Now, look at the graphs of these two functions as shown in Figure 29.10. As you can see, they do intersect in two points. One point we found, $\left(\dfrac{\sqrt{2}}{2}, \frac{\pi}{4}\right)$, the other point is the pole. The reason we did not find the pole when we solved the equations is that for the curve $r = \sin\theta$, the pole is at the points $(0, \pi + n\pi)$. For the curve $r = \cos\theta$, the pole is at $(0, \frac{\pi}{2} + n\pi)$. Both curves pass through the pole but not for the same values of θ. Because they intersect for different values of θ, we cannot find these points by using simultaneous equations. This point is at $(0, 0) = \left(0, \frac{\pi}{2}\right)$.

In order to find the points of intersection of two graphs in polar form, we need to do two things. First sketch a graph of the curves. While your curve does not have to be precise it must be good enough to see how many points of intersection exist. A graphing calculator will be very helpful here.

Second, solve the polar equations simultaneously. Check your solutions to make sure you have all the points of intersection with the other curve and that you do not have any duplicates.

EXAMPLE 29.19

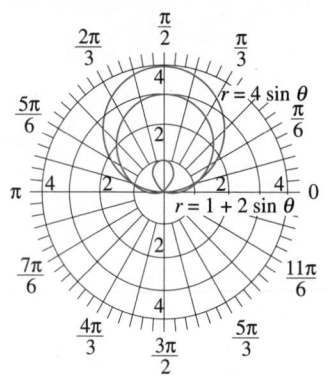

FIGURE 29.11

Find the points of intersection of $r = 4\sin\theta$ and $r = 1 + 2\sin\theta$.

Solution The sketch of these two curves is shown in Figure 29.11. As you can see, they both pass through the pole, so we know that $(0,0)$ is a point of intersection.

There are two other points of intersection. Solving the equations simultaneously, we get

$$4\sin\theta = 1 + 2\sin\theta$$
$$2\sin\theta = 1$$
$$\sin\theta = \frac{1}{2}, \text{ so } \theta = \frac{\pi}{6} \text{ or } \frac{5\pi}{6}$$

The points of intersection are $(0,0)$, $\left(2, \frac{\pi}{6}\right)$, and $\left(2, \frac{5\pi}{6}\right)$. ■

EXAMPLE 29.20

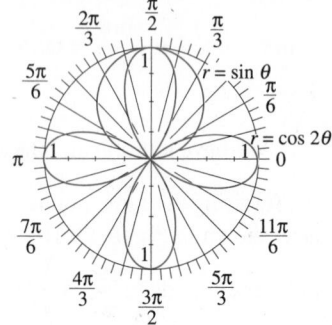

FIGURE 29.12

Find the points of intersection of $r = \sin\theta$ and $r = \cos 2\theta$.

Solution The graphs of these two curves are shown in Figure 29.12. As you can see, there are four points of intersection. Solving the equations simultaneously, we get

$$\sin\theta = \cos 2\theta$$

Using the identity $\cos 2\theta = 1 - 2\sin^2\theta$, we see that

$$\sin\theta = 1 - 2\sin^2\theta$$
$$\text{or} \qquad 2\sin^2\theta + \sin\theta - 1 = 0$$
$$(2\sin\theta - 1)(\sin\theta + 1) = 0$$
$$\sin\theta = \frac{1}{2} \text{ or } \sin\theta = -1$$
$$\text{and} \qquad \theta = \frac{\pi}{6}, \frac{5\pi}{6}, \text{ or } \frac{3\pi}{2}$$

Thus, three points of intersection for these two polar curves are $\left(\frac{1}{2}, \frac{\pi}{6}\right)$, $\left(\frac{1}{2}, \frac{5\pi}{6}\right)$, and $\left(-1, \frac{3\pi}{2}\right) = \left(1, \frac{\pi}{2}\right)$. Checking the pole, we see that $(0,0)$ is also a solution. ■

The methods that we have shown will allow you to find all of the points of intersection for some equations. However, there are other pairs of simultaneous equations that can be used to locate other points of intersection. We will not use these other pairs of equations in this text.

Exercise Set 29.4

In Exercises 1–12, find all points of intersection of the graphs of the two polar equations. Graph all equations.

1. $r = 3\theta, r = \frac{\pi}{2}$

2. $r = \frac{\theta}{4}, r = \frac{\pi}{6}$

3. $r = \frac{1}{2}, r = \cos\theta$

4. $r = 2 - 2\sin\theta, r = 2 - 2\cos\theta$

5. $r = 1 - \sin\theta, r = 1 + \cos\theta$

6. $r = -4 + 4\cos\theta, r = -4 + 4\sin\theta$

7. $r = \sqrt{3}, r = 2\cos\theta$

8. $r = \sin 2\theta, r = \sin\theta$

9. $r = \sin 2\theta, r = \sqrt{2}\sin\theta$

10. $r = 2 + 2\cos\theta, r = \dfrac{1}{1 - \cos\theta}$

11. $r = 1 - \sin\theta, r = \dfrac{1}{1 - \sin\theta}$

12. $r = \sin 2\theta, r = \cos 2\theta$

 **In Your Words**

13. Describe how you would find the intersection of two graphs in polar coordinates.

14. Why do you think it is important to find the points where the graphs of two equations in polar coordinates intersect?

≡ 29.5
AREA IN POLAR COORDINATES

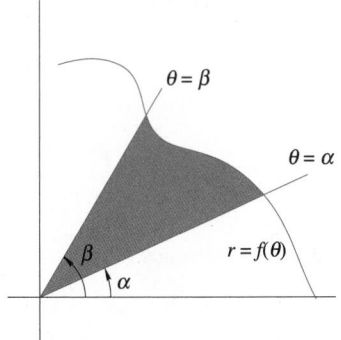

FIGURE 29.13

In Section 29.4, we learned how to find the points of intersection for the graphs of two polar equations. In this section, we will consider the problem of finding the area of a region in a plane that has been enclosed by the graph of a polar equation and two rays that have the pole as a common vertex. Next, we will find the area enclosed by the graph of two polar equations.

Also in this section, we will use the following formula for the area of a sector of a circle.

Area of a Sector of a Circle

If a sector has angle θ (in radians) and radius r, then its area is

$$A = \frac{1}{2}\theta r^2$$

Now, suppose we have a region bounded by $r = f(\theta)$ and the two rays $\theta = \alpha$ and $\theta = \beta$, where $\alpha < \beta$, as shown in Figure 29.13. We want to find the area and so we will subdivide something into n intervals. With rectangular coordinates we divided the interval $[a, b]$. We will do a similar thing here. We will divide the angular interval from α to β into n subintervals where

$$\alpha = \theta_0 < \theta_1 < \theta_2 < \cdots < \theta_n = \beta$$

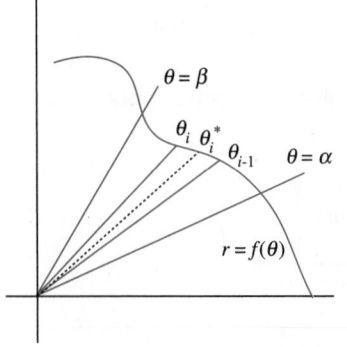

FIGURE 29.14

We will select the ray $\theta_i{}^*$ at the midpoint of the ith interval, as shown in Figure 29.14. The area of this interval is approximately

$$A \approx \frac{1}{2}(\theta_i - \theta_{i-1})[f(\theta_i{}^*)]^2$$

But, what is $\theta_i - \theta_{i-1}$? Nothing more than $\Delta\theta_i$. So, the area becomes

$$A \approx \frac{1}{2}[f(\theta_i{}^*)]^2 \Delta\theta_i$$

Adding all of these intervals and taking the limit, we get the following formula for the area of a region.

Area of a Sector for a Polar Curve

If f is continuous and $f(\theta) \geq 0$ on $[\alpha, \beta]$ with $0 \leq \alpha < \beta \leq 2\pi$, then area A of the region bounded by the graphs of $r = f(\theta)$, $\theta = \alpha$, and $\theta = \beta$ is

$$A = \frac{1}{2}\int_\alpha^\beta [f(\theta)]^2 \, d\theta$$

or

$$A = \frac{1}{2}\int_\alpha^\beta r^2 \, d\theta$$

EXAMPLE 29.21

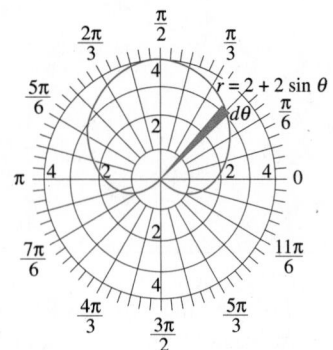

FIGURE 29.15

Find the area enclosed by the cardioid $r = 2 + 2\sin\theta$ from $0 \leq \theta \leq 2\pi$.

Solution The graph of this curve is shown in Figure 29.15. Here, we have $\alpha = 0$ and $\beta = 2\pi$. A typical circular sector has been shaded in the figure and its central angle has been labeled $d\theta$.

$$A = \frac{1}{2}\int_0^{2\pi} (2 + 2\sin\theta)^2 \, d\theta$$

$$= \frac{1}{2}\int_0^{2\pi} (4 + 8\sin\theta + 4\sin^2\theta) \, d\theta$$

Using the half-angle formula $\sin^2\theta = \frac{1}{2}(1 - \cos 2\theta)$, the area becomes

$$A = \frac{1}{2}\int_0^{2\pi} [4 + 8\sin\theta + 2(1 - \cos 2\theta)] \, d\theta$$

$$= \frac{1}{2}(4\theta - 8\cos\theta + 2\theta - \sin 2\theta)\Big|_0^{2\pi}$$

$$= \frac{1}{2}[(8\pi - 8 + 4\pi) - (-8)] = 6\pi$$

The area enclosed by this cardioid is 6π square units.

The following guidelines will help you to find the area of a polar region. Look back over Example 29.21 to see if we followed these guidelines.

Hint

Guidelines for Finding the Area of a Polar Region

1. Sketch the region, labeling the graph of $r = f(\theta)$. Find the smallest value $\theta = \alpha$ and the largest value of $\theta = \beta$ for points (r, θ) in the region.
2. Sketch a typical circular sector and label its central angle $d\theta$.
3. Express the area of the sector in Step 2 as $\frac{1}{2}r^2\, d\theta$.
4. Integrate the expression in Step 3 over the limits α to β.

EXAMPLE 29.22

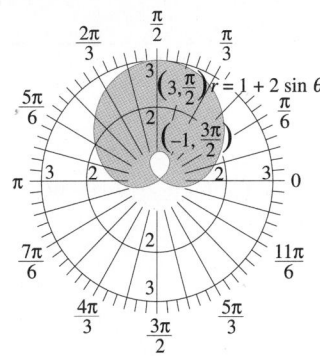

FIGURE 29.16

Find the area enclosed by the outer loop of the limaçon $r = 1 + 2\sin\theta$.

Solution   The graph of this curve is shown in Figure 29.16. The curve intersects the pole when $1 + 2\sin\theta = 0$ or $\sin\theta = -\frac{1}{2}$. This happens twice—at $\theta = \frac{7\pi}{6}$ and $\frac{11\pi}{6}$. Between these two values of θ, we get the inner loop. So, what we want is the area of the curve from 0 to $\frac{7\pi}{6}$ and then again from $\frac{11\pi}{6}$ to 2π. We can get that same area by integrating from $\frac{-\pi}{6}$ to $\frac{7\pi}{6}$, since, on a circle, the angle $\frac{11\pi}{6} = \frac{-\pi}{6}$.

$$A = \frac{1}{2} \int_{-\pi/6}^{7\pi/6} (1 + 2\sin\theta)^2\, d\theta$$

$$= \frac{1}{2} \int_{-\pi/6}^{7\pi/6} (1 + 4\sin\theta + 4\sin^2\theta)\, d\theta$$

$$= \frac{1}{2} \int_{-\pi/6}^{7\pi/6} (1 + 4\sin\theta + 2 - 2\cos 2\theta)\, d\theta$$

$$= \frac{1}{2}(3\theta - 4\cos\theta - \sin 2\theta)\Big|_{-\pi/6}^{7\pi/6}$$

$$= \frac{1}{2}\left[\left(\frac{7\pi}{2} + 2\sqrt{3} - \frac{\sqrt{3}}{2}\right) - \left(\frac{-\pi}{2} - 2\sqrt{3} + \frac{\sqrt{3}}{2}\right)\right]$$

$$= 2\pi + 2\sqrt{3} - \frac{\sqrt{3}}{2}$$

$$= 2\pi + \frac{3}{2}\sqrt{3} \approx 8.88$$

The area enclosed by this region is about 8.88 square units.

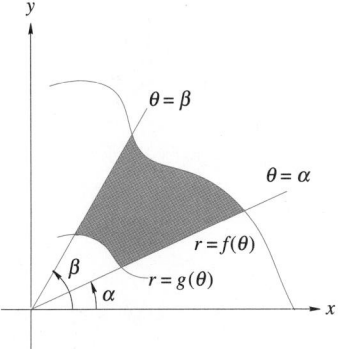

FIGURE 29.17

Now suppose we have two curves, $r = f(\theta)$ and $r = g(\theta)$, with $f(\theta) \geq g(\theta)$. We want to find the area of the region between these curves and bounded by $\theta = \alpha$ and $\theta = \beta$, as shown in Figure 29.17. We do this, as we did with rectangular coordinates, by subtracting the area bounded by the inner curve from the area bounded by the outer curve. This gives us the formula for the area enclosed between two polar curves in the following box.

Area Enclosed Between Two Polar Curves

If f and g are continuous and $f(\theta) \geq g(\theta) \geq 0$ on $[\alpha, \beta]$ with $0 \leq \alpha < \beta \leq 2\pi$, then area A of the region bounded by the graphs of $r = f(\theta)$, $r = g(\theta)$, $\theta = \alpha$, and $\theta = \beta$ is

$$A = \frac{1}{2} \int_{\alpha}^{\beta} [f(\theta)]^2 \, d\theta - \frac{1}{2} \int_{\alpha}^{\beta} [g(\theta)]^2 \, d\theta$$

$$= \frac{1}{2} \int_{\alpha}^{\beta} \left([f(\theta)]^2 - [g(\theta)]^2 \right) d\theta$$

If you let $r_2 = f(\theta)$ and $r_1 = g(\theta)$, then this can be written as

$$A = \frac{1}{2} \int (r_2{}^2 - r_1{}^2) \, d\theta$$

EXAMPLE 29.23

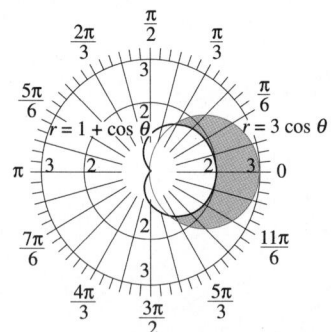

FIGURE 29.18

Find the area inside $r = 3\cos\theta$ and outside $r = 1 + \cos\theta$.

Solution   The area we want to integrate has been shaded in Figure 29.18. We will need to find the points of intersection for these two curves.

Solving the equations simultaneously, we get

$$3\cos\theta = 1 + \cos\theta$$

$$2\cos\theta = 1$$

$$\cos\theta = \frac{1}{2}$$

So, the curves intersect when

$$\theta = \frac{\pi}{3} \text{ and } \frac{-\pi}{3}$$

The area is

$$A = \frac{1}{2} \int_{-\pi/3}^{\pi/3} \left[(3\cos\theta)^2 - (1 + \cos\theta)^2 \right] d\theta$$

$$= \frac{1}{2} \int_{-\pi/3}^{\pi/3} \left[(9\cos^2\theta) - (1 + 2\cos\theta + \cos^2\theta) \right] d\theta$$

$$= \frac{1}{2} \int_{-\pi/3}^{\pi/3} (8\cos^2\theta - 1 - 2\cos\theta) \, d\theta$$

$$= \frac{1}{2} \int_{-\pi/3}^{\pi/3} \left[8 \left(\frac{1 + \cos 2\theta}{2} \right) - 1 - 2\cos\theta \right] d\theta$$

EXAMPLE 29.23 (Cont.)	$$= \frac{1}{2}\int_{-\pi/3}^{\pi/3}(3+4\cos 2\theta - 2\cos\theta)\,d\theta$$ $$= \frac{1}{2}(3\theta + 2\sin 2\theta - 2\sin\theta)\Big|_{-\pi/3}^{\pi/3}$$ $$= \frac{1}{2}\left[(\pi + \sqrt{3} - \sqrt{3}) - (-\pi - \sqrt{3} + \sqrt{3})\right]$$ $$= \pi \text{ square units}$$

We could have made our calculations easier in the last problem if we had taken advantage of the fact that the region was symmetric with respect to the polar axis. Notice that half of the area is above the x-axis and half is below. Thus, we could have found the area from 0 to $\frac{\pi}{3}$ and then doubled it as follows.

$$A = 2 \cdot \frac{1}{2}\int_0^{\pi/3}[(3\cos\theta)^2 - (1+\cos\theta)^2]\,d\theta$$

$$= 3\theta + 2\sin 2\theta - 2\sin\theta\Big|_0^{\pi/3}$$

$$= (\pi + \sqrt{3} - \sqrt{3}) - (0) = \pi$$

Exercise Set 29.5

In Exercises 1–10, sketch the graph of the equation and find the area of the region enclosed by the graph.

1. $r = 4\sin\theta$
2. $r = 5\cos\theta$
3. $r = 1 - \cos\theta$
4. $r = 4 + 4\sin\theta$
5. $r = \sin 2\theta$
6. $r = \cos 3\theta$
7. $r = 4 + \cos\theta$
8. $r = 4 + 3\cos\theta$
9. $r^2 = \sin\theta$
10. $r^2 = 4\cos 2\theta$

In Exercises 11–16, find the area enclosed by the given polar equation and the given rays.

11. $r = 3\cos\theta,\ \theta = 0,\ \theta = \frac{\pi}{4}$
12. $r = 4\sin\theta,\ \theta = 0,\ \theta = \frac{\pi}{3}$
13. $r = \cos 2\theta,\ \theta = 0,\ \theta = \frac{\pi}{4}$
14. $r = \sin 4\theta,\ \theta = 0,\ \theta = \frac{\pi}{8}$
15. $r = e^{2\theta},\ \theta = 0,\ \theta = \frac{\pi}{2}$
16. $r = 5\theta,\ \theta = 0,\ \theta = \frac{\pi}{6}$

In Exercises 17–20, find the area of the region bounded by one loop of the graph of the given equation.

17. $r = 4\cos 2\theta$
18. $r^2 = 4\cos 2\theta$
19. $r = 2\sin 3\theta$
20. $r = \sin 6\theta$

In Exercises 21–26, sketch the graph and find the area of the region described.

21. Inside $r = 1 + \cos\theta$ and outside $r = 1$
22. Inside $r = 2\sin\theta$ and outside $r = 1$
23. Inside $r = 1 - \sin\theta$ and outside $r = 2\cos\theta$
24. Inside $r^2 = 8\sin\theta$ and outside $r = 2$
25. Inside $r = \cos\theta$ and outside $r = 1 - \sin\theta$
26. Inside $r = \sin\theta$ and inside $r = \sqrt{3}\cos\theta$

In Your Words

27. Without looking in the text, describe the guidelines for finding the area of a polar region.

28. Describe how to find the area between two polar curves.

≡ CHAPTER 29 REVIEW

Important Terms and Concepts

Arc length
   Of parametric equations
   Of polar equations
Area
   Between two polar curves
   Of a polar region
   Of a sector of a circle
Parametric equations
   First derivative
   Higher order derivatives

Polar equations
   Derivative
   Intersection of graphs
Sector of a circle
   Area
Surface area
   Generated by parametric equations
   Generated by polar equations

Review Exercises

In Exercises 1–4, (a) find the first and second derivative of the given function and (b) find the slope of the tangent line at the point corresponding to $t = 1$.

1. $x = t^2, y = t^3 + t$

2. $x = \dfrac{1}{t}, y = \dfrac{2}{t}$

3. $x = \cos^2 t, y = \sin t$

4. $x = \sqrt{t}, y = e^{-t}$

In Exercises 5–8, (a) find the first derivative of the given function and (b) find the slope of the tangent line at the point where $\theta = \frac{\pi}{4}$.

5. $r = 3 + 2\cos\theta$

6. $r = 6\cos 2\theta$

7. $r^2 = \sin\theta$

8. $r = \dfrac{8}{3 + \cos\theta}$

In Exercises 9–14, find the arc length of the given curve over the given interval.

9. $x = t^2, y = t^3, 0 \le t \le 2$

10. $x = t^2, y = 2t, 0 \le t \le 3$

11. $x = 4\sin t, y = 4\cos t, 0 \le t \le \frac{\pi}{2}$

12. $r = e^{2\theta}, 0 \le \theta \le \pi$

13. $r = \cos^2\left(\dfrac{\theta}{2}\right), 0 \le \theta \le \pi$

14. $r = 1 - \cos\theta, -\pi \le \theta \le 0$

In Exercises 15–20, find the area of the surface generated by revolving the given curve around the indicated axis.

15. $x = t^2, y = t, 0 \le t \le 2$ around the x-axis

16. $x = t^2, y = t - \dfrac{t^3}{3}, 0 \le t \le 1$ around the x-axis

17. $x = t^2, y = t^3, 0 \le t \le 2$ around the y-axis

18. $x = e^t \sin t, y = e^t \cos t, 0 \le t \le \frac{\pi}{2}$ around the y-axis

19. $r = 6\sin\theta, 0 \le \theta \le \pi$ around the polar axis

20. $r = 4 + 4\cos\theta, 0 \le \theta \le \pi$ around the polar axis

In Exercises 21–24, find the points of intersection of the graph of the given pair of polar equations.

21. $r = 4$, $r = 4 + 4\sin\theta$ 23. $r = 2$, $r = 4 + 4\cos\theta$

22. $r = 3$, $r = 6\sin\theta$ 24. $r = 1$, $r = 2\sin 2\theta$

In Exercises 25–8, find the indicated area and sketch the graph.

25. Inside $r = 1 + \cos\theta$ 28. Inside $r = 6\sin\theta$ and outside $r = 3$

26. Inside $r = 4 + 4\sin\theta$ 29. Inside $r = \sin\theta$ and outside $r = 1 + \cos\theta$

27. Inside $r = 4 + 4\sin\theta$ and outside $r = 4$ 30. Inside $r = 2\sin 2\theta$ and outside $r = 1$

≡ CHAPTER 29 TEST

1. Find the first and second derivative of $x = t^3$, $y = t^2 + 5t$.

2. Find the equation of the tangent line to the graph of $x = \sin t$, $y = \cos^2 t$ at $\left(\dfrac{\sqrt{2}}{2}, \dfrac{1}{2}\right)$, when $t = \dfrac{\pi}{4}$.

3. Find the derivative of $r = 4 - 2\sin\theta$.

4. Find the equation of the normal line to the graph of $r = 8\sin 3\theta$ at the point $\left(4\sqrt{2}, \dfrac{\pi}{4}\right)$.

5. Find the arc length of the curve described by the parametric equations $x = e^t$, $y = 2e^t$ from $0 \le t \le 2$.

6. Find the surface area of the surface generated by revolving the graph of $r = 4\sin\theta$ from $0 \le \theta \le \dfrac{\pi}{2}$ around the polar axis.

7. Find the points of intersection of the curves $r = 4\cos\theta$ and $r = 1 - \cos\theta$.

8. Find the area of the region that lies inside the circle $r = 1$ and outside the cardioid $r = 1 - \cos\theta$.

30

Partial Derivatives and Multiple Integrals

Robots, like the one shown, use a three-dimensional coordinate system. In Sections 30.1 and 30.6, we will study the coordinate systems used by robots like this.

Courtesy of Motoman Inc.

Until now, all of our work has been in the plane. The derivatives and integrals that we have learned have all applied to curves in the plane. We have used two types of coordinate systems—the rectangular and the polar—but we live in a three-dimensional world, so it is important that we study three-dimensional mathematics.

In this chapter, we will learn about measuring and graphing objects in three dimensions. We will learn about three new coordinate systems and see how they are used in robotics, and we will learn how to take derivatives and integrals in three dimensions.

≡ 30.1
FUNCTIONS IN TWO VARIABLES

The functions that we have used until now have always been functions of one variable. The typical function, $y = f(x)$, had x as the variable. Parametric equations of the form $x = f(t)$ and $y = g(t)$, had t as the variable. Polar equations, such as $r = f(\theta)$, had θ as the variable.

We have, however, worked with functions of more than one variable. For example, the formula for the volume of a cylinder, $V = \pi r^2 h$, is a function of two variables, r and h. As you can see, the volume of a cylinder depends on both the radius of the base and the height of the cylinder. To indicate that it is a function of two variables, we could write

$$V = f(r, h) = \pi r^2 h$$

EXAMPLE 30.1

For the function $V = f(r, h) = \pi r^2 h$, find (a) $f(5, 4)$, (b) $f(10, 4)$, and (c) $f(5, 8)$.
Solutions
(a) Here, $r = 5$ and $h = 4$, so we have $f(5, 4) = \pi \cdot 5^2 \cdot 4 = 100\pi$.
(b) In this example, we have $r = 10$ and $h = 4$, and so $f(10, 4) = \pi \cdot 10^2 \cdot 4 = 400\pi$.
(c) Here, $r = 5$ and $h = 8$, with the result $f(5, 8) = \pi \cdot 5^2 \cdot 8 = 200\pi$.

We define a **real function of two variables** as any rule that assigns a unique real number to each ordered pair of real numbers (x, y) in a certain set D of the xy-plane.

We often write a real function of two variables in the form $z = f(x, y)$. Here, z represents a **dependent variable** and x and y are the **independent variables**. The numbers in D of the definition form the **domain** of the function f. The **range** of f is all the real numbers $f(x, y)$, where (x, y) is in the domain D. As you can see, this is very similar to what we did with a function of one variable. Of course, $f(a, b)$ means the value of the function f when $x = a$ and $y = b$.

EXAMPLE 30.2

If $f(x, y) = \dfrac{x^2 - y^2}{x^2 + y^2}$, find $f(2, 1)$

Solution $f(2, 1) = \dfrac{2^2 - 1^2}{2^2 + 1^2} = \dfrac{4 - 1}{4 + 1} = \dfrac{3}{5}$.

EXAMPLE 30.3

If $f(x, y) = e^{x^2 - xy}$, find $f(3, 2)$.
Solution Here, $x = 3$ and $y = 2$, so $f(3, 2) = e^{3^2 - 3 \cdot 2} = e^{9 - 6} = e^3$.

EXAMPLE 30.4

If $f(x, y) = \sqrt{x} + x\sqrt{y}$, find $f(x, 2x) - f(x, x^2)$.

Solution This is a little different because we want to subtract the second value of the function from the first. Also, notice that in both cases the value of x is the same. The value of the y-variable changes.

$$f(x, 2x) = \sqrt{x} + x\sqrt{2x}$$
$$f(x, x^2) = \sqrt{x} + x\sqrt{x^2} = \sqrt{x} + x|x| \qquad \text{Remember, } \sqrt{x^2} = |x|.$$

The difference is

$$f(x, 2x) - f(x, x^2) = (\sqrt{x} + x\sqrt{2x}) - (\sqrt{x} + x|x|)$$
$$= x\sqrt{2x} - x|x|$$
$$= x\left(\sqrt{2x} - |x|\right)$$

Domain

As with functions of one variable, we must be careful that we do not divide by zero. We must also be sure that the values of x and y lead to real number answers. For example, in Example 30.4, if either $x < 0$ or $y < 0$, then $f(x, y)$ would produce an imaginary value. But, this is a *real* value function and imaginary numbers are not allowed. The domain consists of all values of $x \geq 0$ and of $y \geq 0$.

In a similar manner, the domain of Example 30.2 includes all real values of (x, y) except the origin $(0, 0)$, because this gives a zero in the denominator.

Functions of More Than Two Variables

We have been concentrating on functions of two variables. However, we could just as easily have a function of three variables, such as $w = f(x, y, z)$, or of four variables, such as $u = f(x, y, z, w)$, or of as many variables as we like.

EXAMPLE 30.5

If $f(x, y, z, w) = xy^2 + z\tan w - e^{yz}$, find $f\left(3, 1, 2, \frac{\pi}{4}\right)$.

Solution Here $x = 3$, $y = 1$, $z = 2$, and $w = \frac{\pi}{4}$, so

$$f(x, y, z, w) = 3 \cdot 1^2 + 2\tan\frac{\pi}{4} - e^{1\cdot 2}$$
$$= 3 + 2 - e^2$$
$$= 5 - e^2$$

The domain of this function consists of all real numbers for x, y, and z. However, the values of $w \neq \frac{\pi}{2}(2n + 1)$ where n is an integer. Thus, w cannot be an odd multiple of $\frac{\pi}{2}$ such as $\frac{\pi}{2}$, $\frac{3\pi}{2}$, $\frac{5\pi}{2}$, or $-\frac{\pi}{2}$, because the tangent function is not defined at these points.

Example 30.5 shows how to work with functions of four variables. As we saw, these functions are treated in the same manner as those of two variables. For now, we will concentrate on functions of two variables. Using a function of two variables to express a problem is no easier or more different than it is for one variable. No general rules exist that you have not already encountered while working with just one variable. As before, you may originally write the problem with more than two variables. You should then use the relationships between variables to eliminate some of them.

Application

EXAMPLE 30.6

An open rectangular box is constructed out of different materials. The material for the bottom costs $1.25/ft^2$. The material for the sides costs $1.50/ft^2$. The volume of the box is to be $4 ft^3$. Express the cost of the material as a function of the length and width of the box.

Solution   Let's use l for the length, w for the width, and h for the height of this box. Then

$$V = lwh$$

The cost of the material depends on the surface area. The area of the sides is $(2l + 2w)h$. The area of the bottom is lw, so the cost of the material is

$$C = 1.50(2l + 2w)h + 1.25lw$$

This is not a function of the length and width because it includes the variable h. But $V = lwh = 4$, and so $h = \dfrac{4}{lw}$. With this substitution, we get the desired result

$$V = 1.50(2l + 2w)\frac{4}{lw} + 1.25lw$$

$$= \frac{12(l + w)}{lw} + 1.25lw$$

In this section, we will only be writing equations as functions of two variables. Later we will differentiate and integrate these functions in order to solve some problems.

Exercise Set 30.1

Solve Exercises 1–10.

1. Let $f(x, y) = 3x + 4y - xy$. Find: **(a)** $f(1, 0)$, **(b)** $f(0, 1)$, **(c)** $f(2, 1)$, **(d)** $f(x + h, y)$, **(e)** $f(x, y + h)$.

2. Let $g(x, y) = x^2 y + \sin x - \cos y$. Find: **(a)** $g\left(\frac{\pi}{2}, \pi\right)$, **(b)** $g\left(\pi, \frac{\pi}{2}\right)$, **(c)** $g\left(\frac{\pi}{2}, \frac{\pi}{4}\right)$, **(d)** $g(x + h, y)$, **(e)** $g(x, y + h)$.

3. Let $j(x, y) = \sqrt{xy} - x + \dfrac{4}{y}$. Find: **(a)** $j(-1, -2)$, **(b)** $j(-1, -4)$, **(c)** $j(4, 1)$, **(d)** $j(x + h, y)$, **(e)** $j(x, y + h)$.

4. Let $k(x, y) = e^x + e^{xy} - y^2$. Find: **(a)** $k(1, 2)$, **(b)** $k(2, 1)$, **(c)** $k(2, 3)$, **(d)** $k(x+h, y)$, **(e)** $k(x, y+h)$.

5. Let $f(x, y) = \dfrac{2xy - x^2}{y - x}$. Find: **(a)** $f(1, 0)$, **(b)** $f(0, 1)$, **(c)** $f(2, 1)$, **(d)** $f(1, 2)$, **(e)** the domain of f.

6. Let $g(x, y) = \dfrac{\ln(x + y)}{y}$. Find: **(a)** $g(1, 1)$, **(b)** $g(2, 1)$, **(c)** $g(3, 6)$, **(d)** $g(4, -3)$, **(e)** the domain of g.

Solve Exercises 11–24.

11. *Petroleum engineering* The cost of the bottom and top of a cylindrical petroleum storage tank is $200/m² and the cost of the side is $1,000/m². Write the total cost of constructing such a tank as a function of the radius and height.

12. *Wastewater technology* A conical sewage tank is being constructed. The base costs $1,500/ft² and the side costs $200/ft². Write the total cost as a function of the radius r and slant height s.

13. *Medical technology* According to **Poiseuille's law**, the velocity v in cm/min of blood flow r units from the center of an artery of radius R is given by

$$v(R, r) = c\left(R^2 - r^2\right)$$

where c is a constant based on the length of the vessel,

7. Express the volume of a cone as a function of the radius of the base and the height.

8. Express the volume of a prism as a function of the area of its base and its height.

9. Express the lateral surface area of a right circular cylinder as a function of the radius of its base and its height.

10. Express the total surface area of a cone as a function of its slant height and the radius of its base.

the viscosity of the blood, and the blood pressure. If $c = 1$, find $v(0.0075, 0.0045)$.

14. *Meteorology* Under certain conditions, the wind speed in mile per hour of a tornado at a distance d feet from its center can be approximated by the function

$$S(a, V, d) = \frac{aV}{0.51d^2}$$

where a is a constant that depends on certain atmospheric conditions and V is the volume of the tornado in cubic feet. A certain tornado has a volume of 2,400,000 ft³ and $a = 0.62$. Approximate the wind speed **(a)** 100 ft from the center of this tornado and **(b)** 200 ft from the center of this tornado.

15. *Meteorology* Wind speed affects the temperature a person feels and often makes it feel colder than it is. The **Wind Chill Index** (*WCI*) represents the equivalent temperature in degrees Fahrenheit, that exposed skin would feel if there was little or no wind. If F represents the actual temperature in degrees Fahrenheit as measured by a thermometer and v is the velocity of the wind in miles per hour, then

$$WCI(v, F) = \begin{cases} F, & 0 \leq v \leq 4 \\ 91.4 - \dfrac{(10.45 + 6.69\sqrt{v} - 0.447v)(91.4 - F)}{22}, & 4 < v \leq 45 \\ 1.60F - 55, & v > 45 \end{cases}$$

Find the wind chill index to the nearest 1 degree when **(a)** $T = 20°F$ and $v = 20$ mph, **(b)** $T = 10°F$ and $v = 20$ mph, **(c)** $T = 10°F$ and $v = 4$ mph, and **(d)** $T = 20°F$ and $v = 25$ mph.

16. *Meteorology* Humidity affects the temperature a person feels and, in the summer, often makes it feel hotter than it actually is. The **Apparent Temperature Index** (*ATI*) attempts to measure what hot weather "feels like" to the average person under various conditions of temperature and relative humid-

ity. If F represents the actual temperature in degrees Fahrenheit as measured by a thermometer and RH is the relative humidity expressed as a decimal, then

$$ATI(F, RH) = \\ 0.885F - 78.7RH + 1.20F(RH) + 2.70$$

Find the apparent temperature index to the nearest 1 degree, when (a) $T = 95°F$ and $RH = 30\%$, (b) $T = 95°F$ and $RH = 60\%$, (c) $T = 95°F$ and $RH = 90\%$, and (d) $T = 25°F$ and $RH = 75\%$.

17. *Package design*   The packaging department in a company has been asked to design a rectangular box with no top and a partition down the middle, as in Figure 30.1. The dimensions of the box, in inches, are x, y, and z.
   (a) Write a function of three variables for the number of cubic inches in the volume of the box.
   (b) Write a function of three variables for the number of square inches in the material needed to construct the box.

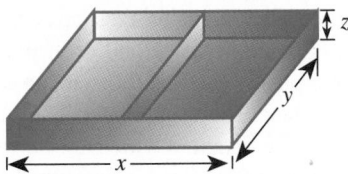

FIGURE 30.1

18. *Construction*   A farmer wants to build a rectangular pen and then divide it with two interior fences, as shown in Figure 30.2. The exterior fence costs \$24/ft, and the interior fence costs \$18.50/ft. Write a function of two variables for the total cost of the material needed to construct the pen.

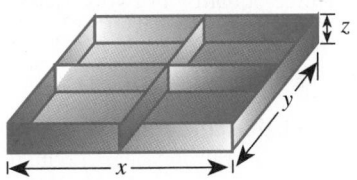

FIGURE 30.2

19. *Police science*   The stopping distance, in feet, for a car after the brakes are applied is given by $S(w, r) = kwr^2$, where k is a constant, w the car's weight in lb, and r its speed in mph. If $k = 0.000\,02$,

determine the stopping distance for a car weighing 2,400 lb and traveling 65 mph.

20. *Recreation*   In using scuba diving gear, a diver estimates the time of a dive according to the equation $t(V, x) = \dfrac{33V}{x + 33}$, where t is the length of the dive in minutes; V the volume of air at sea level pressure, compressed into the tanks, and x the depth of the dive in feet. Find (a) $t(75, 67)$ and (b) $t(60, 32)$.

21. *Medical technology*   The surface area in square meters of a human is approximated by $A(w, h) = 0.18215w^{0.425}h^{0.725}$, where w is the weight of the person in kilograms and h is the height of the person in meters.
   (a) Determine the surface area of a person with a mass of 84 kg and a height of 1.80 m.
   (b) Write a function of two variables to determine the surface area in square inches of a person based on the person's weight in pounds and height in inches.
   (c) Determine the surface area of a person with a weight of 185 lb and a height of 5 ft 11 in.

22. *Advertising*   A company spends x on newspaper advertising and y on television advertising, where x and y are in thousands of dollars per week. It has found that its weekly sales, in thousands of dollars, can be given by $S(x, y) = 5x^2y^3$. (a) Find $S(4, 3)$ and (b) $S(3, 4)$.

23. *Economics*   The **Cobb-Douglas production formula** is used by economists to model the production level of a company. The formula is $P(x, y) = kx^ay^{1-a}$, where P is the total units produced, x is the measure of labor units, y is a measure of capital invested, and k is a constant that varies from product to product. Suppose $P(x, y) = 400x^{0.35}y^{0.65}$ represents the number of units produced by a company, with x units of labor and y units of capital.
   (a) How many units of a product will be manufactured if 300 units of labor and 50 units of capital are used?
   (b) How many units of a product will be manufactured if the units of labor and capital are both doubled?

24. *Agriculture*   The number of cattle that can graze off a certain ranch without causing overgrazing is approximated by $C(x, y) = 9x + 6y - 7$, where x is the number of acres of grass and y is the number of acres of alfalfa.

(a) How many cattle can graze if there are 80 acres of grass and 20 acres of alfalfa?
(b) How many cattle can graze if there are 60 acres of grass and 40 acres of alfalfa?

 In Your Words

25. What are the advantages of expressing something as a function of two variables rather than as a function of one variable?

26. Write an application in your technology area of interest that requires that you describe something as a function of two or three variables.

≡ 30.2
SURFACES IN THREE DIMENSIONS

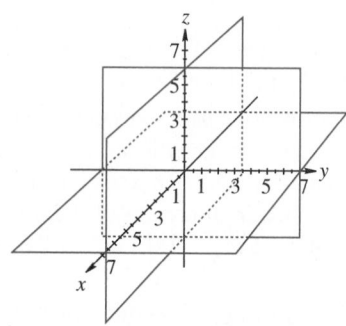

FIGURE 30.3

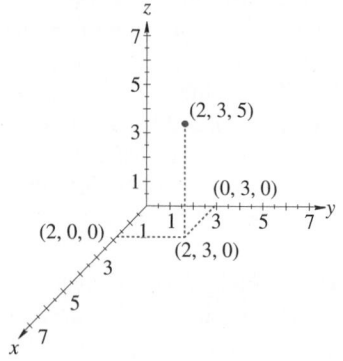

FIGURE 30.4

Graphing on the rectangular coordinate system in two dimensions meant using two perpendicular axes–the x-axis and the y-axis. Graphing in a **rectangular coordinate system in three dimensions** involves using three perpendicular axes–an x-axis, a y-axis, and a z-axis. Each pair of axes forms a plane in space and each plane is perpendicular to the other two planes. The three planes are called by the axes that define them. Thus, we have an xy-plane, an xz-plane, and a yz-plane. The planes divide space into eight regions called the **octants**. The octant in which all the values are positive is known as the **first octant**. The other octants are not numbered. The coordinate system looks something like the one shown in Figure 30.3. Normally, only the axes are drawn and not the planes.

Locating a point is done by the following procedures and is shown in Figure 30.4. Suppose we want to locate the point $(2, 3, 5)$. First we locate the x- and y-coordinates on their respective axes. The x-coordinate is 2, so we locate the point $(2, 0, 0)$. Notice that the x-axis "comes out" towards you. Next we locate the y-coordinate at $(0, 3, 0)$ on the y-axis. From each of these points we draw a line in the xy-plane perpendicular to the axis. This locates the point $(2, 3, 0)$. From this point we draw a line straight up and 5 units long. We have now located the point $(2, 3, 5)$.

In Chapter 15, we showed that a second-degree equation in two variables was of the form

$$Ax^2 + Bxy + Cy^2 + Dx + Ey + F = 0$$

where A, B, and C were not all 0. The graph of this equation was one of four types—circle, ellipse, hyperbola, or parabola. The general first-degree equation in two variables $Ax + By + C = 0$ graphed a straight line.

A second-degree equation in three variables is of the form

$$Ax^2 + By^2 + Cz^2 + Dxy + Exz + Fyz + Gx + Hy + Iz + J = 0$$

This equation graphs one of nine types, called the **quadric surfaces**.

Planes

A general first-degree equation, $Ax + By + Cz + D = 0$, will graph a plane in three dimensions.

EXAMPLE 30.7

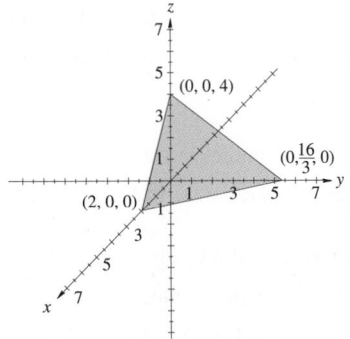

FIGURE 30.5

Sketch the graph of $4x + \frac{3}{2}y + 2z - 8 = 0$.

Solution The intercepts occur where the graph crosses the respective axes. By finding the intercepts, we get three points. It only takes three points to define a plane, so the intercepts are enough to do so.

If we let two of the variables have a value of 0 and solve the remaining equation, we will get the intercepts. If $x = y = 0$, then $z = 4$ and $(0, 0, 4)$ is one point. If $x = z = 0$, then $y = \frac{16}{3}$ and $(0, \frac{16}{3}, 0)$ is the y-intercept. If $y = z = 0$, then $x = 2$ and $(2, 0, 0)$ is the x-intercept. Plotting these three points and connecting them as in Figure 30.5 gives us a representation of the plane. ∎

The shaded region in Figure 30.5 represents only part of the plane. In this particular case, we get the part of the plane that is in the first octant. Figure 30.6 shows the graph of $4x - \frac{3}{2}y - 2z + 8 = 0$. Here again, we only show the part of the plane that is in one octant.

Quadric Surfaces

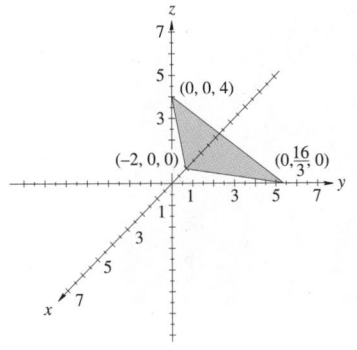

FIGURE 30.6

The previous two graphs are of planes. In general, the graph of an equation in three variables, which is equivalent to a function in two independent variables, is a **surface** in space. So far, we have looked at plane surfaces. In the remainder of this section, we will present the nine quadric surfaces, but before we present them we need to learn some terminology.

The intersection of two surfaces in space is a **curve**. The **traces** of a surface are the curves formed when the surface intersects a coordinate plane. A **section** is formed by the intersection of the surface with a plane other than a coordinate plane. Usually, a section is formed when the surface intersects a plane parallel to a coordinate plane.

In the list of quadric surfaces that follows, we will present the "standard equations." If the surface has a center, it will be at the origin. If the surface has symmetry, it will be symmetric with a coordinate axis. For each surface, we will give the intercepts, a description of the traces, and a sketch of the surface. In all cases, the letters a, b, and c represent positive real numbers.

1. Ellipsoid: $\dfrac{x^2}{a^2} + \dfrac{y^2}{b^2} + \dfrac{z^2}{c^2} = 1$ (Figure 30.7)

The intercepts are $(\pm a, 0, 0)$, $(0, \pm b, 0)$, and $(0, 0, \pm c)$. The traces are ellipses, as are all sections parallel to the coordinate plane.

2. Elliptic Cone: $z^2 = \dfrac{x^2}{a^2} + \dfrac{y^2}{b^2}$ (Figure 30.8)

The only intercept is at the origin $(0, 0, 0)$ and is called the **vertex**. The trace in the xy-plane is the origin. The trace in the yz-plane is the intersecting lines

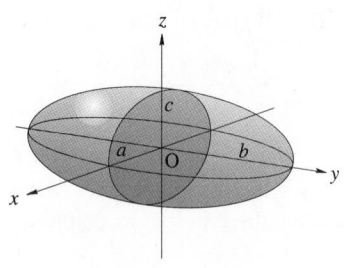

Ellipsoid: $\frac{x^2}{a^2}+\frac{y^2}{b^2}+\frac{z^2}{c^2}=1$

FIGURE 30.7

$z = \pm\frac{y}{b}$. The trace in the xz-plane is the intersecting lines $z = \pm\frac{x}{a}$. Sections parallel to the xy-plane are ellipses. If $a = b$, the elliptic cone become a **circular cone**.

3. Elliptic Paraboloid: $z = \frac{x^2}{a^2}+\frac{y^2}{b^2}$ (Figure 30.9)

The only intercept is at the origin and is called the **vertex**. The trace in the xy-plane is the origin. The trace in each of the other two planes is a parabola—in the xz-plane it is $z = \frac{x^2}{a^2}$ and in the yz-plane $z = \frac{y^2}{b^2}$. Sections parallel to the xy-plane are ellipses and sections parallel to the other two planes are parabolas.

4. Hyperbolic Paraboloid: $z = \frac{y^2}{b^2}-\frac{x^2}{a^2}$ (Figure 30.10)

The only intercept is at the origin. The traces in the xy-plane are the intersecting lines $\frac{y}{b} = \pm\frac{x}{a}$. The traces in the other two planes are parabolas. In the xz-plane it is $z = \frac{-x^2}{a^2}$ and in the yz-plane, $z = \frac{y^2}{b^2}$. Sections parallel to the xy-plane are hyperbolas. Sections parallel to the other two planes are parabolas.

The origin plays a unique role in this figure. It is a minimum point for the trace in the yz-plane and is a maximum point for the trace in the xy-plane. It is known as a **saddle point**.

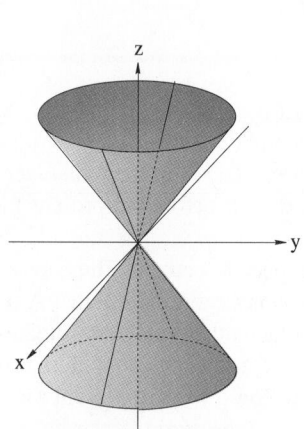

Elliptic cone: $z^2 = \frac{x^2}{a^2}+\frac{y^2}{b^2}$

FIGURE 30.8

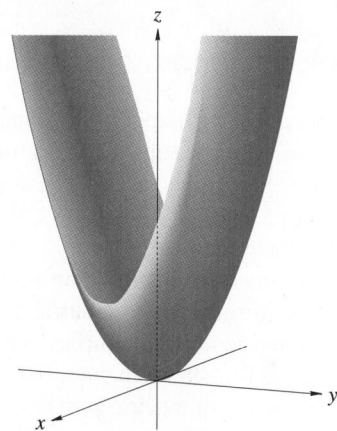

Elliptic paraboloid: $z = \frac{x^2}{a^2}+\frac{y^2}{b^2}$

FIGURE 30.9

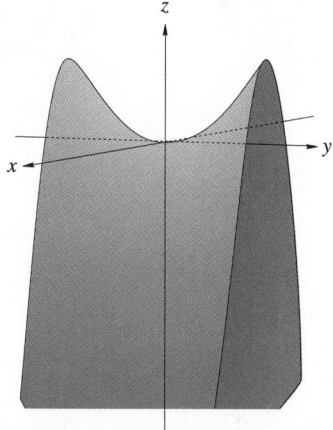

Hyperbolic paraboloid: $z = \frac{y^2}{b^2}-\frac{x^2}{a^2}$

FIGURE 30.10

5. Hyperboloid of One Sheet: $\frac{x^2}{a^2}+\frac{y^2}{b^2}-\frac{z^2}{c^2}=1$ (Figure 30.11)

The intercepts are $(\pm a,0,0)$ and $(0,\pm b,0)$. The trace in the xy-plane is the ellipse $\frac{x^2}{a^2}+\frac{y^2}{b^2}=1$. The traces in the other two planes are hyperbolas. In the xz-plane, the trace is $\frac{x^2}{a^2}-\frac{z^2}{c^2}=1$ and in the yz-plane it is $\frac{y^2}{b^2}-\frac{z^2}{c^2}=1$.

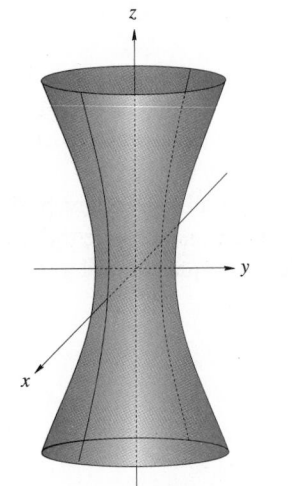

Hyperboloid of one sheet: $\dfrac{x^2}{a^2}+\dfrac{y^2}{b^2}-\dfrac{z^2}{c^2}=1$

FIGURE 30.11

Sections parallel to the xy-plane are ellipses and sections parallel to the other coordinate planes are hyperbolas. If $a=b$, the surface is called a **hyperboloid of revolution**.

6. **Hyperboloid of Two Sheets:** $\dfrac{x^2}{a^2}+\dfrac{y^2}{b^2}-\dfrac{z^2}{c^2}=-1$ (Figure 30.12)

The intercepts are the two points $(0,0,\pm c)$. The trace in the xz-plane is the hyperbola $\dfrac{z^2}{c^2}-\dfrac{x^2}{a^2}=1$ and in the yz-plane, the hyperbola $\dfrac{z^2}{c^2}-\dfrac{y^2}{b^2}=1$. There is no trace in the xy-plane. Sections parallel to the xy-plane are ellipses and sections parallel to the other two planes are hyperbolas.

The last three quadric surfaces are called **cylinders**. A cylinder is generated by a line moving along a curve while remaining parallel to a fixed line. Notice that for each cylinder, one variable is missing.

7. **Parabolic Cylinder:** $x^2=4ay$ (Figure 30.13)

This surface is formed by a line parallel to the z-axis moving along the parabola $x^2=4ay$. One way to visualize this is to take a piece of paper and fold it so that two opposite edges form a parabola.

8. **Elliptic Cylinder:** $\dfrac{x^2}{a^2}+\dfrac{y^2}{b^2}=1$ (Figure 30.14)

This surface is generated by a line parallel to the z-axis moving along the ellipse satisfying the given standard equation. You can visualize this by taking a sheet of paper and folding it until opposite edges meet. The other two edges should be in the shape of an ellipse.

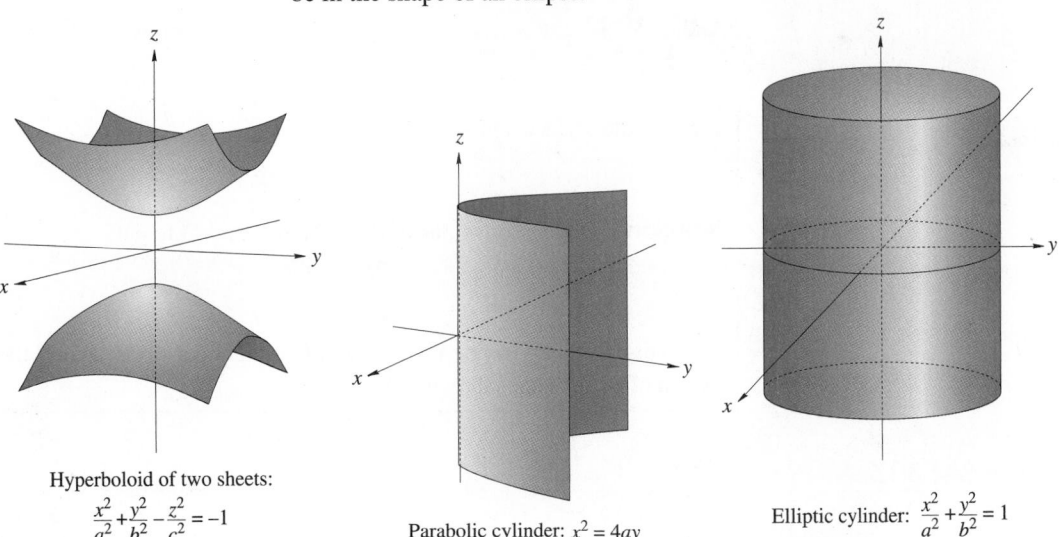

Hyperboloid of two sheets:
$\dfrac{x^2}{a^2}+\dfrac{y^2}{b^2}-\dfrac{z^2}{c^2}=-1$

FIGURE 30.12

Parabolic cylinder: $x^2=4ay$

FIGURE 30.13

Elliptic cylinder: $\dfrac{x^2}{a^2}+\dfrac{y^2}{b^2}=1$

FIGURE 30.14

9. Hyperbolic Cylinder: $\dfrac{x^2}{a^2} - \dfrac{y^2}{b^2} = 1$ (Figure 30.15)

This is generated by a line parallel to the z-axis moving along the hyperbola with the given standard equation. This surface has two parts. You can visualize this surface by folding two sheets of paper where each sheet traces one branch of the hyperbola.

Remember that the nine types of quadric surfaces are all in standard position. It is certainly possible to interchange the variables. This will not change the type of surface, only its orientation. You may have to complete the square in order to get an equation into a recognizable form.

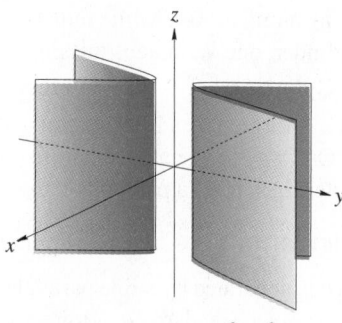

Hyperbolic cylinder: $\dfrac{x^2}{a^2} - \dfrac{y^2}{b^2} = 1$

FIGURE 30.15   **FIGURE 30.16**

EXAMPLE 30.8

Identify the surface and sketch the graph of

$$9x^2 - 36y^2 + 16z^2 = 144.$$

Solution   Divide both sides of the equation by 144 to get

$$\frac{x^2}{16} - \frac{y^2}{4} + \frac{z^2}{9} = 1$$

This is in the form of a hyperboloid of one sheet whose axis is the y-axis. The sketch of this curve is shown in Figure 30.16.

Exercise Set 30.2

In Exercises 1–20, identify each surface and sketch the graph in the rectangular coordinate system in three dimensions.

1. $x + 2y + 3z - 6 = 0$
2. $3x + 2y - z - 6 = 0$
3. $x = 8y^2$
4. $x^2 + y^2 + z^2 = 9$
5. $x^2 + 2y^2 + z^2 = 4$
6. $4x^2 + 9y^2 = 36$
7. $2x^2 + y^2 + 2z^2 = 8$
8. $4x^2 + y^2 - z^2 = 1$
9. $4x^2 + y^2 - z^2 = -1$
10. $4x^2 - 9y^2 = 1$
11. $3x - 4y - 6z = 12$
12. $xy = 4$

13. $z = x^2 + 4y^2 - 16$

14. $2x^2 + y^2 - 2z^2 = -8$

15. $36z = 4x^2 + 9y^2$

16. $36z^2 = 4x^2 + 9y^2$

17. $36z = 4x^2 - 9y^2$

18. $36z^2 = 1 - 4x^2 - 9y^2$

19. $2x = y^2$

20. $3x + 2y - 3z = 6$

 In Your Words

21. Describe how to sketch a three-dimensional rectangular coordinate system.

22. What are quadric surfaces? Name each one and sketch its basic appearance.

☰ 30.3
PARTIAL DERIVATIVES

Derivatives have been a useful tool for us; but the derivatives that we have taken have always been with functions of one variable. Now we are working with functions of two or more independent variables. In this section, we are going to learn a method of taking derivatives that we can use with functions of several variables. This method is called **partial derivatives**.

Suppose we have a function of two variables, such as $z = f(x, y)$. If we treat y as a constant, this acts as a function in just one variable, x. We know how to take the derivative of a function in one variable. If we take the derivative of this "one-variable" function, we will call it a **partial derivative of f with respect to x**. We use the notation $\frac{\partial z}{\partial x}$, $\frac{\partial f}{\partial x}$, or f_x to indicate a partial derivative with respect to x. In the same way, if we take the derivative of $f(x, y)$ with respect to y (by treating x as a constant), then we have taken the **partial derivative of f with respect to y**. The notation $\frac{\partial z}{\partial y}$, $\frac{\partial f}{\partial y}$, or f_y is used for a partial derivative of f with respect to y. The formal definitions for partial derivatives are given in the following box.

Partial Derivatives of a Function of Two Variables

If $z = f(x, y)$, then the **first partial derivatives of f** with respect to x and y are

$$\frac{\partial f}{\partial x} = f_x = \lim_{h \to 0} \frac{f(x+h, y) - f(x, y)}{h}$$
$$\frac{\partial f}{\partial y} = f_y = \lim_{h \to 0} \frac{f(x, y+h) - f(x, y)}{h}$$

provided the limits exist.

 Hint

In these definitions, if $z = f(x, y)$, then to find f_x you need to consider y a constant and differentiate with respect to x. Similarly, to find f_y, consider x a constant and differentiate with respect to y.

EXAMPLE 30.9

If $f(x, y) = x^2 + 2xy$, find $\dfrac{\partial f}{\partial x}$, $\dfrac{\partial f}{\partial y}$, $f_x(1, 2)$, and $f_y(1, 2)$.

Solution In finding the partial derivative of f with respect to x, we treat y as a constant. Thus

$$\frac{\partial f}{\partial x} = 2x + 2y$$

For the partial derivative of f with respect to y, we treat x as a constant and get

$$\frac{\partial f}{\partial y} = 0 + 2x = 2x$$
$$f_x(1, 2) = 2 + 4 = 6$$
$$f_y(1, 2) = 2$$

EXAMPLE 30.10

If $z = \sqrt{x^2 + y^3}$, calculate $\dfrac{\partial z}{\partial x}$ and $\dfrac{\partial z}{\partial y}$.

Solution First we write $z = (x^2 + y^3)^{1/2}$.

$$\frac{\partial z}{\partial x} = \frac{1}{2}(x^2 + y^3)^{-1/2}(2x + 0) = \frac{x}{\sqrt{x^2 + y^3}}$$

$$\frac{\partial z}{\partial y} = \frac{1}{2}(x^2 + y^3)^{-1/2}(0 + 3y^2) = \frac{3y^2}{2\sqrt{x^2 + y^3}}$$

EXAMPLE 30.11

If $z = \dfrac{x}{y}\sin(x^2y^3)$, determine $\dfrac{\partial z}{\partial x}$ and $\dfrac{\partial z}{\partial y}$.

Solution We need to use the product and chain rules of differentiation.

$$\frac{\partial z}{\partial x} = \frac{x}{y}\left[\frac{\partial}{\partial x}\sin(x^2y^3)\right] + \sin(x^2y^3)\left[\frac{\partial}{\partial x}\left(\frac{x}{y}\right)\right]$$

$$= \frac{x}{y}\left[\cos(x^2y^3)\right](2xy^3) + \left[\sin(x^2y^3)\right]\left(\frac{1}{y}\right)$$

$$= 2x^2y^2\cos(x^2y^3) + \frac{1}{y}\sin(x^2y^3)$$

$$\frac{\partial z}{\partial y} = \frac{x}{y}\left[\frac{\partial}{\partial y}\sin(x^2y^3)\right] + \sin(x^2y^3)\left[\frac{\partial}{\partial y}\left(\frac{x}{y}\right)\right]$$

$$= \frac{x}{y}\left[\cos(x^2y^3)\right](3x^2y^2) + \left[\sin(x^2y^3)\right]\left(\frac{-x}{y^2}\right)$$

$$= 3x^3y\cos(x^2y^3) - \frac{x}{y^2}\sin(x^2y^3)$$

Geometric Interpretation of Partial Derivatives

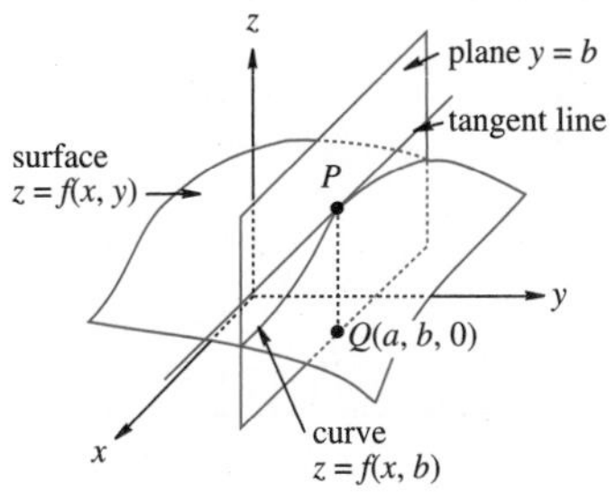

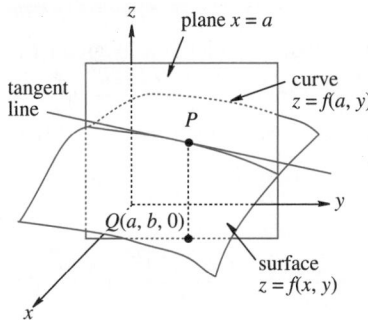

The partial derivatives f_x and f_y are the slopes of tangent lines to certain curves on the surface of $z = f(x, y)$. Consider the point $P(a, b, f(a, b))$ on this surface. This point lies directly above the point $Q(a, b, 0)$ in the xy-plane, as shown in Figure 30.17. The vertical plane $y = b$ is parallel to the xz-plane. (If $b = 0$, the plane $y = b$ is the xz-plane.) The plane $y = b$ intersects the surface in the curve $z = f(x, b)$ through the point P. Because this curve is in the plane $y = b$, the x values can change, but the y values are always b. The partial derivative of z with respect to x represents the slope of a line tangent to this curve $z = f(x, b)$.

In the same way, the partial derivative of z with respect to y represents the slope of a line tangent to the curve $z = f(a, y)$. This curve is formed by the vertical plane $x = a$, which is parallel to the yz-plane, intersecting the surface as shown in Figure 30.18.

FIGURE 30.17

FIGURE 30.18

Hint

To find the slope of a line tangent to surface $z = f(x, y)$ and parallel to the xz-plane at the point $P(a, b, f(a, b))$, calculate $\dfrac{\partial z}{\partial x}$.

Similarly, to find the slope of a line tangent to surface $z = f(x, y)$ and parallel to the yz-plane at the point $P(a, b, f(a, b))$, calculate $\dfrac{\partial z}{\partial y}$.

EXAMPLE 30.12

(a) Find the slope of a line tangent to the surface $z = x^2 + 3y^2$ and parallel to the xz-plane at the point $(2, -1, 7)$. **(b)** Find the slope of the line that is tangent to this surface and parallel to the yz-plane at the same point.

Solutions   **(a)** We want to find $\dfrac{\partial z}{\partial x}$ and then evaluate it at $(2, -1, 7)$:

$$\frac{\partial z}{\partial x} = 2x$$

EXAMPLE 30.12 (Cont.)

So, at $(2, -1, 7)$, $\dfrac{\partial z}{\partial x}\bigg|_{(2,-1,7)} = 4$. **(b)** For the second part, we want $\dfrac{\partial z}{\partial y}$ evaluated at the same point:

$$\frac{\partial z}{\partial y} = 6y$$

So, $\qquad \dfrac{\partial z}{\partial y}\bigg|_{(2,-1,7)} = -6.$

Applications of partial derivatives are found in many ways. The following is but one example. Others are shown in the exercises and in the Section 30.4.

Application

EXAMPLE 30.13

A light source I of a candelabra located r m from a surface inclined at an angle θ provides an illuminance $E = \dfrac{I\cos\theta}{r^2}$ lumens per square meter (lm/m^2). Find the rate at which E varies with respect to I and evaluate it for $\theta = \frac{\pi}{6}$, $r = 2\,\text{m}$, and $I = 100\,\text{cd}$.

Solution $\qquad \dfrac{\partial E}{\partial I} = \dfrac{\cos\theta}{r^2}$

$$\frac{\partial E}{\partial I}\bigg|_{\substack{r=2 \\ \theta=\frac{\pi}{6}}} = \frac{\cos(\frac{\pi}{6})}{2^2}$$

$$= \frac{\sqrt{3}/2}{4} = \frac{\sqrt{3}}{8} \approx 0.217\,\text{lm/m}^2$$

Higher Partial Derivatives

We know that if we have a function in two independent variables, such as $z = f(x, y)$, we can take the partial derivative with respect to either variable: $\dfrac{\partial z}{\partial x}$ and $\dfrac{\partial z}{\partial y}$. We can just as easily differentiate each of these derivatives with respect to either variable. Thus, for $\dfrac{\partial z}{\partial x}$ we find the derivative with respect to x, $\dfrac{\partial}{\partial x}\left(\dfrac{\partial z}{\partial x}\right) = \dfrac{\partial^2 z}{\partial x^2}$, or we can take it with respect to y, $\dfrac{\partial}{\partial y}\left(\dfrac{\partial z}{\partial x}\right) = \dfrac{\partial^2 z}{\partial y\,\partial x}$. In the same way, we can differentiate $\dfrac{\partial z}{\partial y}$ with respect to x, $\dfrac{\partial}{\partial x}\left(\dfrac{\partial z}{\partial y}\right) = \dfrac{\partial^2 z}{\partial x\,\partial y}$, and with respect to y, $\dfrac{\partial}{\partial y}\left(\dfrac{\partial z}{\partial y}\right) = \dfrac{\partial^2 z}{\partial y^2}$. Thus, we have the following second partial derivatives for a function of two variables.

Second Partial Derivatives of a Function of Two Variables

If $z = f(x, y)$, then there are four **second partial derivatives** of f:

1. Differentiate twice with respect to x:

$$\frac{\partial}{\partial x}\left(\frac{\partial z}{\partial x}\right) = \frac{\partial^2 f}{\partial x^2} = f_{xx}$$

2. Differentiate twice with respect to y:

$$\frac{\partial}{\partial y}\left(\frac{\partial z}{\partial y}\right) = \frac{\partial^2 f}{\partial y^2} = f_{yy}$$

3. Differentiate first with respect to x and then with respect to y:

$$\frac{\partial}{\partial y}\left(\frac{\partial z}{\partial x}\right) = \frac{\partial^2 f}{\partial y\,\partial x} = f_{xy}$$

4. Differentiate first with respect to y and then with respect to x:

$$\frac{\partial}{\partial x}\left(\frac{\partial z}{\partial y}\right) = \frac{\partial^2 f}{\partial x\,\partial y} = f_{yx}$$

≡ **Note**

The last two second-order partial derivatives

$$\frac{\partial^2 z}{\partial x\,\partial y} = f_{yx}(x, y) \text{ and } \frac{\partial^2 z}{\partial y\,\partial x} = f_{xy}(x, y)$$

are called **mixed partial derivatives**. Notice the differences in these two equations. The notation f_{xy} means that we should first differentiate with respect to x and then differentiate that result with respect to y.

In Section 30.4, we shall see how to use second partial derivatives to determine relative maximum, relative minimum, concavity, and saddle points, a new property.

EXAMPLE 30.14

Find all second-order partial derivatives of $f(x, y) = \sin(xy)$.

Solution We first find the first-order partial derivatives.

$$f_x = y\cos(xy)$$
$$f_y = x\cos(xy)$$

EXAMPLE 30.14 (Cont.)

Then, we get the following second-order partial derivatives.

$$f_{xx} = \frac{\partial}{\partial x}(f_x) = -y^2 \sin(xy)$$

$$f_{xy} = \frac{\partial}{\partial y}(f_x) = -xy\sin(xy) + \cos(xy)$$

$$f_{yx} = \frac{\partial}{\partial x}(f_y) = -xy\sin(xy) + \cos(xy)$$

$$f_{yy} = \frac{\partial}{\partial y}(f_y) = -x^2 \sin(xy)$$

≡ **Note**   In this example, $f_{xy} = f_{yx}$ for all (x, y). As it turns out, this will be the case for most functions.

Exercise Set 30.3

In Exercises 1–16, find f_x and f_y.

1. $f(x, y) = x^2 y + y^2 x$

2. $f(x, y) = x^2 y + 7y^2$

3. $f(x, y) = 3x^2 + 6xy^3$

4. $f(x, y) = \dfrac{x}{y^2}$

5. $f(x, y) = \dfrac{x^2 + y^2}{y}$

6. $f(x, y) = \dfrac{x + y}{\sqrt{xy}}$

7. $f(x, y) = e^y \cos x + e^x \sin y$

8. $f(x, y) = x^2 \cos y + y^2 \cos x$

9. $f(x, y) = e^{2x+3y}$

10. $f(x, y) = e^{x^2 + y^2}$

11. $f(x, y) = \sin(x^2 y^3)$

12. $f(x, y) = \ln(x^2 y^3)$

13. $f(x, y) = \ln\sqrt{x^2 + y^2}$

14. $f(x, y) = \tan^{-1}\dfrac{y}{x}$

15. $f(x, y) = \sin^2(3xy)$

16. $f(x, y) = e^{\sqrt{x^2 + y^2}}$

In Exercises 17–24, compute all second-order partial derivatives.

17. $f(x, y) = x^3 y^2 - xy^5$

18. $f(x, y) = \sin xy^3$

19. $f(x, y) = 3e^{xy^3}$

20. $f(x, y) = \sin(x + y^2)$

21. $f(x, y) = \dfrac{2x}{y^5}$

22. $f(x, y) = e^x \tan y$

23. $f(x, y) = \ln(x^3 y^5)$

24. $f(x, y) = \sqrt{x^2 y + 3y^2}$

Solve Exercises 25–36.

25. *Physics*   The **ideal gas law** states that $P = \dfrac{nRT}{V}$, where n is the number of moles of an ideal gas, T is the absolute temperature, and R is the universal gas constant. If $n = 20$ and $T = 22°C = 295°K$, what is the rate of change of the pressure with respect to the volume, when the volume is $3\,L$?

26. *Thermodynamics*   The temperature distribution T of a heated plate located in the xy-plane is given by

$$T = \frac{100}{\ln 2}\ln(x^2 + y^2), \quad \text{for } 1 \le x^2 + y^2 \le 4$$

Find the rate of change of T in a direction parallel to the x-axis at the point $(1, 0)$ and at the point $(0, 1)$.

27. *Physics*   The volume V (in cubic centimeters) of one mole of an ideal gas is given by

$$V = \frac{82.06T}{P}$$

where P is the pressure in atmospheres (atm) and T is the absolute temperature. Find the rate of change of the volume of 1 mole with respect to pressure, when $T = 300°K$ and $P = 5$ atm.

28. *Physics*   Find the rate of change of the volume of 1 mole of the ideal gas in Exercise 27 with respect to temperature, when $T = 300°K$ and $P = 5$ atm.

29. Let C be the trace of the paraboloid $z = 16 - x^2 - y^2$ on the plane $x = 1$. Find the slope of the tangent line to C at the point $(1, 3, 6)$.

30. *Physics*   For the ideal gas equation $PV = nRT$, show that

$$\left(\frac{\partial V}{\partial T}\right)\left(\frac{\partial T}{\partial P}\right)\left(\frac{\partial P}{\partial V}\right) = -P$$

31. *Medical technology*   One of **Poiseuille's laws** states that the resistance, R, for blood flowing in a blood vessel depends on the length, L, and radius, r, of the vessel according to the formula $R = k\dfrac{L}{r^4}$, where k is a constant.

(a) What is the rate of change in the resistance with respect to the length?

(b) What is the rate of change in the resistance with respect to the radius?

32. *Economics*   The production of a computer manufacturing company is given approximately by the Cobb-Douglas production formula $P(x, y) = 65x^{0.3}y^{0.7}$, where x is the number of units of labor and y is the number of units of capital.

(a) Find $P_x(x, y)$, the **marginal productivity of labor**.

(b) If the company is currently using 1,500 units of labor and 5,400 units of capital, find the marginal productivity of labor.

(c) Find $P_y(x, y)$, the **marginal productivity of capital**.

(d) If the company is currently using 1,500 units of labor and 5,400 units of capital, find the marginal productivity of capital.

(e) For the greatest increase in productivity, should the management of the company encourage increased use of labor or capital?

33. *Medical technology*   The surface area in square inches of a human is approximated by $A(w, h) = 14.085w^{0.425}h^{0.725}$, where w is the weight of the person in pounds and h is the height of the person in inches.

(a) Find $A_w(w, h)$ and $A_h(w, h)$.

(b) Compute $A_w(w, h)$ and $A_h(w, h)$ for a person with a weight of $185\,lb$ and a height of $5\,ft\ 11\,in$.

(c) Interpret your answers from (b).

34. *Medical technology*   The surface area in square meters of a human is approximated by $A(w, h) = 0.18215w^{0.425}h^{0.725}$, where w is the weight of the person in kilograms and h is the height of the person in meters.

(a) Find $A_w(w, h)$ and $A_h(w, h)$.

(b) Compute $A_w(w, h)$ and $A_h(w, h)$ for a person with a weight of $84\,kg$ and a height of $1.80\,m$.

35. *Police science* The stopping distance for a car after the brakes are applied is given by $S(w, r) = 0.000\,02wr^2$, where w is the car's weight in pounds and r its speed in miles per hour.
   (a) Determine $S_w(w, r)$.
   (b) Calculate $S_w(w, r)$ for a car weighing 2,400 lb and traveling 65 mph.
   (c) Determine $S_r(w, r)$.
   (d) Calculate $S_r(w, r)$ for a car weighing 2,400 lb and traveling 65 mph.
   (e) How much will the stopping distance increase for this car if its velocity is increased from 65 mph to 66 mph?

36. *Meteorology* Wind speed affects the temperature a person feels and often makes it feel colder than it is. The wind chill index (WCI) represents the equivalent temperature, in degrees Fahrenheit, that exposed skin would feel if there were little or no wind. If F represents the actual temperature in degrees Fahrenheit as measured by a thermometer and v is the velocity of the wind in miles per hour, then $WCI(v, F) = 91.4 - \dfrac{(10.45 + 6.69\sqrt{v} - 0.447v)(91.4 - F)}{22}$, where $4 < v \le 45$.
   (a) Evaluate $WCI(v, F)$ for wind velocity of 30 mph and temperature of $-4°$F.
   (b) Determine $WCI_v(v, F)$.
   (c) Calculate $WCI_v(v, F)$ for wind velocity of 30 mph and temperature of $-4°$F.
   (d) Determine $WCI_F(v, F)$.
   (e) Calculate $WCI_F(v, F)$ for wind velocity of 30 mph and temperature of $-4°$F.
   (f) Would you feel more comfortable if the wind velocity increased 1 mph or if temperature dropped $1°$F?

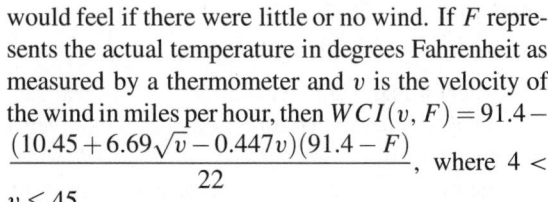

In Your Words

37. Without looking in the text, describe how to find the partial derivatives of a function of two variables.

38. Without looking in the text, **(a)** list all the second partial derivatives of a function of two variables, and **(b)** state how each second partial derivative is found.

☰ 30.4
SOME APPLICATIONS OF PARTIAL DERIVATIVES

Some applications of partial derivatives were discussed in Section 30.3. You may remember that we applied derivatives to such problems as related rates, maxima, and minima. Partial derivatives can be used to solve similar problems. In this section, we will learn how to calculate the differential of a function of two variables and how to apply it. We will also learn how to apply maxima and minima for functions of two variables.

Differentials

If $y = f(x)$ is a function in one variable and if $f'(x)$ exists, then the change in y, Δy, from x to $x + h$, is approximated by the differential of f

$$dy = f'(x)\,dx$$

We discussed this idea in Section 24.7. We can define a differential for a function of two variables in much the same manner.

Differential for a Function of Two Variables

If $z = f(x, y)$ is a function of two variables, then the **differential** dz, usually referred to as the **total differential** of z, is defined as

$$dz = \frac{\partial z}{\partial x} dx + \frac{\partial z}{\partial y} dy$$

EXAMPLE 30.15

Find the differential dz, if $z = x^2 + 3xy - 2y^2$.

Solution We will first compute $\frac{\partial z}{\partial x}$ and $\frac{\partial z}{\partial y}$.

$$\frac{\partial z}{\partial x} = 2x + 3y$$

$$\frac{\partial z}{\partial y} = 3x - 4y$$

Using the formula $dz = \frac{\partial z}{\partial x} dx = \frac{\partial z}{\partial y} dy$, we get the total differential of z.

$$dz = (2x + 3y) dx + (3x - 4y) dy$$

EXAMPLE 30.16

For the function in Example 30.15, estimate the change in z when x changes from 3.0 to 3.1 and y changes from 6.0 to 5.8.

Solution From Example 30.15, we have $dz = (2x + 3y) dx + (3x - 4y) dy$. We have the values $x = 3.0$, $y = 6.0$, $dx = 0.1$, and $dy = -0.2$. Notice that dy is negative because the value of y is decreasing. Substituting these values into the equation for dz, we get

$$
\begin{aligned}
dz &= [2(3.0) + 3(6.0)]0.1 + [3(3.0) - 4(6.0)](-0.2) \\
&= 24(0.1) + (-15)(-0.2) \\
&= 2.4 + 3.0 \\
&= 5.4
\end{aligned}
$$

The answer in Example 30.16 is an approximation to the actual change. We could compute the actual change by finding

$$
\begin{aligned}
f(3.1, 5.8) - f(3.0, 6.0) &= \left[(3.1)^2 + 3(3.1)(5.8) - 2(5.8)^2\right] \\
&\quad - \left[(3.0)^2 + 3(3.0)(6.0) - 2(6.0)^2\right] \\
&= (9.61 + 53.94 - 67.28) - (9 + 54 - 72) \\
&= -3.73 - (-9) = 5.27
\end{aligned}
$$

Application

EXAMPLE 30.17

A bottling company requires cans in the shape of a right circular cylinder of height 11.7 cm and radius 3.1 cm. The can manufacturer claims a percentage error of no more than 0.2% in the height and 0.1% in the radius. What is the maximum error in the volume?

Solution   The volume of a right circular cylinder is given by $V = \pi r^2 h$. The total differential dV is

$$dV = \frac{\partial V}{\partial r} dr + \frac{\partial V}{\partial h} dh$$
$$= 2\pi r h \, dr + \pi r^2 \, dh$$

The relative error of the radius is $\dfrac{|\Delta r|}{r} = \dfrac{|dr|}{r} = 0.001$ since $0.1\% = 0.001$ and the relative error of the height is $\dfrac{|\Delta h|}{h} = \dfrac{|dh|}{h} = 0.002$. Thus, we have $|dr| = 0.001r$ and $|dh| = 0.002h$. The maximum error in the volume is then given by

$$dV = 2\pi r h |dr| + \pi r^2 |dh|$$
$$= 0.002\pi r^2 h + .002\pi r^2 h$$
$$= 0.004\pi r^2 h$$
$$= 0.004V$$

Thus, we see that the change in volume is at most $0.004 = 0.4\%$.
    Since

$$V = \pi r^2 h = (3.1)^2 (11.7)\pi$$
$$\approx 353.23$$

$dV = 0.004V \approx 1.41$ and so the actual volume lies between 351.82 and 354.64 cm^3. ▪

≡ Note

The definition of the total differential can be extended to any number of variables. For example, if you have a function of three variables

$$w = f(x, y, z)$$

then
$$dw = \frac{\partial w}{\partial x} dx + \frac{\partial w}{\partial y} dy + \frac{\partial w}{\partial z} dz$$

Chain Rule

If x and y are parametric equations, where $x = g(t)$ and $y = h(t)$, and you have a function of two variables, $z = f(x, y)$, then you can use the following version of the **chain rule**.

Chain Rule for a Function of Two Variables	$\dfrac{dz}{dt} = \dfrac{\partial z}{\partial x} \cdot \dfrac{dx}{dt} + \dfrac{\partial z}{\partial y} \cdot \dfrac{dy}{dt}$

EXAMPLE 30.18

Find $\dfrac{dw}{dt}$, where $w = e^{-x^2-y^2}$, $x = t$, and $y = \sqrt{t}$.

Solution $\dfrac{dw}{dt} = \dfrac{\partial w}{\partial x} \cdot \dfrac{dx}{dt} + \dfrac{\partial w}{\partial y} \cdot \dfrac{dy}{dt}$

$$= \left(-2xe^{-x^2-y^2}\right)(1) + \left(-2ye^{-x^2-y^2}\right)\left(\frac{1}{2}t^{-1/2}\right)$$

Substituting for x and y, we get

$$= \left(-2te^{-t^2-t}\right) + \left(-2\sqrt{t}e^{-t^2-t}\right)\left(\frac{1}{2\sqrt{t}}\right)$$

$$= -(2t+1)e^{-(t^2+t)}$$

Chain Rule and Related Rates

The chain rule can be used with problems of related rates, such as the one in the next example.

Application

EXAMPLE 30.19

The radius of a right circular cone is increasing at the rate of 3 cm/s, while its height is increasing at the rate of 5 cm/s. How fast is the volume changing, when $r = 18$ cm and $h = 30$ cm?

Solution   We are asked to find $\dfrac{dV}{dt}$, where $V = \frac{1}{3}\pi r^2 h$. Using the chain rule, we obtain

$$\frac{dV}{dt} = \frac{\partial V}{\partial r} \cdot \frac{dr}{dt} + \frac{\partial V}{\partial h} \cdot \frac{dh}{dt}$$

Since $\dfrac{\partial V}{\partial r} = \frac{2}{3}\pi rh$ and $\dfrac{\partial V}{\partial h} = \frac{1}{3}\pi r^2$, we have

$$\frac{dV}{dt} = \frac{2}{3}\pi rh\frac{dr}{dt} + \frac{1}{3}\pi r^2\frac{dh}{dt}$$

When $r = 18$, $h = 30$, $\dfrac{dr}{dt} = 3$, and $\dfrac{dh}{dt} = 5$, we get

$$\frac{dV}{dt} = \frac{2}{3}\pi(18)(30)(3) + \frac{1}{3}\pi(18)^2(5)$$

$$= 1620\pi\ \text{cm}^3/\text{s}$$

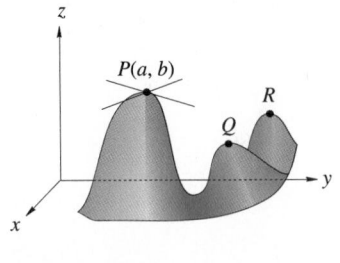

FIGURE 30.19

Maxima and Minima

Our last topic in this section deals with the extrema; that is, the maximum and minimum points on a surface. Suppose we have a surface, $z = f(x, y)$, such as the one in Figure 30.19. We have several points, P, Q, and R, that all seem to be local or relative maximum points. If one of these is a maximum point, then every tangent line is horizontal, as is shown at point P. In particular, we have

$$\frac{\partial z}{\partial x} = 0 = \frac{\partial z}{\partial y}$$

or

$$f_x(a, b) = 0 = f_y(a, b)$$

if P is the point (a, b). The same would be true if P was a minimum point. Thus, in order for $z = f(x, y)$ to have a local maximum or minimum, both partial derivatives of f must exist and have a value of 0. We state this formally as the following first derivative test.

First Derivative Test for Local Extrema

If $z = f(x, y)$ has a local maximum or minimum value at an interior point (a, b) of its domain, where f_x and f_y are both defined, then

$$f_x(a, b) = 0 = f_y(a, b)$$

or

$$\frac{\partial z}{\partial x} = \frac{\partial z}{\partial y} = 0 \text{ at } (a, b)$$

This provides us with a list of critical values but no guarantee that a maximum or minimum exists. The **second derivative test for functions of two variables** provides us with this information. Unfortunately, the second derivative test is fairly complicated.

An important consideration in the determination of relative maxima and minima of $z = f(x, y)$ is the determinant given below, which we have called Δ (delta).

≡ **Note**   Remember that the determinant $\begin{vmatrix} a & b \\ c & d \end{vmatrix} = ad - bc$.

$$\Delta = \begin{vmatrix} \dfrac{\partial^2 z}{\partial x^2} & \dfrac{\partial^2 z}{\partial y\,\partial x} \\[2mm] \dfrac{\partial^2 z}{\partial x\,\partial y} & \dfrac{\partial^2 z}{\partial y^2} \end{vmatrix} = \frac{\partial^2 z}{\partial x^2} \cdot \frac{\partial^2 z}{\partial y^2} - \left(\frac{\partial^2 z}{\partial y\,\partial x} \right)^2$$

(Notice that we are assuming that $\dfrac{\partial^2 z}{\partial x\,\partial y} = \dfrac{\partial^2 z}{\partial y\,\partial x}$.)

Second Derivative Test

If $z = f(x, y)$, $\dfrac{\partial z}{\partial x} = \dfrac{\partial z}{\partial y} = 0$ at (a, b), and

(a) if $\Delta > 0$ and $\dfrac{\partial^2 z}{\partial x^2} > 0$, then $f(a, b)$ is a **relative or local minimum**;

(b) if $\Delta > 0$ and $\dfrac{\partial^2 z}{\partial x^2} < 0$, then $f(a, b)$ is a **relative or local maximum**;

(c) If $\Delta < 0$ at (a, b), then $(a, b, f(a, b))$ is a **saddle point**;

(d) If $\Delta = 0$ at (a, b), the test fails,

$$\text{where} \quad \Delta = \begin{vmatrix} \dfrac{\partial^2 z}{\partial x^2} & \dfrac{\partial^2 z}{\partial y\, \partial x} \\ \dfrac{\partial^2 z}{\partial x\, \partial y} & \dfrac{\partial^2 z}{\partial y^2} \end{vmatrix} = \dfrac{\partial^2 z}{\partial x^2} \cdot \dfrac{\partial^2 z}{\partial y^2} - \left(\dfrac{\partial^2 z}{\partial y\, \partial x} \right)^2$$

It turns out that if $\Delta > 0$, then $\dfrac{\partial^2 z}{\partial x^2} \cdot \dfrac{\partial^2 z}{\partial y^2} > 0$, so these two second-order partial derivatives have the same sign. Thus, you can replace $\dfrac{\partial^2 z}{\partial x^2}$ with $\dfrac{\partial^2 z}{\partial y^2}$ in (a) and (b).

In (c), $\Delta < 0$. This point (a, b) was neither a maximum nor a minimum, but it is a special point called a **saddle point**. An example of a saddle point is at the origin in Figure 30.10. This surface is a minimum for the trace in the yz-plane and is a maximum for the trace in the xy-plane. .

EXAMPLE 30.20

If $f(x, y) = 3x - x^3 - 3xy^2$, find the extrema of f.

Solution We have $f_x(x, y) = 3 - 3x^2 - 3y^2$ and $f_y(x, y) = -6xy$. Critical values will occur when $f_x = f_y = 0$. The fact that $f_y = -6xy = 0$ implies that either $x = 0$ or $y = 0$. The fact that $f_x = 0$ when $x = 0$ implies that $y = \pm 1$, and when $y = 0$, $x = \pm 1$. Thus, we have four critical points: $(1, 0)$, $(0, 1)$, $(-1, 0)$, and $(0, -1)$.

The second-order partial derivatives are $\dfrac{\partial^2 f}{\partial x^2} = -6x$, $\dfrac{\partial^2 f}{\partial y^2} = -6x$, $\dfrac{\partial^2 f}{\partial y\, \partial x} = -6y$, $\dfrac{\partial^2 f}{\partial x\, \partial y} = -6y$, so

$$\Delta = (-6x)(-6x) - (-6y)^2 = 36x^2 - 36y^2$$
$$= 36(x^2 - y^2)$$

at each of the critical points.

Now if $x = 0$, then $x^2 - y^2 = -1$ and so $\Delta < 0$; thus we do not have an extremum. [But we do have a saddle point at $(0, 1)$ and $(0, -1)$.]

If $y = 0$, then $x^2 - y^2 = 1$ and $\Delta > 0$. If $x = 1$, then $\dfrac{\partial^2 f}{\partial x^2} = -6$ and so $(1, 0)$ is a relative maximum. If $x = -1$, $\dfrac{\partial^2 f}{\partial x^2} = 6$ and so $(-1, 0)$ is a relative minimum.

Exercise Set 30.4

In Exercises 1–8, find the total differential of each function.

1. $z = x^2 + y^2$

2. $z = 2x^2 + xy - y^2$

3. $z = 3x^2 + 4xy - 2y^3$

4. $z = xye^{x+y}$

5. $z = \arctan\left(\dfrac{y}{x}\right)$

6. $z = \ln(x^2 + y^2)$

7. $z = x \tan yx$

8. $z = \ln\left(\dfrac{x}{y}\right)$

In Exercises 9–12, find $\dfrac{dz}{dt}$ by using the chain rule.

9. $z = x^2 + y^2,\ x = te^t,\ y = t^2 e^t$

10. $z = \dfrac{1}{x^2 + y^2},\ x = \cos 2t,\ y = \sin 2t$

11. $z = e^u \sin v,\ u = \sqrt{t},\ v = \pi t$

12. $z = x^3 - y,\ x = te^{-t},\ y = \sin t$

In Exercises 13–20, find all critical points and test for relative extrema.

13. $z = x^2 + 4y^2 + x + 8y + 1$

14. $z = x^2 + y^2 - 2x + 4y + 2$

15. $z = 20 + 12x - 12y - 3x^2 - 2y^2$

16. $z = xy + 3x - 2y + 4$

17. $z = x^2 + 2xy - y^2$

18. $z = x^2 - 3xy - y^2$

19. $z = x^3 + x^2 y + y^2$

20. $z = x^3 + y^3 + 3xy$

Solve Exercises 21–36.

21. The base and height of a rectangle are measured as 10 cm, and 25 cm, respectively, with a possible error of 0.1 cm in each. Use differentials to determine the maximum possible error in the area of the rectangle.

22. The radius of the base of a right circular cone is measured as 4 in.; its height is measured as 12 in. A possible error of $\frac{1}{16}$ in. was made in measuring each dimension. Use differentials to estimate the maximum error that might occur in computing **(a)** the volume of the cone and **(b)** the total surface area of the cone.

23. *Electricity* The total resistance R of two resistors R_1 and R_2 connected in parallel is

$$R = \frac{R_1 R_2}{R_1 + R_2}$$

If $R_1 = 100\,\Omega$ and $R_2 = 400\,\Omega$ with a maximum error of 1% in each measurement, what is the maximum error in the value of R?

24. *Physics* The equation relating pressure P, volume V, and temperature T of a confined gas is $PV = kT$, where k is a constant. If $P = 0.5$ psi when $V = 66$ in.^{3}

and $T = 110°F$, approximate the change in P, if V and T change to 70 in.^{3} and 112°F.

25. *Physics* If P and V in Exercise 24 are changing at the rate of $\dfrac{dP}{dt}$ and $\dfrac{dV}{dt}$, respectively, use the chain rule to find a formula for $\dfrac{dT}{dt}$.

26. *Physics* A confined gas satisfies the equation $PV = kT$, as described in Exercises 24 and 25, with $k = 0.4$. If the pressure is increasing at the rate of 0.05 psi/min and the volume is decreasing at the rate of 1 in.^{3}/min, find the rate of change of the temperature at the moment $P = 0.5$ psi and $V = 66$ in.^{3}

27. *Electronics* The energy stored in an inductor is $w = \dfrac{Li^2}{2}$. If L changes from 3.0 to 3.02 H while i changes from 1.0 to 0.9 A, what is the approximate change in w?

28. A vertical line is moving to the right at 2 cm/min, and a horizontal line is moving upward at 3 cm/min. What is the rate of change in the area of the rectangle formed by the coordinate axes and these two lines when $x = 6$ and $y = 7$?

29. *Sheet metal technology*   A manufacturer wishes to make an open rectangular box of volume $V = 500\,\text{cm}^3$. Find the dimensions of the box that will use the least possible amount of material.

30. *Sheet metal technology*   Rework Exercise 29 if the box has a lid.

31. *Sheet metal technology*   An open rectangular box is constructed out of different materials. The material for the bottom costs $1.25/ft². The material for the sides costs $1.50/ft². The volume of the box is to be $4\,\text{ft}^3$. **(a)** What are the dimensions of the box that will cost the least to build? (See Example 30.6.) **(b)** What is this cost?

32. *Thermodynamics*   A flat plate is heated such that the temperature T at any point (x, y) is given by $T = x^2 + 2y^2 - x$. What is the temperature at the coldest point?

33. *Pharmacology*   When x units of drug A and y units of drug B are used in the treatment of one type of cancer, the reaction of the patient 12 h after their administration is given by $R(x, y) = 15xy(2 - x - 4y)$. Find x and y in order to maximize this reaction.

34. *Business*   The total cost in dollars to produce x units of electrical tape and y units of packing tape is given by

$$C(x, y) = 2x^2 + 3y^2 - 2xy + 2x - 125y + 4500$$

(a) Find the number of units of each kind of tape that should be produced so that the total cost is a minimum.
(b) Find the minimum total cost.

35. *Electronics*   Electric charge is distributed uniformly around a thin ring of radius a, with total charge Q. The potential at a point P, a distance x from the center of a ring on a line through the center of the ring and perpendicular to the plane of the ring, is

$$V(x) = \frac{1}{4\pi\epsilon_0} \frac{Q}{\sqrt{x^2 + a^2}}$$

where ϵ_0 is a constant. Find $E_x = -\dfrac{\partial V}{\partial x}$, the rate of change of V for a displacement parallel to the x-axis.

36. *Electronics*   Charge Q is uniformly distributed along a rod with length $2a$. The potential at a point on the perpendicular bisector of such a rod at distance x from its center is

$$V(x) = \frac{1}{4\pi\epsilon_0} \frac{Q}{2a} \ln \frac{\sqrt{a^2 + x^2} + a}{\sqrt{a^2 + x^2} - a}$$

Find $E_x = -\dfrac{\partial V}{\partial x}$.

In Your Words

37. Describe the chain rule for a function of two variables.

38. Explain the first derivative test for local extrema.

≡ 30.5
MULTIPLE INTEGRALS

We have been looking at ways to differentiate functions of two variables and ways to use those derivatives. We will now look at the inverse process—integration of functions of two variables. The process of integrating a function of two variables is very similar to that of differentiating a function of two variables. In each case, an operation (differentiation or integration) is performed while holding one of the independent variables constant.

Let $z = f(x, y)$ be a continuous function over a bounded region R in the xy-plane, as shown in Figure 30.20. We want to find the volume between the surface and R.

We will assume that R is rectangular and that it is bounded by $x = a$, $x = b$, $y = c$, and $y = d$. We will divide the intervals $[a, b]$ and $[c, d]$ into partitions, much

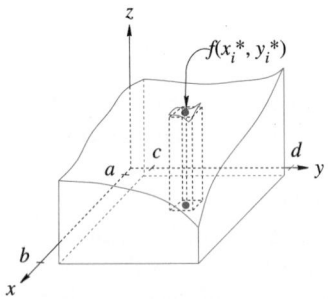

FIGURE 30.20

as we did when we were developing the integral in Chapter 27. We will divide $[a, b]$ into n intervals $a = x_0 < x_1 < x_2 < \cdots < x_i < \cdots < x_n = b$ and we will divide $[c, d]$ into m intervals $c = y_0 < y_1 < y_2 < \cdots < y_j < \cdots y_m = d$. It is possible that $n = m$, but it is not necessary. Thus, we have $\Delta x = \dfrac{b-a}{n}$ and $\Delta y = \dfrac{d-c}{m}$.

Now, select the ith interval on the x-axis, $[x_{i-1}, x_i]$ and the jth interval on the y-axis, $[y_{j-1}, y_j]$. This gives a rectangular region in the xy-plane. From each vertex of this rectangle, erect a vertical line that intersects the surface. What we have approximates a rectangular parallelepiped. We want to get a measure of the volume of that parallelepiped. As before, we will select the midpoint of each interval and call them x_i^* and y_j^*, and this point (x_i^*, y_j^*) is the center of the rectangle. We will erect a vertical line from that point until it meets the surface at $f(x_i^*, y_j^*)$. If we assume that the height of the rectangular parallelepiped has a height of $f(x_i^*, y_j^*)$, then it has a volume of

$$V_{ij} = f(x_i^*, y_j^*), \Delta x \, \Delta y$$

We now add all of the approximate volumes to get an approximation of the total volume. The total volume is

$$\begin{aligned} V \approx V_{11} + V_{12} + V_{13} + \cdots + V_{1m} \\ + V_{21} + V_{22} + \cdots + V_{2m} \\ \vdots \\ + V_{n1} + V_{n2} + \cdots + V_{nm} \end{aligned}$$

If we use summation notation, we need to use two summation signs, because we are summing over two variables i and j.

$$V = \sum_{i=1}^{n} \sum_{j=1}^{m} V_{ij} = \sum_{i=1}^{n} \sum_{j=1}^{m} f(x_i^*, y_j^*) \, \Delta x \, \Delta y$$

When we take the limits of each of these as both Δx and Δy approach zero, we get

$$V = \iint_{R} f(x, y) \, dx \, dy$$

We now give a more formal definition of a double integral.

Double Integral

If the rectangular region, R, is defined by $a \leq x \leq b$ and $c \leq y \leq d$, then we can calculate this double integral as

$$\iint_{R} f(x, y) \, dy \, dx = \int_{a}^{b} \int_{c}^{d} f(x, y) \, dy \, dx$$

$$= \int_{c}^{d} \int_{a}^{b} f(x, y) \, dx \, dy$$

EXAMPLE 30.21

Calculate the volume of the region beneath the surface $z = xy^2 + 2y$ and the rectangle where $0 \le x \le 2$ and $1 \le y \le 4$.

Solution Using the new equation, we have

$$\int_0^2 \int_1^4 (xy^2 + 2y)\, dy\, dx = \int_0^2 \left\{ \int_1^4 (xy^2 + 2y)\, dy \right\} dx$$

We first perform the integration inside the braces. Since this integral is to be taken with respect to y, we treat x as a constant. We will evaluate this definite inner integral:

$$\int_0^2 \left\{ \int_1^4 (xy^2 + 2y)\, dy \right\} dx = \int_0^2 \left(\frac{1}{3}xy^3 + y^2 \right) \Big|_{y=1}^{y=4} dx$$

$$= \int_0^2 \left[\left(\frac{64}{3}x + 16 \right) - \left(\frac{1}{3}x + 1 \right) \right] dx$$

$$= \int_0^2 (21x + 15)\, dx$$

We now have an ordinary single variable integral that we can evaluate easily.

$$\int_0^2 (21x + 15)\, dx = \frac{21}{2}x^2 + 15x \Big|_0^2$$

$$= \frac{21}{2} \cdot 4 + 15 \cdot 2$$

$$= 42 + 30 = 72$$

It is also possible to first integrate this over x and then over y.

EXAMPLE 30.22

Evaluate $\int_1^4 \int_0^2 (xy^2 + 2y)\, dx\, dy$.

Solution $\int_1^4 \int_0^2 (xy^2 + 2y)\, dx\, dy = \int_1^4 \left\{ \int_0^2 (xy^2 + 2y)\, dx \right\} dy$

$$= \int_1^4 \left[\frac{1}{2}x^2y^2 + 2xy \right] \Big|_{x=0}^{x=2} dy$$

$$= \int_1^4 (2y^2 + 4y)\, dy$$

$$= \left(\frac{2}{3}y^3 + 2y^2 \right) \Big|_1^4$$

$$= \left(\frac{128}{3} + 32 \right) - \left(\frac{2}{3} + 2 \right) = 72$$

Average Value Over Rectangular Regions ▬▬▬▬▬

In Section 26.1, we defined the average value of continuous function f over interval $[a, b]$. The **average value**, $\bar{y}$, was

$$\bar{y} = \frac{1}{b-a} \int_a^b f(x)\, dx$$

This definition can be extended to the function of two variables over rectangular regions as shown in the following box. Notice that the denominator in the expression is the area of the rectangle, R.

Average Value Over a Rectangular Region

The **average value** of the function $f(x, y)$ over the rectangular region of width x, with $a \le x \le b$, length y, and with $c \le y \le d$, is

$$\bar{f} = \frac{1}{(b-a)(d-c)} \int\!\!\int_R f(x, y)\, dy\, dx$$

$$= \frac{1}{(b-a)(d-c)} \int_a^b \int_c^d f(x, y)\, dy\, dx$$

$$= \frac{1}{(b-a)(d-c)} \int_c^d \int_a^b f(x, y)\, dx\, dy$$

EXAMPLE 30.23

Find the average value of $f(x, y) = 4 - \frac{3}{2}x - \frac{1}{2}y$ over the rectangular region $R = x \times y$ where $0 \le x \le 2$ and $0 \le y \le 4$.

Solution   Here $a = 0$, $b = 2$, $c = 0$, and $d = 4$.   Using these substitutions, we obtain

$$\bar{f} = \frac{1}{(b-a)(d-c)} \int_a^b \int_c^d f(x, y)\, dy\, dx$$

$$= \frac{1}{(2-0)(4-0)} \int_0^4 \int_0^2 \left(4 - \frac{3}{2}x - \frac{1}{2}y \right) dx\, dy$$

$$= \frac{1}{8} \int_0^4 \left[\left(4x - \frac{3}{4}x^2 - \frac{1}{2}yx \right) \right]_0^2 dy$$

$$= \frac{1}{8} \int_0^4 (5 - y)\, dy$$

$$= \frac{1}{8} \left[5y - \frac{1}{2}y^2 \right]_0^4$$

$$= \frac{1}{8}[20 - 8] = \frac{3}{2}$$

Double Integration Over a Simple Region

Now, suppose that the region is not a rectangle. Then it may be a vertically simple region or a horizontally simple region, as shown in Figure 30.21. A region R is **vertically simple**, if it is described by the inequalities

$$a \leq x \leq b, \quad g_1(x) \leq y \leq g_2(x)$$

where g_1 and g_2 are continuous functions on $[a,b]$. A region R is **horizontally simple**, if it is described by the inequalities

$$c \leq y \leq d, \ h_1(y) \leq x \leq h_2(y)$$

where h_1 and h_2 are continuous on $[c,d]$.

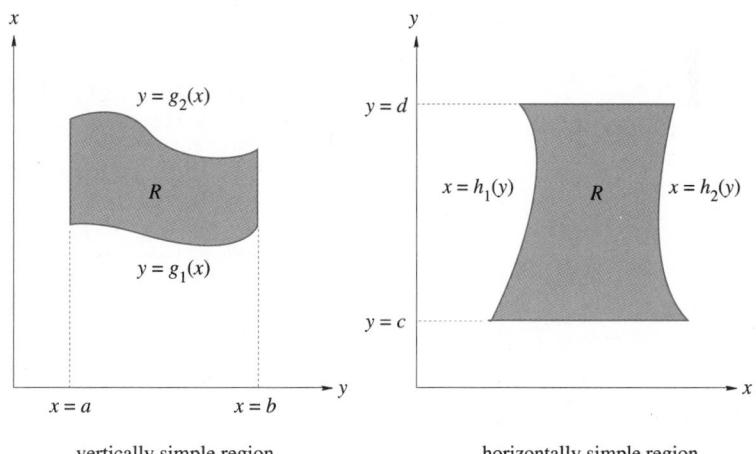

vertically simple region   horizontally simple region

FIGURE 30.21

We can compute a double integral over a region R that is either horizontally or vertically simple, using the following rule.

Double Integration Over a Simple Region

If $z = f(x,y)$ is continuous on the region R and if R is a vertically simple region, then

$$\iint_R f(x,y)\,dA = \int_a^b \int_{g_1(x)}^{g_2(x)} f(x,y)\,dy\,dx$$

If R is a horizontally simple region, then

$$\iint_R f(x,y)\,dA = \int_c^d \int_{h_1(y)}^{h_2(y)} f(x,y)\,dx\,dy$$

The symbol dA represents the change in area of R.

EXAMPLE 30.24

Evaluate $\iint_R (x^3 + 4y)\, dA$, where R is the region in the xy-plane bounded by $y = x^2$ and $y = 2x$.

Solution The region R is shown in Figure 30.22. Notice that this is both horizontally and vertically simple. We can use either rule. Suppose we assume it is vertically simple with $y = x^2$ the lower boundary (g_1) and $y = 2x$ the upper boundary (g_2). We have $0 \le x \le 2$. By the rule for double integration over a simple region,

$$\iint_R (x^3 + 4y)\, dA = \int_0^2 \int_{x^2}^{2x} (x^3 + 4y)\, dy\, dx$$

$$= \int_0^2 (x^3 y + 2y^2)\Big|_{y=x^2}^{y=2x} dx$$

$$= \int_0^2 \left\{ \left[x^3(2x) + 2(2x)^2 \right] - \left[x^3(x^2) + 2(x^2)^2 \right] \right\} dx$$

$$= \int_0^2 (2x^4 + 8x^2) - (x^5 + 2x^4)\, dx$$

$$= \int_0^2 (8x^2 - x^5)\, dx$$

$$= \left(\frac{8}{3} x^3 - \frac{1}{6} x^6 \right)\Big|_0^2$$

$$= \frac{8}{3}(8) - \frac{1}{6}(64)$$

$$= \frac{32}{3}$$

If we assumed that the region was horizontally simple, then we would use $x = \frac{1}{2}y$ as the lower boundary (h_1) and $x = \sqrt{y}$ as the upper boundary (h_2). The integral would then be

$$\iint_R (x^3 + 4y)\, dA = \int_0^4 \int_{1/2y}^{\sqrt{y}} (x^3 + 4y)\, dx\, dy$$

$$= \int_0^4 \left(\frac{1}{4} x^4 + 4xy \right)\Big|_{x=1/2y}^{x=\sqrt{y}} dy$$

$$= \int_0^4 \left[\left(\frac{1}{4} y^2 + 4y^{3/2} \right) - \left(\frac{1}{64} y^4 - 2y^2 \right) \right] dy$$

$$= \frac{32}{3}$$

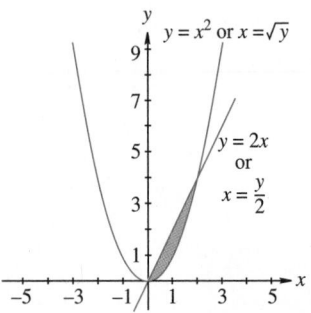

$y = x^2$ or $x = \sqrt{y}$

$y = 2x$
or
$x = \dfrac{y}{2}$

FIGURE 30.22

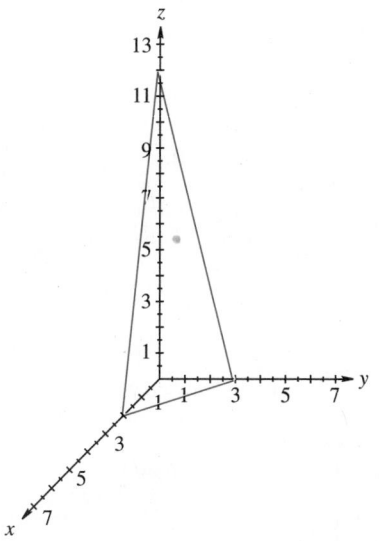

FIGURE 30.23

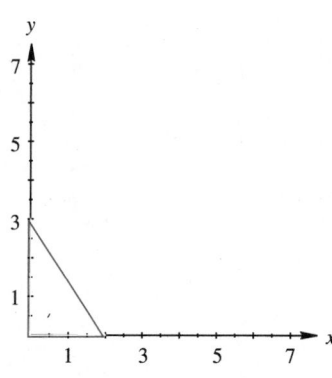

FIGURE 30.24

EXAMPLE 30.25

Find the volume of the solid bounded by the plane $6x + 4y + z = 12$ and the coordinate planes, as shown in Figure 30.23.

Solution When finding a volume, the integrand is always the function that represents the surface. In this case, $z = f(x, y) = 12 - 6x - 4y$. The limits of integration are found by examining the region R in the xy-plane, as shown in Figure 30.24. Notice that the xy-trace in Figure 30.24 is the line $y = -\frac{3}{2}(x - 2)$. Thus, we have $g_1(x) = 0$ and $g_2(x) = -\frac{3}{2}(x - 2)$, and the outer limits are $x = 0$ and $x = 2$. The double integral becomes

$$
\begin{aligned}
V &= \int_0^2 \int_0^{-3(x-2)/2} (12 - 6x - 4y)\,dy\,dx \\
&= \int_0^2 (12y - 6xy - 2y^2)\Big|_0^{-\frac{3}{2}(x-2)}\,dx \\
&= \int_0^2 -18(x-2) + 9x(x-2) + \frac{-9}{2}(x-2)^2\,dx \\
&= \int_0^2 \left(\frac{9}{2}x^2 - 18x + 18\right)dx \\
&= \left(\frac{3}{2}x^3 - 9x^2 + 18x\right)\Big|_0^2 \\
&= 12
\end{aligned}
$$

Double integrals can also be performed in polar coordinates. As we will see after Section 30.6 they can be extended to triple integrals for functions of three variables.

Exercise Set 30.5

In Exercises 1–16, evaluate the double integrals.

1. $\displaystyle\int_0^1 \int_0^{x^2} xy\,dy\,dx$

2. $\displaystyle\int_{-1}^1 \int_{-2}^2 (2xy - 3y^2)\,dy\,dx$

3. $\displaystyle\int_0^\pi \int_{-\pi/2}^{\pi/2} \sin x \cos y\,dy\,dx$

4. $\displaystyle\int_0^1 \int_0^2 \sqrt{x+y}\,dy\,dx$

5. $\displaystyle\int_0^{\ln 2} \int_0^{\ln 5} e^{x+y}\,dy\,dx$

6. $\displaystyle\int_0^1 \int_y^{y^2} (x+y)\,dx\,dy$

7. $\displaystyle\int_0^1 \int_x^{\sqrt{x}} (x+y)\,dy\,dx$

8. $\displaystyle\int_0^4 \int_{-\sqrt{2x}}^{\sqrt{2x}} (2x+y)\,dy\,dx$

9. $\displaystyle\int_0^1 \int_{x^3}^x (y-x)\,dy\,dx$

10. $\displaystyle\int_0^4 \int_y^{\sqrt{y}} (3x+2y)\,dx\,dy$

11. $\displaystyle\int_0^2 \int_0^3 (xy+x-y)\,dy\,dx$

12. $\displaystyle\int_0^1 \int_0^x xy\,dy\,dx$

13. $\displaystyle\int_0^1 \int_0^{1-x} (x^2y+xy^2)\,dy\,dx$

14. $\displaystyle\int_0^2 \int_0^{x^2} (x^2 - y^2)\,dy\,dx$

15. $\displaystyle\int_0^1 \int_{y^2}^y (xy+1)\,dx\,dy$

16. $\displaystyle\int_1^2 \int_y^{y^2} \frac{x}{y}\,dx\,dy$

In Exercises 17–26, evaluate the given integral.

17. $\displaystyle\iint_R (x^2 - y^2)\,dA$, where R is the region bounded by $x = 0$, $y = 1$, and $y = x$.

18. $\displaystyle\iint_R x^2y\,dA$, where R is the region bounded by $y = x^2$, $y = 0$, and $x = 2$.

19. $\displaystyle\iint_R (x^3 - y^3)\,dA$, where R is the region bounded by $y = x^3$, $x = 0$, and $y = 8$.

20. $\displaystyle\iint_R xy\,dA$, where R is the region bounded by $y = x$, and $y = x^4$.

21. $\displaystyle\iint_R (x+y)\,dA$, where R is the region bounded by $xy = 8$ and $x + y = 9$.

22. $\displaystyle\iint_R (x^2 + y^2)\,dA$, where R is the region bounded by $x = 1$, $x = 4$, $y = 0$, and $y = 5$.

23. $\displaystyle\iint_R (x+y)^2\,dA$, where R is the region in the first quadrant bounded by $y = x^3$ and $y = x$.

24. $\displaystyle\iint_R xy\,dA$, where R is the region bounded by $x = y^2$ and $x = y+4$.

25. $\displaystyle\iint_R xy\,dA$, where R is the region bounded by $x = 0$ and $x = y^2 - 9$.

26. $\displaystyle\iint_R \sqrt{xy}\,dA$, where R is the region bounded by $y = 1$ and $y = x^2$.

In Exercises 27–30, find the average value for each of the functions over regions R having the given boundaries.

27. $f(x, y) = x^2 + y^2;\ 0 \le x \le 2, 0 \le y \le 3$

28. $f(x, y) = (x+y)^2;\ 1 \le x \le 5, -1 \le y \le 1$

29. $f(x, y) = \dfrac{x}{y};\ 1 \le x \le 4, 2 \le y \le 7$

30. $f(x, y) = e^{2x+y};\ 1 \le x \le 2, 2 \le y \le 3$

In Exercises 31–34, find the volume of the given solid.

31. The solid bounded by the plane $x + 2y + z = 4$ and the three coordinate planes

32. The solid bounded by the cylinder $x^2 + z^2 = 4$ and the planes $y = 0$ and $y = 2$

33. The solid bounded by the plane $y = 0$, $y = x$, and the cylinder $x^2 + z^2 = 2$

34. The solid bounded by the surface $z = e^{-(x+y)}$ and the three coordinate planes

Solve Exercises 35–38.

35. *Business*   A manufacturing company has two plants. The weekly cost function for the first plant at production level x is $C_1(x) = 0.2x^2 + 40x + 3600$, and at the second plant at production level y the cost is $C_2(y) = 0.4y^2 + 24y + 6,000$.   Find the average weekly cost of the company for $300 \le x \le 400$ and $350 \le y \le 550$, that is, find the average cost of $C_1(x) + C_2(y)$.

36. *Business*   If a company invests x thousand labor-hours and y million dollars in the production of N thousand units of a certain item, then according to the Cobb-Douglas production formula, $N(x, y) = x^{0.75} y^{0.25}$, where $10 \le x \le 20$ and $1 \le y \le 2$. What is the average number of units produced for these ranges of labor-hour and capital expenditures?

37. *Environmental science*   An industrial plant is located in the center of a rectangular-shaped small town

that is 4 mi long and 2 mi wide. The plant emits particulate matter into the atmosphere. The concentration of particulate matter in parts per million at a point d mi from the plant is given by $C = 100 - 15d^2$.
   (a) Express C as a function of x and y, where x and y are the horizontal and vertical distance of d from the plant.
   (b) Determine the average concentration of particulate matter throughout the town.

38. *Police science*   Under ideal conditions, if a person driving a car brakes hard, the car will travel distance L in feet before it stops, where $L = 0.000\,02xy^2$, x is the weight of the car in pounds, and y is the speed of the car in miles per hour. What is the average distance that a car weighing between 2,000 and 3,000 lb will travel at speeds between 50 and 60 mph?

In Your Words

39. Without looking in the text, describe how to determine the double integral of a function defined over a region.

40. How does a simple region differ from a region?

41. Without looking in the text, explain how to determine

the average value of a function over a rectangular region.

42. Describe the geometric meaning of the double integral of a function of two variables.

≡ 30.6
CYLINDRICAL AND SPHERICAL COORDINATES

Until now, all of our three-dimensional work has been done with rectangular coordinates. However, there are several ways to represent points in space. In this section, we will look at two common ways to do this. Both of these methods are used to describe the motion of robots.

Cylindrical Coordinate System

In the cylindrical coordinate system, a point P is identified by $P = (r, \theta, z)$. In this identification, r and θ are the polar coordinates of P as it is projected on the xy-plane. Thus, $r \ge 0$ and $0 \le \theta \le 2\pi$. The value z represents the distance of the xy-plane from P. In Figure 30.25, P is projected onto the xy-plane at Q. Thus, the distance from P to Q is z; that is, $d(P, Q) = z$.

EXAMPLE 30.26

Discuss the graphs of each of the following equations in cylindrical coordinates:
(a) $r = 5$, **(b)** $\theta = \frac{\pi}{3}$, **(c)** $z = 4$, **(d)** $3 \leq r \leq 5$, $\frac{\pi}{4} \leq \theta \leq \frac{\pi}{3}$, $4 \leq z \leq 7$.

Solutions

(a) If $r = 5$ (or any positive constant), then θ and z can assume any value. As a result, we obtain a right circular cylinder with radius r, as shown in Figure 30.26a.

(b) If $\theta = \frac{\pi}{3}$, we obtain a half-plane through the z-axis, as shown in Figure 30.26b.

(c) If $z = 4$, we get a plane parallel to the xy-plane (the polar plane), as shown in Figure 30.26c.

(d) First, the values $3 \leq r \leq 5$ give the region between the cylinders $r = 3$ and $r = 5$. The values $\frac{\pi}{4} \leq \theta \leq \frac{\pi}{3}$ define the "wedge-shaped" region between half-planes $\theta = \frac{\pi}{4}$ and $\theta = \frac{\pi}{3}$. Finally, $4 \leq z \leq 7$ is the "slice" (or "slab") of space between the planes $z = 4$ and $z = 7$. Putting this together, we get the shaded solid shown in Figure 30.26d.

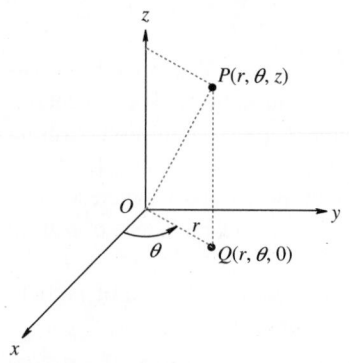

FIGURE 30.25

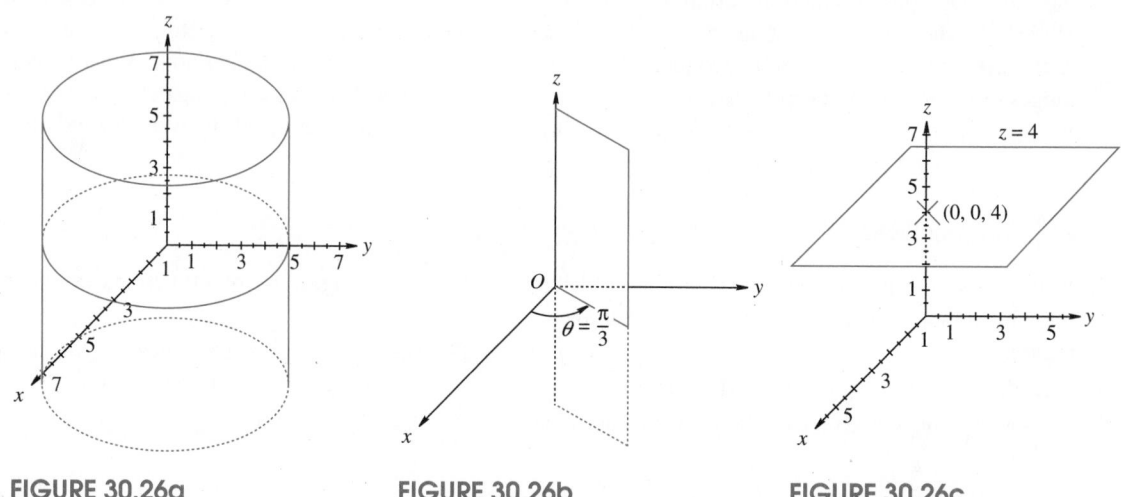

FIGURE 30.26a   **FIGURE 30.26b**   **FIGURE 30.26c**

Changing Between Cylindrical and Rectangular Coordinates

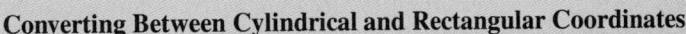

We often find it necessary to change from cylindrical coordinates to rectangular coordinates or vice versa. The following conversion formulas are very similar to the ones we used with polar coordinates.

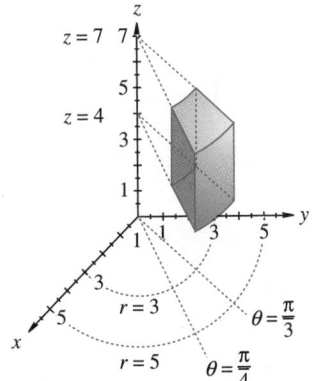

FIGURE 30.26d

Converting Between Cylindrical and Rectangular Coordinates

A point with the cylindrical coordinates (r, θ, z) has the rectangular coordinates (x, y, z), where

$$x = r\cos\theta, \qquad y = r\sin\theta, \qquad z = z \qquad (1)$$

Similarly, a point with the rectangular coordinates (x, y, z) has the cylindrical coordinates (r, θ, z), where

$$r = \sqrt{x^2 + y^2}, \qquad \tan\theta = \frac{y}{x}, \qquad z = z \qquad (2)$$

EXAMPLE 30.27

Convert $P\left(4, \frac{5\pi}{6}, 7\right)$ from cylindrical to rectangular coordinates.

Solution We have $r = 4$, $\theta = \frac{5\pi}{6}$, and $z = 7$. Substituting these into the equations in (1), we obtain

$$x = r\cos\theta = 4\left(\frac{-\sqrt{3}}{2}\right) = -2\sqrt{3} \approx -3.464$$
$$y = r\sin\theta = 4\left(\tfrac{1}{2}\right) = 2$$
$$z = z = 7$$

So in rectangular coordinates, $P = (-2\sqrt{3}, 2, 7)$.

EXAMPLE 30.28

Convert $(5, -12, 2)$ from rectangular to cylindrical coordinates.

Solution Here, we substitute into the equations in (2), with the result

$$r = \sqrt{x^2 + y^2} = \sqrt{5^2 + (-12)^2} = \sqrt{25 + 144} = 13$$

$\tan\theta = \frac{y}{x} = \frac{-12}{5} = -2.4$. So, $\theta = \tan^{-1}(-2.4) = 5.107$ rad [since the point $(5, -12)$ is in the fourth quadrant], and $z = 2$.

Thus, the rectangular coordinates $(5, -12, 2)$ are $(13, 5.107, 2)$ in cylindrical coordinates.

Spherical Coordinates

The other new coordinate system that we are going to learn is the spherical coordinate system. A typical point P in space is represented as

$$P = (\rho, \theta, \phi)$$

where $\rho \geq 0$, $0 \leq \theta \leq 2\pi$, $0 \leq \phi \leq \pi$, and where ρ is the positive distance from P to the origin, θ is the same as in cylindrical coordinates, and ϕ is the angle between $\overrightarrow{OP}$ and the positive z-axis, as shown in Figure 30.27.

EXAMPLE 30.29

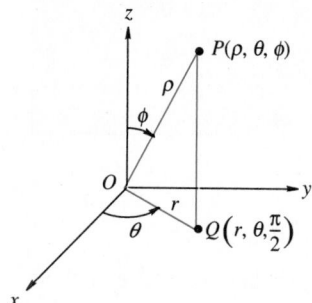

FIGURE 30.27

Discuss the graphs of the following equations in spherical coordinates: **(a)** $\rho = 5$, **(b)** $\theta = \frac{\pi}{3}$, **(c)** $\phi = \frac{\pi}{6}$, **(d)** $3 \leq \rho \leq 5$, $\frac{\pi}{4} \leq \theta \leq \frac{\pi}{3}$, $\frac{\pi}{6} \leq \phi \leq \frac{\pi}{4}$.

Solution

(a) If $\rho = 5$ (or any positive constant), we get all the points in space that are 5 units from the origin. This is a sphere centered at the origin with radius 5, as shown in Figure 30.28a.

(b) If $\theta = \frac{\pi}{3}$, we get a half-plane containing the z-axis, as shown in Figure 30.28b. This is the same graph we would get in cylindrical coordinates, if $\theta = \frac{\pi}{3}$.

(c) If $\phi = \frac{\pi}{6}$, the graph would be obtained by rotating the vector $\overrightarrow{OP}$ around the z-axis. This yields a circular cone, as shown in Figure 30.28c. If $\phi > \frac{\pi}{2}$, the cone would be below the xy-plane, as shown in Figure 30.28d. If $\phi = \frac{\pi}{2}$, we would get the xy-plane.

(d) Here, $3 \leq \rho \leq 5$ gives the region between the spheres $\rho = 3$ and the sphere $\rho = 5$. Next, $\frac{\pi}{4} \leq \theta \leq \frac{\pi}{3}$ is, as in the cylindrical coordinates, a wedge-shaped region between $\theta = \frac{\pi}{4}$ and $\theta = \frac{\pi}{3}$. Finally, $\frac{\pi}{6} \leq \phi \leq \frac{\pi}{4}$ describes the region between the cone $\phi = \frac{\pi}{6}$ and the cone $\phi = \frac{\pi}{4}$. Combining these we would get the **spherical wedge** shown in Figure 30.28e.

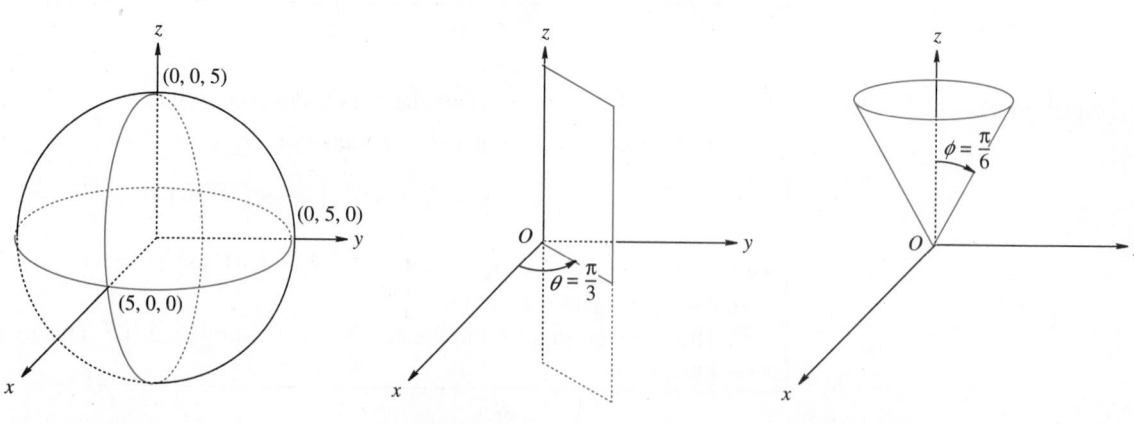

FIGURE 30.28a   **FIGURE 30.28b**   **FIGURE 30.28c**

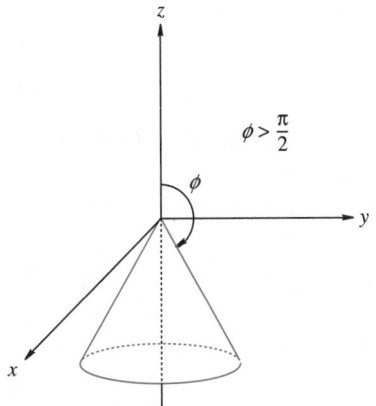

FIGURE 30.28d

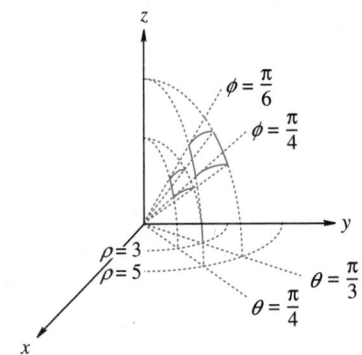

FIGURE 30.28e

Changing Between Spherical and Rectangular Coordinates

To convert from spherical to rectangular coordinates, we use the following formulas:

Converting Between Spherical and Rectangular Coordinates

A point with the spherical coordinates (ρ, θ, ϕ) has the rectangular coordinates (x, y, z), where

$$x = \rho \sin\phi \cos\theta, \qquad y = \rho \sin\phi \sin\theta, \qquad z = \rho \cos\phi \qquad (3)$$

Similarly, a point with the rectangular coordinates (x, y, z) has the spherical coordinates (ρ, θ, ϕ), where

$$\rho = \sqrt{x^2 + y^2 + z^2}, \qquad \tan\theta = \frac{y}{x}, \qquad \cos\phi = \frac{z}{\rho} \qquad (4)$$

EXAMPLE 30.30

Convert $\left(8, \frac{\pi}{6}, \frac{3\pi}{4}\right)$ from spherical to rectangular coordinates.

Solution   We have $\rho = 8$, $\theta = \frac{\pi}{6}$, and $\phi = \frac{3\pi}{4}$. Using the formulas in (3), we have

$$x = \rho \sin\phi \cos\theta = 8 \sin\frac{3\pi}{4} \cos\frac{\pi}{6}$$

$$= 8 \left(\frac{\sqrt{2}}{2}\right) \left(\frac{\sqrt{3}}{2}\right) = 2\sqrt{6}$$

$$y = \rho \sin\phi \sin\theta = 8 \left(\frac{\sqrt{2}}{2}\right) \left(\frac{1}{2}\right) = 2\sqrt{2}$$

EXAMPLE 30.30 (Cont.)

$$z = \rho\cos\phi = 8\left(-\frac{\sqrt{2}}{2}\right) = -4\sqrt{2}$$

So, the spherical coordinates of $\left(8, \frac{\pi}{6}, \frac{3\pi}{4}\right)$ are, in rectangular coordinates, $(2\sqrt{6}, 2\sqrt{2}, -4\sqrt{2})$.

EXAMPLE 30.31

Convert $(1, -\sqrt{3}, \sqrt{5})$ from rectangular to spherical coordinates.

Solution Here, $x = 1$, $y = -\sqrt{3}$, and $z = \sqrt{5}$. Using the formulas in (4) produces

$$\rho = \sqrt{x^2 + y^2 + z^2}$$
$$= \sqrt{(1)^2 + (-\sqrt{3})^2 + (\sqrt{5})^2}$$
$$= \sqrt{1 + 3 + 5} = 3$$
$$\tan\theta = \frac{y}{x}$$
$$\theta_{\text{Ref}} = \tan^{-1}\frac{-\sqrt{3}}{1}$$
$$= -\frac{\pi}{3}$$

Since $x > 0$ and $y < 0$, θ is between $\frac{3\pi}{2}$ and 2π. So,

$$\theta = \frac{5\pi}{3}$$
$$\cos\phi = \frac{z}{\rho} = \frac{\sqrt{5}}{3}$$
$$\phi = \cos^{-1}\frac{\sqrt{5}}{3} \approx 0.730$$

The rectangular coordinates $(1, -\sqrt{3}, \sqrt{5})$ have the spherical coordinates $\left(3, \frac{5\pi}{3}, 0.730\right)$.

Industrial Robots and Coordinate Systems

Industrial robots are currently used to weld, cast, form, assemble, paint, transfer, inspect, and load or unload parts into and out of many different machines. The industrial robot can be thought of as a mechanical arm that is in a fixed location. In fact, it is often bolted to the floor or wall.

The work envelope of a robot consists of all the points in space that can be touched by the end of the robot's arm. This work envelope is a three-dimensional volume of points. Three major axes of motion provide the largest portion of the robot's work envelope. The coordinate system used for these three major axes of motion provides one means of classifying robots. The three major axes of motion are: **(a)** a vertical lift stroke; **(b)** an in-and-out or horizontal reach stroke, and **(c)** a rotational, traverse, or swing motion about the vertical axis.

A cylindrical coordinate robot has a horizontal arm assembled to a vertical axis. The vertical axis is mounted on a rotating base. The horizontal arm can move in and out and move vertically up and down on the vertical axis. This arm assembly can also rotate left and right about the vertical axis. The motions of the three major axes form a portion of a cylinder as the work envelope of the robot, as shown in Figure 30.29.

A spherical coordinate robot consists of an arm that moves in and out in a reach stroke. However, the arm uses a pivoting vertical motion instead of a true vertical stroke and the arm can move left or right about the vertical pivot axis. These motions form a portion of a sphere as a work envelope. (See Figure 30.30.)

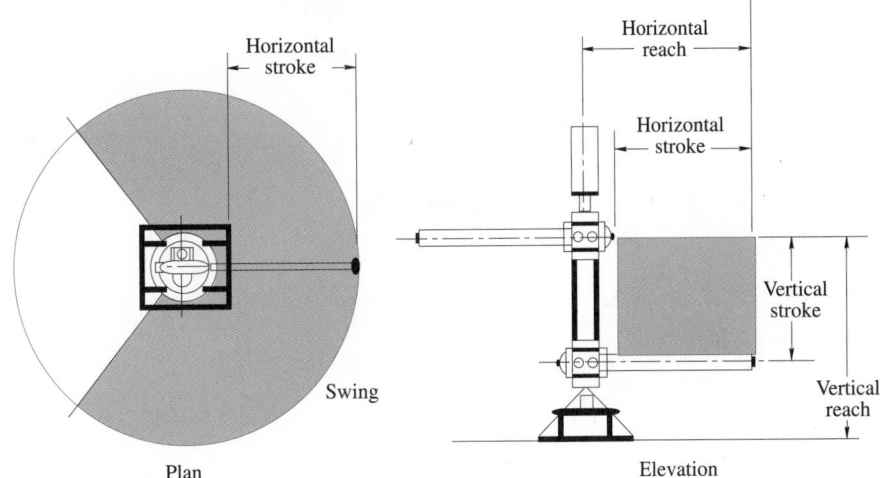

Work Envelope:
Cylindrical Coordinate Robot

FIGURE 30.29

A third work-envelope structure is a rectangular coordinate robot. This type of robot has a horizontal arm assembled to a vertical lift axis. The vertical axis is mounted on a linear traverse base. The work envelope covered by these motions is illustrated in Figure 30.31.

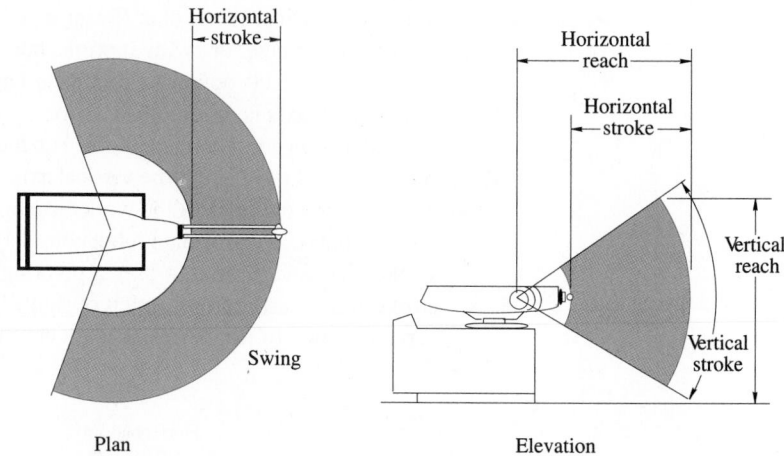

Plan

Elevation

Work Envelope:
Spherical Coordinate Robot

FIGURE 30.30

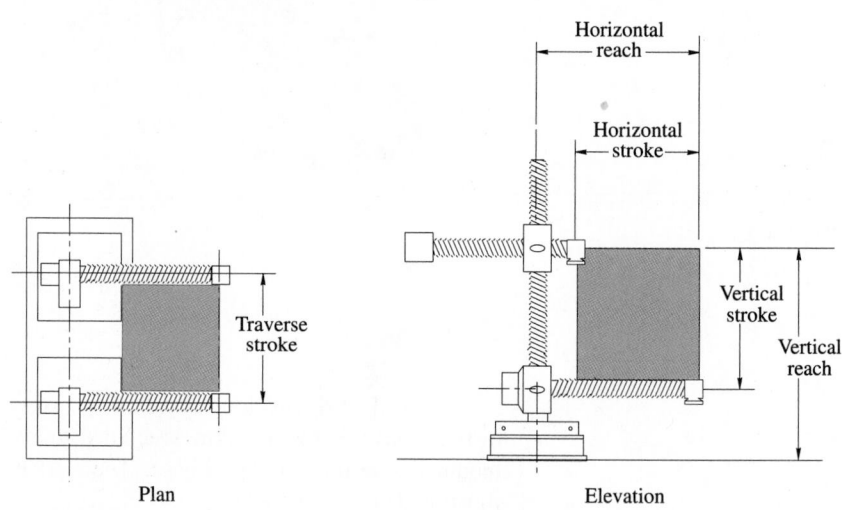

Plan

Elevation

Work Envelope:
Rectangular Coordinate Robot

FIGURE 30.31

Exercise Set 30.6

In Exercises 1–6, convert the cylindrical coordinate to its equivalent rectangular coordinate.

1. $\left(2, \frac{\pi}{4}, 2\right)$

2. $\left(3, \frac{2\pi}{3}, -2\right)$

3. $(2, 0, 4)$

4. $\left(0, \frac{5\pi}{4}, -5\right)$

5. $\left(5, \frac{5\pi}{4}, 0\right)$

6. $\left(4, \frac{5\pi}{3}, 7\right)$

In Exercises 7–12, convert the rectangular coordinate to its equivalent cylindrical coordinate.

7. $(2, 2, 5)$

8. $(3, -4, -5)$

9. $(-4, 3, 2)$

10. $(-1, -\sqrt{3}, 4)$

11. $(12, -5, -3)$

12. $(6, 8, -4)$

In Exercises 13–18, convert the spherical coordinate to its equivalent rectangular coordinate.

13. $\left(4, \frac{\pi}{4}, \frac{\pi}{6}\right)$

14. $\left(1, \frac{\pi}{6}, \frac{\pi}{2}\right)$

15. $\left(3, \frac{\pi}{2}, \frac{5\pi}{3}\right)$

16. $\left(2, \frac{5\pi}{6}, \frac{3\pi}{4}\right)$

17. $\left(5, \frac{7\pi}{6}, \frac{2\pi}{3}\right)$

18. $\left(5, \frac{5\pi}{3}, \frac{\pi}{4}\right)$

In Exercises 19–24, convert the rectangular coordinate to its equivalent spherical coordinate.

19. $(4, 3, 0)$

20. $(4, \sqrt{5}, -2)$

21. $(2, 1, -2)$

22. $(4, -\sqrt{3}, 2)$

23. $(1, 1, \sqrt{2})$

24. $(-\sqrt{3}, \sqrt{3}, -\sqrt{3})$

In Exercises 25–30, describe and sketch the curves and regions determined by the given equations and inequalities.

25. $\theta = \frac{\pi}{4}, \phi = \frac{\pi}{4}$

26. $\theta = \frac{3\pi}{4}$

27. $3 \le \rho \le 5$

28. $\rho = 5, \theta = \frac{\pi}{3}$

29. $\rho \le 4, \phi \le \frac{\pi}{6}$

30. $0 \le \theta \le \frac{\pi}{4}, 0 \le \phi \le \frac{\pi}{4}, 0 \le \rho \le 4$

31. Write a computer program to convert a coordinate in the rectangular, cylindrical, or spherical systems to its equivalent in the other two systems. Check your program on the answers to Exercises 1–24.

 **In Your Words**

32. Explain the cylindrical, spherical, and rectangular three-dimensional coordinate system.

33. Why do we need three different three-dimensional coordinate systems?

☰ 30.7
MOMENTS AND CENTROIDS

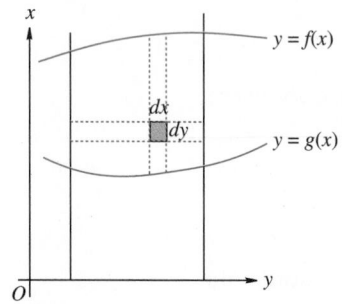

FIGURE 30.32

In Chapter 26, we looked at ways in which integration could be used to determine the centroids and the moments of inertia of a figure. In this section, we will see how to use double integration to find the centroids and moments of inertia of these same figures.

Consider the region shown in Figure 30.32. We can determine the area of this figure by using the integral $A = \int_a^b [f(x) - g(x)]\,dx$. But, $f(x) - g(x) = \int_{g(x)}^{f(x)} dy$, so we can write the area as a double integral:

$$A = \int_a^b \int_{g(x)}^{f(x)} dy\,dx$$

In Section 26.5, we showed that the moment of an area with respect to the y-axis, M_y, is $\rho \int_a^b x\,dA$, where ρ was the mass per unit area. In symbols we wrote this as

$$M_y = \rho \int_a^b x[f(x) - g(x)]\,dx$$

But, since $f(x) - g(x) = \int_{g(x)}^{f(x)} dy$, we can write this as

$$M_y = \rho \int_a^b \int_{g(x)}^{f(x)} x\,dy\,dx$$

Because ρ was constant, its effect was cancelled in locating the coordinates of the centroid. The total mass m was given by ρA.

Thus, the x-coordinate $\bar{x}$ of the centroid is given by

$$\bar{x} = \frac{M_y}{m} = \frac{\rho \int_a^b \int_{g(x)}^{f(x)} x\,dy\,dx}{m}$$

In a similar manner, we can show that the moment of the area with respect to the x-axis is given by

$$M_x = \rho \int_a^b \int_{g(x)}^{f(x)} y\,dy\,dx$$

As a result, the y-coordinate $\bar{y}$ of the centroid is given by

$$\bar{y} = \frac{M_x}{m} = \frac{\rho \int_a^b \int_{g(x)}^{f(x)} y\,dy\,dx}{m}$$

The formulas for determining the mass, moments, and centroid of a region in the xy-plane are summarized in the following box. The examples following the box show how to use these formulas.

Area, Mass, Moments, and Centroid of a Region in the xy-Plane

If ρ is a continuous density function on a plane region R, then the area, mass, moments, and centroid of R are as follows:

Area: $\quad A = \displaystyle\iint_R dy\,dx$

Mass: $\quad m = \displaystyle\iint_R \rho(x, y)\,dy\,dx$

First Moments: $\quad M_x = \displaystyle\iint_R y\rho(x, y)\,dy\,dx$

$$M_y = \iint_R x\rho(x, y)\,dy\,dx$$

Centroid: $\quad (\bar{x}, \bar{y})$, where $\bar{x} = \dfrac{M_y}{m}$ and $\bar{y} = \dfrac{M_x}{m}$

EXAMPLE 30.32

Find the centroid of the region bounded by $y = x$, $y = 6 - 2x$, and the y-axis, as shown in Figure 30.33a.

Solution The two functions intersect at the point $(2, 2)$. So, we have $f(x) = 6 - 2x$, $g(x) = x$, $a = 0$, and $b = 2$.

$$A = \int_0^2 \int_x^{6-2x} dy\,dx = \int_0^2 y\Big|_x^{6-2x} dx$$

$$= \int_0^2 [(6 - 2x) - x]\,dx = \int_0^2 (6 - 3x)\,dx$$

$$= 6x - \frac{3}{2}x^2 \Big|_0^2 = 6$$

$$m = \rho A = 6\rho$$

$$M_y = \rho \int_0^2 \int_x^{6-2x} x\,dy\,dx = \rho \int_0^2 y\Big|_x^{6-2x} x\,dx$$

$$= \rho \int_0^2 (6 - 3x)x\,dx = \rho \int_0^2 (6x - 3x^2)\,dx$$

$$= \rho(3x^2 - x^3) \Big|_0^2 = 4\rho$$

$$M_x = \rho \int_0^2 \int_x^{6-2x} y\,dy\,dx = \rho \int_0^2 \frac{1}{2}y^2 \Big|_x^{6-2x} dx$$

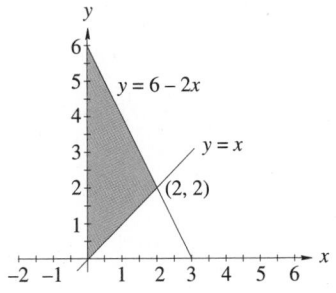

FIGURE 30.33a

EXAMPLE 30.32 (Cont.)

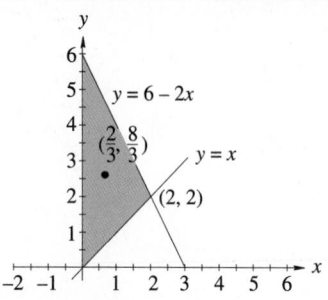

FIGURE 30.33b

$$= \frac{1}{2}\rho \int_0^2 [(6-2x)^2 - x^2]\, dx = \frac{1}{2}\rho \int_0^2 (36 - 24x + 3x^2)\, dx$$

$$= \frac{1}{2}(36x - 12x^2 + x^3)\rho \Big|_0^2 = 16\rho$$

Thus, $\bar{x} = \dfrac{M_y}{m} = \dfrac{4\rho}{6\rho} = \dfrac{2}{3}$ and $\bar{y} = \dfrac{M_x}{m} = \dfrac{16\rho}{6\rho} = \dfrac{8}{3}$.

The centroid is at $\left(\frac{2}{3}, \frac{8}{3}\right)$, as indicated in Figure 30.33b.

EXAMPLE 30.33

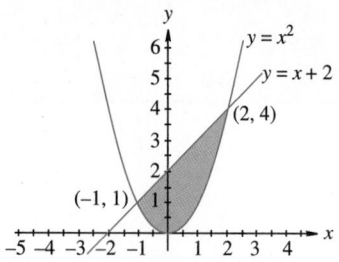

FIGURE 30.34a

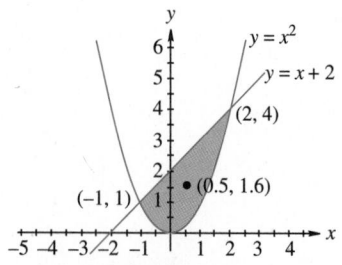

FIGURE 30.34b

Find the centroid of the region bonded by $y = x^2$ and $y = x + 2$.

Solution The region is shaded in Figure 30.34a. The curves intersect at $(2, 4)$ and $(-1, 1)$. We have $f(x) = x + 2$, $g(x) = x^2$, $a = -1$, and $b = 2$.

$$A = \int_{-1}^2 \int_{x^2}^{x+2} dy\, dx = \int_{-1}^2 y \Big|_{x^2}^{x+2} dx = \int_{-1}^2 (x + 2 - x^2)\, dx$$

$$= \left(\frac{1}{2}x^2 + 2x - \frac{1}{3}x^3\right)\Big|_{-1}^2 = 4.5$$

$$m = \rho A = 4.5\rho$$

$$M_y = \rho \int_{-1}^2 \int_{x^2}^{x+2} x\, dy\, dx = \rho \int_{-1}^2 y \Big|_{x^2}^{x+2} x\, dx$$

$$= \rho \int_{-1}^2 \left[(x+2) - x^2\right] x\, dx = \rho \int_{-1}^2 (x^2 + 2x - x^3)\, dx$$

$$= \rho \left(\frac{1}{3}x^3 + x^2 - \frac{1}{4}x^4\right)\Big|_{-1}^2 = 2.25\rho$$

$$M_x = \rho \int_{-1}^2 \int_{x^2}^{x+2} y\, dy\, dx = \frac{1}{2}\rho \int_{-1}^2 y^2 \Big|_{x^2}^{x+2} dx$$

$$= \frac{1}{2}\rho \int_{-1}^2 \left[(x+2)^2 - (x^2)^2\right] dx$$

$$= \frac{1}{2}\rho \int_{-1}^2 (x^2 + 4x + 4 - x^4)\, dx$$

$$= \frac{1}{2}\rho \left(\frac{1}{3}x^3 + 2x^2 + 4x - \frac{x^5}{5}\right)\Big|_{-1}^2 = 7.2\rho$$

$$\bar{x} = \frac{M_y}{m} = \frac{2.25\rho}{4.5\rho} = 0.5, \quad \bar{y} = \frac{M_x}{m} = \frac{7.2\rho}{4.5\rho} = 1.6$$

The centroid is at $(0.5, 1.6)$, as shown in Figure 30.34b.

In Section 26.6, we showed that the **moment of inertia** of a region with respect to the y-axis is

$$I_y = \rho \int_a^b x^2[f(x) - g(x)]\,dx$$

Again, since $\int_{g(x)}^{f(x)} dy = f(x) - g(x)$, we can write

$$I_y = \rho \int_a^b \int_{g(x)}^{f(x)} x^2\,dy\,dx$$

Similarly, we can show that the moment of inertia with respect to the x-axis is

$$I_x = \rho \int_a^b \int_{g(x)}^{f(x)} y^2\,dy\,dx$$

As before, the radius of gyration with respect to the y-axis is $r_y = \sqrt{\dfrac{I_y}{m}}$ and with respect to the x-axis is

$$r_x = \sqrt{\frac{I_x}{m}}$$

These results are summarized in the box on page 1212.

EXAMPLE 30.34

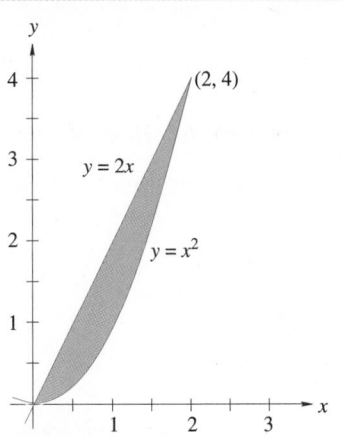

FIGURE 30.35

Find I_x and I_y for the region bounded by $y = 2x$ and $y = x^2$.

Solution The region is shaded in Figure 30.35. Here, we have $f(x) = 2x$, $g(x) = x^2$, $a = 0$, and $b = 2$.

$$I_y = \rho \int_0^2 \int_{x^2}^{2x} x^2\,dy\,dx = \rho \int_0^2 y\Big|_{x^2}^{2x} x^2\,dx$$

$$= \rho \int_0^2 (2x - x^2)x^2\,dx$$

$$= \rho \int_0^2 (2x^3 - x^4)\,dx = \rho \left(\frac{1}{2}x^4 - \frac{1}{5}x^5\right)\Big|_0^2 = 1.6\rho$$

$$I_x = \rho \int_0^2 \int_{x^2}^{2x} y^2\,dy\,dx = \rho \int_0^2 \frac{1}{3} y^3\Big|_{x^2}^{2x}\,dx$$

$$= \frac{1}{3}\rho \int_0^2 (8x^3 - x^6)\,dx$$

$$= \frac{1}{3}\left(2x^4 - \frac{1}{7}x^7\right)\rho\Big|_0^2 = \frac{32}{7}\rho$$

The moments of inertia for the region bounded by $y = 2x$ and $y = x^2$ are $I_x = \frac{32}{7}\rho$ and $I_y = 1.6\rho$.

Moments of Inertia and Radii of Gyration of a Region in the xy-Plane

If ρ is a continuous density function on a plane region R, then the moments of inertia and radii of gyration of R are as follows:

Moments of Inertia (First Moments):

About the x-axis: $\quad I_x = \displaystyle\iint_R y^2 \rho(x, y)\, dy\, dx$

About the y-axis: $\quad I_y = \displaystyle\iint_R x^2 \rho(x, y)\, dy\, dx$

About the origin: $\quad I_0 = \displaystyle\iint_R (x^2 + y^2) \rho(x, y)\, dy\, dx$

Radii of Gyration:

About the x-axis: $\quad r_x = \sqrt{\dfrac{I_x}{m}}$

About the y-axis: $\quad r_y = \sqrt{\dfrac{I_y}{m}}$

About the origin: $\quad r_0 = \sqrt{\dfrac{I_0}{m}}$

where m is the mass of the region.

≡ **Note**   The number I_0 is a measure of the **polar moment of inertia**.

EXAMPLE 30.35

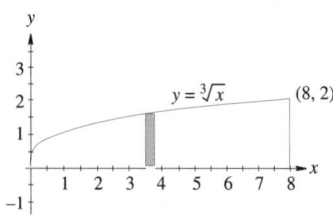

FIGURE 30.36

Find the moments of inertia and the radius of gyration for the region bounded by $y = \sqrt[3]{x}$, $x = 8$, and the x-axis with respect to the y-axis.

Solution   A sketch of this region is shown in Figure 30.36. Here, we have $f(x) = \sqrt[3]{x}$, $g(x) = 0$, $a = 0$, and $b = 8$.

$$A = \int_0^8 \int_0^{\sqrt[3]{x}} dy\, dx = \int_0^8 y \Big|_0^{\sqrt[3]{x}} dx$$

$$= \int_0^8 x^{1/3}\, dx = \frac{3}{4} x^{4/3} \Big|_0^8 = 12$$

$$m = \rho A = 12\rho$$

$$I_y = \rho \int_0^8 \int_0^{\sqrt[3]{x}} x^2\, dy\, dx = \rho \int_0^8 y \Big|_0^{\sqrt[3]{x}} x^2\, dx = \rho \int_0^8 x^{7/3}\, dx$$

$$= \frac{3}{10} \rho x^{10/3} \Big|_0^8 = 307.2\rho$$

$$I_x = \rho \int_0^8 \int_0^{\sqrt[3]{x}} y^2\, dy\, dx = \rho \int_0^8 \frac{1}{3} y^3 \Big|_0^{\sqrt[3]{x}} dx$$

EXAMPLE 30.35 (Cont.)

$$= \frac{\rho}{3} \int_0^8 x \, dx = \frac{1}{6} x^2 \rho \Big|_0^8 = \frac{32}{3} \rho$$

$$r_y = \sqrt{\frac{I_y}{m}} = \sqrt{\frac{307.2\rho}{12\rho}} = \sqrt{25.6} \approx 5.06$$

For this region, the moments of inertia are $I_x = \frac{32}{3}\rho$ and $I_y = 307.2\rho$ and the radius of gyration around the y-axis is 5.06.

EXAMPLE 30.36

Find the moment of inertia and the radius of gyration for the region bounded by $y = \sqrt[3]{x}$, $x = 8$, and the x-axis with respect to the origin.

Solution   This is the same region as that used in Example 30.35, so we know that its mass is $m = 12\rho$.

We first determine its moment of inertia about the origin.

$$I_0 = \rho \int_0^8 \int_0^{\sqrt[3]{x}} (x^2 + y^2) \, dy \, dx$$

$$= \rho \int_0^8 \left(x^2 y + \frac{1}{3} y^3 \right) \Big|_0^{\sqrt[3]{x}} dx$$

$$= \rho \int_0^8 \left(x^{7/3} + \frac{1}{3} x \right) dx$$

$$= \rho \left[\frac{3}{10} x^{10/3} + \frac{1}{6} x^2 \right]_0^8 = \frac{4,768}{15} \rho$$

Using this result, we determine the radius of gyration about the origin

$$r_0 = \sqrt{\frac{I_0}{m}}$$

$$= \sqrt{\frac{4,768\rho/15}{12\rho}} = \sqrt{\frac{1,192}{45}} \approx 5.1467$$

Exercise Set 30.7

In Exercises 1–6, calculate the centroids of the plane region bounded by the given curve and the x-axis over the indicated interval.

1. $y = 2x + 3$, $[0, 3]$
2. $y = x^3$, $[0, 2]$
3. $y = x^{1/3}$, $[0, 8]$
4. $y = x^4$, $[-1, 2]$
5. $y = \sqrt{x + 4}$, $[0, 5]$
6. $y = \sqrt{x^2 + 16}$, $[0, 4]$

In Exercises 7–12, find the centroids of the plane regions bounded by the given curves and lines.

7. $y = x^2$, $y = 4x$, $x = 0$, $x = 2$

8. $y = x^3$, $y = 8x$, $x = 0$, $x = 2$

9. $y = x^{3/2}$, $y = x$

10. $y = x^2$, $y = 18 - x^2$

11. $y = x$, $y = 12 - x^2$

12. $y = x^2$, $y = x^3$

In Exercises 13–20, find the moment of inertia and radius of gyration about the given axis.

13. The region bounded by $y = x^2$, $x = 0$, and $y = 2$, about the y-axis

14. The region in Exercise 13 about the x-axis

15. The region bounded by $x = 0$, $y = 0$, $x = 3$, and $y = 5$, about the x-axis

16. The region in Exercise 15 about the y-axis

17. The region bounded by $y = 4x - x^2$ and $y = 0$, about the x-axis

18. The region in Exercise 17 about the y-axis

19. The region bounded by $y = \dfrac{1}{x}$, $y = x^2$, $x = 2$, and $x = 1$, about the y-axis

20. The region in Exercises 19 about the x-axis

In Exercises 21–24, for each of the plane regions, find (a) the moment of inertia about the origin and (b) the radius of gyration about the origin.

21. The region bounded by $y = x^2$, $x = 0$, $y = 2$. (This is the same region as in Exercise 13.)

22. The region bounded by $x = 0$, $y = 0$, $x = 3$, $y = 5$. (This is the same region as in Exercise 15.)

23. The region bounded by $y = 4x - x^2$, $y = 0$. (This is the same region as in Exercise 17.)

24. The region bounded by $y = \dfrac{1}{x}$, $y = x^2$, $x = 2$, $x = 1$. (This is the same region as in Exercise 19.)

In Your Words

25. Give physical explanations of what is meant by the moments and centroid of a region in the plane.

26. Give physical explanations of what is meant by the radius of gyration of a region in the plane.

≣ CHAPTER 30 REVIEW

Important Terms and Concepts

Centroids
Chain rule for a function of two variables
Curve
Cylindrical coordinate system
Differentials
Functions of three variables
Functions of two variables
   Domain
   Range
Integration
   Double
   Multiple

Moments
Octant
Partial derivatives
   First derivative test
   Second derivative test
Planes
Quadric surfaces
   Ellipsoid
   Elliptic cone
   Elliptic cylinder
   Elliptic paraboloid
   Hyperbolic cylinder

Hyperbolic paraboloid	Intersection of
Hyperboloid of one sheet	Curve
Hyperboloid of two sheets	Section
Rectangular coordinate system	Trace
Related rates	Maxima
Spherical coordinate system	Minima
Surfaces in three dimensions	Saddle point

Review Exercises

In Exercises 1–8, identify and sketch the given surface.

1. $x + 3y + 2z = 6$ 3. $4y^2 + z^2 = 4$ 5. $y = 4x^2$ 7. $x^2 + 4y^2 - z^2 = 16$

2. $4x - 3y - 2z = 12$ 4. $x^2 + y^2 + z^2 = 16$ 6. $9x^2 - 4y^2 = 1$ 8. $36z^2 = 9x^2 + 4y^2$

In Exercises 9–12, find $\dfrac{\partial z}{\partial x}, \dfrac{\partial z}{\partial y}, \dfrac{\partial^2 z}{\partial x^2}, \dfrac{\partial^2 z}{\partial y^2},$ **and** $\dfrac{\partial^2 z}{\partial x \partial y}.$

9. $z = 3x^2 + 6xy^3$

10. $z = \dfrac{x^2 + y^2}{y}$

11. $z = e^x \cos y - e^y \sin x$

12. $z = \ln \sqrt{x^2 + y^3} + \sin^2(3xy)$

In Exercises 13–18, evaluate the given integral.

13. $\displaystyle\int_0^2 \int_0^{x^3} xy\, dy\, dx$

14. $\displaystyle\int_0^{\pi/2} \int_0^{\pi} \sin x \cos y\, dy\, dx$

15. $\displaystyle\int_0^1 \int_0^4 \sqrt{x+y}\, dy\, dx$

16. $\displaystyle\int_0^4 \int_0^9 xy\sqrt{x^2 + y^2}\, dy\, dx$

17. $\displaystyle\int_0^{\ln 4} \int_0^{\ln 10} e^{x+2y}\, dy\, dx$

18. $\displaystyle\int_0^1 \int_0^{x^2} x\, dy\, dx$

In Exercises 19 and 20, convert the cylindrical coordinate to its equivalent rectangular coordinate and equivalent spherical coordinate.

19. $\left(4, \frac{\pi}{6}, 2\right)$

20. $\left(9, \frac{5\pi}{6}, -3\right)$

In Exercises 21–22, convert the spherical coordinate to its equivalent rectangular coordinate and equivalent cylindrical coordinate.

21. $\left(4, \frac{3\pi}{4}, \frac{\pi}{6}\right)$

22. $\left(5, \frac{7\pi}{6}, \frac{2\pi}{3}\right)$

In Exercises 23 and 24, find the average value for each of the functions over regions R having the given boundaries.

23. $f(x, y) = x^2 + y^4, 0 \le x \le 4, -2 \le y \le 2$

24. $f(x, y) = x^3 + 2y, 1 \le x \le 5, 0 \le y \le 4$

Solve Exercises 25–30.

25. *Drafting* An open-topped rectangular box is to have a volume of $300\,\text{in.}^3$ Find the dimensions that minimize its surface area.

26. *Electricity* The total resistance R of two resistances R_1 and R_2 connected in parallel is given by the formula

$$\frac{1}{R} = \frac{1}{R_1} + \frac{1}{R_2} = \frac{R_1 + R_2}{R_1 R_2}$$

Suppose that R_1 and R_2 are measured to be $300\,\Omega$ and $600\,\Omega$, respectively, with a maximum error of 1% in each measurement. Use differentials to estimate the maximum error (in ohms) in the calculated value of R.

27. *Electricity* Ohm's law states that $R = \dfrac{V}{I}$. Measurements are $V = 116\,\text{V}$ and $I = 2\,\text{A}$ with a possible error of $0.2\,\text{V}$ and $0.01\,\text{A}$, respectively. Use differentials to approximate the maximum error in the calculated value of R (in ohms).

28. What is the smallest amount of material needed to build an open-top rectangular box enclosing a volume of $25\,\text{m}^3$?

29. *Physics* According to the ideal gas law, the pressure P, volume V, and absolute temperature T of n moles of an ideal gas are related by

$$PV = nRT$$

where R is a constant. The pressure of 8 moles of an ideal gas is increasing at a rate of $0.4\,\text{N/cm}^2/\text{min}$, while the temperature is increasing at a rate of $0.5°\,\text{K/min}$. How fast is the volume of the gas changing when $V = 1\,000\,\text{m}^3$ and $P = 4\,\text{N/cm}^2$?

30. The radius of a right circular cone is increasing at a rate of $10\,\text{cm/s}$, while its height is increasing at a rate of $15\,\text{cm/s}$. How is the volume changing when $r = 180\,\text{cm}$ and $h = 270\,\text{cm}$?

In Exercises 31–34, find the centroid of the indicated region.

31. $y = 6x,\ x = 1,\ x = 5,\ y = 0$

32. $y = x^4,\ x = 1,\ x = 2,\ y = 0$

33. $y = x^3,\ y = \sqrt[3]{x}$

34. $y = x^2,\ y = 9 - x^2$

In Exercises 35–38, find the moment of inertia and radius of gyration about the given axis.

35. The region bounded by $y = 6x$, $x = 0$, $x = 5$, and $y = 0$, about the x-axis

36. The region bounded by $y = 4 - x$, $x = 0$, $x = 4$, and $y = 0$, about the y-axis

37. The region bounded by $y = \sqrt[3]{x^2}$, $x = 0$, $x = 1$, and $y = 0$, about the x-axis

38. The region in the first quadrant bounded by $y = x^3$ and $y = \sqrt[3]{x}$, about the y-axis

Solve Exercises 39–42.

39. Find (**a**) the moment of inertia and (**b**) the radius of gyration about the origin for the region bounded by $y = 10x$, $x = 0$, $x = 8$, and $y = 0$.

40. Find (**a**) the moment of inertia and (**b**) the radius of gyration about the origin for the region in the first quadrant bounded by $y = 6x$, $y = x^2$.

41. Evaluate $\displaystyle\iint_R (x^2 - y)\,dA$, where R is the region bounded by $y = x$, $x + y = 4$, and $y = 0$.

42. Evaluate $\displaystyle\iint_R (x - 4y)\,dA$, where R is the region bounded by $y = x$ and $y = x^2 + x - 5$.

≡ CHAPTER 30 TEST

In Exercises 1–4, identify and sketch the given surfaces.

1. $2x - y + 4z = 8$

2. $x^2 + y^2 + z^2 = 9$

3. $\dfrac{x^2}{4} + \dfrac{y^2}{9} - \dfrac{z^2}{4} = 1$

4. $z = 9x^2 + 4y^2 - 36$

Solve Exercises 5–14.

5. If $z = 4x^3 - 5x^2y^4$, determine $\dfrac{\partial z}{\partial x}$ and $\dfrac{\partial z}{\partial y}$.

6. If $z = \sqrt{x^2 + 4y^3} + \ln(x^2 y)$, determine $\dfrac{\partial^2 z}{\partial x^2}$.

7. Evaluate $\displaystyle\int_0^4 \int_0^{\pi/4} x \sin y \, dy \, dx$.

8. Evaluate $\displaystyle\int_0^2 \int_0^{x^2} 3xy^2 \, dy \, dx$.

9. Convert the cylindrical coordinates $\left(3, \frac{5\pi}{6}, 8\right)$ to its equivalent rectangular coordinates and equivalent spherical coordinates.

10. Convert the spherical coordinate $\left(4, \frac{7\pi}{6}, \frac{3\pi}{4}\right)$ to its equivalent rectangular coordinates and equivalent cylindrical coordinates.

11. Find the volume of the solid region bounded by the paraboloid $z = 4 - x^2 - 4y^2$ and the xy-plane.

12. Determine the centroid of the region in the xy-plane bounded by the graphs of $y = (x-1)^2 = x^2 - 2x + 1$ and $y = 4$.

13. Find the average value of $f(x, y) = x^2y^3$ for $-1 \le x \le 2, 0 \le y \le 4$.

14. Find **(a)** the moment of inertia and **(b)** the radius of gyration about the origin for the region in the first quadrant bounded by $x = 0$, $y = x$, $y = 6 - x^2$.

CHAPTER

31

Infinite Series

Calculators and computers use infinite series to calculate the values of many functions. In Section 31.1, you will learn how to use the Maclaurin series to perform such calculations.

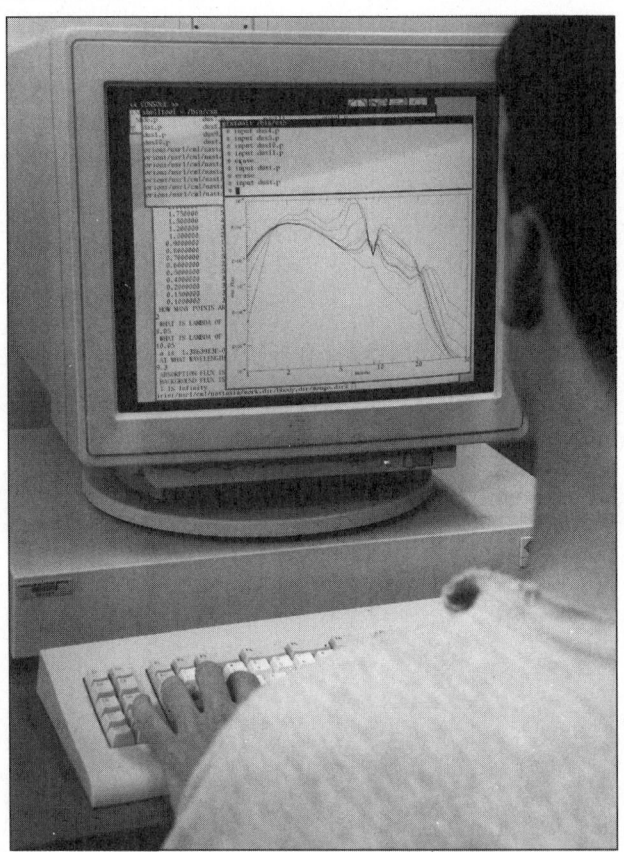

In Chapter 19, we studied sequences and series. An infinite series is an indicated sum of infinitely many terms. In that chapter we looked at some of the cases when an infinite series converges and when it does not. We are able to tell when a geometric series converges and what its sum will be. In this chapter, we will discuss methods for computing the sum of an infinite series. Infinite series arise very naturally in science and mathematics because many functions occur naturally in the form of infinite series.

And, we will see that some functions can be written as infinite series in order to help with numerical computations.

31.1
POWER AND
MACLAURIN SERIES

Until now, the series that we have considered have all had terms that are constants. In this section, the terms of the series will be a function. This means that the series is of the form $\sum f_n(x)$.

Power Series

We will begin by looking at any series of the form

$$\sum_{k=0}^{\infty} a_k x^k = a_0 + a_1 x + a_2 x^2 + a_3 x^3 + \cdots$$

where the coefficients $a_0, a_1, a_2, a_3, \ldots$ are all constants. A series of this type is known as a **power series in** x. In a power series, if $x = 0$, then the series is equal to a_0.

A variation on the power series in x is the power series in $x - c$, which is defined as follows.

Power Series in $x - c$

A series of the form

$$\sum_{k=0}^{\infty} a_k (x - c)^k = a_0 + a_1(x - c) + a_2(x - c)^2 + a_3(x - c)^3 + \cdots$$

is called a **power series in** $x - c$.

Notice that the power series in x,

$$\sum_{k=0}^{\infty} a_k x^k = a_0 + a_1 x + a_2 x^2 + a_3 x^3 + \cdots$$

is a special case of the power series in $x - c$, where $c = 0$.

EXAMPLE 31.1

Write the first five terms of each of the following power series: **(a)** $\sum_{k=0}^{\infty} \dfrac{x^k}{k!}$, **(b)** $\sum_{k=0}^{\infty} \dfrac{kx^k}{3^k}$, and **(c)** $\sum_{k=0}^{\infty} \dfrac{x^k}{k+2}$.

EXAMPLE 31.1 (Cont.)

Solutions

(a) $\displaystyle\sum_{k=0}^{\infty}\frac{x^k}{k!} = \frac{x^0}{0!}+\frac{x^1}{1!}+\frac{x^2}{2!}+\frac{x^3}{3!}+\frac{x^4}{4!}+\cdots$ Recall: $0! = 1$

$$= 1+x+\frac{x^2}{2}+\frac{x^3}{6}+\frac{x^4}{24}+\cdots$$

(b) $\displaystyle\sum_{k=0}^{\infty}\frac{kx^k}{3^k} = \frac{0\cdot x^0}{3^0}+\frac{1\cdot x^1}{3^1}+\frac{2\cdot x^2}{3^2}+\frac{3\cdot x^3}{3^3}+\frac{4\cdot x^4}{3^4}+\cdots$

$$= 0+\frac{x}{3}+\frac{2x^2}{9}+\frac{x^3}{9}+\frac{4x^4}{81}+\cdots$$

(c) $\displaystyle\sum_{k=0}^{\infty}\frac{x^k}{k+2} = \frac{x^0}{0+2}+\frac{x^1}{1+2}+\frac{x^2}{2+2}+\frac{x^3}{3+2}+\frac{x^4}{4+2}+\cdots$

$$= \frac{1}{2}+\frac{x}{3}+\frac{x^2}{4}+\frac{x^3}{5}+\frac{x^4}{6}+\cdots$$

A power series converges for some values of x and diverges for others. The values of x for which the series converges differ from series to series. For example, the series in Example 31.1(a) converges for all values of x, the series in Example 31.1(b) converges if $|x| < 3$, and the series in Example 31.1(c) converges if $-1 \le x < 1$. The values of x for which a series converges is called the **interval of convergence**.

≡ **Note**

It is not our purpose to determine the interval of convergence for a series. We will assume that any series that we work with will converge for all values of x or for some given interval.

Properties of Convergent Series

If a power series has an interval of convergence that is greater than zero, then three properties are true for all values of x within that interval. If $f(x) = \displaystyle\sum_{k=0}^{\infty} a_k x^k = a_0 + a_1 x + a_2 x^2 + a_3 x^3 + a_4 x^4 + \cdots$, then for all values of x within the interval of convergence, **(a)** f is continuous, **(b)** f is differentiable, and **(c)** f can be integrated.

Maclaurin Series

We will immediately use the first two of these properties. Let's assume that for some interval $-R < x < R$, the function f is represented by the power series just given. We may obtain the coefficients $a_0, a_1, a_2, \ldots$ in terms of f and its derivatives by using part (b) from the previous paragraph.

$$f(x) = a_0 + a_1 x + a_2 x^2 + a_3 x^3 + a_4 x^4 + a_5 x^5 + \cdots$$
$$f'(x) = \quad a_1 + 2a_2 x + 3a_3 x^2 + 4a_4 x^3 + 5a_5 x^4 + \cdots$$
$$f''(x) = \quad\quad 2a_2 + 3\cdot 2a_3 x + 4\cdot 3a_4 x^2 + 5\cdot 4a_5 x^3 + \cdots$$

$$f'''(x) = \qquad 3 \cdot 2a_3 + 4 \cdot 3 \cdot 2a_4 x + 5 \cdot 4 \cdot 3a_5 x^2 + \cdots$$
$$f^{(4)}(x) = \qquad 4 \cdot 3 \cdot 2a_4 + 5 \cdot 4 \cdot 3 \cdot 2a_5 x + \cdots$$
$$f^{(5)}(x) = \qquad 5 \cdot 4 \cdot 3 \cdot 2a_5 + \cdots$$

If we let $x = 0$, then we get a value of 0 for each term in these equations, except for the first term. So, we have $f(0) = a_0$, $f'(0) = a_1$, $f''(0) = 2a_2$, $f'''(0) = 3 \cdot 2a_3 = 6a_3$, $f^{(4)}(0) = 4 \cdot 3 \cdot 2a_4 = 24a_4$, $f^{(5)}(0) = 5 \cdot 4 \cdot 3 \cdot 2a_5 = 120a_5$, and so on. Look carefully at each of these terms. In general, we have $f^{(n)}(0) = n(n-1)(n-2)(n-3) \cdots (2) \cdot 1 a_n$. Since $n(n-1)(n-2)(n-3) \cdots 2 \cdot 1 = n!$, we can write $f^{(n)}(0) = n! a_n$. Solving for a_n, we see that

$$a_n = \frac{f^{(n)}(0)}{n!}$$

Caution

Remember that the notation $f^{(n)}(x)$ indicates the nth derivative of $f(x)$ and *not* the nth power of $f(x)$.

Substituting these values into the equation for the function, we get the **Maclaurin series**.

Maclaurin Series
$$f(x) = f(0) + f'(0)x + \frac{f''(0)}{2!}x^2 + \frac{f'''(0)}{3!}x^3 + \frac{f^{(4)}(0)}{4!}x^4 + \cdots$$
$$+ \frac{f^{(n)}(0)}{n!}x^n + \cdots$$
$$= \sum_{n=0}^{\infty} \frac{f^{(n)}(0)}{n!}x^n$$

where $f^{(0)}(x) = f(0)$.

EXAMPLE 31.2

Find the Maclaurin series of $f(x) = e^x$.

Solution   To use our definition, we need to evaluate f and its derivatives at 0.

$$f(x) = e^x \qquad\qquad f(0) = 1$$
$$f'(x) = e^x \qquad\qquad f'(0) = 1$$
$$f''(x) = e^x \qquad\qquad f''(0) = 1$$

EXAMPLE 31.2 (Cont.)

In general, $f^{(n)}(x) = e^x$ and $f^{(n)}(0) = 1$. Thus, the Maclaurin series of $f(x) = e^x$ is

$$e^x = 1 + x + \frac{x^2}{2!} + \frac{x^3}{3!} + \frac{x^4}{4!} + \cdots + \frac{x^n}{n!} + \cdots$$

$$= \sum_{n=0}^{\infty} \frac{x^n}{n!}$$

Notice that this was the same series we used in Example 31.1(a).

EXAMPLE 31.3

Find the Maclaurin series of $f(x) = \cos x$.

Solution Again, we evaluate f and its derivatives at 0.

$$
\begin{array}{ll}
f(x) = \cos x & f(0) = 1 \\
f'(x) = -\sin x & f'(0) = 0 \\
f''(x) = -\cos x & f''(0) = -1 \\
f'''(x) = \sin x & f'''(0) = 0 \\
f^{(4)}(x) = \cos x & f^{(4)}(0) = 1
\end{array}
$$

As you can see, we are back at the beginning: $f^{(5)}(x) = f'(x)$, $f^{(6)}(x) = f''(x)$, and so on. Substituting these values yields the Maclaurin series.

$$\cos x = 1 - \frac{x^2}{2!} + \frac{x^4}{4!} - \frac{x^6}{6!} + \frac{x^8}{8!} + \cdots + (-1)^n \frac{x^{2n}}{(2n)!} + \cdots$$

EXAMPLE 31.4

Find the Maclaurin series of $f(x) = \sin 2x$.

Solution

$$
\begin{array}{ll}
f(x) = \sin 2x & f(0) = 0 \\
f'(x) = 2\cos 2x & f'(0) = 2 = 2^1 \\
f''(x) = -4\sin 2x & f''(0) = 0 \\
f'''(x) = -8\cos 2x & f'''(0) = -8 = -2^3 \\
f^{(4)}(x) = 16\sin 2x & f^{(4)}(0) = 0 \\
f^{(5)}(x) = 32\cos 2x & f^{(5)}(0) = 32 = 2^5
\end{array}
$$

We seem to have a pattern. If n is an even number, $f^{(n)}(0) = 0$. If n is an odd number, then $n = 2k + 1$, and $f^{(n)}(0) = f^{(2k+1)}(0) = (-1)^k 2^{2k+1}$.

The series is

$$\sin 2x = 2x - \frac{8x^3}{3!} + \frac{2^5 x^5}{5!} - \cdots + \frac{(-1)^k (2x)^{2k+1}}{(2k+1)!}.$$

The expansions in the box below are important and have been listed to help you. In Examples 31.2 and 31.3, we developed the first and third expressions. The rest are Exercises 1–4 in Exercise Set 31.1.

Some Important Maclaurin Series

$$e^x = 1 + x + \frac{x^2}{2!} + \frac{x^3}{3!} + \frac{x^4}{4!} + \frac{x^5}{5!} + \cdots \quad \text{, for all } x$$

$$\sin x = x - \frac{x^3}{3!} + \frac{x^5}{5!} - \frac{x^7}{7!} + \frac{x^9}{9!} - \frac{x^{11}}{11!} + \cdots \quad \text{, for all } x$$

$$\cos x = 1 - \frac{x^2}{2!} + \frac{x^4}{4!} - \frac{x^6}{6!} + \frac{x^8}{8!} - \frac{x^{10}}{10!} + \cdots \quad \text{, for all } x$$

$$\sinh x = x + \frac{x^3}{3!} + \frac{x^5}{5!} + \frac{x^7}{7!} + \cdots \quad \text{, for all } x$$

$$\cosh x = 1 + \frac{x^2}{2!} + \frac{x^4}{4!} + \frac{x^6}{6!} + \cdots \quad \text{, for all } x$$

$$\ln(1+x) = x - \frac{x^2}{2} + \frac{x^3}{3} - \frac{x^4}{4} + \frac{x^5}{5} - \frac{x^6}{6} + \cdots \quad \text{, for } -1 < x \le 1$$

Application

EXAMPLE 31.5

A certain type of skin wound heals at the rate of

$$A'(t) = \frac{-50}{t^2 + 20} \text{ cm}^2/\text{day}$$

Suppose that initially a wound had an area of 15 cm². Use a Maclaurin series to approximate the area of the wound after 2 days.

Solution Here, if we let $f(t) = A'(t) = \frac{-50}{t^2 + 20}$, then we have

$$f(t) = \frac{-50}{t^2 + 20} \qquad\qquad f(0) = -\frac{5}{2}$$

$$f'(t) = \frac{100t}{(t^2 + 20)^2} \qquad\qquad f'(0) = 0$$

$$f''(t) = \frac{100(20 - 3t^2)}{(t^2 + 20)^3} \qquad\qquad f''(0) = \frac{1}{4}$$

Thus, we can approximate $A'(t)$ with the Maclaurin series

$$A'(t) \approx -\frac{5}{2} + \frac{1/4}{2!}t^2$$
$$= -\frac{5}{2} + \frac{1}{8}t^2$$

EXAMPLE 31.5 (Cont.)

Since this is the rate at which the wound is healing, the size of the wound in square centimeters at the end of day t is

$$A(t) = \int A'(t)\,dt$$

$$= -\frac{5}{2}t + \frac{1}{24}t^3 + C$$

We are given that the wound was initially $15\,\text{cm}^2$, so $A(0) = 15$. Thus,

$$A(t) = -\frac{5}{2}t + \frac{1}{24}t^3 + 15$$

At the end of day 2, the wound has an area of

$$A(2) = -5 + \frac{1}{3} + 15$$

$$= 10\frac{1}{3}\,\text{cm}^2$$

Exercise Set 31.1

In Exercises 1–16, find the first four nonzero terms of the Maclaurin series expansion for the given function.

1. $\sin x$

2. $\sinh x$ $\left(\text{Remember: } \sinh x = \dfrac{e^x - e^{-x}}{2}.\right)$

3. $\cosh x$ $\left(\text{Remember: } \cosh x = \dfrac{e^x + e^{-x}}{2}.\right)$

4. $\ln(1+x)$

5. e^{3x}

6. $\sin^3 x$

7. $\ln(1+x^2)$

8. e^{-x}

9. $\cos x^2$

10. e^{-x^2}

11. $e^x \sin x$

12. xe^x

13. $x^2 e^{-x^2}$

14. $e^{\sin x}$

15. $x \sin 3x$

16. $e^{-x} \cos x$

Solve Exercises 17–20.

17. *Medical technology* A certain type of skin wound heals at the rate of

$$A'(t) = \frac{-60}{t^2 + 30}\,\text{cm}^2/\text{day}$$

Suppose that initially a wound had an area of $12\,\text{cm}^2$. Use a second-degree Maclaurin series to approximate the area of the wound after 2 days.

18. *Electronics* The current in amperes in a certain circuit is given by

$$i(\alpha) = 2 + \cos\alpha + 2\sin\alpha$$

Find the first four terms of the Maclaurin series expansion of this function.

19. *Hydraulics* The function $E(x) = \sec x$ is used in the study of fluid flow to define the Euler numbers. Find the first three nonzero terms of the Maclaurin series expansion of this function.

20. *Medical technology*   The rate of healing for one type of skin wound in square centimeters per day is given by

$$A'(t) = \frac{-75}{t^2 + 25}$$

Suppose that initially a wound had an area of $16\,\text{cm}^2$.
(a) Use a Maclaurin series to approximate the area of the wound after t days.
(b) What is the approximate area of the wound after 3 days?

In Your Words

21. Describe how to determine a Maclaurin series.

22. Why are series important?   What are some "real world" applications of infinite series?

≡ 31.2
OPERATIONS WITH SERIES

In Section 31.1, we learned how to use Maclaurin series to expand several functions. In this section, we will learn some operations with series that will provide quicker ways to expand these, and other, series. In Section 31.3, we will see how to make use of these series to develop tables.

In Section 31.1, we developed the Maclaurin series for $f(x)$. This same method could be used to expand $f(u)$. Thus, we would have

$$f(u) = \sum_{n=0}^{\infty} \frac{f^{(n)}(0)}{n!} u^n = f(0) + f'(0)u + \frac{f''(0)}{2!}u^2 + \frac{f'''(0)}{3!}u^3 + \cdots$$

By letting u assume values of $2x$, x^2, $-x$, x^{-2}, and so on, we can evaluate $f(2x)$, $f(x^2)$, $f(-x)$, $f(x^{-2})$, and so on.

EXAMPLE 31.6

Find the Maclaurin expansion of e^{3x}.

Solution   From the Section 31.1, we know that

$$e^u = 1 + u + \frac{u^2}{2!} + \frac{u^3}{3!} + \cdots$$

We want the expansion of e^{3x}, so if we let $u = 3x$, we have

$$e^{3x} = 1 + 3x + \frac{(3x)^2}{2!} + \frac{(3x)^3}{3!} + \frac{(3x)^4}{4!} + \cdots$$
$$= 1 + 3x + \frac{9}{2}x^2 + \frac{9}{2}x^3 + \frac{27}{8}x^4 + \cdots$$

EXAMPLE 31.7

Find the Maclaurin expansion of $\cos x^2$.

Solution We know that

$$\cos u = 1 - \frac{u^2}{2!} + \frac{u^4}{4!} - \frac{u^6}{6!} + \cdots$$

Thus, if $u = x^2$, we have

$$\cos x^2 = 1 - \frac{(x^2)^2}{2!} + \frac{(x^2)^4}{4!} - \frac{(x^2)^6}{6!} + \cdots$$

$$= 1 - \frac{x^4}{2!} + \frac{x^8}{4!} - \frac{x^{12}}{6!} + \cdots$$

Compare this to the amount of work that we had to do when we worked Exercise 9 in Exercise Set 31.1.

Basic Operations with Series

The basic operations can be used with series in order to obtain other series. Thus, you can add, subtract, multiply, or divide series in the same way as you do polynomials. The next example will show you how this can be done.

EXAMPLE 31.8

Determine the Maclaurin series expansion of $e^x \sin x$.

Solution We know that if $f(x) = e^x$, then $f(x) = 1 + x + \frac{x^2}{2!} + \frac{x^3}{3!} + \frac{x^4}{4!} + \cdots$, and if $g(x) = \sin x$, then $g(x) = x - \frac{x^3}{3!} + \frac{x^5}{5!} - \frac{x^7}{7!} + \cdots$.

We want $e^x \sin x = f(x) \cdot g(x)$. Thus, we can multiply these two series term by term:

$$e^x \sin x = \left(1 + x + \frac{x^2}{2!} + \frac{x^3}{3!} + \frac{x^4}{4!} + \cdots\right)\left(x - \frac{x^3}{3!} + \frac{x^5}{5!} - \frac{x^7}{7!} + \cdots\right)$$

$$= x + x^2 + \frac{x^3}{3} - \frac{4x^5}{5!} + \cdots$$

Differentiation and Integration of Series

It is also possible to differentiate and integrate functions by differentiating or integrating the terms of their series expansion. We used this procedure to help solve Example 31.5.

EXAMPLE 31.9

Show that $\dfrac{d}{dx}e^x = e^x$.

Solution   We know that $e^x = 1 + x + \dfrac{x^2}{2!} + \dfrac{x^3}{3!} + \dfrac{x^4}{4!} + \cdots$, so

$$\frac{d}{dx}e^x = \frac{d}{dx}\left(1 + x + \frac{x^2}{2!} + \frac{x^3}{3!} + \frac{x^4}{4!} + \cdots\right)$$

$$= 0 + 1 + \frac{2x}{2!} + \frac{3x^2}{3!} + \frac{4x^3}{4!} + \cdots$$

$$= 1 + x + \frac{x^2}{2!} + \frac{x^3}{3!} + \cdots$$

$$= e^x$$

Application

EXAMPLE 31.10

Approximate the area under the graph of $y = \cos x^2$ to four decimal places.

Solution   In order to find this area, we need to evaluate $\displaystyle\int_0^1 \cos x^2\,dx$. In Example 31.7, we found that

$$\cos x^2 = 1 - \frac{x^4}{2!} + \frac{x^8}{4!} - \frac{x^{12}}{6!} + \cdots$$

Integrating the individual terms of this series allows us to obtain the desired integral.

$$\int_0^1 \cos x^2\,dx = \int_0^1 \left(1 - \frac{x^4}{2!} + \frac{x^8}{4!} - \frac{x^{12}}{6!} + \cdots\right)dx$$

$$= x - \frac{x^5}{5\cdot 2!} + \frac{x^9}{9\cdot 4!} - \frac{x^{13}}{13\cdot 6!} + \cdots \Bigg|_0^1$$

$$= 1 - \frac{1}{5\cdot 2} + \frac{1}{9\cdot 4!} - \frac{1}{13\cdot 6!} + \cdots$$

$$\approx 1 - 0.1 + 0.00463 - 0.00010$$

$$= 0.9045$$

The last example of this section will show a relationship between three transcendental functions. This result makes use of the imaginary number $j = \sqrt{-1}$. In this relationship, we will need to assume that the Maclaurin expansions of e^x, $\sin x$, and $\cos x$ are valid for complex numbers.

Look back at the Maclaurin expansions of $\sin x$ and $\cos x$. The expansion of the sine function has only odd powers. The expansion of the cosine function has only even powers. On the other hand, the expansion of e^x contains all powers of x. Now, if we expand e^{jx}, we get

$$e^{jx} = 1 + jx + \frac{(jx)^2}{2!} + \frac{(jx)^3}{3!} + \frac{(jx)^4}{4!} + \frac{(jx)^5}{5!} + \frac{(jx)^6}{6!} + \cdots$$
$$= 1 + jx + \frac{j^2 x^2}{2!} + \frac{j^3 x^3}{3!} + \frac{j^4 x^4}{4!} + \frac{j^5 x^5}{5!} + \frac{j^6 x^6}{6!} + \cdots$$

We need to remember that $j = \sqrt{-1}$, $j^2 = -1$, $j^3 = -j$, $j^4 = 1$, and then the pattern repeats. We get

$$e^{jx} = 1 + jx - \frac{x^2}{2!} - \frac{jx^3}{3!} + \frac{x^4}{4!} + \frac{jx^5}{5!} - \frac{x^6}{6!} - \cdots$$
$$= \left(1 - \frac{x^2}{2!} + \frac{x^4}{4!} - \frac{x^6}{6!}\right) + j\left(x - \frac{x^3}{3!} + \frac{x^5}{5!} - \cdots\right)$$
$$= \cos x + j\sin x$$

This relationship, $e^{jx} = \cos x + j\sin x$, is known as **Euler's identity**.

A complex number in rectangular form is expressed as $a + bj$, where a and b are real numbers. This same number, $a + bj$, is written in polar form as $r(\cos\theta + j\sin\theta) = r\operatorname{cis}\theta$. In exponential notation, $a + bj$ is written as $re^{j\theta}$, where $r = \sqrt{a^2 + b^2}$.

EXAMPLE 31.11	Express $4e^{5.4j}$ in rectangular form.

Solution   We know that $a + bj = re^{j\theta}$. Thus, $r = 4$ and $\theta = 5.4\,\text{rad}$ and we have

$$a + bj = 4(\cos 5.4 + j\sin 5.4)$$
$$\approx 4(0.6347 - 0.7728j)$$
$$= 3.9822 + 0.3764j$$

One final point about Euler's identity: Some of the most familiar and, some people say, interesting numbers in mathematics are $e, \pi, j, 0,$ and -1. Using Euler's identity, we see that

$$e^{j\pi} = \cos\pi + j\sin\pi$$
$$= -1$$

Thus, we have

$$e^{j\pi} = -1$$

or $e^{j\pi} + 1 = 0$

As we will see, Euler's identity is used in differential equations and in the theory of electrical circuits.

Exercise Set 31.2

In Exercises 1–12, find the first four nonzero terms of the Maclaurin expansion of the given function.

1. $f(x) = e^{3x}$
2. $g(x) = e^{-4x}$
3. $h(x) = \cos\dfrac{x}{2}$
4. $j(x) = \sin x^3$
5. $k(x) = \cos x^3$
6. $f(x) = \sin 3x$
7. $g(x) = \sin 2x^2$
8. $h(x) = \ln(1-x)$

9. Show that $\dfrac{d}{dx}\sin x = \cos x$ by using Maclaurin series.

10. Use Maclaurin series to show $\dfrac{d}{dx}\cos x^2 = -2x\sin x^2$.

11. Use Maclaurin series to show that $\dfrac{d}{dx}e^{2x} = 2e^{2x}$.

12. Use Maclaurin series to show that $\dfrac{d}{dx}\sin 3x = 3\cos 3x$.

In Exercises 13–16, evaluate the given integrals by use of three terms of the appropriate series.

13. $\displaystyle\int_0^1 \sin x^2\, dx$

14. $\displaystyle\int_0^{0.1} \frac{e^x - 1}{x}\, dx$

15. $\displaystyle\int_0^{0.2} \sin\sqrt{x}\, dx$

16. $\displaystyle\int_0^1 \frac{\sin x}{x}\, dx$

Solve Exercises 17–20.

17. Find the Maclaurin expansion of $e^{-x}\cos x$ using the method of Example 31.8.

18. Find the Maclaurin expansion of $\frac{1}{2}(e^x - e^{-x})$ by adding the terms of the appropriate series. Compare this to your results to Exercise 2, from Exercise Set 30.1.

19. Find the Maclaurin expansion of $x^2 e^{-x^2}$ using the method of Example 31.8.

20. Find the Maclaurin expansion of $\tan x$ by dividing $\sin x$ by $\cos x$.

In Exercises 21–24, change the given complex number to the indicated form.

21. $5e^{0.8j}$ (rectangular)
22. $5 - 12j$ (polar)
23. $6\operatorname{cis}\frac{4\pi}{3}$ (exponential)
24. $-3 + 4j$ (exponential)

In Exercises 25–28, find the approximate area bounded by the given curves or lines by using the first three nonzero terms of the appropriate Maclaurin series.

25. $y = e^{x^2}$, $x = 0$, $x = 1$, $y = 0$
26. $y = \cos x^2$, $x = 0$, $x = \frac{\pi}{2}$, $y = 0$
27. $y = x^2 e^x$, $x = 0.2$, $x = 0$, $y = 0$
28. $y = e^{-x^2}$, $x = 0$, $x = 1$, $y = 0$

Solve Exercises 29–34.

29. *Electricity*   Use the first five nonzero terms of the Maclaurin expansion of $\cos x$ to obtain an approximation of the voltage $V = 15\cos\Omega t$, when $\Omega t = 0.1$ rad.

30. Use the Maclaurin series expansion of $\sinh x$ and $\cosh x$ to show that $e^x = \sinh x + \cosh x$.

31. *Electricity*   (a) Express as a Maclaurin series the current $i = \sin t^2$ A. (b) What is the amount of charge transmitted by this current from $t = 0$ to $t = 0.02$ s?

32. Find the volume generated by rotating the area bounded by $y = e^{-x}$, $y = 0$, $x = 0$, and $x = 0.1$ about the y-axis, by using the first three terms of the appropriate series.

33. *Environmental technology*   A train derailment has resulted in the spill of a certain pollutant.   The amount of pollutant from the spill in the local groundwater decreases according to the equation $P = P_0 \left(e^{-2t} + e^{-t}\right)$, where t is in weeks and P is measured in ppm.

(a) Find the first four terms of the Maclaurin series expansion of P

(b) Use your answer to (a) to find $\dfrac{dP}{dt}$.

(c) Use your answer to (b) to calculate the value of $\dfrac{dP}{dt}$ at $t = 1.4$.

34. *Quality control*   The manager of a computer manufacturing company guarantees a certain model of computer for two years. The probability that a computer will last t years without service is given by $p(t) = 0.05 e^{-0.05t}$.

(a) Find the first four terms of the Maclaurin series expansion of p.

(b) Use your answer to (a) to find the probability that a computer will need service before it is 2 years old by evaluating $\displaystyle \int_0^2 p(t)\,dt$.

   **In Your Words**

35. What is the advantage of using a series expansion of a function to differentiate or integrate the function?

36. Which of the operations of addition, subtraction, multiplication, and division can be used to combine two series? What precautions do you need to take when you do this? What are the advantages or disadvantages of using these operations to combine two series?

≡ 31.3
NUMERICAL TECHNIQUES USING SERIES

Power series expansions can be used to compute numerical values of transcendental functions. If you use enough terms, you can get the value of the function to any degree of accuracy. Calculations such as these are used to make tables of values.

One question often arises when using a method such as this to approximate a value. That question concerns the accuracy of the approximation.

Alternating Series

Most of the time, we will use a series in which the signs alternate from positive to negative and back to positive. This is called an **alternating series** as defined in the following box.

Alternating Series

An **alternating series** is either of the form

$$\sum_{k=0}^{\infty} (-1)^k a_k = a_0 - a_1 + a_2 - a_3 + a_4 - a_5 + \cdots$$

or of the form

$$\sum_{k=0}^{\infty} (-1)^{k+1} a_k = -a_0 + a_1 - a_2 + a_3 - a_4 + \cdots$$

where each number a_k is positive for all values of k.

If an alternating series converges, then the sum can be obtained to any degree of accuracy by adding the first n terms of the series where the $(n+1)$st term is less than the desired accuracy. This means that if you want an approximation within some value, say E, of the true value, then you need to find the first term of the series, a_{n+1}, when $a_{n+1} < E$. Then $\sum_{n=0}^{n} a_n$ will be within the desired amount of accuracy.

EXAMPLE 31.12

Compute the value of $e^{-0.2}$ with an accuracy within 0.0001.

Solution   Since $e^x = 1 + x + \dfrac{x^2}{2!} + \dfrac{x^3}{3!} + \dfrac{x^4}{4!} + \cdots$, then

$$e^{-x} = 1 - x + \frac{x^2}{2!} - \frac{x^3}{3!} + \frac{x^4}{4!} - \frac{x^5}{5!} + \cdots$$

We want to approximate $e^{-0.2}$ to within 0.0001.

$$e^{-0.2} = 1 - 0.2 + \frac{(0.2)^2}{2!} - \frac{(0.2)^3}{3!} + \frac{(0.2)^4}{4!} - \frac{(0.2)^5}{5!} + \cdots$$
$$= 1 - 0.2 + 0.02 - 0.00133 + 0.000067 - \cdots$$

Since $0.000067 < 0.0001$, we need only add the first four terms to get within 0.0001 of the actual value.

$$e^{-0.2} \approx 1 - 0.2 + 0.02 - 0.0013 = 0.8187$$

The value by using a calculator is approximately 0.81873.

Using Series to Approximate Values

EXAMPLE 31.13

Calculate the value of $\ln 1.1$ with an accuracy within 0.0001.

Solution   We have $\ln(1+x) = x - \dfrac{x^2}{2} + \dfrac{x^3}{3} - \dfrac{x^4}{4} + \cdots$, so

$$\ln 1.1 = \ln(1+0.1) = 0.1 - \frac{(0.1)^2}{2} + \frac{(0.1)^3}{3} - \frac{(0.1)^4}{4} + \cdots$$
$$= 0.1 - 0.005 + 0.00033 - 0.000025 + \cdots$$

When we add the first three terms we get $\ln 1.1 \approx 0.09533$. Since $0.000025 < 0.0001$, we do not need to add the fourth term to get the desired accuracy. The actual value (using a calculator) is approximately 0.09531.

EXAMPLE 31.14

Calculate the value of $\cos 3°$ accurate to within 0.0001.

Solution   We know that

$$\cos x = 1 - \frac{x^2}{2!} + \frac{x^4}{4!} - \frac{x^6}{6!} + \cdots$$

EXAMPLE 31.14 (Cont.)

This is for a value of x in radians. Since $3° = \frac{\pi}{60}$ rad we have

$$
\begin{aligned}
\cos 3° = \cos \frac{\pi}{60} &= 1 - \frac{(\pi/60)^2}{2!} \\
&= 1 - \frac{\pi^2}{7,200} \\
&= 1 - 0.0014 \\
&= 0.9986
\end{aligned}
$$

Since the next nonzero term, $\frac{(\pi/60)^4}{4!} \approx 0.0000003 < 0.0001$, we needed only the first two terms.

EXAMPLE 31.15

Approximate $\int_0^{0.5} e^{-x^2}\, dx$ accurately to within 0.001.

Solution   The Maclaurin series for e^{-x^2} is

$$
1 - x^2 + \frac{x^4}{2!} - \frac{x^6}{3!} + \cdots
$$

Integrating, we get

$$
\begin{aligned}
\int_0^{0.5} e^{-x^2}\, dx &= \int_0^{0.5} \left(1 - x^2 + \frac{x^4}{2!} - \frac{x^6}{3!} + \cdots \right) dx \\
&= x - \frac{x^3}{3} + \frac{x^5}{5\cdot 2!} - \frac{x^7}{7\cdot 3!} + \cdots \Bigg|_0^{0.5} \\
&= \frac{1}{2} - \frac{(1/2)^3}{3} + \frac{(1/2)^5}{5\cdot 2!} - \frac{(1/2)^7}{7\cdot 3!} + \cdots \\
&= \frac{1}{2} - \frac{1}{3\cdot 2^3} + \frac{1}{5\cdot 2!(2)^5} - \frac{1}{7\cdot 3!(2)^7} + \cdots \\
&= 0.5 - 0.04167 + 0.00313 - 0.00019
\end{aligned}
$$

This is an alternating series. The last term, $0.00019 < 0.001$, so the sum of the first three terms gives the desired accuracy:

$$
\int_0^{0.5} e^{-x^2}\, dx \approx 0.5 - 0.04167 + 0.00313 = 0.46146
$$

Application

EXAMPLE 31.16

The current supplied to an initially discharged 1-μF capacitor varies according to $i = \dfrac{0.1(1 - \cos t)}{t}$ A, where t is in seconds. What is the voltage across the capacitor when $t = 0.1\,\text{s}$?

EXAMPLE 31.16 (Cont.)

Solution The voltage across a capacitor at any instant is given by

$$V_C = \frac{1}{C} \int i \, dt$$

where C is the capacitance in farads and V_C is in volts. Here, we have $C = 1 \, \mu F = 10^{-6}$ F, so

$$V_C = \frac{1}{10^{-6}} \int_0^{0.1} \frac{0.1(1 - \cos t)}{t} \, dt = 10^5 \int_0^{0.1} \frac{1 - \cos t}{t} \, dt$$

None of our integration techniques from Chapter 28 and none of the integration formulas in Appendix C show us how to complete this integral. However, using the series for $\cos x$, we can integrate this expression.

We know that $\cos t = 1 - \dfrac{t^2}{2!} + \dfrac{t^4}{4!} - \dfrac{t^6}{6!} + \cdots$, and so $1 - \cos t = \dfrac{t^2}{2!} - \dfrac{t^4}{4!} + \dfrac{t^6}{6!} - \dfrac{t^8}{8!} + \cdots$. Dividing by t produces $\dfrac{1 - \cos t}{t} = \dfrac{t}{2!} - \dfrac{t^3}{4!} + \dfrac{t^5}{6!} - \dfrac{t^7}{8!} + \cdots$.

Substituting this series into the integral, we get

$$V_C = 10^5 \int_0^{0.1} \frac{1 - \cos t}{t} \, dt$$

$$= 10^5 \int_0^{0.1} \left(\frac{t}{2!} - \frac{t^3}{4!} + \frac{t^5}{6!} - \frac{t^7}{8!} + \cdots \right) dt$$

$$= 10^5 \left(\frac{t^2}{2! \cdot 2} - \frac{t^4}{4! \cdot 4} + \frac{t^6}{6! \cdot 6} - \frac{t^8}{8! \cdot 8} + \cdots \right) \Bigg|_0^{0.1}$$

$$= 10^5 \left(\frac{(0.1)^2}{2! \cdot 2} - \frac{(0.1)^4}{4! \cdot 4} + \frac{(0.1)^6}{6! \cdot 6} - \frac{(0.10)^8}{8! \cdot 8} + \cdots \right)$$

$$= 10^5 (0.0025 - 0.000001 + \cdots)$$

$$\approx 250 \, \text{V}$$

There are 250 V across this capacitor when $t = 0.1$ s.

Exercise Set 31.3

In Exercises 1–10, calculate the value of the given function accurately to within 0.0001.

1. $e^{-0.3}$
2. $\sin(0.1)$
3. $\cos(0.1)$
4. $e^{0.2}$
5. $\sin 5°$
6. $\ln 1.5$
7. $\ln 0.97$
8. $\sqrt{e}$
9. $\ln 0.5$
10. $\cos 5°$

Evaluate the integrals in Exercises 11–16 to an accuracy of 0.0001.

11. $\displaystyle\int_0^{0.5} \frac{1-\cos x}{x}\,dx$

13. $\displaystyle\int_0^1 \frac{\sin x - x}{x^2}\,dx$

15. $\displaystyle\int_0^{0.2} \frac{e^x - 1}{x}\,dx$

12. $\displaystyle\int_0^{0.5} \frac{x - \sin x}{x}\,dx$

14. $\displaystyle\int_0^{0.5} \sqrt{x}\cos\sqrt{x}\,dx$

16. $\displaystyle\int_{0.5}^1 \frac{\ln(1+x)}{x}\,dx$

Solve Exercises 17–24.

17. **(a)** Use the Maclaurin expansion to show that, for small values of θ, $\sin\theta = \theta$. **(b)** Determine the values of θ for which this is true.

18. *Electricity*   The current delivered to an initially discharged 1-μF capacitor varies according to $i = \dfrac{0.1(t - \sin t)}{t}$ A, where t is in seconds. What is the voltage across the capacitor when $t = 0.2\,\text{s}$?

19. *Electricity*   The charge q, in microcolumbs (μC), on a certain capacitor is

$$q = \int_0^{0.5} \left(1 - e^{-0.05t^2}\right) dt$$

Determine this charge accurately to within $0.001\,\mu$C.

20. *Electricity*   The charge q, in microcoulombs, on a capacitor is given by

$$q = \int_0^{1.5} \left(1.0 - e^{0.25t^2}\right) dt$$

Use a Maclaurin series to approximate this integral.

21. *Electricity*   The voltage in a circuit is given by $V = 0.75e^{0.75t}$. **(a)** Express the voltage as a Maclaurin series. **(b)** Determine an approximation of the voltage when $t = 0.45\,\text{s}$.

22. *Construction technology*   The **Fresnel integral**

$$c(x) = \int_0^x \cos t^2\,dt$$

is used to describe the longitudinal displacement of a beam subject to a periodic force. This integral cannot be found directly.

(a) Find the first four terms of the Maclaurin series expansion of $\cos t^2$ by substituting t^2 for x in the series for $\cos x$.

(b) Use your answer to (a) to approximate this integral.

(c) Use your answer to (b) to evaluate $c(0.3)$.

23. *Electrical engineering*   The integral

$$V(x) = V_0 \int_0^x e^{-t^2}\,dt$$

is used to calculate the voltage $V(x)$ along a very long transmission line of negligible inductance and capacitance per unit length, where V_0 is the voltage applied to the transmitting end of the line.

(a) Find the first four terms of the Maclaurin series expansion of e^{-t^2} by substituting $-t^2$ for x in the series for e^x.

(b) Evaluate this integral using the first four nonzero terms of the Maclaurin series for this function.

(c) Find $V(1.4)$.

(d) Estimate the maximum error of your approximation in (c).

24. *Electronics*   The charge, q, in millicoulombs (mC), on a certain capacitor is

$$q = \int_0^{1.0} \left(1 - e^{0.1t^2}\right) dt$$

Evaluate this integral using the first four terms of the Maclaurin series for this function.

In Your Words

25. Without looking in the text, describe an alternating series.

26. How do you know when an alternating series will be within a desired amount of accuracy?

☰ 31.4
TAYLOR SERIES

Until now, we have used Maclaurin series to evaluate functions around zero. Let's assume that we have a power series whose interval of convergence is $(c - R, c + R)$, where R is a positive real number. This power series will be of the form

$$f(x) = \sum_{k=0}^{\infty} a_k (x - c)^k = a_0 + a_1(x - c) + a_2(x - c)^2 + \cdots$$

Again, in taking successive derivatives of f, we get

$$f(x) = a_0 + a_1(x - c) + a_2(x - c)^2 + a_3(x - c)^3 + a_4(x - c)^4 + \cdots$$
$$f'(x) = a_1 + 2a_2(x - c) + 3a_3(x - c)^2 + 4a_4(x - c)^3 + \cdots$$
$$f''(x) = 2a_2 + 3 \cdot 2a_3(x - c) + 4 \cdot 3a_4(x - c)^2 + \cdots$$
$$f'''(x) = 3 \cdot 2a_3 + 4 \cdot 3 \cdot 2a_4(x - c) + \cdots$$
$$f^{(4)}(x) = 4 \cdot 3 \cdot 2a_4 + \cdots$$

and, for any positive integer n,

$$f^{(n)}(x) = n!a_n + (n + 1)!\, a_{n+1}(x - c) + \cdots$$

If we evaluate each of these equations when $x = c$, then every term, except the first, will have a value of 0. Thus, we get

$$f(c) = a_0 \qquad \text{or} \qquad a_0 = f(c)$$
$$f'(c) = a_1 \qquad\qquad\qquad a_1 = f'(c)$$
$$f''(c) = 2!a_2 \qquad\qquad\qquad a_2 = \frac{1}{2!} f''(c)$$
$$f'''(c) = 3!a_3 \qquad\qquad\qquad a_3 = \frac{1}{3!} f'''(c)$$
$$\vdots \qquad\qquad\qquad\qquad \vdots$$
$$f^{(n)}(c) = n!a_n \qquad\qquad\qquad a_n = \frac{1}{n!} f^{(n)}(c)$$

Substituting these values into the series expansion for $f(x)$, we get the **Taylor series** for f about c.

Taylor series

The **Taylor Series** for f about c is

$$f(x) = f(c) + f'(c)(x - c) + \frac{f''(c)(x - c)^2}{2!} + \frac{f'''(c)(x - c)^3}{3!} + \cdots$$
$$+ \frac{f^{(n)}(c)(x - c)^n}{n!} + \cdots$$

☰ **Note**   When $c = 0$, the Taylor series is equivalent to the Maclaurin series.

EXAMPLE 31.17

Expand e^x in a Taylor series around 1.

Solution

$$\begin{aligned}
f(x) &= e^x & f(1) &= e \\
f'(x) &= e^x & f'(1) &= e \\
f''(x) &= e^x & f''(1) &= e
\end{aligned}$$

$$\vdots \qquad\qquad \vdots$$

$$\begin{aligned}
f(x) &= e + e(x-1) + \frac{e}{2!}(x-1)^2 + \frac{e}{3!}(x-1)^3 + \frac{e}{4!}(x-1)^4 + \cdots \\
&= e\left[1 + (x-1) + \frac{(x-1)^2}{2!} + \frac{(x-1)^3}{3!} + \frac{(x-1)^4}{4!} + \cdots\right] \\
&= e\left[x + \frac{(x-1)^2}{2!} + \frac{(x-1)^3}{3!} + \frac{(x-1)^4}{4!} + \cdots\right]
\end{aligned}$$

This example gives us a method to evaluate e^x for values of x that are close to 1.

EXAMPLE 31.18

Use a Taylor series to evaluate $e^{0.9}$.

Solution From Example 31.17, we know that for values of x close to 1,

$$e^x = e\left[x + \frac{(x-1)^2}{2!} + \frac{(x-1)^3}{3!} + \frac{(x-1)^4}{4!} + \cdots\right]$$

In this example, $x = 0.9$, so

$$\begin{aligned}
e^{0.9} &= e\left[0.9 + \frac{(0.9-1)^2}{2!} + \frac{(0.9-1)^3}{3!} + \frac{(0.9-1)^4}{4!} + \cdots\right] \\
&= e\left[0.9 + \frac{(-0.1)^2}{2!} + \frac{(-0.1)^3}{3!} + \frac{(-0.1)^4}{4!} + \cdots\right] \\
&= e(0.9 + 0.005 - 0.000167 + 0.000004 - \cdots) \\
&= 0.904837e \\
&\approx 2.4596
\end{aligned}$$

EXAMPLE 31.19

Expand $\sin x$ in a Taylor series about $\frac{\pi}{3}$.

Solution

$$\begin{aligned}
f(x) &= \sin x & f\left(\frac{\pi}{3}\right) &= \frac{\sqrt{3}}{2} \\
f'(x) &= \cos x & f'\left(\frac{\pi}{3}\right) &= \frac{1}{2} \\
f''(x) &= -\sin x & f''\left(\frac{\pi}{3}\right) &= \frac{-\sqrt{3}}{2} \\
f'''(x) &= -\cos x & f'''\left(\frac{\pi}{3}\right) &= -\frac{1}{2}
\end{aligned}$$

$$\sin x = \frac{\sqrt{3}}{2} + \frac{1}{2}\left(x - \frac{\pi}{3}\right) - \frac{\sqrt{3}}{2} \cdot \frac{1}{2!}\left(x - \frac{\pi}{3}\right)^2 - \frac{1}{2} \cdot \frac{1}{3!}\left(x - \frac{\pi}{3}\right)^3 + \cdots$$

EXAMPLE 31.20

Calculate the approximate value of $\sin 61°$ by using a Taylor series.

Solution   We have the Taylor expansion of $\sin x$ around $\frac{\pi}{3} = 60°$ from Example 31.19. We will need to express $61°$ in radians. Since $x = 61° = \frac{61\pi}{180}$ and $c = 60° = \frac{\pi}{3} = \frac{60\pi}{180}$, we have $x - c = \frac{61\pi}{180} - \frac{60\pi}{180} = \frac{\pi}{180}$. Thus, from Example 31.19, we obtain

$$\sin 61° = \frac{\sqrt{3}}{2} + \frac{1}{2}\left(\frac{\pi}{180}\right) - \frac{\sqrt{3}}{2}\cdot\frac{1}{2!}\left(\frac{\pi}{180}\right)^2 - \frac{1}{2}\cdot\frac{1}{3!}\left(\frac{\pi}{180}\right)^3 + \cdots$$
$$\approx 0.8660 + 0.0087 - 0.0001 - \cdots$$
$$= 0.8746$$

Application

EXAMPLE 31.21

The damped current in an electronic circuit has the form

$$i = \frac{\sin \pi t}{t}$$

Calculate the approximate value of this function at $t = 1.9$ by using a Taylor series about $t = 2$.

Solution   Using a procedure similar to the one used in Example 31.19, we determine the following fundamental information needed for the Taylor expansion of $\sin \pi t$ around $t = 2$.

$$f(x) = \sin \pi x \qquad\qquad f(2) = \sin 2\pi = 0$$
$$f'(x) = \pi \cos \pi x \qquad\qquad f'(2) = \pi \cos 2\pi = \pi$$
$$f''(x) = -\pi^2 \sin \pi x \qquad\qquad f''(2) = -\pi^2 \sin 2\pi = 0$$
$$f'''(x) = -\pi^3 \cos \pi x \qquad\qquad f'''(2) = -\pi^3 \cos 2\pi = -\pi^3$$

Using these values, we obtain

$$\sin \pi t = 0 + \pi (t-2) - 0\cdot\frac{1}{2!}(t-2)^2 - \pi^3\frac{1}{3!}(t-2)^3 + \cdots$$
$$= \pi(t-2) - \frac{\pi^3}{3!}(t-2)^3 + \cdots$$

Now dividing by t, we obtain

$$\frac{\sin \pi t}{t} = \pi \left(\frac{t-2}{t}\right) - \frac{\pi^3}{3!}\frac{(t-2)^3}{t} + \cdots$$

and when we let $t = 1.9$, we get

$$\frac{\sin \pi t}{t} = \pi \left(\frac{-0.1}{1.9}\right) - \frac{\pi^3}{3!}\frac{(-0.1)^3}{1.9} + \cdots$$
$$\approx -0.165347 - (-0.002720) + \cdots$$
$$\approx -0.162627$$

A calculator gives $\dfrac{\sin(1.9\pi)}{1.9} \approx -0.162641$.

Exercise Set 31.4

In Exercises 1–12, expand the function in a Taylor series about the given value of c.

1. $f(x) = e^x, c = 2$

2. $g(x) = e^{2x}, c = -1$

3. $h(x) = \sin x, c = \frac{\pi}{4}$

4. $j(x) = \cos x, c = \frac{\pi}{4}$

5. $k(x) = \dfrac{1}{x}, c = 2$

6. $m(x) = \sqrt{x}, c = 4$

7. $f(x) = \dfrac{1}{(x+1)^2}, c = 0$

8. $g(x) = e^{1+x}, c = 1$

9. $h(x) = e^{-x}, c = 1$

10. $j(x) = \ln x, c = 4$

11. $k(x) = \sqrt[3]{x}, c = 8$

12. $m(x) = \tan x, c = \frac{\pi}{4}$

In Exercises 13–24, evaluate the given function by use of the series expansion you used in Exercises 1 through 12.

13. $e^{2.1}$

14. $e^{-0.8}$

15. $\sin 43°$

16. $\cos 43°$

17. $\frac{1}{1.8}$

18. $\sqrt{3.9}$

19. $\dfrac{1}{1.12^2}$

20. $e^{1.9}$

21. $e^{-0.8}$

22. $\ln 4.1$

23. $\sqrt[3]{7.8}$

24. $\tan 43°$

Solve Exercises 25–26.

25. *Electricity*   The damped current in an electronic circuit has the form

$$i = \frac{\sin \pi t}{t}$$

Calculate the approximate value of this function at $t = 2.1$ by using the Taylor series about $t = 2$.

26. *Electricity*   The damped current in an electronic circuit has the form

$$i = \frac{\sin 2\pi t}{t}$$

Calculate the approximate value of this function at $t = 1.05$ by using the Taylor series about $t = 1$.

In Your Words

27. Without looking in the text, describe a Taylor series.

28. How does a Taylor series differ from a Maclaurin series?

≡ 31.5
FOURIER SERIES

So far in this chapter, we have been concerned with functions that can be expanded in power series. In this section, we will consider a different kind of series—a trigonometric series.

Many functions that we use in science and technology are periodic functions. The first periodic functions that we encountered were the trigonometric functions. The **Fourier series** expresses a given function in terms of the sine and cosine functions.

Periodic Function

A **periodic function** is a function f where $f(x+p) = f(x)$. The smallest positive value of p is the period of the function.

The sine and cosine functions both have periods of 2π because $\sin(x+2\pi) = \sin x$ and $\cos(x+2\pi) = \cos x$, and because there is no positive value smaller than 2π for which this is true. In general, $\sin ax$ and $\cos ax$ have periods of $\dfrac{2\pi}{a}$.

Now, suppose that we have a periodic function f that has a period of $2L$. We will represent f by a series of sines and cosines, where

$$f(x) = a_0 + a_1 \cos \frac{\pi x}{L} + a_2 \cos \frac{2\pi x}{L} + \cdots + a_n \cos \frac{n\pi c}{L} + \cdots$$
$$+ b_1 \sin \frac{\pi x}{L} + b_2 \sin \frac{2\pi x}{L} + \cdots + b_n \sin \frac{n\pi x}{L} + \cdots \qquad (*)$$

Now, $\sin\left(\dfrac{\pi x}{L}\right)$ and $\cos\left(\dfrac{\pi x}{L}\right)$ have periods of $\dfrac{2\pi}{\pi/L} = 2L$, $\sin\left(\dfrac{2\pi x}{L}\right)$ and $\cos\left(\dfrac{2\pi x}{L}\right)$ have periods of L, and in general $\sin\left(\dfrac{n\pi x}{L}\right)$ and $\cos\left(\dfrac{n\pi x}{L}\right)$ have periods of $\dfrac{2L}{n}$. All the terms have a common period of $2L$. This series is the **Fourier series**.

We need to find the values of the coefficients a_n and b_n. When we developed the Maclaurin and Taylor series, we used derivatives. To find the coefficients of the Fourier series, we will use integrals. In each case, we will integrate over the interval $(-L, L)$.

To find a_0, we will integrate both sides of equation (*) from $-L$ to L.

$$\int_{-L}^{L} f(x)\,dx = \int_{-L}^{L} a_0\,dx + \int_{-L}^{L} a_1 \cos \frac{\pi x}{L}\,dx + \cdots + \int_{-L}^{L} b_1 \sin \frac{\pi x}{L}\,dx + \cdots$$

Now, $\displaystyle\int_{-L}^{L} f(x)\,dx = a_0 x\Big|_{-L}^{L} = 2La_0$, since the integrals of all the trigonometric

functions are 0. {For example, the integral $\displaystyle\int a_1 \cos \frac{\pi x}{L}\,dx = \frac{a_1 L}{\pi} \sin \frac{\pi x}{L}\Big|_{-L}^{L} =$

$\dfrac{a_1 L}{\pi}[\sin \pi - \sin(-\pi)] = 0.$} Solving for a_0, we get $a_0 = \dfrac{1}{2L}\displaystyle\int_{-L}^{L} f(x)\,dx$.

Next, we will find the value of a_k, $k = 1, 2, 3, \ldots$. To do this, we multiply both sides of equation (*) by $\cos \dfrac{k\pi x}{L}$ and then integrate from $-L$ to L. For example,

$$\int_{-L}^{L} f(x) \cos \frac{k\pi x}{L}\,dx = \int_{-L}^{L} a_0 \cos \frac{k\pi x}{L}\,dx + \int_{-L}^{L} a_1 \cos \frac{\pi x}{L} \cos \frac{k\pi x}{L}\,dx + \cdots$$
$$+ \int_{-L}^{L} a_k \cos^2 \frac{k\pi x}{L}\,dx + \cdots + \int_{-L}^{L} b_1 \sin \frac{\pi x}{L} \cos \frac{k\pi x}{L}\,dx + \cdots$$

$$+ \int b_k \sin \frac{k\pi x}{L} \cos \frac{k\pi x}{L} \, dx + \cdots \qquad (**)$$

From this, you can see that there are two types of integrals: These are

$$\int_{-L}^{L} \cos \frac{m\pi x}{L} \cos \frac{k\pi x}{L} \, dx \quad \text{and} \quad \int_{-L}^{L} \sin \frac{m\pi x}{L} \cos \frac{k\pi x}{L} \, dx$$

where m and k are integers. It can be shown that

$$\int_{-L}^{L} \cos \frac{m\pi x}{L} \cos \frac{k\pi x}{L} \, dx = \begin{cases} 0 & \text{if } m \neq k \\ L & \text{if } m = k \end{cases}$$

and

$$\int_{-L}^{L} \sin \frac{m\pi x}{L} \cos \frac{k\pi x}{L} \, dx = 0$$

Thus, it follows that all the terms on the right-hand side of equation $(**)$ are, with one exception, zero. From this we see that

$$a_k = \frac{1}{L} \int_{-L}^{L} f(x) \cos \frac{k\pi x}{L} \, dx, \ k = 1, 2, 3, \ldots$$

Finally, we will find the values of b_k by using a similar technique. This time we will multiply both sides of (*) by $\sin \dfrac{k\pi x}{L}$ for $k = 1, 2, 3, \ldots$ and integrate from $-L$ to L. When we do this, we find that

$$b_k = \frac{1}{L} \int_{-L}^{L} f(x) \sin \frac{k\pi x}{L} \, dx, \ k = 1, 2, 3, \ldots$$

In summary, we have the following facts about the Fourier series.

Fourier Series

The Fourier series for a function $f(x)$ of period $2L$ is given by

$$f(x) = a_0 + a_1 \cos \frac{\pi x}{L} + a_2 \cos \frac{2\pi x}{L} + \cdots$$
$$+ b_1 \sin \frac{\pi x}{L} + b_2 \sin \frac{2\pi x}{L} + \cdots$$

where the coefficients are given by

$$a_0 = \frac{1}{2L} \int_{-L}^{L} f(x) \, dx$$
$$a_k = \frac{1}{L} \int_{-L}^{L} f(x) \cos \frac{k\pi x}{L} \, dx$$
$$b_k = \frac{1}{L} \int_{-L}^{L} f(x) \sin \frac{k\pi x}{L} \, dx$$

EXAMPLE 31.22

Find a Fourier series for the function

$$f(x) = \begin{cases} -1, & -2 < x < 0 \\ 1, & 0 < x < 2 \end{cases}$$

with a period of 4.

Solution The graph of this function is given in Figure 31.1. From this graph you can see why this is often called a "square wave." Since the period is 4, we have $2L = 4$ or $L = 2$.

$$a_0 = \frac{1}{4}\int_{-2}^{2} f(x)\,dx = \frac{1}{4}\int_{-2}^{0}(-1)\,dx + \frac{1}{4}\int_{0}^{2}(1)\,dx$$

$$= \frac{1}{4}(-x)\Big|_{-2}^{0} + \frac{1}{4}x\Big|_{0}^{2}$$

$$= \frac{1}{4}(0-2) + \frac{1}{4}(2-0) = 0$$

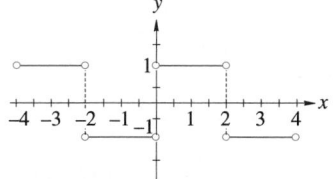

FIGURE 31.1

For $n \ge 1$, we have

$$a_n = \frac{1}{2}\int_{-2}^{2} f(x)\cos\frac{n\pi x}{2}\,dx$$

$$= \frac{1}{2}\int_{-2}^{0}\left(-\cos\frac{n\pi x}{2}\right)dx + \frac{1}{2}\int_{0}^{2}\cos\frac{n\pi x}{2}\,dx$$

$$= \frac{1}{2}\left(\frac{-2}{\pi n}\right)\sin\frac{n\pi x}{2}\Big|_{-2}^{0} + \frac{1}{2}\left(\frac{2}{\pi n}\right)\sin\frac{n\pi x}{2}\Big|_{0}^{2}$$

$$= 0 + \frac{1}{\pi n}\sin(-n\pi) + \frac{1}{\pi n}\sin(n\pi) - 0$$

$$= 0$$

since the sine of a multiple of π is 0.

Finally, for $n \ge 1$, we have

$$b_n = \frac{1}{2}\int_{-2}^{2} f(x)\sin\frac{n\pi x}{2}\,dx$$

$$= \frac{1}{2}\int_{-2}^{0}\left(-\sin\frac{n\pi x}{2}\right)dx + \frac{1}{2}\int_{0}^{2}\sin\frac{n\pi x}{2}\,dx$$

$$= \frac{1}{2}\left(\frac{2}{n\pi}\right)\cos\frac{n\pi x}{2}\Big|_{-2}^{0} + \frac{1}{2}\left(\frac{2}{n\pi}\right)\left(-\cos\frac{n\pi x}{2}\right)\Big|_{0}^{2}$$

$$= \frac{1}{n\pi}[\cos 0 - \cos(-n\pi)] - \frac{1}{n\pi}[\cos(n\pi) - \cos 0]$$

$$= \frac{1}{n\pi}(1 - \cos n\pi) - \frac{1}{n\pi}(\cos n\pi - 1)$$

$$= \frac{2}{n\pi}(1 - \cos n\pi)$$

$$= \begin{cases} 0 & \text{, if } n \text{ is even} \\ \dfrac{4}{n\pi} & \text{, if } n \text{ is odd} \end{cases}$$

EXAMPLE 31.22 (Cont.)

Substituting these values for the coefficients into the series, we get

$$f(x) = \frac{4}{\pi} \sin \frac{\pi x}{2} + \frac{4}{3\pi} \sin \frac{3\pi x}{2} + \frac{4}{5\pi} \sin \frac{5\pi x}{2} + \cdots$$

$$= \frac{4}{\pi} \left(\sin \frac{\pi x}{2} + \frac{1}{3} \sin \frac{3\pi x}{2} + \frac{1}{5} \sin \frac{5\pi x}{2} + \cdots \right)$$

$$= \frac{4}{\pi} \sum_{k=1}^{\infty} \frac{1}{2k-1} \sin \frac{(2k-1)\pi x}{2}$$

It is curious that a series of sine and cosine functions are used to approximate a graph of straight lines. But, if we graph the first term

$$\frac{4}{\pi} \sin \frac{\pi x}{2}$$

we get the graph in Figure 31.2a.

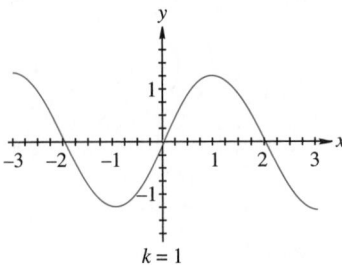

FIGURE 31.2a   FIGURE 31.2b

If we graph the first two terms of this series, $\frac{4}{\pi} \left(\sin \frac{\pi x}{2} + \frac{1}{3} \sin \frac{3\pi x}{2} \right)$, we obtain the graph in Figure 31.2b. The first three terms give the graph in Figure 31.2c. Graphs for the first 4, 5, 10, 20, and 40 terms are shown in Figure 31.2d–h, respectively.

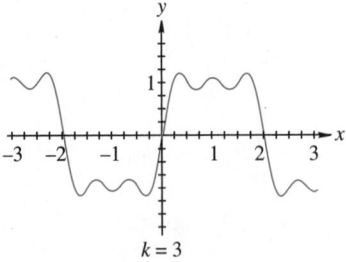

FIGURE 31.2c   FIGURE 31.2d

As you can see, with each additional term the graph gets closer to the graph of the actual function. Notice that at the points of discontinuity, the graph passes through the point midway between the ordinates −1 and 1.

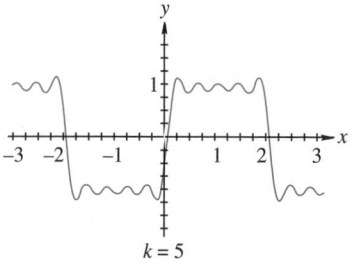

$k = 5$

FIGURE 31.2e

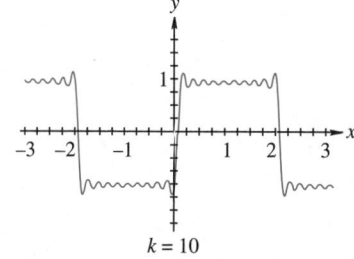

$k = 10$

FIGURE 31.2f

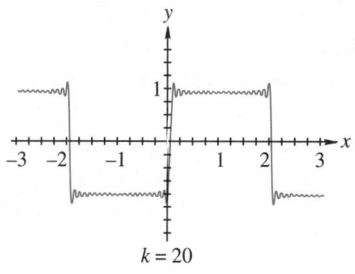

$k = 20$

FIGURE 31.2g

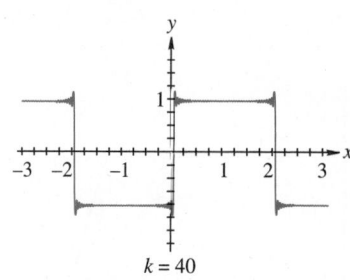

$k = 40$

FIGURE 31.2h

Application

EXAMPLE 31.23

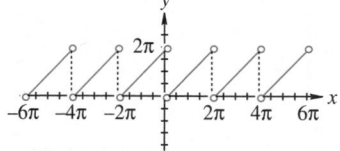

FIGURE 31.3

The electron beam that forms the picture on a television or computer screen is controlled basically by sawtooth voltages and currents. Find a Fourier series for the sawtooth function $f(x) = x$, $0 < x < 2\pi$, with a period of 2π.

Solution   The graph of this function is the saw tooth function shown in Figure 31.3. Since the period is 2π, we have $2L = 2\pi$ or $L = \pi$.

$$a_0 = \frac{1}{2\pi} \int_0^{2\pi} x \, dx = \pi$$

If $n \geq 1$, $a_n = \frac{1}{\pi} \int_0^{2\pi} x \cos \frac{n\pi x}{\pi} \, dx$

$$= \frac{1}{\pi} \int_0^{2\pi} x \cos nx \, dx$$

$$= \frac{1}{\pi} \left(\frac{\cos nx}{n^2} + \frac{x \sin x}{n} \right) \Big|_0^{2\pi}$$

EXAMPLE 31.23 (Cont.)

$$= \frac{1}{\pi} \left[\left(\frac{1}{n^2} + 0 \right) - \left(\frac{1}{n^2} + 0 \right) \right] = 0$$

$$b_n = \frac{1}{\pi} \int_0^{2\pi} x \sin \frac{n\pi x}{\pi} \, dx$$

$$= \frac{1}{\pi} \int_0^{2\pi} x \sin nx \, dx$$

$$= \frac{1}{\pi} \left(\frac{\sin nx}{n^2} - \frac{x \cos nx}{n} \right) \Big|_0^{2\pi}$$

$$= \frac{1}{\pi} \left[\left(0 - \frac{2\pi}{n} \right) - (0 - 0) \right] = \frac{-2}{n}$$

Thus, the Fourier series is

$$\pi - 2 \left(\sin x + \frac{\sin 2x}{2} + \frac{\sin 3x}{3} + \cdots \right)$$

The graph of the first three terms of this Fourier expansion is given in Figure 31.4a and of the first five terms is shown in Figure 31.4b. The graphs in Figure 31.4c–e, are for the first 10, 20, and 40 terms, as indicated. Again, you can see that the larger number of terms gives a better approximation.

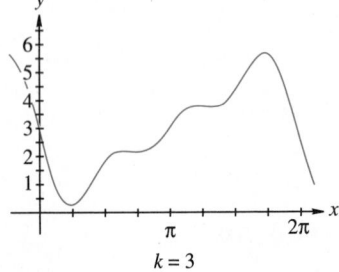

FIGURE 31.4a

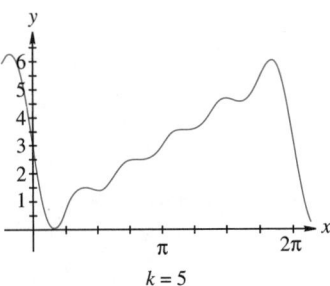

FIGURE 31.4b

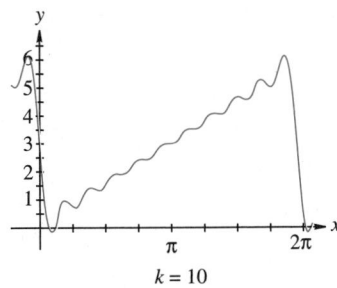

FIGURE 31.4c

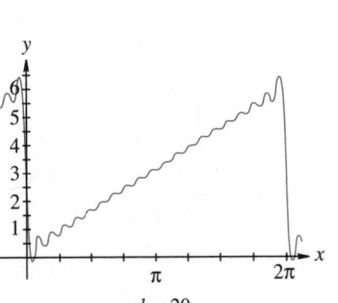

FIGURE 31.4d

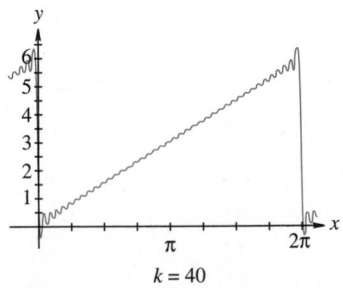

FIGURE 31.4e

Exercise Set 31.5

In Exercises 1–10, obtain Fourier series expansions for the function defined in the indicated interval and with the indicated period outside of the interval. Graph the function.

1. $f(x) = \begin{cases} 2, & -\pi < x < 0 \\ -2, & 0 < x < \pi \end{cases}$, period 2π

2. $g(x) = 1, -2 < x < 2$, period 4

3. $h(x) = \begin{cases} 0, & -3 < x < 0 \\ 2, & 0 < x < 3 \end{cases}$, period 6

4. $j(x) = \begin{cases} 0, & -1 < x < 0 \\ x, & 0 < x < 1 \end{cases}$, period 2

5. $f(x) = \begin{cases} -x, & -\pi < x < 0 \\ x, & 0 < x < \pi \end{cases}$, period 2π

6. $g(x) = \begin{cases} 3, & 0 < x < 4 \\ -3, & 4 < x < 8 \end{cases}$, period 8

7. $h(x) = x, -\pi < x < \pi$, period 2π

8. $j(x) = \begin{cases} 0, & -\frac{\pi}{2} < x < 0 \\ 2, & 0 < x < \frac{\pi}{2} \end{cases}$, period π

9. $f(x) = \begin{cases} 1, & 0 < x < \frac{2\pi}{3} \\ 0, & \frac{2\pi}{3} < x < \frac{4\pi}{3}, \\ -1, & \frac{4\pi}{3} < x < 2\pi \end{cases}$ period 2π

10. $g(x) = \begin{cases} x, & 0 < x \le \pi \\ \pi, & \pi < x < 2\pi \end{cases}$, period 2π

Solve Exercises 11–12.

11. *Electronics* Some electronic devices allow current to pass through in only one direction. When an alternating current is used with one of these devices, the current exists for only half the cycle. The equation for such a **half-wave rectifier** is

$$h(x) = \begin{cases} 0, & -\pi < x < 0 \\ \sin x, & 0 < x < \pi \end{cases}, \text{ period } 2\pi$$

Obtain the Fourier series expansion and graph this function.

12. *Electronics* A **full-wave rectifier** is described by

$$f(x) = \begin{cases} \sin x, & 0 < x < \pi \\ -\sin x, & \pi < x < 2\pi \end{cases}, \text{ period } 2\pi$$

Obtain the Fourier series expansion and graph this function.

 **In Your Words**

13. Without looking in the text, describe a Fourier series.

14. How does a Fourier series differ from either a Maclaurin or a Taylor series?

≡ CHAPTER 31 REVIEW

Important Terms and Concepts

Alternating series
Euler's identity
Fourier series
Infinite series
Maclaurin series

Power series
Series
   Numerical techniques
Taylor series

Review Exercises

In Exercises 1–8, find the Maclaurin series for $f(x)$.

1. $f(x) = \begin{cases} \dfrac{1-\cos x}{x}, & x \neq 0 \\ 0, & x = 0 \end{cases}$

2. $f(x) = xe^{-2x}$

3. $f(x) = \sin x \cos x$

4. $f(x) = \ln(2+x)$

5. $f(x) = (1+x)^{3/4}$

6. $f(x) = e^{-x^4}$

7. $f(x) = \sin x^2$

8. $f(x) = \dfrac{1-e^x}{x}$

Solve Exercises 9–12.

9. Evaluate $\displaystyle\int_0^1 \frac{\sin x}{\sqrt{x}}\,dx$ accurately to 0.0001.

10. Evaluate $\displaystyle\int_0^1 x^3 e^{-x^3}\,dx$ accurately to 0.0001.

11. Evaluate $e^{-0.25}$ accurately to 0.0001.

12. Evaluate $\displaystyle\int_0^1 \cos x^4\,dx$ accurately to 0.0001.

In Exercises 13–16, find the Taylor series expansion about the given value of c.

13. $\cos x$, $c = \frac{\pi}{6}$

14. $\sinh x$, $c = \ln 2$

15. $\sqrt{x}$, $c = 9$

16. $\ln x$, $c = 1$

Use the appropriate Taylor series expansions from the previous four problems to evaluate each of the following.

17. $\cos 29°$

18. $\sinh(\ln 1.9)$

19. $\sqrt{9.1}$

20. $\ln(1.1)$

In Exercises 21–10, determine the Fourier series for the given function. Graph the function.

21. $f(x) = \begin{cases} 1, & -\pi < x < 0 \\ 0, & 0 < x < \pi \end{cases}$, period 2π

22. $g(x) = \begin{cases} 1, & -\pi < x < 0 \\ 2, & 0 < x < \pi \end{cases}$, period 2π

23. $h(x) = x^2$, $-\pi < x < \pi$, period 2π

24. $j(x) = 4 - x^2$, $-\pi < x < \pi$, period 2π

≡ CHAPTER 31 TEST

1. Find the Maclaurin series for $f(x) = \sin x^2$.

2. Find the Maclaurin series expansion for $g(x) = \cos\sqrt{x}$.

3. Find the Taylor series expansion of $h(x) = \dfrac{1}{x}$ about $c = 2$.

4. Use your Taylor series expansion from Exercises 3 to evaluate $\frac{1}{2.1}$.

5. Find the Taylor series expansion of $j(x) = \sqrt[3]{x}$ about $c = -1$.

6. Use your Taylor series expansion from Exercises 5 to evaluate $\sqrt[3]{-0.9}$.

7. Determine the Fourier series for $f(x) = x$, $0 < x < 2\pi$, period 2π.

8. Determine the Fourier series for

$$g(x) = \begin{cases} \cos x, & 0 < x < \pi \\ -\cos x, & -\pi < x < 0 \end{cases}, \quad \text{period } 2\pi.$$

9. Evaluate $\displaystyle\int_0^1 \cos x^3 \, dx$ accurately to 0.0001.

10. The charge q, in microcoulombs, on a capacitor is given by

$$q = \int_0^{0.15} e^{-0.30t^2} \, dt$$

Use a Maclaurin series to approximate this integral.

CHAPTER

32

First-Order Differential Equations

*Building
ventilation
systems can
cause problems
for the
environment.
The systems must
be designed to
ensure efficiency,
safety, and
comfort. In
Section 32.6, we
will solve
problems related
to efficient
ventilation.*

For the last ten chapters, we have worked with equations that involved differentials or derivatives. In some cases, we were required to find the corresponding functions that would produce these equations. An equation that involves an unknown function and its derivative is known as a **differential equation**.

If the unknown function depends on only one independent variable, then it is an **ordinary differential equation**. When we studied derivatives, we learned that a derivative of a function with two or more independent variables is a partial derivative. Similarly, a differential equation in which the unknown function involves two or more independent variables is called a **partial differential equation**.

Differential equations are often used in science and technology. They occur in physics, chemistry, and biology. In this chapter, we will examine certain methods for solving differential equations. We will also show how solving differential equations can help solve applied problems.

≡ 32.1
SOLUTIONS OF DIFFERENTIAL EQUATIONS

The basic types of differential equation that we will consider are those that involve first and second derivatives. As we mentioned in the chapter introduction, differential equations are often divided into ordinary and partial differential equations. Differential equations are also classified by their order and degree.

Order and Degree

The **order** of a differential equation is the order of the highest derivative appearing in the equation. If an equation contains only first derivatives, it is called a **first-order** differential equation. If the equation contains second derivatives and no other higher-order derivatives, it is a **second-order** differential equation. A second-order differential equation can have both first and second derivatives or just second derivatives. The **degree** of a differential equation that can be written as a polynomial is the highest power of the highest-order derivative.

EXAMPLE 32.1

What are the order and degree of each of the following differential equations?

(a) $\dfrac{dy}{dx} = 9x - 7 + y^2$

(b) $\dfrac{d^2y}{dx^2} + 5\left(\dfrac{dy}{dx}\right)^3 = 2$

(c) $6\dfrac{d^3y}{dx^3} + (\cos x)\dfrac{d^2y}{dx^2} + 5x^2y - 3 = y$

(d) $\left(\dfrac{d^2y}{dx^2}\right)^3 + 2y\left(\dfrac{dy}{dx}\right)^4 + y^4\left(\dfrac{dy}{dx}\right)^5 = x$

Solutions   Equation (a) is a first-order differential equation.

Equations (b) and (d) are both second-order differential equations. Equation (c) is a third-order differential equation. Note that in equation (d), the highest derivative is the second, so this is a second-order differential equation even though its first derivatives are raised to the fourth and fifth power and the second derivative is raised to the third power.

Equations (a), (b), and (c) are all of degree one, since in each case the highest-order derivative is only raised to the first power. In equation (d), the degree is 3, because the highest order derivative, $\dfrac{d^2y}{dx^2}$, is raised to the third power.

Caution

Don't confuse order and degree. The symbols $\dfrac{d^2y}{dx^2}$ and $\left(\dfrac{dy}{dx}\right)^2$ do not mean the same thing. The equation $\dfrac{d^2y}{dx^2}=0$ has order 2 and degree 1; the equation $\left(\dfrac{dy}{dx}\right)^2=0$ has order 1 and degree 2.

Solutions of Differential Equations

A **solution** of a differential equation is a relation between the variables that satisfy the equation. This means that when you substitute the solution into the differential equation, you get an algebraic identity. A **particular solution** of a differential equation is any single, or one, solution. The **general solution** of a differential equation is the set of all solutions that satisfy the equation.

EXAMPLE 32.2

Is the equation $y = A\sin 3x + B\cos 3x$, where A and B are constants, a general solution of $y'' + 9y = 0$? What are some particular solutions?

Solution Since $y = A\sin 3x + B\cos 3x$, we need to find y'' and substitute it into the given differential equation.

$$y' = 3A\cos 3x - 3B\sin 3x$$
$$y'' = -9A\sin 3x - 9B\cos 3x$$

Substituting for y'' and y in the given equation, we obtain

$$y'' + 9y = (-9A\sin 3x - 9B\cos 3x) + 9(A\sin 3x + B\cos 3x) = 0$$

and so this is a general solution of the given differential equation.

We can get some particular solutions by substituting values for A and B. If $A = 0$ and $B = 2$, then $y = 2\cos 3x$ is a particular solution of the given equation. If we let $A = 1$ and $B = -4$, then $y = \sin 3x - 4\cos 3x$ is another particular solution.

▪

EXAMPLE 32.3

Show that $y = 2e^{-2x} + xe^{-2x}$ is a solution of the equation $y + 2y' + y'' = xe^{-2x}$.

Solution Differentiating, we get

$$y' = -4e^{-2x} + e^{-2x} - 2xe^{-2x}$$
$$= -3e^{-2x} - 2xe^{-2x}$$
$$y'' = 6e^{-2x} - 2e^{-2x} + 4xe^{-2x}$$
$$= 4e^{-2x} + 4xe^{-2x}$$

EXAMPLE 32.3 (Cont.)

Substituting for y, y', and y'' in the given equation produces

$$y + 2y' + y'' = (2e^{-2x} + xe^{-2x}) + 2(-3e^{-2x} - 2xe^{-2x})$$
$$+ (4e^{-2x} + 4xe^{-2x})$$
$$= 2e^{-2x} + xe^{-2x} - 6e^{-2x} - 4xe^{-2x}$$
$$+ 4e^{-2x} + 4xe^{-2x}$$
$$= (2 - 6 + 4)e^{-2x} + (1 - 4 + 4)xe^{-2x}$$
$$= 0e^{-2x} + xe^{-2x}$$
$$= xe^{-2x}$$

So, $y = 2e^{-2x} + xe^{-2x}$ is a particular solution of $y + 2y' + y'' = xe^{-2x}$.

Earlier we solved some elementary differential equations. For example, when we were looking at some applications of derivatives, we found that if we integrated the acceleration function we got the general solution of the velocity. If we were given the velocity at some particular time, then we were able to find a particular solution. If we integrated the velocity function, we found the general solution for the distance.

EXAMPLE 32.4

Find the function y, if $y' = 6x^2$ and $y(0) = 5$.
Solution Since $y' = 6x^2$, then to find y we need to integrate y'.

$$y = \int 6x^2\,dx = 2x^3 + C$$

We are told that when $x = 0$, $y = 5$. Substituting, we see that $y = 2(0)^3 + C = 5$, and so $C = 5$. The particular solution is $y = 2x^3 + 5$.

Notice that we were given the value of y for a specific value of x. This allowed us to find the particular solution to this differential equation. A differential equation, such as the one in Example 32.4, along with some conditions on the unknown function and its derivatives at the same value of the independent variable is called an **initial value problem**. The conditions are called **initial conditions**. If the conditions are given at more than one value of the independent variable, the problem is a **boundary value problem** and the conditions are called **boundary conditions**.

EXAMPLE 32.5

Solve the boundary value problem $y'' = 4x$, $y(0) = -2$, $y'(1) = 3$.
Solution Since $y'' = 4x$, we can find y' by integrating.

$$y' = \int 4x\,dx = 2x^2 + C_1$$

EXAMPLE 32.5 (Cont.)

We know $y'(1) = 2(1)^2 + C_1 = 2 + C_1 = 3$, so $C_1 = 1$. Thus, we have determined that $y' = 2x^2 + 1$.

Integrating y' in order to find y, we have

$$y = \int y' \, dx = \int \left(2x^2 + 1 \right) dx$$
$$= \frac{2}{3}x^3 + x + C_2$$

Since $y(0) = \frac{2}{3}(0)^3 + 0 + C_2 = -2$, we know $C_2 = -2$, thus the desired solution is $y = \frac{2}{3}x^3 + x - 2$.

Application

EXAMPLE 32.6

A driver in a car traveling 60 mph applies the brakes so that the car's braking system produces a maximum deceleration of 32 ft/s². If the car maintains that deceleration, **(a)** how long will it take the car to stop and **(b)** how far will it travel before coming to a complete stop?

Solutions   We will let s represent the number of feet the car has traveled since the brakes were applied. The velocity of the car at t seconds after the brakes were applied is v and the acceleration a is -32 ft/s². (Notice that the deceleration is really a negative acceleration.) So,

$$a(t) = \frac{dv}{dt} = \frac{d^2s}{dt^2} = -32 \tag{1}$$

Integrating equation (1) produces

$$v(t) = \frac{ds}{dt} = -32t + C_1 \tag{2}$$

We were given $v(0) = 60$ mph, so

$$v(0) = -32(0) + C_1 = 60$$

Thus, $C_1 = 60$ mph. Because the acceleration is in ft/s², we need to convert C_1 to ft/s.

$$C_1 = \frac{60 \, \text{mi}}{1 \, \text{h}} = \frac{60 \times 5,280 \, \text{ft}}{3,600 \, \text{s}} = 88 \, \text{ft/s}$$

So, equation (2) becomes

$$v(t) = -32t + 88 \tag{3}$$

Integrating equation (3), we obtain

$$s = -16t^2 + 88t + C_2$$

EXAMPLE 32.6 (Cont.)

Now at $t = 0$, the car has not traveled, so $s = 0$, so $C_2 = 0$. Thus, the car traveled

$$s(t) = -16t^2 + 88t \text{ ft} \tag{4}$$

t seconds after the brakes were applied.

(a) In order to determine how long it takes to stop the car, we use equation (3). When the car stops, the velocity is 0. So solving $-32t + 88 = 0$ for t, we see that it takes $t = 2.75$ s to stop the car.

(b) Substituting $t = 2.75$ in equation (4), we determine that the car travels $-16(2.75)^2 + 88(2.75) = -121 + 242 = 121$ ft before it stops. ▪

As you can see, in order to solve a differential equation, we need to find some method of changing the equation so that the terms can be integrated. Earlier, with the acceleration and velocity functions, this was an easy task. Not all differential equations are that easy. In the following section we will spend time learning how to transform differential equations so they can be integrated.

Exercise Set 32.1

In Exercises 1–16, show that the each equation is a solution of the indicated differential equation.

1. Equation: $y = e^x$, differential equation: $y'' - y = 0$

2. Equation: $y = e^{2x} + e^x$, differential equation: $y'' - 4y' + 4y = e^x$

3. Equation: $y = 2x^3$, differential equation: $xy' = 3y$

4. Equation: $y = x^3 + 3$, differential equation: $3y - xy' = 9$

5. Equation: $y = 5\cos x$, differential equation: $y' + y\tan x = 0$

6. Equation: $y = e^x - 1$, differential equation: $y' - y = 1$

7. Equation: $y = x^2 + 3x$, differential equation: $xy' - 2y + 3xy'' - 3x = 0$

8. Equation: $y = 4x^2$, Differential Equation: $xy' = 2y$

9. Equation: $y = 2e^x + 3e^{-x} - 4x$, differential equation: $y'' - y = 4x$

10. Equation: $x^2 - y^2 + 2xy = 9$, differential equation: $y' = \dfrac{y+x}{y-x}$

11. Equation: $y = 9x^2 + 6x - 54$, differential equation: $y''' = 0$

12. Equation: $xy^2 = 7x - 3$, differential equation: $x(y')^2 + 2yy' + xyy'' = 0$

13. Equation: $y = x^2 + 4$, differential equation: $y' = 2x$

14. Equation: $2y = \sin x$, differential equation: $y'' + 5y = 2\sin x$

15. Equation: $y = 5e^{-3x} + 2e^{2x}$, differential equation: $y'' + y' - 6y = 0$

16. Equation: $y = 5\cos x$, differential equation: $y'' + y = 0$

In Exercises 17–20, solve the given initial or boundary value problem for y.

17. $y'' = 6x^2$, $y'(1) = -2$, $y(1) = 3$

18. $y'' = \sin x$, $y'(0) = 2$, $y(\pi) = \pi$

19. $\dfrac{dy}{dx} = \sec^2 x$, $y = -6$ when $x = \frac{\pi}{4}$

20. $\dfrac{d^2y}{dx^2} = e^{-x}$, $y'(0) = 0$, $y(1) = -1$

Solve Exercises 21–32.

21. *Automotive technology*   A driver in a car traveling 65 mph applies the brakes so that the car's braking system produces a maximum deceleration of 30 ft/s². If the car maintains this deceleration, **(a)** how long will it take the car to stop and **(b)** how far will it travel before coming to a complete stop?

22. *Automotive technology*   A driver in a car traveling 70 mph applies the brakes so that the car's braking system produces a maximum deceleration of 28 ft/s². If the car maintains this deceleration, **(a)** how long will it take the car to stop and **(b)** how far will it travel before coming to a complete stop?

23. *Highway safety*   Suppose that a car is sitting at a traffic light 151 ft directly ahead of the point where the brakes are applied on the car in Exercise 21. **(a)** How long after the brakes are applied will the cars hit? **(b)** At what speed with the first car strike this stopped car?

24. *Highway safety*   Suppose a car is sitting at a traffic light 184.5 ft directly ahead of the point where the brakes are applied on the car in Exercise 22. **(a)** How long after the brakes are applied will the cars hit? **(b)** At what speed will the first car strike this stopped car?

25. *Physics*   A ball is thrown upward from a point 192 ft above the ground with an initial velocity of 64 ft/s. **(a)** How long will it take for the ball to hit the ground? **(b)** How fast will it be traveling when it hits the ground?

26. *Physics*   A ball is thrown upward from a point 34.3 m above the ground with an initial velocity of 29.4 m/s. **(a)** How long will it take for the ball to hit the ground? **(b)** How fast will it be traveling when it hits the ground? (Use $g = a(t) = -9.8 \, \text{m/s}^2$.)

27. *Medical technology*   A woman with a headache takes a 500-mg pain-relief tablet. The tablets are metabolized (used up) by her body at a constant rate of 75 mg/h. **(a)** Find the formula for the amount left in her body t h after the pills are taken. **(b)** How long will it take for the entire amount of pain reliever to be metabolized if no more tablets are taken?

28. *Medical technology*   After 3 h, the woman in Exercise 27 still has a headache, so she takes another 500-mg pain-relief tablet. There are still 275 mg of the first tablet left in her system. As before, the tablets are metabolized by her body at a constant rate of 75 mg/h. **(a)** Find the formula for the amount left in her body t h after the first pill is taken. **(b)** How long will it take for the entire amount of pain reliever to be metabolized if no more tablets are taken?

29. *Medical technology*   A patient takes a 5-mg dose of cortisone. The patient's body metabolizes the cortisone at the rate of $\dfrac{1}{2\sqrt{t}}$ mg/h. How long will it take for the entire amount of cortisone to be metabolized if no more is taken?

30. *Environmental science*   The population $N(t)$ at time t of a certain species of animal in a controlled environment is assumed to satisfy the logistics growth law

    $$\frac{dN}{dt} = \frac{1}{250} N(1{,}000 - N)$$

    (a) Show that $N(t) = \dfrac{1{,}000}{1 + Ce^{-4t}}$ is a general solution of this general differential equation.
    (b) If $N(0) = 200$, determine the value of C.

31. *Electricity*   In an electric circuit containing constant resistance R and constant inductance L, current i at time t satisfies the differential equation

    $$L\frac{di}{dt} + Ri = V$$

    where V is the constant voltage. Show that a general solution of this differential equation is $i = \dfrac{V}{R} + ke^{-Rt/L}$, where k is a constant.

32. *Pharmacology*   The number $N(t)$ of bacteria in a culture at time t is assumed to satisfy the differential equation

    $$\frac{dN}{dt} = 100 - 0.5N$$

    Show that $N(t) = 200 - Ce^{-0.5t}$ is a general solution of this differential equation.

33. What is a differential equation? How can you tell if a differential equation is ordinary or partial?

34. What is the difference between the order and degree of a differential equation? Illustrate your answer with a second-order differential equation that is not second-degree and with a second-degree differential equation that is not second order.

35. How are particular and general solutions to differential equations alike and how are they different?

≡ 32.2
SEPARATION OF VARIABLES

There is no one method that will allow us to obtain integrable factors of a differential equation. There are some methods that arise often and we will look at these. The first method is the simplest and, as a result, you should always look to see if you can use the method of **separation of variables**.

In Section 32.1, we discussed the order and degree of a differential equation. If a differential equation is of the first order and first degree, then it can be written in the form $\frac{dy}{dx} = f(x, y)$. This can also be written in the differential form

$$M(x, y)\,dx + N(x, y)\,dy = 0 \tag{*}$$

where $M(x, y)$ and $N(x, y)$ may represent constants, functions of either x or y, or functions of x and y.

EXAMPLE 32.7

Write the differential equation $\frac{dy}{dx} = 4x^2$ in the differential form

$$M(x, y)\,dx + N(x, y)\,dy = 0$$

Solution $\frac{dy}{dx} = 4x^2$ can be written as $4x^2\,dx - dy = 0$ by multiplying both sides by dx and then subtracting dy from both sides. Here $M = 4x^2$ and $N = -1$.

EXAMPLE 32.8

Write the differential equation $\frac{dy}{dx} = \frac{x+3y}{x-y}$ in the differential form $M(x, y)\,dx + N(x, y)\,dy = 0$.

Solution Begin by multiplying both sides by the common denominator $dx(x - y)$. Then simplify this result to get it in the desired form.

$$\frac{dy}{dx} = \frac{x+3y}{x-y}$$
$$dy(x - y) = (x + 3y)\,dx$$
$$(x + 3y)\,dx - (x - y)\,dy = 0$$
$$\text{or as}\quad (x + 3y)\,dx + (y - x)\,dy = 0$$

Here $M = x + 3y$ and $N = y - x$.

In order to find the solution to an equation of the form $M(x, y)\,dx + N(x, y)\,dy = 0$, it is necessary to integrate. But in order to integrate M, it must be a function of x alone. In order to integrate N, it must be only a function of the variable y.

As you can see, we have a problem if M is a function of y or if it is a function of both x and y. We also have a problem if N is a function of x or a function of both x and y. We need to use some algebraic means to rewrite $M(x, y)\,dx + N(x, y)\,dy = 0$ in the form

$$A(x)dx + B(y)dy = 0$$

where A is a function of x alone and B is a function of y alone. If this is possible, then we can integrate A with respect to x and B with respect to y.

Separation of Variables

A first order differential equation is **separable** if it can be written in the form

$$M(x, y)\,dx + N(x, y)\,dy = 0$$

or in the equivalent form

$$A(x)dx + B(y)dy = 0$$

When we write the equation in this way, we have then **separated the variables**.

Hint

In using separation of variables, try to get all the terms involving x and dx on one side of the equation and the terms that involve y and dy on the other side.

EXAMPLE 32.9

Solve $\dfrac{dy}{dx} = \dfrac{2y^2}{x^3}$.

Solution First we rearrange the equation so that the terms involving y and dy are on the left-hand side and the terms involving x and dx are on the right-hand side.

$$\frac{dy}{dx} = \frac{2y^2}{x^3}$$

$$\frac{dy}{y^2} = \frac{2\,dx}{x^3}$$

This can be rewritten as

$$\frac{2\,dx}{x^3} - \frac{dy}{y^2} = 0$$

$$\text{or} \quad 2x^{-3}\,dx - y^{-2}\,dy = 0$$

EXAMPLE 32.9 (Cont.)

This is now in the form $A(x)dx + B(y)dy = 0$ with $A(x) = 2x^{-3}$ and $B(y) = -y^{-2}$. Integrating, we get

$$\int 2x^{-3} dx - \int y^{-2} dy = 0$$

$$-x^{-2} + y^{-1} = C$$

or

$$-\frac{1}{x^2} + \frac{1}{y} = C$$

If we can select a function, $I(x, y)$, that we can multiply by the differential equation so that we can integrate the equation, then this function, $I(x, y)$, is known as the **integrating factor**. We need to select the integrating factor so that we remove the x-terms from the coefficient of dy without introducing any factors of y into the coefficient of dx. Similarly, we need to remove the y-terms from the coefficient of dx without introducing any factors of x into the coefficient of dy.

In Section 32.3, we will learn how to find integrating factors, but for now Examples 32.10–32.14 will show the general idea.

EXAMPLE 32.10

Solve $x(y^2 - 3)dx + y(x^2 + 4)dy = 0$.

Solution The integrating factor is $I(x, y) = \dfrac{1}{(y^2 - 3)(x^2 + 4)}$. Multiplying the differential equation by the integrating factor produces

$$\frac{x(y^2 - 3)}{(y^2 - 3)(x^2 + 4)}dx + \frac{y(x^2 + 4)}{(y^2 - 3)(x^2 + 4)}dy = 0$$

$$\frac{x}{x^2 + 4}dx + \frac{y}{y^2 - 3}dy = 0$$

Now we know $\displaystyle\int \frac{x}{x^2 + 4}dx = \frac{1}{2}\ln(x^2 + 4) + C$ and $\displaystyle\int \frac{y}{y^2 - 3}dy = \frac{1}{2}\ln|y^2 - 3| + C$, so we have

$$\int \frac{x}{x^2 + 4}dx + \int \frac{y}{y^2 - 3}dy = 0$$

$$\frac{1}{2}\ln(x^2 + 4) + \frac{1}{2}\ln|y^2 - 3| = C$$

or, using the properties of logarithms,

$$\ln\left|(x^2 + 4)(y^2 - 3)\right| = 2C$$

If we write the constant of integration $2C$ as $\ln C_1$, then

$$\ln\left|(x^2 + 4)(y^2 - 3)\right| = \ln C_1$$

or

$$(x^2 + 4)(y^2 - 3) = C_1$$

This is a general solution to this differential equation. Selecting different values for C_1 will give different particular solutions. Thus, $(x^2+4)(y^2-3) = -10$ and $(x^2+4)(y^2-3) = e^3$ are two particular solutions to the differential equation in Example 32.10.

EXAMPLE 32.11

Solve the differential equation

$$3y\sin x\, dx + 4y^2\cos x\, dy = 0$$

with the initial condition that $y(0) = 5$.

Solution From the note just before Example 32.10, the integrating factor seems to be $\dfrac{1}{y\cos x}$. Multiplying by the integrating factor, we get

$$\frac{3y\sin x}{y\cos x}\, dx + \frac{4y^2\cos x}{y\cos x}\, dy = 0$$

$$\frac{3\sin x}{\cos x}\, dx + 4y\, dy = 0$$

Integrating, we obtain

$$3\int\frac{\sin x}{\cos x}\, dy + 4\int y\, dy = 0$$

$$-3\ln|\cos x| + 2y^2 = C$$

or $\qquad 3\ln|\sec x| + 2y^2 = C$

or $\qquad \ln|\sec^3 x| + 2y^2 = C$

Now, the initial condition states that when $x = 0$, $y = 5$, so

$$\ln\left|\sec^3(0)\right| + 2(5)^2 = C$$

or $\qquad \ln 1 + 50 = C$

And, since $\ln 1 = 0$, we have $C = 50$. Thus, this particular solution is

$$\ln\left|\sec^3 x\right| + 2y^2 = 50$$

Caution

Don't fall into a trap. Not every differential equation needs to be multiplied by an integrating factor. This is shown in Examples 32.12 and 32.13.

EXAMPLE 32.12

Solve: $\dfrac{dy}{dx} = \dfrac{4x^3}{y^2}$.

Solution Rewriting, we get

$$4x^3\, dx - y^2\, dy = 0$$

EXAMPLE 32.12 (Cont.)

This can be integrated as it is:

$$\int 4x^3\,dx - \int y^2\,dy = 0$$

and so

$$x^4 - \frac{1}{3}y^3 = C$$

EXAMPLE 32.13

Solve: $\dfrac{dy}{dx} = \dfrac{y^2}{x^2+9}$.

Solution   We can rewrite this differential equation as

$$\frac{dy}{y^2} = \frac{dx}{x^2+9}$$

Integrating, we get the general solution

$$-\frac{1}{y} = \frac{1}{3}\arctan\frac{x}{3} + C$$

EXAMPLE 32.14

Solve the differential equation

$$y' = \frac{4y}{x(y-5)}$$

Solution   We need to recognize that $y' = \dfrac{dy}{dx}$, so the differential equation can be written as

$$x(y-5)\,dy = 4y\,dx$$

The integrating factor is $\dfrac{1}{xy}$. Multiplying by the integrating factor, the equation becomes

$$\frac{y-5}{y}\,dy = \frac{4}{x}\,dx$$

Now, $\dfrac{y-5}{y} = 1 - \dfrac{5}{y}$, so we can integrate as follows.

$$\int\left(1-\frac{5}{y}\right)dy = \int\frac{4}{x}\,dx$$

$$y - 5\ln|y| = 4\ln|x| + \ln C$$

or

$$y = \ln\left|y^5\right| + \ln x^4 + \ln C$$

$$y = \ln\left|Cy^5 x^4\right|$$

EXAMPLE 32.14 (Cont.)

This can also be written as

$$e^y = Cy^5x^4$$

$$\text{or} \quad C_1e^y = y^5x^4$$

Application

EXAMPLE 32.15

Newton's law of cooling states that the rate of change of the temperature of an object is proportional to the difference between the temperature of the object, T, and the temperature of the surrounding medium, T_m.

(a) Write a differential equation to express Newton's law of cooling.

(b) Use separation of variables to find the general solution to this differential equation.

(c) Find the particular solution if $T = T_0$ at $t = 0$ s.

(d) The temperature of a turkey was $185°$F when it was removed from the oven and placed on a carving tray in a room with a constant temperature of $72°$F. After 10 min, the temperature of the turkey was $178°$F. Find the temperature 15 min after the turkey was removed from the oven.

Solutions

(a) Here T_m is a constant. If $T > T_m$, then $T - T_m$ is positive. However, T is decreasing, so the rate of change of the temperature is negative. Thus, we get $\dfrac{dT}{dt} = -k(T - T_m)$, where k is a positive constant of proportionality.

(b) Rewriting this differential equation, we get

$$\frac{dT}{T - T_m} + k\,dt = 0$$

Integrating produces

$$\int \frac{dT}{T - T_m} + \int k\,dt = \int 0\,dt$$
$$\ln(T - T_m) + kt = C_1$$
$$\ln(T - T_m) = -kt + C_1$$
$$T - T_m = e^{-kt + C_1} = e^{-kt}e^{C_1} = Ce^{-kt}$$
$$T = T_m + Ce^{-kt}$$

The general solution is $T = T_m + Ce^{-kt}$.

(c) Substituting 0 for t and T_0 for T in the general solution, we obtain $T_0 = T_m + Ce^{-k(0)} = T_m + C$. Solving for C produces $C = T_0 - T_m$. Thus, the particular solution is $T = T_m + (T_0 - T_m)e^{-kt}$.

EXAMPLE 32.15 (Cont.)

(d) Here $T_0 = 185$ and $T_m = 72$. When $t = 10$, we have $T = 178$. These values are substituted into the particular solution, and it is solved for k as follows:

$$178 = 72 + (185 - 72)e^{-k(10)}$$
$$106 = 113e^{-k(10)}$$
$$\frac{106}{113} = e^{-k(10)}$$
$$\ln\left(\frac{106}{113}\right) = -k(10)$$
$$0.00639487 \approx k$$

Using this value for k, the given values of T_0 and T_m, and letting $t = 15$, we obtain

$$T = 72 + 113e^{-0.00639487(15)}$$
$$\approx 72 + 102.66 = 174.66$$

So, 15 min after it has been out of the oven, the turkey is about 174.66°F.

Exercise Set 32.2

In Exercises 1–20, solve the given differential equation.

1. $y' = \dfrac{2x}{y}$

2. $\dfrac{dy}{dx} = \dfrac{2y}{x}$

3. $5x\,dx + 3y^2\,dy = 0$

4. $5y\,dx + 3y^2\,dy = 0$

5. $5y\,dx + 3\,dy = 0$

6. $4y^5\,dx - 3x^2\,dy = 0$

7. $4y^5\,dx - 3x^2y\,dy = 0$

8. $xyy' + \sqrt{1+y^2} = 0$

9. $x^3(y^2+4) + yy' = 0$

10. $x^4(y^3-3)\,dx + y^2(x^5-2)\,dy = 0$

11. $ye^{x^2}\,dy = 2x(y^2+4)\,dx$

12. $e^{x^2}\,dy + x\sqrt{4-y^2}\,dx = 0$

13. $2y + e^{-3x}y' = 0$

14. $\sin^2 y\,dx + \cos^2 x\,dy = 0$

15. $(y^2-4)\cos x\,dx + 2y\sin x\,dy = 0$

16. $(1+x)^2\,y' = 1$

17. $y' = \dfrac{x^2y-y}{y^2+3}$

18. $\dfrac{dy}{dx} = \dfrac{y}{x^2+6x+9}$

19. $\sec 3x\,dy + y^2 e^{\sin 3x}\,dx = 0$

20. $2xy + (1+x^2)\,y' = 0$

In Exercises 21–26, find the particular solution of the given differential equation for the given conditions.

21. $y' + y^2x^3 = 0$, $y = -1$ when $x = 2$

22. $x\,dy + y\,dx = 0$, $y = 3$ when $x = 2$

23. $\dfrac{dy}{dx} = \dfrac{3x+xy^2}{y+x^2y}$, $y = 3$ when $x = 1$

24. $\sin^2 y\,dx + \cos^2 x\,dy = 0$, $y = \frac{3\pi}{4}$ when $x = \frac{\pi}{4}$

25. $y' = 3x^2y - 2y$, $y = 1$ when $x = 1$

26. $\sqrt{1+9x^2}\,dy = y^4x\,dx$, $y = -1$ when $x = 0$

Solve Exercises 27–32.

27. *Food science* A roast is removed from a freezer where the temperature is 10°F and placed in an oven with a constant temperature of 325°F. After one hour in the oven, the temperature of the roast is 75°F. The roast will be ready to take out of the oven when its temperature reaches 145°F.
  (a) Write the particular differential equation to fit this situation.
  (b) How long will it take this roast to cook?

28. *Medical technology* Blood plasma is stored at 40°F. Before the plasma can be used, it must be 90°F. When the plasma is placed in a warming oven at 120°F, it takes 45 min for the plasma to warm to 90°F. Assume that Newton's law of cooling applies and that k is not influenced by T_m.
  (a) Write the particular differential equation to fit this situation.
  (b) How long will it take the plasma to warm to 90°F if the oven temperature is set at 150°F?
  (c) How long will it take the plasma to warm to 90°F if the oven temperature is set at 80°F?

29. *Hydrology* One particularly cold morning, the rate of growth of the thickness of a layer of ice on a certain lake is determined by the differential equation $\dfrac{dx}{dt} = \dfrac{5}{x}$. If $x = 2$ in. at 6:00 A.M., at what time will the ice be 6 in. thick? (Hint: Let $t = 0$ at 6:00 A.M.)

30. *Pharmacology* The amount, A, of a certain drug remaining in a body t hours after an injection decreases at a rate proportional to the amount present.
  (a) Write a differential equation to fit this situation.

  (b) Use separation of variables to find the general solution of this differential equation.
  (c) Find the particular solution if the amount of a certain drug decreases at a rate of 2% per hour and an injection of 20 cc is administered.
  (d) Under the conditions in (c), how much of the drug remains in the body 6 hr after the injection?

31. *Medical technology* An influenza epidemic has spread throughout a city of 500,000 people at a rate proportional to the product of the number of people who have been infected and the number who have not been infected. Suppose that 100 people were initially infected and 750 were infected 10 days later.
  (a) Write a differential equation to fit this situation.
  (b) Use separation of variables to find the general solution of this differential equation.
  (c) Use the given infection data to find the particular solution.
  (d) How many will be infected after 20 days?

32. *Advertising* A company decides to use only television commercials to introduce a new product to a city of 250,000 people. Suppose the rate at which people learn about the new product is proportional to the number who have not yet heard of it. Suppose further, that no one knew about the product at the start of the advertising campaign and after 7 days, 40,000 are aware of the product.
  (a) How many people will know about it after 14 days?
  (b) How long will it take for 125,000 people to become aware of the product?

In Your Words

33. Without looking in the text, write the criteria for telling if a first-order differential equation is separable.
34. What is an integrating factor?

≡ 32.3
INTEGRATING FACTORS

In Section 32.2, we often used an integrating factor to allow us to use the separation-of-variables method. An integrating factor is used when separation of variables cannot be done easily. In this section, we will learn how to tell when we need to use an integrating factor and how to determine that factor.

A first-order differential equation of the form $M(x, y)\,dx + N(x, y)\,dy = 0$ is **exact** if there exist a function $f(x, y)$ whose derivative is equal to the left-hand side of the differential equation. Thus,

$$df(x, y) = M(x, y)\,dx + N(x, y)\,dy$$

The general solution of an exact differential equation is $f(x, y) = C$.

Test for Exactness

A first-order differential equation is exact if it passes the following test.

1. Both $M(x, y)$ and $N(x, y)$ are continuous and have continuous first partial derivatives over the defined intervals.
2. The first-order differential equation $M(x, y)\,dx + N(x, y)\,dy = 0$ is exact if and only if

$$\frac{\partial M(x, y)}{\partial y} = \frac{\partial N(x, y)}{\partial x}$$

EXAMPLE 32.16

Test to see if each of the following differential equations are exact.
(a) $(2xy + 4x^2)\,dx + x^2\,dy = 0$
(b) $2y^2 dx - x^3 dy = 0$
(c) $y' = \dfrac{xy^2 - 4}{9 - x^2 y}$
(d) $3x^2 y\,dx + (x^3 + \sin y)\,dy = 0$

Solutions
(a) We can see that $M(x, y) = 2xy + 4x^2$ and $N(x, y) = x^2$. Since M and N are polynomials, they are continuous and have continuous first derivatives. Taking the partial derivatives, we get

$$\frac{\partial M}{\partial y} = \frac{\partial(2xy + 4x^2)}{\partial y} = 2x$$
$$\frac{\partial N}{\partial x} = \frac{\partial x^2}{\partial x} = 2x$$

Since $\dfrac{\partial M}{\partial y} = \dfrac{\partial N}{\partial x}$, this equation is exact.

(b) Here, $M = 2y^2$ and $N = -x^3$. Again, both M and N are continuous and have first partial derivatives that are continuous. Differentiating, we get $\dfrac{\partial M}{\partial y} = 4y$ and $\dfrac{\partial N}{\partial x} = -3x^2$. Since $\dfrac{\partial M}{\partial y} \neq \dfrac{\partial N}{\partial x}$, this equation is not exact.

EXAMPLE 32.16 (Cont.)

(c) $y' = \dfrac{xy^2 - 4}{9 - x^2 y}$ is not in the correct form. Solving, we obtain $(xy^2 - 4)\,dx - (9 - x^2 y)\,dy = 0$, so $M = xy^2 - 4$ and $N = -(9 - x^2 y) = x^2 y - 9$. Both M and N are continuous and have continuous first partial derivatives. Differentiating produces $\dfrac{\partial M}{\partial y} = 2xy$ and $\dfrac{\partial N}{\partial x} = 2xy$. These partial derivatives are the same, so this differential equation is exact.

(d) We have $M = 3x^2 y$ and $N = x^3 + \sin y$. As before, both M and N are continuous and have continuous first partial derivatives. The partial derivatives are

$$\frac{\partial M}{\partial y} = 3x^2 = \frac{\partial N}{\partial x}$$

so the equation is exact.

Integrating Factors

If an equation is not exact, it can often be made exact by the introduction of an integrating factor. In Section 32.2, we introduced the integrating factor. So far, we have been able to find the integrating factor by inspection and through our experience. In the remainder of this section, we develop some procedures for finding integrating factors. First, we will give a more precise definition.

Integrating Factor

A function $I(x, y)$ is an integrating factor for the differential equation $M(x, y)\,dx + N(x, y)\,dy = 0$ if the equation

$$I(x, y)\,[M(x, y)\,dx + N(x, y)\,dy] = 0$$

is exact.

The following table contains combinations of basic differentials, the integrating factor for each, and the resulting exact differential.

 **Note** Some terms have more than one integrating factor. The integrating factor that should be used will depend on the nature of the problem.

TABLE 32.1

	Group of terms	Integrating factor $I(x, y)$	Exact differential $dg(x, y)$
(1)	$y\,dx + x\,dy$	$\dfrac{1}{xy}$	$d(\ln xy)$
(2)	$x\,dx + y\,dy$	1	$d\left[\dfrac{1}{2}\left(x^2 + y^2\right)\right]$
(3)		$\dfrac{1}{x^2 + y^2}$	$d\left[\dfrac{1}{2}\ln\left(x^2 + y^2\right)\right]$
(4)	$y\,dx - x\,dy$	$-\dfrac{1}{x^2}$	$d\left(\dfrac{y}{x}\right)$
(5)		$\dfrac{1}{y^2}$	$d\left(\dfrac{x}{y}\right)$
(6)		$-\dfrac{1}{xy}$	$d\left(\ln\dfrac{y}{x}\right)$
(7)		$\dfrac{-1}{x^2 + y^2}$	$d\left(\arctan\dfrac{y}{x}\right)$
(8)	$ay\,dx + bx\,dy,\ a,\ b$ constants	$x^{a-1}y^{b-1}$	$d\left(x^a y^b\right)$

EXAMPLE 32.17

Solve the differential equation

$$y\,dx - x\,dy + 4x\,dx = 0$$

Solution This equation contains the combination $y\,dx - x\,dy$. We have four integrating factors in Table 32.1 from which to select. If we use integrating factors (5), (6), or (7), the last term in the differential equation, $4x\,dx$, cannot be integrated. We will use integrating factor (4).

Multiplying the given differential equation by the integrating factor produces

$$-\frac{1}{x^2}(y\,dx - x\,dy + 4x\,dx) = 0$$

$$\frac{x\,dy - y\,dx}{x^2} - \frac{4}{x}\,dx = 0$$

According to Table 32.1, the left-hand combination is the derivative of $\dfrac{y}{x}$. The right-hand combination is the derivative of $4\ln|x|$, and so we have

$$\frac{y}{x} - 4\ln|x| = C$$

or $\qquad \dfrac{y}{x} - \ln x^4 = C$

EXAMPLE 32.18

Solve $y^2 dx - y dx + x dy = 0$.

Solution   Here again, we see the combination of $y dx - x dy$. Rewriting, we get

$$y^2 dx - (y dx - x dy) = 0$$

If we multiply by integrating factor (5) of Table 32.1, we obtain

$$dx - \frac{y dx - x dy}{y^2} = 0$$

Integrating this equation results in

$$x - \frac{x}{y} = C$$

or $\frac{x}{y} = x + C_1$, where $C_1 = -C$. Solving for y, we get the general solution

$$y = \frac{x}{x + C_1}$$

EXAMPLE 32.19

Find the particular solution of the differential equation $x dx + x^2 dx + y dy + y^2 dx = 0$ that satisfies the condition that $y = 3$ when $x = 0$.

Solution   This does not appear to contain any of the terms in Table 32.1. If we rearrange terms, we get

$$(x dx + y dy) + (x^2 + y^2) dx = 0$$

The term in the first group of parentheses is like the second group of terms in Table 32.1. Since the second group in the differential equation has a factor of $x^2 + y^2$, we will use the integrating factor in (3). Multiplying by $I(x, y) = \frac{1}{x^2 + y^2}$, gives

$$\frac{x dx + y dy}{x^2 + y^2} + \frac{x^2 + y^2}{x^2 + y^2} dx = 0$$
$$\frac{x dx + y dy}{x^2 + y^2} + dx = 0$$

Integrating this last equation, we get

$$\frac{1}{2} \ln(x^2 + y^2) + x = C_1$$

or $\ln(x^2 + y^2) = -2x + C_2$, where $C_2 = 2C_1$

Taking the exponential of both sides, the equation becomes

$$e^{\ln(x^2 + y^2)} = e^{-2x + C_2}$$

or $x^2 + y^2 = e^{-2x} e^{C_2}$

Since e^{C_2} is a constant, which we will call C, we can rewrite this solution as

$$x^2 + y^2 = C e^{-2x}$$

EXAMPLE 32.19 (Cont.)

This is the general solution. When $x = 0$ and $y = 3$, we obtain

$$0^2 + 3^2 = Ce^{-2 \cdot 0}$$
$$9 = Ce^0 = C$$

Thus $C = 9$, and the particular solution is

$$x^2 + y^2 = 9e^{-2x}$$

Exercise Set 32.3

In Exercises 1–10, test the differential equation to see if it is exact.

1. $(x^2 - y)\,dx + x\,dy = 0$
2. $(x^2 + y^2)\,dx + xy\,dy = 0$
3. $(x^2 + y^2)\,dx + 2xy\,dy = 0$
4. $x\,dx + \sin y\,dx - \sin x\,dy = 0$
5. $x\,dx + y\cos x\,dx + \sin x\,dy = 0$

6. $e^{2x}\,dy + 2ye^{2x}\,dx - dy = 0$
7. $2xy\,dx + y\,dy + x\,dx + x^2\,dy = 0$
8. $4x^3y^2\,dx + (3x^4y + 5x^4)\,dy = 0$
9. $x\,dx + y\,dy = y\,dx - x\,dy$
10. $(y\sin x + x\cos y)\,dx + \left(\frac{1}{2}y^2\cos x + \sin y\right)dy = 0$

In Exercises 11–26, solve the given differential equation.

11. $y\,dx - x\,dy = 0$
12. $y\,dx - x\,dy - x^3\,dx = 0$
13. $y\,dx - x\,dy - 5y^4\,dy = 0$
14. $y\,dx - x\,dy = x^2y\,dx$
15. $y\,dx + y^2\,dx = x\,dy - x^2\,dx$
16. $y\,dx = x\,dy + 3\,dx$
17. $y\,dx + x\,dy = 3x^3y\,dx$
18. $x\,dx + y\,dy = 3x^2\,dy + 3y^2\,dy$

19. $xy^2\,dx + x^2y\,dy = 0$
20. $4y\,dx - 3x\,dy - x^{-3}y^2\,dy = 0$
21. $2y\,dx + x\,dy = 3x\,dx$
22. $x^2\,dx + 2xy\,dy = y^2\,dx$
23. $3x^2(x^2 + y^2)\,dx + y\,dx = x\,dy$
24. $(x^2 + y^2)\,dx = 4x\,dx + 4y\,dy$
25. $\tan(x^2 + y^2)\,dy = x\,dx + y\,dy$
26. $8x\,dy + 4y\,dx + 9x^{-3}y\,dy = 0$

In Exercises 27–30, find the particular solution to the given differential equation that satisfies the given boundary values.

27. $2x\,dy + 2y\,dx = 3x^3y\,dx$, $y = 3$ when $x = 1$
28. $y\,dx - x\,dy = x^2\,dx$, $y = 4$ when $x = 2$

29. $(x^2 + y + y^2)\,dx = x\,dy$, $y = \frac{\pi}{3}$ when $x = \frac{\pi}{3}$
30. $2y\,dx + 3x\,dy = 0$, $y = -2$ when $x = 2$

 **In Your Words**

31. Without looking in the text, describe how to determine if a first-order differential equation is exact.

32. What is an integrating factor? How do you determine if a function is an integrating factor for a differential equation of the form $M(x, y)\,dx + N(x, y)\,dy = 0$?

≡ 32.4
LINEAR FIRST-ORDER DIFFERENTIAL EQUATIONS

At this point, we have solved differential equations by using either separation of variables or inspection and Table 32.1 to find an integrating factor. In this section, we will look at the linear first-order differential equation, because it is always possible to find an integrating factor for an equation of this type.

Linear First-Order Differential Equation

A **linear first-order differential equation** is of the form

$$dy + P(x)y\,dx = Q(x)\,dx$$

$$\text{or} \qquad \frac{dy}{dx} + P(x)y = Q(x)$$

and has an integrating factor of

$$I(x, y) = e^{\int P(x)\,dx}$$

Notice that P and Q are functions of x only, and that this integrating factor does not depend on y.

EXAMPLE 32.20

Find an integrating factor for $dy = 2xy\,dx + x\,dx$.

Solution   We need to write this in the form $dy + P(x)y\,dx = Q(x)\,dx$. Since the term with $P(x)$ also has a factor of y, we will subtract $2xy\,dx$ from (or add $-2xy\,dx$ to) both sides. When we do, we get

$$dy + (-2xy\,dx) = x\,dx$$

which is in the desired form. We can see that $P(x) = -2x$, so the integrating factor is

$$I(x, y) = e^{\int -2x\,dx} = e^{-x^2}$$

The purpose of getting the integrating factor $I(x, y)$ is to multiply the original equation by $I(x, y)$ in order to get an exact differential equation. The left-hand side of the exact differential equation is of the form $\dfrac{du}{dx}$, where $u = yI(x, y) = ye\displaystyle\int P(x)\,dx$. The right-hand side of the exact differential equation is a function of x only, and can be integrated. Remember that when both sides are integrated, $\displaystyle\int \frac{du}{dx}\,dx = u$. Thus, we can make the following statement about the solution of a linear first-order differential equation.

Solution of a Linear First-Order Differential Equation

The solution of a linear first-order differential equation of the form

$$dy + P(x)y\,dx = Q(x)\,dx$$

$$\text{or} \quad \frac{dy}{dx} + P(x)y = Q(x)$$

is

$$ye^{\int P(x)\,dx} = \int Qe^{\int P(x)\,dx}\,dx + C$$

EXAMPLE 32.21

Solve the differential equation in Example 32.20.

Solution The given differential equation was rewritten in the form

$$dy + (-2xy\,dx) = x\,dx$$

with an integrating factor of e^{-x^2}. Multiplying by e^{-x^2}, the equation becomes

$$e^{-x^2}\,dy - 2xye^{-x^2}\,dx = xe^{-x^2}\,dx$$

The left-hand side is the derivative of ye^{-x^2}, so we have

$$\frac{d}{dx}\left(ye^{-x^2}\right) = xe^{-x^2}\,dx$$

Integrating the right-hand side produces

$$\int xe^{-x^2}\,dx = -\frac{1}{2}e^{-x^2} + C$$

so when we integrate the entire equation, we obtain

$$ye^{-x^2} = -\frac{1}{2}e^{-x^2} + C$$

$$\text{or} \quad y = Ce^{x^2} - \frac{1}{2}$$

which is the general solution of this differential equation. ▪

EXAMPLE 32.22

Solve the differential equation $y' - 4y = xe^{4x}$.

Solution Since $y' = \dfrac{dy}{dx}$, this is in the form $\dfrac{dy}{dx} + P(x)y = Q(x)$ with $P(x) = -4$.
Thus, $I(x, y) = e^{\int -4\,dx} = e^{-4x}$. Multiplying by I, we get

$$y'e^{-4x} - 4ye^{-4x} = xe^{4x}e^{-4x}$$

$$\text{or} \quad \frac{d}{dx}\left(ye^{-4x}\right) = xe^{4x-4x} = x$$

EXAMPLE 32.22 (Cont.)

Integrating, we obtain

$$ye^{-4x} = \int x \, dx = \frac{1}{2}x^2 + C$$

So,

$$y = \left(\frac{1}{2}x^2 + C\right)e^{4x}$$

is the general solution of this differential equation.

EXAMPLE 32.23

Solve $\cot x \, dy - y \, dx = (\cot x)3e^{\sin x} \, dx$.

Solution This is not in the correct form. If the equation is divided by $\cot x$, we get

$$dy - \frac{y}{\cot x} \, dx = 3e^{\sin x} \, dx$$

or $\qquad dy - y \tan x \, dx = 3e^{\sin x} \, dx$

Here, $P(x) = -\tan x$ and so the integrating factor is $e^{\displaystyle -\int \tan x \, dx} = e^{\ln|\cos x|} = \cos x$. Multiplying the given equation by the integrating factor, we obtain

$$\cos x \, dy - y \cos x \tan x \, dx = 3(\cos x)e^{\sin x} \, dx$$

Now $\cos x \tan x = \sin x$, so we can rewrite this equation and then integrate the result in order to find the solution:

$$\cos x \, dy - y \sin x \, dx = 3(\cos x)e^{\sin x} \, dx$$
$$\int \frac{d}{dx}(y \cos x) \, dx = 3 \int (\cos x)e^{\sin x} \, dx$$
$$y \cos x = 3e^{\sin x} + C$$

EXAMPLE 32.24

Find the particular solution of $dy + y \, dx = \sin x \, dx$, if $y = 1$ when $x = 0$.

Solution Here, $P(x) = 1$ and so $I(x, y) = e^{\int dx} = e^x$. Thus,

$$ye^x = \int e^x \sin x \, dx$$

We can integrate the right-hand side by using integration by parts twice, or by using formula 87 in the Table of Integrals in Appendix C. Either way, we should see that

$$\int e^x \sin x \, dx = \frac{1}{2}e^x(\sin x - \cos x) + C$$

EXAMPLE 32.24 (Cont.)

So, a general solution is

$$ye^x = \frac{1}{2}e^x(\sin x - \cos x) + C$$

or $$y = \frac{1}{2}\sin x - \frac{1}{2}\cos x + Ce^{-x}$$

To find the particular solution when $x = 0$ and $y = 1$, we see that

$$1 = \frac{1}{2}\sin(0) - \frac{1}{2}\cos(0) + Ce^{-0}$$

$$= \frac{1}{2}(0) - \frac{1}{2}(1) + C$$

Thus, $C = 1 + \frac{1}{2} = \frac{3}{2}$ and the particular solution is

$$y = \frac{1}{2}\sin x - \frac{1}{2}\cos x + \frac{3}{2}e^{-x}$$

Application

EXAMPLE 32.25

The following model has been proposed for weight loss or gain:

$$\frac{dw}{dt} + 0.005w = \frac{1}{3,500}C$$

where $w(t)$ is a person's weight in pounds after t days of consuming exactly C calories/day. Solve this differential equation for $w(t)$.

Solution We first write the differential equation in the form

$$dw + 0.005w\,dt = \frac{1}{3,500}C\,dt$$

In this example w replaces y and t replaces x in the linear first-order differential equation. Here $P(t) = 0.005$ and $Q(t) = \frac{1}{3,500}C$. This differential equation has an integrating factor of

$$I(t, w) = e^{\int 0.005\,dt}$$

$$= e^{0.005t}$$

Multiplying the (rewritten) given equation by the integrating factor, we obtain

$$e^{0.005t}\,dw + 0.005we^{0.005t}\,dt = \frac{1}{3,500}Ce^{0.005t}\,dt$$

Integrating both side of this equation produces

$$we^{0.005t} = \frac{2}{35}Ce^{0.005t} + k$$

EXAMPLE 32.25 (Cont.)

Multiplying this equation by $e^{-0.005t}$, we get the desired result for w.

$$w = \frac{2}{35}C + ke^{-0.005t}$$

Thus, a person's weight in pounds after t days of consuming exactly C calories/day is given by $w = \frac{2}{35}C + ke^{-0.005t}$, where k is a constant.

Exercise Set 32.4

In Exercises 1–20, solve the given differential equation.

1. $dy + 2xy\,dx = 6x\,dx$

2. $y' + y = 2$

3. $y' - 2y = 4$

4. $dy - 3y\,dx = e^{4x}\,dx$

5. $dy + \dfrac{y}{x}\,dx = x^2\,dx$

6. $y' - 6y = e^x$

7. $y' - 6y = 12x$

8. $dy + x^2y\,dx = x^2\,dx$

9. $\dfrac{dy}{dx} = e^x - \dfrac{y}{x}$

10. $\dfrac{dy}{dx} = 6xy$

11. $y' - \dfrac{1}{x^2}y = \dfrac{1}{x^2}$

12. $y' - 4xy = xe^{x^2}$

13. $y\,dy - 2y^2\,dx = y\sin 2x\,dx$

14. $x\,dy - y\,dx = 4x\,dx$

15. $4x\,dy - y\,dx = 8x\,dx$

16. $\sec x\,dy = (y-1)\,dx$

17. $dy = \sec^2 x(1+y)\,dx$

18. $xy' = y + (x^2-1)^2$

19. $x\,dy + (1-4x)y\,dx = 4x^2e^{4x}\,dx$

20. $x\,dy - y\,dx = x^3\sin x^2\,dx$

In Exercises 21–24, find the particular solution for the given equation and conditions.

21. $dy + \dfrac{2y}{x}\,dx = x\,dx$, $y = 3$ when $x = 2$

22. $dy + 3y\,dx = e^{-2x}\,dx$, $y = 2$ when $x = 0$

23. $y' - \dfrac{2y}{x} = x^2\sin 3x$, $y = \pi^2$ when $x = \pi$

24. $\sin x\,dy + (y\cos x - 1)\,dx = 0$, $y = \pi$ when $x = \frac{\pi}{6}$

Solve Exercises 25 and 26.

25. *Health technology* Use the result in Example 32.25, to determine the following for a person weighing 160 lb who goes on a 2,100 cal/day diet.
   (a) What is the particular solution to the differential equation that meets these conditions?
   (b) How much will this person weigh after 30 days on this diet?
   (c) How long will it take for this person to lose 10 lb?
   (d) Find $\lim\limits_{t\to\infty} w(t)$.
   (e) How do you interpret the results in (d)?

26. *Health technology* Use the result in Example 32.25 to determine the following for a person weighing 200 lb who goes on a 2,100 cal/day diet.
   (a) What is the particular solution to the differential equation that meets these conditions?
   (b) How much will this person weigh after 30 days on this diet?
   (c) How long will it take for this person to lose 10 lb?

In Your Words

27. What is a first-order differential equation? What is its integrating factor?

28. How do you solve a linear first-order differential equation?

■ 32.5
APPLICATIONS

As we stated earlier, differential equations have many applications in science and technology. In this section and in Section 32.6, we will look at some of these applications in such areas as growth and decay, cooling, electric circuits, orthogonal trajectories, falling bodies with air resistance, and dilution. In this section and in Section 32.6, we will do these one area at a time in order to help you to understand them better.

Growth and Decay

Suppose that we have a function $N(t)$ representing the amount (or number) of a substance that is either growing or decaying. $N(t)$ represents the amount that is present at any time t. We will assume that $\dfrac{dN}{dt}$, the rate of change of this substance at any time t, is proportional to the amount of the substance present at that time. Then, we have $\dfrac{dN}{dt} = kN$ or

$$\frac{dN}{dt} - kN = 0$$

where k is the constant of proportionality.

Application

EXAMPLE 32.26

A certain radioactive substance is known to decay at a rate proportional to the amount present. An experiment begins with 50 mg of the material. After 8 h it is observed that only 20 mg remain. **(a)** What is the half-life of the substance? **(b)** How much is left after 24 h?

Solutions We begin by determining a formula that describes the amount of radioactive substance present at a given time. Let $\tilde{N}$ denote the amount of the substance present at time t. In particular, when $t = 0$, we know that $N = 50$ and at $t = 8$, we know that $N = 20$.

We have the differential equation $\dfrac{dN}{dt} - kN = 0$. This differential equation is separable as $\dfrac{1}{N}\,dN - k\,dt = 0$. Integrating, we get

$$\begin{aligned} \ln N - kt &= C_1 \\ \text{or} \qquad \ln N &= kt + C_1 \end{aligned}$$

which can be simplified as

$$N = Ce^{kt}$$

EXAMPLE 32.26 (Cont.)

Since $N = 50$ when $t = 0$, we have

$$50 = Ce^{k \cdot 0}$$

and so $\qquad C = 50$

and the solution is now of the form $N = 50e^{kt}$.

Since $N = 20$ when $t = 8$, we have

$$20 = 50e^{8k}$$

or $\qquad \ln 20 = \ln 50 + \ln e^{8k}$

$$= \ln 50 + 8k$$

Thus, $8k = \ln 20 - \ln 50 = \ln \frac{20}{50}$ and $k = \frac{1}{8} \ln \frac{2}{5} \approx -0.1145$. Substituting, we find that the amount of the substance at any time t is given by

$$N = 50e^{-0.1145t}$$

where t is measured in hours. **(a)** The half-life is the amount of time it takes for $\frac{1}{2}$ of the original mass to decay. The original mass was 50 mg and $\frac{1}{2}$ of that is 25 mg. So, when $N = 25$, we have

$$25 = 50e^{-0.1145t}$$

$$0.5 = e^{-0.1145t}$$

$$-0.1145t = \ln 0.5$$

or $\qquad t = \dfrac{\ln 0.5}{-0.1145} \approx 6.05\,\text{h}$

$$= 6\,\text{h}\,3\,\text{min}$$

The half-life of this radioactive substance is 6 h 3 min. **(b)** The amount after 24 h is found from the formula

$$N = 50e^{-0.1145t}$$

when $t = 24$.

$$N = 50e^{(-0.1145)(24)}$$

$$\approx 3.20\,\text{mg}$$

After 24 h, there will be about 3.20 mg of the substance left.

This technique also works on growth problems, particularly those which involve the growth of bacteria as in Exercises 4–6.

Cooling

Newton's law of cooling states that the time rate of change of the temperature of a body is proportional to the temperature difference between the body and the surrounding medium. Let T represent the temperature of the body and T_m represent the temperature of its surrounding medium. The time rate of change of the temperature of the body is $\dfrac{dT}{dt}$. **Newton's law of cooling** can be written as

$$\frac{dT}{dt} = -k(T - T_m)$$

where k is a positive constant.

Application

EXAMPLE 32.27

A metal object at a temperature of 180°C is placed in a room at a constant temperature of 20°C. After 40 min the temperature of the object is 140°C. Find **(a)** the temperature of the metal after 2 h and **(b)** the time it will take to reach 30°C.

Solutions We begin by developing a formula that describes the temperature of the metal object at any time t. With $T_m = 20°C$, we have

$$\frac{dT}{dt} + kT = 20k$$

This differential equation is linear and its solution is

$$T e^{kt} = 20 e^{kt} + C$$

or $\qquad T = 20 + C e^{-kt}$

Since $T = 180$ when $t = 0$, we have $C = 160$. Substituting this value for C, we get

$$T = 20 + 160 e^{-kt} \qquad (*)$$

At $t = 40\,\text{min}$, we are given $T = 140$, so

$$140 = 20 + 160 e^{-40k}$$

$$\frac{120}{160} = e^{-40k}$$

thus, $\qquad k = -\dfrac{1}{40} \ln\left(\dfrac{120}{160}\right) \approx 0.0072$

Substituting this value for k in equation $(*)$, we obtain

$$T = 20 + 160 e^{-0.0072t}$$

where t is in minutes. We now have the necessary formula to answer the questions. **(a)** Since 2 h is 120 min, the temperature after 2 h is given by

$$T = 20 + 160 e^{(-0.0072)(120)} \approx 87.44°C$$

After 2 h, the temperature has dropped to about 87.44°C. **(b)** We want to find t when $T = 30°C$. So,

$$30 = 20 + 160 e^{-0.0072t}$$

or $\qquad 10 = 160 e^{-0.0072t}$

$$\frac{1}{16} = e^{-0.0072t}$$

and so $\qquad -0.0072t = \ln \frac{1}{16}$

EXAMPLE 32.27 (Cont.)

Solving this for t, we find that

$$t = \frac{\ln 1/16}{-0.0072} \approx 385.08\,\text{min}$$

which is about 6 h 25 min 5 s. So, it will take about 6 h 25 min 5 s for the temperature to drop to 30°C.

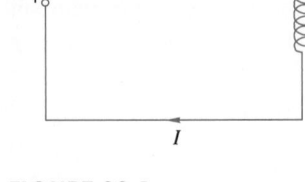

FIGURE 32.1

Electric Circuits

The amount of current I in an RL circuit as shown in Figure 32.1 is governed by the equation

$$\frac{dI}{dt} + \frac{R}{L}I = \frac{E}{L}$$

where R is the resistance in ohms (Ω), L an inductor in henries (H), and an electromotive force (emf) E is in volts (V).

An RC circuit consists of a resistance, a capacitance C in farads (F), an emf, and no inductance, as shown in Figure 32.2. The equation governing the amount of electrical charge q, in coulombs (C), on the capacitor is

$$\frac{dq}{dt} + \frac{1}{RC}q = \frac{E}{R}$$

and the relationship between q and I is

$$I = \frac{dq}{dt}$$

Application

EXAMPLE 32.28

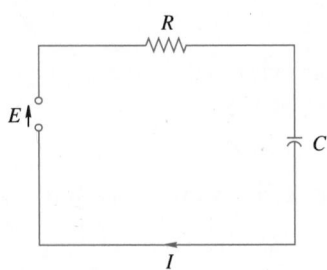

FIGURE 32.2

An RL circuit has an emf of 10 V, a resistance of 75 Ω, an inductance of 1 H, and no initial current. Find the current in the circuit at any time t.

Solution We have $E = 10$, $R = 75$, and $L = 1$, so

$$\frac{dI}{dt} + 75I = 10$$

This equation is linear. Its solution is

$$I = Ce^{-75t} + \frac{10}{75}$$

At $t = 0$, $I = 0$, and so

$$0 = Ce^0 + \frac{10}{75}$$

which means that $C = -\frac{10}{75}$. Thus, the current at any time t is given by

$$I = -\frac{10}{75}e^{-75t} + \frac{10}{75}$$

In the previous equation, the quantity $-\frac{10}{75}e^{-75t}$ is called the **transient current**, since this quantity goes to zero as $t \to \infty$. The quantity $C = \frac{10}{75}$ in the same equation is called the **steady-state current**.

Exercise Set 32.5

Solve Exercises 1–16.

1. *Nuclear physics*   A radioactive substance decays at a rate proportional to the amount present. If there are 100 g initially and 75 g after 15 days, what is the mass at any time t?

2. *Nuclear physics*   What is the half-life of the substance in Exercise 1?

3. *Nuclear physics*   A radioactive substance has a half-life of 1,000 years. How long will it take for 10% of the substance to decay?

4. *Medical technology*   A bacteria culture is known to grow at a rate proportional to the amount present. After 1 h, 2,000 strands of the bacteria are observed in the culture. After 4 h, there are 6,000 strands. Find an expression for the number of strands of the bacteria present in the culture at any time t.

5. *Medical technology*   How many strands were originally in the bacteria culture in Exercise 4?

6. *Medical technology*   How frequently does the bacteria culture in Exercise 4 double?

7. *Archeology*   The half-life of carbon-14 (C-14) is approximately 5,600 years. When an organism dies, its absorption of C-14, by either breathing or eating, stops. Thus, if we compare the proportional amount of C-14 present in a fossil or other object with the constant ratio in the atmosphere, we can get a close estimate of the object's age. If a fossilized bone is found to contain $\frac{1}{500}$ of the original amount of C-14, what is the age of the fossil?

8. *Optics*   When a vertical beam of light passes through a transparent substance, the rate at which its intensity I decreases is proportional to the thickness of the medium in feet. In clear sea water, the intensity 3 ft below the surface of the water is 25% of the initial intensity of the beam. What is the intensity of the beam 18 ft below the surface?

9. *Thermal science*   An object at a temperature of 10°C is placed outdoors where the temperature is 30°C. After 10 min the temperature of the object is 15°C.
   (a) Find a formula for the temperature of the object at time t?
   (b) How long does it take the object to reach a temperature of 22°C?
   (c) What is its temperature after 1 h?

10. *Thermal science*   An object of unknown temperature is placed in a room of constant temperature 40°F. After 10 min, the temperature of the object is 0°F and after 20 min, the temperature is 10°F. What was the initial temperature of the object?

11. *Food science*   A baker removes a cake from a 375°F oven and places it in a 40°F refrigerator to cool. After 15 min, the cake has cooled to 280°F. When the cake cools to 75°F, the icing can be applied. How long will the baker have to wait to apply the icing?

12. *Food science*   An object is removed from a freezer with a temperature of -20°C and placed in a room with temperature 20°C. One minute later the object has warmed up to -16°C. How long will it take the object to reach a temperature of 15°C?

13. *Electricity*   A generator with an emf of 100 V is connected in series with a $10 - \Omega$ resistor and an inductor of 2 H. If the switch is closed at time $t = 0$, determine the current at any time t.

14. *Electricity*   Suppose the 100-V generator in Exercise 13 is replaced with one that has an emf of $20\cos 5t$ V. What is the current at any time t?

15. *Electricity*   An RL circuit has an emf of 5 V, $R = 50\,\Omega$, $L = 1$ H, and no initial current. What is the current in this circuit at any time t?

16. *Electricity*   An RL circuit has an emf of $3\sin 2t$ V, $R = 10\,\Omega$, $L = 0.5$ H, and an initial current of 6 A. What is the current at any time t?

In Your Words

17. Write an application in your technology area of interest that requires you to solve a differential equation that involves growth and decay, cooling, or electric circuits. Give your problem to a classmate and see if he or she understands and can solve your problem using the techniques of this and earlier chapters. Rewrite the problem as necessary to remove any difficulties encountered by your classmate.

18. Write an application in your technology area of interest that requires you to solve a differential equation that involves growth and decay, cooling, or electric circuits. This application should involve a different area than your application in Exercise 17. As before, share your problem with a classmate and rewrite the problem as necessary to remove any difficulties encountered by the classmate.

≡ 32.6
MORE APPLICATIONS

In Section 32.5, we looked at applications of differential equations in the areas of growth and decay, cooling, and electric circuits. In this section, we will look at applications in three-additional areas.

Orthogonal Trajectories

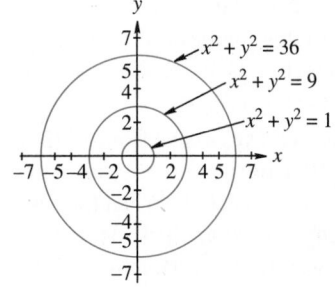

FIGURE 32.3a

A curve that intersects all members of a family of curves at right angles is called an **orthogonal trajectory**. A family of curves is a specified set of curves that satisfy some given conditions.

An example of a family of curves is shown in Figure 32.3a. This is a family of circles that have the origin as their center. The orthogonal trajectories for this family of circles are the family of straight lines through the origin. (Some of the lines are shown as dashed lines in Figure 32.3b). In the same way, the orthogonal trajectories of the family of straight lines passing through the origin are the circles with centers at the origin.

If we have a family of curves of the form $F(x, y, c) = 0$, then to find the orthogonal trajectories we will first implicitly differentiate this equation with respect to x and eliminate the constant c between this derived equation and the original one. We now have an equation in x, y, and y'. Solving this for y', we get an equation of the form

$$\frac{dy}{dx} = f(x, y)$$

The orthogonal trajectories are the solutions of

$$\frac{dy}{dx} = \frac{-1}{f(x, y)}$$

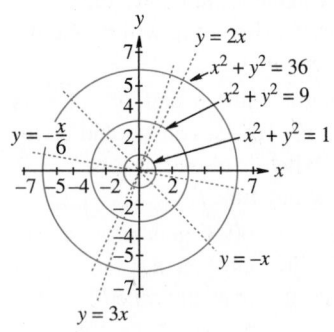

FIGURE 32.3b

Application

EXAMPLE 32.29

Find the orthogonal trajectories for the family of curves formed by the equation $y = cx^2$.

Solution This family is the set of parabolas shown in Figure 32.4a and can be written as $F(x, y, c) = y - cx^2$. It consists of the parabolas symmetric to the y-axis with vertices at the origin. Using implicit differentiation gives the following results.

$$y - cx^2 = 0$$
$$y' - 2cx = 0$$
$$y' = 2cx \quad \text{or} \quad \frac{dy}{dx} = 2cx \qquad (*)$$

Solving $y - cx^2 = 0$ for c, we obtain

$$c = \frac{y}{x^2}$$

Substituting this value for c into $(*)$, we have

$$\frac{dy}{dx} = 2\left(\frac{y}{x^2}\right)x = \frac{2y}{x}$$

Thus, we have $f(x, y) = \dfrac{2y}{x}$. The orthogonal trajectories satisfy the equation $\dfrac{dy}{dx} = \dfrac{-1}{f(x, y)}$, and so

$$\frac{dy}{dx} = \frac{-1}{2y/x} = -\frac{x}{2y}$$
$$\text{or} \quad 2y\,dy + x\,dx = 0$$

The solution of this separable differential equation is

$$y^2 + \frac{1}{2}x^2 = k$$

This is the equation for a family of ellipses. Some members of this family are shown as the dashed lines in Figure 32.4b.

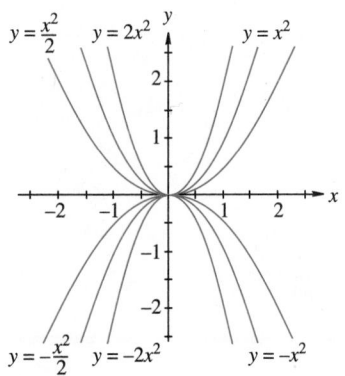

$y = \frac{x^2}{2}$ $y = 2x^2$ $y = x^2$

$y = -\frac{x^2}{2}$ $y = -2x^2$ $y = -x^2$

FIGURE 32.4a

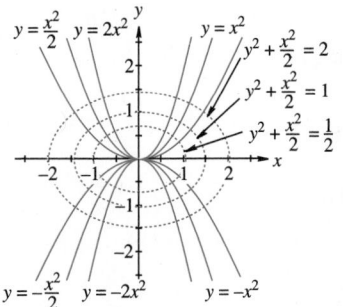

$y = \frac{x^2}{2}$ $y = 2x^2$ $y = x^2$

$y^2 + \frac{x^2}{2} = 2$

$y^2 + \frac{x^2}{2} = 1$

$y^2 + \frac{x^2}{2} = \frac{1}{2}$

$y = -\frac{x^2}{2}$ $y = -2x^2$ $y = -x^2$

FIGURE 32.4b

Falling Bodies with Resistance

The force F acting on a freely falling body in a vacuum is given by $F = mg$ were m is the mass and g is the force of gravity. Most of the time when an object falls it is affected by a resisting medium such as air. The resistance depends on the velocity and the size and shape of the object. At relatively low velocities, the resistance appears to be proportional to the velocity, or kv, where $k > 0$.

Newton's second law of motion states that the net force acting on a body is equal to the time rate of change of the momentum of the body. For an object of constant mass, this means that $F = m\dfrac{dv}{dt}$, where F is the net force of the body and v its velocity, both at time t. Because the resistance opposes the velocity, the net force is $F = mg - kv$. Substituting this into the equation $F = m\dfrac{dv}{dt}$, we obtain $mg - kv = m\dfrac{dv}{dt}$, or

$$\frac{dv}{dt} + \frac{kv}{m} = g$$

Application

EXAMPLE 32.30

A body of mass 10 slugs is dropped from a height of 200 ft with no initial velocity. If the retarding force is equal to 0.2 times the velocity, what is the body's velocity after 5 s?

Solution We have $m = 10$ slugs, $g = 32\,\text{ft/s}^2$, and $k = 0.2$, so

$$\frac{dv}{dt} + \frac{0.2v}{10} = 32$$

or

$$\frac{dv}{dt} = 32 - 0.02v$$

Separating variables, we obtain

$$\frac{dv}{32 - 0.02v} = dt$$

Integrating results in

$$-50\ln(32 - 0.02v) = t + C_1$$

For convenience, we will write $C_1 = -50\ln C$ in order to have

$$-50\ln(32 - 0.02v) = t - 50\ln C$$

or

$$\ln(32 - 0.02v) = -\frac{t}{50} + \ln C$$

Taking the exponential of both side results in

$$32 - 0.02v = Ce^{-t/50}$$

EXAMPLE 32.30 (Cont.)

Now, solving for v, we get

$$v = 50\left(32 - Ce^{-t/50}\right)$$

The object started from rest, so when $t = 0$, $v = 0$. Thus

$$0 = 50(32 - C)$$

and so $C = 32$

As a result, the equation for the velocity is given by

$$v = 50(32 - 32e^{-t/50})$$

or $$v = 1{,}600\left(1 - e^{-t/50}\right)$$

When $t = 5$, we obtain

$$v = 1{,}600\left(1 - e^{-0.1}\right)$$
$$= 1{,}600(1 - 0.9048) = 152.32\,\text{ft/s}$$

So after 5 s, the body's velocity is 152.32 ft/s.

Mixture Problems

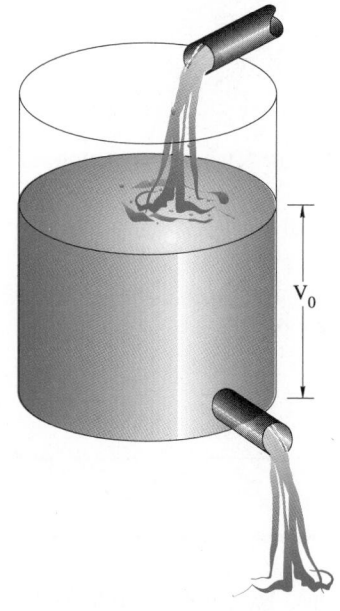

FIGURE 32.5

Suppose you have a tank that originally holds a known quantity V_0 of a solution that contains some quantity a of a dissolved chemical. Another solution containing the quantity b of the same dissolved chemical is added to the tank at the rate of r_1, while at the same time, the well-mixed solution is emptied at the rate r_2, as shown in Figure 32.5. We want to find the quantity of the chemical in the tank at any time t.

If we let $Q(t)$ represent the amount of the chemical at any time, then $\dfrac{dQ}{dt}$ represents the difference between the rate at which the chemical enters the tank and the rate at which it leaves. The chemical enters the tank at the rate of br_1 per unit of time. It leaves at the rate of ar_2 per unit of time. The volume in the tank at any time is $V_0 + r_1 t - r_2 t$, where V_0 represents the volume at $t = 0$. Thus, the concentration of the chemical at any time is $\dfrac{Q}{V_0 + r_1 t - r_2 t}$ and so it leaves at the rate of $r_2\left(\dfrac{Q}{V_0 + r_1 t - r_2 t}\right)$ per unit of time. This means that

$$\frac{dQ}{dt} = br_1 - r_2\left(\frac{Q}{V_0 + r_1 t - r_2 t}\right)$$

or

$$\frac{dQ}{dt} + \frac{r_2}{V_0 + (r_1 - r_2)t}Q = br_1$$

Application

EXAMPLE 32.31

Initially, 50 lb of salt are dissolved into a large tank that holds 300 gal of water. A brine solution of 2 lb/gal is pumped into the tank at the rate of 3 gal/min. A well-stirred solution is pumped out at the same rate. **(a)** How much salt is in the tank at any one time? **(b)** How much salt is there after 1 h?

Solutions

(a) Here we have $r_1 = r_2 = 3$ gal/min, $V_0 = 300$ gal, $a = 50$ lb, and $b = 2$ lb/gal. Thus,

$$\frac{dQ}{dt} + \frac{3}{300 + (3-3)t} Q = 2 \cdot 3$$

$$\frac{dQ}{dt} + \frac{Q}{100} = 6$$

The integrating factor is $e^{t/100}$, and so

$$\frac{d}{dt}\left(e^{t/100} Q\right) = 6e^{t/100}$$

and

$$e^{t/100} Q = 600e^{t/100} + C$$

or

$$Q = 600 + Ce^{-t/100}$$

We are given that $Q = 50$ when $t = 0$, so

$$50 = 600 + C$$

or

$$C = -550$$

Thus, the quantity of salt that is in the tank at any time t, where t is in minutes, is given by

$$Q(t) = 600 - 550e^{-t/100}$$

(b) After 1 h, or when $t = 60$ min, we have

$$Q(60) = 600 - 550e^{-60/100}$$

$$= 600 - 550e^{-0.6}$$

$$\approx 298.154 \text{ lb}$$

Application

EXAMPLE 32.32

An office that measures $15' \times 20' \times 8'$ initially contains 0.1% carbon dioxide. Fresh air containing 0.025% carbon dioxide is pumped into the room at the rate of 100 ft^3/min. The old and new air are well mixed by the ventilation system and the well-mixed air is pumped out at the same rate. **(a)** How much carbon dioxide is in the room at any one time? **(b)** How much carbon dioxide is in the room after 1 h?

EXAMPLE 32.32 (Cont.)

FIGURE 32.6

Solutions

(a) We will let Q represent the amount of carbon dioxide (CO_2) in cubic feet in the room at time t min after beginning to pump fresh air into the room. We are given $r_1 = r_2 = 100\,\text{ft}^3/\text{min}$. The volume of the room is $V_0 = 15 \times 20 \times 8 = 2,400\,\text{ft}^3$, and $a = 0.1\%$ of $2,400 = 0.001 \times 2,400 = 2.4\,\text{ft}^3$ of CO_2. Finally, $br_1 = 0.025\% \times 100 = 0.00025 \times 100 = 0.025\,\text{ft}^3/\text{min}$. Thus, we have the following differential equation.

$$\frac{dQ}{dt} + \frac{100}{2,400 + (100 - 100)t}Q = 0.025$$

$$\frac{dQ}{dt} + \frac{Q}{24} = 0.025$$

$$\frac{dQ}{dt} = 0.025 - \frac{Q}{24}$$

$$= \frac{0.6 - Q}{24}$$

Separating variables, we obtain

$$\frac{dQ}{0.6 - Q} = \frac{dt}{24}$$

EXAMPLE 32.32 (Cont.)

Integration yields

$$-\ln(0.6 - Q) = \frac{1}{24}t + C_1$$

or

$$\ln(0.6 - Q) = -\frac{1}{24}t - C_1$$

So,

$$0.6 - Q = e^{-t/24 - C_1}$$

$$= Ce^{-t/24}$$

We are given that $Q = 2.4\,\text{ft}^3$ when $t = 0$, and so

$$0.6 - 2.4 = C$$

or

$$C = -1.8$$

Thus, the quantity of carbon dioxide in the room at any time t, where t is in minutes, is given by

$$Q(t) = 0.6 + 1.8e^{-t/24}$$

(b) After 1 h, or when $t = 60$ min, we have

$$Q(60) = 0.6 + 1.8e^{-60/24}$$

$$= 0.6 + 1.8e^{-2.5}$$

$$\approx 0.7478\,\text{ft}^3$$

Dividing this by the volume of the room, we find that this amount of CO_2 represents $0.00031 = 0.031\%$ of the air in the room. ▪

Exercise Set 32.6

In Exercises 1–4, find the orthogonal trajectories of the given family of curves. In each case, sketch some of the curves and their orthogonal trajectories. Remember to eliminate the constant from $\dfrac{dy}{dx}$ for the original curves.

1. $x^2 + y^2 = C$ **2.** $xy = C$ **3.** $y = Ce^x$ **4.** $x^2 - y^2 = C$

Solve Exercises 5–16.

5. *Energy science* In heat flow, the family of isothermal curves are the curves that join points of equal temperature. If the family of isothermal curves is given by the ellipses $x^2 + \frac{1}{2}y^2 = C^2$, find the curves along which the heat flows. (In other words, find the orthogonal trajectories of the ellipses.)

6. *Energy science* The lines of equal potential in a field of force are all at right angles to the line of force. In an electric field of force caused by charged particles, the lines of force are given by $x^2 + 2y^2 = C$. Find the equation of the line of equal potential.

7. *Physics* A body of mass of 10 kg (weight 98 N) is dropped from a height of 1000 m with no initial velocity. If the retarding force due to air resistance is 0.2 times the velocity, find a formula for the velocity at time t.

8. *Physics* What is the velocity of the object in Exercise 7 after 10 s?

9. *Physics* If the object in Exercise 7 is dropped with an initial velocity of 1 m/s, find (a) its velocity at any time t, (b) its velocity after 10 s, and (c) if the limiting velocity v_L is defined by $v_L = \frac{mg}{k}$, what is the limiting velocity?

10. *Physics* An object of mass 10 slugs (weight 320 lb) is dropped from a height of 1,000 ft with no initial velocity. The object encounters air resistance proportional to its velocity. The limiting velocity is known to be 320 ft/s. Find (a) the velocity of the object at any time t, (b) the position of the object at any time t, and (c) the time required for the object to reach a velocity of 160 ft/s.

11. *Environmental science* Fifty gallons of brine originally containing 25 lb of salt are in a tank into which 2 gal of water run each minute with the same amount of the well-stirred mixture running out each minute. How much salt remains in the tank after 15 min?

12. *Environmental science* One hundred gallons of brine originally containing 40 lb of salt are in a tank. Water enters the tank at the rate of 5 gal/min. The same amount of mixture from the tank leaves each minute. How much salt is in the tank after (a) 10 min? (b) 30 min? (c) 1 h?

13. *Environmental science* A tank is filled with 8 gal of brine in which 2 lb of salt is dissolved. Brine with a concentration of 3 lb of salt/gal enters the tank at 4 gal/min, and the well-stirred mixture leaves at the same rate. (a) Find the amount of salt at any time t. (b) How much salt is there after 8 min?

14. *Environmental science* A tank has 60 gal of brine with 2 lb of salt per gallon. A solution of 3 lb of salt per gallon enters at 2 gal/min and the well-stirred mixture leaves at the same rate. When will there be 150 lb of salt in the tank?

15. *Wastewater technology* A settling tank contains 10 kg of solid industrial waste dissolved in 200 m³ of water. A solution of 1 kg of solid waste/5 m³ of water is added at a rate of 10 m³/hr and mixed to keep the tank's contents at a uniform concentration. (a) How much waste is in the tank at any time t, where t is in hr? (b) If the contents of the tank is being drained at a rate of 10 m³/hr, how much dissolved solid will the tank contain after 3 hr?

16. *Medical technology* According to the **Gompertzian relation**, a solid tumor grows more slowly with the passage of time. One differential equation that meets these conditions is

$$\frac{dV}{dt} = \lambda e^{-\alpha t} V$$

where $V(t)$ is the volume of dividing cells at time t and λ and α are positive constants. Solve this differential equation.

In Your Words

17. Write an application in your technology area of interest that requires you to solve a differential equation that involves orthogonal trajectories, falling bodies with resistance, or mixture. Give your problem to a classmate and see if he or she understands and can solve your problem using the techniques of this and earlier chapters. Rewrite the problem as necessary to remove any difficulties encountered by your classmate.

18. Write an application in your technology area of interest that requires you to solve a differential equation that involves orthogonal trajectories, falling bodies with resistance, or mixture. This application should involve an area different from your application in Exercise 17. As before, share your problem with a classmate and rewrite the problem as necessary to remove any difficulties encountered by the classmate.

CHAPTER 32 REVIEW

Important Terms and Concepts

Boundary conditions
Boundary value problem
Cooling
Differential equation
 Degree
 Exact
 General solution
 Order
 Ordinary
 Partial
 Particular solution

Electrical circuits
Falling bodies with resistance
Growth and decay
Initial conditions
Initial value problem
Integrating factor
Linear first-order differential equation
Mixture
Newton's law of cooling
Orthogonal trajectories
Separation of variables

Review Exercises

In Exercises 1–16, find the general solution of the given differential equation.

1. $y' = x + y$

2. $x\,dx - y^2\,dy = 0$

3. $y' = y^2 x^3$

4. $y' = 8y$

5. $e^x\,dx - 2y\,dy = 0$

6. $(y^2 - y)\,dx + x\,dy = 0$

7. $(y - xy^3)\,dx + (x - 2x^2y^2)\,dy = 0$

8. $(y+1)\,dx - x\,dy = 0$

9. $y\,dx + (2 - x)\,dy = 0$

10. $(y + x^3 y)\,dx + x\,dy = 0$

11. $y^2\,dx + xy\,dy = 0$

12. $dy = 3y\,dx + 6\,dx$

13. $y' - 4xy = x$

14. $y' + \dfrac{12y}{x} = x^{12}$

15. $y' + y = \sin x$

16. $y' - 7y = e^x$

In Exercises 17–20, find the particular solution of the given differential equation for the given boundary values.

17. $\cos x \, dx + y \, dy = 0$, $y = 2$ when $x = 0$

18. $(x^2 + y + y^2)dx - x \, dy = 0$, $y = \frac{\pi}{3}$ when $x = \frac{\pi}{3}$

19. $y' + \frac{2}{x}y = x$, $y = 0$ when $x = 1$

20. $dy + 6xy \, dx = 0$, $y = 5$ when $x = \pi$

Solve Exercises 21–30.

21. *Nuclear physics*   A certain radioactive substance decays at a rate proportional to the amount present. An experiment finds that after 2 y, 10% of the original mass has decayed. Find **(a)** the expression for the mass at any time t, and **(b)** the half-life of the substance.

22. Find the orthogonal trajectories for the family of curves $y = cx^3$.

23. *Thermal science*   An object at a temperature of $0°F$ is placed in a room with temperature $72°F$. After 15 min, the object has reached a temperature of $20°F$. Find **(a)** the amount of time it will take the object to reach a temperature of $50°F$ and **(b)** the temperature of the object after 1 h.

24. *Electricity*   A 2-H inductor, a 60-Ω resistor, and a 25-V battery are connected in series. Find the current in the circuit at any time t, if the initial current is 0.

25. *Electricity*   An *RC* circuit has an emf of $400\cos 2t$ V, a resistance of $80\,\Omega$, and a capacitance of 10^{-2} F. When $t = 0$, $q = 0$. What is the change in the circuit at any time t?

26. *Physics*   A 192-lb object has a limiting velocity of 16 ft/s when falling in air. The air provides a resisting force proportional to the object's velocity. If $v = 0$ when $t = 0$, what is **(a)** the velocity of the object when $t = 1$ s and **(b)** how long does it take the object to reach a velocity of 15 ft/s?

27. *Environmental science*   A room of volume 2,000 ft^3 contains 0.1% carbon dioxide. Fresh air containing 0.035% carbon dioxide is pumped into the room at the rate of 100 ft^3/min and well-mixed air is pumped out at the same rate. **(a)** How much carbon dioxide

is in the room at any one time? **(b)** How much carbon dioxide is in the room after 1 h? **(c)** How much carbon dioxide is in the room after 3 h?

28. *Environmental science*   A building of volume 25,000 ft^3 contains 0.01% radon. Fresh air containing 0.0005% radon enters the building at the rate of 100 ft^3/min and well-mixed air is pumped out at the same rate. **(a)** How much radon is in the building at any one time? **(b)** How much radon is in the room after 1 h?

29. *Environmental science*   A lake of volume 100 000 m^3 contains a 0.01% concentration of a certain pollutant. Because of a chemical spill, a river flowing into the lake becomes polluted. The polluted river contains 0.05% of the pollutant when it flows into the lake at the rate of 25 m^3/day. **(a)** If the water in the lake is well-mixed before it leaves, and it leaves at the same rate as it enters (25 m^3/day), find the concentration of the pollutant at time t. **(b)** When will the concentration of the pollutant be twice the initial concentration?

30. *Environmental science*   A lake of volume 10 000 000 m^3 contains a 0.01% concentration of a certain pollutant. Because of a mining operation, a river flowing into the lake becomes polluted. The polluted river contains 0.074% of the pollutant when it flows into the lake at the rate of 500 m^3/day. **(a)** If the water in the lake is well-mixed before it leaves, and if it leaves at the rate of 500 m^3/day, find the concentration of the pollutant at time t. **(b)** When will the concentration of the pollutant be twice the initial concentration? **(c)** What is the maximum amount of pollutant that will be in the lake?

≡ CHAPTER 32 TEST

Solve Exercises 1 and 2.

1. Show that $y = 4e^{2x} + 2e^{-3x}$ is a solution of the differential equation

$$\frac{d^2y}{dx^2} + \frac{dy}{dx} - 6y = 0$$

2. Determine if the equation $y^2\,dx + 2xy\,dy = 0$ is exact.

In Exercises 3–8, solve the given differential equation.

3. $y' = \dfrac{5x^4}{y^3}$

4. $(x^2+1)\,dx + x^2y^2\,dy = 0$

5. $(1+y^2)\,dx + (x^2y+y)\,dy = 0$

6. $\sin x \dfrac{dy}{dx} = 1 - 2y\cos x$

7. $x\,dy - y\,dx - x^6\,dx = 0$

8. $4(x\,dy + y\,dx) + 6x^2y\,dx = 0$, given that $x = 1$ when $y = e$.

Solve Exercises 9–12.

9. Find the orthogonal trajectories of the family $y = ce^{-2x}$.

10. What is the half-life of a radioactive substance, if 30% disappears in 40 y?

11. A mass of 5 slugs is falling under the influence of gravity. Find its velocity after 6 s, if it starts from rest and experiences a retarding force of 0.2 times its velocity.

12. An object cools from 50°C to 45°C in 20 min in air that is maintained at 30°C. What is the object's temperature after 1 h?

33

Higher-Order Differential Equations

Objects attached to springs are subjected to several forces including retarding forces, such as friction. In Section 33.4, we will learn how to solve this type of harmonic motion problem by using differential equations.

Courtesy of Monroe Muffler/Brake Inc. Photo by John Myers.

In Chapter 32, we limited our discussion to differential equations of the first order. In this chapter, we will look at higher-order differential equations, particularly at second-order, linear, differential equations with constant coefficients.

We will begin by looking at higher-order homogeneous equations that have constant coefficients. Initially these equations will have distinct real roots, but later we will consider those with repeating real roots or with complex roots. The latter are particularly useful in electrical applications.

Next, we will learn how to solve non-homogeneous differential equations with undetermined coefficients. Finally, we will consider instances where the techniques of this chapter can be applied.

≡ 33.1 HIGHER-ORDER HOMOGENEOUS EQUATIONS WITH CONSTANT COEFFICIENTS

A linear differential equation of order n has the form

$$a_0(x)\frac{d^n y}{dx^n} + a_1(x)\frac{d^{n-1}y}{dx^{n-1}} + a_2(x)\frac{d^{n-2}y}{dx^{n-2}} + \cdots + a_{n-1}(x)\frac{dy}{dx} + a_n(x)y = F(x)$$

where $a_0(x)$, $a_1(x)$, $a_2(x)$, $\ldots$, $a_n(x)$, and $F(x)$ all depend only on x and not on y. It is possible that some of the $a_i(x)$ or $F(x)$ elements may be constants. In fact, if $F(x) = 0$, the equation is called **homogeneous**. If $F(x) \neq 0$, the equation is **nonhomogeneous**. In this chapter, we will look at methods for solving both homogeneous and nonhomogeneous equations. We will, however, restrict our attention to equations with constant coefficients.

Using the above notation, we have the following definition for a second-order, linear, homogeneous equation with constant coefficients.

Second-Order, Linear, Homogeneous Equation with Constant Coefficients

A second-order, linear, homogeneous equation with constant coefficients is of the form

$$a_0\frac{d^2 y}{dx^2} + a_1\frac{dy}{dx} + a_2 y = 0$$

where a_0, a_1, and a_2 are constants.

It is often more convenient to use the notation D and D^2, to indicate the **operations** of taking the first and second derivative. Thus, $Dy = \dfrac{dy}{dx}$ and $D^2 y = \dfrac{d^2 y}{dx^2}$. The symbols D and D^2 are called the **operators** because they define an operation to be performed.

Operator Notation

A second-order, linear, homogeneous equation with constant coefficients is written in **operator notation** as

$$a_0 D^2 y + a_1 D y + a_2 y = 0$$

where a_0, a_1, and a_2 are constants.

This is often written as $(a_0 D^2 + a_1 D + a_0)y = 0$. It could also be written as $a_0 y'' + a_1 y' + a_2 y = 0$.

We now see how operator notation can be used to solve a differential equation.

EXAMPLE 33.1

Solve the differential equation $(D^2 + 5D + 6)y = 0$.

Solution We will begin by factoring the operators as $(D+3)(D+2)y = 0$. If we let $z = (D+2)y$, then $(D+3)z = 0$. We can solve this equation by using separation of variables. Thus, we have

$$\frac{dz}{dx} + 3z = 0$$

or $\qquad \dfrac{dz}{z} + 3\,dx = 0$

and so $\qquad \ln z + 3x = \ln c_1$, when $c = \ln c_1$

Solving for z we get

$$\ln z = \ln c_1 - 3x$$

or $\qquad z = e^{\ln c_1 - 3x}$

$$= e^{\ln c_1} e^{-3x}$$

$$= c_1 e^{-3x}$$

Replacing z by $(D+2)y$, we have

$$(D+2)y = c_1 e^{-3x}$$

This is a linear equation of the first order. Thus, we have

$$dy + 2y\,dx = c_1 e^{-3x}\,dx$$

The integrating factor is $e^{\int 2\,dx} = e^{2x}$ and so

$$ye^{2x} = \int c_1 e^{-3x} e^{2x}\,dx$$

$$= \int c_1 e^{-x}\,dx$$

$$= -c_1 e^{-x} + c_2$$

Multiplying both sides by e^{-2x}, we get

$$y = -c_1 e^{-3x} + c_2 e^{-2x}$$

$$= ke^{-3x} + c_2 e^{-2x}, \text{ where } k = -c_1$$

While we have found a general solution for this given differential equation, we would like to find an easier method for determining this solution. It appears that $y = e^{mx}$ is a solution where m is an unknown constant. We substitute this solution into the original equation of Example 33.1 and get

$$D^2 e^{mx} + 5De^{mx} + 6e^{mx} = 0$$
$$m^2 e^{mx} + 5me^{mx} + 6e^{mx} = 0$$
$$(m^2 + 5m + 6)e^{mx} = 0$$

Now, $e^{mx} > 0$ for all values of m. Thus, the equation is true only if $m^2 + 5m + 6 = 0$ or $(m+3)(m+2) = 0$, which occurs when $m = -3$ or $m = -2$.

The equation $a_0 m^2 + a_1 m + a_2 = 0$ is called the **auxiliary equation** of $a_0 D^2 y + a_1 Dy + a_2 y = 0$. Notice the similarity of the two equations.

There are two roots, m_1 and m_2, of the auxiliary equation and there are two arbitrary constants in the solution of $a_0 D^2 y + a_1 Dy + a_2 y = 0$. These factors direct us to the general solution of the second-order, linear, homogeneous equation with constant coefficients. This general solution is

$$y = c_1 e^{m_1 x} + c_2 e^{m_2 x}$$

where m_1 and m_2 are the solutions to the auxiliary equation. Notice how this corresponds to our solution to Example 33.1.

General Solution of a Second-Order, Linear, Homogeneous Differential Equation

If m_1 and m_2 are distinct real roots of the auxiliary equation $a_0 m^2 + a_1 m + a_2 = 0$, then

$$y = c_1 e^{m_1 x} + c_2 e^{m_2 x}$$

is the **general solution** of the differential equation.

EXAMPLE 33.2

Solve the differential equation $D^2 y - 3Dy + 2y = 0$.

Solution The auxiliary equation is

$$m^2 - 3m + 2 = 0$$

Factoring this equation, we get

$$(m-2)(m-1) = 0$$

and so $\qquad\qquad m_1 = 2$ and $m_2 = 1$

The general solution is

$$y = c_1 e^{2x} + c_2 e^x$$

EXAMPLE 33.3

Solve the differential equation $D^2y - 5Dy = 0$.

Solution The auxiliary equation is

$$m^2 - 5m = 0$$

Solving this, we get $m(m - 5) = 0$, so $m_1 = 0$ and $m_2 = 5$. Thus, the general solution is

$$y = c_1 e^{0x} + c_2 e^{5x} = c_1 + c_2 e^{5x}$$

EXAMPLE 33.4

Solve the differential equation $y'' - y' = 12y$.

Solution We will first rewrite this differential equation using operator notation as follows:

$$D^2y - Dy - 12y = 0$$

and thus the auxiliary equation is

$$m^2 - m - 12 = 0$$

This factors into $(m - 4)(m + 3) = 0$, so the roots are $m_1 = 4$ and $m_2 = -3$. The general solution becomes

$$y = c_1 e^{4x} + c_2 e^{-3x}$$

EXAMPLE 33.5

Solve the differential equation $2\dfrac{d^2y}{dx^2} + 4\dfrac{dy}{dx} - y = 0$ and find the particular solution that satisfies $\dfrac{dy}{dx} = 6$ and $y = 0$, when $x = 0$.

Solution We will first write this differential equation in operator notation:

$$2D^2y + 4Dy - y = 0$$

The auxiliary equation is

$$2m^2 + 4m - 1 = 0$$

This equation does not factor. Using the quadratic formula, we get

$$m = \frac{-4 \pm \sqrt{4^2 - 4(2)(-1)}}{2(2)}$$

$$= \frac{-4 \pm \sqrt{24}}{4} = \frac{-2 \pm \sqrt{6}}{2}$$

Thus, the general solution is

$$y = c_1 e^{\frac{-2+\sqrt{6}}{2}x} + c_2 e^{\frac{-2-\sqrt{6}}{3}x}$$

EXAMPLE 33.5 (Cont.)

Now, let's find the particular solution. When $x = 0$, $y = c_1 + c_2 = 0$, so $c_2 = -c_1$.

Differentiating the general solution, we get

$$y' = \frac{-2+\sqrt{6}}{2}c_1 e^{\frac{-2+\sqrt{6}}{2}x} + \frac{-2-\sqrt{6}}{2}c_2 e^{\frac{-2-\sqrt{6}}{2}x}$$

and when $x = 0$,

$$y' = \frac{-2+\sqrt{6}}{2}c_1 + \frac{-2-\sqrt{6}}{2}c_2$$

Since $c_2 = -c_1$, we have

$$y' = \frac{-2+\sqrt{6}}{2}c_1 + \frac{2+\sqrt{6}}{2}c_1 = \sqrt{6}c_1$$

We are given $\dfrac{dy}{dx} = 6$ and, since $y' = \sqrt{6}c_1$, we have $\sqrt{6}c_1 = 6$ and so $c_1 = \dfrac{6}{\sqrt{6}} = \sqrt{6}$ and $c_2 = -\sqrt{6}$. Thus, the particular solution is

$$y = \sqrt{6}e^{\frac{-2+\sqrt{6}}{2}x} - \sqrt{6}e^{\frac{-2-\sqrt{6}}{2}x}$$

This method works for homogeneous linear differential equations of order higher than two. We summarize these results in the following box.

---

nth-Order Linear Homogeneous Differential Equations

A homogeneous equation of order n with constant coefficients is of the form

$$a_0\frac{d^n y}{dx^n} + a_1\frac{d^{n-1}y}{dx^{n-1}} + \cdots + a_{n-1}\frac{dy}{dx} + a_n y = 0$$

and in operator form is

$$a_0 D^n y + a_1 D^{n-1}y + \cdots + a_{n-1}Dy + a_n y = 0$$

The **auxiliary equation** is of the form

$$a_0 m^n + a_1 m^{n-1} + a_2 m^{n-2} + \cdots + a_{n-1}m + a_n = 0$$

If $m_1, m_2, m_3, \ldots, m_n$ are the roots of this auxiliary equation, then

$$y = c_1 e^{m_1 x} + c_2 e^{m_2 x} + \cdots + c_n e^{m_n x}$$

is the **general solution** of the differential equation.

EXAMPLE 33.6

Solve the differential equation

$$D^3y - 3D^2y + 2Dy = 0$$

Solution   The auxiliary equation is

$$m^3 - 3m^2 + 2m = 0$$

This factors into $m(m-2)(m-1) = 0$, so $m_1 = 0$, $m_2 = 2$, and $m_3 = 1$. The general solution is

$$y = c_1e^{0x} + c_2e^{2x} + c_3e^x$$
$$= c_1 + c_2e^{2x} + c_3e^x$$

■

EXAMPLE 33.7

Solve the differential equation

$$(D^3 - 2D^2 - D + 2)y = 0$$

Solution   This has the auxiliary equation

$$m^3 - 2m^2 - m + 2 = 0$$

The only possible rational roots are ± 1 and ± 2. Using synthetic division, or one of the other methods we used earlier, we get

$$
\begin{array}{r|rrrr}
1 & 1 & -2 & -1 & 2 \\
  &   & 1 & -1 & -2 \\
\hline
  & 1 & -1 & -2 & 0 \\
\end{array}
$$

and so 1 is a root and the equation factors into

$$m^3 - 2m^2 - m + 2 = (m-1)(m^2 - m - 2)$$
$$= (m-1)(m-2)(m+1)$$

So, $m_1 = 1$, $m_2 = 2$, and $m_3 = -1$. Thus the general solution of the differential equation is

$$y = c_1e^x + c_2e^{2x} + c_3e^{-x}$$

■

Exercise Set 33.1

In Exercises 1–20, solve the given differential equation.

1. $(D^2 - 7D + 10)y = 0$

2. $(D^2 + 7D + 12)y = 0$

3. $D^2y - 8Dy + 12y = 0$

4. $6D^2y + Dy = y$

5. $2\dfrac{d^2y}{dx^2} + 9\dfrac{dy}{dx} - 5y = 0$

6. $\dfrac{d^2y}{dx^2} = 9y$

7. $4D^2y - 7Dy - 2y = 0$

8. $4y'' + 11y' - 3y = 0$

9. $y'' - 5y' + 4y = 0$

10. $\dfrac{d^2y}{dx^2} + 4\dfrac{dy}{dx} + 3y = 0$

11. $y'' - y = 0$

12. $D^2y = Dy + 30y$

13. $y'' = 7y$

14. $D^2y + Dy - y = 0$

15. $2\dfrac{d^2y}{dx^2} + \dfrac{dy}{dx} - y = 0$

16. $D^2y + 2Dy - y = 0$

17. $D^3y - 6D^2y + 11Dy - 6y = 0$

18. $D^3y - D^2y - 4Dy + 4y = 0$

19. $y''' + 6y'' + 11y' + 6y = 0$

20. $D^3y - D^2y - 17Dy = 15y$

In Exercises 21–28, find the particular solutions for the given differential equation.

21. $D^2y + 2Dy - 15y = 0$, $y = 2$ and $y' = 6$ when $x = 0$

22. $3D^2y - 14Dy + 8y = 0$, $y = 3$ and $y' = 12$ when $x = 0$

23. $D^2y + 2Dy - 8y = 0$, $y = 0$ and $y' = 6$ when $x = 0$

24. $D^2y + 2Dy - 8y = 0$, $y = 3$ and $y' = -12$ when $x = 0$

25. $y'' + 2y' - 15y = 0$, $y = 5$ and $y' = -1$ when $x = 0$

26. $y'' - 4y' - 32y = 0$, $y = 15$ and $y' = 0$ when $x = 0$

27. $y''' + y'' - 4y' - 4y = 0$, $y = 3$, $y' = 7$, and $y'' = -3$ when $x = 0$

28. $y''' + y'' - 17y' + 15y = 0$, $y = 0$, $y' = -12$, and $y'' = 0$ when $x = 0$

In Your Words

29. Without looking in the text, explain how you can tell if an equation is a second-order, linear, homogeneous differential equation.

30. Without looking in the text, describe how to find the general solution of a second-order, linear, homogeneous differential equation.

≡ 33.2
AUXILIARY EQUATIONS WITH REPEATED OR COMPLEX ROOTS

In Section 33.1, all roots of the auxiliary equation were distinct real numbers. In this section, we will look at repeated and complex roots of auxiliary equations.

Repeated Roots

The following example will show what to do when there are multiple (or repeated) roots.

EXAMPLE 33.8

Solve the differential equation $(D^2 - 6D + 9)y = 0$.

Solution   The auxiliary equation is

$$m^2 - 6m + 9 = (m - 3)^2 = 0$$

This equation has the multiple roots of 3 and 3. Using the method of the last section, we would write $y = c_1 e^{3x} + c_2 e^{3x} = (c_1 + c_2)e^{3x}$ or $y = ce^{3x}$. But, this has only one arbitrary constant.

While we will not show it here, you can check that $y = xe^{3x}$ is also a solution. So, the general solution is given by the linear combination

$$y = c_1 e^{3x} + c_2 x e^{3x}$$

What happens if one root is repeated more than once? A similar thing occurs. For example, if the auxiliary equation is $(m-4)^3 = 0$, then there are three roots that are the same such as 4, 4, and 4. The general solution is given by $y = c_1 e^{4x} + c_2 x e^{4x} + c_3 x^2 e^{4x}$.

General Solution with Repeated Roots

If the auxiliary equation has the real root m, which repeats n times, then

$$y = c_1 e^{mx} + c_2 x e^{mx} + c_3 x^2 e^{mx} + \cdots + c_n x^{n-1} e^{mx}$$
$$= (c_1 + c_2 x + c_3 x^2 + \cdots + c_n x^{n-1}) e^{mx}$$

is the general solution of the equation.

EXAMPLE 33.9

Solve the differential equation $D^6 y + D^5 y - 4D^4 y - 2D^3 y + 5D^2 y + Dy = 2y$.

Solution This differential equation has the auxiliary equation $m^6 + m^5 - 4m^4 - 2m^3 + 5m^2 + m - 2 = 0$. This auxiliary equation factors into $(m-1)^3 (m+1)^2 (m+2) = 0$ and so it has roots of 1, 1, 1, −1, −1, and −2. The general solution of this differential equation is

$$y = (c_1 + c_2 x + c_3 x^2) e^x + (c_4 + c_5 x) e^{-x} + c_6 e^{-2x}$$

Complex Roots

Now, suppose that the auxiliary equation $a_0 m^2 + a_1 m + a_2 = 0$ has real coefficients and complex roots. From the quadratic equation we know that the roots are complex conjugates and are of the form $m = a \pm bj$. From the technique learned in Section 33.1, we can show that the general solution is of the form

$$y = c_1 e^{(a+bj)x} + c_2 e^{(a-bj)x}$$
$$= e^{ax} (c_1 e^{bjx} + c_2 e^{-bjx})$$

The relationship between the exponential and polar form of a complex number states that $re^{j\theta} = r(\cos\theta + j\sin\theta)$. When we apply this to the equation above, we get

$$y = e^{ax} (c_1 [\cos bx + j\sin bx] + c_2 [\cos(-bx) + j\sin(-bx)])$$

But, since $\cos(-\alpha) = \cos\alpha$ and $\sin(-\alpha) = -\sin\alpha$, we can rewrite this as

$$y = e^{ax} [c_1 (\cos bx + j\sin bx) + c_2 (\cos bx - j\sin bx)]$$
$$= e^{ax} (c_1 \cos bx + c_2 \cos bx + jc_1 \sin bx - jc_2 \sin bx)$$
$$= e^{ax} [(c_1 + c_2) \cos bx + j(c_1 - c_2) \sin bx]$$
$$= e^{ax} (c_3 \cos bx + c_4 \sin bx)$$

where $\qquad c_3 = c_1 + c_2$ and $c_4 = j(c_1 - c_2)$

Complex Roots

If the auxiliary equation $a_0 m^2 + a_1 m + a_2 = 0$, with real coefficients, has complex roots of the form $a \pm bj$, the general solution of the differential equation is

$$y = e^{ax}(c_1 \cos bx + c_2 \sin bx)$$

(These are different c_1 and c_2 than we used previously. We use them here simply to indicate two arbitrary constants.)

EXAMPLE 33.10

Solve the differential equation

$$D^2 y + 4Dy + 13y = 0$$

Solution This has the auxiliary equation $m^2 + 4m + 13 = 0$. Using the quadratic formula, we see that

$$m = \frac{-4 \pm \sqrt{4^2 - 4(13)}}{2}$$

$$= \frac{-4 \pm \sqrt{-36}}{2} = \frac{-4 \pm 6j}{2}$$

$$= -2 \pm 3j$$

are the roots of the auxiliary equation. Here, $a = -2$ and $b = 3$, and the general solution is

$$y = e^{-2x}(c_1 \cos 3x + c_2 \sin 3x)$$

EXAMPLE 33.11

Solve the differential equation $\dfrac{d^2 y}{dx} - \dfrac{dy}{dx} + 1 = 0$.

Solution Rewriting this in operator form as

$$D^2 y - Dy + y = 0$$

we see that it has the auxiliary equation

$$m^2 - m + 1 = 0$$

Using the quadratic equation, we get

$$m = \frac{1 \pm \sqrt{1 - 4}}{2} = \frac{1 \pm \sqrt{-3}}{2} = \frac{1}{2} \pm j\frac{\sqrt{3}}{2}$$

and so $a = \frac{1}{2}$ and $b = \dfrac{\sqrt{3}}{2}$.

EXAMPLE 33.11 (Cont.)

The general solution to this differential equation is

$$e^{x/2}\left(c_1\cos\frac{\sqrt{3}}{2}x + c_2\sin\frac{\sqrt{3}}{2}x\right)$$

EXAMPLE 33.12

Solve $y''' - 6y'' + 2y' + 36y = 0$.

Solution This has the auxiliary equation $m^3 - 6m^2 + 2m + 36 = 0$, which has the roots -2 and $4 \pm j\sqrt{2}$. The general solution is

$$y = c_1 e^{-2x} + e^{4x}\left(c_2\cos\sqrt{2}x + c_3\sin\sqrt{2}x\right)$$

Notice how we were able to combine the real and the complex roots into the general solution.

EXAMPLE 33.13

Solve $y'' + 3y' + 4y = 0$, if $y = 2$ and $y' = 4$ when $x = 0$.

Solution The auxiliary equation is $m^2 + 3m + 4 = 0$ and has roots

$$m = \frac{-3 \pm \sqrt{9-16}}{2} = \frac{-3}{2} \pm j\frac{\sqrt{7}}{2}$$

The general solution is

$$y = e^{-3x/2}\left(c_1\cos\frac{\sqrt{7}}{2}x + c_2\sin\frac{\sqrt{7}}{2}x\right)$$

When $x = 0$, this becomes $y = c_1$. Since we were given $y = 2$, then $c_1 = 2$.

The other condition involves y'. If we differentiate the general solution, we get

$$y' = e^{-3x/2}\left(-c_1\frac{\sqrt{7}}{2}\sin\frac{\sqrt{7}}{2}x + c_2\frac{\sqrt{7}}{2}\cos\frac{\sqrt{7}}{2}x\right)$$
$$-\frac{3}{2}e^{-3x/2}\left(c_1\cos\frac{\sqrt{7}}{2}x + c_2\sin\frac{\sqrt{7}}{2}x\right)$$

When $x = 0$, this becomes

$$y' = c_2\frac{\sqrt{7}}{2} - \frac{3}{2}c_1$$

We know that $y' = 4$ when $x = 0$ and that $c_1 = 2$. Making these substitutions, we get

$$4 = c_2\frac{\sqrt{7}}{2} - \frac{3}{2}(2)$$

EXAMPLE 33.13 (Cont.)

$$4 = c_2 \frac{\sqrt{7}}{2} - 3$$

or $\qquad 7 = c_2 \frac{\sqrt{7}}{2}$

and so $\qquad c_2 = \frac{2}{\sqrt{7}} \cdot 7 = 2\sqrt{7}$

The particular solution to this differential equation is

$$y = e^{-3x/2} \left(2\cos \frac{\sqrt{7}}{2} x + 2\sqrt{7} \sin \frac{\sqrt{7}}{2} x \right)$$

Exercise Set 33.2

In Exercises 1–26, solve the given differential equation.

1. $D^2 y + 2Dy + y = 0$
2. $(D^2 - 2D + 1)y = 0$
3. $D^2 y - 4Dy + 4y = 0$
4. $y'' - 10y' + 25y = 0$
5. $9y'' - 6y' + y = 0$
6. $16y'' + 8y' + y = 0$
7. $4y'' + 12y' + 9y = 0$
8. $9y'' - 12y' + 4y = 0$
9. $y'' + 4y' + 5y = 0$
10. $y'' + 4y = 0$
11. $D^2 y + 9Dy = 0$
12. $D^3 y = 0$
13. $D^3 y = y$
14. $\dfrac{d^3 y}{dx^2} + y = 0$

15. $\dfrac{d^3 y}{dx^3} + 8y = 0$
16. $\dfrac{d^3 y}{dx^3} - 27y = 0$
17. $y''' - 6y'' + 11y' - 6y = 0$
18. $D^4 y - 9D^2 y + 20y = 0$
19. $D^4 y + 8D^3 y + 24D^2 y + 32Dy + 16y = 0$
20. $y''' - y'' + y' - y = 0$
21. $(D - 1)^2 (D + 2)^3 y = 0$
22. $(D - 2)^4 (D + 5)y = 0$
23. $(D - 3)^2 (D^2 - 6D - 9)y = 0$
24. $(D + 1)^2 (D^2 + 4D + 9)y = 0$
25. $(3D^2 - 2D + 1)y = 0$
26. $\dfrac{d^2 y}{dx^2} - 2\dfrac{dy}{dx} + 3y = 0$

In Exercises 27–34, find the particular solution of the given differential equation with the given conditions.

27. $y'' - 2y' + 5y = 0$, $y = 4$ and $y' = 7$ when $x = 0$
28. $y'' + 8y' + 16y = 0$, $y = 4$ and $y' = -6$ when $x = 0$
29. $(4D^2 - 12D + 9)y = 0$, $y = 1$ when $x = 0$ and $y = 0$ when $x = 1$
30. $(D^2 - 4D + 5)y = 0$, $y = 1$ when $x = 0$ and $y = 3e^\pi$ when $x = \frac{\pi}{2}$

31. $y'' + 2y' + y = 0$, $y = 1$ and $y' = -3$ when $x = 0$
32. $y'' - 10y' + 25y = 0$, $y = -4$ and $y' = 4$ when $x = 0$
33. $(D^2 - 6D + 34)y = 0$, $y = 4$ and $y' = -3$ when $x = 0$
34. $(D^2 + 10D + 29)y = 0$, $y = 0.5$ and $y' = 1.5$ when $x = 0$

35. Suppose that the auxiliary equation to a higher-order linear homogeneous differential equation has repeated roots. Explain how you find the general solution to this differential equation.

36. Describe how you would find the general solution to a differential equation if its auxiliary equation has real coefficients and nonreal complex roots.

≡ 33.3
SOLUTIONS OF NONHOMOGENEOUS EQUATIONS

In Sections 33.1 and 33.2, we focused on homogeneous linear equations of the form

$$a_0 D^2 y + a_1 Dy + a_2 y = 0$$

where a_0, a_1, and a_2 were constants. Next, we look at nonhomogeneous linear equations of the form

$$a_0 D^2 y + a_1 Dy + a_2 y = f(x)$$

where a_0, a_1, and a_2 are all constants and $f(x)$ is either a constant or a function of x. We will find a way in which to find a general solution, y, which satisfies this equation.

In Sections 33.1 and 33.2, we were able to find a general solution for $a_0 D^2 y + a_1 Dy + a_2 y = 0$. This solution is called the **complementary solution** and is denoted y_c. Thus

$$(a_0 D^2 + a_1 D + a_2) y_c = 0$$

Now, suppose we have found a **particular solution**, y_p, to $a_0 D^2 y + a_1 Dy + a_2 y = f(x)$. This means that $(a_0 D^2 + a_1 D + a_2) y_p = f(x)$. It turns out that $y = y_c + y_p$ is a general solution of the nonhomogeneous linear equation.

Several methods exist for finding y_p. We will use the **method of undetermined coefficients**. This method begins by guessing the form of y_p up to arbitrary multiplicative constants. These constants are then evaluated by substituting the proposed form of y_p into the given differential equation and equating the coefficients of like terms. The guess that we make for y_p is an educated guess. In general, there are the four cases which have been outlined in the following box.

Method of Undetermined Coefficients

Case 1: $f(x) = p_n(x)$, an nth degree polynomial in x. Assume

$$y_p = A_n x^n + A_{n-1} x^{n-1} + \cdots + A_1 x + A_0$$

We will need to determine $A_n, A_{n-1}, \ldots, A_1$, and A_0.

Case 2: $f(x) = e^{ax} p_n(x)$, where a is a known constant and $p_n(x)$ as in case 1. Then, we will assume that

$$y_p = e^{ax}(A_n x^n + A_{n-1} x^{n-1} + \cdots + A_1 x + A_0)$$

As in case 1, we will need to determine the values of $A_n, A_{n-1}, \ldots, A_1$, and A_0.

Case 3: $f(x) = e^{ax} p_n(x) \sin bx$, where a and b are known constants. In this case, we assume that

$$y_p = e^{ax} \sin bx (A_n x^n + A_{n-1} x^{n-1} + \cdots + A_1 x + A_0)$$
$$+ e^{ax} \cos bx (B_n x^n + B_{n-1} x^{n-1} + \cdots + B_1 x + B_0)$$

where $A_n, A_{n-1}, \ldots A_1, A_0$, and $B_n, B_{n-1}, \ldots, B_1, B_0$ are values we will need to determine.

Case 4: $f(x) = e^{ax} p_n(x) \cos bx$, where a and b are known constants. In this case, we make the same assumption for y_p as we made in case 3.

≡ Note

It is possible for a problem to fit more than one of these cases as we will see in Example 33.17.

Now, this is a very lengthy explanation and is somewhat confusing. Let's work several examples to help make it clear.

EXAMPLE 33.14

Solve the differential equation

$$D^2 y - 3Dy + 2y = 4x^2 - 5$$

Solution First we find the complementary solution. In Example 33.2, we found that $y_c = c_1 e^{2x} + c_2 e^x$.

Here we have $f(x) = 4x^2 - 5$, a second-degree polynomial, and so this is a case 1 problem. We assume that $y_p = A_2 x^2 + A_1 x + A_0$. Thus, $y_p' = 2A_2 x + A_1$ and $y_p'' = 2A_2$. Substituting these results into the differential equation, we have

$$(2A_2) - 3(2A_2 x + A_1) + 2(A_2 x^2 + A_1 x + A_0) = 4x^2 - 5$$
$$2A_2 - 6A_2 x + 3A_1 + 2A_2 x^2 + 2A_1 x + 2A_0 = 4x^2 - 5$$

or

$$2A_2 x^2 + (2A_1 - 6A_2)x + (2A_2 - 3A_1 + 2A_0) = 4x^2 + 0x - 5$$

EXAMPLE 33.14 (Cont.)

Thus, $2A_2x^2 = 4x^2$, so $A_2 = 2$. We also have $(2A_1 - 6A_2)x = 0x$. Since $A_2 = 2$, this means that $2A_1 - 6(2) = 2A_1 - 12 = 0$, and so $A_1 = 6$.

Finally, $2A_2 - 3A_1 + 2A_0 = -5$. Since $A_2 = 2$ and $A_1 = 6$, we have $2(2) - 3(6) + 2A_0 = 4 - 18 + 2A_0 = -5$ or $2A_0 = 9$, so $A_0 = \frac{9}{2}$.

Thus $y_p = 2x^2 + 6x + \frac{9}{2}$, and the general solution is

$$y = c_1e^{2x} + c_2e^x + 2x^2 + 6x + \frac{9}{2}$$

EXAMPLE 33.15

Solve the differential equation $y'' - 5y' = 3e^{7x}$.

Solution In Example 33.3, we found that the complementary solution of this differential equation is $y_c = c_1 + c_2e^{5x}$.

Here we have $f(x) = 3e^{7x}$, which is a case 2 problem with $a = 7$ and $p_n(x) = 3$, a constant polynomial. According to case 2, we assume that

$$y_p = A_0e^{7x}$$

So, we have $y_p' = 7A_0e^{7x}$ and $y_p'' = 49A_0e^{7x}$. Substituting these values into the original differential equation gives

$$\begin{aligned} y'' - 5y' &= 3e^{7x} \\ 49A_0e^{7x} - 5(7A_0e^{7x}) &= \\ 49A_0e^{7x} - 35A_0e^{7x} &= \\ \text{or} \quad 14A_0e^{7x} &= 3e^{7x} \end{aligned}$$

Thus $A_0 = \frac{3}{14}$, and the general solution is

$$y = c_1 + c_2e^{5x} + \frac{3}{14}e^{7x}$$

EXAMPLE 33.16

Solve the differential equation

$$(D^2 - 6D + 9)y = \cos 2x$$

Solution In Example 33.8, we found that $y_c = c_1e^{3x} + c_2xe^{3x}$. In this example, $f(x) = \cos 2x$, which is the form in case 4 with $a = 0$, $p_n(x) = 1$, and $b = 2$. Thus, we assume the solution to be

$$y_p = A_0 \sin 2x + B_0 \cos 2x$$

Differentiating, we get $y_p' = 2A_0 \cos 2x - 2B_0 \sin 2x$ and $y_p'' = -4A_0 \sin 2x - 4B_0 \cos 2x$. Substituting into the differential equation, we obtain

$$(-4A_0 \sin 2x - 4B_0 \cos 2x) - 6(2A_0 \cos 2x - 2B_0 \sin 2x)$$
$$+ 9(A_0 \sin 2x + B_0 \cos 2x) = \cos 2x$$

EXAMPLE 33.16 (Cont.)

Collecting terms, we get

$$(-4A_0 + 12B_0 + 9A_0)\sin 2x + (-4B_0 - 12A_0 + 9B_0)\cos 2x = \cos 2x$$

or $(5A_0 + 12B_0)\sin 2x + (5B_0 - 12A_0)\cos 2x = \cos 2x$

Thus, $5A_0 + 12B_0 = 0$ and $5B_0 - 12A_0 = 1$. Solving these simultaneously, we have $A_0 = \frac{-12}{169}$ and $B_0 = \frac{5}{169}$. This means that $y_p = \frac{-12}{169}\sin 2x + \frac{5}{169}\cos 2x$, and the general solution is

$$y = c_1 e^{3x} + c_2 x e^{3x} - \frac{12}{169}\sin 2x + \frac{5}{169}\cos 2x$$

EXAMPLE 33.17

Solve $y'' + 4y = 3xe^x + 5x - 1$.

Solution   From Exercise 10 in Exercise Set 33.3, we see that

$$y_c = c_1 \cos 2x + c_2 \sin 2x$$

In this example, $f(x) = 3xe^x + 5x - 1$. If we write this as $f(x) = g(x) + h(x)$ where $g(x) = 3xe^x$ and $h(x) = 5x - 1$, then h is a case 1 form and g is a case 2 form. Accordingly, we assume that for $g(x)$ we have a solution of the form $e^x(A_1 x + A_0)$ and for $h(x)$ the solution is of the form $B_1 x + B_0$. Thus, we try

$$y_p = A_1 x e^x + A_0 e^x + B_1 x + B_0$$

Differentiating, we get $y_p' = A_1 e^x + A_1 x e^x + A_0 e^x + B_1$ and $y_p'' = 2A_1 e^x + A_1 x e^x + A_0 e^x$. Substituting into the differential equation, we have

$$2A_1 e^x + A_1 x e^x + A_0 e^x + 4(A_1 x e^x + A_0 e^x + B_1 x + B_0) = 3xe^x + 5x - 1$$

or

$$(A_1 + 4A_1)xe^x + (2A_1 + A_0 + 4A_0)e^x + 4B_1 x + 4B_0 = 3xe^x + 5x - 1$$

Thus, $5A_1 = 3$, so $A_1 = \frac{3}{5}$ and $2A_1 + 5A_0 = 0$, and so $A_0 = -\frac{6}{25}$, $B_1 = \frac{5}{4}$, and $B_0 = -\frac{1}{4}$. The general solution of this differential equation is

$$y = c_1 \cos 2x + c_2 \sin 2x + \frac{3}{5}x e^x - \frac{6}{25}e^x + \frac{5}{4}x - \frac{1}{4}$$

There are some exceptions to the four cases just given. However, we will not explore them in this book.

Exercise Set 33.3

Solve each of the differential equations in Exercises 1–20.

1. $(D^2 - 10D + 25)y = 4$

2. $(D^2 - 10D + 25)y = 4x^2$

3. $(D^2 - 10D + 25)y = e^{3x}$

4. $(D^2 - 10D + 25)y = 2xe^{3x}$

5. $(D^2 - 10D + 25)y = 10 + e^x$

6. $(D^2 - 10D + 25)y = 3\sin 2x$

7. $(D^2 - 10D + 25)y = 29\sin 2x$

8. $(D^2 - 9)y = 3\cos x$

9. $(D^2 - 4)y = x^2e^x - 3x$

10. $(D^2 + 4)y = 6x + 3$

11. $(D^2 + 9)y = 4\cos x + 2\sin x$

12. $y'' - 5y' + 6y = 9x + 2e^x$

13. $y'' + y = 2e^{3x}$

14. $(D^2 + 2D + 1)y = 4\sin 2x$

15. $y'' - 4y = 8x^2$

16. $y'' + 4y' + 5y = e^{-x} + 10x$

17. $D^2y + y = 4 + \cos 2x$

18. $3D^2y + 2Dy - y = 4 + 2x + 6x^2$

19. $3D^2y + 2Dy + y = 4 + 2x + 6x^2$

20. $y'' + 9y' - y = x^2 + 6e^{2x}$

In Exercises 21–28, find the particular solution of the given differential equation for the given conditions.

21. $y'' - 2y' + y = x^2 - 1$, $y = 7$ and $y' = 15$ when $x = 0$

22. $y'' - 2y' + y = 10$, $y = 20$ and $y' = 5$ when $x = 0$

23. $y'' - y' - 2y = \sin 2x$, $y = 1$ and $y' = \frac{7}{4}$ when $x = 0$

24. $y'' - y' - 2y = e^{3x}$, $y = 2$ and $y' = 11$ when $x = 0$

25. $y'' + 2y' + y = 3x^2 + 2x - 1$, $y = 3$ and $y' = 5$ when $x = 0$

26. $y'' - 5y' - 6y = e^{3x}$, $y = 2$ and $y' = 1$ when $x = 0$

27. $y'' + 9y = 8\cos x$, $y = -1$ and $y' = 1$ when $x = \frac{\pi}{2}$

28. $y'' - 5y' + 6y = e^x(2x - 3)$, $y = 1$ and $y' = 3$ when $x = 0$

In Your Words

29. What are the differences among a particular solution, complementary solution, and general solution to a nonhomogeneous linear equation?

30. What is the method of undetermined coefficients? When do you use it? How do you use it?

≡ 33.4
APPLICATIONS

In Sections 33.1–33.3, we learned how to solve some higher-order linear differential equations with constant coefficients. In this section, we will look at two applications of second-order equations.

Simple Harmonic Motion ───────────────────────

Perhaps the easiest example of simple harmonic motion is that of a vibrating spring. Consider a spring that is hung from a fixed support as shown in Figure 33.1. Suppose that this spring has an object of mass m suspended from the spring. When the object is at rest we say that it is in the **equilibrium position**. We normally use a vertical coordinate axis with the origin, $y = 0$, at the location of the object in the equilibrium position. This means that y will represent the vertical displacement of the object.

You should remember that, according to Hooke's law, a force F on m due to the stretched or compressed spring is $F = kx$. Newton's second law states that the force acting on a body is equal to its mass times its acceleration. This is equivalent to $F = m\dfrac{d^2x}{dt^2}$. Combining these two values for the force, we have

$$m\frac{d^2x}{dt^2} = -kx$$

where the negative sign is used to indicate that the forces act in opposite directions. An object whose displacement from equilibrium satisfies this equation is said to be executing **simple harmonic motion**.

This, however, is not very practical. The equations that result from solving this differential equation imply that once the object begins to oscillate it will keep moving up and down forever. Our experience tells us that this is not true. What does happen is that the oscillations gradually die down until the object comes to rest at the equilibrium position.

The problem with this new formula is that it ignores the effects of friction. Thus, a reasonable assumption is that there is a retarding force proportional to the velocity. We will represent this force by $-\ell\dfrac{dx}{dt}$, where $\ell > 0$. When this is included, the new formula becomes

$$m\frac{d^2x}{dt} = -kx - \ell\frac{dx}{dt}$$

or $$m\frac{d^2x}{dt} + \ell\frac{dx}{dt} + kx = 0$$

If we divide this last equation by m and let $\dfrac{\ell}{m} = 2b$ and $\dfrac{k}{m} = \omega^2$, then we can write the equation in operator notation as

$$(D^2 + 2bD + \omega^2)x = 0$$

The auxiliary equation $m^2 + 2bm + \omega^2 = 0$ has the roots

$$m = \frac{-2b \pm \sqrt{4b^2 - 4\omega^2}}{2}$$
$$= -b \pm \sqrt{b^2 - \omega^2}$$

There are three possible situations depending on the relative sizes of b^2 and ω^2.

If $b^2 > \omega^2$, we say the motion is **overdamped** and the general solution is of the form

$$x = Ae^{-at} + Be^{-ct}$$

where $a = b + \sqrt{b^2 - \omega^2}$ and $c = b - \sqrt{b^2 - \omega^2}$.

If $b^2 = \omega^2$, we say that the motion is **critically damped** and the general solution is

$$x = (A + Bt)e^{-bt}$$

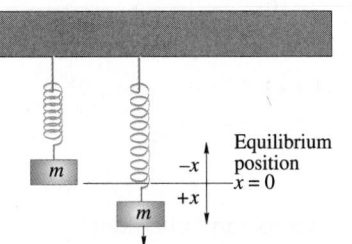

FIGURE 33.1

Equilibrium
position
$x = 0$

$-x$

$+x$

m

m

In both overdamped and critically damped motion, the mass of the object m is subject to such a large retarding force that it slows down and returns to equilibrium. If $b^2 < \omega^2$, then $\sqrt{b^2 - \omega^2}$ is imaginary. If we let $\beta_j = \sqrt{b^2 - \omega^2}$, then the solution is in the form

$$x = e^{-bt}(c_1 \cos \beta t + c_2 \sin \beta t)$$

This motion is known as **underdamped** or **oscillatory motion**. The value $\omega = \sqrt{\dfrac{k}{m}}$ is defined as the **natural frequency** of the system.

If an external force $f(t)$ acts on the spring, then the equation, written in operator notation, becomes

$$(D^2 + 2bD + \omega^2) = f(t)$$

This is an example of **forced oscillation**.

Application

EXAMPLE 33.18

A 10-kg object is attached to a spring stretching it 0.7 m from its natural position. The object is started in motion from its equilibrium position with an initial velocity of 1 m/s downward. Find the position of the object as a function of time, if the force due to air resistance is 90 times the velocity newtons.

Solution   We begin by determining the spring constant. The force stretching the spring is mg. We are given $m = 10\,\text{kg}$ and g, in the metric system, is $9.8\,\text{m/s}^2$. Thus, $F = 10(9.8) = k(0.7)$ and $k = 140\,\text{N/m}$. We are also given ℓ as 90, so $2b = \dfrac{\ell}{m}$ or $b = 4.5$, and $\omega^2 = \dfrac{k}{m} = 14$. Thus, the auxiliary equation is

$$m^2 + 9m + 14 = 0$$

This has the roots 2 and 7, and $b^2 = 20.25 > 14 = \omega^2$, so the general solution is

$$x = Ae^{-2t} + Be^{-7t}$$

Since the object is started in motion from its equilibrium position, we know that when $t = 0$, $x = 0$, and so $A + B = 0$.

The initial velocity is $+1$, so when $t = 0$, $x' = 1 = -2Ae^{-2t} - 7Be^{-7t}$. Thus, $-2A - 7B = 1$. Solving the two equations for A and B, we get $B = -\frac{1}{5}$ and $A = \frac{1}{5}$. Thus,

$$x = \frac{1}{5}(e^{-2t} - e^{-7t})$$

Electric Circuits

There is an electrical situation that parallels the one just described. Consider an RLC circuit similar to the one shown in Figure 33.2. The voltage across an inductor, a

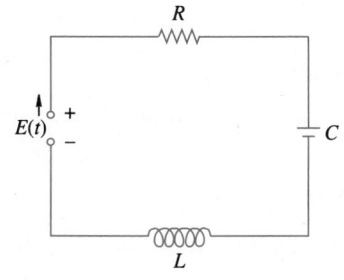

FIGURE 33.2

resistor, and a capacitor is given by

$$L\frac{di}{dt}, \quad Ri, \quad \text{and} \quad \frac{q}{C}$$

respectively. We know that $i = \dfrac{dq}{dt}$ and are then able to rewrite these three expressions as

$$L\frac{d^2q}{dt^2}, \quad R\frac{dq}{dt}, \quad \text{and} \quad \frac{q}{C}$$

If these are connected in series with a generator, then the electromotive force (emf) is expressed as $E(t)$. Kirchhoff's second law states that the sum of the voltage around a closed loop equals the sum of the emf of this loop. Applying Kirchhoff's law to the loop in Figure 33.2, we get the result

$$E(t) = L\frac{d^2q}{dt^2} + R\frac{dq}{dt} + \frac{q}{C}$$

Application

EXAMPLE 33.19

An RLC circuit has $R = 16\,\Omega$, $C = 10^{-2}\,\text{F}$, $L = 1\,\text{H}$, and an emf of $36\sin 10t$. If there is no initial charge on the capacitor but an initial current of $5\,\text{A}$ when the voltage is first applied ($t = 0$), what is the expression for the charge on the capacitor?

Solution   Using the new formula, we have

$$36\sin 10t = \frac{d^2q}{dt^2} + 16\frac{dq}{dt} + 100q \qquad (*)$$

To find the complementary solution q_c, we must solve the auxiliary equation

$$m^2 + 16m + 100 = 0$$

and so,

$$m = \frac{-16 \pm \sqrt{16^2 - 4(100)}}{2}$$
$$= \frac{-16 \pm 12j}{2} = -8 \pm 6j$$

These are complex roots, so

$$q_c = e^{-8t}(c_1\cos 6t + c_2\sin 6t)$$

For this particular solution we have a case 3 situation from the method of undetermined coefficients (see Section 33.3), and so

$$q_p = A\sin 10t + B\cos 10t$$

with $q_p' = 10A\cos 10t - 10B\sin 10t$ and $q_p'' = -100A\sin 10t - 100B\cos 10t$. Substituting these values into $(*)$, we get

$$(-100A\sin 10t - 100B\cos 10t) + 16(10A\cos 10t - 10B\sin 10t)$$
$$+ 100(A\sin 10t + B\cos 10t) = 36\sin 10t$$

EXAMPLE 33.19 (Cont.)

Collecting terms, we obtain

$$(-100A - 160B + 100A)\sin 10t + (-100B + 160A + 100B)\cos 10t = 36\sin 10t$$

Thus, $-160B = 36$ or $B = -0.225$, and $A = 0$.

Then the general solution for $q(t)$ is

$$\begin{aligned} q(t) &= q_c + q_p \\ &= e^{-8t}(c_1\cos 6t + c_2\sin 6t) - 0.225\cos 10t \end{aligned}$$

We are told that there is no initial charge on the capacitor (so $q(0) = 0$) and that the initial current is 5 A (so $q'(0) = 5$). Applying these initial conditions, we obtain

$$q(0) = 0 = c_1 - 0.225 \text{ or } c_1 = 0.225$$

Differentiating produces

$$\begin{aligned} q'(t) &= e^{-8t}(-6c_1\sin 6t + 6c_2\cos 6t) \\ &\quad - 8e^{8t}(c_1\cos 6t + c_2\sin 6t) + 2.25\sin 10t \end{aligned}$$

and when $t = 0$ and $c_1 = 0.225$, we see that

$$\begin{aligned} 5 &= 6c_2 - 8c_1 \\ &= 6c_2 - 8(0.225) \end{aligned}$$

Solving this for c_2, we get $c_2 = \frac{6.8}{6} \approx 1.1333$.

The particular solution for this circuit is

$$q(t) = e^{-8t}(0.225\cos 6t + 1.1333\sin 6t) - 0.225\cos 10t$$

Look more closely at the solution to the last example. The complementary solution, q_c, contains the exponential factor e^{-8t}. As t gets larger this factor approaches 0. After just 2 s, $e^{-8t} = 0.000\,000\,11$. This means that q_c dies out very quickly. In fact, when $t = 2$ s, $q_c = -0.000\,000\,051$. On the other hand, q_p does not contain this factor of e^{-8t}. Thus, since q_c will quickly approach 0, the only significant term left will be q_p. As a result, q_c is called the **transient part** of the solution and q_p is called the **steady-state solution**. The **steady state current** is $i = \dfrac{dq_p}{dt}$.

EXAMPLE 33.20

Find the steady-state current for the RLC circuit in Example 33.19.

Solution The steady-state solution is $q_p = -0.225\cos 10t$. The steady-state current is

$$i = \frac{dq_p}{dt} = 2.25\sin 10t$$

Exercise Set 33.4

Solve Exercises 1–16.

1. *Physics*   A 20-lb weight is suspended from a spring. As a result, the spring stretches 5 in. What is the spring constant?

2. *Physics*   A 15-lb weight stretches a spring 3 in. What is the spring constant?

3. *Physics*   A 5-kg mass is hung from a spring. It stretches the spring 0.2 m. What is the spring constant?

4. *Physics*   An 8-kg mass is suspended from a spring and as a result the spring stretches 0.4 m. What is the spring constant?

5. *Physics*   The 20-lb weight in Exercise 1 is started in motion from the equilibrium position with an initial velocity of 4 ft/s upward. Find the motion of the weight, if the force due to air resistance is twice the instantaneous velocity.

6. *Physics*   The 15-lb weight in Exercise 2 is started in motion from the equilibrium position with an initial velocity of 2 ft/s downward. Find the motion of the weight, if the force due to air resistance is equal to the instantaneous velocity.

7. *Physics*   The object in Exercise 3 is set in motion from the equilibrium position with an initial velocity of 2 m/s downward. Find the equation for the motion of the object, if the force due to air resistance is 40 times the instantaneous velocity newtons.

8. *Physics*   The object in Exercise 4 is started in motion from the equilibrium position with an initial velocity of 3 m/s upward. Find the equation for the motion of the object, if the force in newtons due to air resistance is 80 times the instantaneous velocity.

9. *Electronics*   An RLC circuit with $R = 6\,\Omega$, $C = 10^{-2}\,F$, and $L = 0.1\,H$ has an emf of $50\sin 100t$. Find the charge as a function of time, if $q(0) = 0$ and $i(0) = 0$.

10. *Electronic*   An RLC circuit has $R = 5\,\Omega$, $C = 10^{-2}\,F$, $L = 0.125\,H$, and an emf of $\sin t$. What is

the charge as a function of time, if $q(0) = 0$ and $i(0) = 0$.

11. *Electronics*   A 0.1-H inductor, a 4×10^{-3}-F capacitor, and a generator with emf of $180\cos 60t$ are connected in series. **(a)** What is the charge as a function of time, if $q(0) = i(0) = 0$? **(b)** What is the steady-state current?

12. *Electronics*   An RLC circuit has $R = 10\,\Omega$, $L = 1\,H$, $C = 10^{-2}\,F$, and an emf of $50\cos 10t$. **(a)** What is the charge as a function of time? **(b)** What is the steady-state current?

13. *Medical technology*   Diabetes is usually diagnosed by means of a glucose tolerance test (GTT). In this test the patient comes to the hospital after an overnight fast and is given a large dose of glucose. During the next three to five hours, several measurements are made of the concentration of glucose in the patient's blood, and these measurements are used in the diagnosis of diabetes. If $g(t)$ measures the patient's deviation from an optimum glucose level and if $t = 0$ is the time in minutes at which the glucose load has been completely ingested, then g satisfies

$$\frac{d^2g}{dt^2} + 2\alpha\frac{dg}{dt} + \omega_0{}^2 g = 0$$

where α and ω_0 are positive constants. Solve this differential equation if $\alpha^2 - \omega_0{}^2 < 0$.

14. *Space technology*   The differential equation

$$\frac{d^2z}{d\theta^2} + z = k$$

is used in the study of planetary motion to determine r, the distance from the planet to the sun, where $z = \dfrac{1}{r}$. If k represents a constant and θ is the angle through which the planet has traveled, find the general solution for this differential equation.

15. *Electronics*   Find the equation for $q(t)$ in an RLC series circuit with $L = 1.2$ H, $R = 5.0\,\Omega$, $C = 0.025$ F, and $V(t) = 15$ V.

16. *Electronics*   A capacitor with capacitance is $\frac{1}{505}$ F, and inductor with coefficient of inductance is $\frac{1}{20}$ H,

and a resistor with resistance $1\,\Omega$ are connected in series. If at $t = 0$, $i = 0$ and the charge on the capacitor is 1 C, find **(a)** the charge and **(b)** the current in the circuit due to the discharge of the capacitor when $t = 0.01$ s.

 In Your Words

17. Write an application in your technology area of interest that requires you to solve a differential equation that involves simple harmonic motion, electric circuits, or some area that involves setting up and solving higher-order differential equations. Give your problem to a classmate and see if he or she understands and can solve your problem using the techniques of this and earlier chapters. Rewrite the problem as

necessary to remove any difficulties encountered by your classmate.

18. Write an application in your technology area of interest that requires you to solve a higher-order differential equation in an area different from your application in Exercise 17. As before, give your problem to a classmate and see if he or she understands and can solve your problem using the techniques of this and earlier chapters. Rewrite the problem as necessary to remove any of your classmate's difficulties.

 # CHAPTER 33 REVIEW

Important Terms and Concepts

Auxiliary equation
    Complex roots
    Distinct real roots
    Repeated roots
Electric circuits
Homogeneous equation
Method of undetermined coefficients
Nonhomogeneous equation

Operator
Simple harmonic motion
    Critically damped motion
    Damped motion
    Oscillatory motion
    Overdamped motion
    Undamped motion

Review Exercises

Solve each of the following differential equations.

1. $D^2 y = 0$
2. $(D^2 - 9)y = 0$
3. $(D^2 - 5D + 6)y = 0$
4. $(D^2 - 5D - 14)y = 0$
5. $(D^2 - 6D + 25)y = 0$
6. $(D^2 + 2D - 2)y = 0$
7. $(D^3 - 8)y = 0$

8. $y''' + y'' - y' - y = 0$
9. $(D^2 + 3D - 4)y = 9x^2$
10. $(D^2 - 4D - 5)y = 3e^{2x}$
11. $(D^2 + 7D + 12)y = \cos 5x$
12. $(D^2 - 6D + 8)y = 6x^2 - 2$
13. $(D^2 + 4D + 4) = 4x + e^{2x}$
14. $(D^2 + 4D - 9)y = x\sin 2x$

Solve Exercises 15 and 16.

15. *Physics*   An object of mass $4\,\text{kg}$ stretches a spring $0.2\,\text{m}$. The object is set in motion with an initial velocity of $1\,\text{m/s}$ downward and the force of the air resistance is $10\dfrac{dx}{dt}$. What is the equation that describes the motion of the object?

16. *Electronics*   An *RLC* circuit has $R = 6\,\Omega$, $C = 4 \times 10^{-4}\,\text{F}$, $L = 1\,\text{H}$, and an emf of $16\cos 10t$. **(a)** If $q(0) = i(0) = 0$, what is the charge as a function of time? **(b)** What is the steady-state current?

≡ CHAPTER 33 TEST

What is the auxiliary equation of the differential equation $(5D^2 + 2D - 3)y = 0$?

In Exercises 1–9, solve the differential equation.

1. $(D^2 - 25)y = 0$

2. $(D^2 + 8D - 20)y = 0$

3. $(D - 5)^3(D^2 + 4)y = 0$

4. $(D^2 + 6D + 13)y = 0$

5. $(D^2 + 6D - 7)y = 0$

6. $(D^2 + 6D - 7)y = 6e^{2x}$

7. $(D^2 + 6D - 7)y = 3\cos 4x$

8. $(D^2 + 6D - 7)y = 8x$

9. $(D^2 + 6D - 7)y = e^{4x}$; if $x = 0$, then $y = \frac{4}{11}$ and $Dy = \frac{5}{11}$.

10. Find an expression for the charge and the steady-state current of the *RLC* circuit with $L = 0.5\,\text{H}$, $C = 2 \times 10^{-3}\,\text{F}$, $R = 30\,\Omega$, and an emf of $5\sin 10t$.

CHAPTER
34

Numerical Methods and Laplace Transforms

Laplace transforms are particularly helpful because they allow the use of algebra to solve differential equations. In Section 34.6, we will use Laplace transforms to solve electric circuit problems.

Courtesy of International Business Machines Corporation

In Chapters 32 and 33, we studied some ways to solve differential equations. We looked at a few of the many kinds of differential equations. There are many others, some of which have important applications. For these equations several numerical methods have been developed.

A numerical method for solving a differential equation is a procedure which gives approximate solutions at certain points. Numerical methods use only the four basic operations of addition, subtraction, multiplication and division or the procedure of evaluating functions. A great deal of time and computation is often needed to get the desired solution using numerical methods. For this reason, numerical methods are ideal for the use of calculators and computers.

After we complete our work with numerical methods we will look at Laplace transforms. These transforms provide an algebraic method of obtaining a particular solution to a differential equation with known initial conditions.

34.1
EULER'S OR THE INCREMENT METHOD

The first numerical method for solving differential equations goes by two names. Some books call it **Euler's method** and some call it the **increment method**.

Suppose we have a first-order differential equations of the form $y' = f(x, y)$. Further, we have an initial value for this equation. This means that the graph of this equation would pass through some known point (x_0, y_0). We will find approximate solutions at $x_1, x_2, x_3, x_4, \ldots$, where the difference between any two successive values of x, such as $x_1 - x_0$, $x_2 - x_1$, or $x_n - x_{n-1}$ is the same value $\Delta x = dx$. In keeping with our work with the definition of the derivative, we will let $dx = h$.

Since $y' = \dfrac{dy}{dx} = f(x, y)$, we have $dy = f(x, y)\, dx$ or $dy = f(x, y)h$. Now, in order to find the first approximation, y_1, we find $y_0 + dy_0 = y_0 + f(x_0, y_0)h$. Thus,

$$y_1 = y_0 + f(x_0, y_0)h$$

We know that $x_1 = x_0 + h$, and since we just found the approximation y_1, we can approximate $y_2 = y_1 + f(x_1, y_1)h$. In general, we have the following formula for Euler's method.

Euler's Method for Solving Differential Equations

$$y_{n+1} = y_n + f(x_n, y_n)h$$

This is usually set up in table form, as we will show in Example 34.1.

EXAMPLE 34.1

Find an approximate solution to the differential equation $y' = x - y$, if the curve passes through the point $(0, 2)$.

Solution   For this problem $x_0 = 0$, $y_0 = 2$, and $f(x, y) = x - y$. We select $h = 0.1$ and obtain

$$\begin{aligned} y_{n+1} &= y_n + f(x_n, y_n)h \\ &= y_n + (x_n - y_n)h \end{aligned}$$

Thus,

$$\begin{aligned} y_1 &= y_0 + (x_0 - y_0)h \\ &= 2 + (0 - 2)(0.1) \\ &= 2 + (-0.2) \\ &= 1.8 \end{aligned}$$

EXAMPLE 34.1 (Cont.)

To find y_2, we let $x_1 = x_0 + h = 0 + 0.1 = 0.1$, and so

$$\begin{aligned} y_2 &= y_1 + (x_1 - y_1)h \\ &= 1.8 + (0.1 - 1.8)(0.1) \\ &= 1.8 + (-0.17) \\ &= 1.63 \end{aligned}$$

In this example, we are able to check our work. We have the differential equation $y' = x - y$. Using the technique from Chapter 32, you can determine that this has the particular solution

$$y = 3e^{-x} + x - 1$$

If we check the values when $x = 0.1$, we see that $y = 1.8145$, and when $x = 0.2$, we get $y = 1.6562$. The following table summarizes our results.

n	x_n	y_n	Correct solution $y = 3e^{-x} + x - 1$
0	0	2	2
1	0.1	1.8	1.8145
2	0.2	1.63	1.6562
3	0.3	1.487	1.5225
4	0.4	1.3683	1.4110
5	0.5	1.27147	1.3196

The approximation we get is directly related to the size of h. In order to show that better values are obtained by using smaller values of h, the following table is given. Three values of h were used for this table: 0.1, 0.05, and 0.01. (The first value for h, 0.1, is the same one we just used.) We have left out the intermediate values of x.

	y			
x	$h = 0.1$	$h = 0.05$	$h = 0.01$	Correct solution $y = 3e^{-x} + x - 1$
0	2	2	2	2
0.1	1.8	1.8075	1.81315	1.81451
0.2	1.63	1.64352	1.65372	1.65619
0.3	1.487	1.50528	1.51910	1.52245
0.4	1.3683	1.39026	1.40692	1.41096
0.5	1.27147	1.29621	1.31502	1.31959
0.6	1.19432	1.22108	1.24147	1.24643

EXAMPLE 34.2

Use Euler's method to solve the differential equation $y' = 3x^2 - y^3$, if the curve passes through the point $(1, 2)$.

Solution This example doesn't specify the size of h. We will first let $h = 0.1$ and then let it be 0.05. If $x_0 = 1$, $y_0 = 2$, and $h = 0.1$, then

$$
\begin{aligned}
y_1 = y_0 + f(x_0, y_0)h &= 2 + (3 \cdot 1^2 - 2^3)(0.1) \\
&= 2 + (3 - 8)(0.1) \\
&= 1.5
\end{aligned}
$$

If $h = 0.05$, then

$$
\begin{aligned}
y_1 &= 2 + (3 - 8)(0.05) \\
&= 1.75
\end{aligned}
$$

Notice that when $h = 0.05$, $x_1 = 1.05$, but when $h = 0.1$, $x_1 = 1.10$. Continuing these two situations, we get the following results.

	y	
x	$h = 0.1$	$h = 0.05$
1	2	2
1.05	—	1.75
1.10	1.5	1.64741
1.15	—	1.60536
1.20	1.5255	1.59687
1.25	—	1.60927
1.30	1.60249	1.63526
1.35	—	1.67012
1.40	1.69798	1.71057
1.45	—	1.75431
1.50	1.79643	1.79973

Euler's method is the easiest of many methods that can be used to obtain numerical approximations of the solution to a differential equation. We will look at one other method in Exercise Set 34.1. This second method, called Heun's method, is much more accurate than Euler's method.

Exercise Set 34.1

1. Write a computer program to use Euler's method for solving differential equations. Check your program on Examples 34.1 and 34.2. Your program should allow you to input the size of h and the values of x_0 and y_0.

In Exercises 2–5, solve the given differential equations by Euler's method for the specified value of h. Solve the differential equation exactly and check you approximations against the exact values.

2. $y' = x + 2$, $x_0 = 1$, $y_0 = 4$, $1 \le x \le 2$, $h = 0.1$ and $h = 0.05$

3. $y' = 3y + e^x$, $x_0 = 0$, $y_0 = 0$, $0 \le x \le 1$, $h = 0.1$ and $h = 0.05$

4. $y' = 2xy + 2x$, $x_0 = 0$, $y_0 = 1$, $0 \le x \le 1$, $h = 0.1$

5. $y' = 7y + \sin 2x$, $x_0 = 0$, $y_0 = 1$, $0 \le x \le 0.5$, $h = 0.05$

In Exercises 6–12, solve the given differential equation by using Euler's method for the given initial value of x and y and the given value of h.

6. $y' = xy + 5$, $x_0 = 0$, $y_0 = 0$, $0 \le x \le 1$, $h = 0.1$

7. $y' = x^2 + y^2$, $x_0 = 0$, $y_0 = 1$, $0 \le x \le 1$, $h - 0.1$

8. $y' = x^2 + 4y^2$, $x_0 = 0$, $y_0 = 1$, $0 \le x \le 1$, $h = 0.1$

9. $y' = e^{xy}$, $x_0 = 0$, $y_0 = 0$, $0 \le x \le 1$, $h = 0.1$

10. $y' = \sqrt{4 + xy}$, $x_0 = 0$, $y_0 = 1$, $0 \le x \le 0.5$, $h = 0.05$

11. $y' = \sin(x + y)$, $x_0 = \frac{\pi}{2}$, $y_0 = 1$, $\frac{\pi}{2} \le x \le \frac{3\pi}{4}$, $h = 0.05$

12. $y' = y + \cos x$, $x_0 = \pi$, $y_0 = 2$, $\pi \le x \le \frac{5\pi}{4}$, $h = 0.05$

Solve Exercises 13–16.

13. Heun's method is based on the formula

$$y_{n+1} = y_n + \frac{h}{2}\left[f(x_n, y_n) + f(x_n + h, y_n + n + hy'_n)\right]$$

Write a computer program to use Heun's method for solving differential equations. Check your program on Example 34.1. Your program should allow you to input the size of h and the initial values x_0 and y_0.

14. Use Heun's method to solve Exercises 2–5. Compare your results to the correct solution and to the results you got using Euler's method.

15. *Thermal science* **Stefan's law of radiation** states that the rate of change in temperature of a body at $T(t)$ degrees in a medium at $T_m(t)$ degrees is propor-

tional to $T_m{}^4 - T^4$. Thus, we have

$$\frac{dT}{dt} = k\left(T_m{}^4 - T^4\right)$$

where k is a constant. Let $k = 40^{-4}$ and assume the medium temperature is constant, $T_m(t) = 70°$. If $T(0) = 100°$, use Euler's method with $h = 0.1$ to approximate $T(1)$ and $T(2)$.

16. *Thermal science* According to Newton's law of cooling

$$\frac{dT}{dt} = -k(T - T_m)$$

where k is a positive constant. Let $k = 1$ and assume the medium temperature is constant, $T_m(t) = 70°$. If $T(0) = 100°$, use Euler's method with $h = 0.1$ to approximate $T(1)$ and $T(2)$.

In Your Words

17. What is Euler's method for solving differential equations?

18. Why would you use a numerical technique such as Euler's method for solving differential equations?

☰ 34.2
SUCCESSIVE APPROXIMATIONS

Another method that is used to solve differential equations is based on successive approximations. Using this method, we will develop a function that approximates the actual solution.

As with Euler's method, we begin with a differential equation of the form $y' = f(x, y)$ that has an initial value of (x_0, y_0). Begin by substituting y_0 into

$f(x, y)$. This makes it a function of x only. Then integrate this function, which gives y as a function of x. This we call the first approximation.

If the first approximation is then substituted into the right-hand side of y' and integrated again, a second approximation is obtained. This method can be repeated as long as desired and each time a closer approximation to y is obtained. This method is summarized in the following box.

We will demonstrate this on the differential equations in Examples 34.1 and 34.2 from Section 34.1.

Solving Differential Equations by Successive Approximations

The method of successive approximations to solve the differential equation $y' = f(x, y)$ that has an initial value $f(x_0, y_0)$ uses the following steps:
1. Substitute y_0 into $f(x, y)$ to obtain $f(x, y_0)$.
2. Integrate $f(x, y_0)$ with respect to x.
3. Substitute this result for y in the given equation $y' = f(x, y)$.
4. Integrate this result with respect to x.
5. Repeat steps 3 and 4 to obtain a closer approximation to y.

EXAMPLE 34.3

Obtain the first three approximations to the differential equation $y' = x - y$, if the solution passes through $(0, 2)$.

Solution   We are given $y_0 = 2$. We substitute this for y in the differential equation and get $y' = x - 2$. Integrating, we have

$$\int dy = \int (x - 2)\, dx$$
$$y = \frac{1}{2}x^2 - 2x + C_1$$

We know that this curve passes through $(0, 2)$, and so $C_1 = 2$. Thus we have

$$y = \frac{1}{2}x^2 - 2x + 2$$

as the first approximation.

Next, we substitute this value for y in the differential equation $y' = x - y$ and get

$$y' = x - \left(\frac{1}{2}x^2 - 2x + 2\right)$$
$$= 3x - \frac{1}{2}x^2 - 2$$

Integrating, we obtain

$$y = \frac{3}{2}x^2 - \frac{1}{6}x^3 - 2x + C_2$$

EXAMPLE 34.3 (Cont.)

When $x = 0$, $y = 2$, so $C_2 = 2$ and the second approximation is

$$y = \frac{3}{2}x^2 - \frac{1}{6}x^3 - 2x + 2$$

For the third approximation, we use

$$\begin{aligned}
y' &= x - y \\
&= x - \left(\frac{3}{2}x^2 - \frac{1}{6}x^3 - 2x + 2\right) \\
&= \frac{1}{6}x^3 - \frac{3}{2}x^2 + 3x - 2
\end{aligned}$$

Integrating both sides produces

$$y = \frac{1}{24}x^4 - \frac{1}{2}x^3 + \frac{3}{2}x^2 - 2x + C_3$$

When $x = 0$, $y = 2$, so $C_3 = 2$. The third approximation is

$$y = \frac{1}{24}x^4 - \frac{1}{2}x^3 + \frac{3}{2}x^2 - 2x + 2$$

How accurate is this? When $x = 0.5$ we get $y = 1.31510$. Euler's method with $h = 0.01$ produced $y = 1.31502$ and the actual value is 1.31959. As you can see, this is slightly more accurate than Euler's method with $h = 0.01$.

EXAMPLE 34.4

Obtain the first two approximations to the differential equation $y' = 3x^2 - y^3$, if the curve passes through the point $(1, 2)$.

Solution As before, we begin by substituting the given value for y into the differential equation, getting

$$y' = 3x^2 - (2)^3 = 3x^2 - 8$$

Integrating both sides produces

$$y = x^3 - 8x + C_1$$

Since $y = 2$ when $x = 1$, we can see that $C_1 = 9$, and so the first approximation is

$$y = x^3 - 8x + 9$$

To get the second approximation, we substitute this into the differential equation with the result

$$\begin{aligned}
y' &= 3x^2 - (x^3 - 8x + 9)^3 \\
&= 3x^2 - (x^9 - 24x^7 + 27x^6 + 192x^5 - 432x^4 - 269x^3 \\
&\quad + 1{,}728x^2 - 1{,}944x + 729) \\
&= -x^9 + 24x^7 - 27x^6 - 192x^5 + 432x^4 + 269x^3 - 1{,}725x^2 \\
&\quad + 1{,}944x - 729
\end{aligned}$$

EXAMPLE 34.4 (Cont.)

Integrating both sides and solving for the constant, we have the second approximation

$$y = -\frac{1}{10}x^{10} + 3x^8 - \frac{27}{7}x^7 - 32x^6 + \frac{432}{5}x^5 + \frac{269}{4}x^4 - 575x^3$$
$$+ 972x^2 - 729x + 213.307$$

As you can see, determining the third approximation would be extremely difficult. This method also has the disadvantage that it can be applied only if the resulting differentials can be integrated. But, as we saw in Example 34.3, this often provides a more accurate solution than does Euler's method.

Exercise Set 34.2

In Exercises 1–4, find the third approximation of the solution of the given differential equations whose solution passes through the given point. Evaluate the solution for the indicated value of x. Compare these answers to the exact values and the answers you got using Euler's method in Exercises 2–5 of Exercise Set 34.1.

1. $y' = x + 2$, $(1, 4)$, $x = 2$
2. $y' = 3y + e^x$, $(0, 0)$, $x = 1$
3. $y' = 2xy + 2x$, $(0, 1)$, $x = 1$
4. $y' = 7y + \sin 2x$, $(0, 1)$, $x = 0.5$

In Exercises 5–10, find the second approximation of the solution of the given differential equation whose solution passes through the given point.

5. $y' = x^2 + y^2$, $(0, 1)$
6. $y' = ye^x$, $(0, 1)$
7. $y' = 4 + xy$, $(0, 1)$
8. $y' = x^2 + 4y^2$, $(0, 1)$
9. $y' = \sin 2x + y$, $(\frac{\pi}{2}, 1)$
10. $y' = y + \cos x$, $(\pi, 2)$

Solve Exercises 11–14.

11. Substitute the first few terms of the Maclaurin series for e^{-x} into the expression $3e^{-x} + x - 1$. Compare the results with the solution for Example 34.3.

12. Substitute the first few terms of the Maclaurin series for e^{x^2} into the expression $2e^{x^2} - 1$. Compare the result with the solution for Exercise 3.

13. *Electricity* The current I in an electric circuit at time t is given by

$$I'(t) + 2I(t) = \sin t$$

If $I = 0$ when $t = 0$, calculate I when $t = 0.5$. (Use the third approximation of Euler's method.)

14. *Physics* A 10-kg object is attached to a parachute and dropped from rest. It encounters a force due to air resistance of $4v^2$. The resulting differential equation is

$$\frac{dv}{dt} = 9.8 - 0.4v^2$$

Find the velocity when $t = 0.5$ s by using the third approximation.

In Your Words

15. Describe how the method of successive approximations differs from Euler's method.

16. Is the method of successive approximations or Euler's method easier to use? Why?

≡ 34.3
LAPLACE TRANSFORMS

In Section 34.1, we looked at two numerical methods, Euler's method and Heun's method. In the second section, we looked at the method of successive approximations. In the remainder of this chapter, we will look at a third method of solving differential equations. This last method involves **Laplace transforms**. These transforms give us an algebraic method for finding a particular solution of a differential equation from its initial conditions.

Laplace Transform

The **Laplace transform** of a function $f(t)$ is denoted as either $L(f)$ or $F(s)$ and is defined as

$$L(f) = F(s) = \int_0^\infty e^{-st} f(t)\, dt$$

As we can see from its notation, the Laplace transform is a function of $f(t)$ and a function of s.

You may remember that in Section 27.4 we learned to evaluate an improper integral, such as one whose upper limit is ∞. In general, we use the method

$$\int_0^\infty f(x)\, dx = \lim_{b \to \infty} \int_0^b f(x)\, dx$$

if the limit exists.

EXAMPLE 34.5

Find the Laplace transform of the function $f(t) = 1$.

Solution $L(f) = L(1) = \int_0^\infty 1 \cdot e^{-st}\, dt$

$$= \lim_{b \to \infty} \int_0^b e^{-st}\, dt$$

$$= \lim_{b \to \infty} \left. -\frac{1}{s} e^{-st} \right|_0^b$$

$$= \lim_{b \to \infty} \left(-\frac{1}{s} e^{-sb} + \frac{1}{s} \right) = \frac{1}{s}, \quad s > 0$$

So, $L(1) = \dfrac{1}{s}$, $s > 0$.

Here we assumed $s > 0$ in order to make $\lim_{b \to \infty} e^{-sb} = 0$.

EXAMPLE 34.6

Find the Laplace transform of the function $f(t) = t, t > 0$.

Solution

$$L(f) = \int_0^\infty te^{-st}dt$$

Using integration by parts or Formula 84 from the Table of Integrals in Appendix C, we can evaluate the integral.

$$L(f) = L(t) = \lim_{b \to \infty} \int_0^b te^{-st}\,dt$$

$$= \lim_{b \to \infty} \frac{e^{-st}(-st-1)}{s^2}\Big|_0^b$$

$$= \lim_{b \to \infty}\left[\frac{e^{-sb}(-sb-1)}{s^2} + \frac{1}{s^2}\right] = \frac{1}{s^2}$$

It can be shown that

$$\lim_{b \to \infty} \frac{e^{-sb}(-sb-1)}{s^2} = 0$$

and so we conclude that

$$L(t) = \frac{1}{s^2}, \quad t > 0$$

EXAMPLE 34.7

Find the Laplace transform of the function $f(t) = \sin at$.

Solution $L(f) = L(\sin at) = \int_0^\infty e^{-st}\sin at\,dt$

$$= \lim_{b \to \infty} \int_0^b e^{-st}\sin at\,dt$$

$$= \lim_{b \to \infty}\left[\frac{e^{-st}(-s\sin at - a\cos at)}{a^2+s^2}\right]\Big|_0^b$$

$$= \lim_{b \to \infty}\left[\frac{e^{-sb}(-s\sin ab - a\cos ab)}{a^2+s^2} + \frac{a}{a^2+s^2}\right]$$

$$= 0 + \frac{a}{a^2+s^2} = \frac{a}{s^2+a^2}$$

Thus, $L(\sin at) = \dfrac{a}{s^2+a^2}$.

We could continue in this manner to get more Laplace transforms; in fact, you will be asked to do this as part of the exercises. There are some properties of Laplace transforms that you will find helpful.

In particular, the Laplace transform of a sum of two functions is the sum of their Laplace transforms.

Properties of Laplace Transforms

If $f(t)$ and $g(t)$ are functions and k is a constant, then

1. $L[f(t)+g(t)] = L(f)+L(g)$
2. $L[kf(t)] = kL(f)$

These two properties allow us to say that the Laplace transform is **linear** or has the **linearity property**. A short table of Laplace transforms is given in Table 34.1.

TABLE 34.1   Table of Laplace Transforms

	$f(t)=L^{-1}(F)$	$L(f)=F(s)$		$f(t)=L^{-1}(F)$	$L(f)=F(s)$
1.	1	$\dfrac{1}{s}$	12.	$\sin at - at\cos at$	$\dfrac{2a^3}{(s^2+a^2)^2}$
2.	t	$\dfrac{1}{s^2}$	13.	$t\sin at$	$\dfrac{2as}{(s^2+a^2)^2}$
3.	t^n	$\dfrac{n!}{s^{n+1}}\ (n=1,2,3,\cdots)$	14.	$\sin at + at\cos at$	$\dfrac{2as^2}{(s^2+a^2)^2}$
4.	$\dfrac{1}{\sqrt{\pi t}}$	$\dfrac{1}{\sqrt{s}}$	15.	$t\cos at$	$\dfrac{s^2-a^2}{(s^2+a^2)^2}$
5.	e^{at}	$\dfrac{1}{s-a}$	16.	$\cos at - \cos bt$	$\dfrac{s(b^2-a^2)}{(s^2+a^2)(s^2+b^2)},\ (a^2\neq b^2)$
6.	te^{at}	$\dfrac{1}{(s-a)^2}$	17.	$e^{at}-e^{bt}$	$\dfrac{a-b}{(s-a)(s-b)}$
7.	$t^{n-1}e^{at}$	$\dfrac{(n-1)!}{(s-a)^n}$	18.	$ae^{at}-be^{bt}$	$\dfrac{s(a-b)}{(s-a)(s-b)}$
8.	$\sin at$	$\dfrac{a}{s^2+a^2}$	19.	$e^{-at}\sin bt$	$\dfrac{b}{(s+a)^2+b^2}$
9.	$\cos at$	$\dfrac{s}{s^2+a^2}$	20.	$e^{-at}\cos bt$	$\dfrac{s+a}{(s+a)^2+b^2}$
10.	$1-\cos at$	$\dfrac{a^2}{s(s^2+a^2)}$	21.	$\dfrac{1}{2b^3}e^{-at}\left[\sin(bt)-bt\cos(bt)\right]$	$\dfrac{1}{\left[(s+a)^2+b^2\right]^2}$
11.	$at-\sin at$	$\dfrac{a^3}{s^2(s^2+a^2)}$	22.	$\dfrac{1}{2b^3}e^{-at}\left[abt\cos(bt)+b^2t\sin(bt)-a\sin(bt)\right]$	$\dfrac{s}{\left[(s+a)^2+b^2\right]^2}$

Note: a and b are constants; s and t are variables.

EXAMPLE 34.8

Use Table 34.1 and the linearity property to obtain $L(4\cos 5t + e^{-3t})$.

Solution By the linearity property,

$$L(4\cos 5t + e^{-3t}) = 4L(\cos 5t) + L(e^{-3t})$$

Using transforms 9 and 5 from Table 34.1, we see that

$$4L(\cos 5t) + L\left(e^{-3t}\right) = 4\left(\frac{s}{s^2 + 5^2}\right) + \frac{1}{s+3}$$

$$= \frac{4s}{s^2 + 25} + \frac{1}{s+3}$$

EXAMPLE 34.9

Use Table 34.1 and the linearity property to obtain $L\left(\sin^2 5t\right)$.

Solution We need to use the trigonometric identity

$$\sin^2\theta = \frac{1}{2}(1 - \cos 2\theta)$$

with $\theta = 5t$. This leads to $\sin^2 5t = \frac{1}{2} - \frac{1}{2}\cos 10t$, and so we have

$$L\left(\sin^2 5t\right) = L\left(\frac{1}{2} - \frac{1}{2}\cos 10t\right)$$

By Table 34.1, $L\left(\frac{1}{2}\right) = \frac{1}{2}L(1) = \frac{1}{2}\left(\frac{1}{s}\right)$ and

$$L\left(\frac{1}{2}\cos 10t\right) = \frac{1}{2}L(\cos 10t)$$

$$= \frac{1}{2}\left(\frac{s}{s^2 + 10^2}\right) = \frac{1}{2}\left(\frac{s}{s^2 + 100}\right)$$

Thus,

$$L\left(\sin^2 5t\right) = L\left(\frac{1}{2} - \frac{1}{2}\cos 10t\right)$$

$$= \frac{1}{2}\left(\frac{1}{s}\right) - \frac{1}{2}\left(\frac{s}{s^2 + 100}\right)$$

$$= \frac{50}{s(s^2 + 100)}$$

Exercise Set 34.3

In Exercises 1–4, use the definition to find the Laplace transform of $f(t)$. Check your answer with the transform in Table 34.1.

1. $f(t) = n$ **2.** $f(t) = e^{-at}$ **3.** $f(t) = \cos at$ **4.** $f(t) = te^{at}$

In Exercises 5–12, find the transforms of the given functions by using Table 34.1.

5. $f(t) = t^3$

6. $f(t) = e^{2t}$

7. $f(t) = \sin 6t$

8. $f(t) = e^{2t} \sin 6t$

9. $f(t) = t^3 e^{5t}$

10. $f(t) = e^{-5t} \sin 10t$

11. $f(t) = \sin 4t + 4t \cos 4t$

12. $f(t) = t \cos 7t$

In Exercises 13–22, find the transforms of the given functions by using Table 34.1 and the linearity property.

13. $f(t) = \sin 5t + \cos 3t$

14. $f(t) = e^{3t} \sin 5t + 4t^8$

15. $f(t) = 3t \sin 5t - \cos 4t + 1$

16. $f(t) = t^3 e^{4t} + 4e^{3t}$

17. $f(t) = e^{-2t} \cos 3t - 2e^{5t} \sin t$

18. $f(t) = t \cos 3t - 2t \sin 2t$

19. $f(t) = 4 + 3t + 2e^t$

20. $f(t) = 6 - 4t + 2 \sin 10t + e^{5t}$

21. $f(t) = \cos^2 3t$

22. $f(t) = 5e^{2t} - t^3$

 In Your Words

23. Without looking in the text, give the definition of the Laplace transform.

24. What are the two properties of Laplace transforms? Give an example showing how each is used.

≡ 34.4
INVERSE LAPLACE TRANSFORMS AND TRANSFORMS OF DERIVATIVES

In Section 34.3, we were given a function $f(t)$ and we were able to find its Laplace transform $L(f) = F(s)$. We can reverse this process.

Inverse Laplace Transforms ▬▬▬▬▬

Suppose we have a function $F(s)$. Can we find the **inverse Laplace transform** $L^{-1}(F) = f(t)$? The table of Laplace transforms (Table 34.1) will often help us find the answer.

EXAMPLE 34.10

If $F(s) = \dfrac{1}{s+9}$, find $L^{-1}(F)$.

Solution   If we look in Table 34.1 under the column headed $L(f) = F(s)$, we find the fifth entry is $\dfrac{1}{s-a}$. If $a = -9$, this is the desired form of the formula. The column to the left is headed $f(t) = L^{-1}(F)$ and the fifth entry in that column is e^{at}. We know that $a = -9$, so the answer is

$$L^{-1}\left(\frac{1}{s+9}\right) = e^{-9t}$$

EXAMPLE 34.11

If $F(s) = \dfrac{12}{s^2+16}$, what is $f(t)$?

Solution   Since $f(t) = L^{-1}(F)$, we want to find the inverse Laplace transform. If we look in Table 34.1, we do not find any form that fits. There are two forms that have $s^2 + a^2$ for a denominator. If $a = 4$, the denominator is $s^2 + 16$, which is the one in our problem. One of these, in row 9, has the variable s in the numerator. Our problem does not have an s in the numerator, so this must not be the correct form.

The other form, in row 8 of Table 34.1, has a numerator of a constant, a. We let $a = 4$ and our numerator is $12 = 3 \cdot 4$. We can rewrite this problem as

$$L^{-1}\left(\frac{12}{s^2+16}\right) = 3L^{-1}\left(\frac{4}{s^2+16}\right) = 3\sin 4t$$

EXAMPLE 34.12

Find $L^{-1}\left(\dfrac{4s^2+9s}{s^4+6s^2+9}\right)$.

Solution   We notice that $s^4 + 6s^2 + 9 = (s^2+3)^2$, so

$$\frac{4s^2+9s}{s^4+6s^2+9} = \frac{4s^2+9s}{(s^2+3)^2} = \frac{4s^2}{(s^2+3)^2} + \frac{9s}{(s^2+3)^2}$$

Compare $\dfrac{4s^2}{(s^2+3)^2}$ to transpose 14, $\dfrac{2as^2}{(s^2+a^2)^2}$. If $a^2 = 3$, then $a = \sqrt{3}$ and $4s^2 = \left(\dfrac{4}{2\sqrt{3}}\right)2\sqrt{3}s^2 = \left(\dfrac{2}{\sqrt{3}}\right)2\sqrt{3}s^2$. Thus, we can rewrite $\dfrac{4s^2}{(s^2+3)^2} = \dfrac{2}{\sqrt{3}} \cdot \dfrac{2\sqrt{3}s^2}{(s^2+3)^2}$.

Similarly, compare $\dfrac{9s}{(s^2+3)^2}$ to transform 13, $\dfrac{2as}{(s^2+a^2)^2}$. Again, $a = \sqrt{3}$, so

EXAMPLE 34.12 (Cont.)

$\dfrac{9s}{(s^2+3)^2} = \dfrac{9}{2\sqrt{3}} \cdot \dfrac{2\sqrt{3}\,s}{(s^2+3)^2}$. Thus, we can rewrite the displayed equation as

$$\frac{4s^2+9s}{s^4+6s^2+9} = \frac{2}{\sqrt{3}} \cdot \frac{2\sqrt{3}\,s^2}{(s^2+3)^2} + \frac{9}{2\sqrt{3}} \cdot \frac{2\sqrt{3}\,s}{(s^2+3)^2}$$

By using transforms 14 and 13, respectively, we see that

$$L^{-1}\left(\frac{4s^2+9s}{s^4+6s^2+9}\right) = \frac{2}{\sqrt{3}}L^{-1}\left(\frac{2\sqrt{3}\,s^2}{(s^2+3)^2}\right) + \frac{9}{2\sqrt{3}}L^{-1}\left(\frac{2\sqrt{3}\,s}{(s^2+3)^2}\right)$$

$$= \frac{2}{\sqrt{3}}(\sin\sqrt{3}t + \sqrt{3}t\cos\sqrt{3}t) + \frac{9}{2\sqrt{3}}t\sin\sqrt{3}t$$

Transforms of Derivatives

When we are solving a differential equation, there is another Laplace transform that is important to finding the solution. This involves finding the transform of the derivative of a function.

We will begin by finding the Laplace transform of the first derivative of a function. If $f(t)$ is a function with first derivative $f'(t)$, then by the definition of a Laplace transform

$$L(f') = \int_0^\infty e^{-st} f'(t)\,dt$$

We can integrate this by parts, if we let $u = e^{-st}$ and $dv = f'(t)\,dt$. Then we have the following table:

$u = e^{-st}$	$v = f(t)$
$du = -se^{-st}\,dt$	$dv = f'(t)\,dt$

And so,

$$\int_0^\infty e^{-st} f'(t)\,dt = e^{-st} f(t)\Big|_0^\infty + s\int_0^\infty e^{-st} f(t)\,dt = 0 - f(0) + sL(f)$$

Notice that the integral of the last term on the right-hand side is the Laplace transform of $f(t)$. So, the Laplace transform of $f'(t)$ is

$$L(f') = sL(f) - f(0)$$

If we followed the same procedure on the second derivative of f, $f''(t)$, we would get the Laplace transform

$$L(f'') = s^2L(f) - sf(0) - f'(0)$$

These results are summarized in the following box.

Laplace Transforms of Derivatives

The Laplace transforms of the first derivative, $f'(t)$, and the second derivative, $f''(t)$, are

$$L(f') = sL(f) - f(0) \qquad\qquad (*)$$
$$L(f'') = s^2 L(f) - sf(0) - f'(0) \qquad\qquad (**)$$

EXAMPLE 34.13

If $f(0) = 2$, express the transform of $f'(t) - 5f(t)$ in terms of s and the transform of $f(t)$.

Solution   We will take the Laplace transform of the expression, and by the linearity property we have

$$L(f') - 5L(f)$$

From $(*)$ we have $L(f') = sL(f) - f(0)$, and so, with this substitution, the expression becomes

$$sL(f) - f(0) - 5L(f)$$

We are given the fact that $f(0) = 2$, and, with $L(0) = 0$, the expression becomes

$$sL(f) - 2 - 5L(f)$$
$$\text{or} \qquad (s - 5)L(f) - 2$$

EXAMPLE 34.14

If $f(0) = 1$ and $f'(0) = 4$, express the transform of $f''(t) - f'(t) - 2f(t)$ in terms of s and the transform of $f(t)$.

Solution   Taking the Laplace transform of the expression and using the linearity property, we have

$$L(f'') - L(f') - 2L(f)$$

Using the Laplace transforms of the first and second derivatives for f as given in $(*)$ and $(**)$, we get

$$[s^2 L(f) - sf(0) - f'(0)] - [sL(f) - f(0)] - 2L(f)$$

We are given $f(0) = 1$ and $f'(0) = 4$, so the expression becomes

$$[s^2 L(f) - s(1) - 4] - [sL(f) - 1] - 2L(f)$$

or

$$(s^2 - s - 2)L(f) - s - 4 + 1$$
$$(s^2 - s - 2)L(f) - (s + 3)$$

After Section 34.5, we will combine our knowledge of Laplace transforms from this and Section 34.3 to solve differential equations.

Exercise Set 34.4

In Exercises 1–10, find the inverse Laplace transforms of the given function.

1. $\dfrac{4}{s-5}$

2. $\dfrac{6}{s^2+4}$

3. $\dfrac{16}{s^3+16s}$

4. $\dfrac{4}{(s+9)^2}$

5. $\dfrac{3s}{s^2+9}$

6. $\dfrac{s+6}{(s+16)^2}$

7. $\dfrac{5s}{s^2+9s+14}$

8. $\dfrac{s^2-16}{(s^2+4)^2}$

9. $\dfrac{2s+6s^2}{(s^2+25)^2}$

10. $\dfrac{s+5}{s^2+10s+26}$

In Exercises 11–16, express the transforms of the given expression in terms of s and $L(f)$.

11. $f''(t)+4f'(t)$, $f(0)=1$, $f'(0)=0$

12. $f''(t)-6f(t)$, $f(0)=3$, $f'(0)=2$

13. $2f''(t)-3f'(t)+f(t)$, $f(0)=1$, $f'(0)=0$

14. $y''-4y'$, $f(0)=2$, $f'(0)=1$

15. $y''+6y'-y$, $f(0)=1$, $f'(0)=-1$

16. $y''-3y'+4y$, $f(0)=-2$, $f'(0)=1$

 **In Your Words**

17. Without looking in the text, write the Laplace transforms of derivatives.

18. What is an inverse Laplace transform? How do you find an inverse Laplace transform?

☰ 34.5

PARTIAL FRACTIONS

We are now going to introduce a topic that is often included as an integration technique. Rather than introduce it as an integration method, we chose to let you rely on the table of integrals. But, we have now reached a point where the topic will be valuable and will help us use Laplace transforms to solve differential equations.

It is often necessary to find an inverse Laplace transform, such as

$$L^{-1}\left\{\frac{7}{(s-5)(s+2)}\right\}$$

If you look in Table 34.1 you will not find one for this type of fraction. But, as we will soon show, it is possible to rewrite this fraction as

$$\frac{1}{s-5}-\frac{1}{s+2}$$

We can take the inverse transform of each of these. In fact,

$$L^{-1}\left\{\frac{7}{(s-5)(s+2)}\right\}=L^{-1}\left(\frac{1}{s-5}\right)-L^{-1}\left(\frac{1}{s+2}\right)=e^{5t}-e^{-2t}$$

What we have done is divided the original fraction into **partial fractions**.

If you have an expression of the form $\dfrac{p(x)}{q(x)}$, where $p(x)$ and $q(x)$ are polynomials, and if the degree of p is less than the degree of q, then it is possible to write

$$\frac{p(x)}{q(x)} = F_1 + F_2 + F_3 + \cdots + F_k$$

where each F_i is of the form

$$\frac{A}{(ax+b)^m} \quad \text{or} \quad \frac{Bx+C}{(ax^2+bx+c)^n}$$

where A, B, and C represent constants and m and n are positive integers. One final condition is that $ax^2 + bx + c$ is irreducible in that it has a negative discriminate; that is, $b^2 - 4ac < 0$. The sum on the right-hand side, $F_1 + F_2 + \cdots + F_k$, is called **partial fraction decomposition** of $\frac{p(x)}{q(x)}$ and each F_i is a **partial fraction**.

The process of finding the partial fraction decomposition is rather straightforward; but, it is also quite lengthy. The process involves four steps, as shown in the following box.

Guidelines for Partial Fraction Decomposition

1. If the degree of the numerator $p(x) \geq$ degree of the denominator $q(x)$, divide $p(x)$ by $q(x)$ and work with the remainder.
2. Factor the denominator $q(x)$ into its linear factors $(ax+b)$ and its irreducible quadratic factors (ax^2+bx+c). If any of these are also factors of the numerator, then they should be cancelled. Collect all repeated factors so that the factors are of the form $(ax+b)^m$ and $(ax^2+bx+c)^n$.
3. For each linear factor of the form $(ax+b)^m$, the decomposition contains a sum of m partial fractions of the form

$$\frac{A_1}{ax+b} + \frac{A_2}{(ax+b)^2} + \frac{A_3}{(ax+b)^3} + \cdots + \frac{A_m}{(ax+b)^m}$$

where $A_1, A_2, A_3, \ldots, A_m$ are constants.
4. For each quadratic factor $(ax^2+bx+c)^n$, the decomposition contains a sum of n partial fractions of the form

$$\frac{A_1x+B_1}{ax^2+bx+c} + \frac{A_2x+B_2}{(ax^2+bx+c)^2} + \cdots + \frac{A_nx+B_n}{(ax^2+bx+c)^n}$$

where $A_1, B_1, A_2, B_2, \ldots, A_n, B_n$ are all constants.

EXAMPLE 34.15

Find the partial fraction decomposition of

$$\frac{x+1}{x^3+x^2-6x}$$

EXAMPLE 34.15 (Cont.)

Solution The denominator factors into $x(x+3)(x-2)$, so the partial fraction decomposition is of the form

$$\frac{x+1}{x^3+x^2-6x} = \frac{A}{x} + \frac{B}{x+3} + \frac{C}{x-2}$$

Multiplying both sides by the common denominator x^3+x^2-6x, we get

$$x+1 = A(x+3)(x-2) + Bx(x-2) + Cx(x+3)$$
$$= (A+B+C)x^2 + (A-2B+3C)x - 6A$$

The coefficients of corresponding terms must be equal. This gives the simultaneous equations

$$A+B+C=0 \qquad \text{from the } x^2 \text{ terms}$$
$$A-2B+3C=1 \qquad \text{from the } x \text{ terms}$$
$$-6A=1 \qquad \text{from the constant terms}$$

Solving these simultaneously, we get $A = -\frac{1}{6}$, $B = -\frac{2}{15}$, and $C = \frac{3}{10}$.
The partial fraction decomposition then becomes

$$\frac{x+1}{x^3+x^2-6x} = \frac{-1/6}{x} + \frac{-2/15}{x+3} + \frac{3/10}{x-2}$$

We used simultaneous linear equations to determine the values of the constants A, B, and C. In the next example, we will show a method that can be used to reduce the number of equations that need to be solved.

EXAMPLE 34.16

Find the partial fraction decomposition of

$$\frac{2x+3}{x^3+2x^2+x}$$

Solution The degree of the numerator is less than the denominator. We factor the denominator $x^3+2x^2+x = x(x+1)^2$. These are all linear factors. Thus, we know that the partial fraction decomposition is of the form

$$\frac{2x+3}{x(x+1)^2} = \frac{A}{x} + \frac{B}{x+1} + \frac{C}{(x+1)^2}$$

Notice that the denominator had one factor of x and two factors of $x+1$. The partial fraction decomposition has one fraction with a denominator of x and two with $x+1$ in the denominator.

We need to determine the constants A, B, and C. If we clear the fractions by multiplying both sides of this equation $x(x+1)^2$, we will get

$$2x+3 = A(x+1)^2 + Bx(x+1) + Cx$$

There are a couple of approaches that can be used to solve this for A, B, and C. This equation must be true for all values of x. If $x=0$, two of these terms will

EXAMPLE 34.16 (Cont.)

be 0, and the equation becomes

$$2(0) + 3 = A(0+1)^2 + B(0)(0+1) + C(0)$$

or $\qquad 3 = A$

Since $x + 1 = 0$ when $x = -1$, if we let $x = -1$, then two other terms will become 0, and the equation becomes

$$2(-1) + 3 = A(-1+1)^2 + B(-1)(-1+1) + C(-1)$$

or $\qquad 1 = -C$

and so $C = -1$.

The equation can now be written as

$$2x + 3 = 3(x+1)^2 + Bx(x+1) - x$$
$$= 3(x^2 + 2x + 1) + B(x^2 + x) - x$$
$$= (3+B)x^2 + (6+B-1)x + 3$$

Since there is no x^2 term on the left-hand side, then $3 + B = 0$ or $B = -3$. We now have the values of A, B, and C and can write

$$\frac{2x+3}{x(x+1)^2} = \frac{3}{x} + \frac{-3}{x+1} + \frac{-1}{(x+1)^2}$$

EXAMPLE 34.17

Find the partial fraction decomposition of

$$\frac{x^2+1}{2x^4 - x^3 - x}$$

Solution $\quad \dfrac{x^2+1}{2x^4 - x^3 - x} = \dfrac{x^2+1}{x(x-1)(2x^2+x+1)}$

$$= \frac{A}{x} + \frac{B}{x-1} + \frac{Cx+D}{2x^2+x+1}$$

To clear the fractions, we multiply both sides by $x(x-1)(2x^2+x+1)$ and obtain

$$x^2 + 1 = A(x-1)(2x^2+x+1) + Bx(2x^2+x+1) + (Cx+D)x(x-1)$$

We know that $x(x-1)(2x^2+x+1) = 0$ when $x = 0$ and when $x = 1$. Using each of these in turn, we get the following results:

When $x = 0$:	$1 = -A$, so $A = -1$
When $x = 1$:	$1 + 1 = B \cdot 1(2+1+1)$
And so,	$2 = 4B$ or $B = \dfrac{1}{2}$

The equation can now be written as

$$x^2 + 1 = -1(x-1)(2x^2+x+1) + \frac{x}{2}(2x^2+x+1)$$
$$+ (Cx+D)x(x-1)$$

EXAMPLE 34.17 (Cont.)

If we combine the like powers of x on the right-hand side, we have

$$x^2 + 1 = (-2 + 1 + C)x^3 + \left(2 - 1 + \frac{1}{2} - C + D\right)x^2$$

$$+ \left(1 - 1 + \frac{1}{2} - D\right)x + 1$$

$$= (C - 1)x^3 + \left(\frac{3}{2} - C + D\right)x^2 + \left(\frac{1}{2} - D\right)x + 1$$

Since there is no x^3-term on the left-hand side, we can see that $C - 1 = 0$ and $C = 1$. Likewise, there is no x-term on the left-hand side, so $\frac{1}{2} - D = 0$ and $D = \frac{1}{2}$. Thus, the partial fraction decomposition is

$$\frac{x^2 + 1}{2x^4 - x^3 + x} = \frac{-1}{x} + \frac{\frac{1}{2}}{x - 1} + \frac{x + \frac{1}{2}}{2x^2 + x + 1}$$

If we wanted to take the inverse Laplace transforms of these partial fractions we would easily find the first two in Table 34.1. But there is none in the table like the third. The closest are the two that have $(s + a)^2 + b^2$ as a denominator. We need to write this partial fraction so its denominator is in this form. To do this, we will complete the square.

$$\frac{x + \frac{1}{2}}{2x^2 + x + 1} = \frac{x + \frac{1}{2}}{2\left(x^2 + \frac{1}{2}x + \frac{1}{2}\right)}$$

$$= \frac{1}{2} \cdot \frac{x + \frac{1}{2}}{\left(x^2 + \frac{1}{2}x + \frac{1}{16}\right) + \left(\frac{1}{2} - \frac{1}{16}\right)}$$

$$= \frac{1}{2} \cdot \frac{x + \frac{1}{2}}{\left(x + \frac{1}{4}\right)^2 + \frac{7}{16}}$$

$$= \frac{1}{2}\left[\frac{x + \frac{1}{4}}{\left(x + \frac{1}{4}\right)^2 + \left(\frac{\sqrt{7}}{4}\right)^2} + \frac{\frac{1}{\sqrt{7}} \cdot \frac{\sqrt{7}}{4}}{\left(x + \frac{1}{4}\right)^2 + \left(\frac{\sqrt{7}}{4}\right)^2}\right]$$

Now, if we take the inverse Laplace transform, we obtain

$$L^{-1}\left(\frac{x + \frac{1}{2}}{2x^2 + x + 1}\right) = \frac{1}{2}L^{-1}\left[\frac{x + \frac{1}{4}}{\left(x + \frac{1}{4}\right)^2 + \left(\frac{\sqrt{7}}{4}\right)^2}\right]$$

$$+ \frac{1}{2\sqrt{7}} L^{-1} \left[\frac{\frac{\sqrt{7}}{4}}{\left(x+\frac{1}{4}\right)^2 + \left(\frac{\sqrt{7}}{4}\right)^2} \right]$$

$$= \frac{1}{2} e^{-t/4} \cos \frac{\sqrt{7}}{4} t + \frac{1}{2\sqrt{7}} e^{-t/4} \sin \frac{\sqrt{7}}{4} t$$

$$= \frac{1}{2} e^{-t/4} \left(\cos \frac{\sqrt{7}}{4} t + \frac{1}{\sqrt{7}} \sin \frac{\sqrt{7}}{4} t \right)$$

You should notice that even after we had completed the square and written the denominator in the correct form, we did not have the numerator in the correct form. We had to rewrite the fraction into two fractions so that the numerators were correct.

EXAMPLE 34.18

Find the partial fraction decomposition of

$$\frac{2x^3 + 3}{x(x^2+1)^2}$$

Solution We write

$$\frac{2x^3 + 3}{x(x^2+1)^2} = \frac{A}{x} + \frac{Bx+C}{x^2+1} + \frac{Dx+E}{(x^2+1)^2}$$

Multiplying both sides by $x(x^2+1)^2$ produces the equation

$$2x^3 + 3 = A(x^2+1)^2 + (Bx+C)x(x^2+1) + (Dx+E)x$$

If $x = 0$, we see that $A = 3$, and can write the equation as

$$2x^3 + 3 = 3(x^2+1)^2 + (Bx+C)x(x^2+1) + (Dx+E)x$$

Rewriting and combining like terms changes the equation to

$$2x^3 + 3 = (3+B)x^4 + Cx^3 + (6+B+D)x^2 + (C+E)x + 3$$

From this we see that $B = -3$, $C = 2$, $D = -3$, and $E = -2$. Thus, the partial decomposition is

$$\frac{2x^3 + 3}{x(x^2+1)^2} = \frac{3}{x} + \frac{-3x+2}{x^2+1} + \frac{-3x-2}{(x^2+1)^2}$$

EXAMPLE 34.19

Find

$$L^{-1}\left[\frac{2s^3+3}{s(s^2+1)^2}\right]$$

Solution We found the partial fraction decomposition of this fraction in Example 34.18. Thus

$$L^{-1}\left[\frac{2s^3+3}{s(s^2+1)^2}\right] = L^{-1}\left(\frac{3}{s}\right) + L^{-1}\left(\frac{-3s+2}{s^2+1}\right) + L^{-1}\left[\frac{-3s-2}{(s^2+1)^2}\right]$$

$$= 3 - 3\cos t + 2\sin t - \frac{3}{2}t\sin t - \sin t + t\cos t$$

$$= 3 + (t-3)\cos t + \left(1 - \frac{3}{2}t\right)\sin t$$

Exercise Set 34.5

In Exercises 1–10, find the partial fraction decomposition of the given fraction.

1. $\dfrac{1}{x(x+1)}$

2. $\dfrac{4}{x(x-3)}$

3. $\dfrac{4}{x(x-1)(x+1)}$

4. $\dfrac{5x}{(x-2)(x+3)}$

5. $\dfrac{3x}{x^2-8x+15}$

6. $\dfrac{4x}{(x+1)(x+2)(x+3)}$

7. $\dfrac{x-1}{x^2+2x+1}$

8. $\dfrac{x}{(x+1)^2(x-2)}$

9. $\dfrac{x^3+x^2+x+2}{(x^2+1)(x^2+2)}$

10. $\dfrac{3x^2+5}{(x^2+1)^2}$

Use the method of partial fractions to find the inverse Laplace transforms of the functions in Exercises 11–26. (For Exercises 11–20, use the decomposition from Exercises 1–10.)

11. $L^{-1}\left[\dfrac{1}{s(s+1)}\right]$

12. $L^{-1}\left[\dfrac{4}{s(s-3)}\right]$

13. $L^{-1}\left[\dfrac{4}{s(s-1)(s+1)}\right]$

14. $L^{-1}\left[\dfrac{5s}{(s-2)(s+3)}\right]$

15. $L^{-1}\left(\dfrac{3s}{s^2-8s+15}\right)$

16. $L^{-1}\left[\dfrac{4s}{(s+1)(s+2)(s+3)}\right]$

17. $L^{-1}\left(\dfrac{s-1}{s^2+s+1}\right)$

18. $L^{-1}\left[\dfrac{s}{(s+1)^2(s-2)}\right]$

19. $L^{-1}\left[\dfrac{s^3+s^2+s+2}{(s^2+1)(s^2+2)}\right]$

20. $L^{-1}\left[\dfrac{3s+5}{(s^2+1)^2}\right]$

21. $L^{-1}\left[\dfrac{2}{s^2(s-1)}\right]$

22. $L^{-1}\left[\dfrac{s}{(s+1)(s^2+1)}\right]$

23. $L^{-1}\left[\dfrac{1}{(s+4)(s^2+9)}\right]$

24. $L^{-1}\left[\dfrac{1}{(s^2+1)(s^2-1)}\right]$

25. $L^{-1}\left[\dfrac{1}{(s^2+1)(s-1)^2}\right]$

26. $L^{-1}\left[\dfrac{1}{(s^2+4s+7)(s+1)}\right]$

In Your Words

27. What is a partial fraction?

28. Without looking in the text, outline the guidelines for partial fraction decomposition.

29. The guidelines for partial fraction decomposition only discuss what to do when $q(x)$ is composed of linear factors of the form $ax + b$ and irreducible quadratic factors of the form $ax^2 + bx + c$. What are you supposed to do if q has a factor of degree higher than 2?

≡ 34.6
USING LAPLACE TRANSFORMS TO SOLVE DIFFERENTIAL EQUATIONS

We are now ready to use Laplace transforms to solve differential equations. What we will find are the **particular** solutions of the equation subject to the given initial conditions.

This procedure is relatively direct. It has three parts.

> **Steps to Solving Differential Equations by Laplace Transforms**
>
> 1. Find the Laplace transforms of both sides of the differential equation.
> 2. Solve the resulting algebraic expression for $Y(s)$.
> 3. Find the inverse transform $y = L^{-1}[Y(s)]$.

 **Note**

We have changed our notation. This is more in keeping with the other differential equations that we have solved. We will let $y = f(t)$. Then $L(f) = L(y) = Y(s)$ and $f(0) = y(0)$, $f'(0) = y'(0)$, and so on. Remember from Section 34.4 that $L(y') = sL(y) - y(0)$ and $L(y'') = s^2L(y) - sy(0) - y'(0)$.

EXAMPLE 34.20

Solve the differential equation $y' - y = 2\sin t$, when $y(0) = 0$.

Solution   We take the Laplace transform of both sides:

$$L(y') - L(y) = L(2\sin t)$$

We have $L(y') = sL(y) - y(0) = sY - y(0)$, because $L(y) = Y$. Since $y(0) = 0$, the left-hand side becomes

$$sY - 0 - Y = Y(s - 1)$$

On the right-hand side, $L(2\sin t) = \dfrac{2}{s^2 + 1}$. Thus, we have

$$Y(s - 1) = \frac{2}{s^2 + 1}$$

or $\qquad Y = \dfrac{2}{(s - 1)(s^2 + 1)}$

EXAMPLE 34.20 (Cont.)

and so

$$y = L^{-1} \left[\frac{2}{(s-1)(s^2+1)} \right]$$

To find this inverse transform, we will need to use partial fractions.

$$\frac{2}{(s-1)(s^2+1)} = \frac{A}{s-1} + \frac{Bs+C}{s^2+1}$$

or

$$2 = A(s^2+1) + (s-1)(Bs+C)$$
$$= (As^2+A) + (Bs^2 - Bs + Cs - C)$$
$$= (A+B)s^2 + (C-B)s + (A-C)$$

Thus,

$$A + B = 0$$
$$-B + C = 0$$
$$A - C = 2$$

Solving these simultaneous equations, we get $A = 1$, $B = -1$, and $C = -1$. We conclude that

$$\frac{2}{(s-1)(s^2+1)} = \frac{1}{s-1} + \frac{-s-1}{s^2+1}$$

and

$$y = L^{-1} \left(\frac{1}{s-1} + \frac{-s-1}{s^2+1} \right)$$
$$= L^{-1} \left(\frac{1}{s-1} \right) - L^{-1} \left(\frac{s}{s^2+1} \right) - L^{-1} \left(\frac{1}{s^2+1} \right)$$
$$= e^t - \cos t - \sin t$$

The solution to this differential equation is

$$y = e^t - \cos t - \sin t$$

EXAMPLE 34.21

Solve the differential equation $y'' + 2y' + y = e^{-t}$, where $y(0) = 1$ and $y'(0) = 2$.

Solution   We begin by taking Laplace transforms of both sides, obtaining

$$L(y'') + 2L(y') + L(y) = L(e^{-t}) \tag{*}$$

Now, since

$$L(y'') = s^2 Y(s) - sy(0) - y(0)$$

and

$$L(y') = sY(s) - y(0)$$

we have

$$L(y'') = s^2 Y(s) - s(1) - 2$$

EXAMPLE 34.21 (Cont.)

$$= s^2 Y(s) - s - 2$$

and $\quad L(y') = sY(s) - 1$

Substituting these into equation (∗) produces

$$[s^2 Y(s) - s - 2] + 2[sY(s) - 1] + Y(s) = \frac{1}{s+1}$$

$$(s^2 + 2s + 1)Y(s) - s - 4 = \frac{1}{s+1}$$

or

$$(s^2 + 2s + 1)Y(s) = s + 4 + \frac{1}{s+1}$$

and

$$Y(s) = \frac{s}{s^2 + 2s + 1} + \frac{4}{s^2 + 2s + 1} + \frac{1}{(s+1)(s^2 + 2s + 1)}$$

$$= \frac{s}{(s+1)^2} + \frac{4}{(s+1)^2} + \frac{1}{(s+1)^3}$$

Taking the inverse Laplace transform, we get

$$y = L^{-1}[Y(s)] = L^{-1}\left[\frac{s}{(s+1)^2}\right] + L^{-1}\left[\frac{4}{(s+1)^2}\right] + L^{-1}\left[\frac{1}{(s+1)^3}\right]$$

$$= e^{-t}(1 - t) + 4te^{-t} + \frac{1}{2}t^2 e^{-t}$$

$$= \left(\frac{1}{2}t^2 + 3t + 1\right)e^{-t}$$

From these two examples, you should see that this method of solving differential equations is basically an algebraic method. This is one of the big advantages of Laplace transforms. We can translate a differential equation into an algebraic equation, which is then changed into the solution to the differential equation.

EXAMPLE 34.22

Solve $y'' - 2y' = 4$, where $y(0) = -1$ and $y'(0) = 2$.

Solution
$$L(y'') - 2L(y') = L(4)$$

$$[s^2 Y(s) - s y(0) - y'(0)] - 2[s Y(s) - y(0)] = \frac{4}{s}$$

$$[s^2 Y(s) - s(-1) - 2] - 2[s Y(s) - (-1)] = \frac{4}{s}$$

$$s^2 Y(s) + s - 2 - 2s Y(s) - 2 = \frac{4}{s}$$

$$(s^2 - 2s) Y(s) + (s - 4) = \frac{4}{s}$$

$$(s^2 - 2s) Y(s) = \frac{4}{s} - s + 4$$

$$= \frac{-s^2 + 4s + 4}{s}$$

$$Y(s) = \frac{-s^2 + 4s + 4}{s(s^2 - 2s)}$$

$$= \frac{-s^2 + 4s + 4}{s^2(s - 2)}$$

The partial fraction decomposition of the right-hand side is

$$\frac{-s^2 + 4s + 4}{s^2(s - 2)} = \frac{-3}{s} + \frac{-2}{s^2} + \frac{2}{s - 2}$$

and so

$$y = L^{-1}[Y(s)] = L^{-1}\left[\frac{-s^2 + 4s + 4}{s^2(s - 2)}\right]$$

$$= L^{-1}\left(\frac{-3}{s}\right) + L^{-1}\left(\frac{-2}{s^2}\right) + L^{-1}\left(\frac{2}{s - 2}\right)$$

$$= -3 - 2t + 2e^{2t}$$

As the last example in this chapter, and in this book, we will use an application from electricity.

Application

EXAMPLE 34.23

Recall that for an RLC series circuit, the differential equation for the instantaneous charge $q(t)$ on the capacitor is

$$L\frac{d^2 q}{dt^2} + R\frac{dq}{dt} + \frac{1}{C}q = E(t)$$

Determine the charge $q(t)$ and current $i(t)$ for a series with $R = 20\,\Omega$, $L = 1\,\text{H}$, $C = 0.01\,\text{F}$, $E(t) = 120 \sin 10t$ V, $q(0) = 0$, and $i(0) = 0$.

EXAMPLE 34.23 (Cont.)

Solution Using the given values for R, L, C, and $E(t)$, the differential equation is

$$\frac{d^2q}{dt^2} + 20\frac{dq}{dt} + 100q = 120\sin 10t$$

Taking the Laplace transform of both sides gives

$$L\left(\frac{d^2q}{dt^2}\right) + 20L\left(\frac{dq}{dt}\right) + 100Lq = 120L(\sin 10t)$$

$$[s^2L(q) - sq(0) - q'(0)] + 20[sL(q) - q(0)] + 100L(q) = 120L(\sin 10t)$$

Since $q(0) = 0$ and $i(0) = q'(0) = 0$, we have

$$s^2L(q) + 20sL(q) + 100L(q) = 120\frac{10}{s^2 + 100}$$

$$(s^2 + 20s + 100)L(q) = \frac{1\,200}{s^2 + 100}$$

$$(s + 10)^2L(q) = \frac{1\,200}{s^2 + 10^2}$$

$$L(q) = \frac{1\,200}{(s^2 + 10^2)(s + 10)^2}$$

Using partial fractions, we see that

$$\frac{1\,200}{(s^2 + 10^2)(s + 10)^2} = \frac{0.6}{s + 10} + \frac{6}{(s + 10)^2} + \frac{-0.6s}{s^2 + 10^2}$$

and so

$$q(t) = L^{-1}[L(q)]$$

$$= L^{-1}\left(\frac{0.6}{s + 10}\right) + L^{-1}\left[\frac{6}{(s + 10)^2}\right] + L^{-1}\left(\frac{-0.6s}{s^2 + 10^2}\right)$$

and the charge $q(t)$ is

$$q(t) = 0.6e^{-10t} + 6te^{-10t} - 0.6\cos 10t$$

The current $i(t)$ is $q'(t)$, which is

$$i(t) = q'(t) = -6e^{-10t} + 6e^{-10t} - 60te^{-10t} + 6\sin 10t$$

$$= -60te^{-10t} + 6\sin 10t$$

You can see that the steady-state current is $6\sin 10t$.

Exercise Set 34.6

In Exercises 1–22, use Laplace transforms to solve the given initial value problem.

1. $y' - y = 1$, $y(0) = 0$
2. $y' + 2y = t$, $y(0) = -1$
3. $y' + 6y = e^{-6t}$, $y(0) = 2$
4. $y' - y = \sin 10t$, $y(0) = 0$
5. $y' + 5y = te^{-5t}$, $y(0) = 3$
6. $y' + y = \sin t$, $y(0) = -1$
7. $y'' + 4y = 0$, $y(0) = 2$, $y'(0) = 3$
8. $y'' - 5y' + 4y = 0$, $y(0) = -2$, $y'(0) = 4$
9. $y'' + 4y = e^t$, $y(0) = y'(0) = 0$
10. $y'' + 2y' - 3y = 5e^{2t}$, $y(0) = 2$, $y'(0) = 3$
11. $y'' - 6y' + 9y = t$, $y(0) = 0$, $y'(0) = 1$

12. $y'' + y = 1$, $y(0) = y'(0) = 1$
13. $y'' + 6y' + 13y = 0$, $y(0) = 1$, $y'(0) = -4$
14. $y'' + 2y' + 5y = 8e^t$, $y(0) = y'(0) = 0$
15. $y'' - 4y = 3\cos t$, $y(0) = y'(0) = 0$
16. $y'' - y = e^t$, $y(0) = 1$, $y'(0) = 0$
17. $y'' + 2y' + 5y = 3e^{-2t}$, $y(0) = y'(0) = 1$
18. $y'' - 6y' + 9y = 12t^2 e^{3t}$, $y(0) = y'(0) = 0$
19. $y'' + 2y' - 3y = te^{2t}$, $y(0) = 2$, $y'(0) = 3$
20. $y'' + 2y' - 3y = \sin 2t$, $y(0) = y'(0) = 0$
21. $y'' + 2y' + 5y = 10\cos t$, $y(0) = 2$, $y'(0) = 1$
22. $y'' + 2y' + 5y = 10\cos t$, $y(0) = 0$, $y'(0) = 3$

Solve Exercises 23–29.

23. *Electronics*   Use Laplace transforms to determine the charge $q(t)$, the current $i(t)$, and the steady-state current of an *RLC* series circuit, if $R = 20\,\Omega$, $L = 1\,H$, $C = 0.005\,F$, $E(t) = 150\,V$, and $q(0) = i(0) = 0$.

24. *Electronics*   The differential equation for the current in an *RL* series circuit is $L\dfrac{di}{dt} + Ri = E(t)$, where $E(t)$ is the impressed voltage. If $L = 1\,H$, $R = 10\,\Omega$, $E(t) = \sin t$, and $i(0) = 0$, find the formula for the current $i(t)$.

25. *Electronics*   Determine the charge $q(t)$ and current $i(t)$, of an *RLC* series circuit, if $R = 6\,\Omega$, $L = 0.1\,H$, $C = 0.02\,F$, $E(t) = 6\,V$, and if $q(0) = i(0) = 0$.

26. *Electronics*   An *RLC* series circuit has values of $R = 20\,\Omega$, $L = 0.05\,H$, $C = 100\,\mu F$, and $E(t) = 100\cos 200t$. Determine the charge, current, and steady-state current, if $q(0) = i(0) = 0$.

27. *Physics*   A spring hangs in a vertical position with its upper end fixed. We know that the spring constant is $k = 48$ lb/ft. An object weighing 16 lb is attached to the lower end of the spring. After coming to rest, the object is pulled down 2 in. and released. If the air resistance in pounds is $64v$, describe the motion of the object.

28. *Physics*   If the support of the spring in Exercise 27 is given a motion of $\cos 4t$ ft, describe the motion of the object.

29. *Automotive technology*   A 3,200 lb vehicle rests on a spring-shock absorber system on each of four wheels. The system yields the model

$$y'' + 2y' + 17y = e^{-t}\cos 4t$$

where $e^{-t}\cos 4t$ represents the force in pounds resulting from a bumpy road. If $y = y' = 0$ when $t = 0$, use Laplace transforms to solve this differential equation.

✎ In Your Words

30. Without looking in the text, outline the steps to solving a differential equation by Laplace transforms.

31. Of all the methods used in the text to solve differential equations, which one do you think was the easiest to use? Explain why you think this was the easiest method.

☰ CHAPTER 34 REVIEW

Important Terms and Concepts

Euler's method
Huen's method
Increment method
Inverse Laplace transforms
LaPlace transforms
Numerical methods for solving differential equations

Euler's method
Huen's method
Successive approximations
Partial fractions
Successive approximations

Review Exercises

In Exercises 1–4, solve the given differential equation by using Euler's method for the given initial value of x and y and the given value of h.

1. $y' = -y^2(1+x)$, $x_0 = 0$, $y_0 = 2$, $0 \le x \le 1$, $h = 0.1$, $h = 0.05$

2. $y' = \dfrac{2(x+1)^3}{y}$, $x_0 = 0$, $y_0 = 1$, $0 \le x \le 1$, $h = 0.1$, $h = 0.05$

3. $y' = -0.1y + \dfrac{4}{1+x^2}$, $x_0 = 0$, $y_0 = 60$, $0 \le x \le 1$, $h = 0.05$

4. $y' = \dfrac{6}{\sqrt{1+x^2}}$, $x_0 = 0$, $y_0 = 0$, $0 \le x \le 0.10$, $h = 0.01$

In Exercises 5–8, find the second approximation of the solution of the given differential equations if the solution passes through the given point.

5. $y' = xy + 4$, $(0, 1)$

6. $y' = 2x + y$, $(0, 0)$

7. $y' = y\sqrt{x-9}$, $(10, 1)$

8. $y' = y + \cos x$, $\left(\frac{\pi}{2}, 0\right)$

In Exercises 9–20, solve the given differential equation by use of Laplace transforms, where the solution has the given initial values.

9. $2y' - y = 4$, $y(0) = 1$

10. $3y' + y = t$, $y(0) = 2$

11. $y' + 2y = e^t$, $y(0) = 1$

12. $y' + 5y = 0$, $y(0) = 1$

13. $y'' - y = 0$, $y(0) = y'(0) = 1$

14. $y'' - y = e^t$, $y(0) = y'(0) = 0$

15. $y'' + 2y' + 5y = 0$, $y(0) = 1$, $y'(0) = 0$

16. $y'' + y' + y = 0$, $y(0) = 4$, $y'(0) = -2$

17. $y'' + 2y' + 5y = 3e^{-2t}$, $y(0) = y'(0) = 1$

18. $y'' + 2y' - 3y = 5e^{2t}$, $y(0) = 2$, $y'(0) = 3$

19. $y'' + 4 = \sin t$, $y(0) = y'(0) = 0$

20. $y'' + 4 = t$, $y(0) = -1$, $y'(0) = 0$

Solve Exercises 21 and 22.

21. *Electronics* An *RLC* series circuit has values of $R = 10\,\Omega$, $L = 0.5\,\text{H}$, $C = 0.01\,\text{F}$, and $E(t) = 10\sin t$ V. If $q(0) = 0$ and $i(0) = 10$ A, find the equation for the charge, the current, and the steady-state current.

22. *Physics* A 20-kg mass is attached to a spring with a spring constant of 700 N/m. The mass is started in motion from an equilibrium position with an initial velocity of 1 m/s upward and an applied external force of $5\sin t$. Find the equation for the motion of this mass, if the force in newtons due to air resistance is 90 times the velocity.

CHAPTER 34 TEST

In Exercises 1–4, use Table 34.1 to find the transform of the given function.

1. $f(t) = 5e^{-3t}$

2. $f(t) = 2t^3 + \sin 8t$

3. $f(t) = e^{-5t}\cos 4t$

4. $f(t) = 8t^9 e^{5t}$

In Exercises 5–7, find the inverse Laplace transform of the given function.

5. $F(s) = \dfrac{s-6}{(s-6)^2 + 7}$

6. $F(s) = \dfrac{s}{s^2 - 6s + 13}$

7. $F(s) = \dfrac{5}{(s+3)(s^2 + 1)}$

In Exercises 8–10, use the method of Laplace transforms to solve the given differential equation.

8. $y' + 5y = 0$, $y(0) = 2$

9. $y'' + 2y = e^{-2t}$, $y(0) = 0$, $y'(0) = 0$

10. $y'' + 2y' + y = 0$, $y(0) = 1$, $y'(0) = -1$

Solve Exercises 11 and 12.

11. In an *RLC* series circuit, $R = 8.0\,\Omega$, $L = 0.20\,\text{H}$, $C = 0.002\,\text{F}$, and $E(t) = 12\sin 20t$ V. Find q as a function of time, if $q(0) = 0$ and $i(0) = q'(0) = 0$.

12. Use Euler's method to find an approximate solution to the differential equation $y' = 2x + y$, if the curve passes through the point $(0, 2)$ and $h = 0.2$.

A

The Electronic
Hand-Held Calculator

Until recently you would have had to work all the problems in this book in your head, with pencil and paper, or with a slide rule. There were some calculators available but they were large and bulky—almost the size of a typewriter—and they were very limited in the operations that you could perform with them.

The invention of the integrated circuit made it possible to process and store large amounts of information in a very small space and with very little energy. Later advances in integrated circuits allowed for even more information to be stored on an integrated circuit "chip." The advances in the chip plus the light emitting diode (LED) provided the basis for small, light, inexpensive electronic calculators. More recent calculators use a liquid crystal display (LCD) instead of the LEDs.

≡ A.1
INTRODUCTION

Electronic calculators have quietly and quickly become common tools. People use them at work, at home, and in school. Because we expect that you will use a calculator in your work as a technician, we include this section showing how calculators can and should be used. You are encouraged and expected to use a calculator and/or computer when working problems in this book.

Types of Calculators

What kind of calculator should you get? For the work in this book, you need a "scientific" calculator. This type of calculator naturally has keys for the ten digits, 0 through 9, and for the basic operations ($+$, $-$, $\times$, and $\div$). But, the calculator should also have the keys $\boxed{\sin}$, $\boxed{\cos}$, $\boxed{\tan}$, $\boxed{y^x}$ or $\boxed{x^y}$, $\boxed{\log}$, and $\boxed{\ln x}$. Make sure you have the instruction manual for your calculator. Because not all calculators work the same, the methods described in this book may not work on your calculator. If you have this problem, the instruction manual for your calculator and your teacher should help you learn how to work these problems.

All calculators work in one of two basic ways. Some use **algebraic notation** (**AN**) and the others use **reverse Polish notation** (**RPN**). In the text, most calculator operations are shown using algebraic notation. If it is likely that confusion might result for users of RPN calculators, then an example showing how to use an RPN calculator is provided. It does not matter whether your calculator is an AN or RPN type. Just make sure you read your instruction manual carefully and practice using your calculator.

There are other variations among calculators. Some have one or more addressable memories. Others have the ability to draw graphs.

Another major difference is the order of operations performed by the calculator. A calculator that performs operations according to the usual mathematical order of operations is said to have an **algebraic operating system (AOS)**. Some companies refer to the order in which their calculators perform operations as the **equation operating system (EOS)**. Any calculator that uses the EOS is also using the AOS. If you need to specify that certain operations are performed before others, then you will need to use parentheses.

Caution

Consider the problem $9 + 5 \times 6$. Using the order of operations in Section 1.2, you should get an answer of 39. Work the problem on your calculator. Do you get 39 or 84? Some calculators will give one answer; some will give the other. If your calculator gives the answer 84, then you must be very careful to perform the operations in the correct order and either write the intermediate results or, if your calculator has a memory, store the intermediate results in memory. Some calculators have parenthesis keys that can be used to work problems like the one just described.

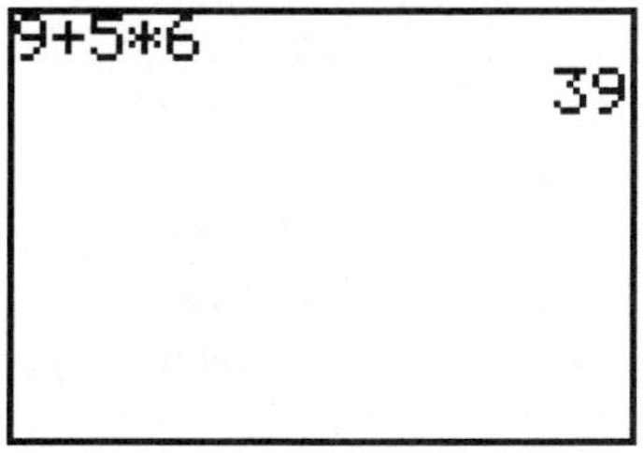

FIGURE A.1

All graphing calculators use an AOS. Figure A.1 shows the result of working $9 + 5 \times 6$ on a TI-82.

A Hewlett-Packard HP-48 gives the same result, but you have two ways, the "stack method" and the "equation editor method" to get an HP-48 to calculate $9 + 5 \times 6$. The keystrokes with the two methods are outlined below.

METHOD	PRESS	DISPLAY
stack	9 ENTER 5 ENTER 6 × +	39
equation editor	⌐ ENTER 9 + 5 × 6 ENTER	'9+5*6'
	EVAL	39

If you have any doubts as to which sequence of operations your calculator will follow when you use combined operations, press the = key after each individual operation and record, or store in the calculator's memory, the intermediate results. Practice and experience will allow you to become familiar enough with your calculator to use it correctly and to take advantage of its built-in capabilities.

Most calculators display 8 to 10 significant digits and store 2 or 3 more digits for rounding off purposes. However, this is not always the case. Some calculators employ a method called **truncation** in which any digits not displayed are discarded. Thus, 489.781 truncated to tenths is 489.7. This same number rounded off to tenths is 489.8. While you will seldom use numbers with enough significant digits for calculator truncation to be a problem, you should be aware of this possibility.

A **graphics calculator** has a larger screen and can draw and display graphs. One advantage of a graphics calculator is that it displays the numbers and operations and allows you to edit or change them.

One other difference between calculators is how they handle the decimal point. Some calculators are designed to show exactly two decimal places with every computation. These calculators have a **fixed decimal point**. Other calculators only show as many decimal places as result from the computation. These calculators have a **floating decimal point**. For example, a fixed decimal point calculator would show 34.50 when a floating decimal point calculator would show 34.5. Similarly, a floating decimal point calculator would show 47.368 when a fixed decimal point calculator would show 47.37. Most scientific calculators have a $\boxed{\text{Fix}}$ key that allows you to set the number of digits to the right of the decimal point that will be displayed.

≡ A.2
BASIC OPERATIONS WITH A CALCULATOR

In Section A.1, we examined the differences between calculators. In this section, we will learn how to use the calculator to solve problems such as those in the first few chapters of the text.

Entering Positive and Negative Numbers

Data are entered by pressing the digits in order from left to right. Thus, the number 48.732 would be entered into the calculator by pressing the keys in this order: $\boxed{4}$ $\boxed{8}$ $\boxed{.}$ $\boxed{7}$ $\boxed{3}$ $\boxed{2}$. If your calculator has a floating decimal point the display should read 48.732; if it has a fixed decimal point, it would read 48.73.

Your calculator probably has a $\boxed{+/-}$, $\boxed{\text{CHS}}$, or $\boxed{(-)}$ key. This key allows you to change the sign of the number in the display. Enter 59.638 in your calculator and press the $\boxed{+/-}$ or $\boxed{\text{CHS}}$ key. The display should now read −59.638. Press the $\boxed{+/-}$ or $\boxed{\text{CHS}}$ key again and you should see the calculator return to the original display of 59.638. Notice that the $\boxed{+/-}$ or $\boxed{\text{CHS}}$ key is pressed *after* the number is entered in the calculator.

Some calculators, such as graphics calculators, use a $\boxed{(-)}$ key to enter a negative number. The use of the parentheses is to help you tell it apart from the $\boxed{-}$ key which is used for subtraction. Here the $\boxed{(-)}$ key is pressed *before* the number is entered. Thus, you would press $\boxed{(-)}$ 59.638 in order to enter the number −59.638.

≡ Note

The $\boxed{+/-}$ or $\boxed{\text{CHS}}$ key should be pressed after the number has been entered. The $\boxed{(-)}$ key should be pressed before the number is entered.

Dual Function Key

Some calculators have a **dual function** key or $\boxed{\text{2nd}}$ key. In order to save space, calculator makers decided to let some keys perform more than one function. The first function is printed on the key; the second is printed just above the key; and a third function may be printed above or below the key.

To use the first function of a key, all you need to do is press the key. To use the second function, you first press the $\boxed{\text{2nd}}$ key and then the key below the desired function. For example, on some calculators there is a key labeled $\boxed{\text{CLR}}$ with CA above it much like this: $\frac{\text{CA}}{\boxed{\text{CLR}}}$. In order to get the CA function, press the $\boxed{\text{2nd}}$ key and then press the $\boxed{\text{CLR}}$ key. We will write this as $\boxed{\text{2nd}}$ $\boxed{\text{CA}}$ to prevent any possible confusion about which function to use. Some calculators indicate the 2nd key by F or 2nd F or a blank key ☐ printed in a solid color such as gold or blue.

Graphics calculators have both a second function key and an alpha function key. For example, on the Texas Instruments TI-81, the x^2 and the LN keys each have two symbols above them as shown here:

$$\sqrt{}\ \text{I} \qquad\qquad e^x\ \text{S}$$
$$\boxed{x^2} \qquad\qquad\quad \boxed{\text{LN}}$$

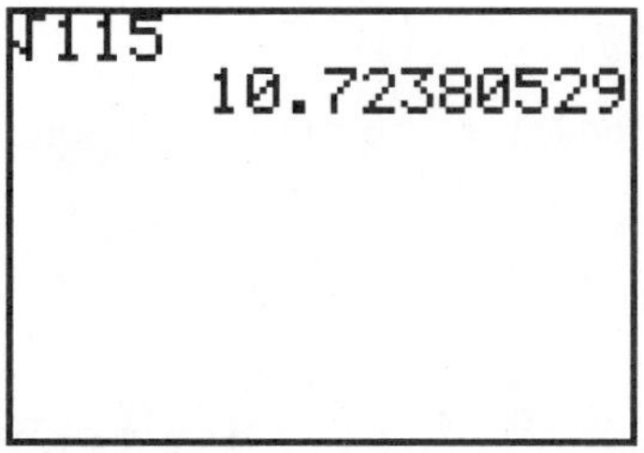

To the left printed in light blue above a key is an operation. For example, to the left above the $\boxed{x^2}$ key is the operation $\sqrt{}$ and to the left above the $\boxed{\text{LN}}$ key is the operation e^x. To use an operation, press the 2nd key and then the key you require. For example, to calculate $\sqrt{115}$, press 2nd $\boxed{\sqrt{}}$ 115 ENTER. The screen displays $\sqrt{115}$ on one line and the result, 10.72380529 on the next line, as shown in Figure A.2.

FIGURE A.2

≡ **Note**

To help you use the correct key, the name of the operation and the 2nd key are in the same color. On a TI-81 and TI-82 calculators, this color is light blue; on a TI-85, it is yellow.

To the right above a key is a letter or other symbol printed in grey. For example, to the right above the $\boxed{x^2}$ key is the letter I and to the right above the $\boxed{\text{LN}}$ key is the letter S. To use these keys, you first press the ALPHA key. So, to enter the letter I you press the ALPHA $\boxed{x^2}$ key. In this book we will indicate that you need to use the ALPHA key by placing the letter inside a square. So, if we want you to use the variable I, we will write ALPHA $\boxed{\text{I}}$.

≡ **Note**

Letters and symbols accessed by pressing the ALPHA key are color-coordinated. On a TI-8x calculator, this color is grey.

Basic Arithmetic Operations ▬▬▬▬▬▬▬▬▬▬▬

The basic arithmetic operations are performed using the operation keys +, −, ×, and ÷, and the =, ENTER, or EXE key. The problem $4 \times 8 + 5 \times 6$ would be worked by keying into the calculator as follows:

Algebraic Calculator	RPN Calculator
4 × 8 + 5 × 6 =	4 ENTER 8 × 5 ENTER 6 × +

In each case, the final display is 62.

To find the sum of 14.32 and 9.37 you press

Algebraic Calculator	RPN Calculator
14.32 + 9.37 =	14.32 ENTER 9.37 +

and the result 23.69 is displayed.

Subtraction, multiplication, and division are handled in much the same manner.

≡ **Note**   From now on, we will indicate what you should enter and what is displayed in the following manner.

PRESS | DISPLAY

14.32 ⊞ 9.37 ⊟   23.69

There will be times when a third column will be added, so we can include some comments or explanations.

EXAMPLE A.1

Use an algebraic or graphing calculator to evaluate **(a)** $72.1 - 83.12$, **(b)** 4.3×6.1, and **(c)** $7.8 \div 2.4$

Solution

	PRESS	DISPLAY
(a)	72.1 ⊟ 83.12 ⊟	−11.09
(b)	4.3 ⊠ 6.1 ⊟	26.23
(c)	7.8 ⊟ 2.4 ⊟	3.25

EXAMPLE A.2

Use an RPN calculator to work the problem $43.7 + 56.2$.

Solution

PRESS	DISPLAY
43.7 ENTER	43.7
56.2 ⊞	99.9

EXAMPLE A.3

Evaluate **(a)** $9.3 + 84.12 - 20.5 - 123.1$ and **(b)** $7.4 \times 2.5 \div 37 \times 8$.

Solution

	PRESS	DISPLAY
(a)	9.3 ⊞ 84.12 ⊟ 20.5 ⊟ 123.1 ⊟	−50.18
(b)	7.4 ⊠ 2.5 ⊟ 37 ⊠ 8 ⊟	4.

These same two problems would be worked on an RPN calculator as follows:

	PRESS	DISPLAY
(a)	9.3 ENTER 84.12 ⊞ 20.5 ⊟ 123.1 ⊟	−50.18
(b)	7.4 ENTER 2.5 ⊠ 37 ⊟ 8 ⊠	4.0

Parentheses

Even if your calculator uses the algebraic operating system (AOS) described earlier, there are times when you will need to specify the exact order in which expressions are to be evaluated. There are several different symbols used for grouping—parentheses ( ), brackets [ ], and braces { }—but calculators and computers primarily use parentheses. The parentheses tell the calculator to perform the operations inside the parentheses first. If you are not sure how your calculator will handle an expression, you should use parentheses.

A calculator does not know about implied multiplication. People often write $(7+3)(8-5)$ when they mean $(7+3) \times (8-5)$. You must tell the calculator when quantities are being multiplied, as shown in Example A.4.

EXAMPLE A.4

Multiply $(7+3) \times (8-5)$.

Solution

PRESS	DISPLAY
$\boxed{(}\ \boxed{7}\ \boxed{+}\ \boxed{3}\ \boxed{)}\ \boxed{\times}\ \boxed{(}\ \boxed{8}\ \boxed{-}\ \boxed{5}\ \boxed{)}\ \boxed{=}$	30.

A slightly more involved use of parentheses is given in Example A.5. Here you want the calculator to evaluate the entire numerator and then divide by the entire denominator. To make sure that this is what the calculator does, extra sets of parentheses must be added.

EXAMPLE A.5

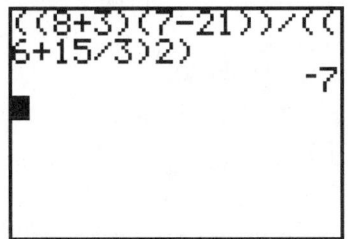

FIGURE A.3

Evaluate $\dfrac{(8+3) \times (7-21)}{(6+15 \div 3) \times 2}$ with your calculator.

Solution   Add another set of parentheses around the numerator and the denominator. The problem now looks like $\dfrac{((8+3) \times (7-21))}{((6+15 \div 3) \times 2)}$. Now use your calculator to evaluate this expression. The result below, and in Figure A.3, is for a TI-82 graphing calculator.

PRESS	DISPLAY
$\boxed{(}\ \boxed{(}\ \boxed{8}\ \boxed{+}\ \boxed{3}\ \boxed{)}\ \boxed{(}\ \boxed{7}\ \boxed{-}\ \boxed{21}\ \boxed{)}\ \boxed{)}\ \boxed{\div}$	
$\boxed{(}\ \boxed{(}\ \boxed{6}\ \boxed{+}\ \boxed{15}\ \boxed{\div}\ \boxed{3}\ \boxed{)}\ \boxed{2}\ \boxed{)}\ \boxed{\text{ENTER}}$	-7

RPN Calculator

Add parentheses to aid in the correct order of operations. It is easiest to work from the inside of parentheses outward, much as you would if you were doing the calculations on paper. Since division is performed before addition, a set of parentheses should be added in the denominator.

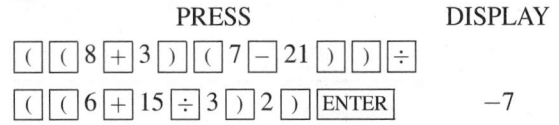

You would then use an RPN calculator in the following manner:

PRESS	DISPLAY	COMMENT
8 ENTER 3 + 7 ENTER 21 − ×	−154	(The numerator)
ENTER	−154	(Stores the numerator)
15 ENTER 3 ÷ 6 + 2 ×	22	(The denominator)
÷	−7	(The result)

 **Note**   Remember that parentheses come in pairs. Whenever you use a left parenthesis, (, you must also use a right parenthesis,), in that same problem.

Example A.6 provides another chance for you to use parentheses with your calculator.

EXAMPLE A.6

Evaluate $\dfrac{9 \times 3 + 7 \times (-8) + (-16)}{((3 + 6 \times 5) + 4.5 - 7.5) \div 1.5}$.

Solution   Notice that there are no parentheses in the numerator. If your calculator does not use the AOS, then you need to place parentheses so that you get the right answer for the numerator. Parentheses should be placed as such:

$$(9 \times 3) + (7 \times -8) + (-16)$$

If your calculator has AOS, then this step is not necessary. You need to rewrite the problem by placing a set of parentheses around the numerator and a set around the denominator. The problem now looks like

$$\frac{(9 \times 3 + 7 \times -8 + (-16))}{(((3 + 6 \times 5) + 4.5 - 7.5) \div 1.5)}$$

Use a TI-8x calculator to evaluate this expression.

PRESS	DISPLAY
(9 × 3 + 7 × (−) 8 + (−) 16) ÷	
(((3 + 6 × 5)) + 4.5 − 7.5) ÷	
1.5) ENTER	−2.25

Reciprocals

Many calculators have a $\boxed{1/x}$ or $\boxed{x^{-1}}$ key that is used to find the reciprocal of a number on display. Remember, your calculator will give the decimal value of the reciprocal. In Example A.7, we use the calculator to evaluate the reciprocals of three numbers.

EXAMPLE A.7

What are the reciprocals of (a) 3, (b) −19, and (c) $\frac{4}{5}$?

Solution

		PRESS	DISPLAY
(a)		3 $\boxed{1/x}$	0.3333333
(b)		19 $\boxed{+/-}$ $\boxed{1/x}$	−0.0526316
(c)	Algebraic calculator:	4 $\boxed{\div}$ 5 $\boxed{=}$ $\boxed{1/x}$	1.25
	RPN calculator:	4 $\boxed{\text{ENTER}}$ 5 $\boxed{\div}$ $\boxed{1/x}$	1.25
	Graphics calculator:	$\boxed{(}$ 4 $\boxed{\div}$ 5 $\boxed{)}$ $\boxed{x^{-1}}$ $\boxed{\text{ENTER}}$	1.25

Notice that in (c), if you knew that the decimal equivalent of $\frac{4}{5}$ was 0.8, then you could have entered 0.8 $\boxed{1/x}$ or 0.8 $\boxed{x^{-1}}$ and the display would have read 1.25. But be careful when you try to use a short cut to solve a problem. Remember that the reciprocal of a reciprocal is the original number. We know that the reciprocal of 3 is $\frac{1}{3}$, so the reciprocal of $\frac{1}{3}$ is 3. Now work Example A.8.

EXAMPLE A.8

Use a calculator to (a) find the reciprocal of the reciprocal of 3, and (b) find the reciprocal of 0.3333333.

Solution

	PRESS	DISPLAY
(a)	3 $\boxed{1/x}$	0.3333333
	$\boxed{1/x}$	3.
(b)	0.3333333 $\boxed{1/x}$	3.0000003

As you can see, the reciprocal of a reciprocal is supposed to return the original number. But, if you enter the decimal equivalent of a number that does not terminate before the eighth place to the right of the decimal point you may not get the exact reciprocal that you want. Sometimes it is worth the few extra steps it takes to enter 1 $\boxed{\div}$ 3 instead of 0.3333333.

Exercise Set A.2

Use a calculator to evaluate Exercises 1–24.

1. $28 + 45$
2. $48 + 27$
3. $93 - 47$
4. $86 - 39$
5. $43 - 192$
6. $57 - 216$

7. 21×15
8. 47×12
9. $18 \div 5$
10. $38 \div 4$
11. $15.32 + -6.71$
12. $18.71 + -7.38$

13. $-14.67 - 8.932$

14. $-21.93 - -6.091$

15. 12.41×-3.4

16. 23.36×-517

17. $-8.35 \div 2.5$

18. $53.72 \div -6.32$

19. $16.37 + 81.4 - 6.437 + -8.5 - -16.372$

20. $-12.4 - 81.37 + -6.9 - 108.4 + -6.43$

21. $8.4 \times 3.7 \div 6.2 \div -4.2 \times 124 \div -0.74$

22. $-165 \div 1.1 \times 2.6 \times -8.5 \div 13 \div -0.015$

23. $\dfrac{(8+7) \times 3 \div ((6.2-5) \times 5)}{18 \div 6 - 3 \times 4.5 - 1.25 \times -8}$

24. $\dfrac{(17.3 + 6.82) \times 4.1 \div (-27.3 \div 3 - 8.9)}{(81.73 - 16.38 \times 220 \div 11 + 40.37) \div (97.6 - 15.4)}$

Use a calculator to determine the reciprocal of the numbers in Exercises 25–30.

25. 4

26. 0.5

27. 4.5

28. 0.1111111

29. 215.3

30. 0.00025

≡ A.3
SOME SPECIAL CALCULATOR KEYS

In Section A.2, we looked at one special key—the $\boxed{1/x}$ or reciprocal key and at the grouping keys $\boxed{(}\ \boxed{)}$. In this section you will learn to use eight more special keys.

Keys for Powers or Roots

The first four special keys are used to find decimal values of powers and roots. Two of these should look familiar. These are the $\boxed{x^2}$ and $\boxed{\sqrt{\ }}$ or $\boxed{\sqrt{x}}$ keys. The $\boxed{x^2}$ will calculate the square (or second power) of any number you enter as shown in Example A.9.

EXAMPLE A.9

Evaluate **(a)** 3^2, **(b)** 5^2, **(c)** 4.2^2, **(d)** 135^2, and **(e)** $(-7)^2$.

Solution

		PRESS	DISPLAY
(a)	$3\ \boxed{x^2}$		9.
(b)	$5\ \boxed{x^2}$		25.
(c)	$4.2\ \boxed{x^2}$		17.64
(d)	$135\ \boxed{x^2}$		18225.
(e)	Algebraic: $7\ \boxed{+/-}\ \boxed{x^2}$		49.
	RPN: $7\ \boxed{CHS}\ \boxed{x^2}$		49.
	Graphic: $\boxed{(}\ \boxed{(-)}\ 7\ \boxed{)}\ \boxed{x^2}\ \boxed{ENTER}$		49.

The $\boxed{\sqrt{}}$ key is used to determine the square root of a number. On many calculators you first enter the number and then press the $\boxed{\sqrt{}}$ key. (This may require pressing the $\boxed{2nd}$ key and then the $\boxed{\sqrt{}}$ key.) On a graphics calculator the $\boxed{\sqrt{}}$ key is pressed before the numbers.

EXAMPLE A.10

Evaluate **(a)** $\sqrt{25}$, **(b)** $\sqrt{2}$, and **(c)** $\sqrt{797}$.

Solution

		PRESS	DISPLAY
(a)	Algebraic or RPN:	25 $\boxed{\sqrt{}}$	5.
	Graphics:	$\boxed{\sqrt{}}$ 25	5.
(b)	Algebraic or RPN:	2 $\boxed{\sqrt{}}$	1.4142136
	Graphics:	$\boxed{\sqrt{}}$ 2	1.4142136
(c)	Algebraic or RPN:	797 $\boxed{\sqrt{}}$	28.231188
	Graphics:	$\boxed{\sqrt{}}$ 797	28.231188

In Example A.10(a), the displayed number is the exact principal square root of the number. The numbers displayed for $\sqrt{2}$ and $\sqrt{797}$ are approximate values. Both of these are irrational numbers. If you evaluate $(1.4142136)^2$ and $(28.231188)^2$, you may not get 2 and 797. For example, on one calculator $(1.4142136)^2 = 2.0000001$ and $(28.231188)^2 = 796.99998$.

Remember, you can only take square roots of positive real numbers. What happens when you enter 2 $\boxed{+/-}$ $\boxed{\sqrt{}}$ into your calculator? Some calculators display 1.4142136 in the flashing mode. The flashing indicates that you have made an error. Other calculators display an "E" or the word "Error." A number, such as 0, may be displayed with the "E".

If you want powers larger than two, use the $\boxed{y^x}$, $\boxed{x^y}$, or $\boxed{\wedge}$ key. This discussion will be written for a calculator with a $\boxed{y^x}$ key. If your calculator has an $\boxed{x^y}$ or $\boxed{\wedge}$ key, then you should make any necessary changes.

Suppose you want to find 2^3. We know that $2^3 = 2 \times 2 \times 2 = 8$. To get this same answer with your algebraic calculator, enter 2 $\boxed{y^x}$ 3 $\boxed{=}$. On an RPN calculator, press 2 $\boxed{ENTER}$ 3 $\boxed{y^x}$ and on some graphing calculators press 2 $\boxed{\wedge}$ 3 $\boxed{ENTER}$.

To evaluate $4^{5/6}$, you need to use parentheses. You should enter 4 $\boxed{y^x}$ $\boxed{(}$ 5 $\boxed{\div}$ 6 $\boxed{)}$ $\boxed{=}$. The result is approximately 3.1748. Now work the problems in Example A.11 to see how to use your calculator.

EXAMPLE A.11

Evaluate: **(a)** 4^3, **(b)** $4.3^{2.7}$, **(c)** $(\sqrt{3})^{\sqrt{2}}$, **(d)** 2^{-3}, **(e)** $\sqrt[4]{2}$, and **(f)** $9^{5/3}$.

Solution

	PRESS	DISPLAY
(a)	4 $\boxed{y^x}$ 3 $\boxed{=}$	64.
(b)	4.3 $\boxed{y^x}$ 2.7 $\boxed{=}$	51.32924
(c)	3 $\boxed{\sqrt{\ }}$ $\boxed{y^x}$ 2.7 $\boxed{\sqrt{\ }}$ $\boxed{=}$	2.466015575
	TI-81: $\boxed{\sqrt{\ }}$ 3 $\boxed{\wedge}$ $\boxed{\sqrt{\ }}$ 2.7 $\boxed{\text{ENTER}}$	2.466015575
(d)	2 $\boxed{y^x}$ 3 $\boxed{+/-}$ $\boxed{=}$	0.125
	Graphics: 2 $\boxed{y^x}$ $\boxed{(-)}$ 3 $\boxed{\text{ENTER}}$	0.125
(e)	2 $\boxed{y^x}$ 0.25 $\boxed{=}$	1.1892071
(f)	2 $\boxed{x^y}$ $\boxed{(}$ 5 $\boxed{\div}$ $\boxed{)}$ $\boxed{=}$	38.9407384

Memory Keys

Memory keys allow you to store numbers that you plan to use later. These numbers are stored in special places in a calculator. The number of memories available in your calculator may range from as few as one for older calculators to 27 for the TI-82 and TI-83. Calculators such as the TI-85 and TI-92 are limited only by the amount of memory, in bytes, that the calculator can use.

To STOre the number on display, you press the $\boxed{\text{STO}}$ key. Some calculators use $\boxed{\text{M}}$ for the Memory key. In calculators with more than one memory, you must tell the calculator in which memory you want to store the number. Calculator memories are numbered. A calculator with ten memories numbers the locations 0–9. If you want to place a displayed number in memory location six, press $\boxed{\text{STO}}$ $\boxed{6}$. To erase the numbers from all memories press $\boxed{\text{2nd}}$ $\boxed{\text{CA}}$.

Calculators that have alphabetical keys have a memory for each letter of the alphabet. For example, to place a number displayed on the TI-81 in memory location B, you press $\boxed{\text{STO(}}$ and then the $\boxed{\text{B}}$ key. Do not press the $\boxed{\text{ALPHA}}$ key. On a Casio fx7700-G, the $\boxed{\rightarrow}$ key acts like the $\boxed{\text{STO(}}$ key, except that you need to press the $\boxed{\text{ALPHA}}$ key. For example, to store a number in memory location B, you press $\boxed{\rightarrow}$ $\boxed{\text{ALPHA}}$ $\boxed{\text{B}}$.

Once you have stored a number in memory, you need to be able to get it back so you can use it. To do this, use the ReCalL key, $\boxed{\text{RCL}}$. On some calculators, this is the Memory Recall, $\boxed{\text{MR}}$, key. If your calculator has more than one memory, then you must indicate which memory location contains the number. To recall the number in memory location five, press $\boxed{\text{RCL}}$ $\boxed{5}$. To recall the number in memory location B, press $\boxed{\text{ALPHA}}$ $\boxed{\text{B}}$.

 **Note**

Recalling a number from memory does not clear the memory. The number is still stored in that memory, so you can use it again.

Instructions for more special calculator keys are included at various places in the text.

Exercise Set A.3

Use a calculator to evaluate Exercises 1–44.

1. 7^2
2. 11^2
3. -25^2
4. 142^2
5. 3.1^2
6. 7.95^2
7. -18.43^2
8. -26.4^2
9. $/\sqrt{36}$

10. $\sqrt{169}$
11. $\sqrt{1296}$
12. $\sqrt{45796}$
13. $\sqrt{10.24}$
14. $\sqrt{34.81}$
15. $\sqrt{151.5361}$
16. $\sqrt{2631.69}$
17. 9^3
18. 7^4

19. 21^5
20. 53^4
21. 4.7^3
22. 0.24^8
23. 14.1^{22}
24. 23.4^{36}
25. $\sqrt[4]{6561}$
26. $\sqrt[5]{32768}$
27. $\sqrt[7]{62748517}$

28. $\sqrt[6]{11390625}$
29. $\sqrt[7]{340.48254}$
30. $\sqrt[5]{656.75808}$
31. $\sqrt[23]{241}$
32. $\sqrt[62]{11562}$
33. 8^{-3}
34. 9^{-7}
35. $1.1^{4.2}$
36. $2.3^{5.7}$

37. 7^3
38. 12^5
39. 25^{-10}
40. 47^{-8}
41. $\sqrt[3.1]{23}$
42. $\sqrt[2.45]{62}$
43. $\sqrt[3]{\sqrt[4]{9}}$
44. $\sqrt[3]{\sqrt[9]{56}}$

Enter each of the numbers in Exercises 45–52 into your calculator in scientific notation.

45. 6.21×10^{14}
46. 3.781×10^{23}

47. 8.92×10^{-21}
48. 7.631×10^{-46}

49. $8,437,100,000,000$
50. $62,790,000,000$

51. 0.00000000000271
52. 0.0000000000472

Evaluate Exercises 53–60 with the help of a calculator.

53. $4.8 \times 10^{12} \times 5.23 \times 10^{25}$
54. $7.912 \times 10^{35} \times 6.39 \times 10^{24}$
55. $1.73 \times 10^{18} \times 8.42 \times 10^{-21}$

56. $2.71 \times 10^{25} \times 4.32 \times 10^{-16}$
57. $4.31 \times 10^{14} \div 2.31 \times 10^{12}$
58. $8.92 \times 10^{19} \div 4.62 \times 10^{23}$

59. $5.42 \times 10^{21} \div 2.47 \times 10^{-5}$
60. $4.72 \times 10^{31} \div 4.93 \times 10^{-34}$

Use the $\boxed{\text{STO}}$ **and** $\boxed{\text{RCL}}$ **keys on a calculator to solve Exercises 61–68.**

61. $\dfrac{45.37 - 82.91}{6.4 \times 3.7 - 8.9 \times 4}$

62. $\dfrac{(9.31 + 8.72)5.4}{7.3 \times 4.6 - 4 \times 9.1}$

63. $\dfrac{3.7 - 91.4}{(18.32 - 6.7)9.2}$

64. $A = (4.3 + 8.1) \div 7.2, B = 5.7A - 34.9, C = \dfrac{B}{A} - A$,
Find C

65. $A = (4.3 + 8.1) \div 7.2, B = 5.7A - 34.9, C = \dfrac{B}{A} - A$,
Find C

66. $A = 12.4 \div (3.7 - 8.9), B = \sqrt{-A}, C = 4B + A$,
Find C

67. $A = 4.3 \times 8.7 - 9.1, B = A^2 + 7, C = (\sqrt{B})A$,
Find C

68. $A = (2.7 + 8.9)^2, B = \sqrt[3]{4A} - \sqrt{A}, C = \dfrac{B^5 + A^3}{AB}$,
Find C

☰ A.4
GRAPHING CALCULATORS AND COMPUTER-AIDED GRAPHING

The introduction of microcomputers has allowed people to use inexpensive computers for writing and performing calculations quickly and accurately; but people discovered that words and numbers were not enough, and so computer graphics were introduced.

Some of these graphing capabilities are now available on calculators. Use of a graphing calculator or a computer removes some of the drudgery of plotting equations by hand. In this text, we show how you can use computer and calculator graphing capabilities. However, we will also try to show you how to interpret the information they give you.

The graphics capabilities of computers and calculators vary widely. Graphics examples in this text were run on either a Casio fx-7700G or a Texas Instruments TI-82 graphing calculator or on a Macintosh computer.

Section 4.5 is an optional section that presents the basics of using a graphing calculator. Additional examples that explain more advanced features of graphing calculators are spread throughout the text.

Exercise Set A.4

Review Exercises

Use a calculator to evaluate Exercises 1–20.

1. $43 + -85$

2. $16 - 34$

3. 25×-81

4. $-62 \div 16$

5. 8.32×4.15

6. $-9.41 + 18.301$

7. $4.32 - 1.011$

8. $15.42 \div 0.002$

9. $8.3 + 9.1 - 4.3 \times 4 + 3 \div 5$

10. $-162 \div 1.2 + 4.3 \div 0.4 - 7$

11. $\dfrac{(3+5) \times -2 \div ((6-4.3) \times 7)}{9 \div 3 - 4 \times 2.3 - 5 \times -7}$

12. $\dfrac{(16.3 - 4) \times 6.2 \div (3.4 \div 18.1 + 3)}{8.1 \div 9 + 10}$

13. 35^2

14. $\sqrt{5}$

15. $(-4)^5$

16. 14^6

17. $\sqrt[7]{813}$

18. $\sqrt[4]{216}$

19. $\sqrt[2.3]{42.1}$

20. $\sqrt[1.3]{25.4}$

In Exercises 21–24, enter the number into your calculator in scientific notation.

21. 4.31×10^{23}

22. 6.42×10^{-16}

23. $34,700,000,000,000,000,000,000,000,000,000$

24. 0.00000000000000000427

Evaluate Exercises 25–26 with the help of a calculator.

25. $2.7 \times 10^{18} \times 3.4 \times 10^{23}$

26. $3.8 \times 10^{-15} \times 4.5 \times 10^{13}$

Appendix

B

The Metric System

The metric system[1] is becoming more important to workers. Almost every major country uses the metric system. Many industries are converting their manufacturing processes and products to metric specifications. In Canada, and every other major industrial country except the United States, the SI metric system is the official measurement system. Competition in the form of international trade has forced many U.S. companies to convert their products to the metric system. The U.S. Congress passed legislation requiring federal agencies to use the metric system in their business activities.

Older technical service publications were printed using the U.S. Customary system, while the latest publications are either in metric or in a combination of both measurement systems. This section gives a brief introduction to the metric system and the different units and their symbols, and tells how to convert within the metric system or between the metric and U.S. Customary system.

Units of Measure

There are seven base units in the metric system and two supplemental units in the SI metric system. All seven base units and the two supplemental units are shown in Table B.1.

All other units are formed from these seven. The base units that you will use most often are those for length (meter), mass or weight (kilogram), time (second), and electric current (ampere). The base unit for temperature is Kelvin, but we will mostly use a variation called the degree Celsius. In addition to the seven base units, there are two supplemental units. We use the supplemental unit for a plane angle (radian).

Each unit can be divided into smaller units or made into larger units. To show that a unit has been made smaller or larger, a prefix is placed in front of the base unit. The prefixes are based on powers of 10. The most common prefixes are shown in Table B.2. For example, 1 kilometer (km) is 1 000 m, 1 milligram (mg) is 0.001 g, 1 megavolt (MV) is 1 000 000 V $=10^6$ V, and 1 nanosecond (ns) is 0.000 000 001 s $=10^{-9}$ s.

---

[1] The official name for the metric system is The International System of Units (or, in French, Le Système International d'Unités). The official abbreviation for the metric system throughout the world is "SI."

TABLE B.1   SI Metric System Base and Supplemental Units

BASE UNITS		
Quantity	**Unit**	**Symbol**
Length	meter	m
Mass	kilogram	kg
Time	second	s
Electric current	ampere	A
Temperature	Kelvin	K
Luminous intensity	candela	cd
Amount of substance	mole	mol
SUPPLEMENTAL UNITS		
Quantity	**Unit**	**Symbol**
Plane angle	radian	rad
Solid angle	steradian	sr

TABLE B.2   Metric Prefixes

Multiple	**Prefix**	**Symbol**	
1 000 000 000	$= 10^9$	giga	G
1 000 000	$= 10^6$	mega	M
1 000	$= 10^3$	kilo	k
1			
0.01	$= 10^{-2}$	centi	c
0.001	$= 10^{-3}$	milli	m
0.000 001	$= 10^{-6}$	micro	μ
0.000 000 001	$= 10^{-9}$	nano	n

Many quantities, such as area, volume, speed, velocity, acceleration, and force require a combination of two or more fundamental units. Combinations of these units are referred to as **derived units** and are listed in Table B.3.

Table B.4 shows some of the metric units most commonly used and their uses in various trade and technical areas. This list is not intended to be exhaustive, but to provide you with some idea as to where the different units are used.

Writing Metrics

You should remember some important rules when you use the metric system.

1. The unit symbols that are used are the same in all languages. This means that the symbols used on Japanese cars are the same as those used on German or American cars.

2. The unit symbols are not abbreviations and a period is not put at the end of the symbol.

TABLE B.3   Derived Units for Common Physical Quantities

Quantity	Derived Unit	Symbol	Alternate Symbol
Acceleration	meter per second squared	m/s^2	
Angular acceleration	radian per second squared	rad/s^2	
Angular velocity	radian per second	rad/s	
Area	square meter	m^2	
Concentration, mass (density)	kilogram per cubic meter	kg/m^3	
	gram per liter	g/L	
Capacitance	farad	F	C/V
Electric current	ampere	A	V/Ω
Electric field strength	volt per meter	V/m	
Electric resistance	ohm	Ω	V/A
Electromotive force (emf)	volt	V	J/C
Force	newton	N	$kg \cdot m/s^2$
Frequency	hertz	Hz	s^{-1}
Illuminance	lux	lx	lm/m^2
Inductance	henry	H	$V \cdot s/A$
Light exposure	lumen second	$lm \cdot s$	
Luminance	candela per square meter	cd/m^2	
Magnetic flux	weber	Wb	$V \cdot s$
Moment of force	newton meter	$n \cdot m$	
Moment of inertia, dynamic	kilogram meter squared	$kg \cdot m^2$	
Momentum	kilogram meter per second	$kg \cdot m/s$	
Power	watt	W	J/s
Pressure	pascal	Pa	N/m^2
Quantity of electricity	coulomb	C	
Speed, velocity	meter per second	m/s	
Volume	cubic meter	m^3	
Volume flow rate	cubic meter per second	m^3/s	
	liter per minute	L/min	
Work, energy, quantity of heat	joule	J	$N \cdot m$

3. Unit symbols are shown in lowercase letters except when the unit is named for a person. Examples of unit symbols that are written in capital letters are joule (J), newton (N), pascal (Pa), watt (W), ampere (A), and coulomb (C). The only exception is the use of L for liter. The symbol L is often used for liter to eliminate any confusion between the lowercase unit symbol (l) and the numeral (1). Some people prefer to use a script ℓ for liter.

4. The symbol is the same for both singular and plural (e.g., 1 m and 12 m).

5. Numbers with four or more digits are written in groups of three separated by a space instead of a comma. (The space is optional on four-digit numbers.) This is done because some countries use the comma as a decimal point.

EXAMPLE B.1

Use 2 473 or 2473 instead of 2,473;
   45 689 instead of 45,689;
   47 398 254.263 72 instead of 47,398,254.26372.

TABLE B.4   Uses of Metric Units in Technical Areas

Quantity	Unit	Symbol	Use
Length	micrometer	μm	paint thickness; surface texture or finish
	millimeter	mm	motor vehicle dimensions; wood; hardware, bolt, and screw dimensions; tool sizes; floor plans
	centimeter	cm	bearing size; length and width of fabric or window; length of weld, channel, pipe, I-beam, or rod
	meter	m	braking distance; turning circle; room size; wall covering; landscaping; architectural drawings; length of pipe or conduit; highway width
	kilometer	km	land distances; maps; odometers
Area	square centimeter	cm^2	piston head surfaces; brake and clutch contact area; glass, tile, or wall covering; area of steel plate
	square meter	m^2	fabric, land, roof, and floor area; room sizes; carpeting; window and/or wall covering
Volume or capacity	cubic centimeter	cm^3	cylinder bore; small engine displacement; tank or container capacity
	cubic meter	m^3	work or storage space; truck body; room or building volume; trucking or shipping space; tank or container capacity; ordering concrete; earth removal
	milliliter	mL	chemicals; lubricant; oils; small liquids; paint;
	liter	L	fuel; large engine displacement; gasoline
Temperature	degree Celsius	°C	thermostats; engine operating temperature; oil or liquid temperature; melting points; welds
Mass or weight	gram	g	tire weights; mailing and shipping packages
	kilogram	kg	batteries; weights; mailing and shipping packages
	metric ton	t	vehicle and load weight; construction material such as sand or cement; crop sales
Bending force, torque, moment of force	newton meter	N · m	torque specifications; fasteners
Pressure/vacuum	kilopascal	kPa	gas, hydraulic, oxygen, tire, air, or air hose pressure; manifold pressure compression; tensile strength
Velocity	kilometers per hour	km/h	vehicle speed; wind speed
	meters per second	m/s	speed of air or liquid through a system
Force, thrust, drag	newton	N	pedal; spring; belt; drawbar
Power	watt	W	air conditioner; heater; engine; alternator
Illumination	lumens per square meter	lm/m^2	intensity of light on a given area

TABLE B.4 (continued)

Quantity	Unit	Symbol	Use
Density	milligrams per cubic meter	mg/m^3	industrial hygiene standards for fumes, mists, and dusts
Flow	cubic meters per second	m^3/s	measure of air exchange in a region; exhaust and air exchange system ratings

6. A zero is placed to the left of the decimal point if the number is less than one (0.52 L, not .52 L).

7. Liter and meter are often spelled litre and metre.

8. The units of area and volume are written by using exponents.

EXAMPLE B.2

5 square centimeters are written as 5 cm^2, not as 5 sq cm;
37 cubic meters are written as 37 m^3, not as 37 cu m

Changing Within the Metric System

Changing units of measurement within a system is called **reduction**. A change from one system to another is called **conversion**. We will now discuss reduction in the metric system, and then conversion between the metric and U.S. customary system afterward.

The metric prefixes in Table B.2 provide a method to help with metric reduction. Using the information in Table B.2, we construct a metric reduction diagram like the one shown in Figure B.1. Starting on the left with the largest prefix shown in Table B.2, we mark each multiple of 10 until we get to the smallest prefix. Notice that not all multiples are labeled. While there is a prefix for each multiple of 10, we have given you only those that you will need.

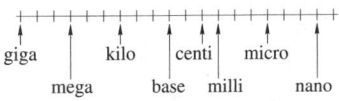

FIGURE B.1

Reduction in the SI Metric System

To change from one metric system unit to another:
1. Mark each unit on the reduction scale in Figure B.1.
2. Move the decimal point as many places as you move along the reduction scale according to the following rules:
   (a) To change to a unit farther to the right, move the decimal point to the right.
   (b) To change to a unit farther to the left, move the decimal point to the left.
3. If you are changing between square units, then move the decimal point *twice* as far as indicated in Step 2.
4. If you are changing between cubic units, then move the decimal point *three times* as far as indicated in Step 2.

EXAMPLE B.3

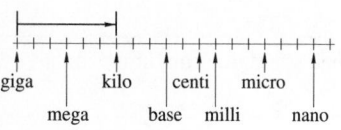

FIGURE B.2

Change 7.35 GV to kilovolts.

Solution Since 7.35 GV represents 7.35 gigavolts, we put a mark at the giga point on the reduction scale in Figure B.2. We want to convert GV to kV, so an arrow is drawn from the first mark to the mark labeled "kilo." The arrow points to the right and is 6 units long, so we move the decimal point 6 units to the right. To do this, we need to insert some zeros, with the result

$$7.35\,\text{GV} = 7\,350\,000\,\text{kV}$$

EXAMPLE B.4

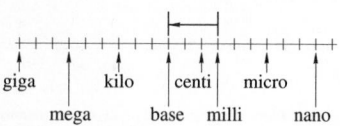

FIGURE B.3

Change 9.6 mm to meters.

Solution Since 9.6 mm represents 9.6 millimeters, we put a mark at the milli point on the reduction scale in Figure B.3. We want to convert mm to m, the base unit, so an arrow is drawn from the first mark to the mark labeled "base." The arrow points to the left and is 3 units long, so we move the decimal point 3 units to the left. Again, we need to insert some zeros, with the result

$$9.6\,\text{mm} = 0.009\,6\,\text{m}$$

EXAMPLE B.5

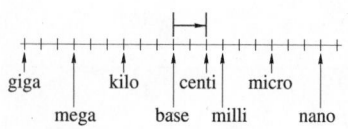

FIGURE B.4

Change 3.7 m^2 to square centimeters.

Solution Since 3.7 m^2 represents 3.7 square meters, we put a mark at the base point on the reduction scale in Figure B.4. We want to convert m^2 to cm^2, so the arrow is drawn from the first mark to the mark labeled "centi." The arrow points to the right and is 2 units long. Since we are converting between square units, we double this number, so we move the decimal point 4 units to the right. Again, we need to insert some zeros, with the result

$$3.7\,\text{m}^2 = 3\,700\,\text{cm}^2$$

To change one set of units into another set of units, you should perform algebraic operations with units to form new units for the derived quantity.

EXAMPLE B.6

Convert a speed of 88.00 km/h to meters per second.

Solution We will use two relationships that will give four possible conversion

EXAMPLE B.6 (Cont.)

factors.

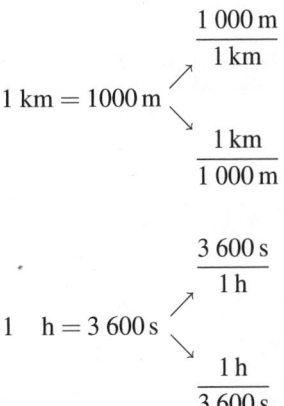

$$1\ km = 1000\ m \begin{cases} \dfrac{1\ 000\ m}{1\ km} \\[2mm] \dfrac{1\ km}{1\ 000\ m} \end{cases}$$

$$1\ h = 3\ 600\ s \begin{cases} \dfrac{3\ 600\ s}{1\ h} \\[2mm] \dfrac{1\ h}{3\ 600\ s} \end{cases}$$

We write the quantity to be changed, then choose the appropriate conversion factors so that all but two of the units "cancel," leaving the units m/s.

$$88\frac{\cancel{km}}{\cancel{h}} \times \frac{1\ 000\ m}{1\ \cancel{km}} \times \frac{1\ \cancel{h}}{3\ 600\ s} \approx 24.44\ m/s$$

Changing Between the Metric and Customary Systems ▬

Many technicians are often called upon to use both the SI metric and the U.S. Customary measurement systems. While it is best not to change from either the metric or the customary system to the other, sometimes a worker has to do so. For example, it is possible that the measuring tools available may be different than the measuring system in which the specifications are given. In such a case, the technician will be required to convert from one system to another. We will show one way to change between the U.S. Customary system and the SI metric system. Table B.5 should help you do this.

EXAMPLE B.7

Express 4.3 kg in pounds.

Solution You are changing from the metric (kg) to the customary system.

Look at Table B.5 under the heading, "From Metric to Customary." You are changing a measurement from kg to lb. Look in the column labeled "To change from" until you find the symbol for kilogram (kg). The symbol in the next right-hand column is the symbol for pound (lb).

Now look in the "Multiply by" column just to the right of these two symbols. The number there is 2.205. Multiply the given number of kilograms (4.3) by this number (2.205).

$$4.3 \times 2.205 = 9.4815$$

This means that 4.3 kg is equivalent to 9.4815 lb (or 4.3 kg ≈ 9.4815 lb).

TABLE B.5   Changing Units Between the Metric and Customary Systems

Quantity	FROM METRIC TO CUSTOMARY			FROM CUSTOMARY TO METRIC		
	To change from	To	Multiply by	To change from	To	Multiply by
Length	μm	mil	0.039 37	mil	μm	25.4
	mm	in.	0.039 37	in.	mm	25.4
	cm	in.	0.393 7	in.	cm	2.54
	m	ft	3.280 8	ft	m	0.304 8
	km	mile	0.621 37	mile	km	1.609 3
Area	cm^2	$in.^2$	0.155	$in.^2$	cm^2	6.451 6
	m^2	ft^2	10.763 9	ft^2	m^2	0.092 9
Volume	cm^3	$in.^3$	0.061	$in.^3$	cm^3	16.387
	m^3	yd^3	1.308	yd^3	m^3	0.764 6
	m^3	gal	264.172	gal	m^3	0.003 785
	mL	fl oz	0.033 8	fl oz	mL	29.574
	L	fl oz	33.814	fl oz	L	0.029 6
	L	pt	2.113	pt	L	0.473 2
	L	qt	1.056 7	qt	L	0.946 4
	L	gal	0.264 2	gal	L	3.785 4
Mass or weight	g	oz	0.035 3	oz	g	28.349 5
	kg	lb	2.205	lb	kg	0.453 6
	t	lb	2205	ton	kg	907.2
Bending moment, torque, moment of force	N · m	lbf · in.	8.850 7	lbf · in.	N · m	0.113
	N · m	lbf · ft	0.737 6	lbf · ft	N · m	1.355 8
Pressure, vacuum	kPa	psi	0.145	psi	kPa	6.894 8
Velocity	km/h	mph	0.621 4	mph	km/h	1.609 3
Force, thrust, drag	N	lbf	0.224 8	lbf	N	4.448 2
Power	W	W	1	W	W	1
Temperature	°C	°F	$\frac{9}{5}(°C) + 32$	°F	°C	$\frac{5}{9}(°F - 32)$

EXAMPLE B.8

Express 6.5 fluid ounces in liters.

Solution   You are changing from the customary (fluid ounces) to the metric (liters) systems.

Look in Table B.5 under the heading "From Customary to Metric." You are changing a measurement from fluid ounces to liters. Look under the column labeled "To change from" until you find the symbol for fluid ounces—fl oz. The fl oz symbol appears twice. Now look in the next right-hand column. Do you see the symbol for liters (L)? It is opposite the second fl oz symbol.

| EXAMPLE B.8 (Cont.) | If you look in the "Multiply by" column to the right of these two symbols (fl oz and L), you find 0.029 6. Multiply the number of fluid ounces (6.5) by this number (0.029 6). |

$$6.5 \times 0.029\ 6 = 0.192\ 4$$

This means that 6.5 fluid ounces is equivalent to 0.192 4 L (or 6.5 fl oz ≈ 0.192 4 L).

Exercise Set

In Exercises 1–16, reduce the given unit to the indicated unit.

27. 347 g to kilograms

28. 0.26 km to meters

29. 7.92 kW to watts

30. 2.3 μs to seconds

31. 0.023 85 Ω to milliohms

32. 0.000 235 47 MW to milliwatts

33. 9.72 mm to centimeters

34. 0.35 mm to nanometers

35. 835 000 cm^2 to square meters

36. 2.34 km^2 to m^2

37. 4.35 mm^3 to cubic centimeters

38. 91.52 m^3 to cubic millimeters

Solve Exercises 13–16.

39. *Machine technology*   A die is 14 mm long. Express this in centimeters.

40. *Law enforcement*   A male suspect is 1.97 m tall. Express this in centimeters.

41. *Environmental science*   A wastewater treatment plant has 92 600 kg of sludge. Express this in metric tons.

42. *Medical technology*   A technician needs 1 125 mL of a sterile solution. Express this in liters.

In Exercises 17–24, change the given units to the indicated unit. (You may want to consult Table B.3.)

43. 45 m/s to km/h

44. 27 mm/s to km/h

45. *Physics*   If a force of 126 N gives a body an acceleration of 9 m/s^2, then the body has a mass of $\frac{126\,\text{N}}{9\,\text{m/s}^2}$. Convert this to kilograms.

46. *Electricity*   An ac electric current that is generated by a source of 120 V and with an impedance of 125 Ω has an effective current of $\frac{120\,\text{V}}{125\,\Omega}$. Convert this to amperes.

47. *Electricity*   The resistance in a heater is $\frac{84\,\text{V}}{8\,\text{A}}$. Convert this to ohms.

48. *Physics*   A technician determines that the absolute pressure in a tank is 112 cm of mercury. The following computation will convert this pressure to kilopascals: $(13\,600\,\text{kg/m}^3)(9.8\,\text{m/s}^2)(112\,\text{cm})$. Convert this to kilopascals.

49. *Automotive technology*
If a 3 250-g piston is held 300 mm above a work surface, then its potential energy relative to the work surface is

$$(3\ 250\,\text{g})(9.8\,\text{m/s}^2)(300\,\text{mm})$$

Express this potential energy in joules.

50. *Construction*   At the instant a 4-kg sledgehammer reaches a velocity of 26 m/s, it has a kinetic energy of $\frac{1}{2}(4\,\text{kg})(26\,\text{m/s})^2$. Express this kinetic energy in joules.

In Exercises 25–40, convert the given measurement to the indicated unit.

51. $4\,\text{m} = $ _____ ft

52. $16\,\text{L} = $ _____ pt

53. $24\,\text{kg} = $ _____ lb

54. $27.5\,\text{N} \cdot \text{m} = $ _____ lbf · ft

55. $105\,\text{km/h} = $ _____ mph

56. $17\,\text{in.} = $ _____ cm

57. $4.5\,\text{qt} = $ _____ L

58. $21\,\text{ft}^2 = $ _____ m²

59. $5.8\,\text{lb} = $ _____ kg

60. $32\,\text{psi} = $ _____ kPa

61. $97\,\text{W} = $ _____ W

62. $98.6°\text{F} = $ _____ °C

63. *Fire safety* A fire extinguisher contains $2\frac{1}{2}$ gal. Convert this to liters.

64. *Law enforcement* An adult female weighs 120 lb. Convert this to kilograms.

65. *Environmental science* A flow-measuring meter at a wastewater treatment plant recorded 9,660,000 gal in one day. Convert this to liters.

66. *Environmental science* A wastewater treatment plant has a daily flow of $59\,500\,\text{m}^3$. Convert this to gallons.

Appendix

C

Table of Integrals

Basic Forms

1. $\displaystyle\int uv = uv - \int vu$

2. $\displaystyle\int u^n\, u = \frac{1}{n+1}u^{n+1} + C, n \neq -1$

3. $\displaystyle\int \frac{du}{u} = \ln|u| + C$

4. $\displaystyle\int e^u\, u = e^u + C$

5. $\displaystyle\int a^u\, u = \frac{1}{\ln a}a^u + C$

6. $\displaystyle\int \sin u\, u = -\cos u + C$

7. $\displaystyle\int \cos u\, u = \sin u + C$

8. $\displaystyle\int \sec^2 u\, u = \tan u + C$

9. $\displaystyle\int \csc^2 u\, u = -\cot u + C$

10. $\displaystyle\int \sec u \tan u\, u = \sec u + C$

11. $\displaystyle\int \csc u \cot u\, u = -\csc u + C$

12. $\displaystyle\int \tan u\, u = \ln|\sec u| + C$
    $\qquad\qquad = -\ln|\cos u| + C$

13. $\displaystyle\int \cot u\, u = \ln|\sin u| + C$

14. $\displaystyle\int \sec u\, u = \ln|\sec u + \tan u| + C$

15. $\displaystyle\int \csc u\, u = \ln|\csc u - \cot u| + C$

16. $\displaystyle\int \frac{du}{\sqrt{a^2 - u^2}} = \sin^{-1}\frac{u}{a} + C$

17. $\displaystyle\int \frac{du}{a^2 + u^2} = \frac{1}{a}\tan^{-1}\frac{u}{a} + C$

18. $\displaystyle\int \frac{du}{u\sqrt{u^2 - a^2}} = \frac{1}{a}\sec^{-1}\frac{u}{a} + C$

19. $\displaystyle\int \frac{du}{a^2 - u^2} = \frac{1}{2a}\ln\left|\frac{u+a}{u-a}\right| + C$

20. $\displaystyle\int \frac{du}{u^2 - a^2} = \frac{1}{2a}\ln\left|\frac{u-a}{u+a}\right| + C$

Forms Involving $\sqrt{a^2 - u^2}$

21. $\displaystyle\int \sqrt{a^2 - u^2}\, u = \frac{u}{2}\sqrt{a^2 - u^2} + \frac{a^2}{2}\sin^{-1}\frac{u}{a} + C$

22. $\displaystyle\int u^2\sqrt{a^2 - u^2}\, u =$
    $\qquad \frac{u}{8}\left(2u^2 - a^2\right)\sqrt{a^2 - u^2} + \frac{a^4}{8}\sin^{-1}\frac{u}{a} + C$

23. $\displaystyle\int \frac{\sqrt{a^2 - u^2}}{u}\, u =$
    $\qquad \sqrt{a^2 - u^2} - a\ln\left|\frac{a + \sqrt{a^2 - u^2}}{u}\right| + C$

24. $\displaystyle\int \frac{\sqrt{a^2 - u^2}}{u^2}\, u = -\frac{1}{u}\sqrt{a^2 - u^2} - \sin^{-1}\frac{u}{a} + C$

25. $\displaystyle\int \frac{u^2\,u}{\sqrt{a^2-u^2}} = -\frac{u}{2}\sqrt{a^2-u^2}+\frac{a^2}{2}\sin^{-1}\frac{u}{a}+C$

26. $\displaystyle\int \frac{du}{u\sqrt{a^2-u^2}} = -\frac{1}{a}\ln\left|\frac{a+\sqrt{a^2-u^2}}{u}\right|+C$

27. $\displaystyle\int \frac{du}{u^2\sqrt{a^2-u^2}} = -\frac{1}{a^2 u}\sqrt{a^2-u^2}+C$

28. $\displaystyle\int (a^2-u^2)^{3/2}\,u =$

$\qquad -\dfrac{u}{8}(2u^2-5a^2)\sqrt{a^2-u^2}+\dfrac{3a^4}{8}\sin^{-1}\dfrac{u}{a}+C$

29. $\displaystyle\int \frac{du}{(a^2-u^2)^{3/2}} = \frac{u}{a^2\sqrt{a^2-u^2}}+C$

Forms Involving $\sqrt{u^2\pm a^2}$

30. $\displaystyle\int \sqrt{u^2\pm a^2}\,u = \frac{u}{2}\sqrt{u^2\pm a^2}\pm$
$\qquad \dfrac{a^2}{2}\ln\left|u+\sqrt{u^2\pm a^2}\right|+C$

31. $\displaystyle\int u^2\sqrt{u^2\pm a^2}\,u = \frac{u}{8}(2u^2\pm a^2)\sqrt{u^2\pm a^2}-$
$\qquad \dfrac{a^4}{8}\ln\left|u+\sqrt{u^2\pm a^2}\right|+C$

32. $\displaystyle\int \frac{\sqrt{u^2+a^2}}{u}\,u =$
$\qquad \sqrt{u^2+a^2}-a\ln\left|\dfrac{a+\sqrt{u^2+a^2}}{u}\right|+C$

33. $\displaystyle\int \frac{\sqrt{u^2-a^2}}{u}\,u = \sqrt{u^2-a^2}-a\cos^{-1}\frac{a}{u}+C$

34. $\displaystyle\int \frac{\sqrt{u^2\pm a^2}}{u^2}\,u =$
$\qquad -\dfrac{\sqrt{u^2\pm a^2}}{u}+\ln\left|u+\sqrt{u^2\pm a^2}\right|+C$

35. $\displaystyle\int \frac{du}{\sqrt{u^2\pm a^2}} = \ln\left|u+\sqrt{u^2\pm a^2}\right|+C$

36. $\displaystyle\int \frac{du}{u\sqrt{u^2+a^2}} = -\frac{1}{a}\ln\left|\frac{a+\sqrt{u^2+a^2}}{u}\right|+C$

37. $\displaystyle\int \frac{du}{u\sqrt{u^2-a^2}} = \frac{1}{a}\cos^{-1}\left|\frac{a}{u}\right|+C$
$\qquad = \dfrac{1}{a}\sec^{-1}\left|\dfrac{u}{a}\right|$

38. $\displaystyle\int \frac{u^2\,u}{\sqrt{u^2\pm a^2}} = \frac{u}{2}\sqrt{u^2\pm a^2}\mp$
$\qquad \dfrac{a^2}{2}\ln\left|u+\sqrt{u^2\pm a^2}\right|+C$

39. $\displaystyle\int \frac{du}{u^2\sqrt{u^2\pm a^2}} = \mp\frac{\sqrt{u^2\pm a^2}}{a^2 u}+C$

40. $\displaystyle\int (u^2\pm a^2)^{3/2}\,u = \frac{u}{4}(u^2\pm a^2)^{3/2}\pm$
$\qquad \dfrac{3a^2 u}{8}\sqrt{u^2\pm a^2}+\dfrac{3a^4}{8}\ln\left|u+\sqrt{u^2\pm a^2}\right|+C$

41. $\displaystyle\int \frac{(u^2+a^2)^{3/2}}{u}\,u = \frac{1}{3}(u^2+a^2)^{3/2}+$
$\qquad a^2\sqrt{u^2+a^2}-a^3\ln\left|\dfrac{a+\sqrt{u^2+a^2}}{u}\right|+C$

42. $\displaystyle\int \frac{(u^2-a^2)^{3/2}}{u}\,u =$
$\qquad \dfrac{1}{3}(u^2-a^2)^{3/2}-a^2\sqrt{u^2-a^2}-a^3\sec^{-1}\dfrac{u}{a}+C$

43. $\displaystyle\int \frac{du}{(u^2\pm a^2)^{3/2}} = \frac{\pm u}{a^2\sqrt{u^2\pm a^2}}+C$

44. $\displaystyle\int \frac{du}{u(u^2+a^2)^{3/2}} =$
$\qquad \dfrac{1}{a^2\sqrt{u^2+a^2}}-\dfrac{1}{a^3}\ln\left|\dfrac{a+\sqrt{u^2+a^2}}{u}\right|+C$

45. $\displaystyle\int \frac{du}{u(u^2-a^2)^{3/2}} = \frac{-1}{a^2\sqrt{u^2-a^2}}-\frac{1}{|a^3|}\sec^{-1}\frac{u}{a}+C$

Forms Involving $a+bu$

46. $\displaystyle\int \frac{u\,u}{a+bu} = \frac{1}{b^2}(a+bu-a\ln|a+bu|)+C$

47. $\displaystyle\int \frac{u^2\,u}{a+bu} =$

$\qquad \dfrac{1}{2b^3}\left[(a+bu)^2-4a(a+bu)+2a^2\ln|a+bu|\right]+C$

48. $\displaystyle\int \frac{du}{u(a+bu)} = \frac{1}{a}\ln\left|\frac{u}{a+bu}\right|+C$

49. $\displaystyle\int \frac{u\,du}{(a+bu)^2} = \frac{1}{b^2}\left(\frac{a}{a+bu} + \ln|a+bu|\right) + C$

50. $\displaystyle\int \frac{du}{u(a+bu)^2} = \frac{1}{a(a+bu)} - \frac{1}{a^2}\ln\left|\frac{a+bu}{u}\right| + C$

51. $\displaystyle\int u\sqrt{a+bu}\,du = \frac{2}{15b^2}(3bu-2a)(a+bu)^{3/2} + C$

52. $\displaystyle\int \frac{u\,du}{\sqrt{a+bu}} = \frac{1}{3b^2}(bu-2a)\sqrt{a+bu} + C$

53. $\displaystyle\int \frac{u^2\,du}{\sqrt{a+bu}} =$
$\displaystyle\frac{2}{15b^3}(8a^2 + 3b^2u^2 - 4abu)\sqrt{a+bu} + C$

54. $\displaystyle\int \frac{du}{u\sqrt{a+bu}}$
$$= \begin{cases} \dfrac{1}{\sqrt{a}}\ln\left|\dfrac{\sqrt{a+bu}-\sqrt{a}}{\sqrt{a+bu}+\sqrt{a}}\right| + C & \text{if } a > 0 \\[2ex] \dfrac{2}{\sqrt{-a}}\tan^{-1}\sqrt{\dfrac{a+bu}{-a}} + C, & \text{if } a < 0 \end{cases}$$

55. $\displaystyle\int \frac{\sqrt{a+bu}}{u}\,du = 2\sqrt{a+bu} + a\int \frac{du}{u\sqrt{a+bu}}$

Trigonometric Forms

56. $\displaystyle\int \sin^2 u\,du = \frac{1}{2}u - \frac{1}{4}\sin 2u + C$
$$= \frac{1}{2}u - \frac{1}{2}\sin u\cos u + C$$

57. $\displaystyle\int \cos^2 u\,du = \frac{1}{2}u + \frac{1}{4}\sin 2u + C$
$$= \frac{1}{2}u + \frac{1}{2}\sin u\cos u + C$$

58. $\displaystyle\int \tan^2 u\,du = \tan u - u + C$

59. $\displaystyle\int \cot^2 u\,du = -\cot u - u + C$

60. $\displaystyle\int \sin^3 u\,du = -\cos u + \frac{1}{3}\cos^3 u + C$

61. $\displaystyle\int \cos^3 u\,du = \sin u - \frac{1}{3}\sin^3 u + C$

62. $\displaystyle\int \tan^3 u\,du = \frac{1}{2}\tan^2 u + \ln|\cos u| + C$

63. $\displaystyle\int \cot^3 u\,du = -\frac{1}{2}\cot^2 u - \ln|\sin u| + C$

64. $\displaystyle\int \sin^n u\,du = -\frac{1}{n}\sin^{n-1}u\cos u + \frac{n-1}{n}\int \sin^{n-2}u\,du$

65. $\displaystyle\int \cos^n u\,du = \frac{1}{n}\cos^{n-1}u\sin u + \frac{n-1}{n}\int \cos^{n-2}u\,du$

66. $\displaystyle\int \tan^n u\,du = \frac{1}{n-1}\tan^{n-1}u - \int \tan^{n-2}u\,du$

67. $\displaystyle\int \cot^n u\,du = \frac{-1}{n-1}\cot^{n-1}u - \int \cot^{n-2}u\,du$

68. $\displaystyle\int \sec^n u\,du =$
$\displaystyle\frac{1}{n-1}\tan u\sec^{n-2}u + \frac{n-2}{n-1}\int \sec^{n-2}u\,du$

69. $\displaystyle\int \csc^n u\,du =$
$\displaystyle\frac{-1}{n-1}\cot u\csc^{n-2}u + \frac{n-2}{n-1}\int \csc^{n-2}u\,du$

70. $\displaystyle\int \sin au\sin bu\,du = \frac{\sin(a-b)u}{2(a-b)} - \frac{\sin(a+b)u}{2(a+b)} + C$

71. $\displaystyle\int \cos au\cos bu\,du = \frac{\sin(a-b)u}{2(a-b)} + \frac{\sin(a+b)u}{2(a+b)} + C$

72. $\displaystyle\int \sin au\cos bu\,du =$
$\displaystyle-\frac{\cos(a-b)u}{2(a-b)} - \frac{\cos(a+b)u}{2(a+b)} + C$

73. $\displaystyle\int u\sin u\,du = \sin u - u\cos u + C$

74. $\displaystyle\int u\cos u\,du = \cos u + u\sin u + C$

75. $\displaystyle\int u^n \sin u\,du = -u^n\cos u + n\int u^{n-1}\cos u\,du$

76. $\displaystyle\int u^n \cos u\,du = u^n\sin u - n\int u^{n-1}\sin u\,du$

77. $\displaystyle\int \sin^n u\cos^m u\,du = -\frac{\sin^{n-1}u\cos^{m+1}u}{n+m}$
$$+ \frac{n-1}{n+m}\int \sin^{n-2}u\cos^m u\,du$$
$$= \frac{\sin^{n+1}u\cos^{m-1}u}{n+m}$$
$$+ \frac{m-1}{n+m}\int \sin^n u\cos^{m-2}u\,du$$

Inverse Trigonometric Forms

78. $\int \sin^{-1}u\,u = u\sin^{-1}u + \sqrt{1-u^2} + C$

79. $\int \cos^{-1}u\,u = u\cos^{-1}u - \sqrt{1-u^2} + C$

80. $\int \tan^{-1}u\,u = u\tan^{-1}u - \frac{1}{2}\ln(1+u^2) + C$

81. $\int u^n \sin^{-1}u\,u =$
$$\frac{1}{n+1}\left[u^{n+1}\sin^{-1}u - \int \frac{u^{n+1}u}{\sqrt{1-u^2}}\right], n \neq -1$$

82. $\int u^n \cos^{-1}u\,u =$
$$\frac{1}{n+1}\left[u^{n+1}\cos^{-1}u + \int \frac{u^{n+1}u}{\sqrt{1-u^2}}\right], n \neq -1$$

83. $\int u^n \tan^{-1}u\,u = \frac{1}{n+1}\left[u^{n+1}\tan^{-1}u - \int \frac{u^{n+1}u}{1+u^2}\right],$
$n \neq -1$

Exponential and Logarithmic Forms

84. $\int u e^{au}\,u = \frac{1}{a^2}(au-1)e^{au} + C$

85. $\int u^2 e^{au}\,u = \frac{1}{a^3}(a^2u^2 - 2au + 2)e^{au} + C$

86. $\int u^n e^{au}\,u = \frac{1}{a}u^n e^{au} - \frac{n}{a}\int u^{n-1}e^{au}\,u$

87. $\int e^{au}\sin bu\,u = \frac{e^{au}}{a^2+b^2}(a\sin bu - b\cos bu) + C$

88. $\int e^{au}\cos bu\,u = \frac{e^{au}}{a^2+b^2}(a\cos bu + b\sin bu) + C$

89. $\int \ln u\,u = u\ln u - u + C$

90. $\int u^n \ln u\,u = u^{n+1}\left(\frac{\ln u}{n+1} - \frac{1}{(n+1)^2}\right) + C,$
$n \neq -1$

91. $\int \frac{1}{u\ln u}\,u = \ln|\ln u| + C$

Hyperbolic Forms

92. $\int \sinh u\,u = \cosh u + C$

93. $\int \cosh u\,u = \sinh u + C$

94. $\int \tanh u\,u = \ln\cosh u + C$

95. $\int \coth u\,u = \ln|\sinh u| + C$

96. $\int \operatorname{sech} u\,u = \tan^{-1}|\sinh u| + C$

97. $\int \operatorname{csch} u\,u = \ln\left|\tanh \frac{1}{2}u\right| + C$

98. $\int \operatorname{sech}^2 u\,u = \tanh u + C$

99. $\int \operatorname{csch}^2 u\,u = -\coth u + C$

100. $\int \operatorname{sech} u\tanh u\,u = -\operatorname{sech} u + C$

101. $\int \operatorname{csch} u\coth u\,u = -\operatorname{csch} u + C$

Answers to Odd-Numbered Exercises

Exercise Set 1.1

1. Natural number, whole number, integer, rational number, and real number   **3.** Irrational number, real number

5. 15   **7.** $\frac{\sqrt{7}}{8}$ **9.** 3

11.

13.

   15. <   **17.** <

19. >   **21.** <   **23.** >   **25.** <   **27.** $-\frac{1}{5}$ **29.** $\frac{3}{17}$
31. $-5, -|4|, -\frac{2}{3}, \frac{-1}{3}, \frac{16}{3}, |-8|$ **33.** (a) $\frac{1}{32}$ inch;
(b) $\frac{1}{8}$ inch; **(c)** 0.0625 inch   **35.** 2.6 V   **37.** $\frac{5}{7}$ (Sheila), $\frac{2}{3}$
(Hazel), $\frac{3}{5}$ (José), $\frac{9}{16}$ (Robert), and $\frac{4}{9}$ (Lamar)   **39.** Mile marker 93 or 221

Exercise Set 1.2

1. Commutative law for addition   **3.** Commutative law for multiplication   **5.** Associative law for addition
7. Identity element for multiplication   **9.** Additive inverse   **11.** Association law for multiplication   **13.** -91
15. $-\sqrt{2}$ **17.** 2   **19.** $\frac{2}{\sqrt{2}} = \sqrt{2}$ **21.** 14   **23.** 13
25. 1   **27.** 31   **29.** 1   **31.** -111 **33.** 270
35. (a) -359 (b) -359 **37.** (a) -32760 (b) -32760

Exercise Set 1.3

1. 50   **3.** 14   **5.** -9 **7.** -24 **9.** 12
11. 15   **13.** $-\frac{38}{4} = -\frac{19}{2}$ **15.** $\frac{1}{8}$ **17.** $\frac{8}{15}$ **19.** $\frac{3}{20}$
21. $\frac{5}{8}$ **23.** $-\frac{47}{30} = -1\frac{17}{30}$ **25.** $\frac{1}{32}$ **27.** $-\frac{8}{15}$ **29.** $\frac{3}{32}$
31. $-\frac{10}{3}$ **33.** $\frac{6}{5}$ **35.** $-\frac{3}{20}$ **37.** $\frac{49}{25} = 1\frac{24}{25}$ **39.** $\frac{7}{2}$
41. $78\frac{1}{2}$ inches $= 6'6\frac{1}{2}''$ **43.** 57 V   **45.** (a) $7\frac{33}{40}$ mi;

(b) $11\frac{1}{5}$ mi; $3\frac{3}{10}$ mi   **47.** $84'9\frac{1}{2}''$ **49.** $\frac{7''}{8}$
51. (a) -21.2 V, (b) The voltage dropped 21.2 V

Exercise Set 1.4

1. 125   **3.** $\frac{3}{2}$ **5.** 16   **7.** $\frac{1}{49}$ **9.** 3^6 **11.** 2^{12}
13. 2^2 **15.** 2^6 **17.** x^{20} **19.** $\frac{a^6}{b^3}$ **21.** $\frac{x^3}{4^3}$
23. $\frac{a^8 b^4}{c^{12}}$ **25.** $\frac{1}{x^7}$ **27.** $\frac{1}{p^3}$ **29.** $\frac{1}{7^5}$ **31.** $\frac{1}{a^3 y^4}$
33. $\frac{y^2}{a^4}$ **35.** $\frac{1}{pr^2}$ **37.** $\frac{5^2}{4y^3}$ **39.** $\frac{y^{15}}{2^3 b^6}$ **41.** b^{24}
43. $51.96\,\Omega$ **45.** $8.5\,\Omega$

Exercise Set 1.5

1. Exact   **3.** Approximate   **5.** Approximate   **7.** 3
9. 3   **11.** 1   **13.** 4   **15.** (a) 6.05, (b) 6.05
17. (a) 5.01, (b) 0.027   **19.** (a) $27,0\tilde{0}0$, (b) $27,0\tilde{0}0$
21. (a) 86, (b) 0.2   **23.** (a) 140.070, (b) 140.070
25. (a) 10, (b) 14, (c) 14.4   **27.** (a) 7, (b) 7.0, (c) 7.04
29. (a) 400, (b) $4\tilde{0}0$, (c) 403   **31.** (a) 300, (b) 310, (c) 305
33. (a) 10, (b) 14, (c) 14.4   **35.** (a) 7, (b) 7.0, (c) 7.04
37. (a) 400, (b) $4\tilde{0}0$, (c) 403   **39.** (a) 300, (b) 310, (c) 305
41. (a) 90, (b) 89.9, (c) 89.899   **43.** (a) 240, (b) 237.3,
(c) 237.302   **45.** (a) 440, (b) 438.0, (c) 437.998
47. (a) 80, (b) 78.7, (c) 78.671
49.

	Absolute error	Relative error
Length	-0.28 mm	-1.17%
Width	0.35 mm	4.38%
Thickness	-0.02 mm	-0.67%

51. 99.37

53. 1618   **55.** 1,020   **57.** 17   **59.** 351.65 m
61. 924.6 mm

Exercise Set 1.6

1. 4.2×10^4 **3.** 3.8×10^{-4} **5.** 9.80700×10^9
7. 9.70×10^{-5} **9.** 4.3×10^0 **11.** 74×10^3
13. 470×10^{-6} **15.** 9.80700×10^9 **17.** 53.10×10^{-6}
19. 5.6×10^0 **21.** 4500   **23.** 40500000   **25.** 0.000063
27. 72   **29.** 75 000   **31.** 47 500 000 000
33. 0.000 039 20   **35.** 83.15   **37.** 1.5504×10^{16}
39. 2.66×10^{-11} **41.** 2.94×10^5 **43.** 2.2528×10^{-1}
45. 1.2×10^5 **47.** 2.5×10^{-3} **49.** 2×10^4
51. 1.8×10^{-3} **53.** $1.373568 \times 10^{25} \approx 1.38 \times 10^{25}$
55. 2.5×10^{22} **57.** 3.9×10^{14} **59.** 675 (Answers may vary.)   **61.** 1.11×10^6 mm   **63.** 2.789808×10^{-19} kg
65. One neutron; approximately 1.84×10^3 times heavier
67. About $282.74 \times 10^3 \, \Omega$

Exercise Set 1.7

1. 5   **3.** 12   **5.** 2   **7.** -3 **9.** 2   **11.** $\frac{2}{3}$
13. 0.2   **15.** -0.1 **17.** 3   **19.** 8.32   **21.** 5
23. $\sqrt[4]{72}$ **25.** 5   **27.** $\frac{1}{2}$ **29.** 7   **31.** $\frac{1}{3}$ **33.** $5^{1/3}$
35. 2   **37.** 8   **39.** 4   **41.** $\frac{1}{4}$ **43.** 81   **45.** $\dfrac{9}{0.2\sqrt{2}}$

47. 0.25   **49.** 2.09 m   **51.** 1449 m/s

Review Exercises

1. (a) Integers, rational numbers, real numbers, **(b)** Rational numbers, real numbers, **(c)** Irrational numbers, real numbers
3. (a) $\frac{3}{2}$, **(b)** $\frac{-1}{8}$, **(c)** -5 **5. (a)** Commutative law for addition, **(b)** Distributive law, **(c)** Additive identity, **(d)** Multiplicative inverse   **7.** -44 **9.** 98   **11.** $-\frac{1}{6}$
13. -3 **15.** $-\frac{2}{5}$ **17.** $\frac{8}{3}$ **19.** $\frac{10}{13}$ **21.** 32
23. -64 **25.** 2   **27.** 2^8 **29.** 2^2 **31.** 4^{15} **33.** 4
35. $\dfrac{a}{b}$ **37.** $\dfrac{1}{a^2 x^4}$ **39. (a)** 2.37, **(b)** 2.37   **41. (a)** both, **(b)** 0.7   **43. (a)** 7.4, **(b)** 7.35, **(c)** 7.4, **(d)** 7.35
45. (a) 2.1, **(b)** 2.05, **(c)** 2.1, **(d)** 2.05   **47.** 3.71×10^{11}
49. 2.4×10^{-11} **51.** 29.08×10^{15} **53.** 75×10^{-24}
55. 12   **57.** 5   **59. (a)** $57.12, **(b)** $5,707.10

Chapter 1 Test

1. (a) $\frac{4}{3}$; **(b)** $\frac{1}{8}$; **(c)** 6   **3.** $\frac{5}{8}$ **5.** -112 **7.** $\frac{22}{3} = 7\frac{1}{3}$
9. $-10\frac{1}{2}$ **11.** $\dfrac{14}{29}$ **13.** 8   **15.** 25   **17.** $\dfrac{b^3}{a}$
19. 0.00051   **21.** $\frac{7}{5}$

ANSWERS FOR CHAPTER 2

Exercise Set 2.1

1. 4 is a constant; x and y are variables.   **3.** 8 and π are constants; r is a variable.   **5.** $3x^3$ and $4x$
7. $\left(2x^3\right)(5y)$, $\sqrt{3}ab$, and $-\dfrac{7a}{b}$ **9.** $-1, 5, x$, and $\dfrac{1}{y}$
11. 47   **13.** $\dfrac{\pi a}{4}$, if a is a constant   **15.** $3x^2 y$ and $17x^2 y$
17. $(x+y)^2$ and $5(x+y)^2$ **19.** $a+b-c$ and $-2(a+b-c)$ **21.** $11x$ **23.** $2z$ **25.** $6x+9x^2$ or $9x^2 + 6x$ **27.** $10w - 7w^2$ or $-7w^2 + 10w$
29. $2ax^2 + a^2 x$ **31.** $11xy^2 - 5x^2 y$ **33.** $12b$
35. $2a^2 + 5b + 4a$ **37.** $2x^2 + 6x$ **39.** $12y^2 + 14x$
41. $-12b + 6c$ **43.** $7a + 7b$ **45.** $x+y$
47. $5a + 5b - c$ **49.** $6x + 6y$ **51.** $5a + 5b$
53. $6a - 2b + 4c$ **55.** $9x + 7y$ **57.** $-6x + 2y - 13z$
59. $-4x + 3y + 25z$ **61.** $6x + 2y$ **63.** $3y + z$
65. $x + 6y$ **67.** $a - 3b$ **69.** $-70a - 8b$
71. $\frac{13}{6}p = 2\frac{1}{6}p$ **73.** $\dfrac{1}{C_T} = \dfrac{C_2 + C_1}{C_1 C_2}$

Exercise Set 2.2

1. $a^3 x^3$ **3.** $6a^2 x^3$ **5.** $-6x^3 w^3 z$ **7.** $-24ax^4 b$
9. $10y - 12$ **11.** $35 - 20w$ **13.** $21xy + 12x$
15. $15t - 5t^2$ **17.** $2a^2 - a$ **19.** $6x^3 - 2x^2 + 8x$
21. $-20y^4 + 8y^3 - 20y^2 + 12y - 24$ **23.** $a^2 + ab + ac + bc$
25. $x^3 + 5x^2 + 6x + 30$ **27.** $6x^2 + xy - y^2$

29. $6a^2 - 7ab + 2b^2$ **31.** $2b^2 + 3b - 5$
33. $56a^4 b^2 + 3a^2 bc - 9c^2$ **35.** $x^2 - 16$ **37.** $p^2 - 36$
39. $a^2 x^2 - 4$ **41.** $4r^4 - 9x^2$ **43.** $25a^4 x^6 - 16d^2$
45. $\dfrac{9}{16}t^2 b^6 - \dfrac{4}{9}p^2 a^4 f^2$ **47.** $x^2 + 2xy + y^2$
49. $x^2 - 10x + 25$ **51.** $a^2 + 6a + 9$ **53.** $4a^2 + 4ab + b^2$
55. $9x^2 - 12xy + 4y^2$ **57.** $12x^3 + 40x^2 - 32x$
59. $x^2 - y^2 - z^2 + 2yz$ **61.** $a^3 + 6a^2 + 12a + 8$
63. $2n^2 + 6n$ **65.** $\dfrac{1}{2}\left(y_2^{\,2} - y_1^{\,2}\right) = \dfrac{1}{2}y_2^{\,2} - \dfrac{1}{2}y_1^{\,2}$
67. $x = 3t^2 + 15t + 12$

Exercise Set 2.3

1. x^4 **3.** $2x^2$ **5.** $3y^2$ **7.** $-3b$ **9.** $11y$
11. $-6ay$ **13.** $18c^2 d^2$ **15.** $\dfrac{-3p}{5n^2}$ **17.** $\dfrac{4bcx}{7y}$
19. $2a^2 + a$ **21.** $4b^3 - 2b$ **23.** $6x^2 + 4x$ **25.** $2x^3 - 3$
27. $-6x^3 + 2x$ **29.** $5x + 5y$ **31.** $2x + 3y$ **33.** $ap - 2$
35. $a + 1$ **37.** $-3xy + z$ **39.** $-b^2 x^2 - b^2$
41. $x + 1 - y$ **43.** $-\frac{3}{2}xy + 2y^2$ **45.** $x + 4$ **47.** $x - 1$
49. $x - 1$ **51.** $2a + 1$ **53.** $4y + 2$ **55.** $x - 2$
57. $2a - 1$ **59.** $2x^2 + x - 1 + \dfrac{3}{2x - 1}$
61. $r^2 - 3r + \dfrac{5}{r + 2}$ **63.** $x^3 + 3x^2 + 9x + 27$
65. $4x^3 - 3x^2 - x + 6$ **67.** $x + y$ **69.** $w^2 + wz + z^2$

71. $x^2 + y^2$ **73.** $c^2d^2 + 2cd + 4$ **75.** $x - y$
77. $p^2r - 2p + 3r^2$ **79.** $a + d + 4 + \dfrac{-4}{a - 3d - 1}$ **81.** af
83. $a + b - c$ **85.** $a^2 - a + 1$ **87.** $\dfrac{1}{R_1} + \dfrac{1}{R_2} + \dfrac{1}{R_3}$
89. $V_2 = \dfrac{V_1 T_2}{T_1}$

Exercise Set 2.4

1. 39   **3.** 12   **5.** -15 **7.** 4.4   **9.** $\dfrac{9}{2}$
11. -8 **13.** $\dfrac{3}{2}$ **15.** $-\dfrac{2}{3}$ **17.** 15   **19.** -24 **21.** 2
23. -4 **25.** $-\dfrac{17}{2}$ **27.** 4   **29.** 4   **31.** -1 **33.** 7
35. $\dfrac{1}{2}$ **37.** $\dfrac{24}{5}$ **39.** 5   **41.** -6 **43.** 9   **45.** -6
47. $-\dfrac{40}{3}$ **49.** 6   **51.** 6   **53.** $-\dfrac{11}{6}$ **55.** $\dfrac{13}{4}$
57. 14   **59.** -18 **61.** 13   **63.** $\dfrac{24}{7}$ **65.** 78
67. $\dfrac{b}{2a}$ **69.** $-\dfrac{x}{x-8}$ or $\dfrac{x}{8-x}$ **71.** $\dfrac{7}{3}$ **73.** 1
75. -3 **77.** -15 **79.** -2 **81.** $a = -12$
83. $a = -2z$ **85.** $C = \dfrac{5}{9}(F - 32)$

Exercise Set 2.5

1. 82   **3.** 60, since no score can be below 60
5. \$1150; \$230   **7.** \$6130   **9.** \$2200 at 7.5% and

\$2300 at 6%   **11.** 7 hours   **13.** 266 miles   **15.** 320 km
17. 2.4 hours or 2 hours 24 minutes   **19.** $\frac{4}{3}$ hours or 1 hour
20 minutes   **21.** 57.5 mL   **23.** 468.75 kg of 75% copper;
281.25 kg of 35% copper   **25.** 350 lb; 8.235 feet from the
500 lb end   **27.** 4.5 ft from the left end   **29.** 6 in from
right end   **31.** 30 cc

Review Exercises

1. $3y$ **3.** $5x - 8$ **5.** $7x^2 + 11$ **7.** $16x + 8$
9. $1 - 3x$ **11.** $9x^2 + x - 2$ **13.** $3a + 9b$ **15.** a^3x^3
17. $27ax^3$ **19.** $20x - 24$ **21.** $8x^2 - 10x$ **23.** $a^2 - 16$
25. $6a^2 - 5ab + b^2$ **27.** $6x^4 - 5x^2 - 6$ **29.** $x^2 + 4x + 4$
31. $30x^3 - 25x^2 - 20x$ **33.** a^3 **35.** $4a$ **37.** $-9ax^2$
39. $18x - 8$ **41.** $x - 4$ **43.** $x^2 + 3x + 9$ **45.** $x - y$
47. $x - 2y$ **49.** 38   **51.** $\dfrac{15}{2}$ **53.** 36   **55.** 5
57. 4.5   **59.** $-\dfrac{3}{2}$ **61.** -7 **63.** -12 **65.** $-\dfrac{1}{5}$
67. 2   **69.** 76   **71.** 3.6 hours or 3 hours 36 minutes
73. 35 kg of lead; 63.64% lead   **75.** 1538.6 N on the left
cable, 1107.4 N on the right cable

Chapter 2 Test

1. $5x^2 - 3x$ **3.** $2x + 8y - 6$ **5.** $4b^2 - 9$
7. $3x^3 + 2x - \dfrac{1}{2x}$ **9.** $\dfrac{y^2 - 13y - 8}{(3y + 2)(y - 1)}$ **11.** $\dfrac{14}{5}$
13. 1.8 qt

ANSWERS FOR CHAPTER 3

Exercise Set 3.1

1. $\frac{\pi}{12} \approx 0.2617994$ **3.** $\frac{7\pi}{6} \approx 3.6651914$
5. $\frac{427\pi}{900} \approx 1.4905112$ **7.** $\frac{109\pi}{120} \approx 2.8536133$
9. $240°$ **11.** $234°$ **13.** $14.323945°$ **15. (a)** $145°$;
(b) $2.5307274 \approx \frac{29\pi}{36}$ **17.** $\angle A = 147°$; $\angle B = 33°$
19. $\angle A = \frac{3\pi}{8}$; $\angle B = \frac{5\pi}{8}$ **21.** $\angle A = 109°30'$;
$\angle B = 70°30'$ **23.** $\frac{24}{5} = 4.8$ **25.** $47.5°$
27. 120π rad/s   **29.** $152°$

Exercise Set 3.2

1. $70°$ **3.** $x = 55°$, $y = 70°$ **5.** $x = y = 60°$
7. $36.87°$ **9.** $A = 23.4$ units^2; $P = 23.4$ units
11. $A = 126$ units^2; $P = 84$ units   **13. (a)** 133.3 m;
(b) 478.5 m^2 **15.** $\sqrt{200} = 10\sqrt{2}$ m ≈ 14.1421 m
17. $\sqrt{5869} \approx 76.6$ ft   **19. (a)** 5261.54 mm and 8123.08 mm;
(b) 9230.77 mm   **21.** $\sqrt{119} \approx 10.9$ m   **23. (a)** 212 m;
(b) 182.8 m   **25.** 73.4 m

Exercise Set 3.3

1. $P = 60$ cm; $A = 225$ cm^2 **3.** $P = 96.5$ in;
$A = 379.5$ in.^{2} **5.** $P = 69$ in; $A = 343.6$ in.^{2}

7. $P = 44$ cm; $A = 168$ cm^2 **9.** $P = 73.8$ mm;
$A = 279.84$ mm^2 **11.** 444 ft^2 **13.** 10 000 mm^2
15. $P = 3.031$ in; $A = 0.663$ in.^{2} **17. (a)** $A = 750$ ft^2;
(b) 94.76 ft   **19.** 2.1875 in.^{2} or 2.19 in.^{2} **21.** 18 638 mm^2

Exercise Set 3.4

1. $A = 16\pi$ cm^2; $C = 8\pi$ cm   **3.** $A = 25\pi$ in.^{2};
$C = 10\pi$ in   **5.** $A = 201.64\pi$ mm^2; $C = 28.4\pi$ mm
7. $A = 146.41\pi$ mm^2; $C = 24.20\pi$ mm   **9. (a)** 576π in.^{2};
(b) $48\pi \approx 151$ in   **11.** 32   **13. (a)** 163.2 mm;
(b) 6935.5 mm^2 **15. (a)** 86.643 in; **(b)** 481.52 in.^{2}
17. (a) $A = 2529.876$ in.^{2}; **(b)** 197.1 in   **19.** 11.91 mm
21. (a) $3\frac{1}{7} - \pi \approx 0.00126$ **(b)** $\dfrac{3\frac{1}{7} - \pi}{\pi}\% \approx 0.040\%$

Exercise Set 3.5

1. $L = 216\pi \approx 678.6$ in.^{2}, $T = 288\pi \approx 904.8$ in.^{2},
$V = 648\pi \approx 2035.8$ in.^{3} **3.** $L = 136\pi \approx 427.3$ mm^2,
$T = 200\pi \approx 628.3$ mm^2, $V = 320\pi \approx 1005.3$ mm^3
5. $L = 735$ in.^{2}, $T = 816.8$ in.^{2}, $V = 859.1$ in.^{3} **7.** A
sphere has no lateral surface area, $T = 7256\pi \approx 804.26$ cm^2,

$V = 682.7\pi \approx 2144.7\,\text{cm}^3$ **9.** $5866.7\,\text{yd}^3$
11. (a) $3600\,\text{ft}^3$; **(b)** $1560\,\text{ft}^2$
13. (a) $103\,455\pi \approx 325\,013.5\,\text{mm}^3$; **(b)** $20\,172.79\,\text{mm}^2$ of
paper **15.** $1333.3\pi \approx 4188.8\,\text{mm}^3$ **17.** $136.1\,\text{yd}^3$
19. $57\,172\,\text{mm}^3$ **21.** $66\pi \approx 207.3\,\text{in.}^2$ **23.** $1208.23\,\text{m}^3$
25. $48\pi \approx 150.8\,\text{ft}^3$

Review Exercises

1. $\frac{3}{20}\pi \approx 0.4712$ **3.** $198°$ **5.** $43°$
7. $\angle A = 152°$, $\angle B = 152°$ **9.** 3.3 **11.** $30°$
13. 33.36 **15.** $A = 294\,\text{units}^2$; $P = 73\,\text{units}$
17. $A = 81\pi \approx 254.5\,\text{units}^2$; $P = 18\pi \approx 56.55\,\text{units}$
19. $A = 924\,\text{units}^2$; $P = 154\,\text{units}$ **21.** $A = 186.75\,\text{units}^2$;

$P = 67.1\,\text{units}$ **23.** $L = 232.32\pi \approx 729.85\,\text{units}^2$,
$T = 525.14\pi \approx 1649.78\,\text{units}^2$, $V = 1405.54\pi \approx 4415.62$
units^3 **25.** $L = 2320\,\text{units}^2$, $T = 3920\,\text{units}^2$, $V = 11{,}200$
units^3 **27.** $L = 520\pi \approx 1633.6\,\text{units}^2$,
$T = 930\pi \approx 2921.68\,\text{units}^2$, $V = 2248.17\pi \approx 7062.84\,\text{units}^3$
29. Spheres have no lateral surface area, $T = 324\pi \approx 1017.88$
sq units, $V = 972\pi \approx 3053.63\,\text{units}^3$ **31.** $430\,\text{ft}$

Chapter 3 Test

1. $\frac{7\pi}{36}$ **3.** $104°$ **5.** 12.6 **7.** $101.4\,\text{cm}$
9. $280.8\,\text{cm}^2$ **11.** about $144.8\,\text{m}$ **13.** $1945.944\,\text{cm}^3$
15. $\approx 201.59\,\text{cm}^2$

≡ ANSWERS FOR CHAPTER 4

Exercise Set 4.1

1. Function **3.** Not a function **5.** Function for $x \geq 0$
7. Function for $x \geq \frac{7}{2}$ **9.** Yes, because for each value of
x there is just one value of y. **11.** No, because when $x = 4$
there are two values of y: -2 and 2, or when $x = 1$, y is 1 or
-1. **13.** Domain: $\{-3, -2, -1, 0, 1, 2\}$; Range:
$\{-7, -5, -3, -1, 1, 3\}$ **15.** Domain:
$\{-3, -2, -1, 0, 1, 2, 3\}$; Range: $\{-25, -6, 1, 2, 3, 10, 25\}$
17. Domain and range are both all real numbers.
19. Domain is all real numbers except 5; that is, the domain is
$\{x : x \neq 5\}$. Range is all real numbers except 1; that is, the
range is $\{y : y \neq 1\}$. **21.** Domain is all nonnegative real
numbers, $x \geq 0$. Range is all real numbers greater than or
equal to -2; that is, the range is $\{y : y \geq -2\}$. **23.** Domain
is all real numbers except 2 and -3; that is, the domain is
$\{x : x \neq 2, x \neq -3\}$. **25.** Domain is all real numbers
except -1 and 5; that is, the domain is $\{x : x \neq -1, x \neq 5\}$.
27. Domain is all real numbers except -4 and 2; that is, the
domain is $\{x : x \neq -4, x \neq 2\}$. **29.** Any function with a
denominator of $x - 5$ will work as long as the numerator is
defined for all real numbers. For example, $y = \dfrac{1}{x - 5}$ is one
answer. **31.** Any function with a denominator of
$(x + 1)(x - 2)$ will work as long as the numerator is defined for
all real numbers. For example, $y = \dfrac{1}{(x + 1)(x - 2)}$ is one
answer. **33.** Any function with an even root will work if the
quantity under the radical sign is positive. For example,
$y = \sqrt{x - 5}$ is one answer. **35.** Any function with $x - 4$ in
the denominator and with an even root where the quantity
under the radical sign is nonnegative when $x \geq -1$. For
example, $y = \dfrac{\sqrt{x + 1}}{x - 4}$ is one answer. **37.** -2 **39.** -11

41. -13 **43.** $3b + 7$ **45.** 0 **47.** -6 **49.** $-\frac{46}{25}$
51. $x^2 - 15x + 50$ **53.** $16m^4 - 20m^2$ **55.** 1
57. $\dfrac{2}{17} \approx 0.1176471$ **59.** $\dfrac{2 - 2x}{4x^2 + 2} = \dfrac{1 - x}{2x^2 + 1}$ **61.** 0
63. -40 **65.** Explicit **67.** Implicit. Solving for y we
get an explicit form of $y = \dfrac{x}{1 - x^2}$. **69.** 7 **71.** -3
73. 4 **75.** 49 **77.** -23 **79.** -3 **81.** If $r = 20$,
$V = 16{,}000\pi\,\text{ft}^3$, $S = 2400\pi\,\text{ft}^2$; if $r = 30$, $V = 36{,}000\pi\,\text{ft}^3$,
$S = 4200\pi\,\text{ft}^2$ **83.** $\$35$ **85.** $4{,}992\,\text{lb/ft}^2$

Exercise Set 4.2

1. $y = x + 5$ **3.** $y = \dfrac{x + 8}{2} = \frac{1}{2}(x + 8) = \frac{1}{2}x + 4$
5. $y = 3x + 2$ **7.** $y = x^2 + 5$
9.

x	-3	-2	-1	0	1	2	3	4
$f(x)$	0	1	2	3	4	5	6	7
$g(x)$	-10	5	0	-1	5	1	-10	7
$(f+g)(x)$	-10	6	2	2	9	6	-4	14

11.

x	-3	-2	-1	0	1	2	3	4
$f(x)$	0	1	2	3	4	5	6	7
$g(x)$	-10	5	0	-1	5	1	-10	7
$(g-f)(x)$	-10	4	-2	-4	1	-4	-16	0

13.

x	-3	-2	-1	0	1	2	3	4
$f(x)$	0	1	2	3	4	5	6	7
$g(x)$	-10	5	0	-1	5	1	-10	7
$(f/g)(x)$	0	$\frac{1}{5}$	undefined	-3	$\frac{4}{5}$	5	$-\frac{3}{5}$	1

15.

x	-3	-2	-1	0	1	2	3	4
$g(x)$	-10	5	0	-1	5	1	-10	7
$(f \circ g)(x)$	-7	8	3	2	8	4	-7	10

17. The domain and range of f are all real numbers. The domain of g is $\{-3, -2, -1, 0, 1, 2, 3, 4\}$. The range of g is $\{-10, -1, 0, 1, 5, 7\}$. **19. (a)** The domain of f/g is $\{-3, -2, 0, 1, 2, 3, 4\}$. The range of f/g is $\{-3, -\frac{3}{5}, 0, \frac{1}{5}, \frac{4}{5}, 1, 5\}$. **(b)** The domain of g/f is $\{-2, -1, 0, 1, 2, 3, 4\}$. The range of g/f is $\{-\frac{5}{3}, -\frac{1}{3}, 0, \frac{1}{5}, 1, \frac{5}{4}, 5\}$. **21.** $x^2 + 3x + 4$
23. $x^2 - 3x - 6$ **25.** $3x - x^2 + 6$
27. $3x^3 + 5x^2 - 3x - 5$ **29.** $\dfrac{x^2 - 1}{3x + 5}$ **31.** $\dfrac{3x + 5}{x^2 - 1}$
33. The domains of both f and g all real numbers
35. $9x^2 + 30x + 24$ **37.** $3x^2 + 2$ **39.** $3x^2 + 4x - 1$
41. $2x - 3x^2 - 1$ **43.** $3x^2 - 2x + 1$ **45.** $9x^3 - x$
47. $\dfrac{3x - 1}{3x^2 + x}$ **49.** $\dfrac{3x^2 + x}{3x - 1}$ **51.** $9x^2 + 3x - 1$
53. $27x^2 - 15x + 2$ **55.** The domains of both f and g are all real numbers. **57. (a)** $C(n) = 15n + 7500$, **(b)** 9000,
(c) $22,000$ **59. (a)** $P(n) = R(n) - C(N) = 60n - 275$
dollars, **(b)** 2725
61. $(P \circ n)(a) = P(n(a)) = 300(7a + 4)^2 - 50(7a + 4)$

Exercise Set 4.3
1.

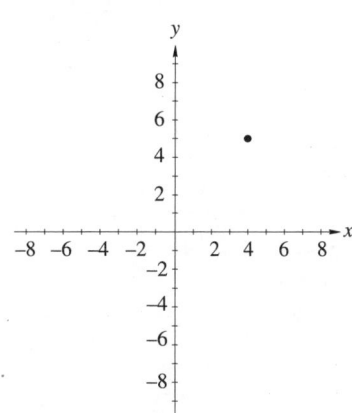

3.

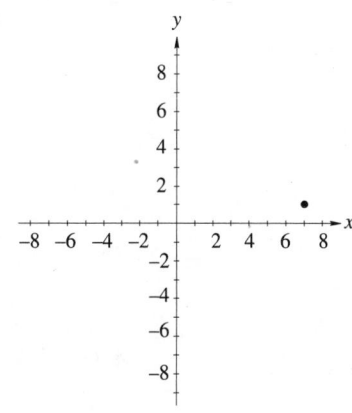

5.

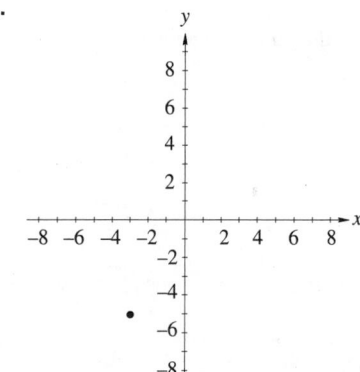

7.

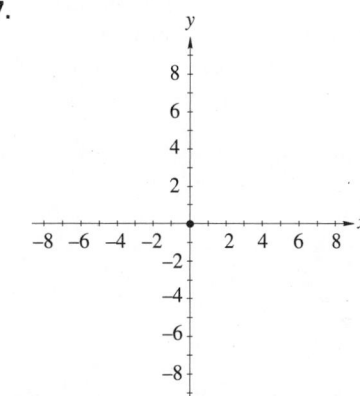

9.

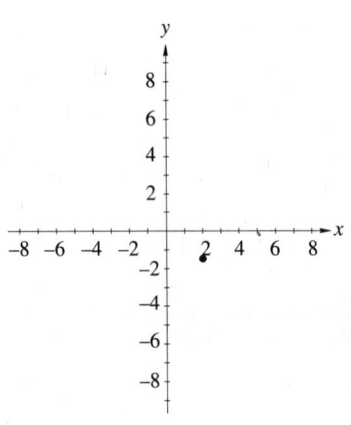

11. $(-1, -4)$

13. They are on a horizontal line and have the same y-coordinate.

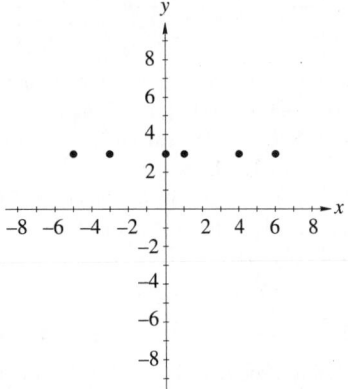

15. The points all lie on the same straight line.

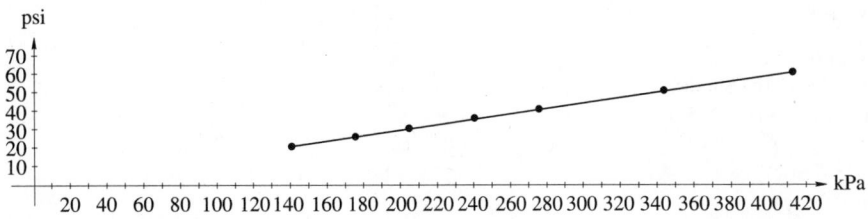

17. On the horizontal line through $y = -2$.   **19.** On the vertical line $x = -5$ **21.** To the right of the vertical line through $x = -3$.   **23.** To the right of the line through the point $(1, 0)$ and below the horizontal line through the point $(0, -2)$.

25. (a)

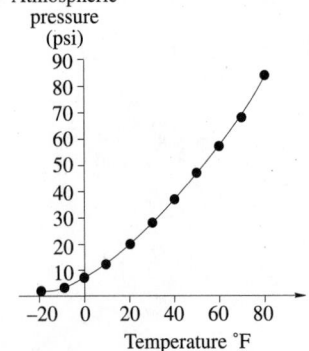

(b) 41.9 psi; **(c)** 99.6 psi

27. (a)

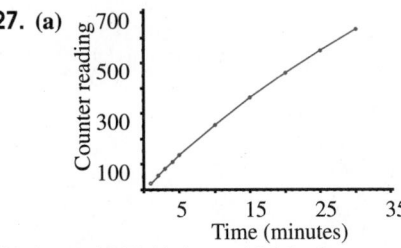

(b) Around 317 **(c)** Around 23.5 min

Exercise Set 4.4

1. (a)

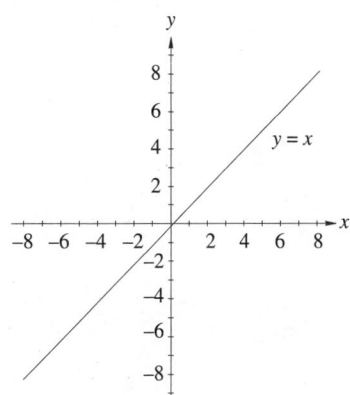

(b) x-intercept is 0; y-intercept is 0; **(c)** $m = 1$

3. (a)

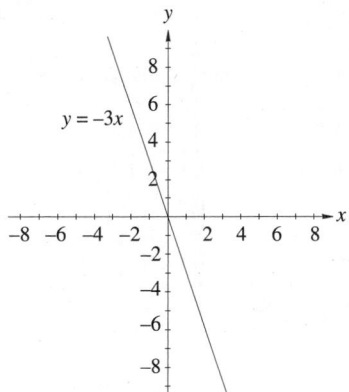

(b) x-intercept is 0; y-intercept is 0; **(c)** $m = -3$

5. (a)

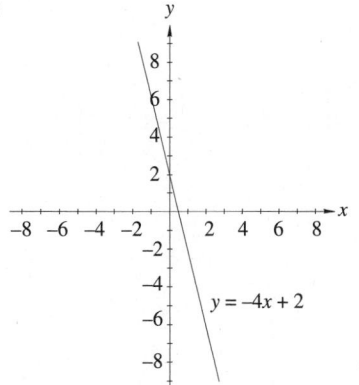

(b) x-intercept is $\frac{1}{2}$; y-intercept is 2; **(c)** $m = -4$

7. (a)

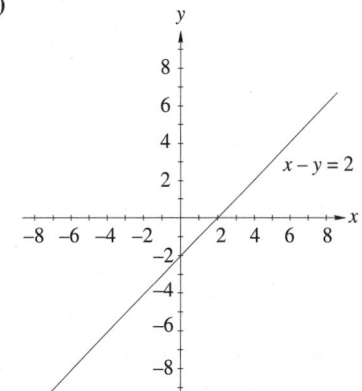

(b) x-intercept is 2; y-intercept is -2; **(c)** $m = 1$

9. (a)

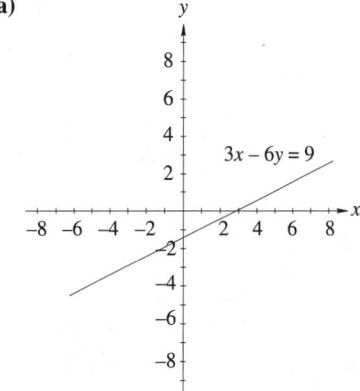

(b) x-intercept is 3; y-intercept is $-\frac{3}{2}$; **(c)** $m = \frac{1}{2}$

11. (a)

x	-3	-2	-1	0	1	2	3
y	9	4	1	0	1	4	9

(b)

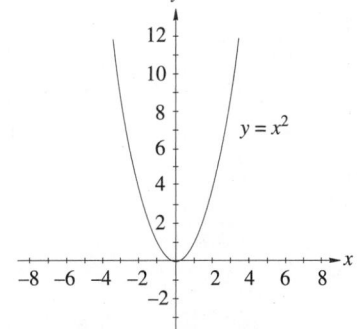

(c) Domain: all real numbers; range: $\{y : y \geq 0\}$

13. (a)

x	-3	-2	-1	0	1	2	3
y	7	2	-1	-2	-1	2	7

(b)

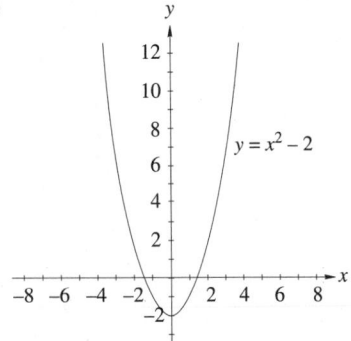

$y = x^2 - 2$

(c) Domain: all real numbers; range: $\{y : y \geq -2\}$

15. (a)

x	-3	-2	-1	0	1	2	3	4	5	6
y	36	25	16	9	4	1	0	1	4	9

(b)

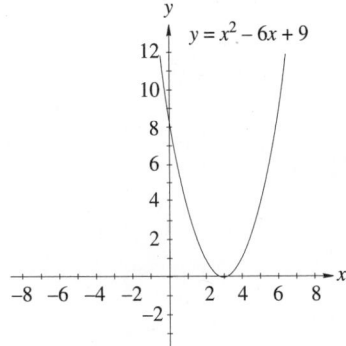

$y = x^2 - 6x + 9$

(c) Domain: all real numbers; range: $\{y : y \geq 0\}$

17. (a)

x	-3	-2	-1	$-\frac{1}{2}$	$-\frac{1}{4}$	0
y	$-\frac{1}{3}$	$-\frac{1}{2}$	-1	-2	-4	undefined

	$\frac{1}{4}$	$\frac{1}{2}$	1	2	3
	4	2	1	$\frac{1}{2}$	$\frac{1}{3}$

(b)

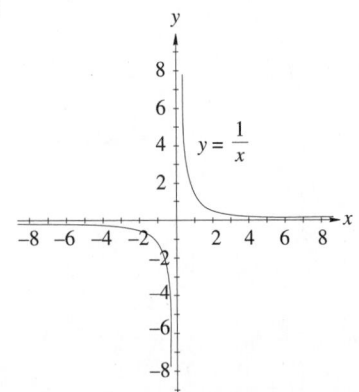

$y = \dfrac{1}{x}$

(c) Domain: $\{x : x \neq 0\}$; range: $\{y : y \neq 0\}$

19. (a)

x	-3	-2	-1	0	1	2	3
y	-27	-8	-1	0	1	8	27

(b)

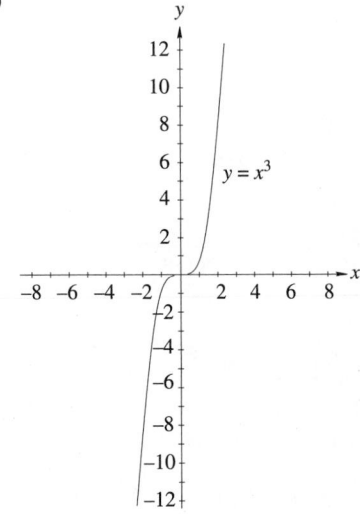

$y = x^3$

(c) Domain: all real numbers; range: all real numbers

21.

x	0	0	5	-5	3	-3	3	-3	4	4	-4	-4
y	5	-5	0	0	4	4	-4	-4	3	-3	3	-3

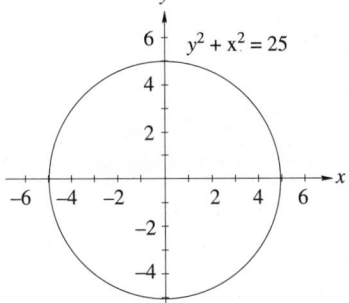

$y^2 + x^2 = 25$

23.

x	-5	-4	-3	-2	-1	0	1	2
y	3	2	1	0	1	2	3	4

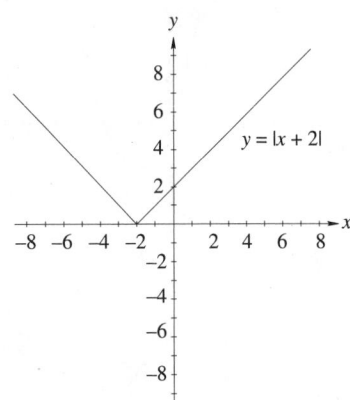

25.

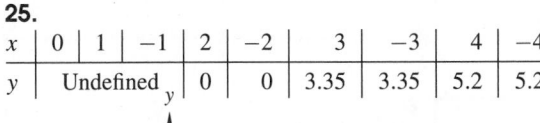

x	0	1	−1	2	−2	3	−3	4	−4
y	Undefined	0	0	3.35	3.35	5.2	5.2		

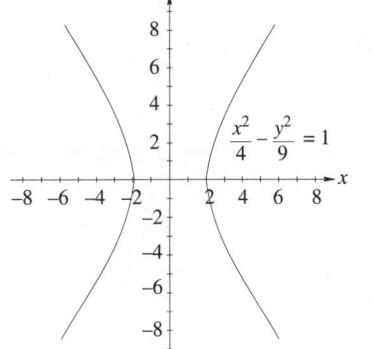

27. Function **29.** Function

31. (a)

k	0	1	2	3	4	5
$p(k)$	25.0	8.33	5.00	3.57	2.78	2.27

k	6	7	8	9
$p(k)$	1.92	1.67	1.47	1.31

(b) yes $p(k)$

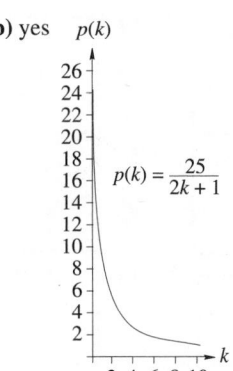

33. (a) $0 < p < \$4.65$

(b)

p	0.25	0.50	0.75	1.00
$d(p)$	10.56	11	11.31	11.50

p	1.25	1.50	1.75	2.00
$d(p)$	11.56	11.50	11.31	11

(c) $d(p)$ **(d)** $1.25

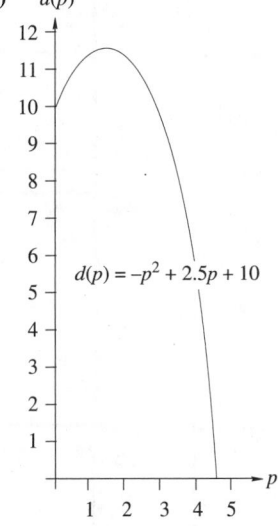

Exercise Set 4.5

1.

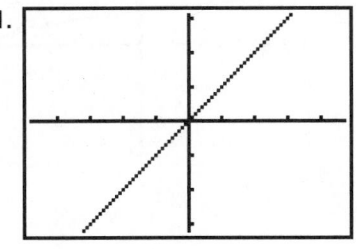

3.

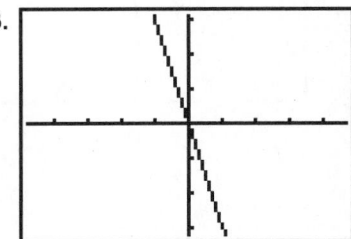

5.

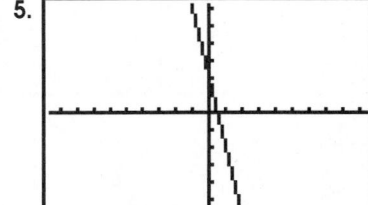

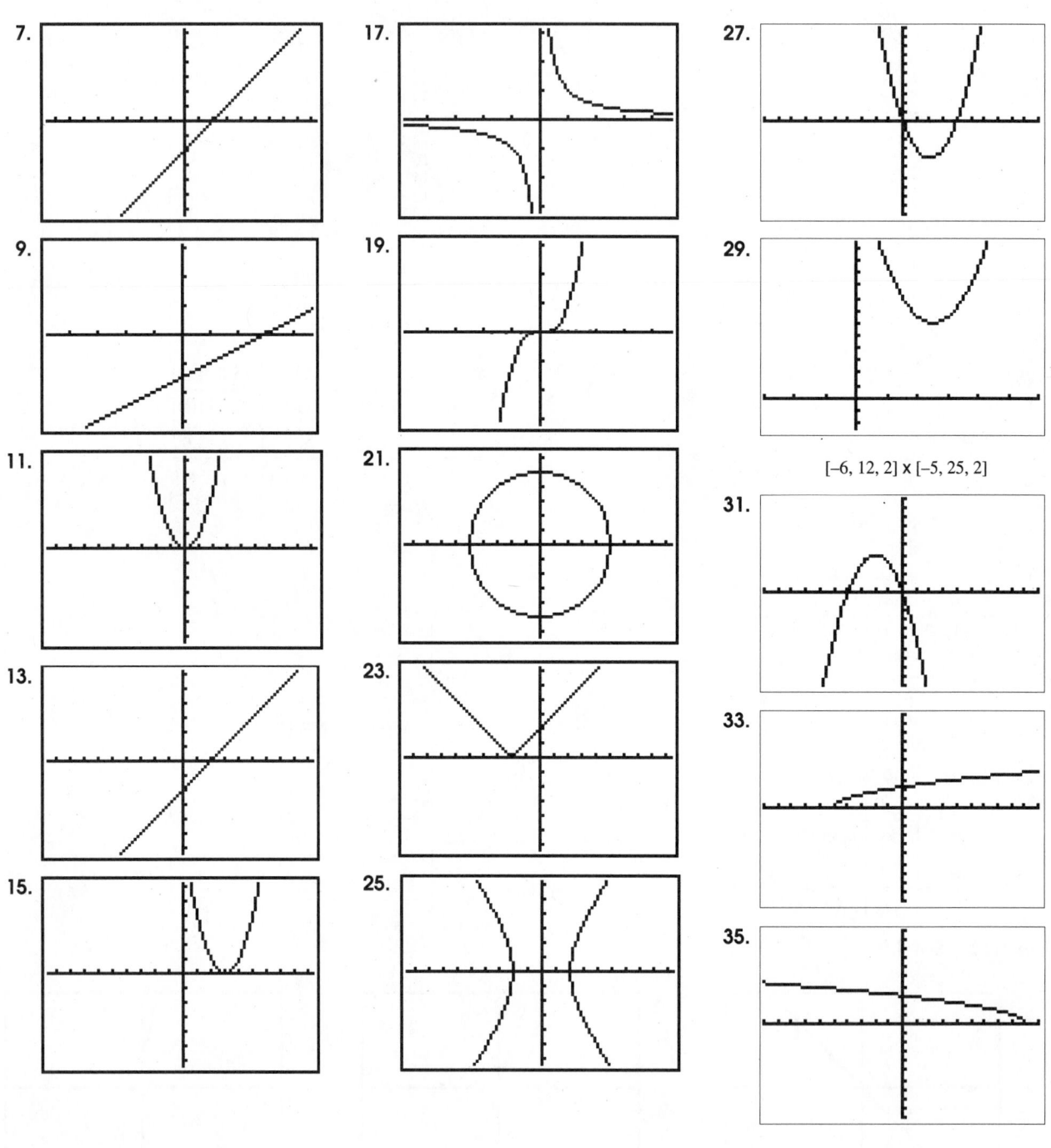

7.

9.

11.

13.

15.

17.

19.

21.

23.

25.

27.

29.

[–6, 12, 2] x [–5, 25, 2]

31.

33.

35.

37.

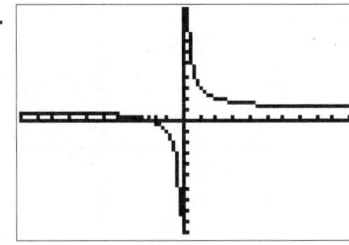

39.

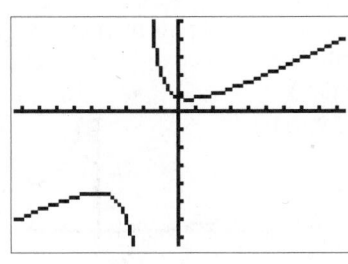

$[-9.4, 9.4, 1] \times [-10, 32, 2]$

41.

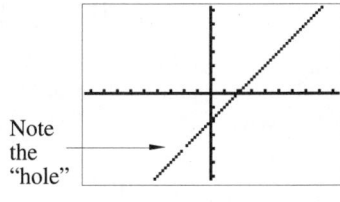

Note the "hole"

$[-9.4, 9.4] \times [-6.2, 6.2]$

Exercise Set 4.6

1. The root is $x = -\frac{5}{2}$

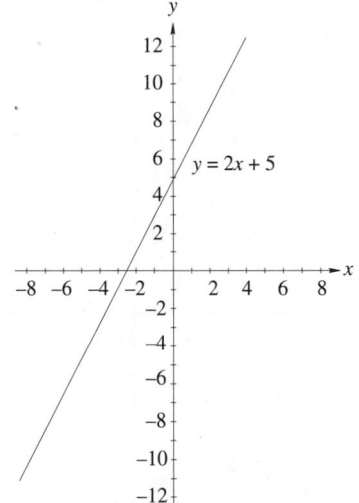

3. The roots are $x = -3$ and $x = 3$

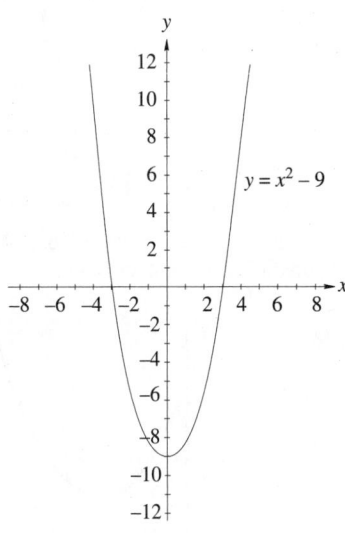

5. The roots are $x = 0$ and $x = 5$

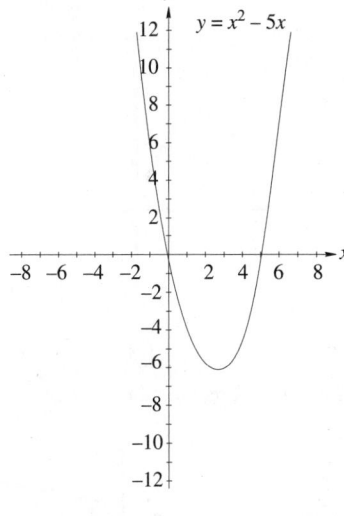

7. The roots are

$$x = -\frac{5}{2} - \frac{\sqrt{37}}{2} \approx -5.5414 \text{ and}$$

$$x = -\frac{5}{2} + \frac{\sqrt{37}}{2} \approx 0.5414$$

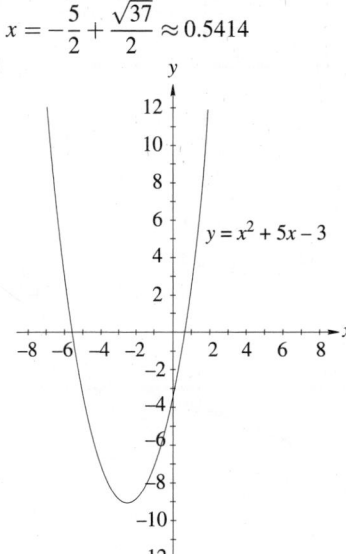

9. The roots are $x = -\frac{9}{4}$ and $x = \frac{6}{5}$

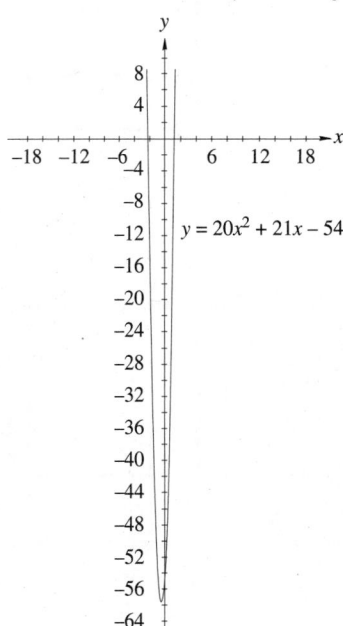

11. The root is $x = -1$.

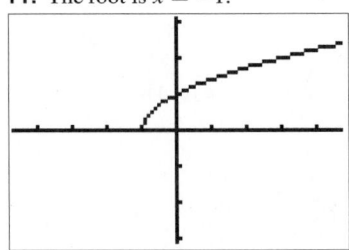

$[-4.7, 4.7] \times [-3.1, 3.1]$

13. The root is $x = 8$.

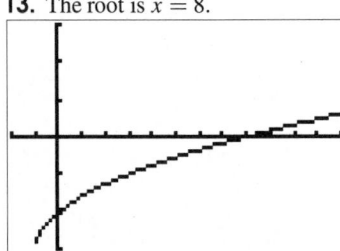

$[-2, 12] \times [-3, 3]$

15. The root is $x = -0.75$.

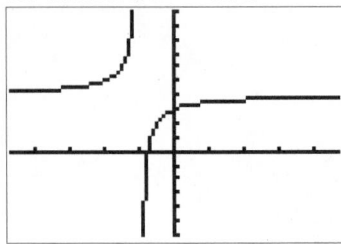

$[-4.7, 4.7] \times [-6, 10]$

17. The root is $x \approx -1.2695$.

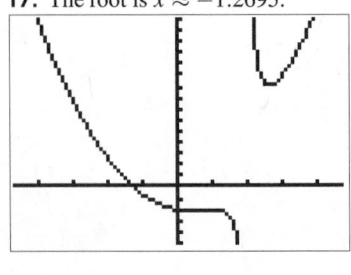

$[-4.7, 4.7] \times [-1, 15]$

19. (a)

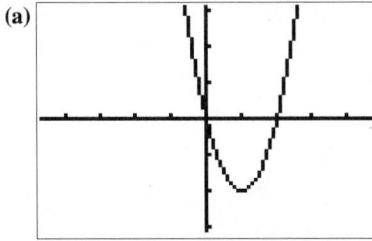

$[-9.4, 9.2, 2] \times [-6.2, 6.2, 2]$

(b) Domain: all real numbers; range: $\{y : y \geq -4\}$

21. (a)

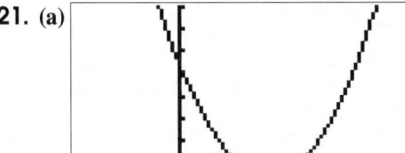

$[-6, 12.8, 2] \times [0, 50, 5]$

(b) Domain: all real numbers; range: $\{y : y \geq 12\}$

23. (a)

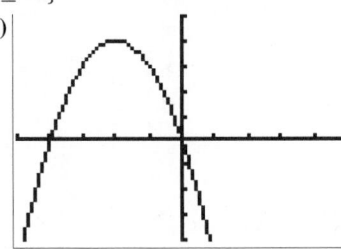

$[-5, 5, 1] \times [-4, 5, 1]$

(b) Domain: all real numbers; range: $\{y : y \leq 4\}$

25. (a)

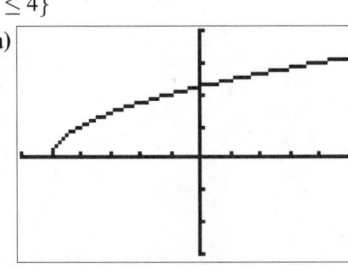

$[-6, 5, 1] \times [-3, 4, 1]$

(b) Domain: $\{x : x \geq -5\}$; range: $\{y : y \geq 0\}$

27. (a)

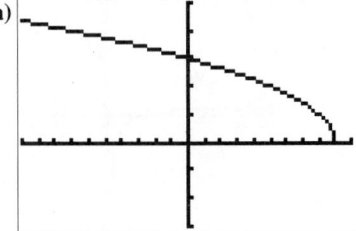

$[-10, 10, 1] \times [-3, 5, 1]$

(b) Domain: $\{x : x \leq 9\}$; range: $\{y : y \geq 0\}$

29. (a)

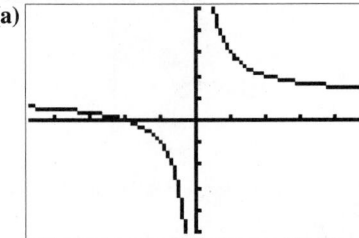

$[-4.7, 4.7, 1] \times [-5, 5, 1]$

(b) Domain: $\{x : x \neq 0\}$; range: $\{y : y \neq 1\}$

31. (a)

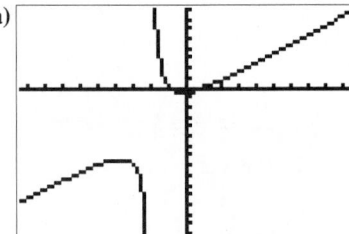

$[-9.4, 9.4, 1] \times [-15, 8, 1]$

(b) Domain: $\{x : x \neq -2\}$; range: $\{y : y < -7.46, y > -0.536\}$ (decimal values are approximate)

33. (a)

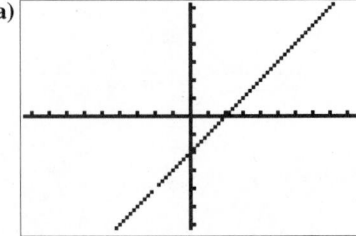

$[-9.4, 9.4, 1] \times [-6.2, 6.2, 1]$

(b) Domain: $\{x : x \neq -2\}$; range: $\{y : y \neq -4\}$

35. Has an inverse function
37. Does not have an inverse function
39.

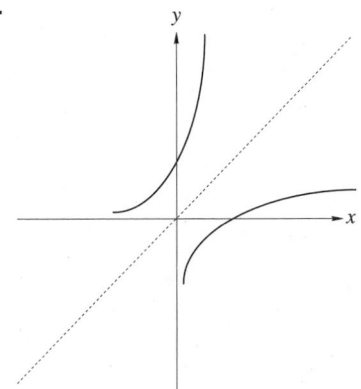

41.

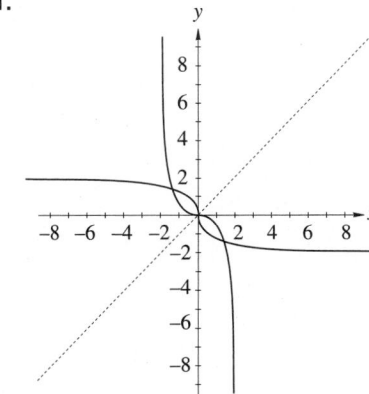

43. (a) 30%; **(b)** 60%; **(c)** 73%
45. No

Review Exercises

1. (a)

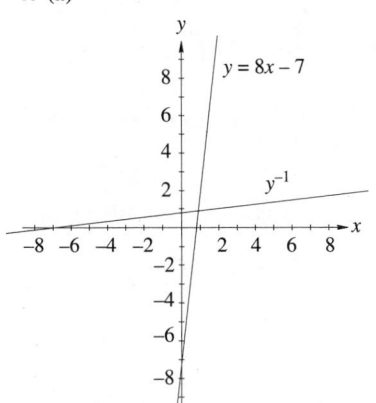

(b) Domain: all real numbers; Range:

all real numbers; x-intercept $\frac{7}{8}$;
y-intercept -7; **(c)** function; **(d)** has an
inverse function
3. (a)

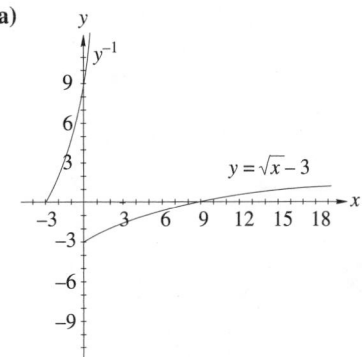

(b) Domain: all non-negative real
numbers, $x \geq 0$; Range: all real
numbers greater than or equal to -3;
x-intercept 9; y-intercept -9;
(c) function; **(d)** has an inverse function
5. (a)

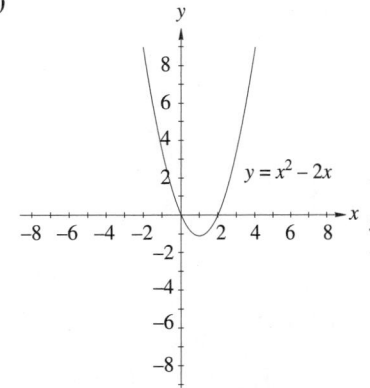

(b) Domain: all real numbers; Range:
all real numbers greater than or equal to
-1; x-intercepts $0, 2$; y-intercept 0;
(c) function; **(d)** does not have an
inverse function
7. -12 **9.** 0   **11.** $4a - 20$
13. -1 **15.** 0   **17.** $\frac{7}{25} = 0.28$

19.

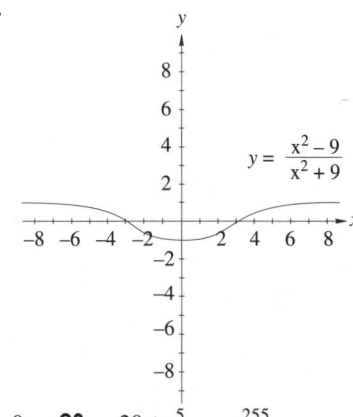

21. 0   **23.** $-20 + \frac{5}{13} = -\frac{255}{13}$
25. 12   **27.** 17   **29.** $\frac{5}{52}$
31. $-\frac{272}{25}$ **33.** -1
35.

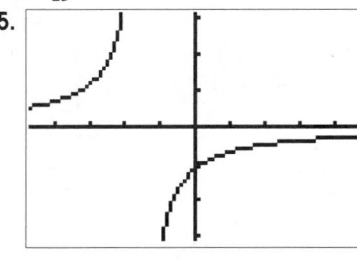

$[-9.4, 9.4] \times [-6.2, 6.2]$

37.

$[-9.4, 9.4] \times [-6.2, 6.2]$

39.

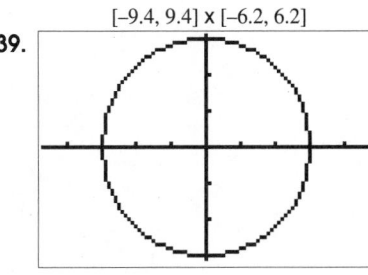

$[-4.7, 4.7] \times [-3.1, 3.1]$
41. $y = -\frac{4}{7}x$; root: $x = 0$

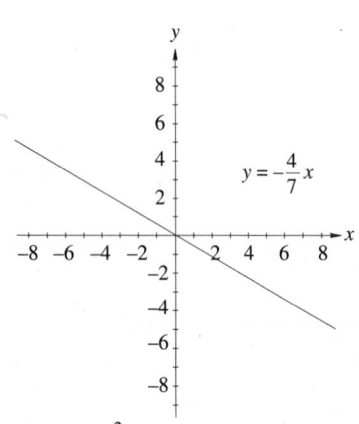

$$y = -\frac{4}{7}x$$

and $x = -1$

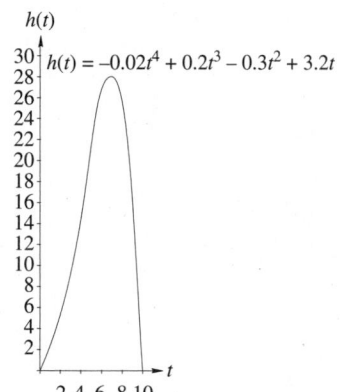

$$y = 2x^2 + 10x + 4$$

45. (a) $29.95, $17.50, $15;
(b) $R(n) = \left(35 - \dfrac{n}{20}\right)n$; **(c)** $3024.95,
$6125, $6000

43. $y = 2x^2 + 10x + 4$; roots: $x = -4$

47.

t	0	1	2	3	4	5
$h(t)$	0	3.08	6.48	10.68	15.68	21.00

t	6	7	8	9	10	
$h(t)$	25.68	28.28	26.88	19.08	2.00	

$$h(t) = -0.02t^4 + 0.2t^3 - 0.3t^2 + 3.2t$$

Chapter 4 Test

1. -19 **3.** -5

5. (a)

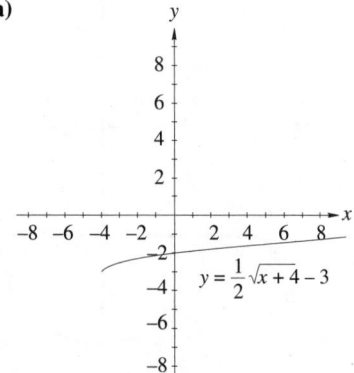

$$y = \frac{1}{2}\sqrt{x+4} - 3$$

(b) $x \geq -4$; **(c)** $y \geq -2$; **(d)** $x = 32$;
(e) $y = -2$ **7.** $\dfrac{3x^2 - x - 70}{x + 5}$
9. $3(x+5) = 3x + 15$

11. $\dfrac{3x - 20}{3x - 10}$

13.

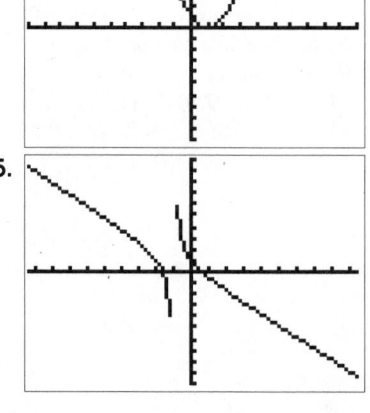

15.

$[-9.4, 9.4] \times [-10, 10]$

17. (a)

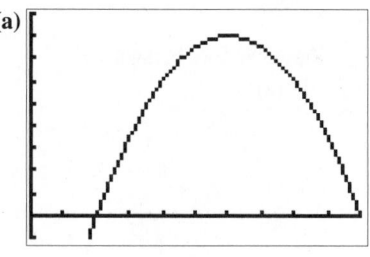

$[0, 10, 1] \times [-2, 18, 2]$
(b) 7, **(c)** 6 ms, **(d)** either 4 ms or 8 ms

ANSWERS FOR CHAPTER 5

Exercise Set 5.1

1. 3   **3.** $-\frac{5}{4}$ **5.** $\frac{10}{7}$ **7.** $m = 0$ **9.** Undefined

11. $y - 3 = 4(x - 5)$ **13.** $y + 5 = \frac{2}{3}(x - 1)$ **15.** $y = -5$

17. $y = -\frac{5}{3}(x - 2)$ **19.** $x = 7$ **21.** $y - 2 = \frac{3}{4}(x + 3)$ or
$y - 5 = \frac{3}{4}(x - 1)$ **23.** $y = 3$ **25.** $y = 2x + 4$

27. $y = 5x - 3$

29. $y = 3x + 6$; $m = 3$, $b = 6$

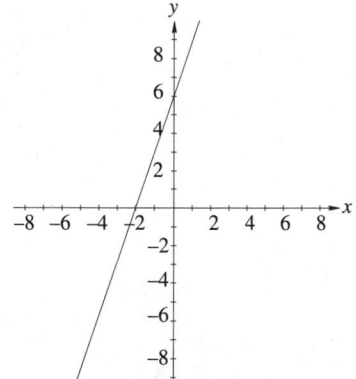

31. $y = \frac{5}{2}x + 4$; $m = \frac{5}{2}$, $b = 4$

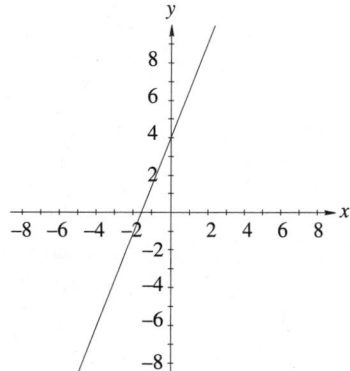

33. $y = \frac{5}{2}x - 5$; $m = \frac{5}{2}$, $b = -5$

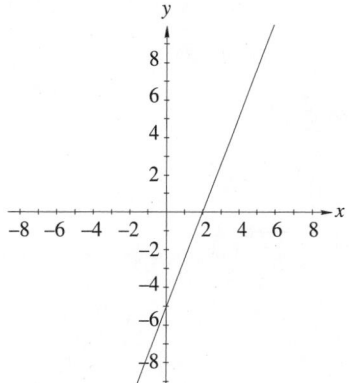

35. $y = -\frac{1}{3}x + \frac{7}{3}$; $m = -\frac{1}{3}$, $b = \frac{7}{3}$

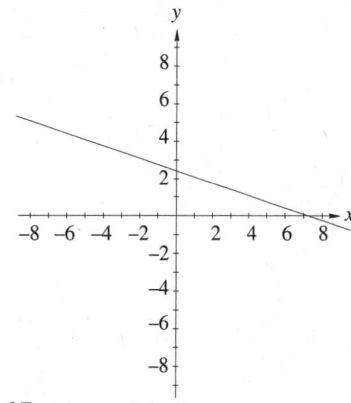

37. (a) $m = 0$
(b)

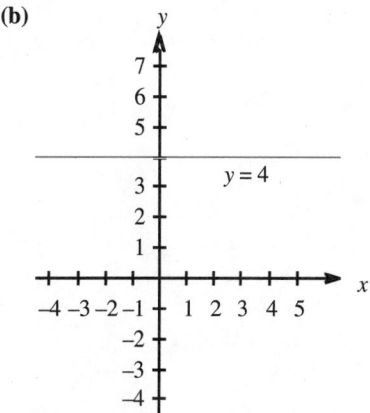

39. (a) The slope is undefined.
(b)

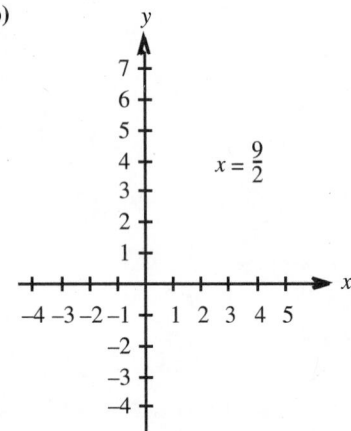

$$x = \frac{9}{2}$$

41. $m = \frac{1780\pi}{60} = \frac{89\pi}{3}$; **43. (a)** $F = kL - kL_0$;
(b) $F = 4.5L - 27$; **(c)**

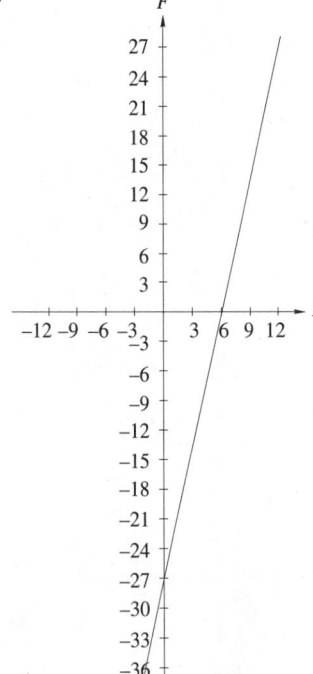

Exercise Set 5.2

1. $(4, 2)$

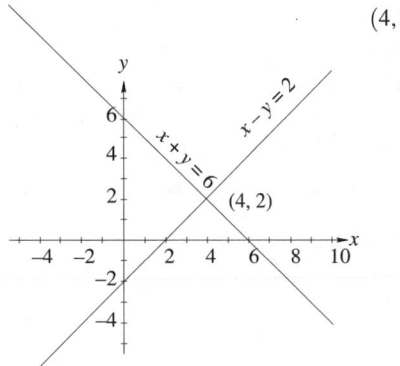

3. $(4.5, 2)$

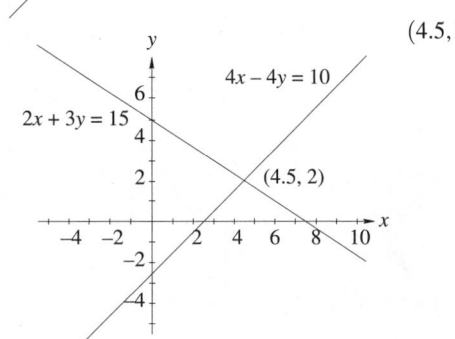

5. $\left(0, \frac{32}{5}\right) = (0, 6.4)$

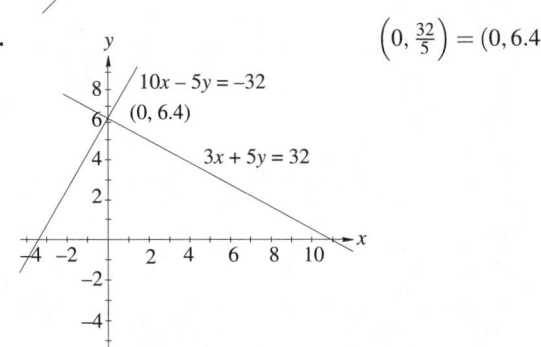

7. $(2.1, -1.1)$

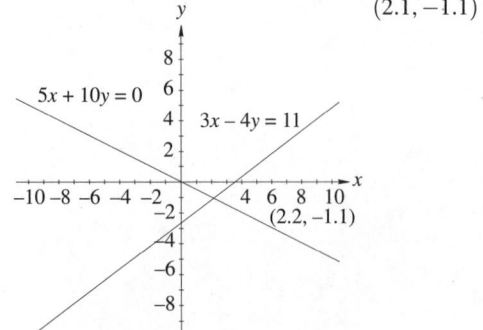

9. $(3,5)$ **11.** $(-4,6)$ **13.** $(8,-2)$ **15.** $(3.3,-1.2)$
17. $(2.6,-3.4)$ **19.** $(-2.1,5.3)$ **21.** $(7,2)$
23. $(-2,1)$ **25.** $(-3,3)$ **27.** $(1.5,-6.5)$
29. $(9.2,-4.6)$ **31.** $(4,-1)$

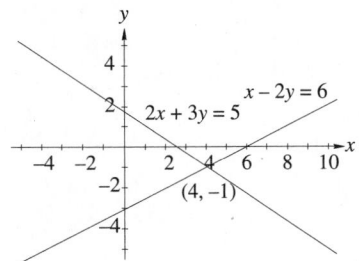

33. $(-1,7)$

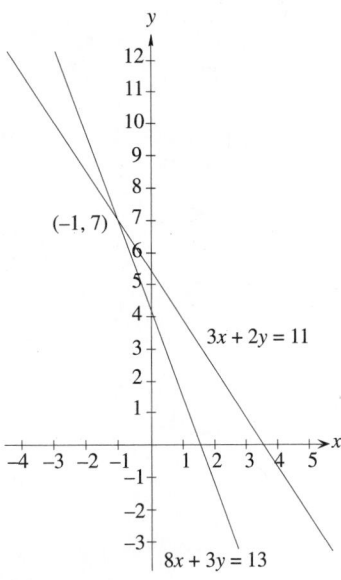

35. $(0.608,-1.324)$

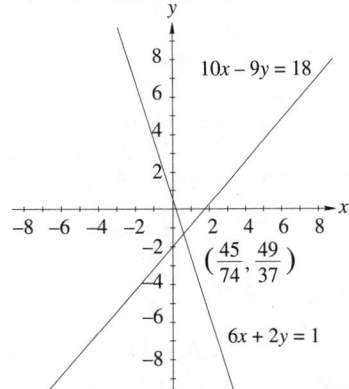

37. $\left(3,\frac{1}{3}\right)$

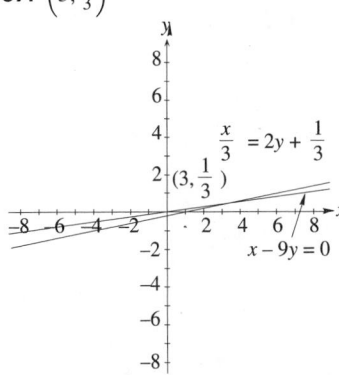

39. $(3.1,-2.7)$ **41.** length $=13\,\text{km}$, width $=5\,\text{km}$
43. length $=16.875\,\text{km}$, width $=5.625\,\text{km}$ **45.** $6250\,\text{L}$ of
5%; $3750\,\text{L}$ of 13% gasohol   **47.** $6.4\,\text{kg}$, $4\,\text{m}$
49. $I_1=1\,\text{A}$, $I_2=1\,\text{A}$, $I_3=0\,\text{A}$ **51.** $I_1=0.224\,\text{A}$,
$I_2=1.052\,\text{A}$, $I_3=0.828\,\text{A}$

Exercise Set 5.3

1. $x=2$, $y=-1$, $z=4$ **3.** $x=-3$, $y=-2$, $z=4$
5. $x=2$, $y=-1$, $z=4$ **7.** $x=-3$, $y=-2$, $z=4$
9. $x=1.3$, $y=2.7$, $z=-2$ **11.** $x=1.25$, $y=3.75$,
$z=-5.5$ **13.** $x=-6$, $y=2$, $z=5$ **15.** $I_1=0.875\,\text{A}$,
$I_2=2.125\,\text{A}$, $I_3=1.25\,\text{A}$ **17.** $2a+6b-c=40$;
$a+7b-c=50$; $a=-4$, $b=6$, $c=-12$ **19.** 4 large
trucks, 2 medium, 3 small   **21.** $40\,\text{lb}$ of the 10-12-15
fertilizer; $240\,\text{lb}$ of the 10-0-5 fertilizer, and $120\,\text{lb}$ of the
30-6-15 fertilizer.

Exercise Set 5.4

1. -14 **3.** 13 **5.** 25 **7.** -10 **9.** (a) 2,
(b) $\begin{vmatrix} 3 & -4 \\ -2 & 1 \end{vmatrix} = -5$, **(c)** 5 **11.** **(a)** 1, **(b)** -13, **(c)** 13
13. **(a)** -2, **(b)** -1, **(c)** -1
15. $3(-3)+7(2)-4(-8)=37$ **17.** 0, rows 1 and 3 are
identical   **19.** 0, row 3 is a multiple of row 1   **21.** 501
23. -120 **25.** 252.86583 **27.** -0.057525 **29.** 5
31. 30   **33.** 0   **35.** Approximately -321.545 **37.** 0
39. See *Computer Programs*   **41.** See *Computer Programs*

Exercise Set 5.5

1. $x=1$, $y=5$ **3.** $x=-2$, $y=4$ **5.** $x=3.5228$,
$y=1.3169$ **7.** no solution, inconsistent system
9. $x=-7.8768$, $y=-17.6891$ **11.** no solution,
inconsistent system   **13.** $x=-2$, $y=4$ **15.** $x=1.2825$,
$y=-1.015$, $z=0.14$ **17.** $x=-2.1$, $y=0.17$, $z=-4.35$,
$w=9.58$ **19.** See *Computer Programs*   **21.** $s_0=200$,
$v_0=18.8$, $a=-9.8$ **23.** $A=530\,\text{km}$, $B=620\,\text{km}$,
$C=750\,\text{km}$ **25.** **(a)** The city should buy from companies

A, B, and C in the proportion $A:B:C = 4:5:2$ units.
(b) Company A: 1,199,012 units, Company B:
1,498,765 units, Company C: 599,506 units

Review Exercises

1. 27 **3.** -67 **5.** 10 **7.** -32 **9. (a)** $\frac{-2}{7}$,
(b) $y - 1 = \frac{-2}{7}(x - 5)$ or $y - 3 = \frac{-2}{7}(x + 2)$ **11. (a)** $\frac{-5}{11}$,
(b) $y - 4 = \frac{-5}{11}(x + 2)$ or $y + 1 = \frac{-5}{11}(x - 9)$
13. $y = 2x - 3$ **15.** $y = \frac{4}{5}x + \frac{8}{5}$
17.

$x = 6$, $y = -14$

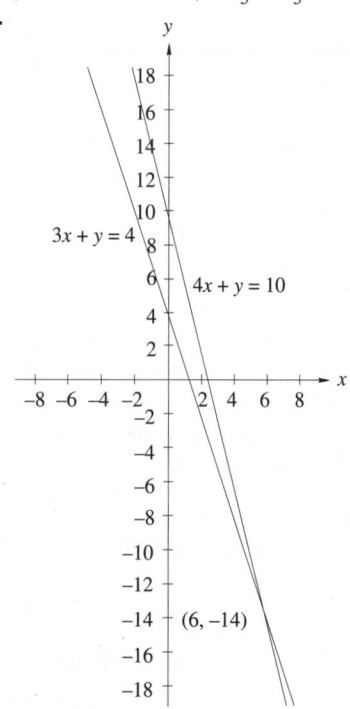

$3x + y = 4$

$4x + y = 10$

$(6, -14)$

19. $x = 3$, $y = -4$ **21.** $x = \frac{19}{11} \approx 1.7273$,
$y = -\frac{1}{11} \approx -0.0909$ **23.** $x = 3$, $y = 5$ **25.** $x = 1$,
$y = 2$ **27.** $x = 5.59375$, $y = 3.71875$, $z = -10.40625$
29. $x = 2$, $y = 1$ **31.** $x = \frac{57}{49} \approx 1.1633$,
$y = \frac{-71}{49} \approx -1.4490$ **33.** $x = 2$, $y = -1$, $z = 1$ **35.** 30
pounds of Colombian Supreme and 20 pounds of Mocha Java
37. 24 large, 32 middle size, and 10 small offices

Chapter 5 Test

1. (a) $\frac{4}{9}$; **(b)** $y - 2 = \frac{4}{9}(x + 4)$ or $y - 6 = \frac{4}{9}(x - 5)$ or, in
slope-intercept form, $y = \frac{4}{9}x + \frac{34}{9}$ **3.** $x = -2$, $y = \frac{5}{3}$
5. -49 **7.** $x = -2$, $y = 3$ **9.** $(-3, 2, 1)$ **11.** $30,
$21, and $9

≡ ANSWERS FOR CHAPTER 6

Exercise Set 6.1

1. $1.38 : 16$ **3.** $86 : 1$ **5.** $4.5 : 1$ **7.** $3.2857 : 1$
9. equal **11.** equal **13.** 28 **15.** 79 **17.** 4.267
19. 14 **21.** 1.78 **23.** $x = 5.6$, $z = 25.2$ **25.** $54 : 13$
27. $28 : 1$ **29.** 1.67 μF **31.** 11.03 cm^3
33. 62,000,000 calories **35.** 3.5 in **37.** 53.7472 L

Exercise Set 6.2

1. $a = 1.5$, $b = 3.75$ **3.** $a = 17$, $b = 42.71$, $c = 47.17$
5. $a = 2.55$, $b = 3.4$, $c = d = 4.25$ **7.** 184,320 in^2,
1280 ft^2 **9.** 5.5 kg **11.** 2143.75 L **13.** 145.56 mm^3
15. (a) 50.27 cm^2; **(b)** 33.51 cm^3

Exercise Set 6.3

1. $R = kl$ **3.** $A = kd^2$ **5.** $IN = k$
7. $r = kt$, $k = \frac{2}{3}$ **9.** $d = kr^2$, $k = \frac{1}{2}$ **11.** 9
13. 180 m/s **15.** 435.6 neutrons **17. (a)** 0.085,
(b) 30.55 m^3 **19.** 106.67 Ω **21.** 160 psi **23.** 0.075 A
25. 142.22 lb

Exercise Set 6.4

1. $K = kmv^2$ **3.** $f = \dfrac{kv}{l}$ **5.** $E = \dfrac{klI}{d^2}$

7. $a = kbc$; $k = 2.5$ **9.** $u = \dfrac{kv}{w}$; $k = 8$ **11.** $r = kst$;
$k = \frac{1}{3}$, $r = 4$ **13.** $a = kxy^2$; $k = 0.12$, $a = 162$

15. $p = \dfrac{kr\sqrt{t}}{w}; k = 1, p = 36$ **17. (a)** $\frac{1}{2}$, **(b)** 25
19. 22.36 V **21.** 17.14 m **23.** 77.5 Btu/hr
25. 1.25 ft^3

Review Exercises
1. 4.5 **3.** 27 **5.** 38 **7.** $x = 36.75, y = 28$
9. 160 **11.** 90 mm; 112.5 mm **13.** 8 : 9 or 88.9%

15. 0.2 m/s **17.** 8.99×10^9 **19.** 1.24×18^8 m/s

Chapter 6 Test
1. $23\frac{2}{3} \approx 22.67$ **3.** 16.8 **5.** $x = 14; y = 10.5$
7. $E = kd^2$ **9.** 27.5 and 66 **11.** 162 m^2

≡ ANSWERS FOR CHAPTER 7

Exercise Set 7.1
1. $3p + 3q$ **3.** $15x - 3xy$ **5.** $p^2 - q^2$
7. $4x^2 - 36p^2$ **9.** $r^2 + 2rw + w^2$ **11.** $4x^2 + 4xy + y^2$
13. $\frac{4}{9}x^2 + \frac{16}{3}xb + 16b^2$ **15.** $4p^2 - 3pr + \frac{9}{16}r^2$
17. $a^2 + 5a + 6$ **19.** $x^2 - 3x - 10$ **21.** $6a^2 + 5ab + b^2$
23. $6x^2 - 7x - 20$ **25.** $a^3 + 3a^2b + 3ab^2 + b^3$
27. $x^3 + 12x^2 + 48x + 64$ **29.** $8a^3 - 12a^2b + 6ab^2 - b^3$
31. $27x^3 - 54x^2y + 36xy^2 - 8y^3$ **33.** $m^3 + n^3$
35. $r^3 - t^3$ **37.** $8x^3 + b^3$ **39.** $27a^3 - d^3$

41. $3a^2 + 12a + 12$ **43.** $5rt + 2r^2 + \dfrac{r^3}{5t}$ **45.** $x^4 - 36$
47. $9x^4 - y^4$ **49.** 0 **51.** $x^3 - 3x^2 - 9x + 27$
53. $r^3 - 3r^2t + 3rt^2 - t^3$ **55.** $125 + 27x^3$
57. $(x + y)^2 - 2(x + y)(w + z) + (w + z)^2$
59. $(x + y)^2 - z^2 = x^2 + 2xy + y^2 - z^2$
61. $z^2 = R^2 + x_L^2 - 2x_L x_C + x_C^2$
63. $a_c = \dfrac{(2t^2 - t)^2}{r} = \dfrac{4t^4 - 4t^3 + t^2}{r}$
65. $d = \left(\dfrac{P}{3EI}\right)\left(l_1{}^3 - l_2{}^3\right)$

Exercise Set 7.2
1. $6(x + 1)$ **3.** $6(2a - 1)$ **5.** $2(2x - y + 4)$
7. $5(x^2 + 2x + 3)$ **9.** $5(2x^2 - 3)$ **11.** $2x(2x + 3)$
13. $7b(by + 4)$ **15.** $ax(3 + 6x - 2) = ax(1 + 6x)$
17. $2ap(2p + 3aq + 4q^2)$ **19.** $(a - b)(a + b)$
21. $(x - 2)(x + 2)$ **23.** $(y - 9)(y + 9)$
25. $(2x - 3)(2x + 3)$ **27.** $(3a^2 - b)(3a^2 + b)$
29. $(5a - 7b)(5a + 7b)$ **31.** $(12 - 5b^2)(12 + 5b^2)$
33. $5(a - 5)(a + 5)$ **35.** $7(2a - 3b^2)(2a + 3b^2)$
37. $(a - 3)(a + 3)(a^2 + 9)$
39. $16(x - 2y)(x + 2y)(x^2 + 4y^2)$ **41.** $2\pi r(r + h)$
43. $\frac{1}{2}d(v_2 - v_1)(v_2 + v_1)$ **45.** $\dfrac{(\omega_f - \omega_0)(\omega_f + \omega_0)}{2\theta}$
47. $W = \frac{1}{2}I(\omega_2 - \omega_1)(\omega_2 + \omega_1)$

Exercise Set 7.3
1. $b^2 - 4ac = 113; \sqrt{113} \approx 10.6$; does not factor using rational numbers **3.** $b^2 - 4ac = 196; \sqrt{196} = 14$; factors

5. $b^2 - 4ac = 169; \sqrt{169} = 13$; factors
7. $(x + 2)(x + 5)$ **9.** $(x - 3)(x - 9)$
11. $(x - 2)(x - 25)$ **13.** $(x - 2)(x + 1)$
15. $(x - 5)(x + 2)$ **17.** $(r + 5)^2$ **19.** $(a + 11)^2$
21. $(f - 15)^2$ **23.** $(6y - 1)(y - 1)$ **25.** $(7t + 2)(t + 1)$
27. $(7b + 1)(b - 5)$ **29.** $(4e - 1)(e + 5)$
31. $(3u + 4)(u + 2)$ **33.** $(9t + 2)(t - 3)$
35. $(3x - 1)(2x + 5)$ **37.** $(5a + 3)(3a - 5)$
39. $(5e + 3)(3e + 5)$ **41.** $(5x - 2)(2x - 3)$
43. $3(r - 7)(r + 1)$ **45.** $7t^2(7t - 1)(t - 2)$
47. $(3x + 2y)(2x - 5y)$ **49.** $(2a + b)(4a - 9b)$
51. $(a - b)(a^2 + ab + b^2)$ **53.** $(2x - 3)(4x^2 + 6x + 9)$
55. $i = 0.7(t^2 - 3t - 4) = 0.7(t - 4)(t + 1)$
57. $(0.009n^2 - 3)(n - 2000) =$
$0.0001(n^2 - 30,000)(n - 2000)$
59. (a) $4x^2 + 32x - 36 = 0$; **(b)** $4(x + 9)(x - 1)$

Exercise Set 7.4
1. $\frac{35}{40}$ **3.** $\dfrac{ax}{ay}$ **5.** $\dfrac{3ax^3 y}{3a^2 x}$
7. $\dfrac{4(x + y)}{x^2 - y^2} = \dfrac{4x + 4y}{x^2 - y^2}$ **9.** $\dfrac{(a + b)^2}{a^2 - b^2} = \dfrac{a^2 + 2ab + b^2}{a^2 - b^2}$
11. $\frac{19}{12}$ **13.** $\dfrac{x}{4}$ **15.** $\dfrac{4}{x - 3}$ **17.** $\dfrac{x - 4}{x + 4}$ **19.** $\dfrac{x}{3}$
21. $\dfrac{x + 3}{x^2 + 5}$ **23.** $\dfrac{2m - m^2}{3 + 6m^2}$ **25.** $\dfrac{x}{x - 3}$ **27.** $\dfrac{2b}{3(b + 5)}$
29. $\dfrac{z + 3}{z - 3}$ **31.** $\dfrac{x + 1}{x + 4}$ **33.** $\dfrac{2x + 1}{x + 5}$ **35.** $\dfrac{y(2y + 1)}{y - 1}$
37. $\dfrac{y^2 + xy + x^2}{2}$ **39.** $x - y$ **41.** $p = \dfrac{2wh}{s + 1}$

Exercise Set 7.5
1. $\dfrac{10}{xy}$ **3.** $\dfrac{12x^2}{5y^3}$ **5.** $\dfrac{3y}{7x}$ **7.** $\dfrac{8x^3}{21y}$ **9.** $\dfrac{5}{6xy}$
11. $\dfrac{5a^3 d}{2b}$ **13.** $\dfrac{8y^2}{25x^2}$ **15.** $\dfrac{y^3}{5p^2}$ **17.** $4y$

19. $5(a - b)$ **21.** $\dfrac{3(x^2 - 100)}{4(x + 5)}$ **23.** $\dfrac{(2x - 1)(x + 3)}{-3x}$

25. $a+2$ **27.** $\dfrac{1}{4a}$ **29.** $\dfrac{1}{x+1}$ **31.** $\dfrac{4}{y}$

33. $\dfrac{x-1}{3(x+2)}$ **35.** $\dfrac{(x-2y)(x-5y)(x-4y)}{(x+4y)(x+2y)(x-7y)}$

37. $\dfrac{3x-5}{x+3}$ **39.** $\dfrac{x+y}{x-y}$ **41.** $\dfrac{x^3+y^3}{4(x^3-y^3)}$ **43.** $\dfrac{x+5}{3x}$

45. $\dfrac{3}{x+3}$ **47.** $\dfrac{a^2+aa'+a'^2}{a(a+a')}$

Exercise Set 7.6

1. 1 **3.** $\dfrac{2}{3}$ **5.** $\dfrac{5}{6}$ **7.** $\dfrac{2}{15}$ **9.** $\dfrac{6}{x}$ **11.** $\dfrac{1}{a}$

13. $\dfrac{5x}{y}$ **15.** $-\dfrac{3r}{2t}$ **17.** $\dfrac{3+x}{x+2}$ **19.** $\dfrac{t-2}{t+1}$

21. $\dfrac{2y}{x+2}$ **23.** $\dfrac{7}{a+b}$ **25.** $\dfrac{2y+3x}{xy}$ **27.** $\dfrac{ad-4b}{bd}$

29. $\dfrac{7x-3}{x(x^2-1)}$ **31.** $\dfrac{6-2x}{(x^2-1)(x+1)}$

33. $\dfrac{x^2+8x-10}{(x^2-36)(x-5)}$ **35.** $\dfrac{11-3x}{(x-3)(x^2-4)}$

37. $\dfrac{1-x-8x^2}{x(3x-1)(x-4)}$ **39.** $\dfrac{3x^2-8x-5}{(x^2-1)(x+4)}$

41. $\dfrac{6y+13-y^2}{(y-2)(y+1)(y+4)}$ **43.** $\dfrac{2x^4-x^3+13x^2+2x-4}{(x^2+3)(x-1)^2(x+2)}$

45. $\dfrac{x+2}{x-3}$ **47.** $\dfrac{x(x-1)}{x+1}$ **49.** $\dfrac{x^2-2xy-y^2}{x^2+y^2}$

51. $\dfrac{x+3}{x+2}$ **53.** $\dfrac{t(t-1)}{t^2+1}$ **55.** $\dfrac{x^2+y^2}{2x}$ **57.** $\dfrac{R_1+R_2}{R_1R_2}$

59. $\dfrac{C_1C_2+C_1C_3+C_2C_3}{C_1C_2C_3}$ **61.** v

63. $\dfrac{V_1R_2R_3+V_2R_1R_3+V_3R_1R_2}{R_2R_3+R_1R_3+R_1R_2}$

Review Exercises

1. $5x^2-5xy$ **3.** $x^3-6x^2y+12xy^2-8y^3$
5. $2x^2-9x-18$ **7.** x^4-25 **9.** $8+12x+6x^2+x^3$
11. $9(1+y)$ **13.** $7(x-3)(x+3)$ **15.** $(x-6)(x-5)$
17. $(x+8)(x-2)$ **19.** $(2x+3)(x-3)$ **21.** $\dfrac{x}{3y}$

23. $\dfrac{x-3}{x+3}$ **25.** $\dfrac{x^2-xy+y^2}{x+y}$ **27.** $\dfrac{3xy}{7}$ **29.** $\dfrac{4}{x}$

31. $\dfrac{7x}{y}$ **33.** $\dfrac{7x^2-27x+2}{(x+2)^2(x-5)^2}$ **35.** $\dfrac{-2x^2}{y^2-x^2}$

37. $\dfrac{2(x^2+36)(x+1)}{(x+2)(x^2-36)}$ **39.** $\dfrac{(x-6)^2}{(x+6)^2}$ **41.** $\dfrac{y-x}{y+x}$

43. -1 **45.** $2x^2-1$

Chapter 7 Test

1. $x^2+2x-15$ **3.** $-15x^3+6x^2+20x-8$
5. $2(x-8)(x+8)$ **7.** $(5x-7)(2x+3)$ **9.** $\dfrac{x-5}{x+1}$

11. $\dfrac{3x(x-1)}{(x+2)^2}=\dfrac{3x^2-3x}{x^2+4x+4}$ **13.** $\dfrac{x^2-2x+6}{x-5}$

15. $\dfrac{(x+1)^2}{x(x+2)}$ **17.** $\dfrac{2r_1r_2}{r_1+r_2}$

▤ ANSWERS FOR CHAPTER 8

Exercise Set 8.1

1. 0.3 **3.** 10 **5.** $-\dfrac{1}{7}$ **7.** 9 **9.** -3 **11.** $\dfrac{15}{8}$
13. $\dfrac{9}{2}$ **15.** 15 **17.** -9 **19.** $\dfrac{2}{5}$ **21.** 1 **23.** $\dfrac{30}{37}$

25. $-\dfrac{3}{7}$ **27.** -6 **29.** 1 **31.** $\dfrac{rt}{r-t}$

33. $\dfrac{R_1R_2R_3}{R_1R_2+R_1R_3+R_2R_3}$ **35.** $\dfrac{V-2\pi r^2}{2\pi r}$

37. $\dfrac{R_1f(n-1)}{R_1-f(n-1)}$ **39.** $C=\dfrac{dR_1R_2}{(9\times10^9)(R_1-R_2)}$

41. $f=\dfrac{r_1r_2}{(n-1)(r_2-r_1)}$ **43.** $D=d+\dfrac{h}{2}$ **45.** $3\frac{3}{4}$ h or

3 h 45 min **47.** 3 hours **49.** $h=\dfrac{28}{5}=5$ hours 36
minutes **51.** 12 h

25. $-\dfrac{3}{2},\ \dfrac{20}{3}$ **27.** $-1,\ 3$ **29.** $\dfrac{6}{5},\ -\dfrac{2}{5}$ **31.** $2,\ -\dfrac{1}{2}$
33. $-12, 7$ **35.** $-3, 7$ **37.** $-1, 24$ **39.** $\dfrac{10}{7}, 5$
41. $7, 24$ **43.** $-\dfrac{1}{7}, 1.5$ **45.** 4 s **47.** width 8 cm,
length 13 cm **49.** 5 m and 12 m **51.** 3.00 in.
53. 12.0 m **55.** Pipe A: 12 h, pipe B: 6 h **57.** 15.00 in.

Exercise Set 8.3

1. $-2, -4$ **3.** $-1, 11$ **5.** $-3\pm\sqrt{6}$ **7.** $\dfrac{5\pm\sqrt{5}}{2}$

9. $\dfrac{3\pm\sqrt{29}}{2}$ **11.** $\dfrac{3\pm\sqrt{29}}{2}$ **13.** $-k\pm\sqrt{k^2-c}$

15. 3.5 s **17.** either 20 or 370 objects were sold
19. There are 2 possible answers. In one, each field is 300 ft $\times$
450 ft; and in the other, each field measures 600 ft $\times$ 225 ft.

Exercise Set 8.4

1. $-4, 1$ **3.** $2, -\dfrac{1}{3}$ **5.** $-1, \dfrac{1}{7}$ **7.** $-1, \dfrac{7}{2}$

9. $-2, \dfrac{4}{3}$ **11.** $-\dfrac{2}{3}, -\dfrac{2}{3}$ **13.** $\dfrac{3\pm\sqrt{17}}{4}$

Exercise Set 8.2

1. ±3 **3.** $-3, 2$ **5.** $-1, 12$ **7.** $2, -4$
9. $0, 5$ **11.** $3, 4$ **13.** $-2, \dfrac{7}{2}$ **15.** $\dfrac{3}{2}, 4$
17. $3, -\dfrac{1}{3}$ **19.** $\dfrac{5}{2}, \dfrac{7}{2}$ **21.** $\dfrac{5}{3}, -\dfrac{7}{2}$ **23.** $\dfrac{1}{5}, \dfrac{3}{2}$

15. $\dfrac{-5 \pm \sqrt{17}}{2}$ **17.** $\dfrac{-3 \pm \sqrt{15}}{2}$ **19.** $1 \pm \sqrt{8}$

21. $\pm \dfrac{\sqrt{6}}{2}$ **23.** no real roots because the discriminant is -48 **25.** no real roots because the discriminant is $-\frac{647}{81}$

27. $\dfrac{-0.2 \pm \sqrt{0.064}}{0.02}$ **29.** $5 \pm \sqrt{13}$ **31.** $\dfrac{2 \pm \sqrt{1.6}}{2.4}$

33. $\dfrac{-\sqrt{3} \pm \sqrt{87}}{6}$ **35.** no real root because the discriminant is negative **37.** $\dfrac{-2 \pm \sqrt{13}}{3}$ **39.** approximately 9.16 s **41.** approximately 11.42 s **43.** approximately 16.38 cm wide and 24.57 cm long **45.** 2.25 inches wide, 7.5 inches long, or 3.75 inches wide and 4.5 inches long **47.** \$20

49. $1062.1\,\Omega$ **51.** $\dfrac{-13 + 3\sqrt{46}}{2} \approx 3.67$ in. **53.** 35.6 in.

55. See *Computer Programs* **57. (a)** $-2, \frac{1}{3}$

Review Exercises

1. 6 **3.** -2 **5.** $\dfrac{2 \pm \sqrt{10}}{3}$ **7.** $\dfrac{A - 2lw}{2(l + w)}$

9. $\frac{pq}{p+q}$ **11.** 1, 7 **13.** 1, 10 **15.** $-7, -8$ **17.** $1, \frac{3}{2}$

19. $-2, \frac{5}{6}$ **21.** $\pm \frac{3}{2}$ **23.** $-5, 1$ **25.** $\dfrac{-19 \pm 9\sqrt{5}}{2}$

27. $\dfrac{-5 \pm \sqrt{193}}{6}$ **29.** $4 \pm \sqrt{11}$ **31.** $1, -\frac{5}{2}$

33. $\dfrac{-1 \pm \sqrt{13}}{3}$ **35.** $\dfrac{4 \pm \sqrt{6}}{5}$ **37.** no answer because the discriminant is negative **39.** $\dfrac{3 \pm \sqrt{17}}{4}$ **41.** 0.0899 A **43.** 3 sec **45.**

Chapter 8 Test

1. 4 **3.** $-1, 5$ **5.** double root, $\frac{3}{4}$ **7.** $\dfrac{-3 \pm \sqrt{5}}{2}$

9. $x = -10$ and $x = 3$ **11.** $H = \dfrac{V + 5}{3D^2}$ **13.** 8.5 m long and 5.5 m wide

ANSWERS FOR CHAPTER 9

Exercise Set 9.1

1. 1.5705 **3.** 1.396 **5.** 2.70475 **7.** 3.75175
9. $114.592°$ **11.** $85.944°$ **13.** $60°$ **15.** $-45°$
17. (c) $510°, -210°$

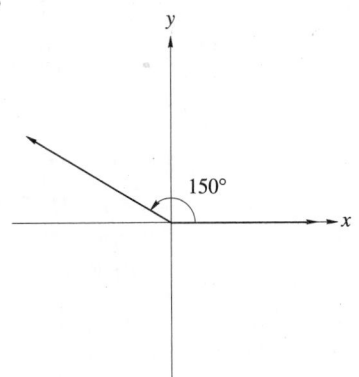

19. (c) $-495°, 225°$

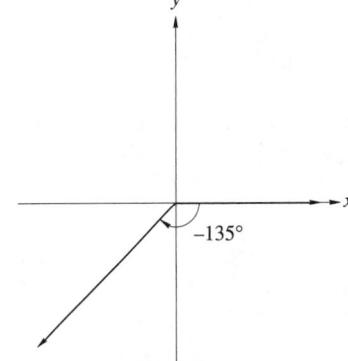

21. $\sin\theta = \frac{3}{5}$; $\cos\theta = \frac{4}{5}$; $\tan\theta = \frac{3}{4}$; $\csc\theta = \frac{5}{3}$; $\sec\theta = \frac{5}{4}$; $\cot\theta = \frac{4}{3}$ **23.** $\sin\theta = \frac{-15}{17}$; $\cos\theta = \frac{8}{17}$; $\tan\theta = \frac{-15}{18}$; $\csc\theta = \frac{-17}{15}$; $\sec\theta = \frac{17}{8}$; $\cot\theta = \frac{-8}{15}$ **25.** $\sin\theta = \dfrac{2}{\sqrt{5}}$; $\cos\theta = \dfrac{1}{\sqrt{5}}$; $\tan\theta = 2$; $\csc\theta = \dfrac{\sqrt{5}}{2}$; $\sec\theta = \sqrt{5}$; $\cot\theta = \frac{1}{2}$

27. $\sin\theta = -\dfrac{4}{\sqrt{41}}$; $\cos\theta = \dfrac{5}{\sqrt{41}}$; $\tan\theta = \frac{-4}{5}$; $\csc\theta = -\dfrac{\sqrt{41}}{4}$; $\sec\theta = \dfrac{\sqrt{41}}{5}$; $\cot\theta = \frac{-5}{4}$ **29.** $\sin\theta = \frac{5}{6}$; $\cos\theta = \dfrac{\sqrt{11}}{6}$; $\tan\theta = \dfrac{5}{\sqrt{11}}$; $\csc\theta = \frac{6}{5}$; $\sec\theta = \dfrac{6}{\sqrt{11}}$; $\cot\theta = \dfrac{\sqrt{11}}{5}$ **31.** $\sin\theta = 0$; $\cos\theta = 1$; $\tan\theta = 0$;

$\csc\theta =$ Does not exist; $\sec\theta = 1$; $\cot\theta =$ Does not exist.
33. $\sin\theta = 0$; $\cos\theta = -1$; $\tan\theta = 0$; $\csc\theta =$ Does not exist; $\sec = -1$; $\cot\theta =$ Does not exist. **35.** $\sin\theta = \frac{4}{5}$; $\cos\theta = \frac{3}{5}$; $\tan\theta = \frac{4}{3}$; $\csc\theta = \frac{5}{4}$; $\sec\theta = \frac{5}{3}$; $\cot\theta = \frac{3}{4}$ **37.** $\sin\theta = \frac{21}{29}$; $\cos\theta = \frac{-20}{29}$; $\tan\theta = \frac{-21}{20}$; $\csc\theta = \frac{29}{21}$; $\sec\theta = \frac{-29}{20}$; $\cot\theta = \frac{-20}{21}$ **39.** $\sin\theta = -\frac{\sqrt{15}}{8}$; $\cos\theta = \frac{-7}{8}$; $\tan\theta = \frac{\sqrt{15}}{7}$; $\csc\theta = -\frac{8}{\sqrt{15}}$; $\sec\theta = \frac{-8}{7}$; $\cot\theta = \frac{7}{\sqrt{15}}$

Exercise Set 9.2

1. $\sin\theta = \frac{5}{13}$; $\cos\theta = \frac{12}{13}$; $\tan\theta = \frac{5}{12}$; $\csc\theta = \frac{13}{5}$; $\sec\theta = \frac{13}{12}$; $\cot\theta = \frac{12}{5}$ **3.** $\sin\theta = \frac{3}{5}$; $\cos\theta = \frac{4}{5}$; $\tan\theta = \frac{3}{4}$; $\csc\theta = \frac{5}{3}$; $\sec\theta = \frac{5}{4}$; $\cot\theta = \frac{4}{3}$ **5.** $\sin\theta = \frac{1.4}{\sqrt{7.25}}$; $\cos\theta = \frac{2.3}{\sqrt{7.25}}$; $\tan\theta = \frac{14}{23}$; $\csc\theta = \frac{\sqrt{7.25}}{1.4}$; $\sec\theta = \frac{\sqrt{7.25}}{2.3}$; $\cot\theta = \frac{23}{14}$ **7.** $\sin\theta = \frac{15}{17}$; $\cos\theta = \frac{8}{17}$; $\tan\theta = \frac{15}{8}$; $\csc\theta = \frac{17}{15}$; $\sec\theta = \frac{17}{8}$; $\cot\theta = \frac{8}{15}$ **9.** $\sin\theta = \frac{3}{5}$; $\cos\theta = \frac{4}{5}$; $\tan\theta = \frac{3}{4}$; $\csc\theta = \frac{5}{3}$; $\sec\theta = \frac{5}{4}$; $\cot\theta = \frac{4}{3}$ **11.** $\sin\theta = \frac{20}{29}$; $\cos\theta = \frac{21}{29}$; $\tan\theta = \frac{20}{21}$; $\csc\theta = \frac{29}{20}$; $\sec\theta = \frac{29}{21}$; $\cot\theta = \frac{21}{20}$
13. $\sin\theta = 0.866$; $\cos\theta = 0.5$; $\tan\theta = 1.732$; $\csc\theta = 1.155$; $\sec\theta = 2$; $\cot\theta = 0.577$ **15.** $\sin\theta = 0.085$; $\cos\theta = 0.996$; $\tan\theta = 0.085$; $\csc\theta = 11.765$; $\sec\theta = 1.004$; $\cot\theta = 11.718$
17. 0.3189593 **19.** 0.3307184 **21.** 4.1760011
23. 0.1298773 **25.** 0.247404 **27.** 0.7291147
29. 0.1417341 **31.** 4.7970857 **33.** 22.61 m **35.** 4.92
V **37.** See *Computer Programs*

Exercise Set 9.3

1. 47.05° **3.** 77.92° **5.** 32.97° **7.** 73.00°
9. 61.68° **11.** 12.12°
13. $B = 73.5°, b = 24.6, c = 25.7$
15. $B = 17.4°, a = 19.1, b = 5.9$
17. $B = 47°, a = 32.3, c = 47.3$
19. $B = 0.65, b = 4.9, c = 8.2$
21. $B = 1.42, a = 2.7, b\,17.8$
23. $B = 0.16, a = 246.6, c = 249.8$
25. $A = 0.54, B = 1.03, c = 17.5$
27. $A = 1.03, B = 0.54, a = 15.4$
29. $A = 0.73, B = 0.84, b = 22.4$ **31.** 32°
33. 1147.3 m **35.** $F_y = 5\,\text{N}, F_x = 8.7\,\text{N}$ **37.** 9.62°
39. 117.2 m from the intersection; 276.2 m from service station **41.** 61.56 ft

Exercise Set 9.4

1. 87° **3.** 43° **5.** $\frac{\pi}{8}$ **7.** 1.36 **9.** 158°
11. 209° **13.** 1.02 **15.** 4.11 **17.** Quadrant II
19. IV **21.** III **23.** II **25.** II and IV **27.** II and III
29. IV **31.** positive **33.** positive **35.** negative

37. negative **39.** 0.6820 **41.** −2.3559 **43.** −0.2830
45. 0.6755 **47.** −0.1853 **49.** 4.9131 **51.** 0.8241
53. −0.0454 **55.** −1.0491 **57.** 0.2151 **59.** −1.8479
61. 0.9090 **63.** 24.26 mA **65.** 440.46 mm
67. 4,546 ft^2

Exercise Set 9.5

1. I, II **3.** II, IV **5.** I, IV **7.** III, IV
9. 30.0°; 150.0° **11.** 153.4°; 333.4°
13. 76.6°; 283.4° **15.** 189.4°; 350.6° **17.** 0.85; 2.29
19. 1.95; 5.09 **21.** 0.23; 2.91 **23.** 1.19; 5.09
25. 57.1° **27.** 76.6° **29.** 18.7° **31.** −32.6°
33. 1.91 **35.** 1.28 rad **37.** 1.25 rad **39.** 0.24 rad
41. −0.6898 rad ≈ −39.52° **43.** 43.57° **45.** See *Computer Programs* **47.** See *Computer Programs*

Exercise Set 9.6

1. 861.525 m **3.** 2261.9 in/m or 37.7 in/s
5. 29.45 cm **7.** 220.80 cm^2 **9.** 0.168 m
11. 24 776.5 mm/min. or 412.9 mm/sec; 24776.5 mm
13. 9.515°; 657.66 miles **15.** 83 (The actual answer is 83.77, but this would be 83 bytes.) **17.** 5100
19. 8.80 m/sec **21.** 7393.3 km **23.** 30°
25. (a) $\frac{\pi}{21600}$ rad/sec ≈ 0.000 145 444 rad/s (b) about 2.175843801 mi/s

Review Exercises

1. $\frac{\pi}{3} \approx 1.047$ **3.** 5.672 **5.** −2.007 **7.** 135°
9. 123.19° **11.** −246.94° **13.** (a) I, (b) 60°, (c) 420° and −300° **15.** (a) IV, (b) 35°, (c) 685° and −35°
17. (a) III, (b) 65°, (c) 245° and −475° **19.** (a) II, (b) $\frac{\pi}{4}$, (c) $\frac{11\pi}{4}$ and $\frac{-5\pi}{4}$ **21.** (a) II, (b) 0.99, (c) 8.43 and −4.13
23. (a) II, (b) 1.17, (c) 1.97 and −10.59 **25.** $\sin\theta = -4/5$; $\cos\theta = 3/5$; $\tan\theta = -4/3$; $\csc\theta = -5/4$; $\sec\theta = 5/3$; $\cot\theta = -3/4$ **27.** $\sin\theta = 21/29$, $\cos\theta = -20/29$, $\tan\theta = -21/20$, $\csc\theta = 29/21$, $\sec\theta = -29/20$, $\cot\theta = -20/21$ **29.** $\sin\theta = 1/\sqrt{50}$, $\cos\theta = 7/\sqrt{50}$, $\tan\theta = 1/7$, $\csc\theta = \sqrt{50}$, $\sec\theta = \sqrt{50}/7$, $\cot\theta = 7$
31. $\sin\theta = 8/17$, $\cos\theta = 15/17$, $\tan\theta = 8/15$, $\csc\theta = 17/8$, $\sec\theta = 17/15$, $\cot\theta = 15/8$ **33.** $\sin\theta = 4/5$, $\cos\theta = 3/5$, $\tan\theta = 4/3$, $\csc\theta = 5/4$, $\sec\theta = 5/3$, $\cot\theta = 3/4$
35. $\sin\theta = 84/91$, $\cos\theta = 35/91$, $\tan\theta = 84/35$, $\csc\theta = 91/84$, $\sec\theta = 91/35$, $\cot\theta = 35/84$
37. $\cos\theta = 24/27.2$, $\tan\theta = 12.8/24$, $\csc\theta = 27.2/12.8$, $\sec\theta = 27.2/24$, $\cot\theta = 24/12.8$ **39.** $\sin\theta = 3.12/4$, $\cos\theta = 2.5/4$, $\tan\theta = 3.12/2.5$, $\csc\theta = 4/3.12$, $\cot\theta = 2.5/3.12$ **41.** $\tan\theta = 0.577$, $\csc\theta = 2$, $\sec\theta = 1.155$, $\cot\theta = 1.732$ **43.** 0.7071 **45.** 0.6619
47. −0.6663 **49.** −0.0585 **51.** 30.0°; 150.0°
53. 138.6°; 221.4° **55.** 2.09; 4.19 **57.** 2.38; 5.52

59. $60.0°$ **61.** $-45.0°$ **63.** $22.6°$ **65.** $22.1°$
67. $6.3A$ **69.** $\theta = 41.3°; V = 66.6\,V$ **71.** $581.2\,ft$
73. $66,705$ miles/hr. **75.** $23.0°$ **77.** 204.2 feet
79. $F_y = 2891.3\,lb; F_x = 1972.3\,lb.$

Chapter 9 Test

1. $\frac{5\pi}{18} \approx 0.8725$ **3. (a)** III; **(b)** $57°$
5. (a) $\frac{9}{23.4} \approx 0.38462$; **(b)** $\frac{9}{21.6} \approx 0.41667$
7. (a) 0.79864; **(b)** 2.47509; **(c)** -1.66164; **(d)** -0.49026
9. $179.15\,ft$

ANSWERS FOR CHAPTER 10

Exercise Set 10.1

1.

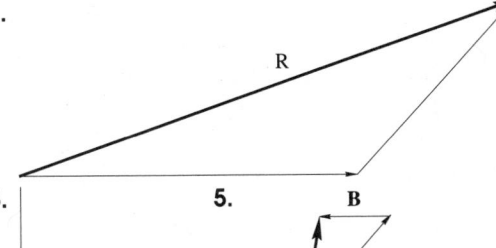

3. **5.**

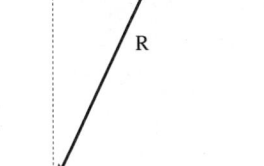

7.

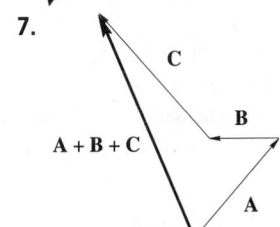

9.

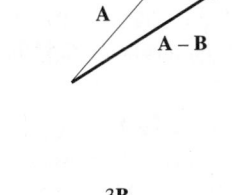

11. **13.**

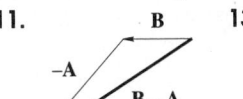
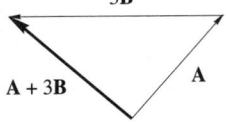

15.

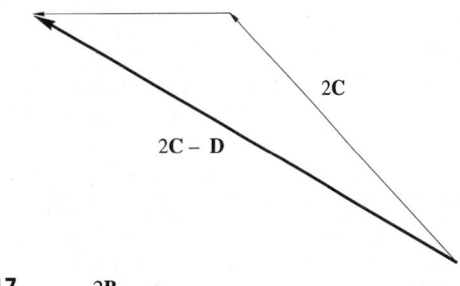

17.

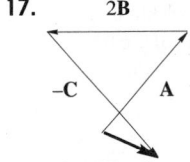

19.

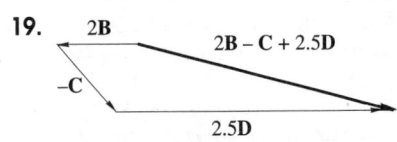

21. $P_x = 5.176$;

$P_y = 19.319$ **23.** $P_x = 4.688; P_y = -17.793$
25. $V_x = 9.372; V_y = 2.687$ **27.** $A = 15$ **29.** $C = 17$
31. $13\,km/h$ **33.** $V_x \approx 295\,N, V_x \approx 930\,N$ **35.** parallel
to the ramp $\approx 31.8\,lb$, perpendicular to the ramp $\approx 149.7\,lb$
37. $V = 17.5\,V, \phi \approx 30.96°$

Exercise Set 10.2

1. $R = 12.5, \theta_R = 180°$ **3.** $R = 34.8, \theta_R = 270°$
5. $R = 73, \theta_R = 131.11°$ **7.** $-89.7, \theta_R = 24.79°$
9. $A = 65, \theta_A = 59.49°$ **11.** $C = 12.5, \theta_C = 20.61°$
13. $E = 6.5, \theta_E = 14.25°$ **15.** $G = 15.1, \theta_G = 56.31°$
17. $R = 12.33, \theta_R = 32.03°$ **19.** $R = 53, \theta_R = 178.11°$
21. $R = 88.50, \theta_R = 223.83°$ **23.** $R = 22.49,$
$\theta_R = 349.18°$ **25.** $R = 13.04, \theta_R = 150.30°$
27. $R = 63.58, \theta_R = 4.6\,rad$ **29.** $R = 13.13, \theta_R = 64.38°$
31. $A = 1213\,N, B = 1832\,N, C = 2107\,N$ **33.** Tension in
the cable: $390.5\,lb$, compression in the boom: $311.9\,lb$
35. ground speed ≈ 488.3 mph, course $\approx 68.2°$, drift angle
$\approx 5.2°$ **37.** $324.9\,V$ **39.** See *Computer Programs*

Exercise Set 10.3

1. $R = 114.50, \theta = 14.49°$ is 70 lb force has direction $0°$
3. (a) 20.48 lb, **(b)** 14.34 lb, **(c)** vertical=16.07, horizontal=19.15 **5.** 1616.8 N forward, 525.33 N sideward
7. 372.16 mi/hr in the compass direction $173.83°$
9. 1112.62 kg perpendicular, 449.53 kg parallel
11. 119.11 lb horizontal, 154.40 lb vertical **13.** 6 A, $38.66°$
15. 10.68 A, $-59.46°$ **17.** 31.24 A, $\theta = -39.81°$
19. $55.84, \theta = 6.52°$ **21.** 126.24 m/s at an angle of $-18.09°$ **23.** 43.1 in. $\times$ 13.5 in. or $3'7\frac{1}{10}'' \times 1'1\frac{1}{2}''$

Exercise Set 10.4

1. $a = 7.59, b = 22.77, C = 75.3°$ **3.** $A = 67.1°$, $C = 15.9°, c = 4.22$ **5.** $C = 73.31°, B = 20.37, b = 6.69$
7. no solution, not a triangle **9.** $B = 57.31°$, $C = 77.69°, c = 22.5$; or $B = 122.69, C = 12.31°, c = 4.91$
11. $A = 148.8°, b = 20.17, c = 23.74$ **13.** $B = 64.62°$, $A = 79.78°, a = 21.13$; or $B = 115.38°, A = 29.02°$, $a = 10.42$ **15.** no solution **17.** $B = 0.44, A = 1.33$, $a = 19.52$ **19.** $B = 1.35, a = 90.67, c = 193.98$
21. From B to C is 305.5 m; From A to C is 369.6 m
23. 34.3 miles **25.** 294.77 m **27.** 444 ft 4 in.
29. 8.49 cm

Exercise Set 10.5

1. $c = 11.31, A = 33.6°, B = 104.1°$ **3.** $b = 68.02$, $A = 40.1°, C = 26.2°$ **5.** $A = 45°, B = 23.2°$, $C = 111.8°$ **7.** $b = 103.25, A = 38.16°, C = 49.40°$
9. $A = 30.34°, B = 44.11°, C = 105.55°$ **11.** $c = 11.77$, $A = 0.56, B = 1.76$ **13.** $A = 0.70\,\text{rad}, b = 74.52$, $C = 0.46\,\text{rad}$ **15.** $A = 0.96, B = 0.59, C = 1.59$
17. $b = 68.56, A = 0.925\,\text{rad}, C = 0.590\,\text{rad}$
19. $A = 1.59\,\text{rad}, B = 0.83\,\text{rad}, C = 0.72\,\text{rad}$
21. 10426 km **23.** 92.25 ft **25.** $30.98\,N, \theta = 39.36°$
27. 76 kΩ

Review Exercises

1.

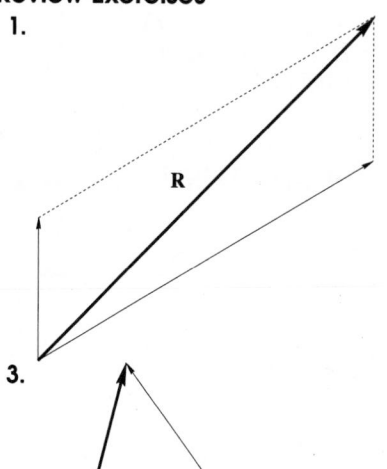

3.

5. $P_x = -13.68, P_y = 32.22$
7. $P_x = -23.29, P_y = 3.09$ **9.** $R = 17.89$, $\theta_R = 333.43°$ **11.** $A_x = 36.71, A_y = 9.84$
13. $C_x = 6.12, C_y = 18.41$ **15.** $R = 50.41, \theta_R = 20.69°$
17. $R = 72.76, \theta_R = 3.13\,\text{rad}$ **19.** $A = 25.7°, B = 97.5°$, $C = 56.8°$ **21.** $A = 145.93°, a = 146.17, C = 14.47°$
23. $B = 0.12, C = 2.90, c = 258.06$ **25.** $A = 0.52$, $B = 0.41, c = 109.27$ **27.** 2831.175 kg at $83.97°$
29. 66.47 lb. parallel; 107.53 lb. perpendicular
31. 195.47 m

Chapter 10 Test

1. $V_x = -21.34, V_y = 41.88$ **3.** $R = 59.69$, $\theta_R = 93.86°$ **5.** $a = 9.38$ **7.** 49.59 in.

ANSWERS FOR CHAPTER 11

Exercise Set 11.1

1. Period $= \pi$, amplitude $= 3$, frequency $= \frac{1}{\pi}$
3. Period $= 2\pi$, amplitude $= 2$, frequency $= \frac{1}{2\pi}$
5. Period $= 1$, amplitude $= 8$, frequency $= 1$
7. Period $= \frac{\pi}{2}$, amplitude $= \frac{1}{2}$, frequency $= \frac{2}{\pi}$
9. Period $= 4\pi$, amplitude $= \frac{1}{3}$, frequency $= \frac{1}{4\pi}$

11. Period 8π, amplitude $= \frac{1}{3}$, frequency $= \frac{1}{8\pi}$ **13.** Period 2π, amplitude $= \frac{1}{2}$, frequency $= \frac{1}{2\pi}$

15.

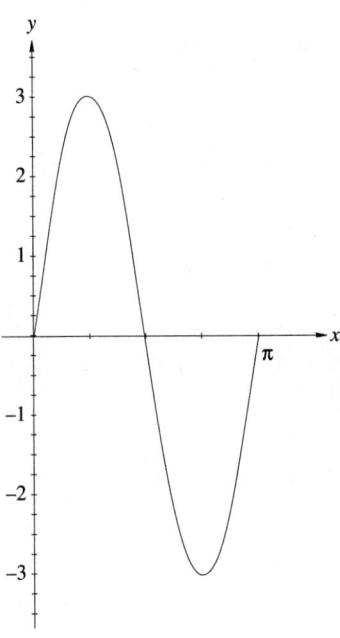

17.

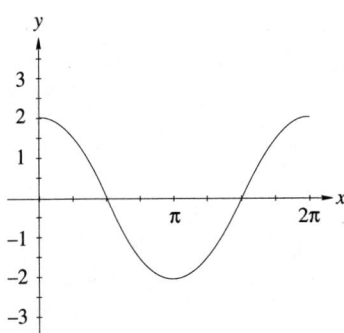

19.

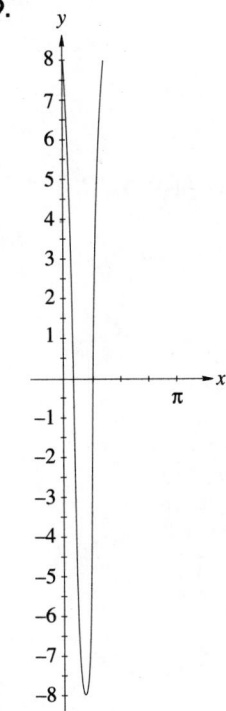

21.

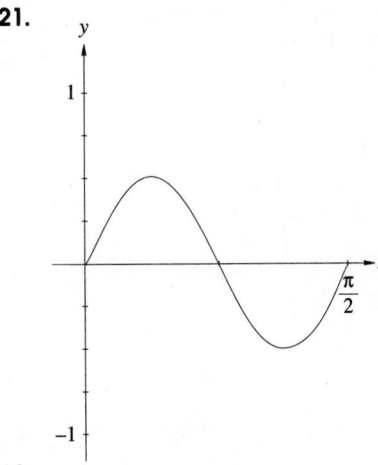

23.

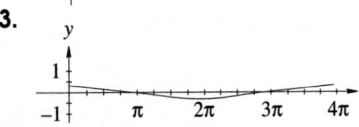

25.

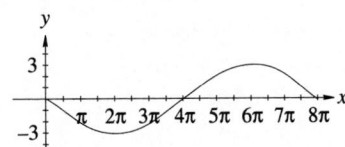

27.

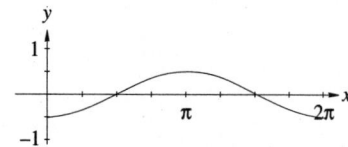

29. (a) Amplitude: 6.5 A, period: $\frac{1}{60}$, frequency: 60; **(b)**

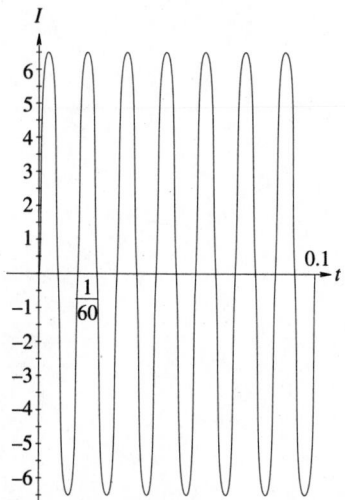

31. (a) Amplitude: 1094, period: 2π, frequency: $\frac{1}{2\pi}$
(b) $G_i = 1094 \ \cos i$

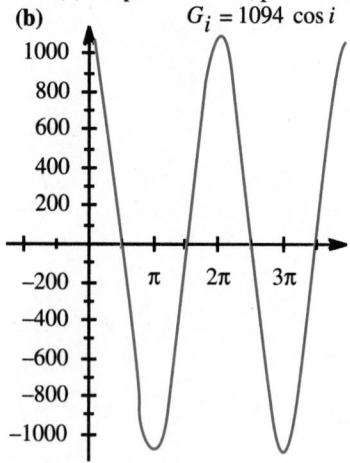

Exercise Set 11.2

1. Amplitude $= 2$, period $= 2\pi$, phase shift $= -\frac{\pi}{4}$ or $\frac{\pi}{4}$ left, vertical displacement $= 0$

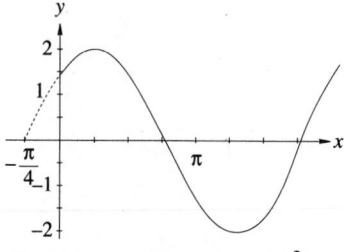

3. Amplitude $= 2.5$, period $= \frac{2\pi}{3}$, phase shift $= \frac{\pi}{3}$ or $\frac{\pi}{3}$ to the right, vertical displacement $= 0$

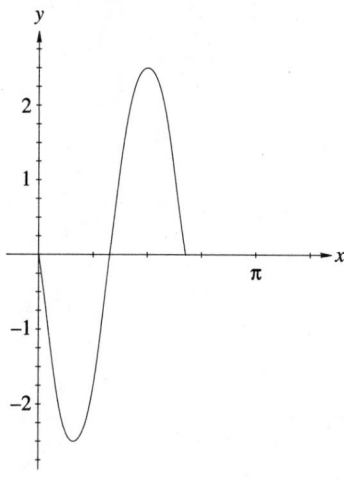

5. Amplitude = 6, period = 240° = $\frac{4\pi}{3}$, phase shift = $-\frac{180°}{1.5} = -120° = -\frac{2\pi}{3}$ to the left, vertical displacement = 0

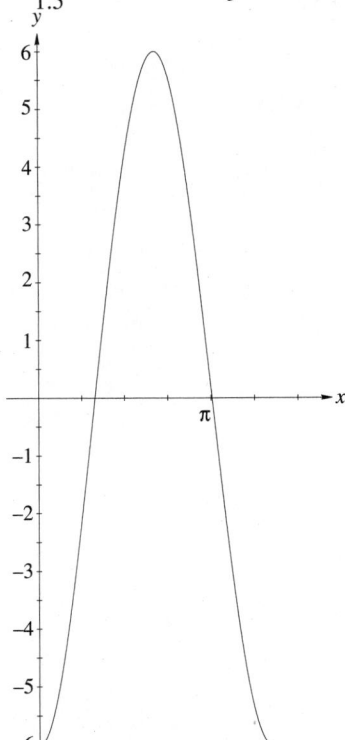

7. Amplitude = 2, period = 120° = $\frac{2\pi}{3}$, phase shift = $\frac{\pi}{6} = 30°$, vertical displacement = 0

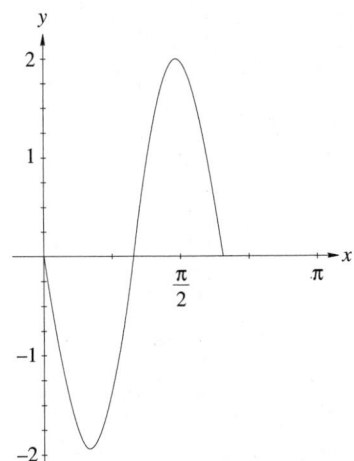

9. Amplitude = 4.5, period = $\frac{\pi}{3}$, phase shift = $\frac{4}{3}$, vertical displacement = 0

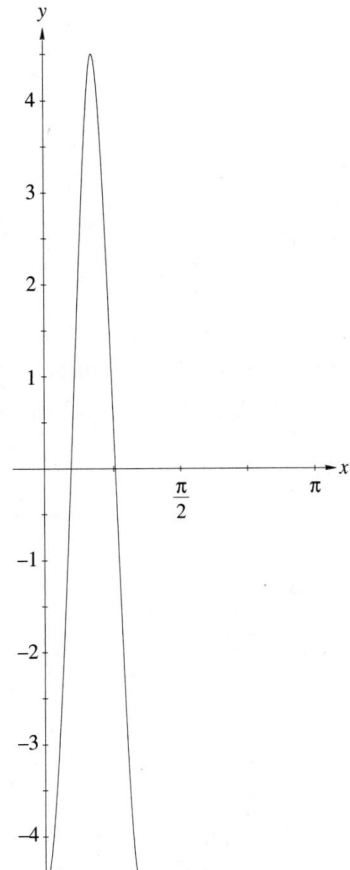

11. Amplitude = 0.5, period = 2, phase shift = $-\frac{1}{8}$, vertical displacement = 0

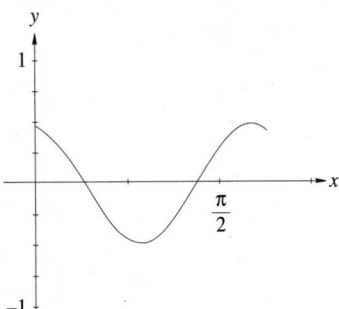

13. Amplitude = 0.75, period = 2, phase shift = $-\frac{\pi}{3}$, vertical displacement = 0

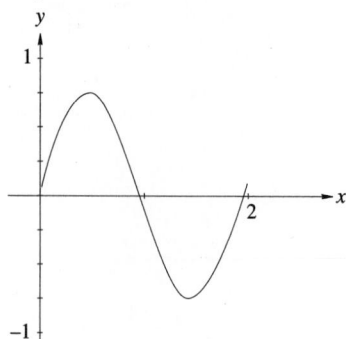

15. Amplitude = 2, period = 2, phase shift = $-\frac{1}{3\pi}$, vertical displacement = 4

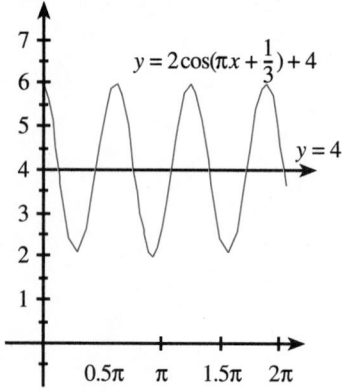

17. Amplitude = $\frac{1}{2}$, period = 2, phase shift = $-\frac{1}{\pi}$, vertical displacement = $\frac{3}{2}$

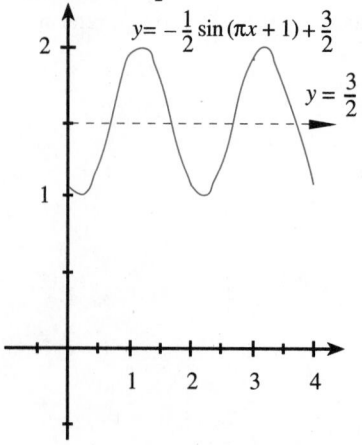

19. Amplitude = 2, period = 180° = π, phase shift = 22.5°, vertical displacement = 3.5

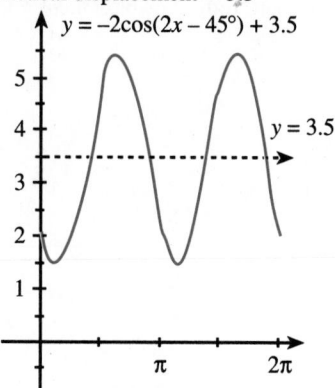

21. (a) Amplitude: 65; Period $\frac{1}{60}$; phase shift: $-\frac{1}{240}$;
(b)

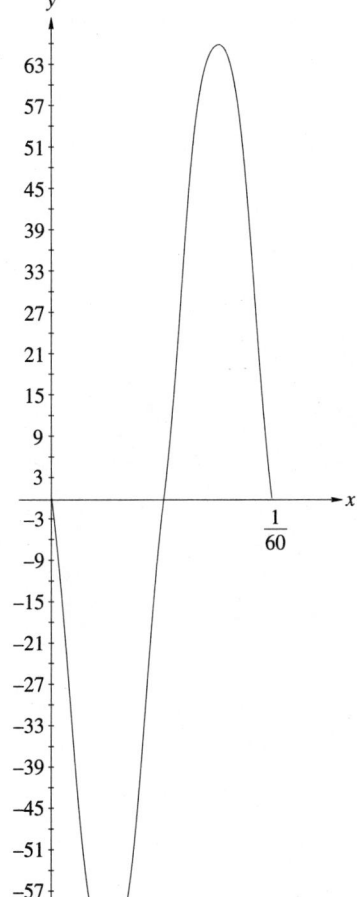

23. (a) $d = 2.8\sin\left(\frac{\pi}{6.2}t - \frac{0.9\pi}{6.2}\right) + 9 = 2.8\sin\left(\frac{5\pi}{31}t - \frac{9\pi}{62}\right) + 9$

(b)

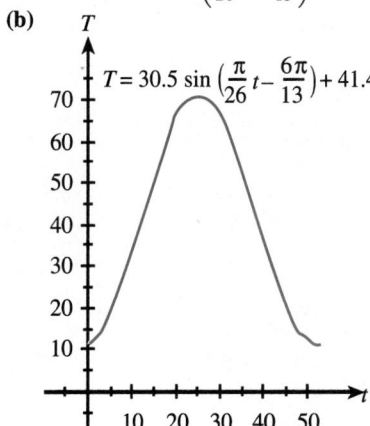

$d = 2.8\sin\left(\frac{5\pi}{3} - \frac{9\pi}{62}\right) + 9$

$y = 9$

25. (a) $T = 30.5\sin\left(\frac{\pi}{26}t - \frac{6\pi}{13}\right) + 41.5$

(b)

$T = 30.5\sin\left(\frac{\pi}{26}t - \frac{6\pi}{13}\right) + 41.4$

Exercise Set 11.3

1.

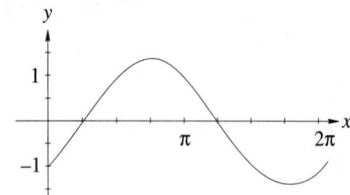

3.

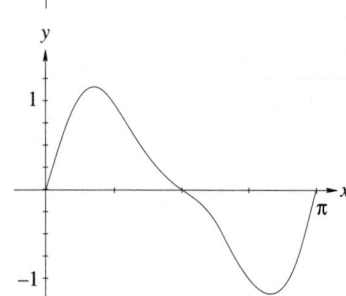

5.

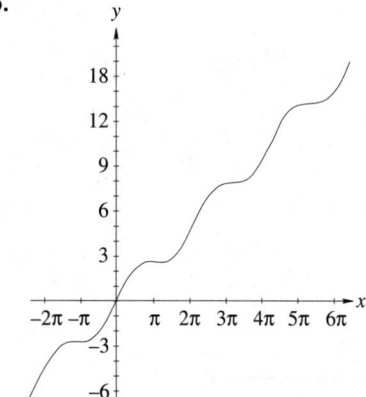

7.

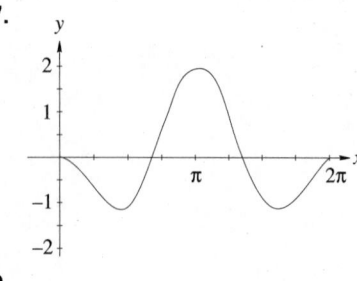

9.

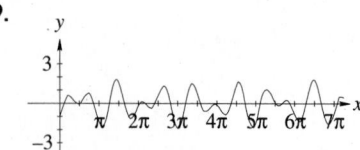

11.

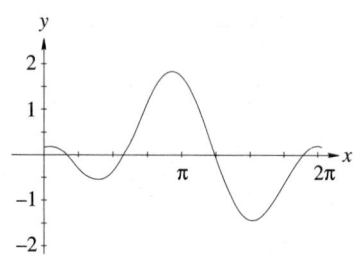

13.

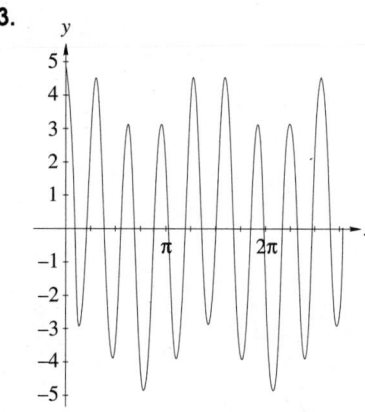

15.

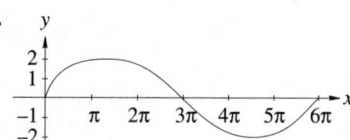

17.

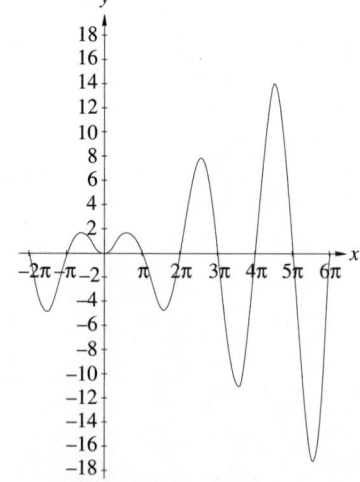

19. (a) Amplitude (maximum): 5, period: π, **(b)** viewing window: one cycle, $\left[0, \pi, \frac{\pi}{4}\right] \times [-3, 7, 1]$; two cycles, $\left[0, 2\pi, \frac{\pi}{4}\right] \times [-3, 7, 1]$, and

(c)

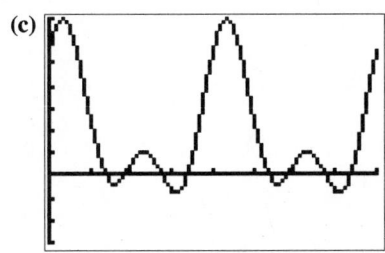

$$\left[0, 2\pi, \frac{\pi}{4}\right] \cdot [-3, 7, 1]$$

21. (a) Amplitude (maximum): 6, period: 2π, **(b)** viewing window: one cycle, $\left[0, 2\pi, \frac{\pi}{4}\right] \times [-9, 3, 1]$; two cycles, $\left[0, 4\pi, \frac{\pi}{4}\right] \times [-9, 3, 1]$, and

(c)

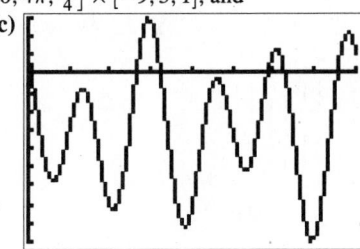

$$\left[0, 2\pi, \frac{\pi}{4}\right] \cdot [-9, 3, 1]$$

23. (a) Amplitude (maximum): 5.5, period: none, not a periodic function, **(b)** viewing window: $\left[0, 4\pi, \frac{\pi}{2}\right] \times [-9, 2, 1]$, and **(c)**

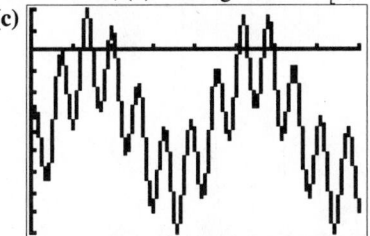

$$\left[0, 4\pi, \frac{\pi}{2}\right] \cdot [-9, 2, 1]$$

25.

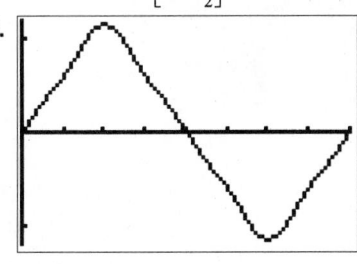

$$\left[0, 2\pi, \frac{\pi}{4}\right] \cdot [0, 3, 1]$$

Exercise Set 11.4

1.

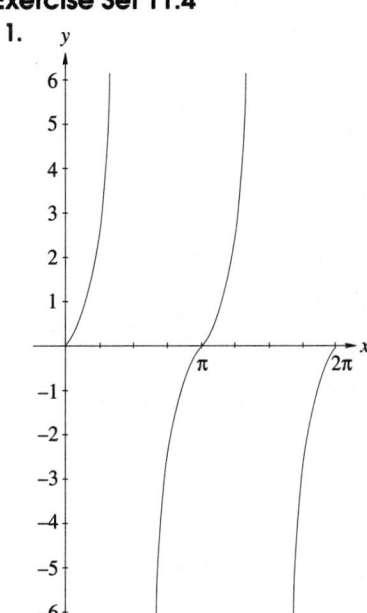

3.

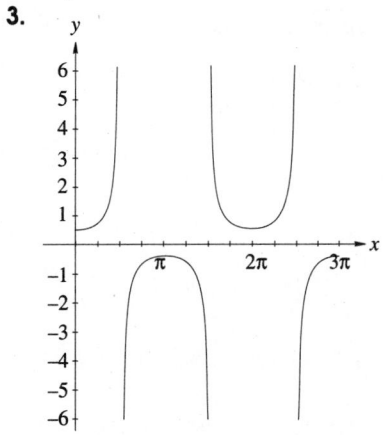

5.

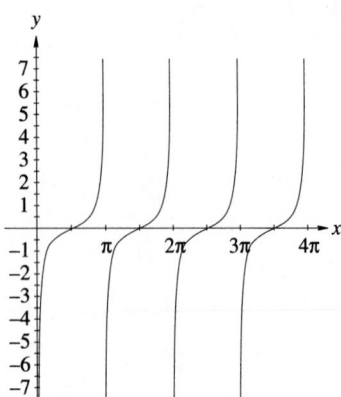

7.

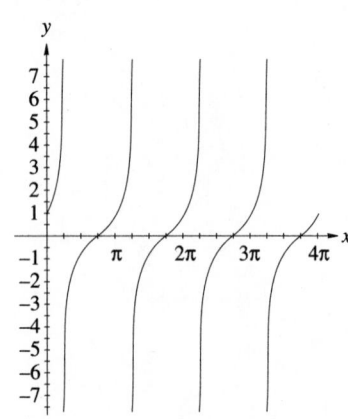

9.

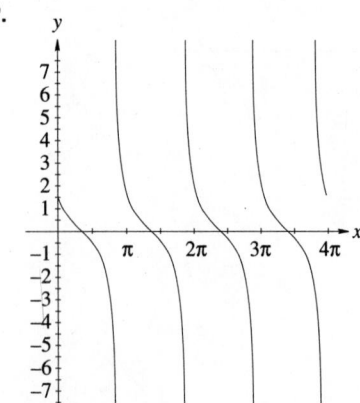

11.

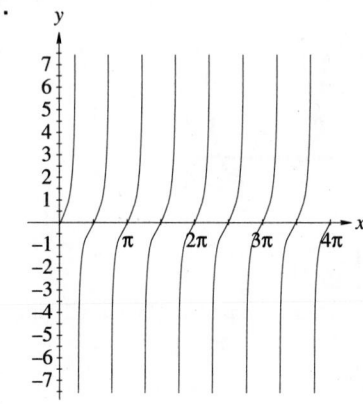

13.

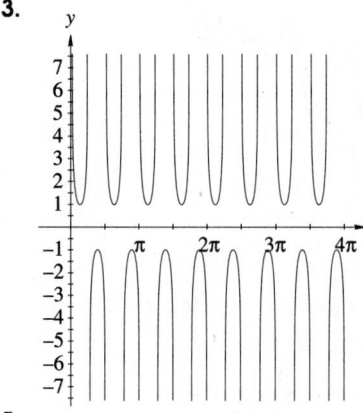

15.

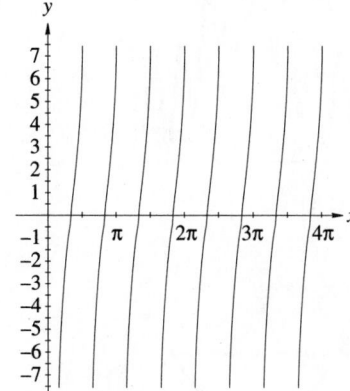

17.

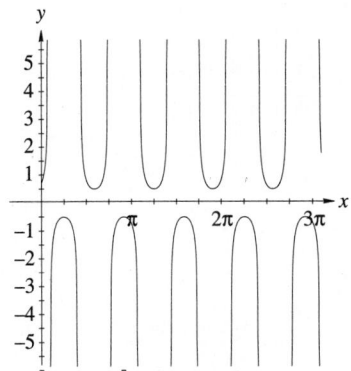

19. (a) $\left[0, 4\pi, \frac{\pi}{2}\right] \times [-8, 6, 2]$,
(b) $\qquad y = 2\tan(0.5x + \frac{\pi}{4}) - 2$

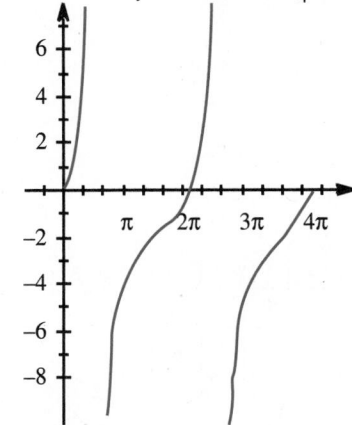

21. (a) about 1.68 cm, **(b)** 15.71°,
(c)

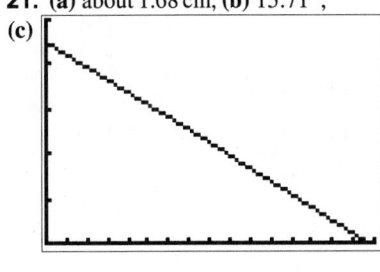

$[0, 16] \times [0, 5]$

Exercise Set 11.5

1. $y = 10\sin 8\pi t$ **3.** $y = 0.8\sin\frac{\pi}{3}t$ **5.** $-2.79\,\text{ft/sec}^2$
7. (a) 8.5 cm; **(b)** $\frac{\pi}{2.8}$; **(c)** $\frac{2.8}{\pi}$

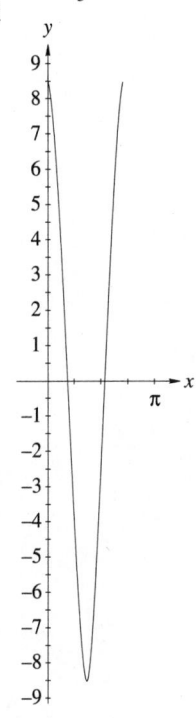

9. amplitude 10 A, period $\frac{1}{60}$ sec/cycle, $f = 60$ Hz,
$\omega = 120\pi$ rad/sec   **11.** $y = 6.8\sin(160\pi t + \pi/3)$
13. $V = 220\sin(80\pi t - \frac{\pi}{3})$ **15.** $V = 220\cos(80\pi t + \frac{\pi}{5})$

17. $f = 10^{23}$, amp $= 10^{-12}$ **19.**

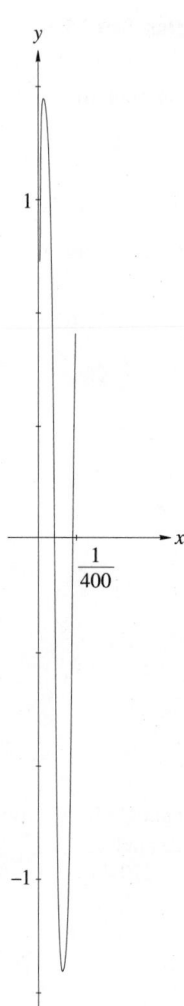

21. (a) Period: $\frac{2\pi}{2\pi \times 10^{18}} = 10^{-18}$ Hz, amplitude: 10^{-10} m,

(b) $y = 10^{-10} \sin(2\pi x \times 10^{18} t + \pi)$

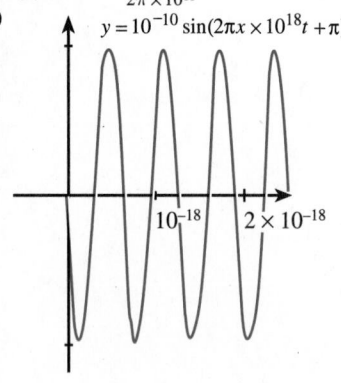

Exercise Set 11.6

1.

t	−3	−2	−1	0	1	2	3; $y = 3x$
x	−3	−2	−1	0	1	2	3
y	−9	−6	−3	0	3	6	9

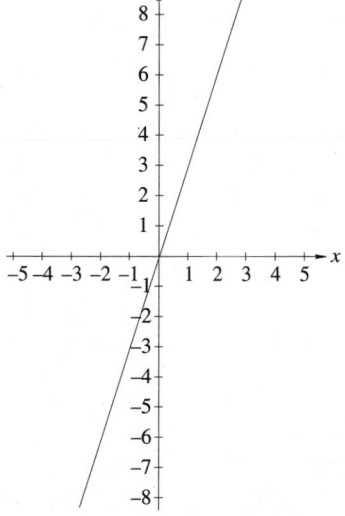

3.

t	−3	−2	−1	0	1	2	3; $y = \dfrac{1}{x}$
x	−3	−2	−1	0	1	2	3
y	$-\dfrac{1}{3}$	$-\dfrac{1}{2}$	−1	*	1	$\dfrac{1}{2}$	$\dfrac{1}{3}$

*Not defined

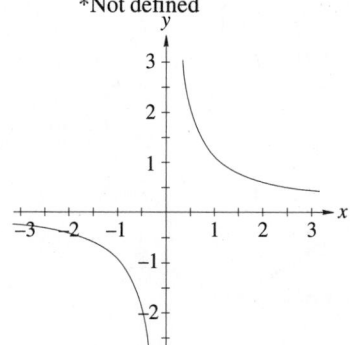

5.

t	−5	−4	−3	−2	−1	0	1	2	3	4	5;
x	8	7	6	5	4	3	2	1	0	−1	−2
y	16	7	0	−5	−8	−9	−8	−5	0	7	16

$y = (x-3)^2 - 9 = x^2 - 6x$

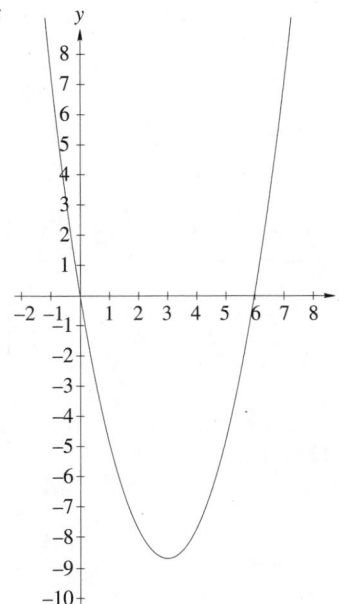

7.

t	0	$\frac{\pi}{4}$	$\frac{\pi}{2}$	$\frac{3\pi}{4}$	π	$\frac{5\pi}{4}$	$\frac{3\pi}{2}$	$\frac{7\pi}{4}$
x	0	$\sqrt{2}$	2	$\sqrt{2}$	0	$-\sqrt{2}$	−2	$-\sqrt{2}$
y	2	$\sqrt{2}$	0	$-\sqrt{2}$	−2	$-\sqrt{2}$	0	$\sqrt{2}$

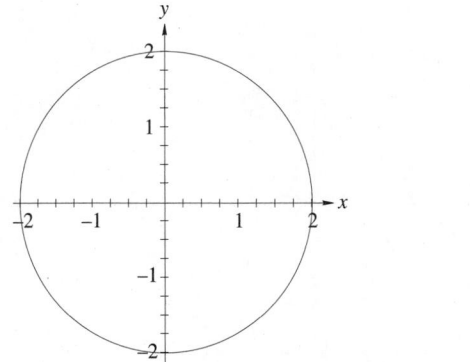

9.

t	0	$\frac{\pi}{4}$	$\frac{\pi}{2}$	$\frac{3\pi}{4}$	π	$\frac{5\pi}{4}$	$\frac{3\pi}{2}$	$\frac{7\pi}{4}$	2π
x	2	$\frac{5\sqrt{2}}{2}$	5	$\frac{5\sqrt{2}}{2}$	0	$-\frac{5\sqrt{2}}{2}$	−5	$-\frac{5\sqrt{2}}{2}$	0
y	0	$\frac{3\sqrt{2}}{2}$	3	$\frac{3\sqrt{2}}{2}$	0	$-\frac{3\sqrt{2}}{2}$	−3	$-\frac{3\sqrt{2}}{2}$	0

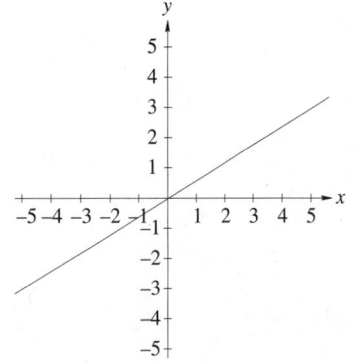

11.

t	-5	-4	-3	-2	-1	0	1	2	3	4	5
x	-5.96	-4.76	-2.86	-1.09	-0.16	0	0.16	1.09	2.86	4.76	5.96
y	0.72	1.65	1.99	1.41	0.46	0	0.46	1.41	1.99	1.65	0.72

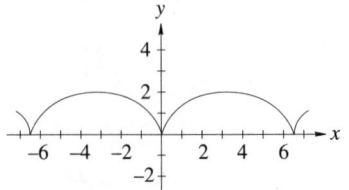

13.

t	0	$\frac{\pi}{6}$	$\frac{\pi}{4}$	$\frac{\pi}{3}$	$\frac{\pi}{2}$	$\frac{2\pi}{3}$	$\frac{3\pi}{4}$	$\frac{5\pi}{6}$	π	$\frac{7\pi}{6}$	$\frac{5\pi}{4}$	$\frac{4\pi}{3}$	$\frac{3\pi}{2}$	$\frac{5\pi}{3}$	$\frac{7\pi}{4}$	$\frac{11\pi}{6}$	2π
x	1	$\frac{2}{\sqrt{3}}$	$\sqrt{2}$	2	*	-2	$-\sqrt{2}$	$-\frac{2}{\sqrt{3}}$	-1	$-\frac{2}{\sqrt{3}}$	$-\sqrt{2}$	-2	*	2	$\sqrt{2}$	$\frac{2}{\sqrt{3}}$	1
y	*	4	$2\sqrt{2}$	$\frac{4}{\sqrt{3}}$	2	$\frac{4}{\sqrt{3}}$	$2\sqrt{2}$	4	*	-4	$-2\sqrt{2}$	$-\frac{4}{\sqrt{3}}$	-2	$-\frac{4}{\sqrt{3}}$	$-2\sqrt{2}$	-4	*

*Not defined

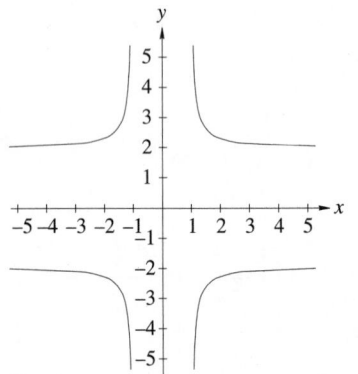

17. (a) $2 : 1$;

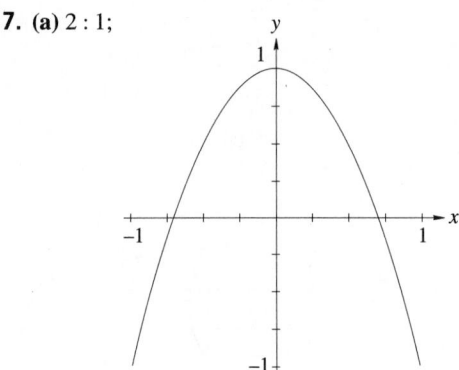

15. (a) $1 : 1$;

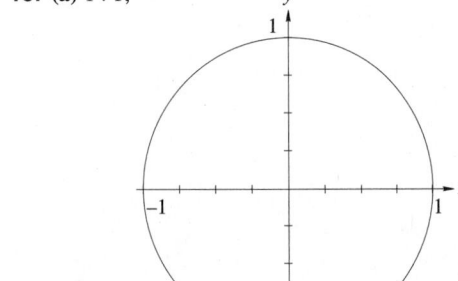

19. (a) $1 : 4$;

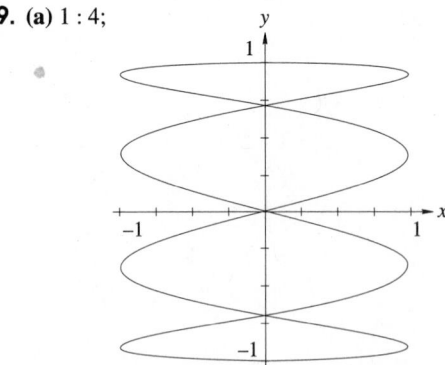

21. **(a)** 2 : 3;

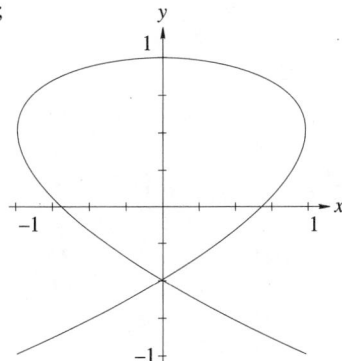

23. **(a)** 2 : 5;

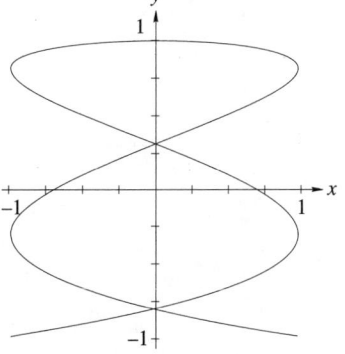

25.

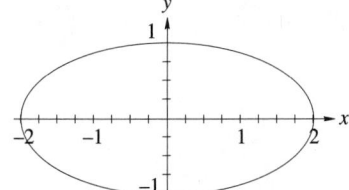

27.

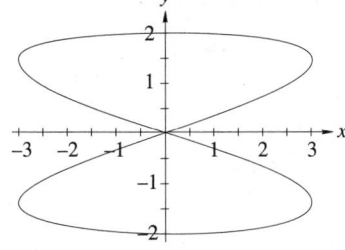

29.

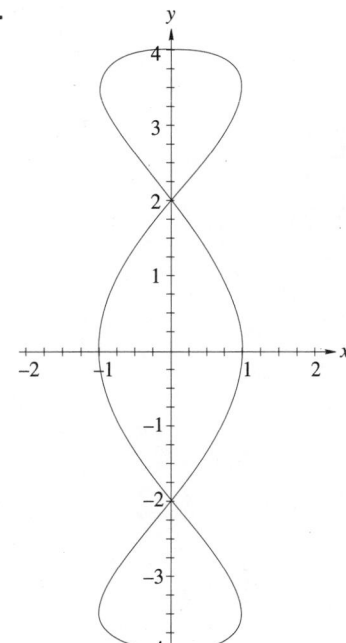

31. **(a)**

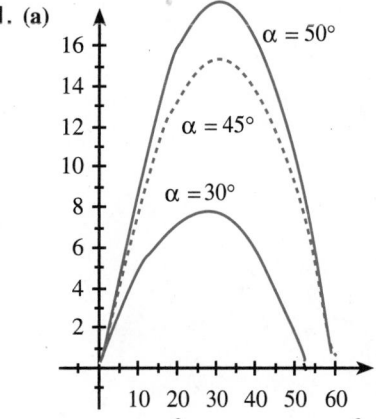

(b) $y = x \tan\alpha - \dfrac{gx^2}{2V^2} \sec^2\alpha$, **(c)** $x = \dfrac{V^2 \sin 2\alpha}{g}$

Exercise Set 11.7

1.

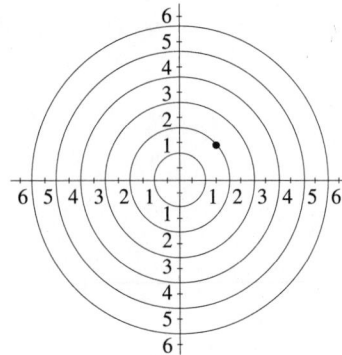

3.

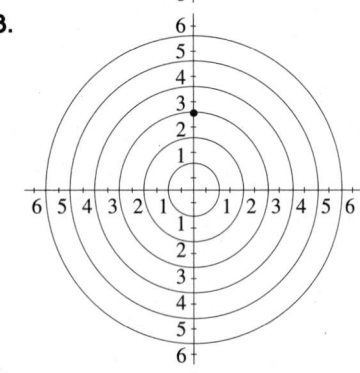

5.

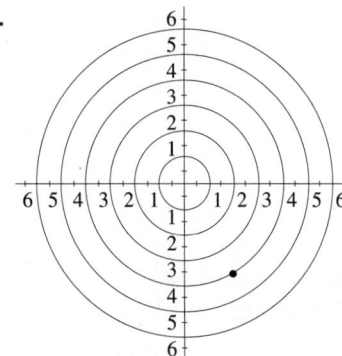

7.

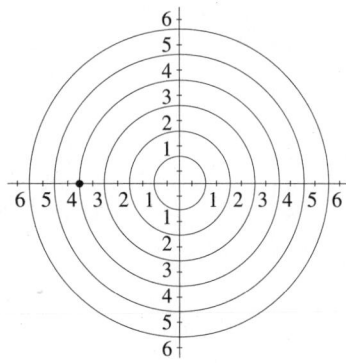

9.

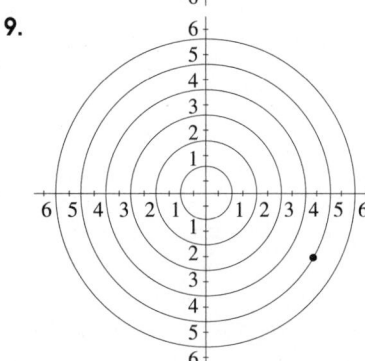

11.

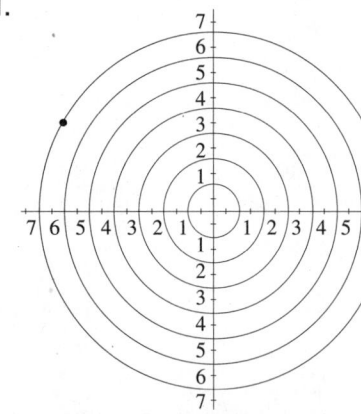

13. $(2, 3.46)$ **15.** $(-1.414, 1.414)$ or $(-\sqrt{2}, \sqrt{2})$
17. $(-5.64, -2.05)$ **19.** $(0.80, 1.83)$ **21.** $(-2.95, 0.52)$
23. $(4.81, 3.59)$
25. $(5.66, 45°) \approx (4\sqrt{2}, 45°) = (4\sqrt{2}, \frac{\pi}{4})$
27. $(5, 35.87°) = (5, 0.64)$
29. $(29, 133.60°) = (29, 2.332)$
31. $(5, 126.87°) = (5, 2.214)$
33. $(12.21, 235.01°) = (12.21, 4.102)$
35. $(9.22, 77.47°) = (9.22, 1.352)$

37.

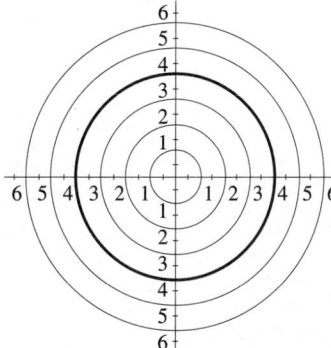

39.

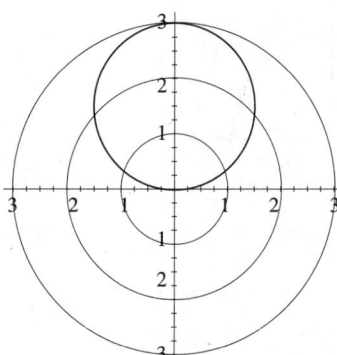

41.

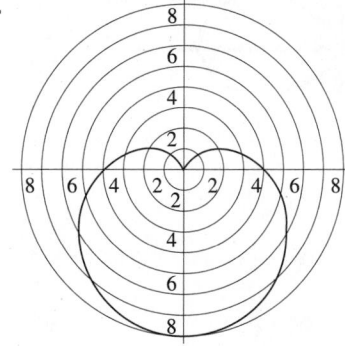

43.

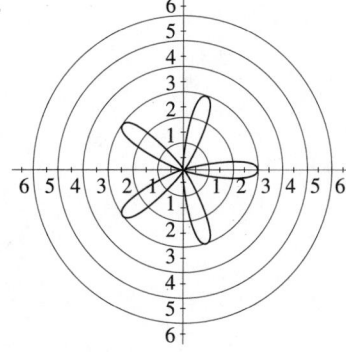

45.

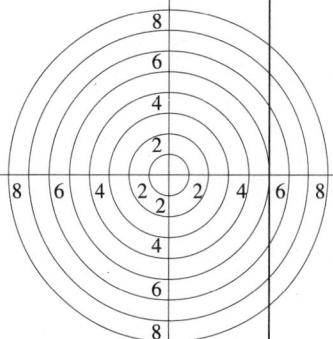

47.

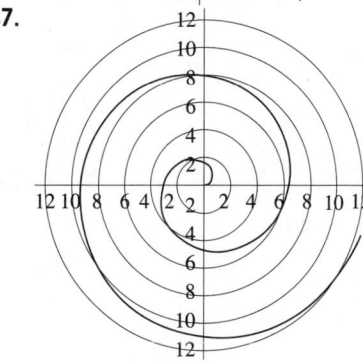

49.

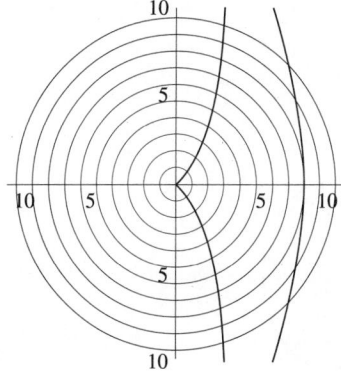

51.

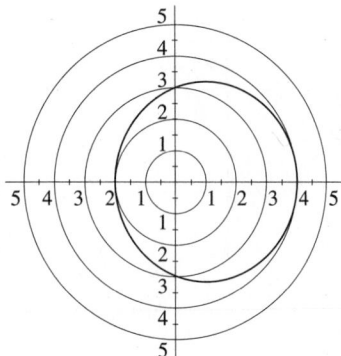

53.

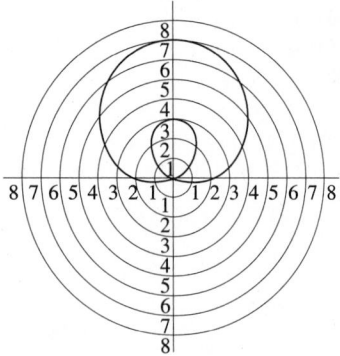

55.

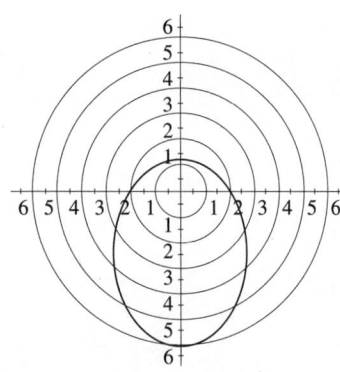

57.

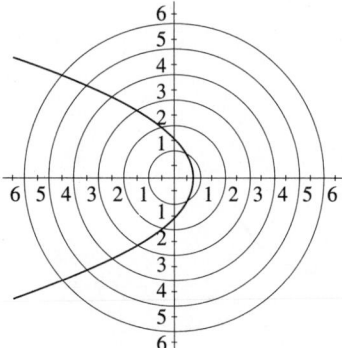

59.

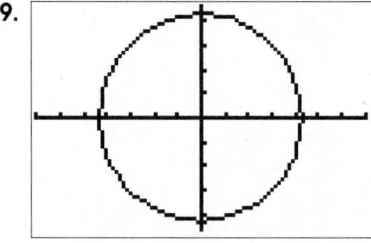

$[-7050, 7050, 1000] \times [-4650, 4650, 1000]$

61.

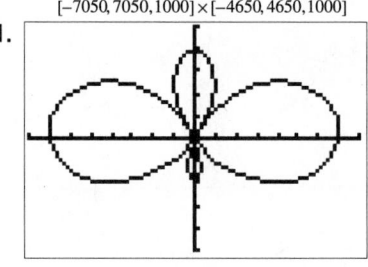

$[-8, 8, 1] \times [-5, 5, 1]$

Review Exercises

1. period $\frac{\pi}{2}$, amplitude 8, frequency $\frac{2}{\pi}$, displacement 0

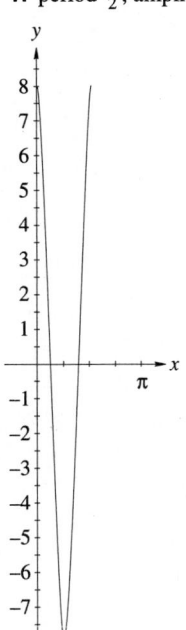

3. period $\frac{\pi}{3}$, amplitude ∞, frequency $\frac{3}{\pi}$, displacement 0

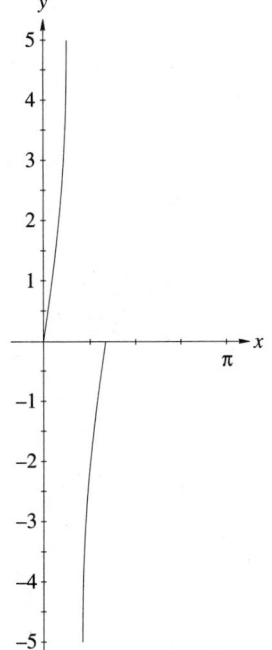

5. period $\frac{2\pi}{3}$, amplitude $\frac{1}{2}$, frequency $\frac{3}{2\pi}$, displacement $\frac{-\pi}{9}$

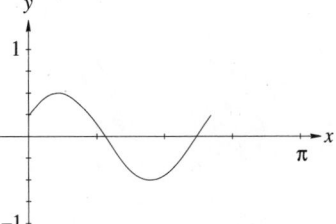

7. period π, amplitude ∞, frequency $\frac{1}{\pi}$, displacement $\frac{\pi}{4}$

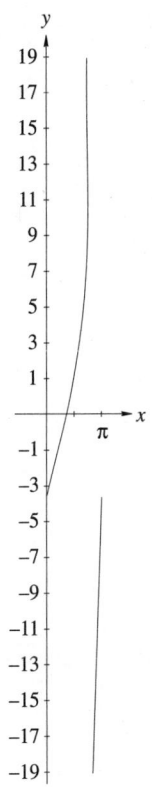

9. period 3π, amplitude 2, frequency $\frac{1}{3\pi}$, displacement $\frac{\pi}{4}$

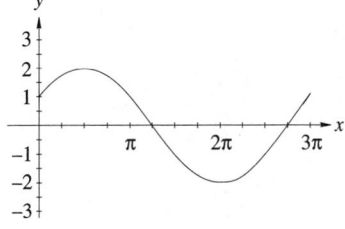

11.

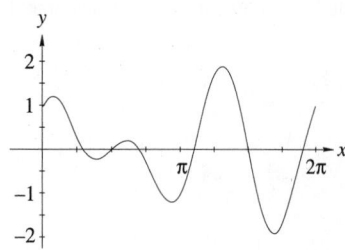

13.

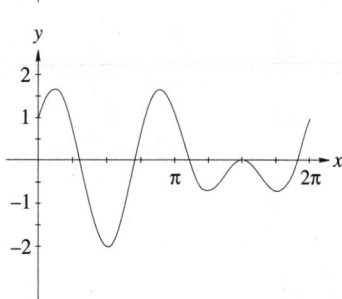

15.

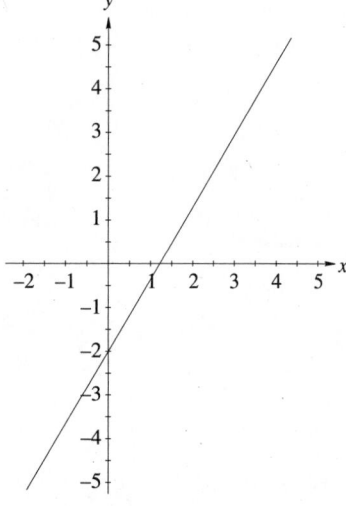

17.

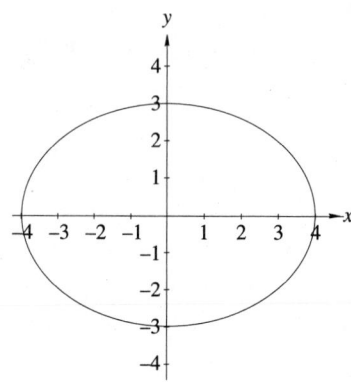

19.

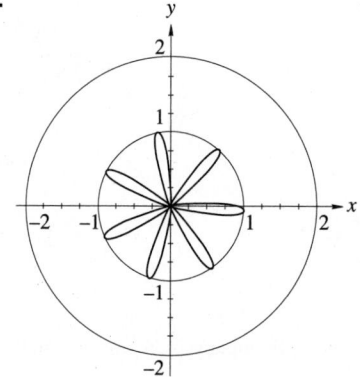

21. $y = 85 \sin 10\pi t$ **23.** $I = 4.8 \sin\left(120\pi t + \frac{\pi}{2}\right)$

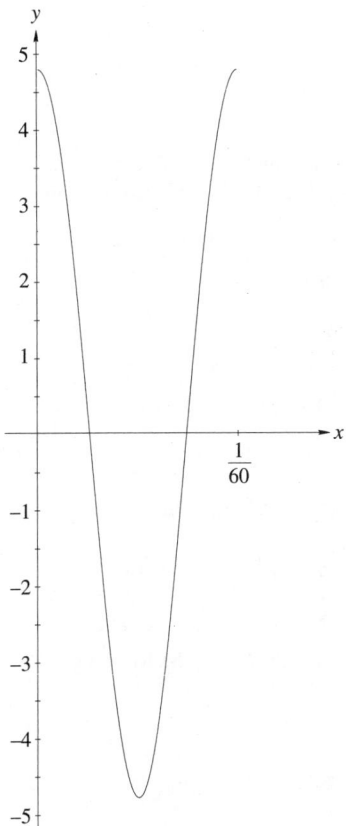

Chapter 11 Test

1. period $\frac{2\pi}{5}$, amplitude 3, frequency $\frac{5}{2\pi}$, displacement 0

period $\frac{2\pi}{3}$, amplitude 2.4, frequency $\frac{3}{2\pi}$, displacement $\frac{\pi}{12}$

period $\frac{\pi}{2}$, amplitude ∞, frequency $\frac{2}{\pi}$, displacement $-\frac{\pi}{10}$

3.

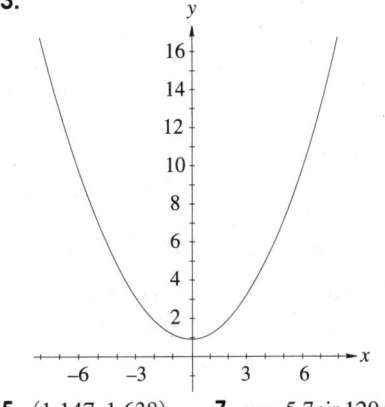

5. $(1.147, 1.638)$ **7.** $y = 5.7\sin 120\pi t$

≡ ANSWERS FOR CHAPTER 12

Exercise Set 12.1

1. 5   **3.** 4   **5.** 1/5   **7.** 1/2   **9.** 9   **11.** 25
13. 1/8   **15.** $-1/2$ **17.** 1/2   **19.** 32   **21.** 3^7
23. 7^4 **25.** x^{10} **27.** y^2 **29.** 9^{10} **31.** x^{21}
33. x^5y^5 **35.** $1/a^5b^5$ **37.** x^8 **39.** $1/x^6$ **41.** x^2
43. $r^{7/4}$ **45.** $a^{5/6}$ **47.** $d^{5/12}$ **49.** b^3/a^3
51. r^2/s^3 **53.** y^6 **55.** a^5/b^{12}

57. $1/yb^{13/2} = \dfrac{1}{yb^{13/2}}$ **59.** p^3/x^2 **61.** $\dfrac{y^6}{z^2}$ **63.** $\dfrac{2y^2}{3x}$

65. $\dfrac{y^{19/15}}{x^{31/15}}$ **67.** $\dfrac{t^{11/6}x^{1/6}}{6}$ **69.** $\dfrac{4y^{11}}{27x^{10}}$
71. 4.0993852   **73.** 25.368006   **75.** 3.5162154
77. 4.5606226   **79.** -0.2253777406 **81.** 221.1125 m
≈ 221 m   **83.** $p = 905\,146.3\,\text{N/m}^2$ $T = 91\,238.747$ K

85. $1 - \left(\dfrac{p_1}{p_2}\right)^{-7/2} = 1 - \left(\dfrac{p_2}{p_1}\right)^{7/2}$

Exercise Set 12.2

1. $2\sqrt[3]{2}$ **3.** $\sqrt[3]{5}$ **5.** y^4 **7.** $a\sqrt[5]{a^2}$ **9.** $xy^3\sqrt{y}$
11. $a\sqrt[4]{ab^3}$ **13.** $2x\sqrt[3]{x}$ **15.** $3x\sqrt{3xy}$ **17.** -2
19. ab^3 **21.** $pq^2r\sqrt[4]{r^3}$ **23.** $2x/3$ **25.** xy^2/z

27. $\dfrac{2x\sqrt[3]{2y^2}}{z^2}$ **29.** $8xy^2\sqrt{x}/3z^2p = \dfrac{8xy^2\sqrt{x}}{3z^2p}$

31. $4\sqrt{3}/3 = \dfrac{4\sqrt{3}}{3}$ **33.** $3\sqrt[3]{2}/2 = \dfrac{3\sqrt[3]{2}}{2}$

35. $5\sqrt{2x}/2x = \dfrac{5\sqrt{2x}}{2x}$ **37.** $3\sqrt[4]{8z}/4z = \dfrac{3\sqrt[4]{8z}}{4z}$

39. $2\sqrt[3]{2y}/x = \dfrac{2x\sqrt[3]{2y}}{x}$ **41.** $-2x\sqrt[3]{by}/3bz = \dfrac{-2x\sqrt[3]{by}}{3bz}$

43. $2 \times 10^2 = 2000$ **45.** $50\sqrt{10}$
47. $2\sqrt{10} \times 10^3 = 2000\sqrt{10}$ **49.** $\sqrt[3]{12.5} \times 10^3$

51. $\sqrt{\dfrac{x^2 + y^2}{xy}} = \dfrac{\sqrt{x^3y + xy^3}}{xy}$ **53.** $a + b$

55. $\sqrt{\dfrac{b+a^2}{a^2 b}} = \dfrac{\sqrt{b^2+a^2 b}}{ab}$ **57.** $\sqrt[18]{27x^2}$

59. $\sqrt[21]{-624.2x^{15}y^{10}z}$ **61.** $f = \dfrac{\sqrt{T\mu}}{2L\mu} = \dfrac{1}{2L\mu}\sqrt{T\mu}$

63. $z = \dfrac{1}{\sqrt{\dfrac{R^2+x^2}{R^2 x^2}}} = \sqrt{\dfrac{R^2 x^2}{R^2+x^2}} = \dfrac{Rx\sqrt{R^2+x^2}}{R^2+x^2} =$

$\dfrac{Rx}{R^2+x^2}\sqrt{R^2+x^2}$ **65.** $1 + \dfrac{\sqrt{m^2+2h}}{m}$ or $\dfrac{m+\sqrt{m^2+2h}}{m}$

Exercise Set 12.3

1. $7\sqrt{3}$ **3.** $5\sqrt[3]{9}$ **5.** $8\sqrt{3}+4\sqrt{2}$ **7.** $3\sqrt{5}$
9. $-\sqrt{7}$ **11.** $5\sqrt{15}/3 = \frac{5}{3}\sqrt{15}$ **13.** $-\sqrt{2}$

15. $3x\sqrt{xy}$ **17.** $(2q+p^2)\sqrt[3]{3p^2 q}$ **19.** $\dfrac{x^2-y^2}{x^2 y^2}\sqrt{xy}$

21. $2\sqrt{3ab}/3$ **23.** $\sqrt{40} = 2\sqrt{10}$ **25.** $x\sqrt{15}$
27. $8x\sqrt{x}$ **29.** $3/4$ **31.** $\sqrt{2x}+2$ **33.** $x+2\sqrt{xy}+y$

35. $\dfrac{\sqrt[3]{5\cdot 7^2}}{7} = \dfrac{\sqrt[3]{245}}{7}$ **37.** $a-b$ **39.** $\sqrt[6]{x^5}$

41. $\sqrt[12]{3^3 49^4 x^{11}} = \sqrt[12]{155{,}649{,}627 x^{11}}$ **43.** 4

45. $\sqrt[3]{2b}/2$ **47.** $\sqrt[6]{\dfrac{x^4}{3}} = \dfrac{\sqrt[6]{243x^4}}{3}$

49. $2\sqrt[12]{\dfrac{2a^5}{b^5}} = 2\dfrac{\sqrt[12]{2a^5 b^7}}{b}$ **51.** $\sqrt[6]{2}$ **53.** $\dfrac{x-\sqrt{5}}{x^2-5}$

55. $\dfrac{(\sqrt{5}-\sqrt{3})^2}{2} = 4-\sqrt{15}$
57. $2x\sqrt{x^2-1}/(x^2-1), x \neq \pm 1$

59. $-\dfrac{\sqrt{x^2-y^2}+\sqrt{x^2+xy}}{y}, \quad y \neq 0$ **61.** $-\dfrac{b}{a}$

63. (a) $R = \dfrac{x^{3/2}}{x+1}$, **(b)** $4.259\,\Omega$ **65.** $\dfrac{c}{a}$

Exercise Set 12.4

1. 22 **3.** 22.5 **5.** no solution **7.** 25 **9.** 16
11. 10 **13.** 2 **15.** $\frac{2829}{296} \approx 14.43$ **17.** no solution

19. no solution **21.** $R = \dfrac{v^2}{\mu_s g}$ **23.** $I_A = 1.419$;
$I_B = 4.691$ **25.** $1{,}183.36\,\text{ft}$

Review Exercises

1. 7 **3.** 125 **5.** $1/27$ **7.** 5^{13} **9.** 2^{32}

11. $x^8 y^{12}$ **13.** $y^5 a^5$ **15.** $x^{5/6}$ **17.** $xy^8/a = \dfrac{xy^8}{a}$

19. -3 **21.** $ab^2\sqrt{b}$ **23.** $\dfrac{-2xy^2}{z}\sqrt[3]{z^2}$ **25.** $\dfrac{\sqrt{a^3-b^3}}{ab}$

27. $8\sqrt{5}$ **29.** $8\sqrt{6}-2\sqrt{3}$ **31.** $(2+3a)\sqrt[4]{a^3 b}$

33. $\dfrac{(a^2-b^2)\sqrt{7ab}}{7ab}$ **35.** $\sqrt{15}$ **37.** $x\sqrt{14}$

39. $a-2b$ **41.** $\dfrac{\sqrt[3]{a}}{2}$ **43.** 82 **45.** no solution
47. 18.14 **49.** 2 **51. (a)** $28.72\,\text{mg}$; **(b)** $16.49\,\text{mg}$
53. $v = \sqrt{2rt^2+4t+5}$

Chapter 12 Test

1. 8^{-2} **3.** $\dfrac{x^5}{y^{15}}$ **5.** -4 **7.** $x^2 y\sqrt{y}$ **9.** $x+3$

11. $2\sqrt{3}x$ **13.** $\dfrac{\sqrt[3]{2y^2}}{y}$ **15.** 12 **17.** 27

☰ ANSWERS FOR CHAPTER 13

Exercise Set 13.1

1. 4.7288 **3.** 5.6164 **5.** 3.4154

7.

x	−3	−2	−1	0	1	2	3	4	5
$f(x)$	$\frac{1}{64}$	$\frac{1}{16}$	$\frac{1}{4}$	1	4	16	64	256	1024

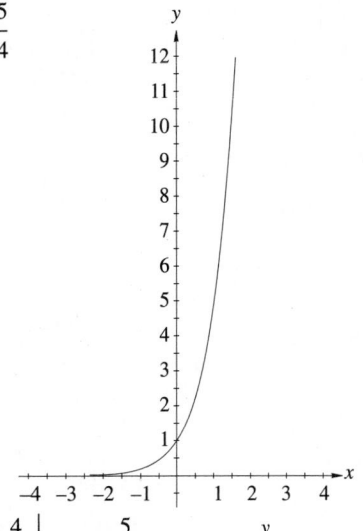

9.

x	−3	−2	−1	0	1	2	3	4	5
$h(x)$	$\frac{8}{27}$	$\frac{4}{9}$	$\frac{2}{3}$	1	1.5	2.25	3.375	5.0625	7.59375

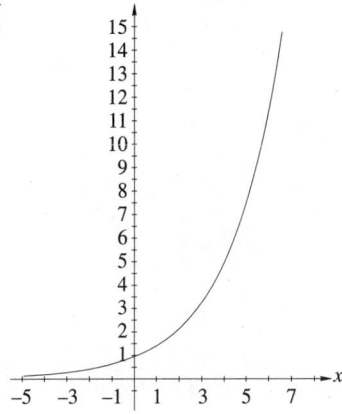

11.

x	−3	−2	−1	0	1	2	3	4	5
$f(x)$	27	9	3	1	$\frac{1}{3}$	$\frac{1}{9}$	$\frac{1}{27}$	$\frac{1}{81}$	$\frac{1}{243}$

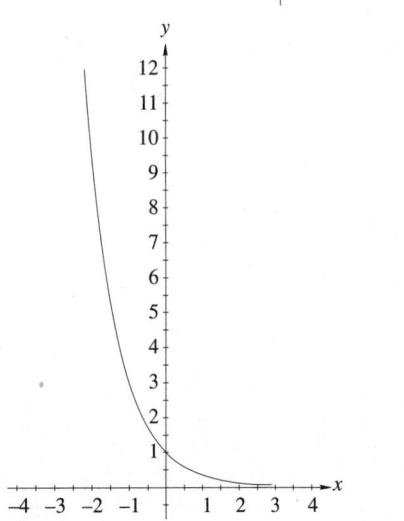

13.

x	-3	-2	-1	0	1	2	3	4	5
$h(x)$	13.824	5.760	2.4	1	0.417	0.174	0.072	0.030	0.0126

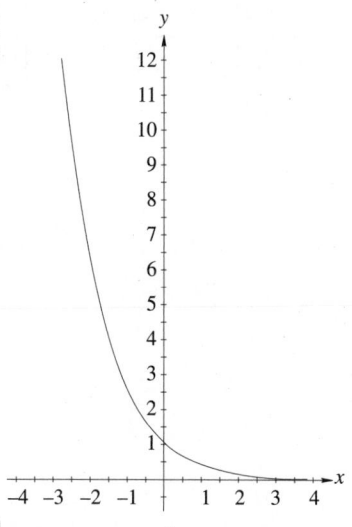

15.

x	-3	-2	-1	0	1	2	3	4
$f(x)$	0.064	0.192	0.577	1.732	5.196	15.588	46.765	140.296

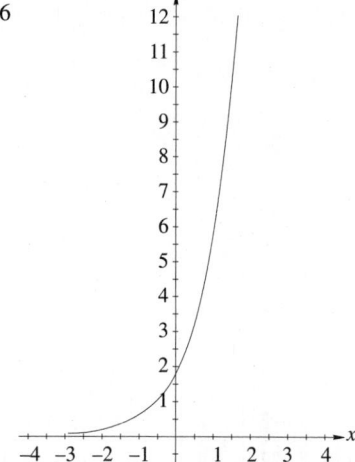

17. (a) \$1338.23; (b) \$1343.92; (c) \$1346.86; (d) \$1348.85
19. \$4.93   **21.** (a) 50,000, (b) 800,000, (c) 1 h, (d) 3 h
23. (a) 400, (b) $\frac{2}{5} = 0.4$, (c) 3,600   **25.** (a) Here $T = 1.75$,
and so $I = 0.88^{1.75} \approx 0.80$.
(b)

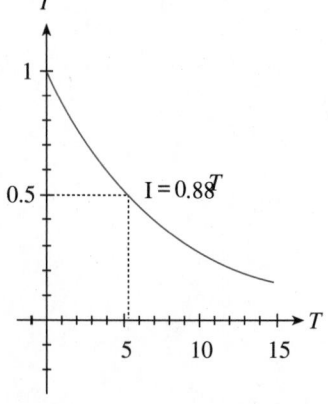

(c) Locate 0.50 on the

I-axis. Draw a horizontal line from there until it intersects the graph of $I = 0.88^{T}$. From there draw a vertical line toward the T-axis. It intersects the T-axis near 5.42, and so we conclude that a thickness of 5.42 mm will produce an intensity of 0.50.

Exercise Set 13.2

1. 20.085 537   **3.** 104.584 986   **5.** 0.018 316
7. 0.063 928

9.

x	-2	-1	0	1	2	3	4
$f(x)$	0.5413	1.4715	4	10.8731	29.5562	80.3421	218.3926

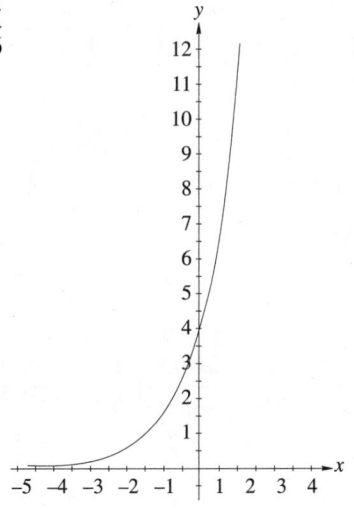

11.

x	-4	-3	-2	-1	0	1	2
$h(x)$	218.3926	80.3421	29.5562	10.8731	4	1.4715	0.5413

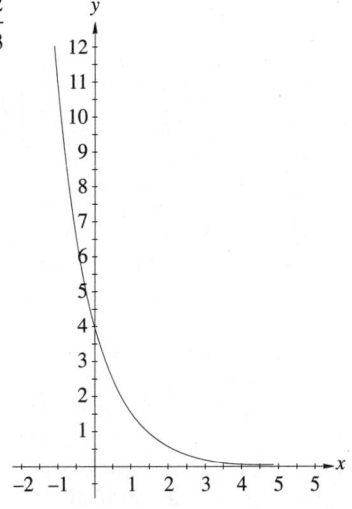

13. (a) 125 mg; **(b)** 85.91 mg; **(c)** 0.31 mg   **15. (a)** 8660;
(b) 25,981; **(c)** 260,980   **17. (a)** 280,510; **(b)** 393,430
19. 110.6°F   **21.** 110.26 min or about 1 hour 50 minutes
23. 0.0066 C   **25.** 25.5 μCi   **27. (a)** 0.05 A, **(b)** 0.048 A,
(c) 0.046 A

Exercise Set 13.3

1. $6^3 = 216$ **3.** $4^2 = 16$ **5.** $\left(\frac{1}{7}\right)^2 = \frac{1}{49}$
7. $2^{-5} = \frac{1}{32}$ **9.** $9^{7/2} = 2{,}187$ **11.** $\log_5 625 = 4$
13. $\log_2 128 = 7$ **15.** $\log_7 343 = 3$ **17.** $\log_5 \frac{1}{125} = -3$
19. $\log_4 128 = \frac{1}{128}$ **21.** 1.609437912
23. 1.558355122   **25.** 0.6020599913   **27.** 1.102776615

29. 1.292029674   **31.** 1.125708821   **33.**

x	0.1	0.5	1	2	4	6	8	10
$\ln x$	-2.30	-0.69	0	0.69	1.39	1.79	2.08	2.30

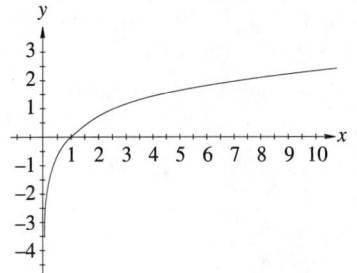

35.

x	0.1	0.5	1	2	4	6	8	10
$\log_2 x$	-3.32	-1	0	1	2	2.58	3	3.32

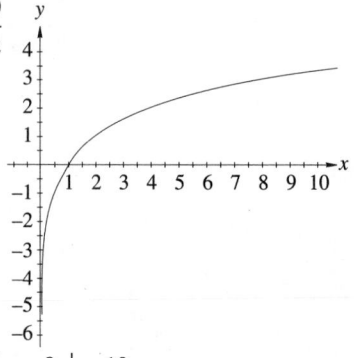

37.

x	0.1	0.5	1	2	4	6	8	10
$\log_{12} x$	-0.93	-0.28	0	0.28	0.56	0.72	0.84	0.93

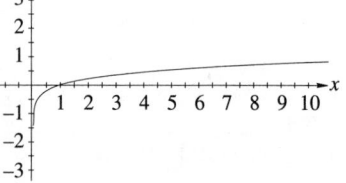

39.

x	0.1	0.5	1	2	4	6	8	10
$\log_{1/4} x$	1.66	0.5	0	-0.5	-1	-1.29	-1.5	-1.66

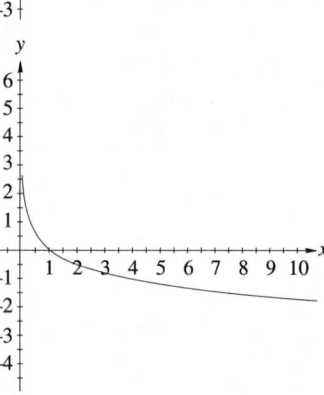

41. (a) $I = I_0 10^{\beta/10}$; (b) 50 dB; (c) 90 dB
43. 3,884 ft³/acre

Exercise Set 13.4

1. $\log 2 - \log 3$ **3.** $\log 2 + \log 7$
5. $\log 2 + \log 2 + \log 3$
7. $\log 2 + \log 3 + \log 5 + \log 5 - \log 7$ **9.** $\log 2 + \log x$
11. $\log 2ax - \log 3y$ **13.** $\log 22$ **15.** $\log \frac{11}{3}$
17. $\log 12$ **19.** $\log \dfrac{4x}{y}$ **21.** $\log(5^2 \cdot 5^3) = \log 4000$
23. $\log \frac{2}{3} \cdot \frac{6}{7} = \log \frac{4}{7}$ **25.** 0.9030   **27.** 1.0791
29. 1.1761   **31.** 1.5741   **33.** 2.3010   **35.** 3.6990
37. 1.5476   **39.** 0.2851   **41.** 6
43. $\log\left(\dfrac{I_2\sqrt{R_2}}{I_1\sqrt{R_1}}\right)^{20} = \log\left(\dfrac{I_2{}^{20}R_2{}^{10}}{I_1{}^{20}R_1{}^{10}}\right)$

Exercise Set 13.5

1. 2.0922   **3.** 0.5746   **5.** -2.4763 **7.** 2.6513
9. 5.5   **11.** 2.6094   **13.** $10^{2.3} \approx 199.53$ **15.** $10^{17} + 5$
17. $\sqrt{\frac{1}{2}e^9} = \dfrac{\sqrt{2}e^{4.5}}{2} \approx 63.6517$ **19.** $\frac{1}{18} \approx 0.056$
21. 101   **23.** 4.040404   **25.** 11.55 yr   **27.** 138.63 yr
29. 5.5%   **31.** 8.15%   **33.** 5.6   **35.** 4418.07 m
37. 0.104 s   **39.** 0.1354 seconds
41. $t = \dfrac{RC}{-2} \ln\left(1 - \dfrac{2E}{U^2 C}\right)$

Exercise Set 13.6

1.

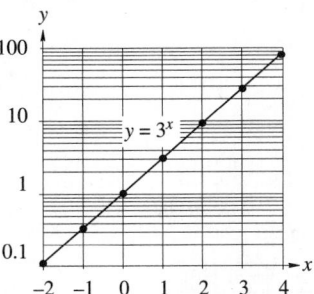

3.

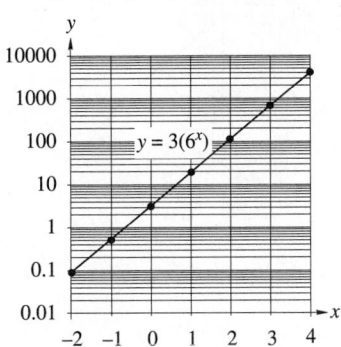

5.

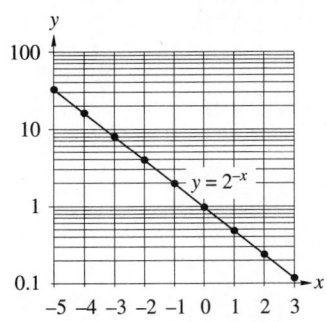

7.

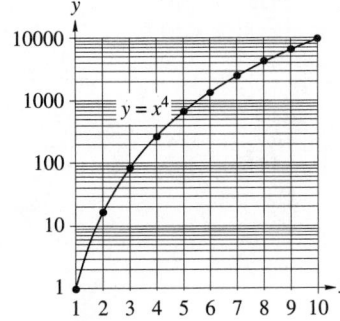

9.

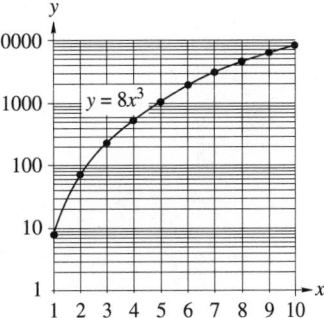

11.

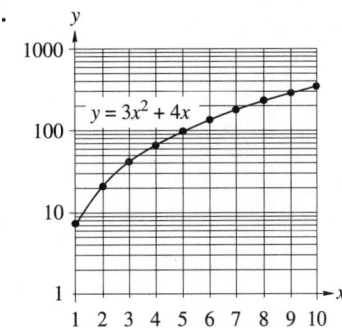

13.

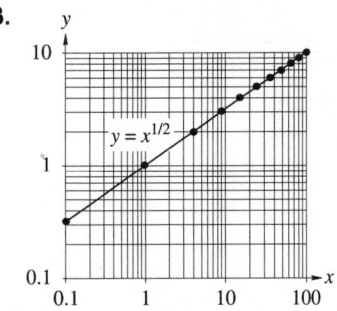

15.

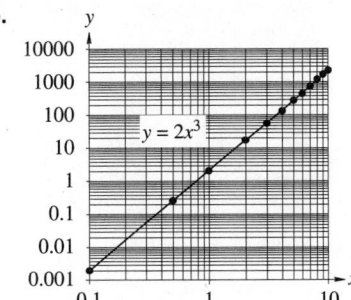

17.

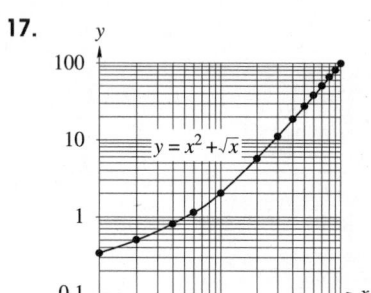

$y = x^2 + \sqrt{x}$

19.

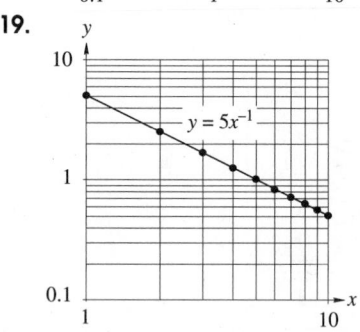

$y = 5x^{-1}$

21.

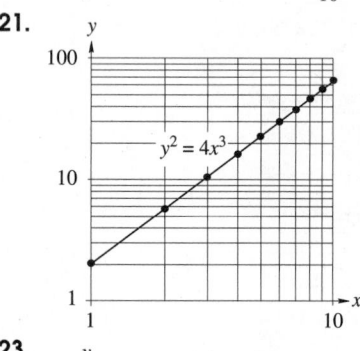

$y^2 = 4x^3$

23.

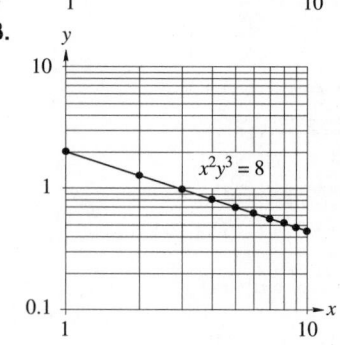

$x^2y^3 = 8$

25.

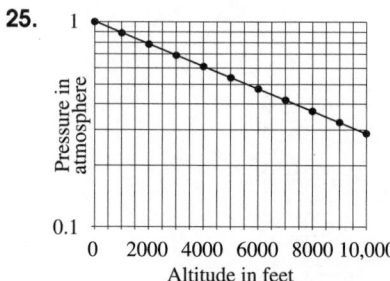

27.

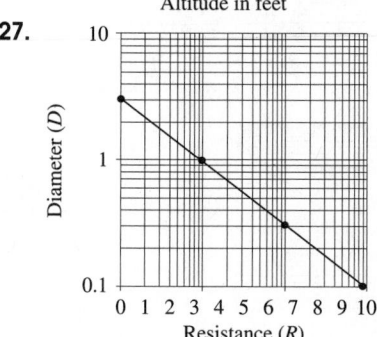

29.

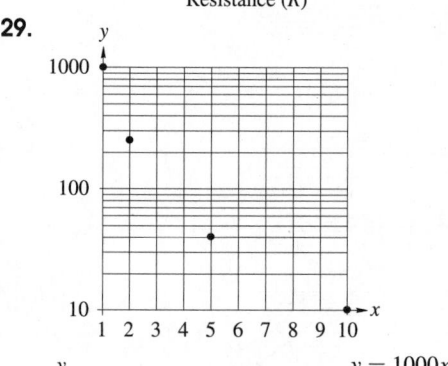

$y = 1000x^{-2}$

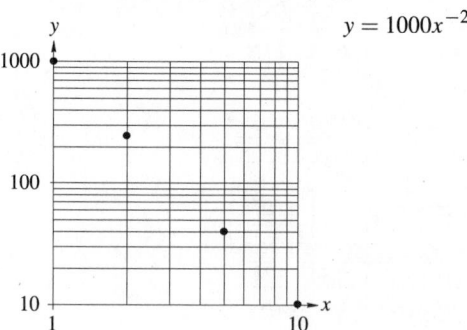

31.

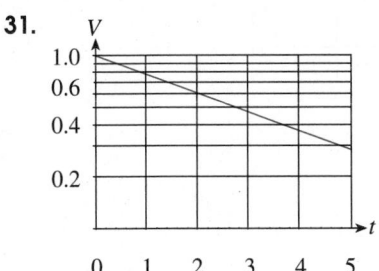

Review Exercises

1. 148.413 16   **3.** 106.697 74   **5.** 0.903 09

7. 4.398 146   **9.**

x	-2	-1	0	1	2	3	4
$5e^x$	0.677	1.839	5	13.59	36.95	100.43	272.99

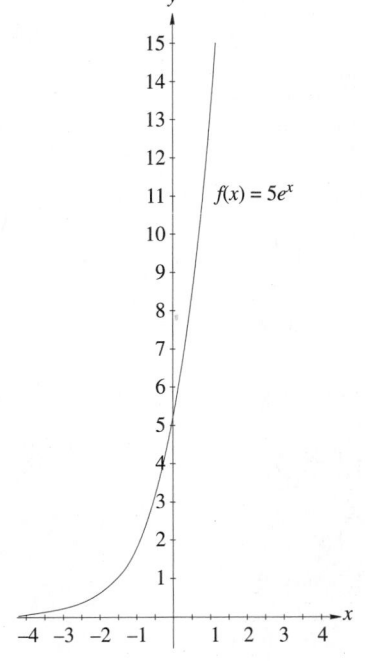

11.

x	-3	-2	-1	0	1	2
5^x	0.008	0.04	0.2	1	5	25

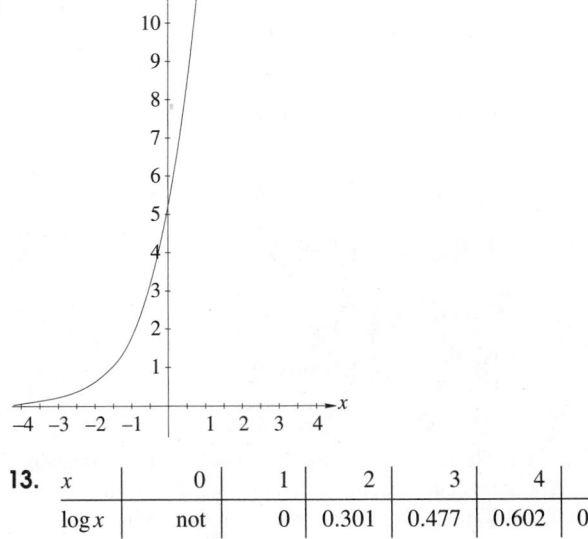

13.

x	0	1	2	3	4	5
$\log x$	not defined	0	0.301	0.477	0.602	0.699

x	6	7	8	9	10
$\log x$	0.778	0.845	0.903	0.959	1

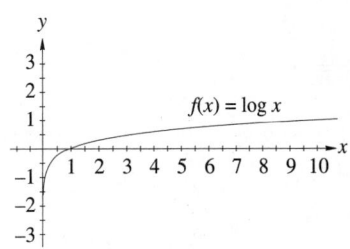

15.

x	$-\frac{1}{2}$	0	1	2	3	4	5	6	7	8	9	10
$\ln(2x+1)$	not defined	0	1.099	1.609	1.946	2.197	2.398	2.565	2.708	2.833	2.944	3.045

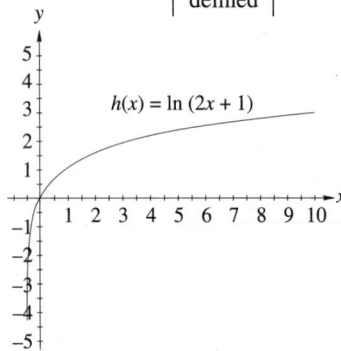

17. $\log 3 - \log 4$ **19.** $\log 5 + \log x$ **21.** $3(\ln 4 + \ln x)$
23. $\frac{1}{2}(\log 6 + \log 5)$ **25.** $\log 45$ **27.** $\log x^5$ **29.** $\log x^8$
31. 2.404   **33.** 0.480   **35.** $e^{9.1} \approx 8955.29$
37. $0.5 \times 10^{50} - 0.5$

39.

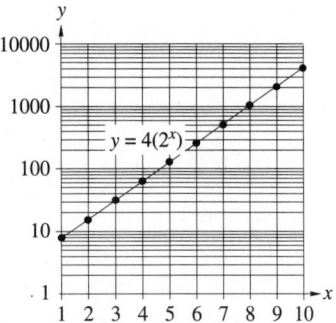

41.

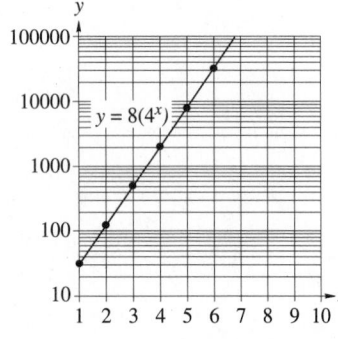

43.

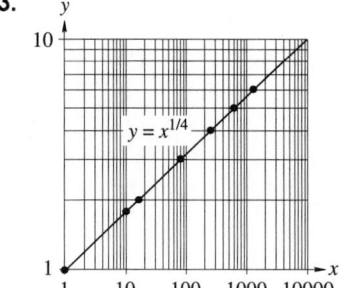

45.

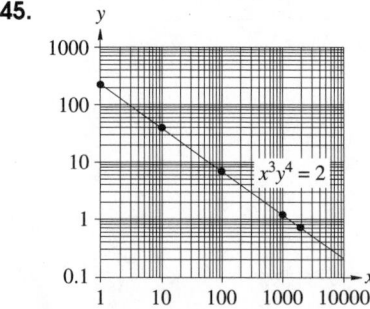

47. **(a)** \$2088.15 in the first bank (compounded semiannually);
(b) \$2112.06 in the second bank (compounded monthly);
(c) \$2117.00 in the third bank (compounded continuously);
(d) 9.24 years   **49.** **(a)** 10,000; **(b)** 156,250
51. $y = 0.4(6.2764)^x$ **53.** **(a)** $I = I_0 \cdot 10^{\frac{6-M}{2.5}}$;
(b) $10^{5.8} \approx 630{,}957.34$ times

Chapter 13 Test

1. 200.33681   **3.** 4.281 515 453
5. $\log 5 + \log 13 + \log x$

7. $\frac{1}{5}(\log 5 + 3\log x) - \log 9 - \log x$ **9.** $\log \dfrac{17}{x^2}$

11. $e^{17.2} \approx 29\,502\,925.92$ **13.** 3.236
15. $10^5 - 2 \approx 99\,998$

17.

x	1	2	3	4	5	6	7
$0.4(3^x)$	1.2	3.6	10.8	32.4	97.2	291.6	874.8

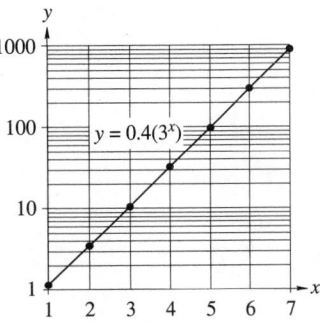

19. $7,200 \, \text{ft}^3/\text{acre}$ **21.** $92.07 \, \text{yr}$

ANSWERS FOR CHAPTER 14

Exercise Set 14.1

1. $5j$ **3.** $0.2j$ **5.** $5j\sqrt{3}$ **7.** $-6j\sqrt{5}$ **9.** $\frac{3}{4}j$
11. $-3j$ **13.** -11 **15.** -18 **17.** -10 **19.** -42
21. $-5j$ **23.** $-j$ **25.** $\frac{2}{3}j$ **27.** $\frac{3}{20}j$ **29.** $-\frac{5}{8}$
31. $0.5j$ **33.** $-2j\sqrt{8.1} = -6j\sqrt{0.9}$ **35.** $-2j\sqrt{15}$
37. $-14j$ **39.** $x = 7, y = -2$ **41.** $x = 10, y = -3$
43. $x = 2, y = -2$ **45.** $x = -\frac{1}{2}, y = -\frac{5}{4}$ **47.** $x = 3.5,$
$y = -4.3$ **49.** $8 + 10j$ **51.** $-10 - 5j$ **53.** $2 + 3j$
55. $-6 + 8j$ **57.** $3 - 4j$ **59.** $1 - 2j$ **61.** $2 + 3j\sqrt{2}$
63. $2 - \frac{1}{2}j\sqrt{6}$ **65.** $\frac{1}{2} - \frac{2}{9}j$ **67.** $\frac{5}{8} - \frac{2}{3}j$
69. $3.6 - 4.5j$ **71.** $7 - 2j$ **73.** $6 + 5j$ **75.** 19
77. $8j$ **79.** $\sqrt{2} - 7.3j$ **81.** $-\frac{1}{2} + \frac{3}{2}j, -\frac{1}{2} - \frac{3}{2}j$

83. $3j, -3j$ **85.** $-\frac{3}{4} + \frac{j\sqrt{47}}{4}, -\frac{3}{4} - \frac{j\sqrt{47}}{4}$

87. $-\frac{1}{5} \pm \frac{2j\sqrt{6}}{5}, -\frac{1}{10} - \frac{j\sqrt{6}}{5}$ **89.** $6.25 + 49.61j\,\Omega$ or

$6.25 - 49.61j\,\Omega$ **91.** $R = \pm\sqrt{\dfrac{X}{B} - X^2}$

Exercise Set 14.2

1. $-1 + 7j$ **3.** $5 - 2j$ **5.** $7 - j$
7. $2 - \sqrt{5} + 11j$ **9.** $8 + 2j$ **11.** $3 - 6j$ **13.** $7 - 2j$
15. $-3 + 6j$ **17.** $10 - 45j$ **19.** $7 + 11j$ **21.** $36 + 8j$
23. $-126 + 48j$ **25.** 9 **27.** $-3 + 4j$ **29.** $48 - 14j$
31. 29 **33.** $1 - 5j$ **35.** $-3j$ **37.** $2j$ **39.** $\frac{1}{13} + \frac{5}{13}j$
41. $-\sqrt{2} - j$ **43.** $4.059 - 0.765j$ **45.** -4

47. $-0.5 + \dfrac{j\sqrt{3}}{2}$ **49.** $(a+bj)+(a-bj) = 2a$, which is a

real number.   **51.** $(a+bj)(a-bj) = a^2 + b^2$, which is a
real number.   **53.** $10.53 - 4.21j$ V   **55.** $5.08 + 2.28j$ V
57. $5.84 + 1.16j\,\Omega$ **59.** $0.4 - 0.3j\,\Omega$
61. $0.74 + 1.12j$ A   **63.** See *Computer Programs*

65. See *Computer Programs*

Exercise Set 14.3

1. $2\sqrt{3} + 2j$

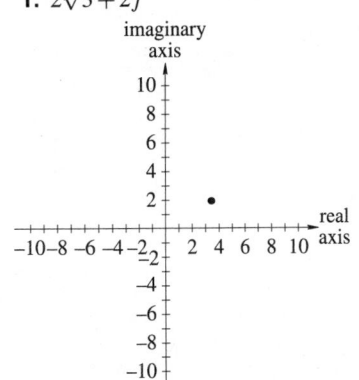

3. $-\dfrac{5\sqrt{2}}{2} + \dfrac{5j\sqrt{2}}{2} \approx -3.5355 + 3.5355j$

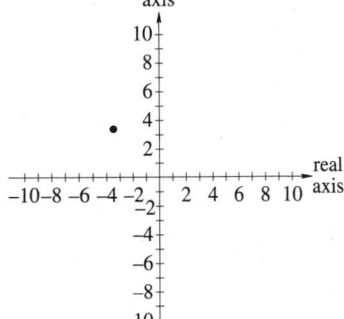

5. $-1.2155 - 6.8937\,j$

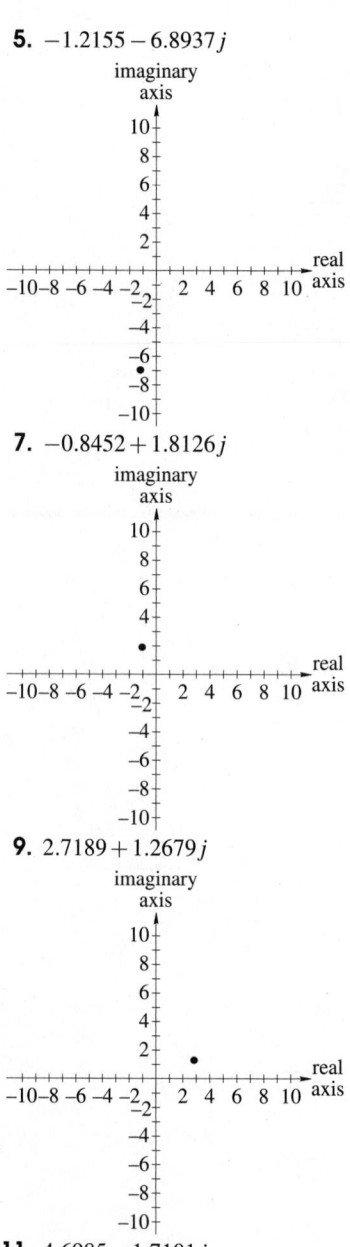

7. $-0.8452 + 1.8126\,j$

9. $2.7189 + 1.2679\,j$

11. $4.6985 - 1.7101\,j$

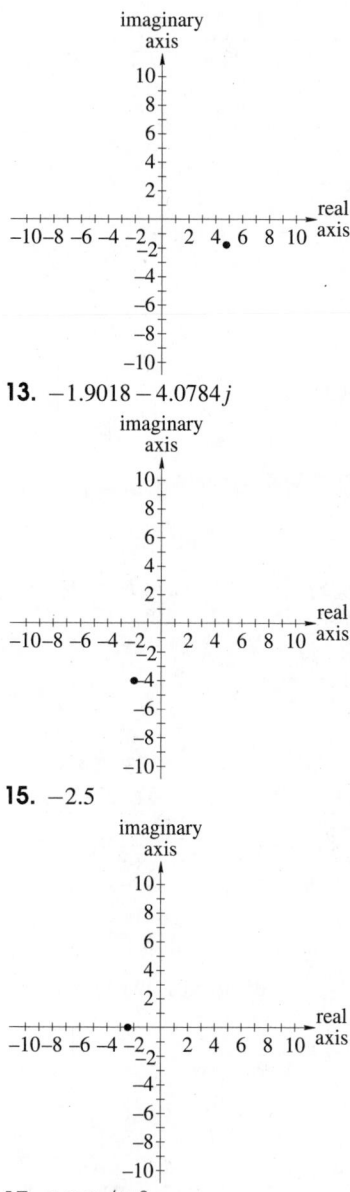

13. $-1.9018 - 4.0784\,j$

15. -2.5

17. $7.071\,\underline{/45^\circ}$

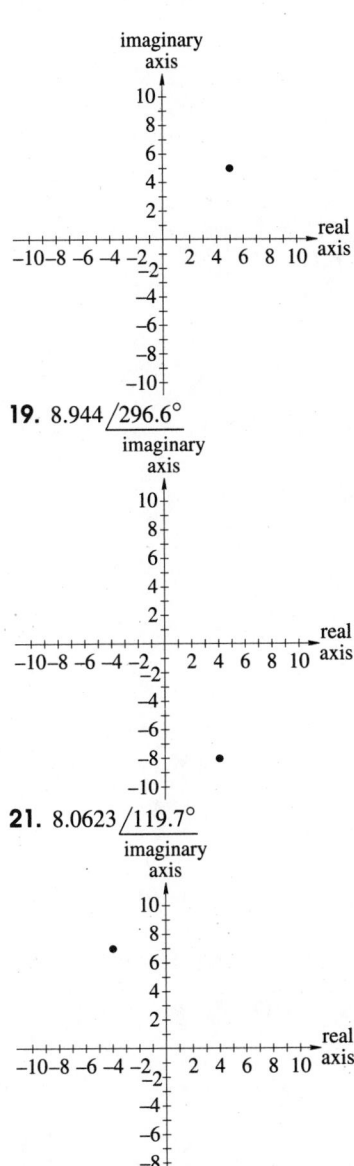

19. $8.944 \underline{/296.6°}$

21. $8.0623 \underline{/119.7°}$

23. $6.3246 \underline{/198.4°}$

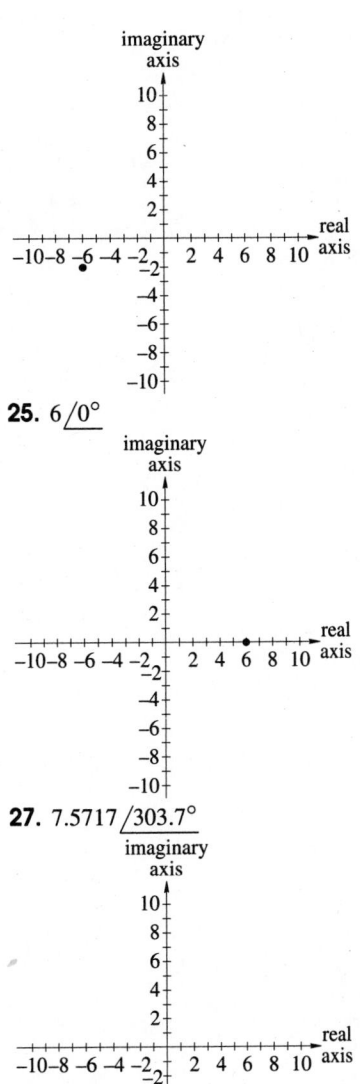

25. $6 \underline{/0°}$

27. $7.5717 \underline{/303.7°}$

29. $5.8034 \underline{/178.0°}$

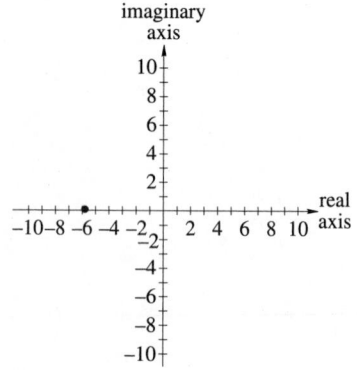

31. $2.7\underline{/180°}$

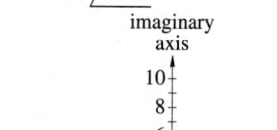

33.

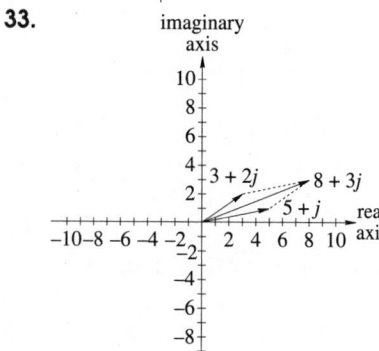

35.

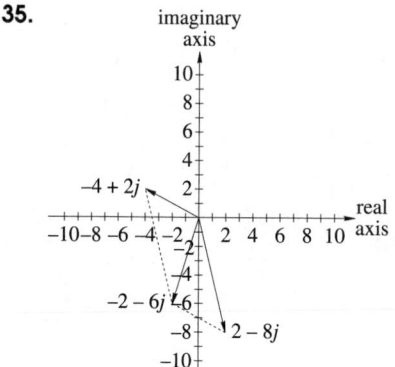

37.

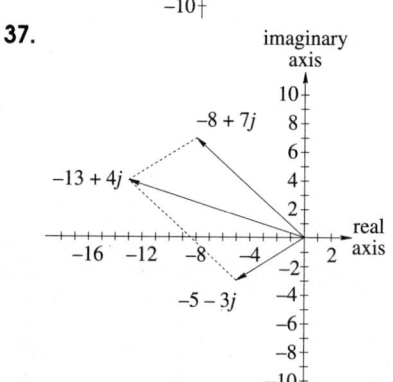

39.

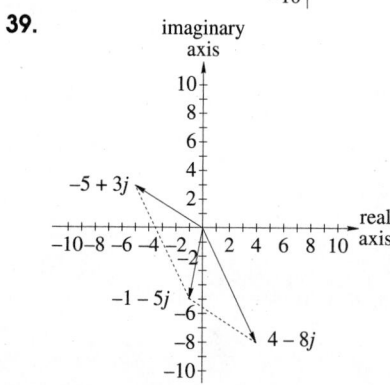

41. $8.0\underline{/-54.1°} = 8.0\underline{/305.9°}$ A   **43.** $119.6\underline{/157.3°}$ V
45. $1.61 - 1.92\,j$ A   **47.** See *Computer Programs*

Exercise Set 14.4

1. $3e^{3\pi/2\,j} \approx 3e^{4.7\,j}$ **3.** $2e^{1.05\,j}$ **5.** $1.3e^{5.7\,j}$
7. $3.1e^{0.44\,j}$ **9.** $10e^{0.64\,j}$ **11.** $15e^{2.21\,j}$
13. $5\,\text{cis}\,0.5 = 4.3879 + 2.3971\,j$
15. $2.3\,\text{cis}\,4.2 = -1.1276 - 2.0046\,j$ **17.** $12e^{5\,j}$
19. $2.4e^{5.9\,j}$ **21.** $28e^{3.72\,j}$ **23.** $4e^{2\,j}$ **25.** $4.25e^{1.5\,j}$
27. $81e^{8\,j}$ **29.** $39.0625e^{6\,j}$ **31.** $2e^{3\,j}$ **33.** $2.5e^{2.1\,j}$
35. $61.4e^{0.4031\,j}$V   **37.** In exponential form
$Z = 135e^{-0.9163\,j}\,\Omega$ or $135e^{5.37\,j}$; in rectangular form

$Z = 82.2 - 107.1j\,\Omega$ **39.** $V = 80e^{0.7029j}$ V
41. $I = 46e^{-0.7937j}$ A   **43.** $3.4e^{1.074j}\,\Omega$

Exercise Set 14.5

1. $15(\cos 69° + j \sin 69°)$ **3.** $10(\cos 4.1 + j \sin 4.1)$
5. $4(\cos 60° + j \sin 60°)$ **7.** $4.5\,\underline{/111°}$ **9.** $12\,\underline{/8.0}$
11. $15625(\cos 144° + j \sin 144°)$
13. $1124.864(\cos 10.26 + j \sin 10.26)$
15. $3.4^4\,\underline{/21.2} = 113.6336$ **17.** $12e^{3.8j}$ **19.** $0.125e^{0.9j}$
21. $1, -0.5 + 0.866j, -0.5 - 0.866j$ **23.** $1.732 - j, 2j,$
$-1.732 - j$ **25.** $1.8478 - 0.7654j, 0.7654 + 1.8478j,$
$-1.8478 + 0.7654j, -0.7654 - 1.8478j$
27. $1.0586 + 0.1677j, 0.1677 + 1.0586j, -0.9550 + 0.5866j,$
$-0.7579 - 0.7579j, 0, 4866 - 0.9550j$ **29.** $0.866 - 0.5j,$
$j, -0.866 - 0.5j$ **31.** $1.9319 + 0.5176j, 0.5176 + 1.9319j,$
$-1.4142 + 1.4142j, -1.9319 - 0.5176j, -0.5176 - 1.9319j,$
$1.4142 - 1.4142j$ **33.** $108\,\underline{/19°}$ **35.** $80\,\underline{/-36.87°}$ V
37. $Z = -0.0448 + 0.1536j = 0.16\,\underline{/106.3°}\,\Omega$
39. **(a)** $0.2137\,\underline{/-20.56°}$ S, **(b)** about $0.200 - 0.075j$ S
41. See *Computer Programs*   **43.** See *Computer Programs*

Exercise Set 14.6

1. **(a)** $3 - 2j$, **(b)** $5 + j$ **3.** **(a)** $1 + 2j$, **(b)** $1.8 - 0.6j$
5. **(a)** $6.803\,\underline{/53.26°}$, **(b)** $1.6611\,\underline{/32.44°}$
7. **(a)** $8.208\,\underline{/0.449}$, **(b)** $1.226\,\underline{/0.45}$ **9.** **(a)** $12 + 5j$,
(b) $1.9207 + 0.2387j$ **11.** **(a)** $6.595\,\underline{/-2.95°}$,
(b) $0.916\,\underline{/29.32°}$ **13.** $-13 - 84j$ V   **15.** 2 V
17. $2.8\,\underline{/23.7°}\,\Omega$ **19.** $X_L = 75.40\,\Omega, X_C = 66.31\,\Omega,$
$Z = 39.07\,\Omega, \phi = 13.45°$ **21.** $X_L = 150.80\,\Omega,$
$X_C = 44.21\,\Omega, Z = 108.45\,\Omega, \phi = 79.37°$
23. $X_L = 12.5\,\Omega, X_C = 100\,\Omega, Z = 91.87\,\Omega, \phi = -72.26°$
25. **(a)** $77.62\,\underline{/14.93°}$, **(b)** 271.67 V   **27.** **(a)** $\sqrt{10}\,\underline{/18.43°}$,
(b) 9.01 V   **29.** $2 - 9j$ **31.** 22.51 Hz
33. $0.16 - 0.12j$ **35.** See *Computer Programs*

Review Exercises

1. $7j$ **3.** $3\sqrt{6}j$ or $3j\sqrt{6}$ **5.** -6 **7.** $9 - 3j$
9. $-1 + 4j$ **11.** $-36 + 8j$ **13.** 25
15. $10.82\,\underline{/-33.69°}$

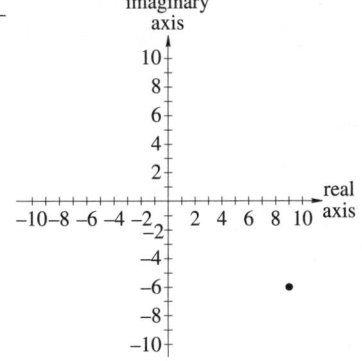

17. $5.657\,\underline{/-45°}$

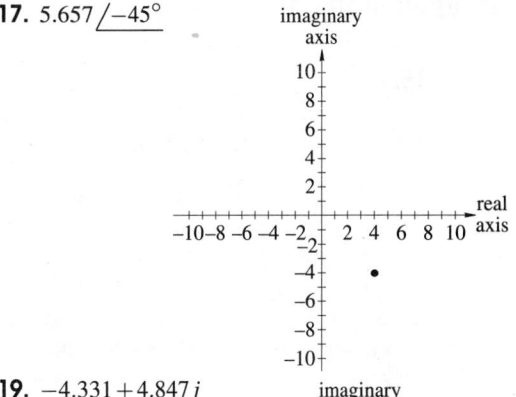

19. $-4.331 + 4.847j$

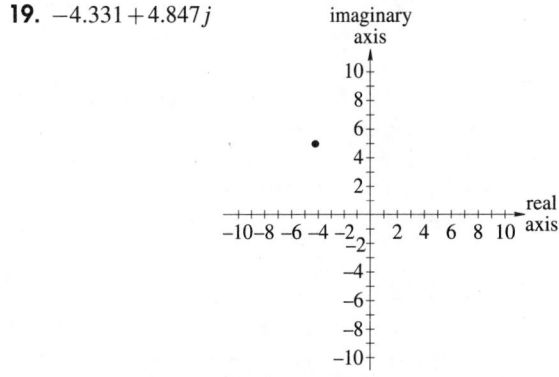

21. -10 **23.** $-4,782,969j$ **25.** $-26.080 + 6.988j$
27. 4096 **29.** $0.866 - 0.5j, j, -0.866 - 0.5j$
31. $2 + 3.464j, -2 - 3.464j$ **33.** $18.385e^{0.391j}$
35. $6740.6e^{2.702j}$ **37.** magnitude: 7.433, direction:
$-10.86°$ **39.** $X_L = 113.10\,\Omega, X_C = 53.05\,\Omega, Z = 81.43,$
$\phi = 47.51°$

Chapter 14 Test

1. $4j\sqrt{5}$ **3.** $-4\sqrt{3} + 4j$ **5.** $-13 + 8j$
7. $-\frac{38}{25} - \frac{9}{25}j = -1.52 + 0.36j$ **9.** $\frac{4}{3}j$
11. $2.1941 + 2.0460j$ **13.** **(a)** $9 - j\,\Omega,$
(b) $\frac{118}{41} - \frac{4}{41}j\,\Omega \cong 2.878 + 0.098j$ **15.** $0.28 + 1.04j\,\Omega$

▤ ANSWERS FOR CHAPTER 15

Exercise Set 15.1

1. $\sqrt{194}$ **3.** $\sqrt{65}$ **5.** $\sqrt{277}$ **7.** 7
9. $\left(\frac{9}{2}, -\frac{5}{2}\right)$ **11.** $\left(1, -\frac{11}{2}\right)$ **13.** $\left(\frac{15}{2}, -6\right)$
15. $\left(\frac{3}{2}, 5\right)$ **17.** $-\frac{13}{5} = -2.6$ **19.** $-\frac{1}{8}$ **21.** $\frac{14}{9}$
23. 0   **25.** $y - 4 = -2.6(x - 2)$ or $y = -2.6x + 9.2$
27. $y + 6 = -\frac{1}{8}(x - 5)$ or $y = -\frac{1}{8}x - 5\frac{3}{8}$
29. $y - 1 = -\frac{14}{9}(x - 12)$ or $y = \frac{14}{9}x - 17\frac{2}{3}$ **31.** $y = 5$
33. 1   **35.** approximately 0.1511352   **37.** 68.19859°
39. 153.43495°   **41.** $-\frac{1}{3}$ **43.** 2   **45.** $y + 5 = 6(x - 2)$
or $y = 6x - 17$ **47.** $y + 2 = -\frac{5}{2}(x - 4)$ or $y = -\frac{5}{2}x + 8$
49. $y + 4 = 1.732(x + 2)$ or $y = 1.732x - 0.536$
51. $y = 0.364x + 3$ **53.** $2x - 3y + 11 = 0$
55. $5x - 2y = 22$ **57.** $m = -\frac{3}{2}$, y-intercept $= 6$,
x-intercept $= 4$ **59.** $m = \frac{1}{3}$, y-intercept $= -3$,
x-intercept $= 9$ **61.** $v = 2.6 + 0.4t$ **63.** $0.8\,\Omega$
65. **(a)** $C = 1,225 + 1.25n$, **(b)** \$26,225
67. **(a)** 362.08 ppm, **(b)** 441.28 ppm, **(c)** Answer varies with
year.

Exercise Set 15.2

1. $(x - 2)^2 + (y - 5)^2 = 9$; $x^2 + y^2 - 4x - 10y + 20 = 0$
3. $(x + 2)^2 + y^2 = 16$; $x^2 + y^2 + 4x - 12 = 0$
5. $(x + 5)^2 + (y + 1)^2 = \dfrac{25}{4}$;
$x^2 + y^2 + 10x + 2y + 19.75 = 0$
7. $(x - 2)^2 + (y + 4)^2 = 1$; $x^2 + y^2 - 4x + 8y + 19 = 0$
9. $C = (3, 4), r = 3$

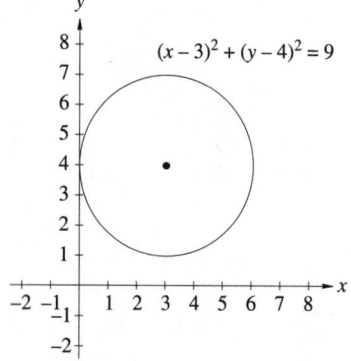

11. $C = \left(-\frac{1}{2}, -\frac{13}{4}\right), r = \sqrt{7}$

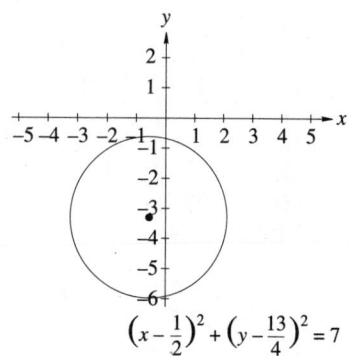

$$\left(x - \frac{1}{2}\right)^2 + \left(y - \frac{13}{4}\right)^2 = 7$$

13. $C = \left(0, \frac{7}{3}\right), r = \sqrt{6}$

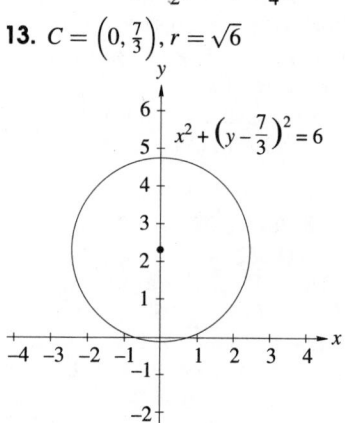

15. Circle, $C = (-2, 3), r = 3$ **17.** Circle, $C = (-5, 3)$,
$r = 9$ **19.** Not a circle since $r^2 = -1$ **21.** Circle,
$C = (-3, 0), r = 5$ **23.** Circle, $C = \left(-\frac{5}{2}, \frac{9}{2}\right), r = 6$
25. Not a circle since $r^2 = -1.3$ **27.** $x^2 + y^2 = 310.03^2$
(assume circle is at the center of a coordinate system.)
29. $\left(x - 1.495 \times 10^8\right)^2 + y^2 = 1.478 \times 10^{11}$;
$x^2 + y^2 + 4x - 12 = 0$ **31.** $x^2 + (y - 0.9)^2 = 0.55^2$
33. **(a)** $x^2 + (y - 5)^2 = 13^2$; **(b)** 28 cm

Exercise Set 15.3

1. $F = (0, 1)$, directrix: $y = -1$, opens upward

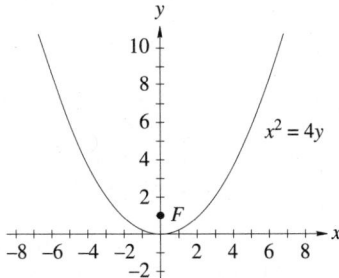

3. $F = (-1, 0)$, directrix: $x = 1$, opens left

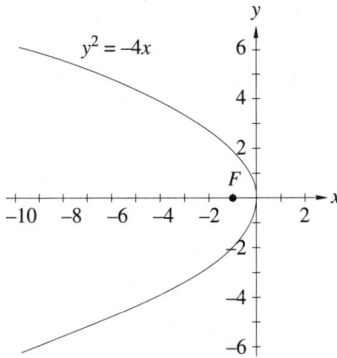

5. $F = (0, -2)$, directrix: $y = 2$, opens downward

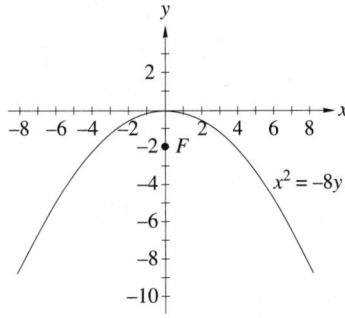

7. $F = \left(\frac{5}{2}, 0\right)$, directrix: $x = -\frac{5}{2}$, opens right

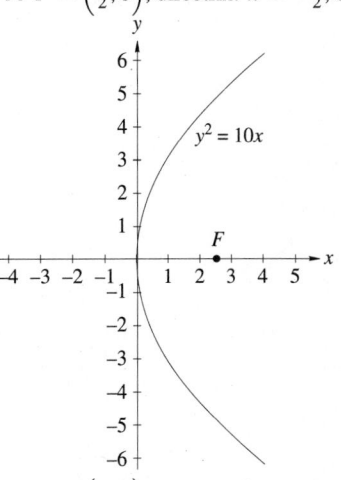

9. $F = \left(0, \frac{1}{2}\right)$, directrix: $x = -\frac{1}{2}$, opens upward

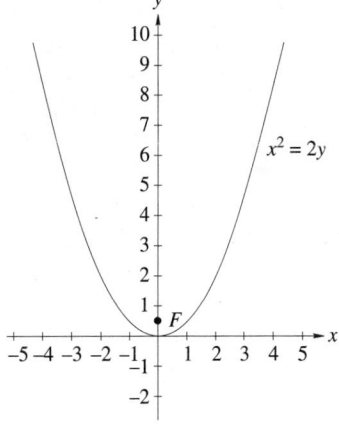

11. $F = \left(-\dfrac{21}{4}, 0\right)$, directrix: $x = -\dfrac{21}{4}$, opens left

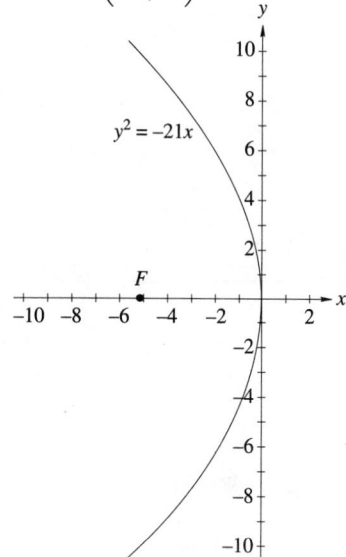

13. $x^2 = 16y$ **15.** $y^2 = -24x$ **17.** $y^2 = 8x$

19. $y^2 = 6x$ **21.** 27.78 m **23.** $\dfrac{6.25}{6} \approx 1.042$ m from vertex **25.**

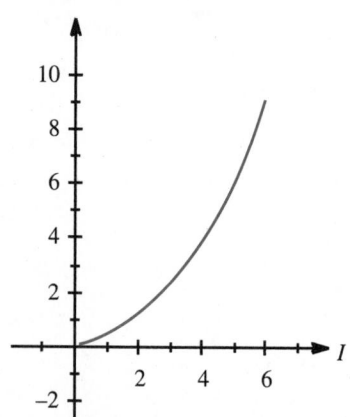

27. 3.125 ft

Exercise Set 15.4

1. $\dfrac{x^2}{36} + \dfrac{y^2}{20} = 1$ **3.** $\dfrac{x^2}{12} + \dfrac{y^2}{16} = 1$ **5.** $\dfrac{x^2}{25} + \dfrac{y^2}{16} = 1$

7. $\dfrac{x^2}{16} + \dfrac{y^2}{9} = 1$ **9.** $V = (0, 3)$, $V' = (0, -3)$,
$F = (0, \sqrt{5})$, $F' = (0, -\sqrt{5})$, $M = (2, 0)$, $M' = (-2, 0)$

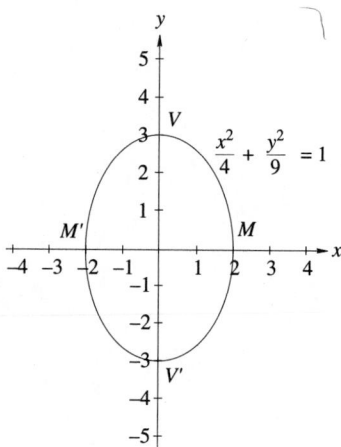

11. $V = (2, 0)$, $V' = (-2, 0)$, $F = (\sqrt{3}, 0)$, $F' = (-\sqrt{3}, 0)$, $M = (0, 1)$, $M' = (0, -1)$

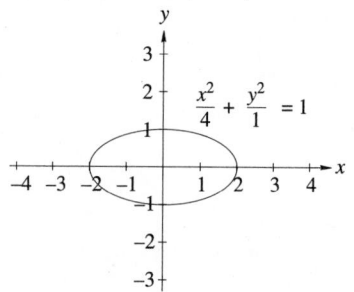

13. $V = (0, 2)$, $V' = (0, -2)$, $F = (0, \sqrt{3})$, $F' = (0, -\sqrt{3})$, $M = (1, 0)$, $M' = (-1, 0)$

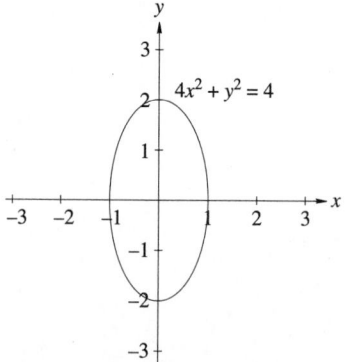

15. $V = (6, 0)$, $V' = (-6, 0)$, $F = (\sqrt{11}, 0)$, $F' = (-\sqrt{11}, 0)$, $M = (0, 5)$, $M' = (0, -5)$

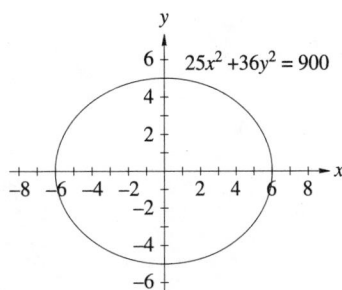

17. 2.992×10^8 km   **19.** major axis $= \sqrt{200} \approx 14.14$; minor axis $= 10$ **21.** 0.9999253 **23.** $8\sqrt{2} \approx 11.3$ in.

Exercise Set 15.5

1. $\dfrac{x^2}{16} - \dfrac{y^2}{20} = 1$ **3.** $\dfrac{y^2}{9} - \dfrac{x^2}{16} = 1$ **5.** $\dfrac{x^2}{9} - \dfrac{y^2}{16} = 1$

7. $\dfrac{x^2}{7} - \dfrac{y^2}{9} = 1$ **9.** $V = (2,0), V' = (-2,0), F = (\sqrt{13},0), F' = (-\sqrt{13},0), M = (0,3), M' = (0,-3)$

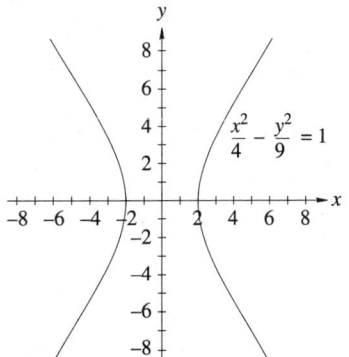

11. $V = (0,2), V' = (0,-2), F = (0,\sqrt{5}), F' = (0,-\sqrt{5}), M = (1,0), M' = (-1,0)$

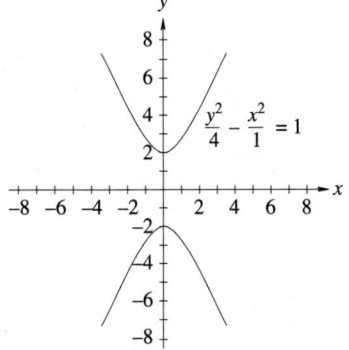

13. $V = (1,0), V' = (-1,0), F = (\sqrt{5},0), F' = (-\sqrt{5},0), M = (0,2), M' = (0,-2)$

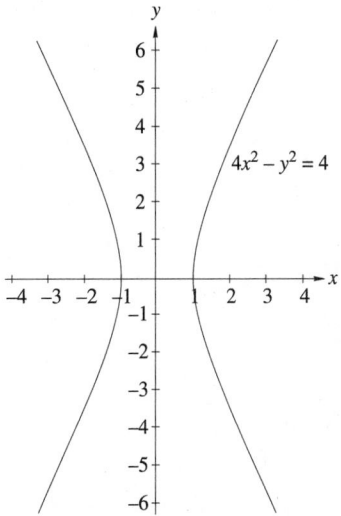

15. $V = (0,5), V' = (0,-5), F = (0,\sqrt{61}), F' = (0,-\sqrt{61}), M = (6,0), M' = (-6,0)$

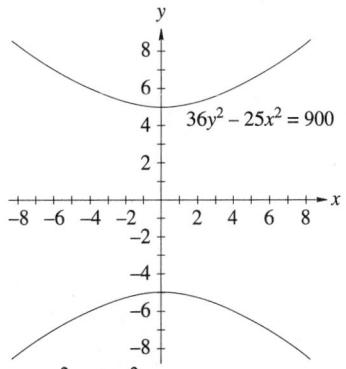

17. $\dfrac{x^2}{49} - \dfrac{y^2}{61.25} = 1$ **19.** $\sqrt{2}$

21.

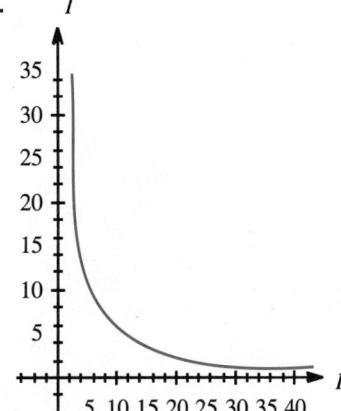

23. $\dfrac{x^2}{1764} - \dfrac{y^2}{337.5} = 1$

25.

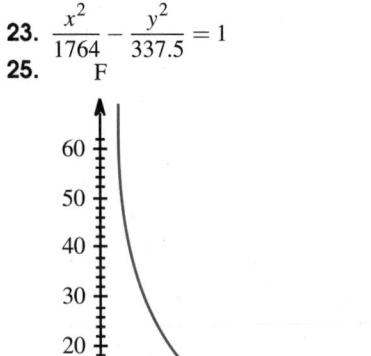

Exercise Set 15.6

1. ellipse;

	$x'y'$-system	xy-system
center	$(0', 0')$	$(4, -3)$
vertices	$(\pm 3', 0')$	$(7, -3), (1, -3)$
foci	$(\pm\sqrt{5}', 0')$	$(\sqrt{5}+4, -3),$ $(-\sqrt{5}+4, -3)$
endpoints of minor axis	$(0', \pm 2')$	$(4, -1), (4, -5)$

$$\dfrac{(x-4)^2}{9} + \dfrac{(y+3)^2}{4} = 1$$

3. hyperbola;

	$x'y'$-system	xy-system
center	$(0', 0')$	$(-4, 3)$
vertices	$(0', \pm 6)$	$(-4, 9), (-4, -3)$
foci	$(0', \pm\sqrt{37}')$	$(-4, \sqrt{37}+3),$ $(-4, -\sqrt{37}+3)$
endpoints of conjugate axis	$(\pm 2', 0')$	$(-3, 3), (-5, 3)$

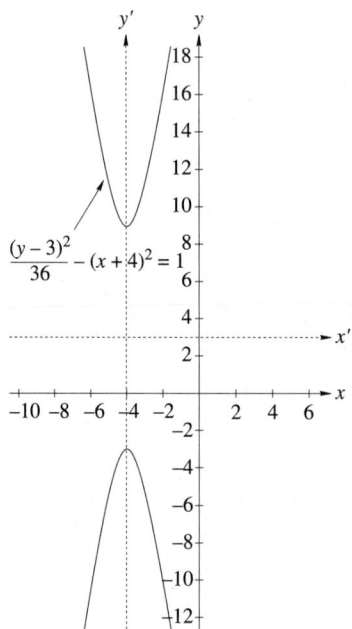

$$\dfrac{(y-3)^2}{36} - (x+4)^2 = 1$$

5. hyperbola;

	$x'y'$-system	xy-system
center	$(0', 0')$	$(-5, 3)$
vertices	$(\pm 2', 0')$	$(3, 0),$ $(-7, 0)$
foci	$(2 \pm \sqrt{26}', 0')$	$(2\sqrt{26}-5, 0),$ $(-2\sqrt{26}-5, 0)$
endpoints of conjugate axis	$(0', \pm 10')$	$(-5, 10),$ $(-5, -10)$

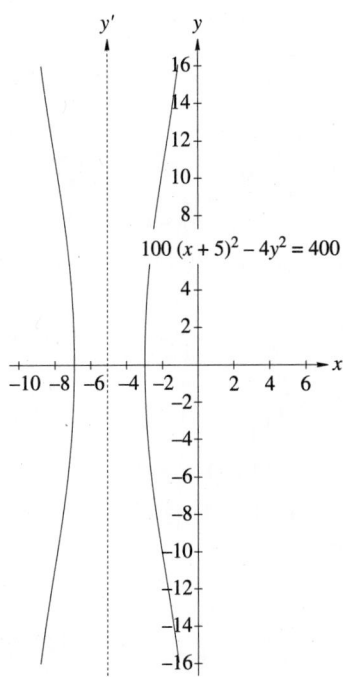

$$100\,(x+5)^2 - 4y^2 = 400$$

7. ellipse;

	$x'y'$-system	xy-system
center	$(0',0')$	$(-2,\frac{3}{2})$
vertices	$(0',\pm 3')$	$(-2,\frac{7}{2})$, $(-2,-\frac{1}{2})$
foci	$(0',\pm\sqrt{3}')$	$(-2,\sqrt{3}+\frac{3}{2})$, $(-2,-\sqrt{3}+\frac{3}{2})$
endpoints of minor axis	$(\pm 1',0')$	$(-1,\frac{3}{2})$, $(-3,\frac{3}{2})$

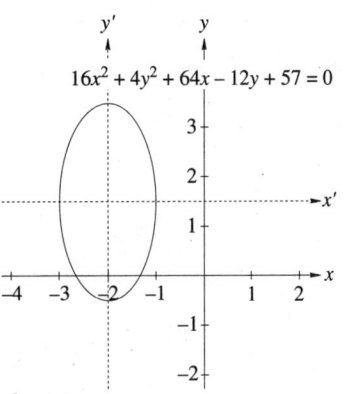

$$16x^2 + 4y^2 + 64x - 12y + 57 = 0$$

9. ellipse;

	$x'y'$-system	xy-system
center	$(0',0')$	$(5,2)$
vertices	$(0',\pm 5')$	$(5,7)$, $(5,-3)$
foci	$(0',\pm\sqrt{21}')$	$(5,2+\sqrt{21})$, $(5,2-\sqrt{21})$
endpoints of minor axis	$(\pm 2',0')$	$(7,2)$, $(3,2)$

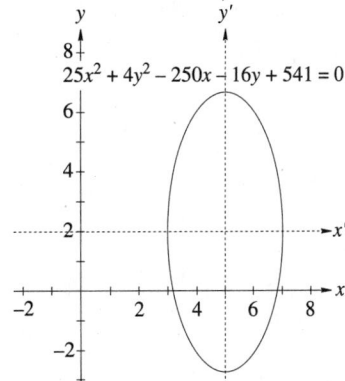

$$25x^2 + 4y^2 - 250x - 16y + 541 = 0$$

11. hyperbola;

	$x'y'$-system	xy-system
center	$(0',0')$	$(4,2)$
vertices	$(\pm\sqrt{2}',0')$	$(4+\sqrt{2},2)$, $(4-\sqrt{2},2)$
foci	$(\pm\sqrt{6}',0')$	$(4+\sqrt{2},2)$, $(4-\sqrt{2},2)$
endpoints of conjugate axis	$(0',\pm 2')$	$(4,0)$, $(4,4)$

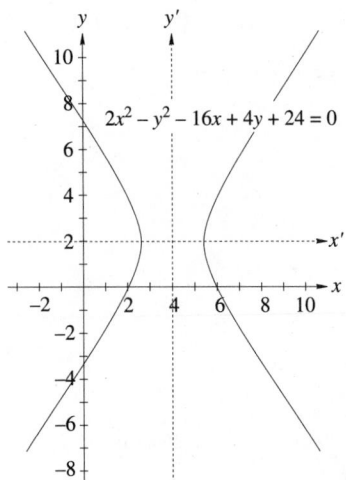

$2x^2 - y^2 - 16x + 4y + 24 = 0$

13. $(x-2)^2 = 32(y+3)$ **15.** $\dfrac{(x-4)^2}{36} + \dfrac{(y+3)^2}{20} = 1$

17. $\dfrac{(y-2)^2}{9} - \dfrac{(x+3)^2}{16} = 1$ **19.** $(y-1)^2 = -16(x+5)$

21. $\dfrac{(x+4)^2}{36} + \dfrac{(y-1)^2}{100} = 1$ **23.** This is a parabola with the equation $-4.9(t-3)^2 = s - 44.1$. Maximum height is

when $t = 3$ and $s = 44.1$ m.

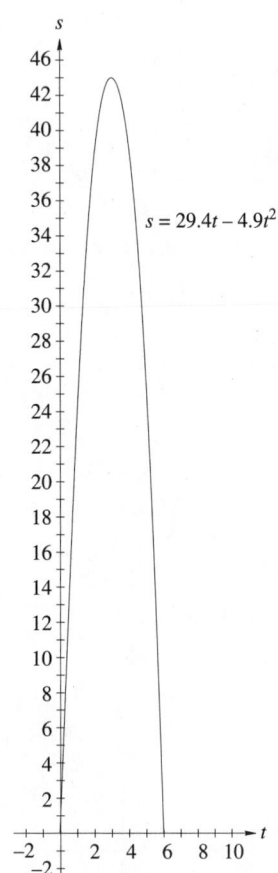

$s = 29.4t - 4.9t^2$

25. If the transverse axis passes through A and B and the conjugate axis passes through the midpoint of $\overline{AB}$, then A is at $(-500, 0)$ and B at $(500, 0)$. The plane is at $(216, 1073.054)$ and lies on the hyperbola $\dfrac{x^2}{8100} - \dfrac{y^2}{241\,900} = 1$

27. (a) $x^2 = -200(y - 0.02)$; (b) 2.00 cm

29. (a) $24(y-125)^2 - x^2 = 15{,}000$, upper branch;

(b) $\left(\dfrac{1{,}200 + 120\sqrt{5}}{19}, \dfrac{2{,}400 + 240\sqrt{5}}{19} \right) \approx (77.3, 154, 6)$

31. (a) $y = 2\sqrt{0.0625 - 0.25x^2}$ and $y = -2.5\sqrt{0.1 - 0.4x^2}$,

(b)

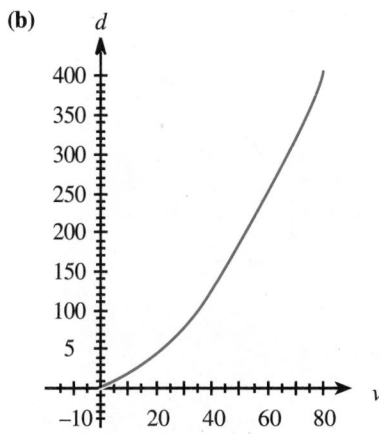

33. (a) $c = 1.14x - 0.01x^2$,

(b)

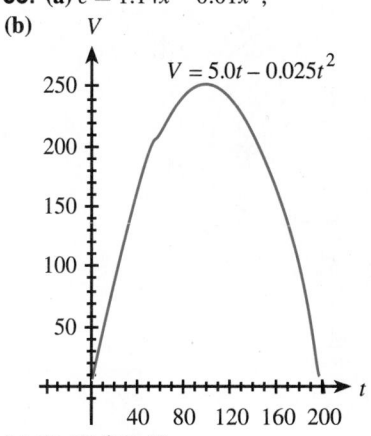

(c) 57, **(d)** $32.49

Exercise Set 15.7

1. (a) discriminant is 1; curve is a hyperbola; **(b)** 45°;

(d) $\dfrac{y''^2}{18} - \dfrac{x''^2}{18} = 1$

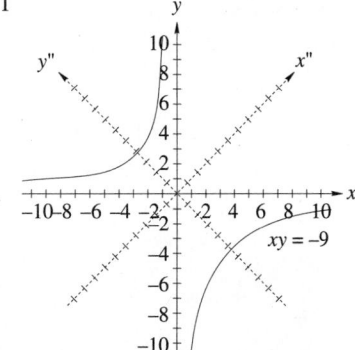

3. (a) discriminant is 32; curve is a hyperbola; **(b)** 45°;

(d) $\dfrac{y''^2}{2} - \dfrac{x''^2}{4} = 1$

(e)

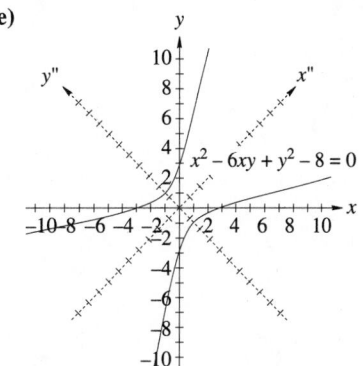

5. (a) discriminant is $-10,000$; curve is an ellipse;

(b) 36.870°; **(d)** $\dfrac{x''^2}{4} + y''^2 = 1$

7. (a) discriminant is 0; curve is a parabola; **(b)** 45°;

(d) $y''^2 = -\dfrac{\sqrt{2}}{2}x''$

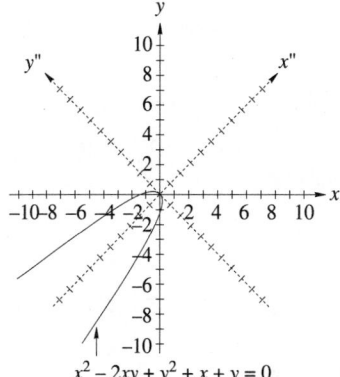

9. (a) discriminant is 168; curve is a hyperbola; **(b)** 33.690°;

(d) $\dfrac{x''^2}{7} - \dfrac{y''^2}{6} = 1$

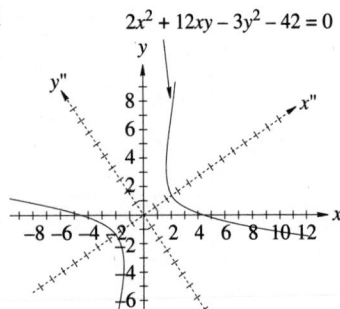

$2x^2 + 12xy - 3y^2 - 42 = 0$

11. (a) discriminant is -119; curve is an ellipse; **(b)** $45°$;
(d) $7x''^2 + 17y''^2 = 52$

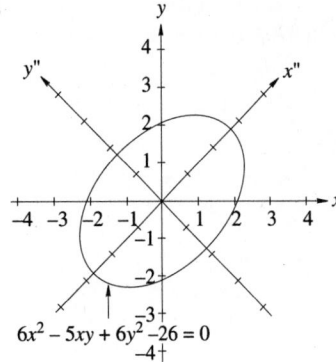

$6x^2 - 5xy + 6y^2 - 26 = 0$

13.

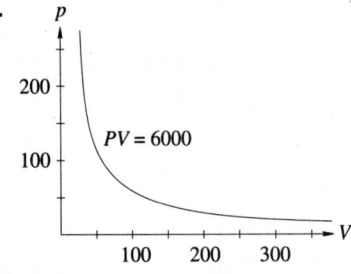

$PV = 6000$

15. Parabola

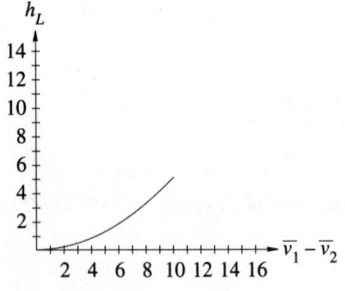

Exercise Set 15.8

1. hyperbola

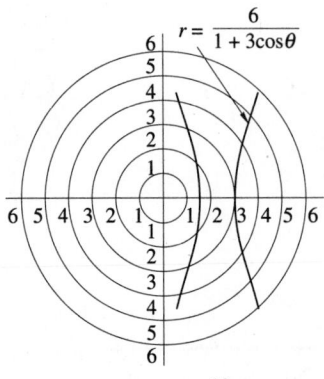

$r = \dfrac{6}{1 + 3\cos\theta}$

3. ellipse

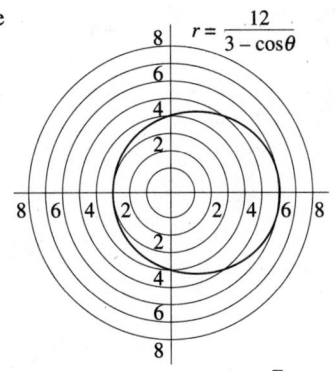

$r = \dfrac{12}{3 - \cos\theta}$

5. parabola

7. ellipse

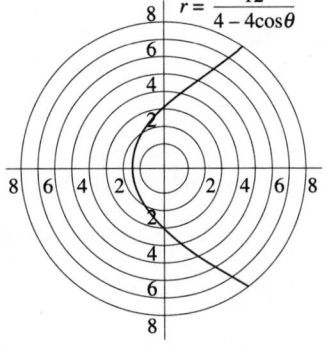

$r = \dfrac{12}{4 - 4\cos\theta}$

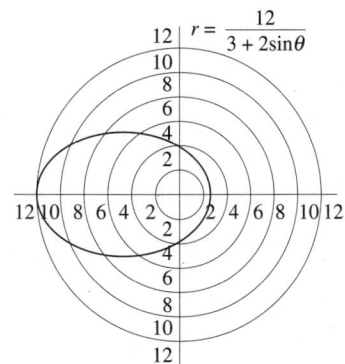

$r = \dfrac{12}{3 + 2\sin\theta}$

9. hyperbola rotated

$\frac{\pi}{3}$ radians

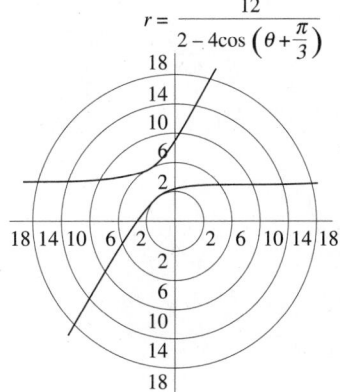

$r = \dfrac{12}{2 - 4\cos\left(\theta + \frac{\pi}{3}\right)}$

11. $r = \dfrac{6}{1 - \frac{3}{2}\cos\theta} = \dfrac{12}{2 - 3\cos\theta}$ **13.** $r = \dfrac{5}{1 - \sin\theta}$

15. $r = \dfrac{\frac{2}{3}}{1 - \frac{2}{3}\cos\theta} = \dfrac{2}{3 - 2\cos\theta}$ **17.** Greatest distance:

4.335×10^7 miles; shortest distance: 2.854×10^7 miles

19.

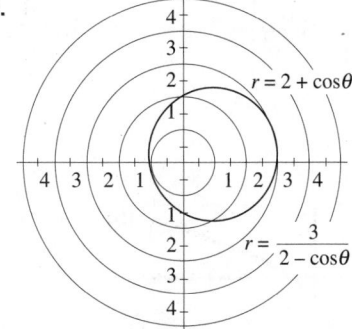

$r = 2 + \cos\theta$

$r = \dfrac{3}{2 - \cos\theta}$

Review Exercises

1. (a) 5; **(b)** $\left(\frac{1}{2}, 7\right)$; **(c)** $-\frac{4}{3}$; **(d)** $y - 5 = -\frac{4}{3}(x - 2)$ or $4x + 3y = 23$ **3. (a)** $\sqrt{104} = 2\sqrt{26}$; **(b)** $(2, 1)$; **(c)** 5; **(d)** $y + 4 = 5(x - 1)$ or $y - 5x = -10$ **5.** line in #1: $\frac{3}{4}$;

line in #2: $\dfrac{12}{5}$; line in #3: $-\frac{1}{5}$; line in #4: 1

7. $y - 5 = -2(x + 3)$ or $y + 2x = -1$

9.

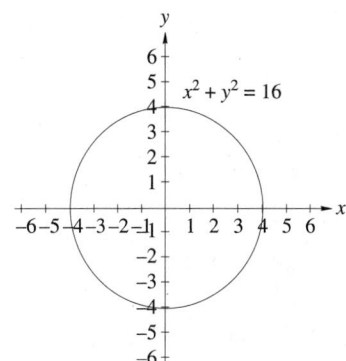

$x^2 + y^2 = 16$

11.

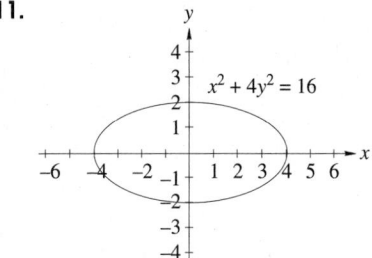

$x^2 + 4y^2 = 16$

13.

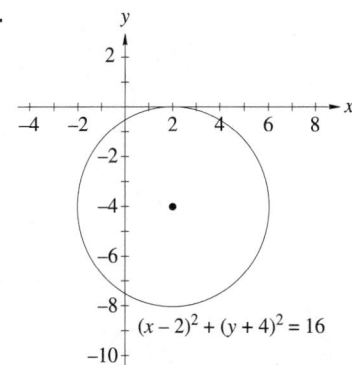

$(x - 2)^2 + (y + 4)^2 = 16$

15.

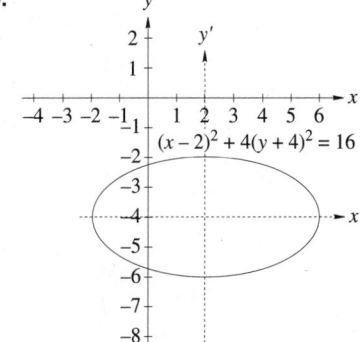

$(x - 2)^2 + 4(y + 4)^2 = 16$

17.

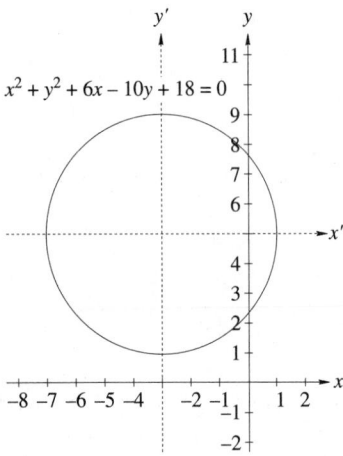

$x^2 + y^2 + 6x - 10y + 18 = 0$

19.

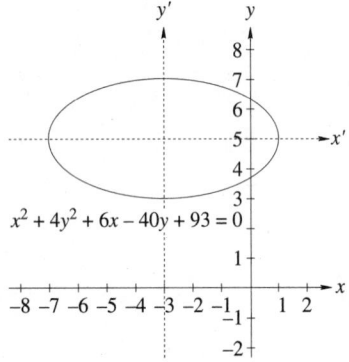

$x^2 + 4y^2 + 6x - 40y + 93 = 0$

21. $2x^2 + 12xy - 3y^2 - 42 = 0$

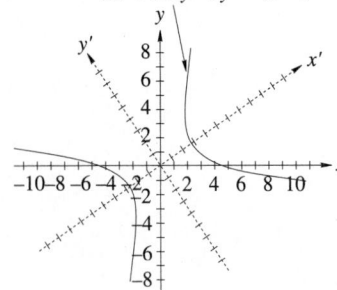

23. $3x^2 + 2\sqrt{3}\,xy + y^2 + 8x - 8\sqrt{3}\,y = 32$

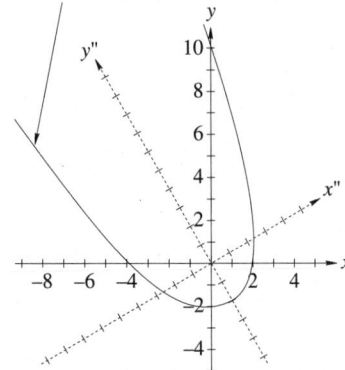

25. $r = \dfrac{16}{5 - 3\cos\theta}$

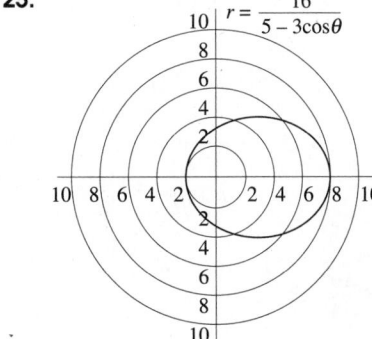

27. $r = \dfrac{9}{3 + 3\sin\theta}$

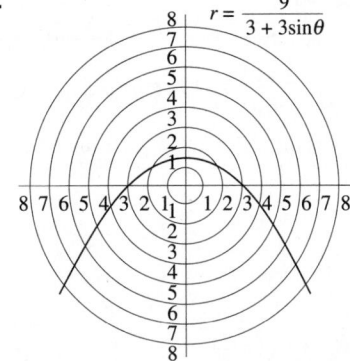

29. (a)

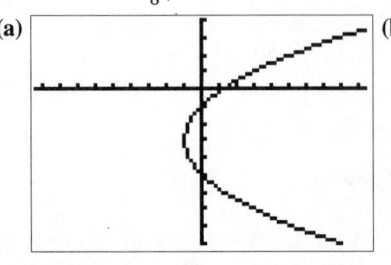

(b) Parabola

$[-9.4, 9.4] \times [-9, 4]$

(c) Horizontal axis, opens to the right, vertex: $(-1, -3)$, directrix: $x = -2$, and focus: $(0, -3)$.

31. $\dfrac{x^2}{50^2} + \dfrac{y^2}{195.84} = 1$ **33.** If the x-axis passes through A and B and the y-axis through the midpoint of $\overline{AB}$, then B has the coordinates $(200, 0)$ and $A = (-200, 0)$. The plane is at $(91.13, -50)$. The plane lies on the hyperbola

$$\dfrac{x^2}{6400} - \dfrac{y^2}{33\,600} = 1.$$

Chapter 15 Test

1. Focus $(-4.5, 0)$; directrix $x = 4.5$

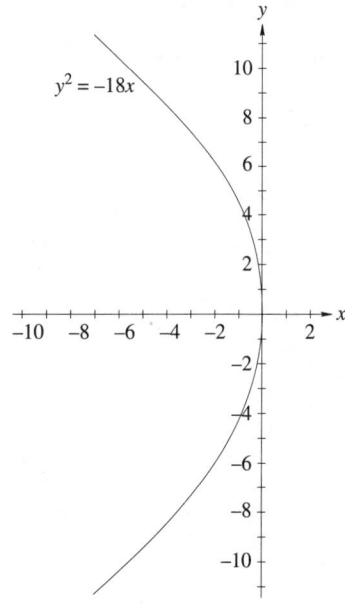

3. (a)

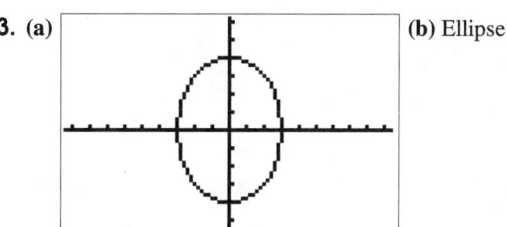

$[-9.4, 9.4] \times [-6.2, 6.2]$

(b) Ellipse

(c) Center: $(0, 0)$, Vertices: $(-3, 0)$ and $(3, 0)$, Foci: $(-\sqrt{7}, 0)$ and $(\sqrt{7}, 0)$; endpoints of minor axis: $(0, 4)$ and $(0, -4)$

5. $y + \dfrac{13}{2} = -\dfrac{2}{3}(x - 3)$ or $6y + 4x + 27 = 0$

7. (a)

(b) Hyperbola

$[-9.4, 9.4] \times [-6.2, 6.2]$

(c) Center: $(-1, 2)$, Vertices: $(-4, 2)$ and $(2, 2)$, Foci: $(-1 - \sqrt{13}, 2)$ and $(-1 + \sqrt{13}, 2)$; endpoints of minor axis: $(-1, 4)$ and $(-1, 0)$ **9. (a)** $\frac{\pi}{4}$; **(b)** an ellipse   **11.** 75 ft above its lowest point.

ANSWERS FOR CHAPTER 16

Exercise Set 16.1

1. $(7, 1)$

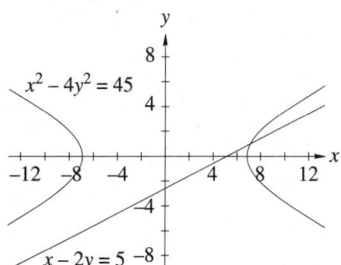

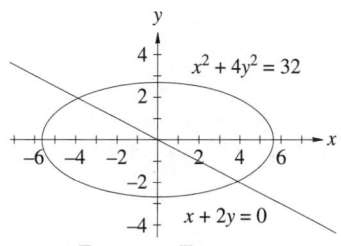

5. $(\sqrt{3}, 1), (-\sqrt{3}, 1)$

3. $(4, -2), (-4, 2)$

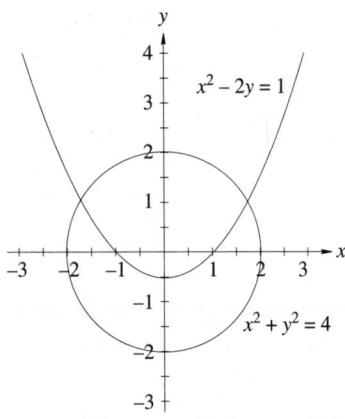

7. $(1, \sqrt{6}), (1, -\sqrt{6}), (-7, j\sqrt{42}), (-7, -j\sqrt{42})$

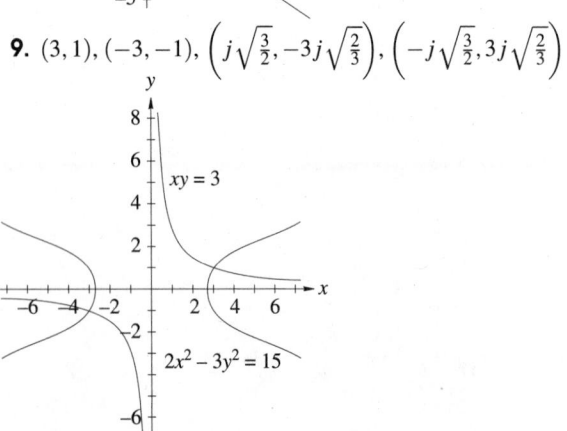

9. $(3,1), (-3,-1), \left(j\sqrt{\frac{3}{2}}, -3j\sqrt{\frac{2}{3}}\right), \left(-j\sqrt{\frac{3}{2}}, 3j\sqrt{\frac{2}{3}}\right)$

11. 42.08 mph for the first part of the trip; 32.08 mph for the last part. **13.** One is $40\,\Omega$ and the other is $60\,\Omega$
15. Either $v_1 = 3\,\text{m/s}$ and $v_2 = 5.5\,\text{m/s}$ or $v_1 = 6\,\text{m/s}$ and $v_2 = 3.5\,\text{m/s}$ **17.** $F = 67.87; \ell = 2.36$

19. $x = 10.16\,\text{mm}, y = 4.57\,\text{mm}$ **21.** $L = 1.336v^2 x$
23. $g = 32, n = 13$

Exercise Set 16.2

1. $x + 5 < 15$ **3.** $3x < 30$ **5.** $\dfrac{-x}{2} > -5$
7. $x^2 < 100$ **9.** $x > 2$

11. $x > -6$

13. $x < 5$

15. $x \geq 6$

17. $x < 18$

19. $x \leq 3\frac{1}{2}$

21. $x \geq -\frac{9}{5}$

23. $-6 < x < 4$

25. $x < -10$ or $x > 2$

27. This is a contradictory inequality. **29.** $-4 \leq x < 7$

31. This is a contradictory inequality **33.** $6357\,\text{km}$ $\leq r \leq 6378\,\text{km}$ **35.** $+0.10° \leq c \leq +1.10°$
37. $266.67\,\Omega < R < 1333.33\,\Omega$ **39.** 43 or more
41. **(a)** $|\epsilon| < 0.6$, **(b)** $|P - 9.0| < 0.6$, **(c)** $8.4 < P < 9.6\,\text{W}$
43. At least 70,001 copies must be sold.

Exercise Set 16.3

1. $x < -1$ or $x > 3$ **3.** $-4 \leq x \leq 1$ **5.** $-1 < x < 1$
7. $-3 \leq x < 5$ **9.** $x \leq -1$ or $x \geq 2$ **11.** $x < 2$ or
$x > 3$ **13.** $-3 < x < -\frac{1}{2}$ **15.** $-\frac{1}{2} < x < 1$
17. Absolute inequality—all real numbers satisfy this
inequality. **19.** $x < -3$ or $-1 < x < 2$
21. $-4 < x < -3$ or $x > \frac{5}{2}$ **23.** $x < -4$
25. $-2 < x < 0$ or $x > 2$ **27.** Contradictory inequality. No
real numbers solve this inequality. **29.** $x < -1$ or
$2 < x < 5$ **31.** $-1 < x < 0$ or $x > 2$ **33.** $x \leq -\frac{1}{3}$ or
$\frac{1}{2} < x \leq 3$ **35.** $x > 9$ or $-1 < x < 1$ **37.** $x > 1$
39. $\dfrac{-3 - \sqrt{17}}{2} \leq x \leq \dfrac{-3 + \sqrt{17}}{2}$ **41.** $1\,\text{m} < d < \sqrt{6}\,\text{m}$
43. $x < 0.123$ or $x > 0.977$ **45.** **(a)** $1\,\text{s} < t < 3\,\text{s}$,
(b) $t > 4.75\,\text{s}$ **47.** $D \geq 35.6\,\text{in.}$

Exercise Set 16.4

1.

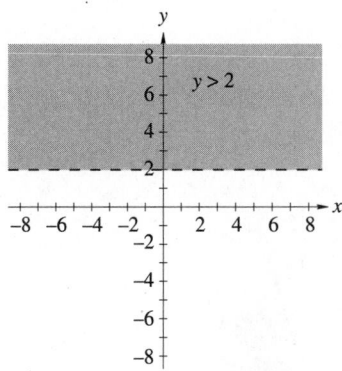

$y > 2$

3.

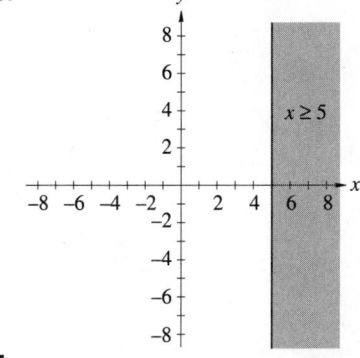

$x \geq 5$

5.

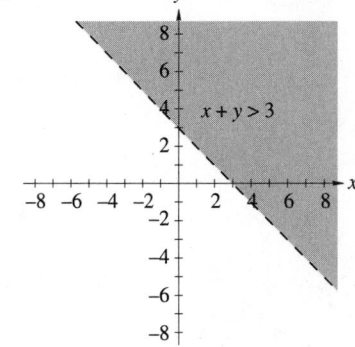

$x + y > 3$

7.

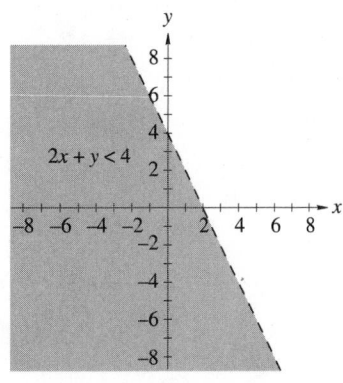

$2x + y < 4$

9.

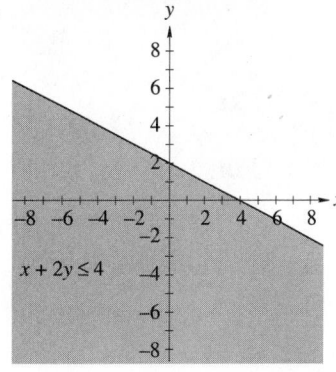

$x + 2y \leq 4$

11.

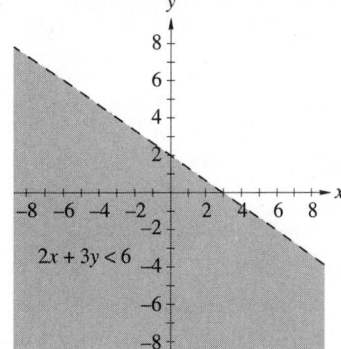

$2x + 3y < 6$

13.

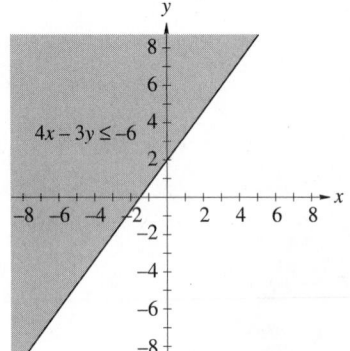

$4x - 3y \leq -6$

15.

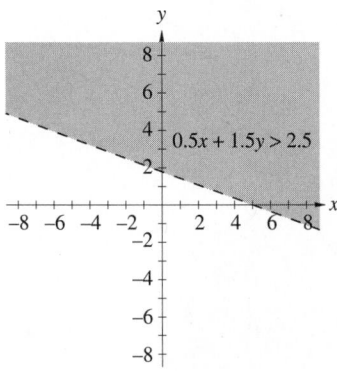

$0.5x + 1.5y > 2.5$

17.

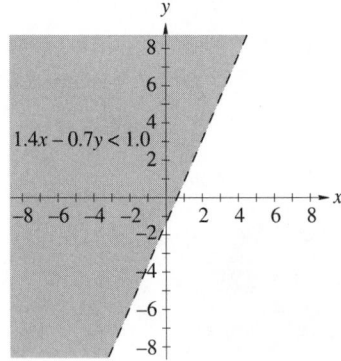

$1.4x - 0.7y < 1.0$

19.

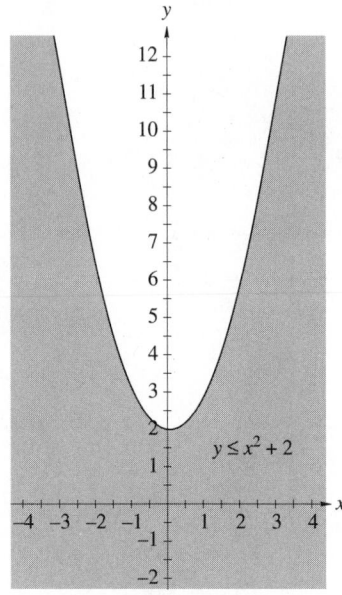

$y \leq x^2 + 2$

21.

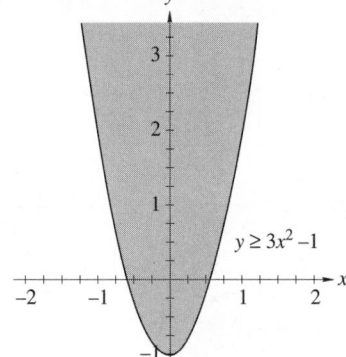

$y \geq 3x^2 - 1$

23.

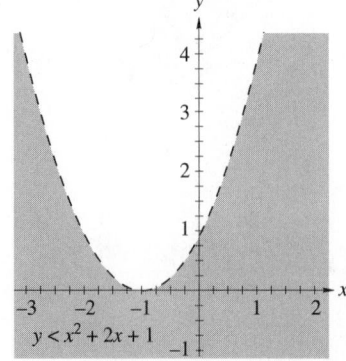

$y < x^2 + 2x + 1$

25.

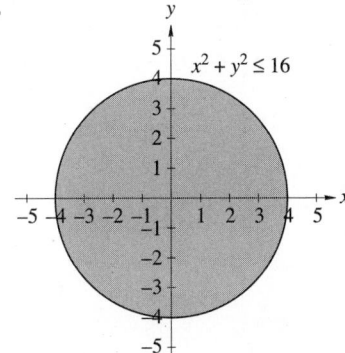

$x^2 + y^2 \le 16$

27.

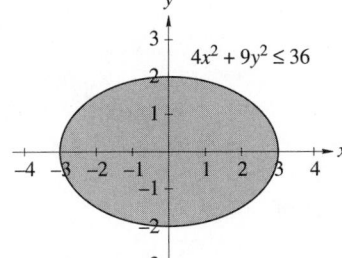

$4x^2 + 9y^2 \le 36$

29.

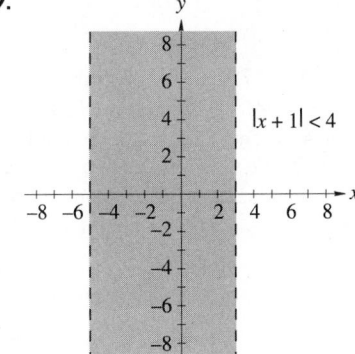

$|x + 1| < 4$

31.

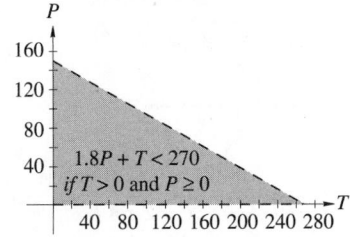

$1.8P + T < 270$
if $T > 0$ *and* $P \ge 0$

33.

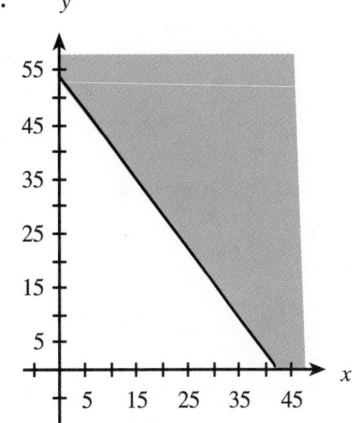

Exercise Set 16.5

1.

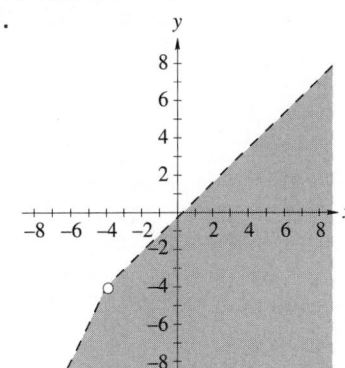

3.

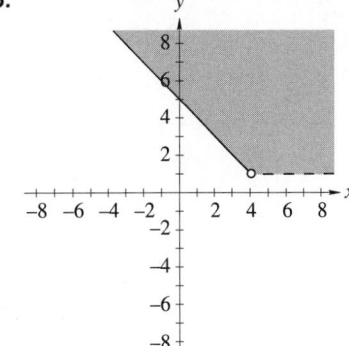

5.

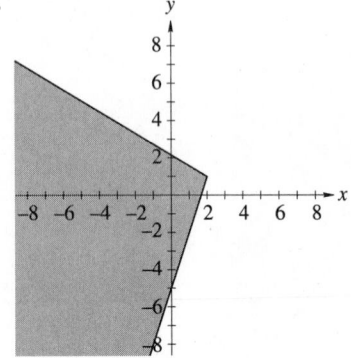

7.

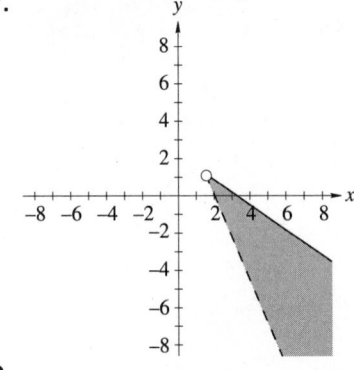

9.

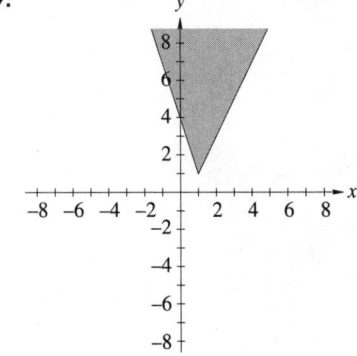

11.

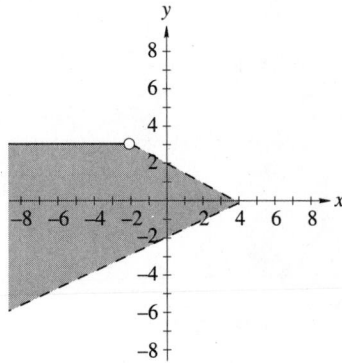

13.

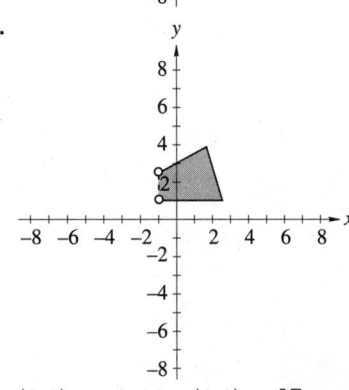

15. $(4, 3)$ max is 11 at $(4, 3)$ **17.** max is 400 at any point on $3x + 2y = 80$ where $20 \le x \le \frac{80}{3}$.   **19.** max is 216 at $(12, 12)$ **21.** 0 of x and 400 of y will produce the most profit: $3,200. $3,008.   **23.** 120 of B and none of A for a profit of $1140.00.   **25.** 37 home loans and 7 commercial loans.   **27.** Buy 10,000 small and 5,000 large, for a total cost of $2,100.

Review Exercises

1. $x < -3$

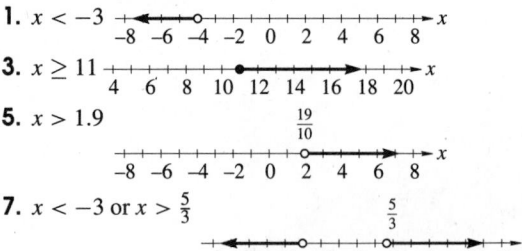

3. $x \ge 11$

5. $x > 1.9$

7. $x < -3$ or $x > \frac{5}{3}$

9.

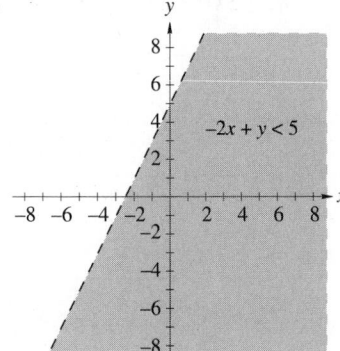

$-2x + y < 5$

11.

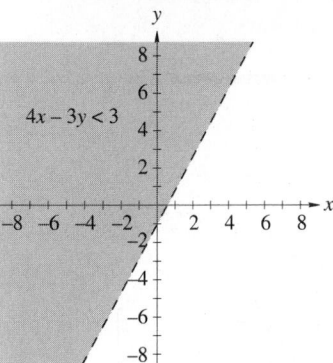

$4x - 3y < 3$

13.

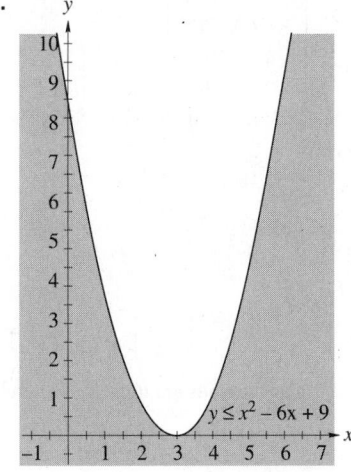

$y \le x^2 - 6x + 9$

15.

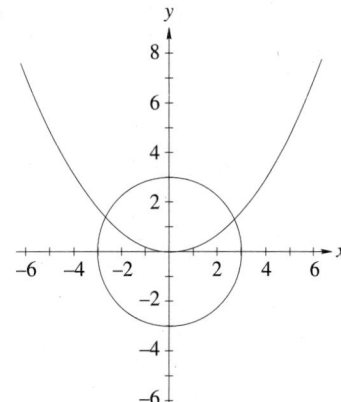

17.

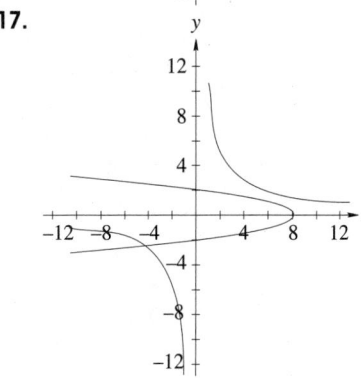

19.

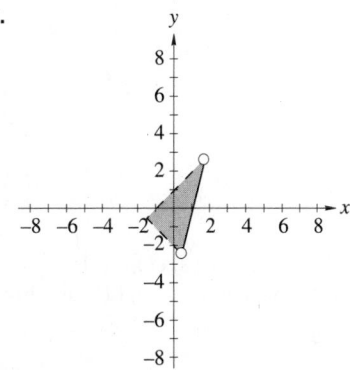

21.

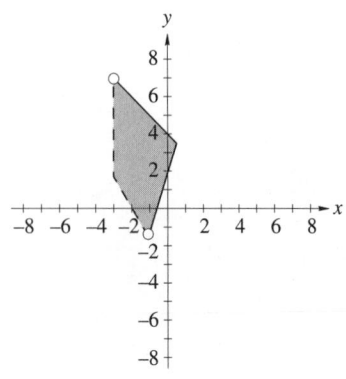

23. $\left(1\frac{1}{6}, 3\frac{2}{3}\right)$ **25.** $R_1 + R_2 \leq 5\Omega$ **27.** 6 of C and 8 of S for a profit of \$20,460.

Chapter 16 Test

1. $x > 6$

3. $x \geq -\frac{11}{2}$

5. $-1 \leq x \leq -\frac{1}{3}$

7. (a) 5 tablets of brand X and 4 of brand Y; **(b)** 44¢

ANSWERS FOR CHAPTER 17

Exercise Set 17.1

1. 2×3 **3.** 3×6 **5.** 4×2 **7.** $a_{11} = 1$; $a_{24} = 13; a_{21} = 8; a_{32} = 6$ **9.** $x = 12, y = -6, z = 2,$ $w = 5$ **11.** $\begin{bmatrix} -3 & 6 & 6 & -8 \\ 12 & 9 & 17 & 0 \\ 9 & 0 & 5 & 10 \end{bmatrix}$

13. $\begin{bmatrix} 2 & -1 & 4 \\ 5 & 9 & -2 \end{bmatrix}$ **15.** $\begin{bmatrix} -1 & 7 & 4 \\ 8 & 1 & 15 \end{bmatrix}$

17. $\begin{bmatrix} 12 & 9 & 6 \\ 15 & 0 & 21 \end{bmatrix}$ **19.** $\begin{bmatrix} -5 & 16 & 8 \\ 12 & 0 & 25 \end{bmatrix}$

21. $\begin{bmatrix} -22 & -1 & -2 \\ -9 & 2 & -5 \end{bmatrix}$ **23.** $\begin{bmatrix} -3 \\ -7 \\ 3 \\ -5 \end{bmatrix}$ **25.** $\begin{bmatrix} 21 \\ 27 \\ -12 \\ 30 \end{bmatrix}$

27. $\begin{bmatrix} 26 \\ 24 \\ -11 \\ 35 \end{bmatrix}$ **29.** $\begin{bmatrix} 0 \\ -22 \\ 9 \\ 25 \end{bmatrix}$ **31.** $\begin{bmatrix} 27 & 12 & 15 \\ 18 & 15 & 16 \end{bmatrix}$

or 27 of computer chip A, 12 of computer chip B, 15 EPROMS, 18 keyboards, 15 motherboards, and 16 disk drives.

33. $\begin{bmatrix} 135 & 82 \\ 65 & 118 \end{bmatrix}$ **35.**

Chip type:	286	386	486	586
Warehouse A	297	2541	3531	292
Warehouse B	1232	4653	3436	83
Warehouse C	352	3439	3017	1113

37. See *Computer Programs*   **39.** See *Computer Programs*

Exercise Set 17.2

1. $\begin{bmatrix} 26 & 9 \\ -20 & -23 \end{bmatrix}$ **3.** $[-23]$

5. $\begin{bmatrix} -3 & 13 & 8 \\ -2 & 16 & 14 \end{bmatrix}$ **7.** $\begin{bmatrix} 5 & 17 & 5 \\ 44 & 67 & 37 \\ -9 & -6 & -7 \end{bmatrix}$

9. $\begin{bmatrix} -12 & 14 & 11 \\ 88 & 18 & 23 \end{bmatrix}$ **11.** $\begin{bmatrix} 7 & 16 \\ 17 & 38 \end{bmatrix}$

13. $\begin{bmatrix} -8 & 2 \\ -12 & 2 \end{bmatrix}$ **15.** $\begin{bmatrix} -68 & 18 \\ -160 & 42 \end{bmatrix}$

17. $\begin{bmatrix} -1 & 18 \\ 5 & 40 \end{bmatrix}$ **19.** $\begin{bmatrix} 42 & 96 \\ 102 & 228 \end{bmatrix}$ **21.** $\begin{bmatrix} 0 & 0 \\ 0 & 0 \end{bmatrix}$,

No   **23.** Warehouse A: \$283,025, warehouse B: \$354,995, and warehouse C: \$364,415

25. $A^2 = \begin{bmatrix} 0 & 1 \\ 1 & 0 \end{bmatrix}\begin{bmatrix} 0 & 1 \\ 1 & 0 \end{bmatrix} = \begin{bmatrix} 0+1 & 0+0 \\ 0+0 & 1+0 \end{bmatrix} = \begin{bmatrix} 1 & 0 \\ 0 & 1 \end{bmatrix} = I$. Similarly, $B^2 = C^2 = I$.   **27. (a)** The commutator of A and B is

$AB - BA = \begin{bmatrix} i & 0 \\ 0 & -i \end{bmatrix} - \begin{bmatrix} -i & 0 \\ 0 & i \end{bmatrix} = 2\begin{bmatrix} i & 0 \\ 0 & -i \end{bmatrix} = 2i\begin{bmatrix} 1 & 0 \\ 0 & -1 \end{bmatrix} = 2iC$. Similar methods are used for (b) and (c).   **29.** See *Computer Programs*

Exercise Set 17.3

1. $\begin{bmatrix} 1 & -1 \\ -5 & 6 \end{bmatrix}$ **3.** $\begin{bmatrix} -1.5 & 2 \\ 4 & -5 \end{bmatrix}$

5. $\begin{bmatrix} -1 & -1.5 \\ -1.5 & -2 \end{bmatrix}$ **7.** $\begin{bmatrix} 0.6 & -2 \\ -0.8 & 3 \end{bmatrix}$

9. $\begin{bmatrix} 0 & 1 & 0 \\ -\frac{1}{3} & 0 & 0 \\ 0 & 0 & \frac{1}{4} \end{bmatrix}$ **11.** Singular matrix because column 2 is twice column 1.   **13.** $\begin{bmatrix} 1 & 3 & -2 \\ -1 & -4 & 3 \\ 0 & -4 & 5 \end{bmatrix}$

15. Singular matrix because row 2 is -2 times row 1

17. $\begin{bmatrix} 0.5 & 0 & 0 \\ -0.5 & 0.5 & 0 \\ 0 & -0.5 & 0.5 \end{bmatrix}$ **19.** $\begin{bmatrix} 3 & 3 & -1 \\ -2 & -2 & 1 \\ -4 & -5 & 2 \end{bmatrix}$

21. All answers should check.

Exercise Set 17.4

1. $x = -1, y = 2$ **3.** $x = -1.5, y = 2.5$ **5.** No solution   **7.** $x = 1.5, y = -4, z = 3.5$ **9.** $x = 5.5, y = -3.5, z = -6.5$ **11.** $x = -2, y = 5$ **13.** $x = -5, y = 7$ **15.** $x = 4.2, y = -2.4$ **17.** $x = -2, y = 4, z = 1$ **19.** $x = -3.4, y = 2.2, z = 1.5$ **21.** $I_1 = 6.05, I_2 = 5.26, I_3 = 0.79$ **23.** $I_1 = 1.22, I_2 = 2.37, I_3 = 1.64$ **25.** See *Computer Programs*   **27.** $A = 530\,\text{km}. B = 620\,\text{km}, C = 750\,\text{km}$ **29. (a)** The city should buy from Companies A, B, and C in the proportion $A : B : C = 4 : 5 : 2$ units. **(b)** Company A: 1,320,000 units, Company B: 1,650,000 units, Company C: 660,000 units   **31.** $A = 12.5, B = 27.25, C = 39.75, D = 20.5$

ANSWERS FOR CHAPTER 18

Exercise Set 18.1

1. 9   **3.** 3   **5.** 130   **7.** 0   **9.** 0   **11.** 11   **13.** 0   **15.** 79   **17.** yes   **19.** yes   **21.** yes   **23.** yes
25. $Q(x) = x^4 + 3x^3 - 8x^2 - 24x + 3; R(x) = 18$
27. $Q(x) = 5x^2 - 8x + 24; R(x) = -63$
29. $Q(x) = 8x^4 + 4x^3 - 2x^2 + 6x + 1; R(x) = \frac{1}{2}$
31. $Q(x) = 2x^3 - 3x^2 - 4; R(x) = 0$

Exercise Set 18.2

1. $1, \dfrac{-5+\sqrt{85}}{10}, \dfrac{-5-\sqrt{85}}{10}$ **3.** $\frac{1}{3}, 3, -3$ **5.** $2, -2, j, -j$ **7.** $3, -3, -1+2j, -1-2j$ **9.** $-1, -1, 2, 3$ **11.** $-1, -1, -3, 1$ **13.** $-\frac{1}{2}, \frac{1}{3}, -2+j, -2-j$ **15.** $1, \frac{2}{3}, \dfrac{-1+j\sqrt{3}}{2}, \dfrac{-1-j\sqrt{3}}{2}$ **17.** $1, 1, 1, -1+j\sqrt{3}, -1-j\sqrt{3}$ **19.** $3, -2, \frac{2}{3}, \dfrac{-1+\sqrt{5}}{2}, \dfrac{-1-\sqrt{5}}{2}$ **21.** $j, -j, 3j, -3j, 7, -1$

Review Exercises

1. $\begin{bmatrix} 4 & 4 & 2 & 7 \\ 9 & 7 & 3 & 4 \\ 9 & 5 & -8 & -3 \end{bmatrix}$ **3.** $\begin{bmatrix} 1 & 5 & 1 & 5 \\ 1 & 8 & -3 & 4 \\ 5 & 7 & -6 & 3 \end{bmatrix}$

5. $\begin{bmatrix} 8 & 3 & 4 & 4 \\ 3 & 14 & -14 & 8 \\ 18 & 35 & -16 & 24 \end{bmatrix}$ **7.** $\begin{bmatrix} 3 & 8 \\ 4 & 13 \\ 1 & 2 \end{bmatrix}$

9. $\begin{bmatrix} 12 & 15 & 3 \\ 8 & 10 & 2 \\ -4 & -5 & -1 \end{bmatrix}$ **11.** $\begin{bmatrix} -2.5 & -1.5 \\ 2 & 1 \end{bmatrix}$

13. $\begin{bmatrix} -0.5 & 0.125 & 0 \\ 0 & 0.25 & 0 \\ 0.5 & -0.125 & 1 \end{bmatrix}$ **15.** $x = 2.5, y = -6.4$

17. $x = 2, y = -3, z = 5$ **19.** $T = 86.47, a = 1.23$
21. $x'' = -x, y'' = -y$

23. $M = \begin{bmatrix} 1 - \dfrac{d}{f_2} & \dfrac{d}{f_1 f_2} - \dfrac{1}{f_1} - \dfrac{1}{f_2} \\ d & 1 - \dfrac{d}{f_1} \end{bmatrix}$

Chapter 17 Test

1. (a) $\begin{bmatrix} 7 & 5 & -7 \\ 19 & -6 & 6 \end{bmatrix}$, and **(b)** $\begin{bmatrix} 26 & -10 & -6 \\ 42 & -18 & -2 \end{bmatrix}$

3. (a) $CD = \begin{bmatrix} -7 & 24 \\ -11 & 16 \end{bmatrix}$, **(b)** $DC = \begin{bmatrix} 2 & 3 \\ -46 & 7 \end{bmatrix}$

5. $x = 1, y = -2, z = 3$ **7.** $I_1 = \frac{20}{11}\,\text{A}, I_2 = \frac{25}{11}\,\text{A}, I_3 = -\frac{5}{11}\,\text{A}$

Exercise Set 18.3

1. Number of Positive Roots: 1; Negative Roots: 0 or 2; Possible Rational Roots: $\pm 1, \pm 2$ **3.** Number of Positive Roots: 0 or 2; Negative Roots: 1; Possible Rational Roots: $\pm 1, \pm 2, \pm 3, \pm 6$ **5.** Number of Positive Roots: 0 or 2; Negative Roots: 1; Possible Rational Roots: $\pm 1, \pm 2, \pm 3, \pm 6$ **7.** Number of Positive Roots: 0 or 2; Negative Roots: 0 or 2; Possible Rational Roots: $\pm 1, \pm 3, \pm \frac{1}{2}, \pm \frac{3}{2}$ **9.** Number of Positive Roots: 0 or 2; Negative Roots: 0 or 2; Possible Rational Roots: $\pm 1, \pm 2, \pm 4, \pm \frac{1}{2}, \pm \frac{1}{3}, \pm \frac{2}{3}, \pm \frac{4}{3}, \pm \frac{1}{6}$ **11.** $1, 3, 5$ **13.** $-1, -1, 2$ **15.** $-\frac{3}{4}, 1 + j\sqrt{3}, 1 - j\sqrt{3}$ **17.** $2, 2, j\sqrt{3}, -j\sqrt{3}$ **19.** $-\frac{1}{2}, -\frac{1}{2}, 3 + \sqrt{3}, 3 - \sqrt{3}$ **21.** no rational roots   **23.** $\frac{1}{2}, 3, 3, j, -j$ **25.** $1, 3, -2, \dfrac{3}{2} + j\dfrac{\sqrt{11}}{2}, \dfrac{3}{2} - j\dfrac{\sqrt{11}}{2}$ **27.** 1 in. or 2 in.   **29.** 9 feet   **31.** $R = 2$ **33.** 3.5 cm

Exercise Set 18.4

1. 0.56   **3.** 0.52   **5.** 0.48   **7.** $-1, 1.35$ (the other two roots are complex numbers)   **9.** $-1, 2.09$

11. $1, -1.6572$ **13.** $-1, 0.24$ **15.** $-1.41, 1.41, -\frac{1}{2}, -\frac{1}{4}$
17. 6.22731 years or about 6 years 83 days
19. (a) $63.87006\,\text{cm}$; **(b)** $521\,101\pi\,\text{cm}^3$ **21.** The
approximate roots are $\pm 0.3536, \pm 0.8654, \pm 1.4639, \pm 1.7321$,
and ± 2.2323 **23.** $3.095\,\text{h} \approx 3\,\text{h}\,5.7\,\text{min}$ and $10.414\,\text{h}$
$\approx 10\,\text{h}\,24.8\,\text{min}$

Exercise Set 18.5

1. vertical asymptote at $x = -2$; horizontal asymptote at
$y = 0$

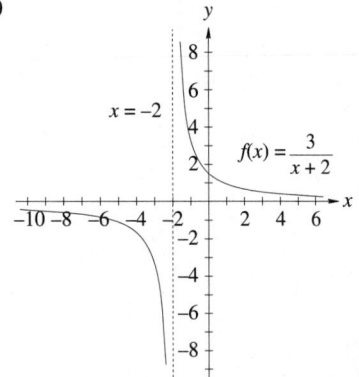

3. vertical asymptote at $x = -4$; horizontal asymptote at
$y = 2$

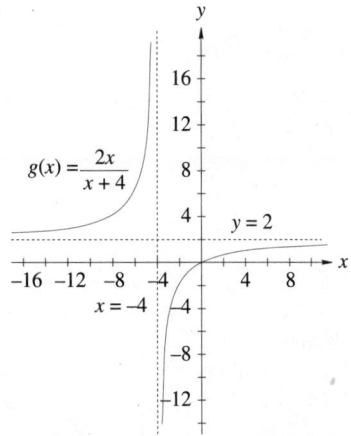

5. vertical asymptotes at $x = -4$ and $x = -2$; horizontal

asymptote at $y = 3$

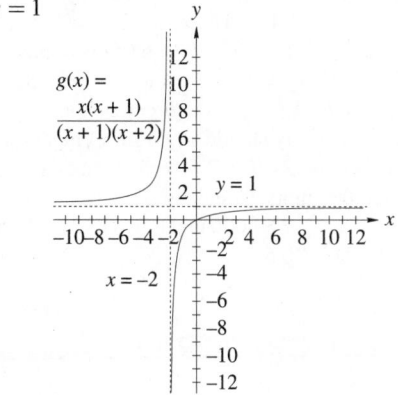

7. vertical asymptote at $x = -2$; horizontal asymptote at
$y = 1$

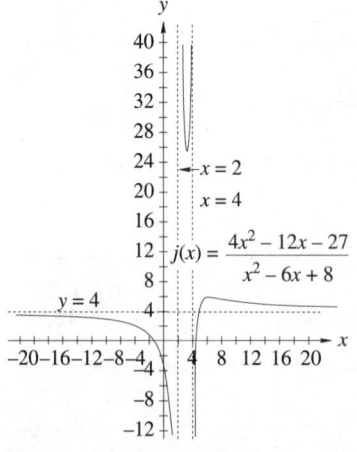

9. vertical asymptote at $x = 4$; horizontal asymptote at $y = 2$

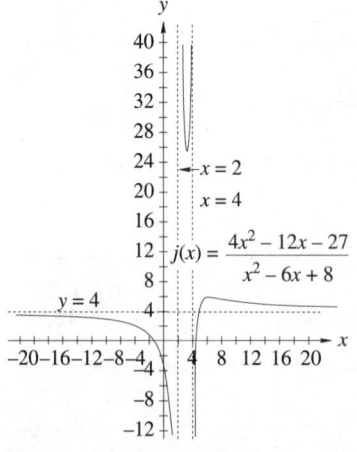

11. $x = -2$ or $x = 3$ **13.** 1 **15.** -5 **17.** $\frac{21}{8}$ **19.** 3
21. 4.49 hours   **23.** $\$73$/set

Review Exercises

1. 30, degree $= 2$ **3.** -13, degree $= 3$
5. $Q(x) = 4x - 18$; $R(x) = 51$
7. $Q(x) = 9x^3 + 4x^2 + 6x - 1$; $R(x) = 0$ **9.** 1, 1, 1
11. $-1, -1, -1, 3$ **13.** $-3.044, -0.328, 0.548, 1.824$
15. $-1, -1, j, j, -j, -j$ **17.** vertical asymptotes at
$x = -4$ and $x = 1$; horizontal asymptote at $y = 0$; solution:
$x = 0$ **19.** vertical asymptotes at $x = 2$ and $x = -2$;

horizontal asymptote at $y = 0$; solution: $x = 18$ **21.** 1
(100%)

Chapter 18 Test

1. 5   **3.** Quotient: $x^3 - 2x^2 - 9x - 10$; Remainder: 15
5. $1, -2, -\frac{2}{3}$ **7.** vertical asymptotes: $x = -3$ and
$x = -2$; horizontal asymptote: $y = 3$; solutions: $x = 1$ and
$x = -\frac{1}{3}$ **9.** 2.48 yr

■ ANSWERS FOR CHAPTER 19

Exercise Set 19.1

1. $\frac{1}{2}$; $\frac{1}{3}$; $\frac{1}{4}$; $\frac{1}{5}$; $\frac{1}{6}$; $\frac{1}{7}$ **3.** 4; 9; 16; 25; 36; 49
5. -1; 2; -3; 4; -5; 6   **7.** $\frac{1}{3}$; $\frac{3}{5}$; $\frac{5}{7}$; $\frac{7}{9}$; $\frac{9}{11}$; $\frac{11}{13}$ **9.** 1; $\frac{2}{3}$;
$\frac{4}{9}$; $\frac{8}{27}$; $\frac{16}{81}$; $\frac{32}{243}$ **11.** 0; $\frac{1}{9}$; $\frac{1}{4}$; $\frac{9}{25}$; $\frac{4}{9}$; $\frac{25}{49}$
13. 1; 2; 6; 24; 120; 720   **15.** 5; 8; 11; 14; 17; 20
17. 2; 2; 4; 64; 16,777,216; 1.329228×10^{36}
19. 1; 2; 9; 262,144; $5^{262,144}$; $6^{(5^{262,144})}$ **21.** 1; $\frac{1}{2}$; $\frac{1}{2}$; $\frac{1}{4}$; $\frac{1}{8}$;
$\frac{1}{32}$ **23.** 0; 2; 2; 0; -2; -2 **25.** 18.75 ft or 18 ft 9 in
27. **(a)** \$30,400; **(b)** \$35,900

Exercise Set 19.2

1. Arithmetic sequence with $d = 8$; $a_9 = 65$
3. Geometric sequence with $r = \frac{-1}{2}$; $a_{10} = \frac{1}{64}$
5. Geometric sequence with $r = -4$;
$a_6 = (-1)(-4)^5 = 1024$ **7.** Arithmetic sequence with
$d = -5$; $a_8 = -32$ **9.** Neither; can't tell from the given
information what a_{10} will be   **11.** Arithmetic sequence with
$d = 0.4$; $a_9 = 3.6$ **13.** Arithmetic sequence with $d = \frac{-1}{6}$;
$a_8 = -\frac{1}{2}$ **15.** Geometric sequence with $r = 0.1$;
$a_6 = 0.00005$ **17.** Arithmetic sequence with $d = 2$;
$a_9 = 17 + \sqrt{3}$ **19.** Neither; $a_6 = 1.2$,
$a_n = -1^{n-1}[4.3 - (1.1)(n-1)]$ **21.** -51 **23.** 243
25. 20   **27.** 9   **29.** -3.5 **31.** $\frac{1}{8}$ miles   **33.** 12.3 cm;
51   **35.** **(a)** about 15.63 miles; **(b)** about 43.05 miles.

Exercise Set 19.3

1. $\frac{3}{2} + \frac{3}{4} + \frac{3}{8} + \frac{3}{16}$ **3.** $1 - \frac{3}{2} + 3 - \frac{27}{4}$ **5.** 0
7. 91   **9.** 225   **11.** $\frac{1,593,342}{67,925} \approx 23.457$
13. Arithmetic; $S_{10} = 165$ **15.** Geometric; $S_8 = 255$
17. Arithmetic; $S_{12} = 192$ **19.** Arithmetic; $S_{20} = 57.5$
21. Geometric; $S_{16} = 0.5625$ **23.** Arithmetic; $S_{14} = 114.8$
25. Geometric; $S_{15} = 17920.253$ **27.** Geometric;
$S_8 = 9840$ **29.** 392.32m when it hits the fourth time,
634.10m when it hits the tenth time.   **31.** 104 ft
33. \$2540.47; \$3.99   **35.** **(a)** \$3,662,000; **(b)** \$7,682,000

Exercise Set 19.4

1. Converges; $\frac{1}{3}$ **3.** Converges; 0.6   **5.** Converges; 4
7. Converges; 1.389   **9.** Converges; 0.0333…
11. Converges; 0.555…   **13.** Diverges; $r = 5/4$
15. Diverges; $r = 10$ **17.** Converges; 2   **19.** Converges;
2.4142   **21.** Diverges; $r = \sqrt{5}$ **23.** 4/9   **25.** $\frac{19}{33}$
27. $\frac{13,520}{9999}$ **29.** $\frac{20,986}{3330} = \frac{10,493}{1665}$ **31.** 45 m   **33.** 1,250
revolutions

Exercise Set 19.5

1. $a^4 + 4a^3 + 6a^2 + 4a + 1$
3. $81x^4 - 108x^3 + 54x^2 - 12x + 1$
5. $\frac{x^6}{64} + \frac{3}{16}x^5 d + \frac{15}{16}x^4 d^2 + \frac{5}{2}x^3 d^3 + \frac{15}{4}x^2 d^4 + 3xd^5 + d^6$
7. $\frac{a^5}{32} - \frac{5a^4}{4b} + 20\frac{a^3}{b^2} - 160\frac{a^2}{b^3} + 640\frac{a}{b^4} - \frac{1024}{b^5}$
9. $a^7 + 7a^6 b + 21a^5 b^2 + 35a^4 b^3 + 35a^3 b^4 + 21a^2 b^5 +$
$7ab^6 + b^7$ **11.** $t^8 - 8t^7 a + 28t^6 a^2 - 56t^5 a^3 + 70t^4 a^4 -$
$56t^3 a^5 + 28t^2 a^6 - 8ta^7 + a^8$
13. $64a^6 - 192a^5 + 240a^4 - 160a^3 + 60a^2 - 12a + 1$
15. $x^{14}y^7 + \frac{7}{2}x^{12}y^6 a + \frac{21}{4}x^{10}y^5 a^2 + \frac{35}{8}x^8 y^4 a^3 +$
$\frac{35}{16}x^6 y^3 a^4 + \frac{21}{32}x^4 y^2 a^5 + \frac{7}{64}x^2 ya^6 + \frac{a^7}{128}$
17. $x^{12} + 12x^{11}y + 66x^{10}y^2 + 220x^9 y^3$
19. $1 + 2a + 3a^2 + 4a^3$ **21.** $1 - \frac{b}{3} + \frac{2b^2}{9} - \frac{14b^3}{81}$
23. 1.464   **25.** 1.049   **27.** 1.008   **29.** 0.983
31. $3003x^{20}y^5$ **33.** $126720x^8 y^4$ **35.** $1001a^{10}b^4$
37. **(a)** $1 + \frac{v^2}{2c^2} + \frac{3v^4}{8c^4}$ **(b)** $mc^2 + \frac{mv^2}{2} + \frac{3mv^4}{8c^2}$
39. $r^3 + \frac{3}{2}\ell^2 r + \frac{3\ell^4}{8r}$

Review Exercises

1. $\frac{1}{3}$; $\frac{1}{4}$; $\frac{1}{5}$; $\frac{1}{6}$; $\frac{1}{7}$; $\frac{1}{8}$ **3.** $\frac{-1}{3}$; $\frac{1}{10}$; $\frac{-1}{21}$; $\frac{1}{36}$; $\frac{-1}{55}$; $\frac{1}{78}$
5. 1; $\frac{1}{2}$; $\frac{1}{6}$; $\frac{1}{24}$; $\frac{1}{120}$; $\frac{1}{720}$ **7.** 1; 3; 4; 7; 11; 18
9. Arithmetic sequence, $d = -3$, $a_{10} = -17$
11. Arithmetic sequence, $d = 4$, $a_7 = 23$ **13.** Geometric

sequence, $r = -\frac{1}{6}$, $a_{10} = -\frac{\overline{}}{3359232} \approx -0.0000003$.
15. -24 **17. (a)** $(a)\frac{4}{3} + \frac{4}{9} + \frac{4}{27} + \frac{4}{81}$ **(b)** $2 + \frac{3}{2} + \frac{4}{3} + \frac{5}{4}$
19. 39 **21.** Arithmetic series; $S_{10} = 265$ **23.** Geometric
series; $S_{14} = 1.4999997$ **25.** Converges; $S = \frac{2}{3}$
27. Diverges **29.** Converges; $S \approx 1.3660$ **31.** $\frac{185}{999} = \frac{5}{27}$
33. $a^5 + 10a^4 + 40a^3 + 80a^2 + 80a + 32$
35. $\frac{x^6}{64} - \frac{9x^5y^2}{16} + \frac{135x^4y^4}{16} - \frac{135x^3y^6}{2} + \frac{1215x^2y^8}{4} -$

$729xy^{10} + 729y^{12}$ **37.** $32,768x^{15} + 245,760x^{14}y +$
$850,160x^{13}y^2 + 1,863,680x^{12}y^3$
39. $1 + 5x + 15x^2 + 35x^3$ **41.** $2,480,640x^{14}y^6$
43. 1.004 **45.** \$727.76 **47.** 65.76 cm **49.** 19.834 sec

Chapter 19 Test
1. $-5, 5, \frac{5}{3}, 1, \frac{5}{7}, \frac{5}{9}$ **3.** arithmetic sequence, $d = -2.5$,
$a_{10} = -7.5$ **5.** 25 **7.** $\frac{435}{999} = \frac{145}{333}$ **9.** 123,000

≡ ANSWERS FOR CHAPTER 20

Exercise Set 20.1
1–29. All answers are proofs and will not be displayed in this
section.
31. This identity is true based on the graphs of $y = 2\csc 2x$
and $y = \sec x \csc x$ shown below.

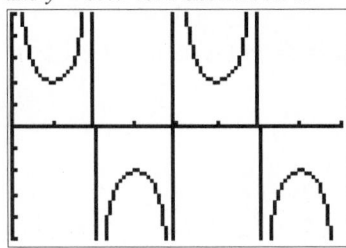

$$y = 2\csc 2x$$

$$\left[0, 2\pi, \frac{\pi}{4}\right] \cdot [-5,5,1]$$

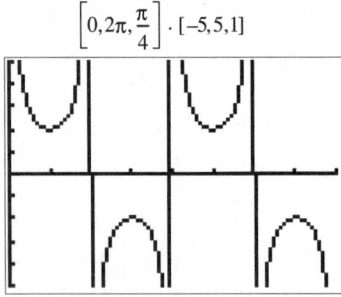

$$y = \sec x \csc x$$

$$\left[0, 2\pi, \frac{\pi}{4}\right] \cdot [-5,5,1]$$

33. This identity is not true based on the graphs of $y = \sin\frac{1}{2}x$
and $y = \frac{1}{2}\sin x$ shown below.

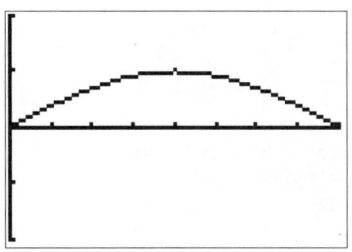

$$y = \sin\frac{1}{2}x$$

$$\left[0, 2\pi, \frac{\pi}{4}\right] \cdot [-2,2,1]$$

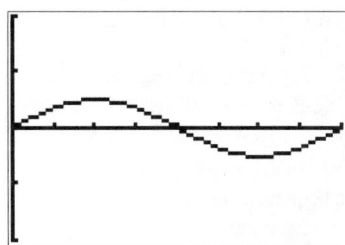

$$y = \frac{1}{2}\sin x$$

$$\left[0, 2\pi, \frac{\pi}{4}\right] \cdot [-2,2,1]$$

35–43. All answers are proofs and will not be displayed in this
section.

Exercise Set 20.2
1. $\frac{\sqrt{6} - \sqrt{2}}{4}$ **3.** $\frac{\sqrt{3}}{2}$
5. $\frac{\sqrt{6} - \sqrt{2}}{\sqrt{6} + \sqrt{2}} = \frac{10 - 4\sqrt{3}}{4} = \frac{5 - 2\sqrt{3}}{2}$ **7.** $\frac{1}{2}$
9. $\cos x$ **11.** $-\sin x$ **13.** $-\cos x$ **15.** $\sin x$
17. $\frac{21 + \sqrt{105}}{32} \approx 0.97647$ **19.** $\frac{21 + \sqrt{105}}{7\sqrt{7} - 3\sqrt{15}} \approx 4.52768$
21. $\frac{7\sqrt{7} + 3\sqrt{15}}{32} \approx 0.94185$ **23.** I

25. $\dfrac{\sqrt{105}-21}{32} \approx -0.33603$

27. $\dfrac{\sqrt{105}-21}{7\sqrt{7}+3\sqrt{15}} \approx -0.35678$

29. $\dfrac{7\sqrt{7}-3\sqrt{15}}{32} \approx 0.21566$ **31.** IV **33.** $\sin 60° = \dfrac{\sqrt{3}}{2}$

35. $\cos 44°$ **37.** $\cos\alpha$ **39.** $\cos^2 x - \sin^2 x$

Exercise Set 20.3

1. $\sqrt{\dfrac{2+\sqrt{3}}{4}} \approx x0.96593$ **3.** $\sqrt{\dfrac{2-\sqrt{3}}{4}} \approx 0.25882$

5. $\sqrt{\dfrac{2+\sqrt{3}}{4}} \approx 0.96593$

7. $\sqrt{\dfrac{1+\sqrt{\dfrac{2+\sqrt{3}}{4}}}{2}} = \sqrt{\dfrac{2+\sqrt{2+\sqrt{3}}}{4}} \approx 0.99145$

9. $\dfrac{\sqrt{2}}{2+\sqrt{2}} \approx 0.41421$ **11.** $\sqrt{\dfrac{2-\sqrt{3}}{4}} \approx 0.25882$

13. $\frac{1}{4}(\sqrt{3}\sqrt{2+\sqrt{2+\sqrt{3}}} - \sqrt{2-\sqrt{2+\sqrt{3}}}) \approx 0.79335$

15. $\sin 2x = -\frac{336}{625}; \cos 2x = \frac{527}{625}; \tan 2x = -\frac{336}{527};$

$\sin\dfrac{x}{2} = \dfrac{7\sqrt{2}}{10}; \cos\dfrac{x}{2} = \dfrac{\sqrt{2}}{10}; \tan\dfrac{x}{2} = 7$ **17.** $\sin 2x = \frac{840}{841};$

$\cos 2x = -\frac{41}{841}; \tan 2x = -\frac{840}{41}; \sin\dfrac{x}{2} = \sqrt{\dfrac{9}{58}}; \cos\dfrac{x}{2} = \sqrt{\dfrac{49}{58}};$

$\tan\dfrac{x}{2} = \frac{3}{7}$ **19.** $\sin 2x = \frac{840}{1369}; \cos 2x = \frac{-1081}{1369};$

$\tan 2x = -\frac{840}{1081}; \sin\dfrac{x}{2} = \sqrt{\dfrac{49}{74}}; \cos\dfrac{x}{2} = -\sqrt{\dfrac{25}{74}}; \tan\dfrac{x}{2} = -\frac{7}{5}$

Exercise Set 20.4

1. $90°, 270°$ **3.** $30°, 210°$ **5.** $228.59°, 311.41°$
7. $51.34°, 231.34°$ **9.** $90°, 270°$ **11.** $180°$
13. $0°, 90°, 180°$ **15.** $0°, 90°, 180°, 270°$
17. $30°, 150°, 210°, 330°$ **19.** $15°, 75°, 195°, 255°$
21. $45°, 225°$
23. $0°, 40°, 60°, 80°, 120°, 160°, 180°, 200°, 240°, 280°,$
$300°, 320°$
25. $45°, 135°, 225°, 315°$ **27.** $0°, 135°, 180°, 315°$
29. $30°, 90°, 150°$ **31.** $20.87°$ **33.** $0.001\,328$ s
35. $0, \pi$

Review Exercises

1. **11.** $139.11°, 319.11°$ **13.** $90°, 120°, 240°, 270°$
15. $0°, 90°, 180°, 270°$ **17.** $0°, 180°$
19. $218.17°, 321.83°$ **21.** $-\frac{120}{169}$
23. $\sqrt{25/26} \approx 0.98058$ **25.** $\frac{-21}{221} \approx -0.09502$
27. $\frac{220}{221} \approx 0.99548$ **29.** $\frac{140}{221} \approx 0.63348$
31. $90°, 210°, 330°$

Chapter 20 Test

1. $-\frac{33}{65}$ **3.** $-\frac{63}{65}$ **5.** $-\frac{24}{25}$ **7.** $\dfrac{2}{\sqrt{5}}$
9. $4\sin 12x$ **11.** **13.** $2r(1-\cos\theta)$

▤ ANSWERS FOR CHAPTER 21

Exercise Set 21.1

1. $\frac{1}{4}$ **3.** $\frac{1}{13}$ **5.** $\frac{3}{13}$ **7.** $\frac{1}{4}$ **9.** $\frac{13}{51}$ **11.** $\frac{1}{6}$
13. $\frac{1}{2}$ **15.** $\frac{1}{36}$ **17.** $\frac{11}{36}$
19.

Die 1 \ Die 2	1	2	3	4	5	6
1	(1,1)	(1,2)	(1,3)	(1,4)	(1,5)	(1,6)
2	(2,1)	(2,2)	(2,3)	(2,4)	(2,5)	(2,6)
3	(3,1)	(3,2)	(3,3)	(3,4)	(3,5)	(3,6)
4	(4,1)	(4,2)	(4,3)	(4,4)	(4,5)	(4,6)
5	(5,1)	(5,2)	(5,3)	(5,4)	(5,5)	(5,6)
6	(6,1)	(6,2)	(6,3)	(6,4)	(6,5)	(6,6)

21. $\frac{1}{36}$ **23.** 0 **25.** $\frac{1}{6}$ **27.** $\frac{1}{6}$ **29.** $\frac{1}{1,296}$ **31.** $\frac{4}{49}$
33. $\frac{4}{7}$ **35.** (a) $\frac{9}{16} = 56.25\%$ (b) $\frac{1}{16} = 6.25\%$ (c) $\frac{3}{8} = 37.5\%$
37. (a) 0.56 (b) 0.06 (c) 0.38 **39.** $0.000\,625$
41. $\frac{1}{10} = 0.10$

Exercise Set 21.2

1.

Number	2	3	4	6	7
Frequency	2	1	4	2	1

3.

Number	73	74	77	80	82	83	84	87	89	92	94	96	100
Frequency	1	1	1	1	1	1	2	1	1	1	3	5	1

5.

Interval	$1-2$	$3-4$	$5-6$	$7-8$
Frequency	2	5	2	1

7.

Interval	$71-75$	$76-80$	$81-85$	$86-90$	$91-95$	$96-100$
Frequency	2	2	4	2	4	6

9.
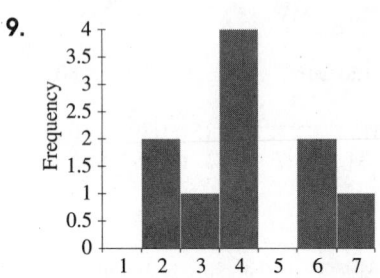

11.
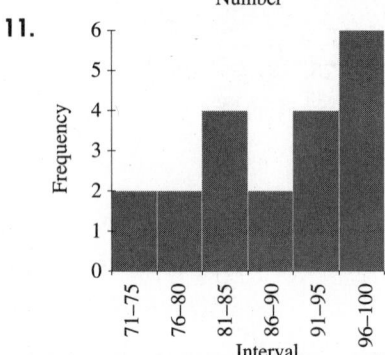

13.
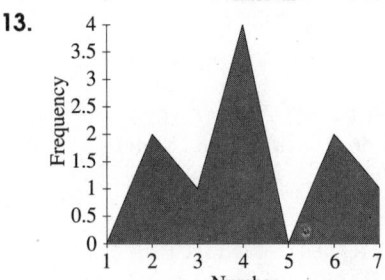

15.
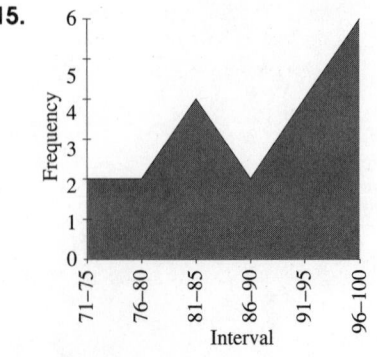

17. Mean: 4.2; Median: 4; Mode: 4; $Q_1 = 3$; $Q_2 = 4$; $Q_3 = 6$

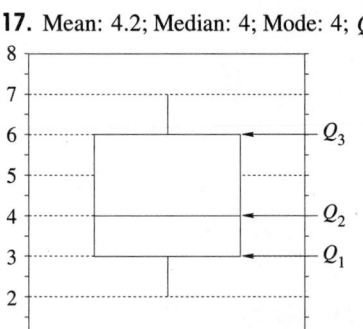
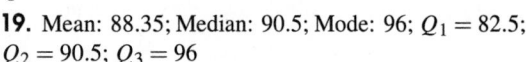

19. Mean: 88.35; Median: 90.5; Mode: 96; $Q_1 = 82.5$; $Q_2 = 90.5$; $Q_3 = 96$

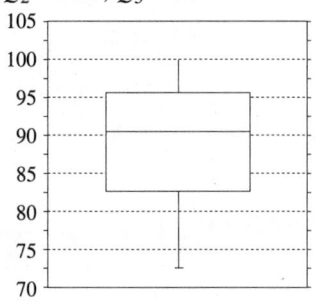

21. See *Computer Programs*

23.
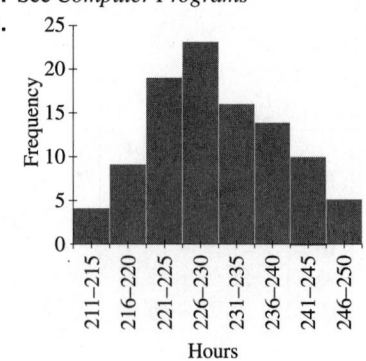

25.

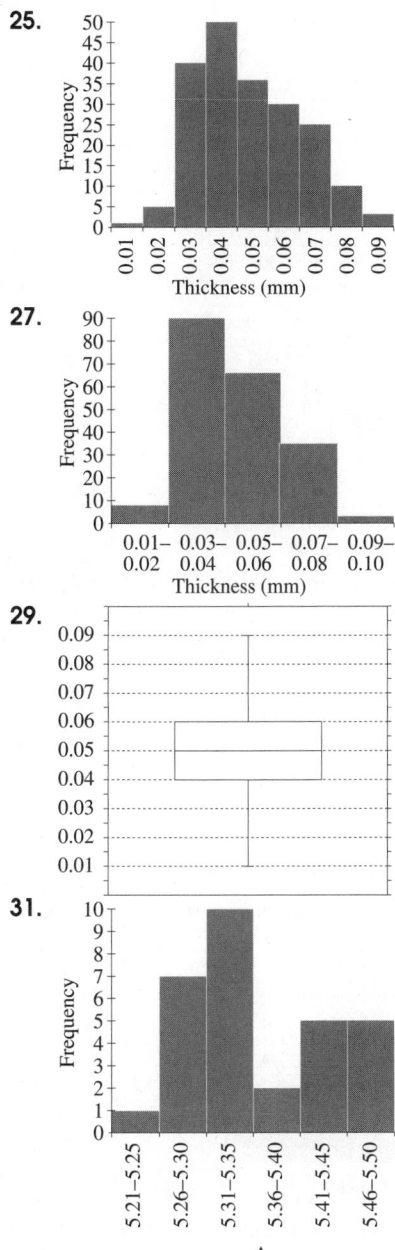

27.

29.

31.

33. $Q_1 = 5.29$; $Q_2 = 5.345$; $Q_3 = 5.43$

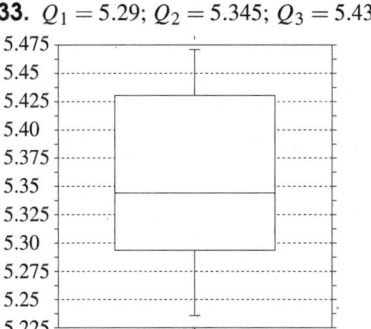

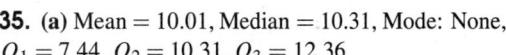

35. (a) Mean $= 10.01$, Median $= 10.31$, Mode: None, $Q_1 = 7.44$, $Q_2 = 10.31$, $Q_3 = 12.36$

(b)

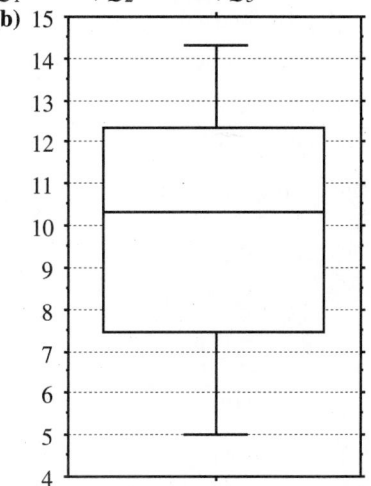

37. Mean $= 0.17$, Median $= 0.16$, Mode $= 0.13$, $Q_1 = 0.12$, $Q_2 = 0.16$, $Q_3 = 0.22$

Exercise Set 21.3

1. 2.84   **3.** 6.62   **5.** 1.69   **7.** 8.23   **9. (a)** 5, **(b)** 10   **11. (a)** 15, **(b)** 20   **13.** See *Computer Programs* **15. (a)** 0.0217, **(b)** 390, **(c)** 480   **17. (a)** 0.0164, **(b)** 116, **(c)** 196   **19. (a)** $\bar{x} = 60.75$, $s = 5.85$, **(b)** 66.7%, **(c)** 100% **21. (a)** $\bar{x} = 659.25$, $s = 174.67$, **(b)** 58.3%, **(c)** 95.8% **23. (a)** $\bar{x} = 25{,}905.65$, $s = 13{,}280.32$, **(b)** 74.2% within one standard deviation, **(c)** 93.5% within two standard deviations.

Exercise Set 21.4

1. $y = 0.833x - 1.333, r = 0.943$

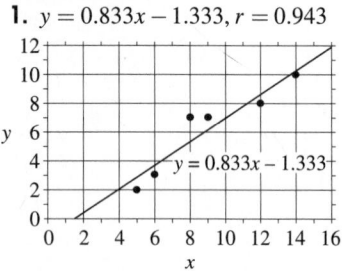

3. $y = 13.55x + 5.1, r = 0.986$

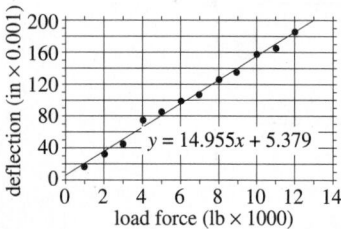

5. $y = 14.955x + 5.379, r = 0.996$

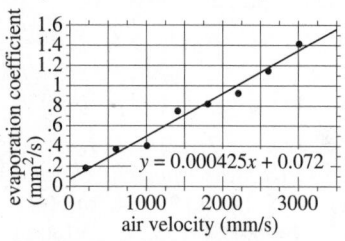

7. $y = 0.000425x + 0.072, r = 0.986$

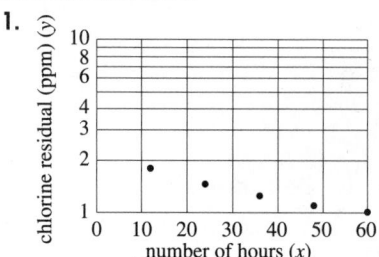

9. See *Computer Programs*

11. $y = -0.26825x + 74.6865, r = -0.70$

13. $y = 5.1x - 10,121, r = 0.98$

Exercise Set 21.5

1.

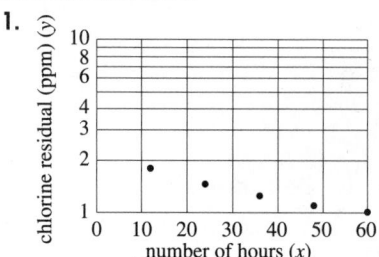

(b) $y = (2.002)(0.989)^x$

3.

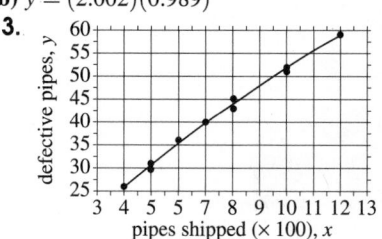

(b) $y = 4.684 - 5.848x + 0.107x^2$; **(c)** Shown by dashed curve in Figure

5.

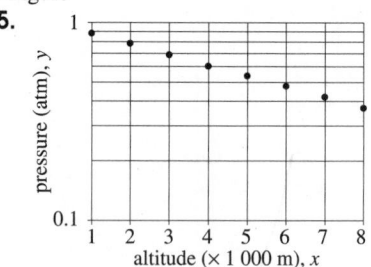

(b) $y = 0.9958e^{-0.12356x}$; **(c)** Shown by dashed curve in Figure

7.

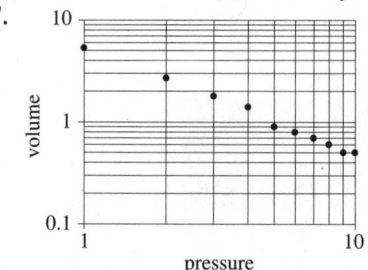

(b) $v = \dfrac{5.41}{p} - 0.064$; **(c)** Shown by dashed curve in Figure.

Review Exercises

1. $\frac{1}{2}$ **3.** $\frac{5}{8}$ **5.** $\frac{1}{3}$ **7.** $\frac{1}{36}$ **9.** $\left(\frac{23}{25}\right)^3 = \frac{12167}{15625}$

11. $Q_1 = 299.7; Q_2 = 300.1; Q_3 = 300.45$ **13.** 300.1

15. 0.931 **17.** 90.9%

19.

(b) $y = 43.94x + 18.94$; (c) See line in graph; (d) $r = 0.996$

21.

(b) $n = -0.686p + 83.714$; (c) Shown by curve in Figure

Chapter 21 Test

1. $\frac{1}{26}$ **3.** 3600.1

5.

7. 3599.994   **9.** 70.0%

11.

(b) $y = -39.774x + 1063.761$; (c) See line in graph;
(d) $r = -0.899$

ANSWERS FOR CHAPTER 22

Exercise Set 22.1

1. 9   **3.** 7.   **5.** 28   **7.** $-1/2$ **9.** 1

11. $\dfrac{x_1^2 + x_1 - 2}{x_1 - 1} = x_1 + 2$ **13.** $2x_1 + h$

15. (a) 29 liters/min; (b) 39 L/m; (c) 34 L/m   **17.** (a) 30,
(b) 27, (c) 25.5, (d) 24.75, (e) 24.3, (f) 24.15   **19.** See

Computer Programs   **21.** (a) 36.5 mi, (b) 43.8 mph

Exercise Set 22.2

1. 84   **3.** 11.9375   **5.** 181   **7.** 72.28125
9. 175.75   **11.** 30.85   **13.** See *Computer Programs*
15. \$2,480   **17.** $1.206\,\Omega$

Exercise Set 22.3

1.

x	0.9	0.99	0.999	0.9999	1.0001	1.001	1.01	1.1
$f(x) = 3x$	2.7	2.97	2.997	2.9997	3.0003	3.003	3.03	3.3

$\lim\limits_{x \to 1} 3x = 3$

3.

x	-1.1	-1.01	-1.001	-1.0001	-0.9999	-0.999	-0.99	-0.9
$h(x) = x^2 + 2$	3.21	3.0201	3.002	3.0002	2.9999	2.998	2.9801	2.81

$\lim\limits_{x \to -1} (x^2 + 2) = 3$

5.

x	-0.1	-0.01	-0.001	-0.0001	0.0001	0.001	0.01	0.1
$f(x) = \dfrac{\tan x}{x}$	1.0033467	1.0000333	1.0000003	1	1	1.0000003	1.0000333	1.0033467

$\lim\limits_{x \to 0} \dfrac{\tan x}{x} = 1$

7.

x	0.9	0.99	0.999	0.9999	1.0001	1.001	1.01	1.1
$h(x) = \dfrac{x}{x-1}$	-9	-99	-999	-9999	10001	1001	101	11

$\displaystyle\lim_{x\to 1}\frac{x}{x-1}$ Does not exist

9. **(a)** 0, both sides of graph near $x = -1$ converge on 0;
(b) Does not exist, graph goes toward -2 when $x < 2$ and goes toward 1 for $x > 2$; **(c)** 4, both sides of graph near $x = 3$ converge on 4.

11.

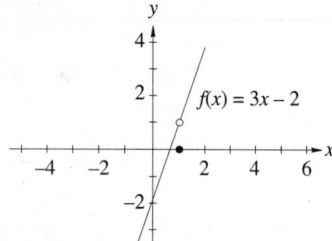

$\displaystyle\lim_{x\to 1} f(x)$ Does not exist

13.

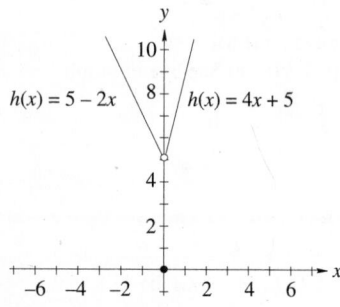

$\displaystyle\lim_{x\to 1} h(x)$ Does not exist

15.

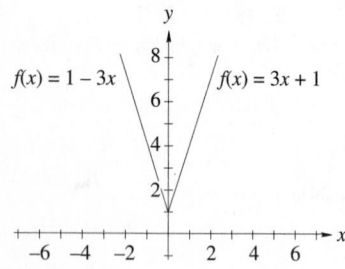

$\displaystyle\lim_{x\to 1} g(x) = 1$

17.

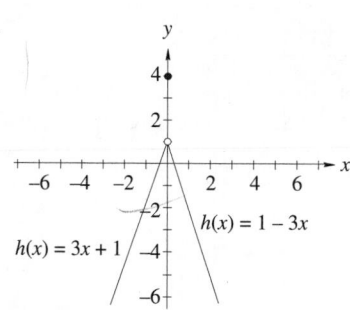

$\displaystyle\lim_{x\to 1} h(x)$ Does not exist

19. -15 **21.** -4 **23.** 8 **25.** 0 **27.** 4

Exercise Set 22.4

1. **(a)** -2; **(b)** 1; **(c)** 4; **(d)** 4 **3.** 3 **5.** ∞ **7.** -4
9. -1 **11.** -1 **13.** 0 **15.** 0 **17.** 0 **19.** 2
21. 0 **23.** $-\infty$ **25.** **(a)** 0.02381, **(b)** 0.0556, **(c)** 0
27. **(a)** \$4.75, **(b)** \$4.75, **(c)** \$4.75

Exercise Set 22.5

1. 8 **3.** 5 **5.** 1 **7.** 6 **9.** 9 **11.** 5
13. 8 **15.** 27 **17.** 1 **19.** $\sqrt{6}$ **21.** 2 **23.** -2
25. 2 **27.** $\frac{13}{9}$ **29.** 8 **31.** $\frac{2}{5}$ **33.** 0

Exercise Set 22.6

1. Continuous **3.** Not continuous,
$\displaystyle\lim_{x\to 1-} h(x) = 4 \neq 3 = h(1)$ **5.** Not continuous, $k(2)$ is
not defined **7.** Not continuous, $\displaystyle\lim_{x\to -2-} g(x) = 4 \neq g(-2)$
9. Not continuous, $\displaystyle\lim_{x\to 2-} j(x) = 0 = j(2)$ **11.** None
13. -1 **15.** $-3, 2$ **17.** None **19.** 2
21. $(-\infty, -3), (-3, \infty)$ **23.** $(-\infty, -2), (-2, 2), (2, \infty)$
25. $[-3, \infty)$ **27.** $(-\infty, -3], [3, \infty)$ **29.** $(-\infty, 0),$
$(0, \infty)$ **31.** **(a)** yes

(b)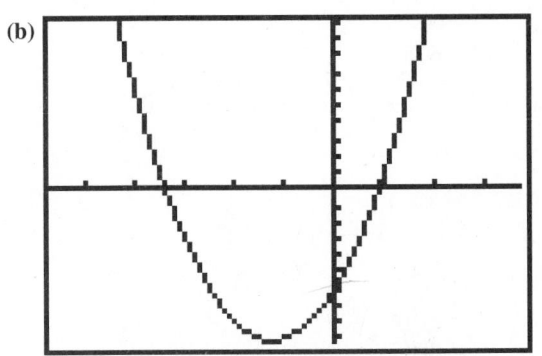

33. (a) \$3,585, **(b)** \$3,585, **(c)** \$3,585, **(d)** Yes, **(e)** \$14,155, **(f)** \$14,155, **(g)** \$14,155, **(h)** Yes

Review Exercises
1. 2 **3.** -10 **5.** $b+1$ **7.** 21 **9.** 65.96875
11. 0 **13.** 6 **15.** 3 **17.** 0 **19.** 0 **21.** 5
23. -3 **25.** $(-\infty, -7)(-7, \infty)$ **27.** $(-\infty, -5), (5, \infty)$

Chapter 22 Test
1. 2 **3.** 13 **5.** 2 **7.** $x = 2$ and $x = 3$

≡ ANSWERS FOR CHAPTER 23

Exercise Set 23.1
1. 3 **3.** -3 **5.** $6x$ **7.** $4x$ **9.** $-8x$
11. $-12x^2$ **13.** $1-9t^2$ **15.** $\dfrac{-1}{(x+1)^2}$ **17.** $\dfrac{8x}{(1-x^2)^2}$

19. $32t - 6$ **21.** $\dfrac{x}{\sqrt{x^2+4}}$ **23.** $\dfrac{1-x}{\sqrt{2x-x^2}}$ **25.** $3; 3$

27. 98, 66, 2 **29.** 0, 3 **31.** $\dfrac{-10}{21^2} = \frac{-10}{441}, \frac{10}{441}$ **33.** $\frac{3}{4}$,

$-\frac{3}{4}$ **35.** 0, $\frac{15}{64}$, $-\frac{20}{27}$

Exercise Set 23.2
1. 0 **3.** 7 **5.** $18x$ **7.** $5x^{14}$ **9.** $-10x^{-3}$
11. $-x^{1/2}$ **13.** $\dfrac{2\sqrt{3}}{\sqrt{x}}$ **15.** $-x^{-4/3} = \dfrac{-1}{x\sqrt[3]{x}}$

17. $18x + 3$ **19.** $x^2 + x - 5 - \frac{4}{3}x^{-3}$

21. $4x^4 - \frac{9}{2}x^2 - x^{-3} + 2x^{-4}$ **23.** $12\sqrt{3}x^3 + \dfrac{3\sqrt{5}}{2\sqrt{x}} + 4x$

25. $32t - 32$ **27.** $-8.0t + t^{-1/2}$
29. $\dfrac{\sqrt{6}}{2}t^{-1/2} - 6t^{1/2} + \frac{2}{3}t^{-1/3}$ **31.** $\frac{35}{3}x^{4/3}$ **33.** $\dfrac{-8}{x^3}$
35. $-\frac{2}{3}x^{-7/6}$ **37.** $\frac{9}{2}x^{1/2} - 7 - 3x^{-5/2}$
39. $f'(x) = 3x^2 + 6x$, $f'(x) = 0$ when $x = 0$ or $x = -2$
41. $g'(x) = 6x^2 + 2x - 4$, $g'(x) = 0$ when $x = -1$ or $x = 2/3$
43. $j(x) = 12t^2 - \dfrac{1}{t^2}$, $j(t) = 0$ when $t = \dfrac{1}{\sqrt[4]{12}} \approx 0.537285$
45. (a) 146,382 people, **(b)** 63.52 increase in population per month **47. (a)** 172 ft, **(b)** 71 ft/s **49. (a)** 133,912 people, **(b)** 1,493 people per year, **(c)** 139,886 people, **(d)** 1,494 people per year **51.** 120 W/A **53. (a)** $P(n) = R(n) - C(n) = (78n - 0.025n^2) - (9,800 + 22.5n) = 55.5n - 0.025n^2 - 9,800$, **(b)** $P'(n) = 55.5 - 0.05n$, **(c)** \$53/telephone **55. (a)** 605 ft above sea level, **(b)** $t = 11$ s, **(c)** $h(2) = 621$ ft, $v(2) = 48$ ft/s, **(d)** $h(4) = 861$ ft, $v(4) = 192$ ft/s, **(e)** $h(6) = 1,325$ ft, $v(6) = 240$ ft/s, **(f)** $h(10) = 1,005$ ft, $v(10) = -720$ ft/s,

(g) $v(11) = -1,320$ ft/s

Exercise Set 23.3
1. $12x - 19$ **3.** $18x^2 - 14x - 8$
5. $-54x^2 + 12x + 36$ **7.** $16t^3 - 72t^2 + 72t$
9. $(3w^3 - 4w^2 + 2w - 5)(2w + w^{-2}) + (9w^2 - 8w + 2)(w^2 - w^{-1}) = 15w^4 - 16w^3 + 6w^2 - 16w + 4 - 5w^{-2}$
11. $\dfrac{(2x+3)(4) - (4x-1)(2)}{(2x+3)^2} = \dfrac{14}{(2x+3)^2}$
13. $\dfrac{(4x-1)(18x) - (9x^2+2)(4)}{(4x-1)^2} = \dfrac{36x^2 - 18x - 4}{(4x-1)^2}$
15. $12s + \dfrac{1}{3s^3}$ **17.** $12t^2 + \dfrac{4}{(t-2)^2}$ **19.** $2x + 1$
21. $\dfrac{(3\phi^2 - 1)(2\phi) - \phi^2(6\phi)}{(3\phi^2 - 1)^2}$
23. $\dfrac{(x^3 - 3x - 1)(9x^2 - 1) - (3x^3 - x + 1)(3x^2 - 3)}{(x^3 - 3x - 1)^2} = $

$\dfrac{-16x^3 - 12x^2 + 4}{(x^3 - 3x - 1)^2}$
25. $\dfrac{t^{1/3}(6t-1) - (3t^2 - t - 1)\frac{1}{3}t^{-2/3}}{t^{2/3}} = \dfrac{5t^2 - \frac{2}{3}t + 1/3}{t^{4/3}} = $

$\dfrac{15t^2 - 2t + 1}{3t^{3/4}}$ **27.** $24w + \dfrac{2}{(w+1)^2}$ **29.** $\dfrac{2x}{(x^2+1)^2}$
31. $\dfrac{(2-x^2)(3x-x^3)(-3x^2) - (4-x^3)(5x^4 - 15x^2 + 6)}{(2-x^2)^2(3x-x^3)^2} = $

$\dfrac{2x^7 - 20x^4 - 12x^3 + 60x^2 - 24}{(2-x^2)^2(3x-x^3)^2}$
33. $\dfrac{(s^2-1)(s-1)(6s^2) - 2s^3[(s^2-1) + (s-1)(2s)]}{(s^2-1)^2(s-1)^2} = $

$\dfrac{12s^5 - 10s^4 - 8s^3 + 6s^2}{(s^2-1)^2(s-1)^2}$

35. $\dfrac{(3x^3+1)(6x^2-22x+14)-(2x^3-11x^2+14x-3)(9x^2)}{(3x^3+1)^2}$

$= \dfrac{33x^4-84x^3+33x^2-22x+14}{(3x^3+1)^2}$

37. $\dfrac{(t^2-1)}{(2t+1)}\left[\dfrac{(2t+1)-(t-1)(2)}{(2t+1)^2}\right]+$

$\dfrac{(t-1)}{(2t+1)}\left[\dfrac{(2t+1)(2t)-(t^2-1)(2)}{(2t+1)^2}\right]=\dfrac{2t^3+3t^2-5}{(2t+1)^3}$

39. 16 **41.** Tangent: $y-2=-\frac{1}{2}(x-3)$ or $2y+x=7$
Normal: $y-2=2(x-3)$ or $y-2x-4$ **43.** Tangent:
$y-3=-\frac{1}{3}(x-1)$ or $3y+x=10$ Normal: $y-3=3(x-1)$
or $y-3x=0$ **45.** Tangent: $y+2=\frac{1}{8}(x+2)$ or
$8y-x=-14$ Normal: $y+2=-8(x+2)$ or $y+8x=-18$
47. 8.2 thousand people per year or 8,200 people per year

49. $V'=\dfrac{6600}{(R+60)^2}$

51. (a) $N'(t)=\dfrac{9t^2+8t^{3/2}-6t-8\sqrt{t}-3}{2\left(3\sqrt{t}+2\right)^2\sqrt{t}}$, (b) 777

bacteria/hr

Exercise Set 23.4
1. $12(3x-6)^3$ **3.** $-25(5x-7)^{-6}$
5. $4(2x+3)(x^2+3x)^3$ **7.** $\dfrac{-4x}{(4x^2+7)^{3/2}}$
9. $3(t^4-t^3+2)^2(4t^3-3t^2)$
11. $3(u^3+2u^{-4})^2(3u^2-8u^{-5})$
13. $2[(x-2)(3x^2-x)]^2[(x-2)(6x-1)+(3x^2-x)]=$
$2[(x-2)(3x^2-x)]^2(9x^2-14x+2)$
15. $6(v^2+1)^2(2v-5)^2+4v(v^2+1)(2v-5)^3$

17. $4\left[\left(x^3-6x^2\right)\left(5x-6x^2+x^3\right)\right]^3\left[\left(x^3-6x^2\right)\left(5-12x+3x^2\right)+\left(5x-6x^2+x^3\right)\left(3x^2-12x\right)\right]=$
$4\left[\left(x^3-6x^2\right)\left(5x-6x^2+x^3\right)\right]^3\left(6x^5-60x^4+164x^3-90x^2\right)$

19. $\dfrac{(3x^2+2)^3(3x^2-7)}{\sqrt{x^3-7x}}+18x(3x^2+2)^2\sqrt{x^3-7x}=\dfrac{243x^8-945x^6-1,710x^4-732x^2-56}{\sqrt{x^3-7x}}$

21. $\dfrac{3(v^3-9)^2(v^2-4v)^2(2v-4)-(v^2-4v)(2)(3v^2)(v^3-9)}{(v^3-9)^4}=\dfrac{12v^7-96v^6+138v^5+540v^4-1728v^3+1728v^2}{(v^3-9)^3}$

23. $6(x^5+4)^5(5x^4)=30x^4(x^5+4)^5$
25. $4x(4x^2-5)^{-1/2}$ **27.** $(14x^2-6)(7x^3-9x)^{-1/3}$
29. $\dfrac{-6}{(x+3)^3}[\dfrac{1}{(x+3)^2}+1]^2$
31. $\frac{1}{3}[(x^3+4)^2-2(x^3+4)]^{-2/3}[2(x^3+4)-2](3x^2)=$
$2x^2\left[\left(x^3+4\right)^2-2\left(x^3+4\right)\right]^{-2/3}(x^3+3)$
33. $\left[16\left(3t^2-4t\right)^3-3\right](6t-4)=2,592t^7-12,096t^6+$
$20,736t^5-15,360t^4+4,096t^3-18t+12$
35. $6(9x^2+4x)^5(18x+4)$
37. $10(11x^5-2x+1)^9(55x^4-2)$ **39.** $\dfrac{-56(18x)}{(9x^2-4)^9}$
41. $\dfrac{3x}{\sqrt[4]{2x^2-5}}$ **43.** $\dfrac{-3072}{x^7}(\dfrac{128}{x^6}-1)$ **45.** $\dfrac{-480}{x^{11}}$
47. $y+8=192(x-2)$ or $192x-y=-394$
49. (a) $P=E'=\dfrac{dE}{dt}=144\left(1+4t^2\right)^2 t$ W, (b) $P=E'=$

$\dfrac{dE}{dt}=\dfrac{144\left(1+4t^2\right)^2 t}{1000}=0.144\left(1+4t^2\right)^2 t$ kW,

(c) 243.36 kW **51.** (a) $P'(t)=\dfrac{250t\left(t^2+30\right)}{\left(t^2+15\right)^{3/2}}$,

(b) 271.8 ppm/day **53.** $R'=\dfrac{4n-4}{(n+1)^3}$
55. (a) $I'=-\dfrac{120X_C}{\left(900+X_C^2\right)^{3/2}}$, (b) -0.0504 A/Ω

Exercise Set 23.5
1. $-4/5$ **3.** $\dfrac{1}{2y}$ **5.** x/y **7.** $-9x/16y$
9. $\dfrac{4x}{2+y}$ **11.** $\dfrac{2x-y-5y^2}{x+10xy}$
13. $(2x+4y)/(1-4x-2y)$ **15.** $\dfrac{1+10x}{3+20y}$ **17.** $\dfrac{-8x}{3y^2}$
19. $\dfrac{y-3x^2}{18y^2-x}$ **21.** $\dfrac{1-2xy}{x^2}$ **23.** $-\dfrac{y^2}{x^2}$
25. $\dfrac{3(x+1)^{-2}-2xy+3(x+1)^{-2}}{x^2}=\dfrac{3-2xy(x+1)^2}{x^2(x+1)^2}$
27. $\dfrac{x^{-2}y-y+1}{x+x^{-1}}=\dfrac{y-x^2y+x^2}{x^3+x}$ **29.** $\dfrac{5+6xy^2-2x^3}{y-6x^2y}$
31. $\dfrac{2y\sqrt{x^2+y^2}+x^3}{2x\sqrt{x^2+y^2}-x^2y}$ **33.** $\dfrac{6x(x^2+y^2)^2}{1-6y(x^2+y^2)^2}$
35. $y+4=\frac{3}{4}(x-3)$ or $4y-3x+25=0$
37. $y-2=\frac{9}{16}(x+2)$ or $9x-16y+50=0$ **39.** tangent:
1; normal: -1 **41.** (a) $\dfrac{ds}{dt}=\dfrac{6\sqrt{st}-s}{8s\sqrt{st}+t}$, (b) 1.464 mi,

(c) 0.399 mi/min $= 23.94$ mi/hr   **43. (a)** 1.48. **(b)** The store should sell 1.48 television sets for each digital satellite that is sold.

Exercise Set 23.6

1. $24x - 12$ **3.** $840x$ **5.** $2x^{-3}$ **7.** $2 + \dfrac{2}{(t+1)^3}$

9. $-6u^{-4} + 24u^{-5} = \dfrac{-6u + 24}{u^5}$ **11.** $-\frac{1}{4}(x+1)^{-3/2}$

13. 48   **15.** $\frac{10}{9}x^{-8/3}$

17. $24x(x+1)^{-5} - 24(x+1)^{-4} = -24(x+1)^{-5}$

19. $216w^2(3w^2+1)^{-4} - 12(3w^2+1)^{-3} = (180w^2 - 12)(3w^2+1)^{-4}$.   **21.** $840x^3 - 480x$

23. $24x^{-5}$ **25.** $\dfrac{18x^3 + 18x^2 + 6x}{(3x-1)^3}$

27. $\frac{3}{4}x(x+1)^{-5/2} - (x+1)^{-3/2} = \dfrac{-x-4}{4(x+1)^{5/2}}$

29. $y' = \dfrac{2x-1}{8y}$, $y'' = \dfrac{16y^2 - (2x-1)^2}{64y^3}$

31. $y' = \dfrac{4}{2xy-1}$; $y'' = \dfrac{-8(4x+2xy^2-y)}{(2xy-1)^2}$

33. $y' = \dfrac{-2y}{x}$; $y'' = \dfrac{6y}{x^2}$ **35. (a)** $N'(t) = 10t - 0.2t^{3/2}$, **(b)** $N''(t) = 10 - 0.3t^{1/2}$ **37. (a)** $h'(t) = 120 - 24t$, **(b)** $h''(t) = -24$

Review Exercises

1. $4x - 1$ **3.** $1 - 6x^2$ **5.** $\dfrac{1}{2\sqrt{x}}$ **7.** $6x + 2$

9. $15x^4 - \dfrac{12}{x^4} + \dfrac{3}{2\sqrt{x}}$ **11.** $\dfrac{3}{(x+3)^2}$

13. $(x-4)(3x^2-3) + (x^3-3x) = 9x^2 - 30x + 12$

15. $\dfrac{1}{2\sqrt{x+1}}$ **17.** $6x^5 - 10x^4 + 12x^3 - 18x^2 - 8x + 8$

19. $\dfrac{4x^3 + 15x^2 + 7}{2\sqrt{(x^3+7)(x+5)}}$

21. $(x^2+5x)^2(-6x^2)(x^3+1)^{-3} + 2(x^2+5x)(2x+5)(x^3+1)^{-2} = \dfrac{-2x^6 - 30x^5 - 100x^4 + 4x^3 + 30x^2 + 50x}{(x^3+1)^3}$

23. $3(x+1)^2(x-1)^2 + 2(x+1)(x-1)^3 = (x-1)^2(x+1)(5x+1)$ **25.** $-30x(x^2+1)^{-4}$

27. $2x(x^2-2)^{-2/3} - 3x^3(x^2-2)^{-5/2} = (x^2-2)^{-5/2}(-x^3-4x)$

29. $-4\left(u^2 + \dfrac{1}{u} - \dfrac{4}{u^2}\right)^{-5}(2u - u^{-2} + 8u^{-3})$

31. $6x^5 + 12x^3 + 4x$ **33.** $-\frac{1}{2}$ **35.** $\dfrac{2y - 2x - y^2}{2xy - 2x}$

37. $\dfrac{y^2 - 2xy^5}{3x^2y^4 - 1}$ **39.** $42x^5 + 24x^2$ **41.** $24 - \dfrac{12}{x^4}$

43. $42 - 420x^{-6}$ **45.** $30 + \frac{3}{8}x^{-5/2}$

47. $\dfrac{6x^2y - 8x + 2y}{(x^2-1)^2}$ **49.** $y' = \dfrac{2x-y}{x-2y}$;

$y'' = \dfrac{2x^3 - 8x^2y + 4x^2 + 8xy^2 + 2xy - 2y^2}{(x-2y)^3}$ **51.** 18 2/9

53. Tangent $\frac{17}{6}$; Normal $-\frac{6}{17}$ **55.** $0, -2$

57. $y - 2 = \frac{5}{4}(x+2)$ or $4y - 5x - 18 = 0$ **59.** $\pm\dfrac{\sqrt{3}}{3}$

Chapter 23 Test

1. $15x^2$ **3.** $2x^5(x^6-2)$ **5.** $\dfrac{2-4x}{(4x^2+1)^{1/2}}$

7. $y' = \dfrac{3x^2 - 2y - y^3}{2x + 3y^2x}$ **9.** $v(t) = 9t + 27$

ANSWERS FOR CHAPTER 24

Exercise Set 24.1

1. $v(t) = 6t - 12$; $a(t) = 6$

t	0	1	2	3	4	5
$s(t)$	5	−4	−7	−4	5	20

3. $v(t) = 3t^3 - 12$; $a(t) = 6t$

t	−4	−3	−2	−1	0	1	2	3	4
$s(t)$	−14	11	18	13	2	−9	−14	−7	18
$v(t)$	36	15	0	−9	−12	−9	0	15	36
$a(t)$	−24	−18	−12	−6	0	6	12	18	24

5. $v(t) = 1 - 4/t^2$; $a(t) = 8/t^3$

t	1	2	3	4
$s(t)$	5	4	17/3	5
$v(t)$	-3	0	0.5555	0.75
$a(t)$	8	1	0.2963	0.125

7. $v(t) = t^{-1/2} - \frac{1}{2}t^{-3/2}$; $a(t) = -\frac{1}{2}(t^{-3/2} - \frac{3}{4}t^{-5/2})$

t	1	2	3	4
$s(t)$	3	3.5355	4.0415	4.5
$v(t)$	0.5	0.5303	0.4811	0.4375
$a(t)$	0.25	-0.0442	-0.0481	-0.0391

9. (a) 9 s; (b) $v = 144$ ft/s downward; $a = 32$ ft/s² downward; (c) 0 ft (ground level)   **11.** (a) 6 s; (b) $v - 29.4$ m/s; $a = -9.8$ m/s²; (c) 0 m   **13.** (a) $\dfrac{dF}{ds} = F(s) = \dfrac{-100}{s^3}$ dynes/cm; (b) $\dfrac{-100}{27}$ dynes/cm   **15.** (a) $\frac{9}{5}$; (b) $\frac{5}{9}$; (c) $\frac{5}{9}$

17. (a) $\dfrac{dQ}{dt} = -50 + t$ gallons/min; (b) -46 gal/min; (c) -42 gal/min   **19.** (a) 4.0 m/s; (b) 8.0 m; (c) 4.0 s

21. (a) $v = -N\dfrac{d\phi}{dt} = -\dfrac{60}{t^{1/2}} + 40t$; (b) -20 V at $t = 1.0$ s and 340 V at $t = 9.0$ s   **23.** 46.24π m²/s   **25.** (a) $P = \dfrac{6.41}{V}$;

(b) $\dfrac{dP}{dV} = -\dfrac{6.41}{V^2}$ **27.** (a) $v(t) = s'(t) = 6t - 0.05t^2$, (b) $a(t) = v'(t) = s''(t) = 6 - 0.1t$, (c) $|v(t)| = |6t - 0.05t^2|$

Exercise Set 24.2

1. Critical values: $x = 4$; increasing on $(4, \infty)$; decreasing on $(-\infty, 4)$; Maximum: none; Minimum: $f(4) = -16$
3. Critical values: $x = 0$; increasing on $(-\infty, 0)$; decreasing on $(0, \infty)$; Maximum: $m(0) = 16$; Minimum: none
5. Critical values: $x = 2$; increasing on $(-\infty, 2)$; decreasing on $(2, \infty)$; Maximum: $j(2) = 31$; Minimum: none
7. Critical values: $x = 2$; increasing on $(2, \infty)$; decreasing on $(-\infty, 2)$; Maximum: none; Minimum: $f(2) = -15$
9. Critical values: $t = \pm 1$; increasing on $(-\infty, -1)$ and $(1, \infty)$; decreasing on $(-1, 1)$; Maximum: $q(-1) = 2$; Minimum: $q(1) = -2$ **11.** Critical values: $x = 0, \pm 6$; increasing on $(-6, 0)$ and $(6, \infty)$; decreasing on $(-\infty, -6)$ and $(0, 6)$; Maximum: $y(0) = 16$; Minimum: $y(-6) = y(6) = -308$ **13.** Critical values: $z = -4$;

increasing on $(-\infty, \infty)$ decreasing nowhere; Max: none; Minimum: none.   **15.** Critical values: $v = -1, 0, 2$; increasing on $(-1, 0)$ and $(2, \infty)$; decreasing on $(-\infty, -1)$ and $(0, 2)$; Maximum: $g(0) = 12$; Minimum: $g(-1) = 7$ and $g(2) = -20$.   **17.** Critical values: none; increasing nowhere; decreasing on $(-\infty, -1)$ and $(-1, \infty)$; Maximum: none; Minimum: none   **19.** Critical values: $x = -1$; increasing on $(-1, 0)$ and $(0, \infty)$; decreasing on $(-\infty, -1)$; Maximum: none; Minimum: $g(-1) = 3$ **21.** Critical values: $x = 0$; increasing on $(0, 1)$ and $(1, 2)$; decreasing on $(-\infty, 0)$ and $(2, \infty)$; Maximum: $f(2) = -4$; Minimum: $f(0) = 0$ **23.** Critical values: $t = \pm 1, \pm\dfrac{\sqrt{2}}{2}$; increasing on $\left(-\dfrac{\sqrt{2}}{2}, \dfrac{\sqrt{2}}{2}\right)$; decreasing on $\left(-1, \dfrac{-\sqrt{2}}{2}\right)$ and $\left(\dfrac{\sqrt{2}}{2}, 1\right)$; Maximum: $f\left(\dfrac{\sqrt{2}}{2}\right) = 0.5$; Minimum: $f\left(-\dfrac{\sqrt{2}}{2}\right) = -0.5$
25. Critical values: $x = 0.6, 1$; increasing on $(-\infty, 0.6)$ and $(1, \infty)$; decreasing on $(0.6, 1)$; Maximum: $(0.6, 0.326)$; Minimum: $(1, 0)$ **27.** (a) 8 s; (b) 1274 ft
29. (a) $P = 0.0$ W; (b) $t = 2.0$ s   **31.** (a) minimum when $R_2 = 0$; no maximum; (b) minimum $R = 0$
33. (a) $P'(n) = -n + 8.6$, (b) 8.6, (c) \$6,498
35. (a) $N'(t) = 2t^{-1/2} + 3t^{1/2} - 2t^{3/2}$, (b) $t = 2$ mo

Exercise Set 24.3

1. Maximum: $(2, 12)$, Concave downward: $(-\infty, \infty)$

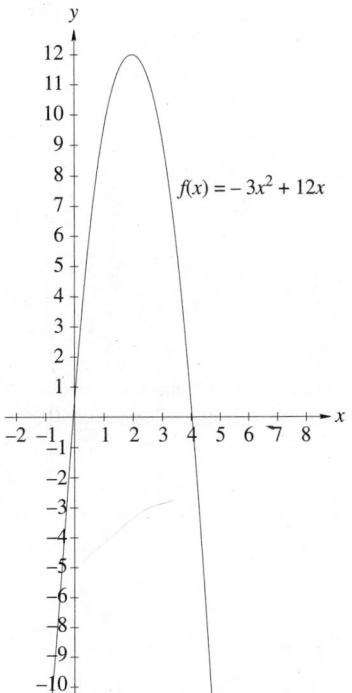

$f(x) = -3x^2 + 12x$

3. Maximum: $(-2, 28)$; Minimum: $(2, -4)$; Concave upward: $(0, \infty)$; Concave downward: $(-\infty, 0)$: Inflection point: $(0, 12)$

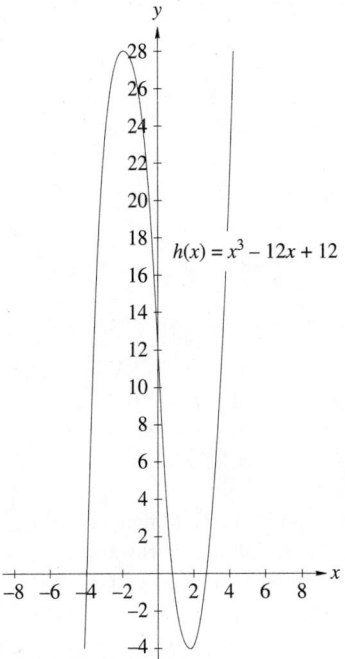

$h(x) = x^3 - 12x + 12$

5. Maximum: none; Minimum: none; Concave upward: $(2, \infty)$, Concave downward: $(-\infty, 2)$: Inflection point: $(2, 4)$

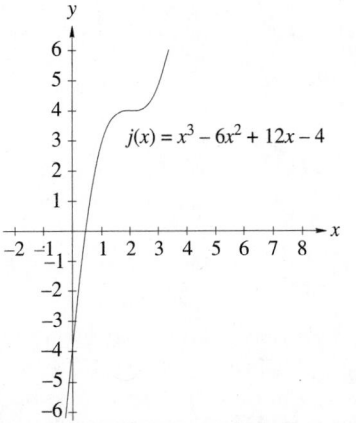

$j(x) = x^3 - 6x^2 + 12x - 4$

7. Maximum: none, Minimum: none; Concave upward: $\left(-\infty, \frac{1}{2}\right)$; Concave downward: $\left(\frac{1}{2}, \infty\right)$; Inflection point: $\left(\frac{1}{2}, -\frac{1}{2}\right)$

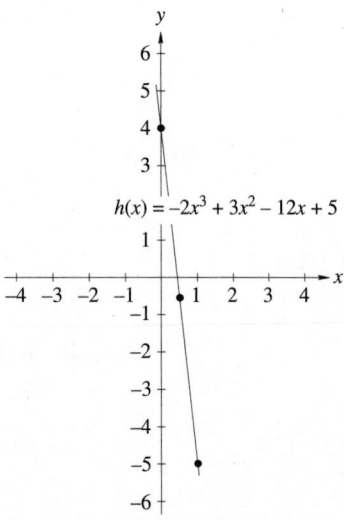

$h(x) = -2x^3 + 3x^2 - 12x + 5$

9. Maximum: $(-2, 10.67)$ and $(1, 1.67)$; Minimum: $(0, 0)$; Concave upward: $(-1.22, 0.55)$; Concave downward: $(-\infty, -1.22)$ and $(0.55, \infty)$; Inflection point: $(-1.22, 6.16)$ and $(0.55, 0.90)$

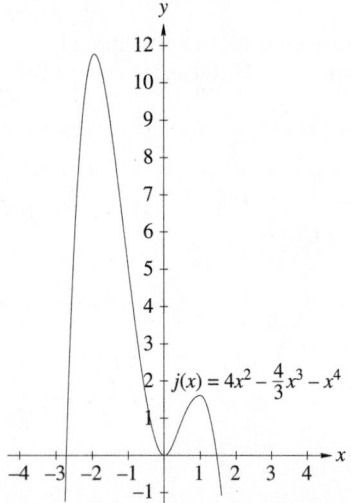

$j(x) = 4x^2 - \frac{4}{3}x^3 - x^4$

11. Maximum: none; Minimum: none; Concave upward: $(-\infty, 1)$; Concave downward: $(1, \infty)$; Inflection point: $(1, 0)$

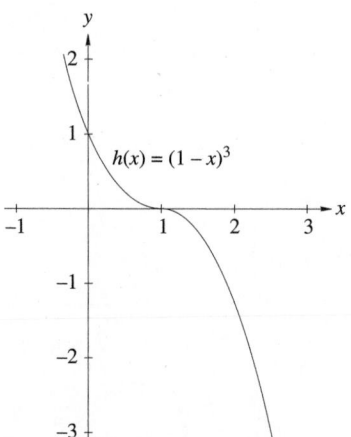

$h(x) = (1-x)^3$

13. Maximum: none; Minimum: none; Concave upward: $(-\infty, 0)$; Concave downward: $(0, \infty)$; Inflection point: $(0, 2)$

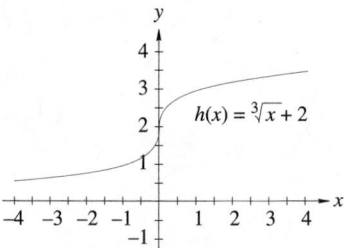

$h(x) = \sqrt[3]{x} + 2$

15. Maximum: $(-2, -4)$; Minimum: $(2, 4)$; Concave upward: $(0, \infty)$; Concave downward: $(-\infty, 0)$; Inflection point: none

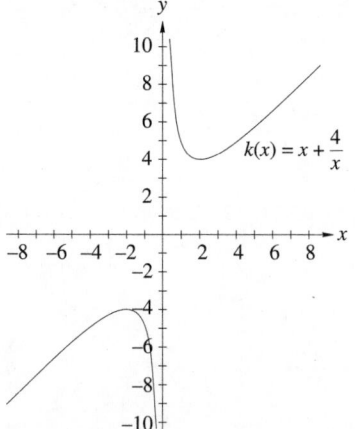

$k(x) = x + \frac{4}{x}$

17. Maximum: none; Minimum: $(-1, 3)$; Concave upward: $(-\infty, 0)$ and $\left(\sqrt[3]{2}, \infty\right) \approx (1.26, \infty)$; Concave downward: $\left(0, \sqrt[3]{2}\right) \approx (0, 1.26)$; Inflection point: $\left(\sqrt[3]{2}, 0\right) \approx (1.26, 0)$

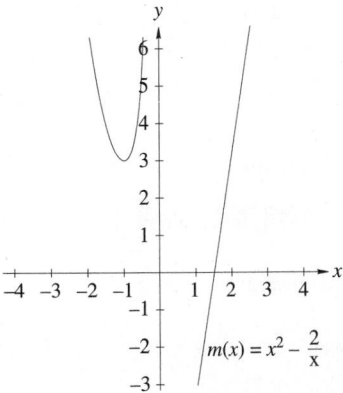

19. Maximum: $(-2, 7.333)$; Minimum: $(3, -13.5)$; Concave upward: $\left(\frac{1}{2}, \infty\right)$; Concave downward: $\left(-\infty, \frac{1}{2}\right)$; Inflection point: $(0.5, -3.083)$

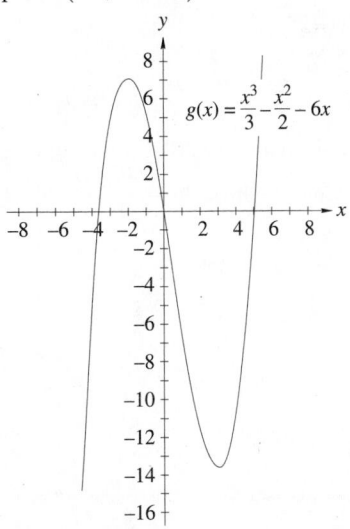

21. Maximum: $(1, 1)$; Minimum: $(0, 0)$; Concave upward: nowhere; Concave downward: $(0, \infty)$; Inflection point: none

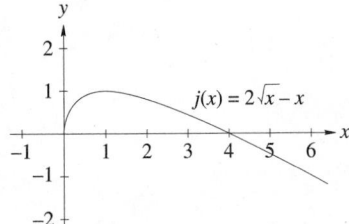

23. Maximum: none; Minimum: $(0, 2)$; Concave upward: nowhere, Concave downward: $(-\infty, \infty)$; Inflection point: none

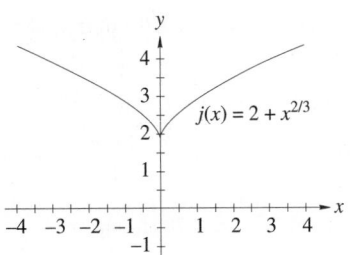

25. Maximum: $(0, 0)$; Minimum: $(4, -15.119)$; Concave upward: $(-\infty, \infty)$; Concave downward: nowhere; Inflection point: none

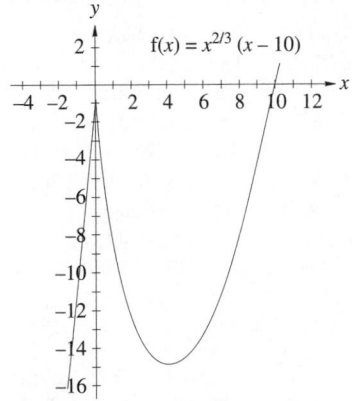

27. Maximum: $\left(1, \frac{1}{2}\right)$; Minimum: $\left(-1, -\frac{1}{2}\right)$; Concave upward: $(-\sqrt{3}, 0)$, $(-\sqrt{3}, 0)$, $(\sqrt{3}, \infty)$; Concave downward: $(-\infty, -\sqrt{3})$, $(0, \sqrt{3})$; Inflection points: $(-\sqrt{3}, -\sqrt{3/4})$, $(\sqrt{3}, \sqrt{3/4})$

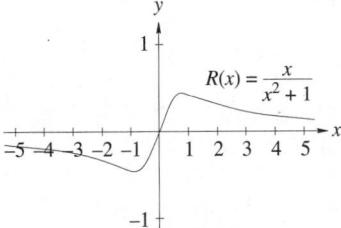

29. $t = 23$ **31.** (a) $N(9) = 116$, (b) $N'(4) = 8$ crimes/mo

Exercise Set 24.4

1. $2500 \, \text{ft} \times 5000 \, \text{ft}$; area $= 12{,}500{,}000 \, \text{ft}^2$ **3.** (a) $2000 \, \text{ft}$ along river $\times 3000 \, \text{ft}$; (b) cost is $\$12{,}000$ **5.** $4 \, \text{cm}$
7. $2\sqrt{2} \approx 2.83$ units horizontally and $\frac{3}{2}\sqrt{2} \approx 2.12$ units vertically   **9.** $4'' \times 4'' \times 2''$ **11.** radius $= 5.05 \, \text{cm}$, height $= 10.1 \, \text{cm}$ **13.** $21.984 \, \text{cm} \times 26.381 \, \text{cm}$
15. (a) $w = \sqrt{300} \approx 17.32 \, \text{in.}$; $d = \sqrt{600} \approx 24.49 \, \text{in.}$
(b) $w = 2r\frac{\sqrt{3}}{3} \, \text{in.}$; $d = 2r\frac{\sqrt{6}}{3} \, \text{in.}$ **17.** $80 \, \text{mm} \times 160 \, \text{mm}$
19. $1.25 \, \text{A}$ **21.** $A = \sqrt{c/b}$ **23.** $2.24 \, \text{m}$

25. (a) $t = 4.5\,s$; (b) $P = 0.0\,W$ **27.** about 3.1 m from the less intense light   **29.** $0.578\,L$ **31.** $\dfrac{h}{d} = \dfrac{3\pi + 2}{4}$
33. $12 - \frac{5}{3}\sqrt{3} \approx 9.11\,\text{mi}$ **35.** (a) $56.2 \times 74.7\,\text{ft}$, (b) $52{,}290$

Exercise Set 24.5

1. $\frac{400}{3} = 133\frac{1}{3}\,\text{cm}^2/\text{min}$ **3.** $\dfrac{\sqrt[4]{5}}{5} \approx 1.789\,\text{ft/min}$
5. $2.9104\,\text{km/min} = 174.6\,\text{km/h}$
7. (a) $\dfrac{5}{8\pi} \approx 0.1989\,\text{ft/min}$; (b) $\dfrac{10}{21\pi} \approx 0.1516\,\text{ft/min}$;
(c) $\dfrac{5}{12\pi} \approx 0.1326\,\text{ft/min}$ **9.** $12.4\,\text{mm/min}$ **11.** $15\,\text{cm}^2/\text{s}$
13. $11.25\,\text{cm}^2/\text{s}$ **15.** $24\,\text{mm}^2/\text{s}$ **17.** -8.3% of the volume per hour   **19.** $387.2\,\text{m/s}$ or $1394\,\text{km/h}$
21. (a) $-18.0\,\text{m/s}^2$; (b) $-18.0\,\text{m/s}^2$ **23.** (a) Either 5.75 mi west and 5.56 mi north of A or 5.75 mi west and 5.56 mi south of A. (b) 14.86 mph (c) $\theta = 51.2°$ north of east or $38.8°$ east of north   **25.** (a) $\frac{50}{7} \approx 7.143\,\text{ft/s}$; (b) $7.143\,\text{ft/s}$; (c) $7.143\,\text{ft/s}$
27. $0.0064\,\Omega/\text{s}$ **29.** $0.86\,\Omega/\text{min}$ **31.** decreasing $0.04725\,\Omega/\text{s}$ **33.** $-2.04\,\Omega/\text{min}$ **35.** (a) $218.47\,\text{Hz}$;
(b) $20.17\,\text{Hz/s}$ **37.** $0.06375\,\text{mm/min}$ **39.** $-1.06\,\Omega/\text{s}$
41. $4.30\,\Omega/\text{s}$ **43.** $0.1\pi\,\text{in.}^2/\text{yr} \approx 0.314\,\text{in.}^2/\text{yr}$

Exercise Set 24.6

1. See *Computer Programs*   **3.** $-2.2143; 0.5392; 1.6751$
5. $0.5760; 2.3810$ **7.** 4.4979 **9.** (b) two solutions: $r_1 = 9.4315\,\text{cm}$, $r_2 = 21.6893\,\text{cm}$

Exercise Set 24.7

1. $dy = (4x^3 - 2x)dx$ **3.** $dy = (15x^2 - 2x + 1)dx$
5. $dy = \frac{-2}{3}(4 - 2x)^{-2/3}dx$ **7.** $dy = 1.0, \Delta y = 1.04$
9. $dy = 3.9, \Delta y = 3.9675$
11. $dy = -0.0074074, \Delta y = -0.0065195$
13. (a) $\frac{16}{3}\pi\,\text{m}^3$; (b) $\frac{16}{300}\pi m^3 = \frac{4}{75}\pi\,\text{m}^3$; (c) 1%
15. (a) $0.96\,\text{in}^2$; (b) $0.00167 = 0.167\%$ **17.** $4\pi\,\text{m}^2$
19. (a) $\pm 24.2\,\text{m}^3$, (b) $\pm 2.42\%$ **21.** $0.08\sigma T^4$
23. $-0.03\,\text{A}$ **25.** $\pm 2.18\,\Omega$ **27.** 1.0%
29. $\pm 0.07\,\text{mm}^3$ **31.** $450\pi \approx 1414\,\text{mm}^3$ **33.** 132.50
35. $19.2\pi \approx 60.3\,\text{cm}^3$.

Exercise Set 24.8

1. $7x + C$ **3.** $2x - x^3 + C$ **5.** $x^4 - x^3 + x^2 + 9x + C$
7. $-\frac{1}{3}x^{-3} - x^{-2} + x^{-1} + 5x + C$
9. $\frac{2}{3}x^{3/2} + \frac{1}{2}x^2 + \frac{1}{x} + C$
11. $-5x^{-0.4} - 0.7x^{-5} - 4x^{-0.3} + C$ **13.** $\frac{1}{3}t^3 + t^2 + C$
15. $21t^2 - 5t + C$ **17.** (a) $6.12\,\text{s}$; (b) $195.96\,\text{ft/s}$
19. (a) $5\,\text{s}$; (b) $525\,\text{ft}$ **21.** (a) $17\,\text{s}$; (b) $1904\,\text{ft}$ **23.** $160\,\text{m}$
25. $-2.5\,\text{m/s}^2$ **27.** (a) $3.9\,\text{rev}$; (b) $4.6\,\text{rev}$
29. (a) $q = 2.2t^2 - 0.7t^3 + 5.0$; (b) $4.6\,\text{C}$

31. (a) $\phi(t) = \frac{-1}{200}(t^2 - 3t^{4/3} - 4.0)$;
(b) $\phi(0.729) = 0.027\,\text{Wb}$ **33.** $f(x) = -2\sqrt{x} + 1$
35. $T = 2875(x + 1)^{-2} + 25$

Review Exercises

1. Critical values: $0, \pm\dfrac{\sqrt{2}}{2}$; Maximum: $(0,0)$; Minimum: $\left(-\dfrac{\sqrt{2}}{2}, -0.25\right)$, $\left(\dfrac{\sqrt{2}}{2}, -0.457\right)$; Inflection point: $\left(\pm\dfrac{\sqrt{6}}{6}, -0.139\right)$; Concave upward: $\left(-\infty, -\dfrac{\sqrt{6}}{6}\right)$, $\left(\dfrac{\sqrt{6}}{6}, \infty\right)$; Concave downward: $\left(-\dfrac{\sqrt{6}}{6}, \dfrac{\sqrt{6}}{6}\right)$

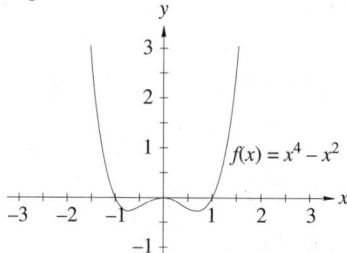

3. Critical values: $0, \pm 3$; Maximum: $(\pm 3, 81)$; Minimum: $(0,0)$; I.P. $\left(\pm\dfrac{\sqrt{3}}{3}, 5.889\right)$; Concave upward: $\left(-\dfrac{\sqrt{3}}{3}, \dfrac{\sqrt{3}}{3}\right)$; Concave downward: $\left(-\infty, -\dfrac{\sqrt{3}}{3}\right)$; $\left(\dfrac{\sqrt{3}}{3}, \infty\right)$

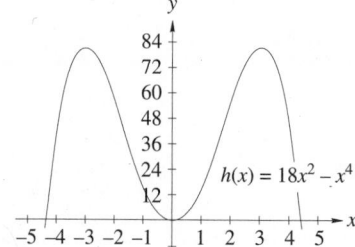

5. Critical value: 0; Maximum: none; Minimum: none; Inflection point: $(0,0)$; Concave upward: $(-\infty, 0)$; Concave downward: $(0, \infty)$

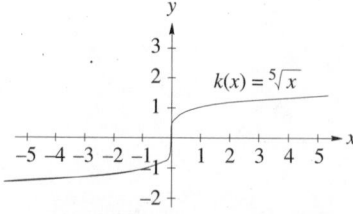

7. Critical value: 0; Maximum: $\left(0, -\frac{1}{4}\right)$; Minimum: none; Inflection point: none; Asymptotes: $x = \pm 2$; $y = 1$; Concave upward: $(-\infty, -2)$, $(2, \infty)$; Concave downward: $(-2, 2)$

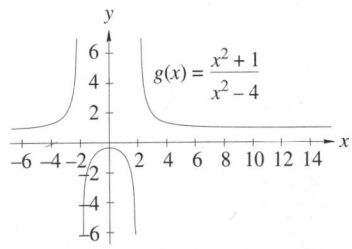

9. Critical values: none; Maximum: none; Minimum: none; Inflection point: $(-0.3275, 0.1475)$; Asymptotes: $x = -2$, $x = 1$, $y = 0$; Concave upward: $(-2, -0.3275)$, $(1, \infty)$; Concave downward: $(-\infty, -2)$, $(-0.3275, 1)$

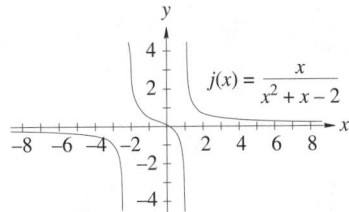

11. distance: Maximum at $(0.056, 0.028)$; Minimum at $(4, -76)$ and $(-4, -212)$; velocity: Maximum at $(4, -23)$ and $(-4, 121)$; Minimum at $(3, -26)$; acceleration: Minimum at $(-4, -42)$ and maximum at $(4, 6)$ **13.** $x^3 - 2x^2 + C$
15. $\frac{2}{3}x^{3/2} + \frac{1}{3}x^3 + \frac{3}{x} + C$ **17. (a)** $v(t) = 288 - 32t$;
$a(t) = -32$; **(b)** height: 896 ft; velocity: 160 ft/s upward; acceleration: -32 ft/s^2; **(c)** 1296 ft; **(d)** 9 s; **(e)** 18 s; **(f)** 288 ft/s downward   **19.** 56.98 mph   **21.** $300 \times 300 = 90000\,\text{m}^2$
23. (a) $0.09\pi \approx 0.2827\,\text{m}^3$; **(b)** $0.12\pi \approx 0.3770\,\text{m}^2$
25. 13.5 V   **27.** $r = \sqrt[3]{\dfrac{10}{\pi}}\,\text{m} \approx 1.4710\,\text{m}$,

$h = 2\sqrt[3]{\dfrac{10}{\pi}}\,\text{m} \approx 2.942\,\text{m}$ **29. (a)** 784 ft; **(b)** 244 ft/s

Chapter 24 Test

1. (a) $x = -2, 0,$ and 2; **(b)** Maximum $(0, 0)$, Minimum $(-2, -16)$ and $(2, -16)$, **(c)** Concave upward: $\left(-\infty, -\sqrt{\frac{4}{3}}\right)$ and $\left(\sqrt{\frac{4}{3}}, \infty\right)$, **(d)** Concave downward: $\left(-\sqrt{\frac{4}{3}}, \sqrt{\frac{4}{3}}\right)$ **3.** 2.646

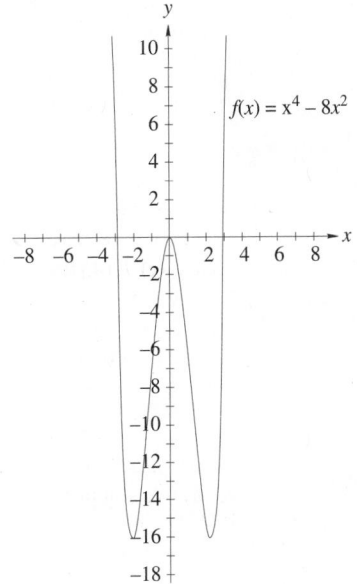

5. (a) $v(t) = 3t^2 - 24t$, $a(t) = 6t - 24$; **(b)** Position: Maximum at $(0, 5)$, Minimum at $(4, -123)$, Velocity: Maximum at $(-2, 60)$, Minimum at $(4, -48)$; Acceleration: Maximum at $(4, 0)$, Minimum at $(-2, -36)$ **7.** 3.38 m

☰ ANSWERS FOR CHAPTER 25

Exercise Set 25.1
1. 4   **3.** 28   **5.** 4   **7.** -6 **9.** $\frac{-2}{3}$ **11.** $\frac{22}{3}$
13. 20   **15.** 72   **17.** $\frac{14}{3}$

Exercise Set 25.2
1. 30   **3.** 24   **5.** $33\frac{1}{3}$ **7.** 6   **9.** 18
11. -18 **13.** -16 **15.** 18   **17.** 20   **19.** -30
21. $\frac{37}{3}$ **23.** $-\frac{37}{3}$ **25.** 2   **27.** 60   **29.** $\frac{26}{3}$
31. $-\frac{7}{8}$ **33.** $\frac{1}{6}$ **35.** $\frac{4}{3}\sqrt{2} \approx 1.8856$ **37.** -2.9501
39. 10.7 m   **41. (a)** 39.4 m/s; **(b)** 74.1 m; **(c)** 5.1258 s
43. (a) $1\,000\,000\,\text{m/s}^2$; **(b)** 5005 m   **45.** 117 moose
47. (a) 2 ft, **(b)** 0.545 ft, **(c)** 3.92 ft

Exercise Set 25.3
1. $9x + C$ **3.** $2x^3 + C$ **5.** $\frac{2}{7}x^{7/2} + C$
7. $\frac{1}{4}t^4 + t + C$ **9.** $\frac{1}{3}y^3 + 2y^2 - 3y + C$
11. $\frac{3x\sqrt[3]{x}}{4} - 9\sqrt[3]{x} + C$ **13.** $\frac{1}{2}(x^2 + 3)^2 + C$
15. $-\frac{(4 - 2x^2)^2}{2} + C$ **17.** $-\frac{1}{4}(3 - x^2)^4 + C$
19. $\frac{1}{3}(x^2 + 4)^{3/2} + C$ **21.** $\frac{1}{2}(\sqrt{x} - 1)^4 + C$
23. $\frac{1}{45}(3x^3 + 1)^5 + C$ **25.** $\sqrt{x^2 + 3} + C$
27. $-\frac{1}{4}(x^2 + 3)^{-2} + C$ **29.** $\frac{9}{5}x^5 - 2x^3 + x + C$
31. $\frac{1}{2}x^2 - 3x + C$ **33.** $\frac{15}{4}(x^2 - 1)^{2/3} + C$
35. $\frac{1}{18}(1 + 3x^2)^3 + C$ **37.** $\frac{3}{20}(4x^3 - 5)^{5/3} + C$

39. $x + 2x^3 + \frac{9}{5}x^5 + C$ **41.** $\frac{1}{3}(4x^2 + 2x)^{3/2} + C$

43. $-\dfrac{1}{x-1} + C$ **45.** $\frac{16}{5}x^{10} - 16x^6 + 36x^2 + C$

47. $\frac{1}{5}(x^3 - 3x)^{5/3} + C$ **49.** $\frac{1}{12}(3x+1)^4 + C$

51. $\frac{2}{5}(x^4 - 2x)^{5/4} + C$ **53.** $\frac{2}{3}(t+7)^{3/2} + C$

55. $\frac{1}{3}y^3 + y^2 + 4y + C$ **57.** -8 **59.** $\frac{9}{2}$ **61.** $156\frac{1}{3}$

63. 139.5 **65.** $6,865,432.5$

67. $L(x) = 0.3x + 0.001x^2 + 0.8$

69. $P(x) = 2x^2 + 35x - 150$

71. $T(x) = 2875(x+1)^{-2} + 25$

Exercise Set 25.4

1. 8

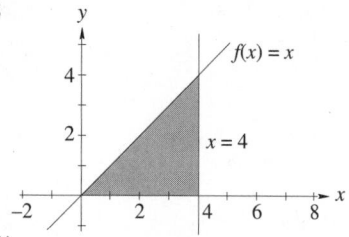

3. $\frac{64}{3}$

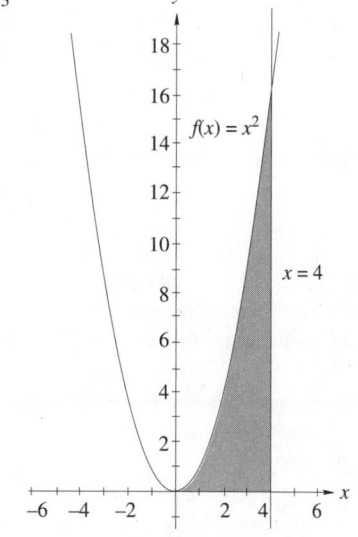

5. $160\frac{1}{6}$

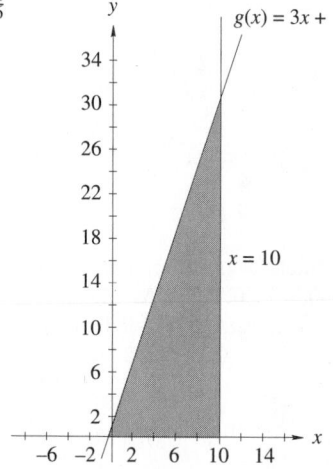

7. 12

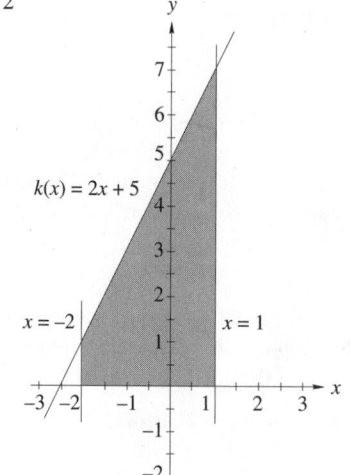

9. $\frac{22}{3}$

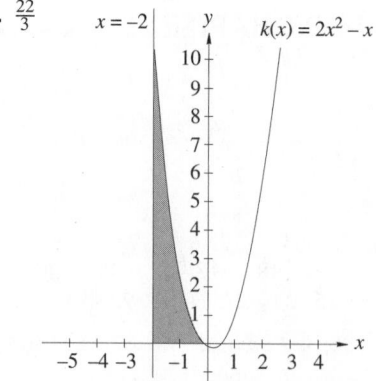

11. $\frac{32}{3}$

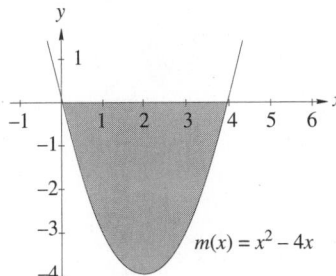

$m(x) = x^2 - 4x$

13. 36

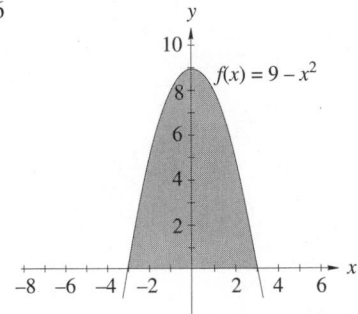

$f(x) = 9 - x^2$

15. 16

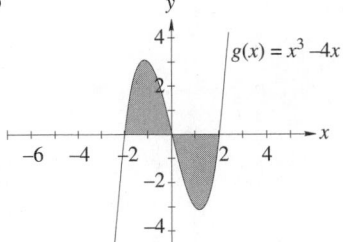

$g(x) = x^3 - 4x$

17. 18

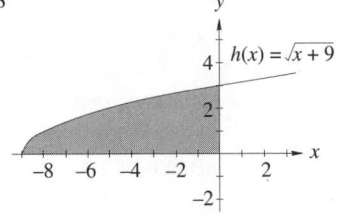

$h(x) = \sqrt{x + 9}$

19. $\frac{4}{3}$

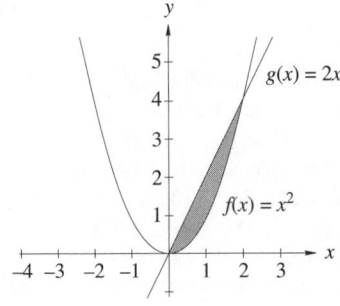

$g(x) = 2x$

$f(x) = x^2$

21. $8\sqrt{6} \approx 19.596$

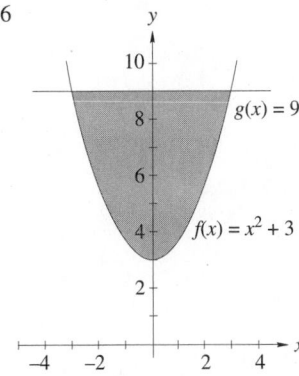

$g(x) = 9$

$f(x) = x^2 + 3$

23. 21.333

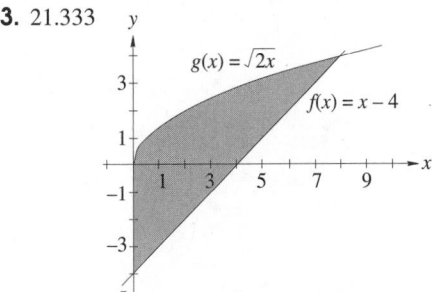

$g(x) = \sqrt{2x}$

$f(x) = x - 4$

25. 20.8333

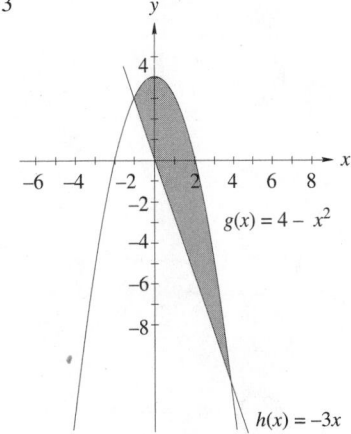

$g(x) = 4 - x^2$

$h(x) = -3x$

27. $\frac{125}{6} \approx 20.833$

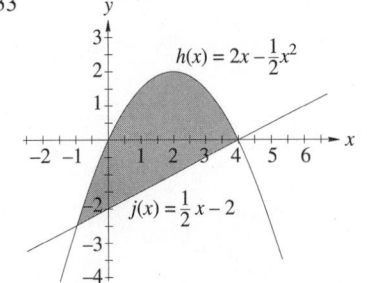

$h(x) = 2x - \frac{1}{2}x^2$

$j(x) = \frac{1}{2}x - 2$

29. 18

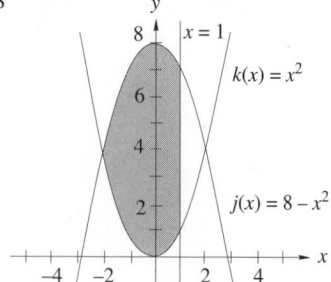

31. 8

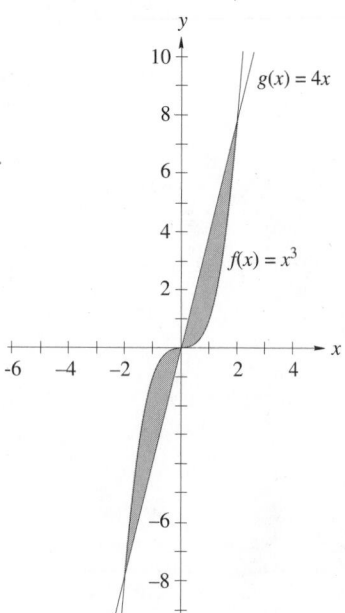

33. $41\frac{2}{3}$

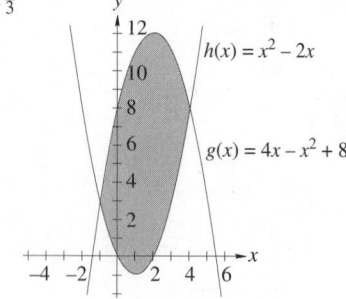

35. 2.25

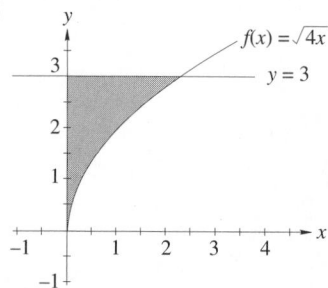

37. $\frac{64}{6} \approx 10.667$

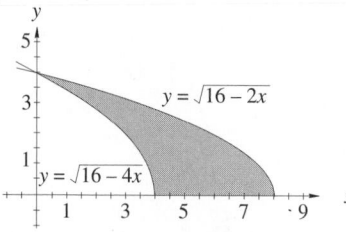

39. 0.375

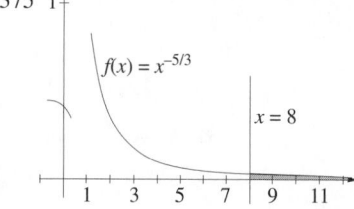

41. $\frac{1}{3}$

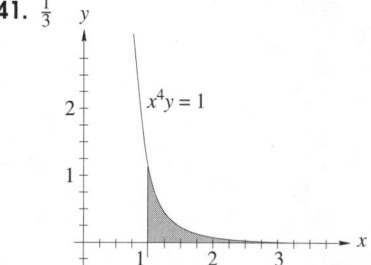

43. White. The blue area is $\frac{5}{12}$ and the white area is $\frac{7}{12}$.
45. $33\,866\frac{2}{3} \approx 33\,866.67\,\text{cm}^3$ **47.** 2630 Btu
49. 139.7 rad/s

Exercise Set 25.5

1. **(a)** 6.264 789 1; **(b)** 6.200 102 86; actual value, 6.2
3. **(a)** 0.643 283; **(b)** 0.656 526 3; actual value $\frac{2}{3}$
5. **(a)** 1.628 968; **(b)** 1.610 847   7. **(a)** 1.103 2107;
(b) 1.098 726   9. **(a)** 3.139 926; **(b)** 3.141 592 6
11. **(a)** 3.251 744; **(b)** 3.241 238   13. **(a)** 7.909 233;
(b) 7.912 321   15. **(a)** 2.141 030; **(b)** 2.170 342
17. **(a)** 30.85; **(b)** 30.833 33   19. **(a)** 24.646 25;
(b) 24.629 167   21. **(a)** 26.76 μC; **(b)** 20.34 μC   23. See
Computer Programs

25. Trapezoidal: 11.264 acres, Simpson: 11.417 acres

Review Exercises

1. $\frac{1}{6}x^6 + C$ **3.** 9   **5.** $\frac{9}{2}$ **7.** $-\frac{1}{3}(t+5)^{-3} + C$

9. $\frac{381}{7} \approx 54.429$ **11.** $\frac{1}{5}u^5 + \frac{1}{3}u^{-3} + C$

13. $\frac{2}{3}(x^2 - 5)^{3/2} + C$ **15.** $-\frac{3}{8}(6-x^2)^{4/3} + C$ **17.** $-\frac{39}{2}$

19. $(y^3 - 5)^{1/3} + C$ **21.** $\frac{16}{35}$

23. $\frac{1}{3}(2x)^{3/2} - (2x)^{1/2} + C$ **25.** $-\frac{1}{4}$

27. $\frac{855}{2312} \approx 0.36981$ **29.** $\frac{1}{3}$ **31.** 53   **33.** $\frac{125}{6}$

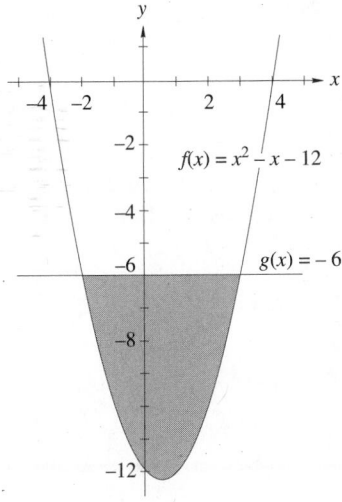

35. $\frac{1}{8}$ **37.** $\frac{9}{16} \approx 0.5625$ **39.** **(a)** 0.776130; **(b)** 0.781752

41. **(a)** 1.613695; **(b)** 1.613656   **43.** **(a)** 1.091812;

(b) 1.086185   **45.** **(a)** 118.5 m; **(b)** $\frac{127}{128} \approx 0.992$ m/s^2

47. 30.8 A

Chapter 25 Test

1. 1   **3.** $\frac{1}{5}x^5 + 3x^3 + 9x + C$ **5.** $x - 2x^{-1} + C$

7. $\frac{2}{3}(10^{3/2} - 2^{3/2}) \approx 19.196$ **9.** $\frac{125}{6}$ **11.** 0.00107 C

ANSWERS FOR CHAPTER 26

Exercise Set 26.1

1. $\bar{y} = 1/3$, $f_{\text{rms}} = \sqrt{1/5} = \dfrac{\sqrt{5}}{5} \approx 0.4472$ **3.** $\bar{y} = 10/3$,

$h_{\text{rms}} = \sqrt{248/14} \approx 4.0661$ **5.** $\bar{y} = 2\frac{38}{15}$,

$\dot{f}_{\text{rms}} = \sqrt{\dfrac{13}{2}} \approx 2.5495$ **7.** $\bar{y} = \frac{28}{3}$,

$f_{\text{rms}} = \sqrt{1648/15} \approx 10.4817$ **9.** $\bar{y} = 6\frac{2}{3}$,

$f_{\text{rms}} = \sqrt{\dfrac{226}{5}} \approx 6.7231$ **11.** $\bar{y} = 10.7083$, $j_{\text{rms}} = 12.8992$

13. 3.7522 cm for 90 days; 5.4755 cm for the year

15. $\bar{s} = 240$ ft; $\bar{v} = -75.8947$ ft/s   **17.** **(a)** 68° F; **(b)** 71° F

19. 2.6547 A   **21.** 23.0309 A   **23.** 27136 W

25. **(a)** 1.3927 V; **(b)** 19.40 W   **27.** 0.02 C   **29.** 2537.5 J

31. 0.6667 V

Exercise Set 26.2

1. 661.33π   **3.** 64π/3   **5.** 30π   **7.** 8π

9. 16π/3   **11.** 768π/7   **13.** $\dfrac{512\pi}{5}$ **15.** 2π/35

17. $\dfrac{10\pi}{3}$ **19.** $32\sqrt{2\pi}/3$ **21.** $V = \frac{4}{3}\pi r^3$

23. **(a)** $x^2 = \frac{2}{3}y$; **(b)** 1.8π   **25.** 16,913,712π ft^3

$\approx 53,135,993$ ft^3 **27.** **(a)** $\frac{800}{3}\pi \approx 837.758$ m^3;

(b) 5760π $\approx 18\,095.574$ m^3

Exercise Set 26.3

1. 8π   **3.** 32π   **5.** 625π   **7.** 768π/7

9. 64π/5   **11.** $\dfrac{8\pi}{35}$ **13.** 10π/7 $\approx 1.42857\pi$

15. 176π/15 $\approx 11.7333\pi$ **17.** $\dfrac{27\pi}{2}$ **19.** $28\pi\sqrt{21}$

21. 99.7 in.^{3}

Exercise Set 26.4

1. 12   **3.** 14/3   **5.** 14/3   **7.** 64.6308π

9. $4\sqrt{2\pi}$ **11.** $32\sqrt{17\pi} \approx 414.5$ **13.** approximately

244.17 m

Exercise Set 26.5

1. $(0.2, 5)$ **3.** $(2.9, 0.2)$ **5.** $\left(\frac{7}{6}, \frac{4}{3}\right)$ **7.** $\left(\frac{1}{74}, \frac{25}{74}\right)$

9. $(1.75, 3.25)$ **11.** $(32/7, 4/5)$

13. $(253/95, 195/152)$ **15.** $(5/4, 17/5)$ **17.** $\left(-\frac{1}{2}, \frac{8}{5}\right)$

19. $(0, 9)$ **21.** $(0.5, 4.4)$ **23.** $(7/4, 0)$ **25.** $\left(0, \frac{3}{5}\right)$

27. $(33/28, 0)$ **29.** $(0, 2.75)$ **31.** $(2.4, 0)$

33. $(5/9, 0)$

35. (a) $\left(\dfrac{5049.5}{199}, \dfrac{5049.5}{199}\right) \approx (25.374, 25.374)$;

(b) $\left(\dfrac{h^2 + hw - w^2}{2(2h - w)}, \dfrac{h^2 + hw - w^2}{2(2h - w)}\right)$ **37.** $m = 4040\,\text{g}$,

$\bar{x} = 26.6$ cm from the less dense end.

Exercise Set 26.6

1. $I_x = 0$; $I_y = 93$; $r_0 = \sqrt{93/8} \approx 3.4095$ **3.** $I_x = 52$;
$I_y = 19$; $r_0 = \sqrt{71/7} \approx 3.1847$
5. $I_y = 8\rho\sqrt{2/15} \approx 0.7542\rho$; $r_y = \sqrt{0.4} \approx 0.6325$
7. $I_x = 125\rho$; $r_x = \sqrt{25/3} \approx 2.8868$
9. $I_x \approx 0.014286\rho$; $r_x \approx 0.414$ **11.** $I_y = 256\,\text{g}\cdot\text{cm}^2$;
$r_y = \sqrt{4.8} \approx 2.1909$ cm **13.** $I_y = 41.6\,\text{g}\cdot\text{cm}^2$;
$r_y = \sqrt{156/55}\,\text{cm} \approx 1.6842$ cm **15.** $I_y = 4\rho\pi/3$;

$r_y = \sqrt{2/3} = 0.8165$ **17.** $I_y = 29.867\pi\rho$; $r_y \approx 1.497$
19. $I_x = 2592\pi\rho$; $r_x = 3\sqrt{2} \approx 4.2426$
21. $I_y = 512\pi\,\text{g}\cdot\text{cm}^2$; $r_y = \sqrt{7.76} \approx 2.7852$ cm
23. $I_x = 158.0952\pi\,\text{g}\cdot\text{cm}^2$; $r_y \approx 1.7436$ cm

Exercise Set 26.7

1. $25/8 = 3.125$ ft·lb **3.** 15 J **5.** 4.0 N
7. 4.8305 ft. **9.** 1125 J **11.** 126 ft·lb. **13.** 127 kJ
15. 179.76 MJ **17.** 99k J **19.** 80,000 mi-lb.
$= 4.224 \times 10^9$ ft·lb. **21.** $1\,238\,720\,\pi$ J $\approx$ J
23. 1 286 250 J **25.** 2.866 MJ **27.** $274\,400\pi$ J
29. (a) $19\,600\,\text{N/m}^2$; **(b)** 1 372 000 N **31.** 137 200 N
33. 39 200 N **35.** 24 578 N

Review Exercises

1. $\bar{y} = 14$, $f_{\text{rms}} = 15.1438$ **3.** $\bar{y} = 16/3$, $h_{\text{rms}} = 6.3666$
5. $V = 1488\pi$, $\bar{x} = 3.7742$ **7.** $V = 511\pi/9$, $\bar{x} = 1.8018$
9. $V = 16\pi/35 \approx 0.4571\pi$, $\bar{y} = 0.5469$ **11.** $V = 20.25\pi$,
$\bar{y} = 4.5$ **13.** $L = 9.0734$, $S = 93.9342\pi$
15. $L = 20.1111$, $S = 411.0124\pi$ **17.** 4.5 N·m **19.** 2 A
21. 35,250 ft·lb **23.** $39\,739\,392\pi$ J

Chapter 26 Test

1. 32 **3.** 227.5π **5.** 14.6238 **7.** $2\,250\,000\pi$ J

ANSWERS FOR CHAPTER 27

Exercise Set 27.1

1. $3\cos 3x$ **3.** $-6\sin 2x$ **5.** $2x\cos(x^2 + 1)$
7. $24\sin 3x \cos 3x$ **9.** $-6x\sin(3x^2 - 2)$

11. $\dfrac{1}{2\sqrt{x}}\cos\sqrt{x}$ **13.** $-3x^2(2x^3 - 4)^{-1/2}\sin\sqrt{2x^3 - 4}$

15. $2x + 2\sin x \cos x$ **17.** $\cos^2 x - \sin^2 x = \cos 2x$

19. $\dfrac{-2\sin 2x \sin x - 4\cos x \cos 2x}{\sin^2 2x} = \dfrac{-2\cos^2 x}{\sin x}$

21. $2x\sin x + x^2\cos x$ **23.** $\frac{1}{2}x^{-1/2}\sin x + \sqrt{x}\cos x$
25. $2(\cos 2x)(\cos 3x) - 3(\sin 2x)(\sin 3x)$
27. $12x^3\sin^2(x^4)\cos(x^4)$
29. $2\sin x \cos x - 2\cos x \sin x = 0$ **31.** $-\sin x$
33. $-\cos x$ **35.** $\sin x$ **37. (a)** $y = -x + 1 + \dfrac{\pi}{4}$,

(b)

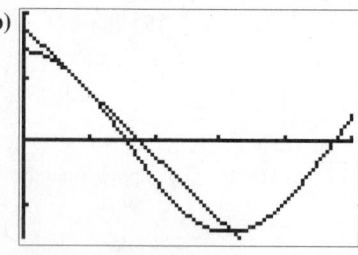

[0, 5, 1] × [−1.5, 2, 1]

39. (a) $y = -\frac{1}{4}\left(x - \dfrac{\pi}{2}\right)$ or $y = -\frac{1}{4}x + \dfrac{\pi}{8}$,

(b)

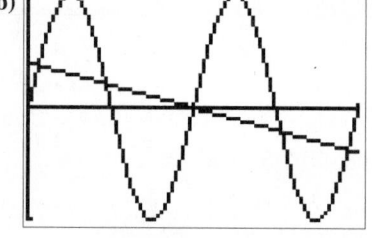

$\left[0, \pi, \dfrac{\pi}{4}\right] \cdot [-1, 1, 1]$

41. −3470 V

43. 0.033 A **45.** $-\dfrac{\sqrt{3}}{2}M \approx -0.866M$ cd/rad
47. maximum 2.24 A; minimum −2.24 A

Exercise Set 27.2

1. $\dfrac{\tan\sqrt{x}\sec^2\sqrt{x}}{\sqrt{x}}$ **3.** $5\sec 5x \tan 5x$

5. $-2\csc(2x - 1)\cot(2x - 1)$
7. $\sin x \sec^2 x + \cos x \tan x = \sin x(\sec^2 x + 1)$
9. $3\tan^2 x \sec^2 x$ **11.** $8x\sec^4(x^2)\tan(x^2)$
13. $\sec^2 x \cot x - \tan x \csc^2 x = 0$

15. $2\sin x\cos x\cot x - \sin^2 x\csc^2 x = 2\cos^2 x - 1 = \cos 2x$

17. $\dfrac{-1}{x^2}\sec^2\dfrac{1}{x}$ **19.** $\dfrac{x\sec^2 x^2}{\sqrt{1+\tan x^2}}$

21. $\dfrac{\sec x(\sec x + \sec^2 x - \tan^2 x)}{(1+\sec x)^2} = \dfrac{\sec x}{1+\sec x}$

23. $\dfrac{\csc x(\csc^2 x - \csc x - \cot^2 x)}{(1-\csc x)^2}$

25. $3(\csc x + 2\tan x)^2(2\sec^2 x - \csc x\cot x)$
27. $\dfrac{6}{5}(\sec^2 2x)(\tan 2x)^{-2/5}$

29. $\dfrac{\sec x(\tan x + \tan^2 x - \sec^2 x)}{(1+\tan x)^2} = \dfrac{\sec x(\tan x - 1)}{(1+\tan x)^2}$ **31.**

33. $8\sec^2 2x\tan 2x$ **35.** $2\sec^2 x(1 + x\tan x)$

37. $\dfrac{\sin y}{1 - x\cos y}$ **39.** $\dfrac{1 - \cos(x+y)}{\cos(x+y) - 1}$ **41.** -6.31

43. $1055.05\,\mathrm{mV/rad}$

Exercise Set 27.3

1. $\dfrac{2}{\sqrt{1-4x^2}}$ **3.** $\dfrac{1/2}{1+x^2/4} = \dfrac{2}{4+x^2}$

5. $\dfrac{2x}{\sqrt{1-(1-x^2)^2}} = \dfrac{2x}{\sqrt{2x^2 - x^4}}$ **7.** $\dfrac{-1}{\sqrt{x-x^2}}$

9. $\dfrac{1-3x^2}{\sqrt{1-(x^3-x^2)^2}}$ **11.** $\dfrac{2}{(2x+1)\sqrt{(4x+2)^2 - 1}}$

13. $\dfrac{x}{\sqrt{1-x^2}} + \sin^{-1}x$ **15.** $2x\cos^{-1}x - \dfrac{x^2}{\sqrt{1-x^2}}$

17. $\dfrac{2}{8x^2 - 4x + 1}$ **19.** $\tan^{-1}(x+1) + \dfrac{x}{1+(x+1)^2}$

21. $\dfrac{-1}{\sqrt{2-x^2}\sqrt{\sin^{-1}(1-x^2)}}$

23. $\cot^{-1}(1+x^2) - \dfrac{2x^2}{2+2x^2+x^4}$

25. $\dfrac{1}{x}\left(\dfrac{1}{\sqrt{1-x^2}} - \dfrac{\sin^{-1}x}{x}\right)$ **27.** $\dfrac{1}{\sqrt[3]{1-x^2}(\arcsin x)^{2/3}}$ 2

$7.1087\,\mathrm{rad/s}$ **31.**

Exercise Set 27.4

5. maxima: $\left(\frac{\pi}{6}, 1.5\right)$ and $\left(\frac{5\pi}{6}, 1.5\right)$; minima: $\left(\frac{\pi}{2}, 1\right)$ and $\left(\frac{3\pi}{2}, -3\right)$; inflection points: $(1.0030, 1.2646)$, $(2.1386, 1.2646)$, $(3.7765, -0.8896)$, and $(5.6483, -0.8896)$

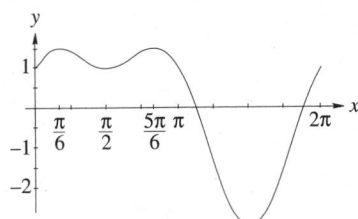

7. maxima: $\left(\frac{7\pi}{4}, \sqrt{3}\right)$; minima: $\left(\frac{3\pi}{4}, -\sqrt{3}\right)$; inflection points: $\left(\frac{\pi}{4}, 0\right)$, $\left(5\frac{\pi}{4}, 0\right)$

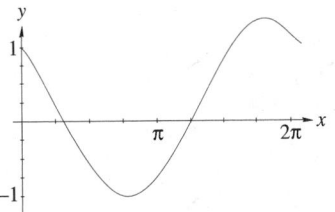

9. maxima: $\left(\frac{\pi}{6}, 1.25\right)$, $\left(\frac{5\pi}{6}, 1.25\right)$; minima: $\left(\frac{\pi}{2}, 1\right)$, $\left(\frac{3\pi}{2}, -1\right)$; inflection points: $(0.7297, 1.2222)$, $(2.4119, 1.2222)$, $\left(\frac{7\pi}{6}, 0\right)$, $\left(\frac{11\pi}{6}, 0\right)$

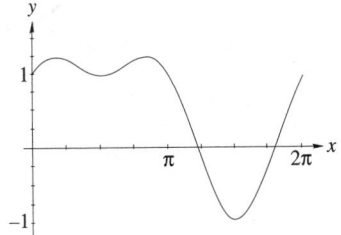

11. $y - 0.5 = \dfrac{\sqrt{3}}{2}\left(x - \dfrac{\pi}{6}\right)$

13. $y - 1 = 6\left(x - \frac{\pi}{4}\right)$

15. $0.0352\,\mathrm{rad/s}$ **17.** $104.71\,\mathrm{m/s}$ **19.** $42.8284\,\mathrm{mi/sec}$

21. (a) $4.0319\,\mathrm{m^3/s}$; **(b)** $4.2602\,\mathrm{m^3/s}$; **(c)** $180°$

23. $1.2364\,\mathrm{A/s}$ **25.** $0.5\,\mathrm{m/s}$ **27.** 2.9883

29. $0, \pm 2.7200$

Exercise Set 27.5

1. $\dfrac{1}{x}\log e = \dfrac{1}{x}\dfrac{1}{\ln 10}$ **3.** $\dfrac{2x+4}{x^2+4x}$

5. $\dfrac{x+2}{x^2+4x}[\ln(x^2+4x)]^{-1/2}$ **7.** $-\dfrac{1}{x}$ **9.** $\sec x\csc x$

11. $\dfrac{1-\ln x}{x^2}$ **13.** $\dfrac{1}{1-x^2}\left[\ln\left(\dfrac{1+x}{1-x}\right)\right]^{-1/2}$ **15.** $\dfrac{2x^2+12}{x(x^2+4)}$

17. $\dfrac{2\ln x}{x}$ **19.** $\dfrac{1}{x\ln x}$ **21.** $\dfrac{-1}{x^2}$

23. $\dfrac{2\ln x - 3}{x^3} = \dfrac{\ln x^2 - 3}{x^3}$ **25.** $\dfrac{3}{2x^2} - \csc^2 x$

27. $-\dfrac{\ln x^2 + 1}{x^2(\ln x^2)^{3/2}}$

29. $2\ln(\sin 2x) + 8x\cot 2x - 4x^2\csc^2 2x$ **31. (a)** about 109 medflies, **(b)** yes, the medflies will be eradicated in about $31.67\,\mathrm{h}$ **33. (a)** about $14.72\,\mathrm{h}$, **(b)** about $0.132\,\mathrm{ppm}$

Exercise Set 27.6

1. $4^x\ln 4$ **3.** $\dfrac{1}{2\sqrt{x}}5^{\sqrt{x}}\ln 5$ **5.** $(\cos x)2^{\sin x}\ln 2$

7. $(2x+1)e^{x^2+x}$ **9.** $4x^3 4^{x^4}\ln 4$ **11.** $e^x(x-2)/x^3$

13. $\dfrac{(x-2)e^x - 2}{x^3}$ **15.** $\sec^2 x e^{\tan x}$ **17.** $2xe^{x^2}\cos e^{x^2}$

19. $3x^2 3^x + (x^3 - 1)3^x \ln 3$

21. $x^{\cos x}\left[\dfrac{\cos x}{x} - (\sin x)(\ln x)\right]$

23. $(\sin x)^x(\ln \sin x + x \cot x)$ **25.** $3e^{3x}\cot e^{3x}$ **27.** e^{-y}

29.   **31.**   **33.** $3(\cosh 3x)e^{\sinh 3x}$

35. $g'(x) = 3x^2 \operatorname{sech}^2 \sqrt{x^3 + 4}$

37. $j'(x) = 3(4x^3 + \cos x)\cosh^2(x^4 + \sin x)\sinh(x^4 + \sin x)$

39. $f'(x) = \dfrac{7}{\sqrt{49x^2 + 1}}$

41. $j'(x) = \dfrac{-|x|}{x\sqrt{1+x^2}} + \sinh^{-1}\dfrac{1}{x}$ **43.** 10.69 people/day

45. (a) $I = q'(t) = -5e^{-t}\cos 2.5t - 12.5e^{-t}\sin 2.5t$,
(b) $q'(0.45) \approx -3.34\,\text{A}$

47. (a) $N'(t) = 10{,}000\,\operatorname{sech}^2(0.1t)$, (b) 6401 bacteria/h,
(c) $699{,}906 \approx 7 \times 10^5$ bacteria

Exercise Set 27.7

1. maxima: none; minima: $(e^{-1}, -e^{-1})$; inflection point: none

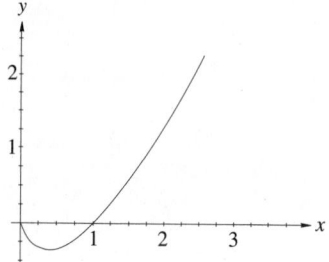

3. maxima: none; minima: $(-1, -e^{-1})$; inflection point: $(-2, -2e^{-2})$

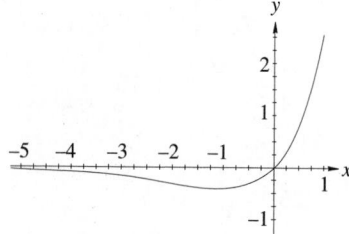

5. maxima: $(0, 0)$; minima: none; inflection points: $(\pm 1, \ln \frac{1}{2}) = (\pm 1, -0.6931)$

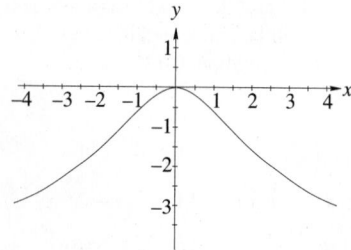

7. maxima: none; minima: none; inflection point: $(0, 0)$

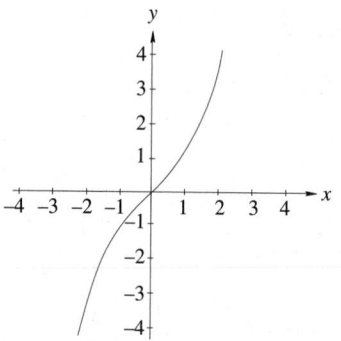

9. maxima: $\left(-\frac{\pi}{4}, 1.5509\right)$, $\left(\frac{7\pi}{4}, 0.0029\right)$; minima: $\left(\frac{3\pi}{4}, -0.0670\right)$ and $\left(-\frac{5\pi}{4}, -35.8885\right)$; inflection points: $(-2\pi, 0.0019)$, $(-\pi, -23.1407)$, $(0, 1)$, $(\pi, -0.0432)$, $(2\pi, 0.0019)$

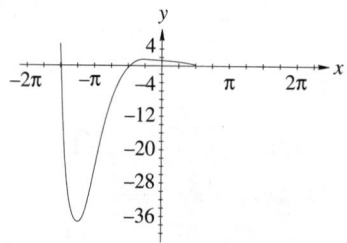

11. $y = x - 1$ **13.** $y - e = 3e(x - 1)$

15. (a) $v(t) = -3e^{-3t}$; (b) never   **17.** (a) $v(t) = e^t\cos(e^t)$;
$a(t) = e^t[\cos(e^t) - e^t\sin(e^t)]$; (b) $t = \ln\frac{\pi}{2} \approx 0.4516\,\text{s}$;
$a(t) = -\left(\frac{\pi}{2}\right)^2 \approx 2.4674\,\text{units/s}^2$ **19.** $x \approx 1.8371\,\text{m}$

21. $t = 0$ **23.** $-0.7085\,\text{W/day}$ **25.** $0.1586; 3.1462$

27. $\pm\left(k\pi + \frac{\pi}{2}\right)$, $k = 0, 1, 2, 3, \ldots$

Review Exercises

1. $2\cos 2x - 3\sin 3x$ **3.** $6\tan 3x \sec^2 3x$

5. $2x\sin x + x^2\cos x = x(2\sin x + x\cos x)$

7. $\dfrac{3}{\sqrt{12x - 9x^2 - 3}}$ **9.** $\dfrac{6x}{1 + 9x^4}$ **11.** $\dfrac{-1}{\sqrt{1-x^2}}$

13. $\dfrac{6x}{3x^2 - 5}\dfrac{1}{\ln 4}$ **15.** $\dfrac{6x^2\ln(x^3 + 4)}{x^3 + 4}$ **17.** $\dfrac{4x - 9}{x(4x - 3)}$

19. $8xe^{4x^2}$ **21.** $2x(\cos x^2)e^{\sin x^2}$

23. $e^{-x}(5\cos 5x - \sin 5x)$

25. maxima: $(-2, 0.5413)$; minima: $(0, 0)$; inflection points: $(-2-\sqrt{2}, 0.3835), (-2+\sqrt{2}, 0.1910)$

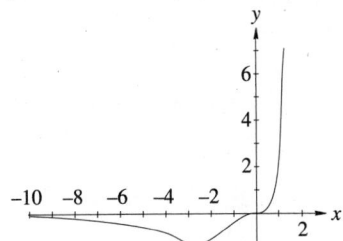

27. maxima: $(0.6629, 0.6964), (3.8045, 0.1448)$; minima: $(2.2337, -0.3175), (5.3753, -0.0660)$; inflection points: $(1.3110, 0.2578), (2.9378, -0.0913), (4.4526, 0.0536), (6.0793, -0.0190)$

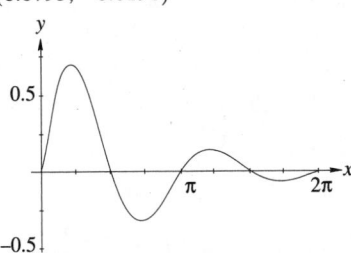

29. maxima: none; minima: $(0.7165, -0.1226)$; inflection

point: $(0.4346, -0.0684)$

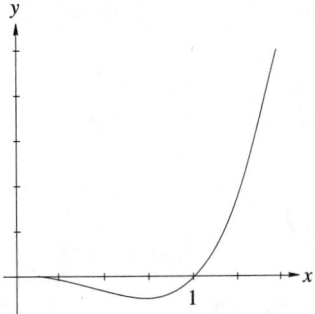

31. **(a)** $y - \frac{\pi}{4} = \frac{1}{2}(x-1)$; **(b)** $y - \frac{\pi}{4} = -2(x-1)$
33. **(a)** $y = x - 1$; **(b)** $y = 1 - x$
35. **(a)** $y + \ln\sqrt{2} = x - \frac{\pi}{4}$; **(b)** $y + \ln\sqrt{2} = \frac{\pi}{4} - x$
37. **(a)** $64.9967°C$; **(b)** $-2.2089°C$ **39.** $19.5959\,\text{ft}$

Chapter 27 Test

1. $7\cos 7x$ **3.** $h'(x) = 2e^{2x}$ **5.** $\dfrac{2xe^{x^2}}{\sqrt{1-e^{2x^2}}}$

7. $(\tan 5x)\left[10(7x^2+3x)\left(\sec^2 5x\right) + (14x+3)\left(\tan^2 5x\right)\right]$
9. $-3.82\,\text{A}$

■ ANSWERS FOR CHAPTER 28

Exercise Set 28.1

1. $\frac{1}{4}\sin^4 x + C$ **3.** $\frac{2}{5}\sin^{5/2} x + C$ **5.** $\frac{1}{2}\tan x^2 + C$
7. $\frac{1}{2}\sec^2 x + C$ or $\frac{1}{2}\tan^2 x + C$ **9.** $-\frac{1}{2}(\arccos x)^2 + C$
11. $-\frac{2}{3}(\operatorname{arccsc} x)^{3/2} + C$ **13.** $\frac{1}{8}(\arcsin 2x)^4 + C$
15. $\frac{1}{5}\ln^5(x+4) + C$ **17.** $-(e^x+4)^{-1} + C$
19. $\frac{1}{8}(e^{2x}-1)^4 + C$ **21.** $1/16$ **23.** 0.8098
25. 0.0689 **27.** $2/3$ **29.** $1.3397\,C$
31. **(a)** $N(t) = \dfrac{1020}{t} - 340t + 9070$, **(b)** $N(10) = 5772$

Exercise Set 28.2

1. $\frac{1}{4}\ln|4x+1| + C$ **3.** $-\frac{1}{3}e^{-3x} + C$
5. $-\ln|\cos x| + C$ **7.** $2\ln|\tan x| + C$ **9.** $\dfrac{4^x}{\ln 4} + C$
11. $e^x - e^{-x} + C$ **13.** $\frac{1}{2}\ln(1+e^{2x}) + c = \ln\sqrt{1+e^{2x}} + C$
15. $-\frac{1}{2}\ln^2(1/x) + C$ **17.** $3e^{\sqrt[3]{x}} + C$
19. $\frac{1}{2}\ln 17 \approx 1.4166$ **21.** $\ln 3 \approx 1.0986$
23. $\frac{1}{4}(e-1) \approx 0.4296$ **25.** 0.1988
27. $e^2 - (e+1/3) \approx 4.3374$
29. **(a)** $s(t) = -15{,}760e^{-0.025t} - 38et + 15{,}760$; **(b)** about

$1.028\,\text{s}$; **(c)** about $5.1196\,\text{m}$ **31.** $\dfrac{\pi}{2}(1-e^{-4}) \approx 1.5420$
33. $(\ln 0.5) \div (-0.15) \approx 4.6210\,\text{mi}$ **35.** $10.574\,\text{ft·lb}$
37. $F(x) = 2 - 2e^{-3x^2}$

Exercise Set 28.3

1. $-2\cos\dfrac{x}{2} + C$ **3.** $-\frac{1}{5}\ln|\cos x| + C = \frac{1}{5}\ln|\sec x| + C$
5. $\ln|\sec x + \tan x| + C$ **7.** $\tan x + C$
9. $-2\ln|\cos\sqrt{x}| + C = 2\ln|\sec\sqrt{x}| + C$
11. $-\frac{1}{4}\cos(4x-1) + C$ **13.** $\sec x + C$
15. $\frac{1}{6}\ln|5+2\sec 3x| + C$ **17.** $\frac{1}{12}\sec^4 3x + C$
19. $\ln|\sin 3x| + C$ **21.** $\tan x + 4\ln|\sec x + \tan x| + 4x + C$
23. $\dfrac{25\pi}{2}$
25. $-\frac{1}{4}(\cot^2 4x + \csc 4x) + C = -\frac{1}{4}\left(\dfrac{\cos 4x+1}{\sin 4x}\right) + C$
27. $4\ln|\sec x/4| + C$ **29.** $\pi/4$ **31.** 0.4765
33. 0.4509 **35.** $\frac{1}{2}\cosh x^2 + C$ **37.** $\left(\sinh x^2\right)^{3/2} + C$
39. $\frac{1}{5}\cosh^5 x - \frac{1}{3}\cosh^3 x + C$ **41.** Delmar did not provide an answer for this exercise **43.** 1 **45.** $.4413$.
47. $2.067\,\text{V}$ **49.** $51.1027\,\text{m}$ **51.** $509.397\,\text{ft}$.

Exercise Set 28.4

1. $\frac{1}{9}\sin^3 3x + C$ **3.** $\frac{1}{4}\sin^4 x + C$ **5.** $\frac{1}{25}\sin^5 5x + C$

7. $\frac{x}{16} + \frac{1}{64}\sin 4x + \frac{1}{48}\sin^3 2x + C$

9. $\frac{5}{16}x + \frac{1}{12}\sin 6x + \frac{1}{64}\sin 12x - \frac{1}{144}\sin^3 6x + C$

11. $\frac{1}{16}\theta - \frac{1}{128}\sin 8\theta + \frac{1}{96}\sin^3 r\theta + C$ **13.** $\frac{1}{3}\tan^3 x + C$

15. $-\frac{1}{2}\cot^2 x - \frac{1}{4}cot^4 x + C$

17. $\frac{2}{9}\sin^{3/2}3\theta - \frac{2}{21}\sin^{7/2}3\theta + C$ **19.** $\csc x - \frac{1}{3}\csc^3 x + C$

21. $\frac{1}{4}\tan^4 x + C$ **23.** $\frac{1}{7}\tan^7 x + C$

25. $-\frac{1}{5}\cot^5 x + \frac{1}{3}\cot^3 x - \cot x - x + C$ **27.** $1 - \pi/4$

29. $\frac{3\pi}{16}$ **31.** $\pi\left(\frac{\pi}{6} + \frac{\sqrt{3}}{8}\right) \approx 2.3251$

33. (a) $I = \frac{mr^2\pi}{4}$, (b) $0.0974\,\mathrm{kg\cdot m^2}$ **35.** $0.5421\,\mathrm{A}$

Exercise Set 28.5

1. $\arcsin\frac{x}{2} + C$ **3.** $\frac{1}{3}\arcsin\frac{3}{2}x + C$ **5.** $\sqrt{9 - x^2} + C$

7. $\frac{1}{2}\operatorname{arcsec}\frac{3x}{2} + C$ **9.** $-\arctan(3 - x) + C$

11. $2\ln(4x^2 + 25) - \frac{3}{5}\arctan\frac{2x}{5} + C$

13. $\arctan(x + 3) + C$ **15.** $\frac{1}{2}\arctan x^2 + C$

17. $-\arcsin(\cos x) + C = x - \frac{\pi}{2} + C$ **19.** 0.6155

21. $\cosh^{-1}\frac{x}{5} + C$ **23.** $\frac{1}{3}\sinh^{-1}\frac{3x}{5} + C$

25. $\frac{\pi}{6} \approx 0.5236$ **27.** 0.36905 **29.** 0.2905 **31.** $\pi/4$

33. $\frac{\pi}{2}(e^2 - 1 - \frac{\pi}{2}) \approx 7.5685$ **35.** $2\pi/3$

37. $y = \operatorname{sech}^{-1}x - \sqrt{1 - x^2} + \operatorname{sech}^{-1}1$

Exercise Set 28.6

1. $-\sqrt{9 - x^2} + C$ **3.** $-\frac{x}{2}\sqrt{9 - x^2} + \frac{9}{2}\sin^{-1}\frac{x}{3} + C$

5. $\frac{1}{5}(9 - x^2)^{3/2}(6 + x^2) + C$

7. $\ln\left|\frac{x}{3} + \frac{\sqrt{x^2 - 9}}{3}\right| + C = \ln|x + \sqrt{x^2 - 9}| + k$ where

$k = C - \ln 3$ **9.** $\frac{1}{2}[x\sqrt{x^2 - 9} - 9\ln(x + \sqrt{x^2 - 9})]$

11. $\sqrt{x^2 + 9} + C$ **13.** $\frac{1}{3}\arctan\frac{x}{3} + C$

15. $(\frac{x^2 - 6}{5})(x^2 + 9)^{3/2} + C$ **17.** $-\frac{1}{12}\frac{(4 - 3x^2)^{3/2}}{x^3} + C$

19. $\frac{1}{243}\sqrt{(9x^2 + 4)^3} - \frac{16}{81}\sqrt{9x^2 + 4} + C$

21. $-\frac{\sqrt{x^2 + 1}}{x} + \ln|x + \sqrt{x^2 + 1}| + C$

23. $\frac{x - 1}{2}\sqrt{4 - (x - 1)^2} + 2\sin^{-1}(\frac{x - 1}{2}) + C$

25. $\ln|x + \sqrt{(x - 1)^2 - 4}| + C$

27. $\frac{1}{2}\Big[(x - 1)\sqrt{x^2 - 2x - 3} -$

$2\ln\left|x - 1 + \sqrt{x^2 - 2x - 3}\right|\Big] + C$

29. $\frac{1}{4}\ln\frac{-4 + \sqrt{x^2 - 6x + 25}}{x - 3} + C$ **31.** 1.59747

33. 0.1355 **35.** $81\pi \approx 254.4690$

37. $\frac{-1}{3}(41\sqrt{7} - 128) \approx 6.5080654$

39. $5 + \ln 3^{9/2} \approx 9.9438$

41. $8\pi + \frac{100\pi}{\sqrt{21}}\sin^{-1}\frac{\sqrt{21}}{5} \approx 104.6074$ **43.** $\approx 0.1829\,\mathrm{A}$

Exercise Set 28.7

1. $\frac{1}{2}x^2(\ln x - \frac{1}{2}) + C$ **3.** $\frac{1}{4}\sin 2x - \frac{x}{2}\cos 2x + C$

5. $\frac{x^3}{3}\ln x - \frac{x^3}{9} + C$ **7.** $\frac{1}{2}[x^2\sin^{-1}x + \sqrt{1 - x}] + C$

9. $e^{2x}(\frac{1}{2}x^3 - \frac{3}{4}x^2 + \frac{3}{4}x - \frac{3}{8}) + C$

11. $\frac{1}{60}(6x - 1)(4x + 1)^{3/2} + C$

13. $e^{x/4}(3x^2 - 32x + 128) + C$ **15.** $\frac{1}{2}e^x(\cos x + \sin x) + C$

17. $\frac{1}{2}x^2\sin x^2 + \frac{1}{2}\cos x^2 + C$

19. $-e^{-x}(x^3 + 3x^2 + 6x + 6) + C$

21. $\frac{1}{2}e^{-x}(\sin x - \cos x) + C$ **23.** $\frac{1}{16}\cos 4x + \frac{x}{4}\sin 4x + C$

25. $\pi^2/4$ **27.** $\sin 8 - 8\cos 8 \approx 2.1534\,\mathrm{C}$

29. $\bar{x} = 1.5972, \bar{y} = 7.8846$ **31.** $\frac{\pi^3}{8} + 3\pi - 6 \approx 7.3006$

33. $1096\,\mathrm{ft}$ **35.** (a) $F = -\frac{e^{-kR}2\pi(kR + 1)P_0}{k^2} + \frac{2\pi P_0}{k^2}$,

(b) $T = -2\pi\mu\left[\frac{1}{k^3}e^{-kR}\left(k^2R^2 + 2kR + 2\right)\right] + 4\pi\mu\left(\frac{1}{k^3}\right)$

37. $3.0776\,\mathrm{ppm}$

Exercise Set 28.8

1. $\frac{1}{3}(2\ln|\sec 3x| + \tan 3x) + C$ **3.** $-\frac{1}{3}\ln\left|\frac{x}{4x - 3}\right| + C$

5. $\frac{\pm\sqrt{16 + x^6}}{48x^3} + C$ **7.** $-\frac{x}{2}\sqrt{9 - x^2} + \frac{9}{2}\sin^{-1}\frac{x}{3} + C$

9. $\sin x - \frac{1}{3}\sin^3 x + C$ **11.** $-\frac{1}{18}\sin^5 3x\cos 3x -$

$\frac{5}{72}\sin^3 3x\cos 3x - \frac{8}{45}\sin 3x\cos 3x + \frac{5}{48}x + C$

13. $\frac{1}{136}e^{10x}(10\cos 6x + 6\sin 6x) + C$

15. $x^8(\frac{\ln x}{8} - \frac{1}{64}) + C$

17. $\frac{1}{4}(4x\sin^{-1}4x + \sqrt{1 - 16x^2}) + C$

19. $\sqrt{9 + x^2} - 3\ln\left|\frac{3 + \sqrt{9 + x^2}}{x}\right| + C$

21. $\frac{1}{2}x^3e^{2x} - \frac{3}{4}x^2e^{2x} + \frac{3}{8}(2x - 1)e^{2x} + C =$

$\frac{1}{8}(4x^3 - 6x^2 + 6x - 3)e^{2x} + C$

23. $-\frac{1}{2}(x^3\cos 2x - \frac{3}{2}x^2\sin 2x + \frac{3}{4}\sin 2x - 3x\cos 2x) + C$

25. $\frac{1}{2.5}(286047.5315) \approx 114\,419.0126\,\mathrm{V}$

27. $\frac{1}{14}\ln 13 \approx 0.1832\,\mathrm{N\cdot m}$

Review Exercises

1. $\frac{1}{9}(3x - 1)e^{3x} + C$ **3.** $-\sqrt{25 - x^2} + C$

5. $\frac{1}{10}\sin^5 2x + C$ **7.** $\frac{1}{9}\ln|9x + 5| + C$

9. $-\frac{1}{7}\cos(7x+2)+C$

11. $-\frac{1}{21}\cos^7 3x+\frac{2}{15}\cos^5 3x-\frac{1}{9}\cos^3 3x+C$

13. $\frac{x}{49\sqrt{4x^2+49}}+C$ **15.** $\frac{1}{3}e^{x^3}+C$

17. $\frac{1}{4}\sqrt{4x^2+49}+C$ **19.** $\frac{1}{8}\ln|9+2\sec 4x|+C$

21. $\frac{1}{12}[\ln(2x+1)]^6+C$ **23.** $\frac{1}{3}\ln|x^3+4|+C$

25. $-5\ln|\cos\frac{x}{5}|+C=5\ln|\sec\frac{x}{5}|+C$ **27.** $-e^{\cos x}+C$

29. $\frac{1}{3}\tan^{-1}(\frac{\sin x}{3})+C$ **31.** $\frac{3e^4}{8}+\frac{1}{8}\approx 20.5993$

33. $\frac{3}{320}(\sin^{4/3}4x)(5\sin^4 4x-16\sin^2 4x+20)+C$

35. $\frac{1}{2}\sin^{-1}(\frac{\sin x}{2})+C$ **37.** $\frac{1}{2}e^{x^2}(x^4-2x^2+2)+C$

39. $\frac{1}{10}(\arctan 2x)^5+C$ **41.** 12.5861

43. $18-9\sqrt{2}\approx 5.2721\approx 5.2721$

Chapter 28 Test

1. $-2\sqrt{9-e^x}+C$ **3.** $4\arctan x+C$

5. $\frac{1}{4}xe^{4x}-\frac{1}{16}e^{4x}+C=\frac{1}{16}e^{4x}(4x-1)+C$

7. $\frac{\pi}{4}(e^2+1)$

ANSWERS FOR CHAPTER 29

Exercise Set 29.1

1. $\frac{dy}{dx}=\frac{1}{2t+1}$, $\frac{d^2y}{dx^2}=\frac{-2}{(2t+1)^3}$, min at $t=-\frac{1}{2}$: $(-\frac{1}{4},\frac{1}{2})$

3. $\frac{dy}{dx}=\frac{1}{2t-6}$, $\frac{d^2y}{dx^2}=\frac{-2}{(2t-6)^3}$, min at $t=3$: $(3,7)$

5. $\frac{dy}{dx}=\frac{2t-1}{2t+1}$, $\frac{d^2y}{dx^2}=\frac{4}{(2t+1)^3}$, min at $t=1/2$: $(\frac{3}{4},-\frac{1}{4})$

7. $\frac{dy}{dx}=1/4$; $\frac{d^2y}{dx^2}=0$, maxima when $t=0,2\pi,\dots$: $(7,0)$;

min when $t=\pi,3\pi,\cdots$: $(-1,-2)$ **9.** $\frac{dy}{dx}=-\tan t$:

$\frac{d^2y}{dx^2}=\frac{-1}{\cos^3 t}$; maxima when $t=\pi,3\pi,\dots$: $(2,-2)$ and when $t=0,2\pi,\dots$: $(2,0)$; min when $t=\pi/2,5\pi/2,\dots$: $(3,-1)$ and $t=3\pi/2,7\pi/2,\dots$: $(1,-1)$ **11.** tangent: $y=3x-1$; normal: $3y+x=7$ **13.** tangent: $9y+2x=-36$; normal: $2y-9x=261$ **15.** tangent: $y=2$; normal: $x=2$
17. horizontal tangent when $t=2$: $(5,-4)$; vertical tangents: none **19.** horizontal tangent when $t=\frac{\pi}{2},\frac{5\pi}{2},\dots$: $(0,5)$ and $t=\frac{3\pi}{2},\frac{7\pi}{2},\dots$: $(0,5)$; vertical tangents when $t=0,2\pi$, $\dots$: $(3,0)$ and $t=\pi,3\pi,\dots$: $(-3,0)$
21. (a) $s_x(5)\approx 12{,}646.57$ ft, $s_y(5)\approx 13{,}645.44$ ft;
(b) $v_x(5)\approx -2809.09$ ft/s, $v_y(5)\approx 2369.32$ ft/s; **(c)** $a_x(5)0$, $a_y(5)=-32$ ft/s² **23. (a)** 100 km at 71.6°,
(b) $v_x=-\frac{100t}{(t^2+1)^{3/2}}$, $v_y=\frac{100}{(t^2+1)^{3/2}}$, **(c)** magnitude:

10 km/s, direction: $-18.4°$, **(d)** $a_x=\frac{200t^2-100}{(t^2+1)^{5/2}}$,

$a_y=-\frac{300t}{(t^2+1)^{5/2}}$, **(e)** magnitude: $\sqrt{37}\approx 6.08$ km/s²,

direction: $-27.9°$ **25. (a)** $v_x=\frac{20t^3}{\sqrt{t^4+1}}$, $v_y=60\sqrt{t}$,

(b) magnitude: $20\sqrt{\frac{t^6}{t^4+1}+9t}$ m/s, direction:

$\tan^{-1}\left(\frac{3\sqrt{t^4+1}}{t^{5/2}}\right)$, **(c)** magnitude: 167.28 m/s, direction:

53.3°

Exercise Set 29.2

1. $\frac{2\cos\theta\sin\theta}{\cos^2\theta-\sin^2\theta}=\tan 2\theta$

3. $\frac{\cos^2\theta-\sin^2\theta+\cos\theta}{-(2\sin\theta\cos\theta+\sin\theta)}=\frac{\cos 2\theta+\cos\theta}{\sin\theta-\sin 2\theta}$

5. $\frac{3\sin 3\theta\sin\theta-\cos\theta-\cos\theta\cos 3\theta}{3\sin 3\theta\cos\theta+\sin\theta+\sin\theta+\cos 3\theta}$

7. $\frac{\sec^2\theta\sin\theta+\tan\theta\cos\theta}{\sec^2\theta\cos\theta-\tan\theta\sin\theta}=(\sec^2\theta+1)\tan\theta$
9. tangent: $y=1$; normal: $x=0$ **11.** tangent:
$y-0.9151=0.6603(x-3.4151)$; normal:
$y-0.9151=-1.5146(x-3.4151)$ **13.** tangent:
$y-3.8971=5.1962(x+2.25)$ or $y-12\sqrt{3}x=36\sqrt{3}$;
normal: $y-3.8971=-0.1925(x+2.25)$ or
$36y+4\sqrt{3}x+45\sqrt{3}$ **15.** tangent: $y+3.7321x=7.7875$;
normal: $y-0.2679x=2.0861$ **17.** Numerator is 0 when
$\cos\theta=0$ or $\sin\theta=\pm\sqrt{6}/6$. Denominator is 0 when $\sin\theta=0$
or $\sin\theta=\pm\sqrt{6}/6$. maxima: $(3,0)$, $(-2,1.1503)$,
$(-2.0543,1.9790)$, $(3,\pi)$; minima: $(2,0.4205)$, $(-3,\pi/2)$,
$(2,2.7211)$, $(2.0543,3.5498)$ **19.** Numerator = 0 when
$\cos\theta=(-1\pm\sqrt{33}/8)$; denominator = 0 when $\theta=0,\pi$,
$\cos^{-1}(-\frac{1}{4})$; maxima: $(3,0)$, $(0.5,1.8235)$, $(-1,\pi)$,
$(0.5,4.4597)$, $(3,2\pi)$; minima: $(2.1862,0.9359)$,
$(-0.6862,2.45738)$, $(-0.6862,3.7094)$, $(2.1862,5.3473)$

21. (a) $\frac{dr}{d\theta}=\frac{30\sin\theta}{(5+3\cos\theta)^2}$, **(b)** $\theta=\pi$

Exercise Set 29.3

1. $\frac{1}{2}(5\sqrt{5}-1)\approx 5.0902$ **3.** 6π **5.** $\frac{1}{2}\pi^2$ **7.** 4
9. 1.3802 **11.** $\frac{\pi}{27}(145^{3/2}-1)\approx 203.0436$ **13.** 2π
15. 2π **17.** $2^5\pi/5=\frac{32}{5}\pi$ **19. (a)** 50.56 cm, **(b)** 10^+ h

Exercise Set 29.4

1. $\left(\frac{\pi}{2}, \frac{\pi}{6}\right)$

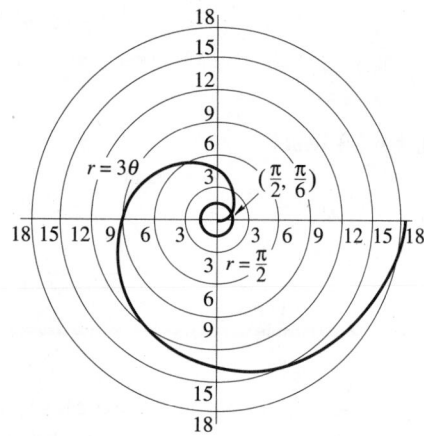

3. $\left(\frac{1}{2}, \frac{\pi}{3}\right), \left(\frac{1}{2}, \frac{5\pi}{3}\right)$

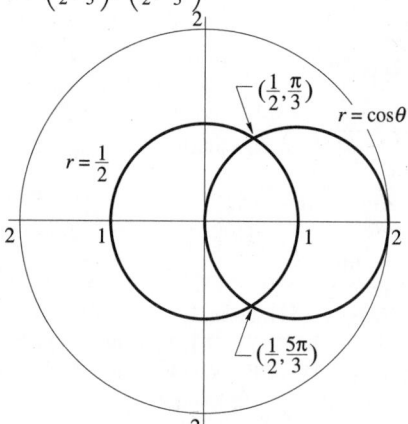

5. $\left(1 - \frac{\sqrt{2}}{2}, \frac{3\pi}{4}\right), \left(1 + \frac{\sqrt{2}}{2}, \frac{7\pi}{4}\right), (0,0)$

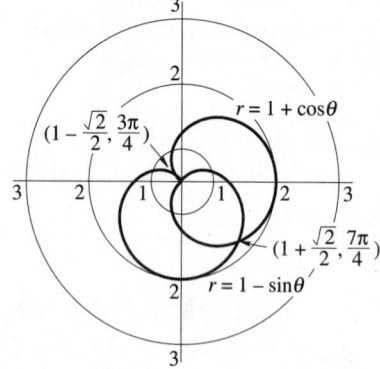

7. $\left(\sqrt{3}, \frac{\pi}{6}\right), \left(\sqrt{3}, \frac{11\pi}{6}\right)$

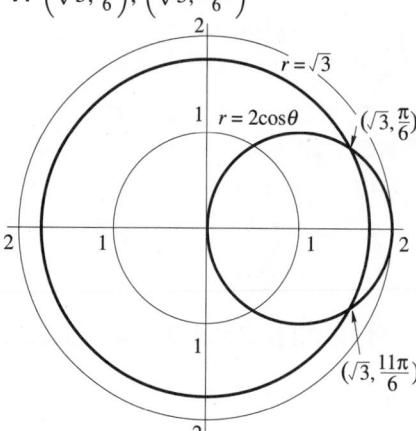

9. $\left(1, \frac{\pi}{4}\right), \left(-1, \frac{7\pi}{4}\right), (0,0)$

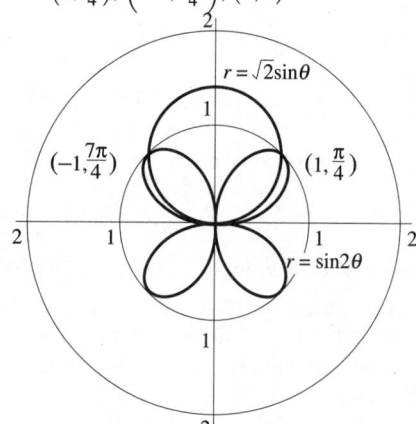

11. $(1, 0), (1, \pi)$

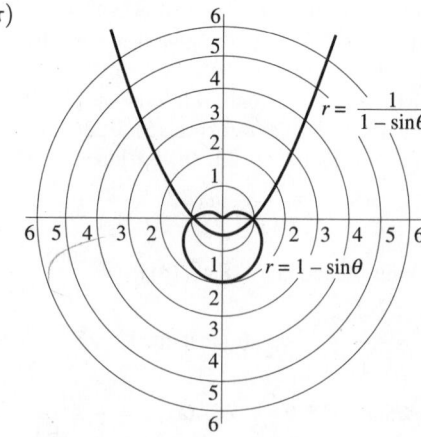

Exercise Set 29.5

1. 4π

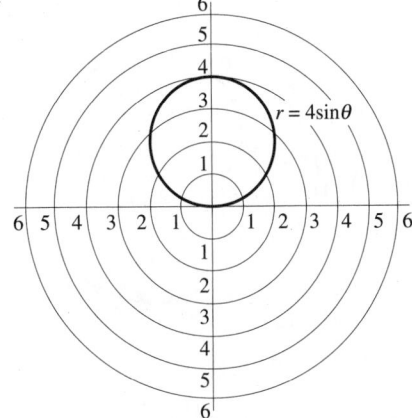

$r = 4\sin\theta$

3. $\frac{3}{2}\pi$

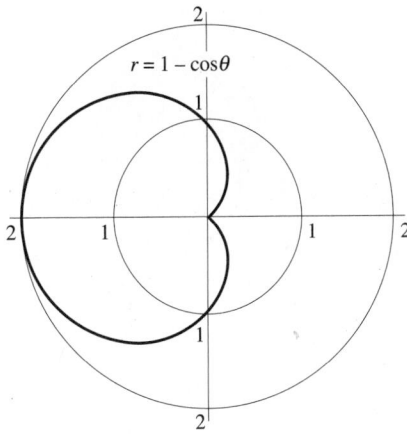

$r = 1 - \cos\theta$

5. $\pi/2$

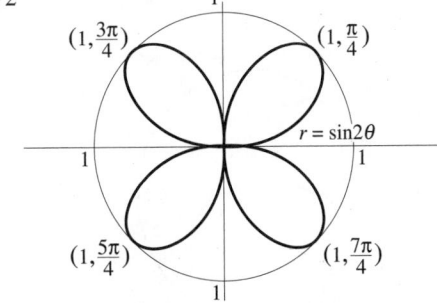

$\left(1, \frac{3\pi}{4}\right)$ $\left(1, \frac{\pi}{4}\right)$ $r = \sin2\theta$ $\left(1, \frac{5\pi}{4}\right)$ $\left(1, \frac{7\pi}{4}\right)$

7. $\frac{33}{2}\pi$

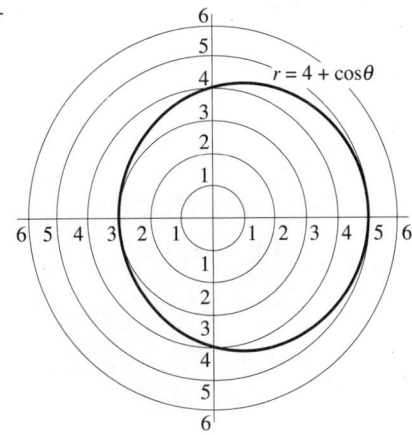

$r = 4 + \cos\theta$

9. 2

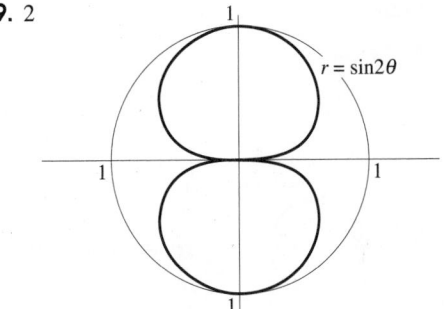

$r = \sin2\theta$

11. $\frac{9}{16}(\pi+2) \approx 2.8921$ **13.** $\pi/16 \approx 0.1963$

15. $\frac{1}{8}(e^{2\pi} - 1) \approx 66.8115$ **17.** 2π **19.** $\pi/3$

21. $2 + \pi/4 \approx 2.7854$

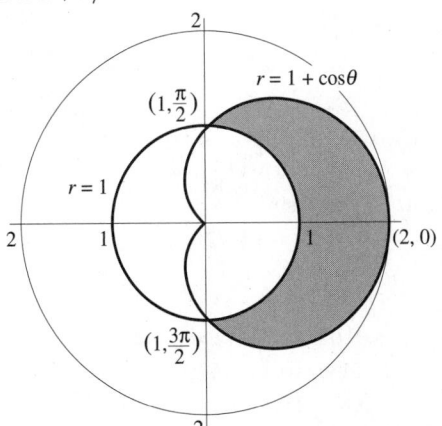

$\left(1, \frac{\pi}{2}\right)$ $r = 1 + \cos\theta$ $r = 1$ $(2, 0)$ $\left(1, \frac{3\pi}{2}\right)$

23. 0.382778

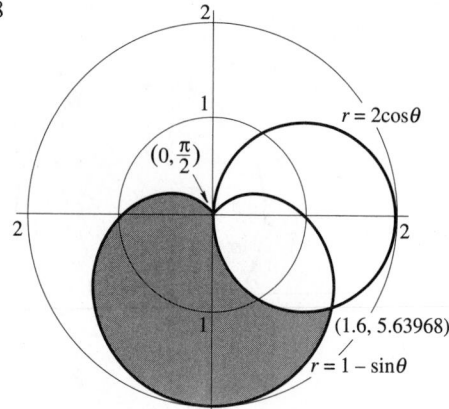

25. $1 - \frac{\pi}{4} \approx 0.2146$

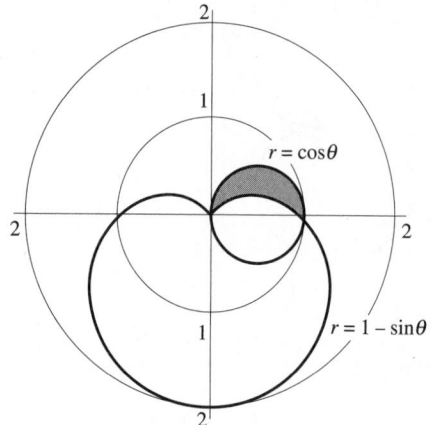

25. $3\pi/2$

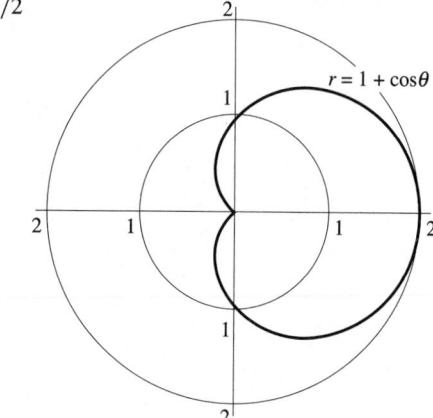

27. $4\pi + 32$

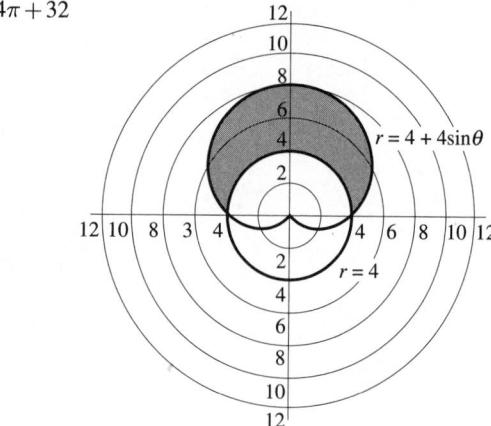

29. $1 - \pi/4 \approx 0.2146$

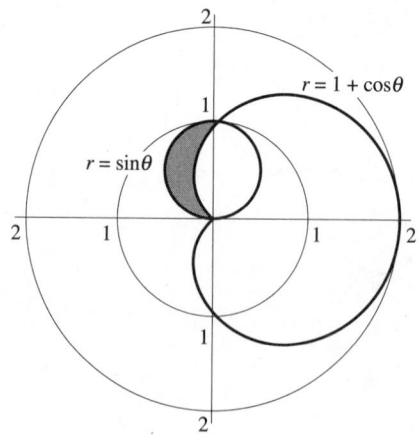

Review Exercises

1. (a) $\dfrac{dy}{dx} = \dfrac{3t^2 + 1}{2t}$; $\dfrac{d^2y}{dx^2} = \dfrac{3t^2 - 1}{4t^3}$; (b) 2

3. (a) $\dfrac{dy}{dx} = \dfrac{-1}{2\sin t}$; $\dfrac{d^2y}{dx^2} = \dfrac{-1}{4\sin^3 t}$; (b) -0.5942

5. (a) $-\dfrac{2\cos 2\theta + 3\cos\theta}{2\sin 2\theta + 3\sin\theta}$; (b) -0.5147

7. (a) $\dfrac{3\cos\theta\sin\theta}{\cos^2\theta - 2\sin^2\theta}$; (b) -3

9. $\frac{1}{27}\left(40^{3/2} - 4^{3/2}\right) \approx 9.0734$ **11.** 2π **13.** 2

15. $\frac{\pi}{6}\left(17^{3/2} - 1\right) \approx 11.5155\pi \approx 36.1769$

17. $41.7485\pi \approx 131.1568$ **19.** $36\pi^2 \approx 355.3058$

21. $0, \pi$ **23.** $2\pi/3, 4\pi/3$

Chapter 29 Test

1. $\dfrac{dy}{dx} = \dfrac{3t^2}{2t+5}$; $\dfrac{d^2y}{dx^2} = \dfrac{6t(t+5)}{(2t+5)^3}$ **3.** $\sin\theta\cos\theta$

5. $\sqrt{5}(e^2-1)$ **7.** $(0.2, 1.3694)$ and $(0.2, -1.3694) = (0.2, -4.9137)$

≡ ANSWERS FOR CHAPTER 30

Exercise Set 30.1

1. (a) 3, **(b)** 4, **(c)** 8, **(d)** $3x+3h+4y-xy-hy$,
(e) $3x+4y+4h-xy-xh$ **3. (a)** $\sqrt{2}-1$, **(b)** 2, **(c)** 2,
(d) $\sqrt{(x+h)y}-(x+h)+\dfrac{4}{y}$, **(e)** $\sqrt{x(y+h)}-x+\dfrac{4}{y+h}$
5. (a) 1, **(b)** 0, **(c)** 0, **(d)** 3, **(e)** all real values of (x,y) except
when $x=y$ **7.** $V=\frac{1}{3}\pi r^2 h$ **9.** $A=2\pi rh+2\pi r^2$
11. $C=200\pi r(2r+h)$ **13.** 0.000036 cm/min
15. (a) $-11°F$, **(b)** $-25°F$, **(c)** $10°F$, **(d)** $-15°F$
17. (a) $V(x,y,z)=xyz$, **(b)** $S(x,y,z)=xy+2xz+3yz$
19. 202.8 ft   **21. (a)** 1.834 m²,
(b) $A(w,h)=14.085w^{0.425}h^{0.725}$, **(c)** 2,847.5 in.²
23. (a) 37,444, **(b)** 74,888

Exercise Set 30.2

1. A plane whose intercepts are the points $(6,0,0)$, $(0,3,0)$,
and $(0,0,2)$

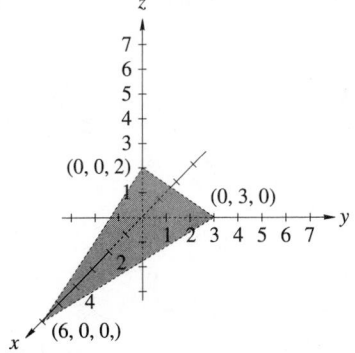

3. Parabolic cylinder

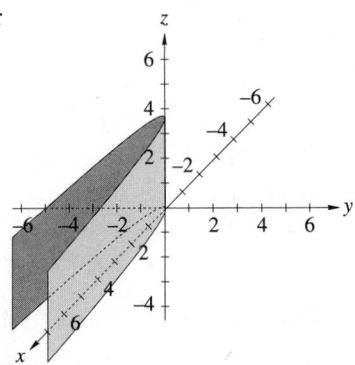

5. Ellipsoid with intercepts at $(\pm2,0,0)$, $(0,\pm\sqrt{2},0)$, and

$(0,0,\pm2)$

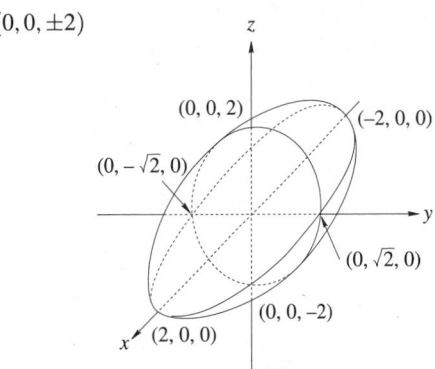

7. Ellipsoid with intercepts at $(\pm2,0,0)$, $(0,\pm2\sqrt{2},0)$, and
$(0,0,\pm2)$

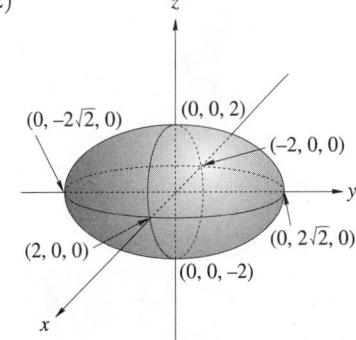

9. Hyperboloid of two sheets with intercepts at $(0,0,\pm1)$

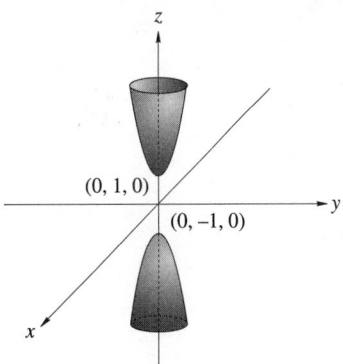

11. Plane with intercepts at $(4,0,0)$, $(0,-3,0)$ and $(0,0,-2)$

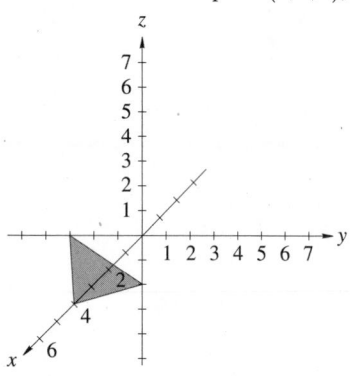

13. Elliptic paraboloid with intercepts $(\pm 4,0,0)$, $(0,\pm 2,0)$, and $(0,0,-16)$

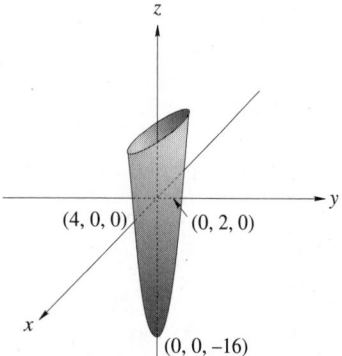

15. Elliptic paraboloid

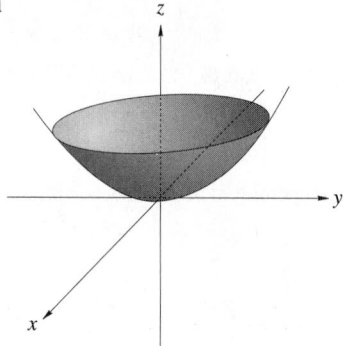

17. Hyperbolic paraboloid

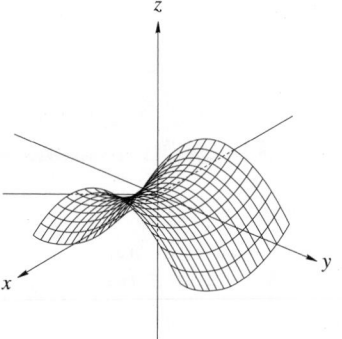

19. Parabolic cylinder

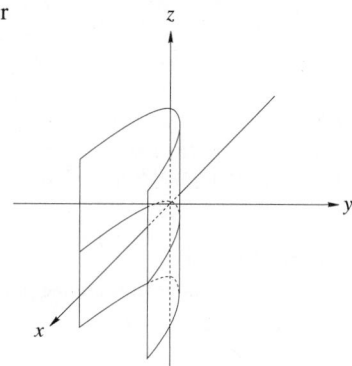

Exercise Set 30.3

1. $f_x = 2xy + y^2$; $f_y = x^2 + 2xy$ **3.** $f_x = 6x + 6y^3$;
$f_y = 18xy^2$ **5.** $f_x = \dfrac{2x}{y}$; $f_y = \dfrac{y^2 - x^2}{y^2}$

7. $f_x = e^x \sin y - e^y \sin x$; $f_y = e^y \cos x + e^x \cos y$

9. $f_x = 2e^{2x+3y}$; $f_y = 3e^{2x+3y}$

11. $f_x = 2xy^3 \cos(x^2 y^3)$; $f_y = 3x^2 y^2 \cos(x^2 y^3)$

13. $f_x = \dfrac{x}{x^2 + y^2}$; $f_y = \dfrac{y}{x^2 + y^2}$

15. $f_x = 6y \sin(3xy) \cos(3xy)$; $f_y = 6x \sin(3xy) \cos(3xy)$

17. $f_{xx} = 6xy^2$; $f_{xy} = f_{yx} = 6x^2 y - 5y^4$; $f_{yy} = 2x^3 - 20xy^3$

19. $f_{xx} = 3y^6 e^{xy^3}$; $f_{xy} = f_{yx} = 9y^2 e^{xy^3}(1 + xy^3)$;
$f_{yy} = 9xye^{xy^3}(2 + 3xy^3)$ **21.** $f_{xx} = 0$; $f_{xy} = f_{yx} = \dfrac{-10}{y^6}$;
$f_{yy} = \dfrac{60x}{y^7}$ **23.** $f_{xx} = \dfrac{-3}{x^2}$; $f_{xy} = f_{yx} = 0$; $f_{yy} = \dfrac{-5}{y^2}$

25. $\dfrac{-5900R}{9}$ kg/m^2 **27.** -984.72 cm^3/atm

29. $\dfrac{dz}{dy} = -2y$. At $(1,3,6)$ the slope is -6. **31.** (a) $\dfrac{k}{r^4}$,

(b) $-\dfrac{4kL}{r^5}$ **33.** (a) $A_w(w,h) = 5.9861w^{-0.575}h^{0.725}$ in.^{2}/lb
and $A_h(w,h) = 10.2116w^{0.425}h^{-0.275}$ in.^{2}/in.,
(b) $A_w(185,71) = 6.54$ in.^{2}/lb; $A_h(185,71) = 29.08$ in.^{2}/in.,

(c) $A_w(185,71) = 6.54 \text{ in.}^2/\text{lb}$ means that for a 185 lb person 5'11" tall, the rate of change in surface area is 6.54 in.^{2} for each pound gained in weight if the height is fixed; $A_h(185,71) = 29.07 \text{ in.}^2/\text{in.}$ means that for a 185 lb person 5'11" tall, the rate of change in surface area is 29.07 in.^{2} for each inch gained in height if the weight remains the same.
35. (a) $S_w(w,r) = 0.000\,02r^2$ ft/lb,
(b) $S_w(2400,65) = 0.0845$ ft/lb,
(c) $S_r(w,r) = 0.000\,04wr$ ft/mph,
(d) $S_r(2400,65) = 6.24$ ft/mph, (e) about 6.24 ft

Exercise Set 30.4

1. $2x\,dx + 2y\,dy$ **3.** $(6x+4y)\,dx + (4x-6y^2)\,dy$
5. $\dfrac{-y}{x^2+y^2}\,dx + \dfrac{x}{x^2+y^2}\,dy$
7. $(\tan yx + yx\sec^2 yx)\,dx + x^2\sec^2 yx\,dy$
9. $2te^{2t}(1+t) + 2t^3e^{2t}(2+t) = 2te^{2t}(1+t+2t^2+t^3)$
11. $\dfrac{e^{\sqrt{t}}}{2\sqrt{t}}\sin\pi t + \pi e^{\sqrt{t}}\cos\pi t$ **13.** minimum at $(-1/2,-1)$
15. minimum at $(2,-3)$ **17.** saddle point at $(0,0)$
19. saddle point at $\left(3,-\frac{9}{2}\right)$; test fails at $(0,0)$ **21.** $3.5\,\text{cm}^2$
23. $0.8\,\Omega$ **25.** $\dfrac{1}{k}(V\cdot\dfrac{dP}{dt} + P\cdot\dfrac{dV}{dt})$ **27.** -0.29
29. $h = 5\,\text{cm}$, $l = w = 10\,\text{cm}$ **31.** (a) $h = 0.89\,\text{ft}$, $l = w = 2.13\,\text{ft}$; (b) \$16.94 **33.** $x = \frac{2}{3}$, $y = \frac{1}{6}$ produces a reaction of $\dfrac{10}{9}$ **35.** $E_x = \dfrac{Qx}{4\pi\epsilon_0(x^2+a^2)^{3/2}}$

Exercise Set 30.5

1. $\frac{1}{12}$ **3.** 4 **5.** 4 **7.** $\frac{3}{20}$ **9.** $\frac{-4}{105}$ **11.** 6
13. $1/30$ **15.** $\frac{5}{24}$ **17.** $\frac{1}{6}$ **19.** -1876.7473
21. $114\frac{1}{3}$ **23.** $\dfrac{31}{120} \approx 0.258$ **25.** 121.5 **27.** $\dfrac{13}{3}$
29. $\frac{1}{2}\ln\frac{7}{2} \approx 0.6264$ **31.** $5\frac{1}{3}$ **33.** $\frac{4}{3}\sqrt{2} \approx 1.8856$
35. \$141,400 **37.** (a) $C(x,y) = 100 - 15(x^2+y^2)$,
(b) 75 ppm

Exercise Set 30.6

1. $(\sqrt{2},\sqrt{2},2)$ **3.** $(2,0,4)$
5. $(-03.5355, -3.5355, 0)$
7. $(2.8284, 0.7854, 5) = (2.8284, \pi/4, 5)$
9. $(5, 2.4981, 2)$ **11.** $(13, 5.8884, -3)$
13. $(\sqrt{2}, \sqrt{2}, 3.4641)$ **15.** $(0, -2.5981, 1.5)$
17. $(-3.75, -2.1651, -2.5)$
19. $(5, 0.6435, 1.5708) = (5, 0.6435, \pi/2)$
21. $(3, 0.4636, 2.3005)$
23. $(2, 0.7854, 0.7854) = (2, \frac{\pi}{4}, \frac{\pi}{4})$ **25.** The graph is a line that makes an angle of $\frac{\pi}{4}$ with the z-axis and whose image on the xy-plane make an angle of $\frac{\pi}{4}$ with the x-axis.

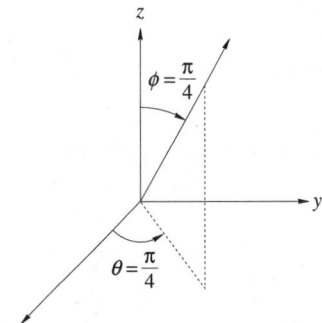

27. The region between the two spheres $\rho = 3$ and $\rho = 5$.

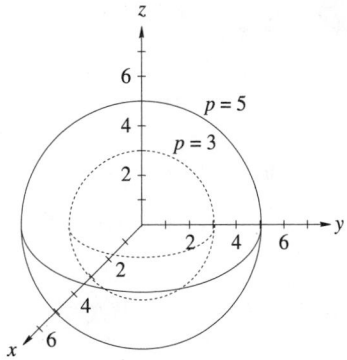

29. A cone shaped figure with spherical base.

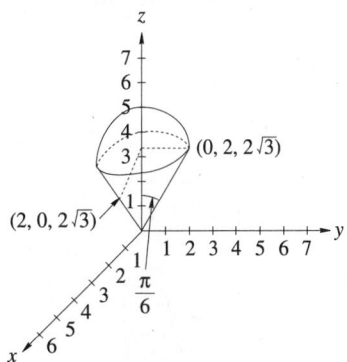

31. See *Computer Programs*

Exercise Set 30.7

1. $\bar{x} = 1.75$, $\bar{y} = 3.25$ **3.** $\bar{x} = \frac{32}{7} \approx 4.5714$, $\bar{y} = 0.8$
5. $\bar{x} = \frac{253}{195} \approx 2.6632$, $\bar{y} = \frac{195}{152} \approx 1.2829$ **7.** $\bar{x} = 1.25$,
$\bar{y} = 3.4$ **9.** $\bar{x} = \frac{10}{21} \approx 0.4762$, $\bar{y} = \frac{5}{12} \approx 0.4167$
11. $\bar{x} = -0.5$, $\bar{y} = 4.4$ **13.** $r_y = \sqrt{2/5} \approx 0.6325$;
$I_y = \dfrac{8\sqrt{2}}{15} \approx 0.7542$ **15.** $r_x = \sqrt{25/3} \approx 2.8868$; $I_x = 125$
17. $r_x = 1.9124$; $I_x = \frac{4096}{105}$ **19.** $r_y = 1.6928$; $I_y = 4.7$

21. (a) $\dfrac{296\sqrt{2}}{105}\rho$, **(b)** $\sqrt{74/35}\approx 1.45$

23. (a) $\dfrac{9472}{105}\rho\approx 90.21\rho$, **(b)** $\sqrt{\dfrac{296}{35}}\approx 2.91$

Review Exercises

1. A plane whose intercepts are $(6,0,0)$, $(0,2,0)$, and $(0,0,3)$

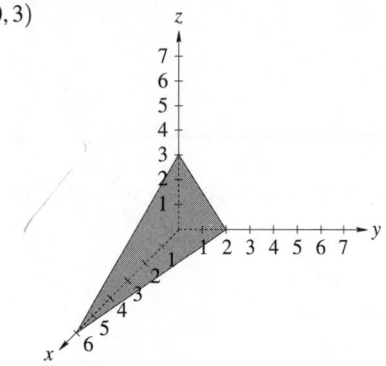

3. An elliptical cylinder whose axis is the x-axis

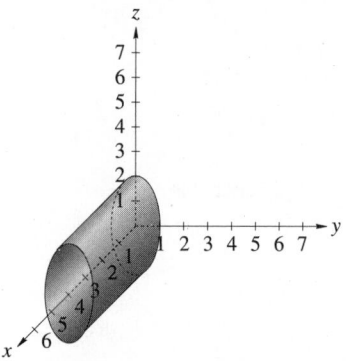

5. A parabola whose axis is the y-axis and has the trace in the xy-plane of $y=4x^2$.

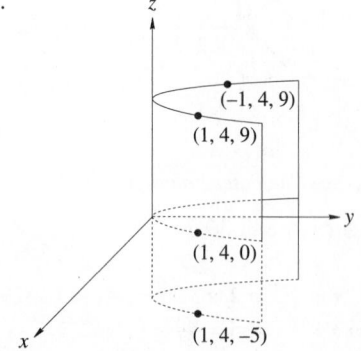

7. Hyperboloid of one-sheet. The trace in th xy-plane is the ellipse $\dfrac{x^2}{16}+\dfrac{y^2}{4}=1$. The trace in the yz-plane is the hyperbola

$\dfrac{y^2}{4}-\dfrac{z^2}{16}=1$ and the trace in the xz-plane is the hyperbola $\dfrac{x^2}{16}-\dfrac{z^2}{16}=1$.

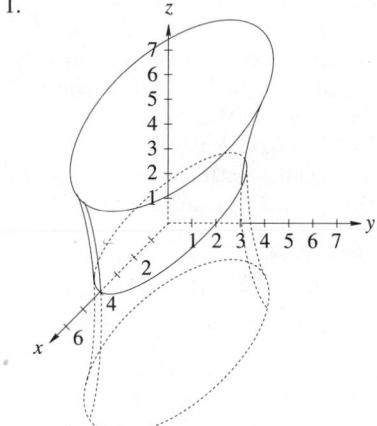

9. $\dfrac{\partial z}{\partial x}=6x+6y^3$, $\dfrac{\partial z}{\partial y}=18xy^2$, $\dfrac{\partial^2 z}{\partial x^2}=6$, $\dfrac{\partial^2 z}{\partial y^2}=36xy$,

$\dfrac{\partial^2 z}{\partial x\,\partial y}18y^2$ **11.** $\dfrac{\partial z}{\partial x}=e^x\cos y-e^y\cos x$,

$\dfrac{\partial z}{\partial y}=-e^x\sin y-e^y\sin x$, $\dfrac{\partial^2 z}{\partial x^2}=e^x\cos y+e^y\sin x$,

$\dfrac{\partial^2}{\partial y^2}=-e^x\cos y-e^y\sin x, =-e^x\sin y-e^y\cos x$ **13.** 16

15. $\dfrac{4}{15}(5^{5/2}-33)\approx 6.1071$ **17.** 148.5 **19.** rectangular:
$(3.4641,2,2)$, spherical: $(4.4721,\pi/6,1.1071)$
21. rectangular: $(-\sqrt{2},\sqrt{2},2\sqrt{3})$, cylindrical:
$(2,3\pi/4,3.4641)$ **23.** $\dfrac{128}{15}\approx 8.53$
25. $l=\omega=\sqrt[3]{600}\approx 8.4343$ in, $h=\tfrac{1}{2}\sqrt[3]{600}\approx 4.2172$ in
27. 0.19Ω **29.** $R-100\,\text{cm}^3/\text{min}$
31. $\bar{x}=\dfrac{248}{72}\approx 3.4444$, $\bar{y}=\dfrac{744}{72}=\dfrac{31}{3}\approx 10.3333$
33. $\bar{x}=\bar{y}=\dfrac{16}{35}\approx 0.4571$ **35.** $I_x=11,250$,
$r_x\approx\sqrt{150}\approx 12.2474$ **37.** $I_x=\tfrac{1}{9}$, $r_x=\sqrt{5/27}\approx 0.4303$
39. (a) $\dfrac{1,054,720}{3}\rho$, **(b)** $\sqrt{\dfrac{3,296}{3}}\approx 33.15$ **41.** 16

Chapter 30 Test

1. A plane whose intercepts are $(4,0,0)$, $(0,-8,0)$, and $(0,0,2)$

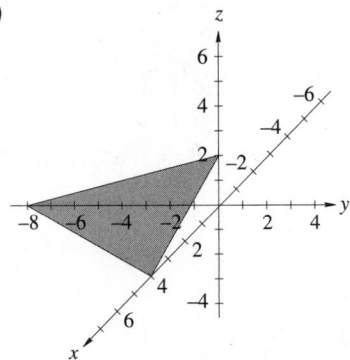

3. Hyperboloid of one-sheet. The trace in the xy-plane is the ellipse $\dfrac{x^2}{4}+\dfrac{y^2}{9}=1$. The trace in the yz-plane is the hyperbola $\dfrac{y^2}{9}-\dfrac{z^2}{4}=1$ and the trace in the xz-plane is the hyperbola

$$\frac{x^2}{4}-\frac{z^2}{4}=1.$$

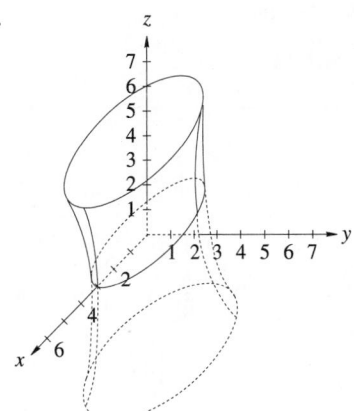

5. $\dfrac{\partial z}{\partial x}=12x^2-10xy^4$; $\dfrac{\partial z}{\partial y}=-20x^2y^3$

7. $8-4\sqrt{2}\approx 2.3431$ **9.** Rectangular coordinates: $\left(-\dfrac{3\sqrt{3}}{2},\dfrac{3}{2},8\right)$; Spherical coordinates: $\left(\sqrt{73},\dfrac{5\pi}{6},0.3588\right)$

11. 4π **13.** 16

■ ANSWERS FOR CHAPTER 31

Exercise Set 31.1

1. $x-\dfrac{x^3}{3!}+\dfrac{x^5}{5!}-\dfrac{x^7}{7!}$ **3.** $1+\dfrac{x^x}{2!}+\dfrac{x^4}{4!}+\dfrac{x^6}{6!}$

5. $1+3x+\dfrac{9}{2}x^2+\dfrac{27}{3!}x^3=1+3x+\dfrac{9}{2}x^2+\dfrac{9}{2}x^3$

7. $x^2-\dfrac{1}{2}x^4+\dfrac{1}{3}x^6-\dfrac{1}{4}x^8$

9. $1-\dfrac{12}{4!}x^4+\dfrac{1680}{8!}x^8-\dfrac{665280}{12!}=1-\dfrac{1}{2}x^4+\dfrac{1}{4!}x^8-\dfrac{1}{6!}x^{12}$

11. $x+x^2+\dfrac{1}{3}x^3-\dfrac{4}{5!}x^5$

13. $\dfrac{2}{2!}x^2-\dfrac{24}{4!}x^4+\dfrac{360}{6!}x^6-\dfrac{6720}{8!}x^8=$
$x^2\left(1-x^2+\dfrac{1}{2}x^4-\dfrac{1}{6}x^6\right)$

15. $\dfrac{6}{2!}x^2-\dfrac{108}{4!}x^4+\dfrac{1458}{6!}x^6-\dfrac{17496}{8!}x^8=$
$3x^2-\dfrac{9}{2}x^4+\dfrac{81}{40}x^5-\dfrac{243}{560}x^8$ **17.** $10\dfrac{4}{15}\approx 10.67\,\text{cm}^2$

19. $E(x)=1+\dfrac{1}{2}x^2+\dfrac{5}{24}x^4$

Exercise Set 31.2

1. $1+3x+\dfrac{9}{2}x^2+\dfrac{9}{2}x^3$ **3.** $1-\dfrac{x^2}{8}+\dfrac{x^4}{384}-\dfrac{x^6}{46,080}$

5. $1-\dfrac{x^6}{2!}+\dfrac{x^{12}}{4!}-\dfrac{x^{18}}{6!}$

7. $2x^2-\dfrac{8x^6}{3!}+\dfrac{32x^{10}}{5!}-\dfrac{128x^{14}}{7!}$ **9.**   **11.**

13. $\dfrac{1}{3}-\dfrac{1}{42}+\dfrac{1}{1320}\approx 0.3103$

15. $\dfrac{2}{3}(0.2)^{3/2}-\dfrac{2}{5\cdot 3!}(0.2)^{5/2}+\dfrac{2}{7\cdot 5!}(0.2)^{7/2}\approx 0.05844$

17. $1-x+\dfrac{2x^3}{3!}-\dfrac{4x^4}{4!}+\dfrac{x^5}{5!}+\cdots$

19. $x^2-x^4+\dfrac{x^6}{2!}-\dfrac{x^8}{3!}+\dfrac{x^{10}}{4!}-\dfrac{x^{12}}{5!}+\cdots$

21. $3.4835+3.5868j$ **23.** $6e^{4\pi j/3}$ **25.** 1.4519

27. $\dfrac{1}{3}(0.2)^3+\dfrac{1}{4}(0.2)^4+\dfrac{1}{5}(0.2)^5\approx 0.0031$ **29.** 14.9251

31. (a) $t^2-\dfrac{t^6}{3!}+\dfrac{t^{10}}{5!}-\dfrac{t^{14}}{7!}+\dfrac{t^{18}}{9!}-\cdots$; (b) 0.0000027

33. (a) $P=P_0\left(2-3t+\dfrac{5}{2}t^2-\dfrac{3}{2}t^3\right)$,

(b) $\dfrac{dP}{dt}=P_0\left(-3+5t-\dfrac{9}{2}t^2\right)$, (c) $-4.82P_0\,\text{ppm/week}$

Exercise Set 31.3

1. 0.7408 **3.** 0.9950 **5.** 0.0872 **7.** -0.0305

9. -0.6931 **11.** 0.0619 **13.** -0.0813 **15.** 0.2105

17. (b) For an accuracy of 0.0001, $x<\sqrt[3]{0.0006}\approx 0.0843$

19. $0.0021\,\mu\text{C}$ **21.** (a) $V=0.75e^{0.75t}=$
$0.75\left(1+0.75+\dfrac{(0.75)^2}{2!}+\dfrac{(0.75)^3}{3!}+\cdots\right)$; (b) 1.05105

23. (a) $e^{-t^2}\approx 1-t^2+\dfrac{1}{2}t^4-\dfrac{1}{6}t^6$

(b) $V(x)=V_0\displaystyle\int_0^x e^{-t^2}\,dt\approx V_0\left(x-\dfrac{1}{3}x^3+\dfrac{1}{10}x^5-\dfrac{1}{42}x^7\right)$

(c) $0.7722V_0$ **(d)** $0.0957V_0$

Exercise Set 31.4

1. $e^x = e^2(x - 1 + \frac{(x-2)^2}{2!} + \frac{(x-2)^3}{3!} + \frac{(x-2)^4}{4!} + \cdots)$

3. $\sin x = $
$\frac{\sqrt{2}}{2}[1 + (x - \pi/4) - \frac{1}{2!}(x - \pi/4)^2 - \frac{1}{3!}(x - \pi/4)^3 + \cdots]$

5. $\frac{1}{x} = \frac{1}{2}(1 - \frac{1}{2}(x - 2) + \frac{1}{2 \cdot 2!}(x - 2)^2 - \frac{1}{8}(x - 2)^3 + \cdots)$

7. $(x + 1)^{-2} = 1 - 2x + \frac{6x^2}{2!} - \frac{24x^3}{3!} + \cdots = $
$1 - 2x + 3x^2 - 4x^3 + \cdots$

9. $e^{-x} = e^{-1}[1 - (x - 1) + \frac{1}{2!}(x - 1)^2 - \frac{1}{3!}(x - 1)^3 + \cdots]$

11. $\sqrt[3]{x} = $
$2 + \frac{1}{12}(x - 8) - \frac{1}{9 \cdot 16 \cdot 2!}(x - 8)^2 + \frac{10}{27 \cdot 2^8 \cdot 3!}(x - 8)^3 + \cdots$

13. $e^2(1.1052) \approx 8.1662$ **15.** 0.6820 **17.** 0.5551
19. 0.7972 **21.** 0.4493 **23.** 1.9832 **25.** 0.14715

Exercise Set 31.5

1. $f(x) = \frac{-8}{\pi}(\sin x + \frac{1}{3}\sin 3x + \frac{1}{5}\sin 5x + \frac{1}{7}\sin 7x + \cdots) = $
$\frac{-8}{\pi}\sum_{k=1}^{\infty}\frac{1}{2k-1}\sin(2k-1)x$

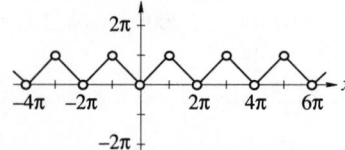

3. $h(x) = 1 - \frac{2}{\pi}\sin\frac{\pi x}{3} + \frac{1}{\pi}\sin\frac{2\pi x}{3} - \frac{2}{3\pi}\sin\frac{3\pi x}{3} + $
$\frac{2}{4\pi}\sin\frac{4\pi x}{3} - \cdots = 1 + \sum_{k=1}^{\infty}(-1)^k\frac{2}{k\pi}\sin\frac{k\pi x}{3}$

5. $f(x) = \frac{\pi}{2} - \frac{4}{\pi}\cos x - \frac{4}{9\pi}\cos 3x - \frac{4}{25\pi}\cos 5x - \cdots = $
$\frac{\pi}{2} - \frac{4}{\pi}\sum_{k=1}^{\infty}\frac{1}{(2k-1)^2}\cos(2k-1)x$

7. $h(x) = 2\sin x - \sin 2x + \frac{2}{3}\sin 3x - \frac{1}{2}\sin 4x + \cdots = $

$2\sum_{k=1}^{\infty}\frac{(-1)^{k+1}}{k}\sin kx$

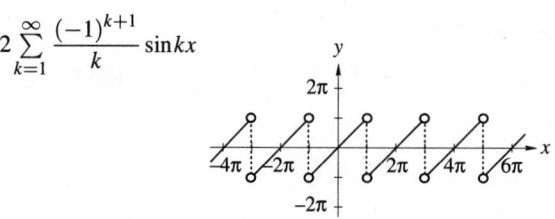

9. $f(x) = \frac{3}{\pi}(\sin x + \frac{1}{2}\sin 2x + \frac{1}{4}\sin 4x + \frac{1}{5}\sin 5x + \cdots) = $
$\frac{3}{\pi}\sum_{k=1}^{\infty}(\frac{\sin(3k-2)x}{3k-2} + \frac{\sin(3k-1)x}{3k-1})$

11. $h(x) = \frac{1}{\pi} + \frac{1}{2}\sin x - \frac{2}{\pi}(\frac{1}{3}\cos 2x + \frac{1}{15}\cos 4x + $
$\frac{1}{35}\cos 6x + \cdots) = \frac{1}{\pi} + \frac{1}{2}\sin x - \frac{2}{\pi}\sum_{k=1}^{\infty}(\frac{1}{4k^2-1}\cos 2kx)$

Review Exercises

1. $\frac{x}{2!} - \frac{x^3}{4!} + \frac{x^5}{6!} - \frac{x^7}{8!} + \frac{x^9}{10!} - \frac{x^{11}}{12!} + \cdots$

3. $x - \frac{5x^3}{3!} + \frac{16x^5}{5!} - \frac{64x^7}{7!} + \frac{247x^9}{9!}$ **5.** $1 + \frac{3}{4}(x' + $
$1) - \frac{s(x'+1)^2}{16 \cdot 2!} + \frac{15(x'+1)^3}{64 \cdot 3!} - \frac{135(x'+1)^4}{256 \cdot 4!} + \cdots$

7. $x^2 - \frac{x^6}{3!} + \frac{x^{10}}{5!} - \frac{x^{14}}{7!} + \frac{x^{18}}{9!} - \cdots$ **9.** 0.6015

11. 0.7788 **13.** $\frac{\sqrt{3}}{2} - \frac{1}{2}(x - \frac{\pi}{6}) - \frac{\sqrt{3}}{2}\frac{(x-\pi/6)^2}{2!} + $
$\frac{1}{2}\frac{(x-\pi/6)^3}{3!} + \frac{\sqrt{3}}{2}\frac{(x-\pi/6)^4}{4!} - \cdots$ **15.** $3 + \frac{1}{6}(x - 9) - $
$\frac{1}{4 \cdot 3^3}\frac{(x-9)^2}{2!} + \frac{3}{8 \cdot 3^5}\frac{(x-9)^3}{3!} - \frac{15}{16 \cdot 3^7}\frac{(x-9)^4}{4!} + \cdots$

17. 0.8688 **19.** 3.0137

21. $\frac{1}{2} - \frac{2}{\pi}(\sin x + \frac{1}{3}\sin 3x + \frac{1}{5}\sin 5x + \cdots) = $
$\frac{1}{2} - \frac{2}{\pi}\sum_{k=1}^{\infty}\frac{1}{2k-1}\sin(2k-1)x$

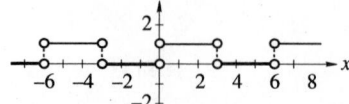

23. $\dfrac{\pi^2}{3} - 4\cos x + \cos 2x -$

$\dfrac{4}{9}\cos 3x + \dfrac{1}{4}\cos 4x - \dfrac{4}{25}\cos 5x + \cdots = \dfrac{\pi^2}{3} + 4\displaystyle\sum_{k=1}^{\infty} \dfrac{(-1)^k}{k^2}\cos kx$

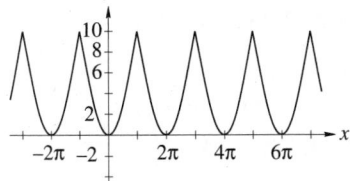

5. $\sqrt[3]{x} = -1 + \dfrac{x+1}{3} + \dfrac{(x+1)^2}{9} + \dfrac{5(x+1)^3}{81} + \cdots$

7. $f(x) = \pi - 2\left(\sin x + \dfrac{\sin 2x}{2} + \dfrac{\sin 3x}{3}\cdots\right)$

9. 18.1818

Chapter 31 Test

1. $\sin x^2 = x^2 - \dfrac{x^6}{3!} + \dfrac{x^{10}}{5!} + \cdots$

3. $\dfrac{1}{x} = \dfrac{1}{2} - \dfrac{x-2}{4} + \dfrac{(x-2)^2}{16} + \cdots$

ANSWERS FOR CHAPTER 32

Exercise Set 32.1

17. $y = \frac{1}{2}x^4 - 4x + 6\frac{1}{2}$ **19.** $y = \tan x - 7$ **21. (a)** about $t = 3.178$ s;**(b)** 151.474 ft   **23. (a)** 3 s;**(b)** $5\frac{1}{3}$ mph
25. (a) $t = 6$ s;**(b)** 128 ft/s downward
27. (a) $p(t) = -75t + 500$; **(b)** $t = 6\frac{2}{3}$ h   **29.** 25 h
31. $L\left(-\dfrac{kR}{L}e^{-Rt/L}\right) + R\left(\dfrac{V}{R} + ke^{-Rt/L}\right) = V$

Exercise Set 32.2

1. $\frac{1}{2}y^2 - x^2 = C$ **3.** $\frac{5}{2}x^2 + y^3 = C$
5. $5x + 3\ln y = C$ **7.** $-4x^{-1} + y^{-3} = C$
9. $\frac{1}{4}x^4 + \ln\sqrt{y^2+4} = C$ **11.** $\ln\sqrt{y^2+4} = -e^{-x^2} + C$
13. $\frac{2}{3}e^{3x} + \ln y = C$ **15.** $\left(y^2-4\right)\sin x = C$
17. $\frac{1}{2}y^2 + 3\ln y = \frac{1}{3}x^3 - x + C$ **19.** $-\dfrac{1}{y} + \frac{1}{3}e^{\sin 3x} = C$
21. $-\dfrac{1}{y} + \frac{1}{4}x^4 = 5$ **23.** $\dfrac{3+y^2}{1+x^2} = 6$
25. $\ln y = x^3 - 2x + 1$
27. (a) $T(t) = 325 - 315e^{-0.231112t}$, **(b)** 2.42 hr $\approx$ 2 hr 25 min
29. 9:24 A.M.   **31. (a)** If i represents the number of people who are infected, then $\dfrac{di}{dt} = ki(500{,}000 - i)$,
(b) $i(t) = \dfrac{500{,}000}{1 + Ce^{-Bt}}$, **(c)** $i(t) = \dfrac{500{,}000}{1 + 4{,}999e^{-0.201620t}}$,
(d) 5,578 people

Exercise Set 32.3

1. No, $\dfrac{\partial(x^2 - y)}{\partial y} = -1$, $\dfrac{\partial x}{\partial x} = 1$ **3.** Yes,

$\dfrac{\partial(x^2 + y^2)}{\partial y} = 2y = \dfrac{\partial 2xy}{\partial x}$ **5.** Yes,

$\dfrac{\partial(x + y\cos x)}{\partial y} = \cos x = \dfrac{\partial\sin x}{\partial x}$ **7.** Yes,

$\dfrac{\partial(2xy + x)}{\partial y} = 2x = \dfrac{\partial(y + x^2)}{\partial x}$ **9.** No, $\dfrac{\partial(x - y)}{\partial y} = -1$,

$\dfrac{\partial(y + x)}{\partial x} = 1$ **11.** $\ln\dfrac{y}{x} = C_1$ or $\dfrac{y}{x} = C_2$ where $C_2 = e^{C_1}$

13. $\dfrac{x}{y} - \frac{5}{3}y^3 = C$ **15.** $\arctan\dfrac{y}{x} - x = C$

17. $\ln xy - x^3 = C$ **19.** $\ln xy = C$ **21.** $x^2y - x^3 = C$
23. $-x^3 + \arctan\dfrac{y}{x} = C$ **25.** $y = \ln|\sin(x^2 + y^2)| + C$
27. $x^2y^2 = e^{x^3} + 9 - e$ **29.** $\arctan\dfrac{y}{x} = x - \dfrac{\pi}{12}$

Exercise Set 32.4

1. $ye^{x^2} = 3e^{x^2} + C$ **3.** $ye^{-2x} = -2e^{-2x} + C$
5. $yx = \frac{1}{4}x^4 + C$ **7.** $ye^{-6x} = \frac{1}{3}e^{6x}(-6x - 1) + C$
9. $yx = e^x(x - 1) + C$ **11.** $ye^{1/x} = -e^{1/x} + C$
13. $ye^{-2x} = \dfrac{e^{-2x}[-2\sin 2x - 2\cos 2x]}{8} + C =$
$-\frac{1}{4}e^{-2x}(\sin 2x + \cos 2x) + C$ **15.** $yx^{-1/4} = \frac{8}{3}x^{3/4} + C$
17. $ye^{-\tan x} = -e^{-\tan x} + C$ **19.** $\dfrac{yx}{e^4x} = \frac{4}{3}x^3 + C$
21. $yx^2 = \frac{1}{4}x^4 + 8$ **23.** $yx^{-2} = -\frac{1}{3}\cos 3x + \frac{2}{3}$
25. (a) $w(t) = 120 + 40e^{-0.005t}$, **(b)** about 154.4 lb, **(c)** about 58 days, **(d)** 120 lb, **(e)** this person's weight will approach

120 lb if this diet is maintained for a long time.

Exercise Set 32.5

1. $N = 100e^{-0.0192t}$, t in days   **3.** 152.0031 years
5. 1387   **7.** 50,208.4 years
9. (a) $T = 30 - 20e^{-0.02877t}$, t in minutes; **(b)** 31.8508 min;
(c) 26.4404°C   **11.** 101.59697 min = 1 h 21 min 35.82 s
13. $I = -10e^{-5t} + 10$ **15.** $I = -0.1e^{-50t} + 0.1$

Exercise Set 32.6

1. $y = kx$

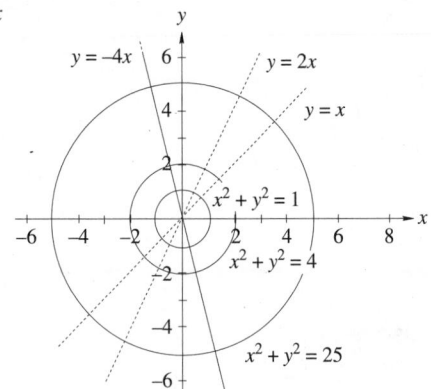

3. $y^2 + 2x = k$

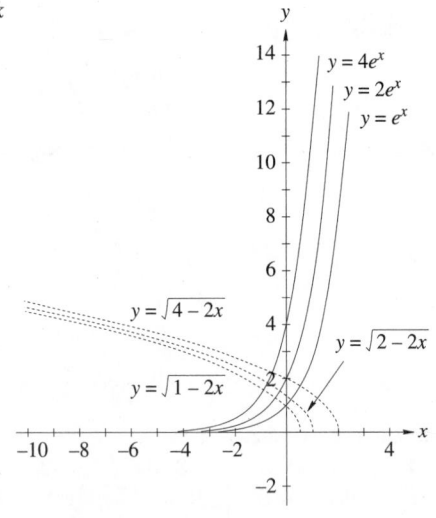

5. $y^2 = kx$ **7.** $v(t) = 490\left(1 - e^{-t/50}\right)$ m/s
9. (a) $v(t) = 490 - 489e^{-t/50}$ m/s; **(b)** 89.64 m/s; **(c)** 490 m/s
11. 13.7203 lb; $Q = 25e^{-t/25}$
13. (a) $Q(t) = 24 - 22e^{-t/2}$; **(b)** 23.5971 lb
15. (a) $Q(t) = 400 - 390e^{-t/200}$, **(b)** about 15.806 kg

Review Exercises

1. $y = Ce^x - (x+1)$ **3.** $x^4 + 4y^{-1} = C$
5. $e^x - y^2 = C$ **7.** $\ln|xy| - y^2 x = C$ **9.** $\dfrac{x-2}{y} = C$
11. $\ln|xy| = C$ **13.** $y = -\frac{1}{4} + Ce^{2x^2}$
15. $y = \frac{1}{2}(\sin x - \cos x) + Ce^x$ **17.** $\sin x + \frac{1}{2}y^2 = 2$
19. $yx^2 = \frac{1}{4}x^4 - \frac{1}{4}$ **21. (a)** $Q(t) = Ce^{-0.0527t}$ where
$C = Q(0)$; **(b)** 13.1576 years ≈ 13 years 57 days 12 hours
48.5 min   **23. (a)** 54.65 min; **(b)** 52.41°F
25. $q(t) = \frac{1}{89}(100\cos 2t + 160\sin 2t - 100e^{-5t/4})$
27. (a) $0.7 + 1.3e^{-t/20}$ ft³; **(b)** 0.7647 ft³; **(c)** 0.7032 ft³
29. (a) $50 - 40e^{-t/4000}$ 4 m³; **(b)** in about 1151 days or 3 years
56 days

Chapter 32 Test

1. Differentiating $y = 4e^{2x} + 2e^{-3x}$ we get
$\dfrac{dy}{dx} = 8e^{2x} - 6e^{-3x}$ and $\dfrac{d^2y}{dx^2} = 16e^{2x} + 18e^{-3x}$. Substituting
these in the given differential equation, we obtain
$\left(16e^{2x} + 18e^{-3x}\right) + \left(8e^{2x} - 6e^{-3x}\right) - 6\left(4e^{2x} + 2e^{-3x}\right) =$
$(16 + 8 - 24)e^{2x} + (18 - 6 - 12)e^{-3x} = 0.$
3. $\frac{1}{4}y^4 = x^5 + C$ **5.** $\arctan x + \frac{1}{2}(1 + y^2) = C$
7. $-\dfrac{y}{x} + \frac{1}{5}x^5 = C$ **9.** $y^2 - x = k$ **11.** ≈ 170.70 ft/s

▤ ANSWERS FOR CHAPTER 33

Exercise Set 33.1

1. $y = c_1 e^{2x} + c_2 e^{5x}$ **3.** $y = c_1 e^{2x} + c_2 e^{6x}$
5. $y = c_1 e^{\frac{1}{2}x} + c_2 e^{-5x}$ **7.** $y = c_1 e^{-\frac{1}{4}x} + c_2 e^{2x}$
9. $y = c_1 e^x + c_2 e^{4x}$ **11.** $y = c_1 e^x + c_2 e^{-x}$
13. $y = c_1 e^{\sqrt{7}x} + c_2 e^{-\sqrt{7}x}$ **15.** $y = c_1 e^{x/2} + c_2 e^{-x}$
17. $y = c_1 e^x + c_2 e^{2x} c_3 e^{3x}$
19. $y = c_1 e^{-x} + c_2 e^{-2x} + c_3 e^{-3x}$ **21.** $y = 2e^{3x}$
23. $y = e^{2x} - e^{-4x}$ **25.** $y = 3e^{3x} + 2e^{-5x}$
27. $y = 5e^{-x} + 2e^{2x} - 4e^{-2x}$

Exercise Set 33.2

1. $y = (c_1 + c_2 x)e^{-x}$ **3.** $y = (c_1 + c_2 x)e^{2x}$
5. $y = (c_1 + c_2 x)e^{x/3}$ **7.** $y = (c_1 + c_2 x)e^{-3x/2}$
9. $y = e^{-2x}(c_1 \cos x + c_2 \sin x)$ **11.** $y = c_1 + c_2 e^{-9x}$

13. $y = c_1 e^x + e^{-x/2}\left(c_2 \cos \dfrac{\sqrt{3}}{2}x + c_3 \sin \dfrac{\sqrt{3}}{2}x\right)$

15. $y = c_1 e^{-2x} + e^x(c_2 \cos \sqrt{3}x + c_3 \sin \sqrt{3}x)$
17. $y = c_1 e^x + c_2 e^{2x} + c_3 e^{3x}$
19. $y = (c_1 + c_2 x + c_3 x^2 + c_4 x^3)e^{-2x}$
21. $y = (c_1 + c_2 x)e^x + (c_3 + c_4 x + c_5 x^2)e^{-2x}$
23. $y = (c_1 + c_2 x)e^{3x} + c_3 e^{(3+3\sqrt{2})x} + c_4 e^{(3-3\sqrt{2})x}$

25. $y = e^{x/3}\left(c_1 \cos \dfrac{\sqrt{3}}{3}x + c_2 \sin \dfrac{\sqrt{3}}{2}x\right)$

27. $y = e^x\left(4\cos 2x + \dfrac{3}{2}\sin 2x\right)$ **29.** $y = (1-x)e^{3/2x}$
31. $y = e^{-x} - 2xe^{-x}$ **33.** $y = e^{3x}(4\cos 5x - 3\sin 5x)$

Exercise Set 33.3

1. $y = (c_1 + c_2 x)e^{5x} + 4/25$
3. $y = (c_1 + c_2 x)e^{5x} + \dfrac{1}{4}e^{3x}$
5. $y = (c_1 + c_2 x)e^{5x} + \dfrac{2}{5} + \dfrac{1}{16}e^x$
7. $y = (c_1 + c_2 x)e^{5x} + \dfrac{21}{29}\sin 2x + \dfrac{20}{29}\cos 2x$
9. $y = c_1 e^{2x} + c_2 e^{-2x} - \left(\dfrac{1}{3}x^2 + \dfrac{4}{9}x + \dfrac{14}{27}\right)e^x + \dfrac{3}{4}x$
11. $y = c_1 \cos 3x + c_2 \sin 3x + \dfrac{1}{4}\sin x + \dfrac{1}{2}\cos x$
13. $y = c_1 \cos x + c_2 \sin x + \dfrac{1}{5}e^{3x}$
15. $y = c_1 e^{2x} + c_2 e^{-2x} - 2x^2 - 1$
17. $y = c_1 \cos x + c_2 \sin x - \dfrac{1}{3}\cos 2x + 4$

19. $y = e^{-x/3}\left(c_1 \cos \dfrac{\sqrt{2}}{3}x + c^2 \sin \dfrac{\sqrt{2}}{3}x\right) + 6x^2 - 22x - 12$

21. $y = (2 + 9x)e^x + x^2 + 4x + 5$
23. $y = e^{2x} + \dfrac{1}{20}e^{-x} - \dfrac{3}{20}\sin 2x + \dfrac{1}{20}\cos 2x$
25. $y = -10e^{-x} + 5xe^{-x} + 3x^2 - 10x + 13$

27. $y = \cos x + \dfrac{2}{3}\cos 3x + \sin 3x$

Exercise Set 33.4

1. $4 \text{ lb/in.} = 48 \text{ lb/ft}$ **3.** 245 N/m

5. $x = \dfrac{-4}{\sqrt{75.2}}e^{1.6t}\sin\sqrt{75.2}t \approx -0.4613e^{1.6t}\sin 8.6718t$

7. $x = \dfrac{2}{\sqrt{33}}e^{4t}\sin\sqrt{33}t \approx 0.3482e^{4t}\sin 5.7446t$

9. $q(t) =$
$e^{-30t}\left(\dfrac{1}{39}\cos 10t + \dfrac{9}{13}\sin 10t\right) - \dfrac{1}{26}\sin 100t - \dfrac{1}{39}\cos 100t$
11. (a) $q(t) = \dfrac{18}{11}\cos 50t - \dfrac{18}{11}\cos 60t$ **(b)** $\dfrac{1080}{11}\sin 60t$
13. $g(t) = e^{-\alpha t}\left(c_1 \cos \sqrt{\omega_0^2 - \alpha^2}\,t + c_2 \sin \sqrt{\omega_0^2 - \alpha^2}\,t\right)$
15. $q = e^{-2.1t}(c_1 \cos 5.4t + c_2 \sin 5.4t) + 0.375$

Review Exercises

1. $y = c_1 + c_2 x$ **3.** $y = c_1 e^{2x} + c_2 e^{3x}$
5. $y = e^{3x}(c_1 \cos 4x + c_2 \sin 4x)$
7. $y = c_1 e^{2x} + e^{-x}(c_1 \cos \sqrt{3}x + c_2 \sin \sqrt{3}x)$
9. $y = c_1 e^x + c_2 e^{-4x} - \dfrac{9}{4}x^2 - \dfrac{27}{8}x - \dfrac{117}{32}$
11. $y = c_1 e^{-3x} + c_2 e^{-4x} - \dfrac{13}{1394}\cos 5x + \dfrac{35}{1394}\sin 5x$
13. $y = (c_1 + c_2 x)e^{-2x} + x - 1 + \dfrac{1}{16}e^{2x}$

15. $x = \dfrac{4}{\sqrt{759}}e^{-5t/4}\sin\dfrac{\sqrt{759}}{4}t$

Chapter 33 Test

1. $5m^2 + 2m - 3 = 0$ **3.** $y = c_1 e^{-10x} + c_2 e^{2x}$
5. $y = e^{-3x}(c_1 \cos 2x + c_2 \sin 2x)$
7. $y = c_1 e^{-7x} + c_2 e^x + \dfrac{2}{3}e^{2x}$
9. $y = c_1 e^{-7x} + c_2 e^x - 8x$ **11.** The charge is given by
$q(t) = e^{-30t}(c_1 \cos 20t + c_2 \sin 20t) + 0.02\sin 10t + \dfrac{1}{75}\cos 10t$;
the steady-state current is $i = 0.2\cos 10t - \dfrac{10}{75}\sin 10t$

ANSWERS FOR CHAPTER 34

Exercise Set 34.1

1. See *Computer Programs*

3.

x	y $h = 0, 1$	$h = 0.05$	Correct solution $y = -\frac{1}{2}e^x + \frac{1}{2}e^{3x}$
0	0	0	0
0.5	–	0.5	0.5528
0.10	0.1	0.1101	0.12234
0.15	–	0.1818	0.2032
0.20	0.2405	0.2672	0.3004
0.25	–	0.3683	0.4165
0.30	0.4348	0.4878	0.5549
0.35	–	0.6285	0.7193
0.40	0.7002	0.7937	0.9141
0.45	–	0.9873	1.1446
0.50	1.0595	1.2138	1.4165
0.55	–	1.4784	1.7369
0.60	1.5422	1.7868	2.1138
0.65	–	2.1459	2.5566
0.70	2.1871	2.5636	3.0762
0.75	–	3.0488	3.6854
0.80	3.0046	3.6120	4.3988
0.85	–	4.2650	5.2337
0.90	4.1805	5.0218	6.2101
0.95	–	5.8980	7.3510
1.00	5.6807	6.9120	8.6836

5.

x	y $h = 0.05$	Correct solution $y = \frac{1}{53}(55e^{7x} - 2\cos 2x - 7\sin 2x)$
0	1	1
0.05	1.35	1.4219
0.10	1.8275	2.0265
0.15	2.4770	2.8904
0.20	3.3588	4.1220
0.25	4.5538	5.8753
0.30	6.1717	8.3686
0.35	8.3600	11.9117
0.40	11.3182	16.9442
0.45	15.3154	24.0898
0.50	20.7149	34.2336

7.

x	y
0	1
0.1	1.1
0.2	1.222
0.3	1.3753
0.4	1.5735
0.5	1.8371
0.6	2.1995
0.7	2.1793
0.8	3.5078
0.9	4.8023
1.0	7.1895

9.

x	y
0	0
0.1	0.1
0.2	0.2010
0.3	0.3051
0.4	0.4147
0.5	0.5327
0.6	0.6633
0.7	0.8121
0.8	0.9887
0.9	1.2092
1.0	1.5062

11.

x	y
$\pi/2 \approx 1.57$	1
1.62	1.0270
1.67	1.0507
1.72	1.0711
1.77	1.0882
1.82	1.1022
1.87	1.1130
1.92	1.1209
1.97	1.1259
2.02	1.1281
2.07	1.1277
2.12	1.1249
2.17	1.1197
2.22	1.1123
2.27	1.1028
2.32	1.0913
2.37	1.0779

13. See *Computer Programs*

15. $T(1) \approx 82.694$, **(b)** $T(2) \approx 76.446$

Exercise Set 34.2

1. $y = \frac{1}{2}x^2 + 2x + \frac{3}{2}$, $y(2) = 7.5$ the exact value

3. $y = \frac{1}{3}x^6 + x^4 + x^2 + 1$, $y(1) = 3\frac{1}{3}$

5. $y = \frac{1}{63}x^7 + \frac{2}{15}x^5 + \frac{1}{6}x^4 + \frac{2}{3}x^3 + x^2 + x + 1$

7. $y = \frac{1}{8}x^4 + \frac{4}{3}x^3 + \frac{1}{2}x^2 + 4x + 1$

9. $y = -\frac{1}{2}\cos 2x - \frac{1}{4}\sin 2x + \frac{1}{2}x^2 + \frac{1-\pi}{2}x + \frac{4-2\pi-\pi^2}{8}$

11. $3(1-x) + \dfrac{x^2}{2!} - \dfrac{x^3}{3!} + \dfrac{x^4}{4!} - \dfrac{x^5}{5!} + x - 1 =$

$2 - 2x + \frac{3}{2}x^2 - \dfrac{x^3}{2} + \dfrac{x^4}{8} - \dfrac{x^5}{40}$. The first four terms are the same. **13.** $I(t) \approx 2\sin t + 3\cos t + 2t^2 - 2t - 3$ is the third approximation $I(0.5) \approx 0.0916$.

Exercise Set 34.3

1. $\dfrac{n}{s}$ **3.** $\dfrac{s}{s^2+a^2}$ **5.** $\dfrac{3!}{s^4}$ **7.** $\dfrac{6}{s^2+36}$

9. $\dfrac{3!}{(s-5)^4}$ **11.** $\dfrac{85^2}{(s^2+16)^2}$ **13.** $\dfrac{5}{s^2+25}+\dfrac{s}{s^2+9}$

15. $\dfrac{30s}{(s^2+25)^2}-\dfrac{s}{s^2+16}+\dfrac{1}{s}=\dfrac{30s}{(s^2+25)^2}+\dfrac{16}{s(s^2+16)}$

17. $\dfrac{s+2}{(s+2)^2+9}-\dfrac{2}{(s-5)^2+1}$ **19.** $\dfrac{4}{s}+\dfrac{3}{s^2}+\dfrac{2}{s-1}$

21. $\dfrac{s^2+18}{s(s^2+36)}$

Exercise Set 34.4

1. $4e^{5t}$ **3.** $1-\cos 4t$ **5.** $3\cos 3t$
7. $-2e^{-2t}+7e^{-7t}$ **9.** $\frac{1}{5}t\sin 5t + \frac{3}{5}(\sin 5t + 5t\cos 5t)$
11. $(s^2+4s)L(f)-(s-4)$
13. $(3s^2-3s+1)L(f)-(2s+3)$
15. $(s^2+6s-1)L(f)-(s+5)$

Exercise Set 34.5

1. $\dfrac{1}{x}+\dfrac{-1}{x+1}$ **3.** $\dfrac{-4}{x}+\dfrac{2}{x-1}+\dfrac{2}{x+1}$

5. $\dfrac{-9/2}{x-3}+\dfrac{15/2}{x-5}$ **7.** $\dfrac{1}{x+1}+\dfrac{-2}{(x+1)^2}$

9. $\dfrac{1}{x^2+1}+\dfrac{x}{x^2+2}$ **11.** $1-e^{-t}$ **13.** $-4+2e^t+2e^{-t}$

15. $-\frac{9}{2}e^{3t}+\frac{15}{2}e^{5t}$ **17.** $e^{-t}-2te^{-t}$ **19.** $\sin t + \cos\sqrt{2}t$

21. Use $L^{-1}\left[\dfrac{-2}{s}+\dfrac{-2}{s^2}+\dfrac{2}{s-1}\right]=-2-2t+2e^t$

23. $L^{-1}\left[\dfrac{1/25}{s+1}+\dfrac{-s/25+4/25}{s^2+9}\right]=$
$\frac{1}{25}e^{-4t}-\frac{1}{25}\cos 3t+\frac{4}{25}\sin 3t$

25. $L^{-1}\left[\dfrac{-1/2}{s-1}+\dfrac{1/2}{(s-1)^2}+\dfrac{1/2s}{s^2+1}\right]=$
$-\frac{1}{2}e^t+\frac{1}{2}te^t+\frac{1}{2}\cos t$

Exercise Set 34.6

1. $y=-1+e^t$ **3.** $y=te^{-6t}+2e^{-6t}$
5. $y=\frac{301}{100}e^{-5t}+(\frac{1}{10}t+\frac{1}{100})e^{5t}$
7. $y=2\cos 2t+\frac{3}{2}\sin 2t$
9. $y=\frac{1}{5}e^t-\frac{1}{5}(\cos 2t+\frac{1}{2}\sin 2t)$
11. $y=\frac{2}{27}+\frac{1}{9}t-\frac{2}{27}e^{3t}+\frac{10}{9}te^{3t}$
13. $y=e^{-3t}(\cos 2t+\frac{1}{2}\sin 2t)$
15. $y=\frac{1}{10}e^{-2t}+\frac{1}{10}e^{2t}-\frac{3}{5}\cos t$
17. $y=e^{-t}(\frac{2}{5}\cos 2t+\frac{13}{10}\sin 2t)+\frac{3}{5}e^{-2t}$
19. $y=-\frac{13}{50}e^{-3t}+\frac{5}{2}e^t-\frac{6}{25}e^{2t}+\frac{1}{5}te^{2t}$
21. $y=2\cos t+\sin t$ **23.** $q(t)=15e^{-10t}\sin 10t$,
$i(t)=150e^{-10t}(\cos 10t-\sin 10t)$, steady state current is 0 A

25. $q(t)=\frac{3}{25}-\frac{3}{20}e^{-10t}+\frac{3}{100}e^{-50t}$ $i(t)=\frac{3}{2}(e^{-10t}-e^{-50t})$
27. $x\approx Ae^{(-64-2\sqrt{1022})t}+Be^{(-64+2\sqrt{1022})t}$ where
$A=1-\dfrac{32}{\sqrt{1022}}$ and $B=1+\dfrac{32}{\sqrt{1022}}$ **29.** $y=\frac{1}{8}te^{-t}\sin 4t$

Review Exercises

1.

| | y | |
x	$h=0.1$	$h=0.05$
0.05	–	1.8
0.10	1.6	1.6299
0.15	–	1.4838
0.20	1.3184	1.3572
0.25	–	1.2467
0.30	1.1098	1.1495
0.35	–	1.0636
0.40	0.9497	0.9873
0.45	–	0.9190
0.50	0.8234	0.8578
0.55	–	0.8026
0.60	0.7217	0.7527
0.65	–	0.7074
0.70	0.6384	0.6661
0.75	–	0.6284
0.80	0.5691	0.5938
0.85	–	0.5621
0.90	0.5108	0.5329
0.95	–	0.5059
1.00	0.4612	0.4809

3.

x	y $h = 0.05$
0.05	59.9
0.10	59.8
0.15	59.6990
0.20	59.5961
0.25	59.4905
0.30	59.3812
0.35	59.2678
0.40	59.1496
0.45	59.0263
0.50	58.8975
0.55	58.7630
0.60	58.6228
0.65	58.4767
0.70	58.3249
0.75	58.1675
0.80	58.0047
0.85	57.8366
0.90	57.6635
0.95	57.4857
1.00	57.3034

5. $y = \frac{1}{8}x^4 + \frac{4}{3}x^3 + \frac{1}{2}x^2 + 4x + 1$

7. $y = \frac{2}{9}(x-9)^3 + \frac{2}{9}(x-9)^{3/2} + \frac{5}{9}$ **9.** $y = -4 + 5e^{t/2}$

11. $y = \frac{2}{3}e^{-2t} + \frac{1}{3}e^t$ **13.** $y = e^t$

15. $y = e^{-t}\cos 2t + \frac{1}{2}e^{-t}\sin 2t$

17. $y = \frac{3}{5}e^{-2t} + \frac{2}{5}e^{-t}\cos 2t + \frac{3}{10}e^{-t}\sin 2t$

19. $y = t - 2t^2 - \sin t$ **21.** $q(t) = \frac{10}{39201}(20\cos t +$
$199\sin t - 20e^{-10t}\cos 10t + 3880.2e^{-10t}\sin 10t)$;
$i(t) = \frac{10}{39201}(-20\sin t + 199\cos t + 39002e^{-10t}\cos 10t -$
$4080.2e^{-10t}\sin 10t)$; steady state current
$= \frac{10}{39201}(199\cos t - 20\sin t)$

Chapter 34 Test

1. $\dfrac{5}{s+3}$ **3.** $\dfrac{s-5}{(s-5)^2 + 4^2}$

5. $L^{-1}(F) = e^{-6t}\sin\sqrt{7}t$

7. $L^{-1}(F) = 0.5e^{-3t} - 0.5\cos t + 1.5\sin t$

9. $y = \frac{1}{6}e^{-2t} + \frac{5}{6}\cos\sqrt{2}t + \frac{1}{3\sqrt{2}}\sin\sqrt{2}t$

11. $q = L^{-1}(Q) =$
$-49.925 + 49.985e^{-40t} - 0.06\cos 20t - 0.03\sin 20t$

Computer Programs and Additional Exercises

This appendix consists of programs written in GWBASIC that serve as answers to each of the computer programming exercises found in this text. Just as the textbook increases in sophistication and complexity as you progress from beginning to end, the computer programs also increase in sophistication and complexity.

The name of each program corresponds to its location in the text. Each name consists of the three letters PRG followed immediately by five digits. The first two digits represent the chapter in the text. Thus, Chapter 9 is 09 and chapter 25 is 25. The next (third) digit denotes the section in the chapter; and the last two digits, the exercise in that section where the program was assigned.

Using the above description, the program for Chapter 9, Section 5, Exercise 43, has the name PRG09543. Similarly, the computer program for Chapter 17, Section 3, Exercise 22, is named PRG17322.

* * * * *

You may wish to write your own versions of these programs. If you do, then use the following programs as guidelines. In some cases, we have added some additional steps to the programs to make them easier to use. You might have some ideas for making the programs even more user-friendly.

However, it is important that anyone programming a computer remember that a computer must be told what to do and how it is done. If the directions are wrong, the computer will give a wrong answer.

There are several special commands, statements, and functions that are used in BASIC. In fact, most of these, with some variation such as SQRT rather than SQR, are used in other computer programming languages. This is not intended as a programming course, and so we assume that you are familiar with BASIC.

Commands, such as LOAD, RUN, LIST, and SAVE are not part of a program, but allow you to tell the computer to do something with a program.

Program *statements* used in these programs include PRINT, PRINT USING, INPUT, REM, GOTO (or GO TO), DEF FN, DIM, FOR ... NEXT, IF ... THEN, GOSUB ... RETURN, CLS, CLEAR, and END.

Functions used in these programs include ABS, ATN, COS, EXP, INT, LEFT$, LOG, SGN, SIN, GQR, TAB, and TAN. You should remember that a computer evaluates trigonometric functions expressed in radians and not degrees. You will need to include conversion factors in any programs where you will enter, or have the computer print, angles given in degrees.

You may also have noticed that BASIC includes only the sine, cosine, tangent, and arctangent functions. If your programs include any other trigonometric formulas, you will have to define them in that particular program.

Naturally, since these are programs involving mathematics, you will need to use the symbols for the mathematical *operations*. The computer uses $+$, $-$, $*$, and $/$ for addition, subtraction, multiplication, and division, respectively. For exponentiation, these BASIC programs use $\wedge$. (Some computers and some languages use either $**$ or $\uparrow$.)

$$* * * * *$$

Not everyone wishes to take the time to write a computer program. Thus, each of these programs is available on a 3.5-inch IBM formatted disk.

Included with these program listings are 30 exercises. Some of the exercises ask you to modify a program so that it will do a better job of performing the desired task or so that it can be used in some additional situations. The remaining exercises ask you to use one of the programs with a given set of numbers or equations.

A solution to each of these additional exercises is given in the *Instructor's Solutions Manual*; solutions to the odd numbered exercises are in the *Student's Solutions Manual*. Whenever possible, each solution contains a copy of the exact wording and format used when the computer printed each solution.

Chapter 5

Section 5.2, Exercise 44

```
  1 REM *** PRG05244 ***
 20 DIM I(20), X(20), Y(20)
 30 CLS
 40 REM THIS PROGRAM USES THE
    GAUSS-SEIDEL METHOD FOR SOLVING
    SYSTEMS OF TWO LINEAR EQUATIONS
 50 PRINT "THIS PROGRAM USES THE
    GAUSS-SEIDEL METHOD FOR SOLVING
    SYSTEMS OF TWO LINEAR EQUATIONS."
 60 REM EQUATION 1 IS OF THE FORM
    AX+BY+C=0
 70 REM EQUATION 2 IS OF THE FORM
    DX+EY+F=0
 80 PRINT"ENTER THE COEFFICIENTS FOR
    EQUATION 1: AX+BY+C=0"
 90 PRINT:INPUT"A = ",A:INPUT"B =
    ",B:INPUT "C = ",C
100 PRINT:PRINT:PRINT"ENTER THE
    COEFFICIENTS FOR EQUATION 2:
    DX+EY+F=0"
110 PRINT:INPUT"D = ",D:INPUT"E =
    ",E:INPUT"F = ",F
120 PRINT:PRINT:PRINT"IF YOU WANT TO
    GUESS A VALUE FOR X ENTER A (1)
    ":PRINT"IF YOU WANT TO GUESS A VALUE
    FOR Y":;INPUT" ENTER A (2) ",Q
130 IF Q <> 1 AND Q <> 2 THEN GOTO 120
140 I = 1
150 IF Q = 2 THEN GOTO 260
160 PRINT:INPUT"ENTER YOUR GUESS FOR X:
    ",X(I)
170 PRINT "I","X(I)","Y(I)"
180 Y(I) = -(D*X(I)+F)/E
190 PRINT:PRINT I,X(I),Y(I)
200 IF I = 1 GOTO 220
210 IF ABS(X(I) - X(I-1))<.01 AND
    ABS(Y(I)-Y(I-1)) < .01 GOTO 400
220 I = I + 1
230 IF I > 20 GOTO 360
240 X(I) = - (B*Y(I-1)+C)/A
250 GOTO 180
260 PRINT:INPUT"ENTER YOUR GUESS FOR Y:
    ",Y(1)
270 PRINT "I","X(I)","Y(I)"
280 X(I) =-(B*Y(I) + C)/A
```

```
290 PRINT:PRINT I,X(I),Y(I)
300 IF I = 1 THEN GOTO 320
310 IF ABS(X(I)-X(I-1))<.01 AND
    ABS(Y(I)-Y(I-1))<.01 THEN GOTO 400
320 I = I + 1
330 IF I > 20 GOTO 360
340 Y(I)=-(D*X(I-1)+F)/E
350 GOTO 280
360 PRINT "YOUR GUESS WAS NOT CLOSE
    ENOUGH." : PRINT
370 INPUT "DO YOU WANT TO TRY ANOTHER
    GUESS? ", Q2$
380 IF Q2$ = "Y" THEN GOTO 120
390 IF Q2$ = "N" THEN GOTO 420
400 PRINT : PRINT : PRINT "THE ANSWER IS"
    : PRINT : PRINT "X = "X(I)", Y= "Y(I)
410 PRINT
420 INPUT "IF YOU WANT TO WORK THIS SAME
    PROBLEM TYPE A 99 ", QU
430 IF QU = 99 THEN GOTO 120
440 END
```

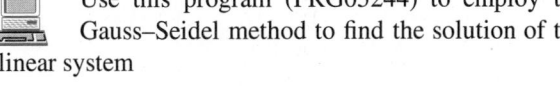 Use this program (PRG05244) to employ the Gauss–Seidel method to find the solution of the linear system

$$\begin{cases} 2x+5y-2=0 \\ 3x-8y+15=0 \end{cases} \quad \text{near } (-2,1).$$

Section 5.4, Exercise 35

```
  1 REM ***PRG05435***
 10 REM THIS PROGRAM WILL ALLOW YOU TO
    INPUT ELEMENTS OF A 2 x 2 DETERMINANT
    OF THE FORM:
 20 REM
 30 REM     A     B
 40 REM
 50 REM     C     D
 60 REM
 70 REM AND EVALUATE IT.
 80 CLS
 90 PRINT " THIS PROGRAM DETERMINES THE
    VALUE OF A 2 X 2 DETERMINANT"
 91 PRINT "OF THE FORM:"
 92 PRINT
 93 PRINT" A B"
 94 PRINT
 95 PRINT" C D"
 96 PRINT
```

```
 98 INPUT "GIVE VALUE FOR A: ",A
100 INPUT "GIVE VALUE FOR B: ",B
110 INPUT "GIVE VALUE FOR C: ",C
120 INPUT "GIVE VALUE FOR D: ",D
130 DET = A*D-C*B
140 PRINT
150 PRINT "THE DETERMINANT IS: ";DET
160 END
```

 Use this program (PRG05435) to evaluate each of the following determinants.

(a) $\begin{vmatrix} 2 & -\frac{1}{2} \\ \sqrt{2} & \frac{1}{\sqrt{7}} \end{vmatrix}$ (b) $\begin{vmatrix} \pi & 0.002 \\ -5 & -\sqrt{19} \end{vmatrix}$

Section 5.4, Exercise 37

```
  1 REM PRG05437
 10 REM THIS PROGRAM DETERMINES THE VALUE
    OF A 3 X 3 DETERMINANT
 20 REM OF THE FORM
 30 REM A1 B1 C1
 40 REM A2 B2 C2
 50 REM A3 B3 C3
 60 REM
 70 CLS
 80 PRINT"THIS PROGRAM DETERMINES THE
    VALUE OF A 3 X 3 DETERMINANT"
 90 PRINT"OF THE FORM:"
100 PRINT"A1 B1 C1"
110 PRINT"A2 B2 C2"
120 PRINT"A3 B3 C3"
130 PRINT
140 PRINT
150 INPUT"GIVE THE VALUE FOR A1: ",A1
160 INPUT"GIVE THE VALUE FOR B1: ",B1
170 INPUT"GIVE THE VALUE FOR C1: ",C1
180 PRINT
190 PRINT
200 INPUT"GIVE THE VALUE FOR A2: ",A2
210 INPUT"GIVE THE VALUE FOR B2: ",B2
220 INPUT"GIVE THE VALUE FOR C2: ",C2
230 PRINT
240 PRINT
250 INPUT"GIVE THE VALUE FOR A3: ",A3
260 INPUT"GIVE THE VALUE FOR B3: ",B3
270 INPUT"GIVE THE VALUE FOR C3: ",C3
275 PRINT
276 PRINT
280 DET = A1*B2*C3 + B1*C2*A3 + C1*A2*B3
```

```
         - C1*B2*A3 - A1*C2*B3 - B1*A2*C3
290 PRINT USING"THE VALUE OF THIS
    DETERMINANT IS:
    ##########.#####";DET
300 END
```

Use this program (PRG05437) to evaluate each of the following determinants.

$$\text{(a)} \begin{vmatrix} 1 & 2 & 3 \\ 0 & 2 & 4 \\ 0 & 0 & -6 \end{vmatrix} \quad \text{(b)} \begin{vmatrix} \pi & \frac{4}{3} & -\frac{17}{6} \\ 1+\sqrt{2} & 7 & 0.4 \\ 2 & \sqrt{11} & 0 \end{vmatrix}$$

Section 5.5, Exercise 17

```
  1 REM PRG05517
 10 CLS
 20 PRINT "THIS PROGRAM WILL USE CRAMER'S
    RULE TO SOLVE A SYSTEM OF LINEAR
    EQUATIONS"
 30 PRINT "TYPE A '2' IF YOU WANT TO
    SOLVE A SYSTEM OF TWO EQUATIONS OR
    TYPE A '3' IF YOU WANT TO SOLVE A
    SYSTEM OF THREE EQUATIONS.": INPUT Q
 40 IF Q <> 2 AND Q <> 3 AND Q <> 0 THEN
    GOTO 30
 50 IF Q = 0 THEN END
 60 IF Q = 3 THEN GOTO 240
 70 PRINT : PRINT : PRINT "THE TWO
    EQUATIONS ARE OF THE FORM:"
 80 PRINT "AX + BY = H"
 90 PRINT "CX + DY = K"
100 PRINT : PRINT "ENTER THE VALUES FOR
    EQUATION 1:"
110 INPUT" A = ",A
120 INPUT" B = ",B
130 INPUT" H = ",H
140 PRINT : PRINT "NOW ENTER THE VALUES
    FOR EQUATION 1:"
150 INPUT" A = ",A
160 INPUT" B = ",B
170 INPUT" K = ",K
180 DET = A * D - B * C
190 IF DET = 0 THEN GOTO 450
200 DX = H * D - K * B:DY = A * K - C * H
210 X = DX/DET:Y = DY / DET
220 PRINT "THE SOLUTIONS ARE:"
230 PRINT : PRINT "X = ";X : PRINT "Y =
    ";Y:GOTO 520
240 PRINT " THIS PROGRAM WILL USE
    CRAMER'S RULE TO SOLVE A SYSTEM OF
```

```
    THREE EQUATIONS OF THE FORM: "
250 PRINT "A(1)X + B(1)Y + C(1)Z = K(1)"
260 PRINT "A(2)X + B(2)Y + C(2)Z = K(2)"
270 PRINT "A(1)X + B(3)Y + C(3)Z = K(3)"
280 PRINT : PRINT "ENTER THE VALUES FOR
    EQUATION 1: "
290 INPUT" A(1) = ",A1
300 INPUT" B(1) = ",B1
310 INPUT" C(1) = ",C1
320 INPUT" K(1) = ",K1
330 PRINT : PRINT "ENTER THE VALUES FOR
    EQUATION 2: "
340 INPUT" A(2) = ",A2
350 INPUT" B(2) = ",B2
360 INPUT" C(2) = ",C2
370 INPUT" K(2) = ",K2
380 PRINT : PRINT "ENTER THE VALUES FOR
    EQUATION 3: "
390 INPUT" A(3) = ",A3
400 INPUT" B(3) = ",B3
410 INPUT" C(3) = ",C3
420 INPUT" K(3) = ",K3
430 DET = A1*B2*C3 + B1*C2*A3 + C1*A2*B3
    - C1*B2*A3 - A1*C2*B3 - B1*A2*C3
440 IF DET <> 0 THEN GOTO 460
450 PRINT : PRINT : PRINT "THE
    DETERMINANT IS ZERO. THE SYSTEM HAS
    NO SOLUTION!!!":GOTO 520
460 DX = K1*B2*C3 + B1*C2*K3 + C1*K2*B3 -
    C1*B2*K3 - K1*C2*B3 - B1*K2*C3
470 DY = A1*K2*C3 + K1*C2*A3 + C1*A2*K3 -
    C1*K2*A3 - A1*C2*K3 - K1*A2*C3
480 DZ = A1*B2*K3 + B1*K2*A3 + K1*A2*B3 -
    K1*B2*A3 - A1*K2*B3 - B1*A2*K3
490 X = DX/DET:Y=DY/DET:Z=DZ/DET
500 PRINT "THE SOLUTIONS ARE: "
510 PRINT "X = ";X : PRINT "Y = ";Y :
    PRINT "Z = ";Z
520 END
```

1. Change this program (PRG05517) to print out either "THIS SYSTEM IS INCONSISTENT" or "THIS SYSTEM IS DEPENDENT", whichever is correct, rather than the statement on line 450.

2. After you have revised PRG05517, as in Exercise 1 above, use it to solve each of the following systems.

(a) $\begin{cases} 2x + y - 3z = 1 \\ x - 4y + z = 6 \\ 4x - 16y + 4z = 24 \end{cases}$

(b) $\begin{cases} 2x + y + z = 0 \\ x + y - z = 0 \\ x + 2y + 2z = 0 \end{cases}$

(c) $\begin{cases} x + 2y - z = 0 \\ 2x - y + z = 0 \\ 8x + y + z = 2 \end{cases}$

Chapter 8

Section 8.4, Exercise 51

```
10 REM ***PRG08451***
20 PRINT"THIS PROGRAM WILL DETERMINE THE
   ROOTS OF A QUADRATIC EQUATION OF THE
   FORM AX^2 + BX + C = 0."
30 PRINT:INPUT"A= ",A:INPUT "B=
   ",B:INPUT"C= ",C
40 D = B * B - 4 *A*C
50 IF ABS(D) <= .00001 THEN D = 0:GOTO
   120
60 IF D > 0 THEN GOTO 80
70 PRINT"THE DISCRIMINANT IS LESS THAN
   ZERO. THERE ARE NO REAL ROOTS.":GOTO
   140
80 Y = SQR(D)
90 X1 = ( - B + Y)/(2*A)
100 X2 = ( - B - Y)/(2*A)
110 PRINT : PRINT"THE ROOTS ARE: ";X1;"
    AND ";X1
120 X1 = -B/(2*A)
130 PRINT : PRINT"THE ROOTS ARE BOTH THE
    SAME: ";X1;" AND ";X2
140 PRINT : PRINT : PRINT"TYPE A 99 IF
    YOU WANT TO SOLVE ANOTHER QUADRATIC
    EQUATION. ":INPUT Q
150 IF Q = 99 THEN GOTO 30
160 END
```

1. Change this program (PRG08451) to print out the nonreal complex roots rather than THERE ARE NO REAL ROOTS.

2. Use your new version of PRG08451 from Exercise 1 above to find the roots of

(a) $4x^2 - 9x + 3.5 = 0,$

(b) $\sqrt{2}x^2 + 4x - \sqrt{12} = 0,$

(c) $1.2x^2 - 5.52x + 6.348 = 0,$

(d) $3.2x^2 - 7.1x + 5.3 = 0.$

Chapter 9

Section 9.2, Exercise 37

```
10 REM *** PRG09237 ***
20 REM THIS PROGRAM COMPUTES AND PRINTS
   THE SINES AND COSINES OF THE ANGLES
30 REM FROM: 0 TO 1.6 RADIANS EVERY 0.1
   RADIAN.
40 PRINT"THIS PROGRAM COMPUTES AND
   PRINTS THE SINES AND COSINES OF THE
   ANGLES"
50 PRINT"FROM: 0 TO 1.6 RADIANS EVERY
   0.1 RADIAN."
60 PRINT"ANGLE","SIN","COS" : PRINT
70 FOR X = 0 TO 1.6 STEP 0.1
80 SX = X - (X^3)/6 + (X^5) / 120 -
   (X^7) / 5040
90 CX = 1 - (X^2)/2 + (X^4) / 24 -
   (X^6) / 720 + (X^8)/ 40320
100 PRINT USING"##.# ##.######
    ##.######";X;SX;CX
110 NEXT X
120 END
```

Section 9.2, Exercise 38

```
10 REM *** PRG09238 ***
20 REM THIS PROGRAM COMPUTES AND PRINTS
   THE SINE AND COSINE OF ANGLES
30 REM*FROM 0 TO 90 DEGREES EVERY 1
   DEGREE
40 PRINT "ANGLE","SIN","COS"
50 PRINT
60 PI = 3.141592653559#
70 FOR I = 1 TO 90
80 X = I * PI / 180
90 SX = X - (X ^ 3) / 6 + (X ^ 5) / 120
   - (X ^ 7) / 5040
100 CX = 1 - (X ^ 2) / 2 + (X ^ 4) / 24 -
    (X ^ 6) /720 + (X ^ 8) / 40320!
110 PRINT USING" ##      ##.####
    ##.####";I;SX;CX
120 NEXT I
130 END
```

 Change these programs (PRG09237 and PRG09238) to print the sines, cosines, and tan-

gents from 0 to 2π radians every 0.1 radian and from 0°
to 180° every 1°, respectively.

Section 9.5, Exercise 45

```
10 REM *** PRG09545 ***
20 CLS
30 REM THIS PROGRAM WILL DETERMINE THE
   ANGLE, IN RADIANS, FROM THE VALUE
40 REM FOR THE TANGENT OF THE ANGLE.
50 PRINT"THIS PROGRAM WILL DETERMINE AN
   ANGLE MEASURE, IN RADIANS, FROM THE
   VALUE"
60 PRINT"ENTERED IN FOR THE ANGLE'S
   TANGENT."
70 INPUT"WHAT IS THE TANGENT OF THE
   ANGLE: ",X
80 Y = ATN(X)
90 PRINT : PRINT"IF ATN(Y) = ";X
100 PRINT"THEN Y IS: ";Y;" RADIANS."
110 END
```

Section 9.5, Exercise 46

```
10 REM PRG09546
20 CLS
30 REM THIS PROGRAM WILL DETERMINE THE
   ANGLE, IN RADIANS, FROM THE VALUE FOR
40 REM THE COSINE OF THE ANGLE.
50 PRINT "THIS PROGRAM WILL DETERMINE
   THE ANGLE, IN RADIANS, FROM THE VALUE
   FOR"
60 PRINT "THE COSINE OF THE ANGLE."
70 PRINT : PRINT:INPUT"WHAT IS THE
   COSINE OF THE ANGLE: ",X
80 IF X = 1 THEN Y = 1:GOTO 140
90 IF X = - 1 THEN Y = 3.1415927#:GOTO
   140
100 IF ABS(X) < 1 GOTO 130
110 PRINT : PRINT : PRINT X;" IS NOT AN
    ACCEPTABLE VALUE. PLEASE ENTER A
    VALUE BETWEEN"
120 PRINT " - 1 AND + 1.":GOTO 70
130 Y = - ATN(X/SQR( - X*X + 1)) +
    1.57079633#
140 PRINT : PRINT"IF ARC COS(Y) = ";X:
    PRINT"THEN Y IS: ";Y;" RADIANS."
150 END
```

Section 9.5, Exercise 47

```
1 REM RPG09547
10 CLS
20 REM THIS PROGRAM WILL ALLOW YOU TO
   ENTER A VALUE AND PRINT ALL POSSIBLE
   INVERSE TRIGONOMETRY FUNCTIONS FOR
   THAT VALUE.
30 PRINT"THIS PROGRAM WILL ALLOW YOU TO
   ENTER A VALUE AND PRINT ALL POSSIBLE
   INVERSE TRIGONOMETRY FUNCTIONS FOR
   THAT VALUE."
40 PI = 3.14159265359#
50 INPUT "WHAT IS THE VALUE OF X: ",X
60 IF ABS(X) > 1 THEN GOTO 140
70 IF X = 1 THEN Y1 = PI /2:Y2 = 0:GOTO
   120
80 IF X = - 1 THEN Y1 = - PI / 2: Y2 =
   PI:GOTO 120
90 Y1 = ATN(X/SQR( - X*X + 1))
100 Y2 = - ATN(X/SQR( - X*X + 1)) + PI /2
110 IF ABS(X) < 1 THEN GOTO 160
120 IF X = 1 THEN Y3 = PI / 2:Y4 =
    PI:GOTO 160
130 IF X = - 1 THEN Y3 = - PI/2:Y4 =
    0:GOTO 160
140 Y3 = ATN(1/SQR(X*X - 1)) + (SGN(X) -
    1)* PI/2
150 Y4 = ATN(SQR(X*X - 1)) + (SGN(X) - 1)
    *PI/2
160 Y5 = ATN(X)
170 Y6 = - ATN(X) + PI/2
180 PRINT : PRINT"IF THE VALUE OF X IS
    ";X" THEN "
190 IF ABS(X) > 1 GOTO 220
200 PRINT : PRINT"ARC SIN(X) = ";Y3;"
    RAD" : PRINT"ARC COS(X) = ";Y2;" RAD"
210 IF ABS(X) < 1 THEN GOTO 240
220 PRINT : PRINT"ARC SIN(X) AND ARC
    COS(X) DO NOT EXIST OR VALUE LARGER
    THAN 1 OR SMALLER THEN - 1"
230 PRINT : PRINT"ARC CSC(X) = ";Y3;"
    RAD" : PRINT"ARC SEC(X) = ";Y4;"
    RAD":GOTO 250
240 PRINT : PRINT"ARC CSC(X) AND ARC
    SEC(X) DO NOT EXIST OR VALUES BETWEEN
    - 1 AND + 1."
250 PRINT : PRINT"ARC TAN(X) = ";Y5;"
    RAD" : PRINT"ARC COT(X) = ";Y6;" RAD"
260 END
```

Section 9.5, Exercise 48

Add the following lines to each program:

To Exercise 45 add
```
43 Y = Y * 180/3.14159254359
45 PRINT "OR"; Y; "DEGREES"
```

To Exercise 46 add
```
83 Y = Y * 180/3.1415927
85 PRINT "OR"; Y; .DEGREES."
```

To Exercise 47 multiply each value of Y1, Y2, etc. by 180/PI

Chapter 10

Section 10.2, Exercise 37

```
1 REM *** PRG10237 ***
10 CLS
20 REM THIS PROGRAM WILL DETERMINE THE
   COMPONENT VECTOR FOR A POSITION
   VECTOR.
30 PRINT "THIS PROGRAM WILL DETERMINE
   THE COMPONENT VECTOR FOR A POSITION
   VECTOR."
40 PI = 3.14159265359#
50 PRINT : PRINT"IF THE DIRECTION OF THE
   POSITION VECTOR IS IN RADIANS A '99'.
   IF THEY ARE GIVEN IN DEGREES THEN
   TYPE A ' - 1'.":INPUT Q
60 PRINT: INPUT"ENTER THE MAGNITUDE OF
   THE VECTOR ",M: PRINT :INPUT" AND ITS
   DIRECTION : ",D
70 IF Q = 99 THEN GOTO 90
80 D = D * PI / 180
90 AX = M * COS(D):AY = M * SIN(D)
100 TX = 0:TY = PI / 2
110 IF AX < 0 THEN TX = PI
120 IF AY < 0 THEN TY = TY + PI
130 PRINT : PRINT : PRINT"THE HORIZONTAL
    VECTOR HAS " : PRINT"MAGNITUDE:
    ";ABS(AX):PRINT" AND DIRECTION:
    ";TX;" RADIANS" : PRINT " OR ";TX *
    180 / PI;" DEGREES"
140 PRINT : PRINT"THE VERTICAL VECTOR HAS
    " : PRINT"MAGNITUDE: ";ABS(AY) :
    PRINT"AND DIRECTION ";TY;" RADIANS" :
    PRINT "OR ";TY * 180 / PI;" DEGREES."
150 END
```

Use this program (PRG10237) to determine the horizontal and vertical components of the given vectors.

(a) Magnitude 20, $\theta = 75°$
(b) Magnitude 16, $\theta = 212°$
(c) Magnitude 18.4, $\theta = 4.97$ rad
(d) Magnitude 23.7, $\theta = 2.22$ rad

Section 10.2, Exercise 38

```
10 REM *** PRG10238***
20 CLS
30 REM THIS PROGRAM WILL DETERMINE THE
   RESULTANT VECTORS WHEN TWO OR MORE
   VECTORS ARE ADDED.
40 PRINT "THIS PROGRAM WILL DETERMINE
   THE RESULTANT VECTORS WHEN TWO OR
   MORE VECTORS ARE ADDED."
50 AX = 0:AY = 0:PI = 3.14159265359#
60 PRINT : PRINT "IF THE DIRECTION FOR
   THE VECTORS ARE IN RADIANS TYPE A
   '99'. IF THEY ARE GIVEN IN DEGREES
   THEN TYPE ' -1'." : INPUT Q
70 PRINT : PRINT:INPUT "ENTER THE NUMBER
   OF VECTORS: ";N
80 PRINT : PRINT : PRINT "FOR THE FIRST
   VECTOR ENTER THE":INPUT" MAGNITUDE:
   ";A(1):INPUT" AND THE DIRECTION:
   ";TA(1)
90 FOR I = 2 TO N
100 PRINT : PRINT "FOR VECTOR NUMBER ";I"
    ENTER THE":INPUT" MAGNITUDE:
    ";A(I):INPUT" AND THE DIRECTION:
    ";TA(I)
110 NEXT I
120 IF Q = 99 THEN GOTO 140
130 FOR I = 1 TO N:TA(I) = TA(I) * PI/
    180:NEXT I
140 FOR I = 1 TO N
150 AX(I)=A(I) * COS(TA(I)):AY(I) = A(I)
    * SIN(TA(I)):AX = AX + AX(I):AY = AY
    + AY(I)
160 NEXT I
170 R = SQR(AX^2 + AY^2)
180 IF AX <> 0 THEN GOTO 220
190 IF AY < 0 THEN GOTO 210
200 ANG = PI/2:GOTO 230
210 ANG = 3 * PI/2:GOTO 230
220 ANG = ATN(AY/AX)
230 IF AX < 0 THEN GOTO 260
240 IF AY < 0 THEN ANG = ANG + 2 * PI
250 GOTO 270
260 ANG = ANG + PI .
```

```
270 PRINT : PRINT : PRINT "THE RESULTANT
    VECTOR HAS" : PRINT : PRINT
    "MAGNITUDE = ";R : PRINT" AND
    DIRECTION = ";ANG;"RADIANS": PRINT
    "OR ";ANG * 180 / PI;" DEGREES"
280 END
```

Use this program (PRG10238) to determine the resultant vector when the given vectors are added.

(a) $A = 109$, $\theta_A = 33.4°$, $B = 125$, $\theta_B = 69.4°$

(b) $C = 137$, $\theta_C = 30°$, $D = 89.4$, $\theta_D = -72.5°$, $D = 164.6$, $\theta_E = 159.8°$

(c) $F = 37.5$, $\theta_F = 0.25$ rad, $G = 49.3$, $\theta_G = 1.92$ rad

Chapter 14

Section 14.2, Exercise 63

```
1 REM *** PRG14263 ***
10 CLS
20 PRINT"THIS PROGRAM WILL ADD OR
   SUBTRACT TWO COMPLEX NUMBERS. THE
   NUMBERS MUST BE IN THE FORM: A + BJ"
30 '
40 '
50 PRINT" ENTER AN 'A' IF YOU WANT TO
   ADD TWO NUMBERS, ENTER AN 'S' IF YOU
   WANT TO SUBTRACT THEM":INPUT A$
60 '
70 '
80 REM **** ENTER COMPLEX NUMBER IN A +
   BJ FORM ****
90 '
100 '
110 CLS : PRINT"ENTER EACH OF THE COMPLEX
    NUMBERS IN THE FORM: A + BJ"
120 IF A$ = "A" THEN GOTO 140
130 PRINT"ENTER THE NUMBER BEING
    SUBTRACTED AS THE SECOND NUMBER"
140 FOR I = 1 TO 2 : PRINT : PRINT"FOR
    NUMBER ";I : PRINT:INPUT"A= ";A(I -
    1) : PRINT:INPUT"AND B = ";B(I -
    1):NEXT I
150 IF A$ = "S" THEN GOTO 280
160 '
170 '
180 REM *** SUBTRACT COMPLEX NUMBERS ***
190 '
200 '
210 A = A(0) + A(1):B = B(0) + B(1)
```

```
220 GOTO 350
230 '
240 '
250 REM *** SUBTRACTING COMPLEX NUMBERS
    ***
260 '
270 '
280 A = A(0) - A(1):B = B(0) - B(1)
290 PRINT : PRINT A;" + ";B;"J"
300 '
310 '
320 REM *** PRINTING ANSWERS ***
330 '
340 '
350 PRINT : PRINT : PRINT"THE ANSWER IS "
360 PRINT : PRINT A;" + ";B;"J"
370 END
```

Use this program (PRG14263) to add or subtract the following complex numbers.

(a) $(7.51 - 2.36j) + (-12.47 + 4.61j)$

(b) $(15.59 - 6.34j) - (-21.3 + 7.21j)$

Section 14.2, Exercise 64

```
1 REM *** PRG14264 ***
10 CLS : PRINT"THIS PROGRAM WILL
   MULTIPLY TWO COMPLEX NUMBERS. THE
   NUMBERS MUST BE ENTERED IN THE FORM:
   A + BJ."
80 '
90 '
100 REM *** ENTER COMPLEX NUMBER IN A +
    BJ FORM ***
110 '
120 '
130 PRINT"ENTER EACH OF THE NUMBERS IN
    THE FORM: A + BJ.
140 FOR I = 1 TO 2 : PRINT : PRINT"FOR
    NUMBER ";I: INPUT"GIVE A =",A(I - 1)
    : PRINT:INPUT"GIVE B = ";B(I -
    1):NEXT I
150 GOTO 220
170 '
180 '
190 REM *** MULTIPLYING COMPLEX NUMBERS
    ***
200 '
210 '
220 A = A(0) * A(1) - B(0)*B(1):B = A(0)
```

```
          * B(1) + A(1) * B(0)
280 '
290 '
300 REM *** PRINT PRODUCT ***
310 '
320 '
330 PRINT : PRINT : PRINT"THE PRODUCT IS:
    "
340 PRINT : PRINT A;" + ";B;"J"
350 END
```

Use this program (PRG14264) to multiply the following complex numbers.

(a) $(5.72 - 4.36j)(-35.87 + 1.61j)$

(b) $(0.09 - 1.74j)(-1.3 + 0.11j)$

Section 14.2, Exercise 65

```
10  REM *** PRG14265 ***
20  CLS : PRINT" THIS PROGRAM WILL DIVIDE
    TWO COMPLEX NUMBERS. THE NUMBER MUST
    BE ENTERED IN THE FORM: A + BJ"
30  '
40  '
50  REM *** ENTER COMPLEX NUMBERS IN THE
    FORM: A + BJ ***
60  '
70  '
80  PRINT"ENTER EACH OF THE COMPLEX
    NUMBERS IN THE FORM A + BJ"
90  PRINT : PRINT"ENTER THE NUMERATOR
    FIRST"
100 PRINT"THE NUMERATOR IS:":INPUT"A =
    ";A(0):PRINT:INPUT"AND B = ";B(0)
110 PRINT:PRINT"AND THE DENOMINATOR
    IS:":INPUT"A = ";A(1):PRINT:INPUT"AND
    B = "; B(1)
120 '
130 ,
140 REM *** DIVIDING COMPLEX NUMBERS ***
150 '
160 DENOM = A(1)*A(1) + B(1)*B(1)
170 A = (A(0) * A(1) + B(0) * B(1)) /
    DENOM:B = (B(0) * A(1) - A(0) *
    B(1))/DENOM
180 '
190 PRINT : PRINT : PRINT"THE QUOTIENT
    IS: "
200 PRINT : PRINT A;" + ";B;"J"
210 END
```

Use this program (PRG14265) to divide the following complex numbers.

(a) $(0.25 + 7j) \div (-9.12 - 0.21j)$

(b) $4j \div (7 + 2j)$

Section 14.3, Exercise 47

```
10  REM *** PRG14347 ***
20  PRINT"THIS PROGRAM WILL CHANGE A
    COMPLEX NUMBER FROM RECTANGULAR FORM
    TO POLAR FORM OR FROM POLAR FORM TO
    RECTANGULAR."
30  PRINT : PRINT : PRINT"IF THE NUMBERS
    YOU ARE ENTERING ARE IN RECTANGULAR
    FORM, THEN TYPE A '1'. IF THEY ARE IN
    POLAR FORM, THEN TYPE A '9'."
40  INPUT A$
50  PI = 3.14159265359#
60  IF A$ = "9" THEN GOTO 200
70  '
80  '
90  REM *** ENTER COMPLEX NUMBERS IN
    RECTANGULAR FORM ***
100 '
110 '
120 CLS:PRINT"ENTER THE COMPLEX NUMBER IN
    THE FORM: A + BJ."
130 INPUT"A = ";A:PRINT:INPUT" AND B =
    ";B
140 GOTO 420
150 '
160 '
170 REM *** ENTER THE COMPLEX NUMBERS IN
    POLAR FORM ***
180 '
190 '
200 PRINT:PRINT"ENTER THE COMPLEX NUMBER
    IN THE FORM R CIS(AN)"
210 PRINT:PRINT"IF ANGLES ARE IN DEGREES
    TYPE '55' IF IN RADIANS TYPE
    '11'":INPUT DR:IF DR <> 11 AND DR <>
    55 THEN GOTO 210
220 INPUT"R = ";R:PRINT:INPUT "AND THE
    ANGLE IS ";AN
230 '
240 '
250 REM *** CONVERT DEGREES TO RADIANS
    ***
260 '
```

```
270 '
280 IF DR = 11 THEN GOTO 350
290 AD = AN:AN = AN * PI / 180
300 '
310 '
320 REM *** CONVERT TO RECTANGULAR FORM
    ***
330 '
340 '
350 A = R * COS(AN):B = R * SIN(AN)
360 GOTO 490
370 '
380 '
390 REM *** PRINTING ANSWER ***
400 '
410 '
420 R = SQR(A*A + B*B)
430 IF A <> 0 THEN GOTO 460
440 IF B > 0 THEN AN = PI / 2:GOTO 490
450 IF B < 0 THEN AN = 3 * PI / 2:GOTO
    490
460 AN = ATN(B/A)
470 IF A = 0 THEN AN = AN - PI
480 IF AN < 0 THEN AN = AN + PI
490 IF AN >= 2 * PI THEN AN = AN - 2 *
    PI:GOTO 490
500 AD = AN * 180/PI
510 '
520 '
530 REM *** PRINTING ANSWER ***
540 '
550 '
560 PRINT:PRINT: PRINT"THE NUMBER IS "
570 PRINT:PRINT A;" + ";B;"J"
580 PRINT:PRINT" OR ":PRINT R;" CIS
    ";AN;" RADIANS" :PRINT R;" CIS ";AD;
    " DEGREES"
590 END
```

Use this program (PRG14347) to convert the following complex numbers from their given form to the indicated form.

(a) Change $-12.4 + 3.7j$ from rectangular form to polar form

(b) Change $4.2 \operatorname{cis} 247°$ from polar form to rectangular form

(c) Change $11.25 \operatorname{cis} 5.13$ from polar form to rectangular form

Section 14.5, Exercise 41

```
10 REM *** PRG14541 ***
20 PRINT" THIS PROGRAM WILL MULTIPLY A
   GROUP OF COMPLEX NUMBERS. THE NUMBERS
   CAN BE ENTERED IN RECTANGULAR FORM OR
   IN POLAR FORM"
30 PRINT:PRINT:PRINT"IF THE NUMBERS YOU
   ARE ENTERING ARE IN RECTANGULAR FORM
   TYP E A '1' IF IN POLAR FORM TYPE
   '9'."
40 INPUT A$
50 PI = 3.14159265359#
60 N = 2
70 PRINT:PRINT:INPUT"ENTER THE NUMBER OF
   COMPLEX NUMBERS YOU ARE MULTIPLYING
   ";N
80 IF A$ = "9" THEN GOTO 220
90 '
100 '
110 REM *** ENTER COMPLEX NUMBERS IN
    RECTANGULAR FORM ***
120 '
130 '
140 CLS:PRINT"ENTER EACH OF THE ";N;"
    COMPLEX NUMBERS IN THE FORM A + BJ"
150 FOR I = 1 TO N:PRINT:PRINT"FOR NUMBER
    ";I:PRINT:INPUT"A =
    ";A(I-1):PRINT:INP UT"AND B =
    ";B(I-1):NEXT I
160 GOTO 430
170 '
180 '
190 REM *** ENTER COMPLEX NUMBER IN POLAR
    FORM ***
200 '
210 '
220 PRINT:PRINT"ENTER THE COMPLEX NUMBER
    IN THE FORM R CIS(AN)"
230 PRINT:PRINT"IF ANGLES ARE IN DEGREES
    TYPE '55' IF IN RADIANS TYPE
    '11'":INPUT DR:IF DR <> 11 AND DR <>
    55 THEN GOTO 230
240 FOR I = 1 TO N:PRINT:PRINT"FOR NUMBER
    ";I:PRINT:INPUT"R =
    ";R(I-1):PRINT:INPUT "AND THE ANGLE
    IS ";AN(I-1):NEXT I
250 '
260 '
270 REM *** CONVERT DEGREES TO RADIANS
```

```
    ***
280 '
290 '
300 IF DR = 11 THEN GOTO 370
310 FOR I = 0 TO N-1:AD(I) = AN(I):AN(I)
    = AN(I) * PI /180:NEXT I
320 '
330 '
340 REM *** CONVERT TO RECTANGULAR FORM
    ***
350 '
360 '
370 FOR I = 0 TO N-1:A(I) = R(I) *
    COS(AN(I)):B(I) = R(I) *
    SIN(AN(I)):NEXT I
380 '
390 '
400 REM *** MULTIPLY COMPLEX NUMBERS ***
410 '
420 '
430 A = A(0) * A(1) - B(0) * B(1):B =
    A(0) * B(1) + A(1) * B(0)
440 IF A$ = "9" THEN AN = AN(0) + AN(1)
450 IF N = 2 THEN GOTO 490
460 FOR I = 3 TO N:A(N) = A:B(N) = B:A =
    A(I-1) * A(N) - B(I-1)* B(N):B =
    A(I-1) * B(N) + A(N) * B(I-1)
470 IF A$ = "9" THEN AN = AN + AN(I-1)
480 NEXT I
490 R = SQR(A*A+B*B)
500 IF A$ = "9" THEN GOTO 560
510 IF A <> 0 THEN GOTO 540
520 IF B > 0 THEN AN = PI/2:GOTO 570
530 IF B < 0 THEN AN = 3 * PI / 2:GOTO
    570
540 AN = ATN(B/A)
550 IF A$ = "1" AND AN < 0 THEN AN = AN +
    PI
560 IF AN < 0 THEN AN = AN + 2*PI:GOTO
    560
570 IF AN >= 2*PI THEN AN = AN - 2 *
    PI:GOTO 570
580 AD = AN * 180 / PI
590 PRINT:PRINT:PRINT"THE TOTAL IS "
600 PRINT:PRINT A;" + ";B;"J"
610 PRINT:PRINT"OR":PRINT R;"CIS ";AN;"
    RADIANS":PRINT R;" CIS ";AD;"
    DEGREES"
620 END
```

 Use this program (PRG14541) to multiply the following complex numbers.

(a) $(5.72 - 4.36j)(-35.87 + 1.61j)$

(b) $(4 \operatorname{cis} 321°)(2.7 \operatorname{cis} 217°)(2 \operatorname{cis} 153°)$

(c) $(6 \operatorname{cis} 2.75)(4.5 \operatorname{cis} 1.25)$

Section 14.5, Exercise 42

```
 10 REM *** PRG14542 ***
 20 PRINT" THIS PROGRAM WILL DIVIDE TWO
    COMPLEX NUMBERS. THE NUMBERS CAN BE
    ENTERED IN RECTANGULAR FORM OR IN
    POLAR FORM"
 30 PRINT:PRINT:PRINT"IF THE NUMBERS YOU
    ARE ENTERING ARE IN RECTANGULAR FORM
    TYPE A '1' IF IN POLAR FORM TYPE
    '9'."
 40 INPUT A$
 50 PI = 3.14159265359#
 60 IF A$ = "9" THEN GOTO 210
 70 '
 80 '
 90 REM *** ENTER COMPLEX NUMBERS IN
    RECTANGULAR FORM ***
100 '
110 '
120 CLS : PRINT"ENTER THE NUMERATOR
    FIRST"
130 PRINT"THE NUMERATOR IS:":INPUT"A =
    ";A(0):PRINT:INPUT"AND B = ";B(0)
140 PRINT:PRINT"AND THE DENOMINATOR
    IS:":INPUT"A = ";
    A(1):PRINT:INPUT"AND B = "; B(1)
150 GOTO 420
160 '
170 '
180 REM *** ENTER COMPLEX NUMBER IN POLAR
    FORM ***
190 '
200 '
210 PRINT:PRINT"ENTER THE COMPLEX NUMBER
    IN THE FORM R CIS(AN)"
220 PRINT:PRINT"IF ANGLES ARE IN DEGREES
    TYPE '55' IF IN RADIANS TYPE
    '11'":INPUT DR:IF DR <> 11 AND DR <>
    55 THEN GOTO 220
230 FOR I = 1 TO 2:PRINT: PRINT"FOR
    NUMBER ";I: PRINT: INPUT"R = ";
    R(I-1): PRINT: INPUT "AND THE ANGLE
    IS "; AN(I-1):NEXT I
```

```
240 '
250 '
260 REM *** CONVERT DEGREES TO RADIANS
    ***
270 '
280 '
290 IF DR = 11 THEN GOTO 360
300 FOR I = 0 TO 1:AD(I) = AN(I):AN(I) =
    AN(I) * PI /180:NEXT I
310 '
320 '
330 REM *** CONVERT TO RECTANGULAR FORM
    ***
340 '
350 '
360 FOR I = 0 TO 1:A(I) = R(I) *
    COS(AN(I)):B(I) = R(I) *
    SIN(AN(I)):NEXT I
370 '
380 '
390 REM *** DIVIDING COMPLEX NUMBERS ***
400 '
410 '
420 DENOM = A(1)*A(1) + B(1)*B(1)
430 A = (A(0) * A(1) + B(0) * B(1)) /
    DENOM:B = (B(0) * A(1) - A(0) *
    B(1))/ DENOM
440 '
450 '
460 '
470 R = SQR(A*A+B*B)
480 IF A$ = "9" THEN AN = AN(0) -
    AN(1):GOTO 540
490 IF A <> 0 THEN GOTO 520
500 IF B > 0 THEN AN = PI/2:GOTO 550
510 IF B < 0 THEN AN = 3 * PI / 2:GOTO
    550
520 AN = ATN(B/A)
530 IF A$ = "1" AND AN < 0 THEN AN = AN +
    PI
540 IF AN < 0 THEN AN = AN + 2*PI:GOTO
    540
550 IF AN >= 2*PI THEN AN = AN - 2 *
    PI:GOTO 550
560 AD = AN * 180 / PI
570 PRINT:PRINT:PRINT"THE TOTAL IS "
580 PRINT:PRINT A;" + ";B;"J"
590 PRINT:PRINT"OR":PRINT R;"CIS ";AN;"
    RADIANS":PRINT R;" CIS ";AD;"
    DEGREES"
```

```
600 END
```

Use this program (PRG14542) to divide the following complex numbers.

(a) $(0.25 + 7j) \div (-9.12 - 0.21j)$

(b) $7.5 \operatorname{cis} 36° \div 2.5 \operatorname{cis} 108°$

(c) $2.8 \operatorname{cis} 5.34 \div 0.7 \operatorname{cis} 1.65$

Section 14.5, Exercise 43

```
  1 REM *** PRG14543 ***
 10 PRINT "THIS PROGRAM WILL USE DE
    MOIVRE'S FORMULA TO FIND THE N-TH
    POWER OF A COMPLEX NUMBER OR THE N
    N-TH ROOTS OF A COMPLEX NUMBER."
 20 PRINT : PRINT : PRINT "IF YOU ARE
    FINDING A ROOT TYPE A '2' IF YOU ARE
    FINDING A POWER TYPE A '9'."
 30 INPUT RP: IF RP <> 2 AND RP <> 9 THEN
    GOTO 20
 40 PRINT : PRINT : PRINT "IF THE NUMBER
    YOU ARE ENTERING IS IN RECTANGULAR
    FORM TYPE '1' IF IT IS IN POLAR FORM
    TYPE A '9'."
 50 INPUT A$
 60 PI = 3.14159265359#
 70 IF A$ = "9" THEN GOTO 330
 80 '
 90 '
100 REM *** ENTER COMPLEX NUMBER IN
    RECTANGULAR FORM ***
110 '
120 '
130 CLS: PRINT "ENTER THE COMPLEX NUMBER
    IN THE FORM A + BJ"
140 PRINT : PRINT : INPUT "A= ";A: PRINT
    : PRINT : INPUT "AND B = ";B
150 '
160 '
170 REM *** CONVERT FROM RECTANGULAR TO
    POLAR FORM ***
180 '
190 '
200 R = SQR(A*A + B*B)
210 IF A <> 0 THEN GOTO 240
220 IF B > 0 THEN AN = PI/2:GOTO 260
230 IF B < 0 THEN AN = 3 * PI/2:GOTO 260
240 AN = ATN(B/A)
250 IF A < 0 THEN AN = AN + PI
260 AD = AD * 180/PI
```

```
270 GOTO 480
280 '
290 '
300 REM ENTER COMPLEX NUMBER IN COMPLEX
    FORM ***
310 '
320 '
330 PRINT : PRINT "ENTER THE COMPLEX
    NUMBER IN THE FORM R CIS(AN)"
340 PRINT : PRINT "IF THE ANGLES ARE IN
    DEGREES TYPE '55' IF": INPUT " IN
    RADIANS TYPE '11'";DR:IF DR <> 11 AND
    DR <> 55 GOTO 340
350 PRINT : PRINT : PRINT : INPUT "R =
    ";R: PRINT : INPUT " AND THE ANGLE IS
    ";AN
350 PRINT : PRINT : PRINT : INPUT "R =
    ";R: PRINT : INPUT " AND THE ANGLE IS
    ";AN
360 '
370 '
380 REM CONVERT DEGREES TO RADIANS ***
390 '
400 '
410 IF DR = 11 THEN GOTO 480
420 AD = AN:AN = AN * PI / 180
430 '
440 '
450 REM *** FIND THE N - TH POWER OF THE
    NUMBER ***
460 '
470 '
480 IF RP = 2 GOTO 630
490 PRINT : PRINT : INPUT "ENTER THE
    POWER";N
500 AN = N * AN
510 R = R^N
520 IF AN >= 0 AND AN < 2*PI THEN GOTO
    570
530 IF AN < 0 THEN GOTO 560
540 AN = AN - 2*PI:IF AN >= 2 * PI GOTO
    540
550 GOTO 570
560 AN = AN + 2 * PI: IF AN <0 GOTO 560
570 PRINT : PRINT : PRINT "THE TOTAL IS "
580 A = R * COS(AN):B = R * SIN(AN)
590 AD = AN * 180/PI
600 PRINT : PRINT A;" + ";B;"J"
610 PRINT : PRINT "OR": PRINT R;" CIS
    ";AN;" RADIANS": PRINT : PRINT R;"
```

```
    CIS ";AD;" DEGREES"
620 GOTO 850
630 PRINT : PRINT : INPUT "ENTER THE
    NUMBER OF ROOTS ";N
640 R = R^(1/N)
650 AG = AN
660 FOR I = 0 TO N - 1
670 AN = (AG + 2 * PI * I)/N
680 IF I = 0 THEN GOTO 740
690 IF I = 1 THEN GOTO 730
700 IF I = 2 THEN GOTO 720
710 PRINT : PRINT : PRINT "THE ";I +
    1;"TH ROOT IS":GOTO 750
720 PRINT : PRINT : PRINT "THE 3RD ROOT
    IS":GOTO 750
730 PRINT : PRINT : PRINT "THE 2ND ROOT
    IS":GOTO 750
740 PRINT : PRINT : PRINT "THE FIRST ROOT
    IS"
750 IF AN >= 0 AND AN < 2*PI THEN GOTO
    800
760 IF AN < 0 THEN GOTO 790
770 AN = AN - 2 * PI: IF AN >= 2*PI THEN
    GOTO 770
780 GOTO 800
790 AN = AN + 2 * PI:IF AN < 0 THEN GOTO
    790
800 A = R * COS(AN):B = R * SIN(AN)
810 AD = AN * 180/PI
820 PRINT : PRINT A ;" + ";B;"J"
830 PRINT : PRINT "OR": PRINT R;" CIS
    ";AN;" RADIANS": PRINT : PRINT R;"
    CIS ";AD;" DEGREES"
840 NEXT I
850 END
```

Use this program (PRG14543) to evaluate the following complex numbers.

(a) $(1.5 + 3.2j)^4$

(b) the four fourth roots of $2 - 2j$

Section 14.6, Exercise 35

```
 1 REM RPG14635
10 PRINT" THIS PROGRAM WILL COMPUTE THE
   TOTAL IMPEDANCE OF A GROUP OF N
   IMPEDANCES IN AN AC CIRCUIT THAT ARE
   CONNECTED IN SERIES OR PARALLEL"
20 PRINT:PRINT:PRINT"THE IMPEDANCE MAY
   BE ENTERED EITHER IN RECTANGULAR OR
```

```
        IN POLAR FORM."                     330 '
 30 FOR I = 1 TO 4000:NEXT I               340 '
 40 PI = 3.14159265359#                    350 REM *** CONVERT TO RECTANGULAR FORM
 50 N = 2                                       ***
 51 PRINT:INPUT"HOW MANY IMPEDANCES ARE    360 '
    IN THE CIRCUIT: ",N                    370 '
 60 DIM A(N),B(N)                          380 A(I - 1) = R(I - 1) * COS(AN(I -
 70 PRINT:PRINT:INPUT"IF THE IMPEDANCES         1)):B(I - 1) = R(I - 1) * SIN(AN(I -
    ARE IN RECTANGULAR FORM TYPE A '1' ;        1)):NEXT I
    IF THEY ARE IN POLAR FORM THEN TYPE A  390 PRINT:PRINT:PRINT"THE TOTAL IMPEDANCE
    '9':",FR:IF FR <> 1 AND FR <> 9 THEN        FOR A"
    GOTO 70                                400 '
 80 IF FR = 9 THEN GOTO 220                410 '
 90 '                                      420 REM *** DETERMINE THE SERIES
100 '                                          IMPEDANCE ***
110 REM **** ENTER RECTANGULAR            430 '
    COORDINATES ****                       440 '
120 '                                      450 FOR I=1 TO N:A(N) = A(N) + A(N -
130 '                                          1):B(N) = B(N) + B(N - 1):NEXT I
140 PRINT:PRINT"ENTER EACH OF THE ";N;"    460 RT = SQR(A(N) * A(N) + B(N) *
    IMPEDANCES IN THE FORM A + BJ"              B(N)):TA =
150 FOR I = 1 TO N:PRINT"FOR IMPEDANCE          ATN(B(N)/A(N)):DA=TA*180/PI
    #";I:PRINT:INPUT"A = ",A(I -           470 PRINT"SERIES CIRCUIT IS" :PRINT:PRINT
    1):PRINT:INPUT "AND B = ",B(I -             A(N);" + ";B(N);"J,":PRINT:PRINT RT;"
    1):NEXT I                                   CIS ";TA;" RADIANS OR ":PRINT:PRINT
160 GOTO 390                                    RT;"CIS ";DA;" DEGREES."
170 '                                      480 '
180 '                                      490 '
190 REM *** ENTER POLAR COORDINATES ***   500 REM *** DETERMINE THE PARALLEL
200 '                                          IMPEDANCE ***
210 '                                      510 '
220 DIM R(N),AN(N)                         520 '
230 PRINT:PRINT"ENTER EACH OF THE ";N;"    530 ZA = 0:ZB = 0
    IMPEDANCES IN THE FORM R CIS(A)"       540 FOR I = 1 TO N:B = A(I - 1) * A(I -
240 PRINT:PRINT"IF ANGLES ARE IN DEGREES       1) + B(I - 1)*B(I - 1):ZA = ZA + A(I
    TYPE '55'; IF":INPUT"IN RADIANS TYPE        - 1)/B:ZB = ZB - B(I - 1) /B:NEXT I
    '11'",DR:IF DR <> 55 AND DR <> 11      550 B = ZA * ZA + ZB * ZB:PA=ZA/B:PB = -
    THEN GOTO 240                              ZB/B
250 FOR I = 1 TO N :PRINT:PRINT"FOR        560 RP = SQR(PA * PA + PB * PB):AR =
    IMPEDANCE NUMBER ";I:PRINT:PRINT:          ATN(PB/PA):AD = AR * 180/PI
    PRINT:INPUT "R = ",R(I - 1):PRINT      570 PRINT :PRINT"AND FOR A":PRINT:PRINT
    :INPUT" AND THE ANGLE IS ",AN(I - 1)       "PARALLEL CIRCUIT IS" :PRINT:PRINT
260 '                                          PA;" + ";PB;"J,":PRINT:PRINT RP;"CIS
270 '                                          ";AR;"RADIANS OR ":PRINT:PRINT
280 REM *** CONVERT DEGREES TO RADIANS         RP;"CIS ";AD;" DEGREES."
    ***                                    580 '
290 '                                      590 '
300 '                                      600 REM *** ASK FOR ANOTHER PROBLEM ***
310 IF DR = 11 GOTO 380                    610 '
320 AN(I - 1) = AN(I - 1) * PI/180         620 '
```

```
630 PRINT:PRINT:PRINT:INPUT"IF YOU WANT
    TO WORK ANOTHER PROBLEM TYPE A '1';
    IF NOT TYPE A '5'.",QU:IF QU <> 1 AND
    QU <> 5 THEN GOTO 630
640 IF QU = 5 THEN GOTO 670
650 CLEAR
660 GOTO 40
670 END
```

Use this program (PRG14635) to find the total impedances if the given impedances are connected **(a)** in series and **(b)** in parallel.

(a) $Z_1 = 2 + 3j$, $Z_2 = 1 - 5j$
(b) $Z_1 = -2 + j$, $Z_2 = 3 - 4j$, $Z_3 = 9 - 2j$
(c) $Z_1 = 2.19\,\underline{/18.4°}$, $Z_2 = 5.16\,\underline{/67.3°}$
(d) $Z_1 = 2.57\,\underline{/0.25}$, $Z_2 = 1.63\,\underline{/1.38}$

Chapter 17

Section 17.1, Exercise 37

```
1 REM *** PRG17137 ***
5 CLS
10 REM ************************
20 REM ************************
30 REM **                    **
40 REM **   MATRIX ADDITION   **
50 REM **   AND SUBTRACTION   **
60 REM **                    **
70 REM ************************
80 REM ************************
100 PRINT : PRINT : PRINT : PRINT
110 PRINT "THIS PROGRAM ADDS AND
    SUBTRACTS TWO": PRINT : PRINT "M X N
    MATRICES": PRINT :
120 GOTO 140
130 PRINT "PLEASE PRINT AN A OR AN S":
    PRINT: PRINT" 140 '
150 INPUT "DO YOU WANT TO ADD OR
    SUBTRACT? ";QU$
160 IF LEFT$ (QU$,1) < > "A" AND LEFT$
    (QU$,1) < > "S" THEN GOTO 130
170 PRINT : PRINT : PRINT : INPUT "ENTER
    THE NUMBER OF ROWS: ";M: PRINT :
    PRINT : INPUT "AND THE NUMBER OF
    COLUMNS: ";N
180 DIM A(M-1, N-1), B(M-1, N-1), C(M-1,
    N-1)
190 CLS : PRINT "ENTER THE ELEMENTS IN
    MATRIX A"
200 PRINT : PRINT
210 FOR I = 0 TO M - 1
220 FOR J = 0 TO N - 1
230 PRINT "A(";I;",";J;") = ";: INPUT
    A(I,J)
240 NEXT J
250 PRINT : PRINT : NEXT I
260 CLS : PRINT "ENTER THE ELEMENTS IN
    MATRIX B"
270 PRINT : PRINT
280 FOR I = 0 TO M - 1
290 FOR J = 0 TO N - 1
300 PRINT "B(";I;",";J;") = ";: INPUT
    B(I,J)
310 NEXT J
320 PRINT : PRINT : NEXT I
330 CLS : PRINT "THE ANSWER IS": PRINT :
    PRINT
340 FOR I = 0 TO M - 1
350 FOR J = 0 TO N - 1
360 IF LEFT$ (QU$,1) = "S" THEN GOTO 390
370 C(I,J) = A(I,J) + B(I,J)
380 GOTO 400
390 C(I,J) = A(I,J) - B(I,J)
400 Z = 8 * J + 1
410 PRINT USING "########.####"; C(I,J);
420 NEXT J
430 PRINT : PRINT : NEXT I
440 GOTO 460
450 PRINT : PRINT "PLEASE REENTER AND
    TYPE A Y OR N": PRINT : PRINT
460 INPUT "DO YOU WANT TO WORK ANOTHER
    PROBLEM? "; QU$
470 IF LEFT$ (QU$,1) < > "Y" AND LEFT$
    (QU$,1) < > "N" THEN GOTO 450
480 IF QU$ = "Y" THEN GOTO 150
490 END
```

Use this program (PRG17137) to find the indicated sum or difference

(a) $\begin{bmatrix} 2 & 4 \\ -5.3 & 1 \\ 3.1 & 4 \end{bmatrix} + \begin{bmatrix} 5.7 & -6 \\ 2.4 & 9 \\ 8 & 2 \end{bmatrix}$ and

(b) $\begin{bmatrix} 3.4 & 2.1 & -1.5 \\ 9.7 & 12.1 & 4 \end{bmatrix} - \begin{bmatrix} 1 & 3.4 & -5 \\ 4 & 3.9 & 6 \end{bmatrix}$

Section 17.1, Exercise 38

```
1 REM *** PRG17138 ***
5 CLS
```

```
 10 REM ******************************
 20 REM ******************************
 30 REM **                          **
 40 REM **   SCALAR MULTIPLICATION   **
 50 REM **        OF MATRICES        **
 60 REM **                          **
 70 REM ******************************
 80 REM ******************************
100 PRINT : PRINT : PRINT : PRINT
110 PRINT "THIS PROGRAM MULTIPLIES A
    SCALAR": PRINT : PRINT " BY AN M X N
    MATRIX : PRINT : PRINT
115 CLEAR
120 PRINT : PRINT : INPUT "ENTER THE
    SCALAR: "; K: PRINT : PRINT
130 PRINT : PRINT : PRINT : INPUT "ENTER
    THE NUMBER OF ROWS:";M: PRINT : PRINT
    : INPUT "AND THE NUMBER OF COLUMNS:
    ";N
140 DIM A(M - 1,N - 1),B(M - 1,N - 1),C(M
    - 1,N - 1)
150 CLS : PRINT "ENTER THE ELEMENTS IN
    THE MATRIX"
160 PRINT : PRINT
170 FOR I = 0 TO M - 1
180 FOR J = 0 TO N - 1
190 PRINT "A(";I;",";J;") = ";: INPUT
    A(I,J)
200 NEXT J
210 PRINT : PRINT : NEXT I
220 CLS : PRINT "THE ANSWER IS": PRINT :
    PRINT
230 FOR I = 0 TO M - 1
240 FOR J = 0 TO N - 1
250 C(I,J) = K * A(I,J)
250 C(I,J) = K * A(I,J)
260 Z = 8 * J + 1
270 PRINT USING "########.####"; C(I,J);
280 NEXT J
290 PRINT : PRINT : NEXT I
300 GOTO 320
310 PRINT : PRINT"PLEASE REENTER AND TYPE
    A Y OR N"; PRINT : PRINT
320 INPUT "DO YOUR WANT TO WORK ANOTHER
    PROBLEM";QU$
330 IF LEFT$ (QU$,1) < > "Y" AND LEFT$
    (QU$,1)< > "N" THEN GOTO 310
340 IF QU$ = "Y" THEN GOTO 115
350 END
```

Section 17.1, Exercise 39

```
  1 REM *** PRG17139 ***
  5 REM *******************************
 10 REM *******************************
 20 REM **                           **
 30 REM **   SCALAR MULTIPLICATION    **
 40 REM **     AND THEN ADDITION      **
 50 REM **       OR SUBTRACTION       **
 60 REM **      OF TWO MATRICES       **
 70 REM **                           **
 80 REM *******************************
 90 REM *******************************
100 PRINT : PRINT : PRINT : PRINT
110 PRINT "THIS PROGRAM ALLOWS YOU TO
    MULTIPLY TWO M X N MATRICES BY A
    SCALAR" : PRINT : PRINT
112 PRINT "AND THEN IT ADDS OR SUBTRACTS
    THE NEW MATRICES."
115 CLEAR
120 PRINT : PRINT : INPUT "ENTER THE
    SCALAR FOR THE FIRST MATRIX: "; K1:
    PRINT : PRINT
130 PRINT : PRINT : INPUT "ENTER THE
    SCALAR FOR THE SECOND MATRIX: "; K2:
    PRINT : PRINT
220 GOTO 240
230 PRINT "PLEASE PRINT AN A OR AN S":
    PRINT: PRINT"
240 '
250 INPUT "DO YOU WANT TO ADD OR
    SUBTRACT? ";QU$
260 IF LEFT$ (QU$,1) < > "A" AND LEFT$
    (QU$,1) < > "S" THEN GOTO 230
270 PRINT : PRINT : PRINT : INPUT "ENTER
    THE NUMBER OF ROWS: ";M: PRINT :
    PRINT : INPUT "AND THE NUMBER OF
    COLUMNS: ";N
290 CLS : PRINT "ENTER THE ELEMENTS IN
    MATRIX A"
300 PRINT : PRINT
310 FOR I = 0 TO M - 1
320 FOR J = 0 TO N - 1
330 PRINT "A(";I;",";J;") = ";: INPUT
    A(I,J)
340 NEXT J
350 PRINT : PRINT : NEXT I
360 CLS : PRINT "ENTER THE ELEMENTS IN
    MATRIX B"
370 PRINT : PRINT
```

```
380 FOR I = 0 TO M - 1
390 FOR J = 0 TO N - 1
400 PRINT "B(";I;",";J;") = ";: INPUT
    B(I,J)
410 NEXT J
420 PRINT : PRINT : NEXT I
430 CLS : PRINT "THE ANSWER IS": PRINT :
    PRINT
440 FOR I = 0 TO M - 1
450 FOR J = 0 TO N - 1
460 IF LEFT$ (QU$,1) = "S" THEN GOTO 490
470 C(I,J) = K1*A(I,J) + K2*B(I,J)
480 GOTO 500
490 C(I,J) = A(I,J) - B(I,J)
500 Z = 8 * J + 1
510 PRINT USING "########.####"; C(I,J);
520 NEXT J
530 PRINT : PRINT : NEXT I
540 GOTO 560
550 PRINT : PRINT "PLEASE REENTER AND
    TYPE A Y OR N": PRINT : PRINT
560 INPUT "DO YOU WANT TO WORK ANOTHER
    PROBLEM? "; QU$
570 IF LEFT$ (QU$,1) < > "Y" AND LEFT$
    (QU$,1) < > "N" THEN GOTO 550
580 IF QU$ = "Y" THEN GOTO 115
590 END
```

Use this program (PRG17139) to determine

$$3\begin{bmatrix} 2 & 4 \\ -5.3 & 1 \\ 3.1 & 4 \end{bmatrix} + \frac{1}{2}\begin{bmatrix} 5.7 & -6 \\ 2.4 & 9 \\ 8 & 2 \end{bmatrix}$$

Section 17.2, Exercise 29

```
  1 REM *** PRG17229 ***
  5 CLS
 10 REM ******************************
 20 REM ******************************
 30 REM **                         **
 40 REM **    MATRIX MULTIPLICATION **
 50 REM **                         **
 60 REM ******************************
 70 REM ******************************
100 PRINT : PRINT : PRINT : PRINT
110 PRINT "THIS PROGRAM MULTIPLIES TWO
    MATRICES": PRINT : PRINT "MATRIX A
    AND MATRIX B": PRINT :
120 CLEAR
130 PRINT : PRINT : PRINT "FOR MATRIX A:"
    : PRINT : PRINT : INPUT "ENTER THE
    NUMBER OF ROWS: " ;M : PRINT : PRINT
    : INPUT "AND THE NUMBER OF COLUMNS:
    ";N
140 PRINT : PRINT : PRINT "FOR MATRIX B:"
    : PRINT : PRINT : INPUT "ENTER THE
    NUMBER OF ROWS: " ;P: PRINT : PRINT :
    INPUT "AND THE NUMBER OF COLUMNS: ";R
150 IF N = P THEN GOTO 210
160 CLS : PRINT : PRINT "YOU HAVE MADE A
    MISTAKE.": PRINT : PRINT "THE NO. OF
    COLUMNS IN MATRIX A MUST BE THE SAME
    AS THE NO. OF ROWS IN MATRIX B"
170 FOR X = 1 TO 4000: NEXT X
180 PRINT: PRINT : PRINT "YOU HAD ";N;"
    COLUMNS IN MATRIX A": PRINT : PRINT "
    AND ";P;" ROWS IN MATRIX B."
190 FOR X = 1 TO 4000: NEXT X
200 PRINT : PRINT : PRINT "PLEASE
    REENTER": FOR X = 1 TO 2000: NEXT X:
    CLS : GOTO 130
210 DIM A(M - 1,N - 1),B(P - 1,R - 1),C(M
    - 1,R - 1)
220 CLS : PRINT "ENTER THE ELEMENTS IN
    MATRIX A":
230 PRINT : PRINT :
240 FOR I = 0 TO M - 1
250 FOR J = 0 TO N - 1
260 PRINT "A(";I;",";J;") = ";: INPUT
    A(I,J)
270 NEXT J
280 PRINT : PRINT : NEXT I
290 CLS : PRINT "ENTER THE ELEMENTS IN
    MATRIX B"
300 PRINT : PRINT
310 FOR I = 0 TO P - 1
320 FOR J = 0 TO R - 1
330 PRINT "B(";I;",";J;") = ";: INPUT
    B(I,J)
340 NEXT J
350 PRINT : PRINT : NEXT I
360 CLS : PRINT "THE ANSWER IS": PRINT :
    PRINT
370 FOR I = 0 TO M - 1
380 FOR K = 0 TO R - 1
390 FOR J = 0 TO N - 1
400 C(I,K) = C(I,K) + A(I,J) * B(J,K)
410 NEXT J
420 Z = 8 * J + 1
```

```
430 PRINT USING "########.####"; C(I,K);
440 NEXT K
450 PRINT : PRINT : NEXT I
460 GOTO 480
470 PRINT : PRINT "PLEASE REENTER AND
    TYPE A Y OR N": PRINT : PRINT
480 INPUT "DO YOU WANT TO WORK ANOTHER
    PROBLEM? "; QU$
490 IF LEFT$ (QU$,1) < > "Y" AND LEFT$
    (QU$,1) < > "N" THEN GOTO 470
500 IF QU$ = "Y" THEN GOTO 120
510 END
```

 Use this program (PRG17229) to find the indicated product:

$$\begin{bmatrix} 3.7 & 4.3 & -2.9 & -17.4 \\ -7.1 & 8.3 & 10 & -5.4 \end{bmatrix} \begin{bmatrix} 0 & 4.2 & 1.7 \\ 1.2 & 0.4 & 2 \\ 8 & -1.9 & 0 \\ -3.5 & -4.1 & 12.7 \end{bmatrix}$$

Section 17.3, Exercise 22

```
  1 REM *** PRG17322 ***
  5 CLS
 10 REM **************************
 20 REM **************************
 30 REM **                    **
 40 REM **   MÁTRIX INVERSION   **
 50 REM **                    **
 60 REM **************************
 70 REM **************************
100 PRINT : PRINT : PRINT : PRINT
110 PRINT "THIS PROGRAM INVERTS AN N X N
    MATRIX":
120 FOR X = 1 TO 1000: NEXT X
130 CLEAR
140 PRINT : PRINT : INPUT "ENTER THE SIZE
    OF THE SQUARE MATRIX ";X
150 DIM A(X - 1,X - 1), B(X - 1,X - 1)
160 CLS : PRINT "ENTER THE ELEMENTS IN
    THE MATRIX"
170 FOR I = 0 TO X - 1
180 FOR J = 0 TO X - 1
190 PRINT "A(";I;",";J;") = ";: INPUT
    A(I,J)
200 NEXT J
210 B(I,I) = 1
220 PRINT : PRINT : NEXT I
230 FOR J = 0 TO X - 1
```

```
240 FOR I = J TO X - 1
250 IF A(I,J) < > 0 THEN 280
260 NEXT I
270 PRINT : PRINT "THIS MATRIX IS
    SINGULAR. IT DOES NOT HAVE AN
    INVERSE.": GOTO 530
280 FOR H = 0 TO X - 1
290 T = A(J,H):A(J,H) = A(I,H):A(I,H) = T
300 M = B(J,H):B(J,H) = B(I,H):B(I,H) = M
310 NEXT H
320 D = A(J,J)
330 FOR H = 0 TO X - 1
340 A(J,H) = A(J,H) / D:B(J,H) = B(J,H) /
    D
350 NEXT H
360 FOR H = 0 TO X - 1
370 IF J = H THEN 420
380 D = A(H,J)
390 FOR K = 0 TO X - 1
400 A(H,K) = A(H,K) - A(J,K) * D:B(H,K) =
    B(H,K) - B(J,K) * D
410 NEXT K
420 NEXT H
430 NEXT J
440 CLS : PRINT : PRINT : PRINT "THE
    INVERSE MATRIX IS": PRINT : PRINT
450 FOR I = 0 TO X - 1
460 FOR J = 0 TO X - 1
470 B(I,J) = INT ( B(I,J) * 10000 + .5) /
    10000
480 PRINT USING "######.####"; B(I,J);
490 NEXT J
500 PRINT : PRINT : NEXT I
510 GOTO 530
520 PRINT : PRINT "PLEASE REENTER AND
    TYPE A Y OR N": PRINT : PRINT
530 INPUT "DO YOU WANT TO INVERT ANOTHER
    MATRIX"; QU$
540 IF LEFT$ (QU$,1) < > "Y" AND LEFT$
    (QU$,1) < > "N" AND LEFT$ (QU$,1) < >
    "y" AND LEFT$ (QU$,1) < > "n" THEN
    GOTO 530
550 IF QU$ = "Y" OR QU$ = "y" THEN GOTO
    130
560 END
```

 Use this program (PRG17322) to find the inverse of the given matrix (if it exists).

(a) $\begin{bmatrix} -20 & 5.1 \\ -7.5 & 1.5 \end{bmatrix}$, **(b)** $\begin{bmatrix} 1 & 2 & -1 \\ 3 & 7 & -10 \\ 7 & 16 & -21 \end{bmatrix}$, and

(c) $\begin{bmatrix} 1.5 & -2.2 & -1 & -2.8 \\ 3.6 & -5.4 & -2 & -7 \\ 2 & -5 & -1 & -5 \\ -1 & 6 & 4 & 10 \end{bmatrix}$

Section 17.4, Exercise 25

```
  1 REM *** PRG17425 ***
  5 CLS
 10 REM ******************************
 20 REM ******************************
 30 REM **                        **
 40 REM **   USING MATRICES TO SOLVE   **
 50 REM **SYSTEMS OF LINEAR EQUATIONS**
 60 REM **                        **
 70 REM ******************************
 80 REM ******************************
100 CLS
110 PRINT "THIS PROGRAM SOLVES A SYSTEM
    OF N EQUATIONS WITH N VARIABLES":
120 FOR X = 1 TO 1000: NEXT X
130 CLEAR
140 CLS : INPUT "ENTER THE NUMBER OF
    EQUATIONS ";X
150 DIM A(X - 1,X - 1),B(X - 1,X - 1),
    C(X - 1),D(X - 1)
160 CLS: PRINT "ENTER THE COEFFICIENTS OF
    THE SYSTEM OF EQUATIONS"
170 FOR I = 0 TO X - 1
180 FOR J = 0 TO X - 1
190 PRINT "A(";I;",";J;") = ";: INPUT
    A(I,J)
200 NEXT J
210 B(I,I) = 1
220 PRINT : PRINT : NEXT I
230 PRINT : PRINT "NOW ENTER THE
    CONSTANTS.": PRINT
240 FOR I = 0 TO X - 1
250 PRINT "C(";I;") = ";: INPUT C(I)
260 NEXT I
270 FOR J = 0 TO X - 1
280 FOR I = J TO X - 1
290 IF A(I,J) < > 0 THEN 320
300 NEXT I
310 PRINT : PRINT "THE MATRIX IS
    SINGULAR. THIS SYSTEM HAS NO
    SOLUTION.": GOTO 650
320 FOR H = 0 TO X - 1
330 T = A(J,H):A(J,H) = A(I,H):A(I,H) = T
340 M = B(J,H):B(J,H) = B(I,H):B(I,H) = M
350 NEXT H
360 D = A(J,J)
370 FOR H = 0 TO X - 1
380 A(J,H) = A(J,H) / D:B(J,H) = B(J,H) /
    D
390 NEXT H
400 FOR H = 0 TO X - 1
410 IF J = H THEN 460
420 D = A(H,J)
430 FOR K = 0 TO X - 1
440 A(H,K) = A(H,K) - A(J,K) * D:B(H,K) =
    B(H,K) - B(J,K) * D
450 NEXT K
460 NEXT H
470 NEXT J
480 CLS : PRINT : PRINT : PRINT "THE
    INVERTED MATRIX IS ": PRINT : PRINT
490 FOR I = 0 TO X - 1
500 FOR J = 0 TO X - 1
510 Y = INT ( B(I,J) * 1000 + .5) / 1000
520 PRINT Y,;
530 NEXT J
540 PRINT : PRINT : NEXT I
550 PRINT : PRINT "THE ANSWER IS": PRINT
560 FOR I = 0 TO X - 1
570 FOR J = 0 TO X - 1
580 D(I) = D(I) + B(I,J) * C(J)
600 NEXT J
610 PRINT "X(";I + 1;") = ";D(I)
620 PRINT : PRINT : NEXT I
630 GOTO 650
640 PRINT : PRINT "PLEASE REENTER AND
    TYPE A Y OR N"; PRINT : PRINT
650 INPUT "DO YOU WANT TO SOLVE ANOTHER
    SYSTEM "; QU$
660 IF LEFT$ (QU$,1) < > "Y" AND LEFT$
    (QU$,1) < > "N" AND LEFT$ (QU$,1) < >
    "y" AND LEFT$ (QU$,1) < > "n" THEN
    GOTO 650
670 IF QU$ = "Y" OR QU$ = "y" THEN GOTO
    130
680 END
```

 Use this program (PRG17425) to solve each of the following systems

$$(a) \begin{cases} 2x + y - 3z = 1 \\ x - 4y + z = 6 \\ 4x - 16y + 4z = 24 \end{cases}$$

$$(b) \begin{cases} -x + y + 3z = 4 \\ 2x + 3y + z = -2 \\ 6x + 4y + 2z = 6 \end{cases}$$

Chapter 18

Section 18.4, Exercise 22

```
  1 *** PRG18422 ***
 10 REM *****************************
 20 REM *****************************
 30 REM **                        **
 40 REM **   LINEAR INTERPOLATION  **
 50 REM **                        **
 60 REM *****************************
 70 REM *****************************
 80 REM
 90 REM
100 REM
110 REM **  DEFINE THE FUNCTION  **
120 REM
130 CLS
140 INPUT "DO YOU WANT TO CHANGE THE
    FUNCTION? (Y/N) "; QU$
150 IF QU$ < > "Y" AND QU$ < > "N" THEN
    GOTO 140
160 IF QU$ = "N" GOTO 220
170 CLS : PRINT : PRINT : PRINT "TO
    CHANGE THE FUNCTION TYPE "
180 PRINT : PRINT "1050 DEF FN F(X) = . .
    . "
190 PRINT : PRINT "WHERE YOU PUT THE NEW
    FUNCTION AFTER THE EQUAL SYMBOL."
200 PRINT : PRINT "AFTER YOU HAVE ENTERED
    THE FUNCTION YOU WILL HAVE TO TYPE
    'RUN' TO GET THE PROGRAM TO START.":
    PRINT
210 END
220 REM
230 REM
240 REM ** SEARCH FOR ROOTS **
250 REM
260 REM
270 GOSUB 1050
280 GOTO 300
290 PRINT : PRINT : PRINT "YOU MADE A
```

```
    MISTAKE. PLEASE ANSWER THE QUESTION
    WITH A Y OR N.": PRINT : PRINT
300 INPUT "DO YOU WANT TO SEARCH FOR
    ROOTS? "; QU$
310 IF LEFT$ (QU$,1) < > "Y" AND LEFT$
    (QU$,1) < > "N" GOTO 290
320 IF LEFT$ (QU$,1) = "N" GOTO 560
330 PRINT : PRINT : INPUT "ENTER THE
    LOWER LIMIT: ";A
340 PRINT : INPUT "NOW ENTER THE UPPER
    LIMIT: ";B
350 CLS : PRINT "THE LOWER LIMIT IS ";A;"
    AND THE" : PRINT "UPPER LIMIT IS ";B
360 PRINT : PRINT : PRINT "X - 1"; TAB(
    8);"F(X - 1)"; TAB(21);"X"; TAB(30);
    "F(X) "
370 FOR X = A TO B - 1
380 P = FN F(X)
390 Q = FN F(X + 1)
400 IF SGN (P) < > SGN (Q) GOTO 420
410 GOTO 440
420 PRINT : PRINT X; TAB( 6);P; TAB( 21);
    X + 1;TAB(26);Q
430 PRINT " THERE IS A REAL ROOT BETWEEN
    ";X;" AND ";X + 1
440 NEXT X
450 PRINT "SEARCH COMPLETED.": PRINT
460 GOTO 480
470 PRINT : PRINT : "PLEASE ANSWER AGAIN.
    TYPE A Y OR N."
480 PRINT : INPUT "DO YOU WANT TO SEARCH
    OVER ANOTHER INTERVAL? ";QU$
490 IF LEFT$ (QU$,1) < > "Y" AND LEFT$
    (QU$,1) < > "N" GOTO 470
500 IF LEFT$ (QU$,1) = "Y" GOTO 330
510 REM
520 REM
530 REM ** USE OF LINEAR INTERPOLATION **
540 REM
550 REM
560 PRINT : PRINT : PRINT "WE WILL NOW
    USE LINEAR INTERPOLATION TO DETERMINE
    APPROXIMATE ROOTS OF F(X).": PRINT
570 PRINT : PRINT "ENTER THE ENDPOINTS OF
    THE INTERVAL": PRINT " WHERE YOU WANT
    THE ROOT.": PRINT
580 INPUT "LEFT ENDPOINT = ";A: INPUT
    "RIGHT ENDPOINT = ";B
590 IF SGN (FN F(A)) = SGN (FN F(B)) GOTO
    610
```

```
600 GOTO 620
610 PRINT : PRINT "THERE IS NO REAL ROOT
    BETWEEN ";A;" AND ";B;".": PRINT
    "PLEASE TRY A DIFFERENT INTERVAL.":
    PRINT : GOTO 570
620 M = (FN F(B)  —  FN F(A))/(B  —  A)
630 C = A  —  (FN F(A)/M)
640 IF SGN (FN F(A)) = SGN (FN F(C)) GOTO
    660
650 B = C: GOTO 670
660 A = C
670 PRINT : PRINT "THE ROOT IS
    APPROXIMATELY ";C : PRINT "THE ACTUAL
    ROOT IS BETWEEN ";A;" AND ";B
680 CHK = ABS (FN F(C))
690 IF CHK < .001 GOTO 710
700 GOTO 620
710 PRINT : PRINT : PRINT "INTERPOLATION
    COMPLETED.": PRINT : PRINT "THE ROOT
    IS APPROXIMATELY ";C: PRINT " AND
    F(C) = "; FN F(C)
720 GOTO 740
730 PRINT : PRINT "PLEASE REANSWER WITH A
    Y OR N."
740 PRINT : INPUT "DO YOU WANT TO SEARCH
    OVER ANOTHER INTERVAL? ";QU$
750 IF LEFT$ (QU$,1) < > "Y" AND LEFT$
    (QU$,1) < > "N" GOTO 730
760 IF LEFT$ (QU$,1) = "Y" GOTO 570
770 END
1000 REM
1010 REM
1020 REM ** SUBROUTINE **
1030 REM
1040 REM
1050 DEF FN F(X) = X ^ 3  +  X ^ 2  —  7
     * X
1060 PRINT : RETURN
```

Use this program (PRG18422) to approximate the root of $f(x) = \cos(x^2) - x$

Chapter 21

Section 21.2, Exercise 32

```
 1 *** PRG21232 ***
15 REM THIS PROGRAM LETS YOU ENTER
   INDIVIDUAL SCORES AND THEIR
   FREQUENCIES TO FIND THE MEAN.
20 CLS
```

```
30 PRINT"ENTER EACH SCORE AND THEN" :
   PRINT : PRINT"ITS FREQUENCY" : PRINT
   : PRINT"WHEN DONE ENTER A NEGATIVE
   NUMBER FOR A SCORE."
40 T = 0:FR = 0
50 PRINT:INPUT "INPUT A SCORE: ";S
60 IF S < 0 THEN GOTO 100
70 PRINT:INPUT "AND IT'S FREQUENCY: ";F
   : PRINT
80 T = T + F*S:FR = FR + F
90 GOTO 50
100 MEAN = T / FR
110 PRINT : PRINT"THE MEAN IS ";MEAN
120 END
```

Section 21.3, Exercise 13

```
 10 REM ** PRG21313 **
 20 CLS
 30 PRINT"THIS PROGRAM GIVES THE MEAN,
    VARIANCE, AND STANDARD DEVIATIONS"
 40 PRINT"OF A NUMBER OF SCORES."
 50 PRINT"ENTER EACH SCORE AND THEN" :
    PRINT : PRINT"WHEN YOU HAVE FINISHED,
    ENTER A NEGATIVE NUMBER FOR THE
    SCORE."
 60 T = 0:FR = 0:A = 0
 70 PRINT:INPUT "HOW MANY SCORES ARE YOU
    GOING TO ENTER? ";N
 80 DIM S(N),F(N)
 90 FOR I = 0 TO N-1
100 PRINT:INPUT"SCORE = ";S(I)
110 PRINT:INPUT "FREQUENCY = ";F(I) :
    PRINT
120 T = T + F(I)*S(I):FR = FR + F(I):A =
    A + F(I) * S(I) * S(I)
130 NEXT I
140 MEAN = T/FR
150 VAR = (FR * A - T * T)/(FR*(FR-1))
160 SD = SQR(VAR)
170 PRINT : PRINT"THE MEAN IS: ";MEAN
180 PRINT : PRINT"THE VARIANCE IS: ";VAR
190 PRINT : PRINT"AND THEN STANDARD
    DEVIATION IS: ";SD
200 END
```

Section 21.3, Exercise 14

```
 1 REM *** PRG21314 ***
10 CLS
20 PRINT "THIS PROGRAM GIVES THE MEAN,
```

```
    VARIANCE, STANDARD DEVIATION, AND THE
    TOTAL NUMBER AND PERCENTAGE OF SCORES
    WITHIN ONE AND TWO STANDARD
    DEVIATIONS OF THE MEAN."
 30 PRINT:PRINT "ENTER EACH SCORE AND
    THEN ITS FREQUENCY" :PRINT
 40 T = 0:FR = 0:A = 0:N1 = 0:N2 = 0
 50 PRINT : INPUT "HOW MANY SCORES ARE
    THERE TO ENTER? ",N
 60 DIM S(N),F(N)
 70 FOR I = 0 TO N - 1
 80 PRINT : INPUT "SCORE = ";S(I)
 90 PRINT : INPUT "FREQUENCY = ";F(I):
    PRINT
100 T = T + F(I)*S(I):FR = FR + F(I):A =
    A + F(I) * S(I)*S(I)
110 NEXT I
120 MEAN = T / FR
130 VAR = (FR * A - T * T) / (FR *
    (FR-1))
140 SD = SQR(VAR)
150 S1 = MEAN - SD : S2 = MEAN + SD:S3 =
    S1 - SD:S4 = S2 + SD
160 FOR I = 0 TO N
170 IF S(I) > S1 AND S(I) < S2 THEN N1 =
    N1 + F(I)
180 IF S(I) > S3 AND S(I) < S4 THEN N2 =
    N2 + F(I)
190 NEXT I
200 PRINT "THE TOTAL NUMBER IN THE SAMPLE
    IS: ";FR
210 PRINT "THE MEAN IS: ";MEAN
220 PRINT "THE VARIANCE IS: ";VAR
230 PRINT : PRINT "AND THE STANDARD
    DEVIATION IS: ";SD
240 PRINT : PRINT "THE NUMBER OF SCORES
    WITHIN ONE STANDARD DEVIATION OF THE
    MEAN I S: ";N1;" OR "
250 PRINT N1 * 100 / FR;" PERCENT OF THE
    TOTAL VALUE."
260 PRINT : PRINT "AND THE NUMBER WITHIN
    TWO S.D. OF THE MEAN IS: ";N2;" OR
    ";N2 *1 00/FR;"PERCENT."
270 END
```

Use this program (PRG21318) on the data in the following table to determine the mean, variance, the standard deviation, the number within one standard deviation of the mean, and those within two standard de-

viations of the mean.

Score	1	2	3	4	5	6	7	8
Frequency	23	51	31	27	12	4	1	1

Section 21.4, Exercise 9

```
  1 REM *** PRG21409 ***
  5 CLS
 10 PRINT "THIS PROGRAM DETERMINES THE
    SLOPE, M, AND Y - INTERCEPTS OF THE
    LEAST SQUARES LINE AND THE
    CORRELATION COEFFICIENT OF A SET OF
    DATA."
 30 PRINT : PRINT : PRINT :INPUT "ENTER
    THE NUMBER OF PAIRS OF POINTS (X,Y)
    IN THE SAMPLE ";N
 40 FOR I = 0 TO N - 1
 50 PRINT : PRINT : PRINT "FOR POINT
    NO.";I + 1
 60 PRINT : INPUT "X = ";X
 70 PRINT : INPUT "Y = ";Y
 80 XX=XX + X : YY=YY + Y : XY=XY + X*Y :
    X2=X2 + X*X : Y2=Y2 + Y*Y
 90 NEXT I
100 XV = (N * X2 - XX * XX)/(N*N) : YV =
    (N * Y2 - YY*YY)/(N*N)
110 XS = SQR(XV):YS = SQR(YV)
120 PRINT : PRINT : PRINT "THE FOLLOWING
    ARE THE TOTALS"
130 M = (XY/N - XX*YY/(N*N))/XV
140 B = YY/N - M*XX/N
150 PRINT : PRINT "M = ";M: PRINT : PRINT
    "B = ";B
160 R = M * XS/YS
170 PRINT : PRINT "R = ";R
180 END
```

Use this program (PRG21409) to determine the slope and y-intercept of the least squares line and the correlation coefficient for the data in the following table.

x	17	21	15	16	15	16	24	27	23	18
y	73	66	64	61	70	71	90	86	84	72

Chapter 22

Section 22.1, Exercise 19

```
   1 REM *** PRG22119 ***
  10 REM THIS PROGRAM WILL FIND THE
     AVERAGE SLOPE, BETWEEN ANY TWO POINTS
     ON A CURVE WHEN YOU GIVE THE
     X-COORDINATES OF THE TWO POINTS.
  20 '
  30 REM *** CHANCE TO CHANGE FUNCTIONS
     ***
  40 '
  50 INPUT "DO YOU WANT TO CHANGE THE
     FUNCTION";AN$
  60 GOSUB 2530
  70 IF LEFT$(AN$,1) = "Y" THEN GOTO 1030
  80 '
  90 REM *** ENTER POINT ***
 100 '
 110 INPUT "ENTER THE FIRST POINT: ";A
 120 PRINT:INPUT "NOW ENTER THE SECOND
     POINT: ";B
 130 '
 140 REM *** COMPUTE SLOPE ***
 150 '
 160 GOSUB 2030
 170 M = (FN F(B) - FN F(A))/(B - A)
 180 '
 190 REM *** PRINT ANSWERS ***
 200 '
 210 PRINT : PRINT : PRINT"POINT 1","POINT
     2","SLOPE"
 220 PRINT A,B,M
 230 '
 240 REM *** ANOTHER POINT? ***
 250 '
 260 PRINT:INPUT "DO YOU WANT TO WORK
     ANOTHER PROBLEM WITH THIS
     FUNCTION";AN$
 270 GOSUB 2530
 280 IF AN$ = "Y" THEN GOTO 110
 290 PRINT:INPUT "DO YOU WANT TO CONTINUE
     WITH A DIFFERENT FUNCTION;"AN$
 300 GOSUB 2530
 310 IF AN$ = "Y" THEN GOTO 1030
 320 END
 340 REM *** ANOTHER POINT? ***
1000 '
1010 REM *** DEFINE FUNCTION DIRECTION ***
1020 '
1030 PRINT : PRINT : PRINT"TO CHANGE THE
     FUNCTION, TYPE:"
1040 PRINT : PRINT" 2030 DEF FN F(X) = . .
     . . ."
1050 PRINT : PRINT"WHERE '. . . .' IS
     WHERE YOU TYPE YOUR FUNCTION."
1060 PRINT : PRINT"AFTER YOU HAVE
     FINISHED, TYPE 'RUN 110' AND PRESS
     <RETURN>"
1070 END
2000 '
2010 REM *** FUNCTION DEFINED ***
2020 '
2030 DEF FN F(X) = 3 * X * X - 5
2040 RETURN
2500 '
2510 REM *** CHECK AN$ ***
2520 '
2530 IF LEFT$(AN$,1) <>"Y" AND
     LEFT$(AN$,1) <> "N" THEN GOTO 2550
2540 RETURN
2550 PRINT:INPUT "PLEASE ANSWER AGAIN. USE
     A 'Y' OR 'N'";AN$:GOTO 2530
```

Section 22.2, Exercise 13

```
   1 REM *** PRG22213 ***
  10 PRINT "THIS PROGRAM WILL APPROXIMATE
     THE AREA UNDER A CURVE USING THE
     RECTANGUALAR METHOD"
  20 '
  30 REM *** CHANCE TO CHANGE THE FUNCTION
     ***
  40 '
  50 INPUT "DO YOU WANT TO CHANGE THE
     FUNCTION? ";AN$
  60 GOSUB 2530
  70 IF LEFT$(AN$,1) = "Y" THEN GOTO 1030
  80 '
  90 REM *** ENTER POINTS ***
 100 '
 110 CLS
 120 PRINT : INPUT "ENTER THE LEFT END
     POINT, A=";A
 130 PRINT : INPUT "NEXT ENTER THE RIGHT
     POINT, B =";B
 140 PRINT : INPUT "AND FINALLY, THE
     NUMBER OF SEGMENTS, N= ";N
 150 DIM Y(N)
 160 '
```

```
170 REM *** COMPUTE AREA ***
180 '
190 DX = (B - A)/ N
200 GOSUB 2030
210 P = A + DX / 2:SUM = 0
220 FOR I = 1 TO N
230 Y(I) = FN F(P)
240 SUM = SUM + Y(I)
250 P = P + DX
260 NEXT I
270 AREA = SUM * DX
280 '
290 REM *** PRINT ANSWERS ***
300 '
310 PRINT "A","B","N": PRINT A,B,N
320 PRINT : PRINT "THE AREA IS ";AREA
330 '
340 '
350 '
420 END
1000 '
1010 REM *** DEFINE FUNCTION DIRECTION ***
1020 '
1030 PRINT : PRINT : PRINT "TO CHANGE THE
     FUNCTION, TYPE:"
1040 PRINT : PRINT " 2030 DEF FN F(X) = .
     . . . ."
1050 PRINT : PRINT "WHERE '. . . .' IS
     WHERE YOU TYPE YOUR FUNCTION."
1060 PRINT : PRINT "AFTER YOU HAVE
     FINISHED, TYPE 'RUN 110' AND PRESS
     <RETURN>"
1070 END
2000 '
2010 REM *** FUNCTION DEFINED ***
2020 '
2030 DEF FN F(X) = 3 * X * X - 5
2040 RETURN
2500 '
2510 REM *** CHECK AN$ ***
2520 '
2530 IF LEFT$(AN$,1) <>"Y" AND
     LEFT$(AN$,1) <> "N" THEN GOTO 2550
2540 RETURN
2550 PRINT : INPUT "PLEASE ANSWER AGAIN.
     USE A 'Y' OR 'N'";AN$:GOTO 2530
```

Use this program (PRG22213) to approximate the area under the function $f(x) = \sin(x^2) + 1$ be-

tween $x = 0$ and $x = \pi$.

Chapter 24

Section 24.6, Exercise 1

```
  1 REM *** PRG24601 ***
  5 CLS
 10 REM
 20 REM ********************
 30 REM ********************
 40 REM **              **
 50 REM ** NEWTON'S METHOD **
 60 REM **              **
 70 REM ********************
 80 REM ********************
 90 REM
100 REM
110 REM ** DEFINE THE FUNCTION **
120 REM
130 CLS
140 INPUT "DO YOU WANT TO CHANGE THE
    FUNCTION(Y/N)";QU$
150 IF QU$<>"Y" AND QU$<>"N" THEN GOTO
    140
160 IF QU$ = "N" THEN GOTO 250
170 CLS: PRINT : PRINT : PRINT "TO CHANGE
    THE FUNCTION TYPE"
180 PRINT : PRINT "1050 DEF FN F(X) = . .
    ."
190 PRINT : PRINT "WHERE YOU PUT YOUR NEW
    FUNCTION AFTER THE EQUAL SYMBOL ON
    LINE 1050"
200 PRINT : PRINT "AND" : PRINT
210 PRINT "1060 DEF FN D(X) = . . . ."
220 PRINT : PRINT "WHERE YOU PUT THE
    DERIVATIVE OF THE FUNCTION AFTER THE
    EQUAL SYMBOL ON LINE 1060"
230 PRINT : PRINT "AFTER YOU HAVE ENTERED
    THE FUNCTION YOU WILL HAVE TO TYPE,
    RUN 270,' TO GET THE PROGRAM TO
    CONTINUE.": PRINT
240 END
250 REM
260 REM
270 REM ** SEARCH FOR ROOT **
280 REM
290 REM
300 GOSUB 1050
310 GOTO 330
320 PRINT : PRINT : PRINT "YOU MADE A
```

```
             MISTAKE. PLEASE ANSWER THE QUESTION
             WITH A 'Y' OR A 'N'." PRINT : PRINT
330 INPUT "DO YOU WANT TO SEARCH FOR
             ROOTS?";QU$
340 IF LEFT$(QU$,1) <>"Y" AND
             LEFT$(QU$,1)<> "N" THEN GOTO 320
350 IF LEFT$(QU$,1) = "N" THEN GOTO 590
360 PRINT : PRINT : INPUT "ENTER THE
             LOWER LIMITS: ";A
370 PRINT : INPUT "NOW ENTER THE UPPER
             LIMITS: ";B
380 CLS: PRINT "THE LOWER LIMIT IS: ";A;"
             AND THE": PRINT "UPPER LIMIT IS: ";B
390 PRINT : PRINT : PRINT "X -
             1";TAB(8);"F(X - 1)";TAB(21);"X";
             TAB(30);"F(X)"
400 FOR X = A TO B - 1
410 P = FN F(X)
420 Q = FN F(X + 1)
430 IF SGN(P) <> SGN(Q) THEN GOTO 450
440 GOTO 470
450 PRINT : PRINT X;TAB(6);P;TAB(21);X +
             1;TAB(26);Q
460 PRINT "THERE IS A REAL ROOT BETWEEN
             ";X;" AND ";X + 1
470 NEXT X
480 PRINT "SEARCH COMPLETE. ": PRINT
490 GOTO 510
500 PRINT : PRINT "PLEASE ANSWER AGAIN.
             TYPE A Y OR N."
510 PRINT : INPUT "DO YOU WANT TO SEARCH
             OVER ANOTHER INTERVAL";QU$
520 IF LEFT$(QU$,1) <> "N" AND
             LEFT$(QU$,1) <> "Y" THEN GOTO 500
530 IF LEFT$(QU$,1) = "Y" GOTO 360
540 REM
550 REM
560 REM ** USE THE NEWTON'S METHOD **
570 REM
580 REM
590 PRINT : PRINT : PRINT "WE WILL NOW
             USE THE NEWTON'S METHOD TO DETERMINE
             APPROXIMATE ROOTS OF F(X).": PRINT
600 PRINT : INPUT "ENTER YOUR GUESS FOR
             THE ROOT ";X: PRINT
610 GOSUB 1050
620 PRINT : PRINT : PRINT
630 PRINT "I"; TAB(9); "X(I)"; TAB(28);
             "F(X(I))"; TAB(47); "F'(X(I))": PRINT
640 PRINT 1; TAB(6);X; TAB(25); FN F(X);
```

```
             TAB(45);FN D(X)
650 FOR J = 0 TO 25
660 XY = X - (FN F(X) / FN D(X))
670 X = XY
680 GOSUB 1050
690 PRINT J+2;TAB(6);X;TAB(26);FN
             F(X);TAB(45);FN D(X)
700 CHK = ABS(FN F(X))
710 IF CHK < .000001 THEN GOTO 740
720 NEXT J
730 CLS
740 PRINT : PRINT "THE ROOT IS
             APPROXIMATELY ";X: PRINT : PRINT
750 GOTO 770
760 PRINT : PRINT "PLEASE REANSWER WITH A
             Y OR N."
770 PRINT : INPUT "DO YOU WANT TO SEARCH
             FOR ANOTHER ROOT";QU$
780 IF LEFT$(QU$,1) <> "N" AND
             LEFT$(QU$,1) <> "Y" THEN GOTO 760
790 IF LEFT$(QU$,1) = "Y" GOTO 600
800 END
1000 REM
1010 REM
1020 REM ** SUBROUTINE **
1030 REM
1040 REM
1050 DEF FN F(X) = X ^ 5 - 111
1060 DEF FN D(X) = 5 * X ^ 4
1070 PRINT :RETURN
```

Use this program (PRG24601) to approximate the root of $f(x) = \cos(x^2) - x$. (Note: This is the same function used in Exercise #24.)

Chapter 25

Section 25.5, Exercise 23

```
  1 REM *** PRG25523 ***
 10 REM ******************************
 20 REM ******************************
 30 REM **                        **
 40 REM **    AREA UNDER A CURVE   **
 50 REM **        USING BOTH       **
 60 REM **      SIMPSON'S RULE      **
 70 REM **          AND            **
 80 REM **    THE TRAPEZOIDAL RULE  **
 90 REM **                        **
100 REM ******************************
```

```
110 REM ****************************
120 REM
130 GOTO 150
140 PRINT : PRINT : PRINT"PLEASE ANSWER
    AGAIN. THIS TIME USE A 'P' OR AN
    'F'."
150 PRINT : PRINT"DO YOU WANT TO FIND AN
    AREA FOR A FUNCTION, Y=F(X), OR FOR A
    SET OF OBSERVED POINTS: ": INPUT QU$
160 IF QU$ <> "F" AND QU$ <> "P" THEN
    GOTO 140
170 IF QU$ = "P" THEN GOTO 2060
180 REM
190 REM
200 REM
210 REM **** DEFINE THE FUNCTION ****
220 REM
230 CLS
240 INPUT "DO YOU WANT TO CHANGE THE
    FUNCTION (Y/N)?";QU$
250 IF QU$ <> "Y" AND QU$ <> "N" THEN
    GOTO 240
260 IF QU$ = "N" THEN GOTO 320
270 CLS: PRINT : PRINT : PRINT "TO CHANGE
    THE FUNCTION, TYPE"
280 PRINT : PRINT : PRINT "1050 DEF FN
    F(X) = ......"
290 PRINT : PRINT "AND PUT YOUR NEW
    FUNCTION WHERE THE DOTTED LINE IS."
300 PRINT : PRINT "WHEN FINISHED, RERUN
    THE PROGRAM BY TYPING: RUN 390"
310 END
320 REM
330 REM
340 REM ** INPUT END POINTS & NO. OF
    INTERVALS **
350 REM
360 REM
370 GOTO 390
380 PRINT : PRINT : PRINT "YOU MADE A
    MISTAKE. PLEASE REENTER THE
    ENDPOINTS." : PRINT : PRINT
390 CLEAR
400 PRINT : INPUT "ENTER THE LEFT
    ENDPOINT: ";A
410 PRINT : INPUT "AND THE RIGHT
    ENDPOINT: ";B
420 IF B - A < = 0 THEN GOTO 380
430 GOTO 450
440 PRINT : PRINT : PRINT : PRINT "YOU
    HAVE MADE A MISTAKE. YOU WANTED ";N;"
    INTERVALS. THIS IS NOT AN EVEN
    NUMBER. PLEASE REENTER."
450 PRINT : INPUT "NOW ENTER THE NUMBER
    OF INTERVALS. THIS MUST BE AN EVEN
    NUMBER ";N
460 CHK = N / 2 - INT ( N / 2) IF CHK <>
    0 GOTO 440
470 REM
480 REM
490 REM ** USING SIMPSON'S AND THE
    TRAPEZOIDAL RULES **
500 REM
510 REM
520 GOSUB 1050
530 DX = (B - A) / N
540 DIM Y(N)
550 FOR I = 0 TO N:X = A + I * DX:Y(I)
    =FN F(X): NEXT I
560 TR = Y(0):SR = Y(0)
570 FOR I = 1 TO N - 3 STEP 2
580 SR = SR + 4 * Y(I) + 2 * Y(I + 1): TR
    = TR + 2 * Y(I) + 2 * Y(I + 1)
590 NEXT I
600 SR = SR + 4 * Y(N - 1) + Y(N): TR =
    TR + 2 * Y(N - 1) + Y(N)
610 SR = SR * DX / 3:TR = TR * DX / 2
620 PRINT : PRINT "THE AREA UNDER F(X)
    OVER THE INTERVAL FROM "A;" TO ";B;"
    IS"
630 PRINT : PRINT SR;"BY SIMPSON'S RULE
    AND
640 PRINT : PRINT TR;"BY THE TRAPEZOIDAL
    RULE
650 END
990 END
1000 REM
1010 REM
1020 REM ** SUBROUTINE **
1030 REM
1040 REM
1050 DEF FN F(X) = 1 / (1 + X)
1060 PRINT : RETURN
2000 REM
2010 REM
2020 REM ** ENTER OBSERVED POINTS **
2030 REM
2040 REM
2050 PRINT : PRINT : PRINT "YOU MADE A
     MISTAKE. YOU NEED AN EVEN NUMBER OF
```

```
       INTERVALS. ";N;" POINTS WILL PRODUCE
       ";N - 1;" INTERVALS. THIS IS NOT AN
       EVEN NUMBER. PLEASE REENTER."
2060 PRINT : INPUT "ENTER THE NUMBER OF
       OBSERVED POINTS ";N:N2 = N - 1
2070 CHK = N2 / 2 - INT (N2 / 2): IF CHK <
       > 0 GOTO 2050
2080 DIM X(N), Y(N)
2090 FOR I = 1 TO N
2100 PRINT : PRINT "X(";I;") = "; TAB (7)
       : INPUT X(I)
2110 PRINT : PRINT "Y(";I;") = "; TAB (7)
       : INPUT Y(I)
2120 NEXT I
2125 A=X(1) : B =X(N)
2130 SR = Y(1):TR = Y(1)
2140 FOR I = 2 TO N - 3 STEP 2:SR = SR + 4
       * Y(I) + 2 * Y(I + 1): TR = TR + 2 *
       Y(I) + 2 * Y(I + 1): NEXT I
2150 DX = (X(N) - X(1))/(N - 1)
2160 SR = SR + 4 * Y(N - 1) + Y(N): SR =
       SR * DX / 3: TR = TR + 2 * Y(N - 1) +
       Y( N): TR = TR* DX / 2
2170 PRINT " PRINT "THE OBSERVED POINTS
       ARE"
2180 PRINT "I","X(I)","Y(I)"
2190 FOR I = 1 TO N: PRINT I,X(I),Y(I):
       NEXT I
2200 GOTO 620
```

Use this program (PRG25523) to approximate the area under the function $f(x) = \sin(x^2) + 1$ from $x = 0$ to $x = \pi$ using 50 intervals. (Note: this is the same function used in Exercise #27. Compare these answers with the one you got in that exercise.)

Chapter 30

Section 30.6, Exercise 31

```
   1 REM *** PRG30631 ***
  10 REM ****************************
  20 REM ****************************
  30 REM **      CONVERSION AMONG       **
  40 REM **        RECTANGULAR,         **
  50 REM **         CYLINDRICAL,        **
  60 REM **       AND SPHERICAL         **
  70 REM **    COORDINATE SYSTEMS     **
  80 REM ****************************
  90 REM ****************************
```

```
 100 :
 110 REM ** DETERMINE ORIGINAL SYSTEM **
 120 :
 130 PRINT "THIS PROGRAM WILL CONVERT FROM
       ONE THREE - DIMENSIONAL COORDINATE
       SYSTEM (THE ORIGINAL) TO ANOTHER (THE
       TARGET)."
 140 GOTO 160
 150 PRINT : PRINT : PRINT "YOU MADE A
       MISTAKE.": PRINT "PLEASE REENTER BY
       TYPING AN 'R','S',OR 'C'."
 160 PRINT : PRINT "TYPE THE FIRST LETTER
       OF THE ORIGINAL SYSTEM.": PRINT "IF
       THE ORIGINAL SYSTEM IS THE": GOSUB
       1030: GOSUB 1040: GOSUB 1050: PRINT:
       PRINT "AND THEN PRESS THE 'RETURN'
       KEY."
 170 INPUT OS$
 180 IF OS$ < > "R" AND OS$ < > "C" AND
       OS$ < > "S" THEN GOTO 150
 190 PI=3.14159265359#
 200 :
 210 REM ** DETERMINE TARGET SYSTEM **
 220 :
 230 PRINT: PRINT: PRINT "TYPE THE FIRST
       LETTER OF THE TARGET SYSTEM.": PRINT
       "IF THE TARGET SYSTEM IS THE"
 240 IF OS$ = "R" THEN GOSUB 1040: GOSUB
       1050: INPUT TS$ : GOTO 330
 250 IF OS$ = "C" THEN GOSUB 1030: GOSUB
       1050: INPUT TS$ : GOTO 530
 260 IF OS$ = "S" THEN GOSUB 1030: GOSUB
       1040: INPUT TS$ : GOTO 730
 300 :
 310 REM ** ENTER RECTANGULAR COORDINATES
       **
 320 :
 330 CLS : PRINT "YOU WILL NOW ENTER THE
       (X,Y,Z) COORDINATES FOR THE
       RECTANGULAR COORDINATE SYSTEM.":
       PRINT
 340 PRINT : INPUT "X= "; X : PRINT :
       INPUT "Y = "; Y: PRINT : INPUT "Z =
       "; Z
 350 IF TS$ = "S" THEN GOTO 400
 360 R = SQR (X * X + Y * Y): GOSUB 1130
 370 PRINT: PRINT "CYLINDRICAL COORDINATES
       OF THE RECTANGULAR POINT ("; X;", ";
       Y;", "; Z;") ARE" : PRINT : PRINT "R
       = ";R: PRINT : PRINT "THETA = "; TH;
```

```
        " = "; TH/PI;"*PI": PRINT: PRINT "Z =
        ";Z :GOTO 930
400 RH = SQR (X * X + Y * Y + Z * Z):
        GOSUB 1130: AC = Z/RH: GOSUB 1230
410 PRINT: PRINT "THE SPHERICAL
        COORDINATES OF THE RECTANGULAR POINT
        (";X;", ";Y ;", ";Z;") ARE": PRINT :
        PRINT "RHO = ";RH: PRINT: PRINT
        "THETA = "; TH;" = "; T H/PI;"*PI" :
        PRINT: PRINT "PHI = " PH;" =
        ";PH/PI;" *PI": GOTO 930
500 :
510 REM ** ENTER CYLINDRICAL COORDINATES
        **
520 :
530 CLS: PRINT "YOU WILL NOW ENTER THE
        (R, THETA, Z) COORDINATES FOR THE
        CYLINDRICAL SYSTEM.": PRINT: PRINT
        "REMEMBER THAT R >= 0.": PRINT: PRINT
        "THETA SHOULD BE IN RADIANS AND
        WRITTEN AS A MULTIPLE OF PI."
540 PRINT "YOU WILL BE ASKED FOR THE
        NUMERATOR AND DENOMINATOR OF THETA."
        : PRINT "FOR EXAMPLE, IF THETA =
        3*PI/4, ENTER 3 AS THE NUMERATOR AND
        4 AS THE DENOMINATOR."
550 PRINT : INPUT "R = "; R: PRINT: PRINT
        "ENTER THETA: ": PRINT: INPUT
        "NUMERATOR = "; TN: PRINT: INPUT
        "DENOMINATOR = ";TD : PRINT: TH=TN *
        PI/TD: PRINT: INPUT "Z= ";Z
560 IF TS$ = "S" THEN GOTO 600
570 X=R*COS(TH):Y=R*SIN(TH)
580 PRINT: PRINT "THE RECTANGULAR
        COORDINATES OF THE CYLINDRICAL
        COORDINATE POINT (";R;",
        ";TN"*PI/";TD;", "Z;") ARE": PRINT:
        PRINT "X = ";X: PRINT: PRINT "Y =
        ";Y: PRINT: PRINT "Z = ";Z: GOTO 930
600 RH=SQR (R * R + Z * Z):AC=Z/RH: GOSUB
        1230
610 PRINT: PRINT "THE SPHERICAL
        COORDINATES OF THE CYLINDRICAL
        COORDINATE POINT (";R;",
        ";TN;"*PI/";TD;", "; Z;") ARE":
        PRINT: PRINT "RHO = ";RH : PRINT: :
        PRINT "THETA = "; TH;" = "; TH/PI;" *
        PI": PRINT: PRINT "PHI = ";PH;" = ";
        PH/PI;" * PI" : GOTO 930
700 :
```

```
710 REM ** ENTER SPHERICAL COORDINATES **
720 :
730 CLS: PRINT "YOU WILL NOW ENTER THE
        (RHO, THETA, PHI) COORDINATES FOR THE
        SPHERICAL SYSTEM.": PRINT: PRINT
        "REMEMBER THAT RHO >=0.": PRINT:
        PRINT "BOTH THETA AND PHI SHOULD BE
        IN RADIANS AND WRITTEN AS MULTIPLES
        OF PI."
740 PRINT "YOU WILL BE ASKED FOR THE
        NUMERATOR AND DENOMINATOR OF THETA
        AND OF PHI.": PRINT "FOR EXAMPLE, IF
        THETA = 3*PI/4, ENTER 3 AS THE
        NUMERATOR AND 4 AS T HE DENOMINATOR."
750 PRINT: INPUT "RHO = ";RH: PRINT :
        PRINT "ENTER THETA: ": PRINT: INPUT
        "NUMERATOR = "; TN: PRINT: INPUT
        "DENOMINATOR = ";TD: PRINT: TH = TN *
        PI/TD: PRINT
760 PRINT "ENTER PHI: ": PRINT: INPUT "
        NUMERATOR = ";PN: PRINT: INPUT "
        DENOMINATOR = "; PD: PH=PN*PI/PD
770 IF TS$ = "C" THEN GOTO 800
780 X= RH * SIN(PH) * COS(TH):Y = RH *
        SIN(PH) * SIN(TH): Z=RH * COS(PH)
790 PRINT: PRINT "THE RECTANGULAR
        COORDINATES OF THE SPHERICAL
        COORDINATE POINT (";RH;",
        ";TN;"*PI/";TD;", "; PN;"*PI/";PD;")
        ARE ": PRINT: PRINT "X= ";X: PRINT :
        PRINT "Y= ";Y: PRINT: PRINT "Z= "Z:
        GOTO 930
800 R = RH * SIN(PH): Z= RH * COS (PH)
810 PRINT: PRINT "THE CYLINDRICAL
        COORDINATES OF THE SPHERICAL
        COORDINATE POINT (";RH;",
        ";TN;"*PI/";TD;", "; PN;"*PI/";PD;")
        ARE ": PRINT: PRINT "R = "; R: PRINT
        : PRINT "THETA = "; TH;" = ";TH/PI;"
        * PI": PRINT : PRINT "Z = "; Z
900 :
910 REM ** ANOTHER PROBLEM **
920 :
930 PRINT: PRINT: PRINT: PRINT "DO YOU
        WANT TO MAKE ANOTHER CONVERSION":
        INPUT " BETWEEN THESE COORDINATE
        SYSTEMS?"; AN$
940 IF AN$ < > "Y" AND AN$ < > "N" THEN
        GOSUB 1330: GOTO 930
950 IF AN$ = "N" THEN GOTO 990
```

```
960 IF OS$ = "R" THEN GOTO 340
970 IF OS$ = "C" THEN GOTO 550
980 IF OS$ = "S" THEN GOTO 750
990 PRINT: PRINT: PRINT: PRINT "DO YOU
    WANT TO MAKE A CONVERSION BETWEEN":
    INPUT "TWO OTHER COORDINATE SYSTEMS?
    "; AN$ : IF AN$ < > "Y" AND AN$ < >
    "N" THEN GOSUB 1330: GOTO 990
995 IF AN$ = "Y" THEN GOTO 160
999 END
1000 :
1010 REM ** SYSTEM NAME SUBROUTINES **
1020 :
1030 PRINT " RECTANGULAR SYSTEM, TYPE
     'R'": RETURN
1040 PRINT " CYLINDRICAL SYSTEM, TYPE
     'C'": RETURN
1050 PRINT " SPHERICAL SYSTEM, TYPE 'S'":
     RETURN
1100 :
1110 REM ** DETERMINING THETA **
1120 :
1130 IF X = 0 AND Y > 0 THEN TH = 0:
     RETURN
1140 IF X = 0 AND Y > 0 THEN TH = PI:
     RETURN
1150 TH = ATN (Y / X)
1160 IF X < 0 THEN TH = PI + TH: RETURN
1170 IF X > 0 AND Y < 0 THEN TH = 2 * PI +
     TH: RETURN
1200 :
1210 REM ** COMPUTE THE ARC COSINE *
1220 :
1230 PH = - ATN (AC / SQR ( - AC * AC +
     1)) + PI / 2: RETURN
1300 :
1310 REM ** INCORRECT RESPONSE **
1320 :
1330 PRINT: PRINT: PRINT: PRINT "YOU MADE
     A MISTAKE. PLEASE REENTER WITH A 'Y'
     OR AN 'N'.": RETURN
```

Use this program (PRG30631) to convert the given coordinates in the rectangular, cylindrical, or spherical system to its equivalent in the other two systems.

(a) the rectangular coordinate $(-2, 5, -4.5)$

(b) the cylindrical coordinate $(3, \frac{\pi}{6}, -2)$

(c) the spherical coordinate $(5, \frac{\pi}{6}, \frac{\pi}{4})$

Chapter 34

Section 34.1, Exercise 1

```
  1 REM *** PRG34101 ***
 10 REM *****************************
 15 REM *****************************
 20 REM **                        **
 30 REM **     EULER'S METHOD     **
 40 REM **       FOR SOLVING      **
 50 REM **       DIFFERENTIAL     **
 60 REM **         EQUATIONS      **
 70 REM **                        **
 80 REM *****************************
 90 REM *****************************
100 PRINT : PRINT : PRINT : PRINT
110 PRINT"THIS PROGRAM WILL GIVE N
    APPROXIMATIONS TO THE DIFFERENTIAL
    EQUATION Y = F(X,Y). THE EQUATION
    CURRENTLY IN THE PROGRAM IS LISTED ON
    LINE 1030"
120 PRINT "WHERE EQ= F(X,Y).": PRINT :
    PRINT
130 INPUT "DO YOU WANT TO CHANGE THE
    DIFFERENTIAL EQUATION?";YN$:IF YN$ <>
    "Y" A ND YN$ <> "N" THEN GOSUB
    2030:GOTO 130
140 IF YN$ = "N" THEN GOTO 170
150 CLS: PRINT "TO CHANGE THE
    DIFFERENTIAL EQUATION TYPE": PRINT :
    PRINT "1030 EQ = .. ..": PRINT :
    PRINT "YOU WILL ENTER F(X,Y) WHERE
    THE DOTS ARE AT THE RIGHT OF THE '='
    SIGN."
160 PRINT : PRINT "WHEN YOU HAVE FINISHED
    TYPING THE NEW EQUATION PRESS THE
    RETURN KEY, TYPE 'RUN 170,'(DON'T
    FORGET THE COMMA), AND AGAIN PRESS
    THE RETURN KEY.":E ND
170 INPUT "HOW MANY APPROXIMATIONS DO YOU
    WANT?";N
180 PRINT : INPUT "ENTER THE SIZE OF
    INCREMENT FOR X: ";H
190 PRINT : INPUT "WHAT IS THE VALUE OF
    X(0)?";X
200 PRINT : INPUT "ENTER THE VALUE OF
    Y(0):";Y: PRINT : PRINT : PRINT
210 PRINT "N","X(N)","Y(N)": PRINT
220 FOR I = 1 TO N
230 PRINT I,X,Y
```

```
240 GOSUB 1030
250 Y = Y + EQ*H
260 X = X + H
270 NEXT I
280 PRINT : PRINT : INPUT "DO YOU WANT TO
    WORK ANOTHER PROBLEM?";YN$:IF YN$
    <>"Y" AN D YN$ <> "N" THEN GOSUB
    2030:GOTO 280
290 IF YN$="Y" THEN GOTO 130
300 END
1000 '
1010 REM ** DIFFERENTIAL EQUATION

     SUBROUTINE **
1020 '
1030 EQ = X + 2
1040 RETURN
2000 '
2010 REM ** MISTAKE SUBROUTINE **
2020 '
2030 PRINT : PRINT "YOU HAVE MADE A
     MISTAKE. PLEASE ANSWER WITH A 'Y' OR
     'N'."
2040 RETURN
```

Index of Applications

This index categorizes the application problems in this book alphabetically. After each topic, the index lists the text page number, followed by the problem number in parentheses. For example, if you are looking for a problem on accounting, you would turn to page 576 and look for problem number 66.

Index